Handbook of Fiber Chemistry

Third Edition

INTERNATIONAL FIBER SCIENCE AND TECHNOLOGY SERIES

Series Editor

MENACHEM LEWIN
Hebrew University of Jerusalem
Jerusalem, Israel
Herman F. Mark Polymer Research Institute
Polytechnic University
Brooklyn, New York

The Editor and Publisher gratefully acknowledge
the past contributions of our distinguished
Editorial Advisory Board

STANLEY BACKER
Fibers and Polymer Laboratory
Massachusetts Institute of Technology
Cambridge, Massachusetts

CHRISTOPHER SIMIONESCU
Romanian Academy of Sciences
Jassy, Romania

SOLOMON P. HERSH
College of Textiles
North Carolina State University
Raleigh, North Carolina

VIVIAN T. STANNETT
Department of Chemical Engineering
North Carolina State University
Raleigh, North Carolina

ELI M. PEARCE
Herman F. Mark
Polymer Research Institute
Polytechnic University
Brooklyn, New York

ARNOLD M. SOOKNE
Burlington Industries
Greensboro,
North Carolina

JACK PRESTON
Research Triangle Institute
Research Triangle Park,
North Carolina

FRANK X. WERBER
Agricultural Research Service
USDA
Washington, D.C.

INTERNATIONAL FIBER SCIENCE AND TECHNOLOGY SERIES

Series Editor: **MENACHEM LEWIN**

1. Handbook of Fiber Science and Technology (I): Chemical Processing of Fibers and Fabrics-Fundamentals and Preparation (in two parts), *edited by Menachem Lewin and Stephen B. Sello*
2. Handbook of Fiber Science and Technology (II): Chemical Processing of Fibers and Fabrics-Functional Finishes (in two parts), *edited by Menachem Lewin and Stephen B. Sello*
3. Carbon Fibers, *Jean-Baptiste Donnet and Roop Chand Bansal*
4. Fiber Technology: From Film to Fiber, *Hans A. Krässig, Jürgen Lenz, and Herman F. Mark*
5. Handbook of Fiber Science and Technology (III): High Technology Fibers (Part A), *edited by Menachem Lewin and Jack Preston*
6. Polyvinyl Alcohol Fibers, *Ichiro Sakurada*
7. Handbook of Fiber Science and Technology (IV): Fiber Chemistry, *edited by Menachem Lewin and Eli M. Pearce*
8. Paper Structure and Properties, *edited by J. Anthony Bristow and Petter Kolseth*
9. Handbook of Fiber Science and Technology (III): High Technology Fibers (Part B), *edited by Menachem Lewin and Jack Preston*
10. Carbon Fibers: Second Edition, Revised and Expanded, *Jean-Baptiste Donnet and Roop Chand Bansal*
11. Wood Structure and Composition, *edited by Menachem Lewin and Irving S. Goldstein*
12. Handbook of Fiber Science and Technology (III): High Technology Fibers (Part C), *edited by Menachem Lewin and Jack Preston*
13. Modern Textile Characterization Methods, *edited by Mastura Raheel*
14. Handbook of Fiber Science and Technology (III): High Technology Fibers (Part D), *edited by Menachem Lewin and Jack Preston*
15. Handbook of Fiber Chemistry: Second Edition, Revised and Expanded, *edited by Menachem Lewin and Eli M. Pearce*
16. Handbook of Fiber Chemistry: Third Edition, *edited by Menachem Lewin*

Handbook of Fiber Chemistry

Third Edition

Edited by
Menachem Lewin

Taylor & Francis
Taylor & Francis Group
Boca Raton London New York

CRC is an imprint of the Taylor & Francis Group,
an informa business

FIRST INDIAN REPRINT, 2010

CRC Press
Taylor & Francis Group
6000 Broken Sound Parkway NW, Suite 300
Boca Raton, FL 33487-2742

© 2007 by Taylor & Francis Group, LLC
CRC Press is an imprint of Taylor & Francis Group, an Informa business

No claim to original U.S. Government works
Printed and bound in India by Replika Press Pvt. Ltd.

International Standard Book Number-10: 0-8247-2565-4 (Hardcover)
International Standard Book Number-13: 978-0-8247-2565-5 (Hardcover)

This book contains information obtained from authentic and highly regarded sources. Reprinted material is quoted with permission, and sources are indicated. A wide variety of references are listed. Reasonable efforts have been made to publish reliable data and information, but the author and the publisher cannot assume responsibility for the validity of all materials or for the consequences of their use.

No part of this book may be reprinted, reproduced, transmitted, or utilized in any form by any electronic, mechanical, or other means, now known or hereafter invented, including photocopying, microfilming, and recording, or in any information storage or retrieval system, without written permission from the publishers.

For permission to photocopy or use material electronically from this work, please access www.copyright.com (http://www.copyright.com/) or contact the Copyright Clearance Center, Inc. (CCC) 222 Rosewood Drive, Danvers, MA 01923, 978-750-8400. CCC is a not-for-profit organization that provides licenses and registration for a variety of users. For organizations that have been granted a photocopy license by the CCC, a separate system of payment has been arranged.

Trademark Notice: Product or corporate names may be trademarks or registered trademarks, and are used only for identification and explanation without intent to infringe.

Library of Congress Cataloging-in-Publication Data

Handbook of fiber chemistry / edited by Menachem Lewin. -- 3rd ed.
    p. cm. -- (International fiber science and technology series ; 16)
   Includes bibliographical references and index.
   ISBN 0-8247-2565-4
   1. Textile fibers. 2. Textile chemistry. I. Lewin, Menachem, 1918- II. Series.

TS1540.H26 2006
677'.02832--dc22
                                                      2006044600

Visit the Taylor & Francis Web site at
http://www.taylorandfrancis.com

and the CRC Press Web site at
http://www.crcpress.com

FOR SALE IN SOUTH ASIA ONLY

Preface

The third edition of the *Handbook of Fiber Chemistry*, expanded from the second edition, contains 13 chapters dealing with the most important natural, human-made, and synthetic fibers, including the additional chapter on the highly important and broadly used Kevlar fiber. Almost a decade has passed since the the second edition, and relatively little change has taken place during this time in the use of the basic fibers. Some important technological advances that have happened during this period are fully discussed in the present volume. Thus, the fibers described in this book maintain their unchallenged economic positions. The technologies used in their production and applications have been greatly improved, as indicated by the considerable number of patents published; thus the current production systems have not been discarded or replaced by inherently new systems. Similarly, the markets for these fibers were not only maintained, but have expanded and diversified.

Fiber science in its present state of development cannot be considered as a mature science. New fibers, including nanofibers and biologically and electronically active fibers, are under development for specific applications at present for relatively limited markets. Several of these fibers are discussed in the four volumes on high-technology fibers included in this series. Their development is, however, derived from the scientific and technological principles of the conventional fibers described in this book. The definitions, morphology, and fine structure, properties, testing, processing methods, and equipment, and the conversions into marketable products are basically similar.

The chapters in this revised and expanded volume, except for the chapters on acrylic and wool fibers, are either new or extensively updated; hence this edition should be considered as entirely a new book. A wide array of new data have become available in the past decade based, to a large extent, on new scientific techniques, instruments, and disciplines. These data have enabled us to gain a better insight into the structure of fibers and structure–property relationships, and have brought about a better understanding of fiber-related phenomena. We have made a serious effort to include the most important developments in fiber science during the last decade in the present volume.

The chapters in this edition are authored by leading experts in the field of fiber science. Many of the chapters (rayon, acetate, silk, polypropylene, polyamide, polyester) are new and written by authors who have not contributed to the previous edition. Other chapters (vinyl fibers, cotton, jute and kenaf, long vegetable fibers) have been fully updated. Of particular importance is the updated comprehensive chapter on cotton fibers. This was prepared by 16 recognized authorities and compiled by P. Wakelyn and R. Bertoniere. It contains a vast amount of up-to-date information presented in a lucid and concise format, and covers all aspects of the science and technology of cotton and cellulose. The recently revived interest in other vegetable fibers is clearly illustrated in the chapters on long vegetable fibers, and on jute and kenaf.

This volume is aimed at a wide audience of scientists, technologists, and engineers in chemistry, physics, biology, medicine, agriculture, materials, textiles, and polymers. We hope that this book will help experts working in these various disciplines to understand the vigorous and complex field of fibers, and as a result, to interact with scientists working on fibers so as to provide new, better routes for developing novel and innovative products and technologies.

I wish to thank all the authors who have contributed to this book, and the editorial staff of the Taylor & Francis Group who helped me in its publication.

Contributors

Subhash K. Batra
North Carolina State University
Raleigh, North Carolina

Noelie R. Bertoniere
Southern Regional Research Center
Agricultural Research Service
U.S. Department of Agriculture
New Orleans, Louisana

Peggy Cebe
Departments of Biomedical
  Engineering,
Chemical & Biological
  Engineering & Physics
Tufts University
Medford, Massachusetts

Anthony J. East
Medical Device Concept Lab
New Jersey Institute of Technology
Newark, New Jersey

J. Vincent Edwards
Southern Regional Research Center
Agricultural Research Service
U.S. Department of Agriculture
New Orleans, Louisiana

Alfred D. French
Southern Regional Research Center
Agricultural Research Service
U.S. Department of Agriculture
New Orleans, Louisiana

Bruce G. Frushour
High Performance Materials
Monsanto Company
St. Louis, Missouri

Vlodek Gabara
Spruance Plant
E.I. DuPont
Richmond, Virginia

Gary R. Gamble
Cotton Quality Research Station
Agricultural Research Service
U.S. Department of Agriculture
Clemson, South Carolina

Wilton R. Goynes, Jr.
Southern Regional Research Center
Agricultural Research Service
U.S. Department of Agriculture
New Orleans, Louisiana

Jon D. Hartzler
Spruance Plant
E.I. DuPont
Richmond, Virginia

Lawrance Hunter
Council for Scientific and Industrial
  Research
Port Elizabeth, South Africa

Michael Jaffe
Department of Biomedical Engineering
New Jersey Institute of Technology
Newark, New Jersey

Leslie N. Jones
Division of Wool Technology
CSIRO
Belmont, Victoria, Australia

David L. Kaplan
Departments of Biomedical Engineering,
Chemical & Biological Engineering & Physics
Tufts University
Medford, Massachusetts

Hyeon Joo Kim
Departments of Biomedical Engineering,
  Chemical & Biological Engineering &
  Physics
Tufts University
Medford, Massachusetts

Raymond S. Knorr
Fibers Division
Solutia, Inc.
Pensacola, Florida

Richard Kotek
College of Textiles
North Carolina University
Raleigh, North Carolina

Herman L. LaNieve
Warren, New Jersey

Kiu-Seung Lee
DuPont Experimental Station
Wilmington, Delaware

Akira Matsumoto
Departments of Biomedical Engineering,
  Chemical & Biological Engineering &
  Physics
Tufts University
Medford, Massachusetts

David D. McAlister
Cotton Quality Research Station
Agricultural Research Service
U.S. Department of Agriculture
Clemson, South Carolina

Takuji Okaya
The University of Shiga Prefecture
Shiga, Japan

Donald E. Rivett
Division of Biomolecular Engineering
CSIRO
Melbourne, Victoria, Australia

David J. Rodini
Spruance Plant
E.I. DuPont
Richmond, Virginia

Marie-Alice Rousselle
Southern Regional Research Center
Agricultural Research Service
U.S. Department of Agriculture
New Orleans, Louisiana

Roger M. Rowell
Modified Lignocellulosic Materials
Forest Products Laboratory
University of Wisconsin
Madison, Wisconsin

Ichiro Sakurada[*]
Institute for Chemical Research
Kyoto University
Kyoto, Japan

Harry P. Stout[*]
British Jute Trade Research Association
Dundee, Scotland

Devron P. Thibodeaux
Southern Regional Research Center
Agricultural Research Service
U.S. Department of Agriculture
New Orleans, Louisiana

Barbara A. Triplett
Southern Regional Research Center
Agricultural Research Service
U.S. Department of Agriculture
New Orleans, Louisiana

Daryl J. Tucker
School of Biological and Chemical
  Sciences
Deakin University
Geelong, Victoria, Australia

Irene Y. Tsai
Departments of Biomedical Engineering,
Chemical & Biological Engineering &
  Physics
Tufts University
Medford, Massachusetts

Phillip J. Wakelyn
National Cotton Council of America
Washington, D.C.

Xianyan Wang
Departments of Biomedical Engineering,
Chemical & Biological Engineering &
  Physics
Tufts University
Medford, Massachusetts

H.H. Yang
Richmond, Virginia

Mei-Fang Zhu
Shanghai, China

[*]Deceased

Table of Contents

Chapter 1 Polyester Fibers 1
Michael Jaffe and Anthony J. East

Chapter 2 Polyamide Fibers 31
H.H. Yang

Chapter 3 Polypropylene Fibers 139
Mei-Fang Zhu and H.H. Yang

Chapter 4 Vinyl Fibers 261
Ichiro Sakurada and Takuji Okaya

Chapter 5 Wool and Related Mammalian Fibers 331
Leslie N. Jones, Donald E. Rivett, and Daryl J. Tucker

Chapter 6 Silk 383
Akira Matsumoto, Hyeon Joo Kim, Irene Y. Tsai, Xianyan Wang,
Peggy Cebe, and David L. Kaplan

Chapter 7 Jute and Kenaf 405
Roger M. Rowell and Harry P. Stout

Chapter 8 Other Long Vegetable Fibers: Abaca, Banana, Sisal, Henequen,
Flax, Ramie, Hemp, Sunn, and Coir 453
Subhash K. Batra

Chapter 9 Cotton Fibers 521
Philip J. Wakelyn, Noelie R. Bertoniere, Alfred D. French, Devron P. Thibodeaux,
Barbara A. Triplett, Marie-Alice Rousselle, Wilton R. Goynes, Jr., J.Vincent Edwards,
Lawrance Hunter, David D. McAlister, and Gary R. Gamble

Chapter 10 Regenerated Cellulose Fibers 667
Richard Kotek

Chapter 11 Cellulose Acetate and Triacetate Fibers 773
Herman L. LaNieve

Chapter 12 Acrylic Fibers 811
Bruce G. Frushour and Raymond S. Knorr

Chapter 13 Aramid Fibers 975
Vlodek Gabara, Jon D. Hartzler, Kiu-Seung Lee, David J. Rodini, and H.H. Yang

Index 1031

1 Polyester Fibers

Michael Jaffe and Anthony J. East

CONTENTS

1.1 Introduction 2
1.2 PET History 3
1.3 PET Polymerization 3
    1.3.1 Monomer Production 3
    1.3.2 Polymerization 4
    1.3.3 Characterization of Poly(ethylene Terephthalate) Chip 5
    1.3.4 PET Processing—Melt Spinning 5
    1.3.5 PET Processing—Drawing 10
    1.3.6 PET Yarn after Processing—Heat-Setting and Bulking 12
    1.3.7 Polyester Yarns for Specific Applications 12
    1.3.8 Physical Properties of PET 13
1.4 Other Polyesters 14
    1.4.1 Polyester Fibers Based on Terephthalic Acid 14
    1.4.2 High-Performance Polyester Fibers—PEN and LCPs 15
    1.4.3 Fibers from Main-Chain Thermotropic Polyesters—LCPs 15
        1.4.3.1 Chemical Structure of LCPs 15
        1.4.3.2 Processing of Thermotropic Polyesters 16
        1.4.3.3 Structure–Property Relationships 17
1.5 Biodegradable Fibers 18
1.6 Modification of Polyester Fibers—Specific Solutions for
Specific Applications 19
    1.6.1 Spin Finishes 19
    1.6.2 Tire Cord 19
    1.6.3 Low-Pill Staple Polyester 19
    1.6.4 Noncircular Cross-Section Fibers 20
    1.6.5 Antistatic and Antisoiling Fibers 20
1.7 Dyeing Polyesters 21
    1.7.1 Introduction 21
    1.7.2 Disperse Dyes 21
    1.7.3 Anionic and Cationic Dyes for Polyester 22
    1.7.4 Mass Dyeing 22
1.8 Bicomponent Fibers and Microfibers 23
    1.8.1 Side–Side Bicomponent Fibers 23
    1.8.2 Core–Sheath Bicomponent Fibers 24
    1.8.3 Multiple Core Bicomponent Fibers 24
    1.8.4 Hollow Fibers 24

1.9   Novel Fiber Forms 25
   1.9.1   Microfibers 25
   1.9.2   Melt-Spinning Microfibers 25
1.10 World Markets and Future Prospects for Polyester Fibers 26
References 26

1.1 INTRODUCTION

Polyester fiber, specifically poly(ethylene terephthalate) (PET), is the largest volume synthetic fiber produced worldwide. The total volume produced in 2002 was 21 million metric tons or 58% of synthetic fiber production worldwide. The distribution of synthetic fiber production by chemistry is shown in Figure 1.1 [1].

If one assumes the total production is a single 5 denier per filament (dpf) (~20 μm in diameter) filament, the total length would be about 0.01 light years (~10^{14} m) or the equivalent of about one million trips to the moon. While other polyesters are commercially produced in fiber form—poly(ethylene naphthalate) (PEN); poly(butylene terephthalate) (PBT); poly(propylene terephthalate) (PPT); and poly(lactic acid) (PLA); thermotropic polyester (liquid crystalline polymer (LCP)—these are of insignificant volume compared to PET. Hence this chapter focuses primarily on PET.

The reasons for the dominating success of PET fiber are:

- Low cost
- Convenient processability
- Excellent and tailorable performance

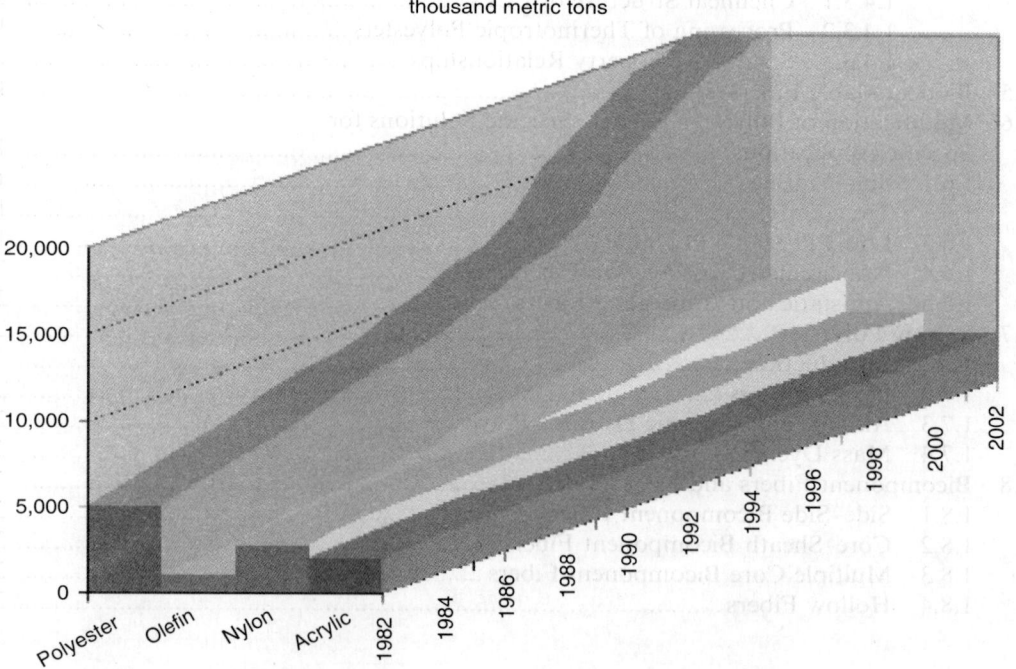

FIGURE 1.1 Worldwide fiber production.

The basis of the low cost lies in the high efficiency of the conversion of mixed xylenes to terephthalic acid, the melting temperature (280°C) being well within the range of commercial heating fluids, and the glass transition temperature (75°C), allowing the convenient stabilization of spinline- or drawline-introduced morphology and molecular orientation. The excellent performance results from the ability to accurately control fiber morphology (distribution and connectivity of crystalline and noncrystalline load-bearing units), allowing the balance of thermal and dimensional stabilities, transport, and mechanical properties to be controlled. Over the decades, since its introduction in the 1960s, polyester technology has evolved into a large number of products that range from cotton-blendable staple to high-performance tire cord. It is likely that PET will continue to dominate as the synthetic fiber of choice in future, although profitability has constantly eroded with time and production has shifted from the United States and Europe to Asia.

Polyester fibers have been reviewed in many publications [2–4], most recently by East [5], and the reader is directed to these publications for additional details. This work provides the reader with an overview of polyester fiber technology, sufficient to allow the vast and detailed open and patent literature related to polyester fibers generally, and PET fiber specifically, to become more meaningful.

1.2 PET HISTORY

The development of PET fiber began with the pioneering work on condensation polymers led by W.H. Carothers of DuPont in the 1930s [6].

Carothers focused on aliphatic polyesters and the resulting properties were poor compared to the aliphatic nylons that were simultaneously explored by his group. Much improved fiber performance was achieved in the early 1940s by the team comprising Whinfield and Dickson [7], Calico Printers Association Laboratory in Great Britain. Their work focused on aromatic–aliphatic polyesters from terephthalic acid (TA) and ethylene glycol. The same studies examined other aliphatic–aromatic polyester compositions, including PBT, PPT, and PEN. Commercialization of PET was rapid after World War II with the introduction of Terylene in Great Britain by ICI and the introduction of Dacron in the United States. Other products soon followed and PET successfully entered the textile market as both filament yarn and staple, and the industrial market as a rubber reinforcement filament yarn, primarily for use in the sidewalls of passenger car tires. Key properties were wash-and-wear characteristics in textiles and high modulus, coupled with excellent modulus retention, in industrial applications. The detailed review of Brown and Reinhart [8] described this history.

1.3 PET POLYMERIZATION

PET is the condensation product of terephthalic acid and ethylene glycol. The key to successful PET polymerization is monomer purity and the absence of moisture in the reaction vessel. PET polymerization has recently been reviewed in detail by East [9].

1.3.1 Monomer Production

The enabling technological breakthrough that allowed for the cost-effective polymerization of PET was the development of low-cost, pure TA from mixed xylenes by the Amoco company in the mid-20th century [10,11]. An alternative to TA, and the monomer of choice before the availability of low-cost TA, is dimethyl terephthalate (DMT). While direct esterification of TA is the preferred method of PET synthesis, ester interchange between DMT and ethylene glycol is still utilized in some PET manufacture, partially because of local choice and partially because DMT is a product of polyester recycling by methanolysis or glycolysis [12]. The second monomer, ethylene glycol, is a major material of commerce, produced by the oxidation of ethylene followed by ring opening with water [13]. The large-scale production of all PET monomers assures low-cost polymers and makes competition from new compositions of fiber-forming polymers very difficult.

1.3.2 Polymerization

The first stage of PET polymerization is, in essence, the production of bishydroxyethylterephthalate (BHET). In the direct esterification of TA, this reaction

$$HOOC-C_6H_4-COOH + 2HOCH_2CH_2OH$$
$$\rightarrow HOCH_2CH_2OCO-C_6H_4-COOCH_2CH_2OH + 2H_2O \qquad (1.1)$$

actually results in a mixture of low amounts of free BHET with a variety of PET oligomers. Water removal is critical to the ultimate achievement of high molecular weights. Similarly, in the first stage of the ester interchange process, BHET is formed along with a mixture of PET oligomers, i.e.,

$$CH_3OCO-C_6H_4-COOCH_3 + 2HOCH_2CH_2OH$$
$$\rightarrow HOCH_2CH_2OCO-C_6H_4-COOCH_2CH_2OH + 2CH_3OH \uparrow \qquad (1.2)$$

The reaction catalysts for the ester interchange reaction have been the subject of intense research for many years and many catalyst compositions are found in the patent literature [14–16]. The introduction of ester interchange catalysts requires the killing of these catalysts later in the polymerization sequence as they are equally effective as depolymerization catalysts.

The next step in the polymerization is the melt polymerization stage. In this reaction step, an ester interchange reaction occurs between two molecules of BHET to split off a molecule of glycol and build polymer molecular weight. The reaction must be catalyzed, and antimony trioxide (Sb_2O_3) is almost universally the moiety of choice. High vacuum is applied to push the reaction to high molecular weights. Typical melt polymerization temperatures are 285°C or higher, and viscosities are on the order of 3000 poise, making uniform stirring and the imparting of a constant shear history across the polymerization mixture difficult, although the power requirement to the stirrer thus becomes a useful QC tool. Recent variations of this method have been patented by DuPont (elimination of vacuum) [19–21] and Akzo (new, nonantimony-based catalyst) [17,18]. As neither DuPont nor Akzo has produced PET fiber in 2005, it is unclear whether these apparent process improvements are actually utilized.

After achieving molecular weight targets, the polymer may be extruded into strands and cut into chips for subsequent melt spinning (batch process) or fed directly into a spinning machine and converted to fiber (continuous process—CP spin-draw). In the case of chipped polymer, the molecular weight can be further increased through solid-state polymerization. In this process, thoroughly dried PET chip is first crystallized at about 160°C to prevent the amorphous as-polymerized chip from sticking together (sintering), and then heated just below the melting point under high vacuum and extreme dryness to advance the molecular weight upward to values of inherent viscosity (IV) of 0.95 (textile grade chip has an IV of about 0.65). [22,23]. The effects of the process thermal history of PET chip and fiber have been extensively studied and are conveniently monitored by thermal analysis techniques. Jaffe et al. [24] have reviewed the thermal behavior of PET and described the expected response of PET to process history in detail.

A variety of side reactions and end-group-induced reactions can lower the thermal stability and cause degradation of PET during spinning. The formation of diethylene glycol through the coupling of two hydroxyl ends from the glycol ends (or BHET ends) by dehydration, forming a diethyleneglycol (DEG) unit in the chain, is especially troublesome. DEG is a foreign unit in the backbone, although it does not directly affect the polymer chain length. This unit reduces crystallinity and lowers the glass transition, thermal stability, and hydrolytic stability of the polymer. It is impossible to completely eliminate DEG formation and about 1.0–1.5 mol% of DEG is always present. Depression of the polymer melting point is easily measured by differential scanning calorimetry (DSC), and this parameter provides an accurate measure of DEG content [24]. Finally, any melt-processed PET always has some cyclic trimer content, which, while not a direct problem for polymer performance, does tend to exude during processing and may cause process upsets.

In reality, commercially produced PET is always made by a continuous process involving a number of linked vessels between which the polymer is continuously pumped until the final product specifications are achieved. While some process descriptions have been published [25], most processing details are highly protected as proprietary information. The process usually involves at least four steps, i.e., an initial esterifier followed by a series of three polymerizers, each designed to further advance the polymer molecular weight. Extreme care is taken to promote within and between batch uniformity, eliminate dead zones where polymer may degrade, and remove all low molar mass reaction products such as glycol or water. A typical PET polymerization process is shown in Figure 1.2.

1.3.3 Characterization of Poly(ethylene Terephthalate) Chip

PET chip or representative samples of CP spin-draw polymer are conveniently characterized as by their molecular weight, cleanliness, and thermal behavior. Molecular weight is characterized by the polymer intrinsic viscosity [η], usually in halogenated solvents; the best halogenated solvents are hexafluoroisopropanol/pentafluorophenol mixtures. Intrinsic viscosity is related to molecular weight by the Mark-Houwink equation, i.e.,

$$[\eta] = KM_v^\alpha$$

where K and α are solvent-dependent, but K is about $1.5 \times 10^{-2} - 1 \times 10^{-1}$ and α is about 0.60–0.85 [26]. High molecular weight or high crystallinity can make polymer dissolution difficult and be responsible for erratic results. Polymer cleanliness is measured microscopically (optical techniques, polarized light microscopy) and is often expressed in units such as the number of black specks or the number of gels per gram of polymer. Acceptable values are determined empirically and are meaningful only in a known process context. Thermal parameters are conveniently monitored by DSC, allowing a quick assessment of DEG content, crystallinity, etc.

1.3.4 PET Processing—Melt Spinning

The melt spinning of PET has been extensively treated in the patent literature, but less in the open literature [27], although the recent chapters by Bessey and Jaffe [28] and Reese [2] are good introductions to the process. We will concentrate here on how changes in the key process variables of spinline stress and temperature profile affect assembly at the molecular level (morphology), and, in turn, how the morphology affects the resulting performance of the yarn. The relationships described here are equally valid for all semicrystalline polymers; LCPs will be treated separately. The average value of key properties and the standard deviation

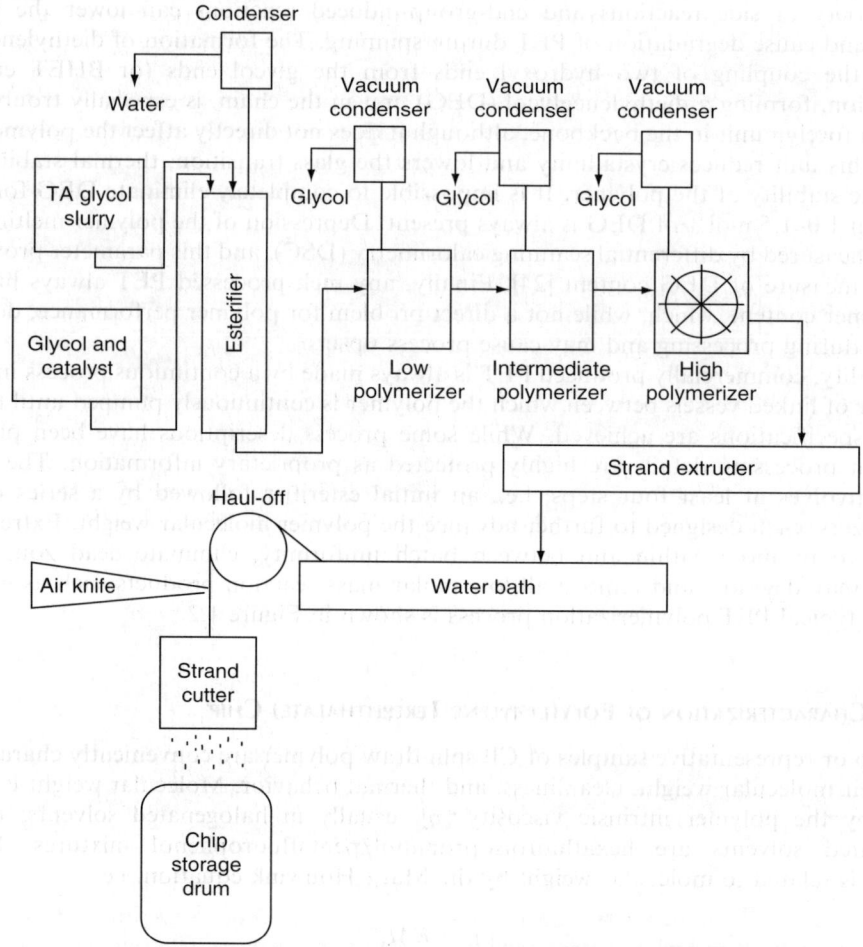

FIGURE 1.2 Typical PET production process.

associated with the mean value must be controlled for fiber products to have commercial value. In general, variation in properties, hence variation in morphology, must be controlled to about 10% for the yarn to be commercially acceptable. Variation means differences between filaments in a yarn or along a given filament. The frequency of variation is also critical; high frequency changes that may be averaged over a critical use length are, in general, more acceptable than a smaller variation along or between filaments that occurs at a lower frequency.

Polymer is introduced into the manifold of the spinning machine either as a dried chip or as produced by the CPU. The manifold may feed as few as one or as many as 200 separate spinnerets and is designed to keep the directed polymer streams as uniform as possible in shear and thermal history. The PET spinning temperature is typically between 280 and 300°C; local shear heating may increase this temperature by as much as 10–15°C. The molten polymer stream is then fed through metering pumps to the spinning pack (assembly that starts with a series of filters and ends at the spinneret—see Figure 1.3). The spinneret consists of five (hosiery yarn) to several thousand holes, typically ranging from 180 to 400 μm in

Polyester Fibers

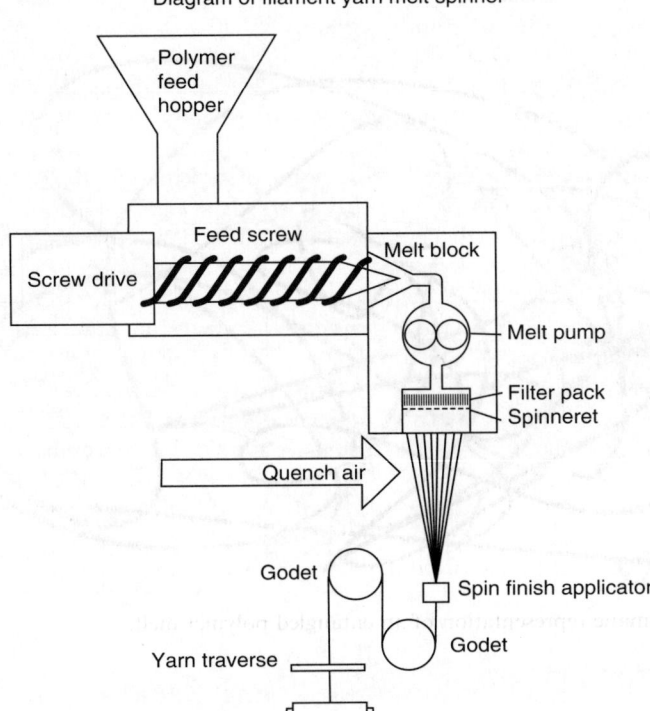

FIGURE 1.3 Key elements of polyester filament yarn melt-spinning machine.

diameter. Pack and spinneret designs are the subject of specialized expertise and the reader is referred to the open and patent literature for the depth of engineering detail available on these subjects [29]. The purpose of pack and spinneret is to insure that filtered (clean) polymer is fed to each hole of the spinneret as uniformly as possible. Passage through the spinneret subjects the polymer to a complex rheological environment (see, for example, the work of Denn [30]), resulting in local increases of molecular orientation and a distribution of orientation between the spinneret wall and center line. On exiting the spinneret, the combined effects of surface tension and relaxation of molecular orientation result in die swell (increase of the filament diameter to greater than the spinneret hole diameter).

From a molecular point of view, the starting polymer melt is best visualized as an entangled network, characterized by the polymer molecular weight, molecular weight distribution, the entanglement density, and the average chain length between entanglements. This is shown diagrammatically in Figure 1.4.

The processes that occur in the spinline, between the exit of the polymer from the spinneret and the point of stress isolation on the first godet or roller at the base of the spin line, involve the changing of this fluid network to the solid-state molecular chain topology of the filament. Within a distance of 3–5 m, and under the influence of an applied force (take-up tension) and quench media, at speeds in excess of 100 miles per hour—less than 0.01 sec residence time—the fiber is transformed from a fluid network to a highly interconnected semicrystalline morphology, characterized by the amount, size, shape, and net orientation

FIGURE 1.4 Diagramatic representation of an entangled polymer melt.

(with respect to the fiber or long axis) of crystalline units, and the orientation of spatial distribution of noncrystalline areas. All of these units are interconnected by molecules that traverse more than one local region (tie molecules) of the load-bearing elements of the fiber structure.

It has been noted [31] that the crystallization rate of polymers increases by up to six orders of magnitude when the crystallization event occurs when the polymer is under an applied stress rather than in a quiescent state. This large increase in crystallization rate is accompanied by a change in crystal habit, the shape of the crystalline phase produced transformed, over a narrow stress regime, from a spherulitic (spherically symmetrical) to a columnar habit (see Figure 1.5).

This transition is surprisingly sharp—occurs at a stress of about 0.1 g/d. Increasing the spinline stress increases the number of rows and decreases the diameter of the fibrillar structure. As the fibrils are stable only in the presence of the spinning stress, they may or may not be visible in the final fiber morphology. A useful way of conceptualizing the process is to divide the spinline into three regions, namely:

- Region 1. Increase local and global molecular orientation
- Region 2. Fibril formation at points of maximum orientation (transient mesogen, mechanical steady state)
- Region 3. Fibril decoration (folded chain crystal growth)

A cartoon of this model of morphology and molecular chain topology development in melt spinning of PET is shown in Figure 1.6.

In Region 1, the spinline stress leads to filament drawdown, causing a net increase of molecular orientation of the molten and amorphous polymers. A consequence of this stress is

Polyester Fibers

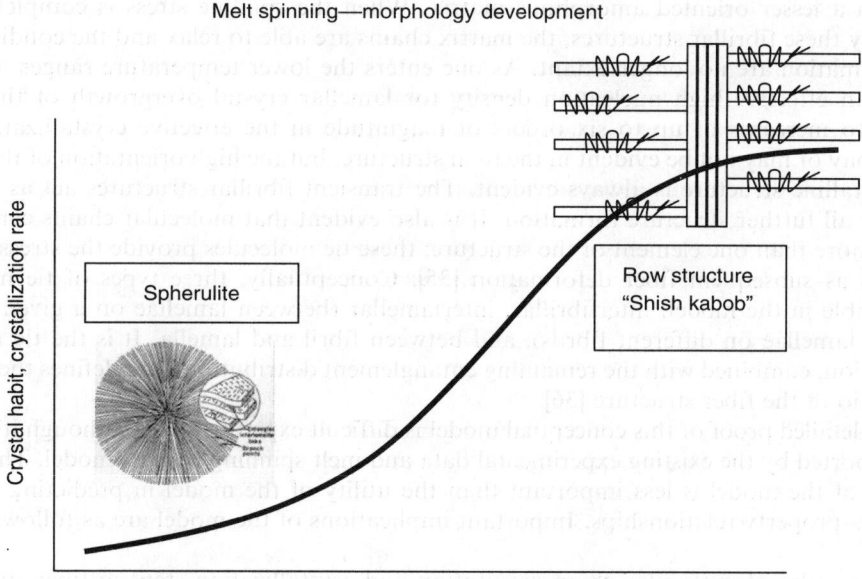

FIGURE 1.5 Morphology development in melt spinning as a function of key spinning parameters.

the disentangling of some of the starting network chains and the increase in the local molecular chain orientation in the proximity of remaining entanglements. As these bundles of locally oriented chains grow in aspect ratio, they satisfy the conditions for nematic phase formation [32–34], leading to a biphasic array comprised of fibrillar mesogenic structures

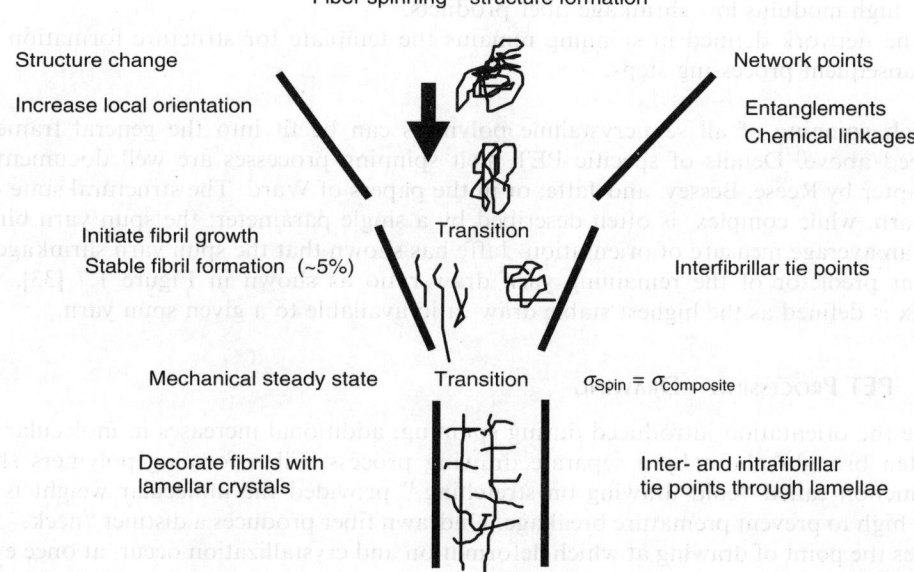

FIGURE 1.6 A cartoon of morphology development in PET melt spinning.

sitting in a lesser oriented amorphous matrix. When the spinline stress is completely supported by these fibrillar structures, the matrix chains are able to relax and the conditions for fibril formation are no longer extant. As one enters the lower temperature ranges, the fibril acts as an effective high nucleation density for lamellar crystal overgrowth of the fibrils, leading to increases of up to six orders of magnitude in the effective crystallization rate. Fibrils may or may not be evident in the final structure, but the high orientation of the wholly semicrystalline structure is always evident. The transient fibrillar structures act as the template for all further structure formation. It is also evident that molecular chains can participate in more than one element of the structure; these tie molecules provide the stress transfer elements as subsequent fiber deformation [35]. Conceptually, three types of tie molecules are possible in the model: interfibrillar, interlamellar (between lamellae on a given fibril or between lamellae on different fibrils), and between fibril and lamella. It is the tie molecule distribution, combined with the remaining entanglement distribution, that defines the residual draw ratio of the fiber structure [36].

The detailed proof of this conceptual model is difficult experimentally, although it is generally supported by the existing experimental data and melt spinning process model. The overall veracity of the model is less important than the utility of the model in predicting process–structure–property relationships. Important implications of the model are as follows:

- The order of molecular chain orientation and crystallization steps in fiber spinning is critical.
  - The formation of a transient fibrillar mesophase is the template for all further morphology development and defines the nucleation density for subsequent crystallization.
- As chain orientation prior to crystallization is increased, the load-bearing aspects of the crystalline network produced also increases, while the noncrystalline load-bearing elements of the structure decrease.
  - Leads to the decoupling of molecular orientation responsible for increased modulus and strength, from oriented chains responsible for entropic shrinkage, allowing for high modulus low shrinkage fiber products.
- The network defined in spinning remains the template for structure formation in all subsequent processing steps.

The melt spinning of all semicrystalline polymers can be fit into the general framework described above. Details of specific PET melt spinning processes are well documented in the chapter by Reese, Bessey, and Jaffe, or in the papers of Ward. The structural state of the spun yarn, while complex, is often described by a single parameter: the spun yarn birefringence, an average measure of orientation. Jaffe has shown that the spun yarn shrinkage is an excellent predictor of the remaining yarn draw ratio as shown in Figure 1.7 [33], where DRmax is defined as the highest stable draw ratio available to a given spun yarn.

1.3.5 PET Processing—Drawing

Despite the orientation introduced during spinning, additional increases in molecular order are often brought about by a separate drawing process. Fiber-forming polymers show a phenomenon called "cold drawing on stretching," provided the molecular weight is sufficiently high to prevent premature breakage. Undrawn fiber produces a distinct "neck," which localizes the point of drawing at which deformation and crystallization occur, at once evident from the change in opacity in the drawn filament due to its optical anisotropy. As-spun PET fibers can be amorphous or crystalline, depending on the spinning conditions (see Figure 1.5).

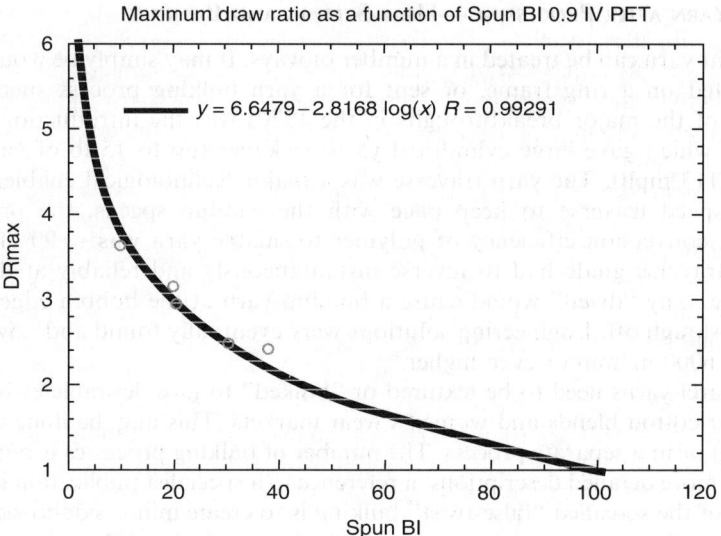

FIGURE 1.7 Variation of maximum draw ratio of high IV PET yarn as a function of molecular orientation imparted during spinning.

All fibers become more crystalline and better oriented when drawn. Faster crystallizing polymers like PBT, PPT, or nylon-6,6 always form crystalline spun fibers, although they often need a drawing stage to induce and complete crystallite orientation. The combination of molecular entanglements and the presence of polymer chain crystallites lock this orientation into place. This, in turn, affects such parameters as tenacity, modulus, elongation at break, and heat-shrinkage. The fiber must be drawn close to its maximum draw ratio for the drawing to be effective. Draw ratio is the ratio of yarn feed velocity to draw-roll haul-off speed: this ranges from about 1.5 to 6.0. The draw point is the actual place where fiber necking takes place and it must be stabilized. In early processes, this was done by a heated metal snubber pin around which the yarn was passed. The pin temperature was set to about 10°C above T_g, i.e., about 85–90°C for PET process. However, this alone was not sufficient and the drawn yarn had an unacceptable degree of heat shrinkage. The latter defect was prevented by heat-setting the fiber by passing it over a long hot plate at about 130–140°C, well above the effective T_g (~125°C) of the drawn, crystallized yarn. This simple system was adequate when draw speeds were low (500 m/min), but, as draw speeds rose considerably, it was necessary to use separately heated feed rolls and draw rolls to achieve the same effect at much higher speeds. The heated rolls allowed for longer yarn contact times for thermal transfer, with the yarn wrapped several times around the roll and over an attendant idler roll. The draw ratio has a major effect on yarn elongation and tenacity. High draw ratios give high-tenacity yarns with higher yarn moduli and lower extensions to break as expected; low draw ratios give lower tenacities with higher extensions. Jaffe [34], Ward [35], and others showed that a consequence of high-speed spinning is to shift the load supporting of the network chains of the fiber structure from noncrystalline to crystalline regions of the fiber morphology. This limits the draw ratio available to fully orient these fibers, resulting in fibers with nearly equivalent tensile properties, but significantly lower shrinkage at an elevated temperature.

1.3.6 PET Yarn after Processing—Heat-Setting and Bulking

Drawn filament yarn can be treated in a number of ways. It may simply be wound onto a yarn package, twisted on a ring frame, or sent for a yarn bulking process such as false-twist bulking. One of the major breakthroughs in the 1970s was the introduction of high-speed yarn winders, which gave large cylindrical yarn packages (up to 15 lb of yarn) and ran at 3000 m/min (113 mph). The yarn traverse was a major technological enabler, as without a reliable high-speed traverse to keep pace with the windup speeds, the process was not runnable (i.e., conversion efficiency of polymer to salable yarn was <~90%). The problem was that the traverse guide had to reverse instantaneously and reliably at the end of each traverse stroke. Any "dwell" would cause a buildup yarn at the bobbin edges and the yarn would simply slough off. Engineering solutions were eventually found and nowadays windup speeds can be 6000 m/min or even higher.

Many apparel yarns need to be textured or "bulked" to give desirable esthetic properties, particularly for cotton blends and women's wear markets. This may be done during drawing (draw-bulking) or in a separate process. The number of bulking processes is numerous and for those wanting more detailed descriptions, a reference to a specialist publication is provided [37]. The principle of the so-called "false-twist" bulking is to create minor side-to-side variations in molecular orientation across a given yarn, causing the yarn to bend during controlled thermal shrinkage to create a 3D structure with a bulky feel. The process entails running a continuous yarn through a device that twists it in the middle. Since no net twist is applied, it is called a "false" twist; the yarn ahead of the machine is wound up and the false twist escapes, but the yarn behind the twister passes through a long tube heated above fiber T_g, so that, as it exits, the false twist is "set" into the yarn. When this twist tries to spring back and unwind, it causes the treated yarn to bulk up into a spiral crimp. The degree of twist is quite high, several hundred twists per meter, so that, if the yarn is running at productive speeds, the rotation of the twister device has to be extremely high, of the order of 1 million rpm. This produces formidable mechanical problems. One ingenious solution is the friction-bulking process, in which the yarn itself is twisted either by running against the internal surface of a rotating friction bush or by contact with the edges of a series of friction disks. Since the yarn diameter is very small compared to that of the bush or the disk, a very high "gear-up" ratio is achieved and the friction device can rotate at far more reasonable speeds. A typical texturing process is shown in Figure 1.8.

Bulked continuous filament (BCF) carpet yarns are heavy decitex bundles of fiber that are bulked by passage through a turbulent blast of steam or hot air well above T_g. The turbulence blows the yarn about and entangles the filaments, and then heat sets them into place, giving them a permanent crimp. Polymers like PET do not have very good resilience as carpet fibers, but PTT ($T_g = 45°C$) lends itself very well to the BCF process and has excellent resilience [38].

1.3.7 Polyester Yarns for Specific Applications

For industrial use, high-tenacity yarns, such as the tire cord, have to be drawn under conditions where low heat shrinkage, low extension, and high modulus products are produced. In fact, a tire cord is a highly specialized product, and complete integrated continuous polymerization spinning and drawing plants (cp-spin-draw) have been developed. The process is little discussed in the open literature and the reader is directed to the patents of DuPont, Fiber Industries, and Allied Chemical Corporation (none of these companies currently exist as fiber producers).

The demands of staple fiber are different from those of filament yarns. Staple fiber is a continuous filament cut into short lengths in centimeters. Staple fibers are discontinuous and are crimped and chopped to the desired staple fiber length to blend at the carding stage with cotton (short staple), wool (long staple), or other natural fibers. The raw polyester fibers are

Polyester Fibers

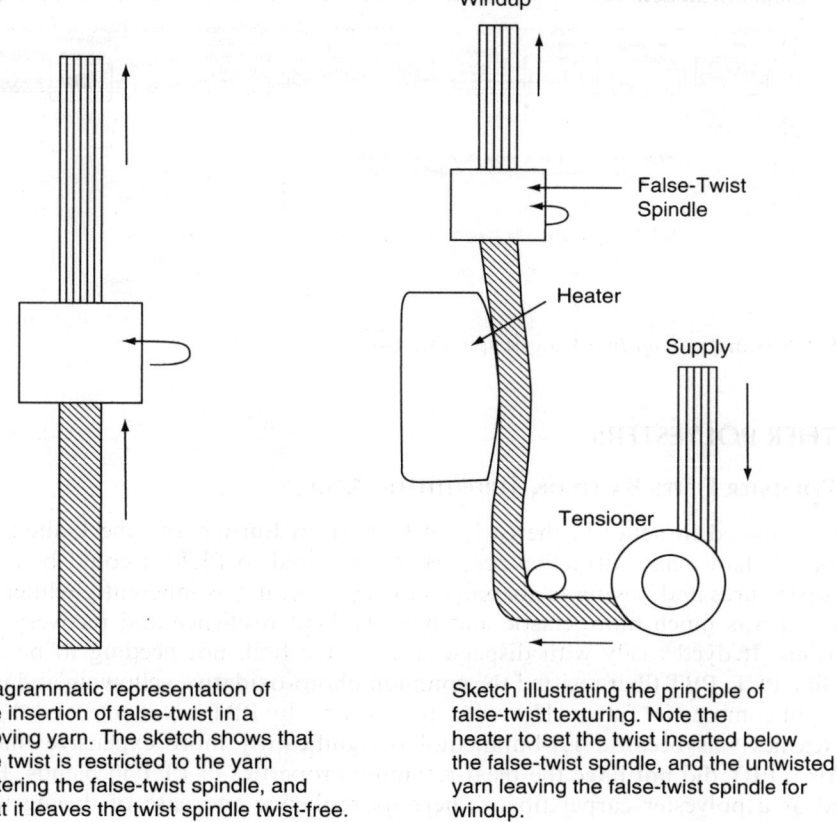

Diagrammatic representation of the insertion of false-twist in a moving yarn. The sketch shows that the twist is restricted to the yarn entering the false-twist spindle, and that it leaves the twist spindle twist-free.

Sketch illustrating the principle of false-twist texturing. Note the heater to set the twist inserted below the false-twist spindle, and the untwisted yarn leaving the false-twist spindle for windup.

FIGURE 1.8 Typical polyester yarn bulking process, false-twist texturing.

melt-spun through many hundreds of holes in a staple spinneret, and hauled off via a godet, but not wound up in the conventional way. Instead, they are deposited via an air ejector loosely into a large drum or "yarn can." When the yarn can is full, fiber bundles from many cans are combined into a thick bundle of fibers called a "tow." These tows may have a yarn count of several million decitex. The thick bundle of fibers is then drawn on a massively constructed drawframe (because the mechanical forces involved are heavy) using multiple sets of draw rolls and feed rolls. The yarn is heat-set in a steam-heated hot box, because this method gives the best thermal transfer to the individual fibers in the tow. The drawn tow then passes to a crimper, often of the stuffer-box type. The tow is overfed into a heated wedge-shaped box with a sprung lid, where it is compressed to form a concertina-type crimp. The bulked tow is finally cut to the desired staple length using a continuous staple cutter. The loose cut fiber is then transferred by an air handler to a bin and compressed into bales. A schematic diagram of a staple line is shown in Figure 1.9.

1.3.8 Physical Properties of PET

PET is a semicrystalline polymer and its physical parameters have been repeatedly determined over many years. The summary of the most recent widely accepted values [39] is shown in Table 1.1.

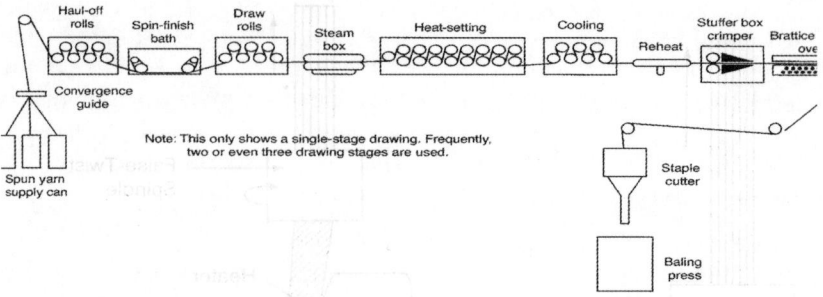

FIGURE 1.9 Schematic diagram of staple spinning line.

1.4 OTHER POLYESTERS

1.4.1 Polyester Fibers Based on Terephthalic Acid

PBT was examined in detail in the early 1950s both in Europe and the United States as a textile fiber. It had many attractive properties compared to PET; it could be melt-spun at lower temperatures and, owing to its polymer chemistry, it was inherently whiter than PET. As a fiber, it was much more elastic and had excellent resilience and recovery from small deformations. It dyed easily with disperse dyes at the boil, not needing to be dyed under pressure like PET. PBT fiber resisted the common photo-oxidative yellowing and it seemed to have a bright commercial future. However, the reason why PBT never achieved the success of PET in textiles was because 1,4-butanediol is significantly more expensive than ethylene glycol. Also, PBT did not have the pleat-retaining properties of PET in blends. However, it succeeded as a polyester carpet fiber, where its resilience and ease of dyeing were assets, although it had to compete against nylon.

Another of the pioneer polyesters was polytrimethylene terephthalate (PTT). This was recognized very early on as a fiber with outstanding resilience. PTT has been known in many ways as an ideal textile fiber for over 60 years. It remained on the shelf until, in the last decade, it became a commercial product owing to two new routes to the crucial intermediate 1,3-propanediol. One route is petrochemically derived (hydroformylation of ethylene oxide), while the other is a fermentation route using corn sugar to make 1,3-propanediol directly using genetically modified bacteria [40].

TABLE 1.1

Crystal habit	Triclinic: one polymer chain per unit cell
Cell parameters	$a = 0.444$ nm; $b = 0.591$ nm; $c = 1.067$ nm, $\alpha = 100°$; $\beta = 117°$; $\gamma = 112°$
Cell density	1.52 g/cm^3
T_m (DSC)	260–265°C
ΔH_f	140 J/g; 33.5 cal/g
T_g (solid chip)	78°C (DSC)
T_g (drawn fiber)	120°C (dynamic loss)
Specific gravity	1.33 (amorphous, undrawn), 1.39 (crystalline drawn fiber)

Eastman Kodak introduced a new polyester as a staple fiber called "Kodel" in 1958. A new diol was introduced to derive a patent-free composition of matter; a mixture of *cis-* and *trans-*1,4- cyclohexanedimethanol made by the exhaustive hydrogenation of dimethyl terephthalate. This polyester had a higher T_g than PET and also a higher melting point, but it was successful enough to find a market. In recent years, the polyester has found use in polyester carpets [41,42].

1.4.2 High-Performance Polyester Fibers—PEN and LCPs

The polyester derived from ethylene glycol and naphthalene-2,6-dicarboxylic acid was first discovered by ICI in the late 1940s [43]. It has a much higher T_g than PET and gives strong, high-modulus fibers, but the inaccessibility of the diacid was an insurmountable problem until recently. Now, firms like Amoco (now Solvay) are able to supply the dimethyl ester of 2,6-NDA, and the polymer (PEN) is increasingly used in high-performance polyester films and for high-softening-point blow-molded bottles and containers. Recently, Honeywell have started producing a high-performance PEN fiber under the name PENTEX. This absorbs UV light owing to the naphthalene ring. In Japan, stretch-blow-molded PEN bottles are used to package vitamins and baby food, which would otherwise be adversely affected by UV light.

1.4.3 Fibers from Main-Chain Thermotropic Polyesters—LCPs

It was recognized early in the development of polymer science that the tensile modulus of polymers should correlate with both the chemical and physical structures, and that maximum property levels would be achieved when all the molecular chain backbone bonds were lined up in the direction of measurement [44,45]. The all-aromatic main chain thermotropic polyesters are semirodlike molecules that naturally organize into nematic liquid crystal domains and many variants are commercially available in resin and fiber form (see literature and websites of Ticona and DuPont). The nematic state can be viewed as similar to "logs floating in a river," leading to ease of flow parallel to the molecular axis (low elongational viscosity) and an extended chain structure in the solid state. Hence, the nematic state in polymers brings both processing ease and high axial tensile properties. Figure 1.10 illustrates the processing of LCPs and the morphology produced, in contrast to conventional polymers such as PET or nylon.

All of the LCPs are composed of stiff, highly aromatic molecules and are characterized by a very high local molecular orientation in the solid state (orientation function >0.95). If processed into fibers, the local orientation is transformed to global. These globally oriented LCP fibers are further characterized by very high specific tensile properties and intrinsically low density when compared with metals, ceramics, and carbon. The highly anisotropic nature of these oriented LCP fibers, causing inherent weakness in shear and compression, limiting their use almost exclusively to applications in tension, is not shown in Figure 1.10. It should also be noted that, in the absence of global orientation, the tensile properties of the thermotropic polymers are similar to filled plastics, and compressive behavior is less of a critical issue.

1.4.3.1 Chemical Structure of LCPs

Thermotropic polyester backbone chemistry is characterized by a high degree of aromaticity, planarity, and linearity in the chain backbone. Most common moieties are *p*-phenylene, 1,4-biphenyl, and 2,6-naphthalyl moieties linked by ester or amide linkages. Polymers that form liquid crystal phases in the melt are thermotropic, whereas those that form liquid crystalline phases in solution are lyotropic. The all-aromatic polyester homopolymers tend to be intractable, decomposing at temperatures well below their melting points and insoluble in most

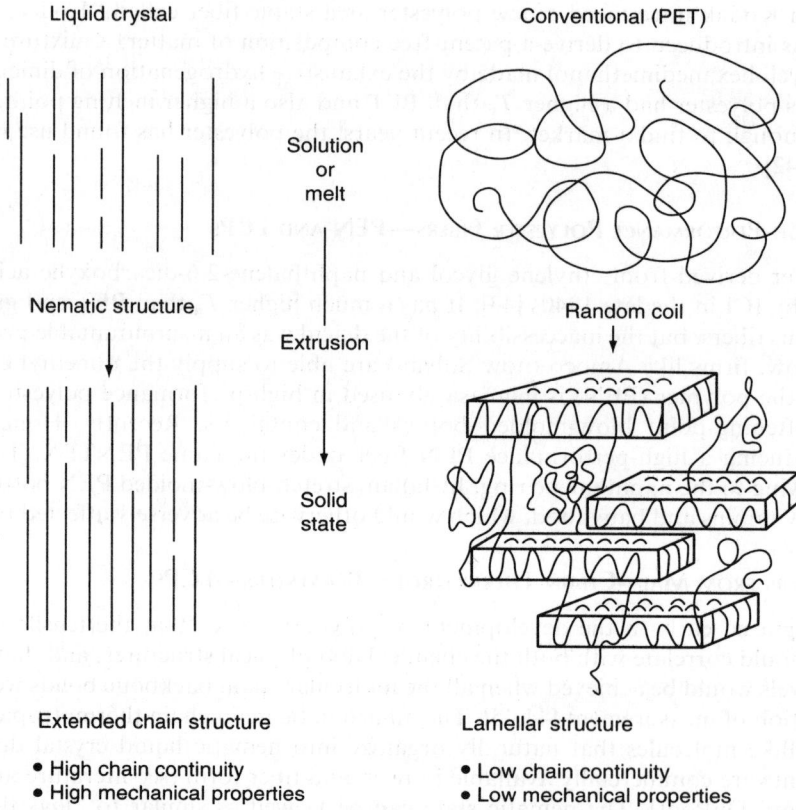

FIGURE 1.10 Structure development during spinning LCP versus PET.

solvents. Successful melting point reduction strategies include incorporation of comonomers to lessen crystal packing, decrease chain linearity, and increase chain-to-chain distance. All these approaches lower the polymer melting point and, when the melting temperature is reduced to below the polymer decomposition temperature, stable melt processing is possible. These approaches have led to large numbers of melt processable thermotropic polyesters. Typical LCP monomer and polymer chemistries of industrial importance are shown in Figure 1.11. Much of the cost of LCP fibers is the result of high monomer cost and limited monomer availability.

1.4.3.2 Processing of Thermotropic Polyesters

Thermotropic polyesters are melt-spun from the nematic phase and orient easily in an elongational flow field (moderate drawdowns/forces are sufficient). In the fiber case, highly oriented fibers form easily with an initial modulus close to theory—typical values range from about 70 to 150 GPa. Ward [46] has shown that the tensile modulus may be described by an "aggregate model," i.e., the modulus is a function of the inherent chain modulus, the molecular chain orientation, and the shear modulus (which described the stress transfer between chains). The tensile strength of LCP fibers follows the prediction of the "lag-shear model" [47]. Both the aggregate model and the lag shear model treat the LCP as though it

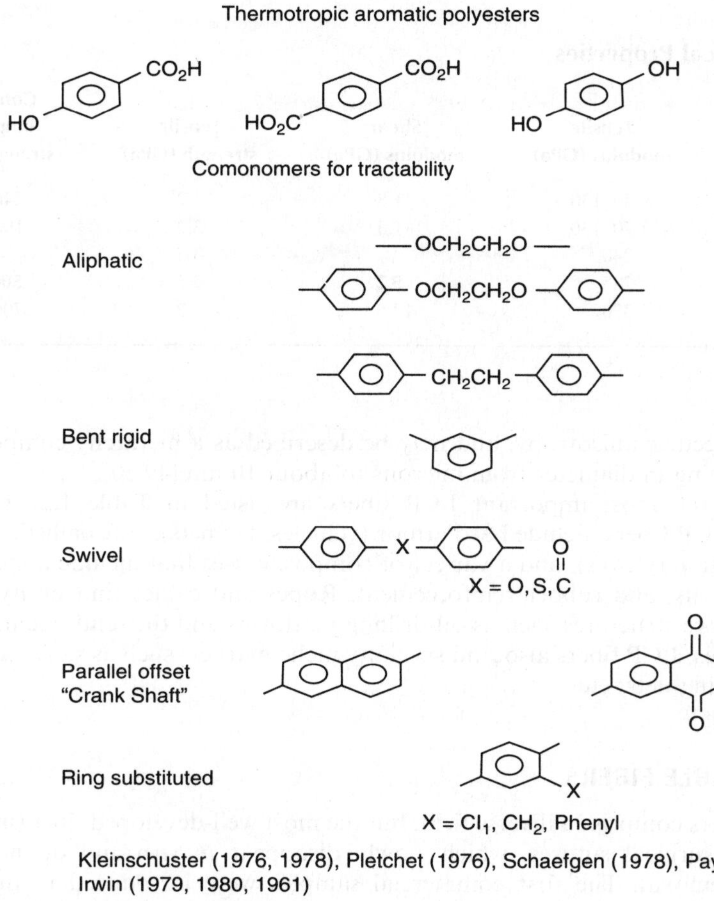

Thermotropic aromatic polyesters

HO—⌬—CO₂H   HO₂C—⌬—CO₂H   HO—⌬—OH

Comonomers for tractability

Aliphatic: —OCH₂CH₂O—
—⌬—OCH₂CH₂O—⌬—
[S ring]
—⌬—CH₂CH₂—⌬—

Bent rigid: [meta-phenylene]

Swivel: —⌬—X—⌬—   X = O, S, C (with C=O)

Parallel offset "Crank Shaft": [naphthalene], [anthraquinone-like]

Ring substituted: [substituted phenyl]—X   X = Cl₁, CH₂, Phenyl

Kleinschuster (1976, 1978), Pletchet (1976), Schaefgen (1978), Payet (1972), Irwin (1979, 1980, 1961)

FIGURE 1.11 Typical LCP monomers.

were a self-reinforced short-fiber-reinforced composite. As-spun tensile strength of the thermotropic copolyesters tends to be on the order of about 1 GPa, and can be advanced to about 3 GPa by annealing free to shrink close to the melting temperature. Kinetics of strength improvement follow those of solid-state polymerization, leading many researchers to associate strength increases with molecular weight increase [48]. The failure of strength and elongation to increase in tandem suggests a mechanism of flaw reduction. Structural perfection and improved intermolecular bonding also play a role in the observed property improvement. Modulus increase during annealing is usually minimal with the thermotropic polyesters, unless structural perfection leads to an increase in overall molecular chain orientation.

1.4.3.3 Structure–Property Relationships

The unifying feature of all fibers spun from LCPs is the very high axial molecular orientation, which leads to extreme anisotropy of microstructure and mechanical properties. In the transverse direction, the strength is only about 20% of the axial strength and the modulus is typically less than 10% of the axial value. The microstructure of LCP fibers reflects the very

TABLE 1.2
Typical Fiber Mechanical Properties

Fiber	Tensile modulus (GPa)	Shear modulus (GPa)	Tensile strength (GPa)	Composite compression strength (GPa)
p-Aramid	70–130	1.8	3.2	240–290
Thermotropic copolyester	70–130	1.3	3.2	100–200
PBO	240		3.5	
M5	285	5.2	3.5	500
Carbon-HM370	370	17.5	2.2	700–900

high orientational molecular anisotropy, and may be described as a hierarchy composed of fibrillar structures ranging in diameter from microns to about 10 nm [49,50].

The properties of the most important LCP fibers are listed in Table 1.2. The key application areas for LCP fibers include hard armor (vehicles, helmets), soft ballistic protection (vests), cut protection (gloves), and a variety of composite uses that include honeycomb structure, pressure vessels, and rubber reinforcement. Ropes and cables find utility in the mooring of huge offshore structures such as oil-drilling platforms and the reinforcement and support of optical cables. LCP fibers also find specialty niche markets such as sails for racing yachts, specialized fishing nets, etc.

1.5 BIODEGRADABLE FIBERS

Biodegradable polyesters comprise a diverse field, but the most well-developed fiber (monofil) market is resorbable surgical sutures, which slowly disappear *in vivo* and do not need subsequent surgical removal. The first commercial samples were introduced in the early 1970s by Ethicon Corporation [51]. These sutures were monofil fibers spun from a copolymer of glycolic acid and D-lactic acid. Such aliphatic hydroxy acids are completely biocompatible and harmless: in U.S. Food and Drug Administration (FDA) terms, these materials are "generally recognized as safe (GRAS)." The properties of polyglycolide and stereochemically pure D- or L-polylactide polymers are quite good, and they form strong, highly crystalline fibers by melt spinning. Other biodegradable polyester fibers have been explored. Synthetic lactones such as ε-caprolactone and 2-dioxanone have been copolymerized with glycolide and lactide [52,53]. ICI began working on poly (3-hydroxybutyric acid) in the 1970s and later developed a copolymer with 3-hydroxyvaleric acid. Both polyhydroxyacids are stereochemically pure and give crystalline polymers, which can be processed into fibers and films. The interesting feature of these polymers is that they are made in very high molecular weight form by bacteria. Certain microorganisms, when cultivated and starved of nitrogen sources, synthesize aliphatic polyesters instead of proteins. The number average molecular weight of the as-harvested polymer can be several million daltons and it must be reduced to allow the polymer to be processed and fabricated. ICI (now Astra-Zeneca) first developed "Biopol" as one product and although others have been introduced by different companies, little has been targeted towards fiber end-use [54]. All the polyhydroxyacids are unstable and degrade on exposure or composting, but the degradation rate is very much governed by the ratio of hydrophobic/hydrophilic properties. While hydrolysis is important, catalyzed degradation by various lipases is also a factor.

Polyester Fibers

1.6 MODIFICATION OF POLYESTER FIBERS—SPECIFIC SOLUTIONS FOR SPECIFIC APPLICATIONS

This wide topic covers both chemical and physical modifications to both the polymer and the fiber. We shall deal with only a few of the more important variations possible on this theme, but all are based on an understanding of polyester chemistry and processing described earlier in this chapter.

1.6.1 Spin Finishes

Fibers need to be treated with surface finishes or lubricants to allow high-speed processing. The various processing steps such as drawing, bulking, and textile processing would be impossible without these spin finishes because so many of them rely on specific frictional properties of the fiber (for example, friction twisting). Spin finishes are often water emulsions of various surface-active agents and lubricant oils; their formulation is a complex process and sometimes more of an art as well as a science. Finish application is made early in the process, before the cooling threadline from the spinner hits the first godet. Earlier, finish was applied from a lick roll rotating slowly in a bath. As spinning speeds increased, the finish was applied directly via a special hollow ceramic yarn guide as a neat oil formulation and metered at precise levels via a metering pump. Staple fiber is sprayed with emulsified finish or the whole tow may be immersed in large baths of finish. Some staple processes use a draw stage in a hot bath of finish.

1.6.2 Tire Cord

During the manufacture of tires (typically radial ply construction for passenger cars), the polyester tire core is subjected to drastic hydrolytic conditions. The rubber is molded into the basic tire shape and rubber vulcanization uses various accelerators, some of which cause severe aminolysis of the polyester chain. The process is run at 175°C in the presence of steam. While PET is fairly resistant to strong aqueous ionic base at moderate temperatures, nonpolar bases like ammonia, hydrazine, and simple aliphatic amines can easily diffuse into the PET structure and cause aminolytic breakdown [55].

To maintain the high strength engineered into the tire cord, it is essential that the IV (molecular weight) drop be minimized. The rate of degradation of the tire cord is directly related to the level of free COOH groups on the chain ends. This reaction is autocatalytic under vulcanization conditions, and reduction from their usual level (about 40 micro equivalents per gram of fiber) improves in-rubber stability. For some years, tire cord manufacturers employed a process in which the yarn was treated with epoxy compounds such as phenyl–glycidyl ether to esterify excess COOH end-groups [56]. This process was convenient because tire cords were treated with various "activating finishes" to improve their rubber adhesion. However, the glycidyl ethers were carcinogens and the process was abandoned in favor of the drastic alternative of melt-injecting ethylene oxide gas under high pressure into the molten polymer during the last stages of polymerization [57]. This reduced the free COOH end-group concentration to about 4–10 μe/g by forming harmless BHET ends, and IV drop at tire molding was significantly reduced.

1.6.3 Low-Pill Staple Polyester

PET staple blends with wool and cotton were highly successful from the very first introduction of PET in the 1950s. However, consumers soon noticed an annoying problem. It was the formation of small fuzzy balls (called "pills") on the surface of fabrics. This phenomenon is known as "pilling" and it is common to all staple fibers, particularly if the level of yarn twist is low, so that the fiber has many loose ends. The pills rub off harmlessly with wool because wool is a weak fiber. However, PET is a strong fiber (tenacity ca. 5 g/decitex) and therefore pills do not rub off; instead, they cling and have a negative impact on fabric esthetics. To reduce pilling, the IV of the polyester is reduced to make weaker fibers. These do not pill so obviously because the pills break away. A polymer of IV = 0.42 was selected as the best compromise for a low-pill PET staple fiber, but it caused many problems. The melt viscosity was so low and the molten polymer so fluid that the process became unstable. A method had to be found to raise the effective melt viscosity of the polymer while maintaining the low-pill properties to give an acceptable melt-spinning process. The method adopted was to introduce branching points into the polymer chain by adding a multifunctional component (either a polyacid or a polyhydric alcohol) so as to produce a star-branched polymer. Such polymers are known to have higher melt viscosities for the same (nominal) polymer IV. The branching agent added (ca. 1 mol%) was usually pentaerythritol. Too much additive would lead to gel formation by forming cross-linked networks, but this is not a problem at low levels [2].

1.6.4 Noncircular Cross-Section Fibers

Synthetic fibers like PET and nylon are normally round in cross section, however no natural fiber has a circular cross section. Wool is irregular, cotton is "dogbone" shaped, and silk is triangular. In the early 1970s, people began to study the effect of noncircular cross-section (NCCS) fibers on yarn and fabric esthetics, which is a subjective topic involving such arcane terms as "feel," "drape," and "handle." Fortunately, a melt-spun fiber lends itself NCCS well to the production of (NCCS) fibers by varying the shape of spinneret orifice, provided the melt viscosity is high enough so that surface tension does not cause the filament to resume a circular shape. Since the holes had to be very small (about 0.015 in. overall), machining a multiplicity of holes at a uniform size and shape was a major engineering problem, particularly in the hard metal alloys used for spinneret plates. Laser etching is one technique used. A hole shaped like a T gave trilobal filaments. In the pioneering days, much of this work was entirely heuristic, but gradually emerged some rules of thumb. Multilobed yarn cross sections (trilobal and octalobal) can give quite different appearances. Trilobal is glittery as the incident light reflects off the fiber surface, while octalobal gives an opaque matte effect, as the light is effectively absorbed by multiple reflections from the many acute angles. Sharp-edged filaments have the prized rustle and high frictional characteristics of pure silk, where it is called "scroop." Flat rectangular filaments give fabrics an unpleasant "slimy" handle. Gradually, these principles were applied to commercial yarns, and many filament yarns for the apparel and BCF carpet markets now use NCCS fibers.

1.6.5 Antistatic and Antisoiling Fibers

These topics are related because the origins of the problems are interrelated. Synthetic fibers in general, and PET in particular, are hydrophobic materials—PET has a moisture regain of 0.4% at 60% RH. PET fibers are difficult to wet and rapidly build up static electrical charges by friction because as water effectively leaks away, voltage is produced. It is possible to build up potentials as high as 50 kV by rubbing a polyester fabric, e.g., by walking on a polyester carpet when the relative humidity is low (5%). Such a potential, discharged by grasping a grounded door handle, would give a very unpleasant electric shock. Static charges also lead to attraction of dust and dirt.

To avoid these problems, the moisture uptake of the polyester should be increased by combining it with hydrophilic materials that are wash-fast. One additive that has been used repeatedly is polyethylene oxide (PEG), a stable, functional, highly hydrophilic, water-soluble, and humectant polymer (see below):

$$HO-CH_2CH_2O[CH_2CH_2O-]_n-H$$

The MW can be from a few hundred to many millions. Copolymers of PET with PEG having a molecular weight of approximately 500–2000 Da were made, and it was possible to incorporate permanently enough PEG without drastic reduction of the PET properties to greatly improve the fiber moisture uptake, but at the expense of severe reduction in the light stability of dyed fibers [58]. Other processes used a PET/PEG block copolymer in aqueous dispersion that was padded and baked onto the fiber as a textile finish. This relied on cocrystallization of the PET segments with the polymer to make the treatment wash-fast [59]. The most satisfactory technique is probably to make a bicomponent fiber with a thin coating of a PET/PEG copolymer on a PET core in a core–sheath configuration. This does not affect fiber properties and minimizes the light fastness issue [60].

1.7 DYEING POLYESTERS

1.7.1 INTRODUCTION

Dyeing synthetic fibers is a huge subject in its own right and the reader is advised to consult one of the many publications that deal with it comprehensively [61]. When PET fibers first appeared, they presented many problems for traditional dyers. PET has no functional groups to give affinity for usual dyestuffs. Natural fibers like wool, cotton, silk, and then later man-made ones like rayon and nylon were well known and had good dye affinities because the fibers had pendant or terminal functional chemical groups such as $-NH_2$, $-COOH$, and $-OH$. These dyes were developed to interact with such groups. The only way to dye polyester was to rely on Van der Waals forces to hold the dye in the fiber. All classic cationic and anionic dyes for wool and silk or direct dyes for cotton had water-solubilizing ionic groups like $-NR_3^+$ and $-SO_3^-$. Such dyes had little or no affinity for PET.

1.7.2 DISPERSE DYES

PET fiber chemistry is in some ways similar to that of cellulose acetate fibers, where the class of dye called "dispersed dyes" were in use. These dyes did not have strongly polar solubilizing groups and were actually dispersed in the aqueous dyebath with a surfactant as a suspension of fine particles in suspension. Such dyes usually had a low molecular weight and this later led to problems with PET due to dye sublimation. Polyester fabrics needed stentering (heat-setting under tension) on a pin-frame to remove creases after dyeing. This became a big problem for PET dyers. It was clear that special dyes were needed for PET. As the polyester fiber market grew, such modified dyes rapidly advanced, and were based on well-understood dye chemistry. Higher molecular weight dyes of the anthraquinone type gave reds, blues, and dark greens, while selected azos were used for yellow and orange shades. These dyes were most effective if they were somewhat water-soluble and the ethanolamino group ($-NRCH_2CH_2OH$) and sometimes its O-benzenesulfonate ester were incorporated, giving a weakly polar nature and bestowing solubility in polyester but without a truly ionic character. The higher molecular weight dyes reduced dye sublimation, but at the cost of slower dyeing and poor dye exhaustions. A breakthrough was achieved in the middle of the 1950s with the development of heterocyclic (nitroaminothiazole) dyes, which gave very stable light-fast azo blues [62]. These had good affinities for PET. Another dye problem that became quite important was gas-fume fading of disperse dyes due to the generation of oxides of nitrogen (NO_x) and even traces of ozone in living rooms, arising from wider use of oil and gas heating systems. It was solved largely by selecting dyes whose chromophores were stable to oxidation by NO_x.

Carriers were introduced to speed up dyeing. These were solubilizing agents that temporarily swelled the fiber and "carried" the dye into the fibrous structure. The carrier was trapped in the amorphous regions of the fiber morphology, since the dense crystalline regions could not be penetrated by the large dye molecules. The carrier itself diffused out again, so it might be regarded as a fugitive plasticizer. Phenols like 2-hydroxybiphenyl (OPP) were widely used and greatly improved the economics of dyeing polyester. An alternative was pressure dyeing, using superheated dye liquor at 135–150°C (well above the T_g of drawn PET fiber), but this was a capital-intensive process since pressure-dyeing vessels were expensive. Eventually, pollution problems with dyehouse liquor waste led to restrictions on the use of carriers. Pressure dyeing is now the norm, although more expensive. This is one reason why non-PET polyester fibers like PBT and PTT are attractive to the dyer. Both have T_gs of about 45°C, so they can be aqueously dyed to heavy shades at the boil at atmospheric pressure. The reluctance of polyester to dye can be turned into a commercial advantage. The argument is that "a fiber that does not readily dye will also not easily stain."

1.7.3 Anionic and Cationic Dyes for Polyester

Since much polyester was originally used in blends with wool, it was natural that attempts should be made to modify PET to make it acid-dyeable with anionic dyes. The most popular theme was to incorporate basic additives by copolymerizing an aminohydroxy compound or aminoacid into the PET structure. All such attempts failed because the copolymers were discolored yellow or brown, and were of low IV. It was found, however, that certain polyamides, containing additional in-chain tertiary amine groups, when melt blended with PET and high-molecular-weight PEG (M_w 20,000) formed a three-phase mixture in which the polyamide was dispersed inside the PEG and this in turn was dispersed inside the PET. Thus, the critical components were prevented from intermixing in the melt. The mixture was melt-spun successfully into fibers at 270°C. Diamond-patterned fabrics were jacquard knitted with mixtures of the dye variant fiber and normal PET. These could be cross dyed to give patterned effects from a single dyebath containing both acid and disperse dyes. However, the process was deemed too complex and expensive for a commercial product and the light stability of the dyes was not adequate [63].

Greater success was achieved by DuPont who copolymerized, the sodium salt of 5-sulfoisophthalic acid into PET to render the polymer dyeable with cationic (basic) dyes. Basic dyeable PET was successfully launched as Dacron 64 in the form of a low-pill staple product [64]. The presence of the sulfonate groups in the polymer chain also acts as an ionic dipolar cross-link and increases the melt viscosity of the polymer quite markedly. Thus, it is possible to melt-spin polymer with IV 0.56 under normal conditions, giving a low-pill fiber variant. The fiber also has a greater affinity for disperse dyes due to the disruption of the PET structure. Continuing this theme, there are "deep dye" variant PET fibers, often used in PET carpet yarns, which are copolymers of PET with chain-disrupting copolymer units like polyethylene adipate. They have less crystallinity and a lower T_g; therefore, they may be dyed at the boil without the use of pressure equipment or carrier at the cost of some loss of fiber physical properties.

1.7.4 Mass Dyeing

Since much polyester staple fiber is dyed to dark, expensive colors (black and navy blue), the fiber is often mass dyed or mass pigmented at the polymerization stage. Clearly, thermally stable pigments and dyes have to be used. Especially fine pigment grades of carbon black are used for black, and this is toned by adding small amounts of navy blue or very dark green melt dyes to remove any traces of brown, which dyers consider unacceptable. Mass dyeing is only economic if the demand warrants it.

A more recent development is "dope dyeing" (a term dating back to the acetate rayon industry), where a range of melt-dyed colors are produced by coloring white polymer immediately before melt spinning by adding calculated mixtures of master-batch pigmented polymer or actual neat dyestuff. This can conveniently be done by adding the dye in the form of pills or granules containing a specific amount of dye at a calculated feed rate to the molten polymer during the melt-spinning process or by adding the coloring agent as a liquid dispersion in a very high boiling point (over 300°C) inert oil, either during polymerization or at melt spinning [65]. The latter process is of particular value in melt coloration of POY feedstock yarns.

1.8 BICOMPONENT FIBERS AND MICROFIBERS

Bicomponent fibers or "heterofil" fibers are filaments made up of different polymers. There are many geometrical arrangements. The three main heterofil geometries are side-by-side, core–sheath (both concentric and eccentric), and the multiple core or "islands in a sea" configuration. The so-called "splittable pie" configurations are used in the production of microfibers (see Figure 1.12).

The two polymer components do not have to differ in "chemical" nature. They can differ only in physical parameters such as molecular weight. Usually, it is desirable that the two components have good mutual adhesion, but not always. Polyolefines do not bond well with polyesters or polyamides and this fact is exploited in the formation of microfibers (see later).

1.8.1 SIDE–SIDE BICOMPONENT FIBERS

Side–side bicomponent fibers can be used to produce self-bulking yarns. Two PET polymers of different molecular weights, spun as a side–side heterofil, produce a self-bulking fiber

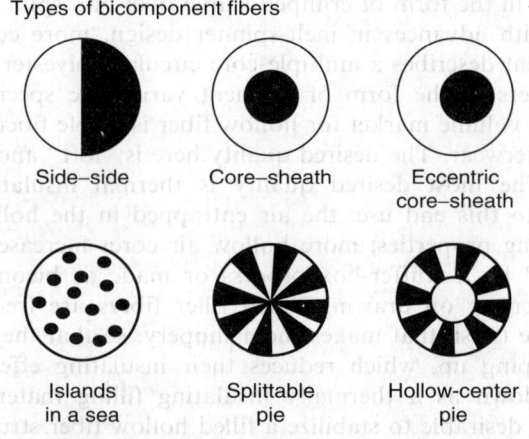

FIGURE 1.12 PET containing biocomponent fiber.

after drawing and relaxing because the spun birefringences differ. The relaxed yarn curls up like a bimetallic strip and results in a spiral crimp. A self-bulking fiber can also be made by cospinning PET with a PET copolyester containing a branching agent in a side–side configuration [66].

1.8.2 Core–Sheath Bicomponent Fibers

The core–sheath (c–s) configuration is adaptable because many different polymers may be applied as a sheath over a solid polyester core, thus giving a variety of modified surface properties while maintaining all the major fiber and textile properties of PET. An early patent by Shima and coworkers uses an eccentric core–sheath configuration to achieve spiral crimp in a yarn [67]. A recent patent by Chang and coworkers discloses the use of side–side or eccentric c–s bicomponent fibers to achieve a self-crimping yarn made from polytrimethylene terephthalate, where one component is a melt-blend of PTT with a small amount of polystyrene [68].

We have already mentioned the antisoil–antistatic fiber made by using a PET–PEG block copolymer coating on a polyester core. A widely used c–s heterofil has a normal PET core with a lower softening-point sheath polymer (typically a PET–isophthalate copolyester). When such fibers are laid randomly in a nonwoven structure and heated to a temperature above the softening point of the sheath polymer, but below the fusion point of the core polymer, the fibers adhere wherever they cross and touch. This may be done either by heated calendar rolls or simply in a forced draught hot air oven. The result is a stable nonwoven fabric [69,70]. A new development is a biodegradable, nonwoven material for disposable fabrics, made by thermally bonding a polylactide core bicomponent fiber with a low melting sheath polymer such as polyethylene [71].

1.8.3 Multiple Core Bicomponent Fibers

The multiple core or "islands in a sea" type of heterofil is mainly of interest in connection with microfibers and this is discussed in the following text.

1.8.4 Hollow Fibers

Hollow fibers are a type of core–sheath heterofil in which the core is composed of air. They are usually made in the form of crimped staple fiber and spun from a modified staple spinning pack [72]. With advances in melt-spinner design, more complex geometries [73] are available. One patent describes a multiple-core circular polyester staple fiber with seven cores [73]. Hollow fibers in the form of filament yarn have specialized uses in medical devices, but the largest volume market for hollow fiber is staple fiberfill for pillows, duvets, quilts, and thermal outerwear. The desired quality here is "loft" and it is better if the fiber is light and bulky. The most desired quality is thermal insulation. Hollow polyester fibers are well suited to this end use: the air entrapped in the hollow cores adds significantly to their insulating properties; more hollow air cores increase this effect. The fibers are frequently crimped by a stuffer-box process or made as bicomponent hollow fibers, which develop spiral crimp on drawing. Such filler fibers are frequently treated with a permanent polysiloxane finish that makes them slippery, so that they slide easily over each other and resist clumping up, which reduces their insulating effect. In this form, they compete with goose down as a thermally insulating filling material. For end-uses like fabric interlinings, it is desirable to stabilize a filled hollow fiber structure by incorporating an additional thermally bondable bicomponent fiber [75].

1.9 NOVEL FIBER FORMS

1.9.1 Microfibers

Microfiber is arbitrarily defined as a filament of less than 1.0 dpf. Normal filament yarn polyester is around 3. 0–5.0 dpf. Microfibers are many times finer than a human hair and finer than the finest silk: diameters are generally less than 10 μm. A typical polyester microfiber has a titer of about 0.5 dpf. Such fine fibers in the form of yarns have excellent textile properties. They are very flexible, giving a soft "hand" and excellent drape to fabrics. The high density of fibers in a typical microfiber fabric makes it inherently windproof and waterproof. There are only tiny gaps for air to blow through, yet the fabrics are largely unwettable, because surface tension effects prevent water from penetrating the interstices in the fabric. These fabrics are comfortable to wear as water vapor from perspiration evaporates easily. Their fabric properties make them ideal for women's wear, sportswear, active, and outdoor wear. They have (radiant) heat-insulating properties because the filaments are of the same dimensional order as the wavelengths of infrared radiation. A 0.5 dpf polyester filament (density ≈ 1.4 g/cm^3) has a diameter of about 7 μm, right in the middle of the IR wavelength range (2–20 μm). Hence, radiation is efficiently scattered by the microfibers and radiation loss of body heat is reduced.

Microfibers lend themselves very well to fabric esthetics. Dyed fabrics appear with solid, bright colors due to the fine size of individual filaments. They are semimatte in appearance, without the need for treatments such as sand washing. The vulnerability to damage from careless ironing is one disadvantage. The thermal capacity of the tiny filaments is so low that it is easy to overheat them. They also snag easily and, as with all fine fabrics, they need to be handled with a degree of care.

1.9.2 Melt-Spinning Microfibers

The first commercial microfibers were produced in Japan [76] in the 1970s and were made by spinning a bicomponent fiber with polyester fibrils dispersed in a matrix polymer in the "islands in a sea" configuration [77]. This was drawn into fibers and processed into fabric and finally the matrix polymer was dissolved, leaving tiny polyester fibrils. These were processed into a synthetic suede material marketed as Ultrasuede. The polyester fibrils were extremely fine, less than 1 μm in diameter. The process was expensive, but the product was successful. At the same time, numerous variations on this theme later followed. One ingenious idea by Sato and coworkers used was a blend of PET in a PET–sulfoisophthalate copolymer rich in SO$_3$Na groups, which dissolved readily in aqueous base leaving the unaffected PET.

There are many patents in the literature, mostly on static devices for melt-spinning multiple fibers with the "islands in a sea" configuration. This is usually done by a multiple series of flow-divider plates that take the initial side–side polymer flow (as in a heterofil spinner) and subdivide it and cross over the flow many times before the spinneret plate, so that each spun filament emerges with the desired structure. Some examples are further ideas of Okamoto [78] and Dugan [79]. More recently, the dissolvable matrix has been made of such materials as polylactic acid (already mentioned), thermoplastic starches, or water-soluble copolyesters. Good review articles on microfibers have been written by Robeson [80], Murata [81], and Isaacs [82].

Another method of making microfibers is the "splittable pie" technique, where a bicomponent fiber of special configurations is spun from two incompatible polymers that adhere poorly. On subjecting these fibers to mechanical stress, as during carding, they split apart to form bundles of microfibers with a wedge-shaped cross section. PET–polypropylene or

PET–nylon-6 are examples of suitable polymer combinations. Such microfibers are frequently used in nonwoven fabrics such as filter materials and specialty fabrics like cleaning cloths for microelectronic components or polishing cloths for lenses and delicate optical instruments.

1.10 WORLD MARKETS AND FUTURE PROSPECTS FOR POLYESTER FIBERS

The total world market for all synthetic fibers in 2002 was around 36,000,000 tons. Of this total, 21,500,000 tons is PET and the rate of consumption is still growing, although apparently slowing. Over the past 15 years, there have been cataclysmic changes in the polyester-producing fiber business. The gradual eclipse of the textile industry in the United States and much of western Europe and its geographical shift to Asia and other places, such as Central America and parts of Eastern Europe, has brought about these changes. Old, firms like ICI, Hoechst, Monsanto, and Eastman have disappeared completely from the fiber-producing scene. The last survivor was DuPont which announced in February 2002 that they would split off all their fiber and textile interests as a separate industry under the name Invista. In November 2003, Koch Industries announced that they would acquire Invista.

The new generation of polyester fiber producers buy polymer in the open market as a commodity item and convert it into fiber and yarn. They have revolutionized the market and superseded the old order. Koch Industries buy PPT polymer from Shell and spin Corterra fibers and market them, although Shell retains the trademark and Koch proposes to build its own polymer plant in Mexico. The emergence of China as a major consumer and producer of polyester fiber (it outstripped the United States in polyester production in 1998) will have a major effect on world markets.

The market for polyester fiber will certainly continue to grow overall, although, as a major commodity item, it is likely to be affected much more than in the past by global economics and trade cycles. Certainly, the price of raw materials like crude oil and natural gas will have an effect on process costs and markets. Nevertheless, there is still a trend to replace other fibers, both natural and synthetic, with polyester. Nylon is still losing markets to polyester. At present, nylon dominates polyester in domestic carpet yarns, but because PET is cheaper, it has a growing share of the contract carpet trade. The new microfibers and newer easy-to-dye polyesters with excellent resilience (like PTT) would be expected to make big inroads into floor coverings and the apparel markets over the next few years. It was confidently expected that PTT would have an immediate impact on the carpet business (one place where PET polyester suffers) in 1999–2000 when Corterra was first launched. So far this has not happened, but fiber price and availability are major factors as always. In the long term, there are new nonoil-based biomass-derived processes in commercial production for making not just intermediates for polyesters but even the polymers themselves. The effect of recycling polyester such as soda bottles into fiberfill and carpet yarns may also have unpredictable effects.

REFERENCES

1. Worldwide Manufactured Fiber Survey, Fiber Economic Bureau, Fiber Organon, June 2002.
2. G. Reese, Polyester fibers, in *Encyclopedia of Polymer Science & Technology*, 3rd edn, Vol. 3, John Wiley & Sons, New York, 2003, pp. 652–678.
3. Y. Murase and A. Nagai, in T. Nakjima (ed.), *Advanced Fiber Spinning*, Woodhouse Publishing, Cambridge, UK, 1994, pp. 25–64.
4. J.E. McIntyre, in J. Scheirs and T.E. Long (eds.), *Chemistry and Technology of Polyesters and Copolyesters*, John Wiley & Sons, New York, 2003, pp. 2–28.

5. A.J. East, Polyester fibres, in J.E. McIntyre (ed.), *Synthetic Fibres: Nylon, Polyester Acrylic, Polyolefine,* Woodhead Publishing Limited, Cambridge, England, 2005, Chapter 3.
6. H. Mark and G.F. Whitby (eds.), *The Collected Papers of W.H. Carothers,* Interscience Publications Inc., New York, 1940.
7. U.K. Patent 578,079 (June 14, 1946), J. Rex Whinfield and J.T. Dickson (to ICI Ltd), also U.S. Patent 2,465,319 (March 22, 1949), J.R. Whinfield and J.T. Dickson (to E I Du Pont de Nemours & Co Inc.).
8. A.E. Brown and K.A. Reinhart, Polyester fiber: from its invention to its present position, *Science,* July 16, 1971, 287–293.
9. A.J. East, *Kirk-Othmer Encyclopedia of Chemical Technology,* 5th edn, to be published.
10. W.S. Witts, *Chem. Proc. Eng.,* 1970, **51**, 55.
11. U.S. Patent 3,584,039 (June 8, 1971), D.H. Meyer (to Standard Oil Co).
12. J. Milgrom, Outlook for plastics recycling in the USA, Decision Resources Inc, 1993, pp. 25–27.
13. G.O. Curme and F. Johnston, *Glycols,* ACE Monograph #114, Reinhold Publishing Corp., New York, 1952, Chapter 2.
14. R.E. Wilfong, *J. Polym. Sci.,* 1961, 54, 388.
15. French Patent 1,169,659 (1959) to ICI Ltd.
16. Canadian Patent 573,301 (1959), W.K. Easley et al. (to Chemstrand Corporation).
17. U.S. Patent 5,684,116 (November 4, 1997), M. Martl et al. (to Akzo-Nobel NV).
18. U.S. Patent 5,789,528 (August 4, 1998), M. Martl et al. (to Akzo-Nobel NV).
19. U.S. Patent 5,552,513 (September 3, 1996), K.K. Bhatia (to E I DU Pont de Nemours & Co Inc.).
20. U.S. Patent 5,677,415 (October 14, 1997), K.K. Bhatia (to E I DU Pont de Nemours & Co Inc.).
21. U.S. Patent 5,865,423 (January 5, 1999), K.K. Bhatia (to E I DU Pont de Nemours & Co Inc.).
22. W. Sorenson, F. Sweeney, and T.W. Campbell, *Preparative Methods of Polymer Chemistry,* 3rd edn, John Wiley & Sons, New York, 2001, pp. 181–185.
23. D.G. Callender, *Polym. Sci. Eng.,* 1985, **25**, 453–457.
24. M. Jaffe, J. Menczel, and W. Bessey, Thermal characterization of polymeric materials, in E.A. Turi (ed.) *Fibers,* 2nd edn., Vol. 1, Academic Press, San Diego, CA, 1997, Chapter 7.
25. G.W. Davis and E. Hill, *Kirk-Othmer Encyclopedia of Chemical Technology Edition III,* Vol. 18, John Wiley & Sons, 1982, p. 535.
26. S. Berkowitz, *J. Appl. Polym. Sci.,* 1984, **29**, 4353–4361.
27. G.W. Davis and E.S. Hill, Polyester fibers, *Kirk-Othmer Encyclopedia of Chemical Technology,* 3rd edn, Vol. 18, John Wiley & Sons Inc., 1982, pp. 531–539.
28. M. Jaffe and W. Bessey, Solid phase processing of polymers, in I. Ward, P. Coates, and M. Dumoulin (eds.), *Solid State Processing of Polymers,* Carl Hanser Verlag, Munich, 2000.
29. A. James, Principles of spin pack design: a guide to selecting components, *Int. Fiber J.,* 1997, **12**, 2.
30. M.M. Denn, Anniversary Article, Fifty years of non-Newtonian fluid dynamics, *AIChE J.,* 2004, **50**(10), 2335–2345.
31. A. Zabicki and A. Wasiak, Effects of molecular deformation and orientation on the crystallization of polymers, *Chemia Stosowana,* 1981, **25**(2), 147–161.
32. I. Onsager, *N Y Acad. Sci.,* 1949, **51**, 627.
33. P.J. Flory, *Proc. R. Soc. London. Ser., Part A, 1956,* London.
34. S.E. Bedford, K. Yu, and A.H. Windle, Influence of chain flexibility on polymer mesogenicity, JACS, *Faraday Trans.,* 1991, **88**(13), 1765–1773.
35. U.S. Patent 4,101,525, H.L. Davis, M. Jaffe, H.L. LaNieve, and E.J. Powers, "Polyester Yarn of High Strength Possessing an Unusually Stable Internal Structure," U.S. Patent 4,195,052, H.L. Davis, M. Jaffe, H.L. LaNieve, and E.J. Powers, "Production of Improved Polyester Filaments of High Strength Possessing and Unusually Stable Microstructure."
36. S.D. Long and I.M. Ward, Tensile drawing behavior of poly(ethylene terephthalate), *J. Appl. Polym. Sci.* 1991, **42**(7), 1911–1920.
37. J.W.S. Hearle, L. Hollick, and D.K. Wilson, *Yarn Texturing Technology,* Woodhead Publishing Co., Cambridge, U.K., 2001.
38. H. Chuah et al., *Chem. Fib. Inter.,* 1996, **46**(6), 424–428.
39. Y. Kitano, Y. Kinoshita, and T. Ashida, *Polymer,* 1995, **36**, 10.

40. H.H. Chuah, Polytrimethylene terephthalate, *Encyclopedia of Polymer Science and Technology*, 3rd edn, Vol. 3, John Wiley & Sons, New York, 2003, pp. 544–557.
41. E.V. Martin and C.J. Kibler, in H.F. Mark, S.M. Atlas, and E. Cernia (eds.), *Man-Made Fiber Science and Technology*, Vol. 3, John Wiley & Sons, New York, 1968, pp. 83–134.
42. S.R. Turner, R.W. Seymour, and T.W. Smith, Cyclohexanedimethanol polyesters, *Encyclopedia of Polymer Science and Technology*, 3rd edn, Vol. 2, John Wiley & Sons, 2003, pp. 127–134.
43. U.K. Patent 604,073 (1948), J.G. Cook, H.P.W. Hugill, and A.R. Low (to ICI Ltd).
44. H. Mark, Phase transitions and elastic behavior of high polymers, *J. Indust. Eng. Chem.*, 1942, **29**, 449–454.
45. Tai-Shung Chung and M. Jaffe, Liquid crystalline polymers, Mainchain, *Encyclopedia of Polymer Science and Technology*, Wiley-Interscience, 2003.
46. I.M. Ward, Mechanical anisotropy of highly oriented polymers, *J. Computer Aided Matl. Design*, 1997, **4**(1), 43–52.
47. H.N. Yoon, *Colloid Polym. Sci.*, 1990, **268**, 230.
48. U.S. Patent 5,945,216, Flint, M. Jaffe, I. Haider, J. Dibiase, and J. Cornetta, "Process for making high denier filaments of thermotropic liquid crystalline polymers and compositions thereof."
49. L.C. Sawyer and M. Jaffe, *J. Mater. Sci.*, 1986, **21**, 1897.
50. L.C. Sawyer et al., *J. Mater. Sci. Lett.*, 1992, **11**, 69.
51. D. Goupil, Sutures, Section 7.9, in B.D. Ratner, A.S. Hoffman, F.J. Schoen, and J.E. Lemons (eds.), *Biomaterials Science*, Academic Press, 1996, pp. 356–360.
52. U.S. Patent 5,869,597 (February 9, 1999), H.D. Newman and D.D. Jamiolowski (to Ethicon Inc.).
53. U.S. Patent 4,841,968 (June 27, 1989), R.L. Dunn and coworkers (to Southern Research Institute).
54. U.S. Patent 4,427,614 (January 24, 1984), P.J. Barham and Paul Holmes (to Imperial Chemical Industries PLC).
55. G. Farrow et al., *Polymer*, 1962, **3**, 17.
56. U.S. Patent 4,016,142 (April 5, 1977), W. Alexander and coworkers (to Millhaven Fibers Ltd, Canada).
57. U.S. Patent 4,442,058 (April 10, 1984), R.L. Griffith and N.A. Favstritsky (to Fiber Industries Inc).
58. D. Coleman, Copolymerization of polyethylene terephthalate with polyoxyethylene glycols, *J. Polym. Sci.*, 1954, **XIV**, 15–27.
59. U.S. Patent 3,557,039 (January 19, 1971), J.E. McIntyre and M.M. Robertson (to Imperial Chemical Industries Limited).
60. U.S. Patent 3,616,183 (October 26, 1971), J.R. Brayford, I.S. Fisher, and M.M. Robertson (to Imperial Chemical Industries Limited).
61. D.R. Waring and G. Hallas (eds.), *The Chemistry and Application of Dyes*, Plenum Press, New York, 1994.
62. J.B. Dickey, E.B. Towne, and G.F. Wright, *J. Org. Chem.*, 1955, **20**, 505.
63. U.S. Patent 3,544,658 (December 1, 1970), A.J. East and W.D. Thackray (to Imperial Chemical Industries Limited).
64. U.S. Patent 3,018,272 (January 23, 1962), J.M. Griffing and W.R. Remington (to E I Du Pont de Nemours & Co Inc.).
65. U.S. Patent 6,110,405 (August 29, 200), C.M. King et al. (to Wellman, Inc.).
66. U.S. Patent 5,723,215 (March 3, 1998), I.A. Hernandez et al. (to E I Du Pont de Nemours & Co Inc.).
67. U.S. Patent 3,520,770 (July 14, 1970), T. Shima et al. (to Teijin Limited, Japan).
68. U.S. Patent 6,641,916 (November 4, 2003), J.C. Chang, J.V. Kurian and R.W. Miller (to E I Du Pont de Nemours & Co Inc.).
69. U.S. Patent 5,082,720 (January 21, 1992), D.J. Hayes (to Minnesota Mining and Manufacturing Company).
70. U.S. Patent 6,441,267 (August 27, 2002), J.S. Dugan (to Fiber Innovation Technology).
71. U.S. Patent 3,772,137 (November 13, 1973), J.W. Tolliver (to E I Du Pont de Nemours & Co Inc.).
72. U.S. Patent 5,344,297 (September 6, 1994), W.H. Hills (to BASF Corporation).
73. U.S. Patent 5,104,725 (April 14, 1992), C.R. Broaddus (to E I Du Pont de Nemours & Co Inc.).
74. U.S. Patent 4,520,066 (May 28, 1985), G. Athey (to Imperial Chemical Industries PLC).

75. M. Okamoto, Ultrafine Fiber and its Applications, Part I, *Japan. Text. News*, November 1977, pp. 94–97; Part II, *Japan. Text. News*, January 1978, pp. 77–81.
76. U.S. Patent 3,705,226 (December 5, 1972), M. Okamoto et al. (to Toray Industries, Inc.).
77. U.S. Patent 4,233,355 (November 11, 1980), Y. Sato and H. Arai (to Toray Industries, Inc.).
78. U.S. Patent 4,350,006 (September 21,1982), M. Okamoto et al. (to Toray Industries Inc.).
79. U.S. Patent 5,366,804 (November 22,1994), J.S. Dugan (to BASF Corporation).
80. Robeson, *J. Appl. Polym. Sci.,* 1994, **52**, 1837–1846.
81. Murata, *Textile World*, 1996, **144**, 42–48.
82. Isaacs, *Textile World*, August 1994, 73–74.

2 Polyamide Fibers

H.H. Yang

CONTENTS

2.1 Introduction 33
   2.1.1 Historic Perspective 33
   2.1.2 Aliphatic Polyamides 34
       2.1.2.1 Definition of Polyamides 34
       2.1.2.2 Examples of Polyamide Compositions 34
2.2 Basic Chemistry of Aliphatic Polyamides 35
   2.2.1 Synthetic Routes 35
   2.2.2 Amidation Reactions 36
   2.2.3 Ring-Opening Reactions 38
   2.2.4 Molecular Weight and Molecular Weight Distribution 40
   2.2.5 Nylon-6,6 Polyamide 45
       2.2.5.1 Synthetic Procedure 45
       2.2.5.2 Kinetics and Thermodynamics 45
       2.2.5.3 Solid-State Polymerization 46
   2.2.6 Nylon-6 Polyamide 47
       2.2.6.1 Synthetic Procedure 47
       2.2.6.2 Kinetics and Thermodynamics 47
   2.2.7 Other Polyamides 56
       2.2.7.1 Nylon-3 56
       2.2.7.2 Nylon-4 56
       2.2.7.3 Nylon-7 56
       2.2.7.4 Nylon-8 56
       2.2.7.5 Nylon-11 58
       2.2.7.6 Nylon-12 58
       2.2.7.7 Nylon-4,2 58
       2.2.7.8 Nylon-4,6 58
       2.2.7.9 Nylon-6,12 58
       2.2.7.10 Nylon-4,1 58
       2.2.7.11 Nylon-6,1 58
       2.2.7.12 Nylon-6,T 59
       2.2.7.13 PACM,12 59
2.3 Polymerization Processes 59
   2.3.1 Monomer Syntheses 59
       2.3.1.1 Caprolactam 59
       2.3.1.2 Adipic Acid 65
       2.3.1.3 Hexamethylene Diamine 67

	2.3.2	Industrial Processes	70
		2.3.2.1 General Operations	70
		2.3.2.2 Nylon-6 Polyamide	72
		2.3.2.3 Nylon-6,6 Polyamide	73
		2.3.2.4 Other Polyamides	73
		2.3.2.5 Nanocomposites	74
		2.3.2.6 Process Simulation	74
2.4	Preparation of Polyamide Fibers		78
	2.4.1	Melt Spinning	78
	2.4.2	Dynamics of Melt Spinning	80
	2.4.3	Quenching	80
	2.4.4	Spin Finish	81
	2.4.5	Drawing	81
	2.4.6	Process Integration	82
		2.4.6.1 Draw-Twisting	83
		2.4.6.2 Spinning–Drawing–Texturing	84
	2.4.7	High-Speed Spinning	84
	2.4.8	Textile Finishing Operations	85
		2.4.8.1 Texturing	85
		2.4.8.2 False-Twist Texturing	85
		2.4.8.3 Crimp Texturing	86
		2.4.8.4 Commingling, Interlacing, and Air-Jet Texturing	87
	2.4.9	Winding and Yarn Package	87
2.5	Structures and Properties of Polyamide Fibers		87
	2.5.1	Polymer Chain Structure	87
	2.5.2	Characteristics of Crystalline and Amorphous Structures	88
		2.5.2.1 Crystalline Forms	88
		2.5.2.2 Crystalline Structures in Fibers	92
		2.5.2.3 Structural Models	93
	2.5.3	Thermal Transitions of Crystalline Phase	94
		2.5.3.1 Melt Temperature	94
		2.5.3.2 Glass Transition Temperature	96
	2.5.4	Characterization of Structural Parameters	98
		2.5.4.1 X-Ray Diffraction	98
		2.5.4.2 Crystallite Size	99
		2.5.4.3 Degree of Orientation	99
		2.5.4.4 Birefringence	100
		2.5.4.5 Crystallinity and Crystalline Indices	101
	2.5.5	Typical Fiber Properties	105
		2.5.5.1 Tensile Behavior	107
		2.5.5.2 Chemical Properties	113
		2.5.5.3 Degradation Behavior	113
		2.5.5.4 Polymer Stabilization	118
		2.5.5.5 Antistatic	118
		2.5.5.6 Flammability	119
		2.5.5.7 Transparency	120
		2.5.5.8 Dye Diffusion	120
	2.5.6	Process–Structure–Property Relationship	120
		2.5.6.1 Effect of Molecular Weight	121
		2.5.6.2 Effect of Water	121

		2.5.6.3	Effect of Spinning and Drawing 121
		2.5.6.4	Effect of Spinning Conditions 122
2.6	Commercialization and Future Outlook 122
	2.6.1	Polyamide Fiber Products 122
		2.6.1.1	Filament Yarn and Staple 122
		2.6.1.2	Bicomponent Fibers 122
		2.6.1.3	Microfibers 123
	2.6.2	End-Uses 124
		2.6.2.1	Textiles 124
		2.6.2.2	Carpets 124
		2.6.2.3	Industrial Applications 124
		2.6.2.4	Specialty Fibers 124
		2.6.2.5	Nonwovens 124
		2.6.2.6	Nonfiber Uses 125
	2.6.3	Worldwide Consumption of Polyamide Fibers 125
	2.6.4	Future Outlook 126
		2.6.4.1	Technology 126
		2.6.4.2	Market Share 126
Acknowledgment 126
References 127

2.1 INTRODUCTION

2.1.1 HISTORIC PERSPECTIVE

The development of polyamide fibers is quite interesting in the history of synthetic fibers. Nylon, the first commercial fiber product of polyamide, was marketed for manufacturing women's hosiery by E.I. du Pont de Nemours & Company in 1938. The immediate success made nylon a household name all over the world. It also provided DuPont a string of 50 years of strong earnings. The success of nylon led to the commercialization of other synthetic fibers such as polyester, polyacrylonitrile, and polyolefin fibers, as well as inorganic fibers such as carbon and boron. These developments have been strongly supported by the progress in polymer and fiber chemistry and material science. In retrospect, therefore, one must credit the chemical industry for its pursuit of new technology and vision of new materials.

In the early 1920s, DuPont embarked on a "pure science" research program for its future business objectives. The company hired Dr. Wallace H. Carothers, a brilliant young chemist, in 1927 to study long-chain polymers, methods of polymerization, and polymer properties. Within three-and-half years, Carothers accomplished the design and syntheses of condensation polymers and the conversion of these polymers to fibers. The properties of such fibers were further enhanced by the discovery of cold drawing. He also discovered a useful synthetic rubber, polychloroprene. In the ensuing years, Carothers attempted to develop useful polyesters and polyamides. His work was hampered by inadequate polymer properties for fiber use, and by his spells of mental depression. Nevertheless, he finally learned to prepare polyamides and was particularly interested in nylon-5,10 from pentamethylene diamine and sebacic acid. This polyamide composition was unsuitable for commercialization because the monomers would always be difficult to find. Rather, he was asked to develop nylon-6,6. The monomers for nylon-6,6 were thought at least theoretically available from benzene. Thus, the first sample of nylon-6,6 was prepared in 1931. The fiber made from nylon-6,6 was coded Fiber 66. For fundamental support, Paul Flory was assigned to study the molecular weight distribution of condensation polymers. DuPont was in the commercial production of nylon-6,6 by February 1935.

Carothers [1] filed for patent on stockings from polyamide yarns on February 15, 1937. He claimed that such stockings would fit properly and would not become baggy like other synthetic fibers. Shortly afterwards, Fiber 66 was evaluated for the full-fashioned women's stockings market. DuPont made an extensive effort to overcome problems of notably knitting, wrinkling, dyeing, and shininess. This resulted in a reasonably reproducible process for stockings by September 1938. Thus, DuPont officially announced the development of a new textile fiber called "nylon." It was described as having "filaments as strong as steel, as fine as a spider's web, yet more elastic than any of the common natural fibers." [2] Full-fashioned nylon stockings then went on sale nationwide on May 14, 1940. When World War II broke out, every ounce of nylon was required for military uses including parachutes and soft body armor. By the 1980s, women's hosiery alone consumed more than 70 million pounds of nylon a year.

Unfortunately, Carothers never saw the completion of the nylon development. He committed suicide in April 1937. Flory later published his distinguished book on polymer chemistry [3]. He went on to become one of America's most revered polymer scientists and a Nobel Laureate.

2.1.2 Aliphatic Polyamides

2.1.2.1 Definition of Polyamides

Aliphatic polyamides are macromolecules whose structural units are characteristically interlinked by the amide linkage —NHCO—. The nature of the structural unit constitutes a basis for classification. Aliphatic polyamides with structural units derived predominantly from aliphatic monomers are members of the generic class of *nylons*, whereas aromatic polyamides in which at least 85% of the amide linkages are directly adjacent to aromatic structures have been designated *aramids*. This chapter is concerned with nylons, especially those of commercial importance. Aramids are discussed in a separate chapter.

The U.S. Federal Trade Commission defines nylon fibers as "a manufactured fiber in which the fiber forming substance is a long chain synthetic polyamide in which less than 85% of the amide linkages (—CO—NH—) are attached directly to two aliphatic groups."

Polyamides that contain recurring amide groups as integral parts of the polymer backbone have been classified as condensation polymers regardless of the principal mechanisms entailed in the polymerization process. Though many reactions suitable for polyamide formation are known, commercially important nylons are obtained by processes related to either of two basic approaches: one entails the polycondensation of difunctional monomers utilizing either amino acids or stoichiometric pairs of dicarboxylic acids and diamines, and the other entails the ring-opening polymerization of lactams. The polyamides formed from diacids and diamines are generally described to be of the AABB format, whereas those derived from either amino acids or lactams are of the AB format.

2.1.2.2 Examples of Polyamide Compositions

Straight-chain aliphatic nylons are commonly identified either as nylon X,Y or nylon Z, where X, Y, and Z signify the number of carbon atoms in the respective monomeric units. The pair X,Y refers to the AABB-type nylons, where the first number X is equal to the number of carbon atoms in the diamine unit and the second number Y represents the number of carbon atoms in the corresponding diacid unit. The number Z refers to the AB-type nylons and is equal to the number of carbon atoms in the amino acid unit. A few examples are:

Nylon-6,10	-[-NH-(CH$_2$)$_6$-NH-CO-(CH$_2$)$_8$-CO-]-
Nylon-6	-[-NH-(CH$_2$)$_5$-CO-]-
Nylon-11	-[-NH-(CH$_2$)$_{10}$-CO-]-
Nylon-6,T	-[-NH-(CH$_2$)$_6$-OCO-(C$_6$H$_4$)-OCO-]-
mXD,6,6	-[-NH-CH$_3$(C$_6$H$_3$)-NH-CO-(CH$_2$)$_4$-CO-]-
Nylon-6,6–6,10 (60:40)	-[-NH-(CH$_2$)$_6$-NH-{-CO-(CH$_2$)$_4$-CO-}$_{60}$/{-CO-(CH$_2$)$_8$-CO}$_{40}$-]-

Thus, nylon-6,10 is the polyamide produced from the 6-carbon hexamethylene diamine and the 10-carbon sebacic acid, whereas nylon-6 is obtained from the 6-carbon caprolactam and nylon-11 from the 11-carbon aminoundecanoic acid. The coding of nylons derived from ring structures usually includes either a single letter or a combination of letters representing the ring-containing unit. Nylon-6,T refers to a polyamide produced from hexamethylene diamine and terephthalic acid, whereas nylon-mXD,6 is derived from *m*-xylylene diamine and adipic acid. Copolyamide compositions are represented by listing the components in the order of decreasing percentages in parenthesis. Thus, nylon-6,6–6,10 (60:40) refers to a product obtained by copolymerizing hexamethylene diamine with 60% adipic acid and 40% sebacic acid.

Among various nylon compositions, nylon-6 and nylon-6,6 are by far the most important polyamides for the commercial production of fibers and resins. This chapter will focus largely on the fiber-related aspects of these two nylons. Other nylons marketed for minor fiber applications will be discussed appropriately. A wealth of information on nylon technology exists in the literature [4–8].

2.2 BASIC CHEMISTRY OF ALIPHATIC POLYAMIDES

2.2.1 Synthetic Routes

Generally speaking, polyamides can be synthesized by direct amidation where an amine reacts with a carboxylic acid with the removal of water. The reactive amine and acid groups may be on a single amino acid molecule:

$$H_2NRCOOH \rightleftharpoons H-(HNRCO)_n-OH + H_2O$$

or in different molecules:

$$RNH_2 + R'COOH \rightleftharpoons RNH_3^+ + R'COO^- \rightleftharpoons RNHCOR' + H_2O$$

where R represents an aliphatic chain segment or aromatic unit. An ester derivative of the carboxyl group may be used in some cases to facilitate the reaction:

$$RNH_2 + R'COOR'' \rightleftharpoons RNHCOR' + R''OH$$

An alternative route involves the reaction of acid chloride and amine at low temperatures to form high-melting polyamides:

$$H_2NRNH_2 + ClCOR'COCl \rightleftharpoons H_2NRNH(COR'CONH\ RNH)_{x-1}COR'COCl$$
$$+ (2R-1)HCl$$

This method of synthesis avoids polymer decomposition or cross-linking at high temperatures. This reaction can be carried out by interfacial polymerization where the diacid is added in a water-immiscible solvent to an aqueous solution of the diamine, an inorganic base, and surfactant. Polymerization will take place in the organic layer at the interface. A single-phase solution polymerization can also be used for this reaction in the presence of an acid acceptor.

For aliphatic polyamides, the method of direct amidation is used predominantly. The syntheses of the commercially important nylons may be represented by the general Equation 2.1 through Equation 2.3. Equation 2.1 refers to the formation of AABB-type nylons:

$$H_2N-(CH_2)_x-NH_2 + HOOC-(CH_2)_{y-2}-COOH$$
$$\rightleftharpoons -[-HN-(CH_2)_x-NH-OC-(CH_2)_{y-2}-CO-]- \quad (2.1)$$

Equation 2.2 and Equation 2.3 pertain to the polycondensation of amino acids and to the ring-opening polymerization of lactams for the synthesis of AB-type nylons, respectively:

$$H_2N-(CH_2)_{z-1}-COOH \rightleftharpoons -[-HN-(CH_2)_{z-1}-CO-]- + H_2O \quad (2.2)$$

$$\overline{HN-(CH_2)_{z-1}-C}=O \rightleftharpoons -[-HN-(CH_2)_{z-1}-CO-]- \quad (2.3)$$

Lactams may be converted to the corresponding nylons by water-initiated polymerization. The reaction may also be initiated by a base that requires anhydrous conditions and proceeds at high reaction rates [9,10]. Initiators used include carbonates, hydrides, alcoholates, and hydroxides of alkali and alkaline earth metals. The conversion of lactams to polyamides is also possible by a cationic process, entailing initiation in an anhydrous medium by either strong protonic acids or their dissociable salts [11].

2.2.2 Amidation Reactions

Nylon-6,6 and other AABB-type aliphatic polyamides are synthesized according to Equation 2.1, while that of aliphatic amino acids forms AB-type polyamides according to Equation 2.2.

These polycondensation reactions proceed by a mechanism that is characterized by carbonyl addition–elimination reactions, which may be catalyzed or uncatalyzed as indicated in the general reaction scheme (2.4)

$$(2.4)$$

Assuming equivalence of all the amide groups formed and independence of the end groups from the molecular chain length, the equilibrium constant K_c, as defined by

Equation 2.5, thus governs the condensation equilibria according to Equation 2.1 and Equation 2.2.

$$K_c = [NHCO][H_2O]/[COOH][NH_2] = K_c/K_c' \qquad (2.5)$$

Thus, the molecular weight of the resulting polyamides will always remain finite and will be affected considerably by any stoichiometric imbalance of the end groups. Nonequivalence of the concentration of functional groups may involve only the bifunctional reactants, or result from the presence of a monofunctional species containing a nonreactive terminal unit, or may be caused by the decomposition of the end groups. Nonequivalence of the concentration of the bifunctional reactants and the addition of monofunctional species such as acetic acid are practiced for molecular weight control. The effect of such nonequivalence on the degree of polymerization (P_n) is represented by Equation 2.6:

$$P_n = \frac{1 + r + q}{(1 + r)(1 - p) + q} \qquad (2.6)$$

where p is the extent of reaction and is defined here by Equation 2.7

$$p = 1 - ([COOH] + [NH_2])/([COOH]_o + [NH_2]_o) \qquad (2.7)$$

and

$$r = [NH_2]_o/[COOH]_o'; \quad q = [COOH]_o'/[COOH]_o$$

where $[COOH]'$ stands for the concentration of a monofunctional structure, here a monocarboxylic acid, and the subscript o refers to initial concentrations.

Applying the concept of equal reactivity of all functional groups, the rate of polycondensation according to the general Equation 2.4 may be expressed in terms of the extent of reaction by the rate equation 2.8.

$$\frac{dp}{dt} = 2k_c[COOH]_o(1 - r)\left\{\frac{r}{(1+r)^2} - \frac{p}{2}\left[1 - \frac{p}{2}\left(1 - \frac{X_w(t)}{K_c}\right)\right]\right\} \qquad (2.8)$$

where K_c is the equilibrium constant as defined by Equation 2.5; k_c is the rate constant that may assume one of the following forms:

1. $k_c = k_c^\circ[\text{cat.}]$, if the reaction is catalyzed by the addition of a catalyst.
2. $k_c = k_c^\circ + [COOH]k_c^\circ$, if the reaction is catalyzed by the carboxyl groups present in the reacting system.

In these expressions, k_c° is the rate constant for the uncatalyzed reaction and [cat.] is the concentration of any added catalyst. The momentary carboxyl group concentration [COOH] may be expressed in terms of the initial concentration and the extent of reaction of Equation 2.9:

$$[COOH] = [COOH]_o\left\{1 - \frac{p(1 + r)}{2}\right\} \qquad (2.9)$$

The term $X_w(t)$ represents the fraction of the water formed by the polycondensation reaction present in the system. Thus, a solution of Equation 2.8 is only possible if an explicit expression for $X_w(t)$ can be developed. The problem obviously entails simultaneous diffusion and reaction since X_W is related to the rate at which water is removed from the system. Approaches for corresponding analytical treatments involving specific reactor geometries [12] and modes of operations utilize differential equations such as Equation 2.10 [13,14].

$$\frac{\partial w}{\partial t} = D_w \left[\frac{\partial^2 w}{\partial z^2} + \frac{g}{z}\frac{\partial w}{\partial z} \right] + \frac{[COOH]_o(1+r)}{2}\frac{dp}{dt} \qquad (2.10)$$

where D_w is the diffusion coefficient, z is distance in the direction of diffusion, and g is a geometrical factor.

2.2.3 Ring-Opening Reactions

The polycondensation equilibrium, as represented by Equation 2.5, is also part of the more complex mechanism of the water-initiated polymerization of lactams (Equation 2.3), which entails hydrolytic ring opening (Equation 2.11), condensation (Equation 2.12), and addition (Equation 2.13), as the principal equilibrium reactions [15].

$$\overline{HN-(CH_2)_{z-1}-CO} + H_2O \underset{k'_h}{\overset{k_h}{\rightleftharpoons}} H_2N-(CH_2)_{z-1}-COOH; \; k_H = k_h/k'_h \qquad (2.11)$$

$$[-CO]_n OH + H[HN-\cdots]_m \underset{k'_c}{\overset{k_c}{\rightleftharpoons}} -[-NHCO-]_{n+m-1} + H_2O; \; K_c = k_c/k'_c \qquad (2.12)$$

$$\cdots -NH]_n H + \overline{OC-(CH_2)_{z-1}-NH} \underset{k'_a}{\overset{k_a}{\rightleftharpoons}} HO\bigl[CO(CH_2)_{z-1}-NH\bigr]_{n+1} H; \; K_A = k_a/k'_a \qquad (2.13)$$

In the anionic polymerization of lactams, the reaction mechanism entails initiation (Equation 2.14) and propagation (Equation 2.15) reactions:

$$(2.14)$$

$$(2.15)$$

The initiation reaction yields an imide moiety, which constitutes a growth center for propagation reaction. Addition of certain imides such as acyl lactams as coinitiators essentially eliminates the initiation reaction and makes possible the polymerization at relatively low reaction temperatures. Mechanistic and kinetic aspects of the anionic polymerization of lactams have been treated quite extensively [16b]. The discussed subjects relate to the various equilibria governing the polymerization process. They comprise equilibria allied to monomer conversion, to the formation of cyclic oligomers, and to the effect of initiator concentrations.

Although the general reaction mechanisms of polymerization are independent of the size of the lactam ring, quantities related to the ring size mainly determine the polymerizability of a particular unsubstantiated lactam containing only carbon atoms.

As any ring-opening polymerization, the polymerization of lactams is characterized by competition between the intermolecular reaction resulting in a linear polyamide and the intramolecular reaction of cyclization. Thus, if thermodynamically feasible, as indicated by a negative value of the free-energy change of polymerization ΔG_p, the conversion of lactam to the linear polyamide can be realized if an appropriate reaction path exists for a reasonable rate in a polymer–monomer equilibrium characterized by a preponderance of linear macromolecules. The corresponding equilibrium monomer concentration $[M]_e$ is related to the standard enthalpy ΔH_p^0 and entropy ΔS_p^0 of polymerization and to temperature by Equation 2.16.

$$\Delta G_p = -RT\ln K_A = \Delta H_p^0 - T\Delta S_p^0 = RT\ln[M]_e \qquad (2.16)$$

It is readily seen that there is a reciprocal relationship between K_A and $[M]_e$. The equilibrium constant K_A is therefore a direct measure of the polymerizability of a particular lactam, and ΔH_p^0 and ΔS_p^0 are thus the corresponding principal parameters.

Polymerization in general is an exoentropic process due to the decrease of translational entropy, resulting from the ordering of individual molecules into a polymer chain. In case of the polymerization of lactams, the decrease in translational entropy is in part compensated by increases in rotational and vibrational entropies resulting from the conversion of the cyclic structures into flexible polymer chain segments. In most lactams, with the probable exception of those with more than 12 ring atoms [17], this compensation does not result in a negative entropy term in Equation 2.16. Therefore, negative values for the overall change of the free energy of polymerization of lactams containing up to 12 ring atoms result from enthalpy changes. The enthalpy change for the polymerization of these lactams is related to the difference between the strain in the particular ring structure and the strain in the corresponding linear polymer segment. Strain in lactams results from molecular deformations, which may be expressed in terms of:

1. Bond stretching (compression), related to the bond length and the motion of bonded nuclei along the internuclear line
2. Bond angle distortion (angle strain, Baeyer strain), related to the radial scissoring motion of the bond angle
3. Bond torsion (bond opposition, Pitzer strain), related to rotational motion around the bond axis and the interaction between substituents on neighboring ring atoms
4. Transannular strain (compression of van der Waals' radii), related to interaction between substituents on nonadjacent ring atoms
5. Conformational strain entailing the amide group

The extent to which each type of strain contributes to the total molecular strain depends on the ring size. Bond angle distortion is the principal source of strain in the three- and four-membered lactams, whereas torsional forces are mainly responsible for the strain in the five- to seven-membered rings. For the five- and seven-membered lactams, there is additional strain due to the inability of the amide groups to assume planar conformation resulting in a decrease of resonance stabilization [18]. Both torsional forces and nonbonded interactions originate strain predominantly in the larger rings, although decreased resonance stabilization of the amide group also contributes to the strain up to the nine-membered lactams. Regardless of any strain energy, however, the polymerization of lactams up to this size is also characterized by an

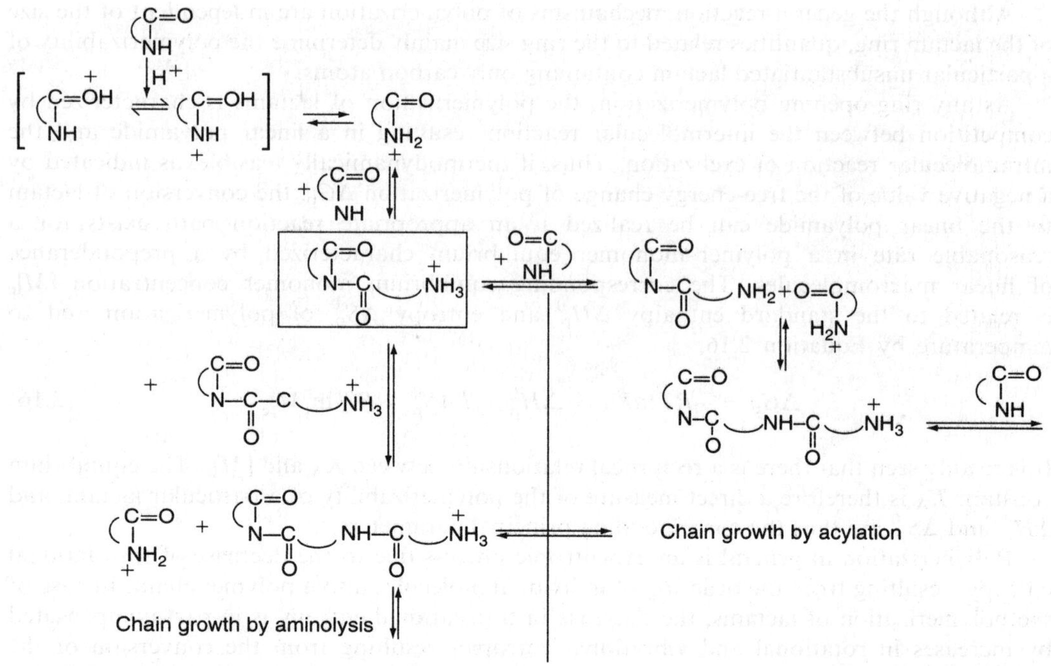

FIGURE 2.1 Mechanism of cationic polymerization of lactams.

additional energy gain of 1.4 kcal/mol due to the *cis–trans* conversion of the amide groups proceeding along with the transition of the ring structures to the linear polyamides.

In addition to the aspects discussed thus far, the polymerizability of lactams is also affected by the presence of heteroatoms in the ring moiety [19–21] and by substituents. The latter affect both enthalpy and entropy of polymerization mainly due to changes in the conformation on conversion of the cyclic structures into linear ones. The overall effects depend markedly on number, size, location, and nature of the substituents [22–26].

Kinetic presentations of the polymerization of lactams may be derived from the general mechanistic schemes presented above. The kinetics of the polymerization of caprolactam has been investigated extensively. It will be reviewed later together with the discussion on nylon-6.

This process for *cationic polymerization* of lactams has been studied extensively. Figure 2.1 shows the principal reactions of monomer conversion and chain growth [27]. Mechanistically, chain growth can commence on both the ammo-terminal end via acylation and the carboxy-terminal end via aminolysis of the polymer molecule. High extents of polymerization are rarely attained because of the occurrence of side-reactions. As shown in Figure 2.2 [27] these side-reactions result in terminal amidine groups that are incapable of adding further lactam. The cationic polymerization process has therefore not attained any practical importance.

An augmented treatment of cationic polymerization has been presented[16d]. Topics regarding initiation, type of initiators, chain growth, and the kinetic particularities of this process were addressed.

2.2.4 Molecular Weight and Molecular Weight Distribution

The polycondensation processes generally produce polyamides that are mixtures of polymer molecules of different molecular weights, the distribution of which usually follows a definite continuous function according to the "most probable distribution" model by Schulz–Flory [3]. This distribution function may, in principle, be derived from the kinetics of polymerization process, but is more readily derived from statistical considerations. In this case, the extent

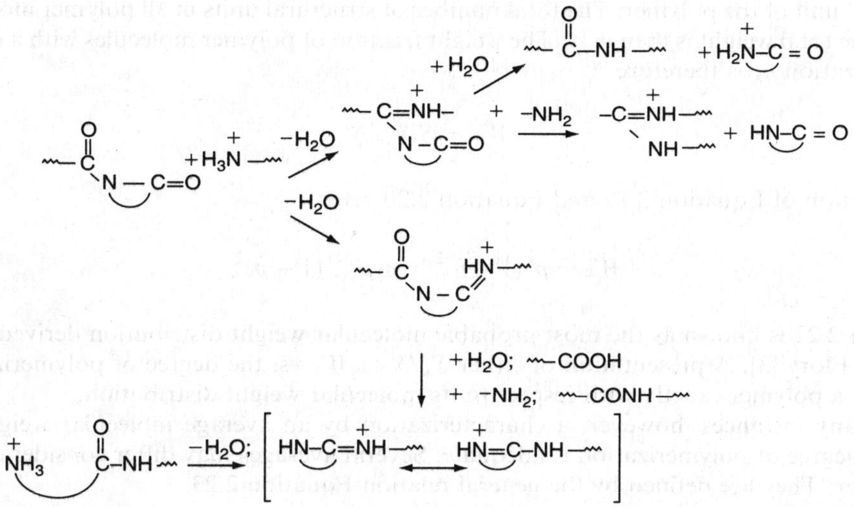

FIGURE 2.2 Formation of amidine functions during cationic polymerization of lactams. (From Reimschuessel, H.K., *J. Polym. Sci., Macromol. Rev.*, 1977, 12, 65. With permission.)

of reaction p (Equation 2.7) is defined as the probability that of all initially present reactive terminal units ($[NH_2]_o + [COOH]_o$), the fraction $2[NHCO]/([NH_2]_o + [COOH]_o)$ has reacted. The fraction of all unreacted units is thus $(1-p)$. Considering a randomly selected polymer molecule, it is now necessary to know the probability that this molecule has exactly λ linkages. The probability of the existence of one linkage is p. The same probability exists independently for any other linkages present in this molecule. Thus, the probability for the existence of λ linkages is p^λ. The probability for the existence of an unreacted function in this molecule is $(1-p)$. The probability for the existence of the complete molecule is equal to the product of these two probabilities:

$$p_\lambda = p^\lambda (1-p) \tag{2.17}$$

This product is equal to the fraction of all molecules that are characterized by λ linkages. The total number of these molecules is given by Equation 2.18 as

$$S_\lambda = S p^\lambda (1-p) \tag{2.18}$$

where S is the total number of polymer molecules. Since $P_n = N/S$, where N is the total number of monomeric species converted to polymer, substitution of S by N/P_n and of P_n by Equation 2.6 yields for $r=1$ and $q=0$

$$S_\lambda = N p^\lambda (1-p)^2 \tag{2.19}$$

This is the number distribution function for a linear step-growth polycondensation at the extent of reaction p. In this instance, λ is given by the relation

$$\lambda = n_\lambda - 1 \tag{2.20}$$

where n_λ is the number of units interlinked by λ linkages. The weight of the polymer molecules consisting of n_λ structural units is $S_\lambda n_\lambda \bar{m}$, and $\bar{m}$ is the average mass of the structural unit of the polymer. The total number of structural units in all polymer molecules is N, and the total weight is then $\bar{m} N$. The weight fraction of polymer molecules with a degree of polymerization n_λ is therefore

$$W_\lambda = S_\lambda n_\lambda / N \qquad (2.21)$$

Combination of Equation 2.19 and Equation 2.20 gives

$$W_\lambda = p^\lambda (1-p)^2 = n_\lambda p_\lambda^{n-1} (1-p)^2 \qquad (2.22)$$

Equation 2.22 is known as the most probable molecular weight distribution derived theoretically by Flory [3]. A presentation of either S_λ/S or W_λ vs. the degree of polymerization n_λ describes a polymer exactly with respect to its molecular weight distribution.

In many instances, however, a characterization by an average molecular weight or an average degree of polymerization is adequate. Several averages may differ considerably from each other. They are defined by the general relation Equation 2.23.

$$\bar{M}_x = \left[\frac{\sum y_i M_i^{\alpha+1}}{\sum y_i M_i} \right]^{1/\alpha} \qquad (2.23)$$

where y_i is the fraction of molecules with a molecular weight of M_i.

For practical use, three molecular weight averages are of importance; they are:

$$\text{Number average } \bar{M}_n (\alpha = -1) \qquad (2.23a)$$
$$\text{Weight average } \bar{M}_w (\alpha = 1) \qquad (2.23b)$$
$$\text{Viscosity average } \bar{M}_v (0.5 < \alpha \geq 1) \qquad (2.23c)$$

The number average may be obtained by end-group titration or osmometry, the weight average by light scattering, and the viscosity average by viscometry of dilute solutions.

Although a relative method, viscometry is the most convenient one and therefore widely used for rapid and reliable characterization of polyamides. It entails the determination of the intrinsic viscosity $[\eta]$, which is a measure of the hydrodynamic volume of the macromolecular coil and depends, for a given solvent, on the molecular weight of the polymer. It is defined by relations such as

$$\lim_{c \to 0} \eta_{\text{red}} = \lim_{c \to 0} \eta_{\text{sp}}/c = [\eta] \qquad (2.24)$$

Therefore, its determination usually entails the extrapolation of the concentration dependence of reduced viscosity quantities to zero polymer concentration. For the determination of $[\eta]$ the most frequently used are the classical Huggins [28] and Kraemer [29] equations:

$$\eta_{\text{sp}}/c = [\eta](1 + k_H [\eta] c) \qquad (2.25)$$
$$\ln \eta_{\text{rel}}/c = [\eta](1 - k_K [\eta] c) \qquad (2.26)$$

where η_{sp} is the specific viscosity, η_{rel} is the relative viscosity of the solution compared to the solvent, c is the concentration in g/dl, and k_H and k_K are the Huggins and Kraemer coefficients, respectively, related by the relationship $k_H + k_K = 0.5$. This relationship may not be applicable for low-molecular-weight polyamides [30,31]. For a given polymer, the intrinsic viscosity is related to the average molecular weight by the equation

$$[\eta] = KM^a \qquad (2.27)$$

which has been associated with the names of a number of scientists: Staudinger, Mark, Kuhn, Houwink, and Sakurada. The parameters a and K are functions of the polymer solvent system. Theoretically, the parameter a may have values from 0.5 to 2.0. The lower value would correspond to a Gaussian molecular coil of unperturbed dimensions (impermeable, no excluded volume, θ-conditions). A value of 2 would indicate a rodlike structure. Aliphatic polyamides are flexible macromolecules for which a has values between 0.5 and 0.9. The higher values are observed in good solvents due to expansion of the molecular coil. Some of the published values for the parameters K and a (Equation 2.27) are listed in Table 2.1 [32–45]. A more complete listing and a detailed discussion on molecular weight determination and dilute solution properties of aliphatic polyamides have been presented [46].

TABLE 2.1
Parameters for Mark–Houwink Equation: $[\eta] = KM^a$

Solvent	Temperature (°C)	Calibration method	MW range (M × 10^{-3})	10^{4}K dl/g	a	Ref.
Nylon-4						
m-Cresol	25	LS (M_w)	10–300	4.0	0.77	32
m-Cresol	25	EG (M_n)	1.7–14	30.0	0.70	33
Nylon-6						
m-Cresol	25	LS (M_w)	5–40	5.57	0.73	34
m-Cresol	25	LS (M_w)	9–335	5.26	0.74	35
m-Cresol	25	EG (M_n)	5–30	18.0	0.65	34
Tricresol	25	OS (M_w)	8–80	2.1	0.90	36
Conc. H$_2$SO$_4$	25	EG (M_w)	4–37	6.3	0.76	37
85% HCOOH	20	V (M_v)	7–120	2.26	0.82	38
CF$_3$CH$_2$OH	25	V (M_v)	13–100	5.36	0.75	38
Nylon-6,6						
m-Cresol	25	LS (M_w)	7–80	24.0	0.61	39
90% HCOOH + 2M KCl	25	LS (M_w)	2.5–50	14.2	0.56	40
90% HCOOH	25	EG (M_n)	6–24	11.0	0.72	41
m-Cresol	20	V (M_v)	10–40	38.0	0.55	42
Nylon-6,10						
m-Cresol	25	EG (M_w)	8–24	1.35	0.96	43
Nylon-8						
m-Cresol	25	EG (M_w)	1–25	7.0	0.76	44
Nylon-12						
m-Cresol	25	LS (M_w)	3–125	8.1	0.70	45
m-Cresol	25	EG (M_n)	1–33	11.8	0.73	45

The average molecular weights as defined by Equation 2.23 correspond to average degrees of polymerization p_λ that are generally used for correlation with the extent of reaction p (Equation 2.7). The most significant averages are the number average $\bar{P}_n$ and the weight average $\bar{P}_w$. They are defined by Equation 2.28 and Equation 2.29.

$$\bar{P}_n = (\sum_\lambda S_\lambda n_\lambda)/(\sum_\lambda S_\lambda) = \Sigma p_\lambda n_\lambda \qquad (2.28)$$

$$\bar{P}_w = (\sum_\lambda S_\lambda n^2)/(\sum_\lambda S_\lambda n_\lambda) = \Sigma w_\lambda n_\lambda \qquad (2.29)$$

Combining with Equation 2.17 and Equation 2.22, respectively, and evaluating the summations

$$\sum_\lambda n_\lambda p_\lambda^{n-1}(1-p)$$

and

$$\sum_\lambda n_\lambda^2 p_\lambda^{n-1}(1-p)^2$$

yields, since $p < 1$:

$$\bar{P}_n = \frac{1}{1-p} \qquad (2.30)$$

$$\bar{P}_w = (1+p)/(1-p) \qquad (2.31)$$

The ratio $\bar{P}_w/\bar{P}_n = 1 + p$ is equal to the ratio $\bar{M}_w/\bar{M}_n$ and is referred to as the polydispersity index, since it is a measure of the polydispersity of the polymer. For linear polyamides, its value approaches 2 with increasing extents of reaction. Nonstoichiometric concentrations of bifunctional reactants or the presence of monofunctional species results in modified expressions for $\bar{P}_n$ and $\bar{P}_w$ (see Equation 2.6), but cause only negligible changes of the ratio $\bar{P}_w/\bar{P}_n$.

Both the molecular weight and the molecular weight distribution are of prime importance in (1) controlling and thus conducting the polymerization process; (2) the fiber forming process, because of the strong effects on the rheological properties of the melt of polyamides; and (3) the final product characteristics, particularly in regard to the tensile properties. Therefore, more than the determination of one molecular weight is necessary for adequate polymer characterization and evaluation.

The literature includes a quite detailed discussion of molecular weight determination, polydispersity, conformational rigidity, and branching [16d]. Subjects associated with end group analysis, membrane osmometry, light scattering, viscometry, turbidimetric titration, gel permeation chromatography, and branching are discussed. The identification of textile fibers by nondestructive interference microscopy also has been reported in a discussion of the quantitative analysis of interferograms in connection with the equirefractive immersion for the classification of textile fibers [47]. The method entails the measurement of the two refractive indices in light polarized parallel and perpendicular to the fiber axis. Computerized evaluation has been used in characterization and identification of chemical fibers by infrared spectrometric methods [48].

2.2.5 NYLON-6,6 POLYAMIDE

Nylon-6,6 is the generic term for poly(hexamethylene adipamide). It is commercially synthesized by polycondensation from hexamethylene diamine and adipic acid according to the amidation reaction of Equation 2.1:

$$n\text{H}_2\text{N}-(\text{CH}_2)_x-\text{NH}_2 + n\text{HOOC}-(\text{CH}_2)_4-\text{COOH}$$
$$\text{Hexamethylene diamine} \qquad \text{Adipic acid}$$
$$\rightleftharpoons \text{H}-[-\text{HN}-(\text{CH}_2)_6-\text{NH}-\text{OC}-(\text{CH}_2)_4-\text{CO}-]_n-\text{OH} + 2n\text{H}_2\text{O}$$
$$\text{Nylon-6,6: poly(hexamethylene adipamide)}$$

2.2.5.1 Synthetic Procedure

Zimmerman et al. [5–7] gave a comprehensive summary of this synthesis. Hexamethylene diamine melts at 40.9°C and is normally used in the form of a concentrated aqueous solution. Adipic acid has a melt temperature of 152.1°C and is used in its pure solid form. A salt solution of about 50% concentration containing precisely stoichiometric quantities of the two intermediates is first prepared. In a typical polymerization reaction, the salt solution is heated to boiling to evaporate water, possibly at elevated pressure, until its salt content reaches ≥60%. The concentrated salt solution is then heated gradually in a reactor as water is evaporated, typically from 212°C to 275°C at 1.73 MPa (250 psi). The polymer molecular weight will reach about 4400 at this point. The pressure is then gradually reduced to atmospheric to allow further reaction for about an hour. The polymer molecular weight is now in the range of 15,000 to 17,000, but is not quite equilibrated. All of the liquid water in the salt solution and nearly all of the potential water of reaction in the form of amine and carboxyl end groups are removed at this point. The loss of hexamethylene diamine, which boils at 200°C, is minimal. The resulting polymer is suitable for melt spinning or chip forming.

2.2.5.2 Kinetics and Thermodynamics

The equilibrium constant K_c is defined in Equation 2.5 as a function of end group and water concentrations: $[\text{NHCO}][\text{H}_2\text{O}]/[\text{COOH}][\text{NH}_2]$. For a typical reaction of high conversion and high polymer molecular weight, the equilibrium constant at 280°C is about 300 ± 50. The amide group concentration [NHCO] at high conversions is almost constant as molecular weight varies. The water concentration in the melt at a given temperature depends only on the water vapor pressure. According to Equation 2.5, therefore, the equilibrium value of $[\text{COOH}][\text{NH}_2]$ is proportional to the water concentration at low steam pressures and high molecular weights. The amidation reaction at high conversions is exothermic with a heat of reaction of about 25–29 kJ mol^{-1} (6–7 kcal mol^{-1})[4]. This is in the same range as the endothermic heat of vaporization of water from nylon based on equilibrium regain data extrapolated to the reaction temperature. For this reason, the decrease in equilibrium constant K_c with increasing temperature is almost exactly offset by the reduction in water content [H$_2$O] at a given steam pressure. The equilibrium value of [COOH][NH$_2$] does not change significantly for some given water vapor pressure as the temperature is varied.

This polycondensation reaction follows a second-order kinetics at conversions up to about 98. It becomes a third-order reaction at higher conversion where it is catalyzed by [COOH] end groups. In the presence of water, the overall rate of reaction is a function of $[COOH]^2[NH_2]$ for amidation and of $[COOH][H_2O]$ for hydrolysis. For nylon-6,6, the activation energy for polyamidation has been estimated at about 88 kJ mol^{-1} (21 kcal mol^{-1}). This is significantly higher than that for diffusion, which is estimated to be 59 kJ mol^{-1} (14 kcal mol^{-1}) at temperatures well above the polymer melting point. The rate of increase is about 40% for a 10°C temperature increase in the normal range of polymerization temperatures. Thus, the reaction equilibrium as manifested in Equation 2.10 is quite favorable for nylon-6,6. Under an atmosphere of steam, the reaction rate is not affected by diffusion (stirring) and by the rate of water removal up to a polymer molecular weight of about 18,000. As with [COOH] carboxyl end groups, hypophosphite salts and phosphonic acids have also been found to exert catalytic effect to the nylon-6,6 reaction at high conversions.

The molecular weight distribution of linear nylon-6,6 follows the most probable distribution of Equation 2.22. The average number polymerization degree $\bar{P}_n = 1/(1-p)$. The number average molecular weight $\bar{M}_n$ can be obtained by multiplying the respective value of $\bar{P}_n$ by the molecular weight of repeat unit. If the end groups are not widely unbalanced, an average value of molecular weight for the two kinds of end groups, 113 g for nylon-6,6, may be used. If 99% of the original end groups react, $p = 0.99$, $\bar{P}_n = 100$, $\bar{M}_n = 11,300$. This appears to be the low limit for most commercial nylon-6,6 today. At higher conversion, for example, $p = 0.993$, $\bar{P}_n = 143$, $\bar{M}_n = 16,140$, and $\bar{M}_w = 32,170$.

For linear polyamides, the viscosity of dilute or moderately concentrated solution can be related closely to $\bar{M}_w$. Thus, the molecular weight of polymer can be evaluated from $\eta_{inh} = \ln \eta_{rel}/c$. This is usually measured at a concentration of 0.5 g of polymer in 100 ml of solvent, e.g., m-cresol. A typical value of η_{inh} of nylon-6,6 is one for $\bar{M}_n$ of about 15,000. Another method of characterization commonly used is to measure the relative viscosity (RV) of an 8.4% solution of polymer in 90% formic acid. Typical values of relative viscosity for nylon-6,6 are in the range of 30 to 70. An RV of 41 corresponds to $\bar{M}_n$ of about 15,000, whereas an RV of 60 corresponds to about 19,000. Polymers in the lower range are used for textile yarns, and those in the higher range for industrial yarns.

2.2.5.3 Solid-State Polymerization

According to a DuPont patent, nylon-6,6 and nylon-6 can be subjected to solid-state polycondensation at temperatures below its melting point [49,50]. As an example, a nylon-6,6 prepolymer with a molecular weight of 2500 was polymerized at 216°C for 4 hr to attain a molecular weight of 16,000. Zimmerman [6] investigated the kinetics of solid-state polymerization of nylon early in 1953 [7]. He observed that the reaction is catalyzed by [COOH] end groups and follows third-order kinetics. Unlike low molecular weight prepolymers having end groups with a high degree of ionization, the polymers for solid-state polymerization are initially at high conversions and their end groups are mostly in the nonionic form. In fact, the third-order reaction rate is not quite four times as fast as expected of the melt reaction at the same temperature. The activation energy for amidation and that for diffusion of reactive ends in solid state are similar—about 84 kJ mol^{-1} (20 kcal mol^{-1}). Thus, it was suggested that the rate of solid-state polymerization is largely diffusion-controlled.

The molecular weight distribution for solid-state polymerization is about normal, if the polymer does not exceed the entanglement point. Branching occurs in the polymer and the ratio of $\bar{M}_w/\bar{M}_n$ will increase to 3.4 beyond this point.

2.2.6 Nylon-6 Polyamide

Nylon-6 is the generic name for polycaprolactam. It is almost exclusively synthesized from ε-caprolactam by a ring-opening reaction: Equation 2.3

$$\overset{\displaystyle\overline{\hspace{3em}}}{HN-(CH_2)_5-C}=O \rightleftharpoons -[-HN-(CH_2)_5-CO-]-$$
$$\text{ε-caprolactam} \qquad\qquad \text{Nylon 6: polycaprolactam}$$

The reaction is essentially an addition polymerization, but can be considered to be the condensation polymerization of AB-type polyamide.

2.2.6.1 Synthetic Procedure

Caprolactam melts at about 69°C. It does not polymerize upon heating to elevated temperatures. However, shortly after Carothers developed nylon-6,6, Schlack [51] of I.G. Farben discovered that the ring-opening reaction occurs readily in the presence of amine and carboxyl groups. Thus, ε-aminocaproic acid, nylon-6,6 salt, or simply water, is employed to hydrolyze lactam to form [COOH] and [NH$_2$] end groups. The [COOH] group catalyzes the addition of [NH$_2$] to the caprolactam ring. This discovery led to the polymerization of caprolactam for nylon-6.

The polymerization of caprolactam is carried out initially in the presence of water at 265°C initially under pressure. It is generally characterized by an induction period to build up the hydrolyzed products. As the end group concentrations increase, the carboxyl-catalyzed amine addition proceeds at an increasing rate and the polymer chain grows. This reaction also produces cyclic oligomers. The [COOH] and [NH$_2$] end groups reach a maximum concentration with time and then decreases as the monomer content depletes to equilibrium. The equilibrium constants for the end groups in nylon-6 is reportedly in the range from about the same as nylon-6,6 to somewhat above, e.g., 428 at 280°C [5–7].

2.2.6.2 Kinetics and Thermodynamics

The heat of amidation is about the same as nylon-6,6. The activation energy for polymerization is about 78 kJ mol^{-1} (18.7 kcal mol^{-1}) for the −COOH-catalyzed reaction and 88 kJ mol^{-1} (21 kcal mol^{-1}) for the uncatalyzed reaction. The activation energy of melt viscosity is about 60 kJ mol^{-1} (14.3 kcal mol^{-1}), which is almost the same as for nylon-6,6. The values of K and a in the Mark–Houwink equation, as shown in Table 2.1, are 18.0×10^4 dl/g and 0.65 in m-cresol at 25°C, respectively.

The molecular weights of nylon-6 are generally in the same range as nylon-6,6. Its molecular weight distribution as prepared under the above conditions is also the same as linear nylon-6,6 and other linear condensation polymers with $\overline{M}_w/\overline{M}_n = 1 + p$, or about two. However, nylon-6 and AB polyamides are unique in that a narrower molecular weight distribution can be attained by adding a bifunctional stabilizer such as the Bb-type dicarboxylic acid [37]. An AB polymer chain can only pick up one BB unit and becomes terminated with B end groups at both ends. For example, if 40 equiv. 10^{-6} g. of BB units are added and the concentration of A ends [NH$_2$] is reduced to 10 equiv. 10^{-6} g, $\overline{M}_w/\overline{M}_n = 1.6$. It would be possible to increase $\overline{M}_n$ by 25% in that case.

Reimschuessel reviewed the water-initiated, ring-opening polymerization of caprolactam in detail [27]. In an attempt to analyze the process from its kinetic and mechanistic aspects, the equilibrium reactions 2.11, 2.12, and 2.13 are conventionally presented as follows:

1. Ring opening

$$\underset{x}{CL} + \underset{w}{H_2O} \underset{k_1'}{\overset{k_1}{\rightleftarrows}} \underset{s_1}{ACA} \quad (K_1)$$

where C denotes the carboxyl group, L the lactam segment, and A the amine group.

2. Polycondensation

$$S_n + S_m \underset{k_2'}{\overset{k_2}{\rightleftharpoons}} S_{n+m} + w$$

$$\underset{a}{-NH_2} + \underset{c}{-COOH} \longrightarrow \underset{z}{-NHCO-} + \underset{w}{H_2O} \quad (K_2)$$

3. Polyaddition

$$\underset{x}{CL} + \underset{s}{S_n} \underset{k_3'}{\overset{k_3}{\rightleftharpoons}} \underset{s-s_1}{S_{n+1}} \quad (K_3)$$

where x is the concentration of caprolactam, w the concentration of water, s the concentration of polymer molecules, the subscripts n and m indicate a degree of polymerization of n or m, S_1 the concentration of ε-aminocaproic acid, c the concentration of carboxyl groups, a the concentration of amine groups, and z the concentration of amide linkages. With $w = w_o - s$, $z = x_o - x - s$, and $s_2 = s$, where w_o is the initial water concentration, s_2 the linear dimer, the following rate equation may be written [9,27]:

$$dx/dt = -k_1[x(w_o - s) - s_1/K_1] - k_3[xs - (s - s_1)/K_3] \quad (2.32)$$

$$ds/dt = -k_1[x(w_o - s) - s_1/K_1] - k_2[s^2 - (x_o - x - s)(w_o - s)/K_2] \quad (2.33)$$

$$ds_1/dt = -k_1[x(w_o - s) - s_1/K_1] - 2k_2[ss_1 - (w_o - s)(s - s_1)/K_2] - k_3(xs_1 - s_1/K_3) \quad (2.34)$$

The kinetic and thermodynamic constants are defined as follows:

$$K_i = k_i/k_i' = \exp[(\Delta S_i - \Delta H_i/T)/R] \quad (i = 1, 2, 3) \quad (2.35)$$

$$k_i = k_i^o + k_i^c c \quad (2.36)$$

$$k_i^j = A_i^j \exp(-E_i^j/RT) \quad (j = o, c) \quad (2.37)$$

where k_i^o is the rate constant for the uncatalyzed reaction and k_i^c is the rate constant for the reaction catalyzed by carboxyl end groups.

Values for all of the kinetic and thermodynamic parameters have been reported in the literature. They were usually obtained from experiments in which caprolactam and definite amounts of water were heated in closed systems for various periods of time. From the mechanism and the corresponding rate equation, it is readily seen that for a given temperature the concentration of water is the principal process parameter. It affects both the rate and the attainable degree of polymerization. If in the kinetic experiment, therefore, any free reactor volume (vapor space) is not essentially eliminated (which may pose some experimental problems), then the effective initial water concentration is lower and consequently a lower rate of polymerization will result. This may be one reason for certain differences in values reported by different investigators. Another reason may entail different analytical approaches. Table 2.2 and Table 2.3 show the kinetic and thermodynamic parameters as reported by two different groups [52,53] for the three principal equilibrium reactions.

Both sets of parameters have been used for simulations of the hydrolytic polymerization process [52,54–56] and both appear suitable to predict general responses in different reactor systems as functions of initial composition and process parameters. Comparisons of observed

TABLE 2.2
Kinetic Parameters

$$k_i^j = A_i^j \exp(-E_i^{0j}/RT)$$

$$(i+1,2,3; j = 0,C)^a$$

i	A_i^0 (kg/mol·h)		E_i^0 (cal/mol)		A_i^c (kg^2/mol^2·h)		E_i^c (cal/mol)	
	I	II	I	II	I	II	I	II
1	1.6940x10^6	5.987x10^5	2.1040x10^4	1.9980x10^4	4.1060x10^7	4.3075x10^7	1.873x10^4	1.8806x10^4
2	8.6870x10^9	1.8942x10^{10}	2.2550x10^4	2.3271x10^4	2.3370x10^{10}	1.2114x10^{10}	2.0674x10^4	2.0670x10^4
3	2.6200x10^8	2.8558x10^9	2.1269x10^4	2.2845x10^4	2.3720x10^{10}	1.677x10^{10}	2.0400x10^4	2.0107x10^4

Note: [a] i = 1, ring opening; i = 2, polycondensation; i = 3, polyaddition.

Source: Reimschuessel, H.K.; Nagasubramanian, K., *Chem. Eng. Sci.*, 1972, 27, 1119; Tai, K.; Teranishi, H.; Arai, Y.; Tagawa T., *J. Appl. Polym. Sci.*, 1980, 25, 77. With permission.

TABLE 2.3
Thermodynamic Parameters

$$K_i = \exp.[(\Delta S_i - \Delta H_i/T)/R] (i = 1,2,3)^a$$

	ΔS_i (e.u.)		ΔH_i (cal/mol)	
	I	II	I	II
1	$-7,8700$	$-7,8846$	$2,1142 \times 10^3$	$1,9180 \times 10^3$
2	$9,3000 \times 10^{-1}$	$9,4374 \times 10^{-1}$	$-6,1404 \times 10^3$	$-5,9458 \times 10^3$
3	$-6,9500$	$-6,9457$	$-4,0283 \times 10^3$	$-4,0438 \times 10^3$

Source: From Tai, K.; Teranishi, H.; Arai, Y.; Tagawa T., *J. Appl. Polym. Sci.*, 1980, 25, 77. With permission.

and calculated data are presented in Figure 2.3 and Figure 2.4. Although the calculated data are in general consistent with the observed ones, certain discrepancies are evident. This is true even for the curves represented by the broken lines, which were calculated using the very same parameters that had been obtained by curve fitting of the experimental data represented by the solid lines [53]. These differences may be a consequence of the definition of the equilibrium constants K_i. Rather than defined by activities, these constants are defined by equilibrium concentrations. Thus they are dependent not only on the temperature but also on the initial composition (that is, the initial water concentration). Consequently, according to Equation 2.35 and Equation 2.37, all of the kinetic and thermodynamic parameters depend on the initial composition. Nevertheless, the kinetic system, as represented by Equation 2.32 through Equation 2.37, has been very useful for predicting responses to changes in process parameters and has been used as a basis for modeling and optimization of the polymerization process [52,54–58].

Both the water concentration and temperature have been identified as the principal process parameters. Their effect on the rate and extent of monomer conversion and on the rate and degree of polymerization is shown in Figure 2.5 through Figure 2.8. The effect of temperature on both reaction rate and equilibrium has been well recognized. As a consequence of the exothermic nature of the addition and condensation reactions, lower temperature is favorable for obtaining higher equilibrium values for both monomer conversion and degree of polymerization.

2.2.6.2.1 Cyclic Oligomers

The formation of low-molecular-weight cyclic polymers during the polymerization of caprolactam is of concern in regard to conversion and polymer molecular weight. These oligomers are soluble in water and in the lower alcohols. Individual members ranging from the cyclic dimer to the cyclic monomer have been identified in polymer extracts. The formation of these ring structures may be explained by intra- or intermolecular transacylation and transamidation, and direct cyclization of the corresponding linear oligomers. The concentration of the cyclic dimer amounts to about 50 mol % of the total concentration of the soluble cyclic oligomers [27]. The formation of these cyclic dimers is represented by reactions in (2.74). A kinetic model postulates that these reactions are governed by the equilibrium reactions (2.32) and (2.33).

$$\overline{NH(CH_2)_5CONH(CH_2)_5CO} + H_2O \rightleftharpoons H[NH(CH_2)_5CO]_2OH$$

$$\overline{NH(CH_2)_5CONH(CH_2)_5CO} + H[NH(CH_2)_5CO]_nOH \rightleftharpoons H[NH(CH_2)_5CO]_{n+2}OH \quad (2.74)$$

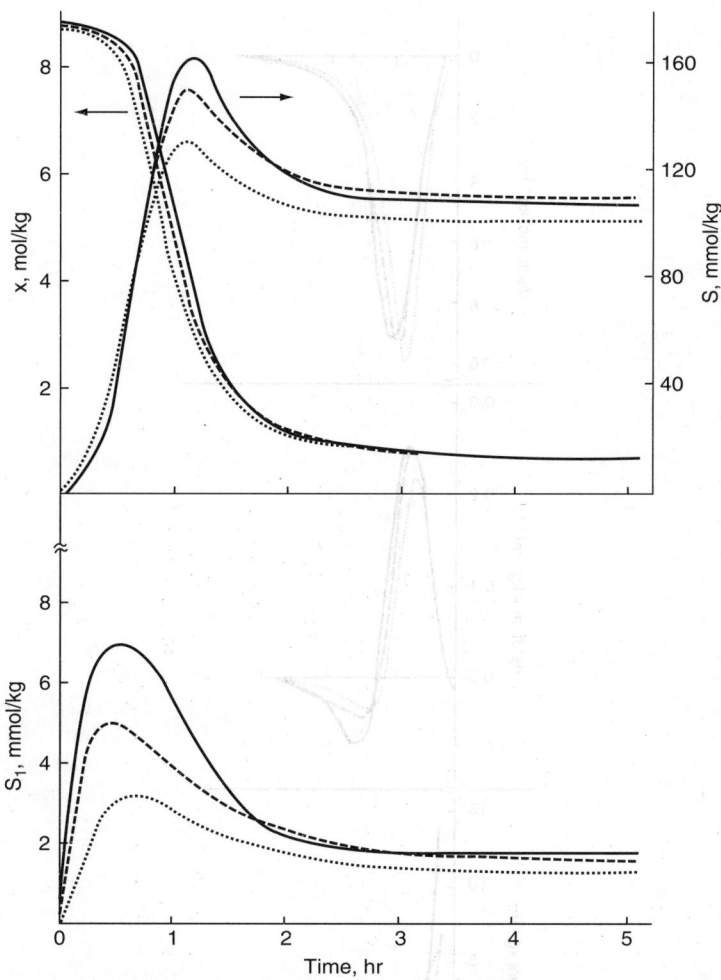

FIGURE 2.3 Comparison of observed and calculated concentration–time relationships for caprolactam (x), polymer chains (s), and amino caproic acid (S_1); $W_o = 0.82$ mol/kg, $T = 259°$C. Solid lines represent experimental data; broken lines calculated using set-II parameters; dotted lines calculated using set I parameters. (From Reirnschuessel, H.K. and Nagasubramanian, K., *Chem. Eng. Sci.*, 1972, 27, 1119. With permission.)

The kinetic and thermodynamic parameters for these equations have already been reported in polymerization kinetics and simulation studies [59–66]. The rate constants have also been obtained by Reimschuessel [52], without considering cyclization, and by Tai et al. [53], with consideration of cyclic dimer formation. Gupta et al. [56] employed these published data in a numerical analysis of rate equations for the above equilibrium reactions to demonstrate the effects of incorporating the formation of cyclic oligomers into the kinetic scheme. There were no discernible differences for the two sets of rate constants for conversion. However, the final degree of polymerization is slightly lower, and the ratio P_w/P_n approaches a value somewhat higher with calculations based upon Arai et al. [66] than the corresponding values based upon Reimschuessel [52].

In a similar effort, Mallon and Ray [67] refined the kinetic model to include the effect of water on nylon-6 equilibria. Variations in equilibrium were attributed to microscopic states

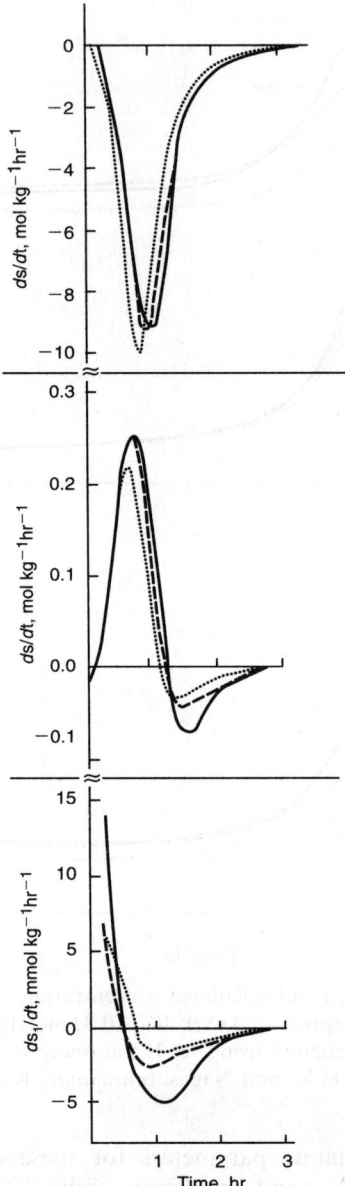

FIGURE 2.4 Comparison of observed and calculated reaction rates dx/dt; ds/dt. $W_o = 0.82$ mol/kg; $T = 259°C$. Solid lines; experimental data; broken lines calculated using set-II parameters; dotted lines calculated using set-I parameters. (From Reimschuessel, H.K. and Nagasubramanian, K., *Chem. Eng. Sci.*, 1972, 27, 1119. With permission.)

of water in nylon melt and were modeled as due to a changing dielectric constant. All reactions were assumed to be acid catalyzed. Results were good for water to polymer ratio ranging from 0.01 to 0.30 and for temperatures from 200 to 280°C. The model framework allowed the calculation of interchange rates and cyclic oligomer concentrations. Figure 2.9 and Figure 2.10 illustrate the results of the proposed model with the model of Tai et al. [68], and Wiloth's

Polyamide Fibers

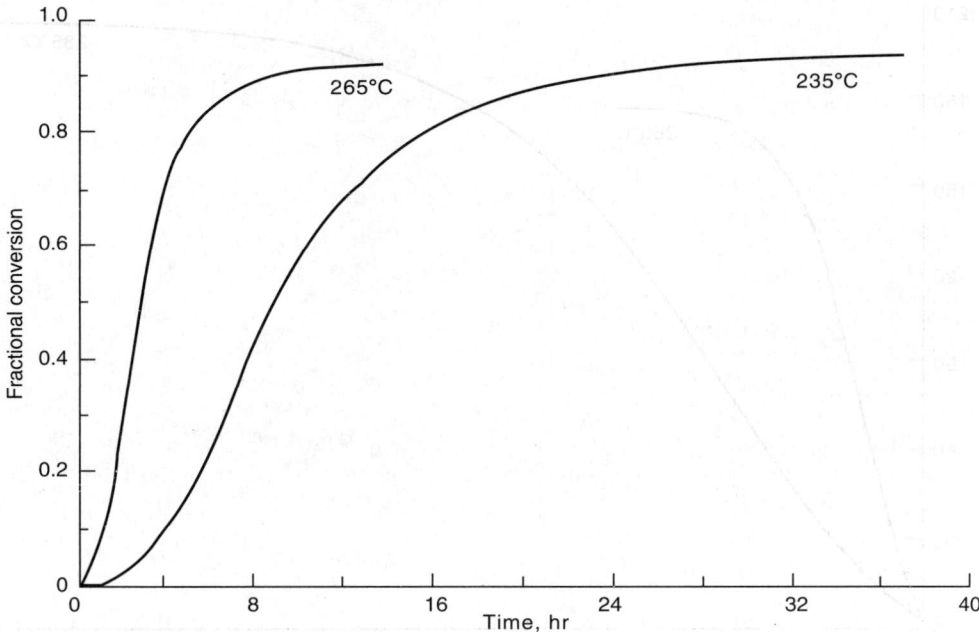

FIGURE 2.5 Effect of temperature on extent of conversion; $[W_o] = 0.02$.

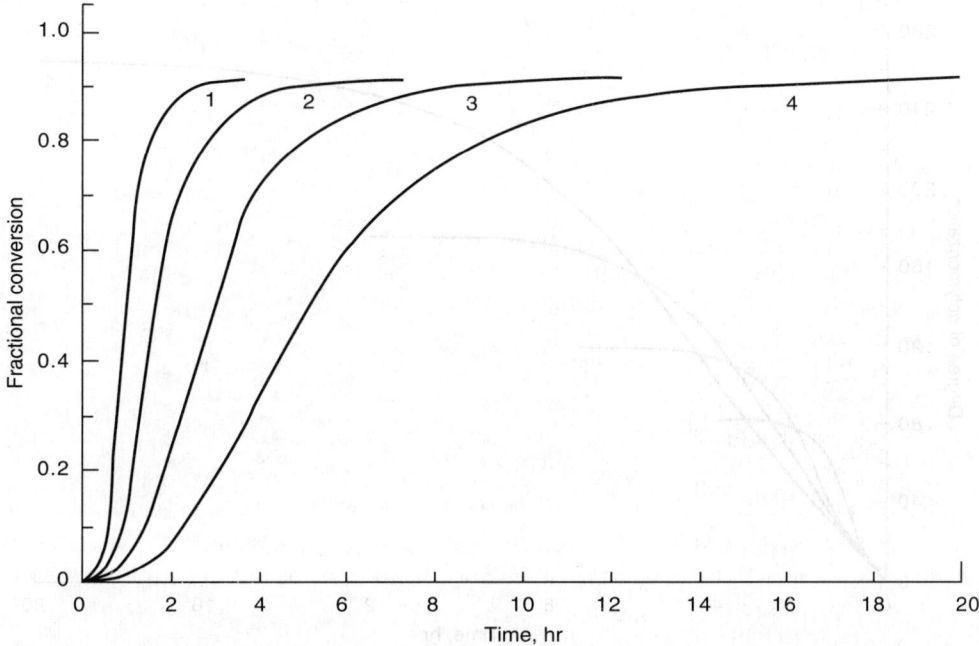

FIGURE 2.6 Effect of initial water concentration $[W_o]$ on extent of conversion; $T = 265°C$; $[W_o]_1 = 0.08$; $[W_o]_2 = 0.04$; $[W_o]_3 = 0.02$; $[W_o]_4 = 0.01$.

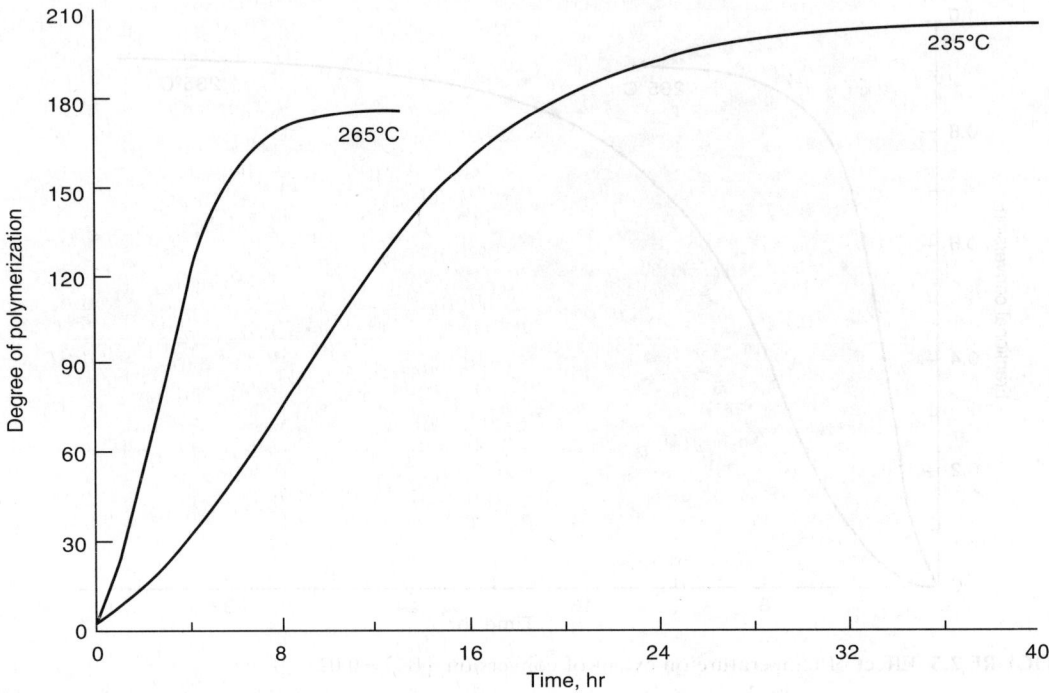

FIGURE 2.7 Effect of temperature on rate and degree of polymerization; $[W_o] = 0.02$.

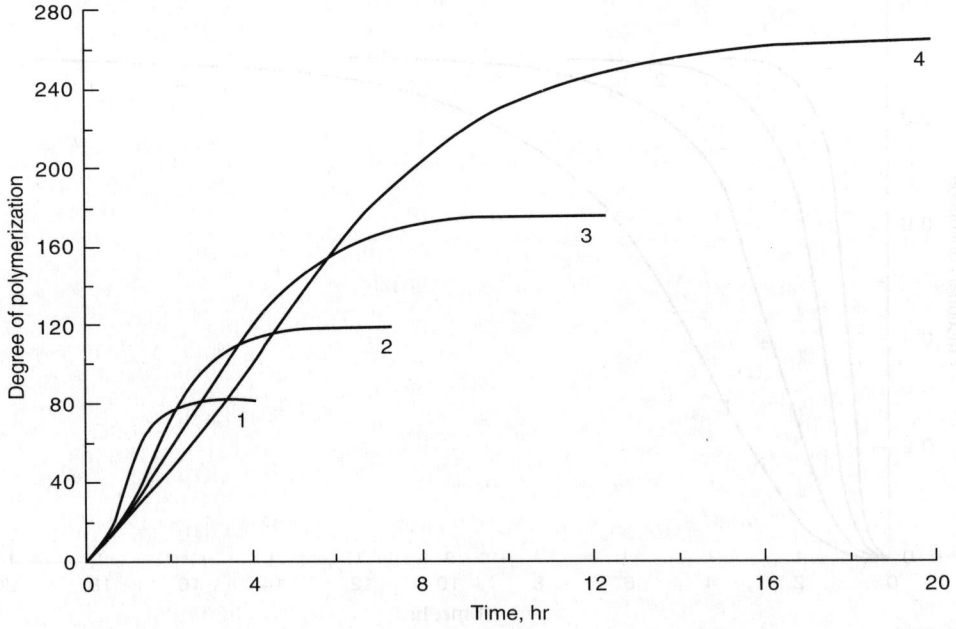

FIGURE 2.8 Effect of initial water concentration $[W_o]$ on rate and degree of polymerization; $[W_o]_1 = 0.08$; $[W_o]_2 = 0.04$; $[W_o]_3 = 0.02$; $[W_o]_4 = 0.01$.

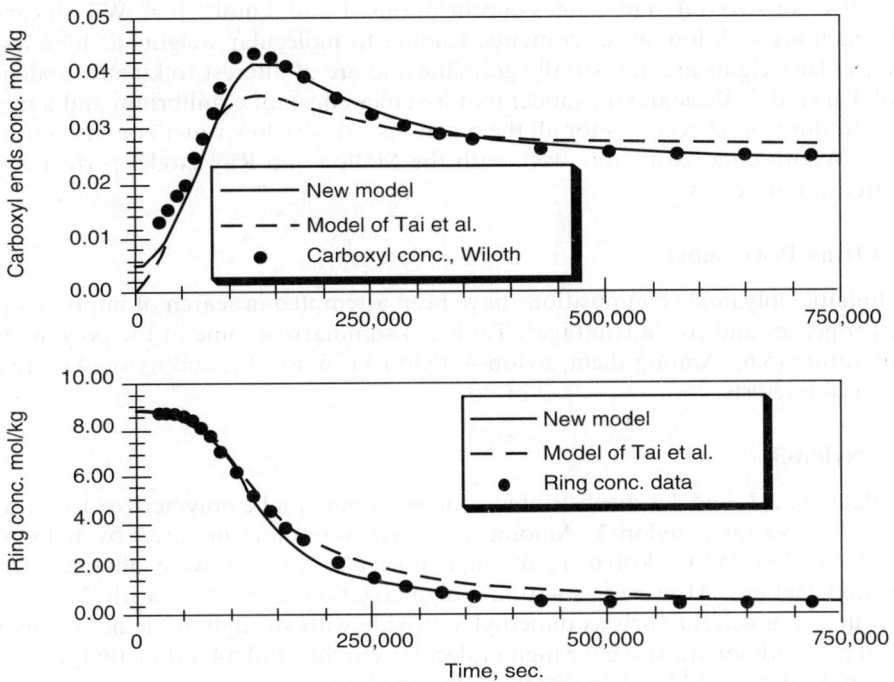

FIGURE 2.9 Polymerization at 220°C, $[W_o] = 1$ mol %. (From Mallon, F.K. and Ray, W.H., *J. Appl. Polym. Sci.*, 1998, 69, 1213–1221. With permission.)

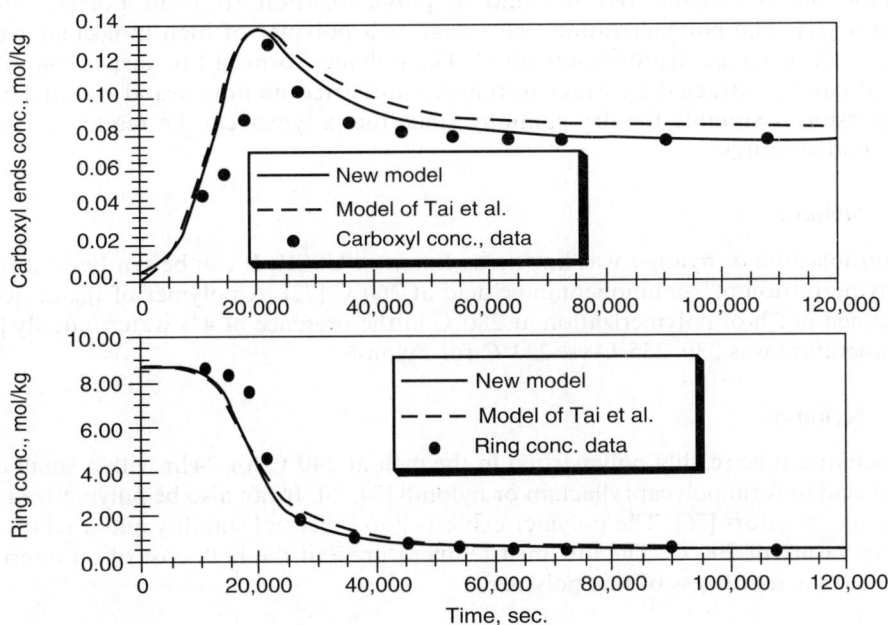

FIGURE 2.10 Polymerization at 220°C,. $[W_o] = 8$ mol %. (From Tai, K.; Tagawa, T.; *Ind. Eng. Chem. Prod. Res. Dev.*, 22, 192, 1983.)

data at 220°C for two different water concentrations: 1 and 8 mol% [69]. Wiloth carried out nylon-6 reactions with low water contents, leading to molecular weights as high as 35,000. Such molecular weights are industrially germane and are of interest to kinetic modeling. The model of Tai et al. is the standard model that assumes constant equilibrium and a mixture of second- and third-order reaction for all the reactions. At very low water content, both models follow the Wiloth data reasonably well, with the Mallon and Ray model performing somewhat better in both cases.

2.2.7 OTHER POLYAMIDES

Many aliphatic polyamide compositions have been attempted in search of improved polymer end-use properties and cost advantages. Table 2.4 summarizes some of the polyamides from recent literature [5,6]. Among them, nylon-4, nylon-11, nylon-12, and nylon-4,6 are notable for industrial interests.

2.2.7.1 Nylon-3

β-Propiolactam, 3,3- and 4,4-disubstituted propiolactams can be polymerized to form unsubstituted and substituted nylon-3. Anionic polymerization was initiated by nylon-6,6 and carried out at 200–250°C. Polymers of high molecular weights were obtained from 3,3-disubstituted lactams. However, solution polymerization at 0–20°C with 25% monomer concentration in a solvent such as dimethyl sulfoxide with strongly basic activators such as potassium pyrrolidinonate also gave high molecular weights. Poly(4,4-dimethylpropiolactam) with η_{inh} of 4.5 dl/g and $\overline{M}_w$ of 50,000 was reported [70].

2.2.7.2 Nylon-4

2-Pyrrolidinone is polymerized by anionic polymerization to form polypyrrolidinone or nylon-4 [71]. The polymerization process gives a polymer of high molecular weight in 80–85% yields at temperatures below 60°C. The polymer forms a fine dispersion in hydrocarbon. It can be extracted by water to remove unreacted monomer and residual catalyst. The dispersion is suitable for dry spinning since the polymer can be readily dissolved at elevated temperatures.

2.2.7.3 Nylon-7

Polyenantholactam or nylon-7 was known as Enant in USSR. It can be synthesized from the melt polymerization of 7-aminoheptanoic acid at 260°C [72]. A polymer of η_{sp} of about 1.2 was obtained in 5 h of polymerization at 280°C in the presence of 4% water initially [73]. Its melt temperature was 230–235°C vs. 223°C for nylon-6.

2.2.7.4 Nylon-8

Capryllactam can be readily polymerized in the melt at 240°C for 24 hr with a small amount of amino acid to form polycapryllactam or nylon-8 [74,75]. It can also be polymerized rapidly with anionic initiators [76]. The polymer exhibits good thermal stability and a relatively low melt temperature of 200°C. The low melt temperature and the high cost of an intermediate have limited the usefulness of this polymer.

Polyamide Fibers

TABLE 2.4
Synthesis of Aliphatic Polyamides

Polymer type	Polymer intermediates	Synthetic route	M_w, M_n, η_{inh}	Melt. temp. (°C)	Major uses	Ref.
AB polyamides						
Nylon-3	β-Propiolactam	Melt or solution polymerization		>320	Fibers	59
3,3-dimethyl 3	3,3-Dimethyl propiolactam			250		
4,4-dimethyl 3	4,4-Dimethyl propiolactam			296		
Nylon-4	2-Pyrrolidinone	Anionic polymerization	η_{inh} 4.5	260	Fibers	60
Nylon-6	Caprolactam	Ring-opening polymerization	M_n 20,000	232	Fibers, resins	61, 62
Nylon-7	ω-Enantholactam, or 7-aminoheptanoic acid	Melt amidation 200°C	η_{sp} 1.2–1.3	233	Fibers	63–65
Nylon-8	Capryllactam	Ring opening		200		6
Nylon-9	ω-Aminopelargonic acid	Amidation		209		6
Nylon-10	Azacycloundecan-2-one	Ring opening		188	Film	8
Nylon-11	ω-Aminoundecanoic acid	Melt amidation 215°C		190	Fibers	8, 66
Nylon-12	Azacyclotridecan-2-one	Ring opening 300°C		170		
AABB polyamides						
Nylon-4,2	Tetramethylene diamine Diethyl oxalate	Amidation	η_{inh} 2.7	380		67
Nylon-4,6	Tetramethylene diamine Adipic acid	Amidation 140°C Solid-phase polym	M_n 32,900	295	Fibers, resins	68, 69
Nylon-6,6	Hexamethylene diamine Adipic acid	Amidation	M_n 20,000	265	Fibers, resins	6
Nylon-6,12	Hexamethylene diamine Dodecanedioic acid	Amidation			Resins	6, 70
Aliphatic–aromatic polymers						
Nylon-4,I	Tetramethylene diamine Isophthalic acid	Amidation				71
Nylon-6,I	Hexamethylene diamine Isophthalic acid	Amidation	M_n 13,000	180–210	Resins	72–74
Nylon-6,T	Hexamethylene diamine Terephthalic acid	Interfacial amidation		370	Fibers	75
PACM,12	Bis(4-aminophenyl)methane Dodecanedioic acid	Amidation		290	Fibers	76–78

2.2.7.5 Nylon-11

Poly(aminoundecanoic acid) or nylon-11 was first synthesized by Carothers in 1935 and commercialized in France in 1955 [8]. The polymer is prepared from ω-aminoundecanoic acid by melt polymerization at 215°C under nitrogen for 3 h. This polymer is hydrophobic and has excellent electrical properties.

2.2.7.6 Nylon-12

ω-Dodecanolactam is polymerized in the melt at temperatures above 300°C with an acid catalyst [8,77]. The polymer has typically low extractable content and low melt temperature of 179°C.

2.2.7.7 Nylon-4,2

Tetramethylene diamine and diethyl oxalate were prepolymerized in a 50/50 phenol/trichlorobenzene mixture at 140°C. The prepolymer, after purification, was subjected to solid-phase polymerization under nitrogen at 250–300°C to form poly(tetramethylene oxalamide) or nylon-4,2 [78]. The polymer exhibited η_{inh} as high as 2.7 dl/g as measured from a 0.5% solution of polymer in 96% sulfuric acid. Interestingly, the polymer was soluble in trifluoroacetic acid, dichloroacetic acid, and 96% sulfuric acid, but not in 90% formic acid. It had a melt temperature of 388–392°C and heat of fusion of 148–154 J/g (35–37 cal/g).

2.2.7.8 Nylon-4,6

Poly(tetramethylene adipamide) or nylon-4,6 is prepared from tetramethylene diamine and adipic acid by polymerization in an organic solvent or by melt polymerization followed by solid-state polymerization [79,80]. The polymer melts at 295°C, about 30°C above nylon-6,6. It is more sensitive to degradation and branching. The volatility of tetramethylene diamine also makes it difficult to control the balance of end groups during polymerization. Excess diamine is added to compensate the losses. Values of $\overline{M}_n$ as high as 32,900 have been reported.

2.2.7.9 Nylon-6,12

Poly(1,6-hexamethylene dodecaneamide) or nylon-6,12 is synthesized from the polycondensation of hexamethylene diamine and dodecanedioic acid. It has been commercially produced by DuPont as a resin, Zytel* 158L [81].

2.2.7.10 Nylon-4,I

The inclusion of a ring unit in an AABB polyamide generally brings about significant changes in polymer chain structures and thermal properties. Thus, poly(tetramethylene isophthalamide) or nylon-4,I and poly(hexamethylene isophthalamide) or nylon-6,I both tend to be amorphous. They can be crystallized with great difficulties in boiling water or swelling agents [82].

2.2.7.11 Nylon-6,I

Poly(hexamethylene isophthalamide) of high molecular weights with $\overline{M}_n = 13,000$ for $\eta_{inh} = 0.74$ dl/g in formic acid was reported. The polymer exhibited melt temperature of 180–210°C and low crystallinity [83–85]. It is used as an engineering resin with good, high transparency and adhesive properties.

---

*Zytel—a registered trademark of E.I. du Pont de Nemours & Co., Inc., Wilmington, Delaware, USA.

2.2.7.12 Nylon-6,T

Poly(hexamethylene terephthalamide) or nylon-6,T has been investigated extensively [86]. The inclusion of a *para*-oriented phenylene ring provides a polymer of higher crystallinity, and higher melt and glass transition temperatures than that of a *meta*-oriented phenylene ring. Because of its high melt temperature of 370°C, it must be polymerized by alternative methods such as interfacial polymerization to avoid polymer degradation. This polyamide is interesting because its intermediates are inexpensive and it offers good product properties.

2.2.7.13 PACM,12

In the 1970s, DuPont commercialized a silk-like Qiana* fiber based on poly(bis[4-aminocyclohexyl]methane dodecaneamide) or PACM,12 [87]. The base polymer is synthesized from bis(4-aminocyclohexyl)methane and dodecanedioic acid. The diamine exists in three isomeric forms: *trans–trans*, *cis–trans*, and *cis–cis*. Its content of *trans–trans* isomer can be controlled by specific catalysts during its preparation. A normal diamine intermediate contained 51% *trans–trans*, 40% *cis–trans*, and 9% *cis–cis* isomers [88,89]. For Qiana fiber, the trans-trans content was about 70%. The polymer had a melt temperature of 290°C.

2.3 POLYMERIZATION PROCESSES

2.3.1 Monomer Syntheses

Nylon-6 and nylon-6,6 are the principal polyamides for commercial production of fibers and resins. The corresponding monomeric intermediates are caprolactam and the salt of hexamethylene diamine and adipic acid (HA-salt). The commercial importance of these two polyamides has stimulated considerable development and optimization of commercial processes for their monomers. In this section, both commercial processes and some of the significant experimental developments will be discussed.

2.3.1.1 Caprolactam

2.3.1.1.1 Overview
Figure 2.11 shows schematically the individual processes for the synthesis of caprolactam. The solid lines indicate processes that have been practiced commercially. As can be seen, all processes start from materials that belong to the group consisting of phenol, benzene, toluene, and cyclohexane. The chemistry of different processes has been reviewed [27,90,91]. Commercially, processes 1, 2, and 3 as shown in Figure 2.11 are important. The principal intermediates are cyclohexanone and cyclohexanone oxime for process 1, cyclohexanone oxime for process 2 [92–95], and cyclohexane carboxylic acid for process 3.

In recent years, BASF disclosed a number of processes for preparing caprolactam by the catalytic reaction of 6-aminocapronitrile (ACN) and water [96–103]. Two of these processes were reported to yield caprolactam and hexamethylene diamine simultaneously. The polymer intermediate ACN is derived from cleaving oligomers and polymers of caprolactam in the presence of a catalyst at high temperatures. Thus, it is possible to use wastes from the polymerization of caprolactam or from fiber spinning as the starting material. AlliedSignal Inc. (now Honeywell International) also reported process technology for depolarizing nylon-containing wastes to form caprolactam [104–107].

Additional synthesis routes include the catalytic aminomethylation of pentenoic acid derivatives to prepare caprolactam [108] and the catalytic conversion of a cyanopentenic

---

*Qiana—a registered trademark of E.I. du Pont de Nemours & Co., Inc., Wilmington, Delaware, USA.

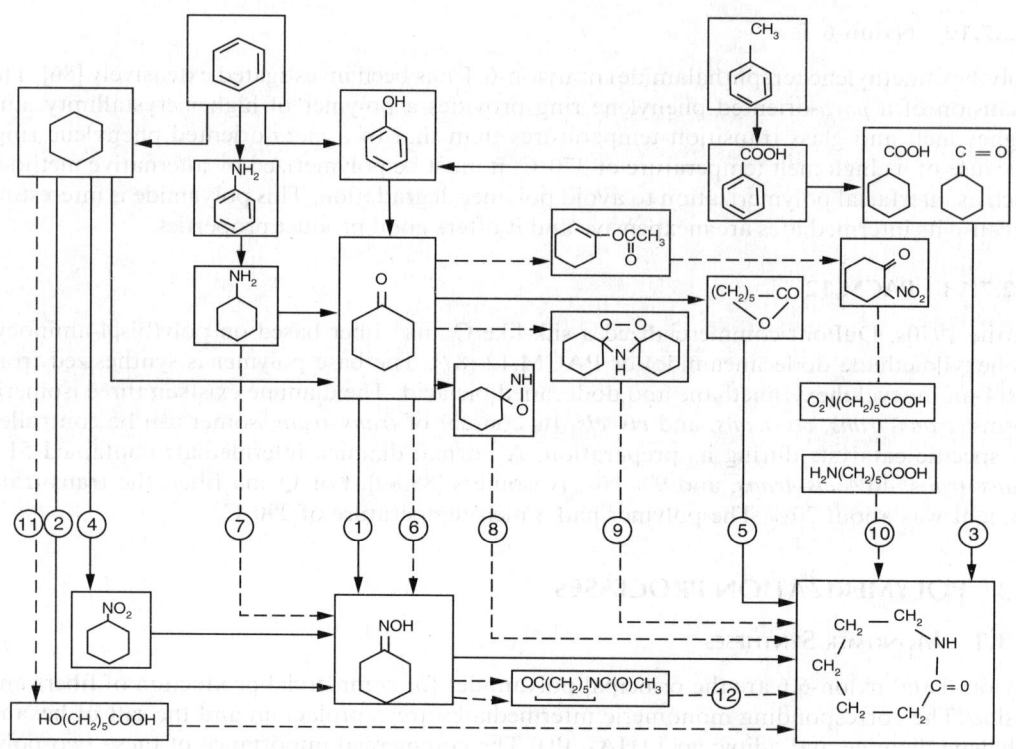

FIGURE 2.11 Block diagram of caprolactam processes.

carboxylic acid or its ester in ammonia [109]. Possible intermediates for the synthesis of caprolactam-entailing products derived from the dimerization of acrylonitril have been reported [110,111].

Procedures for the synthesis of polymer intermediate cyclohexanone:

1. Catalytic hydrogenation of phenol and subsequent dehydrogenation of the resulting cyclohexanol
2. One-step catalytic hydrogenation of phenol using palladium on carbon catalyst
3. Catalytic oxidation of cyclohexane to a cyclohexanol–cyclohexanone mixture
4. Reductive catalytic hydrolysis of cyclohexylamine in a one-stage operation

Though process 1 is still practiced, 2 and 3 are the most significant commercial processes.

Cyclohexanone oxime may be produced by:

1. Reaction of cyclohexanone with hydroxylamine (process 1, Figure 2.11)
2. Photonitrosation of cyclohexane with nitrosyl chloride (PNC process) [112,113] (process 2, Figure 2.11)
3. Hydrogenation of nitrocyclohexane [112,114] (process 4, Figure 2.11)
4. Reaction of cyclohexylamine with H_2O_2 [112,115] (process 7, Figure 2.11)
5. Ammoxydation of cyclohexanone with NH_3 and H_2O_2 [116] (process 6, Figure 2.11)

The hydroxylamine used for the reaction with cyclohexanone is obtained by the Raschig process or by catalytic hydrogenation of either nitric oxide or nitric acid. In the Raschig process, the hydroxylamine is obtained in the form of its sulfate. The raw materials for this process are sulfur dioxide, ammonia, carbon dioxide, and water. A mixture of NO and NO_2, as obtained from the catalytic oxidation of ammonia, is absorbed in an aqueous ammonium carbonate solution to yield ammonium nitrite, which is then reacted with SO_2 in the presence of ammonium hydroxide. The product of this reaction is hydroxylamine disulfonate, which converts upon hydrolysis via the monosulfonic acid hydroxylamine into hydroxylamine sulfate:

$$NH_4NO_2 + NH_4OH + 2SO_2 \xrightarrow{5°C} HON(SO_3NH_4)_2$$
$$\xrightarrow{+2H_2O,\ 100°C} (NH_3OH)HSO_4 + (NH_4)_2SO_4 \qquad (2.38)$$

In 1967, both BASF [117] and Inventa introduced a commercial process for the catalytic hydrogenation of nitric oxide [113]. The reaction is carried out in the presence of sulfuric acid and yields hydroxylamine sulfate according to:

$$NO + 1.5H_2 + H_2SO_4 \rightarrow H(NH_3OH)HSO_4 \qquad (2.39)$$

Therefore, only 0.5 mol of ammonium sulfate per mol of hydroxylamine is then produced in the subsequent neutralization with ammonia. This is half as much as in the Raschig process. Complete elimination of ammonium sulfate formation characterizes a process introduced in 1970 by Stamicarbon [118,119]. This process involves catalytic hydrogenation of nitrate ions on Pd–C in a phosphoric acid ammonium hydrogen phosphate buffering system:

$$HNO_3 + 3H_2 + H_3PO_4 + NH_4H_2PO_4 \rightarrow (NH_3OH)H_2PO_4 + NH_4H_2PO_4 + 2H_2O \qquad (2.40)$$

It is carried out in a gas–liquid contactor in which hydrogen is contacted with a circulating stream containing the nitrate ions, the buffering acid, and the noble metal catalyst. Since the hydroxylammonium ion is unstable, the reaction product is directly contacted with cyclohexanone in a toluene solution to effect the formation of cyclohexanone oxime:

$$C_6H_{10}O + (NH_3OH)H_2PO_4 \rightarrow C_6H_{10} = NOH + H_3PO_4 + H_2O \qquad (2.41)$$

The aqueous $H_3PO–NH_4H_2PO_4$ solution is reacted with a mixture of NO and NO_2, which results in the conversion of any ammonium ions to nitrogen.

$$2NH_4^+ + NO + NO_2 \rightarrow 2N_2 + 2H^+ + 3H_2O \qquad (2.42)$$

Simultaneous introduction of air results in the formation of nitric acid,

$$2NO + 1.5O_2 + H_2O \rightarrow 2HNO_3 \qquad (2.43)$$

which is then recycled into the process.

An interesting process developed and operated since 1962 by Toray for manufacturing cyclohexanone oxime is the photo-nitrosyl chlorination (PNC) process. It is a one-stage process in which cyclohexane is converted to the oxime hydrochloride, and may be explained by the following mechanism [120]:

$$NOCl \xrightarrow{h\upsilon} NO^{\bullet} + Cl^{\bullet} \qquad (2.44)$$

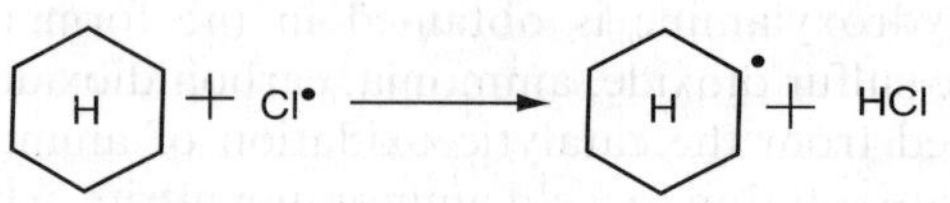

$$C_6H_{11} + Cl^{\bullet} \longrightarrow C_6H_{10}^{\bullet} + HCl \qquad (2.45)$$

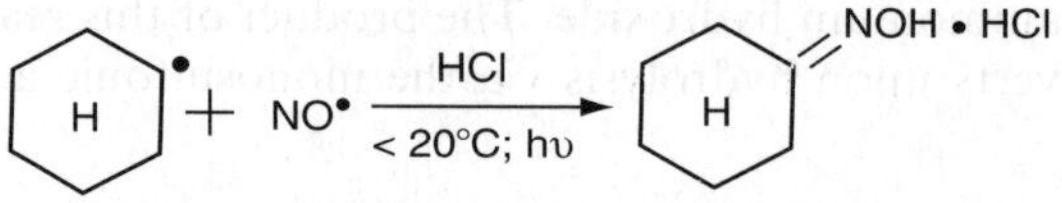

$$C_6H_{10}^{\bullet} + NO^{\bullet} \xrightarrow[< 20°C; h\upsilon]{HCl} C_6H_{10}=NOH \cdot HCl \qquad (2.46)$$

The nitrosyl chloride for this synthesis is obtained by a sequence of reactions that includes:

1. Oxidation of ammonia

$$2NH_3 + 3O_2 \rightarrow NO \cdot NO_2 + 3H_2O \qquad (2.47)$$

2. Formation of nitrosyl sulfuric acid

$$2H_2SO_4 + NO \cdot NO_2 \rightarrow 2NOHSO_4 + H_2O \qquad (2.48)$$

3. Sulfuric acid regeneration

$$NOHSO_4 + HCl \rightarrow NOCl + H_2SO_4 \qquad (2.49)$$

In a process developed to commercial maturity by DuPont, and industrially practiced by this company from about 1963 to 1967 [121], cyclohexanone oxime is obtained by a route that encompasses the nitration of cyclohexane to nitrocyclohexane and the reduction of the latter to the oxime, which is converted to caprolactam by the Beckmann rearrangement. According to U.S. Patent 2,634,269, nitrocyclohexane can be converted directly to caprolactam in the gas phase at 250–450°C using a polyborophosphate catalyst.

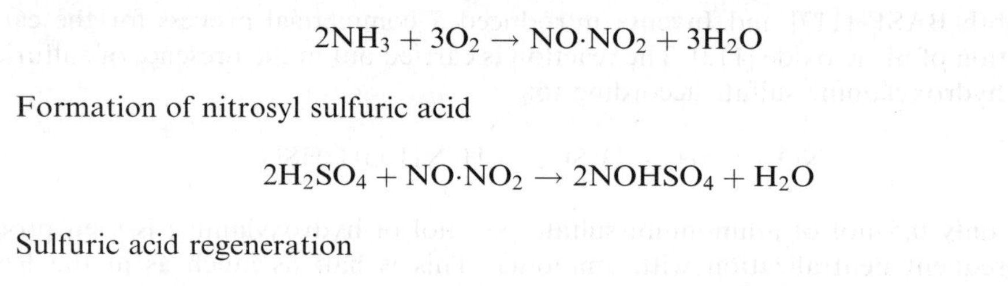

$$C_6H_{12} \xrightarrow{HNO_3\ 120°C, 4\ atm} C_6H_{11}NO_2 \xrightarrow{Ag\ (alkali)} C_6H_{10}=N\begin{smallmatrix}O\\ONa\end{smallmatrix} \qquad (2.50)$$

$$\xrightarrow{Pt/H;(Zn/Cr);\ 140°C, 3\ atm} C_6H_{10}=NOH$$

Cyclohexanone oxime processes that have not been commercially practiced include:

1. The Kahr process [115], in which cyclohexylamine is catalytically oxidized with hydrogen peroxide
2. The ammoxidation process developed by Tao Gosei Chern. Ind. Co. [116], entailing the direct reaction of cyclohexanone, ammonia, and hydrogen peroxide in a liquid phase in the presence of a tungstic acid catalyst:

$$C_6H_{10}O + NiI_3 + H_2O_2 \xrightarrow{(H_3PW_{12}O_4) \cdot 8H_2O} C_6H_{10}NOH + 2H_2O \qquad (2.51)$$

Polyamide Fibers

The cyclohexanone oxime obtained by any of these processes (1, 2, 4, 6, and 7, in Figure 2.11) is converted to caprolactam by the Beckmann rearrangement in oleum. The resulting caprolactam sulfate is neutralized with ammonia and purified. The neutralization process yields about 2 ± 0.2 kg of ammonium sulfate/kg of caprolactam. The lactam purification may entail extraction with organic solvents (toluene, benzene, chlorinated aliphatic hydrocarbons) followed by extraction of the organic solution with water and subsequent isolation of the lactam by either crystallization or distillation. Newer developments are concerned with the gas-phase catalytic rearrangement of cyclohexanone oxime using a boric acid on carbon catalyst in a fluidized-bed operation [122].

Union Carbide Corp. developed a process using cyclohexanone as a principal intermediate and used this process commercially in 1966. According to the following reaction scheme, cyclohexanone is oxidized to caprolactone with peracetic acid, which is obtained by the reaction of acetaldehyde and hydrogen peroxide. The caprolactone is then converted to caprolactam by reaction with ammonia at high temperature and high pressure (process 5, Figure 2.11). The only by-product is acetic acid; the amount of acetic acid obtained is about 1 kg/kg of product [123].

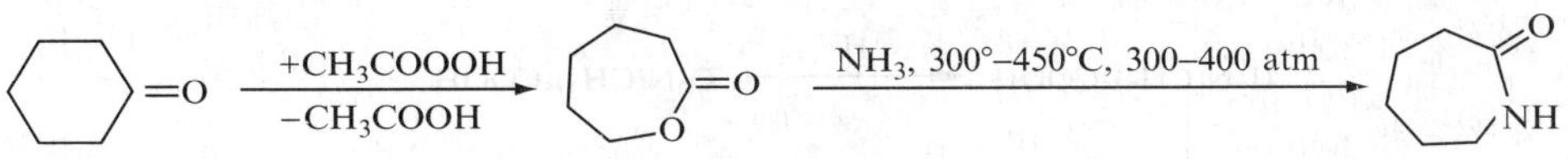

$$\text{cyclohexanone} \xrightarrow[-CH_3COOH]{+CH_3COOOH} \text{caprolactone} \xrightarrow{NH_3,\ 300°-450°C,\ 300-400\ atm} \text{caprolactam} \qquad (2.52)$$

Cyclohexanone is also the starting material for process 8 (Figure 2.11). This process does not appear to be commercially practiced. It is characterized by the formation of cyclohexanonisoxim (3,3'-pentamethylene oxaziridine) upon treatment of cyclohexanone in a toluene solution with ammonia and sodium hypochloride [124]:

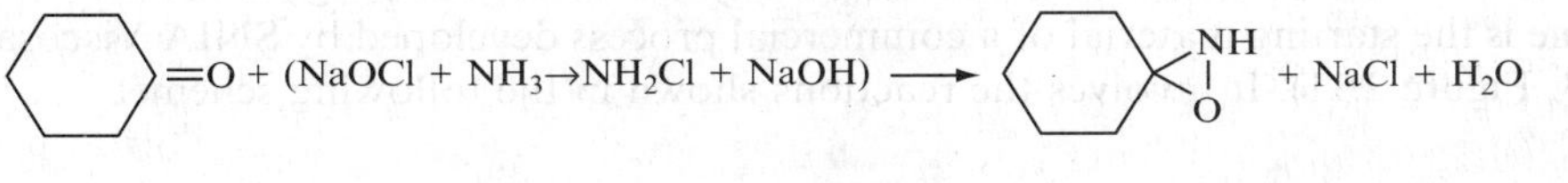

$$\text{cyclohexanone}=O + (NaOCl + NH_3 \rightarrow NH_2Cl + NaOH) \longrightarrow \text{oxaziridine} + NaCl + H_2O \qquad (2.53)$$

Process 9 (Figure 2.11) refers to the reaction of cyclohexanone with hydrogen peroxide and ammonia resulting in 1,1'-peroxy dicyclohexylamine via either of the following two routes:

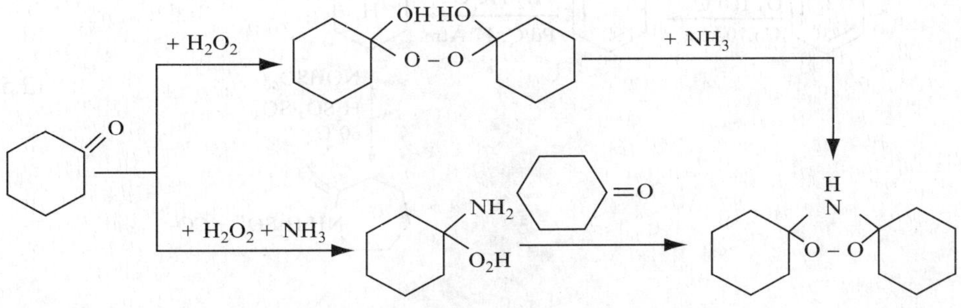

$$(2.54)$$

Cleavage of 1,1'-peroxy dicyclohexyl amine then yields caprolactam with regeneration of one equivalent of cyclohexanone according to the reaction [125,126]:

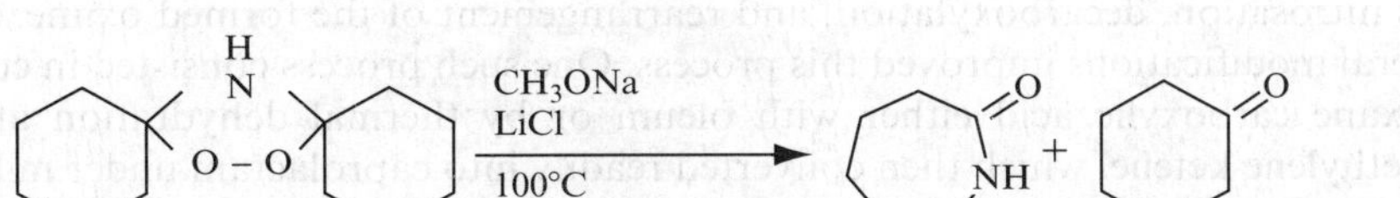

$$\xrightarrow[100°C]{CH_3ONa,\ LiCl} \text{caprolactam} + \text{cyclohexanone} \qquad (2.55)$$

The Techni-Chem. process (process 10, Figure 2.11) that started from cyclohexanone also did not develop beyond pilot plant operations [127]. It is characterized by the following reaction scheme entailing (1) acylation of cyclohexanone with ketene, (2) nitration of the resulting cyclohexenyl acetate with concurrent deacetylation to 2-nitrocyclohexanone, (3) hydrolytic cleavage to ε-nitrocaproic acid, (4) hydrogenation to ε-aminocaproic acid, and (5) cyclization to caprolactam:

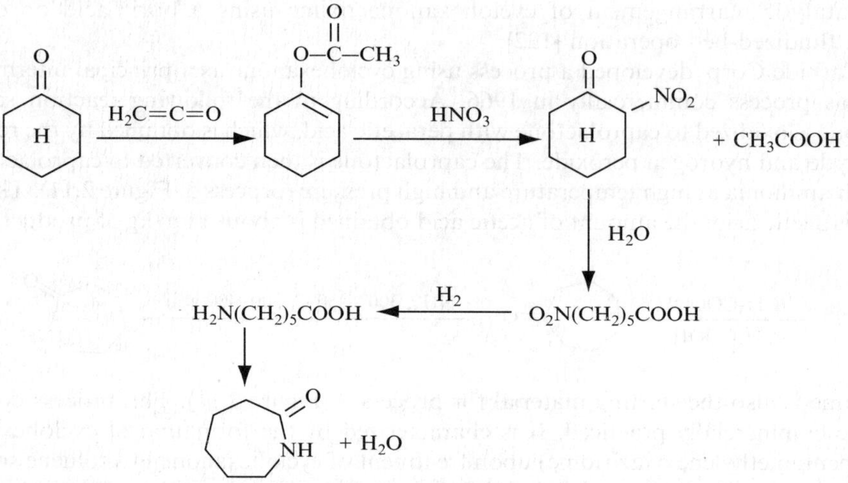

$$\text{cyclohexanone} \xrightarrow{H_2C=C=O} \text{cyclohexenyl acetate} \xrightarrow{HNO_3} \text{2-nitrocyclohexanone} + CH_3COOH \quad (2.56)$$

$$\downarrow H_2O$$

$$H_2N(CH_2)_5COOH \xleftarrow{H_2} O_2N(CH_2)_5COOH$$

$$\downarrow$$

$$\text{caprolactam} + H_2O \quad (2.57)$$

The acetic acid obtained in the nitration step is recycled to the ketene generator.

Toluene is the starting material of a commercial process developed by SNIA Viscosa [128] (process 3, Figure 2.11). It involves the reactions shown in the following scheme:

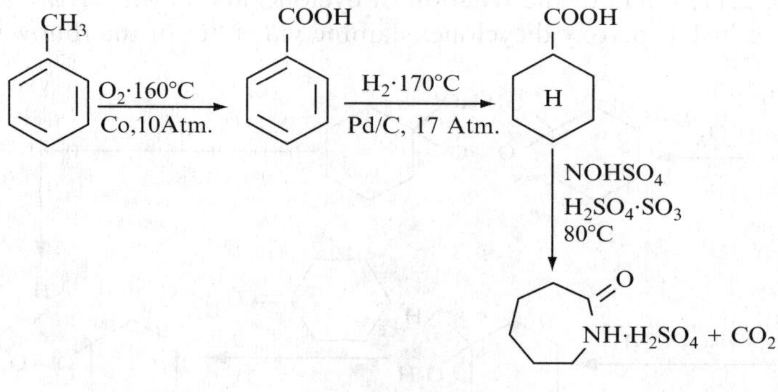

$$\text{toluene} \xrightarrow[\text{Co, 10 Atm.}]{O_2, 160°C} \text{benzoic acid} \xrightarrow[\text{Pd/C, 17 Atm.}]{H_2, 170°C} \text{cyclohexane carboxylic acid}$$

$$\downarrow \begin{array}{c} NOHSO_4 \\ H_2SO_4 \cdot SO_3 \\ 80°C \end{array} \quad (2.58)$$

$$\text{caprolactam} \cdot H_2SO_4 + CO_2$$

Toluene is oxidized in liquid phase to benzoic acid, which is subsequently hydrogenated to cyclohexane carboxylic acid. Reaction of this acid with nitrosylsulfuric acid in oleum then results directly in the formation of caprolactam sulfate by a mechanism that entails simultaneous nitrosation, decarboxylation, and rearrangement of the formed oxime.

Several modifications improved this process. One such process consisted in converting the cyclohexane carboxylic acid either with oleum or by thermal dehydration at 600°C into pentamethylene ketene, which then converted readily into caprolactam under mild conditions on treatment with nitrosylsulfuric acid.

Polyamide Fibers

$$C_6H_{11}COOH \xrightarrow[-H_2O]{H_2SO_4 \cdot SO_3} C_6H_{10}C=O \xrightarrow[-CO_2]{NOHSO_4} \text{caprolactam} \qquad (2.59)$$

In 1974, SNIA revealed a further modification of this process that resulted in eliminating ammonium sulfate formation [129]. In this modified process, the product of the reaction between cyclohexane carboxylic acid and nitrosyl sulfuric acid is slightly diluted with water and treated with an alkylphenol to extract the caprolactam, which is then purified by distillation. The sulfuric acid is mixed with a fuel and thermally cracked to sulfur dioxide, which is recycled into the process.

Processes 11 and 12 (Figure 2.11) also demonstrate approaches for the manufacture of caprolactam without producing ammonium sulfate. Developed by Kanegafuchi [130], process 11 involves oxidation of cyclohexane and reaction of the oxidation products with ammonia and hydrogen in the presence of a copper chromite catalyst. The reaction is conducted at 200–300°C and at pressures up to 3 atm. Process 12 (Figure 2.11) was developed by Kanebo [131]. It entails acetylation of cyclohexanone oxime with either ketene or acetic anhydride, vapor-phase rearrangement over a SiO_2–Al_2O_3 fixed-bed catalyst at 150°C to N-acetyl caprolactam,

$$\text{cyclohexanone N–OAc} \xrightarrow[150°C]{SiO_2-Al_2O_3} \text{N-acetyl caprolactam} \qquad (2.60)$$

and transacetylation according to

$$\text{N-Ac caprolactam} + \text{cyclohexanone oxime (N–OH)} \longrightarrow \text{caprolactam (NH)} + \text{cyclohexanone N–OAc} \qquad (2.61)$$

employing temperatures up to 100°C. The mechanism of the rearrangement may entail coordination of the acetyl carbonyl with the acidic silica–alumina catalyst and thus may be similar to the one that governs the boric oxide catalyzed vapor-phase rearrangement as reported by BASF [132].

2.3.1.2 Adipic Acid

Approaches for the synthesis of adipic acid are shown in Figure 2.12. The basic raw materials are benzene, cyclohexane, phenol, acrylates, and butadiene. The principal commercial processes are based on the oxidation of cyclohexane, which usually proceeds in two stages. The first step entails oxidation with air, yielding either a mixture of cyclohexanone and cyclohexanol (process 1, Figure 2.12) or predominantly cyclohexanol (process 2, Figure 2.12). These reaction products are oxidized in the second stage with nitric acid to adipic acid. Process 1 employs a soluble cobalt oxidation catalyst [133], reaction temperatures in the range of 150–160°C, pressures between 800 and 1000 kPa, and catalyst concentrations of 0.3–3 ppm. At conversions of 5–10%, the selectivity with respect to the cyclohexanone–cyclohexanol mixture is about 70–80 mol%, with an alcohol/ketone ratio of about 2:1. In process 2 the oxidation is carried out in the presence of boric acid or its anhydride. This results in mixtures particularly

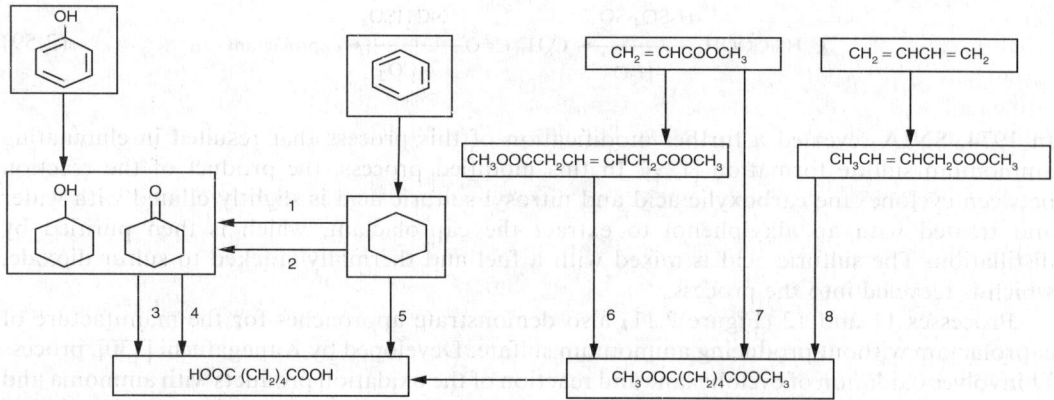

FIGURE 2.12 Block diagram of adipic acid processes.

rich in cyclohexanol. In this process the reaction temperatures are in the range of 170–180°C. At a conversion of about 15%, the selectivity is about 85% for a mixture characterized by an alcohol/ketone ratio of more than 10:1 [134]. The oxidation processes have been largely improved through changes in oxidizing agents and catalyst compositions [135–138].

Hydrogenation of phenol may yield, depending on the type of catalyst and operating conditions, predominantly either cyclohexanol or cyclohexanone. Although ketone is preferred for the production of caprolactam, alcohol is the more desirable product for the manufacture of adipic acid. The hydrogenation is usually carried out in the liquid phase using nickel on silica catalyst, with temperatures at about 140°C, and hydrogen pressures between 0.2 and 1.8 MPa [139]. At 99% conversion, selectivity is higher than 97%.

Although air oxidation of the cyclohexanone–cyclohexanol mixtures on a Cu–Mn catalyst in acetic acid [140] is possible, the principal commercial operations entail oxidation with nitric acid. The reaction is usually carried out at 60–80°C and pressures of 0.1 to 0.4 MPa, employing 50–60% nitric acid and a copper–vanadium catalyst containing between 0.1 and 0.5% Cu and 0.1 and 0.2% V [141]. The yields of adipic acid are in the range of 90–96%. The main by-products are succinic acid and glutaric acid. Their concentration generally increases as the purity of the feed mixture decreases. The adipic acid is isolated by crystallization and purified by recrystallization from water.

Single-step oxidation of cyclohexane to adipic acid (process 5, Figure 2.12) has been demonstrated [142]. This process involves a liquid-phase air oxidation using acetic acid as a reaction medium and cobalt acetate as an oxidation catalyst. The reaction temperatures are in the range of 70–90°C. At residence times of 6–10 h, conversions to about 80% were obtained with selectivities to adipic acid of 70–75%. Several alternate processes have been described for the oxidation of cyclohexane to form adipic acid [143–148].

Routes 6 and 7 in Figure 2.12 are based on methylacrylate, which may be converted to the adipic acid ester by electrolytic coupling. The corresponding technology has been reviewed quite extensively [149–151]. More recent developments have been concerned with catalytic coupling processes [152,153] using catalysts such as $PdCl_2$, $Pd(PPh_3)_2Cl_2$, $RuCl_3$, Ru(acac), $Fe(CO)_5$, and Ca(acac)/AlR_3. Reaction conditions are reflected in the following typical experiment: when a mixture of 100 ml of methylacrylate, 0.2 g of $PdCl_2$, and 0.3 g of benzonitrile under nitrogen was heated at 80°C for 35 min, 45% of methyl acrylate was converted. The reaction product contained 92% dimethyl hexenoate (linear dimer), 3% dimethylmethyl pentenoate (branched dimer), 3% oligomers, and 2% methyl propionate. Hydrogenation of the linear dimer and subsequent hydrolysis of the resulting dimer readily yield adipic acid. An alternate method [154] to process 6 involves the hydrolysis of dimethyl adipate in a 30-tray column at 95–105°C and atmospheric pressure yielded a bottom effluent of 31% adipic acid. This process constitutes a simple continuous reaction operation.

Process 8 in Figure 2.12 indicates an interesting approach. This process involves a two-step carbonylation of butadiene [155]. In the first step butadiene is reacted with carbon monoxide and methanol in the presence of dicobaltoctacarbonyl and a heterocyclic structure containing a tertiary nitrogen moiety (pyridines, picolines, quinoline, isoquinoline):

$$CH_2 = CHCH = CH_2 + CO + CH_3OH \rightarrow CH_3CH = CHCH_2COOCH_3 \qquad (2.62)$$

The reaction is carried out at 120°C and at pressures of 60 MPa. The selectivity with respect to the methyl 3-pentenoate is 98%. About 1% each of the branched isomer and the saturated ester is formed. After removal of unreacted butadiene, the second-step carbonylation is carried out at about 185°C and 3 MPa:

$$CH_3CH = CHCH_2COOCH_3 + CO + CH_3OH \rightarrow CH_3OOC(CH_2)_4COOCH_3 \qquad (2.63)$$

This second step yields about 85% of the dimethyl adipate. The main by-products are about 10% dimethyl methyl glutarate, 3% dimethyl ethyl succinate, 1% dimethyl diethyl succinate, and 1% methyl pentanoates. The dimethyl adipate is isolated by distillation and converted to adipic acid by hydrolysis.

Burke discloses a two-step process for the conversion of butadiene to adipic acid at high yields [156]. The first step is the hydrocarboxylation of butadiene to form 3-pentenoic acid. The second step is the hydrocarboxylation of 3-pentenoic acid with carbon monoxide and water in the presence of a rhodium-containing catalyst, an iodide promoter, and certain inert solvents such as methylene chloride. The first reaction step gives also a significant by-product of γ-valerolactone and a minor by-product of α-methyl-γ-butyrolactone. These lactones can be converted to adipic acid by modified catalyst compositions [157–159]. In a related work, pentenic acids or esters are used as the starting intermediates for conversion to adipic acid [160–166].

2.3.1.3 Hexamethylene Diamine

The block diagram in Figure 2.13 shows the process reactions for syntheses of hexamethylene diamine. This diagram includes processes (8–10) that had been developed to some degree of commercial maturity but do not appear to be practiced commercially. Process 8 is included because it starts from furfural, a material independent of hydrocarbon feedstocks. Process 9, developed by Toray [167], involves first the catalytic gas-phase reaction of caprolactam with ammonia at 300°C to amino capronitrile as the principal product at conversions of up to 75%. The subsequent step is the hydrogenation of the nitrile group. In process 10, developed by Celanese [168], caprolactone is first hydrogenated at 250°C and pressures of 25–30 MPa to hexanediol-1,6 in very good yields using Raney-cobalt or a copper–chromite catalyst. Subsequent catalytic conversion of the hydroxyl groups to amino groups using Raney-nickel proceeds at higher temperatures and pressures [169], to a yield of about 90% with the production of many by-products that complicate further purification.

The commercial processes for the manufacture of hexamethylene diamine entail hydrogenation of adiponitrile. It is a continuous liquid-phase process [170] that is usually conducted at 75°C and 3 MPa pressure in the presence of a chromium-containing Raney-nickel catalyst and aqueous sodium hydroxide:

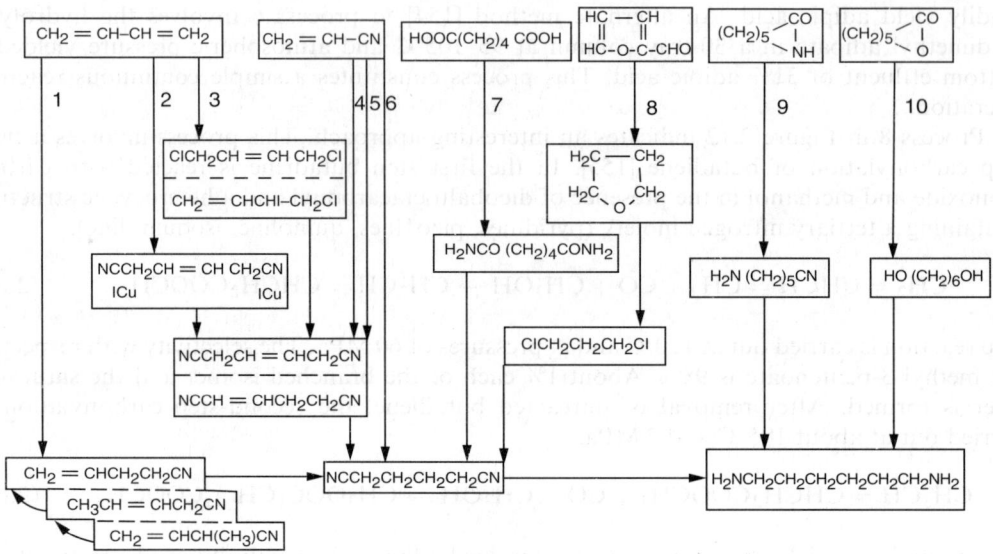

FIGURE 2.13 Block diagram of hexamethylene diamine processes.

$$\text{NC(CH}_2)_4\text{CN} \xrightarrow{\text{H}_2(\text{2MPa}), \ 75°\text{C}, \ \text{Cr/Ni}, (\text{NaOH}, \text{H}_2\text{O})} \text{H}_2\text{N(CH}_2)_6\text{NH}_2 \qquad (2.64)$$

Yields are in the range of 99%.

Adiponitrile in turn is obtained from processes starting from (a) adipic acid, (b) butadiene, and (c) acrylonitrile. The process starting from adipic acid (process 7 in Figure 2.13) is rather well developed and used by most of the major raw material suppliers. It consists of a gas-phase reaction between adipic acid and ammonia at about 270°C at atmospheric pressure. In the presence of dehydration catalysts, such as phosphonic acid or a mixture of phosphoric and boric acids, the reaction proceeds via the formation of adipic acid amide to adiponitrile with about 90% yield.

In the first butadiene process (process 3 in Figure 2.13) developed by DuPont [171], chlorine is reacted with butadiene at about 200°C in a mole ratio of 4:1. At a 95% yield this reaction results in the formation of both 1,4-dichlorobutene and 1,2-dichlorobutene. Treating this mixture with HCN at 130–150°C in the presence of HCl acceptors such as $CaCO_3$ yields 1,4-dicyanobutene-2, which upon treatment with a basic catalyst, isomerizes to 1,4-dicyano-butene-1. Hydrogenation at 250°C and at atmospheric pressure in the gas phase using a palladium catalyst yields hexamethylene diamine directly.

A newer DuPont process [172] is characterized by the direct addition of HCN to butadiene (process 1 in Figure 2.13). It is essentially a two-stage process. The first step is a vapor-phase operation. A gaseous mixture of butadiene, hydrogen cyanide, nitrogen, and hydrogen chloride in a ratio of 1:1:1:0.1 is contacted at about 215°C and at atmospheric pressure with a copper magnesium chromite fixed-bed catalyst for about 50 min. In an essentially quantitative conversion, the reaction product consists of 88% of a mixture of linear pentene-nitrile isomers and 12% of branched 2-methylbutenenitriles:

$$\text{CH}_2\text{=CHCH=CH}_2 \xrightarrow{+\text{HCN}} \begin{array}{l} \text{CH}_2\text{=CH–CH}_2\text{CH}_2\text{CN + isomers} \\ \\ \text{CH}_3\text{CH(CN)CH=CH}_2 \text{ + isomers} \end{array} \qquad (2.65)$$

The second step entails catalytic isomerization of the 2- and 3-pentenenitriles to 4-pentenenitrile and the addition of a second mole of HCN:

$$CH_3CH=CHCH_2CN \rightarrow CH_2=CHCH_2CH_2CN \xrightarrow{HCN} NC(CH_2)_4CN \qquad (2.66)$$

At about 98% conversion, hexamethylene diamine is obtained in 90% yield. The process is a liquid-phase operation, carried out at 120°C. The catalyst is a triphenyl phosphite–Ni(O) complex. It appears that the addition of a promoter such as zinc chloride to the nickel ligand results in a composition that catalyzes the conversion of 2-methylbutenenitrile (the undesired branched isomer) into the linear 3-pentenenitrile [173], which may be recycled into the second process step for isomerization to 4-pentenenitrile.

The approach for process 2 in Figure 2.13 has been developed by Exxon [174]. In this process, butadiene reacts with cuprous cyanide in the presence of iodine, yielding a cuprous iodide complex of 1,4-dicyanobutene-2. This is decomposed with aqueous hydrogen cyanide resulting in the liberation of the dicyanobutene, regeneration of cuprous cyanide, and formation of hydrogen iodide, which on oxidation to iodine can be, together with cuprous cyanide, recycled to the first process stage:

$$\begin{aligned}CH_2=CHCH=CH_2 &\xrightarrow{CuCN/I_2} [ICuNCCH_2CH=CHCH_2CNCuI] \\ &\xrightarrow{(HCN)} NCCH_2CH=CHCH_2CN + CuCN + HI\end{aligned} \qquad (2.67)$$

$$2\,HI \xrightarrow{1/2\,O_2} I_2 + H_2O \qquad (2.68)$$

The isolated 1,4-dicyanobutene-2 is hydrogenated successively to adiponitrile and hexamethylene diamine.

Processes 4, 5, and 6 in Figure 2.13 are based on acrylonitrile. Among them, process 6 is the oldest. Introduced in 1965 as a technical process by Monsanto [175], it involves the electrolytic dimerization of acrylonitrile to adiponitrile. The process is carried out in a system of electrolytic cells containing sulfonated polystyrene membranes. An aqueous solution of tetraethylammonium-p-toluyl sulfonate is used as the catholyte and aqueous sulfuric acid as the anolyte. Using a 40% aqueous solution of acrylonitrile at temperatures of 25–35°C, the following reaction sequences occur:

$$\text{Anode}: H_2O \rightarrow 2H^+ + 1/2\,O_2 + 2e \qquad (2.69)$$

$$\text{Cathode}: 2CH_2=CHCN + 2H_2O + 2e \rightarrow NC(CH_2)_4CN + 2OH^- \qquad (2.70)$$

The reductive dimerization thus proceeds at the cathode, occurs at a pH of 7–9, and yields more than 90% adiponitrile at a conversion of acrylonitrile of about 50%. By-products are 2% bis(cyanoethyl) ether, 0.5% hydroxy propionitrile, 2.5% propionitrile, and about 4% high boilers. The adiponitrile is purified by liquid–liquid extraction and distillation.

Processes 4 and 5 in Figure 2.13 are more recent developments entailing dimerization of acrylonitrile. Although formally analogous to the Monsanto process they are quite different because they proceed via catalysis rather than electrolysis. One of these processes developed by Halcon [176] involves liquid-phase reaction of acrylonitrile at 30°C under a nitrogen pressure of 0.7 MPa in the presence of a two-component catalyst system.

The catalyst system consists of a combination of a tertiary amine, either acyclic or heterocyclic, with a Lewis acid. The reaction yields 2-methylene glutaronitrile:

$$2CH_2=CHCN \xrightarrow{Zn\text{-tosylate}/N(Et)_3} \underset{NCCCH_2CH_2CN}{\overset{CH_2}{\underset{\|}{}}} \quad (2.71)$$

In a second step, the glutaronitrile is isomerized to the desired linear 1,4-dicyanobutene. This reaction is carried out at 255°C in the presence of catalytic amounts of lithium cyanide and appears to proceed via the formation of 1,2,4-tricyanobutane:

$$\underset{NCCCH_2CN}{\overset{CH_2}{\underset{\|}{}}} \xrightarrow{+HCN} \underset{NCCHCH_2CH_2CN}{\overset{CH_2CN}{\underset{|}{}}} \xrightarrow{-HCN} \begin{bmatrix} NCCH=CHCH_2CH_2CN \\ \\ NCCH_2CH=CHCH_2CN \end{bmatrix} \quad (2.72)$$

The same 1,4-dicyanobutenes can be obtained directly by a process disclosed by ICI [177]. According to this process, acrylonitrile is dimerized in an anhydrous proton donor solvent in the presence of a phosphinite, phosphonite, or phosphite catalyst at temperatures of about 60–80°C in an inert atmosphere:

$$2CH_2=CHCN \xrightarrow[\text{t-butyl alcohol}]{CH_3C_6H_4P(OCH_3)_2 \text{ at } 60°C} \begin{bmatrix} NCCH=CHCH_2CH_2CN \\ \\ NCCH_2CH=CHCH_2CN \end{bmatrix} \quad (2.73)$$

At 25% conversion after a reaction time of 1.5 h, yields in excess of 90% have been reported. Conventional processes may readily hydrogenate the linear dicyanobutenes.

2.3.2 Industrial Processes

Although industrial polymerization processes are generally treated as proprietary information, general description of polymerization technology, including flow sheets and schematic diagrams of polymerization processes and reactors, can be found in the literature [27,178–182]. The following sections discuss some general aspects of industrial polymerization processes.

2.3.2.1 General Operations

2.3.2.1.1 Monomer Preparation

The monomer feed preparation section is designed for obtaining carefully controlled monomer composition. This is particularly critical in case of the nylon-6,6 manufacture, where proper stoichiometry must be assured for the preparation of the HA-salt. Nonequivalence of the two monomers may, according to Equation 2.6, significantly affect the attainable molecular weight. The salt may be prepared either by reacting an aqueous dispersion of the diacid with an aqueous solution of the diamine or by mixing alcoholic solutions of the two components. In the latter case, a rather pure salt precipitates that may be used directly for making an aqueous feed solution of about 50% salt concentration. The monomer feed preparation in the nylon-6 process is concerned mainly with producing compositions of molten caprolactam with proper amounts of water and additives such as acids, or bases, pigments, delustrants, antistatic agents, and materials imparting stability against light and heat. Additives of this type, of course, are also used in the nylon-6,6 process.

The monomer compositions are then fed into the polymerization reactor system where the conversion of monomer to polyamide takes place. The polymerization of the HA-salt is either preceded or, in the initial stages, accompanied by evaporation of the water used to dissolve the salt. This evaporation process requires proper temperature and pressure control to prevent solidification of any prepolymer formation and to eliminate or at least minimize any loss of diamine. Since only catalytic amounts of water are used in the caprolactam polymerization process and only a single monomer is present, there is no need for an analogous step for this process. The polymerization of either monomer system is then conducted either in a batch or in a continuous operation using equipment ranging from simple autoclave reactors to multistage flow reactor systems. The process is carried out at temperatures in the range of 250–280°C for periods of about 12 h to more than 24 h. It results in a polymer melt that is either transferred directly to spinning units for filament formation or extruded in the form of ribbons that are subsequently cut into chips. For nylon-6,6, except for drying of the chips, neither the polymer melt nor the chips need further treatment prior to the melt-spinning operation. This is not true for nylon-6 because the polymerization does not result in complete conversion of the caprolactam but in equilibrium.

2.3.2.1.2 Polymer Purification
The quantity of the residual monomer depends on the reaction temperature. At the temperatures used in industrial operations, it amounts to about 8–9%. In addition, there are about 3% oligomers of low molecular weight that are mostly cyclic. Since subsequent processing as well as performance characteristics in many applications is adversely affected by both the residual monomers and the oligomers, it is necessary to remove them from the polymer. This can be accomplished either by hot-water extraction or by vacuum evaporation. The latter is usually part of an integrated continuous operation. In this case, the molten equilibrium polymer is fed into a suitable apparatus, such as a thin-film evaporator, in which most of the monomer and part of the cyclic oligomers are distilled from the melt under vacuum. The efficiency of this operation depends upon the vacuum applied, the residence time, and the design of both the equipment and process, especially of the part that is concerned with condensation and transfer of the cyclic oligomers. Concomitant with the vacuum distillation of the monomer and oligomers is usually an increase of the polymer molecular weight as a result of intercatenary condensation. The polymer melt emerging from the vacuum stage can then be converted directly to filaments or to polymer chips. The chips are usually cylindrically shaped, about 3×3 mm.

If the process does not include a vacuum stage for the removal of residual monomer and oligomers, then hot-water extraction is required. In this case, the molten polymer is extruded in the form of rods of about 3 mm in diameter, which are quenched in water and cut into chips of about 3 mm in length. These chips are then extracted with water at temperatures slightly below the boiling point. The extraction process can be carried out batchwise or continuously in countercurrent operation. The residence time for this diffusion-controlled process is about 8–12 h. The extracted chips must then be dried. This process again can be done batchwise or continuously. The use of vacuum tumble dryers at temperatures of 100–120°C and final pressures of 0.1 torr is most common for batch operation. Rotary dryers or towers are used in continuous processes in which the chips are dried countercurrently with circulating nitrogen. Depending on the particular process and constitutional factors such as molecular weight and type of end groups, residence times in the range of 20–50 h may be required to reduce the water content of the chips to a level suitable for various subsequent melt-processing operations. The residual water content of the chips used in most commercial processes ranges from 0.02 to 0.1%.

Since both extraction and drying steps require long residence times because of the diffusional resistance of the solid nylon, vacuum evaporation appears to be a more attractive mode of operation. However, it is less effective in removing the water-soluble constituents of the equilibrium polymer. The corresponding polyamide usually contains a higher concentration of oligomers when compared with a water-extracted one. Although the total amount of residual water-soluble compounds in extracted commercial nylon-6 is generally below 1%, concentrations of 2.8–3.8% are characteristic of products from which monomer and oligomers are removed by vacuum evaporation. This may be too high for some uses and product requirements. Even if the performance of vacuum evaporation can be improved to meet quality requirements, the route via chip extraction will continue to be practiced, especially in cases where flexibility with respect to subsequent processing steps and product characteristics is of primary concern.

2.3.2.1.3 Monomer Recovery
Although the monomer recovered from the nylon-6 melt by vacuum distillation may be directly recycled into the monomer feed section, the monomer recovered by extraction of the solid chips is usually purified by liquid–liquid extraction and distillation. In case of nylon-6, the corresponding aqueous solutions are generally combined with those leaving the polymer waste recovery section in which the polyamide waste, generated in the various process stages (Figure 2.26), is hydrolyzed to the corresponding monomers with superheated steam.

2.3.2.2 Nylon-6 Polyamide

Although the syntheses of aliphatic polyamides are well established, process modifications are continually developed either to produce specific products or to produce polyamides more economically. According to BASF in Ger. Offen. DE 3,134,716/17 (1983), caprolactam is polymerized in a perpendicular tubular reactor consisting of discrete zones. The resulting polyamide is claimed to be more easily dyed. In a later BASF disclosure [183], a moist inert gas is employed to remove unreacted monomer and oligomers from the unextracted polymer to obtain polycaprolactam of high molecular weight. The preparation of high-molecular-weight polycaprolactam by the removal of water from the melt by nitrogen in the polycondensation stage at normal pressure has also been reported [184]. A similar process also has been described, which entails the removal of water from a flowing film by nitrogen where the flow rate of the nitrogen controls the relative viscosity of polyamide [185]. DuPont disclosed a continuous process that employs a multistage reactive distillation column to produce nylon-6 [186,187]. A feed stream containing a major portion of caprolactam and prepolymer and a minor portion of aminocapronitrile is fed to the column in which the feed stream flows countercurrent to a steam stream. In an another disclosure [188], 6-aminocapronitrile is polymerized to form a prepolymer in a tubular reactor. The liquid prepolymer passes through a flasher at atmospheric pressure to evaporate excess water and volatile products, and is crystallized at 140–160°C.

Unitika patent JP 60,248,730 deals with a process to produce high-molecular-weight polyamides in which molten linear polyamides are treated with adducts of bisoxazolines and dicarboxylic acids. Continuous polymerization resulting in high-molecular-weight nylon-6 entailing the degassing of a thin film is described in the East Ger. DD 234,430 patent. The solid-phase polymerization of polycaprolactam has been reported to increase the polymer molecular weight [189,190]. Barnes and Gottund. [191] discuss the use of certain quaternary ammonium compounds such as methyl tributyl ammonium chloride as accelerator for the polymerization of caprolactam. Such a modification enhances the rate and yield of polymerization significantly at temperatures below 115°C.

2.3.2.3 Nylon-6,6 Polyamide

In the preparation of nylon-6,6 salt, DuPont discloses a process for making highly concentrated solutions of nylon salt at maximum solubility [192]. In the first step of this process, a concentrated salt solution for nylon-6,6 is made with 73.5–77.5 wt% of adipic acid and 22.5–26.5 wt% of hexamethylene diamine at 55–60°C. The solution contains 60–69.5 wt% solute as compared to an ordinarily stoichiometric solution containing 56% diacid and 44% diamine with a maximum solute concentration of about 59 wt%. The second step is to remove water from the solution by evaporation to a solution concentration of 93–96 wt%. When the concentrated solution is ready to be polymerized, addition hexamethylene diamine is added to complete the reaction. A process with similar reaction modifications is developed to prepare an essentially anhydrous mixture of adipic acid and hexamethylene diamine [193]. The reaction mixture is heated to 120–135°C to allow evaporation of water while reacting. The resulting product has a ratio of adipic acid to hexamethylene diamine of 81:19. The molten acid-rich mixture is withdrawn in a continuous process.

As an alternate intermediate, adiponitrile is employed to react with hexamethylene diamine and steam in a multistage distillation column to prepare nylon-6,6 [194]. The process is carried out in the presence of an oxygen-containing phosphorous catalyst at an elevated temperature and pressure. Nylon-6,6 of high molecular weight can be produced by the postpolymerization of lower molecular weight polymer in the molten state in the presence of a phosphonic acid catalyst [195] or in the solid state [196]. In a different approach, a prepolymer is formed in a reactor system.

2.3.2.4 Other Polyamides

The synthesis of a number of polyamides other than nylon-6 and nylon-6,6 has already been discussed in Section 2.2.7. In recent investigations, nylon-12,14 with a long alkane segment was investigated for its crystallization behavior [197]. Copolymers of butylene terephthalate and 1,4-diaminobutylene terephthalate (PB/4-T) were synthesized from the diamides of diaminobutane and dimethyl terephthalate (DMT) with butane diol and more DMT in a concentration range of up to 50% nylon-4,T [198]. The polymerization conditions were similar to those for poly(butylene terephthalate), i.e., melt polymerization followed by solid-state postcondensation. At increasing nylon-4,T content in the copolymers, the melting temperatures increased strongly, heats of fusion decreased slightly, and glass transition temperatures increased linearly. The torsional moduli above the glass transition temperature also increased. Gonsalves and Chen reported the synthesis of copolymers of nylon-2,6,6 and nylon-6,6 by interfacial polymerization of N-glycyl hexane diamine and hexane diamine with adipoyl chloride [199]. The molecular weights of these copolymers were relatively high according to intrinsic viscosity measurements and GPC analysis. The copolymers had similar solubility features as nylon-6,6. The copolymers were semicrystalline. Both melting and glass transition temperatures showed a minimum at around 20–30% nylon-6,6 content. The copolymers with relatively low melting temperatures could conceivably be melt-spun into fibers without appreciable thermal degradation.

New block copolymers of nylon-6-b-polyimide-b-nylon-6 were prepared by first synthesizing a series of imide oligomers end-capped with phenyl 4-aminobenzoate [200]. The oligomers were then used to activate the anionic polymerization of molten caprolactam. In the block copolymer syntheses, the phenyl ester groups reacted quickly with caprolactam anions at 120°C to generate N-acyllactam moieties, which activated the anionic polymerization. In essence, nylon-6 chains grew from the oligomer chain ends. All of the block copolymers gave higher moduli and tensile strengths than those of nylon-6. However, their elongations at break were much lower. The thermal stability, chemical resistance, moisture resistance, and impact strength were dramatically increased by the incorporation of only 5 wt% polyimide in the block copolymers.

2.3.2.5 Nanocomposites

A new composite material was introduced in 1987 with the discovery of a nylon-6/clay hybrid (NCH) [201]. The hybrid was prepared by the *in situ* thermal polymerization of ε-caprolactam with 8% or less montmorillonite, the clay material containing 1-nm thick exfoliated aluminosilicate layers. It exhibited a truly nanometer-sized composite of nylon-6 and layered aluminosilicate. Figure 2.14 depicts conceptually the NCH synthesis and its fine structure. The NCH exhibited high modulus, high strength, and good gas-barrier properties. The unique and superior properties led to the commercialization of NCH. It has also created a new class of nanocomposites and worldwide interest.

Variations in the preparation of nanocomposites have now been investigated extensively. Liu et al. [202] proposed the preparation of nylon-6/clay nanocomposites by a melt-intercalation process. They reported that the crystal structure and crystallization behaviors of the nanocomposites were different from those of nylon-6. The properties of the nanocomposites were superior to nylon-6 in terms of the heat-distortion temperature, strength, and modulus without sacrificing their impact strength. This is attributed to the nanoscale effects and the strong interaction between the nylon-6 matrix and the clay interface. More recently, nanocomposites of nylon-10,10 and clay were prepared by melt intercalation using a twin-screw extruder [203]. The mechanical properties of the nanocomposites were better than those of the pure nylon-10,10.

García et al. [204] prepared composites of nylon-6 polymer with nanometer-sized silica (SiO_2) filler by compression molding. The addition of 2 wt% SiO_2 resulted in a friction reduction from 0.5 to 0.18 when compared with neat nylon-6. This low silica loading led to a reduction in wear rate by a factor of 140, whereas the influence of higher silica loadings was less pronounced.

2.3.2.6 Process Simulation

Advances in the areas of instrumental analysis and numerical computations have made possible meaningful improvement of the polymerization processes.

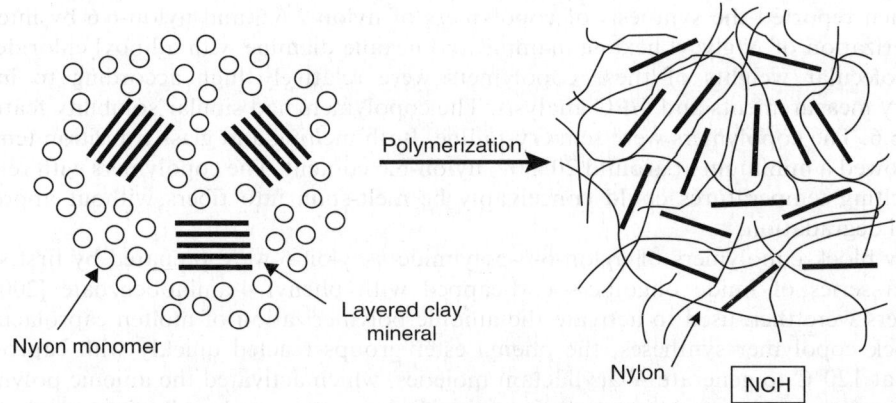

FIGURE 2.14 Conceptual depiction of NCH synthesis and fine structure. (From Kawasumi, M., *J. Polym. Sci. A: Polym. Chem.*, 2004, 42, 4, 819–824. With permission.)

2.3.2.6.1 Staged Polymerization

The optimal process is distinguished by two principal stages. The first entails operations at high water concentrations and high temperatures, resulting in high rates of monomer conversion. In the second stage, low water concentrations and low temperatures characterize the approach to high degrees of polymerization and conversion. The transfer of the reacting mass from the first to the second stage therefore entails the removal of any free water. The additional reactor element to accomplish this may be either a separate unit or a part of the reactor systems in either of the two stages. To realize the effects of a low temperature on conversion and polycondensation equilibrium would necessitate cooling of the polymer mass when conversion equilibrium is approached. Since a highly viscous material is involved, such an additional process step requires a more complex reactor system. Further complications may also result from the concomitant increase of the melt viscosity. Most industrial processes, therefore, operate the essential polymerization stage as much as possible under isothermal conditions employing the highest permissible reaction temperature. The latter is determined by the boiling point of the caprolactam and the extent of undesirable side reactions. In such a case, the optimizing functions depend only on the water concentration; the optimizing criterion being the minimum time required to obtain the desired degrees of polymerization and conversion. The optimum water concentration is either the upper or lower limiting value: the upper with respect to the rate of conversion, the lower with respect to the condensation equilibrium.

A simulation for such a two-stage process is shown in Figure 2.15 in comparison with the course of polymerization for the one-stage process at 265°C. On the basis of the assumption that there are no operational limitations for realizing ideal piston flow and instantaneous removal of water, Figure 2.15 shows that in the two-stage process the desired degree of polymerization is obtained in less than half the time required for the one-stage process.

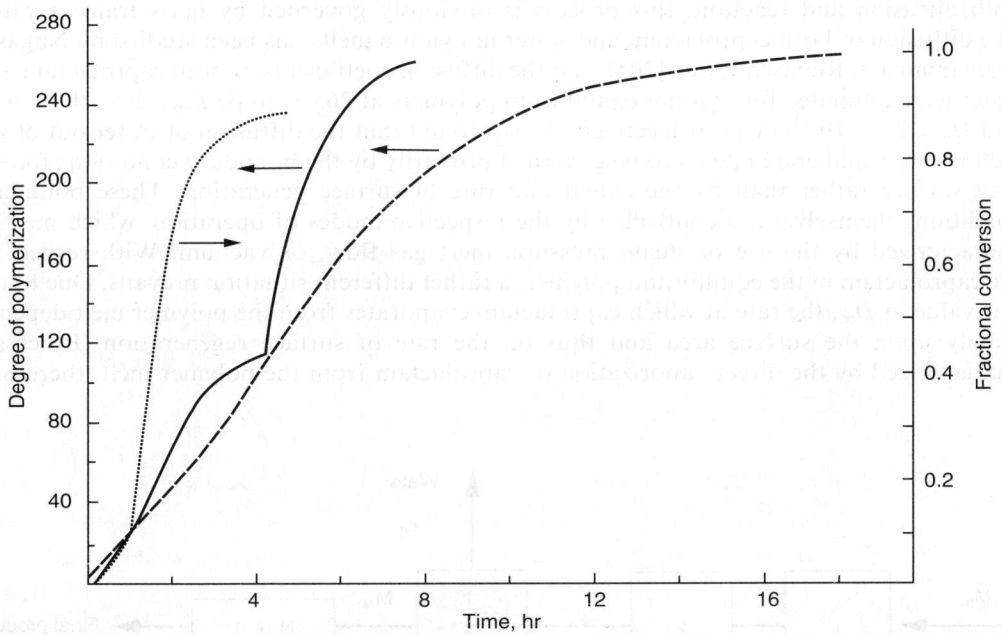

FIGURE 2.15 Course of polymerization for two-stage process. Solid line: two-stage process, initial water concentration; $[W_o]_{init} = 0.04$ mol/mol CL; free water removed at conversion equilibrium; dotted line: conversion at $[W_o]_{init} = 0.04$; broken line: one-stage process, $[W_o] = 0.1$.

Commercial process designs reflect the awareness of the advantage of operating the polymerization process in two distinct stages that are characterized by water content that approaches the upper working limit in the first stage and the lower one in the second stage. The most favorable mode of operation, however, depends on many factors, such as specific product requirements, the availability of reactors that provide the desired performance with regard to heat and mass transfer, and flow patterns.

2.3.2.6.2 Reactor Designs

Industrial processes entailing both batchwise and continuous operations use reactor systems that include autoclaves, with and without internal heat exchangers and stirrers, simple tube reactors, cascades of stirred tank reactors, and combination reactors containing both tubular and back-mix units. Relatively simple reactor systems that meet the considered requirements to some degree have been employed in industrial operations. Most of them are tubular flow reactors that in ideal operations are characterized by true plug flow, which is a condition that cannot readily be realized. Consequently, and also for the achievement of good heat transfer, processes have been proposed and practiced that utilize a cascade of stirred-tank reactors or a combination of such a cascade and tubular plug flow reactor. In this combination, the cascade constitutes the first stage and the tubular reactor the second one.

Principally, this process of polymerization consists of N stirred reactors followed by a tubular reactor as shown schematically in Figure 2.16. To obtain the highest possible rates of conversion, the reactor segment consisting of the stirred tanks 1 to $N+1$ is essentially a closed system and, as such, the first process stage. Removal of the free water is carried out in the Nth tank, from which the polymer melt is fed into the tubular reactor where the final degree of polymerization is obtained. Processes of this type have been quantitatively treated [54,55]. The overall rate of reaction of this multistage process depends, of course, considerably upon the removal of water from the polymer melt after the first process stage. With both diffusion and reaction, this process is obviously governed by mass transfer rates. The diffusion of both caprolactam and water in nylon-6 melts has been studied by Nagasubramanian and Reimschuessel. [205], and the diffusion coefficients of both caprolactam and water were estimated for nylon-6 equilibrium polymers at 265°C to be $D_m = 8 \times 10^{-8}$ cm^2/s and $D_w = 2.5 \times 10^{-4}$ cm^2/s, respectively. It was found that the diffusion of water out of the melt is very rapid and appears to be governed primarily by the boundary conditions for the melt surface rather than by the extent and rate of surface generation. These boundary conditions themselves are controlled by the respective modes of operation, which may be characterized by the use of steam pressure, inert gas flow, or vacuum. With respect to the caprolactam in the equilibrium polymer, a rather different situation prevails. Due to the low value of D_m, the rate at which caprolactam evaporates from the polymer melt depends mainly upon the surface area and thus on the rate of surface regeneration. Processes characterized by the direct vaporization of caprolactam from the polymer melt, therefore,

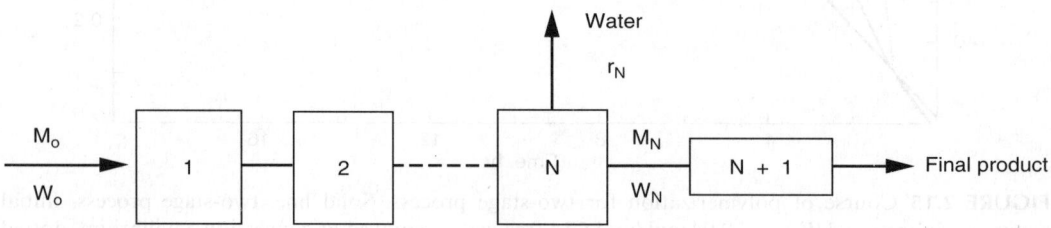

FIGURE 2.16 Combination reactor system.

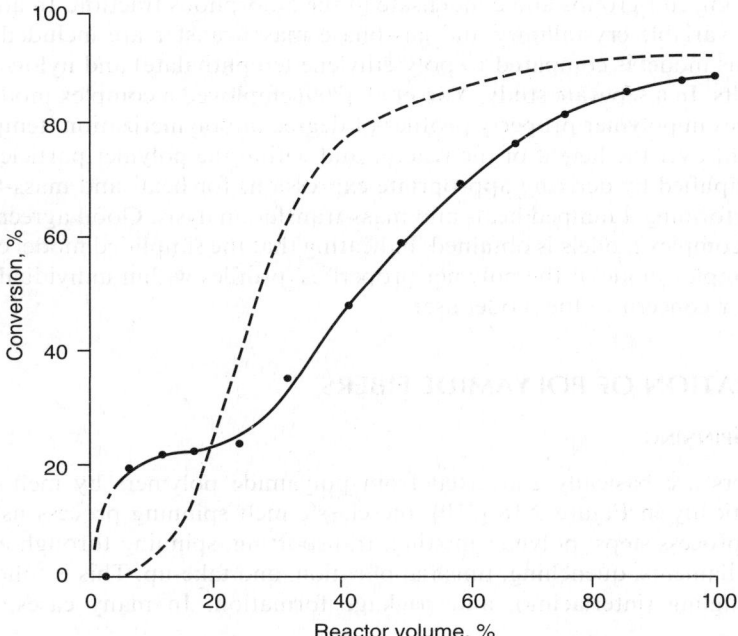

FIGURE 2.17 Conversion as a function of reactor volume for a reactor with back-mix flow in the first stage. (From Bayer, Ger. Patent 2,848,951.)

require more sophisticated operations and equipment than the more conventional ones in which the unreacted monomer is removed from the quenched polymer by hot-water extraction.

The concept of combining back-mix units with tubular plug flow reactors characterizes particularly large capacity reactor systems [181,182]. Compared with ideal plug flow reactors, higher initial conversion rates have even been claimed for a commercial reactor containing a type of back-mix section as the first stage [181]. Figure 2.17 shows the lactam conversion as a function of the reactor volume. The solid curve represents the observed course of conversion, and the broken line the one calculate for plug flow according to data of Reimschuessel and Nagasubramanian [52].

In recent years, twin-screw extruders are increasingly employed as commercial reactors for the polymerization of nylons. Such a reactor system offers the advantages of controlled plug flow, good mixing and heat transfer, and control of reaction time. Nascimento et al. investigated the finishing stage of nylon-6,6 polycondensation in a twin-screw extruder reactor [206,207]. Experimental results from industrial and pilot plants were employed to build various process models. They reported reasonable agreement between their models and industrial data and were able to achieve an increase in industrial production of about 20%.

2.3.2.6.3 Solid-State Polymerization
As discussed earlier, solid-state polymerization reactions are used to increase the degree of polymerization in the production of nylon-6 and nylon-6,6. The solid-state polymerization process has been studied by process simulation. Mallon and Ray [208] developed a comprehensive model to handle the reactions in polymers undergoing polycondensation reactions in the solid state. The polymer crystalline fraction is modeled as containing only repeat units,

thus concentrating end-groups and condensate in the amorphous fraction. In addition, many effects such as variable crystallinity and gas-phase mass transfer are included in a general framework. This model is compared to poly(ethylene terephthalate) and nylon reaction data with good results. In a separate study, Yao et al. [209] employed a complex model to describe dynamic changes in polymer property profiles of degree of polymerization, temperature, and moisture content over the height of the reactor and within the polymer particles. The model was further simplified by deriving appropriate expressions for heat- and mass-transfer coefficients and performing a lumped heat- and mass-transfer analysis. Good agreement between simplified and complex models is obtained, indicating that the simplified model can be used in place of the complex model if the polymer properties' profiles within individual particles are not of particular concern to the model user.

2.4 PREPARATION OF POLYAMIDE FIBERS

2.4.1 MELT SPINNING

Polyamide fibers are basically converted from polyamide polymers by melt spinning. As shown schematically in Figure 2.18 [210], the classic melt spinning process usually encompasses several process steps: polymer melting, transporting, spinning through a spinneret to form multiple filaments, quenching, finish application, and take-up. This is followed by fiber drawing, comingling (interlacing), and package formation. In many cases, the drawing

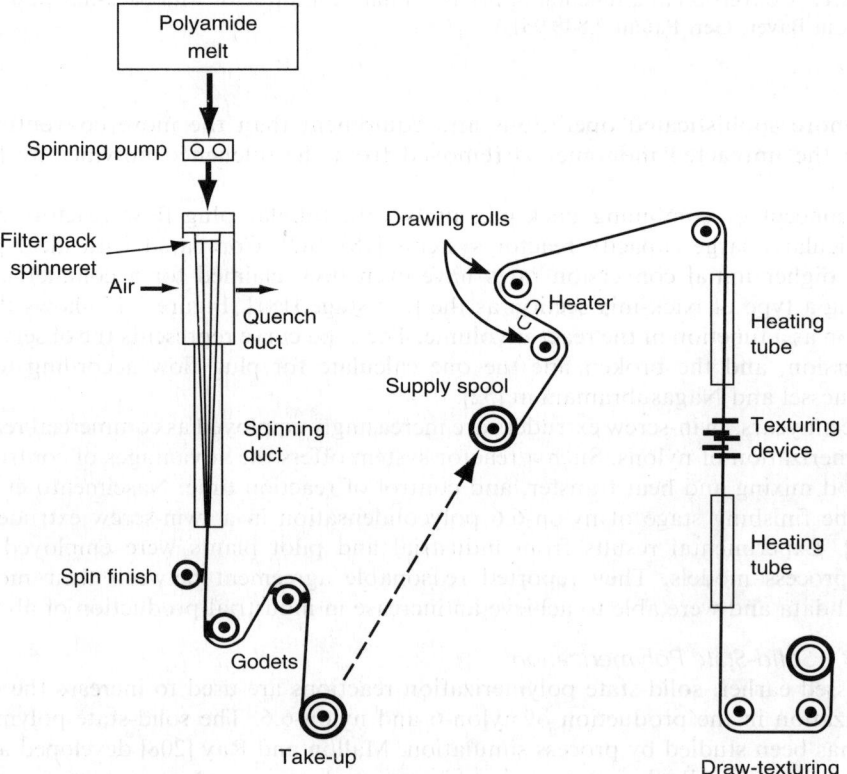

FIGURE 2.18 Traditional two-step melt spinning from chip.

Polyamide Fibers

process is integrated with spinning to form a one-step process. By changing the operating conditions of various process steps in a manufacturing process, different fiber structures and properties can be obtained for different applications.

The molten polyamide obtained either directly from the polymerization vessel or by melting the polymer chips via an extruder or a melt tank is delivered to an accurate metering device called a spin pump. The spinning temperature is usually 30°C higher than the polymer melting point. For nylon-6 with a melting point of 220°C, the spinning temperature is generally targeted at 260–270°C, and around 290°C for nylon-6, with a melting point of 260°C. It is very important to dry the polyamide chip to a consistent moisture level (about 0.12%) for successful spinning. Otherwise, poor spinning yield and poor fiber quality may result from polyamide degradation.

The spin pump (metering pump) delivers the molten polymer to a spin pack, which consists of a top cap and a breaker plate to distribute polymer evenly, then into a filtration media and out of the spinneret. The filtration media contains either different layers of special sand, layers of different size of stainless steel screens, or sintered metals. In addition to removing the foreign particles, gel particles and undesirable conglomerate additives, filtration may also improve the polymer melt homogeneity due to its torturous path and high shear of the filtration media.

The spinneret is similar to a showerhead, which converts the polymer melt into filaments. The orifices of the spinneret are normally round, thus producing filaments with a round cross section. In general, most nylon fiber products requiring high strength in tire cord, rope and cordage, and sling airbag, have round cross sections. However, for textile and carpet applications, fibers with modified cross sections have been developed to achieve different aesthetics such as luster, opacity, and insulation, walk resistance, etc. A fiber with a complex cross section can be used as a filtration medium for air or liquid. Figure 2.19 illustrates examples of nonround filament cross sections, including that of a $T_m = \Delta H_m / \Delta S_m$ TRIAD* fiber [211].

Spinneret orifice length/diameter ratio (L/D) and the orifice layout in the spinneret strongly influence filament uniformity and production yield. In general, an L/D ratio of 3 eliminates the entrance effect of the polymer flow through the orifice, a minimum upstream pressure of >600 psi for uniform distribution of polymer flow across the spinneret, and a shear rate at wall below 5000/s to avoid melt fracture are considered as a "rule of thumb" in spinneret design. Therefore, different polymer types (viscosity), molecular weight, and throughput rate may require different spinneret orifice dimensions for high yield and uniform fibers.

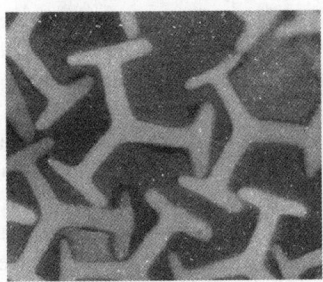
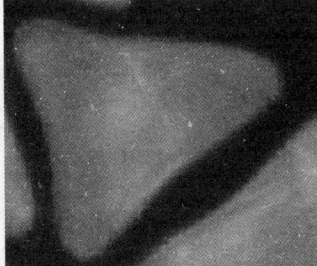
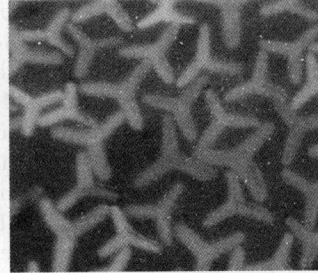

Trilobal fiber　　　　Trilobal fiber　　　　TRIAD^K fiber

FIGURE 2.19 Examples of nonround fibers. (From http://FRAM.com/Honeywell International, FRAM Consumer Products Group, products (accessed September 2004). With permission.)

---

*TRIAD—a registered trademark of FRAM, Honeywell International, Consumer Products Group, Danbury, Connecticut, U.S.A

2.4.2 DYNAMICS OF MELT SPINNING

The melt spinning speed refers to how fast the yarn is taken up on the first godet roll (take-up roll). This is the most important area in fiber spinning and the dynamics is very complicated. Key process parameters such as the throughput rate (jet velocity) through the spinneret, the quenching rate (the rate of cooling to solidify the fiber), the take-up velocity (stretching rate dv/dl), and the dynamic viscosity of the polymer as a function of temperature, are mainly responsible for the physical properties of the resulting fiber. Theoretical studies by computer modeling to relate fiber properties to spinning conditions, heat transfer, polymer rheology, kinetics of crystallization, and molecular orientation have been formulated and resolved in the last 30 years. These results provide a guideline to study the process–structure–properties relationship. However, experimental studies and actual measurements are still the most reliable method to obtain an insight of the fiber properties.

A principle factor for correlating fiber properties to spinning conditions is the velocity gradient (q). It is related to the jet velocity (v_j), the take-up velocity (v_t) and the distance from the spinneret (l) at which the filament cross section becomes constant.

$$q = (v_t - v_j)/l \quad (2.74)$$

For a given Δv (dv/dl), the distance l is a function of the melt viscosity, which depends on the quenching temperature and its gradient. The elongational flow in melt spinning affects the molecular orientation, the extent of which results from the competition between the velocity field and thermal motion, the latter being controlled by temperature and viscosity [212]. Therefore, the temperature profile of the filament bundle in the quenching zone is of great importance, especially for those with a large number of filaments in the bundle. Significant differences in birefringence, diameter, tenacity, and elongation have been reported from the windward and leeward segments of the filament bundle. Filaments from the windward (cooler) part were characterized by larger diameters, lower birefringence, lower tenacity, and higher elongation after draw than the filaments from the leeward (warmer) part of the bundle.

2.4.3 QUENCHING

As the molten filaments emerge from the spinneret, they are cooled by air and sometimes by water. Most melt-spinning operations use air as the cooling medium. The quench air temperature is in the range of 18–20°C and relative humidity in the range of 55–65%. Both temperature and humidity of the air used for quenching must also be controlled because of their effects on orientation and eventual crystallization.

There are at least three major quenching approach systems: cross flow, radial inflow, and radial outflow as illustrated in Figure 2.20 [213]. The key objective is to provide uniform cooling to each individual filament so that all the filaments will have the same morphology and properties within the yarn bundle. If not done correctly, the end-product will be nonuniform, which may result in streaky carpet or streaky fabric because the dye uptake rate is affected by fiber morphology. Therefore, the spinneret layout, the velocity and the flow pattern of the quench air in turbulent or laminar flow, and the take-up speed of the fiber are very important to fiber uniformity. Considerable effort has therefore been directed toward the development of quench stacks designed to achieve uniform air velocities across the filament bundle and to eliminate any turbulence. Additional approaches to reduce filament nonuniformity involved specially designed spinnerets [214]. With the advent of the high-speed computer, flow modeling via computation of flow dynamics by finite element analysis (CFD)

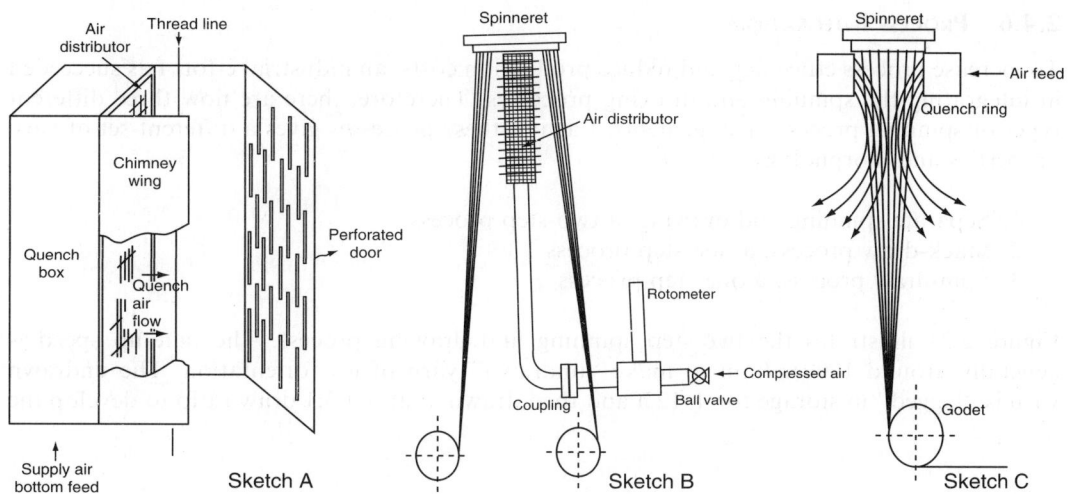

FIGURE 2.20 Quench air systems: A. cross flow; B. inflow; C. outflow. (From James, A., *Intl. Fiber J.*, 1999, 101–103. With permission.)

is currently used to design the quenching hardware by visualizing the interaction between the quench air flow and the filaments to maximize the yarn uniformity.

2.4.4 Spin Finish

As the filaments solidify and travel downstream at high speed, the surface friction through air generates static electricity, which makes the filament bundle less cohesive. To facilitate drawing and downstream processes, a finish emulsion with an antistatic agent, lubricating oils, bactericides, an antisoil agent, etc., are used to "wet" the fiber bundle by a kiss roll or meter finish device. Most of nylon-6 and nylon-6,6 spin finishes are water-based. The amount of finish (water) applied on the yarn may act as a plasticizer to lower the glass transition temperature and change the rate of crystallization once the yarn is wound up on the package. In general, the "wet pick-up" of the finish is about 10% by weight of the fiber to achieve the equilibrium moisture level for the downstream process, and the finish on yarn (FOY) is in the range of 0.2–1% of the fiber weight. Although the effect of the spin finish is an important aspect in fiber production, rather little has been reported in corresponding investigations [215,216].

2.4.5 Drawing

To develop the fiber strength, polymer molecules in both the crystalline and the noncrystalline regions are further oriented by elongating the fiber between two rolls at different speeds. The ratio of the two rolls is called the draw ratio. Higher draw ratio generally leads to higher molecular orientation and thus higher fiber strength. Drawing devices such as heated rolls with different surface roughness, draw pin, or draw point localizer, etc. are used to control the drawing profile. Mostly, increasing the drawing speed (higher elongation rate), drawing at a temperature lower than appropriate, or increasing the orientation of the feeder yarn from the spinning process decreases the fiber uniformity and quality. Therefore, spinning and drawing are best operated as an integral process to control the fiber morphology, physical properties, and end-uses.

2.4.6 Process Integration

To increase process efficiency and reduce production costs, an industrial effort has succeeded in integrating the spinning and drawing processes. Therefore, there are now three different types of spinning process arrangements. Each of these processes gives a different set of yarn properties and morphology.

1. Separate spinning and drawing: a two-step process
2. Stack-draw process: a one-step process
3. Spin-draw process: a one-step process

Figure 2.21 illustrates the two-step spinning and drawing process. The take-up speed is generally around 1000 m/min to make "undrawn" yarn of low orientation. The undrawn yarn is "lagged" in storage for 4–12 h and then drawn at about 3:1 draw ratio to develop the

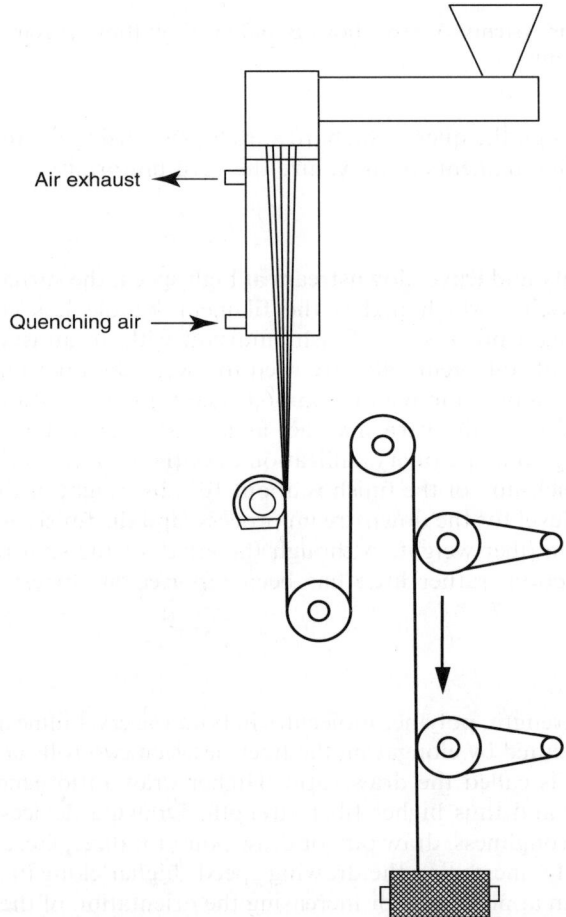

FIGURE 2.21 One-step stack draw process. (From Tam, T.Y. and Lin, C.Y., A unique one-step process of achieving fully oriented yarn properties with <1500 m/min take-up speed, Fiber Society Spring Conference, Princeton, New Jersey, June 1996. With permission.)

fiber strength. The process of this type, as shown in Figure 2.18, yields a fully drawn yarn (FDY). Additional twisting step to provide the filament bundle cohesiveness is generally followed in the same process. Thus, the process is also called a draw-twist process.

Figure 2.21 illustrates the one-step stack-draw process. The take-up speed of this one-step process is greater than 3000 m/min. The product is known as the partially oriented yarn (POY). The combination of drawing and textile finishing seems suited particularly for the production of POY at high spinning speeds.

Figure 2.22 illustrates the one-step spinning followed immediately by drawing and lagging. The take-up speed of this process ranges from 600 to 3000 m/min, coupled with immediate drawing on the panel. The resulting product is a fully drawn yarn.

2.4.6.1 Draw-Twisting

Polyamide filaments produced in high-speed operations that do not combine spinning and drawing processes must be subjected to a subsequent draw-twisting process to develop more useful properties. Since this process is performed in the solid state, the resulting orientation is

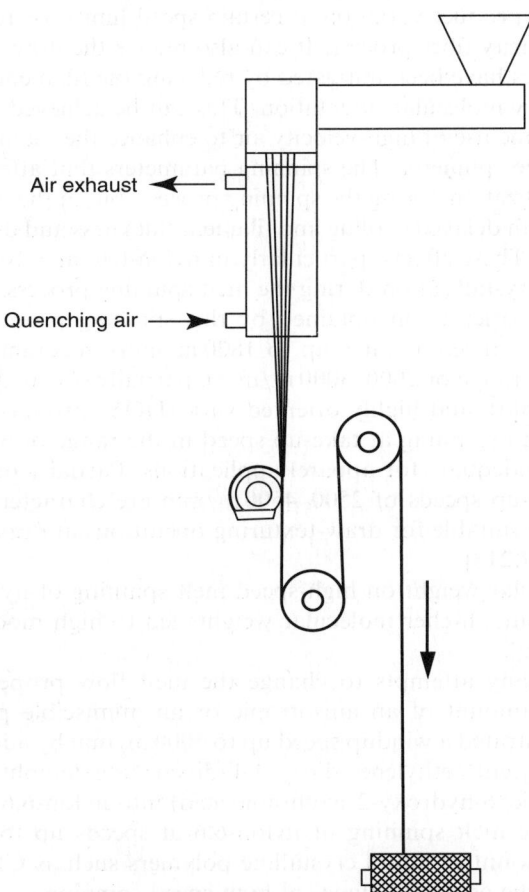

FIGURE 2.22 One-step spin-draw process. (From Tam, T.Y. and Lin, C.Y., Fiber Society Spring Conference, Princeton, New Jersey, June 1996. With permission.)

mainly controlled by the draw (deformation) ratio and is less affected by relaxation factors (molecular motions). Thus the drawing rate and temperature in the draw-twisting process are of less importance than they are in the spin-draw process.

2.4.6.2 Spinning–Drawing–Texturing

BCF production can be converted into a one-step spinning–drawing–texturing process for carpet and other applications. In this process, the yarn is generally spun at a take-up speed of about 2000 m/min, immediately drawn on panel at a draw ratio from 1.1 to 2.5, and passed through a texturing device to develop a crimped fiber. The crimped fiber will continue onto a perforated drum to be cooled before it is wound up on a package. Alternatively, an undrawn yarn spun earlier can be used as a feeder yarn to be drawn and textured as above.

2.4.7 HIGH-SPEED SPINNING

For polyamide fibers, increasing the spinning speed increases productivity as well as the preliminary orientation of polymer molecules with respect to the fiber axis. This leads to a reduction of the extensibility of the spun filaments and the extent to which they can be drawn. It can result in acceptable product variation at certain speed limits, or reduce the fiber tenacity attainable in a high-tenacity fiber process. It can also reduce the draw ratio in the stack-draw process. These difficulties have been mitigated by reducing the filament quenching in spinning to reduce the preliminary molecular orientation. This can be achieved by the use of hot tubes below the spinneret, or the use of high-velocity air to enhance the filament stretching at a high temperature closer to the spinneret. The spinning parameters that affect the molecular orientation also affect crystallization during the spinning process, but in the opposite direction. Thus crystallinity increases with delayed cooling and filament thickness and decreases with increasing take-up velocities [217]. These effects, particularly in nylon-6,6, may be somewhat effaced due to orientation-induced crystallization during the melt-spinning process.

The categorizing of orientation obtained by this spinning process is classified as low-oriented yarn (LOY, produced at rates up to 1800 m/min), medium-oriented yarn (MOY, produced at rates in the range of 2000–3000 m/min), partially oriented yarn (POY, produced at rates 3000–4000 m/min), and highly oriented yarn (HOY, produced at rates higher than 4000 m/min). High-speed spinning at take-up speed in the range of 6000 m/min yields fiber properties that appear adequate for apparel applications. Partially oriented polyamide filaments obtained at take-up speeds of 2500–4500 m/min are characterized by elongations of about 40–100% and are suitable for draw-texturing operations at drawing velocities between 700 and 800 m/min [207,218].

The effect of molecular weight on high-speed melt spinning of nylon-6 has been studied [219]. In as-spun filaments, higher molecular weights led to high modulus and tenacity and lower elongation to break.

There have been many attempts to change the melt flow properties of a polymer by incorporating a small amount of an anisotropic or an immiscible polymer. For example, Brody [220,221] demonstrated a windup speed up to 5000 m/min by adding a small amount of copoly(chloro-1,4-phenylene ethylene dioxy-4,4′-dibenzoate/terephthalate) (CLOTH) or copoly(4-hydroxybenzoic/6-hydroxy-2-naphthoic acid) into nylon-6,6. More recently, Vassilatos [222] disclosed the melt spinning of nylon-6,6 at speeds up to 6000 m/min with the addition of a minor amount of liquid crystalline polymers such as CLOTH. This technique clearly offsets some of the cost advantages of high-speed spinning.

The development of high-speed winding equipment has made the application of the high spinning speeds feasible. Equipment has been developed that realized spinning speed up to

6000 m/min. Modern operations utilize microprocessors to control the winding of the filament on cylindrical bobbins to obtain a "ribbon-free random wind" (RFRW).

2.4.8 Textile Finishing Operations

Textile finishing generally comprises operations such as rewinding, warping, twisting, and texturing. In the warping procedure, an extensive number of filaments are wound concurrently on a warp beam at a rate of 700–1000 m/min. In subsequent steps, the bobbins produced in this way are placed directly on a weaving machine. More recent developments are concerned with combining this process step with a drawing operation. Twisting and texturing, as discussed earlier, are parts of the integrated draw-twisting and spinning–drawing–texturing processes.

2.4.8.1 Texturing

Texturing procedures comprise mechanical distortions of the filaments to improve fiber characteristics such as the apparent volume (bulk), stretching properties (elasticity), shape (appearance), and the perceived feel (touch, handle) of the fiber material. The textured yarns are either fine tex* (1.7–2.2 tex) or heavy tex (110–400 tex) materials. The former are used in apparel applications, mainly for woven and knitted stretch fabrics; the latter are used primarily for carpets. A considerable number of processes and techniques have been developed for the production of textured filaments. A compilation of relevant processes has been presented by Wagner [223].

There are two types of texturing operations: crimping and jet texturing. The classic crimp texturing technology is used in the two-step spinning and drawing process for staple fibers. In this two-step operation, the fiber is first spun into a large container. About 20 to 30 of these tow containers are combined together to make a 30,000-denier tow. This big tow is drawn normally at a 3:1 draw ratio to develop its strength. The drawn tow of about 10,000-denier is continually forced into a stuffer box to be "crimped." The crimp is two-dimensional and the normal intensity is from 9 to 12 CPI (crimp per inch). The "crimped" tow is immediately heat-set and is cut to different lengths of about 7–7.6 cm (2.75–3 in.) or to customer specifications. The typical denier range is from 10 to 15 dpf (denier per filament) for carpet fiber.

Another texturing technology used in conjunction with process (1) is the Sintex texturing technology for three-dimensional texturing. An undrawn yarn is first spun. The undrawn package is then drawn at about 3:1 draw ratio and immediately fed into a texturing jet and wound upon a package. The product is about 1200 denier and typically contains 70 filaments.

Jet texturing is the new technology employed in the one-step spinning–drawing–texturing process (3). In this process, the yarn is spun, immediately drawn on panel without packaging and immediately textured on panel with a texturing jet, cooling drum, and finally wound up on a package.

2.4.8.2 False-Twist Texturing

The most important recent development, applied for the manufacture of about 90% of textured filaments, utilizes a process based on torsion mechanical–thermal procedures. The operation consists of three basic steps: high-twist twisting, fixation of the twist, and detwisting of the twist. In the process known as false-twist texturing, these three procedures are carried out in one process step. Rather detailed descriptions of this procedure are given

---

*1 tex = 0.11 denier = $1 \times 10-4$ kg/m; tex = weight in grams of 1000 meter filament; dtex (decitex) = weight in grams of 10000 meter filament.

in the literature [224,225]. The efficacy of this texturing process has been greatly improved by resolving problems concerning a bearing-free support of the spindle in the texturing aggregate. Adjusting the ratio of the driving wheel diameter to the spindle to about 20:1, revolutions of 106/min and rates to 200–300/min were possible. Twist is defined as the ratio of the spindle revolutions to the delivery rate of the filaments. If the twist of the filament amounts to or exceeds 150 twists/m, the material must be exposed to temperatures ranging from 60 to 80°C to stabilize the induced twist.

Considerable improvement in the production rates resulted from the introduction of frictional texturing [226]. This process uses friction wheels. Essentially, one wheel revolution produces a fiber characterized by a twist number that parallels the ratio of the friction wheel diameter to the fiber diameter. Industrially practiced frictional texturing processes use a variety of wheels and are characterized by a delivery rate that approaches values in the range of $1000 \, m/min^{-1}$. Gall presents a detailed account of this process [224].

Kuznetsov and Usenko have presented some aspects related to the texturizing technology, employing false-twisting operations and discussed ultrahigh frequency heating for thermal fixation of textured fibers [227]. The manufacture of textured yarns of polyamide conjugate fibers (mixtures of nylon-6,6 and nylon-6) is described in a Toray Industries patent [228], which refers to a process operating at 15°C and 3000 turns/m. Conjugate yarns that yield crimpable polyamides are described in a number of patents: a Toyobo patent [229] relates to yarn compositions constituted mainly of nylon-4,6 and nylon-6,6, in which the two polymers were spun into side-by-side conjugate yarns. Toray patents [230] pertain to diverse polyamide compositions yielding self-crimpable composite fibers.

False-twisting is most widely used for texturing fine tex yarns. In the first step, they are heated to about 220°C and 185°C in the cases of nylon-6,6 and nylon-6, respectively. This is followed by twisting to introduce 30–40 turns/cm using spindle or friction devices. The yarn is then cooled and untwisted. The resulting product is characterized by good elasticity. If bulk is required, the yarn is either subjected to a second heat–twist–untwist step or treated with steam or hot water in an autoclave.

2.4.8.3 Crimp Texturing

The edge-crimping process results a textured yarn by passing the filaments over a heated roll and then drawing them across a blunt knife edge at an acute angle. This results in compressed and stretched filament segments, causing the development of a textured structure upon relaxation. The texturing efficacy does not equate the results attained with the false-twisting process.

Used mainly for heavy tex yarns, gear crimping produces a textured material when drawn filaments are passed through sets of meshing gear teeth. The process may or may not involve external heating of the filaments, but heat-setting is necessary for adequate crimp permanence.

In stuffer box crimping (and in the related Spunise process), yarn is heated and forced into a stuffer box (a heated chamber) where it bends, folds, and forms random crimp that is heat-set. In the mechanical process, the yarn is forced into the box through pairs of feeder rolls, whereas in the aerodynamic process, the yarn is transported into the crimping chamber via a jet by a gaseous medium, usually superheated steam [231]. In both cases, the crimped yarn is pulled from the stuffer box after a controlled residence time and at a constant rate. The principal application of this process is the texturing of fibers having titers* in the range of 600 to 4000 dtex. This class includes carpet fibers for which this process is primarily applied.

---

*The titer is the weight of 1000 meter filament and is defined by the relation Titer $= M*1000/(Shy*Nd)$, where $M =$ amount of polymer fed to the spinneret in g/min, Shy $=$ take-up velocity in m/min, and Nd $=$ number of filaments pro spinneret.

2.4.8.4 Commingling, Interlacing, and Air-Jet Texturing

Since a yarn bundle consists of multiple filaments, it is difficult, if not impossible, to further process the yarn into a final product such as carpet or fabric via weaving or tufting processes. The methods to provide filament cohesiveness include twisting in the draw–twist process as mentioned above, commingling, or interlacing. The commingling process is conducted by passing the yarn bundle in and out of a commingling jet where a high pressure, turbulent air inside the jet chamber entangles the filaments together into nodes with uniform spacing. These nodes transform loose filaments into a continuous yarn bundle suitable for weaving, tufting, knitting, and the like.

The design of a commingling jet affects the appearance of an entangled yarn. Some commingling jets produce the node and antinode feature, while some produce a continuous interlaced cohesive yarn bundle without the distinct nodes. In addition, by controlling the feed speed and take-up speed (called overfeed) with a specially designed jet, one can obtain a bulky, air-textured yarn for end uses like luggage fabric, backpack, etc.

2.4.9 Winding and Yarn Package

After the commingling or twisting step, the yarn is finally ready to be wound up on a bobbin. To produce a stable package, winder settings like the wind ratio, helix angle, winder tension, finish or moisture content of the fiber, and the fiber properties must be optimal. As mentioned earlier, in the high-speed spinning of nylon-6, the fiber is not stable at the take-up speed of 1200–3000 m/min for a polymer of 55 FAV (formic acid viscosity). The fiber will grow or elongate as it is being wound up on the package. Good package formation cannot be provided under this circumstance.

2.5 STRUCTURES AND PROPERTIES OF POLYAMIDE FIBERS

2.5.1 Polymer Chain Structure

The structure of polyamide fibers is defined by both chemical and physical parameters. The chemical parameters are related mainly to the constitution of the polyamide molecule and are concerned primarily with its monomeric units, end-groups, and molecular weight. The physical parameters are related essentially to chain conformation, orientation of both polymer molecule segments and aggregates, and to crystallinity.

This characteristic for single-chain aliphatic polyamides is determined by the structure of the monomeric units and the nature of end groups of the polymer molecules. Thus, in accordance with Equation 2.1 through Equation 2.3, these polyamides are represented by either of the two general structures:

1. X,Y polyamides

$$R' - [\,[-HN-(CH_2)_X-NH-CO-(CH_2)_{Y-2}-CO-]\,]_n - R$$

2. Z polyamides

$$R' - [-HN-(CH_2)_{z-1}-CO-] - R''$$

where $R' = H, C(O)R$; $R'' = OH, NHR$; $n \gg 1$; and R = aliphatic or aromatic residue.

For the Z-type polylactams, recent literature has reviewed the molecular and electronic structures of lactams in considerable detail [16a]. Rudolf Puffr's review is concerned with the electronic structure of the amide group and the corresponding responses of molecular parameters. He refers to sp^2-hybridization of the nitrogen and carbonyl carbon atoms, the essential coplanarity of the atoms forming the amine group, the torsion angle about the amide bond, and the spatial electron distribution of the amide group. Conformational characteristics depend on the ring size of the particular lactam and encompass planar structures to puckered rings. In addition to some physical properties, he also addresses acid–base features comprising the acidity of both the NH and CH_2 moieties, characteristics of lactam salts, amide electrophilicity, hydrogen bonding, self-association, amine–water interaction, hydrogen donor complexes, protonation, and inorganic electron acceptor complexes entailing molecular acceptors and metal cation acceptors.

2.5.2 Characteristics of Crystalline and Amorphous Structures

Linear aliphatic homopolyamides are partially crystalline materials. Therefore they are characterized by both an unordered amorphous state and an ordered crystalline state. The latter may exhibit polymorphism. The extent to which each state or specific modification is represented depends, for a given chemical structure, considerably on processing conditions and treatment operations. It affects the properties of the shaped polyamide product. Thus the corresponding structure parameters are of importance for optimizing fiber processes as well as for assessing the performance of fiber products in particular applications.

2.5.2.1 Crystalline Forms

The crystal structure of polyamides results from the conformation of the macromolecules and their lateral packing. Generally, the packing of polymer chains will be such that the occupied volume is at a minimum, thereby minimizing the potential energy of the structure, but maintaining appropriate distances for intermolecular forces between adjacent chain segments. In polyamides, these intermolecular forces are both van der Waal's bonds and hydrogen bonds. The latter, involving the NH and CO moieties, cause the formation of sheet-like arrangements between adjacent chains. The stacking of these hydrogen-bonded sheets controls the size and shape of the unit cell. Since the nature of the bonds in the chain direction is different from that of the lateral bonds, the unit cells of polyamides are generally characterized by monoclinic, triclinic, and rhombic lattices. Hexagonal lattices indicate usually metastable mesomorphic modifications. The general conformation of the chain segments in the unit cell and their mode of packing are the basis for classification with regard to the crystalline forms of various polyamides. One is characterized by fully extended chain segments and is referred to as the α-structure; the other γ-structure consists also of extended segments, which however are twisted or contain some kink, resulting usually from rotation about the $-CH_2NHCO-$ unit [232]. Polyamides of the nylon X,Y-type with even–even carbon atom numbers crystallize mainly in the α-form, whereas those with odd–odd, even–odd, and odd–even numbers crystallize usually in the γ-form. Polyamides of the nylon Z type with even numbers crystallize generally in the γ-form for $Z > 8$. Nylon-4 and nylon-6 can crystallize in either the α- or the γ-form. However, the predominant structure of these two polyamides is the α-form. The two structures are interconvertible. The γ-form can be converted to the α-form by either treatment with aqueous phenol or the application of stress at high temperatures, whereas the α-form can be converted to the γ-form by treatment with aqueous iodine–alkali iodide solutions [233–238].

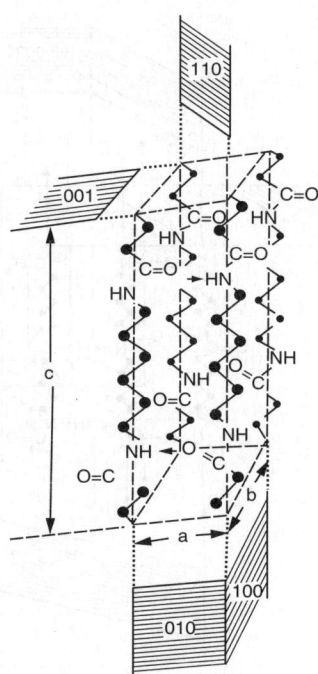

FIGURE 2.23 Unit cell of nylon-6,6.

Nylon-6,6 and nylon-6 are the two polyamides that are of most importance for commercial fiber production. The stable structure of both is the α-form and as such is comprised of stacks of sheets of planar hydrogen-bonded extended-chain segments [238,240]. These sheets are characterized in nylon-6,6 by a parallel alignment of the adjacent extended molecules, which are spaced with a perpendicular chain-to-chain distance of 0.42 nm and which are successively displaced in chain direction by a distance corresponding to one chain atom. The stacking of the hydrogen-bonded sheets entails a perpendicular sheet-to-sheet distance of 0.36 nm and a displacement of each successive sheet of about 0.5 nm in the chain direction. This arrangement results in a triclinic structure. The corresponding unit cell is characterized by identity of the crystallographic period and the chemical repeat unit. There is one extended chemical repeat unit per unit cell [239]. Figure 2.23 shows a schematic drawing of the unit cell and the principal crystallographic planes.

In nylon-6, the hydrogen-bonded sheets of the α-form are characterized by an antiparallel alignment of the extended chain segments. This opposite directionality of the CONH segments makes possible complete formation of unstrained hydrogen bonds. The stacking of the resulting planar sheets is marked by an alternating up-and-down displacement of about 0.37 nm parallel to the chain direction. The resulting structure is monoclinic. The crystallographic repeat unit of the unit cell corresponds to two extended monomer units, and the unit cell consists of four extended-chain segments; it contains, therefore, eight monomer units [240]. Figure 2.24 shows a schematic presentation of the unit cell and the principal crystallographic planes.

The distances between both the hydrogen-bonded chain segments and the sheets linked by van der Waal's forces are about the same as those found for nylon-6,6, namely 0.44 and 0.37 nm, respectively [240]. Although in these structures, due to the relatively strong hydrogen bonds, the chain-to-chain distances do not easily change, sheet-to-sheet distances are more

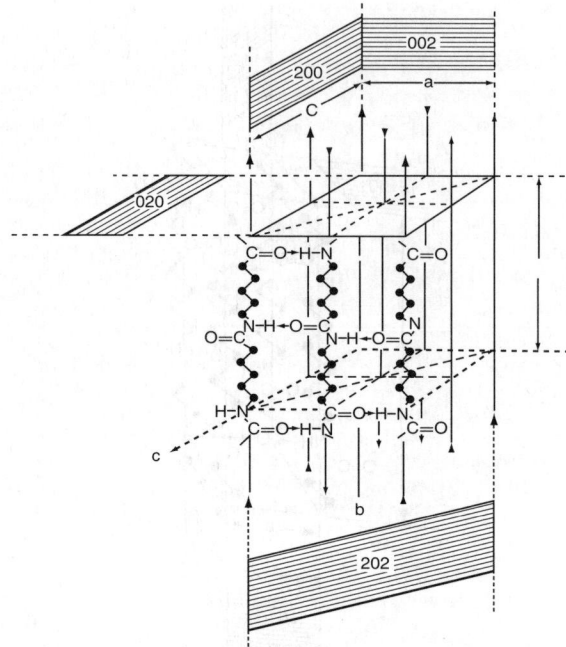

FIGURE 2.24 Unit cell of the monoclinic α-form of nylon-6.

readily affected and susceptible to changes in response to conditions of crystallization and applied external forces. This anisotropy of the intermolecular forces is responsible for the polymorphism that characterizes most aliphatic polyamides.

Stable polymorphic modifications are possible usually in systems distinguished by directionality, which in polyamides is caused by steric polarity resulting from the invariable sequence of the CONH groups. As is shown schematically in Figure 2.25, steric polarity is absent in polyamides of the X,Y-type but is a distinct characteristic of the Z-type polyamides.

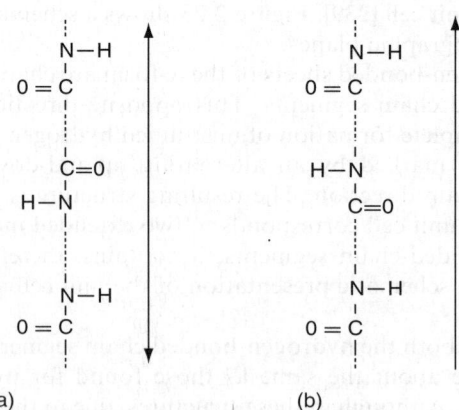

FIGURE 2.25 Steric polarity in polyamides: (a) nylon X,Y; (b) nylon Z.

Polyamide Fibers

Any lattice disorder may correct itself readily due to the lack of directionality of the X,Y-type nylon chain molecules. The x-ray diagram of nylon-6,6 filaments of high axial orientation, for instance, shows a medicinal reflection and layer-line streaks. This was initially explained as a result of an alternating up-and-down displacement of the hydrogen-bonded lattice planes similar to the stacking arrangements, characterizing the α-form of nylon-6 [239]. It was therefore considered a definite polymorph referred to as β-form. Another explanation, however, involves the postulation of rudimentary and intermediate stages in a continuous process of the formation of an ordered α-structure [241,242]. Indeed, upon annealing, this β-form converts readily into the α-form.

Rudimentary organization of chain segments may also be the cause for the formation of a similar metastable mesomorphic modification of nylon-6 upon rapid quenching of the molten polyamide [243]. The resulting structure is pseudo-hexagonal, since the corresponding x-ray diagram shows only a single reflection, but it is in general less well understood. It may be considered a bundle-like arrangement that consists of chain segments of both parallel and antiparallel directionalities with statistically changing equidistant bonding along two crystallographic axes [244,245]. This pseudo-hexagonal structure has been referred to as a γ-form [232,234,246–250], a β-form [251], and a pleated α-form [252]. This structure is formed and is stable within the temperature range of 100–150°C. Upon annealing or any proceeding process of crystallization, adjacent chains organize according to their directionality to form stacks of lattice planes. The resulting structure is characterized by an antiparallel arrangement of adjacent chain segments, either within the individual lattice planes or between successive planes. The antiparallel alignment of extended chain segments within the lattice planes characterizes, as discussed earlier, the α-form, whereas parallel directionality within the lattice planes is now generally referred to as the γ-form. In both forms, hydrogen bonding is complete between the amide functions of adjacent chain segments within the individual lattice planes, as shown by infrared studies [248,253]. Both crystallographic period and chain packing of the γ-form are different from those of the α-form; there is no alternating up-and-down displacement of the hydrogen-bonded sheets in the chain direction, such as the shift of $(3/14)b$ in the α-form. Both forms have the same chain arrangement in the basal plane [236], as can be seen in Figure 2.26.

To realize complete hydrogen bonding between parallel chain segments, the latter cannot be in a planar zigzag conformation, but are, as has been pointed out earlier, slightly twisted due to a rotation by about 60° about the $-CH_2NHCO$ unit. This results in a lowering of the height of the unit cell to 1.688 nm from the 1.724 nm of the α-form.

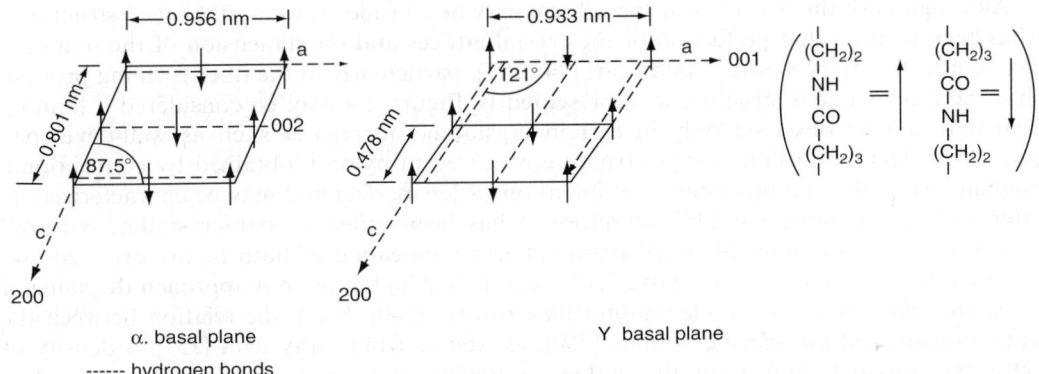

FIGURE 2.26 Schematic presentation of chain arrangement and hydrogen bonds in the basal planes.

2.5.2.2 Crystalline Structures in Fibers

Nylon-6 fibers obtained by melt spinning contain mixtures of the different crystal modifications, the concentrations of which depend upon the conditions of spinning, drawing, and annealing. Thus, fibers produced by conventional operations at moderate spinning speeds contain about equal amounts of α- and γ-forms. Annealing these fibers at temperatures up to about 100°C in the presence of moisture and up to about 150°C in dry conditions results in moderate increases of both forms, with the α-form slightly favored. At annealing temperatures higher than about 100°C, and in the presence of moisture, conversion of the γ-form to the α-form occurs. On drawing at low temperatures, the ratio of the concentrations of the two forms changes, although a mixed structure is retained. Increasing the draw temperature results in an increased conversion to the α-form; in combination with increasing draw ratios, the amount of the γ-form becomes negligible [254,255]. All this indicates that the α-form is energetically favored and is thus the more stable structure. However, fibers spun at high take-up speeds (<3000 m/min) have the γ-form as the predominant structure. Spinning at these very high winding speeds entails an orientation-induced crystallization process; obviously, the kinetics of this process favors the γ-form [254].

A pure γ-form can be obtained by an α-crystal-to-γ-crystal transition, which can be effected by treatment with a solution of iodine in aqueous potassium iodide [232,233, 236,237,248,249,256–267]. The pure γ-form resulting from this process is stable to the extent that it can be annealed and may retain any initial uniplanar axial orientation [236,249]. Although its structure has been considered nearly hexagonal [249,268], pseudohexagonal [237], and orthorhombic [257], it is characterized by a monoclinic geometry [236,249]. The corresponding unit cell has the following dimensions: $a = 0.933$ nm, $b = 1.688$ nm (fiber axis), $c = 0.478$ nm, $\beta = 121°$, and it contains two chain segments each consisting of two monomer chemical units. The crystal-to-crystal transition may be explained by a mechanism that entails interaction of the iodine with the amide group. This interaction results in both the breaking of the hydrogen bonds between antiparallel chain segments within the sheet-like structures and omits twisting the amide moiety out of the plane of the main chain axis by about 60°. This rotation reduces the distance between the amide groups of neighboring sheets to such an extent that after the removal of iodine, hydrogen bonds form between these amide groups. Thus, in accordance with the chain arrangement shown in Figure 2.6, a structure emerges that is composed of hydrogen-bonded sheets of parallel chain segments [249,263,264].

Consistency with x-ray diffraction data has been indicated in a study of iodine-complexed nylon-6 films using polarized resonance Raman spectroscopy [269].

Although both this γ-form and the α-form may be considered well-established structures, it has been realized that perfection of the crystal lattices and the dimension of the unit cells may be affected by processing conditions [245,247], particularly in the fiber-forming process. Thus the monoclinic α-structure as represented in Figure 2.4 may be considered a limiting form that can be observed only in certain crystalline aggregates such as well-developed spherulites. The monoclinic α-type structure observed in nylon-6 obtained by conventional molding, extrusion, and melt-spinning operations is less perfect and may be characterized by differing lattice parameters [245]; therefore, it has been called a "paracrystalline α-form" [247]. Indeed, the densities obtained upon extensive annealing of both highly oriented and nonoriented fibers were slightly above 1.17 g/cm^3 [236,270] but did not approach the value of 1.23 g/cm^3 calculated for the ideal monoclinic α-form [240]. From the relation between the heat of fusion and the specific volume [250], as well as from x-ray data [257], a density of 1.19 g/cm^3 was determined for the perfect monoclinic γ-form as obtained by the iodine treatment; the corresponding experimentally obtainable values were in the range of

1.14–1.15 [271]. The structure of iodide ions in iodinated nylon-6 and the outgrowth of hydrogen bonds between parallel chains have been discussed [272].

The limiting value for the density of the pseudohexagonal structure was found to be 1.13 g/cm^3, whereas a value of 1.084 g/cm^3 has been reported for the amorphous phase [236]. For nylon-6,6, values of 1.233 and 1.12 g/cm^3 for the triclinic crystal [273] and the amorphous phase [274], respectively, were found.

2.5.2.3 Structural Models

The crystalline and noncrystalline phases in polyamide fibers do not appear to be governed by what may be defined as thermodynamic equilibria, nor is there evidence for definite boundaries between a "phase," characterized by a simple or complex state of order and an essentially amorphous "phase." It is therefore quite obvious that the morphological structure of nylons cannot be described adequately in terms of a simple two-phase model according to which ideally ordered crystallites exist together with completely amorphous domains. This model constitutes merely one of the two limiting cases; the other is that of a paracrystal according to which all deviations from the ideal crystal are ascribed to defects and distortions of the crystal lattice [275–277].

The paracrystal model is rather convenient for explaining some observations, such as the broadening of the x-ray diffraction pattern and the essentially linear relationship between the melting temperature and the specific volume of nylon samples that are characterized by rather different thermal and processing histories [278]. On the other hand, , for example, both the existence of a glass transition, which is affected by plasticizing agents, and the observed relationships between density and the diffusion and absorption of water and dyes are less compatible with this model. These phenomena and processes seem to be more readily explained in terms of a two-phase model.

The fact that certain structure-related phenomena tend to support one model, whereas others can be explained adequately only in terms of the other, underlines the limiting nature of either of the two concepts. The actual morphological structure of nylon is more complex and may involve paracrystals of any of the possible polymorphic forms, mesomorphic regions, and essentially amorphous domains. This is also indicated by a multiple melting phenomenon [279] that was investigated by high-temperature x-ray techniques and differential scanning calorimetry, and has been explained based on three phases — crystalline, intermediate, and liquid amorphous [280].

The micromorphology of the ordered phases depends upon the conditions of crystallization and may be characteristic of that of flat dendrite crystals, lamellae, fibrillar and globular aggregates, and rhombohedric single crystals. The thickening of these crystals may be the result of either superimposition of lamellae or spiral growth originating from screw dislocations [281].

The thickness of the lamellae depends on the crystallization temperature and is usually in the order of 6–10 nm. The macromolecules are normal to the lamellae and are folded back and forth on themselves. A single-polymer molecule may belong to more than one lamella, especially in polymers crystallized from the melt. Such interlamellar bonding increases as the molecular weight increases. Chain-folding also exists in some of the fibrous or ribbon-like structures in which nylons may crystallize. These structures are presumed to be the degenerated forms of lamellae. The structure of the observed globular particles appears to be unknown.

The ordered structures discussed thus far are sometimes incorporated into superstructures, which are mainly, but not exclusively, formed upon crystallization from the melt. These superstructures are spherical aggregates, range in sizes from submicroscopic dimensions to millimeters in diameter, and are called spherulites. They can be recognized in the polarized light microscope as circular birefringent areas with a dark Maltese Cross pattern. If they are formed in nylon-6, then they are always "positively birefringent spherulites," which means that the spherulite radius is parallel to the major crystal axis. Conversely, if the spherulite radius is parallel to the minor axis of its component crystal, it is called a "negatively birefringent spherulite." These spherulites appear to consist of ribbon-like lamellae that are twisted together to form sheaf-like centers by electron microscopy.

The spherulite growth may be initiated by either heterogeneous nucleation by foreign particles or homogeneous spontaneous nucleation. The growth rate of spherulites from the polymer melt increases as the temperature decreases. It reaches a maximum in the temperature range of 140–150°C, and then decreases upon further decrease of temperature. The time dependency of the primary crystallization can be described by the known Avrami equation [282].

Drawing of filaments and fibers may result in an orientation of either or both polymer molecules and crystalline aggregates in the direction of draw. In crystalline aggregates, specimen drawing does not change the degree of crystallinity significantly. However, in amorphous or only partially crystallized material, crystallinity is likely to develop and increase.

The drawing orientation step is included in the production of nylon filaments and fibers to impart high mechanical strength and low elongation in the direction of draw, which is the fiber axis in the case of fibers. The applicable draw ratio may vary from 1:3 to 1:6. It depends on the constitution, composition, and the crystalline and morphological structures of the particular undrawn nylon. These factors, in combination with the rate of drawing and temperature, determine significantly the extent and permanence of the orientation. This permanence is directly related to the crystalline structure that results from either or both the drawing-induced crystallization and the transformation of the initially present crystalline form. Such transformation may involve the deformation and destruction of spherulites. Available information on the morphological structure of the drawn fiber is still inconclusive. Transitional states from folded-chain crystals to fringed micelles of extended chains have been considered. It appears that a two-phase model can explain the phenomena involving oriented amorphous and crystalline regions; both folded structures and fibrillar aggregates [283] characterize the latter.

The fibrillar aggregates entail highly extended interfibrillar tie molecules [284], which may be considered as a separate phase that significantly affects the modulus and the strength of the fiber [285,286]. The tie molecules result from chain unfolding during the drawing process, and therefore connect the crystalline blocks from which they originate. They are believed to be located mainly at the outer boundaries of the microfibrils [287].

2.5.3 Thermal Transitions of Crystalline Phase

2.5.3.1 Melt Temperature

The temperature above which all crystalline order disappears is defined as the melting point of a crystalline polymer. This transition is related to the surface-free energy σ, the specific crystal volume V_c, and the heat of fusion ΔH^∞ of an infinite large lamellar crystal involving an infinite large linear polymer molecule, by the following equation [288,289]:

$$T_m = T_m^\infty (1 - 2\sigma V_c / \Delta H^\infty l) \tag{2.75}$$

where T_m^∞ is the limiting or equilibrium melting point and l the thickness of the lamella.

Since the composition with respect to the crystalline modifications, and the micromorphological structure of crystalline nylon-6, and thus the values for V_c, l, and ΔH, depend upon the thermal and processing histories, no definite melting point can be expected. Values reported in the literature for commercial products are generally in the range of 215–228°C.

Knowledge of the equilibrium melting points (T_m^∞) for the various modifications is important for determining corresponding structural and thermodynamic quantities. Extrapolations of experimental data have resulted in values that pertain to the α modification [260,270]. $(T_m^\infty)_\alpha$ was determined to be 260°C. On the basis of the two-phase model, values obtained for σ_α were 47 erg/cm² and for $(\Delta H^\infty)_\alpha$ either 55 or 64 cal/g, using either 0.828 or 0.814 cm³/g for $(V_c)_\alpha$. The stable γ structure as obtained by iodine treatment of monoclinic material was found to melt at 215°C [250]. Assuming a surface-free energy identical to that of the α-crystal, a value of 510 cal/g was estimated for ΔH^∞ [250].

Measurements on single nylon-6,6 crystals yielded a value of 61 cal/g for ΔH^∞ and 69 erg/cm² for σ [290]. The thermodynamic melting point T_m^∞ was found to be 301°C [290,291]. The commercial nylon-6,6 products melt at about 265°C.

The change in free energy (ΔG) during melting is equal to

$$\Delta G = \Delta H_m - T\Delta S_m = 0 \qquad (2.76)$$

The melting temperature therefore is represented by the quotient

$$T_m = \Delta H_m / \Delta S_m \qquad (2.77)$$

The entropy of fusion (ΔS_m) is a function of the chain conformation, whereas the enthalpy of fusion (ΔH_m) is a function of intermolecular forces, such as hydrogen bonding. Thus both are functions of the chemical structure of the polymer. In polyamides, the principal structural element entailed in intermolecular interactions is the amide moiety, which constitutes a barrier to rotation and as such affects chain conformation and chain stiffness. Consequently, the melting point should be related to the concentration of amide groups or the number of CH_2 units interlinking these groups. Figure 2.27 shows this is indeed the case. This figure also indicates that there is an additional dependency on the symmetry of the structural units.

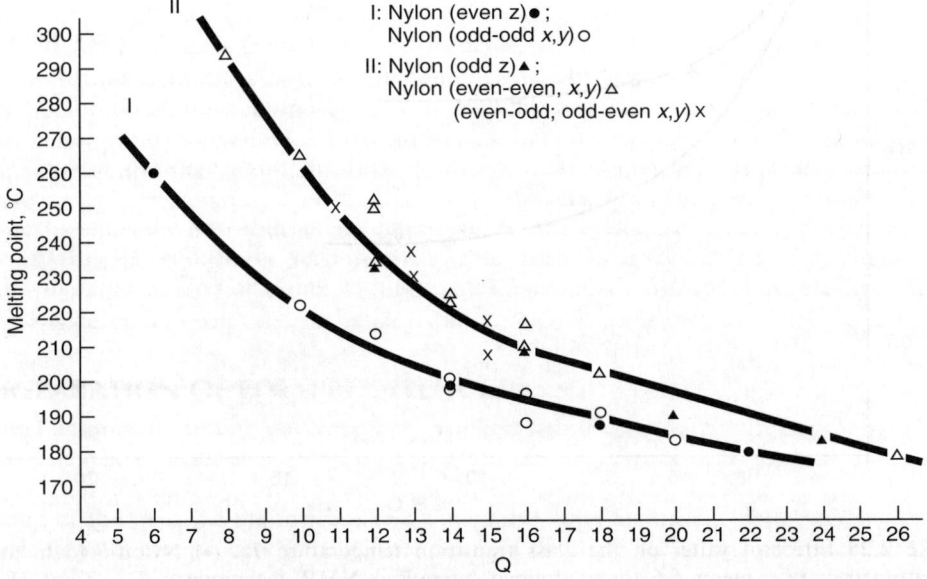

FIGURE 2.27 Melting point of aliphatic polyamides. Q (Nylon Z) = 2 (z−1); Q (Nylon X,y) = x + (y − 2); x,y,z = number of CH_2 groups per monomer unit.

2.5.3.2 Glass Transition Temperature

The most important structural parameter of the noncrystalline (amorphous) phase is the glass transition temperature (T_g) since it has a considerable effect on both processing and properties of the polyamide fibers. It relates to a type of a glass–rubber transition and is defined as the temperature, or temperature range, at which mobility of chain segments or structural units commences. Thus it is a function of the chemical structure; in case of the linear aliphatic polyamides, it is a function of the number of CH_2 units (mean spacing) between the amide groups. As the number of CH_2 units increases, T_g decreases [292]. Although T_g is further affected by the nature of the crystalline phase, orientation, and molecular weight, it is associated only with what may be considered the amorphous phase. Any process affecting this phase exerts a corresponding effect on the glass transition temperature. This is particularly evident in its response to the concentration of water absorbed in polyamides. As shown in Figure 2.28 [409,410] an increase in water content results in a steady decrease of T_g toward a limiting value. This phenomenon may be explained by a mechanism that entails successive replacement of intercatenary hydrogen bonds in the amorphous phase with water. It may involve a sorption mechanism, first suggested by Puffr and Sebenda [293], according to which 3 mol of water interact with two neighboring amide groups, as shown in Figure 2.29.

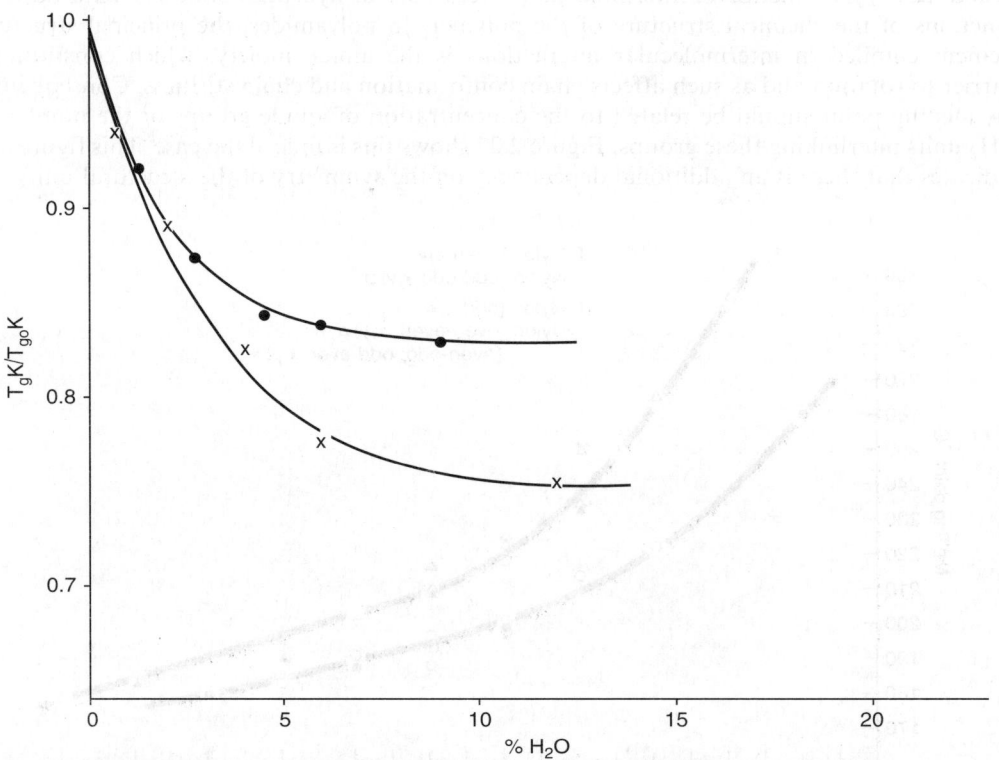

FIGURE 2.28 Effect of water on the glass transition temperature T_g. (•) Nylon-6 (data attained by dilatometry); (x) > nylon 6,6 (data attained by pulsed NMR techniques); T_{go}, T_g at $H_2O = 0$. (From Kaimin, I.F. Apinis, A.P., and Galvanovskii, A.Ya., *Vysokomol. Soedin*, 1975, A17, 41 and E.G. Smith, *Polymer*, 1976, 17, 739.)

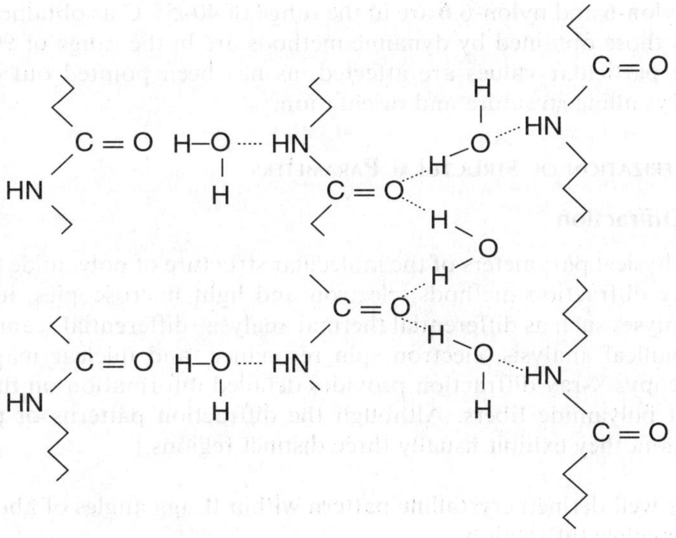

Polyamide Fibers

FIGURE 2.29 Interaction of amine groups with water in nylons.

Equation 2.78 [294] represents the relationship between T_g and water (W) in polyamides

$$T_g = (\Delta T_g)_o \exp\{-[\ln(T_g)_o]W/\tau W_1\} + T_{gl} \qquad (2.78)$$

where $(\Delta T_g)_o = T_{go} - T_{gl}$; $T_{go} = T_g$ at $H_2O = 0$; T_{gl} is the T_g at saturation; W_1 is the saturation water concentration, $dT_g/dW \to 0, = 0.239(1-\alpha)$; α is the fractional crystallinity; and $\tau = W_{(T_{gl}+1)}/W_1 =$ constant $= 0.55-0.58$ for nylon-6 and $0.83-0.88$ for nylon-6,6.

The effect of water on T_g has also been estimated from a mixture theory [292] using expressions that had been derived for plasticizer-containing polymers [295]. It has, however, been shown that water in nylon does not behave as just a simple plasticizer [296]; a finding that supports a more complex interaction of water with polyamides such as shown in Figure 2.29 [293].

Structural changes as result of hydration in nylon-6 showed that water molecules diffuse almost exclusively into the amorphous regions [297]. This diffusion in turn affects the T_g, the modulus, and the crystallinity; the glass transition temperature and the modulus decrease, and the crystallinity increases.

Although the glass transition resembles characteristics of a second-order thermodynamic transition such as changes in the coefficient of expansion and heat capacity, the temperature of the transition is a function of the heating or cooling rate and of the rate of deformation. The methods used to determine T_g are based either on static or dynamic mechanical processes. The former uses volume effects (dilatometry) and heat capacity effects in differential scanning calorimetry (DSC), entailing conditions of very low deformation. The latter utilizes the response to imposed deformation of the system.

With dynamic mechanical methods, a glass transition is observed only for such a combination of deformation temperature and frequency that causes mobility of sufficiently large numbers of chain segments during a given time. Thus, as the frequency increases, more energy in the form of heat is required for the motion of a sufficient number of segments. This is mainly the reason why T_g values obtained by dynamic mechanical methods may be as much as 50°C higher than those resulting from essentially static methods. At 0% water, the T_g

values for both nylon-6 and nylon-6,6 are in the range of 40–55°C as obtained by dilatometry or DSC, whereas those obtained by dynamic methods are in the range of 90–100°C. Within these ranges, the particular values are affected, as has been pointed out earlier, by other factors such as crystalline structure and orientation.

2.5.4 Characterization of Structural Parameters

2.5.4.1 X-Ray Diffraction

Information on physical parameters of the molecular structure of polyamide fibers are usually obtained by x-ray diffraction methods, electron and light microscopies, infrared spectroscopy, thermal analyses such as differential thermal analysis, differential scanning calorimetry, and thermomechanical analysis, electron spin resonance, and nuclear magnetic resonance (NMR) spectroscopy. X-ray diffraction provides detailed information on the molecular and fine structures of polyamide fibers. Although the diffraction patterns of polyamide fibers show wide variation, they exhibit usually three distinct regions:

1. A generally well-defined crystalline pattern within Bragg angles of about 20 to close to 90° by wide-angle diffraction
2. Less well-defined diffuse bands due to scattering from the amorphous phase
3. Diffraction pattern within Bragg angles of less than about 2° of the primary beam by small-angle diffraction, due to longer-range periodicities entailing amorphous and crystalline phases in meridional reflections, particle scattering, and vacuoles that usually cause an equatorial diffraction of continuously declining intensity

"Bragg angle" refers to the well-known relation known as the Bragg equation [298]: $\lambda = 2d \sin \theta$, where λ is the wavelength of the x-rays, d the vertical distance (spacing) of the lattice planes, and θ the angle of the incident x-rays.

The principal features of the diffraction pattern are shown schematically in Figure 2.30 together with the structural information that may be derived from them.

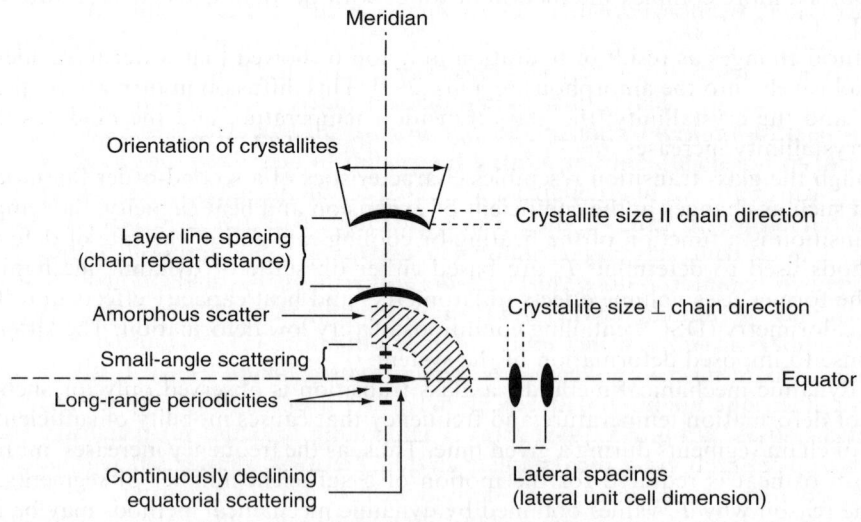

FIGURE 2.30 Schematic diagram of x-ray fiber diffraction pattern.

2.5.4.2 Crystallite Size

The diffraction pattern of polyamide fibers, as well as those of many other fibers, is characterized by a considerable broadening of the wide-angle diffraction maxima. This is usually explained in terms of the crystallites size responsible for the diffraction. The formal relationship between crystallite size and θ is given by the Scherrer equation [299]: $D = K\lambda / \Delta(2\theta) \cos \theta$, where D is the diameter of the crystallite, $\Delta(2\theta)$ is the angular diffraction width of the particular crystal reflection, and K is the shape factor, which is a constant with a value close to unity.

The Scherrer equation is applicable only for defect-free crystals. Drawn filaments, particularly immediately after drawing, usually yield rather diffuse diffraction patterns as a result of considerable lattice disorder. In addition to the crystallite size effect, this lattice disorder causes further broadening of the diffraction maxima [300]. The subsequent sharpening of these reflexes, particularly upon annealing, has therefore been explained as a result of both crystallite growth and "healing" of lattice defects. Since both effects are difficult to separate, the Scherrer equation yields what has been called a "crystallite size equivalent" rather than the actual crystallite size. Conclusive correlations between either of these parameters and fiber processing and performance characteristics are yet to be developed.

2.5.4.3 Degree of Orientation

One of the most important parameters of fiber structure is the orientation entailing both the crystalline and noncrystalline regions. The crystalline orientation relates to the distribution of orientation of all crystal axes relative to the fiber axis. Assuming that the polymer chains are all parallel to the crystallographic axis, the crystallite orientation becomes a function of the angle between the particular crystal axis and the fiber axis, as shown in Figure 2.31. The magnitude of the orientation is reflected in the arc length of the x-ray diffraction pattern and the density distribution within the arc in Figure 2.31. The arc length decreases with increasing orientation toward a minimum length that is determined by the crystallite size effect.

Hermans Orientation Function [301,302] in Equation 2.79 gives a quantitative expression for specifying the degree of axial orientation in crystalline fibers:

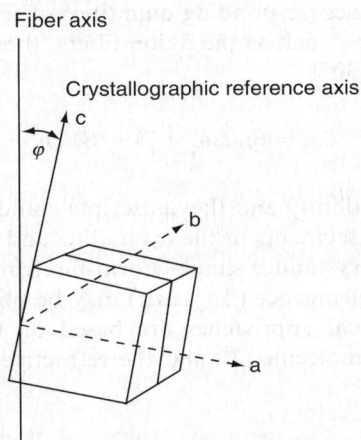

FIGURE 2.31 Crystalline orientation with respect to the fiber axis.

$$f_\Phi = (3 <\cos^2 \Phi > -1)/2 \tag{2.79}$$

where $<\cos^2\Phi>$ represents the mean-square cosine (averaged over all the crystallites) of the angle Φ. The value for $<\cos^2\Phi>$ may be evaluated from the orientation distribution function $I(\phi)$ using the following equation [255,303]:

$$<\cos^2\Phi> = \int_{\Phi_o}^{\pi/2} I(\Phi)\sin\Phi\cos^2\Phi\,d\Phi \Big/ \int_0^{\pi/2} I(\Phi)\sin\Phi\,d\Phi \tag{2.80}$$

The value for f_Φ (Equation 2.79) is within the range $1 \geq f_\phi \geq -0.5$. It is 1 when all polymer chains are parallel to the fiber axis ($\Phi = 0$), 0 for complete isotropy ($\cos^2\Phi = 1/3$), and -0.5 when all polymer chains are perpendicular to the fiber axis ($\Phi = 90°$).

2.5.4.4 Birefringence

Noncrystalline chain segments in fibers may also show orientation in the direction of fiber axis. The extent of this "amorphous orientation" can be estimated from the total orientation in combination with the crystallite orientation as obtained from x-ray analysis. Although spectroscopic, sonic, and NMR methods have been described for the determination of the general orientation, the most facile method is based on optical anisotropy [304]. This fiber property is characterized by the optical birefringence factor Δn, defined by the following equation:

$$\Delta n = n_\| - n_\perp \tag{2.81}$$

where $n_\|$ and $n_\perp$ are the refractive indices of light polarized parallel and perpendicular to the fiber axis.

The optical birefringence is related to the total internal fiber orientation by the optical orientation function, defined by the following equation:

$$f_\phi = (\Delta n/\Delta n°)(d°/d) \tag{2.82}$$

where $\Delta n°$ and $d°$ are, respectively, the birefringence and density of a perfectly oriented (ideal) fiber, whereas Δn and d are the corresponding quantities of the real fiber.

In partially crystalline fibers, such as the nylon fibers, the total birefringence Δn may be represented by Equation 2.83 [305]:

$$\Delta n = \beta f_c \Delta n_c° + (1-\beta) f_a n_a° \tag{2.83}$$

where β is the fractional crystallinity and the subscripts c and a indicate that the particular quantities pertain to the chain segments in the crystalline and amorphous regions. Equation 2.83 neglects the effect of the crystallites shape—form-birefringence.

Values for the intrinsic birefringence ($\Delta n_c°$, $\Delta n_a°$) may be obtained experimentally [306] or analytically [307]. The analytical approaches are based on the relationships between the principal polarizability of the molecule (P) and the refractive indices as represented by the Lorentz–Lorenz equation:

$$(n^2 - 1)/(n^2 - 2) = 4\pi P/3 \tag{2.84}$$

Employing crystal lattice parameters of the α-phase of nylon-6, Δn_c° has been calculated to be about 0.0825 for this polyamide [307].

An analysis of the theoretical methods of calculation of the ideal fiber birefringence has been presented in the literature [308]. Using a modified Lorentz–Lorenz equation, theoretical birefringence was calculated by considering intermolecular interactions. The calculations showed considerable discrepancies between the theoretical and the experimental values.

An expression for estimating the orientation of the chain segments in the amorphous regions is readily obtained by combining Equation 2.79, Equation 2.81, Equation 2.82, and Equation 2.83.

2.5.4.5 Crystallinity and Crystalline Indices

The degree of crystallinity of polyamide fibers may be estimated from density determinations, calorimetric measurements, and infrared and x-ray data. Although not an absolute method, assessment of the degree of crystallinity from the density is a very facile, rapid, and precise procedure. It is independent of orientation or geometry of the sample, but requires dry samples that are free of voids and pigments. This method is based on the assumption that the density ρ or its reciprocal value and the specific volume V are represented by Equation 2.85 and Equation 2.86, respectively.

$$\rho = x_v \rho_c + (1 - x_m) \rho_a \tag{2.85}$$

$$V = x_m V_c + (1 - x_m) V_a \tag{2.86}$$

where x_v is the crystalline volume fraction, x_m is the crystalline mass fraction, and the subscripts c and a indicate crystalline and amorphous quantities. Stipulating that the enthalpy of fusion is a function of only the crystalline fraction, a value for x_m may be obtained from calorimetric measurements according to

$$x_m = \Delta H^\infty / (\Delta H^\infty)_{100} \tag{2.87}$$

where $(\Delta H^\infty)_{100}$ represents the enthalpy of fusion of a 100% crystalline material and is a true value usually obtained by extrapolation [267]. Differential scanning calorimetry may be used to obtain the actual heats of fusion from which, under consideration of the crystallite size effect, values for ΔH^∞ may be estimated.

The density of the crystal (ρ_c) can be obtained from x-ray structural data according to the relationship

$$\rho_c = n M_m / V_u L \tag{2.88}$$

where M_m is the molecular weight of the monomer unit, n the number of monomer units per unit cell, V_u the volume of the unit cell, and L the Loschmidt number $= 6.03 \times 10^{23}$. Values for ρ_a may be obtained by extrapolation of the melt densities to room temperature or by measuring the density of a macroscopic amorphous material as obtained by rapid quenching from the melt. With values for ρ_c and ρ_a, Equation 2.85 yields for the degree of crystallinity

$$x_v = (\rho - \rho_a)/(\rho_c - \rho_a) \tag{2.89}$$

Since polyamides may crystallize in different forms, ρ_c in Equation 2.89 may be substituted by the expression

$$\rho_c = w_\alpha p_{c\alpha} + w_\gamma \rho_{c\gamma} \tag{2.90}$$

where w_α, w_γ, and $\rho_{c\alpha}$, $\rho_{c\gamma}$ are, respectively, the weight fractions and densities of the corresponding crystal forms.

The different crystalline forms of several polyamides have been examined by infrared spectroscopy. The bands at 936 and 1140 cm^{-1} have been used to measure the crystalline and amorphous contents, respectively, in nylon-6,6 [309], whereas bands at 1198 and 1181 cm^{-1} were employed for measuring α and γ crystalline content [310]. Fluctuations in crystallinity in nylon-6,6 were related to the ratios 1430/2910 and 930/2910 [311,312]. All even–even nylons show an α-crystal-related peak at 690 cm^{-1}, which appears at 725 cm^{-1} in odd–odd nylons [313]. In nylon-6, there is a crystallinity band at 1260 cm^{-1}, α-crystalline-sensitive bands at 1028, 960, 950, 930, and 830 cm^{-1}, and γ-crystalline-related bands at 1120, 990, and 970 cm^{-1} [314–316]. A band at 979 cm^{-1} has been attributed to the amorphous phase [315].

Considerable effort has been devoted to obtain information on the relative amounts of crystalline structures and amorphous materials in polyamides from wide-angle x-ray diffraction [263,317–321]. Success depends on adequate separation of the individual contributions of the various structural constituents to the x-ray equatorial diffraction pattern. In the diffraction pattern of polyamide fibers, the most intense region consists of two reflections and relates to the interaction between neighboring chain segments. For the triclinic nylon-6,6, as can be seen in Figure 2.32, this region entails the 100, 010, and 110 reflections. At low crystalline perfection, the equatorial scan shows only one peak in Figure 2.33a. Its width at half-maximum intensity may be considered a measure of order. With increasing crystalline

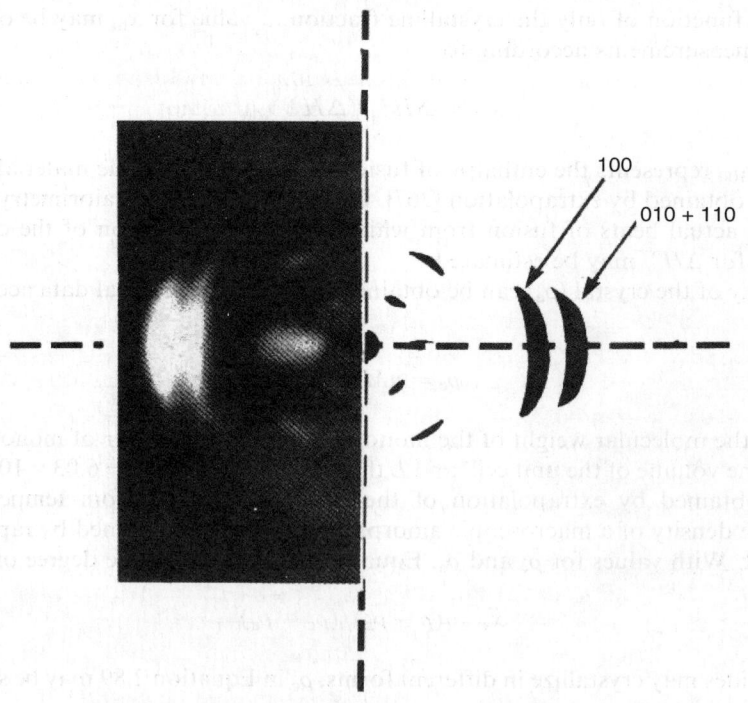

FIGURE 2.32 Wide angle x-ray fiber diagram of nylon-6,6.

Polyamide Fibers

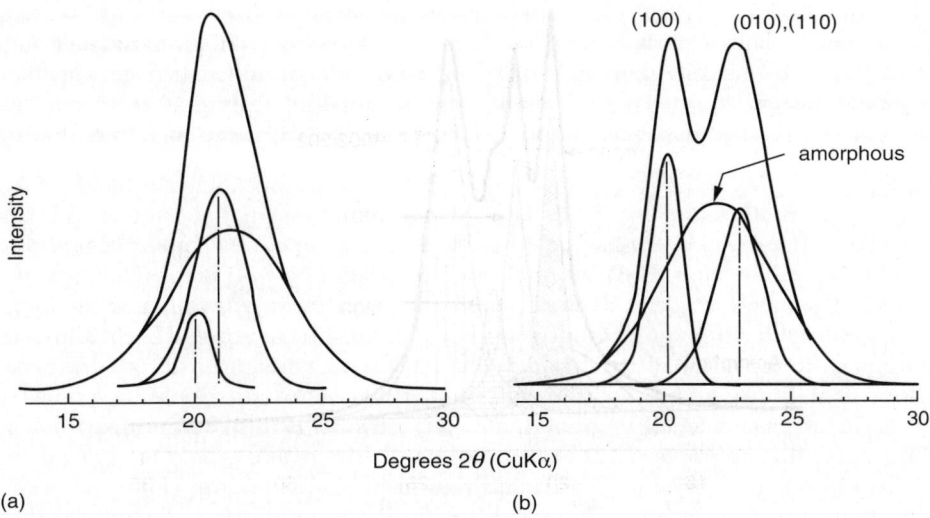

FIGURE 2.33 Wide-angle x-ray equatorial diffraction scans. (a) Low crystalline perfection; (b) high crystalline perfection. (From Clark, E.S. and Wilson, F.C., in *Nylon Plastics*, M. Kohan, Ed., John Wiley & Sons, New York, 1973. With permission.)

perfection, two peaks appear as shown in Figure 2.33b [322] due to decreasing spacing between the hydrogen-bonded lattice planes. The peak separation entails moving of the 010,110 doublet to higher angles. In nylon-6, as was discussed earlier, two crystalline modifications—an α-form and a γ-form—can be found. Their respective x-ray reflections and corresponding schematic presentations are shown in Figure 2.34. Since both the α- and

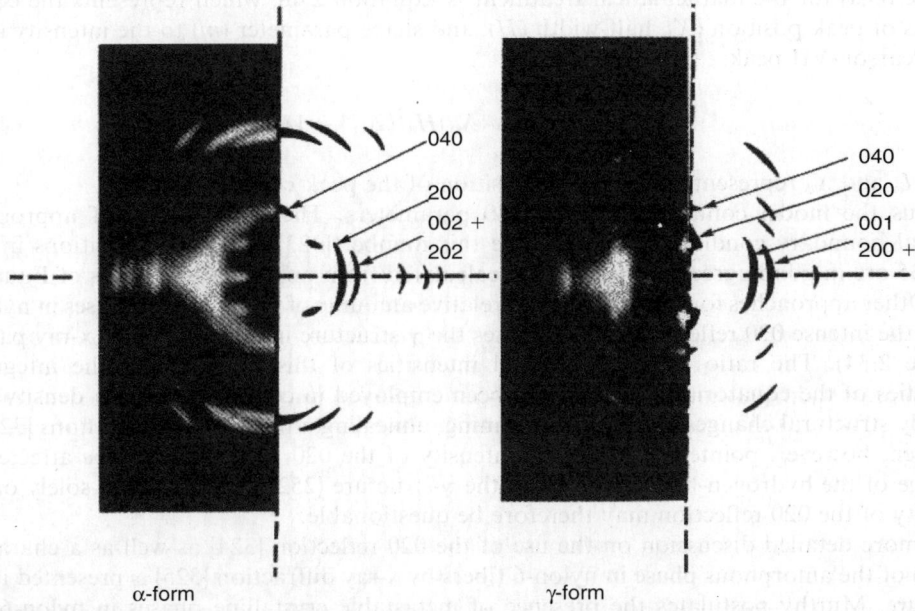

FIGURE 2.34 Wide-angle x-ray diagrams of the α- and γ-forms of nylon-6.

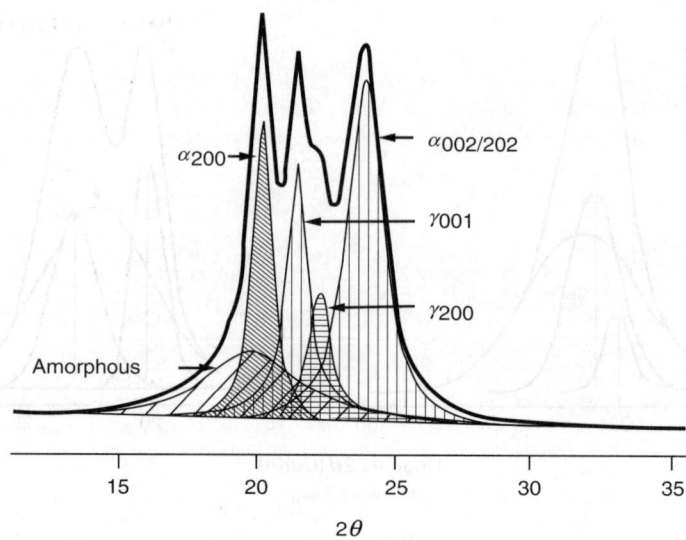

FIGURE 2.35 Experimental and fitted equatorial diffractions.

γ-structures may be present in nylon-6 fibers, the measurement of their relative amounts is more problematic. What appears to be the most successful treatment thus far entails the development of a mathematical model based on a Pearson-VII-type function [320] comprising five curves: two of which are related to the α-form, two to the γ-form, and one to the amorphous part of the fiber. A computer fits these curves to the observed equatorial diffraction scans as shown in Figure 2.35 [323].

The basis for the mathematical treatment is Equation 2.48, which represents the contributions of peak position (X), half-width (H), and shape parameter (m) to the intensity (I) of each Pearson-VII peak:

$$I = I_o\{1 + 4[(X - X_o)H]^2(2^{1/m} - 1)\}^{-m} \tag{2.91}$$

where I_o and X_o represent intensity and position of the peak center.

Thus the model contains a total of 20 parameters. The introduction of appropriate physical boundary conditions could reduce this number [323]. The fitted positions in Figure 2.35 are in good agreement with those calculated from unit cell dimensions of Equation 2.86. Other approaches for determining the relative amounts of the different phases in nylon-6 utilize the intense 020 reflection that indicates the γ-structure in the wide-angle x-ray pattern (Figure 2.34). The ratio of the integrated intensities of this reflection to the integrated intensities of the equatorial reflection has been employed in combination with density data to study structural changes during melt spinning, annealing, and drawing operations [321]. It has been, however, pointed out that the intensity of the 020 reflection may be affected by slippage of the hydrogen-bonded sheets in the γ-structure [252]. Results based solely on the intensity of the 020 reflection may therefore be questionable.

A more detailed discussion on the use of the 020 reflection [324] as well as a characterization of the amorphous phase in nylon-6 fibers by x-ray diffraction [325] is presented in the literature. Murthy postulates the presence of metastable crystalline phases in nylon-6 and discusses the implication on mechanical properties. He has also considered factors associated with the mobility of polymer chains in the crystalline regions [285,326]. Baldrian and Lednicky have reviewed subject matters related to the crystal structures and morphology of lactam-derived nylons [16e]. Their discussion on morphology encompasses topics such as solution-grown crystals, entailing growth from quiescent solutions, and formation of fibrous crystals. It further deals with epitaxy, high-pressure crystallization, deformation, high-modulus fibers, heterophase polyamide systems of blends, and crystallization kinetics. The effect of annealing on the structure and morphology of nylon-6 fibers also has been discussed in the literature [327], as well as changes and effects of void content and free volume in fibers during heat setting [328]. Elsewhere, Reimschuessel has reported polymer–metal halide interactions entailing nylon-6 and zirconium tetrafluoride [329]. ZrF_4 forms stable complexes with both nylon-6 and caprolactam. The incorporation of ZrF_4 constitutes a molecular reinforcement and results in significant increases in both yield stress and tensile modulus. The incorporation of ZrF_4 results in compositions characterized by increased rates of nucleation and a stable, fine crystal structure in the solid state. Genis et al. discuss the effect of plasticizers on the properties of fibrous isotropic lattices of aliphatic polyamides [330].

2.5.5 Typical Fiber Properties

The thermal and mechanical properties, solubility and transport phenomena, and resistance toward chemicals in polyamides are reviewed in considerable detail and discussed in the literature [16f]. Tuzar's discussion on thermal features addresses phenomena related to relaxation and melting. His survey of mechanical properties concerns yielding, orientation, fibrillation, high-modulus and high-strength, molecular fracture, crazing and stress cracking, crack propagation, solubility and transport phenomena, and effects of chemicals.

An extensive discussion of all the important aspects and methods of characterization of physical and mechanical properties of fibers is outside the scope of this chapter. Corresponding information can be found in more comprehensive treatises [331,332]. Some typical properties of nylon-6,6, nylon-6, nylon-11, nylon-6,11 and nylon-6,12 are listed in Table 2.5 and Table 2.6.

To impart and enhance desirable properties, or to remove and reduce undesirable characteristics, procedures have been designed to modify performance attributes of nylon fibers. Either chemical and physical procedures or a combination of both are applied in these operations. As pointed out earlier in this chapter, chemical modification refers to the constitution of the polyamide and relates to the nature of the monomeric units, end groups, and the molecular weight. Although not in accordance with this definition of chemical modification, manipulation used to alter such properties as moisture transport [333], hygroscopic characteristics [334], electrostatic charge, flammability, and soil resistance are usually conceived as chemical modification. Germane approaches utilize addition of suitable modifiers, either to the polymerizing composition or to the polyamide melt. The additives used in these procedures must not interact chemically with the polymer or the polymer-forming materials. Ideally, the additives should have the following properties:

1. They should persist as a fine dispersion in the polymer matrix or polymer-forming composition.
2. They should not affect the rate and extent of polymerization.
3. They should not form or promote the formation of gels.
4. They should not alter the surface characteristic of the final product.
5. They should retain thermal stability to withstand the conditions used in both polymerization and melt processing.
6. They should not be removed by any extraction procedures, which may be applied to the polymer or the final product.

TABLE 2.5
Typical Properties of Nylon-6 and Nylon-6,6 Fibers

Polymer type	Nylon-6			Nylon-6,6			Nylon-11	Nylon-6,12
Product type	Staple	Regular	High strength	Regular	High strength			
Molecular weight	20000–30000			12000–20000				
Glass transition temperature (°C)[b]	45–75			60–80				
Melt temperature (°C)	210–220			255–265			195–219	
Ignition temperature (°C)				530				
Density (g/cm³)	1.14			1.14				1.04–1.05
Moisture regain (%)	4.0–5.0			1.0			1.09	
Tenacity (cN/tex)							4.5–6.8	
Dry	4.2–5.9	4.4–5.7	5.7–7.7	4.9–5.7	5.7–7.7			
Wet	3.5–5.0	3.7–5.2	5.2–6.5	4.0–5.3	4.9–6.9			
Wet/dry tenacity ratio (%)	83–90	84–92	84–92	90–95	85–90			
Elongation at break (%)							15–40	
Dry	38–50	28–42	16–25	26–40	16–24			
Wet	40–53	36–52	20–30	30–52	21–28			
Recovery at 3% elongation (%)	95–100	98–100	98–100	95–100	98–100			
Initial modulus (GN/m²)	0.98–2.45	1.96–4.41	2.75–5.00	3.66–4.38	2.30–3.11		4.0–4.4	

[b] As discussed in Section V.C.2, the glass transition temperature is affected by a number of variables, including the water content of fiber.

TABLE 2.6
Typical Tensile Properties of Nylon Fibers According to End-Uses

	Tenacity (g/d)	Elongation (%)	Modulus (g/d)
Textile draw-twisted	4–6	30–70	40–60
POY	3–5	50–120	25–50
Carpet	3–5	30–60	40–60
Rope and cordage	7–10	15–30	70–100

Fibers with improved dimensional stability are asserted for a process outlined in the Japanese patent 57,193,516 (Teijin, 1982). A process claimed in the Japanese patent 5,898,415 (Teijin, 1983) refers to operations characterized with reduced yarn breakage during draw-spinning, and Japanese patents 5,836,211 and 5,836,212 (Teijin, 1983) assert procedures, resulting in improvements in yarn tenacity. Another Teijin patent, JP 61,132,615, relates to a process yielding fibers distinguished by improved dyeing characteristics. Elsewhere, the role of dye diffusion in the ozone fading of acid and disperse dyes in polyamides has been discussed in a study that deals with the effect of the physical structure of the fiber on the fading by ozone of both acid and dispersed dyes in nylons [335]. The Toray Industries patent JP 60,128,166 (1985) refers to a process that yields fibers with good packaging stability and good dyeing uniformity.

2.5.5.1 Tensile Behavior

Polyamide fibers, as fibrous materials in general, are characterized by anisotropy of physical properties, which is reflected in different values for a given property in the axial and the radial directions of the fiber. This phenomenon is obviously a consequence of orientation of polymer molecules in both the crystalline and the amorphous regions. It is particularly pronounced for important mechanical properties comprising the elastic modulus, and both yield and breaking stresses, or their corresponding elongations. These properties, which constitute responses to applied forces and deformations, largely determine both the behavior of the fibers in processing and their performance characteristics in final applications. Of particular interest are the tensile properties. They represent the behavior of fibers when forces and deformations are applied along the fiber axis. This behavior is completely depicted by stress–strain curves, which are obtained by measuring the elongation of fibers applying gradually increasing force (load), until breakage. The relationships between the specific stress and the tensile strain, the yield point, elastic modulus, and work of rupture are derived from these curves. The specific stress is defined by the following equation:

$$\text{Specific stress} = \text{load/mass per unit length} \tag{2.92}$$

and is expressed in Newton per tex (N/tex). The tensile strain as defined by Equation 2.93 is equal to the ratio of elongation to initial length and is usually expressed as percentage extension:

$$\text{Tensile strain} = \text{elongation/initial length} \tag{2.93}$$

The yield point, characterized by conjugated yield stress and yield strain is that point of the stress–strain curve beyond which at higher strains elastic recovery becomes less complete and permanent deformation starts to take place.

The elastic modulus can be derived directly from the shape of the stress–strain curve and is equal to its initial, constant slope. Its value equals that of the stress that would be required to double the length of the fiber at the initial conditions; it is therefore measured in units of stress or specific stress. The work of rupture, a measure of the toughness of the fiber, is defined as the energy needed to break the fiber and may be represented by the following equation:

$$\text{Work of rupture} = \int_o^{l_b} \text{forces} \times \text{displacement} = \int^{l_b} F \, dl \qquad (2.94)$$

The units are joules. Since the work of rupture is proportional to the fiber mass per unit length, a specific work to rupture may be defined according to the following equation:

$$\text{Specific work to rupture} = \frac{\text{work to rupture}}{(\text{mass/unit length}) \times \text{initial length}} \qquad (2.95)$$
$$= \text{force}/(\text{mass/length})$$

The units are N/ex or kJ/g; Nm/kg. Figure 2.36 is a schematic presentation of a stress–strain curve and indicates some of the features that may be derived from it.

The behavior of a fiber, that is the actual shape of the stress–strain curve and thus a set of important physical properties, depends upon the basic morphology of the particular fiber.

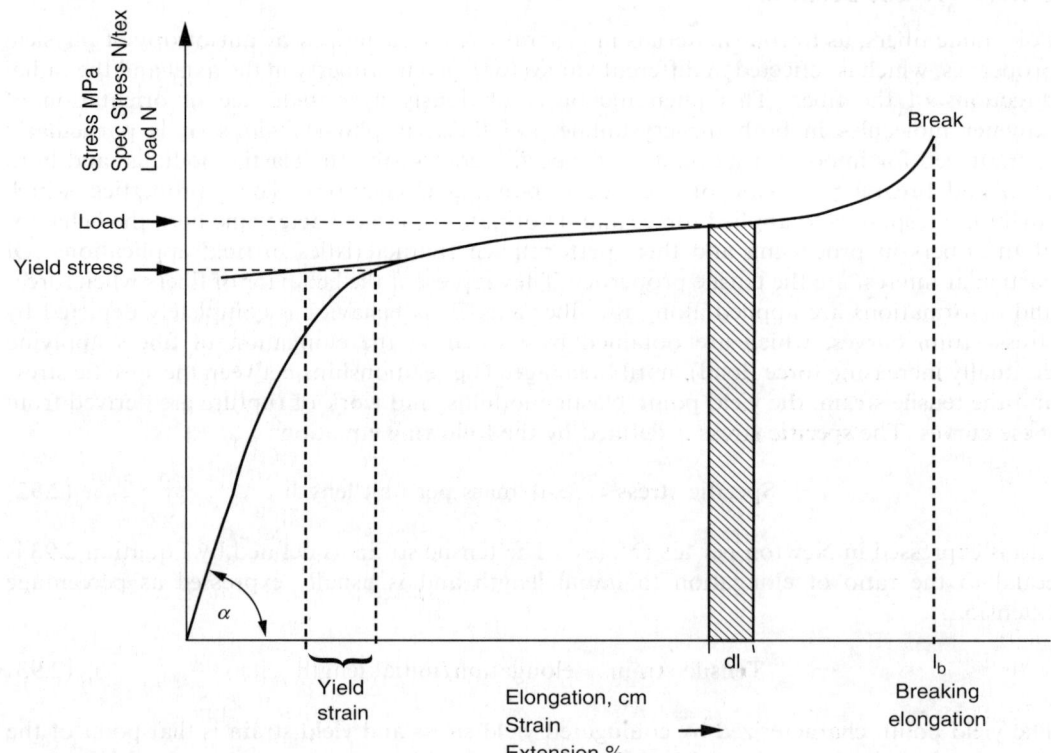

FIGURE 2.36 Schematic stress–strain curve.

Detailed studies of the plastic deformation of semicrystalline fibers have resulted in models for such fibrous structures. They belong to either of two general groups. One assumes a more or less uniform crystal structure for the entire fiber with randomly distributed defects [336,337], whereas the other considers bridging of amorphous layers between crystal blocks by tie molecules that may be either perfectly lax [338,339] or characterized by tautness [340–342].

2.5.5.1.1 Microfibrillar Model

Models that assume a regular alternation of crystalline and amorphous layers within a microfibrillar structure best explain the behavior of polyamide fibers. Such a structure is a result of the extensional flow and plastic deformation to which polyamides are subjected in melt-spinning and drawing operations. The microfibrils in turn are bound tightly together into macrofibrils. Thus any axial force is readily distributed among the microfibrils, thereby homogenizing the corresponding stress field and diminishing the influence of local imperfections. The crystal blocks are connected by both intrafibrillar and interfibrillar tie molecules. Plastic deformation due to fiber drawing causes unfolding of the chains at the outer boundaries of the crystal blocks and results in an increase of the fraction of taut interfibrillar tie molecules. Since this process entails shearing displacement of adjacent microfibrils, it eventually effects the full extension of interfibrillar tie molecules. A schematic presentation of the resulting structure is shown in Figure 2.37.

The increase of the fraction of these extended molecules is proportional to the draw ratio and is paralleled by an increase of the axial elastic modulus toward a limiting value corresponding to the finite concentration of the extended structures. At high draw ratios, such extended molecules can form crystalline bridges between folded chain blocks [283]. This may be reflected in a more than linear increase of the modulus with the draw ratio. No such behavior, however, is indicated for the tensile strength. Proportionality between tensile strength and elastic modulus at any draw ratio would characterize an ideal fiber. In real fibers, however, the increments of any increase in strength decrease for a given extension interval with the draw ratio. This retarded increase in strength and the eventual break is indicative of structural defects. Identification of the nature of the structural defects depends

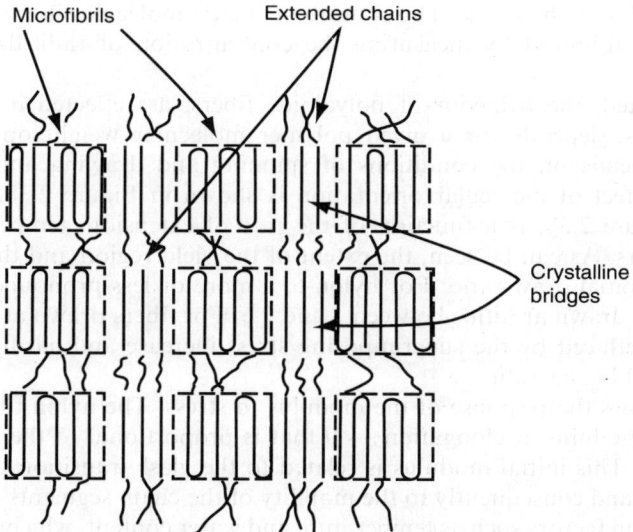

FIGURE 2.37 Microfibrillar structure with noncrystalline extended-chain molecules.

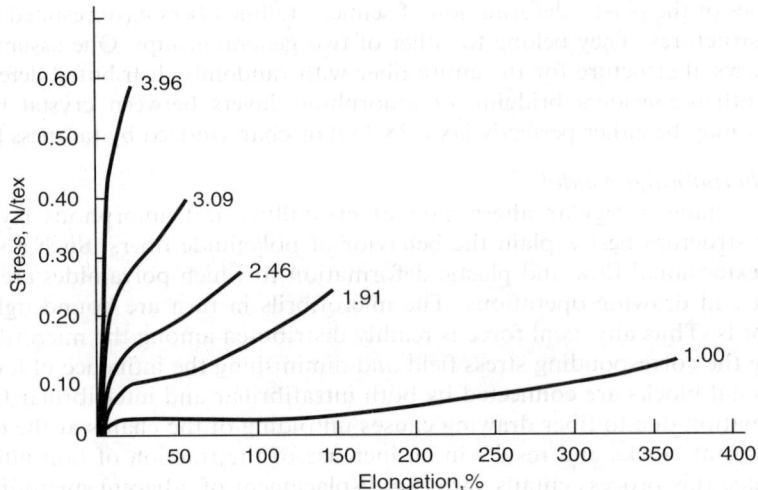

FIGURE 2.38 Stress–strain diagram of nylon-6 filaments drawn at different draw ratios. (From Schultz-Gebhart, F., *Faserf. Textiltechn.*, 1977, 28, 467. With permission.)

upon the particular morphological model. Some models assume as structural defects amorphous layers or domains that are either regularly arranged between subsequent crystal blocks [343] or randomly distributed in a crystal matrix [339,340], whereas, according to another model [344], only the ends of microfibrils constitute structural defects. In any case, tensile failure is indicated by rupture of taut tie molecules, particularly those that are close to areas of structural defects. Such rupture results in the formation of microcracks that, by growth and coalescence, can reach a critical dimension and thereby cause catastrophic failure. The microcracks may grow radially or axially. Radial growth results in the rupture of tie molecules bridging amorphous layers, whereas axial growth separates adjacent microfibrils, and filament failure is then a consequence of the rupture of tie molecules connecting microfibrils on the opposite sides of the crack. The rupture of the tie molecules in polyamide fibers has been detected and followed by measuring the concentration of radicals by electron spin resonance [345].

As already stated, the behavior of polyamide fibers, as reflected in the shape of the stress–strain curves, depends for a given polymer molecular weight on the morphology, which in turn depends on the conditions of spinning and drawing and subsequent heat treatments. The effect of molecular orientation is shown in Figure 2.38. The yield strain, about 10% in Figure 2.38, is a function of the rate of extension, temperature, and water content of the fibers. As can be seen, the extent of the yield region and the total elongation depend upon the initial draw ratio. For nylon-6, a more or less pronounced yield region is indicated for fibers drawn at ratios between 1 and 2.5. For fibers drawn at ratios higher than 2.5, this region is effaced by the superimposing stress increase and hardly recognizable for fibers drawn at still higher ratios.

Figure 2.39 shows the response of the modulus to stress. The nylon fibers are characterized by an initial modulus at elongation $\rightarrow 0$ that is proportional to the ratio at which the fibers were drawn. This initial modulus is related to the glass transition temperature of the amorphous region and consequently to the mobility of the chain segments in these regions. It depends therefore on factors such as temperature and water content, which affect the mobility of the chain segments.

Polyamide Fibers

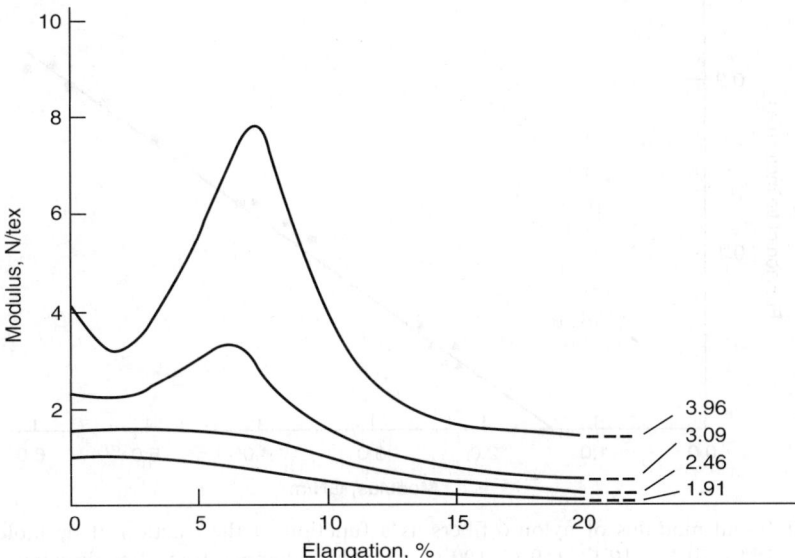

FIGURE 2.39 Modulus–strain diagram of nylon-6 filaments drawn at different draw ratios. (From Schultz-Gebhart, F., *Faserf. Textiltechn.*, 1977, 28, 467. With permission.)

Stress-induced deformation (elongation) results in arrangement of chains in the axial direction. Some disentanglement of tie molecule segments entailed in this process results in a decrease of the modulus with increasing strain. As shown in Figure 2.39, for nylon-6 fibers that had been drawn at ratios higher than about 2.5, this decrease terminates at about 2% elongation. Further increase of the strain results in an increase of the modulus toward a rather pronounced maximum at about 6–8% elongation. The slope of this increase is related to the distribution of the length of the intercrystal tie molecules; its value is inversely proportional to the average length of the tie molecules. The ensuing decrease of the modulus with increasing strain is due to the extended tie molecules and the pulling out of such molecules from the crystal blocks. As mentioned earlier, further increase in strain then results in the formation of both extended interfibrillar tie molecules due to the shearing displacement of adjacent microfibrils and new crystalline bridges in the amorphous layers. This effect may be affected by the length and extensibility of the intrafibrillar tie molecules. The direct proportionality of the modulus of nylon-6 fibers and the concentration of taut tie molecules has been demonstrated [343] using a model entailing mechanical coupling of the crystalline and amorphous regions [346]. Figure 2.40 depicts the results.

The structure of the drawn filaments does not correspond to a thermodynamic equilibrium state. To approach such a state, drawn fibers are generally subjected to a thermal treatment to stabilize their shape and structure. This heat-setting process causes changes in the morphology and may entail recrystallization processes and the formation of new crystalline and intercrystalline structures characterized by lower free enthalpy. Consequently, the drawn fibers shrink on heating. The extent of this shrinkage depends on the existing structure of the fiber, the applied tension, the temperature, the rate of heating, and the concentration of any plasticizing or swelling agent (water). The effect of such heat treatment on the stress–strain relationship is shown in Figure 2.41 for nylon-6 filaments treated with saturated steam at 110° and 130°C, both with and without applied tension. The shift of the maxima in the modulus curves to higher elongations, their flattening in the case of the tension-free filaments, and the modulus decrease

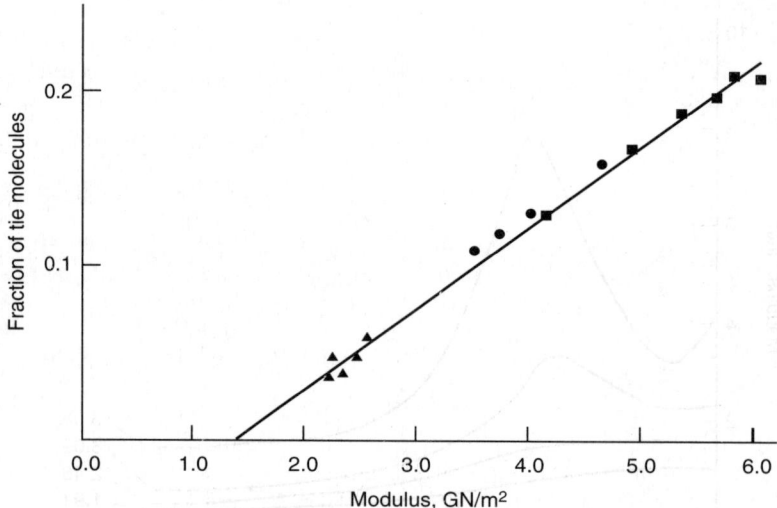

FIGURE 2.40 Initial modulus of nylon-6 fibers as a function of the fraction of tie molecules. Draw temperatures, 35°C, 70°C, 110°C, 150°C, 190°C; quenching temperature after drawing, −16°C; heat seuing temperature, 120°C (20 min); ratios, ▲ = 2.5, ● = 3.5, ■ = 4.5. (From Mishra, S.P. and Deopura, B. L., *J. Appl. Polymer Sci.*, 1982, 27, 3211. With permission.)

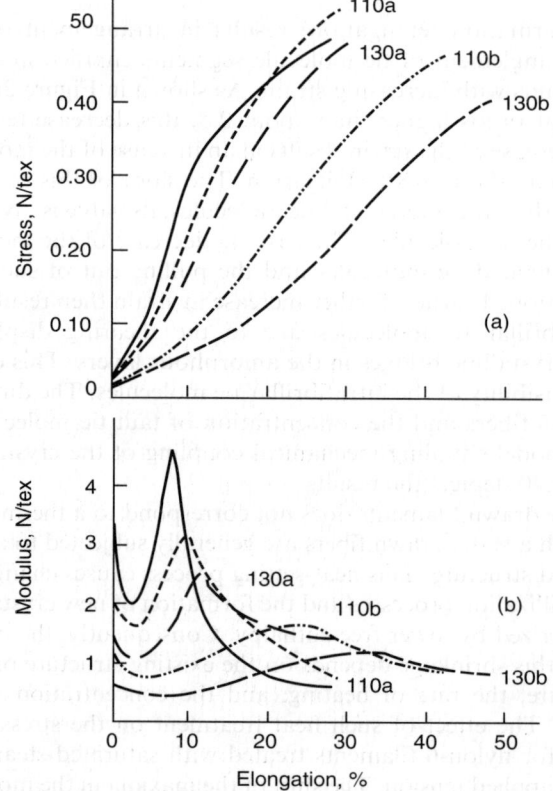

FIGURE 2.41 Stress–strain and modulus–strain curves for steam-treated drawn nylon-6 filaments. (Solid curve.) Untreated filaments; (a) constant filament length; (b) free shrinkage. (From Schultz-Gebhart, F., *Faserf. Textiltechn.*, 1977, 28, 467. With permission.)

may be indicative of both a perfection of the folded crystal blocks and an increase of the length of interfibrillar tie molecules at the expense of less perfect crystalline bridges formed during the drawing process. Disconvolution of these interfibrillar tie molecules due to drawing or tension may then again cause an increase in the modulus.

The effect of temperature and rate of elongation on the characteristic of the stress–strain diagram is important particularly for the examination and characterization of fibers. Increasing the test temperature results in a decrease of the modulus, whereas increasing the rate of elongation yields higher values for the modulus. These aspects have been reviewed in detail [347]. The effect of water is similar to that of the temperature. The decrease of elastic modulus with increasing water concentration parallels that of the glass transition temperature.

2.5.5.2 Chemical Properties

The chemical properties of polyamide fibers are related mainly to the constitutional parameters of the polymer structure and therefore are dependent on the concentrations and reactivates of the amide groups and the particular end groups. Thus polyamides undergo reactions typical for amide groups, such as addition of formaldehyde, resulting in the formation of methyl groups and alkoxymethyl groups. Aqueous acids and bases cause hydrolysis, whereas concentrated acids and phenols are solvents. The same functional groups are effective in some of the dyeing processes employed for nylons.

Different classes of dyes can dye both nylon-6,6 and nylon-6. The most important are acid dyes, metal complex dyes, and dispersion dyes. The acid dyes are mainly sodium salts of mono- and disulfonic acids. The dyeing equilibrium entailing acid dyes has been described by a two-phase model [348] assuming concentration-independent distribution coefficients. Thus far, little has been reported on activity coefficients and dye aggregation. The rate of dyeing depends on the diffusion of the dye into the fiber. It has therefore been described by Fick's equation expressing the temperature dependency of the diffusion coefficient by a WLF-type equation [349]. Models considering deviations from Fick's equation have been suggested [350]. Metal complex dyes are coordination complexes consisting of a metal atom, usually chromium, and either one or two chromophore systems. The latter contain either sulfonic acid groups or nonionic sulfur moieties. These dyes interact first with the basic function of the polyamide fibers by salt formation entailing the dye anion and subsequently by a Nernst distribution of the undissociated dye [351]. Insoluble members of this group act similar to the dispersion dyes. For this group of dyes, the distribution of the dye between the fiber and the dye bath according to Equation 2.96 [352,353] governs the dyeing process:

$$C_F = K_d C_B \text{ for } C_B \geq C_{BS} \tag{2.96}$$

where C_F is the concentration of the dye in the fiber; K_d is the distribution coefficient; C_B is the concentration of dye in the bath; and C_{BS} is the solubility limit.

In addition to the dependence on constitutional parameters, the rate and extent of dyeing also depend significantly on the morphology, crystallinity, and molecular orientation of the fiber, for any of the processes considered [353].

2.5.5.3 Degradation Behavior

Polyamides are susceptible to degradation by heat, oxygen, light, and chemical agents. During melt processing, thermal, oxidative, and hydrolytic processes may be operative that may also contribute, in addition to photochemical degradation, to the deterioration of properties of any finished article. Many approaches have been described in the patent literature to protect polyamide fibers against light-induced degradation. The use of manganese compounds such as the acetates, sulfates, phosphates, and hypophosphites appears to be most effective [354]. Thermo-oxidative degradation of fibers has been reduced by the addition of copper-I salt [355] and the use of antioxidants based on either alkylphenol or aromatic amines in possible combination with trialkylphenylphosphites [356,357].

2.5.5.3.1 Thermal Degradation

At temperatures used in processing and melt-spinning operations, both nylon-6,6 and nylon-6 can undergo thermal degradation as indicated by the formation of carbon dioxide, ammonia, and water [358,359]. Ammonia and carbon dioxide result obviously from deamination and decarboxylation reactions, whereas water results from various condensation and dehydration reactions. In addition, particularly at higher temperatures, low-molecular-weight amines, acids, and nitriles may be formed. Possible mechanisms for the degradation and formation of decomposition products have been proposed [360–363] and reviewed [27,364]. Deamination and decarboxylation can occur as side-reactions under conditions employed in the polymerization processes. The formation of ammonia has been explained by the reaction between two amino end groups:

$$-NH_2 + H_2N- \rightarrow -NH- + NH_3 \tag{2.97}$$

The isolation of di(ω-aminohexyl)amine and di(ω-carboxylpentyl)amine after hydrolysis of extensively heat-treated nylon-6,6 and nylon-6, respectively, supported this explanation. In the case of nylon-6, however, it was found that the amount of the secondary amine structure was inadequate for the quantities of ammonia actually liberated [365]. Additional reactions therefore must be responsible for ammonia production. Thus it was postulated that ammonia release may be a consequence of the formation of either or both of the following two imino-ether moieties [365]:

$$\overline{}OC(CH_2)_5-O-C=N(C^1H_2)_5; \quad \overline{}OC(CH_2)_5-N=C-O(CH_2)_5 \tag{2.98a}$$

by the reaction of an amino end group with, respectively, either caprolactam (lactim form) or the nearest chain amide group. At the temperature range considered, these imine-ether units are susceptible to undergoing a Chapman rearrangement, which for either structure results in the terminal lactam moiety

$$\overline{}OC(CH_2)_5-N-CO(CH_2)_5 \tag{2.98b}$$

that upon hydrolysis of the polymer furnishes the di(ω-carboxypentyl)amine.

The decarboxylation, which appears to be the most significant side-reaction because of the formation of considerable amounts of carbon dioxide, was assumed to be the result of reaction between two carboxyl end groups:

$$-\overset{|}{\underset{|}{C}}-COOH + HOOC-\overset{|}{\underset{|}{C}}- \rightarrow -\overset{|}{\underset{|}{C}}-CO-\overset{|}{\underset{|}{C}}- + CO_2 + H_2O \tag{2.99}$$

Levchik et al. [366] observed further that polylactams like nylon-6, nylon-11, and nylon-12 tend to re-equilibrate to monomeric or oligomeric cyclic products. Diacid–diamine polyamides like nylon-6.6, nylon-6.10, and nylon-6.12 produce mostly linear or cyclic oligomeric fragments and monomeric units. Because of the tendency of adipic acid to fragment with the elimination of CO and H_2O and to undergo cyclization, significant amounts of secondary products from nylon-6.6 have been reported. Many investigators have shown that the primary polyamide chain-scission occurs either at the peptide C(O)—NH or at adjacent bonds, most probably at the alkyl-amide NH—CH_2 bond, which is relatively the weakest in the aliphatic chain. Hydrolysis, homolytic scission, intramolecular C—H transfer, and *cis*-elimination (a particular case of C—H transfer) are all suggested as possible primary chain-scission mechanisms. It was concluded that there are no convincing results or kinetic measurements that tend to generally support one of these degradation mechanisms relative to the others. Rather, it seems that the contribution of each mechanism depends on experimental conditions.

2.5.5.3.2 Hydrolytic Resistance

Hydrolysis of nylon 6,6 that had been extensively exposed to high temperatures [362] did not yield any 5-oxononane-1,9-dicarboxylic acid, which should be expected according to the reaction shown in Equation 2.99. Hydrolysis of nylon-6 revealed the presence of 1,11-diamino-6-oxo-undecane as expected according to the reaction shown in Equation 2.99. However, the amounts were very small in comparison to the observed quantities of carbon dioxide. Subsequent studies of equilibrium polymers in the temperature range 250–290°C indicated that decarboxylation is the result of the interaction of carboxyl end group and an amide group of either a caprolactam molecule or a linear polymer molecule according to the following reaction scheme. In either case, the same cyclic Schiff base structure is the decarboxylation product. Its formation explains the

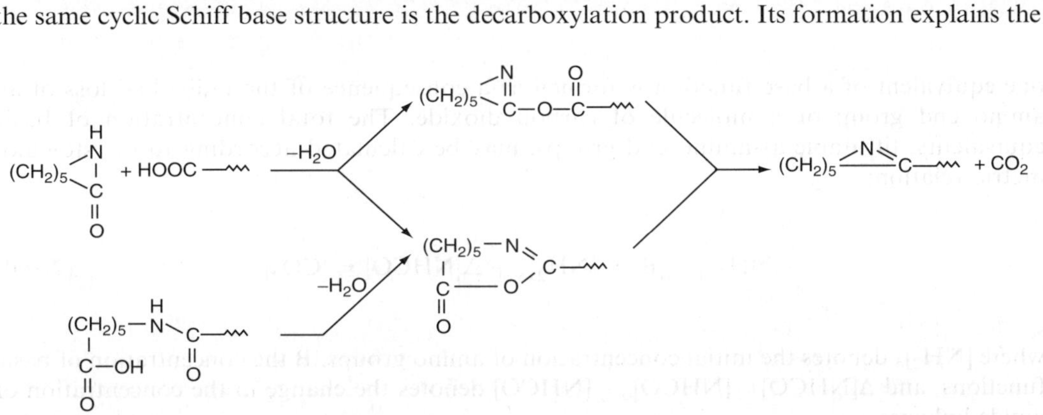

increase of basic functions observed on prolonged heating of nylon-6. Since it was found that the production of carbon dioxide is characterized by an induction period only when no caprolactam was initially present, it is assumed that the interaction between carboxyl groups and caprolactam is the principal decarboxylation reaction. The concentration of caprolactam remains essentially constant upon prolonged heating due to re-equilibration. The rate of decarboxylation is then directly proportional to the concentration of carboxyl groups and inversely proportional to the water concentration [365]. The formation of branched and network structures is the result of secondary reactions that may involve the Schiff base moiety. According to the mechanism on the formation of an imino-ether moiety and a Schiff base structure on the deamination and decarboxylation, respectively,

TABLE 2.7
Measured and Calculated Data for the Concentration of the Basic Functions

Temperature (°C)	Time (hr)	$[NH_2] + [B]$ (meg/g)	
		Found	Calc.
250	0	0.037	—
	24.00	0.036	0.035
	48.17	0.035	0.032
	73.50	0.030	0.029
260	0	0.0379	—
	21.97	0.0336	0.032
	43.17	0.0337	0.033
	71.75	0.0379	0.035
270	0	0.037	—
	23.18	0.036	0.036
	46.68	0.041	0.043
	70.00	0.047	0.051
280	0	0.0384	—
	18.58	0.0467	0.045
	42.58	0.0559	0.057
	69.90	0.0713	0.079

Source: From Reimschuessel, H.K. and Dege, G. J., *J. Polym. Sci.*, 1970, AI, 3265.

one equivalent of a base function is formed as a consequence of the individual loss of an amino end group or a molecule of carbon dioxide. The total concentration of basic equivalents, titratable as amino end groups, may be calculated according to the stoichiometric relation:

$$[NH_2]_o + [B] = [NH_2]_o + \Delta[NHCO] = [CO_2] \qquad (2.100)$$

where $[NH_2]_o$ denotes the initial concentration of amino groups, B the concentration of basic functions, and $\Delta[NHCO] = [NHCO]_o - [NHCO]$ denotes the change in the concentration of amide linkages.

Table 2.7 summarizes the values calculated for the sum $[NH_2] + [B]$ according to Equation 2.99 and the corresponding experimental data as obtained by titration. The rather good agreement between the calculated and the experimental data indicates that the considered reactions may be explained by the proposed mechanisms.

The presence of the 1,11-diamino-6-oxoundecane in hydrolysates of heat-treated nylon-6 [359] thus is obviously the result of the hydrolysis of the Schiff base moiety. The absence of the corresponding oxo-structure in the hydrolysates of gelled nylon-6,6 may be indicative of an analogous decarboxylation mechanism entailing intramolecular acid–amide interaction: Hydrolysis of the Schiff base structure yields, as indicated, hexamethylene diamine and cyclopentanone, both of which have been found.

Polyamide Fibers

2.5.5.3.3 Oxidative Degradation

The sensitivity of polyamides toward oxygen at elevated temperatures has been clearly recognized and has been the subject of intensive studies. The corresponding investigations have been concerned mainly with the elucidation of the mechanisms of the thermal oxidation. In case of nylon-6, it has been shown that the primary attack takes the place of the N-vicinal methylene group whose particular reactivity in turn appears to depend on its conformation with respect to the carbonamide group. Its reactivity, however, is in any case appreciably higher than that of any of the remaining methylene groups, which seem not to be characterized by differentiable activities and are, therefore, attacked according to a statistical pattern as indicated by the composition of hydrolysis products (ω-amino acids and alkyl amines) of extensively thermo-oxidized nylon [365–368]. The formation of water, carbon monoxide and carbon dioxide, acetaldehyde, formaldehyde, and methanol has been attributed to the breakdown of peroxide radicals and peroxides that resulted from the primary attack on the N-vicinal methylene group [369]. Additional products of thermo-oxidative decomposition of nylon-6 are carboxylic acids, amines, nitriles, and cyclopentanone. Essentially the same products are obtained from the thermal decomposition of certain caprolactam oxidation products. It has been shown [370] that at relatively mild temperatures between 70 and 100°C, exposure of caprolactam to air results in the formation of ε-hydroperoxy-ε-caprolactam, which converts easily to adipic acid monoamide. Since caprolactam is always present during polymerization and in the resulting equilibrium polymer, these reactions and the effect of their extent on the molecular weight and the type and concentration of end groups of the corresponding nylon-6 polyamide have been studied in some detail [371]. It was found that the decomposition of the caprolactam hydroperoxide was catalyzed by the adipic acid monoamide and that the latter acts as a chain terminator during the polymerization process. The corresponding polymer exhibited characteristics resembling that of polymers obtained by polymerizing caprolactam in the presence of carboxylic acids. As the extent of peroxidation increases, the molecular weight of the corresponding polyamide decreases and the concentration of acid end groups increases. Exposure of nylon-6,6 and nylon-6 fibers to temperatures in excess of 100°C in the presence of oxygen (air) results in a loss of both strength and breaking elongation. Discoloration and a decrease in amino end groups accompany these changes [372].

Most investigators agree that oxygen first attacks the N-vicinal methylene group, which is followed by the scission of alkyl-amide N–C or vicinal C–C bond. Alternatively, it was suggested that any methylene group that is β-positioned to the amide group methylene can be initially oxidized [365]. There are few mechanisms in the literature that explain discoloration (yellowing) of nylons. UV/visible active chromophores are attributed either to pyrrole type structures, to conjugated acylamides, or to conjugated azomethines. Some secondary reactions occurring during the thermal or thermo-oxidative decomposition lead to the cross-linking of polyamides. Nylon-6.6 cross-links relatively easily, especially in the presence of air, whereas nylon-11 and nylon-12 cross-link very little. Strong mineral acids, strong bases, and some oxides or salts of transition metals catalyze the thermal decomposition of nylons, but minimize cross-linking. In contrast, many fire-retardant additives promote secondary reactions, cross-linking, and charring of aliphatic polyamides.

2.5.5.4 Polymer Stabilization

The question regarding degradation and stabilization of polyamides has been dealt with elsewhere. Lanska et al. [16g] reviewed aging and durability, degradation, stabilization, and resistance to biological effects. The degradation reactions comprise reactions initiated by radiation, light, oxygen, heat, and both mechanical and thermal energies. The extent of these reactions is affected by the chemical and physical structures of the polymer and by certain compounds by either deliberate addition or contamination. Added structures include filler, pigments, dyes, or compounds admixed to affect polymer stability. More recent methods for determining basic groups, acidity, acyl groups, comonomers, molecular weights, delustering agents, and techniques for chemical and physicochemical analyses of polyamide fibers have also been reviewed [373].

Nylon fiber for outdoor uses, such as flags, decorative banners, and personal flotation device covers, must be protected from the ultraviolet light to extend its useful life. Therefore, light stabilizers are generally added during the nylon polymerization process. By reacting a sufficient number of amine or amide-forming functional groups of a hindered amine with the end groups of the polyamide precursor at polymerization temperature, the hindered amine are bound to the polyamide and inhibit migration, leaching, and volatilization of the hindered amine. Thus, a light-stabilized polyamide is formed as illustrated in several recent U.S. patents [374–376]. Hence, the articles manufactured from such a polyamide will retain their breaking strength and ultimate elongation even after many hours of outdoor exposure to ultraviolet light under the sun.

2.5.5.5 Antistatic

The purpose of antistatic modification is to decrease the specific electric resistance to eliminate or reduce the accumulation of static electrical charge. As a result, propensity is decreased for spark discharge and adsorption of dust and dirt. The approaches employed to achieve this comprise conductive coatings [377] and the addition of polar compounds having structural segments containing O, N, P, or S atoms. Representative structures are organic acids, amines, phenols, and amides. Typically, members of these compounds are applied together with poly(ethylene oxide). Inorganic structures with particle dimensions of less than 0.1 μm are generally selected from the group consisting of copper sulfide and the oxides of copper, tin antimony, and the nonconducting titanium dioxide in combination with these metal oxides. Carbon black and fine metal fibers also belong to this group. Ionic structures —both anionic and cationic compounds— including alkylammonium halides, alkaline sulfonates, phosphonium phosphonates and sulfonates, and polyacrylates, have been used often in antistatic fiber finishes.

To impart antistatic properties, specific processes are described in a large number of patents. The use of mixtures of polyethylene glycol monoester with stearylethanolamine and polyethylene glycol stearyldiethanolamine ether in nylon-6 is described in Teijin patent JP 57,139,519. Another approach, which is the subject of Unitika patent JP 57,121,617 (1982), deals with spinning blends of block copolymers containing poly(oxyalkylene) units, a siloxane, and a thermoplastic polymer. Kawaken Fine Chemicals patent JP 5,881,614 (1983) covers the use of polyethylene glycol diglycolate or its alkali metal salt. Asahi Glass patent JP 59,223,781 (1984) is an aerosol-type soil-resistant finishing agent comprising solutions of 3,4-$(MeO_2CNH)MeC_6H_3NHCOC_9F_{19}$ dissolved in trichlorotrifluoroethane and $EtOCH_2$-CH_2OH and combined with dichlorodifluoromethane. Unitika patent JP 60,209,014 (1985) is a composition containing poly(alkylene oxide) and metal sulfonate units. Bayer Japan and Toray Industries patent JP 61,245,379 deals with nylon fabrics treated with hydrophilic polyamides that contain tertiary amino groups and polyalkylene glycol groups. Czech patent CS 215,439 (1985) refers to a process characterized by incorporation of a polymer made from polyoxyethylene diamine and dicarboxylic acids into nylon-6. Czech patent CS 215,440 (1985) claims a process in which caprolactam is polymerized in the presence of polyether–polyamides derived from α,ω-diamine derivatives of polyethylene glycol and C4–12 dicarboxylic acids or esters. Czech patent CS 215,441 (1985) covers the polymerization of caprolactam in the presence of a polyesteramide prepared from diaminopolyoxyethylene and a dicarboxylic acid having 4 to 12 carbon atoms. Czech patent CS 218,606 (1985) describes a process characterized by impregnating polyamide fibers with an aqueous solution or dispersion of a polycaprolactam containing segments of a polyether–polyamide obtained from α,ω–diamine-terminated polyethylene glycol and a dicarboxylic acid having 4 to 8 carbon atoms. Czech patent CS 226,632 (1985) deals with compositions that contain esters or amides derived from amine-terminated polyethylene glycol, polyethylene glycol, and adipic acid. Czech patent CS 231,794 (1986) deals with compositions containing esters derived from α,ω-dihydroxy-poly(alkylene oxide) and dicarboxylic acids. Czech patent CS 246,664 (1987) proclaims improved antistatic properties for nylon-6 compositions that contain poly(alkylene oxide) in cases when the polymerization is initiated with dry H_3PO_4.

Fixation of conducting polymers such as methoxymethylated polyamides and polyethylene glycol may be achieved by the application of their solutions or suspensions to polyamide fibers followed by the evaporation of particular solvents or suspension agents. Other approaches are characterized by encasing fibers with conducting polymers or by applying a filament that contains a conducting core comprising of a mixture of a suitable polymer with high quantities (35 to 80%) of conducting fillers selected from SnO_2, Sb_2O_3, CuI, CuS, or carbon black. In addition to their use as components in antistatic textile compositions, fibers coated with silver, copper, or nickel are used in electronic applications and in uses related to electromagnetic filtration.

2.5.5.6 Flammability

Considerable effort has been applied to reduce the flammability of nylon fibers. Two major approaches are generally employed. The first is to promote the char formation and prevent formation of any dripping polymer melt [378,379]. It requires 20–50% of suitable additives to achieve a nondripping and a self-extinguishing composite. The other approach is to add a high-molecular-weight flame retardant based on organic compositions containing halogen, phosphorus, and nitrogen. Principal halogen compositions are perbromated aromatics; the organic phosphorus compounds are mainly melamine and its derivatives. Most inorganic compounds in use are red phosphorus and antimony trioxide.

An important development in flame retardancy of polymers is the use of ammonium polyphosphate blended with a polyhydric alcohol such as pentaerythritol as a char-forming agent, and a nitrogen derivative such as melamine, guanidine, or urea [380–382]. In general, high-molecular-weight flame retardants are favored because they produce less smoke and may be less toxic during burning. More recently, hexavalent sulfur compounds and particularly sulfamates, metal-based catalysts, and nitrogen derivatives have been incorporated in the flame retardants with or without ammonium polyphosphate [383–387]. The hexavalent sulfur derivatives can be inorganic or organic derivatives of sulfuric and sulfamic acids. Among these materials are sulfamic acid salts, H_2NSO_3, metal condensation products of sulfamic acids, such as imidobisulfonic acid $NH(SO_3H)_2$ and its salts. These additives are less toxic and less corrosive than the formulations used in the art of flame retarding of polymers today. They are also readily available and relatively inexpensive. The effectiveness of flame retardancy is greatly improved so that only small amounts of retardants are required. For example, Lewin et al. [388] reported recently the formulation of 1.5–2.5 wt% of ammonium sulfamate or diammonium imidobisulfonate, together with 0.4–0.85 wt% of pentaerythritol or dipentaerythritol to obtain fully flame-retardant nylon and nylon-6,6.

The function of sulfur derivatives in flame retardancy is not exactly understood. It was suggested that the sulfur derivative appears to be a more effective catalyst for the dehydration, cross-linking, and char formation than ammonium polyphosphate alone [390]. The sulfation and desulfation occur more rapidly than phosphorylation and dephosphorylation. The char is formed both by the sulfation and the phosphorylation routes, but the char obtained appears to be a more effective, more compact, and less penetrable surface barrier. The sulfur compounds may act as synergists of the ammonium polyphosphate, similar to the effect of antimony trioxide in the case of halogen-based additives. It was further noted that the sulfations of nylon-6 and dephosphorylation occur simultaneously and produce cross-links and networks.

2.5.5.7 Transparency

The addition of certain chemicals at concentrations of or below 1% affords processing of essentially transparent filaments. This effect can be attained by adding ethylenbistearamide, ketone-containing higher alkyl groups, certain salts such as Zn-*N*-benzoyl-6-aminohexanoate, and Na-hypophosphite. In addition to the interaction between modifier and polymer, transparency is augmented with increases in both the polymer molecular weight and any branching of the polymer molecule. Transparency decreases with increases in the processing temperature.

2.5.5.8 Dye Diffusion

The relationship between fiber structure and dye diffusion has been reviewed [389]. Dynamic mechanical data are converted into two parameters: an internal viscosity and a mobile noncrystalline fraction. The ratio of these was found to be comparable to dye diffusion.

2.5.6 Process–Structure–Property Relationship

There are significant interactions between process, fiber structure, and properties. A general understanding of these interactions often provides insights into process problems and new product development.

2.5.6.1 Effect of Molecular Weight

The structure and properties of as-spun filaments from high-speed spinning of nylon-6 were characterized using DSC, wide-angle x-ray diffraction, small-angle x-ray scattering, birefringence, and tensile tests [249]. The structural characteristics were found to depend strongly on the molecular weight. The behavior was affected by the crystallinity of the material.

2.5.6.2 Effect of Water

The fine structure and morphology of nylon-6 are significantly influenced by the equilibrium water content as a time-induced phenomenon. The interaction between process and water content is illustrated with the packaging behavior of nylon-6 carpet and textile fibers with molecular weight of 55–60 FAV:

1. At spinning speeds below 1200 m/min, the fiber is amorphous initially. It will absorb moisture from the ambient air during storage and begin crystallizing to form a predominantly α-structure. The wound package will become loose but stable, and further processing is possible.
2. At speeds between 1200 and 3000 m/min, the fiber is increasingly oriented with increasing speed. It will grow rapidly and the wound package may fall off the bobbin during storage; or the fiber grows too rapidly while the package is still on the winder that an undrawn yarn package cannot be obtained.
3. At speeds above 3000 m/min, the fiber becomes increasingly of the γ-structure and is stable on the package again.
4. The speed ranges tend to decrease with increasing polymer molecular weight.

2.5.6.3 Effect of Spinning and Drawing

Crystallinity in the polyamide fibers is generally the result of both primary and secondary crystallizations. The former occurs during the melt-spinning and is therefore more pronounced for nylon-6,6 than for nylon-6, whereas the latter proceeds in the drawing process and may be stimulated by heat, stress, and swelling agents (water). Crystallinity changes associated with the drawing process depend on the morphology of the undrawn material and the drawing conditions. Although heat treatments, either under tension or in the relaxed state, result in increases of both crystallinity and crystal perfection, drawing of highly crystalline fibers at low temperatures can cause an initial decrease in crystallinity due to destructive mechanical forces. Any subsequent crystallinity increase entails an internal temperature increase due to the dissipation of energy. A steady crystallinity increase, on the other hand, characterizes the drawing of fibers of low initial crystallinity. This increase parallels draw ratio and drawing velocity and also depends on self-heating effects. The rate of crystallization during drawing is strongly affected by any previously introduced molecular orientation.

The relationship of take-up speed and processing conditions with fiber properties can be further illustrated by nylon-6 fibers from three different types of spinning operation.

1. The conventional two-stage spin and draw process, as shown in Figure 2.18, yields a fully drawn yarn with a predominantly α structure. This process converts an undrawn yarn to a draw-textured yarn with a predominant α structure, improved dimensional stability, and dye-fading resistance.
2. The partially oriented yarn (POY) from the one-step stack-draw process, as illustrated in Figure 2.21, at a take-up speed faster than 3000 m/min is largely of the γ-structure.

3. In the one-step spin-draw process, as shown in Figure 2.22, there is no lagging for the moisture to change the morphology of the feeder yarn. The yarn morphology resulting from a one-step spin-draw-texturing process may be of the γ- or α-structure, depending on the degree of heat treatment. A fully drawn yarn from this process has a predominantly γ-structure.

2.5.6.4 Effect of Spinning Conditions

Spinning at velocities above 6000 m/min results in yarns that virtually do not retain any residual drawability. This operation has thus been considered to yield fully drawn yarn (FDY) [390]. A study of the effect of spinning conditions relates birefringence with spinning conditions [391]. Interference microscopy showed structural changes of the fibers produced at 260, 265, and 270°C. Cold drawing increased the mean birefringence and the sheath-core structure. Annealing at 190°C for 1 min destroys the sheath-core structure to a large extent. Birefringence measurements also provide information on fiber damage [392] and explain variations in dyeing of woven and knit textiles. Processes concerned with combining both spinning and drawing operations have been described in many patents.

2.6 COMMERCIALIZATION AND FUTURE OUTLOOK

2.6.1 POLYAMIDE FIBER PRODUCTS

The commercial production of polyamide fibers began in the late 1930s in the United States and Europe. They are now produced worldwide. Nylon fiber producers in the United States in 2004 include Fiber Innovation Technology, Inc., Honeywell Nylon Inc., Invista Inc. (formerly DuPont Textiles & Interiors), Nylstar, Inc., Palmetto Synthetics, Polyamide High Performance, Inc. (formerly Acordis), Solutia Inc., Unifi-Sans Technical Fibers, LLC, Universal Fiber Systems LLC, and Wellman, Inc.

Except for fiber production, commercial polyamides are produced in the form of chips for use in molding, composite, resin compounding, and numerous industrial applications. Nylon-6,6 is often in the form of nylon salt for conversion in other regions of the world.

2.6.1.1 Filament Yarn and Staple

Nylon fiber products are produced in five commodity types: textile filament yarns, industrial filament yarns, tows, staple fiber, and carpet BCF filament yarn. A continuous filament yarn may contain multiple filaments or a monofilament. The filament size is in the range of 1.5–28 deniers per filament (dpf). The yarn denier varies in the range of 1200–5000 deniers, while large tows of 10,000–30,000 deniers are also produced, depending on the end-uses. Nylon staple is generally produced by cutting continuous yarn immediately after crimping or texturing. The staple length varies in the range of 6–19 cm (2.75–7.5? in.).

2.6.1.2 Bicomponent Fibers

Bicomponent fiber (BCF) comprises two polymers of different chemical or physical properties. In either case, due to their constitutional variance, the two polymers possess different extendibility characteristics. These differences become evident on heating in air or on exposure to hot water. Both treatments cause the development of bulk. Both polymers are extruded from a common spinneret to form a single filament. Depending on the characteristics of the two polymers, the bicomponent fiber can provide functional properties such as thermal bonding, self bulking, unique cross sections, and achieve functionality of special polymers

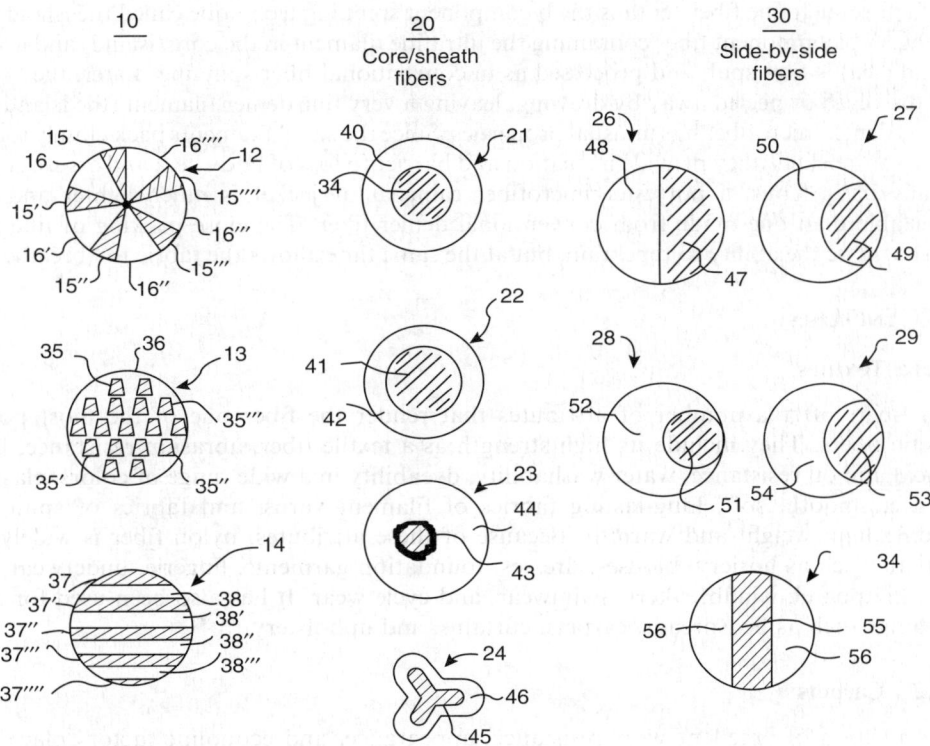

FIGURE 2.42 Examples bicomponent fibers. (From Magill, M.C. and Martmann, M.H., Multi-component fibers having enhanced reversible thermal properties and methods of manufacturing thereof, U.S. Patent Application No. 20030035951, February 20, 2003. With permission.)

or additives for surface adhesion, ultraviolet protection, etc., at reduced cost. Several examples of bicomponent fibers are illustrated in Figure 2.42 [393].

The bicomponent fiber technology began in the 1960s on polyester and polyamide fibers. This technology encompasses a vast variety of polymers, filament shapes, and processes. As recent examples, Howe et al. [394], Wu [395], and Talley et al. [396] have disclosed "spin-texture" processes for the preparation of self-crimped polyamide bicomponent fibers. DuPont [397] disclosed the method for high-speed spinning bicomponent fibers. The use of a bicomponent yarn with another yarn to form a composite yarn bundle has been taught by Stevenson et al. [398] and others.

2.6.1.3 Microfibers

One of the most important developments in recent years has been the technology to extrude extremely fine filaments of less than 1.0 denier while maintaining all of the strength, uniformity, and processing characteristics expected by textile manufacturers and consumers. This product development began with the preparation of conjugated bicomponent filaments that were post-processed to split into ultrafine fibers. The process technology later matured and has been applied to polyesters, polyamides, polypropylene, and polyethylene and polyphenylene sulfide.

Ultrafine fibers are generally called *microfibers*. The definition of microfiber, as accepted in the trade, is a fiber finer than 1.2 dtex for polyester, and finer than 1.0 dtex for polyamide. These fibers are finer than luxury natural fibers such as silk. The most popular technology to manufacture such fine fibers entails the bicomponent spinning technique called the island in the sea [399]. A bicomponent fiber containing the ultrafine filament in the core (island) and a sheath material (sea) is first spun and processed as in conventional fiber spinning. Later, the "sea" is either dissolved or peeled away by drawing, leaving a very fine denier filament (the island). The ultrafine denier microfiber has unusual properties. Since the small filaments pack closely together and trap air pockets, they provide insulation and barrier to loss of body heat and assure comfort on chilly days. Thus, a polyester microfiber raincoat or jacket is much lighter and more comfortable than one made from conventional denier fiber. The close packing of fibers also gives the fabric the ability to repel rain, but at the same time allows the fabric to "breathe".

2.6.2 END-USES

2.6.2.1 Textiles

Nylon fibers offer a number of attributes that render the fiber one of the most popular synthetic fibers. They include its high strength as a textile fiber, abrasion resistance, luster, chemical and oil resistance, water washability, dyeability in a wide range of colors, elasticity, resilience, smooth, soft, long-lasting fabrics of filament yarns, and fabrics of spun yarn, providing light weight and warmth. Because of these attributes, nylon fiber is widely used in apparel such as hosiery, blouses, dresses, foundation garments, lingerie, underwear, raincoats, ski apparel, windbreakers, swimwear, and cycle wear. It has also been used for house furnishings such as bedspreads, carpets, curtains, and upholstery.

2.6.2.2 Carpets

A combination of excellent wear resistance, appearance, and economic factors place polyamide fibers in an excellent position in the *carpet fiber* market, and it can be expected that the nylons will continue to be the most used fibers for carpets.

2.6.2.3 Industrial Applications

Excellent mechanical properties, strength, fatigue resistance, and good adhesion to rubber are reasons for the predominance of polyamide fibers for industrial applications and for use in carcasses of truck tires and airplane tires. Other applications include upholstery fabrics, seat belts, parachutes, ropes, fishing lines, nets, sleeping bags, tarpaulins, tents, thread, monofilament fishing line, and dental floss. Their high strength, toughness, and abrasion resistance are main factors for selecting polyamide fibers for a wide range of military applications.

2.6.2.4 Specialty Fibers

Many specialty fibers have been developed over the years for special end-uses. An example of excellent industrial accomplishment is the development of TRIAD fiber shown in Figure 2.19. Used in FRAM automotive air filter, the TRIAD fiber helps trap dirt inside its microscopic channels that in turn trap more dirt without increasing the air flow restriction. Another example is the preparation of a conductive fiber using the bicomponent fiber technology with a built-in conductive fiber. Figure 2.42, Item 29, shows a fiber cross-section where the dark-colored portion of the fiber is the conductive portion.

2.6.2.5 Nonwovens

Like polyester (PET) fiber, nylon has a high melting point, which provides good high-temperature performance. Nylon fiber is more sensitive to water than PET. However, nylon

Polyamide Fibers

is not considered a comfortable fiber in contact with the skin. The relatively high cost of nylon has somewhat limited use in nonwoven products. It is used as a blending fiber in some cases, because it conveys excellent tear strength. Overall, the resiliency and wrinkle recovery of a nonwoven nylon product is not as good as that from PET fiber.

Because of its toughness, nylon fiber is suitable for nonwoven needle-punched floor-covering products. Nylon nonwoven fiber can be found in garment interlinings and wipes for its strength and resilience. It is also used Ni/H and Ni/Cd batteries as nonwoven separators; in high performance wipes, synthetic suede, heat insulators, battery separators, and specialty papers; and in automotive products, athletic wear, and conveyor belts.

2.6.2.6 Nonfiber Uses

In addition to fiber uses, nylon is a major thermoplastic polymer for film and sheet products, resin compounding, and composite materials. The recently developed nanocomposites, as discussed in Section 2.3.2.5, further broaden the nonfiber uses of nylon. These composites exhibit superior properties such as high modulus, high strength, and good gas-barrier properties. Applications include moisture-barrier film for meat packaging, oxygen-barrier beer bottle, and medicine packaging.

2.6.3 Worldwide Consumption of Polyamide Fibers

In 2002, American Fiber Manufacturers Association published the regional data for the worldwide production of synthetic fibers as shown in Figure 2.43 [400]. The major components of synthetic fibers are nylon and polyester fibers. Asia has experienced the fastest growth

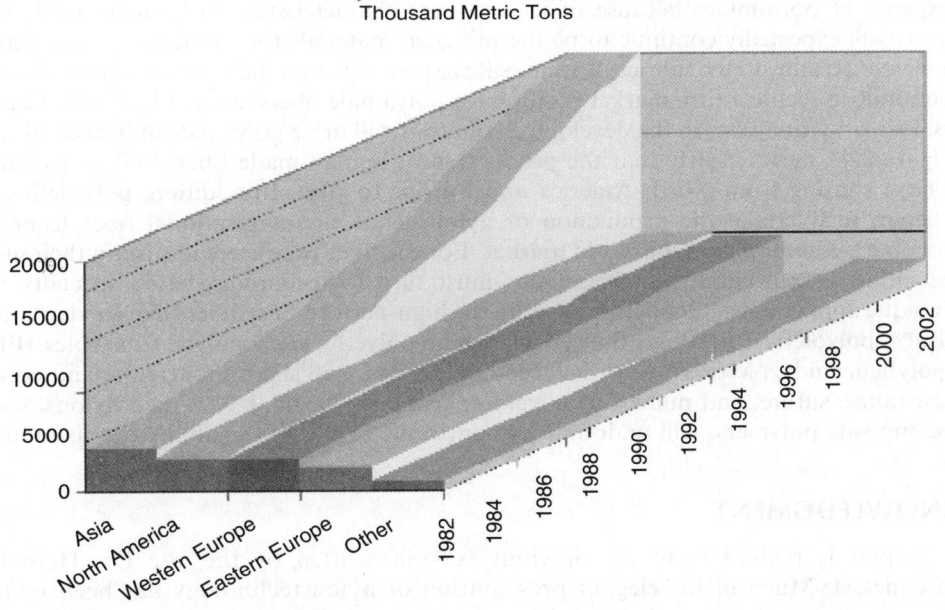

FIGURE 2.43 Worldwide consumption of synthetic fibers by regions. (From http://www.afma.org/American Fiber Manufacturers Association, Fiber Economics Bureau, 2005 World Directory of Manufactured Fiber Producers [associated October 1, 2004]. With permission.)

in synthetic fiber production since 1990 and approached 20M (million) tons in 2002. All the other regions have had much slower growth than Asia; in 2002, North America reached about 5M; Western Europe, 4.5M; Eastern Europe, about 1.5M; and others, 2.5M tons. Thus the total worldwide production of synthetic fibers was in the order of 33.5M tons by 2002. It is further forecasted that global consumption of nonwoven fibers may reach 3.7M tons by 2005 and 4M tones by 2007. The consumption of synthetic fibers was about 8.1% of all textile fibers in 1998. In 2005, it is expected to reach 10% and 10.4% by 2007. The consumption of nylon fiber in nonwovens was 49M tons in 2002, 50M tons in 2004, and is expected to rise to 60M tons in 2007.

2.6.4 FUTURE OUTLOOK

2.6.4.1 Technology

There are some reviews on specific subject matter that address certain aspects of polyamide fibers [401]. High-speed spinning of nylon fibers has been reviewed [402], and goals and problems of the high-speed spinning process have been discussed [403]. Preparation, structure, and mechanical properties of highly oriented polymers with flexible chains also have been reviewed [404] with regard to the preparation of high modulus and high-strength polymers and changes in the molecular structure and morphology during the processes of crystallization. Additional reviews are concerned with progress in the field of high-performance nylon fibers [405] and advances in man-made fibers [406].

2.6.4.2 Market Share

The main uses for both nylon 6,6 and nylon-6 are in apparel, carpets, and industrial applications. In the area of general apparel, polyester has gained considerably in significant market segments at the expense of polyamides because of its easy-care characteristics. Polyamide textile fibers, however, will expectedly continue to be the principal materials for women's hosiery, intimate apparel, and certain stretch fabrics. It seems safe to conclude that the price to performance ratio will continue to secure a firm market position for polyamide fibers, even at increased feedstock prices. Primarily tire yarns in the developing countries will drive polyamide industrial fibers.

Figure 2.43 shows clearly that the production of human-made fibers and its technology have been shifting from North America and Europe to Asia. In addition, polyolefins have also begun to overtake the production of nylon. Thus the conventional fiber technology has become a commodity in the world market. For the fiber producers to sustain their growth in the industrialized countries, their focus must turn to technology-based specialty fibers with unique applications. Good examples are the high-performance fibers like Kevlar*aramid, Spectra** polyolefin (ultrahigh molecular weight polyethylene), polybenzoxazoles (PBO)), and polybenzimidazole (PBI) for advanced composites, ballistic armors, rope and cordage, cut-resistance, suture, and numerous novel applications [407,408]. Modified nylons, such as nanocomposite polymers, will undoubtedly contribute to the sustainability of this industry.

ACKNOWLEDGMENT

This chapter is revised from the previous version written by the late Dr. Herbert K. Reimschuessel. Much of his elegant presentation of nylon technology has been retained,

---

*Kevlar - a registered trademark of E.I. du Pont de Nemours & Co., Inc., Wilmington, Delaware, USA
**Spectra - a registered trademark of Honeywell International, Inc., Morristown, New Jersey, USA

but slightly rearranged to allow updating. The author is indebted to Dr. Joseph Zimmerman, formerly DuPont Fibers, and to Dr. Thomas Y. Tam, Honeywell International, for their kindest guidance.

REFERENCES

1. Carothers, W.H.; Stocking and Method for Producing Same, U.S. Patent 2,157,116, May 9, 1939.
2. Hounshell, D.A.; J.K. Smith, Jr.; Science and Corporate Strategy, DuPont R&D, 1902–1980, Cambridge University Press, 1988, 270.
3. Flory, P.J.; *Principles of Polymer Chemistry*; Cornell University Press, Ithaca, New York, 1953.
4. Sweeny, W.; Zimmerman, J.; Encyclopedia of Polymer Science and Technology, 1969, Vol. 10, 483.
5. Zimmerman, J.; Encyclopedia of Polymer Science and Engineering, 1988, Vol. 11, 315.
6. Zimmerman, J. Polymers for Fibers. In *Comprehensive Polymer Science*, Aggarwal, S.L.; ed., Pergamon Press, Oxford. Vol. 7.
7. Zimmerman, J.; Kohan, M.I.; Polyamides. Nylon—Selected topics, *J. Polym. Sci. Part A, Polym. Chem.*, **2001**, 39, 2565–2570.
8. Kohan, M.I. Polyamides.; In *Ullmann's Encyclopedia of Industrial Chemistry*, VCH, Weinheim, 1992, 179–205; *Nylon Plastic Handbook*, Hanser-Gardner, Munich, 1995.
9. Reimschuessel, H.K.; in *Ring-Opening Polymerization*, Frisch K.C.; Reegen, S.L.; Eds., Marcel Dekker, New York, 1969, 303.
10. Šebenda, J.J.; *MacromoL Sci. Chem.*, **1972**, A6, 1145.
11. Rothe, M.; Bertalan, G.; Mazanek, J.; *Chimia*, **1974**, 28, 527.
12. Secor, R.M.; Beuder, J.A.; *AIChE J.*, **1967**, 13, 15.
13. Higbee, R.; *Trans. Am. Inst. Chem. Eng.* **1935**, 31, 365.
14. Nagasubramanian, K.; Reimschuessel, H.K.; *J. Appi. Polym. Sci.*, 17, 1663.
15. Cubbon, R.; *Makromol. Chem.*, **1964**, 80, 44.
16. Puffr, R.; Kubánek, V.; Eds.; *Lactam-based Polyamides*, Vol. 1, CRC Press, Boca Raton, FL, 1991.
16a. Puffr, R.; 2–21;
16b. Šebenda, J.; 30–67;
16c. Šebenda, J.; 47ff;
16d. Tuzar, Z.; 96–129;
16e. Baldrian, J.; Lednický, F.; 132–185;
16f. Puffr, R.; Raab, M.; Dolezel, M.; I187–260;
16g. Lánská; 261–302.
17. Bonetskaya, A.K.; Skuratov, S.M.; *Vysokomol. Soed.*, **1969**, A11, 532.
18. Šebenda, J.; *Pure Appl. Chem.*, **1976**, 48, 329.
19. Ogata, N.; Asahara, T.; Tohyama, S.; *J. Polym. Sci.*, A-1 **1966**, 4, 1359.
20. Ogata, N.; Tanaka, K.; Inagaki, M.; *Makromol. Chem.*, **1968**, 113, 95.
21. Ogata, N.; Tanaka, K.; Takayama, K.; *Makromol. Chem.*, **1968**, 119, 161.
22. Gol'dfarb, Ya. I.; *Belen'kii. Usp. khim.*, **1960**, 29, 470.
23. Brown, R.F.; van Gulick, N.M.; *J. Org. Chem.* **1956**, 21, 1064.
24. Yumoto , H.J.; *Chem. Phys.* **1958**, 29, 1234.
25. Cubbon, R.; *Makromol. Chem.* **1964**, 80, 44.
26. Reirnschuessel, H.K.; *Angew. Chemie, Int. Ed.*, **1975**, 14, 43.
27. Reimschuessel, H.K.; *J. Polym. Sci., Macromol. Rev.*, **1977**, 12, 65.
28. Huggins, M.L.; *J. Am. Chem. Soc.*, **1942**, 64, 2716.
29. Kraemer, E.O.; *Ind. Eng. Chem.* **1938**, 30, 1200.
30. Šebenda, J.; Králiček, J.; *Colln. Czech. Chem. Commun.*, **1966**, 31, 2534.
31. Lánska, B.; Bohdanecká, M.; Šebenda, J.; Tuzar, Z.; *Eur. Poly. J.*, **1978**, 14, 807.
32. Tuzar, Z.; Kletečková, J.; Králiček, J.; *Colln. Czech. Chem. Commun.*, **1974**, 39, 2201.
33. Sekiguchi, H.; *Bull. Soc. Chim*, France, **1960**, 1831–1839.
34. Reimschuessel, H.K.; Dege, G.J.; *J. Polym. Sci.*, A-1 **1971**, 9, 343.
35. Tuzar, Z.; Kratochvil, P.; Bohdanecký, M.; *J. Polym. Sci. C*, **1967**, 16, 633.

36. Goebel, C.V.; Cefelin, P.; Stehliček, J.; Šebenda, J.; *J. Polym. Sci. Polym. Chem. Ed.*, **1972**, 10, 1411.
37. Schaefgen, J.R.; Flory, P.J.; *J. Am. Chem. Soc.*, **1948**, 70, 2709.
38. Gechele, G.B.; Mattiussi, A.; *Eur. Polym. J.*, **1965**, 1, 47.
39. Burke; J.J.; Orofmo, T.A.; *J. Polym. Sci.*, **1969**, A-2, 7, 1.
40. Elias, H.G.; Schumacher, R.; *Makromol. Chem.*, **1964**, 76, 23.
41. Taylor, G.B.; *J. Am. Chem. Soc.*, **1947**, 69, 635.
42. Liquori, A.M.; Mele, M.; *J. Polym. Sci.*, **1954**, 13, 589.
43. Morgan, P.; Kwolek, S.L.; *J. Polym. Sci.*, **1963**, A-1, 1147.
44. Puffr, R.; Šebenda, J.; *J. Polym. Sci.*, **1973**, 42, 21.
45. Lánská, B.; Bohdanecký, M.; Šebenda, J.; Tuzar, Z.; *Eur. Polym. J.*, **1978**, 14, 807.
46. Tuzar, Z.; Kratochvil, P.; Bohdanecky, M.; *Adv. Polym. Sci.*, **1979**, 30, 117.
47. Heuse, O.; Adolf, F.P.; *J. Forensic. Sci. Soc.*, **1982**, 22(2), 103.
48. Kirret, O.; Lahe, L.; Rajalo, G.; Kirjanen, E.; *Eesti NSV Tead. Akad. Toim.*, **1983**, Keem 32(2), 119.
49. Flory, P.J.; U.S. Patent 2,172,374, September 12, 1939.
50. Monroe, G.C.; U.S. Patent 3,031,433, April 25, 1962.
51. Schlack, P.; U.S. Patent 2,241,321, 1941.
52. Reimschuessel, H.K.; Nagasubramanian, K.; *Chem. Eng. Sci.*, **1972**, 27, 1119.
53. Tai, K.; Teranishi, H.; Arai, Y.; Tagawa T.; *J. Appl. Polym. Sci.*, **1980**, 25, 77.
54. Tai, K.; Arai, Y.; Tagawa, T.; *J. Appl. Polym. Sci.*, **1982**, 27, 731.
55. Nagasubramanian, K.; Reimschuessel, H.K.; *J. Appl. Polym. Sci.*, **1972**, 16, 929.
56. Gupta, S.K.; Kumar, A.; Agrawal, K.K.; *J. Appl. Polym. Sci.*, **1982**, 27, 3089.
57. Hoftyzer, P.J.; Hoogschagen, J.; van Krevelen, D.W.; *Proc. Third Europ. Symp., Chem. React. Eng.*, **1964**, 247.
58. Griscenko, A.Z.; Sachnenko, D.V.; *Faserforsch. Textil-techn.*, **1971**, 22, 362.
59. Tai, K.; Ara, Y.; Teranishi, H.; Tagawa, T.; *J. Appl. Polym. Sci.*, **1980**, 25, 1789.
60. Gupta, S.K.; Naik, C.D.; Tandon, P.; Kumar, A.; *J. Appl. Polym. Sci.*, **1981**, 26, 2153.
61. Gupta, S.K.; Kumar, A.; Tandon, P.; Naik, C.D.; *Polym.*, **1981**, 22, 481.
62. Gupta, A.; Gandhi, K.S.; *J. Polym. Sci.*, **1982**, 27, 1099.
63. Ramagopal, A.; Kumar, A.; Gupta, S.K.; *Polym. Eng. Sci.*, **1982**, 22, 849.
64. Ogata, N.; *Makromol. Chem.*, **1960**, 42, 52; *Makromol. Chem.*, **1961**, 48, 117.
65. Tai, K.; Tagawa, T.; *J. Appl. Polym. Sci.*, **1982**, 27, 791.
66. Arai, Y.; Tai, K.; Teranishi, H.; Tagawa, T.; *Polym.*, **1981**, 22, 273.
67. Mallon, F.K ; Ray, W.H.; A comprehensive model for nylon melt equilibria and kinetics, *J. Appl. Polym. Sci.*, **1998**, 69, 1213–1221.
68. Tai, K.; Tagawa, T.; *Ind. Eng. Chem. Prod. Res. Dev.*, **1983**, 22,192.
69. Wiloth, F.; *KolloidZ.*, **1955**, 143, 129.
70. Graf, R. et al.; *Angew. Chem. Int. Ed. Engl.*, **1962**, 1, 481.
71. Ney, W.O.; Crowther, M.; U.S. Patent 2,739,959, May 27, 1956.
72. Nesmeyanov, A.N. et al.; *Chem. Tech. USSR*, **1957**, 9, 139.
73. Cubbon, R.; *Polym.*, **1964**, 4, 545.
74. Brassat, B.; Brockmann, R.; Process for the production of polyamides, U.S. Patent 3,989,677, November 2, 1976.
75. Puffr, R.; Šebenda, J.; Method for the preparation of lactams, U.S. Patent 4,111,869, September 3, 1978.
76. Dachs, K.; Schwartz, E.; *Angew. Chem. Int. Ed. Engl.*, **1962**, 1, 430.
77. Murabayashi, K.; Resin composition, U.S. Patent 4,716,198, December 28, 1987; 4,839,424, June 13, 1989; U.S. Patent 4,871,804, October 3, 1989.
78. Zimmerman, J. et al.; *J. Appl. Polym. Sci.*, **1973**, 17, 849.
79. Gaymans, R.J.; Baur, E.H.J.P.; Neth. Patent 80, 01,763; 80, 01,764, October 16, 1981.
80. Baur, E.H.J.P.; Warnier, J.M.M.; Eur. Patent 77,106, April 20, 1983.
81. E.I. du Pont de Nemours & Co., in *Zytel Design Handbook*, 1988.
82. Bonner, W.H.; U.S. Patent 3,088,794, May 7, 1963.
83. Jackson, J.B.; *Polym.*, **1969**, 10, 159.
84. Nielinger, W.; Brassat, B.; Neuray, D.; *Angew.Makromol. Chem.*, **1981**, 98, 225.

85. Dolden, J.C.; *Polym.*, **1976**, 17, 875.
86. Sprague, B.S.; Singleton, R.W.; *Text. Res. J.*, **1965**, 35, 999.
87. Clayton, G. et al.; in *Textile Progress: The Production of Man—Made Fibers, Vol.18;* Alvey, P.J., ed.; The Textile Institute, Manchester, UK, 1976, 1, 18.
88. Prince, F.R.; Pearce, E.M.; *Macromolecules*, **1971**, 4, 347.
89. Meyer, R.V.; Dhein, R.; Fahnler, F.; Transparent polyamides from PACM and a mixture of dicarboxylic acids, U.S. Patent 4,205,159, May 27, 1980.
90. Janssen, P.; *Angew. Makromol. Chem.*, **1974**, 40141, 1.
91. RosIer, W.; Lunkwitz, H.; *Chem. Tech.*, **1978**, 30, 67.
92. Danziger, H.; Immel, O.; Kaiser, B.; Schwartz, H.; Bandtel, E.; Abrasion resistant catalyst, U.S. Patent 4,157,986, June 12, 1979.
93. Werther, J.; Fuchs, H.; Brand, U.; Faulhaber, F.R.; Preparation of caprolactam, U.S. Patent 4,268,440, May 19, 1981.
94. Jha, B.; Chhatre, a. S.; Kulkarni, B.; Sivasanker, S.; Process for the preparation of caprolactam, U.S. Patent 5,401,843, March 28, 1995.
95. Jha, B.; Chhatre, a. S.; Kulkarni, B.; Sivasanker, S.; Process for the preparation of caprolactam, U.S. Patent 5,594,137, January 19, 1997.
96. Fuchs, E.; Witzel, T.; Preparation of caprolactam, U.S. Patent 5,495,014, February 27, 1996.
97. Achhammer, G.; Fuchs, E.; Preparation of caprolactam, U.S. Patent 5,495,016, February 27, 1996.
98. Bassler, P.; Luyken, H.; Achhammer, G.; Witzel, T.; Fuchs, E.; Fischer, R.; Schnurr, W.; Simultaneous preparation of caprolactam and hexamethylenedial11ine, U.S. Patent 5,717,090, February 10, 1998.
99. Fuchs, E.; Achhammer, G.; Preparation of caprolactam, U.S. Patent 5,874,575, February 23, 1999.
100. Achhammer, G.; Bassler, P.; Fischer, R.; Fuchs, E.; Luyken, H.; Schnerr, W.; Voit, G.; Hilprecht, L.; Process for slmultaneously preparing caprolactam and hexamethylene diamine, U.S. Patent 6,147,208, November 14, 2000.
101. Bassler, P.; Baumann, D.; Fischer, R.; Fuchs, E.; Melder, J.; Ohlbach, F.; Preparation of caprolactam, U.S. Patent 6,677,449, January 13, 2004.
102. Fuchs, E.; Melder, J.; Schnurr, W.; Fischer, R.; Process for producing caprolactam, U.S. Patent 6,683,177, January 27, 2004.
103. Bassler, P.; Baumann, D.; Fischer, R.; Preparation of caprolactam, U.S. Patent 6,683,179, January 27, 2004.
104. Sifniades, S.; Levy, A.B.; Handrix, J.A.J.; Process for depolymerizing nylon-containing waste carpet to form caprolactam, U.S. Patent 5,929,234, July 27, 1999.
105. Costello, M.R.; Sloan, F.E.; Duffy, E.A.; Babb, W.M.; Ward, A.E.; Process for recovering high value polymer from carpet salvage waste, U.S. Patent 6,059,207, May 9, 2000.
106. Mayer, R.E.; Crescentini, L.; Jenczewski, T.J.; Process for the purification of caprolactam obtained from the depolarization of polyamide—containing carpet, U.S. Patent 6,187,917, February 12, 2001.
107. Sifniades, S.; Levy, A.B.; Handrix, J.A. J.; Process for depolymerizing nylon-containing waste to form caprolactam, U.S. Patent 6,342,555, January 29, 2002.
108. Patols, C.; Spagnol, M.; Preparation of caprolactam, U.S. Patent 5,6,204,379, March 20, 2001.
109. Ohlbach, F.; Ansmann, A.; Fischer, R.; Melder, J.; Method for the production of caprolactam, U.S. Patent 6,683,180, January 27, 2004.
110. Mares, F.; Boyle, W.J. Jr.; Corbo. A.M.; *J. of Catalysis*, **1987**, 107, 407.
111. Mares, F.; Galle, S J.E.; Diamond, E.; Regina, F.J.; *J. Catal.*, **1988**, 112, 145.
112. Paśek, J.; Richter, P.; *Chem. Prumysl.*, **1970**, 20, 88.
113. *Hydrocarbon Process*, **1975**, 4, 119.
114. Sherwood, P.W.; *Ind. Eng. Chem.*, **1963**, 55, 37.
115. Kahr, K.; *Angew. Chem.*, **1960**, 72, 135; *Chem. Ber.*, **1960**, 93, 132.
116. Tsuda, S.; *Chem. Econ. Eng. Rev.*, **1970**, 2, 39.
117. *Nitrogen*, **1967**, 50 (11/12), 27.
118. Layson, S.J.; Nunnink, G.H.J.; *Hydrocarbon Process*, **1972**, 51, 92.
119. *Eur. Chem. News*, **1976**, 28 (743), 20–22; Br. Patent 1,094,221, 1967; *Oil Gas* J., Aug. 24, 1970.
120. Miller, E.; *Melliand Textilber*, **1963**, 44, 484, 596; *Angew. Chem.*, **1959**, 71, 229.

121. Kunststoff-Rundschau, **1968**, 174.
122. Bayer A.G.; Ger. Patents 2,641,389, 2,641,429, 2,641,414, 2,641,449, 2,651,454, 2,651,478, 2,641,381, 2,641,408.
123. Union Carbide; U.S. Patents 3,025,306, 3,064,008, 3,000,880.
124. Schmitz, E.; Heyne, W.U.; Hilgetag, K.P.; Dilcher, H.; Lorenz, R.; *Mitteilungsbl. Chem. Ges. d. DDR*, **1974**, 63.
125. *Eur. Chem. News*, **1969**, 16, (404), 16; Hawkins, E.G.E.; *J. Chem. Soc.*, London, (C), **1969**, 2663.
126. U.S. Patent 3,575,964; Ger. Patent 2,049,560.
127. *Eur. Chem. News*, Caprolactam Supplement, May 2, 1969; *Hydrocarbon Process*, **1971**, 50, 141; Ger. Patent 1,940,809.
128. Taverna, M. and Chiti, M.; *Hydrocarbon Process*, Nov. **1970**, 137; Brev. Ital. Patents 604, 795, 603, 606, 608, 873, 1960.
129. Heath, A.; *Chem. Eng.*, **1974**, July 22, 70.
130. *Nitrogen*, **1970**, 67,38; Br. Patent 1,191,539.
131. U.S. Patent 3,698,477; Ger. Patents 2,142,401, 2,163,539, 1972; 2,307,302, 1974.
132. Ger. Patent 2,120,205, 1972.
133. DuPont; U.S. Patent 2,223,493, 1940.
134. *Eur. Chem. News*, **1969**, 15, 22; Halcon, U.S. Patents 3,243, 449, 3,932,513, 1976.
135. Lyon, J.B.; Stowe, G.T.; Production of cyclohexyl hydroperoxide, U.S. Patent 04,675,450, June 23, 1987.
136. Besmar, U.N.; Lyon, J.B.; Miller, F.J.; Musser, M.T.; Preparation of cyclohexanone and cyclohexanol, U.S. Patent 4,720,59, January 19, 1988.
137. Steinmetz, G.R.; Lafferty, N.L.; Summer, C.E.; Process for the preparation of adipic acid, U.S. Patent 4,902,827, February 20, 1990.
138. Kulsrestha, G.N.; Saxena, M.P.; Gupta, A.K.; Goyal, H.B.; Prasad, R.; Prasada, R.; Turuga, S.R.; Patel, P.D.; Catalyst and a process for preparing carboxylic acids using the catalyst, U.S. Patent 5,547,905, August 20, 1995.
139. Allied; U.S. Patent 2,794,056, 1957.
140. Halcon; Br. Patents 956,779; 956,780; 941,662; 1964.
141. BASF; Can. Patent 789,040, 1968; Br. Patent 1,092,603, 1969; DuPont; U.S. Patent 3,359,308, 1967; ICI; U.S. Patent 3,754,0241974, 1973; ICI; Br. Patent 914,510, 1963; Scholven; Neth. Patent 6,411,575, 1965; USSR Patent 396,315; Zimmer, Ger. Patent 1,518,242, 1973.
142. Tanaka, K.; *Chem. Technol.*, **1974**, 4, 555; *Hydrocarbon Process*, **1974**, 53, 114.
143. DuPont; U.S. Patent 3,306,932, 1967.
144. Monsanto; U.S. Patent 3,654,355, 1972.
145. Kollar, J.; Process for the preparation of adipic acid and other aliphatic dibasic acids, U.S. Patent 5,321,157, June 14, 1994.
146. Kollar, J.; Recycling process for the production of adipic acid and other aliphatic dibasic acids, U.S. Patent 5,463,119, October 31, 1995.
147. Mall, S.; Kumar, S.; Process for preparation of adipic acid, U.S. Patent 6,235,932, May 22, 2001.
148. Srivinas, D.; Chavan, S.A.; Ratnasamy, T.; Process for the preparation of adipic acid, U.S. Patent 6,521,789, February 18, 2003.
149. Tsutomu, N., *Yuki Gosei Kagaku Kyosaishi*, **1977**, 35, 842.
150. Keller, R.E.; *Encyl. Ind. Chem. Anal.*, **1967**, 4, 408.
151. Beck, F.; *Chem. Ing. Tech.*, **1965**, 37, 607.
152. ICI; Neth. Patents 6,515,869, 6,515,603; Fr. Patent 1,627,980, 1968.
153. Nobel, D.; Perron, R.; Denis, P.; Rhodium/iridium catalyzed synthesis of carboxylic acids or esters thereof, U.S. Patent 5,625,094, April 29, 1997.
154. Magnussen, P.; Schumacher, V.; Gebert, W.; Reitz, H.; Praetorius, W.; Continuous preparation of adipic acid, U.S. Patent 4,360,695, November 23, 1982.
155. BASF; Ger. Patents 2,630,086, 2,646,955, 1978.
156. Burke, P.M.; U.S. Patents 4,622,423, 4,788,333,
157. Burke, P.M.; Preparation of adipic acid from lactones, U.S. Patent 5,359,137, October 25, 1994.

158. Atadan, E.M.; Bruner, H.; Process for the preparation of adipic acid or pentenoic acid, U.S. Patent 5,292,944, March 8, 1994.
159. Bruner, H.S.; Lane, S.L.; Manufacture of adipic acid, U.S. Patent 5,710,325, January 20, 1998.
160. Kummer, R.; Merger, F.; Bertleff, W.; Fischer, R.; Preparation of adipic acid, U.S. Patent 4,931,590, June 5, 1990.
161. Denis, P.; Grosselin, J.; Preparation of adipic acid by hydrocarbonylation of pentenic acids, U.S. Patent 5,198,577, March 30, 1993.
162. Denis, P.; Grosselin, J.; Preparation of adipic acid by hydrocarboxylation of pentenic acids, U.S. Patent 5,227,522, July 13, 1993.
163. Denis, P.; Grosselin, J.; Preparation of adipic acid by hydrocarbonylation of pentenic acids, U.S. Patent 5,227,523, July 13, 1993.
164. Delis, P.; Denis, P.; Grosselin, J.; Metz, F.; Preparation of adipic acid by hydrocarboxylation of pentenic acids, U.S. Patent 5,268,505, December 7, 1993.
165. Denis, P.; Metz, F.; Perron, R.; Preparation of adipic acid by hydrocarboxylation of pentenoic acids, U.S. Patent 5,312,979, May 17, 1994.
166. Denis, P.; Grosselin, J.; Metz, F.; Preparation of adipic acid by hydrocarboxylation of pentenic acids, U.S. Patent 5,420, 346, May 20, 1995.
167. Jap. Patent 41-14092.
168. Celanese; Fr. Patent 1,374,807, 1962.
169. Celanese; U.S. Patents 3,215,742, 1960; 3,268,588, 1962.
170. DuPont; U.S. Patent 2,166,151, 1936.
171. DuPont; U.S. Patent 2,532,311, 1948.
172. DuPont; U.S. Patent 3,547,972, 1968.
173. DuPont; USA, Br. Patent 1,210,969, 1968.
174. Baird, W.C.; Surridge, J.H.; *J. Org. Chem.*, **1971**, 36, 2898.
175. *Chem. Eng.*, **1965**, 72, 80, 82.
176. Hakon; Ger. Patents 2,264,932, 1975, 2,559,185, 2,559,186, 2,559,187, 1976; U.S. Patent 3,538,141, 1970.
177. ICI; Ger. Patents 2,720,791, 2,721,808, 2,649,904, 1977.
178. Klare, J.; Fritzsche, E.; Grobe, V.; *Synthetische Fasern aus Polyamiden,* Akademie-Verlag, Berlin, 1963.
179. Jacobs D.B.; Zimmennan, J.; in *High Polym.s,* Schildknecht, C.E.; Skeist, I., Eds.; John Wiley & Sons, New York, 1977, Vol. 29, 424–467.
180. DuPont; Ger. Patent 2,508,566.
181. Bayer; Ger. Patent 2,848,951.
182. BASF; Ger. Patent 2,918,828
183. Martin, W.K.; Ryffel, J.R.; Schuster, H.H.; Wen, C.P.; Continuous process for the hydrolytic production of polycaprolactam by tempering unextracted granular polymer material with moist inert gas, U.S. Patent 4,891,420, January 2, 1990.
184. Belyskov, A.V.; Tyurenkova, E.V.; Kremer, E.B.; Vlasov, I.M.; *Khim. Volokna*, **1983**, 5, 20.
185. Spirin, V.A.; Budko, E.G.; Datsenko, V.I.; Gusakov, M.Ya.; Baranchukova, L.A.; Koleushko, V.P.; *Khim Voloka*, **1985**, 4, 40.
186. Cohen, J.D.; Fergusson, S.B.; Marchildon, E.K.; Marks, D.N.; Mutel, A.T.; Process for the production of nylon 6, U.S. Patent 6,437,089, August 20, 2002.
187. Fergusson, S.B.; Marchildon, E.K.; Mutel, A.T.; A process for making nylon-6 in which caprolactam and water are reacted in a multistage reactive distillation column, U.S. Patent 6,479,620, November 12, 2002.
188. Alsop, A.W.; Blanchard, E.N.; Cohen, J.D.; Iwasyk, J.M.; Lin, C.Y.; Marks, D.N.; Stouffer, J. M.; Process for preparing polyamides, U.S. Patent 6,069,228, May 30, 2000.
189. Kubánek, V.; Šterbáček, Z.; Method for postpolymerization of polyamide granules after polymerization in melt and an equipment for performing the method, U.S. Patent 4,755,590, July 5, 1988.
190. Mielinger, W.; Gittinger, A.; Ostlinning, E.; Idel, K.; Schulte, H.; Process for the production of high molecular weight polyamide 6, U.S. Patent 5,597,888, January 28, 1997.

191. Barnes, A.G.; Gottlund, K.L.; Polymerization of caprolactam with certain quaternary ammonium compounds, U.S. Patent 5,859,178, January 12, 1999.
192. Larsen, H.A.; Preparation of salt solution used for making nylon, U.S. Patent 4,442,260, April 10, 1984.
193. Brearley, A.M.; Lang, J.J.; Marchildon, E.K. A.; Continuous polymerization process for polyamides, U.S. Patent 5,674,974, October 7, 1997.
194. Fergusson, S.B.; Marchildon, E.K.; Mutel, A.T.; Process for making nylon-6,6, U.S. Patent 6,472,501, October 29, 2001.
195. Dujari, R.; Tynan, D.G.; Linear very high molecular weight polyamides and process for producing them, U.S. Patent 5,698,658, December 16, 1997.
196. Dujari, R.; Cramer, G.D.; Marks, D.N.; Method for solid phase polymerization, U.S. Patent 5,955,569, September 21, 1999.
197. Li, Y.; Zhang, G.; Yan, D.; Synthesis and crystallization behavior of nylon-12,14.1. Preparation and melting behavior, *J. Appl. Polym. Sci.*, **2003**, 88, 6, 1581–1589.
198. Gaymans, R.J.; DeHaan, J.L; van Nieuwenhuize, O.; Copolymers of PBT and nylon-4T, *J. Polym. Sci. A: Polym. Chem.*, **1993**, 31, 2, 575–580.
199. Gonsalves, K.E.; Chen, X.; Copolymers of nylon-266 and nylon-66: Synthesis and characterization, *J. Polym. Sci. A: Polym. Chemi.*, **1993**, 31, 3, 701–706.
200. Pae, Y.; Harris, F.W.; Synthesis and properties of novel polyimide/nylon-6 triblock copolymers, *J. Polym. Sci. A: Polym. Chem.*, **2000**, 38, 4247–4257.
201. Kawasumi, M.; The discovery of polymer-clay hybrids, *J. Polym. Sci. A: Polym. Chem.*, **2004**, 42, 4, 819–824.
202. Liu L.; Qi, Z.; Zhu, X.; Studies on nylon-6/clay nano-composites by melt-intercalation process, *J. Appl. Polym. Sci.*, **1999**, 71, 7, 1133–1138.
203. Liu, Z.; Zhou, P.; Yan, D.; Preparation and properties of nylon-1010/montmorillonite nano-composites by melt intercalation, *J. Appl. Polym. Sci.*, **2003**, 91, 3, 1834–1841.
204. García M.; De Rooij, M.; Winnubst, L.; van Zyl, W.E.; Verweij, H.; Friction and wear studies on nylon-6/SiO2 nano-composites, *J. Appl. Polym. Sci.*, **2004**, 92, 3, 1855–1862.
205. Nagasubramanian, K.; Reimschuessel, H.K.; *J. Appl. Polym. Sci.*, **1973**, 17, 1663.
206. Giudici, R.; Nascimento, C.A. O.; Scherbakoff, N.; Modeling of industrial nylon-6,6 polymerization process in a twin screw extruder reactor. I. Phenomenological and parameter adjusting, *J. Appl. Polym. Sci.*, **1998**, 67, 1583–1587.
207. Nascimento, C.A.O.; Giudici, R.; NBeiler, I.C.; Modeling of industrial nylon-6,6 polymerization process in a twin screw extruder reactor. II. Neural networks and hybrid models, *J. Appl. Polym. Sci.*, **1999**, 72, 7, 905–912.
208. Mallon, F.R.; Ray, W.H.; Modeling of solid-state polycondensation, I. Particle models, *J. Appl. Polym. Sci.*, **1999**, 69, 6, 1233–1250.
209. Yao, K.Z.; McAuley, K.B.; 1 ", E, Keith Marchildon, E.K.; Simulation of continuous solid-phase polymerization of nylon-6,6. III. Simplified model, *J. Appl. Polym. Sci.*, **1999**, 69, 14, 3701–3712.
210. Tam, T.Y.; Lin, C.Y.; A unique one-step process of achieving fully oriented yarn properties with <1500 m/min take-up speed, Fiber Society Spring Conference, Princeton, New Jersey, June 1996.
211. http://FRAM.com/, Honeywell International, FRAM Consumer Products Group, products (accessed September 2004).
212. Ziabicki, A.; *Fundamentals of Fiber Formation*, 1976, Wiley, London.
213. James, A.; An overview of quench cabinet systems, *Intl. Fiber J.*, **1999**, 101–103.
214. Kiritchenko, A.M.; *Khimicheskiye*, **1976**, 6, 56.
215. Ziabicki, A;. Wasiak, A.; Proc. Ilnd. Internat. Symp., *Manmade Fibers*, 1977, Kalinin, USSR.
216. Buttner, R.; Jung, A.; Dtsch. Textiltech, **1971**, 21, 493.
217. Bromley, J.E.; McManara, M.M.; Flat nylon-66 yarn having a soft handand process for making same, U.S. Patent 4,093,147, June 6, 1978.
218. Adams, E.B.; Polyhexamethylene adipamide yarn, U.S. Patent 3,994,121, November 3, 1976.
219. Koyama, K.; Suryadevara, J.; Spuriell, J.E.; *J. Appl. Polym. Sci.*, **1986**, 31(7), 2203.
220. Brody, H.; Melt spinning process, U.S. Patent 4,442,057, April 10, 1984.

221. Brody, H.; Process of melt spinning of a blend of a fibre-forming polymer and an immiscible polymer and melt spun fibres produced by such process, U.S. Patent 4,518,744, May 21, 1985.
222. Vassilatos, G.; High speed melt pspinning of fibers, U.S. Patent 6,432,340, August 13, 2002.
223. Wagner, C.-D.; *Chemiefasern/Textilind.*, **1974**, 24/76, 475.
224. Gall, H.; in *Synthesefasem,* B.V. Falkai, Ed.; Verlag Chemie, Weinheim, 1981.
225. Komanický, V.; Cupak, F.; Kvarda, V.; in *Lactam-based Polyamides*, R. Puffr and V. Kubánek, Eds.; 1991, Vol. 2, 91.
226. Klein, W.; Trummer, A.; *Text. Betr.*, **1976**, 94, 87.
227. Kuznetsov, Yu. A.; Usenko, V.A.; *Chem. Vlakna*, **1988**, 38(4), 274.
228. Toray Industries; Jap. Patent 02, 26,932, 1990.
229. Toyobo Co.; Jap. Patent 61,113,823, 1986.
230. Toray Industries; Jap. Patents 61,29,123, -124, -125, -126, -127, 1986.
231. Billica, H.R.; *Chemiefasem/Textilind*, **1977**, 27/79, 328.
232. Kinoshita, Y.; *Makromol. Chem.*, **1959**, 33, 1.
233. Miyasaka, K.; Makishima, K.J.; *Polym. Sci.*, **1967**, A-1 5, 3017.
234. Miyasaka, K.; Ishikawa, K.J.; *Polym. Sci.*, **1968**, A-2 6, 1317.
235. Arimoto, H.; Kobunski Kagaku; **1962**, 19, 204, 212.
236. Arimoto, H.J.; *Polym. Sci.*, **1964**, A-2, 2283.
237. Vogelsong, D.C.; *J. Polym. Sci.*, **1963**, A-1, 1055.
238. Ishikawa, T.; Nagai, S.; *Makromol. Chem.*, **1981**, 182, 977.
239. Bunn, C.W.; Garner, V.; *Proc. R. Soc.*, London 1947, 189A, 39.
240. Holmes, D.R.; Bunn, C.W.; Smith D.J.; *J. Polym. Sci.*, **1955**, 17, 159.
241. Sandemann, I.; Keller, A.; *J. Polym. Sci.*, **1956**, 19, 401.
242. Keller, A.; Maradudin, A.; *J. Phys. Solids*, **1957**, 2, 301.
243. Bankar, V.G.; Spruiell, J.E.; White, J.L.; *J. Appl. Polym. Sci.*, **1977**, 21, 234.
244. Reichle, A.; Prietschk, A.; *Angew. Chem.*, **1962**, 74, 562.
245. Parker, J.P.; Lindenmeyer, P.H.; *J. Appl. Polym. Sci.*, **1977**, 21, 821.
246. Kinoshita, Y.; *Makromol. Chem.*, **1959**, 33, 21.
247. Roldan, L.G.; Kaufman, H.S.; *J. Polym. Sci.*, **1963**, B-1, 603.
248. Miyake, A.; *J. Polym. Sci.*, **1960**, 44 ,223.
249. Arimoto, H.; Ishibashi, M.; Hirai, M.; Chantani, Y.; *J. Polym. Sci.*, **1965**, A-3, 31.
250. Illers, H.K.; Haberkorn, H.; Simak, P.; *Makromol. Chem.* **1972**, 158, 285.
251. Ziabicki, A.; *Kolloid ZZ Polym.*, **1959**, 167, 132.
252. Stepaniak, R.F.; Garton, A.; Carlsson D.J.; Wiles, D.M.; *J. Polym. Sci.*, Polym. Phys. Ed., **1979**, 17, 987.
253. Trifan, D.S.; Terenzi, J.F.; *J. Polym. Sci.*, **1895**, 28, 443.
254. Heuvel, H.M.; Huisman, R.; *J. Appl. Polym. Sci.*, **1981**, 26, 713.
255. Stepaniak, R.F.; Garton, A.D.; Carlsson, J.; Wiles, D.M.; *J. Appl. Polym. Sci.*, **1979**, 23, 1747.
256. West, C.D.; *J. Chem. Phys.*, **1947**, 15, 689.
257. Bradbury, E.M., Brown, L.; Elliott, A.; Parry, D.A.D.; *Polym.*, **1965**, 6, 465.
258. Matsubara, 1.; Magill, J.H.; *Polym.*, **1966**, 7, 199.
259. Takayanagi, M.; *Pure Appl. Chem.*, **1967**, 15, 555.
260. Ishikawa, K.; Miyasaka, K.; Okabe, T.; *Makromol. Chem.*, **1969**, 122, 123.
261. Baldrian, J.; *Czech. J. Phys.*, **1965**, B 15, 838.
262. Abu-Isa, 1.; *J. Polym. Sci.*, **1971**, A-I 9, 199.
263. Stepaniak, R.F.; Garton, A.; Carlsson, D.J.; Wiles, D.M.; *J. Appl. Polym. Sci.*, **1979**, 23, 1747.
264. Frayer, P.D.; Koenig, J.L.; Lando, J.B.; *J. Macromol. Sci. Phys.*, **1972**, B 6, 129.
265. Hoashi, K.; Andrews, R.D.; *J. Polym. Sci.*, **1972**, C-38, 387.
266. Goikhman, A.Sh.; Osokin, G.A.; Konkin, A.A.; *Polym. Sci. USSR*, **1969**, 10, 1903.
267. Nagatoshi, F.; Arakawa, T.; *Polym. J.*, **1970**, 1, 685.
268. Slichter, W.P.; *J. Polym. Sci.*, **1959**, 36, 259.
269. Burzynski, R.; Prasad, P.N.; Murthy, N.S.; *J. Polym. Sci.*, Polymer Physics Edition, **1986**, 24, 133.
270. Illers, K.H.; Haberkorn, H.; *Makromol. Chem.*, **1971**, 142, 31.

271. Illers, K.H.; *Makromol. Chem.*, **1978**, 179, 497.
272. Murthy, N.S.; *Macromolecules*, **1987**, 20, 309.
273. Hinrichsen, G.; *Kolloid Z.*, **1972**, 250, 1162.
274. Illers, K.H.; Haberkorn, H.; *Makromol. Chem.*, **1971**, 146, 267.
275. Hosemann, R.; Bonart, R.; Schoknecht, G.Z.; *Phys.*, **1956**, 146, 588.
276. Bonart, R.; Hosemann, R.; Motzkus, F.; Ruck, N.; Norelco Reporter, **1960**, 7, 81.
277. Hosemann, R.; Bonart, R.; *Kolloid Z.*, **1957**, 152, 53.
278. Arakawa, T.; Nagatoshi, F.; *J. Polym. Sci.*, **1970**, B-8, 41.
279. Weigel, P.; Hirte, R.; Ruscher, Ch.; *Faserforsch. Textil-techn.*, **1974**, 25, 198, 283, 500.
280. Itoh, T.; Mijaji, H.; Asai, K.; *Jpn. J. Appl. Phys.*, **1975**, 14, 206.
281. Ruscher, Ch.; Schulz, E.; *Faserforsch. Textiltechn.*, **1971**, 22, 260.
282. Avrami, M.; *J. Chem. Phys.*, **1941**, 9, 177.
283. Reimschuessel, A.C.; Prevorsek, D.C.; *J. Polym. Sci. Phys. Ed.*, **1976**, 14, 485.
284. Peterlin, A.; *J. Polym. Sci.*, 1971, C-32, 297.
285. Prevorsek, D.C.; Harget,; Sharma, R.K.; Reimschuessel, A.C.; *J. Macromol. Sci. Phys.*, **1973**, B-8, 127.
286. Prevorsek, D.C.; Tirpak, G.A.; Harget, P.; Reimschuessel, A.C.; *J. Macromol. Sci. Phys.*, **1974**, B-9, 733.
287. Kamezawa, M.; Imada, K.; Takayanagi, M.; *J. Appl. Polym. Sci.*, **1979**, 24, 1227.
288. Wunderlich, B.; *Polym.*, **1964**, 5, 611.
289. Hoffman, J.D.; Weeks, J.J.; *J. Res. Natl. Bur. Stand.*, **1962**, A66, 13.
290. Magill, J.H.; Girolamo, M.; Keller, A.; *Polymer*, **1981**, 22, 43.
291. Mitomo, K.; Nakazato, H.; Kuriyama, 1.; *Polymer*, **1978**, 19, 1427.
292. Buchanan, D.R.; Walters, J.P.; *Tex. Res. J.*, **1977**, 47, 398.
293. Puffr, R.; Šebenda, J.; *J. Polym. Sci.*, **1976**, C-16, 79.
294. Reimschuessel, H.K.; *J. Polym. Sci. Chem. Ed.*, **1978**, 16, 1229.
295. Bondi, A.; Tobolsky, A.V.; in *Polymer Science and Materials*, Tobolsky, A.V.; Mark, H.F.; Eds.; Wiley-Interscience, New York, 1971.
296. Kettle, G.J.; *Polym.*, **1977**, 18, 742.
297. Murthy, N.S.; Stamm, M.; Sibilia, J.P.; Krimm, S.; *Macromolecules*, **1989**, 22, 1261.
298. Bragg, W.L.; *Proc. Cam. Phil. Soc.*, **1912**, 17, 43.
299. Scherrer, P.; *Gottinger Nachrichten*, **1918**, 2, 98.
300. Bonart, R.; Hosemann, R.Z.; *Elektrochem.*, **1960**, 64, 314.
301. Hennans, P.H.; Platzek, P.; *Kolloid Z.*, **1939**, 88, 68.
302. Hennans, J.J.; Hennans P.H.; Vennaas, D.; Weidinger, A.; *Rec. Trav. Chim. Pays-Bas*, **1946**, 65, 427.
303. Wilchinsky, Z.W.; *Advances in X-Ray Analysis*, Plenum Press, New York, 1963, Vol. 6., 231; *J. Appl. Phys.*, **1959**, 30, 792.
304. Ward, I.M.; *Structure and Properties of Oriented Polymers*, Applied Science Publishers, London, 1975.
305. Stein, R.S.; Norris, F.H.; *J. Polym. Sci.*, **1956**, 21, 381.
306. Samuels, R.J.; in *Science and Technology of Polymer Films*, Sweeting, 0. J., Ed.; Wiley Interscience, New York, 1968, Chapter 7.
307. Wlochowicz, A.; Rabiej, St.; Janicki, J.; *Acta Polymerica*, **1982**, 33, 117.
308. Wlochowicz, A.; Janicki, J.; Rabiej, S.; *J. Appl. Polym. Sci.*, **1985**, 30(4), 1653.
309. Koenig, J.L.; Itoga, M.; *J. Macromol. Sci. Phys.*, **1972**, B6, 309–327.
310. Bouriot, P.; *Teintex.*, **1969**, 34, 149.
311. Heidemann, G.; *Chemiefasem Text. Anwendungs Tech.*, **1970**, 20, 204.
312. Heidemann, G.; *Text. Chim.*, **1971**, 27, 19.
313. Miyake, A.; *J. Polym. Sci.*, **1960**, 44,223.
314. Koenig, J.L.; Ipekci, M.; Lando, J.B.; *J. Macromol. Sci. Phys.*, **1972**, B6, 713.
315. Simak, P.; *Makromol. Chem.*, **1973**, 28, 75.
316. Tirpak, G.A.; Sibilia, J.P.; *J. Appl. Polym. Sci.*, **1973**, 17,643.
317. Roldan, L.G.; RaW, R.; Paterson, A.R.; *J. Polym. Sd.*, **1965**, C-8, 145.
318. Kyotani, M.; Mitsuhashi, S.; *J. Polym. Sci.*, **1972**, A-2, 10, 1497.

319. Dismore, P.F.; Statton, W.O.; *J. Polym. Sci.*, 1966, C-13, 133.
320. Huisman, R.; Heuvel, H.M.; Lind, K.C.J.B.; *J. Polym. Sci., Polym. Phys. Ed.*, **1976**, 14, 921–941.
321. Gianchandani, J.; Spuiell, J.E.;. Clark, E.S., *J. Appl. Polym. Sci.*, **1982**, 27, 3527.
322. Clark, E.S.; Wilson, F.C.; in *Nylon Plastics*, Kohan, M., Ed.; John Wiley & Sons, New York, 1973.
323. Hauvel, H.M.; Huisman, R.; *J. Polym. Sci.*, **1981**, 19, 121.
324. Murthy, N.S.; Szollosi, A.B.; Sibilia, J.P.; Krimm, S.; *J. Polym. Sci., Polym. Phy. Ed.*, **1985**, 23, 2369.
325. Murthy, N.S.; Hatfield, G.R.; Glans, J.H.; *Macromolecules*, **1990**, 23, 1342.
326. Murthy, N.S.; Correale, S.T.; Moore, R.A. F.; *J . Appl. Polym. Sci.*, Appl. Polym. Sympos., **1991**, 47, 185.
327. Murthy, N.S.; Minor, H.; Latif, R.A.; *J. Macromol. Sci.-Phys.*, **1987**, B26(4), 427.
328. Murthy, N.S.; Reimschuessel, A.C.; Kramer, V.; *J. Appl. Polym. Sci.*, **1990**, 40, 249.
329. Reimschuessel, H.K.; *Colloid. Polym. Sci.*, **1982**, 260(9), 842.
330. Genis, A.V.; Gribanov, S.A.; Khaseleva, L.N.; Fenin, V.A.; *Vysokomol. Soedin*, **1989**, Ser. A 31(7), 1431.
331. Morton and W.E.; Hearle, J.W. S.; *Physical Properties of Textile Fibers*, John Wiley & Sons, New York, 1975.
332. Happey, F.; *Applied Fiber Science*, Academic Press, London, 1979.
333. Komatsu, Y.; Takei, S.; *Sen'i Gakkaishi*, **1982**, 23(9), 359.
334. Acierno, D.; Ciaperoni, A.; La Mantia, F.; Polizzotti, G.; *Acta Polym.*, 1985, 37(6), 351.
335. Moore, R.A. F.; Ruetsch, S.B.; Weigmann, H.-D.; *Textile Chemist and Colorist*, 1984, 16, 250.
336. Fischer, E.W.; Goddar, H.; *J. Polym. Sci.*, **1969**, C16, 4405.
337. Taylor, W.N.; Clark, E.S.; *Polym. Prep.*, ACS Nat. Meeting, August 1977, 332.
338. Kausch, H.H.; Becht, J.; *Rheol. Acta*, **1970**, 9, 137.
339. Kausch, H.H.; DeVries, K.L.; *Int. J. Fracture II*, **1975**,109.
340. Peterlin, A.; *J. Polym. Sci.*, **1965**, C9, 61.
341. Peterlin, A.J. *Mater. Sci.* **1971**, 6, 490.
342. Prevorsek, D.C.; Harget, P.J.; Sharma, R.K.; Reimschuessel, A.C.; *J. Macromol. Sci.*, **1973**, B8, 127.
343. Mishra, S.P.; Deopura, B.L.; *J. Appl. Polym. Sci.*, **1982**, 27, 3211.
344. Peterlin, A.; *J. Macromol Sci.*, **1973**, B7, 705.
345. Johnsen, U.; Klinkenberg, D.; *Kolloid ZZ Polym.*, **1973**, 251, 843.
346. Takayanagi, M.; Proc. 4th Int. Congr., *Rheol.*, **1965**, 161.
347. Schultz-Gebhart, F.; *Faserf. Textiltechn.*, **1977**, 28, 467.
348. McGregor, R.; Harris, P.W.; *J. Appl. Polym. Sci.*, **1970**, 14, 513.
349. Fujita, H.; *Fortschr. Hochpolym.-Forsch.*, **1961**, 3, 1.
350. Hopper, M.E.; McGregor, R.; Peters, R.H.; *J. Soc. Dyers Colour*, **1970**, 86, 117.
351. Back, G.; Zollinger, H.; *Helv. Chim. Acta*, **1958**, 41, 2242; **1959**, 42, 1526–1539.
352. Peters, R.H.; *Textile Chemistry*, Elsevier, Amsterdam, 1975, Vol. III; (b) *The Chemistry of Synthetic Dyes*, Venkataraman, K., Ed.; Academic Press, New York, 1978, Vol. VIII, 133.
353. Warwicker, J.O.; (a) *J. Soc. Dyers Colour*, **1970**, 86,303; (b) *Br. Polym.*, **1971**, 7, 3, 68; (c) *J. Appl. Polym. Sci.*, **1975**, 19, 1147.
354. DuPont; Ger. Patent 1,469,130, 1960.
355. Stamicarpon; Ger. Patent 1,645,457, 1966.
356. Ciba-Geigy; Ger. Patent 2,237,849, 1972.
357. Firestone; U.S. Patent 3,644,280, 1968.
358. Achhamer, B.G.; Reinhardt, F.W.; Kline, G.M.; (a) *J. Appl. Chem.*, **1951**, 1, 301; (b) *J. Res. Nat. Bur. Stds.*, **1951**, 46, 391.
359. Kamerbeck, B.; Kroes, G.H.; Grolle, W.; *Soc. Chem. Ind. Lond.*, Monograph, **1961**, 13, 357.
360. Peebles, L.H.; Huffman, M.W.; *J. Polym. Sci.*, **1971**, AI, 9, 1807.
361. Madorosky, S.L.; *Thermal Degradation of Organic Polymers,* Interscience Publishers, New York, 1964.
362. Straus, S.; Wall, L.A.; *J. Res. Nat. Bur. Stds.*, **1958**, 60, 39; **1959**. 63A, 269.
363. Reimschuessel, H.K.; Dege, G.J.; *J. Polym. Sci.*, **1970**, AI, 3265.

364. Peters, R.H.; Still, R.H.; in *Applied Fiber Science*, Happey, F., Ed.; Academic Press, London, 1979, Vol. 2, 321.
365. Valk, G.; Krussmann, H.; *Angew. Chem. Intern. Ed.*, **1967**, 6, 1001.
366. Levchik, S.V.; Weil, E.D.; Lewin, M.; Thermal decomposition of aliphatic nylons, *Polym. Intl.*, **1999**, 48, 532–557.
367. Valk, G.; Krussmann, H.; Dugal, S.; Gentzsch, Ch.; *Angew. Makromol. Chem.*, **1970**, 10, 127.
368. Valk, G.; Heidemann, G.; Dugal, S.; Krussmann, H.; *Angew. Makromol. Chem.*, **1970**, 10, 135.
369. Levantovskaya, I.I.; Kovarskaya, B.M.; Dralyuk, G.V.; Neiman, M.B.; *Vysokomolekul. Soedin*, **1964**, 6, 1885.
370. Rieche, A.; Schon, W.; *Ber.*, **1966**, 99, 3238; *Kunststoffe*, 1967, 5749.
371. Dege, G.J.; Reimschuessel, H.K. *J. Polym. Sci. Chem. Ed.* **1973**, *11*, 873.
372. Valko, E.I.; Chiklis, C.K. *J. Appl. Polym. Sci.* **1965**, 9, 2855.
373. Wolf, K.; Klee, D.; Zahn, H.; TPI, *Text. Prax. Int.*, **1986**, 41(3), 309; 41(5), 543.
374. Kazmierzak, R.T.; McLeay, RJ. E.; Process for preparing polymer bound hindered amine light stbilizers, U.S. Patent 4,975,489, December 4, 1990.
375. McLeay, R.E.; Myers, T.N.; Multipurpose polymer bound stabilizers, U.S. Patent 4,981,915, January 1, 1991.
376. Lofquist, R.A.; Mohajer, Y.; Light stabilized polyamide substrate and process for making, U.S. Patent 5,618,909, April 8, 1997.
377. Dawczynski, H.; Taeger, E.; Clemens, H.; *Chem. Vlakna.*, **1978**, 28, 1.
378. Stepniczka, H.E.; *Eng. Chem. Prod. Res. Devev.*, **1973**, 12, 29.
379. Vaidya, A.A.; Chattopadhyay, S.; Ravishankar, S.; *Text. Dyer Printer*, **1977**, 10(8), 37.
380. Fukumura, T.; Iwata, M.; Narita, N.; Inoue, K.; Tanaka, M.; Seki, M.; Takahashi, R.; Water insoluble ammonium polyphosphate power for flame-retardant thermolstic polymer composition, U.S. Patent 3,795,930, August 18, 1998.
381. Horacek, H.; Reichenberger, R.; Ritzberger, K.; Prinz, C.; Flameproof glass fiber–reinforced resin composition with melaminemelemphosphoric acid reaction products as flame retardants, U.S. Patent 6,031,032, February 29, 2000.
382. Hirono, M.; Watanabe, N.; Resin composition comprising polyamide resin, U.S. Patent 6,225,383, May 1, 2001.
383. Schleifstein, R.A.; Polybrominated sulfamides, U.S. Patent 4,959,500, September 25, 1990.
384. Schleifstein, R.A.; Flame retardant nylon composition, U.S. Patent 5,264,474, January 1, 1991.
385. Schleifstein, R.A.; Pietrewicz, D.S.; Polymer compositions, U.S. Patent 5,264,474, November 23, 1993
386. Venkataramani, V.S.; Flame resistant polymer compositions, U.S. Patent 5,693,700, December 2, 1997.
387. Levin, M.; Flame retardation of polymeric compositions, U.S. Patent 6,528,558 B2, May 4, 2003.
388. Lewin, M.; Brožek, J.; Martens, M.M; *Polym. Adv. Technol.*, **2002**, 13, 1091–1102.
389. Murthy, N.S.; *Polym. Comm.*, **1991**, 32, 301.
390. Treptow, H.; LOV-MOV-POV-HOV-FOV, *Chemiefasern Text. Ind.*, **1985**, 35, 411.
391. Beier, M.; Bossmann, A.; Heinrichs, C.; Berndt, H.J.; Schollmeyer, E.; *Melliand. Textilber.*, **1986**, 67(3), E73, 154.
392. Nettelnstroth, K.; *Melliand. Textilber.*, **1983**, 64(12), 918.
393. Magill, M.C.; Martmann, M.H.; Multi-component fibers having enhanced reversible thermal properties and methods of manufacturing thereof, U.S. Patent Application No. 20030035951, February 20, 2003.
394. Howe, P.T. Jr.; Wilkie, A.E.; Polyamide spin-texture process, U.S. Patent 4,202,854, May 13, 1980.
395. Wu, W.L.; Spin texture process, U.S. Patent 4,244,907, January 13, 1981.
396. Talley, A.; Wilkie, A.E.; Buchanan, K.H.; A self-set yarn made from bicomponent fibers forms helical crimps that lock in twist and form bulk, U.S. Patent 6,158,204, December 12, 2000.
397. Chang, J.C.; Kurian, J.V.; Nguyen, Y.D.; Van Trump, J.E.; Vassilatos, G.; Method for high-speed spinning of bicomponent fibers, U.S. Patent 6,692,687, February 17, 2004.

398. Stevenson, P.E.; Bruner, J.W.; U.S. Patent 6,020,275, February 1, 2000. Lintecum, B.M.; Shoemaker, R.T.; Anderson, C.R. Jr.; Bicomponent effect yarn and fabrics thereof, U.S. Patent 6,548,429, April 15, 2003.
399. Okanoto, M.; Spinning of ultra-fine fibers. in *Advanced Fiber Spinning Technology*, Nakajima, T., ed; Kajiwara, K.; McIntyre, J.E., English eds.; Woodhead Publishers Ltd., Cambridge, England, 1994.
400. http://www.afma.org/ American Fiber Manufacturers Association, Fiber Economics Bureau, *2005 World Directory of Manufactured Fiber Producers* (accessed October 1, 2004).
401. Saunders, J.H.; in *Kirk-Othmer Encyclopedia Technology*, 3rd ed., 18, 372–405 (1982).
402. Kawaguchi, Tatsuro; *Sen'i Gakkaishi*, **1984**, 40(4–5), 265.
403. Beyreuther, R.; Schoene, A.; Hofmann, H.; Kaufmann, S.; *Textiltechnik*, (Leipzig), **1984**, 34(1),17.
404. Hoff, M.; *Chem. Listy*, **1990**, 84(8), 806.
405. Hirai, T.; Chiba, K.; *Sen'i Gakkaishi*, **1984**, 40(4–5), 256.
406. Tsuji, W.; *Senshoku Kogyo*, **1985**, 33(9), 369.
407. Yang, H.H.; *Aromatic High-Strength Fibers*, Wiley-Interscience, New York, 1989.
408. Yang, H.H.; *Kevlar Aramid Fiber*, Wiley, England, 1993.
409. Kaimin, I.F. Apinis, A.P. and Galvanovskii, A. Ya. *Vysokomol. Soedin*, **1975**; A17, 41.
410. Smith, E.G.; *Polym.*, **1976**, 17, 739.

3 Polypropylene Fibers

Mei-Fang Zhu and H.H. Yang

CONTENTS

3.1 Introduction 141
   3.1.1 Evolution of Polypropylene Fiber 141
   3.1.2 Market Growth 142
      3.1.2.1 Worldwide Growth 143
      3.1.2.2 Regional Productivity 143
      3.1.2.3 Other Synthetic Fibers 144
3.2 Major End-Uses 144
   3.2.1 Carpet and Furnishing 146
      3.2.1.1 Carpet 146
      3.2.1.2 Home Furnishing 147
      3.2.1.3 Automotive Furnishing 147
   3.2.2 Nonwoven Fabrics 147
   3.2.3 Industrial Applications 148
      3.2.3.1 Ropes 148
      3.2.3.2 Civil Construction 148
      3.2.3.3 Cement Reinforcement 148
      3.2.3.4 All Synthetic Paper 149
   3.2.4 Apparel 149
3.3 Preparation of Propylene Polymer 149
   3.3.1 Basic Polymerization Reaction 149
   3.3.2 Reaction Mechanism 150
      3.3.2.1 Catalyst Type and Concentration 152
      3.3.2.2 Monomer Concentration and Hydrogen 152
      3.3.2.3 Polymerization Temperature and Time 152
      3.3.2.4 Polymerization Medium 152
   3.3.3 Catalyst Systems 152
      3.3.3.1 Ziegler–Natta Catalysts 153
      3.3.3.2 Metallocene Catalysts 153
   3.3.4 Polymerization Processes 154
      3.3.4.1 Basic Processes 154
      3.3.4.2 Industrial Operations 156
3.4 Basic Polymer Properties 157
   3.4.1 Molecular Weight and Molecular Weight Distribution 158
   3.4.2 Crystalline Structure 160

		3.4.3	Melt Behavior	160
			3.4.3.1 Phase Transitions	160
			3.4.3.2 Melt Rheology	160
			3.4.3.3 Comparison of ZNPP and MiPP	172

3.5 Additives for Fiber Preparation ... 174
    3.5.1 Stabilizers ... 175
        3.5.1.1 Degradation of Unstabilized Fiber ... 175
        3.5.1.2 Melt Extrusion Stability ... 176
        3.5.1.3 Long-Term Thermal Stability ... 178
        3.5.1.4 UV Stability ... 180
    3.5.2 Pigments ... 184
        3.5.2.1 Fiber-Processing Requirement ... 184
        3.5.2.2 Stability Requirements ... 185
        3.5.2.3 Interaction with Other Additives ... 187
        3.5.2.4 Effects on Fiber Stability ... 188
    3.5.3 Dyeability Modifiers ... 190
        3.5.3.1 Copolymerization ... 190
        3.5.3.2 Grafting ... 191
        3.5.3.3 Polymeric Modifiers ... 191
        3.5.3.4 Fiber Blends ... 192
        3.5.3.5 Organic Metal Salts ... 192
        3.5.3.6 Surface Modification ... 193
        3.5.3.7 Halogen Compounds ... 193
    3.5.4 Flame Retardants ... 193
    3.5.5 Fiber Finishes ... 195

3.6 Preparation of Polypropylene Fibers ... 195
    3.6.1 Long Air-Quench Melt Spinning ... 196
        3.6.1.1 Metering Pump ... 196
        3.6.1.2 Spin Pack ... 197
        3.6.1.3 Spinneret ... 197
        3.6.1.4 Quenching ... 198
        3.6.1.5 Finish Application ... 198
        3.6.1.6 Drawing and Annealing ... 199
    3.6.2 Short Air-Quench Melt Spinning ... 200
    3.6.3 Water-Quench Melt Spinning ... 200
        3.6.3.1 Spinning and Quenching ... 200
        3.6.3.2 Draw Resonance ... 201
        3.6.3.3 Drawing ... 201
    3.6.4 Spun-Bonding Process ... 201
        3.6.4.1 Spun-Bonding ... 202
        3.6.4.2 Melt-Blowing ... 202
    3.6.5 Fibers from Film ... 203
        3.6.5.1 Sheet Film Extrusion ... 203
        3.6.5.2 Drawing and Annealing ... 203
        3.6.5.3 Fibrillation ... 203

3.7 Fiber Properties ... 204
    3.7.1 As-Spun Fibers ... 204
        3.7.1.1 Crystalline Structure ... 204
        3.7.1.2 Spinning Stress ... 205
        3.7.1.3 Mechanical Properties ... 211

    3.7.2 Drawn Fibers 213
        3.7.2.1 Deformation Model 213
        3.7.2.2 Effect of Temperature 215
        3.7.2.3 Effect of Draw Rate 217
        3.7.2.4 Crystalline Structure 220
        3.7.2.5 Structure–Property Relationship 225
    3.7.3 Annealed Fiber 227
    3.7.4 Melting Behavior 229
3.8 New Generation of Polypropylene Fibers 236
    3.8.1 Three-Dimensional Helical Staple 236
    3.8.2 High-Strength, High-Modulus Fiber 238
        3.8.2.1 Rapid Quenching 238
        3.8.2.2 Two-Stage Drawing 239
        3.8.2.3 Fibrillar Crystallization 243
        3.8.2.4 Gel Spinning 243
        3.8.2.5 Current Efforts 244
        3.8.2.6 Commercial Products 246
    3.8.3 High-Shrinkage Fiber 246
        3.8.3.1 Fiber Preparation 246
        3.8.3.2 Fiber Properties 247
        3.8.3.3 Applications 248
    3.8.4 Conductive Fibers 248
    3.8.5 Fibers from Polymer Blends 248
        3.8.5.1 PP–PP Blend 249
        3.8.5.2 PP–PS Blend 249
    3.8.6 Fibers from Nanocomposites 251
        3.8.6.1 Macroscopic Morphology 252
        3.8.6.2 Crystallization Behavior 252
        3.8.6.3 Rheological Behavior 253
        3.8.6.4 Fiber Spinning 253
        3.8.6.5 Dyeability 253
Acknowledgment 254
References 254

3.1 INTRODUCTION

3.1.1 Evolution of Polypropylene Fiber

When propylene was first polymerized in the mid-1930s, the product was a low-melting solid of no possible value as a resin from which synthetic fibers could be prepared. By the mid-1950s, research had led to the preparation of a new kind of polypropylene that had all or most of the pendant methyl groups in a regular position with respect to the polymer backbone. Commercial production of these stereospecific polymers utilized catalysts that were generally based on a combination of $TiCl_3$ and an aluminum alkyl. The stereospecific resin with all the methyl groups in the same position, rather than regular alternating positions, was called isotactic polypropylene. This resin was crystallizable with melt temperatures in the range of 160–174°C, and fibers capable of retaining molecular orientation at normal-use temperatures could be prepared. Since then, catalyst and process improvements have resulted in approximately a tenfold increase in polymer yield per pound of the catalyst and have simplified the polymer purification procedures. The percent noncrystalline resin has been reduced, and clean, gel-free resin required for modern fiber processes is now readily available. Isotactic polypropylene, as originally produced, had a very broad molecular weight distribution compared to most synthetic fiber resins. Polypropylene resin with a narrower molecular weight distribution, which is beneficial to the same fiber processes, is now manufactured.

Polypropylene generally leaves the reactor as a powder, which, after separation from the unreacted propylene or solvent, is quite susceptible to oxygen even at room temperature. Antioxidants are required to protect the polymer during storage; additional antioxidants are needed to permit melt extrusion, either to form pellets or to be spun into fibers. Once formed, the fiber must be protected during its lifetime by a long-term antioxidant. Additional stabilizers may be needed depending on the fiber end-use. Exposure to UV light (sunlight) in the presence of atmospheric oxygen will cause degradation in a chain reaction unless an adequate UV stabilizer is added. In addition to antioxidants for thermal stabilization and a variety of UV stabilizers, certain synergists are used to help the additives function better.

Although there have been several dyeable forms of polypropylene offered commercially, polypropylene fibers are usually not dyeable as the resin does not contain a dye site, like other commercially useful fibers. Therefore, colored polypropylene fibers are manufactured by adding pigments prior to extrusion. The pigments must be properly dispersed, both for optimum economics and for efficient fiber processing. To accomplish this, the pigments are first dispersed in polypropylene at relatively high concentrations (25–50%) to form a pigment concentrate. Carefully measured quantities of several different pigment concentrates are combined with natural polypropylene by the fiber producer to prepare the desired color. The combination of color and resin is accomplished by (1) melt injection, (2) blending of powdered color concentrate and resin, or (3) preparing a letdown prior to spinning. The fiber end-use determines which pigments may be used, just as it establishes stabilizer requirements. Some pigments are not sufficiently stable, or expensive for the end-use, or incompatible with the stabilizers. Achieving fiber color specifications requires not only color control by the fiber producer but also careful attention to specifications from the pigment and color-concentrate manufacturers.

A large variety of textile products in different colors and with different stability properties are available today. The story of modern polypropylene textile products is the story of differences—how they differ from polypropylene products made 20–25 years ago and how these products differ from one another today.

3.1.2 Market Growth

Because polypropylene fiber producers are now able to make a large variety of products and the textile market has learned how to use these products, polypropylene fibers have been able to achieve an increasing share of the textile market and are expected to occupy an even greater part of the market in the future.

Polypropylene is converted to fiber products using different methods. The marketplace was initially penetrated with the monofilament form for use in ropes, cordage, and outdoor furniture webbing, where the strength, light weight, mildew resistance, and economics of polypropylene allowed it to replace other fiber products. Polypropylene is also converted to textile products via fibrillated and ribbon yarns produced by film-making (SF) procedures, by direct fabric-making methods such as spun-bonded (SB) and melt-blown products, and by conventional procedures such as staple (ST) and multifilaments.

TABLE 3.1
Worldwide Production of Polyolefin Fibers[a]

Year	1996	1997	1998	1999	2000	2001	2002	Growth (%) 1999/1998	Growth (%) 2002/2001
Filament (KT)	1850	2033	2168	2292	2406	2677	2694	5.7	+0.6
Staple (KT)	1018	1103	1195	1261	1189	1212	1300	+5.5	+7.3
Film based (KT)	1713	1822	1814	1966	2142	1909	1918	+8.4	+0.5
Total (KT)	4581	4958	5177	5519	5737	5798	5912	+6.6	+2.0

[a]Polyolefin fibers include polyethylene fiber and polypropylene fiber. About 95% of polyolefin fibers is polypropylene fiber. Filament products include single filament and spun-bonded non-woven fabrics.
Source: From Chinese Chemical Fiber Association, *Fiber Organon*, **2003**, 6.

3.1.2.1 Worldwide Growth

The evolution of polypropylene (PP) fiber followed other synthetic fibers, but developed rather rapidly. The production of polypropylene fiber was commercialized in 1959.

In 1981, about one third of the total polypropylene usage was based on conventional staple and multifilament products; about 60 KT (thousand tons) of ST and 54.5 KT of multifilament yarn. The largest textile end-use of polypropylene was in carpet backing, which was made primarily from ribbon yarns. By 1995, its production surpassed aliphatic polyamides (PA) and polyacrylonitrile (PAN), but lagged behind PET. In 2002, the production of polyester fiber was 62%, polypropylene fiber, 17.5%, polyamide fiber, 11.5%, and polyacrylonitrile, 18.1%, of all synthetic fibers.

Table 3.1 presents the worldwide production of polyolefin fibers in 1996–2002, according to Fiber Organon [1]. In 2002, the total production of polyolefin fibers grew to 5.913 million tons (MT). The annual rate of growth was 2% as compared to 8% for polyester, 6.4% for PAN, and 4.4% for polyamides. The production of polypropylene fibers, excluding fibrillated fibers, reached 3.99 MT in 2002 and 4.20 MT in 2003. About 70% of its production was in filament yarn and 30% in staple. As Table 3.1 also shows the portion of polypropylene filaments increased from 40.4% in 1996 to 45.5% in 2002, while fibers from fibrillating film decreased from 35.2 to 32.5%. The portion of ST remained steady at 22.1%.

3.1.2.2 Regional Productivity

Table 3.2 lists the development of polyolefins in various countries in 1996–2002. Polypropylene staple maintained a high growth rate of 7%, particularly in West Europe and Taiwan, where the growth rate reached 20 and 15%, respectively. Filament yarn grew by 1%, which consisted of 4% in China and South Korea, 5% in the remaining Asia region, 3% in West Europe, and 1% in Taiwan and Latin America.

The productivity of polypropylene in West Europe increased from 1.70 MT in 1999 to 1.80 MT in 2002, while it increased from 1.35 to 1.85 MT in 2002 in the United States. After 25 years of developmental efforts, China became the second largest producer of polypropylene after the United States. Its production of polypropylene fibers, excluding fibrillated fibers, was 561 KT. The productivity share was 165% for China and 22.5% for the United States. The growth was fastest in the Middle East and Africa. In the Middle East, the production of polypropylene grew from 0.75 MT in 2000 to 1.95 MT in 2001 due to new investments by Saudi Arabia. The consumption of polypropylene in the Asia Pacific region increased by 10% in 2000 and 5% in 2001, thus draining the export capacity from Saudi and Iran.

TABLE 3.2
Regional Production of Polyolefins

Area		USA	West Europe	Asia[a]	China	Latin America	Total
1996	Production, (KT)[b]	1086 (877)	1160 (576)	1539 (835)	876 (516)	236 (102)	4521 (2869)
	1996/1992 annual growth (%)	4.9	3.7	9.7	10.6	9.1	6.5
	World market share (%)	24.2	25.6	34.0	19.3	5.2	100
1999	Production (KT)	1340 (1009)	1700 (1295)	1630 (897)	916 (544)	328 (145)	5518 (3553)
	World market share (%)	24.2	30.8	29.5	16.5	5.1	100
2002	Production (KT)	1353 (1013)	1346 (1202)	1855 (958)	945 (561)	350 (173)	5913 (3994)
	World market share (%)	22.5	22.7	31.0	16.0	5.9	100

[a] Asia includes China.
[b] Figures in brackets do not include fiber obtained by fibrillating film.
Source: From Chinese Chemical Fiber Association, *Fiber Organon*, **2003**, 6.

3.1.2.3 Other Synthetic Fibers

As shown in Table 3.3, the productivity of all synthetic fibers in 2002 was 34.906 MT at an annual growth rate of 4.2%, and that of other chemical fibers was 349 KT at a growth rate of 4.2%. This includes 240 KT polyurethane (PU), 40 KT polyvinyl chloride (PVC), 13 KT poly(*p*-phenylene benzo*bis*thiazole) (PBT), 10 KT hydrophilic fibers, and 53 KT high-performance fibers, which includes 35 KT aramid fibers.

The past growth of polypropylene and that projected for the future are due to the improvements in the properties of fibers made from the resin and an increased knowledge of how to take advantage of some of its unique characteristics.

3.2 MAJOR END-USES

More than any other synthetic fiber, polypropylene fibers owe their existence to polymer additives that perform various tasks that pure resin is unable to accomplish. When prepared, the polymer is quite unstable, and at melt temperatures it will decompose without stabilizers. Fibers will degrade over longer periods of time even at moderate temperatures if long-term stabilizers are not added. Polypropylene fibers are also very sensitive to UV light, which will promote oxidative degradation unless other stabilizers are added. For these reasons, the earliest polypropylene textile products were made in thick cross sections and put into end-uses that were not very demanding. Through most of its life as a commercial product up to the mid-1970s, polypropylene fiber had a reputation as a product with poor thermal stability and relatively poor UV stability, and as difficult to obtain in desirable colorations. The low melting point was also considered to be a deficiency.

This view of polypropylene, while partially deserved in its early history, has not been true since the early 1980s because of improved polymerization conditions, stabilizers, and pigments. It was recognized that both the chemical inertness that prevents dyeing and the low melting point can be advantageous in many products.

The melting point of polypropylene is an advantage in many new nonwoven products. Polypropylene fibers can be melted sufficiently to bond to one another without destroying fiber properties. Nonwoven fabrics made from polypropylene can, therefore, be fusion-bonded,

TABLE 3.3
Worldwide Production of Synthetic Fibers in 2002

Region fiber type	China Thousand tons	China Annual growth %	USA Thousand tons	USA Annual growth %	West Europe Thousand tons	West Europe Annual growth %	Japan Thousand tons	Japan Annual growth %	Worldwide Thousand tons	Worldwide Annual growth %	Worldwide Share (%)
Chemical fibers[a,b]	9815.5	+20.1	4,162		3647.8		1,525	−9.5	34,906	+6.0	100
Viscose	680	+12.8	47.0	−26	393.9		162.2	−2.1	2715	−0.6	
ST		(6.8)			316.0	+4.3	38.8	−45.0			
F					77.9	−20.6					
Acetate	—		34.0	−14.8			109	−0.7	600	+1.0	(1.9)
Synthetic fibers	9135.5	+20.7			3253.9		1,254		31,591	+6.2	
PET	7721.6	+22.8	1,472		842.3	+2.0	563.8	+8.1	20,950	+8.1	(65.3)
F	4771.6	(78.0)	542	+2.1	391.3	−5.7	323.0	−8.4	12,140	+8.4	
ST	2950		930	+0.5	451.0	+6.6	240.7	−8.7	8,810	+6.8	
PA	474.9	+12.1	1,106		499.3		132.8	+4.4	3,905	+4.4	(11.5)
F		(4.79)	797	+7.1	411.1	−3.5	125.9	−22.5	—		
ST			309	+12.5	88.2	−11.9	64	−21.0	—		
PAN	594.0	+11.3	150		566.3	+2.3	358	−1.9	2,742	+6.4	(8.1)
Subtotal		(5.9)									
PP	299	+6.5	1,353	+1.0	1,346	+7.0	124.4	−14.0	3,994	+2.0	(11.8)
	(561.0)	(5.7)									
F			1,013	−2.0	876	+3.0	75	−17.0	2,694	+1.0	
ST			340	+4.0	570	+15	49	−11.9	1,300	+7.0	
PVC	20.0						32.5	−2.1	—		
Others									349	+4.2	
PU	26.0						42.5	−2.3	240		

[a] Data do not include fiber obtained from fibrillating film.
[b] ST: staple; F: filament yarn; PA: polyamide; PAN: polyacrylonitrile; PVC: polyvinyl chloride; PU: polyurethane.

Source: From Chinese Chemical Fiber Association, *Fiber Organon*, **2003**, 6.

eliminating the need for chemical binders [2]. Such a process is more economical not only because of the increased cost of binders but also because the energy levels for removing water from bonded fabrics are much higher than that required for fusion bonding, and because wastewater problems are eliminated. Use of thermally bonded cover stock in baby diapers and similar products will result in markedly increased use of polypropylene [3]. The fusion characteristics of polypropylene are used not only to bond carded webs but also to improve the dimensional stability of needle-bonded fabrics [4]. A large variety of civil engineering fabrics for road stabilization and dam and dike reinforcement, soil stabilization, and roofing are made from polypropylene fibers. Other major end-uses for nonwoven polypropylene fabrics are in furniture construction and as substrates for vinyl fabrics. The use of polypropylene fibers in nonwovens is expected to grow even more rapidly because of the development of new fibers for that end-use. Special fibers with greater temperature windows for fusions, the gap between the temperatures of fusion bonding and melting, have been prepared [5] especially for nonwovens.

The thermal stability of polypropylene fabrics at temperatures below melting point (120–135°C) has been improved markedly by the addition of stabilizers. This, along with the inherent stability to a wide variety of chemicals, has allowed both woven and nonwoven polypropylene fabrics to be used in filtration and a wide range of other industrial uses. At temperatures below the melting point, the stability of polypropylene fabrics rivals any but the most expensive high-performance fibers.

The use of polypropylene fibers will grow depending on the ability of the textile industry to take advantage of properties that can be built into polypropylene. Some of these properties, such as toughness, low density, chemical stability, etc., are inherent to all polypropylene products. Others, like color stability and UV and thermal stability, are built into the product by additives. Still others, such as improved uniformity due to low-temperature extrusion of modified polymers, open new areas of use. The future growth of polypropylene in textiles depends on the ability of the suppliers and users to take advantage of all the properties to make unique products from polypropylene.

3.2.1 CARPET AND FURNISHING

3.2.1.1 Carpet

The worldwide production of carpet fibers was 2.1 MT in 2002 at an annual growth rate of 6.7%. About 80% of carpet fiber production is attributable to North America and Europe. Polypropylene base fiber for carpets accounted for 1.16 MT and nylon polyamide fibers for 0.94 MT. In 2002, the total production of polypropylene fibers for carpets and furnishings was above 1.75 MT.

The carpet production in the United States in 1998 was 1,471,536,000 sq. m., of which 41% was for domestic furnishing, and 73% of the base fiber was nylon, 39% polypropylene, 7% polyester, and 1% wool. In comparison, the consumption of polypropylene tufted rug in China decreased from 30 million sq. m. in 1995 to 21 million sq. m. in 2000, while nylon carpet fiber increased from 10 million sq. m. to 35 million sq. m. It is expected that the total carpet consumption will increase from 70 million sq. m. in 2000 to 85 million sq. m. in 2005.

The consumption of polypropylene carpet in West Europe is about 0.5 MT, which is about 50% of all flooring materials for 2.6 billion sq.m. It is second to 3.6 billion sq.m. for nylon, whereas polyester is only used for bathroom rugs. Among the flooring materials used in West Europe, tufted carpet accounts for 60%. Among 500,000 tons of polypropylene carpet, tufted carpet accounts for 150 KT, machine woven, 130 KT, needle-punched, 100 KT, and carpet backing, 100 KT.

3.2.1.2 Home Furnishing

It has been pointed out that "inherent defects, such as lack of dyeability and low melting points, were retarding factors" with regard to market growth of polypropylene fibers [6]. Although a great deal of research effort was applied to creating dyeable forms of polypropylene, virtually all colored polypropylene products are prepared with pigments. About two thirds of the multifilament yarn produced in the early 1980s was used in home furnishing, for carpet or upholstery, and most of that was pigmented. Because polypropylene could not be dyed, much effort was applied to producing better pigments for textile end-uses. These improved pigments, along with better thermal and UV stabilizers, have permitted polypropylene fibers to be used in areas where dyed fibers could not perform satisfactorily.

Polypropylene fabric for furnishing has been produced primarily by machine weaving. The current trend is to produce furniture upholstery with polypropylene tufted fabric of medium-denier BCF, air-textured yarn, and draw-textured yarn. More than one fourth of polypropylene upholstery is produced by needle punch. Polypropylene upholstery is widely used for sofa covering. Although the air-textured yarn is most economical, it is also used in combination with the BCF yarn. About 40% of polypropylene fabric is used in furnishing, of which 14% is used in cars. Among the automotive fabrics consumed in West Europe in 2000, 41.7% was polyester, 26.1% nylon, 13.9% polypropylene, and 11.8% viscose. It is estimated that the consumption will increase from 259 KT in 2002 to 302.9 KT in 2006.

3.2.1.3 Automotive Furnishing

In recent years, the quality of polypropylene fiber has been improved to enhance such properties as light weight, moldability, light stability, coloration, recovery feasibility, flame resistance, and antistatic propensity. It is increasingly used for automotive interior furnishing and parts.

Flooring material accounts for one fourth of the automotive interior furnishings. The consumption of carpet has increased from 6 sq.m. per car 7 years ago to 7.3 sq.m. per car presently. For an annual production of 5.56 million automobiles of which 2.5 million are sedans, the total consumption of textile products would exceed 100 KT at 15–20 kg of fibers per car.

Consequently, worldwide consumption of polypropylene and polyester fibers will continue to increase phenomenally, while polyamide fibers may suffer large declines. With the boom in the housing sector, polypropylene BCF, particularly the silicon-treated version, will be increasingly used in hermetically sealed windows at an extra cost. This appears to be an irreplaceable new opportunity for polypropylene fiber in the future.

3.2.2 Nonwoven Fabrics

The worldwide production of nonwoven fabrics in 2002 was 3.625 MT, with an annual growth rate of 10% and a total sales value of $9.3 billion. The United States accounted for 1.282 MT or 41% of the total productivity, West Europe 1.203 MT or 30%, and Japan 0.29 MT or 8%. These fabrics had a total area of 32.2 billion sq.m.

Countries that are major producers of nonwoven products have heavily used these products in health and medical applications. According to 1999 statistics, 38% by weight was devoted to these usages in West Europe, 31% in the United States, and 23% in Japan.

Polypropylene provided 44% of base fiber for the nonwovens in 1990 and increased to 62.7% in 2000, while polyester accounted for 22.5%, nylon, 1.5%, PAN, 2.0%, and viscose, 8%. The share of polypropylene nonwovens was 44.5, 33.0, and 44.5% in West Europe,

United States, and Japan, respectively. According to *Non-Wovens World*, melt-spun nonwoven products increased from 118 to 136 KT worldwide in 1999. About 26%, or 31 KT, of these products was used for filtration. China is rapidly catching up with 10% of the world productivity in this product segment.

It is estimated that the worldwide consumption of nonwoven products will reach 6.3 MT by 2010, of which Europe and the United States will account for 55%. With a rapid annual growth of 10%, China will contribute about 24% of this productivity.

3.2.3 Industrial Applications

Fabrics used in the automotive industry must meet exceptionally high requirements and be subjected to nonstandard tests that simulate the high-temperature and humidity conditions of light exposure [7]. Dyed fibers cannot meet these accelerated exposure specifications, which are met by specially stabilized, pigmented polypropylene fibers. The outdoor use of polypropylene fibers is expected to grow because of improved stabilizers. It has permitted grass-substitute products to be manufactured in a variety of forms. Many of them resemble carpets, which, until 1980, were used only indoors.

Polypropylene accounted for about half of the nonwoven products in industrial uses in 2001. Its share in ropes and nets was 55–60%, and 70–80% in civil construction, where polyester claimed 20–24%. In automotive applications, polypropylene shared 26–30%, nylon almost 50%, and polyester about 20%. Polypropylene contributed to over 86% of agricultural nonwoven, 100% of packaging cloth, 85% of sanitary items, and 64–70% of medical applications. The world consumption of industrial nonwoven products was 1.329 MT in 2000. Polypropylene topped all synthetic fibers with a share of over 40% in this market segment.

3.2.3.1 Ropes

Modern designs and fiber science have segregated the rope and cable industry from the low-priced products. Surpassing the traditional metallic products, ropes and cables of synthetic fibers have now reached a strength–weight ratio of 10:1. Thus, these high-technology products offer a great potential market in oceanography and oil exploitation for the rope industry.

3.2.3.2 Civil Construction

Nonwoven products for use in civil construction amounted to 1.4% in 1985 and increased to 2.3% in 1995. In 1995, about 196 KT of nonwoven construction fabrics were consumed in the United States, 140 KT in West Europe, and 15 KT in Japan. Its market share is estimated to increase to 3.5% by 2005. This growth has been attributed to the wide adoption of nonwoven products in engineering projects on rocky grounds.

3.2.3.3 Cement Reinforcement

The use of synthetic fiber for the reinforcement of road pavement not only enhances strength but also lengthens the useful life from normally 10–15 years to 20–25 years for concrete pavement, and from 3–5 years to 5–10 years for asphalt pavement. In addition, the cost of road maintenance is drastically lowered with fiber reinforcement. Mostly, polypropylene ST is used in the reinforcement of concrete pavement, while polyester ST in asphalt pavement.

Dura fiber that was developed by Hill Brothers Chemical Co. in the United States has been widely adopted for cement reinforcement in the industrialized countries. It has triggered a great deal of interest that led to the development of high-performance fibers for cement reinforcement.

3.2.3.4 All Synthetic Paper

With the growth of high technologies in the 21st century, traditional paper products can no longer meet the ever-increasing requirements for high performance. Faced with this challenge, the paper industry has been searching for high-performance papers with value addition.

Polypropylene paper exhibits a number of attractive properties in addition to those of ordinary papers. These include good appearance and hand, freedom from molding, permeability, high strength, abrasion resistance, tear and penetration resistance, waterproof, and pollution resistance. It can be used for books, maps, accounting books, high-quality money market records, and mailing packages. It can also be used for packaging of food, drinks, and clothes. Pressed polypropylene paper is ideal for use in disposable food containers that can withstand microwave heating and is almost fully recyclable. In the current international market, polypropylene paper has replaced 70% of ordinary paper as the packing material.

There are two types of polyolefin paper at present: polyethylene and polypropylene. Commercial products include the SWP polyolefin papers developed jointly by Mitsui Petrochemicals Inc. and Crown Zellerboch Co., and Pulpex and Perlosa paper products by Hercules Co.

3.2.4 APPAREL

Polypropylene has penetrated the apparel market to a modest degree. Some polypropylene yarn has been used as backing yarn for sliver-knit products. Woven fabrics of polypropylene fiber, particularly of the low-denier (<1.5 dpf), are effective for thermal protection, hand, and wicking. They provide good aeration during perspiration and are waterproof. In apparels, their use can prevent skin dryness and body odor from perspiration, thus providing comfort and personal hygiene. The current production of fine-denier polypropylene fiber is about 4000 tons a year. Its major usage lies in needle-punched military diving suit, winter wear, sports wear for bowling, hiking and skiing, work clothes, underwear, and underpants. A smaller amount of this fiber is used in blends with nylon or acetate fibers for fashion wear. Fine- and superfine-denier polypropylene fibers have been greatly favored in the last 10 years.

The ultimate volume of polypropylene used in apparel depends on the balance of performance and economics relative to nylon and polyester, and this balance is still under evaluation.

3.3 PREPARATION OF PROPYLENE POLYMER

3.3.1 BASIC POLYMERIZATION REACTION

Propylene is catalytically polymerized to form a linear, flexible-chain polymer according to the following equation:

$$n\mathrm{CH}_3\text{—CH}=\mathrm{CH}_2 \xrightarrow{\text{catalyst}} -[-\mathrm{CH}_2-\mathrm{CH}-]_n-$$

Propylene → Polypropylene (with pendant CH_3)

The polymer contains a repeat unit of $-CH_2-HC(CH_3)-$ from the opening of its double bond, and the formation of a pendant methyl group. The reaction is exothermic. Its heat of reaction is in the order of 20.0 kcal/mol or 83.6 kJ/mol:

As will be discussed later, the polymerization of propylene is catalyzed by either Ziegler–Natta catalyst or metallocene catalyst. In the Ziegler–Natta catalyzed polymerization, the propylene monomer accesses the reaction site between the metal and carbon atoms in two ways: the head-to-tail and tail-to-head insertion. The head-to-tail insertion occurs primarily in isotactic polymerization [8–10], while the tail-to-head insertion occurs in the low-temperature catalysis with VCl_4–Et_2AlCl catalyst [11]. There are mostly head-to-tail insertions and very few head-to-head or tail-to-tail inversions in the main chain of polypropylene [12,13].

There are three types of monomer insertions with respect to the pendent methyl groups: the meso, racemic. [14]. Meso insertion produces a polymer with the methyl groups in the same spatial position, which is referred to as *isotactic* polymer; racemic insertion produces a polymer with the methyl groups in alternating locations, referred to as the *syndiotactic* polymer. When the monomer insertion is random and nonstereospecific, a noncrystalline *atactic* polymer is produced. These three forms of polypropylene are schematically represented in the following chain configurations:

1. Isotactic polypropylene

$$-CH_2-CH-CH_2-CH-CH_2-CH-CH_2-CH-CH_2-CH-CH_2-$$
$$\quad\;\;|\qquad\quad\;|\qquad\quad\;|\qquad\quad\;|\qquad\quad\;|$$
$$\quad CH_3\quad\; CH_3\quad\; CH_3\quad\; CH_3\quad\; CH_3$$

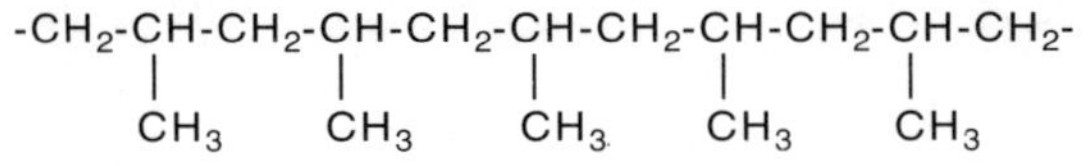

2. Syndiotactic polypropylene

$$\qquad\qquad CH_3\qquad\qquad\;\; CH_3\qquad\qquad\;\; CH_3$$
$$\qquad\qquad\;\;|\qquad\qquad\qquad|\qquad\qquad\qquad|$$
$$-CH_2-CH-CH_2-CH-CH_2-CH-CH_2-CH-CH_2-CH-CH_2-CH-$$
$$\quad\;\;|\qquad\qquad\qquad|\qquad\qquad\qquad|$$
$$\quad CH_3\qquad\qquad\; CH_3\qquad\qquad\; CH_3$$

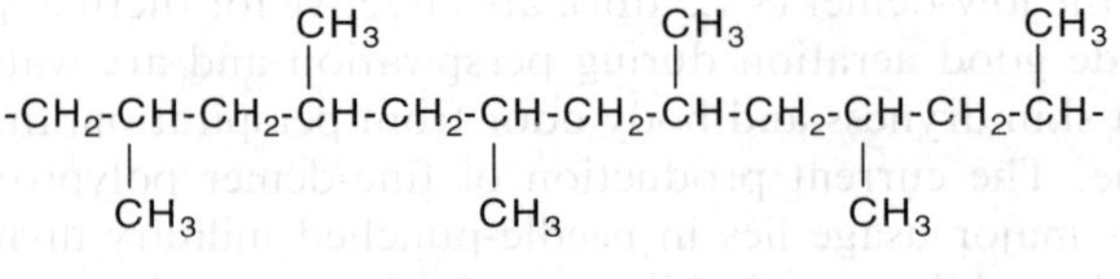

3. Atactic polypropylene

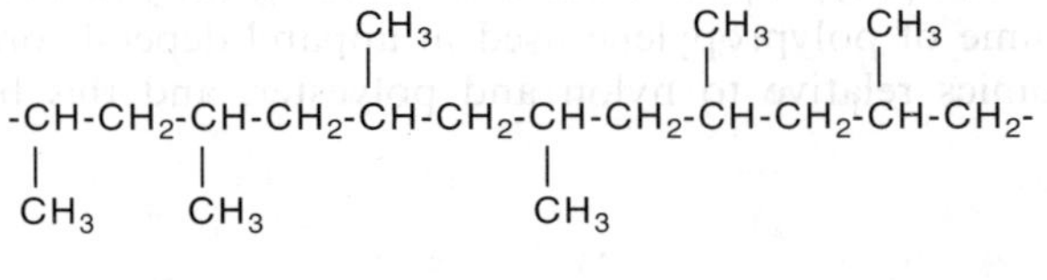

$$\qquad\quad CH_3\qquad\qquad\qquad CH_3\;\; CH_3$$
$$\qquad\quad\;|\qquad\qquad\qquad\quad\;|\qquad\;\;|$$
$$-CH-CH_2-CH-CH_2-CH-CH_2-CH-CH_2-CH-CH_2-CH-CH_2-$$
$$\;\;|\qquad\quad\;|\qquad\qquad\qquad|$$
$$CH_3\quad\; CH_3\qquad\qquad\; CH_3$$

The relative amount of these three types of polymers depends greatly on the catalyst type and its activity. Ziegler–Natta catalyst generally possesses active centers for stereoregular and random insertion [15,16]. The multiactive centers on the surface of the catalyst induce different rates of chain growth during polymerization. This leads to a relatively broad molecular weight distribution with Ziegler–Natta catalysts. In contrast, metallocene catalysts generally give rise to polymers of narrow molecular weight distribution and high stereoregularity. These types of catalysts are thought to have a single active center, or contain a small amount of other active centers [17–19].

3.3.2 REACTION MECHANISM

The mechanism for the catalytic polymerization of propylene is represented by the following equations:

Polypropylene Fibers

1. Initiation

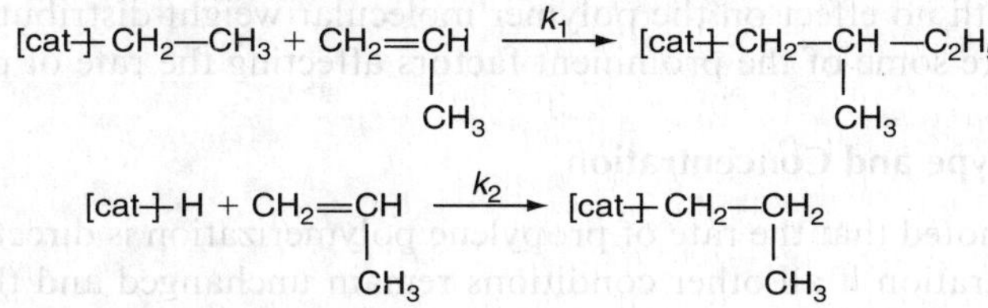

$$[\text{cat}]\text{-CH}_2\text{-CH}_3 + \text{CH}_2=\text{CH}(\text{CH}_3) \xrightarrow{k_1} [\text{cat}]\text{-CH}_2\text{-CH}(\text{CH}_3)\text{-C}_2\text{H}_5$$

$$[\text{cat}]\text{-H} + \text{CH}_2=\text{CH}(\text{CH}_3) \xrightarrow{k_2} [\text{cat}]\text{-CH}_2\text{-CH}_2(\text{CH}_3)$$

2. Chain growth

$$[\text{cat}]\text{-CH}_2\text{-CH}_2(\text{CH}_3) + n\text{CH}_2=\text{CH}(\text{CH}_3) \xrightarrow{k_p} [\text{cat}]\text{-}(\text{CH}_2\text{-CH}(\text{CH}_3))_n\text{-CH}_2\text{-CH}_2(\text{CH}_3)$$

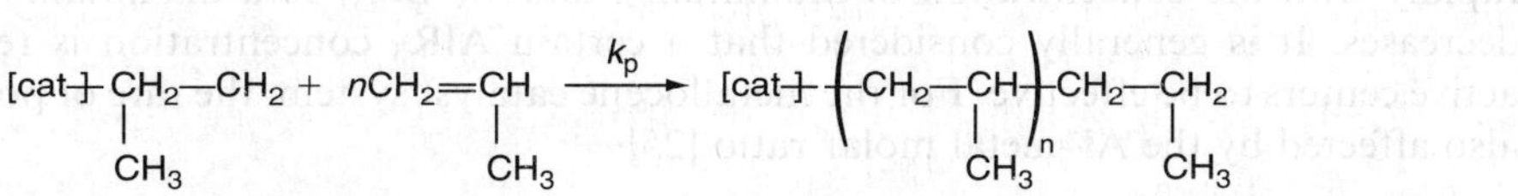

3. Chain transfer and termination

A growing chain generally proceeds through the following reactions to its termination:

(a) Elimination of β-H group

$$\text{Mt}-\text{CH}_2-\text{CH}(\text{CH}_3)-\text{P} \xrightarrow{k\phi-H} \text{Mt}-\text{H} + \text{CH}_2=\text{C}(\text{CH}_3)-\text{P}$$

(b) Elimination of β-CH$_3$ group

$$\text{Mt}-\text{CH}_2-\text{CH}(\text{CH}_3)-\text{P} \xrightarrow{k\phi-\text{Mt}} \text{Mt}-\text{CH}_3 + \text{CH}_2=\text{CH}-\text{P}$$

(c) Transfer to monomer

$$\text{Mt}-\text{CH}_2-\text{CH}(\text{CH}_3)-\text{P} + \text{CH}_2=\text{CH}-\text{CH}_3 \xrightarrow{k_{\text{Mt}}} \text{Mt}-\text{CH}_2-\text{CH}_2-\text{CH}_3 + \text{CH}_2=\text{C}(\text{CH}_3)-\text{P}$$

(d) Transfer to activated hydrogen compound

$$\text{Mt}-\text{CH}_2-\text{CH}(\text{CH}_3)-\text{P} + \text{H}_2 \xrightarrow{k_{\text{Mt}_2}} \text{Mt}-\text{H} + \text{CH}_3-\text{CH}(\text{CH}_3)-\text{P}$$

(e) Transfer to alkyl-metal site

$$\text{Mt}-\text{CH}_2-\text{CH}(\text{CH}_3)-\text{P} + \text{AlR}_3 \xrightarrow{k_{\text{tMt}}} \text{Mt}-\text{R} + \text{R}_2\text{Al}-\text{CH}_2-\text{CH}(\text{CH}_3)-\text{P}$$

where [cat] denotes catalyst, Mt the transition metal, and P the polymer chain.

When a polymer chain stops its growth after chain transfer, an active center is vacated to allow the formation of a new polymer chain. The chain transfer by the elimination of the β-H group is not important for most Ziegler–Natta catalysts, but it is the major chain termination reaction for most metallocene catalysts. The elimination of the β-methyl group does not occur in multiphase catalysis, but is the most important chain termination mechanism for the metallocene catalysts containing Cp$_2$MCl$_2$–MAO, where M is zirconium (Zr) or hafnium

(Hf). With the addition of hydrogen, the chain transfer to the monomer is the main termination reaction for the Ziegler–Natta catalyst system. Thus hydrogen is the most effective chain terminator with no effect on the polymer molecular weight distribution.

The following are some of the prominent factors affecting the rate of polymerization:

3.3.2.1 Catalyst Type and Concentration

Natta et al. [20,21] noted that the rate of propylene polymerization is directly proportional to the catalyst concentration if all other conditions remain unchanged and the effect of monomer diffusion is not considered. The metal alkyls and the Al–metal molar ratio of a catalyst system also affect the rate of polymerization [22–24]. The polymerization rate increases rapidly with the concentration of aluminum trialkyl (AlR_3) to a maximum value and then decreases. It is generally considered that a certain AlR_3 concentration is required for the active centers to be effective. For the metallocene catalyst system, the rate of polymerization is also affected by the Al–metal molar ratio [25].

3.3.2.2 Monomer Concentration and Hydrogen

In a broad range of Ziegler–Natta reaction conditions, the rate of polymerization is a first-order function of monomer concentration, except at very low concentrations [26]. However, the rate of metallocene-catalyzed polymerization is of a higher order of reaction. This is possibly caused by localized heating at high monomer concentrations [27].

The addition of hydrogen regulates the molecular weight of polypropylene while, at the same time, it affects the polymerization rate. These effects vary with different catalyst systems. For the α-$TiCl_3$–$AlEt_3$ catalyst, the reaction rate may decrease on hydrogen addition [28,29]. With the Solvay-type $TiCl_3$ catalyst, however, the reaction rate may increase as much as 50% at a high partial pressure of hydrogen [21]. The addition of hydrogen also increases the reaction rate in the presence of $MgCl_2$ catalyst carrier [30–32] and the metallocene catalyst $EBInZrCl_2$–MAO [33]. Although the addition of hydrogen is primarily aimed at accelerating the initiation reaction, it actually suppresses the reaction rate in the presence of aromatic esters [34].

3.3.2.3 Polymerization Temperature and Time

The rate of polymerization increases with increasing temperature within a given temperature range. At the maximum temperature limit, the catalyst gradually deactivates and the reaction rate declines.

Some catalysts may hold the reaction rate steady for a long period of time after a short acceleration; others may allow the reaction rate to decline with time after the induction period [35,36].

3.3.2.4 Polymerization Medium

The effect of polymerization medium is not obvious with Ziegler–Natta catalysts, but can be quite different with homogeneous metallocene catalysts. Increasing the polarity of the polymerization medium helps increase the activity of metallocene anions [37].

3.3.3 CATALYST SYSTEMS

The growth of polypropylene has largely depended on the development of its polymerization catalysts. Following the invention of Ziegler catalyst, the refinement by Natta and the subsequent introduction of metallocene catalysts brought about significant improvement in catalyst efficiency and polymer quality. A highly stereospecific polypropylene resin, with a high isotactic content, is generally preferred to produce fiber, film, and composite products with high performance. The Ziegler–Natta and the metallocene catalyst systems met this need in two giant steps.

3.3.3.1 Ziegler–Natta Catalysts

In 1953, Ziegler [38] first employed aluminum trialkyl–titanium tetrachloride (R_3Al–$TiCl_4$) catalyst to prepare stereo-irregular polypropylene. Later in 1954, Natta [39,40] refined the Ziegler catalyst with aluminum trialkyl–titanium trichloride (R_3Al–$TiCl_3$) catalyst and successfully prepared highly stereoregular polypropylene. The new catalyst rendered polypropylene the possibility of practical usefulness. In 1957, it led to the construction of the first production facility for polypropylene by Montecatini Co. in Italy.

Since its inception 50 years ago, the Ziegler–Natta catalyst system has been developed to its third and fourth generations of formulation. The new catalysts are many times more reactive than the initial version; they also yield a polypropylene of high stereoregularity. In the Ziegler–Natta formulation, Natta used $AlEt_3$ to reduce $TiCl_4$ in the Ziegler catalyst to form $TiCl_3$ crystals that, in turn, catalyze the polymerization reaction to form polypropylene of 90% isotacticity. The next generation of Ziegler–Natta catalyst added a Lewis base and increased the reactivity four- to fivefold. It is exemplified by a Solvay catalyst developed by Solvay Co. that employed diethyl aluminum chloride (DEAC) as a cocatalyst. This composite catalyst had a specific surface area as high as 150 m^2/g [41,42]. In the third generation of the Ziegler–Natta catalysts, $MgCl_2$ was used as the catalyst support. The preparation of polypropylene was greatly simplified with this catalyst by the elimination of removing the atactic polymer and catalyst residues. Even the palletizing process could be eliminated in some cases. The key to that formulation is the use of activated $MgCl_2$ as the catalyst support [43]. The inclusion of a suitable cationic donor could further increase the stereoregularity of the resulting polypropylene [44]. The fourth generation of the Ziegler–Natta catalyst is the "reactor granule" technology that provides high efficiency and high degree of stereoregularity, and enables the control of polymer particle formation and growth. For example, a spherical catalyst carrier of Himont Co. [45] could allow propylene to copolymerize with another monomer. Mao et al. [46–49] developed a solid catalyst system that consisted of (1) a solid mixture of titanium tetrahalide and an organic aluminum compound, magnesium halide, organic epoxy compound, (2) an organic phosphorous compound, and (3) an organic silicon compound. These appear to be an effort towards a supported catalyst system.

3.3.3.2 Metallocene Catalysts

In the 1980s, Sinn et al. [50], Ewen [51,52], Kaminsky [53,54], and others [55] reported that, using methylaluminumoxane (MAO) as a catalyst activator, certain metallocene complexes not only constitute a highly activated polymerization catalyst but also give highly isotactic or syndiotactic polypropylene. Albizzati et al. [56] at Montell Co. developed a solid catalyst containing active magnesium dichloride that was supported on a titanium tetrachloride and an electron donor ether such as 2,2-diisobutyl-1,3-dimethoxy propane. The new formulation rendered the catalyst very high activity and stereoregularity without the addition of an electron-donor compound. The metallocene compounds have finite, single, active sites that exhibit strong selectivity for the steric conformation of the monomer. Thus, the compounds of transition metals such as titanium (Ti), zirconium (Zr), and hafnium (Hf) have been extensively investigated as the activated catalyst site. [57] This new class of supported catalysts is now known as the *metallocene catalysts*.

TABLE 3.4
Examples of Metallocene Catalyst Systems

Catalyst composition	Effect	Reference
Methylalumoxane (MAO) activated bis(cyclopentadienyl)zirconium dimethyl	Good catalytic efficiency	[50]
Alumoxide and trimethyl aluminum, and bis(cyclopentadienyl)titanium diphenyl	Polypropylene with isotactic-stereoblock units	[52]
Methylalumoxane and bis(cyclopentadienyl)zirconium dichloride	Commercially available starting material for catalyst and high reaction activity	[53]
Methylalumoxane and a bridged metallocene of zirconium dichloride with an unsubstituted and a substituted cyclopentadienyl rings	Highly syndiospecific catalyst and unique polymer microstructure	[58]
Titanium tetrachloride and electron-donor methylcyclohexyldimethoxysilane (MCMS)	High catalyst stability and efficiency	[59]
A solid catalyst of active magnesium dichloride supported on titanium tetrachloride and solid electron-donor such as 2,2-diisobutyl-1,3-dimethoxypropane	Highly active and stereospecific catalyst	[56]
A cationic metallocene ligand with sterically dissimilar cyclopentadienyl rings joined to a positively charged coordinating transition metal atom, and a stable noncoordinating counter anion for the metallocene cation	Highly syndiotactic	[60]
2-Phenyl-6,6-dimethylfulvene and 2-(phenyl cyclopentadienyl)-2-florenyl propane	A stereorigid metallocene compounds as catalyst	[61]
Catalyst activator complexes formed by reacting a $Al(C_6F_5)_3$ with racemic diol	An alternative activator	[62]
Polymer-coated metallocene catalyst and its support	High catalyst surface area	[63]
A mixed catalyst system containing a cyclopentadiene-based ligand as the poor comonomer incorporating catalyst	Preferably used in gas-phase polymerization for copolymers	[64]
Two catalyst systems used separately in a two-stage polymerization process	Mixed polymers with different properties	[65]
A modified catalyst of bis-n-butyl-cyclopentadienyl zirconium dichloride activated by MAO on silica support	Lower polymer melting or softening temperature	[66]

There have been numerous formulations of the metallocene catalysts in the last 20 years. Table 3.4 presents a few examples of metallocene catalyst systems of interest (see all preceding references) [58–66]. It can be seen that earlier efforts were focused on catalyst efficiency and polymer stereospecificity. More recently, some interests seemed to have turned to syndiotactic polypropylene for various polymer and fiber properties.

Table 3.5 compares the key features of Ziegler–Natta and metallocene catalyst systems.

3.3.4 Polymerization Processes

3.3.4.1 Basic Processes

Dozens of polymerization methods have been developed since the commercialization of polypropylene. These methods can be grouped into four classes: slurry, solution, bulk, and

TABLE 3.5
Comparison of Ziegler–Natta and Metallocene Isotactic Polypropylene

Polymer type	ZNiPP	MiPP
Time period	1950s	1980s
Catalyst system	Ziegler–Natta	Metallocene
Catalyst components	Transition metal compounds: Ti, V, Cr, Ni, etc. Alkyl–metal catalyst	Transition-metal organic compounds Zr, Ti, Cr, etc., bonded cyclopentadienyl ring
Distinctive structure	$TiCl_3$, $R_2O/TiCl_3$	$MgCl_2/TiCl_4/Ph(COOiBu)$
Activity	$5 \times 10^3 \sim 1 \times 10^4$ g/gTi	$3 \times 10^4 \sim 6 \times 10^4$ g/gTi
Reaction mechanism	Polycoordination, multiple polymerization centers	Polycoordination, single polymerization center, high catalytic efficiency
Polymer properties		
Degree of isotacticity	~90%	≥98%
Molecular distribution	Wide: $\overline{M}_w/\overline{M}_n = 3-6$	Narrow: $\overline{M}_w/\overline{M}_n = 2$
Melting point	Controllable range: 160–164°C High melting point: 162°C	Controllable range: 130–160°C Low melting point: 148°C
Stereospecificity	Isotactic products (iPP) only	Low atactic content Isotactic or syndiotactic homopolymer, or isotactic/syndiotactic copolymer
Spinning performance	Poor: low elasticity, low viscosity, large degree of nozzle bulking	Good: good spin-stretch, easy processing
Producer	Hoeches, Germany: Low pressure PE Montecatini, Italy: PP	Exxon, USA: PP Hoechest, Germany: PP Knapsack, Germany: PP Mitsui, Japan: PP

gas-phase processes. The process technology has been continually improved with the progress in catalyst chemistry. Thus the earliest postpolymerization process included the removal of catalyst residue and atactic polymers, drying, and palletizing. The catalyst and atactic portions removals have already been eliminated with the new generations of process technology. The next step of process simplification will be to eliminate palletizing.

Slurry polymerization was the earliest method of polymer preparation. Montecatini Co. of Italy first commercialized the Moplen batch slurry process in 1957. A solvent is used in this process to keep crystalline polypropylene particles in suspension but dissolve the amorphous polymers. A highly stereoregular polypropylene is obtained after separating the catalyst by centrifuging. The solvent is separated from the amorphous polypropylene by evaporation. The initial effort by Montecatini was followed by Hercules, Mitsui East Asia Chemical Corp., and Amoco to further refine the slurry method [8,9,67,68]. In addition to using hexane or heptane for catalyst removal, as in the Montecatini process, the Hercules technology involves the addition of hydrogen to the reactor to control the molecular weight of the polymer. As ethanol is generally used in catalyst removal, a considerable amount of process equipment is required for the purification and recovery of ethanol and the slurry solvent. The disadvantage of high capital investment is compounded by relatively low stereoregularity and relatively broad molecular weight distribution of its polymer product. Polypropylene of MFR >15 g/min and copolymers by this process are not suitable for general applications. The slurry process was eventually abandoned due to high cost.

Alpha-olefins are generally used as the solvent in the solution polymerization process. The polymerization temperature can be as high as 140°C. The process is relatively costly because of its complexity and was adopted only by Eastman Chemical Co. commercially. The Ziegler–Natta catalyst had to be replaced by a lithium compound such as lithium–aluminum hydride in this high-temperature process. The unreacted monomer is removed by depressurization and recycling; the catalyst residue is removed by filtration. The polymer solution undergoes multistage evaporation and extrusion to separate solvent from the polymer that is further purified by heptane to remove amorphous polypropylene. The resulting polymer exhibits relatively low melting point. It is used in specialty products of high quality and low market volume [8,10,69].

Rexene Co. and Philips Petroleum Co. first developed the *bulk polymerization* process with the first-generation $TiCl_3$ catalyst [8,11,70]. It was then commercialized by Dart Industries in 1964. The reactor feed contains 10–30% propylene in the liquid phase. A mixture of hexane and isopropanol was employed for the removal of catalyst residue as well as the amorphous polypropylene. The process step of removing residual catalyst was later eliminated after the high-efficiency catalyst was adopted, constituting the so-called "liquid pool process." Subsequently, Philips and Sumitomo companies further developed the liquid-phase polymerization process. This process enhances the reaction rate, catalyst efficiency, monomer conversion, and therefore results in high productivity. It also eliminates the need for solvent recovery and reduces environmental pollution. However, the process is somewhat complicated by the unreacted monomer, which has to be first vaporized and then liquefied before it is reused. The reaction vessel must be designed to operate under high pressures. In most cases, this process employs autoclaves for batch operation and tubular reactors for continuous operation.

The first gas-phase polymerization was first commercialized in Wesseling, Germany by ROW Co. in a joint venture with BASF and Shell companies in 1969. This facility employed the Novolen process for propylene polymerization in the gas phase. UCC and Sumitomo companies later developed fluidized-bed processes for the gas-phase polymerization of propylene. The advantages of this process are its high-efficiency catalysis, elimination of residual removal, and elimination of evaporation or centrifugal separation. Its polymer product can be used in almost all applications [12,13,71,72].

The Spheripol process, which combines the bulk and gas-phase polymerization processes, is probably the most popular method for preparing polypropylene on a commercial scale. This process is based on the fourth-generation Ziegler–Natta catalyst—a technology developed by Montedison Co. and now owned by Basell Co. It employs one or two serial tubular reactors to prepare propylene homopolymer or high-impact copolymers. An advantage of this process is that unreacted monomer can be liquefied under pressure with cooling water and recycled to the reactor; the need for a high-energy compressor is avoided. The resulting polymer is in a uniform, round powder form, thus eliminating the need for palletizing in some applications. It is capable of producing polypropylene of very high molecular weight and narrow molecular distribution for melt spinning or SB products [8,12,13].

3.3.4.2 Industrial Operations

The preparation of crystallizable polypropylene, as practiced in the earlier days of polypropylene, involved the preparation of catalyst, polymerization, purification, solvent recovery, and, finally, compounding. Typically, yields were 500–1000 lb of polymer per pound of catalyst [6]. Each manufacturer practiced one's own version of these process steps in an effort to produce a uniform product with regard to molecular weight and molecular weight distribution, ash content, color and color stability, and atactic, noncrystallizable polymer content. As discussed earlier, some systems used solvents that kept the atactic polymer in solution.

Others, which used propylene under pressure as the solvent, required washing after the unreacted monomer was removed and recycled. The particular process was unique to each of the polymer manufacturers.

As discussed above, there have been major improvements in catalyst efficiency and selectivity. The amount of atactic polymer has been reduced, and the number of pounds of polymer produced per pound of catalyst has been increased from five- to tenfold and more. Polymerization in the gas phase has been improved [73] in order that resin with low atactic content can be produced without solvent removal or polymer washing. The new processes [74] have reduced and, in some cases, eliminated the purification and solvent recovery steps, thereby simplifying the polymerization process and reducing the cost of a polymer plant. In addition, the new processes yield polymers with reduced catalyst residues and fewer gels, resulting in better filterability and improved pack life during fiber melt spinning.

Except for the improvements brought about by metallocene catalysts, there has been very little change in the stereospecificity and molecular weight distribution of resins as they come from the polymer reactor [74]. Although average molecular weight, as indicated by polymer melt flow, can be readily changed, little progress has been made in controlling the polymer molecular weight distribution. However, techniques have been developed to alter molecular weight distribution by further processing following the polymerization reaction.

3.3.4.2.1 CR Resins

The rheological properties and the fiber-processing characteristics of polypropylene polymers are dependent on their molecular weight distribution. Resins with narrow molecular weight distribution, as evidenced by a lower ratio of weight- to number-average molecular weight $\overline{M}_w/\overline{M}_n$, are referred to as controlled-rheology (CR) resins. These CR resins are produced by cracking a resin of high molecular weight to a lower molecular weight by heat, with or without the aid of a peroxide or oxygen. When converted to fibers, CR resins behave differently from normal polypropylene in several ways. Die swell, the amount of melt bulging on exit from a spinneret, is reduced. Higher shear rates can be used with CR resins without excessive die swell when compared to standard resins so that higher spinning rates can be used. The CR resins also permit low-denier fibers to be produced [74]. As maximum drawdowns are increased when using CR resins, low-denier fibers are produced with CR resins than with standard polypropylene [75]. In addition to the spinning improvements, CR resins of high melt flow are extruded to produce excellent fibers at much lower temperatures [74], making possible the use of less thermally stable additives. Some processes were described using extrusion temperatures as low as 177°C [76]. In a broad sense, CR resin is akin to the Ziegler–Natta catalyzed polypropylene. Metallocene-catalyzed polypropylene with high isotacticity behaves more controllably.

At present, polypropylene resins are still produced by the older, lower-yield processes, since all producers have not turned to the catalysts for high productivity. Resins are produced with standard molecular weight distribution as well as with several different versions of narrow molecular weight distribution. With many different average molecular weights offered in the market, the possibility of different molecular weight distributions at each average molecular weight, and the multitude of additives used by the resin and the fiber producer, it is obvious that there are many different resins and fibers that can be made from them. This summary, therefore, should be considered as a status report in the evolution of polypropylene fibers.

3.4 BASIC POLYMER PROPERTIES

Table 3.6 lists the physical properties of polypropylene that do not vary greatly with MFR. The data come from industrial brochures and should be taken as approximate values.

TABLE 3.6
Physical Properties of Polypropylene

Physical properties	Typical values
Moisture regain	<0.1%
Refractive index n_D	1.49
Thermal conductivity	0.95 Btu-in/ft^2·hr·°F
Coefficient of linear thermal expansion	4.0×10^{-5}/°F
Heat of fusion	21 cal/g
Specific heat	0.46 cal/g·C
Density of melt at 180°C	0.769 g/cc
Heat of combustion	19,400 Btu/lb
Oxygen index	17.4
Decomposition temperature range	328–410°C
Dielectric constant (0.1 MHz)	2.25
Dissipation factor (0.1 MHz)	<0.0002
Specific volume resistivity	$>10^{16} \Omega$·cm

3.4.1 Molecular Weight and Molecular Weight Distribution

Polypropylene is a white, semicrystalline polymer. The resins are generally solid and used according to their melt index, which in the case of polypropylene is more commonly called the melt flow rate (MFR). Polymers with MFRs ranging from less than 1 to more than 35 are commercially produced. Most fibers are produced from resins having MFR between 3 and 35.

The procedure for measuring polypropylene MFR is ASTM D1238 Condition L. The temperature of the test is 230°C and the forcing load is 2160 g. The MFR is the die amount of resin extruded through the 0.0825 × 0.315-in. die in 10 min. The MFR is inversely related to the apparent melt viscosity and the end-pressure losses, both of which are strongly dependent on the polymer weight-average molecular weight $\overline{M}_w$. The larger the melt flow, the lower the molecular weight. A curve showing the relation between weight-average molecular weight and MFR is given in Figure 3.1 [77].

The intrinsic viscosity $[\eta]$ is sometimes used as an indicator of molecular weight. The viscosity is often measured in decalin at around 135°C. The relation between the viscosity-average molecular weight $\overline{M}_v$ and $[\eta]$ for this system is:

$$\overline{M}_v = ([\eta] \times 10^4)^{1.25} \quad (137°C)[78] \tag{3.1}$$

$$\overline{M}_v = ([\eta] \times 9091)^{1.28} \quad (135°C)[79] \tag{3.2}$$

In polypropylene, the molecular weight distribution can vary widely. The common measure of dispersity, $\overline{M}_w/\overline{M}_n$, can range from around 2 to 12 or more. In recent years the commercial production of narrow-molecular-weight-distribution polymers, the so-called CR resins, has increased significantly. As discussed earlier, these resins can be spun at higher speeds and at lower temperatures than conventional resins.

Gel permeation Chromatography (GPC) curves for a 12 MFR "normal" resin and for a 12 MFR CR resin are shown in Figure 3.2 [74]. As noted, M_n is higher and M_w is lower for the CR resin. The high-molecular-weight tail is absent from the CR material.

Polypropylene Fibers

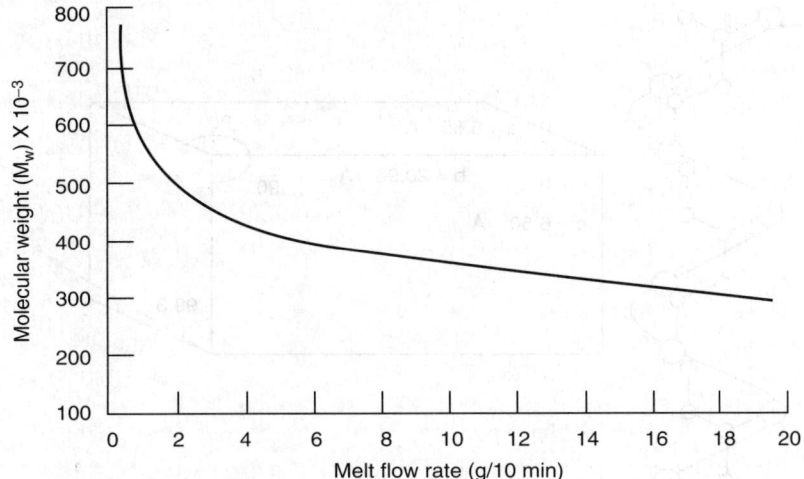

FIGURE 3.1 Polypropylene weight-average molecular weight versus melt flow rate. (From Schroeder, C.W.; Shell Development Co., "Modern Polypropylene Resins," paper presented at Fiber Producer Conference, October 1981.)

The solubles removed by various solvents in recrystallization or extraction procedures are generally taken to be the atactic polymer content. There is no standard procedure for measuring the soluble content. Commonly used solvents include n-heptane, n-pentane, toluene, and xylene. The amount of solubles removed varies according to the measuring technique and the resin manufacturer, but generally ranges from 2 to 5%.

	$M_n \times 10^{-3}$	$M_w \times 10^{-3}$	$M_z \times 10^{-3}$	Q
12 MFR normal	34	261	1580	7.66
12 MFR CR	42	211	596	5.06

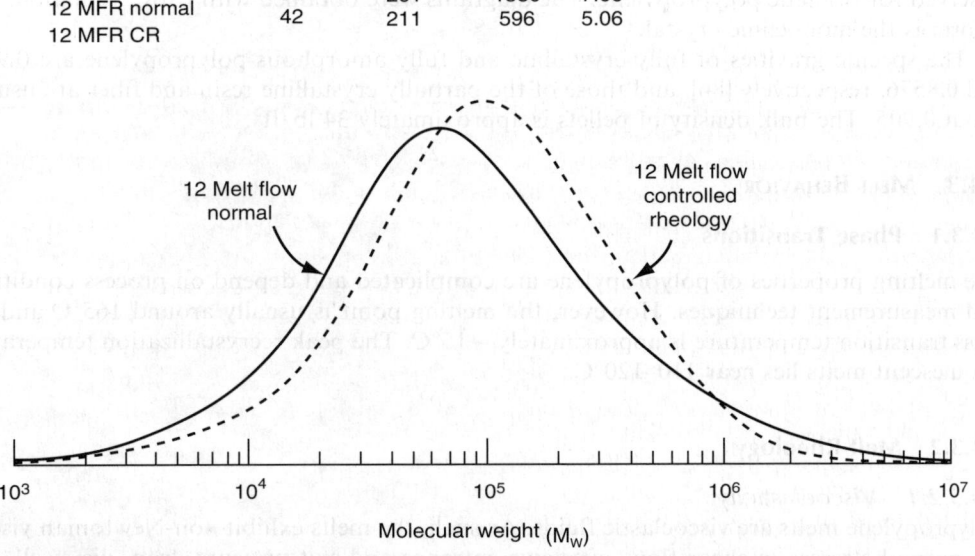

FIGURE 3.2 Molecular weight distribution curves for 12 MFR normal and 12 MFR controlled-rheology polypropylene. (From Spruiell, J.E.; White, J.L.; *Appl. Polym. Symp.*, **1975**, No. 27, 121.)

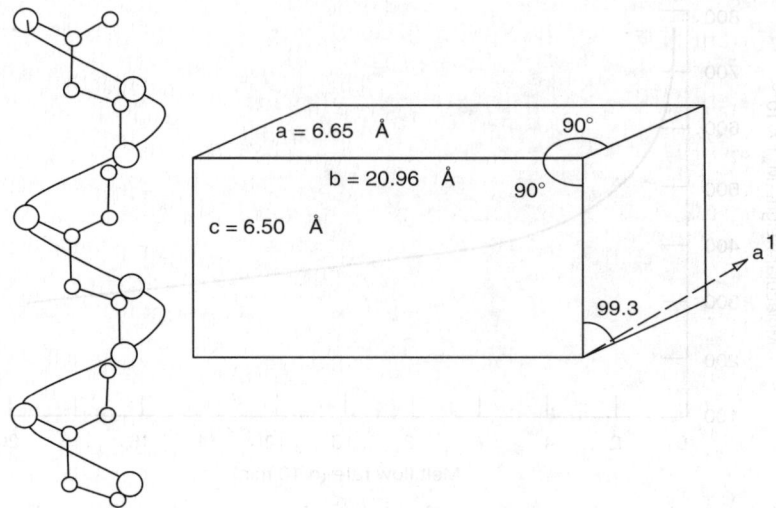

FIGURE 3.3 Unit cell of monoclinic polypropylene. (From Spruiell, J.E.; White, J.L.; *Appl. Polym. Symp.*, **1975**, No. 27, 121.)

3.4.2 CRYSTALLINE STRUCTURE

The degree of crystallinity in polypropylene depends on processing conditions, but is usually between 50 and 65%. Its common crystalline form is monoclinic, but other forms have been observed [80–83]. The monoclinic unit cell is depicted in Figure 3.3 [84]. The polypropylene helix lies along the *c*-axis of the unit cell. The crystals are chain-folded lamellae and are aggregated into spherulites or row-nucleated structures, depending on processing conditions [85]. Figure 3.4 [80] shows the x-ray diffraction diagrams of the various crystalline forms observed for isotactic polypropylene. The diagrams were obtained with CuK_α radiation. The α-form is the monoclinic crystal.

The specific gravities of fully crystalline and fully amorphous polypropylene are 0.9363 and 0.8576, respectively [86], and those of the partially crystalline resin and fiber are usually about 0.905. The bulk density of pellets is approximately 34 lb/ft^3.

3.4.3 MELT BEHAVIOR

3.4.3.1 Phase Transitions

The melting properties of polypropylene are complicated and depend on process conditions and measurement techniques. However, the melting point is usually around 165°C and the glass transition temperature is approximately -15°C. The peak recrystallization temperature in quiescent melts lies near 110–120°C.

3.4.3.2 Melt Rheology

3.4.3.2.1 Viscoelasticity
Polypropylene melts are viscoelastic fluids. As such, the melts exhibit non-Newtonian viscosity, normal stresses in shear flow, excessive entrance-and-exit pressure drop, die swell, secondary entrance flows, melt fracture, and draw resonance. (Newtonian fluids also exhibit draw resonance.) Polypropylene melts are more viscoelastic than melts of nylon and polyester.

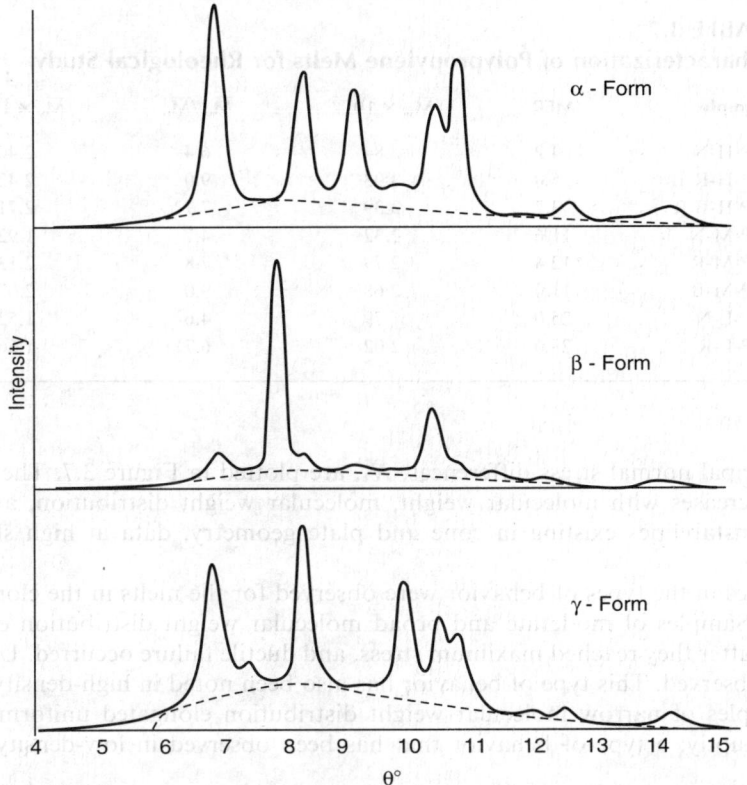

FIGURE 3.4 X-ray diffraction diagrams of isotactic polypropylene crystalline forms. (From Turner-Jones, A.; Aizlewood, Z.M.; Beckett, D.R.; *Makromol. Chem.*, **1964**, *75*, 134–135. With permission.)

For an elementary, unifying review of entrance-and-exit effects in viscoelastic fluids, see Hagler [87]. For a thorough analysis of viscoelastic fluids, see Bird et al. [88].

A systematic study of the basic rheological properties for a wide variety of polypropylene melts has been made by Minoshima et al. [89]. These authors measured shear viscosities at low shear rates in a Rheomatrics mechanical spectrophotometer and at high rates in an Instron capillary rheometer. The principal normal stress difference, N_1, was measured in the mechanical spectrophotometer with a cone and plate device. The elongational viscosity, of special importance to fiber formation, was measured in an apparatus built by Ide and White [90].

The method consists of heating rods of polypropylene on the surface of a hot silicone oil bath and then stretching them. All the experiments were conducted at 180°C. A description of the melts that were studied is given in Table 3.7.

The non-Newtonian shear viscosity, η, versus shear rate is given in Figure 3.5. It is noted that at low shear rates η is a constant, η_0. As shear rate increases, the viscosity decreases—a manifestation of pseudoplastic behavior. The viscosities generally decrease with MFR. At high shear rates, in the range of fiber spinning, the viscosities differ by a factor of only 1.5.

In Figure 3.6, plots of η_0 versus $\overline{M}_w$ and $\overline{M}_v$ are provided. The slope of the $\overline{M}_w$ curve is 3.7. Generally, for flexible polymers η_0 varies with approximately the 3.5th power of the weight-average molecular weight.

TABLE 3.7
Characterization of Polypropylene Melts for Rheological Study

Sample	MFR	$\overline{M}_w \times 10^{-5}$	$\overline{M}_w/\overline{M}_n$	$\overline{M}_v \times 10^{-5}$
PP-H-N	4.2	2.84	6.4	2.40
PP-H-R-B	5.0	3.03	9.0	2.42
PP-H-B-R	3.7	3.39	7.7	2.71
PP-M-N	11.6	2.32	4.7	1.92
PP-M-R	12.4	2.79	7.8	2.13
PP-M-B	11.0	2.68	9.0	2.07
PP-L-N	25.0	1.79	4.6	1.52
PP-L-R-N	23.0	2.02	6.7	1.66

The principal normal stress differences, N_1, are plotted in Figure 3.7. The normal stress difference increases with molecular weight, molecular weight distribution, and shear rate. Because of instabilities existing in cone and plate geometry, data at high shear rates are unobtainable.

Differences in the types of behavior were observed for the melts in the elongational flow experiment. Samples of moderate and broad molecular weight distribution exhibited local necking just after they reached maximum stress, and ductile failure occurred. Usually, several necks were observed. This type of behavior has also been noted in high-density polyethylene [91,92]. Samples of narrow molecular weight distribution elongated uniformly and finally ruptured abruptly; a type of behavior that has been observed in low-density polyethylene [91,92].

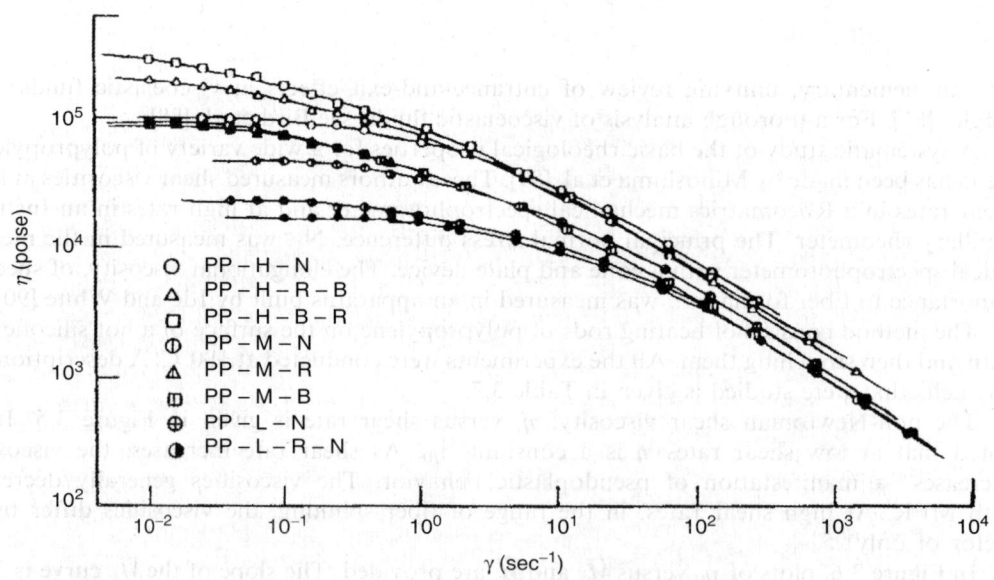

FIGURE 3.5 Shear viscosity–shear rate plots for polypropylene at 180°C. (From Minoshima, W.; White, J.L.; Spruiell, J.E.; *Polym. Eng. Sci.*, **1980**, *20*, 1166. With permission.)

Polypropylene Fibers

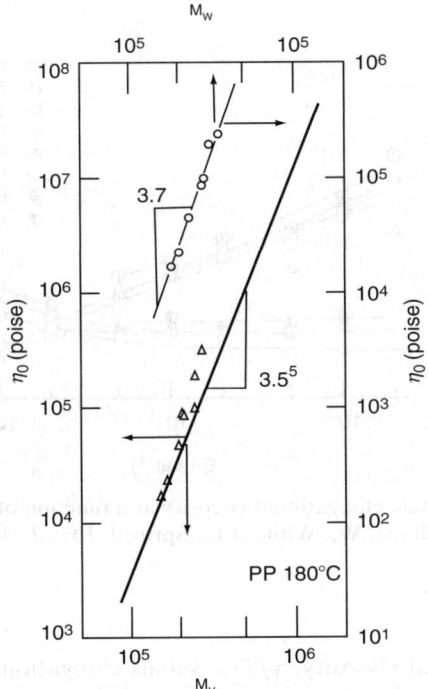

FIGURE 3.6 Zero shear viscosity as a function of weight-average molecular weight and viscosity-average molecular weight for polypropylene. (From Minoshima, W.; White, J.L.; Spruiell, J.E.; *Polym. Eng. Sci.*, **1980**, *20*, 1166. With permission.)

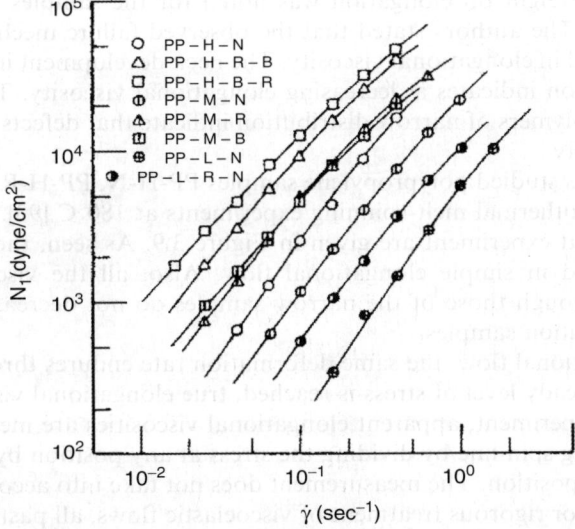

FIGURE 3.7 Principal normal stress difference as a function of shear rate for polypropylene at 180°C. (From Minoshima, W.; White, J.L.; Spruiell, J.E.; *Polym. Eng. Sci.*, **1980**, *20*, 1166. With permission.)

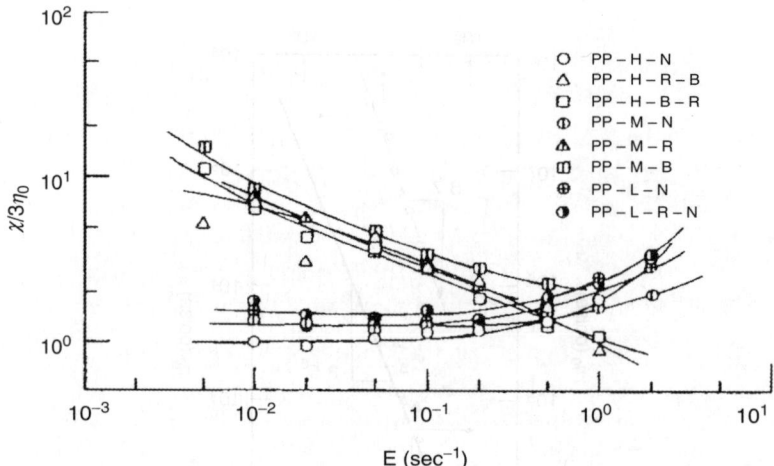

FIGURE 3.8 Reduced steady-state elongational viscosity as a function of elongational rate for polypropylene at 180°C. (From Minoshima, W.; White, J.L.; Spruiell, J.E.; *J. Appl. Polym. Sci.*, **1980**, 25, 287. With permission.)

The reduced elongational viscosity, $\chi/3\eta_0$, versus elongation rate is given in Figure 3.8. (The elongational viscosity of a Newtonian fluid is three times its shear viscosity.) For the samples of narrow molecular weight distribution, a constant χ of approximately $3\eta_0$ was obtained at low elongation rates. At higher rates, the viscosity increased. For the broader samples, the elongational viscosity decreased with elongation rate. The samples of narrow distribution were "strain hardening," whereas the broader samples were "strain softening." It was also found that the elongation to break was larger for the samples of narrow molecular weight distribution and that the elongation increased with decreasing molecular weight. No effect of molecular weight on elongation was noted for the samples of broader molecular weight distribution. The authors stated that the observed failure mechanisms are consistent with the trends found in elongational viscosity. The neck development in samples of moderate and broad distribution indicates a decreasing elongational viscosity. The uniform filaments obtained with the polymers of narrow distribution indicate that defects are healed because of an increasing viscosity.

The same authors studied polypropylene samples PP-H-N, PP-H-R-B, PP-M-N, and PP-M-B in low-speed isothermal melt-spinning experiments at 180°C [93]. The apparent viscosities measured in that experiment are given in Figure 3.9. As seen, the viscosities are larger than those measured in simple elongational flow. Also, all the viscosities decrease with elongation rate, although those of the narrow samples do not decrease as rapidly as those of the broad-distribution samples.

In simple elongational flow, the same deformation rate endures throughout the measurement, and, after a steady level of stress is reached, true elongational viscosity is obtained. In the melt-spinning experiment, apparent elongational viscosities are measured point-to-point along the accelerating spin line by dividing the stress at any position by the velocity gradient existing at the same position. The measurement does not take into account the prior deformation of the melts. For rigorous treatment of viscoelastic flows, all past deformation must be considered. Even flow in the spinneret may affect the apparent elongational viscosity, at least in low-speed spinning.

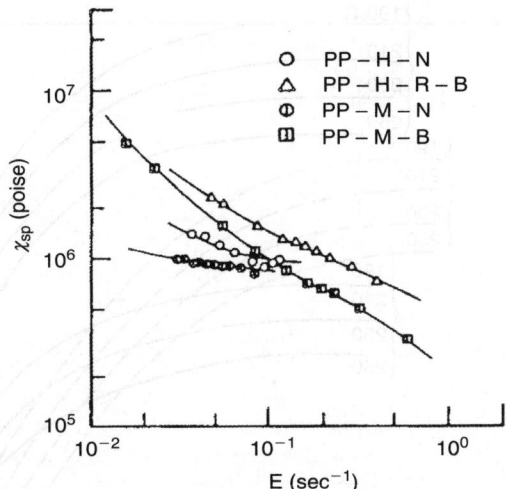

FIGURE 3.9 Apparent melt-spinning elongational viscosity as a function of elongational rate for polypropylene. (From Minoshima, W.; White, J.L.; Spruiell, J.E.; *J. Appl. Polym. Sci.*, **1980**, 25, 287. With permission.)

Van der Vegt [78] studied the shear viscosities of about 40 polypropylenes of different molecular weights and molecular weight distributions. In Figure 3.10 the viscosities of three different grades of polypropylene are plotted with temperature as a parameter. According to the author, the activation energy of polypropylene is 11 kcal/mol. He also found that η_0 is proportional to $\overline{M}_v^{3.6}$.

3.4.3.2.2 Die Swell

Die swell, or the Barus effect, has been observed in Newtonian fluids under special conditions, but the phenomenon is a general characteristic of tube flows of viscoelastic fluids. The more viscoelastic a fluid, the larger is the die swell. The phenomenon is due to elastic strain recovery as the melt exits the restraining die. There is a pronounced entrance effect on die swell; short L/D tubes give larger values of die swell. As the melt enters the tube, it is rapidly accelerated from rest to the tube velocity. The stress generated in the melt by the sudden deformation in the entrance region decays as the material travels through the tube. However, because of normal stresses generated in the melt due to tube shear flow, which also cause the melt to retract when it leaves the tube, the die swell approaches some asymptotic value. Generally, the die swell reaches its limiting value at an L/D of around 40. Die swell is also dependent on the die entrance angle. The view is that gradually tapered dies deform the flowing melt less rapidly and thus generate lower stresses. Also the smaller angle entrances give larger effective L/Ds. Because of gravity and take-up stresses generated in a spinning thread line, die-swell values depend strongly on experimental conditions.

Die swell in polypropylene has been studied by Minoshima et al. [93] for seven of the samples listed in Table 3.13, and also by White and Roman [94] and by Huang and White [95]. Minoshima et al. [93] conducted experiments at 180°C in a die having an L/D of 40. The ratio of extrudate to die diameter, d/D, is plotted as a function of die wall shear rate in Figure 3.11. Figure 3.11 shows that swell for each melt increases with shear rate. Die swell is also highly dependent on molecular weight distribution. Minoshima et al. later heated the solidified extrudates to 180°C in silicone oil for 5 min. The delayed recovery was measured and is plotted as a function of shear rate in Figure 3.12. The samples of broad molecular weight distribution exhibited greater delayed recovery.

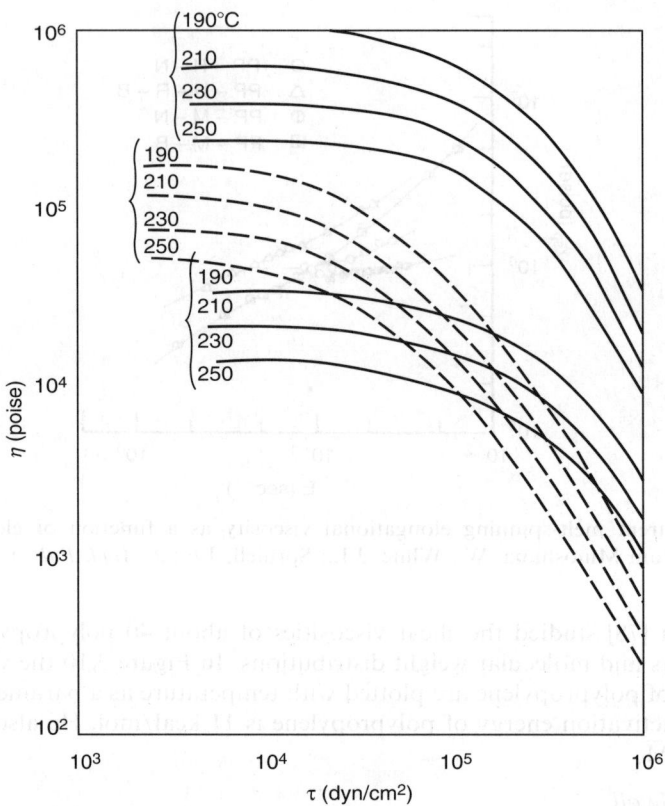

FIGURE 3.10 Effect of temperature on melt viscosity for three different polypropylenes. (From Van Der Vegt, A.K.; *Plastics Inst. [London] Trans. J.*, **1964**, *32*, 165.)

3.4.3.2.3 Capillary Effects

White and Roman [94] measured the effects of L/D and drawdown on die swell of polypropylene. Figure 3.13 shows the die swell as a function of L/D for Hercules Profax 6523 and four other melts at 180°C. The d/D ratios were measured on frozen extrudates. Figure 3.14 shows the die swell as a function of take-up velocity. The relationship between die swell and die entrance angle for a polypropylene was determined by Huang and White [95]. Figure 3.15 presents their data.

Capillary entrance flow patterns have been studied for a polypropylene by Ballenger and White [96]. The material had a viscosity of 100,000 poise at 180°C and a wall shear rate of 5 sec^{-1}. In flow-visualization experiments at 180°C, the authors found that at low shear rate the material flowed into a flat-entry die along Newtonian-like, radially converging stream lines. The naturally occurring entrance angle developed by the fluid remained constant at approximately 150° as shear rates were increased below a critical value. At wall shear rates of approximately 22 sec^{-1}, the emerging extrudate began exhibiting roughness, but the entrance flow remained unperturbed. At shear rates above 45 sec^{-1}, a secondary angular flow began at the entrance and the extrudate began emerging as a helix. Increasing the flow rate caused a violent tornado-like flow at the entrance; the extrudate became more distorted.

These authors also measured the combined entrance-and-exit pressure drops for the same polypropylene at 180°C. Figure 3.16 displays a plot of end losses normalized with wall shear

Polypropylene Fibers

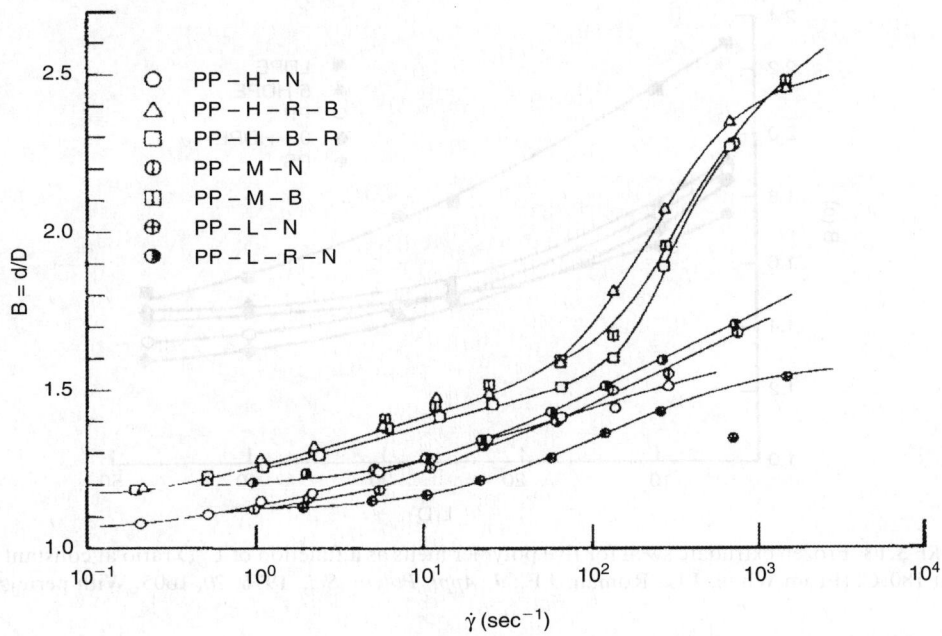

FIGURE 3.11 Extrudate swell as a function of die wall shear rate for polypropylene at 180°C. (From Minoshima, W.; White, J.L.; Spruiell, J.E.; *J. Appl. Polym. Sci.*, **1980**, 25, 287. With permission.)

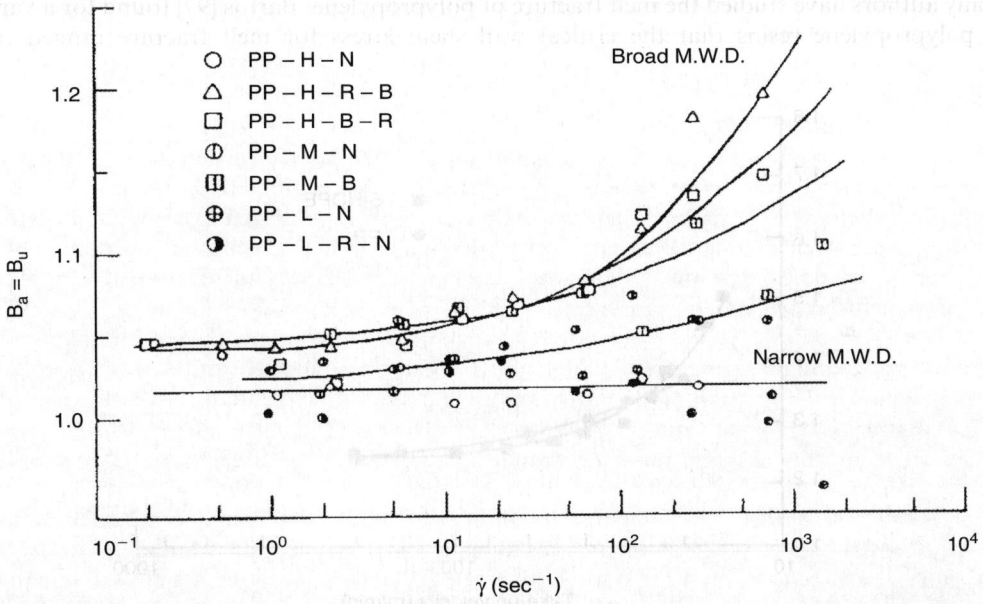

FIGURE 3.12 Delayed recovery as a function of die wall shear rate for polypropylene. (From Minoshima, W.; White, J.L.; Spruiell, J.E.; *J. Appl. Polym. Sci.*, **1980**, 25, 287. With permission.)

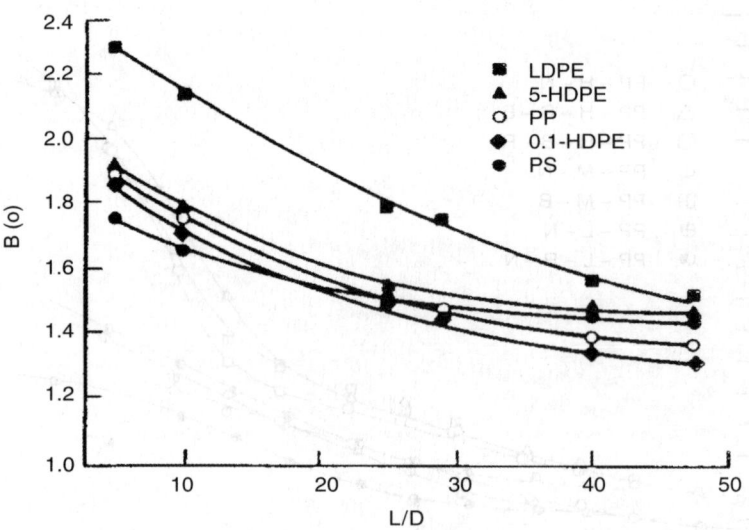

FIGURE 3.13 Frozen extrudate swell for five polymer melts as a function of L/D ratio at constant shear rate at 180°C. (From White, J.L.; Roman, J.F.; *J. Appl. Polym. Sci.*, **1976**, *20*, 1005. With permission.)

rates, σ_w, as a function of shear rate for a number of fluids. The end losses, ΔP_e, are a measure of the first normal stress difference. The ratio $\Delta P_e / \sigma_w$ is viewed as the Weissenburg number. As seen, the normalized end losses for viscoelastic fluids are considerably larger than those for Newtonian fluids.

3.4.3.2.4 Melt Flow Instabilities

Many authors have studied the melt fracture of polypropylene. Bartos [97] found for a variety of polypropylene resins that the critical wall shear stress for melt fracture ranged from

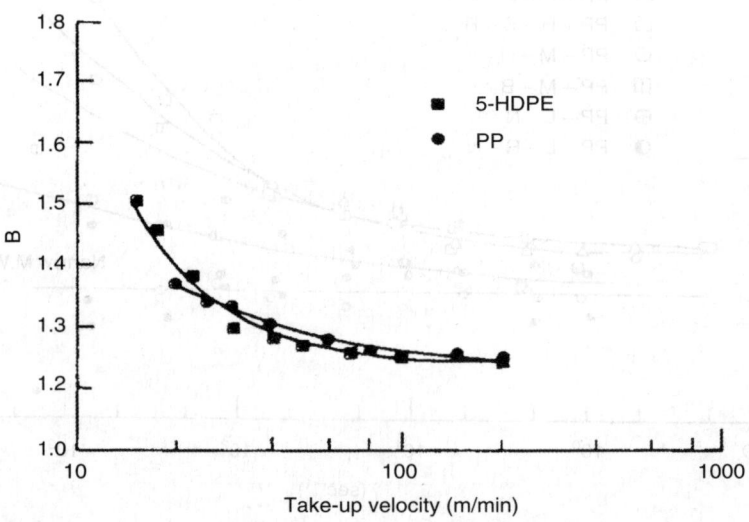

FIGURE 3.14 Die swell as a function of take-up velocity for high-density polyethylene and polypropylene. (From White, J.L.; Roman, J.F.; *J. Appl. Polym. Sci.*, **1976**, *20*, 1005. With permission.)

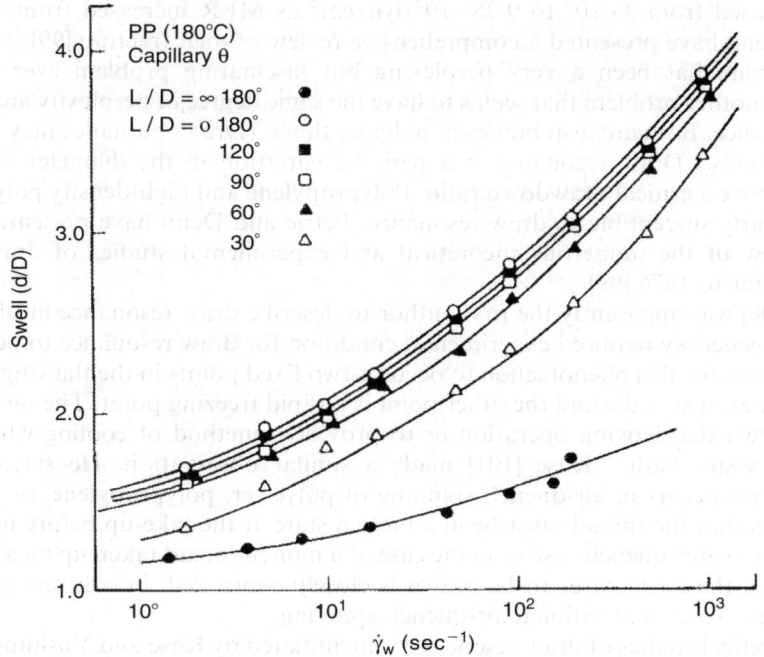

FIGURE 3.15 Effect of capillary entrance angle on die swell for polypropylenes. (From Huang, D.; White J.L.; *Polym. Eng. Sci.*, 1980, *20*, 182. With permission.)

8.9×10^5 to 16×10^5 dyn/cm^2. Melts of many other plastics exhibit melt fracture in the same range of shear stress. Vinogradov et al. [98] studied melt fracture in polypropylenes with MFR ranging from 1.17 to 30.5. He also identified three instabilities. His critical shear

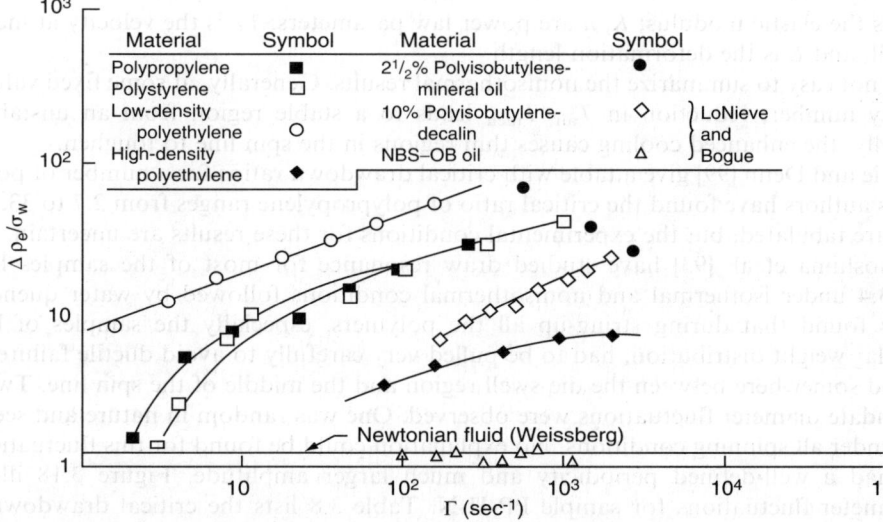

FIGURE 3.16 Reduced end-pressure losses as a function of shear rate. (From Ballenger, T.F.; White, *J. Appl. Polym. Sci.*, 1971, *15*, 1949. With permission.)

stresses increased from 3×10^5 to 9.75×10^5 dyn/cm^2 as MFR increased from 1.17 to 30.5. Petrie and Denn have presented a comprehensive review of melt fracture [99].

Melt fracture has been a very perplexing but fascinating problem ever since it was discovered. Another problem that seems to have the same degree of perplexity and fascination is draw resonance. Both are instabilities in polymer flows. (Draw resonance may also occur in Newtonian fluids.) Draw resonance is a periodic variation in the diameter of a spinning thread line above a critical drawdown ratio. Polypropylene and high-density polyethylene are both particularly susceptible to draw resonance. Petrie and Denn have presented a comprehensive review of the numerous theoretical and experimental studies of draw resonance conducted prior to 1976 [99].

Miller [100] was apparently the first author to describe draw resonance in fibers. He also recognized a generally required experimental condition for draw resonance to occur: "one of the requirements for this phenomenon to occur is two fixed points in the thawing system. One of these is the extrusion die and the other point is a rapid freezing point. The only prevention is to slow down the thawing operation *or* to provide a method of cooling which is not as abrupt as a water bath." Kase [101] made a similar observation. He stated that draw resonance never occurs in air-quench spinning of polyester, polypropylene, or nylon fibers. He also stated that the thread must be in a molten state at the take-up before instability can develop. In the water-quench case or in the case of a molten thread taken up on a chill roll, the point at which the melt ceases to be drawn is closely controlled, in contrast to the lack of control in the case of conventional air-quench spinning.

The theoretical studies of draw resonance were initiated by Kase and Yoshimoto [102] and Pearson and Matovich [103]. These authors found that the critical drawdown ratio for isothermal Newtonian fluids is about 20. Fisher and Denn analyzed both isothermal and nonisothermal flows of spinning of viscoelastic fluids [104,105]. Figure 3.17 gives the results for the isothermal case, where

$$\alpha = \frac{3^{((n-1)/2)} K}{G} \left(\frac{V_0}{L}\right)^n \tag{3.3}$$

and G is the elastic modulus; K, n are power law parameters; V_0 is the velocity at maximum die swell; and L is the deformation length.

It is not easy to summarize the nonisothermal results. Generally, at some fixed value of an elasticity number, reduction in T_{air}/T_{melt} leads to a stable region from an unstable one. Physically, the enhanced cooling causes thin regions in the spin line to toughen.

Petrie and Denn [99] give a table with critical drawdown ratios for a number of polymers. Various authors have found the critical ratio of polypropylene ranges from 2.7 to 33. Higher values are tabulated, but the experimental conditions for these results are uncertain.

Minoshima et al. [93] have studied draw resonance for most of the samples listed in Table 3.4 under isothermal and nonisothermal conditions followed by water quench. The authors found that during string-up all the polymers, especially the samples of broader molecular weight distribution, had to be pulled very carefully to avoid ductile failure, which occurred somewhere between the die-swell region and the middle of the spin line. Two types of extrudate diameter fluctuations were observed. One was random in nature and seemed to occur under all spinning conditions. No explanation could be found for this fluctuation. The other had a well-defined periodicity and much larger amplitude. Figure 3.18 illustrates the diameter fluctuations for sample PP-H-N. Table 3.8 lists the critical drawdown ratios for all the samples studied. Note that critical ratio is higher for the samples of narrow molecular weight distribution and under nonisothermal conditions. The authors argued

Polypropylene Fibers

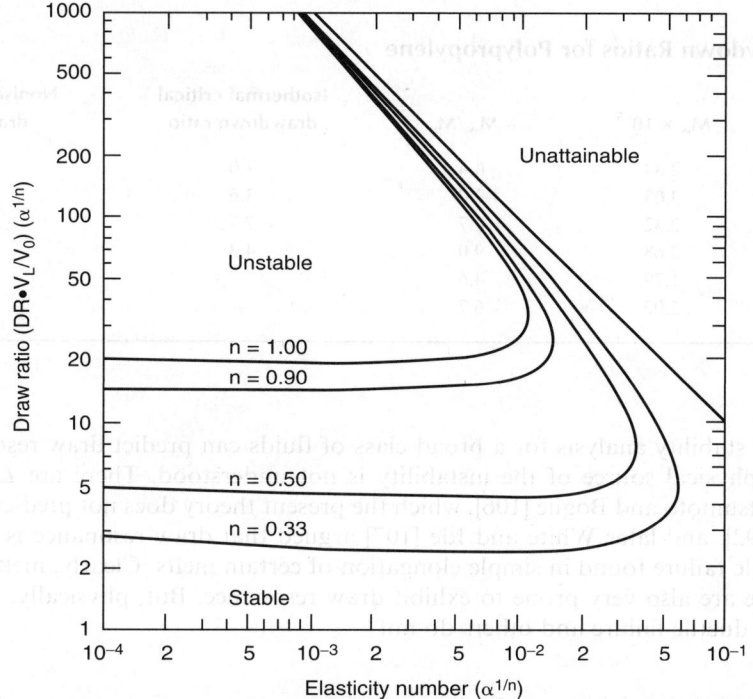

FIGURE 3.17 Theoretical prediction of the effect of viscoelasticity on draw resonance. (From Matsumoto, T.; Bogue, D.C.; *Polym. Eng. Sci.*, **1978**, *18*, 564. With permission.)

that melts that are strain-hardening or those with true or apparent elongational viscosities with a larger slope of the viscosity–elongational rate relationship would develop draw resonance at higher drawdown ratios. This argument is proved in their studies.

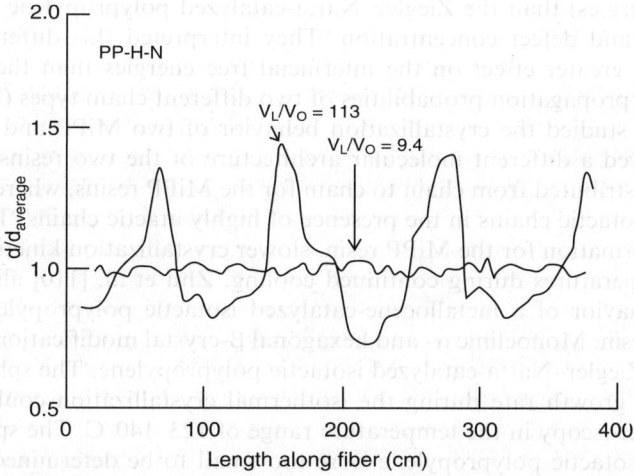

FIGURE 3.18 Diameter fluctuations for polypropylene fibers during spinning: random fluctuation and draw resonance. (From Minoshima, W.; White, J.L.; Spruiell, J.E.; *J. Appl. Polym. Sci.*, **1980**, *25*, 287. With permission.)

TABLE 3.8
Critical Drawdown Ratios for Polypropylene

Sample	$M_w \times 10^{-5}$	$-M_w/M_n$	Isothermal critical drawdown ratio	Nonisothermal critical drawdown ratio
PP-H-N	2.84	6.4	7.0	22.0
PP-H-R-B	3.03	9.0	3.6	10.0
PP-M-N	2.32	4.7	7.7	19.0
PP-M-B	2.68	9.0	4.3	9.0
PP-L-N	1.79	4.6	—	23.0
PP-L-R-N	2.02	6.7	—	19.0

Although stability analysis for a broad class of fluids can predict draw resonance theoretically, the physical source of the instability is not understood. There are L/D effects as noted by Matsumoto and Bogue [106], which the present theory does not predict. Hagler [91], Chen et al. [92], and later White and Ide [107] argued that draw resonance is a continuous form of ductile failure found in simple elongation of certain melts. Clearly, melts that exhibit ductile failure are also very prone to exhibit draw resonance. But, physically, why do some melts exhibit ductile failure and others do not?

3.4.3.3 Comparison of ZNPP and MiPP

3.4.3.3.1 Crystallization Behavior

Galante and Mandelkern have studied the melt behavior of metallocene isotactic polypropylene (MiPP) with the same chain defect concentration and molecular weights ranging from 68,480 to 288,430 [108]. They found that the crystallization rate and the variation of rate with crystallization temperature followed a pattern that is basically independent of molecular weight. The metallocene polypropylenes also showed significantly higher $\sigma_e \cdot \sigma_u$ (product of interfacial free energies) than the Ziegler–Natta-catalyzed polypropylene (ZNPP) of similar molecular weight and defect concentration. They interpreted this difference as the MiPP crystallites having greater effect on the interfacial free energies than the ZNPP, or as the different sequence propagation probabilities of two different chain types (Figure 3.19). Bond and Spruiell [109] studied the crystallization behavior of two MiPP and one ZNPP resins. Their results showed a different molecular architecture of the two resins. The defects were more uniformly distributed from chain to chain for the MiPP resins, whereas the ZNPP resin exhibited highly isotactic chains in the presence of highly atactic chains. This resulted in less perfect lamellae formation for the MiPP resin, slower crystallization kinetics, and crystallization at lower temperatures during continued cooling. Zhu et al. [110] also investigated the crystallization behavior of a metallocene-catalyzed isotactic polypropylene and a Ziegler–Natta-catalyzed resin. Monoclinic α- and hexagonal β-crystal modifications were observed in the conventional Ziegler–Natta-catalyzed isotactic polypropylene. The spherulites were large enough that their growth rate during the isothermal crystallization could be measured by polarized light microscopy in the temperature range of 123–140°C. The spherulites of metallocene-catalyzed isotactic polypropylene were too small to be determined, even during isothermal crystallization for a long time at a temperature very close to their melting point. DSC scans exhibited one melting peak for ZNPP sample, but two melting peaks for MiPP. Wide angle x-ray diffraction (WAXD) results showed the existence of γ-orthorhombic crystal

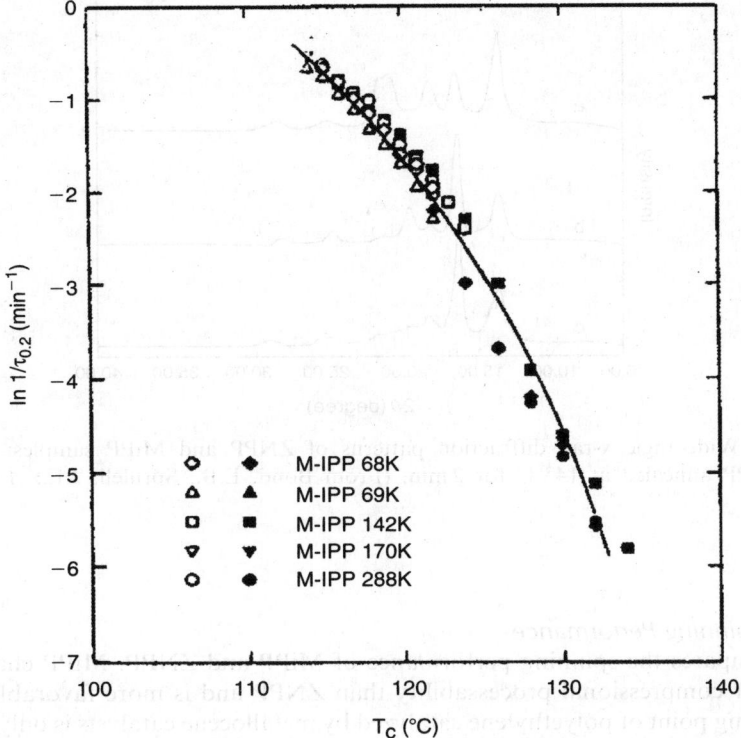

FIGURE 3.19 Plot of the overall rate of crystallization, $1/\tau_{0.20}$ vs. temperature for the listed polypropylenes. Open symbols are data from exotherms; closed symbols from endotherms. (From Galante, M.J.; Mandelkern, L.; Alamo, R.G.; Lehtinen, A.; Paukker, R.; *J. Therm. Anal.*, **1996**, *47*, 913. With permission.)

structure in the MiPP sample, but not in the ZNPP sample. Note the difference in peak height at $2\theta = 14°$ and $17°$ for ZNPP and MiPP, respectively, in Figure 3.20 [7].

3.4.3.3.2 Physical Properties
Referring to Table 3.5, metallocene catalysts, as compared with Ziegler–Natta catalysts, consist only of single catalytic centers and therefore produce polymer chains having much more uniform chain length and distinctly lower atactic content than conventional polypropylene. Thus, metallocene-catalyzed polypropylene exhibits some good properties that are not attainable with Ziegler–Natta-catalyzed polypropylene, such as

1. Lower density: about 0.88 g/cm^3
2. Lower melting point: 130–150°C
3. Low heat of fusion: 15–20 J/g
4. Low atacticity: 0.7–0.9, more narrow molar mass distribution
5. Slow crystallization rate and small crystallite size
6. Good chemical resistance

These improved properties provide new ways to modify polypropylene and its fibers.

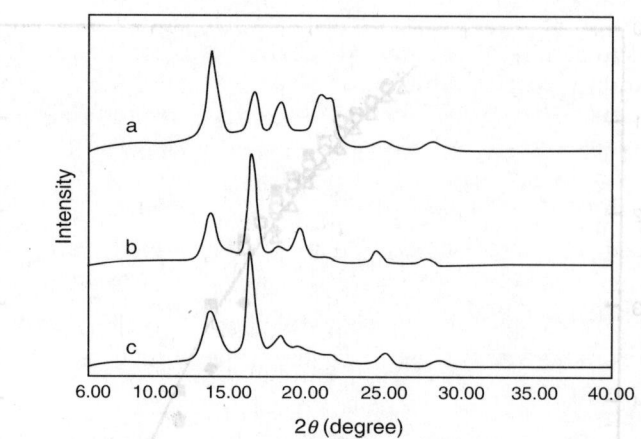

FIGURE 3.20 Wide-angle x-ray diffraction patterns of ZNPP and MiPP samples: (a) ZNPP, (b) MiPP, (c) MiPP annealed at 143°C for 2 min. (From Bond, E.B.; Spruiell, J.E.; *ANTEC*, Europe, **1997**, 388.)

3.4.3.3.3 Spinning Performance

Table 3.9 compares the spinning performance of MiPP and ZNPP. MiPP characteristically exhibits better compressional processability than ZNPP and is more favorable to spinning. The low melting point of polyethylene catalyzed by metallocene catalysts is only 107°C, which is favorable to the preparation of thermal-bonded fiber such as high-shrinkage bonded fiber and fine-denier fiber for SB or melt-blown nonwoven fabrics. These fabrics have good interfusion resistance and gas permeability that are useful for personal hygiene and filter media. In addition to fine-denier fiber production, MiPP is favorable to increasing the spinning velocity. It is potentially useful for the preparation of high-strength fibers or strong films.

3.5 ADDITIVES FOR FIBER PREPARATION

Polypropylene is not very stable at the melt temperature and degrades at room temperature, particularly in the presence of UV light. It is not dyeable: coloration requires the addition of

TABLE 3.9
Spinning Performance of MiPP and ZNPP

Spinning velocity, m/min	Fiber diameter (μm)	ZNiiPP (CR)		MiPP, MI = 18	
		Tensile strength (cN/tex)	Elongation (%)	Tensile strength (cN/tex)	Elongation, (%)
2500	26	22.3	265	36.9	154
3000	26	22.0	265	36.9	142
3500	26	—	—	3.65	145
4000	26	—	—	3.64	137

pigments. Therefore, the preparation of commercially useful polypropylene fibers requires the addition of stabilizers to permit processing and adequate life, and the addition of pigments to provide the color. Additives to modify the resin are essential for polypropylene fibers; the choice of additives is extensive. Hence, the use of additives has resulted in designing a wide variety of properties in polypropylene fibers. The entire additive package must be considered when trying to achieve a particular set of properties, as the additives, both stabilizers and pigments, interact with one another. The mechanism by which unstabilized polypropylene degrades has been studied extensively and provides some understanding to how stabilizers function.

3.5.1 Stabilizers

3.5.1.1 Degradation of Unstabilized Fiber

More than any other polymer used to prepare fibers, polypropylene owes its existence as a fiber-former to stabilizers used to overcome its deficiencies. When prepared, polypropylene is unstable to both heat and light. At the high temperatures used to melt the resin for fiber extrusion, the molecular weight of the polymer and melt viscosity change, creating processing difficulties. During extrusion, polymer reactions occur that result in oxygen-containing groups becoming part of the polymer [111]. As a fiber, unstabilized polypropylene is very susceptible to sunlight-initiated deterioration. Weathering results in a sharp drop in fiber elongation together with an increase in oxygen-containing groups.

The overall oxidation scheme of an unstabilized polypropylene fiber has been the subject of much investigation [112]. Polypropylene, as produced, contains trace impurities which, in the presence of heat or light, permits a polypropylene radical (PP) to be formed. This radical will react with oxygen as shown in the reaction represented by Equation 3.4.

$$PP^{\bullet} + O_2 \rightarrow PPO_2^{\bullet} \tag{3.4}$$

The $PPO_2^{\bullet}$ radical reacts inter- or intramolecularly with polypropylene to form $PP^{\bullet}$ (Equation 3.5), which will react with oxygen as shown in the reaction represented by Equation 3.4, so that the oxidation reaction will repeat itself. The hydroperoxide group formed in the reaction shown by Equation 3.5 reacts with UV light or heat to yield a tertiary alkoxy radical and a hydroxy radical (Equation 3.6). The tertiary alkoxide may react as shown in the reactions represented by Equation 3.7a or Equation 3.7b and the hydroxy radical can also react with polypropylene as shown in the reaction represented by Equation 3.8, in order that, in each case, the product is another polypropylene radical that can react with oxygen as presented in Equation 3.4. Equation 3.7b leads to chain scission and reduction in the molecular weight of the polymer

$$PPO_2^{\bullet} + PPH \rightarrow PPOOH + PP^{\bullet} \tag{3.5}$$

$$PPOOH \rightarrow PCH_2 \underset{\underset{O}{\overset{|}{\underset{\bullet}{|}}}}{\overset{CH_3}{\overset{|}{-}}} CH_2P + HO^{\bullet} \tag{3.6}$$

$$PCH_2-\underset{\underset{O^\bullet}{|}}{\overset{\overset{CH_3}{|}}{C}}-CH_2P \longrightarrow \begin{cases} PCH_2-\underset{\underset{OH}{|}}{\overset{\overset{CH_3}{|}}{C}}-CH_2P + PP^\bullet & (3.7a) \\ \\ PCH_2-\underset{\underset{O}{\|}}{\overset{\overset{CH_3}{|}}{C}} + PCH_2^\bullet & (3.7b) \end{cases}$$

$$HO^\bullet + PPH \rightarrow HOH + PP^\bullet \qquad (3.8)$$

The increase in oxygen concentration accompanies the deterioration of fiber physical properties, but the deterioration does not necessarily relate to the increase directly. Factors such as fiber orientation and the conditions, primarily temperature at which the fiber is drawn, can affect the rate of deterioration [113]. The reduction of elongation that accompanies exposure to UV light is less for a highly oriented fiber because the photooxidation reaction is initially limited to the fiber surface. The photooxidation reaction is accompanied by the formation of cracks on the thin surface [114]. Cold-drawn fiber reacts throughout, degrades faster, but does not form surface cracks. Fiber wettability increases with increased UV exposure [112], which is additional evidence of the formation of surface oxygen bonds.

To avoid or at least reduce the rate of deterioration of polypropylene fiber, a variety of stabilizers are added. These are sometimes categorized [115] as:

1. Radical terminators that react with active polymer radicals to form harmless products
2. Peroxide decomposers that react with the peroxides to form harmless products, preventing the reaction represented by Equation 3.3 from occurring
3. UV absorbers that preferentially absorb UV energy—a mechanism not very effective in thin sections such as fibers
4. Energy quenchers that react with activated polypropylene to prevent decomposition and then degrade themselves to harmless products

Since most stabilizers react in more than one way and combinations of additives are required to prepare stabilized polypropylene fiber, these materials are better discussed in accord with the function they perform for polypropylene rather than the mechanism by which they function. When making fibers, stabilizers are added to polypropylene to (1) provide melt extrusion stability; (2) provide long-term thermal stability at normal use temperatures; and (3) provide stability to exposure to UV light (sunlight). The degree to which these stabilizers effectively accomplish their objectives without introducing other objectionable effects, such as yellowing, determines the value of the resultant fiber product.

3.5.1.2 Melt Extrusion Stability

Control of the fiber extrusion process requires control of the melt viscosity of the polymer within narrow limits. Since unstabilized polypropylene degrades when melted, stabilizers are added to prevent large and uncontrolled changes. Polymer oxidation is inhibited by the use of additives that contain phenolic groups, which react with radicals and are then deactivated by combination. A large variety of hindered phenolic chemicals, containing a t-butyl phenol moiety, have been found to help control polypropylene oxidation. Table 3.10 lists some antioxidants used in 1980–1990s to control polymer melt flow. Polymer melt flow increases as molecular weight decreases and is used as a control measure of molecular weight because polypropylene is insoluble at room temperature in all common solvents. Since fiber extrusion

TABLE 3.10
Effect of Antioxidants on Processing Stability

Antioxidant (0.1%)	Number of extruder passes at 270°C melt flow index[a]		
	1	3	5
Goodrite 3114	5.6	10.7	14.4
Irganox 1010	5.0	9.5	12.2
Topanol CA	5.0	11.0	17.8
Ionox 330	4.1	6.7	9.4
Irganox 1076	5.9	12.0	18.0

[a]ASTM D1328.

is conducted at higher or lower temperatures than the test temperature of 270°C, multiple extrusions provide a measure of the utility of particular agents. Since fiber color control is also important, care must be taken to select a stabilizer or stabilization system that adds as little color as possible to the resultant fiber.

There are numerous formulations of antioxidants and processing stabilizers in the literature. Table 3.11 [116–118] and Table 3.12 present a few examples of historical and current

TABLE 3.11
Exemplary Antioxidants for Processing Stability

General composition	Examples	Reference
Hindered phenolic compounds containing a *t*-butyl phenol moiety	Ciba Irganox HP 2215, 2225	
	Ciba Irganox 3114	
	Octadecyl 3,5-di-*tert*-butyl-4-hydroxyhydrocinnamate	[116]
Thiodipropionaces	Dilauryl thiodipropionate (DLTDP)	
	Distearyl thiodipropionate (DSTDP)	
Organic phosphites	Trilauryl phosphate	[117]
	Tris-nonylphenyl phosphite (TNPP)	[116]
	Weston 618: (distearyl pentaerythritol diphosphite)	
	Weston 626: (2,4-di-5-butylphenyl) pentaerythritol diphosphite	
	Ciba Irgafos 168: tris-(2,4-di-*t*-butylphenyl) phosphite.	
	Ciba SandostabP-EPQ: [tetrakis(2,4-di-*t*-butylphenyl)-4,4′-biphenylenediphosphonite]	
Triazines	Ethanox 314: 1,3,5-tris(3,5-di-tert-butyl-4-hydroxybenzyl)1,3,5-triazine-2,4,6(1H,3H,5H)-trone	[118] [125]
	Hydroxyphenyltriazines	
Peroxide-free hindered hydroxylamine ester	Ciba Irgatec 76	
Phenol-free compounds	Ciba Fiberstab L 112	

TABLE 3.12
Exemplary Long-Term Stabilizers

General composition	Examples	Reference
Thiodiphosphites	Dilauryl thiodipropionate (DLTDP)	
	Distearyl thiodipropionate (DSTDP)	
Organic phosphites	Weston 618: (distearyl pentaerythritol diphosphite)	
Hindered amine light stabilizers (HALS)	Ciba CR 144: poly(2,2,6,6-tetramethyl-4-piperidyl aminotriazine)	[120]

interests. Some additives, occasionally referred to as peroxide decomposers, behave cooperatively and sometimes synergistically with antioxidants to help stabilize polypropylene during melting. Thiodipropionate compounds such as dilauryl (DLTDP) or distearyl (DSTDP) help antioxidants control melt stability. The addition of phosphites such as trilauryl phosphite and tris-nonylphenyl phosphite (TNPP) also help control melt flow. Weston 618 (Table 3.11: distearyl pentaerythritol diphosphite) helps control melt flow with or without DSTDP. Use of a phosphite to help improve processing stability is favored when possible because phosphites help protect color, or at least contribute less to color than do antioxidants.

Thio compounds are less desirable because they contribute to undesirable polymer odors, although they are inexpensive to use. Organic phosphites in general have problems due to hydrolysis and sometimes lose some effectiveness when stored in humid conditions. Sandostab P-EPQ is a phosphonite [tetrakis(2,4-di-*t*-butylphenyl)-4,4'-biphenylenediphosphonite] that is more stable than Weston 618 or TNPP in humid areas [119] and offers excellent melt flow stabilization. Weston 626 is chemically similar to Weston 618 except that the 2,4-di-5-butylphenyl group has replaced the stearyl group esterifying the phosphite. It is a more hydrolysis-resistant phosphite, as is Irgafos 168 [tris-(2,4-di-*t*-butylphenyl) phosphite]. These phosphites offer hydrolysis resistance and can be used advantageously if the higher price is accepted.

3.5.1.3 Long-Term Thermal Stability

Polymer stability at melt temperatures is required to generate fibers. However, such stabilization does not always result in fibers with excellent thermal stability at normal end-use temperatures. Since polypropylene has excellent chemical resistance characteristics, there has been considerable interest in its use for filtration, sometimes at high temperatures. Because the fibers do not absorb dyes readily, the use of polypropylene yarns to produce dye bags or leader fabrics that could be reused many times requires production of fibers with good long-term stability at elevated temperatures. Early polypropylene fibers did not possess the long-term thermal stability to meet such end-uses, and some problems resulted from the decompositions of laundry bags stored in containers at relatively high temperatures. The use of better antioxidants and synergists such as DSTDP and phosphites improved long-term thermal stability. As found for melt stability, DSTDP improves the long-term thermal stability provided by various antioxidants. Unlike melt flow stability, long-term thermal stability is aided more by DSTDP than it is by phosphates. Since phosphates aid color stability and melt flow stability while DSTDP or DLTDP help the long-term thermal stability provided by antioxidants, formulations are frequently made with all three types of additives. The specific additives selected and the amount of each must be determined by the fiber produced with consideration of the end-use intended for the fibers:

Another class of additives for polypropylene stabilizations is compounds containing a 2,2,6,6-tetramethyl piperidyl group (**I**). These compounds are frequently referred to as

TABLE 3.13
Heat Stability of Multifilament Yarn (8000/396)

	Days at 120° to embrittlement
Goodrite 3114(0.10%)	14
Irganox 1010(0.10%)	20
CR 144(0.10%)	47
0.25%CR144+0.1%Goodrite 3114	83
0.25%CR144+0.1% Irganox 1010	86
0.25%CR144+0.1%Weston 618	82

hindered amine light stabilizers (HALS). Although these products are excellent light stabilizers (see below), they are also excellent long-term thermal stabilizers:

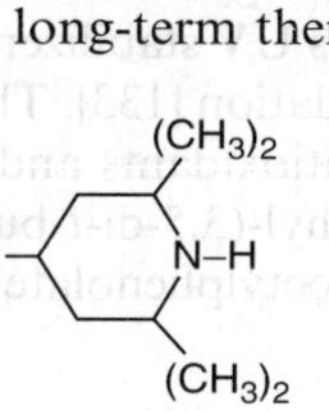

$$\text{structure with } (CH_3)_2, N-H, (CH_3)_2 \tag{I}$$

CR 144 (Table 3.12) is a poly(2,2,6,6-tetramethyl-4-piperidyl aminotriazine) [120] that provides excellent long-term thermal stability by itself or in combination with antioxidants (Table 3.13). It is actually the amine oxide (**II**), which is formed during fiber preparation or exposure that is proposed to be the active stabilizer. It is interesting that CR144 does not provide melt flow processing stability, even in the presence of DLTDP. An antioxidant or a phosphite is required to provide melt processing stability when using CR 144:

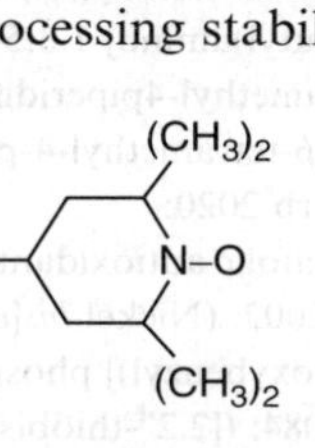

$$\text{structure with } (CH_3)_2, N-O, (CH_3)_2 \tag{II}$$

A triazine compound, Ethanox 314, of the following structure by Albemerie Corp. has a moderate molecular weight of 784 and a melting point range of 218–223°C [118]. It is claimed to retard thermooxidative degradation in various polymers and to enhance their light stability. The antioxidant exhibits high resistance to extraction and low volatility:

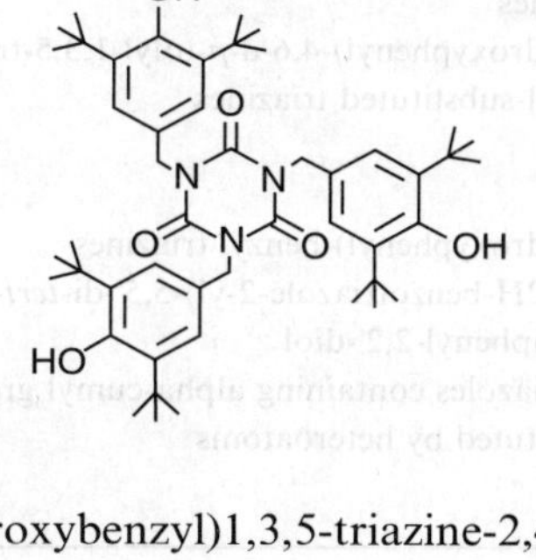

Ethanox 314 antioxidant
1,3,5-tris(3,5-di-*tert*-butyl-4-hydroxybenzyl)1,3,5-triazine-2,4,6(1H,3H,5H)-trone
Molecular weight = 784

3.5.1.4 UV Stability

The sensitivity of polypropylene to heat and light is magnified in fibers by the fiber-processing conditions, because the drawing and annealing processes increase the potential for polymer oxidation [111], which initiates the degradation step. Efforts to stabilize polypropylene fibers to UV light have involved the use of UV absorbers, energy-transfer agents, and, since the 1980s, HALS compounds. Table 3.14 [121–132] shows several representative compositions of UV light stabilizers.

Carbon black is a well-known stabilizer for polyolefins, but has limited utility in fibers unless a black fiber is desired. Even then, the stability of black fibers is not sufficient for many current end-uses. UV absorbers are reasonably effective in products of thick cross section but are of limited value for products of low denier per filament. The small cross section prevents absorption from being effective.

Additives that operate primarily as energy transfer agents dissipate the accumulated energy in a harmless manner. They are effective as UV stabilizers even though they do not absorb light in the wavelengths that cause photodegradation [133]. These stabilizers, known as energy quenchers, provide UV stability, which the antioxidants and the hydroperoxide decomposers cannot equal. Irgastab 2002 (Nickel *bis*[*o*-ethyl-(3,5-di-*t*-butyl-4-hydroxybenzyl)]phosphonate) and Cyasorb UV 1084 ([2,2' [1]-thiobis(4-*t*-octylphenolate)]-*n*-butylamine Nickel II) are the nickel

TABLE 3.14
Exemplary UV Stabilizers

General composition	Examples	Reference
Hindered amine light stabilizers (HALS)	Ciba Chimassorb 944: Poly[[6-[(1,1,3,3-tetramethylbutyl)amino]-1,3,5-triazine-2,4-diyl] [(2,2,6,6-tetramethyl-4piperidinyl) imino]-1,6-hexanediyl[(2,2,6,6-tetramethyl-4-piperidinyl) imino]])	[121]
	Ciba Chimassorb 2020:	
	Mixtures of phenolic antioxidant and hindered amine	
Nickel compounds	Ciba Irgastab 2002: (Nickel *bis*[*o*-ethyl-(3,5-di-*t*-butyI-4-hydroxybenzyl)] phosphonate)	
	Cyasorb UV 1084: ([2,2'¹-thiobis(4-*t*-octylphenolate)]-*n*-butylamine Nickel II)	
N-alkoxy hindered amines (NOR HALS)	*Bis*[1-(2-hydroxy-2-methylpropoxy)-2,2,6,6-tetramethyl piperidin-4-yl] sebacate	[122]
Triazines	2,4-*bis*[2-hydroxy-4-(2-hydroxyethyloxy)-phenyl]-6-(4-chlorophenyl)-1,3,5-triazine	[123]
	2-(2-Hydroxyphenyl)-4,6-diphenyl-triazine trisaryl triazines	[124]
	2-(2-Hydroxyphenyl)-4,6-d-*p*-tolyl-1,3,5-triazine	[125]
	Biphenyl-substituted triazines	[126]
		[127]
		[128]
Benzotriazoles	2-(2'-hydroxyphenyl)-benzyl triazines	[129]
	3,3'-*bis*(2H-benzotriazole-2-yl)-5,5'-di-*tert*-octyl-1,1'-biphenyl-2,2'-diol	[130]
	Benzotriazoles containing alpha-cumyl groups substituted by heteroatoms	[131]
		[132]

stabilizers that were most used to provide UV stability for fibers. Because Irgastab 2002 is a hindered phenol, it is also somewhat effective as a processing and long-term thermal stabilizer, whereas Cyasorb UV 1084 is not effective. One problem with the use of these nickel stabilizers is the green cast resulting from the high stabilizer levels needed to meet some requirements such as automotive interiors or outdoor uses. The stabilizers also resulted in poor color stability in gas-fade testing, and they reacted with certain colorants particularly sulfur-bearing pigments.

The ability of these energy quenchers to stabilize polypropylene fibers to weathering permitted the development of many new end-uses, but their capabilities have been surpassed by a new group of stabilizers that contain a hindered piperidine structure. As shown above, these HALS compounds can be very good long-term thermal stabilizers. Hindered piperidines react with hydroperoxides during polypropylene processing to form nitroxyl radicals (**II**) that are effective polymer radical traps [134]. These nitroxyl radicals react with polymer free radicals to form the polymeric hydroxylamine (**III**).

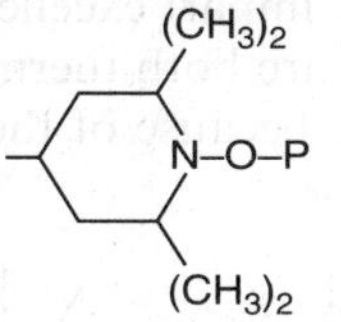

(**III**)

If the reaction of the hindered piperidine were to stop at this point, it would be difficult to explain why these materials are so effective. However, tile polymeric hydroxyl amine is itself an effective stabilizer, and it probably reacts [134], as in the reaction given by Equation 3.9, to form the hydroxyl amine equivalent of the hindered amine (**IV**), which can then react with a peroxyl radical to reform the hindered nitroxyl radical. The regeneration of the nitroxyl radical permits hindered piperidine compounds to be effective in preventing a polymer radical from decomposing in a manner that degrades the polypropylene fiber:

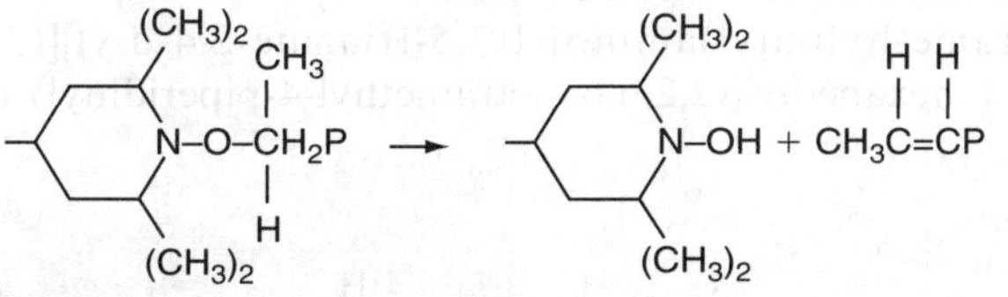

$$\text{N-O-CH}_2\text{P} \rightarrow \text{N-OH} + \text{CH}_3\text{C=CP} \quad (3.9)$$

Tozzi et al. have reported that CR144 is an effective light stabilizer and provides better protection than either an absorber or a quencher at half the concentration [120] (Table 3.15). They also reported that although antioxidants and phosphites did not help UV stability, they did help long-term thermal stability (as in Table 3.13) and were required to provide processing and color stability. It was also shown that the polymeric stabilizer CR144 was much more extraction-resistant, that washing the fiber caused very little property change.

Because substituted hindered piperidine compounds are so effective, there has been much effort by many companies to prepare these compounds. Those commercially available and recommended for fine-denier fibers are made by Ciba-Geigy. Other companies included Cyanamid, Ferro, Goodrich [135], Borg-Warner [136], and Sandoz in the 1980–1990s. The HALS compounds used today are Chimassorb 944, Tinuvin 622, Tinuvin 144, and Tinuvin 770 by Ciba-Geigy. The last compound is not generally recommended for fine-denier fibers.

HALS have been developed for use in polyolefins since the 1980s [137]. They are represented by Chimassorb 944 and Chimassorb 2020 that also contain piperidinyl segment.

TABLE 3.15
Light Stability of Multifilament Yarn (8000/396)

Stabilizer	Hours in weatherometer to 50% tensile strength
Cyasorb 531(0.05%)	410
Cyasorb UV 1084 (0.05%)	470
CR 144(0.25%)	830
CR 144(0.25%)	1100

Their chemical structures, as shown below, are rather bulky with molecular weights of 12,000–3,100 and 2,600—3,400, respectively. The melting point range is 100–135°C and 120–150°C for Chimassorb 2020. They impart excellent light stability and long-term thermal stability to polypropylene fibers. They are both thermally stable, and have low volatility and extractability and a low migration rate because of their polymeric nature.

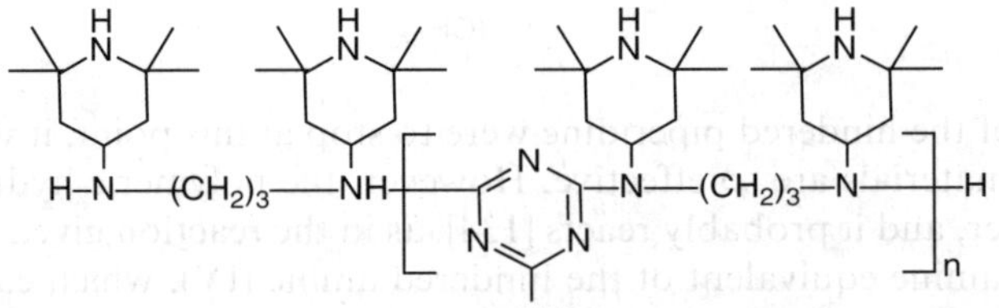

Ciba Chimassorb 944 oligomeric hindered amine.
Poly[[6-[(1,1,3,3-tetramethylbutyl)amino]-1,3,5-triazine-2.4-diyl][(2,2,6,6-tetramethyl-4-piperidinyl) imino]-1,6-hexanediyl[(2,2,6,6 -tetramethyl-4-piperidinyl) imino]]) $M_n = 12,000–3,100$

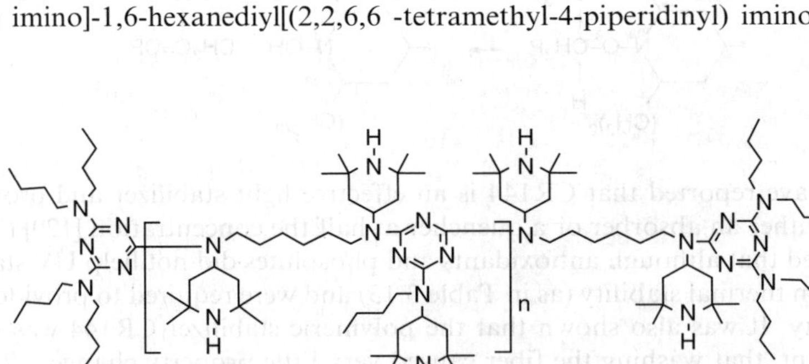

Ciba Chimassorb 2020 block oligomeric hindered amine.
1,6-Hexanediamine,N,N'-bis(2,2,6,6-tetramethyl-4-piperidinyl)-polymer with 2,4,6-trichloro-1,3,5-triazine. $M_n = 2600–3400$.

Table 3.16 demonstrates the effectiveness of a HALS compound like Tinuvin 144 in stabilizing a polypropylene yarn to UV light. Despite the fact that Tinuvin 144 costs much

TABLE 3.16
Light Stability of Multifilament Yarn (120/12)

Stabilizer	Hours in xenotest 1200 to 50% tensile strength
0.1% Antioxidant	200
1.2% Nickel light stabilizer	1150
0.05% Tinuvin 144	700
0.10% Tinuvin 144	1150
0.20% Tinuvin 144	2000

more than a nickel stabilizer does, it is more cost effective (costs less per hour of stability) and can achieve higher levels of stability than are possible with nickel stabilizers.

According to data provided by Ciba-Geigy [138], Tinuvin 622 is even more effective than Tinuvin 144, and Chimassorb 944 provides the most protection of the three. The data provided by American Cyanamid for their new stabilizer, Cyasorb UV 3346 [139], indicate that even better HALS compounds may be developed by the chemical industry. Table 3.17 indicates that Cyasorb UV 3346 may provide even better stability than does Chimassorb 944, particularly in a carbon-arc weatherometer run at 190°C—a test designed for fibers intended for use in automotive fabrics.

Stabilization of polypropylene fibers to meet product end-use specifications requires careful consideration of how the fiber will be used, the cost that can be tolerated, and the properties of the available stabilizers when used together. In many instances stabilizers act synergistically. It is possible to prepare high-UV fibers that are colorless and do not yellow in most end-uses. This is a major advance permitted by some recently developed addictive systems. A combination of heat and light, as is found in automotive end-uses, requires careful formulation to avoid unwanted color changes. Even the selection of a fiber finish is important in meeting color stability requirements in certain end-uses [140]. Multiple formulations are required for multiple products if minimum costs are a consideration. This is true even when color is not part of the formulation. When colorants (mostly pigments) are added, stabilization may become even more complicated because pigment additives may also interact with stabilizers, thereby having effects other than just the addition of color.

TABLE 3.17
Light Stabilization of Blue Multifilament Yarn (612/18)

Stabilizer		Hours to 50% tensile strength	
		CAWOM(190°F)	Xenon WOM(145°F)
Cyasorb UV 3346	0.2%	565	1680
	0.4%	855	2170
	0.6%	1040	2715
	0.8%	1230	3065
Chimassorb 944	0.2%	390	1420
	0.4%	620	2360
	0.6%	615	2635
	0.8%	775	2915

3.5.2 Pigments

Because polypropylene contains no polar groups, it contains no dye sites capable of reacting permanently with dye molecules. Since monomers containing polar groups react with the catalysts used to make polypropylene, it has thus far proven difficult to incorporate dye sites during polymerization. Therefore, coloration of polypropylene fibers has been accomplished either by (1) modification of polypropylene as part of the fiber-manufacturing process to render the resultant fiber dyeable or by (2) the addition of pigments.

Most colored polypropylene products are made from fibers that have been produced by adding pigments to the resin prior to extrusion. In the early days of polypropylene fibers, stabilizers did not provide sufficient stability for the polymer itself to outlast the pigment. Almost any pigment with sufficient thermal stability to be used at melt-spin temperatures was acceptable in terms of color stability. The fiber would have degraded before it lost its color. As discussed above, improvements in stabilization have permitted polypropylene to enter many new markets. Pigmented fibers are used outdoors and in automobiles, where large expanses of glass, sunlight, and relatively high temperatures cause other fibers to fail. There are still many end-uses that are not nearly so demanding. Because of the wide variety of end-uses, there is a need for a wide variety of pigments exhibiting different costs as well as color stability characteristics. Pigments are selected based on various requirements—some related to the fiber-making process and some related to the product. The characteristics considered are (1) suitability for fiber processing, (2) stability requirements of the product, (3) interaction with some additives, (4) color requirements, and (5) cost requirements.

3.5.2.1 Fiber-Processing Requirement

Producer-colored polypropylene fibers are made by bringing together polymer and pigment in such a way that the colored particles are finely dispersed in a polymer matrix in the final fiber. This is required to optimize the color yield and to avoid the spinning and drawing problems that would result if the pigment particles were permitted to form large aggregates. To achieve this file dispersion, the fiber producer uses pigment dispersed previously in polymer at high pigment concentration. Pigment concentrates, usually 25–50% pigment, are made by specialized concentrate manufacturers who use various techniques to break down particles for dispersion, usually in polypropylene but sometimes in other resins. To produce good fiber, it is essential that the pigment manufacturer provides products in sufficiently small particle size and that the concentrate manufacturers prepare pigment concentrates that do not contain aggregates. The procedures used by these manufacturers are a secret and may differ among the producers. However, the end result must be the same. Most fiber producers use special tests to evaluate color-concentrate quality to be certain that the pigment will not interfere with the fiber-making process and that the expected colors will be obtained.

Each fiber producer has developed its own method for modifying the fiber-making process to prepare colored fibers from polymer and concentrate. The color concentrates may be ground to powder and fed in a proper ratio with polymer powder. Alternately, the producer can dilute the colored concentrate pellets in a proper ratio with natural pellets to produce a letdown prior to extrusion. Some systems have been designed to inject melted color concentrates directly into the melt stream during fiber preparation. Such a method proves difficult because for some colors four or more pigments must be added accurately with a concentration ratio difference of 1000:1 or more. The rest of the fiber process is quite conventional, as described elsewhere in this chapter. The extruded colored melt must be quenched, drawn, crimped, and cut to make ST or processed to make filaments with or without texturing.

TABLE 3.18
Time to Crystallize at 121°C

Pigment added	Crystallization time (min)
None—natural resin	8.66
Black (low level)	5.51
Black (high level)	3.78
Red 149 (low level)	3.31
Red 149 (high level)	2.60
Violet 19 (low level)	1.18
Violet 19 (high level)	0.94
Red 194 (low level)	2.05
Red 194 (high level)	1.50
Red 144 (low level)	1.97
Red 144 (high level)	1.26
Yellow 95 (low level)	4.88
Yellow 95 (high level)	1.97

To be useful, a pigment must not interfere excessively with the fiber-making process. The pigment must not only be finely dispersed to begin with but must also not aggregate during processing; otherwise, it would clog processing filters, interfere with the drawing process, and degrade the fiber physical properties. The pigment must also be stable to the thermal conditions used in fiber making. Although thermal stability is inherent to the chemistry of the pigment, filterability, drawability, and physical properties are related to the quality of the 80 pigments' dispersion, which is why the concentrate manufacturer is so important to the fiber producer.

Although it would be most desirable that pigments not interfere with the fiber-making process, it is not possible in actual practice. Even with good-quality color concentrates, final fiber tenacity is frequently reduced and fiber mechanical quality is adversely affected. Because pigment particles may act as nucleating agents [141] and affect crystallization behavior (see Table 3.18), they may influence extrusion and air quenching resulting in the modification of the fiber process, depending on the pigments used.

3.5.2.2 Stability Requirements

If pigments were sufficiently stable to outlast the fiber, there would be no need to consider color stability and all the pigments listed in Table 3.19 would be useful regardless of fiber requirements. This, however, is not the case. Polypropylene fiber extrusion is conducted from 200°C to above 300°C, depending on the process, in such a manner that some pigments used at lower temperatures cannot be used at high temperatures. Since polypropylene fiber products can now be manufactured for uses that require varied levels or stability, some pigments used for indoor home furnishings are not useful in automotive fibers, and still fewer are useful in preparation of fibers for long-term outdoor use.

Table 3.20 lists pigments found suitable during the 1980–1990s for high-temperature extrusion in products requiring good outdoor stability [142]. These pigments were recommended for use in producer-colored polyester as well as polypropylene. Note that, because of the severe end-use requirements, Table 3.20 contains only 11 of the 79 pigments listed in Table 3.19.

Many pigments are quite color-stable when used at high concentrations to give deep colors. Fewer are color-stable at intermediate concentrations and only a few are suitable

TABLE 3.19
Pigments for Polypropylene Fibers

White 6	Red 104	Red 190
	Red 108	Red 194
Red 38	Red 112	Red 199
Red 48:1	Red 113	Red 202
Red 48:2	Red 122	Red 206
Red 48:3	Red 123	Red 207
Red 53	Red 144	Red 208
Red 57:1	Red 149	Red 214
Red 58: 3	Red 166	Red 224
Red 68	Red 175	Red 242
Red 88	Red 177	Red 244
Red 101	Red 185	Red 246
Yellow 17	Yellow 83	Yellow 116
Yellow 24	Yellow 93	Yellow 119
Yellow 34	Yellow 95	Yellow 120
Yellow 37	Yellow 108	Yellow 138
Yellow 53	Yellow 109	Yellow 139
Yellow 81	Yellow 110	Yellow 151
		Yellow 154
Blue 15	Green 7	
Blue 16	Green 17	Violet 19
Blue 22	Green 26	Violet 23
Blue 60	Green 36	Violet 32
	Green 50	
Orange 13		
Orange 20	Orange 34	Orange 48
Orange 23	Orange 36	Orange 49
Orange 31	Orange 38	Orange 61
	Orange 43	Orange 62
	Red 104 (Moly Orange)	

for use at the low concentrations required for tinting to create desired shades. Red 194, for instance, is very stable when used at a concentration of 0.25% or higher [141]. At that level, it can be used in upholstery fibers and even automotive fibers. However, at low levels needed for shading (0.0125%), the pigment is useless because it fades very quickly. In general, high-performance pigments are expensive and low-cost pigments do not perform so well. The use of a high-performance pigment for an end-use that does not require it is not economical,

TABLE 3.20
Pigments for High-Temperature, High-UV Use

White 6	Yellow 37	Orange 20
Red 149	Red 166	Blue 15
Violet 19	Violet 23	Black 7
Green 7	Green 36	

whereas the use of a low-performance pigment that does not meet end-use requirements could result in a disaster. It is important to understand how the pigments in Table 3.19 perform to be able to make a proper choice.

Pigment performance is sometimes a function of the stabilizers used, so the formulator must understand their behavior in his own system. In southern Florida testing, pigment color stability requirements are pushed beyond the point of existing data, because stabilizers can now prolong the life of polypropylene fibers to the equivalent of 2–5 years or more of life. Although these requirements are not new for thick sections, they are new for low-denier fibers. The literature has very little data about color stability of pigments in polypropylene fibers, after 2 years in Florida, because the fibers normally degraded before that time until recently. Using pigments with color stability equivalent to the degradation stability built into the fiber is a commercial requirement. Since some pigments are prodegradants, as will be discussed later, they should be avoided if possible. However, if their color stability is excellent, it may be possible to use them by adding more stabilizers. Obviously, if color stability is not equal to the product requirement, the pigment cannot be used regardless of the type of behavior in the fiber.

Although the final product dictates stabilizer requirements as well as color requirements, it is important to remember that stabilizers and pigments are additives and additives may affect color even if they are not pigments. They may affect thermal and UV stability even though they are considered to be colorants. It has been already mentioned that stabilizers affect color. Modern stabilizers add very little color to fiber products and do not change drastically on exposure; hence, they have little effect on color. However, pigments do interact with stabilizers and cause thermal and UV stability to differ from what is obtained without the pigments.

3.5.2.3 Interaction with Other Additives

Ideally, pigment should provide only color. They are expected to be heat- and light-stable and not have any effect on fiber properties. However interactions do occur. Some are predictable—such as the interaction between some inorganic sulfides with nickel stabilizers. Most interactions are unpredictable and are found only via an Edisonian approach, that is, produce the fiber and test it.

An early study [143] using a 10-mil film showed that pigments contribute favorably to both heat and light stability of unstabilized polypropylene. Presumably, the pigments helped the light stability by acting as an absorber of light. Even iron oxide, known as a severe prodegradant for stabilized polypropylene, served as both a heat and light stabilizer for unstabilized 10-mil films. In stabilized polypropylene, all the pigments acted as thermal prodegradants, presumably because they absorbed the antioxidant stabilizer.

Similar data were found for a 1-mil film containing 0.3% light stabilizer and 0.4% pigment [144], but the data (Table 3.21) varied depending on the stabilizer used. As found in the previous study, light stability improved in all the pigments when added to polypropylene containing no other stabilizer. With a nickel stabilizer, none of the pigments tested, except iron oxide (Red 101), is a prodegradant. With the more effective HALS stabilizer, both Yellow 93 and Red 144 are prodegradants, at least at the 0.4% level.

Steinlin and Saar reported the effect of various concentrations of pigments on the light stability of fine-denier polypropylene yarn (72 denier, 24 filament) containing varied levels of Tinuvin 770 [145]. Although Tinuvin 770 is not a preferred HALS stabilizer for fine-denier fibers, the data presented are of considerable interest. (The authors pointed out that Tinuvin 622 and Chimassorb 944 are preferred for polypropylene fibers.) The amount of energy required to reduce yarn strength by 50% was determined. Tinuvin 770 was added at 0.25,

TABLE 3.21
Interaction of Pigments with Light Stabilizers

	Kilolangleys to 50% tensile strength		
Pigment	Unstabilized	Nickel stabilizer	HALS
None	13	30	80
White 6	23	39	92
Red 101	15	24	—
Yellow 37	17	32	80
Orange 20	18	31	85
Blue 29	13	32	85
Yellow 93	22	31	40
Red 144	15	41	58
Blue 15	16	30	94

0.50, 0.75, and 1.0% l, and the amount of energy to 50% strength loss was determined. Dividing each of these energy values by that for the natural polypropylene fiber provided a value called the protection factor. The protection factor for the sample with no pigment and no Tinuvin 770 is therefore defined as 1. In a similar manner, samples containing different pigments at different Tinuvin 770 levels were tested at 50% strength loss and protection factors determined. Table 3.22 displays the results obtained for 1% pigment at varied Tinuvin 770 levels.

At low levels of Tinuvin 770, most of the pigments are stabilizers as well as colorants. As the amount of Tinuvin 770 increases, the probability that a 1% pigment loading will act as a prodegradant increases. When 1% Tinuvin 770 was added, only 9 of the 26 pigments in Table 3.22 were not prodegradants; i.e., the protection factor was equal to or greater than that of an unpigmented sample.

There are other complicating factors. The data in Table 3.22 were obtained by testing the dry samples. When the samples were wet (as in a weatherometer or exposed outdoors), results differed considerably in some cases. Yellow 37, a cadmium yellow, changed color and caused rapid degradation. These data probably indicate that the Orange 20, a cadmium orange in Table 3.20, as well as Yellow 37, are not suitable for outdoor use. Another complicating factor is that the effect of pigment concentration on degradation is not always the same. The data in Table 3.22 indicate the relative effects of 1.0% pigment at varied Tinuvin 770 levels, but do not necessarily predict what would happen at higher or lower pigment levels. At 0.5% Tinuvin 770, an increased level of Yellow 138 increased the rate of fiber degradation. However, with Yellow 110 and with Violet 37 increasing, the pigment concentration improved fiber stability. Table 3.21 demonstrates that the interaction of pigment and stabilizer is not consistent and depends on the particular stabilizer. Since Tinuvin 770 is not the HALS stabilizer best suited for fiber, these data should be used with caution when applied to other stabilizers.

3.5.2.4 Effects on Fiber Stability

The producer of colored polypropylene fibers for undemanding end-uses can use almost any pigment that fits the process without fear of accelerated degradation, if the base resin has a reasonable stabilizer package. Since the fiber will not last very long in severe conditions, the color stability of the pigment will not create problems. However, as the end-uses become more demanding, the fiber manufacturer must better understand the interaction of the pigments

TABLE 3.22
Pigment Interaction with Tinuvin 770

	Protection factors—concentration of tinuvin 770				
Pigment(1.0g)	0 g	0.25 g	0.50 g	0.75 g	1.0 g
None	1.0	2.4	3.7	4.6	5.4
Yellow 37	1.4	3.2	4.2	4.8	5.4
Yellow 83	1.2	2.0	2.8	3.2	3.5
Yellow 93	1.6	2.2	2.6	3.0	3.2
Yellow 94	1.2	1.6	1.9	2.2	2.4
Yellow 95	1.2	1.6	2.0	2.4	2.6
Yellow 109	1.6	2.4	3.2	3.8	4.4
Yellow 110	1.2	2.8	3.6	4.2	4.6
Yellow 129	1.4	2.8	3.6	4.4	5.0
Yellow 138	1.0	1.8	2.2	2.6	3.0
Yellow 147	1.2	2.8	3.6	4.2	4.8
Orange 31	1.6	2.5	3.0	3.5	4.0
Orange 61	1.3	3.0	3.8	4.8	5.7
Red 48.2	1.2	3.0	3.8	4.5	4.9
Red 88	1.5	2.3	3.0	3.9	4.8
Red 144	1.4	2.6	3.6	4.0	4.8
Red 149	1.2	2.4	3.0	4.5	5.1
Red 166	1.5	2.7	3.6	4.2	4.8
Red 177	2.0	3.2	4.0	4.6	5.2
Red 220	2.0	3.1	4.2	4.8	5.4
Red 224	1.0	1.9	2.8	3.7	4.6
Violet 37	1.5	4.3	5.4	6.1	6.5
Blue 15.3	1.6	5.0	>5.5	>6.0	>7.0
Blue 16	1.2	3.5	4.8	6.1	7.6
Blue 60	1.6	4.8	6.8	7.5	>8
Green 7	1.8	4.1	>5	>5	>5
Black 7	2.5	5.6	>6	>7	>7

with the stabilizer package. The knowledge that the unpigmented fiber meets the stability requirements is not sufficient. The pigment must be sufficiently color-stable, and the stabilizer package must be adjusted so that the end-product meets the required specification. This is possible only if the fiber producer knows how the pigments and stabilizers interact.

Some of the following conclusions were drawn in an excellent review and current knowledge of the influence of pigments on the light stability of pigments [144]:

1. Many pigments, particularly organic yellow, orange, and red, have negative effects on the photostability of polypropylene.
2. Many pigments, such as carbon black, phthalocyanine blue and green, and titanium dioxide, influence light stability favorably by screening, selective absorption of harmful radiation, and deactivation of polymer photo-excited species.
3. Negative influences of pigments on polymer light stability are due to sensitization of singlet oxygen formation, initiation of polymer photooxidation by photo-excited pigment chromophore formation of other photo-excited species with adverse effects on

polymer matrices, adsorption of stabilizing additives, chemical transformation of stabilizing additives, or nucleation of less stable crystalline forms of polymers.
4. No single, simple explanation exists for the many manifestations of unfavorable pigment–polymer interactions. Most organic pigments are structurally complex and many undergo photoisomerization, photorearrangement, photooxidation, and photoreduction reactions. This complexity adds to the complexity of the influence of pigments on polymer light stability, requiring in-depth studies to understand the reasons for negative effects with specific pigments and polymers.
5. Pigment influences on polymer photostability are not completely understood. More work on underlying causes, especially with organic pigments, is needed.
6. Assuming that a particular pigment and concentrate manufactured with that pigment is used in the fiber process, has sufficient color stability for the intended end-use, and will not interact adversely with other additives, the fiber producer may now consider what color the pigment concentrate will provide and how it may be used to provide the color.

3.5.3 DYEABILITY MODIFIERS

Although the mainstream technique of coloration for polypropylene fiber is the addition of pigments before spinning, it still has several disadvantages. This method (1) is suitable only for large-scale manufacturing, (2) is not flexible for changing the chromatogram, and (3) does not allow adjusting and changing the color during weaving. The method of pigmentation limits the diversity of the types of fabric and fails to satisfy the demands of transient customers. Despite the extensive use of pigments to color polypropylene fibers, efforts to develop a generally acceptable dyeable polypropylene fiber have continued. To be successful commercially, a dyeability modification procedure must result in a dyeable fiber at a reasonable cost and produce a product with good dyeability and fastness without disturbing the desirable characteristics of natural polypropylene fibers. A good example here is the addition of dyeable resins into polypropylene. Polypropylene fibers are highly crystalline, and the addition of polymers like phenoxy or epoxy resins results in a disperse-dyeable product as well as decreased crystallinity and orientation. Reduced fiber strength and thermal resistance can also occur. Modifications of extrusion and drawing behavior result in processing difficulties, which add to the cost of the dyeable product.

Table 3.23 illustrates various methods of modifying polypropylene developed over the years [146–149]. They may be classified as:

(1) Copolymerizing with other monomers
(2) Grafting of dye sites on isotactic polypropylene
(3) Adding dyeable polymers before fiber spinning
(4) Adding dyeable filaments during fiber processing
(5) Dissolving or dispersing additives of low molecular weight in the polymer melt
(6) Treating to modify the surface of filaments after extrusion
(7) Adding halogen compounds

3.5.3.1 Copolymerization

A moderately successful method for producing an acid-dyeable polypropylene fiber has been based on the addition of an ethylene–aminoalkyl acrylate copolymer [150].

Researchers at Donghua University (formerly China Textile University), Shanghai, China, have investigated the dyeing modification of polypropylene extensively. By using

TABLE 3.23
Exemplary Dyeability Modification of Polypropylene

Methods	General composition	Reference
Organic metal salts	Nickel complexes	[60]
Copolymers	Ethylene–aminoalkyl acrylate copolymer	[63]
Grafted polymers	Polyetheramine	[151]
	Polyetheramine	[152]
	Polyetherephthalamide	[153]
Polymeric modifiers	Dicarboxylic polymer and polyamine	[61]
	Expoxy oligomers of 4,4′-isopropylidene diphenol and epichlorohyrdrin	[155]
	Propylmethacrylic epoxy	
	Polystyrene	
	N-maleic anhydride and cross-linking agent	
Fiber blends	Cotton and acrylic filaments	[62]
Halogen compound	Ammonium bromide	[162]
Nanocomposites	Organoclay	[265]

polyolefins as an additive and by selecting appropriate compatible agents, they have successfully prepared dyeable polypropylene fiber of <1.2 dpf. This achievement makes polypropylene fiber applicable for the apparel market. In their recent investigation, the condensation polymerization of aminocarbonate was used to copolymerize with polystyrene for blending with polypropylene. Alternately, ethylene–ethyl acetate copolymer was used for blending with polypropylene to obtain a dyeable and printable fiber with tensile strength of 16.5 cN/tex and elongation of 111%. The method of adding maleic anhydride and a cross-linking agent into polypropylene melt and copolymerization was also developed to prepare polypropylene fiber with excellent dyeability and high color durability.

3.5.3.2 Grafting

Dyeable, oligomeric polymers are grafted to polypropylene to provide dye sites. These graft polymers are exemplified by polyetheramine [151], polyetheramine [152], and polyetherephthalamide [153]. Graft modification adds reactive graft groups on the main chain of polypropylene to improve its interfacial properties. Graft copolymerization introduces the dyeable group to polypropylene molecules and thus provides the fiber affinity with dyestuff. For graft-modified polypropylene fiber, the degree of graft is usually limited to 5–10% to avoid the gross increase in polymer molecular weight and possible changes in physical properties. For this reason, catalyst usage is strictly controlled during the graft modification to suppress over-graft. The methods of grafting generally include heating, radiation, and oxidation. All these methods achieve graft by engendering dissociated groups on polypropylene. Because of the complexity of polypropylene graft process and damage to polypropylene fiber, the graft modification method has not been brought to industrial scale.

3.5.3.3 Polymeric Modifiers

An early U.S. patent [154] claims that mixing polypropylene with various resins such as "other polyolefins, epoxy resin [155], polyamides, polyalkenimines, polyesters, polyaminotriazoles, polycaprolactam and mixtures thereof" will result in a dyeable product when spun into filaments. In 1969, a patent claiming acid dyeability due to the addition of a polymer based on a dicarboxylic acid and a polyamine was issued to Hercules [156].

Polymer blending has been applied in preparing new types of polymeric material. Among various methods of dyeability modification, adding dyeable component to polypropylene is most widely used. There are mainly three types of polymers for dyeability modification: low-molecular-weight polymer, high-molecular-weight polymer, and metal salt polymer. Polypropylene fiber manufacturers have already adopted the latter two types. The addition of low-molecular-weight polymer can improve dyeability to a certain degree. Low-molecular-weight polymers are widely available and the cost is relatively low. In comparison, adding dyeable polymer of adequate molecular weight into polypropylene can provide adequate dyeability. Introducing metal salt (mainly organic metal salt) into polypropylene can improve the dyeability of polypropylene fiber. Furthermore, it can provide polypropylene radiation resistance and oxidation resistance.

An important factor to consider on the blending of polymeric materials is that most polymers are incompatible with polypropylene on the molecular scale. This might cause many problems, such as macro-phase separation during blending, low interface adhesion, low tensile transfer rate, and low physical properties, which may be even lower than the unmodified polymer. To sustain good fiber properties, controlling the phase structure and interface adhesion is a necessity.

Recently, adding nonpolar or weak-polar polymers to polypropylene is widely practiced. As an example, blending polyester with polypropylene will form a matrix of two-phase microfibril structure. There is a large amount of interface between the two phases and a great deal of microcrevices for dye diffusion and penetration. With the ester group and phenylene ring in the polyester, polypropylene can absorb the dyestuff and meet the requirements of dyeing. Blending polypropylene with imide has also been developed to attain a dyeable and thermal-expandable polypropylene fiber. Polymers such as polyglycidyl methacrylate and polystyrene are found to be effective with polypropylene.

There are now polymer products in the market for the dyeability modification of polypropylene. Based on journal reports, Eif-Atochem and Centexbel, both EU firms, have developed polymeric additives that are compatible with polypropylene [157]. These compatible polymers imparted outstanding dyeability to spun fibers with selective dispersion dyestuff.

3.5.3.4 Fiber Blends

In 1970, an acid-dyeable polypropylene fiber was described [158]. It was claimed that this fiber could be dyed like nylon and could produce one-batch union shades in blends with cotton and acrylics. The dye site in this fiber was present as a microfilament dispersed within the polypropylene. These microfibers extended to the surface of the fiber and water was readily absorbed in order that acid dyes could penetrate at a reasonable rate to react with the basic dye sites. Acid-dyeable polypropylene fiber products have been available from Phillips [159] and, from Polyolefin Fibres Engineering (PFE) Ltd. [160]. PFE has also introduced a bicomponent fiber readily dyeable with disperse dyestuffs [161].

3.5.3.5 Organic Metal Salts

Nickel complexes used as energy quenchers to provide UV stability can provide dye sites with some dyes designed to form chelate complexes [146]. In recent years, hydrophilic dyestuff with long alkyl graft chain has been developed, but it can only provide light colors. Several kinds of metal salts such as nickel stearate and zinc stearate are added before fiber formation to induce complexes of metal ions in the fiber to be dyed. The addition of accelerating agents and swelling agents during dyeing also promotes dyeability.

3.5.3.6 Surface Modification

There are many methods of surface modification, among which plasma treatment has attracted most attention. It has been observed that polypropylene fiber shows improvement in water absorption and dyeing properties after it is treated with plasma gas such as air and nitrogen. The fact that active groups can react with dye molecules on the fiber surface after treatment is commonly accepted. Examination by x-ray spectroscopy showed that radical groups of N and O, usually hydroxide, carboxyl, amido, and acyl, appeared on the fiber surface. At the same time, weight loss and strength loss were observed because of the etching effect from high-energy plasma on the polypropylene fiber surface. Nevertheless, surface treatment does present many other problems such as high cost, color durability, and abrasion resistance.

3.5.3.7 Halogen Compounds

Lewin et al. [162] disclosed a novel method to prepare dyeable polypropylene or polyethylene fiber, which involves treating the fiber first with dilute bromine water, and then with concentrated ammonia. Both bromine and ammonia are absorbed in the amorphous regions of the polymer, and interact there exothermically without attacking the polymer. The reaction produces gaseous nitrogen and ammonium bromide. The nitrogen expands and creates a porous structure in the amorphous regions and on the surface of the polymer. The polymer becomes dyeable with a variety of dyestuffs, including premetallized dyestuffs. As discussed later, the ammonium bromide treatment also improves the flame retardancy of polypropylene fiber.

There are many options and elegant methods for the dyeability modification of polypropylene fiber. The key to success is the cost, fiber properties, and color durability. With further research and market evaluation, it is quite hopeful that the dyeing problem of polypropylene fiber will be successfully resolved in the near future.

3.5.4 FLAME RETARDANTS

A large variety of flame-retardancy (FR) tests [163] are generally performed on fabrics made from polypropylene fibers and not on the fiber itself. Polypropylene fiber itself does not ignite readily. When exposed to a flame, it shrinks, melts, and pulls away from the flame, thus passing most requirements in fabric form. Mixtures of antimony oxide and certain halogenated materials like decabromodiphenyl oxide have provided improved FR behavior. The thermal degradation of polypropylene has been observed to accelerate when it is mixed with chlorinated hydrocarbons. This has been explained as occurring due to attack on the tertiary hydrogen atom of polypropylene by the chlorine atoms formed during dehydrochlorination of the chlorinated hydrocarbon [164]. Since most FR systems contain halogen-containing molecules and hydrohalogenation is part of the FR mechanism, equipment corrosion during fiber- making is a concern. Researchers have therefore been searching for other ways to prepare FR polypropylene fibers without halogenated compounds.

As listed in Table 3.24, organic phosphorous compounds such as tris(2,4-di-*tert*-butylphenyl)phosphite [165] and tris(tribromoneopentyl) phosphate [166] have been attempted. A new class of *N*-alkoxy hindered amines (NOR HAS) have been pursued by Ciba as fire retardants [167]. An example of these compounds is Flamstab NOR TM 116, the structure of which is shown below. It is a reaction product of N,N'-ethane -1,2-diyl*bis*(1,3-propanediamine),

TABLE 3.24
Exemplary Flame Retardants

General composition	Examples	Reference
Halogen compounds	Antimony oxide and certain halogenated materials like decabromodiphenyl oxide Ammonium bromide	[162, 177]
Organic phosphates	Ciba Irgafos 168; tris(2,4-di-tert-butylphenyl)phosphate,	[165]
	Tris(tribromoneopentyl) phosphate	[166]
N-alkoxy hindered amines	Ciba Flamstab NOR TM 116, a reaction product of N, N'-ethane-1,2-diylbis (1,3-propanediamine). cyclohexane, peroxidized 4- butylamino-2,2,6, 6-tetramethylpiperidine and 2,4,6-trichloro-1,3,5-triazine	[167]
Metallic compounds	Magnesium hydroxide, melamine, and novolac	[172]
Intumescent systems	APP/petol-based intumescent systems catalyzed by divalent and transition metal compounds	[175, 176]

cyclohexane, peroxidized 4-butylamino-2,2,6,6-tetramethylpiperidine, and 2,4,6-trichloro-1,3,5-triazine, with a molecular weight of 2261.

Flamstab NOR TM 116, a reaction product of N, N-ethane-1,2-diylbis (1,3-propanediamine). cyclohexane, peroxidized 4-butylamino-2,2,6,6-tetramethylpiperidine and 2,4,6-trichloro-1,3,5-triazine.

It shows flame retardancy efficacy in polyolefin fibers, at concentrations as low as 1%. Flamstab NOR 116 shows excellent polymer compatibility and high extraction resistance. It also provides superior UV light and thermal stability to the polymer and shows low interaction with acidic species derived from pesticide residues or other halogenated products.

Lewin has led extensive research studies on flame retardants for polymers [168–171]. In one of his works, Weil et al. [172] reported that the combination of magnesium hydroxide, melamine, and novolac at levels as low as 1% surprisingly rendered polypropylene a UL J4 V-O FR rating. Levels of magnesium hydroxide in the formulation were 30–50%, allowing the mixture to be flexible. The novolac provided a useful dimension-stabilizing effect above the melting point of polypropylene. Magnesium hydroxide is a commercially useful flame-retardant additive for polypropylene [173,174]. Melamine is known as a flame retardant in plastics but it has not been very effective in polypropylene. It is hoped that more flexible FR polypropylene formulations may be prepared. Recently, Lewin and Endo [175,176] also reported the catalysis of intumescent flame retardancy of polypropylene by divalent and multivalent metallic compounds. The intumescent system was based on ammonium polyphosphate (APP) and pentaerythritol (petol) in polypropylene. A catalytic effect exerted by the addition of small concentrations of the metallic compounds in the range of 0.1–2.5 wt% of the compositions resulted in an increase in the oxygen index (OI) and UL-94 ratings. This catalytic effect increased with the concentration of the catalyst until a maximum was reached. At higher catalyst concentrations, a decrease in the flame retardancy parameters was observed, accompanied in some cases by a degradation and discoloration of the composition. The catalyst replaced melamine in intumescent systems. The metal compounds included oxides, acetates, acetyl acetonates, borates, and sulfates of bivalent and transition metals. The effect was concentration-dependent and reached a maximum at an optimal catalyst concentration and thereafter decreased. In the case of transition metal compounds at concentrations higher than the optimal, a degradation of the polymer was observed during processing. These observations are interesting in that highly efficient catalysts may evolve from nanosized metals, organometallic compounds as well as macromolecular metal complexes.

Finally, the inclusion of nonreactive ammonium bromide as a flame retardant in polypropylene and foamed polystyrene was known to be effective qualitatively. Ammonium bromide decomposes into ammonium and hydrobromic acid at about 220°C, which is low with respect to the polymer processing temperature. Concern for equipment corrosion has limited its use. Lewin et al. [177] reported a novel method for introducing ammonium bromide into the FR fiber after spinning. This is accomplished by (1) soaking the polymer (fiber, film foam) in an aqueous bromine solution and (2) treating the bromine-containing polymer with an ammonia solution. Bromine is exothermally reduced by ammonia to form ammonium bromide in the amorphous region of the polymer. The resulting fiber exhibited improved dyeability and higher efficacy (increase in OI/% bromine) than organic bromide by direct addition. A polypropylene fabric containing 5.24% ammonium bromide gave an OI of 24.2 as compared to 19.6 for an unmodified polypropylene.

3.5.5 Fiber Finishes

Fiber finishes are normally added as lubricants or as antistats to the surface of fibers to facilitate fiber production and subsequent processing. Finishes are additives that may not achieve exactly what is intended. Low-molecular-weight mineral oils dissolve readily in polypropylene as do other materials added to the surface. It has been proposed that finishes may cause softening of the polymer surface [178], particularly at higher temperatures. Since polypropylene fibers are produced in colored form, the producer finish is not necessarily scoured off because it is for a dyeable fiber. The finish used by the fiber producer most frequently remains as part of the final product—for better or worse. Thus, the fiber finish must be considered not only as a processing aid but also as an additive.

3.6 PREPARATION OF POLYPROPYLENE FIBERS

Polypropylene fibers are produced by a larger variety of processes than other melt-spun fabrics. At one end of the range, the long air-quench process produces high-quality multifilament yarns, and, at the other end, fibrillating slit film produces coarser fibers. The success of the lower-cost polypropylene slit-film fiber is due to the lower price of the polypropylene resin and the unique adaptability of polypropylene to the less expensive slit-film fibrillation process. The water-quench process for monofilament has long been an established technique for producing high-denier fibers. In recent years, the spun-bonding process for producing nonwoven fabrics and, very recently, the short air-quench system for producing fibers in less building space have been developed.

3.6.1 LONG AIR-QUENCH MELT SPINNING

Until short-quench systems were developed in the 1980s, there was no reason to qualify the air-quench spinning process. Now the term "long air quench" is used to distinguish spinning systems that utilize take-up speeds from 300–4000 m/min and a filament freefall length of 3–10 m. The long lengths are needed for sufficiently cooling the high-speed thread line. Yarns with individual filament deniers from around 2 to 25 and total deniers from 75 to several thousand are produced by the tong air-quench process. It is also used to produce staple fibers.

3.6.1.1 Metering Pump

The melt spinning of polypropylene begins at the hopper of an extruder, where the polypropylene, either a uniform blend or a volumetrically or gravimetrically proportioned mixture of natural and colored material (pellets, spheres, or powder), is fed into the throat of the extruder. Resins with MFR of 3–35 are used in the process. As polypropylene is nonhygroscopic, drying of the feed material is not necessary. However, the surface moisture should be removed and nitrogen-blanketing of the extruder hopper is optional. The particulate material is then melted and conveyed through the electrically heated extruder. Normally a single-screw extruder is employed. Extruders of at least 24/1 Length-to-diameter ratio (L/D) are desirable for polypropylene, and some resin suppliers recommend that a portion of the screw be equipped with a mixing zone. Others prefer to mix downstream of the extruder. This mixing is usually accomplished with commercially available static mixers. The comprehensive treatment of single-screw extrusion is in the literature [179]. A monograph treats twin-screw extrusion [180].

The filtration of polymer melt is normally supplied immediately downstream from the extruder. This filtration can range from a coarse single screen supported by a breaker plate to a very fine nonwoven metal filter contained in a separate housing that allows the filter to be changed without interruption of the process. The choice of filtration depends on the process, the feed material, and the product.

After the melt leaves the filter, it travels through the transfer lines or spin manifold to the metering pumps. One extruder may feed as many as a hundred metering pumps. The transfer lines should be designed in such a manner that the material entering each metering pump has the same heat history and residence time. Also, there should be no stagnant regions within the lines. The residence time uniformity is usually accomplished by arranging the transfer lines in a "Christmas tree" pattern. In such a design, the stream from the extruder is divided several times until finally there are as many separate, identical streams as there are pump entrances. The transfer lines are usually enclosed in a space heated by a condensing organic vapor to ensure uniformity of temperature. Electrical heat may be used, but extra caution is necessary in the system design to maintain symmetrical heat flux. The transfer lines must also be designed to ensure that the pressure drop through them is not excessive. If the pressure drop is too high for the desired flow, the metering pumps will be starved. The system housing the transfer lines, metering pumps, etc., is called a pumpblock or spin beam.

The metering pumps are precision-made positive-displacement gear pumps, especially designed for melt spinning. They are used to ensure that a precise proportion of the melt from the extruder is delivered to each thread line and that the thread line denier does not vary with time. Modern pressure-feedback-controlled extenders operate remarkably steadily, but not steady enough for multifilament spinning without meter pumps.

3.6.1.2 Spin Pack

After leaving the metering pumps, the melt travels through short transfer lines and into the spin pack. The design of the spin pack has evolved through the years, and almost every fiber manufacturer has his own specific design. Since the 1980s, top-loading packs have replaced bottom-loading packs for convenience of installation; but when small polymer rates are employed, bottom-loading packs are still preferred because of better thermal contact with the spin beam. Bolted flanges hold the bottom-loading pack in place and melt flows into it vertically. The packs are sealed by gaskets that are crushed when the flange bolts are tightened. Top-loading packs rest on a ledge, and melt flows into them horizontally. During installation, the pack is lowered into its cavity and one horizontal screw crushes the gasket on the opposite side to seal the transfer line pack inlet.

Good pack design ensures that the melt inside the pack flows in a symmetrical manner with respect to the spinneret. If the spinneret is round, the melt flow in areas above the spinneret should be symmetrical with respect to the spinneret axis. If the spinneret is rectangular, the flow should be symmetrical with respect to the planes passing through the center lines in both the length and width directions. Care should also be taken to eliminate stagnant areas.

The pack filter, the breaker plate, and the spinneret are contained in the lower portion of the pack. These components, along with the metering pumps, are the heart of the melt-spinning process. The filter rests on the breaker plate and may be composed of woven wire, nonwoven wire, sintered metal, or sand. The major function of a polypropylene filter is filtration; there seems to be little need for shearing polypropylene. The breaker plate should contain as much open area as possible and should be thick enough to withstand the pressure drop developed through the filter. For flat and relatively thin filters, the only portion of the filter that is used is the portion just above the openings in the breaker plate. The melt flows through these openings, and funnels of flowing melt develop in the filter above them. Regions outside these funnels are relatively stagnant. Rise in pressure is rather rapid in pigmented polypropylene, and longer spin pack life is realized with extended-area filters. There must be a gap between the breaker plate and the rear of the spinneret. This ensures that the pressure drop along melt stream lines from the bottom of the breaker plate to the entrances of spinneret counterbores is negligible in comparison with the pressure drop through the spinneret orifices.

3.6.1.3 Spinneret

The spinneret orifices or capillaries are contained in the lower part of the spinneret and are round for producing round fibers or in almost any shape desired. Triangular, trilobal, and octalobal fiber cross sections are common and are designed for luster and fiber "feel" characteristics. The face of a spinneret is generally coated with a silicone spray to prevent resin from adhering to the spinneret.

There are no general criteria for the design of spinneret orifices. Diameters of round capillaries are usually 200–400 μm. The L/D is normally 2–5; capillaries with large values of L/D are expensive to produce and give unnecessarily high-pressure drops. The entrance angle to the round orifices is usually 40–90°. Orifices of modified shape are generally placed at the center of flat-bottomed counterbores. To produce yarns with good interfilament uniformity, the orifices must be uniform.

Die swell is always present in polypropylene spinning because its melt is viscoelastic. The viscoelasticity also gives rise to melt fracture if temperatures are too low or if orifice shear rates are too high. Melt fracture must be avoided.

3.6.1.4 Quenching

After exiting the spinneret capillaries, the filaments begin their descent to the winder or take-up position. The filaments first enter the quench cabinet, usually 1–2 m in length, where they are solidified. Further cooling takes place in the quench stack between the cabinet and the point of finish application. The quench zone is a very important part of a melt-spinning process. With polypropylene, the emerging filaments are from 70 to 170°C above the melting point (165°C). This difference between spinning temperature and melting point is much higher than it is for nylon or polyester.

Careful consideration should be given to the aerodynamics of the quench cabinet. There is normally a cross-flow of air all along the fiber bundle length. The cross velocity is usually higher in the regions near the spinneret. The thread line, traveling at high speeds, develops an air boundary layer rapidly, and the cross-flow becomes ineffectual after the boundary layer develops sufficiently. Turbulence or unstable flows anywhere along the thread line can cause filament nonuniformity.

In many cases, the spin pack and spinneret are round. The allowable diameter of these spinnerets is limited, because with too large a diameter the quenching of the interior filaments and the filaments downstream of the cross-flow is inadequate. In recent years, quenching of a large number of filaments produced from a circular spinneret has been improved by blowing air radially outward from an air source placed on the spinneret axis. Also, rectangular packs have been introduced for producing large bundles of filaments.

Although some orientation is introduced into the melt during its passage through the spinneret orifices, most of this orientation is lost at the orifice exit because of strain recovery. This unconstrained recovery gives rise to die swell. The orientation in the undrawn yarn is a result of the stretching of the filament between the position of maximum diameter just below the orifice exit and the position of final diameter. However, the development of crystalline orientation in polypropylene continues beyond the position of the final diameter. The resin properties and the spinning conditions determine the amount of orientation in undrawn yarn. Any condition that gives higher stress in stretching the fiber gives higher degree of orientation. Resins of higher molecular weight, larger quench rates, and larger stretch rates give higher spin-line stresses. The take-up speed is one of the most influential parameters affecting orientation.

3.6.1.5 Finish Application

Immediately after exiting the quench stacks (or in some cases at the bottom of the quench cabinet), the yarn receives an application of finish. The finish is applied by contacting the yarn to a rotating roll partially immersed in finish or by metering it onto the running thread line with precision gear pumps. The finish, usually an aqueous emulsion, is normally applied at the 0.5–1.5% level. Functionally, finishes serve as antistatic agents and lubricants. The finish must remain colorless on the yarn and be heat-stable. In addition to the functional ingredients, emulsifiers and antibacterial agents are always present in a finish makeup.

Finish application is considered the last operation common in the spinning process; after this step, the yarn processing can be varied widely, depending largely on the end-product to be made. The undrawn yarn is wound onto a spool and transferred to another area for further processing. These yarns are simply drawn into "flat" yarns with no crimp or bulk, or they are draw-textured on machines similar to those used for draw-texturing polyester; or heavier yarns are drawn and jet-bulked for use in carpets. Instead of being transferred for further processing, yarns are drawn and bulked in a continuous operation with spinning. Staple fibers are normally spun as continuous filaments in the long air-quench process. In modern staple processes, the thread lines formed from a number of spinnerets are combined into a large tow and "piddled" into a large can for subsequent drawing on a horizontal fiber line.

3.6.1.6 Drawing and Annealing

There are many variations of the drawing and annealing process, depending on the yarn denier, spinning speed, etc. However, most processes carry out the same two basic functions: (1) orientation of the filaments to increase the strength and decrease the elongation and (2) annealing of the filaments to relax at least partially the frozen-in stresses and perfect the crystalline structure, and consequently decrease the residual shrinkage. Polypropylene is very viscoelastic, and the short residence time in the in-line annealing zone may not be long enough for sufficient relaxation; further heat treatment may be necessary.

The draw-twister, capable of processing more than 100 yarn packages simultaneously, is the oldest, most common unit for drawing polypropylene yarns. The draw-twister consists of a creel for holding the undrawn packages, two or more sets of rolls, a lay rail that supports the ring with its traveler, and a vertical spindle for holding the pirn onto which the yarn is wound. The yarn is wrapped around the rolls several times. The yarn wraps are separated with the aid of a small offset separator roll, and the drawing roll is rotated faster than the feeding roll. Depending on the orientation of the undrawn yarn, it is stretched, usually two to six times its original length, between the two rolls. Some machines are equipped with a third set of rolls, allowing a two-stage draw to be employed. In the usual case, the feeding roil is heated so that the relaxation times of the yarns will be decreased for better drawing. The drawing roll is usually heated to a temperature higher than that of the feeding roll. This heating (annealing) decreases the residual shrinkage of the yarn. After leaving the drawing roll, the yarn travels under the ring traveler and onto the rotating vertical pirn. Twist is imparted to the yarn because the surface speed of the pirn is higher than the yarn speed and the traveler moves around the ring that surrounds the pirn. The lay rail traverses vertically to lay the yarn along the length of the pirn. A draw-winder is similar to a draw-twister, but the yarn is wound onto a horizontal spool whose surface travels at the yarn speed. The yarn contains no twist.

As mentioned earlier, yarns may be drawn in-line with spinning. The drawing scheme is similar in principle to that of the draw-twister, except that the feeding and drawing rolls must travel at much higher speeds to be economical. In this case too, the yarns have no twist. To provide filament coherence, the yarns are passed through air jets to entangle the filaments at intermittent lengths along the thread line.

Staple is typically drawn on a horizontal fiber line. A fiber line consists of two or more roll stands, perhaps a heating zone between the roll stands, a crimper, and a cutter. There may also be an oven between the crimper and the cutter. Each of the roll stands consists of five to seven large rolls arranged in two rows, which are commonly driven on a large common frame. Circulating liquids or condensing vapors heat the rolls. The heating zones between the roll stands may be hot-air or steam ovens, hot-liquid baths, or hot plates. The crimper is either a common stuffer box crimper or a jet-crimping device, and the cutter is usually of the rotating pressure-cutting type. One or more tows of yarn are drawn simultaneously. The tows travel a serpentine path from one roll to another and are drawn and then annealed between the roll sets. On exiting the last roll set, which rotates with a lower surface speed than the drawing roll to allow the yarn to relax, the yarns are fed directly to the crimper and finally to the cutter. The staple is then transferred to the baler for packaging. In some processes, the yarn may be heat-set in an oven after it is crimped.

Lower-denier polypropylene may be draw-textured on conventional draw-texturing machines, which were developed for polyester. On these machines, undrawn yarns are creeled and fed by a roller system to a heated plate. The yarns are drawn and twisted simultaneously on this hot plate. The twist is supplied by a rotating pin around which the yarn is wrapped or by rotating disks that contact the yarn. On the other side of the twisting device, the twist is removed, but the yarn remembers its deformation on the hot plate and retains a bulk or crimp. The yarn then travels through the second heater and onto the spool. This process is known as false-twist draw texturing.

Heavier-denier BCF yarns are generally spun for tufted or woven carpeting. Typical processes for producing BCF yarns consist of means for drawing the yarn coupled with a stream or hot-air jet-texturing device for imparting bulk. The bulked yarn is removed from the jet-crimping device and wound onto a disposable tube. There are many variations of the BCF process.

There are variations of the spinning process that have not been mentioned here, e.g., bicomponent spinning. Usually, this operation is carried out by employing two extrusion systems that keep the resins separate until just before the streams exit the spinneret. Sheath-core or side-by-side yarns can be spun by this process. Simpler bicomponent yarns are spun with mixed incompatible resins from mixed-pellet blends.

3.6.2 SHORT AIR-QUENCH MELT SPINNING

In the 1980s, a short air-quench process was developed in Europe for polypropylene. The primary purpose of this design is to save building space and reduce labor costs; the entire system can be located on one floor. In principle, the short quench process is basically the same as the long quench process; but, because of the shorter distance available for quenching, no more than a meter or so, the spinning speeds are drastically reduced. Factory literature lists a maximum spinning speed of 150 m/min for these machines, although most operating speeds would be lower than that. To make up for the loss in production due to lower spinning speeds, spinnerets with thousands of orifices are used.

Some of the spinning units are built in modular form with a small extruder for each one or two spinnerets. However, larger extruders are used for feeding larger numbers of spinnerets.

The units are primarily designed for making staple fibers (3–100 denier), but some manufacturers suggest their use for making textured or bulked yarns. In any case, the spinning units are placed in-line with the conventional drawing, annealing, and bulking systems downstream.

To further save space, some units are designed to enable the spinnerets to extrude filaments in a horizontal or vertically upward direction. This allows the extruders and other spinning equipment to be located closer to the floor.

The products produced from some of these machines are apparently not as high in quality as those produced from long air-quench systems. Further information concerning the short air-quench systems is available from the European machinery manufacturers.

3.6.3 WATER-QUENCH MELT SPINNING

The manufacture of large monofilaments, certainly larger than 100 denier, must be carried out in a water-quench process because of the inherent limitation of forced convective heat-transfer with gases. Conventional screw extruders are used for melting and conveying the polypropylene resin, just as in the case of the melt-spinning processes described above. Often, however, there are no metering pumps used in the process.

3.6.3.1 Spinning and Quenching

In the normal polypropylene operation, one large spinneret is employed (may be two or three), with the number of spinneret orifices corresponding to the number of filaments desired. The spinnerets are usually heated electrically. When filament uniformity is important, condensing organic vapors are used for heating. Monofilaments of polypropylene usually go into such products as chair webbing, ropes, etc., and wider limits of uniformity are normally acceptable in these products. If uniformity in size is critical, metering pumps should be used, and extra care must be taken in ensuring that the melt entering the spin pack is uniform, the orifices are uniform and clean, and the temperature is uniform.

After exiting the spinnerets, the filaments travel a short distance and enter a water-quench tank. Guides are arranged within the tank to ensure the filaments travel a proper distance for quenching. For a uniform product, the distance between the spinnerets and the water should be held constant. The water temperature should be maintained within close limits throughout the tank, and care should be taken to ensure the water contains no turbulent regions.

3.6.3.2 Draw Resonance

There are two unique aspects to the water-quenching of polypropylene. As with other thermoplastic melts, this material exhibits draw resonance under certain spinning conditions. Polypropylene apparently is more susceptible to this phenomenon than many other materials. Under certain conditions, the filament will be uniform all along its length. If the take-up speed is increased, a critical value of speed will be reached when the filament diameter begins to oscillate in a regular manner. The amplitude of diameter oscillation can reach 10–50% of the nominal diameter of the filament. The instability seems to be related to a critical value of the ratio V_T/V_E, where V_T is take-up velocity and V_E is extrusion velocity. The critical ratio is affected by resin properties, extrusion temperature, filament nominal diameter, distance between spinneret and bath, and perhaps other process parameters.

When polypropylene is water-quenched under certain conditions, a difference in crystalline order is obtained. In air-quench systems, yarns with a crystallinity of around 55% are obtained. The crystals are usually distributed throughout the yarn as lamellae and are normally in the monoclinic α-form. In the rapidly cooling water-quench process, a less-ordered smectic or paracrystalline structure is produced with a morphology that is basically fibrillar. Under certain drawing conditions, these smectic structures lead to high-strength fibers.

The orientation of the spun monofilaments and, consequently, the draw ratio necessary for producing a satisfactory product depend on the stretch rate during spinning and quenching, just as they do in the air-quench process. The larger the stretch rate, the higher the spin-line stress, the higher the orientation, and the lower the required draw ratio.

3.6.3.3 Drawing

After the filaments leave the quench tank, they must be wiped or blown free of water in preparation for hot-drawing. Wet sections of the yarn will not be heated to the same temperature as the dry sections. The process for drawing monofilaments is very similar to ST. Roll stands and hot ovens, baths, or plates are used for the filaments. The yarns are usually annealed on heated rolls or in an air oven. A two-stage yarn may be used, with a total draw ratio commonly around 6:1. A finish must be applied to the filaments for lubrication and static control. The filaments are wound individually on tubes or flanged bobbins.

3.6.4 SPUN-BONDING PROCESS

The term "spun-bonding process" appears to have several definitions. Perhaps the most general one, and the one used here, is defined as the method wherein the production of fibers and fabrics or webs is combined. The usual spun-bonding process consists of laying down a web or mat of continuous filaments and then bonding the fibers within the fabric either mechanically, thermally, or chemically. The so-called *melt-blowing* process, in which short, fine fibers are laid down, also fits the general definition of a spun-bonding process.

3.6.4.1 Spun-Bonding

The typical spun-bonding process includes a spinneret with a means of conveying the continuous filaments out into the form of a web, or else several spinnerets arranged in a row whose length may approximate the width of the web to be produced. In either case, the process of melting and fiber formation is identical to that used in conventional melt spinning.

The steps beyond filament extrusion are diverse and are usually proprietary. The yarns are pulled from the spinneret, perhaps by air jets, given an electric charge, and are then laid down on a moving screen. Alternatively, the yarns may be pulled away by rolls and drawn between a set of rolls before they are fed to the air jets. In the former case, the yarns are usually only partially oriented, even though they are stretched at very high speeds. After they are formed, the fabrics are thermally bonded by calendering, or needle-punched, or chemically bonded. For information in the patent literature, the reader should see Gillies [181].

3.6.4.2 Melt-Blowing

The melt-blowing process was developed on a small scale at the U.S. Naval Research Laboratories; subsequently, larger-scale development was carried out by Exxon and the Beloit Corporation. The process produces nonwoven fibrous webs from any thermoplastic resin but has been used most extensively with polypropylene, which appears to be particularly suited for melt-blowing.

In this process, a long rectangular die or spinneret containing a large number of capillaries in a single row all along its length is attached to a conventional screw extruder. Molten polymer is fed directly from the extruder through a filter and through the capillaries. Hot air at near-sonic velocities is forced through adjustable slots all along the length of both sides of the die. Fiber attenuation and breakage (usually) occur as the hot melt issues from the capillaries into the convergent high-velocity air streams. Some of the hot air is captured and recirculated. There are also cooling air ducts located along both sides of the die. This air is directed toward the filaments at a position further from the die and aids in quenching. The fibers are collected on a continuous, moving wire screen under which a suction chamber is located to aid in the removal of large volumes of air. The web is fed from the collecting screen to a take-up roll or a hot calendering system. More than one extruder–die combination may be used to give laminates of the same or different materials.

The process produces a web with a width the same as the length of the die. The weight of the fabric is determined by the output of the extruder and the speed of the collecting screen. The average orientation of the fibers relative to the machine direction depends on the air flow and screen speed. The diameter of the fibers can be as small as 0.5 μm, depending on process conditions. The length of the fibers also varies according to process conditions. Obviously, the lower the polymer flow rate and the higher the air flow rate, the finer and shorter are the fibers. Continuous fibers are produced at very low air rates. The temperature and flow rate of the cooling air affect the hand of the web: higher temperatures and lower rates give a softer fabric. A similar effect is obtained by decreasing the distance between the die and collecting screen. As a result of all of the possible operating variations, the process is very versatile. Under extreme conditions, very lightweight webs of very fine fibers are produced and very stiff, almost cardboard-like sheets may also be produced. The major deficiency in the process is that under any condition only undrawn fibers are produced.

3.6.5 Fibers from Film

Since the early 1980s, techniques for producing relatively coarse but very useful fibers or fiber-like tapes from films have been developed. Obviously, the first step in this operation is the production of a film. Films are manufactured in two conformations—sheet and blown films. Because the blown film operation is more complicated and because it produces a less fibrillatable film due to a more ordered crystalline structure and lateral orientation, it is not widely used and will not be discussed further.

3.6.5.1 Sheet Film Extrusion

Sheet films are produced from long, narrow-slit dies, which are usually attached directly to a conventional single-screw extruder. There is normally a filter between the extruder and the die, and sometimes a static mixer is used. Metering pumps are almost never used. The die opening, or a restrictor bar upstream of the opening, can be adjusted along the die length so that the film thickness is maintained uniformly across its width. The film thickness is measured by a gauge and the gauge readout is used for manual die adjustments, or a gauge signal is used for automatic control. Resins with MFR of 1–4 are commonly used for these films.

The film is normally quenched in a water bath but may also be quenched on chilled rolls. The chilled rolls are expensive and their use is restricted. The quench tank for films is very similar to that used for monofilament. The molten film exits the die an inch or so above the water level and film guides route the sheet through the tank so that the tank water temperature is uniformly and constantly controlled and the water is free from turbulence. The film leaves the tank at a speed usually no greater than 45 m/min. The speed is limited because of "water carry-over" by the film. If the film is wet, difficulties will be encountered in drawing. Resin manufacturers control the wettability of films with additives and, consequently, the running speed. Drying methods, such as blowing with air, are applied as an added precaution just after the film leaves the water-quench tank.

3.6.5.2 Drawing and Annealing

Before drawing, the film is usually pulled through a set of knives for slitting to the desired width for drawing. The slitting knives are held in a bar and may be rectangular (similar to razor blades) or round. In some cases, the blades are heated. Drawing and annealing are carried out in ways similar to those used for processing ST and monofilament. Heat is applied with ovens, rolls, plates, etc., and the tapes are drawn and relaxed between large roll stands. Processes for drawing the film before slitting are limited in usage. The drawing is carried out on rolls spaced very closely, in order that the contraction of the film in the width direction is not as great as it is in the more widely used processes. The lower lateral contraction gives a material that more closely resembles blown film with regard to fibrillation tendency. In unrestrained drawing of incompressible tapes, both the width and thickness are reduced by the factor $\sqrt{1/\text{draw ratio}}$.

3.6.5.3 Fibrillation

After the yarn is drawn, it is fibrillated. A large number of devices have been invented for fibrillating tapes in-line to give them a more fiber-like character. In various processes, the tapes are rolled, twisted, brushed, pulled, treated with air jets, subjected to ultrasonics, or contacted with rotating rolls that have cutting or punching devices on their periphery. The most prevalent method is that which uses rotating rolls. This method is simple, easy to control, and gives a more uniform product. The peripheral cutting devices can consist of pins, hacksaw blades, razor blades, or fine threads or serrations cut on the lands of a fluted roll. In operation, the running tape is fed over the rotating roll, which rotates so that its contacting surface travels in the tape direction at speeds of one to three times the tape speed. Oriented polypropylene is very adaptable to this fibrillating process. Strong, very fiber-like tapes with a denier of 300 and up are produced. If the fibrillated tapes are pulled laterally, a network structure is revealed.

There are other, infrequently used processes for producing fibers from films or tapes that should be mentioned. In one of these, the film in the molten state is pressed between two rolls. One of the rolls is smooth and the surface of the other contains closely spaced, shallow peripheral grooves. The film is then embossed all along its length and, in subsequent stretching and perhaps mechanical treatments, it separates along the grooves. Relatively fine monofilaments (6–10 denier) are produced in this manner. Groups of the monofilaments are taken up together to form a heavier yarn. A variation of this process is the direct extrusion of embossed tapes from specially designed orifices, which are subsequently drawn and fibrillated with air jets. In another process, fibers are cut with blade or notched bars at very close intervals to give continuous filaments.

In some instances, chemicals, blowing agents, or other polymers are blended with polypropylene to enhance fibrillation or, in certain cases, to retard it. Additives may also be used for modifying the physical properties of the film fiber. Processes have been suggested for using blowing agents as the major means of fibrillating films, but these methods have not yet proven acceptable. For a detailed review of fibers from film, the reader should see Krassig [182].

3.7 FIBER PROPERTIES

3.7.1 As-Spun Fibers

3.7.1.1 Crystalline Structure

On the basis of the most accepted model of fiber structure, spun polypropylene fiber is considered to consist of two phases: an amorphous phase and a crystalline phase [183,184]. The crystalline phase is composed of discrete lamellae or crystallites. The lamellae, each consisting of many chain-folded molecules, are connected by tie molecules, which make up part of the amorphous phase. Other molecules included in the amorphous phase are those whose tacticity or molecular weight difference prevents their crystallization. In addition, portions of some molecules exist in disordered fold regions on the surface of the crystal. Even within the crystals, imperfections and disorders exist, and these regions contribute to the amorphous content. The crystallites and amorphous regions exist together in large, complicated structures. Quiescent melts crystallize in clusters of spherulites. The spherulites form around a nucleus, which is a chance alignment of molecular chain segments that cause crystallization to begin. When melts solidify under a condition of sufficient stress, as fibers usually do, a different aggregation of crystalline and amorphous regions arises. Instead of point nucleation, row nucleation occurs, and lamellae grow epitaxially along fibrillar molecular structures. This gives rise to the so-called shish-kebab structure. The exact nature of the aggregates existing in the spun fiber depends entirely on tacticity, molecular weight and molecular weight distribution, additives, and spinning conditions.

In spite of the complex nature of the aggregations of the crystalline and amorphous regions in polypropylene fibers, it has been found that the mechanical properties of both spun and drawn fibers depend largely on the relative amounts of the amorphous and crystalline regions present and the molecular orientation of each of the phases [79,183,184].

Polypropylene Fibers

The orientation of the crystalline region is measured by x-ray diffraction techniques. The orientation in the less-ordered amorphous region is determined by x-ray, infrared measurements, sonic modulus technique [183], or by separating the crystalline and amorphous contributions to the fiber birefringence [79]. The average orientation of the crystalline regions is specified by the Hermans–Stein orientation functions:

$$f_a = \frac{3\cos^2\psi_a - 1}{2} \tag{3.10a}$$

$$f_b = \frac{3\cos^2\psi_b - 1}{2} \tag{3.10b}$$

$$f_c = \frac{3\cos^2\psi_c - 1}{2} \tag{3.10c}$$

where a, b, and c refer to the crystal axes of the unit cell, and ψ_a, ψ_b, and ψ_c are the angles between the respective crystal axes and the fiber axis. Polypropylene crystallizes most commonly in the α- or monoclinic form. In rapid quenching, such as spinning into ice water, polypropylene arranges in a paracrystalline or smectic form [13,185,186]. In monoclinic polypropylene, the chains are helices with axes lying along the c-axis. In Equation 3.10, f_i can range between -0.5 and $+1$. An f_i value of -0.5 indicates the crystal axis is perpendicular to the fiber axis, a value of 0 indicates random orientation, and a value of $+1$ indicates the crystal axis is parallel to the fiber axis. For orthogonal crystal axes,

$$f_a + f_b + f_c = 0 \tag{3.11}$$

It should be noted that the orientation functions do not specify the distribution of crystallite orientation. More than one distribution may produce the same orientation functions. In the case of monoclinic polypropylene, the a-axis makes an angle of 99.3° with the c-axis, and Equation 3.11 does not hold. For convenience, it is customary to define an axis, which is not a true crystallographic axis but is one perpendicular to both the b- and c-axes. Then the orientation of the a-axis may be determined from Equation 3.11, f_b, and f_c.

What would seem to be a hopeless task characterizing fibers according to the aggregation of the amorphous and crystalline regions has been greatly simplified by the fact that the fiber properties are related to the amounts and the orientations of the two phases. Another simplification arises with the manner in which these orientations are related to the spinning parameters, at least for the relatively low spinning speeds that have been used in most of the studies presented in the open literature.

3.7.1.2 Spinning Stress

Ziabicki [187] has given an extensive review of studies of the melt-spinning process. Some derivations from this review are discussed briefly. A simple analysis for predicting the effects of process parameters on fiber orientation are also discussed before polypropylene data are presented. To summarize, the properties of spun polypropylene fiber (and other fibers as well) are primarily determined by the stress existing in the spin line at the position of the final diameter, at least at moderate spinning speeds. The stress level is determined largely by:

1. Take-up velocity
2. Molecular weight

3. Extrusion temperature
4. Extrusion velocity
5. Air cross-flow rate and temperature

Following Ziabicki [187], we may assume that the radial variation of velocity is negligible in a spin line. With this assumption, one may integrate the equations of motion to obtain a force balance for any axial position:

$$\pi[D(x)/2]^2 \sigma_{xx}(x) = \pi[D(o)/2]^2 \sigma_{xx}(o) + \rho Q[V(x) - kV(o)]$$
$$+ \pi\alpha[D(o)/2 - D(x)/2] - g\int_0^x (\rho - \rho^\circ \cos\omega)\,\mathrm{d}x \quad (3.12)$$
$$+ \pi\int_0^x \rho^\circ V^2(x) c_\mathrm{f}\,\mathrm{d}x$$

where D is the filament diameter, σ_{xx} the axial stress in spin line, ρ the spin-line density, Q the volumetric flow rate, V the spin-line velocity, k the correction factor for velocity profile, α the surface tension, g the acceleration of gravity, ρ° the density of the surrounding medium, ω the angle between the spin-line axis and the vertical (cos $\omega = +1$ for spinning down, -1 for spinning up), c_f the skin friction coefficient, and χ the spin-line distance measured from zero at the position of maximum spin-line diameter.

Term 1 is the total spin-line force at any spin-line position; term 2 is the spin-line force at position of maximum die swell; term 3 is the force due to acceleration of spin line; term 4 is force due to surface tension; term S is the force due to gravity; and term 6 is force due to air drag. If cross-flow is present, there will be another term. This form of force balance presents a clear picture of the terms contributing to the stress at any position. For making calculations, the integration is carried out from the take-up position, where the force is measured, up to any arbitrary spin-line position. This will eliminate from the calculation the force term at the position of maximum diameter, which is not easily evaluated.

The equation of continuity is integrated to give

$$\pi[D(x)/2]^2 V(x) = Q \quad (3.13)$$

If the absence of a radial temperature distribution is assumed, the following is obtained for the energy balance:

$$QC_p\frac{\mathrm{d}T}{\mathrm{d}x} = C(x)[h(T - T_\mathrm{A}) + \varepsilon k(T^4 - T_\mathrm{S}^4)] + Q\Delta H_\mathrm{f}\frac{\mathrm{d}\Phi}{\mathrm{d}x} \quad (3.14)$$

where Q is the mass flow rate, C_p the heat capacity, C the fiber circumference, T the fiber temperature, T_A the ambient temperature, T_S the temperature of surroundings, ε the emissivity, k the Stefan–Boltzmann coefficient, h the heat transfer coefficient, ΔH_f the heat of fusion, and Φ the crystalline fraction.

The temperature of the fiber at any axial position is of course determined by the extrusion temperature, convective heat transfer, radiative heat transfer, and heat liberation due to crystallization. Heat generated by viscous dissipation is negligible.

These three equations are the general equations for melt spinning in the absence of radial variation of velocity and temperature. Ideally, if a constitutive equation relating stress to deformation and temperature were available, the stress and diameter profiles could be calculated from the equations. Further, if a relation could be found between stress and fiber properties, the fiber properties could be calculated for any set of spinning conditions. Unfortunately, such general relations do not exist now. However, the equations are useful. If the diameter profiles are determined experimentally, the force balance may be used to calculate the stress σ_{xx}, which is a rheological stress, at any thread line position, given approximations for air drag. Ziabicki discusses simple approximations [187]. In addition, with these experimental profiles, estimates of stress level are made with a variety of presently available constitutive equations.

The relative values of the terms in the force balance depend strongly on spinning conditions, and in certain cases some of the terms are negligible. The force due to surface tension is always small. In slow-speed spinning, all terms are negligible except 1, 2, and 5. Ziabicki presented a chart showing the relative magnitudes of the terms for nylon 6 monofilament (Figure 3.21).

Numerous authors have found that, for polypropylene (and other resins), stress in the spin line is the determining factor in the orientation of the molecules prior to solidification, and that this orientation gives rise to rapid crystallization in the spin line. Hagler has given a simple method for estimating qualitatively the effects of various spinning parameters on spin-line stress and consequently the properties of the spun fiber [188]. The stress is related to how rapidly the spin line is stretched and cooled, not how much drawdown exists in the spinning process. Many authors erroneously use drawdown or draw ratio, V_T/V_E, where V_T is the take-up velocity and V_E is extrusion velocity, for explaining variations in spun fiber properties. Draw ratio is important for drawn fiber properties, where the time constant of relaxation is much larger.

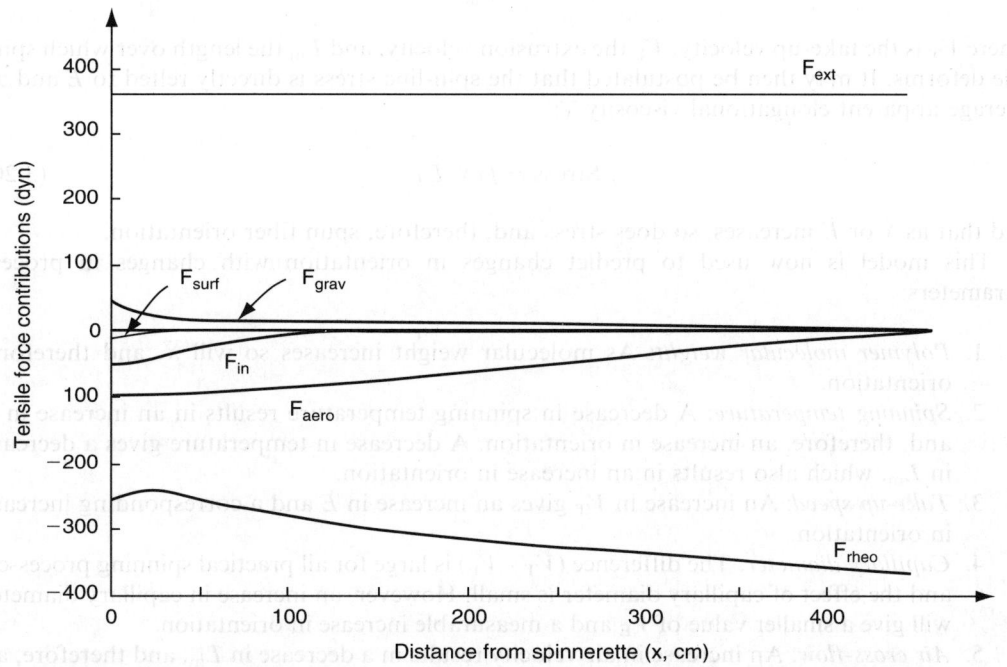

FIGURE 3.21 Components of spin-line tension for 70-denier nylon 6 monofilament spun at 646 m/min. (From Ziabicki, A.; *Fundamentals of Fibre Formation,* John Wiley & Sons, New York, 1976. With permission.)

Following Hagler, if one considers the simple elongation of a rod, the rate of elongation at any time is given in terms of the instantaneous length as

$$E = \frac{1}{\ell}\frac{d\ell}{dt} \quad (3.15)$$

If the rod is a Newtonian fluid, the axial stress is given by:

$$\sigma_{xx} = 3\mu \frac{1}{\ell}\frac{d\ell}{dt} \quad (3.16)$$

where μ is the Newtonian viscosity. For the case of stretching a Newtonian spin line in which the diameter varies:

$$\sigma_{xx}(x) = 3\mu \dot{E}(x) \quad (3.17)$$

For viscoelastic melts or solutions, the above equation has been extended to give

$$\sigma_{xx}(x) = \chi \dot{E}(x) \quad (3.18)$$

The apparent elongational viscosity, x, is not a constant or a function of the instantaneous value of E but depends on the entire history of E. We may define for a spinning process an average elongation rate:

$$\overline{\dot{E}} = \frac{V_T - V_E}{L_m} \quad (3.19)$$

where V_T is the take-up velocity, V_E the extrusion velocity, and L_m the length over which spin-line deforms. It may then be postulated that the spin-line stress is directly relted to E and an average apparent elongational viscosity $\overline{\chi}$:

$$\text{Stress} = f(\overline{\chi}, \overline{\dot{E}}) \quad (3.20)$$

and that as $\overline{\chi}$ or $\overline{\dot{E}}$ increases, so does stress and, therefore, spun fiber orientation.

This model is now used to predict changes in orientation with changes in process parameters:

1. *Polymer molecular weight*: As molecular weight increases so will $\overline{\chi}$, and therefore, orientation.
2. *Spinning temperature*: A decrease in spinning temperature results in an increase in $\overline{\chi}$ and, therefore, an increase in orientation. A decrease in temperature gives a decrease in L_m, which also results in an increase in orientation.
3. *Take-up speed*: An increase in V_T gives an increase in E and a corresponding increase in orientation.
4. *Capillary diameter*: The difference $(V_T - V_E)$ is large for all practical spinning processes, and the effect of capillary diameter is small. However, an increase in capillary diameter will give a smaller value of V_E and a measurable increase in orientation.
5. *Air cross-flow*: An increase in air velocity results in a decrease in L_m, and therefore, an increase in E and orientation.
6. In spinning, the orientation is due to the stretch rate, not the amount of stretch as in the case of drawing.

TABLE 3.25
Polypropylenes Used for Melt Spinning

Sample	Melt flow	M_w before spinning	M_w after spinning
H-0042	0.42	4.3×10^5	3.14×10^5
H-0660	6.6	2.77×10^5	2.74×10^5
H-1200	1.20	2.36×10^5	1.94×10^5
T-0900	9.0	2.39×10^5	2.13×10^5
T-0255	2.55	3.48×10^5	2.88×10^5

Various authors have studied the effects of spinning conditions on the basic properties of spun polypropylene fibers. Notable are the studies of Sheehan and Cole [185,108], Katayama et al. [189,112], Fung et al. [190,113], Kitao et al. [191,114], Anderson and Carr [192,110], Henson and Spruiell [193,116], Spruiell and White [85], Ishizuka and Koyama [194,117], and Nadella et al. [79].

Nadella et al. [79] studied polypropylenes with a broad range of melt flows. Table 3.25 lists the properties of the melts they studied. These melts were spun from a screw extruder at a temperature of 230°C. In the case of H-0060, filaments were also spun at 200 and 260°C. With the exception of H-0042, the polymers were spun through a single capillary at 2.1 g/min. The H-0042 could not be extruded at that rate because of melt fracture; this material was extruded at 0.5 g/min. The filaments were quenched in stagnant air at 25°C and were taken up at speeds of 50, 100, 200, 400, and 550 m/min. X-ray and birefringence measurements were taken along the descending spin-line.

Figure 3.22 shows a plot of spin-line stress versus take-up speed. As may be seen, the stress increases with take-up speed and with molecular weight.

In all of their air-quenched filaments, only the monoclinic crystal form was produced. Crystallization was observed to occur on the spin line, and during crystallization the birefringence rose rapidly. Henson and Spruiell [193], who participated in this study, plotted temperature versus distance from the spinneret for the H-0660 sample (Figure 3.23). Crystallization begins to show on the x-ray at the knee in the plot.

In the Nadella study, the c-axis orientation was observed to increase with spinning velocity. Figure 3.24 shows this increase. An increase in orientation was also obtained with an increase in molecular weight. The H-0042 sample developed higher levels of orientation than the other samples.

These authors found a pronounced effect of temperature on the orientation of the spun fibers. Increasing the spinning temperature with other variables unchanged resulted in a lower orientation.

Figure 3.25 shows a plot of crystalline index versus elapsed time in the spin line for H-0660. As may be seen, crystallization rates generally increase with an increase in take-up speed or stress. However, included in these results are effects due to cooling rate, which also increases with take-up speed. Henson and Spruiell [193] gave a plot of crystallization start temperature versus spin-line stress at several cooling rates (Figure 3.26). As seen, stress increases the start temperatures and increase in cooling rates depresses the start temperatures.

Figure 3.27 shows the degree of crystallinity as a function of spin-line stress. As seen, there is an upward trend in crystallinity with stress.

Figure 3.28 shows the crystalline orientation functions for all the experiments plotted as a function of spin-line stress. The value of f_c increases with stress until the axis is nearly aligned

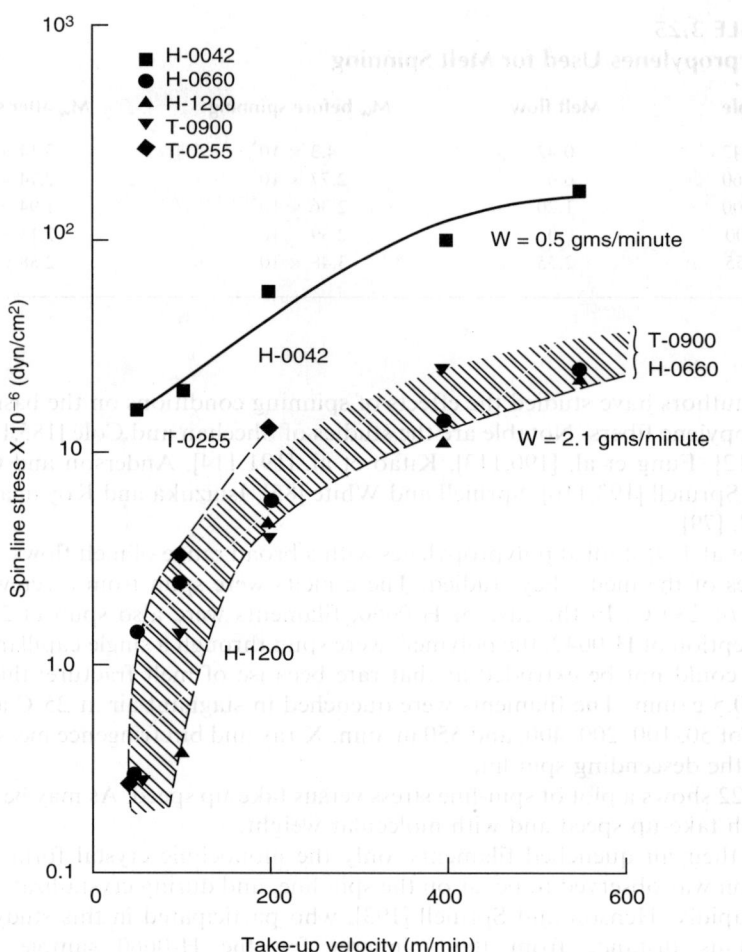

FIGURE 3.22 Effect of take-up velocity on spin-line stress for polypropylene fibers at extrusion temperature of 230°C. (From Nadella, H.P., Henson, H.M., Spruiell, J.E.; White, J.L.; *J. Appl. Polym. Sci.*, **1977**, *21*, 3003. With permission.)

with the fiber axis. The *b*-axis quickly aligns perpendicularly to the fiber axis. The f_a values decline slowly at first but then approach −0.43.

Figure 3.29 gives the birefringence as a function of spin-line stress for all experiments. Although both the crystalline and amorphous orientations increase with stress, the orientation developed in the crystalline regions is always larger than that developed in the amorphous regions. The amorphous orientation functions ranged from slightly less than zero up to almost 0.3. Abhiraman [195] argued that the highly oriented molecules in the melt zone are incorporated into the crystals, and that the molecules remaining in the amorphous region are necessarily less oriented.

The features of the Nadella study are similar to those observed for polypropylene by other authors. The variations in the crystalline orientation functions are also similar to those reported for high-density polyethylene. Dees and Spruiell [196] interpreted these variations for polyethylene to indicate spherulitic structure at low spinning stresses but which undergo a

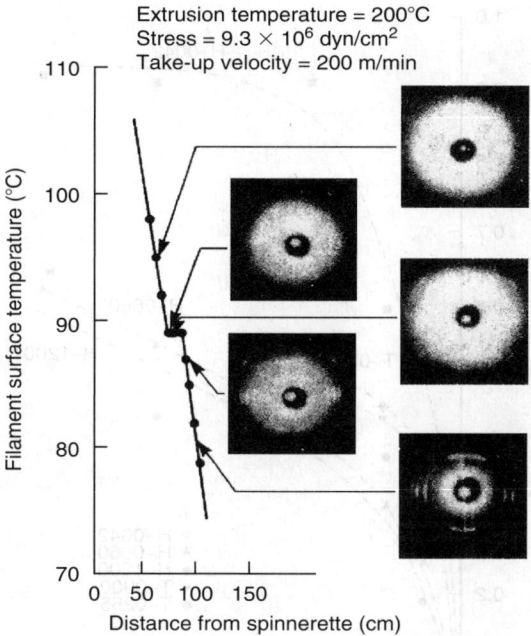

FIGURE 3.23 Fiber temperatures as a function of distance from spinneret for polypropylene. (From Spruiell, J.E.; White, J.L.; *Polym. Eng. Sci.*, **1975**, *15*, 660. With permission.)

transition to a row-nucleated structure as stress increases. Their interpretation was based on the row structure models of Keller and Machin [197]. Meridional, two-point, small-angle x-ray scattering patterns and electron micrographs showing lamellar crystals stacked in a direction parallel to the fiber axis support a row-nucleated lamellar structure. This evidence has been found in the case of high take-up speed for both polyethylene and polypropylene.

Polypropylene exhibits a distinctive bimodal orientation of the unit cells when crystallized from melts undergoing extension. There are crystals whose *c*-axis is oriented parallel with the fiber axis and a smaller population of crystals whose *a*-axis is oriented parallel with the fiber axis. For the H-0660 filaments at high stress, Nadella et al. [79] estimate that about 10–20% of the crystals have an *a*-axis orientation. Anderson and Carr [192] and Clark and Spruiell [198] concluded that the *a*-axis-oriented crystals were very small and were distributed throughout the sample and perhaps grew epitaxially to the *c*-axis-oriented crystals.

3.7.1.3 Mechanical Properties

Nadella et al. [79] also analyzed the mechanical properties of their spun filaments. Figure 3.30 through Figure 3.33 show modulus, yield strength, tensile strength, and elongation, all plotted against spin-line stress. A remarkable phenomenon seen is that of data from all the experiments being well correlated by the spin-line stress.

Shimizu et al. [199] have studied the spinning of a 9 MFR polypropylene at speeds of up to 7000 m/min. Figure 3.34 shows the birefringence of the fiber as a function of take-up speed. The initial modulus is related to birefringence in Figure 3.35. As seen, there is a sharp increase in the slope of the modulus curve at a birefringence of 0.019, which corresponds to a spinning speed of 3000 m/min. Fibers spun at speeds higher than 3000 m/min had no yield point on the stress–strain curve. At lower spinning speeds, density and the birefringence were both

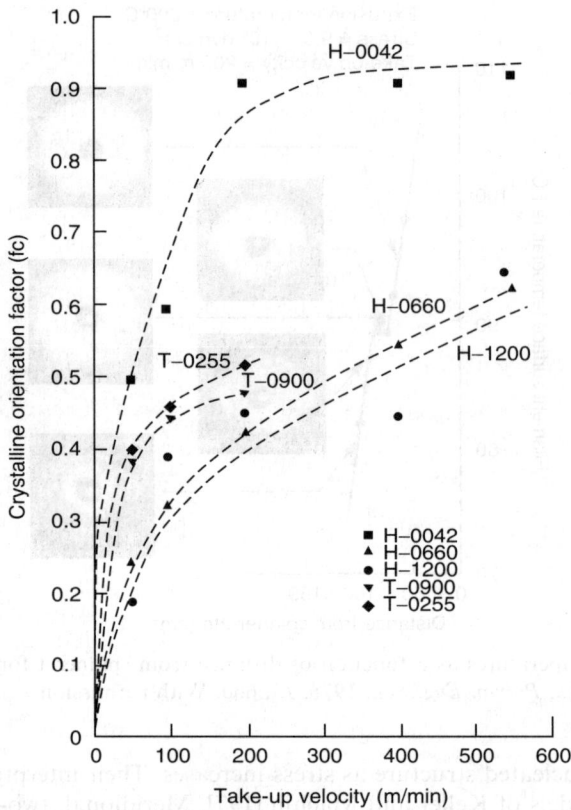

FIGURE 3.24 c-Axis orientation function of take-up velocity showing influence of molecular weight. (From Nadella, H.P., Henson, H.M., Spruiell, J.E.; White, J.L.; *J. Appl. Polym. Sci.*, **1977**, *21*, 3003. With permission.)

dependent on the spinning temperature. However, at high spinning speeds (>3000 m/min) the birefringence did not depend on the temperature. Also at 3000 m/min, f_c, f_a, and f_b leveled out at around 0.8, −0.4, and −0.5, respectively. Figure 3.36 shows the stress–strain curves for fibers formed at 290°C for various spinning speeds.

Minoshimo et al. [93] predicted, based on the data taken at very low take-up speeds, that resins of narrower molecular weight distribution would yield spun fibers with more orientation than resins of broader molecular weight distribution. The data suggested that the apparent elongational viscosity in spinning might be higher at moderate to high extension rates for the materials (Figure 3.9). The higher viscosity, it was predicted, would give larger spin-line stresses and thus higher degrees of orientation. For some applications, this effect would not be beneficial. If a spinning process is speed limited, the increase in orientation results in lower output as the draw ratio must be reduced. If high-speed spun yarns are to be utilized without drawing, the effect will be most useful. Jack and McKinley [200] spun resins of high melt flow, narrow molecular weight distribution (CR resins) at speeds as high as 3500 m/min. Figure 3.37 shows a plot of tenacity and elongation versus take-up speed. At a take-up speed of 3500 m/min, the tenacity is approximately 2.5 g/denier for both the 35 and the 300 MFR resins.

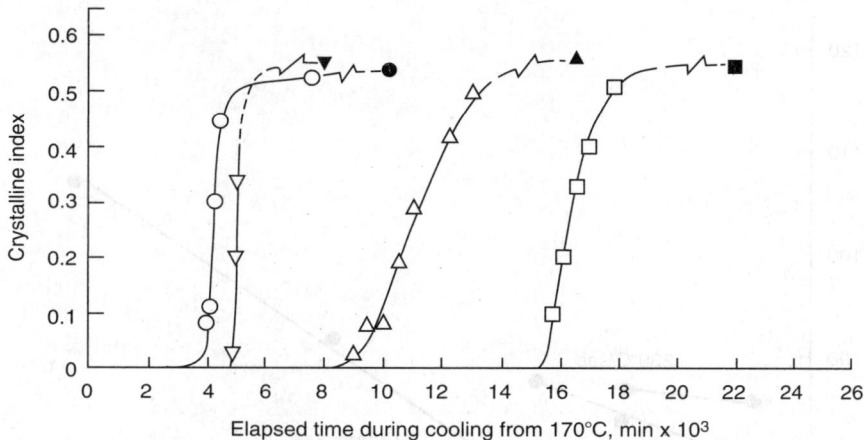

Symbol	Extrusion Temperature, °C	Take-up Velocity, m/min	Spin-line Stress dyn/cm² × 10
□	230	50	1.4
○	230	300	9.0
▽	230	550	22.7
△	200	200	9.2

FIGURE 3.25 Crystalline index as a function of residence time in the spin line for polypropylene H-0660. Open symbols, x-ray; closed symbols, density on bobbin. (From Nadella, H.P., Henson, H.M., Spruiell, J.E.; White, J.L.; *J. Appl. Polym. Sci.*, **1977**, *21*, 3003. With permission.)

Thus far, all the properties presented have been based on average values. For example, the f-values reported in all cases are an average value across the fiber cross section. Fung et al. [190] measured birefringence in 10-μm sections of polypropylene fibers. They found that the birefringence was higher near the surface of the fibers (see Figure 3.38). These higher birefringence values have been postulated to be due to the rapid cooling on the outside and the consequent increase in viscosity and load bearing in the outside layers.

3.7.2 Drawn Fibers

As is well known, melt-spun fibers are not satisfactory for use in most applications until they are drawn. Drawing increases strength and modulus, and decreases elongation. On a macroscopic scale, drawing seems to be a simple flow process. However, on a microscopic scale, profound changes occur to the fiber structure.

3.7.2.1 Deformation Model

Deformation and rearrangement of the morphological structure have been described by Peterlin [201–203] and Samuels [183]. During the initial step of deformation, a crystalline fiber undergoes an affine transformation, i.e., the strains are uniform throughout the material. However, even at still smaller deformations the morphological inhomogeneity of the

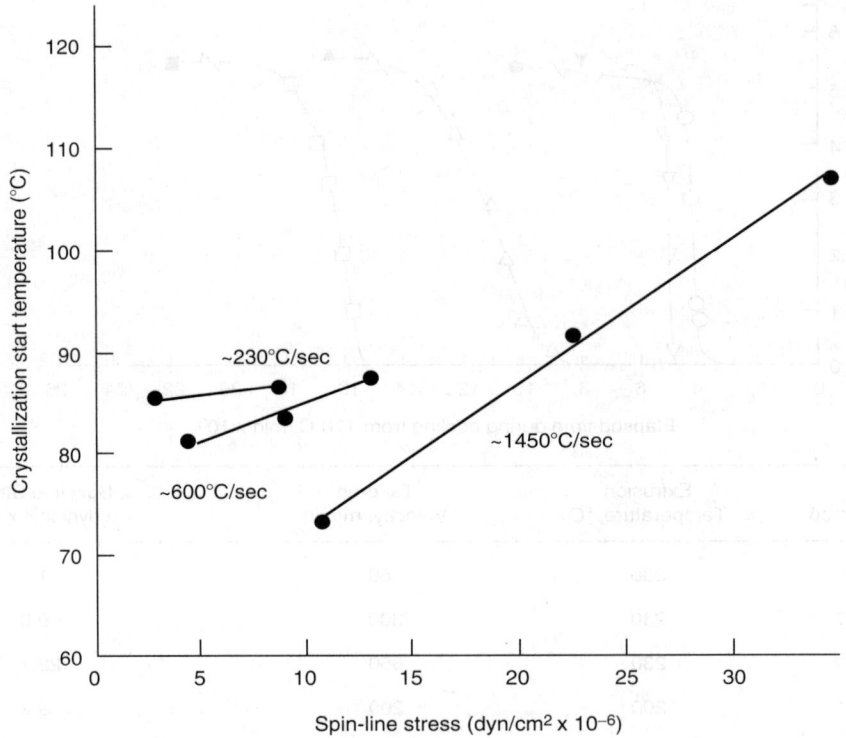

FIGURE 3.26 Crystallization start temperature versus spin-line stress for polypropylene. (From Spruiell, J.E.; White, J.L.; *Polym. Eng. Sci.*, **1975**, *15*, 660. With permission.)

lamellar structure and the small resistance of the lamellae to plastic shear deformation lead to shear, tilt, rotation, and separation. Lamellae parallel to the applied stress may deform in a different manner from lamellae perpendicular to the stress.

With sufficient deformation, the lamellae are broken up into folded chain-blocks approximately 20 nm in width. These blocks subsequently aggregate to form the basic unit of the drawn fiber—the microfibril (Figure 3.39). The microfibrils consist of alternating blocks of folded-chain crystals and amorphous regions that are mostly chain folds. Another essential part of the amorphous region are the tie molecules that connect the blocks in the fiber axis direction and provide the major source of resistance to the deformation and strength to the fiber. The tie molecules are believed to be located mainly on the surface of the microfibrils and to extend along many crystal blocks. The microfibrils are aggregated laterally to form the fibrils. The description of deformation given here applies to commercially produced polypropylene. There are further stages of drawing that have been applied experimentally, which give fibers of very high modulus and strength. These will be discussed later.

The microscopic deformation described above can proceed macroscopically in two different ways. The fiber may deform uniformly or it may deform by the well-known necking process. The mode of deformation depends on the properties of the spun fiber and the conditions of drawing. Shimizu et al. [199] found that fibers spun at speeds greater than 3000 m/min did not exhibit necking when drawn. Nadella et al. [204] found that the tendency to form necks was reduced in fiber spun under high spin-line stress.

Polypropylene Fibers

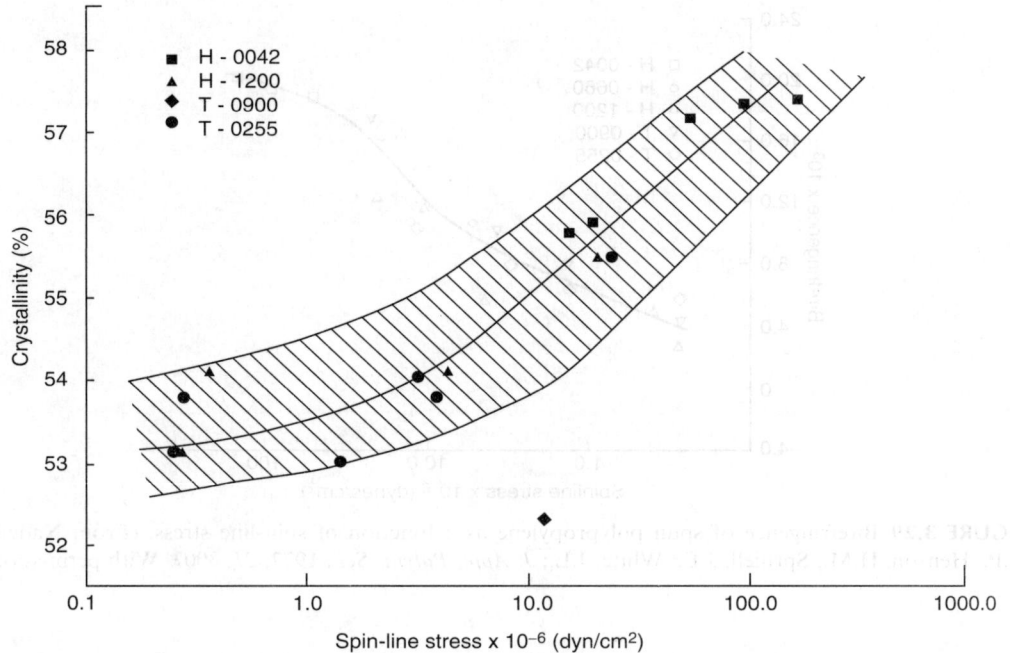

FIGURE 3.27 Crystallinity of polyproylene as function of spin-line stress. (From Nadella, H.P., Henson, H.M., Spruiell, J.E.; White, J.L.; *J. Appl. Polym. Sci.*, **1977**, *21*, 3003. With permission.)

3.7.2.2 Effect of Temperature

Ziabicki [187] shows the effects of temperature on the mode of deformation. Figure 3.40 shows typical stress–strain curves for different temperatures. Below the glass transition temperature, the fiber behaves as a brittle solid with *a*-type deformation. With increases in

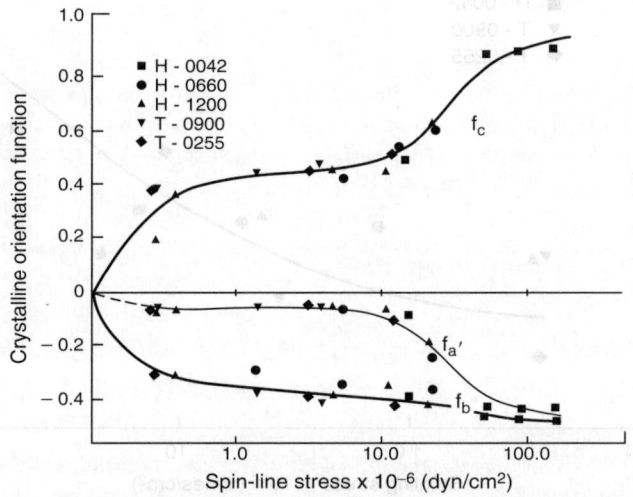

FIGURE 3.28 *a'*-, *b*-, and *c*-axis orientation functions versus spin-line stress. (From Nadella, H.P., Henson, H.M., Spruiell, J.E.; White, J.L.; *J. Appl. Polym. Sci.*, **1977**, *21*, 3003. With permission.)

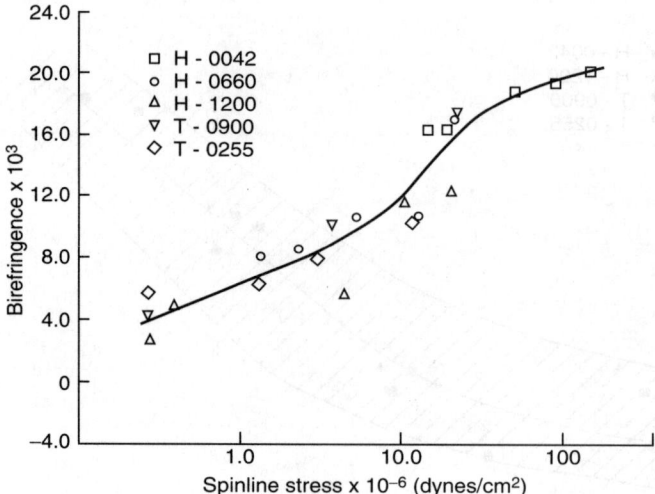

FIGURE 3.29 Birefringence of spun polypropylene as a function of spin-line stress. (From Nadella, H.P., Henson, H.M., Spruiell, J.E.; White, J.L.; *J. Appl. Polym. Sci.*, **1977**, *21*, 3003. With permission.)

temperature, plastic flow becomes more and more pronounced, and the stress–strain curve takes on a sigmoidal shape or *c*-type deformation. With further increases in temperature, the maximum and the horizontal sections disappear, and *b*-type deformation occurs.

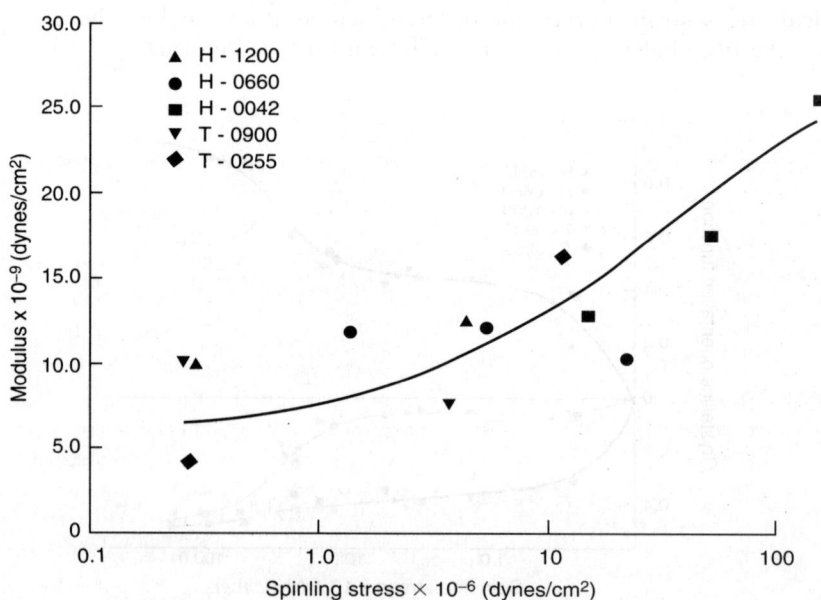

FIGURE 3.30 Modulus versus spin-line stress for spun polypropylene fibers. (From Nadella, H.P., Henson, H.M., Spruiell, J.E.; White, J.L.; *J. Appl. Polym. Sci.*, **1977**, *21*, 3003. With permission.)

Polypropylene Fibers

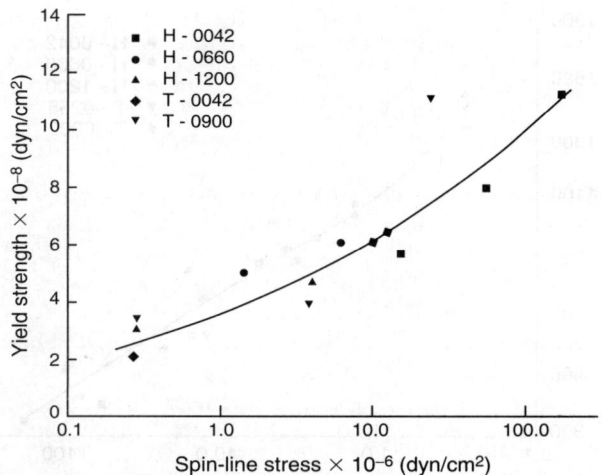

FIGURE 3.31 Yield strength versus spin-line stress for spun polypropylene fibers. (From Nadella, H.P., Henson, H.M., Spruiell, J.E.; White, J.L.; *J. Appl. Polym. Sci.*, **1977**, *21*, 3003. With permission.)

3.7.2.3 Effect of Draw Rate

An increase in the rate of drawing affects stress–strain curves in a manner similar to that of a decrease in temperature. Lazurkin [205] obtained an empirical relationship between yield stress σ^* and extension rate, $\dot{E}$:

$$\sigma^* = a + b \log \dot{E} \tag{3.21}$$

This equation is valid for many polymers. A fit has been determined for polyethylene but not for polypropylene.

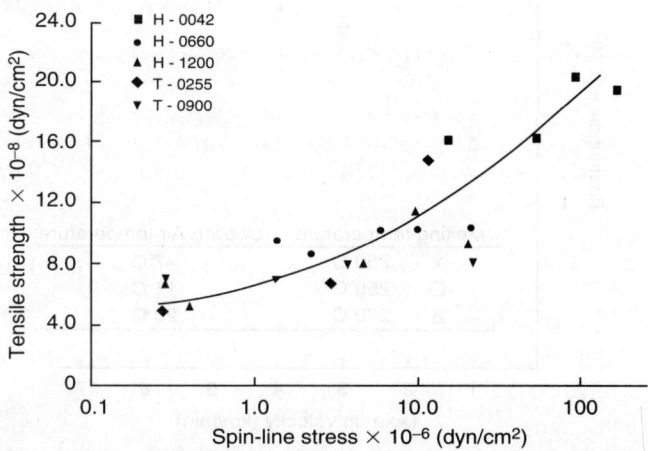

FIGURE 3.32 Tensile strength versus spin-line stress for spun polypropylene fibers. (From Nadella, H.P., Henson, H.M., Spruiell, J.E.; White, J.L.; *J. Appl. Polym. Sci.*, **1977**, *21*, 3003. With permission.)

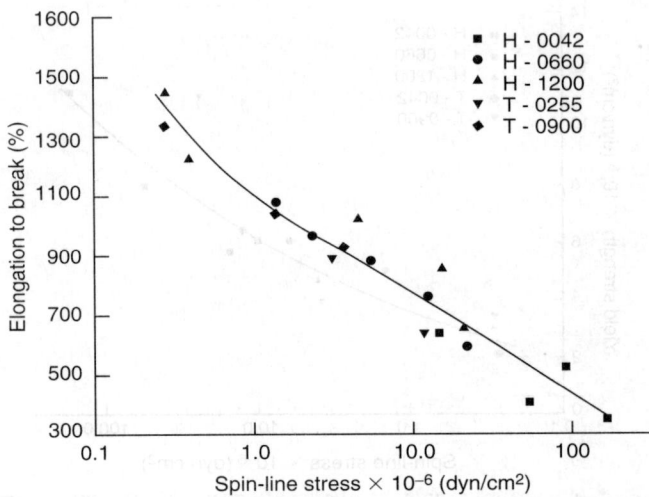

FIGURE 3.33 Elongation versus spin-line stress for spun polypropylene fibers. (From Nadella, H.P., Henson, H.M., Spruiell, J.E.; White, J.L.; *J. Appl. Polym. Sci.*, **1977**, *21*, 3003. With permission.)

Nadella et al. [204] studied the drawing of samples H-0042 and 11-1200, which were produced earlier in their spinning studies [79]. The drawing was carried out on an Instron tensile testing machine. One-inch samples were stretched to three different lengths: 1 in. beyond the natural draw ratio (disappearance of neck), approximately 1 in. prior to fracture, and between the above two. If necking was not observed, the first stage corresponded to a

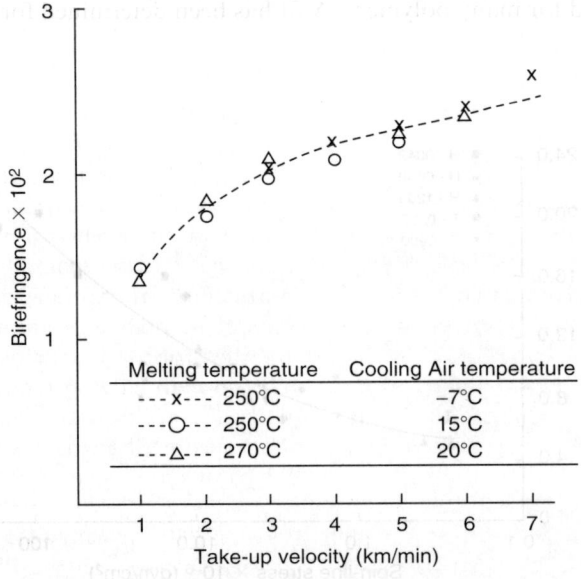

FIGURE 3.34 Relation between birefringence and take-up velocity at various extrusion and cooling air temperatures for polypropylene. (From Shimizu, J.; Toriumi, K.; Imai, Y.; *Sen-i Gakkaishi*, **1977**, *33*, T-255. With permission.)

Polypropylene Fibers

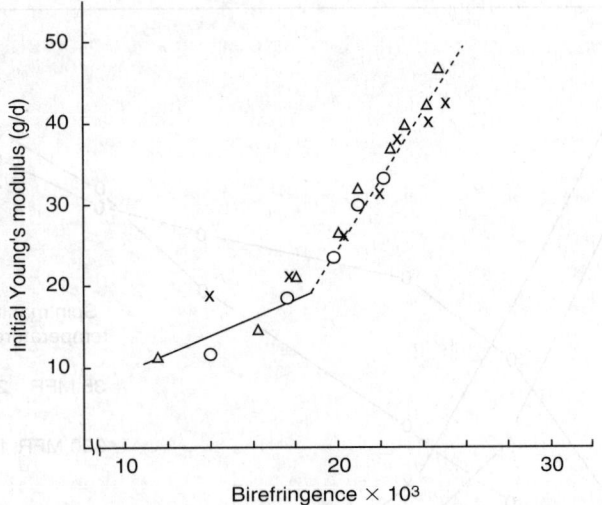

FIGURE 3.35 Plots of initial modulus versus birefringence for polypropylene. See key in Figure 3.34. (From Shimizu, J.; Toriumi, K.; Imai, Y.; *Sen-i Gakkaishi*, **1977**, *33*, T-255. With permission.)

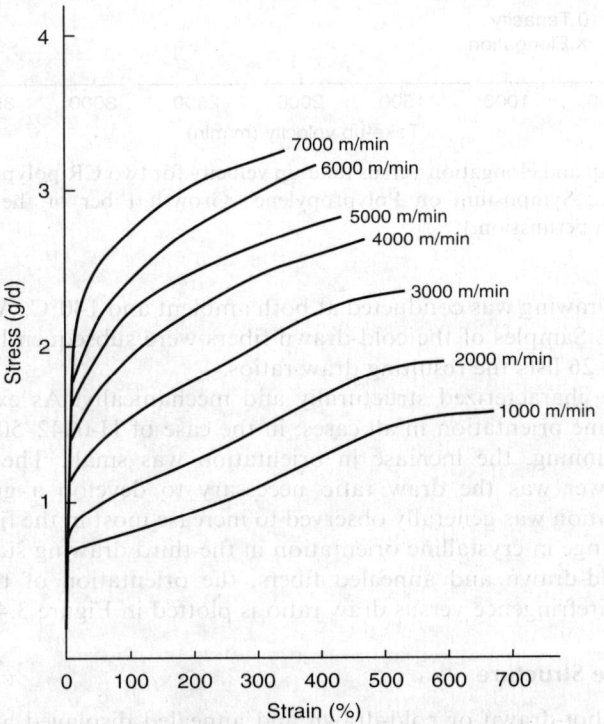

FIGURE 3.36 Stress–strain curves for polypropylene spun at various take-up velocities at 250°C. (From Shimizu, J.; Toriumi, K.; Imai, Y.; *Sen-i Gakkaishi*, **1977**, *33*, T-255. With permission.)

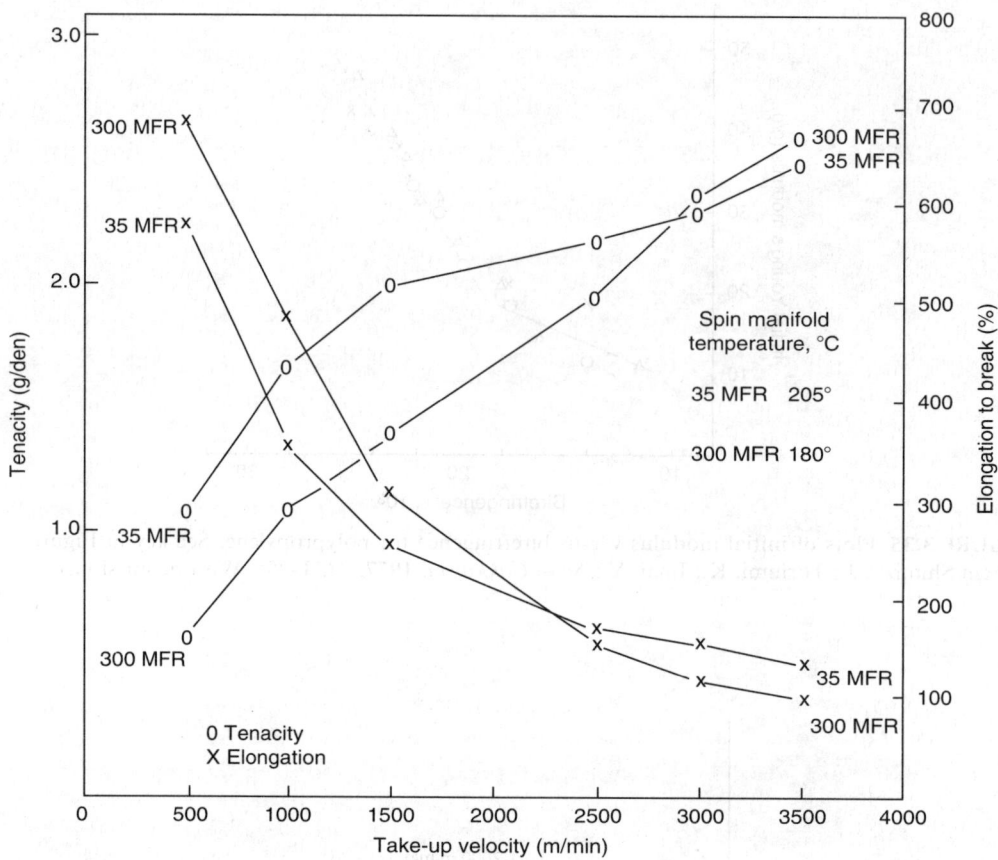

FIGURE 3.37 Tenacity and elongation versus take-up velocity for two CR polypropylenes. (From Jack, H.P.; McKinley, J.R.; Symposium on Polypropylene—Growth Fiber of the Eighties, New York, September 1981. With permission.)

draw ratio of 2.0. Drawing was conducted at both ambient and 140°C. A cross-head speed of 2 in./min was used. Samples of the cold-drawn fibers were subsequently annealed at 140°C for 15 min. Table 3.26 lists the resulting draw ratios.

The fibers were characterized structurally and mechanically. As expected, drawing increased the crystalline orientation in all cases; in the case of H-0042–500, which was highly oriented during spinning, the increase in orientation was small. The higher the as-spun orientation, the lower was the draw ratio necessary to develop a given higher level of orientation. Orientation was generally observed to increase most in the first stage of drawing. There was little change in crystalline orientation in the third drawing stage. In both the hot-drawn and the cold-drawn and annealed fibers, the orientation of the a'-axis gradually disappeared. The birefringence versus draw ratio is plotted in Figure 3.41.

3.7.2.4 Crystalline Structure

Samples that were hot-drawn or cold-drawn and annealed displayed a well-formed monoclinic structure. Cold-drawing transformed the monoclinic form of the spun fibers into a more oriented but disordered structure. The x-ray pattern exhibited a high degree of line

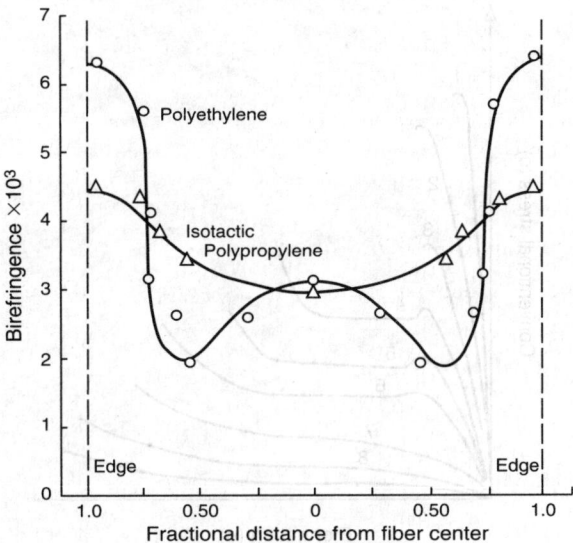

FIGURE 3.38 Birefringence profiles for polyethylene and polypropylene spun fibers. (From Fung, P.Y.F.; Orlando, E.; Carr, S.H.; *Polym. Eng. Sci.*, **1973**, *13*, 295. With permission.)

broadening that was considered to result from a decrease of crystallite size, and from lattice strains and defects.

Densities of the hot-drawn and cold-drawn and annealed fibers were slightly higher than those of the spun fibers. Crystalline fractions of around 61% were observed.

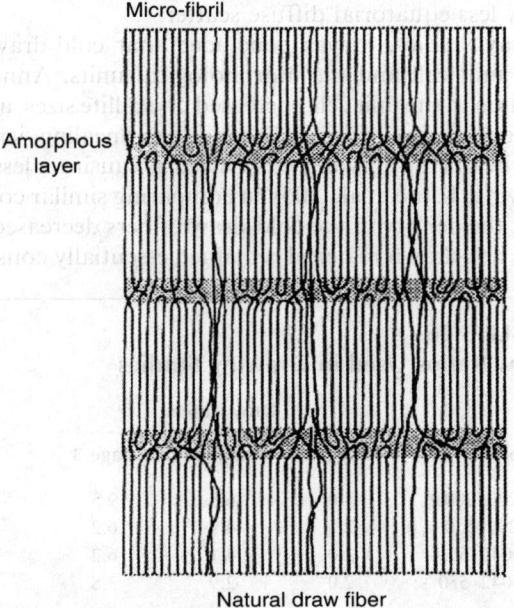

FIGURE 3.39 Proposed morphology of conventionally drawn fiber. (From Taylor, W.N.; Clark, E.S.; *Polym. Eng.*, **1978**, *18*, 518. With permission.)

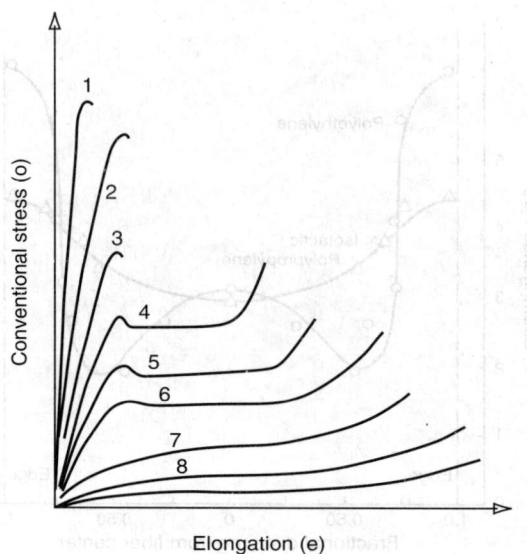

FIGURE 3.40 Stress–strain curves for undrawn fibers. Temperature increase in the order 1–9. (From Ziabicki, A.; *Fundamentals of Fibre Formation*, John Wiley & Sons, New York, 1976.)

Small-angle x-ray analysis was also conducted on the fibers. For the cold-drawn H-1200–200 sample, there was diffuse equatorial scattering. The same fiber, after annealing, showed reduced equatorial scattering and the development of a two-point meridional pattern corresponding to a long period of 147 Å. The small-angle pattern of the hot-drawn fiber was similar, but the scattering corresponded to a long period of approximately 200 Å. The small-angle patterns for the H-0042–550 samples were similar to those of the H-1200–200 sample, except there was slightly less equatorial diffuse scatter.

On the basis of x-ray studies, it was concluded that cold-drawing disrupts the entire structure, including the unit cell and the morphological units. Annealing removed residual strains, increased perfection in the unit cell, increased crystallite sizes, and restructured the axial periodicity with a new long-period spacing. During the annealing, small voids separating the microfibrils appeared to heal. Hot-drawing was viewed as causing a less severe disruption of the structure than cold-drawing. Kitao et al. [206], in comparing similar cold-drawn fibers and hot-drawn fibers, found that the density of the cold-drawn fibers decreased significantly with draw ratio. The density of the hot-drawn material remained essentially constant (see Figure 3.42).

TABLE 3.26
Draw Ratios Used in Drawing Studies

Sample	Draw ratios		
	stage 1	stage 2	stage 3
H -1200–200a[a]	4.0	6.8	9.5
H -1200–550	2.5	4.9	6.2
H - 0042–50	2.0	4.1	6.2
H - 0042–550	2.0	2.9	3.8

[a]Spinning speed, m/min.

Polypropylene Fibers

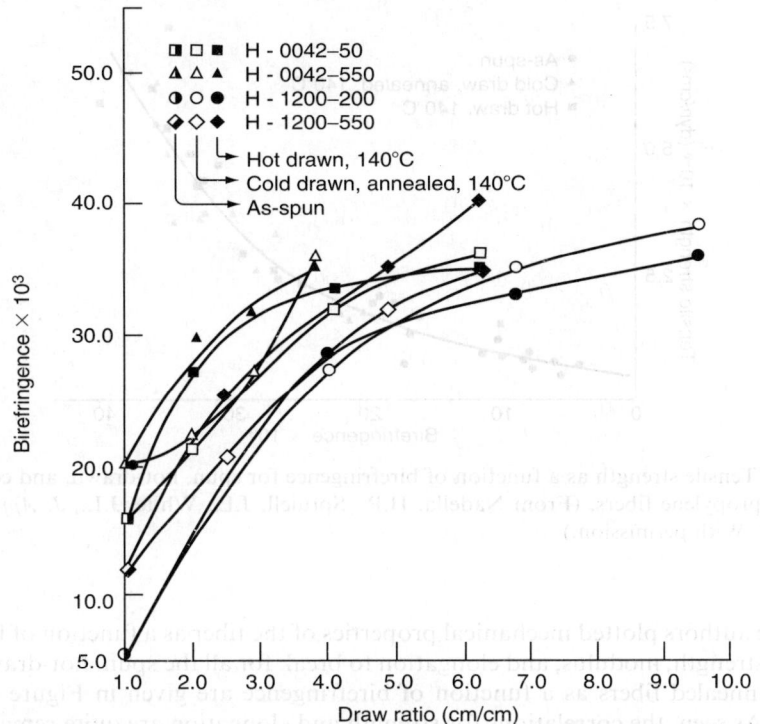

FIGURE 3.41 Birefringence as a function of draw ratio for polypropylene fibers. (From Nadella, H.P.; Spruiell, J.E.; White, J.L.; *J. Appl. Polym. Sci.*, **1978**, *22*, 3121. With permission.)

Though the mechanical properties of the fibers from the Nadella et al. study [204] showed a regular improvement with draw ratio, a general correlation of properties with draw ratio was not possible because of the differences in the orientation of the spun fiber. Following Samuels

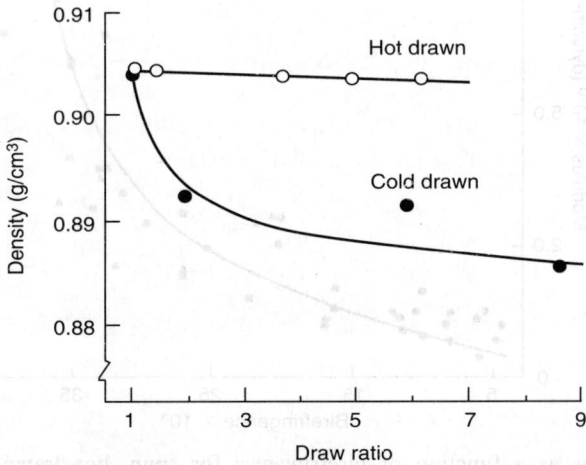

FIGURE 3.42 Density of drawn polypropylene as a function of draw ratio. (From Kitao, T.; Spruiell, J.E.; White, J.L.; *Polym. Eng. Sci.*, **1979**, *19*, 761. With permission.)

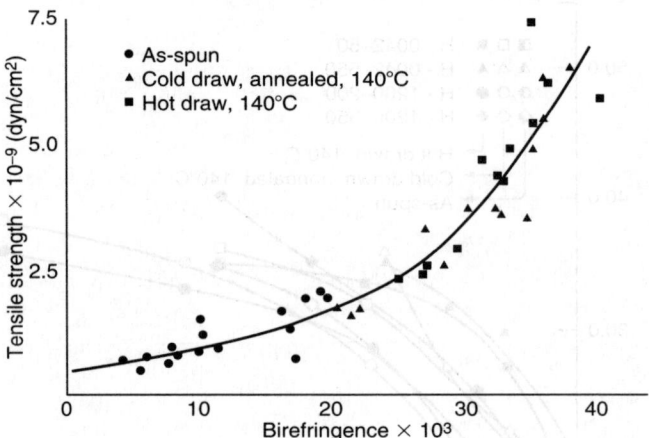

FIGURE 3.43 Tensile strength as a function of birefringence for spun, hot-drawn, and cold-drawn and annealed polypropylene fibers. (From Nadella, H.P.; Spruiell, J.E.; White, J.L.; *J. Appl. Polym. Sci.*, **1978**, *22*, 3121. With permission.)

[183,184], the authors plotted mechanical properties of the fiber as a function of birefringence. The plots of strength, modulus, and elongation to break for all the spun, hot-drawn, and cold-drawn and annealed fibers as a function of birefringence are given in Figure 3.43 through Figure 3.45. As seen, the correlations for strength and elongation are quite remarkable.

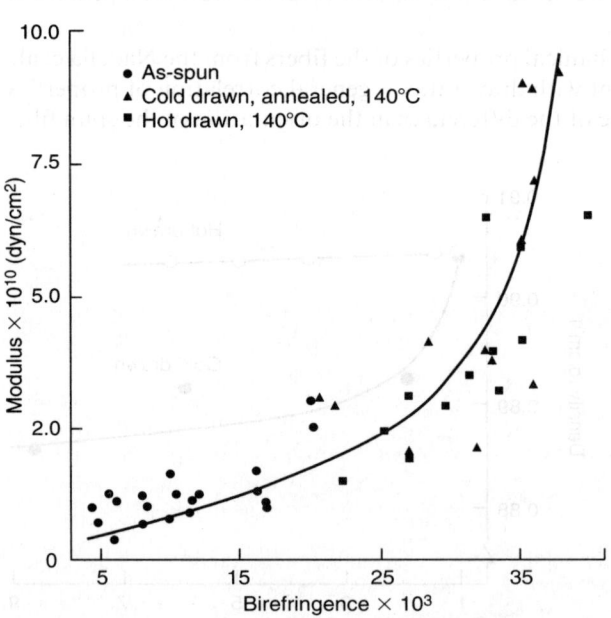

FIGURE 3.44 Modulus as a function of birefringence for spun, hot-drawn, and cold-drawn and annealed polypropylene fibers. (From Nadella, H.P.; Spruiell, J.E.; White, J.L.; *J. Appl. Polym. Sci.*, **1978**, *22*, 3121. With permission.)

Polypropylene Fibers

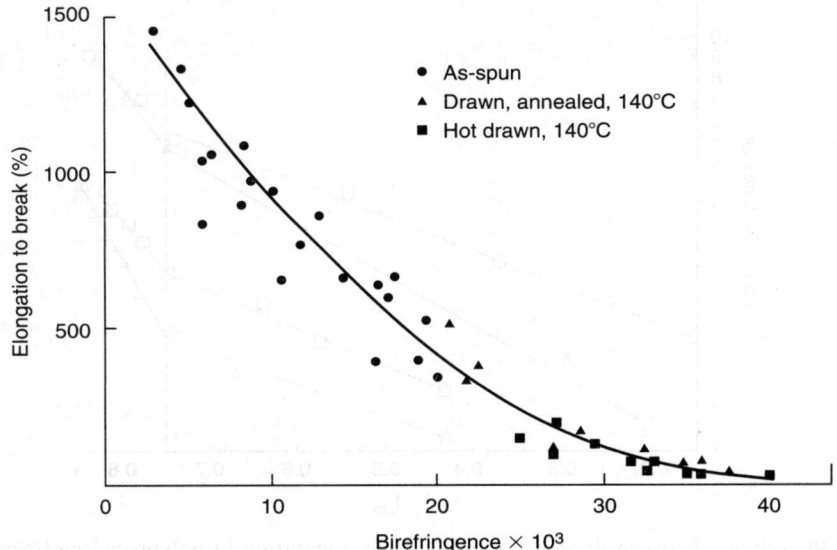

FIGURE 3.45 Elongation as a function of birefringence for spun, hot-drawn, and cold-drawn and annealed polypropylene fibers. (From Nadella, H.P.; Spruiell, J.E.; White, J.L.; *J. Appl. Polym. Sci.*, **1978**, *22*, 3121. With permission.)

3.7.2.5 Structure–Property Relationship

Samuels [183,184] has conducted extensive studies on the relations between structural states and properties. His studies have been carried out on both polypropylene films and polypropylene fibers. According to Samuels, each fabrication process, test, and end-use application involves a deformation, and all these steps cause a change from an initial structural state to a final one. His view was that the initial and final structural states will always lie between that of the undeformed polymer and the final state just before fracture. These states are specified by the relative amounts of crystalline and amorphous materials present and by their orientation and are attainable by different fabrication paths. In addition to f_c and f_{am}, Samuels used an f average for correlating properties. The definition of f_{av} is given by:

$$f_{av} = \beta f_c + (1 - \beta)f_{am} \qquad (3.22)$$

where β is the crystalline fraction.

Figure 3.46 shows a plot of draw ratio versus f_{av} for polypropylene fibers and films. At an f_{av} of 0.75, there was a change in the internal structure of all the drawn materials. They all reached the same internal state, correlated by f_{av}, by different process paths. A similar plot was obtained for draw ratio versus f_c. The slope change occurred at an f_c of 0.9. However, a plot of draw ratio versus $(1-\beta)f_{am}$ showed no such change in slope. Thus, it was concluded that the transition was a result of crystal deformation processes. According to Samuels, this observation has important consequences. Some properties are influenced primarily by one region of the polymer, whereas others are controlled by both. Samuels found the individual contributions of the amorphous and crystalline regions to the total birefringence for the fiber. Figure 3.47 shows a plot showing these contributions.

A plot of tenacity versus f_{am} for all the films and fibers is given in Figure 3.48. As seen, the tenacity correlates very well with the amorphous orientation. Samuels gave many other correlations in his monograph [183].

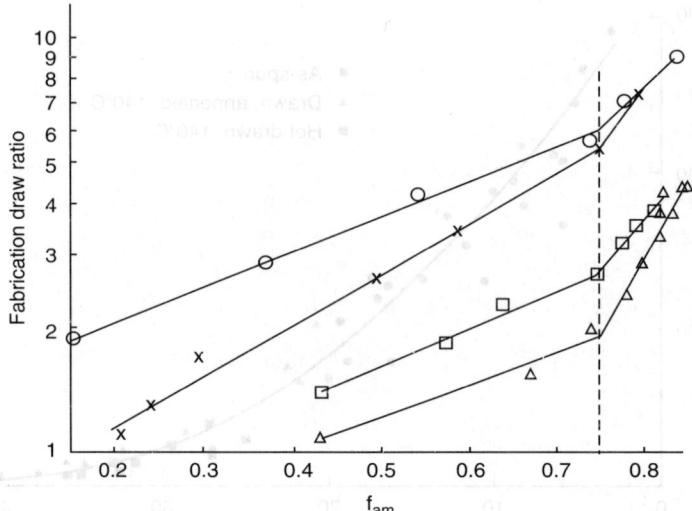

FIGURE 3.46 Relation between draw ratio and average orientation in polypropylene fibers and film: (○) film, draw temperature 135°C; (×) film, draw temperature 110°C; (□) fiber, draw temperature 90°C; (△) fiber, heat-set. (From Samuels, R.J.; *Structured Polymer Properties,* John Wiley & Sons, New York, 1974. With permission.)

Sakthivel [207] has conducted a study of polypropylene spinning and drawing under commercial conditions. A polymer with a $\overline{M}_w$ of 222,000 and $\overline{M}_w/\overline{M}_n$ of 6 was used. The author drew the fibers in a spin-draw process and in a separate conventional drawing process after aging for 48 h. In both processes, the feed rolls and draw rolls were maintained at 100°C. Table 3.27 shows the morphological parameters for fibers spun at 500, 1000, and 2000 m/min and for drawn fibers spun at 2000 m/min. There was a surprisingly small dependence of the parameters on spinning speed. Of course, drawing generally gives a large effect.

Samuels [183] hypothesized that fiber structure and properties lie on a path from generation to utilization. The deformation during physical testing is viewed as the last part of processing that leads to ultimate fiber rupture. According to this picture, if testing proceeds slowly enough, all fibers from a given polymer species should have the same strength at failure if the fibers were formed properly. The strength at failure is calculated by:

$$S_f = \text{Tenacity}(1 + E/100) \tag{3.23}$$

where E is elongation at failure. Hagler [208] prepared polypropylene yarns at take-up speeds from 787 to 2000 m/min and at total spinneret outputs from 8.7 to 19 lb/h. Each of the yarns was drawn at 2×, 2.5×, and at maximum. Tensile tests were conducted at an extension rate of 10%/min. Plots of the yarn tenacity as a function of draw ratio are given in Figure 3.49. The data from all these yarns are plotted in the form of S_f versus elongation in Figure 3.50. As seen, S_f is rather constant for elongations larger than 30%. However, the highly drawn yarns are different from the lesser drawn ones.

Rosenthal [209] has given an empirical expression for the relation between tenacity and elongation:

$$T = CE^{-1/2} \tag{3.24}$$

Polypropylene Fibers

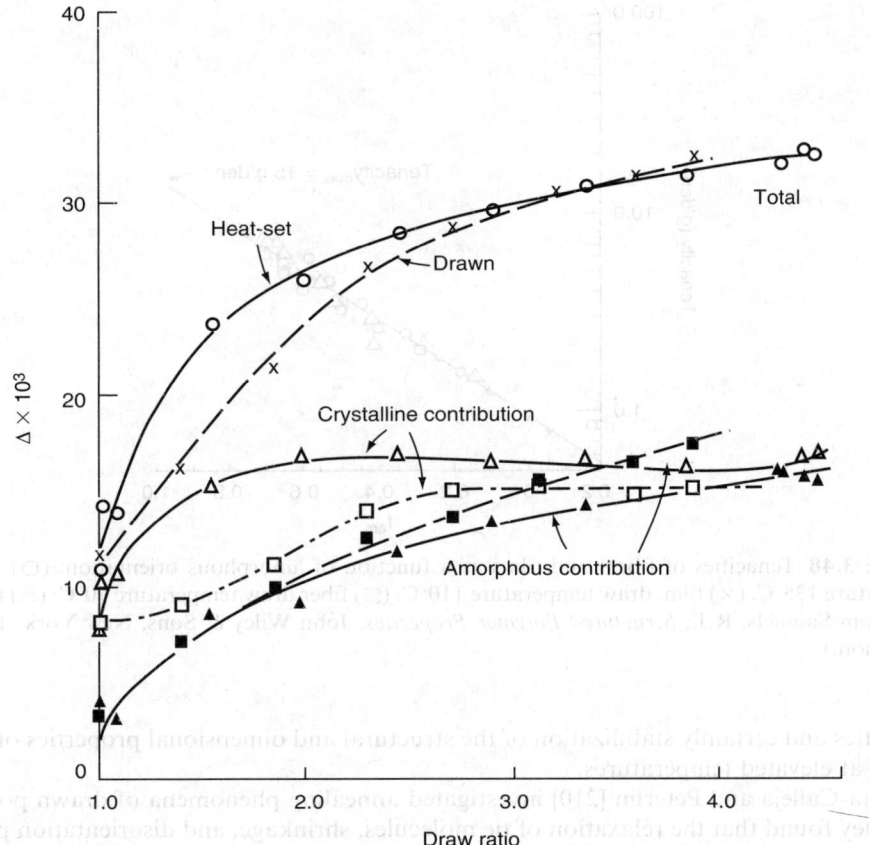

FIGURE 3.47 Birefringence contributions from the crystalline and noncrystalline regions: (□, ■) drawn; (△, ▲) heat-set. (From Samuels, R.J.; *Structured Polymer Properties*, John Wiley & Sons, New York, 1974. With permission.)

He found that this relation holds for fibers from a wide variety of polymers, including polypropylene. The data of Hagler [208] are plotted in Figure 3.51. As seen, Rosenthal's [209] empirical expression fits the data rather well.

3.7.3 Annealed Fiber

For polymers, deformation is followed by either stress relaxation or strain recovery to some thermodynamically stable state. The relaxation or recovery process occurs because polymers are viscoelastic materials. In melts, the process occurs rapidly because the time constants are relatively small. In cooled polymers, relaxation or recovery occurs much more slowly, because the time constants are much larger. The process can be greatly hastened by the application of swelling agents or heat.

Drawing introduces stresses, structural defects, and voids within fibers. Annealing the fibers quickly relaxes the stress, heals the voids and structural defects, and leads to increases in degree of crystallinity, and to perfection and size of crystallites. If heated with free ends, annealing causes the fiber to shrink. Annealing can lead to an improvement in mechanical

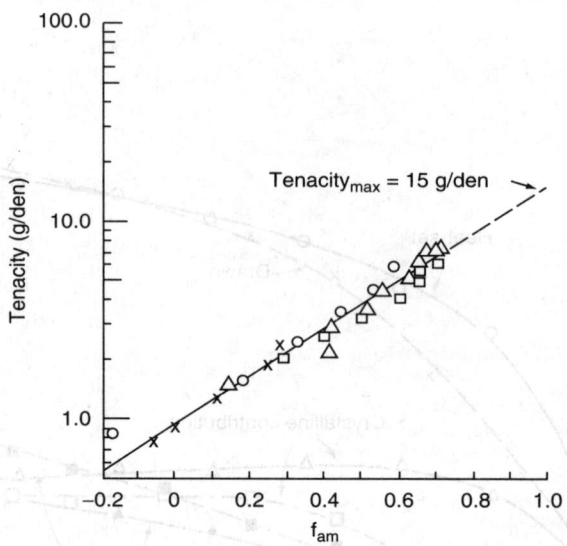

FIGURE 3.48 Tenacities of fibers and films as a function of amorphous orientation: (O) film, draw temperature 135°C; (×) film. draw temperature 110°C; (□) fiber draw temperature 90°C; (△) fiber, heat-set. (From Samuels, R.J.; *Structured Polymer Properties*, John Wiley & Sons, New York, 1974. With permission.)

properties and certainly stabilization of the structural and dimensional properties of the fiber for use at elevated temperatures.

Balta-Calleja and Peterlin [210] investigated annealing phenomena of drawn polypropylene. They found that the relaxation of tie molecules, shrinkage, and disorientation proceeded relatively fast, compared with the long period growth and the increase in density, which continued as a linear function of log time through 1000 min.

Nadella et al. [204] found that cold-drawing of polypropylene caused voids and a decrease of crystallite size and greatly enhanced lattice strains and defects. Annealing the fibers at 140°C restored a well-formed monoclinic structure and healed the voids.

TABLE 3.27
Morphological Parameters of Commercially Produced Fibers

Spin speed = D.R.	Percent Crystallinity		Birefringence	Crystalline orientation function (f_c)	Noncrystalline orientation function (f_{am})	Crystal size (Å)	Percent axially oriented crystals
	X-ray	DSC					
500/UD[a]	65	48	0.0156	0.61	0.20	91	80
1000/UD	62	40	0.0168	0.65	0.22	115	83
2000/UD	59	53	0.0170	0.67	0.22	104	83
2000/1.5X	65	56	0.0216	0.76	0.35	104	91
2000/2.9X	56	68	0.0313	0.86	0.65	109	91

[a]Undrawn.

Polypropylene Fibers

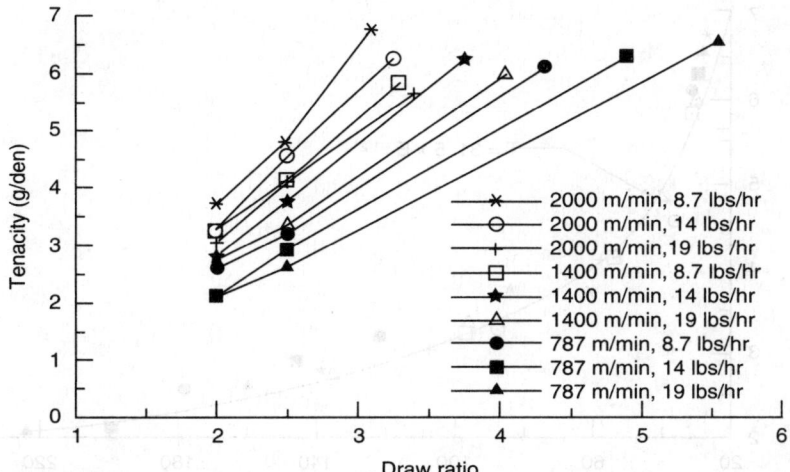

FIGURE 3.49 Tenacity versus draw ratio for polypropylene yarns spun at various spinning speeds and at various spinneret outputs. (From Samuels, R.J.; *Structured Polymer Properties*, John Wiley & Sons, New York, 1974.)

Sakthivel [207] annealed commercially produced fiber and measured changes in crystal size and shrinkage of the fibers during annealing at different temperatures. Table 3.28 gives the crystal sizes and Table 3.29 gives the fiber shrinkage.

3.7.4 Melting Behavior

The melting behavior of polypropylene has been studied by a number of authors [83,211–214]. Jaffe [211] and Samuels [83] have reported results for fibers.

Jaffe studied spun fibers produced at different spin-line stress levels. Figure 3.52 shows DTA traces of fibers spun from Profax 6523 at both a low level and a high level of spin-line stress. Both fibers had been annealed at 140°C prior to melting. The small endotherms at 140°C

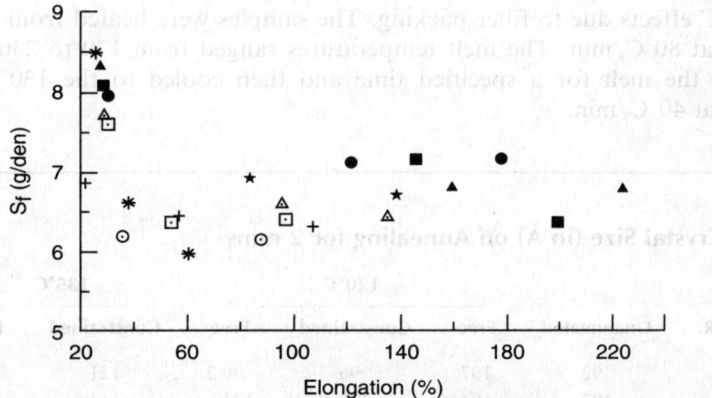

FIGURE 3.50 Tenacity based on denier at break versus elongation for drawn polypropylene yarns spun at various spinning speeds and at various spinneret outputs. (From Hagler, G.E.; unpublished research.)

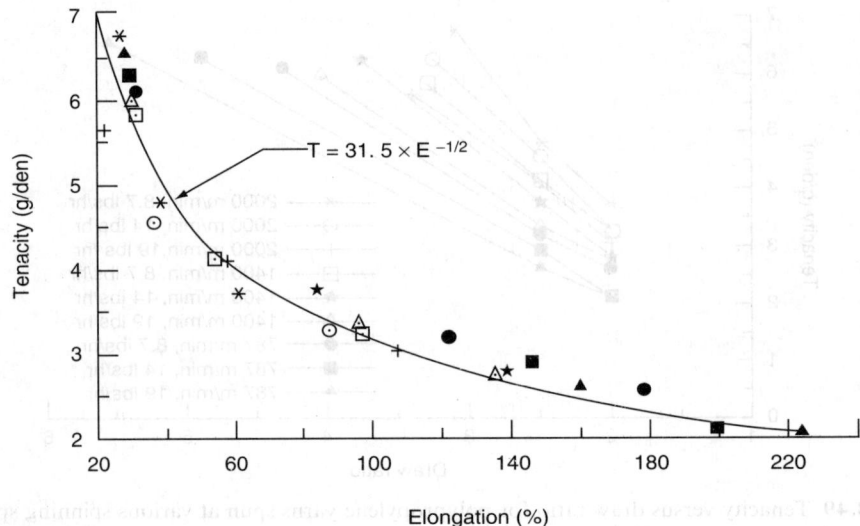

FIGURE 3.51 Tenacity versus elongation for drawn polypropylene yarns spun at various spinning speeds and at various spinneret outputs.

were from materials crystallized during the annealing operation. The high-temperature shoulder present on the main melting peak of the high-stress sample is of importance. It was shown that the sample produced under low spin-line stress contained spherulitic crystalline regions, whereas the high-stress fiber contained row-nucleated structures. Jaffe attributed the high melting shoulder to the melting of the fibrillar crystal nucleating species of the row structure.

Figure 3.53 shows the rate-dependent melting behavior of three fibers spun at different stress levels. In all the cases, the peak melting temperature increases with decreasing heating rate. The high melting shoulder of the sample from high spin-line stress disappeared into the main melting peak at rates greater than 20°C/min.

Jaffe melted and isothermally recrystallized fibers spun under different stress levels. The bulk crystallization kinetics was measured with both DSC and optical microscopy techniques. Table 3.30 lists the samples studied. The film samples were included to ensure the absence of spurious DSC effects due to fiber packing. The samples were heated from 50°C to the melt temperature at 80°C/min. The melt temperatures ranged from 170 to 230°C. The samples were held in the melt for a specified time and then cooled to the 130°C crystallization temperature at 40°C/min.

TABLE 3.28
Changes in Crystal Size (in Å) on Annealing for 2 mins

Spin speed/ D. R.	Unannealed	120°C		135°C		150°C	
		Free	Constrained	Free	Constrained	Free	Constrained
500 MPM/4X	92	107	96	99.2	121	130	108
1000 MPM/3X	102	104	101	123	114	127	—
2000 MPM/2.9X	109	112	105	118	118	127	115

TABLE 3.29
Shrinkage of Drawn Filaments

	Percent shrinkage	
Spin speed/draw ratio	at 120°C	at 135°C
500 MPM/UD[a]	1	1
500 MPM/1.5X (Spin draw)	8.5	9.7
500 MPM/2.8X (Spin draw)	6	7.6
500 MPM/1.5X (Conventional draw)	4.8	7.2
500 MPM/4.0X (Conventional draw)	3.7	4.7

[a]Undrawn.

Figure 3.54 shows actual DSC curves for samples A and B melted at 170°C. The fit of the calculated Avrami parameters to the experimental data is also shown. As seen, the sample spun under high stress crystallized much more rapidly than the sample produced under low stress. Moreover, it was observed that the starting spherulitic and row-nucleated structures were retained in the recrystallized samples.

Figure 3.55 shows the plots of the times to reach maximum crystallization rate and the maximum crystallization rate, $d\varphi/dt$, as a function of the melting temperature. Both plots show that, at higher melting temperatures, the resulting rate of crystallization is diminished, with the implication that crystal nuclei are destroyed. As seen, under none of the conditions utilized did the overall rate of crystallization of the samples under high stress reduce to those observed for the samples under low stress. Jaffe also found that the time spent in the melt had an effect similar to the melt-temperature level. As the melt time increases, crystallization kinetics slow down.

The optical micrographs taken of sample A recrystallized from melt temperatures of 170, 200, and 230°C show that the morphology varied from row to mixed to spherulitic. The line nucleus seemed to break down into shorter lines and finally to a high concentration of points.

Samuels [83] studied the melting behavior of two series of drawn fibers. One series was drawn to different extensions at 90°C. The other was drawn to different extensions at 90°C and then heat-set under tension at 140°C. The structural state parameters were known for all

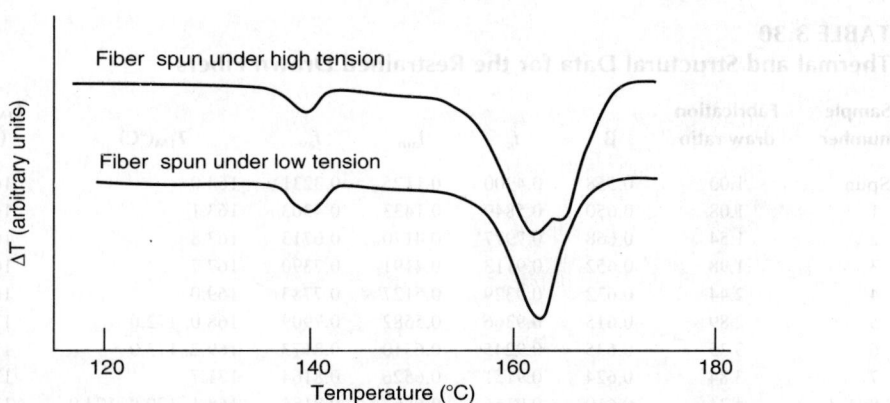

FIGURE 3.52 DTA traces for annealed polypropylene fibers. (From Jaffe, M.; In *Thermal Methods in Polymer Analysis*, Shalaby, S.W.; ed., Franklin Institute Press, Philadelphia, 1977, p. 93. With permission.)

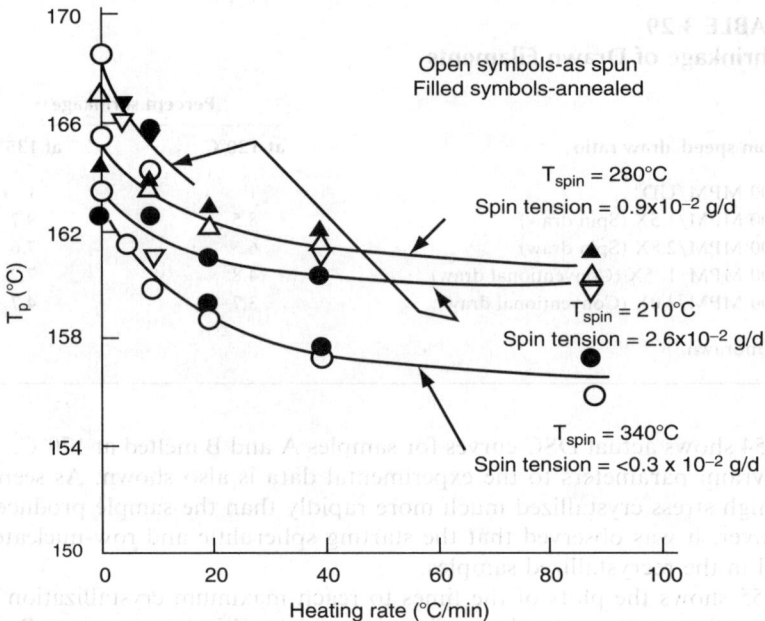

FIGURE 3.53 Melting peak temperature versus heating rate for three polypropylene fibers spun at different spin-line stress levels. (From Jaffe, M.; In *Thermal Methods in Polymer Analysis*, Shalaby, S.W.; ed., Franklin Institute Press, Philadelphia, 1977, p. 93. With permission.)

the fibers. Table 3.30 and Table 3.31 show the draw ratios, the structural parameters, and the endotherm peaks obtained from the fibers. The fibers were restrained and were heated at a rate of 20°C/min. The T_{1M} designates the lower endotherm peak temperature and T_{2M} designates the higher endotherm peak temperature. As seen, the endotherms occurred at higher temperatures the more the fibers were drawn. Figure 3.56 and Figure 3.57 show the

TABLE 3.30
Thermal and Structural Data for the Restrained Drawn Fibers

Sample number	Fabrication draw ratio	β	f_c	f_{am}	f_{av}	T_{1M}(°C)	T_{1M}(av) (°C)	T_{2M}(°C)
Spun	1.00	0.558	0.4900	0.1125	0.3231	164.0	164.0	—
1	1.08	0.650	0.5849	0.1433	0.4303	163.1	163.1	—
2	1.54	0.668	0.7977	0.4170	0.6713	163.8	163.8	—
3	1.98	0.652	0.9113	0.4191	0.7390	167.7	167.7	(173.0)
4	2.44	0.632	0.9329	0.5127	0.7783	169.0	169.0	183.0
5	2.89	0.615	0.9366	0.5582	0.7909	168.0, 172.0	170.0	187.8
6	3.35	0.635	0.9245	0.6310	0.8173	169.2, 173.0	171.0	191.1
7	3.84	0.624	0.9151	0.6526	0.8164	171.7	171.7	193.8
8	4.31	0.610	0.9156	0.6667	0.8185	168.1, 170.8, 174.0	171.0	189.8
9	4.42	0.640	0.9138	0.7163	0.8427	172.0, 174.6, 177.0	174.5	196.3
10	4.48	0.642	0.9281	0.7024	0.8473	172.1, 182.0	177.0	195.5

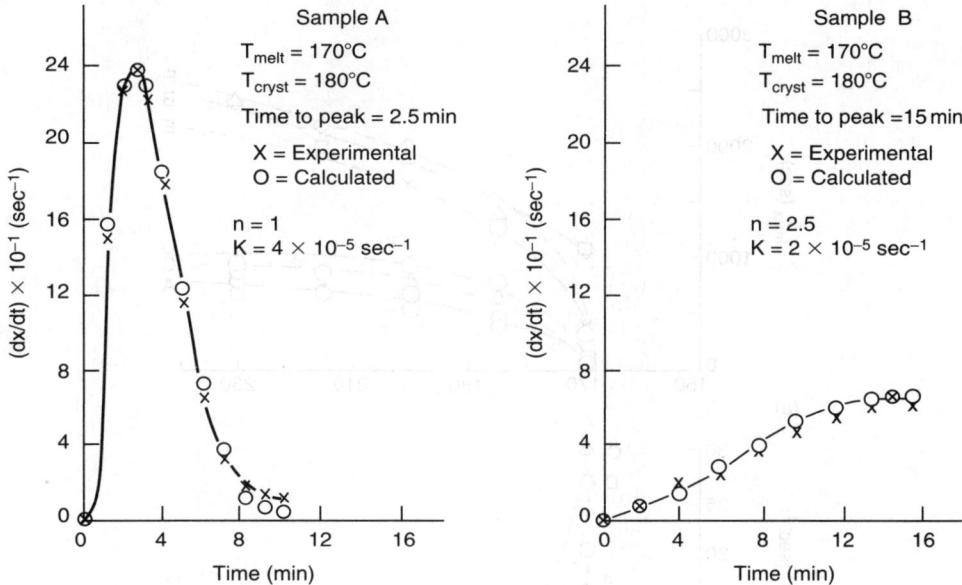

FIGURE 3.54 Crystallization kinetics for polypropylene fibers spun at high (sample A) and low (sample B) spin-line stress levels. (From Jaffe, M.; In *Thermal Methods in Polymer Analysis*, Shalaby, S.W.; ed., Franklin Institute Press, Philadelphia, 1977, p. 93. With permission.)

plots of T_{2M} versus f_c and f_{am}. Clearly, T_{2M} is a function of the initial amorphous orientation in both the drawn and heat-set fibers. If one extrapolates the line through the points in Figure 3.56, a T_{2M} of 220°C is predicted for an $f_{am} = 1.0$—a value also predicted by Cox and Duswalt [215] for isotactic polypropylene. From the data, it appears that a high-temperature peak occurs only in samples that have a minimum f_{am} of 0.40.

Figure 3.58 and Figure 3.59 show the relationships between T_{1M} and f_c and f_{am}. The low-temperature endotherm is generally regarded as the monoclinic melting endotherm. From Table 3.30 and Table 3.31 it is seen that more than two peaks appeared within the main body of the T_{1M} endotherm as draw ratio was increased. The reason for these peaks is unknown. In the curves, an average of these peaks is plotted as T_{1M}. In regions of low orientation, T_{1M} was relatively constant. At an $f_{am} = 0.40$, T_{1M} started increasing. If T_{1M} is extrapolated to $f_{am} = 1.0$, a T_{1M} of 185°C is obtained, again in agreement with Cox and Duswalt [215] (see also ref. [83]).

Samuels also made DSC measurements on a large number of samples of polypropylene resins that were isothermally crystallized in the temperature range of 130–160°C. Two endotherms were also observed for these samples, and when the endotherms were extrapolated to the equilibrium melting point, the high-temperature melting point was predicted to be 220°C, while the low-temperature melting point was predicted to be 185°C. The dual endotherm behavior of the isothermally crystallized polymer seemed to be a manifestation of the same crystal melting process as that observed for restrained fibers.

Samuels' data seem to contradict Jaffe's suggestion that the higher-melting endotherm results from the melting of the fibrillar nucleus of the row structure. The high endotherm is present in samples of isothermally crystallized resins that supposedly contained no row structures. Since x-ray data showed that Samuels' fibers contained no crystal modifications,

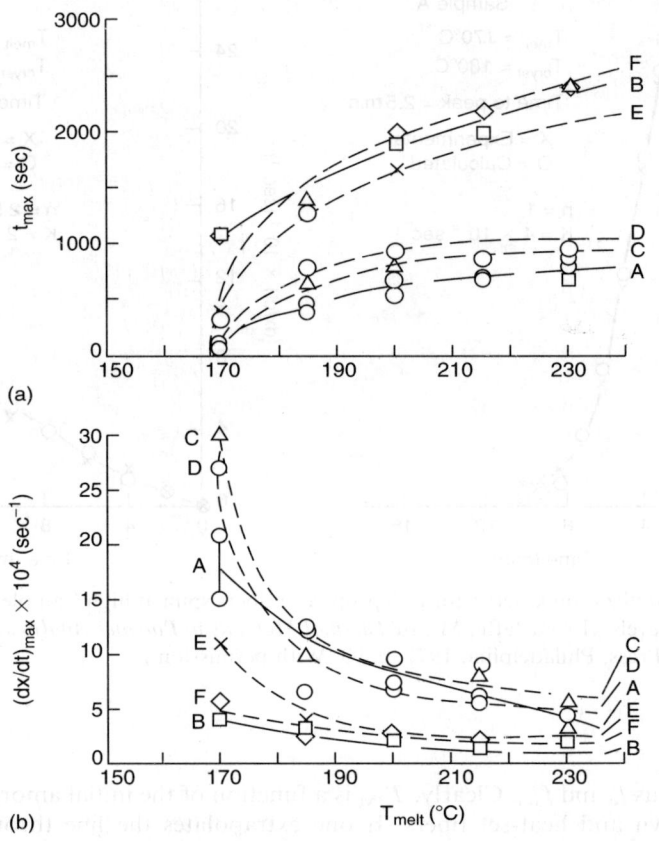

FIGURE 3.55 (a) Time to peak and (b) maximum recrystallization rate versus maximum temperature in melt for polypropylene fibers and films. (From Jaffe, M.; In *Thermal Methods in Polymer Analysis*, Shalaby, S.W.; ed., Franklin Institute Press, Philadelphia, 1977, p. 93. With permission.)

TABLE 3.31
Thermal and Structural Data for the Restrained Heat-Set Fibers

Sample number	Fabrication draw ratio	β	f_c	f_{am}	f_{av}	T_{1M} (°C)	T_{1M} (av) (°C)	T_{2M} (°C)
Spun	1.00	0.558	0.4900	0.1125	0.3231	164.0	164.0	
1	1.08	0.650	0.5849	0.1433	0.4303	163.1	163.1	
2	1.54	0.668	0.7977	0.4170	0.6713	163.8	163.8	
3	1.98	0.652	0.9113	0.4191	0.7390	167.7	167.7	(173.0)
4	2.44	0.632	0.9329	0.5127	0.7783	169.0	169.0	183.0
5	2.89	0.615	0.9366	0.5582	0.7909	168.0,172.0	170.0	187.8
6	3.35	0.635	0.9245	0.6310	0.8173	169.2,173.0	171.0	191.1
7	3.84	0.624	0.9151	0.6526	0.8164	171.7	171.7	193.8
8	4.31	0.610	0.9156	0.6667	0.8185	168.1,170.8,174.0	171.0	189.8
9	4.42	0.640	0.9138	0.7163	0.8427	172.0,174.6,177.0	174.5	196.3
10	4.48	0.642	0.9281	0.7024	0.8473	172.1,182.0	177.0	195.5

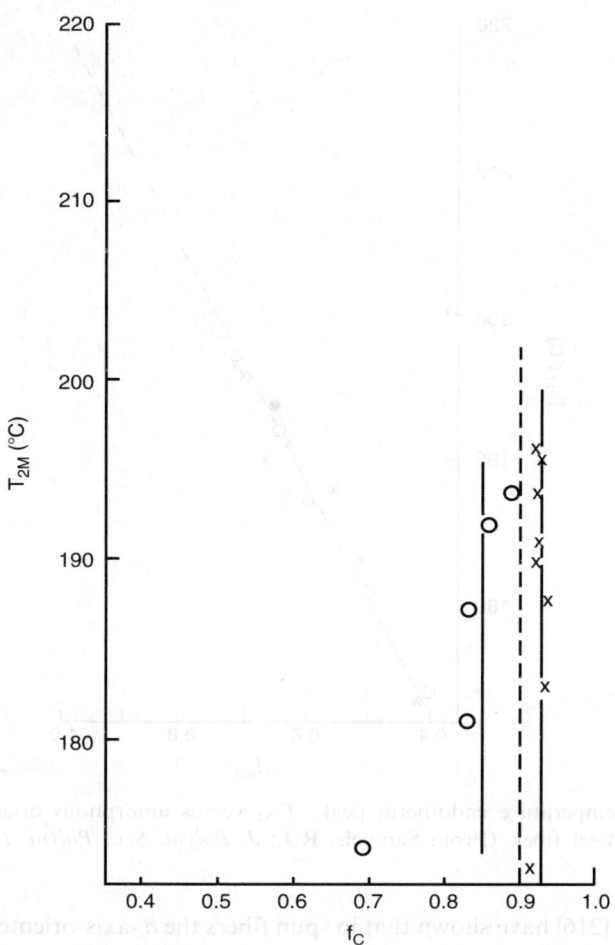

FIGURE 3.56 High-temperature endotherm peak, T_{2M} versus c-axis orientation function: (○) drawn fiber; (×) heat-set fiber. (From Samuels, R.J.; *J. Polym. Sci., Polym. Phys.*, **1975**, *13*, 1417. With permission.)

the presence of different crystal types cannot explain the observed melting behavior. Samuels concluded that this data hinted "that the melting of different sized or disordered crystals is the more likely explanation."

Others have found the presence of multiple endotherms. Lenz et al. [213] observed peaks at about 165 and 174°C for highly stretched films of polypropylene. These authors assigned the low-temperature peak to crystalline domains with a long period of 140–160 Å. The high melting peak appeared during stretching and corresponded to crystalline domains with a long period larger than 200 Å. They claimed the high-temperature fraction to be the paracrystalline building blocks of the microfibrils.

Mucha [214] found multiple endotherm peaks in samples that were isothermally crystallized. The double peaks were observed in samples crystallized at temperatures above 110°C. At lower temperatures, single peaks were observed. Mucha suggested that the two peaks correspond to the melting of the hexagonal and monoclinic crystalline forms.

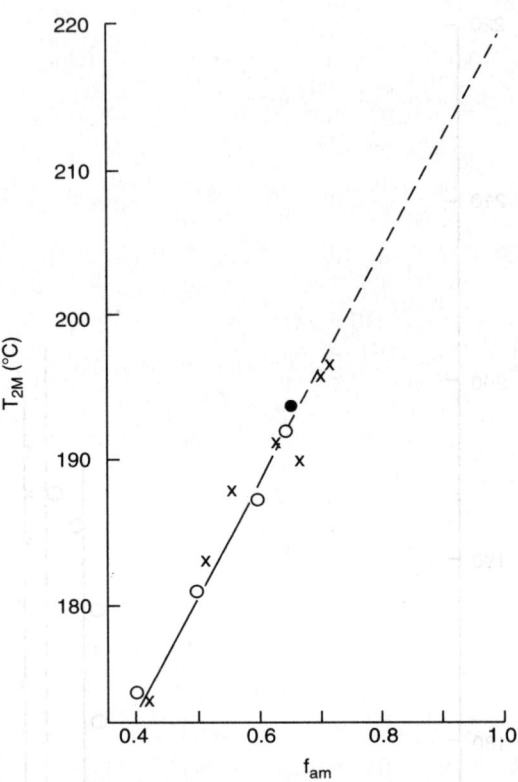

FIGURE 3.57 High-temperature endotherm peak, T_{2M} versus amorphous orientation function: (○) drawn fiber; (×) heat-set fiber. (From Samuels, R.J.; *J. Polym. Sci., Polym. Phys.*, **1975**, *13*, 1417. With permission.)

Katayama et al. [216] have shown that in spun fibers the a'-axis-oriented crystals melted at a temperature lower than the melting temperature for the c-axis-oriented crystals.

3.8 NEW GENERATION OF POLYPROPYLENE FIBERS

3.8.1 THREE-DIMENSIONAL HELICAL STAPLE

Polypropylene staple with permanent three-dimensional helix has been produced by a new processing technology developed by ESL, a British company, in the 1990s [217]. This staple fiber has the same contour as wool. It is particularly useful as a stuffing material for thermal insulation.

There are several process requirements for the preparation of polypropylene staple with permanent three-dimensional helical curvature. Specifically, a rectangular spinneret-pack assembly is used to produce flow perturbation and to impart high internal stress. A specially designed cooling device cools the fiber quickly to form a paracrystalline structure in the fiber. The process principle is that the flow perturbed in the polypropylene melt creates internal stress on one side of the fiber section. Because of the stress memory of polypropylene, the internal stress difference at the interface of streamlined and perturbed flows can sustain in the fiber after it has been cooled and solidified. This leads to different crystal structures and shrink properties, and thus a fiber in the shape of a three-dimensional helix.

Polypropylene Fibers

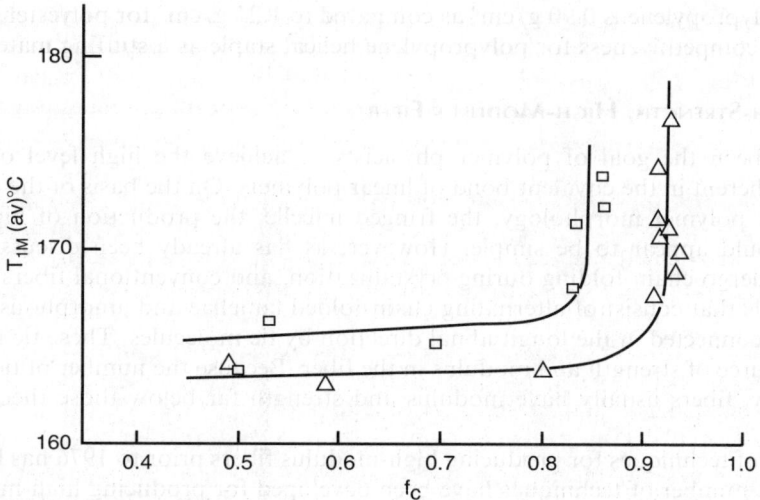

FIGURE 3.58 Low-temperature endotherm peak. $T_{1M}(av)$, versus c-axis orientation function: (□) drawn fiber; (△) heat-set fiber. (From Samuels, R.J.; *J. Polym. Sci., Polym. Phys.*, **1975**, *13*, 1417. With permission.)

The process for preparing the helical polypropylene staple includes: (1) spinning a fiber with differential internal stresses: (2) oven crystallization to control the crystal size through temperature adjustment and thereby to control the helical curvature: (3) multiroller drawing: (4) steam treatment: (5) multiroller drawing: (6) staple cutting: (7) annealing in the relaxed state: and (8) packing. In industrial operations, a typical staple product has a titer of 0.67 dtex, length of 65 mm, helix number of 3.74/cm, compression ratio of 76.5%, compressional resilience ratio of 35.2%, and compression elasticity ratio of 16.6%.

The three-dimensional helix is homogeneous and is of permanent distortion. Thus, the staple has a soft, high fluffiness, good rebound, and good heat preservation properties. The

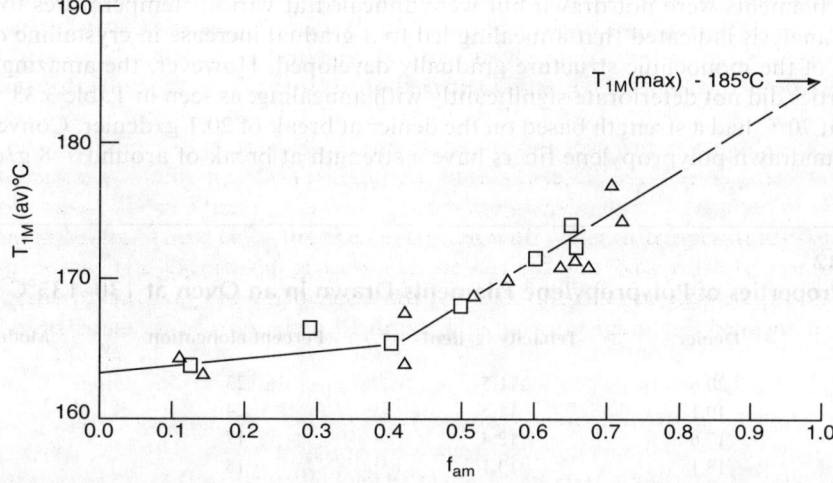

FIGURE 3.59 Low-temperature endotherm peak, $T_{1M}(av)$ versus amorphous orientation function: (□) drawn fiber; (△) heat-set fiber. (From Samuels, R.J.; *J. Polym. Sci., Polym. Phys.*, **1975**, *13*, 1417. With permission.)

density of polypropylene is 0.90 g/cm³ as compared to 1.27 g/cm³ for polyester. This renders considerable competitiveness for polypropylene helical staple as a stuffing material.

3.8.2 High-Strength, High-Modulus Fiber

It has long been the goal of polymer physicists to achieve the high level of mechanical properties inherent in the covalent bond of linear polymers. On the basis of the early concept of crystalline polymer morphology, the fringed micelle, the production of highly oriented materials would appear to be simple. However, as has already been discussed, polymers normally undergo chain-folding during crystallization, and conventional fibers are made up of microfibrils that consist of alternating chain-folded lamellae and amorphous regions. The lamellae are connected in the longitudinal direction by tie molecules. These tie molecules are the major source of strength and modulus in the fiber. Because the number of tie molecules is relatively low, fibers usually have modulus and strength far below those theoretically predicted.

A review of techniques for producing high-modulus fibers prior to 1976 has been given by Bigg [218]. A number of techniques have been developed for producing high-modulus, high-strength fibers; most of the emphasis has been placed on high-modulus fibers. These techniques include: (1) rapid quenching of extruding fiber and subsequent drawing at low rates and high draw ratios; (2) slow, two-stage drawing of a conventionally produced fiber at carefully controlled temperatures; (3) hydrostatic extrusion; (4) inducing fibrillar crystal growth from dilute solutions; and (5) spinning a fiber in the form of a gel and subsequent hot-thawing.

3.8.2.1 Rapid Quenching

Sheehan and Cole [185] prepared monofilaments from polypropylene by extruding them into cold water. This rapid quenching gave a pseudohexagonal or a smectic structure instead of the usual monoclinic crystal form. When these fibers were drawn slowly in an oven at 130–135°C, very strong fibers were attained. The properties of the fibers are shown in Table 3.32.

Noether [186] prepared undrawn filaments in a manner similar to that of Sheehan and Cole. The filaments were not drawn but were annealed at various temperatures for 30 min. The x-ray analysis indicated that annealing led to a gradual increase in crystalline order; the perfection of the monoclinic structure gradually developed. However, the amazing mechanical properties did not deteriorate significantly with annealing, as seen in Table 3.33. The fiber annealed at 70°C had a strength based on the denier at break of 20.1 g/denier. Conventionally produced undrawn polypropylene fibers have a strength at break of around 7–8 g/denier. In

TABLE 3.32
Physical Properties of Polypropylene Filaments Drawn in an Oven at 130–135°C

Draw ratio[a]	Denier	Tenacity (g/den)	Percent elongation	Modulus (g/den)
29	20.6	11.5	23	88
32	19.1	11.5	24	94
33	17.6	12.4	17	106
34	18.1	13.1	18	110

[a] Includes drawdown in spinning of 3X.

TABLE 3.33
Tensile Data of Smectic Polypropylene

Annealing temperature (°C)	Denier	Tenacity (g/den)	Percent elongation	Modulus (g/den)
70	2.6	5.7	253	44
80	2.6	4.4	220	45
90	2.7	4.0	172	47
100	2.9	3.3	190	47
110	2.8	3.7	172	43
120	2.8	4.4	154	43
130	2.7	5.0	171	39

spite of the rather drastic change in crystallinity and morphology, the load-bearing units of the fiber maintained their integrity. Other authors have studied smectic polypropylene fibers [79,219].

3.8.2.2 Two-Stage Drawing

Taylor and Clark [86] have produced high-strength, high-modulus polypropylene fibers by the two-stage drawing process of Clark and Scott [220]. In the first stage, the undrawn spherulitic fiber was drawn to the natural draw ratio by the familiar necking mechanism. The rapid first stage was followed by the slower second stage. Both stages of drawing were conducted in silicone oil at 125–145°C. The first-stage draw rate was 1000%/min, whereas the second stage rate varied from 4 to 400%/min. The drawn filaments were not annealed; however, the filaments remained at the drawing temperature for long periods due to the slow second-stage drawing.

Figure 3.60 and Figure 3.61 show the modulus and strength as a function of draw ratio. The elongation to break versus draw ratio is plotted in Figure 3.62. A drawing temperature of 130°C was considered to be optimum.

The "superdrawn" filaments were very resistant to heat shrinkage, as seen in Figure 3.63. Mechanical property measurements on these filaments after the heat shrinkage showed a 16% loss in modulus and a 10% loss in tensile strength. The fibers began to melt at 155°C, and the peak maximum occurred at 161°C.

The proposed morphology of the superdrawn fiber is shown in Figure 3.64. As seen, the second-stage draw resulted in a substantial increase in the number of tie molecules over those present in a conventionally drawn fiber (see Figure 3.39).

Cansfield et al. [221] have also studied the slow two-stage drawing of polypropylene. A series of polymers with different molecular weights and molecular weight distributions were examined.

The interest in high-strength, high-modulus fibers from flexible polymers stems from the pioneering extrusion studies of Southern and Porter [222]. Most of these studies have been conducted with polyethylene. In one variation of their process, linear polyethylene was extruded between 132 and 138°C and above a critical shear rate. As extrusion proceeded, supercooling, combined with a high elongational velocity gradient in the tapered capillary, promoted massive crystallization in the capillary entrance. At shear rates below the critical level (about 5000 sec^{-1} at 136°C), the polymer extruded as a melt. Temperature was a critical parameter in this technique. At temperatures above 138°C, pressure could not be generated

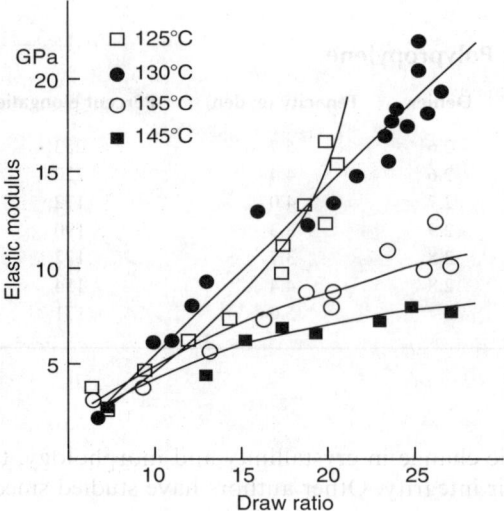

FIGURE 3.60 Modulus versus draw ratio for polypropylene fibers. (From Taylor, W.N.; Clark, E.S.; *Polym. Eng.*, **1978**, *18*, 518. With permission.)

rapidly enough to promote crystallization, and ordinary melt extrusion always occured. At temperatures below 132°C, the polymer is a solid; hence simple hydrostatic extrusion results. Polyethylene fiber with a modulus of approximately 70 GPa (875 g/denier) has been obtained with this method.

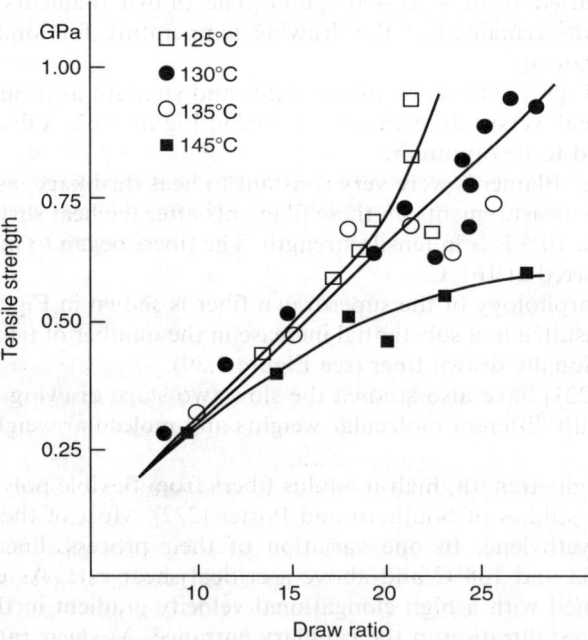

FIGURE 3.61 Tensile strength versus draw ratio for polypropylene fibers. (From Taylor, W.N.; Clark, E.S.; *Polym. Eng.*, **1978**, *18*, 518. With permission.)

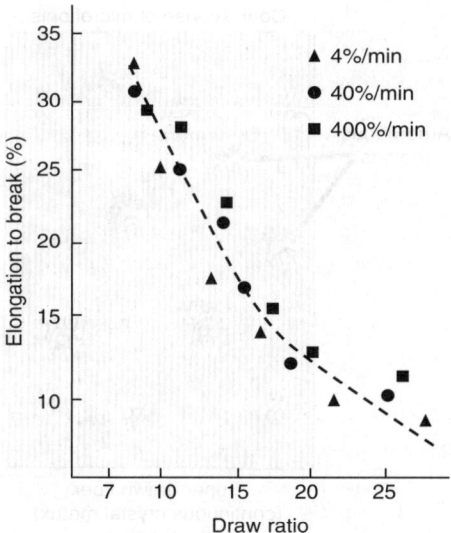

FIGURE 3.62 Elongation versus draw ratio for polypropylene fibers. (From Taylor, W.N.; Clark, E.S.; *Polym. Eng.*, **1978**, *18*, 518. With permission.)

The fibers produced by Porter and Clark are transparent, indicating the absence of structural units larger than about 0.5 μm. In addition, the fibers took a permanent set when bent. The bending caused internal fibrillation.

Hydrostatic extrusion of polypropylene has been conducted by Nakamura et al. [223] and by Williams [224]. Williams extruded cylindrical billets of undrawn polymer with essentially no orientation. The billets were forced through an extrusion die with a 30° entrance angle and a diameter of 0.3 in. Silicone oil was used to transmit pressure to the billet as well as to

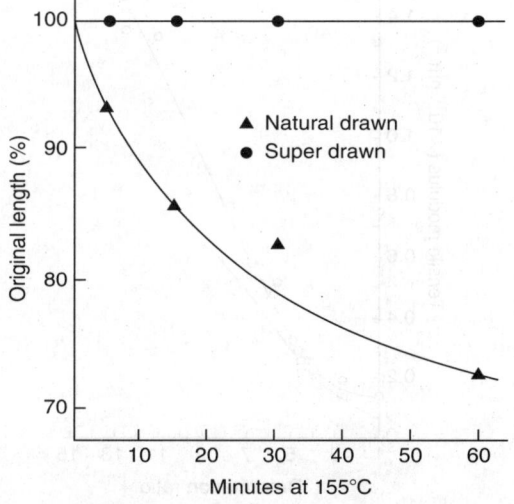

FIGURE 3.63 Shrinkage of natural drawn and superdrawn fibers at 155°C. (From Taylor, W.N.; Clark, E.S.; *Polym. Eng.*, **1978**, *18*, 518. With permission.)

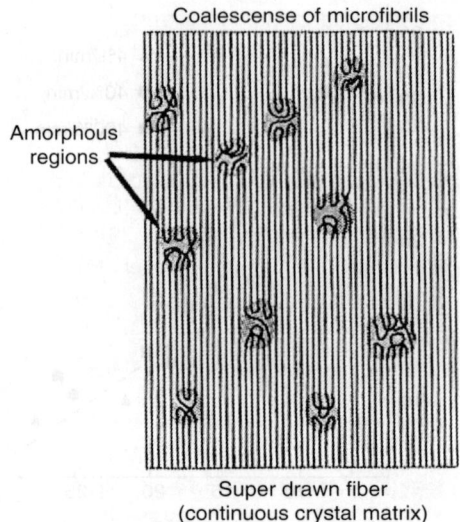

FIGURE 3.64 Proposed morphology of superdrawn fibers. (From Taylor, W.N.; Clark, E.S.; *Polym. Eng.*, **1978**, *18*, 518. With permission.)

lubricate the die walls. The draw ratio was changed by varying the diameter of the billet. The extrusion temperature was 100°C. In addition to the extrusion pressure, weights as large as 130 kg were used to pull on the extrudate. By this technique, draw ratios as large as 15 were achieved. A plot of modulus versus draw ratio for the extruded rods is shown in Figure 3.65.

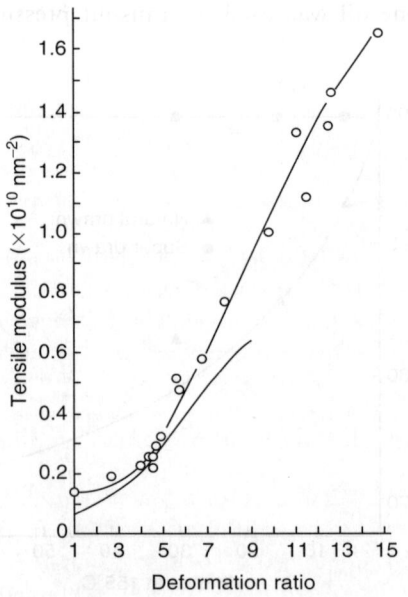

FIGURE 3.65 Modulus versus deformation ratio for hydrostatically extruded polypropylene. (From Williams, T.; *J. Mater. Sci.*, **1973**, *8*, 59. With permission.)

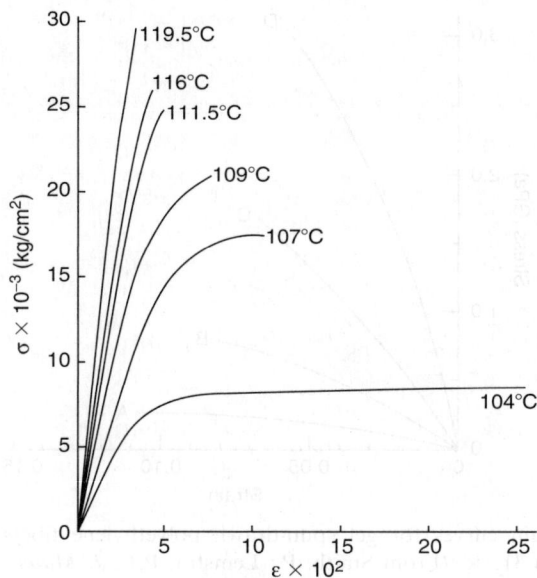

FIGURE 3.66 Stress–strain curves for polyethylene fibers grown from solution in Couette apparatus at various temperatures. (From Zwijnenburg, A.; Pennings, A.J.; *J. Polym. Sci., Polym. Lett.*, **1976**, 14, 339. With permission.)

3.8.2.3 Fibrillar Crystallization

Various techniques have been developed for producing high-performance fiber from solutions of high-molecular-weight polyethylene. These are mentioned here because of general interest and because the methods are probably applicable also to polypropylene. Zwijnenburg and Pennings [225] have studied the production of fibers generated from a dilute solution of polyethylene flowing in an elongational velocity gradient. In one description, high-molecular-weight polyethylene ($M_w = 1.5 \times 10^6$) was used to make a 0.5% (by weight) solution in *p*-xylene. Longitudinal growth of fibrillar crystals was effected in a Couette-type apparatus by pressing seed crystals against the surface of a Teflon rotor. The fibers, which were crystallized at temperatures above 105°C, were prepared at a stirring rate of 20 rpm. Under this condition, growth rates of 20 cm/min could be attained by winding the fibrillar crystals on a variable-speed take-up roll. Figure 3.66 shows the stress–strain curves for the fibers.

Kavesh et al. [226] reported the production of polyethylene filaments having a tenacity of at least 30 g/den by pulling a filament from a curved Teflon surface submerged in a solution of polyethylene. The rate of production of the filament was a least 2 m/min.

3.8.2.4 Gel Spinning

A remarkable phenomenon has been reported by Smith and Lemstra [227] and by Smook and Pennings [228]. Smith and Lemstra dissolved 2% by weight polyethylene ($M_w = 1.5 \times 10^6$) in decalin at 150°C. This highly viscous solution was pumped at 130°C through capillaries, and the extrudate was quenched in cold water to form a gel fiber, which contained almost all the decalin. The fiber was then drawn in a hot-air oven at 120°C at a strain rate of about 1 sec^{-1} to give a solvent-free, highly oriented fiber. Figure 3.67 shows the stress–strain curves for the fibers.

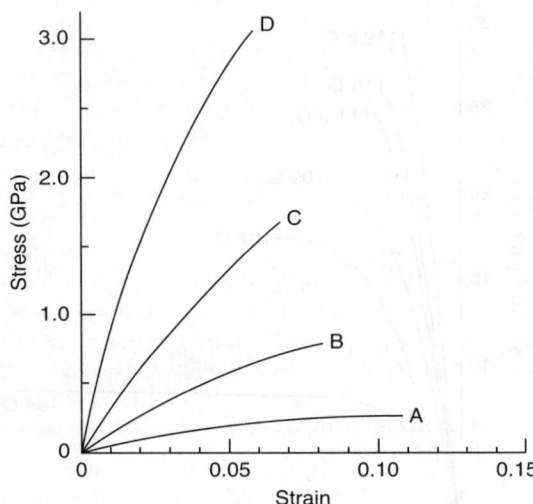

FIGURE 3.67 Stress–strain curves for gel spun-drawn polyethylene fibers. Draw ratios: (A) 2.8×; (B) 8.4×; (C) 15.7×; (D) 31.7×. (From Smith, P.; Lemstra, P.J.; *J. Mater. Sci.*, **1980**, *15*, 505. With permission.)

The technology of gel spinning has not been successful with polypropylene. Unlike polyethylene with a simple linear chain conformation, the helical chain conformation of polypropylene can conceivably forfeit a tight crystal packing required for a high-strength, high-modulus fiber.

3.8.2.5 Current Efforts

The search for a method to produce a high-strength, high-modulus (HSHM) polypropylene fiber has been a continuing effort. In view of its low density, good chemical resistance, and good processing characteristics, polypropylene fiber has a significant potential for industrial applications. According to theoretical calculation, the theoretical strength can reach 4 GPa, while the strength of a corresponding monofilament is about 430 cN/tex. The as-spun fiber typically has an optical dual refractive index of only 8×10^{-4} and strength of only 6–8 cN/tex

The rationality to further increase the fiber strength and modulus is through the spinning, drawing, and heat stabilization processes. These processes embark on increasing the stretch of macromolecular chains, the number of tie molecules, and the control of crystalline morphology and the degree of crystallinity. The goal is to produce a polypropylene fiber with α-monoclinic crystallinity. By conventionally changing the spinning condition, higher melt temperature and lower spinning tension, a fiber with hexagonal crystalline structure can be obtained.

ELS, a British company, developed a two-step process while Fare, an Italian company, pursued a one-step method. Both processes produced high-strength polypropylene filaments with a highly oriented α-monoclinic crystalline structure, an elongation of 16–20% and breaking strength of 75 and 65 cN/tex, respectively.

In the two-step process, polypropylene resin of typically MFR = 35 g/min and isotacticity above 90% is spun at 280°C at a low speed of 200 m/min. The spun fiber, which has a structure of low-oriented hexagonal crystallinity, is drawn at a low temperature of 60°C over seven rolls. The drawn fiber, still having a paracrystalline structure, is drawn again at a higher temperature of 110°C–140°C to change the paracrystallinity into a highly oriented

TABLE 3.34
Tensile Strength of Experimental HSHM Polypropylene Fiber

PP resin no.	Melt flow rate	Extension rate	Strength (cN/tex)
3–29	11.69	5.5	62.6
		6.8	72.3
70218	21.92	5.5	54.7
		6.8	66.2
70218	17.18	5.5	57.3
		6.8	68.8
Z30S	25	5.5	52.9
		6.8	63.5

Source: Mei-Fang Zhu, unpublished data.

monoclinic crystallinity. The resulting HSHM polypropylene fiber is typically 0.3~0.6 tex. The fiber strength reaches 85~116 CN/tex, modulus 580–1175 cN/tex, and elongation at break 24~13%. The one-step process typically produces an HSHM polypropylene fiber with a strength of 65 cN/tex and elongation of 16~18% at break.

Table 3.34 shows the results of preparing HSHM polypropylene fibers from different polymer molecular weights. A polypropylene fiber of high strength may conceivably be useful in ballistic and composite applications where high fiber strength is a primary requirement. However, these experimental polypropylene fibers are still inadequate as compared with other high-strength synthetic fibers. To further increase the fiber strength and modulus, a highly crystallized polypropylene fiber must be sought. To this end, a fiber producer in Japan has reportedly developed a fiber with strength 1.04 GPa, modulus 1300 kg/mm^2, and heat shrinkage rate of 4.5% by processing a highly isotactic polypropylene spun fiber [229]. The spun fiber was drawn at the preheat temperature of 145°C, which is higher than the crystallization temperature, and at the draw ratio above 10. The degree of crystallinity of the resulting fiber reaches 75% and 100% of the crystalline phase consists of α crystals. The physical properties and chemical properties of the HSHM polypropylene fiber are shown in Table 3.35 and Table 3.36, respectively. It can be seen that this fiber has not only good mechanical properties but also high chemical resistance. It also exhibited good wear resistance to aromatic solvents.

Ube-Nitto Chemical Products, Japan, has also developed a high-strength (HS) polypropylene fiber under the brand name Simtex. The technology is principally to optimize the crystallization of fiber by a new orientation technique. The HS fiber exhibits strength up to 88–115 cN/tex versus 40~60 cN/tex for the conventional polypropylene fiber. The heat shrinkage rate is below 50%. It also has high chemical resistance, especially to organic

TABLE 3.35
Physical Properties of HSHM Polypropylene Fiber

Fiber type	Titer (dtex)	Strength (GPa)	Draw ratio (%)	Modulus (GPa)	Heat shrinkage (%) at 140 °C
HSHM PP	2.3	1.04	16	12.7	4.5
Regular PP	7.2	0.43	45	2.5	10

TABLE 3.36
Solvent Resistance of HSHM Polypropylene Fiber

	HS PP fiber		Regular PP fiber	
Solvent	Strength GPa	Modulus GPa	Strength GPa	Modulus GPa
Untreated sample	0.94	10.4	0.51	2.3
Xylene	0.91	9.6	Dissolved	Dissolved
Toluene	0.78	9.3	Dissolved	Dissolved
Chlorobenzene	0.85	9.5	Dissolved	Dissolved

solvents. The commercial production of this fiber is expected to be in the order of 100 tons per month in the immediate future.

3.8.2.6 Commercial Products

The products of HS polypropylene filament are mainly used as bend reinforcements, strong ropes, long-lasting threads, filters for solvents, separator of recharge batteries, and geo-textiles. They also can be used in light-duty ballistic suits. Asota Ltd., an Austrian company, has exploited HS polypropylene staple with a titer of 0.1–0.7 tex, strength of 60cN/tex, and elongation below 30%. The staple can be used to produce outdoor textile goods, insulation material, and geo-textiles. British Bonar Textiles has produced the HS-twisted polypropylene multifilament yarn under the trade name Bonafild with 55.0–330.0 tex in titer, 60 cN/tex in strength, and 20% in elongation. This fiber has good UV stability and is used extensively in high-performance concrete in modern architecture.

The use of polypropylene fiber in concrete has played an increasingly important role in recent years. The polypropylene fiber-reinforced concrete must compete in the marketplace with high strength, good fluidity, excellent wear capability, and high performance–price ratio. Other polypropylene fiber products for concrete uses include the Dura fiber of Hill Brothers Chemical Co. in the United States, and the polypropylene fiber product of SS Co. in South Korea. Dura fiber is produced by the surface modification of 100% polypropylene fiber to impart good cohesion with the concrete matrix and good dispersion capability, so that the polypropylene fibers will disperse into the matrix quickly in the presence of water. The high-strength, acid and alkali resistance, and anti-long-term stability ensure that the polypropylene fiber plays a long efficacy in the concrete.

3.8.3 HIGH-SHRINKAGE FIBER

High-shrinkage polypropylene fiber is an important variety of polypropylene fibers [266,267]. Nobelex high-shrinkage polypropylene staple has been developed in Czech since the 1970s. Up to 1990s, both AKZO and Hoechst companies and Shandong Institute of Synthetic Fibers in China had diligently sought a high-retraction polypropylene fiber.

High-retraction polypropylene fiber has low thermal conductivity for thermal insulation. It also provides fluffiness with high retraction. This fiber is widely used in blanket, cotton products, synthetic leather, nonwoven fabrics, weaving, and needle knitting.

3.8.3.1 Fiber Preparation

Polypropylene is a high polymer that readily undergoes crystallization. Through changes in spinning conditions and the addition of modifying agents, its crystallization behavior and crystalline structure can be changed. As discussed earlier in Section 3.8.1, these changes can

TABLE 3.37
Comparison of Crystalline Structure of Conventional and High-Shrinkage Polypropylene Fibers

Sample type	Crystallinity (%)	Crystallite size (Å)
Ordinary PP fiber	65.5	134.7
High-shrinkage PP fiber	33–34	43–45

lead to a coiling silk-like polypropylene fiber with low crystallinity and low degree of orientation. The spun fiber is highly drawable and can sustain cold drawing at temperatures below 80°C and at draw ratio up to 6×. The resulting drawn fiber, which has small crystallite size, low crystallinity, but a high degree of orientation in the amorphous region, exhibits high shrinkage as compared to an ordinary polypropylene fiber.

3.8.3.2 Fiber Properties

The density of high-shrinkage polypropylene fiber is 0.860 g/cm^3 and that of its amorphous region is 0.85 g/cm^3. Thus the density of the crystalline region is 0.88 g/cm^3 by calculation. Table 3.37 shows the difference in crystallinity and crystallite size between the high-shrinkage and ordinary polypropylene fibers. Table 3.38 shows the birefringence and orientation factors of crystalline and amorphous regions of high-shrinkage polypropylene fiber. As can be seen from these tables, the crystallinity of high-shrinkage polypropylene fiber is lower than that of ordinary polypropylene fiber. The crystallite size of the high-shrinkage polypropylene fiber is quite small, while its overall orientation birefringence and birefringence of amorphous region are higher than the ordinary polypropylene fiber. These structural features give rise to the high shrinkage of this polypropylene fiber.

Table 3.39 lists the typical physical properties of high-shrinkage polypropylene fiber. It can be seen that the hot shrink rate of this fiber is relatively high. Its mechanical processing properties are good and its dyeability is better than ordinary polypropylene fiber. When the high-shrinkage fiber is processed to form fabrics at high temperature, its inherent shrinkage behavior will render the fabric fluffiness, fullness in touch, and high density.

There are two types of high-shrinkage polypropylene staple: the cotton type and the wool type. The cotton staple is typically 0.13 tex/filament and 38 mm long. The wool staple is typically 0.39 tex/filament and 60 mm or 90 mm long. The fracture intensity and crochet strength of high-shrinkage polypropylene staple can be set at 30~40 cN/tex. Its fracture extension rate is 65–86% and crochet extension rate is 50–70%. The hot shrink rate is typically 26–27% at 140°C in 10 min.

TABLE 3.38
Orientation in Crystalline and Amorphous Regions of High-Shrinkage Polypropylene Fiber

	Amorphous region		Crystalline region	
Overall birefringence $\Delta n \times 10^2$	Birefringence, $\Delta n \times 10^2$	Orientation factor (F_{am})	Birefringence, $\Delta n \times 10^2$	Orientation factor (F_c)
2.99	3.05	0.661	2.79	0.901

TABLE 3.39
Physical Properties of High-Shrinkage Polypropylene Fiber

Property index	16.7tex/3f	16.7tex/3f	10.0tex/3f
Linear density(tex)	16.7	12.2	10.07
Unevenness of size (CV %)	1.92	1.24	1.69
Breaking strength (cN/tex)	41.5	36.7	37.5
Breaking elongation (%)	40.6	59.2	61.8
Shrinkage in boiling water (%)	15.70	14.81	15.0
Hot shrinkage in 130°C (%)	31.2	29.8	30.5
Hot shrinkage at 140°C (%)	38.2	36.7	35.8
Dyeability (%)	60–70	60–70	60–70

3.8.3.3 Applications

High-shrinkage polypropylene fiber is mixed with conventional fibers in spinning to yield a fluffy yarn. The proportion of blending depends on the end-use. The blend ratio for mixed spinning generally include the following variations:

(1) 40% high-shrinkage polypropylene staple with 60% ordinary polypropylene staple
(2) 40% high-shrinkage polypropylene staple with 60% wool
(3) 40% high-shrinkage polypropylene staple with 60% viscose fiber
(4) 40% high-shrinkage polypropylene staple with 60% polyester staple
(5) 40% high-shrinkage polypropylene staple with 30% wool and 30% viscose fiber
(6) 35% high-shrinkage polypropylene staple with 65% polyacrylonitrile staple

Knitted fabric of high-shrinkage polypropylene fiber and of its mixed yarn with polyester and PAN staple provide good warmth and ventilative character. They are also pilling- and knot-resistant. The high-density nonwoven fabric for acupuncture uses contains high-shrinkage polypropylene staple. It is ideal for synthetic leather shoes. It can also increase the fluffiness when used in polypropylene carpet.

3.8.4 Conductive Fibers

Mitsubishi Corp. has developed a Rovalpolypropylene fiber with good antistatic, anticorrosive, and color properties. It can be woven into carpet for computer rooms. Courtaulds Corp. invented an electrically conductive polypropylene fiber that has excellent chemical resistance. Unlike surface-treated conductive fibers, this polypropylene fiber provides uniform electrical conductivity across its filament. These conductive fibers are now offered in multifilament and monofilament forms at 12 and 18 tex. An antistatic polypropylene fiber of fine denier was also invented by researchers at East China University of Science and Technology, Shanghai, China. It can be used in domestic polypropylene carpets. With hypothermal agglomerates, it can be prepared in partially oriented polypropylene filament yarn for use in antipollution, antistatic applications such as work clothes for operating personnel in oil, chemical, and electronics industries [230,231].

3.8.5 Fibers from Polymer Blends

Polymer blends are investigated for potentially improved polymer and fiber properties. In the case of polypropylene, a wide variety of conventional polymers have been studied for possible improvement in dyeability, strength, shrinkage, and other physical properties. These polymers include polyethylene, polypropylene [234,235], polystyrene[236–239], polyamide [240–241], polyester [243–245], polyacrylonitrile and even liquid crystalline polymers [247–249]. Polymer dispersion, fibrillar formation, and surface morphology of fibers from these polymer blends were studied extensively. The blend appears to be that of polypropylene with polypropylene [234,235], and that of polypropylene with polystyrene appears to be successful in achieving specific property improvements.

3.8.5.1 PP–PP Blend

Polypropylene can be blended with another polypropylene of different isotactic content or molecular weight. Although the all-polypropylene composite has a strong interfacial adhesion, its crystalline structure, macroscopic structure, and molecular weight distribution can be significantly affected by such blending. Kitayama and Utsumi [250] investigated all-polypropylene composites because of their interfacial, thermal, mechanical, economical, and ecological benefits. It was found that the c-axes of the polypropylene crystal were oriented in the longitudinal direction of the fiber in the *trans*-crystalline layer. This was attributed to the high adhesion strength of the PP–PP composite in the direction perpendicular to the fiber axis.

The concept of PP–PP blending has had some interesting utilities in light of the excellent physical properties of isotactic polypropylene. Fibers made of MiPP exhibit much higher mechanical properties, particularly tenacity, than fibers from ZNPP. Thus Pinoca and Africano [251] employed a polymer blend of highly isotactic polypropylene and a random copolymer of propylene with ethylene or higher olefins in the preparation of polypropylene fiber with high strength. However, it is known that isotactic polypropylene has good thermal resistance and has a very narrow range for thermal bonding in the preparation of thermal-bonded nonwovens. To this end, Demain refined the thermal bonding range with a polymer blend of >50% ZNPP, 5–50% MiPP, and <15% syndiotactic polypropylene [252]. Still another PP–PP blend contained 5–50% syndiotactic polypropylene isotactic polypropylene that gave a yarn with increased shrinkage [253]. The wisdom of blending ZNPP and MiPP was later manifested in the use of MiPP alone for a polypropylene fiber with improved shrinkage [255].

3.8.5.2 PP–PS Blend

When additives and compatibilizers are blended with polypropylene prior to extrusion, the impact resistance is often increased owing to the improvement in stress transmission at the phase interface. Such a mechanism has often been used in the modification for fiber flexibility. Similarly, blending polypropylene with 2~8% of atactic polystyrene (PS) prior to spinning and drawing can result in a modified fiber with good drawability. Since the structure of the as-spun polystyrene fiber is largely spherulitic, polymer blending can improve the plastic deformation of the as-spun fiber. The drawn blend fiber also exhibits improvement in rigidity and creep resistance by the intrinsic rigidity of the polystyrene molecule.

The formation of the fibril is a consequence of dispersed phase deformation and orientation correlating with the viscous force and interfacial tension between the dispersed phase and the matrix [236,237]. For good fibrillation to be achieved, the viscosity of dispersed phase should be lower than that of the matrix (i.e., $p = \eta_d/\eta_m < 1$) [238]. The temperature dependence of p of PS to PP is shown in Figure 3.68. Experimental results showed that p is less than 1 above 210°C and reduces to about 0.5. This indicates that polystyrene is able to form the fibrils in polypropylene matrix when the processing temperature is over 210°C, but finds it difficult below 210°C [239].

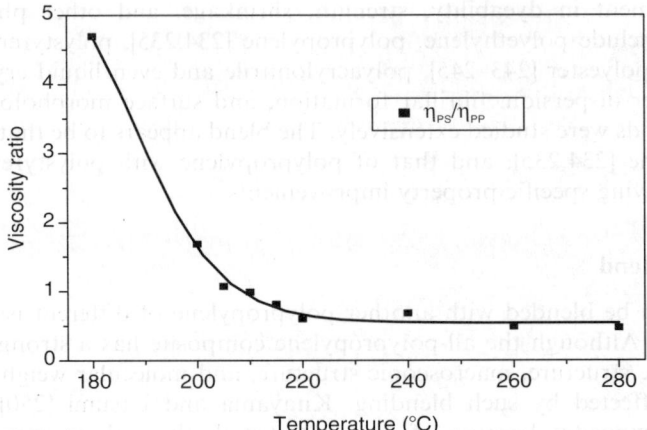

FIGURE 3.68 Viscosity–temperature curve of PP–PS blends. (From Xing, Q.; Wang, Y.H.; Zhu, M.F.; Chen, Y.M.; Pionteck, J.; Adler, H.J.; *IUPC*, 2004, pp. 4.3–206.)

The nature of melt flow plays a decisive role in the dispersed phase. Shear flow tends to break up the polymer domain structure, while extensional flow may consolidate and orient it [255]. In the process of fiber formation, the transformation of polystyrene into fibrils may take place in two stages. First, there is an elongational flow at the spinneret hole entrance; secondly, there exists a strong elongational strain at the spinneret exit. Figure 3.69 and Figure 3.70 present, respectively, the SEM micrographs of cross section and etched longitudinal section of as-spun composite fibers, with 98/2 and 92/8 PP–PS blend compositions at a processing temperature of 260°C.

It is obvious that polystyrene is immiscible with polypropylene and the morphology of polystyrene phase changes with its concentration. In the above PP–PS blends, there was a conversion of dispersion morphology from ellipses at 98/2 and 96/4 to fibrils at 94/6 and 92/8 in the composite fibers. At the lower polystyrene content (2%), the morphology of polystyrene phase was near ellipses along the longitudinal direction with similar short axis at about 200 nm, lower aspect ratio (L/D) from 1.5 to 5, and good dispersion (Figure 3.69a and Figure 3.70a). In comparison, the fibrillar morphology of the 8% polystyrene composite

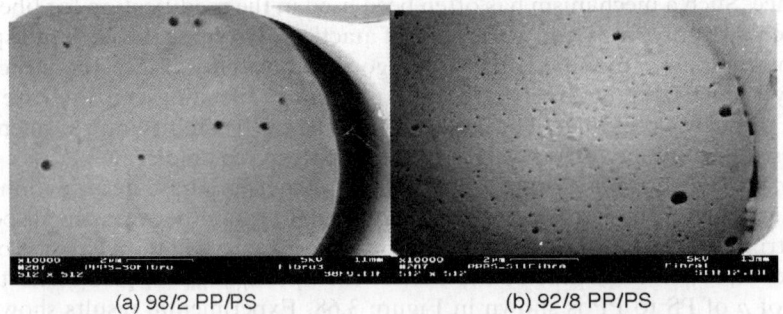

FIGURE 3.69 SEM micrographs of cross section of (a) 92/2 and (b) 92/8 PP–PS blend fibers spun at 800 m/min. (From Wang, Y.H.; Xing, Q.; Zhu, M.F.; Chen, Y.M.; Pionteck, J.; Adler, H.J.; International Symposium on Polymer Physics, Dali, China, June 1–5, 2004, p. 190.)

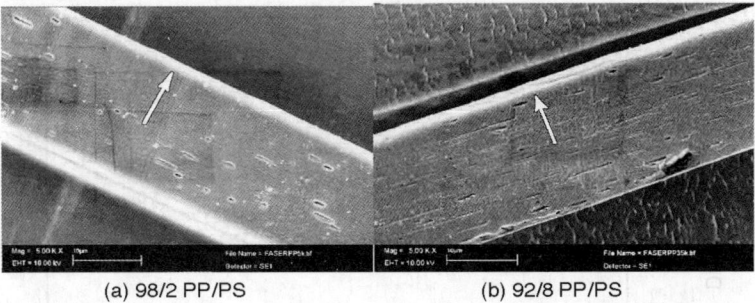

(a) 98/2 PP/PS   (b) 92/8 PP/PS

FIGURE 3.70 SEM micrographs of longitudinal section of (a) 98/2 and (b) 92/8 PP–PS blend fibers spun at 800 m/min: Arrow is in the radial direction, and dark region is the PS phase. (From Wang, Y.H.; Xing, Q.; Zhu, M.F.; Chen, Y.M.; Piontech, J.; Adler, H.J.; International Symposium on Polymer Physics, Dali, China, June 1–5, 2004, p. 190.)

fiber (Figure 3.70b) can be seen distinctly, especially in the gradient polystyrene phase. Along the direction of the arrow from the center to the surface shown in Figure 3.69a and Figure 3.70b, the fibrils became bigger, from 50–70 to 200 nm, and the aspect ratio (L/D) reduced from above 50 to below 10. Thus, the "thin and long" feature evolved to the "fat and short" feature. This was attributed to (1) the "pipe surface" effect on the blend melt flow in the spinneret and (2) the temperature gradient distribution along the radial direction of fiber during spinning. Comparing Figure 3.70a with Figure 3.70b, it can be seen that the fibril formation required a certain minimum polystyrene content, which is similar to thermotropic polymer–thermoplastics *in-situ* composites [254] and there was no gradient phase structure at lower polystyrene content.

Further study by SEM revealed that the surface morphology of as-spun PP–PS composite fibers was rougher than the unmodified polypropylene fiber. This is interpreted as resulting from the polystyrene phase lying on the fiber surface. The surface became rougher with higher polystyrene content. The polystyrene dispersed morphology remained consistent at different blends' ratios. However, the number and size of polystyrene phase increased with increasing polystyrene content [239].

PP–PS drawn fibers were prepared from the as-spun fibers in a heated tunnel above the T_g of polypropylene and polystyrene. This implies the two components can undergo deformation readily and thereby influence the fiber structure.

The diameter of as-spun fiber decreased from around 20–30 μm to about 15 μm after drawing, while the size of the polystyrene morphology reduced about 10%. This indicates the drawing process did not effectively deform the solid dispersed phase, because it is difficult for the draw stress to transfer from the matrix to the dispersed phase through the solid interface, and the free space for the polystyrene phase deformation is limited.

Figure 3.71 shows the change of average polystyrene phase diameter versus distance along the radial direction. The gradient phase structure was intensely related to the polystyrene content. With increasing amount of polystyrene, the size of the dispersed phase in the fiber center decreased continuously to nanosized fibrillar structure [257].

3.8.6 Fibers from Nanocomposites

Polymer nanocomposites based on the incorporation of inorganic or organic filler have attracted considerable attention not only from researchers but also from polymer producers [258–260]. Nanocomposites of polypropylene have been prepared by *in situ* polymerization or

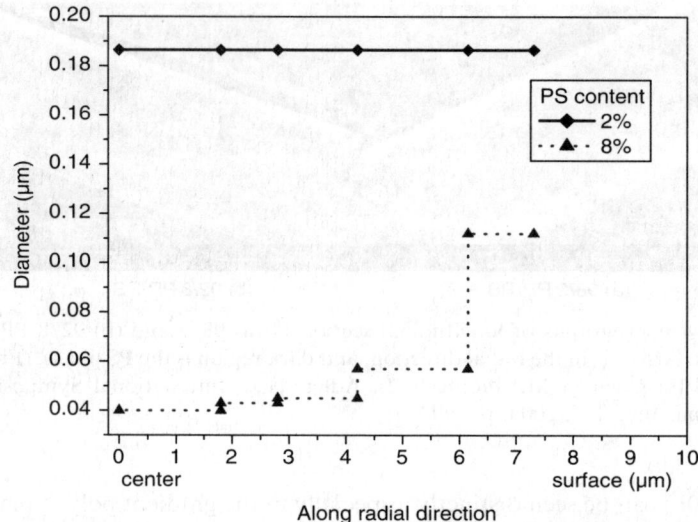

FIGURE 3.71 Plot of average PS phase diameter versus distance along the radial direction in the cross section of PP–PS drawn composite fiber. Diameter is calculated from the area of an equivalent circle of PS phase in the fiber cross section. (From Xing, Q.; Wang, Y.H.; Zhu, M.F.; Chen, Y.M.; Pionteck, J.; Adler, H.J.; *IUPC*, 2004, pp. 43–206. With permission.)

melt intercalation. The nanosized particles are generally fillers such as silica and carbon, organic and inorganic compounds such as montmorillonite, calcium carbonate, and maleic anhydride-modified polymers. By definition, the nanosized particles should be dispersed in the polymer matrix in a layered periodicity of nanometers.

3.8.6.1 Macroscopic Morphology

Pavlíková and Thomann [261] investigated the effect of one-dimensional orientation on the morphology and mechanical properties of polypropylene nanocomposites in the form of fibers. Their microscopic examination of polypropylene/SOMASIF ME C16 composite fibers showed that SOMASIF particles were oriented in the drawing direction of the fibers. Furthermore, a high degree of exfoliation of the SOMASIF particles in the fibers was observed following processing and spinning of polypropylene composites and drawing of fibers. The level of exfoliation was directly proportional to the draw ratio. At a draw ratio of 3, the distance between adjacent layers of the filler was 10 nm. Increasing the SOMASIF filler content and draw ratio of the fibers from 2 to 3 had considerable effects on the tensile strength of the fibers.

3.8.6.2 Crystallization Behavior

In the isothermal crystallization of polypropylene–montmorillonite (PP–MMT) nanocomposites, it was found that the crystallinity of the composites decreased with increasing montmorillonite content, indicating the dispersion of MMT layers in the polypropylene matrices. The nanoparticles confined the polypropylene chains and hindered the polymer crystallization. The spherulites of the PP–MMT nanocomposites were greatly decreased in size as MMT was introduced. On the other hand, the crystallization rate increased dramatically with the increasing of MMT content. The interfacial free-energy per unit area perpendicular to polypropylene chains in PP–MMT nanocomposites decreased with increasing MMT content, suggesting that the MMT layers acted as heterogeneous nuclei in the nucleation of crystallization. The nucleus density increased with the increasing of MMT content, leading to a positive effect on the crystallization [262]. It should be noted that the crystallization kinetics may differ with different nanoparticles [263].

3.8.6.3 Rheological Behavior

As would be expected, polypropylene nanocomposite exhibits somewhat different rheological behavior from the unmodified polymer. In rheological studies, a PP–MMT nanocomposite deviated from the linear viscoelastic behavior at a much lower strain than the polymer matrix. The percolation threshold was near 3 wt % MMT. Having been subjected to steady preshear, the tactoids could be oriented preferentially in the shear direction, and the percolation network was ruptured. The magnitudes of stress overshoots observed in the reverse flow experiments were strongly dependent on the rest time, which indicated that the ruptured network could be reorganized even under quiescent conditions. The composite displayed a strain-scaling stress response to the startup of steady shear. The maxima of stress overshoots appeared at the strain where the deviation of linear viscoelastic behavior started. It might imply that subjected to the deformation below the threshold strain, the network structure could be regard as an elastic one [264].

In nanocomposites of nanosized $CaCO_3$–polypropylene–ethylene copolymer (PPE) and styrene–butadiene–styrene (SBS), the inclusion of nanosized $CaCO_3$ increased the viscosity of polymer matrix significantly [265]. The increased shear force during compounding continuously breaks down SBS particles, resulting in the reduction of the SBS particle size and improving the dispersion of SBS in the polymer matrix. Thus the toughening effect of SBS on the matrix was improved. At the same time, the existence of SBS provides the matrix with a good intrinsic toughness, satisfying the condition that nanosized inorganic particles of $CaCO_3$ efficiently toughen the polymer matrix. The synergistic toughening function of nanosized $CaCO_3$ and SBS on PPE matrix was exhibited.

3.8.6.4 Fiber Spinning

The as-spun fiber from PP–MMT nanocomposite exhibited an increase in crystallization rate over the unmodified polypropylene. This was attributed to the heterogeneous nucleation of nanoclay that gave rise to an increase in nucleation rate. While the composite fiber had much higher crystallinity, it attained much lower orientation than the conventional polypropylene fiber at the same draw ratio. The nanocomposites also exhibited good spinning performance and the moisture absorption of resulting fiber was improved. Enhancement of fiber density and strength was also observed in melt spinning of polypropylene nanocomposite containing single-walled carbon nanotubes [266].

It has also been reported that polypropylene modified with nanosized silicon oxide can distinctively improve the processability of polypropylene while simultaneously increasing its strength. On the basis of the four property indices, specific electrical resistance, water absorption, flex stiffness, and rigidity, the modified fiber has reached or surpassed the indices of polyamide 6.

3.8.6.5 Dyeability

A recent report by Fan et al. [267] claimed that the incorporation of nanosized clay modified with quaternary ammonium salt in polypropylene imparted dyeability. The technique appeared to be effective in aqueous and disperse dyeing. When fully developed, it would afford a low-cost method to prepare dyeable polypropylene fibers.

In summary, there is an impressive amount of research effort on various polypropylene fiber products. The developments of fine-denier spinning, dyeability modification, high fiber strength and modulus, and nanocomposites certainly appear inductive to further growth in market shares and value-in-use for propylene fibers. However, as with other synthetic fibers, the manufacturing process yield and cost, particularly spinning continuity, must not be adversely impacted by any new technology to be commercialized. This is clearly the key to the future success of polypropylene fibers.

ACKNOWLEDGMENT

This chapter is a revision of the previous version by Dr. Marvin Wishman and Dr. Gerald E. Hagler. Much of their excellent presentation has been retained but updated in some technological aspects. The authors wish to acknowledge the kind assistance of Dr. H. Yu, Donghua University, Shanghai, China.

REFERENCES

1. Chinese Chemical Fiber Association, *Fiber Organon*, **2003**, 6.
2. Mordgren, G.; Coldenstein, G.; Textile Industries, **1982**, *146*, 52.
3. Staff Report, Textile Industries, **1982**, *146*, 57.
4. Platt, L.; Wishman, M.; Gentry, D.R.; Williams, J.E.; U.S. Patent 4,042,65, 1977.
5. *Non-wovens Report International*, September, 1982, p. 3.
6. Erlich, V.L.; Honeycutt, E.M.; *Man-Made Fibers*, John Wiley & Sons, New York, 1968, Vol. 3.
7. Park, J.; *Rev. Prog. Coloration*, **1981**, *11*, 19.
8. Natta, G.; Pino, P.; Mantica, E.; *Chim. Ind.* (Milan), **1956**, *38*, 124.
9. Zambelli, A; Locatelli, P.; Sacchi, M.C.; *Macromolecules*, **1982**, *15*, 831.
10. Sacchi, M.C.; Tritto, I.; Locattelli, P.; *Prog. Polym. Sci.*, **1991**, *16*, 331.
11. Zambelli, A.; Longo, P.; Pellecchia, C. et al.; *Macromolecules*, **1987**, *19*, 2035.
12. Zambelli, A.; Giongo, M.; Natta, G.; *Makromol. Chem.*, **1968**, *112*, 183.
13. Ewen, J.A.; *J. Am. Chem. Soc.*, **1984**, *106*, 6355.
14. Pino, P.; Mulhaupt, R.; *Angew. Chem. Int. Ed. Engl.*, **1980**, *19*, 857.
15. Kakugo, M.; Miyatake, T.; Naito, Y.; *Macromolecules*, **1988**, *21*, 314.
16. Kioka, M.; Makyo, H.; Kashima, N.; *Polymer*, **1994**, *35*, 580.
17. Warzelhan, V.; Burger, T.F.; Stein, D.J.; *Makromol. Chem.*, **1982**, *183*, 489.
18. Rishina, L.A.; Vizen, E.T.; *Eur. Polym. J.*, **1980**, *16*, 965.
19. Keii, T,; Doi, Y.; Soga, K.; *Makromol. Chem.*, **1984**, *185*, 1537.
20. Natta, G.; Pasquon, I.; Svab, J.; *Chim. Ind.* (Milan), **1962**, *44*, 621.
21. Goodall, B.L.; *Polypropylene and Other Polyolefins*, Van der Ven, second ed., Elsevier, 1990, pp. 113 and 77.
22. Zakharov, V.A.; Bukatov, G.D.; Yermakov, Yu; *Makromol. Chem.*, **1977**, *178*, 967.
23. Barbe, P.C.; Cecchin, G.; Noristi, L.; *Adv. Polym. Sci.*, **1986**, *81*, 1.
24. Zakharov, V.A.; Makhtarulin, S.I.; Yermakov, Yu; *React. Kinet. Cat. Lett.*, **1978**, *9*, 137.
25. Fischer, D.; Jungling, S.; Munhaupt, R.; *Makromol. Chem., Macromol. Symp.*, **1993**, *66*, 191.
26. Siedle, A.R.; Lamanna, W.M.; Newmark, R.A.; *Makromol. Chem., Macromol. Symp.*, **1993**, *66*, 215.
27. Stehling, U.; Diebold, J.; Brintzinger, H.H.; *Organometallics*, **1994**, *13*, 968.
28. Natta, G.; Mazzanti, G.; Longi, P.; *Chim. Ind.* (Milan), **1959**, *41*, 519.
29. Takahashi, A.; Keii, T.; Japan Chem. Soc. Annual Meeting, 1966.
30. Chadwick, J.C.; Miedemam, A.; Sudmeijer, O.; *Makromol. Chem.*, **1994**, *195*, 167.

31. Chien, J.C.W.; Nozaki, T.; *J. Polym. Sci., Part A: Polym. Chem.*, **1991**, *29*, 205.
32. Albizzati, E.; Giannini, U.; Morini, G.; *Macromol. Symp.*, **1995**, *89*, 73.
33. Tsutsui, T.; Kashiwa, N.; Mizuno, A.; *Makromol. Chem., Rapid Commun.*, **1990**, *11m*, 566.
34. Guyot, A.; Spitz, R.; Dassaud, J.P.; *J. Mol. Catal.*, **1993**, *82*, 37.
35. Galli, P.; Luciani. L.; Cecchin, G.; *Angew. Makromol. Chem.*, **1981**, *94*, 63.
36. Keii, T.; Soga, K.; Saiki, N.; *J. Polym. Sci., Part C: Polym. Symp.*, **1967**, *16*, 1507.
37. Longo, P.; Oliva, L.; Grassi, A.; *Makromol. Chem.*, **1989**, *190*, 2357.
38. Ziegler, K.; Belgian Patent 533,362, 1953.
39. Natta, G.J.; *Polym. Sci.*, **1953**, *16*, 143.
40. Montecatini Co., U.S. Patent 3,112,200,1963; U.S. Patent 3,112,301, 1963.
41. German Patent 2,213.086, 1972.
42. Solay & CIE, U.S. Patent 3,769,233, 1973.
43. Montedision; British Patent 1,286,867, 1968.
44. Montedision; Belgian Patent 785 332, 1972.
45. Himont Inc.; China 1,047,302. 1990; Mitsui Oil Co., Japan Patent 58 138 711.1983.
46. Galli. P.; Barbe, P.C.; Noristi, L.; *Angew. Makromol. Chem.*, **1984**, *120*, 73.
47. Mao, B.; Yang, A.; Zhang, Y.; Li, Z.; China 85,100,997, 1987; Catalyst system for use in olefin polymerization, U.S. Patent 4,784,983, November 15, 1988.
48. Mao, B.; Yang, A.; Zhang, Y.; Li, Z.; China 1,091,748A, 1993.
49. Jing, M.Z.; China 1,215,061A, 1999.
50. Sinn, H.W.; Kaminsky, W.O.; Vollmer, H.J.C.; Woldt, R.O.; Preparing ethylene polymers using Ziegler catalyst comprising cyclodienyl compound of zirconium, U.S. Patent 4,404,344, September 13, 1983.
51. Ewen, J.A.; *J. Am.Chem. Soc.*, **1984**, *106*, 6355.
52. Ewen, J.A.; Isotactic-stereoblock polymers of alpha-olefins and process for producing the same, U.S. Patent 4,522,982, June 11, 1985.
53. Kaminsky, W.; Hahnsen, H.; Kulper, H.; Woldt, R, Process for the preparation of polyolefins, U.S. Patent 4,542,199, September 17, 1985.
54. Kaminsky, W.; Kulper, K.; Brintzinger, H.H.; *Angew. Chem. Int. Ed. Eng.*, **1985**, *24*, 507.
55. Ewen, J.A.; Jones, R.L.; Razavi, A.; *J. Am. Chem. Soc.*, **1988**, *110*, 6255.
56. Albizzati, E.; Barbe, P.C.; Noristi, L.; Scordamaglia, R.; Morini, G.; Components and catalysts for the polymerization of olefins, U.S. Patent 4,971,937, November 20, 1990.
57. Giardollo, M.A.; Eisen, M.S.; Marks, T.J. et al.; *J. Am. Chem. Soc.*, **1995**, 117, 12114.
58. Ewen, J.A.; Process and catalyst for producing syndiotactic polyolefins, U.S. Patent 4,892,851, January 9, 1990.
59. Ewen, J.A.; Catalyst systems for the polymerization of olefins, U.S. Patent 4,927,797, May 22, 1990.
60. Elder, M.J.; Razavi, A.; Even, J.A.; Process and catalyst for producing syndiotactic polyolefins, U.S. Patent 5,225,500, July 6, 1993.
61. Kaminsky, W.; Kueber, F.; Schiemenz, B.; Werner, R.; Schauwienold A.M.; Freidanck, F.; Stereorigid metallocene compounds, U.S. Patent 6,410,661, June 25, 2002.
62. Holtcamp, M.W.; Polymerization catalyst activator complexes and their use in a polymerization process, U.S. Patent 6,806,328, October 19, 2004.
63. Zhang, X.; Yoon, S.C.; Lim, J.G.; Lee, Y.S.; Supported catalyst for producing syndiotactic styrenic polymer with high productivity and significantly reduced reactor fouling, U.S. Patent 6,828,270, December 7, 2004.
64. Vaughan, G.A.; Szul, J.F.; McKee, M.G.; Farley, J.M.; Lue, C.T.; Mixed metallocene catalyst systems containing a poor comonomer incorporator and a good comonomer incorporator, U.S. Patent 6,828,394, December 7, 2004.
65. Kinnan, M.A.; Ehrman, F.D.; Shirodkar, P.P.; Davis, M.B.; Grief-Rust, M.L.; Methods of polymerizing olefin monomers with mixed catalyst systems, U.S. Patent 6,833,416, December 21, 2004.
66. Ernst, E.; Reussner, J.; Roterud, P.; Solli, K.A.; Modified supported catalysts for the polymerization of olefins, U.S. Patent 6,841,634, January 11, 2005.

67. Moore, E.P.; *The Rebirth of Polypropylene: Supported Catalyst*, Munich, Vienna, New York, Hanser/Gardner Publications, 1998.
68. Li, X.G.; ed., *Petrochemical Process Technology IV. Polypropylene*, Chinese Petrochemicals Corp., Beijing, China, 1998.
69. Moore, E.P.; *Polypropylene Handbook*, Munich, Vienna, New York, Hanser/Gardner Publications, 1996.
70. Can, Z.Y.; ed., *Handbook of General-Purpose Resins*, Chinese Petrochemicals Publisher, Beijing, China, 1999.
71. Zhang, T.Z.; ed., *Processing of Polypropylene Derivatives*, Chemical Industry Publisher, Beijing, China, 1995.
72. Chinese Chemical System. Process Evaluation/Research Planning, Polypropylene, **2000**, *1*, 98-99.
73. Angenstein, B.L.; Northern Petrochemical Co., "A Marbting View of Polypropylene Developments," paper presented at Polypropylene Position and Potential Symposium, September, 1977.
74. Schroeder, C.W.; Shell Development Co., "Modern Polypropylene Resins," paper presented at Fiber Producer Conference, October, 1981.
75. Brockschmidt, A.; *Plastics Technol.*, **1982**, *28*, 67.
76. Roberts, J.S.; U.S. Patent 4,347,206, 1982 (to Kling-Tees, Inc.).
77. Phillips Petroleum Company; Training Literature, 1983.
78. Van Der Vegt, A.K.; *Plastics Inst. [London] Trans. J.*, **1964**, *32*, 165.
79. Nadella, H.P., Henson, H.M., Spruiell, J.E.; White, J.L.; *J. Appl. Polym. Sci.*, **1977**, *21*, 3003.
80. Turner-Jones, A.; Aizlewood, Z.M.; Beckett, D.R.; *Makromol. Chem.*, **1964**, *75*, 134–135.
81. Addink, E.J.; Beintema J.; *Polymer*, 1961, *2*, 185.
82. Keith, H.D.; Padden, F.J.; Walter, N.M.; Wyckoff, H.W.; *J. Appl. Phys.*, **1959**, *30*, 1485.
83. Samuels, R.J.; *J. Polym. Sci., Polym. Phys.*, **1975**, *13*, 1417.
84. Spruiell, J.E.; White, J.L.; *Appl. Polym. Symp.*, **1975**, No. 27, 121.
85. Spruiell, J.E.; White, J.L.; *Polym. Eng. Sci.*, **1975**, *15*, 660.
86. Taylor, W.N.; Clark, E.S.; *Polym. Eng.*, **1978**, *18*, 518.
87. Hagler, G.E.; *Fiber Producer*, **1980**, *8*, 8.
88. Bird, R.B.; Armstrong, R.L.; Hassager, O.; *Dynamics of Polymeric Liquids*, John Wiley & Sons, New York, 1977, Vol. 1.
89. Minoshima, W.; White, J.L.; Spruiell, J.E.; *Polym. Eng. Sci.*, **1980**, *20*, 1166.
90. Ide, Y.; White, J.L.; *J. Appl. Polym. Sci.*, **1978**, *22*, 1061.
91. Hagler, G.E.; Ph.D. Dissertation, University of Tennessee, 1972.
92. Chen, I.J.; Hagler, G.E.; Abbott, L.E.; Bogue, D.C.; White, J.L.; *Trans. Soc. Rheol.*, **1972**, *16*, 473.
93. Minoshima, W.; White, J.L.; Spruiell, J.E.; *J. Appl. Polym. Sci.*, **1980**, 25, 287.
94. White, J.L.; Roman, J.F.; *J. Appl. Polym. Sci.*, **1976**, *20*, 1005.
95. Huang, D.; White, J.L.; *Polym. Eng. Sci.*, **1980**, *20*, 182.
96. Ballenger, T.F.; White, J.L.; *J. Appl. Polym. Sci.*, **1971**. *15*, 1949.
97. Bartoš, O.; *J. Appl. Phys.*, **1964**, *35*, 2767.
98. Vinogradov, G.V.; Friedman, M.R.; Yarlykov, B.V.; Malkin, A.Y.; *Rheol. Acta.*, **1970**, 9, 323.
99. Petrie, C.J.S.; Denn, M.M.; *Am. Inst. Chem. Eng. J.*, **1976**, *22*, 209.
100. Miller, J.C.; *SPE Trans.*, **1963**, *3*, 134.
101. Kase, S.; *J. Appl. Polym. Sci.*, **1974**, *18*, 3279.
102. Kase. S.; Yoshimoto, Y.; *Seni Kikai Gakkaishi*, **1966**, *19*, T63.
103. Pearson, J.R.A.; Matovich, M.A.; *Ind. Eng. Chem. Fund.*, **1969**, *8*, 605.
104. Fisher, R.J.; Denn, M.M.; *Am. Inst. Chem. Eng. J.*, **1976**, *22*, 236.
105. Fisher, R.J.; Denn, M.M.; *Am. Inst. Chem. Eng. J.*, **1977**, *23*, 23.
106. Matsumoto, T.; Bogue, D.C.; *Polym. Eng. Sci.*, **1978**, *18*, 564.
107. White, J.L.; Ide, Y.; *J. Appl. Polym. Sci.*, **1978**, *22*, 3057.
108. Galante, M.J.; Mandelkern, L.; Alamo, R.G.; Lehtinen, A.; Paukker, R.; Crystallization kinetics of metallocene type polypropylene, *J. Therm. Anal.*, **1996**, *47*, 913.
109. Bond, E.B.; Spruiell, J.E.; Comparison of the crystallization behavior of Ziegler–Natta and metallocene catalyzed isotactic polypropylene, *ANTEC*, Europe, **1997**, 388.
110. Zhu, M.F.; Li, Tie; Chen, Y.; *PPS-14*, Japan, **1998**, 539.

111. Carlsson, D.J.; Garton, A.; Wiles, D.M.; *J. Appl. Polym. Sci.*, **1977**, *21*, 2963.
112. Carlsson, D.J.; Clark, F.R.S.; Wiles, D.M.; *Text. Res. J.*, **1976**, *46*, 590.
113. Garton, A.; Carlsson, D.J.; Sturgeon, P.Z.; Wiles, D.M.; *Text. Res. J.*, **1977**, *47*, 42,
114. Blais, P.; Carlsson, D.J.; Clark, F.R.S.; Sturgeon, P.Z.; Wiles, D.M.; *Text. Res. J.*, **1976**, *46*, 641.
115. Metk, W.H.; Iverson, R.L.; *Modern Plastics Encyclopedia,* McGraw-Hill, New York, 1976.
116. McLaughlin, K.M.; Townsend, E.B.; DiNardo, V.M.; Stabilization system for improving the melt viscosity of polypropylene during melt processing, U.S. Patent 6.787,006, September 7, 2004.
117. Horsey, D.W.; Andrews, S.M.; Davis, L.H.; Gray, R.E.; Gupta, A.; Hein, B.V.; Puglisi, J.S.; Ravichandran, R.; Shields, P.; Srinivasan, R.; Flame retardant compositions, U.S. Patent 6,599,963, July 29, 2003.
118. http://www.albemerie.com, Albemerie Corp., web site, January 20, 2005.
119. Sandoz Colors & Chemicals; Product Bulletin, 6–267/77.
120. Tozzi, A.; Catatore. G.; Masina, F.; *Text. Res. J.*, **1978**, *48*, 433.
121. Seip, S.D.; Thompson, S.E.; Townsend, E.B.; Polyolefin additive packages for producing articles with enhanced stain resistance, U.S. Patent 6,777,470, August 17, 2004.
122. Galbo, J.P.; Capocci, C.A.; Cliff, N.N.; Detlefsen, R.L.; DiFazio, M.P.; Ravichandran, R.; Solera, P.; Bulliard, C.; Hydroxy-substituted *N*-alkoxy hindered amines and compositions stabilized therewith, U.S. Patent 6,586,507, July 1, 2003.
123. Migdal, C.A.; Heins, J.B.; Kluger, E.W.; Reactive, non-yellowing triazine compounds useful as UV screening agent for polymers, U.S. Patent 4,826,978, May 2, 1989.
124. Valet, A.; Birbaum, J.L.; Slongo, M.; Coating materials stabilized against light-induced degradation, U.S. Patent 5,298,067, March 29, 1994.
125. Schmitter, A.; Burdeska, K.; Slougo, M.; Birbaum, J.L.; Stabilized polymers having heteroatoms in the main chain, U.S. Patent 5,288,778, February 22, 1994.
126. Reinehr, D.; Bacher, J.P.; Process for the preparation of hydroxyphenyl-1,3,5-triazines, U.S. Patent 5,478,935, December 26, 1995.
127. Birbaum, J.L.; Rytz, G.; Van Toan, V.; Valet, A.; Wurms, N.; Latent light stabilizers, U.S. Patent 5,597,854, January 28, 1997.
128. Fletcher, I.J.; Kaschig, J.; Metzger, G.; Reinehr, D.; Bisphenyl-substituted triazines, U.S. Patent 6,841,670, January 11, 2005.
129. Ravichandran, R.; Suhadolnik, J.; Wood, M.G.; Debellis, A.; Detlefsen, JR.L.; Iyengar, R.; Wolf, J.P.; Benzotriazole UV absorbers having enhanced durability, U.S. Patent 6,166,218, December 26, 2000.
130. Valentine, D.H.; Stephen, J.F.; Sassi, T.P.; Mono- and *bis*-benzotriazolyldihydroxybiaryl UV absorbers, U.S. Patent 6,344,505, February 5, 2002.
131. Wood, M.G.; Suhadolnik, J.G.; Ravichandran, R.; Lau, J.; Hendricks-Guy, C.; Bulliard, C.; Benzotriazoles containing alpha-cumyl groups substituted by heteroatoms and compositions stabilized therewith, U.S. Patent 6,451,887, September 17, 2002.
132. Wood, M.G.; Pastor, S.D.; Lau, J.; DiFazio, M.; Suhadoolnik, J.; Benzotriazoles containing phenyl groups substituted by heteroatoms and compositions stabilized therewith, U.S. Patent 6,800,676, October 5, 2004.
133. Ciba-Geigy Corp.; Product Bulletin, A-52A2M81, 1981.
134. Bagheri, R.; Chakraborty, K.B.; Scott, G.; *Polymer Degradation and Stability*, **1982**, 4.1.
135. Goodrich, B.F.; Product Bulletin, GC-75, 1982.
136. Borg-Warner Chemicals, Inc. Product Bulletin, CA-lO8A.
137. http://www.cibasc.com, Ciba Specialty Chemicals Corp., January 20, 2005.
138. Ciba-Geigy, Product Bulletin, A3065M7, 1981.
139. American Cyanamid Co., Preliminary Product Bulletin, 2–2696, 1982.
140. Wishman, M.; Taylor, P.A.; Leininger, J.C.; U.S. Patent 4,291,093, 1981 (to Phillips Petroleum Co.).
141. Wishman, M.; *Fiber Producer*, **1982**, *10*, 50.
142. Handen, S.L.; Graff, J.R.; *Fiber Producer* **1980**, *8*, 35, 66.
143. Uzelmeier, C.; *SPE J.*, **1970**, *26*, 69.
144. Klemchuk, P.P.; *Polym. Photochem.*, **1983**, *3*, 1.

145. Steinlin. F.; Saar, W.; *Melliand Textilberichte*, **1980**, *II*, 941.
146. Shore, J.; *Rev. Prog. Coloration*, **1975**, *6*, 7.
147. Huang, X. et al.; *Synthetic Fiber*, **1998**, *148*(3), 18.
148. Talley, A.; Proc. Clemson University Conference, Polypropylene Technology, **1996**, p. 18.
149. Yu. C. et al.; *Synthetic Fiber Industry*, **2000**, *23*(4), 9.
150. Taniguchi, I. et al.; U.S. Patent 3,395,198, 1968 (to Sumitomo and Toyo Spinning); Hammer, C.F.; U.S. Patent 3,653,804, 1972 (to DuPont); Uzelmeier, C.W.; Schroeder, C.W.; U.S. Patent 3,772,411, 1973 (to Shell).
151. Dominguez, R.J.G.; Henkee, C.S.; Crawford, W.C.; Cummings, G.W.; Hess, K.J.; Clark, R.J.; Evans, R.K.; Dyeable polyolefin containing polyetheramine modified functionalized polyolefin, U.S. Patent 5,985,999, November 16, 1999.
152. Dominguez, R.J.G.; Henkee, C.S.; Crawford, W.C.; Cummings, G.W.; Hess, K.J.; Clark, R.J.; Evans, R.K.; Dyeable polyolefin containing polyetheramine modified functionalized polyolefin, U.S. Patent 6,127,480, October 3, 2000.
153. Li, S.S.; Leggio, A.J.; Ergenc, N.; Dyeable polyolefin fibers and fabrics, U.S. Patent 6,679,754, January 20, 2004.
154. Fior, A.; DelMauro, E.; U.S. Patent 3,137,989, 1964 (to Montecatini).
155. Son, T.W.; Lim, S.G.; Park, J.H.; Disperse dyeable polypropylene fibers and its method of manufacture, U.S. Patent 6,054,215, April 25, 2000.
156. Earle, R.H.; Schmalz, A.C.; U.S. Patent 3,433,853, 1969 (to Hercules).
157. http://www.centexbel.be/Eng/homepage.htm, December 2004.
158. Levine, M.; Weimer, R.P.; *Text. Chem. Colorist*, **1970.**, *2*, 269.
159. Phillips Fibers Corp.; Product Bulletin, 760120–1M.
160. Bradshaw, J.; *Carpet & Rug Ind.*, **1981**, 9, 41.
161. Non-Wovens Report International, September **1982**, p. 4.
162. Lewin, M.; Guttmann, H.; Method of improving the sorption capacity of polymers, U.S. Patent 4,066,387, January 3, 1978; GB 1,543,076, 1979; Japan 1,199,825, 1983; France 7628829, 1981; Canada 26Q,999, 1981; Italy, 1070570, 1985; Israel 46937, 1978.
163. Zoehrer, R.; *Facil. Design Manage.*, **1982**, *1*, 70.
164. Camino, C.; Costa, L.; Trossareli, L.; Polymer Degradation Stability, **1982**, 4, 39.
165. Horsey, D.W.; Andrews, S.M.; Davis, L.H.; Gray, R.E.; Gupta, A.; Hein, B.V.; Puglisi, J.S.; Ravichandran, R.; Shields, P.; Srinivasan, R.; Flame retardant compositions, U.S. Patent 6,800,678, October 5, 2004.
166. Bar-Yakov, Y.; Hini, S.; Fire-retardant polyolefin compositions, U.S. Patent 6,737,456, May 18, 2004.
167. http://www.cibacs.com, Ciba Specialty Chemicals Corps., Kaprimidis, Earhart, N.; Zing, J.; Overview of recent advances in flame retardantcompositions, UV stable flame retardant systems: fully formulated antimony free flame retardant products for polyolefins, January 20, 2005.
168. Lewin, M.; *J. Fire Sci.*, **1999**, *17*, 2–19.
169. Endo, M.; Lewin, M.; in *Recent Advances in Flame Retardancy of Polymer Materials,* Lewin, M.; ed., BCC, 1993, Vol. 4, p. 171.
170. Lewin, M.; Endo, M.; In *Recent Advances in Flame Retardancy of Polymer Materials,*, Lewin, M.; ed, BCC, 1994, Vol. 5, p. 56.
171. Lewin, M.; ed., *Recent Advances in Flame Retardancy of Polymer Materials,* BCC, 1995, Vol. 6, p. 41.
172. Weil, E.D.; Lewin, M.; Lin, H.S.; Enhanced flame retardancy of polypropylene with magnesium hydroxide, Melamine and Novolac, *J. Fiber Sci.*, **1999**, *16*, 383.
173. Molesky, F.; The use of magnesium hydroxide for flame retardant/low smoke polypropylene, *Proc. Int. Cont Fire Safety*, **1991**, *16*, 212–226.
174. Rothon, R.N.; Hornsby, P.R.; Flame retardant effects of magnesium hydroxide, *Polym. Degradation Stability*, **1996**, *54*(2–3), 383–385.
175. Lewin, M.; Endo, M.; Catalysis of intumescent flame retardancy of polypropylene by metallic compounds, *Polym. Adv. Technol. J.*, **2002**, *13*, 1–9.
176. Lewin, M.; Endo, M; In Intumescent systems for flame retarding of polypropylene, *Fire and Polymers,* Nelson, G.; ed., ACS Symp., **1995**, 599, 91–116.

177. Lewin, M.; Guttmann, H.; Sarsour, N.; In A novel system for the application of bromine in flame retarding polymers, *Fire and Polymers*. 1991.
178. Schick, M.J.; *Text. Res. J.*, **1980**, *50*. 675.
179. Tadmor, I.Z.; Klein, I.; In *Engineering Principles of Plasticating Extrusion*, Krieger, R.E.; ed., Huntington, NY, 1970.
180. Janssen, L.P.B.M.; *Twin Screw Extrusion*, Elsevier, New York, 1978.
181. Gillies, M.T.; *Non-Woven Materials in Recent Development*, Noyes Data Corp., Park Ridge, NJ, 1979.
182. Krassig, H.; *J. Polym. Sci., Macromol. Rev.*, **1977**, *12*, 321.
183. Samuels, R.J.; *Structured Polymer Properties*, John Wiley & Sons, New York, 1974.
184. Samuels, R.J.; *Polym. Eng. Sci.*, **1976**, *16*, 327.
185. Sheehan, W.C.; Cole, T.B.; *J. Appl. Polym. Sci.*, **1964**, *8*, 2359.
186. Noether, H.D.; *Prog. Coll. Polym. Sci.*, **1979**, *66*, 804.
187. Ziabicki, A.; *Fundamentals of Fibre Formation*, John Wiley & Sons, New York, 1976.
188. Hagler, G.E.; *Polym. Eng. Sci.*, **1981**, *21*, 121.
189. Katayama, K.; Amano, T.; Nakamura, K.; *Kolloid-Z Z. Polym.*, **1968**, *226*, 125.
190. Fung, P.Y.F.; Orlando, E.; Carr, S.H.; *Polym. Eng. Sci.*, **1973**, *13*, 295.
191. Kitao, T.; Ohya, S.; Furukawa, S.; Yamashita, S.; *J. Polym. Sci., Polym. Phys.*, **1973**, *11*, 1091.
192. Anderson, P.G.; Carr, S.H.; *J. Mater. Sci.*, **1975**, *10*, 870.
193. Henson, H.M.; Spruiell, J.E.; Division of Cellulose, Paper and Textile Chemistry, American Chemical Society, Philadelphia, April, 1975.
194. Ishizuka, O.; Koyama, K.; *Sen-i Gakkaishi*, **1976**, *32*, T43.
195. Abhiraman, A.S., private communication.
196. Dees, J.R.; Spruiell, J.E.; *J. Appl. Polym. Sci.*, **1974**, *18*, 1053.
197. Keller. A.; Machin, M.J.; *J. Macromol. Sci.* (Phys.), **1967**, B1, 41.
198. Clark, E.S.; Spruiell, J.E.; *Polym. Eng. Sci.*, **1976**, *16*, 176.
199. Shimizu, J.; Toriumi, K.; Imai, Y.; *Sen-i Gakkaishi*, **1977**, *33*, T-255.
200. Jack, H.P.; McKinley, J.R.; Symposium on Polypropylene—Growth Fiber of the Eighties, New York, September 1981.
201. Peterlin, A.J.; *Mater. Sci.*, **1971**, *6*, 490.
202. Peterlin, A.; *Polym. Eng. Sci.*, **1977**, *17*, 183.
203. Peterlin, A.; *Polymeric Materials*, American Society for Metals, Metals Park, OH, 1975.
204. Nadella, H.P.; Spruiell, J.E.; White, J.L.; *J. Appl. Polym. Sci.*, **1978**, *22*, 3121.
205. Lazurkin, J.S.; *J. Polym. Sci.*, **1958**, *30*, 955.
206. Kitao, T.; Spruiell, J.E.; White, J.L.; *Polym. Eng. Sci.*, **1979**, *19*, 761.
207. Sakthivel, A., M.S. Thesis, Georgia Institute of Technology, 1983.
208. Hagler, G.E.; unpublished research.
209. Rosenthal, A.J.; Proc. Symp. on Polypropylene Fibers, Southern Research Institute, September 1964.
210. Balta-Calleja, F.J.; Peterlin, A.; *Makromol. Chem.*, **1971**, *141*, 91.
211. Jaffe, M.; In *Thermal Methods in Polymer Analysis*, Shalaby, S.W.; ed., Franklin Institute Press, Philadelphia, 1977, pp. 93.
212. Varga, J.; Menczel, J.; Solti, A.; *J. Therm. Analy.*, **1981**, *20*, 23.
213. Lenz, J.; Wrentschur, E.; Geymayer, W.; *Angew. Makromol. Chem.*, **1983**, *111*, 17.
214. Mucha, M.J.; *Polym. Sci.: J. Polym. Symp.*, **1981**, *69*, 79.
215. Cox, W.W.; Duswalt, A.A.; *Polym. Eng. Sci.*, **1967**, *7*, 1.
216. Katayama, K.; Amano, T.; Nakamura, K.; *Kolloid Z.Z. Polym.*, **1968**, *226*, 125
217. Bai, X.; *Synthetic Fiber*, **1998**, 27(3), 49.
218. Bigg, D.M.; *Polym. Eng. Sci.*, **1976**, *16*, 725.
219. Simizu, J.; Shimazaki, K.; *Sen-i Gakkaishi*, **1974**, *30*, T87.
220. Clark, E.S.; Scott, L.S.; *Polym. Eng. Sci.*, **1974**, *14*, 682.
221. Cansfield, D.L.M.; Capaccio, G.; Ward, I.M.; *Polym. Eng. Sci.*, **1976**, *16*, 72.
222. Southern, J.S.; Porter, R.S.; *J. Macromol. Sci., Phys.*, **1970**, B4, 541.
223. Nakamura, K.; Imada, K; Takayanagi, M.; *Polymer*, **1974**, *15*, 446.
224. Williams, T.; *J. Mater. Sci.*, **1973**, *8*, 59.

225. Zwijnenburg, A.; Pennings, A.J.; *J. Polym. Sci., Polym. Lett.*, **1976**, *14*, 339.
226. Kavesh, S.; Prevorsek, D.C.; Wang, D.G.; U.S. Patent 4,356,138, 1982 (to Allied Chemical Corp.).
227. Smith, P.; Lemstra, P.J.; *J. Mater. Sci.*, **1980**, *15*, 505.
228. Smook, J.; Pennings, A.; *J. Polym. Bull.*, **1983**, *9*, 75.
229. Dan, H.; *Developmental Potential for High-Strength Polypropylene Fiber*, April 24, 1997.
230. Samuels, R.J.; *Macromol. Sci.—Physics, Kep. Fac. Sci.*, **1971**, *6*, 17.
231. Li, H. et al.; Synthetic Fiber Industry, **1997**, *20*(3), 17.
232. Han, F.; Dong G.; *Synthetic Fiber*, **2002**, *9*, 28.
233. Yan, Y. et al.; Synthetic Fiber Industry, **2001**, *2*, 13.
234. Nakamae, K.; Nishino, T.; Shimizu, Y.; Matsumoto, T.; *Polym. J.*, **1987**, *19*, 451.
235. Ota, S.; In *5th Asian Textile Conference*, 1999, Vol. 2, p. 1005.
236. (a) Van Oene H.; *J. Coll. Inter. Sci.*, **1972**, *40*, 448.
237. (b) Hemmati, M.; Nazokdast, H.; Panahi, H.S.; *J. Appl. Polym. Sci.*, **2001**, 82, 1129–1137.
238. Kim B.K., Do I.H.; *J. Appl. Polym. Sci.*, **1996**, *60*, 2207.
239. Xing, Q.; Wang, Y.H.; Zhu, M.F.; Chen, Y.M.; Pionteck, J.; Adler, H.J.; *IUPC*, 2004, p. 4.3–206.
240. Lopez-Manchado, M.A.; Arroyo, M.; *Die Angew. Makromol. Chem.*, **1999**, *265*, 20–24.
241. Tzur, A.; Narkis, M.; Siegmann, A.; *J. Appl. Polym. Sci.*, **2001**, *82*, 661.
242. Wen, B.; Li, Q.C.; Hou, S.H.; Wu, G.; *J. Appl. Polym. Sci.*, **2004**, *91*, 2491–2496.
243. Min, K.; White, J.L.; *Polym. Eng. Sci.*, **1984**, *24*, 1327; Wu, S.; *Polym. Eng. Sci.*, **1987**, *27*, 335.
244. Lin, X.D.; Cheung., W.L.; *J. Appl. Polym. Sci.*, **2003**, *88*, 3100–3109.
245. Taylor, G.I.; *Proc. Royal Soc.* A, **1934**, *146*, 501.
246. Lin, X.D.; Jia, D.; Leung, F.K.P.; Cheung., W.L.; *J. Appl. Polym. Sci.*, **2004**, *93*, 1989.
247. Beery, D.; Kenig, S.; Siegmann, A.; *Polym. Eng. Sci.*, **1991**, *31*, 451.
248. Brinkmann, T.; Höck, P.; Michaeli, W.; *SPE ANTEC.*, **1990**, *37*, 988.
249. Heino, M.T.; Hietaoja, P.T.; *J. Appl. Polym. Sci.*, **1994**, *51*, 259–270.
250. Kitayama, T.; Utsumi, S.; *J. Appl. Polym. Sci.*, **2003**; *88*, 2875.
251. Pinoca, L.; Africano, F.; High tenacity polypropylene fiber and process for making it, U.S. Patent 5,849,409, December 15, 1998.
252. Demain, A.; Polypropylene fibers, U.S. Patent 6,720,388, April 1, 2004; Polypropylene fibers, U.S. Patent 6,730,742, May 4, 2004.
253. Galambos, A.F.; Propylene polymer yarn and articles made therefrom, U.S. Patent 5,455,305, October 3, 1995.
254. Gownder et al.; Reduced shrinkage in metallocene isotactic polypropylene fibers, U.S. Patent 6,505,970, May 20, 2003.
255. Wang, H.M.; Tao. X.M.; Newton, E.; Chung, T.S.; *Polym. J.*, **2002**, *34*, 575–583.
256. Beery, D.; Kenig, S.; Siegmann, A.; *Polym. Eng. Sci.*, **1996**, *36*, 229–236.
257. Wang, Y.H. ; Xing, Q.; Zhu, M.F.; Chen, Y.M.; Pionteck, J.; Adler, H.J.; International Symposium on Polymer Physics, Dali, China, June 1–5, 2004, p. 190.
258. Reichert, P.; Nitz, H.; Klinke, S.; Brandsch, R.; Thomann, R.; Mülhaupt, R.; *Macromol. Mater. Eng.*, **2000**, *275*, 8.
259. Ke, Y.; Lü, J.; Yi, X.; Zhao, J.; Qi, Z.; *J. Appl. Polym. Sci.*, **2000**, *78*, 805
260. Rong, M.Z.; Zeng, H.M.; Schmitt, S.; Wetzel, B.; Friedrich, K.; *J. Appl. Polym. Sci.* **2001**, *80*, 2218.
261. Pavliková, S.; Thomann, R.; Reichert, P.; Mülhaupt, R.; Marcin, A.; Borsig, E.; *J. Appl. Polym. Sci.*, **2003**, *89*, 3, 604–611.
262. Ma, J.; Zhang, S.; Qi, Z.; Li, G.; Hu, Y.; *J. Appl. Polym. Sci.*, **2002**, *83*, 1978–1985.
263. Svoboda, P.; Zeng, C.; Wang, H.; Lee, L.J.; Tomasko, D.L., *J. Appl. Polym. Sci.*, **2002**, *85*, 1562–1570.
264. Li, J.; Zhou, C.; Wang, G.; Zhao, D.; *J. Appl. Polym. Sci.*, **2003**, *89*, 3609–3617.
265. Chen, J.; Wang, G.; Zeng, X.; Zhao, H.; Cao, Jimmy, D.; Yun, J.; Tan, C.K.; *J. Appl. Polym. Sci.*, **2004**, *94*, 796–802.
266. Moore, E.M.; Ortiz, D.L.; Marla, V.T.; Shambaugh, R.L.; Grady, B.P.; *J. Appl. Polym. Sci.*, **2004**, *93*, 2926–2933.
267. http://www.umassd.edu. Fan, Q.; Ugbolae, S.C.; Wilson, A.R.; Dar, Y.S.; Yang, Y.; January 20, 2005.

4 Vinyl Fibers

Ichiro Sakurada[†] and Takuji Okaya[*]

CONTENTS

4.1 Introduction 262
    4.1.1 History 262
    4.1.2 Outline of the Chemistry 263
4.2 Polyvinyl Acetate 265
    4.2.1 Manufacture of Monomer 266
    4.2.2 Polymerization 266
        4.2.2.1 Mechanism of Polymerization 266
        4.2.2.2 Rate of Polymerization 267
        4.2.2.3 Chain Transfer and Degree of Polymerization 268
        4.2.2.4 Chain Transfer to Dead Polymer 269
        4.2.2.5 Industrial Process of Polymerization 273
    4.2.3 Conversion to Polyvinyl Alcohol 274
        4.2.3.1 Methanolysis and Hydrolysis 274
        4.2.3.2 Drop in the Degree of Polymerization 276
        4.2.3.3 Process of Deacetylation 276
4.3 Structure and Properties of Polyvinyl Alcohol 277
    4.3.1 Molecular Structure 277
        4.3.1.1 Chain Configuration 277
        4.3.1.2 End Groups 278
        4.3.1.3 Stereostructure 279
        4.3.1.4 Branching 281
        4.3.1.5 Molecular Weight 281
    4.3.2 Crystal Structure 283
        4.3.2.1 Unit Cell 284
        4.3.2.2 Effect of Water 284
        4.3.2.3 Crystal Structure Model 285
        4.3.2.4 Degree of Crystallinity 285
    4.3.3 Physicochemical Properties 288
        4.3.3.1 Melting Point 288
        4.3.3.2 Glass Transition Temperature 288
        4.3.3.3 Density 289
        4.3.3.4 Orientation and Strength 290
        4.3.3.5 Swelling 291
        4.3.3.6 Aqueous Solution 292

[†] Deceased.
[*] Retired in 2004.

    4.3.4 Chemical Reactions 293
        4.3.4.1 Acetalization 293
        4.3.4.2 Esterification 295
        4.3.4.3 Etherification 295
        4.3.4.4 Complex Formation 296
        4.3.4.5 Grafting 296
        4.3.4.6 Decomposition 296
4.4 Manufacture of Polyvinyl Alcohol Fiber 297
    4.4.1 Traditional Process of Wet-Spinning 297
        4.4.1.1 Fiber Formation 297
        4.4.1.2 Drawing 299
        4.4.1.3 Heat Treatment 300
        4.4.1.4 Acetalization 301
    4.4.2 Different Processes of Spinning 303
        4.4.2.1 Wet-Spinning with Alkali Bath 303
        4.4.2.2 Wet-Spinning of PVA Solution Containing Boric Acid 306
        4.4.2.3 Dry-Spinning 306
        4.4.2.4 Spinning Using Organic Solvent 307
        4.4.2.5 Biodegradability of Polyvinyl Alcohol 309
        4.4.2.6 Miscellaneous 309
    4.4.3 Bicomponent Fiber (PVC/PVA) 310
        4.4.3.1 Emulsion-Spinning 311
        4.4.3.2 Polychlal 312
4.5 Polyvinyl Chloride Fibers 313
    4.5.1 Manufacture of Polyvinyl Chloride 313
        4.5.1.1 Vinyl Chloride Monomer 313
        4.5.1.2 Polymerization 315
    4.5.2 Manufacture of Fiber 316
        4.5.2.1 Dry-Spinning 316
        4.5.2.2 Other Methods of Spinning 320
    4.5.3 Properties of Fiber 321
        4.5.3.1 Flame Retardance 321
        4.5.3.2 Chemical Resistance 321
        4.5.3.3 Triboelectricity 322
        4.5.3.4 Dimensional Stability 323
4.6 Applications and Future Trends 323
Acknowledgments 326
References 326

4.1 INTRODUCTION

4.1.1 HISTORY

Polyvinyl alcohol (PVA) is a polymer that has a polymerized formula of vinyl alcohol, $CH_2=CH-OH$. However, the vinyl alcohol monomer has neither been isolated nor obtained in high concentration. Vinyl chloride or vinyl acetate is obtained by the addition of hydrogen chloride or acetic acid to acetylene. Therefore, vinyl alcohol is formed by the reaction of water with acetylene. This reaction, however, produces acetaldehyde, $CH_3-CH=O$, and not vinyl alcohol. It is probable that vinyl alcohol is formed at a point during the reaction, but it changes instantaneously to acetaldehyde.

Although vinyl alcohol itself is very unstable, its esters are stable. Klatte found in 1912 that vinyl acetate is formed as a by-product in the course of the manufacture of ethylidene diacetate from acetylene and acetic acid [1].

Klatte's reaction is a liquid-phase reaction. In 1921, Baum et al. invented the gas-phase reaction [2]. With this process, it is possible to produce vinyl acetate practically without the formation of ethylidene diacetate.

In 1924, Herrmann and Haenel added alkali to a clear alcoholic solution of polyvinyl acetate to saponify this polymeric ester; from the experience with the saponification of monomeric vinyl acetate, they expected that some impure resinous precipitate would be produced. In contrast to their expectation, however, ivory-colored PVA was obtained [3]. It was remarkable—water-soluble PVA was obtained from hydrophobic polyvinyl acetate.

Independent of Herrmann and Haenel, Staudinger et al. studied PVA and, in a lecture of the Gesellschaft Deuche Naturforscher and Arzte in 1926, reported the reversible change of polyvinyl alcohol ↔ polyvinyl acetate as a strong evidence of their theory of the macromolecule. After consultation between the two groups of researchers, scientific reports on PVA were published simultaneously in 1927 [4,5].

The production of polyvinyl acetate and alcohol was carried out in the early days not only in Germany but also in the United States, France, and United Kingdom. The first practical use of PVA in a large quantity was for warp-sizing of rayon and other synthetic fibers. In emulsion polymerization, it was used as an emulsifier or a stabilizer and also as a thickening agent for aqueous dispersion. All these uses have been continued and expanded.

The idea of making a fiber from PVA was applied by Herrmann and Haenel in 1931 [6]. The main objective of most of the inventions was to produce a water-soluble filament, especially for surgical sutures. "Synthofil," made from PVA fiber, was one such example, which was manufactured by Firma B. Braun and appeared in 1935. There were a few applications that contained textile fiber from PVA in the claim, but none were developed further.

With regard to other vinyl polymers, Klatte applied for a patent in 1913 to make synthetic fiber from polyvinyl chloride [7]. However, it was too early to apply for a patent. In 1932, monofilament was produced for the first time by the thermoplastification of polyvinyl chloride (PVC) by I. G. Since 1934, when chlorinated polyvinyl chloride began to be employed and spun from acetone solution, the fiber has been called "Pe-Ce Faser."

Vinyl polymers and condensation polymers were studied for the production of synthetic fiber. In 1932, Carothers and Hill of Du Pont studied linear aliphatic polyester and showed that fibers of sufficiently good mechanical properties are obtained by melt-spinning and cold-drawing [8]. Polyester fiber was considered unsuitable as a commercial fiber because it has a low melting point and hydrolyzes easily with water. Carothers therefore turned his investigation from polyester to polyamide, and, in 1938, Du Pont announced the success of a new fiber called "nylon." In Japan, studies to produce textile fibers from PVA began in 1938 and were intensively promoted.

4.1.2 Outline of the Chemistry

The process of making water-insoluble PVA fiber was reported in 1939 [9] and is explained here briefly. An aqueous solution of PVA is spun by a wet process using a concentrated aqueous solution of sodium sulfate as a coagulation bath. The spinning proceeds smoothly, but the fiber cannot be washed with water because it is water soluble. When PVA fiber is stretched, it is not soluble in water at room temperature. Its x-ray diagram shows clearly that the fiber is partially crystalline, similar to highly stretched dry rubber. When stretched PVA is relaxed, it is again water soluble.

Although formalization seems to be the easiest way to obtain water-insoluble PVA fiber, it is very difficult to effect a chemical reaction of a water-soluble fiber in an aqueous medium without injuring its fiber structure. It was nevertheless tried with limited success. Fiber obtained by such a procedure is insoluble in boiling water, but it becomes rubbery when the water temperature is raised to 60°C. The hot-water resistance of the fiber reported in 1939 was thus insufficient. Various experiments were subsequently carried out to remove the defect, and it was found that the best method to improve the hot-water resistance was by heat treatment of the PVA fiber before formalization [10]. Thus, the most important problem regarding the suitability of PVA fiber for use as a general-purpose textile fiber was solved.

Researchers at Kyoto University got a hint of the heat treatment from their researches on water-cellulose [11]. Water-cellulose is cellulose with water of crystallization and is obtained from alkali-cellulose (Na-Cell I) by washing Na-Cell I with cold water until it is free from sodium hydroxide and then drying it at a low temperature. When Na-Cell I is washed with hot water or dried at a higher temperature, cellulose II is obtained instead of water-cellulose. Water-cellulose also yields cellulose II after drying at a higher temperature. It is very important that water-cellulose is not formed when cellulose II is immersed in cold water. Water-cellulose contains one molecule of H_2O per one $C_6H_{10}O_5$ unit of cellulose. X-ray diagrams of cellulose II and water-cellulose are similar to one another, but the interplanar spacing of (101) is 7.32 Å for the former and 8.85 Å for the latter. Water is located in the (101) plane and expands the spacing.

In the case of PVA, it was considered that fresh wet fiber taken up from the coagulation bath is in a state similar to that of water-cellulose; therefore, it is likely that water contained in the lattice is driven off by heating at a higher temperature, and crystallization of PVA proceeds to give a fiber of a higher degree of crystallinity and sufficiently good hot-water resistance.

The effect of the heat treatment was studied in detail. Polyvinyl alcohol of degree of polymerization (DP) 1250 was employed to obtain fibers by the usual wet-spinning method, and the coagulated fiber was dried without removing the sodium sulfate from the coagulation bath. The dried fiber was subjected to a 10-min heat treatment in air at a temperature between 130 and 210°C. Hot-water resistance is measured by solubility, swelling, or heat-shrinkage of the heat-treated fibers in hot water at various temperatures. Table 4.1 shows the shrinkage of heat-treated PVA fibers in water at varying temperatures.

The heat treatment has a remarkable effect on the colloid chemical properties of PVA fibers. The fibers treated at 145°C showed a shrinkage of about 50% in water at 45°C. As the temperature of the heat treatment was increased, shrinkage decreased rapidly and became less

TABLE 4.1
Shrinkage of Heat-Treated Polyvinyl Alcohol Fibers in Water of Varying Temperatures

Heat treatment temperature (°C)	Shrinkage (%) of fiber in water at the following temperatures					
	45°C	55°C	65°C	75°C	85°C	95°C
145	49.1	54.4	x^a	x	x	x
170	20.9	42.8	58.8	x	x	x
190	0.48	3.35	14.3	x	x	x
200	0.49	1.46	2.91	38.7	x	x
210	0.0	0.0	0.0	0.0	0.0	x

x^a = Dissolved

TABLE 4.2
Shrinkage of Heat-Treated and Successively Formalized Polyvinyl Alcohol Fibers in Water

Temperature of heat treatment (°C)	Shrinkage (%) of fibers in water at different temperatures					
	45°C	55°C	65°C	75°C	85°C	95°C
145	1.89	5.94	22.9	51.7	65.1	69.7
170	0.0	0.0	4.18	12.2	45.8	63.3
190	0.0	0.0	0.90	2.3	7.7	50.1
200	0.0	0.0	0.99	1.98	3.43	8.8
210	0.0	0.0	0.92	0.44	1.40	2.3

than 1% when the treatment was above 190°C, as shown in Table 4.1. When the heat treatment is carried out at 210°C, the shrinkage is negligible until the temperature of water reaches 85°C; but the fiber is soluble in water at 95°C.

When the fiber is formalized after the heat treatment, the formalization results in the improvement of the hot-water resistance. Table 4.2 shows the shrinkage of heat-treated and successively formalized PVA fibers in water at varying temperatures. Fiber that has been heat-treated at 210°C and successively formalized is no longer soluble in water at 95°C and shows a shrinkage of only 2.3%.

Table 4.1 and Table 4.2 present classical results. Nowadays, the temperature of heat treatment is raised to 230°C or higher, in order that the fibers insoluble in boiling water are obtained even without formalization. In the industry, fibers resistant to boiling water are produced by the combination of heat treatment and formalization.

It was also confirmed by x-ray investigation that the interplanar spacing (101) corresponding to the A_1 interference of PVA decreases from 8.21 Å of the fresh fiber to 7.64 Å of the heat-treated fiber. This is similar to water-cellulose, but the change of the interplanar spacing of PVA is smaller compared to that of cellulose [12].

It is also very important to note that the degree of crystallinity increases with an increase in temperature of the heat treatment. Table 4.3 shows the change of the crystallinity determined by an x-ray diffractometric method. The table shows that the crystallinity increases from 34% of air-dried fiber to 68% of the same fiber after heat treatment at 225°C [12].

4.2 POLYVINYL ACETATE

Polyvinyl alcohol is obtained by the hydrolysis of various polyvinyl esters; in practice, however, mostly polyvinyl acetate is employed as the starting material. Polyvinyl acetate is not only used as a raw material for PVA but is also widely used for various purposes such as

TABLE 4.3
Change in Crystallinity of the PVA Fiber by Heat Treatment

Temperature (°C)	Air dry	80	120	120	160	200	225
Time	—	6 min	100 s	5 s	100 s	100 s	100 s
Crystallinity (%)	34	50	48	49	53	64	68

Source: From Sakurada, I., *Kolloid Z.*, 139, 155, 1954.

adhesives, emulsion coating, and so on. In the following, the emphasis is on polyvinyl acetate, which is used for the manufacture of PVA.

4.2.1 Manufacture of Monomer

Vinyl acetate monomer can be synthesized by the reaction of acetic acid with either acetylene or with ethylene. For the production of vinyl acetate from acetic acid and acetylene, the following process was adopted: a gaseous mixture of acetylene and acetic acid was reacted at about 200°C in the presence of active carbon impregnated with zinc acetate

$$CH \equiv CH + CH_3COOH \rightarrow CH_2 = CH\text{---}OCOCH_3$$

With the development of the petrochemical industry, acetylene was changed to ethylene, as in the case of production of vinyl chloride. The process is explained simply as follows: a gas mixture consisting of ethylene, acetic acid, and oxygen is reacted over a solid catalyst that contains Pd at a temperature between 175 and 200°C under a pressure between 4 and 10 kg/cm^2.

$$CH_2=CH_2 + CH_3COOH + 1/2\, O_2 \rightarrow CH_2=CH\text{---}OCOCH_3 + H_2O$$

4.2.2 Polymerization

4.2.2.1 Mechanism of Polymerization

Vinyl acetate polymerizes very easily by radical mechanism; the technical method of polymerization is also a radical process. First, a typical radical polymerization scheme of the vinyl monomer takes place in the presence of an initiator, I, to yield a pair of free radicals R·:

$$I \xrightarrow{k_d} 2R\cdot \qquad (4.1)$$

The second step is the initiation by addition of a monomer, M, to a primary radical R·, to give a chain radical:

$$R\cdot + M \xrightarrow{k_a} RM_1\cdot \qquad (4.2)$$

The growth of polymer molecules by successive addition of monomer molecules to radicals $RM_1\cdot$ and their successors are presented by

$$RM_1\cdot + M \xrightarrow{k_p} RM_2\cdot \qquad (4.3)$$

$$RM_2\cdot + M \xrightarrow{k_p} RM_3\cdot$$

or, in general,

$$RM_x\cdot + M \xrightarrow{k_p} RM_{x+1}\cdot \qquad (4.3a)$$

The same reaction constant, k_p, is written for each growing step under the assumption that the radical activity is independent of the chain length.

The growth of the polymer molecules is terminated by a mutual reaction of the growing radicals. Two mechanisms are possible for the termination. One is coupling and the other is disproportionation:

$$RM_m\cdot + RM_n\cdot \xrightarrow{k_{tc}} RM_{m+n}R \qquad (4.4)$$

$$RM_m\cdot + RM_n\cdot \xrightarrow{k_{td}} RM_m + RM_n \qquad (4.4a)$$

where k_{tc} and k_{td} are the rate constants of termination by coupling and disproportionation, respectively.

4.2.2.2 Rate of Polymerization

The rate of polymerization, R_p (i.e., the monomer molecules converted to polymer per unit time), is expressed by

$$R_p = -d[M]/dt = k_p[M][M_m\cdot] \qquad (4.5)$$

where $[M_m\cdot]$ is the total concentration of the chain radicals.

The total chain radical concentration $[M_m\cdot]$ is calculated under the assumption of the stationary state. The rate of free radical formation is assumed to be equal to the rate of decomposition of the initiator as shown in Equation 4.1 and is given by

$$\frac{d[R\cdot]}{dt} = k_d[I] \qquad (4.6)$$

Free radicals disappear by the termination reaction shown in Equation 4.4 and Equation 4.4a. Now, we assume that k_{td} is much smaller than k_{tc}, so that the former can be neglected compared to the latter. Writing $k_{tc} = k_t$, the rate of radical disappearance is expressed by

$$-\frac{d[M_m\cdot]}{dt} = k_t[M_m\cdot]^2 \qquad (4.7)$$

The condition of the stationary state is that the rate of free radical formation and disappearance is the same; hence, from Equation 4.6 and Equation 4.7, we have

$$[M_m\cdot] = (k_d/k_t)^{1/2}[I]^{1/2} \qquad (4.8)$$

Putting Equation 4.8 into Equation 4.5, the rate of polymerization is written as

$$R_p = k_p(k_d/k_t)^{1/2}[I]^{1/2}[M] \qquad (4.9)$$

According to Equation 4.9, the rate of polymerization is directly proportional to the concentration of the monomer and to the square root the concentration of the initiator. Experimentation agrees satisfactorily with Equation 4.9, not only in the case of vinyl acetate but also in the case of polymerization of styrene, methyl methacrylate, and methyl acrylate. Table 4.4 shows numerical values reported by Flory for k_p and k_t in the polymerization of the above monomers [13]. As shown in Table 4.4, both the rate constants of polymerization and the termination of vinyl acetate are greater than those of other monomers.

TABLE 4.4
Rate Constants of Polymerization k_p and Termination k_t of Some Common Monomers (at 60°C)

Monomer	k_p (l mol^{-1} s^{-1})	$k_t \times 10^{-7}$ (l mol^{-1} s^{-1})
Vinyl acetate	3700	7.4
Styrene	176	3.6
Methyl methacrylate	367	0.93
Methyl acrylate	2090	0.47

Source: From Flory, P.J., *Principle of Polymer Chemistry*, Cornell Univ. Press, Ithaca, New York, 1953, chap. 4, pp. 1–2.

Studies on the polymerization of vinyl acetate have been hindered by the presence of various kinds of impurities. In the polymerization experiments, an induction period was often observed. It is now believed to be the inhibitors produced by the reaction of impurities such as aldehydes, especially crotonaldehyde or the monomer with O_2.

On the other hand, it is also probable that an impurity combines with O_2 to form peroxides, which act as initiators in the polymerization. These peroxides, which may have initiated the polymerization of vinyl acetate without an initiator, have been observed in early studies [14,15]. Figure 4.1 shows a typical case of such a bulk polymerization that was carried out at 80°C for 26 h. Polymerization is completely inhibited in the initial 7 h and then proceeds smoothly, which can be described by Equation 4.9.

4.2.2.3 Chain Transfer and Degree of Polymerization

Radical transfer is the transfer of a radical from molecule A to molecule B and is written as

$$A\cdot + B \rightarrow A + B\cdot \tag{4.10}$$

In the case of polymerization, generally, the radical transfers from chain radical $M_m\cdot$ to monomer M and the solvent molecule S are important:

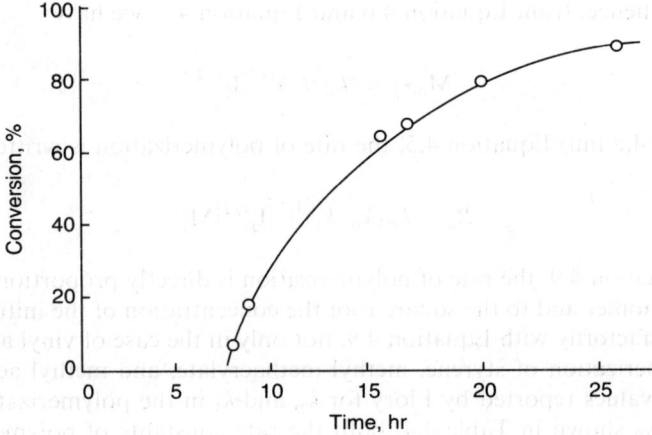

FIGURE 4.1 Time-conversion curve in the polymerization of vinyl acetate in bulk at 80°C without a catalyst.

$$M_m\cdot + M \longrightarrow M_m + M\cdot \qquad (4.11)$$

$$M_m\cdot + S \longrightarrow M_m + S\cdot \qquad (4.12)$$

Equation 4.11 and Equation 4.12 correspond to monomer and solvent transfer, respectively. The free radical S· reacts with the monomer in most cases and grows. Therefore, there is essentially no change in the number of free radicals, hence no change in the rate of polymerization. It is important to note that the growing molecular chain is cut by the transfer reaction to result in a lower DP.

Average DP is calculated by dividing the rate of polymerization by the number of formation of stable polymer molecules per unit time. Stable molecules are formed by termination, monomer transfer, and solvent transfer. The average degree of polymerization $\overline{P}$ is written as

$$\overline{P} = \frac{k_p[M\cdot][M]}{k_t[M\cdot]^2 + k_{trm}[M\cdot][M] + k_{trs}[M\cdot][S]} \qquad (4.13)$$

Equation 4.13 is rearranged into the following form, which is convenient for the evaluation of useful constants from the experimental results:

$$\frac{1}{\overline{P}} = \frac{k_t}{k_p^2} \frac{R_p}{[M]^2} + C_m + C_s \frac{[S]}{[M]} \qquad (4.14)$$

where $C_m = k_{trm}/k_p$ and $C_s = k_{trs}/k_p$ and are called monomer and solvent transfer constants, respectively. Equation 4.14 is written as

$$\frac{1}{\overline{P}} = \frac{1}{\overline{P}_0} + C_s \frac{[S]}{[M]} \qquad (4.15)$$

where $\overline{P}_0$ is the degree of polymerization without the solvent. The numerical value of C_s is easily found by plotting $1/\overline{P}$ against $[S]/[M]$. Figure 4.2 shows a typical example of solution polymerization in methanol.

Solvent transfer constants of various solvents for vinyl acetate are shown in Table 4.5 [16]. Polyvinyl acetate, as the raw material of polyvinyl alcohol for fiber, is produced by solution polymerization of vinyl acetate. In most cases, methanol is employed as the solvent. Solvent transfer constant, C_s, is one of the most important features for the selection of the solvent (see below in Section 4.2.2.5).

4.2.2.4 Chain Transfer to Dead Polymer

In the radical polymerization of vinyl monomers, it is clear that the concentration of dead polymer increases with increasing conversion; therefore, it is likely that chain transfer to dead polymer becomes observable at higher conversion. In the fundamental studies of polymerization, not only in the case of vinyl acetate but also in other monomers, most researchers are interested only in the initial stage of the polymerization (perhaps due to the simplicity of the condition for the theoretical treatment). Nevertheless, there are some reports on the polymerization of vinyl acetate carried out up to higher conversion; it is noteworthy that experimental results that indicate transfer to dead polymer have been reported.

Transfer to dead polymer is written as follows:

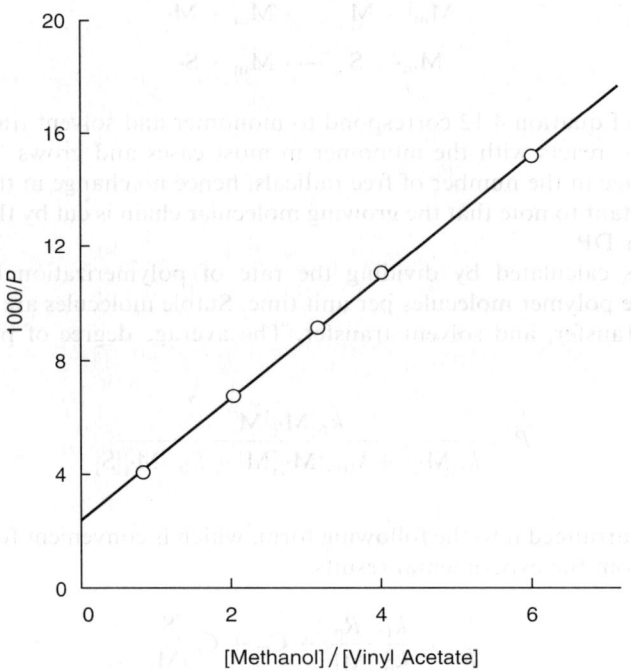

FIGURE 4.2 Effect of solvent/monomer ratio on the DP for vinyl acetate in methanol.

$$M_m\cdot + M_n \rightarrow M_m + M_n\cdot \qquad (4.16)$$

where $M_m\cdot$ and $M_n\cdot$ are chain radicals and M_n and M_m are dead polymers. Equation 4.16 seems to give an impression that the transfer is not so important from the chemical point of view. The most important thing in this transfer is the location of the free radical (the unpaired electron). In the ordinary growing polymer chain (chain radical), the radical is located at the

TABLE 4.5
Transfer Constants of Polyvinyl Acetate Radical to Solvents (at 60°C)

Solvent	$C_s \times 10^4$
Methyl alcohol	2.3–6.0
Ethyl alcohol	25
Butyl alcohol	20
t-Butyl alcohol	0.5–1.3
Ethyl acetate	1.1–3.4
Acetone	11.7–12.0
Chloroform	125–170
Dimethyl sulfoxide	2.0
Benzene	1.1–3.0
Toluene	18–69

Source: From Brandrup, J. and Immergut, E.H., *Polymer Handbook*, 2nd ed., Interscience, New York, 1975, p. II-67.

chain-end of the long molecule. In the transfer of radical to dead polymer, there are a number of sites where the radical can be transferred. The following reaction is written under the assumption that the radical of $M_m\cdot$ reacts with $-CH_3$ of an acetyl group of polyvinyl acetate and dehydrogenates it to form $-CH_2\cdot$.

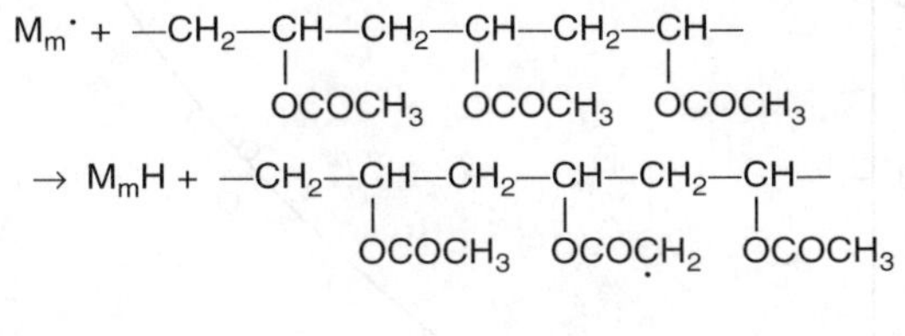

$$M_m\cdot\ +\ -CH_2-CH-CH_2-CH-CH_2-CH-$$
$$|||$$
$$OCOCH_3\ \ OCOCH_3\ \ OCOCH_3$$

$$\rightarrow M_mH\ +\ -CH_2-CH-CH_2-CH-CH_2-CH-$$
$$|||$$
$$OCOCH_3\ \ OCOCH_2\cdot\ \ OCOCH_3$$

The free radical formed on the side chain of polyvinyl acetate grows to form polyvinyl acetate with a long branch as shown below:

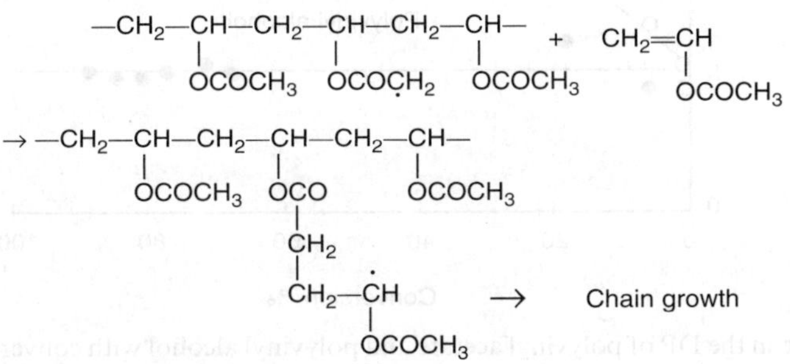

$$-CH_2-CH-CH_2-CH-CH_2-CH-\ \ +\ \ CH_2=CH$$
$$||||$$
$$OCOCH_3\ \ OCOCH_2\cdot\ \ OCOCH_3OCOCH_3$$

$$\rightarrow -CH_2-CH-CH_2-CH-CH_2-CH-$$
$$|||$$
$$OCOCH_3\ \ OCOOCOCH_3$$
$$|$$
$$CH_2$$
$$CH_2-CH\cdot\ \ \rightarrow\ \ \text{Chain growth}$$
$$|$$
$$OCOCH_3$$

The formation of the first branch is shown, but it is clear that the trunk may have more than one branch; it is also likely that the primary branch has second, third, etc., branches. As early as 1940, Blaikie and Crozier reported that PVA obtained from polyvinyl acetate has a lower degree of polymerization than the parent acetate [17]. Further investigation was carried out by McDowell and Kenyon [18] and Staudinger and Warth [19].

Osugi [20], Inoue, and Sakurada [21], and Wheeler et al. [22] pointed out that the drop in the DP of polyvinyl acetate by saponification is due to the cutting of branches at the acetyl groups of polyvinyl acetate. The evidence for this mechanism is found in Figure 4.3, where an example of the dependence of the DP of polyvinyl acetate and alcohol on the conversion is shown.

As shown clearly in the above figure, the DP of polyvinyl acetate increases linearly from 2000 at 0% conversion to 10,000 at 80% conversion. On the other hand, the degree of polymerization of PVA is almost constant from 0 to 80% and has a value of about 2000, which is the same as the initial DP of the polyvinyl acetate. These features of Figure 4.3 are explained as follows: The number of branches by transfer increases in proportion to conversion; hence, the DP of polyvinyl acetate increases with conversion. In the saponification of polyvinyl acetate, all acetyl groups, including the budding points of the branches, are converted to hydroxyl groups, and branches are separated from the trunk. Thus, the branch radicals grow just like the chain radicals, and the separated branches have the same average DP as that of the trunk. The chemical change is shown as follows:

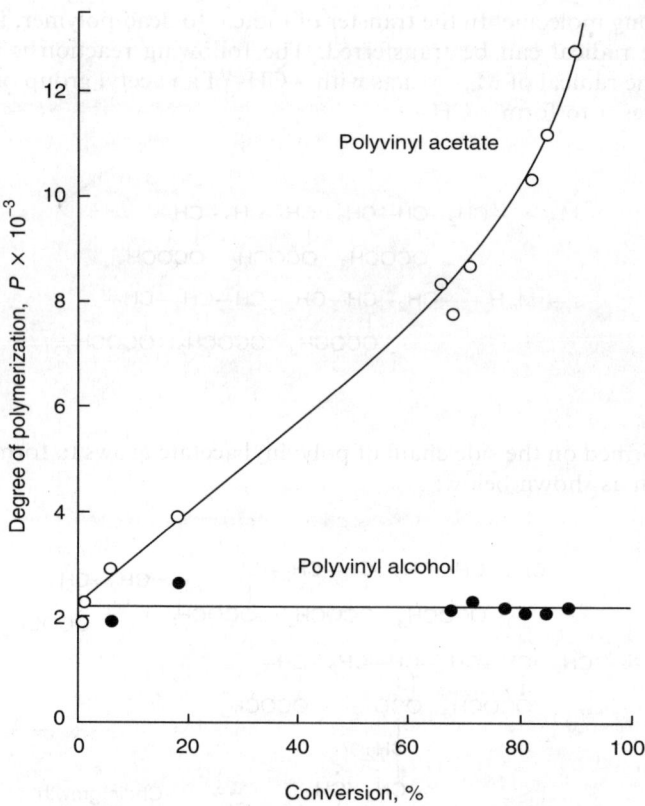

FIGURE 4.3 Change in the DP of polyvinyl acetate and polyvinyl alcohol with conversion in the polymerization of vinyl acetate in bulk. (From Inoue, R. and Sakurada, I., *Kobunshi Kagaku*, **7**, 211, 1950. With permission.)

Also, many researchers pointed out the tendency of decrease in degree of polymerization of PVA with conversion, although this is apart from the essential consideration on the position of branches. Table 4.6 shows an example [23].

As for the position of hydrogen abstraction, chain transfer to dead polymers may occur, in principle, not only at carbon atom (**1**) but also at (**2**) and (**3**) [22]. It is, however, necessary to remember that branches at carbon atoms (**2**) and (**3**) cannot be cut by saponification to convert polyvinyl acetate to PVA.

TABLE 4.6
Change in Number Average DP of Polyvinyl Acetate with Conversion

	Number average degree of polymerization	
Conversion (%)	As polymerized	Saponified and reacetylated
25.0	4380	3930
35.6	(5000)	3900
53.5	(5500)	3630
73.0	(6500)	3380
89.9	—	2350

Polymerization: at 60°C, in bulk.
Numbers in parenthesis: estimated values.
Source: From Graessley, W.W., Hartung, R.D., and Uy, W.C., *J. Polymer Sci.*, **7A-2**, 1919, 1969.

$$\begin{array}{c}(3)\quad(2)\\-CH_2-CH-\\|\\O\\|\\CO\\|\\(1)\ CH_3\end{array}$$

A change in the DP of polyvinyl acetate and alcohol with the progress of saponification shows no indication of the presence of branches that cannot be cut in the saponification. It may be that the transfer to carbon at (**2**) and (**3**) occurs far less frequently than it does at carbon atom (**1**).

4.2.2.5 Industrial Process of Polymerization

There are four kinds of polymerization processes: bulk, solution, emulsion, and suspension polymerization. As Table 4.7 shows [24], the heat of polymerization of vinyl acetate is high compared to other monomers; hence, the control of temperature is difficult in bulk polymerization. In the case of emulsion and suspension polymerization, it is somewhat troublesome to separate dispersed polyvinyl acetate particles from the aqueous medium, and it is necessary to remove the emulsifier and stabilizer completely because these substances induce problems in the process of fiber-making.

TABLE 4.7
Heat of Polymerization of Some Vinyl Monomers ($-\Delta H$ kcal/mol)

Monomer	$-\Delta H$	Monomer	$-\Delta H$
Ethylene	22.7	Acrylonitrile	17.3
Vinyl acetate	21.3	Styrene	16.7
4-Vinyl pyridine	18.7	Methyl methacrylate	13.6
Methyl acrylate	18.6	α-Methylstyrene	8.4

Source: From Joshi, R.M. and Zwolinsky, B.J., *Vinyl Polymerization*, Part I, Ham, G.E., Ed., Marcel Dekker, New York, 1967, p. 461.

Solution polymerization under employment of methanol seems to be the best process from the practical point of view. Methanol is a good solvent of polyvinyl acetate; the solvent transfer constant of methanol is small compared to that of a common solvent such as acetone, so that PVA of a sufficiently high DP is obtained even when a comparatively large amount of methanol is used in polymerization. Methanol and vinyl acetate form an azeotrope at 60°C, the latent heat of which is so large that it is easier to remove the heat of polymerization.

Polymerization is not carried out up to very high conversion and stopped mostly at about 65% conversion. Residual vinyl acetate is recovered, and the methanol solution of polyvinyl acetate is used directly for the saponification; that is, the solvent for the polymerization is utilized for the saponification (methanolysis).

4.2.3 CONVERSION TO POLYVINYL ALCOHOL

Polyvinyl acetate is soluble in many solvents; if polyvinyl acetate is dissolved in methanol or ethanol and caustic alkali is added to the solution, precipitation of polyvinyl alcohol begins in a few minutes. Such a phenomenon was observed for the first time by Herrmann and Haenel, the inventors of PVA [5]. Essentially, the same method is employed for the industrial production of PVA.

Similar to caustic alkali, mineral acid is also effective in catalyzing the reaction. The use of acid, however, does not have any advantage and, often, partly water-insoluble PVA is obtained. Polyvinyl esters of higher fatty acids and aromatic esters were studied, but, from the viewpoint of fiber-making, no advantage in using esters other than acetate has been reported. The tacticity of polyvinyl esters and ethers is discussed in Section 4.3.1.

4.2.3.1 Methanolysis and Hydrolysis

When sodium hydroxide is added to a pure methanol solution of polyvinyl acetate, deacetylation takes place as methanolysis; sodium hydroxide is a catalyst.

$$-CH_2-CH- + CH_3OH$$
$$\quad\quad |$$
$$\quad OCOCH_3$$
$$\xrightarrow{NaOH} -CH_2-CH- + CH_3OCOCH_3$$
$$\quad\quad\quad\quad\quad\quad |$$
$$\quad\quad\quad\quad\quad OH$$

Because NaOH is not consumed in the reaction, it catalyzes methanolysis of a large equivalent amount of acetyl groups. Purification of PVA is simply due to the employment of a very small amount of alkali.

$$-CH_2-CH- + H_2O$$
$$\quad\quad |$$
$$\quad OCOCH_3$$
$$\xrightarrow{NaOH} -CH_2-CH- + NaOCOCH_3$$
$$\quad\quad\quad\quad\quad\quad |$$
$$\quad\quad\quad\quad\quad OH$$

With increasing water content, hydrolysis also takes place. In the production of PVA, methanol of very low water content is used to prevent alkali consumption.

The course of methanolysis and hydrolysis is interesting as a fundamental polymer reaction [25,26]. If we neglect the reactivity of an acetyl group located at the very end of polymer molecules, we may assume that any acetyl group has roughly the same reactivity

Vinyl Fibers

independent of the DP and the location of the acetyl group in the molecule. The rate of deacetylation is calculated as follows:

$$\text{Methanolysis: } dx/dt = k(a - x)b \tag{4.17}$$

$$\text{Hydrolysis: } dx/dt = k(a - x)(b - x) \tag{4.18}$$

where a, b, and x are initial concentrations of the ester group, alkali, and hydroxyl group, respectively, and k is the rate constant of the deacetylation.

The above equations are applicable only in the initial state of the reaction. Figure 4.4 shows the courses of methanolysis in pure methanol (a) and hydrolysis in acetone–water (b). Both curves are sigmoidal, and the courses of the reaction seem to be autocatalytic. The apparent rate constant k in Equation 4.17 and Equation 4.18 increases linearly with increasing conversion. Figure 4.5 shows the relation for the case of hydrolysis in acetone–water; the relation in Figure 4.5 is expressed by

$$dx/dt = k_0[1 + m(x/a)](a - x)(b - x) \tag{4.19}$$

where k_0 is the initial rate constant and m is a constant independent of the conversion. The calculated value of m from the experiment of Figure 4.5 is 42. Such an autocatalytic

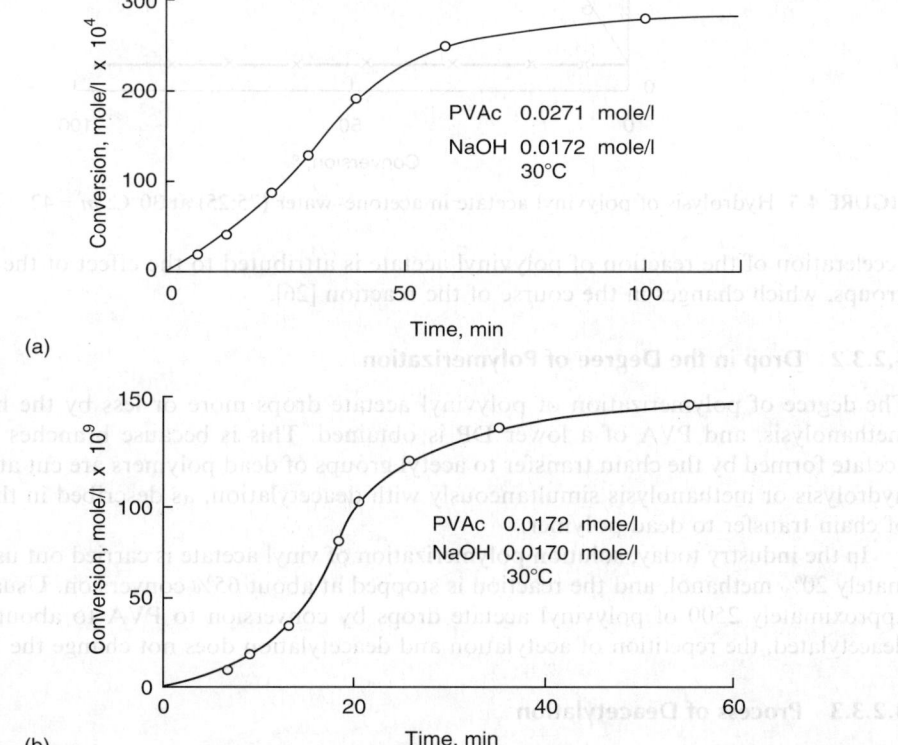

FIGURE 4.4 (a) Methanolysis and (b) hydrolysis of polyvinyl acetate.

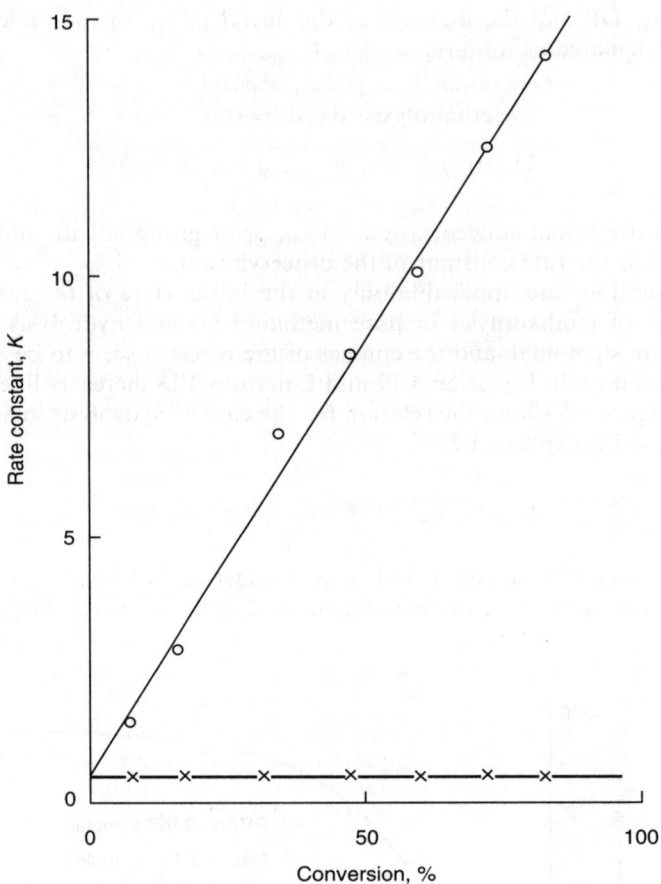

FIGURE 4.5 Hydrolysis of polyvinyl acetate in acetone–water (75:25) at 30°C: $m = 42$.

acceleration of the reaction of polyvinyl acetate is attributed to the effect of the neighboring groups, which changes in the course of the reaction [26].

4.2.3.2 Drop in the Degree of Polymerization

The degree of polymerization of polyvinyl acetate drops more or less by the hydrolysis or methanolysis, and PVA of a lower DP is obtained. This is because branches of polyvinyl acetate formed by the chain transfer to acetyl groups of dead polymers are cut at the point of hydrolysis or methanolysis simultaneously with deacetylation, as described in the discussion of chain transfer to dead polymer.

In the industry today, solution polymerization of vinyl acetate is carried out using approximately 20% methanol, and the reaction is stopped at about 65% conversion. Usually, a DP of approximately 2500 of polyvinyl acetate drops by conversion to PVA to about 1700. Once deacetylated, the repetition of acetylation and deacetylation does not change the DP [27].

4.2.3.3 Process of Deacetylation

Continuous methanolysis is adopted for the deacetylation process. Figure 4.6 shows a schematic flow sheet of deacetylation. The first step of the process is the mixing of a concentrated methanol solution of polyvinyl acetate with a methanol solution of alkali.

Vinyl Fibers

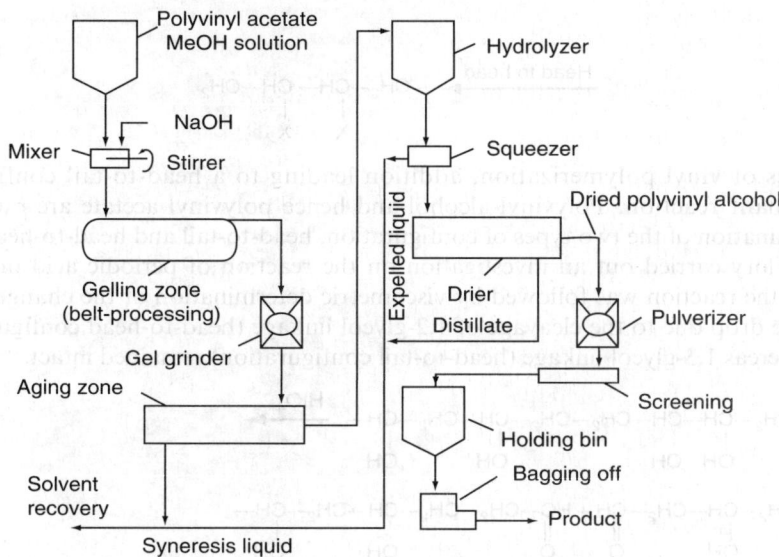

FIGURE 4.6 Schematic of belt-processing for methanolysis of polyvinyl acetate.

When they are mixed without precautions, the reaction begins instantaneously at the contact place of the two solutions to cause flocculation and precipitation. The precipitate picks up the unreacted liquid, and therefore no homogeneous mixing can be expected. This difficulty can be overcome by cooling the liquid before mixing. A screw or belt reactor is employed for the reaction. A typical scheme for the belt-process is shown in Figure 4.6. The main reaction of deacetylation and the subsequent gelling takes place on the belt. The gel is ground and fed to an aging zone, where syneresis takes place, and a liquid consisting of methanol and methyl acetate is separated. Polyvinyl alcohol gel is squeezed to remove the liquid. However, a small amount of water is added in a hydrolyzer before squeezing. The squeezed precipitate is heated in a dryer at 60 to 80°C, and the crystallinity of PVA increases by heating. A slight amount of water contained in the precipitate promotes the crystallization and makes the product insoluble in cold water, so that it can be washed with water to remove residual sodium acetate for the preparation of the spinning solution.

4.3 STRUCTURE AND PROPERTIES OF POLYVINYL ALCOHOL

4.3.1 MOLECULAR STRUCTURE

4.3.1.1 Chain Configuration

The addition of a vinyl monomer to a growing chain radical takes place in either of the two ways:

$$-CH_2-\dot{C}H + CH_2=CH$$
$$\quad\quad\;\; | \quad\quad\quad\;\; |$$
$$\quad\quad\;\; X \quad\quad\quad\; X$$

$$\xrightarrow{\text{Head to tail}} -CH_2-CH-CH_2-\dot{C}H$$
$$\quad\quad\quad\quad\quad\quad\;\;\; | \quad\quad\quad\;\; |$$
$$\quad\quad\quad\quad\quad\quad\; X \quad\quad\quad\; X$$

or

$$\xrightarrow{\text{Head to head}} -CH_2-CH-CH-\dot{C}H_2$$
$$ | |$$
$$ X X$$

In most cases of vinyl polymerization, addition leading to a head-to-tail configuration is the predominant reaction. Polyvinyl alcohol and hence polyvinyl acetate are rare examples where determination of the two types of configuration, head-to-tail and head-to-head, has been performed. Flory carried out an investigation on the reaction of periodic acid on PVA [28]. In his study, the reaction was followed by viscometric determination of the change in the DP. A remarkable drop due to the cleavage of 1,2-glycol linkage (head-to-head configuration) was observed, whereas 1,3-glycol linkage (head-to-tail configuration) remained intact.

$$-CH_2-CH-CH_2-CH-CH-CH_2-CH_2-CH-CH_2-CH- \xrightarrow{HIO_4}$$
$$ | | | | |$$
$$ OH OH OH OH OH$$

$$-CH_2-CH-CH_2-CH + HC-CH_2-CH_2-CH-CH_2-CH-$$
$$ | \| \| | |$$
$$ OH O O OH OH$$

In another investigation, the consumption of the periodate by the reaction was determined by a titration method, and the number of cleavages by the PVA was calculated [29]. Nuclear magnetic resonance (NMR) spectroscopy performed by Amiya and Uetsuki also confirmed the presence of 1,2-glycol linkage [30]. Both results were in good agreement with Flory's method and showed that regular PVA contains about one to two 1,2-glycol linkages per 100 structural units. It follows that 98–99% of the addition reaction leads to a head-to-tail configuration and only 1–2% to head-to-head, tail-to-tail configurations. Even though the content of 1,2-glycol linkage is very small, the effect of the cleavage is remarkable; the DP drops with the treatment with periodic acid from 100 to 50, independent of the initial value.

The most important factor that affects the content of 1,2-glycol linkages is the temperature of polymerization, as Flory first pointed out. The method of polymerization, solvent, and catalyst for the polymerization and conversion have apparently no effect on the configuration [31]. It is also possible that the 1,2-glycol linkage is formed by the coupling reaction in the termination (see Section 4.2.2); however, the contribution of stable polymer molecules produced by termination to the amount of 1,2-glycol linkages is small.

4.3.1.2 End Groups

In the polymerization of common vinyl monomers, such as styrene and methyl methacrylate, most of the polymer molecules are produced by the termination reaction; it is often possible to deduce from the end groups whether the termination has occurred by coupling or by disproportionation. As mentioned above, stable polymer formation by termination in the polymerization of vinyl acetate is not important. Most of the stable polymer molecules are produced by the chain transfer to monomer, polymer, and solvent. As already described, chain transfer consists of the dehydrogenation,

$$-CH_2-\dot{C}H + QH \rightarrow CH_2-CH_2 + Q\cdot$$
$$ | |$$
$$ OCOCH_3 OCOCH_3$$

$$Q\cdot + CH_2=CH \rightarrow\rightarrow Q(CH_2-CH)_nH$$
$$ | |$$
$$ OCOCH_3 OCOCH_3$$

where QH is a chain transfer agent such as monomer, polymer, and solvent. Thus, a chain transfer moiety is introduced at an end group of polyvinyl acetate and PVA. Okaya and his collaborators have performed the synthesis of several PVAs modified at an end group [32–37]. The method consists of using mercaptans as a chain transfer agent with a regulated addition technique. On the basis of the technology, new types of PVAs as well as new block copolymers consisting of PVA and other vinyl polymers such as PVA-*b* polyacrylic acid have been commercialized and utilized in various application fields.

4.3.1.3 Stereostructure

In the early stage of the studies on the stereostructure of polyvinyl esters and PVA, it was observed that polyvinyl formate, trifluoroacetate, and polyvinyl benzyl ether—and especially PVAs derived from these polymers—show properties that are somewhat different from those of common PVAs. It was therefore thought that the stereostructures of these PVAs are different from those of the common ones. In one study, some infrared (IR) absorption bands and the ratio of two absorption bands, D_{916}/D_{849}, were used as measures of the stereostructure [38]. However, the IR method was later found to result in a problem; that is, the absorption bands were dependent on the degree of crystallinity of PVA. Nowadays, the most powerful analytical method to determine the stereostructures is NMR spectroscopy. At first, stereostructures of the reacetylated samples using ^{1}H-NMR spectroscopy were reported [39,40]. But the chemical shifts of the methyl protons arising from the stereospecificity—that is, isotactic, heterotactic, and syndiotactic from low magnetic field—are not so large that there might exist remarkable errors in the calculations of stereospecificity. Later, ^{1}H-NMR [41–43] and ^{13}C-NMR [43,44] spectroscopies of PVA in deuterated dimethyl sulfoxide (DMSO) proved to be useful in terms of sensitivity.

The stereospecificity of PVA derived from various vinyl polymers, where the abovementioned two NMR spectroscopies are utilized, are listed in Table 4.8. Polyvinyl alcohol derived from polyvinyl acetate is almost atactic and independent of polymerization temperature. Polyvinyl alcohol derived from poly-*t*-butylvinylether, polybenzylvinylether, and polyvinyltrimethylsilylether are isotactic-rich, while those derived from polyvinyl formate, polyvinyl trifluoroacetate, polyvinyl pivalate, and polyvinyltrimethylsilylether (polymerized in a polar solvent) are syndiotactic-rich.

Almost 20 years later, polyvinyl pivalate that had been difficult to hydrolyze was found to be fully hydrolyzed in tetrahydrofuran under nitrogen atmosphere [53,54]; a detailed study was carried out [48]. Polyvinyl alcohol derived from polyvinyl pivalate has a high DP, as high as 40,000, due to the absence of acetyl hydrogen in the monomer that is attacked by the propagating radicals in the chain-transfer reaction. Syndiotacticity in the dyad of the PVA derived from the polyvinyl pivalate is not so high from the numerical point of view compared to atactic ordinary PVA and syndiotactic-rich PVA derived from polyvinyl trifluoroacetate, as shown in Table 4.9. However, the properties of the PVA are different from those of ordinary PVAs to a great extent: a 20°C higher melting point, a half degree of swelling of cast film from the DMSO solution, hard to dissolve in water, and a higher melting point and greater strength of the gelled polymer formed by the freeze–thaw method.

As for isotactic PVA, highly isotactic PVA was synthesized rather recently by the cationic polymerization of *t*-butylvinylether using a borotrifluoride–water complex as a catalyst [55,56]. As listed in Table 4.10, the isotactic PVA with 88.6% isotacticity in dyad has a high DP and a melting point at 247°C, and it is not soluble in water even at 120°C. The highest melting point of isotactic PVA ever reported is 235°C (soluble in water at 100 to 110°C). The highly isotactic PVA has a sharp x-ray diffraction pattern that differs from that of ordinary PVA. This suggests that the highly isotactic-rich PVA has a high crystallinity and a different crystal structure.

TABLE 4.8
Tacticity of PVA

Monomer	Temperature (°C)	Tacticity in triad (%)		Reference
		PVAc-NMR (i/h/s)	PVA-NMR (i/h/s)	
$CH_2=CHOC(CH_3)_3$	−78	79/17/4		40
	−78		55/32/13	42
	−78	18/48/34		45
$CH_2=CHOCH_2C_6H_5$	−78	79/17/4		46
	−78	89/8/3		47
$CH_2=CHOCOCH_3$	−40		22/50/28	48
	60		21/50/29	48
$CH_2=CHOCOH$	−78	16/46/38		45
	60	22/48/38		45
$CH_2=CHOCOCF_3$	−78	14/48/38		49
	−78		14/45/41	50
	60		19/48/33	50
$CH_2=CHOCOC(CH_3)_3$	−30	16.5/48.5/35		49
	60		(rr/mm = 2.2)	51
$CH_2=CHOSi(CH_3)_3$	−96	86/10/4		40
	−96		70/23/7	43
	−96	67/25/8		44
	−78	6/40/54		52

Recently, highly syndiotactic polyvinyl acetate was obtained; polymerization of vinyl acetate was carried out in fluoroalcohols [57]. 1,1,1-Tri(trifluoromethyl) methanol is the most effective solvent, where syndiotacticity of the polymers in dyad polymerized at 20 and −78°C are 62.3 and 72.2%, respectively.

TABLE 4.9
Comparison of Syndiotacticity in Dyad of PVAs Derived from Various Polymers

Polymerization temperature (°C)	Syndiotacticity in dyad[a] (%)		
	PVAc[b]	PVTFAc[c]	PVPi[d]
−78	—	63.0	—
−40	53.3	—	—
−30	—	—	64.0
0	53.5	58.7	62.8
60	53.6	57.6	61.5

[a] Equal to S (in triad) + 1/2 H (in triad).
[b] Polyvinyl acetate.
[c] Polyvinyl trifluoroacetate.
[d] Polyvinyl pivalate.

Source: From Fukunishi, Y., Sato, T., Okaya, T., Yamamoto, T., and Kamachi, M., *Report of Poval Committee*, **97**, 59, 1990.

Vinyl Fibers

TABLE 4.10
Characterization of Isotactic PVAs Derived from Poly-*t*-Butylvinylether

Sample number	Tacticity (in triad) (%)			Tacticity (in dyad) (%)		DP[a]	Melting point (°C)
	I	H	S	I	S		
1	79.1	18.9	2.0	88.6	11.4	3540	247
2	77.8	19.6	2.6	87.6	12.4	23800	246

[a]Measured viscometrically after acetylation of polyvinyl alcohol.
Source: From Ohgi H. and Sato, T., *Macromolecules*, **26**, 559, 1993; *ibid.*, **32**, 2403, 1999.

4.3.1.4 Branching

As already mentioned, multibranched polyvinyl acetate is produced by the chain transfer to acetyl groups of dead polyvinyl acetate, but all these branches are cut off by the deacetylation. If chain transfer occurs at a carbon atom on the main chain, it is certain that the branch connected directly to the main chain carbon atom will not be broken to result in the long branches. There are some papers that insist on the chain transfer to the main chain carbon atoms; however, there is no experimental evidence to show the existence of a main chain branch [58–60]. The main chain transfer probably occurs, but its rate is much lower than the rate of transfer to the acetyl group.

The formation of a short branch in polyvinyl acetate, as in the case of high-pressure polymerization of ethylene, by the backbiting intramolecular radical transfer is also operative [61]:

$$-CH_2-CH-CH_2-CH-CH_2-CH\cdot \;\; \rightarrow \;\; -CH_2-\underset{H}{\overset{AcO}{C}}\underset{CH-OAc}{\overset{CH_2-CHOAc}{\diagup\diagdown}}$$
$$|||$$
$$OAcOAcOAc$$

Later, a study using ^{13}C-NMR spectroscopy reported the existence of the short branch. The amount of short branch composed of vinyl alcohol dimer is very small 0.027 per 100 monomer unit [62].

4.3.1.5 Molecular Weight

The molecular weight of PVA is determined by osmometry or light scattering of an aqueous solution of PVA. Often, PVA is acetylated to polyvinyl acetate and the measurement of molecular weight is carried out by means of solution in organic solvents, because acetylation in ordinary conditions does not change the DP.

The simplest method of determining the DP is viscometry. The relation between the degree of polymerization and limiting viscosity number $[\eta]$ is given by the equation

$$[\eta] = KP^a \tag{4.20}$$

where K and a are constants. Numerical values of K and a of PVA in aqueous solution at 20°C were calculated on the basis of the experimental data of Staudinger and Warth [63] and the constants were found [64]:

$$[\eta] = 8.87 \times 10^{-4} \times P^{0.62} \quad (4.21)$$

where $[\eta]$ in Equation 4.21 is expressed in l/g. Since then, Equation 4.21 has been widely used in the field of chemical technology of PVA in Japan. The value of degrees of polymerization of PVA in this book are calculated with this equation, unless otherwise noted.

In the case of fractionated PVA, Nakajima and Furudachi obtained the following relation for aqueous solution at 30°C [65]:

$$[\eta] = 7.50 \times 10^{-4} \times P^{0.64} (l/g) \quad (4.22)$$

It was shown by Matsumoto and Ohyanagi that Equation 4.22 is applicable for samples with the most probable molecular-weight distribution [66]. The value of $[\eta]_t$ measured at temperature t, which is higher than 20°C, is corrected with the following equation to $[\eta]_{20}$ [64]:

$$[\eta]_{20} = [\eta]_t (1.07)^{(t-20) \times 0.1} \quad (4.23)$$

All PVA samples used in the above experiments to derive Equations 4.21 through Equation 4.23 are atactic, so that it is possible that PVA, rich in isotactic or syndiotactic structure, exhibits a slightly different dependence of viscosity on molecular weight. Difference in the amount of 1,2-glycol linkage may also have a small influence on the viscosity. All these problems have to be studied.

Degrees of polymerization of commercially available PVAs are about 300 to 2400. The fiber grade is 1700. Polyvinyl alcohol with a higher DP can be synthesized with bulk polymerization of vinyl acetate at low temperatures by UV light [67] or γ-ray initiation method [68,69]. Because of the high viscosity of the polymerization system, it is difficult to regulate the internal temperature, so that large quantities of the polymer to be used in a spinning test cannot be obtained easily.

Owing to the success of obtaining the high-tenacity polyethylene fiber where high-molecular-weight polymer is utilized, the importance of utilizing high-molecular-weight PVA has been recognized. The viscosity average DP of the polyvinyl alcohol ($\overline{P}_A$) can be expressed as

$$\frac{1}{\overline{P}_A} = \frac{1}{1.82} \left\{ \frac{1.62x}{1-(1-x)^{1.62}} \right\}^{1/0.62} \left\{ C_m + C_s \frac{[S]}{[M]_0} + \frac{1}{2} \frac{k_t}{k_p^2 [M]_0^2} R_{P_0} \right\} \quad (4.24)$$

where x is conversion, $[S]$ and $[M]_0$ denote the concentration of solvent and initial monomer, respectively, C_m and C_s are chain transfer constants of monomer and solvent, respectively, k_t and k_p are rate constants of termination and propagation, respectively, and R_{P_0} denotes initial rate of polymerization [70]. By substituting the values at varying temperatures in the case of bulk polymerization, the relationship of $\overline{P}_A$ at 50% conversion and polymerization temperature was calculated (see Figure 4.7). With decreases in the rate of polymerization, maximal $\overline{P}_A$ increases and the temperature at which maximal $\overline{P}_A$ occurs decreases. Since it is very difficult for industrial producers to adopt the rate of polymerization lower than 1%/h, bulk or solution polymerization to obtain PVA with a DP of 20,000 seems to be impossible.

Vinyl Fibers

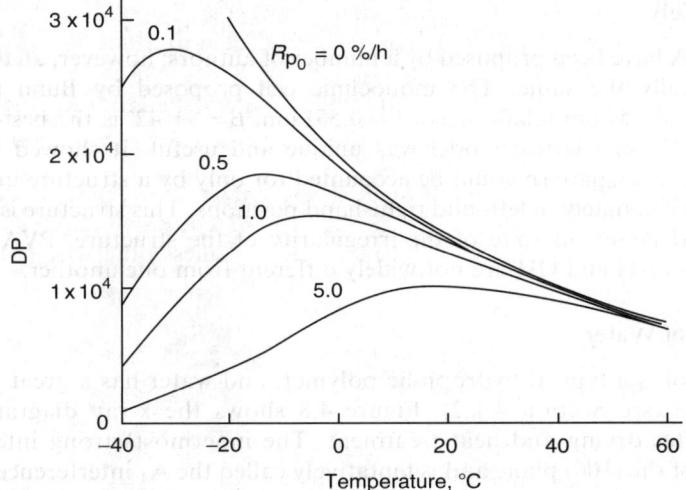

FIGURE 4.7 Relationship between the degree of polymerization of polyvinyl alcohol and polymerization temperature at various rates of polymerization. Conversion: 50%. (From Fujiwara, N., Sato, T., Yuki, K., Yamauchi, J., and Okaya, T., *Report of Poval Committee*, **102**, 22, 1993. With permission.)

By using suspension or emulsion polymerization at low temperatures, a relatively higher rate of polymerization can be utilized to get PVA with a high DP. Table 4.11 shows the comparison of various polymerization methods. As seen in the table, the emulsion polymerization at low temperatures seems to be the best method. Polyvinyl alcohol of DP 18,000 is obtained at −20°C at a polymerization rate of 15%/h.

4.3.2 Crystal Structure

Although polyvinyl acetate is amorphous, PVA derived from the former in most cases gives a crystalline x-ray diffraction pattern. When PVA fiber is drawn, an x-ray fiber diagram is obtained.

TABLE 4.11
Preparation of High-Molecular-Weight PVA from Three Polymerization Methods at Low Temperatures

Polymerization method	Polymerization temperature (°C)	Polymerization time (h)	Conversion (%)	Rate of polymerization (%/h)	Degree of polymerization $\overline{P}_A$
Suspension	5	80	80	1.0	12,000
	0	100	50	0.5	15,000
Emulsion	5	4	62	17	10,000
	−20	5	60	15	18,000
Bulk (UV)	−20	127	26	0.20	24,700
	−30	440	10	0.02	30,000

Source: From Fujiwara, N., Sato, T., Yuki, K., Yamauchi, J., and Okaya, T., *Report of Poval Committee*, **102**, 22, 1993.

4.3.2.1 Unit Cell

Unit cells of PVA have been proposed by a number of authors; however, all the proposed unit cells are essentially the same. The monoclinic cell proposed by Bunn and Peiser with $a = 0.781$ nm, $b = 0.252$ nm (chain axis), $c = 0.551$ nm, $\beta = 91°42'$ is the best-known example [71]. Bunn and Peiser's lattice model was unique and useful. It showed that the relative intensities of the x-ray pattern could be accounted for only by a structure in which hydroxyl groups are indiscriminately in left- and right-hand positions. This structure is atactic. According to Bunn and Peiser, in spite of the irregularity of the structure, PVA is crystallizable because the sizes of H and OH are not widely different from one another.

4.3.2.2 Effect of Water

Polyvinyl alcohol is a typical hydrophilic polymer, and water has a great influence on the crystal structure (see Section 4.1.2). Figure 4.8 shows the x-ray diagrams of polyvinyl alcohol fiber after drying and heat treatment. The innermost strong interference on the equator is that of the (100) plane and is tentatively called the A_1 interference. Films or fibers prepared from an aqueous solution of PVA, after simple air-drying, show very broad and indistinct A_1 interference; with increased drying, A_1 becomes more distinct and the peak of A_1 shifts to the outside on the equator, showing that the interplanar spacing of (100) has decreased. The best method to affect such a contraction of the interplanar spacing is heat treatment.

Table 4.12 shows the change of interplanar spacing of lightly air-dried and heat-treated (60 and 200°C) PVA gel. It is seen that the interplanar spacing of the (101) plane drops from 8.21 to 7.6 Å. It is noteworthy that even after 3 years of air-drying, the interplanar spacing still has a value of 8.0 Å (this result is not given in Table 4.12).

The interplanar spacing of about 7.8 Å does not change through contact with water at room temperature. The relation between water-cellulose and cellulose II is similar to that between PVA swollen with water and heat-treated PVA (see Section 4.1.2).

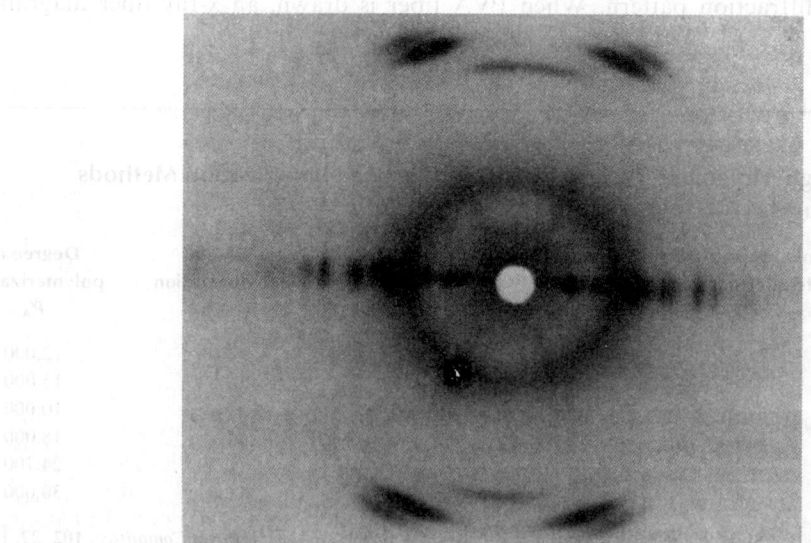

FIGURE 4.8 X-ray diffraction photograph of polyvinyl alcohol.

TABLE 4.12
Effect of Aging and Heat Treatment on the Interplanar Spacing of the (101) Plane of PVA

Temperature (°C)	Time	Water content (%)	Spacing (Å)
No aging	—	42.0	8.21
30	9 days	15.1	7.99
30	25 days	9.0	7.99
60	2.5 h	12.0	7.92
60	1 day	6.3	7.82
200	10 min	5.2	7.64
200	5 min	>5.0	7.64

The mechanism is explained as follows: PVA films or fibers obtained from its aqueous solution still contain a certain amount of hydration water in the (100) plane; such bound water cannot be removed outside of the lattice, and more hydrogen bonds between PVA molecules are formed. Hydrogen bonds between PVA molecules cannot be broken with water at room temperature. If, after heat treatment at 160°C, PVA film is dipped into water of room temperature, a weight increase of about 50% is observed by the swelling. X-ray diffractometric observation shows no change in the interplanar spacing of the percent crystallinity. This is clear evidence that water cannot penetrate into the PVA lattice at room temperature.

4.3.2.3 Crystal Structure Model

A crystal structure model for PVA was proposed by Bunn and Peiser in 1948 [71]; later, a different one was proposed by Sakurada et al. [72]. The unit cells of these models are practically the same as those mentioned above; however, the crystal structures are completely different. Figure 4.9 shows a comparison of these two models. In Bunn and Peiser's model, a double layer of molecules is formed along the c-axis by hydrogen bonds; these two layers stack along the a-axis by weaker van der Waals forces. On the other hand, in the case of the Sakurada et al. model, a single layer of molecules, which is connected along the a-axis by hydrogen bonds, stacks in the c-direction.

It has been shown that the E_t values, i.e., the Young's moduli, perpendicular to the chain axis for various planes for nylon-6 are closely connected with the hydrogen bond direction [73]. The E_t determination of PVA was also carried out; the values are shown in Table 4.13 along with E_1, the Young's modulus in the direction of the fiber axis. The value of 5.2 GPa for the (002) plane is nearly equal to that of polyethylene, whereas the value of about 8 GPa is much larger. Judging from these values, it seems that the hydrogen-bonded layer is along the a-axis, in accordance with the model of Sakurada et al. It is also in accordance with the fact already mentioned that the interplanar distance of (100) of PVA films or fiber prepared from aqueous solution, after simple drying, shows indistinct (100) interference; it becomes more distinct and the interplanar distance becomes shorter by the successive drying.

4.3.2.4 Degree of Crystallinity

Hermans and Weidinger's method [74] has been used for the determination of the degree of crystallinity of PVA in a study by Sakurada et al. [75].

Polyvinyl alcohol fiber was spun from an aqueous solution by a conventional method with moderate drawing and air-drying at room temperature, and, thereby, a starting fiber for a series of fibers with different degrees of crystallinity was obtained. Crystallinity was changed,

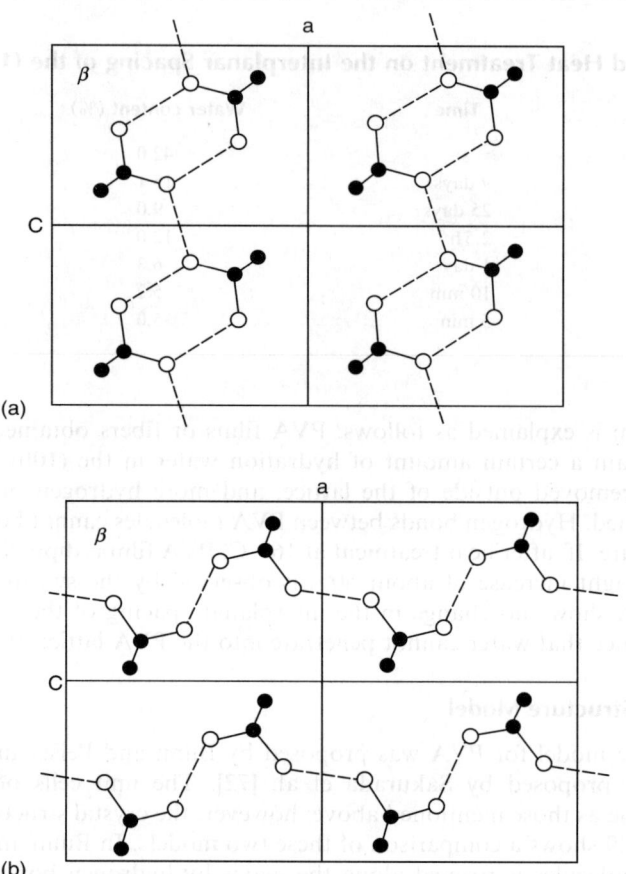

FIGURE 4.9 Crystal structure models of polyvinyl alcohol (*b*-projection): (a) Bunn; (b) Sakurada. Broken lines represent hydrogen bonds.

as shown in Table 4.3, by changing the temperature and the time of heat treatment; after the heat treatment, the fibers were washed with methanol–water or cold water to remove sodium sulfate from the coagulation bath. The degree of crystallinity was calculated from the

TABLE 4.13
Young's Moduli of PVA for the Various Planes of the Crystal Lattice and the Macroscopic Specimen

Crystal modulus	GPa	Macroscopic modulus	GPa
E_t (101)	10.8	Y_t (perpendicular to d.d.)[a]	7.1
E_t (200)	8.0		
E_t (002)	5.2		
E_l (fiber axis)	255.0	Y_l (parallel to d.d.)	7.5

[a]Direction of drawing.

Source: From Sakurada, I., Nukushina, Y., and Mori, N., *Kobunshi Kagaku*, **12**, 302, 307, 311, 1955.

TABLE 4.14
Crystallinity of PVA Films with Different Degrees of Polymerization after Heat Treatment at Various Temperatures

Degree of polymerization	Temperature of heat treatment (°C)				
	40	80	120	160	200
309	0.15	0.20	0.27	0.41	0.54
708	0.25	0.27	0.32	0.39	0.54
1288	0.29	0.32	0.34	0.37	0.53
2317	0.29	0.32	0.34	0.41	0.53
4570	0.30	0.33	0.42	0.45	0.54

Source: From Sakurada, I., Nukushina, Y., and Mori, N., *Kobunshi Kagaku*, **12**, 302, 307, 311, 1955.

diffractometric data. It is seen from the table that higher temperature and longer time of heat treatment is to some extent necessary to affect a higher degree of crystallinity.

As already mentioned, the effect of crystallinity on fiber properties has been useful in the manufacture of PVA fiber that is resistant to hot water and does not shrink or soften in boiling water.

The equation for calculating the degree of crystallinity from the intensities of interference of crystalline and amorphous parts of PVA fibers is also applicable to PVA films. Therefore, the range of experiments could be expanded, and PVAs of various degree of polymerization, from 309 to 4570, were employed in the experiments by Sakurada et al. [75]. The temperature of the heat treatment was varied from 40 to 220°C; the time was always 10 min; the calculated degree of crystallinity is shown in Table 4.14. It is seen that the crystallinity increases with increasing heat treatment temperatures, whereas the DP has no systematic influence on the crystallinity.

Recently, physical properties of PVA with a low degree of polymerization have been reported [76]. By using the method mentioned previously (see Section 4.3.1), PVA having a hydroxyethylthio end group can be obtained with various DPs. As listed in Table 4.15, the degree of crystallinity was found to increase with decrease in the degree of polymerization.

TABLE 4.15
Degree of X-Ray Crystallinity of PVAs with Various Degrees of Polymerization

$\overline{P}_A$ (from VPO)	Degree of crystallinity	
	HT at 40°C	HT at 160°C
26	0.623	0.701
47	0.514	0.623
61	0.508	0.562
88	0.504	0.615
196	0.487	0.587
300	0.480	0.578
1000	0.414	0.537

Source: From Sato, T. and Okaya, T., *Polymer J.*, **24**, 849, 1992.

4.3.3 PHYSICOCHEMICAL PROPERTIES

4.3.3.1 Melting Point

A direct determination of the melting point of PVA is difficult because decomposition begins near the melting point. However, it can be determined directly by measurement of the melting point of the polymer in the presence of a diluent. The following equation for the melting point of polymer–diluent is used in this case [77]:

$$\frac{1}{T_m} - \frac{1}{T_m^0} = (R/\Delta H_2)(V_2/V_1)(v_1 - \chi v_1^2) \tag{4.25}$$

where T_m and T_m^0 are melting points of the polymer in the presence and absence of the diluent respectively and V_1 and V_2 are molar volumes of the polymer repeat unit and diluent respectively, R, ΔH_2, v_1, and χ are gas constant, heat of fusion of the polymer, the volume fraction of the diluent, and the polymer–diluent interaction parameter, respectively. The value of $1/T_m^0$ is easily found by means of the graphical representation of the relation between $1/T_m$ and v_1.

Experimental results when water and some other diluents are used [78] are shown in Table 4.16, along with values of the heat of fusion as calculated with Equation 4.25. It is shown in Table 4.16 that the melting points of PVA calculated from various diluent systems agree satisfactorily with one another. On the other hand, the agreement of the heat of fusion is poor. According to Table 4.16, the melting point of PVA is 267°C. However, much lower melting points are reported in the literature for PVA [79]. It may be that most of the experiments were carried out with PVA that contained some residual water.

4.3.3.2 Glass Transition Temperature

In determining the glass transition temperature, T_g, of PVA, the effect of residual water is great. The effect of the diluent on T_g is given by the following equation [80]:

$$1/T_g = w_1/T_{g1} + w_2/T_{g2} \tag{4.26}$$

where T_g, T_{g1}, and T_{g2} are the glass transition temperatures of the mixture, diluent, and pure polymer, respectively, and w_1 and w_2 are weight fractions of the diluent and pure polymer, respectively.

TABLE 4.16
Melting Point and Heat of Fusion of PVA

Diluent	Water	Ethylene glycol	Dimethyl formamide	Dimethyl acetamide
Melting point (°C)	266	267	267	267
Heat of fusion (kcal/mol)	2.47	2.00	1.43	1.47
Interaction parameter (χ)	0.49	0	0.19	0.40

Source: From Hamada, F. and Nakajima, *Kobunshi Kagaku*, **23**, 395, 1966.

A number of authors have reported on the T_g of dry PVA, which are divided into two groups: one is about 70°C and the other 85°C. It was shown that PVA samples (film or fiber) that have no thermal history above 130°C belong to the first group, whereas those that were treated at 230°C belong to the second group and show a T_g between 85 and 90°C [81]. The T_g of sufficiently wet PVA fiber determined dilatometrically is below 0°C [82], as was assumed in the cases of cotton, rayon, and nylon 66 [83]. A transition temperature of about 120°C [84] has also been reported.

4.3.3.3 Density

The density of PVA can easily be determined by a floating method in an appropriate mixture such as benzene–nitrobenzene–carbontetrachloride. Table 4.17 shows the density of PVA films with different DPs after heat treatment at various temperatures [85]. The range of DP of polyvinyl alcohol was 309 to 4570, and the range of temperature of heat treatment was 40 to 200°C. It is shown in Table 4.17 that the density increases with increasing temperature of the heat treatment.

Solid polymeric substances consist of two parts, crystalline and amorphous. Let ρ_c and ρ_a be densities of the crystalline and amorphous parts, respectively; then the density of ρ of any PVA samples is given by

$$\frac{1}{\rho} = \frac{1}{\rho_c} x + \frac{1}{\rho_a}(1-x) \qquad (4.27)$$

where x is the volume fraction of the crystalline part.

The crystallinities of the same PVA films that are listed in Table 4.17 have already been measured by an x-ray difractometric method and are given in Table 4.14. We can test the applicability of Equation 4.27 to the experimental results by plotting $1/\rho$ against x. Figure 4.10 shows the plots. It is seen that there is a roughly linear relationship between specific volume and crystallinity, independent of the DP of polyvinyl alcohol and the temperature of heat treatment. The intercepts indicate that the density of crystallinities 1 and 0 are 1.3450 and 1.2690, respectively [85]. The former value agrees very well with the value of the density of the PVA crystal calculated from the dimensions of the unit cell containing two CH_2CHOH groups ($a = 7.71$ Å, $b = 2.49$ Å, $c = 5.43$ Å, $\beta = 93°$, ρ_c [calculated] = 1.345).

With Equation 4.27a, derived from Equation 4.27, the crystallinity, x, is calculated from the experimental value of the density ρ.

TABLE 4.17
Density of PVA Films with Different DP after Heat Treatment at Various Temperatures

	Temperature of heat treatment (°C)				
Degree of polymerization	40	80	120	180	200
309	—	—	1.2961	1.3044	1.3094
708	1.2824	1.2893	1.2965	1.3036	1.3062
1288	1.2854	1.2920	1.2987	1.3039	1.3065
2317	1.2879	1.2949	1.3006	1.3050	1.3088
4570	1.2896	1.2954	1.3020	1.3045	1.3088

Source: From Sakurada, I., Nukushina, Y., and Sone, Y., *Kobunshi Kagaku*, **14**, 506, 1957.

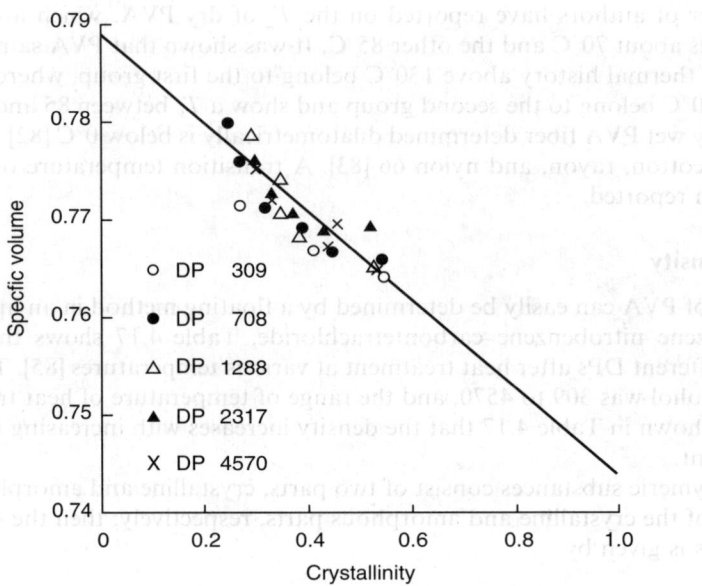

FIGURE 4.10 Specific volume of polyvinyl alcohol films as a function of crystallinity. (From Sakurada, I., Nukushina, Y., and Sone, Y., *Kobunshi Kagaku*, **14**, 506, 1957. With permission.)

$$x = \frac{1/\rho_a - 1/\rho}{1/\rho_a - 1/\rho_c} = \frac{0.7880 - 1/\rho}{0.0445} \tag{4.27a}$$

To use Equation 4.27a for the calculation of crystallinity, it is important that the PVA is completely saponified and has no residual acetyl group.

4.3.3.4 Orientation and Strength

When the fiber is drawn, the crystals are oriented with the *b*-axis parallel to the direction of drawing. Drawing is an important process in the manufacture of PVA fiber.

Table 4.18 shows the effect of the draw ratio (DP = 1700; at 180°C) of the PVA film on tenacity. The tenacity increases with greater ratio than the draw ratio; at a draw ratio of four times, it is observed that the tenacity increases sixfold [86]. These experiments on the drawing of PVA films with various degrees of polymerization were carried out at 180°C to obtain films of the highest draw ratio at each degree of polymerization. The tensile properties of the films are given in Table 4.19 [86]. It is seen from Table 4.19 that the highest possible draw ratio

TABLE 4.18
Effect of Draw Ratio on the Tensile Strength of PVA Fiber

Draw ratio	1	2	3	4
Tensile strength (GPa)	0.10	0.23	0.41	0.66
Elongation (%)	30.2	25.8	17.4	16.5

Source: From Tsuji, W., *Rept. of the Second Symp. PVA*, 1958, p. 100.

TABLE 4.19
Tensile Properties of PVA Films of Various DPs Drawn up to the Highest Draw Ratios

Degree of polymerization	750	1020	1270	1500	1910	3550
Highest possible draw ratio	8	13	16.5	17	16	15.5
Tensile strength (GPa)	0.39	1.03	1.22	1.32	1.38	1.60
Elongation (%)	8.0	9.8	6.5	5.5	8.1	12.1

Source: From Tsuji, W., *Rept. of the Second Symp. PVA*, 1958, p. 100.

increases with increasing degree of polymerization up to a maximum of 1500, and then decreases. Perhaps this is due to the temperature of drawing; 180°C is too low for PVA of a higher degree of polymerization. The highest strength found in this experiment is 1.60 GPa, which is equal to 14 g/den.

4.3.3.5 Swelling

Water is a solvent for PVA, and, when freshly precipitated, PVA is readily soluble in cold water. However, after drying and heat treatment at a higher temperature, its solubility in cold water decreases. Table 4.20 shows the effect of heat treatment of PVA films of various degrees of polymerization on solubility in water at 30°C [87]. Heat treatment temperatures are 40, 80, 120, 160, 200, and 220°C and the time of treatment is 10 min throughout. The solubility is given in fractions of the starting film weight. Table 4.20 shows that water solubility decreases with rising temperatures of heat treatment, and that the effect of the DP is large, especially in the region where the DP is low. The parts of PVA that are insoluble in water at 30°C can be dissolved in boiling water without change in the DP.

TABLE 4.20
Solubility and Swelling in Water at 30°C of PVA Films with Different DPs after Heat Treatment at Various Temperatures

Degree of polymerization	Temperature of heat treatment (°C)				
	40	80	120	160	200
a. Solubility					
309	0.61	0.55	0.43	0.030	0
708	0.45	0.29	0.17	0.025	0
1288	0.23	0.082	0.058	0.013	0
2317	0.10	0.041	0.058	0.015	0
4570	0.038	0	0	0	0
b. Swelling					
309	10.7	8.6	2.7	0.5	0.2
708	5.6	4.4	2.3	0.5	0.2
1288	4.7	3.6	1.9	0.5	0.2
2317	3.6	2.6	1.5	0.5	0.2
4570	3.1	2.2	1.3	0.5	0.2

Source: From Sone, Y. and Sakurada, I., *Kobunshi Kagaku*, **14**, 145, 1957; Sakurada, I. and Nukushina, Y., *Kobunshi Kagaku*, **11**, 472, 1954.

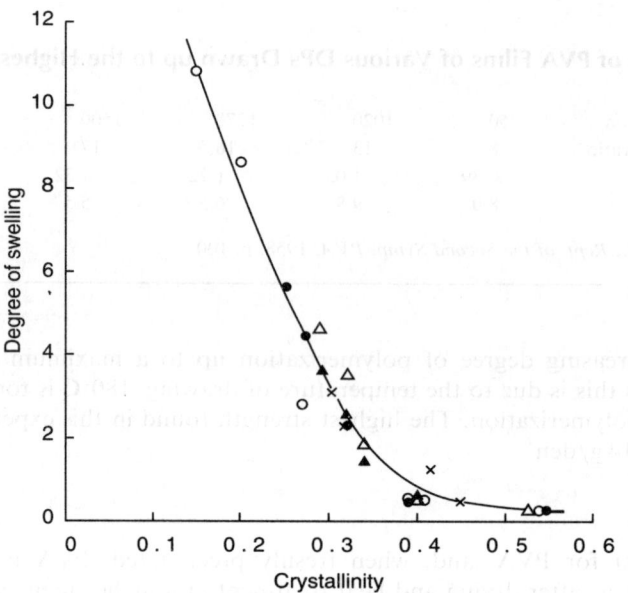

FIGURE 4.11 Degree of swelling as a function of the crystallinity. (Symbols are the same as in Figure. 4.10.) (From Sakurada, I., Nukushina, Y., and Sone, Y., *Kobunshi Kagaku*, **14**, 506, 1957. With permission.)

The insoluble parts in the above experiment are swollen with water. The degree of swelling, based on the weight of the dried film after removal of the soluble part, is calculated using the following equation:

$$\text{Degree of swelling} = \frac{\text{weight of swollen film} - \text{weight of dried film}}{\text{weight of dried film}}$$

As seen in the lower half of Table 4.20, the degree of swelling decreases with increasing heat treatment temperature and increasing DP.

Swelling and dissolution of polymers are due to the same factors. The tendency toward swelling is opposed by a reaction force arising from the network formed by cross-linking. It is assumed that crystalline regions play roles of cross-linkages [88].

X-ray diffractomeric evidence shows that the crystalline regions are not changed by the swelling in water [87]. The degree of swelling is plotted against crystallinity in Figure 4.11. The relationship is represented independently of the DP and the heat treatment temperature with a single curve. However, the degree of swelling is not always determined by the crystallinity. It is also dependent on the method of film formation (i.e., the drying condition of aqueous solution) [89]. This is related to the number of crystallites at the same crystallinity.

4.3.3.6 Aqueous Solution

When a dilute aqueous solution of PVA is allowed to stand at room temperature, the solution becomes somewhat turbid; on heating above 85°C, the solution becomes clear again. Microgel formation in the turbid solution is due to hydrogen bond formation between hydroxyl groups of the PVA molecules. Polyvinyl alcohol is dispersed molecularly in the clear solution after heating [90].

Vinyl Fibers

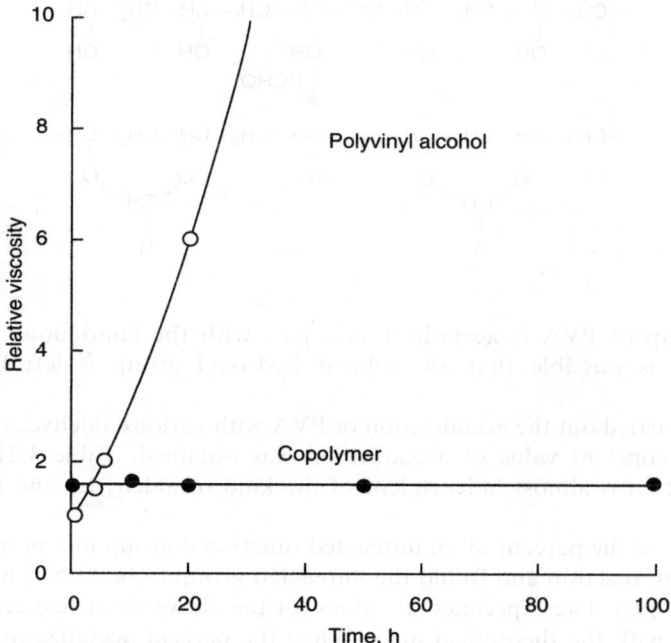

FIGURE 4.12 Change of the viscosity of solutions (conc. 10%) of polyvinyl alcohol and a copolymer of vinyl and allyl alcohol (97:3) with time of standing.

The heat of dilution of aqueous solutions of PVA is negative, and therefore heat is generated when PVA is dissolved in water. It is therefore expected that PVA is dissolved more readily in cold water. However, hot water is used to dissolve PVA. This is because PVA is always partly crystalline, and it is necessary that the crystalline part be changed to an amorphous state before actual dissolution can occur.

In the case of a concentrated solution of PVA, no flocculation is observed on standing, but the viscosity of the solution increases with increase in time of standing. Changes in the viscosity of a 10% solution of PVA with a degree of polymerization of 625 relative to the time of standing in an aqueous solution at 20°C is shown in Figure 4.12. The viscosity increase is so rapid that in 50 h it turns into a hard gel that does not flow. Higher concentrations and lower temperatures of solution result in more rapid gel formation. When polyvinyl alcohol is not pure and contains a small amount—namely, a few percent—of comonomers such as allyl and isopropenyl alcohol, then the solution is much more stable than that of the pure PVA; curve 2 in Figure 4.12 shows such an example.

4.3.4 Chemical Reactions

Polyvinyl alcohol is a typical polyvalent alcohol and undergoes the same chemical reactions that are characteristic of common alcohols. Esterification, etherification, acetalization, complex formation with metallic salts, and many other reactions have been studied in the case of PVA.

4.3.4.1 Acetalization

Acetalization is important not only because it is used in the manufacture of PVA fiber but also because it is a characteristic reaction of PVA.

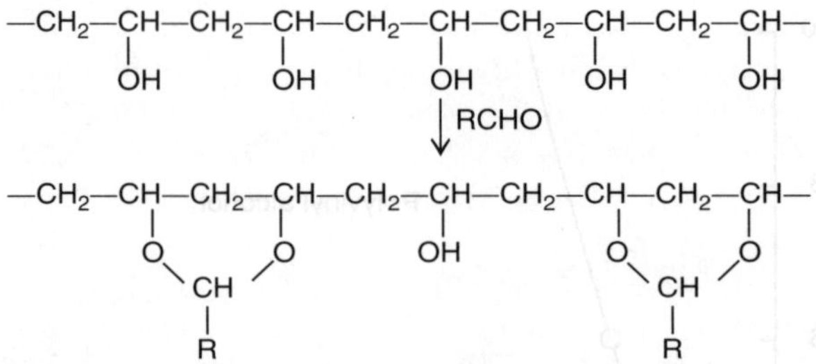

$$-CH_2-CH-CH_2-CH-CH_2-CH-CH_2-CH-CH_2-CH-$$
$$\hspace{1.2em}|\hspace{3.2em}|\hspace{3.2em}|\hspace{3.2em}|\hspace{3.2em}|$$
$$\hspace{1.2em}OH\hspace{2em}OH\hspace{2em}OH\hspace{2em}OH\hspace{2em}OH$$

$$\downarrow RCHO$$

$$-CH_2-CH-CH_2-CH-CH_2-CH-CH_2-CH-CH_2-CH-$$

with acetal rings (O–CH(R)–O) bridging pairs of adjacent carbons, and an isolated OH remaining in the middle.

A hydroxyl group of PVA is acetalized as a pair with the contiguous hydroxyl group; consequently, it is possible that an isolated hydroxyl group is left behind after the reaction.

Noma et al. carried out the acetalization of PVA with various aldehydes and repeated the reaction until a constant value of acetalization was obtained. Table 4.21 shows that the highest acetalization is almost independent of the kind of aldehydes and lies between 83.0 and 85.8% [91].

Flory calculated the percent of an unreacted functional group in a polyvinyl compound after the thorough reaction and found the unreacted group to be 13.5%; hence, the reacted group was 86.5% [92]. The experimental values for the above-described acetalization are in good agreement with the theoretical ones. When the percent acetalization is greater than 70%, all of the above polyvinyl acetals are soluble in organic solvents. This solubility suggests that intermolecular acetalization leading to cross-linking takes place only to a very small extent, if any.

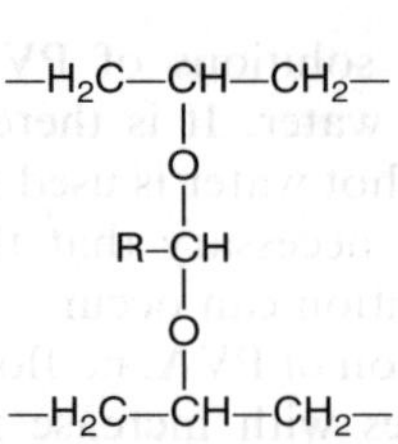

$$-H_2C-CH-CH_2-$$
$$\hspace{3em}|$$
$$\hspace{3em}O$$
$$\hspace{3em}|$$
$$\hspace{2em}R-CH$$
$$\hspace{3em}|$$
$$\hspace{3em}O$$
$$\hspace{3em}|$$
$$-H_2C-CH-CH_2-$$

When a dialdehyde such as glyoxal is allowed to react with PVA, cross-linking occurs easily and an insoluble product is obtained.

TABLE 4.21
Highest Degree of Acetalization

Aldehyde	Acetalization %
Palmitinaldehyde	85.0
Chloracetaldehyde	85.8
o-Chlorobenzaldehyde	84.6
Benzaldehyde	83.0

Source: From Noma, K., Wo, T., and Tsuneda, T., *Kobunshi Kagaku*, **6**, 439, 1949.

Vinyl Fibers

If acetalization is carried out with aldehydes that contain a sulfonic acid group or groups, however, it is impossible to reach such a high degree of acetalization [93]. The maximum degree of acetalization for aldehydes with sulfonic acids is shown in the table below.

Aldehyde	Acetalization %
β-Butyraldehyde sulfonic acid	57.6
o-Benzaldehyde sulfonic acid	44.0
2,4-Benzaldehyde disufonic acid	36.0

The results seem to show that not only isolated but also some other hydroxyl groups are unable to undergo the acetalization reaction due to the repulsion effect of sulfonic acid groups.

4.3.4.2 Esterification

There are two ways to obtain polyvinyl esters. One is polymerization of the corresponding monomer, and the other is by esterification of PVA [94]. In addition to organic esters, some inorganic esters such as nitrate, sulfate, and phosphate also have been synthesized by esterification [94].

4.3.4.3 Etherification

When alkylene oxide is reacted with PVA, poly(vinyl hydroxyalkyl ether) is obtained. In the case of ethylene oxide, the reaction is given as follows:

$$-CH_2-CH(OH)- \;+\; CH_2-CH_2(O) \;\longrightarrow\; -CH_2-CH(O-CH_2-CH_2OH)-$$

Poly(vinyl hydroxyethyl ether) is more easily soluble in water than PVA, and possesses primary hydroxyl groups, whereas PVA possesses secondary ones [95].

Carboxyl groups can be introduced into PVA by the reaction of monochloroacetic acid [96]:

$$-CH_2-CH(OH)- \;+\; ClCH_2COOH \;\longrightarrow\; -CH_2-CH(OCH_2COOH)-$$

Triphenylmethyl chloride is an interesting etherifying reagent with a large molecular volume. When the reaction is carried out under appropriate conditions, it has been found that the highest degree of substitution could not be greater than 42.4%, based on the total hydroxyl groups.

$$-CH_2-CH(OTr)-CH_2-CH(OH)-CH_2-CH(OTr)-CH_2-CH(OH)-CH_2-CH(OH)-CH_2-CH(OTr)-$$

If it is assumed that both the neighboring hydroxyl groups are free from the introduction of the trityl groups into the PVA chain, the simplest case is that the hydroxyl groups should react alternately, and the highest conversion should be 50%. However, in a random reaction, it is possible that the tritylation proceeds so that a pair of unreactive hydroxyl groups remains after the reaction, as shown in the right-hand side of the above structure of a tritylated PVA. Taking into consideration such a probability, a statistical calculation was done, and a value of 42.4% was arrived at as the highest conversion [97]. This is in good agreement with the experimental result. It was further confirmed that the remaining hydroxyl groups could be acetylated smoothly.

4.3.4.4 Complex Formation

Polyvinyl alcohol is liable to form complexes with various inorganic compounds to result in gelation of PVA solution [98]. Complexing with boric acid is the most popular method [99] and is interesting from the industrial point of view because it may be employed for the spinning of PVA fiber (see Section 4.4.2). The monodiol complex shown below is still soluble in methanol. With the addition of alkali, cross-linking is formed between PVA molecules to increase the viscosity and induces gelation. Some titanium compounds show similar gelation effects.

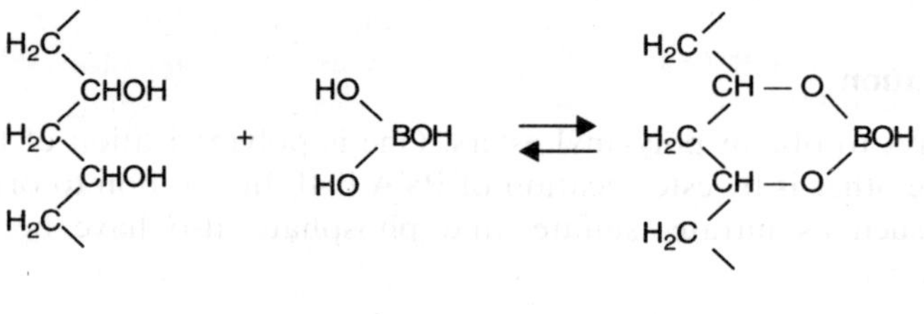

4.3.4.5 Grafting

Grafting of PVA is carried out not only in solution but also in the solid state without losing the original form, such as fiber or film [100]. Cerium salts, peroxides, persulfates, etc., are used as initiators for grafting. These initiators react with PVA to form free radicals on the chain, which initiate grafting in the presence of the appropriate monomer. Radiation-induced grafting has also been studied.

For the modification of fiber properties, grafting of PVA is interesting from the industrial point of view. The reaction can be carried out easily in water with various kinds of monomers to give desired properties to the fiber.

4.3.4.6 Decomposition

When PVA is heated in vacuum to above 200°C, dehydration begins and water is the main volatile decomposition product, although small amounts of acetaldehyde and crotonaldehyde are also detected. The second step of the thermal decomposition occurs at 400°C and generates a small amount of hydrocarbon; the residue is solid carbon [101,102]. By this pyrolytic process, fiber is obtained.

Experiments on the oxidative degradation of PVA with potassium permanganate [103], potassium dichromate [104], hydrogen peroxide [105], and ozone [106] were carried out in detail. It was observed that a random scission of the PVA chain occurs. The scission product was fractionated, and the determination of the DP and end groups was carried out. It was found that the main end groups formed by the scission are either carbonyls or carboxyls. Radiation-induced oxidative degradation has also been studied [107].

Degradation and successive cross-linking of PVA are observed in highly concentrated (20%) hydrochloric acid by heating, although PVA is stable to hydrolysis. It is likely that the reaction is initiated by dehydration [108].

4.4 MANUFACTURE OF POLYVINYL ALCOHOL FIBER

4.4.1 Traditional Process of Wet-Spinning

Polyvinyl alcohol fiber was manufactured by a wet-spinning process using sodium sulfate as the coagulating agent. Other processes, such as wet-spinning with other coagulating agents, dry-spinning, and melt-spinning, are described in the following text. Vinylon is the generic name for synthetic fibers of PVA in Japan. These fibers are called Vinal (or Kuralon) in the United States. Mixed fibers of PVA and polyvinyl chloride can be manufactured by spinning polyvinyl chloride emulsions in an aqueous solution of PVA essentially by the same spinning process so as to produce vinylon. Figure 4.13 shows the outline and Figure 4.14 a flow sheet of the regular process for the wet-spinning with the sodium sulfate coagulation bath.

4.4.1.1 Fiber Formation

Polyvinyl alcohol powder is first washed with cold water to remove impurities such as alkali and sodium acetate, which may cause discoloration of the fiber in the heat treatment, and then dissolved in water under heating with direct steam. In most cases, PVA with a degree of polymerization of 1200 to 1800 is used; the concentration is usually from 14 to 16% and is controlled by the measurement of the viscosity. Table 4.22 shows some examples of the viscosity of the spinning solutions measured mainly at 96°C. It is necessary that the temperature of the spinning solution is kept at least above 70°C [109].

A horizontal spinning machine, as is used in viscose rayon production, was first adopted. However, there is an essential difference in the coagulation mechanism of PVA and viscose solutions. In the case of viscose rayon, hydrolysis of cellulose-xanthogenate to regenerated

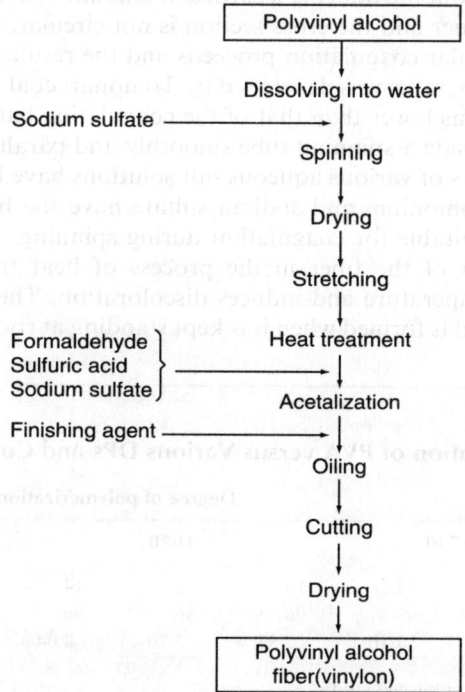

FIGURE 4.13 Outline of the regular process for wet-spinning of polyvinyl alcohol fiber.

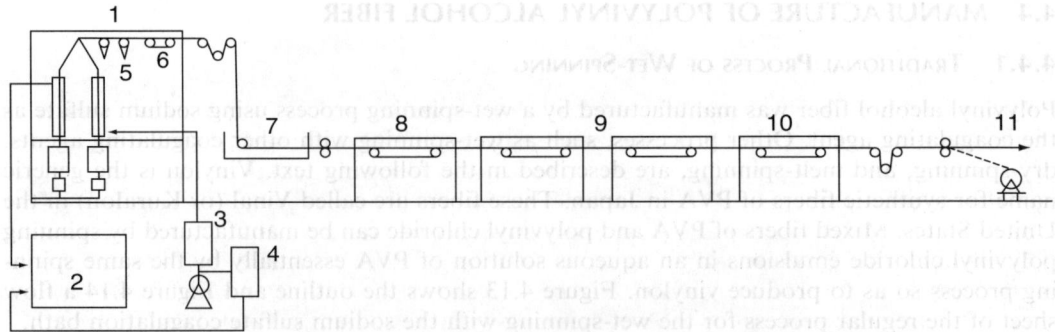

FIGURE 4.14 Flow sheet of the regular process for wet-spinning of polyvinyl alcohol fiber. 1. vertical machine; 2. coagulation bath; 3. bath heater; 4. evaporator; 5. godet rollers; 6. take-up rollers; 7. hot second bath; 8. driers; 9. drawing; 10. heat treatments; 11. winder.

water-insoluble cellulose is the most important chemical reaction for coagulation. In the case of PVA, no fundamental chemical change takes place during coagulation and dehydration; that is, ridding the PVA of bound water is the only reason for coagulation. Consequently, coagulation of PVA proceeds much more slowly, and longer coagulation baths are necessary than in the case of rayon production.

Figure 4.15 shows a typical cross section of PVA fiber obtained by wet-spinning with a coagulation bath of sodium sulfate. The cross section is not circular and consists of skin and granular core. Hence, it is likely that the dehydration that takes place on the surface of the filament forms a densely packed structure to a certain depth and builds up skin. Water goes out from the filament without introducing a sufficient amount of coagulating solution, so that the filament becomes thinner and the cross section is not circular, but kidney-shaped. In the case of the filament, granular coagulation proceeds and the result is a coarse structure.

A vertical spinning apparatus was developed by Tomonari et al. in 1951 [110]. The density of the spinning solution was lower than that of the coagulating bath. Figure 4.16 shows that both the solutions flow inside a spinning tube smoothly and parallel to one another.

The coagulation powers of various aqueous salt solutions have been tested systematically, and it was found that ammonium and sodium sulfate have the highest coagulation power [111]. Both sulfates are suitable for coagulation during spinning. Ammonium sulfate, however, causes discoloration of the fiber in the process of heat treatment because the salt decomposes at a high temperature and induces discoloration. The viscosity of the spinning solution increases and a gel is formed when it is kept standing at room temperature. However,

TABLE 4.22
Viscosity of Spinning Solution of PVA versus Various DPs and Concentration of PVA

	Degree of polymerization							
	1730			1670			1695	
Concentration (%)	16	15	16	15	14	16	15	14
Temperature (°C)	96	96	96	96	96	96	90	90
Viscosity (poise)	9.91	7.10	7.87	5.70	4.04	8.51	10.56	14.96

Source: From Kobayashi, S., unpublished study.

Vinyl Fibers

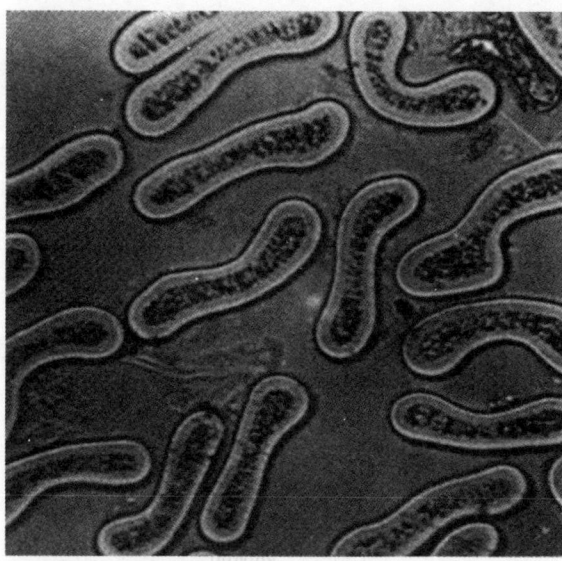

FIGURE 4.15 Microphotogaraph of a typical cross section of polyvinyl alcohol fiber obtained by the regular process of wet-spinning with a sodium sulfate coagulation bath.

the solution is stable at temperatures above 40°C and no viscosity change is observed. The temperature is much higher for the actual spinning, usually above 90°C. The temperature of the coagulation bath is between 40 and 50°C; the solubility maximum of sodium sulfate is in this temperature range, and therefore the solution has the maximum coagulation power [111]. The rate of spinning is rather low, about 50 m/min; as will be seen later, a lower rate of spinning is favorable to the successive drawing.

4.4.1.2 Drawing

Drawing is carried out to affect fiber structure by a parallel orientation of the polymer molecules. When spinning is carried out at a high rate, parallel orientation is effected in the course of the spinning. Successive drawing after the spinning is more natural and reasonable than simultaneous spinning and drawing. In the wet-spinning of PVA fiber, four kinds of drawing are available: (1) guide stretch (GD) (see Figure 4.16), (2) roller stretch (RD) (see Figure 4.16), (3) hot-drawing in wet state (wHD) (see no. 7 in Figure 4.14), and (4) hot-drawing in dry air (dHD) (see no. 9 in Figure 4.14).

In practice, it is seldom that only one kind of drawing is carried out; often, two or more kinds of drawing are done successively. Osugi et al. carried out a systematic study on drawing [112] and found that the highest draw ratios are in the cases of GD, 3–4; RD and wHD, 4–5. They found that dry hot-drawing is the most effective, and achieved a draw ratio of about 10. Even when fibers were predrawn with GD, RD, or wHD, if they were finally hot-drawn in dry air to the highest possible ratio, the total draw ratio that was calculated from the deniers of undrawn and drawn fibers was about 10. These results suggest that the method of preliminary drawing has practically no influence on the total drawing ratio if the final drawing is carried out up to the highest ratio by the dry hot-drawing method.

The relation between the tenacity ratio and the draw ratio of the fiber by various methods of drawing, including combined drawing, is shown in Figure 4.17 [112]. As shown, the

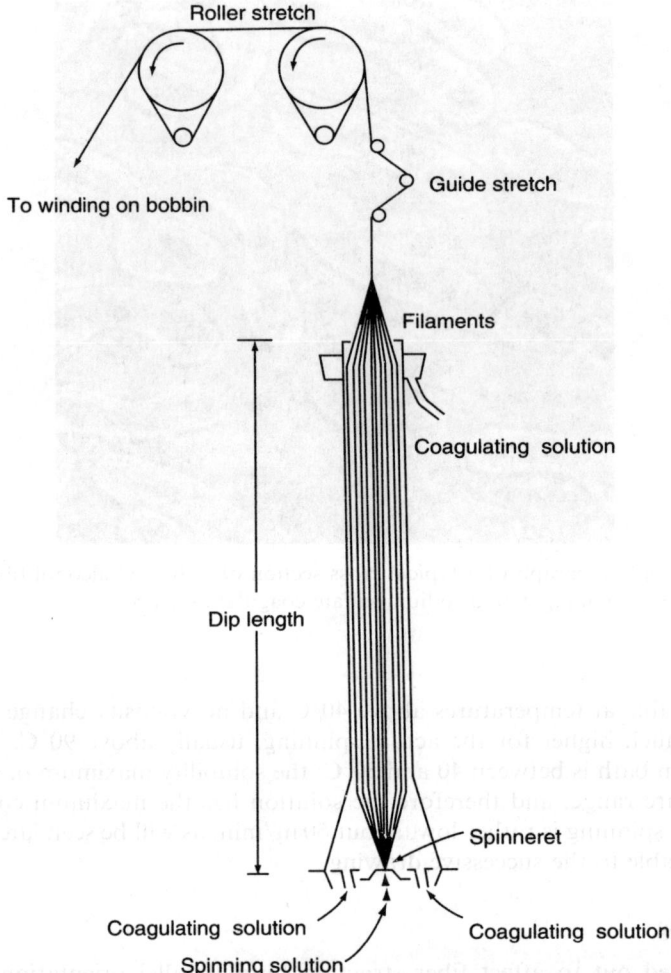

FIGURE 4.16 Schematic of vertical wet-spinning apparatus for PVA with sodium sulfate coagulation bath.

tenacity ratio increases linearly with the draw ratio, independent of the method of drawing. From the inclination of the drawn line, it is found that the average value (of tenacity ratio/draw ratio) is not 1, but 0.82. In any case, it is noteworthy that the tenacity ratio is nearly equal to the draw ratio.

In most cases, hot-drawing in dry air is carried out between 180 and 200°C as a set with heat treatment. The temperature of the heat treatment is somewhat higher than that of the drawing, between 210 and 230°C. Although the hot-drawing temperature is lower than the temperature of normal heat treatment, the effect of heat treatment proceeds to a large extent during the hot-drawing.

4.4.1.3 Heat Treatment

Heat treatment is a process that improves the hot water resistance of the PVA fiber. In the molecular mechanism, the most important change is the increase in crystallinity of PVA through the removal of residual water and the formation of new hydrogen bonds between PVA molecules.

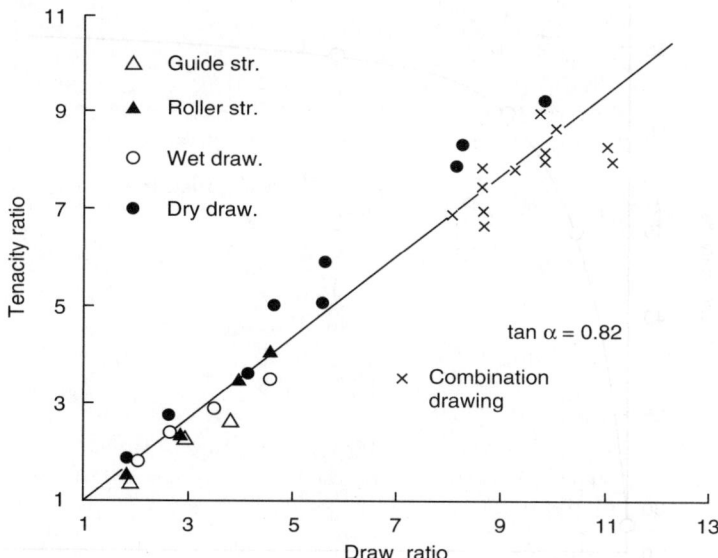

FIGURE 4.17 Relation between tenacity ratio and draw ratio of polyvinyl alcohol fibers drawn by various methods. (From Osugi, T., Tanabe, T., Suda, T., Morimoto, T., and Miyazaki, S., *Sen-i Gakkaishi*, **16**, 155, 1960. With permission.)

Although the effect of heat treatment is a kind of heat setting, it is outstanding because we are dealing with a water-soluble polymer that turns water insoluble by heat treatment.

In experiments on the rate of crystallinity [113], the heat treatment was carried out with an apparatus similar to that used in the industry. Polyvinyl alcohol fiber was subjected to heat treatment at 235°C for various lengths of time, and crystallinity increase was followed by x-ray diffractometry; the results are shown in Figure 4.18. It can be seen that crystallinity increases very rapidly from 1 to 10 sec and then levels off. A second experiment was carried out at varying temperatures, from 120 to 230°C, to discover the effect of temperature on crystallinity; the time of heat treatment was 45 s for each. The results are shown in Figure 4.19. It can be seen that crystallinity increases almost linearly from 120 to 230°C. It was difficult to extend the experiment beyond 235°C due to the melting of the sample.

4.4.1.4 Acetalization

The purpose of acetalization in the manufacture of PVA fiber is to improve hot-water resistance of the heat-treated fiber further. In the production of monofilament, acetalization is somewhat omitted because the denier of monofilament is low and it can be heat-treated homogeneously at sufficiently high temperature. In the case of staple-fiber production, the total denier of the fiber bundle is high, so it is difficult to realize the homogeneous effect of heat treatment, and the process temperature has to be rather low to avoid local overheating. In most cases, consequently, acetalization is necessary. The conditions for acetalization are as follows:

Composition of acetalizing bath: formaldehyde, 60 g/l; H_2SO_4, 250 g/l; Na_2SO_4, 300 g/l
Acetalization: temperature, 60°C; time, approximately 20 min.

In most cases, fibers that are heat-treated at 210°C reach the acetalization stage and are acetalized up to 40 mol%. Such fibers resist up to 10 h heating in boiling water. When the

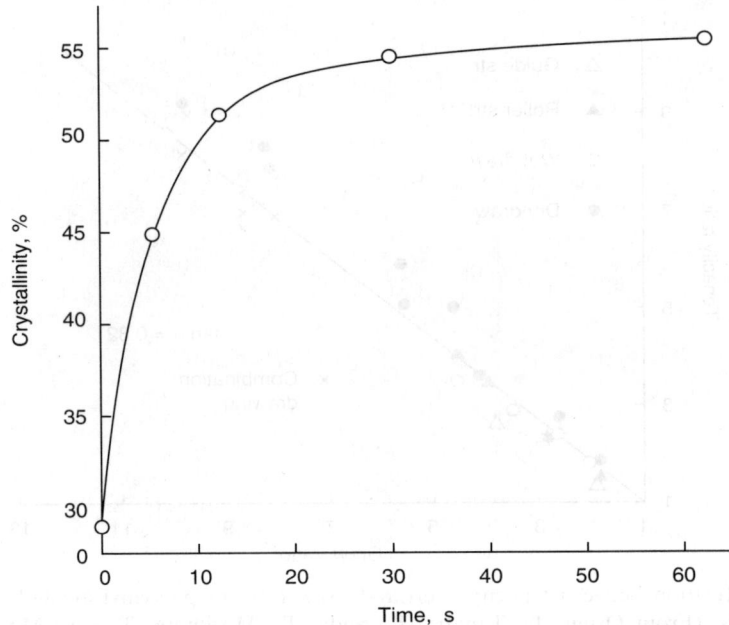

FIGURE 4.18 Increase in crystallinity with the time of heat treatment (at 230°C). (From Kawakami, H., Mori, N., Sato, H., and Miyoshi, A., *Sen-i Gakkaishi*, **16**, 155, 1960. With permission.)

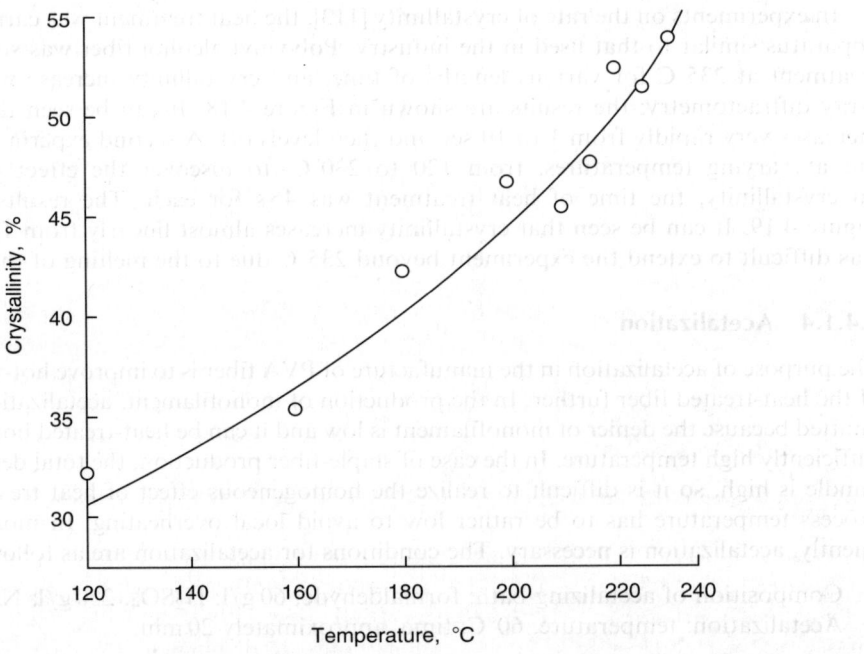

FIGURE 4.19 Increase of crystallinity with the temperature of heat treatment (time: 45 s). (From Kawakami, H., Mori, N., Sato, H., and Miyoshi, A., *Sen-i Gakkaishi*, **16**, 155, 1960. With permission.)

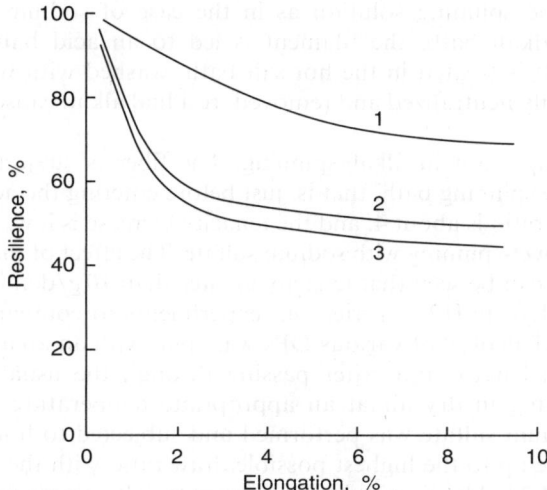

FIGURE 4.20 Influence of formalization on the resilience of fibers: curve 1, 0%; curve 2, 17%; curve 3, 40%. (From Sakurada, I., and Nakamura, N., *Kobunshi Kagaku*, **8**, 476, 480, 484, 539, 1951. With permission.)

fibers are acetalized up to 45%, they resist 1 h heating in water at 110°C. When relaxation is carried out, fibers of 30 mol% acetalization can resist water at 122°C.

Although the most remarkable change by acetalization is the increase in hot-water resistance, the mechanical properties of fibers are also affected. Acetalization has practically no effect on fiber tenacity, but formalization induces brittleness in fibers. For example, a heat-treated fiber shows 70–80% relative knot strength, which drops by formalization to 55–65%. There is a rather large difference of resilience between the formalized and unformalized fibers, as shown in Figure 4.20 [114]. It can be seen that the resilience of the unformalized PVA fibers is rather high, but it drops markedly with an initial formalization of less than 20%; the drop in fiber resilience after further formalization is very small.

If, instead of formaldehyde, acetalization is carried out with hydrophobic aldehydes such as nonylaldehyde [115] or benzaldehyde [116,117], the drop in resilience is not so large [118]; so benzylation was once adopted for commercial production.

It is sometimes said that cross-linking takes place during formalization. However, as has already been mentioned in the discussion on acetalization in Section 4.3.4, intermolecular acetalization leading to cross-linking occurs only to a very small extent, if any. When cross-linking is needed, it is advisable to use a dialdehyde, such as terephthalaldehyde [119], and acetalyl sulfide [120], which enables us to obtain sufficiently heat-resistant PVA fiber without heat treatment, although the fiber is brittle compared to fibers that are acetalized with monoaldehydes.

4.4.2 Different Processes of Spinning

In addition to wet-spinning with the sodium sulfate bath, other processes of spinning are available, some of which have been gaining importance.

4.4.2.1 Wet-Spinning with Alkali Bath

Wet-spinning with an alkaline coagulating bath, such as sodium hydroxide, is important not only because the coagulation mechanism is different from that of sodium sulfate but also because a sodium hydroxide bath gives a smoother filament of better quality, i.e., of higher tenacity.

Practically the same spinning solution as in the case of sodium sulfate is used. After passing through the alkali bath, the filament is led to an acid bath, where the alkali is neutralized, and then it is treated in the hot salt bath, washed with water, and dried. If the alkali is not satisfactorily neutralized and removed, residual alkali causes discoloration during the heat treatment.

Drawing is also important in alkali-spinning. The fiber is subjected to successive spin-drawing as it leaves the spinning bath, that is, just before entering the acid bath for neutralization. The highest draw ratio is about 4, and the tenacity increase is low. Hot-drawing in dry air is carried out as it is in wet-spinning with sodium sulfate. The effect of the hot-drawing is shown in Table 4.23, where it can be seen that tenacity greater than 10 g/den is easily attained [121].

Kawashima and Miyoshi [122] carried out experiments to compare alkali-spinning with salt-spinning. Polyvinyl alcohol of various DPs was spun with an alkali bath, spun-drawn to four times its original length, and, after passing through the usual steps of the process, subjected to hot-drawing in dry air at an appropriate temperature. In a similar manner, salt-spinning with sodium sulfate was performed and subjected to hot-drawing in air. Fiber was drawn in each case up to the highest possible draw ratio with the process. The comparison is shown in Table 4.24. Alkali-spinning is superior to salt-spinning in terms of the tenacity and hot-water resistance of fibers.

Alkali-spinning easily gives filaments of high tenacity; as can be seen in the upper half of Table 4.24, the filaments at all DPs have tenacities higher than 13.9 g/den; the highest tenacity, at a DP of 3000, is as high as 18.2 g/den, which corresponds to 2.1 GPa. It is remarkable that a fiber of such a high tenacity can be produced by a conventional wet-spinning method. The alkali-spinning also produces a PVA filament that is heat resistant in water between 100 and 150°C. Consequently, alkali-spun PVA filament resists boiling water, and acetalization can be omitted in the manufacture.

The above-mentioned characteristics seem to be in close connection with the microscopic structure that is observed in the microphotographs of the fiber cross section, which is shown in Figure 4.21. When we compare Figure 4.21 with the cross section of the fiber produced by wet-spinning with sodium sulfate (see Figure 4.15), the difference is clear. In alkali-spinning, the cross section is circular and quite homogeneous; in the case of the salt-spinning, the cross section is kidney-shaped and consists of a dense skin and a coarse core. According to the dyeing test of the cross sections, the alkali-spun fiber has the same structure as the skin of the salt-spun fiber. The above-mentioned differences in the properties of the two types of

TABLE 4.23
Tensile Properties of PVA Filaments Spun in Sodium Hydroxide Coagulation and Heat-Drawn to Various Ratios

Draw ratio	Denier	Strength (g)	Tenacity (g/den)	Elongation (%)
1.00	2.76	8.81	3.19	24.5
1.37	2.01	9.26	4.61	18.1
1.73	1.58	9.18	5.81	12.4
2.55	1.08	8.39	7.77	8.5
2.84	0.97	9.94	10.20	6.7
3.32	0.83	9.53	10.80	5.8

Source: From Kawakami, H. and Sato, H., *Sen-i Gakkaishi*, **18**, 183, 1962.

TABLE 4.24
Comparison of the Effects of the DP and Draw Ratio on Tensile Properties and Hot-Water Resistance of PVA Filaments Prepared by the Use of Sodium Hydroxide and Sodium Sulfate for Coagulation

	DP of polyvinyl alcohol						
	1000	1700	2200	2400	3000	3500	3800
a. NaOH Spinning							
Drawing temperature (°C)	225	225	225	230	230	230	230
Draw ratio	17.6	18.0	18.6	19.7	19.8	19.8	17.2
Tenacity (g/den)	13.9	14.1	15.7	17.3	18.2	17.8	14.2
Elongation (%)	8.7	8.7	8.5	8.5	7.3	7.3	8.0
Hot-water resistance (°C)	110	110	110	115	115	115	110
b. Na$_2$SO$_4$ Spinning							
Drawing temperature (°C)	220	225	230	230	235	240	240
Draw ratio	14.0	15.0	15.0	14.2	14.0	13.5	12.0
Tenacity (g/den)	8.2	9.2	9.8	9.3	8.4	8.5	8.0
Elongation (%)	8.3	9.0	9.6	9.5	9.3	8.2	9.0
Hot-water resistance (°C)	100	100	100	100	95	95	95

Source: From Kawashima, K. and Miyoshi, A. *Japan Pat.*, 47-8186, 1968, (to Unitica).

fibers can be traced to this structural difference. The salt-spun fiber is always accompanied by a coarse granular structure that is, expressed briefly, dead load to high-tenacity fiber.

The mechanism of coagulation in the process of alkali-spinning is quite different from that of salt-spinning, which has already been explained (Section 4.4.1). According to the

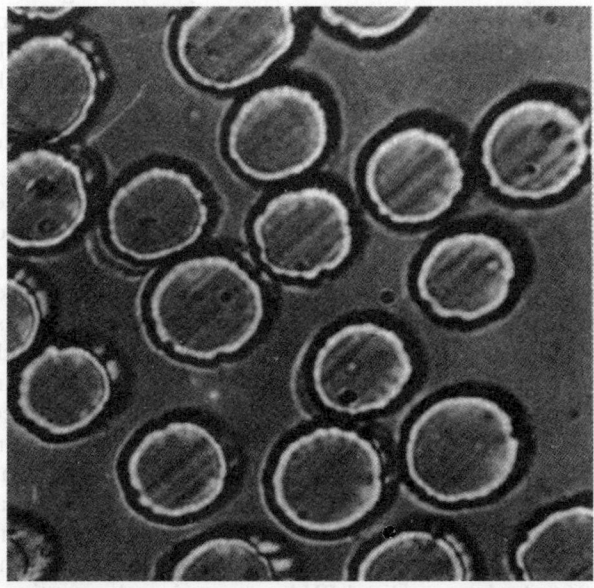

FIGURE 4.21 Micropthograph of cross section of polyvinyl alcohol fiber obtained by wet-spinning with an alkaline coagulating bath.

investigation of Kawakami and Sato [121], the coagulation is a liquid–liquid phase separation, and, throughout the coagulation bath, the water transferred from the fiber to the bath is compensated with the transfer of the alkali solution from the bath into the fiber. Coagulation and fiber formation proceed very slowly, so that the complete coagulation is effected in the hot salt bath.

4.4.2.2 Wet-Spinning of PVA Solution Containing Boric Acid

As described previously, salt-spinning consists of coagulation of the dope, forming a skin-core structure induced by dehydration inside the fiber. On the other hand, alkali-spinning first gives a gelled structure, forming a homogeneous fiber that brings about higher performance in fiber properties. A similar type of gelation is observed when an aqueous solution of PVA containing boric acid is alkalized. On the basis of this observation, a modified spinning method has been developed, with a coagulation bath containing sodium hydroxide and sodium sulfate [123]. The spinning method has been improved and has taken over the alkali-spinning method for high-performance fiber manufacture not only due to its fiber properties but also economy. Although the detailed data have not yet been published, it is believed that fibers produced by the spinning method have a tenacity greater than 11 g/den and they are used in such industrial applications as fiber-reinforced cement (FRC), which is described later.

4.4.2.3 Dry-Spinning

Dry-spinning is essentially simpler than wet-spinning; if water is used, as in the case of PVA, it is necessary to take into account that the latent heat of vaporization is much higher than that of a common solvent. In case it is not necessary to evaporate water quickly from the dope, the situation turns out to be favorable for dry-spinning.

The dry-spinning of PVA is divided into two types, which differ mainly in the rate of the draft, therefore the spinning speed. The first is a low-speed method, which is applied to the production of high-tenacity monofilament and filament yarn of large denier. If the spinning is carried out under high draft, the filament is drawn in the course of spinning, and subsequently the main drawing, that is, hot-drawing in dry air, becomes less effective. This is the reason for low-speed spinning (take-up speed, 10–40 m/min), in which filaments with a tenacity of about 7.5 g/den are produced [124].

In typical spinning conditions, the concentration of PVA in the dope is 40% and the temperature is 130°C. The dope is extruded from a nozzle into the spinning cell under circulation of air at 50°C. Continuous dry-spinning apparently can be effected; however, it is likely that evaporation of water from the dope scarcely occurs in the cell because the heat supply is insufficient, and it is cooled to form soft, solid filaments. The fiber-forming mechanism is the same as it is with conventional melt-spinning, except that the molten substance is a binary mixture. In the present case, the spun-filament is successively led to drying and finally subjected to hot-drawing in dry air, as in the case with the wet-spinning.

The second method of spinning is the high-speed one [125]. The concentration of PVA is little lower than it is in the case of the low-speed-spinning method. In the high-speed method, the liquid filament coming down from the spinneret at high relative humidity is spun under a very high draft (take-up speed, 500–700 m/min). By this method, a fine-denier filament with a tenacity of 5 g/den is produced. This method was developed to produce filament yarn because the wet-spinning of PVA from the sodium sulfate bath was not a suitable method for its production.

Nihon Vinylon Company (now Nitivi) began the production of filament yarn by the dry-spinning process in 1964. Aminoacetalized PVA was used as the starting material to improve the dyeability of the filament. The aim was to obtain a silk-like synthetic fiber, but the resulting filament was not very smooth and was water soluble.

4.4.2.4 Spinning Using Organic Solvent

There were proposals to use solvents other than water for the spinning. It was found by Tsuboi and Mochizuki [126] that single crystals of plate are obtained by the slow cooling of dilute solutions of PVA in polyvalent alcohols such as ethylene glycol, diethylene glycol, triethylene glycol, and glycerin. With the so-called dry-spinning of concentrated solutions of PVA in these polyvalent alcohols, investigators have succeeded in preparing fibers of higher crystallinity and better hot-water resistance [127].

About 40 years ago, PVA in dimethyl sulfoxide was spun into organic solvents such as methanol, acetone, and so on for the first time by Matsubayashi and Segawa [128]. They obtained a high-tenacity and high-modulus fiber. Since the discovery of the high-performance polyethylene fiber by the researchers of DSM using a high-molecular-weight polymer, many other studies have been carried out in the field of PVA. In almost all these studies, an organic solvent such as dimethyl sulfoxide and glycerin and an organic coagulation bath such as methanol have been utilized. The spinning method is classified as "gelation-spinning." The concentration of PVA is rather low compared to that used in the usual spinning method to prevent the polymer molecules from too many entanglements. Hence, PVA of higher molecular weight is needed. The dope is spun into a coagulation bath directly (usual wet-spinning) or via air (dry–wet spinning) to yield PVA gel in the coagulation bath. After the extraction of solvent from the gel and drying the undrawn fiber, drawing is carried out, where a high temperature close to melting point is necessary to melt the microcrystalline parts formed before drawing.

There are much data concerning the tenacities and moduli obtained from gelation spinning [129]. However, these data were derived on the basis of a single filament, and the highest values were reported. From the industrial point of view, average values based on multifilaments should be adopted. According to Mochizuki, PVA fiber of DP 5000 showed a tenacity 20 g/den and a modulus of 480 g/den based on multifilaments [130].

Recently, the gelation-spun PVA fiber was commercialized by Kuraray. According to Ohmori, the fiber is produced by the complete organic solvent-coagulant system. Polyvinyl alcohol is solubilized in dimethyl sulfoxide, and the solution is spun into cold methanol either by dry–wet spinning or wet-spinning. The resulting fiber is washed with methanol and drawn to about 25 times after drying [131,132]. In the coagulation bath, the gel of the dope is formed, which will be mentioned later. There is an air gap between the dope and the coagulation bath in the dry–wet spinning method. At first, this spinning method was tried, but there was a problem in producing the fiber industrially; the number of filaments obtained in a nozzle was small compared to those obtained with usual wet-spinning because there was a tendency of filaments to adhere to one other. Later, the wet-spinning method over the gelation-spinning method was improved, so that multifilaments obtained from over 10,000 holes in a nozzle became available.

Yarn tenacity depends on the degree of polymerization of PVA, as shown in Figure 4.22. The tenacities of the yarn are remarkably high in the case of high DP; 13 and 24 g/den for the DPs of 1,700 and 14,000, respectively. The DPs in the figure are measured for the polymer in the final fibers, and decreases in them are observed to some extent (especially in the case of high DP) from the original polymers during the fiber-making process; for instance, 14,000 of the fiber from 18,000 of the original PVA.

Since tenacity of the fiber produced from gelation-spinning of a conventional PVA with DP 1,700 is sufficiently high for the applications, and the PVA with high DP of 18,000 is still not commercially produced, Kuraray has commercialized the gelation-spun fiber produced

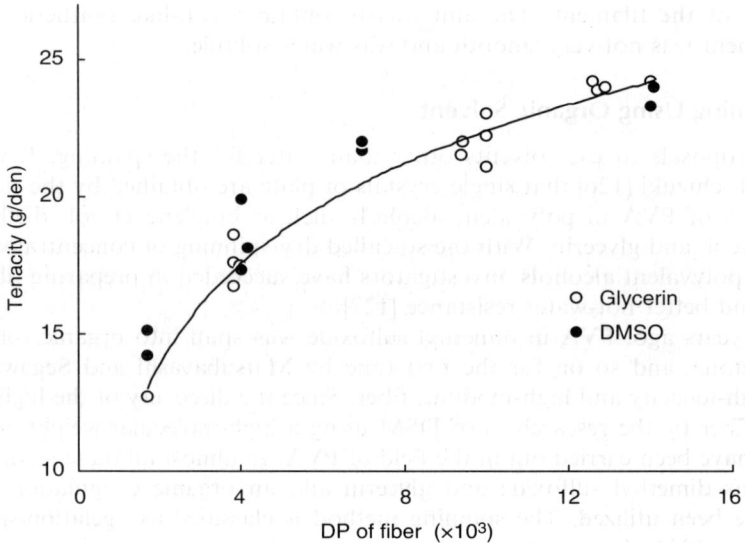

FIGURE 4.22 Relationship between fiber (yarn) tenacity and degree of polymerization of PVA fiber. (From Ohmori, A., Sakuragi, I., and Onodera, M., *Sen-i Gakkaishi*, **55**, 418, 1999.)

from a conventional PVA in the name of K-II Vinylon since 1998; the production capacity is 7,000 tons/year.

The mechanism of gelation-spinning was studied by Kaji et al. using PVA of DP 1,700 and solvents of dimethyl sulfoxide–water and glycerin–water; the study was carried out by cooling the solution [133]. The dope becomes gel either via liquid–liquid phase separation (spinodal decomposition) at a temperature higher than 10°C, or directly at a lower temperature. The cross-linked points are the microcrystals of PVA. In the actual spinning, too many junction points are formed at high temperature to cause difficulty in drawing; on the other hand, the junction points are too few at low temperature to result in the slipping of the molecules. Accordingly, appropriate temperature of the coagulation bath is needed for the industrial production of fibers with high performance.

Gelation-spinning of PVA has produced new types of fibers that were never done by ordinary wet-spinning. In addition to the high tenacity fibers using a conventional PVA with high degree of hydrolysis (DH) of more than 99.8%, Ohmori et al. have succeeded in commercialization of the following three types of new fibers; application fields and the performance of these fibers are shown in Table 4.25 [131,132]. The first are water-soluble fibers by using PVA with low DH (88–99%). Temperatures at which the fiber is soluble in water can be controlled by using different DH. As is seen from Table 4.25, the water-soluble temperatures are selected to be less than 5, 50, 70, and 80°C. The water-soluble fibers have sufficient tenacities (5–8 g/den) and elongations (12–28%) for use. These PVAs with a low DH could not be utilized for fiber production by the usual wet-spinning method already described, because water is used in various processes. The water-soluble fibers thus produced are resistant even at high relative humidity.

The second is a fiber with a thermal bonding ability and water solubility. The fiber is spun by gelation spinning of a mixture of two types of PVA: high DH and low DH. Since both types of PVA are not compatible with each other, phase separation takes place in the fiber. Consequently, fibers with phase separation (sea-island type) are formed; the continuous phase consists of high DP and the noncontinuous one, low DP. When the fiber is heated to a

TABLE 4.25
Type and Properties of "Kuralon[a]" K-II

Type	Grade		Tenacity (g/den)	Elong. (%)	Water-soluble temp. (°C)	Thermal bond. temp. (°C)
High tenacity	Nonwoven	DQ1	11	8	≥100	≥210
	Staple fiber	EQ2	12	8	≥100	—
	Reinforcem.	EQ5	15	6	≥100	—
Water-soluble	Room temp.	WJ2	5	28	<5	≥110
	Medium	WN5	6	20	50	—
	Hot water	WN8	8	10	80	—
Water-soluble and thermal bonding	Hot water	WJ9	8	12	70	≥200
Easy-fibrillation		SA	11	6	≥100	—

[a] English name of Vinylon.

Source: From Ohmori, A., Sakuragi, I., and Onodera, M., *Sen-i Gakkaishi*, **55**, 418, 1999.

temperature higher than the lower melting point of PVA of low DP, but lower than the higher melting point of PVA of high DP, a type of phase inversion takes place; the newly formed continuous phase has a thermal-bonding ability to bond with each other and with other fibers. After the thermal bonding, the fabrics made from the fiber with other fibers are utilized as water-soluble, nonwoven fabrics.

The third is a fiber with an easy fibrillation. The fiber is spun by gelation-spinning of a mixture of PVA and another polymer, such as polyacrylonitrile, which is soluble in dimethyl sulfoxide. Phase-separated fibers are produced; the continuous phase consists of PVA, and the noncontinuous phase consists of polyacrylonitrile. The fiber can be fibrillated with physical strengths such as milling in rubber, stirring in water, and jet-streamed water, to result in fibrillated fibers with diameters less than 1 μm (0.01 den).

4.4.2.5 Biodegradability of Polyvinyl Alcohol

Polyvinyl alcohol is well known as a biodegradable polymer. This is of value to synthetic polymers because it is difficult to decompose almost all of them by microorganisms. Biodegradable polymers must be friendly to the environment. Consequently, it should be clarified that dissolved parts from the water-soluble PVA fibers mentioned above can be decomposed by microorganisms. Figure 4.23 shows the experimental results carried out by the method of Japan Industrial Standard (JIS K6590), where aniline, which is a typical biodegradable compound, is used as a reference. Two kinds of the water-soluble K-II Vinylon show their biodegradability to be close to aniline.

4.4.2.6 Miscellaneous

Phase separation spinning was reported in detail by Zwick [134]. This method allows spinning of fibers at winding speeds of 10–1000 m/min from solutions of polymer contents of 10–25%, with the use of modified melt-spinning equipment. To realize phase separation, the concentration of the solution has to be chosen so that, on the way from the spinneret to the winder, and at a temperature between that of the spinneret and room temperature, the thin streams of polymer undergo phase separation into a concentrated gel, or pure polymer phase, and a solvent phase.

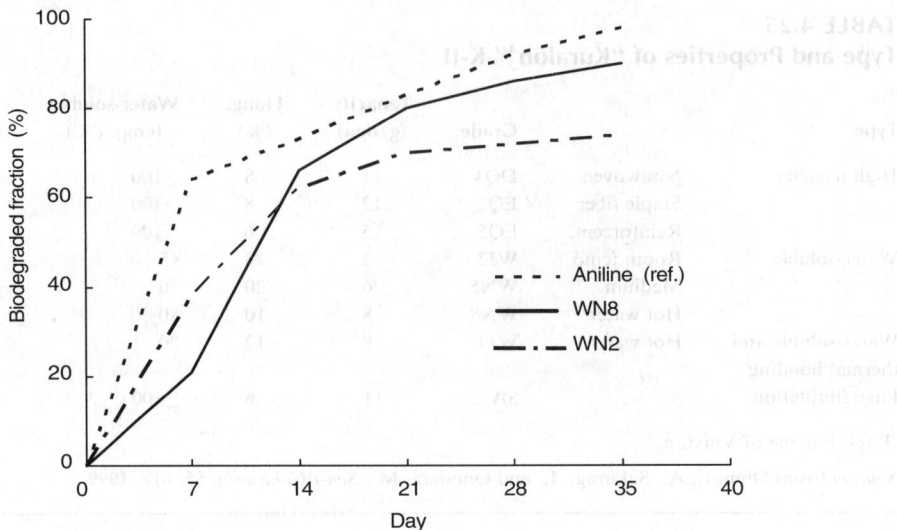

FIGURE 4.23 Biodegradability of water-soluble vinylon K-II fibers after solubilized in water. (From Ohmori, A., Sakuragi, I., and Onodera, M., *Sen-i Gakkaishi*, **55**, 418, 1999.)

The solvents for PVA were investigated in some detail. Single solvents such as benzene sulfonamide, toluene sulfonamide, caprolactam, trimethylol propane, and binary systems such as water–urea, water–thiourea, and dimethyl sulfoxide–pentaerythritol were found suitable for phase separation spinning. Apparatus similar to that used in dry-spinning was employed. Polyvinyl alcohol contents were 13–37%, and the temperature of the spinning solution was 115–195°C. The tenacity of the fibers prepared by this method is rather high and, in most cases, shows values greater than 6 g/den. Phase-separation spinning is a technique that is appropriate for PVA because it does not melt without decomposition.

Spinning of a PVA mixture with some other polymers has been studied by a number of researchers, and properties of the bicomponent fibers are reported mostly in patents. There is a classical report by Kawakami on the mixed-spinning of PVA [135].

It is well known that it is difficult to achieve miscibility of two polymers. However, it is often possible to obtain apparently stable solutions in which two polymers are dissolved without phase separation. Examples of water-soluble polymers that were mixed with PVA and spun by wet-spinning are polyacrylic acid, sodium carboxymethyl cellulose, various proteins, and polyvinyl aminoacetals. The most common objective of mixed-spinning was the modification of dyeing properties, but this method is also applied to modify many other properties of PVA fiber.

Early studies have found that PVA is soluble in liquid ammonia; solutions of 15–55% PVA content were ejected from the spinneret into the spinning cell with its own pressure of about 8 atm and at room temperature. Take-up speed of the fiber was 60–270 m/min [136].

Another interesting method of mixed-spinning of polymer emulsion in PVA solution will be outlined below.

4.4.3 BICOMPONENT FIBER (PVC/PVA)

A bicomponent fiber of PVC and PVA is produced by spinning a mixture of a polyvinyl chloride emulsion and an aqueous solution of PVA. This process is called "emulsion spinning." The fiber is produced by Kohjin Company under the trade name of Cordelan. The fiber's generic name is polychlal.

4.4.3.1 Emulsion-Spinning

Those polymers that are neither soluble in solvents nor melt by heating cannot be converted to fibers or thin films by conventional spinning methods. Emulsion-spinning was first developed to overcome this difficulty [137].

Polytetrafluoroethylene is a well-known polymer for which the emulsion-spinning method is applied. One example of the process is to mix polytetrafluoroethylene emulsion with viscose to obtain a spinning solution containing about 40% polytetrafluoroethylene and 2.3% cellulose. In this example, the spinning was carried out very slowly with a coagulation bath containing sulfuric acid, sodium sulfate, and zinc sulfate. The coagulation bath is suitable not only for viscose spinning but also for the salting out of the polytetrafluoroethylene dispersion. After drying at 190°C, the filament bundle was heated to 389°C to flash off the cellulose, sinter polytetrafluoroethylene, and stretch the bundle of filament. Tensile strength higher than 2 g/den was obtained.

Emulsion-spinning to prepare bicomponent fiber from the emulsified polymer is different from that used to obtain single-component fiber. In the latter case, a polymer dissolved in the dispersed medium during spinning is completely removed from the final fiber; therefore, it cannot affect the properties of the product. In the bicomponent fibers, the two polymers, one in the dispersed medium and the other in the dispersed phase, remain in the final fiber and modify its properties.

Tsuji et al. [138] and Kitamaru et al. [139] published reports on the mixed emulsion-spinning of PVA. In addition to the spinning solution, there is essentially no difference in the process from that of regular wet-spinning.

Typical examples of the mixed emulsion-spinning of polyvinyl chloride, polyethylene, polyvinylidene chloride, and polyvinyl acetate, and some properties of the mixed fibers are shown in Table 4.26. The measurement of the fiber properties is carried out after hot-drawing in dry air at 180°C, heat treatment for 100 sec at 250°C, and acetalization for 40 min at 70°C without tension. The draw ratios shown in the table are the highest possible ratios under the given experimental conditions. In most cases, there is a maximum possible draw ratio at a certain mixing ratio. However, only the results of experiments at mixing ratios of 1:3 and 1:1 are shown in the table.

The fibers show appreciable tenacity, although tenacity decreases with an increase in the proportion of the emulsion polymer. It was confirmed by microscopic investigations that a

TABLE 4.26
Mechanical Properties of Bicomponent Fibers Prepared by Emulsion-Spinning of Mixtures of PVA and Emulsions of Various Polymers

Emulsion polymers	Mixing ratio	Draw ratio	Tenacity (g/den)	Elongation (%)
Polyvinyl acetate/PVA	1/3	3.4	4.33	15.9
	1/1	2.7	1.86	19.2
Polyvinyl chloride/PVA	1/3	3.7	4.20	20.2
	1/1	2.4	2.33	26.4
Polyvinylidene chloride/PVA	1/3	7.0	5.99	15.0
	1/1	5.9	2.27	31.1
Polyethylene/PVA	1/3	2.9	3.09	19.9
	1/1	3.0	2.39	19.9
PVA without additive	0	—	6–8	20–25

Source: From Tsuji, W., Kitamaru, R., Ochi, T., and Mori, N., *Kasen Koen Shu*, **16**, 23, 1959; Kitamaru, R., Ochi, T., Koh, H., and Tsuji, W., *Kasen Koen Shu*, **17**, 39, 1960.

temperature lower than 180°C is sufficient for emulsion particles to coalesce. In the course of the hot-drawing, not only PVA but also emulsion polymers are drawn and microfibrils are formed in macroscopic fibers. It is also important that the mixed fibers show a resistance to boiling water that is similar to that of pure PVA fiber.

4.4.3.2 Polychlal

Fundamental research on the emulsion-spinning of bicomponent fiber of PVC/PVA (polychlal) has been conducted by Okamura et al. [140]. Their first important problem was to prepare a stable mixture of polyvinyl chloride emulsion and aqueous solution of PVA. The problem was solved by forming a graft copolymer of polyvinyl chloride to PVA, which brings about intimate mixing of PVA and emulsion particles of polyvinyl chloride. It was further observed that emulsions of smaller particles give better results in the process of spinning; emulsions of particle diameters of less than 500 Å give lower viscosity and higher stability.

The next problem was that of coalescing emulsion particles and giving oriented fiber structure to polyvinyl chloride. As will be shown below, it was confirmed that the oriented fiber structure is realized in the hot-drawing of the spun filament in dry air.

Polychlal is manufactured essentially by the same process as that used in the manufacture of regular PVA fiber. Total polymer concentration of the spinning solution is about 20%; the ratio of PVC/PVA in most cases is nearly 40/60. The coagulating agent is sodium sulfate, which is suitable not only for the coagulation of PVA but also for the salting out of the emulsion. The filament is drawn about five times its original length at a temperature between 160 and 180°C and subjected to heat treatment between 220 and 230°C. Acetalization with formaldehyde is carried out, and final fiber is obtained.

Properties of PVC/PVA fibers prepared under conditions as mentioned above, and the appearance after extraction with water is shown in Table 4.27. It is seen from the table that the tenacity decreases in proportion to the increase of PVC content. This is in accordance with the assumption that the matrix of the fiber is PVA, at least until its fraction falls to 0.6 of the total polymer. Although polyvinyl chloride apparently does not contribute to tenacity, it shows continuous fiber structure, as indicated in the last column of Table 4.26, when the polyvinyl chloride fraction is greater than 0.40. Some mechanical properties of PVC/PVA fiber, such as higher resilience, are perhaps due to the structure.

As already mentioned, the commercially produced polychlal has a PVC/PVA composition nearing 1:1, and consists of two phases, a PVA phase and a polyvinyl chloride phase. The former is a matrix polymer, and the properties are not the average of the two components. Polychlal shows a softening temperature of 180–200°C, which is much higher

TABLE 4.27
Properties of PVC/PVA Fibers of Various Mixing Ratios

Mixing ratio PVC/PVA	Draw ratio	Tenacity (g/den)	Elongation (%)	Appearance of the residue after extraction of PVA
0/10	5.0	8.4	11.8	Completely dissolved
2/8	5.0	5.7	20.8	Finely dispersed
4/6	5.0	4.0	24.6	Fiber form retained
5/5	5.0	2.8	35.2	Fiber form retained
6/4	5.0	1.7	34.8	Fiber form retained

than that of polyvinyl chloride and not far from that of PVA. Polychlal is also flame-retardant and self-extinguishing, therefore polychlal is widely used in the manufacture of children's clothing.

Although the dyeability of both polyvinyl chloride fiber and regular vinylon is difficult or not satisfactory, polychlal fiber is dyeable with various kinds of dyes. Some physical properties of polychlal, vinylon, and polyvinyl chloride fibers are shown in Table 4.28. All fibers listed in the table are commercial products in Japan [141].

4.5 POLYVINYL CHLORIDE FIBERS

Polyvinyl chloride is a synthetic polymer that was first described by Regnault in 1838 [142]. Although there was already a German patent in 1913, which suggested the formation of new fiber from PVC [143], the chemical industry did not begin to show an active interest in this polymer until the early 1930s.

For fiber formation from a polymer, it is necessary for the polymer to either melt at an elevated temperature or dissolve in a solvent. The first commercially produced PVC fiber (monofilament) was made by a method of melt-spinning [144]. However, melt-spinning of PVC is not an agreeable process because of its high melting point, its high melt viscosity, and its low thermal stability.

Cyclohexanone was the only solvent that was known in early research on PVC. It was often tried as a solvent for wet-spinning, but was not successful. Later, tetrahydrofuran and dimethyl formamide were found to be solvents for PVC; but the former is too expensive and the latter is less attractive compared to mixed solvents.

The progress of PVC fiber production was brought about by the chemical modification of PVC so that it became soluble in common solvents. In the United States, a copolymer of vinyl chloride and vinyl acetate, which was soluble in acetone, was synthesized, and the PVC fiber Vinyon was produced in 1935.

In Germany, a different method was used for the modification of the solubility of PVC, and an acetone-soluble polymer was obtained by the chlorination of PVC. A fiber of this type of polymer was produced for the first time in 1934 and was called the Pe-Ce fiber [145].

The approach in France was quite different. Intensive research resulted in the discovery that a mixture of acetone and carbon disulfide was a suitable mixed solvent for PVC [146]. On the basis of this discovery, the production of Rhovyl began in 1949.

In Japan, Teijin Company found that there are binary solvent mixtures that do not dissolve PVC at room temperature, but dissolve it at higher temperatures. They found that, among several binary mixtures, acetone–benzene is the most suitable one for dry-spinning. The production of the fiber called Teviron of 5 tons/day began in 1956 by this method.

Bicomponent fibers of PVC/PVA are prepared by emulsion-spinning and contain about 40% PVC (see Section 4.4.3).

4.5.1 Manufacture of Polyvinyl Chloride

4.5.1.1 Vinyl Chloride Monomer

The important manufacturing methods of vinyl chloride monomer currently in practice are the oxychlorination method and the mixed-gas method.

Oxychlorination: In this method [147], ethylene and chlorine are reacted to form dichloroethane, and vinyl chloride monomer is prepared by thermal decomposition of dichloroethane in the presence of an appropriate catalyst; hydrogen chloride is liberated as a by-product in this reaction. Hydrogen chloride is reacted with ethylene and oxygen to form dichloroethane,

TABLE 4.28
Some Properties of PVA, Polyvinyl Chloride, and PVC/PVA Fibers

Fiber	Vinylon Staple Regular	Vinylon Staple H.T.[a]	Vinylon Filament Regular	Vinylon Filament H.T.[a]	Polyvinyl chloride Staple Regular	Polyvinyl chloride Staple H.T.[a]	Polyvinyl chloride Filament	PVC/PVA Staple
Tenacity (g/den)								
Standard	4.0–6.5	6.8–10.0	3.0–4.0	6.0–9.5	2.0–2.8	3.3–4.0	2.7–3.7	2.8–3.3
Wet	3.2–5.2	5.3–8.5	2.1–3.2	5.0–8.5	2.0–2.8	3.3–4.0	2.7–3.7	2.8–3.3
Elongation (%)								
Standard	12–26	9–17	17–22	8–22	70–90	15–23	20–25	20–24
Wet	12–26	9–17	17–25	8–26	70–90	15–23	20–25	20–24
Initial modulus (g/den)	25–70	70–130	60–90	70–250	15–25	30–50	30–45	25–35
Resilience at 3% long. (%)	70–85	72–85	70–90	70–90	79–85	80–85	80–90	80–90
Specific gravity (g/cm^3)	1.26–1.30	1.26–1.30	1.26–1.30	1.26–1.30	1.39	—	1.32
Water content (20°C, 65%) (%)	4.5–5.0	—	3.5–4.5	3.0–5.0	0	0	0	2.5–3.5
Softening temperature (°C)	220–230	220–230	220–230	220–230	90–100	60–70	—	180–200

[a]High-tenacity type
Source: From Sakurada, I., J. Pure Appl. Chem. 16, 263, 1968; Society of Chemical Fibers [Japan], Hand Book of Chemical Fibers, Kasen-Kyokai, Osaka, 1981, p. 308, 310, 313.

Vinyl Fibers

and dichloroethane thus formed is also decomposed to vinyl chloride. Water is liberated as a by-product. There are many industrial processes to synthesize vinyl monomer by the oxychlorination method.

$$CH_2=CH_2 + Cl_2 \longrightarrow CH_2Cl-CH_2Cl \longrightarrow CH_2=CHCl + HCl$$

$$CH_2=CH_2 + 1/2 O_2 + 2HCl \longrightarrow CH_2Cl-CH_2Cl + H_2O$$
$$\longrightarrow CH_2=CHCl + HCl$$

Mixed-gas method: In this method [148], naphtha is thermally cracked to form acetylene and ethylene. Dichloroethane is formed by reacting ethylene with chlorine, and vinyl chloride is obtained by thermal decomposition of dichloroethane with simultaneous formation of hydrogen chloride. Further, the reaction between hydrogen chloride and acetylene also gives vinyl chloride.

$$CH_2=CH_2 + Cl_2 \longrightarrow CH_2Cl-CH_2Cl \longrightarrow CH_2=CHCl + HCl$$

$$CH\equiv CH + HCl \longrightarrow CH_2=CHCl$$

4.5.1.2 Polymerization

Methods of polymerization of vinyl chloride currently in practice are suspension, emulsion, and bulk polymerization. The suspension polymerization is adopted in most cases. There is a low-temperature polymerization method to obtain crystalline PVC [149].

Suspension polymerization: In this method, vinyl chloride is polymerized in water containing a suspending agent using an oil-soluble initiator under rather violent mechanical stirring. A water-soluble polymer, such as PVA, is used as a dispersing agent. It is adsorbed on the surface of vinyl chloride droplet, preventing their coagulation. Since PVC is not soluble in vinyl chloride, with increase in conversion, the polymer continues to precipitate in the monomer droplet to form a porous mass. This mass has a very complicated morphology. Much research has been done on the regulation of the morphology of the precipitated polymer mass formed during polymerization. The mass (about 100 μm diameter) contains many smaller masses, and each smaller mass contains tiny several layered masses.

It is important to remove the residual vinyl chloride from the polymer mass as completely as possible for increasing the porosity of the mass. It is also necessary to make thinner skins not only of the large mass but also of the inner smaller masses. On the other hand, the bulk density should not be too small from the viewpoint of handling the polymer. To attain these requirements that contradict with one another, a secondary dispersing agent was studied and used. The secondary dispersing agent is not soluble in water, but is soluble in vinyl chloride, and plays a role in stabilizing and regulating the polymer coagulants that separate in the monomer droplet with the increase in conversion. As the secondary dispersing agent, different types of PVA with very low degree of hydrolysis (less than 50%), low degree of polymerization (less than 500), and modified end groups from the conventional PVAs are utilized industrially.

Low-temperature polymerization: For the preparation of crystalline PVC, a low-temperature polymerization method is used. An example is redox polymerization at −30°C, using a catalyst system comprising hydrogen peroxide, ferrous salt, and ascorbic acid [149]. Syndiotactic PVC, which shows excellent physical characteristics at higher temperature, is obtained by this method.

4.5.2 Manufacture of Fiber

4.5.2.1 Dry-Spinning

Polyvinyl chloride fibers are produced mostly by dry-spinning using a mixed solvent. Rhovyl is spun from an acetone–carbon disulfide solution and Teviron from acetone–benzene. Although there are no significant differences between the two fibers, their manufacturing processes are widely different from one another.

Mixed solvent of acetone–benzene: it was shown by Doty and Zabel [150] that the degree of swelling, Q, is a good measure of the dissolving power of solvents. Q is defined to be the reciprocal of the volume fraction of the polymer in the swollen polymer. Shimeha used a similar method for the evaluation of the dissolving power [151].

The relation between composition of the mixture and Q is shown in Figure 4.24. The maximum value of Q is at the concentration of 40 parts acetone and 60 parts benzene;

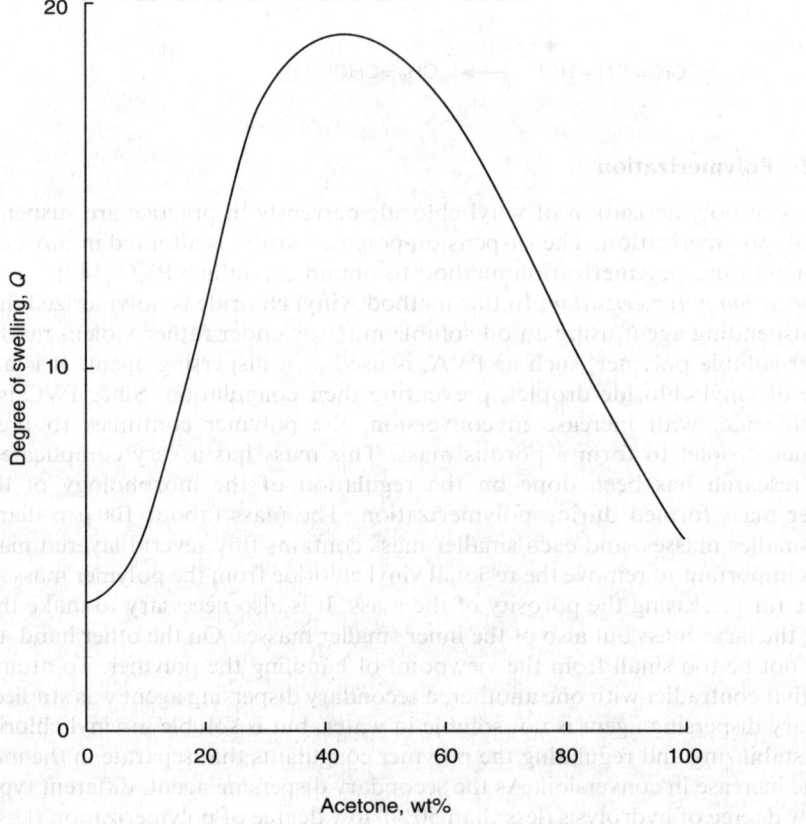

FIGURE 4.24 Relation between Q and acetone content in binary mixture acetone–benzene.

Vinyl Fibers

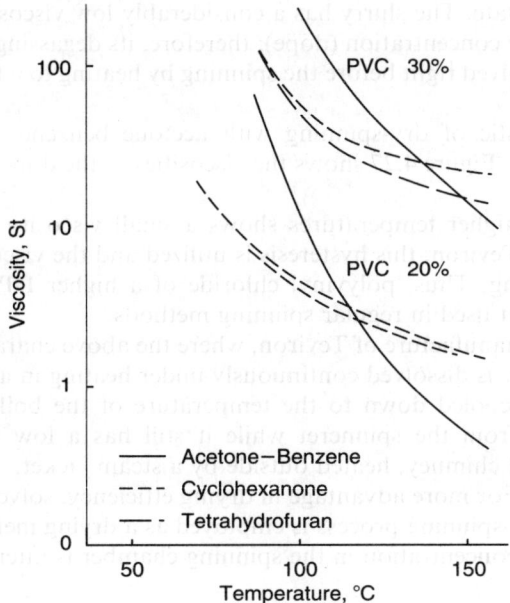

FIGURE 4.25 Temperature dependence of polyvinyl chloride solution viscosity on temperature in three different solvents.

however, even in this composition, the mixture only swells PVC at room temperature and cannot dissolve it.

Figure 4.25 shows the effect of temperature on the viscosity of PVC solutions in acetone–benzene (40:60), tetrahydrofuran, and cyclohexanone. Although the viscosity is very high at lower temperatures, it decreases rapidly with increasing temperature and becomes comparable with solutions in good solvents in a high-temperature range. In the manufacture of Teviron, this characteristic property of acetone–benzene is utilized.

As shown in Figure 4.26, when PVC is mixed with acetone–benzene at room temperature (5–25°C), polymer particles absorb the solvent mixture and swell. The viscosity of the slurry increases with the passage of time, and finally the fluidity is lost. However, if mixing is done at a lower temperature (below 5°C), swelling is suppressed, and the apparent viscosity of the

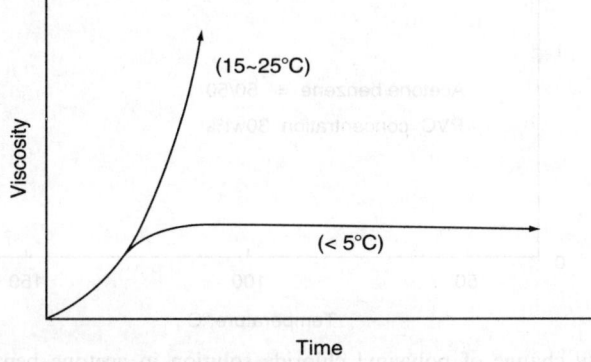

FIGURE 4.26 Schematic of viscosity change of polyvinyl chloride slurry in acetone–benzene with time.

slurry shows a steady state. The slurry has a considerably low viscosity compared to that of the solution of the same concentration (dope); therefore, its degassing and transportation are easy. This slurry is dissolved right before the spinning by heating to a temperature higher than 90°C under pressure.

Another characteristic of dry-spinning with acetone–benzene is the hysteresis of the viscosity of the solution. Figure 4.27 shows the viscosities of the dope in the course of heating and cooling.

Dope prepared at higher temperatures shows a small viscosity increase in the cooling process. In the case of Teviron, this hysteresis is utilized and the viscosity of the dope can be kept low in the spinning. Thus, polyvinyl chloride of a higher DP can be used in higher concentrations than that used in regular spinning methods.

In the commercial manufacture of Teviron, where the above characteristics are taken into consideration, the slurry is dissolved continuously under heating in a closed vessel and, after filtration, it is rapidly cooled down to the temperature of the boiling point of the mixed solvent and extruded from the spinneret while it still has a low viscosity. The spinning chamber is a cylindrical chimney, heated outside by a steam jacket.

Condense-spinning: For more advantage in drying efficiency, solvent recovery, and energy efficiency, the condense-spinning process is employed as a drying method [152]. This method is safe because the gas concentration in the spinning chamber is intentionally fixed at a level

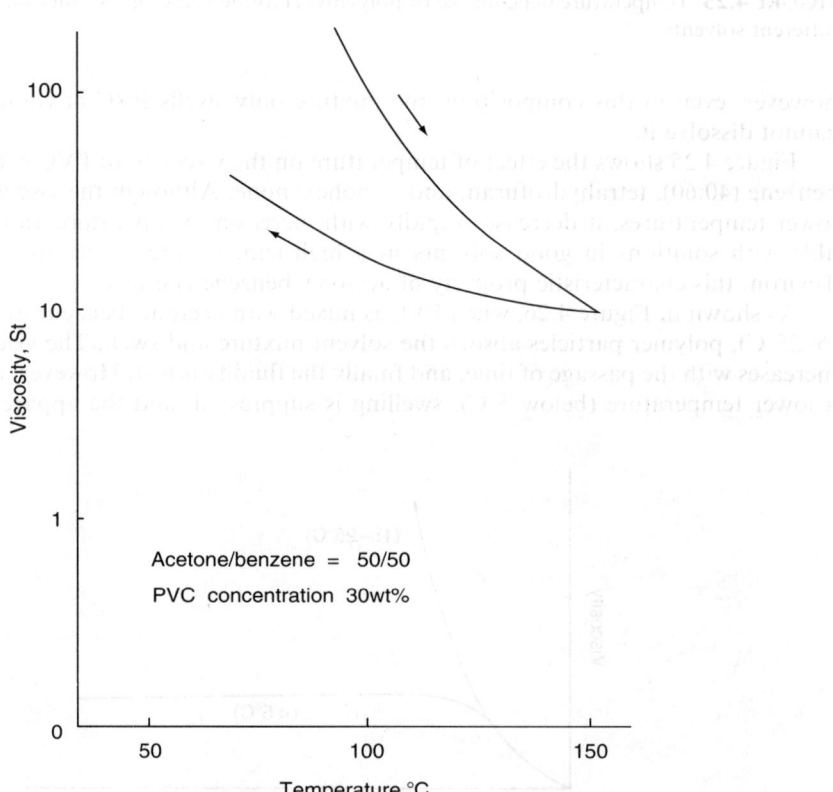

FIGURE 4.27 Viscosity change of polyvinyl chloride solution in acetone–benzene in the course of heating and cooling.

Vinyl Fibers

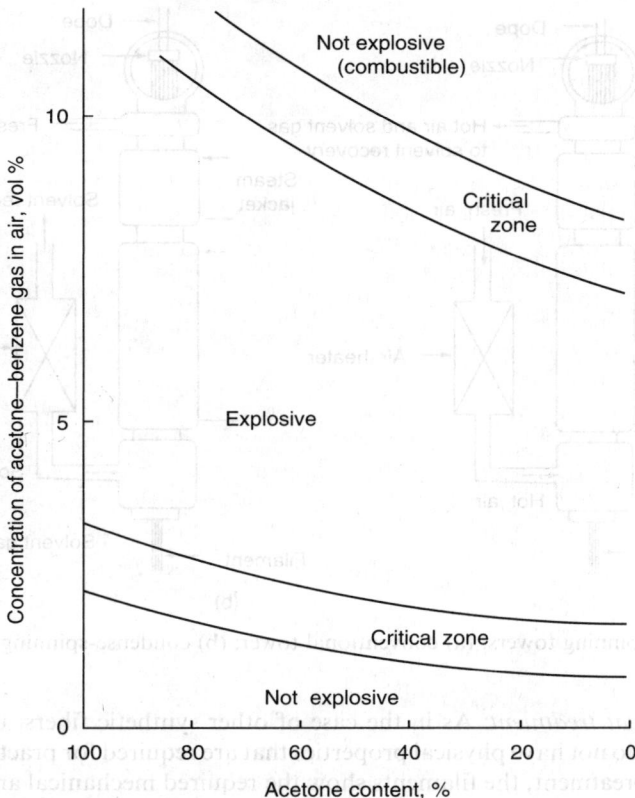

FIGURE 4.28 Explosion limits of acetone–benzene–air mixture.

higher than the upper explosion limit. The explosion limit of the acetone–benzene–air mixture is shown in Figure 4.28, and spinning chambers are shown in Figure 4.29.

In condense-spinning, only a small quantity of air at normal temperatures is fed into the top of the spinning chimney; therefore, gas speed in the chamber is very slow, and the heating of the drying chamber is carried out mainly by the steam-heated jacket. Inside, the spinning chimney is full of the solvent gas of high concentration that has evaporated from the filaments. The gas is introduced into condenser from the bottom of the chamber, where a large portion of the solvent is condensed for recovery.

Condense-spinning has the following advantages over normal dry-spinning:

1. The heat-transfer effect in the spinning chamber and, consequently, the productivity is improved.
2. Yarns of large monofilaments (above 50 den) can be spun.
3. The cross section of the filaments can easily be controlled, so that filaments with intended irregular cross sections can be manufactured.
4. Equipment and power cost can be reduced.

One of the practical applications of condense-spinning is in the manufacture of yarn for wigs. The yarn for wigs has an irregular cross section, its monofilament denier is about 50, and it cannot be manufactured by normal dry-spinning methods.

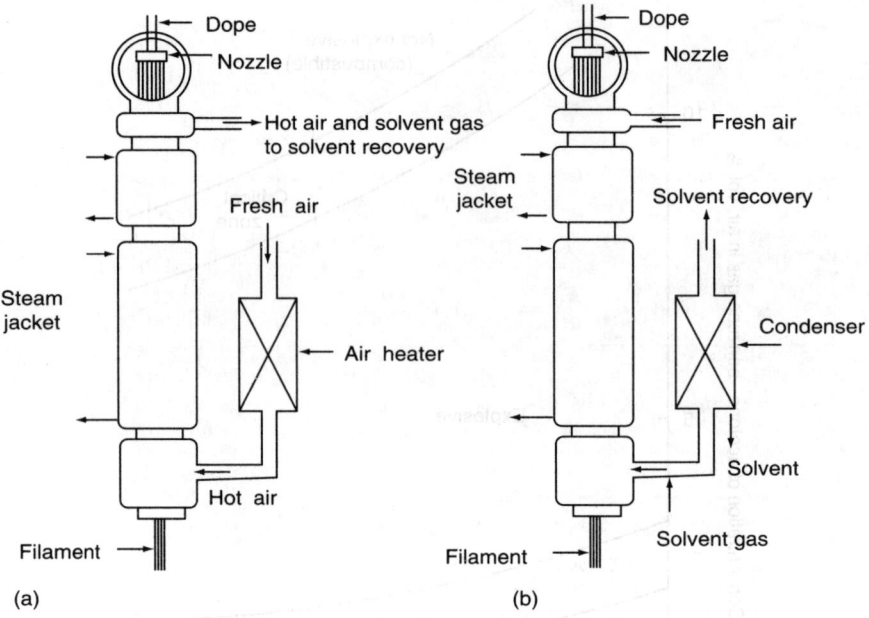

FIGURE 4.29 Dry-spinning towers: (a) conventional tower; (b) condense-spinning tower.

Drawing and heat treatment: As in the case of other synthetic fibers, undrawn polyvinyl chloride filaments do not have physical properties that are required for practical use. Only after drawing and heat treatment, the filaments show the required mechanical and thermal properties. The drawing of PVC fiber is conducted by a wet process using hot water or steam, and, in the case of Teviron, a steam-drawing process is employed. Figure 4.30 shows the change in tensile strength, knot strength, and elongation by drawing. Figure 4.31 shows experimental results of heat-setting, at constant length; however, the effect is not very large.

Mixed solvent of acetone–carbon disulfide: In manufacturing Rhovyl, PVC is dissolved into a mixed solvent of acetone–carbon disulfide and spun. The manufacturing system of Rhovyl and its spinning chamber are described in its patent [146] and explained in a paper by Cord [153].

The temperature of the heating zone of the spinning chamber is 120°C; the solvent is evaporated, and the undrawn filament is taken up at a winding speed of 170 m/min. Air containing a solvent and carbon dioxide–nitrogen gas, which is added to prevent explosion, is cooled down to −17°C in the cooling zone. The solvent condensed at the cooling zone is recovered from the outlet at the bottom. The air and inert gas are circulated to the spinning zone after passing through the reheated room that is kept at 95°C. The undrawn filaments are drawn to a draw ratio of 360% in hot water at 97.5°C. The tensile strength of the drawn fiber is about 3 g/den, and the elongation is 10%.

In the spinning of Rhovyl, the spinning chamber and the solvent recovery process form a closed system, and each spinning position is independent. Because a large quantity of carbon disulfide is used, special attention must be paid to prevent accidents such as explosion.

4.5.2.2 Other Methods of Spinning

Wet-spinning method: Wet-spinning of PVC fiber was used for the manufacture of Leavil by the Montefibre Company. According to their process, syndiotactic PVC is dissolved in cyclohexanone, and the solution is spun into a coagulating bath of a water–alcohol mixture.

Vinyl Fibers

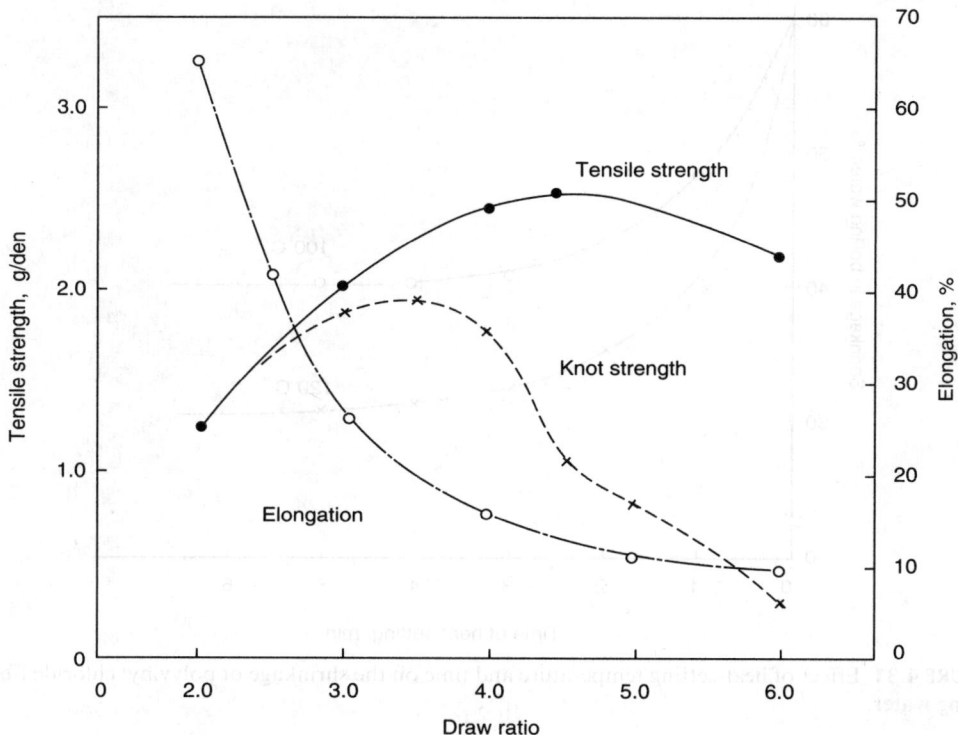

FIGURE 4.30 Tensile, knot strength, and elongation versus draw ratio of polyvinyl chloride fiber.

Melt-spinning method: The melt-spinning of PVC is conducted using an extruder [145]. As PVC easily decomposes and carbonizes when heated for a long time, it cannot be transported in a molten state. Since viscosity of molten PVC is high, gear pumps cannot be used in the melt-spinning of PVC. Also, special attention must be paid to its thermal stability and fluidity.

In the case of the melt-spinning by means of an extruder, the screw measures the amount of the outlet; however, accuracy comparable to that of methods that use gear pumps cannot be expected. Accordingly, in most cases, monofilaments are manufactured by the melt-spinning of PVC.

4.5.3 Properties of Fiber

4.5.3.1 Flame Retardance

When PVC fiber is in contact with flame, it burns with black smoke and an irritating smell; but when the flame is removed, the burning does not continue. This fiber is flame-retardant and self-extinguishing, and has a high limiting oxygen index (LOI) of 37.1 [154], and is the most difficult fiber to ignite among the various fibers shown in Table 4.29.

4.5.3.2 Chemical Resistance

Polyvinyl chloride fiber has a strong resistance to inorganic chemicals, and it swells or dissolves only in a limited number of organic liquids. Weather resistance is also high when it is compared with that of other synthetic fibers such as nylon 6 and vinylon.

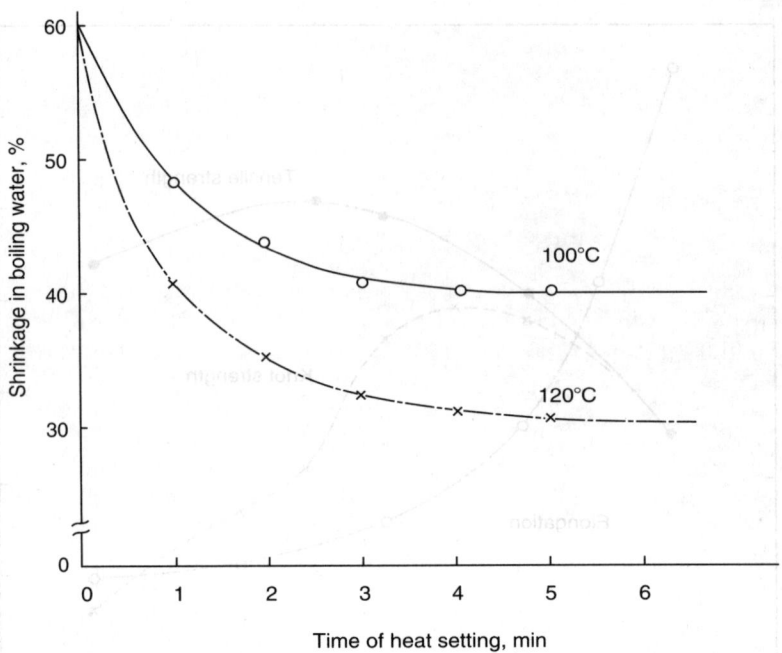

FIGURE 4.31 Effect of heat-setting temperature and time on the shrinkage of polyvinyl chloride fiber in boiling water.

4.5.3.3 Triboelectricity

A unique property of the PVC fiber is that a strong negative static charge is generated on it by frictional force. A triboelectric series is shown in Table 4.30 [151]. Polyvinyl chloride has a very large negative electric charge partly because it is a good insulator.

TABLE 4.29
Limiting Oxygen Indices for Various Fibers

Fiber	LOI
Acrylan	18.2
Arnel triacetate	18.4
Acetate	18.6
Polypropylene	18.6
Vinylon (PVA)	19.7
Rayon	19.7
Cotton	20.1
Nylon	20.1
Polyester	20.6
Wool (dry cleaned)	25.2
Dynel	26.7
Nomex N-4274	28.2
Rhovyl (PVC) "55"	37.1

Source: From Lyons, J.W., *The Chemistry and Uses of Fire Retardants*, John Wiley & Sons-Interscience, New York, 1970, p. 399.

TABLE 4.30
Triboelectric Series

Positive—Glass, Nylon 6,6, Nylon 6, Wool, Silk, Viscose staple, vinylon, Acrylan, Steel, Cotton
Negative—Orlon, Acetate rayon, Dynel, Saran, Rhovyl, Rubber

Source: From Tsuji, W. and Okada, N., *Kasen Koen Shu*, **12**, 42, 1955.

It has been confirmed that when undergarments produced from the PVC fiber are worn, the fibers become negatively charged and the concentration of calcium ions in the blood increases [153]. The fiber also has an excellent heat-retaining property. Because of these factors, it has been shown that the PVC fiber is effective for the treatment of neuralgia or rheumatism.

4.5.3.4 Dimensional Stability

Because of its low crystallinity, the PVC fiber is poor in dimensional stability at high temperature. Shrinkage occurs near the glass transition temperature (85°C), and fiber manufactured below normal heat-treatment conditions shrinks by 40–60% in boiling water. This characteristic may be a defect, but various applications are under development to make use of this property.

Staple fiber previously shrunk at high temperatures does not shrink later in boiling water. A nonshrinking staple thus produced has high elongation and a low Young's modulus.

Polyvinyl chloride fibers with improved thermal dimensional stability in boiling water can be attained by the use of a PVC, rich in syndiotacticity, and by blending with postchlorinated PVC. Further, it is possible to prepare fiber from PVC grafted with polyacrylonitrile, which shows a softening temperature much higher than 100°C.

4.6 APPLICATIONS AND FUTURE TRENDS

One of the characteristic properties of PVA fiber is its high affinity to water. Its strength, modulus, and toughness are not very different from those of any other common synthetic fibers. However, the main weaknesses of PVA fiber are that it is not thermoplastic and shows low elastic recovery, although it is superior in these areas to cotton and rayon.

On the other hand, the most characteristic property of PVC fiber is its nonflammability, which leads to its many applications. The PVC fiber is thermoplastic, highly resistant to water, and shows insulating properties to electricity and heat. Its disadvantages are its low softening temperature and low tenacity.

Polyvinyl alcohol fiber (vinylon) was commercialized in 1950 for use as a textile fiber. In the early years, vinylon was used mainly for office wear and school uniforms because of its good durability to chemicals and sunlight. With an increase in the production of polyethylene terephthalate fiber, the production of vinylon for textile decreased. The reason for the decrease is due to its poor dimensional stability after washing and its poor heat settability. The annual production of vinylon is shown in Figure 4.32. It reached maximum at 75,000 tons in 1971, decreased to about 40,000 tons in 1991, and increased to some extent (45,000 tons) in 2001. Instead of textile uses, industrial applications of vinylon have increased. The ratio of the amount of the industrial applications to total production has been over 90% in the decades from 1970 to 2000, as shown in Figure 4.33.

The spun yarns used in textile fields are prepared mainly by the cotton or rayon staple fiber-spinning systems. The spun yarns used in the industrial field are produced mainly by the

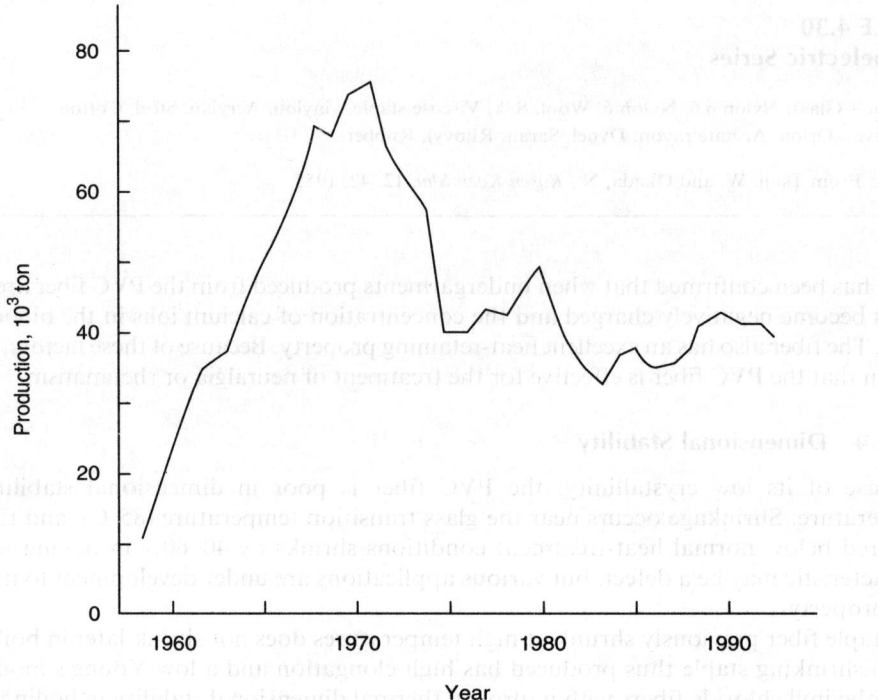

FIGURE 4.32 Annual production of vinylon.

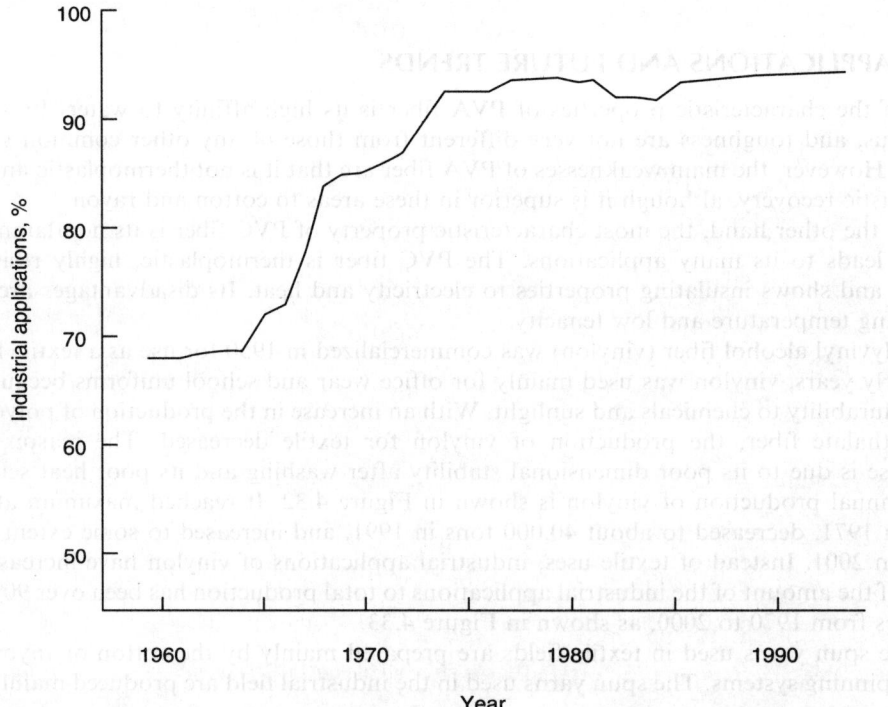

FIGURE 4.33 Change in the ratio of vinylon used for industrial applications.

Vinyl Fibers

stretch-breaking system (perlok or converter system) and partly by the cotton-spinning system. The stretch-breaking system has the advantage of making spun yarns effectively with high strength and toughness from tow, particularly in the case of vinylon.

In the industrial fields, the vinylon spun yarns prepared by the perlok spinning system are used for various purposes such as ropes, fishery materials, industrial sewing threads, bicycle tire cords, hoses (fire hose, braided hose), sheets, industrial heavy fabrics, agricultural materials, etc. The vinylon spun yarns prepared by cotton-spinning systems are used typically in the manufacture of sheets, tents, sporting goods, rubbery-soled cloth shoes, coated fabrics with polyvinyl chloride, fertilizer bags, warp yarns of tatami mats, and so on.

The industrial uses of vinylon filament yarn include laver cultivating nets, conveyor belts, radial tire cords, sewing threads for tatami mats, canvas and duck, base cloths, reinforcing fibers, hoses and so on. Its new applications are in the field of civil engineering industry as reinforcing materials (geotextiles). The water-soluble filament yarns that are prepared by special methods different from regular vinylon are used in the manufacture of base cloths for high-quality lace knitting separation yarn for hosiery fabrics and as reinforcing fibers for pile yarns of towel cloth, and so on. These are the unique applications of vinylon.

Since the completion of the first edition of *Fiber Chemistry*, remarkable progress has taken place in the industrial applications of vinylon. The health hazards of asbestos, which had been used as reinforcement materials in cement, led to the search for an alternative cement-reinforcement fiber. Among the many fibers tested, vinylon showed the most superior performance. Figure 4.34 shows the flexural strengths of the slate in the presence

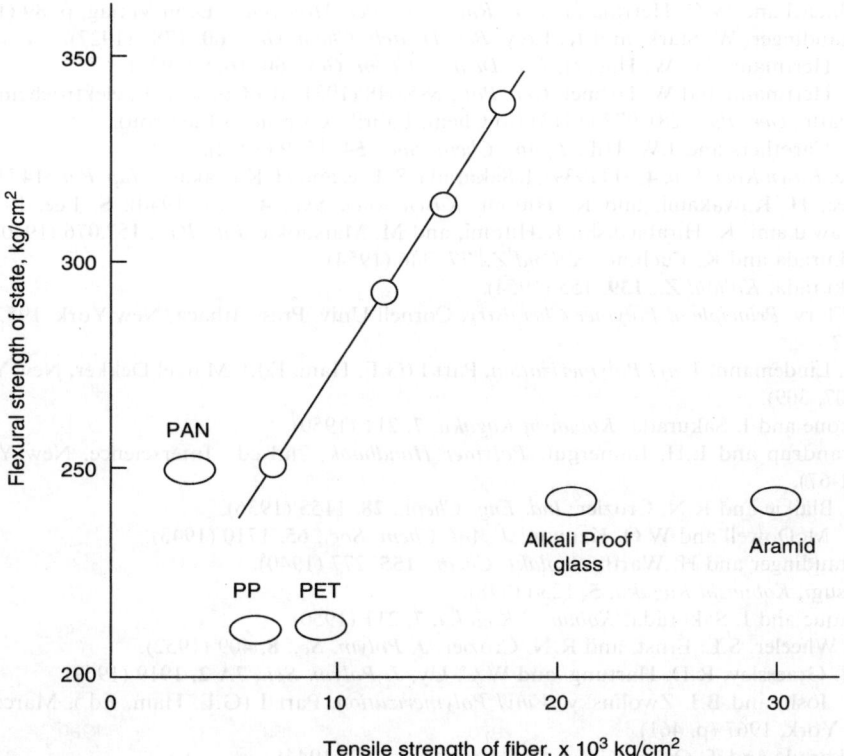

FIGURE 4.34 Relationship between tensile strength of fibers and flexural strength of slates. Fiber/cement: 3 wt%.

of various kinds of fibers. As is clear from the figure, the flexural strength of fiber-reinforced cement (FRC) using vinylon is much higher than that using other fibers such as polypropylene, polyethylene terephthalate, acrylic, alkali-proof glass, and Aramid. There is no relation between the tensile strengths of fibers and the flexural strengths of slates. The high performance endowed by vinylon is the consequence of the excellent affinity of PVA to inorganic materials like cement. The good durability of PVA to the alkali in cement is another important factor. The quantity of vinylon used in FRC application is increasing every year.

As already described in Section 4.4.2, a new series of high-performance and high-functional fibers spun from gelation-spinning have been commercialized, and new application fields using the new type of PVA fiber have been developed.

ACKNOWLEDGMENTS

The authors are grateful to Dr. S. Okamura, Dr. T. Osugi, Dr. M. Matsumoto, Dr. K. Tanabe, and Dr. T. Nakayasu for kindly furnishing research and technological data on PVA. Our thanks are extended to Dr. Sakurai and his colleagues in Teijin Co. for their very kind collaboration in preparing this manuscript on polyvinyl chloride fiber.

REFERENCES

1. F. Klatte, *Ger. Pat.*, 271,381 (1912) (to Chem. Fabrik. Griesheim Elektron).
2. E. Baum, H. Deutch, and W.O. Herrmann, *Ger. Pat.*, 485,271 (1927) (to Consort. F. elektrochem. Ind.).
3. W. Haenel and W.O. Herrmann, *Vom Ringen mit den Molekulen*, Econ-Verlag, p. 89 (1924).
4. H. Staudinger, W. Stark, and K. Frey, *Ber. Deutch. Chem. Ges.*, **60**, 1782 (1927).
5. W.O. Herrmann and W. Haenel, *Ber. Deutch. Chem. Ges.*, **60**, 1658 (1927).
6. W.O. Herrmann and W. Haenel, *Ger. Pat.*, 685,048 (1931) (to Consort. F. elektrochem. Ind.).
7. F. Klatte, *Ger. Pat.*, 281,977 (1913) (to Chem. Fabrik. Griesheim Elektron).
8. W.H. Carothers and J.W. Hill, *J. Am. Chem. Soc.*, **54**, 1579 (1932).
9. S. Lee, *Kasen Koen Shu*, **4**, 51 (1939); I. Sakurada, S. Lee, and H. Kawakami, *Jap. Pat.*, 147,958 (1939).
10. S. Lee, H. Kawakami, and K. Hitomi, *Kasen Koen Shu*, **4**, 115 (1940); S. Lee, I. Sakurada, K. Kawakami, K. Hirabayashi, K.Hitomi, and M. Matsuoka, *Jap. Pat.*, 157,076 (1940).
11. I. Sakurada and K. Fuchino, *Kolloid Z.*, **77**, 346 (1954).
12. I. Sakurada, *Kolloid Z.*, **139**, 155 (1954).
13. P.J. Flory, *Principle of Polymer Chemistry*, Cornell Univ. Press, Ithaca, New York, 1953, Chap. 4, p. 1–2.
14. M.K. Lindemann, *Vinyl Polymerization*, Part I (G.E. Ham, Ed.), Marcel Dekker, New York, 1967 (p. 207, 309).
15. R. Inoue and I. Sakurada, *Kobunshi Kagaku*, **7**, 211 (1950).
16. J. Brandrup and E.H. Immergut, *Polymer Handbook*, 2nd ed., Interscience, New York, 1975 (p. II-67).
17. K.G. Blaikie and R.N. Crozier, *Ind. Eng. Chem.*, **28**, 1155 (1936).
18. W.H. McDowell and W.O. Kenyon, *J. Am. Chem. Soc.*, **65**, 1710 (1943).
19. H. Staudinger and H. Warth, *J. plakt. Chem.*, **155**, 277 (1940).
20. T. Osugi, *Kobunshi Kagaku*, **5**, 123 (1948).
21. R. Inoue and I. Sakurada, *Kobunshi Kagaku*, **7**, 211 (1950).
22. G.L. Wheeler, S.L. Ernst, and R.N. Crozier, *J. Polym. Sc.*, **8**, 409 (1952).
23. W.W. Graessley, R.D. Hartung, and W.C. Uy, *J. Polym. Sci.*, **7A-2**, 1919 (1969).
24. R.M. Joshi and B.J. Zwolinsky, *Vinyl Polymerization*, Part I (G.E. Ham, Ed.), Marcel Dekker, New York, 1967 (p. 461).
25. I. Sakurada and T. Osugi, *Gosei–Seni-Kenkyu*, **2**, 192 (1944).

26. I. Sakurada, *J. Pure Appl. Chem.*, **16**, 263 (1968).
27. I. Sakurada and M. Fujikawa, *Kobunshi Kagaku*, **2**, 143 (1945).
28. P.J. Flory and F.S. Leutner, *J. Polym. Sci.*, **3**, 880 (1948).
29. I. Sakurada and G. Takahashi, *Kasen-Koenshu*, **14**, 37 (1957).
30. S. Amiya and M. Uetsuki, *Analytical Sci.*, **1**, 91 (1985).
31. K. Imai and O. Maeda, *Kobunshi Kagaku*, **16**, 222 (1959).
32. T. Okaya and T. Sato, *Polyvinyl Alcohol-Developments* (C.A. Finch, Ed.), John Wiley & Sons, London, 1992 (p. 105).
33. T. Sato and T. Okaya, *Makromol. Chem.*, **194**, 163 (1992).
34. T. Sato and T. Okaya, *J. Appl. Polym. Sci.*, **46**, 641 (1992).
35. T. Sato, T. Tsugaru, J. Yamauchi, and T. Okaya, *Polymer*, **33**, 5066 (1992); *ibid.*, **34**, 2659 (1993).
36. T. Okaya, H. Kohno, K. Terada, T. Sato, H. Maruyama, and J. Yamauchi, *J. Appl. Polym. Sci.*, **45**, 1127 (1992).
37. T. Sato, K. Terada, J. Yamauchi, and T. Okaya, *Makromol. Chem.*, **194**, 175 (1992).
38. K. Fujii and J. Ukida, *Makromol. Chem.*, **65**, 74 (1963).
39. K. Fujii, Y. Fujiwara, and S. Fujiwara, *Makromol. Chem.*, **89**, 278 (1965).
40. S. Murahashi, S. Nozakura, M. Sumi, H. Yuki, and K. Hatada, *Kobunshi Kagaku*, **23**, 605 (1966).
41. J.R. DeMember, H.C. Haas, and L. McDonald, *J. Polym. Sci., Polym. Lett. Ed.*, **10**, 385 (1972).
42. T. Moritani, I. Kuruma, K. Shibatani, and Y. Fujiwara, *Macromolecules*, **5**, 577 (1972).
43. T.K. Wu and D.W. Ovenall, *Macromolecules*, **6**, 240 (1973).
44. Y. Inoue, R. Chujo, and A. Nishioka, *Polym. J.*, **4**, 244 (1973).
45. K. Fujii, *J. Polym. Sci., Macro. Rev.*, **5**, 431 (1971).
46. S. Murahashi, H. Yuki, T. Sano, U. Yonemura, H. Tadokoro, and Y. Chatani, *J. Polym. Sci.*, **62**, 577 (1962).
47. H. Yuki, K. Hatada, K. Ota, I. Kinoshita, S. Murahashi, K. Ono, and Y. Ito, *J. Polym. Sci.*, **A-1**, **7**, 1517 (1969).
48. Y. Fukunishi, T. Sato, T. Okaya, T. Yamamoto, and M. Kamachi, *Report of Poval Committee*, **97**, 59 (1990).
49. S. Nozakura, M. Sumi, M. Uoi, T. Okamoto, and S. Murahashi, *J. Polym. Sci., Polym. Chem. Ed.*, **11**, 279 (1973).
50. S. Matsuzawa, K. Yamaura, and S. Amiya, Private Communication.
51. K. Imai, T. Shiomi, Y. Tezuka, T. Kawanishi, and T. Jin, *J. Polym. Sci., Polym. Chem. Ed.*, **26**, 1961 (1988).
52. M. Sumi, K. Matsumura, R. Ono, S. Nozakura, and S. Murahashi, *Kobunshi Kagaku*, **24**, 606 (1967).
53. T. Yamamoto, S. Yoda, R. Fukae, and M. Kamachi, *Polym. J.*, **21**, 1053 (1989).
54. T. Yamamoto, S. Yoda, H. Fukase, T. Sato, O. Sangen, and M. Kamachi, *Polymer. J.*, **23**, 185 (1991).
55. T. Sato, *The World of PVA*, Kobunshi Kankoukai, Kyoto, 1992 (p. 1).
56. H. Ohgi and T. Sato, *Macromolecules*, **26**, 559 (1993); *ibid.*, **32**, 2403 (1999).
57. K. Yamada, T. Nakano, and Y. Okamoto, *Proc. Japan Acad.*, **74**, Ser. B, 3, 46 (1998).
58. Y. Morishima, W. Kim, and S. Nozakura, *Polym. J.*, **8**, 196 (1976).
59. M. Kurata, M. Abe, M. Iwama, and M. Matsushima, *Polym. J.*, **3**, 729 (1972).
60. M. Kurata, H. Okamoto, M. Iwama, M. Abe, and H. Homma, *Polym. J.*, **3**, 739 (1972).
61. M. Morishima, Y. Irie, H. Iimura, and S. Nozakura, *Polym. J.*, **7**, 481 (1975); and *J. Polym. Sci., Polym. Chem. Ed.*, **14**, 1267, 1277 (1976).
62. S. Amiya, *Chem. Express*, **1**, 72 (1986); *Prog. Pacif. Polym. Sci.*, **3**, 1 (1993).
63. H. Staudinger and H. Warth, *J. prakt. Chem.*, **155**, 261 (1944).
64. I. Sakurada, *Kogyo Kagaku Zasshi*, **47**, 137 (1944).
65. A. Nakajima and K. Furudachi, *Kobunshi Kagaku*, **6**, 460 (1949).
66. M. Matsumoto and Y. Ohyanagi, *Kobunshi Kagaku*, **17**, 191 (1960).
67. G.M. Burnett, M.H. George, and H.M. Melville, *J. Polym. Sci.*, **16**, 31 (1955).
68. S. Okamura, T. Ikegami, Ko. Hayashi, and T. Natori, *Annual Rep. Japanese Assoc. Rad. Res. Polym.* **91**, 1 (1958/1959).
69. I. Sakurada, Y. Ikada, and Y. Nishizaki, *Bull. Inst. Chem. Res. Kyoto Univ.*, **48**, 1 (1970).

70. N. Fujiwara, T. Sato, K. Yuki, J. Yamauchi, and T. Okaya, *Report of Poval Committee*, **102**, 22 (1993).
71. C.W. Bunn and H.S. Peizer, *Nature*, **161**, 929 (1948).
72. I. Sakurada, K. Fuchino, and A. Okada, *Bull. Inst. Chem. Res. Kyoto Univ.*, **23**, 78 (1950); A. Okada, *Dissertation*, Kyoto Univ. (1957).
73. I. Sakurada and K. Kaji, *Makromol. Chem., Suppl.*, **1**, 599 (1975).
74. P.H. Hermans and A. Weidinger, *J. Polym. Sci.*, **4**, 135, 217 (1949); *ibid.*, **5**, 269, 515 (1950).
75. I. Sakurada, Y. Nukushina, and N. Mori, *Kobunshi Kagaku*, **12**, 302, 307, 311 (1955).
76. T. Sato, and T. Okaya, *Polym. J.*, **24**, 849 (1992).
77. P.J. Flory, *J. Chem. Phys.*, **17**, 233 (1949).
78. F. Hamada and A. Nakajima, *Kobunshi Kagaku*, **23**, 395 (1966).
79. R.K. Tubbs and T.K. Wu, *Polyvinyl Alcohol* (C.A. Finch, Ed.), John Wiley and Sons, London, 1973 (p. 169).
80. T.G Fox, *Bull. Am. Phys. Soc.*, **1**, No. 3, 123 (1956).
81. T. Inoue, Private Communication.
82. I. Sakurada, *Kobunshi Kagaku*, **28**, 316 (1979).
83. G.M. Bryant and A.T. Walter, *Tex. Res. J.*, **29**, 211 (1959).
84. Y. Sone and I. Sakurada, *Kobunshi Kagaku*, **14**, 574 (1957).
85. I. Sakurada, Y. Nukushina, and Y. Sone, *Kobunshi Kagaku*, **14**, 506 (1957).
86. W. Tsuji, *Rept. of the Second Symp. PVA*, p. 100 (1958).
87. Y. Sone and I. Sakurada, *Kobunshi Kagaku*, **14**, 145 (1957); I. Sakurada and Y. Nukushina, *Kobunshi Kagaku*, **11**, 472 (1954).
88. G. Takahashi and I. Sakurada, *Kobunshi Kagaku*, **13**, 502 (1956).
89. I. Sakurada, Y. Sone, and Y. Nukushina, *Kobunshi Kagaku*, **12**, 517 (1955).
90. T. Matsuo and H. Inagaki, *Makromol. Chem.*, **55**, 150 (1962).
91. K. Noma, T. Wo, and T. Tsuneda, *Kobunshi Kagaku*, **6**, 439 (1949).
92. P.J. Flory, *J. Am. Chem. Soc.*, **61**, 1705 (1939).
93. I. Sakurada, Y. Sakaguchi, and Y. Ohmura, *Kobunshi Kagaku*, **21**, 564 (1964).
94. C.A. Finch, *Polyvinyl Alcohol* (C.A. Finch, Ed.), John Wiley & Sons, London, 1973 (p. 183).
95. S.G. Cohen, H.C. Haas, and Slotonick, *J. Polym. Sci.*, **11**, 193 (1953).
96. M. Hida, *Kogyo Kagaku Zasshi*, **55**, 221, 275 (1952).
97. K. Noma and N. Sawagashira, *Kobunshi Kagaku*, **4**, 46 (1947).
98. R.K. Tubbs and T.K. Wu, *Polyvinyl Alcohol* (C.A. Finch, Ed.), John Wiley & Sons, London, 1973 (p. 189).
99. J.G. Pritchard, *Polyvinyl Alcohol: Basic Properties and Uses*, Gordon and Breach, London, 1970 (p. 73).
100. Y. Ikada, *Adv. Polym. Sci.*, **29**, 47 (1978).
101. K. Noma, *Kobunshi Kagaku*, **5**, 190 (1948).
102. Y. Tsuchiya and K. Sumi, *J. Polym. Sci.*, **A-1**, **7**, 3151 (1969).
103. I. Sakurada and H. Matsuzawa, *Kobunshi Kagaku*, **16**, 633 (1959).
104. I. Sakurada and H. Matsuzawa, *Kobunshi Kagaku*, **18**, 252 (1961).
105. I. Sakurada and H. Matsuzawa, *Kobunshi Kagaku*, **16**, 565 (1959).
106. I. Sakurada and H. Matsuzawa, *Kobunshi Kagaku*, **18**, 257 (1959).
107. I. Sakurada and H. Matsuzawa, *Kobunshi Kagaku*, **17**, 268 (1960).
108. I. Sakurada and H. Matsuzawa, *Kobunshi Kagaku*, **20**, 349, 353 (1963).
109. S. Lee, *Kobunshi Kagaku*, **2**, 175 (1945).
110. T. Tomonari, T. Akahoshi, M. Nagai, and T. Osugi, *Japan Pat.*, 188,756 (1951) (to Kurashiki Rayon).
111. E. Nagai, *Kasen Koensyu*, **5**, 5 (1940).
112. T. Osugi, T. Tanabe, T, Suda, T. Morimoto, and S. Miyazaki, *Sen-i Gakkaishi*, **15**, 630 (1959).
113. H. Kawakami, N. Mori, H. Sato, and A. Miyoshi, *Sen-i Gakkaishi*, **16**, 155 (1960).
114. I. Sakurada and N. Nakamura, *Kobunshi Kagaku*, **8**, 476, 480, 484, 539 (1951).
115. I. Sakurada, N. Mori, T. Tanaka, K. Sakurai, and O. Yasutake, *Sen-i Gakkaishi*, **11**, 526 (1955).
116. H. Suyama and E. Kozuki, *Kasen Koenshu*, **6**, 245 (1941).

117. E.T. Cleine, P.S. Pinkney, L. Plambeck, and H.B. Stevenson, *U.S. Pat.* 2,610,300 (appl. 1950) (to Du Pont).
118. K. Tanabe and K. Ono, *Sen-i Gakkaishi*, **12**, 258 (1956).
119. K. Tanabe and K. Ono, *Sen-i Gakkaishi*, **12**, 543 (1956).
120. I. Sakurada, A. Yamamoto, O. Yasutake, and W. Tsuji, *Kasen Koensyu*, **10**, 55 (1954).
121. H. Kawakami and H. Sato, *Sen-i Gakkaishi*, **18**, 183 (1962).
122. K. Kawashima and A. Miyoshi, *Japan Pat.*, 47–8186 (appl. 1968) (to Unitica).
123. T. Ashikaga and T. Kousaka, *U.S. Pat.*, 3,660,556 (appl. 1969) (to Kurashiki Rayon).
124. S. Nakajo and E. Morita, *Japan Pat.*, 36–13112 (appl. 1958) (to Kurashiki Rayon).
125. M. Uzumaki and E. Shimoda, *Sen-i Gakkaishi*, **18**, 397 (1962).
126. K. Tsuboi and T. Mochizuki, *J. Polym. Sci., Polym. Lett.*, **1**, 531 (1963).
127. T. Ashikaga, T. Endo, and H. Kurashige, *Japan Pat.*, 546,553 (appl. 1965).
128. K. Matsubayashi and H. Segawa, *Japan Pat.*, 43–16675 (appl. 1964).
129. H. Narukawa and H. Noguchi, *Sen-i Gakkaishi*, **46**, 466 (1990).
130. M. Mochizuki, *The World of PVA*, Kobunshi Kankoukai, Kyoto, 1992 (p. 63).
131. A. Ohmori, *Report of Poval Committee*, **113**, 65 (1998).
132. A. Ohmori, I. Sakuragi, and M. Onodera, *Sen-i Gakkaishi*, **55**, 418 (1999).
133. Y. Nishikoji, H. Takeshita, K. Nishida, T. Kanaya, and K. Kaji, *Report of Poval Committee*, **113**, 1 (1998).
134. M.M. Zwick, *Appl. Polym. Symposia*, **6**, 109 (1967).
135. H. Kawakami, *Report of the First Symp. PVA*, 1956 (p. 131).
136. K. Yamaguchi and M. Amagasa, *Japan Pat.*, 26–572 (appl. 1949).
137. H.F. Mark and S.M. Atlas, *Man-Made Fibers*, Vol. 1 (H.F. Mark, S.M. Atlas, and E. Cerni, Eds.), Interscience, New York, 1967 (p. 237).
138. W. Tsuji, R. Kitamaru, T. Ochi, and N. Mori, *Kasen Koen Shu*, **16**, 23 (1959).
139. R. Kitamaru, T. Ochi, H. Koh, and W. Tsuji, *Kasen Koen Shu*, **17**, 39 (1960).
140. S. Okamura, T. Yamashita, and H. Asakura, *Japan Pat. Appl.*, 37–12920; H. Asakura, *Gosei Sen-i* (I. Sakurada, H. Sobue, and M. Kushi, Eds.), Asakura-Shoten, Tokyo, 1964 (p. 353).
141. *Society of Chemical Fibers [Japan], Hand Book of Chemical Fibers*, Kasen-Kyokai, Osaka, 1981 (p. 308, 310, 313).
142. V. Regnault, *Ann. Chim. Physique*, **69**, 151 (1838).
143. F. Klatte, *Ger. Pat.*, 281,977 (1913), (to Chem. Fabrik Griesheim-Elektron).
144. H. Rein and F. Davidshoefer, *Chemische Textilfasern Filme and Folien* (R. Pummerer, Ed.), 1952 (p. 583).
145. F. Fourne, *Synthetische Fasern* (F. Fourne, Ed.), Wissenschaftliche, 1964 (p. 174).
146. Rhone-Poulenc, *Fr. Pat.*, 969,363, and *Fr. Pat.*, 980,874.
147. Toyo Soda, *Japan Pat.*, 12602, (1965).
148. Kureha Kagaku, *Japan Pat.*, 7568, (1960).
149. Kureha Kagaku, *Japan Pat.*, 28305, (1964).
150. P. Doty and H.S. Zable, *J. Polym. Sci.*, **1**, 90 (1946).
151. J. Shimeha, *Gosei Sen-i* (I. Sakurada et al., Eds.), Asakura Shoten, Tokyo, 1964 (p. 366).
152. Teijin, *Japan Pat.*, 13008, (1965).
153. L. Cord, *Man-Made Fibers* (H.F. Mark, S.M. Atlas, and E. Cerina, Eds.), Interscience, New York, 1968 (p. 327).
154. J.W. Lyons, *The Chemistry and Uses of Fire Retardants*, John Wiley & Sons-Interscience, New York, 1970 (p. 399).
155. W. Tsuji and N. Okada, *Kasen Koen Shu*, **12**, 42 (1955).

5 Wool and Related Mammalian Fibers

Leslie N. Jones, Donald E. Rivett, and Daryl J. Tucker

CONTENTS

5.1 Background Reading 332
5.2 Fiber Structure and Composition 333
    5.2.1 Introduction 333
    5.2.2 Fiber Cuticle 335
        5.2.2.1 Cellular Features, Function, and Formation 335
        5.2.2.2 Fiber Surface 336
        5.2.2.3 Proteinaceous Resistant Barriers....................................340
        5.2.2.4 Endocuticle...341
        5.2.2.5 Cuticle Cell Isolation and Characterization 342
    5.2.3 Fiber Cortex...342
        5.2.3.1 Cortical Cell Structure, Isolation, and Formation 342
        5.2.3.2 Cortical Cell Types and Ultrastructure 342
        5.2.3.3 Hard Keratin Intermediate Filaments (Microfibrils) 343
        5.2.3.4 Matrix Structure and IF–Matrix Interactions................346
    5.2.4 Medulla..347
    5.2.5 Cell Membrane Complex 348
        5.2.5.1 Definition 348
        5.2.5.2 Staining, Ultrastructure, and Composition 348
        5.2.5.3 Resistant Membranes of the CMC................................349
    5.2.6 Protein Composition of Mammalian Fibers 351
        5.2.6.1 Protein Classes 351
        5.2.6.2 Extraction of Proteins 351
        5.2.6.3 Low-Sulfur Proteins (Properties of SCMKA and Subunits)...........352
        5.2.6.4 Sequence Studies of SCMKA Subunit Polypeptides.....353
        5.2.6.5 Formation of IF Structural Units 353
        5.2.6.6 High-Sulfur Proteins: Isolation and Characterization.....353
        5.2.6.7 Amino Acid Sequences of High-Sulfur Proteins 354
        5.2.6.8 Structure and Conformation of High-Sulfur Proteins.....354
        5.2.6.9 Ultrahigh Sulfur Proteins 355
        5.2.6.10 High-Glycine Tyrosine Proteins 355
    5.2.7 Other Hard Keratin Proteins 356
5.3 Chemical Reactions of Wool 356
    5.3.1 Introduction 356
    5.3.2 Reduction...357

5.3.3 Oxidation ... 358
5.3.4 Alkali ... 359
5.3.5 Acid ... 360
5.3.6 Water, Steam ... 360
5.3.7 Further Reactions of Amino Acid Side-Chains ... 361
    5.3.7.1 Alkylation ... 361
    5.3.7.2 Arylation ... 361
    5.3.7.3 Esterification ... 362
    5.3.7.4 Acylation ... 362
    5.3.7.5 Cross-Linking ... 362
5.3.8 Polymer Applications ... 362
    5.3.8.1 Surface Polymers for Shrink-Proofing ... 362
    5.3.8.2 Internal Deposition of Polymers ... 363
5.3.9 Photodegradation ... 363
5.3.10 Effect of Heat: Flame-Proofing ... 363
5.3.11 Moth-Proofing ... 364
5.3.12 Setting ... 364
5.4 Load–Extension Curve: Conformation of the Protein Chains in Wool at Different Extensions ... 365
  5.4.1 Load–Extension Curve ... 365
  5.4.2 Change in Protein Chain Conformation as Wool is Extended ... 366
  5.4.3 Comparison of the Load–Extension Curves and the X-Ray Diffraction Patterns at Different Extensions ... 367
  5.4.4 Load–Extension Curves Under Different Conditions ... 368
  5.4.5 Effect of Chemical Treatments of Wool on the Load–Extension Curve ... 368
5.5 Sorption ... 368
  5.5.1 Absorption and Desorption Isotherms: Hysteresis ... 368
  5.5.2 Chemical Modification and Moisture Absorption ... 370
5.6 Other Mammalian Fibers ... 370
  5.6.1 Chemical Composition ... 370
  5.6.2 Fiber Structure ... 372
References ... 373

5.1 BACKGROUND READING

A detailed information on wool structure, chemistry, and technology are covered in the published proceedings of eight quinquennial international wool textile research conferences: Australia, 1955; Harrogate, U.K., 1960; Paris, 1965; San Francisco, 1970; Aachen, Germany, 1975; Pretoria, 1980; Tokyo, 1985; and Christchurch, New Zealand, 1990 [1–8]. The proceedings of the Harrogate and San Francisco conferences were published, respectively, in volume 51 of the *Journal of the Textile Institute* (December 1960, number 12) and in Volume 18 of the *Journal of Polymer Science, Applied Polymer Symposia* (1971, parts 1 and 2). The proceedings of the other conferences were published by the organizers of the conferences. Crewther et al. [9] and Bradbury [10] published two important reviews of the investigations of wool structure using biophysical and biochemical techniques in 1965 and 1973; the first came from the CSIRO Division of Protein Chemistry in Parkville, Australia, and the second from the Australian National University in Canberra. *Keratins*, authored by Fraser et al. [11] in 1972, describes the composition, structure, and biosynthesis of epidermal and fibrous keratins. *The Chemistry of Natural Protein Fibres* by Asquith [12] includes chapters on the structure and chemical reactions of wool and other animal fibers. Maclaren and Milligan deal with the chemical reactivity of wool and its relevance to wool technology in their 1981 publication, *Wool Science* [13]. *Wool, Nature's Wonder Fibre*, by Leeder [14], is an easy-to-read reference for the nonspecialist, dealing with wool structure and properties. *The Mechanics of Wool Structure* by Postle et al. [15], deals with the physical properties of wool. *Wool Science Review* is published several times a year by the International Wool Secretariat; its authoritative articles are usually about wool science as related to technology. Two recent reviews by Leeder [16] and Rivett [17] discuss the increasingly important area of the surface and cellular membrane structure of the wool fiber. The most recent review of wool structure, by Rippon, appeared in 1992 in *Wool Dyeing* [18].

The publications mentioned above deal directly with wool; however, there are several books, reviews, and symposia proceedings relating to human hair and speciality fibers, which also contain material relevant to wool structure. Of particular note are the proceedings of the two specialty animal fiber symposia, which were published in Aachen, Germany, in 1987 [19] and 1989 [20], and L. Hunter's *Mohair: A Review of Its Properties, Processing and Application* in 1993 [21].

5.2 FIBER STRUCTURE AND COMPOSITION

5.2.1 INTRODUCTION

Mammalian fibers exhibit wide variations in physical characteristics such as diameter (10–250 μm) and length [22]. These characteristics vary with the species and are influenced by the nutritional and metabolic state of the animal. In addition, fiber length is often dependent on the duration of the anagen or actively growing phase of the follicle. Other hair-fiber characteristics that can vary widely include pigmentation, transverse shape, and general surface contours. It should also be noted that changes in pigmentation, diameter, shape, and scale patterns can occur within a single fiber.

Wool and other mammalian fibers consist of cells, where outer, flattened overlapping *cuticle cells* form a protective sheath around *cortical cells*. The cortical cells in turn surround central vacuolated medullary cells when these are present in coarser fibers.

In fine wools, such as those obtained from Merino sheep, the cuticle is normally one cell thick and usually constitutes about 10% by weight of the total fiber. By contrast, human hair cuticle may contain up to 10 layers of cells and pig bristle cuticle, about 35 layers. Sections of cuticle cells show an internal series of laminations comprising outer sulfur-rich bands known as the *exocuticle* and inner regions of lower sulfur content called the *endocuticle*. On the exposed surface of cuticle cells, a membranelike proteinaceous band (epicuticle) and a unique lipid component form a resistant barrier. These moieties are the functional components of the fiber surface and are significant in fiber protection and textile processing.

The cortex comprises the main bulk and determines many mechanical properties of the mammalian fibers. Cortical cells are long, polyhedral, and spindle-shaped and consist of intermediate filaments (microfibrils) embedded in a sulfur-rich matrix. The filament–matrix texture is organized into larger macrofibrillar units, and these are often observed in cortical cells. In fine Merino wools, for example, the main cortices (ortho and para) are arranged bilaterally (Figure 5.1), and orthocortical cells show different filament–matrix packing arrangements to paracortical cells. The terms *ortho* and *para* are used to describe the cellular texture in the cortex of fine wool fibers. Consequently, it is not strictly correct to apply these terms to the cortical cells of other mammalian fibers such as human hair or cashmere. The arrangement of cell types in fine wools is often thought to affect crimp characteristics.

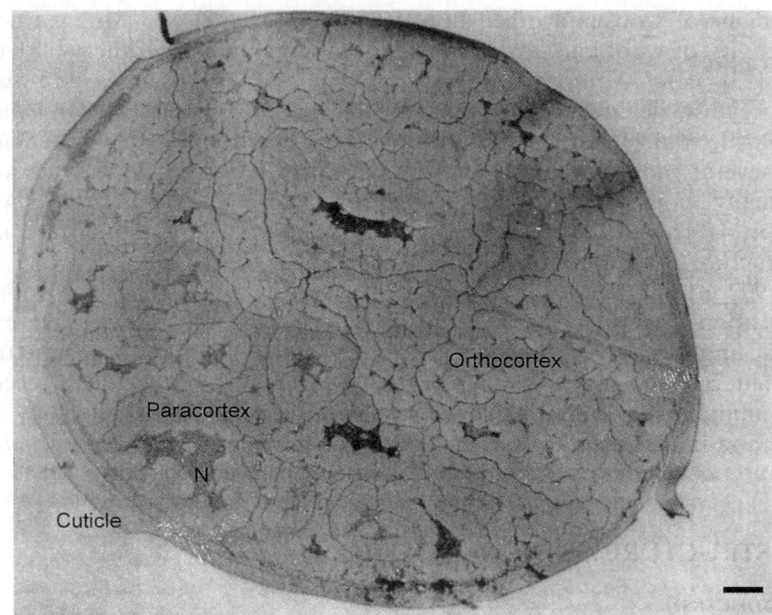

FIGURE 5.1 Transmission electron micrograph showing a transverse section of a Merino wool fiber stained with osmium tetroxide after prior reduction with thioglycolic acid. At low magnification, the outer flattened cuticle cells are observed surrounding cortical cells. Paracortical and orthocortical cells of the cortex show different morphological and staining characteristics and are arranged into roughly bilateral segments. The densely stained material most evident in paracortical cells arises from residual cell nuclei. Bar equals 1 μm.

Between cuticle cells and cortical cells is a continuous intercellular material, which, despite being a relatively minor fraction of the total fiber weight, is a material of increasing interest due to its presumed role in water and reagent penetration of fibers (Figure 5.2). The intercellular material and the apposing cellular "membranes" of cuticle and cortical cells comprise the cell membrane complex of fibers.

The structural components of mammalian fibers result from the complex differentiation processes arising from living germinal epithelial cells located at the base of follicles (Figure 5.3). During differentiation of these cells, the main events involved in fiber formation include synthesis of proteins and other chemical components, assembly of these components into the various structural moieties, and, finally, stabilization of the assembled complexes. Follicle studies have recently assumed increased importance by focusing on the various structural components that can be isolated prior to cross-link formation. Earlier attempts to isolate structural components from fibers were hindered by the need to use potentially damaging reagents. It must be remembered that, although providing an alternative to mature fibers, follicles have complex cellular structures and may be difficult to obtain in suitable quantities for structural and compositional studies.

The major chemical components of mammalian fibers are complex mixtures of proteins, but lipids and carbohydrates are also present in significant quantities and play an important role in fiber function.

Further reading on the structure, composition, and formation of mammalian fibers can be found in the above-mentioned book by Fraser et al. [11] and in the reviews by Leeder [16], Rivett [17], and Rippon [18]. Other reviews are by Montagna and Parakkal [23], Rogers and

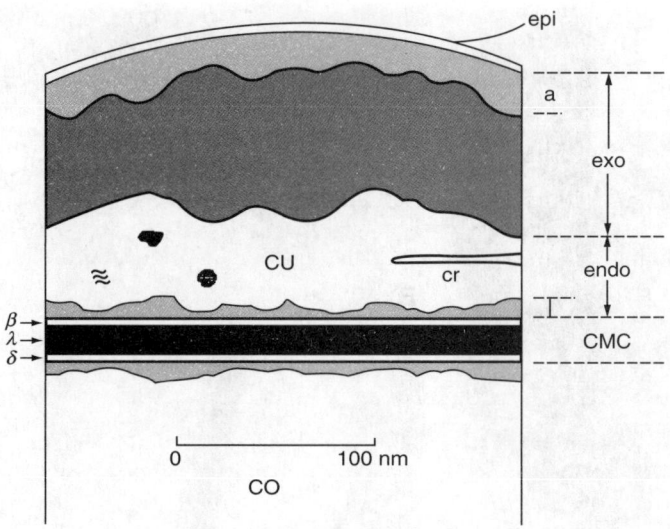

FIGURE 5.2 Diagram showing the main structural features of mature fiber cuticle cells. Cuticle cells (CU) possess an outer surface resistant membrane (epi) and an underlying exocuticle (exo) and its surface associated a-layer (a), which together comprise the main structural barrier components of the fiber surface. Cytoplasmic remnants (Cr) are often observed in the inner endocuticle (endo) layers of cuticle cells. The darker regions (exo, a) are richer in sulfur than the endocuticle. The CMC consists of unstained β-layers (membranes) and a darkly stained intercellular material (δ-layer). An intracellular layer (λ) is associated with the β-layers and plays an important role in reagent penetration of cells. Cortical cells (CO) occupy the fiber bulk and lie within the outer flattened cuticle cells. Cortical cells have a filament–matrix texture when viewed at high magnifications.

Harding [24], Lindley [25], Swift [26], Zahn [27], Orwin [28,29], Fraser and MacRae [30,31], Fraser et al. [32], Speakman [33], and Marshall et al. [34].

5.2.2 Fiber Cuticle

5.2.2.1 Cellular Features, Function, and Formation

Flattened cuticle cells primarily account for most of the surface properties of mammalian fibers. In Merino wool, cuticle cells are approximately $20\,\mu m \times 30\,\mu m \times 0.7\,\mu m$ [35]. Cuticle cells overlap both longitudinally and circumferentially with exposed lips or scale edges pointing toward the distal end of the fiber. The scale edges are thought to aid in the removal of dirt and vegetable matter, but they may also assist in anchoring the fiber to the skin [36].

During their passage up the follicle shaft, presumptive cuticle cells undergo flattening and are organized into an overlapping arrangement [37–39]. While undertaking this process, they become interlocked with apposed cells comprising the cuticle layer of the inner root sheath. Cuticle protein synthesis is also initiated, and these proteins form into a complex arrangement lining the outer cell boundary (exocuticle). An inner cell lining called the endocuticle is also formed [40].

In the mature fiber, cuticle cells essentially perform a protective function; however, they also play an important role in controlling the ingress of water and other chemical substances [16,41]. Hence a detailed understanding of the cuticle structure and reactivity is essential to provide a basis for improving many of the textile properties of wool and other fibers. The nature of the fiber cuticle surface in particular is a high priority of current research activities.

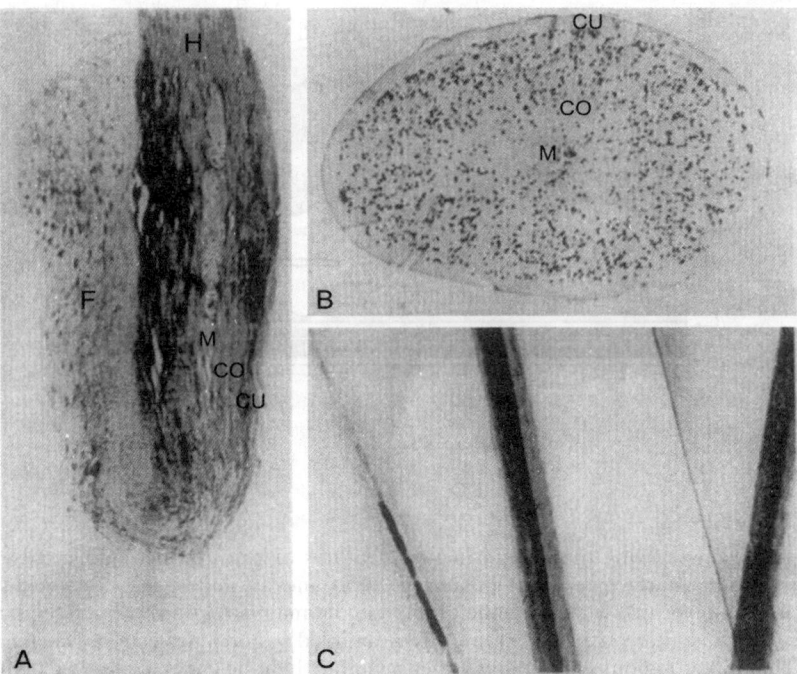

FIGURE 5.3 (A) Longitudinal section of human hair follicle where the emerging hair (H) is derived from developing component cells, cuticle (CU), cortex (CO), and medulla (M). × 150. (B) Transverse section of a human hair fiber. Note the relatively thick cuticle (CU), which surrounds the cortex (CO) and a central medulla (M). The dark structure existing in the cortex are pigment (melanin) granules. × 400. (C) Four human hairs from the same individual demonstrating variation in pigmentation and medulation. × 100. (A) and (B) were stained with toluidine blue.

5.2.2.2 Fiber Surface

Most of our knowledge of cuticle-surface structure and ultrastructure has been derived from the application of physical techniques such as scanning electron microscopy (SEM), transmission electron microscopy (TEM), x-ray photoelectron spectroscopy (XPS), secondary ion mass spectroscopy (SIMS), and infrared spectroscopy (IR). In addition, recent biochemical studies have provided the essential chemical data (lipids and proteins) required for modeling the surface structure [42,43].

External surface features of fibers such as contours, defects and damage, chemical treatments, and polymer coatings are normally observed in the SEM (Figure 5.4). Recent developments in atomic force microscopy (AFM) should expand the potential to provide ultrahigh-resolution data about surface features at the atomic and the molecular levels. Thus, a complementary array of techniques is now available for detailed surface studies.

When ultrathin sections are positively stained and examined by TEM, the fiber cuticle ultrastructure is shown to consist of a laminated arrangement of variously stained components [11,28,44]. At the outermost region of each cuticle cell, a nonstained thin band has been observed after specialized staining procedures [45] or by techniques coating the surface of this band [46, 47]. Estimates of the surface-layer depth have been the subject of many studies and still remain controversial. Despite the wide range in estimates, it is generally agreed that the

Wool and Related Mammalian Fibers

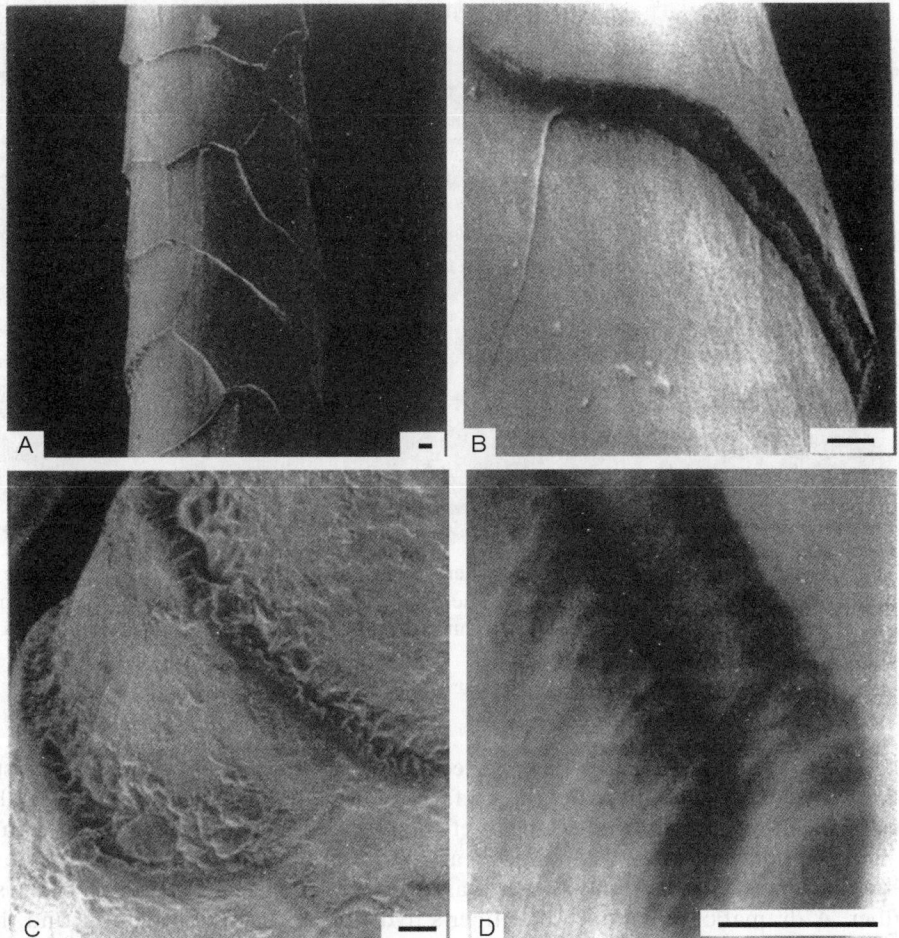

FIGURE 5.4 Scanning electron micrographs showing the surfaces of wool fibers. (A, B, and D) Untreated. (C) After Chlorine-Hercosett treatment.

thickness of this surface layer is between 2 and 7 nm [46,48,49], and Jones, unpublished (1994). The surface membrane plays an important role in wettability, friction, and surface tension in spite of being a relatively minor proportion of the fiber.

The cuticle surface membrane contains a lipid and a proteinaceous component; the term *epicuticle* is sometimes used to describe this surface membrane (Figure 5.5). However, it must be remembered that the epicuticle was originally defined as the membrane raised from the fiber surface. It is highly resistant and is raised as bubbles or sacs from the underlying material after treatment with chlorine and water [50]. After isolation by agitation, as mentioned above, the epicuticle was shown to be predominantly proteinaceous [51]. Its similarity to cuticle composition led King and Bradbury to suggest that the isolated membrane consisted of multilayers derived from the fiber surface [49]. These authors also found lipids to be associated with the epicuticle.

It has long been postulated that the hydrophobic nature of the surface of the wool fiber could be explained if there existed a lipid substance covalently bound to the surface of cuticle

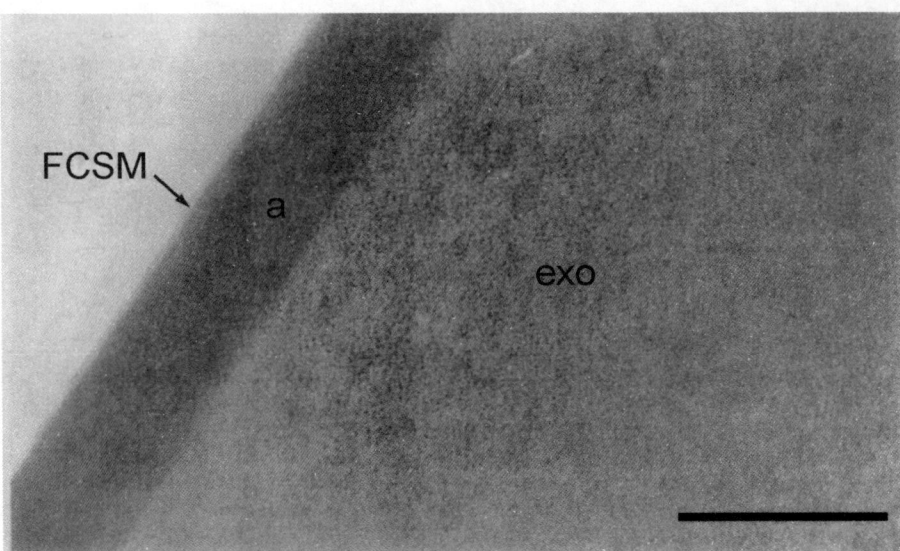

FIGURE 5.5 High magnification TEM of the fiber cuticle surface as it appears in transverse section after bulk fiber staining with osmium tetroxide. The fiber cuticle surface membrane (FCSM) comprising lipids (fatty acids) and proteins is approximately 6 nm thick and is apposed to the a-layer (a) of exocuticle (exo). Bar equals 0.1 μm.

cells [52,53]. Kopke and Nilssen [54] suggested that the surface of wool fibers was studded with long-chain fatty acids linked to the surface by ester linkages. They reasoned that the existence of such linkages would explain the effect observed when wool is treated with alkali. Leeder and Rippon [55] treated wool with anhydrous potassium tertiary butoxide in the nonswelling solvent tertiary butanol, which was assumed to confine the reaction to the surface of the fiber. A dramatic reduction in the fiber's hydrophobicity was observed, and this was attributed to the removal of the postulated lipid layer from the fiber surface, which they named the F-layer. Evans et al. [56] identified the major lipid component, removed by the alkaline treatment, as a methyl-branched 21-carbon fatty acid and suggested that it was bound to the protein of the fiber surface by an ester or thioester bond, as originally proposed by Kopke and Nilssen. Subsequent research reported the finding of this unusual fatty acid in human hair and other mammalian fibers [57–59]. In studies using mass spectrometry and nuclear magnetic resonance spectroscopy (NMR), the fatty acid was conclusively identified as 18-methyleicosanoic acid [57,60].

Proof that the ester-linked fatty acids were predominantly (possibly even exclusively) confined to the cuticle was provided by studies of covalently bound fatty acids from isolated cuticle cells [61,62] and by TEM [63]. The culmination of the chemical research was the presentation of a model for the cuticle cell envelope, which proposed that the cuticle–air interface consisted of long-chain fatty acids covalently linked as thioesters to a heavily cross-linked protein membrane, forming a hydrophobic barrier at the surface of each cuticle cell [42] (Figure 5.6).

A confirmation of the above hypothesis has been provided by static secondary ion mass spectrometry (SSIMS) and x-ray photoelectron spectroscopy (XPS) [64–67]. Ward et al. [67] used atomic ratios from XPS to estimate the thickness of the surface lipid layer as 0.9 nm, which is less than half of that estimated by Negri et al. [42] from the length of a 20-carbon

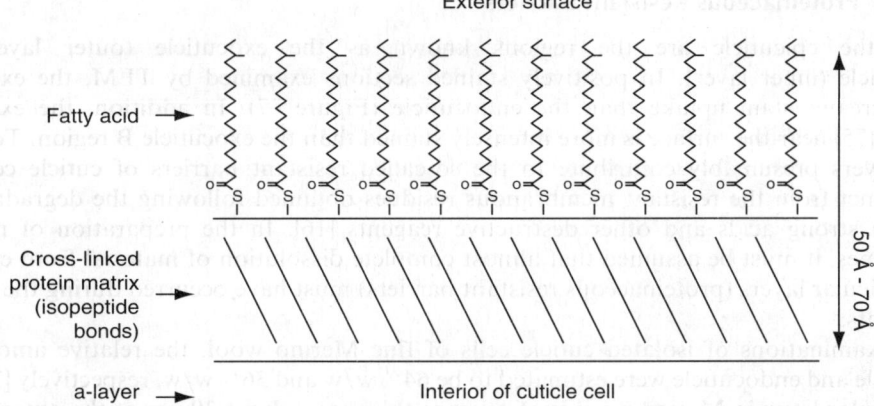

FIGURE 5.6 Model structure proposed for the fiber cuticle surface membrane. The exterior surface of the cuticle consists of a monolayer comprising C_{21} branched-chain fatty acids linked covalently via thioester bonds to the proteins comprising the Allwörden-induced resistant membrane (epicuticle). The epicuticle consists of an inert protein matrix containing isopeptide cross-links and is linked via an unknown mechanism to the underlying a-layer of the exocuticle.

chain. Zahn et al. [43] suggested that this apparent anomaly could be accommodated if the fatty acids are folded back in the direction of the surface, while Peet et al. [68] suggested that the apparent contradiction in the thickness of the lipid layer may be an artifact of the anhydrous, high-vacuum conditions used in the XPS.

The suggestion that the wool fiber surface could be different in vacuum conditions compared to other environments (e.g., in a liquid) may be well founded. Horr [69] has recently studied contact angles and surface energy values previously determined for wool [71] as they related to the surface components that exist on wool such as methyl-, methylene-, keratin, and absorbed vapors. He has suggested that the outermost region on the wool surface cannot consist of methyl groups exclusively as had been indicated in Negri et al.'s model [42].

Owing to the method of contact-angle or surface-energy measurement, the surface of wool necessarily includes the region between cuticle cells in addition to the cuticle itself. Horr has further suggested that vapor adsorption due to capillary condensation may occur at the fiber cuticle scale edges, and that the phenomenon may contribute to the above interpretation that the wool surface is not entirely methyl. Horr also found that the possible composition of the wool fiber surface may even vary depending on the liquid with which it is in contact (e.g., water or methylene iodide).

The cuticle surface membrane of the fiber has recently been shown to be derived partly or completely from intercellular laminae, forming in the intercellular region between apposed fiber-cuticle- and inner root-sheath-cuticle cells [41,72]. In these studies, the original plasma membranes of fiber-cuticle cells were observed to be disrupted during the formation of the underlying exocuticular bands. Earlier suggestions that the fiber cuticle surface membrane [73,74] was derived from a modified plasma membrane were considered unlikely since the original plasma membrane was adapted to function in a physiological environment. The formation of a specialized fiber-cuticle surface membrane, as observed by Jones et al. [41], more likely accounts for the surface properties and for the protective function required by the exposed surface membrane of the mammalian fibers.

5.2.2.3 Proteinaceous Resistant Barriers

Below the epicuticle are the regions known as the exocuticle (outer layer) and endocuticle (inner layer). In positively stained sections examined by TEM, the exocuticle shows greater stain uptake than the endocuticle (Figure 5.7). In addition, the exocuticle A band [75] near the surface is more intensely stained than the exocuticle B region. Together, these layers presumably contribute to the so-called resistant barriers of cuticle cells and are distinct from the resistant membranous residues obtained following the degradation of fibers in strong acids and other destructive reagents [16]. In the preparation of resistant membranes, it must be assumed that almost complete dissolution of material from exo- and endocuticular layers (proteinaceous resistant barriers) must have occurred during the various treatments.

In examinations of isolated cuticle cells of fine Merino wool, the relative amounts of exocuticle and endocuticle were estimated to be 64% w/w and 36% w/w, respectively [76]. The exocuticular layer in Merino wool is of average thickness, about 30 nm at the outer surface region, but appears to continue as a markedly thinner band forming an envelope on the underside of the cuticle cells [11].

A number of studies have suggested that based on their relative affinity for heavy-metal stains, the exocuticle A (a-layer) band is richer in cystine than is the underlying exocuticle B

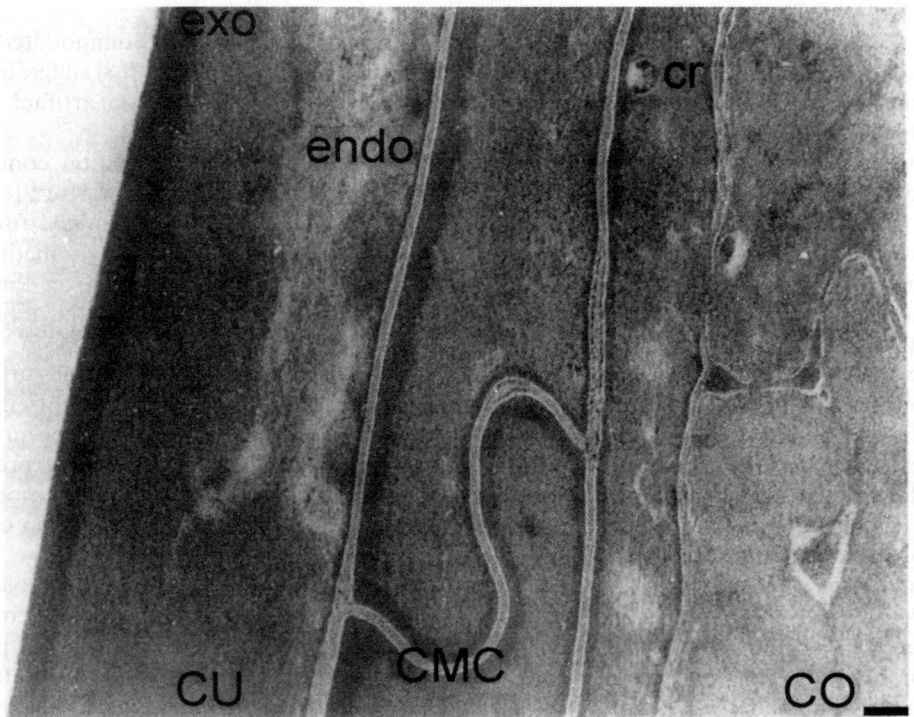

FIGURE 5.7 Transmission electron micrograph showing fiber-cuticle cell (CU) ultrastructure in a transverse section of human hair as it apposes the fiber cortex (CO). Regions associated with the cuticle include the exocuticle (exo) including its densely stained outer band or a-layer (a), the endocuticle (endo), cytoplasmic remnants (Cr), and the CMC. Specimen was reduced and stained with osmium tetroxide, uranyl acetate, and lead citrate. Bar equals 0.1 μm.

band (as it is sometimes known). Prior reduction of fibers appears to enhance heavy-metal uptake by proteins in the a-layer, suggesting that the cleavage of disulfide bonds enhances reaction with osmium [45,77,78]. The problems with osmium and other heavy-metal stains, such as silver, is that chemical specificity can only be assumed, and other factors, such as accessibility, must be considered [79].

More recent developments in microanalytical TEM have provided direct evidence showing the location and relative distribution of sulfur-rich regions in the exocuticular A and B layers of fibers [80–83]. The exocuticle A was shown to contain relatively higher sulfur contents than the exocuticle B and endocuticle (Figure 5.8). Apart from the disulfide bonding in the exocuticle layers, chemical analyses of these structures show relatively high contents of lysine and glutamic acid, suggesting the presence of isopeptide cross-links [ε-γ(glutamyl)-lysine] [84,85]. These types of cross-links are known to exist in various epithelial cells with a particularly insoluble protein complex, such as that found in the hair medulla, inner root sheath, and the envelope proteins of stratum corneum [85–88].

5.2.2.4 Endocuticle

The endocuticle appears to be derived from the developing cuticle cell cytoplasm, nucleus, and cell organelles. This region differs from exocuticle with respect to uptake of heavy-metal stains and susceptibility to attack by proteolytic enzymes [37,79]; as such, the endocuticle is considered to be one of the weakest and most accessible parts of keratin fibers [89]. In high-resolution TEM studies, no ultrastructure has been observed in either endocuticle or exocuticle layers [11]. In various textile processes, the endocuticle may exhibit greater uptake of dyes and other reagents than exocuticular regions. A recent study using microanalytical methods and TEM showed the passage of dyes through the cuticle [82] and clearly demonstrated that dyes were preferentially located in the endocuticle.

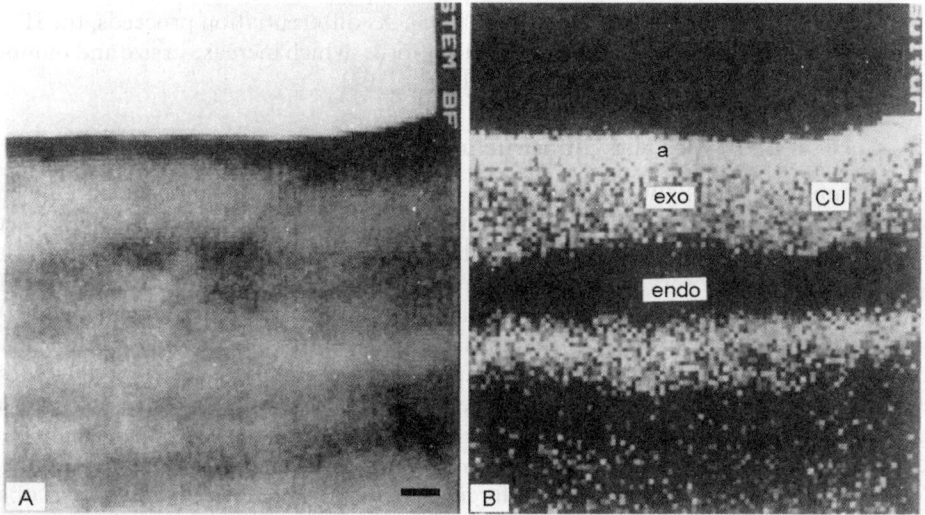

FIGURE 5.8 Bright field (STEM BF) (A) and sulfur x-ray map of the wool fiber cuticle surface (B). In (B), the relative sulfur contents are highest in the outer bands of the cuticle (CU) known as the exocuticle (a, exo). The endocuticle (endo) appears to be devoid of sulfur. Bar equals 0.1 μm.

5.2.2.5 Cuticle Cell Isolation and Characterization

Methods for isolating cuticle cells from wool fibers have been described by Ley and Crewther [90]. After using alkaline reducing conditions for the extraction of cuticle cell polypeptides, they were able to solubilize about 30% w/w of the cells and suggested that the aforementioned isopeptide bonds would account for low yields. In amino acid and electrophoretic studies, these isolated polypeptides were cystine-rich but clearly different from cortical cell high-sulfur proteins [91].

5.2.3 Fiber Cortex

5.2.3.1 Cortical Cell Structure, Isolation, and Formation

Cortical cells comprise the bulk of mammalian fibers and play an important role in their mechanical and physical behavior [11]. In the fiber, these highly elongated cells are aligned in the same direction as the fiber's longitudinal axis [45,92]. They have been isolated by means of enzymes [93] or by treatments with weak acids [94–97]. Newer developments have used a fluorescence cell sorter for separating ortho- and paracortical cell preparations [98]. The length of cortical cells isolated from mammalian fibers are variable, but measurements show they are usually in the range of 80–115 μm [99,100].

Cortical cells form from a central or annular stream of germinal epithelia. During their early differentiation, they undergo elongation and alignment processes. The initial sites for the formation of keratin structural components (arising concurrently with cell elongation) are at the outer boundaries (plasma membranes) of cortical cells, often in association with desmosomes [101,102]. At high magnification, these initial keratin structures appear to consist of 7–8 nm intermediate filaments (IFs) that show appreciable stain uptake when viewed after application of the heavy-metal preparations commonly used for contrast enhancement in TEM studies [102]. After the formation and packing of IFs into lattice structures, a densely stained proteinaceous material appears to occupy the interfilamentous spaces. This proteinaceous material forms a matrix, which is generally considered to be composed of the high-sulfur and high-tyrosine proteins. The formation of these various structural components from their constituent proteins follows this two-stage sequential differentiation [103]. As differentiation proceeds, the IF matrix aggregates to form structures often known as macrofibrils, which increase in size and number and begin to occupy the bulk of presumptive cortical cells [104].

5.2.3.2 Cortical Cell Types and Ultrastructure

In many mammalian fibers, two or more types of cortical cell are distinguished on the basis of relative dye uptake by the internal components of these cells. When Merino wool sections were stained with methylene blue [105,106], bilateral or semicircular arrangements of the two cortical cells types were observed. The well-stained cells were referred to as orthocortical cells; those cells exhibiting relatively little stain uptake were called paracortical cells [106,107]. In later studies, a third cell type (mesocortical cell), intermediate to ortho and paracortical cells in terms of dye uptake, was identified in Merino wool sections [45,89,92,108,109]. Mesocortical cells comprise a relatively minor component between the two segments. The cortical cell arrangements and their staining characteristics in coarser fibers such as human hair are different from those of wool, hence the terms *ortho* and *para* may not be strictly appropriate to all cases. In sections of human hair, differences between cortical cells have been observed [110,111]. More recently, Jones and Pope [112] demonstrated at least two distinct cell types when examining sections of human hair presumptive cortex, and these were described as ortholike and paralike.

The arrangement of cortical cells is highly variable among species, but, as a general rule, finer fibers (less than about 25 μm) tend to exhibit bilateral symmetry [28,113]. In fine wools, the characteristic crimp may be influenced by cortical cell arrangements [105,106,109] where the bilateral symmetry results in the orthocortex following the outer track of the crimp curl. Compared with the bilateral arrangements of ortho-, and mesocortical cells in high-crimp wools, Kaplin and Whiteley [109] have observed a tendency toward ortho–para bilateral symmetry in low-crimp wools. However, relationships among the chemical composition of ortho-, para-, and mesocortical cells, their packing arrangements, and the propensity of fibers for crimp formation are complex, and definitive conclusions are presently not available [22]. In wools where crimp loss is induced by factors such as copper deficiency the normal ortho–para bilateral arrangement was unaffected (Jones, 1986, unpublished). Orwin et al. [114] reached a similar conclusion in studies of Corriedale and Romney wools where fibers were found not conforming to the correlation between bilateral arrangement and crimp.

Fine structure in mature fiber sections was revealed by Rogers [45,92], who combined partial reduction using thioglycolic acid with prolonged staining in osmium tetroxide. Rogers confirmed the filament–matrix texture observed in follicle sections of mammalian fibers [101]. On the basis of high osmium affinity for the interfilament matrix, especially after prior reduction, Rogers put forward the suggestion that the matrix was richer in sulfur (high-sulfur proteins) than the filament component. Other heavy-metal stains demonstrating a high affinity for matrix proteins include silver nitrate [77,78], silver nitrate and silver methenamine [115], and potassium permanganate [116]. The combination of stains developed by Rogers and Filshie [116]—which is now in common usage—involves prior partial reduction of fibers using thioglycolic acid and subsequent bulk fiber staining with osmium tetroxide. Fiber sections are then stained with uranyl acetate followed by lead hydroxide.

The use of these heavy-metal stains has led to other observations concerning the IF–matrix packing arrangements within the various cortical cells. Characteristic whorl-like patterns of filaments appear in orthocortical cells, while paracortical cells exhibit a pseudo-hexagonal filament packing arrangement. Mesocortical cells demonstrate an intermediate stain uptake at the cellular level and, ultrastructurally, the filament patterns are highly ordered in a hexagonal arrangement [109] (Figure 5.9).

Associated with differences in filament arrangements between cortical cell types, variability in the amount of interfilament matrix material is also evident. The matrix component appears to be in greater concentration and more intensely stained in paracortical cells than in orthocortical cells after reduction and osmication of fibers [11,117]. In support of a higher matrix content in paracortex, Kulkarni et al. [93] found a higher proportion of high-sulfur proteins in isolated paracortical cells. Direct evidence for a higher sulfur content in the paracortex than in the orthocortex of Merino wool fiber sections was recently found by Jones et al. [80], who used x-ray mapping techniques in conjunction with TEM (Figure 5.10).

The filament–matrix complexes of mammalian fibers, such as those found in para- and orthocortical cells of wool, appear to exist in characteristically higher ordered structures that are generally referred to as macrofibrils [45,92,101]. In paracortical cells of Merino wool, macrofibrils are not as clearly defined as are the discrete macrofibrillar bundles evident in orthocortical cells [118]. Macrofibrils exhibit wide variability in physical dimensions and appearance among fiber sections from different species.

5.2.3.3 Hard Keratin Intermediate Filaments (Microfibrils)

In modern usage, the observed filaments in fiber sections are strictly called hard keratin IFs. Traditionally, these IFs—known as microfibrils—have probably received more intensive study over the years than other fiber structural components. The range of the various

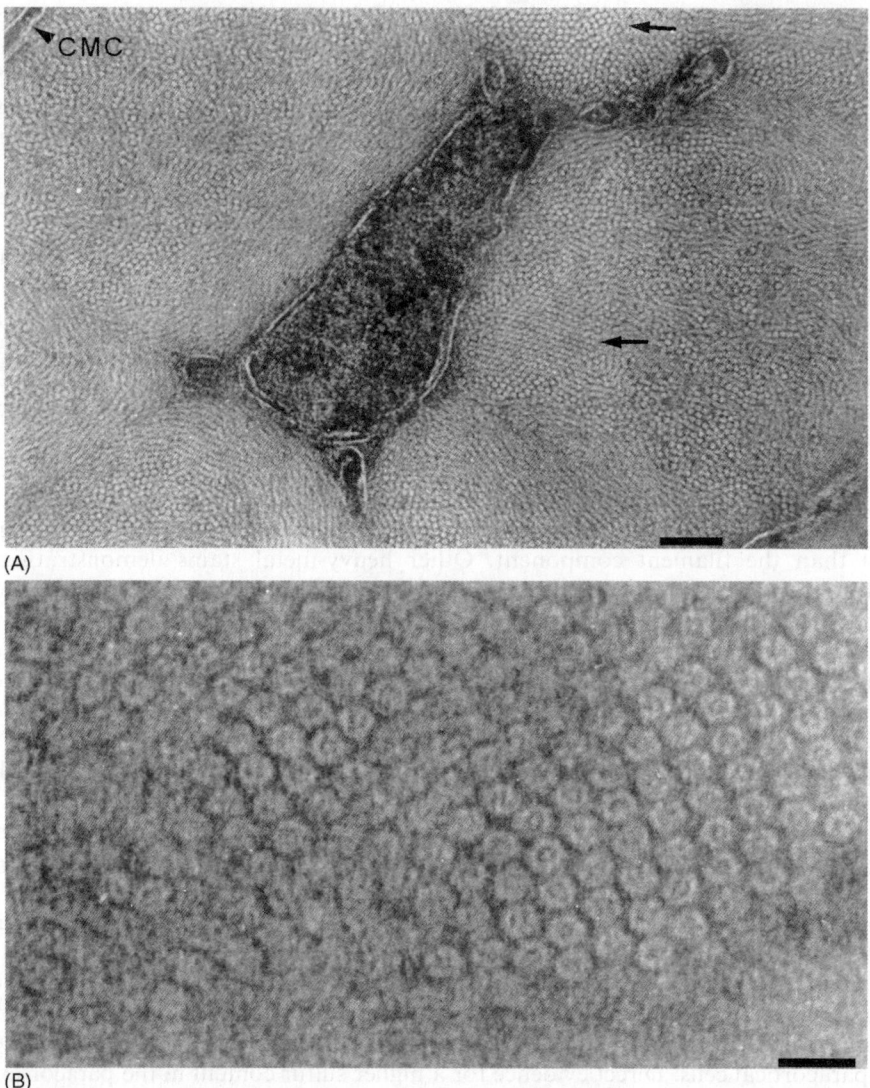

FIGURE 5.9 (A) Ultrastructure of a human hair cortical cell (TEM) in transverse section. Hard keratin IFs (unstained) are embedded in an interfilamentous dark stained matrix. The filament arrangements in human hair are variable but, in some regions, ordered packing is observed (arrowed). The CMC consists of a densely stained intercellular band (δ-layer) surrounded by unstained β-layers. Densely stained material in the center of the micrograph probably comprises remnants of cytoplasmic or nuclear material from cell development processes. Bar equals 0.1 μm. (B) High magnification of intermediate filaments packed into a hexagonal array. This type of packing is often found in transverse sections of paracortical and mesocortical cells of Merino wool. The ultrafine structure of hard keratin IFs (microfibrils) consists of an unstained outer ring, a stained annular zone, and, often, an unstained central core. The diameter of microfibrils is approximately 7 nm and these are surrounded by the dark stained matrix material. In some regions, protrusions appear from microfibril surfaces suggesting the hexagonal lattice may be stabilized by interfilamentous linkages. Bar equals 0.01 μm. In (A) and (B), sections were stained with osmium tetroxide, uranyl acetate, and lead citrate after preliminary reduction with thioglycolic acid. (From I.J. Kaplin and K.H. Whiteley, *Aust. J. Biol. Sci.*, *31*, 231 (1978).)

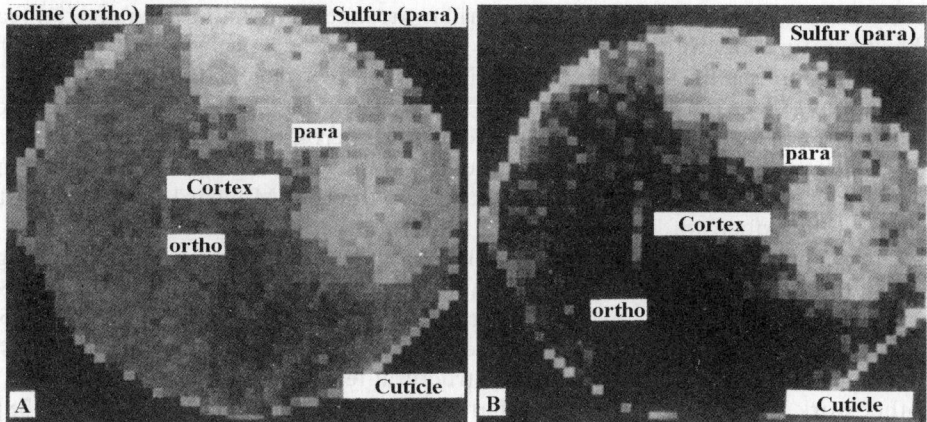

FIGURE 5.10 X-ray maps of wool fiber sections show the relative distributions of (A) iodine and (B) sulfur in cortical and cuticle cells. In A and B, a high sulfur content is present in cells of the paracortex and the cuticle, while in A, iodine is predominant in the orthocortex. Iodine was used to label the high-tyrosine proteins, which appear to be present mostly in the orthocortex. (From L.N. Jones, M. Cholewa, I.J. Kaplin, G.J. F. Legge, and R.W. Ollerhead, *Proc. Eighth Int. Wool Text. Res. Conf.*, Christchurch, *1*, 246 (1990) and L.N. Jones, I.J. Kaplin, and G.J.F. Legge, *J. Computer Assisted Microsc.*, *5*, 85 (1993).)

estimates of the diameters of IFs in sectioned material viewed by TEM was in the range of 7–10 nm [11]. This wide range of results was complicated by the differences in interfilamentous spacing [45,92,117]. From complementary data obtained from studies using a variety of physical techniques [119–121], the diameter generally agreed on for hard keratin IFs is 7–7.5 nm, and this diameter is considered constant across mammalian species.

Hard keratin IFs give rise to an α-type x-ray diffraction pattern [122] similar to the patterns obtained from other fibrous proteins, such as those in muscle, elastin, and fibrin, and are grouped with that class of ubiquitous biological filaments known as the intermediate filaments. Intermediate filaments are widely distributed in eukaryotic cells and thought to perform a mechanical role in the cytoskeletal structure of these cells [123]. The hard keratin IFs have structural and chemical features similar to those of other IFs such as desmin, neurofilaments, vimentin, together with the wide range of IFs found in epidermal cells [124]. The structure and assembly mechanisms of IFs found in hair and wool are considered to be the most complex [112], and attempts to reassemble fully integrated IFs *in vitro* generally have proved unsuccessful.

The early ultrastructural studies of transverse sections of wool and human hair (reduced and osmicated) stained with heavy metals [45,92,125] suggested that IFs consisted of protofibrils—nine peripheral and two central. This interpretation of TEM images of protofibrillar arrangement started the so-called 9 + 2 protofibril controversy. At that time, the equatorial features of x-ray patterns [126] were judged to be consistent with the TEM observations. Protofibrils (2.0 nm in diameter) were considered to contain coiled coils of component α-helices [127,128], with the α-helix part of the individual protein molecules comprising IFs. It was subsequently suggested that coiled-coil units in IFs consisted of two or three protein chains [126]. The protofibril theory was convenient for explaining hard keratin IF structure, especially in view of the ubiquity of such a structure in other biological filaments. Further data in support of the protofibril ultrastructure was put forward by Wilson [129], who calculated Fourier transforms from different combinations of coiled coils accounting for the equatorial x-ray diffraction patterns. However, Fraser et al. [130] subsequently pointed to a relatively poor agreement between protofibrillar arrangements and the equatorial features of the hard keratin x-ray diffraction patterns.

A critical appraisal of variation within individual IF images was subsequently made by Fraser and Millward [131]. Their studies involved using image averaging to reduce the random effects of noise (phase contrast) in electron micrographs. Averaged images of IF transverse sections indicated an unstained outer ring, an annular stained ring, and an unstained central core. In micrographs of hard keratin IFs, Millward [132] demonstrated that increased underfocus of TEM images [133,134] caused increased granularity. This grain structure in the images was due to phase contrast effects [134], which modified images of keratin IFs and had led to the earlier spurious ultrastructural interpretation of IFs. Subsequent studies have used direct imaging electron diffraction of stained specimens [135] and x-ray diffraction of both stained and unstained specimens [136,137], and these studies have indicated that the ring and core are genuine structural features of hard keratin IFs.

In other work relating to the protofibrillar concept in keratin filaments, Dobb [138] used ultrasonication techniques to isolate filaments with diameters of 2.0–2.5 nm from chemically treated keratinous materials. In negatively stained TEM preparations, he further observed an axial repeat of 20-nm spacing in the filaments. Filaments with diameters of approximately 2 nm were also observed by Rogers and Clark [139] and Johnson and Speakman [140,141] in similar preparations.

Questions concerning the origin of these filaments were raised by Millward [142], who demonstrated that cellulosic filaments from plant and animal sources were morphologically indistinguishable from the keratin preparations. The controversy continued [143,144], but serious doubts about the origin of protofibrils were emerging when demonstrations of ultrastronicated laboratory paper tissue preparations yielded similar structures. The question of extraneous artifacts arising in TEM preparations of dispersed specimens needed to be addressed [145]. In subsequent studies, methodology to demonstrate possible contamination and the origin of monodispersed filaments was developed [120,146]. The method devised by Jones [120] involved the isolation of presumptive cortical cells and their subsequent rupture in hypotonic media. By the use of these nondegradative methods he was able to demonstrate definitively the origin of filaments in negatively stained dispersates, and this provided a basis for direct structural and chemical studies of hard keratin IFs [112,121].

Hard keratin IFs have been isolated from the presumptive hair shafts of rat vibrissae [120,121] and human hair [112]. After purification, these filaments in negatively stained preparations were used in high-resolution TEM studies [121]. By means of digital processing, images of individual IFs (full focal series) that had been negatively stained with potassium phosphotungstate were reconstructed, and evidence for a ring ultrastructure was obtained. This internal structure was consistent with the ring-core substructure observed by Millward [132]. Owing to the masking of the core by the ring moiety in these procedures, the unstained core was not observed in the transverse sections. This same procedure was later used by Steinert et al. [147] to demonstrate similar ultrastructural features in reconstituted epidermal and other IFs. Preparations of IFs isolated from developing human hair cortical cells have demonstrated a 22-nm periodicity in individual IFs and in tactoidlike assemblies of these IFs [112]. In the assemblies, IFs appear to associate in a juxtaposed arrangement with the helical domains of IF molecules in register. Accumulations of nonhelical tails in IF assemblies are probably involved in forming interfilamentous bridges between IFs [112].

5.2.3.4 Matrix Structure and IF–Matrix Interactions

In transverse sections of hard keratins, the interfilamentous matrix component appears to be relatively amorphous. A few reports of ordered structures in the matrix have been made [11], but only limited knowledge about the arrangement of polypeptide chains is available [148]. An important observation made in porcupine quill during hydration was that swellings of IFs and matrix moieties were 6 and 13%, respectively [149]. The actual volume of swelling was calculated as 11% for IF and 53% for matrix, which was consistent with the observed mechanical behavior of fibers in wet and dry states [150]. In accounting for the high matrix swelling and other mechanical properties, Crewther and Dowling concluded that disulfide bonding must be largely intramolecular [151].

Since the amount of matrix is variable among hard keratins from fibers of different species [152], it is difficult to explain the modes of assembly in developing cortical cells. As described earlier, the IF moiety initially assembles into an observable orderly arrangement. In the two-stage synthetic model, insertion of matrix proteins in the interfilamentous spaces requires consideration of high-sulfur protein synthesis outside developing IF arrays and the insertion of these proteins into interfilamentous spaces without disrupting the order of the arrays [153]. One possible explanation is that IF arrays are stabilized by linkages between certain low-sulfur proteins protruding from surfaces of juxtaposed IFs [31]. An alternative explanation is that a linking or key high-sulfur protein may stabilize IF arrays [32]. Proposals such as these would not only account for variability of composition and volume of matrix proteins but also for the mechanical behavior of keratinous materials in the wet and dry states [154,155]. The direct imaging of interfilamentous linkages between IFs has so far proved elusive, and a special methodology is required in specimen preparative and staining methods. The recent development of energy filtering [156 and 157] in TEM instruments and its applications to studies of fiber sections should prove to be a valuable technique for these studies.

Evidence to support the existence of interfilamentous linkages recently arose from the work by Jones and Pope [112], where direct visualisation of a 22-nm axial periodicity in negatively stained assemblies of developing IF arrays was observed. In these observations, IFs appeared to be aligned in lateral register, presumably via accumulations of nonhelical "tails" occurring periodically along the surfaces of IFs. These nonhelical tails might perform specialized functions [158], for instance, in the assembly of IF lattices associated with developing cortical cells and the subsequent IF–matrix interactions [112]. The available knowledge of matrix proteins suggests that, together with the IF phase, they play a dynamic role in contributing to a fiber's mechanical and swelling properties. Clearly, the importance of associations between IF and matrix proteins warrants further investigation.

5.2.4 Medulla

A central stream of cells interspersed with vacuoles is termed the medulla, although this medullary cell-type is not present in all mammalian fibers. In fine wools and in human hairs, the medulla is usually absent; it is more prevalent in coarser fibers [11]. The characteristic features of developing medullary cells in the follicle are circular, densely stained granules (medullary granules) that form internal coatings within the membranes of mature cells [159]. In the development of medullary cells, certain arginine residues are converted to citrulline, and isopeptide bonds [ε-γ(glutamyl)lysine] [160] are formed.

In some fibers, very few medullary cells are present and the medullary canal appears as a well-defined air-filled space. Brunner and Comans have classified the medulla on the basis of its absence or interruption [161]. Alternatively, in the forensic sciences, Clement et al. have classified human medullae as either intermediate or fragmental [162]. The latter used the ratio of medulla diameter to hair diameter to obtain the medullary index (MD/HD = MI).

The ultrastructure of material in medullary cells of most mammalian fiber sections appears to be amorphous; however, Clement et al. have claimed to observe a filamentous substructure in human hair medulla, but the possibility of this substructure due to juxtaposed cortical cell remnants needs to be eliminated.

Medullary cells are resistant to attack by normal reagents used for solubilizing keratins [163]. Methods used for the isolation of medulla are based on its inherently inert nature. Thus, medullary isolates result as a residue after preferential dissolution of cuticle and cortical cells [160].

5.2.5 Cell Membrane Complex

5.2.5.1 Definition

In mature mammalian fibers, there are intercellular connections between the main cell types including cuticle–cuticle, cuticle–cortical, cortical–cortical, and, in some fibers, cortical–medullary [32]. Together, these intercellular regions constitute a continuous phase in the fiber and, as such, play an important mechanical and chemical role in fiber function [16]. There is no universal agreement on the question of which structural components comprise the so-called cell membrane complex (CMC) [11,16]. However, the modified plasma membranes (β-layers) and the intercellular material (δ-layers), which are formed in the developing fiber shaft, are considered to be its main components [11,16,26,28].

Based on the early models of Swift and Holmes [46], Leeder described the three major components of the CMC [16]. The intercellular material (δ-layer) is composed mainly of proteinaceous material with low cross-link density. The β-layers are assumed to consist of lipids, possibly as bilayers coupled with inert proteinous (resistant membranes) outer boundaries. Some confusion arises as to whether an additional associated intracellular membrane band (i-layer) should be considered as part of the CMC [11,32]. Although the intracellular band is usually considered as an internal part of the cell, it appears to play an important role in the stabilization of the CMC. A detailed understanding of this i-layer as well as of the other regions of the CMC is important for understanding the transport processes of chemical reagents into keratinized cells.

5.2.5.2 Staining, Ultrastructure, and Composition

In TEM observations of fiber sections stained with heavy metals, the δ-layer typically averages about 15 nm in width and is densely stained, especially between cortical cells [45,92]. The β-layers (approximately 5 nm thick) are usually unstained, although the very thin outer bands do stain in the follicle stage but not in the mature fiber [137]. The intracellular i-band is densely stained and appears similar in some respects to the a-layer of cuticle cells [79].

The staining properties of the δ-layers between apposed cortical cells and apposed cuticle cells are different from each other [164,165]. Studies demonstrate that this difference is enhanced after fibers are treated with formic acid [166,167]. Other studies have indicated differences in chemical composition between cortical–cortical CMC and cuticle–cuticle CMC after histochemical staining [165,168] or through the effects of treatments (with enzymes and reducing agents) used in cell-separation experiments [169,170]. In addition, it has been shown that cuticle CMC is more resistant to modification after formic acid treatment; it has also been shown to contain citrulline [166].

When wool fibers were extracted by means of formic acid and enzymes, partial dissolution of the CMC was assumed to occur [93,97,171]. It is well known that formic acid treatment causes extensive swelling of fibers and subsequent release of lipids and proteins that account for about 1 to 2% of the fiber weight [97]. To support the contention that these chemical moieties were preferentially extracted from the CMC, dispersions of cuticle and cortical cells were obtained after formic acid–utrasonication treatment [97]. Proteins rich in glycine, tyrosine, and phenylalanine obtained after reduction and formamide treatment of wool were shown by TEM to be derived from the CMC [172]. Other agents that appear to specifically disrupt the CMC were dichloroacetic acid [95,173], chloroform, and methanol–water [167]. The wide variation in chemical composition of CMC extracts obtained in various experiments indicates that certain chemical procedures may preferentially remove material from other regions of the fiber, such as the cortical and cuticle cells. Despite the variations in CMC compositions, there has been general agreement about the relatively low levels of cystine, which in turn indicates low contents of disulfide bonding. Hence it is possible to provide a chemical basis for the swelling and accessibility properties of the intercellular region (δ-layer) of the CMC. Consequently, the CMC is often described as nonkeratinous [174,175]. The exact composition of the δ-layer is not known, but it most likely consists of complex mixtures of proteins and lipids.

The lipid moiety found in CMC extracts is thought to arise from the β-layers since these structures appear inert to heavy-metal stains in both mature and developing fiber [28,41,45]. The evidence that the β layer lipids are arranged as bilayers arose from the appearance of preferred fission planes, which appear at the presumed nonpolar junctions after mechanical stress [16].

Recent structural studies have shown that the δ-layer in the CMC often appears to have a laminated ultrastructure. After polymer grafting into thioglycolic-acid-reduced hair [176], the cuticle δ-layer indicated a nonstaining central band surrounded by two densely stained bands. After staining with osmium, the δ-layers in the cuticle CMC of human hair and developing wool fiber cuticle cells have also shown a series of laminations [41,72,177] (Figure 5.11).

Ultrastructure in the δ-layer between cortical cells is not normally observed, except after certain chemical treatments, as described above. Lamina in the δ-layer have been observed after tertiary butoxide (anhydrous) treatment [63], suggesting that the action of this reagent may not be confined to the surface [16]. Hence, it appears that conventional heavy-metal stains used in TEM mask the ultrastructure in CMC components. The recent use of energy-filtered TEM in studies of the CMC has resulted in high contrast and high definition of fine detail after modified staining procedures [41,72]. In these studies, the intercellular material between apposed cuticle cells was seen to consist of stained laminae with a nonstained central band. These intercellular structures are sandwiched between two well-defined nonstained layers derived from the original plasma membranes. It is important to continue to use these new developments in energy filtering with TEM since it is normally difficult to observe the ultrafine structure of the CMC in conventional specimens and TEM instruments.

5.2.5.3 Resistant Membranes of the CMC

By definition, the so-called membranes obtained from the CMC result after the treatment of wool fibers with degradative reagents such as acids, alkalis, proteolytic enzymes, and reducing and oxidizing agents [16,84,178,179]. Hence the resistant membranes of cortical and cuticle cells must be regarded as residues from these treatments or combinations of these treatments. Factors such as preferential extraction during the various treatments and the different experimental parameters of different studies have obviously led to variations in the descriptions of resistant membrane compositions.

The resistant membranes in the CMC originate particularly from intracellular-membrane-associated regions, which form an envelope lining of cortical and cuticle cells [16]. An important difference between resistant membranes in cuticle and cortical cells is the presence of citrulline and ornithine in the cuticle membranes [166]. Some of the common features in the

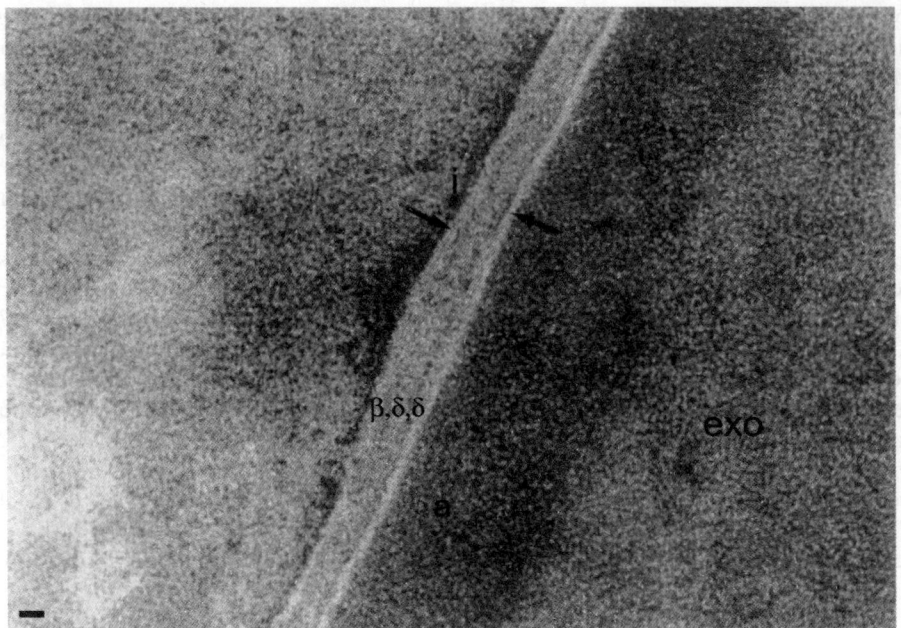

FIGURE 5.11 Transmission electron micrograph of stained transverse section showing the ultrafine structure of the CMC of apposed fiber cuticle cells in human hair. The CMC comprises two unstained modified membranes (β-layers) and an intercellular δ-layer of higher staining intensity. Note also the internal laminae in the δ-layer with thickness approximately 0.005 μm. An intracellular membrane associated layer (i) forms a narrow band on the underside of a fiber-cuticle cell. The dark stained band (a) is the surface a-layer of exocuticle (exo) from an underlying cuticle cell. The section was stained with osmium tetroxide, uranyl acetate, and lead citrate after prior reduction with thioglycolic acid. Bar equals 0.1 μm.

composition of resistant membrane isolates are that cystine contents are similar to those of wool and the amounts of lysine and glutamic acids are elevated [16].

A similar intracellular envelope-like layer has been reported lining the cells of the stratum corneum and shown by Jessen to be rich in cysteine [86]. This inert layer has been called, variously, keratolinin [87] and involucrin [88]. Keratolinin contains proteins cross-linked by intermolecular ε-(γ-glutamyl)-lysine bonds (glu-lys) of the type first identified by Pisano et al. [180].

Similar bonding has been identified in hard keratin fibers [181,182], but it is yet to be established definitively whether the cross-link exists in membrane-associated layers. However, the high contents of glutamic acid and lysine found in resistant membrane preparations of wool would suggest that isopeptide bonds (glu-lys bonds) exist in these membranes.

Previous studies have provided evidence for the existence of glu-lys bonding in the exocuticle, endocuticle, and the CMC [183,184]. In a related study, Röper et al. [185] investigated the presence of glu-lys bonds in membrane proteins but found little evidence of them in enzymatic hydrolysates of wool.

A calcium-dependent transglutaminase existing in the inner root sheath and the medullary cells [186–188] catalyzed the formation of isopeptide bonds. Evidence obtained by Rogers et al. [189] suggests that proteins containing the glu-lys bond, or, more correctly, the δ-(γ-glutamyl)-lysyl cross-link, also contained citulline residues. Their studies indicate that trichohyalin is rich in arginine, glutamic acid, and glutamine residues and was the probable precursor of citrulline-containing proteins. The presence of citrulline in resistant cuticle membrane preparations may provide evidence for isopeptide bonds existing in resistant membrane residues.

5.2.6 Protein Composition of Mammalian Fibers

5.2.6.1 Protein Classes

The main chemical constituents of mammalian fibers are hard keratin proteins, of which three main groups have been identified. Each of these groups in turn consists of complex mixtures of proteins [190,191]. The groups are commonly known as low-sulfur proteins (LSP), high-sulfur proteins (HSP), and high-tyrosine proteins (HTP), and their constituent proteins are arranged within the structural components of mammalian fibers (Section 2.2 through Section 2.5). While there is some evidence that HSP and HTP are located in the matrix [45,191], the only direct evidence linking structure and composition has been the demonstration that the LSPs comprise the IF (microfibril) components of fibers [112,120,121].

5.2.6.2 Extraction of Proteins

The inherently inert nature of hard keratins requires the cleavage of intermolecular bonds prior to attempting protein solubilization. These bonds are primarily disulfide in nature; chemical reactions such as reduction [192,193], oxidation [194], sulfitolysis [196], and oxidative sulfitolysis [197] are required as preliminaries to protein extraction. In most present-day studies, the reductive methods are used in conjunction with the use of protein denaturants. For example, reduction utilizes reagents such as thioglycolic acid, mercaptoethanol, or dithiothreitol in excess at a mildly alkaline pH. Coupled with the presence of swelling and dispersive reagents such as concentrated urea, guanidine hydrochloride, or lithium bromide, the disulfide bonds are made accessible to the reducing agent [191,198,199]. To prevent reoxidation of thiol groups, alkylation with iodoacetic acid (IAA) [194] or $2-^{14}$[C] IAA [199] is used to form an S-carboxymethyl (SCM) kerateine (keratin) derivative. SCM-keratin derivatives are subsequently divided into two fractions by acid precipitation [194] or zinc acetate [191]. These fractions are termed SCMKA and SCMKB and have widely differing contents of SCM-cysteine [9]. In addition, SCMKAs (low-sulfur proteins) contain relatively higher levels of glutamic and aspartic acids, leucine, lysine, and arginine than SCMKB (high-sulfur proteins). The high-sulfur proteins show elevated contents of proline, serine, and threonine but have lower levels of aspartic acid, lysine, alanine, and isoleucine [191].

A third family of hard keratin proteins has been recognized by Gillespie [202]. In early studies, these proteins coprecipitated with low-sulfur proteins [203,204] but were found to represent a distinct family of proteins rich in aromatic residues (tyrosine and phenylalanine), glycine, and serine. These so-called high-tyrosine proteins consisted of two distinct classes, and their proportions varied widely in mammalian keratins (1–30% of total extracts) [205]. Apparently, high-tyrosine proteins are absent in human hair [191].

In Merino wool, the amount of hard keratin proteins that are actually solubilized does not account for the entire fiber weight; however, yields may reach about 80% on a weight (w/w) basis. This level of extraction is considered to be a reasonable representation of the original fiber composition, but extracts of human hair vary widely, from 5% w/w to 75% w/w. Hence comparisons of keratin extracts, especially in poorly solubilized hairs, should be considered carefully in the quantitative context [34].

TABLE 5.1
Amino Acid Compositions of the Main Protein Classes Found in Mammalian Fibers (Residue Percentages)

Amino acid	Low-sulfur SCMKA [200]	High-sulfur SCMKB [200]	Ultrahigh sulfur [191]	High-tyrosine type I [201]	High-tyrosine type II [201]	Wool [200]	Human hair (Jones, 1986, Unpublished)
Alanine	6.9	2.9	2.0	1.5	1.1	5.2	4.6
Arginine	7.3	6.2	6.9	5.4	4.7	6.2	5.8
Aspartic acid	9.0	2.3	0.6	3.3	1.8	5.9	4.9
Half-cystine	6.0[a]	22.1[a]	29.9[a]	6.0[a]	9.8	13.1	17.8
Glutamic acid	15.7	7.9	7.9	0.6	0.7	11.1	11.4
Glycine	7.7	6.2	4.2	27.6	33.6	8.6	6.4
Histidine	0.6	0.7	1.3	1.1	0.1	0.8	0.9
Isoleucine	3.6	2.6	1.7	0.2	0.2	3.0	2.6
Leucine	10.2	3.4	1.3	5.5	5.3	7.2	5.8
Lysine	3.5	0.6	0.9	0.4	0.4	2.7	2.7
Methionine	0.6	nil	nil	nil	nil	0.5	0.6
Phenylalanine	2.5	1.6	0.5	10.3	4.5	2.5	1.6
Proline	3.8	12.6	12.8	5.3	3.0	6.6	8.4
Serine	8.2	13.2	12.7	11.8	10.9	10.8	11.7
Threonine	4.8	10.2	11.1	3.3	1.7	6.5	6.8
Tyrosine	3.6	2.1	1.9	15.0	20.3	3.8	2.0
Valine	6.1	5.3	4.3	2.1	1.4	5.7	5.8

[a] Determined as S-Carboxymethylcysteine.

Solubility of hard keratin proteins may also be lowered markedly with such fiber treatments as mild heating, exposure to sunlight, and mild alkaline solution [206–208]. However, extended exposures to the above treatments presumably causes peptide bond cleavage resulting in higher solubilities. Most soluble proteins originate from cortical IF–matrix components, and minor structural components appeared to have lower solubility [34].

5.2.6.3 Low-Sulfur Proteins (Properties of SCMKA and Subunits)

In mammalian fibers, low-sulfur proteins usually constitute about 50 to 60% of the original fiber weight. These proteins are usually rich in the α-helix favoring amino acid residues. Preparations with very high α-helix contents (approximately 80%) have been obtained in proteolytic-derived fragments obtained in aqueous solution [190]. On the basis of sequence studies, these fragments were distinguished as comprising two distinct entities referred to as type I and type II. SCMKA extracts in turn consist of two major groups or families, one consisting of the polypeptides of component 8 and the other, the polypeptides of components 5 and 7 [190,209–211]. The type I helix-rich fraction originated from the component 8 family, while the type II helix-rich fraction was derived from the component 5 and 7 families.

When SCMKA polypeptides obtained from wool extracts are separated by two-dimensional electrophoresis in polyacylamide gels, two well-defined groups of polypeptides are resolved. The groups that contained the highest molecular weight polypeptides (57 k) have been designated as 5, 7a, 7b, and 7c, while the lower molecular weight group (44 k–48 k) were labeled 8a, 8b, 8c-1, and 8c-2. In wool SCMKA, the intensity of component polypeptides within each family appears very similar, suggesting equimolar amounts; but Gillespie and

Marshall [212] have shown variable stoichiometry in human hair and nail extracts. Dowling et al. [213] have isolated and characterized certain of these polypeptides and shown numerous inter- and intrafamily differences in polypeptides with regard to size, composition, and α-helix contents.

5.2.6.4 Sequence Studies of SCMKA Subunit Polypeptides

The elucidation of amino acid sequences in SCMKA polypeptides has been hindered by the blockage of amino-terminii (N-acetyl), but the available data has indicated that components 5, 7c, 8a, and 8c-1 have a similar molecular framework [214,215]; that is, N-acetyl termini linked with four nonhelical domains interrupted at variable intervals by three regions of α-helical segments terminating in a nonhelical C-terminal domain. The sequences of helical domains contain similar heptapeptide sequences where, characteristically, the first and fourth amino acid residues are mostly nonpolar. The regions of homology between IF components within and between species tend to be restricted to α-helical or rod domains [214–217]. Homology in nonhelical regions of component polypeptides has not been observed [217].

5.2.6.5 Formation of IF Structural Units

The initial stage of filament assembly involves the formation of a polypeptide dimer containing a type I (e.g., 5, 7a, 7b, 7c) and a type II (e.g., 8a, 8b, 8c-1, 8c-2) polypeptide [216,218]. In forming this dimer, two polypeptide chains are aligned in parallel with their α-helical domains in the register. The number of polypeptides comprising the IF structural subunit was unclear until Ahmadi et al. cross-linked low-sulfur proteins with dimethyl suberimidate and demonstrated convincing evidence for a four-chain tetrameric structural unit consisting of two pairs of dimers with component polypeptides derived from each family (5, 7 family and 8 family) [219]. Hence, the structural unit of hard keratin IFs has been shown to be an obligate heteropolymer unlike other IFs [219,220]. Woods and Inglis [220] isolated a four-chain α-helix-rich fragment after limited proteolysis. They showed this tetrameric structural unit originated from the N-terminal helical regions and contained two segments (derived from families 7 and 8) in a parallel alignment.

5.2.6.6 High-Sulfur Proteins: Isolation and Characterization

In extracts of mammalian fibers, the high-sulfur (SCMKB) proteins remain soluble after precipitation of low-sulfur (SCMKA), and high-tyrosine proteins after precipitation with either weak acids or zinc acetate [191]. The SCMKB fraction contains about 6% (by weight) sulfur and consists of a very complex mixture of polypeptides possessing molecular weights in the range of 10,000–30,000 [221]. When extracts are obtained following preliminary oxidation, the so-called γ-keratose fraction (high-sulfur proteins) is soluble after α-keratose precipitation [11].

The extracts (SCMKB, γ-keratose) obtained by both procedures demonstrate marked heterogeneity [222,223]. SCMKB from Merino wool has been shown to contain more than 60 polypeptides [222]. These high-sulfur polypeptides have been found to display heterogeneity with respect to both molecular weight and charge.

The SCMKB fraction consists of four main families:

1. SCMKB 1 with molecular weights (M_r) ranging from 23,000 to 26,000
2. SCMKB 2, M_r 19,000
3. SCMKB IIIA, M_r 16,000
4. SCMKB IIIB, M_r 11,000 [223,244]

An additional group of high-sulfur proteins with SCM-cysteine contents in excess of 30 residues percent of total amino acid residues is the ultrahigh sulfur proteins. These proteins are not resolved in two-dimensional electrophoretograms [225].

The amino acid compositions of SCMKB fractions from a variety of mammalian fibers generally show similarities in being rich in half-cystine, proline, serine, and threonine. These four residues generally account for at least half the total residues. Residues such as lysine, histidine, tyrosine, and phenylalanine are generally low in SCMKB fractions, and methionine is usually absent [226,227]

Estimates of the actual number of high-sulfur proteins are variable. In one study, after fractionation using DEAE chromatography and then subjecting the fractions to gel electrophoresis, the total number of polypeptides was estimated to be between 30 and 40 [228]. Similar numbers have been observed in two-dimensional electrophoretograms [229]; however, no set of electrophoretic conditions that can resolve all components from the various families has yet been developed [191].

In electrophoretic studies aimed at molecular weight estimation, particular problems arise with the SCM high-sulfur proteins (i.e., those of the SCMKB fraction) due to their inherently slower mobilities. The high SCM-cysteine contents (20–30 mol %) affect mobility and hence interpretations of molecular weights [191,230] in SDS polyacrylamide gel electrophoresis.

The large number of high-sulfur proteins indicates an inherent complexity, which can be difficult to explain. However, this complexity may not appear so extreme if the individual components are grouped into the various families as indicated previously. For example, of the six components identified with the SCMKB 2 family, chain lengths range from 151 to 171 residues with an additional 13 amino acid replacements [25]. Hence, the components fall into a limited number of families (five or six) and then exhibit considerable intrafamily heterogeneity due to loss of segments and point mutations [224].

5.2.6.7 Amino Acid Sequences of High-Sulfur Proteins

From the diverse range of mammalian fibers, sequences of high-sulfur proteins have been determined only for a limited number from wool and mohair. From the IIIB family (M_r 11,000), three sequences have been determined, while from the B2 (M_r 19,000) and the IIIA (M_r 16,000), six and eleven proteins, respectively, have been sequenced [25,148,215]. The amino acid sequence of only one ultrahigh sulfur protein has been determined [231].

In the native state, all the proteins contain a half-cystine residue as the carboxyterminal amino acid, and most have the aminoterminal blocked with an acetylalanine (there are a few exceptions in the IIIA family). The components of IIIA and B2 families exhibit a characteristically cystine-rich repetitive pentapeptide (-cys-cys-X-pro-Y-), which is not found in the IIIB family. The X position of the pentapeptide is frequently occupied by glutamine, glutamic acid, or arginine residues, while serine or threonine usually occurs in the Y position. The above sequence pattern has been found to occupy the major portion of the B2 protein [25,224]. In this protein, an intrachain disulfide bond is thought to link cys-2 of the first pentapeptide to cys-1 of the second pentapeptide unit, which would result in a convoluted pattern in the protein backbone [30,31,148,232].

5.2.6.8 Structure and Conformation of High-Sulfur Proteins

Given that many of the high-sulfur proteins are located in the interfilamentous matrix of cortical cells, no ultrastructural texture has been observed in high-resolution TEM studies. It has been suggested that high-sulfur proteins have globular shapes with diameters averaging about 2 nm [229]. The high contents of proline, for example, in position four of the repeating pentapeptides, would argue against either an α-helical or β-sheet formation [148]. Despite the paucity of knowledge about the structure of high-sulfur proteins, it seems likely that they contain tightly folded configurations stabilized with intrachain disulfide bonds [148]. Intrachain disulfide bonding is the most likely explanation for the observed matrix swelling of fibers and other keratins after immersion in water or formic acid [31].

As matrix or interfilamentous proteins, the high-sulfur proteins and the high-tyrosine proteins (see below) are now often referred to as interfilamentous associated proteins or IFAPs, especially when they are considered in the context of intermediate filament systems and terminology [124,192,229,233].

5.2.6.9 Ultrahigh Sulfur Proteins

The ultrahigh sulfur proteins (UHS) are characterized by very rich half-cystine (>30 mol %) [191]. Reis has described these proteins in wool after enrichment of cystine in the experimental sheep [234]. UHS proteins were originally detected in moving boundary electrophoresis [235], but more recently they have been demonstrated as a diagonal wedge of unresolved proteinaceous material in two-dimensional electrophoretograms [236]. Very little is known about UHS proteins, but approximately one third of the amino acid residues are half-cystine. The molecular weights of these proteins are apparently higher than those of the high-sulfur proteins, but the observations made from two-dimensional electrophoretograms need to be carefully assessed [237,238].

No single UHS protein component has been isolated as yet, but sequence data has been obtained using nucleic acid procedures [231], which indicates that UHS proteins are encoded by a separate set of genes. UHS proteins have been identified in sheep, mouse, and human hair but isolation of homogeneous components have been restricted by the extensive heterogeneity and apparent continuous distribution of components [239]. It is of interest that production of UHS proteins was stimulated in regrowth wool after sheep were administered an epidermal growth factor or cyclophosphamide [225,237].

5.2.6.10 High-Glycine Tyrosine Proteins

The group of proteins commonly known as high-tyrosine proteins (M_r less than 10,000) vary widely in their distribution and relative proportions across the various mammalian fiber types and other keratins. These proteins range from virtual absence in Lincoln wool and human hair to about 30–40% in echidna quill [240].

Indirect evidence from x-ray diffraction studies indicates that the space occupied by the matrix (cortex) could accommodate the quantities of high-sulfur and the high-glycine-tyrosine proteins found in extracts of fibers [241]. However, the ultrastructural location of high-tyrosine proteins in the fiber has not been established definitively. A predominance of high-tyrosine proteins in the orthocortex of Merino wool has been demonstrated directly using x-ray mapping microanalytical techniques [81] (Figure 5.10) following iodonization of tyrosine residues, but the spatial resolution of this technique is insufficient to resolve the intermediate filament–matrix texture. Various lines of evidence have suggested that the high-tyrosine proteins originate in the cuticle and the cell membranes [95,172,173]. In keratin materials with a greater content of high-tyrosine proteins, such as Merino wool and echidna quill, however, the membranes would account for only a small proportion of the proteins. The remainder would presumably be located in the matrix of the cortex. Isolated preparations of ortho- and paracortical cells [10], together with microanalytical studies, have shown that the paracortex is enriched in the high-sulfur proteins and that the high-tyrosine proteins are preferentially located in the orthocortex [36,80,81,243].

Two families of high-tyrosine proteins are normally separated (Type I and Type II) according to their relative sulfur contents. Differences in composition show that Type I has a relatively low content of half-cystine but high levels of phenylalanine. The contents of these residues are reversed in Type II proteins [244]. Both of these protein families have shown heterogeneity with respect to charge and molecular size; however, only limited studies have been undertaken on the Type I high-tyrosine proteins [245–247]. The Type I family consists of ten fractions with different amino acid compositions, but these fractions can be separated further into varying numbers of components [247]. Sequence homology between various fractions (Type I C2, C3) [201] has led to the suggestion that the presence of aromatic amino acids (tyrosine or phenylalanine) may contribute to conformational changes and account for the apparent heterogeneity.

In the Type II family, approximately 20 components have been observed following chromatography on QAE-cellulose and subsequent electrophoresis. Despite similarities in the relatively high contents of glycine, tyrosine, half-cystine, and serine [201], many differences were apparent in the other amino acid residues. As more data become available, it is likely that the Type II family of high-tyrosine proteins will be classified into a limited number of multimembered families, as has been done for the other hard keratin proteins.

5.2.7 OTHER HARD KERATIN PROTEINS

Proteins that originate from the inner root sheath and medulla of mammalian fibers are characterized by the presence of isopeptide cross-links (ε-(γ-glutamyl)lysine) [85]. These proteins are insoluble using conventional extraction procedures [248] but can be obtained following proteolysis with enzymes [160]. In protein hydrolysates of medulla and inner root sheath, amino acid compositions indicate a relatively high proportion of citrulline and glutamic acid. Half-cystine is present in low amounts, suggesting that disulfide bonding is not important in medullary and inner root sheath cells.

Isopeptide bonds have been shown to be present in enzymic digests of wool [249] and in protein fractions of medullary and inner root sheath cells [187,250]. The available evidence indicates that citrulline found in the proteins of these digests results from postsynthetic modifications of the arginine-rich precursors existing in the trichohyalin granules of these cell types. *In vitro*, the trichohyalin protein has demonstrated cross-linking activity catalyzed by a calcium-dependent transglutaminase to form a high-molecular-weight component [189]. The conversion of arginine to citrulline is complex but appears to be enzyme-catalyzed by a peptidyl arginine deiminase [251]. The occurrence of citrulline in amino acid hydrolysates of proteins may be a convenient indicator for the existence of isopeptide bonds. Current evidence indicates that the function of citrulline in the proteins of inner root sheath and medullary cells is still unknown [85]. The presence of citrulline occurring in proteins containing isopeptide bonds may be coincidental, but the possibility remains that a mechanistic link exists between the formation of these two moieties.

5.3 CHEMICAL REACTIONS OF WOOL

5.3.1 INTRODUCTION

The wool fiber consists predominantly of proteins; therefore, the reactions of wool are the reactions of the protein backbone (e.g., peptide bond hydrolysis) and the reactions of the side-chains of the 21 different types of amino acid residues of which wool is composed. Six of these amino acids (glycine, alanine, valine, leucine, isoleucine, and proline) are essentially chemically inert so are not normally available for chemical reaction. The reactions of the cystine residues are of particular significance, as the disulfide bond is important in the overall structural integrity of the wool proteins. The modification of the disulfide bond is also the basis for a number of technological processes. The reactivity of the protein backbone and the side-chains are to some degree affected by neighboring groups in the structure. The rate of hydrolysis of the protein peptide bonds, catalyzed by proteolytic enzymes, is affected by the conformation of the protein; it is slower if the chain is in a helical conformation, and faster if it is nonhelical.

There are several reviews of the chemical reactivity of wool. Maclaren and Milligan describe how wool reacts with a complete range of reagents, indicate what changes are caused in wool's physical and chemical properties by each treatment, and highlight the actual and potential industrial relevance of each reaction [13]. A review by Asquith and Leon is also comprehensive in the range of reactions that it covers but is less concerned with the effect of the reactions on the properties of wool and more with the industrial relevance of the reactions [242]. Baumann's review pays particular attention to the differences in the amino acid composition of the peptides (wool gelatins) that are dissolved in the different processes used in the wool textile industry [175]. These are peptide fragments of the intact proteins of wool, produced by protein hydrolysis that occurs during wool processing. Fletcher and Buchanan give a general review of the reactions of proteins that are especially relevant to wool protein chemistry [252], and Ziegler has reviewed wool protein cross-linking reactions [253].

Possibly the most important contribution to the study of the chemical reactions of wool in the last 10 years has been an understanding of the role of fiber structural lipids in wool chemistry and technology. The discovery of covalently bound lipids on the surface of the fiber [42,56,57,61,63,254] has provided an explanation for many empirical observations and has suggested possible new approaches to the chemical processing of wool [255].

5.3.2 Reduction

Reagents containing thiol groups, usually mercaptoethanol or thioglycolic acid, are used to reduce the disulfide cross-links in wool [13,242]. Tributyl phosphine and sulfites are also used. Normally, reduction of wool affects only cystine residues; no other amino acid side-chains react.

On reaction of a cystine residue with a thiol-containing reagent, a cysteine residue and a mixed disulfide are formed first:

$$P-CH_2-SS-CH_2-P + {}^-S-CH_2COO^- \rightleftharpoons P-CH_2-S^- + P-CH_2-SS-CH_2COO^-$$

A second cysteine residue is formed in a further reaction:

$$P-CH_2-SS-CH_2COO^- + {}^-S-CH_2COO^- \rightleftharpoons P-CH_2-S^- + {}^-OOC-CH_2-SS-CH_2COO^-$$

(where P represents the main protein chain.)

Because the thiol anion is the reactive species, the reactions are fastest in alkaline solution.

Thiol or tributyl phosphine are usually used to break cystine residue disulfide cross-links to prepare soluble proteins from wool to be studied in solution. Especially in alkaline solution, cysteine residues in proteins are reoxidized by dissolved oxygen, and cystine residue disulfide cross-links are reformed:

$$P-CH_2-SH + 1/2\,O_2 + HS-CH_2-P \rightleftharpoons P-CH_2-SS-CH_2-P + H_2O$$

Therefore cysteine residues in dissolved proteins are usually blocked to prevent reoxidation (see Section 5.3.7).

When sodium sulfite is dissolved in water, H_2SO_3, HSO_3^-, and S_3^{2-} are in equilibrium, the relative amounts depend on the pH. It is uncertain which of these is the most reactive with cystine residues in wool. A cysteine residue and a cysteine S-sulfonate residue are formed in the reaction

$$P-CH_2-SS-CH_2-P + HSO_3^- \rightleftharpoons P-CH_2-SH + P-CH_2-SSO_3^-$$

At one time, it was thought that the cystine residues could be divided into two groups with distinctly different reactivities toward reducing agents, and that these two groups could each be further divided into two subgroups with different reactivities [13,256]. It was thought that cystine residues in matrix and microfibril, or in orthocortical and paracortical cells, might have different properties, or that intermolecular and intramolecular cystine residues might react to different extents. The present view seems to be that the reactivity of cystine residues is likely to be influenced by the nearby amino acid side-chains, but that there is more likely to be a continuous range of cystine residue reactivities [13,257].

A test to discover whether wool has been damaged in processing is to determine the fraction of the wool that dissolves under standard conditions in a solution containing urea, as a hydrogen bond breaker, and a reducing agent, usually bisulfite. Acid damage increases the solubility of treated, compared with untreated wool, because acid hydrolyzes peptide bonds in the protein chains. Alkali treatment decreases the solubility, because reducible cystine residue disulfide cross-links are slowly replaced by nonreducible lanthionine and lysinoalanine cross-links in alkali (see Section 5.3.4).

Reducing agents are used industrially to put permanent creases or pleats into woolen garments.

5.3.3 OXIDATION

Performic acid and peracetic acid are the reagents usually used to oxidize wool [13,242]. The main reaction is with cystine residues, and the ultimate products are two cysteic acid residues:

$$P-CH_2-SS-CH_2-P + 6[O] \rightarrow 2P-CH_2-SO_3^-$$

Partially oxidized cystine residues are intermediates in the reaction. Some hydrolysis of peptide bonds occurs during oxidation with peracetic acid, less with performic acid. Methionine and tryptophan residues are also oxidized. Oxidized wool is soluble in alkaline solutions. When these solutions of dissolved, oxidized wool proteins are acidified, a low-sulfur protein, α-keratose, is precipitated. A high-sulfur protein, γ-keratose, remains in solution. (β-keratose is the residue of oxidized wool that is not soluble in alkali.)

Hydrogen peroxide is an effective reagant to bleach wool, but damage to the wool is significant when alkaline hydrogen peroxide is used. It is uncertain what is responsible for the natural cream color of wool.

Chlorination is used to shrink-proof wool [258]. The objective is to modify the surface of the cuticular cells that are responsible for wool felting shrinkage and to minimize damage to the interior of the fiber. This is achieved by chlorination in acid solution, in strong salt solutions, in organic solvents, or by gaseous chlorination. Tyrosine and tryptophan residues are also affected by chlorination, and peptide bonds are hydrolyzed. Other oxidizing agents have also been used to shrink-proof wool [13]. Wool is sometimes given a mild chlorination treatment (or oxidation treatment) before the application of a polymer to give a shrink-proof finish [13]. The chlorination permits the polymer to wet the wool and spread on the surface of the fibers.

When wool fibers are immersed in chlorine water, blisters appear on the surface of the fibers within a few seconds. The usual explanation [10] of this reaction, termed the Allworden reaction [50], is that the blisters consist of cuticle membranes (epicuticle) and, probably, underlying material from the surface of the cuticular cells [259]. The chlorine water treatment is thought to break cystine residue disulfide cross-links and peptide bonds just below the semipermeable surface of the cuticular cells. The resulting peptides are assumed to cause osmotic changes under the surface of the cells, permitting water to diffuse in and give rise to the blisters. Recently, it has been shown that acid chlorine solutions liberate a thioester-linked lipid from the surface or the fiber by oxidizing the sulfur [42]. The result is a lipid-free (or lipid-reduced) hydrophilic surface with the cysteine sulfur component oxidized to cysteic acid residues. It is worthy of note that the Allworden blisters only occur under conditions that also lead to cleavage of the protein–lipid bond.

5.3.4 ALKALI

Cystine residues in wool are attacked by alkali [13,242], and two new cross-links—lanthionine and lysinoalanine—are formed. A mechanism for the reaction of cystine residues with alkali has been suggested.

$$P-CH_2-SS-CH_2-P \rightarrow P-CH_2-SS^- + CH_2=P + H^+$$

The formation of lanthionine involves a cysteine residue,

$$P=CH_2 + HS-CH_2-P \rightarrow P-CH_2-S-CH_2-P$$

and the formation of lysinoalanine, a lysine residue.

$$P=CH_2 + H_2N-(CH_2)_4-P \rightarrow P-CH_2-NH-(CH_2)_4-P$$

Lanthionine and lysinoalanine cross-links are not broken by reducing agents and therefore alkali-treated wool is less soluble in urea–bisulfite solution. Alkali treatment degrades the mechanical properties and the feel, or handle, of wool and causes yellowing. Thus for scouring or washing wool, neutral, nonionic detergents are preferred to alkaline soaps. Severe alkaline treatment hydrolyzes asparagine and glutamine residues to aspartic and glutamic acid residues.

$$P-CH_2-CO-NH_2 \rightarrow P-CH_2-COOH$$
$$P-CH_2-CH_2-CO-NH_2 \rightarrow P-CH_2-CH_2-COOH$$

Other amino acid side-chains are attacked, and eventually peptide bonds are hydrolyzed and the wool dissolves completely.

Recent research has shown [42,255] that even mild alkaline treatment, especially if carried out in organic solvents such as ethanol, leads to partial cleavage of the covalently linked fatty acids from the surface of the fiber. The result is a more hydrophilic fiber, the effect increasing with the time and severity of treatment.

$$P-CH_2-S-CO-(CH_2)_n-CH_3 \xrightarrow[ROH]{OH^-} P-CH_2-SH + \text{fatty acid}$$

With the development of a practical procedure, using mild alkaline solutions containing a cationic surfactant [255], wetting times of the fiber and the time for dyeing are significantly reduced without any observable damage to the bulk of the fiber.

5.3.5 Acid

Acid treatment [13,242] of wool hydrolyzes peptide bonds:

$$-NH-CHR-CO-NH-CHR^1-CO- \longrightarrow -NH-CHR-COOH + H_2N-CHR^1-CO-$$

Hydrolysis with 6 M HCl in a sealed tube at 110°C for 24 h converts wool completely to its component amino acids for amino acid analysis. Tryptophan is destroyed by this procedure, and cystine, serine, and threonine are partially degraded. Peptide bonds on both sides of amino acid residues with acidic side-chains (aspartic and glutamic acid residues) are hydrolyzed faster than other peptide bonds [260]. Similarly, peptide bonds on both sides of cysteic acid residues hydrolyze rapidly; therefore, wool oxidized during shrink-proofing is more susceptible to acid damage than untreated wool.

The acid hydrolysis of asparagine and glutamine residues to aspartic and glutamic acid residues is faster than the hydrolysis of peptide bonds. Some N→O peptidyl shift occurs in serine and threonine residues when wool is treated with acid [261]. For example, in serine

$$-NH-CHR-CO-NH-\underset{\underset{\underset{OH}{|}}{\underset{CH_2}{|}}}{CH}-CO- \longrightarrow -NH-CHR-CO-O-CH_2CH(NH_2)-CO-$$

The acid binding capacity of wool at pH 1 is about 0.8 mol/g, which is consistent with the estimated contents of aspartic and glutamic acid residues and carboxyl-terminii.

Vegetable impurities (seeds, burrs, etc.) are removed from scoured wool by carbonizing [175]. In this process, wool is passed through a bath containing about 6% sulfuric acid and then passed through squeeze rollers. If the wool has more than about 10% acid, based on the weight of wool, it is dried at 65°C. It is then baked for 3 min at 150°C. The acid treatment makes the vegetable impurities brittle. The wool is then passed between rollers. The vegetable impurities are crushed and fall out of the wool as dust. Finally, the wool is neutralized by passing through sodium carbonate solution. If carbonizing is carried out correctly, the wool is not damaged. If it is not carried out correctly, some peptide bonds are hydrolyzed, peptides dissolve from the wool, and, therefore, some of the mass of wool is lost and the fibers are weakened.

Wool is usually dyed in acid solution [175], for 1 h at 100°C at a pH value between 2 and 7, for example. At the more acid pH values, peptide bond hydrolysis will occur in the dyebath and the fibers are weakened. Including a cationic surfactant in the dyebath decreases damage.

5.3.6 Water, Steam

Treatment of wool with water or neutral buffer solution [13,242] at temperatures above 50°C causes the formation of lanthionine and lysinoalanine residue cross-links and, therefore, decreases the solubility of wool in urea-bisulfite solutions. When wool is heated in water in a sealed tube, it contracts at temperatures between 128 and 140°C, depending on the rate of heating.

Water or steam treatments are used in the flat pressing and setting of wool fabrics. Resistance to wrinkling [262,263] in wool fabrics can be increased by exposing them to a humid atmosphere. The higher the temperature, the shorter the exposure needed.

5.3.7 Further Reactions of Amino Acid Side-Chains

Wool proteins are composed of amino acid residues with side chains containing hydrophobic groups (alanine, valine, leucine, isoleucine, proline, and phenylalanine), hydroxyl groups (serine, threonine, and tyrosine), carboxylic acid groups (aspartic and glutamic acids), amide groups (asparagine, glutamine), basic groups (lysine, arginine, and histidine), and thiol groups (cysteine), as well as glycine, cystine, methionine, and tryptophan residues. With the exception of glycine and the hydrophobic side-chains, all these residues are reactive, and their reactions have been extensively investigated. They have been investigated in the search for new processes to modify or improve the properties of wool as well as to advance knowledge on the structure of the fiber. For example, modification of almost any side-chains affects the mechanical and sorption properties of wool, and modification of basic or acid side-chains affects the dyeability of wool.

5.3.7.1 Alkylation

When wool proteins are dissolved by alkaline solutions [13,242] of reducing agents, they contain cysteine residues. To prevent reoxidation of the cysteine residues to cystine residue disulfide cross-links by dissolved oxygen, the dissolved reduced wool proteins are usually treated with iodoacetate ions, which react with the cysteine residues.

$$P-CH_2-SH + I-CH_2-COO^- \rightarrow P-CH_2-S-CH_2-COO^- + HI$$

The thioether link is stable during hydrolysis, and the proportion of cysteine residues in the protein before treatment with iodoacetate can be estimated from the proportion of S-carboxymethylcysteine in the hydrolysate.

Other alkylating agents that react with cysteine residue side-chains are methyl iodide, N-ethylmaleimide, and acrylonitrile. Alkylating agents react most rapidly with cysteine residues but more slowly with lysine, histidine, methionine, serine, threonine, tyrosine, aspartic acid, and glutamic acid residue side-chains.

5.3.7.2 Arylation

1-Fluoro-2,4-dinitrobenzene (FDNB) reacts with wool and with proteins dissolved from wool [13,242]. It reacts with the amino group at the end of each protein chain to give a dinitrophenyl (DNP) derivative.

$$-CO-CHR-NH_2 + F-\underset{NO_2}{\underset{|}{C_6H_3}}-NO_2 \longrightarrow -CO-CHR-NH-\underset{NO_2}{\underset{|}{C_6H_3}}-NO_2 + HF$$

FDNB forms DNP derivatives with the side-chain amino groups of lysine residues and, to a lesser extent, the side-chains of histidine, cysteine, tyrosine, serine, and threonine residues.

After acid hydrolysis of wool or proteins that have been treated with FDNB, the DNP derivatives of amino acids from the amino end of protein chains are soluble in ether, whereas amino acids whose side-chains have reacted only with FDNB together with unreacted amino acids are not soluble. Thus the amino acids from the amino ends of protein chains (amino-terminal amino acids) can be separated from other amino acids. The method is used to estimate the proportions of the different amino acids that occupy the amino-terminal positions of proteins in wool fibers and in proteins derived from wool. If peptide bond hydrolysis occurs during wool processing, it introduces more amino-terminal amino acid residues; therefore, FDNB can be used to estimate the amount of peptide bond hydrolysis occurring during processing.

5.3.7.3 Esterification

Aspartic acid and glutamic acid residues in wool can be esterified [13,242] using an alcohol and dilute hydrochloric acid. With methanol, esterification occurs at room temperature, but raised temperatures are needed for esterification with higher alcohols.

$$P-CH_2-COOH + HO-CH_3 \rightarrow P-CH_2-COO-CH_3 + H_2O$$

5.3.7.4 Acylation

Acid chlorides or acid anhydrides have been used to acylate wool [13,242]. The hydroxyl groups of serine, threonine, and tyrosine are acylated.

$$P-CH_2OH + (CH_3CO)_2O \rightarrow P-CH_2-OCOCH_3 + CH_3COOH$$

Arginine, lysine, and cysteine residue side-chains are also acylated.

5.3.7.5 Cross-Linking

Molecules with two reactive groups have been used to cross-link [13,242,253] protein molecules in wool. For example, 1,2-dibromoethane will cross-link reduced wool.

$$P-CH_2SH + BrCH_2CH_2Br + HSCH_2-P \rightarrow P-CH_2SCH_2CH_2SCH_2-P$$

Formaldehyde treatment also cross-links wool. Many pairs of amino side-chains are capable of reacting with formaldehyde, but it has proven difficult to determine which are actually involved. For example, formaldehyde may react with lysine side-chains.

$$2P-(CH_2)_4-NH_2 + HCHO \rightarrow P-(CH_2)_4-NH-CH_2NH-(CH_2)_4-P$$

Djenkolic acid residues are formed from cysteine residues when reduced wool is treated with formaldehyde.

$$P-CH_2SH + HCHO + HSCH_2-P \rightarrow P-CH_2SCH_2SCH_2-P$$

Treatment of wool with carbodiimides catalyzes the formation of amide bonds between lysine residues and aspartic or glutamic acid residues.

$$P-(CH_2)_4-NH_2 + HOOC-CH_2-P \rightarrow P-(CH_2)_4-NH-CO-CH_2-P$$

When wool is treated with alkali or hot water in the absence of added agents, cross-linking also occurs, to form lanthionine and lysinoalanine [253].

The application of cross-linking techniques has the potential to correct damage to wool fibers that is caused by peptide bond hydrolysis occurring during processing.

5.3.8 POLYMER APPLICATIONS

5.3.8.1 Surface Polymers for Shrink-Proofing

Mechanical action during scouring or washing causes fabrics made from untreated wool to shrink or felt. The application [13,258,264] of one of a number of polymers to wool fabrics prevents shrinkage. It is believed that the polymer sticks each fiber to other fibers at many points along its length, thereby preventing relative movement of the fibers and thus shrinkage.

It is possible to treat wool fibers with polymer before they are spun into yarn in a way that will ensure that the resulting fabric is shrink-proof. In such cases, it is necessary to cover a significant portion of the fiber surface with polymer. It is thought that this masks the cuticular cells, or scales, on the fibers and prevents the ratchet effect of the scales, all of which point in the same direction along a fiber and are thought to be responsible for the felting of wool fabrics.

The polymers are applied to the wool fibers or fabrics as solutions or emulsions. The polymers contain reactive side-chains, which form cross-links between the polymer chains during a curing process after application and may form covalent links with the wool protein. For example, Hercosett 57 is a polyaminoamide with reactive azetidinium side-chains, which form cross-links between the polymer chains. They also form covalent links with amino and thiol groups in the wool protein. In the case of some polymers (e.g., Hercosett 57), it is necessary to give the wool a mild chlorination treatment to modify the surface of the fibers before applying the polymer. Such a treatment allows the polymer to spread evenly on the surface.

5.3.8.2 Internal Deposition of Polymers

Vinyl monomers, like styrene or acrylamide, are absorbed by wool fibers and can be polymerized within the fibers using UV radiation or redox initiators [13]. The resulting polymers are covalently linked, or grafted, to the wool protein chains, often at both ends. Cysteine residue is the usual site of the covalent link between polymer and protein. Very large increases in weight due to polymer are possible—up to 50 times the original weight of wool. Even when the weight of wool protein and grafted polymer are of the same order of magnitude, there are considerable changes in the mechanical and sorption properties, as compared with those of untreated wool. However, there have been no commercial applications of wool containing grafted polymer.

5.3.9 Photodegradation

Irradiation causes physical and chemical changes in wool [13,265,266]. Sunlight may cause wool to yellow or bleach, depending on the relative intensities of the radiation at the different wavelengths that compose sunlight. The rate of yellowing or bleaching is increased in the presence of moisture. Wool that has been chemically bleached or treated with fluorescent whitening agents yellows more rapidly than untreated wool does. It has been suggested that the yellow color may be due to degradation products from tryptophan residues, but no degradation products have been isolated from yellowed wool and many other residues are also affected. No satisfactory method of inhibiting wool yellowing has been found.

Sunlight eventually causes loss of fiber strength, which is a serious problem in curtain, carpet, and upholstery fabrics in sunny conditions. The tips of Merino wool fibers may become yellow or become weathered because they are more exposed to sunlight than the rest of the fiber. Exposed tips may become brittle and break off during wool processing. Sunlight converts some cystine residues to cysteic acid residues, so tips that do not break off may dye irregularly, either because of the presence of cysteic acid residues or because of damage to the fiber surface.

5.3.10 Effect of Heat: Flame-Proofing

Dry heat in air causes less damage to wool than wet heat does [13,267,268]. Above 140°C, yellowing or scorching occurs; lanthionine, lysinoalanine, and isopeptide cross-links are formed, and solubility in urea-bisulfite solution decreases. Cysteic acid residues are formed from cystine residues, and many other amino acid residues are affected. The melting point of wool is about 235°C, depending on the rate of heating. Pyrolysis occurs at 250°C.

Wool is much less flammable than are cellulosic fibers, for example, and its flammability can be lowered still more by treatment with zirconium or titanium salts. These are absorbed by the wool and slowly hydrolyzed during fabric use and during laundering. The zirconium treatment has the advantage that, unlike titanium, it does not discolor wool. The mechanism by which these treatments flame-proof wool is unknown.

5.3.11 Moth-Proofing

Clothes-moth [13,175,269,270] larvae attack wool with a mixture of enzymes that catalyze the reduction of cystine residue disulfide cross-links and the hydrolysis of peptide bonds. Wool is usually moth-proofed by treatment with insecticides that are absorbed like dyes and show similar fastness. The wool textile industry applies insect-resistant agents mainly to carpet wools, which account for about 85% of the treated wool. Insect-resistant agents are of two classes: those which have been developed specifically for use on wool, and those which consist of agricultural insecticides that have been specially formulated for use with wool. The former group is usually polychlorinated aromatic compounds, and the latter group is based on synthetic pyrethroid insecticides.

The greatest difficulty facing the continuing insect-proofing of wool is related to environmental concerns. Pyrethroid insecticides are regarded as safe as far as mammalian toxicity is concerned, but their toxicity to fish and invertebrate aquatic life is relatively high. Thus effluent from wool processing plants could be a problem. There have been attempts at the development of noninsecticide methods of moth-proofing wool [271], but the procedures to date are either unreliable or not cost-effective.

5.3.12 Setting

Most fundamental studies on setting [13,272,273] of wool are carried out on single fibers. These fibers are extended (usually by 40%) and then set by immersion in boiling aqueous solutions for up to 2 h. The resulting level of set is generally assessed by releasing the fibers in boiling water for 1 h. The residual extension (percentage of original length) is defined as a permanent set, even though the fiber will contract further on prolonged immersion in boiling water. Fibers that are held extended for shorter times in boiling water (e.g., 5 min) contract to less than their original length and are said to supercontract.

When the fibers are extended 50% in cold water, the original x-ray diffraction pattern (due to the α-helical conformation, approximately 30% of the total length of the protein chains in wool) is partly replaced by the β-pattern due to extended protein chains. The phenomena of supercontaction and permanent set are usually explained by assuming that cystine residue disulfide cross-links are rapidly broken under tension in boiling water. If the tension is released in the first 5 min, the now uncross-linked fiber will contract to less than its original cross-linked state. The protein chains in the supercontracted fiber are in a random conformation. If the fiber is held extended for longer than 5 min in boiling water, new cross-links, lanthionine and lysinoalanine residues, are formed. When the tension is removed from the fiber, the new cross-links hold some of the protein chains in the extended conformation and therefore the fiber is permanently set.

Evidence, which favors this explanation of permanent set and supercontraction, is that lanthionine and lysinoalanine are found in hydrolysates of wool that has been treated in boiling water. The involvement of disulfide exchange in setting is also supported by experiments in which wool fibers have been set in solutions of reducing agents and by experiments that show that oxidized wool sets less readily than untreated wool. Setting occurs more readily in neutral or alkaline solution; it is reduced by esterification and increased by acylation. These latter two effects are attributed to internal pH changes in the fiber.

Studies on setting using single fibers and treatment times of 1–2 h have little relevance to industrial setting processes. Flat finishing, pleats, and creases are generally accomplished by steaming for much shorter times.

5.4 LOAD–EXTENSION CURVE: CONFORMATION OF THE PROTEIN CHAINS IN WOOL AT DIFFERENT EXTENSIONS

5.4.1 Load–Extension Curve

When a wool fiber is extended in water, the load–extension curve (Figure 5.12) shows the following features: For a small extension, up to about 1%, the gradient of the curve increases as the crimp of the fiber is straightened. The gradient is then constant up to about 3% extension. Thus the extension of the fiber is approximately proportional to the load. Hooke's law is approximated; this is the so-called Hookean region of the load–extension curve.

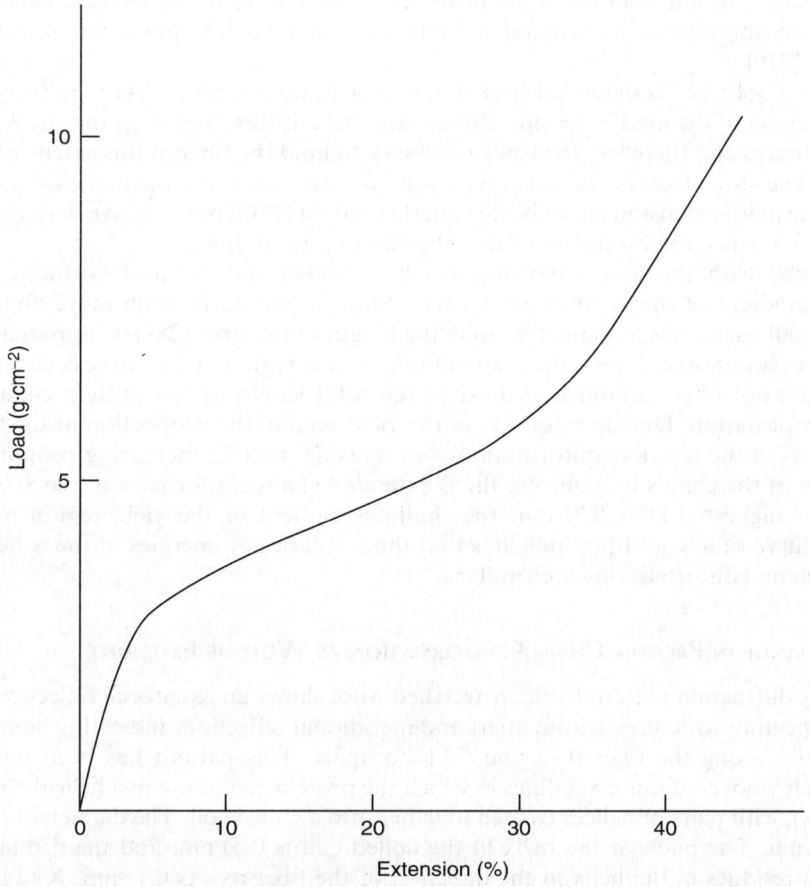

FIGURE 5.12 Load–extension curve of a Merino wool fiber extended in water. (From J.B. Speakman, *J. Text. Inst.*, 18, T431 (1927).)

At about 3% extension, the yield point, there is a sudden decrease in the gradient of the load–extension curve, and the gradient remains small and constant until about 30% extension. After 30% extension, the fiber becomes increasingly resistant to extension and the gradient of the load–extension curve increases continuously until the fiber breaks, at about 40% extension [274–276].

If the fiber is extended by only 30% of its length, or less, and if the load is then removed and the fiber immersed without tension in water for 24 h, the fiber returns to its original length. If the load–extension curve for the fiber is determined for the second time, it follows the path of the first load–extension curve [277]. These recovery properties are unique to wool [272]; no other natural or synthetic fiber exhibits them.

This precise recovery of the fiber's original shape and dimensions and tensile properties appears to be analogous to the denaturation and renaturation, which is possible with soluble proteins in aqueous solution. The protein chain of an enzyme molecule, for example, has a specific, native (natural) conformation in solution. If a hydrogen bond breaker and a reducing agent are added to the solution, the protein chain is denatured and its conformation in solution becomes random. If the hydrogen bond breaker and the reducing agent are now removed from the solution, the protein chain reforms in its original native conformation. Dissolved oxygen will cause the reformation of cystine residue disulfide cross-links between pairs of cysteine residues, in the same positions as they were in the enzyme before denaturation; the enzyme regains its original catalytic activity (which requires the native structure) completely [279].

When a wool fiber is extended in water to a definite extention—say, halfway along the Hookean region of the load-extension curve—and held at this extension, the stress in the fiber gradually decays and therefore the force necessary to hold the fiber at this extension gradually decreases. The slow decrease of stress with time is responsible for the fact that the Hookean region of the load–extension curve is not exactly straight [280], but concave with respect to the extension axis, since the extension of the fiber takes a finite time.

Compared with the load–extension curves of other natural and synthetic fibers, the change in gradient at the yield point is very abrupt; it would be even more abrupt if there were not small variations in diameter along the length of the fiber [281]. Compared with load–extension curves of other fibers, the gradient in the yield region of the curve is very shallow. In unextended wool fibers, about one third of the total length of the protein chains is in the α-helix conformation. During extension in the yield region, the proportion of the total length of the chains in the α-helix conformation is decreasing, and an increasing proportion of the total length of the chains is acquiring the β extended-chain conformation (see Section 5.4.2). It has been suggested [276,282] that the shallow gradient of the yield region in the load–extension curve of a wool fiber indicates that the stabilities or energies of the α-helical and β extended-chain conformations are similar.

5.4.2 Change in Protein Chain Conformation as Wool is Extended

The α x-ray diffraction pattern from unstretched wool shows an equatorial reflection indicating laterally repeating structures 1.0 nm apart and meridional reflections indicating laterally repeating structures along the fiber 0.15 and 0.51 nm apart. This pattern has been interpreted as showing that wool contains crystallites in which the protein chains are in a helical conformation (the α-helix), with pairs of helices twisted together into a coiled coil. The diameter of the α-helix is about 1 nm. The pitch of the helix in the coiled coil is 0.51 nm, and the distance between amino acid residues in the helix in the direction of the fiber axis is 0.15 nm. X-ray diffraction and other evidence indicates that about one third of the total length of the protein chains in wool is in the coiled-coil α-helix conformation (see Section 5.2.3).

The α x-ray diffraction pattern is obtained using bundles of wool fibers. The bundle is held at both ends in jaws that can be moved apart in a metal frame. It is then immersed in water, stretched by a measured length, and dried before it is x-rayed. At 5% extension, there is a measurable decrease in the intensity of the 0.51 nm meridional reflection, indicating that the proportion of the total length of the protein chains in the α-helix conformation has decreased. As the bundle is extended, a new meridional reflection begins to appear at 0.33 nm and a new equatorial reflexion at 0.465 nm. These are the characteristic reflections of the crystallites in the β extended-chain protein conformation. In the β conformation, the protein chains are parallel to the fiber direction and the amino acid residues are 0.33 nm apart in the direction of the fiber axis. The α pattern arises from the unextended microfibrils, in which about half of the total length of the protein chains is in the α-helix confirmation. It is not clear whether the β pattern arises from extended α-helices, from extended nonhelical parts of the protein chains in the microfibril, or from extended protein chains in the matrix. At about 40% extension, before fibers start to break in the bundle, the x-ray pattern shows that both the α and β conformations are present. By extending the bundle in steam, it is possible to reach extensions of about 80%. The pattern then shows that the α reflections have almost disappeared, and by extrapolating graphs of the intensities of α and β reflections versus extension, it appears that the α → β transformation would be complete at about 100% extension [283–286].

5.4.3 Comparison of the Load–Extension Curves and the X-Ray Diffraction Patterns at Different Extensions

In the Hookean region of the load–extension curve, up to about 3% extension, hydrogen bonds between the turns of the α-helices may be strained, but there may be no decrease in the total proportion of the total length of the protein chains in crystallites in the α conformation.

Beyond the yield point at about 3% extension, the proportion of the total length of the protein chains in crystallites of coiled coils of α-helices starts to decrease, and the proportion in crystallites of β extended chains starts to increase. It is not clear from the x-ray diffraction patterns at different extensions why the yield region of the load–extension curve should end at only about 30% extension, or why should there be a steep rise in the gradient of the load–extension curve between about 30% extension and the breaking point at 40% extension. At 30% extension, only about one third of the proportion of the total length of the protein chains in the α-helix conformation has been lost, and only about one third of the estimated final proportion of the total length in β crystallites has been formed.

Two explanations have been put forward to explain the end of the yield region at about 30% extension. In one, it has been pointed out that the matrix and microfibrils in the fiber are extended in parallel [282]. According to this explanation, the load–extension curve of the microfibril, if it could be determined separately from the matrix, might show a yield region from 3 to 100% extension, as expected from the x-ray diffraction patterns from extended bundles of fibers. To account for the steep rise in the gradient of the load–extension curve beyond 30% extension (the change from yield region to postyield region), it is suggested that the load–extension curve of the matrix might be J-shaped with a very low gradient up to about 30% extension followed by a steep rise in the gradient above 30%.

In the alternative explanation of the transition from yield region to postyield region at 30% extension, it was suggested that there might be two types of zones alternating along each microfibril [287,288]. According to this explanation, one of the types of zones might be easy to extend, and these easily extensible zones are all extended when the fiber had been extended 30%. The other type of zone is supposed to be harder to extend (either intrinsically, or because of matrix material associated with it), and these zones that are harder to extend are extended when the fiber is extended beyond 30% extension.

5.4.4 Load–Extension Curves Under Different Conditions

The area under the load–extension curve is the area of work to extend the fiber. The load–extension curve of a fiber can be repeated exactly if the fiber is not extended beyond 30% extension, and if it is allowed to recover for 4 h without tension in water at room temperature (see Section 5.4.1). This means that the work to extend a wool fiber up to 30% ("the 30% index") under different conditions can be compared. Similarly, the work to extend a wool fiber after some chemical treatment can be compared with the work to extend the untreated fiber after it has been allowed to relax without tension for 24 h after the first, calibrating, load–extension experiment.

It is easier to extend wool fibers in acid or alkaline solutions than in neutral solutions [289]. In acid solutions, the carboxylic acid groups of aspartic and glutamic acid residue side-chains lose their negative charges, and in alkaline solution the side-chains of lysine and other basic amino acid residues lose their positive charges. In both cases, there are fewer interactions between positive and negative side-chains ("salt links") to oppose the extension of the fiber.

It is much harder to extend dry wool fibers than it is to extend the fibers in water. The work to extend decreases with increasing humidity and moisture content [276].

5.4.5 Effect of Chemical Treatments of Wool on the Load–Extension Curve

Esterification of the aspartic acid [13] and glutamic acid side-chains, or alkylation or arylation of the side-chains of lysine and other basic amino acid residues, removes salt links from the wool fiber and makes it easier to extend in neutral solution.

Breaking the disulfide cross-links in a wool fiber by oxidation, or by reduction and alkylation, makes it easier to extend.

5.5 SORPTION

5.5.1 Absorption and Desorption Isotherms: Hysteresis

The relationship between relative humidity and the water absorbed at equilibrium by a sample of purified wool fibers, at a fixed temperature, is sigmoid [290]. The relationship shows hysteresis (Figure 5.13). The lower curve represents the equilibrium water content (EWC) of samples of wool that were originally dry and have reached equilibrium at different relative humidities (RH). It is known as the absorption isotherm. The upper curve represents the EWC of originally wet samples that have reached equilibrium at different RHs; it is the desorption isotherm.

The generally accepted explanation of the absorption isotherm at room temperature is as follows [274,291,292]: When a dry sample of wool is exposed to an RH of up to about 5%, water molecules are strongly bound (by hydrogen bonds) onto hydrophilic side-chains and hydrophilic main-chain groups in the protein molecules. This accounts for the steep gradient of the isotherm up to about 5% RH or about 3% EWC. If the dry wool samples are exposed to RHs greater than 5%, then 3% of water is strongly bound and further water molecules are bound less strongly. These further water molecules may be bound to sites on the side-chains or main-chains that bind water molecules with less energy, or they may be bound to the water molecules that are already strongly bound. This less strong binding accounts for the fact that the isotherm is less steep between 5 and 70% RH, or 3 and 15% EWC. This absorbed water causes the fibers to swell. The swelling breaks hydrogen bonds within and between protein chains. In this way, more sites are created to which water molecules can bind, and this accounts for the increase in the gradient of the isotherm between 70 and 100% RH.

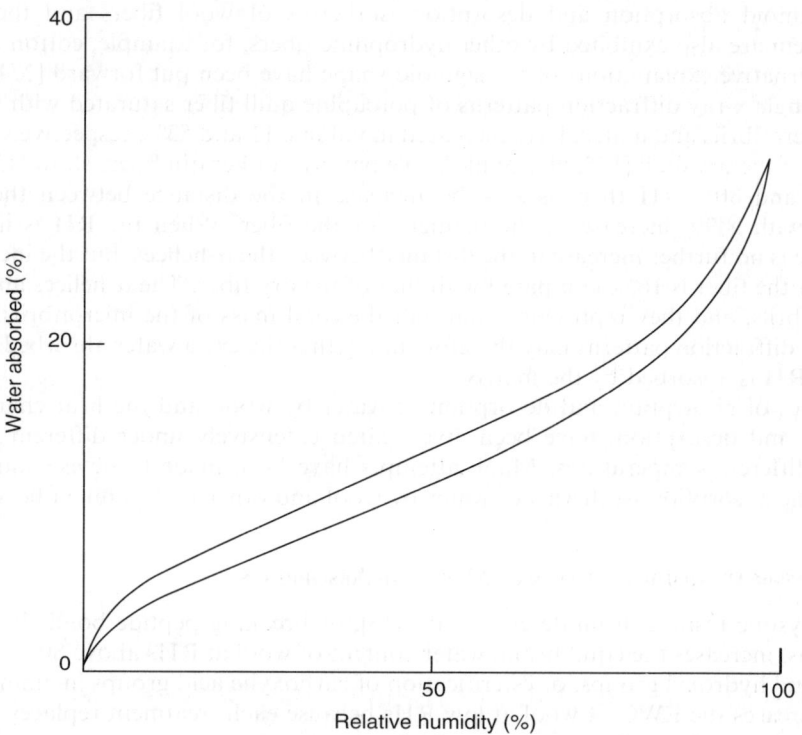

FIGURE 5.13 Percentage water absorbed when wet wool fibers (upper curve) or dry wool fibers (lower curve) are allowed to reach equilibrium at various relative humidities. (From J.B. Speakman, *J. Soc. Chem. Ind.*, 49, 209T (1930).)

The hysteresis between the absorption and desorption isotherms is generally explained in a similar way [274,291,293]. As suggested above, the hydrogen bonds in the structure of wool fibers that are exposed to high relative humidities or immersed in water break, and new sites, to which water molecules can bind, are created. Therefore, a wool sample that has reached equilibrium at a given RH and temperature after immersion in water has a higher EWC than does an originally dry sample that has been equilibrated at the same RH and temperature. It is generally agreed that the two different EWCs at the same RH and temperature, which are indicated on the absorption and desorption isotherms, are true equilibria.

Nuclear magnetic resonance (NMR) experiments indicate that the mobilities of water molecules absorbed by wool are intermediate between the mobilities of water molecules in ice and in liquid water [274]. There are no water molecules with mobilities comparable to those in liquid water, even in wool fibers at equilibrium with atmospheres with very high RHs. There is also no evidence from NMR experiments that only distinct categories of water molecules are absorbed by wool—for example, strongly bound molecules or less strongly bound. The experiments show that water molecules absorbed by wool have a broad, continuous range of mobilities. However, the experiments do show that the water molecules absorbed at low RHs are less mobile than the other water molecules absorbed at higher RHs. This is consistent with the idea that the steep part of the isotherm, up to 5% RH, is due to the absorption of strongly bound water molecules.

The sigmoid absorption and desorption isotherms of wool fibers and the hysteresis between them are also exhibited by other hydrophilic fibers, for example, cotton and viscose rayon. Alternative explanations of the sigmoid shape have been put forward [274,291,293].

Small-angle x-ray diffraction patterns of porcupine quill fiber saturated with water show that the microfibrils and matrix have increased in volume 11 and 53%, respectively, compared with those of the dry quill [135,293]. Wide-angle patterns of keratin fibers show [135,293] that between 0 and 80% RH there is a 4–5% increase in the distance between the α-helices, compared with a 9% increase in the diameter of the fiber. When the RH is increased to 100%, there is no further increase in the distance between the α-helices, but the increase in the diameter of the fiber is 16%, compared with that of the dry fiber. The α-helices are present in the microfibrils, and they represent about half the total mass of the microfibrils. The wide-angle x-ray diffraction patterns may therefore imply that the extra water absorbed between 80 and 100% RH is absorbed by the matrix.

The rates of absorption and desorption of water by wool, and the heat changes during absorption and desorption, have been investigated extensively under different conditions, including different temperatures. Many attempts have been made to devise equations that represent the absorption isotherms of water by wool and other hydrophilic fibers [274,293].

5.5.2 Chemical Modification and Moisture Absorption

Breaking cystine residue disulfide cross-links [13], or breaking peptide bonds in the protein main-chains, increases the equilibrium water content of wool at RHs above 80%. Acetylation of amino and hydroxyl groups, or esterification of carboxylic acid groups in amino acid side-chains, decreases the EWC of wool at low RHs because each treatment replaces hydrophilic groups with less hydrophilic groups.

Cross-linking may affect the EWC of wool by reacting with hydrophilic groups, by introducing hydrophobic groups, or because the cross-links prevent the fiber from swelling. Creating cross-links into dry wool using formaldehyde vapor decreases the EWC at 100% RH. In contrast, creating cross-links into wet wool with aqueous formaldehyde does not decrease the EWC at 100% RH. Thus the extent of cross-linking in a wool fiber cannot be inferred from the EWC under specified conditions. A better method is to measure the swelling in, or the sorption of, formic acid [294], which breaks hydrogen bonds under specified conditions.

5.6 OTHER MAMMALIAN FIBERS

5.6.1 Chemical Composition

Other animal fibers used in the textile industry include mohair, cashmere, alpaca, llama, camel, and angora rabbit.

The only significant differences among the amino acid compositions of the fibers are in their cystine and cysteic acid contents. The genus llama (guanaco, llama, vicuna, and alpaca) has much higher cystine levels than does yak, camel, cashmere, cashgora, and wool [295–297]. The cysteic acid levels of llama, guanaco, vicuna, yak, and camel fibers are higher than those of cashmere, cashgora, and wool [296,297]. The presence of cysteic acid is almost certainly due to photodegradation of the fibers during growth, as it is not a naturally occurring amino acid [296,298]. The variability in amino acid composition observed in wool samples is also apparent with cashmere, with the variation within a large number of cashmere samples as much as that among cashmere, mohair, camel, yak, and wool [296,298]. Even when the diets of the animals are the same, the variations in cashmere are still present, indicating that genetic differences are probably the cause. This variation means that amino acid composition cannot be used to distinguish between the specialty animal fibers.

Keratin protein synthesis is primarily under the control of an animal's genes [225] and therefore should be species-specific. Consequently, differences among the protein compositions of fibers from individual animals of the same species, as well as differences among species, should be demonstrable by high-resolution electrophoresis. Differences among keratin fibers have been shown with two-dimensional polyacrylamide gel electrophoresis (2D-PAGE). The use of acidic gels can show differences among wool, mohair, camel, and alpaca fibers [207]. Acidic gels have been shown to be more useful than alkaline gels showing differences among the specialty animal fibers, although the latter have been used for the analysis of wool–mohair and alpaca–cashmere blends [299,300]. Alkaline gels have been used to compare the protein profiles of mohair, cashmere, and cashgora fibers from individual animals from China and Australia, as well as cashmere and mohair blends from the two countries [301]. The results show that although 2D-PAGE demonstrates differences in protein patterns among fibers from individual angora (mohair), cashmere, and cashgora goats, none of the differences are consistent with any one fiber type. The protein patterns obtained for fiber samples from individual cashmere goats show some differences when compared to those found for commercial blends from the same country of origin, indicating that blending can mask any animal-to-animal variation. The investigation showed that the 2D-PAGE technique using either acidic or alkaline gels does not unequivocally differentiate between cashmere, cashgora, and mohair fibers but does differentiate between wool and goat fiber. One-dimensional PAGE techniques using reduced as well as reduced and alkylated fibers have had limited success in distinguishing between specialty animal fibers [302,303].

Deoxyribonucleic acid (DNA) has been isolated from processed animal fibers [304,305], but not until recently has a goat-specific oligonucleotide been developed that has allowed DNA samples extracted from goat fibers to be distinguished from those of wool [306]. However, it is not yet possible to distinguish between the goat fibers, mohair, cashmere, and cashgora. Species-specific DNA probes have been developed, which enable the identification of yak [307] and angora rabbit [308]. The procedure has been used in the quantitative analysis of yak–cashmere blends [307].

Leeder [16] has shown that the composition of the cell membrane complex (see Section 5.2.5), of which the lipid fraction is one component, has a dramatic influence on fiber and fabric properties. The composition of the internal lipid fractions of a number of specialty animal fibers has been the subject of detailed study [309,310]. Wool, cashmere, cashgora, and mohair contain free cholesterol and desmosterol in the ratio of 1.7–2.6:1 [309]. By comparison, llama, camel, and alpaca fibers contain virtually no free cholesterol or desmosterol. The results for yak vary widely [309,310]. Rabbit and dog hairs have distinctive sterol compositions, which are unlike each other and different from that of wool and goat fibers.

Logan et al. [309] analyzed the free fatty acid composition of wool, mohair, cashgora, rabbit, yak, camel, alpaca, and dog hair and found that palmitic, stearic, and oleic acids accounted for 77–96% of the free fatty acids present. Körner [310] determined the total fatty acid composition for cashmere and yak fibers after saponification and found that the three fatty acids accounted for 50–60% of the fatty acids present. In addition, many fatty acids in the range C7–C26 were present in small amounts. These results confirm that some of the fatty acids are present as esters. When wool, cashmere, and cashgora, from which the surface grease has been removed, are digested with alkali, high yields (16–18%) of 18-methyleicosanoic acid are obtained [309,311]. This fatty acid is covalently bound to the surface of the fibers (see Section 5.2.2).

It may be possible to identify specialty animal fibers based on the basis of the fatty acid composition of their internal lipid fractions. For llama, stearic, palmitic, and oleic acids account for only 50% of the free fatty acids as compared to 77–96% for the other animal fibers [309]. In addition, camel is rich in stearic acid, mohair is high in palmitic acid, and yak has more myristic acid [310]. Cashmere and cashgora fibers contain 5–10% free linoleic acid, whereas mohair contains only trace amounts [309,311]. As a result, ratios of palmitic to linoleic, stearic to linoleic, and oleic to linoleic can be used to differentiate mohair from cashmere and cashgora but not the latter two from each other. Care must be taken in interpreting the results obtained from commercial samples because, if they have been scoured with soap, the fatty acids present in the soap can penetrate the cell membrane complex [309,311].

5.6.2 Fiber Structure

The fine structure of the specialty animal fibers, particularly that of cashmere, cashgora, llama, alpaca, guanaco, vicuna, yak, and camel, has received little attention compared to that given to wool. Prior to the late 1980s, most of the examinations were on longitudinal and transverse fiber sections and used optical microscopy [312–314]. The more recent work arose from the need to differentiate one animal fiber from another [315]. As a result, SEM procedures have been developed. These require the measurement of scale heights [316,317] and can suffer from problems of interpretation due to false scale edges and ill-defined scales.

Early studies on the fine structure of specialty animal fibers involved differential staining using dyestuffs or heavy metals, and subsequent observations were made using light microscopy. By means of these techniques, vicuna fiber was shown to have a bilateral structure [318], and mohair fiber, predominantly ortho- with some paralike material [319] (see Sections 5.2.2 and 5.2.3). Bilateral structures are also observed when wool, cashmere, camel, and alpaca (but not mohair) are treated with sodium hydroxide and examined by means of polarized light [297].

TEM studies of thin transverse fiber sections show that the cortical structure of cashmere is considerably different from that of fine wool [296,311,320]. Australian and Chinese cashmere fibers display both bilateral symmetry and random cell arrangements, not only in cashmere fibers from different samples but also in fibers from the same fleece [296,311], whereas fine wool fiber exhibits bilateral asymmetry only. The variation in cortical structure among fibers from the same cashmere fleece suggests that different mechanisms may be involved in fiber formation. Cashmere cortex is composed predominantly of ortholike and mesolike cells, whereas fine wool is composed predominantly of ortho- and paracortical cells arranged bilaterally. Because of the variations observed, many transverse sections need to be examined before definitive statements can be made about the physical structure of fiber from a given cashmere sample.

The cortical cell distribution for a range of goat fibers is shown in Table 5.2. The comparable figures for fine wool are ortho (66–68%), para (28–32%), and meso (1–4%) [108].

There is an interesting feature of the bilateral segmentation of cashmere when it is considered with respect to that of wool. Fine Merino wool is well crimped and exhibits a bilateral organization of the cortex. (The origin of crimp for a wool fiber is still uncertain; the main theory is that crimp results from the bilateral structure of the fiber, with the orthocortex situated on the outside of the crimp and the paracortex on the inside [105]. Other reports consider the angle of the fiber follicle in the skin to be a factor in crimp formation [192].) Cashmere is devoid of crimp and instead has a wavy macrostructure [312]. Low-crimp fine wools tend to have correspondingly more mesocortical cells than paracortical cells, while the percentage of orthocortical cells remains reasonably constant [109]. The variability of the cortical configuration in cashmere, ranging from bilateral symmetry to complete random orientation, may prevent the formation of any crimp in the fiber; whereas in highly crimped wool, the bilateral structure is uniform along the fiber axis.

TABLE 5.2
Cortical Cell Distribution in Goat Fibers

Sample	Cortical cells (%)		
	Ortho	Para	Meso
Cashmere			
Xinjiang (China)	49	30	21
Liaoning (China)	40	25	35
White 71 (China)[a]	47	16	37
B17 (Australia)	32	28	40
B114 (Australia)	48	28	24
2–012 (Australia)	32	7	61
Cashgora			
P3 (Australia)	53	25	22
P51 (Australia)	37	13	50
3–436 (Australia)	58	20	22
Mohair			
Kid (Australia)	37	8	55
Adult (Australia)	43	10	47

[a]Fibers from a commercial blend, all others from individual fleeces.
Source: From A.H.F. Hudson, The Chemistry and Morphology of Goat Fibers, Ph.D. thesis, Deakin University, Geelong, Australia, 1992.

The structure and chemical composition of mohair has been the subject of a recent review [21]. Mohair fiber is almost 100% orthocortex or predominantly meso- and orthocortex [296,319,321]. Cashgora, which is a crossbred goat from a cashmere doe and an angora buck, can produce fibers that resemble those of cashmere or kid mohair [296,311]. Fibers from the same fleece can exhibit both bilateral symmetry and random arrangements of cortical cells. However, only random cortical cell arrangements have been observed [296, 311] in other cases in which the fibers are also from the one fleece. These observations are not surprising since it is genetically possible for one cashgora goat to have fibers representative of each parent as well as fibers falling between the two extremes. Cashgora, like cashmere, consists mainly of ortho- and mesolike cortical cells.

On the basis of TEM observations of the cortical structure of goat fibers, it is not possible to distinguish between cashmere and cashgora because each can show both bilateral and nonbilateral cortical cell arrangements. It is possible, however, to distinguish between mohair and cashmere fibers because the former does not exhibit bilateral arrangement of cells. Fine wool can be distinguished from cashmere, cashgora, and mohair because it always exhibits bilateral symmetry.

Camel fibers from the same fleece can exhibit both bilateral and random cell arrangements. Similar observations have been made for yak fibers, which tend to consist mainly of ortho- and mesolike cells [296]. Vicuna and guanaco exhibit bilateral structure, whereas llama and alpaca do not [296].

REFERENCES

1. W.G. Crewther, Ed., *Proc. First Int. Wool Text. Res. Conf.*, CSIRO, Australia, vols. A–F (1955).
2. *Proc. Second Int. Wool Text. Res. Conf.*, Harrogate, U.K., published as *J. Text. Inst. Transactions*, 51, no. 12, parts 1 and 2 (1960).

3. *Proc. Third Int. Wool Text. Res. Conf.*, L'Institut Textile de France, Paris, sections 1–4 (1965).
4. *Proc. Forth Int. Wool Text. Res. Conf.*, San Francisco, published as *J. Polymer Science, Applied Polymer Symposia 1971*, 18, parts 1 and 2 (1970).
5. K. Ziegler, Ed., *Proc. Fifth Int. Wool Text. Res. Conf.*, Deutsches, Wollforschungsinstitut, Aachen, Germany, vols. 1–5 (1975).
6. *Proc. Sixth Int. Wool Text. Res. Conf.*, Pretoria, South Africa, vols. 1–5 (1980).
7. M. Sakamoto, Ed., *Proc. Seventh Int. Wool Text. Res. Conf.*, The Society of Fibre Science Technology, Tokyo, vols. 1–5 (1985).
8. G.H. Crawshaw, Ed., *Proc. Eighth Int. Wool Text. Res. Conf.*, Wool Research Organisation of N.Z., Christchurch, N.Z., vols. 1–5 (1990).
9. W.G. Crewther, R.D.B. Fraser, F.G. Lennox, and H. Lindley, In *Adv. Prot. Chem.*, (C.B. Anfinsen, M.L. Anson, J.T. Edsall, and F.M. Richards, Eds.), Academic Press, New York, 1965, pp. 20, 191.
10. J.H. Bradbury, *Adv. Prot. Chem., 27*, 111 (1973).
11. R.D.B. Fraser, T.P. MacRae, and G.E. Rogers, *Keratins, Their Composition Structure and Biosynthesis*, Charles C. Thomas, Springfield, Illinois, 1972.
12. R.S. Asquith, *Chemistry of Natural Protein Fibres*, Plenum Press, New York, 1977.
13. J.A. Maclaren and B. Milligan, *Wool Science, the Chemical Reactivity of the Wool Fibre*, Science Press, Marrackville, N.S.W., Australia, 1981.
14. J.D. Leeder, *Wool, Nature's Wonder Fibre*, Australasian Textiles, Australia, 1984.
15. R. Postle, G.A. Carnaby, and S.de Jong, *The Mechanics of Wool Structure*, Ellis-Horwood, Australia, 1988.
16. J.D. Leeder, *Wool Science Review, 63*, 3 (1986).
17. D.E. Rivett, *Wool Science Review, 67*, 1 (1991).
18. J.A. Rippon, The Structure of Wool, in *Wool Dyeing*, Society of Dyers and Colourists, U.K., 1992, p. 1.
19. A. Körner, Ed., *Proc. of the First Int. Symp. on Speciality Animal Fibres*, Aachen, Germany (1988).
20. A. Körner, Ed., *Proc. of the Second Int. Symp. on Speciality Animal Fibres*, Aachen, Germany (1990).
21. L. Hunter, *Mohair: A Review of Its Properties, Processing and Applications*, CSIR Division of Textile Technology, South Africa International Mohair Association and the Textile Institute, U.K., 1993.
22. I.J. Kaplin and K.J. Whiteley, *Proc. Seventh Int. Wool Text. Res. Conf.*, Tokyo, *1*, 95 (1985).
23. W. Montagna and P.F. Parakkal, in *The Structure and Function of Skin*, 3rd ed., Academic Press, New York, 1974.
24. G.E. Rogers and H.W.J. Harding, in *Biology and Disease of the Hair*, (K. Toda, Y. Ishibashi, Y. Hori, and F. Morikawa, Eds.), University of Tokyo Press, Tokyo, 1975, p. 411.
25. H. Lindley, in *Chemistry of Natural Protein Fibers* (R.S. Asquith, Ed.), Plenum Press, New York, 1977, p. 147.
26. J.A. Swift, in *Chemistry of Natural Protein Fibers* (R.S. Asquith, Ed.), Plenum Press, New York, 1977, p. 81.
27. H. Zahn, *Lenzinger Ber., 42*, 1 (1977).
28. D.F.G. Orwin, in *Fibrous Proteins, Scientific, Industrial and Medical Aspects*, (D.A.D. Parry and L.K. Creamer, Eds.), Academic Press, New York, 1979, pp. 1, 271.
29. D.F.G. Orwin, *Int. Rev. Cytol., 160*, 331 (1979).
30. R.D.B. Fraser and T.P. MacRae, *The Skin of Vertebrates*, Linean Society, London, 1980.
31. R.D.B. Fraser and T.P. MacRae, in *The Mechanical Properties of Biological Materials, Society for Experimental Biology Symposium XXXIV* (J.F.V. Vincent and J.D. Curry, Eds.), Cambridge University Press, London, 1980, p. 211.
32. R.D.B. Fraser, L.N. Jones, T.P. MacRae, E. Suzuki, and P.A. Tulloch, *Proc. Sixth Int. Wool Text. Res. Conf.*, Pretoria, *1*, 1 (1980).
33. P.T. Speakman, Wool Fibers, in *Handbook of Fiber Science and Technology* (M. Lewin and E.M. Pearce, Eds.), Marcel Dekker, New York, 1984, pp. 4, 589.
34. R.C. Marshall, D.F.G. Orwin, and J.M. Gillespie, *Electron. Microsc. Rev., 4*, 47 (1991).
35. J.H. Bradbury and J.D. Leeder, *Aust., J. Biol. Sci., 23*, 843 (1970).

36. S.J. Leach, G.E. Rogers, and B.K. Filshie, *Arch. Biochem. Biophys.*, *105*, 270 (1964).
37. M.S.C. Birbeck and E.H. Mercer, *J. Biophys. Biochem. Cytol.*, *3*, 215 (1957).
38. P.F. Parakkal, In *Advances in Biology of Skin and Hair Growth*, (W. Montagna and R.L. Dobson, Eds.), Pergamon, Oxford, U.K., 1969, pp. 9, 441.
39. J.L. Woods and D.F.G. Orwin, *J. Ultrastruct. Res.*, *80*, 230 (1982).
40. J.L. Woods and D.F.G. Orwin, in *Fibrous Proteins, Scientific, Industrial and Medical Aspects*, (D.A.D. Parry and L.K. Creamer, Eds.), Academic Press, New York, 1980, pp. 2, 141.
41. L.N. Jones, T.J. Horr, and I.J. Kaplin, *Micron.*, *25*, 589 (1994).
42. A.P. Negri, H.J. Cornell, and D.E. Rivett, *Text. Res. J.*, *63*, 109 (1993).
43. H. Zahn, H. Messinger, and H. Höcker, *Text. Res. J.*, *64*, 554 (1994).
44. G.E. Rogers, *Ann. N.Y. Acad. Sci.*, *83*, 408 (1959).
45. I. Stapleton, L.N. Jones, and L.A. Holt, *Proc. Eighth Int. Wool Text. Res. Conf.*, Christchurch, *4*, 117 (1990).
46. J.A. Swift and A.W. Holmes, *Text. Res. J.*, *35*, 1014 (1965).
47. P. Mansour and L.N. Jones, *Text. Res. J.*, *59*, 530 (1989).
48. J. Lindberg, B. Philip, and N. Gralen, *Nature*, *162*, 458 (1948).
49. N.L.R. King and J.H. Bradbury, *Aust. J. Biol. Sci.*, *21*, 375 (1968).
50. K. von Allwörden, *Z. Angew. Chem.*, *29*, 77 (1916).
51. H. Zahn, *Melliand Textilber*, *32*, 419 (1951).
52. R.L. Elliot and B. Manogue, *J. Soc. Dyers Col.*, *68*, 12 (1952).
53. J. Lindberg, *Text. Res. J.*, *23*, 67, 225, 573 (1953).
54. V. Kopke and B. Nilssen, *J. Text. Inst.*, *51*, T1398 (1960).
55. J.D. Leeder and J.A. Rippon, *J. Soc. Dyers Col.*, *101*, 11 (1985).
56. D.J. Evans, J.D. Leeder, J.A. Rippon, and D.E. Rivett, *Proc. Seventh Int. Wool Text. Res. Conf.*, Tokyo, *1*, 135 (1985).
57. P.W. Wertz and D.T. Downing, *Lipids*, *23*, 878 (1988).
58. P.W. Wertz and D.T. Downing, *Comp. Biochem. Physiol.*, *92B*, 759 (1989).
59. D.J. Peet, R.E.H. Wettenhall, D.E. Rivett, and A.K. Allen, *Comp. Biochem. Physiol.*, *102B*, 363 (1992).
60. A.P. Negri, H.J. Cornell, and D.E. Rivett, *Aust. J. Agric. Res.*, *42*, 1285 (1991).
61. U. Kalkbrenner, A. Körner, H. Höcker, and D.E. Rivett, In *Proc. Eighth Int. Wool Text. Res. Conf.*, Tokyo, *1*, 135 (1990).
62. D.J. Peet, *Protein-Bound Fatty Acids in Mammalian Hair Fibers*, Ph.D. thesis, University of Melbourne, Melbourne, 1994.
63. R.I. Logan, L.N. Jones, and D.E. Rivett, in *Proc. Eighth Int. Wool Text. Res. Conf.*, Christchurch, *1*, 408 (1990).
64. G.A. George, H.A. Willis, and R.J. Ward, *Chem. in Aust.*, *59*, 56 (1992).
65. C.M. Carr, I.H. Leaver, and A.E. Hughes, *Text. Res. J.*, *56*, 457 (1986).
66. A. Körner, U. Kalkbrenner, R. Kaufmann, H. Höcker, and D.E. Rivett, *IWTO Report No. 3*, Nice, *1*, (1990).
67. R.J. Ward, H.A. Willis, G.A. George, G.B. Guise, R.J. Denning, D.J. Evans, and R.D. Short, *Text. Res. J.*, *63*, 362 (1993).
68. D.J. Peet, R.E.H. Wettenhall, and D.E. Rivett, *Text. Res. J.*, *65*, 58 (1995).
69. T.J. Horr, *Text. Res. J.*, (in press).
70. B.O. Bateup, J.R. Cook, H.D. Feldtman, and B.E. Fleishfresser, *Text. Res. J.*, *46*, 720 (1976).
71. J.H. Brooks and M.S. Raman, *Text. Res. J.*, *56*, 164 (1986).
72. L.N. Jones, *J. Invest. Dermatol.*, *102*, 559 (1994).
73. J.D. Leeder and J.H. Bradbury, *Text. Res. J.*, *41*, 563 (1971).
74. J.H. Bradbury and J.D. Leeder, *Aust. J. Biol. Sci.*, *25*, 133 (1972).
75. G. Lagermalm, *Text. Res. J.*, *24*, 17 (1954).
76. J.H. Bradbury and K.F. Ley, *J. Biol. Sci.*, *25*, 1235 (1972).
77. P. Kassenbeck, in *Structure de la Laine*, Inst. Text. France, Paris, 1961, p. 50.
78. M.G. Dobb and J. Sikorski, in *Structure de la Laine*, Inst. Text. France, Paris, 1961, p. 37.
79. J.A. Swift, *Proc. Fifth Int. Wool Text. Res. Conf.*, Aachen, *2*, 12 (1976).

80. L.N. Jones, M. Cholewa, I.J. Kaplin, G.J.F. Legge, and R.W. Ollerhead, *Proc. Eighth Int. Wool Text. Res. Conf.*, Christchurch, *1*, 246 (1990).
81. L.N. Jones, I.J. Kaplin, and G.J.F. Legge, *J. Computer Assisted Microsc.*, *5*, 85 (1993).
82. V. Sideris, I.H. Leaver, L.A. Holt, and L.N. Jones, *J. Soc. Dyers Col.*, *108*, 436 (1992).
83. P. Hallegot and P. Corcuff, *J. Microsc.*, *172*, 131 (1993).
84. A. Schwan and H. Zahn, *Proc. Sixth Int. Wool Text. Res. Conf.*, Pretoria, *2*, 29 (1980).
85. G.E. Rogers, In *Biochemistry and Physiology of the Skin* (L.A. Goldsmith, Ed.), Oxford University Press, New York, 1983, pp. 1, 511.
86. H. Jessen, *Histochemie*, *33*, 15 (1973).
87. J.G. Zettergren, L.L. Peterson, and K.D. Wuepper, *Proc. Natl. Acad. Sci. U.S.A.*, *81*, 238 (1984).
88. S. Banks-Schlegel and H. Green, *J. Cell Biol.*, *90*, 732 (1981).
89. D.F.G. Orwin and R.W. Thompson, *Proc. Fifth Int. Wool Text. Res. Conf.*, Aachen, *2*, 173 (1975).
90. K.F. Ley and W.G. Crewther, *Proc. Sixth Int. Wool Text. Res. Conf.*, Pretoria, *2*, 13 (1980).
91. K.F. Ley, R.C. Marshall, and W.G. Crewther, *Proc. Seventh Int. Wool Text. Res. Conf.*, *1*, 152 (1985).
92. G.E. Rogers, *J. Ultrastruct. Res.*, *2*, 309 (1959).
93. V.G. Kulkarni, R.M. Robson, and A. Robson, *J. Polym. Sci. Appl. Polym. Symp.*, *18*, 127 (1971).
94. W.H. Ward and J.J. Bartulovich, *J. Phys. Chem.*, *60*, 1208 (1956).
95. J.H. Bradbury, G.V. Chapman, and N.L.R. King, *Proc. Third Int. Wool Text. Res. Conf.*, Paris, *1*, 359 (1965).
96. J.D. Leeder and J.A. Rippon, *J. Text. Inst.*, *73*, 149 (1982).
97. J.H. Bradbury and N.L.R. King, *Aust. J. Chem.*, *20*, 2803 (1967).
98. K.F. Ley, L.M. Dowling, and R.D. Rossi, *Proc. Eighth Int. Wool Text. Res. Conf.*, Christchurch, *1*, 215 (1990).
99. W.J. Onions, in *Wool: An Introduction to Its Properties, Varieties, Uses and Production*, E. Benn, London, 1962.
100. F. Kidd, *Proc. Third Int. Wool Text. Res. Conf.*, Paris, *1*, 221 (1965).
101. M.S.C. Birbeck and E.H. Mercer, *J. Biophys. Biochem. Cytol.*, *3*, 203 (1957).
102. R.E. Chapman and R.T. Gemmell, *J. Ultrastruct. Res.*, *36*, 342 (1971).
103. K.M. Rudall, *Proc. First Int. Wool Text. Res. Conf. Aust.*, *F*, 176 (1955).
104. D.F.G. Orwin, in *Fibrous Proteins: Scientific, Industrial and Medical Aspects* (D.A.D. Parry and L.K. Creamer, Eds.), Academic Press, New York, 1979, pp. *1*, 271.
105. M. Horio and T. Kondo, *Text. Res. J.*, *23*, 373 (1953).
106. E.H. Mercer, *Text. Res. J.*, *23*, 388 (1953).
107. E.H. Mercer, J.L. Farrant, and A.L.G. Rees, *Proc. First Int. Wool Text. Res. Conf. Aust.*, *F*, 120 (1955).
108. R.M. Bones and J. Sikorski, *J. Text. Inst.*, *58*, 521 (1967).
109. I.J. Kaplin and K.H. Whiteley, *Aust. J. Biol. Sci.*, *31*, 231 (1978).
110. J.A. Swift, *J. Text. Inst.*, *63*, 129 (1972).
111. P. Kassenbeck, in *Hair Research* (C.E. Orfanos, W. Montagna, and G. Stultgen, Eds.), Springer-Verlag, Berlin, 1979, p. 52.
112. L.N. Jones and F.M. Pope, *J. Cell Biol.*, *101*, 1569 (1985).
113. R.D.B. Fraser and G.E. Rogers, *Aust. J. Biol. Sci.*, *8*, 288 (1955).
114. D.F.G. Orwin, J.E. Woods, and S.L. Ranford, *Aust. J. Biol. Sci.*, *37*, 237 (1984).
115. J.A. Swift, *J. Text. Inst.*, *60*, 30 (1969).
116. G.E. Rogers and B.K. Filshie, in *Fifth Int. Congress for Electron Microscopy* (S.S. Breese, Ed.), Academic Press, New York, 1962, pp. 2, 1.
117. M.G. Dobb, *J. Text. Inst.*, *61*, 232 (1970).
118. J.H. Bradbury and D.E. Peters, *Text. Res. J.*, *42*, 471 (1972).
119. R.D.B. Fraser and T.P. MacRae, *Nature*, *233*, 138 (1971).
120. L.N. Jones, *Biochim. Biophys. Acta*, *412*, 91 (1975).
121. L.N. Jones, *Biochim. Biophys. Acta*, *446*, 515 (1976).
122. R.D.B. Fraser, T.P. MacRae, and A. Miller, *J. Mol. Biol.*, *14*, 432 (1965).
123. E. Lazarides, *Nature*, *283*, 249 (1980).

124. P.M. Steinert, C.R. Jones, and R.D. Goldman, *J. Cell Biol.*, 99, 225 (1984).
125. B.K. Filshie and G.E. Rogers, *J. Mol. Biol.*, 3, 784 (1961).
126. R.D.B. Fraser, T.P. MacRae, and G.E. Rogers, *Nature*, 193, 1052 (1962).
127. F.H.C. Crick, *Nature*, 170, 882 (1952).
128. L. Pauling and R.B. Corey, *Nature*, 171, 59 (1953).
129. H.R. Wilson, *J. Mol. Biol.*, 6, 474 (1963).
130. R.D.B. Fraser, T.P. MacRae, and A. Miller, *Nature*, 203, 1231 (1964).
131. R.D.B. Fraser and G.R. Millward, *J. Ultrastruc. Res.*, 31, 203 (1970).
132. G.R. Millward, *J. Ultrastruc. Res.*, 31, 349 (1970).
133. F. Thon, *Nature*, 21a, 476 (1966).
134. F. Thon, in *Sixth Int. Cong. for Electron Microscopy* (R. Vyeda, Ed.), Maruzen Publishers, Tokyo, 1966, pp. 1, 23.
135. R.D.B. Fraser, T.P. MacRae, G.R. Millward, D.A.D. Parry, E. Suzuki, and P.A. Tulloch, *J. Polym. Sci. Appl. Polym. Symp.*, 18, 65 (1971).
136. C.J. Bailey, C.N. Tyson, and H.J. Woods, *Proc. Third Int. Wool Text. Res. Conf.*, Paris, 1, 105 (1965).
137. R.D.B. Fraser, T.P. MacRae, G.E. Rogers, and B.K. Filshie, *J. Mol. Biol.*, 7, 90 (1963).
138. M.G. Dobb, *Nature*, 202, 804 (1964).
139. G.E. Rogers and R.M. Clarke, in *Biology of the Skin and Hair Growth* (A.G. Lyne and B.F. Short, Eds.), Angus and Robertson, Sydney, Australia, 1965, p. 329.
140. D.J. Johnson and P.T. Speakman, *Nature*, 205, 268 (1965).
141. D.J. Johnson and P.T. Speakman, *Proc. Third Int. Wool Text. Res. Conf.*, Paris, 1, 173 (1965).
142. G.R. Millward, *J. Cell Biol.*, 42, 317 (1969).
143. M.G. Dobb and J. Sikorski, *J. Text. Inst.*, 60, 497 (1969).
144. R.D.B. Fraser, T.P. MacRae, and G.R. Millward, *J. Text. Inst.*, 60, 343 (1969).
145. R.D.B. Fraser, T.P. MacRae, and G.R. Millward, *J. Text. Inst.*, 60, 498 (1969).
146. P.G. Whitmore, *J. Cell. Biol.*, 52, 174 (1972).
147. P.M. Steinert, J.M. Starger, and R.D. Goldman, in *Fibrous Proteins: Scientific Industrial and Medical Aspects* (D.A.D. Parry and L.K. Creamer, Eds.), Academic Press, New York 1980, pp. 2, 227.
148. D.A.D. Parry, R.D.B. Fraser, and T.P. MacRae, *Int. J. Biol. Macromol.*, 1, 17 (1979).
149. R.D.B. Fraser, T.P. MacRae, G.R. Millward, D.A.D. Parry, E. Suzuki, and P.A. Tulloch, *J. Polym. Sci., Appl. Polym. Symp.*, 18, 65 (1971).
150. M. Feughelman, *Text. Res. J.*, 29, 223 (1959).
151. W.G. Crewther and L.M. Dowling, *J. Text. Inst.*, 51, T775 (1960).
152. J.M. Gillespie and R.C. Marshall, *Cosmetics and Toiletries*, 95, 29 (1980).
153. R.F. Chapman and K.A. Ward, in *Physiological and Environmental Limitations to Wool Growth* (J.L. Black and P.J. Reis, Eds.), University of New England Press, Armidale N.S.W., 1979, p. 193.
154. E.G. Bendit, *J. Macromol. Sci.-Phys.*, B17, 129 (1980).
155. M. Feughelmann, G.D. Danilatos, and D. Dubro, in *Fibrous Proteins: Scientific, Industrial and Medical Aspects* (D.A.D. Parry and L.K. Creamer, Eds.), Academic Press, London, 1980, pp. 2, 195.
156. R. Castaing, in *Physical Aspects of Electron Microscopy and Microbeam Analysis* (B. Siegel, Ed.), New York, 1975, p. 287.
157. L.N. Jones, T.J. Horr, and I.J. Kaplin, *Micron.*, 25, 589 (1994).
158. W.G. Crewther, L.M. Dowling, P.M. Steinert, and D.A.D. Parry, *Int. J. Biol. Macmol.*, 5, 267 (1983).
159. E.H. Mercer, in *Keratin and Keratinization: An Essay in Molecular Biology*, Pergamon Press, Oxford, U.K., 1961.
160. H.W.J. Harding and G.E. Rogers, *Biochemistry*, 10, 624 (1971).
161. H. Brunner and B.J. Comans, in *Identification of Mammalian Hair*, Inkata Press, Melbourne, 1974.
162. J.L. Clement, R. Hagege, A. Le Pareux, J. Connet, and G. Gustaldi, *J. Forensic Sciences*, 26, 447 (1981).
163. J.H. Bradbury and J.M. O'Shea, *Aust. J. Biol. Sci.*, 22, 1205 (1969).
164. D.F.G. Orwin and R.W. Thompson, *J. Cell. Sci.*, 11, 205 (1972).
165. Y. Nakamura, T. Kansh, T. Kondo, and H. Inagaki, *Proc. Fifth Int. Wool Text. Res. Conf.*, Aachen, 2, 23 (1976).
166. D.E. Peters and J.H. Bradbury, *Aust. J. Biol. Sci.*, 29, 43 (1976).
167. J.D. Leeder, D.G. Bishop, and L.N. Jones, *Text. Res. J.*, 53, 402 (1983).

168. Y. Nakamura, K. Kosaka, M. Tada, K. Hirota, and S. Kinugi, *Proc. Seventh Int. Wool Text. Res. Conf.*, Tokyo, *1*, 171 (1985).
169. A.W. Holmes, *Text. Res. J.*, *34*, 706 (1964).
170. J.A. Swift and B. Bews, *J. Text. Inst.*, *65*, 222 (1974).
171. J.H. Bradbury, G.V. Chapman, N. Hambly, and N.L.R. King, *Nature*, *210*, 1333 (1966).
172. R.A. DeDeurwaerder, M.G. Dobb, and B.J. Sweetman, *Nature*, *203*, 48 (1964).
173. J.H. Bradbury, G.V. Chapman, and N.L.R. King, *Aust. J. Biol. Sci.*, *18*, 353 (1965).
174. H. Zahn, *Proc. Sixth Int. Wool Text. Res. Conf.*, Pretoria, Plenary Lecture (1980).
175. H. Baumann, in *Fibrous Proteins: Scientific, Industrial and Medical Aspects*, (D.A.D. Parry and L.K. Creamer, Eds.), Academic Press, New York, 1979, pp. 1, 299.
176. S. Naito, T. Takahashi, K. Ari, *Proc. Eighth Int. Wool Text. Res. Conf.*, Christchurch, *1*, 276 (1990).
177. L.N. Jones, Ph.D. thesis, University of Melbourne, Melbourne, 1985.
178. J.D. Leeder, M.Sc. thesis, Aust. Nat. University, Canberra, Australia, 1969.
179. J.H. Bradbury, J.D. Leeder, and I.C. Watt, *J. Polym. Sci., Appl. Polym. Symp.*, *18*, 227 (1971).
180. J.J. Pisano, J.S. Finlayson, and M.P. Peyton, *Science*, *160*, 892 (1968).
181. J.L. Abernathy, R.L. Hill, and L.A. Goldsmith, *J. Biol. Chem.*, *252*, 1837 (1977).
182. B. Milligan, L.A. Holt, and J.B. Caldwell, *J. Polym. Sci., Appl. Polym. Symp.*, *18*, 113 (1971).
183. M. Neinhaus, A. Schwan, J. Föhles, and H. Zahn, in *Hair Research, Status and Future Aspects* (C.E. Orfanos, W. Montagna, and G. Stuttgen, Eds.), Springer-Verlag, Berlin, 1981, p. 94.
184. M. Neinhaus and J. Föhles, *Proc. Sixth Int. Wool Text. Res. Conf.*, Pretoria, *2*, 487 (1981).
185. K. Röper, J. Föhles, and H. Klostermeyer, *Methods in Enzymology*, *106*, 58 (1984).
186. S.I. Chung and J.E. Folk, *Proc. Nat. Acad. Sci.*, U.S., *69*, 303 (1972).
187. H.W.J. Harding and G.E. Rogers, *Biochemistry*, 11, 2858 (1972).
188. J.A. Rothnagel and G.E. Rogers, *Mol. Cell. Biochem.*, *58*, 113 (1984).
189. G.E. Rogers, H.W.J. Harding, and I.J. Llewellyn-Smith, *Biochim. Biophys. Acta*, *495*, 159 (1977).
190. W.G. Crewther, *Proc Fifth Int. Wool Text Res. Conf.*, Aachen, *1*, 1 (1976).
191. J.M. Gillespie, In *Biochemistry and Physiology of the Skin*, (L.A. Goldsmith, Ed.), Oxford University Press, Oxford, U.K., 1983, p. 475.
192. P.M. Steinert, W.W. Idler, and S.B. Zimmerman, *J. Mol. Biol.*, *105*, 547 (1976).
193. D.R. Goddard and L. Michaelis, *J. Biol. Chem.*, *106*, 605 (1934).
194. D.R. Goddard and L. Michaelis, *J. Biol. Chem.*, *112*, 361 (1935).
195. P. Alexander and C. Earland, *Nature*, *166*, 396 (1950).
196. C.B. Jones and D.K. Mecham, *Arch. Biochem.*, *3*, 193 (1943).
197. J.M. Swan, *Nature*, *180*, 643 (1957).
198. J.M. Gillespie and F.G. Lennox, *Aust. J. Biol. Sci.*, *8*, 97 (1955).
199. R.C. Marshall, *Text Res. J.*, *51*, 106 (1981).
200. J.M. Gillespie and R.C. Marshall, *Proc. Sixth Int. Wool Text. Res. Conf.*, Pretoria, *2*, 67 (1980).
201. R.C. Marshall, J.M. Gillespie, A.S. Inglis, and M.J. Frenkel, *Proc. Sixth Int. Wool and Text. Res. Conf.*, Pretoria, *11*, 147 (1980).
202. J.M. Gillespie, *Aust. J. Biol. Sci.*, *13*, 81 (1960).
203. B.S. Harrap and J.M. Gillespie, *Aust. J. Biol. Sci.*, *16*, 542 (1963).
204. I.J. O'Donnell and E.O.P. Thompson, *Aust. J. Biol. Sci.*, *17*, 937 (1964).
205. J.M. Gillespie, *Comp. Biochem. Physiol.*, *41B*, 723 (1972).
206. J.E. Kearns and J.A. Maclaren, *J. Text. Inst.*, *79*, 534 (1979).
207. R.C. Marshall, H. Zahn, and G. Blankenburg, *Text. Res. J.*, *84*, 126 (1983).
208. I.H. Leaver, R.C. Marshall, and D.E. Rivett, *Proc. Seventh Int. Wool Text. Res. Conf.*, Tokyo, *4*, 11 (1985).
209. E.O.P. Thompson and I.J. O'Donnell, *Aust. J. Biol. Sci.*, *18*, 1207 (1965).
210. W.G. Crewther, L.M. Dowling, and A.S. Inglis, *Proc. Sixth Int. Wool Text. Res. Conf.*, Pretoria, *2*, 79 (1980).
211. W.G. Crewther, L.M. Dowling, K.H. Gough, R.C. Marshall, and L.G. Sparrow, in *Fibrous Proteins: Scientific, Industrial and Medical Aspects* (D.A.D. Parry, L.K. Creamer, Eds.), Academic Press, New York, 1980, pp. 2, 151.
212. J.M. Gillespie and R.C. Marshall, *Aust. J. Biol. Sci.*, *30*, 401 (1977).

213. L.M. Dowling, K.H. Gough, A.S. Inglis, and L.G. Sparrow, *Aust. J. Biol. Sci.*, *32*, 437 (1979).
214. L.M. Dowling, D.A.D. Parry, and L.G. Sparrow, *Bio. Sci. Rep.*, *3*, 73 (1983).
215. B.C. Powell and G.E. Rogers, in *Biology of the Integument* (J. Bereiter-Hahn, A.G. Matoltsy, and K. Sylvia Richards, Eds.), Springer-Verlag, Berlin, 1986, pp. 2, 695.
216. J.F. Conway, R.D.B. Fraser, T.P. MacRae, and D.A.D. Parry, in *The Biology of Wool and Hair* (G.E. Rogers, P.J. Reis, K.A. Ward, and R.C. Marshall, Eds.), Chapman and Hall, London, 1989, p. 127.
217. L.G. Sparrow, L.M. Dowling, Y.Y. Loke, and P.M. Strike, in *The Biology of Wool and Hair* (G.E. Rogers, P.J. Reis, K.A. Ward, and R.C. Marshall, Eds.), Chapman and Hall, London, 1989, p. 145.
218. D.A.D. Parry, R.D.B. Fraser, T.P. MacRae, and E. Suzuki, in *Fibrous Protein Structure* (J.M. Squire and P.J. Vibers, Eds.), Academic Press, New York, 1987, p. 193.
219. B. Ahmadi, N.M. Boston, M.G. Dobb, and P.T. Speakman, in *Fibrous Proteins: Scientific Industrial and Medical Aspects* (D.A.D. Parry and L.K. Creamer, Eds.), Academic Press, New York, 1980, pp. 2, 161.
220. E.F. Woods and A.S. Inglis, *Int. J. Biol. Macromol.*, *6*, 227 (1984).
221. R.L. Darskus, J.M. Gillespie, and H. Lindley, *Aust. J. Biol. Sci.*, *22*, 1197 (1969).
222. J.F. Joubert and F. MacBurns, *J. Sth. Afr. Chem. Inst.*, *20*, 161 (1967).
223. T. Haylett, L.S. Swart and D. Parris, *Biochem. J.*, *128*, 191 (1971).
224. L.S. Swart, F.J. Jourbert, and D. Parris, *Proc. Fifth Int. Wool Text. Res. Conf.*, Aachen, *2*, 254 (1975).
225. J.M. Gillespie and R.C. Marshall, in *Hair Research* (C.E. Orfanos, W. Montagna, and G. Stuttgen, Eds.), Springer-Verlag, Berlin, 1981, p. 76.
226. J.A. Maclaren and B. Milligan, in *Wool Science, the Chemical Reactivity of the Wool Fiber*, Science Press, Marrickville, N.S.W, Australia, 1981, p. 14.
227. R.C. Marshall and J.M. Gillespie, *Aust. J. Biol. Sci.*, *30*, 389 (1977).
228. R.L. Darskus, *J. Chromatog.*, *69*, 341 (1972).
229. J.M. Gillespie, in *Cellular and Molecular Biology of Intermediate Filaments* (R.D. Goldman and P.M. Steinert, Eds.), Plenum Press, New York, 1990, p. 95.
230. R.C. Marshall, *J. Invest. Dermatol.*, *80*, 519 (1983).
231. P.J. MacKinnon, Ph.D. thesis, University of Adelaide, Adelaide, Australia, 1989.
232. R.D.B. Fraser, T.P. MacRae, L.G. Sparrow, and D.A.D. Parry, *Int. J. Biol. Macromol.*, *10*, 106 (1988).
233. P.M. Steinert, J.M. Starger, and R.D. Goldman, in *Fibrous Proteins: Scientific, Industrial and Medical Aspects* (D.A.D. Parry and L.K. Creamer, Eds.), Academic Press, New York, 1980, pp. 2, 227.
234. P.J. Reis, in *Physiological and Environmental Limitations to Wool Growth* (J.L. Black and P.J. Reis, Eds.), University of New England, Pub Unit New South Wales, 1979, p. 223.
235. J.M. Gillespie and P.J. Reis, *Biochem. J.*, *98*, 669 (1966).
236. J.M. Gillespie and R.C. Marshall, *J. Invest. Dermatol.*, *80*, 195 (1983).
237. J.M. Gillespie, M.J. Frenkel, and P.J. Reiss, *Aust. J. Biol. Sci.*, *33*, 125 (1980).
238. J.M. Gillespie, R.C. Marshall, G.P.M. Moore, B.A. Panaretto, and D.M. Robertson, *J. Invest. Dermatol.*, *79*, 197 (1982).
239. R.C. Marshall and J.M. Gillespie, *Aust. J. Biol. Sci.*, *29*, 1 (1976).
240. J.M. Gillespie and M.J. Frenkel, *Comp. Biochem. Physiol.*, *47B*, 339, (1974).
241. R.D.B. Fraser, J.M. Gillespie, and T.P. MacRae, *Comp. Biochem. Physiol.*, *44B*, 943 (1973).
242. R.S. Asquith and N.H. Leon, in *Chemistry of Natural Protein Fibers* (R.S. Asquith, Ed.), Plenum Press, New York, 1977, p. 193.
243. C.M. Carr, L.A. Holt, and J. Drennan, *Text. Res. J.*, *56*, 669 (1986).
244. J.M. Gillespie and M.J. Frenkel, *Proc. Fifth Wool Text. Res. Conf.*, Aachen, *2*, 265 (1975).
245. H. Zahn and M. Biela, *Eur. J. Biochem.*, *5*, 567 (1968).
246. H. Brunner, A. Brunner, and J. Gerendas, *J. Polym. Sci., Appl. Polym. Symp.*, *18*, 55 (1971).
247. J.M. Gillespie and M.J. Frenkel, *Aust. J. Biol. Sci.*, *27*, 617 (1974).
248. G.E. Rogers, in *The Epidermis* (W. Montagna and W.C. Lobitz, Eds.), Academic Press, New York, 1964, p. 179.

249. R.S. Asquith, M.S. Otterburn, J.H. Buchanan, M. Cole, J.C. Fletcher, and K.L. Gardner, *Biochim. Biophys. Acta*, 427, 315 (1970).
250. H.W.J. Harding and G.E. Rogers, *Biochim. Biophys. Acta*, 257, 37 (1972).
251. K. Sugawara and M. Fujisaki, *Agric. Bio. Chem.*, Japan, 43, 2407 (1979).
252. J.C. Fletcher and J.H. Buchanan, in *Chemistry of Natural Protein Fibers* (R.S. Asquith, Ed.), Plenum Press, New York, 1977, p. 1.
253. K. Ziegler, in *Chemistry of Natural Protein Fibers* (R.S. Asquith, Ed.), Plenum Press, New York, 1977, p. 267.
254. D.J. Peet, R.E.H. Wettenhall, and D.E. Rivett (in preparation).
255. A.P. Negri, H.J. Cornell, and D.E. Rivett, *J. Soc. Dyers Col.*, 296, 109 (1993).
256. H. Lindley, in *Sulfur in Proteins* (R. Benesch, R.E. Benesch, P.D. Boyer, I.M. Klotz, W.R. Middlebrook, A.G. Szentgyorgyi, and D.R. Schwarz, Eds.), Academic Press, New York, 1959, p. 33.
257. R. Cecil and J.R. McPhee, *Adv. Prot. Chem.*, 14, 302 (1959).
258. K.R. Makinson, *Shrinkproofing of Wool*, Marcel Dekker, New York, 1979.
259. C.F. Allen, S.A. Dobrowski, P.T. Speakman, and E.V. Truter, *Proc. Seventh Int. Wool. Text. Res. Conf.*, Tokyo, I, 143 (1985).
260. J. Shultz, *Methods in Enzymol.*, 11, 255 (1967).
261. K. Iwai and T. Ando, *Methods in Enzymol.*, 11, 263 (1967).
262. J.D. Leeder, *Wool Science Rev.* 52, 14 (1977).
263. J.D. Leeder, *Wool Science Rev.* 53, 18 (1977).
264. J. Lewis, *Wool Science Rev.* 54, 2 (1977).
265. C.H. Nicholls, in *Developments in Polymer Photochemistry* (N.S. Allen, Ed.), Applied Science Publishers, London, 1980, pp. 1, 125.
266. *Wool Science Rev.*, 39, 27 (1970).
267. L. Benisek, *Wool Science Rev.*, 50, 40 (1974).
268. L. Benisek, in *Flame Retardant Polymeric Materials* (M. Lewin, S.M. Atlas, and E.M. Pearce, Eds.), Plenum Press, New York, 1975, p. 137.
269. D.M. Lewis and T. Shaw, *Rev. Prog. Coloration*, 17, 86 (1987).
270. G.S. Robinson and E.S. Nielsen, *Tineid Genera of Australia*, CSIRO, Melbourne, 1993.
271. D.E. Rivett, S. Ciccotosto, R.I. Logan, E.S. Nielsen, C.P. Robinson, L.G. Sparrow, and R.M. Traynier, *Proc. Eighth Int. Wool Text. Res. Conf.*, Christchurch, 4, 548 (1990).
272. H.J. Woods, *Physics of Fibers*, Institute of Physics, London, 1955.
273. A. Robson, in *The Setting of Fibers and Fabrics* (J.W.S. Hearle and L.W.C. Miles, Eds.), Merrow, Watford, England, 1971, p. 40.
274. W.E. Morton and J.W.S. Hearle, *Physical Properties of Textile Fibers*, 2nd ed. Heinemann, London, 1975.
275. B.M. Chapman, *J. Text. Inst.*, 60, 181 (1969).
276. E.G. Bendit and M. Feughelman, *Encyclopedia of Polymer Science and Technology*, John Wiley and Sons, New York, 1968, pp. 8, 1.
277. J.B. Speakman, *Proc. Roy. Soc.*, B103, 377 (1928).
278. J.B. Speakman, *J. Text. Inst.*, 18, T431 (1927).
279. C.B. Anfinsen, *New Perspectives in Biology* (M. Sela, Ed.), Elsevier, Amsterdam, 1964, p. 42.
280. E.G. Bendit, in *Fibrous Proteins: Scientific, Industrial and Medical Aspects* (D.A.D. Parry, L.K. Creamer, Eds.), Academic Press, New York, 1980, pp. 2, 185.
281. J.D. Collins and M. Chaikin, *J. Text. Inst.*, 59, 379 (1968).
282. J.W.S. Hearle, B.M. Chapman, and G.S. Senior, *J. Polym. Sci., Appl. Polym. Symp.*, 18, 775 (1971).
283. E.G. Bendit, *Nature*, 179, 535 (1957).
284. E.G. Bendit, *Text. Res. J.*, 30, 547 (1960).
285. A.R.B. Skertchly and H.J. Woods, *J. Text. Inst.*, 51, T517 (1960).
286. A.R.B. Skertchly, *Proc. Third Int. Wool Text. Res. Conf.*, Paris, 161 (1965).
287. M. Feughelman and J.D. Collins, *Text. Res. J.*, 44, 627 (1974).
288. G.D. Danilatos and M. Feughelman, *Text. Res. J.*, 50, 568 (1980).

289. J.B. Speakman and M.C. Hirst, *Trans. Faraday Soc.*, *29*, 148 (1933).
290. J.B. Speakman, *J. Soc. Chem. Ind.*, *49*, 209T (1930).
291. A.R. Urquhart, in *Moisture in Textiles* (J.W.S. Hearle and R.H. Peters, Eds.), Butterworths, London, 1960, p. 14.
292. B.H. Mackay, *Proc. Sixth Int. Wool Text. Res. Conf.*, Pretoria, *1*, 59 (1980).
293. I.C. Watt, *J. Macromol. Sci.—Rev. Macromol. Chem.*, *C18*, 169 (1980).
294. J.B. Caldwell and B. Milligan, *J. Text. Inst.*, *61*, 588 (1970).
295. J. Villarroel, in *Wool Handbook* (W. Von Bergen, Ed.), Interscience, New York, 1963, pp. 1, 315.
296. D.J. Tucker, A.H.F. Hudson, G.V. Ozolins, D.E. Rivett, and L.N. Jones, *Schriftenr. Dtsch. Wollf. Inst.*, *103*, 71 (1988).
297. G. Satlow, M. von S. Cieplik, and G. Fichtner, *Textilechnik*, *16*, 143 (1965).
298. D.J. Tucker, *Schriftenr. Dtsch. Wollf. Inst.*, *106*, 1 (1990).
299. P. Kusch and G. Stephani, *Schriftner. Dtsch. Wollf. Inst.*, *96*, 1 (1984).
300. G. Stephani and H. Zahn, *Proc. Seventh Int. Wool Text. Res. Conf.*, Tokyo, *2*, 195 (1985).
301. D.J. Tucker, A.H.F. Hudson, A. Laudani, R.C. Marshall, and D.E. Rivett, *Aust. J. Agric. Res.*, *40*, 675 (1989).
302. P.T. Speakman and J.C. Horn, *J. Text. Inst.*, *78*, 308 (1987).
303. G. Wortmann and F.J. Wortmann, *Schriftenr. Dtsch. Wollf. Inst.*, *103*, 39 (1988).
304. L.S. Meyer-Stork, J. Kalbe, R. Kuropka, S.L. Sauter, H. Höcker, and H. Berndt, *Textilverdlung*, *23*, 304 (1988).
305. G. Nelson, P.F. Hamlyn, and B.J. McCarthy, *Proc. Eighth Int. Wool Text. Res. Conf.*, Christchurch, *2*, 385 (1990).
306. G. Nelson, P.F. Hamlyn, L. Holden, and B.J. McCarthy, *Text. Res. J.*, *62*, 590 (1992).
307. H. Berndt, J. Kalbe, R. Kuropka, L.S. Meyer-Stork, and H. Höcker, *Schriftenr. Dtsch. Wollf. Inst.*, *106*, 259 (1990).
308. P.F. Hamlyn, G. Nelson, and B.J. McCarthy, *Schriftenr. Dtsch. Wollf. Inst.*, *106*, 259 (1990).
309. R.I. Logan, D.E. Rivett, D.J. Tucker, and A.H.F. Hudson, *Text. Res. J.*, *59*, 109 (1989).
310. A. Körner, *Schriftenr. Dtsch. Wollf. Inst.*, *103*, 104 (1988).
311. A.H.F. Hudson, *The Chemistry and Morphology of Goat Fibers*, Ph.D. thesis, Deakin University, Geelong, Australia, 1992.
312. A.B. Wildman, *The Microscopy of Animal Textile Fibres*, W.I.R.A., Leeds, U.K., 1954.
313. H.M. Appleyard, *Guide to the Identification of Animal Fibres*, 2nd ed., W.I.R.A., Leeds, U.K., 1978.
314. K.D. Langley and T.A. Kennedy, *Text. Res. J.*, *51*, 703 (1981).
315. EEC Working Party on Textile Names and Labelling Analysis, *J. Text. Inst.*, *79*, 155 (1988).
316. F.J. Wortmann and W. Arns, *Text. Res. J.*, *56*, 442 (1986).
317. E. Weidemann, E. Gee, L. Hunter, and D.W.F. Turpie, *IWTO. Report No. 2*, Paris, (1987).
318. G. Laxer, C.S. Whewell, and H.J. Woods, *J. Text. Inst.*, *45*, T482 (1954).
319. R.D.B. Fraser and T.P. MacRae, *Text. Res. J.*, *26*, 618 (1956).
320. D.J. Tester, *Text. Res. J.*, *57*, 213 (1987).
321. V.G. Kulkarni, *Text. Res. J.*, *45*, 183 (1975).

6 Silk

Akira Matsumoto, Hyeon Joo Kim, Irene Y. Tsai,
Xianyan Wang, Peggy Cebe, and David L. Kaplan

CONTENTS

6.1 Introduction 383
6.2 Types of Silk 384
    6.2.1 Silkworm Cocoon Silk 385
    6.2.2 Spider Silks 386
6.3 Mechanical Properties 387
6.4 Thermal Properties 388
6.5 Chemical Composition 388
6.6 Structure 389
    6.6.1 Structural Hierarchy 389
    6.6.2 Crystalline Polymorphs 389
    6.6.3 Structure of the Spider Orb Web 391
6.7 Processing 391
    6.7.1 *In Vivo* Processing 391
    6.7.2 Liquid Crystallinity 392
    6.7.3 Rheology 392
    6.7.4 Mechanisms of Silk Assembly 393
    6.7.5 Biomimetics 394
6.8 Solubilization 394
6.9 Reprocessed Silks for New Materials 394
    6.9.1 Films 395
    6.9.2 Fibers 397
    6.9.3 Hydrogels 397
    6.9.4 Spongy 3D Porous Materials 397
6.10 Degradability 398
6.11 Genetic Engineering 398
    6.11.1 Synthetic Silk Genes 398
    6.11.2 cDNA Expression 399
6.12 Biomedical Materials 400
6.13 Applications 400
Acknowledgments 400
References 401

6.1 INTRODUCTION

Silk is an externally spun fibrous protein secretion formed into fibers, usually resulting in material structures such as cocoons or webs [1–4]. Silks are essentially pure proteins, with only low levels of sugars and minerals in some systems. Of all the natural fibers, silks represent the only ones that are spun. Silk fibers from silkworms have been used in textiles for nearly 5000 years. The primary reasons for this longtime use have been the unique luster, tactile properties, durability, and dyability of silks. Silk fibers are remarkable materials displaying unusual mechanical properties: strong, extensible, and mechanically compressible. Silks also display interesting thermal and electromagnetic responses, particularly in the UV range for insect entrapment and form crystalline phases related to processing. Silk fibers were used in optical instruments as late as the mid-1900s because of their fine and uniform diameter and high strength and stability over a range of temperatures and humidity. Naturalist reports suggest that some spider silks were used in the South Pacific for gill nets, dip nets, and fishing—a testimony to the remarkable mechanical properties and durability of this family of protein polymers. Silks have historically been used in medicine as sutures over the past 100 years and are currently used today in this mode along with a variety of consumer product applications. Commercially, silkworm cocoons are mass produced in a process termed "sericulture". The cocoons are extracted in hot soapy water to remove the sericin glue-like protein. The remaining fibroin or structural silk is reeled onto spools, yielding approximately 300–1200 m of usable thread per cocoon. These threads can be dyed or modified for textile applications. The annual world production of raw silk is about 60,000 tons, with China producing half of the world supply followed by India, Korea, and Japan (http://www.dawn.com/2005/06/27/ebr6.htm).

Silks represent one member of a larger class of fibrous proteins in nature, which include keratins, collagens, elastins, and others [5]. These types of proteins can be considered nature's equivalent of synthetic block copolymers. Aside from their direct use in materials applications, fibrous proteins provide experimentally accessible model systems with simpler and well-controlled genetic template-based protein synthesis. The highly repetitive structure allows key features of the primary sequences of these proteins to be captured in shorter consensus sequences at the corresponding genetic level. Short synthetic genetic variants can then be combined to generate larger genes and thus proteins that represent mimics of the native protein. This technique is useful in simplifying the complex behavior of these proteins to an intelligible level, while retaining their biological relevance and materials function. These shorter genetic variants, when polymerized (multimerized) into longer genes, can be used to explore protein sequence and size relationships.

The novel mechanical and visual features of silk fibers from silkworms and spiders have driven interest in this family of structural protein fibers for centuries. The ability to manipulate silkworms for domesticated production of silk fiber, the opportunity to exploit spider silks via genetic engineering, and future options to mimic the novel features of this family of protein fibers using synthetic approaches, continues to drive strong interest in these protein fibers. With growing applicability of these fibers in biomedical and consumer product applications, this interest is likely to continue to expand.

6.2 TYPES OF SILK

Silkworm cocoon silk from the domesticated silkworm, *Bombyx mori,* is the most well characterized of all silks due to the developed practice of sericulture that has optimized this material over thousands of years in China [6] (Figure 6.1). This silk used in textiles was highly valued during the "silk road" that bridged the East–West trade long ago, and was even considered as valuable as gold during some of this time. Silkworms can be raised in high densities, which permit reasonable production levels for commercialization. This process can also be conducted in local settings; hence, it is very adaptable to rural locations where suitable

FIGURE 6.1 Images of raw silk fibers from commercial courses to show the fibroin fibers before (left) and after (right) extraction. Fibers are approximately 5–7 μm in diameter; insert is lower magnification.

food sources, such as Mulberry leaves, are available. The dragline silk from the orb-weaving spider, *Nephila clavipes*, is probably the most well characterized of the different spider silks. Unlike silkworm silk, spider silk has not been domesticated for textile applications because spiders are more difficult to raise in large numbers due to their solitary and predatory nature, and there are problems associated with multiple silk proteins formed into web structures that result in mixtures of silk proteins instead of purer products. In addition, unlike the cocoon silk from the silkworm, orb webs are not reelable as a single fiber and amounts of silk produced per web are much lower than in a single silkworm cocoon. Silk fibers are formed not only by silkworms and spiders but also by scorpions, mites, and flies. Few of these silks have been characterized, as is the case for many of the silkworms and spiders present in nature. In recent years, the silks from perhaps a dozen of these species have been initially characterized so the understanding of silk sequences, structures, and properties is gradually becoming clarified. Importantly, each silk differs in properties, composition, and morphology.

6.2.1 Silkworm Cocoon Silk

The cocoon silk from *B. mori* contains two structural fibroin filaments coated with a family of glue-like sericin proteins, resulting in a single thread having a diameter of 10–25 μm and consisting of two core fibroin fibers of 5–10 μm in diameter. Wild silkworms generally have larger diameter threads as well as a diverse range of cross-sectional morphologies. The silkworm *B. mori* passes through four different metamorphosizing phases: egg or embryo, larva, pupa, and moth (adult). Smaller quantities of silk are produced at all larval stages except during molts [6]. The total life cycle runs from 55 to 60 days. The majority of silk production occurs during cocoon formation around day 26 in the cycle during the fifth larval instar just before molt to the pupa. The silk is formed and then spun from modified salivary glands near the head of the animal, with silk threads coming out of the mouth of the animal. The fiber is stretched and further aligned postspinning due to the figure-eight movement of the head during the cocoon formation process. Silkworm silk is therefore primarily formed at one stage in the life cycle in cocoon formation, and also consists of basically one type of silk—fibroin—as the major structural element of the fibers. Only a few silkworm fibroin-encoding genes have been sequenced to date (examples include [7–9]), thus the chemical sequence details of most of these sources of protein fibers remains undefined.

6.2.2 SPIDER SILKS

Spider silks in the form of orb webs are synthesized in glands located in the abdomen and spun through a series of orifices or spinerettes [10]. The types and nature of the various silks are diverse and depend on the type of silk and the species of the spider. Some general categories of silks and the gland responsible for their production are listed in Table 6.1 [11–12]. The contrast between silks from spiders and silks from silkworms is instructive; most silks from spiders are produced throughout the life of the animal, whereas the majority of silks from silkworms are produced at a single stage of development of the silkworm. There are different types of spider silks generated by individual animals, whereas there is only one type generated from the silkworm, and silks are spun via the abdomen of spiders instead of the head. Spiders, such as *N. clavipes*, generate a family of silk proteins each of which is evolutionarily "tailored" to perform specific functions [13]. Some serve as a safety line (dragline), adhesives (aggregate) or as barriers against environmental extremes during growth

TABLE 6.1
Mechanical Properties of Silks in Comparison to a Sampling of Other Fibers

Fiber	Elongation (%)	Modulus (GPa)[a]	Strength (GPa)	Energy to break (J/kg)
N. clavipes dragline				
Quasistatic	9–11	22–60	1.1–2.9	3.7×10^4
High strain[b]	10	20	—[c]	1.2×10^5
Other spiders—dragline	10–39	2–24	0.2–1.8	$1–10 \times 10^4$
B. mori fibroin	15–35	5	0.6	7×10^4
Other silkworm fibroins	12–50	2–4	0.1–0.6	$3–6 \times 10^4$
Nylon	18–26	3	0.5	8×10^4
Cotton	5–7	6–11	0.3–0.7	$5–15 \times 10^3$
Kevlar	4	100	4	3×10^4
Steel	8	200	2	2×10^3

[a] Conversion: $1\ \text{GPa} = 10^9\ \text{N/m}^2$.
[b] >500,000%/s.
[c] No data.

Source: From Gosline, J.M.; DeMont, M.E.; Denny, M.W. *Endeavour* 10:37–43, 1986; Cunniff, P.M.; Fossey, S.A.; Auerbach, M.A.; Song, J.W.; Kaplan, D.L.; Adams, W.W.; Eby, R.K.; Mahoney, D.; Vezie, D.L. *Poly. Adv. Technol.* 5:401–410, 1994.

(cocoon), in prey capture, reproduction, as vibrational sensors, safety lines, and dispersion tools. Orb web fibers from some spiders have diameters as low as 0.01 μm. A silking rate of ~1 cm/s is often considered equivalent to natural spinning rates for the spider [14,15]. Some spider silks have been observed to supercontact up to around 50% when unconstrained and exposed to high moisture; other silks such as the silkworm cocoon silk do not contact under similar experimental conditions [16]. A skin core has been reported using light microscopy and electron microscopy and reports of nanoscale subfibrils are also indicated. A variety of silkworm and spider silks have been partially or fully sequenced [17,18,20,21].

6.3 MECHANICAL PROPERTIES

The mechanical properties of silk fibers consist of a combination of high strength, extensibility, and compressibility [22,23,24,25]. Some spider silks exhibit over 200% elongation and others maintain tensile strengths approaching those of high-performance fibers. In terms of energy absorption prior to break, spider silks are unmatched in the world of synthetic or natural fibers (Table 6.2). The mechanical properties of silk fibers are a direct result of the size and orientation of the crystalline domains, the connectivity of these domains to the less crystalline domains, and the interfaces or transitions between less organized and crystalline domains. Studies on dragline silk from *N. clavipes* have included conventional quasistatic Instron tests at 10%/s rates of deformation and high rates of deformation (>500,000%/s). The best properties of the native fibers collected and tested at quasistatic rates were 60 and

TABLE 6.2
Types of Spider Silk Including the Gland Where They Are Synthesized and the General Functions of the Silk

Silk	Gland	Function	Amino Acids
Dragline	Major ampullate safety line	Orb web frame, radii	Glycine (37%), alanine (18%), small side chains (62%), polar (26%)
Viscid	Flagelliform	Prey capture, sticky spiral	Glycine (44%), proline (21%), small side chains (56%), polar (17%)
Glue-like	Aggregate	Prey capture, attachment glue	Glycine (14%), proline (11%), polar (49%), small side chains (27%)
Minor ampullate	Minor ampullate	Orb web frame	Glycine (43%), alanine (37%), small side chains (85%), polar (26%)
Cocoon	Cylindrical (tubuliform)	Reproduction	Serine (28%), alanine (24%), small side chains (61%), polar (50%)
Wrapping	Aciniform	Wrapping captured prey	Serine (15%), glycine (13%), alanine (11%), small side chains (40%), polar (47%)
Attachment	Piriform	Attachment to environmental substrates	Serine (15%), small side chains (32%), polar (58%)

Note: Small side chains = glycine + alanine + serine, polar = aspartic acid + threonine + serine + glutamic acid + tyrosine + lysine + histidine + arginine

Source: From Denny, M.W. In: *Mechanical Properties of Biological Materials*. Vincent, J.F.V.; Currey, J.D.; Eds., Cambridge University Press, Washington DC, 1980; Andersen, S.O. Amino acid composition of spider silks. *Comp. Biochem. Physiol.* 35:705–711, 1970.

2.9 GPa for initial modulus and ultimate tensile strength, respectively. Resistance to axial compressive deformation is another interesting property of the silk fibers. Based on the microscopic evaluations of knotted single fibers, no evidence of kink-band failure on the compressive side of a knot curve was observed [26]. Synthetic high-performance fibers fail by this mode even at relatively low stress levels; this is a major limitation with synthetic fibers in some structural composite applications. Generally, silkworm silks exhibit mechanical properties slightly lower than those of spider silks; however, only a limited range of silks from various spiders and silkworms have been characterized to date.

6.4 THERMAL PROPERTIES

Thermal analysis of *B. mori* silkworm cocoon silk indicates a glass transition temperature (T_g) of 175°C and stability at ~250°C [15]. Spider dragline silk (*N. clavipes*) is thermally stable at about 230°C based on thermal gravimetric analysis (TGA) [15]. Two thermal transitions are observed by dynamic mechanical analysis (DMA): one at −75°C, presumed to represent localized mobility in the noncrystalline regions of the silk fiber, and the other at 210°C, indicative of a partial melt or a glass transition.

6.5 CHEMICAL COMPOSITION

Silkworm cocoon silk contains two structural proteins termed "fibroin heavy chain" with a molecular weight of about 375,000 Da and the "fibroin light chain" with a molecular weight of about 25,000 Da [7]. These two proteins are linked by a single disulfide bond to form a large protein chain that remains linked during protein processing into fibers by the silkworm and may play a role in the regulation of chain-folding and fiber formation. There is also an accessory protein termed "P25" that is involved in the process of protein assembly into fibers [27]. Aside from these core proteins, there is the family of sericin proteins that range in molecular weight between 20,000 and 310,000 Da that bind the fibroin chains together in the silk threads. Some silks, such as those produced by the caddis fly and aquatic midge, which spin silks underwater to form sheltered tubes, have also been characterized and consist of a family of proteins having high cysteine content and protein chains that can be >1,000,000 Da due to the extensive disulfide interactions [28].

The amino acid repeat in *B. mori* silkworm cocoon silk fibroin heavy chain that is responsible for the formation of β sheets in the fibers is the 59mer: GAGAGSGAAG[SGAGAG]$_8$Y (where G = glycine, A = alanine, S = serine, Y = tyrosine) with variations on this general sequence as well as subdomains representing aspects of this sequence [7,29]. These core repeats are surrounded by nonrepetitive less regular repeats. The repetitive structures in the majority of the protein are thought to be the result of genetic level continuous unequal crossovers or genetic recombination events during evolution. These repeats are responsible for the formation of β-sheet crystals during spinning of the solution of protein into fibers. The highly repetitive nature of these repeats distinguishes fibrous proteins such as silks from the family of globular proteins (enzymes, antibodies). These repeats are largely responsible for the unique self-assembly and properties for the formation of materials of this family of protein.

The dragline silk from the spider, *N. clavipes*, originates from the major ampullate gland in the abdomen and contains two proteins or spidroins called major ampullate silk protein 1 (MaSp1) or spidroin 1, with a molecular weight around 275,000 Da, and a second similar protein but enriched with proline [17,30]. There are no sericin-like proteins associated with the dragline fiber. MaSp1 contains amino acid repeats considerably shorter than those found in the silkworm fibroin and not as highly conserved. The repeats consisting of polyalanines of

6 to 12 residues are responsible for the formation of the β-sheet crystals in an analogous fashion to the GA repeats for the silkworm. Regions in the spider dragline silk with GGX repeats, where X = alanine, tyrosine, leucine, or glutamine are involved more in the flexible regions of the proteins when formed into fibers. It is worth noting that the distinctions between silkworm and spider repeat sequences responsible for β-sheet crystal formation are becoming blurred as more species of these types of organisms become sequenced and somewhat similar repeats can be identified both in silkworms and in spiders.

6.6 STRUCTURE

6.6.1 Structural Hierarchy

As outlined above, silk proteins assume complex structural organization from the molecular level up length scales to the macroscopic web or cocoon (Figure 6.2). Intermediate scales of β-sheet crystals, organized supramolecular structures, and liquid crystalline features fill the intermediate scales of the structural organization [31,32]. Nanofibril and fiber organizations complete the next level of ordering. As illustrated with almost any biologically derived material, control of structural hierarchy, initiated from the control of chain chemistry, establishes a template of organization on which complexity ensues and provides the features required to tune mechanical functions.

6.6.2 Crystalline Polymorphs

Most silkworm cocoon and spider dragline silk fibers contain assembled antiparallel β-pleated-sheet crystalline structures [33–35]. Silks are considered semicrystalline materials with 30–50% crystallinity in spider silks, 62–65% in cocoon silk fibroin from the silkworm *B. mori*, and 50–63% in wild-type silkworm cocoons. In the β-sheet crystals the polymer chain axis is parallel to the fiber axis. The extent to which these structures form, as well as their orientation and size, directly impact the mechanical features of silk fibers. Furthermore, the polyalanine repeats or

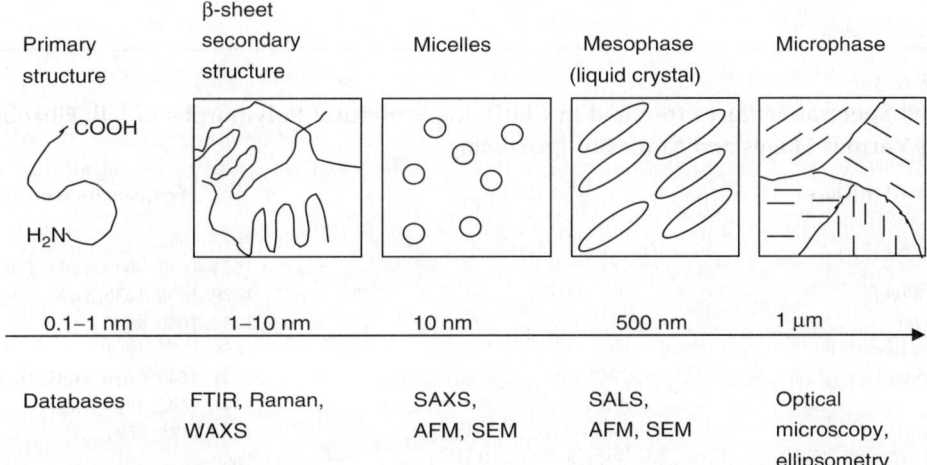

FIGURE 6.2 Schematic showing the hierarchical structures in silks and modes of analytical characterization often used to clarify structures at each of these length scales.

the glycine-alanine repeats are the major primary structure sequences responsible for β-sheet formation. Other silks can also form α-helical structures (such as some bees, wasps, ants) or cross-β-sheet structures (many insects) structures. The cross-β-sheets are characterized by a polymer chain axis perpendicular to the fiber axis. Most silks assume a range of different secondary structures during processing from soluble protein in the glands to insoluble spun fibers. Infrared spectroscopy is often used to distinguish some of the polymorphs (Table 6.3).

The crystalline structure of silks was first described in the 1950s as an antiparallel, hydrogen-bonded β sheet based on characterization of the *B. mori* fibroin. Three crystalline forms of silk have been characterized to date: silk I (prespun), silk II (spun), and silk III (interfacial). The unit cell parameters in the silk II structure (the spun form of silk that is insoluble in water) are 0.94 nm (a, interchain), 0.697 nm (b, fiber axis), and 0.92 nm (c, intersheet). These unit cell dimensions are consistent with a crystalline structure in which the protein chains run antiparallel, with interchain hydrogen bonds perpendicular to the chain axis between carbonyl and amine groups and Van der Waal forces stabilize intersheet interactions (based on the predominance of short side-chain amino acids such as glycine, alanine, and serine in the crystalline regions). The β sheets consisting of the glycine–alanine crystalline repeats in the silkworm fiber are asymmetric, with one surface primarily projecting alanyl methyl groups and the other surface of the same sheet containing hydrogen atoms from the glycine residues. In silk II, these sheets are organized back-to-back such that for every other sheet, the sheet-to-sheet interacting faces are the glycyl side-chains and the alternating interacting faces are the alanyl methyl groups. This arrangement leads to alternating intersheet spacings of 3.70 Å in the glycyl and 5.27 Å in the alanyl interacting intersheet distances. An alternative silk II structure has been proposed by Takahasi (1994) in which the two surfaces of a sheet have both alanyl and glycyl side-groups; thus the intersheet spacing for these now symmetrical sheets is half the c-axis spacing. Solid-state NMR data from spider dragline fibers indicate that all of the crystalline fractions of dragline silk were composed of alanine-rich sequences and these crystalline regions consisted of two orientations: one highly oriented along the fiber axis (~40%) and another less oriented and less dense (~60%) [36,37].

Silk I, the prespun pseudocrystalline form of silk present in the gland in a water-soluble state, remains without a consensus structure in the field and likely represented by many partially stable states. Many models have been proposed and likely reflect many local minima

TABLE 6.3
Infrared Spectral Features to Assist in Clarifying Structural Polymorphs of Silk Fibroin during Various Stages and Modes of Processing

Secondary Structures	Frequencies (cm^{-1})
α helix	1648–1662
β sheet	1624–1642, 1645, 1699, 1703
Antiparallel β	1629, 1630–1636, 1690–1693, 1696
Parallel β	1630, 1640, 1645
Turns and bends	1666–1688, 1691
Random coil	1641, 1649, 1650, 1653, 1656–1660
Silk I	1641, 1645, 1649–1650
Silk II	1624–1625, 1697
Silk III	1662

Note: Amide I bands frequencies (in cm^{-1}) characteristic of the peptide bond in various secondary structures and for silks.

structures during chain-folding and interactions during packing as protein concentrations increase in the gland [38,39]. The silk I structure is unstable and on shearing, drawing, heating, spinning, or exposure in an electric field, or exposure to polar solvents such as methanol or acetone, converts to silk II [40–42]. The change in unit cell dimensions during the transition from silk I to silk II during fiber spinning is most significant in the intersheet plane, with an 18.3% decrease in distance between overlying sheets based on modeling predictions [43]. This change results in the exclusion of water, reducing solubility. Silk II is more thermodynamically stable than silk I and the energy barrier for the transition is low, whereas the return barrier is high and considered essentially irreversible [1,2,14].

A third polymorph, silk III, has also been described based on the interfacial behavior of the silkworm silk fibroin and the partitioning behavior at an air–water interface [44, 45]. Silk III, a structure stabilized at interfaces optimizes the surfactancy of the silk in the core repeats of glycine, alanine, and serine.

6.6.3 Structure of the Spider Orb Web

The construction of the orb web is a remarkable engineering process involving wide area coverage with minimal protein material and formed through a complex set of unknown behavior cues to orchestrate the complexity of web formation. This highly coordinated materials processing, involving control of composition, spatial arrangements, and temporal cues, results in a minimal use of protein and maximal area for prey capture. This process must evolve within the bounds of organismal survival. Thus conservation of energy while fostering survival has provided a driving force toward this complexity in orb web weaving and display. It has been estimated that about 70% of the energy of impact of a flying insect is dissipated through viscoelastic processes when the insect hits a web. Therefore, web construction is designed to balance stiffness and strength against extensibility, both to keep the web from breaking on impact and to ensure that the insect is not ejected from the web by elastic recoil. The ability to dissipate the kinetic energy of a flying insect impacting the web is based on the hysteresis of radical threads and also aerodynamic damping by the web [11,12]. Aside from this remarkable suite of engineering accomplishments, mechanical and structural, some orb webs are also recycled on ingestion by spiders on a daily basis. This may be a conservation tool since some of the amino acids are reused in the construction of new webs. There is also evidence that the composition of the web (amino acid content) changes with diet, suggestive of an energy conservation and evolutionary mechanism for survival depending on food sources available.

6.7 PROCESSING

6.7.1 In Vivo Processing

Silks are synthesized in specialized glands in the head (silkworms) or abdomen (spiders). Initially, some degree of self-organization or self-assembly occurs as a result of protein–protein interactions among the crystalline repeats in the protein chains, driven by both hydrophobic interactions and hydrogen bonding. In the spider gland, changes in physiological conditions such as pH concentrations of salts and content of water accompany the processing and control solubility, assembly, and transitions among solution, gel, and liquid crystalline states. In the silkworm, there are three distinct regions to the glands and two sets of these organs feeding into one final thread [14,15]. The fibroin is synthesized in the posterior region of the gland and the protein moves by peristalsis to the middle region of the gland where it is stored as a viscous aqueous solution or gel until needed for spinning. The protein concentration is 12–15% in the posterior region of the gland where fibroin chain synthesis occurs, increases to around 20–30%

in the middle region of the gland where the fibroin is stored and the sericin is synthesized, and is significantly higher in the anterior region of the gland where spinning begins [15]. During transit, the pH of the processing solution decreases from near neutral to near pH 5 at the point of spinning. These changes appear related to the precipitation of the protein to facilitate high solids content and successful spinning to form mechanically robust fibers. The two lobes of the gland join just before the spinnerettes in the anterior region and the fiber is spun into the air. Aside from binding together the fibroin chains in the final spun fiber, the sericin may function as a reservoir for divalent cations or as a water-holding medium to promote plasticization of the fiber during and after spinning.

6.7.2 Liquid Crystallinity

In the solution state features associated with flexible amphiphiles, including surfactancy, micelle formation, and micelle-related lyotropic liquid crystalline phase behavior would be anticipated for a silk protein chain in which there is a predominant hydrophobic structure with hydrophilic end blocks. There is evidence of liquid state chiral liquid crystals as helicoids precursors for silks as liquid crystalline textures are observed in solutions before crystalline domains form [46,47,31]. This liquid crystalline order can be one mechanism by which long-range order is achieved with these proteins, leading to helicoidal multiscale ordered biological materials. The long-range ordered structures are facilitated by amphiphilic interactions driven by hydrophobic effects due to the repeats described earlier, forming micelle-like structures to enhance or compete with the liquid crystalline ordering. Silk fibroin displays surfactancy and suggests that amphiphilic interactions are largely responsible for structure formation. Folded supersecondary structures are driven by hydrophilic and hydrophobic interactions in aqueous environments. Supersecondary structures and secondary structures are likely important in silk spinning. The surfactancy and the presence of micelle-like structures in solution, and lyotropic liquid crystalline phase behavior may be mechanisms for organizing silk into high-concentration domains, leading to orientation of micelles and possibly enhancing water removal in the gland during processing by providing channels for fluid to flow.

6.7.3 Rheology

During the final spinning step, physical shear plays an important role in many aspects of the process. Shear is likely responsible for alignment of the helicoidal liquid in the late-stage concentrated solution and leads to the formation of the supersecondary structures in the aligned liquid crystal. Rheological experiments indicate that crystallinity in the fiber correlates positively with shear and draw rates. An extrusion rate of around 50 cm/min was found to be a minimum threshold for the appearance of birefringence and the conversion of the soluble silk solution in the gland to the fibers containing β sheets [14,15]. In the posterior region of the gland, which is 0.4–0.8 mm in diameter, the silk solution is optically featureless. A range of secondary structures are present in this region, including random coil and silk I; and the shear rate is low. In the middle region of the gland, the diameter of the gland duct is 1.2–2.5 mm, streaming birefringence is observed, and the shear rate is also low. In the anterior region of the gland, the diameter of the gland duct is narrow (0.05–0.3 mm), the shear rate is high, water appears to be actively transported out of the gland, the pH decreases, and active ion exchange occurs. Viscosity also increases but presumably decreases prior to spinning as a result of the formation of liquid crystalline phases. In the case of spinning dragline silk by the spider, less detail is understood. Protein synthesis occurs in a pair of major ampullate glands in the spider. Generally, a similar process occurs as summarized for the silkworm, although there appear to be differences in terms of the relevant cations. Even less is known about the physiological conditions at different stages in the spinning process when compared with the silkworm fibroin. In the spider, there is no sericin involved in the spinning process.

6.7.4 Mechanisms of Silk Assembly

Regenerated silkworm silk fibroin can be used to mimic this process *in vitro*, and these studies suggest that surfactancy dominates the initial ordering interactions, as hydrophobic and hydrophilic sequence domains separate into different localized regions in solution, resulting in loosely folded chains with some defined secondary and supersecondary structure [48]. A hydrophobic plot of the sequence illustrates the nature of the hydrophobic and hydrophilic segments in this repetitive protein, suggesting types of likely supersecondary transient folded structures present prior to micelle formation. The micelle structures and the process involved are not easily formed unless the protein concentration increases gradually; otherwise premature gelation due to hydrophobic physical cross-links occurs, precluding micellar structures. The large hydrophilic N- and C-termini of the fibroin interacts with the external aqueous environment. These chain-end protein sequence blocks are significantly larger than internal hydrophilic blocks that disrupt the bulk hydrophobic domains in the protein. That soft micelle stage in the processing of silk proteins (native and regenerated) has been identified the suggests this is an important step governing chain interactions (both intra- and interchain) on the path to the formation of lyotropic phases, leading to organized silk structures. These transitions can be duplicated in part using regenerated silkworm silk fibroins and polyethylene oxide to drive the process of water loss (via osmotic stress). This step in silk assembly provides a path toward globular and gel states and leads to liquid crystallinity in the gland. The observations suggest true lyotropic liquid crystallinity, where micelles of flexible molecules grow and alter their curvature and shape in response to changes in concentration and solubility. Typically an amphiphilic flexible surfactant can form small spherical micelles, larger cylinders, aligned cylinders, and lamellar layered structures as the surfactant concentration is increased. Bilayered and multiwalled structures are also a possibility. Given the observed surfactancy of silk proteins and the presence of micelle-like structures in solution, lyotropic liquid crystalline phase behavior suggests itself as a mechanism for organizing silk into high-concentration domains, biasing the orientation at the molecular level through the orientation of micelles. This mechanism possibly enhances water removal in the gland by providing clear channels for fluid to flow. In the final stages of the process, physical shear induces the transition from the hydrated (silk I) state to the crystalline (silk II) β-sheet content. This process induces micelle and globule aggregation and fibril formation.

A comprehensive mapping study of the domain structures in silks from insects and spiders related to protein assembly has been completed [49]. Domain types, sizes, and distributions were distinguished to identify consistent design features that have evolved to meet these requirements. In this study of all silks sequenced to date, silk proteins consist of hydrophilic domains flanking a very long central portion constructed from hydrophobic blocks separated by small hydrophilic features. The large internal highly repetitive hydrophobic blocks are the crystallizable domains that promote intra- and interchain folding with very short intervening hydrophilic blocks to control water content in micellar and globular states. These inclusions serve to prevent crystallization into β sheets until the latter stages of processing into fibers during spinning, and the larger N- and C-terminal hydrophilic blocks interact with water and define micellar partitioning [48]. In addition, sequences of all silks sequenced to date support these "design rules" for this family of hydrophobic proteins [49]. While these triblock designs remain consistent to fit within the aqueous processing environment and spinning requirements, there is also a significant sequence (chemistry) variability permitted in silks. These differences give rise to the variations in mechanical properties related to different functions for different silks. Therefore, an overall template of sequence design is required for the hydrophobic polymer processing, while leaving sufficient room for evolutionary inputs toward different fiber functions. For example, variances in the sequence chemistry in the crystallizable blocks among silks (e.g., polyalanine vs. polyalanine–glycine) are a source of differences in fiber mechanical properties: the polyalanines form a more stable structure. Since silks are the most hydrophobic proteins known yet and are processed in all aqueous environments in spiders and silkworms, these "design rules of nature" provide instructive guidance to synthetic polymer designs in terms of how to optimize processing and functional features of materials formed from these polymers.

6.7.5 Biomimetics

The modified triblock design described above has important subtle differences when compared to traditional synthetic diblock or triblock copolymers; specifically, large internal highly repetitive hydrophobic blocks (crystallizable domains that promote intra- and interchain folding) with very short intervening hydrophilic blocks (to control water content in micellar and globular states). These inclusions serve to prevent crystallization into β sheets until the latter stages of processing (i.e., spinning stage for native silk), and the larger N- and C-terminal hydrophilic blocks interact with water and define micellar partitioning [48]. The above design rules are likely similar in many fibrous proteins due to the "blocky" sequence structures, with long subdomains or primary sequences of predominately hydrophobic or hydrophilic amino acids. This consistency in amino acid patterning gives rise to relatively homogeneous secondary structures in contrast to globular proteins where only small subdomains of secondary structure form. This approach is already being exploited through synthetic block copolymer designs and chemistry to mimic the unusual assembly and materials features found in silks [50], as well as via genetically engineered variants of silks, such as elastin copolymers [51]. This has been a popular avenue to study due to the combined mechanical features of silks and elastomers.

6.8 SOLUBILIZATION

Silks are difficult to resolubilize due to the extensive hydrogen bonding and van der Waals interactions, and the exclusion of water from the intersheet regions. Silk in fiber form from silkworms and spiders are insoluble in water, dilute acids and alkali, and most organic solvents. They are also partially resistant to most proteolytic enzymes, with the exception of chymotrypsin and other V8 protease cocktails [40]. Silkworm fibroin can be solubilized by first degumming or removing the sericin using boiling soap solution or boiling dilute sodium bicarbonate solution, followed by immersion of the fibroin in high-concentration salt solutions such as lithium bromide, lithium thiocyanate, or calcium nitrate. These salt solutions cannot always be used to solubilize spider silk, and high concentrations of propionic acid–hydrochloric acid mixtures and formic acid are often required. After solubilization in these aggressive solvents, dialysis into water or buffers can be used to remove the salts or acids, although premature reprecipitation is a problem unless the solutions are kept at low temperature. Ternary-phase diagrams of silk, water, and chaotropic salt for processing windows have been published for native silkworm silks and for genetically engineered versions of silkworm silk [52].

6.9 REPROCESSED SILKS FOR NEW MATERIALS

Silk from the silkworm *B. mori* has been solubilized as outlined above, and then reformulated into new materials. Formats include films (Figure 6.3), electrospun fibers (Figure 6.4) and

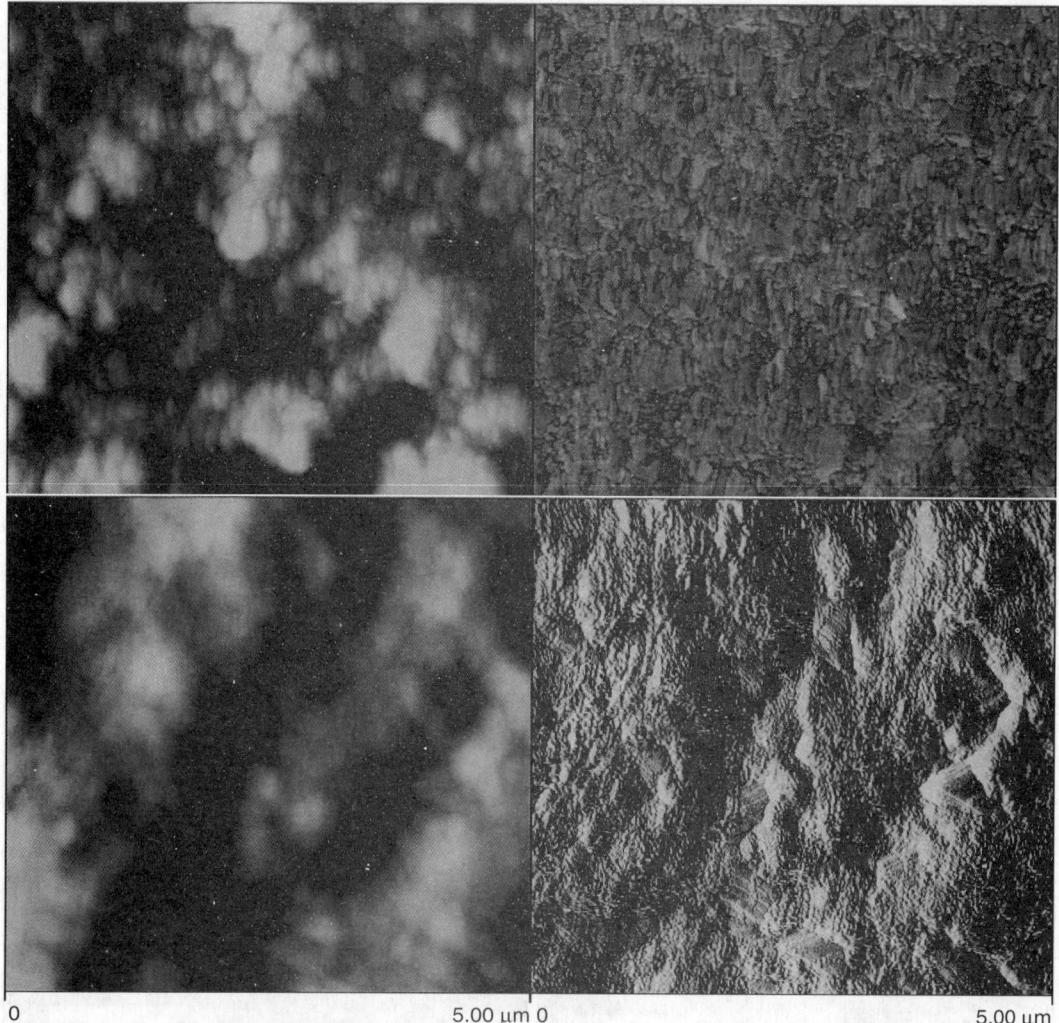

FIGURE 6.3 Atomic force microscopy images of silk films formed from reprocessed fibroin of *B. mori*. Two modes of processing are shown, all aqueous process (top), and methanol-induced β-sheet transition (bottom).

larger diameter fibers, gels, and porous matrices (Figure 6.5). In these studies, usually either aqueous processing or processes involving hexafluoroisopropanol are used, and solids content influence the mechanical properties of the final materials.

6.9.1 Films

Films or membranes formed from reprocessed silkworm silk have been produced by air-drying aqueous solutions prepared from the silk solutions after the salts are removed by dialysis. However, rapid gelation can occur at room temperature, so the solutions must be handled carefully. Maintaining solutions of higher concentrations at 4°C significantly slows

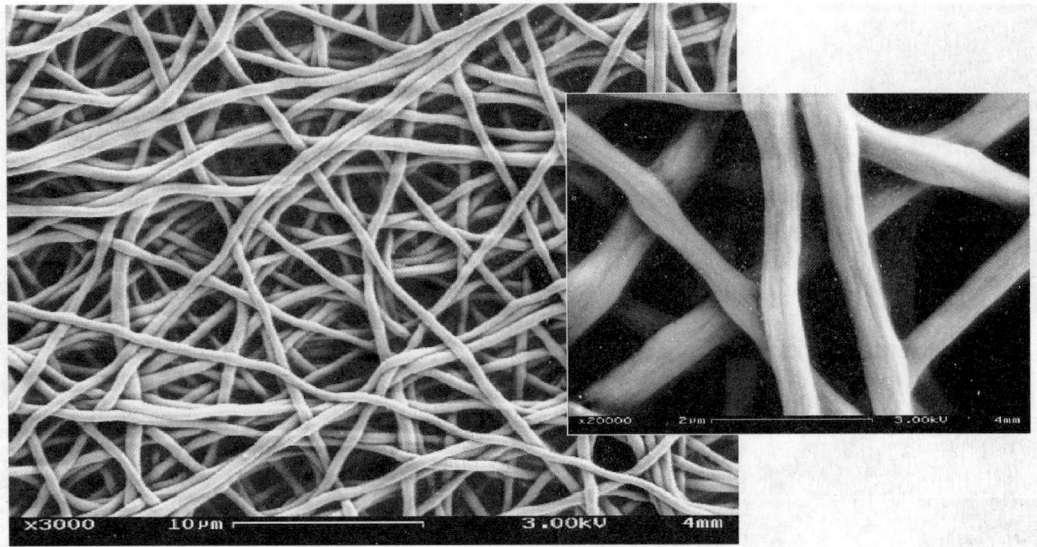

FIGURE 6.4 Images of electrospun fibers from reprocessed fibroin of *B. mori* to illustrate nanoscale diameter fibers with high surface area in nonwoven mat formats, compared to the micron scale diameter native fibers generated by this silkworm. Image on left (10 μm scale bar), on right (2 μm scale bar).

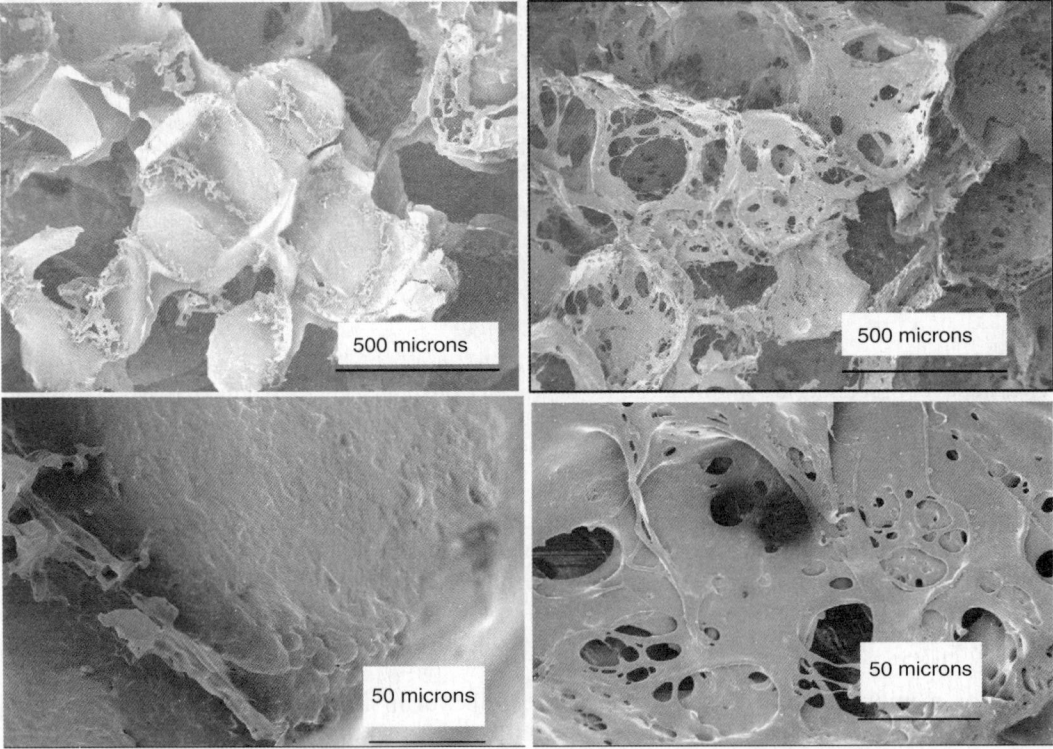

FIGURE 6.5 Scanning electron microscopy images of 3D porous scaffolds formed from reprocessed silkworm fibroin of *B. mori*. Top images are from organic solvent processing (HFIP) and bottom images are from all aqueous processing.

the gelation process and provides wider processing windows. The films formed from the water-soluble protein generally contain a silk I conformation with a significant content of random coil. Many different treatments have been used to modify these films to decrease water solubility by conversion of the protein to the silk II polymorph. Most commonly methanol has been used to induce this structural transition. This process was used successfully to entrap enzymes, although the materials embrittle with time [53]. These types of silk membranes have also been cast from fibroin solutions and characterized for permeation properties. Oxygen and water vapor transmission rates were dependent on the exposure conditions to methanol to facilitate the conversion to silk II [53]. Thin monolayer films have been formed from soloublized silkworm silk using Langmuir techniques to facilitate structural characterization of the protein [54]. In recent studies, methanol treatments were avoided and an all-water annealing process for the films was utilized [54]. This process resulted in a different structural content of the films, reduced silk II content, but the films were insoluble in water and retained flexibility over time unlike the same materials when treated with methanol rapidly embrittled. These approaches have also been used to control drug release from films, since the structural state can be modified by processing conditions, leading to control of release profiles of entrapped drugs.

6.9.2 Fibers

Resolubilized silkworm cocoon silk and genetically engineered variants of silk have also been spun into fibers [56]. These fibers do not exhibit the remarkable mechanical properties of the native materials. Electrospun silk has also been generated from silkworm, spider, and genetically engineered silks [57,58]. These fibers can be formed with diameters in the hundreds to thousands of nanometer diameter size range, depending on spinning conditions (solids content, water or organic solvents). These fibers can also be mineralized to further stiffen the materials to expand their potential applications [59].

6.9.3 Hydrogels

Hydrogels can be formed from reprocessed silkworm silk [60]. These gels are formed with aqueous solutions of the fibroin prepared as outlined earlier. The rate of sol–gel transition is directly dependent on temperature (higher the temperature the more rapid the gelation), pH (lower the pH the more rapid the gelation), and solids content (higher the solids higher the rates of gelation). Cations can also enhance rates of gelation, with the specific salts dependent on the type of silk; potassium plays a role with spider dragline silk and calcium with silkworm silk. The overall rate of gelation has been controlled via osmotic stress, with resulting mechanical, morphological, and structural details dependent on the rate and extent of water removal.

6.9.4 Spongy 3D Porous Materials

3D porous matrices with control of pore size, porosity, and structural content can be formed from reprocessed silk fibroin, either via organic solvent or aqueous processes, and using either salt leaching, gas foaming or freeze drying techniques [61,62]. The nature of the process used to form the porous matrices directly influences mechanical properties, degradability, and cell and tissue formation when used as scaffolds in tissue engineering. Pore sizes in these scaffolds can be controlled from below 100 μm diameter to above 1,000 μm, depending on processing conditions.

6.10 DEGRADABILITY

Silk fibers in textile form degrade over time on exposure to ultraviolet light due to the presence of aromatic amino acids: tyrosine, tryptophan, and phenylalanine [63]. Silk fibers in suture form for biomedical uses also slowly degrade due to proteolytic degradation usually mediated by a foreign body response, with rates dependent on the location *in vivo*. Silk fibers generally lose most of their tensile strength within 1 year *in vivo* [64]. The rate of absorption *in vivo* is dependent on the implantation site, mechanical environment, and variables related to the physiological status of the patient. In recent proteolytic degradation studies *in vitro*, native silkworm fibers lost mass, diameter, and tensile strength in a linear fashion with time, leading to predictable outcomes in terms of fiber function [65].

6.11 GENETIC ENGINEERING

Many attempts have been and are made to clone and express silk and silk-like proteins, either by cloning and expressing cDNAs or more commonly by exploiting the block-like domain features highlighted above through the use of synthetic gene technology. Genetically engineered or recombinant DNA silkworm and spider silks have been produced using either cDNAs isolated from the host animal, or through the use of synthetic genes generated by oligonucleotide synthesis and subsequent cloning into host systems such as bacteria and yeast to generate heterologous silk-like proteins. The aim has been either to overcome the inherent limitations of available quantities of spider silks through heterologous expression or to generate "designer" silks to optimize selective features in the recombinant silk-like proteins for either fundamental study or applied needs [66,67]. An understanding of the genetics of silk production in silkworms and spiders should help in the developing processes for higher levels of silk expression by recombinant deoxyribonucleic acid (DNA) methods. This is of particular interest for spider silks due to the lack of commercial production processes. In addition, the high protein production capability of the silkworm, about 300 μg of silk fibroin per epithelial cell in the silk producing glands, has generated interest from molecular biologists and developmental biologists in elucidating the genetic regulation of this system and then exploiting it for the production of therapeutic and other proteins—transgenic silkworms.

6.11.1 SYNTHETIC SILK GENES

Variations in synthetic gene construction strategy as well as the use of either *Escherichia coli* or the yeast, *Pichia pastoris,* host systems for expression of heterologous proteins have been reported [68–70]. Often, in the above studies, yield and homogeneity of the products of longer genes are limited by premature termination during synthesis. In general, expression levels obtained from synthetic silk genes are low with yields of about 1 to 10 mg/l, representing usually less than 5% of the total protein in the cell, depending on the size of the protein. Precipitation and nonspecific interactions are some of the reasons cited for significant losses of protein during purification in some of the above studies. Tobacco and potato plants have been successfully utilized as transgenic hosts to generate silk-like proteins from synthetic genes [71]. Similarly, variants of these genes have been formed, such as block copolymer systems with elastins, to generate new functions with silk-like proteins. Variations in native silkworm populations indicate a high degree of polymorphism in length and organization of fibroin genes and thus the encoded proteins [72]. A significant degree of variation in protein size, apparently, can be tolerated in native populations of silkworms likely due to the highly repetitive nature of the genes and the encoded proteins, a characteristic of fibrous proteins in general. The result of this genetic organization is that the deletion or addition of protein sequence repeats presumably has little impact on secondary structure and functional performance above a threshold size.

Peptides and genetically engineered proteins with sequence variants from the consensus sequence of the major ampullate spider dragline silk protein from *N. clavipes* were used to assess conformational transitions between soluble and less soluble states. Molecular solubility triggers were incorporated into the silk variants at various positions in relation to the β-sheet-forming polyalanine region, either chemically activated methionine redox or enzymatically activated phosphorylation and dephosphorylation triggers [73,74,75]. The chemical redox reactions were rapid with no significant changes observed beyond 30 min. Biophysical characterization of the protein demonstrated a decrease in β-sheet structure and a corresponding increase in random coil structure on oxidation.

6.11.2 cDNA Expression

The first successful expression of cDNA clones of silks were with partial cDNA clones from *N. clavipes* major ampullate silk gland cDNA [76]. The 1.7-kb fragment was from the 3′-terminus of the major ampullate silk gene and a 43 kDa recombinant silk protein was expressed and characterized. Most recently, partial cDNA clones of the 3′ end of spider silks cDNAs from *Araneus diadematus* were cloned and expressed in mammalian cells as part of an effort to develop transgenic animals that express silk proteins [77]. Transgenic expression of silks in plants (tobacco and potato) and mammalian epithelial cells has been reported [71,77] and may point the way toward more substantive production of these proteins in the future.

TABLE 6.4
Comparison of Mechanical Properties of Common Silks (Silkworm and Spider Dragline) to Several Types of Biomaterial Fibers and Tissues Commonly Used Today

Material	UTS (MPa)	Modulus (GPa)	% Strain @ break	References
B. mori Silk (w/sericin)[a]	500	5–12	19	[25]
B. mori Silk (w/o sericin)[b]	610–690	15–17	4–16	[25]
B. mori Silk[c]	740	10	20	[25]
Spider Silk[d]	875–972	11–13	17–18	[25]
Collagen[e]	0.9–7.4	0.0018–0.046	24–68	[89]
Collagen X-linked[f]	47–72	0.4–0.8	12–16	[89]
PLA[g]	28–50	1.2–3.0	2–6	[90]
Tendon (comprised of mainly collagen)	150	1.5	12	[91]
Bone	160	20	3	[91]
Kevlar (49 fiber)	3600	130	2.7	[91]
Synthetic Rubber	50	0.001	850	[91]

[a]*Bombyx mori* silkworm silk—determined from bave (multithread fibers naturally produced from the silk worm coated in sericin).
[b]*Bombyx mori* silkworm silk—determined from single brins (individual fibroin filaments following extraction of sericin).
[c]*Bombyx mori* silkworm silk—average calculated from data in Cunniff et al. (1994).
[d]*Nephila clavipes* silk produced naturally and through controlled silking.
[e]Rat-tail collagen type I extruded fibers tested after stretching from 0 to 50%.
[f]Rat-tail collagen dehydrothermally cross-linked and tested after stretching from 0 to 50%.
[g]Polylactic acid with molecular weights ranging from 50,000 to 300,000.

6.12 BIOMEDICAL MATERIALS

Silk from the silkworm, *B. mori*, has been used as biomedical suture material for centuries. The unique mechanical properties of these fibers provide important clinical repair options. Biocompatibility problems reported for silkworm silk can result from contamination from residual sericin, thus proper processing to remove these glue-like proteins is important [64]. More recent studies with well-defined silkworm silk fibers suggest that the silk fibroin fibers exhibit biocompatibility *in vitro* and *in vivo* comparable with other commonly used biomaterials [64,78–82,83]. Furthermore, the unique mechanical properties of the silk fibers, the diversity of side-chain chemistries for "decoration" with growth and adhesion factors, and the ability to genetically tailor the protein provide additional rationale for the exploration of this family of fibrous proteins for biomaterial applications. Silk has been used in native fiber form as sutures for wound ligation and became the most common natural suture, surpassing collagen over the past 100 years [64]. Silk sutures are used in ocular, neural, and cardiovascular surgeries, as well as a variety of other tissues in the body. Silk's knot strength, handling characteristics, and the ability to lie low to the tissue surface make it a popular suture in cardiovascular applications. The above features of silk fibroin have led to the recent emergence of silk-based biomaterials for a wide range of cell and tissue studies [84–88]. These efforts demonstrate the importance of silk biomaterials in tissue engineering [78–82].

6.13 APPLICATIONS

With improved analytical techniques, together with the tools of biotechnology, a new generation of silk-related materials is envisioned that depart from traditional textiles and medical sutures, which are currently the focus for these materials. These proteins are already finding broadened applications in medical fields as biomaterials. The unique and remarkable mechanical properties of silks has already prompted studies for their application in high-performance materials, as well as modes to mimic these features through more traditional synthetic organic polymer chemistry routes. Since the specifics of the silks can be modulated through genetic manipulation, through processing and through choice of starting protein, a great deal of control of structure, morphology, and functional attributes in materials derived from these proteins can be obtained. This control is already exploited in the biomaterials arena and suggests that this family of novel proteins can serve as an important blueprint for understanding structure–function relationships as well as for new and novel materials.

A number of consumer products are already promulgated based on silks, including cosmetics, hair replacements and shampoos, among others. Sutures, biomaterials for tissue repairs, wound coatings, artificial tendons, bone repair, and related needs may be possible applications since silks are biocompatible and slowly degrade in the human body. Genetically engineered silks have also been commercialized as cell culture plate coatings to improve cell adhesion. Genetic variants of silks are also actively pursued as controlled release systems to deliver pharmaceuticals in a variety of systems.

ACKNOWLEDGMENTS

The authors wish to thank many postdoctoral fellows, graduate students, undergraduate students, and collaborative scientists who have worked with the group over the years. We also wish to thank the NIH and the NSF for support in the various stages of this effort. We thank Adam Collette (Tufts University) for his pictures of the silk fibers.

REFERENCES

1. Robson, R.M. Silk composition, structure and properties. In: *Fiber Chemistry Handbook of Science and Technology*, Lewin, M., Pearce, E.; Eds., Marcel Dekker, New York, Vol. IV (1985).
2. Kaplan, D.L.; Lombardi, S.J.; Muller, W.S.; Fossey, S.A. Silks. In: *Biomaterials: Novel Materials from Biological Sources*. Byrom, D.; Ed., Stockton Press, New York (1991).
3. Kaplan, D.L.; Adams, W.W.; Viney, C.; Farmer, B. Silks: Materials Science and Biotechnology, *American Chemical Society Symposium Series* 544, (1994).
4. Kaplan, D.L.; Mello, C.; Fossey, S.; Arcidiacono, S.; Muller, W., Silks. In *Protein-Based Materials*. Birkahuser, Boston, (1998).
5. Kaplan, D.L.; Mello, S.M.; Arcidiacono, S.; Fossey, S.; Senecal, K.; Muller, W.; Silk. In: *Protein-Based Materials*. McGrath, K., Kaplan, D.L.; Eds.,. Birkhauser, Boston, (1998).
6. Asakura, T.; Kaplan, D.L. Silk production and processing. In: *Encyclopedia of Agricultural Science*. Academic Press, New York (1995).
7. Zhou, C.Z.; Confalonieri, F.; Medina, N.; Zivanovic, Y.; Esnault, C.; Yang, T.; Jacquet, M.; Janin, J.; Duguet, M.; Perasso, R.; Li, Z.G. Fine organization of Bombyx mori fibroin heavy chain gene. *Nucleic Acids Res.* 28(12):2413–2419 (2000).
8. Zhou, Y.; Wu, S.; Conticello, V.P. Genetically directed synthesis and spectroscopic analysis of a protein polymer derived from a flagelliform silk sequence. *Biomacromolecules* 2:111–115 (2001).
9. Zurovec, M.; Sehnal, F. Unique molecular architecture of silk fibroin in the waxmoth, *Galleria mellonella*. *J. Biol. Chem.* 277: 22639–22647 (2002).
10. Foelix, R.F. *Biology of Spiders*. Harvard University Press, Cambridge, MA, (1992).
11. Denny, M.W. The physical properties of spider's silks and their role in the design of orb-webs. *J. Exp. Biol.* 65:483–506 (1976).
12. Denny, M.W. Silks: Their properties and functions. In: *Mechanical Properties of Biological Materials*. Vincent, J.F.V.; Currey, J.D.; Eds., Cambridge University Press, Washington DC, (1980).
13. Andersen, S.O. Amino acid composition of spider silks. *Comp. Biochem. Physiol.* 35:705–711 (1970).
14. Ilzuka, E. Silk thread: Mechanism of spinning and its mechanical properties. *J. Appl. Polym. Sci.* Japan, 41:173–185 (1985).
15. Magoshi, J.; Magoshi, Y.; Nakamura, S. Crystallization, liquid crystal, and fiber formation of silk fibroins. *J. Appl. Polym. Sci.: Appl. Polym. Symp.* 41:187–204 (1985).
16. Work, R.W. A comparative study of the supercontraction of major ampullate silk fibers of orb-web building spiders (Araneae). *J. Arachnol.* 9:299–308 (1981).
17. Hinman, M.B.; Lewis, R.V. Isolation of a clone encoding a second dragline silk fibroin. *J. Biol. Chem.* 267:19320–19324 (1992).
18. Guerette, P.A.; Ginzinger, D.G.; Weber, B.H.F.; Gosline, J.M. Silk properties determined by gland-specific expression of a spider fibroin gene family. *Science* 272: 112–115 (1996).
19. Hayashi, C.Y.; Lewis, R.V. Evidence from flagelliform silk cDNA for the structural basis of elasticity and modular nature of spider silks. *J. Mol. Biol.* 275:773–784 (1998).
20. Hayashi, C.Y.; Lewis, R.V. Molecular architecture and evolution of a modular spider silk protein gene. *Science* 287:1477–1479 (2000).
21. Hayashi, C.Y.; Lewis, R.V. Spider flagelliform silk: lessons in protein design, gene structure, and molecular evolution. *Bioessays* 23:750–756 (2001).
22. Gosline, J.M., Denny, M.W., DeMont, M.E. Spider silk as rubber. *Nature* 309:551–552 (1984).
23. Gosline, J.M.; DeMont, M.E.; Denny, M.W. The structure and properties of spider silk. *Endeavour* 10:37–43 (1986).
24. Cunniff, P.M.; Fossey, S.A.; Auerbach, M.A.; Song, J.W.; Kaplan, D.L.; Adams, W.W.; Eby, R.K.; Mahoney, D.; Vezie, D.L. Mechanical and thermal properties of dragline silk from the spider *Nephila clavipes*. *Poly. Adv. Technol.* 5:401–410 (1994).
25. Perez-Rigueiro, J.; Viney, C.; Llorca, J.; Elices, M. Mechanical properties of single-brin silkworm silk. *J. Appl. Polym. Sci.* 75:1270–1277 (2000).
26. Mahoney, D.V.; Vezie, D.L.; Eby, R.K.; Adams, W.W.; Kaplan, D. Aspects of the morphology of drag line silk of *Nephia-clavipes*. Silk polymers. 544:196–210 (1994).

27. Inoue, S.; Tanaka, K.; Arisaka, F.; Kimura, S.; Ohtomo, K.; Mizuno, S. Silk fibroin of *Bombyx mori* is secreted, assembling a high molecular mass elementary unit consisting of H-chain, L-chain, and P25, with a 6:6:1 molar ratio. *J. Biol. Chem.* 275: 40517–40528 (2000).
28. Case, S.T.; Powers, J.; Hamilton, R.; Burton, M. *J. Silk Polymers*. 544:80–90 (1994).
29. Mita, K.; Ichimura, S.; James, T.C. Highly repetitive structure and its organization of silk fibroin gene. *J. Mol. Evol*. 38:583–592 (1994).
30. Xu, M.; Lewis, R.V. Structure of a protein superfiber: Spider dragline silk. *Proc. Natl. Acad. Sci. U.S.A.* 87: 7120–7124 (1990).
31. Vollrath, F.; Knight, D.P. Liquid crystalline spinning of spider silk. *Nature* 410:541–548 (2001).
32. Valluzzi, R.; Probst, W.; Jacksch, H.; Zellmann, E.; Kaplan, D.L. Patterned peptide multilayer thin films with nanoscale order through engineered liquid crystallinity. *Soft Mater*. 1:245–262 (2003).
33. Marsh, R.E.; Corey, R.B.; Pauling, L. An investigation of the structure of silk fibroin. *Biochim. Biophys. Acta* 16:1–34 (1955).
34. Lucas, F.; Shaw, J.T.B.; Smith, S.G. Comparative studies of fibroins. I. The amino acid composition of various fibroins and its significance in relation to their crystal structure and toxonomy. *J. Mol. Biol*. 2:339–349 (1960).
35. Fraser, R.D.B.; MacRae, T.P. Silks. In: *Conformation in Fibrous Proteins*. Academic Press, New York, (1973).
36. Simmons, A.; Ray, E.; Jelinski, L.W. Solid-state C-13 Nmr of *Nephia-clavipes* dragline silk establishes structure and identity of crystalline regions. *Macromolecules* 27:5235–5237 (1994).
37. Simmons, A.H.; Michal, C.A.; Jelinski, L.W. Molecular orientation and two-component nature of the crystalline fraction of spider dragline silk. *Science* 271:840–887 (1996).
38. Asakura, T.; Demura, M.; Date, T.; Miyashita, N.; Ogawa, K.; Williamson, M.P. NMR study of silk I structure of Bombyx mori silk fibroin with N-15- and C-13-NMR chemical shift contour plots. *Biopolymers*. 41:193–203 (1997).
39. Monti, P.; Taddei, P.; Freddi, G.; Asakura, T.; Tsukada, M. Raman spectroscopic characterization of Bombyx mori silk fibroin: Raman spectrum of Silk I. *J. Raman Spect*. 32:103–107 (2001).
40. Mello, C.M.; Senecal, K.; Yeung, B.; Vouros, P.; Kaplan, D. Initial characterization of *Nephia-clavipes* dragline protein. *Silk polymers* 554:67–79 (1994).
41. Valluzzi, R.; Winkler, S.; Wilson, D.; Kaplan, D.L. Silk: Molecular organization and control of assembly. *Phil. Trans. R. Soc. Land. B* 357:165–617 (2002).
42. Chen, X.; Knight, D.; Shao, Z.; Vollrath, F. Conformation transition in silk protein films monitored by time-resolve Fourier Transform Infrared Spectroscopy: Effect of potassium ions on Nephila Spidroin films. *Biochemistry*. 41:14944–14950 (2002).
43. Fossey, S.A.; Nemthy, G.; Gibson, K.D.; Scheraga, H.A. Conformational energy studies of β sheets of model silk fibroin peptides. I. Sheets of alanine and glycine. *Biopolymers* 3:1529 (1991).
44. Muller, W.S.; Samuelson, L.A.; Fossey, S.A.; Kaplan, D.L. Formation and characterization of Langmuir silk films. *Langmuir*, 9:1857–1861 (1993).
45. Valluzzi, R.; Gido, S.P.; Zhang, W.; Muller, W.S.; Kaplan, D.L. A trigonal crystal structure of *Bombyx mori* silk incorporating a threefold helical chain conformation found at the air-water interface. *Macromolecules* (1996).
46. Kerkam, K.; Viney, C.; Kaplan, D.L.; Lombardi, S.J. Liquid crystalline characteristics of natural silk secretions. *Nature* 349:596–598 (1991).
47. Knight, D.P.; Vollrath, F. Changes in element composition along the spinning duct in *Nephila* spider. *Naturwissenschaften* 88(4):179–182 (2001).
48. Jin, H.-J.; Kaplan, D.L. Mechanism of processing silk in insects and spiders. *Nature* 424:1057–1061 (2003).
49. Bini, E.; Knight, D.P.; Kaplan, D.L. Mapping domain structures in silks from insects and spiders related to protein assembly. *J. Mol. Biol*. 335:27–40 (2004).
50. Rathore, O.; Winningham, M.J.; Sogah, D.Y.; *J. Polym. Sci. Pol. Chem*. 38:352–366 (2000).
51. Capello, J.; Crissman, J.; Dorman, M.; Mikolajczak, M.; Textor, G.; Marquet, M.; Ferrari, F.A. Genetic engineering of structural protein polymers. *Biotech. Prog*. 6:198–202 (1990).
52. Perez-Rigneiro, J.; Viney, C.; Llorca, J.; Elices, M. Silkworm silk as an engineering material. *J. Appl. Poly. Sci*. 70:2439–2447 (1998).

53. Asakura, T.; Yoshimizu, H.; Kakizaki, M. An ESR study of spin-labeled silk fibroin membranes and spin-labeled glucose-oxidase immobilized in silk fibrion membranes. *Biotechnol. Bioeng.* 35:511–517 (1990).
54. Muller, W.S.; Samuelson, L.A.; Fossey, S.A.; Kaplan, D.L. Formation and characterization of Langmuir silk films. *Langmuir* 9:1857–1861 (1993).
55. Jin, H.J.; Park, J.; Karageorgiou, V.; Kim, U.J.; Valluzzi, R.; Cebe, P.; Kaplan, D.L. Water-stable silk films with reduced β-sheet content. *Advanced Functional Materials*, in press (2005).
56. Hudson, S.M. The spinning of silklike protein into fibers. In: *Fibrous Proteins*. McGrath, K.; Kaplan, D.L.; Eds., Birkhouser (1998).
57. Jin, H-J.; Fridrikh, S.V.; Rutledge, G.C.; Kaplan, D.L. Electrospinning *Bombxy mori* silk with poly(ethylene oxide). *Biomacromolecules* 3:1233–1239 (2002).
58. Jin, H-J.; Park, J.; Kim, U.-J.; Valluzzi, R.; Cebe, P.; Kaplan, D.L. Biomaterials films of *Bombyx mori* silk with poly(ethylene oxide). *Biomacromolecules* 5:711–717 (2004).
59. Li, C.M.; Jin, H.J.; Botsaris, G.; Kaplan, D.L. Silk apatite composites from electrospun fibers. *J. Mater. Res.* 20:3374–3384 (2005).
60. Kim, U.-J.; Park, J.; Li, C.; Jin, H.-J.; Valluzzi, R.; Kaplan, D.L. Structure and properties of silk hydrogels. *Biomacromolecules*. 5:786–792 (2004).
61. Nazarov, R.; Jin, H.-J.; Kaplan, D.L. Porous 3D scaffolds from regenerated silk fibroin. *Biomacromolecules* 5:718–726 (2004).
62. Kim, U.J.; Park, J.; Kim, H.J.; Wada, M.; Kaplan, D.L. Three dimensional aqueous-derived biomaterial scaffolds from silk fibroin. *Biomaterials*, in press (2005).
63. Becker, M.A.; Tuross, N. Initial degradation changes found in *Bombyx mori* silk fibroins. In: *Silk Polymers: Materials Science and Biotechnology*. Kaplan, D.L.; Adams, W.W.; Farmer, B.; Viney, C.; Eds., American Chem. Soc. Symposium Series, 544 (1994).
64. Altman, G.H.; Diaz, F.; Jakuba, C.; Calabro, T.; Horan, R.L.; Chen, J.; Lu, H.; Richmond, J.; Kaplan, D.L.; Silk based biomaterials. *Biomaterials*. 24:401–416 (2003).
65. Horan, R.L.; Antle, K.; Collette, A.L.; Wang, Y.; Huang, J.; Moreau, J.E.; Volloch, V.; Kaplan, D.L.; Altman, G.H. In vitro degradation of silk fibroin. *Biomaterials* 26:3385–3393 (2005).
66. Winkler, S.; Wilson, D.; Kaplan, D.L. Controlling β-sheet assembly in genetically engineered silk by enzymatic phosphorylation/dephosphorylation. *Biochemistry* 39:12739–12746 (2000).
67. Wong Po Foo, C.; Kaplan, D.L. Genetic engineering of fibrous proteins: spider dragline silk and collagen. *Adv. Drug Deliv. Rev.* 54:1131–1143 (2002).
68. Prince, J.T.; McGrath, K.P.; DiGirolamo, C.M.; Kaplan, D.L. Construction, cloning and expression of synthetic genes encoding spider dragline silk. *Biochemistry* 34:10879–10884 (1995).
69. Fahnestock, S.R.; Bedzyk, L.A.; Production of synthetic spider dragline silk protein in *Pichia pastoris*. *Appl. Microbiol. Biot.* 47: 33–39 (1997).
70. Fukushima, Y. Genetically engineered syntheses of tandem repetitive polypeptides consisting of glycine-rich sequence of spider dragline silk. *Biopolymers* 45:269–279 (1998).
71. Scheller, J.; Guhrs, K.H.; Grosse, F.; Conrad, U. Production of spider silk proteins in tobacco and potato. *Nat. Biotechnol.* 19:573–577 (2001).
72. Sprague, K.U.; Roth, M.B.; Manning, R.F.; Gage, L.P. Alleles of the fibroin gene coding for proteins of different lengths. *Cell* 17:407–413 (1979).
73. Valluzzi, R.; Szela, S.; Avtges, P.; Kirschner, D.; Kaplan, D.L. Methionine redox-controlled crystallization of biosynthetic silk spidroin. *J. Phys. Chem.* 103:11382–11392 (1999).
74. Szela, S.; Avtges, P.; Valluzzi, R.; Winkler, S.; Wilson, D.; Kirschner, D.; Kaplan, D.L. Reduction-oxidation control of β-sheet assembly in genetically engineered silk. *Biomacromolecules* 1:534–542 (2000).
75. Winkler, S.; Kaplan, D.L. Molecular biology of spider silk. *Rev. Mol. Biotechnol.* 74: 85–95 (2000).
76. Arcidiacono, S.; Mello, C.; Kaplan, D.L.; Cheley, S.; Bayley, H. Purification and characterization of recombinant spider silk expressed in Escherichia coli. *Appl. Microbiol. Biot.* 49:31–38 (1998).
77. Lazaris, A.; Arcidiacono, S.; Huang, Y.; Zhou, J.F.; Duguay, F.; Chretien, N.; Welsh, E.A.; Soares, J.W.; Karatzas, C.N. Spider silk fibers spun from soluble recombinant silk produced in mammalian cells. *Science* 295:472–475 (2002).

78. Meinel, L.; Karageorgiou, V.; Hoffmann, S.; Fajardo, R.; Snyder, B.; Li, C.; Zichner, L.; Langer, R.; Vunjak-Novakovic, G.; Kaplan, D.L. Tissue engineering of osteochondral plugs using human mesenchymal stem cells and silk scaffolds. *Chem. Ind.* 58:68–69 (2004).
79. Meinel, L.; Kargeorgiou, V.; Hofmann, S.; Fajardo, R.; Snyder, B.; Li, C.; Zichner, L.; Langer, R.; Vunjak-Novakovic, G.; Kaplan, D.L. Engineering bone-like tissue *in vitro* using human bone marrow stem cells and silk scaffolds. *J. Biomed. Mater. Res.* 71A:25–34 (2004).
80. Meinel, L.; Karageorgiou, V.; Fajardo, R.; Snyder, B.; Shinde-Patil, V.; Zichner, L.; Kaplan, D.L.; Langer, R.; Vunjak-Novakovic, G. Bone tissue engineering using human mesenchymal stem cells: effects of scaffold material and medium flow. *Annals Biomed. Eng.* 32:112–122 (2004).
81. Meinel, L.; Hofmann, S.; Karageorgiou, V.; Kirker-Head, C.; McCool, J.; Gronwicz, G.; Zichner, L.; Langer, R.; Vunjak-Novakovic, G.; Kaplan, D.L. Inflammatory responses to silk films *in vitro* and *in vivo*. *Biomaterials* 26:147–155 (2004).
82. Meinel, L.; Hoffmann, S.; Karageorgiou, V.; Zichner, L.; Langer, R.; Vunjak-Novakovic, G., Kaplan, D.L. Engineering cartilage-like tissue using human mesenchymal stem cells and silk protein scaffolds. *Biotechnol. and Bioeng.* 88:379–391 (2004).
83. Panilaitis, B.; Altman, G.H.; Chen, J.; Jin, H.-J.; Karageorgiou, V.; Kaplan, D.L. Macrophage responses to silk. *Biomaterials* 24:3079–3085 (2003).
84. Demura, M.; Asakura, T. Porous membrane of *Bombyx mori* silk fibroins: Structure characterization, physical properties and application to glucose oxidase immobilization. *J. Membr. Sci.* 59:39–52 (1991).
85. Minoura, N.; Tsukada, M.; Nagura, M. Physico-chemical properties of silk fibroin membrane as a biomaterial. *Biomaterials* 11:430–434 (1990).
86. Minoura, N.; Aiba, S.; Gotoh, Y.; Tsukada, M.; Imai, Y. Attachment and growth of cultured fibroblast cells on silk protein matrices. *J Biomedical Materials Research* 29:1215–1221 (1995).
87. Santin, M.; Motta, A.; Freddi, G.; Cannas M. *In vitro* evaluation of the inflammatory potential of the silk fibroin. *J. Biomed. Mater. Res.* 46:382–389 (1999).
88. Sofia, S.; McCarthy, M.B.; Gronowicz, G.; Kaplan, D.L. Functionalized silk-based biomaterials for bone formation. *J. Biomed. Mater. Res.* 54:139–148 (2001).
89. Pins, G.D.; Christiansen, D.L.; Patel, R.; Silver, F.H. Self-assembly of collagen fibers: Influence of fibrillar alignment and decorin on mechanical properties. *Biophys. J.* 73:2164–2172 (1997).
90. Engelberg, I.; Kohn, J. Physiocomechanical properties of degradable polymers used in medical applications: a comparative study. *Biomaterials* 12:292–304 (1991).
91. Aoron, B.B.; Gosline, J.M. Optical-properties of single elastin fibers indicate random protein conformation. *Nature* 287:865–867 (1980).

7 Jute and Kenaf

Roger M. Rowell and Harry P. Stout

CONTENTS

7.1 Introduction ... 406
7.2 Formation of Fiber ... 407
7.3 Separation of Bast Fiber from Core ... 408
7.4 Fiber Structure ... 409
7.5 Chemical Composition ... 411
7.6 Acetyl Content ... 412
7.7 Changes in Chemical and Fiber Properties during the Growing Season ... 414
7.8 Fine Structure ... 419
7.9 Physical Properties ... 420
7.10 Grading and Classification ... 421
7.11 Fiber and Yarn Quality ... 423
7.12 Chemical Modification for Property Improvement ... 424
    7.12.1 Acetylation ... 425
    7.12.2 Cyanoethylation ... 427
7.13 Photochemical and Thermal Degradation ... 428
7.14 Moisture Effects ... 429
7.15 Fastness to Light ... 430
    7.15.1 Undyed Jute ... 430
    7.15.2 Dyed Jute ... 431
7.16 Woolenization ... 432
7.17 Applications and Markets ... 433
    7.17.1 Composites ... 433
    7.17.2 Geotextiles ... 434
    7.17.3 Filters ... 437
    7.17.4 Sorbents ... 438
    7.17.5 Structural Composites ... 438
    7.17.6 Nonstructural Composites ... 438
    7.17.7 Molded Products ... 438
    7.17.8 Packaging ... 439
    7.17.9 Pulp and Paper ... 439
    7.17.10 Pultrusion ... 442
    7.17.11 Combinations with Other Resources ... 442
    7.17.12 Fiber Thermoplastic Blends ... 443
    7.17.13 Fiber Matrix Thermoplasticization ... 447
    7.17.14 Fiber Thermoplastic Alloys ... 449
    7.17.15 Charcoal ... 449

7.18 Future Trends 449
References 450

7.1 INTRODUCTION

Jute is the common name given to the fiber extracted from the stems of plants belonging to the genus *Corchorus*, family *Tiliaceae*, whereas kenaf is the name given to a similar fiber obtained from the stems of plants belonging to the genus *Hibiscus*, family *Malvaceae*, especially the species *H. cannabinus* L. Only two species of *Corchorus*, namely *C. capsular* L. and *C. olitorius* L., are grown commercially, although around 40 wild species are known, whereas other species of *Hibiscus*, particularly *H. sabdariffa* L. are sometimes also marketed as kenaf.

These plants are examples of a number of woody-stemmed herbaceous dicotyledons grown in the tropics and subtropics. Fibers can be extracted from the bast of stems of these plants. Most of the plants cultivated for fiber are grown from seeds annually, as are jute and kenaf, but a few are grown as perennials. Jute is the most important fiber of this type, and it is probable that, in the industrial and engineering uses of textiles, jute is used more than any other single fiber. Kenaf finds use in the domestic market in many countries, but its demand in the international market is much less than that of jute, and estimates of world kenaf production are liable to be erroneous. In many marketing statistics, the production or utilization of "jute and allied fibers" is given to include all the fibers in this group. "Allied fibers" are suitable for processing on jute spinning systems.

Favorable conditions for jute cultivation are found in the deltas of the great rivers of the tropics and subtropics such as the Ganges, the Irrawaddy, the Amazon, and the Yangtze, where alluvial soils and irrigation, often by extensive flooding, are combined with long day lengths to provide an opportunity for considerable vegetative growth before flowering (see Table 7.1). Jute has an optimum growing temperature between 18 and 33°C with a minimum annual moisture requirement of 250 mm in a soil pH between 6.6 and 7.0. Kenaf has an optimum growing temperature between 22 and 30°C with a minimum annual moisture requirement of 150 mm in a soil pH between 6.0 and 6.8. Jute has a growing cycle of approximately 120–150 days with an average yield of 2200 kg/ha, while kenaf has a growing cycle of 150 to 180 days with an average yield of 1700 kg/ha. Since kenaf requires less water to grow than jute, it is now grown in several countries in Europe and South America, and in Mexico, United States, Japan, and China.

Both jute and kenaf grow to 2.5–3.5 m in height at maturity, but kenaf, although it still requires a long day length for vegetative growth, flourishes in drier conditions than jute and can adapt to a wider variety of soils and climates. As a result, it is preferred to jute as a fiber

TABLE 7.1
Climatic Requirements for Growing Jute and Kenaf

Common name	Optimum temperature (°C)	Minimum moisture[a] (mm)	Optimum soil pH	Growing cycle (days)	Fiber yield (kg/ha)
Jute	18–33	250	6.6–7.0	120–150	2200
Kenaf	22–30	120	6.0–6.8	150–180	1700

[a] Water required during the growing season.

crop by many countries in Africa and Latin America, although usually only for internal use. Bangladesh remains the world's principal exporter of this type of fiber, with exports of jute fiber currently running at around 500,000 t/year. This compares with an FAO forecast for world consumption of manufactured jute goods of 4 million tons in 1985.

The commercial use of the base fibers dates back over 150 years, and, although during that time there has been little change in the nature of the technical fiber, considerable developments have taken place in the techniques of conversion to yarn and fabric, and in the end-uses of these products. Scientific studies began around 60 years ago, and although the base fibers did not receive publicity on the scale given to cotton and wool, the broad features of the internal structure and physical characteristics of fibers were elucidated sufficiently long ago for a great deal of common knowledge to be built up. The literature is now extensive, and is contained in a variety of journals. A number of books have become standard for reading, and critical reviews of the literature have appeared from time to time [1–8]. In the description that follows of the structure and properties of jute and kenaf, this common knowledge is presented without critical annotation of references; instead, a list of the principal books and papers considered relevant is appended.

7.2 FORMATION OF FIBER

Jute and kenaf fibers develop in the phloem, or bast, region of the stem of the plants, and they appear as wedge-shaped bundles of cells intermingled with parenchyma cells and other soft tissues (Figure 7.1) in the transverse sections of the stem. In the growing part of the stem, a circumferential layer of primary fibers develops from the protophloem, but, as vertical growth ceases in the lower parts, secondary phloem fibers develop as a result of cambial activity. In mature plants, which reach a height of 2.5–3.5 m and a basal diameter of about 25 mm, the secondary fiber accounts for about 90% of the total fiber bundles.

The plants pass from the vegetative to the reproductive phase when the day length falls below 12.5 h. Vertical growth then ceases and cambial activity declines. The production of cell bundles is much reduced, but, at the same time, the secondary fiber cells begin to mature

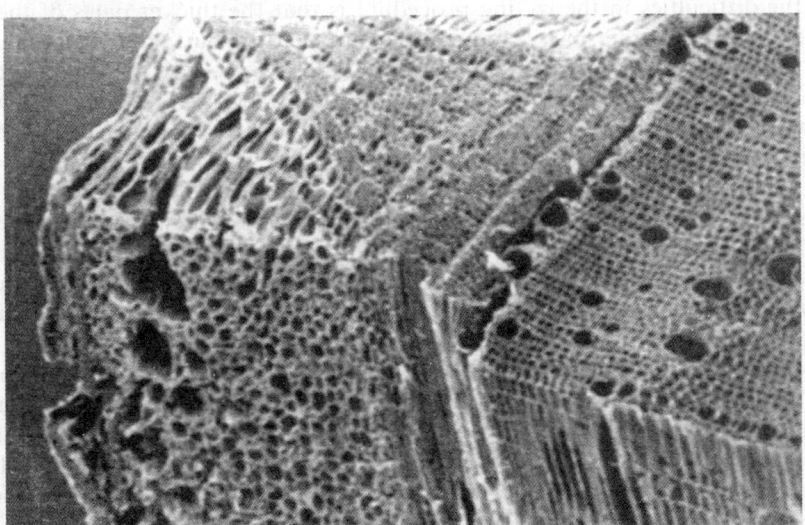

FIGURE 7.1 Jute stem (combined transverse section and longitudinal section). Magnification × 70. (Courtesy of Dr. C.G. Jarman, Tropical Development and Research Institute, London, UK.)

rapidly. Their walls, which have remained thin during the vegetative period, become thicker, and they increase in weight and strength.

Harvesting the plants at the correct time is most important and requires vast experience. For kenaf, the optimum time for harvesting is when about ten flowers are in bloom, and the older flowers have already set their seed. For jute, the optimum time is judged to be when the plants are in the small-pod stage. Harvesting before flowering generally results in lower yields and weaker fiber, whereas, if the seeds are allowed to mature, the fiber becomes harsh and coarse and difficult to extract from the plant.

The plants are harvested by hand with a sickle and cut close to the ground. The cut stems are then tied into bundles, the leaves removed as much as possible, and the bundles submerged in water for retting. This is the process by which the bundles of cells in the outer layers of the stem are separated from the woody core, and from nonfibrous matter, by the removal of pectins and other gummy substances. The action involves water, microorganisms, and enzymes, and takes between 5 and 30 days for completion, depending on the temperature of the water. Constant supervision is required and the time of removal is critical; if the degree of retting is insufficient, the fiber cannot be easily stripped from the woody core and may be contaminated with cortical cells, whereas, if retting proceeds too far, the fiber cells themselves may be attacked and weakened by microorganisms. Stripping the fiber from the stem is done by hand, after which the fibers are washed and dried.

7.3 SEPARATION OF BAST FIBER FROM CORE

The historical removal of the bark and the separation of bast fiber from the core is done by biological retting. Jute has been retted in India and Bangladesh for several hundreds of years by placing the entire plant in a pond and letting the natural decay process remove the bark and separate the long bast fiber from the core or stick. The process takes from 2 to 3 weeks and requires large quantities of water. Since the water contains a mixture of organisms, many biological reactions take place other than retting. The quality of the bast fiber coming from this process is often reduced due to the mixture of organisms and the dirty water. The core is then used for fuel or for fence posts and the bast is sold for use in textiles.

One of the difficulties in the retting procedure is that the thicker parts of the stem take longer to ret than the thinner parts, and, consequently, if the butt ends of the stem are fully retted, the top ends are over-retted and damaged. This can be avoided by stacking the bundles of stems upright with the butt ends in water for a few days, before immersing the whole stem. However, with the fiber intended for export, it is usual to cut off the partly retted butt ends and sell these separately as "cuttings."

Correct retting is an essential first step in the production of good-quality fiber. A comprehensive account of the techniques used, and their effect on fiber quality, has been given by Jarman [9]. Controlling the quality of water along with improving microorganisms used in the process are the keys to improve fiber quality. The use of clean water and specific microorganisms has been shown to greatly improve both the efficiency of the retting process and the quality of the bast fiber.

Extensive research has been done on the mechanical separation of the bast from the core on kenaf. The U.S. Department on Agriculture sponsored a research in mechanical "retting" at the Mississippi State University [10] and with a private firm in Bakersfiled, California [11]. Chopped whole stock was used in a process involving a spiked cylinder and an airline cleaner [12]. Separation efficiencies of 42 to 48% were achieved. It was found that the moisture content was a critical factor in the separation efficiencies and, if controlled, the separation was cleaner and quicker. Fisher [11] used a modified cotton gin and found separation efficiencies of more than 90%.

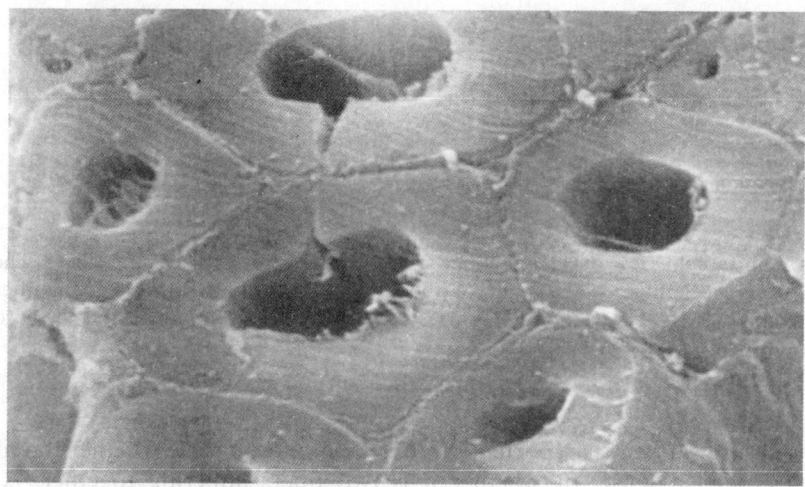

FIGURE 7.2 Part of a fiber bundle of jute as seen in transverse view under the scanning electron microscope. The cementing material between the ultimate fibers can be clearly seen. Magnification × 7600. (Scanning electron micrograph by Mr. A.J. Canning. Courtesy of Tropical Development and Research Institute, London, UK, Crown Copyright, 1982.)

Chemical retting has also been studied using 1, 4, and 7% sodium hydroxide to separate the bast fiber from the core [13].

7.4 FIBER STRUCTURE

In each plant, the rings of fiber cell bundles form a tubular mesh that encases the entire stem from top to bottom. Two layers can usually be distinguished and connected together by lateral fiber bundles, so that the whole sheath is really a lattice in three dimensions [14]. The cell bundles form the links of the mesh, but each link only extends for a few centimeters before it divides or joins up with another link. After extraction from the plant, the fiber sheath forms a flat ribbon in three dimensions.

The jute or kenaf fiber of commerce refers to the sheath extracted from the plant stems, whereas a single fiber is a cell bundle that forms one of the links of the mesh. Staple length, as applied to cotton and wool fibers, has no counterpart in the base fibers, and, as a preliminary to spinning, it is necessary to break up the sheaths by a carding process. The fragments so produced are the equivalent of the staple fibers of the cotton and wool industries.

When a transverse section of a single jute fiber is examined under the microscope, the cell structure is seen clearly. Each cell is roughly polygonal in shape, with a central hole, or lumen, comprising about 10% of the cell area of cross section, as shown in Figure 7.2. In the longitudinal view the fiber appears as in Figure 7.3, which shows the overlapping of the cells along the length of the fiber. The cells are firmly attached to one another laterally, and the regions at the interface of two cells is termed "the middle lamella." Separation of cells can be effected by chemical means, and they are then seen to be thread-like bodies ranging from 0.75 to 5 mm in length, with an average of about 2.3 mm. The cells are 200 times longer than they are broad, and, in common terminology, are referred to as "ultimate" cells. A single fiber thus comprises a bundle of ultimates.

Transverse selections of single fibers show that the number of ultimate cells in a bundle ranges from a minimum of 8 or 9 to a maximum of 20 to 25. Bundles containing up to

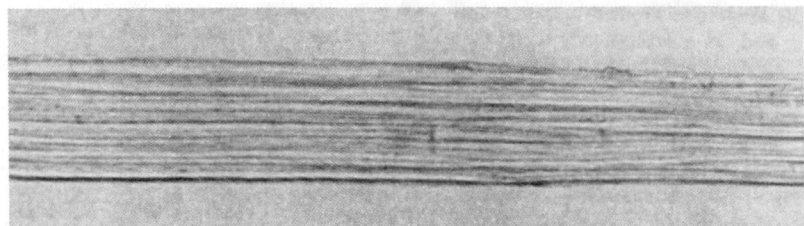

FIGURE 7.3 Longitudinal view of a single fiber strand of jute showing ultimate fibers. The tips of the ultimate can be seen slightly to the right of the center. Stained in safranin. Magnification × 500. (Photomicrograph by Mr. A.J. Canning. Courtesy of Tropical Development and Research Institute, London, UK, Crown Copyright, 1982.)

50 ultimate cells are sometimes reported, but it is then questionable whether the fiber is truly single in the botanical sense, or whether it is two fibers adhering together. A minimum number of cells in the cross section is evidently necessary to provide a coherent and continuous overlapping structure.

Kenaf and many other fiber-bearing dicotyledons have similar ultimate cell dimensions to jute. A distinction must be made between jutelike fibers and flax; however, the ultimate cells in flax are much longer, averaging 20–25 mm, although all are described as base fibers. They are also greater in cross-sectional area, and, because of the longer length, a coherent fiber structure can be built up from only two or three overlapping ultimates. The single fibers of flax are thus much finer than those of jute.

The difference between the bast and core fibers in kenaf is shown in Figure 7.4. The bast fibers have thicker walls (see Figure 7.5) as compared to the core fibers. The longitudinal axis of a kenaf bast fiber is shown in Figure 7.6.

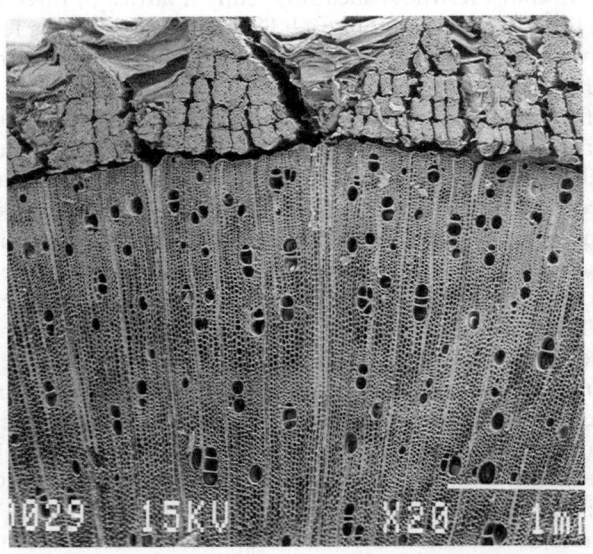

FIGURE 7.4 Cross section of the boundary area between kenaf bast and core. Magnification × 20 (USDA.)

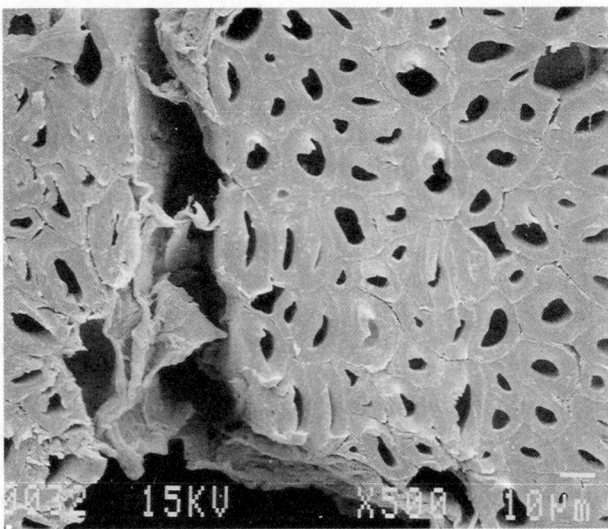

FIGURE 7.5 Cross section of kenaf bast fibers. Magnification × 500 (USDA).

7.5 CHEMICAL COMPOSITION

Retted fibers such as jute and kenaf have three principal chemical constituents, namely, α-cellulose, hemicelluloses, and lignin. The lignin can be almost completely removed by chlorination methods in which a soluble chloro-lignin complex is formed, and the hemicelluloses are then dissolved out of the remaining holocellulose by treatment with dilute alkali. The final insoluble residue is the α-cellulose constituent, which invariably contains traces of sugar residues other than glucose.

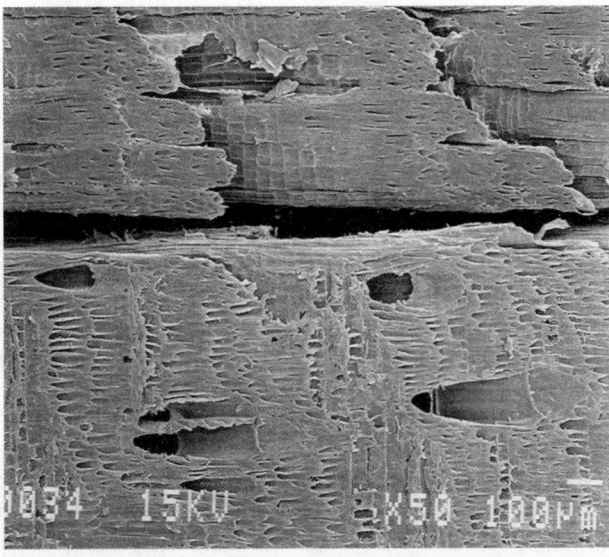

FIGURE 7.6 Longitudinal view of kenaf bast fibers. Magnification × 50 (USDA).

The cellulose has an average molecular weight between 130,000 and 190,000 with an average degree of polymerizatioin of approximately 800 to 1200.

The hemicelluloses consist of polysaccharides of comparatively low molecular weight built up from hexoses, pentoses, and uronic acid residues. In jute, capsularis and olitorius have similar analyses, although small differences occur between different fiber samples. Fiber extracted from jute plants grown in Bangladesh is composed of 12–14% lignin, 58–63% α-cellulose, and 21–24% hemicellulose [15a]. The average molecular weight of the hemicelluloses is in the range of 24,000 to 27,000.

In addition, analysis of the hemicellulose isolated from α-cellulose and lignin gives 8–12.5% xylan, 2–4% galactan, 3–4% glucuronic acid, together with traces of araban and rhamnosan. The insoluble residue of α-cellulose is composed of 55–59% glucosan, 1.8–3.0% xylan, 0.8–1.2% glucuronic acid, together with traces of galactan, araban, mannan, and rhamnosan. All percentages refer to the weight of dry fiber.

Along with the three principal constituents, jute and kenaf contain minor constituents such as fats and waxes, 0.4–0.8%, inorganic matter, 0.3–5%, nitrogenous matter, 0.8–1.5%, and traces of pigments. Totally, these amount to about 2%. Table 7.2 shows the chemical composition of both kenaf and jute reported by different laboratories in the United States, India, and Bangladesh [15b].

The detailed molecular structure of the hemicellulose component is not known with certainty, although, in the isolated material, the major part [3] consists of a straight chain of D-xylose residues, with two side branches of D-xylose residues, whose position and length are uncertain. In addition, there are other side branches formed from single residues of 4-*O*-methyl glucuronic acid, to the extent of one for every seven xylose units.

The third major constituent, lignin, is a long-chain substance of high molecular weight which, like the hemicelluloses, varies in composition from one type of vegetable material to another. The molecular chains are built up from comparatively simple organic units that may differ from different sources, and also in the way in which they are combined.

Most of the studies in lignin have been concerned with wood, and the base fibers have been rather neglected. It seems unlikely that any major differences exist between jute and wood lignin; however, many details of the molecular structure still remain unresolved.

7.6 ACETYL CONTENT

Jute and kenaf, like most vegetable fibers, contain a proportion of acetyl groups that are readily hydrolyzed by dilute alkali to acetic acid. Estimation of the quantity of acetic acid produced per unit weight of fiber then provides an index of the acetyl content.

The acetyl content of any particular type of fiber shows some variation according to where it is grown, and under what conditions, but often these intrafiber variations are small compared to the variations arising between fiber types. This is the case with hibiscus and corchorus fibers, for example, and Soutar and Brydon [16] have reported acetyl contents averaging 110 for hibiscus, 89 for *C. capsularis*, and 76 for *C. olitorius*, all expressed in milliequivalents of acetic acid per 100 g of dry fiber. The higher acetyl content of capsularis than olitorius has since been confirmed by Manzoor-i-Khuda [17].

Soutar and Brydon's results show no significant difference between *H. cannabinus* and *H. sabdariffa*, which is, perhaps, surprising in view of the difference between the two jute varieties, but the acetyl content does appear to offer a means of differentiating between jute and kenaf. For such a comparison to be valid, of course, there must have been no prior treatment of the fiber with alkali, which occasionally happens in chemical retting experiments.

Jute and Kenaf

TABLE 7.2
Chemical Composition of Kenaf and Jute Reported by Different Laboratories

Botanical name	Common name	Location	Cross and Bevan cellulose	α-Cellulose	Lignin	Pentosans	Alcohol benzene	Hot water	1% NaOH	Ash
Hibiscus cannabinus	Kenaf, hurds	IL	53.8d	34.7	15–18	—	3.7	—	30.9	—
Hibiscus cannabinus	Kenaf	FL	52.2d	34.0	10.5	21–23	3.4	—	29.4	—
Hibiscus cannabinus	Kenaf	—	47–57	31–39	12.1	18.3	—	—	—	2–5
Hibiscus cannabinus	Kenaf, stem	MD	53.1cc	36.5	13.2	22.7	4.3	11.2	33.0	—
Hibiscus cannabinus	Kenaf, stem	GA	58.0cc	40.2	7.7	19.7	3.3	7.4	28.4	4.1
Hibiscus cannabinus	Kenaf, whole	—	54.4cc	37.4	8.0	16.1	—	—	—	5.5
Hibiscus cannabinus	Kenaf, bast	—	57.2cc	42.2	17.4	16.0	—	—	—	2.9
Hibiscus cannabinus	Kenaf, core	—	51.2cc	33.7	13.4	19.0	—	—	—	—
Hibiscus cannabinus	Kenaf, bottom	—	53.9d	35.3	—	20.1	3.4	9.7	32.4	—
Hibiscus cannabinus	Kenaf, top	—	46.4d	29.8	—	—	5.5	12.8	39.6	—
Hibiscus cisplantinus	Kenaf	NC	46.1d	30.5	—	—	2.5	—	33.2	—
Hibiscus eelveldeanus	Kenaf	FL	51.9d	36.3	—	21.6	7.5	—	31.0	—
Corhorus capsularis	Jute	FL	56.3d	39.1	—	—	3.5	—	28.6	—
Corhorus capsularis	Jute	—	57–58	—	21–26	18–21	0.6	—	—	0.5–1
Corhorus capsularis	Jute	—	58.0	—	26.8	—	—	1.4	25.9	1.9
Corhorus capsularis	Jute	—	71.5c	—	8.1	21.6	—	—	—	—
Corhorus capsularis	Jute	India	—	60.7	12.5	15.6	0.9	—	—	0.8
Corhorus olitorius	Jute	India	—	61.0	13.2	15.9	—	—	—	0.5
Corhorus capsularis	Jute	India	—	58.9	13.5	17.0	—	—	—	0.5
Corhorus capsularis	Jute	India	57.6c	—	21.3	—	—	1.1–1.8	—	0.3–5
Corhorus capsularis	Jute	Bangladesh	21–24m	58–63	12–14	18.8	—	—	—	—

Source: From Man, J.S., and Rowell, J.S., *Paper and Composites from Jute and Kenaf Resources*, CRC/Lewis Publishers, Boca Raton, FL 1997, MS 83.

An interesting feature of this study is the measurement of the acetyl content for other fibers covering a wide range of hemicellulose and lignin contents. The authors conclude that the acetyl content shows a steady increase with increase in hemicellulose content, as the latter ranges from about 4 to 26%, but the correlation with varying lignin content is not marked.

An alternative method of distinguishing between jute and kenaf is by means of the crystals in the ash of the fiber after incineration. These crystals are present in the parenchyma and retain their original form during asking. In kenaf, cluster crystals are commonly found in the ash, whereas they are relatively uncommon in the case of jute. Jarman and Kirby [18], however, have shown that jute can be distinguished by the fact that the ash contains solitary crystals occurring in chains. Solitary crystals may occur in kenaf, but not in chains.

7.7 CHANGES IN CHEMICAL AND FIBER PROPERTIES DURING THE GROWING SEASON

Different parts of a plant have different chemical and physical properties. That is, the chemical composition and fiber properties of the plant tissue taken from the roots, stem, trunk, and leaves are different. The chemical composition and fiber properties of the plant tissue are different at different stages of the growing season.

The University of Manchester, the Shirley Institute, and the British Textile Technology Group in the United Kingdom have spent years working on jute. While some of the research has been published, the results relating to the changes in the properties of jute fiber as a function of the growing season were done for the International Jute Organization in Bangladesh, but were not published [19]. The research records are stored in Bangladesh, and attempts to gain access to them have failed. Personal communications concerning these results indicate that juvenile jute fiber looks and feels like silk, but this has never been documented in print.

Chatterjee, working at the Technological Jute Research Laboratories in Calcutta, India, first reported the changes in chemical composition at different stages of jute plant growth [20]. Table 7.3 shows a summary of his results. These results show that there is little difference in cellulose, holocellulose, and the lignin content, but the content of xylan, ash, and iron decreases as the plant matures. The aggregate fiber length increases as the growing season progresses. Without defining what is meant by "best," Chatterjee reports that the best fiber is obtained at the bud stage.

Later, Mukherjee et al. while working at the Indian Jute Industries' Research Association in Calcutta, studied characteristics of the jute fiber at different stages of growth [21]. They

TABLE 7.3
Changes in Chemical Composition of Jute at Different Stages of Plant Growth

Component	Stage of plant growth reported on 100 g of dry material				
	Pre-bud	Bud pod	Flower pod	Small	Large
α-Cellulose	58.3	57.6	59.4	58.7	59.1
Holocellulose	86.8	87.8	87.3	87.1	86.8
Xylan	15.5	14.8	14.4	13.7	13.9
Lignin	12.7	12.1	12.4	12.0	12.0
Ash	0.57	0.53	0.47	0.67	0.47
Iron	0.020	0.018	0.009	0.011	0.008
Reed length (mm)	198	273	279	288	321

TABLE 7.4
Changes in Chemical Composition of Kenaf at Different Stages of Plant Growth

Data from the top 0.66 m of the plant, % by weight

Component	90 days after planting	120 days after planting	138 days after planting	147 days after planting	158 days after planting	244 days after planting
Hot water extractives	37.4	39.0	35.2	31.6	30.6	12.8
Lignin	4.5	3.9	6.4	7.2	7.4	11.4
α-cellulose	10.6	14.5	18.5	18.1	20.6	29.8
Pentosan	5.0	12.5	16.1	17.0	16.7	20.1
Protein	25.0	17.9	16.1	13.3	14.9	11.1

found that, at the early stages of growth, there was an incomplete formation of the middle lamella in the cell wall and the parallel bundles of fibrils were oriented at an angle with respect to the fiber axis that gradually decreased with growth. After about 35 days of growth, the fibrils run parallel to the fiber axis. In the mature plant, a few helically oriented fibrils were observed in the Z-direction just below the primary cell wall layer.

Clark and Wolff carried out the first studies on the changes in the chemical composition of kenaf as a function of the growing season [22]. They also studied the chemical differences along the stem and between leaves and stem. This data showed that the pentosans, lignin, and α-cellulose content increases with age, while the protein and hot water extractives content decreases with age. Data taken from the top part of the plant shows similar trends; however, the top part has less cellulose, pentosans, and lignin, but higher hot water extractives and protein than the bottom part of the plant (Table 7.4).

TABLE 7.5
Changes in Fiber Properties of Kenaf at Different Stages of Plant Growth

	Stage of plant growth			
Component	90 days after planting	120 days after planting	150 days after planting	180 days after planting
Bast fiber				
Length (mm)	3.34	2.28	2.16	2.42
Width (microns)	18.3	14.5	13.6	15.1
Lumen width (microns)	11.1	5.4	6.8	7.7
Cell wall				
Thickness (microns)	3.6	4.6	3.4	3.7
Core fiber				
Length (mm)	0.55	0.54	0.45	0.36
Width (microns)	36.9	31.2	32	31.6
Lumen width (microns)	22.7	14.8	18.6	18.7
Cell wall				
Thickness (microns)	7.1	8.2	6.7	6.4

TABLE 7.6
Crystallinity Values (dap = Days after Planting)

after planting	C-108		T-1		E-41		45–9	
	fiber	core	fiber	core	fiber	core	fiber	core
56	84.25	78.38	80.77	87.72	78.4	81.08	80.53	82.35
84	78.87	72.13	76.43	73.85	81.48	72.22	77.62	76.92
112	78.91	78.87	73.94	70.63	80.43	71.79	78.32	75.41
140	80.71	72.31	80.29	79.39	78.17	66.93	78.10	68.18
168	73.57	66.93	77.77	65.19	70.90	64.00	74.64	67.19
196	72.34	68.18	72.86	73.11	70.63	69.53	68.38	70.83

Clark et al. also studied the changes in fiber properties during the growing season [23]. Table 7.5 shows that the bast single fibers are longer than core fibers and both decrease in length with age. Core single fibers are twice as wide and have twice the cell wall thickness as bast single fibers and both dimensions decrease with age. Finally, the lumen width is wider in pith fibers as compared to bast single fibers and both decrease with age.

In a recent study, Han et al. reported changes in kenaf as a function of the growing season [24]. Their data do not necessarily agree with that of Clark and Wolff. The most critical difference between Han et al. and Clark and Wolff was the difference in fiber lengths. The average length of a bast and core (stick) fiber increased as the plant aged in contrast to that of Clark and Wolff.

Han et al. studied changes in the chemical composition during the growing season for four varieties of kenaf: C-108, Tainung-1, Everglade 45–49, and Everglade 41 [24]. Samples were collected weekly starting from about 50 days after the planting (dap) to the end of growing at about 170 days after planting.

X-ray diffraction of kenaf samples were used for crystallinity values (Table 7.6). Cystallinity values decreased as the plant matured.

Ash contents of fibers and cores were determined before and after extraction (Table 7.7). Ash content decreased as the plant reached maturity.

The protein content of fibers and cores was determined before and after the extraction (Table 7.8). The protein content decreased as the plant reached maturity.

Extractives, lignin, and sugar contents were also determined (Table 7.9). The values are averages of four different cultivars. Klason lignin analysis was done after the extraction. The Klason lignin values increased from ca. 4% at the beginning to 10% at the end of the growing season (Bagby et al. reported about 10% using Florida kenaf) [25]. This value is significantly lower than that of softwood (26–32%) and hardwood (20–28%). The actual value of Klason

TABLE 7.7
Ash Content (T-1) (% Dry Basis)

Growth days after planting	Fiber-(Unext)	Fiber-(Ext)	Core-(Unext)	Core-(Ext)
49	13.08	9.54	14.52	7.77
98	7.52	6.83	5.10	4.21
147	6.24	6.00	4.55	4.44
175	3.76	3.84	2.84	2.39

TABLE 7.8
Protein Content (T-1) % (Dry Basis)

Growth days after planting	Fiber-(Unext)	Fiber-(Ext)	Core-(Unext)	Core-(Ext)
49	10.6	7.65	12.75	8.00
98	4.25	3.25	4.85	4.65
147	2.05	3.35	3.40	3.20
175	1.43	1.23	4.15	3.28

lignin could be lower than it appears to be due to the presence of protein in the kenaf. Kjeldahl determination of protein was performed. (Han et al. combined several batches of Klason lignin samples and measured the amount of protein in the Klason lignin.) [24]. The protein content of kenaf was between 4 to 14% of the Klason lignin, depending on the age of the plant. Only 38% of the protein was found in the Klason lignin and the rest was found in the hydrolysate (unpublished FPL Data). In general, the protein content decreased with plant age.

The solvent extractive content varied as a function of growth. In general, it was high in the beginning, decreased during the first part of the growing time, and then increased again. L-Arabinose, L-rhamnose, L-galactose, and D-mannose content decreased as a function of growth, while D-glucose and D-xylose content increased over this same period of time.

The fiber length increased as a function of growth (Table 7.10). The core fiber lengths were ca. 0.8 mm at the end of 84 growing days with an average diameter of ca. 0.5 mm.

The weights ratio between the fiber and the core (core/fiber) increased as the growing days advanced (Table 7.11). A maximum of 1.8 was reached at 175 days after planting in T-1.

The holocellulose content was measured after the extraction (Table 7.12). The juvenile samples had low holocellulose values and gradually increased as the plant aged.

The height of the plant increased with the age of the plant at an even rate. This is a function of the growing conditions and would change with different moisture and sun

TABLE 7.9
Chemical Composition of Kenaf Fiber (% Oven Dry Basis)

Growth days after planting	Extractives	Klason lignin	Polysaccharides content (% anhydro sugars on oven dry basis)					
			arabinan	rhamnan	galactan	glucan	xylan	mannan
35	14.87	4.32	3.95	2.72	0.78	28.86	6.54	1.76
42	8.80	6.00	3.18	1.82	0.62	33.20	7.31	1.63
57	5.13	8.32	2.21	1.46	0.55	35.45	8.08	1.59
63	4.34	7.74	2.43	1.48	0.62	37.08	8.61	1.53
70	4.63	8.70	2.02	1.25	0.46	40.53	9.37	1.47
77	4.99	9.23	2.05	1.36	0.39	40.52	9.16	1.34
84	5.07	8.33	2.27	1.63	0.49	39.88	9.39	1.53
91	5.68	9.38	1.91	1.35	0.42	42.82	9.98	1.31
98	2.42	8.81	2.13	1.43	0.48	41.60	9.69	1.35
133	8.03	8.94	1.67	1.15	0.48	41.98	9.72	1.31
155	7.83	9.99	1.27	0.87	0.38	46.39	11.20	1.19
161	11.51	10.22	2.54	1.52	0.56	39.22	9.75	1.33
168	12.31	9.74	2.18	1.37	0.47	41.41	10.36	1.39
175	8.23	9.69	1.40	0.87	0.36	49.33	12.29	1.02

TABLE 7.10
Fiber Length (Unit = mm)

Growth days after planting	C - 108		Tainung - 1		45–49	
	Bast	Core	Bast	Core	Bast	Core
50	2.3	0.7	2.2	0.7	2.2	0.7
60	2.7	0.7	2.8	0.7	2.7	0.7
77	2.9	—	3.0	—	3.0	—
84	3.4	0.8	3.1	0.8	3.7	0.8

conditions. The diameter of the stalk was also increased gradually with age until 160 days after planting. At the end of 160 days after planting, the rate of the growth became more significant. However, this dramatic increase in volume is indicative of an increase in the core and not the bast fiber. A maximum weekly growth of 30 cm was achieved during high temperature and a good rainfall.

Scanning electron microscopy (SEM) studies indicate that the bast fiber bundles are thin walled at 63 days after planting and are in the process of thickening. The middle lamella is not well formed, as suggested by the weak bonding. The fibers' tendency to gelatinize may be due to the wind effect that often results during the bending of plants. At this stage, the parenchyma bands separating the bast fiber bundles are well formed and occupy a considerable area of tissue system.

At 71 days after planting, the plants become comparatively stronger as bast fiber bundles occupy more area, the fiber wall thickens, and lignification of middle lamella becomes

TABLE 7.11
Weight Ratio of Fiber versus Core

Growth days after planting	C-108	T-1	45–49	E-41	C/F Ave
53	0.88	0.90	0.87	0.74	0.85
57	0.93	1.17	1.02	1.03	1.03
63	1.02	1.10	1.11	1.05	1.07
70	1.29	1.10	1.31	1.14	1.21
77	1.19	1.03	1.15	1.31	1.17
84	1.10	1.33	1.12	1.25	1.20
91	1.42	1.23	1.22	1.41	1.32
98	1.14	1.33	1.03	1.22	1.18
105	1.23	1.94	1.09	1.22	1.37
112	1.16	1.36	1.03	1.80	1.34
120	1.23	1.30	1.01	1.29	1.21
126	1.21	1.62	1.19	1.41	1.36
133	1.28	1.70	1.25	1.30	1.38
140	1.29	1.82	1.41	1.78	1.57
147	1.31	1.31	1.45	1.24	1.32
155	1.40	1.71	1.39	1.98	1.62
161	1.57	2.12	1.69	1.64	1.76
175	1.31	1.81	1.33	1.83	1.57

TABLE 7.12
Holocellulose Content

Growth days after planting	Holocellulose (%)	Growth days after planting	Holocellulose (%)
35	58.97	42	64.47
49	68.24	57	69.02
63	73.91	70	75.69
77	76.52	84	74.15
91	76.13	98	74.87
105	73.65	126	75.40
133	75.27	140	78.50
147	76.18	161	73.88
168	76.27	175	78.60

apparent. Parenchyma cells tend to crash due to the development of fiber bundles, thus allowing more area to be occupied by fiber bundles. At 84 to 108 days after planting, the bast fiber bundles comprising of primary and secondary phloem fibers tend to show more thickening, and the separation of primary and secondary phloem fibers becomes obvious. The secondary phloem fibers start thickening, but with a somewhat weak middle lamella. At this stage of development, in addition to wall thickening, a deposition of silica on the wall surface is seen. The fibers are long and broad and are mainly comprised of an S_2 layer that is encrusted with amorphous silica. At 112 days after planting, bast fiber bundles comprising of primary phloem fibers and secondary phloem fibers are thickened with prominent middle lamella formation. The cells are compact with thickened cell walls and decreased lumen width. The middle lamella is not well lignified at this stage of maturity. The fibers are long and broad with a well-formed S_2 layer [24].

Similar sequential development is seen in secondary phloem fibers. At 63 days after planting, there is little thickening of the fiber wall and the fibers thicken gradually with maturity from 73 days to 112 days after planting.

7.8 FINE STRUCTURE

The location of the three main chemical components of the fibers are reasonably well established. Alpha cellulose forms the bulk of the ultimate cell walls, with the molecular chains lying broadly parallel to the direction of the fiber axis. The hemicellulose and lignin, however, are located mainly in the area between neighboring cells, where they form the cementing material of the middle lamella, providing strong lateral adhesion between the ultimates. The precise nature of the linkages that exist between the three components, and the role played by the middle lamella in determining the fiber properties, are not completely understood. Lewin [26], some years ago, in an interesting literature survey on the middle lamella of base fiber, brought together a great deal of relevant information that highlighted many of the problems, but a thorough understanding of the intercell structure is still awaited.

X-ray diffraction patterns show the basic cellulose crystal structure, but, in jute and kenaf, although the crystallite orientation is high, the degree of lateral order is relatively low in comparison with, for example, flax. There is also considerable background x-ray scattering arising from the noncellulosic content of the fiber.

The cellulose molecular chains in the secondary walls of ultimate cells lie in a spiral around the fiber axis. The effect of this is to produce double spots in the x-ray diffraction patterns, the centers of the spots separated by an angular distance of twice the Bragg angle. For large angles, such as those that occur in coir fiber, and some leaf fibers such as mauritius hemp, the two spots are visibly separated, but, for the small angles found in jute and kenaf, the spots overlap. In this case, the distribution of intensity across the width of the spots, instead of reaching a peak at the center of each, is spread out into a single, flatter, peak. The [002] equatorial reflection shows these effects particularly well, and the analysis of the intensity distribution allows calculation of the Hermans RMS spiral angle. A wide range of base and leaf fibers have been examined in this way [27], with results showing the Hermans angle to range from about 8° for jute and kenaf to up to 23° for sisal. Coir fiber, *Cocos nuciferos*, is an exception, having a Hermans angle of about 45°.

The leaf fibers cover a wide range of ultimate cell dimensions along with covering a good range of spiral angles. The results indicate that, among this group of fibers, the spiral structure averages a constant number of turns per unit length of cell, about 10 per mm, and, with this arrangement, the spiral angle then depends solely on the breadth of the cell. Whether this constancy of turns applies to individual cells, or whether as in wood, the longer cells tend to have steeper spirals, was not, however, investigated.

For the secondary base fibers, the cell dimensions show little variation between plant species, but the number of spiral turns per unit length of cell averages only about four per millimeter, appreciably less than for the leaf fibers.

The importance of the spiral angle measurements lies in the control that the spiral structure exercises on the extension that the fiber can withstand before breaking. Regarding the structure as a helical spring, the extension necessary to straighten a spring of initial angle q, to the axis is (sec $q - 1$) × 100%. A 10° spring will thus extend by 1.54%, a 20° spring by 6.4%, and a 30° spring by 15.5%.

The coconut fiber coir has a spiral of about 45° and its helical spring extension is 41.4%. Such a large extension is easily measured and has been shown to be reasonably correct. Moreover, it is possible to carry out the extension in stages and to measure the angle whereas the fiber is stretched and under tension. X-ray measurements showed the angle to decrease with the extension as predicted by the spring structure, and it was concluded that the extensibility of coir fiber is almost entirely due to the spiral structure of the ultimate cells [28]. This has been confirmed by other studies in which the spirality of the cell wall was investigated microscopically using replica and ultrathin sectioning techniques [29].

It is interesting to note that when coir fiber relaxes after stretching, it shortens in length and the spiral angle increases according to the spring theory. There is usually a semipermanent set left in the fiber after relaxation, but this can be removed by steaming and the fiber can be restored virtually to its original unstretched length.

It is difficult to carry out similar measurements of the extension–spiral angle relationship for low-angle fibers such as jute and kenaf because, as the changes in angle are small, the overlapping of the spots in the x-ray diffraction pattern could introduce significant errors. With coir, the angles over most of the extension range are measurable to a higher degree of accuracy.

Assuming, therefore, that the coir results are of general applicability to fiber cells, it appears that the helical spring theory could be used to calculate the order of magnitude of the extensibility of the fiber, and to rank fibers accordingly.

7.9 PHYSICAL PROPERTIES

Jute and kenaf are strong fibers, exhibiting brittle fracture, but having only a small extension at break. They have a high initial modulus, but show very little recoverable elasticity. Tenacity measurements recorded in the literature vary widely, and, although some of this variation is due to differences in the methods of measurement, a major part arises from variations in the linear density of the fibers themselves. All linear densities are given in tex units of grams per kilometer.

Taking account of all the available evidence, a tenacity of 70 g/tex is a reasonable middle value for a wide range of jute fibers, based on single fiber test lengths of 10 mm or less, and a time to break of 10 s. This value of tenacity is appropriate to fibers of linear density 1.8 tex, and it is important to state the linear density, because, statistically, an increase of 0.1 tex reduces the tenacity by about 1.5 g/tex. This inverse dependence of tenacity on linear density is common to most fibers and also to fine metal wires.

The elongation at which a fiber breaks is a more invariant and fundamental property than the load at which it breaks. It is neither affected significantly by changes in linear density nor by changes in the method of loading. The length of the test specimens does have an effect, however, as irregularities in diameter prevent all sections of a long fiber from being elongated equally. For test lengths of 10 mm, the elongation is generally between 1 and 2% of the initial length, but is difficult to measure accurately with such short lengths. In one particular case, 500 fibers from a bulk of medium-quality jute had a mean elongation of 1.60% (of the 10-mm test length) with a coefficient of variation (CV) of 25%. The breaking load of the fibers, however, had a much higher CV of 40% [30]. It may be noted that 1.6% elongation corresponds to a spiral angle of $10°12'$, which, although slightly greater than the Hermans angle reported, is still within the uncertainty of the comparison.

The initial Young's modulus of the fibers, calculated from the slope of the load–elongation curve, has a mean value of about 4×103 g/tex/100% extension. The value for any particular group of fibers will, of course, be dependent on the linear density, to some extent, owing to the dependence of tenacity values on this factor.

The bending of jute fibers has been studied by Kabir and Saha, who calculated the Young's modulus from measurements of the force required to deflect the free ends of a fringe of fibers arranged in a cantilever fashion [31]. For this calculation, it is necessary to know the fiber diameter instead of the linear density, and this causes difficulty because the cross section of the fibers is irregular in outline and often far from circular. The authors assumed an elliptical configuration, and measured minimum and maximum diameters of a number of cross sections microscopically for insertion in the appropriate formula. Their calculations showed that, over a wide range of commercial fiber qualities, Young's modulus decreased over 60% between an average diameter of 46 mm to a diameter of 68 mm. These values correspond to 3050 and 815 g/tex/100% extension, respectively, and again demonstrate the marked effect of variations in fiber dimensions. Extrapolations of Kabir and Saha's data to smaller diameters show that the tensile value for the modulus of 4000 g/tex/100% extension would be reached at a mean diameter of about 40 mm.

Kabir and Saha also examined the effect of delignification on the bending modulus of jute, using the fringe technique, and showed that successive extractions of lignin on the same fibers resulted in increasing flexibility and decreasing Young's modulus [32]. The delignification method was treatment with sodium chlorite solution followed by extraction with sodium bisulfate, and removal of 10% of lignin reduced the modulus by more than 30%. At the same time, however, the diameter of the fibers was reduced significantly, and this may have affected the flexibility.

7.10 GRADING AND CLASSIFICATION

The grading and classification of base fibers such as jute and kenaf for commercial purposes has a long history, but is still done subjectively by hand and eye. Official standards have been formulated, but these are purely descriptive and no quantitative values are assigned to the stated criteria. Nevertheless, a surprising degree of consistency is achieved, particularly for export purposes, and experienced buyers and sellers do not find it too difficult to find out whether or not the grade assigned to a particular consignment of fiber is correct.

For jute fiber exported from Bangladesh, for example, the current grading system first separates *C. capsularis* and *C. olitorius* into white and tossa categories, respectively, and then further classifies each into five grades denoted by the letters A to E. The highest prices are paid for Grade A, although sometimes a special grade is introduced for which a higher price can be demanded.

The principal criteria used are color, luster, strength, cleanliness, and freedom from retting defects. From a spinning point of view, color is irrelevant, but certain end-users traditionally prefer particular colors of fiber for the sake of appearance. Luster is commonly an indication of strength, for if, for example, the fiber has been over-retted so that the cellulose, or middle lamella, has been attacked and weakened, the surface appears dull. A lack of luster thus downgrades the fiber, although occasionally this same effect may result from inadequate washing, without any loss of strength. The strength of the fiber is also assessed by snapping a few strands by hand—a qualitative procedure that gives a useful indication to an experienced operator.

Cleanliness and freedom from nonfibrous matter is an important feature, and, in this respect, the physical imperfections that may result from improper retting can have a profound effect on the allotted grade. Adhering bark in any form results in downgrading, irrespective of the intrinsic value of the fiber, and, in the case of plants grown on flooded land, which stand in water, the bark becomes so difficult to remove that, for export, the root ends are cut off and sold separately as "cuttings" to be used in heavy yarns of low quality.

The linear density of the individual fibers making up the network is given little consideration, despite the importance of this characteristic in staple fibers, where it is a major factor controlling the levelness of the spun yarn. Adhering bark increases the linear density of the fiber and makes subjective assessment difficult.

Manzoor-i-Khuda et al. [17] have studied the variation in chemical constituents of jute fiber taken from different grades of both white and tossa and concluded that certain correlations exist between the analytical results and the commercial grade. Thus, it is claimed that the lignin content increases as the grades go from higher to lower, and that the ash content and copper number show similar negative correlations.

Although it might be expected that variations in the chemical composition would result in variations in physical characteristics, a correlation with grade is surprising. The chemical composition is that of the fiber itself, and can scarcely take account of the physical imperfections resulting from inadequate retting, which are so important in commercial grading.

The essential feature of any system of grading is that it be self-consistent in the sense that buyers and sellers can mutually agree on the attributes of fiber placed in a particular grade. However, it does not follow that a subjective system based on appearance and feel will classify fiber in a similar manner to an objective system based on measurement. Both systems may be valid, but in different ways, and there is no need to seek a close correlation between them except, perhaps, for the top and bottom grades.

Commercial buying and selling takes place by a subjective system. A buyer selects a range of fiber grades from which blends are made appropriate to the different yarn qualities required. If these fiber grades can now be measured for quality on an objective system, more precision in blending will be possible.

Any system of objective grading based on measurable characteristics must, in fact, be concerned with the fiber as it is, including nonfibrous matter, and not merely with the single fibers themselves. With this precondition in view, Mather [33], in work extending over a decade at the British Jute Trade Research Association laboratories in Scotland, studied the classification of a bulk of jute for its "spinning quality."

7.11 FIBER AND YARN QUALITY

The principal outlets for jute yarns are for industrial purposes in which, to give a satisfactory performance, adequate strength is essential. Appearance and color are of little significance, and so for jute yarns "quality" relates specifically to tensile properties. An objective classification of fiber in bulk thus requires the identification of those attributes of the raw fiber that affect yarn strength. Each grade of fiber bought commercially must then have these attributes measured and the grades assessed for corresponding yarn quality. By blending together fibers having different values of these attributes, the average value serves to predict the tensile strength properties of the yarn spun from the blend.

From an extensive series of correlations between fiber properties and yarn properties, Mather concluded that the tensile properties of a yarn could be predicted from two measurements only on the raw fiber, namely, the linear density and the ballistic work of rupture of uncarded strands of fiber.

The linear density was measured by an air-flow method using a modification of apparatus designed for cotton and wool. A sample of fiber weighing 27 g was used, and care was taken to include in the sample a similar amount of nonfibrous matter as was contained in the bulk. The nonfibrous component effectively increased the linear density, and a less regular yarn resulted when spun to a fixed count.

The ballistic work of rupture was measured by stretching strands of fiber, of known linear density, transversely across the path of a falling pendulum and recording the energy lost by the pendulum in breaking the strands. The energy lost per tex is then a measure of the specific work of rupture and is related to the product of tensile strength and extensibility. The particular feature of the work of rupture is that it appears to control the average length of fiber after carding. Staple length has no meaning in the bulk fiber, and it is only after the mesh has been fragmented by carding that average length becomes meaningful.

The yarns were spun on a standard system of carding, drawing, and spinning frames, programmed to produce yarn of linear density 275 tex. Different spinning systems and different linear densities also affect yarn strength, and this must be taken into account. It is inappropriate to discuss the technology of jute spinning in this article, but a detailed account of the experimental work on which Mather's conclusions are based has been compiled by Stout [34].

Quantitatively, Mather concludes that for jutes exported from Bangladesh (or the erstwhile East Pakistan), the range of linear density is about 1.3–2.4 tex, whereas work of rupture ranges from 4.0 to 8.3 g/cm/tex. For kenaf, although work of rupture is little different, the linear density is often higher than for jute, and a survey of *H. sabdariffa* grown in Thailand showed a range of 1.9–3.0 tex.

Moreover, in the overlapping region of linear density 1.9–2.4 g/cm/tex, it was noticeable that the kenaf fibers were intrinsically coarse but free from nonfibrous matter, whereas the jute fibers were intrinsically much finer but carried a significant amount of adhering bark.

It was also concluded from the statistical correlations that the change in the tenacity of a jute yarn, resulting from a certain percentage change in fiber linear density, is about three times greater than that resulting from a similar percentage change in ballistic work of rupture. Moreover, no correlation was found between linear density and work of rupture, so that these two parameters must exercise their effects quite separately.

The fiber linear density is a measure of the average number of fibers in the cross section of a given yarn, and this controls the yarn irregularity. The more fibers in the cross section, the more uniform is the yarn thickness from point to point. As yarns break at their thinnest points, the breaking load is greater, irrespective of the intrinsic fiber strength.

The high modulus of jute has, in turn, made jute materials a partial substitute for glass fiber as a reinforcement for polyester or epoxy resins in resin transfer technologies. It has not, however, found general acceptance in this reinforcement field, partly because it provides lower impact strength than glass and partly because the economic advantages are not sufficiently attractive. Jute and kenaf have found success as reinforcement fillers in thermoplastic composites. This is discussed in another section in this chapter.

7.12 CHEMICAL MODIFICATION FOR PROPERTY IMPROVEMENT

The performance of any lignocellulosic fiber composite is restricted by the properties of the fiber itself. Jute and kenaf composites change dimensions with changes in moisture content, are degraded by organisms, are degraded by ultraviolet radiation, and burn. If these negative properties of the natural fiber can be improved, all types of jute and kenaf composites can have a greatly improved performance. To understand how jute and kenaf fiber can be used in property-enhanced applications, it is important to understand the properties of the components of the cell wall and their contributions to fiber properties.

Jute and kenaf, like all agro (lignocellulosic) fibers, are three-dimensional polymeric composites, primarily made up of cellulose, hemicelluloses, lignin, and small amounts of extractives and ash. The cell wall polymers and their matrix make up the cell wall and are, in general, responsible for the physical and chemical properties of the jute and kenaf fiber. Properties such as dimensional instability, flammability, biodegradability, and degradation caused by acids, bases, and ultraviolet radiation are a result of the environment trying to convert the natural composites back into their basic building blocks (carbon dioxide and water).

Jute and kenaf fibers change dimensions with changing moisture content because the cell wall polymers contain hydroxyl and other oxygen-containing groups that attract moisture through hydrogen bonding. The hemicelluloses are mainly responsible for moisture sorption, but the accessible cellulose, noncrystalline cellulose, lignin, and surface of crystalline cellulose also play major roles. Moisture swells the cell wall and the fiber expands until the cell wall is saturated with water. Beyond this saturation point, moisture exists as free water in the void structure and does not contribute to further expansion. This process is reversible and the fiber shrinks as it loses moisture.

Jute and kenaf fibers are degraded biologically because organisms recognize the carbohydrate polymers (mainly the hemicelluloses) in the cell wall and have very specific enzyme systems capable of hydrolyzing these polymers into digestible units. Biodegradation of the high-molecular-weight cellulose weakens the fiber cell wall because crystalline cellulose is primarily responsible for the strength of the cell wall. Strength is lost as the cellulose polymer undergoes degradation through oxidation, hydrolysis, and dehydration reactions. The same types of reactions take place in the presence of acids and bases.

Jute and kenaf fibers exposed outdoors undergo photochemical degradation caused by ultraviolet light. This degradation takes place primarily in the lignin component, which is responsible for the characteristic color changes. The lignin acts as an adhesive in the cell walls, holding the cellulose fibers together. The surface becomes richer in cellulose content as the lignin degrades. In comparison to lignin, cellulose is much less susceptible to ultraviolet light degradation. After the lignin has been degraded, the poorly bonded carbohydrate-rich fibers erode easily from the surface, which exposes new lignin to further degradative reactions. In time, this "weathering" process causes the surface of the composite to become rough and can account for a significant loss in surface fibers.

Jute and kenaf fibers burn because cell wall polymers undergo pyrolysis reactions with increasing temperature to give off volatile, flammable gases. The hemicellulose and cellulose polymers are degraded by heat much before the lignin is degraded. The lignin component contributes to char formation, and the charred layer helps insulate the composite from further thermal degradation.

Because the properties of the jute and kenaf fiber result from the chemistry of the cell wall components, the basic properties of a fiber can be changed by modifying the basic chemistry of the cell wall polymers.

Dimensional stability can be greatly improved by bulking the fiber cell wall either with simple bonded chemicals or by impregnation with water-soluble polymers. For example, acetylation of the cell wall polymers using acetic anhydride produces a fiber composite with greatly improved dimensional stability and biological resistance. The same level of stabilization can also be achieved by using water-soluble phenol–formaldehyde polymers followed by curing.

Biological resistance of fiber-based materials can be improved by several methods. Bonding chemicals to the cell wall polymers increases resistance due to the lowering of the equilibrium moisture content point below that needed for microorganism attack and by changing the conformation and configuration requirements of the enzyme–substrate reactions. Toxic chemicals can also be added to the composite to stop biological attack. This is the basis for the wood preservation industry.

Resistance to ultraviolet radiation can be improved by bonding chemicals to the cell wall polymers, which reduces lignin degradation, or by adding polymers to the cell matrix to help hold the degraded fiber structure together so that water leaching of the undegraded carbohydrate polymers cannot occur. Fire retardants can be bonded to the fiber cell wall to greatly improve the fire performance. Soluble inorganic salts or polymers containing nitrogen and phosphorus can also be used. These chemicals are the basis of the fire-retardant wood-treating industry.

The strength properties of fiber-based composites can be greatly improved in several ways. The finished composites can be impregnated with a monomer and polymerized *in situ* or impregnated with a preformed polymer. In most cases, the polymer does not enter the cell wall and is located in the cell lumen. By using this technology, mechanical properties can be greatly enhanced. For example, composites impregnated with acrylates, methacrylate, epoxy, or melamine monomers, and polymerized to weight gain levels of 60 to 100% show increases (compared to untreated controls) in density from 60 to 150%, compression strength from 60 to 250%, and tangential hardness from 120 to 400%. Static bending tests show 25% increase in modulus of elasticity, 80% in modulus of rupture, 80% in fiber stress at proportional limit, 150% in work to proportional limit, and 80% in work to maximum load, and at the same time a decrease in permeability of 200 to 1200%.

Many chemical reaction systems have been published for the modification of agrofibers. These chemicals include ketene, phthalic, succinic, maleic, propionic and butyric anhydrides, acid chlorides, carboxylic acids, many types of isocyanates, formaldehyde, acetaldehyde, difunctional aldehydes, chloral, phthaldehydic acid, dimethyl sulfate, alkyl chlorides, beta-propiolactone, acrylonitrile, ethylene, propylene, and butylene oxide, and difunctional epoxides [35,36].

7.12.1 Acetylation

By far, maximum research has been done on the reaction of acetic anhydride with cell wall polymer hydroxyl groups to give an acetylated fiber. Jute [37–39] and kenaf [40,41] have been reacted with acetic anhydride. Without a strong catalyst, acetylation using acetic anhydride alone levels off at approximately 20 weight percent gain (WPG). The equilibrium moisture content (EMC) and thickness swelling at three relative humidities for fiberboards made from these fibers is shown in Table 7.13.

The rate and extent of thickness swelling in liquid water of fiberboards made from control and acetylated fiber are shown in Table 7.14. Both the rate and extent of swelling are greatly

TABLE 7.13
Equilibrium Moisture Content (EMC) and Thickness Swelling (TS) of Fiberboards Made from Control and Acetylated Fiber

Fiber	Weight percent gain	EMC and TS at 27°C					
		30%RH		65%RH		90%RH	
		TS	EMC	TS	EMC	TS	EMC
Kenaf	0	3.0	4.8	9.6	10.5	33.0	26.7
	18.4	0.8	2.6	2.4	5.8	10.0	11.3
Jute	0	3.7	5.8	8.6	9.3	17.4	18.3
	16.2	0.6	2.0	1.7	4.1	7.3	7.8

reduced as a result of acetylation. At the end of 5 days of water soaking, control boards swelled 45%, whereas boards made from acetylated fiber swelled 10%. Drying all boards after the water-soaking test shows the amount of irreversible swelling that has resulted from water swelling. Control boards show a greater degree of irreversible swelling as compared to boards made from acetylated fiber.

Table 7.15 shows the results of jute cloth acetylated to different levels of acetylation in a fungal cellar test. The fungal cellar is made using unsterilized soil that contains a mixture of white-, brown-, and soft-rot fungi. Control cloth shows a fungal attack at 2 months and is completely destroyed at 6 months. The acetylated cloth at 7.4 PWG shows a slight attack at 3 months and is destroyed at 12 months. At a level above 16% weight gain, the acetylated cloth is not attacked at 36 months [42].

The modulus of rupture (MOR), modulus of elasticity (MOE) in bending, and tensile strength (TS) parallel to the board surface are shown in Table 7.16 for fiberboards made from control and acetylated kenaf fiber. Acetylation results in a small decrease in MOR, but about

TABLE 7.14
Rate and Extent of Thickness Swelling in Liquid Water of Kenaf Fiberboards Made from Control and Acetylated Fiber and a Phenolic Resin [Resin content of boards: 8%]

Fiber	Thickness swelling at–							
	minutes			hours				
	%							
	15	30	45	1	2	3	4	5
Control	15.5	17.1	21.1	22.6	24.7	26.8	31.1	32.6
18.4 WPG	6.7	6.8	6.8	7.0	7.0	7.0	8.0	8.1

Fiber	Days					Oven drying	Weight loss after test
	%						
	1	2	3	4	5		
Control	37.7	41.5	42.6	43.5	44.5	19.0	2.0
18.4 WPG	8.5	8.5	8.7	8.8	9.0	0.7	2.8

TABLE 7.15
Fungal Cellar Tests of Jute Cloth Made from Control and Acetylated Fiber[a]

Weight percent gain	Rating at intervals (months)[b]							
	2	3	4	5	6	12	24	36
0	2	3	3	3	4	—	—	—
7.4	0	1	1	2	3	4	—	—
11.5	0	0	0	1	2	3	4	—
13.3	0	0	0	0	0	1	2	3
16.8	0	0	0	0	0	0	0	0
18.3	0	0	0	0	0	0	0	0

[a]Nonsterile soil containing brown-, white-, and soft-rot fungi and tunneling bacteria.
[b]Rating system: 0 = no attack; 1 = slight attack; 2 = moderate attack; 3 = heavy attack; 4 = destroyed.

equal values in MOE and TS. All strength values given in Table 7.16 are above the minimum standard as given by the American Hardboard Association [43]. The small decrease in some strength properties resulting from acetylation may be attributed to the hydrophobic nature of the acetylated furnish, which may not allow the water-soluble phenolic or isocyanate resins to penetrate into the flake. The adhesives used in these tests have also been developed for unmodified lignocellullsics. Different types of adhesives may be needed in chemically modified boards [44].

7.12.2 CYANOETHYLATION

Jute can be made to react with acrylonitrile in the presence of alkali under conditions that do not reduce the tensile strength of the fibers to any important extent. The properties of cyanoethylated cotton have been known for some time [45], and this particular chemical modification is claimed to provide increased stability against degradation by acids and by heat. Cotton containing more than 3% nitrogen is also said to show high resistance to microbiological deterioration [46].

Experiments with jute yarn at the British Jute Trade Research Association have shown that although untreated yarn subjected to hydrolysis with 0.2 N sulfuric acid at 100°C for 60

TABLE 7.16
Modulus of Rupture (MOR), Modulus of Elasticity (MOE), and Tensile Strength (TS) Parallel to the Board Surface of Fiberboards Made from Control or Acetylated Kenaf Fiber and 8% Phenolic Resin

Board	MOR	MOE	TS
	MPa	GPa	MPa
Kenaf			
Control	47.1	4.6	31.0
18.4 WPG	38.6	5.1	27.1
ANSI Standard	31.0	—	10.3

min retained only 20% of its initial strength, a yarn cyanoethylated to 4.6% nitrogen content retained 80% of its strength under similar conditions.

Jute yarns cyanoethylated to different extents showed increasing resistance to degradation by heating, and whereas untreated yarn heated at 150°C for 24 h retained only 55% of its initial strength, and similar yarn with a nitrogen content of 4.9% retained 90%.

Resistance to rotting was also examined by incubation of yarns under degrading conditions, which caused complete breakdown of strength after 2 weeks. Cyanoethylation up to 1.5% nitrogen showed little improvement, but for 2.8% nitrogen and more, even 16 weeks incubation reduced strength by only 10%. A copper naphthenate treatment with 1.2% copper, for comparison, retained only 30% of strength under similar conditions of exposure. Thus, cyanoethylation gives effective protection against rotting, provided the nitrogen content approaches about 3% [46].

7.13 PHOTOCHEMICAL AND THERMAL DEGRADATION

All cellulose-containing fibers lose strength on prolonged exposure to sunlight. This effect is mainly attributable to the ultraviolet component of the radiation, and its scale is such that, in cotton, about 900-h exposure reduces the strength to 50% of the initial value. In jute, however, a similar strength reduction occurs after about 350-h exposure, and so, although the exposure times are not precise, it is clear that jute loses strength at more than twice the rate for cotton.

In both fibers, there is a loss in strength due to primary bond breakages in the cellulose constituent, but, when seeking an explanation for the difference in behavior, the important question is whether it arises entirely from a greater rate of bond breakage in jute than in cotton, or whether the cohesion between the ultimate cells in jute is also reduced as a result of changes in the middle lamella.

The rate of breakage of cellulose bonds in cotton is readily found from the changes in the degree of polymerization (DP) as exposure continues, using the cuprammonium fluidity as a measure of the DP. In jute, however, this method is not always satisfactory because it is difficult to achieve a complete dissolution of the cellulose component in cuprammonium hydroxide because of interference from the lignin in the fiber. Moreover, preliminary removal of lignin is not advisable, as whatever the process used, it is always liable to cause some degradation of the cellulose.

Nitration techniques that do not degrade the cellulose component have been used successfully to determine the DP of wood cellulose [47], and similar methods are equally satisfactory for jute or other lignified materials [48,49]. In one study carried out in the laboratories of the British Jute Trade Research Association [50], the nitrated lignin and hemicellulose components were first removed by solvent extraction and fractional precipitation, and the DP of the residual cellulose nitrate then determined from viscosity measurements in acetone solution. The viscosities have to be referred to a standard rate of shear and the whole procedure is rather lengthy, but the results showed that after the same exposure conditions jute and cotton had similar DPs within experimental limits of error. Moreover, a plot of 1/DP against time of exposure in standard sun hours was linear, suggesting that the kinetic equation for random breakdown of a polymer chain, namely $1/(DP)_t = 1/(DP)_o * kt$ applies in this case. $(DP)_o$ and $(DP)_t$ are the DPs measured before exposure and after exposure for time t, whereas k is a constant representing the rate of bond breakage.

Exposure to sunlight for periods up to 600 h gave values of k equal to 15.4×10^{-7} and 13.3×10^{-7} for jute and cotton, respectively, in units of reciprocal DP per hour exposure. Exposure to artificial sources of UV light such as a mercury arc lamp or a xenon arc lamp gave lower values of k than for sunlight, but again jute and cotton were similar. With the mercury lamp, the k values were 8.0×10^{-7} and 8.9×10^{-7} units for jute and cotton, whereas the xenon lamp gave 5.9 and 6.4×10^{-7} units, respectively, for jute and cotton.

The rate of photochemical breakdown of cellulose thus appears largely independent of whether lignin is present or not, and, contrary to views that have been expressed in the past, lignin does not act as a photosensitizer for the breakdown. The greater loss in strength of jute compared to cotton must therefore be related to photochemical changes taking place in the middle lamella, which reduce the cohesion between ultimate cells.

Cellulose-containing fibers also lose strength on prolonged exposure to elevated temperatures, but, in this case, cotton and jute show only minor differences in strength losses under similar heating conditions. At 140°C, both fibers lose 50% of strength after 80–85 h exposure, whereas at 160°C only about 10-h exposure is required for the same fall in strength. Thus, although the temperature is a major factor determining the rate of loss in strength, cotton and jute behave similarly and there is no suggestion that the cohesion of the middle lamella is changed by exposure to heat.

Measurement of the change in DP in heating presents difficulties, as the cellulose nitrate now becomes insoluble in acetone or other solvents. This may be due to cross-linking between reactive groups produced in the cellulose molecules by the thermal exposure. In any case, it appears that the chemical changes taking place in thermal degradation are different from those occurring in light-induced degradation.

The nitration techniques used in the measurement of the DP of α-cellulose merit further discussion, particularly in relation to the effect of time of nitration on the cellulosic constituents of the fiber. Nitration of lignin results in products soluble either in the nitrating acids or in methanol, and, by a suitable extraction procedure, the lignin component of the fiber can be completely removed.

After removal of lignin, the nitrated cellulosic products can be separated into three fractions, which are designated A, B, and C, of which Fraction A is insoluble in acetone; Fraction B is soluble in acetone, but insoluble in water; and Fraction C is soluble in both acetone and water. Analysis shows the acetone-soluble fractions B and C to consist of nitrated α-cellulose of DP about 4450, and nitrated hemicellulose, respectively. Both these products are also found in the acetone-insoluble fraction, A.

The amount of α-cellulose that is released as the acetone-soluble Fraction B increases as the time of nitration is increased, and, although small at first, finally reaches the analytical value of about 60% of the whole fiber. The time required for reaching the analytical value is temperature dependent, and, although Lewin and Epstein [49] report that at 3°C more than 24-h nitration is required, they point out that Timell [48] obtained a similar result in only 1 h at 17°C.

As the acetone-soluble Fraction B increases, the acetone-insoluble Fraction A decreases. Interpolation in Lewin and Epstein's results suggests that the two fractions become equal after 11- or 12-h nitration, and that, at this point of equality, their value is about 50% of the maximum value achieved by Fraction B, namely the analytical value.

This pattern of behavior is considered by Lewin and Epstein to indicate the presence of chemical linkages between α-cellulose and the hemicelluloses in jute that hold these components together in Fraction A and render the complex insoluble in acetone. The release of increasing amounts of acetone-soluble nitrated α-cellulose in Fraction B then arises from breakage of the links by the nitrating acids, with more breakages occurring as the time of nitration is increased.

7.14 MOISTURE EFFECTS

The equilibrium moisture held by jute when exposed to an atmosphere of different relative humidity (RH) shows appreciable hysteresis according to whether there is absorption from low humidities or desorption from high humidities. Thus, at 65% RH and 20°C, the equilibrium moisture regain is about 12.5% for absorption by dry fiber and 14.6% for desorption of wet fiber, whereas exposure to 100% RH gives an equilibrium regain of 34–35%. These are average values, and different samples of fiber may show minor differences. It is noted that at 65% RH the equilibrium regain of jute is about 6% higher than that of cotton.

Jute swells in water to an extent of about 22%, a value similar to that of cotton, despite a greater proportion of noncrystalline material in jute. Delignification has a pronounced effect, and it is reported that when the lignin content has been reduced to 0.78%, the swelling may reach almost 40% [51].

Apart from swelling, delignification also affects the equilibrium regain of jute fiber, and Kabir et al. have shown that when delignified by 10%, using a chlorite treatment followed by sodium bisulfite solution extraction, the absorption and desorption regains at 65% RH are each increased by about 1% [52].

7.15 FASTNESS TO LIGHT

7.15.1 Undyed Jute

A major practical difficulty affecting the performance of dyed or bleached jute materials is the change in color that occurs when jute fiber is exposed to sunlight. In the UV region of the spectrum, exposure to light of wavelengths between 3000 and 3600 Å results in yellowing of the fiber, whereas exposure to wavelengths between 3800 and 4000 Å, on the fringe of the visible spectrum, has a bleaching effect. The final color is the resultant of the two processes, and, in general, the initial color change is an obvious yellowing, or darkening, of the fiber, but on longer exposure this color slowly gets lighter and less intense.

Bleaching before exposure generally accentuates the discoloration of the fiber compared to unbleached jute, although part of this is due to the heightened contrast between the nearly white bleached fiber and the exposed fiber. The onset of yellowing varies considerably with different bleaches. Alkaline or neutral hypochlorite, a cheap bleaching medium, gives a product with a rather rapid yellowing tendency, whereas alkaline hydrogen peroxide gives a good white color and a less marked yellowing than hypochlorite. Sodium chlorite, applied under acid conditions, shows the least yellowing tendency, but care must be taken that in obtaining the best conditions to prevent yellowing no drastic loss of strength takes place.

A bleaching process developed in the United States and patented for jute by Fabric Research Laboratories involves treatment with hydrogen peroxide and acid permanganate and gives a better resistance to yellowing than chlorite bleaching. Treatment with acid permanganate alone leaves the natural color of the jute almost unchanged and also provides a higher resistance to yellowing. This improvement probably represents a true reduction in yellowing, although part of it may be due to the smaller contrast between the original bleached color and the exposed color than found with the whiter bleaches.

Improvements in the stability of jute to light exposure result from acetylation or methylation. Treatment with acetic anhydride in xylene solution, for example, combined with a reduction process using sodium borohydride may confer virtually complete stability, whereas methylation with diazomethane confers a marked improvement without preventing yellowing entirely.

Color changes in jute are associated with the lignin content of the fiber, the isolated α-cellulose and hemicellulose fractions being unaffected by exposure to UV light of the correct wavelength band. The importance of lignin has also been demonstrated by irradiating cellulosic fibers of different lignin contents, and, for a series of fibers covering the range of 0% (cotton) to 13% (*Phormium tenax*), it was evident that the intensity of yellowing became more pronounced as the lignin content increased [53].

The formation of colored products from irradiated lignin involves complex reaction chains that are difficult to elucidate fully. It is probable that orthoquinone groups are responsible for the yellow color, formed from orthophenol groups as intermediates. Acetylation blocks both phenolic and aliphatic hydroxyl groups and prevents the objectionable reactions from taking place. The less effective methylation, however, blocks only the phenolic hydroxyl groups.

7.15.2 Dyed Jute

Jute can be dyed with a wide range of dye stuffs. All those generally used for cellulosic fibers, such as direct, vat, and reactive dyes, can be used successfully on jute, but, in addition, jute has a strong affinity for both acid dyes and basic dyes, which normally have little or no dyeing capacity for cotton or rayon but are used extensively for wool.

Many acid and basic dyes give strong, bright colors on jute, but performance is disappointing in regard to color fastness on exposure to sunlight. Some of the dyes used have intrinsically poor light fastness, but it has long been apparent that many acid dyes that give excellent light fastness on wool became fugitive when applied to jute. Yellowing of the jute background causes an apparent change in the color of dyed jute, and although poor fastness to light usually means fading of color, any change in color is in fact regarded as a lack of fastness.

Although systematic studies of dye stuffs on jute have not been frequent, the comprehensive studies carried out at the British Jute Trade Research Association merit discussion [53]. In these studies, a number of dyes were taken from each of several different classes and used to dye a standard jute fabric, both natural and after bleaching. The fastness to light of these dyed samples was then assessed by exposure to xenon light, alongside a series of light fastness standards, and the results compared with the known fastness value of the dye stuff on cotton. There are eight standards in all; No. 1 is the most fugitive and No. 8 the most resistant, and the experimental conditions for assessment are well standardized [54].

The results indicated that with vat dyes, accelerated fading of the dye stuff on jute compared with cotton was largely absent and that the yellowing was the main factor on which the apparent light fastness depended. With acid and basic dyes, however, accelerated fading appeared to be the predominant effect, although the balance with yellowing varied with both color and chemical structure of the dye. A number of acid dye stuffs known to give fastness ratings of 6 or more on wool were rated only 3–4 on jute, with loss of dye color the main cause, whereas a group of basic dyes with ratings of 6 or more on acrylic fibers were reduced to ratings of 2–3 on jute. Again, although yellowing was evident, accelerated fading was the principal cause.

A selection of 200 direct dyes, representative of the range of chemical types in this class and all having fastness ratings of 4 or more on cotton, were used for test dyeing on jute, both natural, chlorite bleached, and peroxide bleached. On cotton, 66% of dyes had fastness grade 6 or more, but on natural jute only 17% retained this grade; the number fell further to 12% on chlorite bleached jute and 5% on peroxide bleached. The average grading was 5.7 on cotton and 4.8 on natural jute, 4.7 on chlorite bleached, and 4.6 on peroxide bleached. Thus, on average, the grade on jute was about 1.0 lower than that on cotton.

The drop in grade was, however, far from regular for different dyes, and the balance between yellowing and accelerated fading did not follow a predictable pattern. Dye color played an important role, for yellow dyes dropped only about 0.5 grade on jute against cotton, whereas for blue colors the average drop was 1.4 grades.

In the case of reactive dyes, test dyeing was done on natural and chlorite-bleached jute. Of the dyes used, 55% were graded 5–6 and more than 6, but on jute only 2% were retained in this category. The mean grade for cotton was 5.3, compared to 4.2 on natural jute and 4.1 on chlorite bleached.

In general, therefore, few dye stuffs retain the same light fastness on jute, natural or bleached, as on cotton. Reduction of the underlying yellowing is helpful in many cases, but there are examples of accelerated fading on jute. Acetylation and methylation can improve the fastness considerably, by preventing the background yellowing, and possibly this prevention may also affect the accelerated fading. However, these treatments are expensive and not simple to use, and alternative methods of obtaining light stability are needed if the standard of jute dyeing is to be raised.

7.16 WOOLENIZATION

When jute fiber is treated with strong alkali, profound changes occur in its physical structure. Lateral swelling occurs, together with considerable shrinkage in lengths, as a result of which the fiber is softened to the touch and develops a high degree of crimp or waviness. The crimp gives a wool-like appearance to the fiber, and much attention has been given to assessing the commercial possibilities for this chemical modification.

On stretching the fibers to break, the crimp is straightened and thereby the extensibility of the fiber is increased. The effect is small at alkali concentrations up to about 10%, but the extensibility increases rapidly at concentrations of 15% and upward and may reach 8 or 9%. At the same time, however, the tensile strength of the fiber decreases with increasing alkali concentration, but the product of extensibility and tensile strength, the breaking energy, appears to pass through a maximum at 15–20% concentration [55]. This has a beneficial effect on spinning because the carded fiber has a longer average length than normal and this results in a more uniform yarn.

The rapid change in extensibility in the vicinity of 15% concentration is similar to the effect of slack mercerization on cotton. The nature of the chemical changes occurring in jute on mercerization have been discussed by Lewin [26], especially in regard to the role played by lignin in the fiber structure. The sheathing of the ultimate cells by a lignified membrane affects the free swelling of the cells and produces tension, whereas the irregular shape of fibers in cross section leads to folding under tension once the middle lamella material is weakened by the treatment.

The crimp statistics have been studied in detail at the Institute for Fibers and Forest Products Research in Jerusalem, and much information has been brought together by, for example, Lewin et al. [56]. Two parameters are measured to define the crimp, namely the RMS value of the width (D) and the number of crimps per unit length of the stretched fiber (n). As the crimp is three dimensional, the fiber is rotated during the measurements. Typical values for jute fibers immersed in 12.5% NaOH, for 1 h at a temperature of 2°C are reported to be about 1.6 mm for D, with a standard deviation of 0.55 mm, and about 0.098 mm^{-1} for n, with a standard deviation of 0.035 mm^{-1}. The extension of the fibers at break was 15% relative to the initial length of the crimped fiber under a load of 10 mg, and the crimp disappeared for loads of about 2000 mg. The energy required to uncrimp the fiber was equivalent to about 3.9 g per 1% of extension.

The above figures refer specifically to an alkali concentration of 12.5%. At concentrations below 6%, no crimp is formed, whereas at 9% alkali, D reaches a maximum value of about 1.9 mm. At concentrations of 15% and above, D takes up a reasonably constant value of about 1.35 mm. The value of n, however, is scarcely affected by changes in alkali concentration.

It is said that the optimum temperature for crimp formation is about 2°C and that at higher temperatures the crimp parameters are reduced, becoming zero at 40°C. An immersion time of at least 0.5 h is necessary for the crimp to be formed.

Banbaji [57] has examined the tensile properties of jute fibers before and after alkali treatment and has shown that the tenacity decreases with increasing concentration: an initial value of 3.6 g/den falling to 2.5 g/den at 9% alkali and to 1.5 g/den at 24% alkali, at 2°C and 1-h immersion. The extension at break, referred to the fiber length before immersion, increased from 1.2% without alkali treatment to 3.6% at 9% alkali and then fell slightly to 2.4% at 24% alkali.

The tenacity changes are no doubt linked with the losses in weight that occur with alkali treatment, but there may be more profound changes taking place internally within the ultimate cells. Such changes are at present imperfectly understood, but, if useful commercial developments are to be made, further investigation of structural changes appears essential. Moreover, the crimp is a "once only" effect, and to be really useful a small degree of elasticity must be introduced into the fiber.

The stability of the crimp is poor, and, once the fiber has been straightened under tension, there is no tendency to revert to the crimped state when the tension is removed; that is, the woolenizing treatment does not confer elasticity on the fiber.

Under mercerizing conditions, the fibers lose considerable weight (15% or more) and give the appearance of being opened up. It is commonly said that there is a considerable reduction in diameter, which implies a lower linear density and hence the production of more regular yarns. However, just as in natural jute, there appears to be a limit below which the diameter does not fall as with mercerized fiber.

The physical effects of the mercerizing process are different when the jute material is kept under tension, instead of being slack. Experiments reported from the Bangladesh Jute Research Institute with treated jute yarns [58] show that the shrinkage is greatly reduced by tension, falling from 11–12% when slack to 1.5–2.5% under 3-kg tension. The loss in weight of 12–13% when slack was reduced by a few percent under 3-kg tension. The effect of temperature change from 30 to 60°C was small in all cases.

The appearance and feel of jute fabrics is much improved by the woolenizing process, and bleached and dyed fabrics appear to have commercial possibilities. The problem is the cost of the treatment, and to achieve similar effects more cheaply may require a deeper knowledge of the internal changes that take place within the fiber.

7.17 APPLICATIONS AND MARKETS

The large historic markets for jute in sacking, carpet backing, cordage, and textiles have decreased over the years and have been replaced by synthetics. Fiber from jute and kenaf can be used in handicraft industries, to make textiles, to make paper products, or to produce a wide variety of composites. A great deal of research is presently underway in each of these fields; however, the largest potential markets are in composite products. These composites range from value-added specialty products to very large volume commercial materials. These markets are potentially larger than the past markets for jute and kenaf and could lead to new dynamic uses for these and other natural fibers.

7.17.1 COMPOSITES

A composite is any combination of two or more resources held together by some type of mastic or matrix. The mastic or matrix can be as simple as physical entanglement of fibers to more complicated systems based on thermosetting or thermoplastic polymers. The scheme shown below gives possible processing pathways that lead to the composite products identified in this report that can come from each fraction of the plant. The entire plant (leaves, stock, pith, roots) can be used directly to produce structural and nonstructural composites such as particleboards or fiberboards. By using the entire plant, processes such as retting, fiber separation, fraction purification, etc. can be eliminated, which increases the total yield of plant material and reduces the costs associated with fraction isolation. This also gives the farmers a different option in their crop utilization; that is, bringing in the entire plant to a central processing center and not having to get involved in plant processing [59].

Another option is to separate the higher value long fiber from other types of shorter fibers and use it in combination with other materials to make value-added structural composites. When the long fiber is separated, the by-product is a large amount of short fiber and pith material that can be used for such products as sorbents, packing, light-weight composites, and insulation. By utilizing the by-product from the long fiber isolation process, the overall cost of long fiber utilization is reduced.

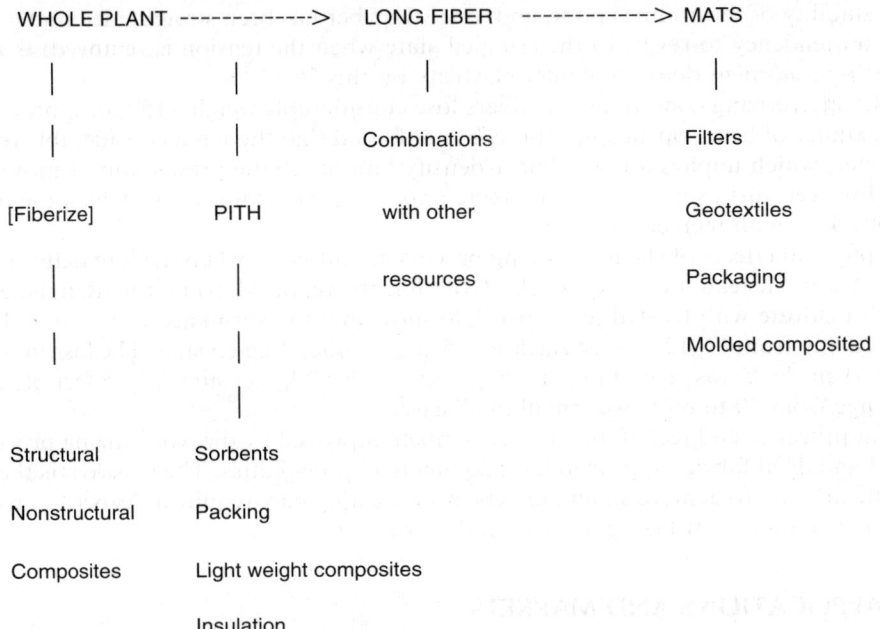

The isolated long fiber can then be used to make mats that have value-added applications in filters, geotextiles, packaging, molded composites, and structural and nonstructural composites. Composites can be classified in many ways as follows: by their densities, by their uses, by their manufacturing methods, or other systems. For this report, they will be classified by their uses. Eight different classes are covered: geotextiles, filters, sorbents, structural composites, nonstructural composites, molded products, packaging, and combinations with other materials. There is some overlap between these areas. For example, once a fiber web has been made it can be directly applied as a geotextile, filter, or sorbent, or can further be processed into a structural or nonstructural composite, molded product, used in packaging, or combined with other resources. Within each composite made there are opportunities to improve the performance of that composite by improving the performance of the fiber used in the composite.

7.17.2 GEOTEXTILES

The long bast or leaf fibers can be formed into flexible fiber mats, which can be made by physical entanglement, nonwoven needling, or thermoplastic fiber melt matrix technologies.

The two most common types are carded and needle-punched mats. In carding, the fibers are combed, mixed, and physically entangled into a felted mat. These are usually of high density, but can be made at almost any density. In the mid-1960s, a mechanical system was developed to process long synthetic fibers for use in medium density fiberboard (Figure 7.7). Section A in Figure 7.1 is where the kenaf or jute bast fiber is fed into the system. Section B is a fiber opener where fiber bundles are separated and can be mixed with other fibers. Between A and B, the fibers are formed into a continuous mat, which is fully formed at C. At D, the web can go on through a needle board where the web is "needled" together in a nonwoven process. Another option at D is to run the mat through a heated chamber or heated metal rollers to melt a plastic fiber that was blended into the web at stage C. Other similar systems have been made using the same principles. Figure 7.8 shows a web that has been made using the needed system.

Work has been done that demonstrates how additives, such as super absorbent powders and binders, can be added to the web during the forming process. In the case of super absorbents, one advantage of this approach is that the super absorbent powder when near the area of maximum void space in the web can absorb liquids faster and in greater quantity than if added to a finished web as part of a laminate in an off-line process. Also, because of their uniform dispersion, powdered binders can perform in much the same manner to insure maximum strength with a minimum add-on. Medium- to high-density fiber mats can be used in several ways. One is for the use as a geotextile. Geotextiles derive their name from the two words geo and textile and, therefore, mean the use of fabrics in association with the earth.

Geotextiles have a large variety of uses. These can be used for mulch around newly planted seedlings (Figure 7.9). The mats provide the benefits of natural mulch; in addition, controlled-release fertilizers, repellents, insecticides, and herbicides can be added to the mats as needed. Research results on the combination of mulch and pesticides in agronomic crops have been promising.

The addition of such chemicals could be based on silvicultural prescriptions to ensure seedling survival and early development on planting sites where severe nutritional deficiencies, animal damage, insect attack, and weed problems are anticipated. Medium-density fiber mats can also be used to replace dirt or sod for grass seeding around new homesites or along highway embankments (Figure 7.10). Grass or other type of seed can be incorporated in the

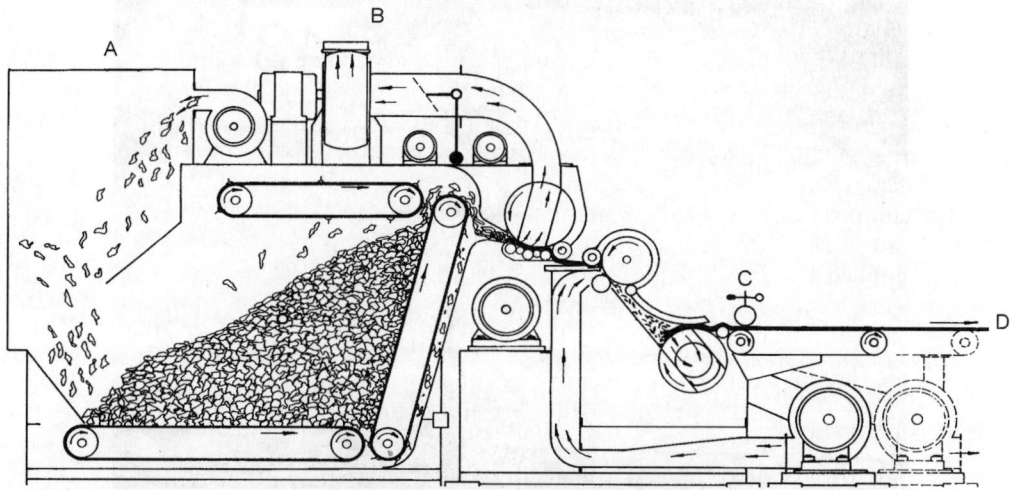

FIGURE 7.7 Schematic diagram of a web making machine (USDA).

FIGURE 7.8 Fiber web (USDA).

fiber mat. Fiber mats promote seed germination and good moisture retention. Low- and medium-density fiber mats can be used for soil stabilization around new or existing construction sites. Steep slopes, without root stabilization, lead to erosion and loss of top soil.

Medium- and high-density fiber mats can also be used below the ground in road and other types of construction as a natural separator between different materials in the layering of the back fill. It is important to restrain slippage and mixing of the different layers by placing

FIGURE 7.9 Mulch mat used to plant tree seedlings (USDA).

FIGURE 7.10 Geotextile used to stabilize a steep slope (USDA).

separators between the various layers. Jute and kenaf geotextiles have been shown to work very well in these applications, but the potential exists for any of the long jute and kenaf fibers.

7.17.3 Filters

Medium- and high-density fiber mats can be used as air filters. The density of the mats can be varied, depending on the size and quantity of the material being filtered and the volume of air required to pass through the filter per unit of time. Air filters can be made to remove particulates and can be impregnated or reacted with various chemicals as an air freshener or cleanser.

Medium- to high-density mats can also be used as filtering aids to take particulates out of waste and drinking water or solvents. Figure 7.11 shows a filter unit that is in place to remove metal ions from water that has come from an abandoned coal mine. Jute and kenaf fibers can also be modified to become more efficient in removing a wide variety of contaminates from water.

FIGURE 7.11 Filter unit containing filters made of kenaf fiber (USDA).

7.17.4 Sorbents

Tests are presently underway to use jute and kenaf sorbents to remove heavy metals, pesticides, and oil from rain water run off in several cities in the United States. Medium- and high-density mats can also be used for oil spill clean up pillows. It has been shown that the core material from kenaf preferentially sorbs oil out of seawater when saturated with water. There are many other potential sorbent applications of agrofiber and core resources such as removal of dyes, trace chemicals in solvents, and in the purification of solvents.

It is also possible to use core materials as sorbents in cleaning aids such as floor sweep. While this is not a composite, it does represent another way in which jute and kenaf resources can be used as sorbents.

7.17.5 Structural Composites

A structural composite is defined as one that is required to carry a load in use. In the housing industry, for example, these represent load-bearing walls, roof systems, subflooring, stairs, framing components, furniture, etc. In most, if not all, cases, performance requirements of these composites are spelled out in codes and in specifications set forth by local or national organizations.

Structural composites can range widely in performance from high-performance materials used in the aerospace industry down to wood-based composites, which have lower performance requirements. Within the wood-based composites, performance varies from multilayered plywood and laminated lumber to low-cost particleboard. Structural wood-based composites intended for indoor use are usually made with a low-cost adhesive, which is not stable to moisture, while exterior-grade composites use a thermosetting resin that is higher in cost but stable to moisture. Performance can be improved in wood-based as well as jute and kenaf composites by using chemical modification techniques, fire retardant, and decay control chemicals, etc.

7.17.6 Nonstructural Composites

As the name implies, nonstructural composites are not intended to carry a load in use. These can be made from a variety of materials such as thermoplastics, textiles, and wood particles, and are used for such products as doors, windows, furniture, gaskets, ceiling tiles, automotive interior parts, molding, etc. These are generally lower in cost than structural composites and have fewer codes and specifications associated with them.

7.17.7 Molded Products

The present wood-based composite industry mainly produces two-dimensional (flat) sheet products. In some cases, these flat sheets are cut into pieces and glued or fastened together to make shaped products such as drawers, boxes, and packaging. Flat-sheet wood fiber composite products are made by making a gravity formed mat of fibers with an adhesive and then pressing. If the final shape can be produced during the pressing step, then the secondary manufacturing profits can be realized by the primary board producer (Figure 7.12). Instead of making low-cost flat-sheet-type composites, it is possible to make complex-shaped composites directly using the long bast fiber.

In this technology, fiber mats are made similar to the ones described for use as geotextiles; except, during mat formation, an adhesive is added by dipping or spraying of the fiber before mat formation or added as a powder during mat formation. The mat is then shaped and densified by a thermoforming step. Within certain limits, any size, shape, thickness, and

FIGURE 7.12 Three-dimensional composites made using a fiber web (USDA).

density is possible. These molded composites can be used for structural or nonstructural applications as well as packaging, and can be combined with other materials to form new classes of composites. This technology is described later.

7.17.8 Packaging

"Gunny" bags made from jute have been used as sacking for products such as coffee, cocoa, nuts, cereals, dried fruits, and vegetables for many years. Although there are still many applications of long fibers for sacking, most of the commodity goods are now shipped in containers. These containers are not made of agrofibers nowadays, but there is no reason why they cannot be made. Medium- and high-density jute and kenaf fiber composites can be used for small containers, for example, in the tea industry and for large sea-going containers for commodity goods. These composites can be shaped to suit the product by using the molding technology described previously or made into low cost, flat sheets and made into containers.

Jute and kenaf fiber composites can also be used in returnable containers where the product is reused several times. These containers can range from simple crease-fold types to more solid, even nestable, types. Long agrofiber fabric and mats can be overlayed with thermoplastic films such as polyethylene or polypropylene to be used to package such products as concrete, foods, chemicals, and fertilizer. Corrosive chemicals require the plastic film to make them more water resistant and reduce degradation of the jute and kenaf fiber. There are many applications for jute and kenaf fiber as paper sheet products for packaging also. These vary from simple paper wrappers to corrogated, mutifolded, multilayered packaging.

7.17.9 Pulp and Paper

Using trees for the production of pulp and paper is much easier than using kenaf or jute as a source of fiber. Kenaf and jute must be harvested at a set time, collected, stored, cleaned, separated, and transported to a pulp mill. A tree can stand in the forest until needed, cut, transported, debarked, chipped, and then pulped. For some countries, however, trees are not available for pulping and so kenaf or jute are logical options. Kenaf can be harvested and put into piles that can be stored for 1 to 2 years without significant loss of quality. Trials have even been done to use a biopulping approach that reduces both the energy and quantity of chemicals needed in a later chemical pulping process [60].

Kenaf and jute contain a lower amount of lignin; therefore, less pulping chemicals are needed and they have more accessible cell wall structures that allow easier access to pulping chemicals compared to wood [61–63]. The stalk contains more hemicelluloses that result in faster hydration.

A great deal of research has been done to use kenaf as a source of pulp and paper [64]. Kenaf, like jute, contains both an outer layer of long bast fibers and a short fiber core. The bast, on a dry weight basis, contains about 20% of the whole stem with an average fiber length of approximately 2.6 mm. The 80% core fibers are much shorter with an average fiber length of only 0.6 mm. The bast gives a higher yield of pulp and the pulp produced has much higher strength properties compared to the pulp produced from the core.

Pulp has also been produced using the entire unseparated plant using chemical, chemithermomechanical (CTMP), chemimechanical (CMP), thermomechanical (TMP), and mechanical pulping processes [64]. Chemical pulping can be done using either soda or kraft processes. Chemical pulping of kenaf has been studied using a variety of pulping systems including keaft, soda, soda-anthraquinone, acidic sulfite, nitric acid, neutral sulfite, and organosolv [64].

The first commercial kenaf pulp mill was in Khon Kaen, Thialand, which started in 1982 with an annual capacity of 70,000 t [65].

The primary pulping process for kenaf is a CMP process using cold soda. The kenaf is steeped in a caustic soda solution for a short period of time and then fiberized in a disk refiner [64]. A 10% caustic soda cook of whole chopped and washed kenaf stalk cooked at 170°C for 3.5 h produces a good bleachable pulp [66,67]. Kenaf stalk can also be pulped using a slightly modified kraft process to give a good pulp with good drainage, freeness, and strength properties similar to a softwood pulp [68,69]. Whole stalk kraft pulping has also been done as reported by Mittal and Maheshwari [70]. They found a high percentage of bast fiber in the pulp resulting in a higher average fiber length and good physical properties in the paper. Table 7.17 shows the properties of kenaf pulped using either a soda or kraft process.

Thermomechanical pulping of whole kenaf was done, but the resulting pulp had very low strength properties [71]. Chemithermomechanical pulping has also been done using alkaline hydrogen peroxide [72]. Table 7.18 shows the properties of paper made from whole kenaf using either TMP and CTMP.

Han et al. pulped core and bast components separately using sodium sulfide in sodium hydroxide [73]. Table 7.19 shows the results of this work.

TABLE 7.17
Properties of Kenaf Paper Produced by Soda or Kraft Processes

Property	Soda	Kraft
Yield, %	62	55
Cellulose (alpha)	68	71
Pentosans, %	19	20
Kappa number	45	27
Yield bleached, % (Cl_2)	53	48
Burst factor, g/cm^2/gsm	54	50
Tear factor, g/gsm	102	93
Breaking length, m	9,600	10,300

Source: From Touzinsky, G.F. Laboratory paper machine runs with Kenaf thermochemical pulp, TAPPI, 1980, 63(3), 109; Touzinsky, G.F. Kenaf, In *Pulp and Paper Manufacturing*, Vol. 3., Secondary fibers and non-wood pulping, Chapter 8, TAPPI Press, Atlanta, GA; 1987, 106.

TABLE 7.18
Properties of Paper Made Using Whole Kenaf Stalk Using Either TMP or CTMP

Property	TMP	CTMP
Brightness, %	67	70
Burst index, MN/kg	1.4	1.7
Tear index, Nm^2/kg	8.1	8.0
Breaking length, km	3.5	4.3
Apparent density, kg/m^3	318	388
Long fiber, %	36	33
Fines, %	49	53
Opacity, %	95	90

Source: From Touzinsky, G.F. Laboratory paper machine runs with Kenaf thermochemical pulp, TAPPI, 1980, 63(3), 109; Touzinsky, G.F. Kenaf, In *Pulp and Paper Manufacturing*, Vol. 3., Secondary fibers and non-wood pulping, Chapter 8, TAPPI Press, Atlanta, GA; 1987, 106.

CMP produced from kenaf core using alkaline hydrogen peroxide gave a pulp yield of 80%, breaking length of over 4 km, brightness of 60%, and opacity of 92% [74].

Chemical and semichemical kenaf pulps are easy to bleach using a three-stage process including chlorination, caustic extraction, and hypochlorite stages [62]. Bleaching is done after removing shives and fines to reduce the consumption of bleaching chemicals.

Studies have been done on the recycling of kenaf paper [75]. The zero-span breaking length was not affected, but the freeness was significantly reduced. Tear strength increased on the first two recycles, but then decreased after the third cycle.

Jute has also been used to make paper, although the entire plant is rarely used [64]. Pulp mills generally buy old jute sacks, cuttings, and waste wrapping material that are mainly bast fiber. Jute is usually pulped using either a chemical or by one of several chemimechanical processes. The Jute Technological Research Laboratory (JTRL) in Calcutta, India, has done

TABLE 7.19
Hand Sheet Paper Made from Wither Bast Fiber or Core Fiber Using Sodium Sulfate and Sodium Hydroxide

Test	Bast paper	Core paper
Density, kg/m^3	571	906
Freeness, CSF (mL)	631	279
Caliper, mm	0.121	0.075
Strain (elongation), %	2.29	2.38
Tensile strength, kN/g	5.36	7.22
ISO brightness, %	25.3	20.5
Printing opacity, %	97.0	94.4
Burst strength, kPa	333.8	381.8
Burst index, kPa.m^2/g	4.92	5.71
Tear resistance, mN	1446.3	263
Tear index, mN.m^2/g	20.9	3.9
Smoothness, sheffield units	329.8	72.6
Fiber length, Kajaani, mm	2.8	0.81

TABLE 7.20
Properties of Pulp and Paper from Jute Using Three Different Processes

Fiber	Pulping Process	Yield UB	B	Breaking Length (km)	Burst Factor	Tear Factor	Fold
Bast	Chemi-mech	8690	8285	7.3	30	130	250
	Soda	6567	6264	8.5	38	135	900
	Kraft	6870	6365	8.8	40	150	1077
Stick	Chemi-mech	7678	6870	5.0	20	51	78
	Soda	45	42	6.5	29	70	200
	Kraft	48	44	7.1	29	78	241
Whole plant	Chemi-mech	8082	7680	6.0	20	60	250
	Soda						
	Kraft	68	63	7.5	30	101	375

UB = unbleached, B = bleached (Young 1997).

the most research on pulping jute [8]. Using caustic soda (10 to 15%) prior to mechanical disintegration in a disc refiner produces good quality pulp. Jute bast and core can also be pulped using pure soda or a kraft process. Table 7.20 shows the properties of pulp and paper produced using a chemimechanical, soda, or kraft process. Jute can also be pulped using fungal treatment prior to an alkaline pulping process and the pulp has higher strength properties than the pulp produced without the fungal pretreatment.

One mill in India uses a two-stage kraft process, where the first stage is run at low pressure and the second stage at high pressure [64]. The resulting pulp is washed and run through a beater. Jute can also be pulped using an alkaline sulfite or neutral sulfite anthraquinone process [76]. The process is carried out on jute bast fiber using sodium sulfite and sodium carbonate. Jute stick can also be pulped this way, but the strength properties are lower than when bast fiber is used.

Jute pulps are generally bleached using a 5 to 10% solution of sodium or calcium hypochlorite in a two-stage process. The process gives a brightness of 50 to 60. Jute pulp is used in cigarette papers, printing, bond and writing papers, but almost always in combination with other pulps.

7.17.10 PULTRUSION

Jute and kenaf bast fibers can be used to substitute for glass finer in pultrusion technology [77,78]. The long bast fiber can be pulled through a bath of phenolic, polyester, or other thermosetting resin and molded to make a wide variety of stiff, strong profiles. After curing, the profiles can be cut to any length desired. Door frames, U-channels, and sports equipment have been successfully made using this procedure.

7.17.11 COMBINATIONS WITH OTHER RESOURCES

It is possible to make completely new types of composites by combining different resources. It is possible to combine, blend, or alloy leaf, bast and stick fiber with other materials such as glass, metals, plastics, and synthetics to produce new classes of materials. The objective is to combine two or more materials in such a way that a synergism between the components results in a new material that is much better than the individual components.

Jute and kenaf fiber–glass fiber composites can be made using the glass as a surface material or combined as a fiber with other lignocellulosic fibers. Composites of this type can have a very high stiffness to weight ratio. The long bast fibers can also be used in place of glass fiber in resin injection molding (RIM) or used to replace, or in combination with, glass fiber in resin transfer molding (RTM) technologies. Problems of dimensional stability and compatibility with the resin must be addressed, but this could also lead to new markets for property-enhanced jute and kenaf materials.

Metal films can be overlayed on to smooth, dimensionally stabilized fiber composite surfaces or applied through cold plasma technology to produce durable coatings. Such products could be used in exterior construction to replace all aluminum or vinyl siding—markets where jute and kenaf resources have lost market share.

Metal fibers can also be combined with stabilized fiber in a matrix configuration in the same way as metal fibers are added to rubber to produce wear-resistant aircraft tires. A metal matrix offers excellent temperature resistance and improved strength properties, and the ductility of the metal lends toughness to the resulting composite. Application for metal matrix composites could be in the cooler parts of the skin of ultrahigh-speed aircrafts. Technology also exists for making molded products using perforated metal plates embedded in a phenolic-coated fiber mat, which is then pressed into various shaped sections.

Bast or leaf fiber can also be combined in an inorganic matrix. Such composites are dimensionally and thermally stable, and they can be used as substitutes for asbestos composites. Inorganic bonded bast fiber composites can also be made with variable densities that can be used for structural applications.

One of the biggest new areas of research in the value-added area is in combining natural fibers with thermoplastics. Since the price of plastic has risen sharply over the past few years, adding a natural powder or fiber to plastics provides cost reduction to the plastic industry (and in some cases increases performance as well), but, to the jute and kenaf industry, this represents an increased value for the jute and kenaf component.

7.17.12 Fiber Thermoplastic Blends

Before 1980, the concepts of blends and alloys were essentially unknown in the plastic industry. Today, there are more than 1000 patents relating to plastic blends and alloys, and it is estimated that 1 out of every 5 kg of plastic sold in the United States is a blend or an alloy [79]. Blends and alloys have revolutionized the plastic industry, as they offer new materials with properties that were not available before and materials that can be tailored for specific end-uses. The jute and kenaf industries have the same opportunity to follow this trend and greatly expand markets for new materials based on blends and alloys with other resources.

Newer materials and composites that have both economic and environmental benefits are considered for applications in the automotive, building, furniture, and packaging industries. Mineral fillers and fibers are used frequently in the plastic industry to achieve desired properties or to reduce the cost of the finished article. For example, glass fiber is used to improve the stiffness and strength of plastics, although there are several disadvantages associated with the use of the fiber. Glass fibers need a great deal of energy to produce since processing temperatures can exceed 1200°C. They tend to abrade processing equipment and also increase the density of the plastic system. Jute and kenaf fibers have received a lot of interest for use in thermoplastics due to their low densities, low cost, and nonabrasive nature. The inherent polar and hydrophilic nature of the jute and kenaf fibers and the nonpolar characteristics of the polyolefins can lead to difficulties in compounding and result in inefficient composites. Proper selection of additives is necessary to improve the interaction and adhesion between the fiber and matrix phases.

Recent research on the use of jute and kenaf fiber suggests that these fibers have the potential use as reinforcing fillers in thermoplastics and a brief preliminary account was published earlier [80]. The annual growth of agricultural crop fibers such as kenaf has resulted in significant property advantages as compared to typical wood-based fillers and fibers such as wood flour, wood fibers, and recycled newspaper [81–85]. The results indicate that kenaf fiber–polypropylene (PP) composites have significant advantages over conventional inorganic filled and reinforced PP systems for certain applications. The low cost and densities and the nonabrasive nature of the fibers allow high filling levels, thereby resulting in significant cost savings. The primary advantages of using these fibers as additives in plastics are the following: low densities, low cost, nonabrasive nature, high-filling levels possible, low energy consumption, high specific properties, renewable, widely distributed, biodegradable, and improvement in the rural or agriculture-based economy.

The two main disadvantages of using jute and kenaf fibers in thermoplastics are the high moisture absorption of the fibers and composites [80] and the low processing temperatures permissible. The moisture absorbed by the composite and the corresponding dimensional changes can be reduced dramatically if the fibers are thoroughly encapsulated in the plastic and there is good adhesion between the fiber and the matrix. If necessary, moisture absorption of the fibers can be significantly reduced by the acetylation of the hydroxyl groups present in the fiber [86], although this is possible with some increase in cost. The disadvantage of the high moisture absorption of the composite can be minimized by selecting applications where the high moisture absorption is not a major drawback. For example, polyamide and its composites absorb large amounts of water, but applications are such that this deficiency is not of prime importance. The processing temperature of the lignocellosic fibers in thermoplastics is limited due to potential fiber degradation at higher temperatures. The plastics that can be used are limited to low-melting-temperature plastics. In general, no deterioration of properties due to fiber degradation occurs when processing temperatures are maintained below about 200°C for short periods.

Kenaf bast fibers with a filament length longer than 1 m are common. These filaments consist of discrete individual fibers, generally 2 to 6 mm long, which are themselves composites of, predominantly, cellulose, lignin, and hemicelluloses. Filament and individual fiber properties can vary, depending on the source, age, separating techniques, and history of the fiber. Furthermore, the properties of the fibers are difficult to measure, so we have made no attempt to measure the properties of kenaf.

Kenaf filaments, about 15 to 20 cm long, a maleic anhydride grafted polypropylene (MAPP used as a coupling agent to improve the compatibility and adhesion between the fibers and matrix), and polypropylene were compounded in a high-intensity kinetic mixer where the only source of heat is generated through the kinetic energy of rotating blades. The blending was accomplished at 4600 rpm that resulted in a blade tip speed of about 30 m/s and then automatically discharged at 190°C.

The mixed blends were then granulated and dried at 105°C for 4 h. Test specimens were injection molded at 190°C. Tensile tests were conducted according to ASTM 638–90, Izod impact strength tests according to ASTM D 256–90, and flexural testing using the ASTM 790–90 standard. The cross-head speed during the tension and flexural testing was 5 mm/min. Although all the experiments were designed around the weight percent of kenaf in the composites, fiber volumes fractions can be estimated from composite density measurements and the weights of dry kenaf fibers and matrix in the composite. The density of the kenaf present in the composite was estimated to be 1.4 g/cc. The results are shown in Table 7.21.

To develop sufficient stress transfer properties between the matrix and the fiber, two factors need to be considered. Firstly, the MAPP present near the fiber surface should be

TABLE 7.21
Properties of Kenaf and Jute Reinforced Polypropylene Composites

Filler/reinforcement in PP	ASTM standard	None	Kenaf	Jute	Talc	CaCO$_3$	Glass	Mica
% filler by weight		0	50	50	40	40	40	40
% filler by volume (estimated)		0	39	39	18	18	19	18
Tensile modulus, GPa	D638	1.7	8.3	7.8	4	3.5	9	7.6
Specific tensile modulus, GPa		1.9	7.8	7.2	3.1	2.8	7.3	6.0
Tensile strength, MPa	D638	33	68	72	35	25	110	39
Specific tensile strength, MPa		37	58	67	28	20	89	31
Elongation at break, %	D638	≫10	2.2	2.3	×	×	2.5	2.3
Flexural strength, MPa	D 790	41	91	99	63	48	131	62
Specific flexural strength, MPa		46	85	92	50	38	107	49
Flexural modulus, GPa	D 790	1.4	7.8	7.7	4.3	3.1	6.2	6.9
Specific flexural modulus, GPa		1.6	7.3	7.1	3.4	2.5	5.0	5.5
Notched izod impact, J/m	D256A	24	32	31	32	32	107	27
Specific gravity		0.9	1.07	1.08	1.27	1.25	1.23	1.26
Water absorption %—24 h	D570	0.02	1.05	×	0.02	0.02	0.06	0.03
Mold (linear) shrinkage cm/cm		0.028	0.003	×	0.01	0.01	0.004	×

strongly interacting with the fiber surface through covalent bonding and acid–base interactions. This means sufficient MA groups should be present in the MAPP so that interactions can occur with the –OH groups on the fiber surface. Secondly, the polymer chains of the MAPP should be long enough to permit entanglements with the PP in the interphase. Polar polymers that can develop hydrogen bonding between chains tend to reach mechanical integrity at lower molecular weights.

A small amount of the MAPP (0.5% by weight) improved the flexural and tensile strength, tensile energy absorption, failure strain, and unnotched Izod impact strength. The anhydride groups present in the MAPP can covalently bond to the hydroxyl groups of the fiber surface. Any MA that has been converted to the acid form can interact with the fiber surface through acid–base interactions. The improved interaction and adhesion between the fibers and the matrix leads to better matrix to fiber stress transfer. There was little difference in the properties obtained between the 2 and 3% (by weight) MAPP systems. The drop in tensile modulus with the addition of the MAPP is probably due to molecular morphology of the polymer near the fiber surface or in the bulk of the plastic phase. Transcrystallization and changes in the apparent modulus of the bulk matrix can result in changes in the contribution of the matrix to the composite modulus and are discussed later. There is little change in the notched impact strength with the addition of the MAPP, while the improvement in unnotched impact strength is significant. In the notched test, the predominant mechanism of energy absorption is through crack propagation as the notch is already present in the sample. The addition of the coupling agent has little effect in the amount of energy absorbed during crack propagation. On the other hand, in the unnotched test, energy absorption is through a combination of crack initiation and propagation. Cracks are initiated at places of high stress concentrations such as the fiber ends, defects, or at the interface region where the adhesion between the two phases is very poor. The use of the additives increases the energy needed to initiate cracks in the system and thereby results in improved unnotched impact strength values with the addition of the MAPP. Entanglement between the PP and MAPP molecules results in improved interphase properties and the strain to failure of the composite. There is a plateau after which further addition of a coupling agent results in no further increase in the ultimate failure strain. There is little difference in the tensile strength of uncoupled composites compared to the unfilled PP, irrespective of the amount of fiber present. This suggests that there is little stress transfer from the matrix to the fibers due to incompatibilities between the different surface properties of the polar fibers and nonpolar PP. The tensile strength of the coupled systems increased with the amount of fiber present and strengths of up to 74 MPa were achieved with higher fiber loading of 60% by weight or about 49% by volume. As is the case with tensile strength, the flexural strength of the uncoupled composites was approximately equal for all fiber-loading levels, although there was a small improvement as compared to the unfilled PP. The high shear mixing using the thermokinetic mixer causes a great deal of fiber attrition. Preliminary measurements of the length of fibers present in the composite after injection molding show that few fibers are longer than 0.2 mm. The strength obtained in our composites was thus limited by the short fiber lengths. Higher strengths are likely if alternate processing techniques are developed that reduce the amount of fiber attrition while at the same time achieve good fiber dispersion.

The specific tensile and flexural moduli of 50% by weight kenaf coupled composites were about equivalent to or higher than the typical reported values of 40% by weight coupled glass–PP injection-molded composites [87]. The specific flexural moduli of the kenaf composites with fiber contents greater than 40% were extremely high and even stiffer than a 40% mica–PP composite. Table 7.21 shows some typical data of commercially available injection-molded PP composites and the comparison with typical jute and kenaf–PP composites. Data on the talc, mica, calcium carbonate, and glass composites were compiled from the Resins and Compounds (*Modern Plastics Encyclopedia*) [88], and Thermoplastic Molding Compounds (*Material Design*) [89]. The properties of kenaf-based fiber composites have properties superior to typical wood (newspaper) fiber–PP composites. The specific tensile and flexural moduli of 50% by weight of kenaf–PP composites compares favorably with the stiffest of the systems shown, that of glass–PP and mica–PP. This technology has been used to make many products including decking shown in Figure 7.13.

FIGURE 7.13 Extruded kenaf thermoplastic products (USDA).

The failure strain decreases with the addition of the fibers. Addition of a rigid filler and fiber restricts the mobility of the polymer molecules to flow freely past one another, and thus causes premature failure. The addition of MAPP followed a similar trend to that of the uncoupled system, although the drop in failure strain with increasing fiber amounts was not as severe. There is a decrease in the failure strain with increasing amounts of kenaf for a coupled system. The stress–strain curve is not linear, which is due to the plastic deformation of the matrix. The distribution of the fiber lengths present in the composite can influence the shape of the stress–strain curve since the load taken up by the fibers decreases as the strain increases; detailed explanations are available elsewhere [90]. The tensile energy absorption and the integrated area under the stress–strain curve up to failure behave in roughly the same manner as the tensile failure strain. The difference between the coupled and uncoupled composites increases with the amount of fibers present, although the drop in energy absorbed for the coupled composites levels off after the addition of about 35 vol.% of fiber.

The impact strength of the composite depends on the amount of fiber and the type of testing, i.e., whether the samples were notched or unnotched. In case of notched samples, the impact strength increases with the amount of fibers added until a plateau is reached at about 45% fiber weight, irrespective of whether MAPP is used or not. The fiber bridge cracks and increases the resistance of the propagation of the crack. The contribution from fiber pullout is limited since the aspect ratio of the fibers in the system is well below the estimated critical aspect ratio of about 0.4 mm [91]. In case of the unnotched impact values of the uncoupled composites, the presence of the fibers decreases the energy absorbed by the specimens. the addition of the fibers creates regions of stress concentrations that require less energy to initiate a crack. Improving the fiber–matrix adhesion through the use of MAPP increases the resistance to crack initiation at the fiber–matrix interface and the fall in impact strength with the addition of fibers is not as dramatic.

The two main disadvantages of using kenaf–PP as compared to glass–PP are the lower impact strength and higher water absorption. The lower notched impact strength can be improved by using impact modified PP copolymers and the use of flexible maleated copolymers, albeit with some loss in tensile strength and modulus, which will be discussed in a later paper. Care needs to be taken when using these fibers in applications where water absorption and the dimensional stability of the composites are of critical importance. Judicious use of these fibers makes it possible for jute and kenaf fibers to define their own niche in the plastic industry for the manufacture of low-cost, high-volume composites using commodity plastics.

An interesting point to note are the higher fiber volume fractions of the jute and kenaf composites compared to the inorganic filled systems. This can result in significant material cost savings as the fibers are cheaper than the pure PP resin, and far less expensive than glass fibers. Environmental and energy savings by using an agriculturally grown fiber instead of the high energy utilizing glass fibers or mined inorganic fillers are benefits that cannot be ignored, although a thorough study needs to be conducted to evaluate the benefits.

7.17.13 Fiber Matrix Thermoplasticization

There have been many research projects over the years studying ways to thermoform lignocellulosics. Most of the efforts have concentrated on film formation and thermoplastic composites. The approach most often used involves the chemical modification of cellulose, lignin, and the hemicelluloses to decrystallize and modify the cellulose and to thermoplasticize the lignin and hemicellulose matrix to mold the entire lignocellulosic resource into films or thermoplastic composites [92–97].

Jute and kenaf fibers are composites made up of a rigid polymer (cellulose) in a thermoplastic matrix (lignin and the hemicelluloses). If a nondecrystallizing reaction condition is used, it is possible to chemically modify the lignin and hemicellulose, but not the cellulose. This selective reactivity has been shown to occur if uncatalyzed anhydrides are reacted with wood fiber [98]. The goal is to only modify the matrix of jute and kenaf fibers allowing thermoplastic flow, but keeping the cellulose backbone as a reinforcing filler. This type of composite should have reduced heat-induced deformation (creep), which restricts thermoplastic-based composites from structural uses.

The modification of the kenaf bast fibers using succinic anhydrides (SA) was performed using either solution reactions with xylene or solid-state reactions using SA in a melt state [99]. Since xylene does not swell the fiber, it is only a carrier for the reagent. The rate of reaction is fastest at higher concentration of SA in xylene and at temperatures above 140°C. The rate of reaction in the melt state has not been determined. It has also not yet been determined what level of modification is needed to give the desired thermoplasticity so it is not known what optimum reaction time is needed.

Thermal analysis (DSC) showed the first glass transition temperature decreased from 170°C to about 133°C [98].

Samples of reacted fiber were pressed into pellets using a powder pressing die consisting of a heavy-walled steel cylinder with a separate bottom and a ram (diameter 10.4 mm) to compress the fibers. Fiber was placed in the preheated cylinder and then compressed to a pellet thickness of 8.7 mm (target density 1.5 g/cm^3, target volume 0.736 cm^3) for 10 min at 190°C.

Scanning electron micrographs (SEM) were taken of the pressed control and SA-reacted fiber specimens using a Jeol 840 scanning electron microscope [99]. Figure 7.14 shows the SEM of the hot pressed control and esterified kenaf fiber. The SA fiber is derived from a reaction done according to the solid-state reaction method and pressed at 190°C for 10 min. The weight percent gain due to esterification is 50. The control fiber (A) shows little tendency to thermally flow under the pressure of the hot press, whereas the esterified fiber (B) shows thermal flow at this temperature. Views A, B, and D are taken from the top of the compressed

FIGURE 7.14 Scanning electron micrographs of pressed kenaf fiber: A, Control (30X), B, SA reacted (50 WPG, 30X), C, SA reacted (50 WPG, 50X), D, SA reacted (50 WPG, 100X) (USDA).

pellet, while C is taken from the side of the pellet. The side view (C) shows a definite layering of the fiber has occurred and view D shows that fiber orientation is still evident.

The research done so far in this area shows that kenaf fiber can be reacted with SA to give high weight gains of esterification of the cell wall polymers either by solution or solid-state chemistry. The esterified fiber shows a reduced transition temperature from about 170°C down to about 135°C, regardless of the weight gained. Electron microscopy of hot pressed fiber indicates matrix thermoplasticity with a rigid fiber structure still in existence.

7.17.14 FIBER THERMOPLASTIC ALLOYS

Research to develop jute and kenaf fiber thermoplastic alloys is based on first thermoplasticizing the fiber matrix as described above, followed by grafting of the modified fiber with a reactive thermoplastic. This type of composite has the thermoplastic bonded onto the jute or kenaf so there is only one continuous phase in the molecule. This is done in one of two ways. In one case, the matrix is reacted with maleic anhydride that results in a double bond in the grafted reacted molecule. This can then be used in vinyl-type additions or in free radical polymerization to either build a thermoplastic polymer or graft one onto the jute or kenaf backbone. In the second method, the matrix is reacted with a bonded chemical and then reacted with a low-molecular-weight thermoplastic that has been grafted with side-chain anhydride groups.

The anhydride functionality in the compatibilization research described before may react with the lignocellulosic, but there is no evidence to support that at this time. A higher level of grafted anhydride on the polypropylene would be required for the alloy reactions, and it would be expected that the reaction between grafted thermoplastic and jute or kenaf would take place both on the matrix polymers (lignin and hemicelluloses) and in the cellulose backbone. Some decrystallization of the cellulose may be desired to give more thermoplastic character to the entire composite.

Preliminary results indicate that maleic anhydride reacts with the jute or kenaf matrix, both in liquid- and solid-state reactions, to similar weight gains as given for SA. Research in this area continues.

Combining jute and kenaf fibers with thermoplastics provides a strategy for producing advanced composites that take advantage of the enhanced properties of both types of resources. It allows the scientist to design materials based on end-use requirements within a framework of cost, availability, recyclability, energy use, and environmental considerations. These new composites make it possible to explore new applications and new markets in such areas as packaging, furniture, housing, and automotive.

7.17.15 CHARCOAL

Jute stick or core is often compressed and pyrolyzed into charcoal for cooking in India and Bangladesh. After the core has been compressed, it is heated for 2 h at 500°C in the presence of an inorganic salt to give a 35 to 40% yield of high-grade charcoal. The charcoal can also be used as a filler in vulcanized rubber and in the production of carbon disulfide [8].

7.18 FUTURE TRENDS

The main commercial developments in the jute industry have been concerned with the spinning and weaving technology, and considerable improvements in productivity have also taken place. However, it is time now to consider what new innovations would assist the spread of jute materials into textile uses outside the traditional fields of packaging and carpets.

Agricultural developments to breed Corchorus or Hibiscus plants containing fibers of significantly lower linear density would allow yarns of lower count to be spun than is feasible at present, and therefore enable lightweight fabrics to be produced. Such fabrics could have increased potential for decorative and furnishing use, especially if the constraint of instability of color could first be removed and some process then devised to produce additional elasticity on a more permanent basis than is done by the woolenizing process.

Kenaf is now being grown in several countries where the bast fiber is used for geotextiles and the pith is going into sorbents for oil spill clean up and animal litter. The production of pulp and paper from kenaf is growing, but it is only used for limited types of papers at present. The utilization of the whole plant of both jute and kenaf is under consideration for structural and nonstructural composites. Automotive interior door panels are now produced in Germany and the United States out of jute and kenaf bast fiber in combination with thermoplastics.

REFERENCES

1. R.H. Kirby, *Vegetable Fibers*, Leonard Hill (Books) Ltd., London (1963).
2. B.C. Kundu, K.C. Basak, and P.B. Sarkar, *Jute in India*, Monograph, Indian Central Jute Committee, Calcutta (1959).
3. W.A. Bell, *Sci. News*, **54**, 39 (1960).
4. R.P. Mukheriee and T. Radhakrishnan, *Tex. Progr.*, **4**, 1 (1972).
5. J.N. Mather, *Carding—Jute and Similar Fibers*, Iliff, London (1969).
6. R.M. Rowell, J.A. Young, and J.K. Rowell, eds., *Paper and Composites from Jute and Kenaf Resources*, CRC Lewis Publishers, Boca Raton, FL (1997).
7. T. Sellers, Jr and N.A. Reichert eds., *Kenaf Properties, Processing and Products*, Mississippi State University Press, Mississippi State, MS (1999).
8. S.N. Pandey and S.R. Anantha Krishnan, *Fifty Years of Research 1939–1989*, Jute Technological Research Laboratories, Hooghly Printing Co. Ltd, Calcutta, India (1990).
9. C.G. Jarman, FAO Agricultural Services Bulletin No. 60, Rome (1985).
10. E.P. Columbus and M.J. Fuller, *Kenaf Properties, Processing and Products*, Mississippi State University Press, Mississippi State, MS 83–89 (1999).
11. G. Fisher, *Proc. Sixth Annual International Kenaf Conf.*, New Orleans, LA 8–12 (1994).
12. G.N. Ramaswam, *Kenaf Properties, Processing and Products*, Mississippi State University Press, Mississippi State, MS 91–96 (1999).
13. E.P. Columbus and W.S. Anthony, *U.S. Cotton Ginning Laboratory Annual Report—Crop Year 1994*, Stoneville, MS 198–201 (1995).
14. S.C. Barker, British Association for Advancement of Science, *J. Text. Inst.*, **30**, 273 (1939).
15a. A. Islam and A.M. Bhuyian, *Jute and Jute Fabrics No. 8*, Bangladesh (1978).
15b. J.S. Han and J.S. Rowell, *Paper and Composites from Jute and Kenaf Resources*, CRC Lewis Publishers, Boca Raton, FL, MS 83 (1997).
16. T.H. Soutar and M. Bryden, *J. Text. Inst.*, **46**, T521 (1965).
17. M. Manzoor-i-Khuda, A.S.M. Serajuddin, M.M.A. Islam, N. Am in, M. Bose, A.A. Khan, and Md. Shahjahan, *Pakistan J.* Sci. Ind. Res., *13, 153, 316, 321 (1970)*.
18. C.G. Jar man and R.H. Kirby, *Colon. Plant Animal Prod.*, **5**, 281 (1955).
19. V. Ozsanlav, British Textile Technology Group, Cheshire, UK, Personal communication (1992).
20. H. Chatterjee, *J. Sci. and Ind. Research*, **18C**, 206 (1959).
21. A.C. Mukherjee, A.K. Mukhopadhyay, and U. Mukhopadhyay, *Textile Res. J.*, **56(9)**, 562 (1986).
22. T.F. Clark and I.A. Wolff, *TAPPI*, **52(11)**, 2606 (1969).
23. T.F. Clark, S.C. Uhr, and I.A. Wolff, *TAPPI*, **50(11)**, 2261 (1967).
24. J.S. Han, W. Kim, and R.M. Rowell, International Kenaf Association Conference Proceedings, March 9–10, Irving, TX, (1995).
25. M.O. Baggy, *TAPPI*, **54**, 11 (1971).

26. M. Lewin, *TAPPI*, **41**, 403 (1958).
27. F. Stern and H.P. Stout, *J. Text. Inst.*, **45**, T896 (1954).
28. F. Stern, *J. Text. Inst.*, **48**, T21 (1957).
29. C.G. Jarman and V. Laws, *J.R. Microbiol. Soc.*, **84**, 339 (1965).
30. I.G. Cumming, D.F. Leach, and G.M. Smith, *Bull. Br. Jute Trade R. A.*, **10**, 74 (1964).
31. M. Kabir and N.G. Saha, *Pakistan J. Sci. Ind. Res.*, **14**, 162 (1971).
32. M. Kabir and N.G. Saha, *Bangladesh J. Sci. Ind. Res.*, **12**, 91 (1977).
33. J.N. Mather, *Bull. Br. Jute Trade R. A.*, **12**, 63 (1968).
34. H.P. Stout, *Fibre and Yarn Quality in Jute Spinning*, The Textile Institute, Manchester (1988).
35. R.M. Rowell, Commonwealth Forestry Bureau, Oxford, England, **6(12)**, 363 (1983).
36. R.M. Rowell, *Handbook on Wood and Cellulosic Materials*, D.N.-S. Hon and N. Shiraishi, eds., Marcel Dekker, Inc., New York, NY, (1991).
37. H.J. Callow, *J. Indian Chem. Soc.*, **43**, 605 (1951).
38. M. Andersson and A.M. Tillman, *J. Appl. Polym. Sci.*, **37**, 3437 (1989).
39. R.M. Rowell, R. Simonson, and A.M. Tillman, European Patent 0213252 (1991).
40. R.M. Rowell, International consultation of jute and the environment, Food and Agricultural Organization of the United Nations, ESC:JU/IC 93/15, 1 (1993).
41. R.M. Rowell and S.E. Harrison, *Proceedings, Fifth Annual International Kenaf Conference*, M.S. Bhangoo, ed., California State University Press, Fresno, CA, 129 (1993).
42. R.M. Rowell, G.R. Esenther, J.A. Youngquist, D.D. Nicholas, T. Nilsson, Y. Imamura, W. Kerner-Gang, L. Trong, and G. Deon, *Proceedings: IUFRO Wood Protection Subject Group*, Honey Harbor, Ontario, Canada. Canadian Forestry Service, 238 (1988).
43. American National Standard. Basic hardboard. ANSI/AHA 135.4, American Hardboard Association, Palatine, IL (1982).
44. C.B. Vick and R.M. Rowell, *Internat., J. Adhes. and Adhesives*, **10(4)**, 263 (1990).
45. J. Compton, W.H. Martin, and R.P. Barber, *Text. Res. J.*, **25**, 58 (1955).
46. J. Compton, *Text. Res. J.*, **27**, 222 (1955).
47. T.E. Timell, *Pulp Paper Mag.*, Canada, **56**, 104 (1955).
48. T.E. Timell, *Text. Res. J.*, **27**, 854 (1957).
49. M. Lewin and J.A. Epstein, *Text. Res. J.*, **30**, S-20 (1960).
50. H. Muir (with W.A. Bell), Doctoral thesis, Edinburgh (1964).
51. M.M. Roy and I.K. Sen, *J. Text. Inst.*, **43**, T396 (1952).
52. M. Kabir, M.S. Rahman, and M. Shahidullah, *Bangladesh J. Jute Fiber Res.*, **2**, 45 (1977).
53. W.A. Bell, *Bull. Br. Jute Trade R. A.*, **12**, 154 (1969).
54. British Standard 1006, British Standards Institution, London.
55. I.G. Cumming, *Bull. Br. Jute Trade R. A.*, **8**, 240 (1961).
56. M. Lewin, M. Shiloh, and J. Banbaji, *Text. Res. J.*, **29**, 373 (1959).
57. J. Banbaji, *Text. Res. J.*, **30**, 798 (1960).
58. A.M. Bhuiyan, M.A. Salam, A. Sukur, and A.S.M.N. Ahsan, *Bangladesh J. Jute User Res.*, **2**, 85 (1977).
59. R.M. Rowell, Food and Agriculture Organization of the United Nations, ESC: JU/EGM 94/3 (1994).
60. H.S. Sabharwal, M. Akhtar, R.M. Blanchette, and R.A. Young, *TAPPI*, **77(12)**, 1 (1994).
61. T.F. Clark, *Pulp and Paper Manufacturing*, Vol. 2, 2nd ed., R.O. MacDonald, ed., McGraw-Hill, New York, NY 1 (1969).
62. D.K. Misra, Pulp and Paper, *Chemistry and Chemical Technology*, Vol. 1, 3rd ed., J.P. Casey, ed., Wiley-Interscience, New York, NY 504 (1980).
63. A.M. Hunter, Non-wood plant fiber pulping, Progress Report No. 19, TAPPI Press, Atlanta, GA, 49 (1991).
64. R.A. Young, *Paper and Composites from Agro-Based Resources*, R.M. Rowell, J.A. Young, and J.K. Rowell, eds., CRC Lewis Publishers, Boca Raton, FL 135–248 (1997).
65. V.P. Leehka and S.K. Thapar, Experience in kenaf pulping in Thailand, non-wood plant fiber pulping. Progress Report No. 14, TAPPI Press, Atlanta, GA (1983).
66. T.F. Clark, G.H. Nelson, H.J. Niechlag, and I.A. Wolff, *TAPPI*, **45(10)**, 780 (1962).

67. T.F. Clark and I.A. Wolff, *TAPPI*, **45(10)**, 786 (1962).
68. M.O. Bagby, Non-wood plant fiber pulping, Progress Report No. 9, TAPPI Press, Atlanta, GA 75 (1978).
69. A.J. Watson, G. Gartside, D.F. Weiss, H.G. Higgins, H. Mamers, G.W. Davies, G.M. Irvine, I.M. Wood, A. Manderson, and E.J.F. Crane, *Div. Chem. Technol*, Paper No. 7, CSIRO, Australia (1976).
70. S.K. Mittal and S. Maheshwari, *Proceedings Pulping Conference TAPI*, San Diego, 105 (1994).
71. G.F. Touzinsky, Laboratory paper machine runs with kenaf thermomechanical pulp, *TAPPI*, **63(3)**, 109 (1980).
72. G.F. Touzinsky, Kenaf, In *Pulp and Paper Manufacturing*, Vol. 3., Secondary fibers and non-wood pulping, Chapter 8, TAPPI Press, Atlanta, GA 106 (1987).
73. J.S. Han, E.S. Miyashita, and S.J. Spielvogel, *Kenaf Properties*, Processing and Products, Mississippi State University Press, 267–283 (1999).
74. Y. Shuhui, H. Zhrivei, and L. Yorsen, *Proc. Seventh International Symposium on Wood and Pulping Chem.*, Vol. 3, Beijing, PR China, 317 (1993).
75. H. Pande and D.N. Roy, *Kenaf Properties, Processing and Products*, Mississippi State University Press, 315–320 (1999).
76. M. Shafi, A.F.M. Akhtaruzzamarn, and A.J. Mian, *Holzforschung* **47**, 83 (1993).
77. A.K. Rana and K. Jayachandran, *Mol. Cryst. and Liq. Cryst.*, **353**, 35 (2000).
78. R.M. Rowell, Final Report on Composites to the UNDP (1998).
79. V. Wigotsky, *Plast. Eng.*, Nov., 25 (1988).
80. A.R. Sanadi, D.F. Caulfield, and R.M. Rowell, *Plast. Eng.* April, 27 (1994).
81. R. Woodhams, T.G. Thomas, and D.K. Rodgers, *Polym. Eng. Sci.* **24**, 1166 (1984).
82. C. Klason and J. Kubat, *Composite Systems from Natural and Synthetic Polymers*, L. Salmen, A. de Ruvo, J.C. Seferis, and E.B. Stark, eds., Elsevier Science, Amsterdam (1986).
83. G. Myers, E.C.M. Clemons, J.J. Balatinecz, and R.T. Woodhams, *Proc. Annual Technol Conf.*, Society of Plastics Industry, 602 (1992).
84. B.V. Kokta, R.G. Raj, and C. Daneault, *Polym. Plast. Technol. Eng.*, **28**, 247 (1989).
85. A.R. Sanadi, R.A. Young, C. Clemons, and R.M. Rowell, *J. Rein. Plast. Compos.*, **13**, 54 (1994).
86. R.M. Rowell, A.M. Tillman, and R. Simonson, *J. Wood Chem. Tech.*, **6**, 427 (1986).
87. A.R. Sanadi, D.F. Caulfield, R.E. Jacobson, and R.M. Rowell, *Ind. and Eng. Chem. Res.*, **34**, 1889 (1995).
88. Resins and Compounds, *Modern Plastics Encyclopedia*, McGraw-Hill, NY, 269 (1993).
89. Thermoplastic Molding Compounds, *Material Design*, Material Selector Issue, Penton Publishing, OH, 184, (1994).
90. D. Hull, *An Introduction to Composite Materials*, Cambridge University, Cambridge (1981).
91. A.R. Sanadi, R.M. Rowell, and R.A. Young, AICHE Summer National Meeting, paper 24f (1993).
92. D.N.-S. Hon and N.-H. Ou, *J. Appl. Polym. Sci.: Part A: Polymer Chemistry*, **27**, 2457 (1989).
93. D.N.-S. Hon and L.M. Xing, Viscoelasticity of Biomaterials, W.G. Glasser, ed., Am. Chem. Soc., Washington, D.C. 118 (1992).
94. H. Matsuda, *Wood Sci. Technol.*, **21**, 75 (1987).
95. H. Matsuda and M. Ueda, *Mokuzai Gakkaishi*, **31(3)**, 215 (1985).
96. N. Shiraishi, *Wood and Cellulose Chemistry*, D.N.-S. Hon and N. Shiraishi, eds., Marcel Dekker, Inc., New York, NY, 861, (1991).
97. M. Ohkoshi, N. Hayashi, and M. Ishihara, *Mokuzai Gakkaishi*, **38(9)**, 854 (1992).
98. R.M. Rowell, R. Simonson, S. Hess, D.V. Plackett, D. Cronshaw, and E. Dunningham, *Wood and Fiber Sci.*, **26(1)**, 11 (1994).
99. R.M. Rowell, J. O'Dell, and T.G. Rials, Proceedings, Second Pacific Rim Bio-Based Composites Symposium, Vancouver, Canada (1994).

8 Other Long Vegetable Fibers*: Abaca, Banana, Sisal, Henequen, Flax, Ramie, Hemp, Sunn, and Coir

Subhash K. Batra

CONTENTS

8.1 Source and Classification of Vegetable Fibers 454
8.2 Nature of Plants and Fiber Extraction 456
    8.2.1 Abaca (Manila Hemp) and Banana 456
    8.2.2 Sisal and Henequen 460
    8.2.3 Flax 463
        8.2.3.1 Plant 463
        8.2.3.2 Extraction: Retting 465
        8.2.3.3 Extraction: Scutching 466
    8.2.4 Ramie, Rhea, or China Grass 467
    8.2.5 Hemp and Sunn 470
    8.2.6 Coir 472
8.3 Fiber Morphology and Chemical Composition 474
    8.3.1 Morphology 474
    8.3.2 Chemical Composition of the Fiber 475
        8.3.2.1 Abaca 479
        8.3.2.2 Flax 479
        8.3.2.3 Sisal, Coir, and Hemp 479
        8.3.2.4 Results of Nitration and Other Studies 480
8.4 Physical Properties 481
    8.4.1 Structure of Fibers and Ultimates 481
    8.4.2 Degree of Polymerization 488
    8.4.3 Specific Gravity and Porosity 489
    8.4.4 Optical Properties 489
    8.4.5 Moisture Absorption, Desorption, and Swelling 489
    8.4.6 Thermal Properties 492
    8.4.7 Electrical Properties 493

---

*To my mentors: Professors Stanley Backer (MIT), Frederick F. Ling (RPI) and Edward A. Fox (RPI) who shaped my career in many significant ways.

8.5	Mechanical Properties	495
	8.5.1 Load–Extension Characteristics	495
	8.5.2 Load Deformation in Other Modes of Deformation	502
	8.5.3 Recovery from Deformations	503
	8.5.4 Time Effects	505
8.6	Chemical Properties	506
	8.6.1 Action of Alkalis and Other Swelling Agents	506
	8.6.2 Action of Mineral Acids	509
	8.6.3 Action of Organic Acids	510
	8.6.4 Bleaching	510
	8.6.5 Stability to Light	511
	8.6.6 Finishing Treatments	511
	8.6.7 Action of Microorganisms and Enzymes	512
	8.6.8 Action of Dyestuffs	514
8.7	Concluding Remarks	514
Acknowledgments		514
Appendix		515
References		515

8.1 SOURCE AND CLASSIFICATION OF VEGETABLE FIBERS

In a broad sense, the source of vegetable fibers, as the name suggests, is the rich plant life on planet Earth. In a narrower sense, different fibers come from different parts of different plants. In general, fibrous assemblies of various types are the structural components of plant life. The load–deformation characteristics of these assemblies are determined by the functional role played by the plant part in the total architecture of the plant, and the soil–climate environment in which it grows.

Commercially useful fibers come primarily from the leaves or stems or seed coverings of specific plants. The functional role of fibers in the plant clearly need not be as structural components. Fibers such as cotton and coir serve to protect the seed or fruit from mechanical and perhaps pest or microbial damage. Cotton and other similarly attached seed fibers are also expected, by nature, to play an extremely important role in plant propagation; the low mass-to-volume ratio of the seed–fiber ensemble makes it possible for the aerodynamic or buoyancy forces to transport it over large distances.

A widely accepted classification of fibers is based on their location in the plant. Accordingly, the three principal categories are seed fibers, bast fibers, and leaf fibers. Fibers that do not belong to one of these three categories are classified as miscellaneous fibers.

Such an approach to fiber classification appears to be systematic and objective. Unfortunately, plants do not always follow the pattern of roots, trunk, leaves, seed, or fruit. There are anomalies. For example, the banana-like plants, yielding abaca, do not have the woody trunk conventionally associated with other plants. Instead, its stem consists of layers of thick, crescent-shaped (in cross section) sheaths wrapped around each other; they reduce to spindly growths that unfurl from the stem and become the thick central stems of fronds (see Section 8.2.1). Botanically, the sheath-frond system is called a leaf; consequently, the fiber extracted from the sheath is classified as leaf fiber. On the other hand, sisal or henequen fibers come from the swordlike leaves of their respective plants. Bagasse (from sugarcane) fiber, used for paper or fiberboard, comes from the stem, which is neither a leaf nor a woody trunk. Thus, the classification of fibers as seed, bast, leaf, or miscellaneous fibers is somewhat arbitrary.

Furthermore, traditionally, agronomists and botanists have developed the vegetable fiber classification system. Therefore, a great deal of emphasis has been placed on their botanical names and classifications according to family names, genus, and species. In the Appendix, we include the traditional classification tree. The botanical classification of each category of fibers, based largely on Kirby's [26] superb tome on the agriculture and botany of vegetable fiber plants, was given by Batra and Bell [22] in 1975. Such a classification, however, does not assist the fiber technologist in any meaningful way. The existence of several names for the same fiber (e.g., sunn, also called Banates, black hemp, Bombay hemp, Madras hemp, etc.) further adds to the confusion of the technologist. Batra and Bell [22] prepared an alphabetical listing of fiber names, showing the interrelationship of names, geographical sources, as well as usage. In this listing, 26 fibers are represented by 340 different names. Batra and Bell also rearranged the listing to show the availability of fibers in different geographical regions and countries of the world as known in 1974.

The literature on vegetable fibers suggests the existence of two other subtle, but implicit, classifications. The first is based on the technological processes (systems) used to convert the fibers to yarns. Most bast and leaf fibers are processed on a system that can deal with fiber lengths greater than 120–150 mm. Parts of this process are modifications of the woolen system. All fibers processed in this system are called long-staple fibers. In contrast, the useful cotton fibers range from about 20–60 mm in staple length; the cotton processing system is designed to cope with this range. In keeping with this rationale, therefore, the fibers in this staple length range are called short-staple fibers.

The second classification is based on functional criteria. Fibers with very small cross-sectional areas tend to be low in bending and torsional rigidities. Products made from these are, therefore, soft to touch. The fibers, therefore, are called soft fibers. Typical examples are cotton and jute. On the other hand, coarser fibers have higher bending and torsional rigidities. Products made from these are harsh to the touch. These fibers are, therefore, called hard fibers. Typical examples are sisal and abaca. Despite these rationales, the fiber classifications remain somewhat arbitrary, albeit often sufficiently useful.

Until the latter part of the 20th century, the interest in natural fibers, other than cotton and wool, was largely confined to developing nations; for many of them, these fibers are a natural resource, which, if exploited commercially, could provide sustained livelihood to farmers and to the labor involved in their conversion to useful products. Many of these people live on the margin. During the latter part of the 20th century, the "green" revolution that spread through both the developed and developing countries reignited interest in natural fibers other than cotton and wool. A simple search of only two databases, using Science Finder of the American Chemical Society, reveals 4263 patents and 2936 journal articles during 1990–2004 related to natural fibers, which include cotton and wool. The abstract of a paper by Peijs [132] states this trend succinctly:

"Environmental legislation and the demand for continuous sustainability, including recyclability, in manufacturing has increased interest in wood *fibers* and such bast *fibers* as flax, hemp, and kenaf for use as fillers for polymer composites. In addition to these environmentally friendly composites, researchers have developed recyclable single component polymer composites for the automotive and construction industries. EIN Engineering of Japan developed composites from waste wood and recycled plastics for outdoor applications, including crash barriers and sound absorbing panels. Tech-Wood International's Tech-Wood composite contains 70 percent pine wood *fibers* and 30 percent compatibilized polypropylene and has applications in the manufacture of hurricane resistant housing. Daimler-Chrysler developed natural *fiber*-reinforced composites for underbody panels, engine and transmission covers, and sound insulation. Audi developed interior door trim panels made from polyurethane reinforced with flax and *sisal* nonwoven mats. Ford developed injection moldable flax/polypropylene composites for radiator grills, front ends, and engine shields."

In the same vein, the development of products using natural fibers is reflected in another report, thus [62]:

"Nonwoven fabrics for automotive applications consist of 50/50 blends of natural fibers (kenaf, *hemp*, and *flax*) and manmade fibers (polypropylene and polyester). Natural fiber nonwoven fabrics feature low fabric weights, high flex modulus, high tensile modulus, excellent impact properties, dimensional stability, and good noise and vibration properties. The manufacture of natural fiber nonwoven fabrics for automotive applications offers shorter cycle times, lower capital costs, and rapid and inexpensive prototyping. Current applications include door panels and inserts; center consoles; A,B,C pillars; package trays; rear quarter panels; seat backs; and trunk trim. Annual growth in the use of natural fiber reinforced plastics for automotive applications will average 38 percent through 2006. Expanding markets in the immediate future include recreational vehicles, modular housing, aerospace and marine applications, packaging, and office interiors."

In addition, lignocellulosic natural fibers and even their waste, generated as a by-product of the production, and their use is likely to find high-value outlets. For example, Lee and Rowell [87] report on studies to assess their potential to remove copper, nickel, and zinc ions from aqueous solutions as a function of their lignin content. Williams and Reed [193] report efforts to develop technology to convert hemp and flax waste into activated carbon mats, which could be used for protection against gaseous chemical hazards. At the same time, flax and hemp crops have been found to be effective in removing heavy metals from soil polluted by nonferrous metal works [92].

Finally, production data on vegetable fibers, fortunately, can now be found online from the FAO statistics database easily. The annual yields of commercially important fibers are given in Table 8.1.

8.2 NATURE OF PLANTS AND FIBER EXTRACTION

Conversion of fibers into useful products is the concern of the textile technologist. Technologies used to convert them into useful products are based on the physical, mechanical, and chemical characteristics of the fibers. These characteristics are determined during growth of the plant and subsequent fiber extraction. It is useful, therefore, to examine the nature of various plants and the methods of fiber extraction. Discussion in the sequel is largely drawn from the excellent publications by Weindling [187], Himmelfarb [63], Wilson [194], Kirby [76], Cook [44], Dujardin [49], Mauersberger [108], Allison [7,8], Allison et al. [4,6], Jarman [68], and many others. It is necessarily restricted to fibers of significant commercial and economic (development) interest (past, present, and future). Batra and Bell [22] give a more detailed treatment of the subject, including agriculture and botany of the plants, and manufacturing technologies employed to convert them into useful products.

8.2.1 Abaca (Manila Hemp) and Banana

The abaca plant was originally found in the Philippine Islands; the Malays used the extracted fiber for ropes, fishing nets, and even woven cloth. Locally, the fiber is known by the term "abaca," presumably Spanish in origin; the name "Manila hemp" was given to it by the English and Americans after they became aware of its commercial usefulness in the early 1880s. The term "Manila hemp" is a misnomer because it is a "hard" fiber, whereas hemp is a "soft" fiber; furthermore, it was not grown around Manila. Thus, to eliminate any confusion, henceforth the fiber will be referred to as "abaca." In 2003, the plant was grown in Philippines

TABLE 8.1
Annual Production Estimates of the Commercially Important Fibers

Fiber name	Annual production (1000 t)	Year(s) of estimates
	Seed/fruit associated	
Cotton	19,417	2003 (a)[a]
Coir	637	2003 (a)
Kapok	125	2003 (a)
	Leaf associated	
Agave fibers nes[b]	54	2003 (a)
Sisal	269	2003 (a)
Henequen	24	2003
Abaca	100	2003 (a)
	Bast associated	
Jute	2,750	2003 (a)
Jutelike fibers[c]	405	2004 (a)
Kenaf and allied fibers	500	2003/2004
Flax (fiber and tow)	751	2003 (a)
Hemp	83	2003 (a)[d]
Sunn	(Included in jutelike fibers)	2003 (a)
Ramie	269	2003 (a)
	All fibers from crops	
Total crop based fibers	28,912	2003 (a)

[a]FAO database
(http://apps.fao.org/faostat/form?collection = Production.Crops.   Primary&Domain = Production&   servlet = 1& hasbulk = 0& version = ext&language = EN)
[b]AGAVE FIBRES NES Including inter alia: Haiti hemp *(Agave foetida)*; henequen *(A. fourcroydes)*; ixtle, tampico *(A. lechequilla)*; maguey *(A. cantala)*; pita *(A. americana)*; Salvador hemp *(A. letonae)*.
[c]JUTELIKE FIBRES Including inter alia: China jute *(Abutilon avicennae)*; Congo jute, malva, paka *(Urena lobata; U. sinuata)*; Indian flax *(Abroma augusta)*; kenaf, meshta *(Hibiscus cannabinus)*; rosella hemp *(H. sabdariffa)*; sunn hemp *(Crotalaria juncea)*.
[d]USA data not reported.

(72,000 MT), Ecuador (26,000 MT), Costa Rica (1100 MT), Indonesia (600 MT), Equatorial Guinea (500 MT), Kenya (30 MT), and perhaps Malaysia [FAO database].

The abaca plant is a perennial and looks like a banana plant as shown in Figure 8.1a. At maturity, the plant consists of 12–30 stalks that radiate from the same root system. The stalks can be 3–7.5 m tall and 120–300 mm in diameter at the stem. The stem is composed of 90% water and sap and 2–5% fiber; the rest is soft, cellular tissue. The stem structure consists of tightly packed, long, crescent-shaped sheaths that grow from a central core; the inner sheaths envelope the contained layers almost entirely. In cross section, the stalk-stem appears as shown in Figure 8.2a. At the top, the sheaths narrow down to thick spindles, which unfurl from the stalk and constitute the central stem of leaflike foliage (fronds). The fronds may be 1–2 m in length and nearly 0.3 m wide. Beyond full maturity, the plant flowers, followed by the appearance of inedible banana-like fruit, about 80 mm long.

In each sheath, there are three distinct layers. The outer layer, including the epidermis, contains the bundles of fibers dispersed in a soft tissue matrix. The middle layer consists of water-transporting fibrovascular tissue. The inner layer consists of soft, cellular tissue. The quantity and quality of fiber in each sheath depends on its width and its location in the stem. The sheaths, about 20 per stem, are separated into four groups: outside sheaths, adjacent to the

FIGURE 8.1 (a) The abaca plant. (From http://www.bar.gov.ph/abaca/abaca2.htm. With permission.) (b) Extracted abaca fibers. (From http://www.peral.net/fiber.htm. With permission.)

outside sheaths, middle, and inner sheaths. Each group yields different grades of fibers. To extract the fiber, the stem is cut at the bottom at an angle and its fronds are removed. At this stage, one of two different procedures may be followed to obtain the commercially usable fiber.

The first procedure, used extensively in the Philippines, is largely or completely manual and is called stripping. On the plantation site, the plant stems are de-sheathed, the sheaths

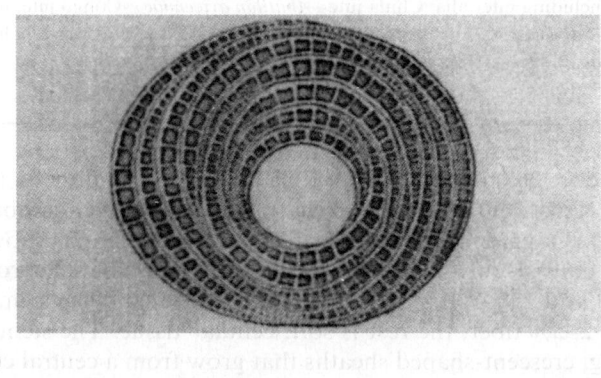

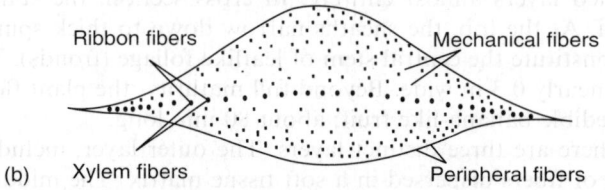

FIGURE 8.2 (a) The cross section of stalk stem of abaca plant. (From Thieme J.G., *The Textile Quarterly*, **5 (1)**, 56–65 1955.) (b) The three zones of the sisal leaf. (From Thieme J.G., *The Textile Quarterly*, **5 (1)**, 56–65, 1955.) [170]

flattened, a knife inserted between the outer and the middle layer, and a 50–80 mm wide strip is separated and pulled off all along the length. The strip is called a *tuxy* and the separation procedure is called *tuxying*. All the fiber is removed in tuxies from each sheath. The tuxies are then scraped by pulling them through or between a wooden block and a serrated knife (400–2000 serrations/m or no serration) under considerable pressure. The manual effort, which is considerable, decreases with decreasing serration density. To minimize the effort and to improve productivity, the scraping process has been mechanized. The machine called Hagotan is designed to allow the cleaning of six to eight tuxies (otherwise implying as many workers) simultaneously. The fineness of the stripped fiber (see Figure 8.1b) increases with increasing serration density of the blade and the pressure used; a nonserrated knife (or infinite serrations/m) and 20,000 kN/m^2 pressure gives best results.

The second procedure called *decortication* is used in Central America, Indonesia, and parts of the Philippines; the resulting fiber is called Deco. The machines (Krupp or Robey) consist of a long (30 m) conveyor belt, a crushing press, a pair of crushing rollers, a rope belt, and the decorticator. The stalks are transported by the conveyor belt into the crushing press, followed by a pair of crushing rollers. The crushed stalks are carried forward on the rope conveyor to the decorticator. First, one-half of the stalk is decorticated and then the other, as in the case of sisal. Decorticated stalks pass through another pair of squeeze rollers into a brushing machine, where the fiber strands are opened up. The fibers are then examined and sorted according to color, etc.; scissors are used to cut off the dirty and tangled ends. They are then passed through a hot-air dryer (100°C), dried, and baled. Such decortication operations can process 16,000–35,000 kg/h of fresh stalks.

Machine-decorticated fiber is reported to sustain processing damage that reduces its strength. In contrast, the stripped fiber is stronger and more lustrous. Fiber loss as waste, however, is higher in the stripping method. The yield of decorticated fiber is 2–3% of the plant weight; 200–800 kg fiber is produced per hectare of land [109]. Grading of the fiber is carried out based on color, color uniformity, cleanliness, method of drying and handling, and the position of its parent sheath in the stalk.

The principal use of the fiber is in industrial and marine cordage, twines, tea and coffee bags, sausage casing and other forms of paper; small amounts are also used as place or floor mats, carpet backing, and other products of the craft industry in the Philippines.

The plant belongs to the Muscacease (banana) family: its genus is *Musa*. As many as 100 species (varieties) have been identified. Of these, 20 are commercially important; only three or four are grown extensively in the Philippines; *Musa textilis* is one of them.

Banana fiber is obtained from the edible-fruit-bearing plant, species *M. cavendishi* and *M. sapientum*. While India and Brazil are said to be the largest producers of bananas, data with regard to fiber production are scarce. The fibers are obtained mainly from the stem of the plant. About 37 kg (average weight) of stem yields about 1 kg of good quality fiber; the yield is 1–1.5% of the dry fiber [39]. The fiber obtained from the central core is of lower quality. The fiber has not been exploited much commercially hitherto, as it is considered inferior to abaca and other available hard fibers. The fiber can be extracted by hand scraping, retting, or by using raspadors; it can also be extracted chemically, for example, by boiling in NaOH solution [39]. Extraction of the fiber for local use (in cordage) or for cottage industries in India has been through manual means, similar to those discussed for abaca. Owing to its high cellulose and low lignin content, the use of fiber in the paper industry (tissue, filter, specialty nonwovens, document, printing, surgical and hygienic applications, coffee bags, meat casings, etc.) has been reported [39]. In Okinawa, the tradition of making banana fiber cloth, dating back to the 13th century, is being preserved by the Kijoka Banana Fiber Cloth, which "was made an important intangible cultural property by the nation" [69]. Over the years, there has been considerable interest in exploiting it [23,39,129,159,160] for a variety of household and industrial uses on a commercial scale. For instance, the use of banana fiber as reinforcement with autoclaved cement mortar, with air-cured cement or with air-cured plaster was investigated [12]. More recently, its use as a reinforcing fiber in polymer matrix composites [135] as well as in ropes [56] is investigated. The potential yield of the fiber in India has been estimated to be on the order of 2.2 million tons annually [14].

8.2.2 Sisal and Henequen

The sisal plant is indigenous to the Western Hemisphere, particularly Mexico, where the ancient Mayas used it along with related plants to produce ropes, mats, and some articles of clothing. The Spanish conquerors also found it useful for the cordage needs of their vessels. Outside Mexico, interest in sisal emerged in the early 1880s. The plant was gradually introduced to other countries with suitable tropical climates. The name "sisal" comes from the Yucatan port of Sisal from which the fiber was originally exported. The fiber was produced in 21 countries in 2004. The leading producers were Brazil (191,771 MT), Tanzania (23,500 MT), Kenya (20,000 MT)...down to Malawi (90 MT) [FAO database].

The plant grows as tall as 2 m with a short, stocky trunk, 0.15–0.23 m in diameter, as shown in Figure 8.3a. At various stages of growth, it has 100–150 dark-to-pale green leaves formed in a rosette on the trunk. The mature leaves, covered with a waxy bloom, resemble a two-edged sword 1–2 m long, 0.1–0.15 m wide, and 6 mm thick (at the center) with a black, 25-mm long terminal spine (thorn) [187]. The fibers lie longitudinally in the leaf, and, when extracted, range from 1 to 2m in length; short, broken fibers are known as "tow." A mature leaf containing about 1000 fibers weighs about 1 kg, of which 3–4% is the weight of the (dry) fiber. Within each leaf, three main fiber zones are identified: peripheral, median, and ground tissue, as shown in Figure 8.2b.

The peripheral zone contains two or three rows of mechanical fibers, which are nearly round in cross section. The "ribbon" fibers making up one row in the central portion of the leaf (median zone) are coarser than the mechanical fibers and have a crescent-shaped cross section that is not shown. They run the full length of the leaf, from butt to the tip. Ribbon fibers are split in the longitudinal direction relatively easily during the extraction and brushing processes. Mechanical fibers do not split; therefore, they determine the maximum diameter of the commercial fiber. The ground tissue zone, situated between the median and the peripheral zones, contains a combination of mechanical and ribbon fibers.

The thickness, length, and strength of fiber in the leaf depend on the maturity of the leaf at the time of cutting and its position along the leaf (fiber being thickest at the butt end). Initially, all leaves grow vertically on the plant; gradually, with age, they fan out to an angle of about 45°. The mature leaves are those closest to the ground; they contain the coarsest and the longest fibers. Fiber extracted from immature leaves is finer, shorter, and probably weaker than the mature fiber. A normal plant yields about 4.5–7 kg of white-to-pale-yellow sisal fiber in its lifetime.

When ready for harvesting, the lower two to three rows of leaves (15–20 leaves) are cut from each plant manually, leaf by leaf, using machetes. Cut leaves, with spines trimmed off, are bundled in groups of 20–50, tied and prepared for transport to the decortication factory. It is important to extract the fiber from the leaf within 24–48 h after cutting; otherwise, drying of the leaf makes extraction very difficult due to hardening of various gums therein.

Fiber can be extracted by microbiological retting, hand scraping, or by using raspador machines. Fiber extraction procedures in various countries, other than Brazil, are quite similar. A sisal plantation or estate consists of a number of large fields, usually arranged so they are connected to a central fiber extraction (or decortication) factory by roads or narrow-gauge railway. In Brazil and India, with some exceptions, sisal is grown on small holdings;

A: feed carrier
B: first beater drum
C: gripping of fiber
D: releasing of grip
E: second beater drum
F: discharge of fiber
I up to 7 inch. gripping ropes

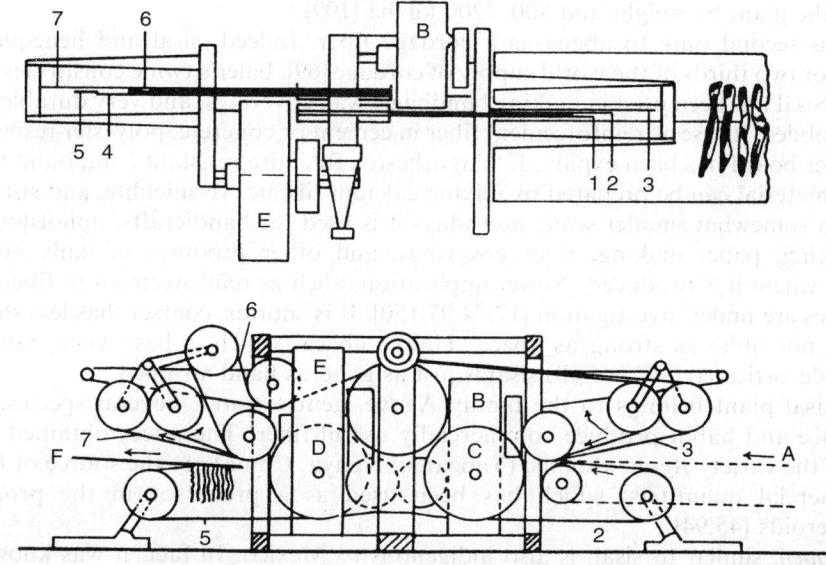

FIGURE 8.3 (a) The sisal plant. (From http://vesmir.msu.cas.cz/Madagaskar/fs17.html. With permission.) (b) Plantation scale decorticator. (From Jarman, C.G., Canning, A.J., and Mykoluk, S., *Trop. Sci.*, 20(2), 91–116, 1978.)

they are not large enough to warrant decortication factories on site. A double-retting process, reportedly developed in India, makes extraction of the fibers easier. The leaves are removed from the tank when retting is half complete, dried, and retted again after a few months [39]. The cost of mechanical (Raspador) extraction of 60 kg fiber in India was estimated to be Rs 110 in 1982 [39].

Large sisal decortication machines (Stork, Robey, and Corona) are also used to extract the fiber. The machines are similar in their operation. In each case, an endless conveyor feeds 200–300 leaves/min into the machine (see Figure 8.3b). They are first crushed by a pair of corrugated metal rollers, then grabbed near the center of the leaf (between two moving brass chains), carried along, and subjected to the scraping action of a bladed drum (similar in principle to the Raspador drum) rotating at 200–500 rpm. A different pair of chains then grips the leaves at a different point so that the unscraped portion of the leaf can be cleaned by a second rotating bladed drum. During the scraping stages, water (about 35,000 L/h) is sprayed onto the leaves to assist in separation of the fleshy plant material from the fiber. Water and waste material is discarded or sent through a flume tow recovery machine (a large rotating wire drum), which screens out the usable fiber tow. Wet decorticated fiber emerges on a chain at the other end of the machine and is dropped into a tank of water for final washing. Occasionally, as in Java, large centrifugal washing machines are used; these are capable of handling about 5 tons of dry fiber per day.

The fiber is removed from the washing tanks and dried, either outdoors in the sun, draped over wires, or artificially by spin-dry centrifuges or steam-heated chambers. The centrifuge machines (Haitre, Reinevald, and Krantz) used in Indonesia dry 1.8–2.3 tons of (dry) fiber in 20–25 min at 800–900 rpm down to about 50% moisture. Final moisture content is below 15% after about 20 min in 80–110°C temperatures. After drying, the fibers are brushed mechanically to remove the clinging dust and other matter and to bring out the luster. Before baling, the fiber is graded subjectively, according to cleanliness and freedom from pulp; bale weights vary from 200–300 kg, depending on the country of origin. Yield of sisal fiber is reported to be 2–5% of the plant by weight and 800–2200 kg/ha [109].

Sisal is second only to abaca as a cordage fiber. Indeed, sisal and henequen used to account for two thirds of the world supply of cordage [69]; baler's twine constitutes 70% of the cordage. Sisal has been used in making handbags, wall coverings, and very durable floor mats in automobiles. Its use as reinforcement fiber in cement or concrete, polyester resins, bitumen, and plaster board has been explored. "An asbestos-free, fire resistant compound for use as a building material can be prepared by mixing calcium silicate, vermiculite, and sisal fiber..." [39]. On a somewhat smaller scale, nowadays it is used for handicrafts, upholstery padding, sack making, paper making, floor coverings, and other products of daily needs in the countries where it is produced. Newer applications such as reinforcement in fiber-reinforced composites are under investigation [17,74,97,150]. It is shorter, coarser, has less sheen, and is generally not quite as strong as abaca. Unlike abaca, which is best when hand-stripped, machine-decorticated (95% of all) sisal is just as good as hand-stripped.

The sisal plant belongs to the family Agave, genus *Agave*. Several species, similar in appearance and habit, produce commercially useful fiber. The juices obtained in decortication of the variety *Agave sisalana* (Tanzania, Kenya, China) are the source of Hecogenin, in commercial quantities, which has been used as a precursor in the production of corticosteroids [45,94].

Henequen, similar to sisal, is also indigenous to Mexico. In fact, it was known as sisal. Owing to the commercial importance of fibers from the two plants, differences between the two are recognized, and the name henequen (of Mayan derivation) has emerged. The henequen plant has a taller trunk (1–2 m), has leaves with spines (thorns) along the edges (3–5 mm long), and the fibers usually have a reddish or yellow tinge. The decorticating machines used for henequen are similar to or sometimes even the same as the sisal machines. Following decortication, the henequen fiber is generally not brushed. The henequen plant produces 2–3% decorticated fiber by weight and about 2600 kg of fiber may be produced on 1 hectare of land [109]. About 90% of the total henequen fiber is produced in Mexico and the remaining perhaps in Australia and Cuba [109]. It is used principally in baler's twine (cordage) and, to some extent, in handicrafts, mattress and furniture stuffing, sack making, floor covering, paper making, industrial ropes, hammocks, etc. [69]. The specie of the fiber producing plant is said to be *Agave fourcroydes* [109].

8.2.3 FLAX

Of the five most ancient textile fibers (others being cotton, hemp, sunn, and ramie), flax was used in Egypt as early as 3400 B.C. for coarse linens. Linen curtains, made in about 1250 B.C., were found well preserved in King Tutankhamun's tomb, which was opened in 1922 [44]. From Egypt, the use of the fiber spread to the area encompassing present day Israel, Jordan, Iraq, etc. during the Biblical period, and to England and France during the Roman period. Gradually, it spread to all of Europe. By the latter part of the middle ages, it was the most widely used textile fiber. "Flax, having a lower lignin content in the fiber, was for centuries the premier fiber for apparel in western cultures"[190].

It is grown in temperate, moderately moist climates. During 2004, flax was produced in 11 countries. China was the largest producer (500,500 MT), followed by France (84,000 MT), Belarus (25,000 MT), etc. [FAO database]. There appears to be significant interest in bringing flax or linen back into fashion apparel in Europe [2,31,43]. In addition, the use of flax and hemp in thick insulation slabs to compete with glass and mineral wool building insulation is being explored [15]. Finally, there appears to be a good deal of interest in using flax as a reinforcing fiber in fiber-reinforced composites [50].

8.2.3.1 Plant

The flax plant, as shown in Figure 8.4, grows to a height of about 0.5–1.25 m and has a stem diameter between 1.6 and 3.2 mm. Seeds in the plant are contained in small spherical balls, or pods, at the top of the stalks; this is the linseed of commerce from which linseed oil is

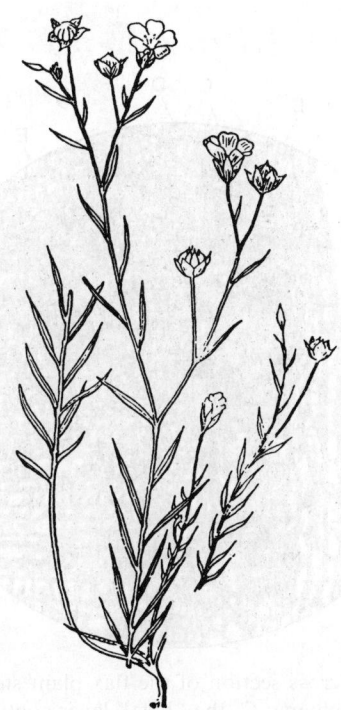

FIGURE 8.4 The ancient flax plant. (From Mauersberger H.R., ed., *Mathews Textile Fibers,* 5th ed., John Wiley and Sons, New York, 1947 [108].)

produced. Flax is usually grown for both fiber (one tenth to one fourth of the weight of the stalk) and seed. Indeed, special varieties of flax are grown for linseed and for fiber. For fiber production, a high planting density is preferred to avoid branching of the stem. The fiber diameter averages 14–15 μm for the customary planting density of 1800–2000 plants/m^2 as against about 18 μm for 300 plants/m^2 at the nursery stage [119]. These differences can be attributed directly to the cross-sectional size of the constituent ultimate cells. When extracted, they are between 0.25 and 0.75m long.

Five distinct regions have been identified in the cross section of the flax plant stem, as shown in Figure 8.5. The outer surface of the first epidermis layer is covered with a thin layer of wax, which prevents excessive moisture evaporation and protects the plant. Bacteria enter the stem through this layer during retting. The next layer, the cortex, consists of circular cortical cells that are not lignified, but contain pectic substances and coloring matter, which must eventually be bleached. In the third "bast" layer, fiber bundles are found surrounded by parenchyma. Each fiber bundle consists of 10–40 ultimates, or cells; there may be 15–40 bundles in the ring, depending on variety and agricultural variables. The ultimates are 14–30 μm in diameter, and 25–30 mm long; they are thicker near the root, and thinner and longest near the apex of the stem. The primary walls of the cells contain pectic substances with a trace of lignin; the secondary wall contains mainly cellulose. The fourth, the cambium layer, consists of tender growth tissue composed of thin-walled cells that separates the fiber layer from the fifth layer. The fifth layer is woody tissue consisting of thick-walled cells first, then thin-walled ones surrounding the pith cavity. This cavity is an air chamber that extends through the length of the stalk.

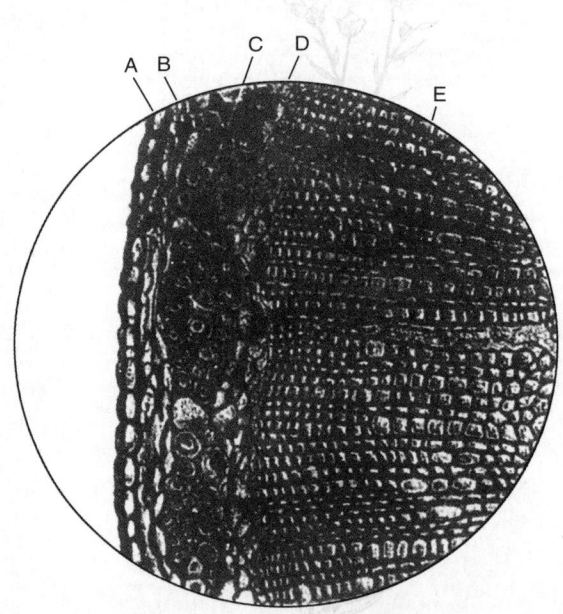

FIGURE 8.5 The five zones in the cross section of the flax plant stems: A, layer of cuticular cells; B, intermediate layer of cortical parenchyma; C, the "bast" layer containing the flax fibers; D, cambium layer E, woody tissue. (From Mauersberger H.R., ed., *Mathews Textile Fibers,* 5th ed., John Wiley and Sons, New York, 1947; *J. Adhes. Sci. Tech.*, **18(9)**, 1063–1076, 2004.)

8.2.3.2 Extraction: Retting

After harvesting by mechanical pulling, the stalks are left on the ground for 1 or 2 days; then bundles are set up in ventilated piles (stocks) and allowed to dry for 10–14 days. Val'ko et al. [176] report that spontaneous heating caused by pectin-degrading bacteria of wet flax straw may occur during storage if the ambient temperature is high enough. Before retting, the stalks are deseeded by crushing the seed bolls (if very dry), or by rippling. The dried seeds are crushed for oil.

The fibers are extracted from the stalks by retting and degumming in one or two ways: (a) biological or natural retting, in which bacteria (water retting) or fungi (dew retting) are the active ingredients; (b) chemical retting or degumming, in which dilute acids or bases are the active ingredients. Natural retting is a biological phenomenon. The pectic substances in the soft cells are dissolved by means of microorganisms, which free the fiber bundles and make it possible to separate them from the woody core [49]. The parts of fiber bundles that contain lignin resist the action of the microorganisms.

During retting, the organisms feed on the following [49]:

1. Pectic substances (free from nitrogen). (the pectic substances of the outer middle lamella—the encrusting, binding material holding together the fiber bundles in the bast layer and gluing this layer to the rest of the stem—are more easily decomposed than those of the lignified inner middle lamella of the ultimate)
2. Proteins of the protoplasm of the cells
3. Sugars
4. Starch
5. Fats and waxes
6. Tannin
7. Mineral substances such as calcium, potassium, magnesium, iron, sulfur, and phosphorous

Early methods of (cold) water retting, which are still used, involved placing of the stalk bundles into available nearby bodies of water, or into vats or ditches dug for the purpose, and holding them under the water surface with stones. Later methods of water retting used tanks, which may be open, closed, or in a cascade arrangement. In the cascade arrangement (three or four interconnected tanks), the water flows continuously at the rate of one (water) change every 2 days. Warm water retting (30–35°C) is considered more efficient commercially than previous methods. Water retting normally requires 8–14 days, depending on temperature and softness of water, to complete the process; warm water retting requires 3–4 days.

The completion of retting [49] is judged by the gray-blue color, as against yellow-brown, over the entire length of the stem. Alternatively, pH of the retting liquor decreases to about 4.6–4.9, where it may remain for 6–10 h and then rise again. Rise in pH implies over-retting and deterioration of the fiber strength. Warm water retting in large tanks is employed in Belgium, Holland, Hungary, Poland, and Romania. The straw is dried artificially after completion of the retting operation. Relatively large amounts of energy are required, both for retting and for drying. In addition, large quantities of waste liquor resulting from the process have to be disposed of [134].

Thus, warm water retting requires high capital expenditure and high water consumption. To combat the problem of high water consumption, and subsequent disposal, especially on plantation scale, aerobic bacilli are added (Rossii process) and the tanks are aerated. This results in lower concentration of volatile acids and a more alkaline solution, which enables the water to be reused with no limit. It also reduces the offensive odor otherwise given off during retting. The problem of over-retting is eliminated because the bacilli do not attack the middle lamella of the fiber.

Dew retting is carried out in spring, under conditions of moderate humidity, warmth, and freedom from wind. After harvesting and deseeding, the stalks are spread evenly in a field reserved for the purpose. Once a week, the stalks are turned over to ensure even retting. Mold and other fungi are the active retting agents. The process takes anywhere from 3 to 7 weeks, depending on the fungi and bacteria present in the soil and on the weather conditions.

Dew retting is employed in France, Czech Republic, Slovakia, and (to the extent of 80% of its crop) in the former USSR. The drying of retted stalks is accomplished in the field. The straw is collected in early autumn, at the latest, to avoid difficulties in drying. A part of the crop is not mature (fully retted) and chemical after-treatments are often necessary for complete extraction of the fibers. The homogeneity of the fibers thus obtained is not high. Steaming of the straw for short periods (former USSR) may be substituted for chemical after-treatments; the quality of fiber thus obtained is inferior to that obtained by other methods [137]. Bleaching of dew-retted flax is difficult and costly due to the dark color of the fibers sometimes encountered; the latter is associated with a fungus called "Altanaria" [161].

Mukherjee and Radhakrishnan [124], in their review, point out that, while both bacteria and fungi act as retting agents, isolation of the most effective individual strains has been less than fully successful. "It is believed that, besides pectolytic and hemicellulolytic activities, association of proteolytic organisms is necessary for the promotion of retting."

They also cite the work of Rosamberg and DeFranca [143] related to an analytic method for establishing termination time of retting of flax. The pH tended to remain between 4.0 and 5.0 and did not signify any specific level of loosening of the fibers. On the other hand, measurement and monitoring of the galacturonic acid (GA) content showed that GA was generated by the bacterial action on the pectinous matter. When the retting was complete (pectinous matter was used up), the bacteria began to use GA. Thus, GA concentration increased to a maximum and then decreased. The point of zero concentration gradient signified the completion of retting.

Chemical degumming substitutes the action of chemicals for that of bacteria or fungi. In the Peufaillit process, the flax straw is degummed 6–10 h in kiers, under a pressure of 1–2.5 atm, in a mixture of water plus 4% naphtha [49]. In the Baur process, the straw is treated with 0.5% sulfuric acid solution for about 3 h at 90–115°C. It is then washed with soda solution to neutralize excess acid.

More recently, the use of ultrasonic energy in the retting of flax and hemp has been investigated [186] as an eco-friendly process. Also, the progress in retting of flax has been reviewed by Akin et al. [2,3].

8.2.3.3 Extraction: Scutching

The retted stalks must be thoroughly dried and the retting action stopped, before the final process of breaking and scutching is undertaken [187]. This is accomplished either by artificial means or by methods similar to those used prior to retting, but this time by spreading the stalks evenly instead of bundling them. If the retting is complete, the fibers can be easily separated by first breaking the surrounding woody matter into small pieces of shive, which remain clinging to the fibers, and scraping them clean in the scutching process. The hand methods used earlier for breaking and scutching have been mechanized. Fluted rollers are used to smash and break the core into pieces of clinging shive as the stalks are fed into the machine and rotating bladed-wheels, operating at close tolerances against curved plates, to scutch the fiber clean. Manual labor is needed only for feeding straw in the machine direction and for removing fibers to the baling machine.

The machines are quite similar in operation to sisal-decorticating machines. Continuing interest in improving the effectiveness of the process is manifest in the studies of Savinovskii [151], and Lavitskii and Ipatov [86]; improvements sought relate to productivity, quality of scutched fiber, and prevention of the loss of long fiber.

During the scutching operation, some fibers are broken and removed in the scutching waste, together with pieces of stalk and shive. These short fibers are separated from the waste and the adhering shive by means of a special rescutching machine. They are known as the "scutched tow"; their percentage decreases with thoroughness or effectiveness of the scutching operation from 30–5% of the overall fiber content of the stalk. Their length ranges between 0.1 and 0.5 m, and they are baled and sold separately for use in coarse fabrics and ropes. The long flax fibers are known as "scutched flax" or "line" and are sorted into lots according to weight (density), fineness, softness, strength, color, uniformity, silkiness or oiliness, length, cleanliness, and handling.

The line fibers undergo a special combing operation, called "hackling," before spinning. The long fibers are separated from the short fibers and from the remaining shive in this operation. Unfortunately, some of the long fibers are broken during hackling and yield the "hackling line" fibers with lengths up to 1 m, and the "machine tow" with lengths between 0.1 and 0.5 m.

A mechanical process for decortication of retted and unretted flax and hemp fibers has been described by Munder et al. [125].

Flax is one of the strongest of the vegetable fibers, even stronger when wet. Color of the fiber varies somewhat depending on the retting conditions. A flax plant produces 12–16% of line fiber by weight with a yield of 480–720 kg/ha [109].

Finer flax is used mainly for damask (twilled table linen), sheeting, lace, and apparel; coarser grades are used in twines, canvas, bags, fishnets, sewing thread, fire hoses, and sail cloth. Flax tow (short fiber) is used for such products as towels, cigarette, and other specialty papers. Attempts to use rotor spinning for flax and hemp are being made [43].

The flax plant belongs to the family Flax, genus *Linum*. Of the 150 species of this genus, only one, *Linum usitatissimum,* produces commercially useful fiber.

8.2.4 RAMIE, RHEA, OR CHINA GRASS

Egyptian mummies from 5000 to 3300 B.C. were found wrapped in China grass fabric. Ramie was also used by the early inhabitants of America. Native Americans used ramie twine to attach blades of knives and spears to handles and shafts [44]. Ramie fabrics were also found in the Mawangdui tomb in Changsha, PRC, dating back to 206 B.C. [130]. Ramie is a perennial native to China, Japan, and the Malay Peninsula, where it has been used as a textile fiber for centuries. The plant is also found in Taiwan, India, and Algeria, and was reintroduced to the United States (and probably Mexico) around 1855. Ramie plant produces about 3.5% of fiber by weight and a one hectare field yields 880–2200 kg [109]. During 2003, significant productions were reported only for China (265,000 MT), Laos (1700 MT), and Brazil (1100 MT).

Over the years, ramie fiber has attracted much attention as a textile fiber because of its excellent properties, but problems in degumming the fiber efficiently (as will be discussed below) have blocked the way for its more widespread use, in general. In China, however, due to economic factors indigenous to the country, attempts continue to be made to increase utilization of the fiber in textile products. In particular, technological research is focused on (1) studying the structure of the fiber [188]; (2) reducing the cost of degumming (through use of specially developed enzymes, coupled with recycling and heat exchange techniques) [130]; (3) how the fiber may be processed optimally on modern spinning, weaving, and knitting machines; and (4) how apparel fabrics for modern consumer markets may be developed [99].

The ramie plant (Figure 8.6) is cheap to grow requiring about 45–55 days in hot (30°C) and humid climates [13]. It grows 1–2.5 m high, with broad, heart-shaped leaves on a nearly branchless stem. The fibers are located in the cortex layer of the stem just underneath the thin outer bark. The cortex layer, in turn, surrounds a thick, woody layer surrounding the soft, pithy core. The core constitutes about two thirds of the 10–20 mm diameter of the stem. The stem and leaves of the plant contain about 80% moisture, 2–4% fiber ribbons (cortex and outer bark), and 1–1.5% degummed fiber, depending on various agricultural and extraction factors. The rest is leaves and woody pith. The structure of the stem is similar to that of flax and other bast fiber plant stems.

In China, ramie is cut by hand in a manner similar to that of high land jute. The stalks can be cut close to the ground as they mature in a year-round process in a small operation, or two to three times a year in larger operations. Since the stalks are prone to bacterial attack soon after cutting, they are either dried or ribboned without delay.

To ribbon, the stalk is first broken in the middle, the shive pushed out, and the undamaged ribbon layer is peeled off on both sides manually. Alternatively, the stem is broken near the ground [76], carefully, so that the ribbon layer is not damaged; then the strips are easily pulled away from the base of the plant. The strips are then scraped by hand, using a blunt knife blade and a block, or other simple hand-held tools.

Due to the gummy pectinous matter in the bark, ramie fiber is difficult to extract completely from the bark. Retting of the ramie stalks in water is not self-induced, as in the case of other bast fibers. This is perhaps due to high concentration of gums in the plant, which protect the strips from microbial fermentation. The scraped fiber, therefore, is only partially separated and requires further processing to be spinnable. The partially separated undegummed fiber strips are hung to dry and bleach in the sun. Often, the fumes from burning charcoal or sulfur are used to bleach the fiber.

Other methods of ramie extraction include beating the stems against a stone surface or with a wooden mallet. A method used in parts of China, as well as in Indonesia, involves allowing the plants to become overmature before cutting. After cutting, the outer bark is

FIGURE 8.6 Photograph of a ramie plant. (From http://www.aboutjoel.com/archives/000147.html. With permission.)

scraped from the plant, leaving the fiber exposed. Then, after the stalk has been washed and dried, the fiber is peeled off in strands as with jute. This fiber is suitable for twines, but is too coarse to be used for fabrics.

In Japan, small decorticating machines are usually used to strip and scrape the stalks. The machines resemble the Raspador, originally developed for sisal, each consisting of a single-bladed decorticating drum. The stalks, after removing leaves, are decorticated by feeding one end into the machine, about half way, withdrawing it, then feeding the other end. In another case, several stalks are passed through fluted crushing rollers at the entrance to the machine. The linear speed of the feed rollers is slower than that of the bladed drum; the stalks are thereby constrained as their full length is decorticated in one pass through the machine. It is subsequently passed through brushing machines (resembling Raspador), which, with two operators, convert 75 kg of rough fiber strips to 45 kg of brushed fiber strips in a day. The fiber constitutes about 5% of the weight of the plant, with leaves.

Before the fiber can be spun into the fine yarns necessary to produce high-quality ramie fabrics, it must be given a degumming treatment. The thick fiber bundles are cemented together by natural gums, which constitute, normally, 20–30% of the weight of the fiber strips. Usually, degumming is completed in spinning mills prior to spinning. The process involves soaking in alkali baths for prescribed periods of time at prescribed temperatures. The most common chemical used is caustic soda. The others are trisodium polyphosphate, sodium sulfite, sodium carbonate, sodium pyrophosphate, sodium silicate, and sodium citrate, which can be used alone or in combination with NaOH. Alkalis are used for degumming (usually in concentrations of less than 1%) because they break down pectins without harming cellulose. Sometimes, though, a mild acid pretreatment or "scouring" stage is added to the process. Each mill seems to have its own degumming process, the details of which are usually proprietary information. These often include several separate soakings (2–4 h each) in alkali baths at near-boiling temperatures, with intermediate washings. The processes are either batch (using large vats) or continuous (with conveyor belts carrying trays of fiber through alkali baths). After degumming, the fiber is first dried (in squeeze-rollers, centrifuges, or hot-air dryers), and then sprayed with soap, oil, and glycerin solution (to keep fibers from drying extensively and becoming brittle).

Because of the need for an extensive degumming process prior to spinning, ramie is not as popular in world markets as other bast fibers (jute, flax, etc.), although it is very similar to flax in appearance and properties. Wang et al. [181] report that a rapid degumming process for ramie has been developed. Adding sodium phosphate, which has a strong chelating power with multivalent ions and good hydrolytic stability, results in rapid degumming of ramie. Liu et al. [102], on the other hand, used Bacillus D 773-22 and mutant Nf-9 strains, which contained pectinase for degumming ramie. The mutant Nf-9 strain increased the fiber's dispersability by 13–17%. High pectinase activity was reported at pH 9 and 55°C. Enzyme activity was decreased by calcium ions and increased by potassium orthophosphate. The effects of different carbon and nitrogen sources in the culture media on enzymatic activity have also been studied. Jarman et al. [70] provide an excellent review of cultivation, extraction, and processing of ramie.

More recently, Bhattacharya and Das [25] seem to have "determined the optimal conditions for degumming ramie fibers with sodium metasilicate alone or in combination with such other *alkali* solutions as sodium carbonate and trisodium phosphate. The weight loss, whiteness index, and color strength of the degummed fibers were better than those of *ramie* fibers conventionally degummed with sodium hydroxide. The sodium metasilicate degumming process eliminated the need for subsequent conventional bleaching and yielded fibers with improved luster and hand properties at a cost equal to that of conventional degumming methods."

In the past, ramie has been found useful for fishing nets, sewing thread, canvas, filter cloth, upholstery, fire hoses, marine packaging, and clothing [84]. The short fibers from processing wastes have been used for specialty paper, for example, currency bills, bank notes, and cigarette paper [67]. Blends of ramie with other fibers for apparel fabrics, with improved functional properties, have been proposed [131,1] from time to time. According to Greenhalgh [56], in Brazil, ramie is used as a textile substitute for flax, sacking, and carpet backing, whereas in Japan it is used for mosquito netting and matt edge cloth, but mostly in clothing in the form of blends with other fibers. In 1989 [130], Japan was reportedly the leading market for ramie in automotive upholstery fabrics, ramie-worsted wool fabrics, table cloths, napkins, curtains, special bank notes, cigarette and certificate paper pulp, and as top-grade packing material for industrial and marine journal boxes and seal-lubricant moving parts. In Switzerland, ramie yarns are used in combination with polypropylene yarns to produce fabrics for reconstruction of river banks. With the increasing interest in sustainability and ecological issues, the use of ramie in fiber-reinforced composites is under investigation [126]. It can be used as a source for pure cellulose as it is very rich in cellulose and essentially free from lignin.

Ramie belongs to the nettle family, genus *Boehmeria nivea*. The variety grown in China, Taiwan, and India has silvery-white hair on the underside of its leaves and yields China grass, or white ramie. The variety grown in Malaysia, Africa, and Mexico has leaves with green undersides and yields rhea, or green ramie. Strips from both are very similar, except that the white ramie fibers are considered to have a finer texture.

8.2.5 Hemp and Sunn

Hemp (not to be confused with Mauritius "hemp", Manila "hemp", and other plants that have no botanical relationship with true hemp) is considered the oldest cultivated fiber plant. Originating in central Asia, the plant has been cultivated in China for about 5000 years[1]. Today (2004), hemp, *Cannabis sativa*, grown for industrial fiber, is found in nearly all temperate regions of the world; principally, it is cultivated in China (38,000 MT), Spain (15,000 MT), North Korea (12,800 MT), Russian Federation (7000 MT), and so on. In the United States, it has long been a subject of sociopolitical controversy [189] because of the plant's seeming similarity to the plant of the same genus that yields high quantities of the drug *marijuana*. The psychoactive narcotic ingredient in Cannabis is delta-9 tetrahydrocannabinol (THC) [110]. The recent development of new strains of non-narcotic varieties, however, has led to the approval of experimental cultivation of industrial hemp in several countries, such as England in 1993, Australia in 2004, Germany... in the fall of 1995 [32]. In the United States, in a January 2000 report, USDA concludes that the market for hemp, fiber, and oil in the United States is too small for its cultivation to be economically viable [175].

Mignoni gives a broader view of this issue in EU countries [116]. He also reviews much other useful information regarding history, agriculture, extraction, processing, properties, and uses of hemp. In the same vein, development of technology to "establish *hemp* textile production in Great Britain with *hemp* garments and soft furnishings" has been reported [116]. In the United States and elsewhere, its use in fiber-reinforced composites [65] and in fiber-reinforced cement [98] is under investigation. Finally, hemp seeds can be the source of as many as 19 fatty acids, yielding oil 26–37% by weight [79].

---

[1]"... introduced by the Emperor Shen Nung in the twenty-eighth century BC. The wild *Cannabis* ancestor is believed to have grown somewhere in a general area between western China and the eastern Caucasus, north of the Hindu Kush. This wild ancestor is not found today." (http://www.gametec.com/hemp/IndHmpFrmfr.html)

Other Long Vegetable Fibers

When cultivated densely for fiber, the plants grow up to 4.5 m high, with a stem diameter between 4 and 20 mm. Varieties grown for seed, however, are short and shrublike (as is also the variety grown for drug, hashish, cultivated in tropical areas). The best stalk dimensions, from the fiber production point of view, are 2 m in length by 5 mm in diameter. The structure of hemp stem is similar to flax stem, with the pith cavity one half the diameter of the stalk. The pith cavity of the narcotic-containing plant specie is much smaller, as shown in Figure 8.7 [162].

After harvesting, the stalks are often laid on the ground to dry for 4–6 days before retting. Retting is accomplished, as in flax, in water, snow, or dew; water retting takes about 10–15 days at 15–20°C. Following retting, the stalks are dried, then subjected to breaking, and scutching operations on machines similar to sisal decorticating machines.

Hemp plant produces 3.5% of line fiber by weight; fiber yields as high as 900–2600 kg/ha are possible [148]. Hemp fiber is graded based on color and luster, although no generally recognized standards of grading exist. The fiber is longer, but coarser and less flexible than flax, and bleaching is difficult. The finest hemp is Italian hemp, which is soft and silky, with a pale gray color and strands about 2 m long.

Hemp has been replaced in its use in ropes largely by hard fibers (sisal and abaca) and synthetic fibers. It is used today mainly as a substitute for flax in the production of yarns and twine. Other uses include fine and coarse household clothing, bags, industrial and marine rope, cables, twines, nets, sailcloth, and canvas; the "hurd" from the core of the stem is pulped for paper making. Detailed information on European varieties of hemp, subjected to different methods of fiber extraction, is given by Leupin [61].

The true hemp belongs to the family Mulberry, genus *Cannabis*. The fiber-producing species is called *Cannabis sativa*.

Sunn has been grown in India since prehistoric times [44]. The plant is used not only for fiber but also for green manure. The fiber is used locally to make fishing twines, thick canvas, cordage, ropes, and other items; a fair portion of the fiber is also exported. The use of sunn for chemical pulp and for viscose rayon has been explored [39]. Attempts are also made to blend it with other fibers to produce ring spun yarns, presumably for clothing applications [48]: India (186,624 MT), Thailand (57,000 MT), China (44,000 MT), and the former Russian Federation (48,000 MT) in 2003 were the largest producers of the so-called jutelike fibers as well as sunn [FAO database].

(a)

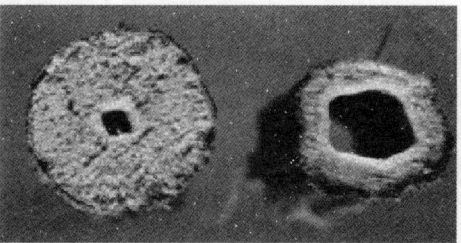
Hollow stem of non fiber (left) vs. fiber Cannabis [162]
(b)

FIGURE 8.7 (a) The hemp plant; (b) stem cross sections of two species. (From http://www.naihc.org/hemp_information/hemp_defined.html. With permission.) (From Small. E., The Species Problem in Cannabis. Corpus, Canada, 1979. With permission.)

When closely spaced, the plants grow to a height of 3 m, with a stem diameter of up to 25 mm; when widely spaced, they grow to be shrublike (short, with many branches). At harvest, the plants are often laid out in the field to dry for 2–3 days for leaves to drop off. Roots and tops are cut, the sunn stalks are retted, much like jute stalks, for 3–15 days (October through December). Fiber is stripped from the retted stalks by beating them against the surface of water to wash away extraneous matter and to separate the fibers on the surface.

After washing and drying, the fiber is pulled off and tied into small hanks (or handfuls). At a later stage, they are cleaned and "dressed" by whipping against and drawing through upright pins mounted in a board (pin hackling). The fiber is then graded according to strength, color, texture, length, extent of tangling, condition of the butt ends, cleanliness, and finally pressed into "Pucca" (final) bales.

The sunn plant produces 2–4% of dry fiber by weight. The sunn fiber is white to gray in color, has a shiny luster and fine texture. It has high tensile strength, but is coarser than jute and kenaf and not as pliable. Therefore, it is unsuitable for mixing with jute in making jute hessian. Its other uses include canvas, sailcloth, industrial ropes, nets, and twines.

The sunn plant is a member of the Pea family, its genus is *Crotalaria*, which has 200 species, of which three (white, green, dewghuddy) are grown for fiber in India. The botanical name of the fiber is *Crotalaria juncea*. [109]

8.2.6 COIR

The coir (from Malayalam *kayaru*—cord) fiber is extracted from the husk of coconuts (see Figure 8.8), native to islands of the Indian Ocean. It was the first hard fiber introduced to European rope makers, and was found to be superior to hemp because of its bulkiness, resilience, and resistance to degradation by seawater. Until recently, coir ropes were used for marine towing, presumably because they could endure sudden pulls that would snap the otherwise much stronger ropes made from hemp or other hard fibers. Coir has been used for making bags, either by itself or in combination with cantala yarn as the warp. During 2003, it was produced in significant quantities in India (450,000 MT), Sri Lanka (127,250 MT), Malaysia (28,000 MT), and so on [FAO database].

Coir fiber comes in two categories: white and brown. In a coconut, the fibrous husk constitutes the outer covering of the hard, nearly spherical shell containing copra. The husk, itself covered by a stiff covering called exocarp, weighs about 0.25–0.43 kg, with the fiber

FIGURE 8.8 Photograph of a coconut tree; a pile of fiber containing husk and exocarp removed from the fruit. (From http://www.bio.ilstu.edu/armstrong/syllabi/coir/coir1.htm. With permission.)

content ranging from 32–44% by weight. Coconuts are picked from the trees by climbing; ripeness is judged by tapping the nut.

To extract the fiber, the coconut is dehusked manually or with the aid of machines (for brown fiber only). Manual dehusking is accomplished either by striking the husk with a machete or by striking the coconut against a sharp stake or spike, followed by hand stripping. The hand-stripped husks are shown in Figure 8.8. Mechanically, Downs dehusking machine or a bursting machine may be used. A dehusking machine, developed by the Tropical Products Institute (London), cuts through the husk longitudinally at four locations, 90° apart, and peels off the husk instantaneously.

The white coir, finer and lighter in color, is obtained from immature coconuts. For as long as 9 months, retting softens the husk segments; the remainder of the extraction process is presumably similar to the brown fiber case.

The brown fiber is produced from ripe coconuts. The husks require soaking in water, which softens the vegetable matter and facilitates fiber extraction. The soaking is carried out in natural or man-made pits with stagnant water, from 2–6 weeks, or soaking in concrete tanks (up to 200,000 husks; 10–14 days uncrushed, 4–5 days crushed), where warm water is changed periodically. The fiber from the latter is cleaner and superior. Rajan, Senan, and Abrahms [139] suggest that phenolic substances present in the husk retard the retting process; during retting in coastal regions, the tidal action leaches away the retardants. Indeed, they give a comprehensive review of microbiology and biochemical aspects of retting as well as that of efforts to *soften* the fiber.

The fiber is extracted from soaked husks by wet milling, or defibering, on special machines called drums, which are arranged in pairs on the same axle. The first drum (breaker), consisting of a wooden cylinder equipped with coarse, widely spaced iron nails (40–50 mm apart), is enclosed in a casing; the casing has a front opening used for inserting (feeding) the husk segments. First, one end is fed part-way and held (by a pair of fluted rollers); the pins tear away at the connective tissue (pith) and the outer shell (exocarp); the shorter mattress fibers are also removed and delivered through a chute at the bottom of the casing. Then the segment is withdrawn, reversed in direction, and reinserted; the second end is thus cleaned and the longer bristle fibers (at least 0.18 m long) are left in the hand of the operator. Next, the bristle fibers from three or four husks are held, first one end and then the other, against the second, the cleaner drum, which is similar to the breaker drum, but with finer and denser (25–35 mm apart) nails. The remaining mattress fibers are extracted by the cleaner drum; the long bristle fibers are left in the hand of the operator. The pair of drums (at 150–250 rpm and about 10 hp) can handle 2000 husks in 8 h, producing 100 kg of bristle and 250 kg of mattress fibers.

The extraction of fiber may also be carried out by dry milling or decortication of smaller husks, or where soaking facilities are not feasible. The decorticators (Downs or Neugeng) consist of two sections: the bursting mill and the sieve or sifter. The bursting mill disintegrates (and partially opens) the husk segments, using metal beater bars attached to a metal disk rotating at up to 2500 rpm. Small amounts of water are added frequently to facilitate disintegration. The husk segments are next conveyed to the sieve or sifter; it completes the cleaning and opening by further beating by steel beater blades attached to a revolving (up to 900 rpm) horizontal (Downs) or coaxial vertical (Neugeng) disk. The daily production varies from 700 to 1300 kg; the fiber so obtained may be further cleaned of dust in the turboscreens (not used in Sri Lanka). The decorticated fiber looks very much like the mattress fiber, but is cleaner and much stronger, because in this case the bristle fiber has not been separated. It is also more resilient and is, therefore, used exclusively in twisting.

The brown coir fibers may receive several additional preparatory treatments, specific to the anticipated end-use, before they are graded, packaged, and shipped or exported.

The above extraction procedures for brown fiber are used predominantly in Sri Lanka. In India, it is extracted by a somewhat different wet-milling procedure. Different types of equipment are used. In the Fehrer equipment, the husk segments are crushed by five fluted rollers, stored for 24–72 h in a tank filled with fresh water, then fed to the defibering machine by an automatic feeding disk. The husks, gripped at one end, are taken past a revolving picker drum whose sharp pins comb through the exposed tuft, remove pith, exocarp fragments, and the mattress fiber. The tufts are then transferred to the periphery of another drum; the unclean, exposed segments are then taken past the same picker drum (on its other side); again, the mattress fibers, pith, and exocarp pieces are removed. The clean bristle fiber is thus dropped in a separate collection. Such machines can process 4000 husks, producing 320–360 kg of fiber (80% mattress, 20% bristle) in 8 h.

The Ennor- and Nakano-type equipment perform the same functions with somewhat different mechanisms. Their respective productivity is 1.5 and 4 times that of the Fehrer equipment, and they produce mattress and bristle fiber in a 70:30 ratio.

The mattress fiber from any of the above equipment used in India is passed through sifting machines or rotating screeners to remove the pith and very short fibers. All fibers are sun-dried, as in Sri Lanka.

In India, too, the brown fiber is produced in the mattress, bristle, and decorticated variety. Indian bristle fiber is not long enough for brush-making; mixed with decorticated and mattress fiber, it is used for twisting. The decorticated fiber and mattress fiber are also used for stuffing mattresses and upholstery.

The white fiber is fine and flexible, and therefore can be spun into yarn, twine, marine cordage, mats, tufted floor mats, and woven carpets; it is produced almost exclusively in the Kerala region of India. Brown fiber is coarse and stiff and finds usage in mattresses, furniture stuffing, automobile seats, brushes, brooms, nets, air filters, reinforcement of river embankments, as geotextiles to prevent soil erosion [89], covering of land-based drainage pipes, etc. It is used as a precursor for granular and powdered activated charcoal. Its potential as reinforcement in polymeric matrix composites and fibrous active carbon has been investigated [64,81,82,138,177] and continues to be investigated [55,107]. Even a use for coir pith is being investigated. Namasivayam and Kavitha report on investigation wherein carbonized pith was used "as adsorbent for the removal of 2,4-dichlorophenol (2,4-DCP) from water [127]."

The coconut belongs to the family Palm and genus *Coco*; its botanical name is *Cocos nucifera*.

8.3 FIBER MORPHOLOGY AND CHEMICAL COMPOSITION

8.3.1 MORPHOLOGY

Vegetable fibers as obtained from nature are single and multiple cellular systems. Cotton and kapok are examples of single-cell systems; all others are usually found as bundles (aggregates) of multiple cells, also called "ultimates." The ultimates in the aggregate bundle are bound together by natural polymers, variously called resins, gums, cementing materials, encrusting materials, and middle lamella [95,96]. Lewin [95,96] states that in bast fiber plants the bast layer "originates and develops from the terminal meristem close to the growing tip and is in its first stage of development, multinucleate and rich in protoplasm. When differentiating from a single cell, it dissolves some of the middle lamella and develops simultaneously in length and diameter." It is proposed here that this mechanism leads the ultimates to pack together tightly in a polygonal form in cross section and to perfectly oriented (lengthwise) longitudinal bundles with tapered ends (at least partially a consequence of the tight packing).

Owing to the tight packing, the middle lamella substance between the cells form a thin, mostly uniform, but unbroken film (layer or lamina) that surrounds each ultimate completely. Microscopic examinations of the morphological structure of the leaf (sisal, abaca, etc.) and multicellular seed (coir) fibers reveal very comparable structures. Thus, a comparable mechanism of formation of multicellular strands may also be postulated.

The dimensions of cells (length, diameter) are highly variable, dependent on species, maturity, location (position) within the plant, and extraction procedures. Meaningful comparisons can be made only in terms of orders of magnitude. Harris [58] collected the most comprehensive set of data from different sources. Table 8.2 is a composite of these, and other data, for fibers of interest and some for purposes of comparison. From these data, it is quite clear that, on average, the unit cells of ramie, flax, and hemp are longer and coarser than the cotton fiber; jute, abaca, and hemp yield the longest fibers, and abaca, henequen, and sisal yield the coarsest fibers.

The morphological structure of the multicellular fibers makes them analogous to the modern-day fiber-reinforced, rigid-matrix composites; materials in which (a) the fibers are all aligned in the same direction (unidirectional), (b) the volume fraction of fibers is very high, and (c) the fibers have discrete (discontinuous filaments) lengths. As in composites, the properties of multicellular fibers are determined by the physical, mechanical, and chemical properties of the morphological constituents.

According to this picture, the middle lamella in bast fibers plays the role of the matrix. It is, however, a complex role. First, the middle lamella holds the ultimates together in the fiber bundle or strand. This is termed the "inner" middle lamella. Because of its close packing and its possible penetration into the fiber walls, as well as because of the possibility of its being chemically linked to the cellulose of the cell wall, it is relatively stable to chemical and microorganism attack.

A second function of the middle lamella is to glue together the fiber bundles in the bast layer and this layer to the other layers of the stem. This is termed the "outer" middle lamella. It is less closely packed than the inner middle lamella and is probably less chemically linked to the other layers. Consequently, it is more easily attacked by the retting bacteria, fungi, and chemical reagents. The possibility of selective attack on the outer middle lamella by microorganisms had been recognized by our ancient ancestors and is used in practice to this day in the retting and degumming operations.

8.3.2 Chemical Composition of the Fiber

Because of the immense importance in usefully exploiting the fibers in a vast variety of end-uses, the chemical composition of vegetable fibers attracted the attention of many investigators in the past. Consequently, it is generally agreed that vegetable fibers contain one or more (usually more) of the following [173]:

1. Fats and waxes, which are found mostly on the surface and can be extracted with benzene.
2. Water solubles, which are extracted by boiling the dewaxed fibers in boiling water.
3. Pectin, which exists in water-insoluble form as calcium, magnesium, and iron salts of pectic acid. The different forms of "pectin" are all short-chain polyuronides, derived from galacturonic acid. During biological retting, these substances are transformed into acetic and butyric acids. Pectic acid can also be removed by boiling in 0.1 N alkali. Boiling in 12% hydrochloric acid yields CO_2. Pectin dissolves in boiling ammonium oxalate or citrate and can be precipitated with calcium ions. Its molecular weight is between 35,000 and 100,000.

TABLE 8.2
Dimensions of Fibers and Ultimate Cells in Typical Vegetable Fibers

Fiber	Length (cm) range	Diameter (mm) range	Fineness (denier) range	Cell length (mm) range	Cell length (mm) mean	Cell width/diameter (µm) range	Cell width/diameter (µm) mean	Shape
Abaca		0.01–0.28	38–400	3–12		6–46	9.9 (minor) 28.1 (major)	Round/oval
Long	200 or more							
Normal	100–200							
Short	60–100							
Tow	Under 60							
Banana		0.011–0.034	54–68	0.9–5.5		18–30		Cylindrical
Sisal		0.1–0.46	9–406	0.8–8	3.3	7–47	21	Cylindrical
Long	100 or more							
Normal	60–100							
Short	40–60							
Tow	Under 40							
Henequen						8–33	12 (minor) 22 (major)	Polygonal
Flax	20–140	0.04–0.62	1.7–17.8	4–77	33	5–76		
Ramie				40–250	—	16–126		Hexagonal/oval
Ex. Long	150 and more	0.06–0.904	4.6–6.4					
Very long	125–150							
Long	100–125							
Normal	80–100							
Short	40–80							
Tow	Under 40							
Jute	150–360	0.03–0.14	13–27	0.8–6	—	15–25	115	Polygonal/oval
Kenaf	—	—	50	1.5–11	3.4	12–36	24	Cylindrical
Hemp (true)	100–300	—	3–20	5–55	25	10–51	25	Polygonal
Sunn	—	—	—	2–14	7	8.3–61	31	Irregular
Coir	—	0.1–0.45	1–3.3	0.3–1.0	0.7	12–24	20	Round/oval
Cotton	Same as the cell dimensions			15–56	—	—	—	

Composite data given by Bhatia and Gupta [23], Harris [58], Himmelfarb [63], Kasewell and Platt [75], Kulkarni et al. [80–82], McGovern [109], Sinha [159,160], N. Chand et al. [39].

4. Hemicelluloses: Buschle-Diller et al. [32] describe them as "matrix polysaccharides", usually composed of a backbone of homo- or heteroxylans or -glucans with substituents in the O-2 and O-3 positions. The side-chains may be single units or di-, tri-, or tetrasaccharides, with the major components being xylose, galactose, arabinofuranose, and glucuronic acid residues with some esterification and cross-linking. They are insoluble in hot water and primarily hydrogen-bonded to cellulose. The hydrogen bonding to cellulose fibrils is considerably stable, so that even with alkaline extraction a certain amount of hemicellulose residues can be expected to remain in the structure.
5. Cellulose, which is the principal constituent of the fiber ultimates. Its chemical structure has been discussed in detail elsewhere in this book.
6. Lignin, which is a short-chain isotropic and noncrystalline (DP 60) made up of units derived from phenylpropane. Lignin is found in the middle lamella of the fiber bundle, as well as in the woody core and the epidermal and cortical cells of the plant stem. It is also often found in the walls of the ultimate cells of the fibers; the nature of lignin in different parts of the plant can be different. To determine the lignin content, the fiber sample, from which waxes, water solubles, pectic matter, and hemicellulose have been extracted, may be first chlorinated and then extracted with sodium sulfite solution; what remains is cellulose. The lignin can also be dissolved in sodium chlorite. In the Klason method, the cellulose is dissolved in 72% H_2SO_4 and the remaining lignin is weighed [167].
7. Coloring matter, which in bast fiber, is located within the cortical cells, some of which remain attached to fiber bundles after scutching. It consists of chlorophyll, xanthophyll, carotene, as well as their alteration products [188]; they are associated with complex compounds of the tannin type. Another source of coloring matter is in the cambium cells; this is usually connected with proteinic materials containing aromatic groups [41,42,57,93,108].

Several investigators have determined the chemical composition of vegetable fibers. Independent determinations of the composition of the same fiber by two or more investigators have often given different results. Turner [173] attributes this to, among others, the following factors: (a) the number of closely related substances in the fiber, some of which are very similar in their chemical reactions; (b) the variations within one type of fiber caused by levels of maturity and different geographical sources. Comparison of values reported by McGovern [109], suggest a third factor: inclusion or exclusion of moisture in the cellulose content.

Lewin [95] makes a very important point in that "the analyses reported in the literature usually refer to the fibers without differentiating as to the degree of their purification by retting or by chemical methods. It may, therefore, be assumed that most of these analyses, in many cases carried out with inadequate analytical methods, actually refer not only to the ultimate fibers and the inner middle lamella, but also to the varying parts of the outer middle lamella."

In Table 8.3, analytical data published by Turner [173], Timell [172], Bhatia and Gupta [23], and McGovern [109] are given. These data differ significantly from the data of Whitford [191] and others cited by Harris [58]. The procedures for determination of chemical composition have been standardized and are described in numerous publications, including those of Turner [173], Wise, and Jahn [195].

In the same vein, Wilson [194] gives composition of the oven-dried sisal as 62% true cellulose, 18% pentosans, 10% other carbohydrates, 8% lignin, 2% wastes, and 1% ash. Such differences in composition can be attributed to reasons cited earlier.

TABLE 8.3
Chemical Composition of Bast and Leaf Fibers (Percentages)

	Cellulose	Hemi-celluloses	Pectin	Lignin	Water solubles	Fat and wax	Moisture	Ash
A. Flax (unretted)	56.5	15.4	3.8	2.5	10.5	1.3	10.0	
Flax (retted)	64.1*(71.2)*	16.7*(18.6)*	1.8*(2.0)*	2.0*(2.2)*	3.9*(6.0)*	1.5	10.0*(−)*	
Jute	64.4*(71.5)*	12.0*(13.4)*	0.2*(0.2)*	11.8*(8.1)*	1.1*(1.8)*	0.5	10.0*(−)*	
Hemp	67.0*(74.9)*	16.1*(17.9)*	0.8*(0.9)*	3.3*(3.7)*	2.1*(3.1)*	0.7	10.0*(−)*	
Ramie	68.6*(76.2)*	13.1*(14.6)*	1.9*(2.1)*	0.6*(0.7)*	5.5*(6.4)*	0.3	10.0*(−)*	
Sunn hemp	67.8	16.6	0.3	3.5	1.4	0.4	10.0	
Sisal	65.8*(73.1)*	12.0*(13.3)*	0.8*(0.9)*	9.9*(11.0)*	1.2*(1.6)*	0.3	10.0	
Manila hemp	63.2*(70.1)*	19.6*(21.8)*	0.5*(0.6)*	5.1*(5.7)*	1.4*(1.8)*	0.2	10.0*(−)*	
Phormium	45.1*(71.3)*	30.1	0.7	11.2	2.2	0.7	10.0	
Cotton	82.7*(92.9)*	5.7*(2.6)*	*(2.6)*	—	1.0*(1.9)*	0.6	10.0	
B. Hemp	78.3	5.47	2.5	2.9	—	—		0.53
Jute	59.4	18.92	4.47	12.9	—	—		0.62
Kapok	43.2*(64.0)*	32.40*(23.0)*	6.61*(23.0)*	15.1*(13.0)*	—	—		0.76
C. Banana *	67.6			5.4	2.4	8.7	1.2	
**	63–64	19		5		10–11		
***	50–60	25–30	3–5	12–18				
D. Coir	36–43	0.15–0.25	3–4	41–45	Some			2.22
E. Flax	60–81	14–18.6	1.8–2.3	2–3				
Jute	51–72	12–20.4	0.2	5–13				
Abaca	60.8–64*(70.1)*	21*(21.8)*	0.8*(0.6)*	12*(5.7)*	*(1.8)*			
Sisal	43–88	10–13	0.8–2	4–12				
Kenaf	36*(72.8)*	21	2	18				
Ramie	68.6–76	13.1–15.0	1.9–2	0.6–1				
Hemp	70–78	17.9–22	0.9	3.7–5				
Cotton	82.7–92	2–5.7	5.7	0.5–1				
Coir	43	0.3	4.0	4.5				
Banana	60–65	6–19	3–5	5–10				
Henequen	60–78	4–28	3–4	8–13				
Baggase	40	30	10	20				
Pineapple	80–81	16–19	2–2.5	12				
Wood	45–50	23		27				

Note: Data for some fibers does not equal 100%.
Sources: Section A: From Turner, A.J. *The structure of Textile Fibers*, A.R. Urauhart and F.O. Howitt, eds., The Textile Institute, manchester, V.K., pp. 91–117, 1953; Section B: Timell, T.E., Text. Res. J., **27**, 854–859, 1957; Section C:* Bhatia, P., and Gupta, K.C. The Indian Text. J., **101(10)**, 60–62, 1991 (report 29.4% compounds soluble in 1% NaOH); ** Chand, N., Sood, S., Rohatgi, P.K., and Satyanarayana, K.G., J. Sc. Ind. Res. (India), **43**, 489–499, 1984; *** u,c....[196]; Section D: Varma, D.S., Varma, M., Varma, I.K. Tex. Res. J., **54**, 12, 827–832, 1984; Section E: Biagiotti, J., Puglia, D., and Kenny José, M. A Review on Natural Fibre-Based composites-Part I: Structure, Processing and Properties of vegetable Fibres, J. Nat. Fibers, **1(2)**, 37–68, 2004; Italicized data from McGovern, J.N. Fibers, vegetable, In Polymers—Fibers and Textiles. A compendium, University of Wisconsin, Madison, WI, 1990.

Lewin [95], in his review, also highlighted several important features of the chemistry of substances in the middle lamella of jute and flax. His observations regarding flax are discussed briefly along with some particular features of the chemical composition of specific fibers in the following text.

8.3.2.1 Abaca

Recently, the presence of hydroxycinnamate esters consisting of ferulic and p-coumaric acids esterified to long-chain fatty alcohols (C_{20} to C_{28}) and ω-hydroxyfatty acids (C_{22} to C_{28}) in abaca fiber has been reported by del Río et al. [72]. del Rio et al. [73] also report that native lignin in kenaf, jute, sisal, and abaca is at least partially acetylated.

8.3.2.2 Flax

From the examination of Turner's data [173] on the chemical composition of retted and unretted flax (see Table 8.3), it could be argued that during retting 2% pectin and a small amount of lignin are removed. Whiting's work [192] suggested that the outer middle lamella of flax contains a polyuronide hemicellulose, composed of arbinose, xylose, galactose, glucose, mannose, rhamnose, and another unknown uronic acid. This hemicellulose is attacked during retting and is not contained in the pectin. The work carried out at the Israel Fiber Institute revealed that lignin and pectin of the inner middle lamella decrease with increasing severity of the cottonization of flax. The idea behind cottonization is to break down the inner middle lamella and to release the individual ultimate cells from the fiber bundle. The ultimate cells seemed to be unaffected by the cottonization treatments. Lignin was found to be both in the inner middle lamella and the ultimate itself. Couchman [46] suggested the presence of two types of lignin in the flax fiber:

1. A fraction (2% based on oven-dried weight) of Klason lignin soluble in 0.7% sodium chlorite solution and containing nearly all (5.3%) the methoxyl groups was associated with the middle lamella just outside the surface of the fiber.
2. The lignin insoluble in sodium chlorite solution (3% based on oven-dried weight) and containing 0.26% methoxyl groups was thought to be included in the fiber wall. The latter fraction was extremely resistant to acids, but disintegrated partly in sodium hydroxide; other evidence suggested that it may be a lignin–carbohydrate complex.

The pectin of the middle lamella of flax appears partly in the form of calcium and magnesium salts, resulting in its low solubility. It is not pure polygalacturonic acid; its hydrolysate contains methoxyl groups and up to 3% arbinose, linked with polygalacturonic acid. Felsar [53] and Luedtke and Felsar [103–105] identified two pectins, A and B, in the outer and inner middle lamella of flax, respectively; the latter being much more stable to chemical and biological attack than the former. Bock and Einsele [28] suggested that the degradation of pectin A in the retting process occurs through a chemical or enzymatic action and its molecular weight decreases from 20,000–30,000 to 12,000–13,000. While a mild hydrolysis of flax straw, with 1% acid, removes pectin A as much as it is during retting, pectin B is destroyed only in the severe cottonization process. Even so, the studies of Rath and Angster [140] on the reactivity of isolated pectin A and B lead to the conclusion that both pectins have the same structure and properties; the higher stability of pectin B in the fiber is attributed to the likely existence of specific linkages, such as "ester bridges" [165], with cellulose of the cell walls.

8.3.2.3 Sisal, Coir, and Hemp

In the case of sisal, Wilson [194] states that 85% of the oven-dried fiber consists of lignocellulose; the nature of this material, however, is not quite clear as sometimes it behaves as though lignin and cellulose are distinctly separate and sometimes as though they are chemically combined. Of course, they can be separated by chemical treatments. The lignin content in sisal is the highest in the middle lamella; in the ultimates, it is maximum in the cells at about 100 mm from the growth point, but falls off exponentially along the fiber toward the tip. The primary wall of the ultimates contains low-molecular-weight polysaccharides (hemicellulose).

Varma et al. [178] report elemental analysis of treated and untreated coir fiber as shown below; the samples were dried for 18 h in vacuum at 60°C.

Treatment	Carbon (%)	Hydrogen (%)	Moisture-regain (%)
Untreated	49.58	6.07	8.0
Desalted	50.52	5.97	7.0
10% NaOH	49.22	6.30	9.0
10% HCl	50.14	6.07	16.0
10% Acetic acid	50.48	6.07	11.0

IR spectra of treated and untreated bristle coir fibers have also been studied [178]. No significant change was noted in the acetic acid–treated fibers; in the alkali-treated fibers, a small absorption peak at 1740 cm^{-1} (perhaps due to carbonyl group) disappeared. With the alkali treatment, the absorption band of 910–1200 cm^{-1} of the untreated fibers changed to a strong absorption peak at 1020 cm^{-1}. The HCl-treated fiber exhibited a light shift in the absorption peak from 1600 cm^{-1} to 1620 cm^{-1}.

Chemical composition of coir fiber reported by Varma et al. [178] is given in Table 8.3. [124] had suggested the lignin content to be about 35%, most of it in the middle lamella and the primary wall; as such, it is said to protect the cellulose from chemical and physical attack [139]. An estimated moisture content of 20% has also been reported [109]. Timell [172], based on extractive-free material, gives the carbohydrate content of hemp and jute products, as shown in Table 8.4.

8.3.2.4 Results of Nitration and Other Studies

Lewin and Epstein [92], through nitration of flax and jute, demonstrate that cellulose in flax is not chemically linked to other carbohydrate components, whereas in jute it is. The linkage appears to be an ester linkage between hydroxyl groups of the cellulose and carboxyl groups of polyuronide fraction. It can be opened gradually from nitration as a whole fiber at 3°C. The opening of this linkage proceeds according to first-order kinetics and the rate constant for jute was found to be 3.6×10^{-4}/min, whereas for *Eucalyptus camaldulenis rostrata* and for *Pinus halepensis* rate constants of 0.276×10^{-4}/min and 1.76×10^{-4}/min, respectively, were

TABLE 8.4
Carbohydrate Composition of Various Hemp and Jute Products

	Hemp			Jute		
Components	Native material	α-Cellulose	Nitrate	Native material	α-Cellulose	Nitrate
Galactan	1.5	Nil	Nil	Trace	Nil	Nil
Glucan	83.3	76.5	80.3	65.6	58.2	60.6
Mannan	5.7	0.9	1.7	2.7	Nil	0.4
Araban	1.0	0.4	Nil	0.4	Nil	Nil
Xylan	1.3	0.5	Nil	9.5	1.2	Nil

Source: From Timell T.E. *Text. Res. J.*, **27**, 854–859, 1957.

obtained. Owing to the existence of this linkage, analytical results on the percentages of α-cellulose and hemicellulose in plant materials and the DP of the cellulose obtained by direct nitration method may be erroneous [91,92,96]. Although it is expected that a similar linkage might exist in other vegetable fibers, no information on its existence and its rate of decomposition appear to have been published.

The existence of a lignin–uronic acid ester linkage in jute was proposed earlier by Sarkar et al. [147,148]. They showed that the acid (carboxyl) value of jute increases progressively on delignification as well as on treatment with 1% NaOH. The maximum value of carboxyl content obtained corresponded closely to the value of uronic acid–carbon dioxide content obtained on distillation of the jute with 12% hydrochloric acid. The rate of delignification of jute with sodium chlorite increased after treating the fiber with cold dilute alkali. A first-order rate constant was found for saponification of the linkage with excess alkali [24]. The opening of this linkage, which is highly unstable in alkalis, increases the accessibility of the fiber to chemicals and microorganisms.

Ester linkages between cellulose and lignin, which are stable to alkali, were also considered by Mukherjee and Woods [123] as an explanation for the complete mercerization of jute and for the difficulty of dissolving wood cellulose in cuprammonium hydroxide [24].

Bhatia and Gupta [23] give the reaction to coloring agents for four different fibers, which can be useful in fiber identification. These are listed below:

Reagent	Banana	Agave	Jute	Sunn
Zinc chloride plus iodine reagent	Golden yellow	Dark brown	Dark brown	Dark brown
Para-nitroaniline reagent	Bright orange	Bright red	Red	Bright red
Potassium permanganate	Pink	Bright red	Bright red	Bright red
Malachite green	Green	Green	Green	Green
Iodine and sulfuric acid reagent	Yellow	Light brown	Dark brown	Dark brown

8.4 PHYSICAL PROPERTIES

8.4.1 Structure of Fibers and Ultimates

The range of physical dimensions of ultimates (ultimate cells) and fibers (including their fineness) listed in Table 8.2 has been discussed as part of the morphology of fiber bundles, albeit, in passing. To repeat, the principal constituent of ultimates is native or α-cellulose (also called cellulose I); it may be intermingled with hemicellulose (polysaccharides) and even lignin to some degree. The cells have lumens that may contain nitrogenous residue of protoplasm or organic and inorganic salts; typically, in the case of sisal fiber the lumen is 18–25 mm and contains calcium oxalate crystal. In the walls of the cells, the spirally wound fibrils (or microfibrils) contain cellulose chains organized in the form of ordered (crystallites) and disordered regions. The fibrils vary in length from 50 Å to several microns (μm) [108]. In the ultimates, the spirally wound fibrils exist in a series of layers; the direction of the spirals changing from z-spiral to s-spiral to z-spiral, etc., in successive layer. The experimental evidence of Frey-Wyssling (see, [108]) on native cellulose suggests that fibrils in the fiber form a three-dimensional open network with empty spaces (cavities) ranging from 10Å to greater than 100Å in diameter. The organization of the ultimates and other aspects of the physical structure of specific fibers are, henceforth, discussed individually.

Microscopic examination of the cross section of the *abaca* fiber reveals bundles of polygonal-shaped ultimates (with or without rounded corners) having sufficiently large oval to circular lumen and, sometimes, relatively thin walls, as shown in Figure 8.9. Scanning electron

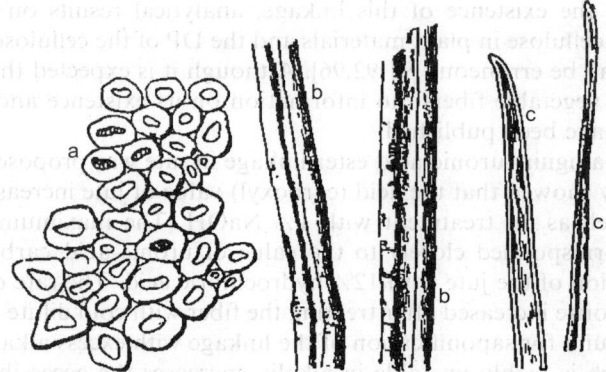

FIGURE 8.9 Abaca fibers: (a), cross section; (b), longitudinal views; (c), ends. (From Mauersberger H.R., ed., *Mathews Textile Fibers,* 5th ed., John Wiley and Sons, New York, 1947.)

micrographs (SEM) obtained by Tayag et al. [178] reveal almost the same features, albeit more elegantly. Longitudinally, the ultimates are smooth and lustrous, with long tapering ends [58] as shown in Figure 8.9. The abaca fiber bundles are characteristically surrounded by a layer of stegma cells, filled with silica (which is left as a residue on burning). The spiral angle for abaca has been estimated to be 22.5° [35] from x-ray diffraction patterns of the fibers in the natural state. The x-ray diffraction spots of Musa fibers (abaca, banana) are well dispersed, indicating predominance of helical fibrils.

The *banana* fiber is relatively shorter in length, but finer. Its fineness can be closer to kenaf and jute. It consists of four types of cells (xylem, phloem, schlerenchyma, and parenchyma) arranged in a particular fashion; the cell shapes are circular to polygonal, with rounded corners. The cell walls are thin and uniform. X-ray analysis reveals that cellulose crystallites are arranged in helices with a helix angle of 11–12° [81]. Chand et al. [39] report a spiral angle of 30°. Banana fibers like jute and ramie have been found to be highly crystalline [179]; delignification does not change the structure of its cellulose.

Examination of the *sisal* fiber by microscopy reveals either round or crescent-shaped cross sections [171], consisting of sharply polygonal ultimates containing large, rounded lumens as shown in Figure 8.10. The longitudinal view of the ultimates shows no cross markings. Accessory cells containing crystals of calcium oxalate are often present in the fiber and aid

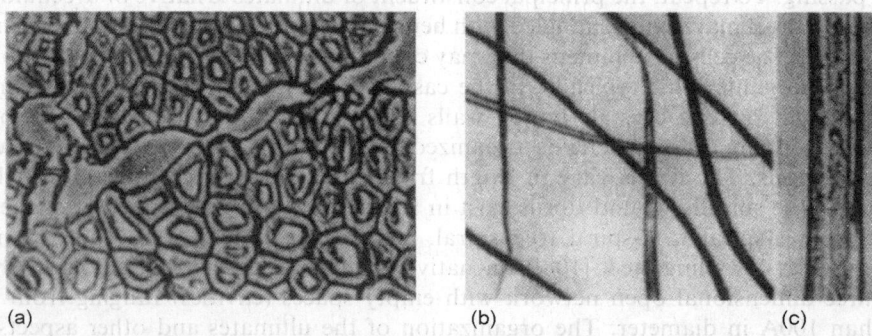

FIGURE 8.10 Sisal fiber: (a) cross section (275×, approximately); (b, c) longitudinal views (275× and 57×, approximately). (From Harris M., *Harris's Handbook of Textile Fibers,* Harris Res. Lab., Inc., Washington, D.C., 1954. (Now Gillette Research Laboratory, Bethesda, MD.))

in identification. If the spiked leaf of sisal is cut before it unfurls, the extracted fiber is finer and tapered at both ends; this allows for uniform packing in a yarn. The fiber extracted from the furled leaves is longer, coarser, and untapered at the butt end; however, it is stronger and therefore preferred for cordage applications. Treatment with 20% NaOH swells the fiber, makes the lumens circular, and reveals as many as 10 layers (1 week per layer) in the lamellar structure [194]. The fiber cross sections turn yellow when treated with iodine–sulfuric acid reagent and show evidence of the middle lamella [108].

The x-ray diffraction patterns of the sisal fiber [58] indicate a high degree of orientation, together with a reasonably good crystalline order. The presence of lignin makes the patterns somewhat diffuse; delignification of the fibers sharpens the pattern to that of true cellulose. Although lignin is generally assumed to be amorphous, evidence has been cited which indicates that it can exist in crystalline form [194]. The crystallites (or micelles) of cellulose are oriented along spirals that are, in turn, oriented either at 10° [194] or 20–25° [39] to the axis of the ultimates. Dhyani et al. [47], using x-ray analysis, found the fiber to be moderately crystalline. Balashov et al. [18] demonstrated the fibrillar organization of the secondary wall. The spiral angle of the fibrils in swollen fiber was measured to be 36° (average). The angle of orientation of the crystallites in the ultimate cells in which the secondary wall deposits had not yet formed is reported to be 65°. By inference, that is the spiral angle of the fibrils in the primary wall. The degree of orientation of the cellulose molecules in the primary wall is stated to be relatively poor.

The morphological and microscopical features of *henequen* are similar to those of sisal. While the fine structure of henequen does not appear to have been studied explicitly, it is believed to be identical to that of sisal.

The cross sections of *flax* fibers show ultimates strongly polygonal in shape, with thick secondary walls enclosing a small round or slitlike lumen, see Figure 8.11. The longitudinal views of the ultimates reveal smooth surfaces with tapered ends. In transmitted light, in addition to the lumen along the length of the fiber, faint cross marks called "nodes or dislocations" often in the form of an "X" are observed. The nodes become very observable in polarized light when the fiber is stained with methyl violet or chlor-iodide of zinc solution; the lumen in the center appears as a narrow yellow line and is completely filled with protoplasm. The number of nodes is high and can reach up to 800 in a fiber strand. On extension of the fiber, the nodes disappear; on relaxation, they reappear. Peters [133] suggested several explanations for the cause of nodes: they may be due to minute fissures or local separation of the fibrils that make up the cell walls, or could be points at which the direction

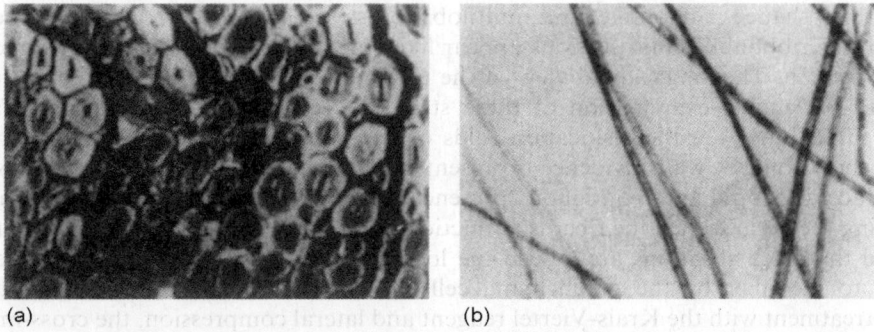

FIGURE 8.11 Flax fiber: (a) cross section (288×, approximately); (b) longitudinal view (57×, approximately). (From Harris M., *Harris's Handbook of Textile Fibers,* Harris Res. Lab., Inc., Washington, D.C., 1954. (Now Gillette Research Laboratory, Bethesda, MD.))

of the fibrils is modified. Both these postulates are supported by the more recent [10] SEM examination of the flax fiber. At 1100×, the fibers show longitudinal fissures, as well as localized transverse, large compression or dislocation bands, giving rise to thicker diameter knots, interspersed with finer compression or shear surface kinks inclined at a slight angle to the transverse direction. Disappearance of the nodes under tension would imply the straightening out of the fibrillar dislocations at these bands or kinks. The reappearance of nodes on release of tension would imply the existence of a certain number of fibrils in the internal structure that are straight in the unstretched fiber and provide the necessary strain energy to restore the fiber to its unstretched configuration.

Sloan [161] reports the belief that nodes represent low-density accessible regions of the flax strand and are responsible for its ease of bending, relatively poor abrasion resistance, and ease of creasing. The very rapid absorption and desorption of water, responsible for the unique cool handle of flax, is also attributed to the nodes, as is the very high reactivity to cross-linking agents. The cross-linking takes place, preferentially, in the nodal regions, bringing about a high concentration of cross-links and consequently an embrittlement of these regions. With the exception of absorption and desorption of water and the "cool hand" attributes, most of these beliefs are plausible with the structure of the nodes postulated above, to a greater or lesser degree.

Boiled-off, but unbleached, linen, or bleached, but unboiled, flax show the following characteristics: when treated with iodine and sulfuric acid, the cross sections show the external yellow layer of lignin, whereas the rest of the pure cellulose turns pure blue [108]. Boiled linen fibers, upon treatment with Schweitzer's reagent, become swollen, but do not dissolve away completely; the fiber blisters irregularly and the shriveled up, insoluble cuticle of the lumen canal is revealed floating in the reagent.

The x-ray diagram of purified flax (linen) is much more distinct; that is, better resolution of [101] and [10$\bar{1}$] spots [124], than that of unpurified flax [58]. The outermost layer of flax has a *Z-spiral*, followed by an *S-layer* in the thick middle wall, and another *Z-layer* toward the center [113]. The spiral angle is estimated to be 6.5° [35]. Unlike cotton, there are no reversals in the directions of spirals along the fiber length.

On treatment with NaOH solution of mercerizing strength, the cellulose I pattern of purified flax completely transforms to the cellulose II pattern. In other natural cellulosic fibers (except ramie), this transformation is only partial [124]. The degree of crystallinity of flax is estimated to be 70% [113].

Microscopic examination reveals that *ramie* fiber is a bundle of ultimate cells. The cells are rounded polygons with a thin, well-defined lumen, both of irregular shapes. The cells also have a tendency to develop radial cracks. More recent SEMs [13] of the fibers (ultimates) reveal kidney-shaped, ribbonlike, and multilobal cross sections with or without lumens; the lumens in the ribbonlike cross sections appear like cracks oriented in the longer direction of the cross section. The longitudinal view of the ultimates shows cross striations, as shown in Figure 8.12. Detailed examination of these striations reveals that they are not cracks or fissures; rather, they are the dislocation folds or constrictions in the straight continuity of fibrils in the secondary wall. Evidence is presented [88] to suggest that cross-markings on the fiber surface develop due to contiguous presence of cross-walls of the adjacent parenchyma cells during the period when the fiber is a functioning member of the plant stem. As such, it is suggested that such striations are indeed the location of paths through which nutrients are supplied to the fiber by the parenchyma cells. It is further demonstrated that, through swelling treatment with the Krais-Viertel reagent and lateral compression, the cross-markings can be made to disappear. Under lateral compression, the fiber is shown to be composed of a series of well-defined, parallel bundle of fibrils, nearly parallel to the fiber axis.

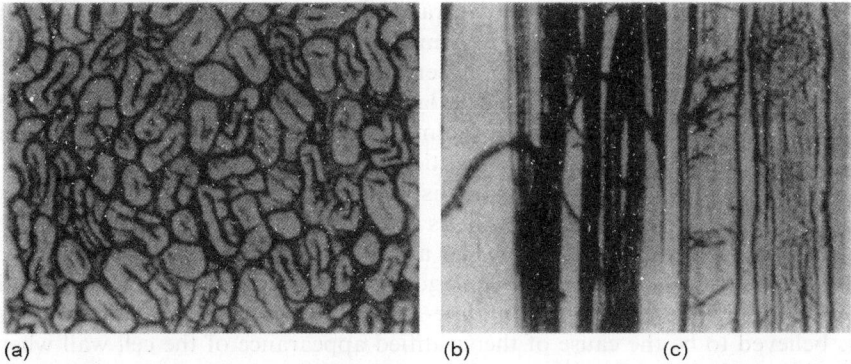

FIGURE 8.12 Ramie fiber: (a) cross section (240×, approximately); (b, c) longitudinal views (240×, 48×, approximately). (From Harris M., *Harris's Handbook of Textile Fibers*, Harris Res. Lab., Inc., Washington, D.C., 1954. (Now Gillette Research Laboratory, Bethesda, MD.))

Ramie fibers consist of a very high proportion of pure cellulose. Therefore, treatment with iodine and sulfuric acid turns the fiber pure blue, but treating it with aniline sulfate gives no color. Treatment with zinc chlor-iodide reagent gives the fiber a blue color, while the fiber color is rose red when treated with calcium chlor-iodide [108].

X-ray diffraction photographs of ramie show the sharp characteristic patterns of cellulose I. The degree of crystallinity of ramie (native cellulose) by x-ray methods is generally estimated to be about 70% [19,113]. In other studies, the estimates range from 74% in the dry state to 54% in the moist state [124]. The treatment with NaOH solution of mercerizing strength completely transforms the cellulose I pattern to cellulose II [169] and reduces the degree of crystallinity to 50% [113]. On the other hand, treatment with dilute NaOH (about 5% solution) increases the crystallinity [185]; the same is true for acid hydrolysis [19].

The fine structure of decorticated and degummed ramie fibers has been studied by Ray et al. [141]. They conclude that while crystallinity increases, the crystallite orientation decreases with increasing (gradual) removal of gum; this may be attributable to the swelling action of the reagents. Optical orientation, in contrast, falls at a more rapid rate. A two-stage degumming method, which can reduce the gum content to 2–3% in *ramie* fiber, improves its surface properties; however, degummed and bleached ramie fiber loses bundle strength [1].

Wenbang [188] reports on the crystalline structure and optical properties of three different types of ramie fibers and different parts of the fibers. The author finds some variation in crystallinity in different parts of the fiber. He also finds tenacity, breaking elongation, and work of rupture of the fibers somewhat correlated to crystallinity and orientation factors of the fibers.

Kulshreshtha et al. [83], as well as Mitra and Mukherjee [118], postulate the existence of paracrystallinity in ramie, jute, and hemp. The latter claim to have developed refinements to the techniques of measurement of parameters of the three-phase model for cellulose I [117]. Subsequently, by measuring the structural parameters relative to [002], [101], and [101] reflections, they show that the degrees of crystallinity and paracrystallinity determined for each of the three reflections are different (anisotropy) for each of the three fibers (ramie, jute, and hemp) [122]. The results also confirm that the greater the paracrystalline distortion, the smaller the paracrystallite size.

The orientation by birefringence, f_0 (Hermans' factor) [61], representing an average value for both crystalline and less-ordered regions and the x-ray-derived crystallite orientation, f_x, were 0.94 and 0.96, respectively, for native ramie fibers. For hydrolyzed ramie, the values increased to 0.97 and 0.98 [19]. Cotton, in comparison, yielded corresponding values of 0.90 and 0.95 for f_0 and f_x [60]. From similar observations, the spiral angle was estimated to be 7° [35]. It has been shown that the outside layer of ramie has a *Z-spiral*, followed by an *S-spiral* layer, and a core of indeterminate, almost parallel orientation [113]. The unit cell in the crystallites of ramie is believed to be identical to that of cotton.

Under the microscope, the *hemp* fiber is generally uneven in diameter and exhibits frequent joints and longitudinal as well as cross striations and fissures, as shown in Figure 8.13. The fiber ends are thick-walled and blunt; in contrast to flax, they may even seem forked. The cross-sections show polygonal cells with rounded edges, as shown in Figure 8.13. Vasculose present in the middle lamella is also present in the cellulose in the cell and is believed to be the cause of the stratified appearance of the cell wall when treated with an iodine–sulfuric acid reagent [108]. The microchemical reactions render the fiber (longitudinal views) bluish green (with iodine-sulfuric reagent), rose red with traces of yellow (with calcium chloride), blue or violet with traces of yellow (with zinc chloride), yellowish-green (with aniline sulfate), and pale red (with ammoniacal fuchsine solution) when treated with the indicated reagents. With Schweitzer's reagent, the fiber dissolves almost completely [108]. When treated with ammoniacal copper oxide solution, the cell membrane is bluish green and swells like a blister; the inner cell walls remain intact in the form of spirally wound tubes contained within a swollen mass of fiber.

X-ray analysis reveals that the hemp fiber structure is very similar to the other fibers (ramie, flax), which have low lignin content. The spiral structure of hemp is comparable to that of ramie [58] and the spiral angle has been estimated to be to be 7.5° [35].

The cross-sectional shapes of the *sunn* fiber ultimates are highly irregular and oblong, separated by thick layers of lignin in the middle lamella, when observed as part of the aggregate bundles. The lumen is thick, long, and irregular. The secondary walls appear to show a secondary structure of different optical density. The longitudinal view of the separated ultimates show cells of varying diameter and an occasional thick spot along the cell. There are partial or complete transverse markings at irregular intervals [171]. X-ray studies revealed that crystallites are well oriented; the crystals are about 600 Å long and 50 Å wide with a spiral angle of 9.8° [164]. The treatment with iodine and strong sulfuric acid swells the fiber; the outer layer becomes a yellow mass, over which a blue semiliquid mass of cellulose flows, leaving a greenish-yellow inner tube as residue.

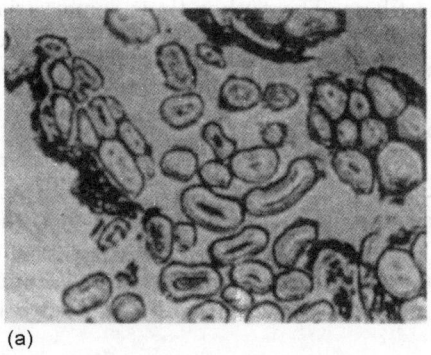 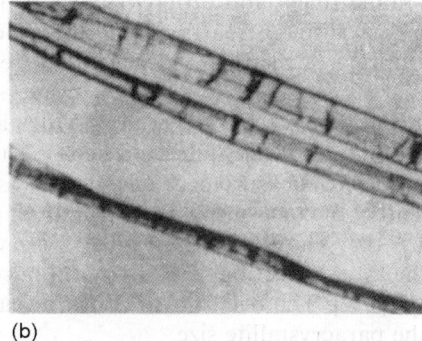

(a) (b)

FIGURE 8.13 Hemp fiber: (a) cross section (200×, approximately); (b) longitudinal view (100×, approximately). (From TI: *Identification of Textile Materials,* 7th ed. The Textile Institute, Manchester, U.K., 1975.)

The *coir* fiber is composed of spindle shaped (tapered ends) ultimate cells, which are contained in a porous sheath. In cross sections, the outline of the fiber (bundle of 30–300 cells) may be elliptical, oval, or nearly circular, sometimes with a small hollowness in the center (see Figure 8.14). The cell lumen may have a significant diameter; its shape may vary from irregular polygonal to nearly round. The EDAX examination reveals that some of the pores in the sheath are plugged with silica-rich material [64]. The longitudinal view of the ultimate cells shows the existence of lumen and oblique striations (at an angle to the cell axis), indicating spiral–fibrillar fine structure. Sometimes, pores may be observed on the lumen surface [108].

The fine structure of coir [70] ultimate cells consists of a primary layer, followed by S^1, S^2, and S^3 (going from outside in) secondary layers. The outer surface of the primary layer is essentially featureless except for small cellular pit openings. The secondary S^1 and S^2 layers are fibrillar in structure. The fibrils in the S^1 layers are organized as helices, oriented at 50–55° to the cell axis. The mean thickness of the S^1 layer (Z-helices) has been measured to be 1.1 ± 0.3 μm and that of the secondary layer (S-helices) S^2 as 1.24 ± 0.4 μm. The inner secondary wall S^3 does not reveal the fibrillar structure. The pit openings observed on this wall are in the form of slits, oriented at 48° to the cell axis.

The x-ray diagram [58] of coir confirms its two spiral-structured S^1 and S^2 layers. The degree of crystallinity of the coir does not appear to be very high. Harris [58], Chakravarty, and Hearle [35] report the spiral angle to be 45°. Varma et al. [178] report a more extensive X-ray analysis of treated and untreated fibers. The present crystallinity and Hermans' orientation factor, f_0, obtained are tabulated below:

Treatment	f_0	% Crystallinity index
Untreated	0.909	25.72
Desalted	0.888	22.53
10% NaOH treatment	0.906	28.96
10% HCl treatment	0.880	23.37
10% Acetic acid treatment	0.881	22.82

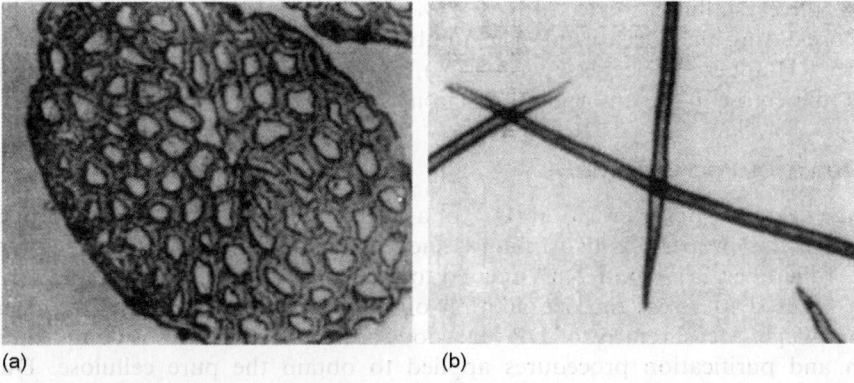

(a) (b)

FIGURE 8.14 Coir fiber: (a) cross section view of a bundle of ultimates (165×, approximately); (b) longitudinal view of ultimate (100×, approximately). (From TI: *Identification of Textile Materials*, 7th ed. The Textile Institute, Manchester, U.K., 1975.)

The Hermans' orientation factor of cellulose I crystallites appears to change with acid treatments. The increase in the crystallinity of the coir with alkali treatment is attributed to the removal of cementing material. The surface morphology of untreated fibers showed globular protrusions identified as silicated stegmata. Desalting the fibers significantly reduces the size of the protrusions. Alkali treatment eliminates the protrusions giving rise to visible voids, but acid treatment only reduces their size slightly.

Coir, when treated with iodine and sulfuric acid, turns golden yellow; and, with aniline sulfate, intense yellow. Schweitzer's reagent does not attack the fiber; the latter reaction indicates that it is a lignified fiber [108].

Tayag et al. [188] reported a comparative study on five fibers (abaca, banana, kenaf, maguey, ramie), three of which are of interest here. The study looked at the influence of boiling of fibers at 95°C for 1 to 2 h in 17.5% (by weight) solution of NaOH. The sorption curves (% regain vs. % rh at 23°C, starting with vacuum-dried fiber) showed a marked difference between treated and untreated fibers. The regain of treated fibers was lower in each case; the decrease was maximum in the case of ramie followed by maguey, kenaf, abaca, and banana in descending order. The BET analysis [30] of the sorption isotherm data showed "that the maximum volume of the mono-layered absorbed water per unit mass of dry cellulose also decreased" with treatment.

The x-ray diffractograms showed the presence of cellulose I structure in all raw fibers. In the treated banana fiber, cellulose II structure was dominant. While cellulose I ([002] peak) is found in all five raw fibers, the [002] peak cellulose II is found as a shoulder in kenaf and ramie. The measured crystallinity ($x\%$) in raw and alkali-treated fibers are shown in the table below:

Samples	$x\%$ — Raw	$x\%$ — Treated
Abaca	52	62
Banana	55	62
Kenaf	47	62
Maguey	55	60
Ramie	61	66

Clearly, the crystallinity increased by 5–15%; it is believed to be due to loss of lignin and hemicellulose during alkali treatment. The number of water molecules sorbed on each glucose unit at 66% RH appear to decrease with x (%), suggesting an increase in the accessibility of water due to increase in the amorphous fraction of cellulose.

8.4.2 Degree of Polymerization

The average degree of polymerization (DP) of a fiber serves as a characteristic property. The importance of this property is two fold: (a) the mechanical properties of the ultimate, and therefore of the fiber, are significantly related to the DP of a fiber [33,78,93]; and (b) the DP value can be used to assess the degradation of cellulose caused by physical, chemical, or radiation damage. Measurement of DP for wood, bast, and leaf fibers is complicated by the extraction and purification procedures applied to obtain the pure cellulose. During the process, the cellulose is usually degraded to varying degrees, depending on the reaction conditions. Furthermore, the cellulose is often chemically linked to lignin or hemicellulose as mentioned previously. It will, therefore, not dissolve in solvents used for viscometric DP

TABLE 8.5
Degree of Polymerization of Cellulose in Vegetable Fibers

Fiber	Source		
	Harris [58]	Timell [145]	Chand et al. [39]
Abaca	1990		
Sisal	2160		
Flax (three varieties)	2190, 2390, 2420	4700	
Ramie	2660[a]	5800	
Hemp (three varieties)	2200, 2200, 2300	4800	
Cotton lint	2020	4700	
Jute	1920	4700	
Nettle	2280		
Kapok		3300	
Milkweed floss		5700	
Sunn			1289–1614

[a]Cabradilla and Zeronian [33] give a value of 2150; Basch and Lewin [19] give a value of 3300.

determinations [91,92,96]. The values derived, therefore, can only by considered as best estimates. Such values from several sources in the literature are given in Table 8.5. Cotton, jute, and some other fibers have been included for the sake of comparison. Of the fibers listed, ramie appears to have the highest DP, while jute has the lowest.

8.4.3 Specific Gravity and Porosity

Several sources report the specific gravity of some vegetable fibers. They are listed in Table 8.6. The range of values given suggests that attainment of precision in this measurement is difficult. Part of the problem may be in the porous nature of the fibers. Sinha [159,160] gives the explicit value for porosity, as well as apparent and true density of some vegetable fibers; these are summarized in Table 8.7.

8.4.4 Optical Properties

Birefringence, the difference between refractive index parallel ($n_{\parallel}$) to the fiber axis and that of normal ($n_{\perp}$) to the fiber axis is a measure of the orientation of molecular chains in a polymeric fiber. The values of refractive indices ($n_{\parallel}$ and $n_{\perp}$) and birefringence of several vegetable fibers have been measured. These are summarized in Table 8.8.

8.4.5 Moisture Absorption, Desorption, and Swelling

Moisture sorption is a physicochemical phenomenon [120]. The hydroxyl groups of the carbohydrate fraction attract water molecules, and attachment occurs due to hydrogen bonding. In a bone-dry fiber, water molecules can diffuse in, initially, to form such bonds. Late-arriving molecules can attach themselves to either existing water molecules (indirect attachment) or to the other hydroxyl groups (direct attachment). Diffusion of water

TABLE 8.6
Average Specific Gravity of Fibers

Fibers	Harris [58]	T.I.[a] [171]	Chakravarty [38]	Mackie [105]
Hemp	1.48	1.48–1.50		1.45
Jute	1.48	1.48–1.50		1.40
Flax	1.50	1.50	1.53–1.55	1.43–1.5
Cotton	1.50	1.50–1.55	1.54–1.55	
Ramie	1.51	1.51–1.55	1.54–1.56	1.50

[a] T.I., The Textile Institute.

molecules does not take place in the crystalline regions of native cellulose in the fiber because of the relatively tight packing of molecules. Thus, absorption of water by the fiber is proportional to the extent of their noncrystalline, less-oriented regions; this observation is affirmed by the work of Tayag et al. [178]. Since hemicelluloses and polyuronides are not crystalline, their presence in the fiber increases its regain. On the other hand, the presence of lignin decreases moisture absorption, since lignin is hydrophobic. In addition, layers of lignin in the inner middle lamella and close to the fiber surface hinder the penetration of moisture into the cellulosic cell wall. Thus, the moisture sorption phenomenon in seed, bast, and leaf fibers is more complicated than in the case of cotton; it resembles, to some extent, the behavior of wood. It is the middle lamella, therefore, which is responsible for the 12% moisture-regain in the case of flax and hemp and 13.8% in jute, as compared to 8.5% in cotton, at 22°C and 65% RH [157]. Assuming that 30% of the α-cellulose and all the hemicellulose are not crystalline, Sen and Hermans [157] found by calculation that the ratio of noncrystalline constituents of jute and cotton is 1.56 as compared to the moisture sorption ratio of 1.6. Sarkar et al. [146,149] measured and compared the moisture-regain values of isolate α-cellulose, hemicellulose, and lignin to that of raw jute. The regain value of

TABLE 8.7
Porosity, Apparent, and True Density of Some Vegetable Fibers

Fiber	Porosity (%)	Apparent density (g/cc)	True density (g/cc)
Banana	35–53	0.86–0.62	1.33–1.31
Mesta	18	1.21	1.47
Aloe	21	1.17	1.47
Abaca	17–21	1.20, 1.10	1.45, 1.40
Sisal	17	1.20	1.45
Roselle	15	1.24	1.46
Jute	14–15	1.23	1.44
Sunn	12	1.35	1.53
Ramie	7.5	1.44	1.56
Flax	10.7	1.38	1.54

Source: From Sinha M.K., J. Text. Inst., 65(1), 27–33, 1974; Sinha M.K., J. Text. Inst., 65(11), 612–615, 1974; Chakravarty A.C., Text. Res. J., 41(4), 318–321, 1971.

TABLE 8.8
Refractive Indices of Some Vegetable Fibers

Fiber	$n_\parallel$	$n_\perp$	Δ_n
Ramie (native)	1.596	1.528	0.068
Flax (native)	1.596	1.528	0.068
Flax (mercerized under tension)	1.571	1.517	0.054
Flax (mercerized without tension)	1.556	1.518	0.038
Cotton (native)	1.578	1.532	0.046
Cotton (mercerized under tension)	1.586	1.522	0.044
Cotton (mercerized without tension)	1.554	1.524	0.030

Source: From Harris M., *Harris's Handbook of Textile Fibers*, Harris Res. Lab., Inc., Washington, D.C., 1954. (Now Gillette Research Laboratory, Bethesda, MD.)

hemicellulose was found to be much higher than that of raw jute, and more than twice as high as that of α-cellulose. According to Lewin [95]:

"The small affinity of the lignin for moisture was, according to these investigators [146,149], in accordance with the small changes in swelling behavior of the jute fiber after a considerable part of the lignin was removed. This is also the obvious explanation of the high moisture absorption in flax in spite of its well known high degree of crystallinity. If it is assumed that the regain of nonlignin middle lamella components in jute and flax is similar, it follows that, in the light of the much lower lignin content of flax, a higher regain should be expected for flax than jute. The lower crystallinity of α-cellulose in jute explains this discrepancy. The imperfect crystallinity of jute, arising from the hindrance due to close packing of the crystallite in the secondary wall due to the extension of the lignin membrane between them, will, therefore, be an additional factor causing the relatively high moisture absorption of jute."

The information on the contribution of the middle lamella, and its constituents, toward moisture absorption of bast and leaf fibers, other than flax and jute, is very scarce and limited.

Conventionally, the moisture sorption characteristics of fibers are depicted by absorption–desorption isotherms. In practice, it is difficult to compare isotherms of different fibers over the whole range of RH and temperature values. It is often sufficient to compare their regains at the standard conditions of 65% RH and 21°C (70°F). Table 8.9 gives absorption regains for a few fibers, together with the difference between absorption and desorption regains.

Textile fibers that absorb moisture also change their axial and transverse dimensions. This phenomenon has important practical consequences: for example, the swelling of fibers could effectively close the pores in a tightly woven fabric and alter its mechanical characteristics.

The swelling behavior can be evaluated in terms of fractional (or %) change in length S_l, diameter S_D, cross-sectional area S_A, or volume S_V. The anisotropic orientation of the molecules makes the swelling of the fibers anisotropic in character.

Data on swelling in water of some fibers of interest are available and shown in Table 8.10. Through a very careful, intensive study, Chakravarty [37] demonstrates that percent volumetric swelling (S_V) of vegetable fibers at different relative humidities is strongly related to their corresponding moisture-regain (γ): indeed, the relationship is given by the equation:

$$S_V = 0.88\gamma + 0.011\gamma^2$$

TABLE 8.9
Regains at Standard Conditions

Fiber	Absorption Regain (%) at 65% RH 70°F	Difference between Desorption and Absorption Regains 65% RH, 70°F
Abaca	9.5	—
Sisal	11.0	—
Cotton	7–8	0.9
Cotton (mercerized)	8–12	1.5
Cotton (scoured)	6	—
Hemp	8	—
Hemp (bleached)	8	—
Jute	12	1.5
Jute (bleached)	10	—
Kapok	10	—
Ramie (bleached)	6	—
Wool (scoured)	14	20
Coir	10	—
Banana	15.2	—

Source: From Sinha M.K., *J. Text. Inst.*, **65(1)**, 27–33, 1974; Sinha M.K., *J. Text. Inst.*, **65(11)**, 612–615, 1974; Harris M., *Harris's Handbook of Textile Fibers*, Harris Res. Lab., Inc., Washington, D.C., 1954. (Now Gillette Research Laboratory, Bethesda, MD.); Morton W.E. and Hearle J.W.S., *Physical Properties of Textile Fibers*, The Textile Institute, Manchester, U.K., 1975; Prabhu G.N., *Coir*, **1(1)**, 17, 1956.

The heat evolved in the process is related to the absorption of moisture in a fiber. The differential heat of sorption Q is the heat evolved when 1 g of water vapor is absorbed by an infinite mass of the material for a specific value of moisture-regain. The heat of sorption Q_l from liquid water is also called "heat of swelling." Alternatively, heat of wetting W (or the integral heat of sorption) is the heat evolved when the specimen of a material at a given regain, whose mass is 1 g, is completely wetted. The measurement of Q_l and W available, for vegetable fibers other than cotton, are very limited. The heat of wetting of flax at different values of regain is given by Harris [58] as follows:

Regain (%)	0.0	0.51	0.73	2.5	4.71	4.93	7.38	7.47
W (cal/g)	13.0	11.6	11.2	7.34	4.91	4.59	2.45	2.55

8.4.6 THERMAL PROPERTIES

The heat capacity of a few fibers of interest has been measured. The values listed in Table 8.11 show that the specific heat does not vary much from fiber to fiber.

Other thermal properties of fibers include the decomposition temperature and the loss of strength due to prolonged exposure to heat. Data for cotton, ramie, and linen due to Illingsworth [66] are given in Table 8.12. The decomposition of cotton occurs at 150°C, and ignition at 390°C.

Other Long Vegetable Fibers

TABLE 8.10
Swelling of Fibers in Water

Fiber	Transverse swelling (%)		Axial swelling (%)	Volume swelling (%)
	Diameter	Area		
Cotton	20, 23, 7	40, 47, 21		
	15		1.2	
Cotton (mercerized)	17	46, 24	0.1	
Sisal				39.5
Abaca				42.2
Jute	20, 21	40	0.37	44.3
Sunn				45.4
Flax		47	0.1, 0.2	29.5
Flax (raw)		65		
Ramie				32.0

Source: From Harris M., *Harris's Handbook of Textile Fibers*, Harris Res. Lab., Inc., Washington, D.C., 1954. (Now Gillette Research Laboratory, Bethesda, MD.); Meredith R., *Mechanical Properties of Textile Fibers*, Inter-science Publishers, New York, 1956; Morton W.E. and Hearle J.W.S., *Physical Properties of Textile Fibers*, The Textile Institute, Manchester, U.K., 1975; Chakravarty A.C., *Text. Res. J.*, **41(4)**, 318–321, 1971.

8.4.7 Electrical Properties

The electrical resistance of fiber mass is defined as mass-specific resistance (R_S), the resistance in ohms between the ends of a specimen 1 m long having a mass of 1 kg; its units are ohm-kg.m^2. The resistance R of an arbitrary specimen is derived from

$$R = \frac{R_s l}{NT} \times 10^5$$

TABLE 8.11
Specific Heat of Fibers

Fiber	Specific heat (cal/g/°C)
Cotton	0.292
Cotton (mercerized)	0.295
Sisal	0.317
Abaca	0.322
Hemp	0.323
Jute	0.324
Flax (linen)	0.322
Kapok	0.324

Source: From Harris M., *Harris's Handbook of Textile Fibers*, Harris Res. Lab., Inc., Washington, D.C., 1954. (Now Gillette Research Laboratory, Bethesda, MD.)

TABLE 8.12
Loss of Strength on Prolonged Exposure to High Temperature

	Percentage strength retained			
	After 20 days		After 80 days	
Fiber	At 100°C	At 130°C	At 100°C	At 130°C
Cotton	92	38	68	10
Flax (linen)	70	24	41	12
Ramie	62	12	26	6

Source: From Illingsworth J.W., *J. Text. Inst.*, **44**, 328, 1953.

where l is distance between ends of the specimen, N is the number of ends of yarn or fiber, and T is count of yarn or fiber in tex units. The electrical resistance of textile fiber mass is strongly influenced by the ambient RH. The three quantities are found to be related by

$$R_s \cdot M^n = K$$

(where M is moisture content in percentage; n, and K are constants depending on fiber type) for most hygroscopic fibers for RH between 30 and 90%. However, experimental evidence suggests that the relationship between R_S and (RH) H can be expressed fairly well as

$$\log R_s = -aH + b$$

where a and b are constants [120].

TABLE 8.13
Values of n and Log K at Medium Humidities (20°C)

Material	n	Log K	(Log K − n) = Log R$_s$ at M = 10%	Log R$_s$ at 65% RH
Cotton (1) and (2)	11.4	16.6	5.25	6.8
Purified cotton (1)	10.7	16.7	5.97	7.2
Ramie	12.3	18.6	6.26	
Purified ramie	11.7	18.5	6.76	
Ramie	9.7	14.4	4.78	7.5
Ramie 10% KCL	8.8	12.0	3.20	
Mercerized cotton	10.5	17.3	6.82	7.2
Hemp	10.8	17.8	7.02	7.1
Purified jute	10.5	18.9	8.40	
Flax	10.6	16.4	5.78	6.9
Purified flax	11.2	18.0	6.82	

Source: From Morton W.E. and Hearle J.W.S., *Physical Properties of Textile Fibers*, The Textile Institute, Manchester, U.K., 1975; Hearle J.W.S., *J. Text. Inst.*, **44**, T117, 1953.

Table 8.13 gives the measured values of n and log K for cottons, ramie, jute, and hemp [59].

This reviewer finds power–law relationships between dimensional quantities very troublesome, as they can lead to erroneous interpretations. This difficulty can be avoided if the quantity raised to some power can be defined as a nondimensional ratio.

8.5 MECHANICAL PROPERTIES

In principle, mechanical properties of fibers should be treated as part of the overall physical properties. In most cases, however, they provide the primary motivation for their utilization in industrial and other applications. Consequently, their study constitutes a substantial portion of investigation into the properties of fibers. As such, they merit a separate discussion of their own.

8.5.1 Load–Extension Characteristics

Textile structures are useful members of mechanical systems where substantial load-bearing capacity is required along the yarn length or in the plane of the fabric, without concomitant high stiffness in bending or shearing. This is accomplished by geometric arrangements of fibers in such a way that their load–elongation characteristics are utilized to best advantage, collectively, in the desired directions.

Load–elongation, or stress–strain properties, of several vegetable fibers have been measured. Figure 8.15 gives typical stress–strain curves[2] of abaca, henequen, and flax; the curves for cotton, nylon, and polyester have been included for comparison [63]. Similar data have been obtained for ramie, flax, hemp, jute, New Zealand hemp (phormium), sisal, abaca, coir, and aloe by Chakravarty and Hearle [35]; for abaca, sansevieria, sisal, and henequen by Kasewell and Platt [75]; for banana, sisal and sunhemp by N. Chand et al. [39]; and for coir (bristle fiber) by Varma et al. [178], etc. The useful parameters obtained from such data include tenacity (breaking stress), extension at break, (initial) elastic modulus, and work of rupture. The values of these parameters obtained by different investigators reflect the inherent variability of these materials (referred to earlier), variability in testing procedures, and precision of data analyses. Table 8.14A through Table 8.14C gives a composite listing of these data.

Varma et al. [178] report the influence of chemical treatments on the stress–strain characteristics of coir (bristle) fiber. Their influence on tenacity, initial modulus, and breaking strain are listed in Table 8.14A. The shape of the curves in all these cases might be described as elastic–plastic with nearly constant strain hardening. The influences of alkali treatment are attributed to the (a) rupture of alkali-sensitive bonds and partial removal of hemicellulose, (b) formation of new hydrogen bonds between certain cellulose chains due to removal of hemicellulose, and (c) dissolution and leaching out of the fatty acids that form the waxy cuticle layer of the fiber. Hydrolysis of the cellulose is associated with acidic treatments, which is presumed responsible for the degradation of the mechanical properties. Exposure to UV light for 100 h leads to slight yellowing of the fiber, indicating photo degradation of its components.

---

[2] The value of stress in g/den or g/tex refers to specific stress. To convert specific stress to engineering (or nominal) stress in unit of dynes/cm^2, multiply g/den by $8.829 \times 10^8 \times \rho$ (density of the fiber, g/cm^3), and multiply the g/tex value by $0.981 \times 10^8 \times \rho$.

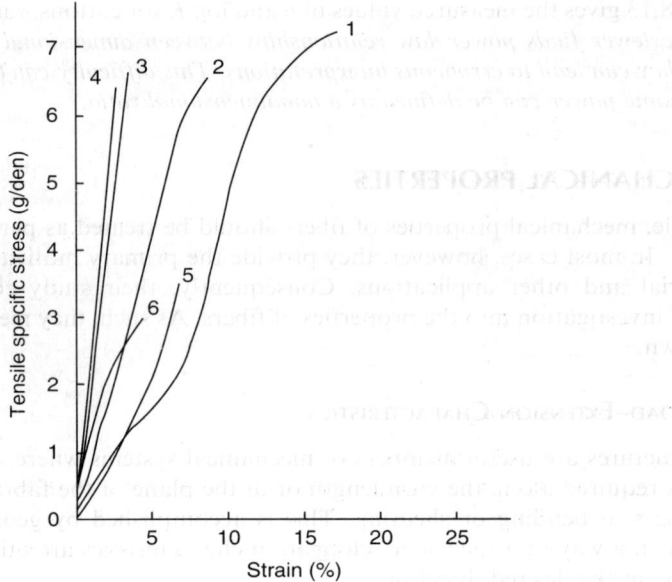

FIGURE 8.15 Tensile stress–strain curves of several fibers: 1. nylon; 2. polyester; 3. flax; 4. abaca; 5. cotton; 6. henequen. (From Himmelfarb, *The Technology of Cordage Fibers and Ropes*, Leon and Hill, London, 1957.)

It has been mentioned earlier that vegetable fibers other than cotton and kapok are multicellular. The mechanical properties of these fibers, therefore, reflect an interaction between their ultimates and their middle lamella substances; ramie is an exception because the size of its ultimates is large enough to act as a fiber.

Chakravarty and Hearle [35] obtained stress–strain property data for ultimates of ramie, flax, hemp, jute, phormium, sisal, abaca, aloe, and coir. From these data, it appears that the initial elastic modulus of ultimates of the fibers may be ranked, in descending order as follows: flax, hemp, jute, phormium, abaca, ramie, sisal, aloe, and coir. The ranking of breaking stress, in descending order, is flax, jute, sisal, hemp, phormium, aloe, ramie, and coir. The extension to break may be ranked in ascending order as: flax, hemp, jute, phormium, ramie, abaca, sisal, aloe, and coir. While the authors explored dependence of initial modulus (tensile) of the ultimates on the average spiral angle (measured from x-ray defraction photographs) and on the modulus of crystallites in the fibril, the precise theoretical relationship remains elusive. It was, however, demonstrated that a one-to-one correspondence between tensile properties of the ultimates and those of the fiber strand should not be expected; this is an acknowledgement of the role played by properties of the middle lamella and structural arrangement of the ultimates in the fiber.

The influence of various components of the middle lamella of flax fiber on its tensile strength was studied by Lindberg [101] and Lindberg and Lang [100]. They assumed that the tensile strength of wet flax strands, when determined at clamp distances greater than the length of the ultimates, is a measure of the strength of the middle lamella. The results are shown in Figure 8.16. It should be stated here, parenthetically, that the tensile strength of the wet flax so obtained is a measure of the contribution of the middle lamella material and its chemical composition, to the strength of the composite fiber strand (ultimates serving as reinforcing elements in the middle lamella matrix).

TABLE 8.14A
Mechanical Parameters from the Stress–Strain Data of Some Fibers

Fiber	Fineness (d-tex)	Tenacity (g = d-tex)	Initial modulus (g = d-tex)	Extension at break %	Work of rupture	References
Flax	5.5		183	3.0	0.082	[120]
	18.0	5.2		1.6	0.0	[152]
(line)	2.1	5.8	182	3.3	0.097	[111,112]
(tow)	3.0	5.1	184	2.8	0.070	[111,112]
	18.0	5.7		1.8		[153]
	10–40	3–5.5		2.0		[106]
Ramie			270		0.06	[163]
		6.0	149	3.7	0.108	[120]
	6.0	6.6		4.6		[152]
	7.0	6.1	152	3.6	0.103	[111,112]
	5.0	6.1	145	3.8	0.108	[111,112]
			167		0.08	[163]
	7.0	5.8		2.3		[153]
	6–12	5–6.0		3.0		[106]
Hemp		4.8	221	2.2	0.054	[120]
	14.0	5.6		2.0		[152]
	3.5	4.7	183	2.6	0.059	[111,112]
			200		0.04	[153]
	12–60	2.9–3.5		2.0		[106]
	22.0	6.1		1.7		[153]
Abaca	42.0	3.6		2.5		[153]
	319.0	6.8		2.6	0.077	[75]
	414.0	5.5		2.7	0.09	[75]
			175		0.07	[163]
		3.5–4.5		2.9–3.0		[159,160]
Banana		1.7–7.9		1.5–9.0		[39]
Sisal	10.0	48.0	257	1.9	0.044	[111,112]
	46.0	3.8		1.9		[153]
		4–4.5		2.5–4.5		[159,160]
		3.7		2.2		[75]
Henequen			127		0.05	[163]
		2.9		4.0		[75]
Coir	1.8		43	16.0	0.16	[114]
Bristle		2.03[a]	34.47	28.8		[178]
Desalted		1.99[a]	36.09	28.6		-do-
10% NaOH treat.		1.87[a]	39.24	27.4		-do-
10% HCl treat.		1.07[a]	30.24	21.8		-do-
10% Acet. Acid	1.70[a]		31.77	27.3		-do-
Untr. exp. to UV		1.21[a]	29.25	19.4		-do-
Desalted exp. to UV		1.30[a]	31.95	20.0		-do-
Sunn	80–100	1.17–1.9		5.5		[39]

[a]The coefficient of variation in these data range from 23.3% to 30%.

TABLE 8.14B
Mechanical Properties of Vegetable Fibers

Fiber	Fineness (km/kg)	Breaking length (km)	Elongation (%)	Modulus of elasticity (N = tex)[a]	Modulus of rupture (mN = tex)
Abaca	32	32–69	2–4.5		6
Sisal	40	36–45	2–3	25–26	7–8
Henequen	32	20–42	3.5–5		
Flax		24–70	2–3	18–20	8–9
Ramie		32–67	4.0	14–16	11
Jute	489	25–53	1.5	17–18	2.7–3
Kenaf	180	24	2.7		
Hemp	139	38–62	1–6	18–22	6–9
Coir		18	16	4.3	16

[a] To convert N/tex to g/den, multiply by 11.33.

Source: From McGovern J.N., Fibers, Vegetable, In *Polymers—Fibers and Textiles. A Compendium*, University of Wisconsin, Madison, WI, 1990.

It is well known that the tensile properties of hygroscopic fibers are affected by their moisture content, and that fiber products are often used in a water or moist environment. It is important, therefore, to determine tensile properties of fibers in the dry (65% RH, 21°C) and the wet state. In the wet state, flax and ramie fibers show greater values of breaking extension (2.2% wet vs. 1.8% dry and 2.4% wet vs. 2.3% dry, respectively) in breaking stress (5 and 18% greater, respectively) [153]. In the case of abaca, sisal, henequen, and sansevieria, the wet breaking extension is almost always greater than that in the dry state, but the extent of

TABLE 8.14C
Physical and Mechanical Parameters of Various Natural and Other Fibers

Fiber	Density (g/cc)	Diameter (μm)	Extension at break (%)	Tensile strength (MPa)	Young's modulus (GPa)	Specific modulus (GPa-cm^3 = g)	Price (Euro/kg)
Flax	1.40–1.50	40–620	2.7–3.2	343–1035	27–80	19–53	2.29–11.47
Jute	1.30–1.50	30–140	1.4–3.1	187–773	3–55	2–37	0.12–0.35
Abaca	1.5	17–21	10–12	980	72	481	0.81–0.92
Sisal	1.30–1.50	100–300	2.0–2.9	507–855	9.0–28.0	7–19	0.70–1.02
Kenaf	1.22–1.40	40–90	3.7–6.9	295–930	22–53	18–38	0.53–0.61
Ramie	1.5	40–60	3.6–3.8	400–938	44–128	29–85	1.44–2.40
Hemp	1.40–1.50	16–50	1.3–4.7	580–1110	3–90	2–60	0.57–1.73
Cotton	1.50–1.60	16–21	2.0–10.0	287–597	5.5–12.6	4–8	1.61–4.59
Banana	1.30–1.35	50–280	3–10	529–914	7.7–32.0	6–24	0.7–0.9
Henequen	1.49	20–500	3.0–5.0	430–580	10.1–16.3	7–11	0.38–0.67
Bagasse	0.55–1.25	200–400	0.9	20–290	2.7–17.0	5–14	0.15
Pineapple	1.52–1.56	200–8800	0.8–3.0	170–1627	6.21–82	4–53	0.36–0.72
E-glass	2.50–2.55	10–20	2.5	2000–3500	73.0	29	1.25
Aramid	1.4–1.45	12	3.3–3.7	3000–3150	63.0–67.0	45–48	7.2
Carbon	1.40–1.75	5.5–6.9	1.4–1.8	4000	230.0–240.0	164–171	12.0

Source: From Biagiotti J., Puglia D. and Kenny José M. A Review on Natural Fibre-Based Composites—Part 1: Structure, Processing and Properties of Vegetable Fibres, *J. Nat. Fibers*, (1)2, 37–68, 2004.

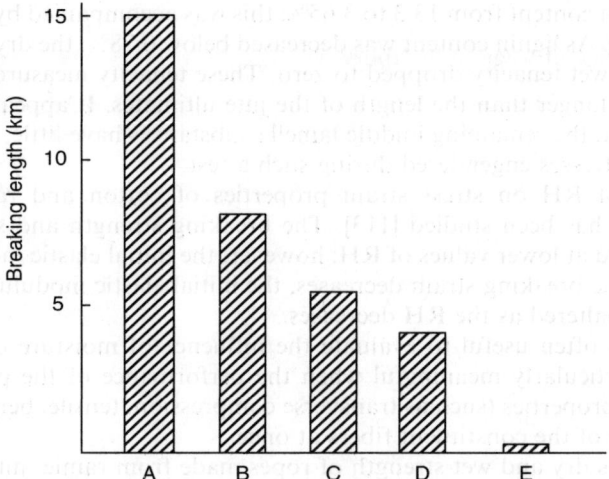

FIGURE 8.16 Tensile strength of wet green (unretted) flax fiber strands treated with : (A) hot water; (B) boiling ammonium oxalate for the removal of pectin; (C) ammonium oxalate and sodium chlorite for the removal of pectin and lignins; (D) ammonium oxalate, hot alkali (for the removal of pectin B of the inner middle lamella); (E) as in D with the addition of sodium chlorite for the removal of lignin. (From Lewin M., *TAPPI*, **41(8)**, 403–415, 1958.)

increase and decrease in the wet strength depends on the specie of the fiber [75]. The wet strength of coir fiber appears to decrease by about 20% compared to its dry strength [191].

In a somewhat similar vein, Kulkarni et al. [80] report a 50–58% decrease in the ultimate elongation and 20–45% increase in the tenacity of the coir fiber as it loses 9–10% of its moisture. While they attribute these changes to the increase of hydrogen bonds in the cellulosic part of the fiber, the changes could also more likely be attributed to the plasticizing effect of moisture on middle lamella substances (polyuronides and hemicellulose).

An earlier, ingenious, study by Roy [145] revealed the role of middle lamella substances on the strength of multicellular fibers quite dramatically. Jute served as the model. His results are shown in Table 8.15. A slow decrease in the dry and the wet tenacity was observed with

TABLE 8.15
Effect of Delignification on the Mechanical Properties of Jute

	Air dry			Wet		
Residual lignin (%)	Tenacity (g = den)	Loss in strength (%)	Extension at break (%)	Tenacity (g = den)	Loss in strength on wetting (%)	Extension at break (%)
13.26	4.31	Nil	1.17	3.17	13.9	1.09
8.89	3.74	13.2	1.03	—	—	—
3.65	3.68	14.6	1.05	3.28	10.9	1.01
0.78	3.02	29.9	0.89	Nil	100.0	—

Temp. = 70° ± 2°F; Humidity = 65% ± 2% RH; Length of test specimen = 4 cm.
Source: From Roy M.M., *J. Text. Inst.*, **44**, T44, 1953.

decrease in the lignin content from 13.3 to 3.65%; this was accompanied by a small decrease in elongation-at-break. As lignin content was decreased below 0.78%, the dry tenacity continued to decrease but the wet tenacity dropped to zero. These tenacity measurements were carried out at gage lengths longer than the length of the jute ultimates. It appears that, at or below 0.78% lignin content, the remaining middle lamella substances have little, if any, resistance to intercellular shear stresses engendered during such a test.

The influence of RH on stress–strain properties of cotton and ramie at a constant temperature (21°C) has been studied [113]. The breaking strength and strain of cotton are considerably reduced at lower values of RH; however, the initial elastic modulus increases. In the case of ramie, the breaking strain decreases, the initial elastic modulus increases, and the strength remains unaltered as the RH decreases.

In practice, it is often useful to evaluate the influence of moisture on a fiber assembly directly. This is particularly meaningful when the performance of the product depends on several mechanical properties (such as transverse compression, tensile, bending, and torsional characteristics, etc.) of the constituent fibers at once.

Table 8.16 shows dry and wet strength of ropes made from ramie, jute, sunn, kenaf, and roselle. In each case, wet strength shows some improvement. Prolonged storage of sunn and Italian hemp ropes in water deteriorates their strength (see Table 8.17); the deterioration in sea water is much more drastic. Deterioration due to prolonged storage may be attributable to microbial action on fibers.

Table 8.18 shows the influence of "dry" storage (2.5 years) on the tenacity and the extensibility of sisal, henequen, sansevieria, and abaca. There appears to be a considerable reduction in the breaking strain of fibers, but less than $\pm$ 5% change in tenacity. Some varieties of sisal and sansevieria show a 6–9% increase in tenacity.

El-Naggar et al. [52] report considerable reduction in mechanical properties (tenacity, breaking elongation) of pure and alkali-treated sisal due to radiation. Similarly, tenacity of the fiber decreases with increasing degree of grafting with 10% styrene and 90% ethylacrylate. In contrast, dye uptake with basic dyes increases at low levels of grafting, but deteriorates at higher levels.

Similarly, Table 8.19 shows the influence of low temperatures ($-70°F$) on the tenacity and rupture elongation of oiled and combed abaca fiber with and without wet-out [75]. Both

TABLE 8.16
Strength of Ropes from Various Cordage Fibers

	Strength (kg)	
Fiber	Dry	Wet
Ramie	110	126
Jute (*C. capsularis*)	65	66
Jute (*C. olitorius*) (Tossa)	51	56
Sunn	51	72
Kenaf	52	60
Roselle	41	53

Ropes from which the above values were obtained were the same size in cross section and 1.2 m in length.

Source: From Whitford A.C., *Textile Age*, **8**, 50, 58, 70, 74, 1944.

TABLE 8.17
Breaking STRENGTH of sunn and hemp cordage

After exposure to fresh and salt water

		Breaking strength (lb)	
Condition	Time	Sunn	Italian hemp
Dry		126.0	
	91.8	126.0	
Sea water	1 month	66.1	96.4
	2 months	35.9	50.5
	3 months	3.2	7.1
Fresh water	40 days	83.0	103.0
	80 days	77.1	101.9
	160 days	72.1	102.0
	282 days	70.3	92.2

Source: From Whitford A.C., *Textile Age*, **8**, 50, 58, 70, 74, 1944.

parameters appear to drop at low temperatures. The drop is less if the wet-out (saturated) specimens are subsequently frozen. The influence of a wet-out of the fiber, frozen to $-70°F$ temperature and back to standard conditions does not appear to deteriorate it significantly.

TABLE 8.18
Effect of Storage for 2.5 Years; All Fibers Tested at 70°F, 65% RH

	Tenacity (g/den)			% Elongation to rupture		
Fiber	1948[b]	1950–1951[a]	% change	1948[b]	1950–1951[a]	% change
Sisal						
African	4.15	4.08	−1.7	2.77	2.22	−20.0
Brazilian	4.74	5.03	+6.1	3.01	2.65	−11.9
Haitian	4.40	4.45	+1.4	2.77	2.22	−20.0
Java	4.46	4.40	−1.3	2.33	1.94	−16.6
Henequen						
Mexican	3.37	3.21	−4.8	4.93	4.00	−18.9
Tampico	3.26	3.18	−2.5	4.25	3.47	−18.5
Victoria	3.17	3.21	+1.3	5.12	4.28	−16.5
Sansevieria						
Longiflora	4.63	4.45	−3.9	2.93	2.43	−17.5
Lorentii	4.81	5.06	+5.2	2.55	2.36	−7.5
Trifasciata	4.02	4.39	+9.2	2.35	2.00	−15.0
Abaca						
Guatemalan	6.96	6.66	−4.3	2.73	2.04	−25.1
Honduras	7.60	7.29	−4.1	2.92	2.20	−24.8

[a]All values average of 45 tests.
[b]All values average of 125 tests except Sansevierias, which are average of 25.

Source: From FRL (Now Albany International Research Co.) Technical Report No. 4, Mansfield, Massachusetts, February, 1951.

TABLE 8.19
Mechanical Properties of Oiled and Combed Abaca Fiber at Low Temperatures

	Tenacity (g/den)	Rupture elongation (%)
Control (+ 70°F, 65% RH)	6.90 (27)[a]	3.26 (19)
Equilibrium at −70°F	4.46 (34)	1.86 (24)
Wet-out and frozen at −70°F	5.01 (46)	1.96 (35)
Wet-out, frozen (−70°F) and reconditioned to +70°F, 65 RH	6.65 (26)	2.71 (26)

[a] Figures in parentheses are coefficients of variation.

Source: From Kasewell E.R. and Platt M.M., *Text. Res. J.*, **21**, 263–276, 1951.

8.5.2 Load Deformation in Other Modes of Deformation

The anisotropic character of textile fibers is well known. The load–deformation properties in the transverse direction are, therefore, generally different from those in the longitudinal direction. Chakravarty [36] estimated transverse moduli of jute, hemp, ramie, abaca, sisal, and coir fibers. The theoretical model on which the estimations are based is very simplistic; the results cited in Table 8.20 and Table 8.21 should therefore be used with a great deal of caution.

The shear modulus of fibers in torsional deformation has been measured for cotton, jute, ramie, and sisal, etc., as shown in Table 8.22. The data for ramie appear to be contradictory; the reason for the contradiction may lie in the differences in the theoretical models (hence assumptions) and the experimental techniques used by the two sets of investigators. Therefore, caution is advised in the use of such data.

Table 8.23 shows the knot tenacity and the knot efficiency for abaca, sisal, henequen, and sansevieria fibers. The efficiency drops to 22% for abaca and 41% for sansevieria. The abaca

TABLE 8.20
Transverse Modulus of Some Plant Fibers

	Fiber	Mean width ($\times 10^{-2}$ cm)	Transverse Modulus ($\pm$ Std Error 10^9 dynes/cm^2)
1.	Jute, tossa (*Corchorous olitorius*)	8.6	0.77 ± 0.081
2.	Jute, white (*Corchorus capsularis*)	9.0	0.59 ± 0.071
3.	Mesta (*Hibiscus cannabinus*)	8.4	0.73 ± 0.093
4.	Hemp (*Cortalaria juncea*)	10.8	0.61 ± 0.079
5.	Ramie (*Bohemeria nivea*)	6.2	0.98 ± 0.109
6.	Sisal (*Agave sisalana*)	21.5	0.28 ± 0.022
7.	Manila (*Musa textilis*)	18.0	0.28 ± 0.028
8.	Aloe (*Furacrea gigantica*)	20.0	0.28 ± 0.022
9.	Coir (*Cocos mucifera*)	27.5	0.86 ± 0.069
10.	Brush fiber	35.0	0.45 ± 0.040

Source: From Chakravarty A.C., *Text. Res. J.*, **39**, 878–881, 1969.

TABLE 8.21
Transverse Modulus of Some Plant Fibers after Different Treatments

	Transverse modulus $M_T \pm$ standard error (10^9 dynes/cm^2)			
	Air-Dry	Wet	Oiled	Batched
Jute, tossa	0.77 ± 0.081	0.69 ± 0.087	0.66 ± 0.073	0.72 ± 0.080
Mesta	0.73 ± 0.093	0.71 ± 0.092	0.70 ± 0.089	0.75 ± 0.097
Ramie	0.98 ± 0.109	0.87 ± 0.096	1.06 ± 0.138	0.89 ± 0.115
Sisal	0.28 ± 0.022	0.23 ± 0.018	0.28 ± 0.002	0.25 ± 0.020
Coir	0.86 ± 0.069	0.46 ± 0.037	0.66 ± 0.053	0.68 ± 0.054

Source: From Chakravarty A.C., *Text. Res. J.*, **39**, 878–881, 1969.

fibers, once knotted and unknotted, do not lose their breaking strength significantly. Similarly, the effect of initial twisting and untwisting of fibers results in some loss of tensile strength; abaca, sisal, and henequen all lose tenacity when stretched under pretwisted conditions.

8.5.3 RECOVERY FROM DEFORMATIONS

In their normal usage, textile structures are loaded–unloaded to levels well below their breaking tenacity. These deformations may be single or repeated excursions. In either case, it is of practical consequence to know whether (a) the fibers or the structures made therefrom recover partially or completely from the imposed deformation, and (b) to what extent are the subsequent load–deformation characteristics modified. Both these questions relate to the dimensional stability and effective long-term performance of the fiber based structures.

Generally, if the initial extension is small, the extent of recovery of the fibers is quite high. Table 8.24 illustrates this in the case of cotton, flax, hemp, jute, and ramie. From a fixed strain

TABLE 8.22
Shear Modulus in the Cross-Sectional Plane of the Fiber

Fiber	Mass per unit length (μg/cm)	RH (%)	Approximate porosity incl. lumen (%)	Shear modulus (10^9 dynes/cm^2)
Jute[a]	16.2	79	38.2	4.2
(*C. olitorius*)	20.3	79	39.9	3.8
Jute[a]	11.6	79	34.9	5.1
(*C. capsularis*)	15.5	79	40.3	4.0
Ramie[a]	8.5	73	19.3	6.9
(*B. nivea*)	12.5	73	25.9	4.8
Sisal[a]	45.0	73	25.7	4.7
(*Agave*)	57.9	73		5.0
Ramie[b]	6.9	65		17.0
Flax[b]	1.8	65		14.0
Cotton[b]	2.4	65		25.0

Source: From [a]Sen K.R. and Bose S.K., *Text. Res. J.*, **24**, 754–755, 1954; [b]Meredith R., *Mechanical Properties of Textile Fibers*, Inter-science Publishers, New York, 1956.

TABLE 8.23
Influence of Knots on the Mechanical Performance
A. Knot Tenacity of Unoiled Hard Fibers

Fiber	Tenacity Knot[a] (g/den)	Tenacity Straight (g/den)	Knot efficiency[b] (%)	Elongation (straight break) (%)
Abaca	1.48 (25)[c]	6.71	22.1	2.65
Sisal	1.36 (22)	4.40	30.8	2.72
Henequen	1.36 (20)	3.30	41.1	4.77
Sensevieria	1.32 (27)	4.50	29.4	2.70

[a] Each value is the average of 125 breaks.
[b] Defined as the ratio of knot tenacity to straight tenacity.
[c] Figures in parentheses are coefficients of variation (%).

B. Effects of Tying a Knot on the Tensile Strength of Abaca Fiber (Preloaded to Approximately 0.3 g/den)

	Before knotting	After knotting and unknotting[a]
Tenacity (g/den)	6.71	6.62[b]
Elongation to rupture (%)	2.65	2.70[b]

[a] Knot efficiency is 22.1%.
[b] Each value is the average of 50 breaks.

C. Effects of the Twisting Operation on Mechanical Properties of Abaca Fiber (Fibers Taken from Standard Yarn: 11.0 turns/ft; 300 ft/lb)

	Original oiled and combed fiber	Fiber from inside of yarn	Fiber from outside of yarn
Tenacity (g/den)	6.90 (27)[a]	6.70 (25)	5.90 (30)
Elongation to rupture (%)	3.26 (19)	3.13 (24)	2.96 (29)
Number of tests	100	350	350

[a] Values in parentheses are coefficients of variation (%).

D. Effect of Twist on the Tenacity[a] of Unoiled Hard Fibers

Fiber twist (turns/in)	Tenacity of Abaca (g/den)	Tenacity of Sisal (g/den)	Tenacity of Henequen (g/den)
0	6.71 (22)[b]	4.40 (21)	3.30 (20)
0.6	6.50 (24)	4.20 (23)	3.24 (20)
1.2	6.60 (25)	4.40 (23)	3.20 (19)
1.8		4.40 (25)	3.24 (16)
2.4	6.70 (24)	4.32 (22)	3.15 (19)
3.0		4.35 (21)	3.15 (16)
3.6	5.80 (32)	4.20 (17)	3.18 (18)

[a] Each value, except those for zero twist, is the average of 50 breaks.
[b] Figures in parentheses are coefficients of variation (%).

Continued

TABLE 8.23 (continued)
Influence of Knots on the Mechanical Performance
E. The Effect of Oiling and Combing on Mechanical Properties of the Abaca Fiber

	Unoiled fiber	Oiled and combed fiber
Tenacity (g/den)	6.71 (21)[a]	6.90 (27)
Rupture elongation (%)	2.65 (22)	3.26 (19)
Energy absorption (g.cm/cm.den)	78×10^{-3} (19)	100×10^{-3}
Knot tenacity (g/den)	1.48 (25)	1.78 (20)
Knot efficiency (%)	22.1	25.6
Repeated-stress performance (90% of ultimate load):		
permanent set (%)	0.62 (21)	0.54 (22)
elastic performance	0.53 (15)	0.60 (20)
coefficient residual elongation to rupture (%)	2.38 (8)	2.61 (10)
tenacity (g/den)	7.61 (9)	7.12 (14)
energy absorption to rupture (g.cm/cm.den)	89×10^{-3} (17)	100×10^{-3}
Tenacity when twisted:		
0 turns/in	6.71 (22)	6.90 (27)
0.6 turns/in	6.50 (24)	6.60 (20)
1.2 turns/in	6.60 (25)	6.70 (27)
2.4 turns/in	6.70 (24)	6.60 (31)
3.6 turns/in	5.80 (32)	6.70 (30)

[a]Values in parentheses are coefficients of variation (%).
Source: From Kasewell E.R. and Platt M.M., *Text. Res. J.*, **21**, 263–276, 1951.

point of view, cotton shows greater elastic recovery than the rest. From a fixed stress point of view, flax shows by far the best recovery followed by jute and ramie. It is useful to note that the elastic recovery often also implies greater mechanical work recovery and smaller work loss.

Table 8.25 gives some data from repeated loading–unloading experiments on sisal, abaca, and henequen [75]. After five cycles, sisal shows the smallest amount of permanent set and hysterisis loss while retaining significant levels of tenacity. Abaca was next best in this respect.

8.5.4 Time Effects

Vegetable fibers also show time-dependent properties associated with other polymeric materials. There is, however, only limited data available, which analyzes creep, relaxation, or strain-rate behavior of vegetable fibers.

Figure 8.17 shows stress relaxation of cotton and flax yarns from different loads associated with different levels of initial strain. It appears that both yarns (fibers), within the range of strains evaluated, obey the laws of linear viscoelasticity. Typically, viscoelastic materials show higher elastic modulus if the tests are carried out at high strain rates; such values are usually referred to as dynamic elastic moduli. Meyer and Lotmar [115] studied the influence of RH on the dynamic modulus of ramie and hemp, among other cellulosic fibers. For fibers in wet, air-dry, and bone-dry states, their values for viscose rayon were as follows: 1.3, 11.5, 14.7 (10^{10} dynes/cm^2), respectively; for raw ramie: 18.7, 51.0, 59.0 (10^{10} dynes/cm^2), respectively; for raw hemp: 34.3, 57.0, 69.0 (10^{10} dynes/cm^2). In other words, moisture lowers the resistance of these fibers to deformation significantly, even at high strain rates. This is attributable to the rupture of some of the hydrogen bonds during wetting.

TABLE 8.24
Elastic Recovery of Various Fibers

Fiber	Approximate fineness (denier)	stress (g/den)					Elastic recovery at: strain (%)				
		1	2	3	4	5	0.5	1	1.5	2	3
Flax I	1.7	0.80	0.73	0.68	0.64	0.61	0.83	0.76	0.70	0.66	0.60
Flax II	2.6	0.76	0.70	0.64	0.60	0.57	0.77	0.71	0.66	0.63	—
Hemp	3.1	0.64	0.55	0.50	0.49	0.50	—	0.55	0.51	0.50	—
Jute	12.6	0.72	0.74	0.75	—	—	0.73	0.74	0.75	—	—
Ramie	6.2	0.76	0.56	0.48	0.43	0.41	0.82	0.67	0.58	0.52	0.44
Cotton (sakel)	1.8	0.65	0.45	0.34	0.27	—	—	0.90	—	0.73	—

Source: [111,112]

Meredith [113] reports a similar influence of temperature. The viscose rayon static (at 20°C) and dynamic (at 20°C and −190°C liquid air) moduli are 8.4, 14.7, and 13.3 (10^{10} dynes/cm^2), respectively; the corresponding values for ramie are 14.7, 59, and 47 (10^{10} dynes/cm^2). Figure 8.18 shows the effect of prestrain on the dynamic modulus of abaca, henequen, viscose, and acetate [113].

8.6 CHEMICAL PROPERTIES

8.6.1 ACTION OF ALKALIS AND OTHER SWELLING AGENTS

The swelling of bast and leaf fibers in alkaline solutions is a complex phenomenon. Their structure is more complicated than in the case of pure cellulose. Different components of the middle lamella react in different ways, and rates, in relation to each other and to the cellulose. In addition, the composition of the fiber changes with the time of treatment. A part of the

TABLE 8.25
Elastic Properties[a] of Unoiled Hard Fibers (Repeated Stress at 50% and 90% of Breaking Strength)

Fiber	Conditioning load (% of ultimate load)	Permanent set (%)	Elastic performance coefficient	Residual elongation (%)	Tenacity (g/den)	Energy absorption (g.cm/cm.den)
Henequen, Victoria	50	0.78 (33)[b]	0.52 (12)	4.38 (20)	3.44 (9)	95×10^{-3} (30)
Sisal, Java	50	0.22 (57)	0.63 (27)	2.28 (12)	4.94 (11)	61×10^{-3} (21)
Henequen, Victoria	90	1.91 (30)	0.39 (12)	3.31 (9)	3.75 (10)	74×10^{-3} (24)
Sisal, Java	90	0.57 (24)	0.54 (11)	2.32 (12)	5.06 (10)	64×10^{-3} (22)
Abaca, Manila	90	0.62 (21)	0.53 (15)	2.38 (8)	7.61 (9)	89×10^{-3} (17)

[a] Values for henequen and sisal are the averages of 10 tests; values for abaca, the averages of 15 tests.
[b] Figures in parentheses are coefficients of variation (%).

Source: From Kasewell E.R. and Platt M.M., *Text. Res. J.*, 21, 263–276, 1951.

Other Long Vegetable Fibers

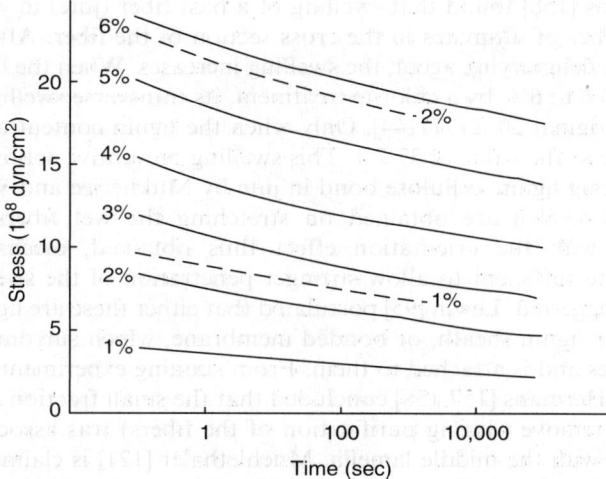

FIGURE 8.17 Stress relaxation of cotton and flax yarns at standard conditions (21°C, 65% RH).(———) cotton (- - - - -) flax (From Meredith R., *Mechanical Properties of Textile Fibers*, Inter-science Publishers, New York, 1956.)

fiber (hemicelluloses) is dissolved; thereby some of the chemical linkages are severed. The structure is opened for deeper penetration of the alkali.

Dilute alkali penetrates only accessible regions of the fiber causing intercrystalline or interfibrillar swelling. The latter is much more profound in that it brings about changes in crystal structure of the fibers, as evidenced by mercerization (in the case of alkali) and the formation of ammonia cellulose (in the case of liquid ammonia).

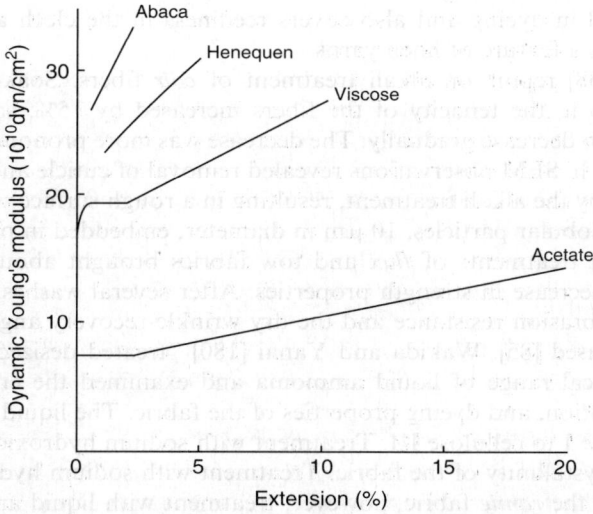

FIGURE 8.18 Influence of pre-strain on the dynamic Young's modulus at standard conditions (65% RH, 21°C). (From Meredith R., *Mechanical Properties of Textile Fibers*, Inter-science Publishers, New York, 1956.)

Sen and Hermans [158] found that swelling of a bast fiber (jute) in water decreases with increase in the number of ultimates in the cross section of the fiber. After treating the fiber with alkali or with a delignifying agent, the swelling increases. When the lignin content of jute is decreased from 13.6 to 6%, by a chlorite treatment, its transverse swelling does not increase much beyond the original 20–21% [144]. Only when the lignin content drops to 0.78% does the swelling increase to the value of 37.7%. This swelling preventive action of lignin led to the postulation of a strong lignin–cellulose bond in jute by Mukherjee and Woods [123]. Higher swelling values (38–43.6%) are obtained on stretching the wet fibers after the alkaline treatments. Along with the orientation effect thus obtained, mechanical rupturing of some bonds at a rate sufficient to allow stronger penetration of the swelling agent into the cellulose has been suggested. Lewin [95] postulated that either these are lignin–cellulose bonds or they are part of a lignin sheath, or bonded membrane, which surrounds and encases cell walls of the ultimates and is attached to them. From staining experiments with Victoria Blue (C.I. 729), Sen and Hermans [157,158] concluded that the small fraction of lignin in jute that is most difficult to remove (during purification of the fibers) was associated both with the secondary wall and with the middle lamella. Muehlethaler [121] is claimed to have obtained electronic micrographs of a skeleton of lignin, in the form of sheets, after saccharification of the cellulose of sisal fibers with concentrated sulfuric acid.

The mercerization of bast fibers (*jute*) by alkali under the usual conditions does not reach completion; along with the newly formed cellulose II, an appreciable amount of cellulose I remains. The total mercerization of jute can be obtained after complete delignification [156] or after pretreatment with sulfuric acid [123]. In both cases, the restraining effect of lignin membrane is removed: in the first case by dissolving it and in the second case by hydrolyzing the chains in the noncrystalline regions that are linked to the lignin of the membrane.

In mercerization of linen (*flax*), the whole fiber, including the nodes, swells uniformly [161]. A homogenization of the structure is thus reached, which is believed to be responsible for a significant increase in abrasion resistance. Furthermore, it is believed to decrease the "embrittlement" brought about by crease-resistance treatment with cross-linking agents, presumably due to a more homogeneous distribution of chemicals in the fiber. Mercerization improves color yield in dyeing and also covers reediness in the cloth associated with yarn unlevelness, which is a feature of linen yarns.

Prasad et al. [138] report on alkali treatment of *coir* fibers. Soaked in 5% NaOH at $28 \pm 1°C$ for 72–76 h, the tenacity of the fibers increased by 15%; soaking beyond 76 h caused the tenacity to decrease gradually. The decrease was more pronounced if the alkali was replenished every 24 h. SEM observations revealed removal of cuticle and "tyloses" from the surface of the fiber by the alkali treatment, resulting in a rough surface with regularly spaced pits; "tyloses" are globular particles, 10 μm in diameter, embedded in pits in the cell walls.

Liquid ammonia treatments of *flax* and tow fabrics brought about a decrease in DP, accompanied by a decrease in strength properties. After several washes, the tensile strength increased; the wet abrasion resistance and the dry wrinkle-recovery angles of the ammonia-treated goods increased [85]. Wakida and Yanai [180] "treated desized and scoured *ramie* fabric with a practical range of liquid ammonia and examined the crystallinity, moisture regain, water absorption, and dyeing properties of the fabric. The liquid ammonia treatment transformed cellulose I to cellulose III. Treatment with sodium hydroxide and liquid ammonia decreased the crystallinity of the fabric. Treatment with sodium hydroxide increased the water absorption of the *ramie* fabric, however, treatment with liquid ammonia and sodium hydroxide and liquid ammonia decreased water absorption. Treatment with sodium hydroxide increased the early dyeing rate of the fabric. Treatment with liquid ammonia decreased the early dyeing rate of the *ramie* fabric. Treatment with liquid ammonia and sodium hydroxide/liquid ammonia decreased the shear modulus, bending hysteresis width 2HG, bending modulus B, and bending hysteresis width 2HB."

The appearance of dyed liquid ammonia-treated linen was found to be cleaner, and dye penetration superior, than in the alkali-treated, dyed linen fabrics [161].

According to Warwicker [184], sisal swells best in 5.0–10.0 N NaOH solution at room temperature. Sisal, preswollen in $7N$ NaOH, acquires a golden yellow color; on washing out of NaOH, it regains its creamy-white color. Washed sisal is soft in water and develops crimp during drying unless fibers are held under tension. NaOH-treated sisal loses strength by about 40%. El-Naggar et al. [52] report that etching treatment with 8% and 30% NaOH solution causes the fiber to fibrillate into its ultimates.

Swelling of cellulose is brought about by various alkalis and inorganic salts (e.g., $ZnCl_2$); their ability to do so is a direct consequence of the supposed degree of hydration of their ions. In the case of *sisal*, Warwicker [184] reports that, at room temperature, $ZnCl_2$ solution does not have any effect; however, when heated to 55°C in 75% $ZnCl_2$ solution, sisal disintegrates and turns brown. Swelling and "mercerization" of cotton has been carried out successfully with liquid ammonia at –40°C, whereas sisal does not swell in liquid ammonia [184].

In the case of *banana*, Bhatia and Gupta [23] report no reaction with dilute (1%) NaOH even on heating; with concentrated (20%) NaOH, the fiber swells only after heating and boiling for several minutes.

Varma et al. [178] report change in color of the *coir* fiber from pale yellowish brown to dark brown, on treatment with 10% NaOH. The fibers develop crimp and their diameter decreases; the latter effect is more pronounced in fibers that are initially coarser. "The size of the central lacuna of the fiber and the lumen of the cell also decreased;" the solution turned dark brown. A weight loss of about 9.3% was reported [178].

Wang et al. [183] explore the role of alkaline boil in the removal of pectins and lignin in the *hemp* fiber. They conclude that (a) the lignin content of the fiber differs from bottom to middle to top; (b) pectins are easier to remove than lignin; (c) alkaline boil (sodium hydroxide) is more effective than acid scouring; (d) sodium sulphite addition accelerates the process of removal; and (e) the concentration of the alkali and sodium sulphite and the treatment time can be adjusted to remove all of the pectins, but some lignin still remains.

8.6.2 Action of Mineral Acids

Swelling action due to mineral acids is more complicated. In sulfuric acid, swelling begins at concentrations above 50%; at 60%, the cellulose degrades into small molecular weight fractions and dissolves. A very short time of contact converts the cellulose to a gel which, when dry, is transparent and gives a parchment-like surface. Similarly, other mineral acids, such as HCl, HNO_3, and phosphoric acid, can swell or dissolve cellulose at specific optimum concentrations.

According to Warwicker [184], sisal in 60% sulfuric acid solution swells after about a 4-min exposure; in addition, the fibers tend to split longitudinally. In 65% solution, the fibers turn yellow (in 15 s) and start disintegrating (3.5 min). In 70% solution, the fibers turn yellow and split longitudinally, but do not disintegrate. At 75% concentration, both dry and wet fibers split longitudinally and form a brown deposit on the surface. The brown deposit can be removed to yield creamy-white fiber, which does not react with sulfuric acid any further. At 80% concentration, results are similar to the 75% case. Sisal is fully resistant to concentrated sulfuric acid (~98%) over short durations. The brown deposit formed can be washed off, leaving behind creamy-white, lustrous, strong yarn, resistant to further attack by sulfuric acid. The brown deposit is expected to be the hydrolyzed part of lignin. It is conceivable that, with lignin removed, the fiber may not yellow any more and may be dyed more easily.

Treatment of cellulose with dilute mineral acids brings about reduction in the chain length of the cellulose molecule by random hydrolysis of the 1–4 glycosidic bond. Hydrocellulose (or hydrolyzed cellulose) retains its fibrous structure, but loses its mechanical strength significantly. The degree of modification depends on temperature, concentration, and duration of contact. Pure cellulosic fibers, such as *cotton* and *ramie*, give colloidal solutions of low-molecular-weight compounds when treated with strong aqueous sulfuric acid. Fibers with significant lignin content, such as *jute*, *kenaf*, and *coir*, remain fibrous at such concentrations and do not reduce to colloidal solution, unless subjected to severe mechanical action [124].

In the case of *banana* fiber, Bhatia and Gupta [23] report no reaction with dilute and concentrated HCl and with dilute HNO_3. With concentrated HNO_3, on heating, the fiber yellows and then disintegrates. With dilute H_2SO_4, on heating, the fiber swells slightly and then disintegrates; in concentrated H_2SO_4, the fiber dissolves completely.

Treatment of *coir* (bristle) fiber with 10% HCl did not change its color, although about 7.5% weight loss was reported [178].

8.6.3 Action of Organic Acids

Treatment of cellulose fibers, such as cotton, by organic acids (formic, acetic, oxalic, tartaric, etc.) is much less degrading. Volatile acids (formic, acetic) are removed from the fiber and do not affect it. Nonvolatile acids in aqueous solutions have no appreciable effect on the fiber. With increasing temperature, however, a gradual conversion of cellulose to hydrocellulose and xylose to furfural occurs. The banana fiber is reported to become yellow and soft in acetic acid, but is unaffected by formic acid and other reagents such as chloroform, petroleum ether, cuprammonium, and acetone.

Treatment of *coir* fiber with 10% acetic acid caused a slight brownish tinge and a weight loss of about 6.5% [178].

8.6.4 Bleaching

Bleaching of bast and leaf fibers is strongly influenced by the high percentage of their noncellulosic components. Retted flax contains only about 70% α-cellulose. The hemicelluloses, which are the main constituent of the middle lamella, are of low degree of polymerization and are highly soluble in alkali; losses in weight of up to 25% may be obtained in alkaline bleaching treatments [140].

A major objective of bleaching of bast and leaf fibers is to remove the lignin with which most of the coloring matter is associated. The 2% of dark-colored lignin in retted flax must be removed completely to achieve a high level of whiteness. The removal of woody matter remaining in the fibers after the mechanical processing stages, if any, is also essential. A two-stage combination bleaching is often applied, usually with sodium chlorite, followed by hydrogen peroxide [124,161]. The more severe the conditions used in bleaching, the higher the degree of whiteness obtained, and the greater the loss in weight.

The degree of bleaching is usually determined according to the nature of the product desired. For tablecloths and similar uses, a hard hand is desired; therefore, little or no mild extraction with sodium carbonate is performed to minimize removal of hemicellulose. On the other hand, for apparel products, a soft hand is needed and a strong alkaline extraction is carried out to remove most of the noncellulosic materials [93,134]. In fully bleached linens, the strand structure that gives flax fibers their characteristic handle, stiffness, and high tearing strength is almost completely altered; the linen is composed almost entirely of ultimates, as in the cottonization processes.

The bleaching processes used for bast fibers are usually milder, with longer treatment times than those used for cotton, to preserve the strand structure as much as possible. Details of bleaching processes for flax and jute with hypochlorite, hydrogen peroxide, sodium chlorite, peracetic acid, sodium dichlorisocyanurate, and cyanuril chloride have been reviewed by Lewin [93].

More recently, Wang and Postle [182] found that increasing the concentrations of sodium hydroxide and sodium sulphite in the alkaline boiling process, X, Y, Z, and WIE (whiteness) of the hemp fiber increase linearly. The alkali boil facilitates further improvements in the fiber color during bleaching. Peroxide bleaching of the fiber can greatly improve its color in terms of X, Y, Z, and whiteness.

8.6.5 Stability to Light

Fibers with significant lignin content have a tendency to develop yellow or brown coloration on the surface on prolonged exposure to light. The color is because of oxidation of lignin due to photosensitivity. According to Mukherjee and Radhakrishnan [124]:

> "From the available evidence, it has been suggested that the lignin component undergoes complex degradative changes leading to the formation of such compounds as syringaldehyde, vanillin, syringic acid, vanillic acid, and other phenolic substances, together with hydrogen peroxide. The phenolic compounds appear to be trans-hemicellulosic components, being reduced in turn to hydroquinone derivatives. These latter components can be oxidized back to quinone in the presence of hydrogen peroxide. In this process of degradation, lignin suffers some loss of methoxyl groups. It has also been suggested that the degradation of lignin is associated with the production of simpler aromatic compounds having two phenolic functional groups, or a phenolic and an aldehydic functional group attached to the nucleus. These compounds then appear to undergo condensation reactions involving a free radical mechanism. In this process, chromophore groups are formed with quinonemethide types of structures, and these are responsible for the yellowing. The blocking of reactive phenolic groups by etherification or esterification would be expected to arrest the yellowing process. Methylation or cyanoethylization has been found to provide little protection against discoloration, while acetylation under suitable conditions has been found to effect the prevention of yellowing."

Warwicker [184] cites the work of Wilson [194] to make much the same point for yellowing of sisal and its prevention by acetylation or methylation.

The prevention of discoloration of jute occurs when the acetyl content reaches 14–19%. Lignin acquires the acetyl groups more easily than cellulose; similarly with hemicellulose, which is much more accessible to acetylating reagent [34].

The light-fastness of bleached bast (*flax*) fibers is of special interest. Since removal of lignin during bleaching is undesirable, as it causes excessive loss in strength, bleaching processes with sodium chlorite are applied; thus, most of the lignin is retained along with a high degree of whiteness. *Jute* fibers bleached in such a way undergo a gradual reversion to the original color (photo yellowing). Light-fast bleaching processes have been described in the literature in which only the surface of the fiber is delignified and the inner portions are not affected appreciably. This can be attained by a pretreatment with chlorine and water or sodium hypochlorite of pH 6 and subsequent extraction with sulfite or alkali [124,174].

8.6.6 Finishing Treatments

The chemistry of the cross-linking of bast fibers, in some ways, is essentially similar to that of cotton. It is, however, more complicated due to specific structural features. For example, the cross-linking is of particular interest to linen, which is used for both clothing and household tablecloths and bedsheet applications. The main difficulties in cross-linking linen are caused by the nodal regions of the fiber, wherein most of the chemical reagents tend to concentrate.

The migration of the cross-linking agents during drying, along with easy movement of moisture, is much more pronounced in linen than in cotton. The acidic catalyst, which is usually a small molecule, migrates freely and concentrates at high spots of the yarns (crowns), causing local embrittlement, whereas the migration of the cross-linking agent can be limited either by homopolymerization or by reaction with cellulose. In an effort to overcome this problem, the use of an oligomeric polyester prepared from triethyleneglycol and citric acid has been suggested. This was applied together with a carbamate-based resin, but resulted in only slight improvement [161].

Oh and colleagues [128] investigated nonformaldehyde treatment of ramie to obtain creases resistance. They conclude that ramie and ramie-blended fabrics can be treated with glyoxal and polyethylene glycol (PEG) to obtain superior performance properties.

8.6.7 Action of Microorganisms and Enzymes

Excellent reviews of the subject were carried out by Mukherjee and Radhakrishnan [124] in 1972 and earlier by Lewin [95] in 1958. According to Mukherjee and Radhakrishnan, "Depending upon the nature of carbon constituents, utilized for their metabolic function, the action of micro-organisms on vegetable fibers can be useful or detrimental." For example, in the retting of flax, jute, kenaf, etc., the microorganisms decompose the pectins and gums to help extract fibers from the surrounding woody and other materials. Anaerobic bacteria of the *Clostridium* genus have been found to be active agents; in particular, *Clostridium felsineum* constitutes the basis of the patented Corbone process for the retting of flax. In addition, aerobic bacteria, particularly those belonging to genus *Bacillus*, have been found to be active; the *Bacillus comesii* constitutes the basis of the previously mentioned Rossi process for retting of flax. In addition to bacteria, fungi belonging to the genera such as *Aspergillus, Penicillium, Chaetomium,* and *Mucor* have varying degrees of retting ability, depending on various ambient conditions.

Both bacteria and fungi exhibit cellulolytic activity. The fungi *Alternaria, Chaetomium, Aspergillus, Curvularia, Fusarium,* as well as bacteria belonging to the *Cytophage, Cellulomonas, Cornybacterium,* and Vibrios groups can decompose cellulose at suitable moisture levels: bacterial growth requires greater moisture content than fungal growth. Microbial growths on fibers cause loss of the tensile strength and develop a musty odor and several types of stains on the fiber substrate (e.g., brownish stains (rust) on flax).

The chemical composition of the fibers influences their susceptibility to microorganism growth. Lignin content offers some protection, as evidenced by the high resistance of coir fibers, which contain 35% lignin. The lignin–hemicellulose ratio, the crystalline content of cellulose fraction, and the presence of micronutrients are other factors that determine the extent of microorganism activity on the fibers.

Elkin [51], by using Searles' retting techniques [154], showed that the resistance of flax to microbial decay increases with the severity of the alkaline boil, because the latter treatment removes more accessible noncellulosic constituents of the middle lamella. Basu [21] found jute to be much more resistant to fungal attack than cellulose; indeed, the relative importance of a given fungus may even be opposite in the two cases. Lignin is the most resistant constituent of the fiber; jute, delignified with sodium chlorite solution, becomes much more susceptible to fungal attack. In contrast, according to Basu [20], hemicellulose is highly susceptible to fungal attack, but, when associated with α-cellulose, it enhances its decomposition considerably.

Chaudhury and Ahmed [40] isolated several fungi and bacteria during the softening of hard root cuttings of jute by the action of microbes. In subsequent laboratory tests, of these, three bacteria and one fungus were found to be effective in softening the hard roots. The isolated fungus *Sclerotium rolfsii* produced the most beneficial effect in softening and improving the color of the cuttings, without requiring any additional nutrients. The fungus, however, did not survive temperatures greater than 45°C, or water content less than 45%. In a similar study [11], mixed cultures of bacteria isolated from rhizosphere of ramie plants and that of cultivated legume were found capable of degumming ramie fibers to the extent of 8–9%, on incubation for 10 days; at the same time, they reduced fiber tenacity by 20%, suggesting cellulolytic activity.

A modification of the constituents of middle lamella by physical and chemical agents strongly influences the microbiological stability of the jute fibers. After 180-h exposure to sunlight, the fibers were found to be much more prone to attack by 20 different fungi known to attack cellulose [95], especially *Aspergillus fumigatus* (Fresenius), *Penicillium vermiculatum* (Dangeard), *Aspergillus sydowi*, and *Penicillium cytrinum* (Thom). Acid or alkali treatments, which removed up to 20% of the hemicellulose, 0.4 to 2% lignin, but no α-cellulose, increased the potency of fungal attack on jute considerably [95]. The effect of alkali treatment, even a mild one (15 min with 0.5% NaOH, at room temperature), was much more pronounced than that of the acid treatment. Lewin [95] suggested that the effect of alkali treatment may be partly due to removal of the acetyl group in the untreated jute, which is known to have a stabilizing effect. Alternatively, the rupture of lignin polyuronide linkages and lignin carbohydrate linkages may be responsible for the observed effect. In the unruptured state, these bonds may cause steric hindrances or barriers to the penetration of fungi into the otherwise amorphous matter of the middle lamella and the noncrystalline portions of the fiber.

The formation of new cross-links in the middle lamella by suitable agents or blocking of the reactive groups of lignin and hemicellulose by acetylation, phosphorylation, cyanoethylation, or other similar treatments may greatly improve biological stability of lignocellulosic materials [95,124]. These methods are, however, relatively costly. Alternatively, the fibers (yarns and fabrics) may be treated by toxic substances. Compounds of Cu, Zn, Pb, Cr, and Zr are important inhibitors [124]. In this connection, chlorophenols and their derivatives may also be used for the purpose. Impregnation and polymerization, *in situ*, of fabrics with resins, such as urea–formaldehyde and melamine–formaldehyde, can also give substantial protection against microbial growth. Mercuration of lignin containing bast fiber fabrics (jute, kenaf, flax) with an aqueous solution of mercuric acetate has been found to be an effective antimicrobial process. The strength of the bond between mercury and lignin remains stable even at 100°C for 4 h.

Buschle-Diller et al. [32] studied the enzymatic hydrolysis of hemp fabrics using cellulose, hemicellulase, and b-glucosidase. They concluded that "the largest total porosity and the highest number of small pores occur when using just cellulase. The hemicellulase admixture helps to generate smaller-sized pores initially, but appears to promote the formation of larger pores for longer treatment times. Cellobiase, however, seems to assist in the creation of bigger pores from the start of the hydrolysis reaction." Similarly, Jin and Maekawa [71] have treated *ramie* and linen fabrics with a commercially available pectinase and determined the effects of temperature, enzyme concentration, and treatment time on fabric degradation rate.

Bhattacharya and Shah [26] studied the influence of environment-friendly enzyme treatment of flax fabrics using BGLU enzyme with hemicellulase and pectinase activities. Under optimal conditions of enzymalysis, weight loss was in the region of 12%, similar to that obtained by conventional caustic soda treatment (10–16%). In addition, the fabrics had improved absorbency, whiteness, and dyeability with tolerable loss of tensile strength.

Sugai et al. [166] suggest that chemically modified cellulase in the presence of borate buffer might be useful for biopolishing of cellulosic fibers. They found that the decline in the breaking strength of ramie yarns was improved, when the modified enzyme was used.

8.6.8 ACTION OF DYESTUFFS

Teri et al. [169] report the highest depth of dyeing for sisal fibers relative to jute and coir in the case of direct, acid, and metal complex dyes, where, as in the case of "cationic dyes, coir exhibited the highest depth of shade, followed by jute and *sisal*, due to absorption of the dye at localized sites in the *fiber*."

In the case of ramie, Wakida and Yanai [180] report "treatment with sodium hydroxide increased the early dyeing rate of the fabric. Treatment with liquid ammonia decreased the early dyeing rate of the ramie fabric."

More recently, Zhou et al. [198]. studied the dyeing of mercerized and unmercerized ramie fabrics cross-linked with 1,2,3,4-butanetetracarboxylic acid using a typical direct dye CI Direct Red 81. They concluded that mercerization improved the equilibrium adsorption and dyeing rate of the fabrics, while reducing the maximum dye adsorption and equilibrium constant. "Cross-linking with butanetetracarboxylic acid decreased most of these parameters related to the dyeing profile, although the activation energy of dyeing was increased." The same treatment also appears to decrease the tensile strength of ramie fabrics, although such losses in mercerized fabrics are less [197].

Bhattacharya and Shah [26] found enzymatic treatment to be more beneficial compared with the conventional caustic soda treatment for removing noncellulosics from flax fabrics. The dye uptake of both direct (124.7%) and reactive dyes (106.2%) appears to increase with enzymatic treatment relative to conventional caustic soda treatment.

8.7 CONCLUDING REMARKS

The interested reader is advised to read other reviews of the subject. For instance, Biagiotti, Puglia, and Kenny [27] provide a good comprehensive review of the structure, processing, and properties of vegetable fibers.

ACKNOWLEDGMENTS

The author takes this opportunity to express his deep indebtedness to (1) Prof. Dr. Menachem Lewin, Director of the Israel Fiber Institute, Jerusalem, who motivated and encouraged the writing of this chapter, provided the institutional facilities to accomplish its completion, and generously spared many hours of his time, as a friend and an editor, for fruitful discussions and refinement of the manuscript; (2) Mrs. Florence Alexander, Mrs. Naomi Levy, Mr. Joseph Netaf, Dr. Joseph Banbaji, and Prof. Shay Armon, whose diligent assistance, constant support, and friendship were invaluable in overseeing the preparation of the manuscript; and (3) the School of Textiles, NCSU, and N.C.-Israel Scholar Exchange Program, which together made possible the sabbatical leave (January–June, 1984) spent in Jerusalem, during which this chapter was completed.

These acknowledgements were written nearly 20 years ago. Time has not in any way diminished the sentiments expressed then.

APPENDIX

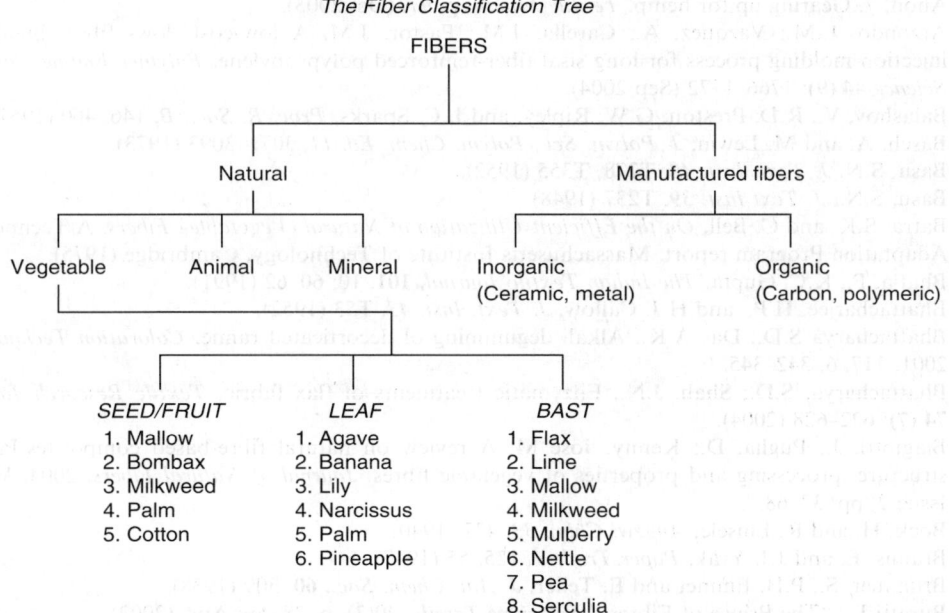

REFERENCES

1. Ahmed, M.; Chattopadhyay, S.K.; Gaikwad, R.S.; Dey, S.K. Characteristics of degummed ramie fibre and its cotton blended yarns., *Indian Journal of Fibre & Textile Research* 29 (3): 362–365 (2004).
2. Akin D.E.; McAlister D.D. III; Foulk J.A.; Evans J.D., Cotton fibres: Properties and interaction with flax fibres in blends., International Cottontest Conference. Bremen; 26, 189, 8p, Mar. 13–16, (2002).
3. Akin, Danny E.; Henriksson, Gunnar; Evans, Jeff D.; Adamsen, Anders Peter S.; Foulk, Jonn A.; Dodd, Roy B. Progress in enzyme-retting of flax. *Journal of Natural Fibers*, 2004, Vol. 1 Issue 1, pp. 21–47, (2004).
4. Allison, R.V. and V.A. Boots, Soil and Crop Science Society of Florida, Proceedings, 22, 176–185 (1962).
5. Allison, R.V. and V.A. Boots, Annual Meeting, Crop Science Society of Florida, Proceedings, 30, 100–112 (1970).
6. Allison, R.V., G.E. Tradwell, and G.E. Vernie, Crop Science Society of Florida, Proceedings, 32, 66–73 (1972).
7. Allison, R.V., *Chemurgic Digest*, March 14–16 (1952).
8. Allison, R.V., Soil and Crop Science Society of Florida, Proceedings 24, 377–383 (1964).
9. Angelova, V.; Ivanova, R.; Delibaltova, V.; Ivanov, K., Bio-accumulation and distribution of heavy metals in fibre crops (flax, cotton and hemp), *Industrial Crops and Products* 19 (3): 197–205 (2004).
10. Anon. 1, *Textiles* **11**, No. 1, 16–17 (1982).
11. Anon. 2, Indian Council of Agricultural Research, Jute Technological Research Laboratory, *Annual Report*, p. 18 (1970).
12. Anon. 3, *High Performance Textiles*, 7–8, April (1991).

13. Anon. 4, *Textiles* **17**, No. 1, 26–28 (1988).
14. Anon. 5, *Asian Textile Journal*, **12** (3), p. 20 (2003).
15. Anon. 6, *Nonwovens Report International*, Issue 373, 66 (Apr. 2002).
16. Anon. 7, Gearing up for hemp. *Textile Month*; p. 97, (Oct 2003).
17. Arzondo, L.M.; Vazquez, A.; Carella, J.M.; Pastor, J.M. A low-cost, low- fiber -breakage, injection molding process for long sisal fiber-reinforced polypropylene, *Polymer Engineering and Science* 44 (9): 1766–1772 (Sep 2004).
18. Balashov, V., R.D. Preston, G.W. Ripley, and L.C. Sparks, *Proc. R. Soc., B*, 146, 460 (1957).
19. Basch, A. and M. Lewin, *J. Polym. Sci., Polym. Chem. Ed. 11*, 3071–3093 (1973).
20. Basu, S.N. *J. Text. Inst. 43*, T278, T355 (1952).
21. Basu, S.N., *J. Text Inst. 39*, T237 (1948).
22. Batra, S.K. and C. Bell, *On the Efficient Utilization of Natural (Vegetable) Fibers*. A Technology Adaptation Program report, Massachusetts Institute of Technology, Cambridge (1975).
23. Bhatia, P., K.C. Gupta, *The Indian Textile Journal*, **101**, 10; 60–62 (1991).
24. Bhattacharjee, H.P., and H.J. Callow, *J. Text. Inst. 43*, T53 (1952).
25. Bhattacharya S.D.; Das A.K., Alkali degumming of decorticated ramie. *Coloration Technology*, 2001, 117, 6, 342–345.
26. Bhattacharya, S.D.; Shah, J.N., Enzymatic treatments of flax fabric, *Textile Research Journal* **74 (7)**: 622–628 (2004).
27. Biagiotti, J.; Puglia, D.; Kenny, José M. A review on natural fibre-based composites-Part I: structure, processing and properties of vegetable fibres. *Journal of Natural Fibers*, 2004, Vol. 1 Issue 2, pp. 37–68.
28. Bock, H. and R. Einsele, *Angew. Chim. 33*, 432 (1940).
29. Brauns, E. and J.J. Yrak, *Paper Trade J.* 125, 55 (1947).
30. Brunauer, S., P.H. Emmet and E. Teller, *J. Am. Chem. Soc.*, **60**, 309 (1938).
31. Buratti L., "The Prince of Fibres," *Selezione Tessile*, 40(3), p. 28, 5p; Apr. (2002).
32. Buschle-Diller G.; Fanter C.; Loth F. Structural changes in hemp fibers as a result of enzymatic hydrolysis with mixed enzyme systems, *Textile Research Journal*, 69(4), 244–252 (1999).
33. Cabradilla, K.E. and S.H. Zeronian, *J. App. Polym. Sci., 19*, (2), 503–518 (1975).
34. Callow, H.J. and J.B. Speakman, *J. Soc. Dyers Color. 65*, 650 (1949).
35. Chakravarty, A.C. and J.W.S. Hearle, *J. Text. Inst. 58*, 651–656 (1967).
36. Chakravarty, A.C., *Text. Res. J. 39*, 878–881 (1969).
37. Chakravarty, A.C., *Text. Res. J.* 41, 4, 318–321 (1971).
38. Chakravarty, A.C., *The Ind. Text. J.*, 81–89, July (1974).
39. Chand, N., S. Sood, Rohatgi, P.K., K.G. Satyanarayana, *J. Sc. Ind. Res. (India)*, **43**, 489–499 September (1984).
40. Chaudhury, S.D. and Q.A. Ahmed, *Jute and Jute Fabrics 7*, 230 (1968).
41. Chilikan, M., Bull. *Feder. Int. Assoc. Chem. Text. Color. 5*, 367 (1935).
42. Chilikan, M., *Mell. Textilber*. 18, 365 (1937).
43. Cierpucha W.; Mankowski J.; Wasko J.; Mankowski T.; Zareba S.; Szporek J. Application of flax and hemp cottonised fibres obtained by mechanical method in cotton rotor spinning, *Fibres & Textiles in Eastern Europe*, Vol. 10 Issue 2 (37), 32 (2002).
44. Cook, J.G., *Handbook of Textile Fibers*. W.S. Coswell, Ltd., UK (1964).
45. Coppen, J.J.W., *Trop. Sci. 21* (3), 163–170 (1979).
46. Couchman, J.F., *J. Text. Inst. 46,* T 735 (1955).
47. Dhyani, K.C., N.G. Paul *Jute Bull.*, **36**, 1 (1974).
48. Doke, S.S.; Sarkar, M.K. Study of sunnhemp & sunnhemp blended DREF II spun yarn, *Man-Made Textiles in India*, Feb 2004, 47, Issue 2, 57.
49. Dujardin, A., *Retting of Flax*. Flax Development Committee, Belfast, Ireland (1948).
50. Dweiba, M.A., B. Hub, O'Donnella, A., Shenton, H.W. and Wool, R.P. All natural composite sandwich beams for structural applications, *Composite Structures* 63 (2): 147–157 (2004).
51. Elkin, H.A., *J. Text. Inst. 26*, P201 (1935).
52. El-Naggar, A.M., M.B. El-Hosamy, A.H. Zaharan, M.H. Zohdy, *Am. Dyestuff Reporter,* **81**, NO. 1: 40–44 (1992).

53. Felsar, H., *Bastfaser 1*, 148 (1941).
54. FRL (Now Albany International Research Co.) *Technical Report* No. 4, Mansfield, Massachusetts, February (1951).
55. Geethamma, V.G.; Pothen, Laly A.; Rhao, Bhaskar; Neelakantan, N.R.; Thomas, Sabu. Tensile stress relaxation of short-coir- fiber -reinforced natural rubber composites, *Journal of Applied Polymer Science*, **94(1)**, 96–104 (2004).
56. Ghosh S.K. Blending Jute with Natural and Manmade Fibers. *Textile Trends* (India), Nov. 1981, Vol. 24, p. 41.
57. Greenhalgh, P., *Trop. Sci. 2* (1), 1–9 (1979).
58. Harris, M., *Harris's Handbook of Textile Fibers*. Harris Res. Lab., Inc., Washington, D.C. (1954).
59. Hearle, J.W.S., *J. Text. Inst. 44,* T117 (1953).
60. Hebert, J.J., R. Giardina, D. Mitchum, and M.L. Rollins, *Text. Res. J.* **40**, 126 (1970).
61. Hermans, P.H., *Physics and Chemistry of Cellulose Fibers,* Elsevier Publishing Co., Amsterdam (1949).
62. Hickey H., Natural fiber composites in automotive applications, *Conference Proceedings: Needlepunch 2002 International Conference* (INDA); Mar. 19–21, (2002)., 14 pages.
63. Himmelfarb, *The Technology of Cordage Fibers and Ropes*, Leon and Hill, London (1957).
64. Hitchcock, S.J., B. McEnaney, and S.J. Watling, *J. Chem. Tech. and Biotechnol. 33A*, 157–163 (1983).
65. Hokens, D.; Mohanty, A.K.; Misra, M.; Drzal, L.T. The influence of surface modification and compatibilization on the performance of natural fiber reinforced biodegradabale thermoplastic composite, *Polymer Preprints*, **43(1)**, 482–483 (2002).
66. Illingsworth, J.W., *J. Text. Inst. 44*, P328 (1953).
67. Jarman, C.G., A.J. Canning, and S. Mykoluk, *Trop. Sci. 20* (2), 91–116 (1978).
68. Jarman, C.G., *Report G29*, Tropical Product Institute, London (1971).
69. Jarman, C.G., *Trop. Sci. 21* (3), 163–170 (1979).
70. Jarman, C.G., V. Laws, *J. Roy. Microbiol. Soc. 84*, (3), 339–346 (1965).
71. Jin C.; Maekawa M. Calculation of degradation rate of pectic substances in ramie and linen fabrics. (English), *Sen'i Gakkaishi* (Journal of the Society of Fiber Science & Technology, Japan); 58 (2), 59, 4p. (Feb. 2002).
72. José C. del Río, Ana Gutiérrez, Ángel T. Martínez, Identification of intact long-chain p-hydroxycinnamate esters in leaf fibers of abaca (Musa textilis) using gas chromatography/mass mass spectrometry, *Rapid Communications in Mass Spectrometry* (2004), 18(22), 2691–2696.
73. José C. del Río, Ana Gutiérrez, Ángel T. Martínez, Identifying acetylated lignin units in non-wood fibers using pyrolysis-gas chromatography/mass spectrometry *Rapid Communications in Mass Spectrometry* (2004), 18(11), 1181–1185.
74. Kalaprasad, G.; Francis, Bejoy; Thomas, Selvin; Kumar, C. Radhesh; Pavithran, C.; Groeninckx, G.; Thomas, Sabu, Effect of fibre length and chemical modifications on the tensile properties of intimately mixed short sisal/glass hybrid fibre reinforced low density polyethylene composites, *Polymer International*, 53(11), 1624–1638 (2004).
75. Kasewell, E.R. and M.M. Platt, *Text. Res. J.* 21, 263–276 (1951).
76. Kirby, R.H. *Vegetable Fibers*. Interscience. New York (1963).
77. Kornreich, E., *Mell. Textilber*. 19, 61 (1938).
78. Krassig, H. and W. Kitchens, *J. Polym. Sci. C. 51*, 123 (1961).
79. Kriese, U., E. Schumann, W.E. Weber, M. Beyer, L. Brühl and Matthäus, Oil content, tocopherol composition and fatty acid patterns of the seeds of 51 *Cannabis sativa* L. genotypes, *Euphytica* 137 (3): 339–351 (June 2004).
80. Kulkarni, A.G., K.A. Cheriyan, K.G. Satyanarayane, and P.K. Rohatgi, *J. App. Polym. Sci. 28*, 625–632 (1983).
81. Kulkarni, A.G., K.G. Satyanarayane, P.K. Rohatgi, and K. Vijayan, *J. Mat. Sci. 18,* 2290–2296 (1983).
82. Kulkarni, A.G., K.G. Satyanarayane and K. Sukumaran, *J. Mat. Sci. 16*, 905–914 (1981).
83. Kulshreshtha, A.K., N.E. Dweltz, and T. Radhakrishnan, *Ind. J. Pure Appl. Phys. 9*, 986 (1971).
84. Kundu, B.C., *Sci. Cult.* 23 (9), 461–470 (1958).

85. Lambrinou, I., *Melliand Textilber*. 56, 277 (1975).
86. Lavitskii, I.N. and A.M. Ipatov, *Tekhnol. Tekstil. Prom. No. 3* (106), 22–26 (1975).
87. Lee, Beom-Goo; Roger M., Rowell, Removal of heavy metal ions from aqueous solutions using lignocellulosic Fibers. *Journal of Natural Fibers*, (2004), 1, 1, 97–108.
88. Lee, H.N., *Supplement No. 1. FRL* (Now Albany International Research Co.) *report*, Dedham, Massachusetts (1948).
89. Lekha, K.R., Field instrumentation and monitoring of soil erosion in coir geotextile stabilised slopes - A case study, *Geotextiles and Geomembranes* 22 (5): 399–413 (2004).
90. Leupin, M., From the plant hemp to textiles http://www3.itv-denkendorf.de/itv2/downloads/d0004920/WG3BucLeupin.pdf (2005).
91. Lewin, M. and J.A. Epstein, *TAPPI 42*, 549 (1959).
92. Lewin, M. and J.A. Epstein, *Text. Res. J. 30*, 520 (1960).
93. Lewin, M., *Chemical Processing of Fibers and Fabrics, Vol. 1 B*. (M. Lewin and S.B. Sello, eds.). Marcel Dekker, New York, pp. 91–256 (1984).
94. Lewin, M., O. Elsner, M. Mielcharek, and T. Bernstein, (a) Israel Pat. 16,324 (1961), U.S. Pat. 3,277,129 (1966); (b) Israel Pat. 16,340 (1961), U.K. Pat 974,878 (1962); (c) Israel Pat. 16,362 (1961), U.K. Pat. 974,876 (1962); (d) Israel Pat. 16,363 (1961), U.K. Pat 974,877 (1961), U.S. Pat. 3.172,884 (1965), Holland Pat. 120,204, (1965), Mexico Pat. 85,181 (1967).
95. Lewin, M., *TAPPI 41* (8), 403–415 (1958).
96. Lewin, M. and J.A. Epstein, *TAPPI 41*, 240 (1958).
97. Li, Y., The investigation of fracture properties of sisal textile reinforced polymers, *Acta Mechanica Solida Sinica* 17 (2): 95–103 (June 2004).
98. Li, Z. Wang, L. Wang, X., Compressive and flexural properties of hemp fiber reinforced Concrete, *Fibers and Polymers* 5 (3): 187–197 (2004).
99. Lin, T., X. Huang, *Journal of China Textile University*, (Eng. Ed.) **11**, 1, 8–14 (1994).
100. Lindberg, G. and P.W. Lange, *Ing. Henlinder Proc.*, 198 (1948).
101. Lindberg, G., *Experientia* 4, 476 (1948).
102. Liu, R.Z., S.G. Zhang, Y.L. Zhao, H.R. Wu, and L.M. Hao, *Weishenguxue Tongbao 8*, No. 5, 209–212 (1981).
103. Luedtke, M. and H. Felsar, *Annalen 549* (1944).
104. Luedtke, M. and H. Felsar, *Bastfaser 1*, 97, 141 (1941).
105. Luedtke, M. and H. Felsar, *Cellulose Chemie 32*, 86 (1943).
106. Mackie, G., Hemp: *Cannabis sativa, Proceedings of the Flax and Other Bast Plants Symposium*, 30 September and 1 October 1997, Poznan, Poland, pp. 50–58.
107. Mahdi, E., Hamouda, A.S.M.; Sen, A.C. Quasi-static crushing behaviour of hybrid and non-hybrid natural fibre composite solid cones, *Composite Structures* 66 (1–4): 647–663 (Oct-Dec 2004).
108. Mauersberger, H.R., Ed., *Mathews Textile Fibers,* 5th ed., John Wiley and Sons, New York (1947).
109. McGovern, J.N., Fibers, vegetable, in *Polymers – Fibers and Textiles. a Compendium,* University of Wisconsin, Madison, WI (1990).
110. McNulty, Sara (Editor), *Report to the Governor's Hemp and Related Fiber Crops Task Force* June 1995: www.globalhemp.com/Archives/Government_Research/USDA/report_to_governor_of_ky.html.
111. Meredith, R., *J. Text. Inst. 36*, T107 (1945).
112. Meredith, R., *J. Text. Inst. 36*, T147 (1945).
113. Meredith, R., *Mechanical Properties of Textile Fibers*. Inter-science Publishers, New York (1956).
114. Meredith, R., *J. Text. Inst. 37*, P469 (1946).
115. Meyer, K.H. and W. Lotmar, *Helv. Chim. Acta 19*, 68 (1936).
116. Mignoni, G., Cannabis as a licit crop: recent developments in Europe. http://www.globalhemp.com/Archives/Government_Research/UN/03_odccp_bulletin.html?print=yes#3.
117. Mitra, G.B. and P.S. Mukherjee, *J. Appl. Cryst. 14*, 421–431 (1981).
118. Mitra, G.B. and P.S. Mukherjee, *Polymer 21*, 1403 (1980).
119. Monrocq, R. *Bull. Inst. Text. de France 4*, No. 3, (15), 209–218 (1975).
120. Morton, W.E. and J.W.S. Hearle, *Physical Properties of Textile Fibers*. The Textile Institute, Manchester, U.K. (1975).
121. Muehlethaler, A., *Biophys. Biochim. Acta 3*, 15, (1949).

122. Mukherjee, P.S. and G.B. Mitra, *Polymer 24*, 525–528 (1983).
123. Mukherjee, R.R. and H.J. Woods, *J. Text. Inst. 41*, T422 (1950).
124. Mukherjee, R.R. and T. Radhakrishnan, *Text. Prog. Text. Inst. 4* (4), (1972).
125. Munder, F.; Fürll, Ch.; Hempel, H. Advanced decortication technology for unretted bast fibers. *Journal of Natural Fibers*, 2004, 1, 1, 49–66.
126. Nam, Sunghyun; Netravali, Anil N. "Characterization of ramie fiber/soy protein concentrate (SPC) resin interface" *Journal Of Adhesion Science and Technology* **18 (9)**: 1063–1076 (2004).
127. Namasivayam, C.; Kavitha, D. "Adsorptive Removal of 2,4-Dichlorophenol from Aqueous Solution by Low-Cost Carbon from an Agricultural Solid Waste: Coconut Coir Pith," *Separation Science and Technology*, 39(6), 1407–1425. (2004).
128. Oh K.W.; Jung E.J.; Choi H.-M., Nonformaldehyde crease-resistant finishing of ramie with glyoxal in the presence of a swelling agent. *Textile Research Journal*, 2001, 71, 3, 225–30.
129. Okinawa Prefecture Cooperative Union, *Kijoka Banana Fiber Cloth,* http://www.kougei.or.jp/english/crafts/0130/f0130.html.
130. O'Shea, M. *Textile Horizon,* **9**, 10, 41 (1989).
131. Pal B.K.; Sharma S.K.; Sen Gupta P. A Study of the Properties of Polyester-Ramie Blended Yarns. *Textile Research Journal*, 52(6), 415 (1982).
132. Peijs T. Composites Turn Green!, *E-polymers*, Feb 11, 2002 Issue T 002, p1, 12p.
133. Peters, R.H., *Textile Chemistry,* vol. 1. Elsevier Publishing Co., Amsterdam (1963).
134. Peters, R.H., *Textile Chemistry,* vol. 2 Elsevier Publishing Co., Amsterdam (1967).
135. Pothan, Laly A.; Neelakantan, N.R.; Rao, Bhaskar; Thomas, Sabu. "Stress relaxation behavior of banana fiber-reinforced polyester composites," *Journal of Reinforced Plastics and Composites*, **23(2)**, 153–165 (2004).
136. Prabhu, G.N., *Coir 1* (1), 17 (1956).
137. Prackowiak, A., *Przegl Wlok*. 35, 471–477 (1981).
138. Prasad, S.V., C. Pavithram, and P.K. Rohatgi, J. Mat Sci. 18, 2043–1454 (1983).
139. Rajan, A., Abraham, T.E., Coir fiber—process and opportunities, *The Journal of Natural Fibers,* To be published.
140. Rath, H. and A. Angster, *Mell. Textilber. 27*, 227 (1946).
141. Ray, P.K., S.C. Bag, and A.C. Chakravarty, *J. App. Polym. Sci. 19*, 99–1004 (1975).
142. Reichert, G. 1994. Hemp (Cannabis sativa 1.). Bi-weekly Bulletin. Agriculture and Agri-Food Canada.
143. Rosamberg, J.A. and F.P. DeFranca, *Appl. Microbiol*. 15, 484 (1967).
144. Roy, M.M. and M.K. Sen, *J. Text. Inst. 43*, T396 (1952).
145. Roy, M.M., *J. Text. Inst. 44*, T44 (1953).
146. Sarkar, P.B. and A.K. Mazumdar, and K.B. Pal, *Text. Res. J.* 22, 529 (1952).
147. Sarkar, P.B., H. Chatterjee, A.K. Mazumdar, and K.B. Pal, *J. Text. Inst. 39*, T1 (1948).
148. Sarkar, P.B., H. Chatterjee, and A.K. Mazumdar, *J. Text. Inst. 38*, T318 (1947).
149. Sarkar, P.B. and A.K. Mazumdar, *Science and Culture 15*, 328 (1950).
150. Savastano, Holmer, Jr.; Warden, Peter G.; Coutts, Robert S.P. Evaluation of pulps from natural fibrous material for use as reinforcement in cement product. *Materials and Manufacturing Processes* **19(5)**, 963–978 (2004).
151. Savinovskii, V.I., *Tekhnol. Tekstil. Prom. No. 2* (105), 35–38 (1975).
152. Schiefer, H.G., L. Fourt, and R.T. Kopf. *Text. Res. J. 17*, 689 (1947).
153. Schmidhauser, O., *Melliand Textilber. 17,* 905 (1936).
154. Searle, G.O., *J. Text. Inst. 20*, T162 (1929).
155. Sen, K.R. and S.K. Bose, *Text. Res. J. 24*, 754–755 (1954).
156. Sen, M.K. and H.J. Woods, *Biophys. Biochim. Acta 3,* 510 (1949).
157. Sen, M.K. and P.H. Hermans, *Nature 164,* 628 (1949).
158. Sen, M.K. and P.H. Hermans, *Rev. Trav. Chim. Pays-Bas 68,* 1079 (1949).
159. Sinha, M.K., *J. Text. Inst.* 65 (1), 27–33 (1974).
160. Sinha, M.K., *J. Text. Inst.* 65 (11), 612–615 (1974).
161. Sloan, FRW, *Rev. Prog. Coloration 5*, 12–15 (1974).
162. Small. E. 1979. *The Species Problem in Cannabis.* Corpus, Toronto.

163. Smith, H.D., *Textile Fibers*, p. 22 (1944).
164. Stout, H.P., J.A. Jenkins, *Ann. Sci. Textiles Belges*, No. 4/12/55, 231 (1955).
165. Stroink, J.B., H.F. Bendel, and D.A. Beerens, *Mell. Textilber, 27*, 139 (1946).
166. Sugai, Jitsuo; Joko, Kyohei; Hayashi, Toshio; Arai, Motoo. Sen'i Gakkaishi (Journal of the Society of Fiber Science & Technology, Japan); (2004), Vol. 60 Issue 1, p. 16, 5p, 7
167. *TAPPI* Standard T13M (1955).
168. Tayag, C.E., Y. Watanabe, T. Hatakeyama, *Sen-i Gakkaishi,* **47**, 8, 434–438 (1991).
169. Teli M.D.; Adivarekar R.V.; Pardeshi P.D. Dyeing of sisal, jute & coir fibres with various dyes. *Indian Textile Journal*, 113, 2, 13, 6p.
170. Thieme, J.G., *The Textile Quarterly 5* (1), 56–65 (1955).
171. TI: *Identification of Textile Materials,* 7th ed. The Textile Institute, Manchester, U.K. (1975).
172. Timell, T.E. *Text. Res. J. 27*, 854–859 (1957).
173. Turner, A.J., *The Structure of Textile Fibers* (A.R. Urquhart and F.O. Howitt, eds.). The Textile Institute, Manchester, U.K., pp. 91–117 (1953).
174. U.S. Pat. 3,521,991 (1967).
175. USDA report: *Industrial Hemp in the United States: Status and Market Potential*, (January 2000).
176. Val'ko, N.I., S.I. Antonov, and A.N. Gavrilova, *Tekhnol. Tekstil. Prom.* No. 4 (148), 22–23 (1982).
177. Varma, D.S., M. Varma, I.K. Varma, *Proc. ACS Conference*, April 13–16 NY, NY (1986)
178. Varma, D.S., M. Varma, I.K. Varma, *Tex. Res. J.* **54**, 12, 827–832 (1984).
179. Venkateswarlu, K. and K.G. Menon, *Acta. Phys. Polon.* **33**, 605 (1968).
180. Wakida T.; Hayashi A.; Lee M.S.; Lee M.; Doi C.; Okada S.; Yanai Y. Dyeing and mechanical properties of ramie fabric treated with liquid ammonia. (English) *Sen'i Gakkaishi* (Journal of the Society of Fiber Science & Technology, Japan); **57 (5)**, p. 148, 5p (May 2001).
181. Wang, D. et al., *J. China Eng. Assn.* **4,** No. 2, 83–86 (1983).
182. Wang, H.M.; Postle, R. Improving the color features of hemp fibers after chemical preparation for textile applications. *Textile Research Journal*, Sep 2004, Vol. 74 Issue 9, pp. 781–6.
183. Wang, H.M.; Postle, R.; Kessler, R.W.; Kessler, W. Removing pectin and lignin during chemical processing of hemp for textile applications. *Textile Research Journal*, 2003, 73, 8, 664–669.
184. Warwicker, J.O., *FAO Report*, HF RS No. 14, (June 1974).
185. Watanabe, S., K. Kimura, and T. Akahori, *Hokkaido Daigaku Kogakubu Kenkyu Hokaku* **47**, 121 (1968).
186. Watzl, Alfred., Ultrasonic flax retting process, *International Textile Bulletin*, Vol. **49** Issue 5, p. 42, 5p (Oct 2003).
187. Weindling, L., *Long Vegetable Fibers*. Columbia University Press, New York (1947).
188. Wenbang, Z., *J. China Textile University,* (Eng. Ed.) 1, 106–117 (1986).
189. West, David P., *Fiber Wars: The Extinction of Kentucky Hemp*: http://www.globalhemp.com/. Archives/Essays/Fiber/fiber_wars.html.
190. West, David P., *Industrial Hemp Farming:* http://www.gametec.com/hemp/IndHmpFrmfr.html.
191. Whitford, A.C., *Textile Age* **8**, pp. 50, 58, 70, 74 (1944).
192. Whiting, G.C., *Nature 168*, 833 (1951).
193. Williams, Paul T.; Reed, Anton R. High grade activated carbon matting derived from the chemical activation and pyrolysis of natural fiber textile waste, *Journal of Analytical and Applied Pyrolysis*, **71 (2)**, 971–986. (2004).
194. Wilson, P., *FAO Report*, HFRS. NO. 8 (1971).
195. Wise, L.E. and E.C. Jahn, *Wood Chemistry*. Rheinholt Publishing Corp., New York (1952).
196. Yu C., Zhang Y., Zheng L. Properties and processing of pineapple and banana fiber. International Textile Clothing & Design Conference: Magic World of Textiles, Book of Proceedings, Oct. 6–9, 2002, p. 147, 5p.
197. Zhou, L.M.; Yeung, K.W.; Yuen, C.W.M.; Zhou, X. Tensile strength loss of mercerized and crosslinked ramie fabrics. *Textile Research Journal*, 73 (4), 367–72 (2003).
198. Zhou, L.M.; Yeung, K.W.; Yuen, C.W.M.; Zhou, X. Effect of mercerisation and crosslinking on the dyeing properties of ramie fabric. *Color Technol.*, 119, 3, 170–176 (2003).

9 Cotton Fibers

*Philip J. Wakelyn, Noelie R. Bertoniere, Alfred D. French,
Devron P. Thibodeaux, Barbara A. Triplett, Marie-Alice
Rousselle, Wilton R. Goynes, Jr., J. Vincent Edwards,
Lawrance Hunter, David D. McAlister, and Gary R. Gamble*

CONTENTS

9.1 General Description of Cotton 523
9.2 Biosynthesis of Cotton 530
9.3 Chemical Composition of Cotton 536
9.4 Solvents for Cotton 540
9.5 Structural Properties of Cotton 542
    9.5.1 Structure Overview 542
    9.5.2 Cellulose Molecule—Constitution and Molecular Weight Distributions 546
        9.5.2.1 Constitution of Chain 546
        9.5.2.2 Molecular Weight Distributions 548
    9.5.3 Three-Dimensional Structures of Cellulose Molecules,
        Crystallites, and Fibers 551
        9.5.3.1 Describing Molecular Shape 551
        9.5.3.2 Experiment- and Theory-Based Modeling Studies 552
        9.5.3.3 Experimental Studies on Cellulose Structure 558
        9.5.3.4 Electron Microscopy and Lattice Imaging 565
        9.5.3.5 Small Angle X-Ray Scattering 566
        9.5.3.6 Diffraction Studies of Crystallite Size 566
        9.5.3.7 Crystallite Orientation by Diffraction 566
    9.5.4 Crystal and Microfibrillar Structure by Chemical Methods 567
        9.5.4.1 Sorption 567
        9.5.4.2 Acid Hydrolysis 571
        9.5.4.3 Formylation 571
        9.5.4.4 Periodate Oxidation 572
        9.5.4.5 Chemical Microstructural Analysis 572
        9.5.4.6 Summary of Average Ordered Fraction
               Values Determined by Chemical Methods 573
    9.5.5 Fiber Structure 574
        9.5.5.1 Morphology 574
        9.5.5.2 Pore Structure 583
9.6 Chemical Properties of Cotton 584
    9.6.1 Swelling 587
        9.6.1.1 Water 587
        9.6.1.2 Sodium Hydroxide 587
        9.6.1.3 Liquid Ammonia 589

- 9.6.2 Etherification 591
  - 9.6.2.1 General 591
  - 9.6.2.2 Wrinkle Resistance 591
  - 9.6.2.3 Reactive Dyeing 592
  - 9.6.2.4 Flame Resistance 593
  - 9.6.2.5 Multifunctional Properties 597
- 9.6.3 Esterification 597
  - 9.6.3.1 General 597
  - 9.6.3.2 Acetylation 597
  - 9.6.3.3 Formaldehyde-Free Wrinkle Resistance 598
  - 9.6.3.4 Esters of Inorganic Acids 599
- 9.6.4 Degradation 600
  - 9.6.4.1 General 600
  - 9.6.4.2 Oxidation 601
  - 9.6.4.3 Acid 603
  - 9.6.4.4 Alkali 604
  - 9.6.4.5 Biodegradation 608
  - 9.6.4.6 Pyrolysis or Thermal Degradation 610
  - 9.6.4.7 Visible, Ultraviolet, and High-Energy Radiation 614
- 9.6.5 Weather Resistance 615
- 9.6.6 Enzymatic Modification 616
- 9.6.7 Corona 619
- 9.6.8 Dyeing 620

9.7 Physical Properties of Cotton 621
- 9.7.1 Maturity and Fineness 621
- 9.7.2 Tensile Strength 623
- 9.7.3 Elongation, Elasticity, Stiffness, Resilience, Toughness, and Rigidity 625
- 9.7.4 Electrical Properties 627
- 9.7.5 Advanced Fiber Information System 628

9.8 Cotton Fiber Classification and Characterization 628
- 9.8.1 U.S. Classification System 630
  - 9.8.1.1 Sampling 630
  - 9.8.1.2 Grade 630
  - 9.8.1.3 Length 633
  - 9.8.1.4 Strength 633
  - 9.8.1.5 Linear Density 633
  - 9.8.1.6 Color (Measurement of Reflectance and Yellowness) 634
  - 9.8.1.7 Trash Content (Measurement) 634
  - 9.8.1.8 Reporting Cotton Classification Results 635
- 9.8.2 Classification Systems of Some Countries Other than the United States 635
  - 9.8.2.1 China 635
  - 9.8.2.2 Central Asia Republics 635
  - 9.8.2.3 Francophone Africa 636
  - 9.8.2.4 Sudan 637
  - 9.8.2.5 South Africa 637
  - 9.8.2.6 Egypt 637
  - 9.8.2.7 India 638
  - 9.8.2.8 Pakistan 638
  - 9.8.2.9 Australia 638

           9.8.3   Impact and Future of HVI Systems 638
    9.9   Production, Consumption, Markets, and Applications 639
           9.9.1   Production, Consumption, and Markets 639
           9.9.2   Applications 641
           9.9.3   Future Trends 644
    9.10 Environmental, Workplace, and Consumer Considerations 645
           9.10.1   Environment 645
                    9.10.1.1   Environmental Stewardship 645
                    9.10.1.2   Emissions to the Environment 645
           9.10.2   Workplace 645
                    9.10.2.1   Inhalation or Respiratory Disease 645
                    9.10.2.2   Skin Irritation or Dermatitis 646
                    9.10.2.3   Formaldehyde 647
           9.10.3   Consumer 647
                    9.10.3.1   Formaldehyde 647
References 647

9.1 GENERAL DESCRIPTION OF COTTON

Cotton (Figure 9.1) is the most important natural textile fiber, as well as cellulosic textile fiber, in the world, used to produce apparel, home furnishings, and industrial products. Worldwide about 38% of the fiber consumed in 2004 was cotton [1]. Cotton is grown mostly for fiber but it is also a food crop (cottonseed)—the major end uses for cottonseeds are vegetable oil for human consumption; whole seed, meal, and hulls for animal feed; and linters for batting and chemical cellulose.

Its origin, development, morphology, chemistry, purification, and utilization have been discussed by many authors [2–12]. The chemistry, structure, and reaction characteristics of cellulose, the carbohydrate polymer that forms the fiber, are thoroughly treated in a number of excellent works [8,9,12–19]. This chapter is intended to provide an overview of the current state of knowledge of the cotton fiber. Much of the information reported here is taken from the references cited at the end of the chapter, which should be consulted for a more in-depth treatment.

FIGURE 9.1 Mature cotton in the field ready for harvesting. (Courtesy of the National Cotton Council of America, Memphis, TN.)

Cotton fibers are seed hairs from plants of the order Malvales, family Malvaceae, tribe Gossypieae, and genus Gossypium [2–5,10,11]. Botanically, there are four principal domesticated species of cotton of commercial importance: *hirsutum, barbadense, aboreum, and herbaceum*. Thirty-three species are currently recognized; however, all but these four are wild shrubs of no commercial value. Each one of the commercially important species contains many different varieties developed through breeding programs to produce cottons with continually improving properties (e.g., faster maturing, increased yields, and improved insect and disease resistance) and fibers with greater length, strength, and uniformity.

Gossypium hirsutum, a tetraploid, has been developed in the United States from cotton native to Mexico and Central America and includes all of the many commercial varieties of American Upland cotton. Upland cottons now provide over 90% of the current world production of raw cotton fiber. The lengths, or staple lengths, of the Upland cotton fiber vary from about $\frac{7}{8}$ to $1\frac{1}{2}$ in. (22–36 mm), and the micronaire value (an indicator of fiber fineness and maturity but not necessarily a reliable measure of either; see Section 9.8) ranges from 3.8 to 5.0. If grown in the United States, *G. hirsutum* lint fibers are 26–30 mm (1 to 1–3/16 in.) long [20]. Fiber from *G. hirsutum* is widely used in apparel, home furnishings, and industrial products.

Gossypium barbadense, a tetraploid, is of early South American origin and provides the longest staple lengths. The fiber is long and fine with a staple length usually greater than $1\frac{3}{8}$ in. (35 mm) and a micronaire value of below 4.0. If grown in the United States., *G. barbadense* lint fibers are usually 33–36 mm ($1\frac{5}{6}$ to $1\frac{1}{2}$ in.) long [20]. Commonly known as extra-long-staple (ELS), it supplies about 8% of the current world production of cotton fiber. This group includes the commercial varieties of Egyptian, American–Egyptian, and Sea Island cottons. Egypt and Sudan are the primary producers of ELS cottons in the world today. Pima, which is also ELS cotton, is a complex cross of Egyptian and American Upland strains and is grown in the western United States (mainly California with some in Arizona, southwestern Texas, and New Mexico), as well as in South America. Pima has many of the characteristics of the better Egyptian cottons. This fiber from *G. barbadense* is used for the production of high-quality apparel, luxury fabrics, specialty yarns for lace and knitted goods, and sewing thread.

The other commercial species—*Gossypium aboreum* and *Gossypium herbaceum*, both diploids—are known collectively as "Desi" cottons, and are the Asiatic or Old World short staple cottons. These rough cottons are the shortest staple cottons cultivated (ranging from $\frac{3}{8}$ to $\frac{3}{4}$ in. (9.5–19 mm)) and are coarse (micronaire value greater than 6.0) compared with the American Upland varieties. Both are of minor commercial importance worldwide but are still grown commercially in Pakistan and India. *G. aboreum* is also grown commercially in Burma, Bangladesh, Thailand, and Vietnam [3,10].

Varietal development programs were once confined to classical methods of breeding that rely on crossing parents within species. Currently, in addition to the conventional breeding methods, research is underway on hybrids to produce new varieties, and modern biotechnology (recombinant DNA technology) to produce biotech or transgenic cottons [21,22], which enhance production flexibility. Biotech cottons deliver high-tech options to farmers and consumers without compromising environmental quality. Since the introduction of *Bacillus thuringiensis* (Bt) biotech cotton in 1996, cotton has been one of the lead crops to be genetically engineered, and biotech cotton has been one of the most rapidly adopted technologies ever. The current varieties of commercial importance address crop management or agronomic traits that assist with pest management (insect resistant) or weed control (herbicide tolerance). Nine countries, representing over 59% of world cotton area, allow biotech cotton to be grown: Argentina, Australia, China (Mainland), Colombia, India, Indonesia, Mexico, South Africa, and United States. Other countries (e.g., Pakistan, Brazil, Burkina Faso, Egypt, and the Philippines) are considering approving the cultivation of biotech cotton. In

2003–2004, about 21% of the world's cotton acreage and about 30% of the cotton produced in the world was biotech cotton; in 2004–2005, about 35% of world's cotton production was biotech cotton; and within five years, world biotech cotton production could be close to 50%.

The initial biotech efforts have been centered on insect resistance and herbicide tolerance. Insect resistance has been conferred through the incorporation of genes from Bt that produce Bt δ-endotoxin, a naturally occurring insect poison for bollworms and budworms. The reduction in the use of insecticides minimizes the adverse effects on nontarget species and beneficial insects. Herbicide tolerance enables reduced use of herbicides and encourages use of safer, less persistent materials to control a wide spectrum of weeds that reduce yield and lint quality of cotton. Transgenic herbicide tolerance moves cottonweed management away from protective, presumptive treatments toward responsive, as-needed treatments. While insect resistance and herbicide tolerance are the only traits currently available in biotech cottons, a broad range of other traits are under development using modern biotechnology. These may impact the agronomic performance, stress tolerance, fiber quality, and yield potential directly. However, in 2005 few of these traits are close to commercialization. As soon as new developments in bioengineered cotton for insect resistance, herbicide tolerance, stress tolerance, yield potential, improved fiber quality, etc., are available, they will be incorporated into conventional cotton varieties.

The cottons of commerce are almost all white (creamy yellow to bright white). In recent years, there has been a renewed interest in naturally pigmented and colored cottons, which have existed for over 5000 years [23,24]. These cotton varieties are spontaneous mutants of plants that normally produce white fiber. The availability of inexpensive dyes and the need for higher-output cotton production worldwide caused the naturally colored cottons to almost disappear about 50 years ago. Yields were low and the fiber was essentially too short and weak to be machine spun. Breeding research over the last 15 years reportedly has led to improvement in yields, fiber quality, fiber length and strength, and color intensity and variation [25]. Naturally colored cottons are a very small niche market. The cottons available today are usually shorter, weaker, and finer than regular Upland cottons, but they can be spun successfully into ring and rotor yarns for many applications [26]. They can be blended with normal white cottons or blended among themselves. For a limited number of colors, the use of dyes and other chemicals can be completely omitted in textile finishing, possibly generating some savings, which can compensate for the higher raw material price. The color of the manufactured goods can intensify with washing (up to 5 to 10 washings), and colors vary somewhat from batch to batch [26]. Naturally colored cottons are presently grown in China, Peru, and Israel. The amount available in 2005 is very small, perhaps 10,000 U.S. bale equivalents (about 2270 metric tons). Shades of brown and greens are the main colors that are available. Other colors (mauve, mocha, red) are available in Peru in a very limited supply and some others are under research. The color for brown and red-brown cottons appears to be in vacuolar tannin material bodies in the lumen (Figure 9.2a). The different colors of brown and red-brown are mostly due to catechin–tannins and protein–tannin polymers [27]. The green color in cottons (Figure 9.2b) is due to a lipid biopolymer (suberin) sandwiched between the lamellae of cellulose microfibrils in the secondary wall [27–29]. The brown fibers (and white lint) do not contain suberin. Green cotton fibers are characterized by high wax content (14–17% of the dry weight) whereas white and brown fibers contain about 0.4–1.0% wax [27].

Cotton grown without the use of any synthetically compounded chemicals (i.e., pesticides, fertilizers, defoliants, etc.) is considered as "organic" cotton [30–34]. It is produced under a system of production and processing that seeks to maintain soil fertility and the ecological environment of the crop. To be sold as organic it must be certified. Certified organic cotton was introduced in 1989–1990 and over 20 countries have tried to produce organic cotton. Since 2001, Turkey has been the largest producer of organic cotton. There are small projects

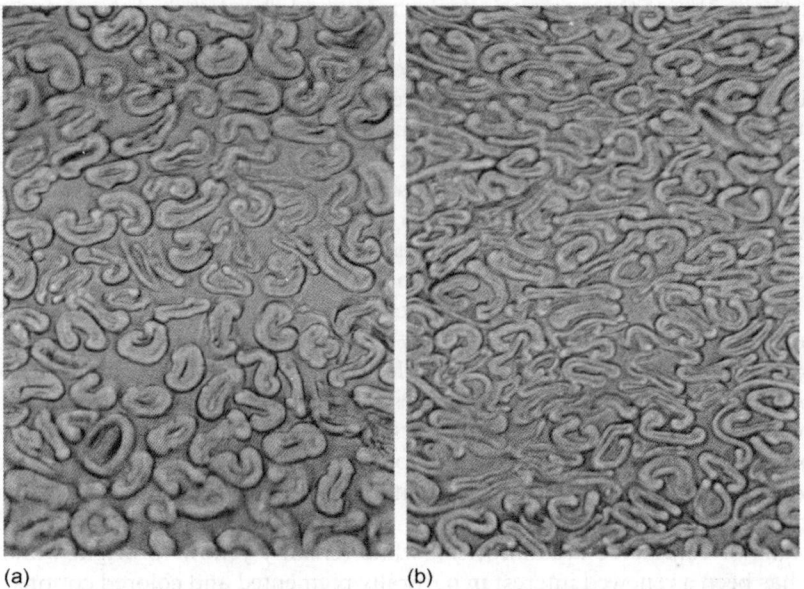

(a)   (b)

FIGURE 9.2 Fiber bundle cross-sections obtained with transmitted light microscopy (a) for natural brown cotton where the presence of material bodies are visible within the lumens of some fibers and (b) for natural green cotton where the fibers are quite immature and are chacterized by the presence of suberin in the fiber walls and not material bodies in the lumen.

in Mali, Kyrgyzstan, and some other developing countries [35]. In 2001, about 14 countries in the world produced about 27,000 U.S. bale equivalents (5700 metric tons) of organic cotton, with Turkey, the United States, and India accounting for about 75% of production [36]. In 2003, in the United States, 4628 U.S. bale equivalents of organic cotton were produced; in 2004, 4674 acres were planted and about 5000 U.S. bale equivalents were produced [37]. In 2005, the world's production of organic cotton was about 100,000 U.S. bale equivalents (about 25,000 metric tons), which is less than 0.12% of world's total cotton production [38].

    Unlike synthetic fibers, which are spun from synthetic or regenerated polymers in factories, cotton fiber is a natural agricultural product. The United States and some other countries use the newest and latest tested technology to produce the cotton crop [8,9]. Cotton production is scattered among hundreds of thousands of small, independent farms (about 20,000–25,000 in the United States), whereas the synthetic fiber industry is concentrated in a few corporations that have production plants worldwide. Cotton is grown in about 80 countries in the world; in 2004–2005, 59 countries cultivate at least 5000 ha (12,350 acres (1 ha = 2.47 acres)) [39]. In many developing countries, cotton farms range in size from less than 1 to about 10 ha. In 2004, developing countries accounted for about 75% of the world's production, and China, the United States, India, Pakistan, Uzbekistan, Turkey, and Brazil account for over 81% of the world's cotton production [40]. Cotton production in the United States is usually about 18–20 million U.S. bale equivalents (8.6–9.6 billion pounds (3.9–4.4 million metric tons)) annually and worldwide production is about 90–105 million bales (45.0–52.5 billion pounds (20.4–23.8 million metric tons)) annually. Raw cotton is exported from about 57 countries and cotton textiles from about 65 countries.

    Cotton production, harvesting, and ginning are described in more detail in other sources [2–5,8,9,41,42]. The cotton plant is a tree or a shrub that grows naturally as a perennial, but for commercial purposes it is grown as an annual crop. Botanically, cotton bolls are fruits. Cotton is a warm-weather plant, cultivated in both hemispheres, mostly in North and South America, Asia, Africa, and India (in tropical latitudes). Mostly it is cultivated in the Northern Hemisphere. It is primarily grown between 37°N and 32°S but can be grown as far north as 43°N latitude in Central Asia and 45°N in mainland China. Planting time for cotton varies with locality, i.e., from February to June in the Northern Hemisphere [2–5,8–11]. The time of planting in the Northern Hemisphere is the time of harvest in the Southern Hemisphere. Seedlings emerge from the soil within a week or two after planting, 5–6 weeks later flower buds or squares form, and white (Upland cotton) or creamy yellow (Pima cotton) to dark-yellow blossoms appear in another 3–4 weeks. The time interval from bloom to open boll is about 40–80 days. The open boll lets air in to dry the white, clean, fiber, and fluff it for the harvest (Figure 9.3).

Each cotton fiber is a single, elongated, complete cell that develops in the surface layer of cells of the cottonseed. The mature cotton fiber is actually a dead, hollow, dried cell wall [4,5,43]. In the dried out fiber, the tubular structure is collapsed and twisted, giving cotton fiber convolutions, which differentiate cotton fibers from all other forms of seed hairs and are partially responsible for many of the unique characteristics of cotton. The biosynthesis and morphology of the cotton fiber are discussed in more detail later (see Section 9.2).

The relatively long fiber lengths (about 1 in. (25.4 mm) or longer) on the cottonseed relate to the fiber that is used by the textile industry. This raw cotton fiber, which can be spun into textile yarns, is called lint. However, another type of fiber, *linters* or *fuzz fibers*, which are very short, is also produced on the seed along with the lint [44]. The distribution of the lint and fuzz fibers over the seed surface is neither uniform nor random. The base of the seed mostly produces lint fibers, whereas cells near the tip of the seed mostly produce fuzz fibers.

The long lint fibers are removed at the cotton gin, but the short linter fibers are still attached to the seed after ginning for *G. hirsutum* (Upland cotton); *G. barbadense* seed does not contain linters. The linter fibers are removed at the cottonseed oil mill by the delintering process prior to the oil extraction process, producing fibers of different lengths. In general, ginned cottonseed is composed of about 8% linters. Linters can be distinguished from lint by several characteristics:

1. Length (commercial lint averages 1 in. (25.4 mm) and first-cut linters average about 0.5 in. (12–15 mm)
2. Pigmentation (linters are often light brown or highly colored)

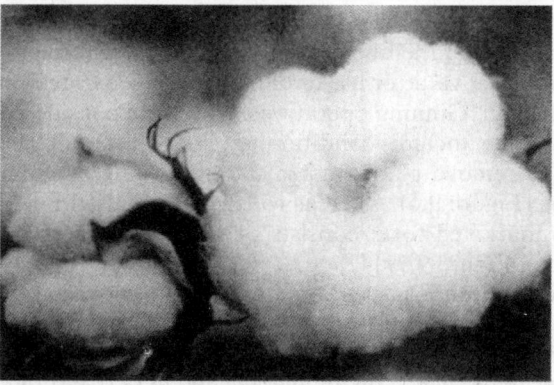

FIGURE 9.3 Open cotton boll. (Courtesy of the National Cotton Council of America, Memphis, TN.)

3. Strength of adherence to seed (linters are more tightly held to the seed)
4. Chemical and physical properties

Linters are also much coarser than lint fibers; linters have a coarse, stiff form without the flexibility or convolutions of lint fibers; and the tips of linter fibers are somewhat tapered but not to the same extent as those of lint fibers. Linter fibers have hardly any lumen. The diameter of the linter fiber is usually about twice the diameter of the lint fiber. In lint fibers, the growth of the primary cell wall starts on the day of flowering and within 20 days they exhibit a 3000-fold increase in length, whereas fuzz fibers (linters) initiate elongation about 4 days after flowering. The secondary cell wall, formation of which begins about 20 days after flowering, is different in lint and in fuzz fibers; for example, the secondary cell wall at the base of the fiber is much thicker in linters than in lint. Although not suitable for textile processing, the linters are used in other applications, for example, second-cut linters (the shortest linter fiber) are used as a chemical feedstock for manufacturing plastics and rayons, whereas first-cut linters (the longer fiber) can be used for batting and padding in bedding, upholstered furniture, automotive applications, and paper (most currencies are made of cotton paper).

Harvesting time for cotton varies with locality (Table 9.1).

The operations of harvesting [11] and ginning the fiber [41], as well as cultural practices during the growing season, are very important to the quality of the cotton fiber [5,11]. Harvesting is one of the final and most important steps in the production of a cotton crop [5,9,11,12], as the crop must be harvested before the inclement weather can damage the quality and reduce the yield. Because of economic factors, virtually the entire crop (>99%) in the United States and Australia is harvested mechanically (Figure 9.4). In rest of the world (~75%) hand harvesting of cotton, one boll at a time, is still quite prevalent, particularly in the less developed countries and in countries where the labor is cheaper [45].

Mechanically harvested cotton, either with cotton picker machines (cotton burr remains attached to the stalk) or with stripper machines (cotton burr is removed along with the seed cotton), can contain more trash and other irregularities than hand-harvested cotton. However, according to "Cotton contamination surveys" by the International Textile Manufactures Federation (ITMF), the most contaminated cottons originate from some of the countries where cotton is hand-picked, whereas some of the cleanest can be sourced in the USA where cotton is machine harvested [34]. Most of the mechanically harvested cotton is harvested with cotton pickers (~75% in the United States and all in Australia).

After harvesting, the seed cotton (consisting of cotton fiber attached to cottonseed and plant foreign matter), a raw perishable commodity, is transported to the ginning plant in trailers or modules, or is stored in the field in modules. In the United States, module storage is used for almost the entire crop. Field storage in modules maximizes efficiency at the gin.

Ginning is the separation of the fibers from seed and plant foreign matter [41,42]. Cotton essentially has no commercial value or use until the fiber is separated from the cottonseed and the foreign matter at the gin. Ginning operations, which are considered a part of the harvest, are normally considered to include conditioning (to adjust moisture content), seed–fiber separation, cleaning (to remove plant trash), and packaging (Figure 9.5). Upland cottons are ginned on saw gins (Figure 9.6), whereas roller gins are used for ELS cottons.

Raw cotton in its marketed form consists of masses of fibers in densely packed bales (>22 lb/ft^3 (352 kg/m^3), Figure 9.7) [42].

The bales into which cotton is packed are of varying dimensions, volumes, densities, and weights (see Table 9.1) and are mainly covered with woven polypropylene, polyethylene film, burlap, or cotton fabrics (Figure 9.8).

In the United States, bales weigh on an average 490 lb (222 kg). A single pound of cotton may contain 100 million or more individual fibers (about 50 to 55 billion fibers in a 480-lb

TABLE 9.1
Typical Weight and Densities of Cotton Bales in Various Countries and Harvest Date

Country	Harvest	Bale weight avg. kg	Bale weight avg. lb	Bale weight range kg	Bale weight range lb	Density kg/m³	Density lb/ft³
North and Central America							
U.S.	July–Jan.	225	495	217–230	477–506	427.1	26.7
Mexico	June–Jan.	227	499	180–240	396–528	380.1	23.8
Guatemala	Nov.–Mar.	227	499	220–236	484–519	251.4	15.7
South America							
Brazil	Aug.–Jan. or Feb.–May	140	308	110–180	242–396	243.1	15.2
Argentina	Feb.–June	220	484	195–250	429–550	346–429	21.6–26.8
Paraguay	Feb.–June	198	436	160–225	352–495	420.9	26.3
Columbia	July–Sept. or Dec.–Mar.	225	495	220–240	484–528	516.2	32.3
Peru	Feb.–Oct.	240	528				
Venezuela	Feb.–May						
Europe							
Greece	Sept.–Nov.	234	516				
Spain	Sept.–Nov.	225	495	200–250	440–550	396.8	24.8
Asia–Oceania							
Uzbekistan	Sept.–Nov.	200	440	190–220	418–484	490.2	30.6
China Bale I	Sept.–Nov.	85	187	80–90	176–198	442.7	27.7
Bale II	Sept.–Nov.	200	440	190–210	418–462	439.8	27.5
India	July–Jan. or Dec.–May	170	375	165–175	363–385	379.8	23.7
Pakistan	Sept.–Feb.	170	375	160–185	352–407	547.5	34.2
Turkey	Sept.–Dec.	217	478	190–245	418–539	342.3	21.4
Australia	Apr.–June	227	499	150–240	330–528	449.6	28.1
						327.6	20.4
Iran	Oct.–Dec.	200	440	170–240	375–528	226.8	14.1
Syria	Sept.–Nov.	205	451	192–212	422–466	425–434	26.6–27.1
Africa							
Egypt	Sept.–Oct.	327	720	290–345	638–759	541.6	33.8
Sudan	Jan.–Apr. or Sept.–May	188	414	185–191	407–420	361–397	22.6–24.8
South Africa	Apr.–May	206	453	170–240	375–528	437.4	27.3
Ivory Coast	Oct.–Jan.	215	474				
Tanzania	May–July	183	403				
Nigeria	Dec.–Feb.	185	407				

Source: Bale weights—From Munro, J.M., *Cotton*, 2nd ed., Longman Scientific and Technical, Essex, England and John Wiley & Sons, New York, 1987, p. 333; Bale survey—From International Cotton Advisory Committee, Oct. 1995; and Harvest date—From Volkart Brothers Holding, Ltd., Switzerland, 1991.

bale). In the rest of the world, bales usually weigh 375–515 lb (170–233 kg), depending on the country where they are produced (Table 9.1) [19,41]. The *International Organization for Standardization* (ISO) specifies (ISO 8115 [46]) that bale dimensions and densities should be: length 1400 mm (55 in.), width 530 mm (21 in.), height 700–900 mm (27.5–35.4 in.), and

FIGURE 9.4 (a) Mechanical harvesting by means of a cotton picker. (b) Mechanical harvesting by means of a cotton stripper. (Courtesy of the National Cotton Council of America, Memphis, TN.)

density of 360–450 kg/m^3 (22.5–28.1 lb/ft^3). Typical bale dimensions for the U.S. universal density (UD) bales are 1400 mm (55 in.) length × 533 mm (21 in.) width × 736 mm (29 in.) height, resulting in a bale density of approximately 448 kg/m^3 (28 lb/ft^3).

At the gin, baled cotton is sampled so that grade and other quality parameters can be determined and the cotton is classed at the U.S. Department of Agriculture (USDA) classing offices. Classification is a way of measuring the fiber quality and physical attributes of this natural product that affect the manufacturing efficiency and quality of the finished product (see Section 9.8). Cotton bales are usually stored in warehouses prior to going to the textile mill [42].

Cotton is merchandized and shipped by a cotton merchant prior to arriving at the textile mill, where it is manufactured into products for the ultimate consumer. The marketing of cotton is a complex operation that includes transactions involving the buying, selling, or reselling from the time the cotton is ginned until it reaches the textile mill. In the United States, after the cotton is ginned and baled, growers usually sell their cotton to a merchant or store it in a government-approved warehouse and borrow money against it. In some other countries, the cotton is sold as seed cotton and the buyer has it ginned and then sells the lint cotton.

When cotton is grown and processed in a responsible manner, it does not have adverse effects on the environment, the workplace, and the consumer [47] (see Section 9.10).

9.2 BIOSYNTHESIS OF COTTON

The cotton fibers used in textile commerce are the dried cell walls of formerly living cells. Botanically, cotton fibers are trichomes or seed coat hairs that differentiate from epidermal cells of the developing cottonseed. The cotton flower blooms only for one day and quickly becomes senescent thereafter. On the day of full bloom, or anthesis, the flower petals are pure white in most *G. hirsutum* varieties. By the day after anthesis, the petals turn bright pink in color and, usually by the second day after anthesis, the petals fall off the developing carpel (boll). The day of anthesis serves as a reference point for all subsequent events in the seed and fiber development.

Typically, there are 20–30 ovules in a boll containing three to five segmented compartments (locules). The ovules are attached to the plant via a connection called the funiculus. Fertilization of the ovules is essential for subsequent development of the seed and fiber. When there is a failure in the fertilization of a seed, development is aborted and the resulting mote may cause problems during later fiber and fabric processing steps. It is possible to remove unfertilized ovules from the boll and mimic the conditions for seed and fiber growth in tissue

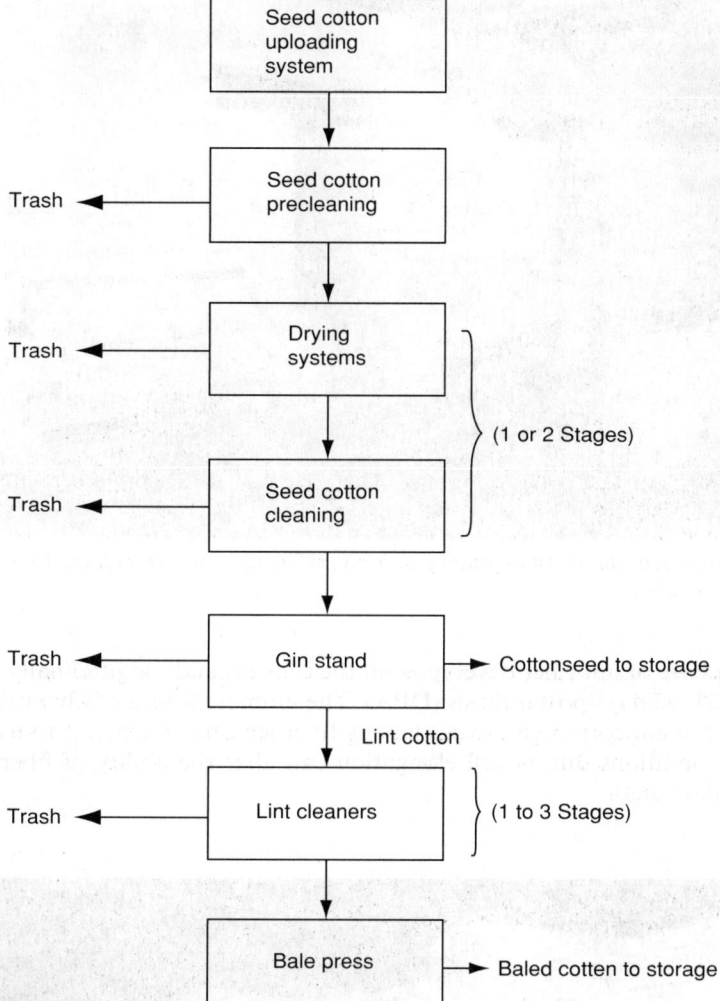

FIGURE 9.5 Simplified flow chart of ginning systems. (Courtesy of the National Cotton Council of America, Memphis, TN.)

culture [48]. Two phytohormones, namely auxin and gibberellic acid, are required in ovule culture to stimulate seed and fiber development.

On, or slightly before the day of anthesis, the morphological events marking the initiation of fiber development are evident. Lint fiber development initiates first at the more rounded end (chalaza) of the seed and proceeds around the seed surface to the micropyle. The fiber cells first assume a fairly rounded or bulbous appearance and are visible above the formerly smooth ovule epidermal surface. Approximately one out of every four epidermal cells begins this cellular differentiation process [49]. The physiological and biochemical factors that regulate which epidermal cells will become fiber cells are unknown at present. Approximately 6–7 days later, a second type of fiber cells called fuzz fibers (linters) begin growing. Fuzz fibers are distinguished from lint fibers by their larger perimeter, shorter length, and final chemical composition [50].

FIGURE 9.6 Cotton gin stand, saw-ginning process. (Courtesy of the National Cotton Council of America, Memphis, TN.)

In the next phase of lint fiber development, the cells expand longitudinally, reaching their final lengths in 21–35 days postanthesis (DPA). The ultimate length of fiber cells is controlled genetically by different cotton genotypes having fiber lengths of nearly 1 to 6 cm. Abnormal environmental conditions during cell elongation can alter the ability of fiber cells to reach their full potential length.

FIGURE 9.7 Packaged bales (universal density) ready to be shipped to the textile mill. (Courtesy of the National Cotton Council of America, Memphis, TN.)

FIGURE 9.8 A newly bound bale comes off the press at conclusion of ginning process. (Courtesy of the National Cotton Council of America, Memphis, TN.)

During the elongation phase of fiber development, the cell is delimited by a primary cell wall and covered by a waxy layer or cuticle (Figure 9.9).

Recently, genotypic variation in mature cotton fiber surface components, presumably from the cuticle, has been reported [51]. In the cytoplasm, smaller vacuoles coalesce into one large central vacuole leaving only a thin ribbon of cytoplasm between the vacuole membrane and the cellular membrane (plasmalemma). Organelles are distributed in the cytoplasm in a manner that is consistent with a model for cell expansion occurring by intercalation rather than tip growth [52,53]. The random orientation of cell wall polymers in the primary cell wall leads to an interwoven network of carbohydrate and protein macromolecules. New primary cell wall material must be coordinately deposited into the exocellular matrix, while at the same time the integrity of the primary cell wall must not be broken. Cotton fiber primary cell wall structure changes during development as demonstrated by changes in sugar composition and molecular mass distribution of the noncellulosic polysaccharides, including xyloglucan [54,55]. The biochemistry of primary cell wall expansion is an active area of investigation in plant biology [56,57]. As in other plant cell types, the biosynthesis and modification of most of the primary cell wall polymers are believed to occur in Golgi bodies with subsequent transport to the plasmalemma in small vesicles. In contrast, cellulose microfibrils of the secondary wall are synthesized by multisubunit enzyme complexes associated with the cell membrane.

There has been little direct characterization of enzymes that are responsible for the synthesis of cotton fiber cell wall components; however, application of molecular genetic techniques in recent years has greatly increased our knowledge about the structure of genes coding for these enzymes. Several large-scale efforts to characterize expressed sequence tags (ESTs) from elongating cotton fiber have resulted in an enormous increase in gene sequence information for the genus *Gossypium*. ESTs are partial or incomplete DNA sequences corresponding to the genes that are transcribed (active) in a particular tissue or stage of development. In public databases, such as GenBank at the National Center for Biotechnology Information (http://www.ncbi.nlm.nih.gov), comparisons of cotton fiber ESTs can be made with genes from other organisms. These searches predict the most likely match based on nucleotide sequence similarity or amino acid sequence similarity from deduced proteins. Estimates are that >2500 genes or over 14% of the currently available EST sequences for *G. arboreum* are related to primary cell wall metabolism and another 12% of the EST sequences are related to secondary cell wall biogenesis [58].

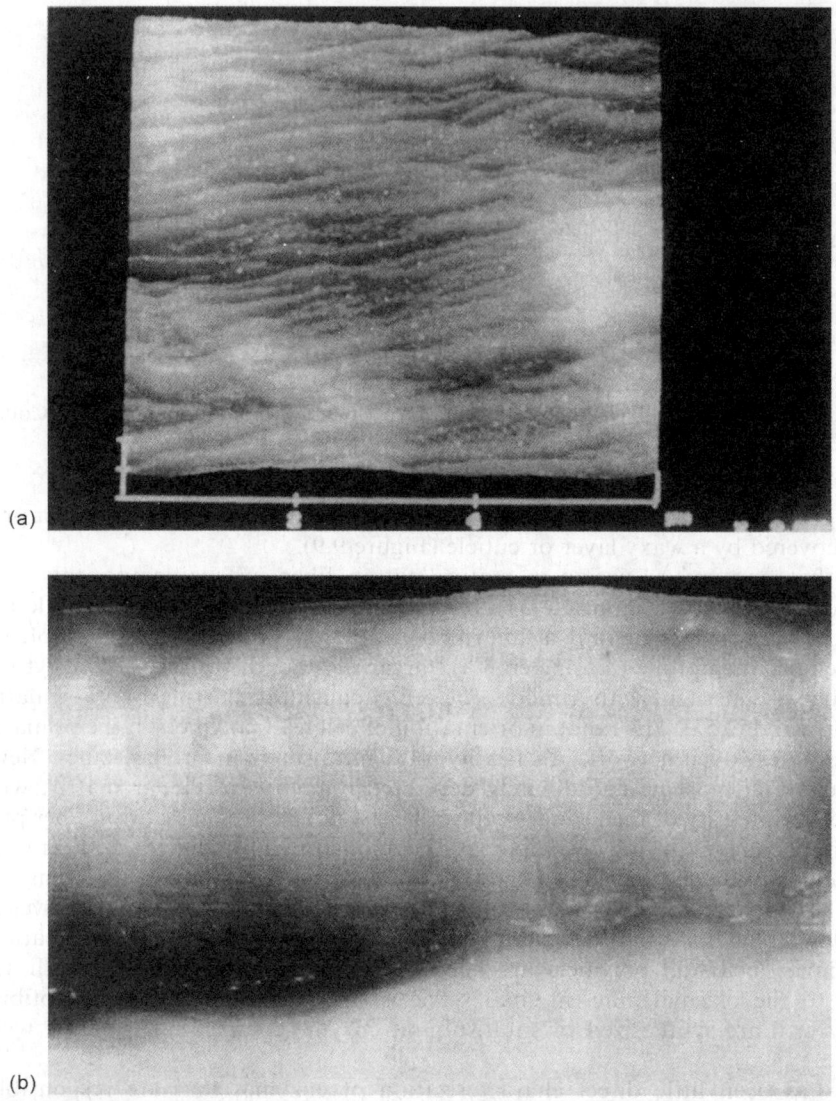

FIGURE 9.9 Atomic force micrographs of the cotton fiber cuticular layer on the surface of unprocessed fiber. (a) Cuticular material is deposited in undulating waves (16-μm scan). (b) Irregular deposition of cuticular material at higher magnification (550-nm scan). (Courtesy of T. Pesacreta, University of Southwestern Louisiana, and L. Groom, USDA-Forest Service.)

At approximately 15 to 19 DPA, lint fiber cells begin producing a secondary cell wall composed of highly crystalline, nearly pure cellulose, a β-(1→4)-D-glucan. The nature of the signal responsible for triggering secondary cell wall biosynthesis is unknown. Cellulose synthesis continues for 30–45 days until the cell wall is 2–6 μm thick. By the time the fiber is mature, over 90% of the dry weight of the fiber is cellulose. Callose, a β-(1→3)-D-glucan polymer is also transiently synthesized during the transition phase of development [59–61] and is deposited between the thickening secondary wall and the plasmalemma [62]. The function of callose in cotton fiber remains an enigma; however, recent evidence suggests that the cotton gene coding for callose synthase is not identical to the gene for cellulose synthase as had been proposed for many years [63].

In the secondary cell wall, cellulose microfibrils are deposited in a very ordered fashion. Concentric lamellae of cellulose microfibrils are subsequently deposited in a helical manner with a gradual increase in pitch. Periodically, the gyre of the helical microfibril orientation changes. These changes in the gyre are called reversals and have been shown to occur at or near where fiber breakage is likely [64,65]. Cellulose microfibril orientation angle is an important factor in fiber strength [66]. The orientation of cellulose microfibrils in the cotton fiber secondary cell wall is regulated by the orientation of structures in the cytoplasm collectively called the cytoskeleton [67]. Principal components of the cytoskeleton are the filamentous structures known as microtubules and microfilaments. Microtubules are composed of two proteins, α- and β-tubulins, and microfilaments are composed of the protein, actin. Regulation of the formation, organization, and disassembly of the cytoskeleton is believed to be controlled by other, less abundant proteins associated with the major cytoskeletal proteins. The characterization of these cytoskeletal components, their regulatory molecules, and how they might dictate cellulose microfibril organization in cotton fiber is an ongoing research effort [68–72].

Although sucrose is the primary transport sugar in the developing boll, the primary substrate for the enzymatic synthesis of cellulose is UDP–glucose [73]. Particulate forms of the enzyme, sucrose synthase [SuSy; EC 2.4.1.13], are believed to synthesize most of the UDP–glucose in the reversible reaction, sucrose+UDP ↔ UDP–glucose+fructose [74]. The case for SuSy involvement in cellulose biosynthesis is enhanced by the finding that SuSy localizes in the same regions of the cell where cellulose biosynthesis is predicted to occur [75]. β-sitosterol-D-glucoside, a lipid-linked sugar, has been proposed to function as a primer for cellulose biosynthesis [76,77]; however, *Arabidopsis* mutants that are genetically impaired in sitosterol synthesis can still produce cellulose [12].

Structures associated with cellulose biosynthesis in higher plants were first characterized [78] by freeze-fracture microscopy. Arranged in the plasma membrane, these hexagonal structures, called rosettes, are assembled in the Golgi apparatus [79]. The six particles making up a rosette are composed of four to six multisubunit complexes, each one of which synthesizes a molecule of cellulose. Due to numerous difficulties in purifying enzyme complexes from higher plants capable of synthesizing cellulose [80–82], attention has focused away from the enzymology of cellulose biosynthesis and diverted toward characterization of genes encoding the enzymes. A breakthrough was achieved when ESTs from cotton fiber were shown to be somewhat similar to short regions in the DNA sequence of a cellulose synthase from the bacterium, *Acetobacter xylinum* [83]. The regions in the bacterial gene that corresponded to two cotton ESTs share sequence similarity with the binding site and catalytic site of the glycosyltransferase family of genes, of which cellulose synthase is a member. The genes from cotton were named *CesA1* and *CesA2*, and were the first clones ever characterized for a higher plant cellulose synthase (*CesA*; EC 2.4.1.12). An antibody to cotton *CesA1* was used to demonstrate directly that the catalytic subunit for cellulose biosynthesis colocalizes with rosettes [84].

Since discovery of the catalytic subunit gene in cotton, *CesA* genes have been cloned from many different plants, most notably from the model organism, *Arabidopsis thaliana*, whose genome has been fully sequenced. Genomic analysis of *Arabidopsis* indicates that there are at least 10 genes in the true *CesA* subfamily and a multitude of other related genes in six subfamilies of cellulose synthase-like (*Csl*) [85]. Multiple forms of the *Arabidopsis CesA* genes must be expressed in the same cell [86]. Current evidence suggests that, in *Arabidopsis,* three distinct *CesA* genes code for the enzyme that produces cellulose for the primary wall [87,88] and three other genes code for the enzymes that produce cellulose in the secondary cell wall [89,90].

The first two higher plant cellulose synthase genes cloned, *GhCesA1* and *GhCesA2,* are transcribed during the secondary wall thickening phase of cotton fiber development [25].

A putative third cellulose synthase gene that is expressed during the same developmental period has recently been cloned and characterized [91]. Measurements of relative transcript abundance for the cotton *CesA* genes indicate that some members of this gene family are transcribed during the cell elongation stage of fiber development and other *CesA* genes are transcribed during secondary cell wall formation [55]. Cotton is an allotetraploid with part of the genome contributed by one diploid progenitor (A subgenome), and the remainder of the genome contributed by another diploid progenitor (D subgenome). Interestingly, cotton *CesA* genes coded by both subgenomes appear to be transcribed [55,92].

The molecular weight of primary cell wall cellulose is less than the molecular weight of cellulose in the secondary cell wall [93,94]. Based on the temporal regulation of *CesA* gene expression, this difference in cellulose molecular weight may be due to the two different types of cellulose synthase complexes formed during the two phases of fiber development. Difficulty in reconstituting enzyme activity from cloned plant genes has prevented a direct resolution of this issue. Another complication is that in addition to *CesA*, several other genes are important for cellulose production as demonstrated by genetic approaches in *Arabidopsis*. These additional genes include those coding for a β-(1→4)-glucanase gene (*KOR*), a serine-rich protein (*KOB1*), three enzymes involved in N-linked glycan formation (*KNF, rsw3, CYT1*), three sterol biosynthesis genes (*FK, HYD1, SMT1/CPH*), a GPI-anchored COBRA protein (*COBL1*), and a chitinase-like gene (*CTL1*). How these gene products are involved in cellulose biosynthesis remains unknown at present. Genomic approaches to identify orthologs of these genes from cotton are ongoing [95].

Cotton bolls dehisce at maturity, leaving the fibers fully exposed to air and sunlight. The water content of the fiber decreases rapidly, the cytoplasm dries against the inner surface of the wall, and a large lumen is left where the central vacuole was once located. The formerly tube-shaped cell collapses and assumes a twisted ribbon conformation with a kidney-like cross-sectional pattern. These twists, or convolutions, permit the spinning of fiber cells into yarns.

9.3 CHEMICAL COMPOSITION OF COTTON

Raw cotton fiber, after ginning and mechanical cleaning, is approximately 95% cellulose (Table 9.2) [96–99]. The structure of cotton cellulose, a linear polymer of β-D-glucopyranose, is discussed in Section 9.5 and chemical properties in Section 9.6.

TABLE 9.2
Composition of Typical Cotton Fibers

	Composition (% dry weight)	
Constituent	typical %	range %
Cellulose	95.0	88.0–96.0
Protein (% N × 6.25)[a]	1.3	1.1–1.9
Pectic substances	0.9	0.7–1.2
Ash	1.2	0.7–1.6
Wax	0.6	0.4–1.0
Total sugars	0.3	0.1–1.0
Organic acids	0.8	0.5–1.0
Pigment	trace	–
Others	1.4	–

[a] Standard method of estimating percent protein from nitrogen content (% N).

The noncellulosic constituents of the fiber are located principally in the cuticle, in the primary cell wall, and in the lumen. Cotton fibers that have a high ratio of surface area to linear density generally exhibit a relatively higher noncellulosic content. The noncellulosic constituents include proteins, amino acids, other nitrogen-containing compounds, wax, pectic substances, organic acids, sugars, inorganic salts, and a very small amount of pigments. Variations in these constituents arise due to differences in fiber maturity, variety of cotton, and environmental conditions (soil, climate, farming practices, etc.). After treatments to remove the naturally occurring noncellulosic materials, the cellulose content of the fiber is over 99%.

The noncellulosic materials can be removed by selective solvents. The wax constituent can be removed selectively with nonpolar solvents, such as hexane and chloroform, or nonselectively by heating in a 1% sodium hydroxide solution. Hot nonpolar solvents and other water-immiscible organic solvents remove wax but no other impurity, hot ethanol removes wax, sugars, and some ash-producing material but no protein or pectin, and water removes inorganic salts (metals), sugars, amino acids and low-molecular-weight peptides, and proteins. Most of the nonpolymeric constituents including sugars, amino acids, organic acids, and inorganic salts may be removed with water. The remaining pectins and high-molecular-weight proteins are removed by heating in a 1% sodium hydroxide solution or by appropriate enzyme treatments. All of the noncellulosic materials are removed almost completely by boiling the fiber in hot, dilute, aqueous sodium hydroxide (scouring or kier boiling), then washing thoroughly with water.

Of all the noncellulosic constituents, the nitrogen-containing compounds constitute the largest percentage when expressed as percent protein (1.1–1.9%) [6,7,99,100]. The nitrogen content of scoured fiber is about 0.035% (about 0.22% protein). Most of the nitrogenous material occurs in the lumen of the fiber, most likely as protoplasmic residue [101], although a small portion is also extracted from the primary wall [102]. The nitrogen-containing compounds located in the lumen may be removed using water, while those located in the primary cell wall are removed by heating in a 1% sodium hydroxide solution (a mild alkali scour such as that used to prepare cotton fabrics for dyeing and finishing). Cotton fiber and its primary wall both contain proteins and peptides, free amino acids, and most likely nonprotein nitrogen [102,103]. The free amino acids that have been detected are glutamic acid, aspartic acid, valine, serine, and threonine [103].

Cotton wax (about 0.4 to 1.0% of fiber dry weight) comprises the cuticle on the outer surface of the fiber. The quantity of wax increases with the surface area of the cotton, and the finer cottons tend to have a larger percentage of wax. The wax is a mixture of high molecular weight, primarily long-chain saturated fatty acids and alcohols (with even numbers of carbon atoms, C_{28} to C_{34}), resins, saturated and unsaturated hydrocarbons, sterols, and sterol glucosides [104,105], including montanyl triacontanoate (10–15%), montanol (25%), 1-triacontanol (18%), and β-sitosterol (10%). A mutant variety (*G. hirsutum L.*) of cotton with natural green color related to a suberin-like wax biopolymer is located between cellulose lamellae in the secondary wall [28], with a 14–17% wax composition relative to fiber weight. This waxy polymer is composed predominantly of ω-hydroxydocosanoic acid with which glycerol and the phenolic compound, caffeic acid, are associated [29,106]. Naturally colored brown-lint cottons and other colored cottons seem to have normal wax contents. Other phenolic compounds may be responsible for color formation in the other colored fiber varieties, but the identity of these compounds is not known at present.

The natural wax content serves as a protective barrier both to water penetration and to microbial degradation of the underlying polysaccharides. The wax serves as a lubricant that is essential for proper spinning of cotton fiber into yarn. Once the yarn is spun, however, the wax does reduce the tensile strength of the yarn as well as hinders dyeing and finishing of the fiber. Wax is detrimental in the chemical processing and finishing of the cotton yarns and fabrics because it interferes with wetting of the fiber and penetration of the reagents. If the wax layer is disrupted, for example, by the microbiological action due to weathering, it can affect the sizing operation. The sizes will penetrate too far into the fiber and the fiber will pick up too high a level of size. The cotton wax, therefore, is removed (saponified) by a mild scouring treatment using sodium hydroxide solution in the normal preparation of cotton yarns and fabrics for dyeing and finishing. After the wax is removed from yarn in the scouring operation, the fullest development of strength is attained in the end product.

The content and texture of the wax are important in predicting the processing potential of the cotton. Mild scouring treatments of cotton, like those acceptable to qualify cotton as washed cotton under the U.S. cotton dust standard, sometimes can alter the wax surface and adversely affect textile processing without lowering the amount of wax on the fiber [107].

Underlying the waxy cuticle is the primary cell wall, which is composed of two distinct layers [108]. The outermost layer is comprised primarily of pectin substances (usually designated as pectin) in the form of free pectic acid (linear polymer of (1→4)-D-galacturonic acid) and its insoluble calcium, magnesium, and iron salts, and constitutes approximately 0.7–1.2% of the dry fiber weight. The innermost layer is comprised of hemicelluloses, primarily in the form of xyloglucan, and cellulose. Pectin is removed either by the same scouring process that removes wax as part of preparation for dyeing and finishing, or by a combination of pectinase enzymes. The method of pectin analysis has much to do with the percent pectin reported. Removal of pectin does not significantly alter the tensile strength of the fiber and has little effect on the yarn and the fabric properties.

Soluble sugars (about 0.1 to 1.0% of fiber dry weight) found on cotton originate from two sources: metabolic residues (plant sugars) located within the dried lumen and the outer fiber surface and insect sugars (insect "honeydew" excretion) found on the outer surface of the fiber [109]. Plant sugars or metabolic residues occur as a result of the normal growth process and are composed primarily of the monosaccharides, glucose and fructose, and to a lesser degree the disaccharide, sucrose [110]. They may range from about 0.1–1.0% of the dry fiber weight and vary in concentration depending upon cotton fiber maturity and environmental factors (i.e., area of growth, weathering, and microbial activity). The levels of these sugars on cotton are determined by one of the several simple sugar test methods [111].

Insect sugars, commonly known as honeydew, most often come from aphids and whiteflies [110]. They are found intermittently on cotton and are generally a problem with cottons grown in arid regions because rainfall in less arid regions serves to wash these impurities off the lint. Insect sugars usually are randomly deposited as spots or specks on the cotton, causing stickiness [112]. Stickiness from high levels of either plant or insect sugars on cotton lint may cause serious processing problems in textile mills, causing fibers to stick to draft rolls or other processing equipment. The sugar profiles for aphid and whitefly honeydews are quite distinct from one another. Aphid honeydew consists primarily of glucose, fructose, melezitose [113,114], maltose, and a dozen or more longer chain oligosccharides, most of which have yet to be characterized. Whitefly honeydew, by comparison, exhibits a less complex profile than the primary sugars, which are glucose, fructose, and trehalulose [115–118], a disaccharide with a high hygroscopicity. The two honeydew types are generally distinguished from one another by the presence or absence of trehalulose, as aphid honeydew contains little if any of this sugar. Arabitol and mannitol (monosugar alcohols), which are products of fungal activity, can sometimes be detected and are indicators of microbial damage to cotton [119]. The sugars are readily removed by water. They are removed by the normal scouring and bleaching processes that are used for the preparation of the fiber for dyeing and finishing.

Organic acids (0.5–1.0% of fiber dry weight) in the raw fiber, exclusive of pectic acid, are primarily l-malic acid (up to 0.5%) and citric acid (up to 0.07%), are present in the lumen as metabolic residues, and are removed during the normal scouring and bleaching due to their high water solubility. Analyses indicate that other acids are also present, totaling some 0.3% but these have not been identified. Organic acids are removed during normal scouring.

Inorganic cations are also present as metabolic residues, again primarily in the lumen, as salts of organic acids or inorganic anions. The inorganic salts (phosphates, carbonates, and oxides) and salts of organic acids present in the raw fiber are reported as percent ash (about 1.2% of fiber dry weight) and expressed as the oxides of the elements present (excluding chlorine, which is expressed as such). The amounts of these cations present on the cotton fiber vary considerably [120] because of maturity differences, environmental factors (e.g. rainfall), and agricultural practices, as well as the field and the handling procedures that affect deposition of material (plant parts and soil) on the fiber. As is the case with all growing plants, mineral salts are necessary for the development of the cotton plant. During the production of cotton, the plant absorbs potassium and other metals as normal nutrients. Metals are incorporated from the soil into plants as natural constituents. In addition to metals absorbed by plant tissue, soil and plant parts may be deposited directly onto the lint, especially during harvesting. Ca, P, S, K, and Fe are plant part elements and Mg, Al, Si, Fe, Cr, Se, Hg, Ni, Cu, K, and Ca are soil elements [121].

The most abundant cation is potassium (2000–6500 ppm), accounting for approximately 70% by weight of total cationic content, followed by magnesium (400–1200 ppm), which accounts for approximately 14% [120]. Calcium (400–1200 ppm) is found both in the lumen and in the pectin fraction, where it serves as a cross-linking agent, and accounts for approximately 14% of the total cationic content. Sodium (100–300 ppm), iron (30–90 ppm), zinc (1–10 ppm), manganese (1–10 ppm), and copper (1–10 ppm) are also present in relatively small quantities. Lead and cadmium were not detected (Table 9.3). In untreated cotton, arsenic levels are usually less than 1 ppm [122,123]. Silicon, phosphorus, chlorine, sulfur, and boron are detected sometimes in trace amounts.

A secondary source of inorganic content on cotton fiber is the deposition of wind-borne particles onto the outer surface of the cotton fiber. Although such particles may be present

TABLE 9.3
Metal Content of Cotton

Metal	ppm
Potassium	2000–6500
Magnesium	400–1200
Calcium	400–1200
Sodium	100–300
Iron	30–90
Manganese	1–10
Copper	1–10
Zinc	1–10
Lead	n.d.[a]
Cadmium	n.d.
Arsenic	trace (<1)[b]

[a] n.d. = not detected.
[b] From Refs. [122,123].
Source: From Brushwood, D.E. and Perkins, H.H., Jr., *Text. Chem. Color.*, 26, 32, 1994.

only in trace amounts, their presence in cotton is of importance to processors because they can contribute to problems in yarn manufacturing, bleaching, and dyeing. Silicon as silica and other metals as oxides can cause frictional problems in rotor spinning and needle wear in knitting [120]. Iron and copper metal particles, introduced to the fiber through deposition from machinery parts, can cause problems in the peroxide bleaching process as well as contribute to a permanent coloration that may affect dyeing. Peroxide bleaching also can be affected by magnesium salts [124]. Insoluble calcium and magnesium salts can interfere with dyeing [125] and copper and iron can contribute to yellowness of the finished denim goods [126]. Iron can contribute to the permanent brown or pink color of the fiber, which affects dyeing [124]. The metals of potential concern in wastewater effluents from textile dyeing and finishing are copper and zinc. The levels of these metals in cotton fiber are low enough so that they do not contribute significantly to effluent problems [47]. The metals are removed for the most part by proper scouring and bleaching processes that are used to prepare the fiber and fabric for dyeing and finishing.

Also of potential concern is the presence of arsenic-containing compounds, which are introduced primarily through agricultural practices such as harvest aid products (arsenic acid) and postemergent herbicides (e.g., cacadylic acid). Arsenic acid is no longer registered for use in the United States [127], and organic arsenic containing postemergent herbicides are used in less than 4% of the U.S. cotton production and are being phased out. While these compounds are generally removed through the scouring process, their presence may be of some concern for both health and marketing reasons. For solid waste, such as textile mill fiber waste (e.g., undercard and pneumafil waste), the U.S. Environmental Protection Agency (EPA) has established limits for the metals leachable from the waste. If these levels are exceeded, the waste has to be treated as hazardous [47]. Cotton does not normally contain any metals in sufficient quantity to be of concern and therefore, if the cotton fiber is not recycled, it can be disposed of in normal municipal landfills or lined landfills [47]. Some textile mill carding and other yarn manufacturing wastes are presently used as animal feed, which indicates that yarn manufacturing fiber wastes have no or very low toxicity and are generally regarded as safe.

9.4 SOLVENTS FOR COTTON

Cellulose is soluble only in unusual and complex solvent systems. The subject has been reviewed [128–131]. Solvents for cellulose are central to the rayon and cellulose film industries, but are also necessary for solubilizing cotton for the determination of molecular weight and degree of polymerization (DP) by chromatographic methods. These solvents fall into several categories. The solvents discussed do not include processes where cellulose is converted to a derivative that is subsequently dissolved in another medium. For example, cellulose acetate is soluble in acetone, but this is not a solution of cellulose. However, the viscose process that forms a cellulose xanthate derivative, from which cellulose is readily regenerated, is generally considered to use a cellulose solution because solvation and derivatization occur simultaneously. The viscose process is the most important method for making cellulose solutions for industrial use [132]. Alkali cellulose (pulp swollen in NaOH) is pressed and aged to reduce molecular weight. Xanthation (a reaction with CS_2) takes place in a vessel that contains an inert atmosphere (CS_2–air mixtures are explosive). The orange xanthate is subsequently dissolved in aqueous alkali to make the spinning dope. The dope is pumped through spinnerets in which there are from 14 to 40,000 holes. The spun dope is converted back to cellulose by the sulfuric acid in the coagulating bath. Another system with simultaneous derivitization and dissolution uses dimethyl sulfoxide and formaldehyde [133].

Several solvents, such as cupriethylenediamine (CUEN) hydroxide, depend on the formation of metal–ion complexes with cellulose. While not as widespread in use as the viscose process, CUEN and its relatives with different metals and ammonium hydroxide find substantial industrial use [131]. The cadmium complex, cadoxen, is now the solvent of choice in laboratory work [134].

Aqueous salt solutions such as saturated zinc chloride or calcium thiocyanate can dissolve limited amounts of cellulose [131]. Two nonaqueous salt solutions with a lengthy history are ammonium thiocyanate/ammonia and dimethylacetamide/lithium chloride (DMAc/LiCl). Solutions up to about 15% can be prepared with these solvents. DMAc–LiCl has been used for molecular weight determinations of cotton [135] (see Section 9.5.2).

Trifluoroacetic acid–methylene chloride and N-methyl morpholine N-oxide monohydrate (NMMO) [136–138] are two other solvent systems that have been studied [139]. The new generic class of regenerated cellulose fibers, lyocell (e.g., Tencel [Courtaulds Fibres Limited, London, England]), is spun from aqueous solutions of NMMO [140]. Lyocell is an alternative to the generic name "rayon" for a subcategory of rayon fibers where the fiber is composed of cellulose precipitated from an organic solution in which no substitution of the hydroxyl group takes place and no chemical intermediates are formed. Lyocell may have a different crystalline structure (a mixture of cellulose II and cellulose III [141]) than other rayons and cotton cellulose. No information has been published on cotton molecular weight determinations using NMMO as the solvent.

A lengthy multiday procedure can be required to produce complete dissolution of the high-molecular-weight cellulose of cotton in the DMAc/LiCl solvent system. Other problems have been found with the DMAc/LiCl solvent system, including incomplete dissolution of some celluloses and degradation of cellulose caused by heating in the solvent system [142–145]. Despite 20 years of use of the DMAc/LiCl solvent system, a recent review [146] further documented the problems with incomplete dissolution and possible aggregation of some celluloses in DMAc/LiCl, and highlighted the need for continuing basic research on this solvent.

A new cellulose solvent system, 1,3-dimethyl-2-imidazolidinone/lithium chloride (DMI/LiCl), was recently reported [147] with the advantages of rapid complete dissolution of cellulose and reduced health risks. This solvent system was adapted for gel permeation chromatography (GPC) analysis of cotton cellulose [148] and was found to more completely, rapidly, and easily dissolve the high-molecular-weight cellulose of cotton than the DMAc/LiCl solvent system, yielding weight average molecular weights of 542400, 232900, 230700, and 114800 for cotton, ramie, linen, and Tencel, respectively. A detailed examination of the suitability of DMI/LiCl for SEC–MALLS (size-exclusion chromatography coupled to multiangle laser light scattering) analysis of cellulose found that it is a true solvent for cellulose. Cellulose molecules dissolved in this solvent are separated by their molecular mass or root-mean-square radius by SEC, no aggregates are produced, and the cellulose solutions were stable over several months at room temperature [149]. It remains to be seen if this solvent will prove to be more effective for measurements of molecular weight distributions of cotton cellulose than the DMAc/LiCl solvent.

Aqueous 9% sodium hydroxide at -5 to $+5°C$ can dissolve steam-exploded chemical wood pulps [150]. Isogai and Atalla [151] adapted this procedure for the dissolution of microcrystalline cellulose, and compared the solubility of several other celluloses. Microcrystalline cellulose is suspended in water, sodium hydroxide is added, and the mixture is shaken at room temperature to dissolve the sodium hydroxide and suspend the cellulose in the solution. The suspension is cooled to $-20°C$ and held at that temperature until a solid frozen mass is formed. On thawing, a gel-like mass results and with the addition of water and gentle shaking, a clear solution of 2% cellulose in 5% aqueous NaOH forms. Low-DP cellulose and

Avicels (either original, regenerated, ethylenediamine (EDA)-treated, or mercerized) are 100% soluble; linters and mercerized linters are about 32% soluble. High DP fibrous plant celluloses are soluble only up to 37% even after swelling treatments.

Organic salts that are liquids at or near room temperature are referred to as ionic liquids. Ionic liquids that contain strong hydrogen-bond acceptors, such as Cl^-, Br^-, or SCN^- can be used to dissolve cellulose [152]. The authors compared the free chloride ion concentration in a typical 10% DMAc/LiCl solution, ca. 6.7%, with the chloride concentration in an ionic liquid consisting of 1-butyl-3-methylimidazolium cation and a chloride anion; in the ionic liquid, the chloride concentration is almost three times as high. They speculated that the high chloride concentration and activity in the ionic liquid is highly effective in breaking the hydrogen-bonding network of cellulose, and leads to faster dissolution and higher concentrations of cellulose than traditional solvent systems. Ionic liquids are heated with high efficiency by microwaves, and such heating speeds up the dissolution. Regenerated celluloses prepared from the above ionic liquid did not have significant changes in DP or polydispersity. Studies of the use of ionic liquids for extrusion of cellulose fibers are in progress [153].

9.5 STRUCTURAL PROPERTIES OF COTTON

9.5.1 Structure Overview

Although there is extensive practical knowledge of cotton processing, a more thorough understanding of cotton fiber structure will improve exploitation of today's fiber. Still, the ultimate reason to seek such understanding is to be able, through genetics, biochemistry, or chemistry, to tailor fiber to have new or improved properties. The initial structure of a cotton fiber is determined by biosynthesis, a series of processes that are subject to substantial influence during fiber growth. After the boll opens, there are many factors that affect the structure, from the weather before the fiber is harvested to the industrial processes such as mercerization. Cotton fibers are composed mostly (e.g., 95%) of the long-chain carbohydrate molecule, the cellulose (the sugar of cell walls). In this overview and Section 9.5.2 through Section 9.5.5, we are concerned with the physical structure of cellulose and the fibers, as revealed by various methods.

Other commercial sources of cellulosic fibers include hemp, jute, flax (linen), and ramie. Wood fibers are used in papermaking and as a feedstock for rayon. Of these sources, cotton provides the purest cellulose. From an experimentalist's point of view, algae and even animals (the tunicates) are also interesting sources of cellulose. Bacteria such as *Acetobacter xylinium* make extracellular cellulose, but in higher plants and algae, cellulose occurs in the walls of individual cells.

Some workers refer to cotton lint (the normal fibers) as cellulose to distinguish it from seed cotton (fiber still on the seed) or the entire plant. Herein, the word cellulose has only the strict chemical meaning: linear β-(1→4)-D-glucan. In the cell wall, cellulose occurs in small, crystalline microfibrils that are arranged in multilayer structures (see Figure 9.10). An especially important layer is the primary wall (see Figure 9.11) although it is a small fraction of the mature, fully developed fiber.

Detailed structures of many crystalline materials can be determined by diffraction methods. However, because of the complex hierarchy of the cotton fiber and its very small crystallites, diffraction experiments on cotton fibers cannot provide fine details of molecular structure. Instead, the best data on cellulose structure comes from other sources. One of the major points of interest is the finding that cellulose has many different crystalline forms, or polymorphs, depending on the sources and subsequent treatments. Historically, there are four polymorphs or allomorphs, I to IV, and subclasses have been identified for all but cellulose II.

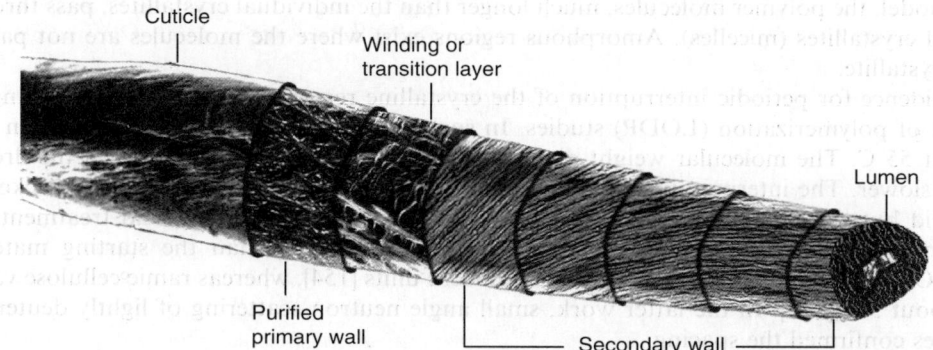

FIGURE 9.10 A computer-generated montage of a fiber segment constructed from individual transmission electron micrographs (TEMs) and scanning electron micrographs (SEMs) of layers of the fiber to show an overview of the layered structures of the fiber. Fiber surface, primary wall, and secondary layers have been shown at different magnifications to better visualize fibrillar structures and the various fiber layers from surface to lumen. The surface marked cuticle is an SEM view of a scoured and bleached fiber surface and was used as the skeleton of the montage. All other segments are taken from transmission micrographs and are shown at higher magnifications. No fibrils are visible in the SEM of the cuticle because of its relatively low magnification as well as the presence of noncellulosic materials. The fiber surface at the cut end of the fiber shows fibrils that have been separated by swelling. Although the montage uses actual pictures of cotton fiber layer structures, it is intended to represent a possible fiber morphology rather than to indicate an exact cotton fiber structure. (Credit to Wilton Goynes.)

Some cotton cellulose is noncrystalline or amorphous in the sense of "lacking definite crystalline form." One reason is that cotton cellulose has a broad molecular weight distribution, making high-crystalline perfection impossible. The small crystallites constitute deviations from ideal crystals that are infinite arrays. The remaining amorphous character of most polymers is often thought to arise from the "fringed micelle" model of the solid structure. In

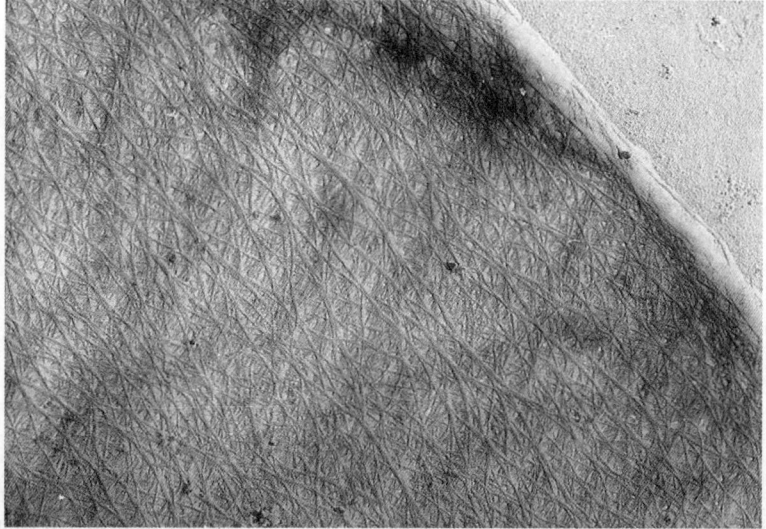

FIGURE 9.11 Transmission electron micrograph of the loose tangle of cellulose microfibrils in a cotton fiber cell wall. (Credit to Wilton Goynes.)

that model, the polymer molecules, much longer than the individual crystallites, pass through several crystallites (micelles). Amorphous regions exist where the molecules are not part of any crystallite.

Evidence for periodic interruption of the crystalline regions comes from the leveling-off degree of polymerization (LODP) studies. In such work, the cellulose is hydrolyzed in 1 M HCl at 55°C. The molecular weight drops rapidly for about a day, after which the drop is much slower. The interpretation is that periodic amorphous regions are readily attacked by the acid but the crystalline regions are much more resistant. After the acid treatment, the remaining material, called hydrocellulose, is more crystalline than the starting material. The LODP values for cotton are about 175 glucose units [154], whereas ramie cellulose values are about 300 [155]. In the latter work, small angle neutron scattering of lightly deuterated samples confirmed the spacing.

Other amorphous contributions arise from imperfections in the crystals. These defects include molecular ends inside the microfibrils, causing discontinuities, disorder on the crystallite surfaces (which may comprise as many as half of the molecules in these small crystals), and mechanical bending and twisting of the crystallites [156]. These bases for the amorphous character are embodied in Figure 9.12. This representation is attractive because of the high density of fibrous cotton (e.g., 1.55 g/cm^3) when purely crystalline cellulose is 1.62 g/cm^3. The high density suggests that the chains are more orderly and compactly arranged than indicated by most depictions of fringed micelles.

Given a DP of about 20,000 glucose residues, a stretched out molecule is perhaps 100,000 Å (10^{-5} m) long. Thus, compared to the fiber's length of about 30 mm (3×10^{-2} m), a cotton fiber is some 3000 times longer than a molecule. Coincidentally, the molecular length of 10 μm is roughly the same as the diameter of the cotton fiber. In fibers such as linen, the molecules are aligned parallel to the fiber axis to within a few degrees. In cotton fibers, however, alignments are distributed over a much larger range. Early in the fiber development, the fibrils are parts of, or near, the primary wall and some have angles nearly perpendicular to the fiber axis (See Figure 9.11). As the secondary cell wall develops on the interior of the

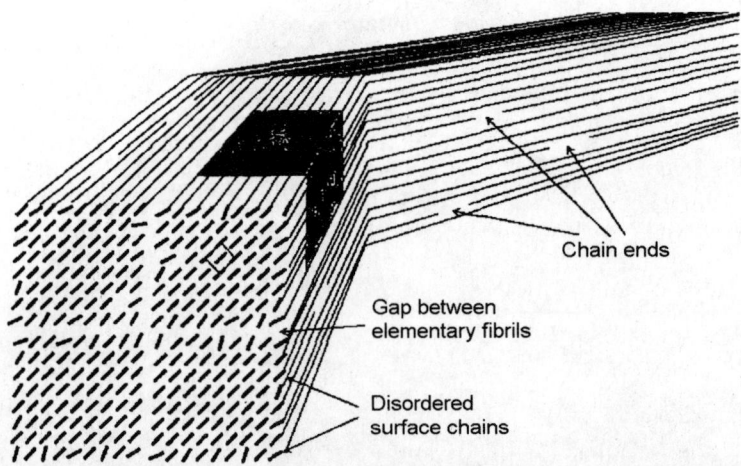

FIGURE 9.12 Cellulose microfibril composed of four elementary fibrils, viewed along the chain axes. The unit cell is shown, and the chains on the elementary fibril surfaces are disordered. Chain ends within the elementary fibrils account for more disorder. The surfaces of the crystallites are related to the crystallographic planes shown in gray.

primary wall, the microfibrils are thought to be progressively aligned more closely with the fiber axis (Figure 9.10). Within a given layer of secondary wall, the direction of the molecules may be roughly constant until it suddenly changes, causing a reversal to appear. Cellulose is birefringent and these reversals are visible under a polarizing optical microscope as well as in the scanning electron microscope (SEM) (see Section 9.5.5).

The mature cotton fiber has a noncellulosic covering called the cuticle that contains waxes, pectins, and proteins left over from biosynthesis (see Section 9.3). This cuticle is intermingled with the primary wall. Figure 9.13 shows the location of immunolabeled pectin within the primary wall. The structure of the primary wall, which changes substantially during fiber development, is not well understood. It is responsible for maintaining the integrity of the fiber and may account for much of the strength of the cotton fiber. Most of the cuticle is dissolved and removed by industrial scouring of fabric, but it has important functions during spinning of the fibers into yarn and during weaving the yarn into fabric. One reason for scouring is that the waxes block access to the interior of the cotton fiber for molecules such as dyes.

Thus, a complete description of fiber structure requires knowledge of the structure of the cellulose molecule, the structure and perfection of its crystalline arrays, the packing of these arrays (elementary fibrils) into microfibrils, and then the arrangement of these microfibrils in the primary and secondary cell walls. The noncrystalline material is also important. The large-scale structures must also be understood. Fibers have various pores or voids that are important in textile finishing reactions. After the cotton boll opens, the cellular fluid dries, leaving a cavity, the lumen, which contains biological material. That material constitutes a small percentage of the total dry weight. With drying of the fiber, the lumen collapses, leaving fiber cross-sections with irregular, kidney-bean shapes, as shown in Figure 9.14. Section 9.7

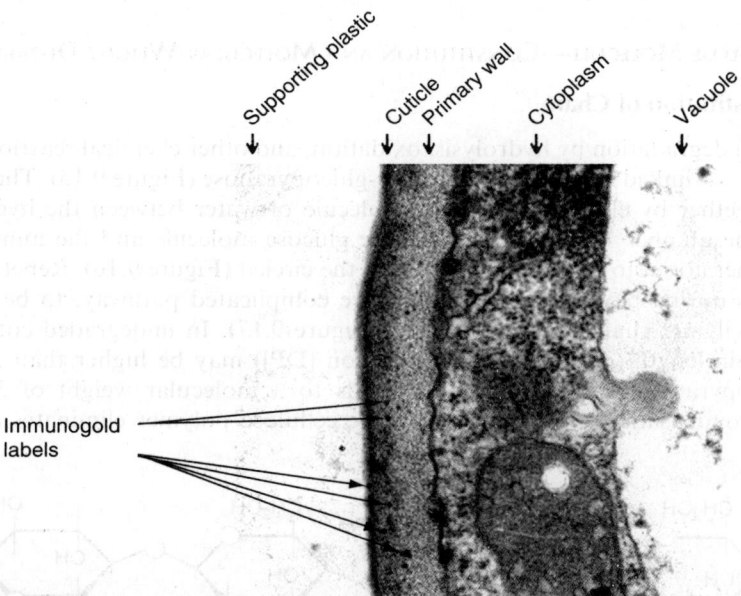

FIGURE 9.13 Transmission electron micrograph of a section of primary cell wall from very immature cotton fiber. The black dots are immuno-labels of pectin molecules. (Credit to Kevin Vaughn, USDA-ARS, Stoneville, MS.)

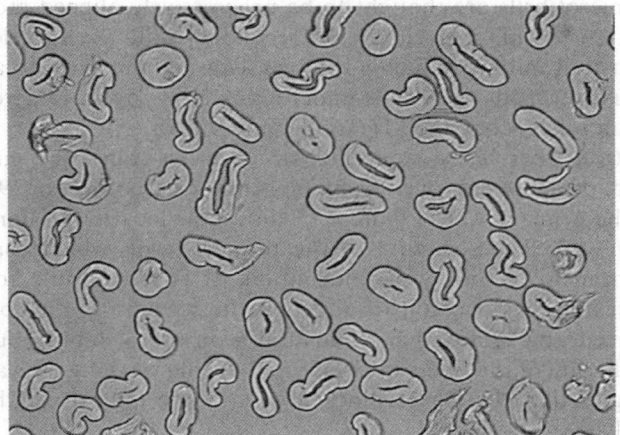

FIGURE 9.14 Cross-sections of cotton fibers showing their variable but generally kidney-bean shapes and lumens. These fibers are from a lot of cotton that, on average, was fully mature. (Credit to Devron Thibodeaux.)

discusses how the degree of development (thickness and shape) of the secondary wall (the maturity) affects many performance characteristics.

Treatments of cotton fiber often alter the structure at more than one level. Therefore, it is not always obvious that the effect of the treatment should correlate with a change at a particular level of structure. For example, cotton is often treated with NaOH in a process called mercerization. This treatment can alter the crystal structure, but it also reduces the density of the whole fiber and increases its luster. For example, the change in crystal structure is probably not directly responsible for the change in fiber luster.

9.5.2 Cellulose Molecule—Constitution and Molecular Weight Distributions

9.5.2.1 Constitution of Chain

Evidence from degradation by hydrolysis, oxidation, and other chemical reactions shows that cellulose is a 1→4-linked linear polymer of β-D-glucopyranose (Figure 9.15). These monomers are linked together by elimination of one molecule of water between the hydroxyl groups attached to the number 1 carbon atom of one glucose molecule and the number 4 carbon atom of another about to be joined with loss of the circled (Figure 9.16). Repetitions of these condensations during biosynthesis (with a more complicated pathway, to be sure) lead to unbranched polymer chains of great lengths (Figure 9.17). In undegraded cotton fiber, the molecular chain length (degree of polymerization [DP]) may be higher than 20,000 monomeric D-glucopyranosyl units. This corresponds to a molecular weight of 3,240,000 Da. Because the condensation reaction to make the cellulose polymer eliminates a molecule of

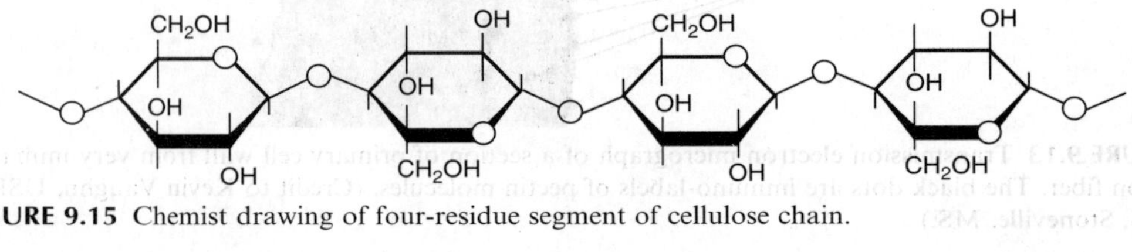

FIGURE 9.15 Chemist drawing of four-residue segment of cellulose chain.

Cotton Fibers

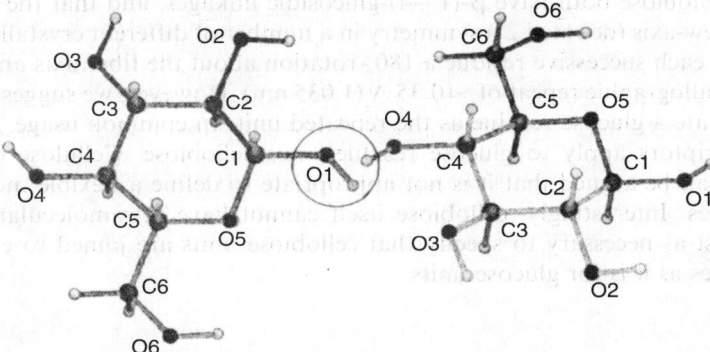

FIGURE 9.16 Two glucose residues in the $^4C_1$ (chair) conformation about to be joined, losing the circled hydroxyl group and hydrogen atom. Drawings were made with Chem-X, formerly developed, and distributed by Chemical Design, Ltd.

water per glucose, the monomers are often referred to as anhydroglucose units or as glucose residues. Each cellulose chain has a reducing end (O1—H) and a nonreducing end (O4—H). Reducing ends are especially reactive, but they are present in such small amounts in cellulose that they are often ignored.

It is often stated that the repeating unit of cellulose is a cellobiose residue, a hydrolysis product of cellulose that is composed of two glucose residues. This acknowledges that

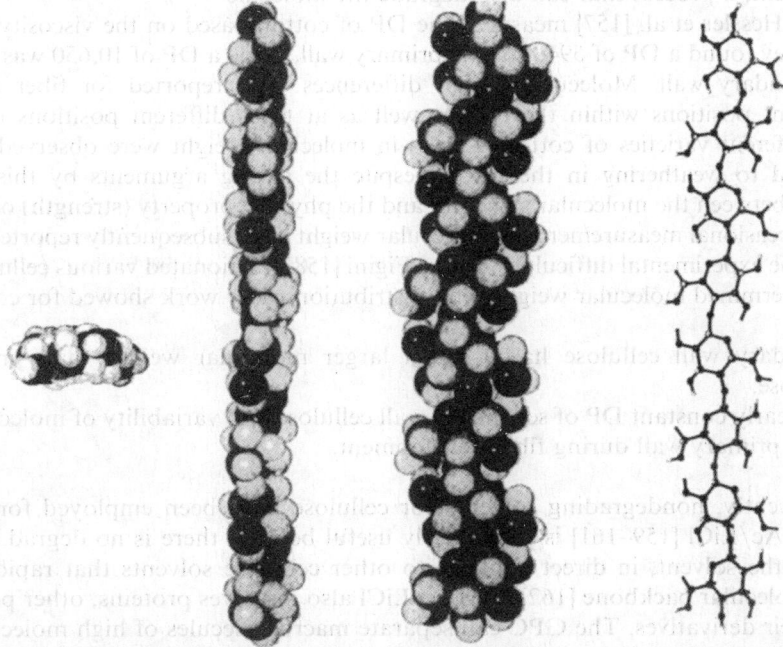

FIGURE 9.17 Projections of short cellulose chains. Leftmost: along chain axis. Left-center: along ribbon edge. Right-center: Maximal width view. Rightmost: Ball and stick model in the same orientation as the space-filling model to its left.

cellulose and cellobiose both have β-(1→4)-glucosidic linkages, and that the cellulose chain has twofold screw-axis (denoted 2_1) symmetry in a number of different crystalline forms. This symmetry gives each successive residue a 180° rotation about the fiber axis and a translation of half the crystallographic repeat of ~10.35 Å (1.035 nm). However, we suggest that it is more useful to designate a glucose residue as the repeated unit. In common usage, DP values and symmetry descriptors apply to glucose residues, not cellobiose. Cellulose can take other shapes, and it can be argued that it is not appropriate to define a flexible molecule by only one of its shapes. Interestingly, cellobiose itself cannot have intramolecular 2_1 symmetry. Finally, it is just as necessary to specify that cellobiose units are joined to each other with β-(1→4)-linkages as it is for glucose units.

9.5.2.2 Molecular Weight Distributions

It has long been thought that cotton fiber strength is influenced by the structural organization of the cellulose chains [157]. Molecular weight of a polymer is one of the most important influences on its physical properties, and the determination of molecular weight distribution is critical for predicting performance of a polymer. Vastly differing distributions can produce the same averages but physical properties will reflect the disparities. For polymers, higher molecular weight and narrower molecular weight distribution are positively correlated with increased strength. Unfortunately, polymer characterization techniques generally depend upon dissolving of the polymers. Attempts to identify the true molecular weight of native cellulose have been limited mostly because cellulose is difficult to dissolve (see also Section 9.4). In addition, cellulose solutions are often unstable and sometimes undergo rapid oxidative degradation. Alternatively, cellulose can be converted to a derivative that is soluble in an organic solvent, a process that can also degrade the molecule.

In 1948, Hessler et al. [157] measured the DP of cotton based on the viscosity of nitrated cellulose. They found a DP of 5940 for the primary wall, while a DP of 10,650 was determined for the secondary wall. Molecular weight differences were reported for fiber taken from three different positions within the boll as well as at three different positions on the seed for three different varieties of cotton. Losses in molecular weight were observed for cotton fiber exposed to weathering in the field. Despite the strong arguments by this group for relationship between the molecular structure and the physical property (strength) of the cotton fiber, only occasional measurements of molecular weight were subsequently reported, probably because of the experimental difficulties. Marx-Figini [158] fractionated various cellulose derivatives and determined molecular weights and distributions. Her work showed for cotton fiber:

1. Secondary wall cellulose has a much larger molecular weight than primary wall cellulose.
2. The nearly constant DP of secondary wall cellulose and variability of molecular weight in the primary wall during fiber development.

More recently, nondegrading solvents for cellulose have been employed for characterization. DMAc/LiCl [159–161] is particularly useful because there is no degradation of the polymer by the solvent, in direct contrast to other cellulose solvents that rapidly degrade the macromolecular backbone [162]. DMAc/LiCl also dissolves proteins, other polysaccharides, and their derivatives. The GPC can separate macromolecules of high molecular weight with the advantage that the molecular weight distribution of a polymer can be obtained in a relatively short time using automated, computer-based data acquisition and calculations. Multiple detectors bypass the need for cellulose standards (which are not available).

Cotton fibers dissolved directly in the solvent DMAc/LiCl have been analyzed by GPC [94]. This procedure completely solubilizes wall polymers without extraction or derivatization for GPC separation processes that could degrade high-molecular-weight components. This approach permits the characterization of the entire array of cell wall polymers, not previously possible, and represents a step toward solving the major problem of precise analysis. The molecular weight distribution for mature, field-grown cotton fiber from a genetic standard variety (*Gossypium hirsutum*, Texas Marker-1) is shown in Figure 9.18. Again, the identified locations of the primary and secondary walls showed:

1. Lower molecular weight for the primary compared with the secondary walls
2. The larger weight fraction of material found in the secondary wall, confirming limited previous reports [157,158]

During development, the composition of the cell wall of the cotton fiber changes continuously, ending with the cessation of the fiber's metabolic activity [163]. Cell wall polymers from cotton fibers under developmental stages were characterized via GPC analysis of DMAc/LiCl solutions [94]. Primary and secondary wall compositions of cotton fiber polymers were monitored from 8–60 DPA. As expected, cell wall polymers from fibers at primary cell wall stages had lower molecular weights than the cellulose from fibers at the secondary wall stages. However, the high-molecular-weight cellulose characteristic of mature cotton was detected as early as 8 DPA. High-molecular-weight material decreased during the period of 10–18 DPA with a concomitant increase in lower-molecular-weight wall components, possibly indicating hydrolysis during the later stages of elongation. During the maturation stages (past about 20 DPA), the high-molecular-weight components of the secondary wall increased dramatically to give the fiber profile shown in Figure 9.18. These observations are consistent with a picture of fiber development that starts with the construction of the full-sized primary

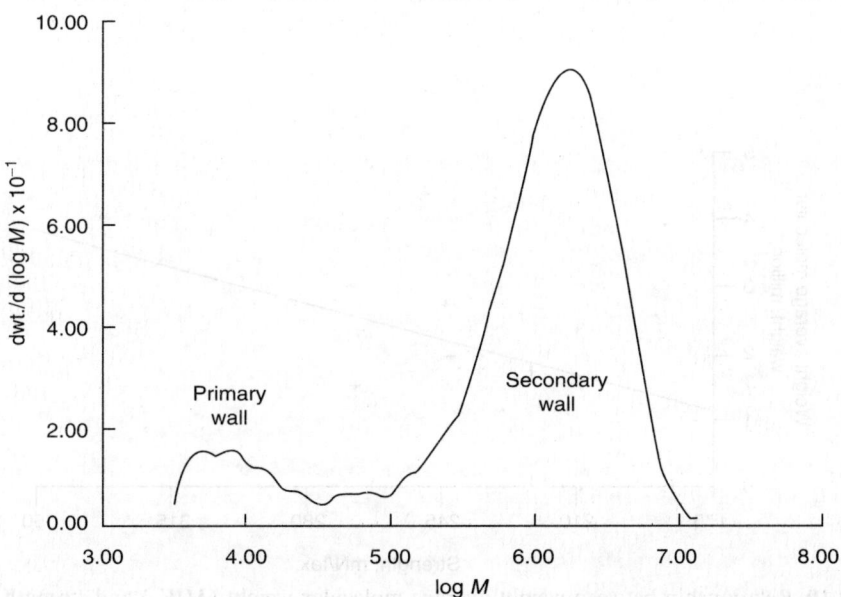

FIGURE 9.18 Molecular weight distribution for cotton printcloth, desized, scoured, and bleached.

wall and proceeds through the addition of secondary wall material on the inside of the primary wall.

Molecular weight distributions were determined for three commercial varieties of American Upland cotton that were similar in all physical properties except strength as measured by the high volume instrument (HVI) [164]. (It is an unsolved mystery why traditional Stelometer strength measurements gave almost identical values for all three varieties.) Fiber samples from the three varieties had different molecular weight distributions, different locations of peaks for the secondary wall fraction, and different weight average molecular weights. Ranking by HVI strength corresponded with ranking by peak molecular weight and weight average molecular weight. Fiber classification standards (for HVI) representing a range of lengths and strengths were sampled and assessed by GPC analysis [165]. The shortest fiber (0.903 in.) with a Stelometer strength of 21.4 g/tex had DP_w (Weight averaged degree of polymerization) = 15,000. The longer fiber (1.236 in) with 40% higher strength (31.0 g/tex) had DP_w = 23,700. The relation of the average DP_w to strength is shown in Figure 9.19.

The general correlation is evident between molecular weight and strength whereby the weight average molecular weight accounts for ~38% of the variability in strength. Thus, molecular compositional profiles indicate correlation of higher average molecular weight with greater strength of the cotton fiber. However, exceptions were evident. Two samples with equivalent lengths and strengths had differences in weight-average molecular weights. The sample with the higher molecular weight had, however, a broader-molecular-weight distribution than the lower-molecular-weight sample. The increased molecular weight in one sample apparently was offset by the narrowness of the molecular weight distribution in the other, giving equivalent strengths.

Molecular weight distributions of cotton fibers were evaluated with variables of the variety and the growth environment [166]. GPC analysis of cotton fiber samples demonstrated consistency of the secondary wall cellulose peak for the variety, alterations in the composition of the cellulose chains as a response to dryland conditions, and differences in molecular weight distributions according to fruiting zones for different irrigation methods.

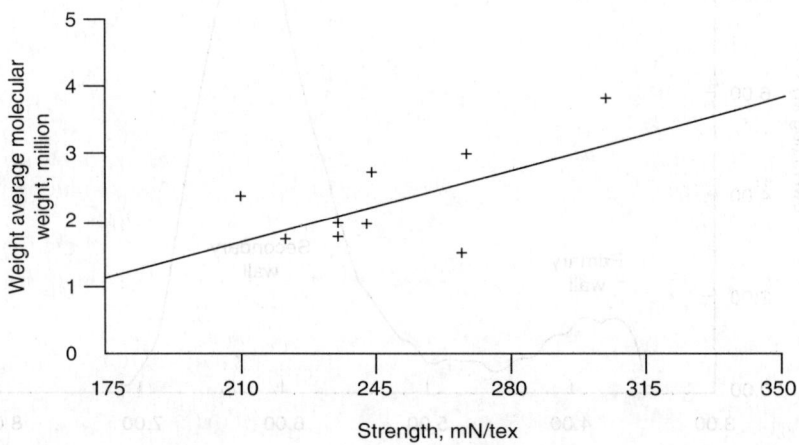

FIGURE 9.19 Relationship between weight average molecular weight (MW_w) and strength for cotton fibers with a range of lengths and strengths. Line shown is least squares fit with R of 0.62.

After cross-linking cotton fabrics with dimethylolethyleneurea to impart durable-press (nonwrinkling) properties, the cellulose was evaluated for strength losses and structural changes [167]. Cellulose nitrate derived from the cotton fabric before and after cross-linking treatment was dissolved in tetrahydrofuran for analysis by GPC. In this case, nitration is advantageous because the cross-links are broken during the derivatization thereby facilitating the characterization. Molecular degradation and cross-link embrittlement were measured as a function of treatment conditions. In fabrics with shorter cure times, the predominant strength loss (which dropped rapidly) came from cross-link embrittlement. Molecular degradation became the major source of strength loss at longer cure times. Acid-catalyzed treatment of the fabric with the substituted ureas substantially reduced the DP of the cotton cellulose.

Molecular weight distributions of cotton fabrics were compared before and after treatment with a total cellulase [168]. A lengthy multiday procedure was required to completely dissolve the cotton in the DMAc/LiCl solvent system. Despite considerable weight loss and breaking load reduction produced by the cellulase treatment, GPC analysis using DMAc/LiCl did not show reduction in molecular weights of the cellulose. This result supports a hypothesis that exoglucanases rapidly cleave cellobiose units from cellulose chains once the chains have been clipped by endoglucanases, resulting in a total removal of cellulose chains and a reduction in microfibrillar size, but with the remaining chains not yet degraded by the total cellulase. GPC analysis of DMAc/LiCl solutions of enzymatically treated lyocell made possible a clarification of depilling mechanisms [169]. The relatively low-molecular-weight cellulose of lyocell was rapidly and easily dissolved, in contrast to the multiday dissolution of the high-molecular-weight cotton cellulose noted above.

Incomplete dissolution of a sample results in an overestimation of the molecular weight. GPC analysis of cotton fabric using the DMI/LiCl solvent system appears to yield a faster, more complete dissolution of high-molecular-weight cellulose than does DMAc/LiCl. It gave a weight average molecular weight of $542,400 \pm 23,200$ for cotton printcloth [148].

9.5.3 THREE-DIMENSIONAL STRUCTURES OF CELLULOSE MOLECULES, CRYSTALLITES, AND FIBERS

Near the end of Section 9.5.1, there was a lengthy list of attributes needed to specify the three-dimensional structure of cellulose. The present section is concerned with descriptions of the smaller-scale structures. The information on shape comes from both theoretical and experimental methods, with the classic advantages and disadvantages of each. Theoretical methods can be used to fill the gaps between experiments and help interpret ambiguous results, but in the end, they are still "educated guesses." On the other hand, laboratory experiments on cellulose give real data, but they are done on single, specific situations. These results may not apply in general and are often ambiguous. It is much more powerful to combine both approaches.

The shapes of cellulose chains are consequences of the slightly variable shape of the glucose ring and the more variable geometry of the β-(1→4) linkage, along with some environmental influence. Almost all D-glucose rings have the chair shape ($^4C_1$) that places the bonds to the hydroxyl and hydroxymethyl groups, including O1 and O4, in equatorial (nearly parallel to the main plane of the pyranosyl ring) orientations (Figure 9.20). We are accustomed to the chair form of the ring, but the preferred linkage geometry is a more involved question, discussed below.

9.5.3.1 Describing Molecular Shape

When all monomeric units and their linkages in a polymer have identical geometries, the molecule will meet the mathematical criteria for a helix. A molecule need not have a central

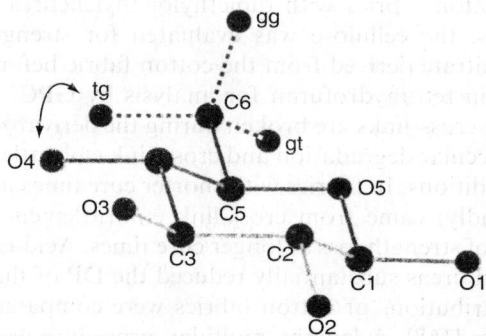

FIGURE 9.20 β-D-Glucopyranose with all three staggered orientations of O6 shown (*gg*, *gt*, and *tg*).

cavity to be helical; it suffices to have intramolecular screw-axis symmetry. Helices are described in terms of the number n of monomeric units (glucose residues) per turn of the helix, and whether the screw thread is left- or right-handed. If left-handed, n is given a minus sign. In case of $n = 2$, however, the helix can be considered to be of either chirality. Also, the advance along the helix axis per residue h and the pitch P the distance between turns of the helix, signify how extended or collapsed the helix is. The cellulose chain is mostly extended, even when it is not in a crystal (see Figure 9.17). For common crystalline cellulose, these parameters are $n = 2$ and $h \approx 5.18$ Å. The degree of extension can be understood by comparing h with the length of the glucose unit in small-molecule crystals, 5.42 to 5.57 Å [170]. Twofold screw helices have symmetry such that for every atom at x, y, and z, there is an identical atom at $-x$, $-y$, $(z + 0.5c, c = P)$. The x and y values are distances from the helix axis, and z is the distance along the crystal unit cell of dimension c and the fiber axis. This definition in Cartesian coordinates is an alternative but equivalent definition to the one in Section 9.5.2.1 that was in cylindrical polar coordinates.

If not in a crystal, cellulose chains would deviate from the ribbon shape in a random way, dictated by the lowest free energy. Deviations from the shape that has the lowest possible enthalpy do not increase the enthalpy very much, but variation increases the entropy. Therefore, deviations are favored. Still, even in solution or noncrystalline regions, the cellulose chain is expected to be rather extended, with the occasional kink [171]. In any case, helix nomenclature can still apply, but only in approximation for local segments.

Crystalline cellulose derivatives take other shapes, such as the trinitrocellulose helix with five residues repeating in two turns with $n = |2.5|$ [172]. The nitromethane complex of cellulose triacetate has eight residues in three turns ($n = -2.67$) [173]. Along with numerous cellulose derivatives that have $n = |3|$, the soda cellulose II complex is a threefold helix that is probably left-handed [174,175]. Values of h are still greater than 5 Å, indicating substantial extension. Greater deviation from the twofold structure is also possible. According to energy calculations, loop structures (Figure 9.21) are also possible, although these are usually somewhat higher in energy than the simple extended models. These other shapes raise the question as to whether the 2_1 shape observed in pure cellulose crystals is the favored form or whether it is distorted so it can pack most efficiently. This is a good opportunity for input from modeling.

9.5.3.2 Experiment- and Theory-Based Modeling Studies

One way to study the shapes of cellulose chains is to construct models that accommodate the available experimental data. There are many approaches to modeling, and comprehensive studies require extensive computations. The first computer model of a carbohydrate was a

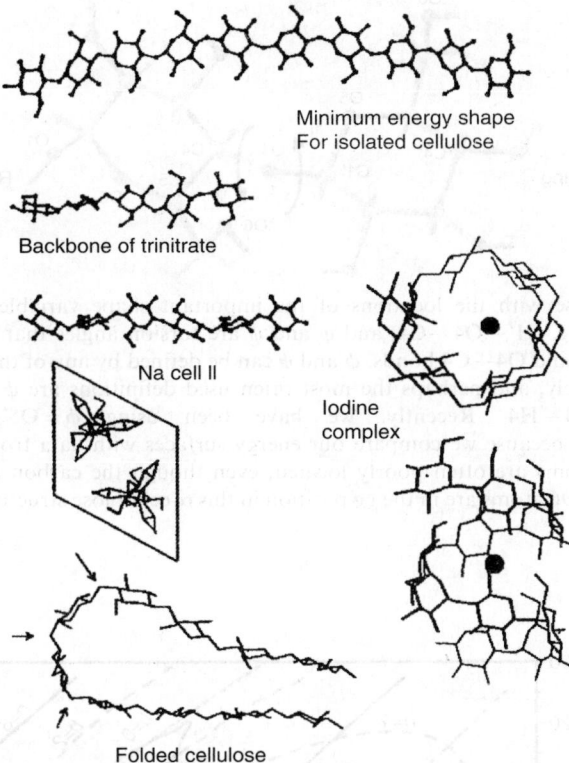

FIGURE 9.21 Various shapes of cellulose backbones proposed [18]. The arrows near the hairpin structure point to glycosidic linkages having approximate conformations of $\phi = 0°$ and $\psi = 180°$.

part of the experimental diffraction studies of cellulose [176]. Since then, there have been substantial improvements in both computers and their representations of molecules.

9.5.3.2.1 Extrapolated Experimental Models

The simplest models are extrapolations [170] of the experimentally determined structures of small molecules that are related to cellulose, such as cellobiose. Many such structures have been determined and are conveniently available in large databases [177,178]. The models consist of the x, y, and z coordinates of the atoms. Given the coordinates, it is simple to calculate the bond lengths, bond angles, and torsion angles. Purely geometric manipulation is used to repeat the geometry of a single experimentally observed glucose unit, connected by an observed linkage geometry. Instead of constructing models from each combination of glucose and linkage geometry, it is useful to take a shortcut. This shortcut consists of a generalized conversion from the ϕ and ψ torsion angle values to n and h values, using an average glucose ring geometry and glycosidic angle, τ (C1′—O4—C4). Figure 9.22 shows the ϕ and ψ torsion angles and valence angle, τ, which are used to characterize the linkage geometry. The values of ϕ and ψ range from 0 to 360° and are calculated from the O5′—C1′—O4—C4 and C1′—O4—C4—C5 torsion angles. Our conversion uses a constant value of 116° for τ. A conversion chart from ϕ and ψ to n and h is shown in Figure 9.23; its axis values are chosen so that the $n = 2$ line is a centered diagonal.

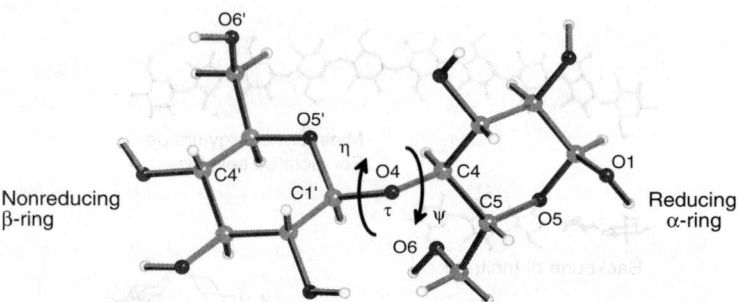

FIGURE 9.22 Cellobiose with the locations of the important shape variables ϕ, ψ, and τ. τ is a conventional bond angle, C1'—O4—C4, and ϕ and ψ are torsion angles that specify the amount of twist about the C1'—O and O4—C4 bonds. ϕ and ψ can be defined by any of the three atoms attached to C1' or C4, respectively, and perhaps the most often used definitions are $\phi =$ H1'—C1'—O4—C4 and $\psi =$ C1'—O4—C4—H4. Recently, we have been using $\phi =$ O5'—C1'—O4—C4 and $\psi =$ C1'—O4—C4—C5 because we compare our energy surfaces with data from crystal structures in which the hydrogen atoms are often poorly located, even though the carbon and oxygen atoms are accurately determined. O6 atoms are in the *gg* position in this α-cellobiose structure taken from Peralta–Inga et al. [177].

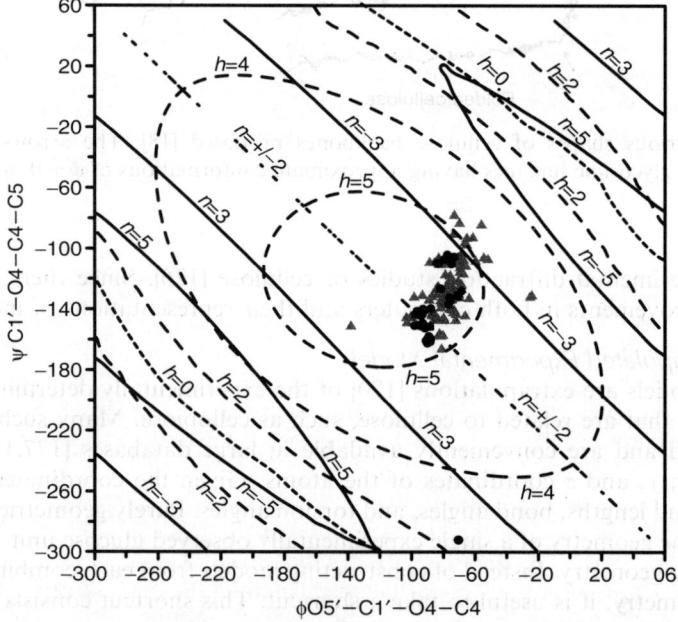

FIGURE 9.23 Contours of *iso-n* and *iso-h* values in ϕ–ψ space for model cellulose helices composed of the nonreducing residue of crystalline cellobiose [179] and a value of $\tau = 116°$. Also shown as dots are the experimentally determined values of ϕ and ψ from crystal structures of small molecules related to cellulose. With one exception at the bottom of the map, all of the dots correspond to extended helices with the numbers of residues per turn between 2 and 3. Geometries found in complexes of cellulose fragments and proteins are shown as triangles. They have a similar distribution but the range is expanded. Left-handed helices are indicated with negative values of *n*.

For example, take the atomic coordinates of the nonreducing ring of β-cellobiose [179] and the linkage geometry of that same structure. Exact extrapolation gives a helix with $n = -2.38$ and $h = 5.13$ Å [170]. Similarly, the hydrated sodium iodide complex of α-cellobiose [180] leads to helices with $n = -2.82$ and $h = 5.04$ Å, and one crystalline form of polymorphic methyl-4-O-methyl-β-cellobioside [181] yields helices with $n = 2.04$ and $h = 5.14$ Å. Review of the small filled circles in Figure 9.23 shows that the latter two values of n cover most of the range of extrapolations of small-molecule crystal structures, with one exception. That is a heavily substituted cellobiose [182] with $\phi = -71°$ and $\psi = -292°$. It would lead to helices with $n \approx +4.5$ and $h \approx 3.1$ Å. With that exception, the extrapolated chains all have $h = 5$ Å or greater and $n = 2$ to 3, with most of the chains left-handed. Approximate twofold conformations exist even when no O3′—H···O5 hydrogen bond or internal symmetry is possible, such as for acetates of xylobiose [183] and cellotriose [184].

Somewhat larger variations in model cellulose shapes are found if geometries are extrapolated from the less-accurately determined crystals of proteins that are complexed with fragments of cellulose or related molecules. There are more than 100 values of the linkage geometry for such complexes in the literature, shown as triangles in Figure 9.23. As described above for the small-molecule crystals, there is a heavy concentration of observed values of ϕ and ψ that correspond to n and h values between 2 and −3. Unlike the results from the simple crystals, some of the points on Figure 9.23 could be in error. Other points are intriguing because they may indicate that the protein is causing distortions that have a beneficial purpose, such as increased catalytic activity.

9.5.3.2.2 Theoretical Models

The above extrapolated shapes for cellulose depend on the availability of observed structures. They also depend on the assumption that a geometry that is observed in one environment could exist under other conditions. A different modeling approach is to calculate the internal, intrinsic energy of all possible structures. The probability would be maximal for the structure with lowest energy and then would decrease according to the higher energy values for the remaining structures. This can be understood by comparing the molecule to a spring. If a spring is stretched or compressed from its lowest energy (relaxed) shape, then it will have potential energy to restore itself to the lowest energy form when the compression or stretching force is removed.

Energy calculations are based on two fundamentally different methods. One of the methods, the electronic structure theory describes a molecule by approximating its electron orbitals with probability functions and makes use of very fundamental physical constants. Such *ab initio* quantum mechanics (QM) calculations improve the accuracy, as more complete theory is used. However, the advanced levels of theory require a great deal of computer time. Alternatively, the energy can be calculated with empirical force fields, also called molecular mechanics (MM) [185]. Here, components of molecular structure such as bond lengths, bond angles, torsion angles, van der Waals forces, and electrostatic forces, are described with simple equations that can be evaluated quickly. For example, MM calculations of energy for changes of bond length can be modeled with Hooke's law for stretching a spring. There are varied approximations used in these force fields, and variations in results from different methods and implementations are the basis of continuing controversy.

Either MM or QM can be used to carry out energy minimization. For example, a molecule can be drawn and its crude coordinates can be submitted for geometry optimization. The modeling software would systematically shift the atoms until the calculated energy was minimized. However, there is no way to know that this local minimum is the lowest possible (global) minimum. The method of conformational analysis systematically puts the molecule into all possible shapes and, in recent times, minimizes the energy at each increment of change.

Besides energy minimization, molecular dynamics (MD) can be employed. In models using molecular dynamics, the individual atoms have kinetic energy commensurate with their temperatures. The atomic motion resulting from the kinetic energy is restrained by the potential energy, depicting variations in molecular shape over time. The individual structural components, such as a particular torsion angle, may be monitored and this record of the time-based variation is called a trajectory. The MD is also used for conformational searching of complex molecules, especially at higher temperatures. It permits explicit inclusion of solvent in the model, and is advantageous because solutions are inherently dynamic. Understandably, the computer time needed increases substantially when hundreds or thousands of solvent molecules are included.

Cellobiose, the shortest cellulose chain, has been extensively modeled with conformational analysis and MD. When solvent water is present, it competes with the cellobiose hydroxyl groups [186]. The Chemistry at Harvard Macromolecular Mechanics (CHARMM) program found a new region especially favored by solvated cellobiose [187]. QM has also been used to find the lowest energy form of cellobiose [188], which, if extrapolated as above for a few residues, would lead to folding. A conformational analysis using QM of a cellobiose analog that is missing all of the hydroxyl and hydroxymethyl groups is reasonably predictive of the observed shapes (see Figure 9.24), even though many of the observed shapes are from structures that have interring hydrogen bonds.

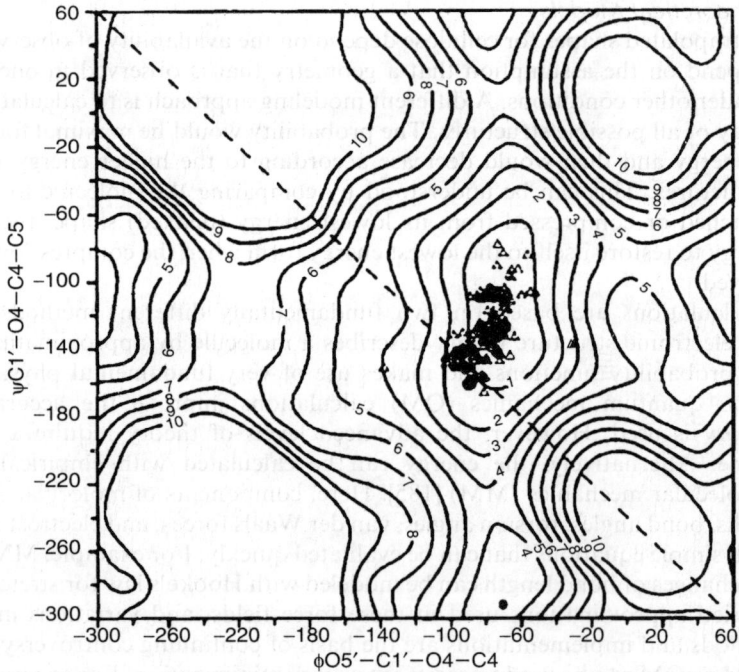

FIGURE 9.24 Quantum mechanics (B3LYP/6-311++G**) energy surface for a cellobiose analog composed of tetrahydropyran residues (cellobiose with all of the exocyclic groups replaced by hydrogen atoms). Contours are shown in 1 kcal/mol increments up to 10 kcal/mol of relative energy. Dots indicate the ϕ and ψ values of experimentally determined structures of small molecules and triangles represent comparable values of structures of molecules related to cellulose that are in complexes with crystalline proteins. The diagonal line represents conformations that would have twofold screw-axis symmetry.

Hybrid modeling studies of cellobiose that are based on both the QM calculations (Figure 9.24) and MM4 for the hydroxyl and hydroxy methyl groups are even more predictive of the observed shapes of the cellobiose linkage [189]. That was true only as long as the electrostatic forces are substantially reduced compared to the full strength. Nonexperimental molecular conformations were apparently favored when using potential energy functions that have strong hydrogen bonding [188,190].

Longer cellulose oligomers have been modeled with MD as well. Two studies have attempted to discuss the molecular shapes in aqueous solution as well as the solvent–cellodextrin and cellodextrin–cellodextrin interactions [191,192]. The first of these studies showed that chains were heavily solvated and not fully extended or in contact with other cellulose fragments. The simulation was proposed as a model for freshly prepared cellophane. The latter study was more in agreement regarding chain shapes with the results in Figure 9.25. In addition, that work showed, unlike the simulated cellotetraose molecules in the same study, that cellohexaose fragments stayed in contact with each other during the simulation. That was in agreement with the low solubility of longer chain crystals. Earlier, MD studies of cellooctaose showed that the central residues in such long chains might undergo major changes in ring shape [193]. Such deformations might be found in the amorphous regions of fibrous materials. Work to model the mechanical properties of cellulose was based on MM, with proposed differences in intramolecular hydrogen bonding for cellulose I and cellulose II leading to computed moduli similar to experimental values of about 130 and 80 GPa (GN/m^2) [194].

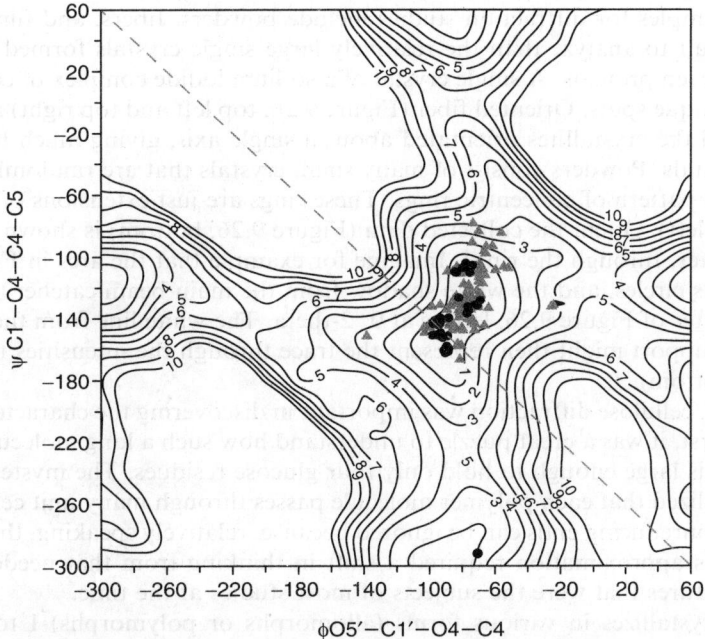

FIGURE 9.25 Energy map over $\phi-\psi$ space for cellobiose based on the quantum mechanics energies from Figure 9.24 for the backbone and empirical force field (MM4) energies for the hydroxyl and hydroxymethyl groups.

9.5.3.3 Experimental Studies on Cellulose Structure

Now that the range of likely shapes has been defined by experiments on related molecules and by energy calculations, we focus on the details of specific structures that have been observed for real, crystalline cellulose molecules, primarily by x-ray, neutron, and electron diffraction studies. A number of landmark concepts have been established with electron microscopy, as well. Infrared (IR), Raman, and nuclear magnetic resonance (NMR) spectroscopy have all also been important in the quest for understanding cellulose structure. Such data, while so far not able to provide complete definitive structures themselves, constitutes additional criteria that any proposed structure must be able to explain. In addition, unlike crystallography, the resolution of spectroscopic methods is not directly affected by the dimensions of the crystalline units, but only by the degree of order within the crystallites [195]. A comprehensive review of the contributions of spectroscopy to cellulose structure has been provided by Rajai Atalla [196].

9.5.3.3.1 Basic Information on Diffraction

Diffraction occurs when a beam of radiation with a wavelength comparable to atom–atom distances interacts with the periodic arrays of molecules in a crystal. Diffracted rays are recorded on imaging systems of many different types, originally photographic film. The positions of the diffracted rays indicate the dimensions of the unit cell, the smallest unit of the crystal structure that can be repeated by simple translation to generate the entire crystal. The intensities of the diffracted rays indicate the types and relative positions of the constituent atoms, and the reciprocal of the width of the diffracted ray indicates the size and degree of perfection of the crystal. Radiation wavelengths of around 1 Å are provided by x-ray generators, electron microscopes (electron beams), and nuclear reactors (neutron beams). Synchrotrons produce very intense x-ray beams.

Cellulose samples for diffraction studies include powders, fibers, and films, all of which are more difficult to analyze than the relatively large single crystals formed by many small molecules and even proteins. A single crystal of a sodium iodide complex of cellobiose gave a total of 9474 unique spots. Oriented fiber (Figure 9.26, top left and top right) and film sample usually have all the crystallites orientated about a single axis, giving much less information than single crystals. Powders consist of many small crystals that are randomly oriented, and therefore give a pattern of concentric rings. These rings are just extensions of the arcs shown in Figure 9.26 (left). Often, the collected data (Figure 9.26, bottom) is shown as a trace from the pattern center, through the rings. Imagine for example that the arcs in Figure 9.26 (top) were continuous circles, and the white shadow from the main beam catcher (the small white circle at the center of Figure 9.26, left) is at 0° 2-theta. The white line from the shadow of the beam catcher support might then represent the trace through the intensities that is shown in Figure 9.26 (bottom).

Historically, cellulose diffraction was important in discovering the characters of polymeric materials. At first, it was a great puzzle to understand how such a long molecule could fit into a unit cell that is large enough to hold only four glucose residues. The mystery was resolved when it was realized that each polymer molecule passes through many unit cells, and that the reducing and nonreducing ends can be ignored because, relatively speaking, their numbers are very small. This approximation required a shift in thinking from that needed for the nonpolymeric structures that were the subjects of most studies at the time.

Cellulose crystallizes in various forms (allomorphs or polymorphs) I to IV, with subclasses. Recently, our knowledge of these different crystal structures has advanced because of studies of highly crystalline cellulose films with synchrotron x-ray and neutron diffraction. The neutron data enabled the determination of the hydrogen bonding systems. The molecular

Cotton Fibers

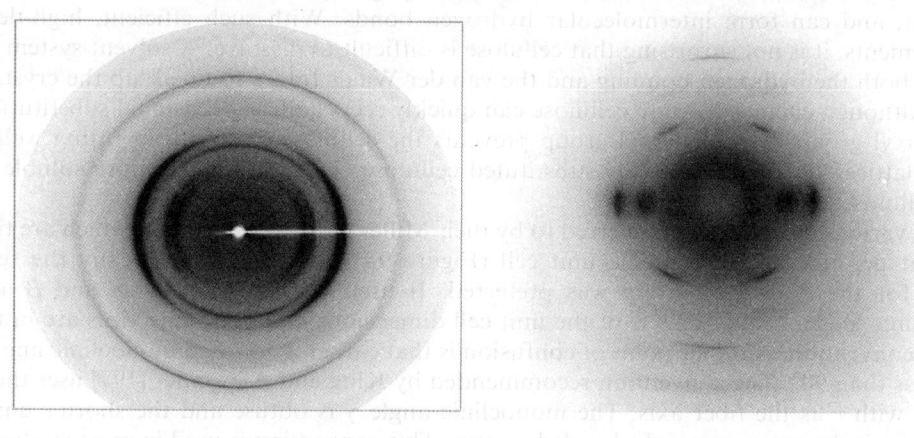

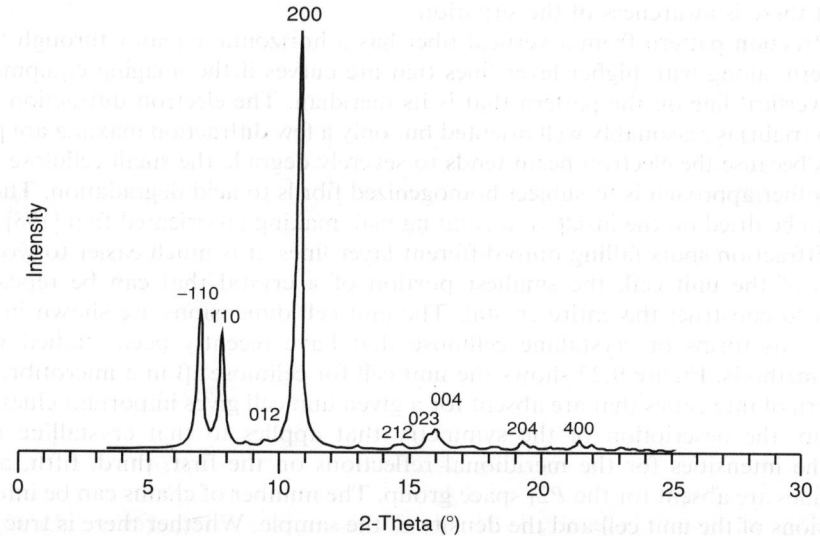

FIGURE 9.26 Cellulose diffraction patterns. Top: synchrotron radiation x-ray diffraction pattern for cotton fiber bundle. The fiber was vertical and the white circle and line correspond to a shadow from the main beam catcher and its support. (Credit to Zakhia Ford.) Middle: electron diffraction pattern of fragments of cotton secondary wall. The much shorter arcs are in top right figure are due to the good alignment and small number of crystallites in the electron beam. (Credit to Richard J. Schmidt.) Bottom: a synthesized powder pattern for cellulose, based on the unit cell dimensions and crystalline coordinates of Nishiyama et al. [209]. (Credit to Zakhia Ford.) Also shown are the *hkl* values for the Miller indices. The 2-theta values are for molybdenum radiation instead of the more commonly used copper radiation.

shape and general mode of packing in these different forms is similar. As a result of crystal packing, the chains take shapes that are flat ribbons in cross section, which can be packed with high density (as high as 1.62 g/cm^3). This packing results in low free energy because of the extensive, intermolecular interactions. The wider (about 9 Å), flat sides of the ribbons are hydrophobic while the thin (about 4 Å) edges of the ribbons are covered with hydroxyl groups. These chains are placed with their flat sides next to each other, allowing extensive interchain van der Waals interactions. The hydroxyl groups on the thin edges are similarly adjacent, and can form intermolecular hydrogen bonds. With such efficient, high-density arrangements, it is not surprising that cellulose is difficult to dissolve. A solvent system must disrupt both the hydrogen bonding and the van der Waals forces to break up the crystalline solid. Although chemically pure cellulose can quickly recrystallize, occasional substitution of a hydroxyl group by a methoxyl group prevents the cellulose chains from fitting well in a crystal lattice. Therefore, partially substituted cellulose chains are much more soluble than pure cellulose.

The various data peaks are referred to by their Miller indices (h, k, and l), which are tied to different periodic spacings in the unit cell (Figure 9.26, bottom). Historically the second setting for the $P2_1$ space group was preferred. It used b as the fiber axis and β as the monoclinic angle. Many reports of the unit cell dimensions and crystallite sizes are in terms of that convention. Another point of confusion is that earlier work used monoclinic angles (β or γ) less than 90°. The convention recommended by Klug and Alexander [197] uses the first setting, with c as the fiber axis. The monoclinic angle γ is obtuse and the shorter unit cell dimension is the a-axis in a right-handed system. This convention is used in most studies that report atomic coordinates (and in Figure 9.26, bottom). These differences are not big problems if there is awareness of the situation.

The diffraction pattern from a vertical fiber has a horizontal equator through the middle of the pattern, along with higher layer lines that are curves if the imaging equipment is flat. There is a vertical line on the pattern that is its meridian. The electron diffraction pattern in Figure 9.26 (right) is reasonably well oriented but only a few diffraction maxima are present. In part, this is because the electron beam tends to severely degrade the small cellulose crystals in cotton. Another approach is to subject homogenized fibrils to acid degradation. The resulting particles can be dried on the inside of a rotating vial, making an oriented film [198].

With diffraction spots falling onto different layer lines, it is much easier to work out the dimensions of the unit cell, the smallest portion of a crystal that can be repeated in all dimensions to construct the entire crystal. The unit cell dimensions are shown in Table 9.4 for the various forms of crystalline cellulose that have recently been studied with high-resolution methods. Figure 9.27 shows the unit cell for cellulose Iβ in a microfibril.

A pattern of intensities that are absent for a given unit cell gives important clues about the space group, the description of the symmetry that applies to that crystalline form. For example, the intensities for the meridional reflections on the first, third, fifth, and higher odd layer lines are absent for the $P2_1$ space group. The number of chains can be inferred from the dimensions of the unit cell and the density of the sample. Whether there is true symmetry in the cellulose crystallites has always been controversial, with weak odd-order meridional reflections [199] often appearing, thus failing to meet the strict requirement of absence. These

TABLE 9.4
Unit Cell Dimensions

Allomorph	Å			Degrees		
	a	b	c	α	β	γ
Cellulose Iα	6.717	5.962	10.400	118.08	114.80	80.37
Cellulose Iβ	7.784	8.201	10.380	90.00	90.00	96.50
Cellulose II	8.10	9.03	10.31	90.00	90.00	117.10
Cellulose III$_I$	4.45	7.85	10.31	90.00	90.00	105.10

Cotton Fibers

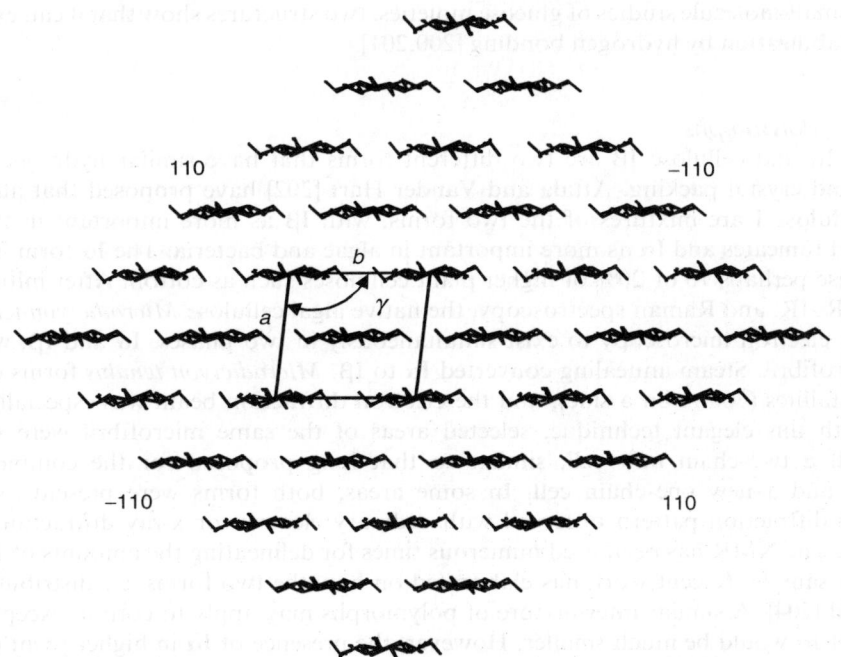

FIGURE 9.27 Cross-sectional view of cellulose Iβ crystallite with 36 molecules. The chain direction is toward the viewer. Also indicated are the *a* and *b* dimensions and the monoclinic angle γ of the unit cell as well as the +110 and −110 faces of the crystallite. A total of two chains pass through the unit cell: the entire central chain and approximately one-fourth of each of the four corner chains.

odd-order meridional intensities may be artifacts, and in any case, any deviations from symmetry have not been shown to indicate meaningful differences in structure.

When there is only one chain per unit cell, then all of the chains must be parallel. The reducing ends will all be at one end of the crystal and the nonreducing ends at the other. However, if there are two chains per cell, then the chains could be either packed parallel or antiparallel. Antiparallel chains have alternating reducing and nonreducing ends at each end of the crystal. A knowledge of the chain-packing mode is critical to understanding the biosynthetic processes. A much less obvious distinction is whether the parallel chain structures are "parallel-up" or "parallel-down." The convention used to describe the unit cell is important in this definition. A model is said to be parallel-up if the *z*-coordinate of O5 is greater than that of C5. The difference in packing can be illustrated by the following: the cellulose I model unit cell can be changed from up to down by shifting the adjacent sheets of chains parallel to the *b*-axis so that the monoclinic angle is changed to acute. Turning the model over then restores the unit cell to the correct, obtuse convention, but the chains are pointed in the opposite direction and have different interchain contacts.

Another issue in structural studies of cellulose is the orientation of the primary alcohol group (Figure 9.20). The O6 atom could be oriented so that it is *gauche* to O5 and *trans* to C4. This position is referred to as *gt*, (*gauche–trans*) or is described as having an O6—C6—C5—O5 torsion angle of approximately 60°. The other orientation found in about equal amounts to *gt* structures in various small-molecule studies is *gauche* to both O5 and C4. This is called the *gg* (*gauche–gauche*) structure (O6—C6—C5—O5 torsion angle ≈ −60°). A third staggered orientation is called *tg* (*trans–gauche*, O6—C6—C5—O5 torsion angle ≈ 180°). The latter is rarely found in small-molecule studies of glucose moieties: two structures show that it can exist, given enough stabilization by hydrogen bonding [200,201].

9.5.3.3.2 Polymorphs

Cellulose Iα and cellulose Iβ are two different forms that have similar hydrogen bonding systems and crystal packing. Attala and Vander Hart [202] have proposed that all types of native cellulose I are mixtures of the two forms, with Iβ as more important in the higher plants and tunicates and Iα as more important in algae and bacteria. The Iα form is thought to comprise perhaps 10 or 20% of higher plant celluloses such as cotton. After initial studies with NMR, IR, and Raman spectroscopy, the native algal cellulose *Microdictyon tenuius* was shown by electron microscopy to exist simultaneously in two phases, Iα and Iβ, within the same microfibril. Steam annealing converted Iα to Iβ. *Microdictyon tenuius* forms especially large crystallites (300 Å on a side), and the electron diffraction beam was especially narrow [203]. With this elegant technique, selected areas of the same microfibril were shown to have both a two-chain unit cell, similar to that long proposed for the common native cellulose, and a new one-chain cell. In some areas, both forms were present, giving the combined diffraction pattern observed with ordinary electron or x-ray diffraction of algal celluloses. The NMR has been used numerous times for delineating the amounts of Iα and Iβ in a given sample. Recent work has elaborated on how the two forms are distributed in the microfibril [204]. A similar intermixture of polymorphs may apply to cotton, except that the fraction of Iα would be much smaller. However, the presence of Iα in higher plant celluloses has been challenged [205,206]. If present in cotton, varying quantities of Iα could lead to variation in the powder diffraction spacings [207] that are routinely ascribed to slightly different unit cell sizes.

The recent cellulose Iα [208] and Iβ [209] crystal structures were obtained with *Glaucocystis nostochinearum* and *Halocynthia roretzi* (tunicin), respectively. They share similar intra- and intermolecular hydrogen bonding systems, with the shifting of the adjacent sheets of chains that contain the intermolecular O—H···O hydrogen bonding systems as the major difference. These sheets are parallel to the crystallographic b axis that is shown in Figure 9.27. In both structures, the crystal packing appears similar when viewed along the chain axes. In both structures, the second sheet is elevated about 2.58 Å relative to the first, and in the Iα structure the third sheet is elevated another 2.58 Å. In Iβ, the third sheet is shifted back down relative to the second sheet, putting it at the same height as the first sheet.

Both cellulose I structures have similar, complex hydrogen bonding systems that contain the O3—H···O5 hydrogen bond that is very typical for β-(1→4)-linked molecules (Figure 9.28). There are variations in the detailed geometries of those interactions. More importantly, there is substantial variation in O6···O2 intramolecular and O6···O3 intermolecular hydrogen bonds in each polymorph. The hydrogen bonds are described as disordered or having fractional occupancy. Both forms have O6 in the *tg* position, otherwise a rarity for glucose residues. The *tg* orientation of O6 permits a second intrachain, interresidue hydrogen bond, either O2—H···O6' or O6—H···O2, both of which occur at either various times or various places in the disordered systems. One of the several disordered systems, however, does not have either of the O6···O2 bonds. Similarly, there are weak O6—H···O1 or O2—H···O1 bonds in some but not all of the systems. Perpendicular to the hydrogen-bonded sheets, C—H···O hydrogen bonds have been found, and there are also strong van der Waals interactions. These attractions do not stabilize the structure as much as the hydrogen bonds within sheets, as indicated by the expansion during heating [210]. Upon heating Iα to 220 or 230°C, the structure converts to Iβ [211]. At this temperature, the intersheet spacing increases enough to allow a sliding shift of the chains to the Iβ form.

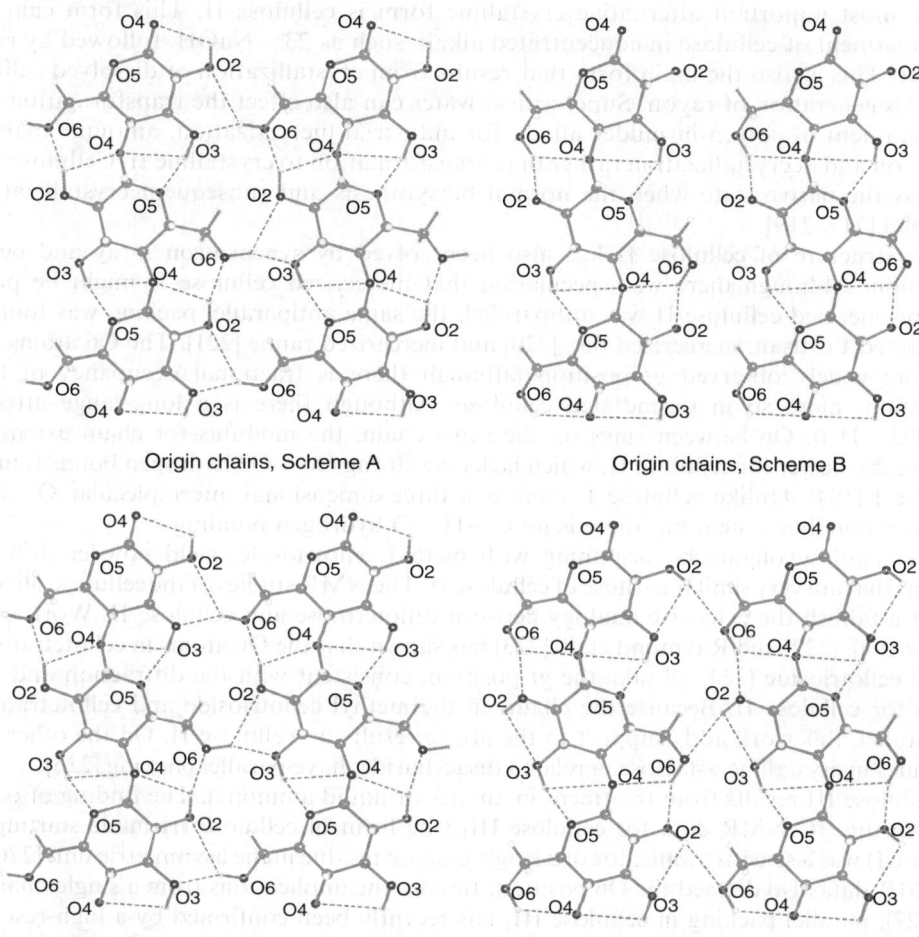

FIGURE 9.28 Nishiyama et al. reported four different hydrogen bonding schemes that arise in cellulose Iβ because of fractional occupancy of different orientations of the different hydroxyl groups and the slightly different internal geometries of the origin and center chains in the unit cell [209]. Note that the hydroxyl hydrogen atom was not located in the origin B scheme. The hydrogen bonds to O4 were not reported by Nishiyama et al. because they appear only with more liberal definition of a hydrogen bond. Even with the more liberal criteria, no interchain hydrogen bonds are found for the origin chains, scheme B.

Atomic force microscopy (AFM) has been applied to surfaces of *Valonia macrophysa*. The images were compared with surfaces generated from molecular models, and close correspondence was found. The data can be subjected to Fourier transformation to yield a diffraction pattern that can be compared with diffraction patterns made from the model surfaces [212]. The initial images are similar to the lattice images from electron microscopy.

Another AFM study of Valonia cellulose I showed O6 to be in the *gt* orientation [213]. Work with NMR spectrometry has gone further. Difference spectra show that the surfaces have extensively disordered regions as well as *gt* and *gg* O6 orientations [214]. When cellulose I is in water, NMR studies have indicated that its surface hydroxyl groups are also in *gt* positions [215].

The most important alternative crystalline form is cellulose II. This form can result from treatment of cellulose in concentrated alkali, such as 23% NaOH, followed by rinsing in water. This is also the main form that results from crystallization of dissolved cellulose, such as regeneration of rayon. Supercritical water can also effect the transformation [216]. The treatment of cotton in milder alkali, for industrial mercerization, amounts mainly to disruption and decrystallization rather than transformation to crystalline II. Cellulose II can occur as the native state when the normal biosynthesis and subsequent crystallization is disrupted [217–219].

The structure of cellulose II has also been solved by synchrotron x-ray and neutron diffraction. Although there was speculation that mercerized cellulose II might be parallel while regenerated cellulose II was antiparallel, the same antiparallel packing was found for regenerated Fortisan, mercerized flax [220] and mercerized ramie [221]. The O6 atoms are in the more widely observed *gt* position although there is fractional occupancy of the *tg* orientation, more so in regenerated cellulose. Although there is a long-range attraction from O3—H to O6 between rings on the same chain, the modulus for chain extension is considerably lower for cellulose II, which lacks the strong O6···O2 hydrogen bonds found for cellulose I [194]. Unlike cellulose I, there is a three-dimensional intermolecular O—H···O hydrogen bonding system but there is no C—H···O hydrogen bonding.

The cellulose oligomers, beginning with methyl cellotrioside, yield powder diffraction patterns that are very similar to those of cellulose II. The NMR studies of the cellulose oligomers further establish the extensive analogy between cellotetraose and cellulose II. Work in both Gessler et al. [222] and Raymond et al. [223] has shown that the O6 atoms in cellotetraose and methyl cellotrioside [224] all take the *gt* position, consistent with the diffraction and NMR results for cellulose II. Because the chains in the methyl cellotrioside and cellotetraose are antiparallel, this work adds support to the above results on cellulose II. On the other hand, molecules in crystalline α-lactose, a related disaccharide, have parallel packing [225].

Cellulose III results from treatment in amines or liquid ammonia. The finding of only six peaks in the ^{13}C NMR data for cellulose III_I (the form of cellulose III made starting with cellulose I) was a sensitive indicator of a single glucose residue in the asymmetric unit [226], and the NMR data also defined the O6 position. Besides the implications from a single-chain unit cell [227], parallel packing in cellulose III_I has recently been confirmed by a high-resolution study that located all the atoms, including hydrogen [228]. The sample was *Cladophora* seaweed cellulose, treated with supercritical ammonia, and the *gt* O6 positions allow an extensive cooperative hydrogen bonding network. Except for its one-chain unit cell, and therefore parallel chains, the packing and hydrogen bonding are reasonably similar to that of cellulose II.

Cellulose IV results from treatments at high temperatures, such as in glycerol at 260°C. The early cell wall material from some cells was reported to be cellulose IV [229]. The latest information is that cellulose IV made from cellulose I is actually a less crystalline cellulose I [226]. The NMR was instrumental in developing that idea. This is consistent with the finding of cellulose IV in the primary cell wall [229], but more developments are expected on this subject. (I just heard a paper at the ACS meeting in which the NMR work showed that the cellulose IV sample used in the x-ray study may not have been properly prepared.)

Cellulose Iβ and cellulose II have monoclinic, two-chain unit cells. Cellulose III_I has a monoclinic one-chain cell [227], and the one-chain unit cell of Iα is triclinic with no 2_1 symmetry. Still, all of the chain shapes are very similar to each other. It had been speculated that cellulose chain linkage geometries would alternate between the quite different linkages found in crystalline β-cellobiose and in methyl β-cellobioside [196]. That idea is now obsolete. Such a departure from symmetry would be far greater than indicated by the above high-resolution studies. When molecules from the high-resolution structures for all of the polymorphs are superimposed, differences in their backbone structures are barely visible.

Solid-state cellulose can also be noncrystalline, sometimes called amorphous. Intermediate situations are also likely to be important but not well characterized. One example, "nematic ordered cellulose" has been described [230]. In most treatments that produce amorphous cellulose, the whole fiber is severely degraded. For example, decrystallization can be effected by ball milling, which leaves the cellulose as a fine dust. In this case, some crystalline structure can be recreated by placing the sample in a humid environment. Another approach uses phosphoric acid, which can dissolve the cellulose. Precipitation by dilution with water results in a material with very little crystallinity. There is some chance that the chain may adopt a different shape (a collapsed, sixfold helix) after phosphoric acid treatment. This was concluded because the cellulose stains blue with iodine (see Figure 9.21), similar to the sixfold amylose helix in the starch–iodine complex.

One of the most special aspects of cellulose polymorphy is the transformation from I to II. The conversion of the parallel-packed cellulose I structures to an antiparallel cellulose II structure is interesting because it can occur without loss of the fibrous form. This transformation is widely thought to be irreversible, although there are several reports [231–233] of regenerated cellulose I. The observation that there are two different forms of cellulose III and of IV is also remarkable. The two subforms of each allomorph have essentially identical lattice dimensions and at least similar equatorial intensities. Other intensities are different, particularly the meridional intensities, depending on whether the structures were prepared initially from cellulose I or II. The formation of the III and IV structures is reversible and the preceding polymorph (I or II) results.

Work in the 1970s proposed the apparently correct explanation for the above reversible and irreversible transformations, i.e., that the chains in cellulose I are packed parallel and in cellulose II are antiparallel. However, there were problems with the work that left room for genuine doubts. These doubts were gradually erased by many separate studies. Now, it appears that conversion from I to II with the retention of the fibrous form is unique to cellulose in secondary cell walls in which the adjacent crystallites are tightly packed and constrained by a primary wall [234,235]. Extracellular celluloses and loosely packed, parenchymal or primary wall celluloses do not retain the fibrous form after treatment in strong NaOH. Instead, they contract into lumps that exhibit cellulose II diffraction patterns. These lumps could contain folds that would account for the antiparallel chains; a model fold is shown in Figure 9.21. The explanation for the ability to convert from parallel to antiparallel chain packing with retention of fibrous form is that each fibril is composed of parallel molecules in the original fibers, but that the fibrils themselves are antiparallel. During treatment with NaOH, the fibers swell and the molecules from the adjacent antiparallel fibrils interdigitate. This allows the formation of the lower-energy crystalline form.

Besides the cellulose structures I–IV and their subclasses, cellulose forms a variety of crystalline complexes. Soda celluloses were mentioned above, and there is an extensive array of complexes with amines [236]. Soda cellulose IV [237] is actually a hydrate of cellulose and contains no sodium (historically, cellulose hydrate meant cellulose II, which is now known to contain no water!). Many cellulose derivatives such as the nitrate (see above) and the triacetate [238] also give diffraction patterns. The most recent analysis of triacetate I shows a single-chain unit cell [239].

9.5.3.4 Electron Microscopy and Lattice Imaging

Work with electron microscopes showed that there is preferential enzymatic activity at only one end of the native microfibrils. This indicates that the reducing ends are all at one end of the microfibril and thus the chains are parallel, not antiparallel [240]. Electron microscopy and diffraction work on algal and bacterial cellulose confirmed the parallel-up nature of the chain orientation in the unit cell and the addition of new glucose residues to the cellulose chain at the nonreducing end [241]. Similar attempts with ramie fibers were not successful.

Lattice images of cellulose can be obtained from cellulose samples in the electron microscope and subjected to the same Fourier transformations as AFM images. Both of these techniques confirm the idea that cellulose chains are very extended in crystalline microfibrils and emphatically do not undergo folding within linear microfibrillar structures, as had been proposed by some authors.

Lattice images of algal, bacterial, and ramie cellulose have been obtained. These images show the individual molecular chains and the sizes of microfibrils, which vary in size and shape according to the source of cellulose [242,243]. There is also some variation within a given source. For example, microfibrils of Valonia ranged from 150 to 250 Å (15 to 25 nm). No evidence of elementary fibrils was seen.

9.5.3.5 Small Angle X-Ray Scattering

Although conventional x-ray diffraction equipment does not permit large (>100 Å) structures to be studied, a special apparatus can be configured to detect the behavior at very small scattering angles. Such devices are often used to study synthetic polymers. One such experiment on various cellulose samples was able to detect the pores of cotton fibers [244]. The results for normal cotton depended on a novel fractal analysis of the data rather than on the classical Guinier and Porod analyses. The void volume fraction ranges from 0.7 to 3.4% in the cotton samples and was 17% in Valonia. Dewaxing, scouring, and bleaching increased the void volume, where NaOH mercerization and ammonia treatments decreased not only the packing efficiency but also the void volume. In hydrocellulose II, an acid-hydrolyzed cotton cellulose II, the average pore size was 85 Å and the specific inner surface area was 15.3 m^2/cm^3.

9.5.3.6 Diffraction Studies of Crystallite Size

Cellulose powders can be created by cutting fibers into small particles, perhaps with a Wiley mill (Arthur H. Thomas Company, Swedesboro, New Jersey). On a laboratory x-ray system, powder diffraction patterns take 30 min. The positions of the peaks indicate the polymorphic form (I–IV); the powder diffraction pattern is often used as a fingerprint for comparison with the known pattern for a given crystalline form [207]. The breadth of the peaks is related to the extent of crystallinity (Figure 9.26, bottom). Using the Scherrer formula [245,246] and assuming no other distortions, the crystallite size can be calculated. Values for cotton perpendicular to the molecular axis are around 40 Å. That corresponds to a 6×6 array of cellulose chains (Figure 9.28). (Note that 20 of these 36 chains constitute the surfaces of the crystallite.) The crystallinity index (CI) can be simply calculated by comparing the minimum intensity just before the largest peak with the peak height [247]. By this simple method, the crystallinities for various varieties of cotton are in the range of 80%. Other methods for measuring crystallinity are based on different physical phenomena and the absolute results can be quite different. The rank ordering is usually similar, however.

9.5.3.7 Crystallite Orientation by Diffraction

Another important aspect of cotton structure, the crystallite orientation, can be determined with diffraction analysis of bundled fibers [248]. The degree of arcing of the spots on fiber patterns (e.g., Figure 9.26, left) indicates the extent to which the long axes of the crystallites are not completely parallel to the fiber axis. The deviation from perfect alignment is an important indicator of fiber strength (see Section 9.7.2). In the case of cotton fiber bundles, the spots on the diffraction pattern are large arcs, resulting from the spiral arrangement of the microfibrils, with reversals (see Section 9.5.5). The molecular axes in cotton fibers are not parallel to the fiber axis. Instead, different samples have average deviations from about 25 to 45°. Cotton is considerably weaker than ramie or linen, both of which have much smaller arcs. If a slurry of cotton fibers in water is cut into small particles in a tissue homogenizer, the spiral structure is broken up and the electron diffraction pattern of the individual particles (Figure 9.26, right) is similar to that for a ramie fiber bundle [249]. That work showed that the arc lengths on an electron diffraction pattern from cut up ramie particles are even shorter.

9.5.4 Crystal and Microfibrillar Structure by Chemical Methods

Much information about the structure of cellulose can be gleaned from data using sorption and chemical reactivity techniques. Information is provided on the accessibility and reactivity of the cellulose sample. An important aspect of such data is the fact that it must be interpreted as accessibility and reactivity to the specific agent and the test conditions used. This subject has recently been reviewed [250,251].

9.5.4.1 Sorption

Deuterium, moisture, iodine, and bromine sorption have been utilized for investigating the supramolecular structure of cotton and mercerized cotton. The methods have been described elsewhere [251]. Average ordered fractions are given in Table 9.5.

In some instances, nonaccessibility, or the so-called average ordered fraction, is measured rather than crystallinity. Values vary depending on the size of the probe molecule and its ability to penetrate and be adsorbed in all the disordered regions. It will be noted that the

TABLE 9.5
Average Ordered Fraction (AVF)[a,b] and Crystallinity (XAL)[b] of Cotton and Mercerized Cotton Determined by Sorption

Technique	Cotton		Mercerized cotton		Ref.
	AVF	XAL	AVF	XAL	
Deuteration	0.58	—	0.41	—	[251]
Moisture Regain					
Valentine	0.615	—	0.45[c]	—	[252]
Hailwood-Horrobin	0.67	—	0.50	—	[251]
Jeffries	0.58	—	0.465	—	[253]
Zeronian	0.63	0.91	0.48	0.79	[261]
Iodine	0.87	—	0.68	—	[251]
Bromine	0.46–0.80[d]	—	—	—	[255]

[a] AVF can be considered the nonaccessible fraction of the sample since the measurements consider crystallite surfaces to be part of the low-order regions.
[b] Expressed as a fraction.
[c] Calculated from Valentine's relation [251] using a value of 1.43 for the sorption ration of mercerized cotton [260].
[d] Depending on cotton variety.

average ordered fraction is relatively close for the deuteration and moisture regain methods. In addition, the average ordered fraction is decreased about 25% by mercerization.

Deuteration of the accessible hydroxyl groups is accomplished with saturated deuterium oxide vapor at room temperature. The extent of deuteration and therefore accessibility can be estimated gravimetrically by infrared spectroscopy. Accessibility rather than crystallinity is measured because deuteration of the hydroxyl groups on crystallite surfaces can also occur.

Water vapor at room temperature will not penetrate well-defined crystallites but will be adsorbed in the amorphous regions. Consequently, moisture sorption measured gravimetrically at a given relative vapor pressure and temperature has been used to determine order in cellulosic materials. In the case of Valentine [252] and Jeffries [253], the fraction of ordered material was obtained by correlating moisture sorption with values obtained by the deuterium oxide method. Hailwood and Horrobin [254] developed an equation for water sorption of cellulose based on a solution theory that allowed the calculation of the fraction of the sample inaccessible to water.

Lewin and coworkers [255–260] developed an accessibility system based on equilibrium sorption of bromine, from its water solution at pH below 2 and at room temperature, on the glycosidic oxygens of the cellulose. The size of the bromine molecule, its simple structure, hydrophobicity, nonswelling, and very slow reactivity with cellulose in acidic solutions, contribute to the accuracy and reproducibility of the data obtained. The cellulose (10 g/l) is suspended in aqueous bromine solutions of 0.01–0.02 mol/l for 1–3 h, depending on the nature of the cellulose, to reach sorption equilibrium. The diffusion coefficients of bromine in cotton and rayon are 4.6 and 0.37×10^{-9} cm^2/min, respectively. The sorption was found to strictly obey the Langmuir isotherm, which enables the calculation of the accessibility of the cellulose as follows:

At equilibrium,

$$C_B = KC_f(C_{B(s)} - C_B) \tag{9.1}$$

where C_B is the concentration of bromine in mol/kg of cellulose (e.g., the number of occupied sites), $C_{B(s)}$ the saturation concentration of bromine in mol/kg, $C_{B(s)} - C_B$ the number of unoccupied sites; C_f the equilibrium concentration of bromine in mol/l of solution; and K the equilibrium constant.

If we put the accessibility $A = 100/n$ and $n = m/C_{B(s)}$, where $1/n$ is the number of anhydroglucose units (AGUs) available for sorption, assuming that one AGU is accessible to one bromine molecule, and $n = 1000/162$, e.g., the number of mole AGUs in 1 kg of cellulose, we obtain:

$$\left(\frac{K}{n}\right)\left(\frac{m}{C_B}\right) - K = \frac{1}{C_f} \tag{9.2}$$

The value of n is calculated from the extrapolation of the straight line obtained by plotting m/C_B against $1/C_f$ to the value of $1/C_f = 0$ (see Figure 9.29).

Straight-line relationships were obtained between the accessibilities by the bromine method and the IR crystallinity indices and the wide angle x-ray scattering (WAXS) indices for 16 native and regenerated celluloses with accessibilities ranging from 70 to 6% (See Table 9.6 and Figure 9.30 and Figure 9.31).

The accessibilities determined by the bromine method involve only the noncrystalline, less-ordered regions (LORs) and do not include the surfaces of the crystalline regions. The glycosidic oxygen on these surfaces is buried half a molecule deep within the crystallite [260]

Cotton Fibers

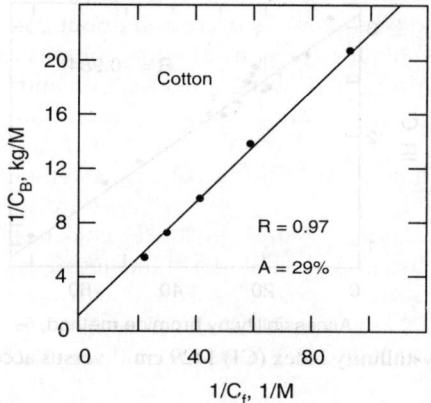

FIGURE 9.29 Langmuir sorption isotherm of Br_2 on cotton fibers.

and is inaccessible both to H_3O^+ which is responsible for hydrolysis, as well as to bromine molecules. Unlike the glycosidic oxygen, the hydroxyl groups protrude from the crystallite surfaces and are, therefore, accessible to deuteration and moisture absorption, which yield similar accessibilities [260]. Substitution reactions similarly occur on the hydroxyl groups.

TABLE 9.6
Accessibilities and Crystallinity Indices (CI) of the Cellulose Samples

Sample	$n = \dfrac{1}{C_{B(s)}} \times m$[a]	S[a]	R[b]	Accessibility by Br method, $A = 100/n$	CI by X-ray	CI by IR $\dfrac{1429\,\text{cm}^{-1}}{893\,\text{cm}^{-1}}$	CI by IR $\dfrac{1372\,\text{cm}^{-1}}{2900\,\text{cm}^{-1}}$
1. Pay master cotton fibers	4.54	0.14	0.97	22	0.785	2.56	0.59
2. Amsark cotton fibers	3.7	0.12	0.97	27	0.740	2.3	0.55
3. Pima cotton fibers	5.0	0.10	0.96	20	0.780	2.44	0.66
4. Acala 4-420 cotton fibers	4.34	0.06	0.98	23	0.750	2.5	0.64
5. Bradley cotton fibers	3.4	0.11	0.97	29	0.740	2.2	0.57
6. Moores cotton fibers	1.85	0.11	0.97	54		1.6	0.45
7. Pima cotton fibers	3.1	0.16	0.98	32	0.700	2.2	0.51
8. Cross-linked Pima cotton fabric	9.09	0.07	0.92	11	0.830	2.74	0.65
9. Hydrolyzed Pima cotton fabric	7.14	0.11	0.97	14	0.860	2.72	0.6
10. Ramie fabric	5	0.11	0.98	20	0.800	2.46	0.59
11. Hydrolyzed ramie fabric	15.2	0.11	0.98	6.5	0.880	3.07	
12. Evlan fibers	3.7	0.10	0.98	27	0.720	2.26	0.53
13. High modulus rayon fibers	4.46	0.10	0.97	22.4	0.730	2.6	0.54
14. Vincel 28 fibers	2.11	0.11	0.97	47	0.630	1.76	0.48
15. Vincel 66 fibers	3.25	0.10	0.97	30.7	0.740	2.35	0.53
16. Cellulose triacetate	1.42	0.037	0.97	70		1.07	0.37

[a]The plot of $1/C_B$ against $1/C_f$ is a straight light with the slope $S = n/Km$.
From the intercept $1/C_{B(s)}$, $n = (1/C_{B(s)}) \times m$ is calculated.
[b]R = degree of correlation.
Source: From Lewin, M., Guttman, H., and Saar, N., *Appl. Polym. Symp.*, 28, 791, 1976.

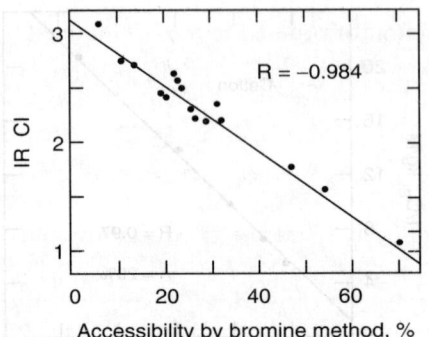

FIGURE 9.30 Infrared (IR) crystallinity index (CI) 1429 cm^{-1} versus accessibility by the Br$_2$ method for different cellulose fiber.

Zeronian et al. [261] hypothesized that if microcrystalline cellulose is prepared that is a facsimile of the crystalline regions present in the fiber then the fraction of amorphous material (F) of the fiber can be obtained from the relation

$$F = \frac{(M_s - M_c)}{(M_a - M_c)} \tag{9.3}$$

where M_s, M_c, and M_a are the monomolecular moisture regains of the sample, its microcrystalline counterpart, and amorphous cellulose, respectively. F is a measure of the disordered cellulose that does not include crystalline surfaces. Accessibility (A_s) is given by

$$A_s = \frac{M_s}{M_a} \tag{9.4}$$

The fraction of crystalline material (X) is given by

$$X = 1 - F \tag{9.5}$$

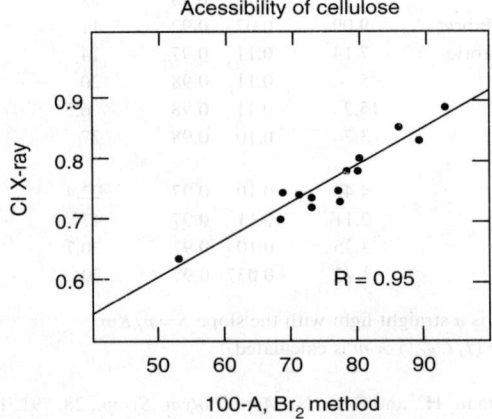

FIGURE 9.31 X-ray crystallinity index (CI) versus accessibility by the Br$_2$ method.

Cotton Fibers

The values of X obtained by Zeronian et al. [261] for cotton and mercerized cotton agree well with the values for fraction of ordered material obtained by acid hydrolysis.

9.5.4.2 Acid Hydrolysis

Acid hydrolysis is usually carried out with mineral acids at elevated temperatures. A portion of the cellulose reacts much faster than the remainder under these conditions. It is believed that the initial reaction occurs in the disordered regions and later extends to the ordered regions. The chain cleavage occurs and the products are glucose, soluble oligosaccharides, and an undissolved residue designated hydrocellulose. The weight of hydrocellulose is plotted against time [262] (Figure 9.32). Resolution of the plot into two components and extrapolation of the data at the slower rate to zero time gives an estimate of the ordered fraction. The sum of the initial percentages of crystalline and amorphous cellulose is always less than 100%. This may be due to incomplete analysis of the weight-loss data. Estimates of the degree of order by this technique may be high. Here the chemical agent is quite small and the acidic medium might disrupt the ordered regions that would permit the reaction to continue.

9.5.4.3 Formylation

The formylation method is based on the determination of the ratio of the extent of esterification of cellulose by formic acid after a given time interval to that of soluble starch for the same time interval. It is assumed that the starch is fully accessible to the reagent; thus a measure of the accessible fraction of the cellulose can be calculated [263,264]. By extrapolating the plot of this ratio against time to zero time, the initial accessible fraction of the sample can be determined [263]. The complement of this value is the ordered fraction. Other workers had arbitrarily measured accessibilities after 16 h of esterification [264–266]. However, much

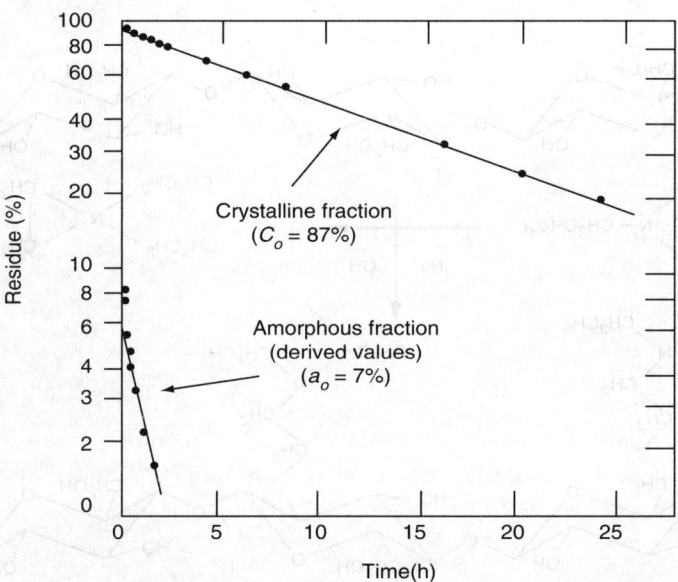

FIGURE 9.32 Relationship between residue weight and time of hydrolysis at 100°C with 6 M HCl for cotton linters. (From Nelson, M.L., *J. Polym. Sci.*, 43, 351, 1960.)

closer agreement with acid hydrolysis values is obtained if the initial ordered fraction is obtained. Nickerson's estimates of the initial nonaccessible fractions of cotton, mercerized cotton, and viscose rayon are 0.91–0.92, 0.85–0.86, and 0.70–0.72, respectively [263]. The advantages of this method are that the reaction is autocatalytic; the formic acid molecule is small, polar, and water miscible; the reagent is a relatively strong swelling agent for cellulose but does not penetrate the ordered regions [263]. A disadvantage is some chain scissions may occur [264] and result in crystallization.

9.5.4.4 Periodate Oxidation

This method is based on the preferential oxidation of the disordered regions by sodium metaperiodate [266,267]. Conditions are selected so that the reaction is confined as far as possible to the Malaprade course resulting in the formation of 2,3-dialdehyde units. The course of the reaction is followed by measuring the oxidant consumption from the amount of periodate consumed. From plots of log oxidant consumption against time, a measure of the fraction of ordered material can be calculated analogous to that of the acid hydrolysis method.

9.5.4.5 Chemical Microstructural Analysis

Chemical microstructural analysis (CMA) method is based on reactivity of the cellulosic hydroxyl groups with diethylaminoethyl chloride under very mild basic conditions, which, to the best of our knowledge, does not further disrupt ordered regions. The cotton cellulose is reacted with diethylaminoethyl (DEAE) chloride as shown in Figure 9.33.

The DEAE cellulose is then hydrolyzed to substituted glucoses and glucose, which are silylated. The relative quantities of OH-2, OH-3, and OH-6 DEAE glucose are determined via gas chromatography. The fraction of total reactivity under these conditions for each of these hydroxyls is given in Table 9.7.

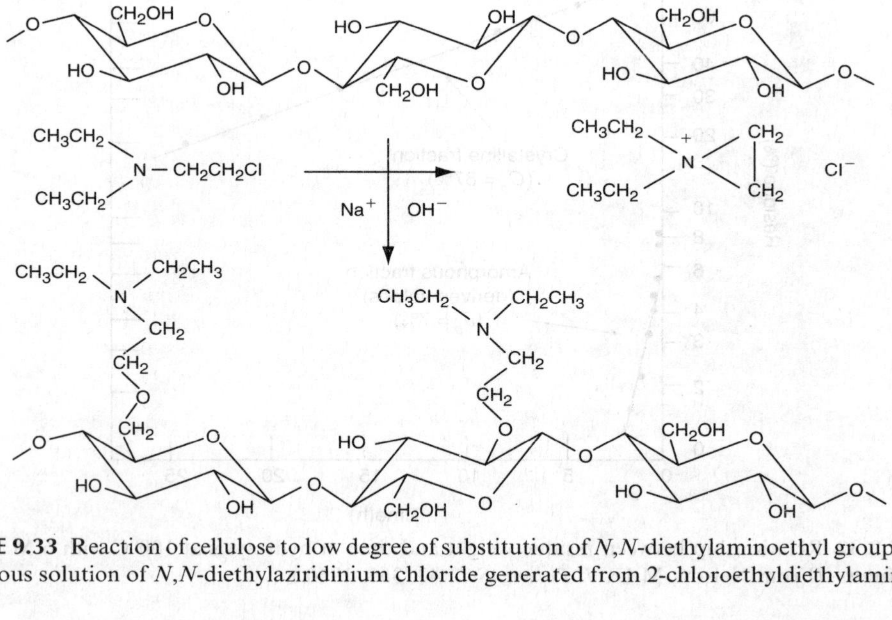

FIGURE 9.33 Reaction of cellulose to low degree of substitution of N,N-diethylaminoethyl groups occurs in aqueous solution of N,N-diethylaziridinium chloride generated from 2-chloroethyldiethylamine.

TABLE 9.7
Fraction of Total Reactivity

Hydroxyl	Fraction
OH-2	0.655
OH-3	0.059
OH-6	0.286

Note that under these nondisruptive, basic conditions the most reactive hydroxyl is OH-2. In this sample, the fraction of total reactivity of OH-2 is 11 times that of OH-3 and more than double that of the primary hydroxyl OH-6. This is quite different from relative reactivity observed for soluble glucans in which OH-6 is usually the most reactive hydroxyl. The CMA technique has been discussed in detail elsewhere [250,251]. Because OH-2 is the most reactive in this reaction, data is reported relative to this hydroxyl. Representative results for several cottons are given in Table 9.8.

The relative availability of each hydroxyl group of totally amorphous cellulose is 1.00 by definition. Mercerization converts the native cellulose I to cellulose II. Hydrocellulose is the unreacted cellulose after acid hydrolysis, which removed the more disordered regions as discussed above. The native cellulose for which the data is presented is cotton that has been dried. Drying alters the relative availability of the OH groups [268]. In agreement with the predictions (Figure 9.22) based on the unit cell structure, the OH-3 to OH-2 ratio has been shown to approach a limiting value of zero in a never-dried cotton. It is assumed that this represents almost perfect crystallinity. This value increases upon desiccation of the fiber on boll opening [268] (Figure 9.34).

This is the first introduction of stress and disorder into the fiber. Greige mill processing further increases this microstructural disorder as evidenced by increasing availability of the OH-3 as shown in Figure 9.35 [269].

9.5.4.6 Summary of Average Ordered Fraction Values Determined by Chemical Methods

The values for the average ordered fractions in cotton and mercerized cottons determined by the different chemical methods are summarized in Table 9.9.

These data for cotton and mercerized cotton are comparable to those determined by the sorption techniques (Table 9.5), by x-ray diffraction (0.73 and 0.51, respectively), and density (0.64 and 0.36, respectively) [250].

TABLE 9.8
Relative Availabilities of Hydroxyls

Cotton	OH-3/OH-2	OH-6/OH-2
Native	0.27	0.82
Mercerized	0.79	0.86
Hydrocellulose	0.23	0.74
Amorphous	1.00	1.00

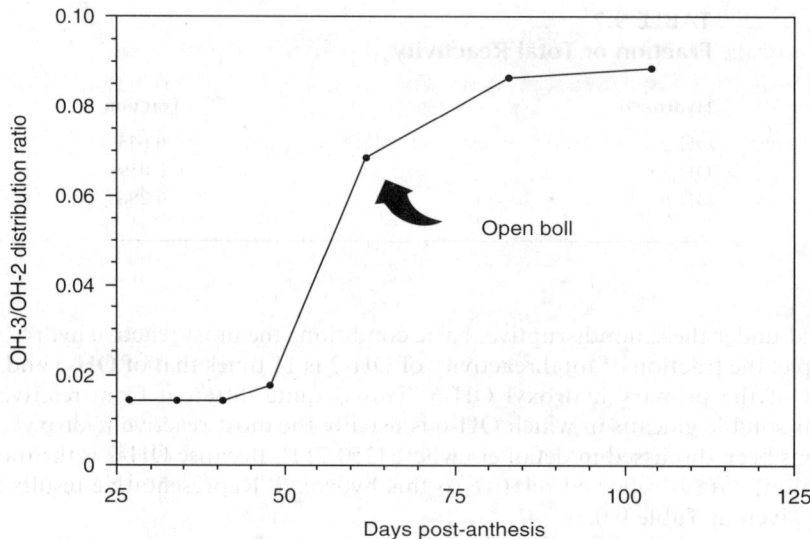

FIGURE 9.34 OH-3/OH-2 distribution ratios for cotton fibers as a function of days post-anthesis (DPA).

9.5.5 FIBER STRUCTURE

9.5.5.1 Morphology

Cotton is a unique textile fiber because of the interrelationships of its subunits. From its multicomponent primary wall, through the pure cellulose secondary wall to the lumen, the organization of subfiber units provides the fiber with characteristic properties that make it a processible, strong, and comfortable textile fiber. The outer skin of the fiber, the cell wall (cuticle-primary wall), is composed of an inner network of microfibrils randomly organized within a mixture of waxes, pectins, proteins, and other noncellulosic materials [270]. In the

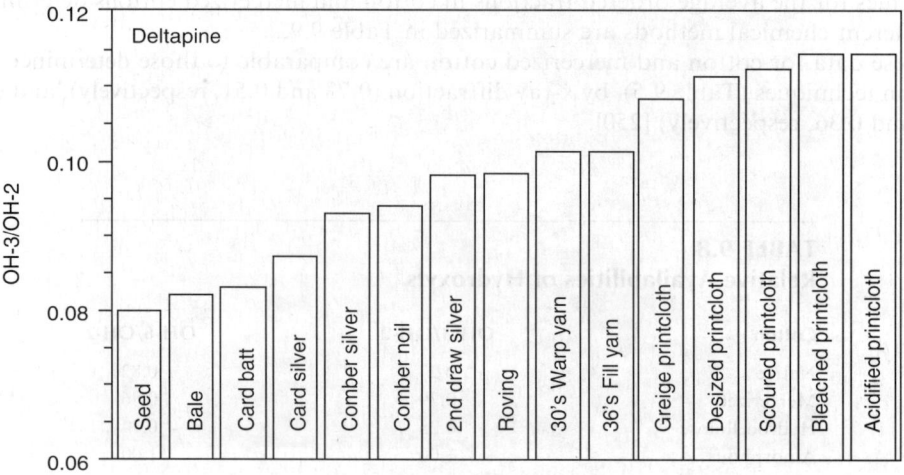

FIGURE 9.35 Effect of processing on the OH-3/OH-2 distribution ratio in cotton fibers.

TABLE 9.9
Average Ordered Fraction in Cotton and Mercerized Cotton Determined by Chemical Methods

Method	Cotton	Mercerized cotton
Acid hydrolysis	0.90	0.80
Formylation	0.79	0.65
Periodate oxidation	0.92	0.90
Chemical microstructural analysis[a]	0.73	0.29

[a] Based on availability of 3-OH.

dried fiber, the microfibrils are present in a network of crisscrossing, threadlike strands that encase the entire inner body of the fiber. The noncellulosic components of the cell wall give the fiber surface a nonfibrillar appearance, and provide both a hydrophobic protection in the environment, and a lubricated surface for processing (Figure 9.36 and Figure 9.37).

It is the waxy component of the primary wall that must be partially removed to allow the finishing and dyeing chemicals to access the body of the fiber. Inside the primary wall, a thin layer, called the winding layer (Figure 9.38), consists of bands of helical microfibrils that are laid down in a lacy network, which has been associated both with the primary wall [271] and with the secondary wall [99,272].

During fragmentation of the fiber, the winding layer often separates with the primary wall; however, it is believed that the winding layer cellulose is deposited at the time of decreasing elongation when secondary wall synthesis begins, and therefore may be more

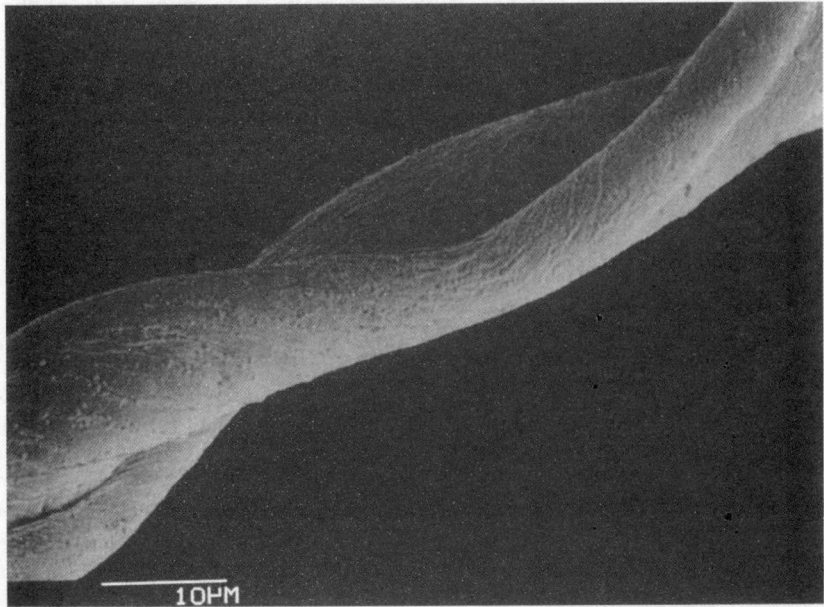

FIGURE 9.36 Typical dimensional structure variation in cotton fiber after drying (scanning electron microscope, SEM).

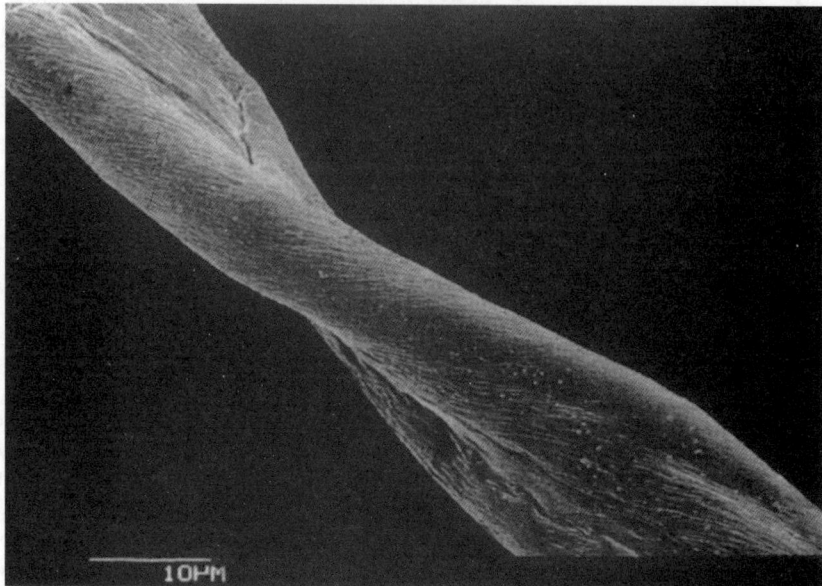

FIGURE 9.37 Less convoluted dimensional structural variation in cotton fiber after drying (SEM).

closely connected chemically to the secondary wall. The intermeshed fibrillar network of the primary wall and the woven mat of fibers of this winding layer just beneath it provide a dynamic casing that allows limited swelling within the secondary wall, which has its microfibrils oriented more along the fiber axis. The casing protects the secondary wall fibrils from lateral separation forces during swelling. As long as the primary wall-winding layer is intact, the inner portion of the fiber is less accessible to damage. The main body of the fiber, the

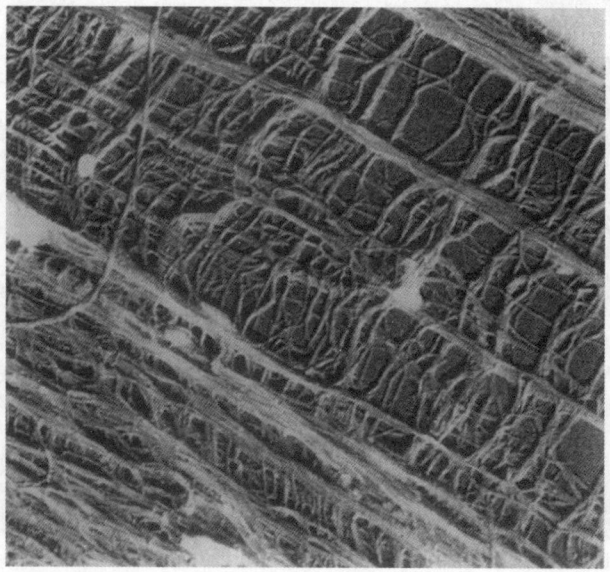

FIGURE 9.38 Cross-hatch structure of winding layer (transmission electron microscope, TEM).

Cotton Fibers

FIGURE 9.39 Parallel microfibrils of secondary wall (TEM).

secondary wall, consists of layers of nearly parallel fibrils laid down concentrically in a spiral formation (Figure 9.39).

Secondary wall fibrils closer to the primary wall lie at an approximate 45° angle to the fiber axis (Figure 9.40), while this orientation becomes aligned more closely with the fibrillar axis as the fiber core, or lumen, is approached (Figure 9.41).

The direction of the spiral around the axis of the fiber reverses at random intervals along the length of the fiber. These fibrillar directional changes are called reversals, and can be detected by following the direction of wrinkles on the fiber surface (Figure 9.42).

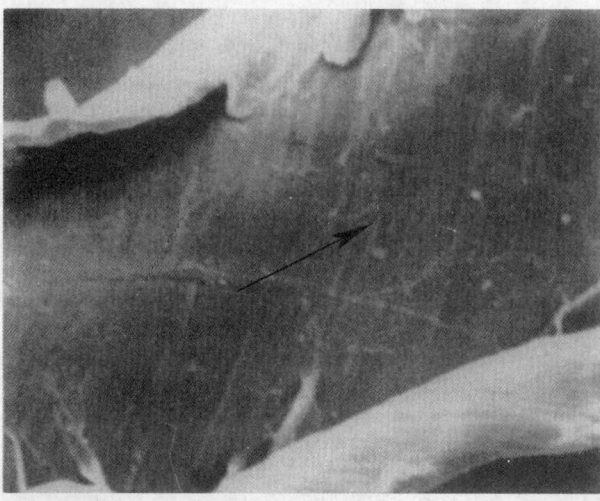

FIGURE 9.40 Fiber with peeled primary wall revealing underlying secondary fibrils at 45° to fiber axis (SEM). Arrow indicates fiber axis.

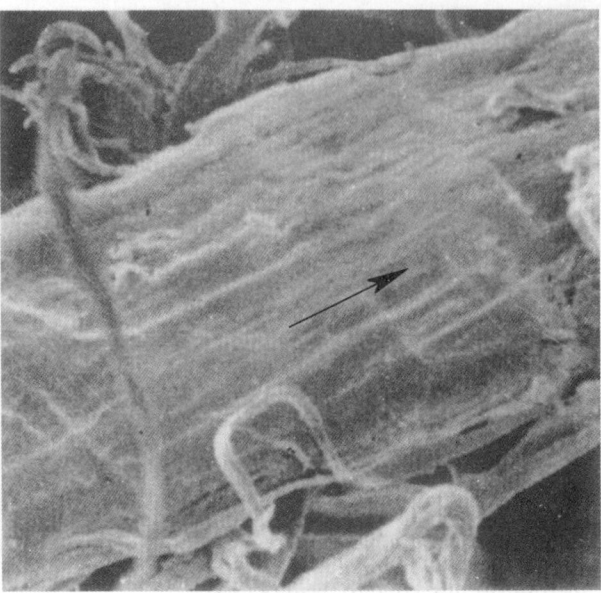

FIGURE 9.41 Fiber with primary and outer secondary walls peeled to show secondary wall fibrils lying almost parallel to fiber axis (SEM). Arrow indicates fiber axis.

Reversals represent zones of variations in breaking strength. The areas immediately adjacent to either side of a reversal are more likely to break under stress than are other fiber areas [65,273]. Figure 9.43 shows cracking of a fiber damaged by fungus on each side of a reversal, as indicated by wrinkles in the fiber's primary wall.

The densely packed fibril layers of the secondary wall are considered to be pure cellulose. The thickness of this wall, from primary wall to lumen, is closely related to gravimetric fineness

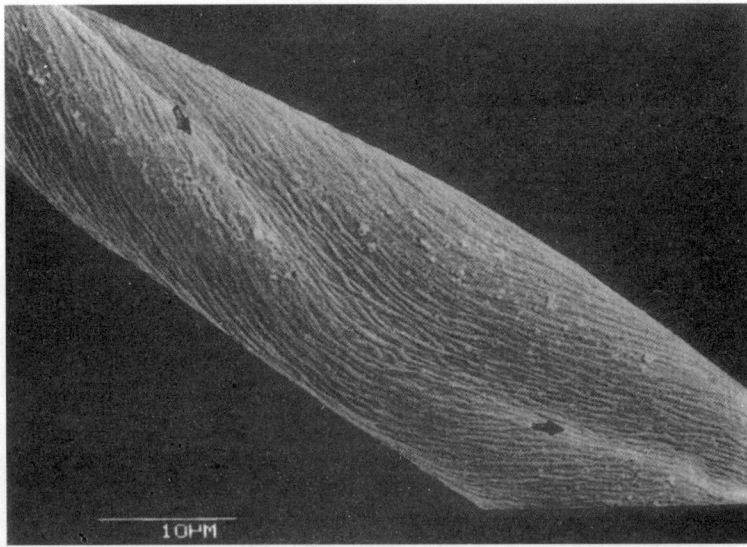

FIGURE 9.42 Fiber surface wrinkles showing underlying reversal in fibrillar direction. Arrows indicate direction of fibril wrap (SEM).

Cotton Fibers

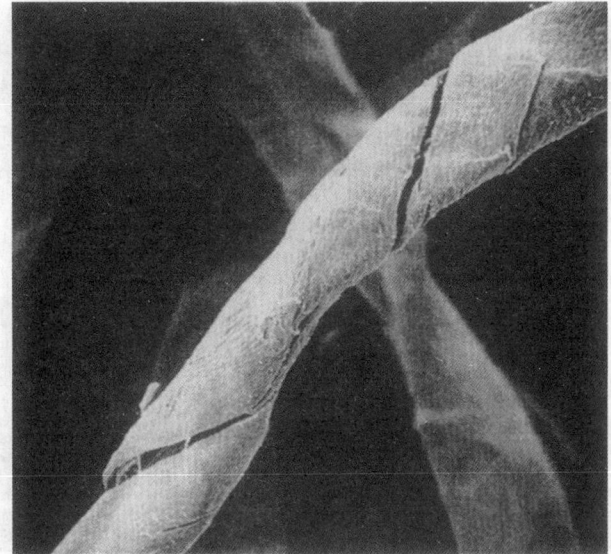

FIGURE 9.43 Fiber showing cracks developed on either side of a reversal (SEM).

(mass per unit length). Fibers with no secondary wall development exhibit no individual fiber integrity and can exist only in clumps [274]. Figure 9.44 illustrates cross section of a bundle of fibers with primary wall, but no secondary wall. Development of the secondary wall provides the fiber with rigidity and body.

Figure 9.45 shows intermediary secondary wall development (immature fibers) in cross section, while Figure 9.46 illustrates cross sections of mature fibers.

These sections were harvested live and processed in the wet state so they are more rounded, and have not assumed the characteristic Kidney-bean-shaped fiber cross-sectional shape of the

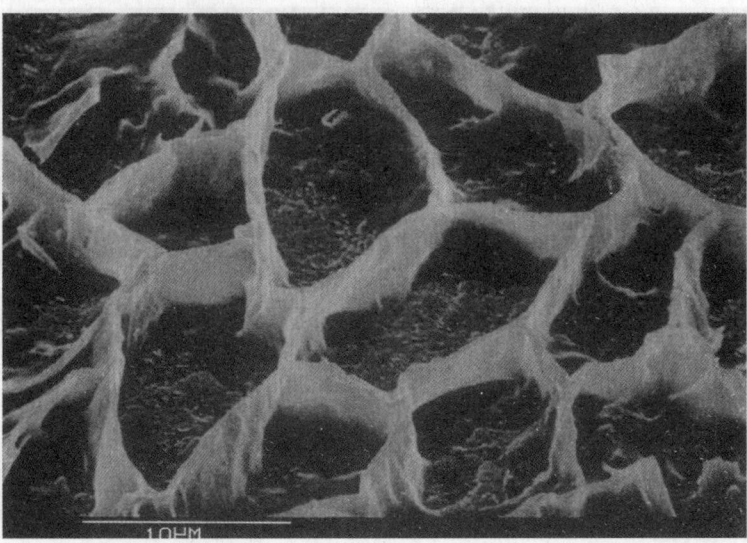

FIGURE 9.44 Cross-section of fibers showing primary wall but no developed secondary wall (SEM).

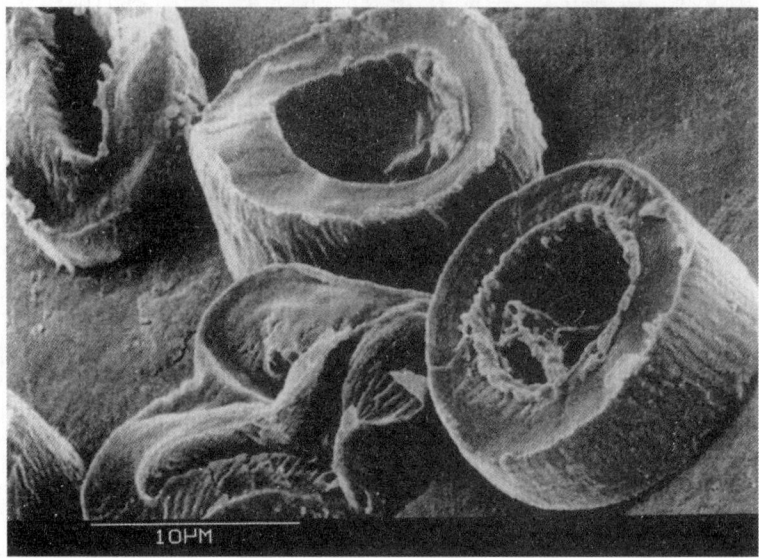

FIGURE 9.45 Cross-section of fibers showing partially developed primary walls, immature fibers (SEM).

dried fiber. Fibers with thinner secondary walls are known as immature, while those with walls at or approaching their maximum thickness are called mature. Thus, maturity is a relative term that is difficult to measure objectively (see Section 9.7.1). Secondary wall thickness is directly related to fiber properties such as strength, dyeability, and reactivity. Figure 9.47 is a cross section of a typical fiber bundle, showing a mixture of both mature and immature fibers.

When the fiber is dried, no differentiation between successive layers of secondary wall fibers can be distinguished. However, when fibers are swelled and viewed at higher magnifications in cross section, lacey, layered patterns become apparent [275]. Figure 9.48

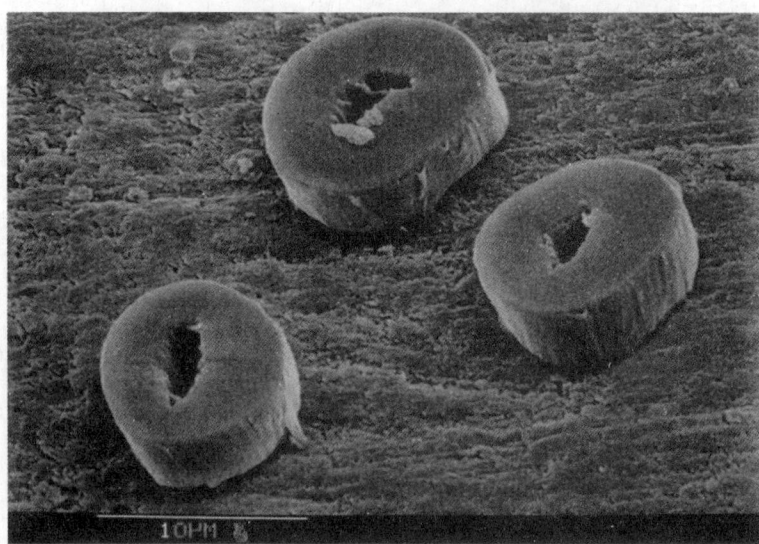

FIGURE 9.46 Cross-sections of fibers showing fully developed primary walls, mature fibers (SEM).

Cotton Fibers

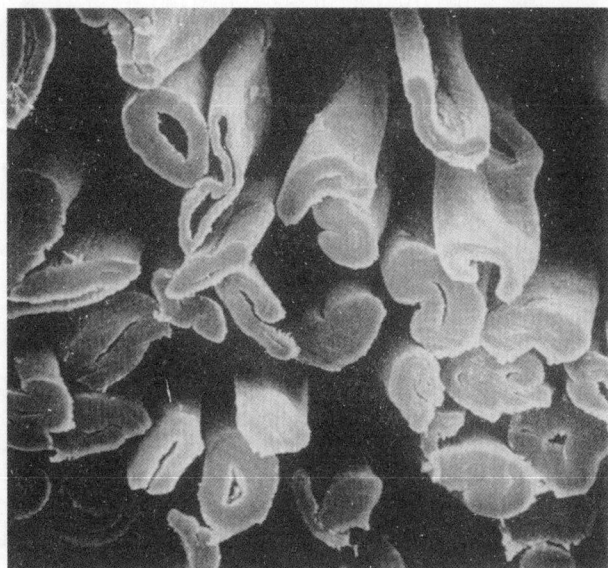

FIGURE 9.47 Cross-section of typical fiber bundle showing both mature and immature fibers (SEM).

illustrates layering that occurs when fibers wet with water or other liquids such as lower alcohols, ethylene glycol, or glycerin are embedded by polymerization of methyl and butyl methacrylates.

The open spaces within the fiber represent sites of entry for the liquids, and thus accessibility of the fiber to the liquid [276]. Figure 9.49 shows this layering at higher magnification, revealing the fibrils that compose the layers of the secondary wall.

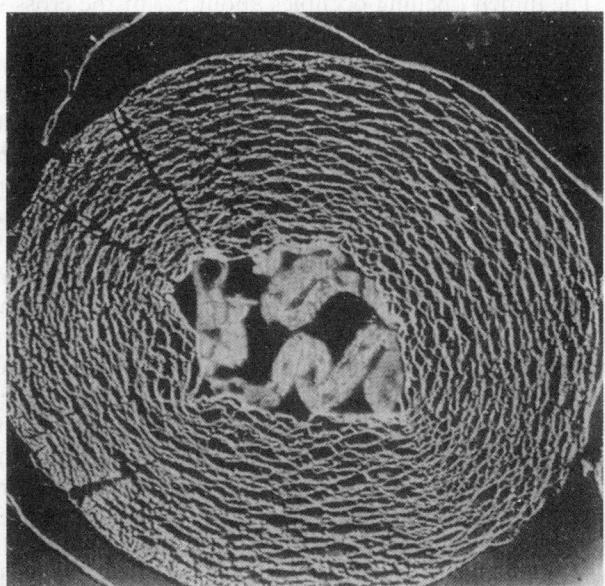

FIGURE 9.48 Thin cross-section showing layering in a swelled fiber at low magnification (TEM).

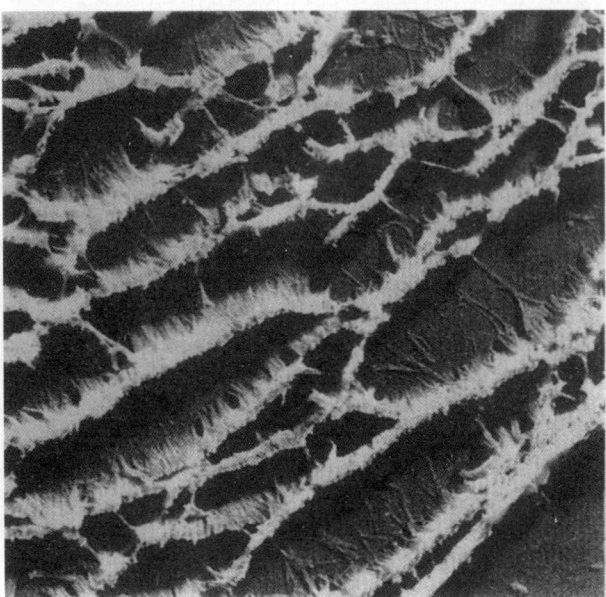

FIGURE 9.49 Portion of thin cross-section showing fibrils in layers of a swelled fiber at higher magnification than Figure 9.48 (TEM).

The lamellar pattern is believed to be due to the differential rates of fibrillar formation during daily growth cycles [277,278]. Differential compaction of the fibrils provides variations in fibrillar bonding, and determines accessibility or permeability into the inner areas of the fiber. The lumen is the central opening in the fiber that spans its length from base nearly to the tip. It contains the dried residues of cell protoplasm, the only source of noncellulose materials in the fiber other than those in the primary wall. A thin cell wall (lumen wall) provides the inner cell boundary. The lumen opening occupies about 5% of the cross-sectional area of the mature fiber.

The live, hydrated fiber exists in a tubular configuration that conforms to available space within the boll locule. When the boll opens, removal of water causes the internal layers of the fiber to twist and collapse, and the primary wall, which is less able to shrink because of its network structure, wrinkles and molds to the underlying fiber layers, producing folds and convolutions (twists), and compression marks (Figure 9.50). Fibers often collapse in a nonuniform elliptical pattern, whose cross section has a convex and a concave side (see Figure 9.36 and Figure 9.47).

This pattern is more pronounced in fibers of low maturity. Even in mature fibers, the lumen cross section assumes an elongated shape on drying, thus giving the fiber cross section a long and a short axes. This asymmetric structure indicates that there may be differences in fibrillar packing densities around the perimeter of the fiber. Such zones would present different areas of accessibility in the fiber. It is not known whether these zones of variations in fibrillar density are due to inherent differences in fibrillar structure at different areas of the cross section, or whether physical forces during drying compress the structure in some areas and expand it in others [279,280]. Dried fibers with relatively thick secondary walls produce the thick, bean-shaped cross section usually associated with the structure of the cotton fiber. This drying and shrinking process produces the nonuniform, convoluted cotton textile fiber, which, although it has its basis in the living biological fiber, is structurally quite different in the dried state than in the living, hydrated state.

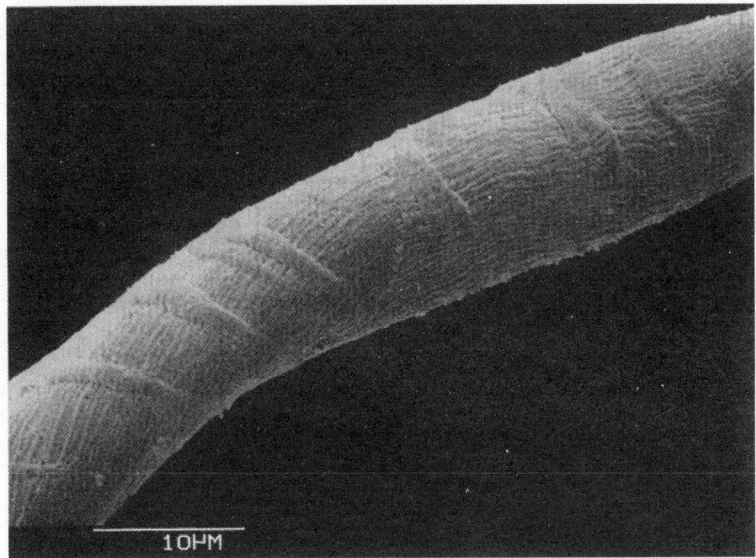

FIGURE 9.50 Surface of a cotton fiber showing compression marks due to removal of water on drying (SEM).

9.5.5.2 Pore Structure

The cotton fiber is hydrophilic and porous. Upon immersion in liquid water, the cotton fiber swells and its internal pores fill with water. Pure cotton holds a substantial percentage of its dry weight in water under conditions of centrifugation. The amount of liquid water held depends upon the severity of the centrifugation used in testing. This is approximately 30% for the water of imbibition [281] or 50% for the water retention value [282]. Centrifugation conditions are less severe in the latter case.

Pores accessible to water molecules are not necessarily accessible to chemical agents. Chemical modification is required to impart many desired properties to cotton fabric. These include color, permanent press, flame resistance, soil release, and antimicrobial properties to name a few. Thus, a knowledge of cotton's accessibility under water-swollen conditions to dyes and other chemical agents of various sizes is required for better control of the various chemical treatments applied to cotton textiles.

The accessibilities of cotton fibers have been measured by solute exclusion. A simplified mechanism is shown in Figure 9.51.

This subject has recently been reviewed [283]. Both static techniques [284], glass column chromatography [285] and liquid chromatography [286,287], have been used. Series of water-soluble molecules of increasing size are used as molecular probes or "feeler gauges." The molecules used as probes must penetrate the cellulose under investigation and not be absorbed on the cellulosic surfaces. These include sugars of low molecular weight, ethylene glycols, glymes, and dextrans. Their molecular weights and diameters are given in Table 9.10. The molecular diameters of the sugars have been reported by Stone and Scallan [284]. Estimates of the molecular diameters of the lower-molecular-weight ethylene glycols are based on extrapolations from measurements of Nelson and Oliver [288]. Measurements of molecular diameters were not available for the glymes but have been approximated by assuming that molecular sizes of the hydrated molecules are the same as those of the parent glycols at the same molecular weight.

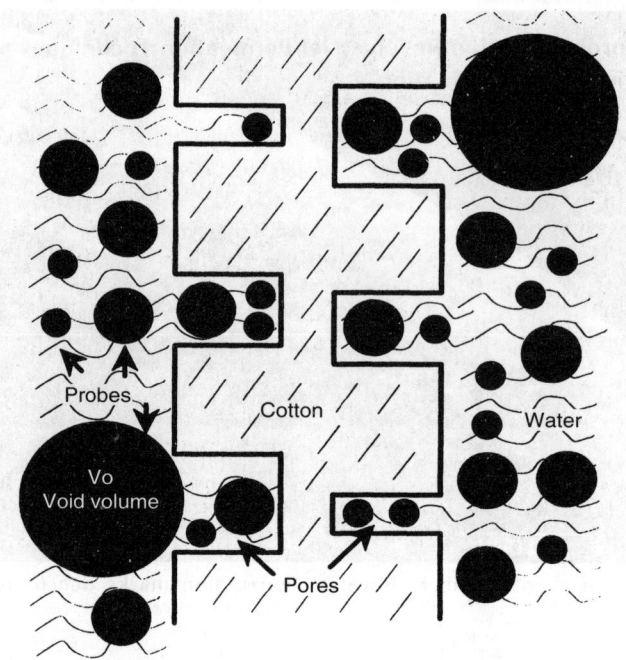

FIGURE 9.51 Simplified illustration of gel permeation mechanism.

The static measurement is based on the addition of a water-swollen cellulose to a solution of the molecular probe. Water in pores accessible to the solute dilutes the solution. In the chromatographic techniques, either glass or standard liquid chromatography columns were packed with cellulose in various forms. The elution volumes of the molecular probes used were determined. Data is generally plotted as internal volume accessible to individual solutes against their molecular sizes. This is illustrated in Figure 9.52.

The degree to which the internal volume has been expanded or contracted is deduced. Similar information is obtained from the static and chromatographic methods. Cellulose has been evaluated in the forms of decrystallized, ball-milled cotton, chopped cotton, cotton batting, and whole fabric.

These techniques have been used to elucidate the effects of variety [289], temperature [287], scouring–bleaching [290], caustic mercerization [290–292], liquid ammonia treatment [290,292,293], cross-linking with different agents under varying conditions [294–298], dyeability [296,299], and treatment with cellulases [300–302], on the cotton. The trends observed are summarized in Table 9.11.

9.6 CHEMICAL PROPERTIES OF COTTON

The cotton fiber is predominantly cellulose and its chemical reactivity is the same as that of the cellulose polymer, a β-(1→4)-linked glucan (Figure 9.53).

The chemical structure shows that the 2-OH, 3-OH, and 6-OH sites are potentially available for the same chemical reactions that occur with alcohols. If the glucan were water soluble, the primary 6-OH, for steric reasons, would be the most available hydroxyl for the reaction. However, as discussed earlier, the chains of cellulose molecules associate with each other by

TABLE 9.10
Molecular probes Used in Reverse Gel Permeation Techniques for Pore Size Distribution Determination

Molecular probe	Molecular weight	Molecular diameter, Å
Dextran T-40 (Void Volume)	40,000	
Sugars		
Stachyose	666.58	14
Raffinose	504.44	12
Maltose	342.30	10
Glucose	180.16	8
Ethylene Glycols		
Degree of polymerization:		
6	282.33	15.6
5	238.28	14.1
4	194.22	12.7
3	150.17	10.8
2	106.12	8.4
1	62.07	5.5
Glymes		
1	222.28	13.8
2	178.22	12.1
3	134.17	9.9
4	90.12	7.4

forming intermolecular hydrogen bonds and hydrophobic bonds. These coalesce to form microfibrils that are organized into macrofibrils. The macrofibrils are organized into fibers.

The cotton fiber is subjected to many treatments that affect swelling and change its crystal structure. The agents employed must be able to interact with and disrupt the native crystalline structure in order to change it to different polymorphs. The chemical reactions of commerce generally involve the water-swollen fiber, which retains a highly crystalline

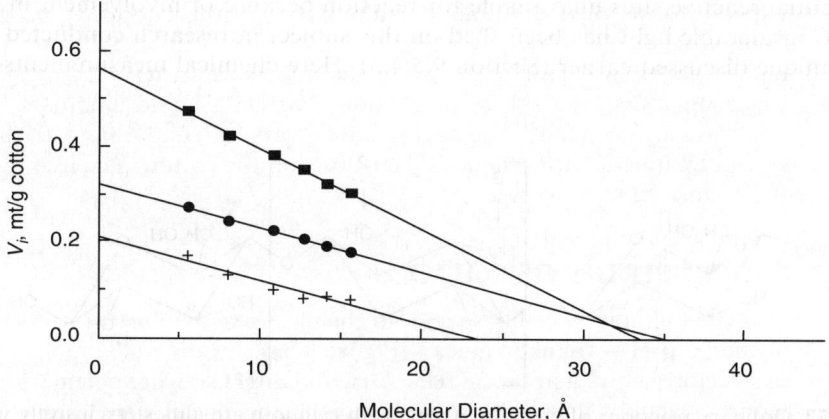

FIGURE 9.52 Internal water (V_i) that is accessible to scoured/bleached (•), caustic mercerized (■), and cross-linked with 4% DMDHEU (+) cotton cellulose. Ethylene glycols were used as molecular probes.

TABLE 9.11
Summary of Changes in Internal Pore Volume

Effect	Internal pore volume	Refs.
Variety	DP-90 > NX-1[a]	[289]
Temperature	Same from 30–60°C	[287]
Scouring and bleaching	Slightly increased[b]	[290]
Caustic mercerization	Substantially increased[c]	[290–292]
Liquid ammonia treatment	Moderately decreased[c] or increased[d]	[290,292,293]
Removal techniques	Moderate increases[c] from:	[293]
	Water immersion >	
	Dry, 25°C >	
	Dry, 95°C	
Cross-linking	Substantially reduced[c]	[295,297,298]
DMDHEU[e]	Reduction progressive with add-on	[295]
Formaldehyde-free agents	Less reduction than with DMDHEU	[297]
Polycarboxylic acid catalysis	More effective catalysts (e.g., NaH_2PO_2) cause greater decreases	[298]
Dyeability	DHDMI[f] cross-linking reduces internal volume but cotton remains accessible to small, but not large, direct dyes	[296]
Cellulase treatment	Mixed results	
8–14 Å pores	No change	[300]
<60 Å pores—small	Decrease in size	[301]
>60 Å pores—large	No change	[302]

[a]DP-90 is a common upland variety; NX-1 is a hybrid of Upland and Pima cottons.
[b]Relative to the original greige cotton.
[c]Relative to scoured and bleached cotton.
[d]Relative to starting purified medical cotton batting.
[e]Dimethyloldihydroxyethyleneurea.
[f]4,5-dihydroxy-1,3-dimethyl-2-imidazoildinone.

structure. Reactions with this highly crystalline, water-insoluble polymer are therefore heterogeneous. Chemical agents that have access to the internal pores of the fiber find many potential reactive sites unavailable for reaction because of involvement in hydrogen bonding. Considerable light has been shed on this subject in research conducted using the CMA technique discussed earlier (Section 9.5.4.5). Here chemical measurements based on

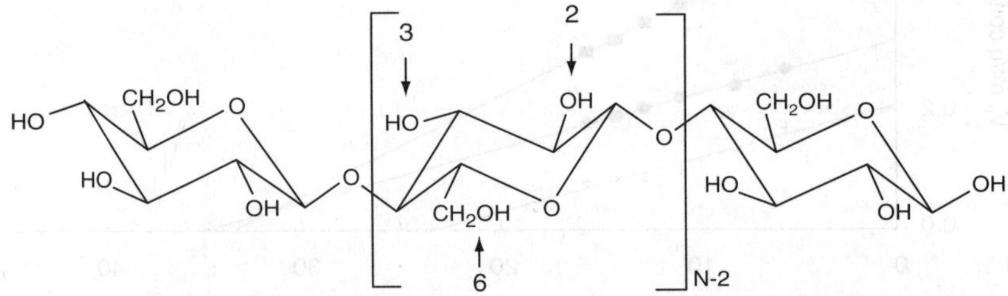

FIGURE 9.53 Ordinary reactions of chemical agents with cellulose are almost exclusively with the 2,3 and 6 hydroxyl groups that are not involved in formation of the linear polymer consisting of D-anhydroglucose units joined via β-1,4-linkages.

reaction with DEAE chloride under mild conditions showed that decreasing availability of the hydroxyl groups in cotton is 2-OH > 6-OH ≫ 3-OH. The total reactivity of the hydroxyls of cellulose and the relative reactivities of the 2-OH, 3-OH, and 6-OH differ depending on the swelling pretreatment, the reagent, and the reaction conditions. These have not been delineated for all systems.

9.6.1 Swelling

9.6.1.1 Water

Cellulose is hydrophilic and swells in the presence of water. Normally cellulose–water interactions are considered to occur either in intercrystalline regions or on the surfaces of the crystallites and the gross structures.

Water vapor adsorption isotherms have been obtained on cotton from room temperature up to 150°C [303,304]. Theoretical models for explaining the water vapor sorption isotherms of cellulose have been reviewed [303]. Only adsorption theories will be discussed here at ambient temperatures. The shape of the isotherm indicates that multilayer adsorption occurs and thus the Brunauer, Emmett and Teller (BET) or the Guggenheim, Anderson and deBoer (GAB) theory can be applied. In fact, the BET equation can only be applied at relative vapor pressures (RVPs) below 0.5 and after modification up to a RVP of 0.8 [305]. The GAB equation, which was not discussed in the chapter in the book *Cellulose Chemistry and Its Applications* [303], can be applied up to RVPs above 0.9 [306]. Initially as the RVP increases, a monomolecular layer of water forms in the cellulose. By a RVP of 0.19–0.22 the monomolecular layer is complete [303], and the moisture regain, when a monomolecular layer has just formed, for cotton and mercerized cotton is 3.27 and 4.56%, respectively [261,303]. By a RVP of 0.83–0.86, about three layers of water molecules are formed, and at higher RVPs it is thought that condensation occurs in the permanent capillary structure of the sample [307].

It is well known that at low moisture uptakes, the water associated with the cellulose exhibits properties that differ from those of liquid water and it has been called by such terms as "bound water," "nonsolvent water," "hydrate water," and "nonfreezing water." From a review of the literature, which included determinations by such techniques as NMR and calorimetry, Zeronian [303] concluded that between 0.10 and 0.20 g/g of the water present in the fiber cell wall appeared to be bound. Such regains are obtained at RVPs between 0.85 and 0.98.

The fiber saturation point (FSP) of cotton is the total amount of water present within the cell wall expressed as a ratio of water to solid content. It is equivalent to the water of imbibition of the fiber, also called its water retention value. The FSP has been measured using solute exclusion, centrifugation, porous plate, and hydrostatic tension techniques. It occurs at RVP greater than 0.997 and from the review of the papers, it has been concluded that the studies have yielded a value for FSP in the range of 0.43 to 0.52 g/g [303].

At equilibrium and at a particular RVP, the amount of adsorbed water held by a cellulose generally will be greater if it has been obtained following desorption from a higher RVP and not by adsorption from a lower RVP. The cause of this hysteresis is not fully established [303]. One explanation is based on the internal forces generated when dry cellulose swells, limiting the amount of moisture adsorbed whereas when swollen cellulose shrinks, stress relaxation occurs since the cellulose is plastic and permits a higher uptake of moisture.

9.6.1.2 Sodium Hydroxide

The swelling of cotton with an aqueous solution of sodium hydroxide is an important commercial treatment. It is called mercerization after its discoverer, John Mercer, who took a patent on the process in 1850 [308]. Other alkali metal hydroxides, notably lithium hydroxide and potassium hydroxide, will also mercerize cotton, but normally sodium hydroxide is used. Mercerization is utilized to improve such properties as dye affinity, chemical reactivity, dimensional stability, tensile strength, luster, and smoothness of the cotton fabrics [309]. The treatment is normally applied either to yarn or to the fabric itself either in the slack state to obtain, for example, stretch products, or under tension to improve such properties as strength and luster. The interaction of alkali metal hydroxides and cellulose has been extensively reviewed. Earlier reviews can be traced from relatively recent ones [99,310,311].

The term mercerization has to be used with care. One of the changes that occur to the treated cotton is that its crystal structure can be converted from cellulose I to II. To a researcher the term implies that the caustic treatment has induced close to complete, or full, conversion of the crystal structure to cellulose II. On the other hand, the industrial requirement is improvement in the properties described above and these changes can be produced without full conversion of the crystal structure. For a given temperature and concentration of sodium hydroxide, the amount of swelling that occurs depends on the form of the sample. Swelling deceases in the order fiber > yarn > fabric. In addition, properties are affected by whether the material is treated in slack condition or under tension. Finally, depending on the processing time, the material, and other conditions (e.g., caustic concentration, temperature, slack or tension treatment) mercerization, as defined by researchers, might not extend beyond peripheral regions. In mercerizing fabrics industrially, the following variables need to be considered: caustic strength, temperature, time of contact, squeeze, framing and washing, and the use of penetrants. Abrahams [309] has provided the following guidelines. The caustic concentration should preferably be in the range of 48–54°Tw (approximately 6.8 to 7.6 M), although if improved dye affinity is the objective 30–35°Tw (roughly 4.0 to 4.7 M) can be used. Temperature may vary in the range of 70–100°F (21.1 to 37.8°C) at higher concentrations but has to be monitored more closely when the concentration is 30°Tw. A contact time of 30 sec can be used normally. A penetrant is essential if the fabric is in the greige state to permit wetting. Washing on the frame is enhanced by using a mercerizing penetrant that is an active detergent over a wide caustic range.

Hot mercerization allows better penetration of the alkali into the fibers than the ambient temperatures used normally [312]. However, to obtain optimum improvement in properties the caustic has to be washed out after the fabric is cooled.

During mercerization, the swelling induced by the caustic is inhibited from outward expansion by the presence of the primary wall of the cotton fiber. The changes observed in fiber morphology by mercerization include deconvolution, decrease in the size of the lumen, and a more circular cross section.

Changes in the fine structure that occur when cotton is mercerized include a conversion of the crystal lattice from cellulose I to II, a marked reduction in crystallite length, a marked increase in moisture regain, and a reduction in degree of crystallinity [99,311]. A higher concentration is required to induce the optimum changes as the temperature is increased from subambient to room temperature. The conversion from cellulose I to II is substantially complete in cotton yarn treated at 0°C for 1 h, with 5 M LiOH, NaOH, or KOH [313]. In the case of the sample treated with 5 M NaOH, the following changes were noted: the extent of swelling, measured by the 2-propanol technique, roughly tripled; the moisture regain increased by about 50%; and the crystallite length decrease by approximately 40% [313]. An estimate of the loss in crystallinity on mercerization, determined by moisture regain measurements, can be found in Table 9.5.

The effect of mercerization on tensile properties depends on the type of cotton tested. In one study six *G. barbadense* samples were slack mercerized and the breaking forces and tenacities of the fibers relative to their nonmercerized counterparts ranged from 88 to 122% and from 80 to 114%, respectively [314]. A larger change was found in the case of a *G.*

hirsutum Deltapine Smoothleaf sample. In this case, the relative breaking force and tenacity were 186 and 134%, respectively. Relative breaking strains ranged from 160 to 189% for the *G. barbadense* samples and 150% for the *G. hirsutum*. The increased strength and extensibility of slack mercerized cotton have been attributed partly to the deconvolution that has occurred and partly to the relief of internal stresses [315]. The reduction in crystallinity and crystallite length that results from mercerization contributes to the relief of stresses in the fiber as well as in giving a product of higher extensibility.

There is some evidence that the degree of hydration of alkali hydroxide ions affects their ability to enter and swell cellulose fibers [310]. At low concentrations of sodium hydroxide, the diameters of the hydrated ions are too large for easy penetration into the fibers. As the concentration increases, the number of water molecules available for the formation of hydrates decreases and therefore their size decreases. Small hydrates can diffuse into the high order, or crystalline regions, as well as into the pores and low-order regions. The hydrates can form hydrogen bonds with the cellulose molecules.

Ternary complexes called soda celluloses can form between cellulose, sodium hydroxide, and water [310]. In these complexes, some of the water molecules of the sodium hydroxide hydrates are replaced by the hydroxyl groups of the cellulose [310]. The x-ray diffraction diagrams have been obtained for five soda celluloses as intermediates in the formation of cellulose II from cellulose I [310,316].

9.6.1.3 Liquid Ammonia

Another swelling reagent for cotton cellulose, which is also used industrially, is liquid ammonia. This treatment has been extensively reviewed and discussed [311,312,317–321]. Anhydrous ammonia penetrates the cellulose relatively easily and reacts with the hydroxyl groups after breaking the hydrogen bonds. The reaction occurs first in the LORs, and gradually later in the crystalline regions of the fibers. An intermediate ammonia–cellulose (A–C) complex, held together by strong hydrogen bonds, is formed. This complex can decompose in several ways, and yield different products, depending on the condition of the removal of the ammonia. Lewin and Roldan [319] developed a phase diagram (Figure 9.54) of the four major phases represented in the four corners of a tetrahedron [319].

The directions of the transitions between the various phases are indicated by the arrows, i.e., a transition from D to III is possible on application of dry heat. A transition from III to D is impossible unless a strong swelling agent like ammonia is used. A transition from III to I is possible by the application of water and heat or by a prolonged application of water at ambient conditions. The reverse transition is impossible without an intermediate swelling step. The transitions are usually not complete, especially in industries, and a wide range of products can be obtained as indicated by the phase diagram. The ammonia–cellulose complex and cellulose III can also be obtained from cellulose II. There is, however no reversion to cellulose I.

The CI decreases upon liquid ammonia treatment and rinsing with water with or without heating, from 79 to 30–40 Å, and the crystallite size decreases from 54 to 37–34 Å [319]. The circularity and homogeneity are also increased. The tensile strength is greatly increased and the elongation is decreased upon stretching the ammonia-treated fibers. The accessibility and consequently the dyeability of the fibers are also greatly increased.

The interactions between cellulose and ammonia have attracted industrial attention. Applications to wood [321–325] by treating it with liquid ammonia and with gaseous ammonia under pressure [326] to cotton fabrics [320,325–327], and to sewing treads [327] have been described.

It has been suggested that the great depth of color or dye yield found with mercerized cotton is due to the caustic treatment inducing an abundance of large pores in the fiber. In

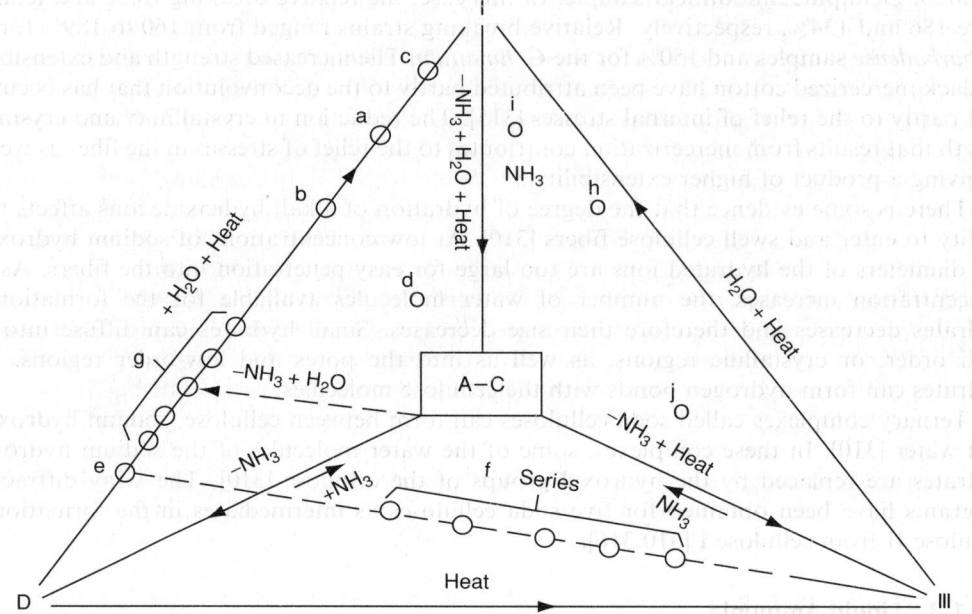

FIGURE 9.54 Phase diagram of ammonia–cellulose (A–C), disordered cellulose (D), Cellulose I (I) and Cellulose III (III). A–C is the vertex of a tetrahedron and is placed above the plane of the paper. The various samples studied are placed in the basal plane. Arrows show the transition directions.

contrast, the high level of resilience associated with liquid ammonia treatment has been ascribed to a low level of large pores in the fiber [292].

9.6.2 Etherification

9.6.2.1 General

Cellulose ethers generally are very stable. Many etherified cottons are highly resistant to hydrolytic removal of substituent groups under both acidic and alkaline conditions. Because of this stability, many of the most practical chemical treatments of cotton are based on etherification reactions [9,328–331]. These treatments provide cotton products with useful, durable properties including wrinkle resistance, water repellency, flame resistance, and antimicrobial action.

Treatments based on condensation reactions (such as in the classical Williamson synthesis) produce the most stable cotton derivatives. On the other hand, treatments based on addition reactions (such as in the Michael reaction) yield cellulose ethers that are somewhat less stable. This lower level of stability is because of the equilibrium nature of the addition reaction. Typical examples of these two types of cellulose etherification are carboxymethylation [9,329,330] and cyanoethylation [9,329,330,332], respectively, both of which proceed in the presence of alkali.

In carboxymethylation of cotton, the fibers are impregnated with aqueous sodium hydroxide and then treated with chloroacetic acid:

$$\text{Cell} + 2\text{NaOH} + \text{ClCH}_2\text{COOH} \rightarrow \text{Cell}-\text{OCH}_2\text{COONa} + \text{NaCl} + 2\text{H}_2\text{O} \qquad (\text{R9.1*})$$

*R denotes reaction.

Cotton Fibers

The equation for the cyanoethylation of cotton with acrylonitrile can be represented as:

$$\text{Cell—OH} + CH_2{=}CHCN \rightleftharpoons \text{Cell—OCH}_2\text{—CH}_2\text{—CN} \qquad \text{(R9.2)}$$

Cotton etherification can be carried out so that the cotton fiber retains its fiber structure and textile properties. The latter, however, often are greatly modified, even at low degrees of substitution (DS). It is possible, through etherification, depending upon the ether group substituted and the DS on the cellulose, to obtain fibrous products with widely diverse properties. For example, there are cellulose ethers that have very high levels of moisture absorbency or even water solubility (carboxymethylated cotton) and others that have high levels of water repellency (long-chain alkyl ethers and stearamidomethyl ethers).

A generalized representation of cellulose ethers is Cell—OR, wherein the ether group (R) is alkyl, aromatic, heteroalkyl, heterocyclic, or other substituent, including ether groups bearing other functional groups. Cellulose ethers with mixed ether substituents also have been prepared by treatment with two or more reactants, either in combination or in sequence.

The extent to which the cellulose molecule is modified is denoted as the DS. Thus, the DS is the average number of alcohol (OH) groups of the anhydroglucose unit of the cellulose molecule that have been substituted. The DS can range to a value of 3.0, which indicates that all the OH groups of the anhydroglucose units that make up the polymeric cellulose molecule have been etherified. A DS of 0.25 indicates that, on an average, there is 0.25 ether group per anhydroglucose unit, or alternatively stated, there is an average of one ether group for every four anhydroglucose units.

It must be noted that in cotton not all of the OH groups are accessible for reaction and substitution because of the unique polymeric structure of the cotton fiber. Thus, the average DS of the etherified cotton usually signifies a higher DS in the accessible regions and no substitution in the inaccessible crystalline regions of the fiber. It must be remembered that because of the heterogeneous nature of cotton we are always speaking of the average DS of the cotton product.

The reaction between cotton and ethylene oxide to give hydroxyethylated cotton appears to be simple and straightforward:

$$\text{Cell—OH} + \underset{\underset{O}{\diagdown\;\diagup}}{CH_2CH_2} \rightarrow \text{Cell—OCH}_2CH_2OH \qquad \text{(R9.3)}$$

However, once introduced on the cellulose molecule, the hydroxyethyl group is highly reactive and capable of further reaction with another molecule of ethylene oxide. This type of reaction, called graft etherification, can build up relatively long-chain substituents on the cellulose molecule without involvement of many additional cellulosic OH groups:

$$\text{Cell—O}(CH_2CH_2O)_nH \qquad \text{(R9.4)}$$

In these products, simple analysis gives the average molar substitution (MS) per anhydroglucose unit of the cellulose rather than the DS. Specialized analytical techniques must be employed to ascertain the DS of the grafted derivatives.

9.6.2.2 Wrinkle Resistance

The most important cotton etherification treatments are those that produce wrinkle resistance in fabrics [331,333,334]. The aldehydes, formaldehydes, and glyoxals, react with the OH groups of two cellulose chains as well as those of one chain. Reaction in which a bond is established between the two cellulose molecules is called cross-linking and is the basis for profound changes in the cotton fiber. Cross-linking produces resiliency in the fiber to give the needed dimensional stabilization, wrinkle resistance, and crease retention for modern durable-press cellulosic fabrics. Cross-links based on etherification reactions traditionally have been used because of their durability to repeated laundering and wear.

Methylenation of cotton by treatment with formaldehyde has been an elusive objective of textile finishers for almost 100 years:

$$2\text{Cell}-\text{OH} + \text{HCHO} \leftrightarrow \text{Cell}-\text{OCH}_2\text{O}-\text{Cell} + \text{H}_2\text{O} \qquad (R9.5)$$

Although formaldehyde is inexpensive, readily available, highly reactive, and ideally would be the simplest ether cross-link between cellulose chains, there has been only limited successful usage of this reagent to produce wrinkle resistant cotton. A treatment based on gaseous- or vapor-phase application of formaldehyde to cotton under rigidly controlled conditions has gained some acceptance.

Methylolamide agents are most commonly used to produce wrinkle resistance and dimensional stabilization in cotton [331,333–337]. These agents, formaldehyde adducts of amides or amide-like nitrogenous compounds, introduce ether cross-links between cellulose molecules of the cotton fiber. Chemical processing of cotton with methylolamide agents is the most widely practiced textile finishing treatment throughout the world.

The first methylolamide agent for cotton was the urea–formaldehyde adduct. Today, most finishing of cotton with methylolamide agents uses cyclic urea–formaldehyde adducts. The most commonly used agent is dimethyloldihydroxyethyleneurea (DMDHEU):

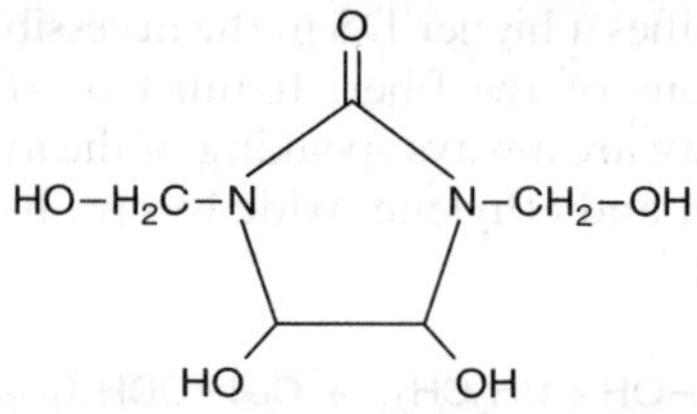

The chemical name of this reagent is 1,3-bis(hydroxymethyl)-4,5-dihydroxy-imidazolidinone-2 but it is usually called DMDHEU or the glyoxal reactant because it is prepared from glyoxal, urea, and formaldehyde. Other methylolamide agents that have been used for producing wrinkle resistance in cotton include the aforementioned urea–formaldehyde, dimethylolurea, dimethylolethyleneurea, and formaldehyde adducts of melamines (triazines), acetylenediurea, propyleneurea, uron, triazones, and alkyl carbamates. Reactions between methylolamides and cellulose occur in the presence of acid (or Lewis acid) catalysts and are very fast at elevated temperatures—sufficiently so that they are adaptable to the requirements of rapid, commercial processing of cotton fabrics.

9.6.2.3 Reactive Dyeing

Another important commercial utilization of cotton etherification is in coloration of fabrics with reactive dyes [338–340]. Reactive dyes contain chromophoric groups attached to moieties that have functions capable of reaction with cotton cellulose by nucleophilic addition or nucleophilic substitution to form covalent bonds. In the nucleophilic addition reaction, an alkaline media transforms the reactive dye to an active species by converting the sulfatoethylsulfone to the vinyl sulfone, which reacts with cellulose to form an ether bond. In the nuceophilic substitution reaction, a halogen atom on the reactive dye molecule is replaced by an oxygen cellulose ester covalent bond. Reactive dyes that have been commercialized include monofunctional dyes with the following reactive groups: mono- and dichlorotriazine, sulfatoethylsulfone, sulfatoethylsulfonamide, trichlorpyrimidine, dichloroquinoxaline, difluoropyrimidine, and difluorochloropyrimidine; and bifunctional dyes (contain two different reactive moieties on the dye molecule) with the following reactive moieties: bis aminochlorotriazine, bis aminonicotinotrazine, aminochlorotriazine–sulfatoethylsulfone, and aminoflorotriazine–sulfatoethylsulfone [341].

Reactive dyes with methylolamide-like groups were used on cotton at one time [342]. Bonding to cellulose was similar to that in etherification treatments to produce wrinkle resistance. However, because of technical problems in their application, usage of these formaldehyde-based reactive dyes has essentially ceased. Fixatives are used, which act through methylol groups, to improve color fastness of direct and other dyes on cotton. Their mechanism includes bonding (etherification) between dye and cellulose as well as between dye molecules.

9.6.2.4 Flame Resistance

Improved flame resistance is an important and useful property that can be imparted to cotton fibers and cotton textiles. The flammability and flame retardancy of cotton have been studied extensively and several comprehensive reviews are available [343–346].

Once ignited, virtually all common textile fabrics will burn. Textile fabrics burn by two distinctly different processes: flaming combustion and smoldering combustion. As the fibers that make up fabrics are composed of large, nonvolatile polymers, flaming combustion (e.g., that caused by an open-flame source, such as a match) requires that the polymer undergo decomposition to form the small, volatile organic compounds that constitute the fuel for the flame. The combustion of polymers is a very complex, rapidly changing system that is not yet fully understood. For many common polymers, this decomposition is primarily pyrolytic with little or no thermo-oxidative character. Smoldering combustion (e.g., that caused by a cigarette) on the other hand involves direct oxidation of the polymer and chars and other nonvolatile decomposition products. Since smolder ignition and open-flame ignition are different mechanisms, they usually require different flame retardant (FR) treatments and treatments to control open-flame ignition can adversely affect smolder resistance of 100% cellulosic and predominately cellulosic fabrics [347].

The flame resistance of a textile is test method dependent, i.e., it should be specified what test the material passes when making a claim of flame resistance. There are currently many mandatory U.S. federal and state and international standards, as well as voluntary national and international standards (smolder or cigarette resistance; small and large open flame), for the flammability of textiles. There are component standards and composite or full scale standards. The current smolder test methods use a standard cigarette. Large open-flame standards for mattress and box springs use burners that mimic burning bedclothes; other large flame sources use, e.g., an 18 kW flame for 3 min, or a trash can full of burning paper. Small open-flame sources usually are matches, cigarette lighters, and candles (e.g., a butane or a propane flame for 15 or 20 s). For general wearing apparel, there is a 45° angle test with a small open flame and burn rate is measured. For children's sleepwear, the test is a vertical flame test with a small open flame and the distance the flame travels is measured.

Whenever cotton is ignited in the presence of oxygen and the temperature is high enough to initiate combustion, untreated cotton will either burn (flaming combustion) or smolder (smolder combustion means burning and smoking or wasting away by a slow and suppressed oxidation without flame) until carbonaceous material or combustion gases result (see Section 9.6.4.6).

Ignition (defined as initiation of self-sustaining flaming combustion for an observable time) occurs as a result of exothermic reactions between volatile decomposition products and oxygen [348]. The major factors that influence ignition of cellulosic materials like cotton are air flow, relative humidity of the fabric, the amount of oxygen available, physical factors (geometry, density, thickness, etc.) [349,350], chemical factors (e.g., inorganic impurities) [351–354]), heat source, and how fast the cotton is heated [350]. Because of these variables, it is not really correct to talk about a specific ignition temperature or autoignition temperature (defined as the temperature under controlled conditions in air when there is a runaway weight loss and flaming combustion resulting from contact with heated air in the absence of any spark or flame) for cotton. Thermal analysis studies in air and in 8.4% oxygen indicate that cotton and pure cellulose ignite at about 360 to 425°C [348,355,356]. Cellulose does not melt but is reported to decompose (char) at 260–270°C [357] and cotton at 338°C [348].

Numerous end uses for cotton depend on its ability to be treated with chemical agents that confer flame resistance. The chemical FR treatments used to confer flame resistance to untreated cotton depend on many factors [358]: is the finish intended to be durable or nondurable; is the treatment used is to prevent burning or smoldering; what is the construction of the textile to be treated; and is the textile 100% cotton or does it contain some percentage of thermoplastic human-made fibers, i.e., polyester? In addition, problems of toxicological and ecological concerns have more recently assumed significance [359]. Thus, it is clear that the process of treating and finishing cotton fabrics to make them flame resistant is complex and may be relatively expensive.

9.6.2.4.1 Nondurable Finishes

A number of nondurable finishes have been developed, which are in use in industrial cotton fabrics. Most are based on borax ($Na_2B_4O_7 \cdot 10H_2O$), boric acid (H_3BO_3), diammonium phosphate ((NH_4)$_2HPO_4$), or sodium phosphate, dodecahydrate ($Na_3PO_4 \cdot 12H_2O$) [360]. These agents are applied to the textile from water solutions, followed by squeezing to reduce the wet pick up and finally by drying. It should be realized that although such treatments are removed by conventional laundering, most of them could withstand several nonaqueous launderings with dry cleaning solvents and remain effective. These nondurable finishes are recommended for 100% cotton textiles. For cotton–polyester blend textiles, a reagent such as ammonium bromide, which decomposes on heating and becomes active in the gaseous phase, should be added to the agents referred to above for greater effectiveness.

9.6.2.4.2 Durable Finishes

Many durable flame retardants for cotton have been developed to convey open-flame resistance [344,346,360,361]. The vertical flame test for determining the U.S. children's sleepwear flammability (16 CFR 1615 and 1616) is a rather severe test and cotton fabrics require a FR treatment to pass the test. The test method requires treatments that are durable to 50 hot water wash and dry cycles. Currently there are relatively few commercially available FR chemistries that are durable under these conditions required today. Some of the reasons include low commercial availability of the chemicals, costs, safety concerns, process control issues, and difficulty in application.

Treatments with more limited durability would be more appropriate for the U.S. general wearing apparel (16 CFR 1610), which has a 45° angle test because the test and the cleaning requirements are less severe (see Section 9.6.2.4.4). Currently only resistance to dry cleaning and hand washing are required.

The main durable FR finishes used on cotton to meet more severe open-flame resistance requirements are phosphorus based [343,358]. One of the problems with typical phosphorus-based FR treatments on fleece, which only requires a mild treatment to pass the 45° angle test, is that the often-required levels alter the esthetic properties of the fleece, resulting in a fabric that is stiff or matted and often has unpleasant odors. Most common types of dyes used on cotton are affected by pH or oxidation–reduction procedures that are used during the FR treatments.

The most successful durable FR finish for open-flame resistance is based on the neutralized moiety of tetrakis(hydroxymethyl)phosphonium chloride (THPC), i.e., (THPOH). When the THPOH is applied to cotton fabric as described previously, the dried fabric can be cured with ammonia gas to form the water-insoluble THPOH–NH_3 polymer [362]. The resin polymerizes within the cotton fibers to form an insoluble polymer that withstands many repeated aqueous launderings. A THPOH precondensate applied to cotton fibers along with ammonia (precondensate–NH_3 process) is currently the commercial system used successfully to impart durable flame resistance to cotton garments (e.g., safety apparel). Durability and freedom from fabric odor is greatly improved by peroxide oxidation of the ammoniated fabric [360]. The tendency of THPOH to release formaldehyde during drying can be a potential health problem for textile workers during finish applications (see Section 9.10). This has been largely solved by the reaction of urea with the phosphonium salt prior to application to the cotton textile and by more efficient hooding of the drying section of the process [363].

Other phosphorus-containing flame retardants achieve their durability by reacting with the cellulose molecules and polymerizing within cotton fibers. Reactive phosphorus-based flame retardants (e.g., phosphonic acid ester) are typically applied by a pad–dry–cure method in the presence of a phosphoric acid catalyst. Examples of such flame retardants are N-methylolamides of phosphines and phosphine oxides, and vinyl phosphonates. Nitrogen-containing agents used with these phosphorus-containing agents include urea, N-methylolpropionamide, and methylolated melamines.

Although the precondensate–NH_3 process has been the most successful FR treatment for 100% cotton textiles, it is not completely effective when the cotton textile contains appreciable amounts of thermoplastic synthetic fibers like polyester [364,365]. Extensive research has been carried out on the treatment of cotton–polyester blends with flame-retardant finishes and it has been found that a dual treatment (i.e., one for the cotton component and one for the polyester) is the most successful [360]. The best finish for the cotton component is precondensate–NH_3. The best agent for rendering the polyester component flame retardant was tris(2,3-dibromopropyl)phosphate (Tris) [366]. This chemical was found to be a mutagen–carcinogen, and in 1977, the Consumer Product Safety Commission banned its use on children's sleepwear. The children's sleepwear industry responded by marketing 100% thermoplastic fabrics (i.e., polyester and nylon) without any chemical treatment. These materials melt at relatively low temperatures before the fabric ignition temperature is reached. There is still a problem of skin injury from the contact of the molten synthetic component with the skin, but this was deemed less serious than the possible exposure to a known carcinogen, i.e., Tris. The U.S. children's sleepwear flammability standard test method (16 CFR 1615 and 1616) was amended in 1978 so that fabrics no longer had to be treated to control for the melt drip phenomenon. These standards were further amended in 1996 to not require the vertical flame test for tight-fitting garments and for infant wear (0–9 months). Currently the most successful FR finish for cotton–polyester fabric involves the application of antimony–halogen finishes [367]. The finish is expensive and difficult to apply, but the results are good, even when the fabric is intended for outdoor use.

Research has also been done on the FR treatment of cotton blend fabrics containing a flame-resistant synthetic fiber, such as a FR-modacrylic, or an aramid. A FR finish must still be applied to the cotton component. Thus so far, research has proved that these treated blends are more successful with heavier-weight goods, such as those intended for military applications. The observation that heavier-weight goods are easier to treat for flame resistance than their lightweight counterparts holds true for almost all FR finishes on almost all cotton containing fabrics. The best finish for durable fire retardant properties on lightweight cotton goods (i.e., 4 oz/yd^2 or less) is precondensate–NH$_3$ process.

Backcoating of upholstery fabrics, using decabromodiphenyl ether (deca-BDE) or hexabromocyclododecane, antimony trioxide, and an acrylic latex, is effective for flame resistance, but there are toxicity concerns. The use of synthetic barrier fabrics (i.e., interior fire-blocking) may remove the need for chemical backcoating.

Another approach to flame-resistant cotton containing fabrics involves the use of core spun yarns [368–372]. There are two components in these specialized yarns. One component is a central core usually made from a human-made polyester or nylon, or a nonflammable core like fiberglass. The other component is a cotton cover that is wound around the central core to form the core yarn. The core yarn is woven or knitted into an appropriate textile, then treated with a finish to make the flame-resistant cotton cover. When the core yarns are spun to restrict their synthetic content to 40% or less, the FR treatment of the cotton component alone will frequently make the array flame resistant. The need for a separate FR treatment of the polyester or nylon component is no longer required.

9.6.2.4.3 Cotton Batting or Filling

For upholstered furniture, mattresses and foundations, and filled bedclothes, there is a need for both smolder-resistant and open flame-resistant cotton batting. Alkali metal ions on fabrics and batting (natural in raw cotton and from finishing agents) induced smoldering of cotton [351,352]. Postrinsing of upholstered fabrics or washing cotton batting does not completely remove the alkali metal ions and could cause a multitude of technical problems [373]. Thermoplastic fiber in blended cotton fabrics and batting appears to convey cigarette or smolder resistance until the cotton content exceeds about 80%. Most treatments that control for flame resistance do not control for smolder resistance [347]. Nondurable agents such as the boric acid, borax, and ammonium phosphate have been used to inhibit the smolder of cotton and to pass open-flame mattress and upholstered furniture flammability requirements. Research into converting these treatments into more durable finishes also has been investigated. Incorporation of trimethylolmelamine (TMM) into a borax-type finish seemed promising [374], but ecological problems with TMM slowed its widespread application.

Engineered cotton batting properly treated with boric acid (~10% or greater) and blended with inherently flame-resistant fibers (e.g., enhanced FR-modacrylic, FR-PET, Visil (silica containing cellulosic), etc.) is cigarette (smolder) resistant and open flame-resistant and can be used like any other padding material [375]. It is an improved product that can be used as a drop-in component fire-blocking barrier in the mainstream soft furnishings (i.e., mattresses, beddings, and upholstered furnitures). Ammonium phosphate-treated cotton, blended with FR-polyester or FR-modacrylic, is also used to make flame-resistant and smolder-resistant engineered batting to meet the flammability requirements of the mattresses and upholstered furnitures. Cotton batting—boric acid or ammonium phosphate treated or blended—should be helpful in meeting the various cigarette resistance and open-flame resistance regulations for mattresses, futons, and upholstered furniture.

9.6.2.4.4 Polycarboxylic Acids as Flame Retardants

Carpet materials larger than 1.2×1.8 m must pass a standard test method (referred to as the pill test; 16 CFR 1632) to be used commercially. Some untreated high-density cotton-containing materials pass this test, but the risk of failure is too high without further modification to improve the flame resistance. To be of commercial use chemical modification of cotton-containing carpets must be devoid of formaldehyde or other harmful substances, must not discolor the surface, must not cause dye shade change, must undergo reaction at less than 150°C to avoid synthetic backing deformation, must be applied at low moisture contents, and must be cost effective. An effective solution has come from an unexpected place. Polycarboxylic acids, such as 1,2,3,4-butanetetracarboxylic acid (BTCA), and citric acid catalyzed by sodium hypophosphite or sodium phosphate that have been shown to be effective durable-press agents for cotton fabrics [376], are also effective for improving the flame resistance of carpets and other cotton textiles that require a mild FR treatment to pass the appropriate flammability test for that product. These are not considered FR agents but these polycarboxylic acid systems were shown to be capable of providing cotton-containing carpeting with flame suppression properties sufficient to pass the methenamine pill test [377–379]. This is attributed to increased char formation upon combustion. A similar approach with polycarboxylic acids has been used for raised-surface apparel [380].

9.6.2.5 Multifunctional Properties

Etherification of cellulosic fibers by N-methylol groups in cross-linking resins usually occurs directly by reaction with the various hydroxyl groups in the anhydroglucose unit. However, when polyethyleneglycols are present, the semicrystalline polyols are preferentially etherified by the N-methylol groups at very low curing temperatures (as low as 80°C) and some grafting occurs on the cellulosic hydroxyl groups. The net effect is a flexible composite structure of the cross-linked polyol as an elastomeric coating on cotton fibers or fabrics.

The resultant fabrics are unique in that they have many functional property improvements: thermal adaptability due to the phase change nature of the bound polyol, durable press or resiliency, soil release, reduction of static charge, antimicrobial activity, enhanced hydrophilicity and improved flex life, and resistance to pilling. Because of the different molecular weights of polyols, resins, acid catalysts, and fabric constructions, there are numerous modified fabrics that can be produced with sets of improved attributes. Each fabric must be carefully evaluated for optimum curing conditions and formulations to produce the desired product. Several licenses have been granted for this process. Various types of apparel, healthcare items, and industrial fabrics are currently evaluated for commercial production [381,382].

9.6.3 Esterification

9.6.3.1 General

Hydroxyl groups of cotton cellulose can be reacted with carboxylic acids, acyl halides, anhydrides, isocyanates, and ketenes to produce cellulose esters retaining the original fiber, yarn, and fabric form of the textile material. Prior to the reaction, native cotton exhibits a high degree of crystallinity, unusually strong and extensive hydrogen bonds, and a cellulose molecular weight exceeding 1 million. These properties greatly limit accessibility of the cotton fiber's interior to penetration by chemical agents. Tightness of yarn and fabric construction also hinders esterification to high DS. Pretreatment with swelling agents is usually required for esterification with monofunctional carboxylic acids and anhydrides. Strong acid catalysts may be used just as in esterifying simple alcohols. However, acid-catalyzed esterification of cotton must be at or below room temperature to prevent chain cleavage of the cellulose, and as a result, 1–3 h may be needed to reach the desired DS.

9.6.3.2 Acetylation

Acetylation of spun or woven cotton to an acetyl content of 21% results in a DS of 1.0 acetyl group per anhydroglucose unit (AGU), and imparts rot resistance as well as heat or scorch resistance to yarn or fabric. The partially acetylated cotton has been used commercially as a covering material for ironing boards and hothead laundry presses. Fully acetylated cotton has a DS of 2.7–3.0 acetyl groups per AGU. It has even greater rot and heat resistance, and is sufficiently thermoplastic to permit permanent creasing and pleating at conventional ironing temperatures. It is highly resistant to organic solvents, in contrast to low-molecular-weight cellulose triacetate, but readily accepts the disperse dyes used with acetate rayon.

9.6.3.3 Formaldehyde-Free Wrinkle Resistance

The high-temperature cross-linking of cotton cellulose by polycarboxylic acids, having three to four carboxyls per molecule, has been extensively investigated as a method of formaldehyde-free durable-press finishing. In 1963, Gagliardi and Shippee [383] showed that polycarboxylic acids are capable of imparting wrinkle and shrinkage resistance to cellulosic fabrics such as cotton, viscose rayon, and linen. Citric acid was the most effective agent tested but it did cause more fabric discoloration than other polycarboxylic acids. In 1963, progress was made with the use of alkaline catalysts such as sodium carbonate or triethanolamine [384]. Acids having four to six carboxylic groups per molecule were usually more effective than those having two or three carboxyls. These finishes were recurable as the ester cross-links appeared to be mobile at high temperature. Acids investigated were cis-1,2,3,4-cyclopentanetetracarboxylic acid and BTCA.

Later a breakthrough came when a series of weak base catalysts were discovered that are more active than sodium carbonate or tertiary amines. Alkali metals salts of phosphoric, polyphosphoric, phosphorous, and hypophosphorous acids were proven effective [376,385–389].

Subsequent to this breakthrough, the subject was extensively investigated and was subsequently reviewed [390]. The acids most effective are BTCA, tricarballylic acid, and citric acid. Owing to the low cost and wide availability of citric acid, it is undergoing widespread commercial development for esterification cross-linking of cotton and paper products, often with minor amounts of BTCA as an activator, together with suitable catalysts.

Pad-dry-heat curing technology suitable for use in continuous mill processing of cotton fabric has been demonstrated with BTCA and citric acid, using weak bases as high-speed curing catalysts. In order of increasing effectiveness as catalysts are the sodium salts of phosphoric, polyphosphoric, phosphorous, and hypophosphorous acid, most of which are superior to sodium carbonate or tertiary amines in catalytic activity. Heat curing is carried out at 160–215°C for 10–120 s. Smooth drying properties so imparted are durable to repeated laundering with alkaline detergents at 60°C (140°F) and pH 10. The surprising degree of resistance of the ester groups to alkaline hydrolysis is apparently due to the presence of unesterified carboxyl groups in the fabric finish, which are converted to negatively charged carboxylate ions by reaction with alkaline detergent. These negatively charged fibers tend to repel hydrolytic attack on ester groups by carbonate, phosphate, and hydroxide ions of the detergent. The carboxylate ions of the finish strongly adsorb cationic dyes, and impart soil release properties.

The mechanism of high-temperature, base-catalyzed esterification involves cyclic anhydride formation by the polycarboxylic acid, followed by base-catalyzed reaction of anhydride groups with cellulosic hydroxyls. Fumaric acid, which cannot form a cyclic anhydride because of the *trans* arrangement of the two carboxyl groups, fails to react with cotton in the presence of weak bases and heat.

As formaldehyde-free, durable-press finishing agents, the polycarboxylic acids are apparently free of adverse physiological effects noted for diepoxides, vinyl sulfone derivatives, and polyhalides previously proposed for this purpose. In addition to imparting shrinkage resistance, smooth drying, and durable-press properties, ester-type cross-linking has undergone commercial development for treatment of woven and nonwoven cellulosic materials used in diapers, sanitary napkins, and other multilayer personal care products where resiliency and recovery from wet compression are required.

9.6.3.4 Esters of Inorganic Acids

Esters of cotton cellulose with inorganic acids have also been prepared in fiber, yarn, or fabric form. Nitric acid can be reacted with cotton linters to produce explosive cellulose nitrates known as guncotton, which contains more than 13% nitrogen, and is used in smokeless powder. The FR properties can be imparted to cotton fabric by oven curing with mono- or diammonium phosphate at high temperature in the presence of urea as a catalyst and coreactant. The resulting ammonium cellulose phosphate is flame- and glow-resistant, but ion exchange with sodium salts or especially calcium salts, during fabric laundering suppresses these properties. Aftertreatment of phosphorylated fabric in its ammonium or sodium ion form with aqueous titanium sulfate produces a finish having increased resistance to ion exchange with calcium or magnesium salts. The flame and glow resistance obtained are fairly durable to repeated laundering. Sulfation of cotton with ammonium sulfamate (AS) in the presence of urea or urea-based cross-linking agents imparts an excellent flame resistance to cotton that is durable to over 50 alkaline launderings in soft as well as in hard water [391–393]. When bis(hydroxymethyl)uron was used as cross-linking agent together with AS, the durable-press rating increased from 1.0 for the untreated fabric to 2.5–3.0 after 25 launderings. The severe afterglow of the sulfated fabrics could be overcome either by an aftertreatment with diammonium phosphate, which was not durable, or by a combined and simultaneous sulfation and phosphorylation using AS and phosphorus triamide (PA) or methyl substituted PA in the ratio of P:S of 1.3:1.0. The tensile and tear strengths of the phosphorylated and sulfated fabrics are decreased by about 30–40%, whereas the stiffness and air permeability remain nearly the same. The treated fabrics stood up to 25 hard water launderings with chlorine bleaches. Sulfamide, $SO_2(NH_2)_2$, cross-links cotton and imparts flame resistance when cured on fabric at high temperatures in the presence of urea. The flame resistance is durable to laundering with ionic detergents, but resistance to afterglow is lacking unless nonionic phosphorus-containing flame retardants are also applied [376,388,389,391,394–401].

9.6.3.4.1 Bioconjugation of Esterified Cotton Cellulose
Biologically active conjugates of cotton cellulose have been synthesized using glycine esterification as a starting point in the synthesis (Figure 9.55).

Peptides and proteins were synthesized or immobilized on cotton to create biologically active fibers with protease sequestering and antibacterial activity. Glycine esterified to cotton cellulose was employed as a linker to assemble peptide sequences and immobilize enzymes on fabrics [402]. Esterification of cotton cellulose with glycine was accomplished using a dimethylaminopyridine-catalyzed, carbodiimide–hydroxybenzotriazole acylation reaction. When the base-catalyzed esterification of cotton cellulose was compared between cotton twill fabrics and cellulase-treated cotton twill fabrics dimethylaminopyridine-catalyzed esterification of cotton gave higher levels of glycine esterification in the cellulase-treated samples. Peptide substrates of human neutrophil elastase (HNE) were synthesized on glycine esterified cotton cellulose to demonstrate the use of HNE substrates as protease sequestering agents of HNE covalently attached to cotton fibers. The HNE substrate peptide conjugate on cotton cellulose has served as a route to demonstrating the efficacy of removing proteases from the chronic wound. This approach is also a model for the design of peptidocellulose analogs in wound dressing fibers for chronic wounds. It has lead to the development of interactive cotton-based wound dressings that redress the proteolytic imbalance, resulting from the inflammatory pathophysiology of the chronic wound [403,404]. In analogous approaches, glycine esterified cotton cellulose was employed to immobilize lysozyme on cotton twill fabric [405]. Lysozyme was immobilized on glycine-bound cotton through a carbodiimide reaction, resulting in more robust antibacterial activity of the enzyme when bonded to cotton fibers than in solution. Retention of biological

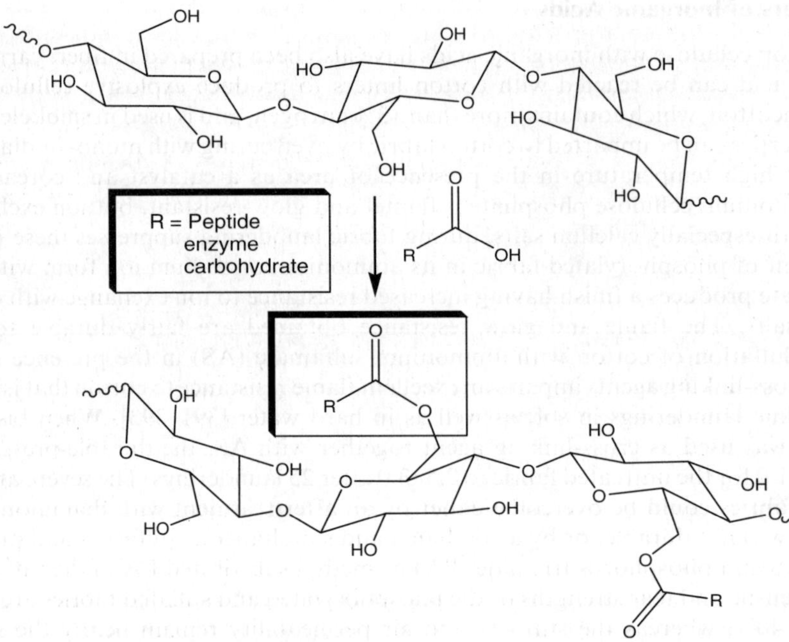

FIGURE 9.55 Peptide, enzyme, and carbohydrate esters of cotton cellulose that form bioconjugates creating biologically active cotton textile surfaces.

activity of the lysozyme-conjugated cotton-based fabric was an impetus for covalent attachment of organophosphorus hydrolase enzymes to cotton wipes as a method of wound-surface removal and decontamination of organophosphorous nerve agents [406].

Carbohydrate conjugates of cotton cellulose have also been prepared in the form of cotton-based wound dressings using a modification of the esterification of cotton cellulose with polycarboxylic acids outlined in Section 9.6.3.3. When citric acid was applied to cotton with the monosaccharide fructose and glucose using the pad-dry-heat curing technology, an esterified cellulose-citrate-linked ester of the monosaccharide was formed [407]. This same approach was also applied to grafting alginate onto cotton wound dressings to create a combination cotton-alginate wound dressing with enhanced properties of absorbency and elasticity. The conjugates formed during these esterification reactions linking cotton cellulose alginate have been termed "algino-cellulose" [408]. The alginate–citrate finishes of cotton gauzes were applied in various formulations containing citric acid, sodium hypophosphite, and polyethylene glycol and confer gelation properties upon hydration of the fabric.

9.6.4 Degradation

9.6.4.1 General

The prime position of cotton among textile fibers is due to its capacity to withstand chemical damage during processing. Nevertheless such damage can occur, under certain circumstances, in which a study of degradation is so important. It is necessary to understand degradation to be able to prevent it. The agents most likely to degrade cotton during processing or subsequent use are acids, alkalis, oxidizing agents, heat, radiations, and enzymes, all of which will be considered in this section. A recent exhaustive review of textile degradation in general contains 827 references, of course many to fibers other than cotton [409].

Native cotton is nearly pure cellulose; the approximately 6% of minor constituents are usually removed during preparation for wet processing. The chemistry of cotton is therefore the chemistry of cellulose, but the effects of chemical changes on its textile properties depend on its supramolecular structure as well as on the nature of the changes themselves. The term degradation has been used in more than one sense. Originally, it implied a loss of tensile strength sufficient to render a fiber or fabric unfit for use. When it became known that one of the main causes of such loss was the severance by hydrolysis of glycosidic bonds in the cellulose chain, it was applied to this process. When this reaction occurs, it yields glucose. However, from a chemical point of view, a complete degradation of cellulose would yield the substances from which it was originally photosynthesized, namely carbon dioxide and water. Thus partial degradation, which interests the textile chemist most, may include chain scission by the hydrolysis of glycosidic bonds and the partial oxidation of AGUs without chain scission. It might be supposed that the latter process could occur without loss of tensile strength, but this has never been observed. Usually chain scission accompanies oxidation, but loss of strength can sometimes be attributed to other causes, such as cross-linking or changes of supramolecular structure.

The mechanical properties of cotton are very sensitive to some types of chemical changes. Quite small changes may affect them profoundly, sometimes even reducing the cotton to powder. However, small variations in the constitution of a large linear macromolecule are not easy to investigate by the usual methods of organic chemistry. However, during the period 1920–1960, several tests, originally devised for monitoring industrial production, were developed into quantitative measurements for the characterization of slightly degraded cellulose [410]. The terms hydrocellulose and oxycellulose were already in common use for the water-insoluble products of acid hydrolysis and oxidation, respectively [411–413]. To study a particular degrading reagent, a series of progressively modified materials was prepared and characterized by measuring such properties as content of acid groups, reducing power, fluidity in cuprammonium hydroxide (CUAM), and tensile strength, the latter two are related to the DP. Valuable insights into the nature of various types of degradation were gained and the methods themselves were studied and improved. Stoichiometric methods are now available for measuring carboxy, aldehyde, and ketone groups, as well as terminal hemiacetals. IR spectroscopy has sometimes been used for the determination of these groups, but its application is limited by the small amounts of substituents usually present. It has of course been found more useful in the study of pyrolysis, along with other instrumental methods [414]. In many laboratories, CUAM for DP measurements has been replaced by the more stable and conveniently handled CUEN, cadoxen (cadmium ethylenediamine) or the Fe(III)–tartrate complex, FeTNa (EWNN in the German literature) [134,415].

The DP results rendered by most of these methods are low and inaccurate, especially when the celluloses are alkali-sensitive, i.e., contain carbonyl groups bringing about chain cleavage in alkaline solutions or active carbonyls, which initiate the stepwise depolymerization of the cellulose chains (known as the peeling reaction) according to the carbonyl elimination mechanism of Isbell et al. [416–420]. The most accurate method for DP determination in all celluloses is the nitration method, in which the cellulose is nitrated in a solution of nitric and phosphoric acids and phosphorus pentoxide, and dissolved in butyl acetate [420–422].

9.6.4.2 Oxidation

Stoichiometric methods for measuring total carboxyl content depend on the exchange between a cation in solution and the solid cellulose substrate in its free-acid form:

$$\text{RcellCOOH} + M^+ \rightleftharpoons \text{RcellCOOM} + H^+ \qquad (R9.6)$$

The equilibrium in the above reaction must be established as far to the right as possible and the main difference between the numerous methods that have been proposed is the way in which this is achieved [423]. The simplest of all is to measure the fall in concentration of a standard sodium hydroxide solution in which a known mass of material has been steeped. The solution should contain enough sodium chloride to equalize the concentrations of alkali in the solid and liquid phases by suppressing the Donnan effect [424]. This method is only valid if the cellulose contains no carbonyl groups since they cause the production of extra carboxy groups in the presence of alkali. The calcium acetate method [425] employs a dilute solution of this salt buffered to achieve a final pH ≥ 6.5, the fall in calcium concentration is measured by ethylenediaminetetraacetic acid titration. Results may be spuriously low if any carboxyl groups have formed lactones, because these are stable at pH 6.5. Methylene blue absorption [426] has two advantages over other methods because of the very high affinity of the dye compared with other cations: it does not require the material to be in its free-acid form and a degree of precision is obtainable even with materials of low carboxyl content. Virtually complete exchange is assured by establishing the following conditions in the final solution: $[\text{dye}] = 10^{-4}$ M, $[\text{Na}^+]/[\text{dye}] \leq 4$, pH = 8. The change in the concentration of dye in the bath is measured spectrophotometrically. The methylene blue absorption of purified cotton is about 0.5 mmol per 100 g, which, if it is assumed to arise from acidic end groups, gives a number-average DP of about 1200.

Lactones in acid-washed samples of modified cotton are not hydrolyzed in methylene blue solution. They may, however, be determined along with free carboxyl groups by steeping the material in a solution containing potassium iodide and iodate, sodium chloride, and an excess of sodium thiosulfate [427,428]. Hydrogen ions from the material liberate iodine according to the reaction

$$6H^+ + 5I^- + IO_3^- \rightarrow 3I_2 + 3H_2O \qquad (R9.7)$$

The iodine does not appear but reacts immediately with the thiosulfate, the consumption of which is determined by titration with the standard iodine solution.

Uronic acid groups (carboxyl groups at C6 in the AGUs) can be estimated from the yield of carbon dioxide on boiling with 12% hydrochloric acid for 8 to 10 h [429–431].

Two types of carbonyl groups may be introduced into cotton during degradation: aldehyde groups (including hemiacetals) and ketone groups. The former have strong reducing powers, but the latter do not. Some analytical methods measure aldehyde groups only, and some total carbonyl. The earliest measure of aldehyde content was the copper number, which is defined as the mass in grams of Cu(II) reduced to Cu(I) by 100 g of dry material on boiling with an alkaline copper solution of specified composition for 3 h [432,433]. The method is empirical, the amount of copper reduced per aldehyde group depends on the position of the group in the chain molecule. For example, a hemiacetal group reduces about 22 atoms of copper [434], but the two aldehyde groups in a periodate oxycellulose (see below) reduce 1.6 and 8.9 atoms, respectively. Thus, the copper number is a very sensitive measure of hemiacetal end groups and the fact that it is close to zero for scoured cotton is strong evidence of the absence of such groups in this material. The best available method of aldehyde determination consists of measuring the increase in carboxyl content of a material when it is treated with chlorous acid at pH 3 [435]. The copper number should be reduced to nearly zero in the process. If it is not, a plot of copper number against carboxyl generated at various times must be extrapolated to zero copper number, but the results should be treated with caution [436].

Total carbonyl content can be measured by means of one of the well-known condensation reactions such as the formation of oximes, phenylhydrazones, or cyanohydrins. Some of the methods proposed have serious limitations, but the reaction with sodium cyanide is usually considered to be satisfactory [437,438].

$$R'COR'' + NaCN + H_2O \rightarrow R' - \underset{\underset{CN}{|}}{\overset{\overset{OH}{|}}{C}} - R'' + NaOH \qquad (R9.8)$$

The consumption of cyanide is accurately determined by argentometric titration. The method was found to apply both to oxidized starch [439] as well as to cellulose [418]. The method is accurate and recently the coefficient of variance and standard deviation for an oxidized cellulose containing 5.60 mmol per 100 g of ketone groups, were found to be 0.0046 and 0.0068, respectively. The corresponding values for the carboxyl groups by the methylene blue method of the same samples were 0.001 and 0.031 [440].

The cyanide method is presently the only method for the determination of ketone groups in the polymers and was highly instrumental in the chemical characterization of degraded and oxidized celluloses. The use of this method enabled the development of the first two systems for the preparation of keto-cellulose, namely by mild oxidation with aqueous bromine at low pH values at room temperature [441,442] and by mild oxidation with hydrogen peroxide at pH 10 and 80°C [420,443].

If the carbonyl content is very low, it is better to use ^{14}C-labeled cyanide and determine the radioactivity of the cyanohydrin [444].

9.6.4.3 Acid

The degradation of cotton by acids consists of the hydrolysis of glycosidic linkages in accordance with Scheme 9.1.

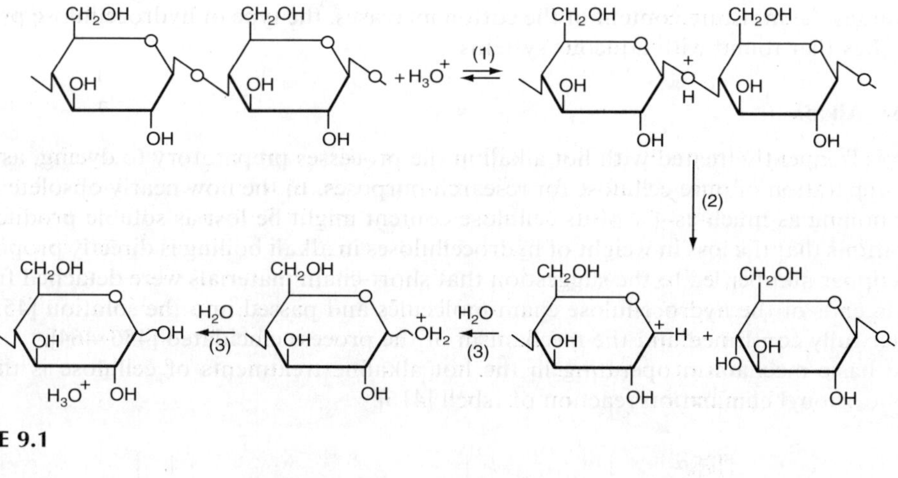

SCHEME 9.1

The reaction takes place in three steps, the formation of a carbonium ion as the glycosidic bond breaks (step 2) is the rate determining step. It is exactly analogous to the hydrolysis of simple glycosides [445]. In homogeneous hydrolysis, for example in 72% sulfuric acid, a yield of D-glucose as high as 90% can be obtained. In a textile context, only dilute aqueous acids come into contact with cotton, which on drying become concentrated, and hydrolysis is confined to the accessible regions of the fiber. Hydrocelluloses, materials of reduced DP having one reducing and one nonreducing end group per macromolecule, are formed. Their properties depend solely on the number and distribution of glycosidic linkages broken. This is governed by the concentration of hydrogen ions, the temperature, and the time of treatment; the nature of the anion is irrelevant. Hence, the tensile strength, copper number, and fluidity in CUAM of hydrocelluloses are uniquely related to one another irrespective of how they have been produced. The products of the early stages of hydrolysis are fibrous, but those with fluidities above about 44 rhe (1 rhe = $10 \text{ m}^2/(\text{N s})$) are usually powders. After some time, the fluidity ceases to rise and a rate plot reaches a plateau. With native cotton, this occurs at a fluidity of 50 rhe, which corresponds to a DP of about 220. This is known as the leveling-off DP (LODP); with mercerized cotton, it is close to 180. The particles of an LODP hydrocellulose are generally identified with the crystallites in the original cotton.

Of the three stages in the acid hydrolysis of cotton, the first is very brief with a reaction rate 10,000 times greater than in the second stage. This is due to the presence of so-called weak bonds [446,447]. The nature of the weak bonds has been the subject of controversy, but it is now generally accepted that they are ordinary glycosidic bonds under abnormal physical stress arising during the original formation of the fibers [448,449]. The second stage represents the random hydrolysis of glycosidic bonds in the accessible regions of the fibers. The third stage consists of the end-wise attack of acid on the otherwise impervious crystallites. Thus, small soluble fragments are progressively removed, causing a continuous loss in weight but no significant fall of DP [450]. This is expected if there is an exponential distribution of crystallite lengths and all crystallites contain the same number of chain molecules [451]. Recent work has suggested that, while this may be true for mercerized cotton, it is only an approximation for native cotton [452,453].

When cotton is treated with very dilute solutions of hydrogen chloride in an aprotic solvent such as benzene it suffers severe degradation. This is because the small amount of hydrogen chloride in the solvent is redistributed in the water adsorbed on the cotton, forming a very concentrated aqueous solution of hydrochloric acid [454]. At low moisture contents, the sites of the consequent hydrolysis are near the ends of the cellulose chains. The relation between DP and copper number therefore differs from that for normal aqueous hydrolysis. However, as the moisture content of the cotton increases, the type of hydrocellulose produced approaches that found with aqueous systems.

9.6.4.4 Alkali

Cotton is frequently treated with hot alkali in the processes preparatory to dyeing, as well as in the preparation of pure cellulose for research purposes. In the now nearly obsolete process of kier boiling as much as 4% of its cellulose content might be lost as soluble products. The observations that the loss in weight of hydrocelluloses in alkali boiling is directly proportional to the copper number led to the suggestion that short-chain materials were detached from the reducing ends of the hydrocellulose chain molecules and passed into the solution [455]. This was later fully confirmed and the mechanism of the process elucidated [456–464].

The basic mechanism operating in the hot alkaline treatments of cellulose is the beta-alkoxyl carbonyl elimination reaction of Isbell [416]:

$$R_1O-\overset{|\beta}{\underset{|}{C}}-\overset{|\alpha}{\underset{|}{C}}-R_2 + OH^- \rightarrow [R_1O-\overset{|}{\underset{|}{C}}-\overset{|}{\underset{|}{C^-}}-R_2] + H_2O \rightarrow R_1O^- + \overset{|}{\underset{|}{C}}=\overset{|}{\underset{|}{C}}-R_2 \quad \text{(R9.9)}$$

According to this reaction any strongly negative group in the position β of an ether, where the α-carbon is carrying a hydrogen, will render the ether bond sensitive to alkali. The hydrogen on the α-carbon atom adjacent to a ketone or aldehyde group will become acidic enough to be removed by a base. This will be followed by an elimination of the alkoxyl group from the β-carbon atom so that an unsaturated product will be formed along with the cleavage of the etheric bond [465]. The upper equation in Scheme 9.2 illustrates the case when the negative carbonyl group is located on C-6. The double bond containing AGU was eliminated from the chain R—OH, which contains a reducing C-1 end. By the Lobry de Bruyn–Van Eckenstein transformation, the C-1 aldehyde is converted into a C-2 ketone, which will be in the position β to the ether bond and thus will cause another chain cleavage and elimination. This stepwise depolymerization, which is called the peeling reaction, continues until the whole length of the chain in the LORs will be depolymerized or until it reaches the crystalline region when a stopping reaction sets in. The 50% of the cleaved chains beginning with C-4 will remain unchanged (see Figure 9.56) [420,441,442,465,466].

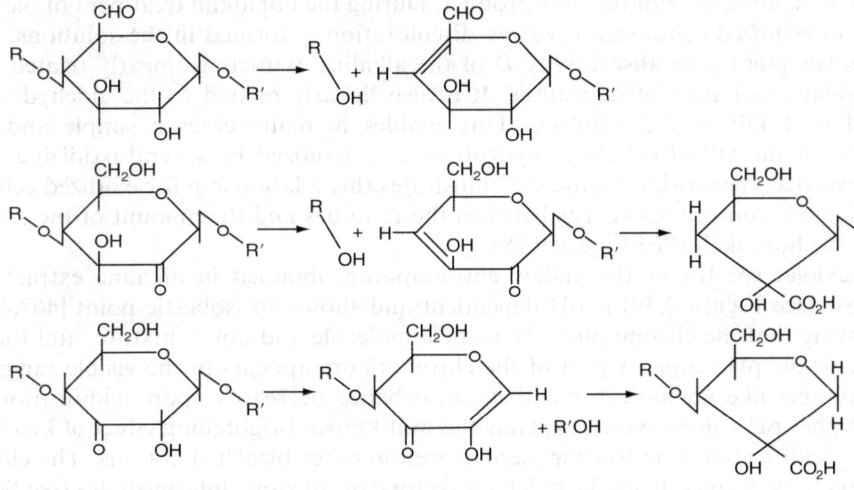

SCHEME 9.2

When the carbonyl group is located on C-2 (see Scheme 9.2 bottom equation) the chain scission will occur on C-4, producing a new chain with a reducing end at C-1. This new chain will behave

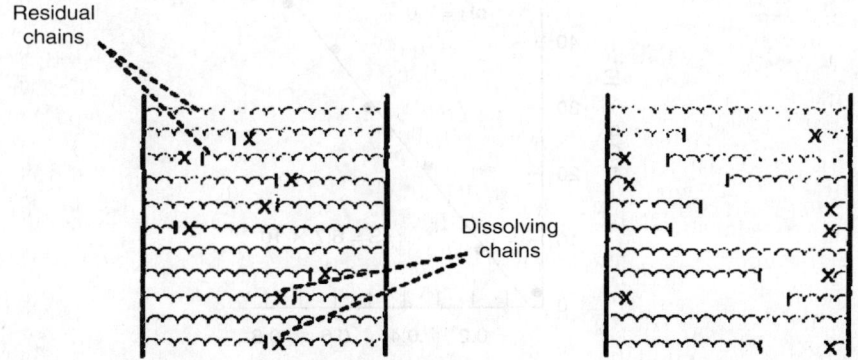

FIGURE 9.56 (X) Reducing end of chain. (|) Nonreducing end of residual chain. Left: after oxidation. Right: after alkaline extraction and peeling.

according to the upper equation in Scheme 9.2, which means that the depolymerization reaction will propagate along the chain and consequently the hot alkali solubility will be high. With a C-3 ketone group (see equation at the bottom in Scheme 9.2), the scission will occur at C-1, producing a nonreducing chain and a diketo derivative, which may be transformed into a saccharinic acid. There will be no continuing depolymerization and the alkali solubility of such celluloses will be insignificant. The ketocelluloses produced by acidic bromine [420,441,442] and by alkaline peroxide [443], discussed above, behave in this manner. Aldehyde groups on C-2 and C-3, as obtained by periodate oxidation, will produce a rupture of the pyranose ring but no peeling reaction will ensue [442]. Hypochlorite oxidized celluloses may, depending on the pH of oxidation, contain carboxyl groups (pH > 10), aldehyde and ketone groups (pH 5–10).

The simplified Scheme 9.3 shows that the diketo moiety formed in the hot alkaline extracts of hydrocellulose can undergo the benzylic rearrangement and produce a colorless isosaccharinic acid, which dissolves in the extract and constitutes the major product of alkaline degradation. It is, however, not the only product. During the hot alkali treatment of bleached, hydrolyzed, or oxidized celluloses, a yellow discoloration is formed in the solutions, which obeys Beer's law [466]. The absorptivity D of the alkaline extracts is linearly related to the extent of oxidation or time of hydrolysis. It is also linearly related to the aldehyde group content and to $1/DP$ of the cellulose. This enables in many cases a simple and rapid determination of the DP of celluloses hydrolyzed and oxidized by several oxidizing agents (excluding hydrogen peroxide). Figure 9.57 illustrates this relationship for oxidized cellulose. A straight-line relationship also exists between the D values and the amount of the cellulose dissolved in the hot alkali (See Figure 9.58).

The ultraviolet spectra of the yellow chromophore obtained in alkaline extracts of a hydrocellulose (see Figure 9.59) is pH dependent and shows an isobestic point [467,468] at 270 nm, proving that the chromophore is a single molecule and not a mixture, and that it is stable in the whole pH range. A part of the chromophore appears in the visible range. The chromophore acts like an indicator and its absorbance decreases upon acidification, and darkens at higher pH values, which explains the well-known brightening effect of kier boiled fabrics upon acidification (the souring step) as well in other bleached cottons. The chromophore appears to be formed from the aldehydo-ketone or diketone intermediates (see Scheme 9.3). Its formation competes with the formation of isosaccharinic acid. The ratio of the rates

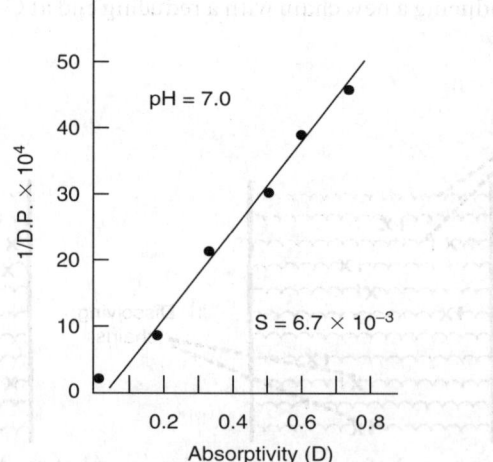

FIGURE 9.57 $1/DP$ versus D for cotton oxidized with hypochlorite at pH 7.

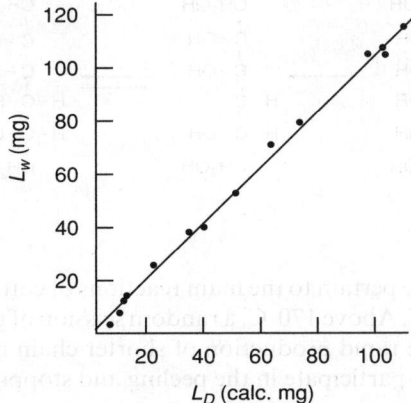

FIGURE 9.58 Gravimetrically determined loss in weight (L_w) on bicarbonate extraction versus loss in weight calculated from the D values on the basis on D of glucose, L_D. Pima cotton, hydrolyzed with 5N HCl at 25°C.

of these two reactions is influenced by the presence of cations, their nature, and valencies. Most effective in the decrease of the chromophore concentration is the calcium ion [469,470]. The peeling reaction takes place in substituted and cross-linked cottons, as determined by measuring the yellowing of alkaline extracts of such celluloses. A schematic representation is shown in Scheme 9.4 [442].

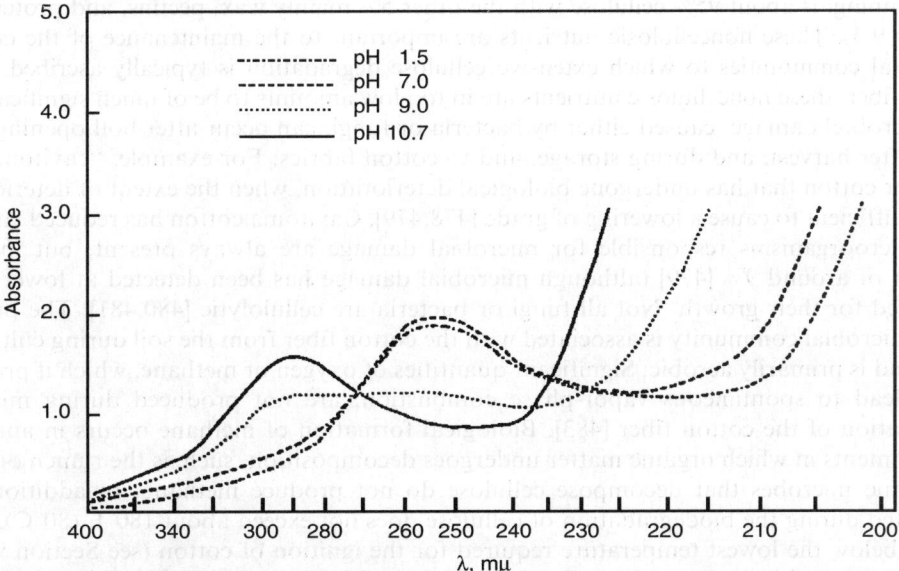

FIGURE 9.59 Ultraviolet (UV) spectra of hydrocellulose, boiled for 1 h in 5% bicarbonate at several pH values. Hydrocellulose prepared from Deltapine cotton fiber by shaking in 5N HCL for 72 h at 20°C. Amount dissolved in bicarbonate corresponds approximately to 3.7 mmol anhydroglucose units (AGUs).

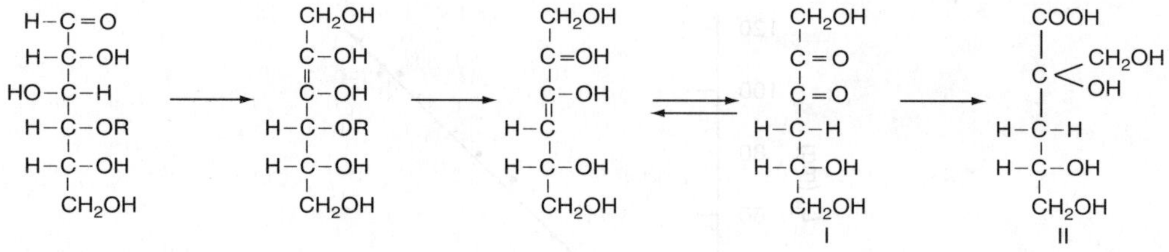

$$
\begin{array}{c}
H-C=O \\
H-C-OH \\
HO-C-H \\
H-C-OR \\
H-C-OH \\
CH_2OH
\end{array}
\longrightarrow
\begin{array}{c}
CH_2OH \\
C-OH \\
C-OH \\
H-C-OR \\
H-C-OH \\
CH_2OH
\end{array}
\longrightarrow
\begin{array}{c}
CH_2OH \\
C=OH \\
C-OH \\
H-C \\
H-C-OH \\
CH_2OH
\end{array}
\rightleftharpoons
\begin{array}{c}
CH_2OH \\
C=O \\
C=O \\
H-C-H \\
H-C-OH \\
CH_2OH \\
I
\end{array}
\longrightarrow
\begin{array}{c}
COOH \\
C{<}{CH_2OH \atop OH} \\
H-C-H \\
H-C-OH \\
CH_2OH \\
II
\end{array}
$$

SCHEME 9.3

The processes discussed above pertain to the main reactions of cotton with alkaline solutions up to temperatures of 120–130°C. Above 170°C, a random scission of glycosidic bonds in accessible regions occurs, leading to the rapid production of shorter chain molecules with new reducing end units. These immediately participate in the peeling and stopping reactions just described.

9.6.4.5 Biodegradation

The biodegradation of cellulose is caused by enzymes known as cellulases [471–475]. Cellulases are produced by many microorganisms (bacteria and fungi). The most widely studied cellulases are of fungal origin, e.g., *Trichoderma* [471,473]. The cellulose-digesting bacteria of the rumen are a complex anaerobic community [476,477].

The primary reaction in the enzymatic degradation of the cellulose is hydrolysis, and degradation is a function of the available surface area and crystallinity of the cellulose. The biological degradation of the cellulose is discussed thoroughly by others [471,472]. The intentional use of enzymes in textile applications for desizing and finishing and some of the mechanisms of their actions is described in Section 9.6.6. Cotton fiber is a highly crystalline form of cellulose and the ability to degrade this form of cellulose is relatively rare among microbes, and, due to the solid nature and insolubility of the substrate, is not very rapid. The cotton fiber, after ginning, is about 95% cellulose with the other 5% mainly wax, pectins, and proteins (see Section 9.3). These noncellulosic nutrients are important to the maintenance of the complex microbial communities to which extensive cellulose degradation is typically ascribed, but in cotton fiber, these noncellulosic nutrients are in too low amounts to be of much significance.

Microbial damage, caused either by bacteria or fungi, can occur after boll opening, prior to or after harvest, and during storage, and to cotton fabrics. For example, "cavitoma" is a term for cotton that has undergone biological deterioration, when the extent of deterioration is not sufficient to cause a lowering of grade [478,479]. Cavitoma cotton has reduced strength. The microorganisms responsible for microbial damage are always present, but moisture content of around 9% [479] (although microbial damage has been detected at lower levels) is needed for their growth. Not all fungi or bacteria are cellulolytic [480,481]. The predominant microbial community is associated with the cotton fiber from the soil during cultivation [482] and is primarily aerobic. Significant quantities of oxygen or methane, which if produced could lead to spontaneous vapor-phase combustion, are not produced during microbial degradation of the cotton fiber [483]. Biological formation of methane occurs in anaerobic environments in which organic matter undergoes decomposition, such as the rumen of cattle. The same microbes that decompose cellulose do not produce methane. In addition, heat generated during the biodegradation of cellulose does not exceed about 180°F (80°C), which is well below the lowest temperature required for the ignition of cotton (see Section 9.6.4.6) and the autoignition temperature of cellulose (about 750°F (400°C)). Therefore, contrary to what has been suggested by some, the raw cotton fiber, unless contaminated with 5–10% oil and wet cotton fiber, is not capable of spontaneous combustion.

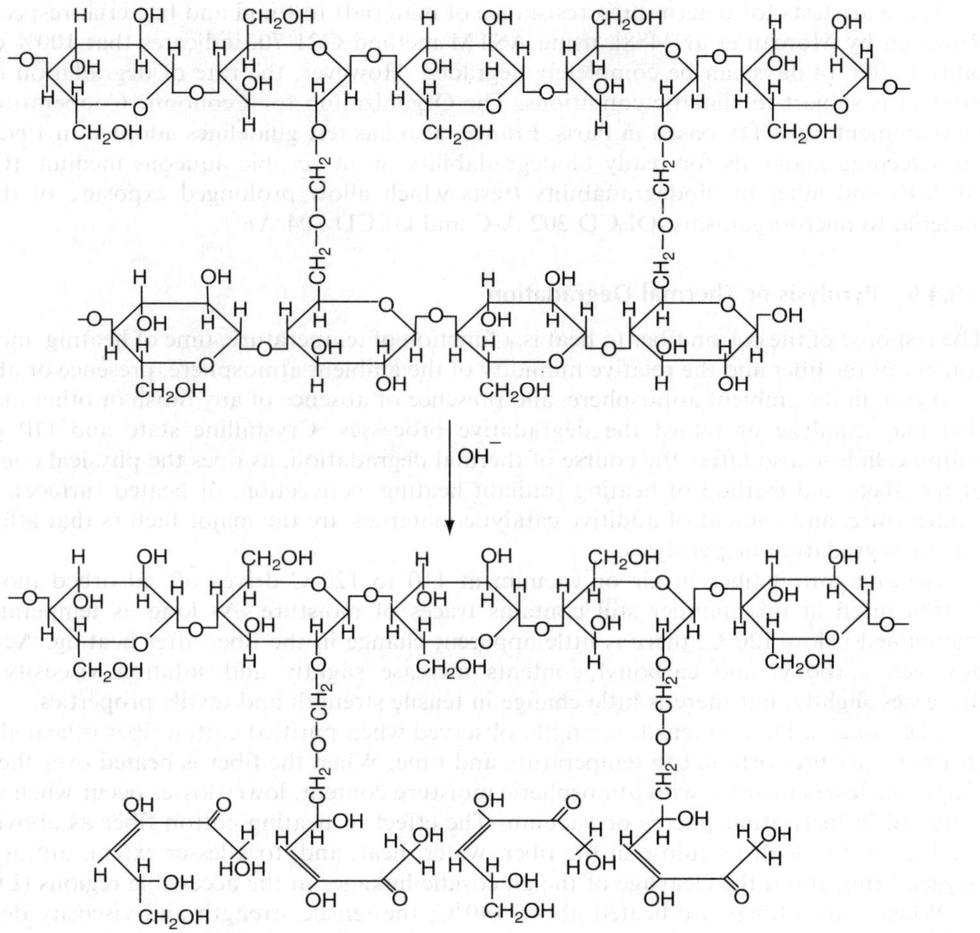

SCHEME 9.4

Tests currently available to identify and quantify microbial damage are reviewed by Allen et al. [484]. These tests include CUAM fluidity tests, pH determinations, reducing sugar determinations, microscopic techniques, and staining methods.

The U.S. Federal Trade Commission (FTC) has requirements [485] for products that make environmental marketing claims of biodegradability. A product claim should be substantiated by competent and reliable scientific evidence that the entire product will completely breakdown and return to nature, i.e., decompose into elements found in nature within a reasonably short time after customary disposal. Claims should be qualified to the extent necessary to avoid consumer deception about the products' ability to degrade in the environment, where it is customarily disposed, and the rate and extent of degradation.

The U.S. EPA considers cellulosic materials clearly biodegradable [486] and lists acceptable tests for biodegradable materials:

1. ASTM Method G21-70 (1984a)
2. ASTM Method G22-76 (1984b) [487]

These are tests for determining resistance of materials to fungi and bacteria, respectively. Research by Moreau et al. [488], using ASTM method G21-70, indicates that 100% cotton fabrics after 14 days can be completely degraded. However, the rate of degradation of any product is subject to climatic conditions. The Organization for Economic Cooperation and Development (OECD), based in Paris, France, also has test guidelines adopted in 1992 [489] for screening materials for ready biodegradability in an aerobic aqueous medium (OECD 301A-F) and inherent biodegradability (tests which allow prolonged exposure of the test material to microorganisms; OECD 302 A-C and OECD 304 A).

9.6.4.6 Pyrolysis or Thermal Degradation

The response of the cotton fiber to heat is a function of temperature, time of heating, moisture content of the fiber and the relative humidity of the ambient atmosphere, presence or absence of oxygen in the ambient atmosphere, and presence or absence of any finish or other material that may catalyze or retard the degradative processes. Crystalline state and DP of the cotton cellulose also affect the course of thermal degradation, as does the physical condition of the fibers and method of heating (radiant heating, convection, or heated surface). Time, temperature, and content of additive catalytic materials are the major factors that affect the rate of degradation or pyrolysis.

Heating cotton fiber in air or vacuum at 110 to 120°C drives off adsorbed moisture. Cotton dried in this manner still contains traces of moisture. As long as temperature is maintained below 120°C, there is little apparent change in the fiber after heating. Actually, however, carboxyl and carbonyl contents increase slightly and solution viscosity (DP) decreases slightly, but there is little change in tensile strength and textile properties.

The losses in DP and tensile strength, observed when purified cotton fiber is heated in air to 150°C, are proportional to temperature and time. When the fiber is heated over the same range, the losses increase with atmospheric moisture content; lower losses occur when cotton is heated in inert atmospheres or vacuum. The effect of heating cotton fiber as above is to produce cotton hydrocellulose in the fiber; water, heat, and, to a lesser extent, atmospheric oxygen bring about the cleavage of the glucosidic linkages in the accessible regions (LORs).

When cotton fibers are heated above 140°C, the tensile strength and viscosity decrease while carbonyl and carboxyl contents increase. Distinct discoloration of the fiber is seen; first, a yellowing (scorching) and then a deepening to brown as thermal degradation escalates. In air, the oxidation of cotton cellulose occurs at these temperatures. Complete loss of tensile strength develops after a few hours at or above 200°C in the absence of oxygen. Above 200°C, cotton cellulose is at the verge of thermal decomposition, and depolymerization begins. Between 200 and 300°C, primary volatile decomposition evolves. The early work on cellulose pyrolysis products was reviewed by Shafizadeh in 1968 [490]. These products can be trapped and analyzed in vacuum pyrolysis. The further decomposition to secondary products occurs when the primary products are heated at ambient pressure. The secondary products separate into a gaseous phase and a distillate (liquid) phase. The most recent work on cellulose pyrolysis is directed at optimizing the distillate fraction for conversion of celluloses or lignocellulosics to renewable fuels. A char is the only remaining solid after heating cellulose above 400°C. The proportions of gas, distillate, and char depend on the conditions of heating and the amount of inorganic catalyst or flame retardant present in the original cotton cellulose. When pure cotton cellulose is heated in vacuum, the gaseous phase constitutes about 20% of the total pyrolysis products, the liquid phase about 65%, and the char about 15%. Differential thermal analysis (DTA), differential scanning colorimetry (DSC), thermogravimetric analysis (TGA), gas–liquid chromatography-Fourier transform infrared spectroscopy (GC-FTIR), and direct pyrolysis-mass spectrometry (Py-MS, also referred to as evolved gas analysis, EGA) are techniques that have played important roles in following the pyrolysis of cotton fiber.

Several kinetic studies of the pyrolysis of cellulose have been reported [491–495] and, possibly due to the complications arising from impurities, have not led to any agreement on the kinetic order or the mechanism. However, Chatterjee found experimentally [496] and showed [497] from the results of Lipska and Parkera [492] a linear relationship between the square root of the weight loss versus time. This relationship was explained based on a two-step reaction sequence: initiation (i.e., random scission of glucosidic bonds producing reactive molecules) and propagation (i.e., decomposition with weight loss of the reactive molecules). This square root relationship was confirmed by Lewin and coworkers [498–500], who found, upon isothermal pyrolysis (at 251°C) of a series of carefully purified cottons and other celluloses of various origins, that the rate increases with decreasing crystallinity and increasing orientation and is inversely proportional to the square root of the DP (correlation factor of 0.940). The energy of activation increases linearly with crystallinity and extrapolated values of 30 and 65 kcal/mol are obtained for 100 and 0% LORs, respectively [498].

The isothermal pyrolysis in the presence of air proceeds at a much faster rate and higher weight losses are obtained as compared to vacuum pyrolysis at the same temperature. The first order rate constant obtained is linearly related to the expression: $[\%LOR + \sigma$ (% crystallinity)$]/f_0$ with a degree of correlation $r = 0.923$, where σ is the accessible surface fraction of the crystalline regions according to Tyler and Wooding [501], and f is the orientation factor. No correlation could be found with DP due to very rapid depolymerization. The fact that the rate is inversely proportional to the orientation and that it decreases with the increase in the thickness of the fibers indicates that the rate of the diffusion of the oxygen into the fibers controls the kinetics and that oxidation is the predominant process in air pyrolysis.

Pure and flame-retardant cotton fibers were studied by TGA and Py-MS [502] to measure the pyrolysis kinetics and primary pyrolysis products in order to propose mechanisms of cellulose degradative processes in the absence of oxidative processes or further degradation of the primary products. At a heating rate of 5°C/min, cellulose pyrolysis occurs in three consecutive stages, each with its own kinetic and product profiles [503]. The first stage, which occurred at 250–290°C, was characterized by a low energy of activation (114 kJ/mol) and a large negative entropy of activation (−137 J/(mol °C)). The volatile products from this stage consisted of only carbon dioxide, carbon monoxide, and water and accounted for only about 2% of the dry weight of cellulose. It was proposed that this stage consists of a random chain scission in the LORs of the cellulose followed by relaxation of the broken chains and dehydration, decarboxylation, or decarbonylation of AGUs previously affected by oxidization in the preparation of cotton. The chain scission was proposed to occur by a conversion of an AGU in a LOR from the normal $^4C_1$ to the $^1C_4$ conformation followed by a rate-determining attack of O6 on the acetal carbon (C1) of the affected AGU to leave a cellulose chain terminated in a 1,6-anhydro-AGU and another chain with a free C_4OH at the non-reducing end. This mechanism would account for the observed kinetic parameters and the strength loss observed at or below this temperature range.

The second pyrolysis stage follows the first stage at 290–310°C and is characterized by a higher energy of activation (165 kJ/mol) and a less negative entropy of activation (−40J/(mol °C)). The volatile products of this stage account for about 5% of the cellulose weight and include anhydroglucoses [e.g., levoglucosan (1,6-anhydro-β-D-glucopyranose), 1,6-anhydro-β-D-glucofuranose, and 1,4:3,6-dianhydro-α-D-glucopyranose]. The mechanism of this stage was proposed to be characterized by cellulose chain unzipping processes in the LORs from the chain ends formed in the first stage. The levoglucosan is formed by a scission process in the penultimate AGU similar to that in the first stage, resulting in a free levoglucosan and another 1,6-anhydro-AGU at the end of the chain. The cellulose chain with the free C_4OH terminal AGU is proposed to unzip by converting from the $^4C_1$ to the $^1C_4$ conformation and rear-side attack of the C_4 to displace the C_1—O—cellulose bond. This displacement may occur before or after a displacement of the C_6OH by the C_3OH oxygen. These processes give the observed anhydroglucofuranose and dianhydroglucose. These processes are expected to have the observed intermediate activation energy and less negative entropy of activation. The attribution of the second pyrolysis stage to the LORs is supported by the behavior of ball-milled cotton cellulose, which shows no crystallinity by x-ray diffraction. Ball-milled cotton shows the same first pyrolysis stage as unmodified cotton cellulose, but the second pyrolysis stage of ball-milled cotton covers a temperature range of 290–354°C and consumes 80% of the original weight. A LODP has also been reported in the pyrolysis of unmodified cotton at about the same DP, weight, and increased crystallinity as observed in the acidic hydrolysis of unmodified cotton cellulose.

The third stage in the vacuum pyrolysis of pure cotton cellulose follows the second stage at 310–350°C and is characterized by a still higher energy of activation (260 kJ/mol) and a positive entropy of activation (123 J/(mol °C)). This stage accounts for approximately 80% of the original cellulose weight. The volatile products include, in addition to those from the first and the second stages, products formed by dehydration of the anhydroglucoses: 5-hydroxymethyl-2-furfural, 2-furyl hydroxymethyl ketone, and levoglucosenone (1,6-anhydro-3, 4-dideoxy-β-D-glycero-hex-3-enopyranos-2-ulose).

The higher energy of activation and the relatively large entropy of activation are consistent with the rate of this stage, determined by disruption of the crystal structure of the native cellulose. The higher crystallinity of the remaining cellulose, the LODP, and the absence of the third stage in decrystallized cotton are consistent with such a mechanism.

The x-ray diffraction studies show that there are no changes in the crystal structure of cotton heated in sealed tubes below 250°C. Heating to 250–270°C causes loss of crystallinity and on further heating (280–300°C) the cellulose pattern disappears. The lowest temperature for self-ignition in air is 266°C, with a heating time of 2.5 h. What ignites, however, is not the cotton fiber but an organic residue, as the fiber has lost 60–70% of its original weight. Changes in the molecular structure can be followed readily by IR as the fiber is heated. The changes observed in the IR spectrum arise from the transformation of cellulosic hydroxyl groups into carbonyl and carboxyl groups, reduction in hydrogen bonding, breaking of glucosidic linkages, etc. As the fiber is heated to higher and higher temperatures, the characteristic cellulose bands disappear.

The practical aspects of the pyrolysis of cotton fibers are in the flame retardant cottons and in the ironing of cotton fabric. The subject of flame-retardant cotton fabrics and the means of producing them, which is discussed in Section 9.6.2.4, will only be touched upon lightly; extensive treatment of the subject can be found elsewhere [361]. Flaming combustion is a gas-phase oxidation of volatile fuels. These fuels are produced by pyrolysis of the solid fuel (cotton cellulose) as a result of the heat from the gas-phase combustion. Phosphorus-based flame retardants used in durable finishes for cotton textiles act by reducing the amount of volatile fuels formed in the pyrolysis, reducing the pyrolysis temperature, and increasing the amount of carbon-containing char remaining after the pyrolysis step. The durable phosphorus-containing flame retardants have specific effects on the kinetics and pyrolysis products of each stage of the pyrolysis [504]. Halogen-containing finishes act by generating volatile halides that act as free-radical traps and interrupt the gas-phase oxidation of volatile fuels. The phosphorus-containing, durable flame retardants for cotton textiles also suppress afterglow, which is the glowing combustion (a direct gas–solid oxidation) of the char remaining after pyrolysis or flaming combustion of cotton fabrics. A number of inorganic salts, such as potassium carbonate or magnesium chloride [505] suppress flaming combustion of cotton textiles, but they are not durable to washing and do not prevent afterglow in the residual char from heating the textile. Other inorganic compounds, such as ammonium phosphates, prevent both flaming and glowing combustion but are not durable to washing. Borax, boric acid, and other compounds of boron dried into cotton fibers inhibit combustion, but are not durable to washing.

Smoldering combustion can occur by sparks or lighted cigarettes falling on unprotected mattresses or cushions, and can lead to catastrophic flaming combustion involving furniture, etc. Smoldering propensity of cotton (e.g., upholstery fabrics and batting) has been found to be induced by alkali metal ions such as potassium, sodium, or calcium, as well as salts of iron, chromium, and lead [351,352]. Such ions are found naturally in raw cotton (which is used for batting and many upholstery fabrics) or as residues of dye assistants, softeners, detergents, etc. These compounds can effect an increase in the yield and reactivity of char formed during the pyrolysis of cellulosics [351]. Chemicals that inhibit the smolder of cellulosics appear to be compounds containing boron or phosphorus (e.g., boric acid and ammonium phosphates). These smolder inhibitors appear to intervene chemically in the oxidation reactions on char surfaces [351]. Boric acid, deposited on cotton batting fibers by way of the volatile methyl borate, is useful for preventing smoldering combustion in cotton batting. Other agents that are effective for cotton fabrics are not effective in cotton batting for preventing smoldering combustion, which is an oxygen-limited process controlled by the rate of diffusion of air through the batting to the fire.

Scorching of the cotton fabric begins in the ironing of cotton when flatiron temperatures reach about 250°C. Wet fabric scorches very much faster than dry fabric and, peculiarly, ease of scorching varies with cotton variety. The temperature of iron, the pressure of ironing, and the time of heating during ironing all have a relationship on the appearance of fabric scorching. Except for the effect of pressure, the effects of degree of temperature and time of heating are, as would be expected, that the degradation occurs with low temperature and long periods of heating or high temperature and shorter periods. In ironing, cotton fabric is in contact with the heated surface for only a short length of time, usually not more than 1 min. The appearance of scorching is a visual indicator that thermal degradation is occurring and can be taken as an end point for setting limits on ironing conditions. Tests have shown that for constant heating time the scorch temperature is reduced when ironing pressure is increased from 1.3 to 4 lb. When pressure is held constant and heating time is lengthened from 2.4 to 30 s, scorch temperature is again reduced. Scorch temperature of the fabric for these conditions ranged from 255 to 332°C. When cotton is exposed to temperatures of this magnitude for several hours, severe thermal degradation results, as has already been pointed out. Some of the original durable-press finishes for cotton cellulose promoted scorching in cotton fabrics, which had been exposed to chlorine bleaches, but this is no longer a problem with modern durable-press finishes.

Recent work on the thermal degradation of cellulose has shifted away from flame retardancy as the regulatory climate has changed. Areas of current research on thermal degradation of cotton include effects of heating during processing. The Acala 4-42 cotton was found to crystallize upon heating in a continuous curing oven for 10 min at 180, 190, and 200°C [259]. This thermal crystallization proceeds by first-order rate kinetics with an energy of activation of 32.5 kcal/mol. The accessibility of cotton decreased from the original 32.2 to 21.7% at 180°C, to 10.9% at 190°C, and to 4.6% at 200°C. The corresponding decreases in moisture regain from the original 7.92% to 7.01, 6.40, and 5.28%, respectively, were obtained. Rushnak and Tanczos [506] also obtained a decrease in moisture regain and dyeability of cotton upon heating. Back and coworkers [507,508] suggested that the increase in crystallinity is caused mainly by cross-linking reactions between aldehyde groups or between hydroxyl groups of adjacent chains in the LORs, producing interchain ether linkages. A similar assumption was made to explain the initial rapid weight loss in the isothermal pyrolysis at

250–275°C [498]. In addition, hydrogen bonded water molecules are removed from the polymer by the heating and enable the formation of new strong hydrogen bonds between hydroxyls of adjacent chains. These considerations explain the decrease in moisture regain and the increase in wet strength of heated paper.

Heating conditions during drying of cotton lint at the gin have been studied [509] to determine effects of drying conditions on cotton fiber quality. Excessive heating causes discoloration, strength loss, and reduced spinning quality, possibly from effects of heat on the noncellulosic components of raw cotton fibers. There has been considerable work on pyrolysis of cellulose and lignocellulosic materials to generate gaseous or liquid fuels from biomaterials and by-products. Richards [510] has done considerable work on pyrolysis mechanisms and useful materials from forestry by-products.

9.6.4.7 Visible, Ultraviolet, and High-Energy Radiation

Photochemistry involves the interaction of visible and ultraviolet light with cellulose whereas radiation chemistry involves its interaction with high-energy radiations, such as from ^{60}Co γ-radiation. Light promotes the deterioration of cellulosic products, particularly cotton fabrics, and certain dyes or other additives greatly accelerate this process known as phototendering. Ionizing radiation is used to sterilize medical and bioproducts many of which are cellulosic. Several authoritative reviews are available [511–514].

The regions of the electromagnetic spectrum that are of interest in photochemistry are the visible (400–800 nm), the near-ultraviolet (200–400 nm), and the far-ultraviolet (10–200 nm) regions. To effect a chemical change directly the radiation must be absorbed by cellulose. Pure cotton cellulose, a saturated compound, lacks the structural features required to absorb light in the visible region. Even absorption in the near ultraviolet remains uncertain. Light with wavelengths greater than 310 nm (Pyrex glass filter) is not able to photolyze cellulose directly. Mercury emission at predominantly 254 nm can cause photodegradation and produce free radicals. It is not clear what structural feature is responsible for the absorption of radiation [514]. Irradiated carbohydrates do form a species absorbing at 265 nm [515] that exhibits an autocatalytic influence. The direct photolysis of cellulose with 254-nm radiation results in degradation and is independent of the presence of oxygen [516,517]. Changes are increase in the solubility, the reducing power, the formation of carboxyl groups, and lowering in the DP.

Although light of wavelengths greater than 310 nm cannot degrade cellulose directly, some other compounds such as dyes and some metallic oxides can absorb near-ultraviolet radiation or visible light and in their excited states can induce the degradation of cellulose. These reactions are designated photosensitized degradation but do not have a common mechanism. The photochemical tendering of cotton fabrics containing vat dyes has been recognized for many years [518–520]. Emphasis has been placed on practical problems involving textile performance rather than on mechanistic studies. It is known that this sensitized deterioration depends on the presence of oxygen in contrast to direct photolysis. The cotton cellulose is also subjected to degradation upon exposure to high-energy radiation. The radiation sources have included cobalt-60, fission products in spent fuel rods, and high-voltage electrons generated by linear acceleration [521]. High-energy radiation causes oxidative depolymerization of cellulose in the presence or absence of oxygen to yield hydrogen, one-carbon gaseous products, and multicarbon residues. At low doses, fibrous properties are retained. The effects of radiation dosages ranging up to 833 kGy on some of the physical and chemical properties of cellulose are shown in Table 9.12. The oxidative depolymerization of cellulose is shown by the increase in carbonyl and carboxyl groups and by the solubility of the irradiated residues.

TABLE 9.12
γ-Irradiation Effect on Some of the Physical and Chemical Properties of Cotton Cellulose[a]

Dosage (kGy)	DP of cellulose[b]	Carbonyl groups (mol = g)	Carboxyl groups (mol = kg)
0.00	4400	0.00	0.002
0.83	3100	0.01	0.002
8.3	880	0.05	0.005
42	320	0.22	0.015
83	210	0.36	0.023
417	79	1.30	0.070
833	56	2.66	0.139

[a]Purified cotton fibers irradiated in nitrogen at 298 K.
[b]Precision of these determinations is low.

The high-energy radiation forms macrocellulosic radicals that are stable in the crystalline areas of cellulose. These radicals can initiate reactions with vinyl monomers to yield grafted polyvinyl-cellulosic fibers with desired properties [522–524].

9.6.5 WEATHER RESISTANCE

Most of the cotton fabrics, although capable of maintaining their strengths for many years indoors in a dry environment, deteriorate readily when kept in an outdoor environment. This is primarily because of bacterial attack and sunlight's actinic degradation (see Section 9.6.4.5 and Section 9.6.4.7). Although bottom weight (i.e., 8.0 oz/yd^2 and greater) cotton fabric has been used for years for tentage and awning material, its poor resistance to the above-mentioned outdoor hazards has limited its usable life, especially in rainy, damp climates. An entire industry has grown up in an effort to protect cotton fabric intended for outdoor use. Many treatments with agents that repel or kill microorganisms and that shield cotton fabric from actinic degradation have been proposed over the last half century. Formulations containing inorganic salts of chromium, copper, zinc, mercury, iron, titanium, and zirconium, as well as a number of organic compounds of sulfur, chlorine, and nitrogen, have found use as antimicrobial and sunlight-resistant agents for cotton fabrics intended for use outdoors [525,526].

Although damage from moisture and mildew are insignificant in most indoor environments, sunlight damage can be quite severe, for products like cotton drapes. Those formulations containing chromium or titanium, or vat or pigment dyes in the blue or grey shades have been the most successful. If titanium is used, it is usually used as TiO_2 in its tetragonal (anatase) form and is usually applied to that side of the drape, which faces the sunlight. If the drape is to be laundered, the treated side must also be coated with a vinyl or acrylic film. Environmental considerations and concerns for heavy metal ions in processing plants have begun to limit the choices open to the finishers in the United States.

For bacterial and mildew resistance, inorganic materials are applied in such a way as to permit them to be slowly released from some soluble form, thereby providing a constant minimal presence of the toxic agent over a long period of time. Agents such as chromium oxide (as in the pearl grey finish) [527,528], copper and zinc naphthenate, copper-8-hydroxyquinolate, organomercury compounds, mixed agents containing both zirconium and chromium (i.e., the "Zirchrome" finish) [527,528], iron salts, and numerous organic compounds such as thiazolylbenzimidazole, octyl isothiazolinone, chlorinated phenols (e.g., pentachlorophenyl laureate), and the quaternary naphthenates have historically been employed by cotton fabric finishers to provide rot resistance. An excellent study comparing the effectiveness of a number of the commercially available fungicides and sunlight protectors over a 3-year period was reported in 1978 [526]. This study compared both inorganic and organic materials with and without a protective coating at several weathering sites. Of all the agents evaluated, those containing chromium salts or octyl isothiazolinone resisted the mildew and retained the strength best. Coated treatments always outperformed uncoated ones [526]. Many of the inorganic agents resulted in the deposition of insoluble inorganic pigments on and into the fabric yarns and interstices. These materials are frequently opaque, and although all of the above-mentioned chemical treatments were used primarily for bacterial and algae protection, it was understood that some of them would also provide a substantial resistance to sunlight degradation. This is especially true of the chromium, the titanium, and to a lesser extent, the zirconium compounds. For more extended usable life outdoors, these fabrics were often treated with additional fungicides and either a water-repelling wax or a vinyl or acrylic coating.

9.6.6 Enzymatic Modification

Enzymes are used in textile applications [529,530] for the hydrolysis of starch after desizing of cotton, for the removal of surface fibers, for giving cotton fabric a softer hand, and for enhancing the color of the dyed fabric. What differentiates enzymes from other catalytic systems is their remarkable specificity. As an example, starch polysaccharide is hydrolyzed by α- or β-amylase at the α-(1→4)-glucoside linkage, whereas a cellulase has no effect at this site. This specificity is attributed to the active-site geometries of these proteins that are dictated by the distinctive three-dimensional shapes in aqueous solution. These complex three-dimensional configurations comprise alpha-helix designs, beta-pleated sheets, and hairpin turns as part of the unit structure. The enzymes that hydrolyze cellulose are known as cellulases or 1,4-β-glucanases [472]. Endoglucanases attack in the middle of the chain whereas exoglucases, also called cellobiohydrolases, remove cellobiose units of the chain end. The resulting cellobiose units are hydrolyzed to glucose by cellobiose hydrolases. The molecular weights of these proteins range from approximately 10,000 to 80,000 Da or more in the fermented broth. They are obtained from both bacterial and fungal sources. The three-dimensional structures of many of these proteins have been determined by x-ray crystallography. Enzymes that must attack insoluble substrates such as crystalline cellulose usually contain a domain that binds tightly to the cellulose and a catalytic domain attached to it by a flexible peptide. Hydrolysis of the glycosidic bond of cellulose (Figure 9.60) can occur with either retention or inversion of the anomeric configuration depending on the cellulase.

The endo-1,4-β-glucanases, exo-1,4-β-glucanases (cellobiohydrolases), and cellobiose hydrolases work synergistically. The catalytic sites of the endo-1,4-β-glucanases are generally located in a cleft whereas those of the exo-1,4-β-glucanases are in a tunnel-like structure. Examples of endoglucanase [531] and exoglucanase [532] structures are given in Figure 9.61.

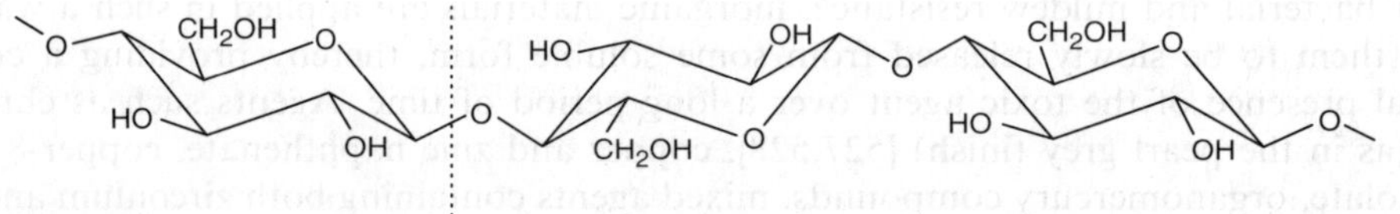

FIGURE 9.60 Glucosidic bond in the cellulose polymer that is hydrolyzed by cellulase (indicated by broken line).

Cotton Fibers

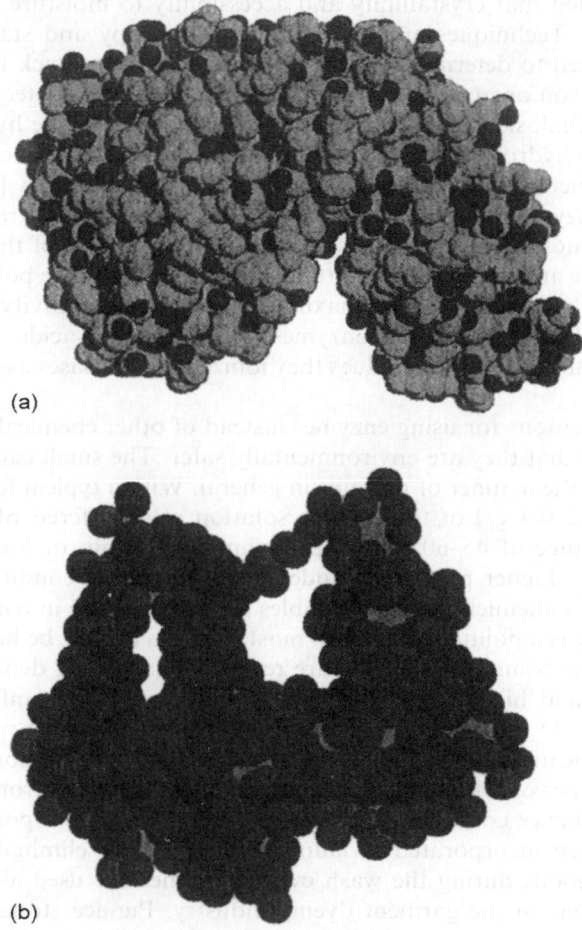

FIGURE 9.61 Molecular structures of a 1,4-β-endoglucanase (a) and 1,4-β-exoglucanase (b) showing the cleft and tunnel features, respectively. These drawings were prepared from files downloaded from the Brookhaven Protein Data Bank (http://pdb.pdb.bnl.gov). (a) (1CEC) and (b) (3CBH, Cα skeleton) were rotated to reveal the cleft and tunnel, respectively, with RasMol, Version 2.5, by Roger Sayle, freeware available from the Brookhaven National Laboratory web site.

The cleft in the endoglucanase permits it to attach to the crystalline microfibril (see Figure 9.12) and effects hydrolysis of a glucosidic bond within the polymer chain, creating new terminal sites. The polymer chain is then passed through the tunnel containing the active site of the exo-1,4-β-glucanase (cellobiohydrolase) and the disaccharide cellobiose is liberated. The cellobiose is hydrolyzed to glucose by a cellobiose hydrolase. After the polymer chain is fragmented by attack of the endoglucanase, the smaller chain segments are rapidly removed from the microfibril by the exoglucanase. The microfibril is swiftly planed by the enzymes, leaving behind a crystalline material structurally similar to the original. The subjects of the structures and mechanisms of glycosyl hydrolases, such as cellulases, have recently been reviewed by Davies and Henrissat [533]. Modification of the cellulase structures and the composition of the mixtures are pursued as a means of tailoring their activity to better meet the practical objectives of the cotton textile industries.

It has been reported that crystallinity and accessibility to moisture do not change after enzymatic hydrolysis. Techniques employing light microscopy and staining of fibers with Congo Red can be used to determine the location of enzymatic attack for cotton and linen, but not for viscose rayon or ramie. In addition, SEM is useful for detecting changes in fiber morphology for all cellulosic fibers [534]. An important aspect of the hydrolysis of cellulose by cellulase is that the hydrolysis reaches a point where the hydrolytic activity levels off. This is reported to be a function of particle size, hydratability, and porosity [535].

The stability of enzymes is dependent on the hydrogen ion concentration of the environment. Cellulases are most stable at their isoelectric point (the sum of the cationic charges is equal to the sum of the anionic charges), but this is not necessarily the point at which catalytic activity of the particular enzyme is at its maximum. The greatest activity may be at higher or lower pH levels. At higher pH values enzymes ionize as weak acids and are precipitated by cationic agents, while at lower pH values they ionize as weak bases and are precipitated by anionic agents [536].

One of the main reasons for using enzymes instead of other chemicals as finishing agents for cotton cellulose is that they are environmentally safer. The small catalytic quantities will eventually degrade in the manner of proteins in general. With a typical formulation, cellulose is hydrolyzed with 0.2–0.4 g/l of the enzyme solution at a buffered pH of 4.5–5 and at a temperature in the range of 45–60°C for approximately 60 min or longer. Other enzymes designed to operate at higher pH values under neutral or basic conditions are also useful. Generally, the rate of a chemical reaction doubles for each 10° rise in temperature, but in the case of enzymes, there is a point above which most enzymes cannot be heated without loss of catalytic activity. Most common cellulases are rendered inactive or denatured by heating to approximately 65°C and higher. This inactivity is associated with unfolding of the three-dimensional structure. In contrast, cellulose is hydrolyzed much more rapidly under harsher acidic conditions by heating at 80°C in the presence of 2.5 N hydrochloric acid.

One of the major uses of cellulase on cotton is for removing fibers from the surface of dyed textile products to enhance color. This operation is referred to as biopolishing, and cellulase preparations have been incorporated in laundry detergents to eliminate or reduce surface fuzziness on cotton goods during the wash cycle. Enzymes are used also for biostoning of dyed garments, as done in the garment dyeing industry. Pumice stones have been used in recent years to soften garments and to remove color and produce unique washed-down appearances through abrasive action. However, such processes are not only environmentally unsound because of the disposal problem of ground-down stone particles after processing, but the stones are also very damaging to the processing equipment as well. The use of enzymes eliminates or substantially reduces the need for treatment with pumice stones [537,538]. The effectiveness of any of these operations for the removal of the material from the substrate is dependent on the type of mechanical action during processing. This includes the abrasive action of fabric-to-fabric contact or the cascading effect of aqueous solution on the cellulosic substrate [539].

The effects of cellulase enzymes on cotton substrates that were dyed with various classes of dyes have been reported. It was found that vat dyes do not inhibit cellulase activity on cotton, and hydrolysis of the substrate with subsequent color removal transpires. In contrast, cellulase activity on cotton was inhibited by the presence of some direct and reactive dyes. There is the probability that a dye–enzyme complex is formed with less activity than that of the free enzyme. The increased weight loss from enzymatic treatment on untreated cellulose is associated with a slight reduction in dye sorption, and this may be because of a reduction of amorphous regions in cellulose where dye molecules are usually sorbed. Cellulase activity is more prominent on mercerized cotton than on unmercerized cotton. This is because mercerization is associated with a decrease in the crystallinity of the cellulosic structure [540]. One major negative aspect of cellulase treatment of cotton fabric is the strength loss associated with the approximately 3–5% weight reduction under normal treating conditions. Of course, this is to be expected after cellulose is hydrolyzed. Future research efforts may lead to processes in which enzymes would be useful for total fabric preparation including not only the desizing of cotton, but also scouring and bleaching of cotton, as well as for the elimination of imperfections on the cellulosic substrate. The possibility also exists for the use of enzymes as catalysts for chemical modification of cellulose by esterification and etherification reactions.

9.6.7 Corona

The phenomenon of *corona discharge* is defined as a nondisruptive discharge of the air between two electrodes separated by a small space and maintained at high electrical potential with respect to each other. The discharge creates a plasma (a partially ionized gas) containing numerous tiny thermal sparks that can alter the surface properties of materials placed in the discharge zone. A considerable amount of research was devoted to study the effects of corona treating wool and mohair fibers with both air and air–chlorine mixtures to improve the strength, shrink-proofing, and processibility of these fibers [541]. Subsequently these treatments were applied to cotton fibers with increases in fiber friction, yarn strength, abrasion resistance, and spinnability [542]. The corona treatment apparatus for cotton, basically developed by Thorsen, consisted of a sliver drafting system that converted six slivers into a thin, 18-cm-wide web that is conveyed between parallel glass-covered electrode plates separated by approximately 1 cm [543]. The treatment zone was maintained at 95°C with 15 kV at a frequency of 2070 Hz applied across the electrodes. Thorsen conducted a rather definitive study with cotton demonstrating that the treatment did not alter the basic fiber strength but did increase yarn strength by the order of 25%. He further demonstrated that the treatment had many potential practical applications including:

1. Improved mercerization with measurable increases in conversion to cellulose II
2. Better yarn stretch and setting properties
3. Increased dyeability
4. Increased production of free radicals making possible the induction of enhanced polymerization reactions and treatments

Thibodeaux and Copeland [544] reported further enhancements of the corona treatment of cotton webs by introducing a higher-frequency, high-energy (3300 Hz, 4.2 kVA) power supply. The most significant finding of this research was that an approximate 50% increase in frequency of the corona discharge lowered the cell temperature requirements from 95 to 65°C for optimum treatment. The increase in fiber friction measured by cohesion test on roving, processed from treated sliver, indicated that drafting tenacity could be more than doubled and thus the twist necessary to be put into the roving for cohesion could be significantly reduced (by as much as 20%). Finally, the study demonstrated that ring-spun yarn could be produced with significantly lower twists than feasible with untreated cotton, with higher or equal strength than control yarns with higher twist. This translated into an increased production rate between 15 and 20%.

Further improvements in the corona processing of cotton were made by Australian researchers at CSIRO. The basic approach of treating drafted web in a wide cell was improved by developing a corona machine that treated whole sliver passing between three 5-cm-diameter rollers mounted circumferentially on the surface of a 15-cm-diameter roller and spaced 4 cm apart. A spacing (0.5 mm) is maintained between the rollers as the sliver pass through and high voltage is maintained between the large and small rollers, leading to a corona discharge in the nip zones. The experiments on treating cotton–viscose blended yarns [545] indicated that corona treatment increased sliver tenacity, cohesion, and wettability. For a given level of power, treatment effectiveness is proportional to residence time, and for constant residence time, effectiveness is proportional to power. Further studies [546] with the same apparatus, using cotton slivers, produced similar results for the effects of power and residence time and further indicated that the effect was enhanced by temperature, diminished by moisture content, and was independent of four different gases (Air, CO_2, N_2, and Ar) used in the reactor. These studies were then extended to study the effects of corona on yarn strength, fabric strength, and processing efficiency [547]. The strengths of yarns spun from corona-treated cotton slivers increased, especially for lower twists and coarser counts. Strength was also increased by repeated treatments. There was no difference in fabric strengths when comparing treated and untreated cottons, despite the fact that the treated cottons had been spun at lower twists. The hand of fabrics from treated cottons was somewhat harsher than fabrics from untreated cottons spun at the same twist level. However, this harshness virtually disappeared with fabrics from the lower twist yarns from the treated cottons.

9.6.8 Dyeing

Cotton requires some pretreatment (referred to as preparation) prior to dyeing and printing [548]. The preparation processes of cotton textiles include singeing, desizing, scouring, bleaching, and mercerization. These treatments remove natural and human-induced impurities, i.e., noncellulosic constituents and other unwanted substances, and increase the affinity of cellulose for dyes and finishes. In addition, color enhancement can be accomplished through the treatment of cotton with cellulase enzymes either before or after dyeing. The increased enhancement occurs because of the removal of fibers, which give the surface of cotton fabric a frosted appearance [549].

The dyeing of cotton has been extensively reviewed. The composition of the dye molecule, the dyeing conditions, and the nature of the cellulosic substrate influence the kinetics and equilibria in dyeing cellulosic fibers. Cotton is dyeable in fiber form (raw stock), yarn, or fabric form with an extensive number of dye classes, including, azoic, direct, indigo, pigment, reactive, sulfur, and vat dyes [339,550–557]. The selection of the appropriate dyes and dyeing processes for cotton is based on numerous factors depending on the application for which the dyed cotton fibers are to be used. A detailed review of these considerations and the properties of the dyes are beyond the scope of this chapter.

Anionic dyes used on cotton are essentially water soluble because the dye molecule contains sulfonic acid groups. The most widely used dye classes for coloring cotton are the direct and reactive ones. The dyeing of cellulosic fibers with direct dyes is relatively simple, as the dyes are affixed by hydrogen bonding and van der Waals forces. More elaborate procedures are employed for dyeing cotton with sulfur, vat, and reactive dyes. Sulfur and vat dyes are generally water insoluble, but they are solubilized as part of the dyeing process by converting the dye to its leuco form as the soluble sodium salt. After diffusion of the dye into the fiber during the dyeing process, the water-insoluble keto form is formed again upon oxidation. Thus, the insoluble dye remains trapped within the cellulosic fibers.

Most dyes do not chemically react with the cellulose molecule to affix the color. The roles of interchain hydrogen bonding and van der Waals forces in the application of azoic, direct, sulfur, and vat dyes are the physical and chemical effects and are not classical chemical reactions. True chemical reaction between cellulose and the dye molecule occurs with reactive dyes, which comprise different chemical types (see Section 9.6.2.3). Such chemical reaction results in covalent bond formation between the dye molecules and the C6 hydroxyl groups of the cellulosic chains. In many cases, these covalent bonds are formed at temperatures ranging from approximately 25°C (over an extended period of time) to 80°C for 30–60 min at pH 11–12. The reactive dyes, such as di- or monochlorotriazine or other suitable derivatives, react with cellulosic hydroxyl groups by splitting out HCl and forming ether linkages (etherification). Sulfatoethylsulfone reactive dyes react with cotton cellulose hydroxyl groups in alkali solution forming ester bonds (esterification). Cottons dyed with these dyes have excellent colorfastness to washing. Of course, any laundering of the dyed cotton should be conducted without bleaching agent, such as sodium hypochlorite, which will destroy the chromophore and also cause cleavage of the covalent bond [558,559].

Cotton also can be colored with pigments, which are water-insoluble colorants of either inorganic or organic composition. As such, pigments have no affinity for cellulose and are used with a binder, such as a polymeric resin, that secures the colorant to the substrate. Their lack of water-solubilizing groups and solubility in the aqueous medium distinguishes pigments from dyes. Pigments are applied to textiles by padding, printing, or batch exhaustion application methods [560].

Normally, cotton is not dyeable after it has been cross-linked with N-methylolamide agents, such as DMDHEU. This is because the fibers are cross-linked in a collapsed state at elevated temperatures and they cannot swell adequately in aqueous solution to accommodate the relatively large dye molecules. However, cross-linked cotton is dyeable with anionic dyes under acidic pH conditions if reactive alkanolamines or hydroxyalkyl quaternary ammonium salts are incorporated in the finishing formulation. In addition, cotton cross linked with polycarboxylic acid, such as BTCA, or citric acid, is dyeable in similar fashion by using the same methods. Thus, these cross-linked cottons have affinity for acid, direct, and reactive anionic dyes at the pH range of 2.5–6.5, depending on the exact chemical composition of the substrate [561–564].

Basic dyes are not normally used to color cotton textiles, but they do react with the few carboxyl groups in the regular cotton or with the greater number of groups in the oxidized cotton (methylene blue for staining oxidized cotton). In addition, cotton that has been cross linked with polycarboxylic acid has affinity for basic dyes because of the presence of free carboxyl groups. However, the basic dyes are not strongly bonded to the substrate, the color is rather easily removed [565], and has poor lightfastness. All other dyeing processes produce dyed cotton of varying degrees of fastness, to washing and to light. Color and brightness of shade vary with the dye class and dyeing process, as does the cost of dyeing.

9.7 PHYSICAL PROPERTIES OF COTTON

9.7.1 Maturity and Fineness

The term fiber maturity is generally understood to refer to the degree of development or thickening of the fiber secondary wall [566,567]. Fiber maturity is a function of the growing conditions that can control the rate of wall development and of catastrophic occurrences such as premature termination of growth due to such factors as insect infestation, disease, or frost. As we have seen, the fiber develops as a cylindrical cell with a thickened wall. As the diameter of the fiber cylinder is largely genetic or species-dependent, a simple absolute measure of the thickness of the fiber secondary wall is not sufficient to define maturity. Probably the best definition of cotton fiber maturity has been proposed by Raes and Verschraege [568] who state "...the maturity of cotton fibers consists in defining it as the average *relative* wall thickness." What is implied in this statement is that maturity is the thickness of the cell wall relative to the diameter or perimeter of the fiber.

A term that has more recently gained popularity is the degree of thickening, Θ, defined as the ratio of the area of the cell wall to the area of a circle having the same perimeter as the fiber cross section, or $\Theta = 4\pi A/P^2$ where A is the wall area (μm^2) and P the fiber perimeter (μm).

In selecting a cotton to be used in a manufacturing process, it is important to have a knowledge of the maturity of cotton, which will determine the ultimate quality of the product as related to dyeability and ease of processing. Immature cottons tend to not dye uniformly, and result in large processing wastes in large numbers of spinning and weaving breaks and faults.

Fiber maturity can be measured directly or indirectly. In general, the direct methods are more accurate and precise, but are much slower and more tedious than the indirect methods. In practice, direct methods are used to calibrate or standardize the indirect methods.

Three of the most significant direct methods include:

1. The caustic swelling [569] test, in which whole fibers are swollen in 18% caustic soda (NaOH) and examined under the light microscope with a specific assessment of the relative width of the fiber versus its wall thickness used to identify a fiber as mature, immature, or dead.
2. The polarized light test [569–571], in which beards of parallel fiber are placed on microscope slides on a polarized light microscope using crossed polars and a selenite retardation plate. The interference color of the secondary wall will be a direct measure of its thickness and thus maturity. Generally, mature fibers appear orange to greenish yellow whereas immature fibers appear as blue-green to deep blue to purple.
3. The absolute reference method of *image analysis* of fiber bundle cross sections [572–574], wherein an image analysis computer system is used to automatically measure the area and perimeter of several hundred fiber sections and statistically analyzed to measure the average Θ and perimeter.

The indirect methods are characterized by the need to be rapid as well as accurate and be reliable enough to be used in the cotton marketing system. These methods may be divided into the double compression airflow approach and near infrared reflectance spectroscopy (NIRS). Examples of the former (airflow) method are the Shirley Developments Fineness and Maturity Tester (FMT) [575], and the Spinlab Arealometer [576]. The NIRS approaches have been developed and discussed by Ghosh [577] and Montalvo et al. [578].

The term fiber *fineness* has had many interpretations and understanding in fiber science. Some of the most important parameters used to define fineness include:

1. Perimeter
2. Diameter
3. Cross-sectional area
4. Mass per unit length
5. Specific fiber surface

Of all of these five parameters, the perimeter has proven to be the least variable with growing conditions and is essentially an invariant property with respect to genetic variety. For this reason perimeter has become recognized by many as inherent or intrinsic fiber fineness. Because of the irregularity of cotton fiber cross sections it is very difficult to measure a real diameter (this would presuppose that the fiber was circular in cross section). Similarly, cross-sectional area, mass per unit length, and specific fiber surface are dependent on maturity and thus are not real independent variables that we desire. However, from the standpoint of the spinner, the most important of the possible fineness parameters listed above is mass per unit length. A knowledge of this parameter allows the selection of fibers based on the minimum numbers of fibers required to spin a certain size yarn, i.e., the finer the yarn, the finer the fiber required.

Fiber fineness or mass per unit length can be measured both directly and indirectly. The direct method [579] consists of selecting five separate bundles from the sample. In each bundle or tuft, the fibers are combed straight and each is cut at the top and bottom to leave 1-cm-long bundles. The fibers from each bundle are laid out on a watch glass beneath a low magnification lens and 100 fibers are counted out from each of the bundles, compacted together, and weighed separately on a sensitive microbalance. For each bundle, we obtain the fiber fineness in 10^{-8} g/cm. The closest thing to an indirect method for measuring fineness is the micronaire test [580]. However, as we will show, the micronaire test actually measures the product of fineness and maturity. It is based upon measurement of airflow through a porous plug of cotton fibers. In the standard micronaire test, 50 gr (3.24 g) of fiber are loosely packed into a cylindrical holder. The cylinder and the walls enclosing it are both perforated to allow the flow of air under pressure that compresses the fiber into a 1-in. diameter by 1-in.-long porous plug that will offer resistance to the flow of air under 6 lb/in.^{2}. Research has shown that the flow through the cotton is given by $Q = aMH$ where a is a constant, M is maturity, and H is fineness. These results imply that for a constant maturity, a micronaire instrument will be nearly linearly dependent on fineness. However, for samples of various finenesses and maturity, it has been demonstrated that there is a quadratic relationship between the product of fineness and maturity (MH) and micronaire. This relationship is best expressed by the quadratic $MH = aX^2 + bX + c$, where X is micronaire and a, b, and c are constants [566]. Thus, given that any two of the parameters (fineness, maturity, or micronaire) are known, the third can be determined, and the processor will have a much more complete picture of the quality of the cotton being processed.

9.7.2 Tensile Strength

An accurate knowledge of the tensile behavior of textile fibers (their reaction to axial forces) is essential to select the proper fiber for specified textile end-use applications. However, to have meaningful comparisons between fibers, experience has shown that it is necessary to conduct measurements under known, controlled, and reproducible experimental conditions [581]. These include mechanical history, relative humidity and temperature of the surrounding air, test or breaking gauge length, rate of loading, and degree of impurities. Mechanical history is important because fiber could be annealed, extended beyond its elastic limit, or otherwise affected by mechanical manipulation. Cotton is the only significant textile fiber whose strength increases with humidity while most others are weakened by increased moisture. Textile fibers universally lose strength with increasing temperature. It seems logical that the longer the breaking or gauge length, there is more chance for an imperfection to occur that will cause a failure. Likewise, the presence of impurities in a material will tend to lead to disorder and weakness.

One definition of strength is the power to resist force. In the case of an engineering material such as textile fibers, this can be translated as a breaking strength or the force or load necessary to break a fiber under certain conditions of strength. Although textiles will be forced to endure a wide variety of forces and stresses, experience is that tensile breaking load is an excellent benchmark of fiber strength. The general approach to relative ranking of the strength of materials is with breaking stress, i.e., the tensile load necessary to break a material normalized for its cross-sectional area. This is expressed by the equation, $\sigma = T_b/A$ where T_b is the breaking tension, and A is the cross-sectional area. The CI units for σ are N/m^2 (or Pa) [315]. However, when dealing with textiles it is often more convenient to think of strength in terms of force per mass rather than per area. When dealing with fibers this translates into force per mass (m) per fiber length (l) that defines *specific stress* (tenacity) as given by $\sigma_{sp} = T_b/(m/l)$ (N m/kg or Pa m^3/kg). Because of the magnitude of quantities dealing with textiles, it has been found to be more convenient to go to the tex system for linear density (1 tex = 1 mg/m) and for single fibers to measure load in millinewtons with the resultant stress units of mN/tex or gf/tex. Here *gf* refers to grams force or grams weight, which is the force necessary to give 1 gram mass an acceleration of 980 cm/s^2.

The values for the tensile strength of single fibers range from about 13 gf/tex (127.5 mN/tex) to approximately 32 gf/tex (313.7 mN/tex) [582]. Calculations based purely upon bond strengths of the cellulose molecule would predict much higher strengths for cotton, but other factors also contribute significantly to determine ultimate fiber tenacity. These include crystallite orientation, degree of crystallinity, fiber maturity, fibrillar orientation, and other features of the fiber structure. Although single-fiber testing is quite tedious and time-consuming, considerable studies in the past show rather consistent findings including that:

1. Fiber breaking load increases with fiber coarseness, though not in direct proportion to fiber cross-section.
2. Breaking tenacity correlates well with fiber length and fineness.
3. There is a correlation between fiber breaking load and fiber weight within any single variety [583].

Although, as mentioned above, the procedures for measuring single fiber breaks have made this type of investigation nearly prohibitive, especially in the case of quality control testing, a new instrument has been designed that shows potential for making single-fiber testing more feasible.

The most significant advance in the technology of fiber strength measurements has been the development of the Mantis, a single-fiber tensile tester [584]. Mantis's unique design allows for rapid loading of single fibers between the breaker jaws without use of glue and has computer system to control single-fiber breaks and record both stress–strain curves and other pertinent data. The single fiber test used by the Mantis consists of two measurement modes: mechanical and optical. Fiber mounting is semiautomatic, i.e., the operator places a fiber across the jaw faces and the fiber is straightened by a transverse airflow caused by two lateral vacuum pipes. Small jaws clamp the fiber ends and a slight stress (< 0.2 g) is applied to remove crimp. Optical measurements are accomplished by detection of the attenuation of infrared radiation by the presence of the fiber. The degree of attenuation is proportional to the fiber's projected profile (ribbon width). A uniform stress is applied, causing the elongation of the fiber until it breaks. A plot of grams force versus elongation is provided until the fiber breaks. In addition to the stress–strain curve, Mantis supplies the force to break, T_b (g), the fiber ribbon width, RW (μm), and the work of rupture (J).

The main purpose of testing raw cotton's strength is to predict the strength of yarn spun from the fiber. As yarn strength is determined by not only fiber strength but also by fiber to fiber interactions as induced by length, friction, and degree of twist, it has been found that breaking bundles of parallel fibers give a better predictor of yarn strength by simulating the combination of fiber tenacity and interaction. The two most commonly used bundle testers are the Pressley and the Stelometer testers [585]. The Pressley operates with a flat bundle of parallel fibers clamped between a set of jaws that may be operated such that there is essentially no gap (zero gauge) between the clamps holding the fibers or that there is a 0.125 in. spacing (eighth-inch gauge) between the clamps. The loading on the jaws is initiated by a weight rolling down an inclined plane. When the breaking load is reached, the bundle breaks, the jaws separate, and the breaking force is read from an attached scale. The force is determined by the length of travel of the carriage weight and is thus applied to the bundle specimen at a nearly constant rate of loading. The mass of the broken bundle is also measured and the breaking tenacity is calculated. There are, however, some mechanical problems with the Pressley and as a result, an alternative bundle tester has been developed that is almost universally accepted as the method of choice. The Stelometer operates on the principle of a pendulum and by proper adjustment can be set to operate at a constant rate of loading of 1 kg/s. The fiber bundle to be tested is mounted between Pressley clamps set to in. The clamps are mounted between a pendulum and beam that are set to allow the pendulum to rotate about a fulcrum. When the beam is released, it rotates about a point, causing the pendulum to rotate in such a fashion as to produce a constantly increasing rate of loading of the bundle under test.

Cotton bundle strengths range from a low of about 18 gf/tex (176.5 mN/tex) for short coarse Asian cottons to a high of approximately 44 gf/tex (431.5 mN/tex) for long fine Egyptian cottons. The degree of crystallite orientation, the fine-structural parameter obtained by x-ray diffraction, is directly related to the bundle strength [586]. This parameter is a measure of the angle of the fibrils spiraling around the fiber axis. The angle varies from about 25° for Egyptian cottons to about 45° for the coarser and weaker species. The degree of crystallite orientation increases with decreasing spiral angle. In general, the more highly oriented fibers are stronger and more rigid. In general, cotton strength increases with moisture content and decreases with temperature.

The reaction of a material to a tensile stress is to stretch or elongate. This is measured as *tensile strain*, defined as the elongation or increased length per initial length. Strain is a dimensionless unit that is usually expressed as a percent (percent elongation).

9.7.3 Elongation, Elasticity, Stiffness, Resilience, Toughness, and Rigidity

In the previous section, we discussed the evaluation of the strength of cotton under conditions of static or steady loading. In practice, textile fibers are exposed to a variety of dynamic forces and their response to "stress in motion" will better characterize their performance during processing or in end use. To quantify a material's dynamic response, it is necessary to record the elongation corresponding to increasing load [582]. A typical load–elongation or stress–strain curve is shown in Figure 9.62.

This represents the response of cotton loaded to the point of breaking. It begins with a curvilinear region (AB) in which fiber crimp or kinkiness is removed as load is applied. The crimp of a cotton fiber is a minor parameter compared with the other factors in the stress–strain properties of cotton, and consequently little quantitative consideration has been given to it [587]. However, it has long been recognized that the crimp of a fiber plays a major role, leading to the phenomenon of fiber cohesion, a property that causes materials to cling together. Without cohesion, it would literally be impossible to spin yarn from staple fibers. In the case of the synthetic fibers, crimp must be artificially induced by elaborate texturing schemes.

With further loading of the fiber, the curve (BC) becomes linear. In this region, stress is proportional to strain with the ratio referred to as the Hookean slope (i.e. follows Hook's law). The point C at which the curve becomes nonlinear is referred to as the yield point where the loaded elements begin to deform in a nonelastic or irreversible fashion and redistribute the stresses. From here to the breaking point (D), the curve becomes essentially linear. The stress–strain curve is strongly influenced by any of several conditions as the rate of loading, sample moisture content, specimen length (where structural imperfections in the fiber come into effect), and extent of mechanical preconditioning.

The elongation of cotton is expressed as percent elongation taken at the point of breaking, hence the term elongation at break [581]. For most cotton, elongation at break, or just elongation, is in the range 6–9%. The effect of moisture is most pronounced on elongation. An elongation of about 5% at low relative humidity will increase to about 10% when the

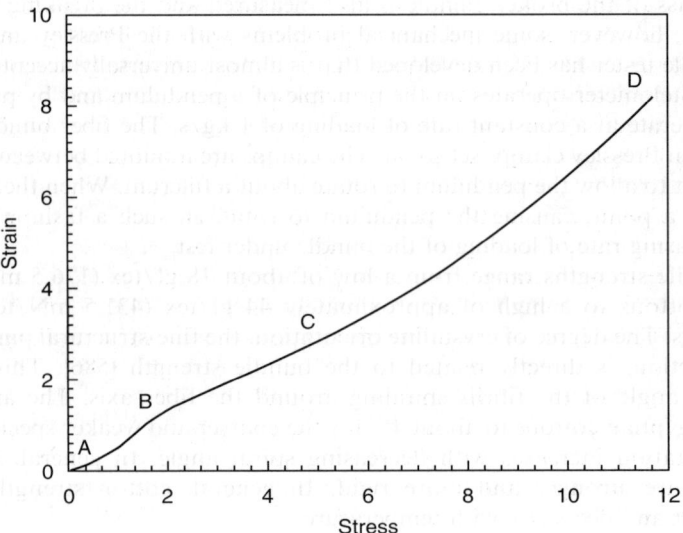

FIGURE 9.62 A typical stress–strain curve for cotton.

relative humidity is almost at the saturation point. The adsorption of water in the pores and amorphous regions of the fiber serves to reduce interfibrillar cohesion and to relieve internal fiber stresses. A more uniform distribution of applied stresses is thereby realized. The internal fibrillar orientation can be seen with x-ray diffraction. Swelling treatments, such as mercerization, ethylamine, and liquid ammonia treatments, affect the fiber elongation in a far greater fashion than water does.

Whereas the parameter elongation is the ability of the fiber to undergo deformation, elasticity is fiber's ability to return to its original shape when the loading is released. This property is highly time-dependent. Young's modulus, the ratio of stretching stress per unit cross-sectional area to elongation per unit length, may be calculated from data taken at either the beginning of elongation (initial modulus) or at the breaking point. Initial Young's moduli for cotton range from about 80 g/den for Sea Island down to 40 g/den for Asian cottons. The elasticity of cotton is imperfect because it does not return to its original length after stretching. When a fiber is stressed and allowed to recover, Young's modulus is found to be approximately one third of the initial value, indicating that some permanent deformation is now present in the fiber.

Elastic recovery of the fiber can be estimated by two methods. In one method, fibers are preconditioned mechanically by cyclic loading to specific levels. The recoveries after stretching and releasing to zero stress give elastic recovery, the ratio of recoverable elongation after a specified cycle to total elongation at the end of the cycle. In this method, elastic recovery is a function of percent elongation; elastic recovery drops curvilinearly from 0.9 at about 1% elongation to about 0.4 at the elongation at break (approximately 6%). In the second method, elastic recovery is resolved into three components: immediate elastic recovery, delayed elastic recovery (5 min after removal of load), and permanent set (or stretch). There is no mechanical preconditioning; a new sample is tested at each successive cycle to higher loadings. Percent elongation at each load is plotted against percent of total elongation to give a plot resembling a phase diagram.

The resilience of a fiber is the ratio of the energy absorbed to the energy recovered when the fiber is stretched and then released. To obtain this index, the areas under the stress–strain curves for the extension phase as well as the recovery phase are measured. The ratio of extension to recovery areas is the index of resilience. As would be expected cotton is not a very resilient fiber. Another parameter from the stress–strain curve is toughness or the energy to rupture. Toughness is determined from the area under the stress–strain curve measured up to the point of break. It may be closely approximated by the product of the breaking stress and the strain at break divided by 2. The units of toughness are therefore also gf/tex or mN/tex (CI). Values for cotton's toughness range from about 5 to 15 mN/tex. Compared to unswollen fibers, toughness is increased considerably by swelling treatments that end in drying without tension. However, swelling of fibers followed by drying under tension decreases toughness.

Rigidity of the fiber is another elastic parameter that is of great significance in describing a fiber's resistance to twisting. Thus, rigidity will obviously have applications to the spinning of textile fibers. Sometimes referred to as "torsional rigidity," it is defined as torque necessary to impart unit twist or unit angular deflection between the ends of a specimen of unit length [581]. In analogy to Young's modulus, the shear modulus or modulus of rigidity is defined as the ratio of shear stress to shear strain or as the ratio of torsional force per unit area to the angle of twist (displacement) produced by the torque. The finer varieties show less rigidity than coarser fiber: fine Egyptian cottons are in the range 1.0–3.0 mN/m^2; American cottons, 4.0–6.0 mN/m^2; and coarse Indian cottons, 7.0–11.0 mN/m^2. It may be convenient to introduce a specific torsional rigidity of unit linear density (tex) that is independent of fiber fineness. This may be defined as $R_t = \varepsilon n/\rho$, where R_t is the specific torsional rigidity (mN mm^2/tex^2), ε is a shape factor that equals approximately 0.7 for cotton, n is the shear modulus, and ρ is the density. Rigidity will vary with growth conditions and fiber maturity. Fiber rigidity increases with temperature and decreases with moisture content. Difficulties in fiber rigidity during spinning are thus eased by maintaining a reasonably warm and humid atmosphere.

The elastic properties discussed so far relate to stresses applied at relatively low rates. When forces are applied at rapid rates, then dynamic moduli are obtained. The energy relationships and the orders of magnitude of the data are much different [570]. Because of the experimental difficulties, only little work at rapid rates has been carried out with cotton fiber compared to that done with testing at low rates of application of stress. In contrast, cotton also responds to zero rate of loading, i.e., the application of a constant stress. Under this condition the fiber exhibits creep that is measured by determining fiber elongation at various intervals of time after the load has been applied. Creep is time-dependent and may be reversible upon removal of the load. However, even a low load applied to a fiber for a long period of time will cause the fiber to break.

9.7.4 Electrical Properties

Electrical properties of fibers were first considered important because of the effects associated with the build up of static charges that could hinder mechanical processing and with certain discomfort and hazards associated with electrical charges in clothing, carpets, upholstery, etc. The degree to which a material is susceptible to electrical charging is referred to as its electrical permittivity or dielectric constant. Another factor, closely related to dielectric constant is the electrical resistivity, which describes the degree to which electrical charges can be conducted through a material to which an electrical potential or voltage differential is applied [581].

In practice, the dielectric constant, ε_r, is determined from the measurements of electrical capacitance as the ratio C_p/C_o, where C_p is the capacitance with the material between the plates of the condenser and C_o is the capacitance of the empty space. The capacitance and thus dielectric constant of cotton is dependent on three parameters including electrical frequency, moisture content, and temperature. The dielectric constant of all materials decreases with increasing frequency. As the moisture content increases, dielectric increases. Thus, in the case of cotton at 0% RH, the dielectric constant ranges from 3.2 at 1 kHz to 3.0 at 100 kHz. At 65% RH, cotton's dielectric constant decreases from 18 at 1 kHz to 6.0 at 100 kHz [588].

The electric resistance of a material is defined as the ratio of the voltage applied across a material to the current (measured in amperes) that flows as a result. The unit of resistance is Ω (V/A). It is more convenient to use specific resistance r, defined as the resistance between opposite faces of a 1-m cube. However, as was the case for fibers, it is more convenient to base the measurement on the linear density rather than cross-sectional area. This leads to the mass specific resistance, $R_S = rd$, where d is mass density. R_S is usually expressed in Ω g/cm^2. Under standard conditions, R_S for raw cotton is approximately 0.5×10^6 Ω g/cm^2. As raw cotton is washed and otherwise purified, its resistance increases at least 50-fold [589]. Moisture content has even a greater role in resistance with $R_S = 10^{11}$ Ω g/cm^2 at 10% RH decreasing a million-fold to $R_S = 10^5$ Ω g/cm^2 at 90% RH [590].

9.7.5 Advanced Fiber Information System

A recent development that incorporates several fundamental measures into one system is the advanced fiber information system (AFIS) [591–593]. The AFIS measures several fiber properties that are key to predicting the ease of spinning and quality of the finished product. Included in the measurements are fiber neps (small tangles of fiber), dust, trash, fiber length including short fiber, and maturity. The measurements are unique in that, as will be explained below, the technology is such that it automatically counts and sizes individual fibers and particles (neps and dust). The principle of operation is that a fiber individualizer aeromechanically opens and separates the sample into single fibers that are injected into an airstream. Dust and trash particles are diverted to a filter while the airstream transports the fibers and neps past an electro-optical sensor that is calibrated to measure the specific size characteristics of the fibers and neps. With regard to neps, AFIS determines their average size and size distribution. The measurements of dust and trash include the distribution of the size of dust and trash particles along with their number per gram and average size. The AFIS length measurements include percent of short fiber content (<0.50 in./12.7mm) by number and by weight; average length by number and by weight; coefficient of variation of fiber length by number and by weight; upper quartile length (i.e., the length for which 75% of fibers are shorter) by weight; and the 5.0% length (i.e., the length for which 95% of the fibers are shorter) by number. Fiber maturity measurements include fiber fineness (linear density measured in millitex), and the immature fiber content (percentage of immature fibers by number).

9.8 COTTON FIBER CLASSIFICATION AND CHARACTERIZATION

Cotton bale weights vary from about 375 to 515 lb (170 to 233 kg) depending on the country in which they are produced (see Table 9.1). A pound of cotton contains 100 million or more individual fibers and each individual fiber varies in properties along its length from one end to the other. Cotton classification is a practical, cost-efficient way of measuring the general quality and physical attributes of bales of cotton fiber that affect the quality of the finished product and manufacturing efficiency and allows a market value to be established for the cotton to facilitate utilization.

Many varieties of the four domesticated species of cotton are produced commercially in widely varying locations and growing conditions throughout the world by thousands of farmers on millions of acres of land. The varying locations and conditions result in many different grades of cotton that vary widely in quality and fiber properties because of different varieties, soils, weather (rainfall, temperature, etc.), farming cultural practices (irrigation, fertilizers, crop protection products, etc.), insect damage, length of growing season, exposure of open cotton before harvest, methods of harvesting and ginning, storage, and a host of other variables [3,6,10,42,594–599]. Variations can exist not only between farms but also on the same farm and within a single bale. These considerations do not apply to synthetic fibers.

The quality of cotton is important to cotton farmers because it is related to the price and yield of cotton produced and provides farmers information to evaluate production, harvesting, and ginning practices and to market their cotton effectively. Quality is even more important to the users of cotton because it is highly related to efficiency in manufacturing and to the quality and utilization of yarns and fabrics produced, as well as to effecting the dyeing and finishing processes and outcome. For example, the color grade of raw cotton can affect dyeability if the cotton is not scoured and bleached adequately [600]. It is, therefore, of commercial importance that producers and users know the quality of cotton.

Because of the size and method of packaging, it is not practical for a buyer to inspect actual bales of cotton at the time of purchase. In general practice, market transactions are made on the basis of representative samples drawn from each bale. A tag or coupon showing the bale number or other identification normally accompanies the sample. This sample furnishes the specimen for classification.

There are two methods for estimating fiber quality, expert appraisal by a trained classer, and instrumentation. Many different classification systems are used worldwide in the various countries where cotton is grown. For example, the United States [601], China [3,594,602], Uzbekistan [603], and Egypt [594] all have different systems. All systems measure cotton quality using similar parameters, even though the actual classification scheme and classification designations may be different. The determinations, either manually or by instrument, may be different for individual parameters also. The quality parameters are the following:

1. Amount of foreign matter
2. Preparation (destructive effects of ginning)
3. Color and luster (discoloration)
4. Fiber length and regularity (feel and appearance)
5. Fiber fineness
6. Strength
7. Growth area, and how ginned (in some countries these are considered the most important parameters)

Saw-ginned cottons (Upland cottons) are classed on a different system than roller-ginned cottons (longer-staple fibers and the very short Asiatic "desi" cottons).

The U.S. HVI system for the classification of fiber quality and instrumentation for measuring other fiber properties are becoming more widely used all over the industrialized world, as well as in some developing countries for determining the proper laydown and mix for processing cotton in textile mills. Therefore, the U.S. system will be described in more detail than the other classing systems. Information also is given on how the various quality parameters can affect cotton fiber chemistry.

The present HVI system as developed in the United States has been used principally for marketing of the U.S. cotton to the U.S. textile industry. With the disappearance of the U.S. textile industry there has been an impetus toward the standardized testing of cotton on a global basis [604]. The concern is that HVI measurements of length, strength, fineness, color, and trash were probably sufficient for marketing Upland cotton to an industry concentrating on production of medium weights primarily with rotor spinning, but are not satisfactory for handling the global market that is still primarily ring spinning of fine counts. To adequately support the marketing of cotton produced from around the world (wide diversity of varieties, growth conditions, and harvesting and ginning protocols), it would seem that several other fiber properties would be necessary for satisfactory fiber characterization. These would probably include short fiber content, fineness and maturity, neps, and stickiness. The establishment and management of such a global system offer several challenges including

1. Developing a system that contains a minimum number of measurements that are truly independent and are of demonstrated value for predicting process and product quality for global textile manufacturing.
2. Maintaining all testing stations at comparable testing levels.
3. Establishing a worldwide database of fiber properties

The USDA Agricultural Marketing Service has established such a marketing system for the U.S. cotton and should be the model and leader in establishing such a global marketing system. One of the immediate challenges is that many cotton-testing systems around the world are unable to afford to maintain exact environmental conditions for testing (70°F and 65% RH). One of the proposals to handle this is to supply HVI systems with built-in sensors that monitor the moisture content of the sample under test and automatically correct the test results to what they would be when testing at standard conditions.

9.8.1 U.S. Classification System

In the United States, cotton classification is the process of describing the quality of cotton in terms of color grade, leaf grade, length, strength, and micronaire reading according to the official cotton standards [601]. The USDA Cotton Program in 2004 operated 240 HVI instruments for classifying the U.S. cotton crop. These instruments measure micronaire, length, length uniformity, strength, reflectance (Rd), yellowness (+b), and trash percent area. Grade has been shifted into color grade and leaf grade. The HVI color Rd and color +b values are combined into an official HVI color grade via the official USDA color chart. Work is in progress to replace the classer's visual grade with an HVI determined leaf grade. In recent years, USDA has classed about 98% of the cotton produced on the U.S. farms.

9.8.1.1 Sampling

A classification sample normally consists of two parts of about 85 g each taken from opposite sides of a bale. To obtain the sample, cuts 15-cm wide are made in the side of the bale before or after they are wrapped, deep enough to provide a representative specimen from the bale.

9.8.1.2 Grade

The official cotton standards of the United States for the grade of Upland cotton are also called the universal standards. Leading cotton associations in major cotton-consuming countries meet periodically to establish a continuing consensus of cotton classification. International conferences are held every 3 years in the United States to consider revisions and to ensure accurate reproduction of the standards. By this method, the U.S. cotton classification system maintains sensitivity and responsiveness to cotton consumer needs.

Grade represents an assessment of two factors, color grade and leaf grade, which are now assessed and reported separately, the former derived from the HVI values for color Rd and color +b. Color grade and leaf grade standards, prepared and maintained by Cotton Division, Agricultural Marketing Service (AMS), provide a basis for maintaining consistency in the assessment of these grades. Color grade and leaf grade designations are shown in Table 9.13 through Table 9.15.

Cotton Fibers

The Pima and the Upland grade standards differ. The Pima cotton is naturally of a deeper yellow color than the Upland cotton. The leaf contents of Pima standards are peculiar to this cotton and do not match that of Upland standards. Pima cotton is ginned on roller gins and has a stringy and lumpy appearance. Pima cotton grades range from the highest grade of 1 to the lowest grade of 10.

9.8.1.2.1 Color
The Upland cotton is naturally white. Continued exposure in the field to weathering and the action of microorganisms can cause the cotton to become darker. Under extreme conditions of weather damage, the color can become a very dark bluish gray. If growth is stopped prematurely by frost, drought, etc., the lint is characterized by increased yellowness. Cotton can also become discolored or spotted by the action of insects, microorganisms, and soil stains. Any departure from the bright color of normally opened cotton is interpreted to indicate deterioration in quality. In the classification of Upland cotton, these color differences are recognized, divided into categories, and described in terms of color grades (see Table 9.13).

9.8.1.2.2 Leaf and Other Trash
Cotton usually becomes contaminated by leaf and other trash in various amounts through exposure in the field and in harvesting. The amount of foreign matter remaining in the lint after ginning is largely dependent upon the trash content, the condition of the cotton at the time of harvest, and the number of cleaning and drying machinery used by the gin. Much of the foreign matter is removed by the cleaning and drying processes during ginning, but it is impractical to remove all of it.

Leaf includes dried and broken plant foliage of various kinds and may be divided into two general groups:

1. Large leaf
2. "Pin" or pepper leaf

Large leaf is usually less objectionable because large particles are more easily removed by the textile manufacturing process. Leaf and other foreign matter in cotton must be removed

TABLE 9.13
Color Grades of the U.S. Upland Cotton[a]

Color grades	White	Light spotted	Spotted	Tinged	Yellow stained
Good middling (GM)	11*	12	13	—	—
Strict middling (SM)	21*	22	23*	24	25
Middling (M)	31*	32	33*	34	35
Strict low middling (SLM)	41*	42	43*	44	—
Low middling (LM)	51*	52	53*	54	—
Strict good ordinary (SGO)	61*	62	63*	—	—
Good ordinary (GO)	71*	—	—	—	—
Below grade (BG)	81*	82	83	84	85

[a]The first number of the 2-digit code for color grade is for GM (1) through BG (8) and the second number is for white (1) through yellow-tinged (5); e.g., 11 is good middling white, 33 is middling spotted.
*Physical standards, all others are descriptive.
Source: From The Classification of Cotton, in *Agricultural Handbook Number 566*, Agricultural Marketing Service, U.S. Department of Agriculture, Washington, D.C., April 2001.

as waste in the textile manufacturing process either in mechanical or in wet processing. Small particles of foreign matter that are not removed in manufacturing detract from the quality and appearance of the manufactured yarn and fabric. Cottons with the smallest amount of foreign matter, other properties kept equal, have the highest value. In the classification of the U.S. cotton, these differences are described in terms of leaf grade (see Table 9.14). Bark, parts of seeds, shale (lining of the bur), motes (immature undeveloped seeds), grass, sand, oil, and dust are also sometimes found in ginned cotton. Research is in progress to replace the classer's visually measured leaf grade with an instrument (HVI) measured leaf grade.

9.8.1.2.3 Preparation

Preparation describes the relative appearance of the ginned lint in terms of smoothness or roughness. Various methods of harvesting, handling, and ginning cotton produce differences in roughness or smoothness of preparation that sometimes are very apparent. Laboratory tests do not show that these differences in preparation in the raw cotton result in important differences in spinning quality; however, generally, cotton with a smooth appearance introduces less waste than cotton with a rough appearance. Longer cottons normally will have a rougher appearance after ginning than shorter cottons, but that does not necessarily mean that yarns made from such cottons will be relatively poorer.

9.8.1.2.4 Neps

Neps are small, entangled clumps of fibers caused by mechanical processing. They are visible as dots or specks when a thin web of fibers is held to the light or against a dark background. Neps in lint are undesirable because they will appear as defects in yarns and fabrics. The removal of neps from the lint is difficult, costly, and sometimes impossible. The longer, finer cottons tend to have more neps than the shorter, coarser cottons. Lint having a high percentage of thin-walled, immature fibers is especially likely to be neppy, which can lead to whitish flicks or specks on the fabric after dyeing. Neps are difficult for the classer to detect or evaluate in classification.

9.8.1.2.5 Special Conditions

The USDA cotton classing regulations provide for special designations because of the presence of extraneous matter or other irregularities that cause the usefulness of the cotton

TABLE 9.14
Relationship of Trash Measurement to Classer's Leaf Grade for the U.S. Cotton[a]

Trash measurement (4-year average) (% area)	Classer's leaf grade
0.13	1
0.20	2
0.34	3
0.51	4
0.72	5
1.00	6
1.25	7
1.57	8

[a] The surface of the cotton sample is scanned by a video camera and the percentage of the surface area occupied by trash particles is calculated. The trash determination and classer's leaf grade are not the same but there is a correlation between the two as shown in this table.

Source: From The Classification of Cotton, in *Agricultural Handbook Number 566*, Agricultural Marketing Service, U.S. Department of Agriculture, Washington, D.C., April 2001.

to be below that normally expected. When such a designation is made, it is reported as a comment in addition to the normal grade information. Notations are normally made on classification memoranda and provide a basis for segregating bales with special conditions from normal bales. Conditions that require special comments include rough preparation, gin-cut or reginned cotton, repacked, false packed, mixed packed, or water-packed cotton, and excessive grass, bark, or extraneous matter.

9.8.1.3 Length

Fiber length is determined by instrument measurement of a tuft of fibers prepared automatically by the mechanical sampler. The length distribution of the fibers in the specimen is determined by scanning the tuft of fibers to produce a fibrogram. Statistical properties of the fibrogram are then used to determine two length parameters for use in assigning length values to the bale from which the specimen was selected. The upper half mean length (UHM) is the average length of the longest half by number of the fibers in the specimen. The mean length (ML) is the mean length of all the fibers. Values actually reported for classification are UHM and length uniformity index (UI), which is the percentage ratio of ML and UHM. These two values, UHM and UI (measured by closely calibrated instruments), are used for length classification and reported as staple length.

9.8.1.4 Strength

Fiber strength measurements are made on the same specimen used in making the length measurement. After the length measurement is completed, the specimen is repositioned to a point of constant mass as determined by the instrument, clamped in two sets of jaws, and broken by transverse motion of one of the sets of jaws. A sensitive transducer attached to the jaws provides a very accurate measurement of the force required to break the specimen. Since, theoretically, the mass of the specimen at the breaking point is held constant, the grams force required to break the specimen is directly related to the tenacity of the specimen. The tenacity measurement is expressed as grams force (gf) per unit mass (tex), where tex is the weight in grams of a specimen having 1000 m of length. Tenacity of the Upland cottons ranges from approximately 20 to 30 gf/tex.

9.8.1.5 Linear Density

The linear density of the fibers is an important factor in determining the processing performance, notably the spinning performance, and the quality of cotton yarns. Linear density is typically reported in millitex (mass in milligrams of 1000 m of fiber). For a complete definition of linear density and fiber fineness characteristics, measurement of both fineness and maturity are required. Although methods for measuring these two factors separately are studied, no practical method has been accepted for use in cotton classification. Fineness and maturity in combination are measured by resistance to airflow. A specimen of specific weight is compressed to a specific volume in a porous chamber. Air is forced through the specimen and the resistance to airflow is related to the specific surface area of the fibers and is a function of both the fiber linear density (fineness) and maturity. The property so measured, called micronaire, has been used for a number of years and is an integral part of the U.S. classification system, and provides a measure of, but is not equivalent to, fiber linear density. The micronaire reading is simple, low cost, and fast and is one of the most useful quality measurements in cotton classification. Typical micronaire readings for good-quality Upland Cottons range from 3.5 to 5.0 but vary from year to year due to growing conditions, the most desirable range between about 3.5 and 4.1. When instruments are developed that are capable of measuring fineness and maturity independently and rapidly, they will likely be adopted into the classification system quickly.

9.8.1.6 Color (Measurement of Reflectance and Yellowness)

The Upland cotton is naturally white in color when it opens under normal growing conditions in the field. A number of factors influence changes in its color after the cotton has opened. Whatever the source of the change, color reduction from the characteristic white observed for newly opened cotton is likely to be an indicator of deterioration in quality. Color differences in cotton can be a source of variations in dye shades in finished fabric. Color in raw cotton is measured by the cotton colorimeter. Two characteristics are measured—the reflectance (Rd) and the yellowness (+b)—under standard illumination and density. In the U.S. system, these two values are used to determine the color grade, which was previously determined visually.

9.8.1.7 Trash Content (Measurement)

Trash content, or foreign matter, includes all materials in the cotton sample except the lint. Types of trash possibly present in cotton have been listed previously. Instrument systems used for cotton classification include image analysis components consisting of high-resolution video cameras and computer hardware and software for interpreting video image scans of cotton samples. The dark contrast of the trash particles against the lighter background of the cotton lint allows these instruments to count the number of particles, measure their sizes, and determine what percentage of the specimen surface area is covered by the trash particles. This percentage area is related to the gravimetric weight of trash in the cotton sample. Percent area

TABLE 9.15
Percentage Distribution of Color Grade and Leaf Grade, the U.S. Cotton—2003 Crop

Color grade	Leaf grade					
	1&2	3	4	5	6	7
11 & 21	12.5	9.6	0.8	*	*	*
31	3.3	29.8	11.1	0.6	*	*
41	0.3	13.0	11.0	1.1	0.1	*
51	*	0.4	0.6	0.1	*	*
61	*	*	*	*	*	*
71	*	*	*	*	*	*
12 & 22	0.3	0.5	0.41	0.06	*	*
32	0.1	0.8	2.17	0.68	*	*
42	*	0.6	4.23	1.14	*	*
52	*	0.1	1.01	0.28	*	*
62	*	*	*	*	*	*
13 & 23	*	0.03	*	*	*	*
33	*	0.18	0.1	*	*	*
43	*	0.35	0.1	*	*	*
53	*	0.20	*	*	*	*
63	*	0.02	*	*	*	*

*Less than 0.01% of crop.
Source: From Cotton Quality Crop of 2003. U.S. Department of Agriculture, Agricultural Marketing Service, Cotton Division, Memphis, TN, Vol. 77, No. 7, pp. 10–11, May 2004.

measurements of trash content range from 0.1 for very clean cotton to 0.9 for cotton with very high trash content (see Table 9.14). These measurements are reported in addition to the leaf grade assigned from visual observations, work in progress to replace the classer's leaf grade with an instrument (HVI) leaf grade.

9.8.1.8 Reporting Cotton Classification Results

Cotton classification services are provided to the U.S. cotton growers by Cotton Division, AMS, USDA on a fee basis. The classification data belongs to the cotton growers and is reported to them by various mechanisms (including a card showing all test values or electronic data transfer and printout). The classification data is maintained in a national data bank by the AMS. As the cotton moves through marketing channels, classification data is made available to the owner of the cotton by electronic data transfer. To obtain the data, the proof of ownership and a small fee are required. As measurements become more precise and accurate and industry develops confidence in the test results, it is anticipated that all cotton users will rely on test information from the national data bank for marketing and utilization of the U.S. cotton.

9.8.2 CLASSIFICATION SYSTEMS OF SOME COUNTRIES OTHER THAN THE UNITED STATES

There is a worldwide movement toward the increased use of instrument measurements such as the HVI in characterizing cotton quality and in cotton classification. This is encouraged by international bodies such as the International Cotton Advisory Committee (ICAC) and the International Textile Manufacturers Federation (ITMF).

9.8.2.1 China

In China [3,10,594,602], cotton is grown on small plots, handpicked, and harvested in sacks. The seed cotton is weighed, sampled, and graded by the government classer (first classing). Moisture, staple length, and lint turnout are determined, and similar lots—by one of the seven grades—are combined and ginned according to class. The saw-ginned cotton is classed again with a system very similar to the U.S. manual system previously used. The Chinese use a three-digit numbering system to classify lint. The first digit is for the grade, which is determined by color, maturity, and ginning quality of the cotton. The second and the third digits are the staple length in millimeter. For example, Type 327 would be grade 3 (under the U.S. system this would be middling color, about leaf grade 4 (see Table 9.13 and Table 9.15), 27 mm (equivalent to 1 1/32 in.). In addition, moisture content, trash content, and bale weight are also determined in classification. In 2003, China unveiled a cotton classification reform plan aimed at transitioning from the current manual classification system to an instrument-based classification system within 5 years. The reform plan includes transitioning to UD bales, two-sided bale sampling, and HVI inspection of all Chinese cotton (bale form). The China Fiber Inspection Bureau, responsible for cotton classification, has acquired 75 HVI systems in 2004 and 2005 as part of a process to shift from a hand-class system, based on grades and staple length to the U.S. style HVI system. A color chart for grading Chinese cotton and a method to determine the gravimetric trash content are under development.

9.8.2.2 Central Asia Republics

In the former Soviet Union [603], there were two classification systems. One was applied when supplying cotton to the domestic market and other socialist countries, the other for cotton sold to hard currency markets. The domestic system was GOST 3279-76. It had no scientific basis and could not estimate satisfactorily fiber characteristics for yarn production. The export version had no officially recognized physical standards. So a new system somewhat approaching the American system that could be used in international markets was developed.

In 1993, a new classification system for cotton fiber quality, RST Uz 604-93, was introduced in Uzbekistan [603]. Since 1995, this standard has also been used in Kazakhstan and Tadjikistan [603]. In Azerbaijan, attempts are made to classify its cotton in accordance with international standards [605].

The Uzbekistan classification of cotton fiber is based on three quality levels: types, sorts, and grades, which are determined with instruments (including HVI type) and classer's methods, they have some 35 HVI systems in place at gins and at Tashkent, all operated by GOST. Their cotton is still traded, however, on the traditional hand-class standards basis established by GOST. Types of cotton are based on length, linear density, and strength (g/tex). The basis of division into types is fiber length in millimeter. There are nine types: 1a to 3 are long staple cottons; 4 to 7 are medium stable. Cotton sort is defined on color and maturity. There are five sorts bearing the names of ordinal numbers in Uzbeki: Birinchi, Ikinchi, Uchinchi, Turchinchi, and Beshinchi. Birinchi sort represents medium staple cotton with white color, no spots, and maturity greater than or equal to 1.8 (CIS method; about 3.6 micronaire, maturity index 0.85). Ikinchi sort has pale yellow spots (like very light spotted) and lower maturity (up to 1.6). In Uchinchi sort, the yellow spots (similar to spotted) are more intense and maturity lower to 1.4. Turchinchi is creamy with brown stains (like tinged) and the maturity lower to 1.2. Beshinchi is mat-grey with brown stains and maturity is very low. Cotton grade is based on preparation and trash content. Within Birinchi and Ikinchi sorts, there are five grades, within Uchinchi and Turchinchi four grades, and within Beshinchi three grades. Altogether, there are 21 grades, which are called Oliy, Yakshi, Urta, Oddiy, and Iflos in Uzbeki (i.e., high, good, middle, ordinary, and contaminated, respectively). The new cotton fiber classification in Uzbekistan is comparable to USDA classification.

Uzbekistan classification and differentiation of cotton is effected immediately after the harvest. At present, cotton quality evaluation is carried out by cotton mill laboratories. This is felt to be undesirable (because of the effect on competitiveness and prices). Cotton is classified as seed cotton before ginning and as fiber after ginning (as described above). The picked raw cotton is taken to collecting points where the classers and laboratories carry out evaluation of raw cotton fiber quality. They sort fiber according to uniform fiber quality features: origin (according to farms), type and sort, trash content, and moisture content. To regulate receiving and classing of raw cotton in points of preliminary preparation, there is a special government standard for raw cotton (RST Uz 615-94 standard: Raw Cotton, Technical Conditions, introduced in 1994). Grade 1 is manually picked cotton; grade 2 is mechanically harvested cotton and also manually picked but contains more trash than grade 1; grade 3 is cotton picked from the ground and contaminated cotton of manual and mechanical picking, which contains more trash than grade 2. The seed cotton classing is meant to stimulate producers to adopt practices to lower the contamination of raw cotton. Also ginning of large quantities of picked cotton of uniform quality produces uniform fiber, which in some cases is sold without fiber quality evaluation.

9.8.2.3 Francophone Africa

For most of the former French colonies that are now independent countries in Africa, as well as in many other African countries, cotton cultivation management, ginning, and fiber marketing are operated by one company [606]. Cotton classing is also operated by the company's classing office, and is based on types and grades. Some HVI lines are in operation but are not used for classing *per se*. Samples of each bale sorted by gin plant are conditioned over several hours.

Classing is performed by trained classers visually and micronaire measurements are made by instrument. Length is estimated with the pulling method and expressed in 1/32 in. Grade is visually estimated according to the color, trash, and preparation. Color and trash are not considered separately. There is a specific grade standard system for cottons of French-speaking countries in Africa. The system is based on six grades 0, 1, 2, 3, 4, 5, from best to worst. These standards are the base for determination of sales types specific to each country. Bales' marking is done at the gin according to a codification for length and another for grade.

9.8.2.4 Sudan

In Sudan [607], cotton is classified twice, as seed cotton before the gin and as lint at the port of shipment. The handpicked cotton is piled up on covers where trash and other contaminants are removed. Grades are assigned to the piles and the graded cotton is transported to the gin. The field classification system is felt to be key to implementing quality control measures such as timely and early picking, which can reduce stickiness. The ginned cotton is baled and transported to Port Sudan where the lint is reclassified and samples are taken for fiber testing.

9.8.2.5 South Africa

In South Africa [608] prior to ginning, producers submit samples of seed cotton both handpicked and machine picked for classification processes by using RSA Seed Cotton Standards for grade to obtain information on the expected quality of their crop. When buying seed cotton, it also determines the price paid to the farmer whether buying or contract ginning; accurate preselection of seed cotton and homogenous blending of lots prior to ginning is viewed by ginners as important to ensure that the lots of cotton lint are uniform in grade and quality. Ginners are currently extensively using HVI evaluation, and the seed cotton grading function is viewed by ginners as a control measure.

The Quality Control Division of Cotton, South Africa, is mainly responsible for the grading and classification of the South African cotton crop but this is not a compulsory service. Ginners who participate submit samples from each bale to receive a grading certificate that is used for the local marketing of their crop. Ginners who are responsible for their own grading and classification are required to use the same grade and quality specifications.

The Quality Control Division uses HVI for fiber analysis. The system measures color, trash, fineness, length, strength, elongation, and length uniformity of cotton. The determination of grade is done visually in accordance with the South African lint standards, namely Deal, Dirk, Doly, Duns, and Lfy, which are comparable to U.S. lint standards good middling, strict middling, middling, strict low middling, and low middling, respectively. (As the U.S. system now separates grade into color grade and leaf grade, these grades would be equivalent to the U.S. color grade with about a leaf grade 3 (See Table 9.13 and Table 9.15).) The South African lint standards for grade with regard to color and trash are similar to the (white) Universal Standards for Grade of the American Upland Cotton, except that provision is made for soil stained and yellow spot in the lower South African standards (Duns and Lfy). As a certain percentage of the South African crop is handpicked, South African cotton tends to be a brighter white than the U.S. standards.

9.8.2.6 Egypt

The Egyptian cotton [594] is classed by the appraisal of experts based on their own standards. The standards are meant to illustrate color, lightness, immature and dead fiber content, foreign matter content, and preparation of the cotton. There is no separate classing of length: length is deduced from the variety. An exact comparison of the Egyptian and the American systems of classing is impossible because of their differing principles. However, when only foreign matter content, preparation, and to some extent color are taken into account, the grades can be related.

9.8.2.7 India

The Indian cotton [3,594] is classed in a way resembling the Egyptian system and not based on the American system. The system takes into account species, variety, grade, and length. Place of cultivation and ginning method are also mentioned. East India Cotton Association prepares and maintains grades and staple standards. There are five principal grades: extra superfine, superfine, fine (basis), fully good, and good with staple lengths varying from 17 to 40 mm 22/32 to 1 18/32 in.) and are roughly comparable to the U.S. standards of strict low middling to good ordinary (see Table 9.13 and Table 9.15).

9.8.2.8 Pakistan

The situation in Pakistan [3,609] is similar to that in India; in Pakistan, trading is based upon hand-classed variety and grade. Little attention has been paid to the cleanliness of the product. The crop is not graded, and cleaning at the gin (there are few seed cotton and lint cleaners) has been largely neglected [590]. However, this is changing. HVI systems have been introduced for conducting tests and preparation of standards for cotton grades [609].

9.8.2.9 Australia

The cotton classing in Australia [610] encompasses elements of the old and revised USDA system. Qualified cotton classers classify the cotton by examining three characteristics: grade, staple, and micronaire, and HVI is widely used as a guide by classers. Micronaire is determined with an airflow instrument. There are 44 different grades of cotton based on color of the fiber, trash content, preparation, and staple length (which has 3 different standards). Length is reported as 1/32 of an in. with color and leaf grades determined by means of USDA standard grade boxes, and color and leaf grades are reported separately. Other cotton fiber properties (strength, length uniformity, elongation, stickiness, nep count, and moisture content) are tested to determine the fiber manufacturing potential, many of which are measured on HVI systems.

9.8.3 Impact and Future of HVI Systems

Hunter [611] has presented an excellent review of literature relevant to theoretical or empirical relationships and fiber quality indices (FQIs) that have been developed to predict yarn quality, especially strength and evenness, from the measurement of raw fiber properties. Theoretical approaches are based upon fundamental principles of physics and mechanics but due to the complexity of the mechanisms of yarn formation, doubt is generally raised as to the practicality of such approaches to reliable predictions of textile quality. Empirical relationships are primarily based on linear- or multiple-regression analysis approaches, which work well for the particular cottons in the data set but have limited usefulness when attempting to expand beyond the particular range of cottons covered. More recently, many approaches including artificial neural networks and fuzzy logic have been used for modeling of textile quality from fiber properties . However, Hunter concludes that, as of yet, there is no viable solution. Hunter's review shares an interesting breakdown of the relative contributions of HVI fiber properties to predicting the strengths of rotor versus ring yarns. From Table 9.16 we see that strength, elongation, and color play a more significant role in rotor yarn strength, whereas length and length uniformity play a more significant role in ring yarn strength.

TABLE 9.16
Relative Contributions of HVI Fiber Properties in Predicting the Strengths of Rotor versus Ring Yarns

HVI property	% of property contributing to rotor yarn strength	% of property contributing to ring yarn strength
Strength	24	20
Length uniformity	17	20
Length	12	22
Micronaire	14	15
Elongation	8	5
Color or Reflectance	6	3
Unexplained	13	12

Suh and Sasser [612] have reviewed the impact of HVI systems on both the domestic and world cotton textile industries. HVI has totally revolutionized the way in which cotton is marketed and processed in the modern textile industry. By purchasing cotton bales based on HVI measurements the textile industry is now able to systematically store, retrieve, and form multibale input laydowns designed for uniformity of process and product quality. In addition to product uniformity, HVI data allows for selection of appropriate raw materials at reduced costs while maximizing product quality and profit. Despite the obvious assets of HVI, at this time, it still does not produce information of certain fiber properties that are critical to process optimization and control. Included among these are true short fiber content, fineness and maturity, seed coat fragments, and sugar. It is hoped that research now carried out, especially in the United States, will lead to further enhancement of HVI systems.

In December 2003, the ICAC formed an Expert Panel on Commercial Standardization of Instrument Testing of Cotton (CSITC). The purpose of the CSITC is to promote instrument testing of cotton on a global basis. The driving force behind this is to enhance the competitiveness of cotton with synthetic fibers in the global market place. At a meeting at Mumbai, India, in 2004, the panel set for itself several goals to encourage worldwide testing of cotton with standardized instruments. These included:

1. Definition of specifications for cotton trading
2. Definition of international test rules
3. Implementation of test rules
4. Certification of testing laboratories
5. Definition and provision of calibration standards
6. Specification of commercial control limits for trading
7. Establishment of arbitration procedures

The two primary agencies proposed to help implement this program are the Bremen Fiber Institute and the USDA's AMS Cotton Program [613].

9.9 PRODUCTION, CONSUMPTION, MARKETS, AND APPLICATIONS

9.9.1 Production, Consumption, and Markets

Cotton is the most important natural vegetable textile fiber used in the world for spinning to produce apparel, home furnishings, and industrial products [614,615]. It continues to be one of the most dominant textile fibers in much of the world and accounted for over 38% of the worldwide textile fiber consumption in 2005 [1] (see Table 9.17). China, India, Pakistan, Turkey, the United States, and Brazil are the major cotton-consuming countries [40,614] (see Table 9.18). Consumption is measured by the amount of raw cotton fiber, the textile mills purchase and use to manufacture textile materials. China, the United States, India, Pakistan, Uzbekistan, Turkey, and Brazil are the major cotton producing countries, accounting for over 81% of world cotton production in 2004 [40,614] (Table 9.19). Organic cotton, which is a very small niche market, was produced in at least 16 countries in 2003, with Turkey as the leading producer (Table 9.20).

In addition to the various markets for cotton lint, there are also markets for cottonseed and its products [616]. Cottonseed represents about 15–20% of the total value of cotton. Vegetable oil for human consumption, whole cottonseed, meal, hulls for animal feed, and linters for batting and chemical cellulose are the major end uses for cottonseed [616].

Since 1988, the trend has returned to natural fibers in the United States and Europe. National surveys of U.S. consumer attitudes about cotton versus synthetic or manufactured fibers show that consumers think cotton has substantial advantages over polyester regarding functional and quality attributes. The difference is greatest regarding comfort, but substantially more

TABLE 9.17
World Production of Textile Fibers (Million Lbs.)

Fiber	2000	2001	2002	2003	2004
Cotton[a]	42,628.3	47,404.8	42,376.8	45,653.3	57,528.5
Man-made fibers	62,685.3	62,446.4	66,526.7	69,893.1	75,409.9
Synthetics	57,801.7	57,855.3	61,842.8	64,886.0	69,916.7
multifilament and monofilament yarns	32,564.1	32,997.3	35,257.7	37,219.1	40,259.1
staple, tow, & fiberfill[b]	25,237.6	24,858.0	26,585.0	27,666.9	29,657.5
acrylic	5,806.9	5,647.5	5,981.9	5,937.0	6,047.8
multifilament & monofilament yarns	11.0	11.0	11.0	11.0	5.9
staple, tow, and fiberfill	5,795.9	5,636.5	5,970.9	5,926.0	6,041.9
nylon	9,075.6	8,341.5	8,688.3	8,728.0	8,996.5
multifilament & monofilament yarns	7,944.9	7,404.1	7,694.7	7,747.8	8,034.0
staple, tow, & fiberfill	1,130.7	937.4	993.6	980.2	962.5
polyester	42,228.5	43,127.5	46,402.0	49,289.7	53,804.6
multifilament & monofilament yarns	24,148.1	25,077.3	27,024.2	28,835.9	31,489.0
staple, tow, and fiberfill	18,080.4	18,050.2	19,377.8	20,453.8	22,315.6
other (except olefin)	690.7	738.8	770.5	931.3	1,067.7
multifilament & monofilament yarns	460.1	504.9	527.8	624.4	730.2
staple, tow, and fiberfill	230.6	233.9	242.7	306.9	337.5
Cellulosics	4,883.6	4,591.1	4,683.9	5,007.1	5,493.2
multifilament and monofilament yarns	1,109.1	1,089.3	1,022.7	1,080.7	1,063.3
staple, tow, & fiberfill[b]	3,774.5	3,501.8	3,661.2	3,962.4	4,429.9
Wool (scoured or cleaned basis)	3,042.0	3,000.0	2,884.0	2,659.0	2,687.0
Silk	N/A	N/A	N/A	N/A	N/A
Total	108,355.6	112,851.2	111,787.5	118,205.4	135,625.4
% Cotton	39.3%	42.0%	37.9%	38.6%	42.4%

[a] 2003 Forecast.
[b] Excludes acetate cigarette filter tow.
Source: Fiber Organon, 76(7), July 2005 (Fiber Economics Bureau, Inc., Arlington, VA, USA)
N/A-no data available

TABLE 9.18
World Cotton Consumption (1000 U.S. 480 lb bales)

Country	1990/91	1995/96	2000/01	2004/05	2005/06
China	19,987	19,388	23,485	38,476	45,471
India	8,956	11,969	13,535	14,791	16,490
Pakistan	5,644	7,195	8,095	10,743	11,743
Turkey	2,478	4,360	5,164	7,095	6,896
United States	8,652	10,640	8,856	6,689	5,996
Brazil	3,319	3,757	4,197	4,197	3,997
Indonesia	1,492	2,135	2,448	2,249	2,299
Thailand	1,505	1,424	1,649	2,149	2,124
Bangladesh	446	551	999	1,874	2,074
Mexico	767	1,099	2,099	2,099	1,999
Russia	5,466	1,149	1,599	1,424	1,499
Korea, South	2,000	1,666	1,449	1,324	1,199
Taiwan	1,588	1,397	1,199	1,199	1,199
Egypt	1,456	1,009	750	949	999
Uzbekistan	839	872	1,024	874	799
Italy	1,469	1,538	1,329	924	750
Vietnam	188	170	430	675	750
Japan	3,025	1,528	1,189	814	725
Syria	250	375	550	700	725
Argentina	615	459	350	620	620
Other	15,328	13,054	11,704	8,963	8,504
Total	**85,470**	**85,736**	**92,100**	**108,828**	**116,855**

Source: USDA, Foreign Agricultural Service, April 2006. Crop year runs from August 1 to July 31.

consumers also believe that cotton offers better overall quality. This has created an increased demand for fabrics and garments of 100% cotton and blends with high cotton content. Previously, most blended fabric contained about 65% polyester; now the trend is toward reverse blends (60% cotton and 40% polyester or 85% cotton and 15% polyester) with higher cotton content. However, in the less developed countries polyester usage is increasing faster than cotton.

The textile industry in most European countries and the United States has become smaller or almost totally disappeared. China, India, Pakistan, and Turkey are the major textile producers. In both the United States and Europe, textile imports have greatly increased. The textile and apparel industries appear to be moving from the highly industrial countries to the developing countries. The high imports do show that demand for cotton textiles in the United States and Europe continues to be strong at the consumer level.

9.9.2 Applications

Cotton fabrics combine durability with attractive wearing qualities and comfort [617]. Cotton is inherently strong because the convolutions create friction within the fabric that prevents fibers from slipping. The wet cotton fabric is stronger than the dry cotton fabric. Cotton can withstand repeated washings and is ideal for household goods and garments that must be laundered often. A cotton garment on laundering may shrink because of the tensions introduced by spinning and weaving, but the cotton fiber itself is dimensionally stable and does not contribute significantly to the shrinkage. Cotton fabric will wrinkle and crease. However, cotton fabric can be treated to impart wrinkle resistance and dimensional stability

TABLE 9.19
World Cotton Production (1000 U.S. 480 lb bales)

Country	1990/91	1995/96	2000/01	2004/05	2005/06
China	20,687	21,886	20,287	28,982	26,183
United States	15,495	17,889	17,177	23,236	23,885
India	9,129	13,242	10,924	18,988	18,288
Pakistan	7,517	8,195	8,195	11,293	9,744
Uzbekistan	7,312	5,736	4,397	5,197	5,596
Brazil	3,291	1,883	4,309	5,896	4,697
Turkey	3,005	3,909	3,598	4,147	3,548
Australia	1,988	1,969	3,698	2,998	2,598
Greece	965	2,066	2,034	1,799	1,974
Syria	666	1,009	1,674	1,599	1,549
Burkina	354	294	525	1,179	1,324
Mali	527	775	480	1,099	1,149
Turkmenistan	2,006	1,149	824	919	974
Egypt	1,377	1,087	919	1,331	949
Kazakhstan	468	340	400	680	675
Mexico	856	973	394	625	635
Tajikistan	1,175	550	485	799	625
Tanzania	220	377	188	525	575
Argentina	1,354	2,089	758	675	550
Iran	547	799	735	615	550
Other	8,147	7,448	6,795	7,839	7,442
Total	**87,086**	**93,663**	**88,794**	**120,420**	**113,510**

Source: USDA, Foreign Agricultural Service, April 2006. Crop year runs from August 1 to July 31.

as was discussed earlier in the chapter. Cotton can be dyed easily with a wide range of colors [618]. Cotton cellulose is not affected unduly by moderate heat, so that cotton fabrics can be ironed with a hot iron without damage.

Cotton fabrics are cool in hot weather. Although cotton is used as a fabric for hot-weather wear, it is also able to provide warmth. The warmth of the garment depends very largely on the pockets of air that are entrapped between the fibers in the fabric and the resistance of the fabric to the wind. Cotton fibers make good insulators when made into padded or quilted garments. Much of the comfort of a textile material depends upon its ability to absorb and desorb moisture. Moisture from perspiration collects in and passes through clothing as worn, and the properties of fabrics influence both the collection and passage of this moisture. Dynamic surface wetness of fabrics correlate with skin contact comfort in wear for a variety of fabric types, suggesting that mobility of thin films of condensed moisture is an important element of wearing comfort. Dynamic surface wetness of fabrics can be used to show why the cotton clothing is considered more comfortable than the clothing made of synthetic fibers [619]. A garment that does not absorb moisture, like garments made from synthetic fibers, will tend to feel clammy as perspiration condenses on it from the skin. Cotton fibers are able to absorb appreciable amounts of moisture, and having done so will get rid of it readily to the air. Cotton garments are, therefore, comfortable and cool, passing on the perspiration from the body into the surrounding air. No matter how tightly woven a cotton fabric may be, it will permit the body to breathe in this way. In addition, the absorbency of cotton makes it an excellent material for household fabrics such as sheets and towels. Static electricity is not a problem with cotton so the clothes do not cling to the wearer or to each other.

TABLE 9.20
World Organic Cotton Production [metric tons]

Country	1990/91	1991/92	1992/93	1993/94	1994/95	1995/96	1996/97	1997/98	1998/99	1999/00	2000/01	2001/02	2002/03	2003/04	2004/05
Argentina															67
Australia						75								25	45
Benin								300				38	46	1,601	1,870
Brazil				500	500	400	300						596	122	240
Burkina Faso								5							6,320
China (PRC)				1	5	1		1	5	20	30	106	122	2,231	436
Egypt	14	45	50	153	600	650	625	500	360	200	200	200	855	380	2
Greece					300	150	125	100	75	50	50				65
India				250	400	925	850	1,000	825	1,150	1,000	696	390	35	296
Israel										5		<425			
Kenya								5	5				19		
Kyrgyzstan						100	75	50							600
Mali					20	20	20	20							70
Mozambique													256	400	813
Nicaragua					75	50	50	50	650	500	550	300	9	60	27
Pakistan				100				10	50	146	200		300	404	1,213
Paraguay				675	900	900	900	650	230	190	180	400	6	6	10,460
Peru			200			1	1						380	600	900
Senegal						30	30	100	835	7,840	7,697	5,504	12,865	11,625	1,968
Tanzania			789	200	463	725	933	1,000	250	246	248	250	500	740	2
Turkey		820	2,155		25	75	75	450	1,878	2,955	1,860	2,227	1,571	1,041	
Uganda	330			4,274	5,365	7,425	3,396	2,852		5		2	3		
USA								1	5						
Zambia															
Zimbabwe															
Total	344	865	3,894	6,153	8,978	11,527	7,382	7,094	5,188	13,317	12,035	10,148	19,270	17,645	25,394

Source: Organic Exchange, 2005–06 and ICAC, 2005–06 for most of the data; and Baird Garrott, Paul Reinhart AG, for Turkey, Israel, Mali and Burkina Faso in 2004/05. Crop year August 1 to July 31.

Cotton can be used in making rainwear fabrics. It can be woven tightly to keep out the driving wind and rain, yet the fabric will allow perspiration to escape. Special rainwear materials are woven in such a way that water swells the cotton fibers and closes up the interstices in the cloth.

The versatility of cotton has made it into one of the most valuable and most widely used of all textile fibers. Wherever a fabric is needed that is strong, hardwearing, and versatile, cotton can be used. There are literally thousands of actual uses (about 100 major uses) for cotton in textile items, ranging from baby diapers to the most fashionable dresses, coats, and jackets [617]. These uses can be classified into three main categories: apparel, home furnishings, and industrial.

In the apparel market, men's and boys' trousers and shorts use the largest amount of cotton, followed respectively by shirts worn by men and boys and men's and boys' underwear. Cotton is the major fiber for jeans, men's and boys' underwear and shirts. Denim fabrics utilize more cotton than any other single apparel item for trousers worn by men and boys. [617].

In the home furnishings, towels and washcloths account for the largest amount of cotton used, and sheets and pillowcases account for the second largest amount. The dominant fiber in towels and washcloths is cotton, supplying better than 90% of the market. It also holds almost half of the market for sheets and pillowcases [617].

Industrial products containing cotton are as diverse as tarpaulins, bookbindings, and zipper tapes, and cotton products can be used for clean up of agrochemical spills [620] and oil spills [621]. Some of the major industrial markets for cotton are medical supplies and industrial thread.

As a raw material for military items, cotton equips and helps support the military of every country. Many textile items are required by the armed forces in war and peace. It has been recognized by leading authorities on textiles and the national defense that too great a dependence upon fibers drawn from petrochemical feedstocks could present undesirable hazards to the military services from a supply standpoint [622,623]. A strong, viable cotton industry is essential to the defense capabilities of every country.

9.9.3 Future Trends

Future demand in the world should remain high for cotton. The near term estimates of production and consumption are good [614,624] (Table 9.18 and Table 9.19). There is increased production in India, Pakistan, Turkey, and Brazil, and the world consumption in 2005 is expected to exceed production [614,625].

Much research on cotton is carried out in the United States and the rest of the world. The quality improvements in cotton continue to be significant. In the United States, for example, cotton breeders, farmers, and processors have responded to the demands of new technology by developing, producing, and delivering the longest, whitest, finest, and strongest cotton fiber ever grown. These trends should continue with new cottons, developed through biotechnology for insect resistance, herbicide tolerance, and improved fiber properties (e.g., stronger, finer, whiter fibers) as well as hybrid cottons and better new cotton varieties by conventional breeding (see Section 9.1). These new cottons are expected to have better fiber properties and require less crop protection products. Also, shorter-season, high-yielding plants are expected in the future. With these cottons, better yarns and fabric will be able to be produced. Cotton finishing will continue to be improved and open up new markets for cotton (e.g., rugs, carpets, nonwovens [626]). With respect to fiber quality, improved color fastness and evenness or levelness of dyeing, durability, and easy care characteristics look promising. Higher strength cotton can open new opportunities in industrial textile applications. New technology virtually has the potential to eliminate fiber contaminants before they reach customers. All of this research should help the cotton industry meet the projected increase in worldwide demand for cotton.

In summary, cotton's future is positive. Cotton use should benefit from consumer demand stemming from favorable economic growth prospects and because of research. On the production side, global output should continue to provide an adequate supply for mill demand. Finally, cotton, one of the most important textile fibers and one of the world's important oilseed crops, should continue to be recognized as a significant commodity in world trade and the consumption of this important fiber, food, and feed crop will continue to grow but at a slower rate than synthetic fibers.

9.10 ENVIRONMENTAL, WORKPLACE, AND CONSUMER CONSIDERATIONS

9.10.1 Environment

9.10.1.1 Environmental Stewardship

Cotton, as grown in the United States and most other countries, is an environmentally responsibly produced and managed product [47,627]. Cotton production faces many challenges. Cotton can be affected by insects [628], weeds [629], diseases [630], nematodes [631], and mycotoxins [632]. The newest and latest tested technology is used to produce the crop [9,47]. Modern cotton production minimizes soil erosion by using conservation tillage and other practices, nutrient loss with nutrient management programs, and ground water contamination. At least 90% of the U.S. cotton production uses crop protection products with a wide array of integrated pest management (IPM) programs as well as computer programs and other technologies (e.g., global positioning systems) to apply only the crop protection products that are needed and only where they are needed. A biocontrol (competitive exclusion) method for managing aflatoxin (a mycotoxin that can be a serious food safety hazard on cottonseed) on cotton has been developed for use in Arizona, Texas, and California [633].

9.10.1.2 Emissions to the Environment

The cotton fiber does not contribute anything that causes hazardous air emissions from textile operations processing cotton [9,47]. Cotton production and ginning can be sources of particulate matter (PM) emissions that are regulated by the U.S. EPA. Neither is a major source under EPA regulations. Cotton production can also be a minor source of volatile organic chemicals that are precursors of ozone and of oxides of nitrogen.

For environmental effluent and solid fiber waste concerns and processing of all kinds, it is beneficial for textile mills to know the concentrations of noncellulosic constituents of cotton fiber and what is leachable and removable from the fiber. Then mills know how to handle their water and fiber waste and how to dye and finish the cotton [47].

9.10.2 Workplace

Workers handle and process cotton through many work operations from harvesting and ginning through yarn and fabric manufacturing.

9.10.2.1 Inhalation or Respiratory Disease

Inhalation of cotton-related dust, generated during the textile manufacturing operations where cotton fiber is converted into yarn and fabric, has been shown to cause an occupational lung disease, byssinosis, in a small number of textile workers [634,635]. The U.S. Occupational Safety and Health Association (OSHA) regulations for cotton dust apply to textile processing and weaving but do not apply to handling or processing of woven or knitted materials, harvesting, ginning, warehousing, classing or merchandizing, or knitting operations. The OSHA permissible exposure limits (PELs) for cotton dust measured as an 8-h time-weighted-average with the vertical elutriator cotton dust sampler are as follows:

1. Yarn manufacturing, 200 $\mu g/m^3$
2. Textile mill waste house, 500 $\mu g/m^3$
3. Slashing and weaving, 750 $\mu g/m^3$
4. Waste processing (waste recycling and garneting), 1000 $\mu g/m^3$ [636]

It usually takes 15 to 20 years of exposure to higher levels of dust (above 0.5 to 1.0 mg/m^3) for workers to become reactors. Cotton dust, an airborne PM released into the atmosphere as cotton, is handled or processed in textile processing and is a heterogeneous, complex mixture of botanical trash, soil, and microbiological material (i.e., bacteria and fungi), which varies in composition and biological activity [637]. The etiological agent and pathogenesis of byssinosis are not known [638–640]. However, control studies in experimental cardrooms suggest that, in today's world, appropriate engineering controls in cotton textile processing areas, along with work practices, medical surveillance, and personal protective equipment for the most part, can eliminate incidence of workers' reaction to cotton dust [641]. Cotton plant trash associated with the fiber and the endotoxin from Gram-negative bacteria on the fiber, plant trash, and soil are thought to be the causative or to contain the causative associated with workers' reaction to dust [634,635]. The cotton fiber, which is mainly cellulose, is not the causative, because cellulose is an inert dust that does not cause respiratory disease. In fact, cellulose powder has been used as an inert control dust in human exposure studies.

A mild water washing of cotton by batch kier washing systems [642,643] and continuous batt systems [107] reduces the residual level of endotoxin in both lint and airborne dust to below levels associated with a zero percentage change in acute reduction in pulmonary function as measured by forced expiratory volume in 1 sec (FEV_1). Levels of endotoxin from Gram-negative bacteria generated during the processing of cotton are associated with the occupational respiratory disease that affects some textile workers [634,635,643]. Washed cotton is determined by levels of potassium and water-soluble reducing substances (WSRS) in the washed lint [107,642,643]. The OSHA has accepted several mild washing systems as qualifying as washed cotton exemptions under the cotton dust standard [643,644]:

1. Mild washing by the continuous batt system or a rayon rinse system is with water containing a wetting agent, at not less than 60°C, with water-to-fiber ratio not less than 40:1, with bacterial levels in the wash water controlled to limit bacterial contamination of the cotton.
2. The batch kier washing system is with water containing a wetting agent, with a minimum of one wash cycle followed by two rinse cycles for each batch, using fresh water in each cycle, and with bacterial levels in the wash water controlled to limit bacterial contamination of the cotton.
    a. For low temperature, at not less than 60°C, with water-to-fiber ratio not less than 40:1; or
    b. For high temperature, at not less than 93°C, with a water-to-fiber ratio not less than 15:1.

9.10.2.2 Skin Irritation or Dermatitis

Handling or processing conventional U.S. cotton does not cause skin irritation or dermatitis. Cellulose is essentially an inert substance and nothing on the fiber surface is known that could cause dermatitis problems. However, it is remotely possible that some very atypical cottons that have been treated with substances that are not approved for use or that are off-grade and perhaps are highly microbiologically damaged might cause skin irritation. These rare atypical cottons should be evaluated on a case-by-case basis, if they are to be used in a conventional way.

9.10.2.3 Formaldehyde

Cotton dyeing and finishing operations can expose workers to formaldehyde in concentrations that exceed the U.S. OSHA workplace PEL for formaldehyde of 0.75 ppm of air as an 8-h time weighted average concentration and 2 ppm of air as a 15-min short-term exposure limit (STEL) [645].

Formaldehyde, a component of resins used to impart easy care and durable press and other properties to cotton fabrics, was classified in 1987 by U.S. EPA as a probable human carcinogen (an animal carcinogen and limited evidence that it is a carcinogen in humans) under conditions of unusually high or prolonged exposure [646]. Since then, studies of industrial workers have suggested that formaldehyde is associated with nasopharyngeal cancer and possibly leukemia. In June 2004, the International Agency for Research on Cancer (IARC) reclassified formaldehyde as a known human carcinogen [647]. Sensory irritation of the mucous membranes of the eyes and the respiratory tract, and cellular changes in the nasal cavity are noncancer effects of exposure to low airborne concentrations of formaldehyde.

9.10.3 CONSUMER

By the time cotton textiles reach the ultimate consumer, there should be nothing known on or extractable from the original cotton fiber that would cause any health concerns to consumers [47]. However, various dyeing and finishing treatments that cotton fabrics are subjected to can leave residues on the fabric or release substances that could cause irritation to consumers, if the treatments are not properly applied.

9.10.3.1 Formaldehyde

Exposure to formaldehyde from cotton textiles is controlled by the chemical technology on low-emitting formaldehyde resin technology and nonformaldehyde finishes (discussed earlier in Section 9.6.2 and Section 9.6.3) and by increased ventilation. It should be noted that in the 1980s, the U.S. Consumer Product Safety Commission (CPSC) studied the effects of various dyeing and finishing treatments, including durable-press finishing of cotton [648,649], and found no acute or chronic health problems of concern to consumers due to exposures to formaldehyde or other finishing chemicals from textiles [650].

In summary, when cotton is grown and processed (including yarn manufacturing and wet processing) in a responsible manner, the production of cotton should not have any adverse effects on the environment (through external emissions, wastewater effluents, and solid wastes), the workers (because of acute or chronic effects), and the consumers (because of acute or chronic effects).

REFERENCES

1. *Fiber Organon*, 75(7), Fiber Economic Bureau, Inc., Arlington, VA, 2004.
2. Basra, A.S., Ed., *Cotton Fibers, Developmental Biology, Quality Improvement, and Textile Processing*, Food Products Press, The Haworth Press, Binghamton, New York, 1999.

3. Bell, T.M. and Gillham, F.E.M., *The World of Cotton*, ContiCotton, EMR, Washington, D.C., 1989.
4. Chaudhry, M.R. and Guitchounts, A., *Cotton Facts*, International Cotton Advisory Committee (ICAC), Washington, D.C., 2003.
5. Hake, S.J, Kerby, T.A., and Hake, K.D., Eds., *Cotton Production Manual*, Pub. 3352, University of California, Division of Agriculture and Natural Resources, Oakland, CA, 1996.
6. Hamby, D.S., Ed., *The American Cotton Handbook*, 3rd ed., Vols. 1 and 2, Interscience, New York, 1965.
7. Cook, J.G., Natural fibres, in *Handbook of Textile Fibres*, 4th ed., Vol. 1, Merrow, Watford, England, 1968.
8. Wakelyn, P.J., Cotton, in *Encyclopedia of Polymer Science and Technology*, 3rd ed., Vol. 5, John Wiley & Sons, New York, 2004, pp. 721–759.
9. Wakelyn, P.J., Bertoniere, N.R., French, A.D., Thibodeaux, D.P., Tripplett, B.A., Goynes, W.R., Jr, Hughs, S.E., Knowlton, J.L., Norman, B.M., and Lanclos, D.K., Cotton, in *Kirk–Othmer Encyclopedia of Chemical Technology*, 5th ed., Vol. 8, John Wiley & Sons, New York, 2004, pp. 1–40.
10. Gillham, F.E.M., Bell, T.E., Ryan, P.D., and Gilson, S.R., *Cotton from Field to Fabric*, Vol. 1, Tom Bell Associates, Falls Church, VA, 1993.
11. Kohel, R.J. and Lewis, C.F., Eds., *Cotton*, Agronomy Monograph No. 24, American Society of Agronomy, Crop Sciences Society of America and Soil Science Society of America, Madison, WI, 1984.
12. French, A.D. and Bertoniere, N.R., Cellulose, in *Kirk–Othmer Encyclopedia of Chemical Technology*, 4th ed., Vol. 5, Interscience, New York, 1993, pp. 476–496.
13. Ward, K., Jr., Ed., *Chemistry and Chemical Technology of Cotton*, Interscience, New York, 1955.
14. Bikales, N.M. and Segal, L., Eds., *Cellulose and Cellulose Derivatives*, Parts IV and V, Wiley Interscience, New York, 1971.
15. Ward, K., Jr., and Morak, A.J., *Chemical Reactions of Polymers*, Fettes, E.M., Ed., Interscience, New York, 1964, pp. 321–365.
16. Nevell, T.P. and Zeronian, S.H., Eds., *Cellulose Chemistry and Its Applications*, Ellis Horwood Ltd., Chichester, 1985, pp. 1–551.
17. Bertoniere, N.R., The chemical nature of cellulose, Proceedings of the Beltwide Cotton Conferences, National Cotton Council, Memphis, TN, 1993, pp. 1413–1416.
18. French, A.D., Molecular arrangements in cellulose, Proceedings of the Beltwide Cotton Conferences, National Cotton Council, Memphis, TN, 1993, pp. 1417–1420.
19. Munro, J.M., *Cotton*, 2nd ed., Longman Scientific and Technical, Essex, England and John Wiley & Sons, New York, 1987.
20. Rayburn, S.T., Bitton, R., and Keen, E., *National Variety Test: Yield, Boll, Seed, Spinning and Data*, USDA, ARS, Stoneville, MS, 1992.
21. Fitt, G.P., Wakelyn, P.J., Stewart, J.M., Roupakias, D., Pages, J., Giband, M., Zafar, Y., Hake, K., and James, C., Report of the Second Expert Panel on Biotechnology in Cotton, International Cotton Advisory Committee (ICAC), Washington, D.C., November 2004.
22. Wakelyn, P.J., May, O.L., and Menchey, E.K., Cotton and biotechnology, in *Handbook of Plant Biotechnology*, Christou, P. and Klee, H., Eds., John Wiley & Sons, Chichester, West Sussex, UK, 2004, chap. 57, pp. 1117–1131.
23. Vreeland, J.M., Jr., Naturally colored and organically grown cottons: anthropological and historical perspectives, Proceedings of the Beltwide Cotton Conferences, National Cotton Council, Memphis, TN, 1993, pp. 1533–1536.
24. Vreeland, J.M., Jr., *Sci. Am.*, 280, 112, 1999.
25. Wakelyn, P.J. and Gordan, M.B., *Text. Horizons*, 15(1), 36–38, 1995.
26. Kimmel, L.B. and Day, M.P., *AATCC Rev.*, 1(10), 32, 2001.
27. Ryser, U., Cotton fiber initiation and histodifferentiation, in *Cotton Fibers, Developmental Biology, Quality Improvement, and Textile Processing*, Basra, A.S., Ed., Haworth Press, Binghamton, New York, 1999, chap. 1, pp. 21–29.
28. Ryser, U., Meier, H., and Holloway, P.J., *Protoplasma*, 117, 196, 1983.
29. Schmutz, A., Jenny, T., Amrhein, N., and Ryser, U., *Planta*, 189, 453, 1993.

30. The National Organic Program, available at http://www.ams.usda.gov/nop/indexIE.htm, Agricultural Marketing Service, USDA.
31. Chaudhry, M.R., *The ICAC Recorder*, Vol. XVI, No. 4, December 1998; Chaudhry, M.R., *The ICAC Recorder*, Vol. XXI, No. 1, March 2003.
32. Guillou, G. and Scharpé, A., *Organic Farming—Guide to Community Rules*, Directorate General for Agriculture, European Commission, ISBN 92-894-0363-2, 2000.
33. Myers, D. and Stolton, S., *Organic Cotton—from Field to Final Product*, Intermediate Technology Publications, Intermediate Technology Development Group, 103–105 Southhampton Row, London, WC1B 4HH, UK., 1999.
34. Wakelyn, P.J. and Chaudtry, M.R., Organic Cotton, in *Cotton: Science and Technology*, Gordon, S. and Heich, L., Eds., Woodhead Publishing Limited, UK, 2006.
35. Kyrgyzstan: organic cotton tested in the south, Reuters Alert Net, Dec 28, 2004. (http://www.alertnet.org/).
36. Marquardt, S., Organic cotton: production and market trends in the United States and Canada—2001 and 2002, Proceedings of the 2003 Beltwide Cotton Conference, National Cotton Council, Memphis, TN, 2003, pp. 362–366.
37. U.S. Organic Cotton Production Drops Despite Increasing Sales of Organic Cotton Products, Organic Trade Association, News Release, December 28, 2004. (http://www.ota.com).
38. Organic Exchange, http://organicexchange.org, 2005.
39. Variation in yields among countries, *ICAC Recorder*, Vol. XXII, No. 3, September 2004, pp. 3–7.
40. U.S. Department of Agriculture, Foreign Agriculture Service, June 2005. (http://www.fas.usda.gov/psd/).
41. Anthony, W.S. and Mayfield, W.D., Eds., *Cotton Ginners Handbook*, U.S. Dept. of Agriculture, *Agricultural Handbook 503*, Washington, D.C., 1994.
42. Wakelyn, P.J., Thompson, D.W., Norman, B.M., Nevius, C.B., and Findley, D.S., *Cotton Gin and Oil Mill Press*, 106(8), 5, 2005.
43. O'Connor, R.T., *Instrumental Analysis of Cotton Cellulose and Modified Cotton Cellulose*, Marcel Dekker, New York, 1972.
44. Berlin, J.D. and Watson, M., Fine structural differentiation of lint and fuzz fibers, Proceedings of the Beltwide Cotton Production Research Conferences, National Cotton Council of America, Memphis, TN, 1974, p. 52.
45. Supak, J.R. and Snipes, Eds., *Cotton Harvest Management: Use and Influence of Harvest Aids*, The Cotton Foundation Reference Book Series, No. 5, The Cotton Foundation, Memphis, TN, 2001; Chaudhry, M.R., Harvesting and ginning of cotton in the world, Proceedings of the Beltwide Cotton Production Research Conferences, Vol. 2, National Cotton Council of America, Memphis, TN, 1997, pp. 1617–1619.
46. ISO 8115, Cotton bales—dimensions and density, International Organization for Standardization, Switzerland, 1986.
47. Wakelyn, P.J., Cotton: environmental concerns and product safety, Proceedings of the 22nd International Cotton Conferences, Bremen, Harig, H. and Heap, S.A., Eds., Faserinstitut Bremen eV, Bremen, Germany, 1994, pp. 287–305.
48. Beasley, C.A. and Ting, I.P., *Am. J. Bot.*, 61, 188, 1974.
49. Stewart, J.McD., *Am. J. Bot.*, 62, 723, 1975.
50. Temming, H., *Linters: Technical Information on Cotton Cellulose*, Glückstadt, West Germany, 192, 1973.
51. Gamble, G., *J. Agric. Food Chem.*, 51, 7995, 2003.
52. Seagull, R.W., *Tip Growth in Plant and Fungal Cells*, Heath, I.B., Ed., Academic Press, San Diego, CA, 1990, pp. 261–284.
53. Tiwari, S.C. and Wilkins, T.A., *Can. J. Bot.*, 73, 746, 1995.
54. Tokumoto, H., Wakabayashi, K., Kamisaka, S., and Hoson, T., *Plant Cell Physiol.*, 43, 411, 2002.
55. Tokumoto, H., Wakabayashi, K., Kamisaka, S., and Hoson, T., *J. Plant Physiol.*, 160, 1411, 2003.
56. Reiter, W.D., *Curr. Opin. Plant Biol.*, 5, 536, 2002.
57. Scheible, W.R. and Pauly, M., *Curr. Opin. Plant Biol.*, 7, 285, 2004.

58. Arpat, A.B., Waugh, M., Sullivan, J.P., Gonzales, M., Frisch, D., Main, D., Wood, T., Leslie, A., Wing, R.A., and Wilkins, T.A., *Plant Mol. Biol.*, 54, 911, 2004.
59. H.R. Huwyler, G. Franz, and H. Meier, *Plant Sci. Lett.*, 12, 55, 1978.
60. Huwyler, H.R., Franz, G., and Meier, H., *Planta*, 46, 635, 1979.
61. Maltby, D., Carpita, N.C., Montezinos, D., Kulow, C., and Delmer, D.P., *Plant Physiol.*, 63, 1158, 1979.
62. Waterkeyn, L., *Protoplasma*, 106, 49, 1981.
63. Cui, X., Shin, H., Song, C., Laosinchai, W., Amano, Y., and Brown, R.M., Jr., *Planta.*, 213, 223, 2001.
64. Wakeham, H. and Spicer, N., *Text. Res. J.*, 25, 585, 1955.
65. Hearle, J.W.W. and Sparrow, J.T., *Text. Res. J.*, 41, 736, 1971.
66. Moharir, A.V., van Langenhove, L., van Nimmen, E., Louwagie, J., and Kiekens, P., *J. Appl. Polym. Sci.*, 72, 269, 1999.
67. Seagull, R.W., *J. Cell Sci.*, 101, 561, 1992.
68. Dixon, D.C., Seagull, R.W., and Triplett, B.A., *Plant Physiol.*, 105, 1347, 1994.
69. Andersland, J.M., Dixon, D.C., Seagull, R.W., and Triplett, B.A., *In Vitro Cell Develop. Biol. Plant*, 34, 173, 1998.
70. Whittaker, D.J. and Triplett, B.A., *Plant Physiol.*, 121, 181, 1999.
71. Dixon, D.C., Meredith, W.R., Jr., and Triplett, B.A., *Int. J. Plant Sci.*, 161, 63, 2000.
72. Andersland, J.M. and Triplett, B.A., *Plant Physiol. Biochem.*, 38, 193, 2000.
73. Delmer, D.P., Heiniger, U., and Kulow, C., *Plant Physiol.*, 59, 713, 1977.
74. Amor, Y., Haigler, C.H., Wainscott, M., Johnson, S., and Delmer, D.P., *Proc. Natl. Acad. Sci. U.S.A.*, 92, 9353, 1995.
75. Haigler, C.H., Ivanova-Datcheva, M., Hogan, P.S., Salnikov, V.V., Hwang, S., Martin, K., and Delmer, D.P., *Plant Mol. Biol.*, 47, 29, 2001.
76. Peng, L., Kawagoe, Y., Hogan, P., and Delmer, D.P., *Science*, 295, 147, 2002.
77. Schrick, K., Fujioka, S., Takatsuto, S., Stierhof, Y.D., Stransky, H., Yoshida, S., and Jurgens, G., *Plant J.*, 38, 227, 2004.
78. Meuller, S.C. and Brown, R.M., Jr., *J. Cell Biol.*, 84, 315, 1980.
79. Haigler, C.H. and Brown, R.M., Jr., *Protoplasma*, 134, 111, 1986.
80. Hayashi, T., Read, S.M., Bussell, J., Thelen, M.T., Lin, F.-C., Brown, R.M., Jr., and Delmer, D.P., *Plant Physiol.*, 83, 1054, 1987.
81. Okuda, R., Li, L., Kudlicka, K., Kuga, S., and Brown, R.M, Jr., *Plant Physiol.*, 101, 1131, 1993.
82. Lai-Kee-Him, J., Chanzy, H., Müller, M., Puteaux, J.-L., Imai, T., and Bulone, V., *J. Biol. Chem.*, 277, 36931, 2002.
83. Pear, J.R., Kawagoe, Y., Schreckengost, W.E., Delmer, D.P., and Stalker, D.M., *Proc. Natl. Acad. Sci. U.S.A.*, 93, 12637, 1996.
84. Kimura, S., Laosinchai, W., Itoh, T., Cui, X., Linder, C.R., and Brown, R.M., Jr., *Plant Cell*, 11, 2075, 1999.
85. Richmond, T. and Somerville, C.R., *Plant Mol. Biol.*, 47, 131, 2001.
86. Taylor, N.G., Howells, R.M., Huttly, A.K., Vickers, K., and Turner, S.R., *Proc. Natl. Acad. Sci. U.S.A.*, 100, 1450, 2003.
87. Arioli, T., Peng, L., Betzner, A.S., Burn, J., Wittke, W., Herth, W., Camilleri, C., Höfte, H., Plazinski, J., Birch, R., Cork, A., Glover, J., Redmond, J., and Williamson, R.E., *Science*, 279, 717, 1998.
88. Schieble, W.-R., Eshed, R., Richmond, T., Delmer, D., and Somerville, C., *Proc. Natl. Acad. Sci. U.S.A.*, 98, 10079, 2001.
89. Holland, N., Holland, D., Helentjaris, T., Dhugga, K.S., Xoconostle-Cazares, B., and Delmer, D.P., *Plant Physiol.*, 123, 1313, 2000.
90. Taylor, N.G., Laurie, S., and Turner, S.R., *Plant Cell*, 12, 2529, 2000.
91. Kim, H.J. and Triplett, B.A., Regulation of gene expression in the transition from cell elongation to secondary wall formation in cotton fiber, Proceedings of the Beltwide Cotton Conferences, National Cotton Council, Memphis, TN, 2005, 1043.
92. Adams, K.L., Cronn, R., Percifield, R., and Wendel, J.F., *Proc. Natl. Acad. Sci. U.S.A.*, 100, 4649, 2003.
93. Marx-Figini, M., *Nature*, 210, 755, 1966.

94. Timpa, J.D. and Triplett, B.A., *Planta*, 189, 101, 1993.
95. Haigler, C.H., Zhang, D., and Wilkerson, C.G., *Physiol. Plant.*, 124, 285, 2005.
96. Tripp, V.W., Moore, A.T, and Rollins, M.L., *Text. Res. J.*, 21, 886, 1951.
97. Guthrie, J.D., The chemistry of lint cotton, in *Chemistry and Chemical Technology of Cotton*, Ward, K., Jr., Ed., Interscience, New York, 1955, p. 2.
98. McCall, E.R. and Jurgens, J.F., *Text. Res. J.*, 21, 19, 1951.
99. Warwicker, J.O., Jeffries, R., Colbran, R.L., and Robinson, R.N., A Review of the Literature on the Effect of Caustic Soda and Other Swelling Agents on the Fine Structure of Cotton, Pamphlet No. 93, Shirley Institute, Manchester, UK, 1966.
100. Rollins, M.L., *For. Prod. J.*, 18(2), 91, 1968.
101. Catlett, M.S., Giuffria, R., Moore, A.T., and Rollins, M.L., *Text. Res. J.*, 21, 880, 1951.
102. Tripp, V.W. and Rollins, M.L., *Anal. Chem.*, 24, 1721, 1952.
103. Wakelyn, P.J., *Text. Res. J.*, 45, 418, 1975.
104. Goldthwaite, C.F., Kettering, J.H., and Guthrie, J.D., Chemical properties of cotton fiber, in *Matthews' Textile Fibers*, 5th ed., Mauersberger, H.P., Ed., Wiley Interscience, New York, 1947, pp. 266–304 [see also 6th ed., 1954].
105. Ferreti, R.J., Merola, G.V., Marsh, P.B., and Simpson, M.E., *Cotton Grow. Rev.*, 52, 136, 1975.
106. Scmutz, A., Buchala, A.J., and Ryser, U., *Plant Physiol.*, 110, 403, 1996.
107. Wakelyn, P.J., Jacobs, R.R., and Kirk, I.W., Eds., *Washed Cotton: Washing Techniques, Processing Characteristics, and Health Effects*, USDA, Washington, D.C., 1986.
108. Vaughn, K.C. and Turley, R.B., *Protoplasma*, 209, 226, 1999.
109. Brushwood, D.E. and Perkins, H.H., Jr., Variations in cotton insect honeydew composition and the related effects on test methods and processing quality, Proceedings of the Beltwide Cotton Conferences, National Cotton Council, Memphis, TN, 1995, p. 1178.
110. Hector, D.J. and Hodkinson, I.D., Stickiness in cotton, *ICAC Review Articles on Cotton Production Research*, No. 2, International Cotton Advisory Committee, 1901 Pennsylvania Avenue, Washington, D.C., 1989.
111. Perkins, H.H., Jr., A survey of sugar and sticky cotton test methods, Proceedings of the Beltwide Cotton Conferences, National Cotton Council, Memphis, TN, 1993, 1136.
112. Brushwood, D.E. and Perkins, H.H., Jr., Characterization of sugar from honeydew contaminated and normal cottons, Proceedings of the Beltwide Cotton Conferences, National Cotton Council, Memphis, TN, 1994, p. 1408.
113. Bourely, J., *Cotton Fibers Tropicale*, 35(2), 189, 1980.
114. Cheung, P.S.R., Roberts, C.W., and Perkins, H.H., Jr., *Text. Res. J.*, 50, 55, 1980.
115. Hendrix, D.L., Wei, Y.A., and Leggett, J.E., *Comp. Biochem. Physiol.*, 101B, 23, 1992.
116. Tarczyski, M.C., Byrne, D.N., and Miller, W.B., *Plant Physiol.*, 98, 753, 1992.
117. Bates, R.B., Byrne, D.N., Kane, V.V., Miller, W.B., and Taylor, S.R., *Carbohydr. Res.*, 201, 342, 1990.
118. Byrne, D.N. and Miller, W.B., *J. Insect Physiol.*, 36, 433, 1990.
119. Roberts, C.W., Cheung, P.S.R., and Perkins, H.H., Jr., *Text. Res. J.*, 48, 91, 1978.
120. Brushwood, D.E. and Perkins, H.H., Jr., *Text. Chem. Color.*, 26(3), 32, 1994.
121. Matsumura, R.T., Ashbaugh, L., James, T., Carvacho, O., and Flocchini, R., Size distribution of PM-10 soil dust emissions from harvesting crops, Proceedings of the International Conferences on Air Pollution from Agricultural Operations, Kansas City, MO, February 7–9, 1996.
122. Columbus E.P. and Morris, N.M., *Trans. Am. Soc. Ag. Eng.*, 27, 546, 1984.
123. Perkins, H.H., Jr. and Brushwood, D.E., *Text. Chem. Color.*, 23(2), 26, 1991.
124. Greenwood, P.F., *Text.*, 2, 23, 1993.
125. Cook, F.C., *Textile World*, May, 84, 1991.
126. Rucker, J.W., Freeman, H.S., and Hsu, W.-N, *Text. Chem. Color.*, 24(9), 66, 1992.
127. Wakelyn, P.J., Supak, J., Carter, F.C., and Roberts, B., Public and environmental issuses, in *Cotton Harvest Management: Use and Infuence of Harvest Aids*, The Cotton Foundation Reference Book Series, No. 5, Supak, J.R. and Snipes, C.E., Eds., The Cotton Foundation, Memphis, TN, 2001, chap. 10, pp. 275–302.
128. Turbak, A.F., Hammer, R.B., Davies, R.E., and Hergert, H.L., *Chem. Tech.*, 10, 51, 1980.
129. Hudson, S.M. and Cuculo, J.A., *J. Macromol. Sci.*, C18, 1, 1980.

130. Turbak, A.F., Newer cellulose solvent systems, Proceedings of the 1983 International Dissolving and Specialty Pulps Conferences, *TAPPI*, Atlanta, U.S., 1983.
131. Johnson, D.C., Solvents for cellulose, in *Cellulose Chemistry and Its Applications*, Nevell, T.P. and Zeronian, S.H., Eds., Ellis Horwood Ltd., Chichester, 1985, chap. 7, pp. 181–201.
132. Treiber, E.E., Formation of fibers from cellulose solution, in *Cellulose Chemistry and Its Applications*, Nevell, T.P. and Zeronian, S.H., Eds., Ellis Horwood Ltd., 1985, chap. 18, pp. 455–479.
133. Johnson, D.C., Nicholson, M.D., and Haigh, F.L., *Appl. Polym. Symp.*, 28, 931, 1976.
134. Jayme, G., Investigations of solutions: new solvents, in *High Polymers*, Vol. V, *Cellulose and Cellulose Derivatives*, Part IV, Bikales, N.M. and Segal, L., Eds., Wiley Interscience, New York, 1971, p. 381.
135. Timpa, J.D., *J. Agric. Food Chem.*, 39, 270, 1991.
136. Graenacher, C. and Sallmann, R., U.S. Patent 2179181, November 7, 1939.
137. McCorsley, C.C, III, and Varga, J.K., U.S. Patent 4142913, March 6, 1979.
138. McCorsely, C.C., III, U.S. Patent 4246221, January 20, 1981.
139. Augustine, A.V., Hudson, S.M., and Cuculo, J.A., Direct solvents for cellulose, in *Steam Explosion Techniques, Fundamentals and Industrial Applications*, Focher, B., Marzetti, A., and Crescenzi, V., Eds., Gordon & Breach, Philadelphia, PA, 1991, p. 251.
140. Rules and regulations under the textile fiber products identification act, Federal Register, 60, pp. 62352–62354, December 6, 1995.
141. Mitzutani, C., Wartell, L.H., Bertoniere, N.R., and French, A.D., unpublished data, 1995.
142. Westermark, U. and Gustafsson, K., *Holzforschung*, 48, 146, 1994.
143. Sjoholm, E., Gustafsson, K., Pettersson, B., and Colmsjo, A., *Carbohydr. Polym.*, 32, 57, 1997.
144. Dupont, A.L., *Polymer*, 44, 4117, 2003.
145. Potthast, A., Rosenau, T., Sixta, H., and Kosma, P., *Tetrahedron Lett.*, 43, 7757, 2002.
146. Strlic, M. and Kolar, J., *J. Biochem. Biophys. Methods*, 56, 265, 2003.
147. Takaragi, A., Minoda, M., Miyamote, T., Liu, H.Q., and Zhang, L.N., *Cellulose*, 6, 93, 1999.
148. Rousselle, M.A., *Text. Res. J.*, 72(2), 131, 2002.
149. Yanagisawa, M., Shibata, I., and Isogai, A., *Cellulose*, 11, 169, 2004.
150. Kamide, K. and Okajima, K., U.S. Patent 4634470, 1987.
151. Isogai, A. and Atalla, R.H., *Cellulose*, 5, 309, 1998.
152. Swatloski, R.P., Spear, S.K., Holbrey, J.D., and Rogers, R.D., *J. Am. Chem. Soc.*, 124, 4974, 2002.
153. Broughton, R., Wang, W., Farag, R., Rogers, R., and Swatloski, R., Cellulose Fibers Extruded from Ionic Liquids, paper presented at INTC 2004, Toronto, Canada, September 20–23, 2004.
154. Krassig, H. and Kappner, W., *Makromol. Chem.*, 44–46, 1, 1961.
155. Nishiyama, Y., Kim, U.-J., Kim, D.-Y., Katsumata, K.S., May, R.P., and Langan, P., *Biomacromolecules*, 4, 1013, 2003.
156. Ziderman, I.I. and Perel, J., *J. Macromol. Sci. Phys.*, B24, 181, 1985–1986.
157. Hessler, L.E., Merola, G.V., and Berkley, E.E., *Text. Res. J.*, 18, 628, 1948.
158. Marx-Figini, M., *Cellulose and Other Natural Polymer Systems: Biogenesis, Structure, and Degradation*, Brown, R.M., Ed., Plenum Press, New York, 1982, p. 243.
159. McCormick, C.L., Callais, P.A., and Hutchinson, B.H., *Macromolecules*, 18, 2394, 1985.
160. Turbak, A.F., *Wood and Agricultural Residues*, Soltes, E.J., Ed., Academic Press, New York, 1983, 87.
161. Kennedy, J.F., Rivera, Z.A., White, C.A., Lloyd, L.L., and Warner, F.P., *Cellul. Chem. Technol.*, 24, 319, 1990.
162. Dawsey, T.R. and McCormick, C.L., *Rev. Macromol. Chem. Phys.*, C30, 403, 1990.
163. Meinert, M.C. and Delmer, D.P., *Plant Physiol.*, 59, 1088, 1977.
164. Timpa, J.D. and Ramey, H.H., *Text. Res. J.*, 59, 661, 1989.
165. Timpa, J.D. and Ramey, H.H., *Text. Res. J.*, 64, 557, 1994.
166. Timpa, J.D. and Wanjura, D.F., *Cellulose and Wood: Chemistry and Technology*, Schuerch, C., Ed., John Wiley & Sons, New York, 1989, p. 1145.
167. Segal, L. and Timpa, J.D., *Text. Res. J.*, 43, 185, 1973.
168. Rousselle, M.A. and Howley, P.S., *Text. Res. J.*, 68, 606, 1998.
169. Morgado, J., Cavaco-Paulo, A., and Rousselle, M.A., *Text. Res. J.*, 70, 696, 2000.

170. French, A.D. and Johnson, G.P., *Cellulose*, 11, 5, 2004.
171. Brant, D.A. and Christ, M.D., Realistic conformational modeling of carbohydrates: applications and limitations in the context of carbohydrate-high polymers, in *Computer Modeling of Carbohydrate Molecules*, French, A.D. and Brady, J.W., Eds., ACS Symposium Series, No. 430, American Chemical Society, Washington, D.C., 1990, p. 42.
172. Meader, D., Atkins, E.D.T., and Happey, F., *Polymer*, 19, 1371, 1978.
173. Zugenmaier, P., Konformations- und packungsanalyse von polysacchariden, in *Polysaccharide*, Burchard, W., Ed., Springer–Verlag, Berlin, 1985, p. 271.
174. Whitaker, P.M., Nieduzynski, I.A., and Atkins, E.D.T., *Polymer*, 15, 125, 1974.
175. Sarko, A., Nishimura, H., and Okano, T., Crystalline alkali–cellulose complexes as intermediates during mercerization, in *The Structures of Cellulose*, Atalla, R.H., Ed., ACS Symposium Series, No. 340, American Chemical Society, Washington, D.C., 1987, p. 169.
176. Jones, D.W., *J. Polym. Sci.*, 42, 173, 1960.
177. Allen, F.H., *Acta Crystallogr., Sect. B*, 58, 380, 2002.
178. Berman, H.M., Battistuz, T., Bhat, T.N., Bluhm, W.F., Bourne, P.E., Burkhardt, K., Feng, Z., Gilliland, G.L., Iype, L., Jain, S., Fagan, P., Marvin, J., Padilla, D., Ravichandran, V., Schneider, B., Thanki, N., Weissig, H., Westbrook, J.D., and Zardecki, C., *Acta Crystallogr., Sect. D*, 58, 899, 2002.
179. Chu, S.S.C. and Jeffrey, G.A., *Acta Crystallogr., Sect. B*, 24, 830, 1968.
180. Peralta-Inga, Z., Johnson, G.P., Dowd, M.K., Rendleman, J.A., Stevens, E.D., and French, A.D., *Carbohydr. Res.*, 337, 851, 2002.
181. Rencurosi, A., Röhrling, J., Pauli, J., Potthast, A., Jäger, C., Pérez, S., Kosma, P., and Imberty, A., *Angew. Chem., Int. Ed. Engl.*, 41, 4277, 2002.
182. Ernst, A. and Vasella, A., *Helv. Chim. Acta.*, 79, 1279, 1996.
183. Leung, F. and Marchessault, R.H., *Can. J. Chem.*, 51, 1215, 1973.
184. Perez, S. and Brisse, F., *Acta Crystallogr., Sect. B*, 33, 2578, 1977.
185. Burkert, U. and Allinger, N.L., Molecular mechanics, in *ACS Monograph 177*, American Chemical Society, Washington, D.C., 1982.
186. Hardy, B.J. and Sarko, A., *J. Comput. Chem.*, 14, 848, 1993.
187. Schmidt, R., Trojan, C., Tasaki, K., and Brady, J.W., Proceedings of the Second Tricel Symposium on Trichoderma Reesei Celulases and other Hyrdrolayses, Suominen, P. and Reinikainen, T., Eds., Foundation for Biotechnical and Industrial Fermentation Research 8, 1993, p. 41.
188. Strati, G.L., Willett, J.L., and Momany, F.A., *Carbohydr. Res.*, 337, 1833, 2002.
189. French, A.D., Johnson, G.P., Kelterer, A.-M., and Csonka, G.I., *Tetrahedron: Asymmetry*, 16, 577, 2005.
190. Kroon-Batenburg, L.M.J., Kroon, J., Leeflang, B.R., and Vliegenthart, J.F.G., *Carbohydr. Res.*, 245, 21, 1993.
191. Tanaka, F. and Fukui, N., *Cellulose*, 11, 33, 2004.
192. Umemura, M., Yuguchi, Y., and Hirotsu, T., *J. Phys. Chem. A*, 108, 7063, 2004.
193. Hardy, B.J. and Sarko, A., *J. Comput. Chem.*, 14, 831, 1993.
194. Kroon-Batenburg, L.M.J., Kroon, J., and Northolt, J.G., *Polym. Commun.*, 27, 290, 1986.
195. Sturcova, A., His, I., Apperley D.C., Sugiyama, J., and Jarvis, M.C., *Biomacromolecules*, 5, 1333, 2004.
196. Atalla, R.H., *Comprehensive Natural Products Chemistry*, Barton, D., Nakanishi, K., and Meth-Cohn, O., Eds., Vol. 3, *Carbohydrates and Their Derivatives Including Tannins, Cellulose, and Related Lignins*, Pinto, B.M., Ed., Elsevier, Amsterdam, 1999, p. 169.
197. Klug, H.P. and Alexander, L.E., *X-Ray Diffraction Procedures*, 2nd ed., John Wiley & Sons, New York, 1974, p. 13.
198. Nishiyama, Y., Kuga, S., Wada, M., and Okano, T., *Macromolecules*, 30, 6395, 1997.
199. Paralikar, K.M., Betrabet, S.M., and Bhat, N.V., *J. Appl. Crystallogr.*, 12, 589, 1979.
200. Jeffrey, G.A. and Park, Y.J., *Acta Crystallogr., Sect. B*, 28, 257, 1972.
201. Jeffrey, G.A. and Huang, D.-B., *Carbohydr. Res.*, 222, 47, 1991.
202. Atalla, R.H. and van der Hart, D.L., *Science*, 223, 283, 1984.
203. Sugiyama, J., Vuong, R., and Chanzy, H., *Macromolecules*, 24, 4168, 1991.

204. Imai, T. and Sugiyama, J., *Macromolecules*, 31, 6275, 1998.
205. Atalla, R.H. and van der Hart, D.L., *Solid State Nucl. Magn. Reson.*, 15, 1, 1999.
206. Wickholm, K., Larsson, P.T., and Iversen, T., *Carbohydr. Res.*, 312, 123, 1998.
207. Wada, M., Okano, T., and Sugiyama, J., *J. Wood. Sci.*, 47, 124, 2001.
208. Nishiyama, Y., Sugiyama, J., Chanzy, H., and Langan, P., *J. Am. Chem. Soc.*, 125, 14300, 2003.
209. Nishiyama, Y., Langan, P., and Chanzy, H., *J. Am. Chem. Soc.*, 124, 9074, 2002.
210. Wada, M., Kondo, T., and Itoh, T., *Polym. J.*, 35, 155, 2003.
211. Wada, M., *J. Polym. Sci., Part B: Polym. Phys.*, 40, 1095, 2002.
212. Kuutti, L., Peltonen, J., Pere, J., and Teleman, O., *J. Microsc.*, 178, 1, 1995.
213. Baker, A.A., Helbert, W., Sugiyama, J., and Miles, M., *J. Biophys.*, 79, 1139, 2000.
214. Vietor, R.J., Newman, R.H., Ha, M.-A., Apperley, D.C., and Jarvis, M.C., *Plant J.*, 30, 721, 2002.
215. Newman, R.H. and Davidson, T.C., *Cellulose*, 11, 23, 2004.
216. Sasaki, M., Adschiri, T., and Arai, K., *J. Agric. Food Chem.*, 51, 5376, 2003.
217. Roberts, E.M., Saxena, I.M., and Brown, R.M., *Cellulose and Wood—Chemistry and Technology*, Schuerch, C., Ed., Wiley Interscience, New York, 1989, p. 689.
218. Kuga, S., Takagi, S., and Brown, R.M., *Polymer*, 34, 93, 1993.
219. Hirai, A., Tsuji, M., and Horii, F., *Cellulose*, 9, 105, 2002.
220. Langan, P., Nishiyama, Y., and Chanzy, H., *J. Am. Chem. Soc.*, 121, 9940, 1999.
221. Langan, P., Nishiyama, Y., and Chanzy, H., *Biomacromolecules*, 2, 410, 2001.
222. Gessler, K., Krauss, N., Steiner, T., Betzel, C., Sarko, A., and Saenger, W., *J. Am. Chem. Soc.*, 117, 11397, 1995.
223. Raymond, S., Heyraud, A., Tran Qui, D., Kvick, A., and Chanzy, H., *Macromolecules*, 28, 2096, 1995.
224. Raymond, S., Henrissat, B., Tran Qui, D., Kvick, A., and Chanzy, H., *Carbohydr. Res.*, 277, 209, 1995.
225. Noordik, J.H., Beurskens, P.T., Bennema, P., Visser, R.A., and Gould, R.O., *Z. Kristallogr.*, 168, 59, 1984.
226. Wada, M., Heux, L., and Sugiyama, J., *Biomacromolecules*, 5, 1385, 2004.
227. Wada, M., Heux, L., Isogai, A., Nishiyama, Y., Chanzy, H., and Sugiyama, J., *Macromolecules*, 22, 3168, 2001.
228. Wada, M., Chanzy, H., Nishiyama, Y., and Langan, P., *Macromolecules*, 37, 8548, 2004.
229. Chanzy, H., Imada, K., Mollard, A., Vuong, R., and Barnoud, F., *Protoplasma*, 100, 303, 1979.
230. Kondo, T., Togawa, E., and Brown, R.M., Jr., *Biomacromolecules*, 2, 1324, 2001.
231. Atalla, R.H. and Nagel, S.C., *Science*, 185, 522, 1974.
232. Whimore, R.E. and Atalla, R.H., *Int. J. Biol. Macromol.*, 7, 182, 1985.
233. Atalla, R.H., Ellis, J.D., and Schroeder, L.R., *J. Wood Chem. Technol.*, 4, 465, 1984.
234. Shibazaki, H., Kuga, S., and Okano, T., *Cellulose*, 4, 75, 1997.
235. Dinand, E., Vignon, M., Chanzy, H., and Heux, L., *Cellulose*, 9, 7, 2002.
236. Blackwell, J., Kurz, D., Su, M.-Y., and Lee, D.M., X-ray studies of the structure of cellulose complexes, in *The Structures of Cellulose*, Atalla, R.H., Ed., ACS Symposium Series, No. 340, American Chemical Society, Washington, D.C., 1987, p. 199.
237. Lee, D.M. and Blackwell, J., *Biopolymers*, 20, 2165, 1981.
238. Stipanovic, A.J. and Sarko, A., *Polymer*, 19, 3, 1978.
239. Sikorski, P., Wada, M., Heux, L., Shintani, H., and Stokke, B.T., *Macromolecules*, 37, 4547, 2004.
240. Kuga, S. and Brown, R.M., *Carbohydr. Res.*, 180, 345, 1988.
241. Koyama, M., Helbert, W., Imai, T., Sugiyama, J., and Henrissat, B., *Proc. Natl. Acad. Sci. U.S.A.*, 94, 9091, 1997.
242. Kuga, S. and Brown, R.M., Jr., *J. Electron Microsc. Tech.*, 6, 349, 1987.
243. Kuga, S. and Brown, R.M., Jr., *Polymer Commun.*, 28, 311, 1987.
244. Lin, J.S., Tang, M.-Y., and Fellers, J.F., Fractal analysis of cotton cellulose as characterized by small-angle x-ray scattering, in *The Structures of Cellulose*, Atalla, R.H., Ed., ACS Symposium Series, No. 340, American Chemical Society, Washington, D.C., 1987, p. 234.
245. Statton, W.O. and Godard, G.M., *J. Appl. Phys.*, 28, 1111, 1957.
246. Klug, H.P. and Alexander, L.E., *X-Ray Diffraction Procedures*, John Wiley & Sons, New York, 1954, chap. 9.

247. Segal, L., Creely, J.J., Martin, A.E., Jr., and Conrad, C.M., *Text. Res. J.*, 29, 786, 1959.
248. Moharir, A.V. and Vijayraghavan, K.M., *J. Appl. Polym. Sci.*, 48, 1869, 1993.
249. Hebert, J.J. and Muller, L.L., *J. Appl. Polym. Sci.*, 18, 3373, 1974.
250. Rowland, S.P. and Bertoniere, N. R., Chemical methods of studying supramolecular structure, in *Cellulose Chemistry and Its Applications*, Nevell, T.P. and Zeronian, S.H., Eds., Ellis Horwood Ltd., Chichester, England, 1985, chap. 4, p. 112.
251. Bertoniere, N.R. and Zeronian, S.H., Chemical characterization of cellulose, in *The Structures of Cellulose*, Atalla, R.H., Ed., ACS Symposium Series, No. 340, American Chemical Society, Washington, D.C., 1987, p. 255.
252. Valentine, L., *Chem. Ind. (London)*, 47, 1279, 1956.
253. Jeffries, R., *J. Appl. Polym. Sci.*, 8, 1213, 1964.
254. Hailwood, A.J. and Horrobin, S., *Trans. Faraday Soc.*, 42B, 84, 1946.
255. Lewin, M., Guttman, H., and Saar, N., *Appl. Polym. Symp.*, 28, 791, 1976.
256. Lewin, M., New chemical approaches to the structure of cellulose in *Cellulose and Its Compounds*, Kennedy, J.F., Phillips, G.O., Wedlock, D.J., and Williams, P.A., Eds., Ellis Harwood, Ltd., Chichester, England, 1985, pp. 27–36.
257. Lewin, M. and Ben Bassat, A., SIRTEC, First International Symposium on Cotton Research, Paris, Institute Textiles de France, 1969, pp. 535–556.
258. Lewin, M., Guttmann, H., and Shabtai, D., *Appl. Polym. Symp.*, 31, 163, 1977.
259. Lewin, M., Guttmann, H., and Derfler, D., *J. Appl. Polym. Sci.*, 27, 3199, 1982.
260. Lewin, M., Guttmann, H., Knoll, A., and Derfler, D., *J. Polym. Sci., Polym. Chem. Ed.*, 20, 929, 1982.
261. Zeronian, S.H., Coole, M.L., Alger, K.W., and Chandler, J.M., *J. Appl. Polym. Sci., Appl. Polym. Symp.*, 37, 1053, 1983.
262. Nelson, M.L., *J. Polym. Sci.*, 43, 351, 1960.
263. Nickerson, R.F., *Text. Res. J.*, 21, 195, 1951.
264. Marchessault R.H. and Howsmon J.A., *Text. Res. J.*, 27, 30, 1957.
265. Rowland, S.P. and Pittman, P.F., *Text. Res. J.*, 35, 421, 1965.
266. Jeffries, R., Roberts, J.G., and Robinson, R.N., *Text. Res. J.*, 38, 234, 1968.
267. Cousins E.R., Bullock, A.L., Mack, C.H., and Rowland, S.P., *Text. Res. J.*, 34, 953, 1964.
268. Rowland, S.P. and Howley, P.S., *J. Polym. Sci., Poly. Sci. Ed.*, 23, 183, 1985.
269. Bertoniere, N.R., Howley, P.S., Ruppenicker, G.F., Anthony, W.S., and Hughs, S. F., Effect of ginning, greige mill, and wet processing on the microstructure of cotton fiber, Proceedings of the Beltwide Cotton Conference, 3, 1267–1269, 1992.
270. Guthrie, J.D., The chemistry of lint cotton, in *Chemistry and Chemical Technology of Cotton*, Ward, K., Ed., Interscience Publishers, New York, 1955, pp. 3–13.
271. Rollins, M.L., *Text. Res. J.*, 15, 65, 1945.
272. Flint, E.G., *Biol. Rev.*, 25, 414, 1950.
273. Warwicker, J.O., Simmens, S.C., and Hallam, P., *Text. Res. J.*, 40, 1051, 1970.
274. Goynes, W.R., Ingber, B.F., and Triplett, B.A., *Text. Res. J.*, 65, 4008, 1995.
275. Rollins, M.L., Moore, A.T., Goynes, W.R., Carra, J.H., and deGruy, I.V., *Am. Dyest. Rep.*, 54, 36, 1965.
276. Tripp, V.W., Moore, A.T., deGruy, I.V., and Rollins, M.L., *Text. Res. J.*, 30, 140, 1960.
277. Gipson, J.R. and Ray, L.L., *Cotton Grow. Rev.*, 47, 257, 1970.
278. Haigler, C.H., Rao, N.R., Roberts, E.M., Huang, J.-Y., Upchurch, D.R., and Trolinder, N.L., *Plant Physiol.*, 95, 88, 1991.
279. Kassenback, P., Incidence of Bilateral Structure of Cotton Fibers, in *First International Symposium on Cotton Textile Research*, Paris, 1969, p. 455.
280. Boylston, E.K. and Hebert, J.J., *Text. Res. J.*, 53, 469, 1983.
281. Welo, W.H., Ziffle, H.M., and McDonald, A.W., *Text. Res. J.*, 22, 261, 1952.
282. Jayme, G., *Tappi J.*, 41, 180A, 1958.
283. Bertoniere, N.R., *Modern Textile Characterization Methods*, Raheel, M., Ed., Marcel Dekker, New York, 1996, chap. 5, pp. 265–290.
284. Stone, J.E. and Scallan, A.M., *Cellul. Chem. Technol.*, 2, 343, 1968.

285. Martin, L.F. and Rowland, S.P., *J. Chromatogr.*, 28, 139, 1967.
286. Bredereck, K. and Blüher, A., *Melliand Textilberichte*, 73, E279 (English), 652 (German), 1992.
287. Ladisch, C.M., Yang, Y., Velayudhan, A., and Ladisch, M.R., *Text. Res. J.*, 62, 361, 1992.
288. Nelson, R. and Oliver, D.W., *J. Polym. Sci.*, Part C, 36, 305, 1968.
289. Bertoniere, N.R., King, W.D., and Hughs, S.E., *Lignocellulosics—Science, Technology, Development and Use*, Kennedy, J.F., Phillips, G.O., and Williams, P.A., Eds., Ellis Horwood Ltd., Chichester, England, 1992, p. 457.
290. Bertoniere, N.R. and King, W.D., *Text. Res. J.*, 59, 114, 1989.
291. Blouin, F.A., Martin, L.F., and Rowland, S.P., *Text. Res. J.*, 40, 809, 1970.
292. Rowland, S.P., Wade, C.P., and Bertoniere, N.R., *J. Appl. Polym. Sci.*, 29, 3349, 1984.
293. Bertoniere, N.R., King, W.D., and Rowland, S.P., *J. Appl. Polym. Sci.*, 31, 2769, 1986.
294. Blouin, F.A., Martin, L.F., and Rowland, S.P., *Text. Res. J.*, 40, 959, 1970.
295. Bertoniere, N.R. and King, W.D., *Text. Res. J.*, 60, 606, 1990.
296. Bertoniere, N.R. and King, W.D., *Text. Res. J.*, 59, 608, 1989.
297. Bertoniere, N.R. and King, W.D., *Text. Res. J.*, 62, 349, 1992.
298. Bertoniere, N.R., King, W.D., and Welch, C.M., *Text. Res. J.*, 64, 247, 1994.
299. Ladisch, C.M. and Yang, Y., *Text. Res. J.*, 62, 481, 1992.
300. Rousselle, M.A., Bertoniere, N.R., Howley, P.S., and Goynes, W.R., Jr., *Text. Res. J.*, 72, 963, 2002.
301. Rousselle, M.A., Bertoniere, N.R., and Howley, P.S., *Text. Res. J.*, 73, 921, 2003.
302. Li, C., Ladisch, C.M., and Ladisch, M.R., *Text. Res. J.*, 71, 407, 2001.
303. Zeronian, S.H., Intercrystalline swelling of cellulose, in *Cellulose Chemistry and Its Applications*, Nevell, T.P. and Zeronian, S.H., Eds., Ellis Horwood Ltd., Chichester, England and Halsted Press, New York, 1985, chap. 5, pp. 138–158.
304. Jeffries, R., *J. Text. Inst.*, 51, T441, 1960.
305. Babbett, J.D., *Can. J. Res.*, A20, 143, 1943.
306. Zeronian, S.H. and Kim, M.S., Proceedings of the 1987 International Dissolving and Specialty Pulps Conference, TAPPI Press, Atlanta, GA, 1987, 125.
307. Weatherwax, R.C., *J. Colloid Interface Sci.*, 49, 40, 1974.
308. Mercer, J., British Patent 13296, 1850.
309. Abrahams, D.H., *Am. Dyestuff Reptr.*, 83(9), 78, 1994.
310. Freytag, R. and Donzé, J.-J., Alkali treatment of cellulose fibers, in *Handbook of Fiber Science and Technology*, Vol. 1, Part A, Lewin, M. and Sello, S.B., Eds., Marcel Dekker, New York, 1983, pp. 93–165.
311. Zeronian, S.H., Intracrystalline swelling of cellulose, in *Cellulose Chemistry and Its Applications*, Nevell, T.P. and Zeronian, S.H., Eds., Ellis Horwood Ltd., Chichester, England and Halsted Press, New York, 1985, chap. 6, pp. 159–180.
312. Vigo, T.L., *Textile Processing and Properties*, Elsevier, Amsterdam, 1994.
313. Zeronian, S.H. and Cabradilla, K.E., *J. Appl. Polym. Sci.*, 17, 539, 1973.
314. Aboul-Fadl, S.M., Zeronian, S.H., Kamal, M.M., Kim, M.S., and Ellison, M.S., *Text. Res. J.*, 55, 461, 1985.
315. Zeronian, S.H., *J. Appl. Polym. Sci., Appl. Polym. Symp.*, 47, 445, 1991.
316. Sarko, A., Recent x-ray crystallographic studies of celluloses, in *Cellulose: Structure, Modification and Hydrolysis*, Young, R.A. and Rowell R.M., Eds., John Wiley & Sons, New York, 1986, pp. 29–49.
317. Stevens, C.V. and Roldán-González, L.G., Liquid ammonia treatment of textiles, in *Handbook of Fiber Science and Technology*, Vol. 1, Part A, Lewin, M. and Sello, S.B., Eds., Marcel Dekker, New York, 1983, pp. 167–203.
318. Bredereck, K., *Mell. Textilber.*, 72(6), 446, 1991.
319. Lewin, M. and Roldan, L.G., *J. Polymer Sci.*, Part C, 36, 213, 1971.
320. Lewin, M., Rau, R.O., and Sello, S.B., *Text. Res. J.*, 44, 680, 1974.
321. Schuerch, C., *For. Prod. J.*, 14, 377, 1964.
322. Schuerch, C., *Ind. Eng. Chem.*, 55, 139, 1963.
323. Schuerch, C., U.S. Patent 3282313, 1966.

324. Schuerch, C., Budrick, M.P., and Mahdalik, M., *I&E.C. Prod. Res. Dev.*, 5, 101, 1966.
325. Pentoney, R.C., *I&E.C. Prod. Res. Dev.*, 5, 105, 1966.
326. Norwegian Patent 152995, 1966.
327. Coats, J. and Coats, P., British Patent 1136417, 1967.
328. Majewicz, T.G. and Podlas, T.J., Cellulose ethers, in *Encyclopedia of Chemical Technology*, 4th ed., Vol. 5, John Wiley & Sons, New York, 1993, pp. 541–563.
329. Savage, A.B., Klug, E.D., Bikales, N.M., and Stanonis, D.J., Cellulose ethers, in *Encyclopedia of Polymer Science and Technology*, Vol. 3, Interscience, New York, 1966, pp. 459–549.
330. Nicholson, M.D. and Merritt, F.M., Cellulose ethers, in *Cellulose Chemistry and Its Applications*, Ellis Horwood Ltd., Chichester, England, 1985, chap. 15, pp. 363–383.
331. Vail, S.L., Crosslinking of cellulose, in *Cellulose Chemistry and Its Applications*, Ellis Horwood Ltd., Chichester, England, 1985, chap. 16, pp. 384–422.
332. Bikales, N.M., Cyanoethylation, in *Encyclopedia of Polymer Science and Technology*, Vol. 4, Interscience, New York, 1966, pp. 533–562.
333. Petersen, H., Crosslinking chemicals and the chemical principles of the resin finishing of cotton, in *Chemical Aftertreatment of Textiles*, Mark, H., Wooding, N.S., and Atlas, S.M., Eds., Wiley Interscience, New York, 1971, pp. 135–233.
334. Ryan, J.J., Wash-and-wear fabrics, in *Chemical Aftertreatment of Textiles*, Mark, H., Wooding, N.S., and Atlas, S.M., Eds., Wiley Interscience, New York, 1971, pp. 417–464.
335. Heap, S.A., Hunt, R.E., Rennison, P.A., and Tattersall, R., Polycondensation products: urea-formaldehyde resin in the treatment of textiles, in *Chemical Aftertreatment of Textiles*, Mark, H., Wooding, N.S., Atlas, S.M., Eds., Wiley Interscience, New York, 1971, pp. 267–317.
336. Herbes, W.F., O'Brien, S.J., and Weyker, R.G, Polycondensation products: melamine-formaldehyde, in *Chemical Aftertreatment of Textiles*, Mark, H., Wooding, N.S., and Atlas, S.M., Eds., Wiley Interscience, New York, 1971, pp. 319–329.
337. Weyker, R.G., O'Brien, S.J., and Herbes, W.F., Polycondensation products: methylol compounds, in *Chemical Aftertreatment of Textiles*, Mark, H., Wooding, N.S., and Atlas, S.M., Eds., Wiley Interscience, New York, 1971, pp. 331–355.
338. Dolby, P.J., *Text. Chem. Color.*, 9(11), 32, 1977; Hilderbrand, D.R., *CHEMTECH*, 8, 224, 1978.
339. Aspland, J.R., Reactive dyes and their application, *Text. Chem. Color.*, 24(5), 31–36, 1992; Aspland, J.R., Practical application of reactive dyes, *Text. Chem. Color.*, 24(6), 35–40, 1992.
340. Reactive dyes, in *Cotton Dyeing and Finishing: A Technical Guide*, Cotton Incorporated, Cary, NC, 1996, pp. 76–92.
341. Shore, J., Dyeing with reactive dyes, in *Cellulose Dyeing*, Shore, J., Ed., The Society of Dyers and Colourists, Bradford, UK, 1995, chap. 4.
342. Siegel, E., Reactive dyes: reactive groups, in *The Chemistry of Synthetic Dyes*, Vol. 6, Venkataraman, K., Ed., Academic Press, New York, 1972, pp. 182–194.
343. Horrocks, A.R., *J. Soc. Dyers Colour.* 16, 62, 1986.
344. Lewin, M. and Sello, S.B., *Flame Retardant Polymeric Materials*, Vol. 1, Lewin, M., Atlas, S.M., and Pierce, E.M., Eds., Plenum Press, New York, 1975, pp. 19–136.
345. Lewin, M. and Bash, A., *Flame Retardant Polymeric Materials*, Vol. 2, Lewin, M., Atlas, S.M., and Pearce, A.M., Eds., Plenum Press, New York, 1978, pp. 1–42.
346. Lewin, M., *Handbook of Fiber Science and Technology*, Vol. 2, *Chemical Processing of Fibers and Fabrics, Part B: Functional Finishes*, Lewin, M. and Sello, S.B., Eds., Marcel Dekker, New York, 1984, pp. 1–141.
347. Wakelyn, P.J., Adair, P.K., and Barker, R.H., *Fire Mater.*, 29, 15, 2005.
348. Miller, B., Martin, J.R., and Turner, R., *J. Appl. Polym. Sci.*, 28, 45, 1983.
349. Ohlemiller, T.J., *Combust. Sci. Technol.*, 26, 89, 1981.
350. Miller, B., Martin, J.R., and Meiser, C.H., Jr., *J. Appl. Polym. Sci.*, 17, 629, 1973.
351. McCarter, R.J., *J. Consumer Prod. Flammability*, 4, 346, 1977.
352. Krasney, J.F., *Text. Chemist and Colorist*, 24(11), 12, 1992.
353. Hshieh, F.Y. and Richards, G.N., *Combust. Flame*, 80, 395, 1989.
354. Hshieh, F.Y. and Richards, G.N., *Combust. Flame*, 76, 49, 1989.
355. Shafizadeh, F. and Sekiguchi, Y., *Combust. Flame*, 55, 171, 1984.

356. Koenig, P.A., Neumeyer, J.P., Knoepfler, N.B., and Vix, H.L.E., Monograph, Organic Coatings and Plastics Chemistry, Vol. 33, No. 1, Presented at the 165th American Chemical Society Meeting, April 1973, pp. 476–483.
357. Weast, R.C., Ed., *Handbook of Chemistry and Physics*, 46th ed., Chemical Rubber Company., Cleveland, OH, 1965, p. C-247.
358. Wakelyn, P.J., Rearick, W.A., and Turner, J., *Am. Dyestuff Reptr.*, 87(2), 13, 1998.
359. Hooper, N.K. and Ames, B.N., *Regulation of Cancer-Causing Flame-Retardant Chemicals and Governmental Coordination of Testing of Toxic Chemicals*, Serial No. 95-33, U.S. Government Printing Office, Washington, D.C., 1977, 42; Studies show flame retardants break down, data said to refute previous industry studies, *BNA Daily Report for Executives*, November 24, 2003, p. 24.
360. Calamari, T.A., Jr. and Harper, R.J., Jr., Flame retardants for textiles, in *Kirk–Othmer Encyclopedia of Chemical Technology*, 4th ed., Vol. 10, Kroschivitz, J.I. and Howe-Grant, M., Eds., John Wiley & Sons, New York, 1993, p. 998.
361. Weil, E.D., *Flame Retardant Polymeric Materials*, Vol. 2, Lewin, M., Atlas, S.M., and Pearce, E.M., Eds., Plenum Press, New York, 1978, p. 123.
362. Beninate, J.V., Boylston, E.K., Drake, G.L., and Reeves, W.A., *Am. Dyest. Rep.*, 57(25), 981, 1968.
363. A performance comparison—Indura Proban cotton vs. Nomex, Westex, Inc., Chicago, IL, 1991.
364. Tesoro, G.C. and Meiser, C., Jr., *Text. Res. J.*, 40, 430, 1970.
365. Kruse, W., *Melliand Textilber*, 50, 460, 1969.
366. Baer, E., Proceedings of the Symposium on Textile Flammability, Lelanc Research Corp., East Greenwich, RI, 1973, p. 117.
367. Mischutin, V., Proceedings of the Symposium on Textile Flammability, LeBlanc, R.B., Ed., LeBlanc Research Corp., East Greenwich, RI, 1975, p. 211.
368. Harper, R.J., Jr., Ruppenicker, G.F., Jr., and Donaldson, D.J., *Text. Res. J.*, 56, 80, 1986.
369. Ruppenicker, G.F., Jr., Harper, R.J., Jr., Sawhney, A.P.S., and Robert, K.Q., *Text. Technol. Forum*, 90, 71, 1990.
370. Sawhney, A.P.S., Ruppenicker, G.F., and Price, J.B., *Text. Res. J.*, 68, 203, 1998.
371. Sawhney, A.P.S., Ruppenicker, G.F., Calamari, T.A., and Parachuru, R., A fire barrier of predominately-cotton content, Proceedings of the Beltwide Cotton Conference, 1365, 1999.
372. Ruppenicker, G.F. and Sawhney, A.P.S., *Text. Sci.*, 93, 642, 2001.
373. Berkley, R., *Text. Chem. Color.*, 25(5), 18, 1993.
374. Donaldson, D.J. and Harper, R.J., Jr., *J. Consum. Prod. Flam.*, 7(1), 40, 1980.
375. Wakelyn, P.J., Adair, P.K., and Wolf, S., Cotton and cotton modacrylic blended batting fire-blocking barriers for soft furnishings to meet federal and state flammability standards, Proceedings of the 2004 Beltwide Cotton Conferences, National Cotton Council, Memphis, TN, 2004, pp. 2829–2842.
376. Welch, C.M. and Andrews, B.K., U.S. Patent 4820307, 1989.
377. Blanchard E.J., Graves, E.E., and Salame, P.A., *J. Fire Sci.*, 18, 151, 2000.
378. Blanchard, E.J. and Grawes, E.E., *Colourage Annu.*, 49, 17, 2001.
379. Blanchard, E.J. and Grawes, E.E., *Text. Res. J.*, 72, 39, 2002.
380. Rearick, W.A., Wallace, J.L., Martin, V.B., and Wakelyn, P.J., *AATCC Rev.*, 2, 12, 2002.
381. Vigo, T.L. and Bruno, J.S., Applications of fibrous substrates containing insolubilized polymers, Proceedings Technology 2002: Third NASA Natl. Tech. Trans. Conf., 2, 307, 1992.
382. Vigo, T.L. and Bruno, J.S., Fibers with multifunctional properties: a holistic approach, in *Handbook of Fibers and Science Technology*, Vol. III, High Technology Fibers, Part C, Lewin, M. and Preston, J., Eds., 1993, pp. 3–356.
383. Gagliardi, D.D. and Shippee, F.B., *Am. Dyestuff Reptr.*, 52, 300, 1963.
384. Rowland, S.P., Welch, C.M., Brannan, M.A.F., and Gallagher, D.M., *Text. Res. J.*, 37, 933, 1967.
385. Welch, C.M., *Text. Res. J.*, 58, 480, 1988.
386. Welch, C.M. and Andrews, B.A.K., U.S. Patent 4936865, 1990.
387. Welch, C.M. and Andrews, B.A.K., U.S. Patent 4975209, 1990.
388. Welch, C.M. and Andrews, B.A., *Text. Chem. Color.*, 21(2), 13, 1989.
389. Welch, C.M., *Text. Chem. Color.*, 22(5), 13, 1990.

390. Welch, C.M., *Surface Characteristics of Fibers and Textiles*, Pastore, D.M. and Kiekens, P., Eds., Marcel Dekker, New York, 2000, chap. 1, pp. 1–32.
391. Isaacs, P., Lewin, M., Sello, S.B., and Stevens, C.V., *Text. Res. J.*, 44, 700, 1974.
392. Lewin, M., Isaacs, P., Sello, S.B., and Stevens, C., *Textilveredlung*, 8, 158, 1973.
393. Lewin, M. and Isaacs, P., German Patent 2127188, 1972.
394. Buras, E.M., Cooper, A.S., Keating, E.J., and Goldthwait, C.F., *Am. Dyestuff Reptr.*, 43(7), P203, 1954.
395. Buras, E.M., Hobart, S.R., Hamalainan, C., and Cooper, A.S., *Text. Res. J.*, 27, 214, 1957.
396. Andrews, B.A.K., *Text. Chem. Color.*, 22(9), 63, 1990.
397. Welch, C.M., *Am. Dyestuff Reptr.*, 83(9), 19, 1994.
398. Yang, C.Q., *Text. Res. J.*, 61, 433, 1991.
399. LeBlanc, R.B. and LeBlanc, D.A., *Text. Chem. Color.*, 6(10), 29, 1974.
400. Little, R.W., Ed., *Flameproofing Textile Fabrics*, Series, No. 104, ACS Monograph, Reinhold, New York, 1947, pp. 196 and 198.
401. LeBlanc, R.B. and Symm, R.H., U.S. Patent 3409463, to Dow Chemical Company, 1968.
402. Edward, J.V., Batiste, S.L., Gibbins, E.M., and Goheen, S.C., *J. Peptide Res.*, 54, 536, 1999.
403. Edwards, J.V., Yager, D.R., Cohen, I.K., Diegelmann, R.F., Montante, S., Bertoniere, N., and Bopp, A.F., *Wound Rep. Reg.*, 9, 50, 2001.
404. Edwards, J.V., Cohen, I.K., Diegelmann, R.F., and Yager, D., Wound Dressings with Protease-Lowering Activity, U.S. Patent 6627785, 2003.
405. Edwards, J.V., Sethumadhavan, K., and Ullah, A.H.J., *Bioconjugate Chem.*, 11, 469, 2000.
406. Grimsley, J.K, Singh, W.P., Wild, J.R., and Giletto, A., A novel enzyme–based method for the wound–surface removal and decontamination of organophosphorus nerve agents, in *Bioactive Fibers and Polymers*, Edwards, J.V. and Vigo, T.L., Eds., ACS Symposium Series, No. 792, American Chemical Society, Washington D.C., 2001, pp. 35–49.
407. Edwards, J.V., Eggleston, G., Yager, D.R., Cohen, I.K., Diegelmann, R.F., and Bopp, A.F., *Carbohydr. Polym.*, 50, 305, 2002.
408. Edwards, J.V., Bopp, A.F., Batiste, S.L., and Goynes, W.R., *J. Biomed. Mater. Res.*, 66A, 433, 2003.
409. Slater, K., *Text. Prog.*, 23/1/2/3, 1, 1991.
410. Whistler, R.L. and BeMiller, J.N., Eds., *Methods in Carbohydrate Chemistry*, Academic Press, New York, Vol. III, 1963, p. 31274; Vol. V, 1965, p. 249; Vol. VI, 1972, p. 76.
411. Girard, A., *Compt. Rend.*, 81, 1105, 1875.
412. Witz, G., *Bull. Soc. Ind. Rouen*, 10, 416, 1982; 11, 169, 1982.
413. Cross, C.F. and Bevan, E.J., *J. Soc. Chem. Ind.*, 3, 206, 291, 1884.
414. Shafizadeh, F., *Cellulose Chemistry and Its Applications*, Nevell, T.P. and Zeronian, S.H., Eds., Ellis Horwood Ltd., Chichester, England, 1985, chap. 11, pp. 266–289.
415. Jayme, G. and Verburg, W., *Reyon, Zellwolle, Chemiefasern*, 32, 193, 275, 1951.
416. Isbell, H.S., *J. Res. NBS*, 32, 54, 1944; *Ann. Rev. Biochem.*, 12, 205, 1943.
417. Davidson, G.F., *J. Text. Inst.*, 43, T291, 1952; *J. Text. Inst.*, 29, T195, 1938.
418. Lewin, M. and Epstein, J.A., *J. Polym. Sci.*, 58, 1023, 1962.
419. Lewin, M. and Albeck, M., *Mild Oxidation of Cotton*, Final Report FG-IS-101-58, Submitted to ARS, USDA, Jerusalem, 1963.
420. Lewin, M., *Handbook of Fiber Science and Technology*, Vol. 1, *Chemical Processing of Fibers and Fabrics, Fundamentals and Preparation*, Part B, Lewin, M. and Sello, S.B., Eds., Marcel Dekker, New York, 1984.
421. Mitchel, R.L., *Ind. Eng. Chem.*, 38, 843, 1946.
422. Cumberbirch, R.J.E. and Harland, W.G., *Shirley Inst. Mem.*, 31, 199, 1958.
423. Samuelson, O., *Methods in Carbohydrate Chemistry*, Whistler, R.L. and BeMiller, J.N., Eds., Academic Press, New York, 1963, p. 31.
424. Samuelson, O. and Wennerblom, A., *Svensk. Papperstidn.*, 58, 713, 1955.
425. Sobue, H. and Okubo, M., *Tappi J.*, 39, 415, 1956.
426. Davidson, G.F., *J. Text. Inst.*, 39, T65, 1948.
427. Ludtke, M., *Angew. Chem.*, 48, 650, 1935.

428. Nabar, G.M. and Padmanabhan, C.V., *Proc. Indian Acad. Sci.*, 31A, 371, 1950.
429. Lefevre, K.V. and Tollens, B., *Ber. Dtsch. Chem. Ges.*, 40, 4513, 1907.
430. Andersen, D.M.W., *Talanta*, 2, 73, 1959.
431. Nevell, T.P., *Methods in Carbohydrate Chemistry*, Vol. III, Whistler, R.L. and BeMiller, J.N., Eds., Academic Press, New York, 1963, p. 49.
432. Swalbe, C.G., *Ber. Dtsch. Chem. Ges.*, 40, 1347, 1907.
433. Clibbens, D.A., *J. Text. Inst.*, 45, P173, 1954.
434. Colbran, R.L. and Davidson, G.F., *J. Text. Inst.*, 52, T291, 1961.
435. Davidson, G.F. and Nevell, T.P., *J. Text. Inst.*, 46, T407, 1955.
436. Davidson, G.F. and Nevell, T.P., *J. Text. Inst.*, 48, T356, 1957.
437. Ellington, A.C. and Purves, C.B., *Can. J. Chem.*, 31, 801, 1953.
438. Lewin, M., *Methods in Carbohydrate Chemistry*, Vol. VI, Whistler, R.L. and BeMiller, J.N., Eds., Academic Press, New York, 1972, p. 76.
439. Schmorak, J. and Lewin, M., *Anal. Chem.*, 33, 1403, 1961.
440. Godsay, M. and Lewin, M., Cellulose and wood chemistry and technology; Proceedings of the 10th Cellulose Conference, Schuerch, C., Ed., Wiley Interscience, Syracuse, 1990, pp. 1059–1084.
441. Albeck, M., Ben-Bassat, A., and Lewin, M., *Text. Res. J.*, 35, 935, 1965.
442. Lewin, M., *Mild Oxidation of Cotton*, Project FG-IS-109, Final Report, submitted to the ARS USDA, Jerusalem, 1968.
443. Lewin, M. and Ettinger, A., *Cellul. Chem. Technol.*, 3, 9, 1969.
444. Isbell, H.S., *Methods in Carbohydrate Chemistry*, Vol. V, Whistler, R.L. and BeMiller, J.N., Eds., Academic Press, New York, 1965, p. 249.
445. BeMiller, J.N., *Adv. Carbohydr. Chem.*, 22, 25, 1967.
446. Sharples, A., *J. Polym. Sci.*, 14, 95, 1954.
447. Daruwalla, E.H. and Narsion, M.G., *Tappi J.*, 49, 106, 1966.
448. Michie, R.I., Sharples, A., and Walter, A.A., *J. Polym. Sci.*, 51, 85, 1961.
449. Sippel, A., *Das Papier*, 13, 413, 1959.
450. Nelson, M. and Tripp, V.W., *J. Polym. Sci.*, 10, 577, 1953.
451. Batista, O.A., *Microcrystal Polymer Science*, McGraw-Hill, New York, 1975.
452. Wood, B.F., Conner, A.H., and Hill, C.G., Jr., *J. Appl. Polym. Sci.*, 37, 1373, 1989.
453. Lin, C.H., Conner, A.H., and Hill, C.G., Jr., *J. Appl. Polym. Sci.*, 42, 417, 1991.
454. Nevell, T.P. and Upton, W.R., *Carbohydr. Res.*, 49, 163, 1976.
455. Davidson, G.F., *J. Text. Inst.*, 25, T174, 1934.
456. Richards, G.N. and Shepton, H.H., *J. Chem. Soc.*, 64, 4492, 1957.
457. Machell, G.N. and Richards, G.N., *J. Chem. Soc.*, 64, 4500, 1957.
458. Machell, G.N. and Richards, G.N., *J. Chem. Soc.*, 69, 1932, 1960.
459. Lobry de Bruyn, C.A. and Alberda van Ekenstein, W., *Recl. Trav. Chim.*, 14, 201, 1895; 16, 262, 1897; *Ber. Dtsch. Chem. Ges.*, 28, 3078, 1895.
460. Speck, J.C., *Adv. Carbohydr. Chem.*, 13, 63, 1958.
461. Corbett, W.M., *J. Soc. Dyers Col.*, 76, 265, 1960.
462. Haas, D.W., Hrutfiord, B.F., and Sarkanen, K.V., *J. Appl. Polym. Sci.*, 11, 587, 1967.
463. Lai, Y-Z. and Outto, D.E., *J. Appl. Polym. Sci.*, 23, 3219, 1979.
464. Johansson, M.H. and Samuelson, O., *J. Appl. Polym. Sci.*, 22, 615, 1978.
465. Albeck, A., Ben-Bassat, A., Epstein, J.A., and Lewin, M., *Text. Res. J.*, 35, 836, 1965; *Isr. J. Chem.*, 1(3a), 304, 1963.
466. Lewin, M., *Text. Res. J.*, 35, 979, 1965.
467. Ziderman, I., Bel-Aiche, J., Basch, A., and Lewin, M., *Carbohydr. Res.*, 43, 255, 1975.
468. Lewin, M., Ziderman, I., Weiss, N., Basch, A., and Ettinger, A., *Carbohydr. Res.*, 62, 393, 1978.
469. Ziderman, I., *Cellul. Chem. Technol.*, 14, 703, 1980.
470. Lewin, M. and Ziderman, I., *Cellul. Chem. Technol.*, 14, 743, 1980.
471. Gascoigne, J.A. and Gascoigne, M.M., *Biological Degradation of Cellulose*, Butterworth & Co., London, England, 1960.

472. Finch, P. and Roberts, J.C., Enzymatic degradation of cellulose, in *Cellulose Chemistry and Its Applications*, Nevell, T.P. and Zeronian, S.H., Eds., Ellis Horwood Ltd., Chichester, England, 1985, chap. 13, pp. 312–343.
473. Griffin, D.H., *Fungal Physiology*, Wiley–Liss, New York, 1994, pp. 160–163.
474. Misaghi, I.J., *Physiology and Biochemistry of Plant–Pathogen Interactions*, Plenum Press, New York, 1982, p. 25.
475. Leschine, S.B., *Annu. Rev. Microbiol.*, 49, 399, 1995.
476. Weimer, P.J. and Odt, C.L., Cellulose degradation by ruminal microbes: physiological and hydrolytic diversity among ruminal cellulolytic bacteria, in *Enzymatic Degradation of Insoluble Carbohydrates*, Saddler, J.N. and Penner, M.H., Eds., ACS Symposium Series, No. 618, American Chemical Society, Washington, D.C., 1995, chap. 18.
477. Stanier, R.Y., Adelberg, E.A., and Ingraham, J.L., *The Microbial World*, Prentice Hall, Englewood Cliffs, NJ, 1976, pp. 780–781.
478. Hamby, D.S., Ed., *The American Cotton Handbook*, 3rd ed., Vol. I, Interscience, New York, 1965, chap. 2, p. 35.
479. Hall, L.T. and Elting, J.P., *Text. Ind.*, 117, 100, 1953.
480. Hegn, A.N.J., *Text. Res. J.*, 120, 137, 1956.
481. Marsh, P.B., Guthrie, L.R., and Butler, M.L., *Text. Res. J.*, 21, 565, 1951.
482. Hegn, A.N.J., *Text. Res. J.*, 27, 591, 1957.
483. Chun, D., Use of high cotton moisture content during storage to reduce stickiness, Proceedings of the Beltwide Cotton Research Conference, Vol. 2, 1997, pp. 1642–1648.
484. Allen, S.J., Auer, P.D., and Pailthorpe, M.T., *Text. Res. J.*, 65, 379, 1995.
485. Interpretation and substantiation of environmental claims, U.S. Code of Federal Regulations, 16, Part 260, p. 5.
486. U.S. EPA, Federal Register, Vol. 57, November 19, 1992, p. 54456.
487. *1988 Book of ASTM Standards*, American Society for Testing and Materials, Philadelphia, PA, 1988.
488. Moreau, J.P., *J. Nonwoven Res.*, 10, 14–22, 1990; Goynes, W.R., Delucca, A.J., and Moreau, J.P., Structural evaluation of biodegradability of nonwoven fabrics, Proceedings of the 49th Annual Meeting of Electron Microscopy Society of America, 1991.
489. OECD guidelines for testing of chemicals, Organization for Economic Co-operation and Development, Paris, France, 1992.
490. Shafizadeh, F., *Advances in Carbohydrate Chemistry* Vol. 23, Wolfrom, M.C. and Tipson, R.S., Eds., Academic Press, New York, 1968, p. 419.
491. Madorsky, S.L., Hart, V.E., and Strauss, S., *J. Res. Natl. Bur. Stand.*, 56, 343, 1956; 58, 343, 1958.
492. Lipska, A.E. and Parker, W.J., *J. Appl. Polym. Sci.*, 10, 1439, 1965.
493. Akita, K., *Rep. Fire Res. Inst. Jpn.*, 9, 10, 1959.
494. Stamm, A.J., *Ind. Eng. Chem.*, 48, 413, 1956.
495. Tang, W.K. and Neil, W., *J. Polym. Sci.*, C, 6, 65, 1964.
496. Chatterjee, P.K. and Conrad, C.M., *Text. Res. J.*, 36, 487, 1966.
497. Chatterjee, P.K., *J. Appl. Polym. Sci.*, 12, 1859, 1968.
498. Basch, A. and Lewin, M., *J. Polym. Sci., Polym. Chem. Ed.*, 11, 3077–3093, 3095–3107, 1973; 12, 2053–2063, 1974; 13, 493–499, 1975.
499. Lewin, M., Basch, A., and Roderig, C., Proceedings of the International Symposium on Macromolecules, Rio de Janeiro, Mano, E., Ed., Elsevier, Amsterdam, 1975, pp. 225–250.
500. Lewin, M. and Basch, A., *Encyclopedia of Polymer Science and Technology*, Suppl. 2, Wiley, New York, 1977, pp. 340–363.
501. Tyler, D.N. and Wooding, N.S., *J. Soc. Dyers Col.*, 74, 283, 1958.
502. Franklin, W.E. and Rowland, S.P., *J. Macromol. Sci., Chem.*, A19, 165, 1983.
503. Franklin, W.E., *J. Macromol. Sci., Chem.*, A19, 619, 1983.
504. Franklin, W.E., *J. Macromol. Sci., Chem.*, A21, 377, 1984.
505. Franklin, W.E., Proceedings of the Eleventh North American Thermal Analysis Society Conference, Vol. 11, 1981, p. 471.

506. Rushnak, I. and Tanczos, I., Preprints of papers presented at IUPAC Symposium on Macromolecules, Helsinki, Vol. 5, 1972, p. 127.
507. Back, F.L. and Klinga, L.O., *Svensk Papperstidn*, 66, 745, 1965.
508. Back, F.L. and Didriksen, E.I., *Svensk Papperstidn*, 72, 687, 1969.
509. Brushwood, D.E., *Text. Res. J.*, 58, 309, 1988.
510. Manley-Harris, M. and Richards, G.N., *Carbohydr. Res.*, 254, 195, 1994.
511. Phillips, G.O. and Arthur, J.C., Jr., Photochemistry and radiation chemistry of cellulose, in *Cellulose Chemistry and Its Applications*, Nevell, T.P. and Zeronian, S.H., Eds., Ellis Horwood Ltd., Chichester, England, 1985, chap. 12, pp. 290–311.
512. Baugh, P.J. and Phillips, G.O., *Cellulose and Cellulose Derivatives*, High Polymers Part V, Bikales, N.M. and Segal, L., Eds., John Wiley & Sons, New York, 5, 1971, p. 1047.
513. McKellar, J.F., *Radiat. Res. Rev.*, Elsevier, Amsterdam, 3, 141, 1971.
514. Phillips, G.O., *The Carbohydrates*, 2nd ed., Vol. IB, Pigman, W. and Horton, D., Eds., Academic Press, New York, 1980, pp. 1217–1299.
515. Phillips, G.O., Baugh, P.J., McKellar, J.F., and Von Sonntag, C., Interaction of radiation with cellulose in the solid state, in *Cellulose Chemistry and Technology*, Arthur, J.C., Jr., Ed., ACS Symposium Series, No. 48, American Chemical Society, Washington, D.C., 1977, p. 313.
516. Launer H.F. and Wilson, W.K., *J. Am. Chem. Soc.*, 71, 958, 1949.
517. Egerton, G. S., *J. Soc. Dyers Colour.*, 65, 764, 1949.
518. Appleby, D.K., *Am. Dyestuff Reptr.*, 38, 149, 1949.
519. Egerton, G.S., *Text. Res. J.*, 18, 659, 1948.
520. Bernard, W.N., Gremillion, S.G., Jr., and Goldthwait, C.F., *Text. Res. J.*, 26, 81, 1956.
521. Blouin, F.A. and Arthur, J.C., Jr., *J. Chem. Eng. Data*, 5, 470, 1960.
522. Arthur, J.C., Jr., Free-radical initiated graft polymerization of vinyl monomers onto cellulose, in *Graft Copolymerization of Lignocellulosic Fibers*, Hon, D.N.-S., Ed., ACS Symposium Series, No. 187, American Chemical Society, Washington D.C., 1982, p. 21.
523. Arthur, J.C., Graft polymerization onto polysaccharides, in *Advances in Macromolecular Chemistry*, Pasik, W.M., Ed., Academic Press, London, England, 1970, p. 1.
524. Arthur, J.C., Jr., Special properties of block and graft copolymers and applications in fiber form, in *Block and Graft Copolymers*, Burke, J.J. and Weiss, V., Eds., Syracuse University Press, Syracuse, New York, 1973, p. 295.
525. Blumenthal, W.B., *Ind. Eng. Chem.*, 42, 640, 1950.
526. Conner, C.J., Brysson, R.J., Walker, A.M., Harper, R.J., Jr., and Reeves, W.A., *Text. Chem. Color.*, 10(4), 17, 1978.
527. Matthews, J.M., *Application of Dyestuffs*, John Wiley & Sons, New York, 1920, p. 521.
528. Conner, C.J., Danna, G.S., Cooper, A.S., Jr., and Reeves, W.A., *Text. Res. J.*, 37(2), 94, 1967.
529. Kirk, O., Damhus, T., Borchert, T.V., Fuglsang, C.C., Olsen, H.S., Hansen, T.T., Lund, H., Schiff, H.E., and Nielsen, L.K., Enzyme applications (industrial), in *Kirk–Othmer Encyclopedia of Chemical Technology*, 5th ed., Vol. 9, John Wiley & Sons, New York, 2004.
530. Betrabet, S.M., *Colourage*, 41(5), 21, 1994.
531. Dominguez, R., Souchon, H., Spinelli, S., Dauter, Z., Wilson, K.S., Chauvaux, S., Begium, P., and Alizari, P.M., *Nat. Struct. Biol.*, 2, 569, 1995.
532. Rouvinen, J., Bergfors, T., Teeri, T., Knowles, J.K.C., and Jones, T.A., *Science*, 249, 380, 1990.
533. Davies, G. and Henrissat, B., *Structure*, 3, 853, 1995.
534. Buschle-Diller, G., Zeronian, S.H., Pan, N., and Soon, M.Y., *Text. Res. J.*, 64(56), 270, 1994.
535. Reese, E.T., Segal, L., and Tripp, V.W., *Text. Res. J.*, 27(8), 626, 1957.
536. Gascoigne, A. and Gascoigne, M.M., *Biological Degradation of Cellulose*, Butterworth & Co., London, Great Britain, 1960, p. 72.
537. Tyndall, M., *Text. Chem. Color.*, 24(6), 23, 1992.
538. Ajgaonkar, S.B., *Colourage*, 42(1), 35, 1995.
539. Cox, T., Hawks, P.E., and Klahorst, S.A., U.S. Patent 5232851, August 3, 1993.
540. Koo, H., Ueda, U., Wakida, T., Yoshimura, Y., and Igarashi, T., *Text. Res. J.*, 64(2), 70, 1994.
541. Thorsen, W.J., *Text. Res. J.*, 41, 331, 1971.
542. Thorsen, W.J., *Text. Res. J.* 41, 455, 1971.

543. Thorsen, W.J., *Text. Res. J.*, 44, 422, 1974.
544. Thibodeaux, D.P. and Copeland, H.R., Corona treating cotton: its relationship to processing performance and quality, *Proceedings of the 15th Textile Chemistry and Processing Conference*, New Orleans, 1975, 36.
545. Belin, R.E., *J. Text. Inst.*, 67, 249, 1976.
546. Abbott, G.M., *Text. Res. J.*, 47, 141, 1977.
547. Abbott, G.M. and Robinson, G.A., *Text. Res. J.*, 47, 199, 1977.
548. Anon., Preparation of cotton yarns, knits, and woven fabrics, *Cotton Dyeing and Finishing: A Technical Guide*, Cotton Incorporated, Cotton Incorporated, Cary, NC, 1996, pp. 1–26.
549. Rivlin, J., *The Dyeing of Textile Fibers—Theory and Practice*, Philadelphia College of Textiles and Science, Philadelphia, PA., 1992, 72.
550. Aspland, J.R., *Text. Chem. Color.*, 23(11), 41, 1991.
551. Aspland, J.R., *Text. Chem. Color.*, 24(3), 21, 1992.
552. Aspland, J.R., *Text. Chem. Color.*, 24(1), 22, 1992.
553. Horne, C.M., *Text. Chem. Color.*, 27(12), 27, 1995.
554. Aspland, J.R., *Text. Chem. Color.*, 24(8), 26, 1992; Aspland, J.R., *Text. Chem. Color.*, 24(9), 74, 1992.
555. Gregory, P., Dyes and dye intermediates, in *Kirk–Othmer Encyclopedia of Chemical Technology*, 4th ed., Vol. 8, John Wiley & Sons, New York, 1993, p. 542.
556. McGregor, R., *Text. Chem. Color.*, 12(12), 19, 1980.
557. Anon., Dyes, in *Cotton Dyeing and Finishing: A Technical Guide*, Cotton Incorporated, Cary, NC, 1996. pp. 28–132.
558. Bide, M., Dyes, in *Kirk–Othmer Encyclopedia of Chemical Technology*, 5th ed., Vol. 8, John Wiley & Sons, New York, 2004.
559. Dolby, P.J., *Text. Chem. Color.*, 9(11), 264, 1977.
560. Aspland, J.R.,*Text. Chem. Color.*, 25(10), 31, 1993.
561. Blanchard, E.J. and Reinhardt, R.M.,*Text. Chem. Color.*, 24(1), 13, 1992.
562. Harper, R.J., Jr., Blanchard, E.J., Allen, H.A., Reinhardt, R.M., Cheek, L., English, S., Etters, J.N., Hsu, L.H., and Roussel, L., *Text. Chem. Color.*, 20(5), 25, 1988.
563. Reinhardt, R.M. and Blanchard, E.J., *Am. Dyestuff Reptr.*, 79(6), 15, 1990.
564. Blanchard, E.J., Reinhardt, R.M., Andrews, B.A.K, *Text. Chem. Color.*, 23(5), 25, 1991.
565. Andrews, B.A.K., Blanchard, E.J., and Reinhardt, R.M., *Am. Dyestuff Reptr.*, 79(9), 48, 1990.
566. Lord, E. and Heap, S.A., *The Origin and Assessment of Cotton Fiber Maturity*, International Institute for Cotton, Manchester, England, 1988.
567. Pierce, F.T. and Lord, E., *J. Text. Inst.*, 30, T173, 1939.
568. Raes, G.T.J. and Verschraege, L., *J. Text. Inst.*, 72, 191, 1981.
569. Standard Test Method for Maturity of Cotton Fibers (Sodium Hydroxide Swelling and Polarized Light Procedures), Vol. 07.01, ASTM Designation D1442-00, December 2000.
570. Smith, J.C., McCracken, F.L., Schiefer, H.F., and Stone, W.K., *Text. Res. J.*, 26, 281, 1956.
571. Calkins, E.W.S., *Text. Res. J.*, 6, 441, 1946.
572. Thibodeaux, D.P. and Evans, J.P., *Text. Res. J.*, 56, 130, 1986.
573. Boylston, E.K., Thibodeaux, D.P., and Evans, J.P., *Text. Res. J.*, 60, 80, 1993.
574. Thibodeaux, D.P. and Price, J.B., *Melliand Textilber.*, 70, 243, 1989.
575. Lunenschloss, J., Gilhaus, K., and Hoffman, K., *Melliand Textilber.*, 61, 5, 1980.
576. Hertel, K.L. and Craven, C.J., *Text. Res. J.*, 21, 765, 1951.
577. Ghosh, S., *Text. World*, 28, 45, 1985.
578. Montalvo, J.G., Faught, S.E., Buco, S.M., Saxton, and A.M., *Appl. Spectrosc.*, 41, 645, 1987.
579. Linear Density of Cotton Fibers (Array Sample), ASTM Designation D1769–77, 1982.
580. Lord, E., *J. Text. Inst.*, 47, T16, 1956.
581. Morton, W.E. and Hearle, J.W.S., *Physical Properties of Textile Fibers*, John Wiley & Sons, New York, 1976.
582. Meredith, R., *J. Text. Inst.*, 36, T107, 1945.
583. Hearle, J.W.S., *J. Appl. Polym. Sci., Appl. Polym.Symp.*, 47, 1, 1991.
584. Hebert, J.J., Thibodeaux, D.P., Shofner, F.M., Singletary, J.K., and Patelke, D.B., *Text. Res. J.*, 65, 440, 1995.

585. Lord, E., *Manual of Cotton Spinning, Vol. II, part 1—The Characteristics of Raw Cotton*, Textile Book Publishers, Inc., New York, 1961, p. 214.
586. Meredith, R., *J. Text. Inst.*, 37, T205, 1946.
587. Alexander, E., Lewin, M., Musham, H.V., and Shiloh, M., *Text. Res. J.*, 26, 606, 1956.
588. Hearle, J.W.S., *Text. Res. J.*, 24, 307, 1954.
589. Walker, A.C. and Quell, M.H., *J. Text. Inst.*, 24, T123, 1933.
590. Hearle, J.W.S., *J. Text. Inst.*, 44, T117, 1953.
591. Davidonis, G.H., Johnson, A., Landivar, J.A., and Hood, K.B., *Text. Res. J.*, 69, 754, 1999.
592. Zurek, W., Greszta, M., and Frydrych, I., *Text. Res. J.*, 69, 804, 1999.
593. Krowicki, R.S., Hinojosa, O., Thibodeaux, D.P., and Duckett, K.E., *Text. Res. J.*, 66, 70, 1996.
594. Doberczak, A., Dowgielewicz, St., and Zurek, W., *Cotton, Bast, and Wool Fibers*, (translated from Polish) published for U.S. Department. of Agriculture and the National Science Foundation, Washington, D.C. by Centralny Instytut Informacji Naukowo-Technicznej I Ekonomiczncj, Warszawa, Poland, 1964, pp. 16–160.
595. Hunter, L., *Textiles: Some Technical Information and Data V: Cotton*, South Africa Wool and Textile Research Institute Special Publication, Port Elizabeth, Republic of South Africa, 1981.
596. Cotton, *Technical Monograph*, No. 3, Ciba–Geigy Agrochemicals, Ciba–Geigy, Ltd., Basel, Switzerland, 1972.
597. Elliot, F.C., Hoover, M., and Porter, W.K., Jr., *Cotton: Principles and Practices*, The Iowa State University Press, Ames, IA, 1968.
598. Reeves, B.G. and Garner, W.E., *CRC Handbook of Transportation and Marketing in Agriculture*, Vol. II, *Field Crops*, The Chemical Rubber Company., Cleveland, OH, 1982, pp. 353–384.
599. Jordan, A.G. and Needham, D.K., Cotton bale weight, density and size standardization, Proceedings of the International Cottontest-Conference, Bremen, Faserinstitut Bremen eV, Bremen, West Germany, 1982.
600. Hinojosa, O. and Thibodeaux, D.P., *Text. Chem. Color.*, 25(1), 27, 1993.
601. The Classification of Cotton, in *Agricultural Handbook Number*, 566, Agricultural Marketing Service, U.S. Department. of Agriculture, Washington, D.C., 1993.
602. Anon., Cotton management in China, Indian Cotton Mill Federation, June 1989, pp. 79–81.
603. Ustyugin, V.E., Development of standards and certification of cotton fibre in Uzbekistan, Fourth International Cotton Conference, Gdynia, Poland, September 14–15, 1995, pp. 33–40.
604. Macdonald, A.G., Expert panel on commercial standardization of instrument testing of cotton (CSITC), Proceedings of the International Cotton Conference, Bremen, Faserinstitut Bremen eV, 2004.
605. Belhia, A., The Cotton industry in Azerbaijan, *Cotton International*, 1992, p. 192.
606. Personal communication, Eric Hequet, Head Cotton Technology Laboratory, Centre de Cooperation Internationale en Recherche Agronomique Pour le Development (CIRAD), Paris, France, 1995.
607. Mursal, I.E., Field classification restores Sudan cotton quality, *Cotton International*, 1994, p. 219.
608. Personal communication, Hein Schroder, Quality Control Manager, South Africa Cotton Board, 1995.
609. Assal, A., and El, S., Adoption of new instrumental grading system for Pakistan's cotton, Proceedings of the Internatonal Committee on Cotton Testing Methods, Bremen, Germany, March 1–2, 1994.
610. *Natural Threads, The Australian Cotton Story*, Australian Cotton Foundation, Ltd., Waterloo, Australia, 1993.
611. Hunter, L., Proceedings of the 27th International Cotton Conference, Bremen, Faserinstitut, Bremen, eV, 2004, pp. 62–70.
612. Suh, M.W. and Sasser, P.E. *J. Text. Inst.*, 87(3), 43, 1996.
613. Townsend, T.P., Commercial standardization of instrument testing of cotton: How soon a reality?, Proceedings of the 2005 Beltwide Cotton Conference, National Cotton Council, Memphis, TN, 2005, pp. 2386–2389.
614. Adams, G., Slinsky, S., Boyd, S., and Huffman, M., *The Economic Outlook for U.S. Cotton*, 1996 edition, Economic Services, National Cotton Council of America, Memphis, TN, 2005.

615. Meyer, L., MacDonald, S., and Skinner, R., Cotton and wool outlook, USDA, CWS-05e, June 13, 2005 [Electronic Outlook Report from the Economic Research Service, www.ers.usda.gov].
616. Anon., *Cottonseed and Its Products*, 9th ed., National Cottonseed Products Association, Memphis, TN, 1990; O'Brien, R.D., Jones, L.A., King, C.C., Wakelyn, P.J., and Wan, P.J., Cottonseed Oil, in *Bailey's Industrial Oil and Fat Products*, 6th ed., Vol. 2, Shadida, F., Ed., John Wiley & Sons, New York, 2005, chap. 5, pp. 173–279.
617. Huffman, M., *Cotton Counts Its Customers (The Quality of Cotton Consumed in Final Uses Produced in the United States)*, 2004 edition, National Cotton Council of America, Memphis, TN, 2004.
618. Anon., *Cotton Dyeing and Finishing: A Technical Guide*, Cotton Incorporated, Cary, NC, 1996.
619. Scheurell, D.M., Spivak, S.M., and Hollies, N.R.S., *Text. Res. J.*, 55, 394, 1985.
620. Choi, H., Moreau, J.P., and Srinivasan, M., *J. Environ. Sci. Health*, A29(10), 2151, 1994.
621. Choi, H. and Cloud, R.M., *Environ. Sci. Technol.*, 26, 772, 1992.
622. *Fibers as Renewal Resources for Industrial Materials*, National Academy of Sciences, Washington, D.C., 1976.
623. *Renewal Resources for Industrial Materials*, National Academy of Sciences, Washington, D.C., 1976.
624. *Long-Term Agricultural Baseline Projections 1995–2005*, U.S. Department of Agriculture, World Agricultural Outlook Board, Staff Report, WAOB-95-1, 1995.
625. Robinson, E., Cotton Analysts to Come out Swinging in July, Delta Farm Press, June 29, 2005.
626. Strahl, W.A., Cotton in rugs and carpets, Proceedings of the 1996 Beltwide Cotton Research Conference, National Cotton Council, Memphis, TN, 1996, pp. 1501–1507.
627. Wakelyn, P.J., Menchey, K., and Jordan, A.G., Cotton and environmental issues, in *Cotton— Global Challenges and the Future*, Papers Presented at a Technical Seminar at the 59th Plenary Meeting of the International Cotton Advisory Committee (ICAC), Cairns, Australia, November 9, 2000, pp. 3–11.
628. King, E.G., Phillips, J.R., and Coleman, R.J., Eds., *Cotton Insects and Mites: Characterization and Management*, The Cotton Foundation Reference Book Series, No. 3, The Cotton Foundation, Memphis, TN, 1996.
629. McWorter, C.G. and Abernathy, J.R., Eds., *Weeds of Cotton: Characterization and Control*, The Cotton Foundation Reference Book Series, No. 2, The Cotton Foundation, Memphis, TN, 1996.
630. Kirkpatrick, T.L. and Rothrock, C.S., Eds., *Compendium of Cotton Diseases*, 2nd ed., APS Press, St. Paul, MN, 2001.
631. *Cotton Nematodes, Your Hidden Enemies: Identification and Control*, The Cotton Foundation, National Cotton Council, and Aventis Crop Science, 2002.
632. Cotty, P.J., Cottonseed losses and mycotoxins, in *Compendium of Cotton Diseases*, 2nd ed., Kirkpatrick, T.L. and Rothrock, C.S., Eds., APS Press, St. Paul, MN, 2001, pp. 9–13.
633. Cotty, P.J., *Phytopathology*, 84, 1270, 1994.
634. Castellan, R.M., Olenchock, S.A., Kingsley, K.B., and Hankinson, J.L., *N. Engl. J. Med.*, 317, 605, 1987.
635. Jacobs, R.R. and Wakelyn, P.J., Assessment of toxicology (respiratory risk) associated with airborne fibrous cotton-related dust, Proceedings of the 1998 Beltwide Cotton Conference, National Cotton Council, Memphis, TN, 1998, pp. 213–220.
636. Cotton dust, U.S. Code of Federal Regulations, 29CFR1910.1043, (a) Scope and application and (c) Permissible exposure limits and action levels.
637. Wakelyn, P.J., Greenblatt, G.A., Brown, D.F., and Tripp, V.W., *Am. Ind. Hyg. Assoc. J.*, 37, 22, 1976.
638. Pickering, C.A.C., The search for the aetiological agent and pathogenic mechanisms of byssinosis: a clinician's view of byssinosis, Proceedings of the 15th Cotton Dust Research Conference, Jacobs, R.R., Wakelyn, P.J., and Domelsmith, L.N., Eds., National Cotton Council, Memphis, TN, 1991, pp. 298–299.
639. Rohrbach, M.S., The search for the aetiological agent and pathogenic mechanisms of byssinosis: a review of *in vitro* studies, Proceedings of the 15th Cotton Dust Research Conference, Jacobs, R.R., Wakelyn, P.J., and Domelsmith, L.N., Eds., National Cotton Council, Memphis, TN, 1991, pp. 300–306.

640. Nichols, P.J., The search for the aetiological agent and pathogenic mechanisms of byssinosis: *in vivo* studies, Proceedings of the 15th Cotton Dust Research Conference, Jacobs, R.R., Wakelyn, P.J., and Domelsmith, L.N., Eds., National Cotton Council, Memphis, TN, 1991, pp. 307–320.
641. Glindmeyer, H.W., Lefants, J.J., Jones, R.N., Rando, R.J., Kader, H.N.A., and Weill, H., *Am. Rev. Respir. Dis.*, 144, 675, 1991.
642. Perkins, H.H., Jr., and Olenchock, S.A., *Annu. Agric. Environ. Med.*, 2, 1, 1995.
643. The Task Force for Byssinosis Prevention, Washed cotton, a review and recommendations regarding batch kier washed cotton, *Current Intelligence Bulletin*, 56, U.S. Department of Health and Human Service, NIOSH, Aug. 1995.
644. Occupational exposure to cotton dust, Final Rule, Federal Register, 50, pp. 51120–51179, December 13, 1985; Washed cotton, 29 CFR 1910.1043(n); Occupational exposure to cotton dust, Direct Final Rule, Federal Register, 65, pp. 76563–76567, December 7, 2000.
645. Formaldehyde, U.S. Code of Federal Regulations, 29 CFR 1910.1048.
646. Assessment and control of indoor air pollution, Report to Congress on Indoor Air Quality, Vol. 1, Office of Air and Radiation, U.S. Environmental Protection Agency, 1989; Formaldehyde risk-assessment update, Office of Toxic Substances, U.S. Environmental Protection Agency, Washington, D.C., June 11, 1991.
647. IARC classifies formaldehyde as carcinogenic to humans, International Agency for Research on Cancer, June 2004.
648. Robbins, J.D., Norred, W.P., Bathija, A., and Ulsamer, A.G., *J. Toxicol. Environ. Health*, 14, 453, 1984.
649. Robins, J.D. and Norred, W.P., Bioavailability in Rabbits of Formaldehyde from Durable Press Textiles, Final Report on CPSC IAG 80-1397, USDA Toxicology and Biological Constituents Research Unit, Athens, GA, 1984.
650. *Status Report on Formaldehyde in Textiles Portion of Dyes and Finishes Project*, U.S. Consumer Product Safety Commission, Washington, D.C., January 3, 1984.

10 Regenerated Cellulose Fibers

*Richard Kotek**

CONTENTS

10.1 Introduction 668
10.2 Cellulose Solvents 668
    10.2.1 N-Methylmorpholine-N-Oxide and Water 669
    10.2.2 N,N-Dimethylacetamide and Lithium Chloride 669
    10.2.3 Trifluoroacetic Acid and Chlorinated Alkanes 670
    10.2.4 Calcium Thiocyanate and Water 671
    10.2.5 Ammonia/Ammonium Thiocyanate 671
    10.2.6 Steam Explosion and Aqueous Solutions of Sodium Hydroxide 673
    10.2.7 Ionic Liquids 673
    10.2.8 Phosphoric Acid 674
10.3 Nonviscose Rayon Processes 675
    10.3.1 N-Methylmorpholine-N-Oxide or Lyocell Process 675
        10.3.1.1 Preparation and Properties of N-Methylmorpholine-N-Oxide 676
        10.3.1.2 Preparation and Properties of Cellulose–N-Methylmorpholine-N-Oxide Solutions 678
        10.3.1.3 Side Reactions and By-Product Formations in N-Methylmorpholine-N-Oxide–Cellulose Solutions 688
        10.3.1.4 Fiber Formation and Properties of Lyocell Fibers 691
    10.3.2 Phosphoric Acid Process (Akzo Process) 698
    10.3.3 Carbamate Process 703
    10.3.4 Cuprammonium Process 708
    10.3.5 Other Potential Processes 709
        10.3.5.1 Amine–Salt Process 709
        10.3.5.2 Cellulose Carbonate Process 710
10.4 Rayon Process 711
    10.4.1 Introduction 711
    10.4.2 History 712
    10.4.3 Viscose Rayon 716
        10.4.3.1 Early Production 717
    10.4.4 Chemistry of Viscose Rayon Process 720
        10.4.4.1 Steeping 720
        10.4.4.2 Shredding 722
        10.4.4.3 Aging 722
        10.4.4.4 Xanthation 723
        10.4.4.5 Mixing 725

---

*The authors of the original chapter published in the first and second editions are John Dyer and George C. Daul, Eastern Research Division, ITT Rayonier, Inc., Stanford, Connecticut.

		10.4.4.6	Filtration	725
		10.4.4.7	Ripening	725
		10.4.4.8	Spinning	726
	10.4.5	Production of Viscose Rayon		732
	10.4.6	Types of Rayon		744
	10.4.7	Rayon Structure		748
	10.4.8	Rayon Properties and Uses		756
		10.4.8.1	Industrial Yarns	757
		10.4.8.2	Textile Rayons	758
Reference				764

10.1 INTRODUCTION

Cellulose, which is found in plant walls, is the most abundant raw material on Earth. Millions of pounds of this biorenewable polymer are produced every year. The total worldwide consumption of cellulosic fibers in 1998 was 4817 million pounds [1]. Cellulose is plentiful, inexpensive, and biodegradable. It is capable of producing a number of fibrous products with excellent properties whose utility extends into numerous end uses and industries. Cellulose is an excellent source of textile fibers, for both the commodity and the high-end, fashion-oriented markets. A common example is rayon. In addition, cellulose provides fibers for industrial end uses requiring strong, tough fibers. A common example is fibers used in tire cord.

Because of the strong intermolecular bonds, cellulose does not melt and does not dissolve readily in ordinarily available solvents; chemists have resorted to the derivatization of cellulose to render it soluble and processable. Specifically, the viscose process was developed. It converted cellulose into sodium cellulose xanthate, which was soluble in a caustic solution, making it possible to wet-spin the polymer into a fiber or film. This technique was accepted worldwide and has prospered. The process, however, consists of multiple steps and causes pollution. As a result, end users have looked for alternate methods of processing cellulose.

In recent years cellulose dissolution has been researched quite extensively and new solvents have been discovered that are more environmentally friendly. Several new processes that rely on these solvents have been developed for manufacturing fibers. Furthermore, research has also been focused on cellulose derivatization processes that pollute less and are more economical.

In this chapter we will elaborate on recent advances in dissolution and derivatization of cellulose and follow up with a description of new processes that lead to regenerated cellulose fibers. Finally, we will describe viscose processes and rayon fiber properties.

10.2 CELLULOSE SOLVENTS

The discussion of cellulose dissolution must recognize that cellulose can exist in four polymorphic forms: native cellulose known as cellulose I polymorph; cellulose II obtained by regeneration of cellulose I; cellulose III, which is derived from the liquid ammonia treatment of cellulose I or cellulose II; and cellulose IV, which refers to the thermal treatment of cellulose I or cellulose III [2]. It is important to recognize these distinctions because the respective cellulose polymorphs can have different solubility characteristics in particular solvents, as will become evident further in this chapter.

As suggested by Łaszkiewicz [3], cellulose dissolution may be divided into four groups as specified in Table 10.1.

TABLE 10.1
Cellulose Solvents

Cellulose as	Solvents
Base	Phosphoric, sulfuric, nitric acids
	Zinc chloride, thiocyanates, iodides, bromides
Acid	Organic amines, aminoxides, CH_3NH_2
Complex	Inorganic complexes of cadmium, copper, iron
	Organic complexes: $CH_3NH_2/DMSO$
Cellulose derivatives	Stable compounds: esters, ethers
	Unstable derivatives of
	Sulfur: xanthates, SO_2/amines-sulfites
	Nitrogen: N_2O_4/DMF
	Carbon: $DMSO/(CH_2O)_x$

Source: Adapted from Łaszkiewicz, B., *Manufacture of Cellulose Fibers without the Use of Carbon Disulfide*, ACGM LODART, SA, Łódź, Poland, 1997. With permission.

A polymer is generally dissolved in a solvent to enable its manipulation into a usable, profitable, and marketable product. Hence, dissolution *per se*, is not the ultimate objective. In addition to rendering the polymer soluble, the resulting solution must have certain desirable characteristics, such as chemical and thermal stability, proper viscoelastic properties, environmental friendliness, and a general ease of manipulation, including ease of recovery. The first five solvents discussed exhibit some of these desirable characteristics. The steam explosion process is mentioned more for completeness than as a potential contender in the cellulose fiber production market.

10.2.1 N-Methylmorpholine-N-Oxide and Water

The dissolution and spinning of cellulose (Figure 10.1) in the N-methylmorpholine-N-oxide (NMMO, Figure 10.2) system recently has been developed commercially by Courtaulds. The preparation of the solution involves adding cellulose to a mixture of aqueous NMMO and n-propyl gallate (PG). The PG is used as an antioxidant to stabilize the degree of polymerization (DP) of the cellulose. The mixture is placed in an airtight vessel and then stirred and heated at 130°C. Much care is taken in heating the mixture because temperatures above 150°C can cause undesirably rapid decomposition of the solvent and lead to explosions [5]. Complete dissolution normally occurs within 30 min at 130°C, and is even faster under optimum conditions. The temperature and time of complete dissolution, however, varies with the composition of the solvent and other process conditions. The complete dissolution of cellulose at given concentrations depends directly on the DP of cellulose. The NMMO solvent system is of particular interest because very high cellulose concentrations (ca. 50%) can be attained leading to the formation of anisotropic solutions. The work of Chanzy et al. [5] is widely regarded as the basis for the development of the NMMO or Lyocell process.

10.2.2 N,N-Dimethylacetamide and Lithium Chloride

McCormick [6] discovered that N,N-dimethylacetamide (DMAc) (Figure 10.3) and lithium chloride (LiCl) would dissolve the cellulose. He and his coworkers also observed cholesteric lyotropic mesophases of cellulose in this solvent system [7,8], which formed at cellulose

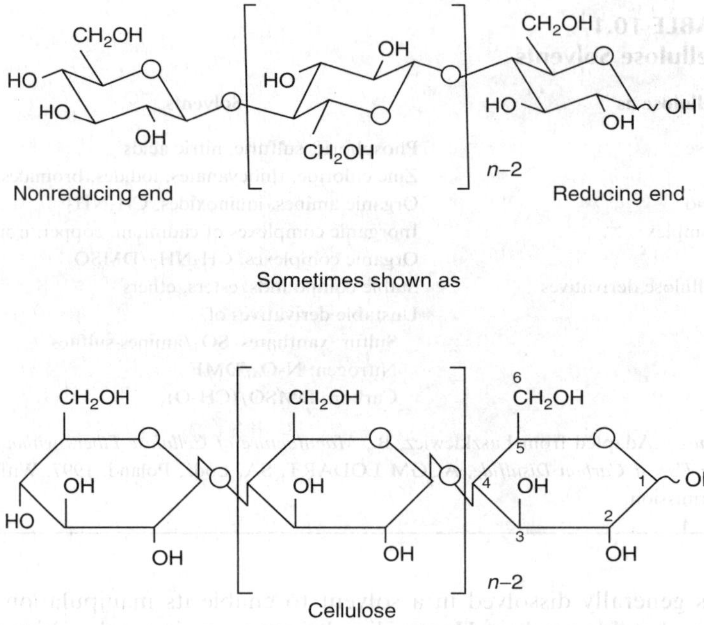

FIGURE 10.1 Chemical structure of cellulose (http://www.fibersource.com/f-tutor/cellulose.htm#chemistry as viewed on June 6, 2004).

concentrations above 15% (w/w) in 9% LiCl–DMAc [9]. Terbojevich et al. reported that LiCl–DMAc neither degraded nor reacted with cellulose. The mechanism by which dissolution actually occurs is believed to proceed through the formation of complexes between the solvent and the cellulosic hydroxyl groups. The LiCl–DMAc complex has been referred to as macrocation [7]. Terbojevich et al. indicated that the best method for dissolving cellulose in this solvent involves prewetting the cellulose with DMAc [9]. A known weight of DMAc is added to a weighed amount of dried cellulose. The mixture is refluxed at ca. 165°C, in a nitrogen atmosphere for 20–30 min. The mixture is then cooled to ~100°C and a predetermined amount of LiCl is added while stirring. Continued stirring at 80°C for 10–40 min ensures complete dissolution. Complete dissolution can be obtained with polymer concentrations up to 15% (w/w) of cellulose (DP = 130) [6]. Above this critical concentration, however, undissolved, swollen particles of cellulose are detected in the viscous solutions. Properly prepared solutions then can be spun into useful fibers.

10.2.3 Trifluoroacetic Acid and Chlorinated Alkanes

Patel and Gilbert demonstrated that mixtures of trifluoroacetic acid (TFA, Figure 10.4) and chlorinated alkanes, such as 1,2-dichloroethane and methylene chloride, are also good solvents for cellulose (and cellulose triacetate) [10,11]. It was discovered early that these solvents

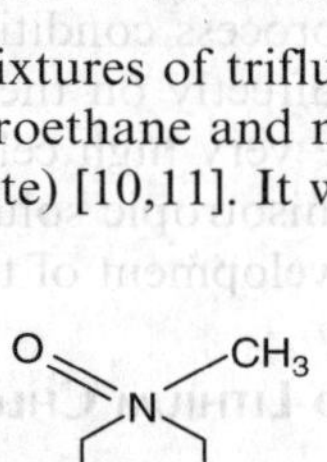

FIGURE 10.2 Chemical structure of N-methylmorpholine-N-oxide (NMMO).

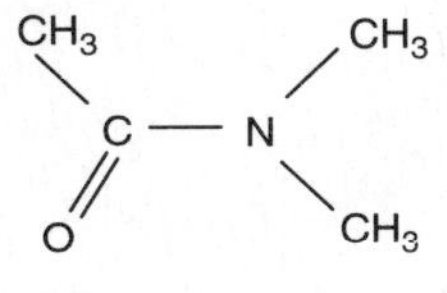

FIGURE 10.3 Chemical structure of N,N-dimethylacetamide.

require rather long dissolution times and degrade the cellulose quite severely. The rate of degradation decreases with a decrease in the TFA:CH_2Cl_2 ratio. Excessive degradation, together with costly, corrosive, environmentally unfriendly, and other noxious characteristics, impeded serious long-term interest in this solvent system. The above authors were interested mainly in the formation of lyotropic cellulose solutions for the purpose of extruding high-tenacity, high-modulus fibers. It is interesting that one author speculated that the cellulose polymorph IV might be the preferred conformation for fibers made with this solvent system [12].

10.2.4 Calcium Thiocyanate and Water

Dubose found that mixtures of calcium thiocyanate and water would dissolve the cellulose [13]. This was the first solvent system for cellulose. Little work has been done on this solvent system because of its high propensity for thermal degradation of cellulose. Despite this problem, the solvent demonstrates many properties common to other direct solvent systems.

To prepare the solution, a known weight of dried cellulose is added to $Ca(SCN)_2 \cdot 3H_2O$, with a solvent composition of 34.4–68.3 wt% of salt and 17.5–54.6 wt% of water [14]. The cellulose concentrations generally used are in the range of 10–30 wt%. The mixture is then heated to 120–140°C until a clear solution is obtained. Dissolution can be obtained within 30–40 min depending upon solution concentration and DP of cellulose. Generally one observes some undissolved cellulose. Prolonged heating can lead to discoloration, from clear to yellow, and then, finally brown. This indicates cellulose decomposition in the solution. Dissolution is accompanied with cellulose degradation of up to 40% loss in DP [13]. The solution forms a clear gel that melts reversibly at 80°C. Fibers prepared from this solution were of very poor quality.

Recently, dissolution of cellulose in aqueous calcium and sodium thiocyanate solutions has been investigated by Hattori et al. [15,16]. The sodium thiocyanate solvent was found ineffective, but the hydrated $Ca(SCN)_2$ dissolved the cellulose because of the complex formation shown in Figure 10.5.

10.2.5 Ammonia/Ammonium Thiocyanate

Hudson and Cuculo discovered that ammonia/ammonium thiocyanate (NH_3/NH_4SCN) is an excellent solvent for cellulose [17]. They showed that the solvent has several practical advantages, including low cost and readily available components. The boiling point of the

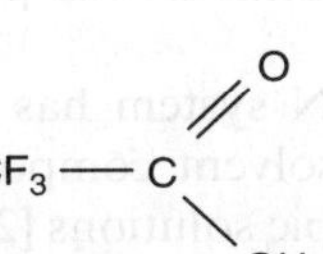

FIGURE 10.4 Chemical structure of trifluoroacetic acid.

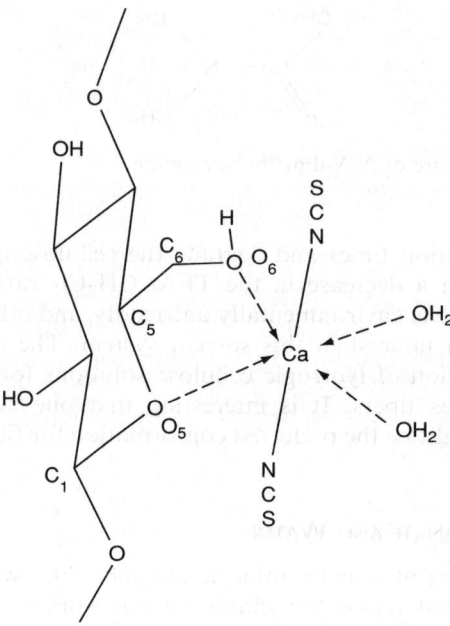

FIGURE 10.5 Schematic representation of interaction between cellulose and calcium thiocyanate–water complex. (Reprinted from Hattori, M., Koga, T., Shimaya, Y., and Saito, M., *Polym. J.*, 30, 43, 1998. With permission.)

solvent is ~70°C; as a result, it can be handled easily. There seems to be no cellulose degradation or reaction between the cellulose and the solvent [18,19]. Mesophases are formed readily in the solvent. Solvent preparation is quite simple. It involves placing a known weight of salt in a chilled polymer kettle that contains a stirrer and ammonia, which is condensed by the kettle using a cold finger. Dissolution of the salt occurs easily and the solvent is ready for use.

In the course of research on this solvent, a novel and powerful technique was developed for the rapid and convenient dissolution of cellulose, which the authors called the *rapid temperature cycling* technique. This technique has served remarkably well in the laboratory. The rapid temperature cycling technique is conducted as follows. Respective known weights of cellulose and solvent are placed in a sealable polyethylene bag and are homogenized with hand mixing. The bag is then subjected to the temperature cycling process, which involves placing the bag in a cold bath for a few minutes at ca. −33°C and then placing the bag at a higher temperature of ca. 40°C. In the interim, before the next subjection to the low and the high temperatures, the mixture is subjected to shearing forces by passing the bag forward and backward through a rolling pin. For difficult dissolutions with, for example, cellulose of very high DP and high crystallinity, multiple treatments are effective. This rapid cycling technique evolved gradually over several years until it was perfected and proved to be reliable and convenient.

Over the years, the NH_3/NH_4SCN system has been extensively studied. The research topics covered include the effects of solvent composition, mesophase formation, and fiber formation from isotropic and anisotropic solutions [20,21]. The authors' primary interest was in determining the intimate mechanism of dissolution of cellulose in the solvent. In the two papers involving NMR in its various forms, this matter was studied and the results were reported [22,23]. It appears that the process of dissolution proceeds via the transformation of the cellulose polymorphs from I to II to III to IV and finally to amorphous as dissolution occurs. This sequence comes about, of course, through the breakage of intra- and intermolecular hydrogen bonds.

10.2.6 Steam Explosion and Aqueous Solutions of Sodium Hydroxide

Kamide et al. [24] and Yamashiki et al. [25–28] dissolved regenerated cellulose ($M_w = 8 \times 10^4$ g/mol) in 10-wt% aqueous NaOH at 4°C to give a 5% (w/w) cellulose solution. Cellulose was pretreated by steam explosion prior to its dissolution in aqueous alkali solution. The steam explosion process causes the breakdown of intramolecular hydrogen bonds, which strongly affect the dissolution of the cellulose [27]. The process of steam explosion is carried out at steam pressures in the range of 1.0–4.9 MPa, which corresponds to steam temperatures of 183–252°C. Cellulose is in contact with the steam for $\approx$15–300 s. The cellulose is then cooled to room temperature, washed, and dried prior to dissolution in NaOH solution.

Steam explosion brings about not only the breakdown of intramolecular hydrogen bonds, but also a reduction of DP and a change in cellulose polymer morphology. The sensitivity of cellulose to mild steam explosion, in terms of the change in polymorph, the reduction of DP, and the solubility, is as follows: cellulose I > cellulose III > cellulose II. Kamide et al. observed that under severe steam explosion treatment, cellulose II attained the lowest level-off DP without a complete change in the polymorph [24]. Cellulose III, however, changes its polymorph almost completely to cellulose I. The degree of solubility in aqueous NaOH increases remarkably for cellulose I and III, subjected to steam treatment, but increases only slightly for cellulose II. The authors indicated that the increase in solubility for cellulose I and III in aqueous NaOH after steam treatment was caused by an increase in the breakdown of intramolecular hydrogen bonds at the hydroxyl group at the carbon-3 position in the glucopyranose ring, as estimated by solid-state cross-polar–magic-angle sample spinning (CP–MAS)^{13}C NMR.

Yamashiki et al. dissolved 50 g (water content $\approx$ 8–12%) of steam-treated cellulose (DP = 331) in 950 g of 9.1-wt% aqueous NaOH solution (precooled at 4°C) [28]. The mixture was left for 8 h with intermittent stirring. The resultant solution was centrifuged at 10,000 rpm for 1 h at 4°C to remove remaining undissolved cellulose and air bubbles. Yamashiki et al. attempted spinning these solutions [28]. The fibers exhibited the cellulose II polymorph.

10.2.7 Ionic Liquids

Salts that melt at temperatures below 100°C form useful ionic liquids (ILs) [29]. Because they lack measurable vapor pressure (even up to 300°C), the ILs can be used as green solvents or reaction media. As a result, they have recently received a lot of attention [30]. Simple ILs usually contain imidazolium, pyridinium, or organic ammonium cations. The anions could be chloride, bromide, or more complex structures such as hexafluorophosphate, trifluoromethylsulfonate, bis(trifluoro-methylsulfonyl)imide [31].

Room temperature ILs (RTILs) have more complex structures [32]; however, they may have greater potential as green solvents particularly for cellulose. Huddleston et al. [33] investigated a series of hydrophilic and hydrophobic 1-alkyl-3-methylimidazolium ILs. Based on these studies, Rogers and coworkers showed that 1-butyl-3-methylimidazolium chloride ([C_4mim]$^+$Cl$^-$, Figure 10.6) [34], which melts around 60°C, easily dissolves the cellulose (Table 10.2). The polymer can be regenerated by precipitation with water. The authors used cellulose from Lyocell and rayon production lines as well as fibrous cellulose having DP $\approx$ 1000. They were able to achieve a very high polymer concentration of up to 25%,

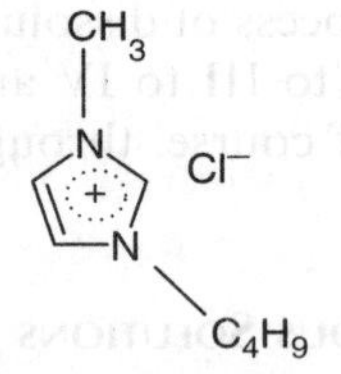

FIGURE 10.6 Structure of 1-butyl-3-methylimidazolium chloride.

and the fiber and film formations were documented. Rogers and coworkers suggested that the high chloride concentration is responsible for breaking the hydrogen network, thus allowing the dissolution of cellulose. It is important to mention that the presence of more than 1-wt% water in $[C_4mim]^+$ Cl^- deteriorates cellulose dissolution thus showing that the degree of ion hydration is a key factor in this process.

Molten salt hydrates such as $LiCO_4 \cdot 3H_2O$, $LiSCN \cdot 2H_2O$, or $ZnCl_2 \cdot 4H_2O$ form complexes and were found to dissolve nonactivated cellulose (DP of 950) [35]; however, the regenerated cellulose had much lower molecular weight, particularly for the latter salt. Heinze et al. [35] also show that cellulose with a DP up to 650 can be dissolved in dimethyl sulfoxide containing 10–20% (w/v) tetrabutylammonium fluoride trihydrate without pre-pretreatment within 15 min at room temperature. Although the cellulose concentration was only 2.9%, the authors were able to conduct acetylation of cellulose in high yields.

10.2.8 PHOSPHORIC ACID

Aqueous solutions of phosphoric acid (H_3PO_4), hydrochloric acid (HCl), or sulfuric acid (H_2SO_4) are known to dissolve cellulose [3]. However, in the first case, the cellulose solutions are very viscous even at a low concentration of 2%. On the other hand, solutions of the other acids, undergoing rapid hydrolysis, are not stable.

TABLE 10.2
Dissolution of Cellulose Pulp (DP ≈ 1000) in Ionic Liquids

Ionic liquid	Method	Solubility (wt%)
[C₄mim]Cl	Heat (100°C)	10
[C₄mim]Cl	Heat (80°C) + sonication	5
[C₄mim]Cl	Microwave heating 3–5-s pulses	25 (clear viscous solution)
[C₄mim]Br	Microwave	5–7
[C₄mim]SCN	Microwave	5–7
[C₄mim][BF₄]	Microwave	Insoluble
[C₄mim][PF₆]	Microwave	Insoluble
[C₆mim]Cl	Heat (100°C)	5
[C₈mim]Cl	Heat (100°C)	Slightly soluble

Source: Reprinted from Swatloski, R.P., Spear, S.K., Holbrey, J.D., and Rogers, R.D., *J. Am. Chem. Soc.*, 124, 4974, 2002. With permission.

A group of Dutch researchers [36] undertook extensive studies on dissolution of cellulose in H_3PO_4, despite its high viscosity. The authors found that phosphoric acid free of water easily dissolves cellulose and that this stable solution can be used for making high-modulus, high-strength, regenerated cellulose fibers. The strength of phosphoric acid is usually determined by the concentration of phosphorus pentoxide (P_2O_5). When the concentration reaches 72.4% by weight, the phosphoric solution contains no water. The authors determined that anisotropic cellulose solutions are easily formed at a cellulose concentration of 8% (w/w) in solutions containing 72.4–76% P_2O_5. Such conditions can be obtained by mixing 99% phosphoric acid with pyrophosphoric acid ($H_4P_2O_7$), polyphosphoric acid, or phosphorus pentoxide. The authors reported highly oriented cellulose II fibers with modulus of 45 GPa and strength of 1.3 GPa.

10.3 NONVISCOSE RAYON PROCESSES

Among the new nonviscose processes, the NMMO process appears to be the most promising replacement for conventional viscose processes. It basically involves the use of polar, aqueous NMMO solvent for dissolution of cellulose. The fibers produced by using this approach are called NMMO fibers with the generic name Lyocell [37], and the process for making these fibers is sometimes called the Lyocell process. Other nonviscose processes exist as well. The H_3PO_4 process developed by Akzo is very simple; however, anhydrous phosphoric acid must be used as solvent to make the cellulose solutions. The carbamate process is still under extensive development and may play an important role in the future for making regenerated cellulose fibers. The cuprammonium process has been known for a long time but it is only used for making little regenerated cellulosics such as cuprosilk (continuous filament) and cuprophane (membranes) [37].

10.3.1 N-METHYLMORPHOLINE-N-OXIDE OR LYOCELL PROCESS

Tertiary amine oxides were first used in 1938 by Graenacher and Sallmann [38] to dissolve cellulose; however, Johnson [39] of Eastman Kodak is credited with the use of NMMO for cellulose dissolution. McCorsley and Varga [40] of Akzona showed that highly concentrated solutions of cellulose in NMMO of up to 23% can be prepared by treating cellulose with aqueous NMMO solution and subsequently removing water. Franks and Varga [41] also demonstrated that cellulose can be easily coagulated with an excess of water. Researchers from Courtaulds continued this fundamental research till the company began the commercial production of staple regenerated cellulose fibers, trademarked Tencel, in Mobil, Alabama, in May 1992. In the meantime, Lenzig also developed Lenzig Lyocell staple fibers using the NMMO technique.

The production of regenerated cellulose fibers with NMMO as a cellulose solvent is a modern, highly efficient, nonpolluting process. Fink et al. [37] listed the following steps for production of regenerated cellulose fibers:

1. Preparation of a homogeneous solution (dope) from cellulose pulp in NMMO–water solution
2. Extrusion of the highly viscous spinning dope at elevated temperatures through an air gap into a coagulation bath (dry-jet wet-spinning process)
3. Coagulation of the cellulose fibers
4. Washing, drying, and posttreatment of the cellulose fibers
5. Recovery of NMMO from coagulation baths

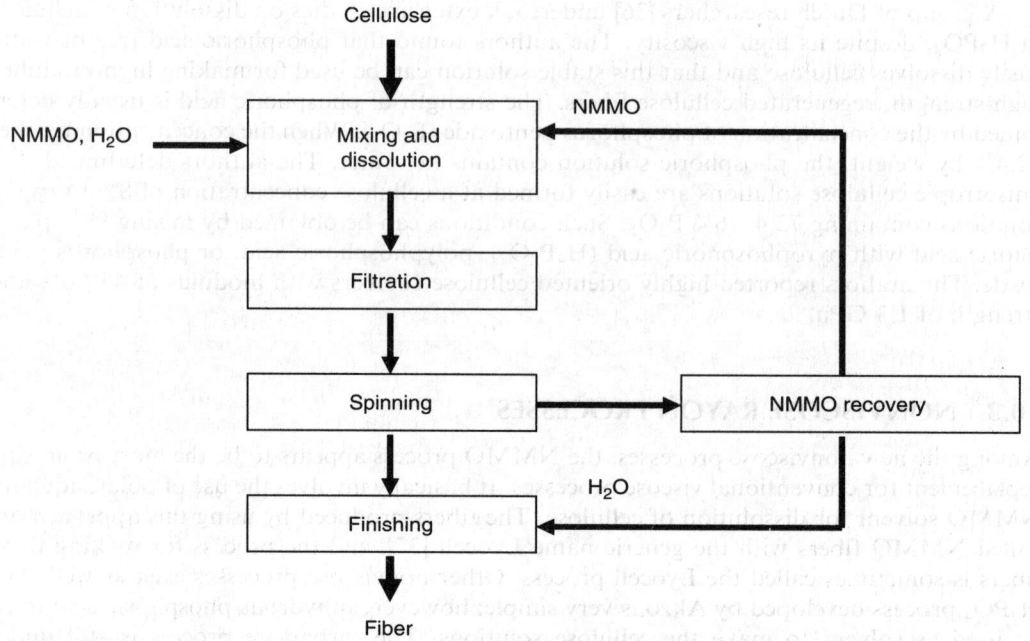

FIGURE 10.7 Schematic of NMMO process. (Adapted from Łaszkiewicz, B., *Manufacture of Cellulose Fibers without the Use of Carbon Disulfide*, ACGM LODART, SA, Łódź, Poland, 1997. With permission of Professor B. Łaszkiewicz.)

As shown in Figure 10.7, NMMO process is a closed loop process in which the NMMO is recycled. The production cycle is relatively short and does not exceed 8 h. On the other hand, the conventional viscose process is very long (Figure 10.8), exceeding 40 h. The viscose process requires handling toxic gas by-products, namely H_2S and CS_2. The NMMO fibers are spun at higher spinning speeds than the rayon fibers. These high-quality fibers have excellent mechanical properties; however because of a high degree of crystallinity, the wet Lyocell fibers have increased susceptibility to fibrillation.

10.3.1.1 Preparation and Properties of *N*-Methylmorpholine-*N*-Oxide

The NMMO is a highly polar compound with the N—O group having the dipole moment (Figure 10.9) that can form hydrogen bonds with the hydroxyl groups of cellulose [3,37]. Consequently the aqueous solutions of NMMO dissolve the cellulose [41]. NMMO can be produced from morpholine (1-oxa-4-azacyclohexane). Morpholine (M) is a colorless liquid having boiling point of 129–130°C at 760 torr. The compound is a weak organic base ($pK_b = 9.25$) and can be synthesized from ammonia and ethylene oxide (Figure 10.10). M has been used extensively for manufacturing pharmaceuticals; however, in recent years this raw material has also been used for making NMMO according to the reaction shown in Figure 10.11. The oxidation of *N*-methylmorpholine (NMM) is accomplished by using a 35% excess of H_2O_2 and carrying out the reaction at 67–72°C for 4–5 h. $NaHCO_3$ and $Na_4P_2O_7 \cdot 10H_2O$ can be used as oxidation catalysts [3,39,42]. NMMO can be crystallized from acetone, a colorless crystalline compound that can be dissolved in water and in

Regenerated Cellulose Fibers

a variety of solvents. The basic properties of a 50% aqueous NMMO solution are given below [43]:

Appearance	Clear to light yellow liquid
Freezing point	−20°C
Boiling point	118.5°C
Viscosity at 50°C	7.4 cP
Density at 90°C	1.084 g/cm^3
50°C	1.113 g/cm^3
25°C	1.130 g/cm^3

A typical composition of aqueous NMMO solution is as follows:

NMMO content	50.0%
NMM content	1 wt%
H$_2$O$_2$	100 ppm

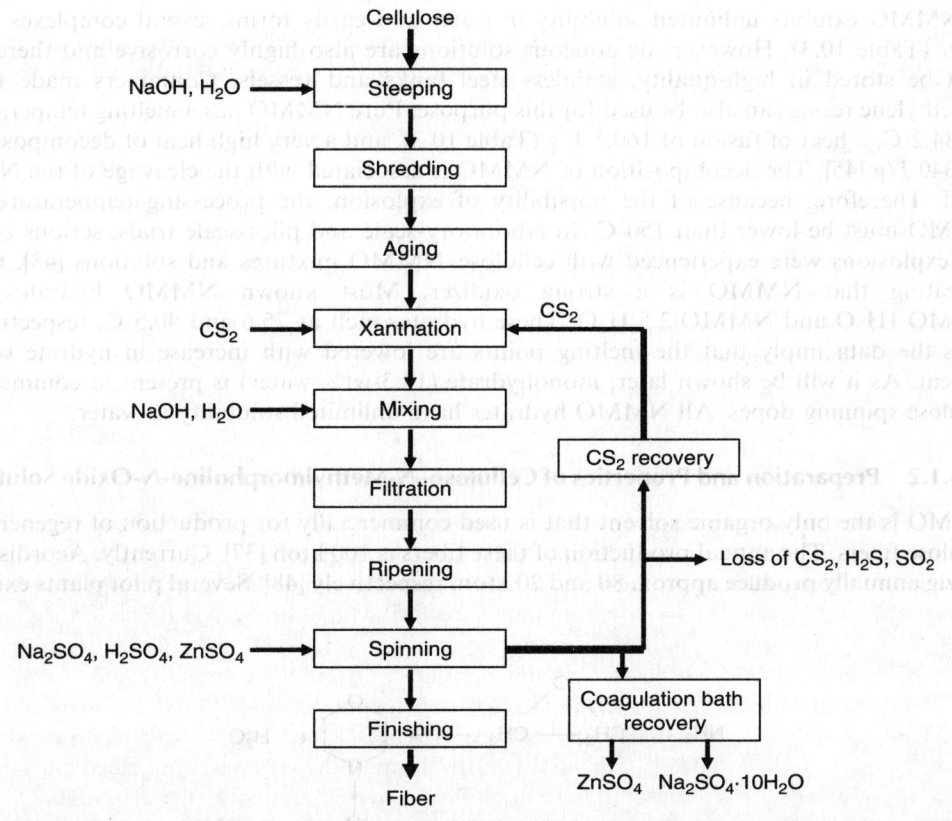

FIGURE 10.8 Schematic of viscose process. (Adapted from Łaszkiewicz, B., *Manufacture of Cellulose Fibers without the Use of Carbon Disulfide*, ACGM LODART, SA, Łódź, Poland, 1997. With permission.)

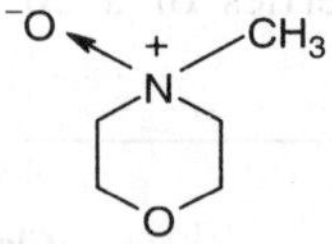

FIGURE 10.9 Polar structure of N-methylmorpholine-N-oxide.

NMMO is manufactured by Degussa, BASF AG, Texaco U.K., and a number of other companies that supply the solvent in small quantities. NMMO is an excellent solvent for other polymers such as poly(ethylene oxide), poly(vinyl pyrrolidone), poly(vinyl acetate), silk fibroin, wool, aliphatic polyamides, polyacrylonitrile, starch, cellulose diacatate, and many other cellulose derivatives [3].

The main advantage of the NMMO solvent is the lack of toxicity. Recent studies by the Lenzig researchers [44] proved that NMMO is biodegradable. The experiments showed that the activated sludge can be adapted to NMMO within 15–20 days. The adapted sludge can degrade the solvent, metabolize it to concentrations below detection levels, and retain this ability even during limited periods without the solvent being present in the wastewater. Degradation takes place in several steps. In the first step NMMO is converted to NMM followed by the demethylation to M. Once M is formed, the adaptation proceeds very quickly until none of the intermediates can be detected any longer.

NMMO exhibits unlimited solubility in water and easily forms several complexes with water (Table 10.3). However, its aqueous solutions are also highly corrosive and therefore must be stored in high-quality, stainless steel tanks and vessels. Containers made from polyethylene resins can also be used for this purpose. Pure NMMO has a melting temperature of 184.2°C, a heat of fusion of 160.5 J/g (Table 10.3), and a very high heat of decomposition of 1340 J/g [45]. The decomposition of NMMO is associated with the cleavage of the N—O bond. Therefore, because of the possibility of explosion, the processing temperature for NMMO must be lower than 150°C. In laboratory-scale and pilot-scale trials, serious blasts and explosions were experienced with cellulose–NMMO mixtures and solutions [48], thus, indicating that NMMO is a strong oxidizer. Most known NMMO hydrates are NMMO·1H$_2$O and NMMO·2.5 H$_2$O. These hydrates melt at 75.6 and 40.5°C, respectively. Thus the data imply that the melting points are lowered with increase in hydrate water content. As it will be shown later, monohydrate (13.3-wt% water) is present in commercial cellulose spinning dopes. All NMMO hydrates have unlimited solubility in water.

10.3.1.2 Preparation and Properties of Cellulose–N-Methylmorpholine-N-Oxide Solutions

NMMO is the only organic solvent that is used commercially for production of regenerated cellulose fibers. The annual production of these fibers is 100 kton [37]. Currently, Acordis and Lenzig annually produce approx. 80 and 20 kton, respectively [48]. Several pilot plants exist at

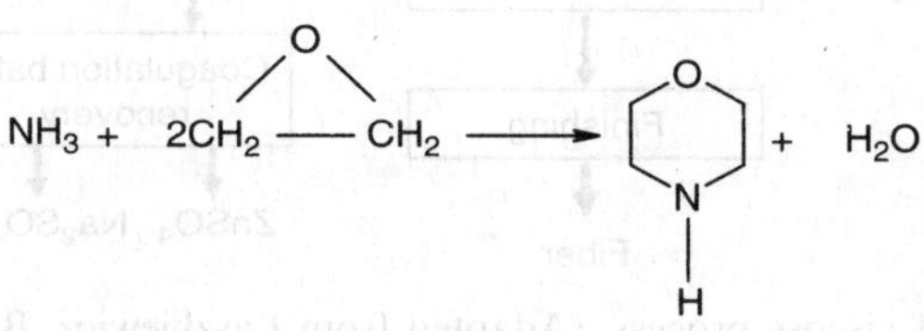

FIGURE 10.10 Synthesis of morpholine.

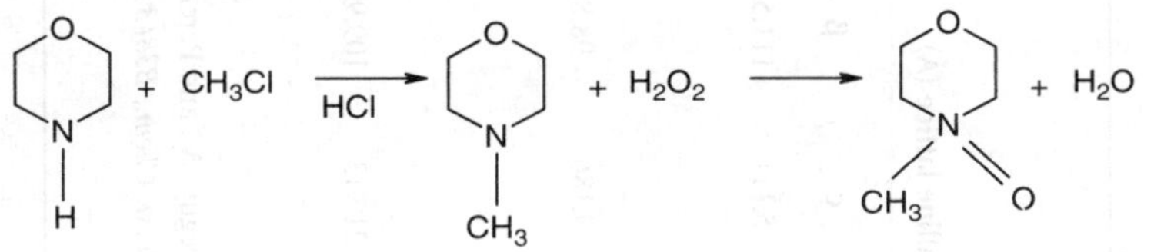

FIGURE 10.11 Synthesis of *N*-methylmorpholine-*N*-oxide.

TITK (Germany), FCFC (Taiwan), Hanil (Korea), and Grasim (India) [48]. The generic term "Lyocell" is commonly used to designate the industrial process (Figure 10.12), the fiber production therein, and the NMMO–cellulose mixture [48]. Depending on the DP, highly concentrated spinning dopes can be prepared with a cellulose concentration up to 35%. NMMO technology has only a few disadvantages namely [3]:

1. NMMO is an oxidizer and a highly corrosive solvent; hence a stainless steel equipment for processing must be used.
2. At temperatures higher than 150°C, NMMO can undergo highly exothermic decomposition that can be catalyzed by copper or iron ions and some other catalysts that induce N—O cleavage [48]. All necessary precautions must be undertaken to avoid any undesired explosion.

As indicated by Fink et al. [37], nowadays, the NMMO process is well established, overcoming initial difficulties such as high investments costs, recovery of the expensive solvent, and other technical problems. Łaszkiewicz [3] listed the following advantages for NMMO technology:

1. NMMO process has a short production cycle. The dissolution and spinning process is usually done within 5 h. Currently, full continuous dissolution of cellulose is implemented in the Lyocell process.
2. NMMO process is eco-friendly. No toxic gases or by-products are formed. The recovery of NMMO can be as high as 99.5%.
3. NMMO process can be utilized to obtain highly concentrated cellulose solutions (25–35%).
4. NMMO fibers can be spun at a relatively high spinning speed in the range of 150–300 m/min.
5. NMMO technology allows regenerated cellulose fibers to be modified by spinning fibers from the solutions containing cellulose and some other polymers including cellulose acetate, cellulose derivatives, polyvinyl alcohol, and some other natural polymers.
6. NMMO technology can be utilized to make cellulose membranes and films [37].

As Rosenau et al. [49] indicated that paper grade pulp, unbleached chemical pulp, cotton and rayon fiber wastes, or even paper wastes can be used as raw materials for Lyocell fiber production, even though problems with spinnability may be encountered in some cases. In preparation for spinning dope, a 50–60% aqueous NMMO is used with the addition of 0.01–0.10% antioxidant to prevent cellulose degradation. A typical antioxidant is PG [50]. In a typical Lyocell industrial process, the slurry is produced from cellulose pulp and an aqueous NMMO solution. Typical compositions are 50–60% NMMO, 20–30% water, and 10–15% pulp [48]. Subsequently excess water is efficiently evaporated at temperatures lower than 150°C and

TABLE 10.3
Properties of N-Methylmorpholine-N-Oxide and Its Hydrates

Chemical formulae	CAS number	Molecular weight (g = mole)	DSC melting temperature (°C)	Heat of fusion (J = g)	Density (g/cm³) exp.	Density (g/cm³) calc.	Dimension of crystalline lattice (Å) a	b	c	β
$C_5H_{11}NO_2$	7529-22-8	117.1	184.2 [45] 184 172	160.5	1.25	1.25	9.884	6.621	5.111	111.54
$C_5H_{11}NO_2 \cdot xH_2O$ $0 < x < 1$	—	—	102 [45]	95	—	—	—	—	—	—
$C_5H_{11}NO_2 \cdot 1H_2O$	70187-32-5	132.5	75.6 [45] 78 76 75 74	160.5	1.29	1.29	25.481	6.045	9.186	98.88
$C_5H_{11}NO_2 \cdot 2.5H_2O$	80913-65-1	162.2	40.5 [45] 39 36	143.4	1.22	1.257	12.803	6.500	21.913	109.99

Source: From Laszkiewicz, B., *Manufacture of Cellulose Fibers without the Use of Carbon Disulfide*. ACGM LODART, SA, Łodz, Poland, 1997; Maia, E., Peguy, A., and Perez, S., *Acta Crystallogr.* Sect. B: *Struct. Crystallogr. Cryst. Chem.*, B37(10), 1858, 1981; Maia, E; and Perez, S., *Acta Crystallogr.* Sect. B: *Struct. Crystallogr. Cryst. Chem.*, B38(3), 849, 1982.

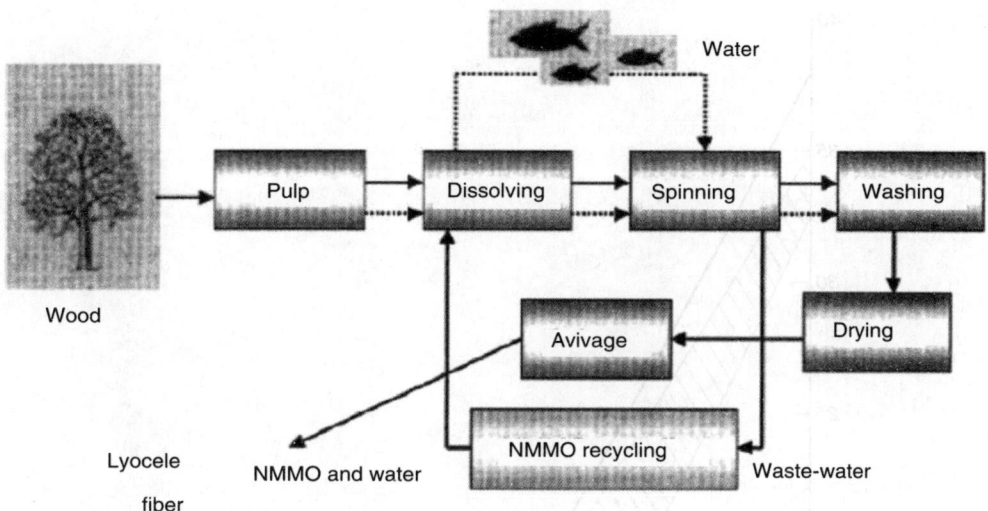

FIGURE 10.12 The Lyocell process in general. (From Rosenau, T., Potthast, A., Sixta, H., and Kosma, P., *Prog. Polym. Sci.*, 26, 1763, 2001. Reprinted with permission of Elsevier B.V.)

under reduced pressure till the cellulose solution is formed. Online control of process parameters such as temperature, cellulose concentration, pressure, and shear stress are practiced. A typical isotropic spinning dope composition contains 14% cellulose, 10% water, and 76% NMMO. As shown in Figure 10.13 through Figure 10.15, the solubility of cellulose in the binary NMMO solution definitely depends on water content. Usually aqueous NMMO solutions containing more than 15–17 wt% water do not dissolve cellulose. One or two hydrogen bonds can be formed between the oxygen of the N—O bond in NMMO and the hydroxyl group of water, an alcohol, or cellulose [46,47]. At high water concentrations, generally greater than 17 wt% [48], hydrogen bonds between NMMO and water dominate, thus preventing cellulose dissolution. At lower concentrations, the oxygen of the N—O bond in NMMO forms hydrogen bonds with the cellulose hydroxyl groups and dissolution can be observed. At water concentrations lower than 4%, dissolution temperatures are very close to the decomposition point of NMMO and therefore $c_{water} = 4$ wt% is considered as the lower cellulose dissolution limit [48]. Typical, safe processing temperatures are in the range of 80–130°C. The effect of various process parameters on cellulose dissolution is shown in Figure 10.16.

Anisotropic cellulose solutions cannot be obtained if the solvent system contains monohydrate of NMMO or 13.3-wt% water and 20-wt% cellulose. To produce such solutions, the water content should be below 11% thus indicating that some water from hydrated NMMO molecules must be released [3,51]. The conditions that assure mesophase formation at 5-wt% water are shown in Table 10.4. These results show that anisotropic cellulose solutions can be formed at critical concentrations greater than 20 wt% for cellulose having DP of 600. At a low DP of 35, a mesophase was observed when the critical polymer concentration was greater than 40 wt%. Although cellulose is a semirigid polymer, the fundamental work of Chanzy et al. [5] clearly showed the possibility of making high-modulus, high-strength fibers from concentrated anisotropic cellulose dopes with a molar ratio of NMMO to water of less than one. Obviously, the viscosity of cellulose anisotropic solutions is strongly dependent on cellulose concentration and DP.

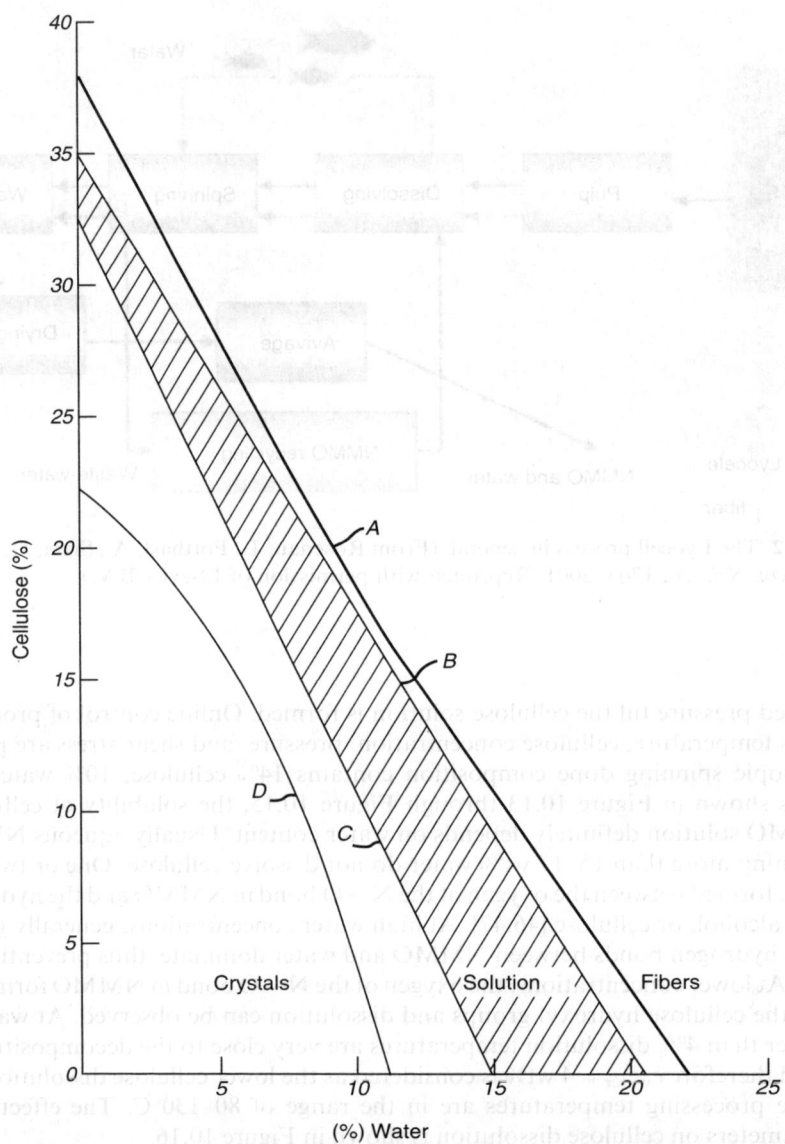

FIGURE 10.13 Solubility of cellulose with medium DP at 100°C in aqueous NMMO. (From McCorsely, C.C. and Varga, J.K., U.S. Patent 4,142,913, to Akzona Inc.; Franks, N.E. and Varga, J.K., U.S. Patent 4,196,282, to Akzona Inc.)

LIST AG has developed a continuous process for cellulose dissolution in NMMO by using mixing and kneading technology [52,53] for producing Lyocell fibers. This new NMMO process comprises two steps, namely the continuous conditioning (prewetting) of the cellulose pulp in a corotating processor and the continuous dissolution of the cellulose in a dissolver. Figure 10.17 demonstrates this concept and also shows that NMMO recycling is part of the process. The conditioner, a continuous mixer or kneader, mixes and homogenizes cellulose pulp with aqueous NMMO under specified conditions. The homogenized slurry serves as a feed stream for the dissolver. The dissolution of cellulose involves, as was indicated earlier, the evaporation of water and the dissolution of the polymer in aqueous NMMO containing

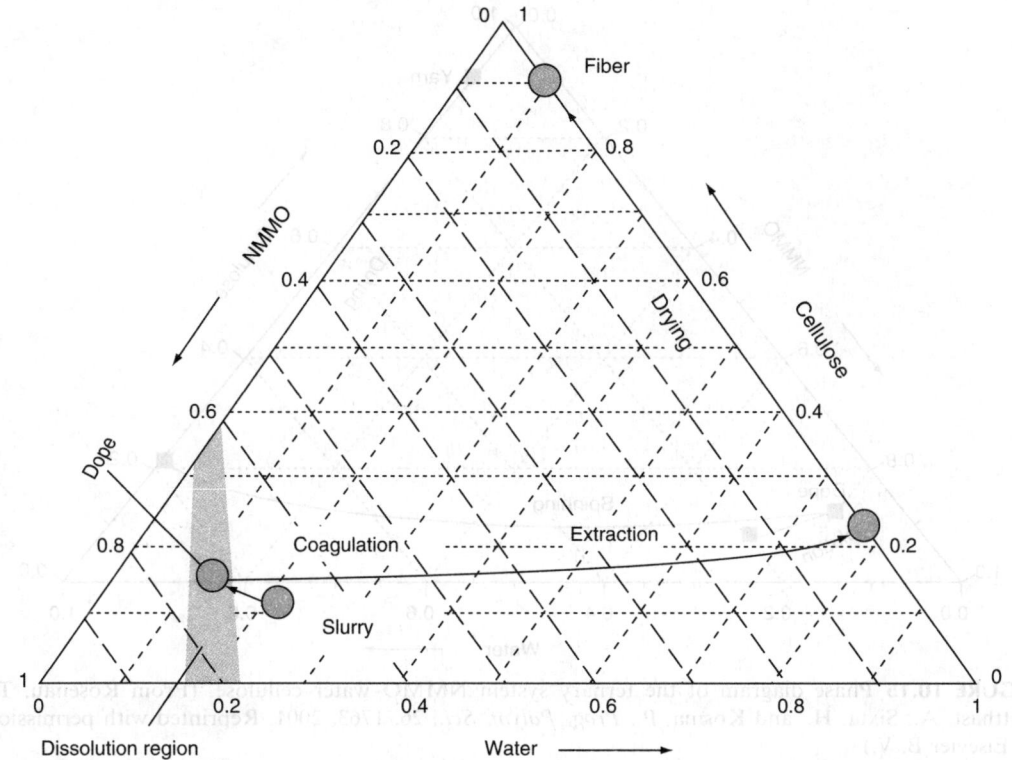

FIGURE 10.14 Ternary phase diagram of NMMO, cellulose, and water. Medium cellulose DP at 100°C. (From Wachsman, U. and Diamantoglou, M., *Das Papier*, 51(12), 660, 1997. Reprinted with permission of dpw-Verlagsgesellschaft GmbH.)

less than 17-wt% water. Homogeneous cellulose solutions containing up to 18 wt% are formed quickly because of the high shear rates used in the process chamber of the dissolver.

This highly efficient technology is utilized for the production of NMMO fibers at Acordis and Lenzig. In mid-1998, Alceru Schwarza GmbH (Rudolstadt, Germany) and Grasim Industries (Nagda, India) started the operation of a pilot plant with annual capacity in the range of 300 to 400 t. At present, LIST AG provides whole integrated production units with annual capacities of 1,000, 5,000, 10,000, and 15,000 t [53]. Mixing and kneading technology has been developed with TITK (Institute for Textile and Plastic Research), Rudolstadt, Germany. For safety reasons, the LIST dissolving technology operates at low temperatures from 80 to 120°C. The LIST technology is comparable to mixing and kneading equipment used for the thermal processing of highly viscous materials. The technology employs equipments to handle large volumes of cellulose and provides the dual benefits of efficient mixing and kneading and self-cleaning of the surfaces that serve as heat exchangers.

The benefits of LIST mixers or kneaders [53] are as follows:

1. Processing of heterogeneous product mixtures in a single unit
2. High interface renewal rates assuring effective mass transfer
3. Efficient combination of static and dynamic elements in the process chambers that leads to a high degree of shear homogenization

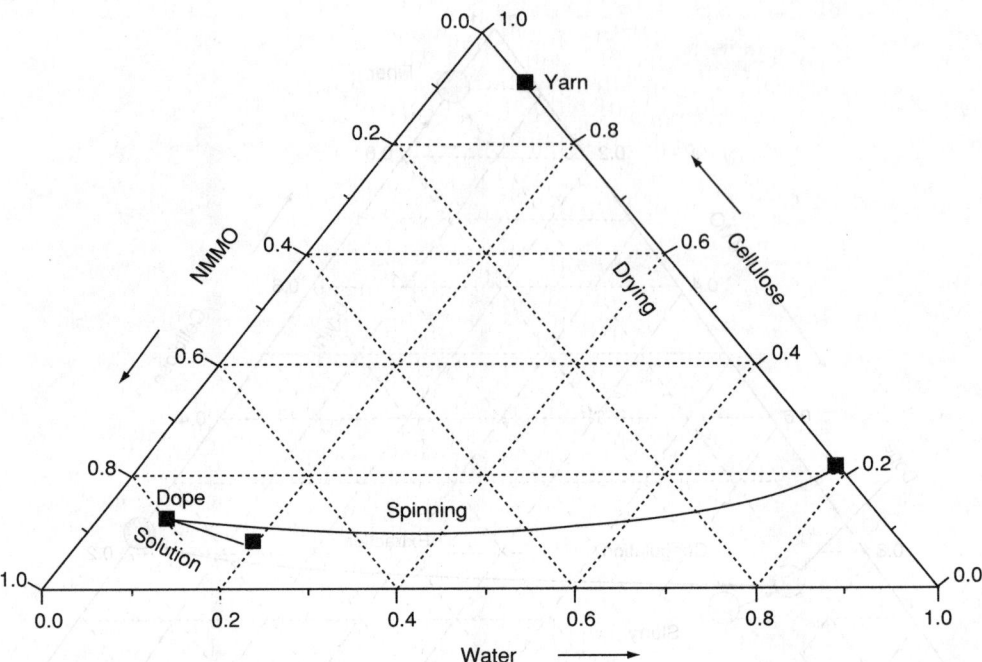

FIGURE 10.15 Phase diagram of the ternary system NMMO–water–cellulose. (From Rosenau, T., Potthast, A., Sixta, H., and Kosma, P., *Prog. Polym. Sci.*, 26, 1763, 2001. Reprinted with permission of Elsevier B. V.)

4. High-energy flux and close control of product temperature
5. Large capacities in a single production unit
6. Closed, contained design that guarantees low operational risk when handling dangerous materials
7. Wide range of retention time from 10 min to several hours
8. Large area to volume ratio

The LIST Corotating Processor (CRP, see Figure 10.18) continuously conditions feed materials and allows for [53] the following:

1. Mixing and homogenization of cellulose pulp with an aqueous NMMO in the low range of temperatures from 80 to 85°C
2. Blending of additive such as antioxidants or dyestuffs and quick, effective thermal conditioning of the mixture, thus avoiding the thermal degradation of the feed material

The LIST Discotherm B processor accomplishes continuous dissolution of cellulose (Figure 10.19) and allows for [53] the following:

1. Water evaporation till the desired water content and product temperature has been reached
2. Dissolution of cellulose and homogenization of the solution under vacuum, which is enhanced by high shear rates at temperatures from 90 to 120°C

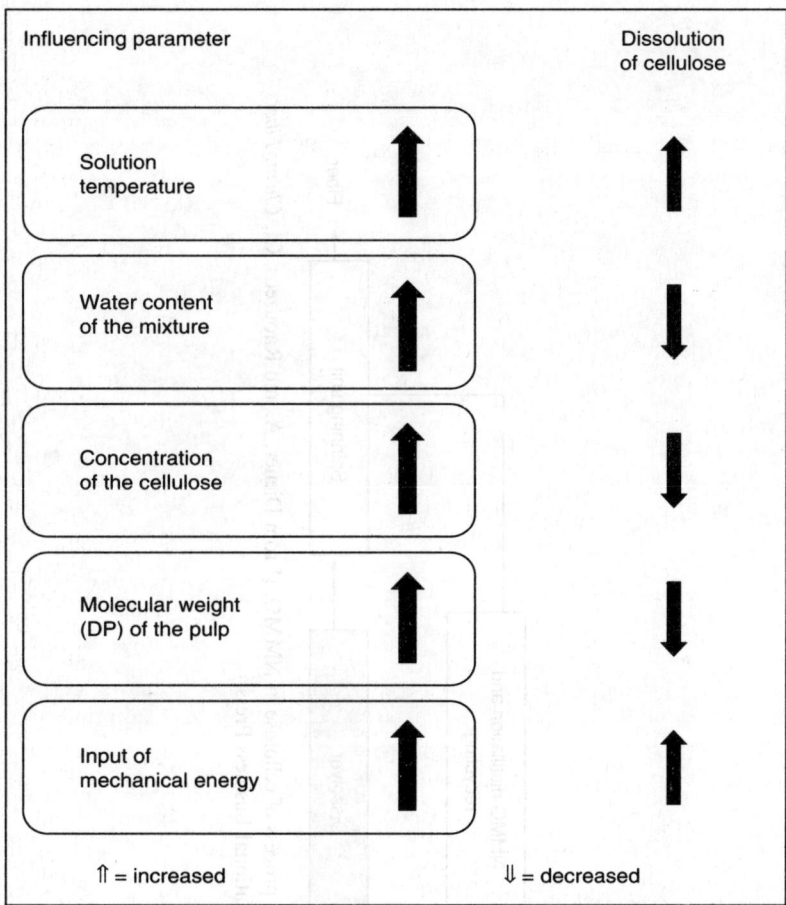

FIGURE 10.16 Effects of various factors on cellulose dissolution in NMMO–water. (From Rosenau, T., Potthast, A., Sixta, H., and Kosma, P., *Prog. Polym. Sci.*, 26, 1763, 2001. Reprinted with permission of Elsevier B. V.)

TABLE 10.4
Influence of Cellulose Concentration and Degree of Polymerization on Formation of Anisotropic Solutions of Cellulose in NMMO–Water Mixture[a]

Cellulose material	Concentration range[b]
Cotton linters (DP = 600)	20–35
Dissolving pulp (DP = 600)	20–35
Avicell PH 101 (DP = 130)	25–35
Microcrystalline cellulose from rayon (DP = 35)	45–55

[a] The mole ratio of water to anhydrous NMMO was kept below unity and 0.4 in optimum cases.
[b] The upper limit corresponds to the limit of solubility.
Source: Reprinted from Chanzy, H. and Peguy, A., *J. Polym. Sci., Polym. Phys. Ed.*, 18, 1137, 1980. With permission.

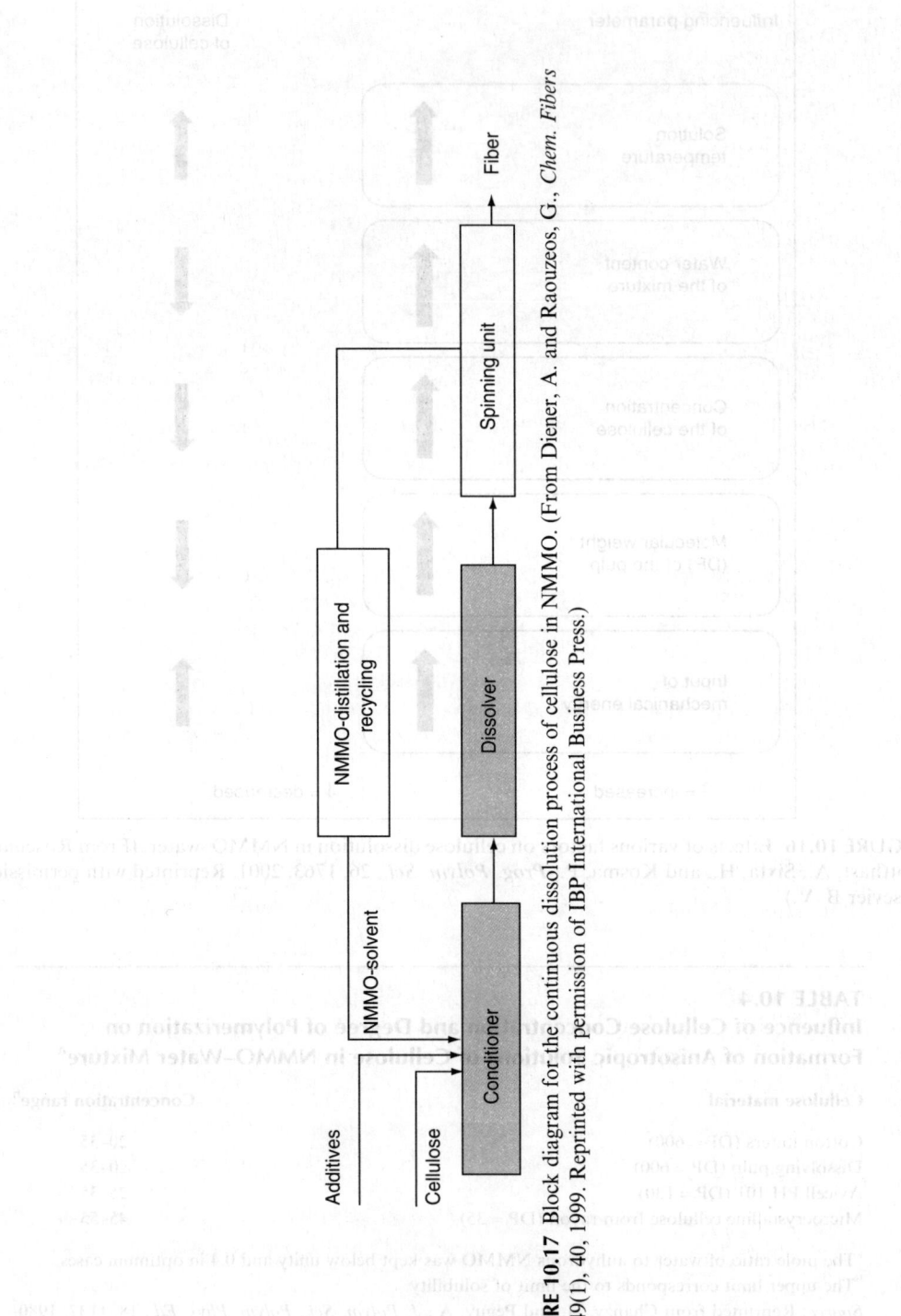

FIGURE 10.17 Block diagram for the continuous dissolution process of cellulose in NMMO. (From Diener, A. and Raouzeos, G., *Chem. Fibers Int.*, 49(1), 40, 1999. Reprinted with permission of IBP International Business Press.)

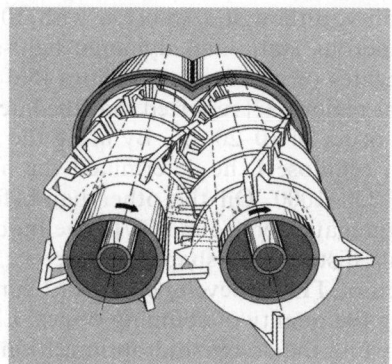

FIGURE 10.18 LIST Corotating Processor. (From Diener, A. and Raouzeos, G., *Chem. Fibers Int.*, 49(1), 40, 1999. Reprinted with permission of IBP International Business Press.)

3. High cellulose content up to 17 wt%, thus providing low energy consumption for water evaporation
4. Minimization of NMMO use for cellulose dissolution, thus lowering the cost of its recycling

Figure 10.20 shows the basic process flow diagram for the continuous dissolution of cellulose in NMMO by means of LIST mixing and kneading technology. As can be seen in the figure, an agitated buffer tank is placed between the conditioner and the dissolver. This intermediate vessel is necessary to ensure smooth continuous operation between the conditioner that operates under atmospheric pressure and the dissolver that works under vacuum. Gravity feeds the conditioned cellulose–NMMO–water mixture to the buffer. The twin piston valve separates the buffer from the dissolver. A twin-screw extrudes a highly viscous spinning dope from the dissolver that can be used for making fibers, films, or membranes.

Studies of NMMO spinning dopes reveal some undissolved cellulose particles. Therefore, for assessing the quality of spinning dope, TITK developed and defined the qualitative parameter *filter value* (F_p) [53]. The filter value is derived from particle analysis of the spinning solution. It is defined as the quotient of the largest particle diameter (X_m) and the logarithm of the number of particles (N_{10}) greater than 10 µm. Spinning dopes with $F_p < 50$

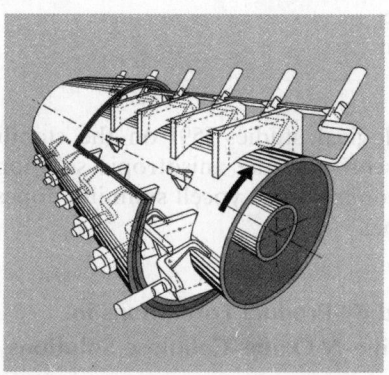

FIGURE 10.19 LIST Discotherm B Fiber. (From Diener, A. and Raouzeos, G., *Chem. Fibers Int.*, 49(1), 40, 1999. Reprinted with permission of IBP International Business Press.)

exhibit excellent spinning quality, although a solution with $50 < F_p < 100$ exhibits only adequate spinning quality. Numerous static and dynamic light-scattering studies also show [37,54,55] that cellulose aggregates or cellulose crystallites [56] remain in the NMMO solutions. Fink et al. in an excellent review [37] reported that the lateral dimension of undissolved cellulose microfibrils is in the order of 10–20 nm. By using the average cross section of 0.32 nm^2 of one cellulose chain in cellulose I, he determined the number of polymer chains in cellulose I microfibrils to be 250–1000. On the other hand, for cellulose xanthate in 1M NaOH only ten chains per aggregate were found [57]. The model for cellulose aggregates is given in Figure 10.21 and has a distinct star-shaped character.

LIST, with the cooperation of TITK, developed a third generation technology (see Figure 10.22) aimed at optimizing the quality of spinning dopes. LIST also developed a process simulation program for supporting the design and optimization of pilot plants and commercial processes. The major advantages of this new process are as follows:

1. Using an operating temperature less than 100°C, therefore lowering the degree of NMMO discoloration
2. Ability to use raw cellulose of different origins
3. Minimization of gel formation in the spinning filter

The degradation of cellulose in NMMO–water solution was one of the major obstacles in the early stages of Lyocell development process. Łaszkiewicz studied the degradation of 5% cellulose solution in an NMMO–water solution at 80°C. He clearly showed that if there is no antioxidant present, the DP rapidly decreases [3]. As shown in Figure 10.23, DP decreased from 700 to around 140 after 120 min at 80°C. The use of antioxidants, particularly PG (see Figure 10.24) that is commonly used in the Lyocell process, prevents cellulose degradation.

The rheological properties of cellulose solutions in NMMO greatly affect the spinnability and even the properties of the resulting Lyocell fibers. At high concentrations greater than 20%, these solutions can be anisotropic. Navard and Haudin [58] were the first researchers to use capillary rheometry to determine the isotropic–anisotropic temperature. As shown in Figure 10.25), for cellulose with DP of 600 in solutions containing 24% cellulose, the critical temperature, T^*, ranges from 87 to 92°C. Although the cellulose solutions in NMMO were not stabilized, the authors were able to show that the cellulose anisotropic solutions in NMMO–water could be formed but are strongly affected by four parameters:

DP of cellulose
Concentration of cellulose
Temperature of the system
Water content

Most recent, extensive rheological studies [59] on the stabilized NMMO–water solution proved that the apparent viscosity of the anisotropic solutions is relatively high (see Figure 10.26). Therefore, most commercial Lyocell spinning dopes usually have concentrations less than 17% and are isotropic.

10.3.1.3 Side Reactions and By-Product Formations in N-Methylmorpholine-N-Oxide–Cellulose Solutions

The dissolution of a polymer in a solvent is usually considered as a process that does not involve any chemical reactions. It is known that in a few cases some polymer solvents may

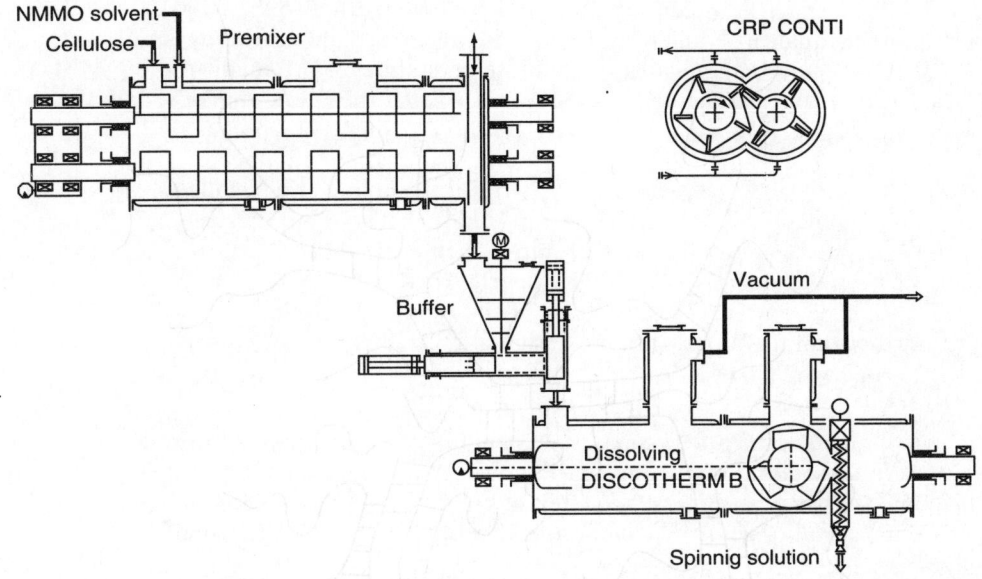

FIGURE 10.20 Process flow diagram for continuous dissolution process of cellulose by means of LIST mixing and kneading technology. (From Diener, A. and Raouzeos, G., *Chem. Fibers Int.*, 49(1), 40, 1999. Reprinted with permission of IBP International Business Press.)

undergo some side reactions and eventually affect the polymers' properties. The dissolution of cellulose in NMMO–water as a solvent for cellulose is far from being a true physical process, as shown in the excellent review by Rosenau et al. [49]. Indeed, NMMO can undergo distinct chemical reactions that involve highly reactive free radicals; NMM, M, and formaldehyde (HCHO) have been recognized as the main by-products. Figure 10.27 shows the complex interactions that could lead to the gradual decomposition of NMMO, the formation of chromophores, the degradation of cellulose, and the occurrence of exothermic reactions, blasts, and explosions. Most of these undesired processes can be avoided by using suitable stabilizers. As it was mentioned before, PG (see Figure 10.24) has been found to be the most suitable stabilizer for the Lyocell process. Recently, a mixture of PG and a novel oxa-chromanol stabilizer (Figure 10.28) showed large improvements.

As shown in Figure 10.29, the main radical species formed in the Lyocell system are the primary radical cation **6** and the carbon-centered radicals **7** and **8**, which are derived from NMMO by one-electron reduction [60]. Obviously, a good stabilizer has to act as a radical scavenger to prevent undesired free radical side reactions that can usually occur, leading to the dramatic degradation of the cellulose polymer (see Figure 10.23). Transition metal ions such as iron and copper can further induce free radical reactions. The action of copper is by far more severe than iron and is related to small amounts of Cu (I) present in the polymer [50]. These copper ions are highly unstable in aqueous solution but become more stable in NMMO systems. As a result, copper is used in numerous reactions that involve NMMO (Figure 10.30). Interestingly, Rosenau et al. [49] state that all transition metal ions that undergo valency changes can affect the stability of NMMO.

In addition to the free radical reactions, Rosenau et al. [61] proposed two main heterolytic reactions draw on N—O bond cleavage. These are the Polonovski (or often quoted

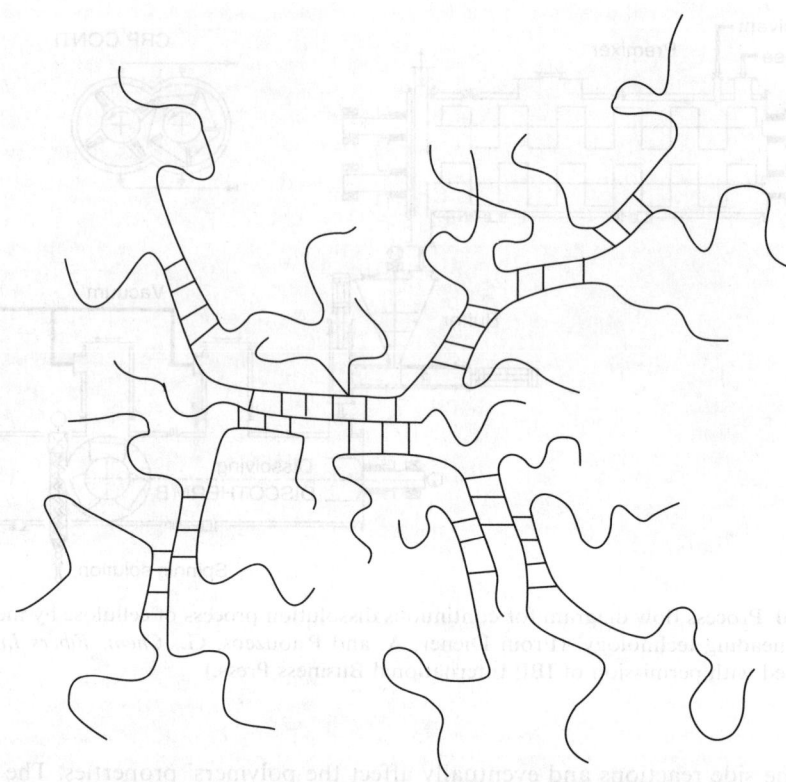

FIGURE 10.21 Model of cellulose aggregate in a cellulose solution. (From Schulz, L., Seger, B., and Burchard, W., *Macromol. Chem. Phys.*, 201, 2008, 2001. Reprinted with permission of Wiley-VCH Verlag GmbH & Co. KGaA and Professor W. Burchard.)

as Polonovski [62]) reactions and the autocatalytic decomposition of NMMO catalyzed by N-(methylene)morpholinium ions (see compound **5** in Figure 10.29). The Polonovski reactions are induced by O-acetylation of the amine oxides. Any acid halide or acid anhydride can induce nonhydrated NMMO Polonovski reactions. Some other acetylating agents can be carboxylic groups present in cellulose, as well as formic or gluconic acid. These are extremely exothermic processes [49] because the intermediate products undergo further rearrangements. As a result of the Polonovski reactions, two main by-products form, mainly M and HCHO. Due to the presence of the highly reactive carbonyl group, the latter can react with hydroxyl groups in the cellulose causing hydroxymethylation and the formation of semiacetals and acetals, respectively.

The ideal stabilizer for the Lyocell system should have the following functions [60]:

1. Prevention of hemolytic reactions by trapping free radicals
2. Transition metal ion complexation to minimize metal-induced hemolytic reactions
3. Adjustment of pH above 7 of the Lyocell solution to minimize Polonovski reactions that are induced by acids
4. Prevention of heterolytic degradation reactions by trapping carbonium–iminium ions (**5**)
5. Scavenging formaldehyde to minimize the formation of carbonium–iminium ions and prevent a reaction of formaldehyde with cellulose

FIGURE 10.22 The third generation dissolving unit developed by LIST. Reprinted with permission of IBP International Business Press.)

PG seems to be the best stabilizer for the Lyocell system; however, it is often used in combination with some other undisclosed additives. It satisfies most of the requirements listed by Rosenau et al. and can react with up to six radicals [60] present in the Lyocell system, namely **6**, **7**, and **8** (see Figure 10.29), to generate the phenoxyl radical **9**; two phenoxyl radical **9** undergo carbon–carbon coupling to form ellagic acid. The latter is an effective antioxidant as shown in Figure 10.31 [60]. The main disadvantage is that the compound **11** is deeply discolored and it is a major cause of the discoloration of Lyocell solutions. As shown in Figure 10.32, PG is also an excellent trap for formaldehyde as well as for N-(methylene)morpholine ions (**5**) (see Figure 10.33).

10.3.1.4 Fiber Formation and Properties of Lyocell Fibers

Lyocell fibers are produced by using a dry-jet wet-spinning process for cellulose in an NMMO–water solution (see Figure 10.34). The molar ratio of NMMO to water is close to 1:1. The wet-spinning technique is relatively slow with a spinning speed less than 100 m/min. The coagulation (or the mass transport) of liquid jets is a critical stage that determines the final speed of the process. Mortimer and Peguy [63] reported that for a 50-μm fiber it takes 1–2 s to accomplish full phase separation. Water or aqueous NMMO solutions are used as coagulants in commercial Lyocell processes. Usually fibrillar cellulose II is regenerated. A higher speed tends to improve final fiber orientation and depends on the air-gap length. The polymer chain orientation takes place mainly in the air gap. Generally, a longer air gap leads to chain relaxation and a lower degree of orientation. Air gaps vary from 20 to 250 mm [63]. The spinning temperature varies from 90 to 120°C. The molecular weight of the cellulose pulp that is used in the Lyocell process is lower than that used in the viscose process [37].

The final properties of Lyocell fibers depend on a number of variables that are grouped in Figure 10.35. As shown in the figure, the final fiber strength will depend on the properties of

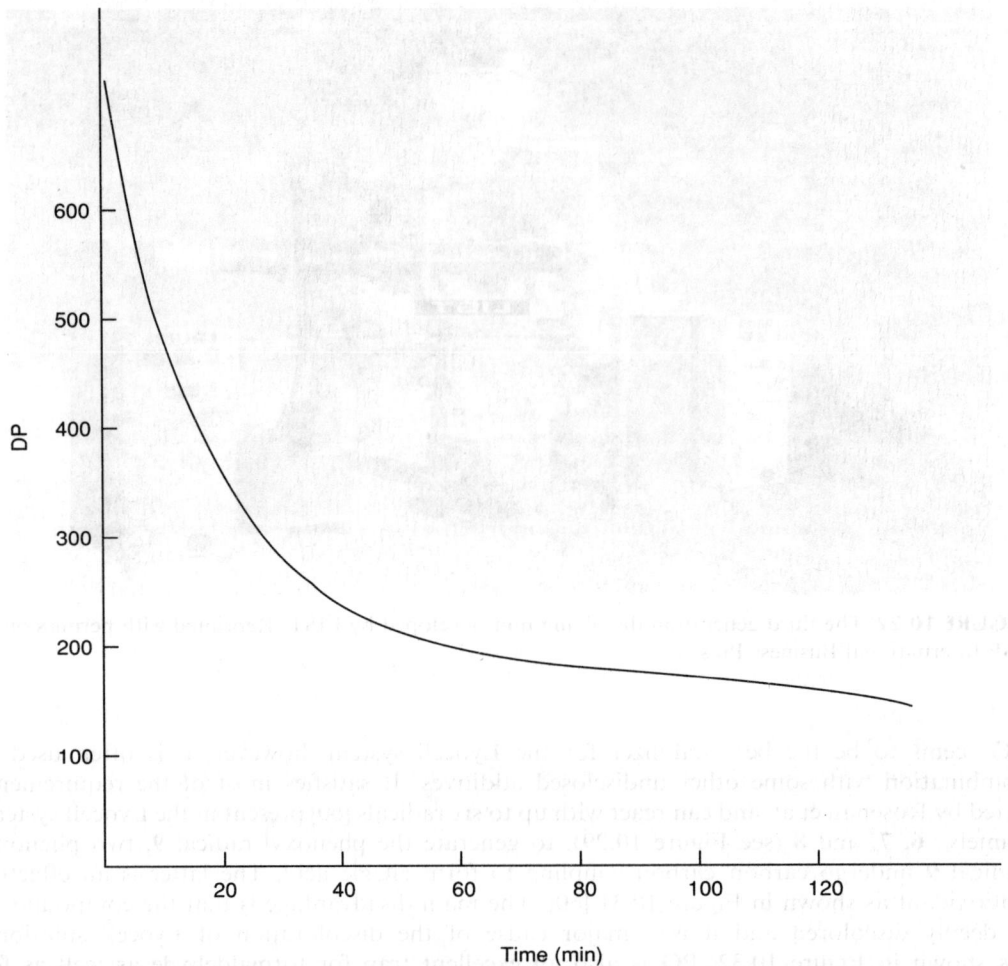

FIGURE 10.23 Degradation of cellulose in NMMO–water at 80°C. The cellulose was dissolved at 85°C and quickly cooled to 80°C before determination of degree of polymerization (DP). (Adapted from Łaszkiewicz, B., *Manufacture of Cellulose Fibers without the Use of Carbon Disulfide*, ACGM LODART, SA, Łódź, Poland, 1997. With permission of Professor B. Łaszkiewicz.)

spinning dope, spinning conditions, coagulation condition, and posttreatment conditions. Numerous studies have reported in the literature that covers the subject [63–69]. The work of Mortimer and Peguy [63–65] is the most elegant and complete. Figure 10.36 shows the effect of the air-gap length (or distance from spinnerette) and draw ratio (DR) on birefringence of Lyocell filaments. They conclude that DR of 4 to 6 is high enough to produce Lyocell fibers with excellent properties. Their work also showed that short air gaps give a dramatic rise in fibrillation. Mortimer and Peguy demonstrated that relatively high air humidity reduces or eliminate fibrillation particularly for long air gaps.

Although, heat setting generally affects the degree of crystallinity of synthetic fibers such as PET, Nylon 6, or Nylon 66, drying of these fibers has little effect. Fink et al. [37] recently discovered that extensive drying for Lyocell fibers under small tension leads to an increase in

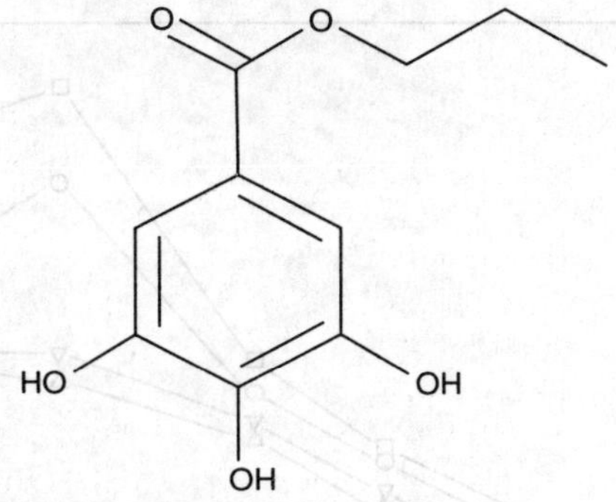

FIGURE 10.24 Structure of *n*-propyl gallate.

orientation of the (1$\bar{1}$0) planes. This phenomenon is attributed to the crystallization of oriented cellulose chains.

Lyocell fibers are more crystalline and more oriented than viscose fibers. They have an oval or round shape (Figure 10.37) and tend to be highly fibrillar (Figure 10.38). The skin–core morphology for these fibers can be obtained only if coagulation is done with liquids other than water [70]. On the other hand, viscose fibers exhibit skin–core morphology and are more porous (see Figure 10.37b). A high degree of orientation leads to improved tensile properties for Lyocell fibers (Figure 10.39). Most of these fibers are manufactured as staple

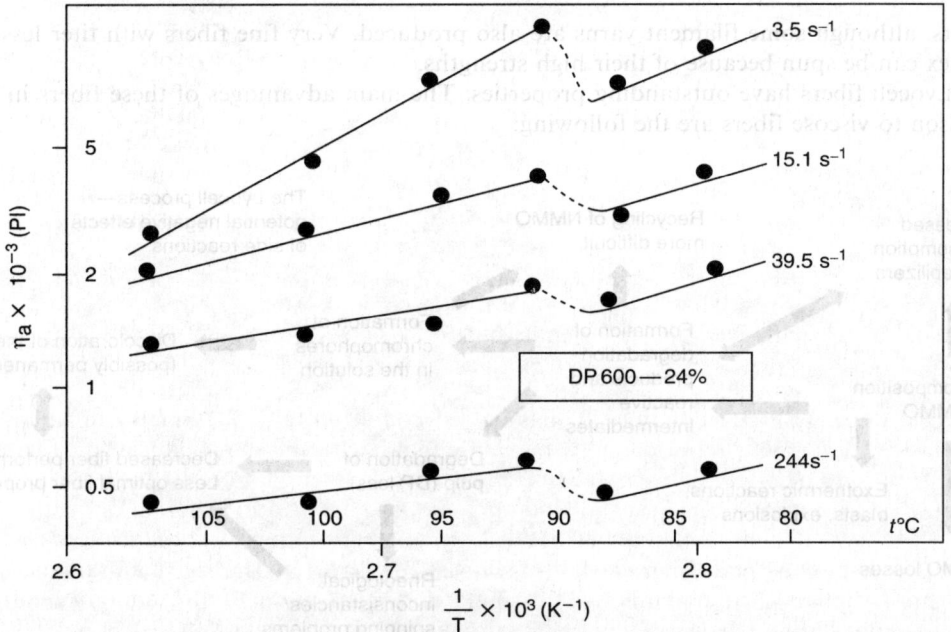

FIGURE 10.25 Temperature dependence of capillary apparent viscosity, η_a of a 24% cellulose–NNMO solution (7.2% water content) at different shear rates. (From Navard, P. and Haudin, J.P., *Br. Polym. J.*, 12(4), 174, 1980. Reprinted with permission from John Wiley & Sons on behalf of SCI.)

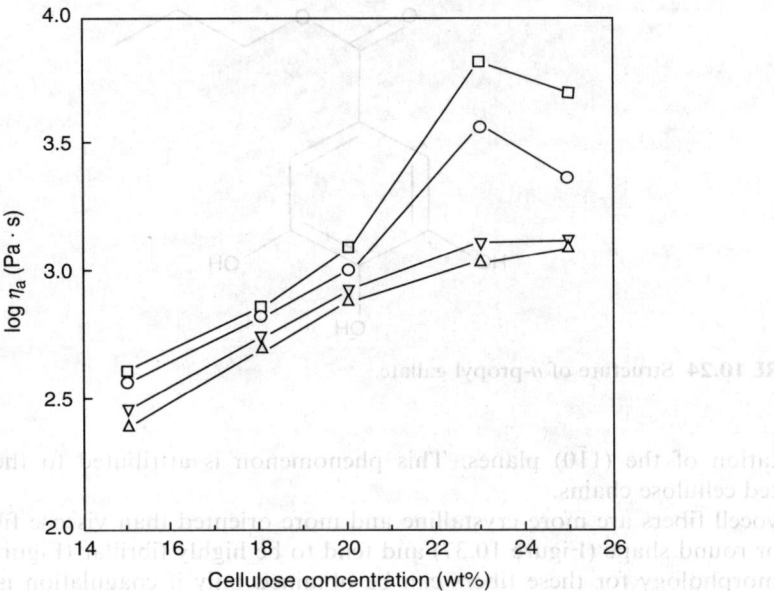

FIGURE 10.26 Concentration dependence of apparent viscosity, η_a for cellulose NMMO–water solution at various temperatures (□) 85°C; (o) 90°C; (▽) 100°C; (△) 110°C at shear rate of 50 s^{-1} and DP of 1180. The molar ratio of water to NMMO was 0.8. The Rabinowitsch correction was used for all viscosity values. (From Kim, S.O., Shin, W.J., Cho, H., Kim, B.C., and Chung, I.J., *Polymer*, 40(23), 6443, 1999. Reprinted with permission of Elsevier B.V.)

fibers, although some filament yarns are also produced. Very fine fibers with titer less than 1 dtex can be spun because of their high strengths.

Lyocell fibers have outstanding properties. The main advantages of these fibers in comparison to viscose fibers are the following:

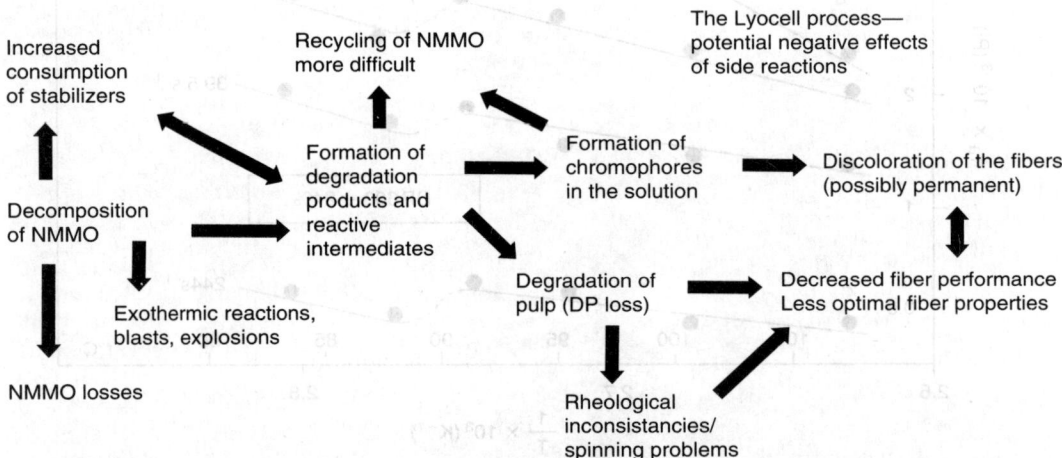

FIGURE 10.27 Side reactions and by-product formation in the Lyocell system. (From Rosenau, T., Potthast, A., Sixta, H., and Kosma, P., *Prog. Polym. Sci.*, 26, 1763, 2001. Reprinted with permission of Elsevier B.V.)

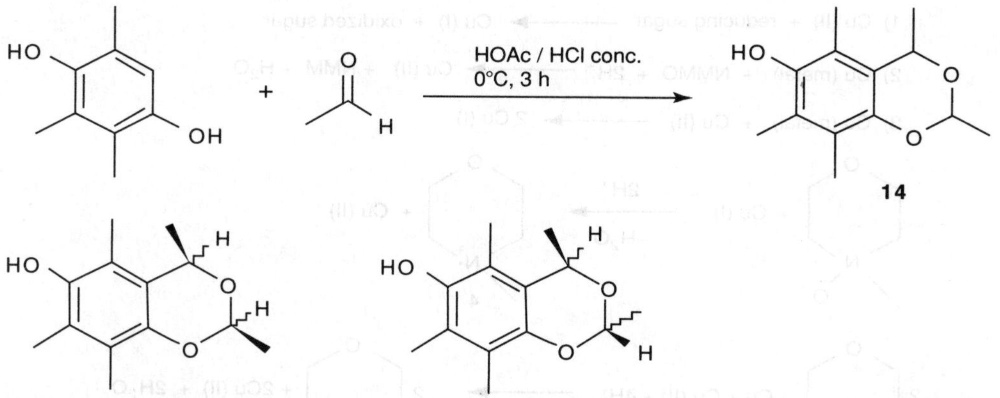

Regenerated Cellulose Fibers

FIGURE 10.28 Synthesis of 2,4,5,7,8-pentamethyl-4H-1,3-benzodioxin-6-ol (PBD). (From Rosenau, T. et al., *Cellulose*, 9, 283, 2002. Reprinted with permission from Springer.)

1. Much higher dry and wet strength
2. Much higher dry and wet Young's modulus
3. Higher knot- and loop-strength

Typical properties of Lyocell fibers are summarized in Table 10.5. Tensile properties can be improved further by introducing ammonium chloride and other additives [72]. One of the disadvantages of Lyocell fibers is the tendency for fibrillation (see Figure 10.38). Interestingly, the fibers exhibit very unique hand.

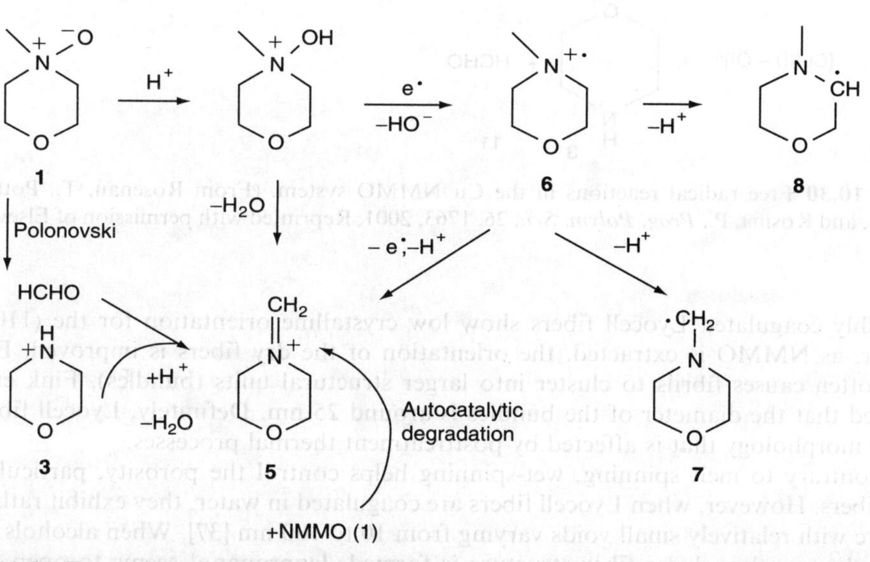

FIGURE 10.29 Formation of reactive intermediates by homolytic (radical) and heterolytic (ionic) side reactions of NMMO. (From Rosenau, T. et al., *Cellulose*, 9, 283, 2002. Reprinted with permission from Springer.)

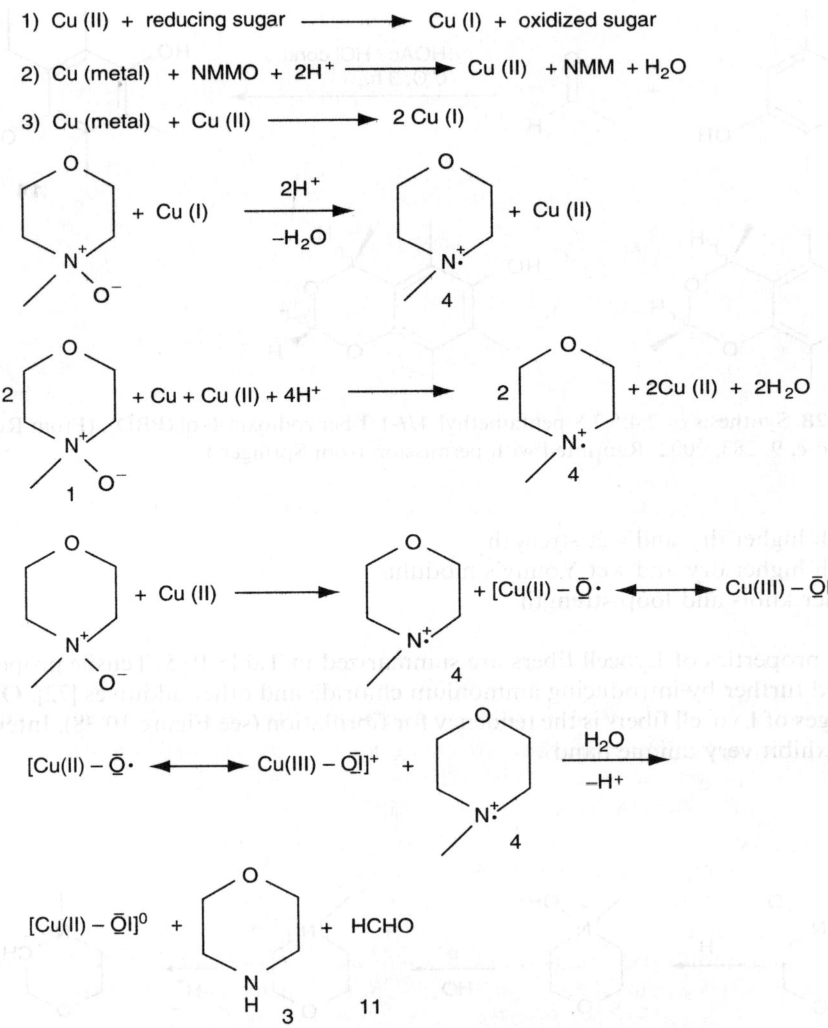

1) Cu (II) + reducing sugar ⟶ Cu (I) + oxidized sugar

2) Cu (metal) + NMMO + 2H$^+$ ⟶ Cu (II) + NMM + H$_2$O

3) Cu (metal) + Cu (II) ⟶ 2 Cu (I)

[morpholine-N-oxide] + Cu (I) →(2H$^+$, –H$_2$O)→ [morpholine-N-methyl, 4] + Cu (II)

2 [morpholine-N-oxide, 1] + Cu + Cu (II) + 4H$^+$ ⟶ 2 [morpholine-N-methyl, 4] + 2Cu (II) + 2H$_2$O

[morpholine-N-oxide] + Cu (II) ⟶ [morpholine-N-methyl, 4] + [Cu(II) – Ō·] ⇌ [Cu(III) – ŌI]$^+$

[Cu(II) – Ō·] ⇌ [Cu(III) – ŌI]$^+$ + [morpholine-N-methyl] →(H$_2$O, –H$^+$)→

[Cu(II) – ŌI]^0 + [morpholine-NH, 3] + HCHO (11)

FIGURE 10.30 Free radical reactions in the Cu–NMMO system. (From Rosenau, T., Potthast, A., Sixta, H., and Kosma, P., *Prog. Polym. Sci.*, 26, 1763, 2001. Reprinted with permission of Elsevier B. V.)

Freshly coagulated Lyocell fibers show low crystalline orientation for the (110) plane; however, as NMMO is extracted, the orientation of the dry fibers is improved. Extensive drying often causes fibrils to cluster into larger structural units (bundles). Fink et al. [37] estimated that the diameter of the bundles is around 25 nm. Definitely, Lyocell fibers have fibrillar morphology that is affected by posttreatment thermal processes.

In contrary to melt spinning, wet-spinning helps control the porosity, particularly, for rayon fibers. However, when Lyocell fibers are coagulated in water, they exhibit rather dense structure with relatively small voids varying from 10 to 100 nm [37]. When alcohols are used as coagulants, a less dense fiber structure is formed. Isopropanol seems to open the fiber structure [37] but other alcohols such as *n*-pentanol and *n*-hexanol dramatically alter the coagulation process, thus lowering the fiber's orientation. Larger pores of up to 300 nm were

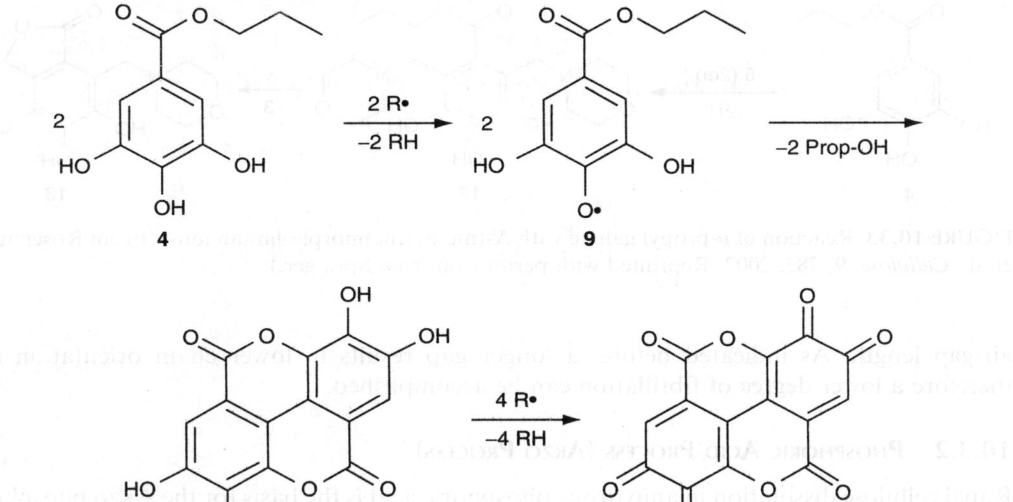

FIGURE 10.31 Reactions of *n*-propyl gallate as a free radical trap. (From Rosenau, T. et al., *Cellulose*, 9, 283, 2002. Reprinted with permission from Springer.)

observed in the core of the fiber. Lyocell fibers that are formed using higher alcohols show skin–core morphology, where the skin is denser than the core. As shown in Figure 10.40, such fibers obviously have much lower strength but also exhibit a much lower tendency for fibrillation (Figure 10.41). Lower fiber strength is generally an undesired fiber property, therefore Fink et al. [37] proposed a clever two-stage coagulation technique for Lyocell fibers (Figure 10.42). Higher alcohols such as *n*-pentanol and *n*-hexanol are not miscible with water and are lighter than water. The double coagulation bath, therefore, has the higher alcohol on the top layer and water on the bottom. Consequently, a liquid jet slowly coagulates first at the outer jet boundary, followed by fast coagulation in water. As a result, the fiber surface is more porous (less dense) and the core of the fiber is highly dense and oriented. The two-stage coagulation technique produces fibers with good strength and high fibrillation resistance [37]. It is not clear if this method is commercially used. Fibrillation of Lyocell fibers can also be controlled by adding various modifiers [63,72,73] to the spinning dope or by controlling the

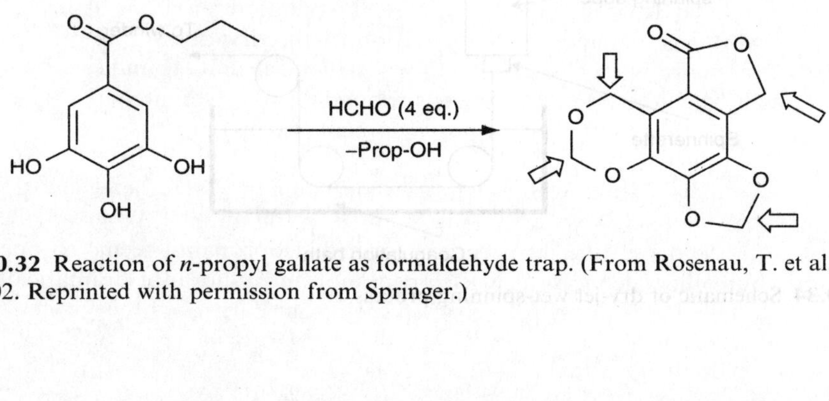

FIGURE 10.32 Reaction of *n*-propyl gallate as formaldehyde trap. (From Rosenau, T. et al., *Cellulose*, 9, 283, 2002. Reprinted with permission from Springer.)

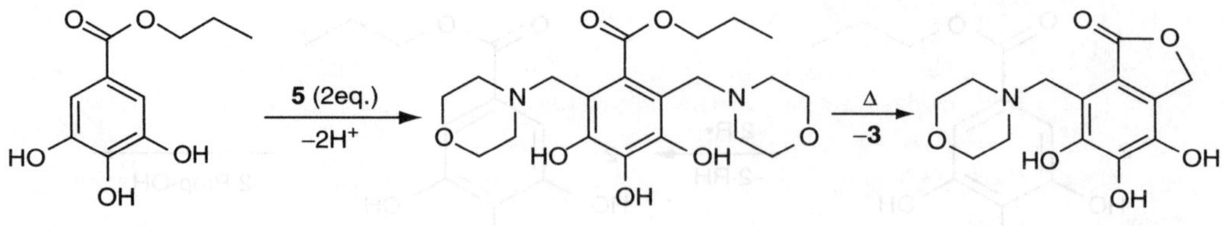

FIGURE 10.33 Reaction of *n*-propyl gallate with *N*-(methylene)morpholinium ions. (From Rosenau, T. et al., *Cellulose*, 9, 283, 2002. Reprinted with permission from Springer.)

air-gap length. As indicated before, a longer gap results in lower chain orientation and therefore a lower degree of fibrillation can be accomplished.

10.3.2 PHOSPHORIC ACID PROCESS (AKZO PROCESS)

Rapid cellulose dissolution in anhydrous phosphoric acid is the basis for the Akzo phosphoric acid process [36,74]. A similar concept has been practiced for making Kevlar fibers, in which anhydrous sulfuric acid is used to dissolve poly(*p*-phenylene terephthalamide). However, anhydrous sulfuric acid is highly corrosive and is produced by mixing concentrated sulfuric acid with small amounts of fuming sulfuric acid.

As phosphoric acid can form oligomers such as dimer, trimer, tetramer, or even polyphosphoric acid, water-free phosphoric acid solutions of different compositions can be easily obtained by mixing together two or more components, such as orthophosphoric acid (H_3PO_4), pyrophosphoric acid ($H_4P_2O_7$), polyphosphoric acid ($H_6P_4O_{13}$), phosphorus pentoxide (P_2O_5), and water. However, it takes considerable time before the new equilibrium distribution of acids is reached. Optimal cellulose dissolution conditions were obtained when the P_2O_5 concentration is in the range of 72.4 to 76%. The melting point of these compositions is below room temperature. Orthophosphoric acid contains 72.4% w/w P_2O_5.

Boerstoel [74] found that an IKA-Duplex kneader or a twin extruder can be utilized to form 17.1% w/w birefringent cellulose solutions for cellulose having DP of 700 to 800 within 2 to 3 min at 28°C (Figure 10.43). Lower temperatures tend to increase the dissolution times, but even at 8°C it takes only 20 min to achieve an anisotropic cellulose solution. Such an

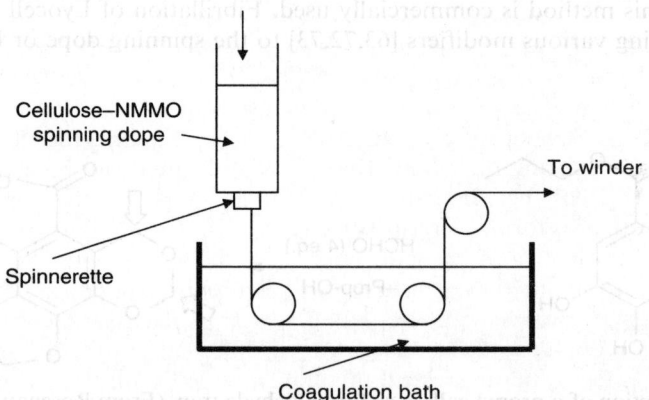

FIGURE 10.34 Schematic of dry-jet wet-spinning process.

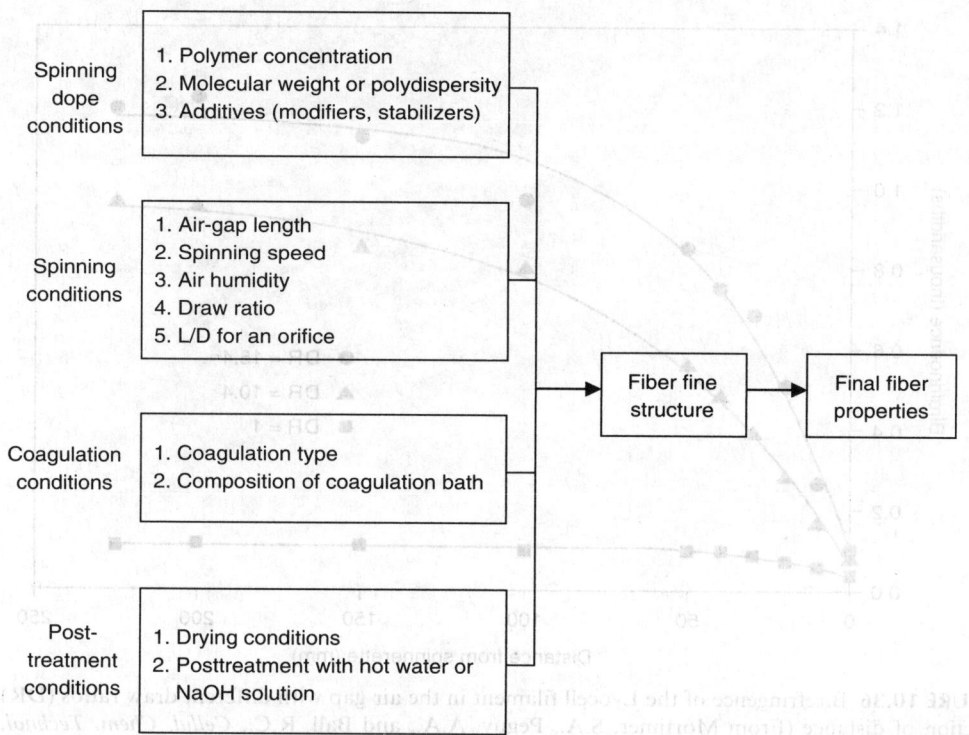

FIGURE 10.35 Lyocell process variables.

extremely short dissolution time is a great achievement and a strong commercial advantage. Usually, it takes close to 48 h to produce viscose fibers.

Boerstoel [74] was first to report anisotropic cellulose solutions in anhydrous phosphoric acid. The effect of cellulose concentration in phosphoric acid (74.4% w/w P_2O_5) on clearing temperature is given in Figure 10.44. The temperature at which a solution becomes isotropic is usually called the clearing temperature. As can be seen in the figure, anisotropy can be seen at concentrations higher than 7.5% w/w at ambient temperature for cellulose having DP of 800. Such a phenomenon clearly demonstrates that phosphoric acid is a powerful solvent for cellulose polymers. Clearing temperature tends to increase with increasing cellulose concentration, up to 38% w/w; however, cellulose degradation starts above 60°C. Figure 10.44 also shows that clearing temperatures for poly(p-phenylene terephthalamide) are much higher than those for cellulose. The open square in Figure 10.44 represents the clearing temperature for chitin, thus demonstrating that anhydrous phosphoric acid can be used to make anisotropic solutions of this polymer. Boerstoel et al. [76] also reported that anisotropic cellulose acetate solutions were also formed in this solvent.

It was found that water has detrimental effects on the anisotropic properties of cellulose solutions, as evidenced by a significant decrease in clearing temperature [36,74]. The authors showed that if the P_2O_5 content is greater than 72.4% w/w, the clearing temperature remains constant for a given cellulose solution concentration; however, if the concentration of P_2O_5 is less than 72.4% w/w, i.e., water is present, this transition temperature decreases rapidly. This phenomenon can be caused by the interaction of water molecules with the hydroxyl groups of cellulose [74].

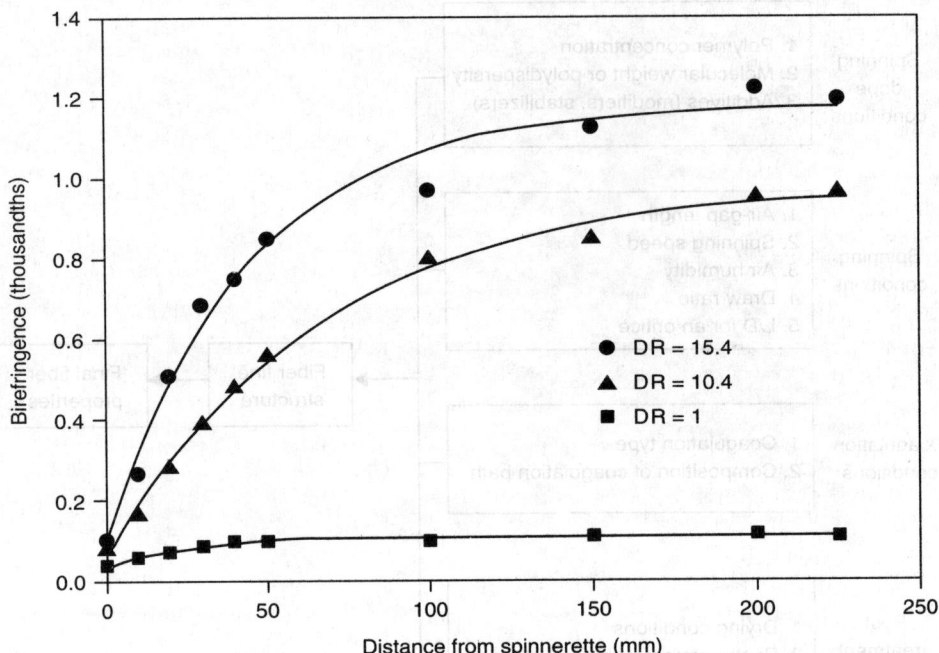

FIGURE 10.36 Birefringence of the Lyocell filament in the air gap with different draw ratios (DR) as a function of distance (From Mortimer, S.A., Peguy, A.A., and Ball, R.C., *Cellul. Chem. Technol.*, 30, 251, 1996.)

As anisotropic cellulose solutions in anhydrous phosphoric acid can be easily formed, the Akzo researchers [36,74] developed experimental high-performance (high-modulus and high-tenacity) fibers. In a typical process [36] (see Figure 10.45), powdered cellulose with an equilibrium content of about 5% w/w moisture and anhydrous phosphoric acid (72.4–74% w/w P_2O_5) were thoroughly mixed in a ZSK 30 twin-screw extruder. The solution, containing 19% w/w of dry cellulose, was filtered, heated, and extruded through a spinning pack equipped with a spinnerette containing 1500 capillaries with 65 μm diameters. The filaments passed through air, where they were subjected to stretching, and coagulated in the vertical

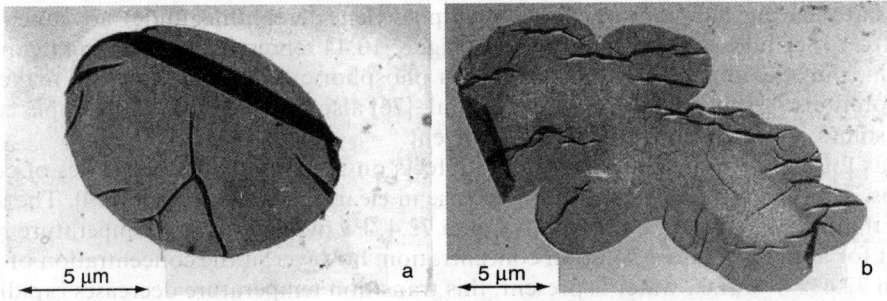

FIGURE 10.37 Fiber cross section of (a) Lyocell fiber and (b) Viscose fiber. (From Fink, H.P., Weigel, P., Purz, H.J., and Ganster, J., *Prog. Polym. Sci.*, 26, 1473, 2001. Reprinted with permission of Elsevier B. V.)

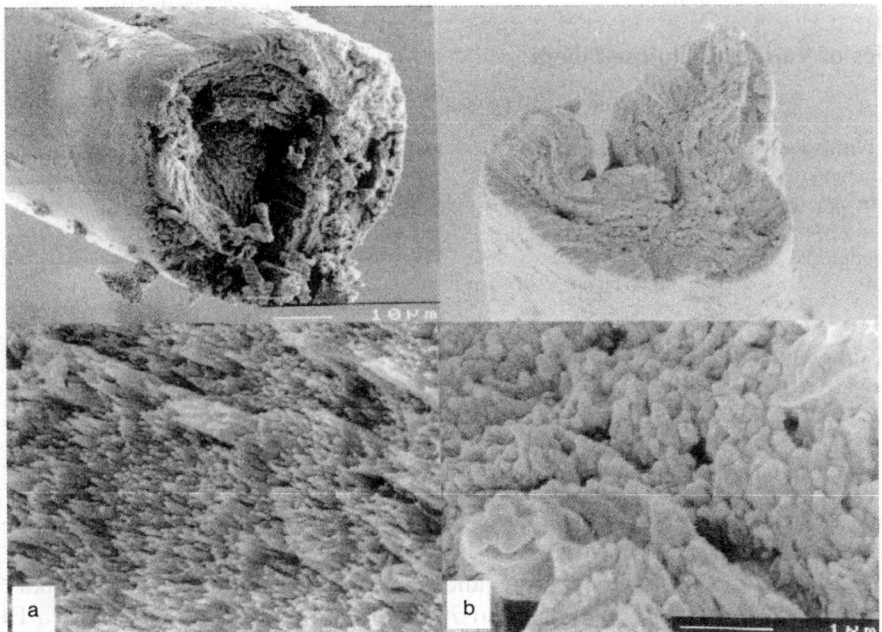

FIGURE 10.38 SEM micrographs of fractured surfaces for (a) NMMO fibers and (b) Viscose fibers. (From Fink, H.P., Weigel, P., Purz, H.J., and Ganster, J., *Prog. Polym. Sci.*, 26, 1473, 2001. Reprinted with permission of Elsevier B. V.)

acetone bath at 5°C. Phosphoric acid was removed from the yarn by subsequent washing with water and finally with 2% w/w Na_2CO_3, and then was dried on a heated godet at a winding speed of 100 m/min. The Akzo researchers indicate that the drop of DP of cellulose in phosphoric acid during processing is relatively small; whereas the starting material had DP of 800, the fiber DP was 620.

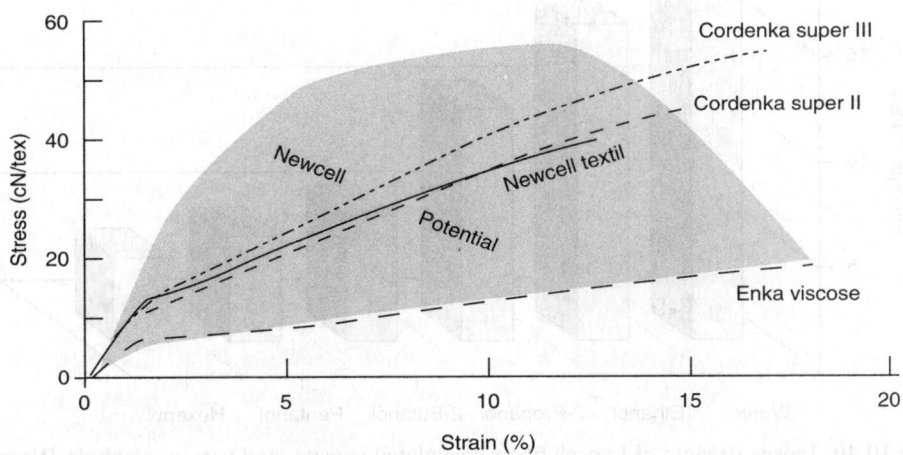

FIGURE 10.39 Stress–strain curves of NewCell (based on NMMO) and Cordenka (viscose-based) fibers. (From Wachsman, U. and Diamantoglou, M., *Das Papier*, 51(12), 660, 1997. With permission.)

TABLE 10.5
Properties of Various Cellulose Fibers

Property	Cotton	Lyocell	Polynosic	HWM	Viscose	Cupro
Degree of fibrillation[a]	2	4–6	3	1	1	2–3
Dry tear strength (cN/tex)	22	42	38	35	22	20
Wet tear strength (cN/tex)	28	36	30	20	12	10
Strength ratio (wet/dry)	~1.25	~0.85	~0.70	~0.60	0.55	0.50
Water retention (%)	50	65	55–70	75	90–100	100–120
Average DP	1600–2000 (bleached)	~600	~500	~400	~300	~500

[a] 0 = minimal, 6 = maximum degree of fibrillation.
Source: Adapted from Breier, R., *Lenzinger Berichte*, 76, 108, 1997.

Tensile stress–strain curves for Enka Viscose (textile yarn) and high-performance cellulose fibers are given in Figure 10.46. As shown in the figure, the new Akzo experimental fiber, called Fiber B, has much higher modulus and tenacity than commercial Cordenka 660 and 700 tire cord yarns. The Cordenka 660 and 700 and the high-modulus Cordenka EHM are produced with the commercial viscose process. As Fiber B has an initial modulus of 45 GPa, strength of 1.3 GPa, and elongation at break of 5.1% (Table 10.6), it can be considered as a high-modulus, high-tenacity, or high-performance fiber. The corresponding tensile properties for Enka Viscose textile yarn are 9.3 GPa, 0.26 GPa, and 23.5%.

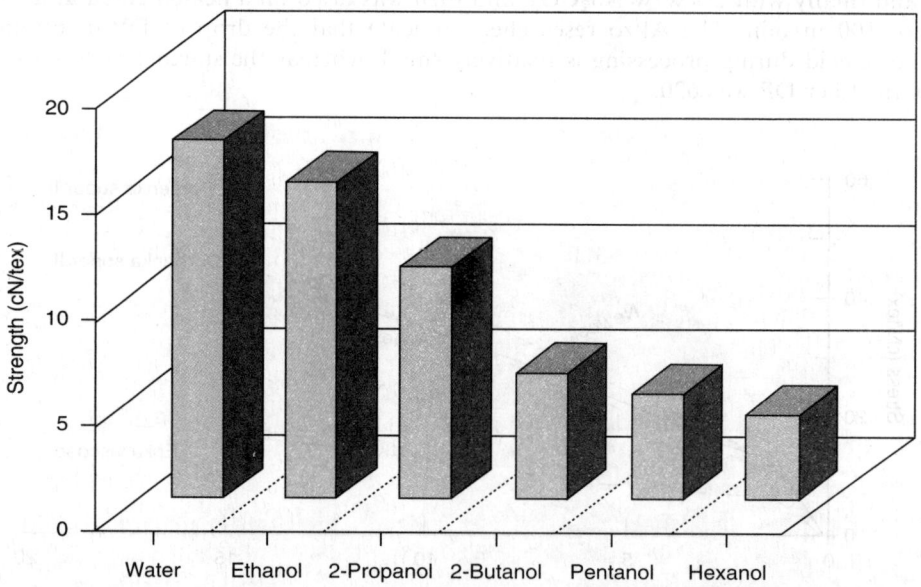

FIGURE 10.40 Tensile strength of Lyocell fibers coagulated in water and various alcohols. (From Fink, H.P., Weigel, P., Purz, H.J., and Ganster, J., *Prog. Polym. Sci.*, 26, 1473, 2001. Reprinted with permission of Elsevier B. V.)

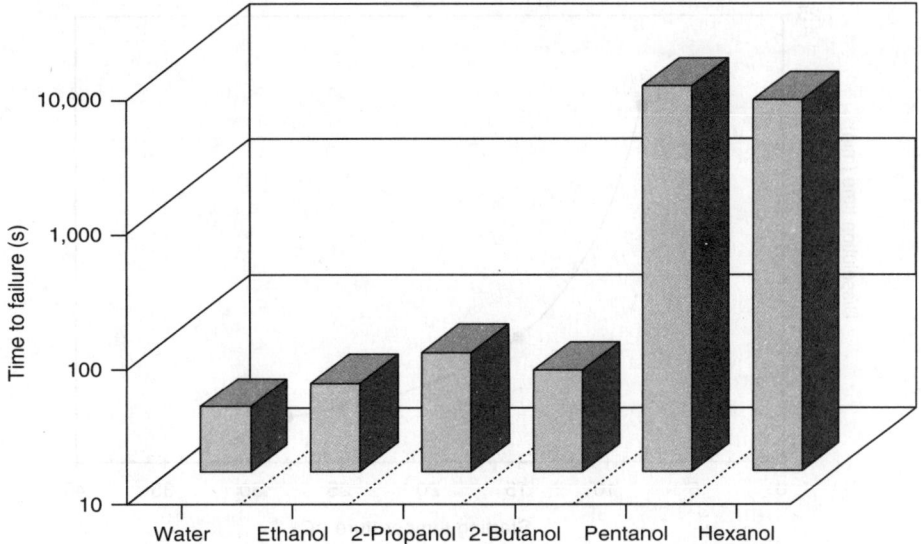

FIGURE 10.41 Fibrillation resistance of Lyocell fibers that were coagulated in water and various alcohols. (From Fink, H.P., Weigel, P., Purz, H.J., and Ganster, J., *Prog. Polym. Sci.*, 26, 1473, 2001. Reprinted with permission of Elsevier B. V.)

10.3.3 CARBAMATE PROCESS

Cellulose carbamate was discovered and patented by Hill and Jacobson [77]. A polymer usually can be formed by reacting cellulose with isocyanic acid (IA) (Figure 10.47). In most reactions, the hydroxyl group present at C6 position in the repeating cellulose unit reacts with

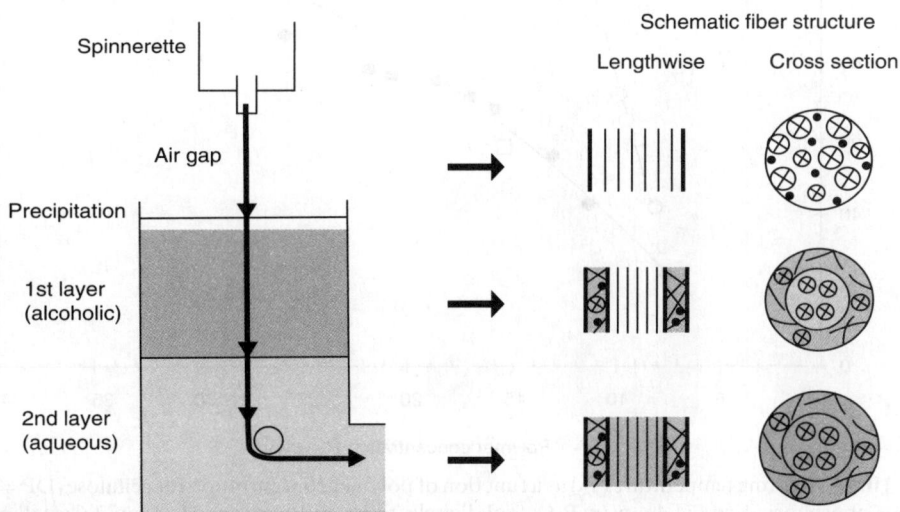

FIGURE 10.42 Formation of skin–core Lyocell fibers by using a two-stage coagulation technique. (From Fink, H.P., Weigel, P., Purz, H.J., and Ganster, J., *Prog. Polym. Sci.*, 26, 1473, 2001. Reprinted with permission of Elsevier B. V.)

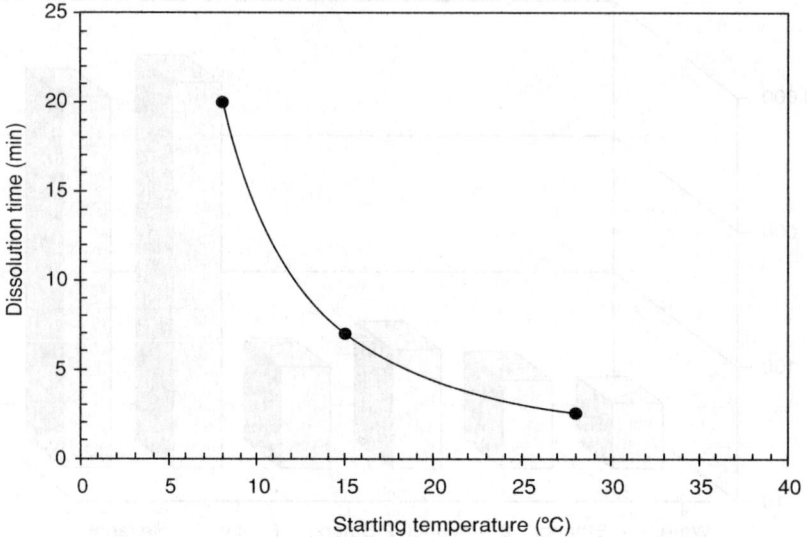

FIGURE 10.43 Cellulose dissolution time in and IKA-Duplex kneader as a function of temperature. $C_{cellulose} = 17.1\%$ w/w; $DP_{cellulose} = 800$; $C_{P_2O_5} = 74.4\%$ w/w. (From Boerstoel, H., Liquid Crystalline Solutions of Cellulose in Phosphoric Acid for Preparing. Cellulose Yarns, Ph.D. dissertation, University of Groningen, 1998.)

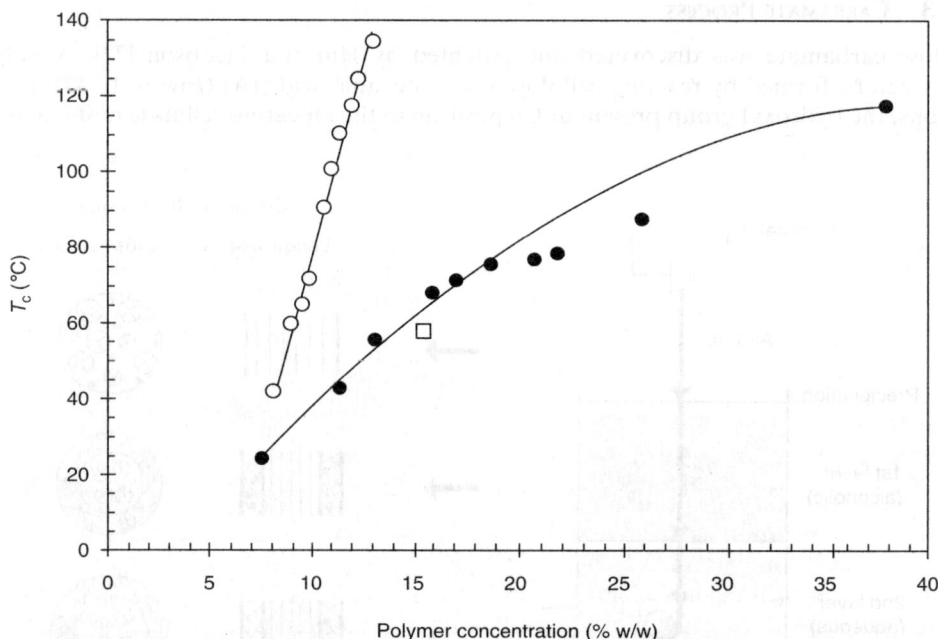

FIGURE 10.44 Clearing temperature (T_c) as a function of polymer concentration for cellulose (DP = 800) in anhydrous phosphoric acid (74.4% w/w P_2O_5, solid circles). (From Boerstoel, H., Liquid Cystalline Solutions of Cellulose in Phosphoric Acid for Preparing. Cellulose Yarns, Ph.D. dissertation, University of Groningen, 1998.). Open circles represent the clearing temperature for poly(p-phenylene terephthalamide) ($M_w = 31,000$ g/mol) in sulfuric acid (From Picken, S.J., Macromolecules, 22, 1766, 1989.). The square represents the clearing temperature for chitin ($M_w = 400,000$ g/mol) solution at the polymer concentration of 15.5% w/w.

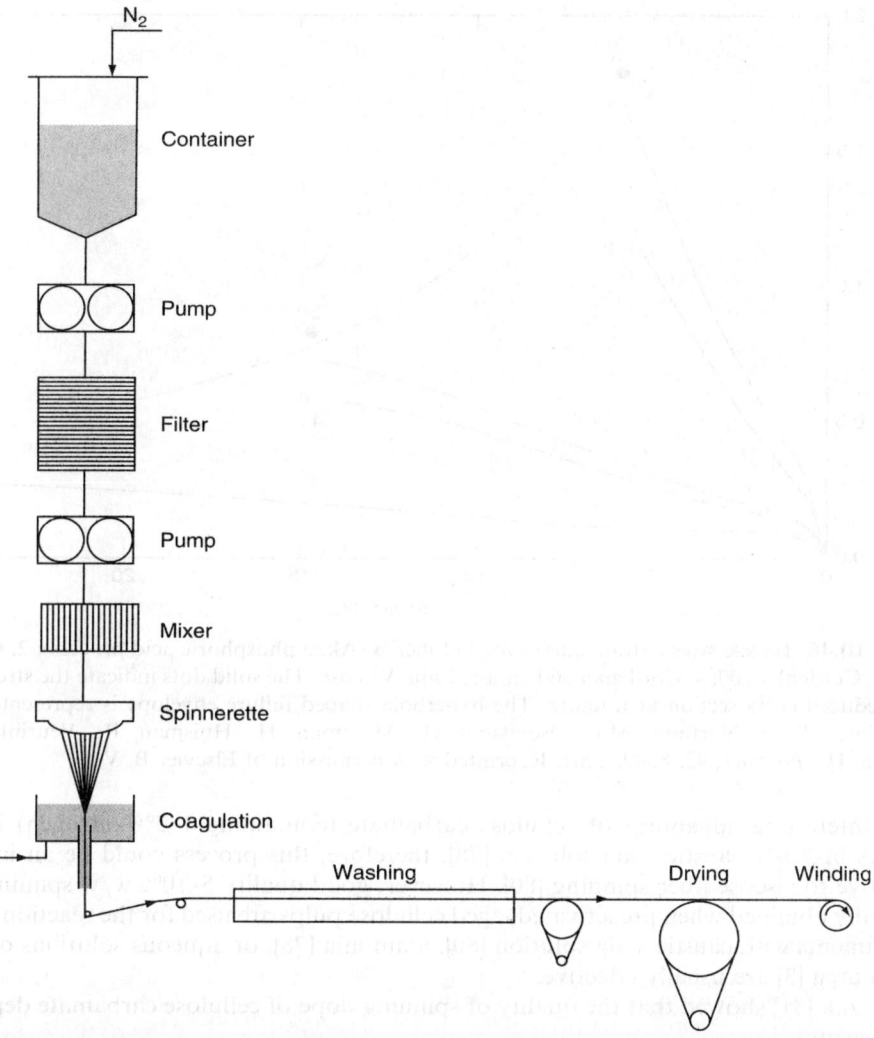

FIGURE 10.45 Dry-jet wet-spinning process for cellulose solutions in anhydrous phosphoric acid (Akzo process). (From Boerstoel, H., Liquid Crystalline Solutions of Cellulose in Phosphoric Acid for Preparing. Cellulose Yarns, Ph.D. dissertation, University of Groningen, 1998.)

IA. There is also some substitution of hydroxyl groups at C2 and C3 [78]. The overall degree of cellulose substitution is relatively low and varies from 0.15–0.25. The corresponding polymer nitrogen content is 1.2–2.0% [3].

The least expensive source of IA is urea. Petropavlovskii and Zimina [79] reviewed reactions of cellulose with urea and the preparation of alkali-soluble cellulose carbamates. IA and ammonia usually form when urea is heated above 135°C [3]; however, as shown in Figure 10.48, IA undergoes some side reactions and various nontoxic, water-soluble by-products can form. Because cellulose carbamate is a stable solid that is not soluble in water, the by-products can be easily separated by extraction with water. The dried polymer can be stored for a prolonged period of time.

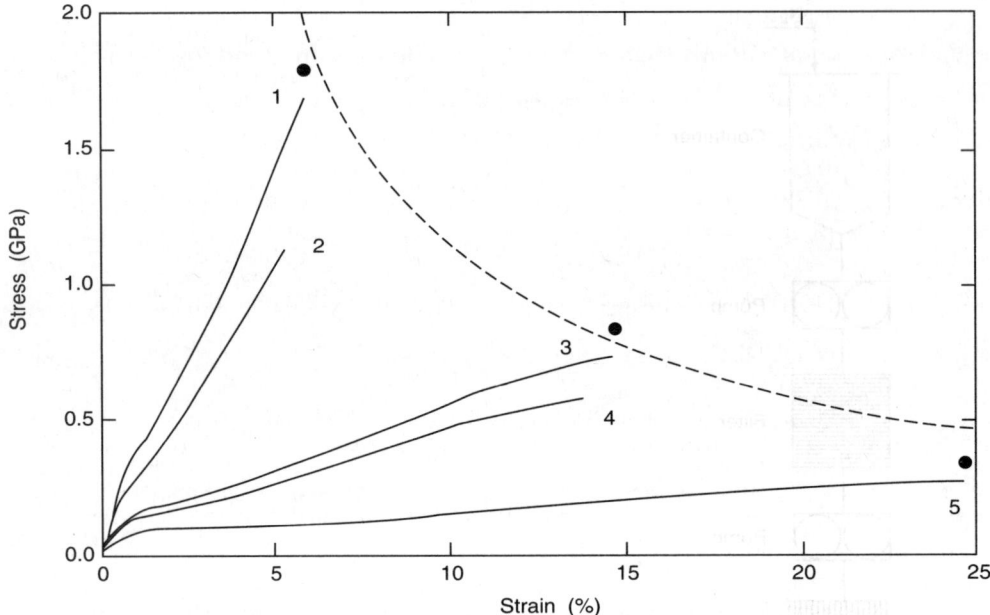

FIGURE 10.46 Tensile stress–strain curves for 1, Fiber B (Akzo phosphoric acid process); 2, Cordenka EHM; 3, Cordenka 700; 4, Cordenka 660; and 5, Enka Viscose. The solid dots indicate the stress related to the reduced cross section at fracture. The hyperbola-shaped failure envelope is represented by the dashed line. (From Northolt, M.G., Boerstoel, H., Maatman, H., Huisman, R., Veurink, J., and Elzerman, H., *Polymer*, 42, 8249, 2001. Reprinted with permission of Elsevier B. V.)

An interesting advantage of cellulose carbamate (containing 1–3% nitrogen) is that it dissolves in 7–9% caustic soda solution [78]; therefore, this process could be an important alternative to viscose fiber spinning [80]. However, good quality 5–10% w/w spinning dopes are usually obtained when preactivated, aged cellulose pulps are used for the reaction with IA. Pretreatments with caustic soda solution [80], ammonia [78], or aqueous solutions of ammonia and urea [3] are usually effective.

Sobczak [81] showed that the quality of spinning dope of cellulose carbamate depends on the following:

1. polymer concentration
2. DP of cellulose
3. degree of substitution (DS) of cellulose
4. concentration of undissolved cellulose carbamate particles

The last factor is extremely important because it determines the spinning efficiency.

It is very important to mention that at ambient temperature carbamate groups tend to hydrolyze under alkaline conditions more easily. Therefore, solutions of cellulose carbamate must be chilled off [3]. Struszczyk [82,83] demonstrated that such solutions are stable below 0°C. All commercial processes, namely dissolution of cellulose carbamate and solution storage must be handled at low temperatures to avoid changes in viscosity. Some additives also can be used to control the solution's viscosity and to reduce gelation. The addition of 1–3% zinc oxide to a spinning dope tends to increase the viscosities of these solutions and improves the spinnability [3]. This additive also extends the solution's storage life.

TABLE 10.6
Tensile Properties of Various High-Performance Cellulose Fibers and Viscose Textile Fibers

Fiber	Initial modulus (GPa)	Tenacity (GPa)	Elongation at break (%)
Fiber B (Akzo phosphoric acid process)	45	1.3	5.1
Cordenka EHM	38	0.9	4.6
Cordenka 700	18.9	0.6	12
Cordenka 660	17.3	0.54	12
Enka Viscose	9.3	0.26	23.5

Note: The samples were conditioned at 21°C and 65% relative before testing. The gauge length was 10 cm. The strain rate was 10%/min.
Source: Adapted from Northolt, M.G., Boerstoel, H., Maatman, H., Huisman, R., Veurink, J., and Elzerman, H., *Polymer*, 42, 8249, 2001.

Fiber spinning of cellulose carbamate solution usually consists of three steps:

1. Degassing of cellulose carbamate solutions
2. Coagulation of cellulose carbamate liquid jets
3. Hydrolysis of cellulose carbamate fibers

In the second step a sulfuric acid solution is generally used as a commercial coagulant; other coagulating baths such as methanol or aqueous solution of aluminum sulfate or sodium carbonate [3] can be utilized to form cellulose carbamate fibers. The third step is conducted in a separate bath that contains a diluted solution of caustic soda. Łaszkiewicz [3] indicates that a high temperature of 80–90°C is used to hydrolyze the carbamate groups. Although, most of the groups are hydrolyzed, regenerated cellulose fibers always contain a small amount of nitrogen.

Struszczyk and coworkers [84] of the Institute of Chemical Fibers (Instytut Włókien Chemicznych, Łódź, Poland) were instrumental in developing a Polish carbamate process for making regenerated cellulose fibers. The best cellulose carbamate solutions were obtained with 8 wt% of α-cellulose and 8.0–8.8% w/w NaOH. These solutions were characterized by a ripeness degree ranging from 10 to 14°H and showed the best spinning properties for the preparation of fibers with good mechanical properties. The Polish process relies on a special cellulose preactivation process that includes the use of aqueous urea solutions containing ammonia or its urea salts [85] or activation of cellulose or its mixtures with urea with NaOH. This is followed by neutralization with CO_2 or SO_2 or other acids or anhydrides [86].

Researchers from the Finish company, Neste Oy, discovered their own cellulose process by activating cellulose with liquid ammonia at −35°C [87,88]. They then treated the activated cellulose with urea at 135–145°C which led to the formation of cellulose carbamate with the DS of 0.15–2.0. The polymer can be readily dissolved in diluted caustic soda solution. The main disadvantage of this technology is its relatively high cost.

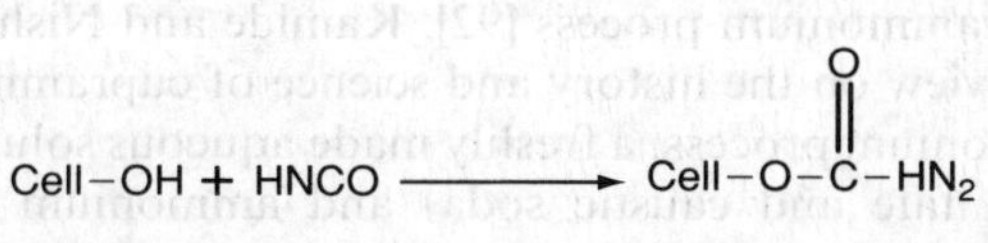

$$\text{Cell-OH} + \text{HNCO} \longrightarrow \text{Cell-O-C(=O)-HN}_2$$

FIGURE 10.47 Reaction of isocyanic acid with cellulose.

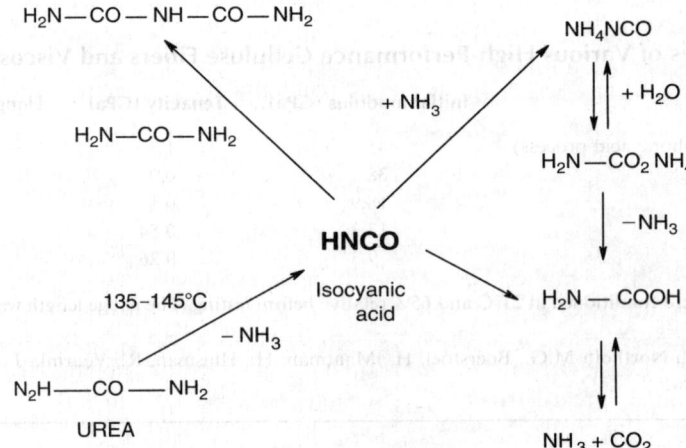

FIGURE 10.48 Thermal decomposition of urea and side reactions of isocyanic acid. (Adapted from Łaszkiewicz, B., *Manufacture of Cellulose Fibers without the Use of Carbon Disulfide*, ACGM LODART, SA, Łódź, Poland, 1997. With permission of Professor B. Łaszkiewicz.)

In recent years, tremendous progress has been made by researchers from the Institute for Man-made Fiber Research, Denkendorf, the Fraunhofer IAP, and Zimmer AG. A new patented carbamate technology, Carbacell, has been developed [80]. The mains steps of this process involve the following [89]:

1. Steeping wood pulp in caustic soda solution for chemical activation and degradation
2. The treatment of alkali cellulose with urea in xylene, which leads to the formation of cellulose carbamate
3. Dissolution of cellulose carbamate in dilute aqueous NaOH
4. Filtering and degassing of the solution
5. Wet-spinning into filament or staple fibers

Carbacell fibers have a smooth surface at the fracture thus indicating a lack of the fibrillar morphology that is usually associated with Lyocell fibers. The property profile of Carbacell fibers largely matches that of viscose fibers [90].

Although the new technology is more expensive, it can be expected that this environmentally friendly carbamate process will find wider commercial success in the near future.

10.3.4 Cuprammonium Process

In 1937, Schweizer [91] discovered that cellulosic fibers such as cotton and hemp readily dissolve in copper hydroxide and ammonium hydroxide solutions. His system is recognized as the Schweizer reagent. The Bemberg Rayon Industry later utilized this solvent for the industrial production of cuprammonium fibers (or cuprammonium rayon) and developed the Bemberg process or cuprammonium process [92]. Kamide and Nishiyama [93] have recently published an excellent review on the history and science of cuprammonium technology.

In a typical cuprammonium process, a freshly made aqueous solution of copper hydroxide (formed from copper sulfate and caustic soda) and ammonium hydroxide is used as a direct solvent for the dissolution of activated cellulose pulp. The ammonia concentration is usually between 124 and 250 g/l and the copper concentration must be greater than 25 g/l

[92,94]. It is generally accepted that cuprammonium ions namely, $Cu(NH_3)_4^{+2}$ form a complex with the hydroxyl groups of cellulose. The complex is highly sensitive to light and oxygen and may decompose quickly if it is not protected. A clear polymer solution free of oxygen is usually degassed and filtered before extruding into a slightly alkaline coagulation bath. Cuculo [92] describes that liquid jets travel into a specialized funnel-type coagulating bath where they are attenuated and coagulated. As the velocity along the funnel increases, the extrudate is stretched up to 400%. Coagulated cuprammonium fibers are usually washed with 5% H_2SO_4.

The main advantage of the Bemberg process over the viscose process is that the process uses cellulose pulp with a higher DP. The cuprammonium solvent is quite powerful and dissolves cellulose pulps with DP as high as 550. Polymer concentrations greater than 10% can also be produced, thus achieving greater productivity. The physical properties of cuprammonium fibers are slightly lower than those of viscose fiber (Table 10.5) but they are fully accepted as valuable textile fibers.

Asahi Chemical Industry has been very active in developing this process at high spinning speeds of 1000 to 2000 m/min [92]. The resulting process is referred to as the Asahi Bemberg process [93].

The main disadvantage of the cuprammonium process is the toxicity of copper sulfate. Hence, it must be fully recovered from the process; as a result large-scale production is limited. In 1990, annual production of cuprammonium yarns was close to 30,000 t [74,94,95]. Although, the cuprammonium process is used for making fibers, other products can also be produced. For example Akzo Nobel is using the cuprammonium process to make dialysis membranes (Cuprophane).

10.3.5 OTHER POTENTIAL PROCESSES

10.3.5.1 Amine–Salt Process

Hattori et al. have recently discovered a two-component system (amine–salt) consisting of hydrazine (NH_2—NH_2) [96] or ethylenediamine (NH_2—CH_2—CH_2—NH_2, EDA) [97–99] and various thiocyanate salts such as LiSCN, NaSCN, or KSCN that dissolves cellulose pulp. However, a high 40–50% salt concentration is generally required to obtain high concentration (up to 18–20% w/w) spinning dopes. The authors showed that cellulose I, II, and III can be dissolved in an amine–salt solvent.

The authors found that isocyanates readily dissolve in both diamines. Interestingly, neither diamine dissolves cellulose; however, hydrazine is a good swelling agent for cellulose as reported by Trogus and Hess [100]. Hydrazine is a molecule similar to ammonia. Its boiling point is 113.5°C, which is much higher than that of ammonia (−33.4°C) [96]. High boiling point of hydrazine (113.5°C) or EDA (118°C), offers some potential advantages for investigating the solubility and behavior of cellulose under ambient temperature and pressure.

The most important part of the amine–salt process developed by Cuculo and Hattori is the temperature cycling step [97]. Although, no cellulose preactivation is necessary, temperature cycling must be used to dissolve the pulp. Generally, dissolution does not occur if the cellulose is placed in an amine–salt solution. In a typical process, cellulose pulp is combined with a thiocyanate salt and amine (hydrazine or ethylenediamine) solution in a mixer in presence of nitrogen. The mixture is chilled to −10°C for about an hour and then warmed up to 50°C with intermittent shearing for 30 min. The process is then repeated several times till cellulose dissolution occurs. Studies have shown that cellulose (DP of 210) solution with concentration of 18 to 20% w/w can be easily prepared by using this technology.

The Cuculo technique, which subjects the cellulose–solvent system to repeated cooling and heating cycles, may be rationalized thermodynamically by changing the balance of entropy (ΔS_{mix}) and enthalpy (ΔH_{mix}), resulting in changes in the types of interactions between each molecule in the system. As dissolution occurs if the free energy $\Delta G_{mix} < 0$ (Equation 10.1), the entropy component, namely $T\Delta S_{mix}$ must be relatively large or the enthalpy of mixing should be minimized to less than 0.

$$\Delta G_{mix} = \Delta H_{mix} - T\Delta S_{mix} \qquad (10.1)$$

The calculated values of $T\Delta S_{mix}$ for NH_2-NH_2 (52 g) with cellulose (5 g) at -10 and $50°C$ are correspondingly 0.8 and 1.0 cal [96]. Furthermore, the authors also report that even at room temperature the dissolution of cellulose in the hydrazine system is slightly exothermic ($\Delta H_{mix} < 0$), thus indicating that hydrogen bonds form quite easily. Therefore, it appears that the temperature cycling effect is relatively small for a system consisting of cellulose, hydrazine, and sodium thiocyanate. The corresponding changes of free energy at various amounts of hydrazine associated with cellulose at various heats of mixing are shown in Figure 10.49.

The authors suggest that amine–salt solvent systems based on ethylene diamine, rather than hydrazine, [99] have greater potential for fiber and film formation of regenerated cellulose. Indeed, they reported that stable, anisotropic solutions form at cellulose (DP of 210) concentrations greater than 10% w/w if EDA and KSCN are used.

Frey et al. [101,102] have shown that an amine–salt solution (EDA with 56% KSCN) at 8% w/w cellulose can be converted into nanofibers (Figure 10.50) by using an electrospinning process. Interestingly, the fiber resembled cotton fibers. Thus, so far these recent, positive achievements indicate that the amine–salt process should be further developed before commercialization.

10.3.5.2 Cellulose Carbonate Process

In 1968, the reaction of carbon dioxide with sodium salts of alcohols was first reported by Nazarov and Chirkina [103]. The authors showed that alkyl carbonate can be obtained by passing dry CO_2 through a sodium salt of an alcohol in ethanol. Several years ago Okuda [104] reported preparation of aliphatic carbonates from alcohols and carbon dioxide using organic bases and sulfonate esters.

Recently, a new eco-friendly, patented process has been reported by Yoo and coworkers [105,106]. It utilizes a similar idea and involves the reaction of alkali cellulose with carbon dioxide (Figure 10.51). The Yoo process consists of the following:

1. Pretreatment of cellulose pulp with caustic soda
2. Carbonation of alkali cellulose with CO_2
3. Dissolution of cellulose carbonate in diluted caustic soda
4. Regeneration of cellulose

The authors obtained cellulose carbonate by reacting carbon dioxide with cellulose pulp having DP of 850 and dissolving the polymer in a 10% sodium hydroxide solution. The molecular weight of cellulose did not change during carbonation. The best results were obtained if cellulose pulp was pretreated with 20% $ZnCl_2$, acetone, or ethyl acetate. Clear 5 to 5.5% w/w polymer solutions that were stable at low temperatures of -5 to $0°C$ could be produced. The authors reported that the addition of 1 to 3% zinc oxide in sodium hydroxide solution greatly enhanced the solubility of cellulose carbonate. The use of a very high DP for cellulose in the Yoo cellulose carbonate process is highly encouraging. Definitely, this clever approach should be explored in greater detail in the near future. Perhaps some other more efficient pretreatment processes for cellulose could also be explored.

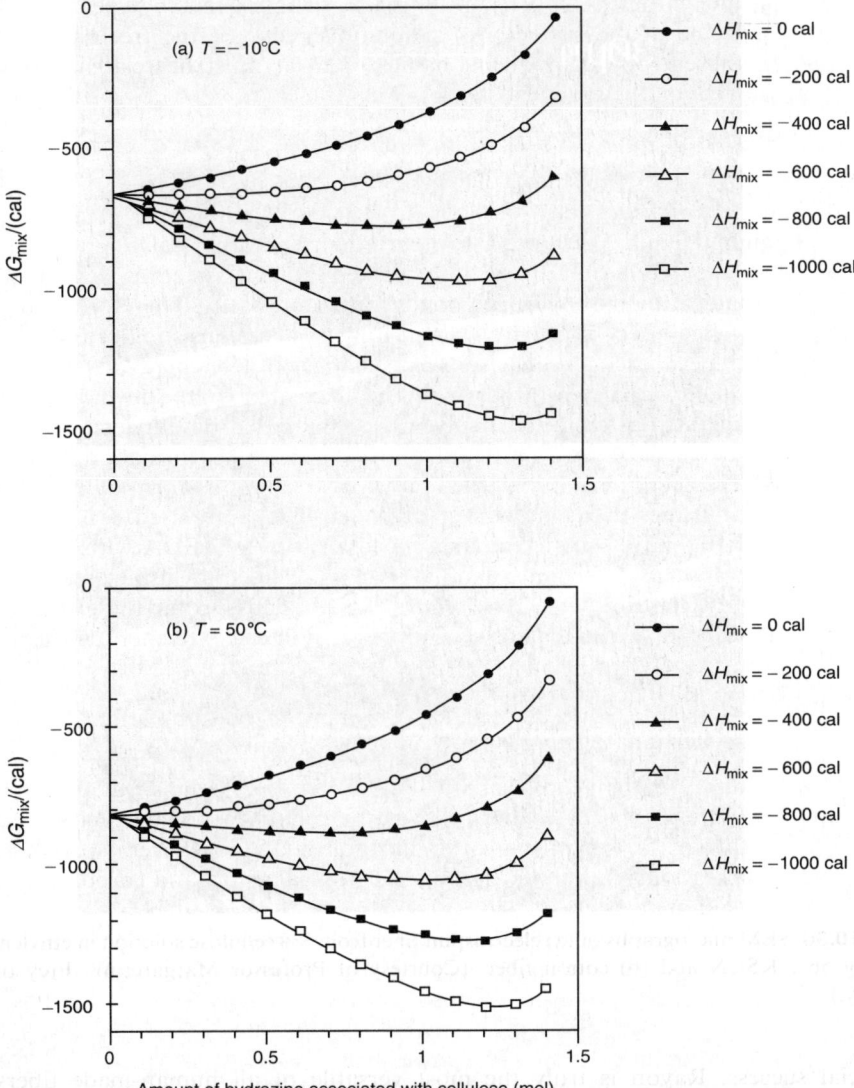

FIGURE 10.49 ΔG_{mix} dependence on the amount of hydrazine associated with cellulose in the cell I (DP = 210) NH_2—NH_2 NaSCN system (5, 52.2, 47.8 g) at (a) −10°C and (b) 50°C. (From Hattori, K., Cuculo, J.A., and Hudson, S.M., *J. Polym. Sci., Part A: Polym. Chem.*, 40(4), 601, 2002. Reprinted with permission of John Wiley & Sons.)

10.4 RAYON PROCESS

10.4.1 INTRODUCTION

Of all the human-made fibers, none has a more fascinating or interesting history than the family of fibers called rayon. The story has been recorded in several small volumes [107–109]. Of particular interest is the work done by Beer who lived through the early struggles and accomplishments and worked with those who contributed so much to make rayon a

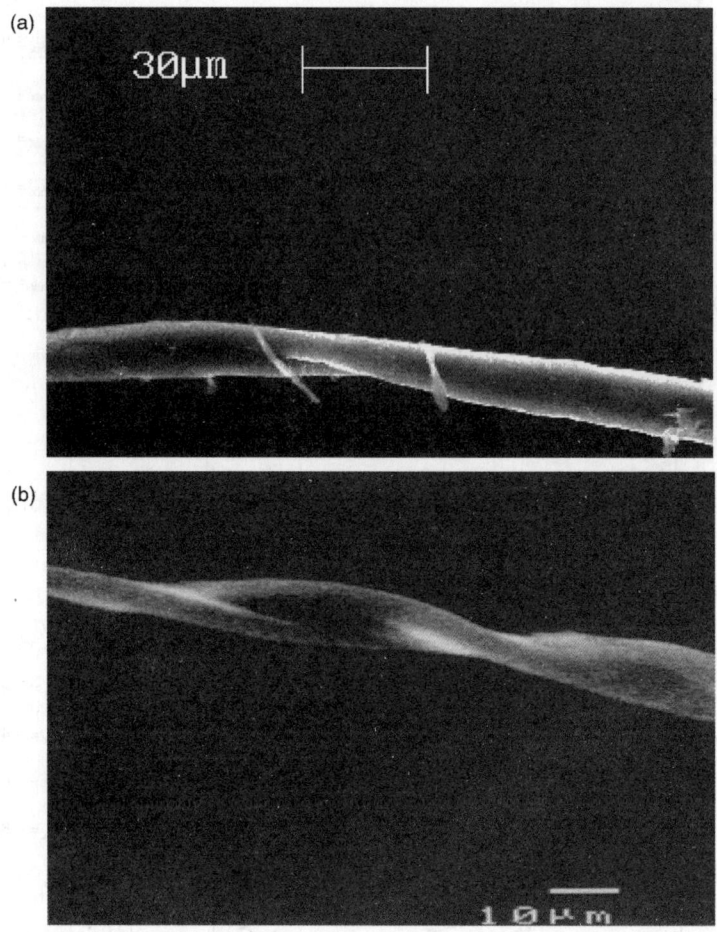

FIGURE 10.50 SEM micrographs of (a) electrospun fiber from 8% cellulose solution in ethylenediamine containing 56% KSCN and (b) cotton fiber. (Courtesy of Professor Margaret W. Frey of Cornell University.)

commercial success. Rayon is truly the most versatile of all human-made fibers, partly because it is the subject of experimentation and development for more than 100 years. Today, the production of rayon represents an example of the most complex case of fiber formation by a wet-spinning process. Although the production of synthetic fibers has overtaken rayon since the 1960s, the prime attributes of rayon-comfort, attractiveness, ease of processing, price, and raw material availability will assure its viability for many years to come.

10.4.2 History

The first recorded idea of a human-made fiber was attributed to Dr. Robert Hooke of London, who, in 1664, wrote in his book *Micrographia* [110]

> And I have often thought that probably there might be a way found out, to make an artificial glutinous composition, much resembling if not full as good, nay better, than that Excrement, or whatever other substance it be out of which, the silkworm wire-draws his clew. If such a

$$\text{Cell—OH} \xrightarrow{\text{NaOH}} \text{Cell—ONa} \xrightarrow{CO_2} \text{Cell—OCO}_2\text{Na} \xrightarrow{\text{Acid}} \text{Cell—OH}$$

FIGURE 10.51 Schematic of the Yoo cellulose carbonate process.

composition were found, it were certainly an easier matter to find very quick ways of drawing it out into small wires for use. I need not mention the use of such an Invention, nor the benefit that is likely to accrue to the tinder, they being sufficiently obvious. This hint therefore may, I hope, give some Ingenious, inquisitive Person an occasion for making some trials, which if successful, I have my aim, and I suppose he will have no occasion to be displeased.

Young Dr. Hooke was a brilliant inventor of such items as a telegraph, balance springs for watches, numerous microscopes, a barometer, gear-cutting machine, and many others that are, too numerous to mention. Unfortunately, he was unable to put his fiber idea into practice due to his preoccupation with preparing a plan for the rebuilding of London after a devastating fire.

Silk threads from cocoons and even filaments from spider webs had long been used to make exquisite fabrics for hundreds of years. However, only the very wealthy could afford such articles. Hooke's idea intrigued inquisitive men for many years, and around 1734–1742 René de Reaumur, a French naturalist, wrote in his *Memoirs pour Servir a l'Histoire des Insects* a suggestion that liquid varnish might be drawn into threads to resemble the silk made from the gum of silkworms.

The basic aim of Hooke and de Reaumur was to find some chemical and mechanical means to produce a less expensive substitute for natural silk. Two essential ingredients were missing, namely the spinning solution and a spinnerette.

One hundred years after de Reaumur speculated about the production of artificial silk, the first practical solution capable of being spun was invented by a Swiss chemist named Schoenbein. In 1846, he invented guncotton (nitrocellulose) which, when dissolved in alcohol and ether, produced collodion, a thick, viscous solution. Another Swiss chemist, George Audemars, took out the first known patent [111] that was granted in 1855 for production of rayon in England.

Audemars used an ancient Chinese method for producing threads from silkworm excretion by dipping the point of a needle into the solution and drawing it out. He used collodion made from mulberry bark and combined it with a gummy rubber solution. In this way, he was able to painstakingly produce fine filaments. He was, however, not able to produce a commercially practicable fiber.

Sir Joseph Swan, an English chemist who invented a successful electric lamp at the same time as Thomas Edison, was interested in fiber production for carbonized filaments to be used in his electric lamps. In 1883, he invented a new method for filament production that consisted of forcing nitrocellulose dissolved in acetic acid through a small hole into alcohol and drawing out a continuous filament of indefinite length. The dried filaments were denitrated with ammonium sulfide and the cellulose was regenerated. Thus, the combination of spinning solution and spinnerette came together. Sir Joseph soon recognized that his filaments could be used as textile fibers. His wife crocheted some of his finest filaments into doilies and table mats which were exhibited at the inventors' Exhibition of London in 1885.

The first practical, commercial production of rayon was accomplished by a French chemist known as the father of the rayon industry, Count Hilaire de Bernigaud de Chardonnet. Chardonnet studied in Paris under Louis Pasteur, who at that time was investigating a silkworm disease that threatened the important French silk industry. Chardonnet assisted Pasteur in this research and learned a great deal about how the silkworm produced silk. One of his future goals was to develop a human-made fiber that would protect the French silk industry against the silkworm disease and its economic consequences.

Years later, while working in his laboratory in Besançon, he accidentally dropped a bottle of collodion on a worktable. The next day, when scooping up the partially dried, sticky collodion, he noticed that it formed long threads resembling silk. This rekindled his interest, and, after 29 years of painstaking research, he was awarded a French patent on May 12, 1884 [112].

After building a plant at Besançon to produce his artificial silk, Chardonnet was able to display a brilliant collection of fabrics at the 1889 Paris Exhibition. As a result of this success and the excitement that followed, Chardonnet was awarded the Grand Prix and was made a knight of the Legion of Honor.

In 1891, Chardonnet's plant had an output of 100 lb of nitrocellulose rayon yarn per day. Fabrics made from this yard were known as "Chardonnet," or "Besançon silks" and became immediately popular. In 1899, the Chardonnet Silk Mill of Tubize, Belgium, was organized by Chardonnet and the silk firm of Wardle and Davenport, a company that was later called the Tubize Company. In 1900, Chardonnet rayon sold for $3.00/lb.

In 1920, the Tubize Company built a plant to produce the yarn in the United States. By 1934, however, other types of superior rayon had been developed, so the nitrocellulose plant was sold to a company in Brazil. Several incidents of explosions and fires caused by the incompletely denitrated cellulose resulted in setbacks to the Chardonnet silk process, but, fortunately, the simultaneous development of cuprammonium and viscose solutions for spinning rayon rapidly replaced the more dangerous nitrocellulose fibers.

Chardonnet, however, has the distinction of being the first to produce multifilament yarns by forcing a cellulose solution through very small holes in a spinnerette and to produce and market rayon. He was awarded the Perkin medal in 1914 for this achievement.

A different type of solution, made from cellulose dissolved in Schweitzer's reagent [91] (ammoniacal copper hydroxide solution), was used by the French chemist Louis Henry Despeissis to produce filaments experimentally. He was awarded a French patent on the process in 1890 [113]. A few years later, three German chemists, Drs. Emile Bronnert, Pauly, and Max Fremery, and an engineer, J. Urban, succeeded in producing cuprammonium rayon on a commercial scale by a process patented in 1897 [114]. Within two years, a series of patents was recorded in the names of Bronnert, Urban, and Fremery [115]. In 1898, a company called Vereingte Glanzstoff Fabriken (VGF) was formed to commercialize these inventions. VGF built a plant at Oberbruch. The company took advantage of the unfortunate experiences Chardonnet encountered with his explosive product and did not hesitate to use this deficiency to promote its own system. In 1900, the plant was employing over 1000 people and producing 1200 lb/day of fiber, which they called "Silkimit," at a selling price of $2.50/lb.

Cuprammonium yarn became so successful that several other companies were formed to use the process, among them were a French company, La Soie Artificielle, and a British company, United Cellulo Silk Spinners. Dr. Edmund Thiele of the latter company developed a stretch-spinning process that gave finer denier filaments and greatly improved the physical properties of the yarn. This yarn closely resembled silk and was used in combination with real silk for production of figured fabrics by differential dyeing. The British Glanzstoff Company started production in 1910, and by the end of the following year was producing 84,000 lb of heavy-denier single-filament yarn for artificial hair, brushes for brushes, and fiber for furniture stuffing.

All of these cuprammonium fiber plants went out of business for economic or other reasons, and it was not until 1925 that the process was introduced into the United States with the establishment of American Bemberg Corporation. In 1931, British Bemberg began operations at its plant in Doncaster. It was affiliated with the German Bemberg Company, which acquired rights to the patents owned by Glanzstoff. Just before World War I, Glanzstoff converted its cuprammonium plants to viscose rayon. Although cuprammonium rayon is still manufactured by a few companies, total production represents but a very small fraction of worldwide rayon production.

The development of viscose was largely the work of two English cellulose chemists, Cross and Bevan, who, with Beadle, received a patent on the process in 1892 [116]. In 1893, they sent a sample of viscose solution to Switzerland, where Charles H. Stearn had been working with Charles F. Topham on carbon filaments for electric lamps. They had worked with nitrocellulose rayon for this purpose and had a small lamp factory for utilizing the carbon filaments.

The viscose obtained from Cross and Bevan was so successful for production of lamp filaments that Stearn asked Topham to try to spin it for use in textiles. The first experiments failed dismally. After several years of painstaking work, Topham made several discoveries essential to the spinning of yarn from viscose: aging (ripening) of the solution, filtration to remove particles, multiple-hole platinum spinnerettes, and a circular, centrifugally operated yarn collecting device that twisted the yarn and packaged it in convenient cake form [117]. The "Topham box," as it is still called, or variations of it are still on many of the continuous-filament rayon machines today.

In 1899, Cross and Stearn formed the Viscose Spinning Syndicate in Kew, near London, where Topham developed the platinum spinnerettes.

Several attempts at viscose rayon production were made by companies in the United States with the patent rights from Cross and Bevan, Steam, and Topham. None of these efforts were successful at producing fiber, and most of their viscose production went into sheets (films) and molded forms. Well-known American names were involved in these early struggles—A.D. Little, Daniel C. Spruance, Willard Saulsbury, Carleton Ellis, and T.S. Harrison [108].

In 1900, representatives of the British silk firm Samuel Courtauld and Company saw an impressive display of viscose rayon yarn at the Paris Exhibition. When the company heard of the progress made by Topham and Stearn, they conducted a careful study of the process, compared it with the other processes, and decided to get into the viscose rayon business. In 1904, Courtauld purchased the British rights to manufacture textile yarn from viscose, the patents of Cross, Bevan, and Stearn, and the rights to Topham's spinning box. A plant was built at Coventry, and the first yarn was produced in November 1905. The yarns were accepted with enthusiasm by the textile trade, and, by 1909, Courtauld was producing 150,000 lb/year. In 1910, expansion with new equipments raised the total to 2,000,000 lb. Viscose rayon was thus established as a viable human-made component of the textile trade.

Also in 1910, Courtauld formed an American subsidiary, The American Viscose Company, and began construction of a plant at Marcus Hook, Pennsylvania. Subsequently, other companies—DuPont, Tubize Comany, Belamore (later known as Hartford Rayon), Industrial Rayon, American Cellulose and Chemical Manufacturing (later called Celanese), etc.—were organized to produce rayon.

The name rayon was officially adopted in 1924 by the National Retail Dry Goods Association. Prior to this, the fiber was called artificial silk, wood-silk, or viscose silk. On October 26, 1937, the Federal Trade Commission (FTC) officially defined rayon as "a textile fiber or yarn produced chemically from cellulose or with a cellulose base." This definition covered cuprammonium and viscose rayon as well as acetate fiber. To avoid confusion in the trade, FTC rules were adopted on December 11, 1951, which defined rayon as "man-made textile fibers and filaments composed of regenerated cellulose." A separate definition was adopted for acetate, "man-made textile fibers and filaments composed of cellulose acetate."

During World War II and subsequently, major advances in viscose preparation and spinning technology were made that vastly improved the properties of rayon for textile as well as industrial uses. Rayon became a family of fibers with widely different performance characteristics. Several companies began calling their improved fibers by different trade names to distinguish them from the older type of filament rayon, which, because of some property deficiencies such as poor wet strength, had developed a bad reputation in the trade. Petitions to the FTC for new names for the improved fibers were denied, and a new, more restrictive definition was adopted: "a manufactured fiber composed of regenerated cellulose in which substitutes have replaced not more than 15% of the hydroxyl groups."

This definition did nothing to classify the various types available, and only through extensive promotion and advertising was the public informed of the wide differences in properties and performance of the many viscose-derived fibers that were all called rayon.

Basically, all methods for producing rayon filaments or fibers depend on solubilizing cellulose, then reshaping it into long-fibered products by extrusion through the small holes of a spinnerette, immediately followed by conversion into solid cellulose. Although there are a number of ways in which this can be done, the viscose rayon process is by far the most important and widely practiced.

10.4.3 Viscose Rayon

Viscose rayon enjoys a unique position as the most versatile of all human-made fibers in end-use applications. This has resulted from the ability to engineer the fiber chemically and structurally in ways that take advantage of the properties of the cellulose from which it is made. Process technology for converting cellulose into rayon has evolved over a period of more than 100 years. The first attempts to produce the fiber were empirical, hit-or-miss experiments, so during the first 50 years there was a paucity of knowledge about the chemical and physical interactions between pulp (cellulose), viscose, and fiber. Production during that period was more art than science. Although there is still some art in the development of improved rayons, much knowledge has been acquired that delineates causes and effects.

The viscose process is composed of several steps, all of which must be carefully controlled to produce the desired end product. Changes in operating parameters in single or multiple steps result in a wide variety of rayons that affords the versatility for which rayon is so well known.

An outline of the viscose process is shown in Figure 10.52. By this process, short-fibered cellulose (wood pulp) is converted in a series of controlled and coordinated steps to a spinnable solution and then into longer filaments, which may be precisely controlled as to length, denier, cross-sectional shape [118], and other physical properties.

The pulp is first steeped in an aqueous solution of sodium hydroxide (17–18%), which causes the fibers to swell and converts the cellulose to sodium cellulosate, commonly called alkali cellulose or white crumb. After steeping, the swollen mass is pressed to obtain a precise ratio of alkali to cellulose and then shredded to provide adequate surface area for uniform reaction in subsequent process steps. The alkali cellulose is aged under controlled conditions of time and temperature to depolymerize the cellulose by oxidation to the desired DP prior to reacting with carbon disulfide to form sodium cellulose xanthate. The xanthate, which is a yellow to orange crumb, is dissolved in dilute sodium hydroxide to yield a viscous orange-colored solution called viscose. The solution is filtered, deaerated, and ripened to the desired coagulation point (called salt index) appropriate for spinning.

The rayon filaments are formed when the viscose solution is extruded through very small holes of a spinnerette into a spin-bath consisting basically of sulfuric acid, sodium sulfate, zinc sulfate, surfactant, and water. Coagulation of the filaments occurs immediately upon

Regenerated Cellulose Fibers

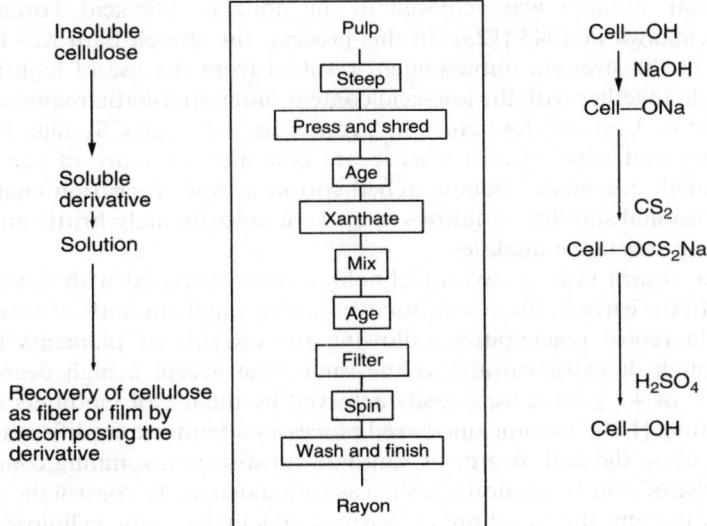

FIGURE 10.52 Outline of the viscose process.

neutralizing and acidifying the cellulose xanthate followed by simultaneous controlled stretching and decomposition of the cellulose xanthate to cellulose. These latter steps are important for obtaining the desired tenacity and other properties of the rayon. Finally, the newly formed rayon is washed and chemically treated (desulfurized), either in the form of continuous filament (yarns or tow) or cut into staple, to remove impurities before applying a processing finish and packaging.

10.4.3.1 Early Production

Modern technology for spinning viscose rayon has developed along two paths: (a) zinc-based processes and (b) nonzinc-based processes. The first practical spin-baths for the production of rayon were developed by Muller [119] and contained sulfuric acid and sodium sulfate. Around 1911, Napper [120] found that about 1% zinc sulfate in the spin-bath improved coagulation and permitted stretching. The Muller bath-spun fibers, produced with practically no stretching, were relatively weak, 1.2 g/den (1.3 g/dtex). Commercial introduction of stretching, with the filaments drawn in air between driven rolls (godets), occurred in the 1920s. The stretch achieved at that time was no more than about 30%, but this was sufficient to almost double the strength of the fiber from 1.2 to 2.2 g/den (2.4 g/dtex).

Surprisingly, one of the strongest rayon fibers ever made was developed in 1925 by Lilienfeld, who patented a nonzinc-based process using 64% sulfuric acid as the spin-bath [121]. He obtained fibers as strong as 7 g/den (7.8 g/dtex) as a result of high (~200%) stretch that could be achieved with this system. His fiber resembled the high-strength polynosic rayons developed much later by the Japanese. However, Lilienfeld's process did not achieve commercial importance because of the hazards involved in using such strong acid.

Further progress was made in the discovery that stretch could be increased by using hot water or steam to plasticize the newly formed filaments during stretching. Together with an increased amount of zinc sulfate (3–4%) in the spin-bath, the use of hot-water stretch-baths as patented by Givens et al. [122] in 1937 advanced the development of rayon tire yarn to about

3.5 g/den. Similar strength was achieved in the nonzinc, low-acid Toramomen process patented by Tachikawa in 1943 [123]. In this process, the stretch-bath was the same as the spin-bath. Part of the strength improvement resulted from the use of high-DP cellulose in the viscose, which, together with the low-acid content of the spin-bath, required changes in the viscose composition. Low cellulose and alkali concentration, high CS_2, high DP, high viscosity, and spinning underripe (green) viscose are common to many of the polynosic-type processes. Although polynosic fibers matched cotton in low-elongation characteristics and gave good dimensional stability in fabrics, they were unfortunately brittle and showed poor wear and abrasion-resistance qualities.

The next advance in viscose rayon technology was associated with the development of super tire yarn. In the early 1950s, it was discovered that small amounts of alkyl amines added to viscose would retard regeneration, allowing the coagulated filaments to deswell and consolidate through dehydration and at the same time accept a high degree of orienting stretch. Strengths of 4.4 g/dtex were easily achieved by the use of modifiers as described by Cox in a 1950 patent [124]. The nonzinc-based processes do not use modifiers added to viscose but rely instead upon the high degree of xanthate substitution, spinning conditions, and, in some cases, the use of spin-bath additives such as formaldehyde to control the rate of cellulose regeneration. Although the reaction of various aldehydes with cellulose xanthate was patented in 1938 [125], it was not until about 1962 [126] that commercial processes using formaldehyde were introduced to make rayon. Polynosic fibers produced in this way can have tenacities in excess of 9 g/den (10 g/dtex).

In the period 1955–1965, many improvements in viscose modification were made. The most notable of these was the development of high wet-modulus (HWM) rayon. A synergistic effect between amine and polyglycol modifiers, in conjunction with the action of zinc, enabled higher stretching during spinning, This modifier system, attributed to Mitchell et al. [127], when used with appropriate spinning conditions such as reduced primary spin-bath temperature and slower spinning speed, resulted in an increase in the important property of wet-modulus.

Although the nonzinc-based processes also yielded a fiber with high wet-modulus, the growth of zinc-based HWM rayon relative to polynosics is associated with simpler and more economic manufacturing methods. By 1965, small amounts of zinc sulfate (1 g/l) were often used in polynosic-type spin-baths to enable faster spinning speeds at higher temperatures. Since the initial Japanese work, many processes have been developed for producing high-tenacity, high-modulus rayons, and means have been found to attain a compromise in dimensional stability at higher elongation levels that help to avoid fibrillation and brittleness [128].

In the period 1965–1980 a wide variety of new, stronger, and more durable rayon fibers were developed. Rayon variants are now produced which utilize the comfort and aesthetic qualities of cellulose to compliment synthetic fibers in many textile applications. Considerable emphasis has been placed on the economics and ways to meet environmental and safety standards. Special effects, such as crimp or hollow filaments, may be obtained by appropriate viscose formulations, point-of-stretch applications, spin-bath compositions, and modifiers. Flame-retardant (FR), acid-dyeable, and superabsorbent rayons are typical of the properties that can be attained by incorporating various materials in the fiber structure. Rayon is unique in the respect that the fiber can be permanently modified for a wide variety of end uses simply by adding the appropriate material to viscose.

In the early days, viscose composition was around 7.0% cellulose, 6.0% NaOH, and 34% CS_2 (based on cellulose), with a viscosity of about 50 P. Wood pulp of about 88% α-cellulose content was the typical cellulose source for ordinary rayon, although purified cotton linters were used for higher-strength rayons. With commercialization of the zinc-based process, higher α-cellulose-content pulp was required to obtain full benefit of this improved high-stretch-spinning practice. Improvements in wood pulp quality, α-cellulose content, uniformity of chain length, reactivity, and processability, as well as lower cost, resulted in the gradual replacement of cotton linters.

Review of the progressive change in purity for viscose-grade cellulose indicates that the long-chain α-cellulose content increased from below 90% in 1930 to more than 98% in some pulps available today. The average DP and cellulose chain-length distribution were identified as significant properties in the production of rayon [129]. With the advent of more sophisticated spinning processes, wood pulps were required to yield adequate DP in the finished rayons that could range from about 260 in regular rayons to about 450 in HWM rayons and even as high as 800 for some of the polynosics and high-tenacity rayons. Another benefit derived from the increased degree of purification was the narrower chain-length distribution in the pulp (Figure 10.53) which, when used with the high-stretch-spinning process, contributed to improved rayon properties. The improvements were the result of removal of the short chain-length cellulose fraction by cold caustic extraction as well as a reduction in the natural contaminating resins [130].

An essential requirement for a textile fiber is that the polymer from which it is made comprises long-chain molecules [131]. This is necessary to achieve physical properties such as strength and extensibility. Functional groups that may be part of the polymer backbone or attached to it confer specific characteristics and properties to the fiber. The extent to which these two contributing factors, molecular size and chemical functionality, are manifest in the fiber properties is determined by the structure of the fiber.

Unlike most human-made fibers where the polymers must first be synthesized from the monomers, rayon is made from cellulose, nature's most abundant polymer. In growing

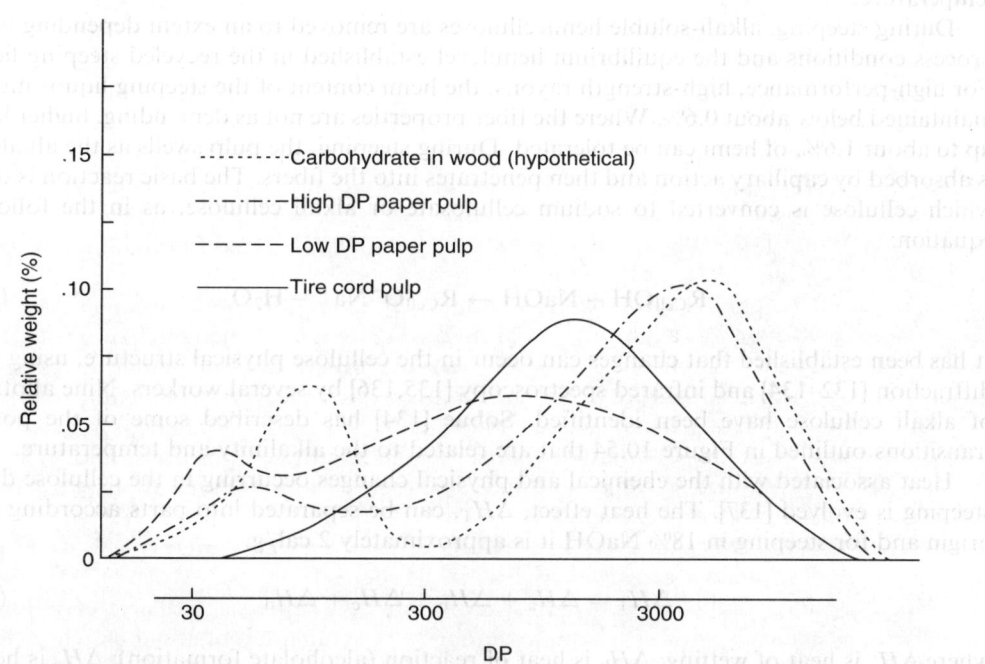

FIGURE 10.53 Chain-length distribution of cellulose by GPC.

plants, glucose synthesized from carbon dioxide and water under the action of sunlight becomes the basic monomer unit of cellulose. The D-glucose monomer units are joined by 1,4 β-glucosidic bonds with the elimination of water to form the ether linkage. In this structure (Figure 10.1), the anhydroglucose units (AGUs) in the preferred chair conformation are arranged such that they alternate about the 1,4 β-glucosidic bonds with every other unit rotated at 180°. As a consequence, the cellulose molecule is relatively stiff. Although the polymer contains numerous hydroxyl groups, normally conferring water solubility, it is insoluble in most common solvents. This is attributed to strong hydrogen bonding between the molecules in the structure as isolated from wood by chemical pulping processes. Such hydrogen bonding limits penetration of the solid by solvents. In addition, the arrangement of the cellulose molecules, which because of their length cannot assume a completely crystalline state characteristic of its polar monomer units, will exhibit variation in the degree of order of the fiber structure on both macroscopic and microscopic levels. Because of the poor solubility, the cellulose is converted to a soluble derivative to prepare a solution for spinning. The structure and accessibility of cellulose is of profound importance to the reactivity in the conversion of wood pulp to rayon. This is done through the various chemical and physical reactions in the multistage viscose process for fiber production, which consists essentially of two parts: (a) preparing the viscose solution and (b) forming the rayon fiber.

10.4.4 Chemistry of Viscose Rayon Process

10.4.4.1 Steeping

In commercial practice, pulp is steeped (soaked) in aqueous sodium hydroxide of about 18% concentration at ambient or slightly higher temperature (~25°C). Because swelling of cellulose is temperature-dependent, less caustic is used at lower temperatures and more at higher temperature.

During steeping, alkali-soluble hemicelluloses are removed to an extent depending on the process conditions and the equilibrium hemilevel established in the recycled steeping liquor. For high-performance, high-strength rayons, the hemi content of the steeping liquor must be maintained below about 0.6%. Where the fiber properties are not as demanding, higher levels, up to about 1.6%, of hemi can be tolerated. During steeping, the pulp swells as the alkali first is absorbed by capillary action and then penetrates into the fibers. The basic reaction is one in which cellulose is converted to sodium cellulosate or alkali cellulose, as in the following equation:

$$R_{Cell}OH + NaOH \rightarrow R_{Cell}O^-Na^+ + H_2O \qquad (10.2)$$

It has been established that changes can occur in the cellulose physical structure, using x-ray diffraction [132–134] and infrared spectroscopy [135,136] by several workers. Nine allotropes of alkali cellulose have been identified. Sobue [134] has described some of the possible transitions outlined in Figure 10.54 that are related to the alkalinity and temperature.

Heat associated with the chemical and physical changes occurring in the cellulose during steeping is evolved [137]. The heat effect, ΔH_T, can be separated into parts according to its origin and for steeping in 18% NaOH it is approximately 2 cal/g.

$$\Delta H_T = \Delta H_a + \Delta H_b + \Delta H_c + \Delta H_d \qquad (10.3)$$

where ΔH_a is heat of wetting; ΔH_b is heat of reaction (alcoholate formation); ΔH_c is heat of possible lattice transition; and ΔH_d is heat of other reactions (degradation, oxidation, etc.).

Regenerated Cellulose Fibers

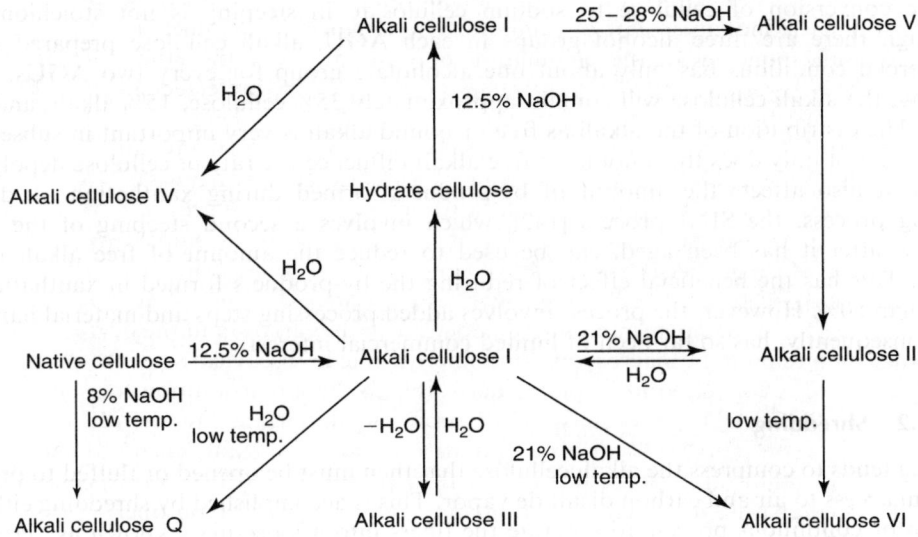

FIGURE 10.54 Alkali cellulose transitions.

Dissipation of this heat is a variable influenced during steeping by conditions such as fill rate and may cause local temperature variations of several degrees.

An initial decrease in the average chain length (DP) is observed when pulp is steeped. This has been attributed to oxidative degradation involving dissolved and absorbed air and alkali-sensitive groups such as carbonyl on the cellulose. After this initial loss of DP, which consumes the available oxygen and entails random scission of the cellulose chains, the degradation proceeds more slowly by alkaline hydrolysis. The mechanism for this involves end-group attack to form low-molecular-weight degradation products. Effective removal of hemicellulose and the alkali-soluble material facilitates improvements in subsequent reactions where the soluble fraction would scavenge reactants. Because alkaline hydrolysis produces short-chain material, it is desirable to continue the steeping for the shortest time needed to obtain uniform distribution of alkali in the pulp. Surfactants are often used to facilitate wetting and generally result in improved viscose quality [138].

Extraction of alkali-soluble material, hemicellulose, and degradation products from the pulp depends on a number of variables. Extensive information has been obtained and reported from numerous studies, including those of Sihtola and Nizovsky [139], Treiber [140], and Kleinert [141]. Besides alkali concentration, temperature, and time, the separation between the pulp sheets is important in the batch process; whereas in the slurry system, the pulp sheets are disintegrated, facilitating removal of the soluble fraction. Low temperature favors swelling and dissolution, so, to avoid excessive material loss, slurry steeping is usually conducted at a higher temperature (around 50°C) than sheet steeping (30°C).

At the end of the steeping cycle, the liquor is drained from the swollen cellulose mass, which is then pressed until the alkali cellulose weight is approximately 2.8 times that of the original pulp. The draining is returned to the supply system for reuse while the pressing, rich in hemicellulose, is sometimes used to mix regular viscose or must be purified by dialysis if the caustic is to be recycled as steep liquor for high performance rayon. In this way, equilibrium hemicellulose levels are attained in the steeping liquors where different types of rayon are produced.

The conversion of cellulose to sodium cellulosate in steeping is not stoichiometric. Although there are three alcohol groups in each AGU, alkali cellulose prepared under commercial conditions has only about one alcoholate group for every two AGUs. After pressing, the alkali cellulose will contain approximately 35% cellulose, 15% alkali, and 50% water. The distribution of the alkali as free or bound alkali is very important in subsequent reactions. Not only does the amount of free alkali influence the rate of cellulose depolymerization, it also affects the amount of by-products formed during xanthation. A double steeping process, the SINI process [142], which involves a second steeping of the alkali cellulose after it has been aged, can be used to reduce the amount of free alkali in the crumb. This has the beneficial effect of reducing the by-products formed in xanthation by more than 50%. However, the process involves added processing steps and material handling and, consequently, has so far been of limited commercial interest.

10.4.4.2 Shredding

Pressing tends to compress the alkali cellulose that then must be opened or fluffed to provide uniform access to air and carbon disulfide vapor. This is accomplished by shredding either in a batch or continuous process to separate the fibers into a loose mass known as "crumb." Time, temperature, and mechanical work done in shredding are controlled variables that must be adjusted for each different type of pulp processed. If the alkali cellulose crumb is too wet from underpressing, or if condensation in the equipment causes localized wetting, it will not xanthate properly. Overpressing, drying out, and carbonate formation must also be avoided. Alkaline oxidation and hydrolysis of the cellulose, which begin when the pulp is steeped, continue during shredding. The shredding temperature and mechanical work are often used as adjuncts to aging to control the extent of depolymerization.

10.4.4.3 Aging

The viscosity of the viscose, an important processing parameter, and the final rayon properties are dependent on the average chain length or DP of the cellulose. Control of this variable is achieved by aging the alkali cellulose crumb under conditions yielding the appropriate extent of depolymerization for the type of rayon produced.

There are at least two chemical reactions that will result in reducing the chain-length hydrolysis and oxidation. The oxidation can be represented as a reaction of zero order. Numerous studies of the reaction mechanism, since the classical work of Entwhistle et al. [143], have confirmed that cellulose chains are randomly cleaved by alkaline oxidation. On the other hand, hydrolysis involves end-group attack, which removes single units from the ends of the cellulose chains [144]. Depolymerization by this mechanism occurs much more slowly than with oxidation and yields alkali-soluble degradation products. The rates of depolymerization for alkali cellulose aged under oxygen and nitrogen are compared in the aging curves shown in Figure 10.55.

The depolymerization is accelerated at elevated temperature by catalysts such as Fe, Mn, and Co [145] and by oxidants like persulfate, perchlorate, and peroxide. In these ways, the time required for aging crumb can be reduced from more than 24 h to as little as 1 or 2 h.

Another way to achieve rapid depolymerization is to irradiate the cellulose with a beam of accelerated electrons [146–148]. In this case, the time required for the depolymerization is of the order of seconds. This technology has been investigated as a means to eliminate the need to age alkali cellulose in the viscose process, and rayon has been successfully made from irradiated pulp [149,150]. The process, however, has not been adopted by the industry because, like double steeping, material handling poses special requirements.

Regenerated Cellulose Fibers

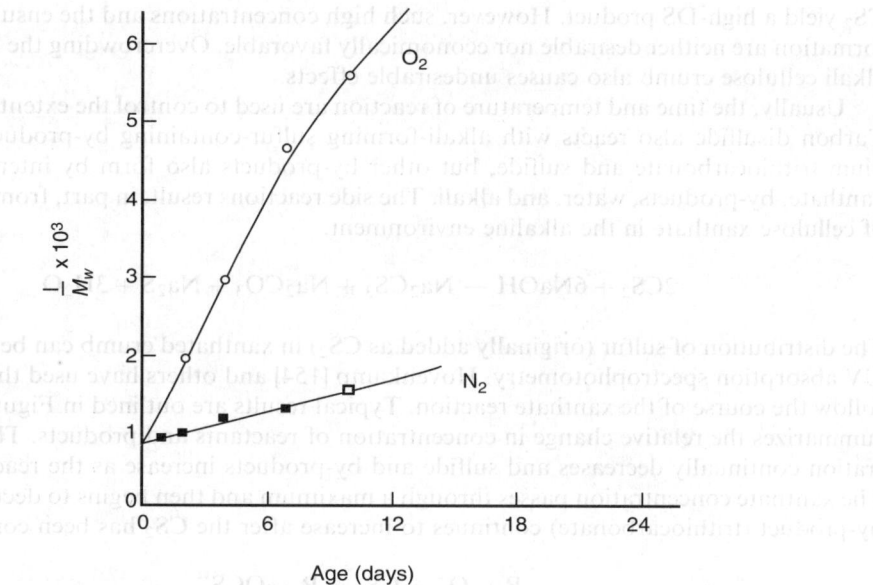

FIGURE 10.55 Alkali aging under oxygen and nitrogen environment.

10.4.4.4 Xanthation

The soluble derivative, sodium cellulose xanthate, is formed by reacting alkali cellulose with carbon disulfide. The reaction has been studied for both cellulose and simple model systems [151,152]. Xanthation is normally conducted by placing alkali cellulose crumb in a reactor, pulling a vacuum, and then introducing CS_2. As the reaction proceeds, CS_2 is consumed and the vacuum is regained. The extent of vacuum regain is used to follow the reaction.

In the commercial process, it is believed that the carbon disulfide, which is added to the reaction mass as a liquid, must first dissolve in the aqueous alkali adsorbed on the alkali cellulose (AC) so that it can be transferred to the reaction sites on the cellulose molecules. There the hydrated CS_2 adds onto the alcoholate ion. The basic reaction is summarized in the following equations:

$$R_{Cell}OH + OH^- \rightarrow R_{Cell}O^- + H_2O \tag{10.4}$$

$$R_{Cell}O^- + CS_2 \rightarrow R_{Cell}OCS_2^- \tag{10.5}$$

The mechanism is reported to be a bimolecular nucleophilic substitution. Kolosh and Eriksson have examined the penetration of CS_2, into alkali cellulose sheets [153] and demonstrated conditions leading to nonuniform reaction and poor viscose quality.

There are $3n + 2$ alcohol groups in each cellulose molecule ($n =$ DP) located at carbons 2, 3, and 6 of the AGUs and one at each end of the chain. Because n is large, the end-groups can be neglected. In freshly xanthated cellulose, the DS seldom exceeds 1 and is usually about 0.7, i.e., there is incomplete substitution of the available alcohol groups, but this is generally sufficient for viscose production. The extent of xanthation is influenced by the composition of the alkali cellulose crumb and the amount of CS_2 used. High alkalinity and large amounts of

CS_2 yield a high-DS product. However, such high concentrations and the ensuing by-product formation are neither desirable nor economically favorable. Overcrowding the xanthator with alkali cellulose crumb also causes undesirable effects.

Usually, the time and temperature of reaction are used to control the extent of xanthation. Carbon disulfide also reacts with alkali-forming sulfur-containing by-products such as sodium trithiocarbonate and sulfide, but other by-products also form by interaction of CS_2, xanthate, by-products, water, and alkali. The side reactions result, in part, from the instability of cellulose xanthate in the alkaline environment.

$$2CS_2 + 6NaOH \rightarrow Na_2CS_3 + Na_2CO_3 + Na_2S + 3H_2O \tag{10.6}$$

The distribution of sulfur (originally added as CS_2) in xanthated crumb can be determined by UV absorption spectrophotometry. Hovenkamp [154] and others have used this technique to follow the course of the xanthate reaction. Typical results are outlined in Figure 10.56, which summarizes the relative change in concentration of reactants and products. The CS_2 concentration continually decreases and sulfide and by-products increase as the reaction proceeds. The xanthate concentration passes through a maximum and then begins to decrease, while the by-product (trithiocarbonate) continues to increase after the CS_2 has been consumed.

$$R_{Cell}O^- + CS_2 \rightleftharpoons R_{Cell}OCS_2^- \tag{10.7}$$

Xanthation is an equilibrium reaction, and as the CS_2 is used to form by-products, it is replaced by decomposition of the xanthate. There is an induction period in the formation of trithiocarbonate. At elevated temperatures, xanthation is accelerated but by-product formation and xanthate decomposition are also affected in the same way. Also, the reaction is exothermic. For these reasons, the temperature is usually maintained below about 32°C by the use of cooling water, but excessive xanthation times are avoided.

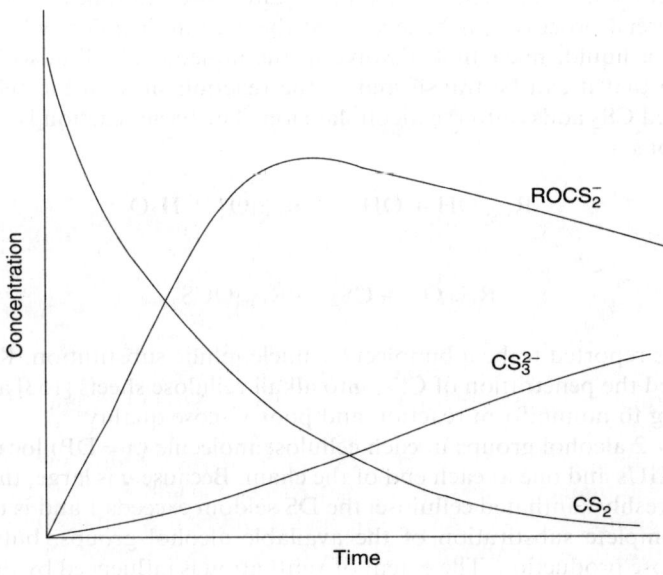

FIGURE 10.56 Xanthation reaction.

10.4.4.5 Mixing

The xanthate crumb is dissolved in a dilute aqueous solution of sodium hydroxide of about 5–8% concentration. During this step, reactions of xanthation and by-product formation continue and, in some processes, more CS_2 can be added to the mix to obtain improved viscose quality. Modifiers, additives, and delustrants can be included at this stage but can also be added later by injection into the viscose when it is pumped to the spinning machines. A low mixing temperature is preferred because this will minimize xanthate decomposition and by-product formation. The viscous, orange-colored solution obtained, viscose, will have the desired composition, i.e., cellulose and caustic content depending on the intended end use.

10.4.4.6 Filtration

Following the dissolution step, the viscose is carefully blended to achieve the best uniformity of properties according to the process used, continuous or batch. From the blender, the viscose is filtered in a number of steps to remove undissolved particles and contaminants. Durso and Parks have reviewed the early work on viscose filtration [155]. The primary objective of filtration is to remove particles that would plug spinnerette holes and cause fiber defects. Because of the deformable nature of many viscose particles, it is impossible to remove all of them by filtering; some will be extruded and others break up into smaller particles [156,157]. Research on the effects of such particles that might be extruded is expected to lead to significant advances in the control of fiber quality [158].

10.4.4.7 Ripening

In 1897, Topham [117] first discovered that ripening is necessary in the preparation of a good spinning viscose and to obtain desired rayon properties.

During ripening, chemical and physical changes occur in the viscose. The solution darkens with an increase in by-products affecting the ease with which the viscose coagulates in subsequent spinning. The viscosity also changes, first decreasing as the solution uniformity improves and then increasing, slowly at first and then quite rapidly, as the cellulose xanthate decomposes and eventually forms a gel. The reactions involved in ripening have been studied by several groups of researchers [159–166]. It is generally accepted that a number of reactions are involved, which result in changes in the distribution and form of sulfur in the viscose:

$$\text{Xanthate decomposition: } R_{Cell}OCS_2^- \rightarrow R_{Cell}O^- + CS_2 \qquad (10.8)$$

$$\text{Rexanthation: } R_{Cell}O^- + CS_2 \rightarrow R_{Cell}OCS_2^- \qquad (10.9)$$

$$\text{Transxanthation: } R_{Cell}OCS_2^- + R'_{Cell}OH \rightarrow R'_{Cell}OCS_2^- + R_{Cell}OH \qquad (10.10)$$

$$\text{By-product formation: } 2CS_2 + 6NaOH \rightarrow Na_2CS_3 + Na_2CO_3 + Na_2S + 3H_2O \qquad (10.11)$$

Most important is the number and distribution of xanthate groups on the cellulose chains. When cellulose is xanthated, the reaction is heterogeneous and the most accessible hydroxyl groups react first, resulting in a very nonuniform distribution. Then, as the cellulose xanthate is dissolved, the conditions for the ripening reactions become homogeneous. Andersson and Samuelson [166] demonstrated the reversible nature of the xanthation reaction by bubbling nitrogen through a solution of viscose and measuring the amount of CS_2 removed. Under normal conditions of viscose ripening, as CS_2 is released from the cellulose xanthate it is consumed in reactions, forming by-products and rexanthating the cellulose. The extent of these reactions is influenced by viscose composition and ripening conditions [160]. Also, under the homogeneous conditions in solution, the reactivity of the different cellulose alcohol groups at C2, C3, and C6 becomes an important factor affecting the redistribution of xanthate groups. Besides rexanthation, it has been suggested that a transesterification of the cellulose in which xanthate groups redistribute without involving free CS_2 can occur as well [167].

The degree of ripening is controlled by time and temperature until the required ripeness is obtained. Ripeness is usually expressed as a salt index (SI) or Hottenroth number, which is the amount of sodium chloride solution or ammonium chloride solution, respectively, needed to coagulate the viscose [168]. The tests measure the ease of forming a coagulated viscose filament in spinning. Regular rayon is usually spun when the SI is from 4.5 to 5.0 for a given viscose. Hottenroth numbers are higher and could be 11–14 for the same viscose. The cellulose DP has an inverse relationship to the SI reflecting solubility–molecular weight relationship: the higher the DP, the lower the SI. Hemicellulose in the viscose will cause a high SI. Also, the distribution of xanthate groups has an effect. A freshly mixed viscose will often have a lower SI than one that has been ripened for a few hours. Improved solubility resulting from the redistribution of xanthate groups more than compensates for the lower degree of xanthation in the ripened viscose [169].

10.4.4.8 Spinning

In spinning, the alkaline viscose solution is exposed to an acidic spin-bath under carefully controlled conditions to form filaments of rayon. The process as described by Vroom involves a complex series of chemical and physical reactions that take place almost simultaneously [170]. Initially, the action of the acid or salt spin-bath causes the viscose to coagulate and forms a skin around the filament. Then, as the acid penetrates into the viscose filament, neutralization of the alkali occurs. At the same time, cellulose xanthate is decomposed regenerating cellulose and carbon disulfide from which it was made (Equation 10.12).

$$2R_{Cell}OCS_2Na + H_2SO_4 \rightarrow 2R_{Cell}OH + 2CS_2 + Na_2SO_4 \qquad (10.12)$$

Diffusion is a principal mechanism by which material transfer, necessary for the reactions between the viscose and spin-bath components, occurs [171]. Conditions affecting diffusion will influence the rates at which the various chemical reactions proceed within the forming fiber. Consequently, the extent of reaction at any given time after extrusion will vary across the filament diameter.

Transformations that occur in the initial stages of the viscose-spinning process are shown schematically for a single filament in Figure 10.57. These have been studied by adding indicators to the viscose and observing the distance from the spinnerette at which color changes occur [172]. The principal reactions beginning at the spinnerette are coagulation, neutralization, and regeneration. Coagulation, in which the viscose sets up as a gel, occurs quite rapidly and can be regarded as leading to primary structure formation, which terminates when the coagulation front reaches the fiber axis. Continuing after this are what may be called secondary structure-formation processes including dehydration, densification, orientation, regeneration, and crystallization. In this way, various cross-sectional structural characteristics

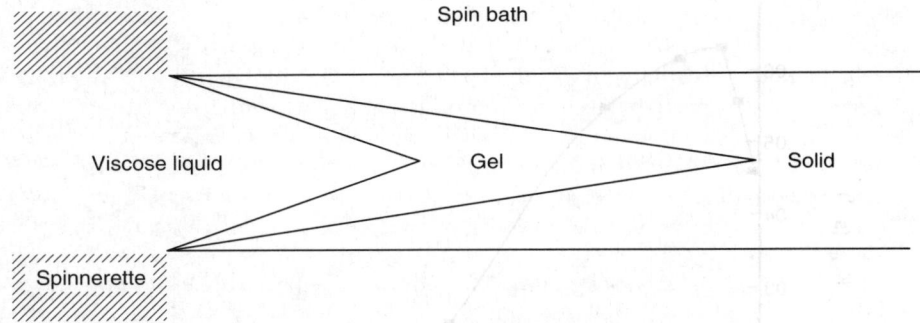

FIGURE 10.57 Changes occurring during initial fiber formation.

become established by the chemical and physical conditions imposed upon the fiber during spinning [173,174].

The decomposition of cellulose xanthate in acidic environment has been studied extensively, and much of the work has been referred to by Tornell [175]. With the aid of model compounds, such as alcohol xanthates, a mechanism for the complex reaction has been established [176], as shown in Figure 10.58. Compounds I, II, and III have been observed by UV spectrophotometry.

The influence of acidity on the decomposition of cellulose xanthate is illustrated in Figure 10.59. In strong acid such, as the Lilienfeld-type spin-bath, the xanthic acid (II) is protonated and stabilized, leading to a slower regeneration of the cellulose. Steric effects associated with the group R also affect the course of the reaction. Thus when R is a bulky t-butyl group, protonation is hindered and t-butyl xanthate decomposes very rapidly, even in strong acid [177].

There are three positions in the basic structural unit of cellulose at which the xanthate groups can be located. Each has a different stability, and, consequently, the distribution of xanthate on the polymer will affect the rate of decomposition. The actual distribution of xanthate groups at the time of spinning depends upon the conditions used to make and

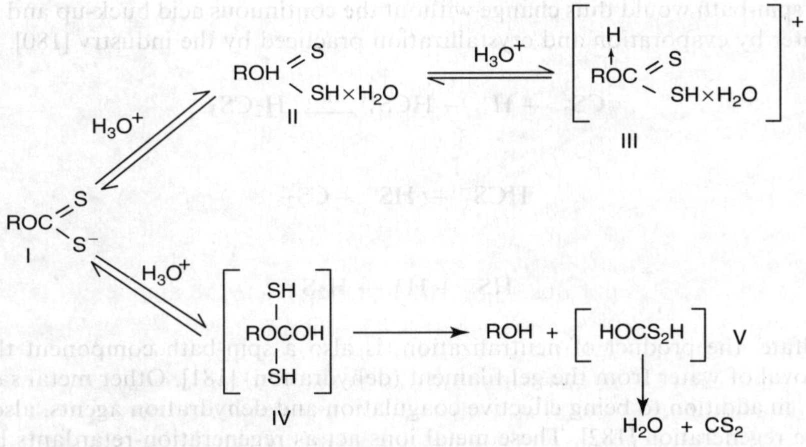

FIGURE 10.58 Xanthate decomposition by acid.

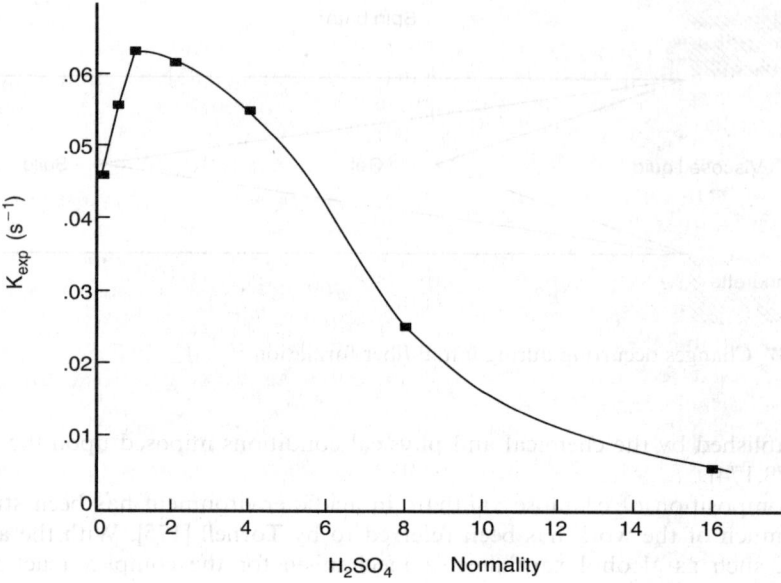

FIGURE 10.59 Influence of acidity on the decomposition of cellulose xanthate.

ripen the viscose. It has been demonstrated that redistribution of xanthate groups by trans-xanthation also occurs in acidic solution at a rate of the same magnitude as the decomposition [167]. Under normal spinning conditions, there are 45 individual reaction rate constants in the decomposition of pure cellulose xanthate by acid alone [176].

Viscose and spin-bath compositions have very important effects on the cellulose regeneration kinetics [178]. Besides the ionic concentration being a driving force for the diffusion processes, the salts formed within the filament have a buffering action on the pH. By-products in the viscose are also decomposed during spinning (Equation 10.13 through Equation 10.15), sodium trithiocarbonate being the main source of hydrogen sulfide in the viscose rayon process [179]. The products of these reactions, together with those from the neutralization of the alkali as the acid penetrates the forming fiber, diffuse into the spin-bath. The composition of the spin-bath would thus change without the continuous acid buck-up and removal of salt and water by evaporation and crystallization practiced by the industry [180].

$$CS_3^{2-} + H^+ \rightarrow HCS_3^- \xrightarrow{H^+} H_2CS_3 \qquad (10.13)$$

$$HCS_3^- \rightarrow HS^- + CS_2 \qquad (10.14)$$

$$HS^- + H^+ \rightarrow H_2S \qquad (10.15)$$

Sodium sulfate, the product of neutralization, is also a spin-bath component that mainly affects removal of water from the gel filament (dehydration) [181]. Other metal salts such as zinc sulfate, in addition to being effective coagulation and dehydration agents, also influence the cellulose regeneration [182]. These metal ions act as regeneration-retardants by forming derivatives of cellulose xanthate that are more resistant to decomposition [183,184]. Several forms of zinc cellulose xanthate have been proposed [185]: $(CellOCS_2)_2Zn$ (I) and $CellOCS_2Zn^+X^-$ (II), the most frequently mentioned is a cross-linked zinc cellulose xanthate (I) [186]. However, not only would such a structure require favorable disposition of two xanthate groups, but also the cross-linking would be expected to reduce the ability to stretch the filament contrary to known behavior. A more plausible structure involving zinc would be a zinc cellulose xanthate cation associated with an anion X^-, such as HCS_3^- (II).

With zinc in the spin-bath, the nature of the Donnan membrane established at the interface between the viscose filament and the spin-bath, and across which the diffusion processes and mass transfer must occur, is changed [170,187,188]. Although it has been proposed that the membrane retards acid diffusion into the filament, it is equally probable that the observed effect is caused by the structure suppressing diffusion of water from the filament. Tornell's study of viscose spinning has shown that zinc reduces deswelling of the filament [189].

High-performance rayons can be obtained in several ways, notably through the use of the following:

1. Special spinning conditions and viscose compositions
2. Viscose modifiers
3. Spin-bath additives

All of these approaches achieve the desired fiber properties by increasing the time required to regenerate the cellulose completely. Also, the way in which the cellulose molecules come together to form the fiber structure is affected.

An example of the first type is the use of low-acid, low-salt, and low-temperature spin-baths, which slow down the cellulose regeneration sufficiently to yield HWM polynosic rayon [190]. Another example is the Lilienfeld process for which the viscose, made from unaged alkali cellulose with excess carbon disulfide and only a short ripening, is spun into a cold spin-bath containing 50–85% of sulfuric acid. This is a case of stabilizing the xanthic acid. Rayon produced in this way has tenacities greater than 5 g/den.

Addition of chemical modifiers to viscose, first patented by Cox [124], represents a most important and widely practiced route to the high-performance rayons in use today [191]. Many organic compounds will function as regeneration-retardants, e.g., amines, quaternary bases, polyoxyalkylene derivatives, polyhydroxypolyamines, and dithiocarbamates [173,192]. Their effectiveness as viscose modifiers appears to depend on an ability to associate with other components in the spinning system to provide control over fiber structure development [193]. By the combined action of viscose modifiers with spin-bath and viscose components, a semipermeable membrane or structure is formed around each filament, which slows diffusion rates for acid penetration and water removal [194]. Besides retarding the rate of cellulose regeneration and changing the crystallite size, the amount of orienting stretch that can be applied to the fiber is increased by the use of modifiers. Although many different modifiers are known, most systems used are based on compounds containing nitrogen or oxygen [195].

Nitrogen-containing modifiers such as DMA, discussed by Deshmukh [196], are generally ineffective without zinc in the spin-bath. It is known that DMA reacts with carbon disulfide in viscose to form dimethyldithiocarbamate (Equation 10.16), which is an effective agent in modifying viscose. The mechanism by which cellulose xanthate decomposition is retarded is believed to involve association of the thiocarbamate with the xanthate group by a bridging zinc atom (Equation 10.17).

$$\text{H}_3\text{C} \diagup \text{NH} + \text{CS}_2 \longrightarrow \text{H}_3\text{C} \diagup \text{N}-\text{C} \diagdown \overset{\text{S}}{\underset{\text{S}^-}{}} \qquad (10.16)$$

$$\text{R}_{\text{cell}}-\text{O}-\text{C} \diagdown \overset{\text{S}}{\underset{\text{S}^- \text{Zn}^{++} \text{-S}}{}} \diagup \text{C}-\text{N} \diagdown \overset{\text{CH}_3}{\underset{\text{CH}_3}{}} \qquad (10.17)$$

Charles [197] proposed that a semipermeable membrane was formed by combined action of the modifier with zinc and trithiocarbonate ions. However, in his idealized structure of the membrane, xanthate groups could be substituted for the trithiocarbonate ions with similar effects. Sisson [174] had claimed earlier that the membrane or cuticle retarded penetration of hydrogen ions into the filament but did not affect the transfer of zinc ions or water. Smith [173] suggested a different explanation, in that the modifier action enhanced penetration of zinc into the filament by forming a complex solubilizing zinc over a wide pH range. Still other investigators claimed that colloidal precipitates were formed at the filament surface that functioned as the semipermeable membrane [194]. Levine and Burroughs also suggest formation of a semipermeable membrane in which the pores are initially blocked by compounds formed from modifiers, trithiocarbonate, and zinc [193]. This is consistent with the finding of Klare and Grobe [198] and Grobe et al. [199], but the conversion of sodium cellulose xanthate to zinc cellulose xanthate occurs to a limited extent, and mostly at the filament surface.

Polyoxyalkylene derivatives are typical of the oxygen-containing viscose modifiers. In this type of modifier, association involving the ether oxygen of the polyoxyalkylene chain is believed to be part of the mechanism by which acid diffusion and cellulose regeneration are retarded. Two possibilities exist, (I) protonation of the ether oxygen [200] retards hydrogen ions from penetration into the filament and facilitates formation of zinc cellulose xanthate and (II) formation of chelate compounds with zinc and zinc salts that increase the stability of the semipermeable membrane [201].

$$\underset{\text{I}}{-\text{R}-\overset{\overset{\text{H}^+}{\uparrow}}{\text{O}}-\text{R}-} \qquad \underset{\text{II}}{\text{R} \diagup \overset{\text{O}}{\underset{\text{O}}{}} \diagdown \text{ZnS} \diagup \overset{}{\underset{}{}} \text{R} \qquad \text{R} \diagup \overset{\text{O}}{\underset{\text{O}}{}} \diagdown \text{ZnCS}_3 \diagup \overset{}{\underset{}{}} \text{R}}$$

It is known that the effectiveness of this class of modifier depends on the basic structural unit R [202]. Also, the length of the polyoxyalkylene chain, although not a factor in the modifier action, is quite important to the practical aspect of mixing it with viscose. Among the most important developments was the use of mixed modifier systems consisting of amine and polyoxyalkylene glycols [126], which provide the basis for the major part of present-day technology. In a typical mixed modifier system, polyethylene glycol of about 1500 molecular weight is most effective when used with dimethylamine [203]. An important aspect of the modifier function, besides regeneration-retardance, is the lower gel swell of the filaments, which has an important effect on subsequent structure formation [204] and affects the shape of the filament cross section. Together with zinc, the modifiers influence crystallite formation in the fiber and the resultant skin–core structure [174].

Adducts of alkylene oxide and amines are also effective modifiers. Typical of this class containing both oxygen and nitrogen described in the patent literature are such products as alkylene oxide adducts of aniline [205], p-aliphatic anilines [206], N-fatty alkyl alanines [207], amino fatty acyl amides [208], triethanolamine [209], the combination of epoxyalkane polymers with amines [210] or nitrogen-containing polyoxyalkylene compounds [211], oxyalkylated ammonia [212], and poly(epoxy ethane)-substituted fatty acid amine [213]. These materials are generally available containing different amounts of polyoxyalkylene derivative. Other patents claim polyhydroxy amines as regeneration-retardants [175]. Although a clear distinction has been made between the actions of modifiers based on either amines or glycols [202], a mechanism for modifiers containing both functionalities has not been developed.

Many organic sulfur compounds, such as thiocarbamates, thioureas, and thioalcohols, can be used as modifiers. However, because compounds of these types are formed in viscose by interaction of amine- or glycol-type modifiers with carbon disulfide and sulfur-containing viscose by-products, they are generally not used in commercial practice. The important role of the viscose by-product sodium trithiocarbonate has been described by Phillip [215]. High-strength, high-elongation rayon was produced by Cox [216] by adding sodium trithiocarbonate to a modifier-free viscose, but it was not possible to achieve equivalent fiber properties using modifiers in a by-product-free viscose. Butkova et al. [217] have reported that fiber properties and cross-sectional shape similar to those of regular rayon are obtained when the by-products are removed by ion-exchanging viscose before spinning. Murakami [202] claimed that the viscose by-products function as modifier assistants yielding improved control over the spinning process and fiber quality.

Formaldehyde added to the spin-bath is a useful regeneration-retardant [125] that, in acidic solution, forms an S-methylol derivative (Equation 10.18) of cellulose xanthate [218].

$$R_{cell}-O-C\begin{matrix}\\S\\\\S^-\end{matrix} + HCHO \xrightarrow{H^+} R_{cell}-O-C\begin{matrix}\\S\\\\SCH_2OH\end{matrix} \quad (10.18)$$

The very slow decomposition of this derivative and the strong membrane formation permits extremely high stretch [126] and is the basis of a number of high-strength rayons [219–221]. In contrast to other types of viscose modifiers, formaldehyde reacts not only with the labile cellulose groups [222,223] but also with the cellulose backbone to form permanent cross-links. Disadvantages associated with this modifier include difficulty in completing regeneration and competing reactions (Equation 10.19 through Equation 10.21) with viscose and spinning by-products such as sodium trithiocarbonate [224] and hydrogen sulfide [225]. These side reactions cause practical problems in spinning and spin-bath reclaim [226]. Efforts to use other compounds that provide the functionality of formaldehyde but avoid its disadvantages have not been successful in achieving equivalent fiber properties.

$$CS_3^{2-} + 2HCHO \xrightleftharpoons{H^+} S=C(SCH_2OH)_2 \quad (10.19)$$

$$S=C(SCH_2OH)_2 \xrightarrow{H^+} HOCH_2SH + CS_2 + HOCH_2OH \quad (10.20)$$

$$3HOCH_2SH \longrightarrow \begin{array}{c} S-CH_2 \\ CH_2 \\ S-CH_2 \end{array} S + 3H_2O \qquad (10.21)$$

Prior to spinning, the processes are concerned with preparing the viscose. At the time of spinning, basic structural units already exist in the viscose solution. The latent structural characteristics of the rayon fiber, however, are established by the way the cellulose molecules are brought together in the spinning process. Control of this step depends on properly balancing the rates of the various chemical and physical reactions. As viscose composition and structure has already been established at this point, control must be achieved through the spinning conditions selected to produce the fiber [227]. Bath temperature and composition and rate and distance of filament travel are the variables that are usually adjusted at this stage of the process. Before the regeneration is completed, the filaments are stretched to obtain molecular orientation or lining-up of the structural units parallel to the fiber axis [228]. In this way, higher tenacity and lower extensibility of the fiber at break can be obtained. This is illustrated by the tensile properties of rayon produced from the same viscose at different stretch levels (Figure 10.60). The maximum tenacity derived from stretching is achieved at some spinning stretch below the maximum or the break stretch, at which the filament bundle breaks under the particular spinning conditions. Stretching beyond this optimum condition disrupts the fiber structure and adversely affects strength. This condition is recognized in spinning by thick-thinning and roughening of the filament bundle as it emerges from the stretch-bath.

The change in molecular orientation caused by stretching the filaments is accompanied by an increase in the tension in the bundle. Tension measurements taken on a filament bundle (tow) leaving the stretch-bath are plotted as a function of the fiber wet tenacity in Figure 10.61. Considerable energy must be expended to achieve a relatively small increase in strength at stretch levels close to break. Under such conditions, it is easy to envisage how particles in the viscose could disrupt the fiber structure, causing defects and even spinning breaks. From this point on, the processes are mainly concerned with washing the fiber in various solutions to remove processing chemicals and acid, completely desulfuring, and sometimes bleaching. A lubricant or spin finish is usually applied to the fiber to facilitate subsequent processing before it is dried and conditioned to about 11% moisture and packaged for shipment. The thoroughness with which these final steps are completed is very important to quality and performance of the rayon fiber in subsequent conversion processes and end uses.

10.4.5 PRODUCTION OF VISCOSE RAYON

Today, viscose rayon represents the product of a mature industry. As described in Section 10.2, most production facilities are based on technology and equipment that was developed many years ago. The major developments in recent years have been focused on the installation of facilities to protect the environment. Considerable emphasis has also been placed on process technology improvements that would enable lower production costs using existing equipment [229]. These developments have seen changes in viscose composition to use less chemicals in the process and equipment modifications, such as the cluster jets allowing increased production capacity.

The expansion of rayon production in undeveloped countries, such as Taiwan, Indonesia, and South America, where new rayon plants have been constructed, has in many cases incorporated the newer technology. Typical of this would be the continuous belt xanthator

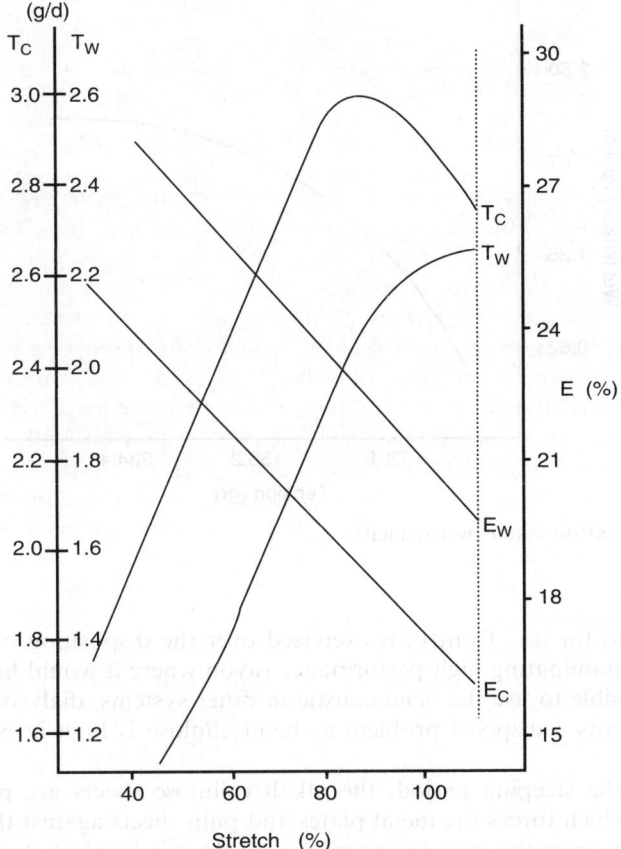

FIGURE 10.60 Influence of molecular orientation (stretch) on tensile properties (c = conditioned fiber, w = wet fiber).

of the Chemtex process [230] and the Maurer–Buss Contisulf continuous process [231]. It is in these areas that viscose rayon production will benefit most from the experience and knowledge gained from the many years of extensive research and development on viscose technology.

The widely different and varied design of present-day viscose rayon plants makes it impractical to describe all of the facilities for making rayon. However, because the chemistry of the process is common to all viscose fiber production, a brief outline of the principal methods will help to consolidate the understanding of how rayon is made. The basic steps of viscose rayon manufacture are outlined schematically in Figure 10.62. Basically, there are two types of process technology, batch and continuous. Pulp is supplied normally in bales of a predetermined weight and sheet size or in rolls.

For conventional batch steeping, the rectangular pulp sheets are arranged in books consisting of up to about 20 sheets placed between perforated metal plates in a large trough or steeping press. Caustic soda is let in from the bottom of the press at a predetermined fill rate, depending on the physical characteristics (mainly swell rate) of the pulp used. The pulp is allowed to soak (steep) in the alkali for a length of time, depending on the manufacturing process specifications; 20 min to 1 h is the usual soak time, after which the caustic solution, containing alkali-soluble hemicellulose (short-chain material removed from the pulp), is

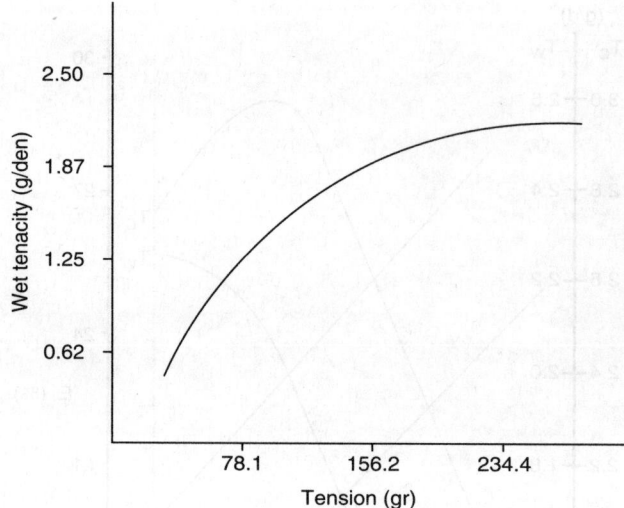

FIGURE 10.61 Tow tension versus wet tenacity.

drained and recycled for use. Control is exercised over the disposition of the hemicellulose caustic to avoid contaminating high-performance rayon where it would have adverse effects. Where it is not possible to use the hemi-caustic in other systems, dialysis is sometimes used [232], but this presents a disposal problem as hemicellulose is high in biochemical oxygen demand (BOD).

At the end of the steeping period, the alkali cellulose sheets are pressed under high pressure by a ram which forces the metal plates and pulp sheets against the opposite wall of the press. The caustic from this pressing operation is higher in hemicellulose content than that

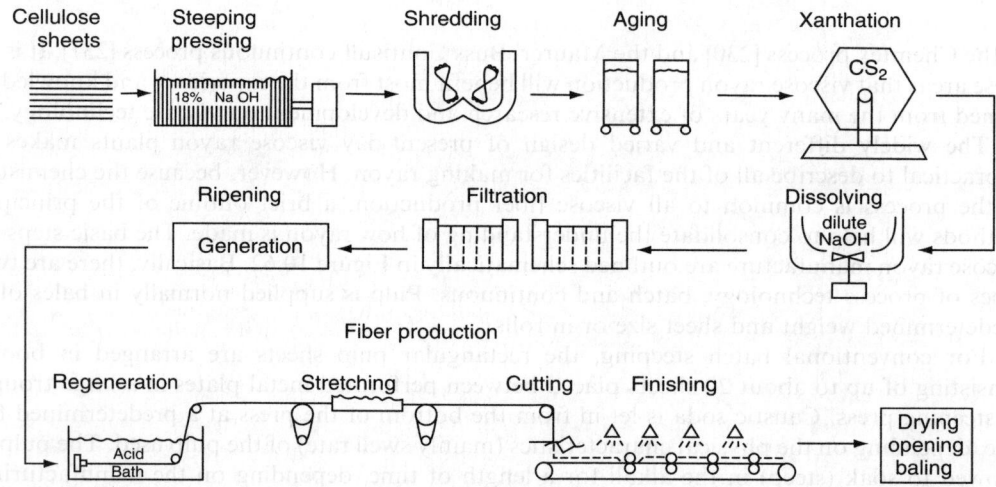

FIGURE 10.62 Conventional batch process for viscose rayon manufacture (schematic). (Courtesy of John Wiley & Sons, New York.)

FIGURE 10.63 Pulp feeding to slurry-steeping tank. (Courtesy of Chemiefaser Lenzing, A.G. Lenzing, Austria.)

which had been drained by gravity [233]. This enables the rayon manufacturer to segregate the high hemi-pressings for control of the hemi-balance in the system. The amount of pressure applied is calculated to give a precise press weight ratio (PWR) corresponding to a cellulose: caustic soda ratio of around 33:15.

In the continuous process or slurry-steeping system, the pulp sheets are fed at a controlled rate into a large tank filled with caustic (Figure 10.63). A paddle continually stirs the mixture, and the slurry consistency is maintained at the desired level, about 4–7%, depending on the type of equipment used, by balancing the rate of feed and takeoff. In the slurry system, pressing can be done by passing the slurry between perforated rolls, using a vacuum box and pressure to pull off the liquid (Figure 10.64), or in a screw press. Again, the PWR achieved is about 2:8.

Next, the alkali cellulose is shredded into a crumb to provide adequate surface area for subsequent reactions. For batch-shredding, the alkali cellulose sheets are transferred to a shredder resembling a dough mixer, except that the curved blades have blunt teeth (serrations) that rotate against a serrated saddle bar to disintegrate the pulp sheets. In this operation, time, temperature, blade clearance, and revolutions per minute (speed) must be carefully controlled. Blade clearance is also important in this operation to avoid compacting the crumb. For optimum processing, each pulp type requires different conditions, which are usually determined by the fiber length of the wood species involved [234]. Long-fibered pine, for example, gives a less dense, bulkier, white crumb than short-fibered hardwoods such as eucalyptus or gum. Continuous shredders, such as the Sprout Waldron, operate by centrifugal action, which throws the alkali cellulose mass from the press and between a series of closely spaced teeth on two discs, one rotating, the other stationary (disc refiner).

Shredders are usually jacketed to permit cooling or heating as required. Mechanical action and the heat involved will cause depolymerization, which affects subsequent aging requirements. It is important to avoid excessive drying, carbonation, or localized wetting from condensation of the crumb within the shredder. Carbonation, due to reaction of CO_2 from air with the alkali on the crumb, forms sodium carbonate, which blocks access of CS_2 to the

FIGURE 10.64 Alkali cellulose from a Sund-Impco Press. (Courtesy of Chemiefaser Lenzing, A.G. Lenzing, Austria.)

cellulose in xanthation, while wet spots reduce the critical concentration of caustic in portions of the crumb, causing poor reactivity and filtration problems as described by Treiber [235].

After shredding, the alkali cellulose crumb is aged to obtain the DP for the type of rayon to be produced. Not only is the DP important to fiber properties, but it also affects the processing characteristics of the viscose. Solution viscosity is proportional to the DP and polymer concentration. Consequently, those fibers requiring a relatively high DP for reasons of strength must be made from viscose containing less cellulose [236].

Aging to depolymerize the cellulose is done in a number of ways. In all processes, temperature, time, and alkali cellulose composition are controlling variables. In addition, catalysts such as salts of manganese and cobalt [237] can be added to the pulp sheets and the steeping liquor to accelerate the aging.

In one aging process, the shredded white crumb is transferred to steel cans, each holding the crumb from one complete steep of the batch process. The cans are covered and stored in a temperature-controlled room for a period of time calculated to give the correct DP. Aging time may vary from a few hours to 24 h or more at a temperature from 20 to 35°C. When an aging catalyst is used, cooler temperatures and shorter times are possible.

Other process equipment for aging the white crumb includes silos and aging drums (Figure 10.65), large storage containers lined with shelves, and platforms through which the crumb is passed by wiping action of rotating blades or rotating action of the drum. Aging tunnels similar to tunnel dryers are used at some rayon manufacturing facilities. In this case, the crumb is conveyed through the tunnel on a belt; the temperature (about 50–60°C) used is considerably higher than that used in crumb-can aging to facilitate the depolymerization in a residence time of 3–4 h. When the alkali cellulose is aged to the DP appropriate for the rayon to be produced, it is ready for reaction with CS_2.

Just as with alkali cellulose aging, xanthation can be carried out in several ways. The most usual methods are batch processes using dry or wet churns. The dry churn, sometimes called a baratte, is a hexagonal, jacketed vessel with mixer blades and a hollow perforated shaft through the center. This center tube is for evacuating the churn and admitting CS_2. In operation, a weighed amount of white crumb is placed into the churn through a hinged

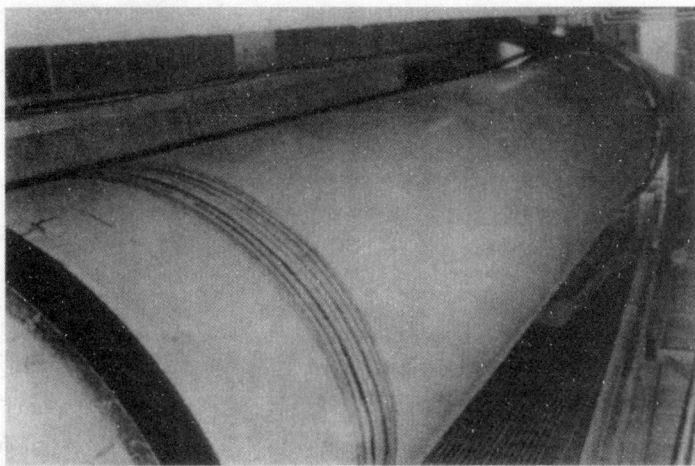

FIGURE 10.65 Alkali cellulose aging drum. (Courtesy of Chemiefaser Lenzing, A.G. Lenzing, Austria.)

door on one side. The door is bolted securely prior to starting the process, and, because of the explosive nature of CS_2, churns are made of nonsparking material with explosion-proof motors. For safety, churn rooms have blowout walls and the operators observe the reaction from a remote, protected station. The operation begins by evacuating the churn to a given negative pressure, then admitting CS_2, either slowly over a period of time or as fast as it will go in. The xanthation reaction is exothermic and must be controlled by chilled water running through the jacket. The entire operation can be controlled automatically, and in modern rayon plants it can be observed remotely by closed-circuit TV. As CS_2 enters the churn, the pressure rises; but as the reaction proceeds, the CS_2 is exhausted by reaction with the alkali cellulose and vacuum is restored. By watching the chart of a pressure recorder, the operator can tell when the reaction is complete and ready for the next step. The time of reaction depends on the temperature and the amount of CS_2 used, and may vary from about 30 min to an hour or more. It has been reported [238] that 30°C gives optimum results with the dry churn.

The wet churn differs from the dry churn in size and in the way in which it is used. Simplex churns (Figure 10.66) of various sizes will hold the alkali cellulose crumb made from 1 to 12 bale of pulp; the most usual sizes hold from 2 to 4 bale (3000 to 6000 lb of white crumb). In operation, the charging of the churn with crumb and CS_2 is similar to that for the dry churn, but the difference arises when part of the mixer caustic charge is introduced shortly after vacuum regain begins. This may occur about 30–40 min after the addition of CS_2 began. Reaction continues for about 10–20 min in the wet churn before the contents of the churn are discharged into a mixer, where the remaining portion of the mixer caustic charge is added. Low mixing temperatures, generally below about 16°C, are used to produce quality viscose; the blowdown temperature from the wet churn is also controlled.

A continuous belt xanthator (CBX) [239] is used at some facilities. In this equipment, the white crumb is fed through a vacuum gate where air is removed and then onto a conveyor belt where it forms a bed that can be several feet thick. The belt moves slowly through a chamber as the white crumb is sprayed with CS_2. As there is no stirring of the reaction mass, the bed is usually dropped onto a second belt traveling in the opposite direction in the chamber. In this way, the xanthation is made more uniform. The xanthated crumb is discharged from the reactor through a sump filled with alkali.

FIGURE 10.66 Simplex wet churn. (Courtesy of Chemiefaser Lenzing, A.G. Lenzing, Austria.)

Another continuous process, the Maurer–Buss Contisulf process [240], uses a series of screw kneaders not only to prepare the xanthate but also to be in mixing (Figure 10.67). Alkali cellulose can also be prepared in this type of equipment, and two or more steps can be done in the same screw. This is possible by using a long barrel and arranging the screws such that alkali, CS_2, and the mixing caustic charge can be introduced at the appropriate points. In this case, the reaction is controlled by temperature, screw size, and the rate of screw feed, which determines the reaction time in each zone. The crudely mixed xanthate is discharged into a tank where it is further mixed and blended as necessary to obtain a uniform viscose.

Several other machines for producing viscose continuously have appeared in the patent literature. Since Richter obtained a patent in 1932 [241], inventors such as Seaman [242], Yasui et al. [243], Von Kohorn [244], and Treiber [245] all claim success in making the several stages in the viscose process on a continuous basis. (Except for the CBX and Contisulf units mentioned before, successful full-rayon-plant operation of the others has not yet been attained.) Most rayon producers, however, prefer batch xanthation in large churns.

Viscose mixing is done in a jacketed vessel with propeller blades, which is contained in an interior vessel, sometimes called a dissolver. Xanthate crumb from the dry churn, or the slurry from the wet churn, is forced in the sides of the vessels in a draft tube and then flows down the center and through the propeller blades continuously. The heat of mixing is controlled by chilled water flowing through the outside jacket of the dissolver. As cellulose xanthate dissolves best in cold caustic, temperatures of 10–15°C are normally employed to prevent premature ripening and by-product formation. However, for economic reasons, higher temperatures of 18–25°C or more can be used for rayons intended for applications where strength and modulus are not critical.

The composition of the viscose is determined by the amount of cellulose xanthate and caustic soda used. For regular rayon, the cellulose content may vary from 8 to 9.5% and the caustic soda from 5 to 6%, depending on process specifications and the efficiency of the equipment at specific facilities. Viscoses required for high-performance rayon, HWM, and tire yarn use a higher DP cellulose and richer (more alkali) compositions with 6.0–7.5% cellulose and 6.0–7.5% NaOH. Generally, from 27.5% to 29.5% CS_2, based on the cellulose in

Regenerated Cellulose Fibers

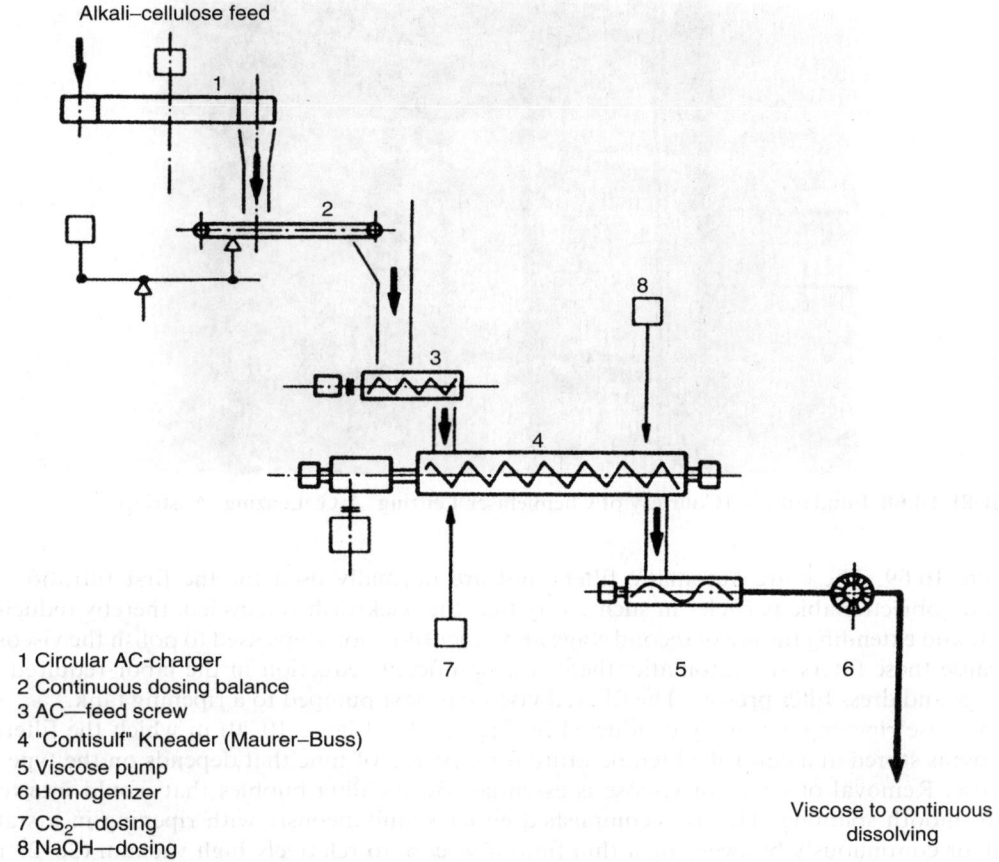

1 Circular AC-charger
2 Continuous dosing balance
3 AC—feed screw
4 "Contisulf" Kneader (Maurer–Buss)
5 Viscose pump
6 Homogenizer
7 CS_2—dosing
8 NaOH—dosing

FIGURE 10.67 Maurer–Buss Contifulf process (diagram). (Courtesy of ING. A. Maurer, SA, Berne, Switzerland.)

the white crumb, is used for regular rayon, 30–34% CS_2 for HWM rayon, and 34–38% CS_2 for tire yarn, with even higher amounts used for some polynosic and specialty fibers.

Modifiers that retard regeneration are required for HWM and high-strength industrial rayons and may be added during the dissolving stage or later in the process, just prior to spinning, by injection into the viscose stream. For production of variant fibers, delustrants (dulling agents) such as finely ground titanium dioxide with a dispersing agent, pigments (vat dyes, carbon black, or metal oxides) for making spun-dyed fibers, or other additives such as flame-retardants and components for alloy fibers may also be added during the mixing stage or by injection.

Filtration of viscose is usually done using plate and frame filter presses [232]. Filter materials vary among manufacturers and may be cotton batting with a tightly woven fabric of cotton, pulp sheets, nylon, polyolefin, metal, or various nonwoven materials. The important criteria are that the filter material is strong enough to withstand the pressure applied to force the viscose through the interstices and tight enough to hold back the undissolved fibers, partially dissolved gels, and particles that would clog the fine spinnerette holes during the spinning operation. Developments in this area have included the Fundafilter [246], shown in Figure 10.68, which uses a PVC filter aid, and the Viscomatic filter [247,248], shown in

FIGURE 10.68 Fundafilters. (Courtesy of Chemiefaser Lenzing, A.G. Lenzing, Austria.)

Figure 10.69. These are automatic filters and are normally used for the first filtration to remove objectionable particles in such a way that the backwash is recycled, thereby reducing waste and extending the life of second stage and other filtration steps used to polish the viscose. Because these filters are automatic, there is a significant reduction in the labor required to change and dress filter presses. The filtered viscose is next pumped to a ripening tank.

Viscose ripening is usually conducted in large tanks (Figure 10.70) in which the filtered viscose is stored at a controlled temperature for a period of time that depends on the type of viscose. Removal of air from viscose is essential for avoiding bubbles that would interfere with smooth spinning. This is accomplished either simultaneously with ripening in a static tank or continuously by exposing a thin film of viscose to relatively high vacuum (28–29 in. Hg) immediately prior to pumping into the ripening tank. For static deaeration, vacuum is applied to the ripening tank and air is removed gradually during the process. Some tanks have an outboard high-shear mixture (attritor), which circulates the viscose out of the tank and back to obtain a uniform, homogenous solution. After the required ripening time, when the viscose has attained the appropriate SI, it is ready for spinning.

Basically, a spinning machine to produce viscose rayon consists of several essential components (Figure 10.71): a metering pump that delivers a precise flow of viscose to the spinnerette or jet; troughs (and sometimes also a tube) to contain the spin-bath and other processing baths as needed; a means to propel the filament bundle through the machine, complete with appropriate guides and drive; and a mechanism for collecting the fibers (Figure 10.72) and pass them on to the next steps of cutting, washing, and processing [249]. Spinning may be done in a horizontal or vertical arrangement. The variations on the basic design of viscose spinning machines depend not only on the type of product (cut staple or continuous filament) made, but also on the physical layout of the particular facility.

In some rayon plants, spinning is done from both sides of the machine, and machines are arranged in long rows on each side of aisles from which the machines are accessible to the workers. Normally the spinning machines are enclosed so that the gases generated can be safely removed in a stream of air (Figure 10.73).

One of the most important pieces of equipment used in rayon manufacture is the spinnerette or jet used to give form to the filaments (Figure 10.74). Spinnerettes are thimble-shaped and made of corrosion-resistant metal, such as platinum or gold, or glass [250]. Typical

FIGURE 10.69 Viscomatic filter. (Sund). (Courtesy of Chemiefaser Lenzing, A.G. Lenzing, Austria.)

spinnerettes are illustrated in Figure 10.75. They range in diameter from 0.5 in. to several inches and can contain from 1,000 to more than 20,000 holes of precisely controlled size and configuration. In most cases, the holes are round and range in size from 0.001 to 0.01 in. The rate at which viscose is delivered to the spinnerette, the number of holes, and the degree to which the filaments are stretched determine the filament denier and tow size. Cellulose content of the viscose is another parameter influencing the finishing fiber denier.

Hydrodynamic conditions of viscose spinning must be balanced carefully so that each filament is exposed to the same acidic environment. Although viscose extrusion causes a pumping action drawing the spin-bath to the center of the filament bundle, a condition known as "acid-starvation" can occur if appropriate pump speed, spin-bath composition, and spinnerette hole spacings are not selected. A recent development avoiding these difficulties entails the use of cluster jets. These consist of assemblies of smaller spinnerettes arranged to produce as many as 70,000 filaments from a single assembly (Figure 10.76). Increased

FIGURE 10.70 Viscose ripening tanks. (Courtesy of Chemiefaser Lenzing, A.G. Lenzing, Austria.)

production capacity as well as improved fiber quality are cited as advantages of using cluster jets, especially for the production of HWM fibers where slow spinning speeds are necessary [231].

At the prescribed distance from the spinnerette, the fiber bundle is drawn around a rotating wheel (godet), which may serve as a holding point for application of stretch (Figure 10.77). A second wheel, or several wheels driven at a higher speed, provides the stretching tension. Depending on the particular process, stretch may range from 20 to 200% or more. Often, the total stretch is obtained in multiple stages involving more than one bath, the last of which is usually low in acid and maintained at high temperature to fix the stretch. This is done as the regeneration of the cellulose is completed.

Following spinning, after the fiber leaves the stretching, it must be carefully washed and processed to remove acid, salts, and occluded sulfur. Although the processes are essentially

FIGURE 10.71 Viscose rayon spinning machine. (Courtesy of Kemira Oy Sateri, Valkealkoski, Finland.)

Regenerated Cellulose Fibers

FIGURE 10.72 Spinning machine stretch rolls. (Courtesy of Kemira Oy Sateri, Valkealkoski, Finland.)

the same for continuous filament and staple fiber, the mechanics of the final steps in rayon production differ.

Continuous filament is collected either in a rapidly rotating centrifugal pot machine, on a bobbin (spool or roller) on the bobbin machine, or on a spool or reel on the continuous machine. The yarn packages from the pot and bobbin machine are wound onto perforated cylinders for washing and processing. This is done in batches. Subsequent steps may involve bleaching, finishing (addition of lubricant and antistat), drying, and winding onto a cone. With a continuous machine, all of these steps are performed successively on the advancing yarn, and the yarn coming off the machine represents a finished package.

Rayon staple fiber is usually obtained by collecting the filaments from a large number of spinnerette assemblies (often 100 or more) into a tow, which is then cut to the desired length.

FIGURE 10.73 Spinning machine with enclosure for pollution control. (Courtesy of Chemiefaser Lenzing, A.G. Lenzing, Austria.)

FIGURE 10.74 Filament formation, 12,000-hole spinnerette. (Courtesy of Kemira Oy Sateri, Valkealkoski, Finland.)

The total denier of a tow for staple often exceeds several million and considerable power is required to apply stretch. Immediately after stretching, the tow passes into a cutter (Figure 10.78). There, the relative speed of the tow and the cutter blades determine the length of the staple. The rayon fiber is washed from the cutter into a sluice and formed into a bed on a moving screen for processing (Figure 10.79). The first washing normally uses hot water, containing a small amount of sulfuric acid. The blanket then moves into other wash stations where solutions of controlled chemical composition and temperature are alternately applied and removed by squeezing between rubber rolls. In this way, the staple fiber is washed, desulfured, neutralized, sometimes bleached, and treated with processing chemicals (lubricants, etc.). A final heavy squeeze removes excess liquid before the blanket is opened (fluffed) and passed on a moving belt through a tunnel dryer (Figure 10.80). The staple fiber is conditioned to about 11% moisture, opened, and then baled for shipment (Figure 10.81).

10.4.6 Types of Rayon

The classification of viscose rayon fibers into different types is done mostly on the basis of physical and chemical properties. Fibers produced by nonviscose processes are usually identified separately. It has already been described how the fiber can be produced to have almost any desired structure, and it is considered to be the most versatile of all human-made fibers. It is available in various cross-sectional shapes, from multilobed, serrated, and round to flat; longitudinally, it may be straight or curled (crimped). It comes in fine deniers

FIGURE 10.75 Rayon spinnerette, 1100 holes. (Courtesy of ITT Rayonier, Stamford, CT.)

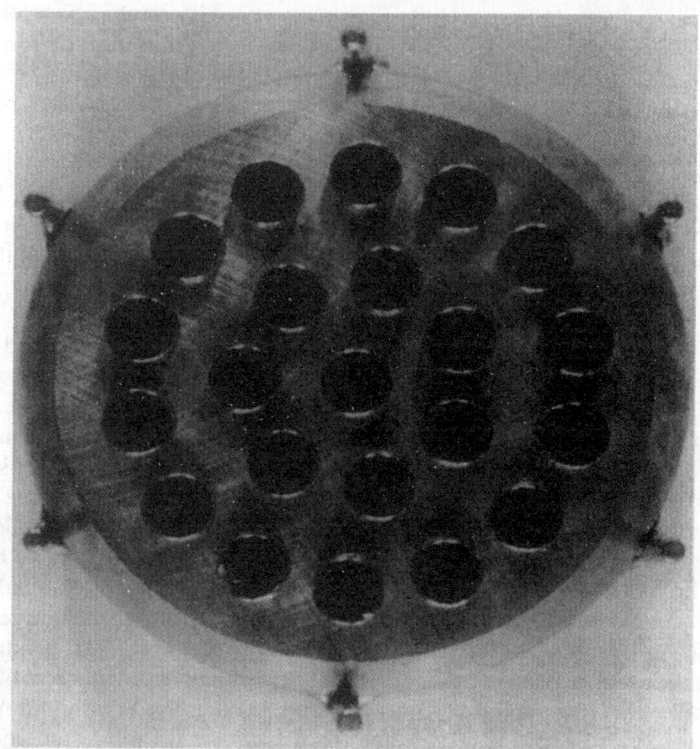

FIGURE 10.76 Cluster of small spinnerettes. (Courtesy of ITT Rayonier, Stamford, CT.)

FIGURE 10.77 Spinning machine. (Courtesy of Courtaulds, Ltd.)

resembling silk and fine cotton, or coarse as wool, horsehair, or ribbon. The product may be in the form of continuous filament or cut staple [251]. The major types of rayon, together with subtypes, are listed in Table 10.7.

The hydrophilic properties of cellulose confer moisture absorption capacity to fabrics made from blends of rayon and synthetics. This property of moisture absorption is one of the reasons why rayon is used extensively in nonwoven, disposable applications, as the cross-sectional swelling of regular rayon is about three times that for cotton. The main disadvantage of regular rayon in textiles is its low wet-modulus; in the wet state, fabrics made from rayon will be weak and easy to stretch and deform.

The high-performance rayons overcome this disadvantage. The HWM fiber has cotton like mechanical properties and a caustic resistance that allows mercerization. It is compatible in blends with all grades of cotton where it adds strength, improved luster and appearance, and a softer hand. In blends with nylon, polyester, acrylics, and triacetate, it has good strength retention after resination, and the blended fabrics have superior wash-and-wear performance and resistance to pilling.

High-tenacity rayons find application where strength, toughness, and durability are required. These fibers have a cross-sectional structure that is mostly thick skin to all-skin. Although they shrink in hot water, these fibers are dimensionally stable when used as reinforcement in tires, conveyor belts, drive belts, and hoses, where the yarns are embedded in rubber or plastic. Besides tire cord and reinforcing yarns, other applications for this type of rayon include industrial sewing thread, tent fabrics, tarpaulins, and some staple used as reinforcing fabric in sheeting. For those applications where shrinkage would be a problem in product fabrication and end use, low-shrinkage, high-strength fibers have been developed.

Although cellulose is not thermoplastic, and rayon cannot be mechanically crimped and textured in the same way as synthetic fibers, a crimped fiber can be made by causing an imbalance in the uniformity and thickness of the skin during fiber production. Greater shrinkage of the skin in hot water, compared with shrinkage of the core, will cause the fiber to crimp. The skin effect in crimped rayon has been described by Sisson and Morehead [252]. Crimped rayon is wool like and is used alone or in blends with polyester, nylon, acetate, or wool to make carpets, upholstery, bedspreads, and clothing fabrics.

FIGURE 10.78 Tow feeding into cutter. (Courtesy of Chemiefaser Lenzing, A.G. Lenzing, Austria.)

Polynosic rayons also are characterized by high strength and high wet-modulus, but they differ from the rayons that are produced using modifiers in several ways. Important in end use are the low extensibility and high caustic-resistance of most polynosic rayons. The fibers possess a hand resembling cotton but have a structure that is brittle and prone to fibrillation, although this deficiency has been largely overcome in recent years. Polynosic fibers are used in most of the applications common to other high-performance rayons.

A large number of rayon variants were developed and commercialized during the 1960s and 1970s to meet specialized markets and provide required properties when these could not be obtained from available fibers. These developments have been described in reviews by Braunlich [253], Daul [254], Treiber [191], and Dyer and Daul [229]. The most significant variants include FR rayons and superabsorbent fibers. For both of these types, an additive incorporated in the viscose and included in the fiber structure provides the basis of the property enhancement or change. Other rayon variants produced on a limited scale include acid-dyeable and high-denier crimped fibers. Technology is reportedly available for

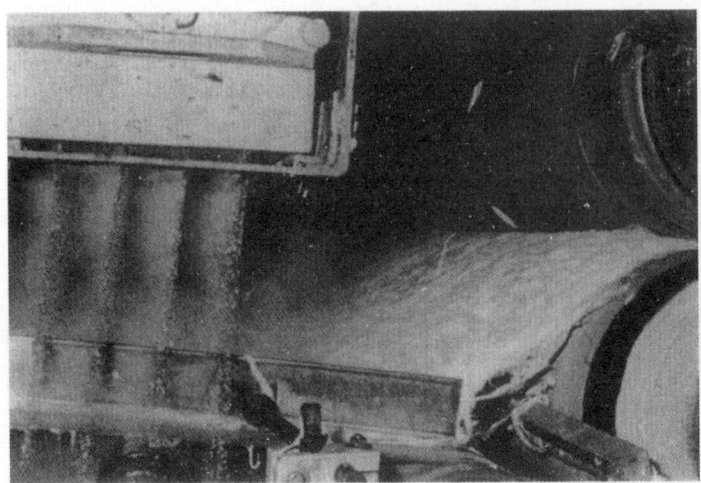

FIGURE 10.79 One section of processing machine. (Courtesy of Chemiefaser Lenzing, A.G. Lenzing, Austria.)

producing a wide variety of rayons with properties suitable for application in ion exchange, hemostats, bacteriostats, and as precursors for ceramic and carbon fibers.

The properties of cuprammonium rayon are sufficiently different from those of viscose rayon that today it is produced as a specialty fiber for several applications. Apart from its use as a substitute for silk in scarves, ties, fine dresses, and linings, the use of hollow cuprammonium fibers for hemodialysis in artificial kidneys has become important.

10.4.7 RAYON STRUCTURE

The basic chemical unit of rayon is the anhydroglucose unit of the cellulose molecule. Through the viscose process, cellulose becomes engineered into the many different types of

FIGURE 10.80 Tunnel drier. (Courtesy of Chemiefaser Lenzing, A.G. Lenzing, Austria.)

Regenerated Cellulose Fibers

FIGURE 10.81 Rayon baled, ready for shipment. (Courtesy of Chemiefaser Lenzing, A.G. Lenzing, Austria.)

rayon suitable for a wide variety of end uses. The fiber properties will depend on (a) how the cellulose molecules are arranged and held together in the fiber and (b) the average size and size distribution of the molecules. It is usual to consider anhydrocellobiose, a diglucopyranose consisting of two AGUs, as the basic structural, as distinct from chemical, unit in rayon because of the molecular conformation about the 1,4 β-glucosidic bond.

Early work on the fine structure of cellulose has been reviewed extensively by Sisson [255] and by Hermans [256]. The observations of Herzog and Jancke [257] who, in 1920, recognized that natural cellulosic material gave identical x-ray diagrams and led to the conclusion that these materials should have identical crystalline structure. The unit cell structure proposed by Meyer and coworkers [258–260] is still used to describe cellulose. The structure of the unit cell for cellulose is shown in Figure 10.82.

TABLE 10.7
Types of Rayon

Regular rayons
   High-tenacity (high-performance) rayons
      Low wet-modulus (LWM) rayons
      Intermediate wet-modulus (IWM) rayons
      High wet-modulus (HWM) and modal rayons
      High-strength, high-elongation (HSHE) rayons
Polynosic rayons
   Modified (special) rayons
      Flame-retardant rayons
      High absorbency (alloy rayons)
      Hollow rayons
      Cuprammonium rayons

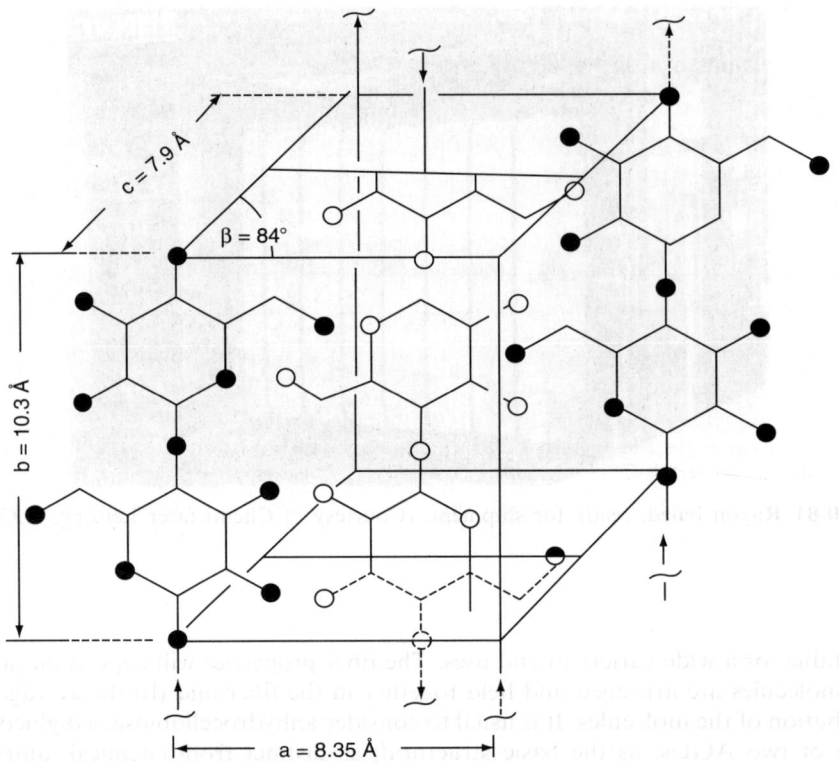

FIGURE 10.82 Structure of cellulose unit cell.

In regenerated cellulose, sometimes called cellulose hydrate, the unit cell is an allotropic modification of cellulose I, designated cellulose II. This form is general for all rayons, cellophanes, and mercerized cellulose. Other allotropic modifications are known. Proposed dimensions for cellulose I, II, III, and IV are listed in Table 10.8.

In native cellulose, the structure develops under conditions of thermodynamic equilibrium and occurs very slowly. For regenerated cellulose, however, not only must the structure be formed rapidly, but also the organization of the macromolecules by crystallization is constrained by the extent of tangling present in the solution. It was suggested by Baker [261] that the structure of cellulose derivatives could be represented by a continuous range of states of local molecular order rather than by definite polymorphic forms of cellulose. This view is supported by the observation that the x-ray diffraction pattern of rayon often reveals both cellulose II and IV components to an extent, depending on the conditions used to make the fiber. Hindeleh and Johnson [262] have described an x-ray diffraction procedure to measure crystallinity and crystallite size in cellulose fibers by which the relative proportions of cellulose II and IV in rayon can be determined.

The lattice of cellulose IV is similar to that of cellulose I. The formation of cellulose IV is generally associated with the production of high-performance rayons [2]. Both cellulose IV and cellulose I have densities in the range of 1.545–1.562 g/cm^3, compared with those of cellulose II and III, which have lower densities of 1.515–1.523 g/cm^3. A difference in the hydrolyzable fraction of these polymorphs has also been reported, 20% for cellulose II and III compared with about 10% for cellulose I and IV. Transformation into cellulose IV reportedly

TABLE 10.8
Unit Cell Dimensions of Cellulose Polymorphs

Cellulose Polymorph	a (Å)	b (Å)	c (Å)	Angle β (°)
Cellulose I	8.35	10.3	7.9	84
Cellulose II	8.1	10.3	9.1	62
Cellulose III	7.74	10.3	9.9	58
Cellulose IV	8.11	10.3	7.9	90

has no effect on the tensile strength of regenerated cellulose; however, the swelling and sorptive capacities of rayon are reduced.

Hearle [263] has given a comprehensive review of the many proposals describing ways in which the cellulose molecules might be arranged to form the fiber fine structure. Essentially, they all entail the formation of crystallites or ordered regions, together with regions described as disordered or amorphous. Although there are no regular dimensions to these regions, which can range in size over several orders of magnitude in terms of the number of molecules involved, they can be oriented in the fiber direction. This is accomplished by a combination of the orienting forces in the spinning process and stretching the filament, while it is still in a plastic deformable state as regeneration is completed. In this way, desirable properties of strength and stability can be obtained [264].

The old popular concepts of fiber fine structure are the fringed micelle, which Abitz et al. [265] suggested for gelatin and collagen in 1930, and, the fringed fibril proposed by Hearle [266] (Figure 10.83). Some attention has also been given to lamellar [267] and folded-chain structures [268]. Examination of rayon by the electron microscope has provided ample evidence for fibrillar structure in rayon. Although the fringed fibril structure shown in Figure 10.83 appears to fit best with the tendency of some rayons to fibrillate in the wet state under certain conditions; the fringed micelle structure can also account for observed properties.

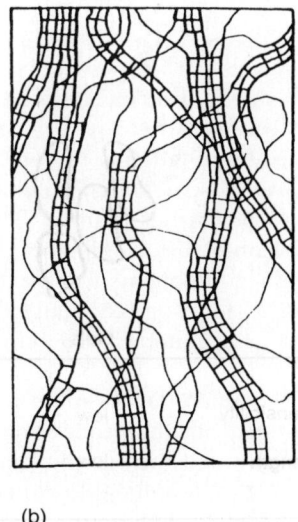

(a) (b)

FIGURE 10.83 Cellulose structure: (a) fringed micelle and (b) fringed fibril.

In 1926, Berle and Lange [269] observed structure formation during viscose ripening. It is generally accepted that ripened viscose has some structural organization before it is extruded. The cellulose xanthate molecules have begun to associate as a consequence of the increasing fraction of underivatized cellulose, which results from xanthate decomposition caused by ripening. Various types of structure formation and their relationships to fiber tensile properties are illustrated schematically in Figure 10.84. If the viscose is extruded into a spin-bath containing only acid, the viscose will rapidly coagulate and begin to form crystallites based on the order that exists in the solution as the cellulose regenerates. Without the use of regeneration-retardants or Lilienfeld-type spin-baths, only limited orientation can be achieved by stretching. As a consequence, the fiber structure is relatively weak.

When zinc and viscose modifiers are used, the chemical reactions are slowed and a different type of structure results. Using staining techniques, first done by Preston [270], the existence of two structurally different areas in the fiber cross section can be detected. They are referred to as skin and core. As the name implies, skin is normally found at and near the surface of the filament and surrounds the core. An improved staining technique was later developed by Morehead and Sisson after their studies of the skin effect in viscose rayon [186,252]. Typical examples of stained cross sections showing skin as the dark areas and core as the lighter areas are given in Figure 10.85. It is possible to vary the structure from all-skin to all-core depending on the conditions used to make the fiber. The fibers thus formed have quite different properties [271]. It has been established that skin contains numerous small crystallites, and the core has fewer but larger crystallites [174]. Because the molecular-weight distribution (chain length) is the same throughout the filament cross section, the molecules in the skin pass through many more ordered regions than in the core. Skin is stronger than core as a result of the greater number of ordered regions with which each cellulose molecule is associated. The arrangement of cellulose molecules between numerous small ordered and disordered regions produces a structure that is more extensible than the core. Not only are the smaller crystallites more able to slip past one another (because, with relatively small contact surface area, there will be fewer hydrogen bonds) but also the ties between the crystallites will be less likely to restrict their movement relative to their length.

Skin swells less than core does. Because the cellulose molecules are fixed in the crystallites at more frequent intervals, there will be less freedom for the small amorphous areas to swell.

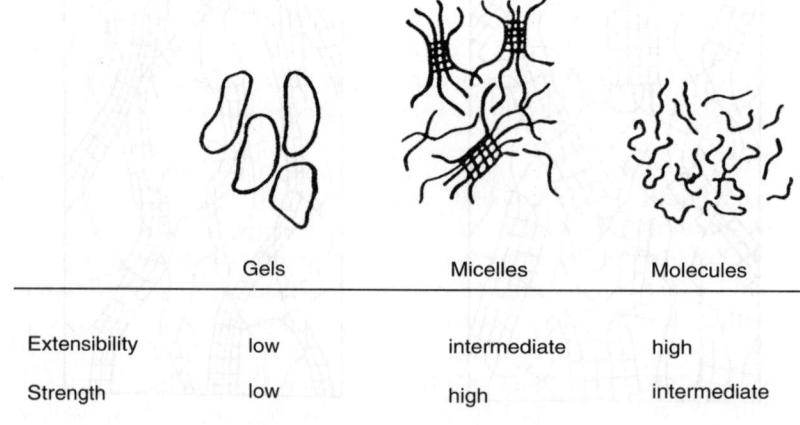

	Gels	Micelles	Molecules
Extensibility	low	intermediate	high
Strength	low	high	intermediate

FIGURE 10.84 Structure and fiber formation; influence of structural units on fiber tensile properties.

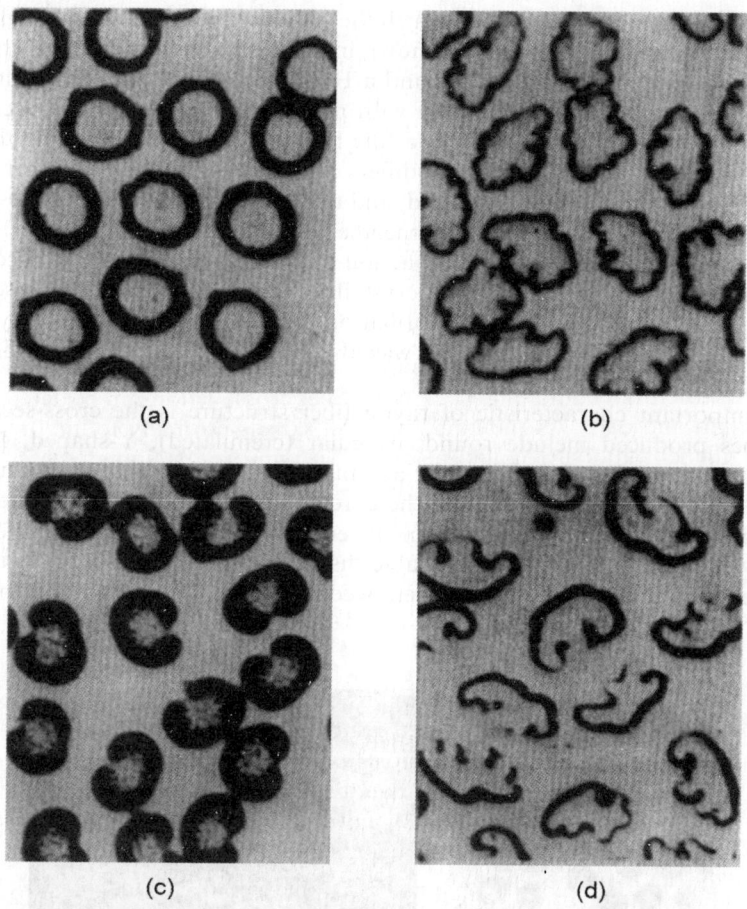

FIGURE 10.85 Stained rayon cross section showing skin and core: (a) high wet-modulus; (b) regular; (c) tire cord; and (d) crimped.

As it is in the more open amorphous regions of the structure that water will be retained on centrifuging, water retention is lower in skin than in the core. On the other hand, moisture regain is higher. This is explained by an increased number of hydroxyl groups available for bonding with water, as a result of the larger total surface area of the more numerous smaller crystallites.

The ratio of wet to oven-dry tenacity has been used to estimate the proportion of accessible hydrogen bonds in the amorphous regions [272]. This ratio also qualitatively reflects the relative amount of skin and core in the fiber structure when the fibers are prepared from the same cellulose. In the wet state, swelling disrupts the structure at the surface of the ordered regions and causes chain ends to become detached, reducing the number of ties between the crystallites. Core swells more than skin, and, because there are fewer but larger ordered regions, the loss of tenacity is much greater for core than for skin. It is the ordered regions that provide strength to the structure, and, generally, these are not disrupted by moisture absorption. When rayon fibers are worked in the wet state [273] or swollen with acid or alkali [274] and then pressed, the filament structure can be made to disintegrate into a fibrillar texture. The extent to which this occurs reflects the order that exists in the fiber structure as a consequence of the way in which the cellulose molecules are brought together in spinning. Thus, the three different fibers shown in Figure 10.86 illustrate the fibrillation in a polynosic fiber, a crimped HWM rayon, and a HWM rayon with uniform skin–core and a round cross section as a result of treating with nitric acid. Crystallization occurring in the highly oriented polynosic rayon produces a fine fibrillar structure, according to Drisch and Priou [275] and Herrbach [276]. When modifiers are used, considerable structural organization exists before the filaments are stretched, and the numerous small crystallites thus formed comprise a fringed micelle structure that becomes oriented into much thicker fibrils by the stretching action imposed in spinning. The action of swelling on the fringed micelle structure, because of the large number of separate crystallites with which each cellulose molecule is associated, is much less likely to cause fibrillation than with a fringed fibril-type structure. The fine structure of regenerated cellulose was also revealed by chemical swelling in a study reported by Dlugosz and Michie [277].

Another important characteristic of rayon fiber structure is the cross-sectional shape. Various shapes produced include round, irregular (crenulated), Y-shaped, E-shaped, U-shaped, T-shaped, and flat. It is generally accepted that the crenulated structure is formed by greater shrinkage of the skin than of the core. All-core and all-skin filaments generally retain a smooth round shape. Viscose filaments emerging from the jet are uniform cylinders that remain well separated some considerable distance into the spin-bath. Eventually, they come together and contact as they are removed from the spin-bath. The most probable

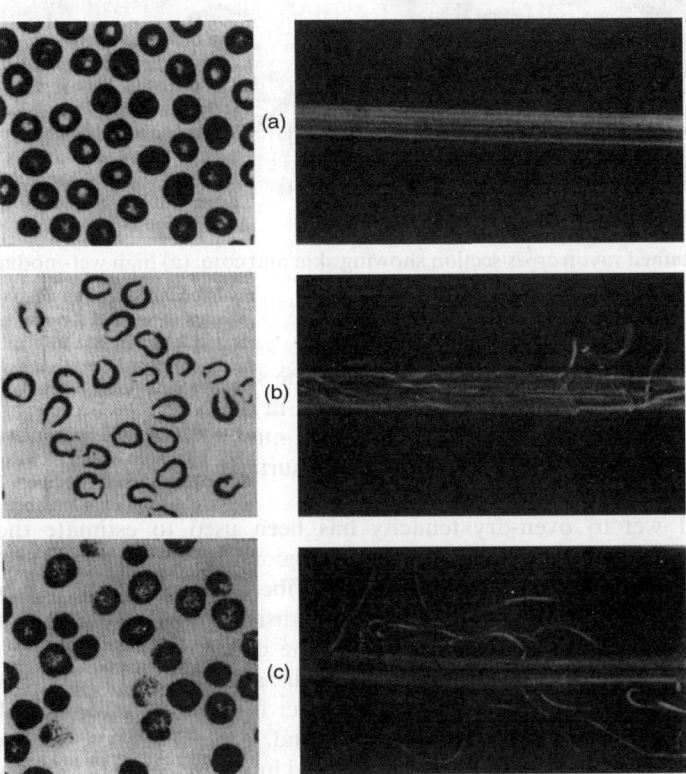

FIGURE 10.86 Rayon fibers showing fibrillation on treatment with nitric acid: (a) high wet-modulus; (b) crimped HWM; and (c) polynosic.

configuration for the filaments at this stage is a closely packed hexagonal one, and this arrangement can be retained in the fiber bundle. With incomplete regeneration, the coagulated gel filaments will deform to conform to the stresses exerted on the system. The forces that act to shape the filament have been classified as (a) osmotic pressure developing as a result of diffusion processes; (b) tensile and compressive forces from the mechanical action of spinning and stretching; and (c) collapse of the structure by dehydration and shrinkage. Close examination of freshly-sectioned fiber bundles reveals that the shape of the filaments in the hexagonal, close-packed arrangement conforms to the surface of adjacent filaments. The actual shape retained by the fiber depends on the thickness and uniformity of the skin, as it is removed form the spin-bath. It is reasonable to expect that fibers accepting as much as 150% stretch in the regeneration bath can be deformed on the godet. Deformation on the godet, where the fiber bundle can contain many thousand filaments, occurs mostly in a direction perpendicular to the fiber axis. The majority of filaments in a fiber sample will have similar cross-sectional appearance. In high-performance rayons, where a tough skin is usual, fibers from the hexagonal arrangement can have up to six lobes. Typical examples of fiber cross sections, flat and three-lobed, originating from deformation are shown in Figure 10.87. Because the skin forms while the filaments are still in a swollen state, the fiber must shrink as water is removed. This can result in several effects. First, there will be an orientation of the crystallites parallel to the fiber radius due to the lateral pressures developing as a result of the shrinkage. Then, depending on the skin's toughness, thickness, and uniformity, the surface will become lobed or crenulated as shrinkage occurs. In the absence of skin, shrinkage is uniform and a round shape is retained.

The round cross-sectional shape of some HWM fibers suggests that there is sufficient free space in the completely regenerated core to accommodate the shrinkage that must occur in

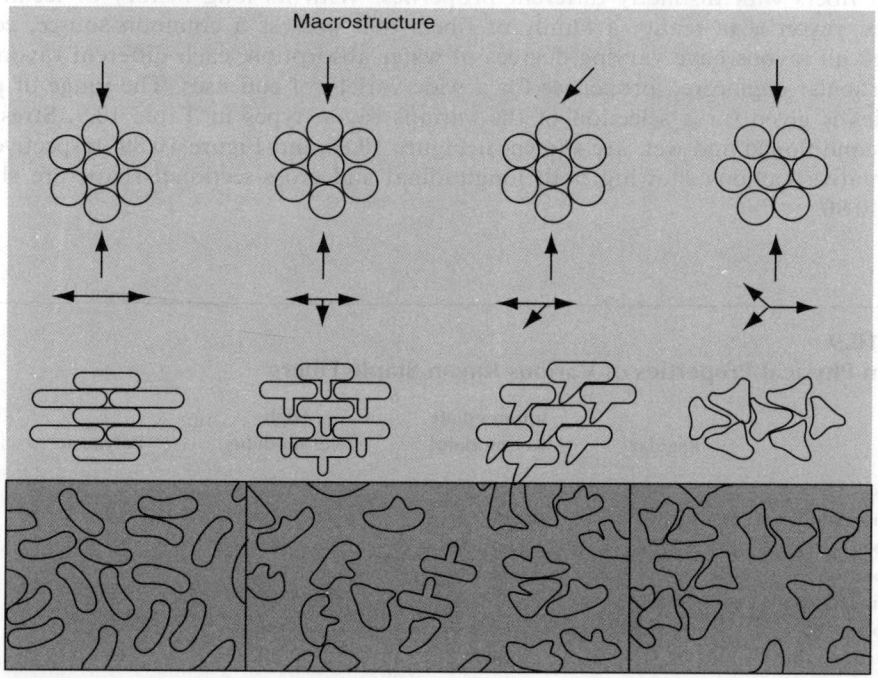

FIGURE 10.87 Flat and three-lobed rayon cross sections.

drying. Collapse of the structure while the core is only partly regenerated is more difficult because, in such a state, the core will resist compression. It is this property that causes the skin to rupture in the hot regeneration bath, enabling the production of crimped fibers. The early types of crimped rayon exhibited stained cross sections that were highly crenulated, with skin on about two thirds of the section, and little or no skin on the remaining portion. The crenulated structure was characteristic of regular rayon, and these crimped fibers were indeed relatively weak. The HWM rayons such as Prima or Avrill III are lobed rather than crenulated. Absence of skin on the outer periphery of lobes indicates that rupture occurred during stretching, because the hot regeneration bath did not contain the zinc ions necessary for skin formation. These fibers crimp due to differential shrinkage of the skin and core.

Examination of green or never-dried viscose filaments sampled at various stages of spinning, as reported by Morehead [278], reveals the formation of a network-type structure in the core resembling microfibrils. As formation of the filament proceeds, water is removed by the dehydrating action of the spin-bath. Together with the strain caused by stretching the filament, the network structure collapses. Viewed in the electron microscope after drying, the cross section appears as a solid matrix containing voids of round to elongated shapes where stretching the filament has reduced the voids to slits. The microfibrillar texture is again observed when the sample is swollen in dilute sodium hydroxide. Microfibrils can be obtained by mechanical action using a Waring blender to disintegrate rayon or by the action of ultrasonic vibrations. The extensive work of Morehead [186] and of Grobe [279] amply illustrate the characteristics of rayon cross sections as observed under the electron microscope.

10.4.8 Rayon Properties and Uses

Relatively minor changes in the chemical and physical parameters of the spinning system can result in fibers with distinctly different properties. With its long history of technological advances, rayon is in reality a family of fibers that possess a common source, cellulose. Although all rayons have varying degrees of water absorption, each different rayon has its own particular engineered properties for a wide variety of end uses. The range of physical properties is given for a selection of the various rayon types in Table 10.9. Stress–strain curves, conditioned and wet, are shown in Figure 10.88 and Figure 10.89, respectively, and representative sections showing both longitudinal and cross-sectional shape are shown in Figure 10.90.

TABLE 10.9
Range in Physical Properties of Various Rayon Staple Fibers

Property	Regular	Intermediate wet-modulus	High wet-modulus	Polynosic	Modal
Tenacity (g/den)					
Conditioned fiber	1.2–3.2	3.5–4.5	3.5–5.0	3.5–5.5	3.5–4.5
Wet fiber	0.7–1.8	2.0–3.0	2.2–3.2	2.7–4.0	2.5–3.5
Elongation (%)					
Conditioned fiber	15–30	12–18	10–15	5.6–10	8–12
Wet fiber	20–40	15–25	12–18	7.0–12	9–15
Water retention (%)	90–110	60–75	60–80	60–70	60–80

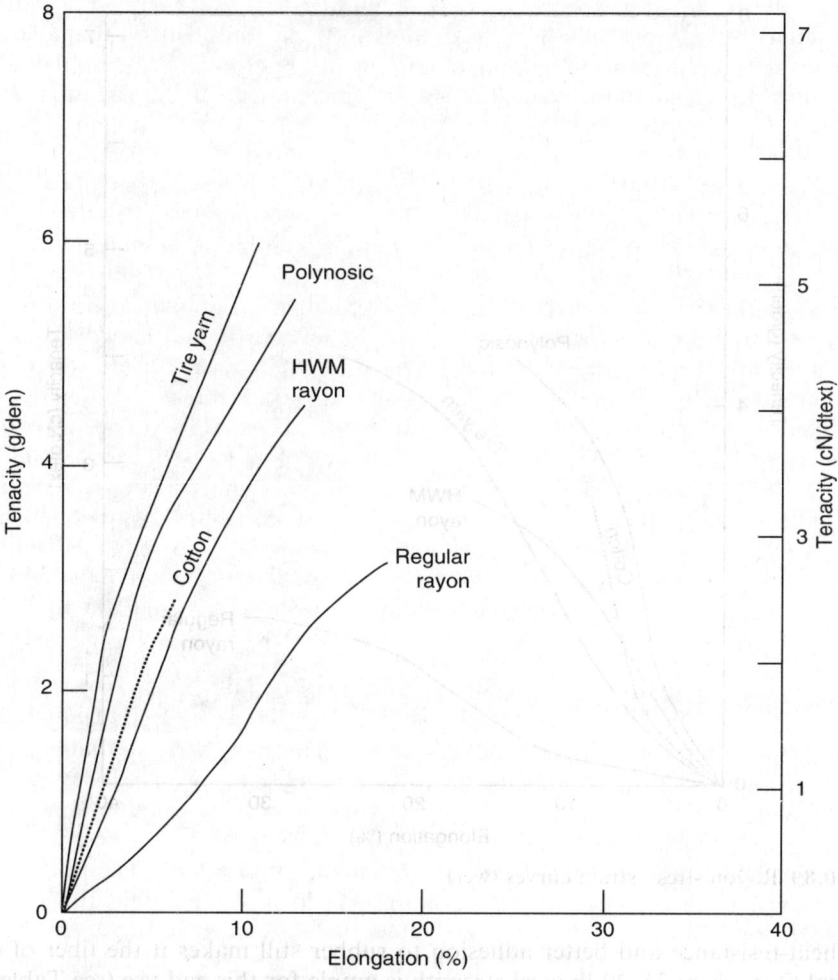

FIGURE 10.88 Regular rayon stress–strain curves after conditioning.

10.4.8.1 Industrial Yarns

With the discovery by Cox [124] of viscose modifiers to retard regeneration and the synergistic action of zinc in the spin-bath, high-strength continuous-filament rayon became a reality by 1950. This development and other improvements utilizing mixed modifiers, notably alkyl amines and polyglycols, by Mitchell et al. [126] almost doubled the strength of rayon, as it had been known. These developments paved the way for the complete replacement of cotton in the industrial field for reinforcement of rubber articles such as tires, conveyor belts, and hoses. Improvements in tenacity were accompanied by significant increases in fatigue-resistance, abrasion-resistance, and work-to-rupture.

Yarns for tire cords are mostly spun 1650 den/1100 filaments twisted into two-ply cords, although other weight yarns are used for various purposes. Since 1965, rayon has gradually lost its prominence in this field, first to nylon and then to polyester. However, rayon's

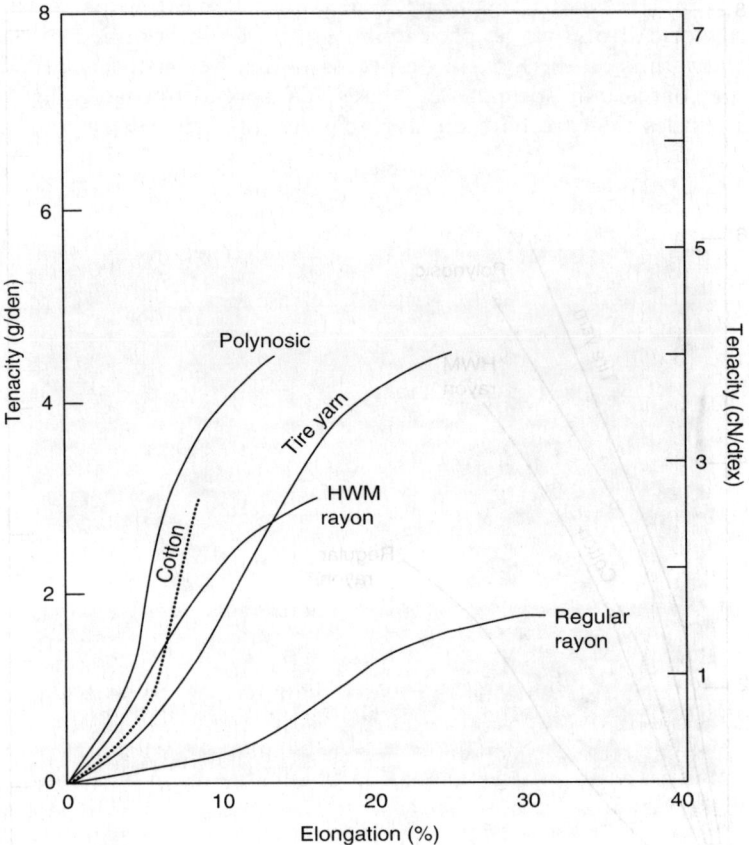

FIGURE 10.89 Rayon stress–strain curves (wet).

superior heat-resistance and better adhesion to rubber still makes it the fiber of choice for some radial tires where 36–40-lb cord strength is ample for this end use (see Table 10.9).

Continuous-filament industrial rayons are produced from high-purity wood pulps. They are mostly highly oriented, all-skin fibers, with round, or nearly round, cross sections. Typical stress–strain curves are shown in Figure 10.88 and Figure 10.89. Individual filaments have tenacities in the range of 4.0–5.5 g/den (3.6–5.0 cN/dtex) and elongations of 12–20%.

Other uses for continuous-filament industrial yarns include reinforcement cords for plastics, as a precursor for carbon and ceramic filaments, and for tents, awnings, strappings, etc.

10.4.8.2 Textile Rayons

10.4.8.2.1 Continuous Filament

Continuous-filament (regular) rayon is produced by either the cuprammonium or viscose process. Both are supplied as yarn having a high sheen resembling silk. End uses include those applications where durability and dimensional stability are not particularly important. Cuprammonium fibers can be spun in finer deniers than viscose fibers, and they find a market for ladies' shawls, scarves, blouses, coat linings, etc. The amount of cuprammonium fiber produced is quite small relative to viscose rayon.

Continuous-filament textile rayon, produced from viscose, while representing only a fraction of the world's total rayon production, is still an important article of commerce. Although its strength is on the low side, especially when wet, its other aesthetic properties make it unique for certain specialty end uses. Of special appeal is lightweight, sheer, richly colored textiles that are both comfortable and attractive. Coat linings are a major market.

FIGURE 10.90 SEM micrographs of various rayons showing cross section and longitudinal shape: (a) regular; (b) crimped.

Continued

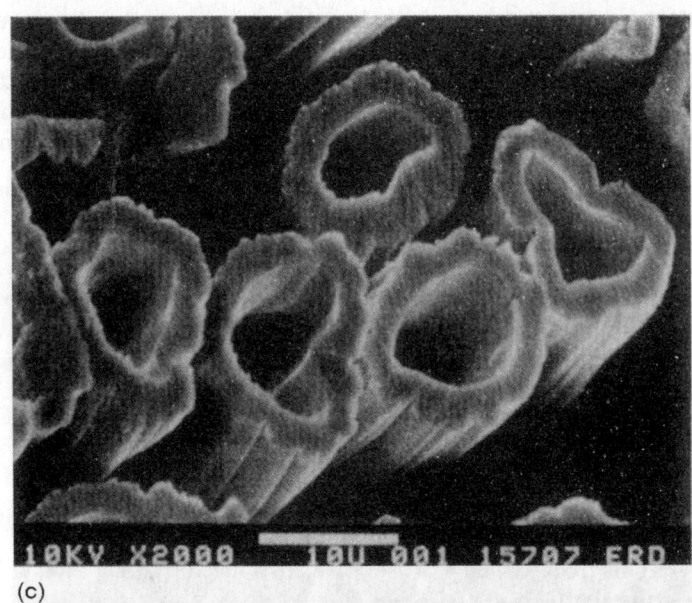

(c)

(d)

FIGURE 10.90 (Continued) SEM micrographs of various rayons showing cross section and longitudinal shape: (c) hollow; and (d) tire cord.

10.4.8.2.2 Staple Fiber

10.4.8.2.2.1 Regular Rayon

Regular viscose rayon staple supplied in all the common deniers, staple lengths, lusters, and even dope-dyed is also an important fiber for the textile industry. The most widely used fabrics made from regular rayon are the lightweight challis for soft, drapable garments and heavier-weight linens. Blends with cotton, polyester, and flax with after treatments to give durable-press qualities often overcome the major inherent defects in properties of this fiber, namely, poor dimensional stability and rather limp, mushy hand.

Because of regular rayon's high water absorption (100%) and lateral swelling and longitudinal shrinkage when wet, unless it is adequately finished with cross-linking agents, fabrics made from regular rayon do not have the dimensional stability required for repeated washing. Also, due to the high wet-elongation properties of the fiber, untreated fabrics tend to stretch; accordingly, profit-oriented fabric finishers can gain 8–10% in fabric yardage by stretching during the finishing operations. Such fabrics have the lean hungry look that has given rayons, in general, an undeserved bad reputation. To avoid this, regular rayon fabrics must be dyed and finished in a relaxed state and after treated with a permanent finishing agent that will not disappear after a few washes.

10.4.8.2.2.2   High Wet-Modulus Rayons

Two fibers with exceptionally high wet-modulus were developed years ago, but the importance of this property to dimensional stability in fabrics was not recognized at the time. Surprisingly, one of the stronger rayon fibers with conditioned tenacity of 6–7 g/den (5.4–6.3 cN/dtex) and wet-modulus over 1 g/den (0.9 cN/dtex) was Lilienfeld rayon [121], developed in 1928. The other was not a viscose rayon but saponified acetate that had similar physical properties. This fiber, called Fortisan, was produced by Celanese during World War II for use as parachute cords and for other industrial uses where strength was important. Unfortunately, both these fibers suffered from a low work-to-rupture, were brittle, and had low knot- and loop-strength. Accordingly, neither could be considered for the textile industry.

10.4.8.2.2.3   Polynosic Rayon

In 1952, in Japan, Tachikawa [128] patented a spinning system for producing strong, HWM rayon. Nonmodified viscoses with a high viscosity (DP) and high degree of xanthation were spun into a simple spin-bath containing small amounts of acid and low salt concentrations maintained at lower temperatures than those used for normal rayon spinning. The result was that very high stretches could be applied (up to 300%), and a fiber with a high degree of orientation could be produced.

The fibers that were produced were given wide recognition, especially in France where the name "polynosics" originated. They had a unique fibrillar structure, high strength wet and dry, low elongation (8–11%), relatively low water retention, and very high wet-modulus of 1 g/den (0.9 cN/dtex). Their resistance to caustic soda was exceptional for rayon; in fabrics, they had the firm, crisp hand of cotton.

Polynosic fibers were produced for a time in several countries, including France, the United States, and Japan, but it soon became evident that they were too brittle for the textile market. Research workers soon began to modify the viscose and spinning conditions to produce several versions of hybrid fibers.

One of the more interesting modifications was the use of small amounts of formaldehyde in the spin-bath. The research laboratories of Courtaulds NA [219] developed a fiber called W-63 with unusually high tenacity (7–10 g/den (6–9 cN/dtex)) and modulus. A yarn called Tenex was produced for a while by Courtaulds for industrial uses. However, this fiber also suffered from brittleness, and the problems associated with spin-bath recovery (formaldehyde reaction products) were not commercially solvable.

10.4.8.2.2.4   Modifier Types

About the same time when the polynosic fibers were developed in Japan, the producers of high-strength tire yarn successfully adapted the process to produce staple fiber. These fibers had good dry and wet strength and wet-moduli sufficient to overcome the stretching and shape loss that can occur with low-modulus regular rayon. There are many forms of HWM fibers, but all are much stronger than regular rayon (both conditioned and wet) and have lower elongations; thus the energy required to stretch the fabric out of shape is much higher than that needed for similar fabrics made from regular rayon. Lower water retention (70 versus 100%) of HWM fibers results in less swelling and therefore less initial and progressive shrinkage. It was found by Lund and Waters [280] that a wet fiber tenacity of about 0.5 g/den (0.45 cN/dtex) is necessary to prevent progressive shrinkage of cellulosic fabrics. This represents the approximate wet-modulus of cotton, whose dimensional stability is unquestioned.

The characteristics and properties of HWM rayon have been described by Goldenberg [281]. The HWM fiber production is increasing worldwide, mainly by conversion of regular rayon production lines but in some cases by new plants or spinning lines built specifically for the purpose.

The HWM staple fibers have essentially all of the best attributes of regular rayon except for a few important differences. They swell less in water, are somewhat stiffer due to higher cellulose DP (IV) and orientation, and are almost twice as strong and resist dimensional change. Fabrics made from HWM fibers can be dyed and later finished (cross-linked) by much the same techniques as those used for cotton fabrics. In 50–50 blends of polyester and HWM rayon, fabrics can be made that are virtually indistinguishable from some cotton counterparts [282].

Another important advantage of true HWM fibers is their better resistance to caustic soda. Although garments made of rayon are not normally subjected to exposure to strong caustic solutions, fabric preparation, such as dyeing, bleaching, etc., sometimes requires caustic treatment. In HWM blends with cotton, fabrics are sometimes mercerized or treated with 12–16% sodium hydroxide, or they are Kier-boiled in dilute caustic. Fabrics made from blends of HWM fiber and polyester are usually treated with 3–5% NaOH to permit easy dyeing of the polyester component. A simple test, developed in ITT Rayonier laboratories, determines to a large degree the resistance of fibers to the adverse reaction of caustic soda. The solubility in 6.5% NaOH at 20°C ($S_{6.5}$) is an indication of how the internal structure of rayon will resist such damage. For example, regular rayon will lose 25% or more in weight when subjected to this test. The better grades of HWM fiber will lose 12% or less, indicating more integrity in the fiber. In the $S_{6.5}$ test, cotton loses only 1 or 2%. The loss is attributed to short-chain cellulose, which adds little to the physical properties except bulk. Research by Szego [283], of Chatillon, has studied the behavior of HWM fibers after treatment with 5–23.5% NaOH.

While HWM fibers demand and get a premium (they require better purity wood pulp), the added expenses of modifiers and production rates are much lower. As a result, in some instances due to economic pressures, compromises in quality have been made. Unfortunately, some rayon fibers are sold as HWM that have actual wet-moduli somewhat lower than the 0.5 g/den (shown by Szego) to be necessary for dimensional stability. Such fibers also have reduced resistance to caustic soda, as is evident by higher $S_{6.5}$ values.

In general, fabrics made from HWM fibers can be prepared, dyed, and finished in the same manner as those made from regular rayon with a few exceptions. For durable-press effects, less resin or cross-linking agents may be used than for regular rayon but slightly more than required for 100% cotton fabrics.

10.4.8.2.2.5 Modal

An attempt has been made by European rayon producers to get the International Organization for Standardization (ISO) and others to adopt a new generic name for those fibers that are shown to be much superior to regular, low-modulus rayons. This is intended to remove

Regenerated Cellulose Fibers

the stigma associated with the name rayon and the poor-quality fabrics that have contributed to this stigma.

Several proposals have been made, all specifying a minimum wet-modulus around 0.5 g/den (0.45 cN/dtex) for 1.5-denier fiber. Human-made cellulosic fibers that meet this criterion could then legally be called Modal.

10.4.8.2.3 Variants to Compete with Cotton

A number of new cotton like rayons shown in Figure 10.91 was introduced to the textile market during the 1970s. One of these developments in HWM technology was the result of research at ITT Rayonier Research Laboratories in Whippany, New Jersey [284–287]. A chemically crimped fiber called Prima, with an unusual lobed cross section, HWM properties, low $S_{6.5}$, and more than adequate strength gives more bulk and cover to fabrics. It can be blended with cotton or polyester or in heavier denier with wool. It is also used 100% in the tube-sock hosiery business in direct competition with cotton [288].

Courtaulds' approach to the cotton-substitute rayon was the development of a hollow fiber called Viloft [289]. It has a hollow core (like cotton), which gives opacity. Because Viloft is less dense than regular rayon fibers of the same diameter, it gives added bulk to fabrics. It is promoted for use in knit underwear and towels. This fiber more closely resembles regular rayon in dry tenacity but is weaker in the wet state.

10.4.8.2.4 Specialty Rayons
10.4.8.2.4.1 Flame-Retardant Fiber

As a result of the 1953 regulation of U.S. Department of Commerce on the flammability of clothing textiles, a number of FR rayons were developed. Some resulted from research that had been done years before but had not been economical to produce. The regulations, however, altered the economics sufficiently to encourage commercialization. The rationale

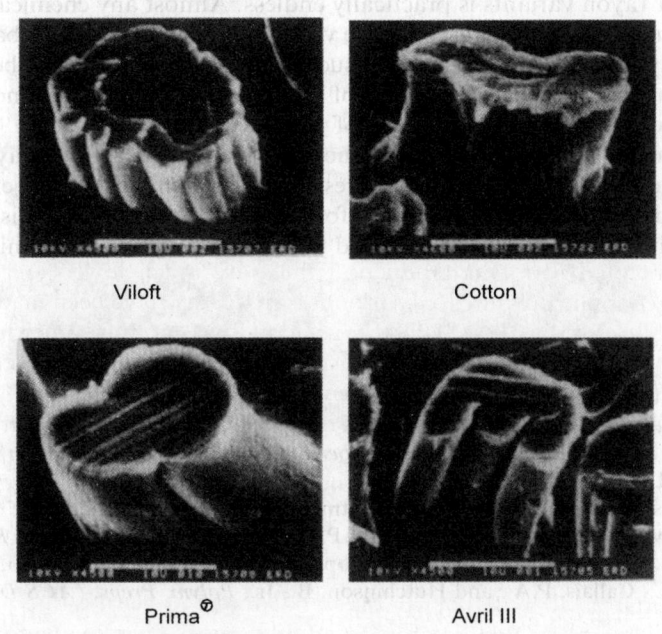

FIGURE 10.91 SEM micrographs of cotton like rayons.

for FR rayon has been discussed by Portnoy and Daul [290]. Flame-retardance of rayon is relatively simple to obtain, providing the correct FR chemical is selected for addition to the viscose. Examples of additives are alkyl, aryl, and halogenated alkyl or aryl phosphates, phosphates, phosphazenes, phosphonates, and polyphosphonates

When 18–25% of the chemical, usually in finely divided powder form or polymeric liquid, is added to the viscose (on cellulose weight) the result is a fiber that, when woven into fabric, can pass the government's standard. The FR rayons have an advantage over FR cotton as the latter must be after treated and most of the flame-retardant remains on the surface of the fibers. FR rayons on the other hand have the agent distributed uniformly throughout the interior of the fiber.

10.4.8.2.4.2 Superabsorbent Rayon

Although regular rayon, with water retention as much as 100% of its weight, is the most absorbent of the human-made fibers, variants have been produced that far exceed this capacity [291]. The demand for superabsorbent fibers arises from the growing use of rayon in surgical and medical supplies, sanitary napkins, disposable diapers, and other nonwovens [292]. The rayon industry has responded with several versions of modified fibers.

One approach is to include water-holding polymers such as sodium polyacrylate or sodium carboxymethylcellulose in the viscose prior to spinning. The result is a fiber that will soak up and retain up to as much as 150–200% of its weight in water. Such absorbent fibers are known as alloy fibers and were produced by American Enka (Absorbit) [293] and Avtex (Maxisorb) [294], in the United States. Courtaulds in England has taken a different route by actually modifying the spinning conditions to produce a superinflated, all-cellulose, hollow fiber [295]. Indications are that a gas producing agent, probably sodium carbonate, is added to the viscose, which when acidified liberates CO_2. The voids caused by the gas evolution inflate the fiber and increase its water-holding capacity.

10.4.8.2.4.3 Others

The possibility for rayon variants is practically endless. Almost any chemical compound that can resist short-term exposure to alkali (in the viscose) or acid (in the spin-bath) is eligible for inclusion in the rayon fiber. The demand for such variants, however, must be large enough to justify commercial production. Short runs of a specialty fiber are just not economical in plants, which produce hundreds of millions of pounds yearly.

Some of the variants that have had some success have been spun-dyed rayons in which dye pigments are included; they have excellent resistance to laundering and exposure to light. Another is an acid, dye-receptive rayon [296] for blending with wool. In this case, proteins or polymers containing NH_2 groups are included in the viscose prior to spinning.

REFERENCE

1. *Fiber Organon*, July, 1999, p. 107.
2. Kulshreshtha, A.K., *J. Text. Inst.*, 1979, *70*, 13.
3. Łaszkiewicz, B., *Manufacture of Cellulose Fibers without the Use of Carbon Disulfide*, ACGM LODART, SA, Łódź, Poland, 1997.
4. http://www.fibersource.com/f-tutor/cellulose.htm#chemistry as viewed on June 6, 2004.
5. Chanzy, H., Nawrot, S., Peguy, A., and Smith, P., *J. Polym. Sci., Polym. Phys. Ed.*, 1982, *20*, 1909.
6. McCormick, C.L., U.S. Patent 4,278,790, to Hopkins Agricultural Chemical Co., July 14, 1981.
7. McCormick, C.L., Callais, P.A., and Hutchinson, B., Jr., *Polym. Prepr. (ACS Div. Polym. Chem.)*, 1983, *24(2)*, 271.
8. McCormick, C.L., Callais, P.A., and Hutchinson, B., Jr., *Macromolecules*, 1985, *18*, 2394.

9. Terbojevich, M., Cosani, A., Conio, G., Ciferri, A., and Bianchi, E., *Macromolecules*, 1985, *18*, 640.
10. Patel, D.L. and Gilbert, R.D., *J. Polym. Sci., Polym. Phys. Ed.*, 1981, *19*, 1231.
11. Patel, D.L. and Gilbert, R.D., *J. Polym. Sci., Polym. Phys. Ed.*, 1981, *19*, 1449.
12. Roche, E.J., O'brien, J.P., and Allen, S.R., *Polym. Commun.*, 1986, *27*, 138.
13. Dubose, *J. Bull. Rouen*, 1905, *33*, 318.
14. Bechtold, H. and Werntz, J., U.S. Patent 2,737,459, 1956.
15. Hattori, M., Koga, T., Shimaya, Y., and Saito, M., *Polym. J.*, 1998, *30*, 43.
16. Hattori, M., Shimaya, Y., and Saito, M., *Polym. J.*, 1998, *30*, 49.
17. Hudson, S.M. and Cuculo, J.A., U.S. Patent 4,367,191, to Research Corporation, January 4, 1983.
18. Hudson, S.M. and Cuculo, J.A., *J. Polym. Sci., Polym. Chem. Ed.*, 1982, *20*, 499.
19. De Groot, A.W., Carroll, F.I., and Cuculo, J.A., *J. Polym. Sci., Polym. Chem. Ed.*, 1986, *24*, 673.
20. Liu, C.K., Cuculo, J.A., and Smith, C.B., *J. Polym. Sci., Part B: Polym. Phys.*, 1990, *28*, 449.
21. Aminuddin, N.M., *Studies on the Dissolution and Fiber Formation of Cellulose in Ammonia/Ammonium Thiocyanate*, M.Sc. thesis, North Carolina State University, NC, 1993.
22. Cuculo, J.A., Smith, C.B., and Sangwatanaroj, U., *J. Polym. Sci., Part A: Polym. Chem.*, 1994, *32*, 229.
23. Cuculo, J.A., Smith, C.B., Sangwatanaroj, U., Stejskal, E.O., and Sankar, S.S., *J. Polym. Sci., Part A: Polym. Chem.*, 1994, *32*, 241.
24. Kamide, K. and Okajima, K., U.S. Patent 4,634,470, to Asahi Kasei Kogyo Kabushiki Kaisha, January 6, 1987.
25. Yamashiki, T., Matsui, T., Saitoh, M., Okajima, K., and Kamide, K., *Br. Polym. J.*, 1990, *22*, 73.
26. Yamashiki, T., Matsui, T., Saitoh, M., Okajima, K., and Kamide, K., *Br. Polym. J.*, 1990, *22*, 121.
27. Yamashiki, T., Matsui, T., Saitoh, M., Matsuda, Y., Okajima, K., Kamide, K., and Sawada, T., *Br. Polym. J.*, 1990, *22*, 201.
28. Yamashiki, T., Saitoh, M., Yasuda, K., Okajima, K., and Kamide, K., *Cellul. Chem. Technol.*, 1990, *24*, 237.
29. Wilkes, J.S., *Green Chem.*, 2002, *4(2)*, 73.
30. Earle, M.J. and Seddon, K.R., *Pure Appl. Chem.*, 2000, *72(7)*, 39.
31. Endres, F., *Zeitschrift fur Physikalische Chemie*, 2004, *218(2)*, 255.
32. Marsh, K.N., Boxall, J.A., and Lichtenthaler, R., *Fluid Phase Equilib.*, 2004, *219(1)*, 93.
33. Huddleston, J.G., Visser, A.E., Reichert, W.M., Willauer, H.D., Broker, G.A., and Rogers, R.D., *Green Chem.*, 2001, *3(4)*, 156.
34. Swatloski, R.P., Spear, S.K., Holbrey, J.D., and Rogers, R.D., *J. Am. Chem. Soc.*, 2002, *124*, 4974.
35. Heinze, T., Dicke, R., Kischella, A., Kull, A.H., Klohr, E.A., and Koch, W., *Makromol. Chem. Phys.*, 2000, *201*, 627.
36. Northolt, M.G., Boerstoel, H., Maatman, H., Huisman, R., Veurink, J., and Elzerman, H., *Polymer*, 2001, *42*, 8249.
37. Fink, H.P., Weigel, P., Purz, H.J., and Ganster, J., *Prog. Polym. Sci.*, 2001, *26*, 1473.
38. Graenacher, G. and Sallmann, R., U.S. Patent 2,179,181, to Society of Chemical Industry, Basel.
39. Johnson, D.J., U.S. Patent 3,447,939, to Eastman Kodak.
40. McCorsely, C.C. and Varga, J.K., U.S. Patent 4,142,913, to Akzona Inc.
41. Franks, N.E. and Varga, J.K., U.S. Patent 4,196,282, to Akzona Inc.
42. Albert, H.E. and Haines, P.G., U.S. Patent 3,333,000, to Pennsalt Chemical Corporation.
43. Technical bulletin for NMMO provide by Huntsman Corporation as viewed on July 3, 2004 at http://www.huntsman.com/performance_chemicals/Media/NMMO.pdf
44. Meister, G. and Wechsler, M., *Biodegradation*, 1998, *9(2)*, 91.
45. Navard, P. and Haudin, J.M., *J. Therm. Anal.*, 1981, *22(1)*, 107.
46. Maia, E., Peguy, A., and Perez, S., *Acta Crystallogr., Sect. B: Struct. Crystallogr. Cryst. Chem.*, 1981, *B37(10)*, 1858.
47. Maia, E. and Perez, S., *Acta Crystallogr., Sect. B: Struct. Crystallogr. Cryst. Chem.*, 1982, *B38(3)*, 849.
48. Wachsman, U. and Diamantoglou, M., *Das Papier*, 1997, *51(12)*, 660.

49. Rosenau, T., Potthast, A., Sixta, H., and Kosma, P., *Prog. Polym. Sci.*, 2001, *26*, 1763.
50. Buijtenhuijs, F.A., Abbas, M., and Witteveen, A.J., *Das Papier*, 1986, *40(12)*, 615.
51. Chanzy, H. and Peguy, A., *J. Polym. Sci., Polym. Phys. Ed.*, 1980, *18*, 1137.
52. Diener, A. and Raouzeos, G., *Chem. Fibers Int.*, 1999, *49(1)*, 40.
53. Diener, A. and Raouzeos, G., *Chem. Fibers Int.*, 2001, *51(4)*, 260.
54. Burchard, W., *Papier*, 1994, *48*, 755.
55. Schurz, J., *Cellul. Chem. Technol.*, 1977, *11*, 3.
56. Röder, T. and Morgenstern, B., *Polymer*, 1999, *40*, 4143.
57. Schulz, L., Seger, B., and Burchard, W., *Macromol. Chem. Phys.*, 2000, *201*, 2008.
58. Navard, P. and Haudin, J.P., *Br. Polym. J.*, 1980, *12(4)*, 174.
59. Kim, S.O., Shin, W.J., Cho, H., Kim, B.C., and Chung, I.J., *Polymer*, 1999, *40(23)*, 6443.
60. Rosenau, T. et al., *Cellulose*, 2002, *9*, 283.
61. Rosenau, T., Poothaust, A., Hofinger, A., Sixta, H., and Kosma, P., *Holzforschung*, 2002, *56*, 199.
62. Polonovski, M., *Bull. Soc. Chim.*, 1927, *41*, 1190.
63. Mortimer, S.A. and Peguy, A.A., *Cellul. Chem. Technol.*, 1996, *30*, 117.
64. Mortimer, S.A., Peguy, A A., and Ball, R.C., *Cellul. Chem. Technol.*, 1996, 30, 251.
65. Mortimer, S.A. and Peguy, A.A., *J. Appl. Polym. Sci.*, 1996, *60*, 1747.
66. Coulsey, H.A. and Smith, S.B., *Lenzinger Berichte*, 1996, *75*, 51.
67. Berger, W., Keck, M., and Philipp, B., *Cellul. Chem. Technol.*, 1988, *22*, 387.
68. Loubinoux, D. and Chaunis, S., *Text. Res. J.*, 1987, *57*, 61.
69. Navard, P. and Haudin, J.M., *Polym. Proc. Eng.*, 1985, *3(3)*, 291.
70. Fink, H.P., Purz, H.J., and Weigel, P., *Das Papier*, 1997, *51*, 643.
71. Breier, R., *Lenzinger Berichte*, 1997, *76*, 108.
72. Chanzy, H., Paillet, M., and Hagege, R., *Polymer*, 1990, *31*, 400.
73. Firgo, H., Eibl, M., and Meister, G., *Lenzinger Berichte*, 1994, *74(9)*, 81.
74. Boerstoel, H., *Liquid Crystalline Solutions of Cellulose in Phosphoric Acid for Preparing Cellulose Yarns*, Ph.D. dissertation, University of Groningen, 1998.
75. Picken, S.J., *Macromolecules*, 1989, *22*, 1766.
76. Boerstoel, H., Maatman, H., Picken, S.J., Remmers, R., and Westerink, J.B., *Polymer*, 2001, *42*, 7363.
77. Hill, I.W. and Jacobson, R.A., U.S. Patent 2134825, to E.I. du Pont de Nemours & Company.
78. Ekman, K., Eklund, V., Fors, J., Huttenen, J.I., Selin, J.F., and Turunen, O.T., *Cellulose Structure, Modification and Hydrolysis*, John Wiley & Sons, New York.
79. Petropavlovskii, G.A. and Zimina, T.R., *Zhurnal Prikladnoi Khimii (Sankt-Peterburg)*, 1994, *67(5)*, 705.
80. Fink, H.P., Gensrich, J., Rihm, R., and Hanemann, O., Abstracts of Papers, 225th ACS National Meeting, New Orleans, LA, March 23–27, 2003.
81. Sobczak, M., *Manufacture of Cellulose Carbamate*, Thesis, Man-Made Politechnika Lodzka, Poland, 1988.
82. Struszczyk, H., *Wlokna Chemiczne*, 1988, *14(4)*, 365.
83. Struszczyk, H., *Wlokna Chemiczne*, 1989, *15(1)*, 7.
84. Mikolajczyk, W., Wawro, D., and Struszczyk, H., *Fibres Text. East. Eur.*, 1998, *6(2)*, 53.
85. Struszczyk, H. et al., Polish Patent 163049, to Instytut Włókien Chemicznych, Łódz, Poland.
86. Starostka, P. et al., Polish Patent 165916 B1, to Instytut Włókien Chemicznych, Łódz, Poland.
87. Eklund, V. et al., Finish Patent 67562, to Neste Oy.
88. Selin, J.F. et al., Finish Patent 66624, to Neste Oy.
89. Voges, M., Brück, M., Fink, H.P., and Gensrich, J., The Carbacell process: an environmentally friendly alternative for cellulosic man-made fibre production, Proceedings of the Akzo Nobel Man-Made Fibre Seminar, Stenungsund, June 13–15, 2000.
90. Fraunhofer IAP Annual Report 2000.
91. Schweizer, E., *J. Prakt. Chem.*, 1857, *72*, 109.
92. Cuculo, J.A., Aminudin, N., and Frey, M.W., *Solvent Spun Cellulose Fibers, Structure Formation in Polymeric Fibers*, Salem, D.R., Ed., Hanser Publishers, Munich, 2001, pp. 296–328.
93. Kamide, K. and Nishiyama, K., *Cuprammonium Processes, Regenerated Cellulose Fibres*, Woodings, C., Ed., Woodhead Publishing, Cambridge, U.K., 2001.

94. *Kirk–Othmer Encyclopedia of Chemical Technology*, 4th ed., Vol. 10, John Wiley & Sons, New York, 1993, p. 696.
95. Browning, B.L., Sell, O.L., and Abel, W., *Tappi J.*, 1954, *37*, 273.
96. Hattori, K., Cuculo, J.A., and Hudson, S.M., *J. Polym. Sci., Part A: Polym. Chem.*, 2002, *40(4)*, 601.
97. Cuculo, J.A. and Hattori, K., U.S. Patent 6,827,773, to North Carolina State University, December 7, 2004.
98. Hattori, K., Yoshida, T., and Cuculo, J.A., New solvents for cellulose using amine/salt system, Abstract of Paper, 224th ACS National Meeting, Boston, MA, August 18–22, 2002.
99. Hattori, K., Yoshida, T., and Cuculo, J.A., Dissolution and solution properties of cellulose in the amine/thiocyanate salt system, Abstracts of Papers, 227th ACS National Meeting, Anaheim, CA, United States, March 28–April 1, 2004.
100. Trogus, C. and Hess, K., *Z. Phys. Chem. B*, 1931, *14*, 387.
101. Frey, M.W. and Song, H., Cellulose fibers formed by electrospinning from solution, Abstracts of Papers, 225th ACS National Meeting, New Orleans, LA, March 23–27, 2003.
102. Frey, M.W., Joo, Y., and Kim, C-W., New solvents for cellulose electrospinning and preliminary electrospinning results, Abstracts of Papers, 226th ACS National Meeting, New York, September 7–11, 2003.
103. Nazarov, V.N. and Chirkina, G.D., *Trudy Instituta—Moskovskii Khimiko-Tekhnologicheskii Institut Imeni D.I. Mendeleeva*, 1968, *57*, 33–36.
104. Okuda, F., Jpn. Patent 09208530, to Idemitsu Kosan Co., Ltd., 1997.
105. Oh, S.Y., Yoo, D.I., Shin, Y., Lee, W.S., and Jo, S.M., *Fibers Polym.*, 2002, *3(1)*, 1.
106. Oh, S.Y., Yoo, D.I., Shin, Y., and Seoc, G., *Carbohydr. Res.*, 2005, *340*, 417.
107. Hard, A.H., *The Romance of Rayon*, Whittaker and Robinson, The Albert Press, Manchester, U.K., 1933.
108. Leeming, J., *Rayon: The First Man-Made Fiber*, Chemical Publishing, New York, 1950.
109. Beer, E.J., *The Beginning of Rayon*, Phoebe Beer, Paignton, U.K., 1962.
110. Hooke, R., *Micrographia*, Octavo, London, 1964.
111. Audemars, G., Br. Patent 283, 1855.
112. Chardonnet, H.B., Fr. Patent 165,349, 1884.
113. Despaissis, L.H., Fr. Patent 203,741, 1890.
114. Pauly, H., Ger. Patent 98,642, 1897.
115. Urban, J., Bronnert, H., Bronnert, E., and Fremery, M., Ger. Patents 109,996; 111,313; 119,230; 121,429; 121,430.
116. Cross, C.F., Bevan, E.J., and Beadle, C., Br. Patent 8700, 1892.
117. Topham, C.F., Br. Patents 12,157–12,158, 1902.
118. Moncrieff, R.W., *Man-Made Fibers*, John Wiley & Sons, New York, 1957, pp. 17–23.
119. Muller, M., U.S. Patent 836,452, 1906.
120. Napper, S.S., U.S. Patent 1,045,731, 1912.
121. Lilienfeld, L., U.S. Patent 1,683,199, 1928.
122. Givens, J.H., Biddulph, H.W., and Rose, L., Br. Patent 467,500, 1937.
123. Kothari, R.P., *Handbook of Rayon*, Century Rayon, Data Press, Bombay, India, 1970, p. 357.
124. Cox, N.L., U.S. Patent 2,535,045, 1950.
125. Alles, F.P., U.S. Patent 2,123,493, 1938.
126. Mitchell, R.L., Berry, J.W., and Wadman, W.H., U.S. Patent 3,018,158, 1962.
127. Mitchell, R.L., Berry, J.W., and Wadman, W.H., U.S. Patent 2,942,931, 1960.
128. Tachikawa, S., U.S. Patent 2,592,355, 1952; 2,732,279, 1956; 2,946,782, 1960.
129. Hermans, P.H., *J. Phys. Chem.*, 1941, *45*, 827.
130. Alexander, W.J. and Muller, T.E., *J. Polym. Sci.*, 1971, *C36*, 87.
131. Hearle, J.W.S., *Skinners Rec.*, 1964, *11*, 1027.
132. Susich, G. Von and Woff, W.W., *Z. Phys. Chem.*, 1930, *B8*, 221.
133. Hess, K. and Trogus, C., *Z. Phys. Chem.*, 1929, *B4*, 321; 1930, *B11*, 381.
134. Sobue, H., *J. Soc. Chem. Ind. Jpn.*, 1940, *43*, B24.

135. Stepanov, B.I., Zhbankov, R.G., and Rosenberg, A.Y., *Russ. J. Phys. Chem.*, 1959, *33(9)*, 215.
136. Bouriot, P., *Bull. Inst. Text. Fr.*, 1962, *103*, 1197.
137. Ranby, B.G., *Acta Chem. Scand.*, 1952, *6*, 101.
138. Lyubova, T.A., Kovaleva, S.A., Tokareva, L.G., Serebryakova, Z.G., and Kartashova, T.P., *Khim. Volokna*, 1977, *19(5)*, 66.
139. Sihtola, H. and Nizovsky, B., *Paperi Puu*, 1963, *45*, 299.
140. Treiber, E., *Das Papier*, 1959, *13*, 253.
141. Kleinert, T.N., *Tappi J.*, 1956, *39*, 807, 813, 818.
142. Sihtola, H., *Cellul. Chem. Technol.*, 1972, *6(1)*, 71.
143. Entwhistle, D., Cole, E.H., and Wooding, N.S., *Text. Res. J.*, 1949, *19*, 527, 609.
144. Davidson, G.F., *J. Text. Inst.*, 1934, *25*, T174.
145. Keil, A., Philipp, B., and Jacob, R., *Faserforschung u. Textiltechnik*, 1962, *13*, 540.
146. Saeman, J.F., Millett, M.A., and Lawton, E.J., *Ind. Eng. Chem.*, 1952, *44*, 2848.
147. Neal, J.L. and Kraessig, H.A., *Tappi J.*, 1963, *46*, 70.
148. Imamura, R., Ueno, T., and Kurakami, K., *Kyoto Diagaku Nippon Kagakuseni Kenkyosho Koensitct*, 1959, *16*, 131.
149. Fr. Patent 1,192,359, 1959; Br. Patent 830,820, 1960; Ger. Patent 1,151,494, 1963.
150. Ueno, T., Murakami, M., and Imamura, R., *Jpn. Tappi*, 1972, *26(4)*, 164.
151. Philipp, B., Dautzenberg, H., and Schmiga, W., *Faserforschung u. Textiltechnik*, 1969, *20(3)*, 111.
152. Philipp, B., Dautzenberg, H., and Schmiga, W., *Faserforschung u. Textiltechnik*, 1969, *20(12)*, 573.
153. Kolosh, F. and Eriksson, M., *Tappi J.*, 1969, *52(5)*, 930.
154. Hovenkamp, S.G., *J. Polym. Sci.*, 1963, *C(2)*, 341.
155. Durso, D.F. and Parks, L.R., *Svensk Pupperstidn.*, 1961, *64(23)*, 853.
156. Samuelson, O., *Svensk Pupperstidn.*, 1949, *52*, 465.
157. Treiber, E., *J. Polym. Sci.*, 1961, *51*, 297.
158. Dyer, J. and Smith, F.R., *ACS Symposium, Series No. 49*, 3, 1977.
159. Matthes, A., *Faserforschung u. Textiltechnik*, 1952, *3*, 127.
160. Sihtola, H., Kaila, E., and Nizovsky, B., *Paperi ja Puu*, 1956, *4a*, 174.
161. Easterwood, M. and Mueller, W.A., *J. Appl. Polym. Sci.*, 1960, *4(10)*, 16.
162. Lyselius, A. and Samuelson, O., *Svensk Pupperstidn.*, 1961, *64*, 735; 1961, *64*, 815.
163. Dunbrant, S. and Samuelson, O., *Tappi J.*, 1963, *46*, 520.
164. Dolby, I., Dunbrant, S., and Samuelson, O., *Svensk Pupperstidn.*, 1964, *67(3)*, 110.
165. Dolby, I. and Samuelson, O., *Svensk Pupperstidn.*, 1965, *68*, 136; 1966, *69*, 305.
166. Andersson, L. and Samuelson, O., *Svensk Pupperstidn.*, 1967, *70*, 567; 1968, *71*, 727.
167. Dyer, J. and Phifer, L.H., *Svensk Papperstidning*, 1968, *71(9)*, 385.
168. Hottenroth, V., *Chemiker-Ztg.*, 1915, *39*, 119.
169. Lyselius, A. and Samuelson, O., *Svensk Pupperstidn.*, 1961, *64(5)*, 145.
170. Vroom, R.A., *Kinetic Aspects of the Viscose Rayon Spinning Process*, U. Waltman, Delft, 1963.
171. Vermaas, D., *Text. Res. J.*, 1962, *32(5)*, 353.
172. Grobe, A., Maron, R., and Klare, H., *Faserforsch.-Textiltech.*, 1961, *12*, 196.
173. Smith, D.K., *Text. Res. J.*, 1959, *29*, 32.
174. Sisson, W.A., *Text. Res. J.*, 1960, *30(3)*, 153.
175. Tornell, B., Decomposition of Cellulose Xanthate During Spinning of Viscose in Solutions of Low Acid Content, Thesis, Goteborg, 1967; *Svensk. Pupperstidn.*, 1964, *67*, 756; 1966, *69*, 658, 695; 1967, *70*, 1, 268, 303, 449, 489.
176. Phifer, L.H. and Dyer, J., *Macromolecules*, 1969, *2*, 118.
177. Dyer, J. and Phifer, L.H., *Macromolecules*, 1969, *2*, 111.
178. Morimoto, S. and Murakami, E., *Sen-i Gakkaishi*, 1964, *20(2)*, 87.
179. Hovenkamp, S.G., *Faserforsch.-Textiltech.*, 1966, *17(7)*, 305; 1966, *17(8)*, 370; 1966, *17(9)*, 400.
180. Chakravarty, S., *Handbook of Rayon*, Century Rayon, Data Press, Bombay, India, 1970, p. 189.
181. Murakami, E., *Sen-i Gakkaishi*, 1964, *20(3)*, 149.
182. Horio, M., *Text. Res. J.*, 1950, *20*, 373.
183. Phifer, L.H. and Plummer, H.K., *Tappi J.*, 1965, *48(5)*, 290.
184. Grobe, A., Klare, H., and Riedel, E., *Faserforsch.-Textiltech.*, 1960, *11*, 113.

185. Svensson, S. and Tornell, B., *J. Appl. Polym. Sci.*, 1972, *16*, 2185.
186. Morehead, F.F. and Sisson, W.A., *Text. Res. J.*, 1945, *15*, 443.
187. Grobe, A., Klare, H., and Jost, H., *Faserforsch.-Textiltech.*, 1960, *11*, 209.
188. Murakami, E., *Sen-i Gakkaishi*, 1964, *20(3)*, 289.
189. Tornell, B., *Svensk Pupperstidn.*, 1967, *70*, 489.
190. Schappel, J.W. and Bockno, G.C., *High Polymers: Cellulose and Cellulose Derivatives*, Vol. 5, Bikales, N.M. and Segal L., Eds., Wiley Interscience, New York, 1971, p. 1131.
191. Treiber, E., *Faserforsch.-Textiltech.*, 1978, *29(9)*, 605.
192. Hilgers, F., *Chemiefasern*, K. Gotze Springer-Verlag, Berlin, 1967, p. 635.
193. Levine, M. and Burroughs, R.H., *J. Appl. Polym. Sci.*, 1959, *11(5)*, 192.
194. Murakami, E., *Sen-i Gakkaishi*, 1964, *20(2)*, 94.
195. Klare, H., *Deutsche Textiltech*, 1966, *16(2)*, 73.
196. Deshmukh, R.M., *Colourage*, 1980, *27(9)*, 9.
197. Charles, F.R., *Can. Text. J.*, 1966, *83(16)*, 37.
198. Klare, H. and Grobe, A., *Osterr. Chem. Ztg.*, 1964, *65*, 209.
199. Grobe, A., Maron, R., Jost, H., Paul, D., and Klare, H., *Faserforsch.-Textiltech.*, 1965, *16(1)*, 33.
200. Greenwald, H.L. and Brown, G.L., *J. Phys. Chem.*, 1954, *58*, 825.
201. Murakami, E., *Sen-i Gakkaishi*, 1964, *20(5)*, 301.
202. Murakami, E., *Sen-i Gakkaishi*, 1964, *20(3)*, 155.
203. Mitchell, R.L. and Daul, G.C., *Encyclopedia of Chemical Technology*, 2nd ed., Vol. 17, John Wiley & Sons, New York, 1968, p. 186.
204. Murakami, E., *Sen-i Gakkaishi*, 1964, *20(5)*, 206.
205. Thumm, B.A., U.S. Patent 2,975,020, 1961, to American Viscose Co.
206. Lytton, M.R., U.S. Patent 2,908,581, 1959, to American Viscose Co.
207. Lytton, M.R., U.S. Patent 2,961,329, 1960, to American Viscose Co.
208. Hollihan, J.P., Howsman, J.A., and Sisson, W.A., U.S. Patent 2,971,816, 1961, to American Viscose Co.
209. Sisson, W.A. and Thumm, B.A., U.S. Patent 2,971,817, 1961, to American Viscose Co.
210. Limburg, P.C., Buurnan, A., and Vroom, R.A., U.S. Patent 2,978,292 1961, to American Enka Co.
211. Br. Patent 857,170, 1960, to N.V. Onderzoe-Kinginstituit Res. Corp.
212. Br. Patent 846,533, 1960, to VGF, AG.
213. Vosters, H.L., U.S. Patent 3,016,305, 1962, to Kunstzijdespinnerij Nyma.
214. Lytton, M.R., U.S. Patent 2,792,279; 2,792,280; 2,792,281, 1957.
215. Phillip, B., *Faserforsch.-Textiltech.*, 1957, *8*, 21.
216. Cox, N.L., U.S. Patent 2,581,835, 1951, to DuPont.
217. Butkova, N.I., Petrova, N.I., Sofronova, I. S. Pakshver, A.B., and Finger, G.G., *Khim Volokna*, 1978, *20(5)*, 44.
218. Thumm, B.A. and Tryon, S., *J. Org. Chem.*, 1964, *29*, 2999.
219. Klein, E., Wise, H., and Richardson, W.C., U.S., Patent 3,109,698, 1963, to Courtaulds North America.
220. Richardson, W.C., U.S. Patent 3,109,699, 1963, to Courtaulds North America.
221. Klein, E., Nelson, D.S., and Bingham, B.E.M., U.S. Patent 3,107,700, 1963.
222. Phifer, L.H. and Ticknor, L.B., *J. Appl. Polym. Sci.*, 1965, *9*, 1055.
223. Phifer, L.H., Dux, J.P., Ticknor, L.B., and O'Shaughnessy, M.T., *J. Appl. Polym. Sci.*, 1965, *9(3)*, 1067.
224. Dyer, J., *Tappi J.*, 1966, *49(10)*, 447.
225. Bogdanski, J. and Chrzqszcewski, J., *Acta Chim.*, 1959, *4*, 37.
226. Credali, L. Mortillaro, L., Galiazzo, G., delFante, N., and Carazzalo, G., *J. Appl. Polym. Sci.*, 1965, *9*, 2895.
227. Murakami, E., *Sen-i Gakkaishi*, 1964, *20(8)*, 519.
228. Budnitskii, G.A., Rogovina, A.A., Veretnikova, T.P., and Serkov, A.T., *Khim. Volokna*, 1971, *13(3)*, 39.
229. Dyer, J. and Daul, G.C., *Ind. Eng. Chem.*, Product R&D, 1981, *20*, 222.

230. Dewey, F.J., *Fiber Producer*, 1975, July, 40.
231. Fueg, W., *Chemiefaser Textil-Ind.*, 1973, *29(6)*, 450.
232. Mauersberger, H.R., *American Handbook of Synthetic Textiles*, Textile Book Publishers, New York, 1952, p. 141.
233. Gotze, K., *Chemiefasern*, Springer-Verlag, Berlin, 1967, p. 780.
234. Treiber, E., *Lenzinger Berichte*, 1964, *19*, 5.
235. Treiber, E., *Svensk. Pupperstidn.*, 1968, *71(4)*, 99.
236. Treiber, E., Abrahamson, B., and Lundin, H., *Holzforschung*, 1964, *18*, 33.
237. Sihtola, H. and Boestrom, *Paperi ja Puu*, 1952, *34*, 23.
238. Treiber, E. and Wangberg, L., *Tappi J.*, 1969, *52(2)*, 305.
239. von Bucher, H.P., *Tappi J.*, 1978, *61(4)*, 91.
240. Meister, W., U.S. Patent 3,438,969, 1969; Br. Patent 1,202,476, 1970, to Ing. A. Maurer.
241. Richter, G.A., U.S. Patent 1,842,688, 1932, to Brown Company.
242. Seaman, S.E., U.S. Patent 2,530,403, 1950.
243. Yasui, K., Yamamoto, K., Itami, K., Fugisawa, K., and Sasaki, T., U.S. Patent 3,671,279, 1972; Br. Patent 1,289,223, 1972; Jpn. Patent 4,949,038, 1974, to Asahi Kasei Kogyo.
244. Von Kohorn, O., U.S. Patent 2,985,647, 1959.
245. Treiber, E., U.S. Patent 3,385,845, 1968, to Sund's Verkstader Aktiebolag.
246. Treiber, E., *Lenzinger Berichte*, 1966, *21*, 7.
247. Durson, D.F., Benning T.C., and Goode, J.R., *Svensk Pupperstidn.*, 1967, *70*, 837.
248. Ivnas, L. and Svensson, L., *Tappi J.*, 1974, *57(8)*, 115.
249. Von Bucher, H.P., *Handbook of Rayon*, Century Rayon, Data Press, Bombay, India, 1970.
250. Schwab, M., *Fiber Prod.*, 1978, *6*, 42.
251. Ford, J.E., *Textiles*, 1980, *9(1)*, 2.
252. Sisson, W.A. and Morehead, F.F., *Text. Res. J.*, 1953, *23(3)*, 152.
253. Braunlich, R.H., *Am. Dyestuff Reptr.*, 1965, *54(5)*, 38.
254. Daul, G.C., *Am. Dyestuff Reptr.*, 1965, *54(22)*, 48.
255. Sisson, W.A., *High Polymers; Cellulose and Cellulose Derivatives*, Vol. 5, Ott, E., Ed., Wiley Interscience, New York, 1943, pp. 203–285.
256. Hermans, P.H., *Physics and Chemistry of Cellulose Fibers*, Elsevier, New York, 1949.
257. Herzog, R.O. and Jancke, W., *Z., Zeitsehr. f. Phys.*, 1920, *3*, 196.
258. Meyer, K.H. and Mark, H., *Z., Physik. Chem.*, 1929, *B2*, 115.
259. Meyer, K.H. and Misch, L., *Ber. Dtsch. Chem. Ges.*, 1937, *70B*, 266.
260. Meyer, K.H. and Misch, L., *Helv. Chim. Acta*, 1937, *20*, 232.
261. Baker, W.O., *Ind. Eng. Chem.*, 1945, *37*, 246.
262. Hindeleh, A.M. and Johnson, D.J., *Polymer*, 1974, *15*, 697.
263. Hearle, J.W.S., *Fiber Struct.*, Hearle, J.W.S. and Peters, R.H., Eds., Butterworth and Textile Institute, London, 1963, pp. 209–234.
264. Nikonovich, G.V., Shoskina, V.I., and Usmanov, K.U., *Cellul. Chem. Technol.*, 1974, *8*, 509.
265. Abitz, W., Gerngross, O., and Herrmann, K., *Naturwissenschaften*, 1930, *18*, 754.
266. Hearle, J.W.S., *J. Polym. Sci.*, 1958, *28*, 432.
267. Hess, K., Mahl, H., and Gutter, E., *Kolloid-Z.*, 1957, *155*, 1.
268. Chang, M., *J. Polym. Sci.*, 1974, *12*, 1349.
269. Berl, E. and Lange, A., *Cellul. Chem.*, 1926, *7*, 145.
270. Preston, J.M., *J. Soc. Chem. Ind.*, 1931, *50*, T199.
271. Dyer, J., *Cellulose Technology Research ACS Symposia*, Serial No. 10, Turbak, A., Ed., American Chemical Society, Washington, D.C., 1975, pp. 181–194.
272. Bingham, B.E.M., *Makromol. Chem.*, 1964, *77*, 139.
273. Schappel, J.W., *Tappi J.*, 1963, *46(10)*, 18A.
274. Ychida, Y., *Pure Appl. Chem.*, 1967, *14*, 461.
275. Drisch, N. and Priou, R., *Bull. Inst. Text. Fr.*, 1962, *101*, 667.
276. Herrbach, P., *Lenzinger Berichte*, 1964, *17*, 13.
277. Dlugosz, J. and Michie, R.I.C., *Polymer*, 1960, *1*, 41.

278. Morehead, F.W., *High Polymers: Cellulose and Cellulose Derivatives*, Vol. 5, Bikales, N.M. and Segal, L., Eds., Wiley Interscience, New York, 1971, pp. 213–265.
279. Grobe, A., *Svensk Pupperstidn.*, 1968, *71(15)*, 636.
280. Lund, G.V. and Waters, W.T., *Text. Res. J.*, 1959, *29*, 950.
281. Goldenberg, D.Z., *Text. Chem. Color.*, 1969, *1(19)*, 374.
282. Howarth, J., *Text. Ind., South. Afr.*, 1980, *3(5)*, 3.
283. Szego, L., *Faserforsch.-Textiltech.*, 1970, *21(10)*, 442.
284. Daul, G.C. and Barch, F.P., U.S. Patents 3,632,468, 1972; 3,793,136, 1974.
285. Stevens, H.D. and Muller, T.E., U.S. Patent 3,720,743, 1973.
286. Bellano, A., *Textilia*, 1978, *54(6)*, 11.
287. Muller, T.E., Batch, F.P., and Daul, G.C., *Text. Res. J.*, 1976, *46(3)*, 184.
288. Newcomb, W.B., *Text. World*, 1980, *10*, 105.
289. Lane, M. and McCombes, J.A., *ACS Symposia, Series No. 58*, 197, 1977.
290. Portnoy, N.A. and Daul, G.C., *Repr. Nat. Tech. Conf. AATCC*, 269, 1978.
291. Palfreyman, T., *Text. Mon.*, 1980, *6*, 23.
292. Welch, M.J. and McCombes, J.A., *Repr. INDA 8th Tech. Symp. Nonwovens*, 3, 1980.
293. American Enka Co., *Nonwovens Report*, 1978, *78(10)*.
294. Avtex Fibers, Inc., Br. Patent 1,517,398, 1978.
295. Courtaulds Ltd., *Nonwovens Report*, 1979, *97(5)*, 3.
296. Ward, J.S. and Hill, R., *Text. Inst. Ind.*, 1969, *7*, S274.

11 Cellulose Acetate and Triacetate Fibers

Herman L. LaNieve

CONTENTS

11.1 Introduction 774
11.2 Cellulose 775
11.3 Cellulose Characterization 777
    11.3.1 Cellulose Purity 777
    11.3.2 Intrinsic Viscosity 777
    11.3.3 Cellulose Reactivity 777
    11.3.4 Sheet Density...778
    11.3.5 Moisture Content 778
11.4 Cellulose Triacetate and Cellulose Acetate Processes 778
    11.4.1 Basic Principles 778
    11.4.2 Cellulose Acetate Unit Operations..779
        11.4.2.1 Shredding 780
        11.4.2.2 Pretreatment 780
        11.4.2.3 Acetylation 781
        11.4.2.4 Hydrolysis 782
        11.4.2.5 Precipitation 783
        11.4.2.6 Flake Washing 783
        11.4.2.7 Flake Drying 783
        11.4.2.8 Flake Blending 783
    11.4.3 Acetic Acid Recovery 784
    11.4.4 Triacetate Process 784
11.5 Cellulose Acetate and Cellulose Triacetate Characterization....................................784
    11.5.1 Acetyl Value 784
    11.5.2 Solubility 784
    11.5.3 Intrinsic Viscosity 785
        11.5.3.1 Solution and Dope Properties for Cellulose Acetate...................785
11.6 Dope Preparation 786
11.7 Spinning of Cellulose Acetate and Triacetate 788
    11.7.1 Dry-Spinning 788
        11.7.1.1 Dope Preheating...790
        11.7.1.2 Dope Metering 790
        11.7.1.3 Spinneret Design 790
        11.7.1.4 Spinning Column or Cabinet 790
        11.7.1.5 Volatility of the Solvents...791

		11.7.1.6	Spinning Parameters versus Yarn Properties	791
		11.7.1.7	Yarn Lubrication and Finish	791
		11.7.1.8	Packaging	791
		11.7.1.9	Solvent Recovery	791
		11.7.1.10	Staple and Tow	792
	11.7.2	Wet-Spinning		792
	11.7.3	Melt-Spinning		792
11.8	Yarn Types and Packages			792
11.9	Cellulose Acetate and Triacetate Properties			793
	11.9.1	Overview		793
	11.9.2	Fiber Cross Sections		793
	11.9.3	Fine Structure of Cellulose Acetate and Triacetate Fibers		795
		11.9.3.1	Cellulose Acetate	795
		11.9.3.2	Cellulose Triacetate	795
	11.9.4	Absorption and Swelling Behavior		796
	11.9.5	Thermal Behavior		798
	11.9.6	Tensile Properties		798
		11.9.6.1	Tenacity Units of Measurement	798
		11.9.6.2	Stress–Strain Curves	799
		11.9.6.3	Wet Strength	801
		11.9.6.4	Tensile Strain Recovery Behavior	802
		11.9.6.5	Fiber Toughness	802
		11.9.6.6	Fiber Bending	802
		11.9.6.7	Abrasion Resistance	803
	11.9.7	Dyeing Characteristics		803
	11.9.8	Light Stability		805
11.10	End-Uses			806
	11.10.1	Textiles		806
	11.10.2	Cigarette Tow		806
	11.10.3	Other Applications		807
11.11	Prospects for Future Applications			807
Acknowledgments				808
References				808

11.1 INTRODUCTION

Camille and Henry Dreyfus developed the first commercial process to manufacture cellulose acetate in 1905 and commercialized the spinning of cellulose acetate fibers in 1924 in the United States. At that time, the only other human-made fiber was viscose rayon, which was still in its early stages of commercialization. The main textile fibers were natural fibers: cotton, wool, silk, and flax. Cellulose triacetate textile fiber was commercialized later in the 1950s. The tremendous technical effort by the Dreyfus Brothers resulted in more than 300 patents describing such significant inventions as the dry-spinning process and disperse dyeing.

Cellulose acetate fiber was first marketed as "artificial silk" and found applications as tricot knits and woven fabrics in blouses, dresses, apparel linings, velvets, and decorative ribbons. The combined cellulose acetate and triacetate textile fiber production continued to grow until in 1971 it peaked at 426,000 MT worldwide. The impact of synthetic fibers, namely polyester and nylon, during the 1970s was significant and has gradually taken market share

TABLE 11.1
World Production of Cellulose Acetate and Triacetate Fibers (1000 MT)

Year	1970	1980	1990	1995	2002
Cellulose acetate/Triacetate textile fibers	426	344	200	109	104
Cellulose acetate filter tow	154	335	465	552	596
All man-made fibers	8,136	13,718	17,664	20,942	36,000

Source: From Textile Economics Bureau, *Textile Organon*, (June, 2003).

from cellulose acetate and triacetate. Cellulose acetate tow was introduced to the market in 1952 and has had major success worldwide for cigarette filter media. Cellulose acetate tow production has steadily grown to 552,000 MT in 1995. World production of cellulose acetate textile fiber and tow is compared with total human-made fiber production in Table 11.1. The production of cellulose acetate and triacetate for plastics and films represents another important end-use [1].

Cellulose acetate and triacetate fibers have survived in the marketplace because they have certain unusual properties that demonstrate significant advantages over other polymeric materials. Cellulose acetate and triacetate textile fibers are luxurious. Fabrics made from them have an excellent hand, dye to brilliant, attractive shades, and are soft and comfortable. Regarding cellulose acetate and triacetate plastics and films, no other polymers can match the sparkling clarity possessed by these. For cigarette-smoke filtration, cellulose acetate offers a unique balance of properties including smoke removal efficiency and contribution to taste that makes it the standard of the industry.

11.2 CELLULOSE

Cellulose is obtained from wood, cotton, and other plants and is a natural linear polymer. The basic structural repeating unit for cellulose is the cellobiose unit, represented by the following structure:

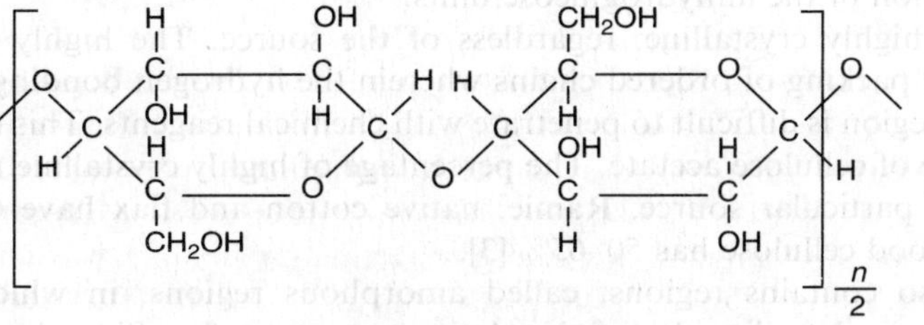

Cellulose

The second ring in the cellobiose unit is chemically identical to the first ring. The important difference is that the second ring is inverted over from the general plane of the first ring. The oxygen atom connecting the rings is referred to as a β-glycoside linkage. The rings are not planar but puckered. The following structure for cellulose has appeared in the literature to emphasize the point that rings are not planar:

$$\left[\text{...cellulose structure with HO, OH, H, CH}_2\text{OH groups...} \right]_{\frac{n}{2}}$$

Each ring is called an anhydroglucose unit. Since each one is identical chemically, it is more convenient to express cellulose as a polymer with the anhydroglucose unit as the repeating unit. This has the advantage of describing the chemistry involved in cellulose acetate in a simpler manner. The structure using the anhydroglucose repeating unit is as follows:

$$\left[\text{...anhydroglucose repeating unit with numbered carbons 1-6, CH}_2\text{OH, OH, H groups...} \right]_n$$

There are three hydroxyl groups in each anhydroglucose unit located in the 2, 3, and 6 positions. These sites are available for acetylation to prepare the acetates.

The symbol n represents the average degree of polymerization (DP). The DPs for undegraded cotton and wood cellulose are very high, and estimates of 6000–8000 for undegraded cotton cellulose and 5000–8000 for spruce wood cellulose have been reported [2]. However, a large decrease in DP occurs during processing and purification to produce an acetate-grade cellulose; the average DP for the latter is about 1000–2000. (This DP range and other DPs quoted in this chapter are based on the 8000 DP for undegraded cotton. For many years, this value was commonly held to be the best estimate. Now, the estimate is considered to be 12,000–17,000 for undegraded cotton.) Although cellulose is a linear polymer, it is complicated by the fact that there is strong intermolecular bonding between adjacent cellulose chains. Because of the preponderance of hydroxyl groups, cellulose has extensive hydrogen bonding between chains. The hydrogen bonding is enhanced by the continual alternating steric configuration of the anhydroglucose units.

Cellulose is highly crystalline, regardless of the source. The highly crystalline regions represent a tight packing of ordered chains wherein the hydrogen bonding is at a maximum. The crystalline region is difficult to penetrate with chemical reagents. This fact is important in the manufacture of cellulose acetate. The percentage of highly crystalline regions in cellulose depends on the particular source. Ramie, native cotton and flax have 60–87% crystalline cellulose, and wood cellulose has 50–65% [3].

Cellulose also contains regions, called amorphous regions, in which there can be a variation from complete disorder of the chains to some order. Since interchain bonding is minimal in the amorphous region, chemical reagents diffuse or penetrate more easily in this region. It is believed that the amorphous regions are acetylated first in the acetylation of cellulose. An important point is that, together with the cellulose (or α-cellulose) in wood and in cotton linters, there are hemicelluloses. Hemicelluloses are shorter-chain polysaccharides that occur in the native cellulose. A greater percentage of hemicellulose is associated with wood cellulose than with cotton cellulose. Acetylation grade cellulose contains 95–98%

α-cellulose, but strong economic incentives have led to the commercialization of 92–93% or lower α-cellulose wood pulps. The main hemicelluloses are mannan and xylan. Cotton linters are generally purer than wood pulp and are used for special plastics grade acetates requiring low color and high clarity. The presence of hemicelluloses in wood pulp affects the solution properties of the cellulose acetate. The hemicelluloses contain hydroxyl groups, so they too are acetylated. However, they have different solubility characteristics from those of cellulose acetate.

11.3 CELLULOSE CHARACTERIZATION

The cellulose acetate manufacturer receives the cellulose in rolls or bales. For wood pulp, the bales consist of sheets of pulp. However, for cotton linters the bales can consist of sheets or cellulose in bulk form. The most important chemical properties are cellulose purity, intrinsic viscosity (IV), and cellulose reactivity. The most important physical properties are sheet density and moisture content.

11.3.1 Cellulose Purity

Cellulose purity can be determined by R10, S10 – S18, and S18. The R10 value, which is the percentage of the cellulose sample insoluble in 10% sodium hydroxide solution at 20°C, represents long-chain cellulose. The S10 value is the percentage of the cellulose sample soluble in 10% NaOH. Of course, R10 equals 100 – S10. The S18 value is the percentage of the cellulose sample soluble in 18% NaOH at 20°C. The S10 – S18 value represents short-chain cellulose. The S18 value represents mostly hemicellulose. The terms R10, S10 – S18, and S18 are similar in value to α-, β-, and γ-cellulose, respectively [4].

Acetate-grade celluloses have high R10 and low S18 values. High-grade cotton linters are the purest, with R10 generally around 99% and S18 less than 0.5%. Wood-pulp celluloses for acetate application generally have R10 values ranging from 95 to 98% and S18 values from about 1 to 3.5%. In general, the purer the grade, the higher the price. Therefore, the bottom line is price–performance, and this depends to a great extent on the cellulose acetate manufacturer's particular process, equipment, and end-use products.

11.3.2 Intrinsic Viscosity

The IV of cellulose is determined by measuring the viscosity of a very dilute solution (such as 0.5%) of the cellulose in cupriethylenediamine hydroxide (abbreviated cuene) or cuprammonium hydroxide (abbreviated cuam). A relationship has been developed between IV and DP so that the two can be used interchangeably.

As pointed out previously, the DP range for acetate-grade celluloses is about 1000–2000. This DP range corresponds to a cuene IV range of about 6–11 dl/g. It should be noted that the DP is a "number" average DP determined by the nitrate method. Acetate grades from cotton linters generally have higher IV or DP values than those from wood pulp.

11.3.3 Cellulose Reactivity

The wood species and the type of pulp manufacturing process relate to the reactivity of the cellulose in acetylation. For example, a highly purified acetate-grade wood pulp from hardwoods manufactured by the prehydrolyzed kraft (PHK) process has about the highest reactivity. Within the category of softwood pulps, there are different levels of reactivity depending on the wood species and the extent of purification. The importance of cellulose reactivity is that cellulose acetate manufacturers must adjust their particular acetylation conditions to optimize the processing of a given cellulose.

11.3.4 Sheet Density

The physical property of the cellulose sheet that relates mostly to the shreddability of the sheet is the sheet density. The lower the density, the easier the sheet is to shred or open up. Opening up the sheet to enhance the diffusion of chemical reagents into the bundles of cellulose fibers is of utmost importance. The purer the pulp and the lower the density, the softer the sheet generally. Sheet densities for acetate grades usually range from about 0.43 to 0.55 g/cm^3 based on bone-dry cellulose.

11.3.5 Moisture Content

Cellulose always contains some moisture. In fact, it is not good to overdry cellulose because this tends to impair its reactivity, which subsequently affects cellulose acetate solution properties. When high α pulps (high R10) have been dried to very low moisture content, the resulting cellulose acetate solutions were entirely unacceptable because of poor filterability, even though the solutions appeared clear and relatively free of insolubles [5]. The natural moisture content of acetate-grade cellulose is about 7%. The cellulose manufacturer carefully controls the moisture content in the range of 6–7%. Extra moisture represents direct dollar cost because the water consumes acetic anhydride.

11.4 CELLULOSE TRIACETATE AND CELLULOSE ACETATE PROCESSES

11.4.1 Basic Principles

Cellulose triacetate is obtained by esterification of cellulose with acetic anhydride. The anhydroglucose unit contains three OH sites for acetylation. Therefore, the maximum degree of acetylation of cellulose is a triacetate; however, the acetylation does not quite reach the maximum of three units per glucose unit. The extent to which the hydroxyl groups are substituted is called the degree of substitution (DS). For cellulose acetate, also called secondary cellulose acetate or cellulose diacetate, the DS is about 2.4.

The overall reaction in the conversion of cellulose to cellulose triacetate is:

$$\text{Cell} \begin{array}{c} \diagup \text{OH} \\ - \text{OH} \\ \diagdown \text{OH} \end{array} + 3\text{Ac}_2\text{O} \xrightarrow{\text{Acid Catalyst}} \text{Cell} \begin{array}{c} \diagup \text{OAc} \\ - \text{OAc} \\ \diagdown \text{OAc} \end{array} + 3\text{HOAc} \quad (11.1)$$

where Cell is the anhydroglucose ring (without -OH groups) and Ac is an abbreviation for acetyl, $COCH_3$. Equation 11.1 shows that 3 mol of acetic anhydride react with 1 mol of cellulose to give 1 mol of cellulose triacetate and 3 mol of acetic acid.

During hydrolysis, some of the acetate groups are hydrolyzed. This is represented approximately by the equation:

$$\text{Cell} \begin{array}{c} \diagup \text{OAc} \\ - \text{OAc} \\ \diagdown \text{OAc} \end{array} + \text{H}_2\text{O} \xrightarrow{\text{Acid Catalyst}} \text{Cell} \begin{array}{c} \diagup \text{OH} \\ - \text{OAc} \\ \diagdown \text{OAc} \end{array} + \text{HOAc} \quad (11.2)$$

Equation 11.2 shows the formation of a diacetate. Actually, the DS is higher at 2.4, which means that less than one acetyl group per anhydroglucose unit is hydrolyzed on an average, or three acetyl groups are hydrolyzed for every five anhydroglucose units.

A series of simultaneous, complex reactions occurs during acetylation of cellulose and during hydrolysis of cellulose triacetate. Reactions include acetylation of the cellulose, some sulfation of the cellulose, degradation of the cellulose chain and reaction to chain-ends, acetylation of the hemicellulose components, and then random hydrolysis of the acetylated or sulfated product [6–12].

There is no satisfactory commercial means to directly acetylate to the 2.4 acetyl level and obtain a secondary acetate that has the necessary solubility for fiber preparation. Since cellulose is highly crystalline and its polymer chains are held tightly together in an ordered manner through extensive hydrogen bonding, it is insoluble in the reaction medium until almost complete acetylation is achieved. Thus, commercially, cellulose is fully acetylated to triacetate and then hydrolyzed back to secondary acetate of 2.4 DS. Careful hydrolysis is nearly random yielding uniform polymer.

The accessibility of the cellulose is very important, because most of the acetylation is a heterogeneous reaction. The greater the accessibility, the easier it is for the reactants to diffuse into the interior of the cellulose. Therefore, adequate shredding of the cellulose sheet and adequate swelling of the cellulose through a pretreatment step are necessary operations prior to the acetylation. The basic principle behind pretreatment is that the penetrating medium, which always contains some water, disrupts the hydrogen bonding in the bundles of cellulose chains causing swelling and enhancing diffusion of reactants.

Although monomeric alcohols can be acetylated with acetic acid using a mineral acid catalyst, acetic anhydride is needed to speed the acetylation of cellulose to reduce a competing depolymerization reaction. There will be some chain depolymerization, but it can be controlled so that the end products will have adequate average chain length.

Acetylation of cellulose is a highly exothermic reaction using acetic anhydride and acid catalyst. Therefore, it is necessary to control the temperature during acetylation, or the rate of depolymerization will become excessive resulting in a loss of target viscosity. However, the extent of cooling needed depends on the particular process.

A strong mineral acid catalyst is needed for the acetylation. Sulfuric acid is generally used for commercial processes. Sulfuric acid catalyst actually combines with the cellulose, forming sulfate linkages; however, most of these are removed during acetylation via an exchange with acetyl groups [13]. It is important that the final cellulose triacetate or cellulose acetate contains only a very small residual amount of sulfate groups generally measured in parts per million because the residual sulfate groups adversely affect the properties, especially the color. After acetylation, residual sulfate groups are removed in a special desulfation step. First excess acetic anhydride is hydrolyzed by the addition of an aqueous acetic acid solution. A slow rate of addition of water for killing the excess anhydride lowers the sulfate content and has resulted in cellulose triacetate with a DS in the range of 2.91–2.96 [6].

Perchloric acid is an excellent catalyst for the acetylation of cellulose. It does not combine with the cellulose; therefore, a full cellulose triacetate (DS of 3) can be made with this catalyst. However, sulfuric acid is preferred over perchloric acid for commercial processes because sulfuric acid is lower in cost and it does not present a potential hazard associated with the buildup of perchlorates in the acid recovery system.

The production of secondary cellulose acetate is done in a separate operation, wherein the cellulose triacetate is hydrolyzed under carefully controlled conditions to a DS of about 2.4.

11.4.2 Cellulose Acetate Unit Operations

A schematic flowsheet for the cellulose acetate process is shown in Figure 11.1. The cellulose sheet is broken up or disintegrated in the shredder. The shredded cellulose is conveyed to the pretreater, where acetic acid is added to pretreat the cellulose. The pretreated cellulose is

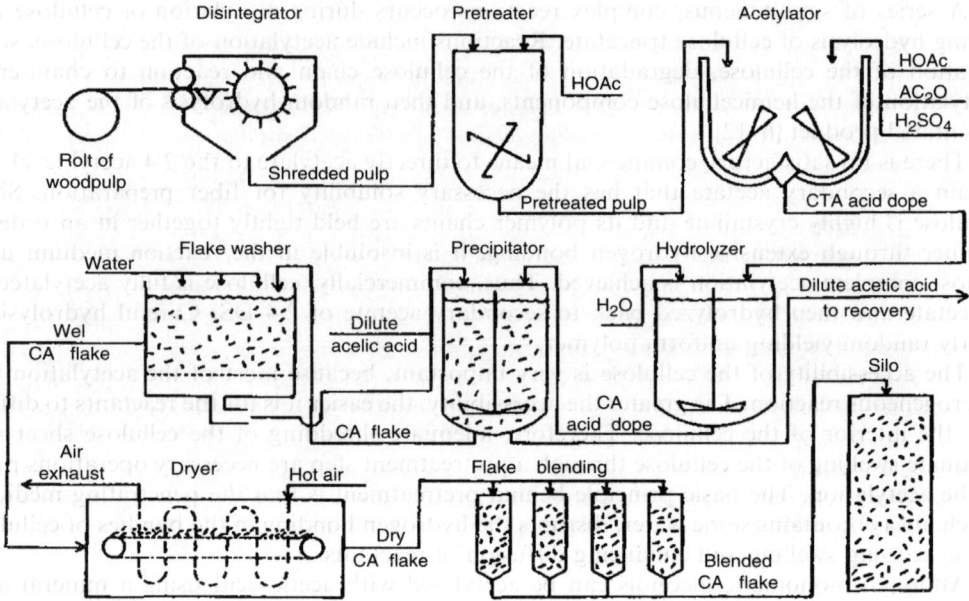

FIGURE 11.1 Cellulose acetate process.

charged into the acetylator and acetylated with acetic anhydride and sulfuric acid catalyst. After acetylation, the excess anhydride is killed with water and the acid dope pumped to the hydrolyzer, in which the cellulose triacetate is hydrolyzed to secondary cellulose acetate. The cellulose acetate flakes or pellets are precipitated from the acid dope in the precipitator. Then they are washed with water in a continuous countercurrent washer. The washed flakes or pellets are pressed or centrifuged to remove excess wash liquor and then dried in a continuous dryer and conveyed to storage bins. The flakes or pellets from many bins are blended by means of a computerized system. The blended material is then conveyed to silos for storage.

11.4.2.1 Shredding

Three basic types of shredders are used commercially to disintegrate the cellulose sheets, namely, hammer mills, disc refiners, and pinpickers. The hammer mill consists of hammers that rotate at high speed and pulverize the continuously fed sheet of cellulose. A cellulose shred is obtained. The disc refiner consists of circular plates having teeth or nobs on the inner surface of each plate. One plate is stationary and the other rotates at a high speed adjacent to the stationary one with a certain, predetermined clearance. Cellulose shreds are fed continuously into the disc refiner and a cellulose fluff is obtained. The combination of a hammer mill and a disc refiner used in sequence represents optimum shredding. The pin-picker also produces a cellulose fluff. It consists of a rotating drum having sharp pins protruding from the surface of the drum at close intervals. A set of continuous sheets of cellulose, usually four, are fed horizontally in layer fashion to touch the rotating drum at right angles. The result is a continuous shredding of the edges of the sheets, producing a light cellulose fluff.

11.4.2.2 Pretreatment

The pretreatment step can be vapor phase or slurry. A vapor-phase pretreatment involves a small amount of acetic acid based on the weight of cellulose. This is done in a closed stainless steel vessel with minimum agitation at about 25 to 50°C for usually 1 h. The small amount of acetic acid hardly wets the bulky cellulose. However, there is a continual vaporization and condensation of the acetic acid within the interstices of the fibers, resulting in uniform swelling throughout the cellulose. If the cellulose contains 7% moisture and 35% glacial acetic acid is used based on the weight of bone-dry cellulose, the composition of the pretreating medium would be 82/18 acetic acid and water—a composition that is excellent for swelling cellulose.

A slurry pretreatment involves a comparatively high ratio of liquid to cellulose. The cellulose is dispersed in water or in aqueous acetic acid with vigorous agitation to disperse the cellulose uniformly; then the water is exchanged with acetic acid by a series of continuous alternate operations of concentrating the cellulose by pressing or centrifuging the slurry and treating with a higher and higher percentage of acetic acid. The final dispersion is in substantially acetic acid with only a small amount of water present. The advantage of slurry pretreatment is that the shredding step is eliminated; that is, sheets of cellulose are fed directly to the vessel equipped with an agitator and containing the dispersing medium.

In some processes, there is an activation step following pretreatment, which is sometimes referred to as a second-stage pretreatment. Activation consists of adding a small amount of sulfuric acid dissolved in acetic acid to the already pretreated cellulose. The result is the sorption of sulfuric acid into the interior of the cellulose, which results in a faster acetylation as shown by shorter clearing time [14].

11.4.2.3 Acetylation

There are three systems for the commercial production of cellulose triacetate and cellulose acetate: acetic acid, methylene chloride, and the heterogeneous systems. Most cellulose acetate is manufactured by the acetic acid system. In the acetic acid system, the acetic acid serves as the solvent for the cellulose triacetate as it is formed during acetylation of the cellulose. Acetylations are completed at moderately elevated temperatures. Acetic acid has better solvent power for cellulose triacetate at elevated temperatures than it does at lower temperatures.

The shredded, pretreated cellulose is added to a mixture of acetic acid, acetic anhydride, and sulfuric acid catalyst in a stainless steel acetylator equipped with an agitator and a jacket for cooling. An excess of acetic anhydride (typically 5–15 wt% excess) assures complete reaction. The esterification of cellulose with acetic anhydride and the reaction of acetic anhydride with water from the pretreatment are both highly exothermic reactions so that cooling is necessary. In high-catalyst processes, initial chilling of the acetylation mixture in a separate vessel is necessary to the point of freezing crystals of acetic acid to provide a heat sink for the process. The acetic anhydride reacts with the moisture in the pretreated cellulose, giving acetic acid and a completely anhydrous reaction medium. Then the cellulose begins to react with the acetic anhydride; the initial reaction is mainly in the amorphous regions of the cellulose. The early stages of the acetylation consist of a heterogeneous reaction wherein the cellulose or partly acetylated cellulose is dispersed in the reaction medium. Good agitation of the reaction mixture is necessary to ensure that the acetylation is proceeding as uniformly as possible throughout the dispersion. As the degree of acetylation increases, the viscosity of the reaction medium increases. The dispersion changes to a uniform opaque viscous mixture. As acetylation proceeds further, the opaque viscous mass begins to clear. When acetylation is virtually complete, the reaction mixture is viscous and clear. It is important that proper clearing is obtained. First there is a grainy stage, that is, the acid dope appears grainy. It is necessary to proceed beyond the grainy stage to attain good filtration characteristics for both the cellulose triacetate and the subsequent cellulose acetate. This is done by "holding" the reaction mixture until graininess disappears, giving a clear acid dope. "Holding" assures that the highly crystalline regions of the cellulose have been penetrated by the reactants and acetylated. At this point, a substantially full cellulose triacetate is dissolved in the acetic acid and excess acetic anhydride, although there are also some residual cellulose sulfate linkages present. It is important that target viscosity is attained for the clear acid dope.

Excess acetic anhydride is then killed by adding stop acid (aqueous acetic acid). It is important to note that the addition of the stop acid has three purposes: it kills excess anhydride; it helps to desulfate the residual sulfate linkages, especially when added slowly [6]; and it provides some water in the reaction mixture so that the latter is no longer anhydrous. This last item is important because chain degradation of the cellulose triacetate is much slower in an aqueous acetic acid system than it is in an anhydrous acetic acid system, especially at elevated temperatures. Therefore, stop acid helps to maintain the target viscosity because in many processes not all of the sulfuric acid catalyst is neutralized at this point.

There are numerous modified conditions of the acetic acid acetylation system practiced commercially. But the basic principles are the same. The objective in all of the commercial processes is to produce a clear, gel-free, acid dope having target viscosity. Some of the main processing factors are catalyst level, acetic acid to anhydride ratio, initial temperature of the A mix (A mix refers to a mixture of acetic acid and acetic anhydride.), amount of excess acetic anhydride, the equipment used to agitate the reaction medium, the extent of jacket cooling to control the exotherm, peak temperature (the maximum temperature reached during the acetylation), and the temperature–time profile for the acetylation.

Most acetylations in the acetic acid system are of the batch type. However, there are a few commercial continuous processes. The pretreated cellulose, usually slurry-pretreated resulting in swollen cellulose in substantially acetic acid, is fed continuously to an acetylation kneader, which is usually a horizontal stainless steel vessel equipped with a unique central-operating agitator and is jacketed to control the temperature. The acetic anhydride and sulfuric acid catalyst are generally added concurrently, mostly at the beginning. According to Serad and Sanders [15], the continuous acetylator is a specialized high-energy input reactor of materials-handling capability. The acetylator must provide (a) thorough mixing and kneading action and positive transport of material through the reactor, (b) a plug-flow residence time distribution to ensure complete reaction and uniform flake properties, and (c) good heat-transfer capability to control the exothermic acetylation reaction.

The second most common process is the methylene chloride system. Methylene chloride, an excellent solvent for cellulose triacetate even at low temperatures, replaces the acetic acid as solvent. Perchloric acid is frequently used as catalyst. Acetic acid is formed as a byproduct of the acetylation; therefore, the solvent is a mixture of methylene chloride, acetic anhydride, and acetic acid during and at the end of the acetylation [16].

There is also a seldom used heterogeneous process that uses an organic solvent as medium, and the cellulose acetate produced never dissolves. The heterogeneous system is used for producing only a cellulose triacetate [17].

A truly homogeneous process is attractive from a technical standpoint, although it has not been commercialized. Several solvents permit nondegradative dissolution of the cellulose, e.g., dimethyl formamide–dinitrogen tetroxide, dimethyl sulfoxide–formaldehyde, dimethylacetamide–lithium chloride, and N-methylmorpholine oxide. Cellulose in solution can be esterified uniformly and directly to the desired degree of substitution [18–22].

11.4.2.4 Hydrolysis

The acid dope is transferred from the acetylator to the hydrolyzer, either by gravity or by pumping. The hydrolyzer is generally a vertical, jacketed, cylindrical vessel equipped with a heavy-duty central agitator. The acid dope is heated by means of hot water or steam in the jacket, or by the use of direct steam. Although conditions will vary from one hydrolysis process to another, there are some general similarities.

The hydrolysis of cellulose triacetate to secondary cellulose acetate is acid-catalyzed. Therefore, chain cleavage also takes place simultaneously with the hydrolysis of the acetate groups. The sulfuric acid catalyst level is usually 4–6%, based on the weight of the starting cellulose. This is equivalent to less than 1%, based on the weight of the acid dope. Temperatures in the range of 60–80°C are generally used. The water concentration in the acid dope is usually 10–15%. Temperature and catalyst level may be purposely varied throughout the hydrolysis. Hydrolysis time depends on the specific catalyst level and the temperature range. The higher the temperature, the lower the catalyst level to prevent degradation. Excessive chain degradation must be avoided during the hydrolysis step. After hydrolysis, the catalyst is usually neutralized with a suitable base such as magnesium acetate. Patents [23,24] have been issued on high-temperature hydrolysis. Temperatures of 120–160°C were reported. In high-temperature hydrolysis, the catalyst is completely neutralized during hydrolysis, or there would be severe chain degradation.

11.4.2.5 Precipitation

The important factors in precipitation are to obtain a uniform flake that is easily washed and to obtain acetic acid for recovery at maximum concentration. A flake is produced by the addition of a dilute acetic acid solution to the acid dope with agitation in a very controlled manner. A common practice is to extrude the dope into a knife-like stream of dilute acetic acid. This tends toward a coarse particle that is easy to wash, and it also minimizes fines. It cannot be overemphasized that the precipitation must be carefully controlled in all cases to obtain a uniform flake that is easily washed. The acetic acid stream that goes to recovery consists of about 30% aqueous acetic acid.

11.4.2.6 Flake Washing

Flake washing is done in horizontal, slowly rotating, cylindrical vessels or in vertical towers equipped with a central agitator. Almost all processes employ a countercurrent wash pattern with fresh water entering where the flake exits from the washer and dilute acetic acid leaving where the flake enters the washer. It is necessary to remove all but traces of acetic acid from the flake, otherwise the latter could char during drying.

11.4.2.7 Flake Drying

Following washing, the wet flake is pressed to remove excess wash liquor and then dried in a continuous dryer. This is generally the moving-belt type in which hot air passes through the moving bed of flake. There are usually zones in the dryer having different temperatures, the maximum being about 100°C. The flake is dried to about 2–5% moisture content.

11.4.2.8 Flake Blending

Since most cellulose acetate processes are of the batch type, it is desirable to blend flakes from many batches. This is done by computer program to obtain a most uniform blended flake on a continuous basis.

11.4.3 Acetic Acid Recovery

The acetic acid recovery system is vital to the economics of cellulose acetate manufacture. Approximately 4–4.5 kg of acetic acid is recovered per kg of cellulose acetate: about 0.5 kg is consumed by the product and the remaining 4 kg is recovered as aqueous acetic acid at about 30% concentration. This aqueous acetic acid is extracted using a liquid–liquid continuous extraction system. Organic solvents are used such as methyl ethyl ketone or ethyl acetate. The upper organic phase containing the extracted glacial acetic acid is fractionally distilled to recover the solvents for recycling. The residual acetic acid from the distillation goes to storage. The bottom aqueous layer from the extraction, which contains salts mostly from neutralizing the sulfuric acid catalyst, is usually discarded.

Part of the acid recovered must be converted to acetic anhydride. This is done efficiently by catalytic pyrolysis. An anhydride process from coal synthesis gas is also available [25].

11.4.4 Triacetate Process

In the manufacture of cellulose triacetate using sulfuric acid catalyst in the batch process, there is a stabilization step following the acetylation. The residual sulfate groups are hydrolyzed (called desulfation) at an elevated temperature by slow addition of dilute aqueous acetic acid solution containing magnesium or sodium acetate or other suitable base to neutralize the liberated sulfuric acid [26,27].

11.5 CELLULOSE ACETATE AND CELLULOSE TRIACETATE CHARACTERIZATION

11.5.1 Acetyl Value

Degree of acetylation or substitution (DS) is specified by two terms: acetyl value (%) and combined acetic acid (%). The ratio of these terms is determined by the molecular weight ratio of acetyl group to acetic acid, $(CH_3CO)/(CH_3COOH)$, or 43/60.

$$\text{Acetyl value (\%)} = 4304 * DS/(159.1 + 43.04 \times DS)$$

Commercial cellulose triacetate has a DS of about 2.91–2.96, which corresponds to an acetyl value of 44.0–44.4% acetyl and a combined acetic acid of 61.4–62.0%. Note that a full triacetate with DS of 3 has an acetyl value of 44.79% and a combined acetic acid of 62.50%. Commercial cellulose acetate or secondary cellulose acetate has a DS of about 2.4, which corresponds to an acetyl value of 39.3% and a combined acetic acid of 54.8%. The values could be slightly higher or slightly lower, depending on the particular manufacturer.

11.5.2 Solubility

Cellulose triacetate is soluble in several organic solvents. Methylene chloride or the mixed solvent of 9/1 methylene chloride–methanol by weight is used for preparing dopes of cellulose triacetate for spinning. Methanol enhances the solubility of cellulose triacetate in methylene chloride. Cellulose triacetate is also soluble in chloroform, sym-tetrachloroethane, trichloroethanol, dimethylformamide, trioxane, sulfane, dimethylacetamide, formic acid, and acetic acid.

The universal solvent for cellulose acetate is acetone. Spinning dopes are prepared by dissolving the flake in an acetone–water solvent composition of about 95/5. The small amount of water significantly reduces the viscosity of a concentrated solution of cellulose acetate such as for a 25% solution. Other solvents for cellulose acetate are ethyl methyl ketone, dioxane, pyridine, nitroethane, dimethylformamide, dimethylacetamide, formic acid, and acetic acid. Although methylene chloride alone is not a good solvent for cellulose acetate, 9/1 methylene chloride–methanol is a good solvent.

11.5.3 Intrinsic Viscosity

The IV of cellulose triacetate is determined by measuring the viscosity of a dilute solution in 9/1 methylene chloride–methanol. The IV is usually in the range of 2.0–2.4 dl/g. The number average DP for cellulose acetate and cellulose triacetate is about 400–500. The IV of cellulose acetate is determined by measuring the viscosity of a dilute solution in acetone. The IV is usually in the range of 1.5–1.7 dl/g. The number average DP for cellulose acetate is about 200–250.

11.5.3.1 Solution and Dope Properties for Cellulose Acetate

The major solution and dope properties that are evaluated when studying flake properties are solution haze and color, the false viscosity effect, and filterability of the acetone dopes.

Haze refers to the turbidity of the acetone solutions. Haze is conveniently measured on a 12% solution of cellulose acetate in acetone using the Hunter colorimeter (Hunter Lab Model D25D2P Color/Color Difference Meter). Neal [28] has demonstrated that haze is caused mainly by xylan acetate present in the cellulose acetate. Conca et al. [29] and, later, Wilson and Tabke [30] studied the more specific hemicelluloses that cause haze in greater detail. They found that, for sulfite wood pulps, 4-O-methylglucuronxylan and glucomannan cause haze, with glucomannan predominating. In PHK wood pulps, xylan and glucomannan cause haze, with xylan predominating.

Solution color refers to the yellowness of the cellulose acetate solution. The most meaningful measurement is the yellowness index (YI) for an 18% solution of cellulose acetate in 9/1 methylene chloride–methanol (by weight). Measurement is made by the Hunter colorimeter described above. Hemicelluloses in the pulp affect the color of the solution of cellulose acetate, but there are other factors that relate to solution color including brightness of the wood pulp and cellulose acetate processing conditions.

False viscosity effect, or anomalous viscosity, refers to viscosities of concentrated cellulose acetate solutions, or dopes. False viscosity effect is measured by the ball fall method using 18% cellulose acetate solutions in acetone containing 1.35% water. False viscosity effect is defined as the difference in percent between the observed viscosity and the predicted viscosity had cotton linters been used as the acetate furnish [31]. A cellulose acetate made from cotton linters gives a lower viscosity in 95/5 acetone–water (by weight) than a cellulose acetate made from wood pulp of the same IV. Some factors other than average chain length contribute to the viscosities of the dopes. The additional viscosity over that contributed by the average chain length is called the false viscosity effect, or anomalous viscosity. Malm et al. [32] and also Lohmann [33] have attributed anomalous viscosity of wood-pulp acetate in highly concentrated solutions to high carboxyl content and subsequent cross-linking through the formation of salts with bivalent cations such as calcium and magnesium ions. Steinmann and White [34] showed that mannan in wood pulp also increases the anomalous viscosity of the

TABLE 11.2
Comparison of Cellulose Acetate Solution and Dope Properties for an Acetate-Grade Softwood Sulfite Pulp and an Acetate-Grade PHK Hardwood Pulp

Cellulose catalyst level during acetylation	Acetate-grade softwood sulfite pulp		Acetate-grade hardwood PHK pulp
	Low catalyst (4.2% H_2SO_4)	High catalyst (14.0% H_2SO_4)	High catalyst (14.0% H_2SO_4)
Key cellulose analyticals			
R10 (%)	95.2	95.2	97.7
S18 (%)	2.9	2.9	1.2
Elrepho brightness (%)	95.0	95.0	94.5
Solution properties			
Hunter haze in 12% acetone solution[a]	23.3	16.0	20.1
YI in 9/1 $MeCl_2$/MeOH solution[b]	14.1	15.0	14.6
Acetone dope properties			
False viscosity effect (%)	150	80	50
Filterability (g/cm^2)	360	320	330

[a] ASTM Method D 1003.
[b] ASTM Method D 1925.

cellulose acetate, whereas xylan does not. Neal [28] confirmed that mannan acetate causes anomalous viscosity. Wilson and Tabke [30] showed specifically that glucomannan in either sulfate or PHK pulps causes false viscosity and that 4-O-methylglucuronoxylan in sulfite pulps or xylan in PHK pulps has practically no effect on false viscosity.

Filterability is measured by determining the plugging value of an 18% solution of cellulose acetate in acetone in which the solution contains 1.35 water. The plugging value test is conducted at constant pressure (40 psig) through 5 μm polypropylene filtering medium at room temperature. Reduced plugging is related to improved spinning performance, fewer spinneret changes, and fewer filter media changes.

A comparison of cellulose acetate solution properties as a function of wood-pulp source and acetylation catalyst level is shown in Table 11.2.

11.6 DOPE PREPARATION

The basic principles are the same for the preparation of both the cellulose acetate dope and the cellulose triacetate dope with the exception of the particular solvent mixture used for each. The flake, the solvent mixture, and a filtering aid are added to a heavy-duty mixer. The solution is prepared in a fully enclosed system to minimize solvent losses and also to meet strict exposure levels regarding the workers. In the case of cellulose acetate, the main solvent is acetone, which is highly flammable. Therefore, the vapor–air ratio must be maintained at a level that meets safety regulations. Strict fire codes are maintained in the dope-preparation department as well as in the fiber-spinning department.

The typical solvent composition for cellulose acetate is about 95% acetone and 5% water, and the typical solids concentration in the extrusion solution is about 20–30%, depending on the polymer molecular weight. The viscosity of the solution at room temperature was reported to be about 100–300 Pa·s (1000–3000 poise). Concentration, temperature, and mixing uniformity are closely controlled through several mixing stages [15,35].

Variations in false viscosity are reduced when the solution is delivered to the spinneret at spinning pressure. Sprague and Homer [36] reported that the anomalous viscosity of dopes from acetate-grade wood pulps is not a problem because under a shear stress of greater than 3000 dynes/cm^2 the solution from an acetate-grade wood pulp and that from cotton linters have identical viscosities and behave identically in plant operations.

Cellulose triacetate is most often spun from about a 20% solution in methylene chloride containing 5–15% methanol [35–38]. Concerns with oral toxicity of methylene chloride have led to the termination of the only commercial manufacture of cellulose triacetate fiber in the United States although manufacture exists elsewhere in the world [39,40].

Both bright and dull yarns for cellulose acetate and cellulose triacetate are produced. In the case of bright yarn, the spinning dope contains cellulose acetate or triacetate and the solvent mixture. For dull yarn, about 1–2% finely ground titanium dioxide pigment is added to the dope. It is necessary that the pigment be milled to a very small particle size, 1–2 μm, or even smaller, so that it can pass through the filtration system as well as the small jet holes of the spinneret. There are two methods that can be used to prepare the spinning dope containing the delustrant (titanium dioxide). One is to add the TiO_2 in the initial mixing stage, a procedure that has the advantage of giving a very uniform dispersion of the pigment in the dope. However, a disadvantage is that completely separate dope-preparation systems from the very initial stages of dope preparation are necessary. Greater flexibility to the dope-preparation department is achieved by injecting a pigment slurry after filtration [35]. However, it is necessary that thorough mixing as well as strict control of the pigment slurry is obtained by the latter method. The delustering solution is prepared by adding flake and titanium dioxide to the solvent in a ball or pebble mill and then blending to exact specification [38].

Besides dull yarn, black fiber is also produced. It is produced in a similar manner to that of dull-yarn production by adding 2% (based on the weight of cellulose acetate) of very finely dispersed carbon black with an average particle size of 1–2 μm to the extrusion dope [35].

The production of acetate and triacetate fibers in solution-dyed colors has been described. This method of coloration is accomplished by adding suitable coloring agents, usually organic or inorganic pigment, to the spinning dope prior to extrusion, which is sometimes called "dope dyeing." It was also reported that in Europe, where the dyeing of acetate and triacetate fiber is relatively more expensive than in the United States, extensive use was made of soluble dyes instead of pigments in the solution-dyeing process. These solvent soluble dyes have the advantage of low cost and ease of handling since they do not have to be dispersed in the acetate or triacetate before use. The most important parameter about spun-dyed yarns is precise control of color from batch to batch for a given dyed fiber [36].

Extensive filtration of the dope is needed to remove mechanical impurities, unacetylated particles, gels, and dirt. The filtration must be thorough enough so that the dope passes through the fine holes in the spinneret without causing jet blockage. The fine jet holes range in size from 30 to 80 μm in diameter [15]. There are generally three stages of dope filtration using plate and frame filters. Cotton batting and cotton fabric or filter paper are generally used as the filtering media. Between filter stages are holding tanks that also serve to deaerate the dope. The dope viscosity is maintained as low as is practical by warming. Lower viscosity also facilitates deaeration and increases the flow rate of the dope. Just prior to spinning, there are two more filters to prevent jet blockage. For the removal of any small particles that may have passed through the primary stages of filtration, a filter is placed in the fixture to which the spinneret is fastened. Another filter is placed in the spinneret assembly over the top of the spinneret itself [35].

11.7 SPINNING OF CELLULOSE ACETATE AND TRIACETATE

11.7.1 Dry-Spinning

Cellulose acetate and triacetate fibers are produced by dry-spinning. A schematic diagram of the extrusion process is shown in Figure 11.2 [41]. The preheated dope is extruded through the fine holes of the spinneret into a heated column or cabinet. Each of the emerging fine dope streams encounters a flow of hot air that brings about an instantaneous evaporation of the solvent from the surface of each stream, resulting in immediate filament formation with a solid skin forming over the still plastic interior of the filament. As the filaments pass down the column, solvent is increasingly removed from the interior of the filaments, thereby continually solidifying and strengthening them. Figure 11.3 shows a typical dry-spinning column with the

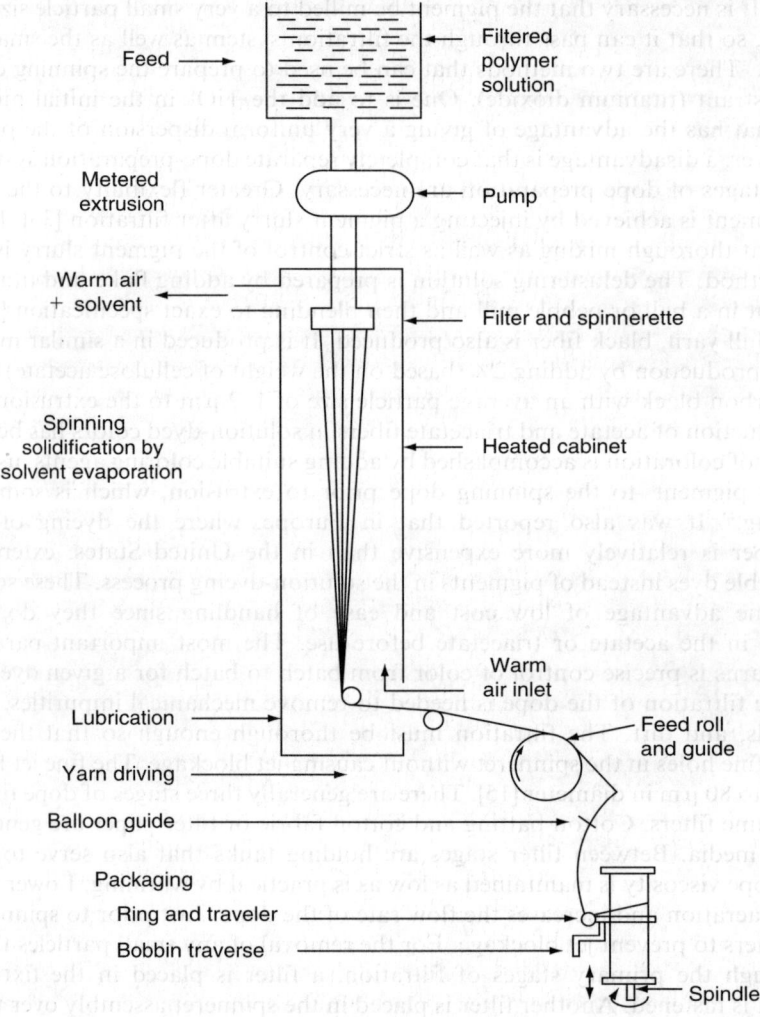

FIGURE 11.2 Dry extrusion–dry-spinning of cellulose acetate fibers. (From C.E. Schildknecht, ed., *Polymer Processes*, Interscience, New York, 1956, p. 841.)

Cellulose Acetate and Triacetate Fibers

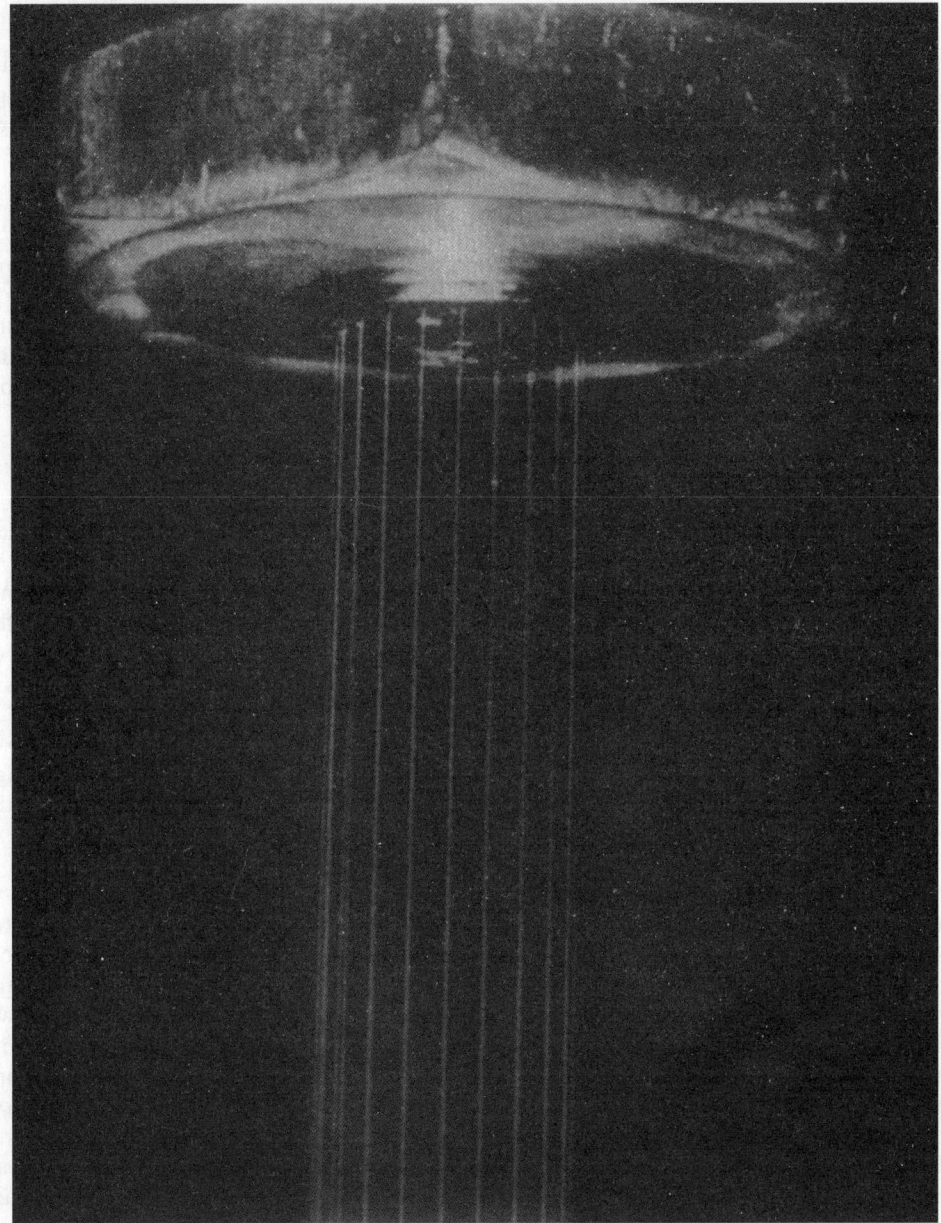

FIGURE 11.3 Dry extrusion filaments emerging from spinner. (From G.M. Moelter and R. Steele, In J. McKetta, ed., *Encyclopedia of Chemical Processing and Design*, vol. 1, Marcel Dekker, New York, 1976, p. 171.)

filaments emerging from the spinneret. Near the bottom of the column, the individual filaments converge to form a yarn, which is a bundle of parallel filaments. The yarn passes over a lubrication wheel and is pulled at constant speed by the feed roll. It then passes through a balloon guide and is taken up on a ring and traveler and wound on a bobbin. There are certain fundamental aspects of dry-spinning worth detailing as described below.

11.7.1.1 Dope Preheating

The dope is preheated before extrusion to lower its viscosity and thus reduce the extrusion pressure. Preheating the dope also supplies part of the heat required for evaporating the solvent during extrusion. The dope is heated by a thermostatically controlled preheater or by locating the final filter and spinneret assembly inside the heated cabinet [35].

11.7.1.2 Dope Metering

The deaerated dope is metered by positive displacement pumps to hundreds of individual yarn extrusion positions from the manifold. It is necessary to have a separate pump for each spinneret position to ensure uniform fiber formation and denier. Good control and maintenance of the metering pumps are necessary to produce uniform, high-quality yarn products. Moelter and Steele [35] commented that, although the principal of extrusion is simple in conception, the precision of the operation reflects the ingenuity of its engineering.

11.7.1.3 Spinneret Design

The spinnerets can have a variety of designs, depending on the yarn products produced. The spinneret, or jet, consists of a certain number of precisely machined holes or orifices. The number of holes can vary from 13 to several hundreds [35]. Each hole diameter is the same for a given spinneret. However, the hole diameter can range from 30 to 80 μm for different spinnerets [15]. The design of the spinneret hole relates to the flow pattern of the dope through the hole and also to the cross section of the yarn produced. Spinneret holes that are round in cross section give yarn having a roughly round, crenulated cross section. The crenulation is the result of the rapid setting-up of a thin, highly viscous skin and the subsequent volume reduction due to solvent evaporation through the rigid skin. A triangular cross-sectional hole in the spinneret produces yarn with cross section in the form of a three-pointed star [35]. The various yarn cross sections govern important features of the final fabrics such as covering power and hand or the surface area and packing of fibers in a tow. Jet or spinneret design is a basic fundamental of dry-spinning.

In commercial processes, the dope flow in the spinneret hole reaches wall shear rates of over 1 million reciprocal seconds. Dope rheology studies show that significant shear thinning occurs at these shear rates, reducing the viscosity and the pressure required for spinning. Viscoelastic effects occur at these shear rates giving rise to such phenomena as die-swell where the diameter of the dope extruding from the spinneret hole expands by a factor of nearly 2, requiring more draw-down in the spin-line to get to the target filament size [42–44]. Spinneret designs have been described to offer improved spinning performance by streamlining flow in the spinneret to reduce the elastic effects [45–47].

The spinnerets are made of stainless steel or other suitable metals. Selection of material for spinnerets is made on the basis of resistance to corrosion by the solvent system and ease of machinability. The spinneret material must be readily machinable to close tolerances because each hole of a spinneret is precision-made to the specific size and shape.

11.7.1.4 Spinning Column or Cabinet

The air column or cabinet height is 2 to 8 m, depending on the extent of drying required and the spinning speed [15]. More than 80% of the solvent can be removed from the filaments during the brief residence time of less than one second in the hot-air column. The air flow may be concurrent or countercurrent with the direction of the travel of the filaments. In some cases, air enters at the top and at the bottom of the cabinet and is withdrawn at the middle

[35]. The fiber properties are contingent on the solvent removal rate, and precise air flow and temperature control are necessary [15].

11.7.1.5 Volatility of the Solvents

The high volatility of acetone for cellulose acetate spinning and, likewise, the high volatility of methylene chloride and methanol for cellulose triacetate spinning greatly enhance the ease of fiber formation and solvent removal at moderate temperatures, resulting in reasonably high spinning speeds.

11.7.1.6 Spinning Parameters Versus Yarn Properties

In the spinning process, the denier of the yarn is determined by the dope concentration, the metering pump output, and the spinning speed. However, the properties of the yarn and even the denier which can be produced in a given process are dependent on many interrelated spinning parameters including dope rheological properties, temperature, concentration, feed-roll or spinning speed, spinneret hole diameter and design, number of holes, column height, and conditions of air temperature and flow. The feed-roll applies tension to the yarn to draw it down to the desired denier and to withdraw it from the cabinet. Mathematical modeling of the dry-spinning process based on consideration of the fundamental force, heat and mass transfer has been successful in relating these process parameters and has assisted in the design of dry-spinning processes for cellulose acetate. Fiber strength and elongation as well as specific surface area and cigarette filter performance are related to stress, concentration, and temperature profiles down the spin-line and across the radius of the fiber predicted by the model [48–50].

11.7.1.7 Yarn Lubrication and Finish

Yarns must be lubricated to impart frictional and antistatic properties necessary for further successful processing. The lubricant is applied precisely, in amounts ranging from 1 to 5%, from a wick or wheel positioned just beyond the point where the yarn emerges from the cabinet. Different lubricants are used depending on the type of application intended for the yarn.

11.7.1.8 Packaging

When the fibers are spun, they contain residual spinning solvent, and yarn strength is at a minimum. Therefore, it is necessary to impart some twist to the substantially parallel bundle of filaments or to entangle the filaments to hold the yarn together and to build-in sufficient strength for packaging. The yarn taken up on a ring twister obtains enough twist to be wound, without difficulty, onto the bobbin. Instead of the application of twist, the yarn filaments can also be entangled via impingement with air jets for the ease of handling [51]. A metier is an array of individual extrusion positions on one common machine. There are usually 100–200 such positions [15]. Each position produces a full bobbin or tube of yarn on a time-schedule basis, thereby maintaining a constant production per package for a given yarn product.

11.7.1.9 Solvent Recovery

For acetate spinning, it is necessary to recover 3 kg of acetone for each kg of fiber produced. The solvent-laden air exiting from the cabinet is passed to adsorption beds of activated carbon. When the beds are saturated with solvent, they are stripped of the solvent with steam. The resulting vapors go to a fractionating column where the solvent is recovered. Sometimes, the solvent-laden air exiting from the cabinet is refrigerated to condense and recover the solvent. In any case, it is necessary to have an efficient recovery of the solvent. Recovery is reported to be about 99% efficient [15]. Acetone offers the advantage of ready adsorption for recovery and having a low boiling point for enhancing its removal from the beds of activated carbon by steam; it is also noncorrosive.

11.7.1.10 Staple and Tow

The production of cellulose acetate or triacetate staple follows the same basic principles as those for the production of continuous filament yarn. Dry-spinning by extruding the dope through a spinneret into a hot-air column and over the feed-roll is still done; usually there are more holes per spinneret for tow compared with continuous filament. Instead of collecting and winding the yarn from a single spinneret, the yarn-filament ends from a number of spinnerets are gathered together into a ribbon-like strand or tow. Staple is produced by cutting the tow, which may be crimped, into short lengths (about 4–5 cm) and baling the staple fiber. Both acetate and triacetate staple fibers are produced.

Cigarette tow is produced from acetate fibers. Cigarette tow is a continuous band composed of several thousand filaments held loosely together by crimp—a wave configuration set into the band during manufacture. A tow is formed by combining the output of a large number of spinnerets and crimping the collection of filaments to create an integrated band of continuous fibers. The tow is then dried and baled [15].

11.7.2 WET-SPINNING

Although wet-spinning of cellulose acetate or triacetate is feasible, it is not practiced to any large extent because the economics are not as favorable due to the low linear speeds necessitated by passing the yarn through the liquid bath.

11.7.3 MELT-SPINNING

Neither acetate or triacetate has good stability in the melt. Severe discoloration and considerable decomposition of the melt occur, especially if held long as a melt. Therefore, melt-spinning of cellulose acetate or triacetate is not considered to be an attractive method for producing fibers. However, a relatively small amount of triacetate yarn has been made by melt-spinning. Special techniques are necessary to prevent degradation at high temperatures at which triacetate melts (>300°C). Therefore, a primary requirement is to hold the polymer in the molten state for only a short time [36].

11.8 YARN TYPES AND PACKAGES

Yarns are available in a wide range of weights and thicknesses. There are two common systems for describing th yarns: denier and tex. Denier is the weight in grams of 9000 m of yarn. Tex is the weight in grams of 1000 m of yarn. Decitex is the weight in grams of 10,000 m of yarn. Thus, a yarn having a denier of 144 has a tex of 16 and a decitex of 160. Since the yarn thickness also depends on the number of holes in the spinneret, yarns are also defined as the denier per filament (dpf) or tex per filament. Thus, a spinneret having 40 holes producing a denier of 144 has a dpf of 3.6 or 0.4 tex per filament. Cellulose acetates are produced from 36 to >90 denier or 4 to >100 tex. Individual filament deniers are usually 2–4 dpf. Common continuous filament yarns have total deniers of 55, 60, 75, and 150 [15].

The principal package types used by the textile industry are tubes, cones, and beams. Tubes contain 3 to 18 lb of yarn. They are about 4 in. height with a diameter of 7–10 in. [35]. The yarn is traversed on the package to provide stability and ease of removal. Zero-twist entangled yarn is usually packaged on tubes for use in circular knits and on tricot and section beams for warp knits [15]. Cones may contain from about 2–18 lb of yarn. During the conversion of bobbins to cones, further twist of the yarn is generally applied. The yarn twist in the bobbins is only about 0.2 turn per inch (tpi), whereas, during coning, the yarn twist is increased to the range of 0.8–20 tpi [35]. Inserting twist in the yarn increases the yarn strength. Also, during coning most of the residual acetone is removed from the acetate yarn. Both entangled and twisted yarns are packaged on cones. Beams are large spools for packaging yarn. They are constructed of aluminum alloy. The most common size of beam used by the tricot knitting trade measures 42 in. width and 21 or 30 in. diameter. Section beams for weaving are usually 54 in. wide and 30 in. diameter. Hundreds of individual yarn-ends are continuously fed parallel to one another and wound on the beam in a very precise manner. A beam may contain as many as 2400 individual yarn-ends, and, when filled, a beam may hold 250–1500 lb of yarn. The length of yarn on beams varies with the yarn denier, the beam capacity, and the intended use. Yarn lengths ordinarily are in the range of 12,000–85,000 yards [35].

11.9 CELLULOSE ACETATE AND TRIACETATE PROPERTIES

11.9.1 OVERVIEW

The interactions among fibers in a fabric array are complex but the performance of the fabric reflects in part the inherent properties of the fiber itself as well as how the fibers are assembled. An overview of the fiber properties for cellulose acetate and triacetate is given in Table 11.3. These fibers are not particularly strong like polyester and nylon, although they have ample strength for their intended end-uses. The key words for cellulose acetate and triacetate fibers and fabrics are silk-like softness, pleasing appearance, and comfort. These fabrics are characterized in the trade as having superb, soft hand, and good draping quality. The overall characteristic for both fibers is simply an excellent balance of properties at economic cost.

Cellulose acetate is not too hydrophobic nor is it hydrophilic. This balance provides a moisture-regain that is responsible for its excellent comfort factor. This balance also results in easy dyeability of the fibers and the resistance of these fibers to microorganisms and insects. Cellulose triacetate is more hydrophobic than cellulose acetate but is not as hydrophobic as polyester. It has enough hydrophobicity that the heat-treated fabrics possess "wash-and-wear" and "ease-of-care" characteristics.

11.9.2 FIBER CROSS SECTIONS

The fiber cross section affects fabric appearance, hand, drape, flexibility, and moisture transport. Cellulose acetate and cellulose triacetate (Arnel) fibers from Celanese Corporation were analyzed by scanning electron microscopy (SEM). The SEM micrographs are shown in Figure 11.4. These were identified as follows: cellulose acetate (CA) 75/25/20 Lot 2575A and cellulose triacetate (Arnel) 150/2M/22 Lot 15163K. The micrographs A and B apply to cellulose acetate, and magnifications are 540× and 2000×, respectively. Micrographs C and D apply to cellulose triacetate, and magnifications are 540× and 1800×, respectively. The main features shown by the micrographs are: (1) both the CA and Arnel fibers have irregular crenulated cross sections; the generally lighter edges may be due to contrast variation between

TABLE 11.3
Overview of Fiber Properties for Cellulose Acetate and Triacetate

Property	Cellulose acetate	Cellulose triacetate
Fiber crosssection	Both are irregular crenulated with skin from round jet holes.	
Fine structure	Very low order of crystallinity and fiber orientation	Heat-treated fiber has high order of crystallinity and orientation
Appearance and color	Both bright yarns have outstanding luster; dull yarns are very white and uniform; yarns have soft silk-like appearance.	
Absorption and swelling behavior	Moisture-regain,[a] 6.5% water imbibition,[b] 24%	Moisture-regain,[a] 3.5% water imbibition:[b] Nonheat-treated, 16% Heat-treated, 10%
Tenacity and elongation	Similar for both. Typical range: 1.13–1.36 g/den (65% RH, 21°C) Similar for both. Typical range: 25–45% (65% RH, 21°C)	
Specific gravity	1.32	1.30
Refractive index	Parallel to fiber axis, 1.478 Perpendicular to fiber axis, 1.473 Very little birefringence	Parallel to fiber axis, 1.472 Perpendicular to fiber axis, 1.471 Practically no birefringence
Thermal behavior	Softening point, 190–205°C Melting point, ca. 260°C	Softening point: Nonheat-treated, 190–205°C Melting point, ca. 305°C
Electrical behavior	Very high electrical resistivity. Typical value, 1.27 ohm·cm	Very high electrical resistivity Typical values: Nonheat-treated, 3.81 ohm·cm Heat-treated, 15.2 ohm·cm
Dyeability	Both can be spun-dyed. For both dyeing of fibers and fabrics mostly disperse dyes are used. Attractive colors and shades with excellent color fastness are obtained	
Light resistance	Good. Both are similar to cotton and better than rayon, better than pigmented nylon and silk, but less resistant than acrylics and polyester. Relative humidity and temperature are important	
Resistance to microorganisms and insects	High; must be more resistant than cotton	Very high, approaches polyester, acrylics, and nylon in resistance

[a]Moisture-regain according to ASTM D 1909–68.
[b]Water imbibition values represent equilibrium established between fiber and air at 100% RH while the fiber is being centrifuged at forces up to 1000g.

the outer skin and inner core typical of acetate fibers; (2) because of the depth of the field, one can see not only the cross sections but also the longitudinal surfaces of the fibers; the "column" effect confirms the crenulated cross section; and (3) there do not appear to be any basic differences in the SEMs between CA and Arnel. It should be recognized that fiber

Cellulose Acetate and Triacetate Fibers

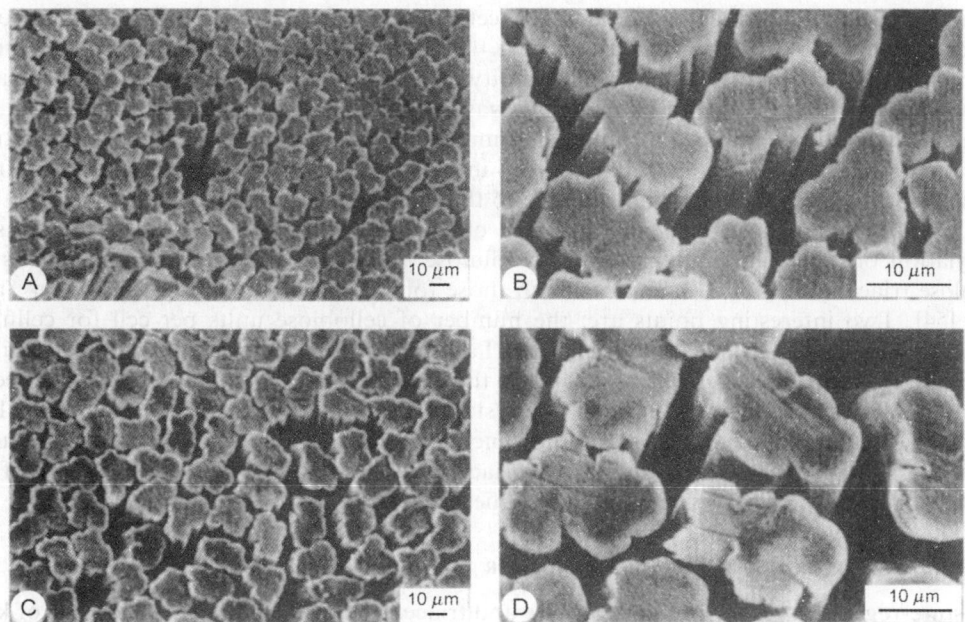

FIGURE 11.4 SEM micrographs.

cross sections depend on the particular geometry of the spinning jet orifices. Therefore, fibers will vary in appearance from one particular jet orifice to another. In fact, SEM is an excellent diagnostic tool for distinguishing between cross sections of the same polymeric fiber spun from different jet orifices and also between cross sections of different polymeric fibers spun from the same type of jet orifice.

11.9.3 Fine Structure of Cellulose Acetate and Triacetate Fibers

11.9.3.1 Cellulose Acetate

Although cellulose from either cotton linters or wood pulp is highly crystalline, dry-spun cellulose acetate fiber has very little crystallinity and fiber orientation. It is apparent that the acetylation process followed by partial hydrolysis of the triacetate to a secondary acetate destroys the fine structure of the original cellulose. Even annealing the fibers or dry-stretching the yarn just short of the breaking point did not increase the crystallinity of cellulose acetate. Only orientation of the already existing small amount of crystallinity was increased by these treatments. The basic reason for the lack of crystallinity and order in cellulose acetate is the heterogeneity in the molecular segments along the chain molecules. Regeneration of cellulose by saponification of cellulose acetate fiber done under stretch results in a very highly crystalline and highly orientated cellulose having unusually high tensile strength (Fortisan fiber) [52].

11.9.3.2 Cellulose Triacetate

Cellulose triacetate dry-spun fiber before any heat treatment has very low crystallinity. However, annealing the fiber at elevated temperatures markedly increased the crystallinity [53]. The development of high crystallinity in the fiber has beneficial effects on the fiber and fabric properties. Heat-treated cellulose triacetate yarns possess a higher softening point, thereby increasing their resistance to heat. Heat treatment also lowers the moisture-regain, thereby improving dimensional stability. The practical result of heat treatment is to impart the so called "ease-of-care" and "wash-and-wear" characteristics to fabrics and garments made from cellulose triacetate. An important point is that, although the moisture-regain is lower, it is not too low. This results in outstanding comfort for fabrics made from cellulose triacetate, which is much superior to that for polyester.

Detailed studies have been made of the effects of cellulose source and conditions of acetylation on the crystalline structure of cellulose triacetate. The unit cell dimensions for cellulose triacetates I and II compared with those for celluloses I and II are shown in Table 11.4 [54]. Two interesting points are: the number of cellobiose units per cell for cellulose triacetates I and II is 4, versus 2 for celluloses I and II; and the measured density for cellulose triacetate II was 1.315 g/cc, which is less than the calculated density of 1.348 g/cc as expected because cellulose triacetate is not 100% crystalline. The above studies on the crystalline structure of cellulose triacetate lead to the conclusion that commercial heat-treated cellulose triacetate is expected to have the cellulose triacetate II crystalline structure. Analysis of the crystal structure of cellulose triacetate continues [55].

11.9.4 Absorption and Swelling Behavior

Moisture-regain curves for cellulose acetate and triacetate fibers compared with those of kier-boiled cotton are shown in Figure 11.5 [56,57]. The curves for acetate fiber are positioned only slightly below those for cotton and actually mesh for intermediate values of relative humidity. Since moisture regain relates to the comfort factor of fabrics, the curves explain why cellulose acetate, like cotton, possesses this important characteristic.

The moisture-regain curves for triacetate fiber are positioned below those of acetate fibers, as expected, because triacetate is more hydrophobic. However, the curves for triacetate show substantial moisture regain at high relative humidity. Therefore, triacetate fabrics also possess the comfort factor, even though they have the "wash-and-wear" and "ease-of-care" characteristics.

A comparison of the commercial moisture-regain values for several fibers is shown in Table 11.5 [58]. In general, cellulose acetate occupies an intermediate position, indicating a good balance. Its value of 6.5% is close to the value of 7.0% for natural cotton yarn. The values of 3.5% for triacetate and 4.5% for nylon indicate that these fibers are not nearly as hydrophobic as acrylics at 1.5% and polyester at 0.4% commercial moisture-regain.

TABLE 11.4
Unit Cell Dimensions of Cellulose Triacetates I and II Compared to those of Celluloses I and II

Dimension	Triacetate I	Triacetate II	Cellulose I	Cellulose II
a	22.6 Å	25.8 Å	8.35 Å	8.1 Å
b	10.5	10.5	10.3	10.3
c	11.8	11.45	7.9	9.1
	79	66.4	84	62
Number of cellobiose units	4	4	2	2
Calculated density	1.39 g/cc	1.348 g/cc	—	—

Source: From B.S. Sprague, J.L. Riley, and H.D. Noether, *Text. Res. J.*, 28, 275 (1958).

Cellulose Acetate and Triacetate Fibers

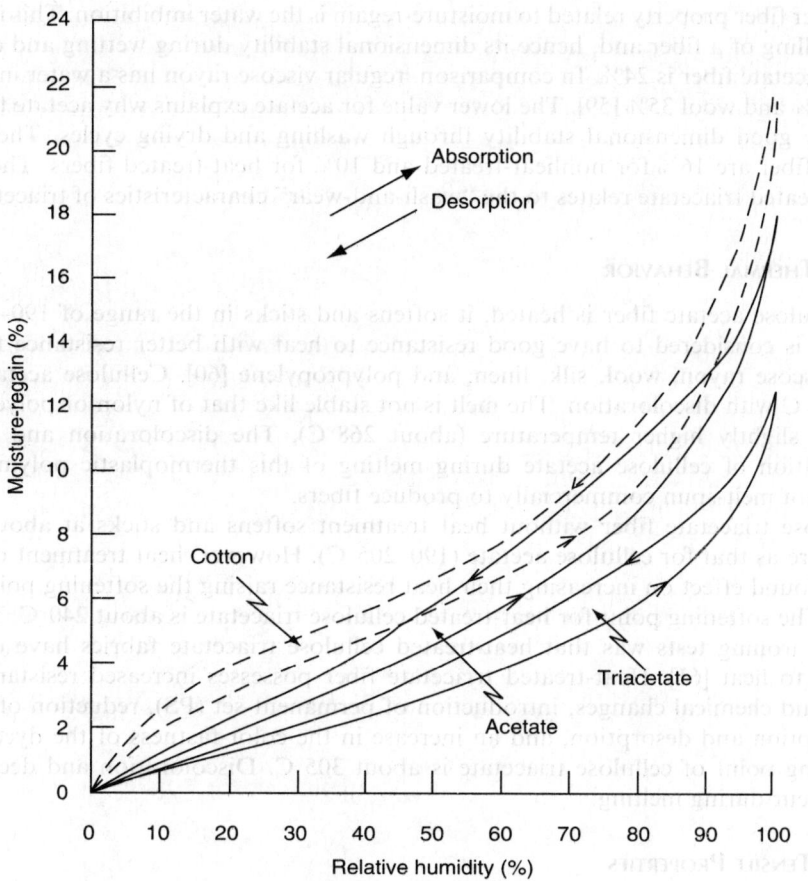

FIGURE 11.5 Comparison of moisture-regain curves for cotton, cellulose acetate, and triacetate. (From Beever, D.K. and Valentine, L. *J. Text. Inst.*, 49, T95, 1958; *Kirk Othmer Encyclopedia of Chemical Technology*, 2nd ed., vol. 4, John Wiley, NewYork, pp. 599–600, 1991.)

TABLE 11.5
Commercial Moisture-Regain Values for Various Fibers

Fiber	(%) Moisture-regain
Wool (all forms)	13.6
Silk	11.0
Rayon (regenerated cellulose)	11.0
Linen	8.75
Natural cotton yarn	7.0
Cellulose acetate	6.5
Nylon	4.5
Cellulose triacetate	3.5
Acrylic	1.5
Polyester	0.4
Olefin	0.0

Source: From ANSI/ASTM D 1909–77, Table of Commercial Regains for Textile Fibers.

Another fiber property related to moisture-regain is the water imbibition. This is a measure of the swelling of a fiber and, hence its dimensional stability during wetting and drying. The value for acetate fiber is 24%. In comparison, regular viscose rayon has a water imbibition of about 120% and wool 35% [59]. The lower value for acetate explains why acetate fabrics have reasonably good dimensional stability through washing and drying cycles. The values for triacetate fiber are 16% for nonheat-treated and 10% for heat-treated fibers. The low value for heat-treated triacetate relates to the "wash-and-wear" characteristics of triacetate fabrics.

11.9.5 THERMAL BEHAVIOR

When cellulose acetate fiber is heated, it softens and sticks in the range of 190–205°C [36]. This fiber is considered to have good resistance to heat with better resistance than nylon, cotton, viscose rayon, wool, silk, linen, and polypropylene [60]. Cellulose acetate melts at about 260°C with discoloration. The melt is not stable like that of nylon or polyester, which melt at a slightly higher temperature (about 268°C). The discoloration and progressive decomposition of cellulose acetate during melting of this thermoplastic polymer explains why it is not melt-spun commercially to produce fibers.

Cellulose triacetate fiber without heat treatment softens and sticks at about the same temperature as that for cellulose acetate (190–205°C). However, heat treatment of the fibers has a profound effect on increasing their heat resistance raising the softening point by 90°F, or 50°C. The softening point for heat-treated cellulose triacetate is about 240°C. The conclusion from ironing tests was that heat-treated cellulose triacetate fabrics have outstanding resistance to heat [61]. Heat-treated triacetate fiber possesses increased resistance to both physical and chemical changes, introduction of permanent-set (PS), reduction of the rate of dye absorption and desorption, and an increase in the color-fastness of the dyed fiber [62]. The melting point of cellulose triacetate is about 305°C. Discoloration and decomposition usually occur during melting.

11.9.6 TENSILE PROPERTIES

11.9.6.1 Tenacity Units of Measurement

A cellulose acetate or triacetate yarn contains a number of filaments, depending on the number of holes in the spinneret. Therefore, it has a certain dpf, and the total denier equals the dpf times the total number of filaments making up the yarn. Yarn tenacity refers to the tenacity for the total denier of the yarn. This is usually expressed as grams/denier (g/den) or grams-force/denier (gf/den) and in the Standard International (SI) units of newtons/tex (N/tex). Other methods of expressing tenacity are centinewtons/decitex (cN/tex), millinewtons/tex (mN/tex), and millinewtons/denier (mN/den). In the first system, the breaking load is expressed in grams. In the SI system, it is expressed in N, cN, or mN. Since 1 kg equals 9.81 N and 1 tex equals 9 denier, the following conversion factors can be calculated:

Conversion	Conversion factor
N/tex to gf/den	Multiply by 11.33
cN/dtex to gf/den	Multiply by 1.133
mN/tex to gf/den	Multiply by 0.0113
mN/dtex to mN/den	Multiply by 1.11

11.9.6.2 Stress–Strain Curves

A perfectly elastic material obeys Hooke's law, which gives a linear stress–strain curve where the stretch is proportional to the stress and the slope is the modulus of elasticity. Most textile fibers are not completely elastic. They may be linear or Hookean under low stresses, but when higher stresses are applied, they elongate out of proportion to the amount of stress applied. This causes the stress–strain curve to become nonlinear. Materials that show this behavior are called viscoelastic. Acetate and triacetate fibers are viscoelastic.

Figure 11.6 shows the typical stress–strain curve for either cellulose acetate or triacetate yarn [35]. The linear part of the curve under low stresses is not as steep as that for cotton or rayon. At about 0.9 g/den stress, the curve starts to show large increments of strain for very small increments of stress. This continues until the fiber breaks at about 1.36 g/den and 26% strain or elongation. Tenacities and elongations for various acetate and triacetate fibers may be slightly different, depending on the particular manufacturing conditions.

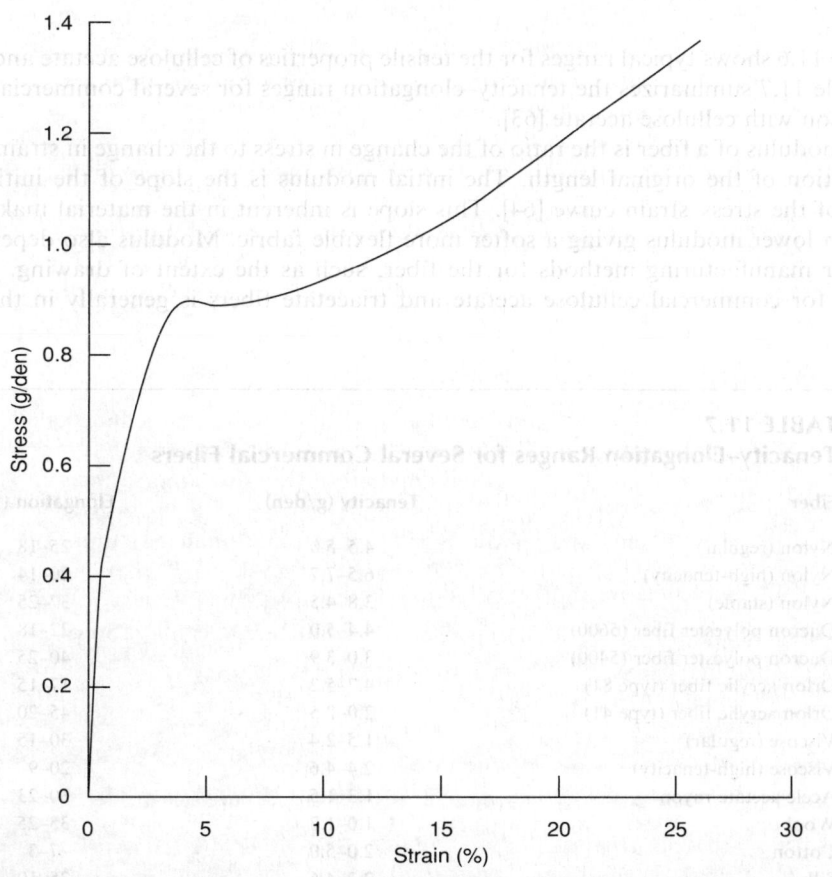

FIGURE 11.6 Stress–strain curve typical of either an acetate or triacetate yarn (Instron Tensile Tester at 60%/min, rate of extension). (From G.M. Moelter and R. Steele, In J. McKetta, ed., *Encyclopedia of Chemical Processing and Design*, vol. 1, Marcel Dekker, New York, 1976, p. 171.)

TABLE 11.6
Typical Tensile Properties for Commercial Cellulose Acetate and Triacetate Fibers

Tenacity	(g/den)	(N/tex)
Standard conditions[a]	1.2–1.4	0.106–0.124
Wet	0.8–1.0	0.071–0.088
Bone dry	1.4–1.6	0.124–0.141
Knot, standard conditions	1.0–1.2	0.088–0.106
Loop, standard conditions	1.0–1.2	0.088–0.106
Elongation at break		Percent
Standard conditions		25–45
Wet		35–50

[a]Standard conditions, also referred to as conditional, represent yarn equilibrated at 65% RH and 70°F.
Source: G.M. Moelter and R. Steele, In J. McKetta, ed., *Encyclopedia of Chemical Processing and Design*, vol. 1, Marcel Dekker, New York, 1976, p. 171.

Table 11.6 shows typical ranges for the tensile properties of cellulose acetate and triacetate [35]. Table 11.7 summarizes the tenacity–elongation ranges for several commercial fibers for comparison with cellulose acetate [63].

The modulus of a fiber is the ratio of the change in stress to the change in strain expressed as a fraction of the original length. The initial modulus is the slope of the initial straight portion of the stress–strain curve [64]. This slope is inherent in the material making up the fiber with lower modulus giving a softer more flexible fabric. Modulus also depends on the particular manufacturing methods for the fiber, such as the extent of drawing. The initial modulus for commercial cellulose acetate and triacetate fibers is generally in the range of

TABLE 11.7
Tenacity–Elongation Ranges for Several Commercial Fibers

Fiber	Tenacity (g/den)	Elongation (%)
Nylon (regular)	4.5–5.0	25–18
Nylon (high-tenacity)	6.5–7.7	20–14
Nylon (staple)	3.8–4.5	37–25
Dacron polyester fiber (5600)	4.4–5.0	22–18
Dacron polyester fiber (5400)	3.0–3.9	40–25
Orlon acrylic fiber (type 81)	4.7–5.2	17–15
Orlon acrylic fiber (type 41)	2.0–2.5	45–20
Viscose (regular)	1.5–2.4	30–15
Viscose (high-tenacity)	2.4–4.6	20–9
Acele acetate rayon	1.3–1.5	30–23
Wool	1.0–1.7	35–25
Cotton	2.0–5.0	7–3
Silk	2.2–4.6	25–10

Source: From L.G. Ray, Jr., *Text. Res. J.*, 22, 144 (1952).

25–45 gf/den [15]. Most yarn properties are reported for yarn equilibrated at standard conditions of 65% relative humidity (RH) and 21°C (70°F).

11.9.6.3 Wet Strength

Wetting a hydrophilic fiber with water causes the tenacity to decrease and the elongation to increase decreasing the initial modulus. These results are expected because the water destroys hydrogen bonding between chains by acting as a wedge between the chains. Therefore, it is easier for the chains to slide over one another as a fiber is put under stress while wet. There is less resistance to deformation so there will be much larger increments of elongation per increment of stress applied. Fiber with lower moisture absorption has less reduction in tensile properties when wet. The wet tenacity curves for cellulose acetate and triacetate fibers tested in water at 20°C are shown in Figure 11.7 [65]. The wet tenacity and elongation for cellulose acetate (Dicel) are 0.7 g/den and 48%, respectively. The initial wet modulus was reported to be 6 g/den. The wet tenacity and elongation for cellulose triacetate (Tricel) are 0.8 g/den and 38% elongation. The initial wet modulus was reported as 20 g/den. The triacetate fiber has a significantly higher wet modulus than secondary acetate fiber, as expected. Higher fiber wet modulus means that a fabric would be expected to have better dimensional stability during washing and drying cycles [65].

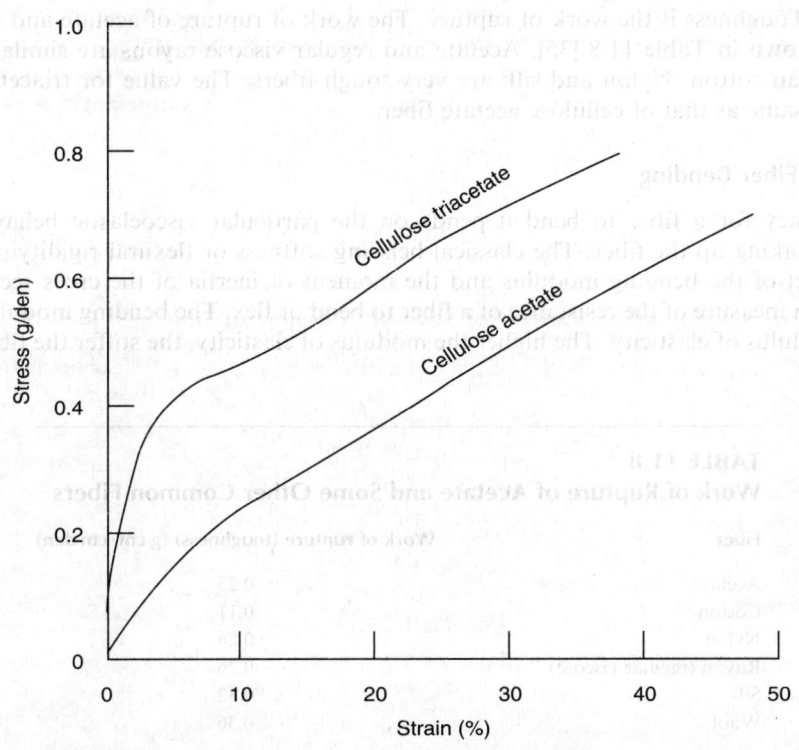

FIGURE 11.7 Comparison of stress–strain curves for cellulose acetate and triacetate yarns tested in water at 20°C. (From Cellulose (Secondary) Acetate (Dicel) and Cellulose Triacetate (Tricel), Technical bulletins published by British Celanese, Coventry, England, 1965.)

11.9.6.4 Tensile Strain Recovery Behavior

Detailed analysis of the viscoelasticity of the fiber as measured by the stress–strain curve, shown in Figure 11.6, has been related to end-use performance as well as to fiber processability. The area under the stress–strain curve represents the work absorbed by the fiber as stress is applied until fiber rupture. But not all of this work is recoverable. When a fiber is stretched under stress and then the stress removed, there are three responses in the behavior of the fiber. These are called immediate elastic recovery (IER), delayed elastic recovery (DR), and PS. DS is also called primary creep and PS secondary creep. The degree of any of the three responses for a given fiber depends on the stress applied. For very low stresses, IER predominates. As the stress increases, IER decreases and DR and PS increase. The PS is at a maximum near or at the breaking load. Molecular mechanisms are discussed to explain the behavior of acetate and triacetate for each of these responses. The PS represents unrecoverable work, that is, that part of the work absorbed by the fiber that is not recoverable. The ratio of the work recovered to the total work absorbed (measured by the respective areas under the stress–strain and stress–recovery curves) is designated as work recovery. Most end-uses and processes should be designed to keep the fiber in the region with no PS. Acetate fiber has zero PS up to 50% of breaking tenacity [36,66–70].

11.9.6.5 Fiber Toughness

Fiber toughness is associated with the area under the stress–strain curve. The larger the area, the tougher the fiber. Toughness represents the ability of a fiber to absorb energy during straining. Toughness is the work of rupture. The work of rupture of acetate and some other fibers is shown in Table 11.8 [35]. Acetate and regular viscose rayons are similar; both are tougher than cotton. Nylon and silk are very tough fibers. The value for triacetate fiber is about the same as that of cellulose acetate fiber.

11.9.6.6 Fiber Bending

The tendency for a fiber to bend depends on the particular viscoelastic behavior of the material making up the fiber. The classical bending stiffness or flexural rigidity of a fiber is the product of the bending modulus and the moment of inertia of the cross section of the fiber. It is a measure of the resistance of a fiber to bend or flex. The bending modulus depends on the modulus of elasticity. The higher the modulus of elasticity, the stiffer the fiber, with all

TABLE 11.8
Work of Rupture of Acetate and Some Other Common Fibers

Fiber	Work of rupture (toughness) (g·cm/cm·den)
Acetate	0.25
Cotton	0.11
Nylon	0.86
Rayon (regular viscose)	0.26
Silk	0.82
Wool	0.36

Source: From G.M. Moelter and R. Steele, In J. McKetta, ed., *Encyclopedia of Chemical Processing and Design*, vol. 1, Marcel Dekker, New York, 1976, p. 171.

other factors the same. The moment of inertia depends on the cross-sectional area of the fiber. Of course, the greater the cross-sectional area, the greater the moment of inertia and, hence, the greater the stiffness. Therefore, the flexural rigidity varies directly with the denier of the fiber. The flexural rigidity or bending stiffness for cellulose acetate fibers as a function of denier is shown in Table 11.9 [35].

11.9.6.7 Abrasion Resistance

The abrasion resistance of cellulose acetate is lower compared with that of other fibers. Abrasion resistance was measured by the wet-flex abrasion determined with the Stoll Abrasion Tester. Abrasion resistance of several fibers was rated in the following decreasing order: nylon, polyester fiber, acrylic fiber, wool, cotton, viscose rayon, and acetate. It was suggested that the abrasion resistance of fabrics is related to the strength and the recovery properties of fibers. The fact that acetate is not a particularly strong fiber probably accounts in part for its inferior abrasion resistance. Heat-treated cellulose triacetate fabrics have both higher tensile strength and abrasion resistance than secondary acetate fabrics for the conditions of dry, wet, and hot wet (80°C) [53,63].

11.9.7 DYEING CHARACTERISTICS

Cotton and rayon are dyed readily with dyes that are soluble in the aqueous dye-bath such as the class of direct dyes. The apparent reason for the success of water-soluble dyes with cotton or rayon is that cellulose swells appreciably in the aqueous dye-bath, thereby enhancing diffusion of the large dye molecules into the interior of the fibers. Cellulose exhausts the dye from the bath rapidly, resulting in deep color shades for cotton and rayon. For the most part, the classes of soluble dyes are not applicable to the more hydrophobic cellulose acetate and triacetate fibers and fabrics [37,71].

The preferred dyes for cellulose acetate and triacetate are the disperse dyes. These consist mainly of azo, anthraquinonoid, and nitrodiphenylamine dyes that are insoluble or at best slightly soluble in hot aqueous dye-bath systems. These crystalline dyes must be ground to a very small particle size, e.g., 2 μm. They are then pasted with water and a wetting agent and added to the dye-bath through a fine mesh screen. A dispersing agent and a carrier or accelerant may be present in the dye-bath to enhance diffusion of the dye molecules into

TABLE 11.9
Flexural Rigidity of Cellulose Acetate Fibers as a Function of Denier

Fiber denier	Flex rigidity (g·cm^2 × 10^4)
3.75	0.35
5	0.56
10	2.5
15	7.4
20	8.9
30	27
60	83

Source: From G.M. Moelter and R. Steele, In J. McKetta, ed., *Encyclopedia of Chemical Processing and Design*, vol. 1, Marcel Dekker, New York, 1976, p. 171.

the fibers. For cellulose acetate, dyeing carriers are not necessary and a temperature range of 70–80°C is used. Higher temperatures, such as 85°C, should be avoided because they could cause delustering of the yarn.

The relative dyeing rate for acetylated cellulose was shown to decrease with increasing acetyl value. The curve for dye index versus acetyl value is shown in Figure 11.8 [72]. For cellulose triacetate (Arnel), carriers and a higher temperature, in the range of 96–100°C, are used [37,72–76]. The rate of dyeing increases markedly with increasing temperature, which is shown in Figure 11.9 [72]. Cellulose triacetate may also be dyed with disperse dyes at temperatures up to 130°C, where suitable pressure equipment is available [62,72].

The requirements for the dyed fibers and fabrics are very demanding. The most important are: attractive colors and shades, good wash-fastness, good light-fastness, resistance to gas fading (oxides of nitrogen), resistance to O-fading (ozone), and fastness to sublimation when fabrics undergo thermal finishing operations such as pleating. These features depend on dye selection, dyeing conditions as described above, and after-treatment of the dyed fibers or fabrics. An example of an after-treatment on a dyed fiber to improve the dye fastness and other characteristics is the heat treatment of dyed cellulose triacetate. Heat-treating dyed Arnel fiber at 425°F for 10 s caused the dye to penetrate the interior of the fiber completely. In other words, there was further migration of the dye molecules into the interior of the fiber during the heat treatment. Such a heat treatment imparts extremely high washability

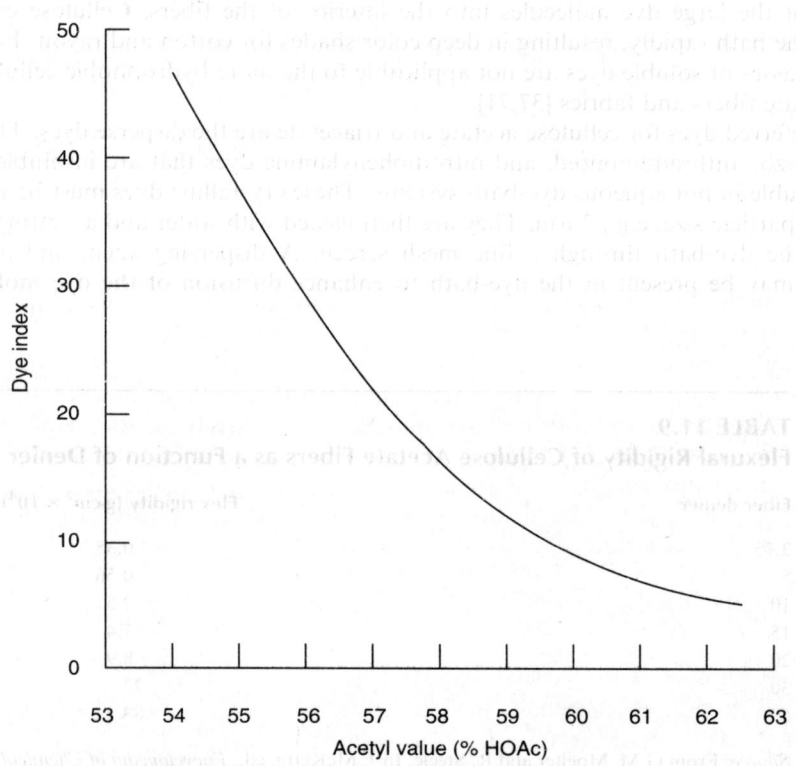

FIGURE 11.8 Relative dyeing rate as a function of the acetyl value (% HOAc). (From F. Fortess and V.S. Salvin, *Tex. Res. J.*, 28, 1009 (1958).)

Cellulose Acetate and Triacetate Fibers

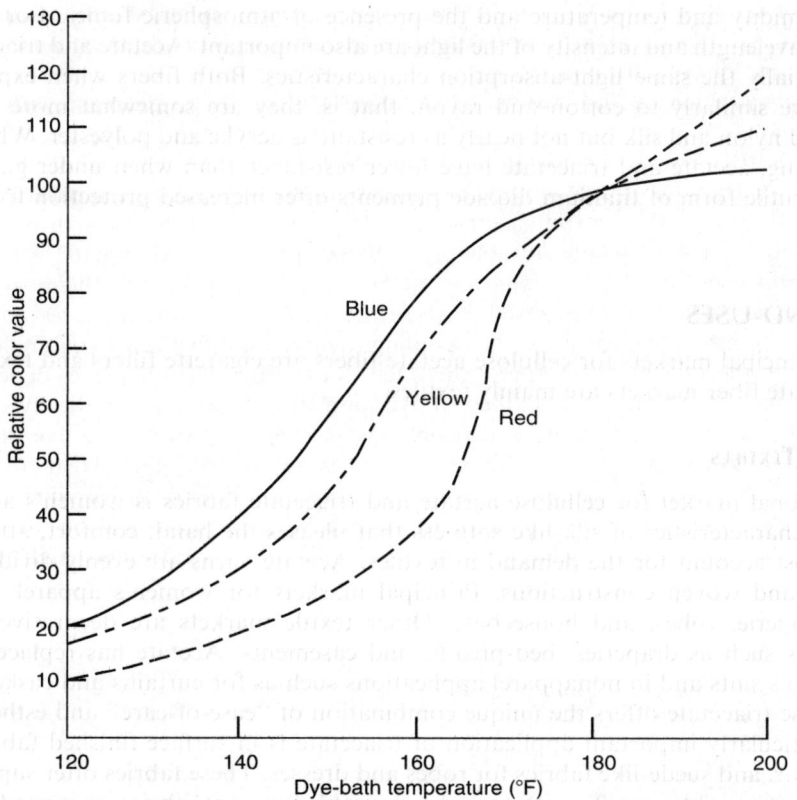

FIGURE 11.9 Slow dyeing, high-temperature dyes. Relative rates of color exhaustion from 1% dyeings at 30:1 liquor ratio using 1 h at 180°F as standard rate. (From F. Fortess and V.S. Salvin, *Tex. Res. J.*, 28, 1009 (1958).)

performance for the dyed fabrics. There is also improvement in gas-fading resistance, O-fading resistance, and crocking resistance [61,73].

For cellulose acetate–viscose or cotton fabric blends, cross-dyeing can be done, because cellulose acetate is unstained by many direct dyestuffs and because disperse dyes used for acetate do not dye viscose or cotton. Two-color and resist effects can be obtained on such blends from one dye-bath [37,77].

Cellulose acetate and triacetate can be spun-dyed. Spun-dyed acetate and triacetate yarns are extremely colorfast to washing, dry cleaning, sunlight, perspiration, sea water, and crocking and are very resistant to gas or fume fading [15].

11.9.8 Light Stability

When textile fibers are exposed to direct sunlight and indirect sunlight (under glass), degradation occurs, resulting in strength loss. The amount of degradation and strength loss depends on many factors. The important ones are: the particular polymer making up the fiber; the absence or presence of pigments, including the titanium dioxide delustrant; the particular dyestuff used in the case of dyed fibers; and the conditions of exposure to sunlight, such as relative humidity and temperature and the presence of atmospheric fumes. For accelerated tests, the wavelength and intensity of the light are also important. Acetate and triacetate fibers have essentially the same light-absorption characteristics. Both fibers when exposed under glass behave similarly to cotton and rayon, that is, they are somewhat more stable than unstabilized nylon and silk but not nearly as resistant as acrylic and polyester. When exposed to weathering, acetate and triacetate have lower resistance than when under glass. Carbon black and rutile form of titanium dioxide pigments offer increased protection from sunlight [71,78–81].

11.10 END-USES

The two principal markets for cellulose acetate fibers are cigarette filters and textiles. Cellulose triacetate fiber markets are mainly textiles.

11.10.1 TEXTILES

The traditional market for cellulose acetate and triacetate fabrics is women's apparel. The appealing characteristics of silk-like softness that pleases the hand, comfort, attractiveness, and low cost account for the demand in textiles. Acetate yarns are evenly divided between tricot-knit and woven constructions. Principal markets for women's apparel are dresses, blouses, lingerie, robes, and housecoats. Other textile markets are decorative household applications such as draperies, bedspreads, and casements. Acetate has replaced rayon as liner in men's suits and in nonapparel applications such as for curtains and caskets [82].

Cellulose triacetate offers the unique combination of "ease-of-care" and esthetic properties. A particularly important application of triacetate is in surface-finished fabrics such as fleece, velour, and suede-like fabrics for robes and dresses. These fabrics offer superb esthetic qualities at reasonable cost. Triacetate is also desirable for print fabrics, as it produces bright, sharp colors [15].

With addition of chemicals, both cellulose acetate and triacetate yarns can pass the U.S. government flame-retardant fabric regulations (e.g., DOC FF 3–71).

Cellulose acetate and triacetate fibers are blended with nylon and polyester for numerous end-uses. The synthetic fibers contribute strength and durability inasmuch as acetate and triacetate fibers have lower strength and abrasion resistance. Combination yarns can be prepared by twisting or by air-entanglement and bulking. Yarns prepared by air-entanglement and bulking have unique characteristics and esthetics that permit their use in casement and upholstery fabric markets [15]. Triacetate–nylon-blended filament yarn, giving light weight with strength and bulk, is particularly useful in producing woven fabrics of the traditional "silk" type, such as plain taffetas, printed foulards, fine crepes, and delicate jacquard figured styles for blouse and lingerie wear [83]. Cellulose triacetate–polyester blends exemplified by Lanese for use in warp-knitted casements having high bulk and low weight represent the unique combination of properties in which the outstanding characteristics of both polyester and triacetate fibers are realized in the same fabric [83].

11.10.2 CIGARETTE TOW

Cellulose acetate tow is the standard for cigarette filters. Acetate is unique for this application. Its high surface-to-volume and moisture-regain help to give it high smoke-removal efficiency with low-pressure drop. The crimped acetate tow has the weight and feel that the customer expects. The acetate fiber and its plasticizer actually improve the taste of the cigarette. Acetate is not toxic and not a risk for lung disease [84].

Tow properties are influenced by the shape of the filament cross section, dpf, and total denier, and the crimp imparted to the tow. The tow properties in turn control the properties of the cigarette filter. Important filter properties of weight, firmness, pressure drop, smoke-removal efficiency and cost are controlled by the fiber and tow properties [15,85,86].

The tow is a continuous band of thousands of fibers held loosely together by crimp—a wave configuration set into the tow during processing. The crimp holds the tow together for ease of processing and imparts bulk to the filter. Acetate tow for filters is shipped to the customer in bales. The tow can be opened either by mechanical methods or by air pressure. Once opened, a plasticizer can be added to the tow band to bond the fibers together and to add firmness to the final filter; however, unbonded filters can also be produced. The tow goes from the plasticizer addition process through a garniture that shapes the opened tow into a cylindrical rod. The rods are covered with paper and cut into the desired lengths. From the rodmaker, the finished rods are put on a cigarette machine that cuts the rods to the desired filter length and attaches the filter to the cigarette [35].

11.10.3 OTHER APPLICATIONS

Other large volume applications for cellulose acetate fibers are filament yarns for ribbons for decorative packaging and tows for ink dispensers for felt tip pens. Cellulose acetate is used in nonwovens and in paper processes as staple and short-cut [87,88].

Semipermeable membranes and hollow fibers are produced from cellulose acetate. Dry-jet wet-spinning techniques are described to provide asymmetric and homogeneous hollow fiber membranes. Manipulation of spinning conditions leads to morphologies that permit higher rejection and higher fluxes. The excellent balance of the hydrophobic–hydrophilic characteristics for cellulose acetate makes this polymer useful for reverse osmosis [89–93]. Cellulose acetate membranes and hollow fiber membranes are commercially available for hemopurification. [94], for ultrafiltration [95], and for other commercial separation processes.

Water soluble cellulose acetate polymers and fibers are produced at DS in the range of 0.5–1 [96].

11.11 PROSPECTS FOR FUTURE APPLICATIONS

R&D efforts continue for cellulose acetate and triacetate. More publications are now from Southeast Asian countries. There is continued interest in acetylation technology for lower grade pulps and even biomass and wood fibers [97–99]. Through the years, research efforts have been directed toward improving certain deficiencies such as abrasion resistance, shrinkage resistance, and strength. Prospects for future applications are viewed as related to the ability to add new performance features to cellulose acetate including thermal processability, water dispersibility, and the ability to interact with other polymers on the molecular level [100].

Approaches to modify acetate include polymer blending. No miscible polymer systems have been identified for acetate. To improve compatibility, graft copolymer additives and cross-linking prepolymer additives have been investigated, and some potentially useful compatible blends have been identified [101–106]. Extraction of one blend component has provided voids in the acetate fiber for holding useful additives or for gas extraction in chromatography [107]. Cellulose acetate fiber has been treated to become a substrate for ion exchange and for immobilization of catalyst [108–110].

Segmented block copolymers have been demonstrated by degrading cellulose triacetate and coupling it to low-molecular-weight polyesters or polyethers capped with isocyanate end groups and spinning from methylene chloride to make Spandex type elastomeric fibers [111,112].

High-tenacity fibers have been demonstrated by dry-jet wet-spinning of anisotropic solutions of cellulose triacetate, but a product has not yet been commercialized [113,114]. Nanofiber composites have been demonstrated by electro-spinning dilute cellulose acetate and triacetate solutions. These composites have exceptionally high specific surface area, and exciting new applications are envisioned [115]. There are many new specialty applications, and research is continually finding new applications that take advantage of the unique balance of properties of cellulose acetate [107,109,110,116].

ACKNOWLEDGMENTS

The author wishes to acknowledge the work of Henry W. Steinmann, who, with help from the ITT Rayonier Research staff, prepared the initial chapter in 1982, which remained unchanged through two revisions of the handbook. Dr. Steinmann's work provided the base for this present update and revision. Contributions from George A. Serad are acknowledged and appreciated [117].

REFERENCES

1. Textile Economics Bureau, *Textile Organon* (June, 2003).
2. J.P. Casey, *Pulp and Paper*, vol. 1, Interscience Publishers, New York, 1960, p. 8.
3. J.P. Casey, *Pulp and Paper*, vol. 1, Interscience Publishers, New York, 1960, p. 19.
4. TAPPI Method, T-203 OS-74.
5. J.R. Proffitt, H.M. Graham, E.R. Purchase, and R.C. Blume, *TAPPI*, 37, 28 (1954).
6. C.J. Malm, L.J. Tanghe, and B.C. Laird, *Ind. Eng. Chem.*, 38, 77 (1946).
7. L. Laamanen and H. Sihtola, *Paper and Timber*, 46, 159 (1946).
8. E.L. Akim, *Pure Appl. Chem.*, 14, 475 (1967).
9. A.J. Rosenthal, *Pure Appl. Chem.*, 14, 535 (1967).
10. L. Segal, In N.M. Bikales and L. Segal, eds., *Cellulose and Cellulose Derivatives*, High Polymer Series, vol.5, Wiley-Interscience, New York, Chapter XVII-A (1971).
11. G.D. Hyatt, W.J. Rebel, In N.M. Bikales and L. Segal, eds., *Cellulose and Cellulose Derivatives*, High Polymer Series, vol. 5, Wiley-Interscience, New York, Ref. 10, Chapter VII-B (1971).
12. C.L. Smart, C.N. Zellner, In N.M. Bikales and L. Segal, eds., *Cellulose and Cellulose Derivatives*, High Polymer Series, vol. 5, Wiley-Interscience, New York, Ref. 10, Chapter XIX-C (1971).
13. S. Shimamoto, T. Kohmoto, and T. Shibata, *ACS Symp. Series*, 688, 194–200 (1998).
14. A.J. Rosenthal and B.B. White, *TAPPI*, 43, 69 (1960).
15. G.A. Serad and J.R. Sanders, *Kirk-Othmer Encyclopedia of Chemical Technology*, 3rd ed., vol. 5, John Wiley and Sons, New York, p. 89, (1979).
16. C.J. Malm and G.D. Hiatt, In E. Ott, H.M. Spurlin, and M.W. Graffin, eds., *Cellulose and Cellulose Derivatives*, High Polymer Series, 2nd ed., vol. 5, Wiley-Interscience, New York, Pt.II (1954).
17. J. Geurden, *Pure Appl. Chem.*, 14, 507 (1967).
18. B. Phillip, H. Schleicher, and W. Wagenknecht, *Chem. Tech. Leipzig*, 7 (11), 79 (1977).
19. S.M. Hudson, and J.A. Cuculo, *J. Macromol. Sci. Chem.*, 18 (1), 1 (1980).
20. A.L. Turbak, A. ElKafrawy, T.W. Snyder, and A.B. Auerbach, U.S. Pat. 4, 302, 252 (1981).
21. A. Isogai, A. Ishizu, and J. Kanako, *Cellul. Chem. Technol.*, 17, 123 (1983).
22. J. Wu, J. Zhang, H. Zhang, J. He, Q. Ren, and M. Guo, *Biomacromolecules*, 5 (2), 266–268 (2004).
23. K.C. Campbell, J.M. Davis, and R.E. Woods, Jr., U.S. Pat. 3,767,642 (1973) (to Celanese Corporation).
24. H. Yabune, Y. Ikemoto, Y. Kato, and Mr. Uchida, U.S. Pat. 4,439,605 (1984) (to Daicel Chemical Industries).
25. *Chem. Eng. News*, 6 (Jan. 14, 1980).
26. C.L. Fletcher, U.S. Pat. 2, 259, 462 (1941).

27. H. Dreyfus, Brit. Pat. 566,863 (Feb. 26, 1945).
28. J.L. Neal, *J. Appl. Polym. Sci.*, 9, 947 (1965).
29. R.J. Conca, J.K. Hamilton, and H.W. Kirchner, *TAPPI*, 46, 644 (1963).
30. J.D. Wilson and R.S. Tabke, *TAPPI*, 57, 77 (1974).
31. F.M. Tedesco, R.J. Conca, J.P. Thelman, and A.B. Auerbach, *TAPPI Conference Papers*, Fourth International Dissolving Pulp Conference, p. 85, (1977).
32. C.J. Malm, L.J. Tanghe, and G.D. Smith, *Ind. Eng. Chem.*, 42, 730 (1950).
33. H. Lohmann, *J. Prakt. Chem.*, 115, 299 (1940).
34. H.W. Steinmann and B.B. White, *TAPPI*, 37, 225 (1954).
35. G.M. Moelter and R. Steele, In J. McKetta, ed., *Encyclopedia of Chemical Processing and Design*, vol. 1, Marcel Dekker, New York, p. 171, (1976).
36. B.S. Sprague and L.T. Homer, *Encyclopedia of Polymer Science and Technology*, vol. 3, John Wiley and Sons, New York, p. 419, (1965).
37. R.W. Moncrieff, *Man-Made Fibers*, 6th ed., Interscience, New York, p. 232, (1975).
38. K.C. Laughlin, In H.F. Mark, ed., *Man-Made Fibers: Science and Technology*, vol. 2, Interscience, New York, p. 103, (1968).
39. *Text. Marketing*, 5 (June, 1986).
40. J. Amsel, K.J. Soden, R.L. Sielken, and C. Valdez-Flora, *Am. J. of Ind. Med.*, 40 (2), 180–191 (2001).
41. C.E. Schildknecht, ed., *Polymer Processes*, Interscience, New York, p. 841, 1956.
42. I. Brazinsky, A.G. Williams, and H.L. LaNieve, *Polym. Eng. Sci.*, (15), 834 (1975).
43. D. Hardwick, Ph.D. Dissertation, The University of Tennessee (Aug. 1969).
44. A. Idris, A.F. Ismail, S.A. Gordeyev, and S.J. Shilton, *Polym. Test.*, 22 (3), 319–325 (2003).
45. J.A. Manning, C.E. Bishop, and V.G. Kight, U.S. Pat. 3,277,009 (1966).
46. J.G. Santangelo, U.S. Pat. 3,537,135 (1970).
47. H.L. LaNieve, U.S. Pat. 4,015,924 (1977).
48. Z. Gou, and A.J. McHugh, *Inter. Polym. Process.*, 19 (3), 244–253 (2004).
49. Z. Gou, and A.J. McHugh, *J. Non-Newtonian Fluid Mech.*, (118), 121 (2004).
50. Y. Sano, *Drying Technology*, 19 (7), 1335–1359 (2001).
51. W.W. Bunting, Jr., and T.L. Nelson, U.S. Pat. 2,985,995 (1961).
52. R.W. Work, *Text. Res. J.*, 19, 381 (1949).
53. R.G. Stoll, *Text. Res. J.*, 25, 650 (1955).
54. B.S. Sprague, J.L. Riley, and H.D. Noether, *Text. Res. J.*, 28, 275 (1958).
55. P. Sikorski, M. Wada, L. Heux, H. Shintani, and B.T. Stokke, *Macromolecules*, 37 (12), 4547–4553 (2004).
56. D.K. Beever and L. Valentine, *J. Text. Inst.*, 49, T95 (1958).
57. *Kirk–Othmer Encyclopedia of Chemical Technology*, 2nd ed., vol. 4, John Wiley, New York, pp. 599–600, (1991).
58. ANSI/ASTM D 1909-77, Table of Commercial Regains for Textile Fibers.
59. R.H. Braunlich, *Am. Dyest. Rep.*, 54, 38 (1965).
60. H.M. Fletcher, *Am. Dyest. Rep.*, 38, 603 (1949).
61. A.F. Tesi, *Am. Dyest. Rep.*, 45, 512 (1956).
62. J. Boulton, *J. Soc. Dyers Colour.*, 71, 45 (1955).
63. L.G. Ray, Jr., *Text. Res. J.*, 22, 144 (1952).
64. ANSI/ASTM D 3822-79, Standard Test Method for Tensile Properties of Single Textile Fibers.
65. Cellulose (Secondary) Acetate (Dicel) and Cellulose Triacetate (Tricel), Technical bulletins published by British Celanese, Coventry, England, (1965).
66. G. Susich and S. Backer, *Text. Res. J.*, 21, 482 (1951).
67. R. Meredith, *J. Text. Inst.*, 37, T107 (1945).
68. G. Susich and S. Backer, *Text. Res. J.*, 21, 482 (1951).
69. W. Zurek, I. Sobieraj, and T. Trzesowska, *Text. Res. J.*, 49(8), 438 (1979).
70. E.R. Kaswell, *Textile Fibers, Yarns, and Fabrics*, Reinhold Publishing Corp., New York, p. 57 (1953).
71. G.S. Egerton, *Am. Dyest. Rep.*, 38, 608 (1949).

72. F. Fortess and V.S. Salvin, *Tex. Res. J.*, 28, 1009 (1958).
73. F. Fortess, *Am. Dyest. Rep.*, 44, 524 (1955).
74. C.L. Bird, *J. Soc. Dyers Colour.*, 70, 68 (1954).
75. G.S. Hartley, *J. Chem. Soc.*, 1972 (1938).
76. R.K. Fourness, *J. Soc. Dyers Colour.*, 72, 513 (1956).
77. N. Caleshu, U.S. Pat. 5,476,518 (1995).
78. G.S. Egerton, *J. Soc. Dyers Colour.*, 65, 764 (1949).
79. M. Fels, *J. Text. Inst.*, 51, 648 (1960).
80. R.C. Harrington, Jr. and C.A. Jarrett, *Mod. Text.*, 4, 67 (1963).
81. C.H. Giles, *Text. Res. J.*, 49, 724 (1979).
82. I.B. Lawson, *Textiles*, 8, 30 (1979).
83. A.R. Van Landingham, *Knitting Times*, 45, 25 (1976).
84. H. Collazo, W.A. Crow, L. Gardner, B.L. Phillips, W.M. Dyer, V.A. Marple, and M.J. Utell, *Inhal. Toxicol.*, 14 (3), 247–262 (2002).
85. Acetate Tow Production and Characterization, Filter Products Division, Technical Bulletin FPB-4, Hoechst Celanese Corp., Charlotte, N.C. (1989).
86. B.M. Lee, and C.S. Winebarger, U.S. Pat. 5,269,996 (1993).
87. S.F. Nielsen, and C.E. Johnson, U.S. Pat. 5,167,764 (1992).
88. M.J. Mitchell, and L.R. Partin, U.S. Pat. 5,505,888 (1996).
89. R.L. Leonard, Ger. Offen. 2,338,786 (1974) (to Monsanto Co.).
90. H. El-Saied, A.H. Basta, B.N. Barsoum, and M.M. Elberry, *Desalination*, 159 (2), 171–181 (2003).
91. W.S. Stephen, U.S. Pat. 3,763,299 (1973).
92. K. Arisaka, K. Watanabe and K. Sasazima, U.S. Pat. 4,127,625 (1978).
93. D.T. Chen and R.D. Mahoney, Brit. Pat. 2,065,546 (July 1, 1981).
94. S.H. Ye, J. Watanabe, and K. Ishihara, *J. of Biomaterials Science*, Polymer Edition, 15 (8), 981–1001 (2004).
95. J. Qin, Y. Li, L. Lee, and H. Lee, *J. Membr Sci.*, 218 (1–2), 173–183 (2003).
96. T.C. Bohrer, U.S. Pat. 3,482,011 (1969).
97. S. Saka, K. Takanashi, and H. Matsumura, *J. Appl. Poly. Sci.*, 69 (7), 1445–1449 (1998).
98. S. Saka and K. Takanashi, *J. Appl. Poly. Sci.*, 67 (2) 289–297 (1998).
99. X.F. Sun, R.C. Sun, L. Zhao, and J.X. Sun, *J. Appl. Poly. Sci.*, 92 (1), 53–61 (2004).
100. W.G. Glasser, *Macromol. Symp, 208, 371–394 (2004)*.
101. S. Lepeniotis, B. Feuer, and J.M. Bronk, *Chemometrics Intell. Labo. Syst.*, 44 (1–2), 293–306 (1998).
102. N. Mikhailov et al., U.S.S.R. 285,156 (1970).
103. N.V. Mikhailov et al., *Khim. Volokna*, 15, 72 (1973).
104. A.F. Turbak, J.P. Thelman, and A.B. Auerbach, U.S. Pat. 40,118,350 (1978).
105. C. Coolidge and J.S. Reese, U.S. Pat. 2,375,838 (1945) (to E.I. DuPont).
106. G. Afnana et al., *Zhurnal Prikladnoi Khimii*, 49, 2103 (1976).
107. M.A. Farajzedeh and M. Hatami, *J. of Sep. Sci*, 26 (9–10), 802–808 (2003).
108. Y. Ikeda, Y. Kurokawa, K. Nakane, and N. Ogata, *Cellulose*, 9 (3–4), 369–379 (2002).
109. C.A. Bargo, A.M. Lazarin, Y. Gushikem, R. Landers, and Y.V. Kholin, *J. Braz. Chem. Soc.*,15 (1), 50–57 (2004).
110. A.M. Lazarin, R. Landers, Y.V. Kholin, and Y. Gushikem, *J. Colloid Interface Sci.*, 254 (1), 31–38 (2002).
111. H.W. Steinmann, U.S. Pat. 3,386,932 (1968) (to Celanese Corporation).
112. J.G. Santangelo and H.W. Steinmann, U.S. Pat. 3,386,930 (1968) (to Celanese Corporation).
113. M. Panar and O.D. Willcox, Ger. Pat. 2,705,382 (1977) (to DuPont).
114. J. Bheda, J.F. Fellers, and J.L. White, *J. Appl. Polym. Sci.*, 26, 3955 (1981).
115. H. Liu and Y.L. Hsieh, *J. Poly. Sci., B: Poly. Phys.*, 40 (18), 2119–2129 (2002).
116. J.C. Chen, and K.J. Soden, U.S. Pat. 6,500,539 (2002).
117. G.A. Serad, *Kirk–Othmer Concise Encyclopedia of Chemical Technology*, 4th ed., John Wiley, New York (1998).

12 Acrylic Fibers

Bruce G. Frushour and Raymond S. Knorr

CONTENTS

12.1	Introduction	812
12.2	History	813
12.3	Polymer Manufacture	816
	12.3.1 Preparation of Acrylonitrile	816
	12.3.2 Acrylonitrile Polymerization	817
	12.3.2.1 Solution Polymerization	818
	12.3.2.2 Bulk Polymerization	822
	12.3.2.3 Emulsion Polymerization	825
	12.3.2.4 Aqueous Dispersion Polymerization	826
	12.3.3 Copolymerization Kinetics	828
	12.3.3.1 Homogeneous Copolymerization	828
	12.3.3.2 Heterogeneous Copolymerization	833
	12.3.4 Commercial Polymerization Methods	833
	12.3.4.1 Semibatch Polymerization	834
	12.3.4.2 Continuous Aqueous Dispersion Processes	834
	12.3.4.3 Solution Polymerization	837
	12.3.4.4 Bulk Polymerization	837
12.4	Solid-State Structure of Acrylic Polymers	838
	12.4.1 Stereoregularity and Chain Conformation	840
	12.4.1.1 Stereoregularity	840
	12.4.1.2 Chain Conformation	841
	12.4.2 Diffraction Studies of the Crystalline Structure	843
	12.4.2.1 Analysis by X-Ray Diffraction	843
	12.4.2.2 Diffraction Analysis of Single Crystals and Fibers	846
	12.4.2.3 Evidence for Two-Phase Morphology and Determination of Crystallinity	847
	12.4.3 Thermal Properties: Melting, Gelation, and Crystallization	852
	12.4.3.1 Evidence for Melting of Acrylic Polymers	852
	12.4.3.2 Effect of Diluents and Comonomers on Melting Behavior	853
	12.4.4 Glass Transition and Dynamic–Mechanical Properties	861
12.5	Fiber Manufacturing	864
	12.5.1 Polymer Solubility and Spinning Dope Preparation	864
	12.5.2 Fiber Formation	869
	12.5.2.1 Fiber Extrusion and Coagulation	870
	12.5.2.2 Structure of the Coagulated Fiber	874
	12.5.3 Tow Processing	881

            12.5.3.1 Orientational Drawing (Wet Stretching) 883
            12.5.3.2 Drying and Collapsing 888
            12.5.3.3 Relaxing, Crimping, and Bulk Development 890
    12.5.4 Special Topics 891
            12.5.4.1 Special Wet-Spinning Processes 891
            12.5.4.2 Melt Extrusion 892
            12.5.4.3 Optimization of Process Spinnability 893
            12.5.4.4 Modification of Cross Section 899
            12.5.4.5 Dyeing of Acrylic Fibers 900
12.6 Fiber Properties 905
    12.6.1 Physical Properties 907
    12.6.2 Chemical Properties of Acrylic Fibers 914
    12.6.3 Flammability 916
    12.6.4 Modifications of Properties 917
            12.6.4.1 Modification of Handle 917
            12.6.4.2 Wear Comfort 919
            12.6.4.3 Reduced Pilling 921
            12.6.4.4 Improved Hot–Wet Properties 921
            12.6.4.5 Improved Abrasion Resistance and Increased Fiber Density 922
            12.6.4.6 Improved Whiteness and Thermal Stability 923
            12.6.4.7 Antistatic Fibers 923
            12.6.4.8 Antisoiling Fibers 924
12.7 Analysis and Identification 924
    12.7.1 Fiber Identification 924
    12.7.2 Instrumental Analysis 925
            12.7.2.1 Polymer Characterization 926
            12.7.2.2 Fiber Characterization 929
12.8 Commercial Textile Products 930
    12.8.1 Standard Staple and Tow 930
    12.8.2 Acrylic Filament Yarns 934
    12.8.3 Fibers with High-Bulk and Pile Properties 935
    12.8.4 Flame-Resistant Fibers 936
    12.8.5 Specialized Products 936
12.9 New Products and Applications 939
    12.9.1 Carbon and Graphite Fibers 939
    12.9.2 Asbestos Replacement Fibers and Other Composite Applications 945
    12.9.3 Fibers for Reverse Osmosis, Ion Exchange, and Filtration 947
    12.9.4 Use as Moisture-Absorbent Synthetic Paper 948
    12.9.5 Electrically Conducting Fibers 948
    12.9.6 Metallized Fibers 950
    12.9.7 Oxidized PAN 950
12.10 Future Trends 951
References 958

12.1 INTRODUCTION

The acrylic fiber business has matured in the sense that most of the growth in traditional markets have been realized. The U.S. shipments of acrylic staple products peaked in the late 1970s, and recent growth has taken the form of expansion to developing countries. However, innovative technology has created many new nontextile applications for acrylic fibers. Some of these, such as precursor fiber for the manufacture of high-strength graphite fiber, will support acrylic fiber production for many years into the future.

This review covers the science and technology of acrylic fibers in the general sense of the terminology. The term "acrylic fiber," however, specifically refers to a fiber containing at least 85% acrylonitrile (AN) comonomer. A similar class of fibers is the modacrylics, in which the percentage of acrylonitrile must be less than 85% but greater than 35%. Modacrylic fibers typically contain substantial levels of halogen-containing comonomers and are intended primarily for applications where superior flame resistance is required. The number of modacrylic fibers on the market today is very small. Therefore, the topic of modacrylic fibers is not treated separately in this review.

In line with the chemistry and physics background of the authors, the emphasis in this review is on polymerization chemistry and fiber physics. We have attempted to emphasize the fundamental science behind the acrylic fiber technology as often as possible. The reader, wishing to learn more about other topics, such as analysis and testing or yarn processing, may wish to consult some of the excellent reviews that have been published on acrylic fibers [1–9].

12.2 HISTORY

The first reported synthesis of acrylonitrile and polyacrylonitrile (PAN) was by the French chemist Moreau [10]. In 1893, Moreau reported two methods for synthesizing acrylonitrile and a year later, he reported the polymerization of acrylonitrile. The polymer received little attention for a number of years because there were no known solvents and because the polymer decomposes before reaching its melting point. This made processing of the polymer nearly impossible. The first breakthrough in developing solvents for PAN came shortly before World War II in Germany. Rein of I.G. Farbenindustrie was successful in spinning fibers from a solution of the polymer in aqueous solutions of quaternary ammonium compounds, such as benzyl peritoneum chloride, or of metal salts, such as lithium bromide, sodium thiocyanate, and aluminum perchlorate [11].

Early interest in acrylonitrile polymers was not based on its potential use in synthetic fibers. Instead, most interest in these polymers was for their use in synthetic rubber. In 1937, I.G. Farbenindustrie introduced the first acrylonitrile–butadiene rubber. Synthetic rubber compounds based on acrylonitrile were developed in the United States during the early 1940s in response to wartime needs. American Cyanamid, however, was the sole U.S. producer of acrylonitrile at that time. In addition to acrylonitrile–butadiene rubber, polyblends of acrylonitrile–butadiene with acrylonitrile–styrene copolymers were developed by the United States Rubber Co. After the war, the demand for acrylonitrile dropped sharply, and American Cyanamid was still the sole U.S. producer.

The situation changed dramatically when DuPont introduced the first commercial acrylic fiber under the trade name of Orlon. This commercial development took place shortly after DuPont [12] and I.G. Farbenindustrie [13] simultaneously reported solvents suitable for spinning acrylonitrile fibers in 1942. Based on this solvent breakthrough, DuPont was able to develop a commercial process for producing acrylic fibers. The DuPont process was based on dry spinning with N,N-dimethylformamide (DMF) as the solvent. The product was introduced in 1944 as Orlon. Shortly thereafter Chemstrand (later to become Monsanto Fibers and Intermediates Company) introduced Acrilan, Süddeutsche Chemiefaser (Hoechst) introduced Dolan, and Bayer introduced Dralon. Developments in this fledgling industry occurred rapidly from that time on.

Major technical problems were solved by these companies to keep their commercial efforts alive. Difficulties in dyeing were overcome by developing cationic dyes and by modifying the fiber morphology with comonomers, such as methyl acrylate (MA) and vinyl acetate (VA).

Dye sites were incorporated by using free radical initiators and chain transfer agents, which leave sulfate and sulfonate end groups on the polymer chains. Where a greater degree of dyeability was required, sulfonated monomers were incorporated in the polymer composition.

Chemstrand's Acrilan process was based on a wet-spinning technology, which produces a fibrillar microstructure. As a result, early acrylic fiber products suffered from problems with abrasion originating with a lack of coherence in the fibrillar surface of the fibers. This was overcome by adding a steam-annealing step, which, combined with the presence of vinyl acetate as comonomer, makes the fibrils that compose each filament fuse together.

The first modacrylic fiber was developed by Union Carbide in 1948 under the trade name Vinyon N. It was a continuous filament yarn based on 60% vinyl chloride and 40% acrylonitrile. The staple form was introduced in 1949 under the trade name of Dynel [14,15]. Interest in modacrylics began slowly, with Tennessee Eastman Co. introducing a version based on vinylidene chloride, known as Verel, in 1956. In 1962, Courtaulds introduced the continuous filament yarn known as Teklan and later in 1966 switched to a staple form in 1966 in response to prevailing market conditions. Other modacrylics were made from combinations of three halogenated monomers, vinyl chloride, vinylidene chloride, and vinyl bromide. Monsanto's SEF is a prime example, formulated from five separate monomer components, including two halogenated monomers, a dye receptor, and a morphology modifier. Vinyl chloride was eliminated as a component of acrylic fibers in the 1970s when it was identified as a carcinogen. Today, vinyl bromide is phased out for similar reasons.

The earliest polymerization processes were either batch mode or semibatch. The semibatch method was used for products, where the two monomers differed greatly in reactivity, as in Union Carbide's early Dynel, acrylonitrile–vinyl chloride, process. Bulk, solution, and emulsion polymerization processes have also been developed for acrylonitrile and its copolymers. However, in recent years nearly every major acrylic fiber producer has used a continuous aqueous suspension process, employing a redox catalyst, followed by a series of steps, which includes slurry filtration and polymer drying.

Modified melt-spinning processes were developed by DuPont and American Cyanamid, but were never commercialized to any significant extent. Solution spinning accounts for virtually all world production of acrylic fibers. Three major solution-spinning processes are used. These are dry spinning using DMF as the solvent and MA comonomer to modify the morphology in order to increase the dyeability, wet spinning with dimethylacetamide (DMAC) as the solvent and VA as the comonomer, and wet spinning with aqueous sodium thiocyanate solution as the solvent and MA as the comonomer.

The acrylic fiber industry experienced a spectacular growth in the 1950s with at least 18 companies introducing acrylic fiber products during that period. Europe had I.G. Farbenindustrie, Bayer and Hoechst in Germany, Courtaulds in England, Rhone-Poulenc in France, and Montefibre and Snia-Viscosa in Italy. By 1957, the majority of the acrylic fiber produced was staple. Filament processes were much more costly than staple and the cotton and wool replacement markets favored staple. In 1960, DuPont produced a bicomponent acrylic fiber that was designed to more closely simulate the desirable characteristics of wool. These bicomponent fibers were designed to have a natural helical crimp, either by incorporating side-by-side components of differing melting points and shrinkage levels or, in the case of so-called water-reversible crimp bicomponents, one component was more hydrophilic than the other. The differential moisture absorption between the two components causes a differential expansion thus generating the crimp.

By 1960, annual worldwide production had risen to over 200 million pounds. In the 1950s and 1960s, world production was concentrated in Western Europe and the United States. Early products were largely filament yarns for industrial and outdoor applications. However, staple processes were soon developed and acrylic fibers became a major competitor in markets held primarily by woolen fibers. Much of the initial demand for acrylic fibers can be attributed to the development of high-bulk processes for sweater yarns and heavy-denier carpet yarns, with the intention of replacing wool sweaters and carpets. By 1963, these two markets accounted for almost 50% of the total acrylic market.

From 1960 to 1970, acrylic fiber consumption continued its rapid growth with a 19% increase in consumption worldwide, peaking at almost 2.1 billion pounds in 1970. Acrylics found wide use as a wool replacement fiber in carpets, home furnishings, and knitted apparel products. The world market share claimed by both acrylics and polyester increased sharply over this period primarily at the expense of wool and nylon. For acrylic, this increase was from about 14% of the world market in 1960 to approximately 23% in 1969. In the next decade, however, the growth rate decreased to around 8%. This was due primarily to the maturing of the wool replacement market in the United States. In addition, nylon became the dominant carpet fiber, reducing the acrylic market share from 25% at its peak to just 6% by 1976. Fibers and blends, such as polyester-cotton, also cut into the acrylic share of the synthetic fibers market. By 1980, the world market share held by nylon had fallen to approximately 30% from over 60% in 1960. Polyester now accounts for roughly 50% of the world market, while acrylics continue to hold approximately 20%.

A major development in the 1970s and 1980s was the rapid rise in worldwide production capacity relative to total worldwide consumption. During the 1970s, there was rapid growth in Japan, Eastern Europe, and the developing countries. Production reached 800 million pounds per year in Japan in 1980. China, India, and South Korea added another 1.4 billion pounds in Asia. South and Central America producers added another 200 million pounds by 1976. By 1981, an estimated overcapacity of approximately 21% had developed. This overcapacity decreased through the 1980s with continued increases in world production balanced by markets opening in China, Eastern Europe, Russia, and the Americas.

Compounding this growth in worldwide production capacity was a decline in overall synthetic fiber consumption beginning in the early 1970s. Before this time, the major growth in synthetic fiber consumption was in the industrialized countries. This was sparked by low raw material costs, general improvement in per capita income, and the rapid development of major new applications, such as easy-care synthetics and carpet fibers from nylon and acrylic fibers. The world economy changed this favorable situation. Beginning with the oil crisis in the mid-1970s, raw materials and energy costs rose sharply and the world economy entered a recession that has persisted into the 1980s. Acrylic fiber consumption was affected more than nylon and polyester. The most important factor in this decline was probably the maturing of the wool replacement market. Next is the continuing dominance of nylon in the carpet sector. In apparel, the use of polyester by itself in filament form or in blends with cotton cut deeply into markets that might otherwise have gone to acrylics. This is particularly true in the United States, which is very much polyester oriented. Another factor contributing to the leveling off of growth in acrylic consumption is the fact that acrylics are widely used in home-furnishing fabrics. This restricts consumption to the high-income countries.

In 1981, worldwide demand was 4.6 billion pounds, while worldwide capacity was 5.8 billion pounds, 1.2 billion pounds higher demand [16]. Consequently, prices have been soft since the mid-1970s. Over the past decade, we have seen the natural consequences of this situation. Except in the developing countries where there is a natural advantage in the cost of labor, investment in plants and equipment has been minimal. Research and development work, which yielded so much progress in the 1960s and the early 1970s, has been essentially frozen since cutbacks in the mid-1970s. Crucial research into long-range technology was cut sharply. Most of the research and development work today is aimed at holding market share or maintaining slim profit margins in the face of stiff competition. Growth in acrylic fiber consumption was approximately 4–5% through 1987. Much of this growth occurred in the

Soviet Union, Eastern Europe, the Far East, and the Americas. Recession in Western Europe, Japan, and the United States lowered demand in those regions and many producers withdrew from the market.

By 1990, there were 68 plants producing a total of 5.1 billion pounds of fiber in Eastern and Western Europe, Asia, and the Americas. However, production in the United States and Western Europe has declined, while production in the underdeveloped countries has increased. In the 1990s, the slow growth of acrylic fiber consumption has been aggravated by the resurgence of cotton as a preferred fiber for apparel, a trend that began to emerge in the 1980s. Growth in the underdeveloped countries is driven in part by the need for cotton replacement fibers, since in these countries with large populations, especially China, cotton crops compete with food crops for available agricultural land. These large populations provide an inexpensive source of labor, which encourages investment in new plant startups [17].

By 1991, after closing its plants in Waynesboro, Virginia and Camden, New Jersey, DuPont had finally withdrawn from the acrylic fiber business, leaving Monsanto and Cytec as the only major U.S. producers. The same trend has occurred in Western Europe and the developed countries of Asia. New production capacity in apparel markets continues to be added, but this is restricted to the developing countries. The rest of the world is focusing on restructuring their product mix, targeting new, nontraditional markets. Recent history and trends in the acrylic fiber industry have been reviewed in more detail by Matzke [17].

12.3 POLYMER MANUFACTURE

12.3.1 Preparation of Acrylonitrile

In 1893, the French chemist Moreau described two routes for the synthesis of acrylonitrile that were based on the dehydration of either acrylamide or ethylene cyanohydrin [10]. There was very little interest in acrylonitrile until 1937 when synthetic rubber based on acrylonitrile–butadiene copolymers was first developed in Germany. A process based on the addition of hydrogen cyanide to acetylene was developed at that time and in the 1950s, the acrylic fiber industry provided the stimulus for further process developments. Today acrylonitrile is made commercially by one of three possible methods: (a) from propylene, (b) from acetylene and hydrogen cyanide, and (c) from acetaldehyde and hydrogen cyanide.

In the ethylene cyanohydrin process, which was developed for commercial use by Union Carbide [18], ethylene oxide is reacted with hydrogen cyanide in aqueous alkaline solution. The process is carried out at 50–60°C for up to 10 h in the presence of a catalyst, such as diethylamine. The resulting ethylene cyanohydrin decomposes spontaneously into acrylonitrile and water when heated to 200°C in the presence of magnesium carbonate.

The acetylene process was developed in Germany in the early 1940s to supply the synthetic rubber industry [19]. Acetylene is reacted with hydrogen cyanide in an aqueous medium in the presence of catalytic amounts of cuprous chloride. The reaction is maintained at 80–90°C at a pressure of 1–2 atm. The reaction is highly exothermic forming a gaseous reactor effluent. This crude product is water-scrubbed and the pure acrylonitrile product is recovered from the resultant 1–3% aqueous solution by fractional distillation. The major drawbacks of this process are the large number of by-products formed by hydration, the loss of catalyst activity from hydrolysis reactions, and the buildup of ammonium chloride and tars.

The acetaldehyde process was developed in 1958 by Knapsack-Griesheim [20]. In the first stage, acetaldehyde is reacted with hydrogen cyanide to form lactonitrile, which in turn is mixed with catalytic phosphoric acid and heated rapidly to 600–700°C. The mixture is then chilled rapidly to 50°C in a solution of 30% phosphoric acid to form the crude acrylonitrile product.

Today the most cost-effective processes are those based on propylene as the starting material. There are three major variations of propylene processes, the Distillers process [21–23], the Sohio process [24], and the DuPont process [25,26]. All three processes are based on the ammonoxidation of propylene. The Distillers process is carried out in two stages. In the first, propylene is oxidized in air to form acrolein and water. These intermediate products are allowed to react in the second stage with ammonia in the presence of molybdenum oxide and air to form crude acrylonitrile. The pure monomer is recovered by a series of azeotropic distillations. The Sohio process is carried out in just one stage. Ammonoxidation of propylene takes place in air at 2–3 atmospheric pressure and 425–510°C. With catalysts, such as concentrated bismuth phosphomolybdate or other oxides of molybdenum and cobalt, the reaction takes place with over 50% yield in a reaction time of only about 15 s. In the DuPont version of this process, the ammonoxidation is brought about with nitric oxide at 500°C using silver on silica catalyst. The chemistry of acrylonitrile monomer has been reviewed by a number of authors [27–30].

12.3.2 Acrylonitrile Polymerization

The homopolymer, PAN, is rarely used in fiber manufacturing with the exception of industrial applications where resistance to chemical attack is of prime importance. PAN is a homopolymer that is difficult to spin and dye, and therefore virtually all commercial acrylic fibers are made from acrylonitrile combined with at least one other monomer. The reasons for this will be discussed in more detail in Section 12.4 and Section 12.5. The comonomers most commonly used are neutral comonomers (in the sense that they are not intended to participate in any chemical reaction), such as MA and VA to increase the solubility of the polymer in spinning solvents, modify the fiber morphology, and improve the rate of diffusion of dyes into the fiber. Sulfonated monomers, such as sodium styrene sulfonate (SSS), sodium methallyl sulfonate (SMAS), and sodium sulfophenyl methallyl ether (SPME) are used to provide dye sites or to provide a hydrophilic component in water-reversible crimp bicomponent fibers. Halogenated monomers, usually vinylidene chloride, vinyl bromide, and vinyl chloride, impart flame resistance to fibers used in the home furnishings, awning, and sleepwear markets. Modacrylic compositions are used when the end use requires high flame resistance. Almost all of the modacrylics are flame-resistant fibers with very high levels of halogen monomers.

Molecular weight and the distribution of molecular weight are also key properties. Typical acrylic polymers have number average molecular weights in the 40,000–60,000 g/mol range or, roughly, 1,000 repeat units. The weight average molecular weight is typically in the range of 90,000–140,000, with a polydispersity index between 1.5 and 3.0. The solution properties of the polymer and rheological properties of the dope must be precisely defined for compatibility with dope preparation and spinning. The molecular weight of the polymer must be low enough that the polymer is readily soluble in spinning solvents, yet high enough to give a dope of moderately high viscosity (again to be discussed in Section 12.4 and Section 12.5).

Fiber dyeability is critically dependent on the molecular weight distribution of the polymer because most acrylic fibers derive their dyeability from sulfonate and sulfate initiator fragments at the polymer chain ends. Thus, the dye site content of the fiber is inversely related to the number average molecular weight of the polymer and very sensitive to the fraction of low-molecular-weight polymer. A critical balance must be maintained between the molecular weight distribution required for good rheological properties and the distribution required for good fiber dyeability. Where such a balance cannot be achieved, it is usual practice to incorporate one of the sulfonated monomers as a means of establishing the required fiber dyeability. DuPont's Orlon 42, for example, is believed to contain a small amount of SSS as a supplemental dye receptor. The very dense fiber structure, produced by DuPont's dry-spinning process, results in very low dye-diffusion rates. The addition of a sulfonated monomer, therefore, compensates by increasing the total dye site content of the fiber.

Acrylonitrile and its comonomers can be polymerized by any of the well-known free radical methods. Bulk polymerization is the most fundamental of these, but its commercial use is limited by its autocatalytic nature. Aqueous dispersion polymerization is the most common commercial method, while solution polymerization is used in cases where the spinning dope can be prepared directly from the polymerization reaction product. Emulsion polymerization is used primarily for modacrylic compositions where a high level of a water-insoluble monomer is used or where the monomer mixture is relatively slow-reacting. The subject of acrylonitrile polymerization was reviewed by Goldfein and Zyubin [31]. Details of these polymerization methods are described below.

12.3.2.1 Solution Polymerization

Solution polymerization is widely used in the acrylic fiber industry. The reaction is carried out in a homogeneous medium by using a solvent for the polymer. Suitable solvents can be highly polar organic compounds or inorganic aqueous salt solutions. DMF [32–35] and dimethyl sulfoxide (DMSO) are the most commonly used commercial organic solvents, although polymerization in alpha-butyrolactone, ethylene carbonate [32–34], and DMAC is reported in the literature. Examples of suitable inorganic salts are aqueous solutions of zinc chloride [38] and aqueous sodium thiocyanate solutions [39].

The homogeneous solution polymerization of acrylonitrile follows the conventional kinetic scheme developed for vinyl monomers [36,40,41]. This kinetic scheme can be presented as follows:

Radical formation—depends on the type of initiation; thermal and redox initiators generally are used.

Initiation of chain growth

$$I^* + M \rightarrow P_1^* \tag{12.1}$$

Chain propagation

$$M + P_n^* \rightarrow P_{n+1}^* \tag{12.2}$$

Free radical transfer reactions (to monomer, solvent, additives)

$$P_n^* + X \rightarrow P_n + X^* \tag{12.3}$$

Termination by radical recombination

$$P_n^* + P_m^* \rightarrow P_{n+m} \tag{12.4}$$

Termination by radical disproportionation

$$P_n^* + P_m^* \rightarrow P_n + P_m \tag{12.5}$$

Termination by metal ion

$$P_n^* + Fe(III) \rightarrow P_n + Fe(II) \tag{12.6}$$

Acrylic Fibers

The parameters are defined as follows:

Initiator	I
Initiator radical	I*
Initiation efficiency	f
Monomer	M
Polymer	P
Polymer radical length, n	P_n^*
Transfer agent	X
Metal ion	T
Solvent	S
Rate of initiation	R_i
Rate of polymerization	R_p
Degree of polymerization	D_p

The rate of initiation is normally written

$$R_i = 2fK_i [I] \quad (12.7)$$

The rate of polymerization is normally written

$$R_p = \frac{K_p}{K_t^{1/2}} (fK_i)^{1/2} [I]^n [M]^m \quad (12.8)$$

where n and m are exponential factors that characterize the rate dependence on the initiator and monomer concentrations, respectively.

The degree of polymerization is normally written as follows:

$$\frac{1}{D_p} = \frac{K_t R_p}{K_p [M][M]} + C_m + C_s \frac{[S]}{[M]} + C_x \frac{[X]}{[M]} \quad (12.9)$$

where K_p is the rate constant for chain growth, K_i is the rate constant for initiation, K_t is the rate constant for termination, C_m is the chain transfer constant for monomer, C_s is the chain transfer constant for solvent, C_x is the chain transfer constant for additive X, [S] is the solvent concentration, [M] is the monomer concentration, and [I] is the initiator concentration.

Thermally activated initiators, such as azobisisobutyronitrile (AIBN), ammonium persulfate, or benzoyl peroxide, can be used in solution polymerization but these initiators are slow acting at the temperatures required for textile-grade polymer processes. Half-lives for this type of initiator are in the range of 10–20 h at 50–60°C [42]. Therefore, these initiators are used mainly in batch or semibatch processes, where the reaction is carried out over an extended period.

Polymerization in solution follows conventional kinetics except for certain solvent-specific side reactions. At monomer concentrations above 2–2.5 M, the reaction order with respect to monomer and initiator has been found to be 1.0 and 0.5, respectively [35]. In DMF, however, a monomer reaction order greater than expected was explained by chain transfer to solvent followed by slow reinitiating by the DMF radical [43]. At higher monomer concentrations, however, the monomer has the effect of adding a nonsolvent to the reaction mixture. Under these conditions, the reaction orders with respect to initiator and monomer can deviate from the expected values. Vidotto et al. [35] found that the reaction became heterogeneous at monomer concentrations above $4\,M$ in DMF and at $6\,M$ in ethylene carbonate. In ethylene carbonate, the rate and molecular weight went through a maximum when the monomer concentration was increased above $6\,M$. Owing to the high chain transfer to solvent activity of DMF, however, the molecular weight in that solvent increased monotonically at monomer concentrations above $4\,M$. The reaction order in monomer also deviates from unity at very low monomer concentrations. This was explained by taking into account chain propagation through the nitrile group.

Chain transfer is an important consideration in solution polymerizations. As we have pointed out, chain transfer to solvent may reduce the rate of polymerization as well as the molecular weight of the polymer. Other chain transfer reactions may introduce dye sites, branching, chromophoric groups, and structural defects that reduce thermal stability. Table 12.1 shows the values for transfer constants for common solvents and comonomers [42]. Table 12.1 also gives values for transfer to polymer, a side reaction that can be very significant at the high conversion levels required for commercial polymerizations. Many of the solvents used for acrylonitrile polymerization are very active in chain transfer. DMAC and DMF, at 4.95×10^{-4}–5.10×10^{-4} and 2.70×10^{-4}–2.80×10^{-4}, respectively, both have very high chain transfer constants when

TABLE 12.1
Chain Transfer Constants for Solvents, Monomers, and Polymers Used in Acrylic Fiber Manufacture

Solvent	$C_s \times 10^4$	Reaction temperature	Reaction conditions
Chain transfer to solvent in solution polymerization			
Benzene	2.46	60	Dispersion polymerization
α-Butyrolactone	0.66	50	
	0.74	50	
Copper(II) chloride	190,000	35	Aqueous solution (NaCl)
Dimethylacetamide	4.95	50	
	5.05	50	
Dimethylformamide	2.8	50	
	2.7	50	
Dimethylsulfoxide	0.11	50	
	0.29	50	
	0.80	50	
Diphenylamine	700	60	DMF solution
Ethylene carbonate	0.33	50	
	0.474	50	
	0.50	50	
Iron(III) chloride	33,300	60	DMF solution
Magnesium perchlorate	<0.05	50	Aqueous solution
Sulfur dioxide	0	50	Anhydrous
Triethylamine	5,900	60	DMF solution, polymerization retarded
Trimethylamine	790	60	Polymerization retarded
Tripropylamine	4,280	60	DMF solution
Zinc chloride	0.006	50	Aqueous solution
Transfer to monomer			
Acrylamide	0.2	25	Gamma-ray initiation, 25°C
	0.6	60	Gamma-ray initiation, 60°C
Acrylonitrile	1.50	30	DMSO solution
	0.05	50	Zinc chloride solution

continued

TABLE 12.1 (Continued)
Chain Transfer Constants for Solvents, Monomers, and Polymers Used in Acrylic Fiber Manufacture

Solvent	$C_s \times 10^4$	Reaction temperature	Reaction conditions
	0.27	50	Magnesium perchlorate solution
	8.20	50	SO_2 solution
Methyl acrylate	0.275	55	Peroxide initiation
	0.036	60	Hydroperoxide initiation
	0.325	60	Peroxide initiation
Methacrylonitrile	5.81	60	
Methyl methacrylate	0.10	50	Benzene solution
	0.15	50	
	0.103	50	Peroxide initiation
Styrene	0.40	50	Benzene solution
	0.50	50	Peroxide initiation, emulsion
	0.60	50	Peroxide initiation
Vinyl acetate	1.29	50	
	4.55	50	Peroxide initiation
	2.4	60	
Vinyl acetate	1.91	60	
	1.93	60	
Vinyl chloride	6.4	50	Butyl acrylate solution
	7.8	50	Butyl chloride solution
4-Vinylpyridine	6.4	25	
N-Vinylpyrrolidone	4.0	20	
Transfer to polymer in homopolymerization			
Acrylonitrile	4.7	50	Magnesium perchlorate solution
	3.5	60	Estimated
N,N-Dimethylacrylamide	0.61	50	
Methyl acrylate	0.5	60	Estimated
	1.0	60	Estimated
Methyl methacrylate	1.5	50	
	350	50	Transfer to end groups
Styrene	4.5	50	
	16.6	50	
Vinyl acetate	3.0	50	
	10.2	50	
Vinyl chloride	5.0	50	Estimated

Source: From Brandrup, J. and Immergut, E.H., *Polymer Handbook*, 2nd ed., Interscience, New York, 1975.

compared to a value of only 0.05×10^{-4} for acrylonitrile itself. DMSO and aqueous zinc chloride, in contrast, have relatively low transfer constants. Hence, the relative desirability of these two solvents is above than that of the former. DMF, however, is used by several acrylic fiber producers as a solvent for solution polymerization.

Between the two most common comonomers, MA and VA, MA is the least active in chain transfer while VA is as active in chain transfer as DMF and DMAC. Vinyl acetate is also known to participate in the chain transfer to polymer reaction [44]. This occurs primarily at high conversion, where the concentration of polymer is high and monomer is scarce. PAN can also participate in branching reactions. Ulbricht [45], in a study of acrylonitrile polymerization in magnesium perchlorate, suggested that branch formation may occur by a reaction in which a growing radical chain abstracts a hydrogen atom from the α-carbon, thereby starting the formation of a side chain by monomer addition. It was shown in this study that branch formation occurs when the ratio of polymer to monomer concentration exceeds one or a conversion of 80%. At this condition, one branch occurs for every 2000 growth steps. Thus, at a molecular weight of 10^5, each molecule shows only one branch on the average. PAN branching was also studied by Peebles [46]. Peebles, however, suggested that PAN branching occurs by polymerization through the nitrile group. Branch formation can also occur as a result of radical termination by disproportionation or other chain transfer reactions. Any reaction that leaves a terminal double bond can lead to long-chain branching if the double bond subsequently reacts with a growing polymer radical.

The advantage of solution polymerization is that the polymer solution can be converted directly to spinnable dope by removing the unreacted monomer. However, it is more difficult to achieve high molecular weight. The solvents required are often chain transfer agents and chain termination is more rapid. Incorporation of nonvolatile monomers, such as the sulfonated monomers commonly used as basic dye acceptors, can also be a problem. The sulfonated monomers, in particular, have poor solubility in organic solvents and must be solubilized by converting them to a soluble form such as the amine salt form. Nonvolatile monomers are also difficult to recover from the reaction medium since the usual distillation techniques are unsuitable. Monomer recovery systems based on carbon adsorption have been developed. However, the usual practice is to maximize the single-pass utilization of these monomers.

12.3.2.2 Bulk Polymerization

The bulk polymerization of acrylonitrile is very complex. Even today, after many investigations into the kinetics of the polymerization, it is still not completely understood. This complexity arises because the polymer precipitates from the reaction mixture barely swollen by its monomer. This heterogeneity has led to kinetics that deviate from the normal and that can be interpreted in several ways.

When initiator is first added, the reaction medium remains clear while particles 100–200 Å in diameter are formed. As the reaction proceeds, the particle size increases, giving the reaction medium a white milky appearance. An example of polymer particles from bulk, aqueous dispersion, and emulsion polymerization is shown in Figure 12.1a through Figure 12.1c. The dense, spherical morphology is typical of high-conversion bulk polymerization (Figure 12.1d), while the broad range of particle sizes is the result of the nucleation and agglomeration processes.

When a thermal initiator, such as AIBN or benzoyl peroxide, is used the reaction is autocatalytic. This contrasts sharply with normal homogeneous polymerizations in which the rate of polymerization decreases monotonically with time. Studies by Bamford and Jenkins [47–49] and others [35] show that three propagation reactions occur simultaneously to account for the anomalous autoacceleration. These are chain growth in the continuous monomer phase; chain growth of radicals that have precipitated from solution onto the particle surface; and chain growth of radicals within the polymer particles [50,51].

The polymerization of radicals within the core of the polymer particles is considered to be very slow since it is limited by the diffusion of monomer from the particle surface to the poorly swollen core. Bamford and Jenkins showed that the swelling of the polymer is so poor that the radicals within the particle core are highly restricted in mobility. These radicals react with available monomer faster than it can diffuse into the core. As a result, the core radicals become trapped within the particle matrix. This effect is best demonstrated with ultraviolet (UV) or gamma-ray-induced polymerization because the source of free radicals can be easily removed. The persistence of free radicals, can be demonstrated directly using electron spin

Acrylic Fibers

resonance (ESR) techniques. Bamford et al. [48], for example, studied the effect of added solvent and temperature on the population of trapped radicals. When the concentration of DMF was increased systematically from 0 to 60 mol%, in a photoinduced polymerization of acrylonitrile at 20°C, it was found that the number of trapped radicals reached a maximum at approximately 10 mol% DMF and became negligible at approximately 50 mol% DMF. Apparently, the additional swelling resulting from DMF concentrations up to 10 mol%

FIGURE 12.1 (a) Scanning electron micrograph of AN–VA polymer particle prepared by aqueous dispersion polymerization at 3.5 water-to-monomer ratio. (b) Scanning electron micrograph of AN–VA polymer particle prepared by aqueous dispersion polymerization at 2.0 water-to-monomer ratio.

Continued

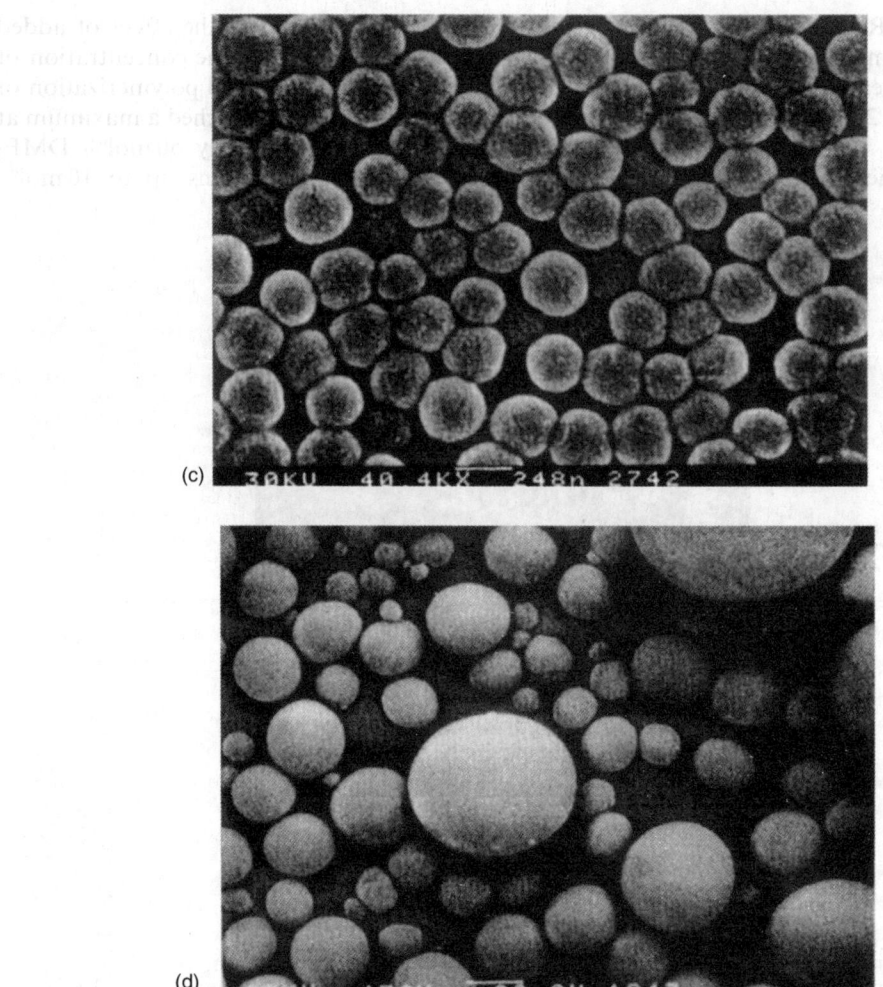

FIGURE 12.1 (Continued) (c) Emulsion polymerization of AN–VA copolymer. (d) Bulk polymerization of AN–VA copolymer.

increases chain mobility enough to allow radicals to diffuse more easily into the particle core but not enough to allow unrestricted radical recombination. When the effect of temperature was considered, it was found that the trapping effect decreased rapidly as the reaction temperature was increased from 20 to 60°C. At 60°C, chain mobility increased sufficiently so that the concentration of trapped radicals was almost negligible.

Polymerization in the continuous monomer phase is also limited. This is because the polymer precipitates at a very low degree of polymerization. In practice, the soluble radical oligomers may either coalesce with similar unprecipitated chains to form particle nuclei or they may collide with an existing particle and become adsorbed on the particle surface. The nucleation process is favored at low conversion when the overall particle population is very low. The particle population increases with conversion, however, so that ultimately the majority of oligomeric soluble radicals are adsorbed by existing particles. At that stage, polymerization occurs primarily on the particle surface and polymerization in the continuous phase then becomes a secondary rate process. The autocatalytic effect is due to two factors. First, as the particle population increases, the total particle surface available as a reaction phase increases giving the effect of a larger reaction volume. Second, there is a reduction in the chain mobility of radicals in the particle surface layer. Radical termination is consequently restricted while chain growth is affected relatively little.

Although bulk polymerization of acrylonitrile seems adaptable, it is rarely used commercially. This is because the autocatalytic nature of the reaction makes it difficult to control. This, combined with the fact that the rate of heat generated per unit volume is very high, makes large-scale commercial operations difficult to engineer. Finally, the viscosity of the medium becomes very high at conversion levels above 40–50%. Therefore, commercial operation at low conversion requires an extensive monomer recovery operation. The subject of heterogeneous radical polymerization of acrylonitrile was reviewed by Guyot [52].

12.3.2.3 Emulsion Polymerization

Emulsion polymerization is another method that is used to polymerize acrylonitrile. The mechanism of emulsion polymerization was first developed qualitatively by Harkins [53] and later quantitatively by Smith and Ewart [54,55] and Gardon [56,57]. It was shown that the emulsifier disperses a small portion of the monomer in aggregates of 50–100 molecules, approximately 50 Å in diameter, called micelles. The majority of the monomer stays suspended in droplet form. These droplets are typically 10,000 Å in diameter, much larger than the micelles. Since a water-soluble radical initiator is used, polymerization begins in the aqueous phase. The micelle concentration is normally so high that the aqueous radicals are rapidly captured [58]. The micelle is essentially a tiny reservoir of monomer. Therefore, polymerization proceeds rapidly, converting the micelle to a polymer particle nucleus. The ability of emulsion polymerization to segregate radicals from one another is of great importance commercially. The effect is to minimize the rate of radical recombination, allowing high rates of polymerization to be achieved along with very high molecular weight. The particle growth mechanism differs from bulk and aqueous dispersion in that agglomeration is minimized by the presence of surfactant at the surface. The result, shown in Figure 12.1, is that many particles are produced at the micelle stage and these grow relatively uniformly with increasing conversion, with little nucleation or agglomeration at the latter stages of the polymerization.

In practice, many commercial process employ a chain transfer agent to control molecular weight at a reduced level. Processes of commercial importance are the copolymerization of butadiene with styrene or acrylonitrile to produce synthetic rubber and the polymerization of acrylic esters, vinyl chloride, vinylidene, and vinyl acetate to produce latexes for adhesives and paints.

The use of emulsion polymerization to make fiber-forming polymers for the textile is generally limited to the manufacture of modacrylic compositions, though it is employed extensively in other areas of polymer manufacturing. There are many examples of commercial emulsion polymerization processes in the literature. One notable example of an emulsion process is the old Union Carbide process for Dynel [14,15]. Acrylonitrile particle formation in emulsion and heterogeneous solution polymerization was reviewed by King et al. [59]. Comprehensive reviews of emulsion polymerization technology have been published by Blackley [60], Bassett and Hamielic [61], Piirma [62], and Eliseeva et al. [63], and a review of emulsion polymerization reactor modeling has been published by Min and Ray [64].

12.3.2.4 Aqueous Dispersion Polymerization

By far, the most widely used method of polymerization in the acrylic fibers industry is aqueous dispersion. It is generally agreed that when inorganic compounds, such as persulfates or perchlorates or hydrogen peroxide, are used as radical generators, the initiation and initial radical growth steps occur mainly in the aqueous phase. Chain growth is limited in the aqueous phase, however, because the monomer concentration is normally very low and the polymer is insoluble in water. The two most likely outcomes are illustrated schematically in Figure 12.2. Nucleation occurs when aqueous chains aggregate or collapse after reaching a threshold molecular weight. At low conversion, when the aqueous monomer concentration is highest and the existing particle population is low, nucleation is favored.

At high conversion, as in the case of commercial continuous polymerizations, the monomer concentration is low and many polymer particles are present. Aqueous phase polymerization is slow and the aqueous radicals are likely to be captured on the particle surface by a sorption mechanism. The particle surface may be richer in monomer than the aqueous phase. Therefore, the polymerization continues in this monomer-rich layer and the sorption becomes irreversible as the chain end grows into the particle.

As polymer swelling is poor and the aqueous solubility of acrylonitrile is relatively high, the tendency for radical capture is very limited, compared to an emulsion polymerization of a monomer such as styrene. Consequently, the rate of particle nucleation is high throughout the course of the polymerization and particle growth occurs predominantly by a process of agglomeration of primary particles. Unlike emulsion particles of a readily swollen polymer, such as polystyrene, the acrylonitrile aqueous dispersion polymer particles are massive agglomerates of primary particles that are approximately 1000 Å in diameter. This can be seen in Figure 12.1, where scanning electron micrographs of aqueous dispersion particles are shown in comparison to emulsion and bulk polymerization particles.

The kinetics of aqueous dispersion polymerization differ very little from acrylonitrile bulk or emulsion polymerization. Redox initiation is normally used in commercial production of polymers for acrylic fibers. This type of initiator can generate free radicals in an aqueous medium efficiently at relatively low temperatures. The most common redox system consists of ammonium or potassium persulfate (oxidizer), sodium bisulfite (reducing agent), and ferric or ferrous iron (catalyst). This system gives the added benefit of supplying dye sites for the fiber.

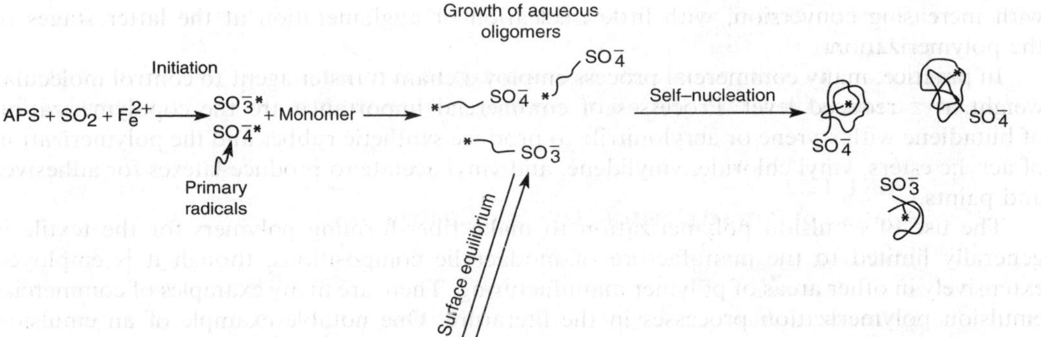

FIGURE 12.2 Schematic depiction of particle nucleation and radical absorption that occur during the aqueous dispersion polymerization of acrylonitrile. APS, ammonium persulfate.

This redox system works at pH levels in the range 2–4, where the bisulfite iron predominates. Two main reactions account for radical production. These are the oxidation of ferrous iron by persulfate, and the reduction of ferric iron by SO_2 in the bisulfite form.

$$S_2O_8^{2-} + Fe^{2+} \rightarrow Fe^{3+} + SO_4^{2-} + SO_4^{*-} \quad (12.10)$$

$$HSO_3^- + Fe^{3+} \rightarrow Fe^{2+} + HSO_3^* \quad (12.11)$$

The sulfate and sulfonate radicals thus produced react with monomer to initiate rapid chain growth. Termination generally occurs by radical recombination though in most commercial processes chain transfer agents are used to control molecular weight and impart acid dye sites. Bisulfite ion, the most widely used chain transfer agent, is believed to react according to the following mechanism [65]:

$$P_n^* + HSO_3^- \rightarrow P_nH + SO_3^{*-} \quad (12.12)$$

The above reaction is apparently very rapid since the bisulfite feed has a very pronounced effect on polymer molecular weight with virtually no effect on the overall rate of polymerization. Studies by Peebles and coworkers [65,66] showed that the ratio of bisulfite to persulfate in the reaction mixture has a strong effect on the dye site content of the polymer. In the absence of chain transfer reactions, all dye sites are derived from initiator radicals. Thus, if termination occurs exclusively by radical recombination, then each polymer chain contains a dye site at each end. Sulfate and sulfonate radicals are produced at equal rates, so the total dye site content must be an equimolar mixture of these two distributed among the chain ends at random. Chain transfer to bisulfite, however, terminates one chain with a hydrogen atom while starting another with a sulfonate radical. This increases the total dye site content of the polymer by reducing the polymer molecular weight. But, at the same time, this reaction produces chains with just one dye site. At a given molecular weight the dye site content of the polymer can, in theory, vary from two per chain at low bisulfite levels to one per chain at very high bisulfite levels.

Data taken from the cited work are shown in Table 12.2. The dye site contents of polymer prepared at sodium bisulfite to potassium persulfate weight ratios of 3.0 and 0.33 are shown. Although the viscosity average molecular weights (M_v) of the two polymer samples are nearly the same, the number average molecular weight (M_n) is much lower in the low bisulfite case. The total dye site content of the low bisulfite polymer is 50% higher, corresponding not only to the lower M_n, but also to a higher total number of dye sites per chain. The distribution of sulfate and sulfonates is approximately as expected on a theoretical basis. The total number of dye sites per chain is between 1 and 2, approaching unity as the bisulfite level is increased. The

TABLE 12.2
Analysis of Polyacrylonitrile Dye Site Content

Bisulfite/persulfate	3/1	1/3
Sulfates/molecule	0.16	0.24
Sulfonates/molecule	0.93	1.03
Total/molecule	1.09	1.27
SAG microequivalents/gram	21.6	31.2
$M_v \times 10^{-4}$	12.5	12.8
$M_n \times 10^{-4}$	5.06	4.08
M_v/M_n	2.5	3.1

Note: M_n = Number average molecular weight.

sulfonate content is close to the theoretical value of unity while the sulfate content is only 0.24 at the low bisulfite level. The theoretical maximum of unity would not be expected unless the bisulfite levels were so low that no chain transfer could occur. The subject of polymer end-group kinetics has been discussed recently by Ebdon et al. [67,68].

In commercial practice, reducing agent to oxidizing agent ratios are used that are equivalent to ratios of bisulfite to potassium persulfate ranging from 8 to 15. These high ratios give narrower molecular weight distributions and the combination of high activator and low oxidizer gives a relatively low-conversion reaction. Low conversion is an effective means of minimizing branching and color-producing side reactions.

A comprehensive review of the literature on aqueous polymerization has been published by Palit et al. [69]. Reviews of acrylonitrile polymerization have been published by Stueben [28], the American Cyanamid Co. [29], Jenkins [30], and Thomas [70].

12.3.3 Copolymerization Kinetics

12.3.3.1 Homogeneous Copolymerization

When carried out in a homogeneous solution, the copolymerization of acrylonitrile follows the normal kinetic rate laws of copolymerization. The controlling variables are the relative reactivities of the component propagation reactions. In a binary copolymerization of monomers 1 and 2, these reactions would be as follows.

$$M_1^* + M_1 \xrightarrow{K_{12}} M_1^* \tag{12.13}$$

$$M_1^* + M_2 \xrightarrow{K_{12}} M_2^* \tag{12.14}$$

$$M_2^* + M_2 \xrightarrow{K_{21}} M_2^* \tag{12.15}$$

$$M_2^* + M_1 \xrightarrow{K_{21}} M_1^* \tag{12.16}$$

Both the rate and copolymer compositions can be characterized by the absolute values of the four possible rate constants. The absolute values of the four possible rate constants are required to specify the overall rate of polymerization, while the reactivity ratios, defined below, determine the polymer composition and monomer sequence distribution for any given monomer mixture. These ratios, R_1 and R_2, are defined

$$R_1 = K_{11}/K_{12} \tag{12.17}$$

TABLE 12.3
Kinetic Characteristics Possible in Copolymerization

Reactivity ratios	Copolymerization characteristics
$R_1 = 0$, $R_2 > 0$	Monomer 1 incapable of homopolymerization—blocking of monomer 1 impossible, mole fraction of monomer 1 cannot be greater than 0.5
$R_1 < 1$, $R_2 < 1$	Alternating copolymerization—monomers tend to alternate along polymer chain
$R_1 > 1$, $R_2 \sim 0$	Kinetically incompatible monomer pair; monomer 1 adds in blocks while monomer 2 adds only by alternating with monomer 1
$R_1 > 1$, $R_2 < 1$	Blocking tendency—monomer 1 tends to add to the polymer in blocks
$R_1 = 1$, $R_2 = 1$	Azeotropic copolymerization—polymer composition same as monomer mixture
$R_1 > 1$, $R_2 > 1$	Simultaneous homopolymerization of both monomers—no case is known

and

$$R_2 = K_{22}/K_{21} \qquad (12.18)$$

Reactivity ratios express the relative tendency for monomer blocking versus monomer alternation along the polymer chain. A value of R_1 greater than 1 indicates a tendency for monomer 1 to incorporate in blocks, while a value of R_1 less than 1 indicates a tendency for monomer 1 to alternate with monomer 2 along the polymer chain. The kinetic behavior of monomer pairs can be classified into the categories described in Table 12.3. Typical reactivity ratios for monomers commonly used in acrylic fiber production are listed in Table 12.4 [71,72].

TABLE 12.4
Reactivity Ratios for Acrylonitrile Polymerizations

Monomer 2	R_1	+/−	R_2	+/−	Temperature	Reaction medium
Acrylamide	0.94	0.16	1.04	0.27	40	Redox initiation
Acrylic acid	0.35		1.15		50	Aqueous solution peroxide
Allyl sulfonic acid, sodium salt	1.85	0.01	0.43	0.01	45	DMSO solution
	1.00	0.01	0.38	0.02	45	DMSO solution
	1.25	0.01	0.28	0.02	45	DMSO solution 94/6 solution
1,3-Butadiene	0.05	0.01	0.35	0.01	50	
1-Butene	8.0		0.10		60	
2-Butene, cis or trans	14.0		0.0		60	
Isobutylene	1.02		0.00		60	
Isoprene	0.03	0.03	0.45	0.05	50	
Methyl acrylate	1.5	0.10	0.84	0.05	50	Peroxide initiation
	1.17		0.76		50	
	1.02	0.02	0.70	0.02	50	DMSO solution
Butyl acrylate	1.00	0.01	1.01	0.01	60	
Methacrylonitrile	0.35	0.05	2.80	0.40	50	Benzene solution
Butyl methacrylate	0.31		1.08		60	Peroxide initiation
Methyl methacrylate	0.15	0.07	1.20	0.14	60	Solution polymer, peroxide
Sodium styrene sulfonate	0.05	0.02	1.50	0.20	40	
Styrene	0.05	0.02	0.37	0.02	50	
	0.07	0.006	0.37	0.03	50	
Vinyl acetate	4.05	0.3	0.061	0.01	60	
	4.2		0.05		50	DMF solution, peroxide
Vinyl bromide	2.25		0.055		60	
Vinyl chloride	3.6	0.2	0.052	0.009	50	
	3.7		0.074		50	Acetone solution
Vinyl fluoride	44.0	0.2	0.005	0.2	25	Gamma initiation, benzene solution
Vinylidene chloride	0.44		0.40		40	Various solvents
	0.70		1.8			
	0.91	0.10	0.37	0.37	60	Peroxide initiation
Vinyl formate	3.0	0.05	0.04	0.005	60	Solution polymer, peroxide
2-Vinylpyridine	0.113	0.002	0.47	0.03	60	
4-Vinylpyridine	0.113	0.005	0.41	0.09	60	
N-Vinylpyrrolidone	0.27	0.16	0.041	0.02	60	
Vinyl stearate	4.2	0.02	0.064	0.005	60	DMF solution
Vinyl sulfonic acid	4.52		0.22		60	ZnCl$_2$/water solution

Source: From Brandrup, J. and Immergut, E.H., *Polymer Handbook*, 2nd ed., Interscience, New York, 1975.

The three main types of copolymerization behaviors are ideal, alternating, and incompatible. These are illustrated in Figure 12.3 and Figure 12.4. The product $R_1 \times R_2$ is considered to be a measure of the ideality of a monomer pair. An ideal copolymerization is considered to be one in which the attacking radical shows no preference for one monomer or the other. The product of $R_1 \times R_2$ in such cases is unity. The monomer pair, acrylonitrile–methyl acrylate, with $R_1 = 1.02$ and $R_2 = 0.70$ at 50°C, is very close to be an ideal monomer pair. Both monomers are similar in resonance, polarity, and steric characteristics. The acrylonitrile radical shows approximately equal reactivity with both monomers and the methyl acrylate radical shows only a slight preference for reacting with acrylonitrile monomer. The effect, with component propagation rate constants shown in Table 12.5, is that the acrylonitrile radical shows approximately equal reactivity with both monomers and the methyl acrylate radical shows only a slight preference for reacting with acrylonitrile monomer. This makes the acrylonitrile–methyl acrylate monomer pair an excellent candidate for synthetic fibers where chemical homogeneity is important.

Most acrylonitrile monomer pairs fall into the nonideal category. One such nonideal monomer pair is acrylonitrile–vinyl acetate, with $R_1 = 4.05$ and $R_2 = 0.061$ at 60°C. This is an example of a nonideality sometimes referred to as "kinetic incompatibility." Acrylonitrile, because of the potential resonance stabilization offered by the nitrile group, is a reactive monomer but a relatively unreactive radical. On the other hand, vinyl acetate offers little possibility for resonance stabilization, so it can be categorized as a relatively unreactive monomer but highly reactive radical. The effect, shown in Table 12.6, is that the reaction between the very reactive acrylonitrile monomer with the highly reactive vinyl acetate radical has an extremely high rate constant.

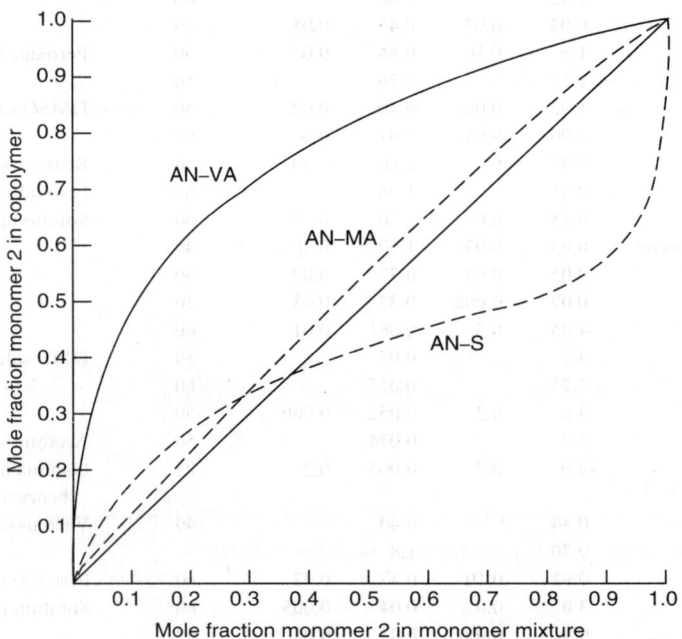

FIGURE 12.3 Incremental copolymer composition versus monomer composition for three commonly observed reaction patterns: acrylonitrile–vinyl acetate (AN–VA, $R_1/R_2 = 4.05/0.061$); acrylonitrile–methyl acrylate (AN–MA, $R_1/R_2 = 1.5/0.84$); and acrylonitrile–styrene (AN–S, $R_1/R_2 = 0.04/0.03$).

Acrylic Fibers

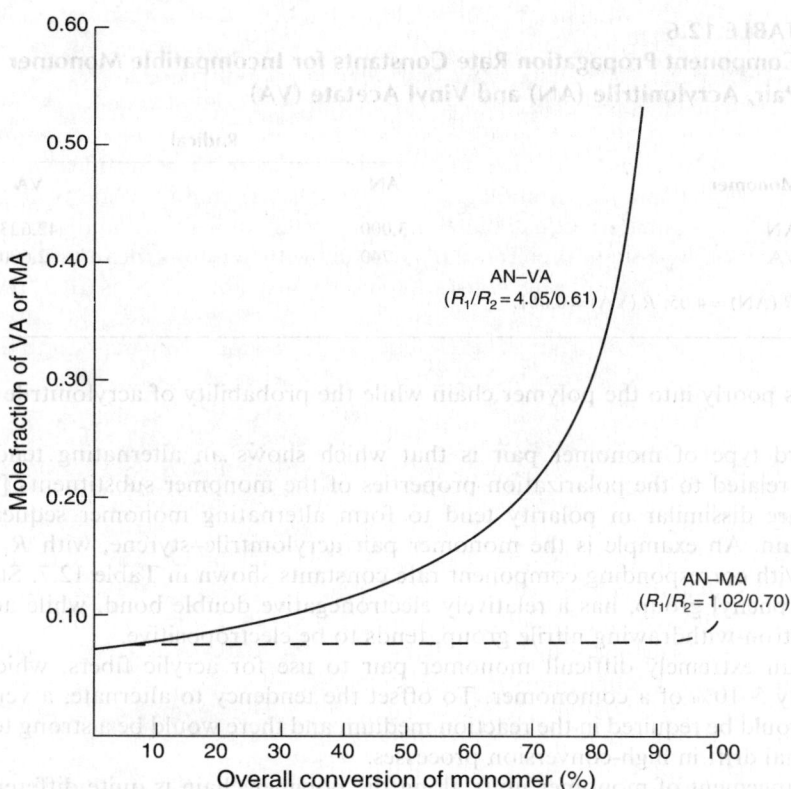

FIGURE 12.4 Illustration of drift in copolymer composition with conversion in a batch reactor for two commercially important comonomer pairs. VA content in copolymer increases sharply with conversion in acrylonitrile–vinyl acetate copolymerization ($R_1/R_2 = 4.05/0.061$) and content of MA in copolymer changes very little with conversion in acrylonitrile–methyl acrylate copolymerization ($R_1/R_2 = 1.02/0.70$).

In this case, the acrylonitrile radical chain end reacts with an acrylonitrile monomer unit much faster than it reacts with a vinyl acetate monomer unit at the same concentration. At the same time, a vinyl acetate radical chain end reacts much faster with an acrylonitrile monomer unit than with another vinyl acetate monomer unit. The net result is that vinyl acetate

TABLE 12.5
Component Propagation Rate Constants for Ideal Monomer Pair, Acrylonitrile (AN) and Methyl Acrylate (MA)

	Radical	
Monomer	AN	MA
AN	3000	1800
MA	2941	1260

R (AN) = 1.02; R (MA) = 0.7.

TABLE 12.6
Component Propagation Rate Constants for Incompatible Monomer Pair, Acrylonitrile (AN) and Vinyl Acetate (VA)

Monomer	Radical	
	AN	VA
AN	3,000	42,623
VA	740	2,600

$R(AN) = 4.05$; $R(VA) = 0.061$.

incorporates poorly into the polymer chain while the probability of acrylonitrile blocking is very high.

The third type of monomer pair is that which shows an alternating tendency. This tendency is related to the polarization properties of the monomer substituents [73]. Monomers that are dissimilar in polarity tend to form alternating monomer sequences in the polymer chain. An example is the monomer pair acrylonitrile–styrene, with $R_1 = 0.07$ and $R_2 = 0.37$, with corresponding component rate constants shown in Table 12.7. Styrene, with its pendant phenyl group, has a relatively electronegative double bond, while acrylonitrile, with its electron-withdrawing nitrile group, tends to be electropositive.

This is an extremely difficult monomer pair to use for acrylic fibers, which typically requires only 5–10% of a comonomer. To offset the tendency to alternate, a very low level of styrene would be required in the reaction medium and there would be a strong tendency for compositional drift in high-conversion processes.

The arrangement of monomer units along the polymer chain is quite different for these three types of polymerizations. The monomer units incorporate randomly along the chain since the growing radical has no preference for either monomer. The incompatible monomer pair incorporates in blocks of the more reactive monomer, particularly when the more reactive monomer is present in excess. And, as might be expected, the monomer units alternate in the third case, except when one of the monomers is in great excess.

Incompatible and ideal copolymerizations behave very differently in high-conversion batch polymerizations. This is illustrated in Figure 12.4 for the AN–VA and AN–MA monomer pairs discussed above. In the AN–VA system, the AN is consumed more rapidly than the VA. Therefore, the unreacted monomer mixture becomes richer in VA as conversion increases. This results in a drift in copolymer composition with conversion. The AN–MA system, in contrast, shows little or no drift, and consequently a more uniform copolymer

TABLE 12.7
Component Propagation Rate Constants for Alternating Monomer Pair, Acrylonitrile (AN) and Styrene (S)

Monomer	Radical	
	AN	S
AN	3,000	486
Styrene	42,857	180

$R(AN) = 0.07$; $R(S) = 0.37$.

composition can be made. The problem of compositional heterogeneity is dealt with commercially by using semibatch or continuous stirred tank reactors.

Copolymer composition can be predicted for copolymerizations with two or more components, such as those employing acrylonitrile plus a neutral monomer and an ionic dye receptor. These equations are derived by assuming that the component reactions involve only the terminal monomer unit of the chain radical. This leads to a collection of $N \times N$ component reactions and $N \times (N-1)$ binary reactivity ratios, where N is the number of components used. The equation for copolymer composition for a specific monomer composition was derived by Mayo and Lewis [74], using the set of binary reactions, rate constants, and reactivity ratios described in Equation 12.13 through Equation 12.18. The drift in monomer composition, for bicomponent systems was described by Skeist [75] and Meyer and coworkers [76,77]. The theory of multicomponent polymerization kinetics has been treated by Ham [78] and Valvassori and Sartori [79]. Comprehensive reviews of copolymerization kinetics have been published by Alfrey et al. [80] and Ham [81,82], while the more specific subject of acrylonitrile copolymerization has been reviewed by Peebles [83]. The general subject of the reactivity of polymer radicals has been treated in depth by Jenkins and Ledwith [84].

12.3.3.2 Heterogeneous Copolymerization

When copolymer is prepared in a homogeneous solution, the kinetic expressions described above can be used to predict copolymer composition. Bulk and dispersion polymerization are somewhat different since the reaction medium is heterogeneous and polymerization occurs simultaneously in separate loci. In bulk polymerization, for example, the monomer-swollen polymer particles support polymerization within the particle core as well as on the particle surface. In aqueous dispersion or emulsion polymerization, the monomer is actually dispersed in two or three distinct phases—a continuous aqueous phase, a monomer droplet phase, and a phase consisting of polymer particles swollen at the surface with monomer. This affects the ultimate polymer composition because the monomers are partitioned such that the monomer mixture in the aqueous phase is richer in the more water-soluble monomers than the two organic phases. Where polymerization occurs predominantly in the organic phases, these relatively water-soluble monomers may incorporate into the copolymer at lower levels than expected. Fordyce et al. [85–87], for example, in studies of the emulsion copolymerization of acrylonitrile and styrene, found that the copolymer was richer in styrene than copolymer made by bulk polymerization, using the same initial monomer composition. Analysis of the reaction mixtures by Smith [88] showed that nearly all of the styrene was concentrated in the droplet and swollen particle phases. The acrylonitrile, on the other hand, was distributed between both the aqueous and organic phases. The monomer compositions in the droplet and particle phase were found to be essentially the same. The effect of monomer partitioning on copolymer composition is strongest with the ionic monomers since this type of monomer is usually soluble in water and nearly insoluble in the other monomers.

Reviews of emulsion copolymerization kinetics and the effects of reaction heterogeneity on reaction locus have been published by Ley and Fowler [89,90] and Eliseeva et al. [91].

12.3.4 COMMERCIAL POLYMERIZATION METHODS

Aqueous media, such as emulsion, suspension, and dispersion polymerization, are by far the most widely used in the acrylic fiber industry. Water acts as a convenient heat transfer and cooling medium and the polymer is very easily recovered by filtration or centrifugation. Fiber producers that use aqueous solutions of thiocyanate or zinc chloride as the solvent for the polymer have an additional benefit. In such cases, the reaction medium can be converted directly to dope to save the costs of polymer recovery. Aqueous emulsions are less common. This type of process is used primarily for modacrylic compositions, such as Dynel. Even in such processes the emulsifier is used at very low levels, giving a polymerization medium with characteristics of both a suspension and a true emulsion.

The most common reactor type today is the continuous stirred tank. However, in the early years of acrylic fiber production, during the 1950s and the early 1960s, the semibatch polymerization process was commonly used for the commercial production of acrylonitrile copolymer. In this type of process, the reaction vessel is charged with a portion of the reactants and the reaction is induced by using radical initiators, such as potassium persulfate. Control of copolymer composition is difficult in this type of reactor because the comonomers most frequently used have vapor pressures, solubilities, or reactivities that differ greatly from that of acrylonitrile. Acid-dyeable monomers, such as SSS, are nonvolatile solids with high water solubility. Monomers used to impart flame retardancy, such as vinyl chloride, vinyl bromide, and vinylidene chloride, are nearly insoluble in water and have high vapor pressures. As a result, the various monomers employed in a given reaction mixture often react at widely differing rates and the copolymer formed in the early stages of the reaction has a different composition and molecular weight than that formed at the later stages.

12.3.4.1 Semibatch Polymerization

An example of a commercial semibatch polymerization process is the early Union Carbide process for Dynel, one of the first flame-retardant modacrylic fibers [14,15]. Dynel, a staple fiber, which was wet-spun from acetone, was introduced in 1951. The polymer is made up of 40% acrylonitrile and 60% vinyl chloride. The reactivity ratios for this monomer pair are 3.7 and 0.074 for acrylonitrile and vinyl chloride in solution at 60°C. Thus, acrylonitrile is much more reactive than vinyl chloride in this copolymerization. In addition, vinyl chloride is a strong chain transfer agent.

To make the Dynel composition of 60% vinyl chloride, the monomer composition must be maintained at 82% vinyl chloride. Since acrylonitrile is consumed much more rapidly than vinyl chloride, if no control is exercised over the monomer composition, the acrylonitrile content of the monomer will decrease to approximately 1% after only 25% conversion. The low acrylonitrile content of the monomer required for this process introduces yet another problem. That is, with an acrylonitrile weight fraction of only 0.18 in the unreacted monomer mixture, the low concentration of acrylonitrile becomes a rate-limiting reaction step. Therefore, the overall rate of chain growth is low and, under normal conditions, with chain transfer and radical recombination, the molecular weight of the polymer is very low.

The low rate of copolymerization and tendency for low-molecular-weight polymer are overcome by using emulsion polymerization. The rate of polymerization and polymer molecular weight are then controlled by varying the rate of initiation and the surfactant concentration. The copolymer composition is controlled by adding acrylonitrile monomer to the reactor at a rate that maintains a constant pressure of 75–76 psi at 40°C. This pressure is produced by a free monomer phase consisting of 18% acrylonitrile. Thus, as long as there is a free monomer phase, the 18% acrylonitrile level can be maintained by holding the reaction pressure constant at 76 psi. The Union Carbide process requires 77 h and 19 additions of acrylonitrile. The yield of copolymer is 65.8% and the final vinyl chloride content of the polymer is 60.5%.

12.3.4.2 Continuous Aqueous Dispersion Processes

Processes using a continuous stirred tank reactor have replaced the semibatch process except where low-volume specialty products are made [92,93]. For start-up, the reactor is charged with a certain amount of the reaction medium, usually solvent or pH-adjusted water. In more

Acrylic Fibers

sophisticated processes, the start-up period may be minimized by filling the reactor with overflow from a reactor already operating at steady state. The reactor feds are metered in at a constant rate for the entire course of the production run, which normally continues until equipment maintenance is needed. A steady state is established by taking an overflow stream at the same mass flow rate as the combined feed streams. The main advantage of this process over the semibatch process is that control of molecular weight, dye site level, and polymer composition is greatly improved.

An example of a continuous aqueous dispersion process is shown in Figure 12.5 [92]. A monomer mixture composed of acrylonitrile and up to 10% of a neutral comonomer, such as methyl acrylate or vinyl acetate, is fed continuously. Polymerization is initiated by feeding aqueous solutions of potassium persulfate (oxidizer), sulfur dioxide (reducing agent), ferrous iron (promoter), and sodium bicarbonate (buffering agent). The aqueous and monomer feed

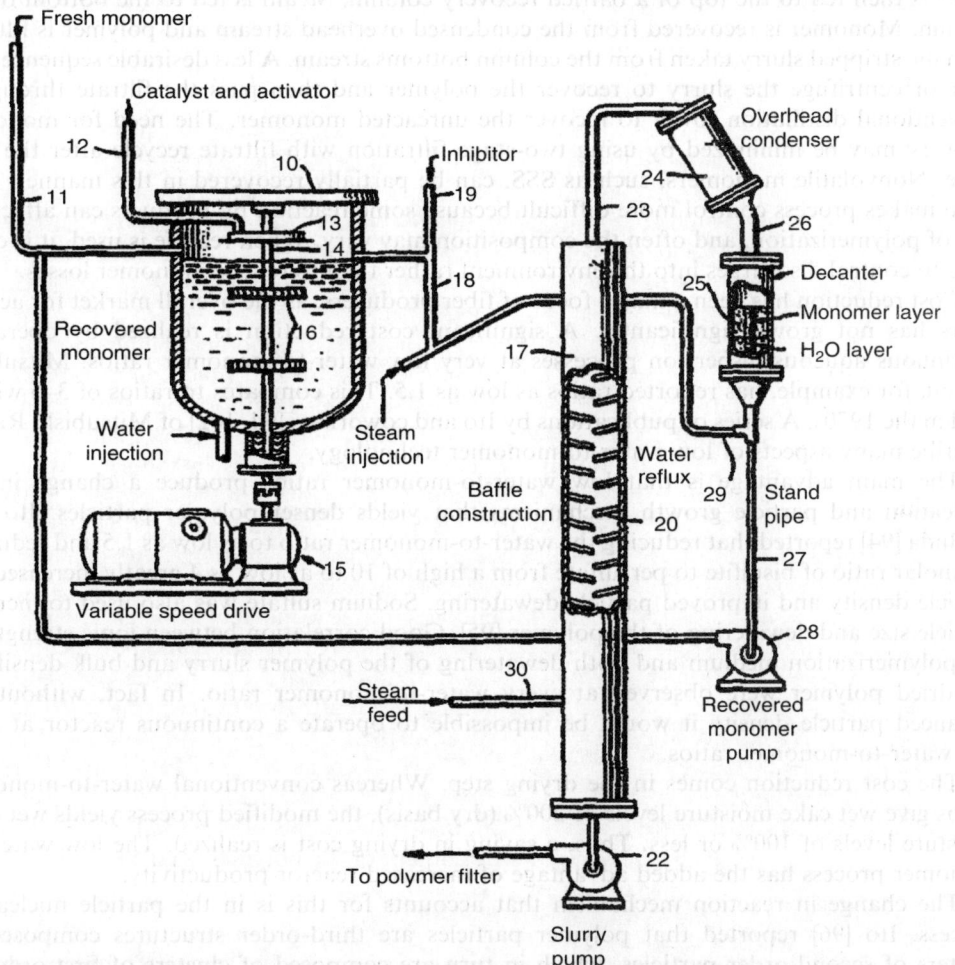

FIGURE 12.5 Schematic diagram of aqueous dispersion polymerization reactor and slurry stripping column. (From Cheape, D.W. and Eberhardt, W.R., U.S. Patent 3,454,542, July 8, 1969.)

streams may be fed at rates that give a reactor dwell time of 40 to 120 min and a feed ratio of water to monomer in the range 2–5. The product stream, an aqueous slurry of polymer particles, is mixed with ethylene diamine tetraacetic acid (iron chelate) or oxalic acid to stop the polymerization. This slurry is then fed to the top section of a baffled monomer separation column. The separation of unreacted monomer can be affected by contacting the slurry with a countercurrent flow of stream introduced at the bottom of the column. The monomer is condensed from the overheads stream and separated from the resulting water mixture using a decanter. The stripped slurry is taken from the column bottoms stream and separated from the water using a continuous vacuum filter. After filtration and washing, the polymer is pelletized, dried, ground, and then stored for later spinning.

The monomer recovery process may vary in commercial practice. An effective way of doing this is to strip the volatile monomers directly from the reactor product stream. This process, known as slurry stripping, is used with aqueous dispersion polymerization in continuous stirred tank reactors. A polymerization inhibitor, normally a chelating agent that deactivates the catalytic iron in the reaction mixture, is added to the reactor overflow stream. The product slurry is then fed to the top of a baffled recovery column. Steam is fed to the bottom of the column. Monomer is recovered from the condensed overhead stream and polymer is filtered from the stripped slurry taken from the column bottoms stream. A less desirable sequence is to filter or centrifuge the slurry to recover the polymer and then pass the filtrate through a conventional distillation tower to recover the unreacted monomer. The need for monomer recovery may be minimized by using two-stage filtration with filtrate recycle after the first stage. Nonvolatile monomers, such as SSS, can be partially recovered in this manner. This often makes process control more difficult because some reaction by-products can affect the rate of polymerization, and often the composition may vary. When recycle is used, it is often done to control discharges into the environment rather than to reduce monomer losses.

Cost reduction has been a major focus of fiber producers, as the overall market for acrylic fibers has not grown significantly. A significant cost reduction is realized by operating continuous aqueous dispersion processes at very low water-to-monomer ratios. Mitsubishi Rayon, for example, has reported ratios as low as 1.5. This compares to ratios of 3–5 widely used in the 1970s. A series of publications by Ito and coworkers [94–102] of Mitsubishi Rayon describe many aspects of low water-to-monomer technology.

The main advantage is that low water-to-monomer ratios produce a change in the nucleation and particle growth mechanisms that yields denser polymer particles. Ito and Yoshida [94] reported that reducing the water-to-monomer ratio to as low as 1.5 and reducing the molar ratio of bisulfite to persulfate from a high of 10 to as low as 4 greatly increased the particle density and improved particle dewatering. Sodium sulfate was also used to increase particle size and dewatering of the polymer [95]. Good correlation between ionic strength of the polymerization medium and both dewatering of the polymer slurry and bulk density of the dried polymer were observed at every water-to-monomer ratio. In fact, without the enhanced particle density it would be impossible to operate a continuous reactor at such low water-to-monomer ratios.

The cost reduction comes in the drying step. Whereas conventional water-to-monomer ratios give wet cake moisture levels of 200% (dry basis), the modified process yields wet cake moisture levels of 100% or less. Thus, a saving in drying cost is realized. The low water-to-monomer process has the added advantage of increased reactor productivity.

The change in reaction mechanism that accounts for this is in the particle nucleation process. Ito [96] reported that polymer particles are third-order structures composed of clusters of second-order particles, which in turn are composed of clusters of first-order or primary particles. The primary particles form directly from polymerization and nucleation in the aqueous phase and the rate of formation increases with decreasing water-to-monomer ratio. The primary particles, therefore, are similar and give a more compact packing of the second-order particles with decreased water-to-monomer ratios. A kinetic model of C_m and M_w proposed by Ito [97] is consistent with the concept of aqueous phase polymerization and primary particle formation in the W/M range from 1.75 to 4.0. Ito [98] suggests aqueous initiation with chain growth in the aqueous phase until termination. This may be inconsistent with observations by Fordyce and Ham [86,87] and Smith [88] on copolymer composition in heterogeneous aqueous copolymerizations.

In other studies of low water-to-monomer polymerization, Ito et al. [99] showed that particle size roundness, uniformity, and bulk density improved with increasing agitator speed and impeller size. Molecular weight distribution (MWD) was studied at W/M from 1.75 to 4.0. The value for M_w/M_n varied between 4 and 6 for all of the conditions studied [100,101]. However, there was little effect of water to monomer observed. One of the main factors affecting M_w/M_n was found to be steady-state conversion. The value of M_w/M_n increased with conversion from 2 at low conversion to about 4 at high conversion. Also, studies of end-group content found one SAG per chain regardless of M_w. When A/C was increased, there were proportionately more sulfonate end groups [102]. The authors concluded that chain transfer was the main process of chain termination.

12.3.4.3 Solution Polymerization

The only other commercial polymerization process used for acrylic fibers is solution polymerization. This type of process can be implemented by feeding the monomers to a continuous mixing tank along with a solvent for the polymer. The overflow stream from this tank is then routed to a form of continuous reactor where the polymerization is carried out in a homogeneous solution. Monomer is removed from the product stream and the resulting polymeric solution is used directly for spinning. An obvious advantage of this process is that considerable cost savings can be achieved by eliminating the filtration, drying, and dope-making steps required in the aqueous dispersion process. There are two major drawbacks associated with solution polymerization. First, it is difficult to produce dopes of high solids, particularly with the organic solvents. Second, most of the effective solvents have very high chain transfer constants, making it difficult to produce polymer of high molecular weight. Solvents suitable for this type of commercial polymerization are DMF and DMSO. Solution polymerizations based on polymer solutions of aqueous zinc chloride and thiocyanate salt have also been reported. DMAC, a popular wet-spinning solvent, is not suitable for solution polymerization, however, because of the powerful chain transfer activity of this solvent.

12.3.4.4 Bulk Polymerization

Although bulk polymerization of acrylonitrile seems adaptable, it is rarely used commercially. This is because the autocatalytic nature of the reaction makes it difficult to control. This, combined with the fact that the rate of heat generated per unit volume is very high, makes large-scale commercial operations difficult to engineer. Finally, the viscosity of the medium becomes very high at conversion levels above 40–50%. Therefore, commercial operation at low conversion requires an extensive monomer recovery operation.

However, a commercially feasible process for bulk polymerization in a continuous stirred tank reactor has been developed by Montedison Fibre [103,104]. The heat of reaction is controlled by operating at relatively low-conversion levels and supplementing the normal jacket cooling with reflux condensation of unreacted monomer. Operational problems with thermal stability are controlled by using a free radical redox initiator with an extremely high decomposition rate constant. Since the initiator decomposes almost completely in the reactor,

TABLE 12.8
Illustration of Thermal Stability in a Continuous Process with Normal Initiator Activity

Reaction temperature °C	Rate (%/h)	Notes
40	0.6	Reaction unstable
50	6.0	Reaction unstable; temperature control impossible
60	NA	Autocatalytic condition

the polymerization rate is insensitive to temperature and can be controlled by means of the initiator feed rate. Polymer molecular weight and dye site content are controlled by using mercapto compounds and oxidizable sulfoxy compounds.

It is claimed in the cited patent that when the product of the initiator decomposition rate constant (K_d) and the reactor dwell time is greater than or equal to unity, the reaction becomes insensitive to temperature. The data in Table 12.8 illustrate this by comparing two continuous processes with 60-min dwell times that differ only in the activity of the initiator used. The first process, represented in Table 12.8, employs AIBN at 2% based on monomer fed. With a K_d of only 0.01 h at 50°C, this initiator yields a polymerization process that is highly sensitive to temperature perturbations. When the temperature was increased from 40 to 50°C, for example, the rate of polymerization increased tenfold. The improved thermal stability offered by the Montedison process is illustrated in Table 12.9, where an extremely active redox initiator is used.

In this second set of data, polymerization was initiated with 0.16% (based on monomer) cumene hydroperoxide (CHP) plus an equimolar equivalent of sodium methyl sulfite (SMS). The decomposition rate constant for this redox pair is 2.4 h at 50°C. Consequently, the overall rate of polymerization is much higher and the change in rate with temperature is very small. In this case, thermal stability is excellent and commercial operation becomes feasible.

The complete process set is not described in the cited reference. However, the polymer can easily be recovered by simple vacuum filtration or centrifugation of the polymer slurry. This can be followed by direct conversion of the filter cake to dope by slurrying the filter cake in chilled solvent and then passing the slurry through a heat exchanger to form the spinning solution and a thin film evaporator to remove residual monomer.

A summary of the commercial processes reported by the major acrylic fiber manufacturers is given in Table 12.10.

12.4 SOLID-STATE STRUCTURE OF ACRYLIC POLYMERS

The majority of textile fibers have a morphology that can be described by the classical two-phase model for semicrystalline polymers [105]. In this model, discrete crystalline domains on

TABLE 12.9
Illustration of Thermal Stability in a Continuous Process with an Extremely Active Initiator

Reaction temperature °C	Rate (%/h)	Notes
40	43	Stable reaction; fluid reaction medium
50	48	Stable reaction; fluid reaction medium
60	48	Stable reaction; fluid reaction medium

TABLE 12.10
Polymerization Processes Reported by Major Acrylic Fiber Producers

Fiber producer	Trade name	Major comonomer	Dye sites	Solvent
Fiber producers using aqueous dispersion polymerization				
Monsanto	Acrilan	VA	Ionic end groups	
DuPont	Orlon	MA	Sodium styrene sulfonate	
Fabel	Acribel	MA	Aliphatic sulfonate	
Hoechst	Dolan	MA	Ionic end groups	
Mitsubishi Rayon	Vonnel	VA	Sulfoethyl acrylate	
	Finel	MA	Ionic end groups	
Rhone-Poulenc	Crylor	MA	Ionic end groups	
Bayer	Dralon	MA	Ionic end groups	
Asahi	Cashmilon	MA	Ionic end groups	
Montefibre	Leacril	VA	Ionic end groups	
Asahi	Cashmilon	MA		
Bayer	Dralon	MA		
Courtaulds	Courtelle	MA		
Cydsa	Crysel	VA		
Cytec (formerly American Cyanamid)		(Previously MMA)		
Fabetta	Acribel	MMA		
Hoechst	Dolan	MA		
Japan Exlan	Exlan	MA		
Kanagafuchi	Kanecaron	VCl		
Mitsubishi Rayon	Vonnel	VA		
	Finel	MA		
Monsanto	Acrilan	VA		
Montefibre	Leacril	VA		
Fiber producers using solution polymerization				
Snia-Viscosa	Velicren	MA	Ionic end groups	DMF
Somes	Finacryl	MA	Ionic end groups	DMF
Toray	Toraylon	MA	Aliphatic sulfonate	DMSO
Cortaulds	Courtelle	MA	Acrylic acid	Thiocyanate
American Cyanamid	Creslan	MA, MMA	Ionic end groups	Thiocyanate
Nippon Exlan	Exlan	MA, MMA	Ionic end groups	Thiocyanate
Badische	Zefron	MA	Sulfoethyl	Zinc chloride
Chloride		Methacrylate		
Toho-Belson Chloride	Belson	MA	Ionic end groups	Zinc chloride
Toho Rayon	Beslon	MA		Zinc chloride
Toray	Toraylon	MA		DMSO

Source: From Masson, J.C., Ed., *Acrylic Fiber Technology and Applications*, Marcel Dekker, New York, 1995.

the order of several hundred angstroms are mixed with amorphous domains of similar size. The individual polymer chains have end-to-end distances on the order of 1000–2000 Å and a single chain can span two or more adjacent crystallites. Such a chain is called a tie molecule, and their assembly in the intercrystalline regions forms the amorphous phase. In a sense, then, the fiber morphology can be viewed as a microcomposite structure of the crystalline and amorphous domains, and the tie molecules provide the connectivity that supports tensile loads on the fiber.

The two-phase model is the framework for formulating structure–property relationships of fibers [106]. The descriptive parameters of the model can be precisely defined and measured. Physical characterization techniques have been developed for determining the volume fractions of the crystalline and amorphous domains, the size and perfection of the crystalline domains, and the degree of orientation of the polymer chains with respect to the fiber axis. These parameters are very sensitive to the fiber spinning and processing conditions. It is well known that the tensile properties and dyeing behavior of synthetic fibers, such as nylon and polyester, can be manipulated by controlling the degree of drawing and relaxation after initial fiber formation.

Developing useful structure–property relationships for acrylic fibers has proven difficult because the polymer is not semicrystalline in the conventional sense, and discrete crystalline and amorphous phases are not clearly seen. The PAN or acrylic morphology has been described as "amorphous with a high degree of lateral bonding," or as a "two-dimensional liquid crystalline-like" structure with many defects. Fortunately, the recent application of newer polymer characterization techniques has revealed a high degree of structural detail and provides the basis for a better understanding of acrylic fiber properties. The current views on acrylic morphology are summarized in this section. We begin with the chemical structure of the isolated PAN chain and focus on the highly polar nitrile group and its influence on the chain conformation. The current models of the chain conformation are followed from discussions on x-ray and electron diffraction studies of polymer, fibers, and solution-grown single crystals. Various models of the acrylic morphology are described at this point utilizing additional information gleaned by structural-sensitive techniques such as solvent swelling and wide-line NMR measurements. The last part of the chapter deals with the thermal properties of PAN and how they are influenced by addition of comonomers. Included under thermal properties are the melting point, glass transition, and molecular relaxations. The effect of comonomers on the structure and properties of acrylics is emphasized.

12.4.1 STEREOREGULARITY AND CHAIN CONFORMATION

12.4.1.1 Stereoregularity

The stereoregularity of a vinyl polymer influences its crystallizability [107,108]. A good example of this is polypropylene. Isotactic polypropylene is highly crystalline, whereas the atactic polymer is completely amorphous. To form a crystalline unit cell, the backbone chain of a polymer must first be able to assume some regular chain conformation such as a planar zigzag or helical structure, and then the individual chains can pack to form the cell. A regular chain conformation allows maximization of intermolecular and intramolecular interactions, e.g., van der Waals forces, hydrogen bonding, and dipolar bonding in the crystal. Highly perfect crystalline packing is not possible if the chain is atactic. It may not be possible for the chain to form the same regular structure as the corresponding isotactic or syndiotactic chain because steric hindrance between substitute groups on adjacent monomer units may occur when the backbone attempts to assume this structure. Even if a regular conformation is found, one may still encounter steric hindrance between groups on adjacent chains. Most commercial vinyl polymers, such as polystyrene and polymethyl methacrylate, are atactic and form amorphous glasses that have no long-range crystalline order. Fibers can be melt-spun from these glassy polymers, but they cannot be used for textile applications because of the lack of crystallinity. As soon as these fibers are heated near the glass transition temperature, which is approximately 100°C for the latter two polymers, they begin to shrink. There are, however, exceptions. Some atactic polymers are not truly amorphous and they possess a high degree of order or pseudocrystallinity. Acrylics fall into this class of polymers. Common among them is a substitute group capable of forming strong interactions through secondary bonding, which is the nitrile group for the acrylics. Other examples of commercial importance are polyvinyl alcohol and polyvinyl chloride (PVC).

All commercial PAN and attendant fiber-forming acrylic copolymers are manufactured using free radical polymerization processes (see Section 12.3). Free radical polymerization generally produces polymers with little or no stereoregularity, and this is confirmed for acrylics by nuclear magnetic resonance (NMR) analysis. Schaefer [109] has shown that the ^{13}C NMR spectrum of the nitrile carbon can be interpreted in terms of steric triads and pentads. Three NMR lines are clearly resolved and are attributed to the three possible steric triad configurations, i.e., the hetero, syndio, and isotactic triad configurations. A completely atactic polymer should have concentrations of the hetero, syndio, and isotactic triads in the ratio of 2:1:1 [110]. Schaefer observed a ratio of 5:2:3, which is not markedly different from expectations for the atactic case, and therefore concluded that stereoregularity was low.

Attempts have been made to increase the stereoregularity of acrylic polymers with the hope of producing a more crystalline polymer with improved fiber tensile properties, especially under hot–wet conditions. Most of these efforts have either given only marginal improvements or involved approaches that could not easily be commercialized. The application of Ziegler-Natta catalysis to produce stereoregular polymers, while successful with polypropylene, apparently cannot be used to make polar polymers including acrylic and PVC [111]. Chiang [112] claimed to have produced a PAN with enhanced stereoregularity by using a proprietary organometallic catalyst. The basis for the claim was an increase in the temperature required to dissolve the polymer in propylene carbonate (PC). The solubility temperature was increased from 135°C for the conventional free radical polymer to 175°C, though no corresponding improvement in the crystalline perfection as measured by x-ray diffraction was reported. The theoretical basis for using dissolution temperature as a means for characterizing the degree of stereoregularity and crystalline perfection was also described by Chiang et al. [113].

12.4.1.2 Chain Conformation

The fact that commercial fibers can be made from acrylic polymers suggests that some degree of crystallinity must be present. All textile fibers, whether they are synthetic or natural, have a crystalline phase. This crystallinity imparts good tensile properties and prevents the fiber from shrinking when it is heated above the glass transition of the amorphous phase, and that can occur during common textile-processing operations including dyeing, washing, and ironing. We now consider how PAN, an atactic polymer, could have a crystalline phase, and for this purpose it is constructive to compare PAN with atactic polypropylene. Both are monosubstituted vinyl polymers with carbon–carbon backbones as shown in Figure 12.6. In polypropylene, the carbon is bonded to three hydrogen atoms to form the methyl group and in PAN, it is bonded to a nitrogen atom to form the nitrile group. These groups have similar molar volumes (32.3 cm^3/mol for the methyl group and 27.2 cm^3/mol for the nitrile) [114]. The distinguishing feature of the nitrile group is the very large dipole moment with a magnitude of 3.9 D [114]. The methyl group has little or no dipole moment, so when we compare these two polymers we are seeing directly the influence of the large dipole moment, and the way an assembly of them can interact. The glass transitions of the atactic polypropylene and PAN are approximately −20 and 95°C, respectively [114]. The glass transition is the temperature at which an amorphous polymer undergoes a transition from the glassy to the rubbery state and is attributed to the onset of large-scale segmental motion. The atactic polypropylene is a waxy material readily soluble in many solvents, while PAN is soluble only in very polar solvents. Billmeyer [115] contends that the differences in T_gs can be expected given that the solubility parameters of polypropylene and PAN are 16.6 and 31.5 (J/cm^3)$^{1/2}$,

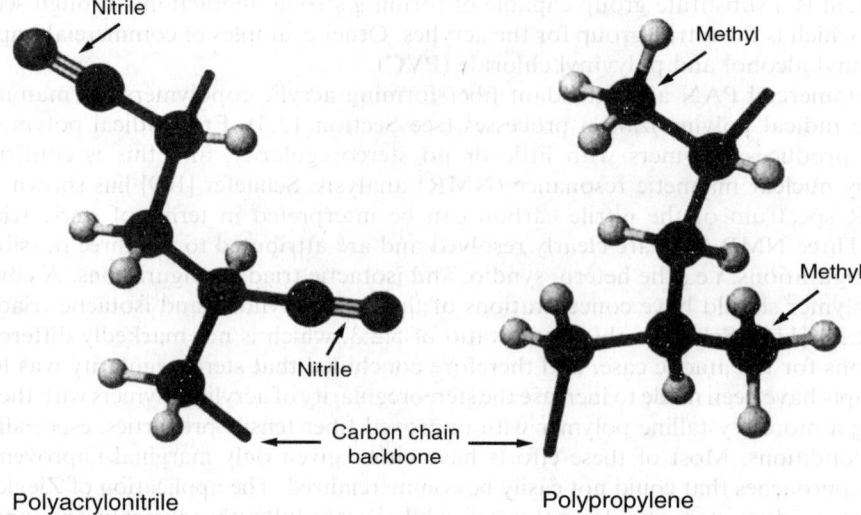

FIGURE 12.6 Dimer segments of polyacrylonitrile and polypropylene polymer chains showing the relative sizes of the nitrile group in polyacrylonitrile versus the methyl group in polypropylene. The small, light-colored atoms are hydrogen.

respectively. The magnitude of the solubility parameter is a measure of the intermolecular attraction (cohesive energy density) and can be calculated from molar attraction constants that are available for various functional groups that comprise the polymer structure. The value of these constants for the methyl and nitrile groups is 303 and 725 $(J/cm^3)^{1/2}/mol$, respectively [115].

A key factor controlling the acrylic fiber structure is the dipolar interaction energy between two nitrile groups; this has been described by Hinrici-Olive' and Olive' [116]. The interaction energy can be either attractive or repulsive, depending upon the spatial orientation of the nitriles, while the magnitude of the interaction depends upon the distance of separation. Possible modes of interaction between two nitrile groups are shown in Figure 12.7. We have to consider both intrachain and interchain dipolar interactions in order to understand the solid-state structure of PAN. In the isolated chain, the preferred conformation will be one that minimizes the chain potential energy [117]. If the backbone chain were placed in a planar zigzag arrangement, then the adjacent nitrile groups would fall into a parallel alignment, thus giving a net repulsion. The chain potential energy can be lowered by placing the adjacent nitrile groups as far apart as possible and this will require that the backbone chain become helical. In a helical conformation, all of the nitrile groups will be pointing away from the axis of the helix, and if several chains are grouped together, some of the nitriles on adjacent chains will be positioned in the antiparallel orientation to produce an attractive interaction. Thus we can begin to see how chain conformation leads to a solid-state structure, where the repulsive interactions are minimized and attractive interactions are maximized.

Krigbaum and Tokita [118] performed potential energy calculations on both the syndiotactic- and isotactic-isolated PAN chain to determine the net dipolar interactions among the nitrile groups, and in both cases the net interaction was always repulsive. They also cited the melting behavior of PAN gels and solution properties of PAN in concluding that the chain conformation must be extended and highly irregular. The above-mentioned work supports the concept of strong dipolar interactions controlling the individual chain conformation and solid-state structures. In the next section, we discuss the x-ray and electron diffraction studies

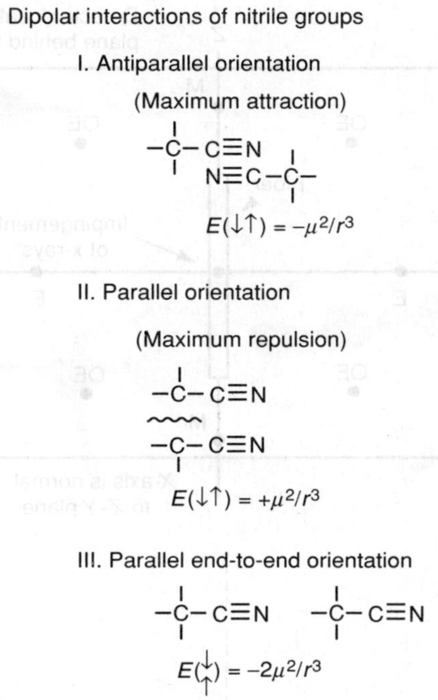

Dipolar interactions of nitrile groups

I. Antiparallel orientation

(Maximum attraction)

$$-\overset{|}{\underset{|}{C}}-C\equiv N \quad \underset{|}{\overset{|}{N\equiv C-\underset{|}{\overset{|}{C}}-}}$$

$$E(\downarrow\uparrow) = -\mu^2/r^3$$

II. Parallel orientation

(Maximum repulsion)

$$-\overset{|}{\underset{|}{C}}-C\equiv N$$
$$-\overset{|}{\underset{|}{C}}-C\equiv N$$

$$E(\downarrow\uparrow) = +\mu^2/r^3$$

III. Parallel end-to-end orientation

$$-\overset{|}{\underset{|}{C}}-C\equiv N \quad -\overset{|}{\underset{|}{C}}-C\equiv N$$

$$E(\overset{\downarrow}{\uparrow}) = -2\mu^2/r^3$$

FIGURE 12.7 Types of dipolar interactions between nitrile groups: (E) dipolar interaction; (μ) dipole moment; and (r) vector between dipoles. (From Hinrici-Olive, G. and Olive, S., *Adv. Polym. Sci. 32*, 128, 1979.)

of PAN fibers to see how a structure having crystalline order could emerge from the packing of these highly polar atactic chains.

12.4.2 Diffraction Studies of the Crystalline Structure

12.4.2.1 Analysis by X-Ray Diffraction

The crystalline structure of a fiber is determined by analysis of the x-ray diffraction pattern. In textile fibers, the polymer chains are always oriented parallel to the fiber axis so that the tensile forces lie along the covalent bonds of the polymer chains. The diffraction pattern is generated by mounting the fiber vertically and impinging the x-ray beam at 90° to the fiber, as shown in Figure 12.8a, and Figure 12.8b. Scattered x-rays are detected using film or an x-ray detector mounted behind the fiber. If crystallites are present, discrete reflections off of the Bragg mirror planes will generate a pattern of spots on the film. Proper analysis of the film pattern yields the unit cell structure and conformation of the individual chains.

The relationship between the directionality of crystalline order and the diffraction pattern can best be seen by reference to the following axes in Figure 12.8: let axis X be colinear with the incident x-ray beam (horizontal), axis Z be colinear with the fiber (vertical), and finally axis Y be mutually normal to X and Z, and hence normal to the fiber.

Now consider how various reflection planes within the fiber might reflect the x-rays. The incident x-ray beam is in the $X-Y$ plane. Any Bragg planes that are parallel to Z, the fiber axis, would reflect the x-rays such that they remain in this plane. These are called the equatorial reflections. For the x-rays to be reflected above or below the $X-Y$ plane, the corresponding Bragg planes must be tilted with respect to the fiber axis, and these reflections out of the $X-Y$ plane are termed off-equatorial. There is a major distinction between

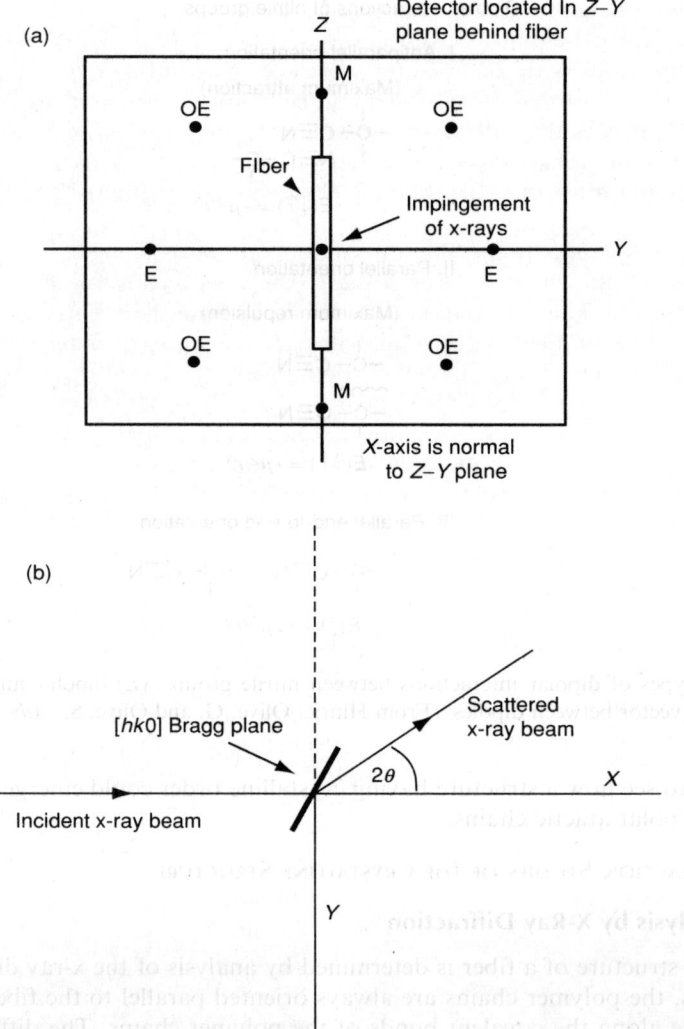

FIGURE 12.8 (a) Geometry of fiber-scattering experiment showing $Z-Y$ plane. The fiber is mounted along the Z-axis and the incident x-ray beam arrives along the X-axis and is normal to the $Z-Y$ plane. The polymer chains are oriented in the direction parallel to the long axis of the fiber, hence they are also oriented along the Z-axis. The x-ray detector, either film or electronic, is mounted in the $Z-Y$ plane behind the fiber. The three possible types of Bragg reflections for a fiber are shown: E, equatorial; OE, off-equatorial; and M, meridional. (b) Geometry of the scattering experiment showing the $X-Y$ plane. The $[hk0]$ Bragg planes (1) are parallel to the fiber direction, which is mounted along the Z-axis. Therefore all scatterings fall within the $X-Y$ plane, and results in equatorial reflections as shown in (a). Similarly, for the meridional reflections in (a), the Bragg planes must be normal to the Z-axis and hence be of the $[001]$ type, and for the off-equatorial reflections in (a), the planes must be inclined with respect to the Z-axis and hence be of the $[hkl]$ type. The angle 2Θ is half of the Bragg angle Θ.

equatorial and off-equatorial reflections. The Bragg planes responsible for the equatorial reflections can be uniquely defined by a two-dimensional unit cell arising from a high degree of packing in the direction normal to the fiber axis [105]. In other words, a regular packing in the X–Y plane is sufficient to define Bragg planes that will produce the equatorial reflections. The requirement for off-equatorial reflections is stricter, for now the Bragg planes will be defined by intersections with all three unit cell axes, including now the axis parallel to the fiber. The existence of a three-dimensional unit cell requires a high degree of order along the fiber axis, which means that all of the polymer chains must assume the identical regular conformation, whether it happens to be planar zigzag or some helical conformational. If these requirements are not quite met, then the off-equatorial reflections will begin to broaden and eventually vanish.

It is instructive to compare the fiber diffraction patterns of polyethylene terephthalate and acrylic fibers, which are shown in Figure 12.9. The fiber axis is vertical and the equatorial direction is horizontal. The pattern for polyester shows both the equatorial and off-equatorial reflections indicative of three-dimensional order, whereas only equatorial reflections are found for the acrylic fiber. The unit cell of the polyester is triclinic [119]. The polymer chain is planar zigzag and is parallel with the c-axis of the unit cell. Bohn et al. [120] published the first detailed model of the PAN fiber morphology. They observed several sharp equatorial reflections, but none that were off-equatorial. These reflections could be indexed in a two-dimensional hexagonal lattice such that the intense 5.2-Å reflection (the most intense reflection) would correspond to an interchain distance of 6 Å. The hexagonal lattice is thought to arise from the packing of adjacent chains. It is assumed that a single chain will, on the average, occupy a cylinder with a diameter of 6 Å, and the equatorial reflections then arise from the packing of these cylinders somewhat like the packing of sticks of chalk in a box. The conformation of the chain within the cylinder would be extended in a highly irregular fashion, and therefore true crystallographic order in the direction of the fiber axes is highly unlikely.

This model explains the origin of the intermolecular bonding among neighboring chains. The proposed extended, kinked chain will not fit into a 6-Å cylinder unless some of the nitrile groups extend beyond the confines of the cylinder and is shown in Figure 12.10. These outlying groups are potentially available for intermolecular bonding because they will be oriented in the antiparallel direction that produces a net attraction (see Figure 12.7). In

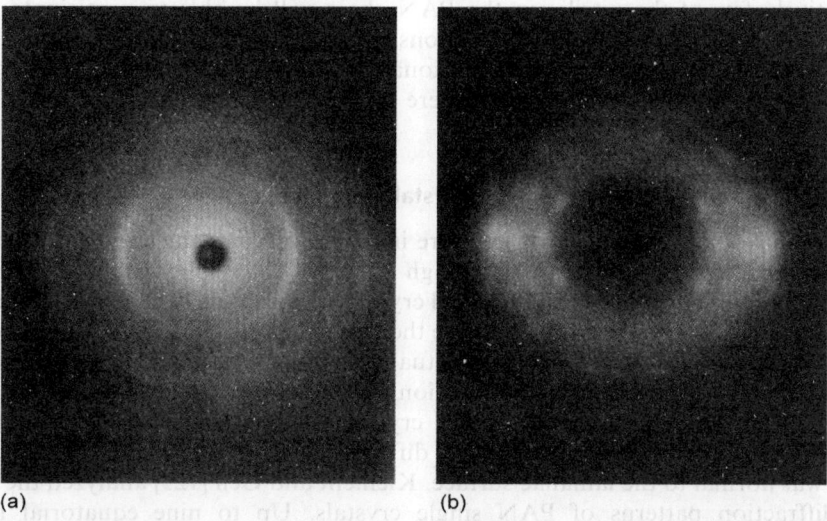

FIGURE 12.9 X-ray diffraction patterns of (a) commercial acrylic and (b) polyethylene terephthalate fibers. The acrylic fiber was a commercial sample containing 7% vinyl acetate comonomer. In both cases, the fiber axis is vertical.

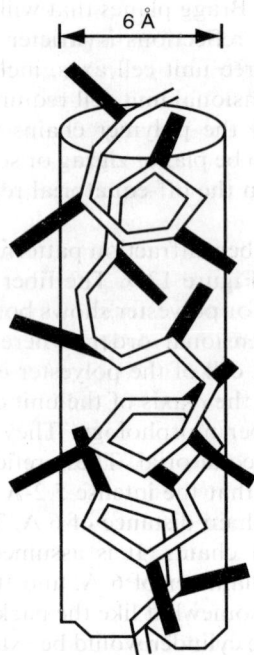

FIGURE 12.10 Model of assumed rigid, irregular helical conformation of the polyacrylonitrile chain as it would exist in the solid-state polymer. (From Hinrici-Olive, G. and Olive, S., *Adv. Polym. Sci. 32*, 128, 1979.)

support of this model, one can cite the work of Saum [121], who has shown that dipolar interactions of the antiparallel type in simple nitriles can lead to the formation of dimers. While relatively few of the nitriles in the PAN chain will be able to participate in perfect antiparallel bonding pairs, still many attractions will be present. The term "laterally bonded" has been used to describe this two-dimensional crystalline morphology. It has also been compared to the liquid-crystalline state where stiff, rod-like molecules will pack together, giving two-dimensional order [122].

12.4.2.2 Diffraction Analysis of Single Crystals and Fibers

Many efforts have been made to extract more information from the acrylic fiber diffraction pattern by analyzing samples with very high polymer chain alignment. One method of obtaining this alignment is to grow individual crystals of polymers isothermally from solution [123]. The level of crystallinity is high because the polymer concentration is kept very low (less than 1%) during the crystallization, thus virtually eliminating the chain entanglements that limit crystalline perfection during crystallization from the melt. Holland et al. [124] were able to grow elliptically shaped, lamellar, single crystals of PAN from PC. The crystals were approximately 100-Å thick and the electron diffraction patterns indicated that the polymer chain axis was normal to the lamallae surface. Klement and Geil [125] analyzed the x-ray and electron diffraction patterns of PAN single crystals. Up to nine equatorial reflections were reported indicating that the lateral packing of the chains in these single crystals is much improved over the packing in the fibers discussed earlier, but still no off-equatorial reflections were observed. The equatorial reflections were indexed as a two-dimensional orthorhombic unit cell with the following dimensions: $a = 21.18 \pm 0.04$ Å and $b = 11.6 \pm 0.02$ Å. Patel and coworkers [126–128] prepared single crystals from PC using a film formation method. The electron diffraction spots could be indexed using the unit cell proposed by Klement and Geil [125]. Kumamaru et al. [129] prepared single crystals during solution polymerization. They were able to obtain a few weak and diffuse off-equatorial reflections in addition to the equitorial reflections observed by Klement and Geil [125]. Using the diffuse off-equatorial reflections, they proposed an orthorhombic unit cell, where the c-axis is parallel to the fiber and polymer chain axes.

Evidence of three-dimensional crystallinity in very highly oriented PAN fibers was reported by Colvin and Storr [130]. The fibers were spun from an aqueous sodium thiocayanate solution, drawn 10× in steam, and then after drying they were given an additional 1.5× stretch by drawing over a hot surface. They observed four sharp off-equatorial reflections in addition to the equatorial reflections discussed previously. All of the reflections could be indexed in an orthorhombic unit cell having the following dimensions: $a = 21.48 \pm 0.02$ Å, $b = 11.55 \pm 0.03$ Å, and $c = 7.096 \pm 0.03$ Å. The a and b dimensions are identical to those reported by Klement and Geil [125]. The new off-equatorial reflections indicate an increase in perfection along the chain axis. This unit cell will accommodate 24 monomer units with six located on the base plane formed by the a- and b-axes and then four monomer units in the c-direction. The c-axis repeat distance of 7.096 Å and the appearance of an intense [004] reflection indicates a monomer-to-monomer translation distance of 1.774 Å, which gives a structure reminiscent of syndiotactic polypropylene [131]. This structure locates the nitriles in a manner that will minimize the repulsions among adjacent groups. The calculated density is $1.199 \pm 0.002 \, \text{g/cm}^3$. This agrees well with the experimental values of 1.197–1.20 g/cm^3 reported for fibers in the absence of microvoids [132].

Bashir and coworkers [133–138] have analyzed acrylic polymer prepared by heating in the presence of polar organic liquids that depress the melting point. They demonstrate that the appearance of the orthorhombic unit cell is caused by residual solvent, such as PC, becoming associated with polymer chain [133–135]. In the orthorhombic unit cell, the distance between chains is on the order of 5 Å, and this spacing would be adequate to allow polar liquids to fit into the lattice. Subsequent treatments that remove any traces of the solvent cause the structure to revert to the hexagonal form. Accordingly, all dry and solvent-free samples of the polymer, such as fibers, films, powder, or molded sheets, show the hexagonal polymorph. The orthorhombic cell can also be induced to form by using water as the melt plasticizer [136,137]. Bashir and coworkers [134,135,138] have also studied thermoreversible gelation of PAN and observed that solvated polymer crystals may be formed when PAN solutions prepared from certain solvents are cooled; i.e., the solvent and polymer crystallize together. It is thought that this polymer–solvent complex is accommodated in the orthorhombic unit cell.

12.4.2.3 Evidence for Two-Phase Morphology and Determination of Crystallinity

A controversy regarding the applicability of the two-phase morphology for PAN and acrylic copolymers has existed since the original work by Bohn et al. [120], and the issue has still not been entirely resolved. These authors argued that the laterally bonded crystalline structure exists throughout the fiber, and that there was no evidence for an amorphous phase that would be expected if the two-phase model applied. They noted the absence of the characteristic amorphous scattering halo in the fiber patterns, but there was diffuse scattering throughout the pattern. This suggested that the laterally bonded structure was essentially a single phase that contained many defects. One can see in Figure 12.9 that the amorphous scattering halo is quite pronounced in the polyester fiber diffraction pattern, but is largely absent in the corresponding pattern of the acrylic fiber. Instead, one observes diffuse scattering located in a narrow band at a Bragg spacing of approximately 3.4 Å. Bohn et al. [120] also compared the polymer volumetric expansion coefficient with the temperature dependence of the 5.2 Å Bragg spacing and observed an increase in both parameters at 85°C. This comparison, shown in Figure 12.11, is inconsistent with the classic two-phase model, since the glass transition of the amorphous phase should have little effect on the chain packing within the crystalline phase. These observations are taken in support of a hybrid single-phase morphology that has both crystalline and amorphous polymer properties.

Lindenmeyer and Hoseman [139] applied the theory of paracrystallinity to the acrylic fiber diffraction pattern and concluded that the diffuse scattering could arise as the net result of having different conformations distributed along the chain, with no single conformation persisting over a long distance.

More recently, Liu and Ruland [140] analyzed two-dimensional x-ray diffraction intensity distributions of a PAN fiber and concluded that the predominant chain configuration in PAN is the planar zigzag. Given the atactic nature of the chain, based on NMR, the authors contend that the continuity of the zigzag is interrupted by defect kinks related to sterically unfavorable sequences of CN groups. No evidence was found for the existence of a two-phase structure with ordered and disordered domains. Furthermore, the rotational disorder about the chain axes produces a diffuse scattering maximum in the equatorial scattering that has been previously interpreted as the amorphous halo commonly seen for more typical semi-crystalline polymers [105] and incorrectly used to evaluate a crystalline fraction of PAN. Hobson and Windle [141] also proposed a mixed tacticity chain conformation, containing both syndiotactic and isotactic sequences.

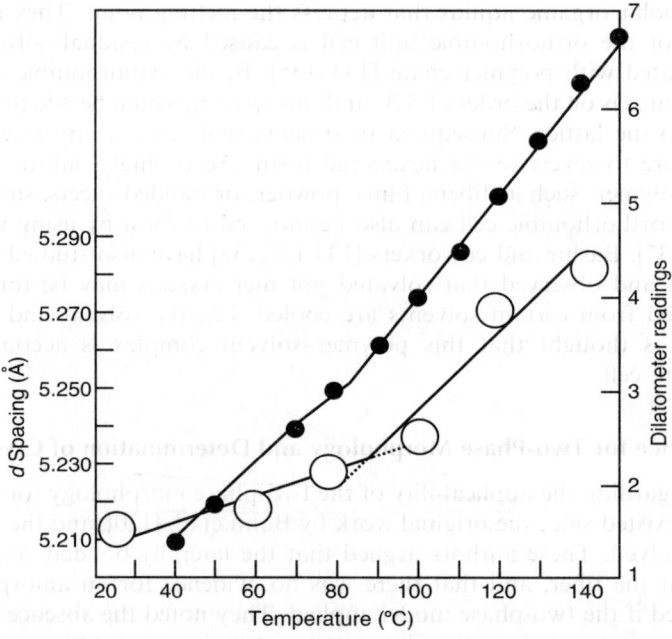

FIGURE 12.11 Thermal expansion of polyacrylonitrile. Open circles refer to diffraction d spacings, closed circles to dilatometer readings. (From Bohn, C.R., Schaefgen, J.R. and Statton, W.O., *J. Polym. Sci.* 55, 531, 1961.)

Despite the strong evidence for the single-phase model, numerous groups have attempted to determine a degree of crystallinity for acrylic polymers and fibers [142–145] by the classic analysis of the x-ray diffraction patterns [105]. The degree of crystallinity by this method is defined as the ratio of the intensity under the crystalline peaks divided by the total scattering intensity. Gupta and Singhal [144] have reviewed methods for defining a crystallinity index for acrylic polymers and they are compared in Figure 12.12. The methods of Hinrichsen [143] (coefficient of crystallinity) and Bell and Dumbleton [142] (crystalline index) are shown along with the authors' preferred method (degree of order). Matta et al. [145] have utilized an amorphous standard in determining the crystallinity of acrylic precursors for manufacturing carbon filters. Using this method, crystallinities from 64 to 66% were reported for two commercial PAN precursors. Modification by stretching an additional 60% in nitrogen increased the crystallinity to 70% and improved the physical, mechanical, and thermal properties [146]. PAN that is amorphous by x-ray analysis has been reported by Imai et al. [147] and Joh [148].

A number of investigators have performed drawing and annealing experiments on acrylic polymers, copolymers, and fibers that provide support for a limited two-phase model of the structure [143,149–153]. Hinrichsen and Orth [149], for example, have observed periodicities on the order of 120–105 Å in the small-angle x-ray pattern of drawn films of the PAN homopolymer. These observations strongly suggest the presence of a two-phase structure, perhaps with lamellar crystalline units, separated by less-ordered material of the chain-folded lamellae [154]. Hinrichsen [143] has observed a similar behavior in drawn PAN fibers.

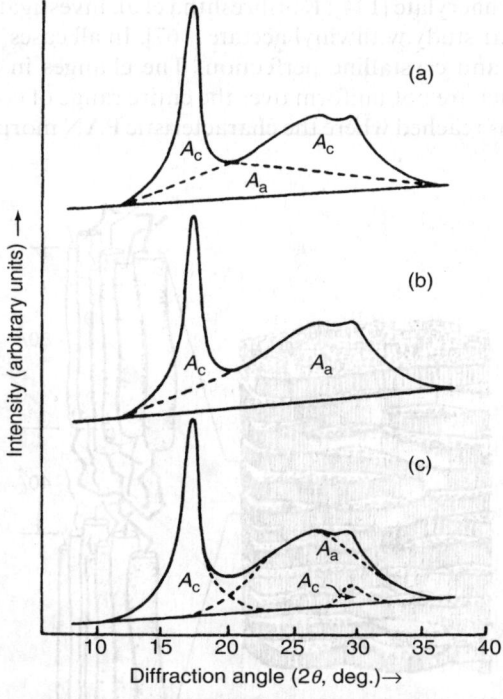

FIGURE 12.12 Schematic representation of extrapolation of crystalline and amorphous components in the x-ray diffraction pattern of PAN: (a) Hinrichsen's method. (From Hinrichsen, G., *J. Polym. Sci.* **38**, 303, 1972.), (b) Bell and Dumbleton's method. (From Bell, P. and Dumbleton, J.H., *Textile Res. J.* **41**, 196, 1971.) (c) Method of Gupta and Singhal. (From Gupta, A.K. and Singhal, R.P., *J. Polym. Sci. Polym. Phys. Ed.* **21**, 2243, 1983.)

A two-phase model for the structure of acrylic fibers has been proposed by Warner et al. [152], who studied the oxidative stabilization of these fibers during the preparation of graphite fibers. In this model, the fiber is composed of fibrillar subunits that contain distinct regions of amorphous and partially ordered material. The model utilizes the laterally bonded concept of Bohn et al. [120] for the partially ordered phase. A drawing of this model is shown in Figure 12.13. The ordered or laterally bonded crystalline phase is thought to have a lamellar texture oriented perpendicular to the fiber axis. Small-angle x-ray diffraction patterns by Warner [153] suggest that rows of lamallae can be formed that are oriented at an angle with respect to the fiber axis.

Gupta and coworkers [144,155–163] investigated both thermal treatment and the incorporation of comonomers on the morphology. Their analysis utilized dielectric relaxation measurements, x-ray diffraction, and infrared (IR) spectroscopy. Measurements were made on PAN disks that were heated at 120 and 160°C for 24 h in air. The heating increased the crystallinity and crystalline perfection. The glass transition temperature was derived from a William–Landel–Ferry (WLF) analysis [164,165] of the dielectric loss curves and it increased continuously for up to 16 h and then decreased. The study included the effects of heat treatment on the structure and properties of acrylic fibers [160]. Drawn PAN fibers were heated at 110 and 150°C for 24 h. Measurement of the tensile properties showed that the fibers broke at higher loads with heat treatment and the breaking elongation decreased slightly; all of this is consistent with an increase in crystalline order.

The effect of comonomers on the morphology has been the subject of several investigations. Gupta and coworkers investigated 2-hydroxyethyl methacrylate [155,156], methacrylonitrile [144,162], and methyl methacrylate [144]; Kulshreshtha et al. investigated methyl acrylate [166], and Frushour did a similar study with vinyl acetate [167]. In all cases, addition of comonomer reduces the crystallinity and crystalline perfection. The changes in the structure caused by addition of the comonomer are not uniform over the entire range of composition, but instead a critical comonomer level is reached where the characteristic PAN morphology begins to rapidly

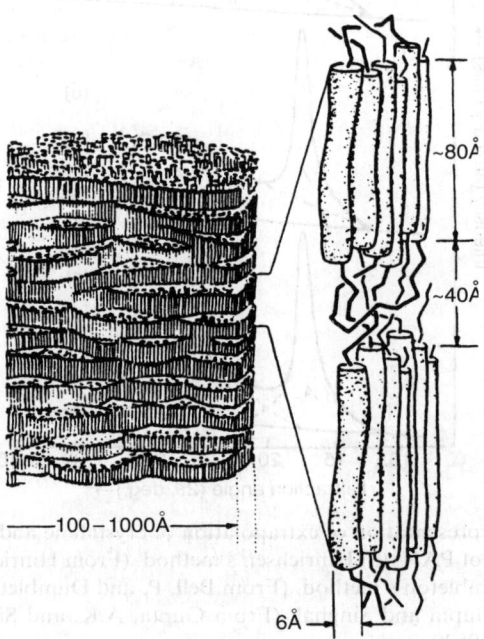

FIGURE 12.13 Schematic diagram of molecular structure of highly oriented acrylic fibers. (From Warner, S.B., Uhlmann, D. and Peebles, L., *J. Mater. Sci. 10*, 758, 1975.)

disappear. This can be seen in Figure 12.14 where the line width of the intense 5.2-Å reflection is plotted against the mole fraction of comonomer for the three copolymers. Recall that this reflection measures the average interchain distance in the crystalline phase. The increase in peak width becomes very apparent at approximately 8–10 mol% for the vinyl acetate and methyl methacrylate copolymers, but significantly more methacrylonitrile, e.g., approximately 20 mol%, is required to achieve approximately the same level of disorder. Vinyl acetate and methyl methacrylate are similar in size, and both are larger than the methacrylonitrile group, hence expected to be more effective in disrupting the crystalline morphology. This effect of comonomers on the morphology will be discussed again in Section 12.4.4.

Complex swelling behavior observed by Andrews and coworkers [168–170] and Grobelny and coworkers [171–175] can also be interpreted in terms of a multiphase morphology. The equilibrium weight gain in aqueous solutions of iodine–potassium iodide was measured as a function of concentration. Multiple steps in the absorption curve were observed and interpreted as the penetration of the solution into domains of increasingly higher order within the polymer. Lewin et al. [176] have studied the interactions of bromine with PAN copolymers and observed both a reversible and irreversible absorption of bromine. The former is removable by treatment with water, but the latter irreversible sorption requires aqueous ammonia solution for removal. The effects of the bromine treatment on tensile properties, glass transition (wet and dry), and swelling were interpreted as evidence for two phases in the less-ordered regions of the PAN.

In summary, one can conclude that a strong case can be made for the existence of at least a limited two-phase morphology for acrylic fibers. The differences in order between the two phases may be much less when compared to conventional melt-spun fibers. Furthermore, the two phases may be highly coupled since the chains in the less ordered region may be rather stiff and extended owing to the presence of the intermolecular dipolar bonds.

We close this section with a few comments regarding the ultrahigh-modulus acrylic fibers. It is estimated that the theoretical crystalline modulus for polyethylene is 240 GPa; hence there has been an intense effort over the last 10–15 years to develop spinning processes to exploit the high-modulus potential. This goal has been achieved by gel-spinning techniques [177]. Allen et al. [178] have estimated the theoretical modulus that might be obtained for

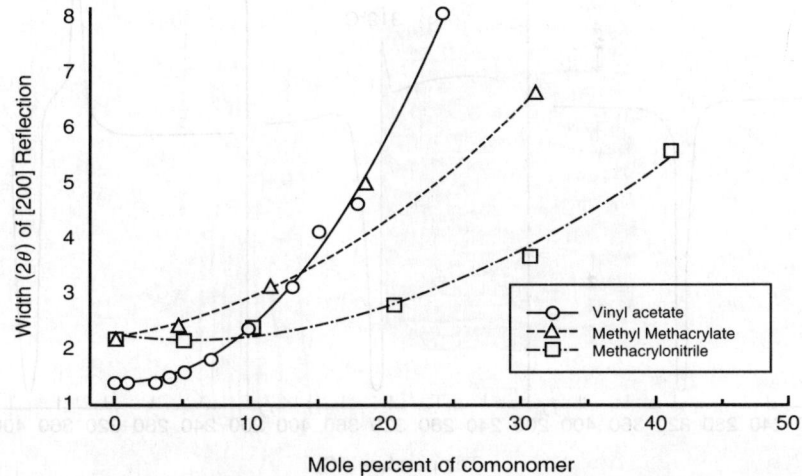

FIGURE 12.14 Response of the major equatorial reflection to incorporation of the following comonomers in copolymers with acrylonitrite: vinyl acetate, methyl methacrylate, and methacrylonitrile. (From Bell, J.F. and Murayama, T., *J. Appl. Polym. Sci.* 12, 1795, 1968.)

PAN by use of gel-spinning or other techniques to create a completely extended chain. Their estimate, based on a comparison of the rod-like PAN conformation with helical chain conformations of other polymers, is that the maximum tensile modulus of atactic PAN would be about 55 GPa. They conclude that PAN with ultrahigh modulus cannot be made by gel spinning, or by any other means, due to the intrinsic chain properties. It is proposed that the strong intramolecular nitrile repulsions that cause the PAN to adopt a rod-like, semiextended conformation do not allow the chain to unravel completely from its semiextended conformation. Attempts to develop gel-spinning processes in acrylic fibers are discussed in Section 12.5.4.

12.4.3 Thermal Properties: Melting, Gelation, and Crystallization

12.4.3.1 Evidence for Melting of Acrylic Polymers

Obtaining good melting data for acrylics is not straightforward, since they belong to the class of polymers where the melting point is higher than the thermal degradation temperature. PAN undergoes a degradation reaction at elevated temperatures in which the adjacent nitrile groups on the polymer chain form six-membered rings [179]. This reaction obviously precludes normal melt processing for acrylics, and happens to be the basis for production of graphite fibers from PAN.

When PAN is analyzed for melting in a differential thermal analyzer (DTA) or differential scanning calorimeter (DSC) using normal scanning rates of 20°C/min, the cyclization reaction produces a large exotherm and no melting is observed. (Readers not familiar with these thermal analytical techniques should consult sources such as Turi [180].) However, if the heating rate is sufficiently high, then some melting will occur before the polymer degrades, and this melting endotherm can be detected in the thermogram. This can be seen in Figure 12.15, where DSC thermograms of PAN are compared at heating rates of 20, 40, and

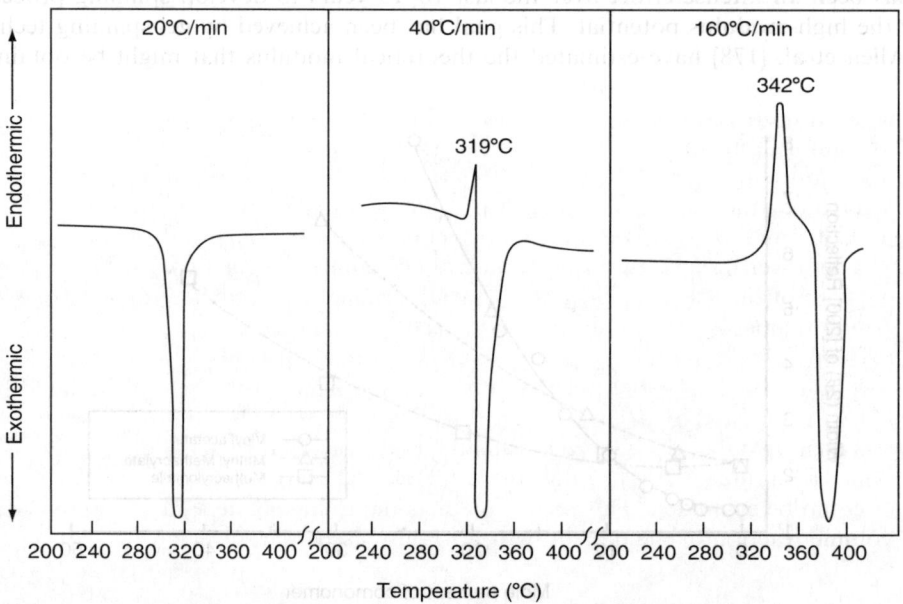

FIGURE 12.15 Differential scanning calorimetric scans of polyacrylonitrile homopolymer in nitrogen atmosphere. Exothermic peak arises from the cyclic degradation reaction.

160°C/min. At the lowest heating rate, the exothermic reaction appears at 315°C and no trace of a melting endotherm is seen. Increasing the heating rate to 40°C/min shifts the exotherm to 330°C and now the beginning of an endotherm near 319°C can be detected. The melting endotherm becomes quite evident at a heating rate of 160°C/min, because the exotherm has now shifted to 390°C. We conclude that PAN undergoes a melting-like transition in the range of 320–330°C.

12.4.3.2 Effect of Diluents and Comonomers on Melting Behavior

One objective in studying the melting behavior of a polymer is to derive basic thermodynamic parameters for the pure crystalline polymer. These parameters include the melting point, T_m, and ΔH_m and ΔS_m, the change in the heat and entropy that occurs during melting. Their values can give insight into the structure of the crystalline polymer and the nature of the crystalline bonding forces. Fortunately, it is not necessary to work exclusively with the pure PAN homopolymer to obtain this information. Diluents can be used to lower the melting point of the polymer, and the specific dependence of T_m on the diluent concentration can be analyzed using theories that yield the fundamental thermodynamic parameters. In the case of acrylics, this approach circumvents the problem of thermal degradation.

One effective diluent for a polymer is a solvent. When a crystalline polymer is dissolved in a solvent at a specific temperature, the melting of the polymer has, by definition, been depressed at least as low as that temperature. When PAN solutions of appropriate concentration are allowed to stand at room temperature, the viscosity begins to increase, and eventually a rubbery gel is formed. This gelation is thermoreversible, i.e., the solution will return to the original viscosity when heated. The gel is formed because some of the polymer chains phase separate and form very small-ordered regions, or microcrystallites, that serve as reversible cross-links. The melting of the gel upon heating corresponds to the melting or redissolution of these crystallites.

Krigbaum and Tokita [118] studied the melting behavior and gelation of PAN gels prepared from concentrated solutions of DMF and γ-butyrolactone. A recording dilatometer was used to follow the volume change during gelation and subsequent gel melting. The PAN solution will contract upon gelation and the gel will then expand upon melting. This follows from the general observation that the density of the crystalline polymer phase is greater than that of the amorphous phase [105]. Polymer solutions were prepared over a wide concentration range and then allowed to gel by cooling to a temperature, where the gelation rate was determined to be at a maximum. The gelled solution was then transferred to the dilatometer and heated. In most cases, two transitions in the thermal expansion coefficient could be detected. The lower temperature transition was the smaller of the two and attributed to the glass transition of the gel and higher temperature transition was gel melting. In Figure 12.16, the transition temperatures are plotted as a function of polymer volume fraction for the two solvent systems. The gel melting points increase rapidly with polymer concentration, and extrapolation to pure polymer gives a PAN melting point of 317°C, which is in good agreement with the directly measured values discussed previously.

The gel melting data were analyzed using the Flory theory for the melting point depression of a polymer by a diluent [181] so that the fundamental thermodynamic parameters referred to earlier could be evaluated. This theory predicts the following dependence of melting point on the volume fraction of the diluent, which in this case is the solvent:

$$1/T_m - 1/T_m^\circ = RV_u/(\Delta H_u V_1)(V_f - \chi V_f^2) \qquad (12.19)$$

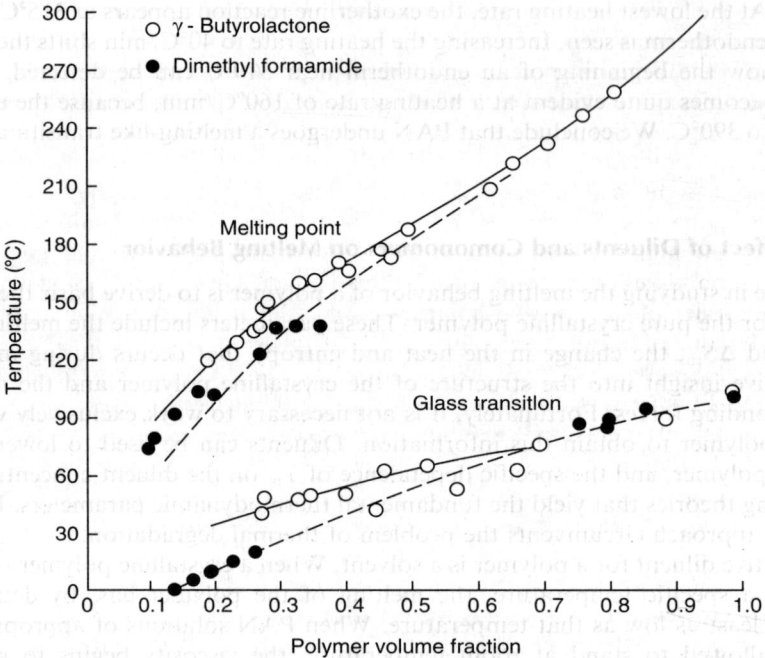

FIGURE 12.16 Melting and glass transition temperatures of polyacrylonitrile gels plotted against the volume fraction of polymer. (From Krigbaum, W.R. and Tokita, N., *J. Polym. Sci.* 43, 467, 1960.)

where T_m and T_m° are the melting points in kelvin of the polymer with diluent present and the pure polymer, respectively, R is the gas constant, ΔH_u is the heat of fusion per mole of crystalline repeat unit, V_u and V_1 are the molar volumes of the repeat unit and diluent, V_f is the volume fraction of diluent, and χ is the polymer–diluent interaction parameter. The theoretical heat of fusion of pure crystalline PAN polymer, ΔH_u, is derived from the gel-melting data by plotting the reciprocal melting point against the diluent volume fraction and substituting the initial slope of that line into Equation 12.19. The other fundamental parameter we need to evaluate is the entropy of fusion, ΔS_m, and this is easily calculated from ΔH_u and T_m (recall that T_m for the pure homopolymer is determined by extrapolating the experimental gel melting point to 100% polymer volume fraction). To calculate ΔS_m, we take advantage of the fact that the free energy of melting, ΔG_m, of a polymer will be equal to zero at the melting point, and we can write the following equation [182]:

$$\Delta G_m = 0 = \Delta H_m - T_m \Delta S_m \qquad (12.20)$$

We can see from this equation that the melting point equals the ratio of the enthalpy and entropy changes that occur upon melting. The entropy change, ΔS_m, can be easily calculated from the other two known quantities. The enthalpy change upon melting reflects the strength of the intermolecular bonds in the crystalline phase that must be broken in order for melting to occur, while entropy change is primarily a measure of the increase in chain flexibility upon melting. All polymer chains are rigid in the crystalline state, but those that are inherently flexible in the melt undergo a much larger change in entropy than do those polymers that are relatively inflexible. Thus, a polymer may have a high melting point for two possible reasons.

First, the melting enthalpy can be large, and second, the change in entropy can be small. The ratio determines the melting point.

In Table 12.11, the melting point, heat, and entropy of fusion derived from the gel-melting experiments are given and compared with those of other semicrystalline polymers. Krigbaum and Tokita [118] point out that the heat of fusion of PAN is about the same as that of polychlorotrifluoroethylene and is considerably less than that of a polyethylene or polyethylene oxide. Since it is difficult to reconcile strong attractive forces with the chemical structures of these three polymers, it appears unlikely that strong intramolecular interactions are responsible for the high melting point of PAN. For many polymers, the entropy of fusion per single chain bond lies in the range of 1.5–2 cal/degree. This quantity is related in a general way to the flexibility of the polymer chain in the melt or in solution. For example, cellulose trinitrate has a value of only 0.75 cal/degree and this polymer is known to be highly extended in solution [183]. Similarly, the entropy per bond of the PAN chain lies below the normal range of values. Furthermore, the unperturbed dimensions estimated for this polymer from dilute solution measurements indicate a highly extended conformation [184], thus supporting the above arguments. This analysis of the melting behavior is also consistent with the model of the structure based on the x-ray diffraction fiber patterns discussed in Section 12.4.2. The rather sporadic nature of the dipolar bonding between the nitrile groups on adjacent chains would not be expected to yield large values for the heat of fusion.

Up to this point, we have discussed the structure and melting behavior of the PAN homopolymer. However, only rarely is the homopolymer used for fiber spinning, and virtually all commercial acrylic fibers are spun from acrylonitrile polymers containing 5–10% comonomer. The comonomers are introduced for several reasons. A fiber spun from the PAN homopolymer is very difficult to dye because the laterally bonded structure is not easily penetrated by the relatively large dye molecules. Incorporation of comonomers such as methyl acrylate and vinyl acetate disrupts the laterally bonded structure in a manner that allows for more rapid diffusion of the dye molecules during dyeing. Several other benefits accrue from their use. The polymer becomes much more soluble, thus making the preparation and storage of the spinning dopes much easier, and the resulting fiber is also more extensible and therefore less prone to fibrillation, albeit the hot–wet strength and modulus are decreased in comparison with fiber spun from the homopolymer.

TABLE 12.11
Thermodynamic Parameters Derived from Melting Point Measurements

Solvent	T_m° (°K)	ΔH_u (kcal)	ΔS (cal/deg.)	ΔS per Bond
Polyacrylonitrile				
γ-Butyrolactone	590	1.25	2.1	1.0
Dimethylformamide	(590)	1.16	2.0	1.0
Other polymers				
Polyethylene oxide	339	1.98	5.85	1.95
Polyethylene	410	1.57	3.80	1.90
Polychlorotrifluoroethylene	483	1.20	2.50	1.25
Cellulose 2.44-nitrate	890	1.35	1.51	0.76
Cellulose trinitrate	970	0.9–1.5	1.5	0.75

Source: From Krigbaum, W.R. and Tokita, N., *J. Polym. Sci.*, 43, 467, 1960.

Incorporation of another monomer into an otherwise crystalline homopolymer will generally reduce the crystallinity and lower the melting point in a continuous manner. Several theoretical models have been developed addressing the reduction of crystallinity and melting point. In discussing these models, we will designate the major comonomer as the monomer in the base homopolymer, and any others will be minor comonomers.

In a model of the crystalline copolymer developed by Flory, the assumption is made that the copolymer crystallizes in a manner that replaces the minor comonomer in the amorphous domain. The crystalline phase is formed by parallel alignment of chain segments containing uninterrupted sequences of the major monomer. The presence of the minor comonomer in the chain reduces the average sequence length of the major monomer, which in turn reduces the number and average size of the crystallites. This reduction in size increases the surface-to-volume ratio of the crystallites, resulting in a higher free energy and hence lower melting point for the crystalline phase.

The following equation derived by Flory gives the theoretical dependence of the melting point on the amount of minor comonomer added to the crystalline homopolymer [185]

$$1/T_m - 1/T_m^\circ = (R/\Delta H_u) \times X_B \qquad (12.21)$$

where T_m and T_m° are the melting points of the copolymer and homopolymer, respectively, R is the gas constant, ΔH_u is the heat of fusion per mole of crystalline repeat unit, and X_B is the mole fraction of minor comonomer. The theory implies that the melting point depression will be independent of the size, or any other structural feature of the minor comonomer. The only role played by the comonomer is to reduce the average sequence length of the major monomer.

An alternative model for a crystalline copolymer is the defect crystal model proposed by Eby [186], where the minor comonomers are incorporated into the crystalline lattice as defects. These defects decrease the melting point and heat of fusion by disrupting the intermolecular bonding within the crystalline lattice. The efficacy of a particular comonomer to depress the melting point is to a first approximation proportional to the molar volume of the comonomer. The slope of the reciprocal melting point versus comonomer mole fraction curve is no longer inversely proportional to the heat of fusion of the homopolymer, as in the Flory theory. This slope can be interpreted as an estimate of the degree to which the comonomer disrupts the lattice. We will now review several studies on the melting behavior of acrylic copolymers and compare the results to the predictions based on the above two theories.

Slade [187] determined the melting point of a series of acrylonitrile–vinyl acetate (AN–VA) copolymers using DTA. The VA content of the copolymer covered the range from 0 to 38.5% by weight. The extrapolated value of the homopolymer melting point was 322°C. The heat of fusion of the homopolymer calculated from the Flory theory (Equation 12.21) was 573 cal/mol, which is much lower than the values obtained from analysis of the gel melting points. The melting behavior of acrylonitrile–styrene (AN–S and acrylonitrile–isobutylene (AN–I) copolymers was studied by Berger et al. [188] using DSC. The calculated heat of fusion for the PAN homopolymer, again using the Flory theory, was found to be dependent on the particular comonomer. Values based on the AN–S and AN–I copolymers were 650 and 1260 cal/g, respectively. The authors suggested that the disparity in heat of fusion values might be interpreted as support for the Eby defect model. The molar volume of the styrene comonomer is significantly larger than that of isobutylene, and from the Eby model we would expect it to disrupt the crystalline lattice to a greater degree. Direct measurement of the experimental heat of fusion of the copolymers was given as additional support for the defect model.

A novel thermal analytical technique was developed by Frushour [167,189,190] that utilizes water to depress the melting point of AN copolymers. This allows melting and melt crystallization studies to be carried out on the PAN homopolymer and copolymers below the temperature, where the exothermic cyclization reaction begins. Information obtained by this technique gives additional insight into the morphology of the PAN homopolymer and copolymers. In this technique a polymer, or fiber, is mixed with water and sealed in a high-pressure DSC capsule [167]. As the water content is increased, the polymer melting point will decrease continuously until a critical water concentration is reached, where a pure water phase is formed [189]. The maximum melting point depression will have occurred at this point and excess water simply goes into the pure water phase.

This critical water concentration is dependent upon the acrylonitrile level of the polymer. For PAN homopolymer it is 33% water, based on polymer weight, and at this concentration and for all higher concentrations the PAN melting point remains at 185°C. This represents an 135°C depression in the melting point, assuming a dry-polymer melting point of 320°C. The region of excess water is referred to as the constant melting plateau. Incorporation of comonomers decreases both the critical water concentration required to reach the constant melting plateau and the corresponding plateau melting point. This can be seen in Figure 12.17, where the melting point is plotted against water concentration for the homopolymer and two AN–VA copolymers. It is the presence of the constant melting plateau that makes the analysis of the melting and crystallization behavior so straightforward because if excess water is used, then the measurement becomes sensitive only to the polymer structure.

The melting endotherms that can be obtained by this procedure are shown in Figure 12.18. The peak temperatures, T_m, and endothermic areas, $\Delta H_{f(exp)}$, can be easily and reproducibly

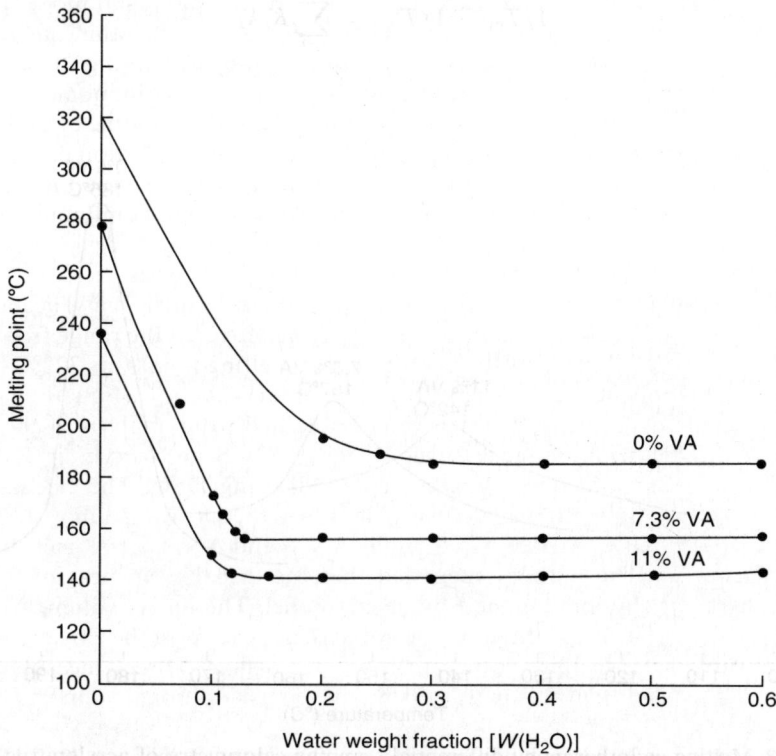

FIGURE 12.17 Dependence of the polymer melting point on water content for acrylonitrile–vinyl acetate copolymers. (From Frushour, B.G., *Polym. Bull.* 7, 1, 1982.)

measured. The technique was used to study the melting behavior of copolymers and higher order copolymers [190]. The effect of comonomer incorporation on reduction of the melting point and experimental heat of fusion (endotherm area) is apparent in Figure 12.18. The reciprocal melting point is plotted against mole fraction of comonomer for a number of different copolymers both in the dry state and with excess water in Figure 12.19. The slopes of the curves for the dry and wet polymers are essentially identical, and the magnitude of the slope increases as the size of the comonomer becomes larger. The first observation means that for virtually any acrylic polymer the wet-polymer melting point, which is easily measured, can be used to calculate the dry-polymer melting point, which probably cannot be easily measured due to thermal degradation.

The correlation between the comonomer size and the slope was established by calculating the molar volume of the comonomer side chain and comparing this to the slope. These values are given in Table 12.12 along with the results of the least-squares correlation between the molar volumes and the slopes. This reasonably strong correlation was taken as justification for interpreting the value of the slope for each comonomer as a parameter that measured the degree to which the comonomer disrupted the crystallite lattice. This parameter was termed the melting point depression constant. The correlation between slope and comonomer volume was taken as support for the Eby model of the copolymer, where the minor comonomers become incorporated in the crystalline domains as defects.

The melting point depression constants can be used in a generalized melting point equation for acrylic copolymers of any order:

$$1/T_m - 1/T_m^\circ = \sum_{i=1}^{n-1} K_i X_i \qquad (12.22)$$

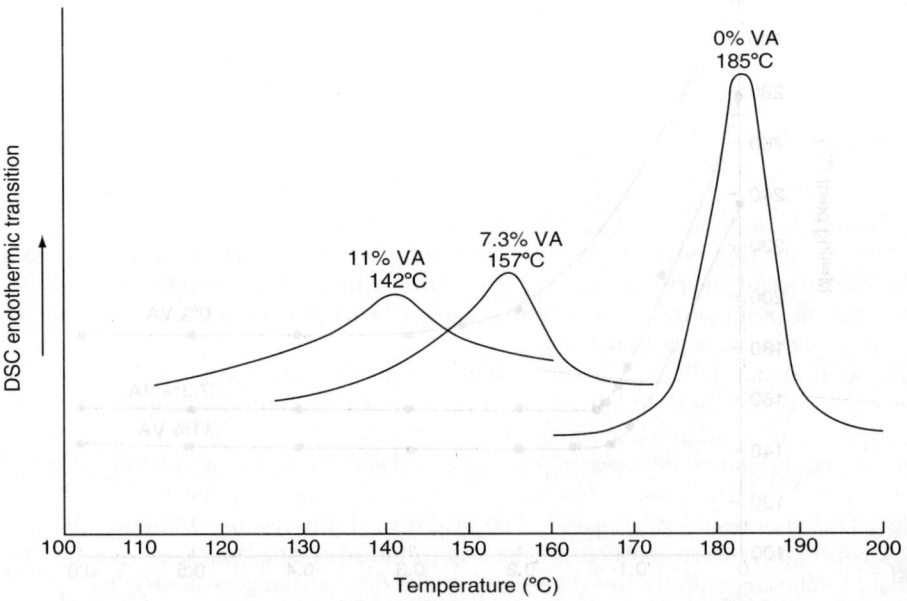

FIGURE 12.18 Melting endotherms by differential scanning calorimetry of acrylonitrile–vinyl acetate copolymer. The polymers were mixed with water (one part polymer: two parts water) and sealed in a special high-pressure capsule. (From Frushour, B.G., *Polym. Bull.* 7, 1, 1982.)

Acrylic Fibers

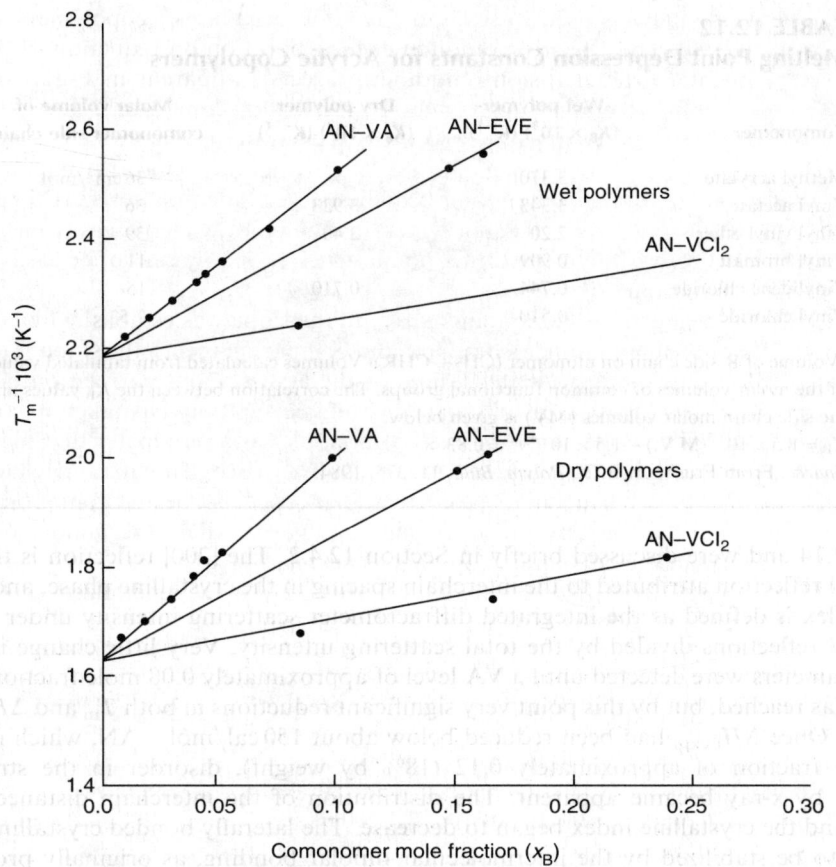

FIGURE 12.19 A comparison of the dependence of the wet and dry-polymer melting points on comonomer type and mole fraction: VA, vinyl acetate; EVE, ethyl vinyl ether; and VCl$_2$, vinylidene chloride. (From Frushour, B.G., *Polym. Bull. 11*, 375, 1984.)

where T_m and T_m° are the melting points of the copolymer and homopolymer, respectively, K_i and X_i are the melting point depression constants and mole fraction of the ith comonomer, and n is the order of the polymer ($n = 2$ for a copolymer, 3 for a terpolymer, etc.).

The melting point of numerous terpolymers and tetrapolymers could be accurately predicted using the above equation [190].

The wet-polymer DSC technique is much more sensitive than x-ray diffraction to the structural changes brought about by the incorporation of comonomers into the PAN chain [167]. The melting point and experimental heat of fusion directly measured the regularity and strength of the dipolar bonding network that stabilized the two-dimensional laterally bonded structure. DSC and x-ray are complementary techniques that are sensitive to different levels of order. This conclusion was based on comparison of the results obtained when a series of AN–VA copolymers were analyzed by the two techniques. This comparison is summarized in Table 12.13. The structurally sensitive x-ray diffraction parameters (half-width of the intense [200] equatorial reflection and the crystalline index) and DSC parameters (wet-polymer melting point and experimental heat of fusion, $\Delta H_{f(exp)}$) for a VA range of 0–0.24 mole fraction, or 34% by weight, are given. The half-width data in Table 12.13 also appear in

TABLE 12.12
Melting Point Depression Constants for Acrylic Copolymers

Comonomer	Wet polymer $(K_B \times 10^3)$ (K^{-1})	Dry polymer $(K_B \times 10^3)$ (K^{-1})	Molar volume of comonomer side chain[a]
Methyl acrylate	3.370	—	36 cm^3/mol
Vinyl acetate	3.343	3.923	36
Ethyl vinyl ether	2.20	2.40	39
Vinyl bromide	0.909	—	11
Vinylidene chloride	0.744	0.710	15
Vinyl chloride	0.510	—	7.5

[a] Volume of R side chain on monomer (CH$_2$—CHR). Volumes calculated from tabulated values of the molar volumes of common functional groups. The correlation between the K_B values and the side-chain molar volumes (MV) is given below:
$K_B = 8.5 \times 10^{-5}$(M.V.) $- 1.5 \times 10^{-4}$ $r^2 = 0.83$
Source: From Frushour, B.G., Polym. Bull., 11, 375, 1984.

Figure 12.14 and were discussed briefly in Section 12.4.2. The [200] reflection is the intense equatorial reflection attributed to the interchain spacing in the crystalline phase, and the crystalline index is defined as the integrated diffractometer scattering intensity under the sharp equatorial reflections divided by the total scattering intensity. Very little change in the two x-ray parameters were detected until a VA level of approximately 0.08 mole fraction (12% by weight) was reached, but by this point very significant reductions in both T_m and $\Delta H_{f(exp)}$ had occurred. Once $\Delta H_{f(exp)}$ had been reduced below about 150 cal/mol × AN, which required a VA mole fraction of approximately 0.12 (18% by weight), disorder in the structure as measured by x-ray became apparent. The distribution of the interchain distances became broader and the crystalline index began to decrease. The laterally bonded crystalline phase is assumed to be stabilized by the intermolecular dipolar bonding, as originally proposed by Bohn et al. [120], though the well-defined melting endotherms may indicate that this bonding

TABLE 12.13
Comparison of the Response of Thermal and X-Ray Parameters to Vinyl Acetate Comonomer Level

Vinyl acetate (mole fraction)	Wet polymer $(T_m, °C)$	Wet polymer $\Delta H_{f(exp)}$ (cal/mol AN)[a]	Half-width [200] reflection (θ)	Crystallinity index
0	185	1150	1.42	0.39
0.0092	177	860	1.4	0.40
0.0294	166	550	1.4	0.37
0.0404	158	500	1.5	0.37
0.0509	153	420	1.6	0.37
0.0708	142	300	1.8	0.38
0.0981	125	205	2.4	0.33
0.130	108	70	3.1	0.29
0.150	102	50	4.1	0.20
0.178	None	None	4.6	0.12
0.240	None	None	8.0	—

[a] $\Delta H_{f(exp)}$ is the experimentally determined heat of fusion.
Source: Frushour, B.G., Polym. Bull., 4, 305, 1981.

is more regular than had been previously suspected. The VA comonomers are incorporated into the crystalline phase as defects at the expense of the dipolar bonds. The crystalline structure is sufficiently imperfect to tolerate a loading of up to 12% by VA by weight before significant disruption on the scale measured by the x-ray technique is detected. These results are not specific to VA. The molar concentration required to effect a given level of crystalline disruption should increase as the molar volume of the comonomer decreases.

Min et al. [191–193] also studied water-induced melting behavior as part of their efforts to develop a plasticized melt-spinning process, and extended Frushour's work to include the effect of adding polar materials (solvents and hydrophilic polymers) and the stability of the hydrated melt at elevated temperatures. Addition of small amounts of ethylene carbonate [191] and DMF to the water further depressed both the melting and crystallization temperatures.

The hydrated melt is sufficiently stable for processing. Samples containing various water contents were annealed at temperatures between 160 and 180°C for up to 60 min. An absence of solution viscosity increase indicated that cross-linking did not occur, and a slight melting point increase was attributed to intramolecular cyclization of nitrile groups. A number of patents and papers have explored the possibility of spinning fibers or making shaped articles from these hydrated melts. These will be described in Section 12.5.

12.4.4 Glass Transition and Dynamic–Mechanical Properties

PAN and the fiber-forming acrylic copolymers do have glass transitions, but with some unusual characteristics. The glass transition is the temperature range over which a glassy polymer becomes rubbery. Polymer chain segmental mobility, or chain Brownian motion, becomes activated and for typical glassy polymers such as atactic polystyrene the modulus of the glass decreases by about three and a half decades over a 15°C range. Other changes that occur at the glass transition are pronounced increases in specific volumes, heat capacity, and diffusion rate of absorbed molecules. The glass transition is a very important thermal property of a fiber, because physical and tensile properties will change as the fiber is heated through the transition. The fiber modulus will decrease and the fiber will become more extensible. In dyeing operations, it is necessary to be above the glass transition of the wet fiber so that the dye molecules can diffuse into the fiber and reach the dye sites.

The glass transition temperature, T_g, can be determined in a number of ways. In general, one monitors the change in a physical property as a function of temperature, and then defines T_g as either the beginning or the midpoint of the change in the property. Each technique may give a slightly different value owing to the fact that the glass transition is a viscoelastic phenomenon [194] and the measured T_g will be a function of the heating rate and measurement frequency. The physical properties that are most often monitored to measure T_g include thermal expansion coefficient, heat capacity, diffusion coefficient, and dynamic modulus. The T_g values based on the first three measurements are referred to as static measurements because the polymer sample is at rest during the measurement. In dynamic measurements, the modulus or dielectric constant is measured many times per second as the sample is heated, and the observed T_g will increase with the measurement frequency.

Kimmel and Andrews [195] have summarized measurements of the PAN T_g, up to the year 1965. The values fall into two ranges. For static measurements, distinct changes in the temperature dependence of the measured property are observed around 85–95°C. Higher values of T_g, 105–140°C, are obtained using dynamic–mechanical and dielectric measurements, T_g values of 85 and 87°C, based on the linear expansion coefficient, were reported by Bohn et al. [120] and Howard [196]. Howard determined the T_g for a series of (AN–VA) copolymers by measuring the linear expansion coefficient of copolymer discs. The T_g for the PAN homopolymer was determined to be 87°C. Upon incorporation of VA, the value of T_g

remained constant until the VA level exceeded 27 wt% and then began to decrease and approach the literature value of 30°C for the pure polyvinyl acetate. At VA levels of 27% and greater the data fit a straight line in accordance with the classic behavior described by the Loshack and Fox equation [197]. The constant T_g region was attributed to the interfering effect of the crystalline phase, and extrapolation of the line to zero VA gives a value of 107°C, which was interpreted as the T_g, for completely amorphous PAN.

DSC measurements give a similar dependence of T_g on VA level [198]. The DSC also gives quantitative values of the change in heat capacity that occurs as the polymer is heated through the glass transition.

The effect of water on the glass transition of acrylic polymers and fibers is quite pronounced. Water lowers T_g from 100 to 64°C. It is not surprising that water would plasticize the acrylic structure. Water has a high dielectric constant and the water molecules would be effective in decreasing the magnitude of the dipolar interactions between nitrile groups. The large plasticization effect of water has important technological consequences for acrylic fibers. It means that when these fibers are placed in very hot water, such as near-boiling conditions of dyeing, they will become highly plasticized and soften.

Several experimental approaches have been applied for determining the fiber T_g under hot–wet conditions [199–203]. Aiken et al. [199] compared the T_g of a commercial acrylic yarn in the dry state and in water using dynamic–mechanical analysis, and observed a reduction from 92 to 72°C. Bell and Murayama [200] observed that the T_g of a commercial AN–NA copolymer decreased from 128°C when dry to 80°C in a 100% relative humidity atmosphere. Gur-Arieh and Ingamells [201] related the extension in length of Acrilan filaments to a T_g reduction and showed a shift from a 90°C in air to 57°C in water. Finally, Hori et al. [202] used DSC to show that the T_g of four kinds of acrylic fibers decreased with increasing water content and approached an almost constant value for all four fibers.

Dynamic–mechanical measurements of polymers are very sensitive to molecular transitions such as the glass transition and the more localized transitions that are found at temperatures below T_g [203,204]. Dynamic–mechanical analysis continuously gives the real and loss moduli as a fiber sample is heated at a constant rate. The ratio of the moduli is referred to as the tan δ. When a polymer begins to undergo a molecular transition such as the glass transition, the real modulus will begin to drop and the tan δ will go through a maximum. This represents a sudden absorption of energy as the polymer transition becomes activated. Dynamic–mechanical measurements are useful because a wide spectrum of transitions can be detected. The glass transition is by far the most prominent transition. Less prominent transitions are found in many glassy polymers at temperatures below T_g that arise from rather specific main chain or pendant side-group molecular motions. The combination of the real modulus and the tan δ is especially beneficial, because tan δ is able to resolve distinct transitions that underlay the inflection in the real modulus.

Analysis of the dynamic–mechanical properties of acrylic polymers has been the subject of numerous investigations [169,205–214]. In most of these investigations, the measurements were made using a Rheovibron, which utilizes the forced-resonance principle [203].

The dynamic–mechanical spectrum of a PAN fiber obtained by Minami [205] is shown in Figure 12.20. The real modulus at room temperature is approximately 4×10^9 MPa. The modulus begins to decrease near 75°C and drops to 2×10^8 MPa, which is a decrease of slightly over one order of magnitude, and corresponds to the onset of the glass transition. The T_g based on the midpoint of the decrease in real modulus is approximately 115°C. This is higher than the values obtained using thermal expansion coefficient (85°C) or DSC (100°C), and simply reflects the temperature shift characteristic of dynamic techniques mentioned earlier.

Most glassy polymers such as polystyrene exhibit a much larger decrease in the real modulus upon heated through the glass transition temperature. A decrease of approximately

Acrylic Fibers

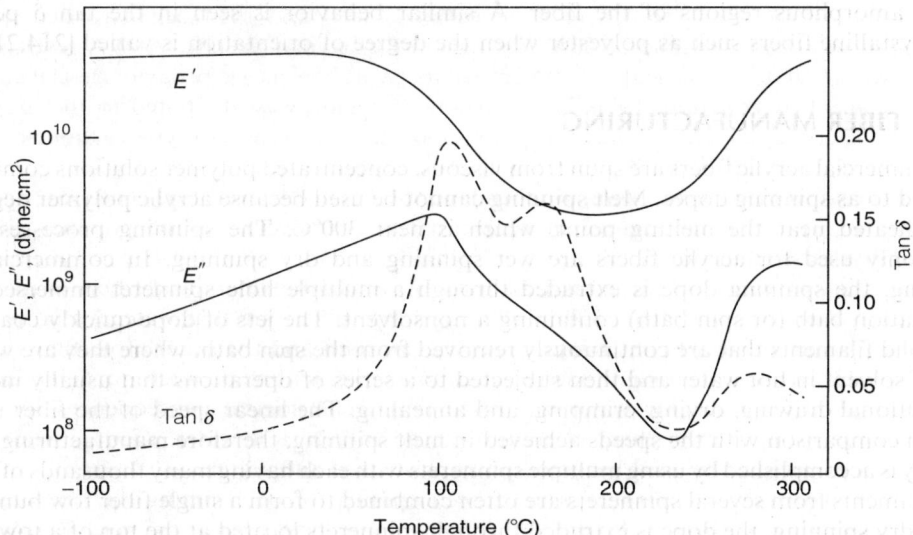

FIGURE 12.20 Dynamic–mechanical properties of undrawn polyacrylonitrile fiber measured with a Rheovibron at 110 Hz. (From Minami, S.J., *Appl. Polym. Sci. Appl. Polym. Symp.* 25, 145, 1974.)

three to four orders of magnitude would be considered typical behavior [215]. Keyon and Rayford [212] interpret the small drop in the real modulus as evidence of crystallinity in PAN. Using an empirical relationship between the decrease in modulus of a polymer and the degree of crystallinity developed by Nielsen [216], a crystalline fraction of 0.47 was calculated.

The tan δ curve of PAN in Figure 12.20 has two distinct transitions appearing near 110 and 160°C at a measurement frequency of 110 Hz. We will use the convention of Minami and refer to the 110 and 160°C transitions as a_{II} and a, respectively. The interpretation of these two peaks has not yet been entirely resolved. Some investigators have taken their appearance as evidence of a double glass transition for this polymer [168–170,213,217]. The PAN structure is assumed to have two phases, both essentially amorphous, but with different levels of intermolecular bonding, and both capable of undergoing a glass transition, though at different temperatures with the more ordered phase having the higher T_g. Padhye and Karandikar [213] treated PAN fibers in various solvents and then observed the effects of these solvents on the tan δ peaks. They found that aromatic solvents (phenol, aniline, and resorcinol) seem to make the lower transition disappear and reduce the temperature for the higher one, whereas nonaromatic solvents (methanol, amyl amine, dimethyl amine, ethylene glycol, and acetonitrile) shift the lower temperature transition to lower temperatures and the higher temperature transition appears to move even higher.

Dynamic–mechanical analysis is very sensitive to the structural changes that are imposed on the acrylic polymer during fiber manufacturing, and the technique derives much of its practical value from this sensitivity [218]. Minami and coworkers [205,206,209,211] have studied the effects of chain orientation and relaxation on the loss peaks (a_{II} and a_I at 110 and 160°C, respectively) and arrive at a somewhat different assignment of the transitions. They associate a_{II} with an unspecified motion within the laterally bonded crystalline phase and a_I, the higher temperature transition, to the glass transition of a lesser ordered or amorphous phase. They observe a decrease in the intensity of a_I upon stretching and a subsequent increase upon relaxation. Little change is observed in a_{II} during this procedure, and it is concluded that the intensity of a_I is sensitive to the orientation of the polymer chains in the amorphous regions of the fiber. A similar behavior is seen in the tan δ peak in semicrystalline fibers such as polyester when the degree of orientation is varied [214,218].

12.5 FIBER MANUFACTURING

All commercial acrylic fibers are spun from viscous, concentrated polymer solutions commonly referred to as spinning dopes. Melt spinning cannot be used because acrylic polymer degrades when heated near the melting point, which is near 300°C. The spinning processes most commonly used for acrylic fibers are wet spinning and dry spinning. In commercial wet spinning, the spinning dope is extruded through a multiple hole spinneret immersed in a coagulation bath (or spin bath) containing a nonsolvent. The jets of dope quickly coagulate into solid filaments that are continuously removed from the spin bath, where they are washed free of solvent in hot water and then subjected to a series of operations that usually includes orientational drawing, drying, crimping, and annealing. The linear speed of the fiber is very slow in comparison with the speeds achieved in melt spinning; therefore manufacturing productivity is accomplished by using multiple spinnerets with each having many thousands of holes. The filaments from several spinnerets are often combined to form a single fiber tow bundle.

In dry spinning, the dope is extruded through spinnerets located at the top of a tower. As the uncoagulated filaments flow down the tower, they are brought into contact with an inert gas heated above the boiling point of the dope solvent. The solvent evaporates from the filaments as they pass down the column and solidify. Conceptually, dry spinning can be considered to be a special case of melt spinning, in which the polymer solidification or crystallization temperature has been depressed by the solvent. The solidification temperature of the filaments will continuously increase as the solvent evaporates until solid filaments are formed. The filaments are continuously removed from the bottom of the tower, washed free of solvent, and then for the most part processed like the wet-spun fibers.

The physical processes underlying fiber formation will be discussed in this section. Particular emphasis will be placed on the links between the fiber formation process, structure, and properties, but details of the actual manufacturing process will not be covered. Excellent review articles covering wet spinning and dry spinning of acrylic fibers have been written by Capone [219] and von Falkai [220], respectively, and the theoretical basis of solution spinning has been described by Tsurumi [221]. The following topics will be discussed in this section:

1. Polymer solubility and spinning dope preparation
2. The wet and dry fiber-spinning processes with emphasis on the fiber formation mechanism
3. Treatment of the tow after fiber formation, e.g., washing, orientational stretching, drying, crimping, and dyeing

Several special topics will be discussed briefly, including gel spinning, which is a process to make very high-strength acrylic fibers, plasticized melt spinning, and in-line (producer) dyeing, where the dye is incorporated during the spinning process.

12.5.1 POLYMER SOLUBILITY AND SPINNING DOPE PREPARATION

Acrylic polymers and fibers are not soluble in most common organic solvents. Good solvents used to make the spinning dopes are polar molecules with large dipole moments and low molecular weight. It has been proposed that the polar groups of the solvent form dipolar bonds with the nitrile groups, thereby disrupting the crystalline structure and allowing dissolution [222]. Solvents used commercially include DMF, DMAC, and DMSO. It is also possible to dissolve the polymer in concentrated aqueous solutions of certain inorganic salts including sodium thiocyanate and zinc chloride. A typical spinning dope will have a polymer concentration of approximately 25 wt%, which gives a dope viscosity in the 500 P range. The viscosity and spinning performance is extremely sensitive to molecular weight changes [223,224], and there are times when the molecular weight must be adjusted to optimize some specific property. A good example is the control of dyeability. Most acrylic fibers utilize end groups as dye sites, and when a shift in dye level is required a molecular weight change must be made despite the concomitant effect on spinning performance. Compensating adjustments can be made, however, such as changing the polymer concentration to recover the initial value of the dope viscosity. The range over which this trade-off is feasible is restricted by another property of the dope, and that is the gelation rate. Under certain conditions, acrylic polymer dopes will form an elastic gel upon standing [225,226]. Gelation of the spinning dope is accompanied by a large increase in viscosity and must be avoided. It is a physical phenomenon caused by the formation of small crystallites that cross-link the chains in a manner analogous to the chemical cross-linking of rubbers, and has been observed with concentrated solutions of other polar polymers [227]. It is reversible, and when the gelled dope is heated above the gel melting point the crystallites melt and the original viscosity is recovered. The rate of gelation will increase with any of the following three conditions: cooling below the gel melting point, increasing the polymer concentration, and increasing the level of acrylonitrile comonomer in the polymer [225–227]. Dopes made from PAN homopolymer, for example, will gel very quickly and the addition of comonomers will slow down the gelation rate and make the dope easier to handle. In summary, no single formula can be given for the preparation of a good spinning dope. For each solvent and fiber production system, the interrelated variables of polymer concentration and the coagulation conditions must be optimized.

A variety of solvents are used in commercial spinning. Seven are listed in Table 12.14 along with the approximate capacity of world fiber production based on each solvent [219,228]. DMF is the most frequently used solvent and is employed in both dry and wet spinning. Aqueous solutions of sodium thiocyanate, zinc chloride, and nitric acid are the most commonly used inorganic solvents. Typical solvents and polymer concentrations are given in Table 12.15 [6].

The manner in which the spinning dope is prepared depends primarily on the polymerization method and form of the polymer. In solution polymerization, the same solvent that is used in the polymerization can be carried into spinning; therefore, it is not necessary to recover the polymer as a separate intermediate. The end result of the polymerization is a solution of the polymer in the spinning solvent, and to prepare the dope it is only necessary to adjust the polymer concentration. In heterogeneous aqueous polymerization, the polymer is usually recovered by filtration in the form of a wet-polymer cake. If the spinning dope solvent is an aqueous salt solution, then the wet polymer may be dissolved directly without drying [221,229].

For organic solvents, at least two methods have been described for conversion of the wet polymer into a solvent-based spinning dope. In both methods, the wet polymer is first slurried with solvent. In one method, the slurry is stripped of water in a wiped film evaporator [230], while in the other method the slurry is passed into a tubular heat exchanger to dissolve the polymer and then on through a vaporization zone to remove the water [231]. The wet cake may also be dried, in which case the spinning dope is prepared in a continuous process by metering the solvent and dry polymer into a mixing device to form a slurry that is converted into the dope by passing through a heat exchanger [232].

The properties of acrylic fibers are enhanced by the incorporation of additives in the spinning dope. Dull or semidull fibers are delustered by adding TiO_2. Colored fibers may be

TABLE 12.14
Acrylic Fiber Production

Country	Producer	Solvent	Spinning	Capacity (t[spi43]/year, 1993)	Gel-dyed	Pigmented	Technology/comments
North and South America							
United States	Cytec (American Cyanamid)	NaSCN	Wet	60,000	×		
	Mann Group	ZnCl$_2$	Wet	136,000			Dow-Badische—Shutdown 1993
	Monsanto	DMAC	Wet	196,000			
Total							
Latin America							
Argentina	Hisisa	HNO$_3$	Wet	19,000		×	Asahi—Shutdown 1992
Brazil	Celbras	DMF	Dry	18,000			Mitsubishi Rayon
Rhodia		DMF	Wet	75,000			Rhone-Poulenc
Mexico	Cydsa	DMF	Wet	34,000			Rhone-Poulenc
	Fibras Nacionales de Acrilica	DMF	Wet	30,000	p		Snia
	Fibra Sinteticas	HNO$_3$	Wet	31,000	×		Asahi
Peru	s.d.F.	DMF	Dry	207,000		×	Bayer
Total							
Western Europe							
Germany	Bayer, Dormagen	DMF	Dry	120,000		×	
	Bayer, Lingen	DMAC	Wet	50,000		×	
	Hoechst, Kehlheim	DMF	Dry	8,000			
	Hoechst, Kehlheim	DMF	Wet	35,000			
	Markische Faser	DMF	Wet	48,000			
Great Britain	Courtaulds	NaSCN	Wet	80,000	×		
Greece	Vomvicryl	DMF	Wet				Snia—Shutdown 1990
Italy	Enichem Fibre	DMAC	Wet	140,000			Monsanto
	Montefibre	DMAC	Wet	120,000			Shutdown
	Snia	DMAC	Wet				
Ireland	Asahi Cashmilon	HNO$_3$	Wet	25,000	×		
Spain	Courtaulds	NaSCN	Wet	62,000	×		
	Montefibre	DMAC	Wet	70,000	×		

Acrylic Fibers

Portugal	España Fisipe	DMAC	Wet	36,000	Mitsubishi Rayon
Total				794,000	
Eastern Europe					
Bulgaria	Bulana Burgas	DMF	Wet	15,000	Snia p
Hungary	Magyarviscosa	DMF	Wet	11,000	Snia x
	Magyarviscosa	DMF	Dry	20,000	DuPont/Simon-Carves
Poland	ZWS Anilina	NaSCN	Wet	15,000	Courtaulds x
Romania	Savinesti	Ethylene carbonate	Wet	65,000	Modacrylic Kanegafuchi p
Russia	Novopolatsk	NaSCN	Wet	10,000	Courtaulds x
	Navoiazot	NaSCN	Wet	40,000	Courtaulds
	Saratov	NaSCN	Wet	40,000	
	Mapan Novopolatsk	DMF	Wet	35,000	Snia x
Yugoslavia	Ohis	NaSCN	Wet	30,000	Courtaulds x
Total				281,000	
Middle East					
India	Ind. Petro Chem	HNO_3	Wet	24,000	Chemtex
	Ind. Synth. Co.	DMAC	Wet		Planned Montefibre
	Ind. Acrylic Ltd.	DMF	Dry	12,000	DuPont/Chemtex
	Reliance Text.	DMF	Dry		Planned DuPont
	Consolidated	NaSCN	Wet	Under constr.	Exlan 15,000 T/A
	J.K. Synthetics	DMAC	Wet	24,000	Montefibre
	Pasupati	DMF	Wet	15,000	Snia
Iran	Polyacryl Corp.	DMF	Dry	30,000	DuPont
Turkey	Aksa	DMAC	Wet	150,000	Montefibre x
	Yalova	DMF	Wet	15,000	Snia x
Total				270,000	
Far East					
China	Daqing	NaSCN	Wet	50,000	Cytec p
	Anqing	NaSCN	Wet	Under constr.	Cytec 50,000 T/A
	Qinhuangdao	DMF	Dry	30,000	DuPont/Chemtex
	Gao Qiao	NaSCN	Wet	2,300	x
	Ningbo	DMF	Dry	Under constr.	DuPont/Chemtex 30,000 T/A

Continued

**TABLE 12.14 (Continued)
Acrylic Fiber Production**

Country	Producer	Solvent	Spinning	Capacity (t[spi43]/year, 1993)	Gel-dyed	Pigmented	Technology/comments
	Jin Shan	NaSCN	Wet	52,000	x		
	Jin Yang	NaSCN	Wet	20,000			
	Mao Ming	DMF	Dry	Under constr.			DuPont/Chemtex 30,000 T/A
	Lanzhou	NaSCN	Wet	14,000			Courtaulds
	Fushun	DMF	Dry	30,000			DuPont/Chemtex
	Zibo	DMF	Dry	45,000			DuPont/Chemtex
Total China				243,300			
Indonesia	Hamparan Rajecki	DMF	Dry	40,000			DuPont/Chemtex
Japan	Asahi	HNO$_3$	Wet	101,000			13,000 T/A filament
	Kanebo	DMF	Wet	29,000			
	Kanegafuchi	Acetone	Wet	41,000		x	Modacrylic
	Mitsubishi Rayon	DMF	Dry	5,000			
	Mitsubishi Rayon	DMAC	Wet	98,000		x	Monsanto
	Nippon Exlan	NaSCN	Wet	60,000			
	Toho Beslon	ZnCl$_2$	Wet	43,000			
	Toray	DMSO	Wet	43,000			
Total Japan				420,000			
North Korea	Anilon	Ethylene carbonate	Wet	4,000			
South Korea	Hamil	HNO$_3$	Wet	118,000	x		
	Taekwang	NaSCN	Wet	79,000			
Taiwan	Formosa Plastic	DMF	Wet	102,000			
	Tong Hua	NaSCN	Wet	55,000			
Thailand	Thai Acrylic Fibers	NaSCN	Wet	17,000			
Total				1,741,600			

Source: From Capone, G.C., in *Acrylic Fiber Technology and Applications*, Masson, J.C., Ed., Marcel Dekker, New York, 1995, pp. 69–104.

TABLE 12.15
Typical Solvents and Polymer Concentrations

Solvent	Polymer concentration (%)
Dimethylformamide	28–32
Dimethylacetamide	22–27
Dimethyl sulfoxide	20–25
Ethylene carbonate	15–18
Aqueous NaSCN, (45–55%)	10–15
Aqueous HNO_3 (65–75%)	8–12
Aqueous $ZnCl_2$	8–12

Source: From Hobson, P.H. and McPeters, A.L., *Kirk–Othmer: Encyclopedia of Science and Technology*, 3rd ed., Vol. I, Wiley, New York, 1978.

produced by the incorporation of either insoluble pigments in the dope or addition of soluble cationic dyestuffs that react with dye sites within the fiber during coagulation. Flame resistance of fibers is increased by the use of additives containing chlorine, bromine, or phosphorous. Antimony compounds [233,234] are used for their synergistic effect on flame resistance in fibers containing a halogen. Heat stabilizers are added to minimize discoloration of the fiber. Typical stabilizers include organic acids and sequestering agents [235], organophosphorous compounds [236,237], tin compounds [238], zinc compounds [239], and zirconium salts [240]. Other additives may be used to improve fiber light stability. It is especially important to use a proper additives package with modacrylics, which can have 35–85% copolymerized AN. The balance of the composition will contain high levels of halogen-containing comonomers such as vinyl chloride, vinylidene chloride, and vinyl bromide, which can render the fiber extremely sensitive to thermal and photodegradation.

12.5.2 Fiber Formation

In this section, the fiber formation process will be described with emphasis on the development of the optimum fiber structure for commercial applications. By fiber formation, we mean the process by which the spinning dope is converted into fiber in the spin-bath or dry-spinning column. The subsequent operations of tow processing will be covered in Section 12.5.3. Wet spinning is the most widely used spinning process and accounts for approximately 85% of the worldwide production, with the remainder accounts for dry spinning [219]. Each spinning system has evolved over the past 50 years to produce commercially viable products. The strategy for developing a commercial spinning process usually includes the following goals: (1) optimizing fiber structure and properties; (2) increasing production rate as determined by linear speed, filament tex (or denier), and tow size; and (3) improving process stability and fiber uniformity. The combination of process conditions required to achieve all of these features simultaneously is strongly influenced by the desired filament size. This is usually described as the denier per filament (weight in grams per 9000 m), or, if the SI units are preferred, the tex per filament (weight in grams per 1000 m). The filament denier in all spinning operations is controlled by the material balance given by the following relationship:

$$\text{Denier/filament} = \frac{K(d \times W_2 \times Q)}{(S \times V_1)} \qquad (12.23)$$

where K is a proportionality constant, d is the spinning dope density in grams per liter, W_2 is the polymer concentration in the dope expressed as the weight fraction, Q is the volumetric flow rate per spinneret hole in liters per second, V_1 is the first roll speed (linear speed in meters per second at which the filaments leave the spin-bath or dry-spinning column), and S is the overall stretch ratio including any relaxation.

The meaning and usefulness of the above equation can be best conveyed by recognizing that the parameters in the equation can be approximately factored into three groups representing three distinct stages of fiber production. These are dope preparation (d, W_2), fiber formation in the spin-bath or dry-spinning column (Q/V_1), and tow processing (S). Now suppose that an increase in the denier per filament is desired, e.g., if a spinning machine is to be converted from producing a three denier per filament textile fiber to a 15 denier per filament carpet fiber, or vice versa. It is likely that the manufacturer will not want other key properties to change along with the denier (and would prefer to be able to independently change any variable). Holding the fiber properties constant means that S cannot be changed, because any change here would affect the degree of polymer chain orientation in the fiber. This will directly affect tensile properties and dyeing rate, among others. The parameters related to the spinning dope, d and W, also should remain largely unchanged because their optimization depends primarily upon the solvent type, polymer molecular weight, and polymer composition. This leaves the ratio of Q/V_1 as the lever for controlling the filament size. In practice, one would normally increase Q, the flow rate per hole, to increase the denier per filament. This may require an increase in the spinneret capillary diameter to prevent excessive dope shear rates, which causes spinning instabilities such as dope fracture. The alternative of decreasing V_1 has the disadvantage of reducing the total machine productivity. In reality, all of the parameters in the equation are interrelated to various degrees, and some compromises must be made when fine-tuning a new spinning process. The goal is to arrive at the optimum set of spinning conditions that produces the desired product in the most economical manner.

12.5.2.1 Fiber Extrusion and Coagulation

In wet spinning, the polymer dope is extruded through a spinneret capillary into the spin bath containing a nonsolvent or coagulant. The polymer is not soluble in the coagulant, but the coagulant is miscible with the spinning dope element. Fiber formation occurs rapidly as the filament, or threadlike component, of dope coagulates and is drawn out of the bath for subsequent tow processing. All of the wet-spinning solvent systems have several features in common. In each case, the spin-bath coagulant is an aqueous solution of the spinning dope solvent. The use of these aqueous solutions as coagulants has advantages of cost and simplicity, and environmental problems are minimized. All of the process steps that utilize aqueous solutions of the solvent in one way or another, such as polymer slurrying and dope preparation, spin-bath solvent concentration control, and tow washing, can be tied together in a common solvent recovery and distribution system. Another common feature is the spinning of many filaments from each spinneret. Spinnerets may have from 10,000 to 60,000 holes with hole sizes from 0.05 to 0.38 mm (2–15 mils). The holes are often divided into segments to improve the diffusion of the coagulant across the spinneret face. Several spinnerets can be located in a single coagulation bath and the filaments from each spinneret are combined to form a single large tow upon exiting the bath on the first take-up roll [6]. The various acrylic spinning methods in use throughout the world are listed in Table 12.14. In addition to the solvent and spinning method, we also give the capacity (as in 1993), indicating the manufacturers who produced gel-dyed or pigment-dyed fiber (in both cases, the dyeing operation is integrated into the spinning process), and the origination of the technology.

A brief description of the dry-spinning operation was given previously. Now we shall give some of the details of a commercial spinning operation that are based primarily on the patent literature. Recall that fiber formation occurs as the polymer filaments are drawn down into a tower filled with an inert hot gas. The tower consists of a vertical jacketed cylinder with a spinneret positioned in the center of the top and the fibers exiting from the bottom [241]. In one mode of operation, the heated gas is drawn into the top of the column where it flows downward with the filaments and is withdrawn near the bottom of the tower [241]. In another approach, the column is divided into three-heated zones with the inlet gas flowing counter-current to the filaments and withdrawn above each zone [242]. The solvent is recovered from the gas exiting each zone. Finish may be applied below the exit by bringing the filaments in contact with a roll wetted with the finish solution. The filament bundle is next wrapped around a takeout roll for several turns and then directed to subsequent process rolls.

As in wet spinning, process optimization and production capacity are greatly influenced by the filament denier, and the relationships among the spinning parameters given in Equation 12.23 are applicable. The spinneret may have 300–900 holes arranged in concentric circles. The number of holes is constrained by the need to maintain filament separation. Turbulence in the tower is also minimized so that the filaments will not touch and fuse. Other important manufacturing considerations include gas tow rate, gas temperature, solvent-to-gas ratio and the explosive limits, heat input from the walls of the tower, tower length, and solvent boiling point [6]. The principal solvent used for dry-spinning acrylic fibers is DMF, which boils at 153°C, and complete removal of the solvent is not possible at typical manufacturing spinning speeds. Therefore, the remainder of the solvent is washed out during the processing of the tow [241]. Typical exit speeds are 200–400 m/min compared with 3–10 m/min in wet spinning. Despite the higher speeds, dry-spinning productivity is lower than in wet spinning because the number of holes per spinneret is much lower. This translates into higher manufacturing costs for dry spinning. One reason for the success of dry spinning is that the filament cross sections have a high-aspect, dog-bone-shaped cross section resulting in a lower bending modulus and more supple feel in the fabrics in comparison to the more rounded cross-sectional characteristic of wet spinning. We return to this subject later in Section 12.6.

It is impractical to carry out fundamental process studies on plant-scale equipment. Most of the investigations to be discussed here were done using small, laboratory-scale spinning lines. A schematic diagram of a typical laboratory wet-spinning line used by McPeters and Paul [243] is shown in Figure 12.21. A polymer dope with a viscosity of about 500 P is pumped to a spinneret immersed in a coagulating bath containing an aqueous solution of DMAC.

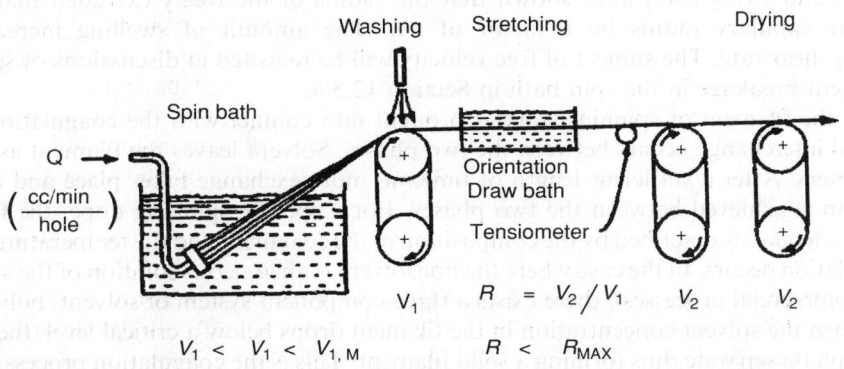

FIGURE 12.21 Schematic diagram of laboratory wet-spinning line. (From McPeters, A.L. and Paul, D.R., *Appl. Polym. Symp.* 25, 159, 1974.)

A positive displacement gear pump is used to meter the dope at a precise volumetric flow rate of Q liters per second per hole. Without take-up on the first roll, the filaments emerge from the spinneret with a free velocity, V_f, on the order of 1–6 ft/min depending upon Q, the volumetric flow rate per spinneret [244]. For fiber production, the filaments are taken up at a velocity V_1 by the first roll shown in Figure 12.21 and the fiber is washed with pure water. There is a maximum velocity of take-up $V_{1,m}$ that the filaments can undergo without breaking in the bath. The fiber is next pulled through a water bath by a second roll at a velocity of V_2 to obtain an orientation draw ratio of $R = V_2/V_1$. The temperature of the water in the draw bath can be adjusted but is most often maintained close to the boiling point. For fixed conditions up to the first roll, there is a well-defined draw ratio, R_{max}, where the filaments begin to break. The measurement of threadline stress is often desired for studies of the rheological forces involved in wet spinning. The tension can be readily measured by placing an in-line tensiometer either before the first roll for measurement of the fiber tension in the coagulation bath or before the drawing roll to measure the drawing tension.

Coagulation in the spin bath produces filaments that are extremely porous for reasons to be discussed later. Even after orientational drawing of the fiber, the void volume fraction can be over 50%. The wet-porous fiber is taken over two pairs of steam-heated rolls, where the voids collapse and the porosity is eliminated in the course of drying. At this point, the fiber density approaches the polymer density of 1.17 g/cm^3.

We will now discuss the mechanism by which the fiber is formed in the coagulation bath. The spinning dope is pumped through the spinneret capillary hole at a velocity (V), which can be calculated from the following equation:

$$(V) = \frac{Q}{(\pi/4)D^2} \quad (12.24)$$

where Q is the volumetric flow rate per hole and D is the capillary hole diameter. The ratio of V_1 to (V) is referred to as the jet stretch ratio. The magnitude of the jet stretch is almost always less than one, but this does not mean that the filament is not stretched while in the spin bath. It is true that the dope velocity while inside the capillary does exceed V_1 but immediately upon exit from the spinneret the normal stresses act to swell the filament and its velocity decreases [245]. Under the conditions of free extrusion (where the filaments emerge freely from the spinneret under zero tension), the so-called free velocity, V_f, may be considerably less than V_1. This swelling arises because the elasticity of the dope permits it to store energy arising from the shearing forces that act upon entry and travel through the spinneret capillary. Fitzgerald and Craig [245] have shown that the radius of the freely extruded filament can exceed the capillary radius by a factor of ten. The amount of swelling increases with increasing shear rate. The subject of free velocity will be revisited in discussions of spinability and filament breakage in the spin bath in Section 12.5.4.

When the filament of spinning dope is brought into contact with the coagulation bath, a diffusional interchange occurs between the two phases. Solvent leaves the filament as the nonsolvent enters. After a sufficient length of time, no more exchange takes place and a state of equilibrium is achieved between the two phases. For a specific spinning dope, the final state should be adequately described by the composition of the coagulant and the temperature at which the coagulation occurs. In the case where the nonsolvent is an aqueous solution of the solvent (as in most commercial processes), there exists a three-component system of solvent, polymer, and water. When the solvent concentration in the filament drops below a critical level, the polymer begins to phase separate thus forming a solid filament. This is the coagulation process.

Two conceptually distinct phase transitions are thought to occur during coagulation: gelation and phase separation (or precipitation). Gelation is the gradual transition of the polymer solution into a single-phase elastic gel, whereas phase separation implies that the polymer phase separates from the solvent. Ideally it is desirable to have some degree of gelation preceding precipitation [246]. If coagulation were to occur without gelation, then the filament would have very little strength because interconnectivity of the polymer chains would be absent. In gelation, polymer microcrystallites are formed that serve as physical cross-links between chains such that all polymer chains are part of a continuous single phase–phase network [225–227]. Now, when the gel undergoes phase separation forming the solid filament, the polymer chains remain interconnected by the physical cross-links, thus giving the internal cohesion needed for the drawing step upon emergence from the bath.

Paul [225,247] has studied the diffusional processes that occur during coagulation. A copolymer of acrylonitrile and vinyl acetate (7% VA by weight) was used with DMAC solvent. The coagulation bath consisted of various mixtures of DMAC and water. The equilibrium composition of a gelled rod was measured as a function of bath composition. These gelled rods were also used to study the process of diffusion during coagulation with the aim of elucidating the mechanism of diffusion. By following the composition of the bath and the weight of the polymer phases, the amounts of DMAC removed and water added at any given time could be determined. A very distinct moving boundary is associated with coagulation [247,248]. One side of the boundary is hard, coagulated polymer whereas the other side is a soft, elastic gel. The rate at which the boundary progresses inward decreases as the DMAC concentration in the bath is increased. The data were analyzed by means of various diffusion models: equal flux, constant flux ratio, and variable flux ratio. The equal flux model assigns equal flux for coagulant inward and solvent outward. The solution to Fick's law yields a boundary movement within the coagulated filament proportional to the square root of the time. However, the model prediction deviates from experimental data as the coagulation time increases. The variable flux model assumes the flux ratio of solvent to coagulant is constant at the coagulation boundary and changes from the coagulated filament surface to the center of the filament. The constant flux ratio model, in which the solvent to nonsolvent flux ratio was assumed constant at the boundary and within the filament, was reported by Paul to be most representative of the experimental data and is a reasonable physical description of the diffusion process. As the water content at a point within the gelled rod increases, the solids level also increases. Both of these factors decrease the solubility of the polymer in the surrounding liquid. A fibrillar network is thus formed during this period of densification just ahead of the moving boundary. We shall see in the following section that such a fibrillar structure can be found within the coagulated filaments. It is the nature of this fibrillar structure that determines to a large degree the range of properties that one can achieve in wet spinning.

Capone [219] has summarized more recent analysis of the diffusion behavior, and an example is the work by Baojin et al. [249]. The rate of diffusion is modeled from cylindrical coordinates again based on Fick's law. The composition of actual filaments from the spin bath was analyzed, and the coagulant was a DMF–water system. Correlations are presented for diffusion coefficients and flux ratios as functions of jet stretch, polymer solution concentration, and coagulation temperature. The flux ratios, they reported, are similar to those reported in Paul's data, 20 years earlier. The diffusion coefficients are in the same range of $4-10 \times 10^{-6}$ cm^2/s that Paul found for DMAC–H$_2$O systems.

Jian et al. [250] and Terada [251] studied the diffusion relationships in DMF, DMSO, and NaSCN solvents with water as the nonsolvent. Similar models were used, and the reported data show diffusion coefficients again in the same range as reported by Paul and Baojin. If the diffusion coefficients are similar for different solvents, then what, if any, structural effects can be attributed to solvents? Grobe and Heyer [248] argue that the thickness at the boundary is accurately predicted by diffusion models, but the pore size of the boundary is different and is the controlling mechanism for solvent diffusion. The formation of a fibrillar network creates these pores, and is a function of the polymer–solvent–nonsolvent interaction and phase behavior.

The phase behavior in wet spinning is the same as in other ternary systems that are controlled by temperature, concentration, and bonding energy between solvents, nonsolvents, and polymers. The mechanism of phase separation has been developed by Cohen et al. [252] for cellulose acetate membranes (cellulose acetate–acetone–water). Both theoretical models and laboratory measurements have been reported by other investigators [253–258]. Ziabicki [246] presented a qualitative picture of wet-spinning systems using ternary phase diagrams and phase separation models. He concluded that thermodynamic and kinetic aspects of phase separation are controlling the wet-spinning process.

Given the phase separation and diffusion models for wet-spinning acrylic fiber, there are fiber structure implications that dictate the physical properties of wet-spun acrylic and determine fiber property differences brought about by different polymer solvent–nonsolvent-spinning systems. The coagulation bath produces a porous gel network formed by separation of a polymer solution into a polymer-rich phase and a solvent-rich phase. The polymer-rich region phase is composed of an interconnecting network of bonded polymer chains. Control of the pore size of this network is a powerful factor for influencing the tensile, dyeing, and transport of wet-spun acrylic fibers.

The major difference between dry and wet spinning is the absence of precipitation in dry spinning. The filament is thought to remain essentially single-phased as the solvent evaporates and the polymer gels. In the absence of a coagulant diffusing into the filament, the fibrillar network so characteristic of wet-spun filaments is largely absent, and the dry-spun filament closely resembles that which hypothetically would be produced by melt spinning. The pore size within the polymer network for wet-spun acrylic polymers is large compared to the single-gel phase generated by a dry-spinning mechanism. This is a result of the relatively high rate of phase separation with systems using nonsolvents and having a counterdiffusion mechanism. The structure of the wet- and dry-spun filaments will now be discussed in greater detail.

12.5.2.2 Structure of the Coagulated Fiber

The structure of the freshly coagulated fiber has a major influence on the tensile properties, abrasion strength, and most other mechanical properties of the finished fiber. In this section, we discuss the characteristics of the coagulated fiber structure that influence these properties and show how the structure may be optimized by proper choice of the coagulation conditions and polymer composition. Two important features of the structure can be determined by looking at cross sections of the filaments under low magnification. The first is the gross cross-sectional shape of the filaments and the second is the presence or absence of large tear-shaped voids, known as macrovoids, that begin near the outer edge of the filament and extend to the center. In Figure 12.22, cross sections are shown for fibers spun into DMAC–water baths at temperatures of 10, 40, 55, and 70°C. With increasing temperature, the cross-sectional shape undergoes a transition from kidney bean to round, and macrovoids begin to appear at the higher temperatures. The cross-sectional shape is a manifestation of the relative volumetric transfer rates of the spin-bath liquid (nonsolvent) into the fiber versus the outward transfer of the solvent [261,262]. Since the fiber skin is the first portion to undergo coagulation, it becomes, in effect, the limiting volume that the fiber can occupy. If there is less volume of nonsolvent diffusing in that solvent diffusing out, the shape will become noncircular and progress toward the kidney-bean shape seen at the lower temperatures. The dry-spinning process is, in a sense, the limiting case of this phenomenon. Here there is no inflow of nonsolvent, and as the solvent diffuses out, the skin collapses around the solidifying polymer core, thus giving the characteristic dog-bone, cross-sectional shape.

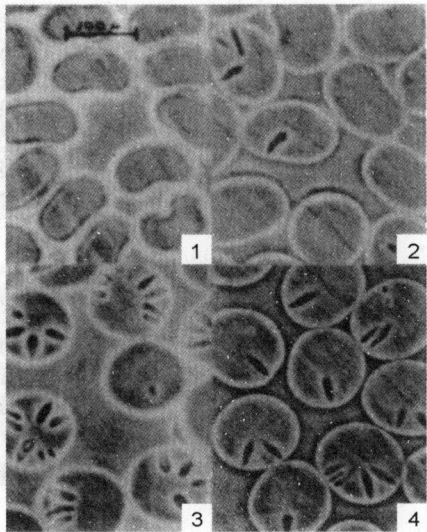

FIGURE 12.22 Cross sections of acrylic fibers taken under a light microscope in the uncollapsed and unoriented state. Spin-bath temperatures: (1) 10°C, (2) 40°C, (3) 55°C, and (4) 70°C. (From Craig, J.P., Knudsen, J.P. and Holland, V.F., *Text. Res. J. 32*, 435, 1962.)

The cross-sectional shape is useful in identifying the manufacturer of commercial acrylic fibers. Each manufacturer uses a different spinning process, which gives rise to a characteristic cross section. Scanning electron micrographs of these cross sections are shown in Figure 12.23. They range from almost perfectly round to dog-bone-shaped. Because of its dog-bone shape, the dry-spun fiber has a lower bending modulus or stiffness relative to a round or kidney-bean-shaped fiber [263]. The dependency of fiber stiffness on cross section is shown in Figure 12.24 for a round, kidney-bean, and dog-bone cross section. These results predict softer yarn or fabric aesthetics will be obtained at a comparable fiber denier with the dry-spinning process.

The dog-bone-shaped cross section also prevents the fibers from packing as closely in the yarn assembly leading to a less dense or bulkier product. Measurements of yarn thickness by Onions et al. [264] confirm this theory. Dog-bone-shaped fibers were 30 and 15% thicker than round and kidney-bean-shaped fibers, respectively, in comparable yarn structures. Studies by Lulay [263] also indicate that the lower packing ability of dry-spun fibers contributes to increased comfort in garments during high-stress activities, where rapid removal of perspiration from the skin is essential. The basket sink test, one measure of water transport, confirms the superiority of dry-spun acrylics compared to wet-spun acrylics, cellulosics, polyester, and polypropylene [265]. This is shown in Figure 12.25, where the amount of water transported in a gram of fiber per second is plotted against fiber type.

Various reasons have been given for the formation of the tear-shaped macrovoids seen in Figure 12.22. Grobe and Meyer [261] have argued that rupture of the skin followed by penetration of the nonsolvent into the interior of the filament is responsible. These voids are definitely undesirable. Even though they collapse during the drying process the surfaces are not eliminated; therefore, they can serve as internal flaws and adversely affect the lateral strength of the fibers. Good lateral strength is needed for resistance to fibrillation, which is the tendency of individual filaments to split.

The density of the coagulated fiber is typically near 0.40–0.50 g/cm^3, which is well below the 1.17 g/cm^3 value for the polymer and indicates that the filaments are extremely porous.

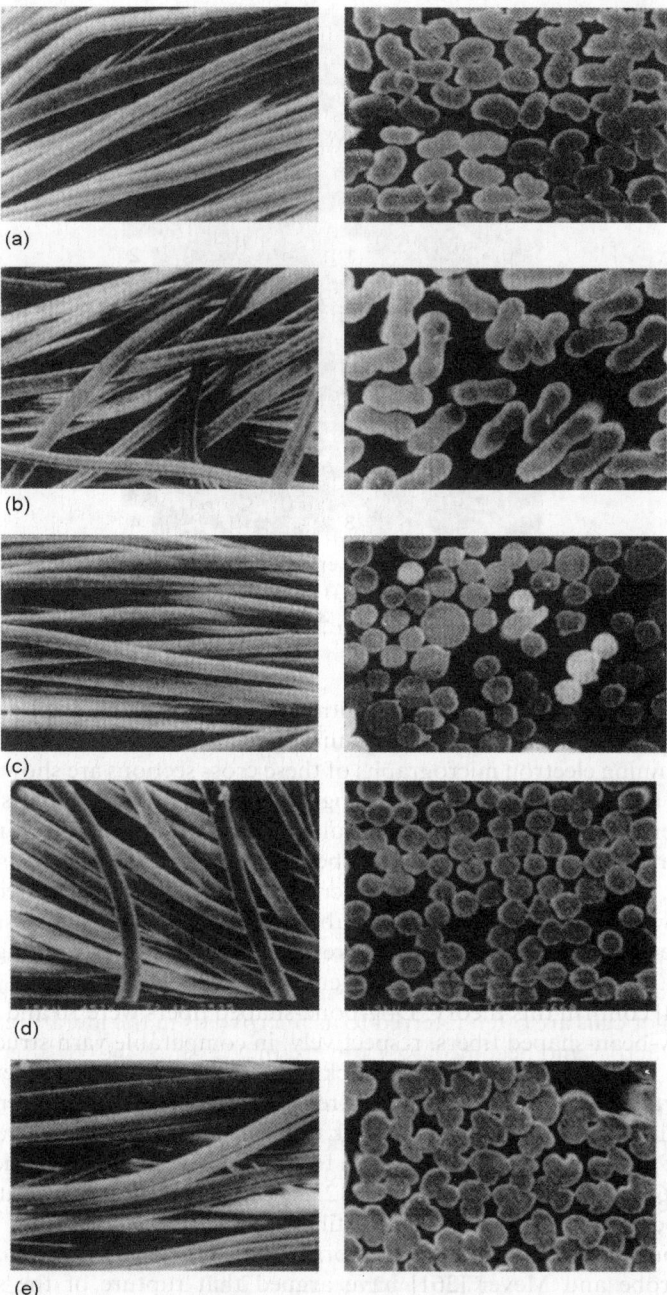

FIGURE 12.23 Scanning electron micrographs of five commercial acrylic fibers. The Orlon fibers are dry-spun from dimethylformamide. Examples of both monocomponent and bicomponent Orlon fiber are shown, with the former exhibiting the characteristic dog-bone-shaped cross section. The other three fibers are wet-spun. Acrilan is spun from dimethylacetamide into an aqueous solution of that solvent and exhibits a pronounced bean-shaped cross section. The Courtelle and Creslan fibers are thought to be spun from aqueous inorganic salt solutions and have round cross sections. (a) Acrilan 410× and 790×. (b) Orlon 250× and 780×. (c) Courtelle 334× and 495×. (d) Creslan 452× and 617×. (e) Orlon 21 bicomponent 318× and 469×.

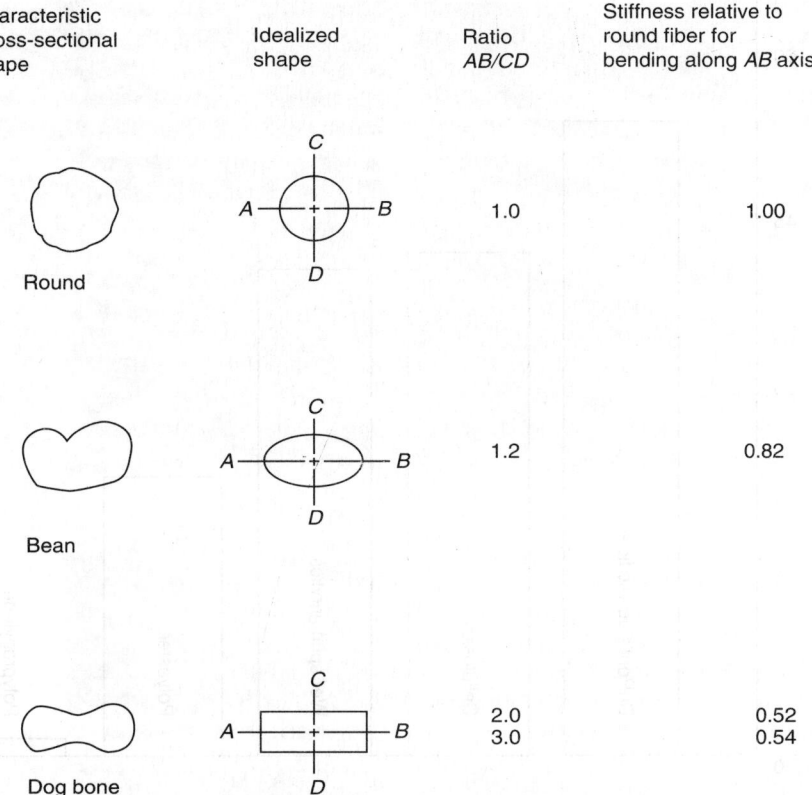

FIGURE 12.24 Dependency of fiber stiffness on cross section. (From Lulay, A. in *Acrylic Fiber Technology and Applications*, J.C. Masson, Ed., Marcel Dekker, New York, pp. 314–315, 1995.)

The nature of this porosity can be seen in Figure 12.26, where a transmission electron micrograph of a cross section is shown. The micrograph reveals a structure that can be described as cellular or microfibrillar in which small rods are interconnected by nodal regions. These open spaces or cells are often referred to as microvoids in the literature. They are on the order of 0.1–1 μm across and become larger toward the center of the cross section. Craig et al. [262] were able to characterize the microvoid structure using a variety of physical techniques. (The general problem of characterizing the void structure of synthetic fibers has been described by Quynn [266].) The microvoid structure was analyzed by Craig et al. [262], first, as a parallel array of rods and voids, and second, as spherical voids in a solid polymer matrix. Equations were derived for calculating fibril and void sizes and the number of voids.

Parallel array of fibrils and voids

$$\text{Fibril radius}: R_p = 2Ad_1 \tag{12.25}$$

$$\text{Void radius}: R_c = 2(d_1 - d_2)/Ad_1d_2 \tag{12.26}$$

where d_1 and d_2 are the polymer density and fiber mercury density, respectively, and A is the total surface area.

Spherical voids in a solid polymer matrix

$$\text{Void radius}: R_s = 3(d_1 - d_2)/Ad_1d_2 \tag{12.27}$$

$$\text{Number of voids}: N_s = A/4R_s^2 \tag{12.28}$$

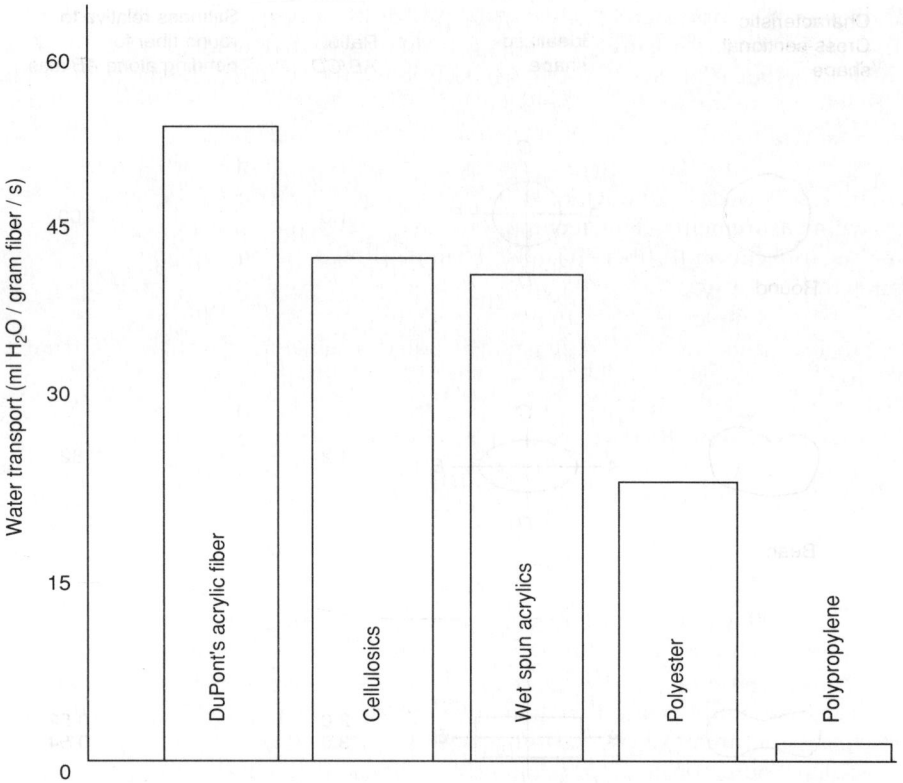

FIGURE 12.25 Water transport of various fibers, basket sink test (staple). (From *Manufactur's Journal*, Absorption Rate, Imbibition Test, EP-SAP 2301 (Basket Sink Test), June 18, 1980.)

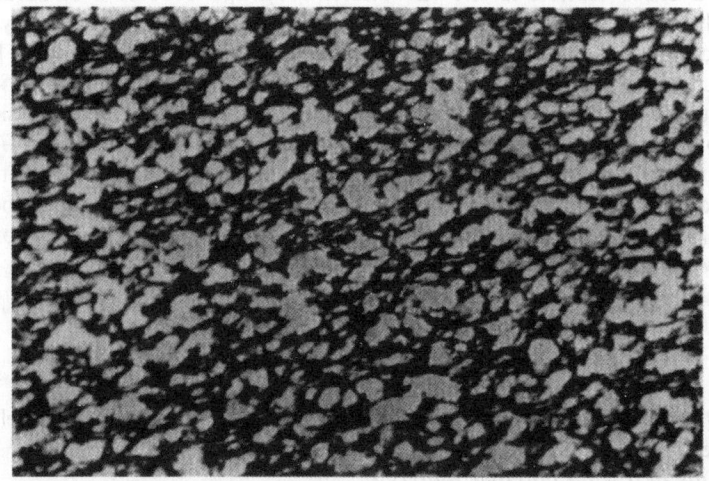

FIGURE 12.26 Transmission electron micrograph of a cross section of acrylic fiber removed from a 55°C dimethylacetamide water spin bath and freeze-dried. Magnification, 20,000×.

The parallel array of fibrils and voids may be the more appropriate model for the fiber after orientational drawing, for example. In any event, fibers that tend to have relatively small values of R, whether it be R_r, R_c, or R_s, and large values of N_s can be described as having a fine microvoid structure, and vice versa: large R values and small N_s values are characteristic of a coarse microvoid structure.

Transmission electron microscopy (TEM) photographs usually reveal a distribution of the microvoid sizes. A quantitative description of this distribution can be derived from mercury porosimetry measurements. This technique is an extension of the mercury density measurement and its application to fibers has also been described by Quynn [266]. The experimental procedure consists of measuring the cumulative volume change of freeze-dried, uncollapsed, coagulated fiber completely surrounded by mercury at some initial low pressure and then subjected to increasing pressure. The total mercury volume change is the total pore volume filled by mercury over the pressure range. The pore radius and the pressure at which a given volume of pore space is filled are connected through a reciprocal relationship, the Kelvin equation [264], and from the shape of the pressure–volume curve one can obtain information about the pore size distribution, i.e., the differential volume as a function of the pore radius.

One can also use small-angle x-ray scattering to measure microvoids. The intensity of the diffuse scattering is measured as a function of the angle between the scattered and incident beams, and this intensity is then converted into the void size distribution by use of the Gunier method of analysis [267]. This method appears to be suitable for only the very small pore range (less than 200 Å in radius) and is more useful in determining the existence and shape of the pores rather than the size and number [266].

The microvoid structure is strongly affected by the spin-bath conditions under which the fiber is coagulated, and the specific comonomer composition. Craig et al. [262] have studied the microvoid structure of fibers spun into DMAC–water baths over a range of temperatures as well as fibers spun from the aqueous NaSCN and $ZnCl_2$ solvent spin-bath systems. Dry-spun fibers covering a wide range of residual solvent content were included for comparison with the wet-spun fibers. The results of the study are shown in Table 12.16. The density measured in toluene, which freely penetrated the fiber interior, remained at $1.17\,g/cm^3$ for all the fibers. This is the accepted value for the polymer density, thus indicating that the degree of crystallinity is not strongly affected by the coagulation conditions. The first four fibers in Table 12.16 represent progressively lower spin-bath temperatures. Reducing the temperature results in a significantly higher mercury density and total internal surface area. The microvoid structure becomes less coarse as evidenced by the decrease in fibril radius or void radius depending upon the model that is used. TEM photographs of the cross sections confirmed that lowering the spin-bath temperature produces a denser and finer structure. A similar temperature dependence of the microvoid structure was observed for the fibers spun from aqueous solution of NASCN and $ZnCl_2$. The dry-spun fibers were found to have very dense structures, as might be expected from a system in which precipitation does not occur.

The last three items in Table 12.16 represent dry-spun fibers in which the residual solvent upon exiting the dry-spinning tower was allowed to increase from 1 to 20%. At the 1% level, the fiber emerges from the tower completely collapsed. Some porosity is introduced at a residual solvent level of 20%. The electron micrographs confirmed that the cross section is very dense and the microfibrillar structure characteristic of the wet-spun fibers is completely absent at this stage of the dry-spinning process though, as we shall see, the microfibrillar structure is formed during the orientational drawing.

The structure of the fiber formed in the spin bath strongly influences the tensile properties of the final product. Knudsen [268] has demonstrated that the influence of the coagulation bath variables on the fiber tensile properties can be rationalized in terms of the uncollapsed fiber structure. As the spin-bath temperature is lowered and a denser and finer structure is

TABLE 12.16
Structural Parameters for Unoriented Acrylic Fibers Spun by Various Techniques

Sample number	Solvent	Coagulant composition	Coagulation temperature (°C)	Mercury density (g/cm³)	Toluene density (g/cm³)	Area ratio	Surface area (m²/g)	R_p (Å)	R_s (Å)	N_s (pores/g × 10^{14})
A-1	DMAC	DMAC/H₂O	70	0.42	1.16	2.8	64	265	717	10
A-2	DMAC	DMAC/H₂O	55	0.41	1.17	2.9	105	160	454	40
A-3	DMAC	DMAC/H₂O	40	0.46	1.17	2.5	187	91	212	323
A-4	DMAC	DMAC/H₂O	10	0.58	1.17	2.0	206	83	122	1100
B-1	55% NaSCN in H₂O	10% NaSCN in H₂O	0	0.52	1.17	2.3	292	58	109	1960
B-2	55% NaSCN in H₂O	55% NaSCN in H₂O	30	0.21	1.17	5.6	140	122	835	16.0
C-1	60% ZnCl₂ in H₂O	35% ZnCl₂ in H₂O	7	0.51	1.17	2.3	288	59	114	1767
C-2	60% ZnCl₂ in H₂O	60% ZnCl₂ in H₂O	42	0.16	1.00	7.3	112	151	1450	4.3
D-1	DMAC	Hot air, 1% residual solvent in fiber at tower exit		1.17	1.17	a	a	a	a	a
D-2	DMAC	Hot air, 10% residual solvent in fiber at tower exit		1.00	1.15		a	a	a	a
D-3	DMAC	Hot air, 20% residual solvent in fiber at tower exit		0.83	1.11	1.42				

[a] Surface area is less than 1 m²/g.
Source: From Craig, J.P., Knudsen, J.P., and Holland, V.F., Text. Res. J., 32, 435, 1962.

formed, one observes an increase in fiber tenacity, modulus, and abrasion resistance. Other coagulation variables examined were polymer concentration in the spinning dope, coagulation bath composition, and jet stretch. Some of the salient effects are summarized in Table 12.17. Changes in coagulation conditions that lead to a denser and more homogeneous structure in the fiber emerging from the spin bath tend to bring about improvements in the tensile properties of the finished fiber.

Another way to obtain the desired dense-coagulated fiber structure is to add an ionic comonomer to the chain. Terada [269] has used mercury porosimetry to follow the change in the microvoid size distribution that occurs when comonomers of differing polarity and ionic structure are added to PAN. When hydrophobic comonomers such as methyl acrylate (MA) are used, the peak of the pore size distribution is shifted in the direction of increased size. However, ionic comonomers such as SSS or sodium allyl sulfonate (SAS) shift the peak in the direction of decreased size. With a three-component composition containing both the hydrophobic and ionic comonomers, as in poly(AN–MA–SAS), the distribution has both characteristics. Examples of void size distributions are shown in Figure 12.27. The 5% MA comonomer has a fairly sharp peak near 850 Å and the maximum size is extended to 200 Å. Moreover, there is only a slight proportion of pores with small sizes present.

The 1% SA copolymer shows two broad peaks near 300 and 55 Å. Other void size characterization methods were used to confirm these results and show that the effect also carries over into the fibers that have gone through the orientational drawing step. These results suggest that the use of ionic comonomers could lead to improved fiber properties.

12.5.3 Tow Processing

When acrylic fibers emerge from the drying tower or spin bath, they are processed to remove residual solvent and then given the orientational drawing and relaxation treatments that are

TABLE 12.17
Summary of Responses to Coagulation Variables

	Protofiber structure parameters[a]			
Variable	Protofiber density	Protofiber surface area	Cross-sectional shape	Homogeniety
Dope solids	(+)	0	0	++
DMAC concentration of spin bath	(+)	(+)	0	+
Spin-bath temperature	––	––	–	
Jet stretch	0	(–)	0	

	Finished fiber properties[a]			
Variable	Tenacity	Elongation	Initial modulus	Abrasion resistance
Dope solids	+	0	0	+
DMAC concentration of spin bath	0	0	0	0
Spin-bath temperature		+	–	

[a] Code: (+), slight positive response; +, strong positive response; ++, very strong positive response; 0, no effect; (–), slight negative response; –, strong negative response; ––, very strong negative response.
Source: From Knudsen, J.P., Text. Res. J., 33, 13, 1963.

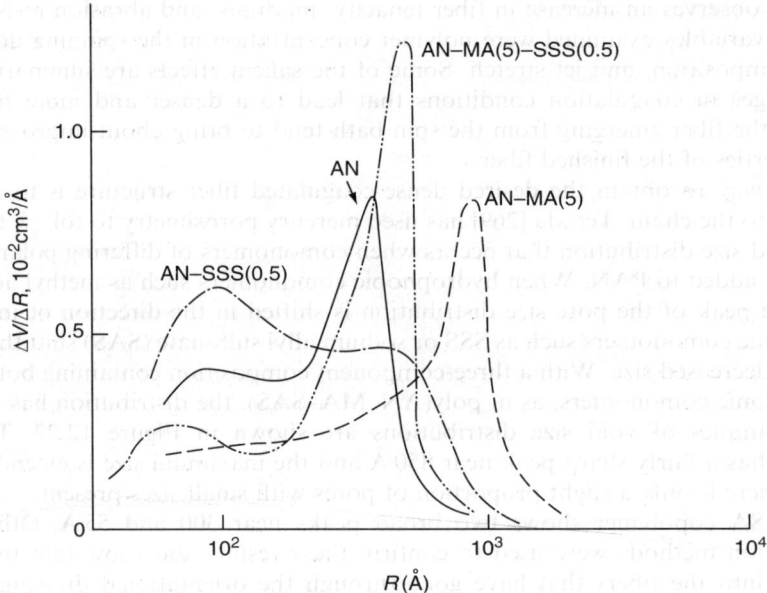

FIGURE 12.27 Void-size distribution in freeze-dried uncollapsed and unoriented fibers as determined by mercury porosimetry. The weight fraction of comonomers methyl acrylate (MA) and sodium styrene sulfonate (SSS) are in parentheses. (From Terada, K., *Sen'i-Gakkaishi 29*, 12, T451, 1973.)

needed to develop a desirable balance of tensile properties, abrasion resistance, and dyeing behavior. The complexity of tow processing for dry-spun fibers is determined to some extent by the level of residual solvent upon exiting the tower. If the solvent level is low, i.e., only several percent, then the bundle of fibers can be immediately drawn between heated rolls since no washing is required [242]. Much higher throughput in the dry-spinning operation can be achieved when 10–20% residual solvent is left in the fiber. But with exit speeds of 200–400 m/min, direct introduction into an aqueous tow-washing bath is not feasible because residence time in the bath may not be sufficient to remove the solvent. It is more typical to combine the solvent laden tows from several columns into a rope, which is then piddled into a can [270], and then multiple ropes are reeled together for tow processing. Wet-spun filaments from several spinnerets may be combined after leaving the coagulation bath and drawn directly into the tow line. The weight of water and solvent in the tow can range from 100 to 300% of the fiber weight, depending upon the filament denier and porosity.

The essential steps of washing, orientational drawing, drying, and relaxation are common to all acrylic fiber-spinning processes, although the sequence in which they are carried out may differ. In Figure 12.28, we show a simplified flow diagram that illustrates the features of several tow processes described in the patent literature [6]. After the steps of washing, stretching, and finish application, at least three different procedures have been described. In one, the fibers are dried and collapsed with no shrinkage permitted and in another controlled shrinkage is allowed by overfeeding the drying rolls. In a third, the fibers are dried and collapsed on a belt without restraint so that shrinkage may occur. Crimp may also be developed at this point. In some cases, the dried fiber will then undergo an additional stretching and relaxation step. Finally, the fibers are either packaged as tow or cut into staple and packed out into bales.

In the following sections, we will describe the various operations of tow processing with regard to the development of the fiber structure. The operations of orientational drawing,

Acrylic Fibers

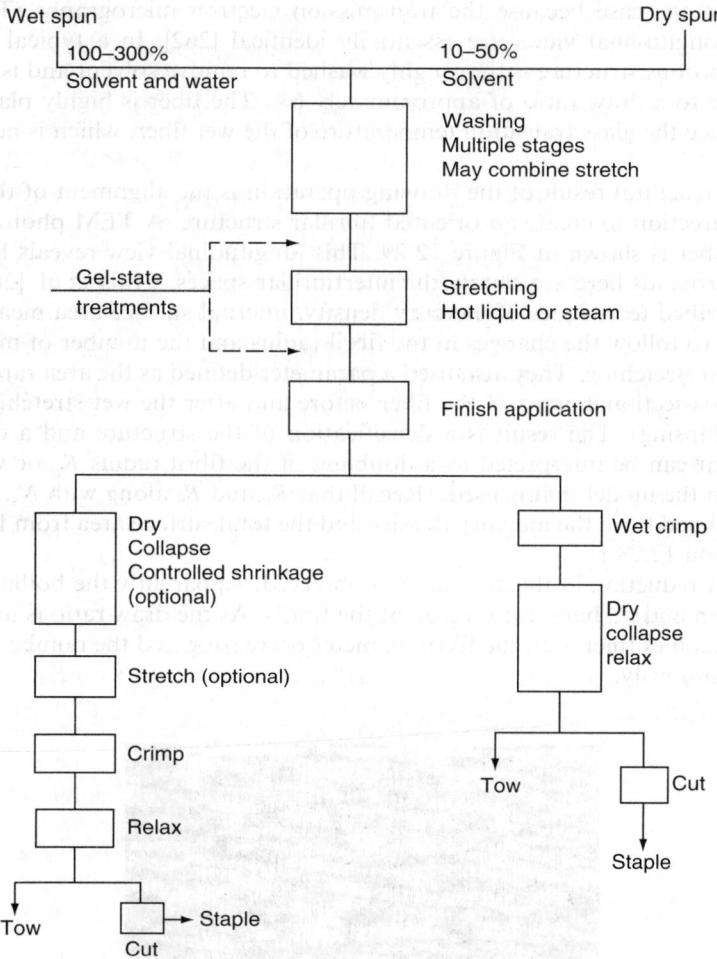

FIGURE 12.28 Tow processing of acrylic and modacrylic fibers. (From Hobson, P.H. and McPeters, A.L., *Kirk–Othmer: Encyclopedia of Science and Technology*, 3rd ed., Vol. I, Wiley, New York, 1978.)

drying and collapsing, and relaxation will be treated separately, so that within a given operation we can describe how the choice of polymer composition and coagulation conditions affects the structural development.

12.5.3.1 Orientational Drawing (Wet Stretching)

The filaments emerging from either the dry-spinning tower or wet-spinning coagulation bath are nonoriented and have little strength. Development of the anisotropic mechanical properties required of textile fibers is accomplished largely in the orientational drawing operation. The first roll fibers are usually washed free of solvent and then stretched in water that is held close to the boil. The stretching causes the polymer chains to become oriented along the fiber axis, resulting in high modulus and breaking strength. Recall from the previous section that the structure of a coagulated wet-spun fiber consists of small fibrils that are separated by void spaces called microvoids. The fibrils are randomly oriented with respect to the fiber axis. We know this to be the case because the transmission electron micrographs (TEA) of cross-sectional and longitudinal views are essentially identical [262]. In a typical wet-stretching operation, the porous structure is thoroughly washed to remove solvent and is then stretched in boiling water to a draw ratio of approximately 6×. The fiber is highly plasticized and is easily drawn since the glass transition temperature of the wet fiber, which is nearly 75°C, has been exceeded.

The major structural result of the drawing operation is the alignment of the fibrils along the fiber axis direction to create an oriented fibrillar structure. A TEM photograph of a 6× wet-stretched fiber is shown in Figure 12.29. This longitudinal view reveals highly oriented fibrils. The microvoids here are clearly the interfibrillar spaces. Craig et al. [262] utilized the previously described techniques of mercury density, internal surface area measurement, and toluene density to follow the changes in the fibril radius and the number of macrovoids that are formed upon stretching. They also used a parameter defined as the area ratio, which is the ratio of the cross-sectional areas of the fiber before and after the wet stretching (but before drying and collapsing). The result is a densification of the structure and a decrease in the surface area that can be interpreted as a doubling of the fibril radius R_p or void radius R_s, depending upon the model that is used. (Recall that R_p and R_s along with N_s, the number of voids, are calculated from the mercury density and the total surface area from Equation 12.25 through Equation 12.28.)

Also a slight reduction in the area ratio is observed. Apparently the boiling water causes some contraction and perhaps even fusion of the fibrils. As the draw ratio is increased to 5×, the structure becomes finer with the fibril diameter decreasing and the number of microvoids increasing continuously.

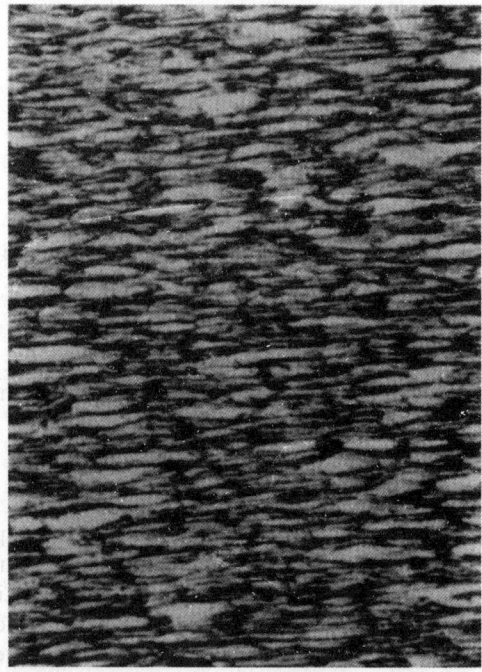

FIGURE 12.29 Transmission electron micrograph of an uncollapsed fiber after a 6× orientation stretch. This longitudinal section reveals the orientation of the fibrils along the fiber axis. Magnification, 20,000×. (From Holmes, D.F., *Am. Assoc. Tex. Technol. 353*, 431, 1952.)

The changes on the molecular level associated with the drawing operation have been studied as well. McPeters and Paul [243] used sonic modulus to follow the total molecular orientation of the fibrils before and after the drying and collapse, and in both cases a continuous increase in orientation was observed as a result of the drawing operation. The use of sonic modulus to measure the polymer chain orientation has been described by Mosely [271] and Desper [272]. The sonic modulus is proportional to the velocity at which sound waves are propagated along the fiber. This velocity becomes higher as the degree of chain alignment increases. The sound pulses propagate more efficiently along the chain axis than in the transverse direction. The sonic modulus data are shown in Figure 12.30. In both the roll-dried (collapsed) and air-dried (uncollapsed) fibers, the orientation begins to reach a limiting value near a draw ratio of 5×.

One can also use wide-angle x-ray diffraction to follow the orientational process during stretching. Bell and Dumbleton [273] showed that the crystalline regions become aligned along the fiber axis upon cascade stretching, albeit no significant increase in the crystallinity, as measured by x-rays or density, is observed. This distinguishes acrylics from other synthetic textile fibers where an increase in the degree of crystallinity is usually observed upon orientation. The argument has been made that the predominant orientation mechanism during drawing is the alignment of the fibrils by simple rotation about the nodal regions and that below a draw ratio of about 4× very little actual molecular deformation takes place. A geometrical model, the gel network mechanics model [274,275], which is based on the concept of an assembly of rigid rods (the fibrils) attached to flexible junctions (the nodule

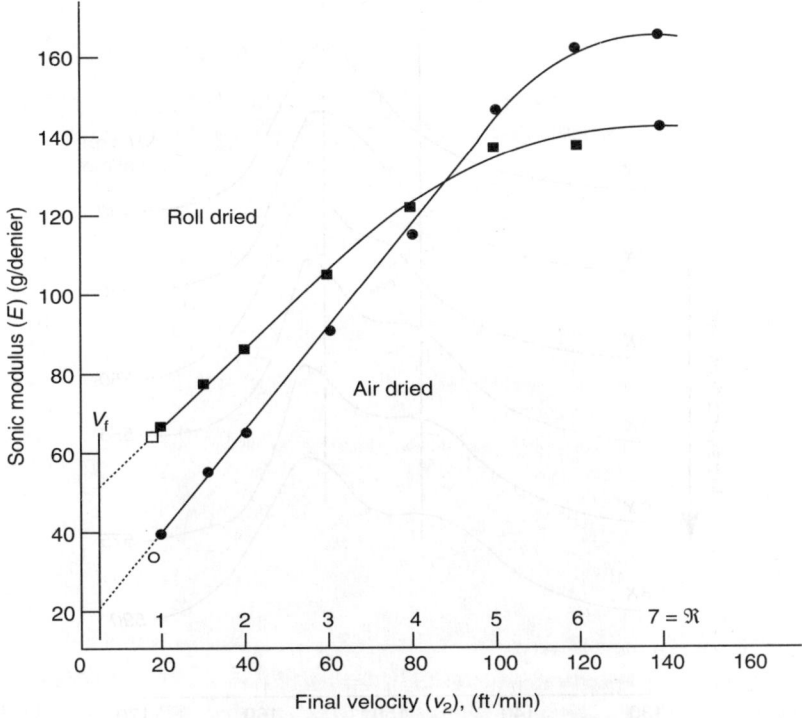

FIGURE 12.30 Comparison of sonic modulus for roll-dried (collapsed) and air-dried (uncollapsed) fibers at different stretch ratios. Open points are for fibers that were relaxed by 10% in the draw bath. Line marked V_f corresponds to observed free velocity for these extrusion conditions. (From McPeters, A.L. and Paul, D.R., *Appl. Polym. Symp.* 25, 159, 1974.)

regions), was proposed and does a good job of accounting for the functional dependence of the crystalline orientation on the draw ratio.

We have seen previously in our discussion of the x-ray diffraction analysis of the acrylic polymer morphology (see Section 12.4) that this technique is sensitive to the amount of lateral bonding present, but it is not particularly sensitive to the strength and regularity of the dipolar bonding networks that stabilize the structure. A technique that is sensitive to the dipolar bonding is the DSC approach described in Section 12.4 whereby water is added to the polymer or fiber to depress the melting point. Application of this technique to acrylic fibers shows that the heat of fusion increases continuously as a function of draw ratio in boiling water, and furthermore there are changes in the shape of the melting endotherm that indicate significant molecular deformation at draw ratios as low as 2× [167]. These DSC endotherms are shown in Figure 12.31 along with the dependence of the heat of fusion (endotherm area) on the draw ratio. The increase in heat of fusion indicates that as the chains become extended and aligned along the fiber axis they are able to participate more fully in the intermolecular bonding process that creates networks of dipolar bonds. These networks of intermolecular bonds should contribute to the increase in modulus and tenacity, and perhaps more importantly, improve the lateral strength of the fiber.

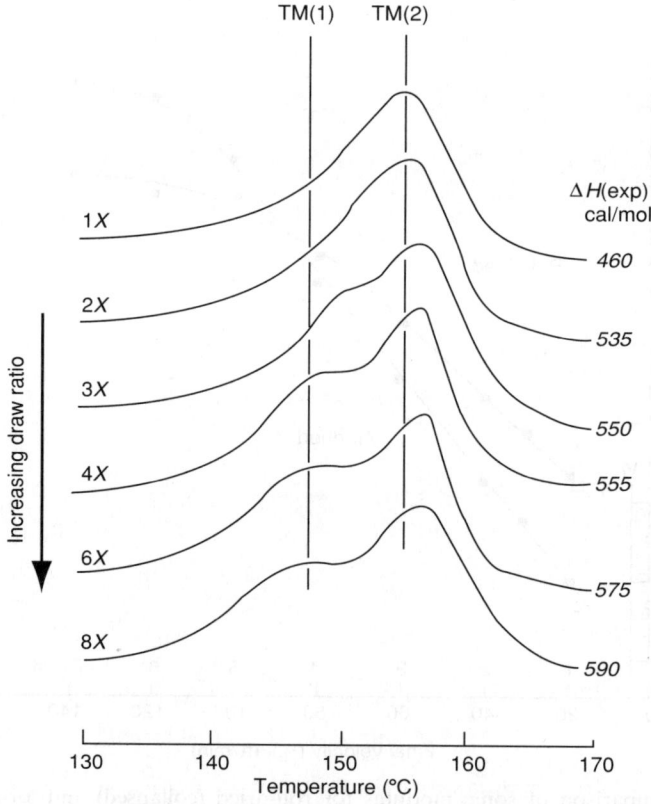

FIGURE 12.31 Differential scanning calorimeter endotherms of oriented and collapsed fiber at different levels of orientation stretch. Water was added to the fiber to depress the melting point and prevent thermal degradation. (From Frushour, B.G., *Polym. Bull. 4*, 305, 1981.)

The dimensions of the fibrils and microvoids seen in the uncollapsed drawn fiber can be traced back to the structure of the freshly coagulated fiber. Knudsen [268] has shown that the reduction of the spin-bath temperature gives a denser and finer structure in both the first roll and drawn uncollapsed fiber, and furthermore, the tensile properties are improved. In Figure 12.32, the breaking tenacity and elongation are shown as a function of the draw ratio for different spin-bath temperatures. The lower spin-bath temperature produces a stronger and stiffer fiber. The lateral properties of the finished fiber are also improved by having the denser structure. The number of cycles-to-fail in a multifilament flex-abrasion test increases by a factor of 2–3× when the spin-bath temperature is lowered from 55 to 0°C. This is attributed to an increase in the number of lateral interconnections between fibrils in the drawn fiber as the bath temperature is reduced.

The presence of the ionic comonomer influences the drawing behavior [276]. The copolymers containing nonionic comonomers such as AN–methyl acrylate display a moderate decrease in the mercury density upon drawing, but a proportionately larger increase in the internal surface area gives rise to a reduction in the fibril diameter and microvoid size. The

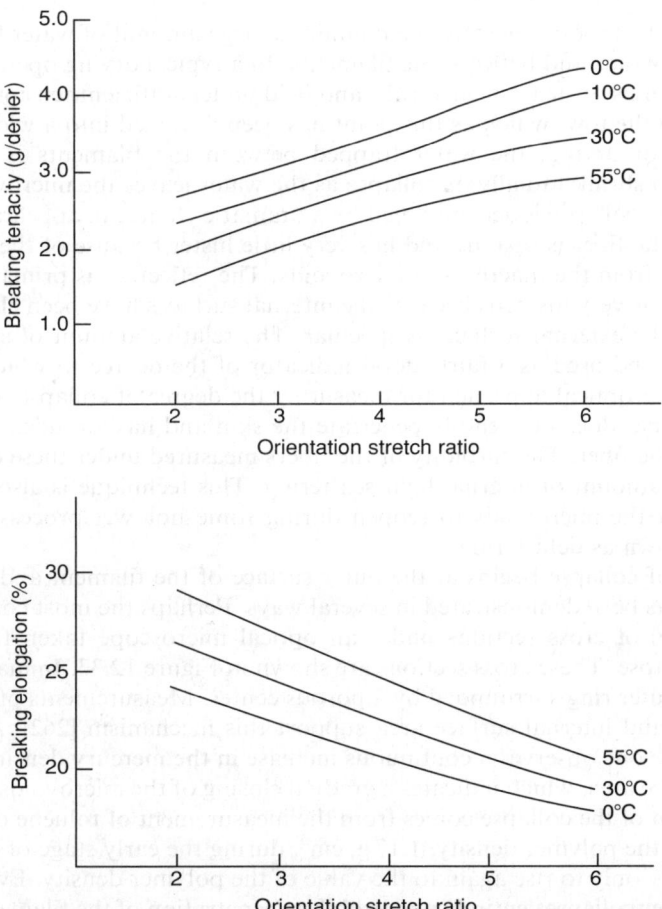

FIGURE 12.32 Dependence of fiber-breaking tenacity (top) and elongation (bottom) on orientation stretch ratio for different spin-bath temperatures. (From Knudsen, J.P., *Text. Res. J.* **33**, 13, 1963.)

presence of the ionic comonomer radically affects the structure of the drawn fiber. While the mercury density increases, the internal surface area usually decreases, implying a partial collapse and fusion of the fine fibrils into larger ones.

There is very little published work, describing the influence of the orientational drawing on the morphology of the dry-spun fiber. Some experiments by Craig et al. [262] indicate that the level of porosity in the drawn fiber is highly dependent on the residual solvent content of the fiber as it emerges from the dry-spinning tower. Recall from Table 12.16 that microvoids are absent when the residual solvent content is 1%, and even at a 20% solvent level the fibers have a surface area approaching the geometrical area. The low residual solvent fiber gives rise to an extremely dense structure after drawing with a mercury density approaching the polymer density. However, upon drawing the fiber that had a 10% residual solvent before washing, fibrillar structure can be clearly seen in TEM photographs that is not unlike that seen in the wet-spun fibers. The simplest consistent explanation for the appearance of this structure is to postulate the generation of an unoriented fibrillar structure in the dry-spun fabric in the tower. This structure is then simultaneously opened and oriented in the drawing step.

12.5.3.2 Drying and Collapsing

After stretching, the fabric is porous and contains a large amount of water located within the macro- and microvoids and between the filaments. In a typical drying operation, the wet tow is passed over a series of steam-heated rolls and held under sufficient tension to provide good heat transfer into the tow, which as this point has been flattened into a wide ribbon. During the initial stage of drying, the water trapped between the filaments is evaporated. The filaments begin to shrink radially or collapse as the water leaves the microvoids and diffuses to the surface. The collapse is accompanied by a dramatic change in appearance of the fibers. Before collapse, the fiber is opaque and has very little luster because of the multiple internal scattering arising from the macro- and microvoids. The reflection is primarily diffuse. After collapse, the fiber is very lustrous because the internal surfaces have been eliminated, and the reflection off of the external surfaces is specular. The relative amount of specular reflection can be measured and used as a fairly good indicator of the degree to which the fibers have collapsed. Another optical approach for measuring the degree of collapse is to submerge the fibers in a fluid that does not readily penetrate the skin and has an index of refraction that matches that of the fiber. The turbidity of the fibers measured under these conditions will be indicative of the amount of internal light scattering. This technique is also used to measure the propensity of the microvoids to reopen during some hot–wet processing treatments, a phenomenon known as delustering.

The process of collapse begins at the outer surface of the filament and proceeds inward [262,277]. This has been demonstrated in several ways. Perhaps the most convincing evidence is the appearance of cross sections under an optical microscope taken from samples just beginning to collapse. These cross sections are shown in Figure 12.33. In many filaments, one can see a dense outer ring surrounded by a porous center. Measurements of mercury density, toluene density, and internal surface area support this mechanism [262]. As the process of collapse proceeds, one observes a continuous increase in the mercury density and decrease in the internal surface area, which indicates a gradual closing of the microvoids. Evidence for the radial progression of the collapse comes from the measurement of toluene density. The latter remains equal to the polymer density, $1.17 \, g/cm^3$, during the early stage of collapse and then decreases abruptly only to rise again to the value of the polymer density. Evidently, the outer portion of the fiber collapses entirely and prohibits penetration of the filament by the toluene. If the interior is porous, then the result is a drop in the density at this point, but as the collapse

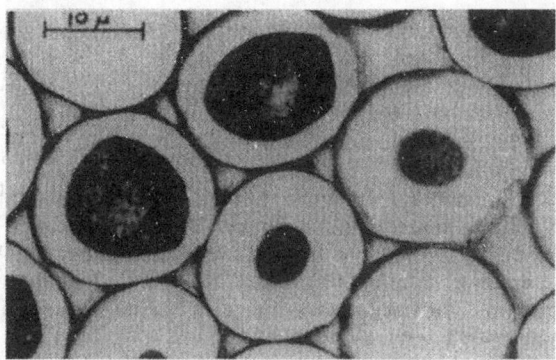

FIGURE 12.33 Cross sections in light microscope showing varying stages of collapse. (From Craig, J.P., Knudsen, J.P. and Holland, V.F., *Text. Res. J. 32*, 435, 1962.)

front moves into the porous interior, the density begins to rise and eventually approaches the polymer density. McPeters and Paul [243] utilized sonic modulus measurements to show that a significant amount of molecular disorientation occurs during the collapse. This was attributed to relaxation of the polymer chains in the amorphous regions of the fiber since the crystalline orientation is not markedly affected by the drying and collapse. The polymer composition has a strong influence on the manner of collapse following the previously described relationship between the comonomer type and the microvoid size [277,278]. Fibers having a relatively fine microfibrillar structure, such as those spun from polymers containing 1% of an ionic comonomer, will collapse fully at temperatures as low as 100°C, whereas temperatures close to 170°C are required for fibers that have a coarser structure. The use of hydrophobic comonomers such as methyl acrylate and the absence of the ionic comonomer produce the coarser structure.

Most aspects of collapse can be explained by considering the action of capillary attraction forces on a highly plasticized microfibrillar structure [277,278]. Before collapse, most of the water absorbed in the filaments is located between the parallel fibrils. The fibrils are highly plasticized by the water; therefore, their resistance to bending will be minimal. As the water begins to leave the interfibrillar spaces, menisci will be formed. These will increase the free energy of the system because water–air interfaces are created at the expense of the more favorable water–polymer interfaces. However, the network of fibrils can respond through bending and lateral translation to reduce the volume of the interfibrillar spaces and eliminate the menisci. This proceeds continuously during the drying until the interfibrillar spaces are eliminated. Once this is completed, the only water that remains is the dissolved water that is plasticizing the fiber, and this leaves during the last stage of the drying process. The temperature of the fiber begins to rise at this point and approaches the temperature of the drying roll. Further evidence in support of this mechanism comes from the observation that a dry-porous fiber will not collapse at any temperature, but upon wetting the fiber can be made to collapse readily in the normal manner. The fiber also will not collapse very readily if the water is replaced with a liquid that does not plasticize the fiber and does not have a high surface tension [279]. The radially directed collapse force arising from the menisci effect is inversely proportional to the effective radius of the pore, or microvoid formed by two adjacent fibrils [277,278]. The magnitude of the force expressed as a pressure can be calculated for various microvoid sizes, and these calculations show that the reduction in pore size and concomitant increase in collapse pressure arising from the addition of ionic comonomers to the polymer composition account for the relative ease with which these fibers collapse.

12.5.3.3 Relaxing, Crimping, and Bulk Development

The processes of stretching and drying leave the fiber with an inadequate balance between tensile and lateral properties. The breaking strength is more than sufficient for most textile applications, but breaking elongation is low and the lateral properties are poor. The fiber tends to fibrillate readily under abrasion. Also, the fibers are difficult to dye at this point. The fibers must be given some type of relaxation or annealing treatment that will make them more extensible. Usually this annealing involves some type of heat treatment in the presence of moisture that causes the fiber to shrink. The ensuing molecular disorientation lowers the modulus and tenacity, but this is more than offset by an increase in the breaking elongation and resistance to fibrillation. The serious commercial consequences of fibrillation problems have been described [280].

The relaxation step can be accomplished using either the wet or dry fiber. If the fiber has been stretched and then collapsed and dried under tension, the relaxation step must involve the reintroduction of water into the fiber for plasticization. The relaxation may be carried out online or in batches using an autoclave [228]. Relaxation of the dry fiber by heat alone is unacceptable since the high temperatures that are required cause discoloration. It is sometimes possible to combine the steps of drying, collapse, and relaxation into a single operation where the wet fiber is piddled onto a belt and dried unconstrained. The method of relaxation chosen by a manufacturer will depend upon the type of spinning process, e.g., dry spinning versus wet spinning, the specific process used for coagulation, and the polymer composition. The manufacturing details are proprietary and little information is available in either the patent or open literature. The few published studies on relaxation do indicate that the major structural response is the relaxation of the amorphous chains while the crystalline orientation remains largely unchanged [282,283]. This conclusion is based on the continuous decrease of the sonic modulus and birefringence with shrinkage while the wide-angle diffraction pattern shows that the crystalline orientation remains constant. This result would seem to lend strong support to the concept of a two-phase structure for acrylic fibers (see Section 12.4). Structural reorganization during wet and dry relaxation treatments have also been followed by Sotton et al. [284] using a silver sulfide staining technique that can reveal cracks and fissures within the filament. The fibers are first treated with hydrogen sulfide and then soaked in aqueous silver nitrate solution, which penetrates into the cracks and fissures and reacts with the out-diffusing hydrogen sulfide to form silver sulfide deposits that are opaque and can be clearly seen under an optical microscope. A characteristic staining pattern is observed for each specific relaxation condition. Autoclaving the fibers renders the interior of the fiber virtually crack-free when the temperature approaches 140–150°C. This is not observed for the dry, thermally treated fibers. It is proposed that the water plasticizes the molecules to such a degree that internal surfaces resulting from cracks, fissures, and even adjacent fibril–fibril contacts are fused together. This mechanism may be one means by which the relaxation process improves the abrasion resistance and reduces fibrillation.

Crimp is put into the fiber to impart bulk and cover to the spun yarn. The crimp also provides the necessary interfilament cohesion for conversion of the staple into the yarn. Stability of the crimp during the yarn conversion and during dyeing is very important insofar as the fiber crimp in the final fabric is often dominated by this characteristic. Moreover, control of crimp permanency is sometimes desirable because the crimp level required during processing may differ from that required in the final product. Usually, the most drastic loss in crimp occurs at the dyeing stage in processing when the load on the filaments can simply pull out the crimp. Acrylics suffer from a very severe loss in modulus and strength when placed in boiling water so that the forces arising from the turbulence of the water may be sufficient to reduce the crimp level. For a given acrylic fiber, the stability of crimp has been found to increase with an increase in the following: (1) time–temperature history during crimping and crimp setting, (2) extent of crimp (angle of crimp curvature), (3) reciprocal of the radius of curvature, and (4) the degree of plasticization by water or steam during crimping [285].

If a high degree of bulk is desired in the spun yarn for applications such as sweaters and craft-knitting yarns, then the manufacturer must turn to bicomponent fibers. These fibers are composed of two polymers in separate areas of the cross section. The polymers are coextruded through the spinneret and therefore each is continuous along the fiber length and permanently joined at the interface. If the two components have different shrinkage potential, then the fiber will develop crimp when given thermal treatments such as dry air or steam annealing. This crimp will be permanent. The two components will usually have slightly different contents of a nonionic comonomer such as methyl acrylate or vinyl acetate. The amount of shrinkage at a given temperature for a monocomponent fiber will increase as the level of comonomer is increased. Therefore, in the case of the bicomponent fibers, the differential shrinkage causes the fiber to crimp.

Fitzgerald and Knudsen [286] have described the mechanics of crimp development in bicomponent yarns. The extent of crimp development will depend on (1) the shrinkage differential between the components, (2) the distribution of components in the fiber, and (3) the translational restraints that may inhibit crimp development. Obviously, the maximum crimp would be developed when the fibers were comprised of equal parts of each component and these components were separated and located on opposite sides of the fiber. On the other hand, if the cross section revealed a distribution in which one polymer was separated into two regions, each on one side of the center slice, then any differential shrinkage forces would be neutralized and no crimp would be developed. In practice, many different distributions of the two components may be observed depending upon the technique used to mix the two polymer dope streams and deliver them to the spinneret hole [287–290].

A nonpermanent, water-reversible crimp may be produced if different levels of ionic comonomers are placed in the two compositions. The result is a differential swelling during hot–wet treatments such as dyeing. Crimp develops when the fiber is subsequently dried and the differential swelling reverses and becomes a differential shrinkage. This approach circumvents the problem of crimp loss during dyeing because the final crimp is developed after dyeing and during drying. Each approach to developing the bicomponent crimp has certain advantages in processing, aesthetics, and product care. The properties of each can be modified by changes in the spinning process and polymer compositions [291–293].

12.5.4 Special Topics

The material discussed up to this point has been intended to give the reader an understanding of how the majority of acrylic fibers are manufactured. In this section, we discuss some of the less conventional spinning processes that could be used to produce fibers for special applications. We also include a section on optimization of spinnability in which some of the more fundamental aspects of the rheological behavior of the fiber in the spin bath are discussed. Completing this section is a discussion of the dyeing of acrylic fibers.

12.5.4.1 Special Wet-Spinning Processes

Several modified processes have been described that are based on coagulation conditions that optimize the microvoid structure. The use of nonpenetrating fluids in the coagulation bath will lead to a dense fiber with very fine fibrils and superior tensile properties in comparison with conventional fibers. Coagulation baths consisting of polyalkylene glycol and solvent have been described that produce this desired effect [294,295]. The high-molecular-weight glycols diffuse into the fiber very slowly and a fine microvoid structure is formed as the solvent diffuses out. In essence, one is attempting to create a structure in wet spinning in which the diffusional and gelation characteristics are more likely to those in dry spinning. The use of hexanetriol [296], low-molecular-weight alcohols [297], and cumene–paraffin mixtures [298] has also been reported.

Another approach to optimize the microvoid structure is gel-state spinning, which is a hybrid between the dry- and wet-spinning processes. A hot-concentrated polymer solution is extruded downward through air and the filaments are then allowed to gel by cooling before they pass into an aqueous bath for solvent extraction [299]. Homogeneous, macrovoid-free fibers are claimed. Apparently, the establishment of a true gel state before phase separation also produces a finer microvoid structure. This approach is different from the dry-jet wet-spinning process [300,301] in which the spinneret is positioned a few centimeters above the coagulation bath surface. Much higher speeds may be achieved because the threadline stress is not transmitted back into the spinneret. (This threadline stress will cause the filaments to break at the surface of the spinneret in conventional spinning if the speed at which the filaments are taken out of the bath exceeds an upper limit. This phenomenon will be discussed later under the topic of spinnability.) The high-spinning speeds achieved with the above processes could be particularly useful for making continuous filament products [302,303].

In the last 10 years, gel spinning has been associated with the spinning process used to make ultrahigh orientation polyethylene fibers that achieve very high tenacities in the range of 30–45 g/denier. These fibers are spun from gelled solutions of very high molecular weight in a manner that creates parallel extended chains that are defect-free [304]. Polyethylene is normally melt-processed, but melt spinning does not yield the proper defect-free structure required for the high molecular weight. Based on the success with polyethylene, gel spinning has been applied to two polymers that are normally solution spun, polyvinyl alcohol and PAN. Representative information from patents shows that a tenacity around 20 g/denier has been achieved by dissolving PAN having a molecular weight more than 1,000,000 g/mol in $NaSCN–H_2O$ to form a 5–10% solution, coagulating in a lower temperature bath to form a gel-like fiber, and drawing first in hot water or glycerol and then in the dry state [305]. Commercially, tenacities around 13–15 g/denier have been reported.

12.5.4.2 Melt Extrusion

It was stated at the outset of this section that the polymers used for acrylic and modacrylic fibers cannot be melt-spun because they degrade when heated near their melting point. Melt spinning would offer great economic advantages because the costly step of solvent recovery could be eliminated. One approach toward this objective has been high solids spinning in which the polymer melting point is reduced by adding 10–60% weight of certain plasticizers [306–309] or a solvent [310] and the resulting high-velocity melts can be extruded as in melt spinning. Part of the economic advantage is lost if the plasticizer or solvent must be extracted and recovered. The continuous plasticized melt extrusion of PAN was reported by Atureliya and Bashir [309] using PC as the plasticizer. The molten extrudate solidified on-line as a result of cooling, without the need for coagulation. In this regard, PAN–PC solutions behaved differently when compared with other solutions commonly used in the wet spinning of PAN fibers. Spontaneous solidification upon cooling means that the filament could be wound up on a take-up system without the need to pass it through a coagulation bath. The solidification occurred as a result of the rapid crystallization of the PAN from the PC solution. The x-ray diffraction pattern of the solidified filament (which contained 40–50% solvent) was different from that of the dry PAN powder, indicating that a different polymorph had been formed. However, upon drawing the x-ray pattern reverted to the normal hexagonal polymorph.

A PAN–PC (50:50 by weight) melt had a shear viscosity comparable with that of a conventional thermoplastic such as extrusion-grade polyethylene and showed thin-thinning behavior. The work demonstrates the feasibility of such a process, albeit a system of solvent recovery would be required for large-scale spinning processes.

We saw in Section 12.4 that the melting point of acrylic polymers can also be depressed by the addition of water, though this must be done under pressure. This is the basis for a number of patents describing water–PAN melt-spinning processes. When the polymer–water mixture contains sufficient water to hydrate a certain percentage of nitrile groups, the melt can be extruded to produce textile-denier fibers [311–315]. The key is to use sufficient amount of water to depress the melting point to a temperature where thermal degradation mechanisms are not dominating, but still maintain a single-phase system. Recall that the melting point of PAN will decrease continuously upon addition of water until it reaches a limiting value of 185°C, and then a second phase forms and the melting point will not go further lower. If the temperature is maintained above 185°C, the system will be single-phased. The limiting value for acrylic copolymers will be lower than the value for PAN, since the initial melting point is lower. The acceptable levels of water are approximately between 15 and 25%.

This spinning process has been recently described by von Falkai [316], and a phase diagram for the PAN–water system is given showing the effects of temperature and pressure. A pressure of 30–70 bar is required for extrusion of the hydrate melt. Pressure-resistant spinnerets with 60-mm apertures and an L/D ratio of two should be used for spinning. Since the PAN hydrate melt is stable only under pressure, in the spinning process it becomes imperative to release the bound water without destroying the filament. The spinning tube should therefore be under pressure to provide the atmosphere to evaporate the water. Fibers produced by this have a tenacity of about 4.5 g/denier and a breaking elongation of 40%. The cross section is round with a core–sheath structure and the surface has a corrugated appearance. Fabrics produced from these fibers have a somewhat harsher and crisper hand [316].

Bashir [317] studied the interaction of water and PAN under conditions similar to those created by the melt-spinning conditions described above. PAN powder that had been blended with water could be compression molded at 210°C like a thermoplastic material. The water-plasticized films were flexible and could be uniaxially drawn. A comparison of the x-ray diffraction patterns before and after addition of the water showed the formation of a new polymorph attributed to hydrogen bonding of the water with the nitrile groups. On drawing the plasticized film, the water was mostly expelled from the film and the x-ray patterns reverted to the original hexagonal structure.

Min et al. [318–320] studied the formation of these melt hydrates using mixtures of solvent and water in anticipation of lowering the melting point. Combined use of water and solvent (ethylene carbonate) exhibited a synergistic effect on the plasticization of PAN. However, water was found to be more effective than solvent in reducing the melt viscosity [318]. The thermal stability of the PAN molecule in the hydrated melt was investigated by monitoring the melting and crystallization temperatures, in the presence of water, as a function of annealing time, and both were observed to increase slightly over a period of 1 h at annealing temperatures of 160–180°C. This was attributed to chain rigidity mainly due to intramolecular cyclization of nitrile groups to form the random imine structure rather than intermolecular cross-linking [320].

12.5.4.3 Optimization of Process Spinnability

Spinnability of a polymer is the ease with which the polymer can be converted to fiber without encountering filament breakage and other problems related to fiber formation. Filament breakage usually occurs near the face of the spinneret. The breakage is often encountered when attempts are made to modify an existing process for a number of reasons, including increasing the linear speed to improve productivity and modifying the polymer composition or the coagulation conditions to alter the fiber properties. Often the problem will occur during scale-up from the laboratory-spinning machine to the plant. Even though every attempt may be made to simulate the plant process in the laboratory, the coagulation conditions can never be completely equivalent because the plant spinnerets may have up to 60,000 holes compared with 500–1000 for laboratory-spinning lines. Therefore, gradients in spin-bath composition in the plant-spinning baths will be encountered that are more severe than in laboratory spinning.

This section describes some of the basic studies of the rheology of wet spinning that were undertaken to clarify the principles that govern spinnability. Han and Segal [321] have approached the problem of spinnability by endeavoring to characterize the rheological behavior of the coagulating threadline and relate these properties to the spinning conditions. Elongational viscosities were measured using a laboratory-spinning line similar to the one described earlier. The solvent and spin-bath nonsolvent were aqueous solutions of NaSCN. The tension of the fiber emerging from the spin bath was measured and decomposed into component forces. The measured value is the total take-up force, $F(tot)$, and equals the sum of several forces, the primary components of which are the rheological force, $F(rheo)$, and the drag force, $F(drag)$. $F(rheo)$ is the force required to deform the threadline when it resides in the spin bath and it is related to the axial stress component, σ_{xx}, and to the elongational viscosity, η_E through the following equation:

$$F(\text{rheo})/A(x) = \sigma_{xx}(x) = \eta_E dV(x)/d(x) \qquad (12.29)$$

where $dV(x)/d(x)$ is the axial velocity gradient and $A(x)$ is the cross-sectional area of the threadline. It is seen that $F(rheo)$ exists only in the region where the axial velocity gradient is nonzero. Furthermore, $F(rheo)$ is considered a constant through the length of the fiber, where the latter condition holds and $F(drag)$ is considered to be dependent upon the distance x from the spinneret face. Other external forces such as the accelerating forces are lumped together with the drag forces in the following force balance equation:

$$F(\text{tot})(x) = F(\text{rheo}) + F(\text{coag})(x) \qquad (12.30)$$

where $F(coag)$ is the sum of all the external force components.

At a fixed distance from the spinneret, $F(tot)$ is observed to increase with the jet stretch. (Recall that the jet stretch is the ratio of the speed at which the fiber leaves the spin bath to the theoretical speed of the dope within the spinneret capillary.) Furthermore, at constant jet stretch $F(tot)$ will also increase as the distance from the spinneret is increased. Extrapolation to $x=0$ gives $F(rheo)$. In Figure 12.34, $F(rheo)$ is plotted against the jet stretch for different coagulation bath conditions. It is seen that $F(rheo)$ increases as the temperature is decreased from 40 to 0°C, and also as the bath concentration is decreased from 20 to 10%. The change in area of the filament as a function of the distance from the spinneret is $A(x)$, and it can be measured by photographic determination of the filament diameter.

The attenuation of the filament diameter as a function of the distance from the spinneret is shown in Figure 12.35 for the combination of jet stretch and spinneret capillary diameter. The change in diameter with distance is greater for larger jet stretch and larger spinneret capillary diameter. From the diameter profile and the volumetric flow rate, the linear velocity of the filament $V(x)$ can be calculated and the derivative of the velocity becomes the rate of elongation. In Figure 12.36, the rate of elongation is shown to decrease very significantly as the distance from the spinneret face is increased. One can now evaluate a quasi-extensional viscosity at any fixed position x by simply taking the ratio of the tensile stress σ_x to the rate of

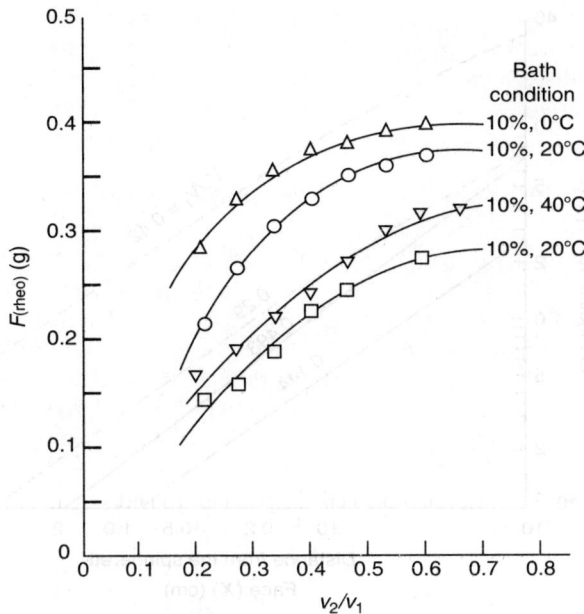

FIGURE 12.34 Rheological force versus jet stretch at different spin-bath conditions. Spinneret diameter was 0.0127 cm (5 mils) and the plug flow velocity within the spinneret was 16.62 cm/s. The concentration of sodium thiocyanate (NaSCN) in the bath and the bath temperature are given. (From Han, C.D. and Segal, L., *J. Appl. Polym. Sci. 14*, 2973, 1970.)

elongation as defined by Equation 12.29. Either an increase in bath concentration or a decrease in bath temperature will decrease the extensional viscosity. The effect of bath temperature at different jet-stretch ratios is illustrated in Figure 12.37. The slopes of these lines were interpreted by Han and Segal as an activation energy for the molecular processes

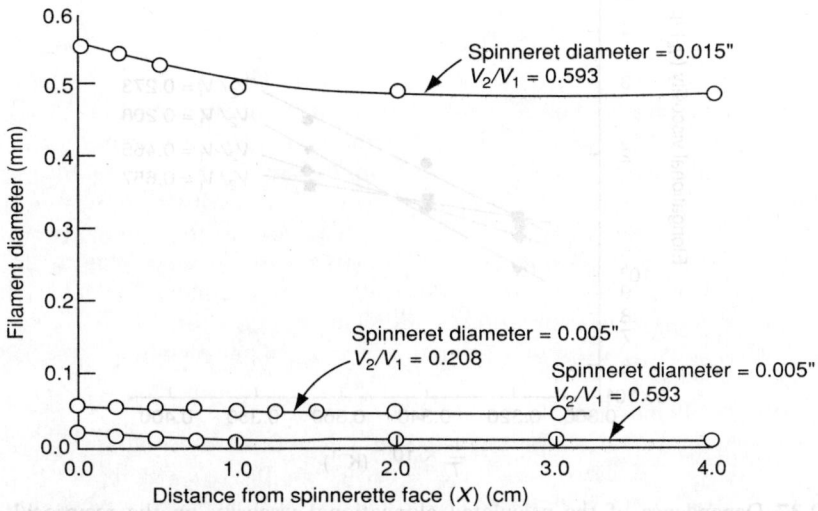

FIGURE 12.35 Filament diameter versus distance in centimeters from the spinneret face. Bath conditions: 10% aqueous NaSCN, bath temperature 20°C, shear rate, $10^4 \, s^{-1}$. (From Han, C.D. and Segal, L., *J. Appl. Polym. Sci. 14*, 2973, 1970.)

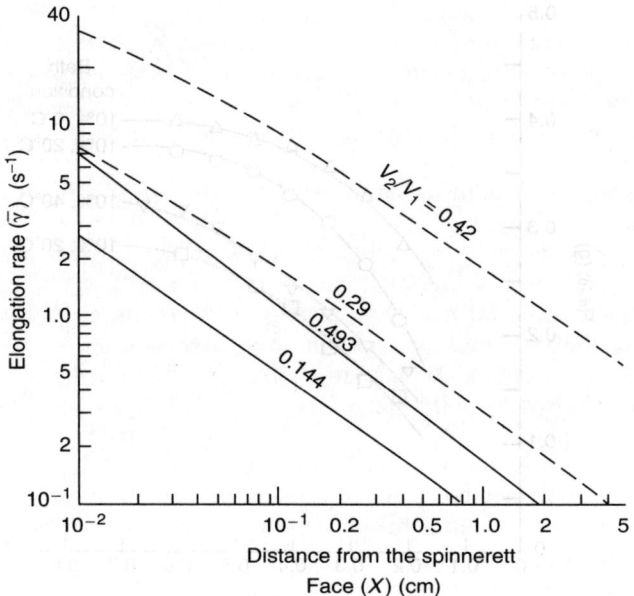

FIGURE 12.36 Rate of elongation versus distance in centimeters from the spinneret face. Bath conditions are same as in Figure 12.35. Shear rate was $10^4 \, s^{-1}$. Solid lines for 0.0127 cm (5 mil) and dashed lines for 0.0381. (From Han, C.D. and Segal, L., *J. Appl. Polym. Sci. 14*, 2973, 1970.)

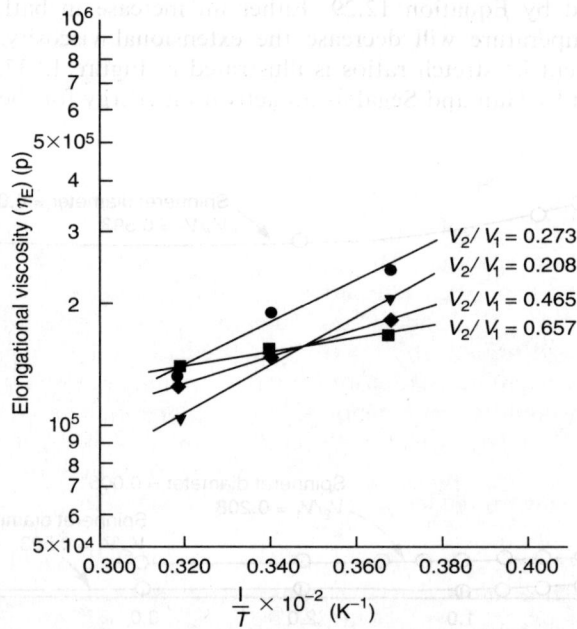

FIGURE 12.37 Dependence of the calculated elongational viscosity on the reciprocal of the bath temperature. Data shown for four jet stretch ratios: $V_2/V_1 = 0.273$, 0.208, 0.465, and 0.657. Spin-bath concentration was 10% NaSCN. (From Han, C.D. and Segal, L., *J. Appl. Polym. Sci. 14*, 2973, 1970.)

that accompany filament attenuation. A maximum is observed when the extensional viscosity is plotted against the rate of elongation. This is consistent with previous observations made on polystyrene rods [322]. The magnitude of the extensional viscosity compared well with values predicted from theoretical models. Furthermore, the results indicate that the influence of coagulation variables on the extensional viscosity is far more important than the rate of elongation.

For any set of spinning conditions in which the volumetric flow rate is held constant, there exists a well-defined maximum in the velocity at which the fiber can be removed from the bath. When this maximum, $V_{1,m}$ is exceeded, the filaments will begin to break [323]. The filament breakage is postulated to occur close to the spinneret face where a skin is beginning to form as the spin-bath liquid begins to penetrate the filament. A filament cross section would show an uncoagulated fluid core surrounded by a coagulated solid sheath, or skin, with the core diameter decreasing as the plane of the cross section moves further away from the spinneret. If the skin were to rupture due to excessive tensile stress, the filament would break since the fluid core would be unable to support the load [244].

The maximum take-up velocity is very reproducible and depends essentially on every process variable, e.g., polymer composition and molecular weight, spinning dope composition, bath composition, bath temperature, spinneret capillary diameter, and flow rate. A parameter that seems to determine the upper limit in take-up velocity is the free velocity, V_f [244,323]. Recall that the free velocity is the velocity at which the filament leaves the spinneret under zero tension. The free velocity will increase in an approximately linear manner with the volumetric flow rate through the capillary, and at a fixed flow rate it will increase as the capillary diameter is decreased. When the take-up velocity, V_1, is increased while V_f is held constant, the magnitude of the rheological force, $F(rheo)$, will increase until the filament breaks at the spinneret face.

The magnitude of V_f is directly related to the swelling of the filament upon emerging from the spinneret. This is the well-known Barus effect [324], which arises in this case from the release of stored elastic energy originating from the deformation of the polymer dope upon entering the capillary plus that generated by shear within the capillary. The diameters and velocities within and immediately outside the capillary are related by the following equation:

$$D_f/D = (<V>/V_f)^{1/2} \qquad (12.31)$$

where D and D_f are the diameters within and immediately outside the capillary and $<V>$ and V_f are the corresponding velocities. Thus D_f/D should depend on the spinneret hole design, the shear rates, and the rheological properties of the spinning dope.

The dependence of $V_{1,m}$, V_f, and the spinning line tension on the volumetric flow rate is shown in Figure 12.38. Initially both velocities increase with the flow rate, but then a maximum appears in each curve, followed by a minimum. The maximum corresponds to the onset of a dope instability phenomenon known as dope fracture. The curve suggests that $V_{1,m}$ should correlate well with V_f, which in fact it does. Paul [244] was able to convert the free velocities into the swelling ratio, D_f/D, by using Equation 12.31 and to demonstrate an excellent correlation between the jet swell and the shear rate. This is shown in Figure 12.39. At the higher shear rate, the swelling ratio begins to increase rapidly and goes through a maximum that probably accounts for the corresponding maximum in V_f. In other words, the maximum in V_f and also in $V_{1,m}$ and the onset of dope fracture are all shear-dependent phenomena.

To maintain a stable-spinning process V_f must be kept below the critical value at which the instability is observed. As the coagulation bath conditions are changed, the corresponding influence on $V_{1,m}$ essentially mirrors the behavior of V_f. The maximum in V_f occurs at lower

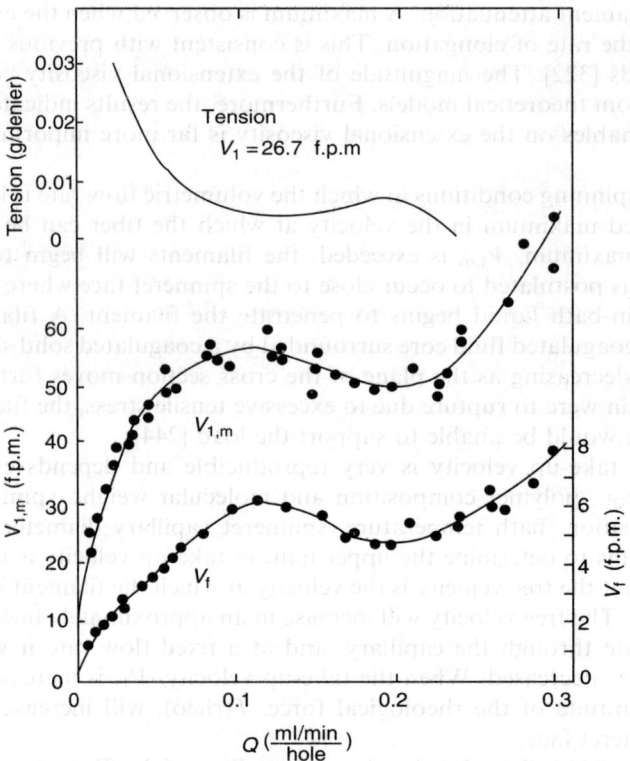

FIGURE 12.38 Dependence of maximum first roll take-up speed, $V_{1,m}$; and free velocity, V_f, on the flow rate per hole, Q. Spin bath was 55% dimethylacetamide and spinning dope concentration was 25% polymer. (From Paul, D.R., *J. Appl. Polym. Sci.* 12, 2273, 1968.)

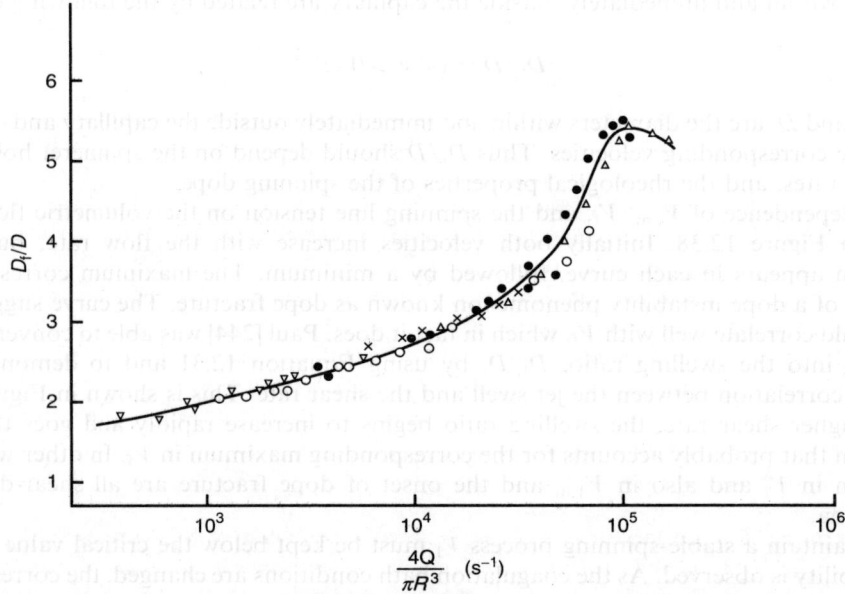

FIGURE 12.39 Shear rate dependence of jet swell. Spin-bath conditions are same as in Figure 12.38. Symbols indicate different spinneret sizes. See original reference for specific values. (From Paul, D.R., *J. Appl. Polym. Sci.* 12, 2273, 1968.)

flow rates as the bath temperature is decreased and $V_{1,m}$ decreases accordingly. Therefore, one way to increase the spinnability is to increase the bath temperature. The effect of coagulation rate on both V_f and $V_{1,m}$ is complex and deserves some comment. In general, a faster coagulation rate will increase V_f. The initial formation of a skin limits the amount of die swell and hence the filament issues at a faster rate than if the skin were not present.

This factor becomes more pronounced as the coagulation rate is increased. The effect of coagulation rate on $V_{1,m}$ is more complicated. At constant V_f, an increase in the coagulation rate will decrease $V_{1,m}$ because the thicker skin resulting from the faster coagulation rate will cause the fiber to be less extensible. The magnitude of F(rheo) will reach the critical value for filament breakage more quickly. Therefore, it would appear that there are two competing mechanisms as the coagulation rate is increased, and as a result, $V_{1,m}$ goes through a minimum when the coagulation rate is increased at constant V_f [244].

12.5.4.4 Modification of Cross Section

Having the ability to manipulate the cross-sectional shape of a fiber is advantageous because this offers a route to improved properties. For example, a trilobal cross section has long been used in nylon carpet yarns because it imparts a greater stiffness to the fiber in comparison to the round cross section of the same denier. In melt spinning, the cross-sectional shape generally does not severely deviate from the shape of the spinneret capillary. If one can quench the molten filament into the solid state by rapid cooling or crystallization, then the shape of the capillary is retained, although some distortion occurs due to the normal stresses that tend to swell the filament when it emerges from the capillary. The situation is more complicated in the case of wet spinning because the cross-sectional shape is strongly influenced by the diffusional characteristics of the particular spinning system. In dry spinning, for example, the cross section is dog-bone-shaped even though the capillary hole is round and, as we shall see, in wet spinning one can choose coagulation conditions that will give rise to filaments that retain the shape of the capillary hole surprisingly well.

Han and Park [325,326] have discussed the rheology of shaped fiber formation for wet- and melt-spinning fibers, and described the optimum-spinning conditions for each case. Experiments in spinning polypropylene through rectangular spinneret capillaries from the melt show that the deviation from the capillary dimensions occurs relatively close to the spinneret face. The magnitude of the die swell can be related to the distribution of the normal stresses. There is a more pronounced die swell on the long side than on the short side of the rectangular capillary hole that arises from the nonuniform distribution of normal wall stresses along the two different sides. It is possible to retain the sharp corners of the rectangular shape in the filament because, first, the magnitude of the normal stress is zero in the corners of the capillary and second, the magnitude of the surface tension is not sufficient to round off the corners. The result of the above factors is a filament rectangular in shape, but with the sides bowed outward.

The same rheological principles apply to wet spinning, but now the surface tension forces are no longer insignificant owing to the lower extrudate viscosity. Consequently, the effect of jet stretch and coagulation conditions must be considered. Han and Park [326] were able to spin acrylic fibers from noncircular capillaries using a spinning dope of 10% PAN in aqueous NaSCN. The shapes of the spinneret holes were rectangular, trilobal, and round with lugs, and fibers having corresponding shapes could be spun. The key to success is in obtaining the proper balance between the aspect ratio of the capillary and the jet stretch. The jet stretch tends to offset the die swell and therefore to minimize the degree of shape rearrangement in the filament before coagulation. However, the jet stretch can only be increased to a point at which the filaments begin to break, and if the desired shape is not achieved at that point, the only recourse is to choose a capillary with a larger aspect ratio.

12.5.4.5 Dyeing of Acrylic Fibers

We begin this part with a brief review of the commercial methods by which acrylic fibers, yarns, and fabrics are dyed. This is followed by a discussion of the dyeing mechanism, which includes the chemistry of dyeing, and a description of the dye diffusion process and its dependence on the fiber morphology. The entire subject has been reviewed by Peters [327].

Acrylic fibers can be dyed with either cationic (basic) dyes, anionic (acidic) dyes, or disperse dyes. The bright, fast colors that can be achieved with cationic dyes make this the method of choice. The fiber must have acid group dye sites that will accept the cationic dyestuff. The dye sites are often sulfonate and sulfate end groups arising from the persulfate–bisulfite-type redox initiator system or are intentionally added through copolymerization with vinyl monomers having acidic pendant groups (see Section 12.3.4). A commonly used monomer for this purpose is SSS. Disperse dyes can also be used to achieve a good range of colors, though not as bright as in the latter case. Much less commonly used are the anionic (acidic) dyes.

The most commonly used commercial dyeing processes for acrylics for stock, package, skein, and piece dyeing [327]. These operations are usually carried out at the boil, where the fibers must be handled very gently owing to their extremely low hot–wet modulus. The modulus of the fiber drops from 45 g/denier (4 N/tex) when dry at 200°C to about 0.6 g/denier (0.05 N/tex) at the boil [328]. The fiber or fabric can be easily distorted if loads arising from excessive water turbulence of handling are not minimized. In stock dyeing, the fiber is dyed in staple form before yarn spinning. In package dyeing, the yarn is wound onto cones and the dye liquor is forced around the outside of the package and into the center of the cone. Complete penetration of the package is necessary to achieve uniform dye uptake. Much of the yarn intended for knit goods is dyed by skein dyeing. Skeins of yarn are hung onto rods and placed into the dye vat where the dye liquor is circulated around the skeins. It is essential that the skeins be allowed to hang freely during the dyeing so that the proper bulk level can be developed. Many of the craft yarns used in hand knitting are dyed this way. This includes much of the high-bulk bicomponent yarn. In piece dyeing, an entire piece of goods such as a sweater blank is dyed. The fabric is placed into a bag that is put into the dye vat and agitated with paddles. It is necessary to cool the dye bath slowly after dyeing until the fiber is well below the wet glass transition temperature, which is approximately 75°C, to achieve uniform dyeing of the fabric. Lowering the temperature too rapidly will produce marks where the fabric is folded in the bag.

Producer dyeing of acrylics, or gel dyeing, is a relatively new process in which the dye is incorporated during the spinning process. The final product is colored and difficulties of handling the fiber under hot–wet dyeing conditions are largely circumvented. However, the new process does require very good process control and good inventory planning. The manufacturers who practice this technology are indicated in Table 12.14. This process takes advantage of the fact that during the wet-spinning process the acrylic fiber is very porous up to the drying and collapsing stage (recall previous discussion on the structure of the coagulated fiber in this section). Producer dyeing is not applicable to dry spinning since the required porous structure is not available. Before drying, wet-spun fibers have a porous structure, which may be 50% or more void volume. This sponge-like structure is ideal for dyeing. Penetration of dyestuff is rapid and complete—perhaps as little as 1 s is required [329]. This means that the process can be operated without a reduction of the normal wet-processing speeds using a dye applicator or dip bath [330].

Dye can be applied to a wet-spun fiber at almost any point during the processing: in the coagulation bath, during washing and stretching, and after stretching [331]. The dyeing rate is related to the surface area of the coagulated gel, and it changes through the processing steps. Until the fiber is dry, however, the rate is sufficient to support a short-exposure-time process. The preferred option for dye application is in the washing–stretching operation. The dye applicator is positioned so there is sufficient washing after the dye application to remove unfixed dye and other impurities such as solubilizing agents introduced with the dye. Uptake of dye is even more rapid if it is applied before the fiber is oriented. Mullinax [332] revealed Monsanto's system in block form, shown in Figure 12.40.

The preferred dyestuffs for producer application are the liquid dyes—concentrated aqueous solutions of cationic dyestuffs. These can be fed to the applicator either as individual streams of one to four dyes or as a mixed solution of the desired components. In either case, the dye concentration may be adjusted from that of the received dye by addition of water. Color control is achieved in the first case by adjusting the individual rates, and in the second by adjustment of the ratio of the components in the mixed solution to adjust hue plus changes in feed rate to adjust chroma. Many fiber producers are able to guarantee color matching from one production to the next, even on the most critical applications. Producer-dyed colors exist for one reason only—lower systems cost to the mill compared to alternative dyeing methods. Because they require a colored fiber and yarn inventory, the products are most economical on large-volume colors, and mills often use traditional dyeing methods to produce specialty colors. Even so, the number of individual colored products (considering color,

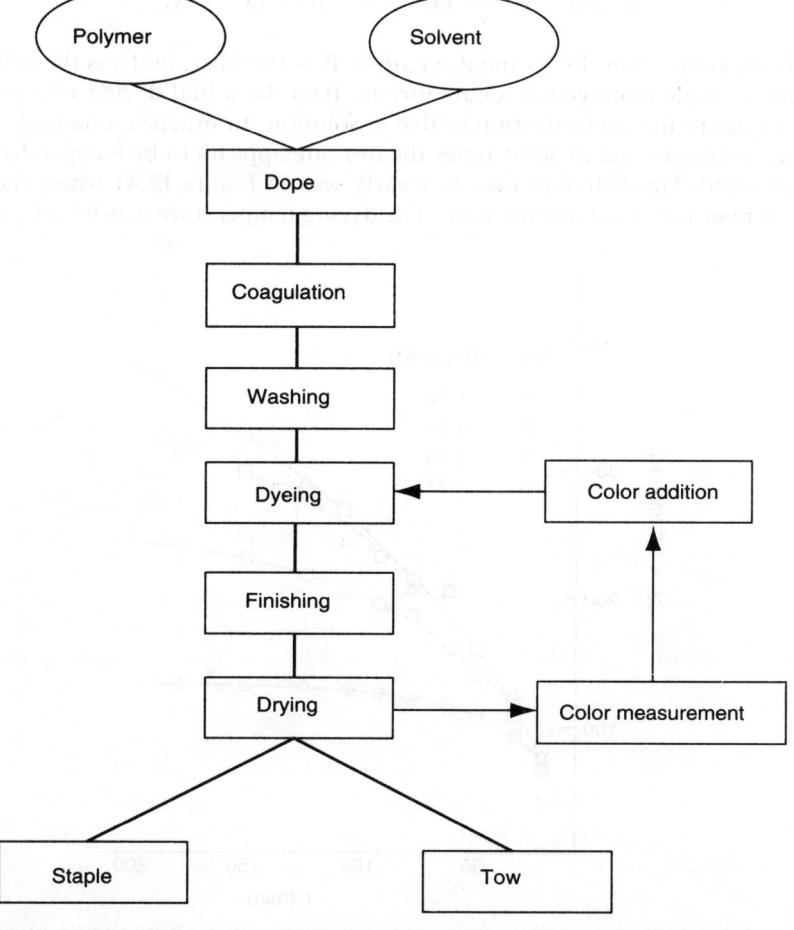

FIGURE 12.40 Dyestuff addition in a producer-dyed fiber process. (From Mullinax, G.B., New Developments in Colored Acrylic Fibers, Textile Research Institute Presentation, Charlotte, NC, April 13, 1988.)

luster, and denier) produced is large. In 1987, Monsanto reported [333] to offer over 250 colored products, and in 1982, Courtaulds reported [334] running 100 shades per week with 10,000 in their computer.

When first introduced commercially, acrylic fibers were regarded as difficult to dye. These difficulties were attributed to a lack of specific dye sites in the fiber and to the high level of intermolecular dipolar bonding that severely limited the diffusion rate of dye molecules. Eventually the manufacturers learned that these barriers could be overcome through copolymerization [335]. The use of ionic comonomers as dye sites was described previously. Incorporation of vinyl monomers with bulky side chains, such as methyl acrylate and vinyl acetate, was found to disrupt the intermolecular bonding (or pseudocrystallinity) and greatly increase the level of polymer chain segmental mobility, thus penetrating the dye to diffuse into the fiber.

The fundamental aspects of acrylic fiber dyeing have been reviewed by Peters [327]. The dyeing process has been described as an ion-exchange equilibrium that follows a Langmuir isotherm [336–338]. The saturation uptake corresponds to a limited number of anionic dye sites that exchange their counterions for dye cations of higher affinity. The ion-exchange reaction may be described by the following equation:

$$R-SO^{3-} M^+ + D_s^+ = R-SO^{3-} D^+ + M_s^+ \tag{12.32}$$

where M^+ is the counterion, D^+ is the dye cation, R is the fiber, and s is the solution. If the reaction were a simple homogeneous equilibrium, then the initial dyeing rate would be first order with respect to the concentration of dye in solution. In practice, one finds the reaction order to be close to zero and at short times the dye rate appears to be independent of the dye concentration [336]. This behavior may be clearly seen in Figure 12.41 where the concentration of dye on fiber is plotted against time. The dyeing temperature was 97.3°C, which is well

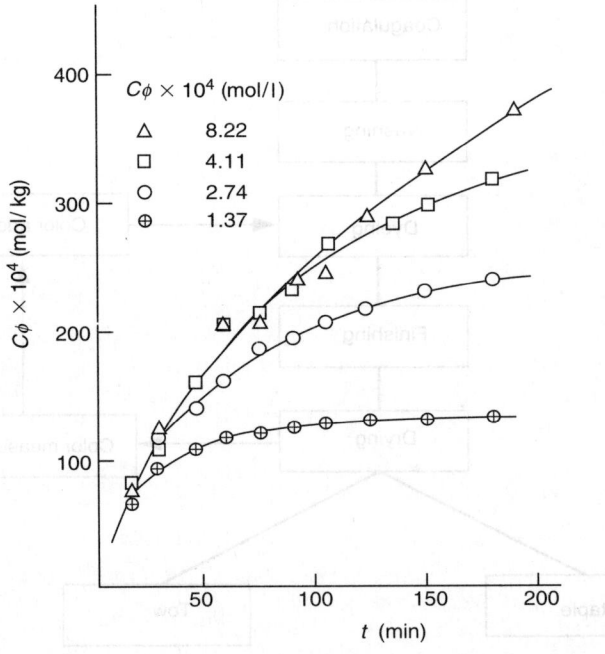

FIGURE 12.41 Dependence of dyeing rate on concentration of basic dye. C_0 is the initial concentration of dye in the bath. (From Rosenbaum, S., *J. Appl. Polym. Sci.* 7, 1225, 1963.)

above the polymer glass transition temperature. The dyeing behavior is best understood in terms of very fast saturation of the outer fiber surface followed by slower diffusion within the fiber from the saturated surface layer.

The dye rate will increase markedly when the fiber denier is decreased, simply because the specific surface area also increases. However, the dye diffusion coefficient may, in fact, decrease in the above case if the reduction in denier is brought about by a process change that increases the molecular orientation in the fiber skin. The rate of dye diffusion within the fiber is controlled by the degree of polymer chain segmental mobility. This follows from the observation that the dye diffusion rate increases precipitously when the dye bath temperature reaches the vicinity of the wet-polymer glass transition temperature. This can be seen in Figure 12.42 where the diffusion rate is plotted against reciprocal temperature for Orlon acrylic fiber. Other way of increasing the segmental mobility is to increase the level of nonionic comonomer or to increase the size of the comonomer side chain so that it disrupts the PAN crystalline structure to a greater degree. In either case, the dye diffusion rate will increase.

The dyeing rate of acrylics can also be enhanced by the use of additives both in the spin bath and in the dying operation. Bajaj and Munukutla [339] showed that adding to the spinning dope polymeric materials such as hydrolyzed acrylonitrile terpolymer, secondary cellulose acetate, and poly(vinylpyrrolidone) would increase dye diffusion rate by generally increasing the disorder in the acrylic fiber morphology. Addition of solvents into the dye bath will increase the dyeing rate if the solvents are plasticizers for the acrylic fiber. Shukla et al. [340] evaluated several solvents and found benzyl alcohol to be especially effective at 2% level in the dye bath. Gur-Arieh and Ingamells [341] also reported on the effectiveness of benzyl alcohol and attributed this to the plasticization effect, i.e., the increase in polymer segmental mobility [341].

If the dye diffusion rate is related to the degree of polymer chain segmental mobility, it follows that one should be able to correlate the diffusion rate with the dynamic–mechanical

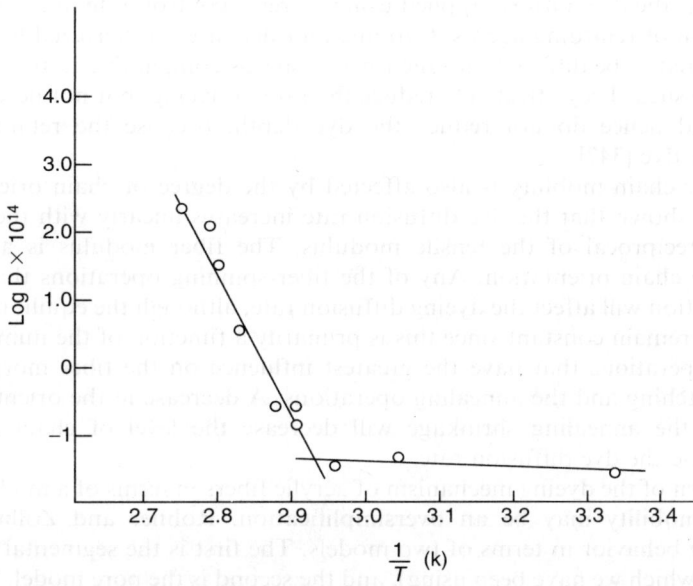

FIGURE 12.42 Dependence of the dye diffusion constant, D, on the dyeing temperature for Orlon 42 dyed with basic dye. (From Rosenbaum, S., *J. Appl. Polym. Sci.* 7, 1225, 1963.)

properties. We have seen in Section 12.4.3 that the loss modulus and tan δ are measures of the increase in polymer chain segmental mobility that occurs when the polymer is heated through the glass transition temperature. Bell and Murayama [342] determined the loss modulus of a series of acrylic fibers at 100°C in water and found a linear relationship between the log of the loss modulus and the log of the dye diffusion coefficient. The data could be fitted to the following equation:

$$\ln(D/R) = C - B_d \ln E'' \qquad (12.33)$$

where D is the dye diffusion coefficient, R is the gas constant, T is the temperature, C and B are empirical constants, and E'' is the dynamic loss modulus.

Bryant and Murayama [343] showed that the above concept could be extended to acrylic fibers composed of polymer blends. The usefulness of able to predict the dyeing behavior from the dynamic–mechanical properties was emphasized. Peters and Wang [344] developed a correlation between the saturation dye uptake and the area under the tan curve from 20°C up to the dyeing temperature. This relationship assumes that the ability of the fiber to absorb dye must be associated with the proportion of the chain segments that are mobile or in a rubber-like condition. This mobile fraction becomes significant when the temperature approaches the glass transition region and continuously increases as the temperature passes through the region ultimately reaching a maximum. The tan δ was measured using a Rheovibron modified so that the fiber could be immersed in water. A linear relationship was observed between the integrated tan δ area and the dye uptake measured at several temperatures, both with and without the presence of carriers, thus establishing the general validity of the relationship.

The strong dependence of the dye uptake rate on temperature introduces some problems in obtaining uniform or level dyeing of large batches. Since it is impossible to insure complete temperature uniformity in many commercial dyeing machines, an unevenness in dye uptake may result from temperature gradients in the bath. Since the affinity of the dye sites for the dye is so high, adequate leveling cannot be achieved through migration alone [345,346]. Hence, the dye must be applied using extreme control of temperature or by dyeing with the addition of retarding agents. Cationic retarders are quarternized long-chain amines and are considered to be diffused into the fiber at speeds comparable to the dyes and occupy some of the dye sites. They effectively reduce the rate of dyeing, but not necessarily the total dye uptake, and hence do not reduce the dye depth, because the retarder is gradually displaced by the dye [347].

The polymer chain mobility is also affected by the degree of chain orientation. Rosenbaum [336] has shown that the dye diffusion rate increases linearly with the tensile compliance, i.e., the reciprocal of the tensile modulus. The fiber modulus is a fairly accurate indicator of the chain orientation. Any of the fiber-spinning operations that alter the level of chain orientation will affect the dyeing diffusion rate, although the equilibrium value of the dye uptake will remain constant since this is primarily a function of the number of dye sites. The spinning operations that have the greatest influence on the fiber morphology are the orientation stretching and the annealing operations. A decrease in the orientation stretch or an increase in the annealing shrinkage will decrease the level of chain orientation and therefore increase the dye diffusion rate.

Interpretation of the dyeing mechanism of acrylic fibers in terms of a model based entirely on segmental mobility may be an oversimplification. Rohner and Zollinger [348] have analyzed dyeing behavior in terms of two models. The first is the segmental mobility or free volume model (which we have been using), and the second is the pore model. The free volume model describes dyeing kinetics as the diffusion of dye molecules or ions through amorphous parts of the polymeric matrix. The rate of diffusion is determined by the mobility of segments of the polymer chains. While there can be no doubt that segmental mobility plays a major role in the dyeing kinetics, there may be contributions from other structural factors that must be considered if a complete understanding of the phenomenon of acrylic fiber dyeing behavior is a goal. One such structural factor is the size and distribution of pores. We have seen that acrylic fibers go through a porous intermediate structure in their manufacturing process (see Section 12.5.3). These pores are normally collapsed at some point during the spinning operation (which includes drawing and annealing steps), but in some variants a significant degree of porosity is allowed to remain.

The pore model of dyeing was first suggested for dyeing cellulose fibers in the thirties [349]. It is based on the hypothesis that, under conventional dyeing conditions (in an aqueous dye bath), the fibers are solids containing networks of channels or pores filled with solution. The dissolved dye molecules diffuse into these pores and will be adsorbed onto the pores' surfaces. Rohner and Zollinger [348] compared the dyeing behavior of two different acrylics that had essentially identical polymer composition, but significant differences in porosity. It was thought that the dependence of the segmental mobility on temperature for these two fibers should be very similar, since their glass transitions in water should be very close. Then, any differences in dyeing behavior could be attributed to the pores. Extensive analysis led to the conclusion that neither model alone would explain all aspects of the mechanism of dyeing acrylics with cationic dyes above the glass transition temperature of dyeing. Pore diffusion and matrix diffusion are present simultaneously, but in varying ratios depending on the type of fiber. In the more porous fiber, pore diffusion is kinetically dominant and matrix diffusion is relatively low. However, for the less porous regular acrylic fiber, the overall rate is slower and therefore the ratio of matrix diffusion to pore diffusion lies more on the side of matrix diffusion, even at relatively short-dyeing times.

The role of fiber porosity on dyeing behavior as discussed by Zollinger and coworkers also included the influence of dye size [349] and the relative amounts of bound and free water [350]. In the water-saturated state, the more porous fibers contain much more free water than do the normal acrylics, while both types of fibers contain approximately the same level of bound water.

12.6 FIBER PROPERTIES

This section will cover physical properties, chemical properties, and flammability. Included in physical properties are tensile properties and dimensional stability at various temperatures in dry and moist environments, shrinkage properties, moisture uptake, and durability or wear resistance. Chemical properties cover resistance to sunlight, chemical and biological agents, and heat. The subsection on modification of properties addresses these property deficiencies and discusses progress made toward their improvement.

The sustained commercial success of acrylic fibers over the past 50 years can be attributed largely to their desirable balance of properties. Apparel goods and carpets made from acrylics have appealed to the consumer because they are aesthetically pleasing in comfort and appearance, easy to care for, and reasonably durable. The physical properties of the fiber are by no means remarkable and some obvious deficiencies such as a poor hot–wet strength and only modest tensile strength and abrasion resistance have prevented penetration of acrylic into some markets.

Lulay [351] has compared acrylic fiber properties against other fibers, and the comparisons are shown in Table 12.18. A scale of five (highest or best) to one (lowest or poorest) was used to assess properties. The basic properties of the various fibers have been translated to end-use performance, and the importance of these performance properties to the consumer for apparel has been segregated into three categories: highly desired, somewhat desired, and

TABLE 12.18
Qualitative Comparisons of the Properties of Textile Fibers

Properties	Polyester	Nylon	Acrylic	Wool	Cellulosics
Category 1: Highly desired					
Abrasion resistance (durability)	4	5	3	3	3
Strength (durability)	4	5	3	2	3
Wash–wear performance	5	3	3	1	1
Wrinkle resistance	5	3	3	4	1
Pill resistance	3	1	3	3	5
Category 2: Somewhat desired					
Bulk (cover)	3	3	5	4	1
Water transport (HS comfort)	3	2	5	3	4
Static resistance	2	1	4	5	5
Speed of drying	5	5	5	1	1
Category 3: Relatively Unimportant					
Resistance to burning	5	5	5	5	1
Resistance to degradation by dry heat	5	3	5	5	3
Resistance to degradation by wet heat	3	4	5	2	4
Resistance to degradation by sunlight					
Outdoors Brt. yarn or nat.	3	3	5	2	3
Beh. glass Brt. yarn or nat.	5	3	5	2	3
Resistance to insects	5	5	5	1	1
Resistance to microorganisms	5	5	5	1	1
Avg. ranking, category 1	4.2	3.4	3.0	2.6	2.6
Avg. ranking, category 2	3.2	2.7	4.7	3.2	2.7
Avg. ranking, category 3	4.4	4.0	4.7	2.6	2.3
Avg. ranking, overall	4.1	3.5	4.2	2.8	2.5

Source: From Lulay, A., in *Acrylic Fiber Technology and Applications*, Masson, J.C., Ed., Marcel Dekker, New York, 1995, pp. 314–315.

relatively unimportant. Polyester fibers have the highest performance qualities in the highly desired category followed by nylon, acrylics, cellulosics, and wool. Acrylic fibers have intermediate properties in the highly desirable category, poorer than polyester or nylon but superior to wool and the cellulosics. Acrylic fibers are deficient to polyester in strength, abrasion resistance, wash–wear performance, and wrinkle resistance.

Acrylic fibers rank highest in the somewhat desired and relatively unimportant categories. Some of the properties in these two categories are considered more important for nonapparel markets, such as good resistance to sunlight degradation for drapery or good resistance to microorganisms for sandbags. In summary, acrylic fibers exhibit a well-rounded combination of performance properties with no major deficiencies. However, for woven apparel, the key properties of good wash–wear performance, abrasion, and wrinkle resistance favor polyester fibers.

Masson [352] has described how the acrylic's balance of properties matches well with the overall requirements for home-furnishing applications. The acrylics offer ability to dye to brilliant shades, good light fastness, good abrasion resistance, excellent property retention in environmental exposure, high resistance to staining, and good cleanability. Acrylic's weaknesses compared to other synthetics are deformation under hot–wet properties, poor high-temperature color stability, and higher cost.

12.6.1 Physical Properties

In this section, we will describe the physical properties of acrylic fibers and, whenever possible, the relationship among structure, properties, and applications will be emphasized.

Acrylic fibers are mostly sold as stable and tow, though a small amount of continuous filament is also produced. Acrylics are restricted to staple yarn processes because a good method for texturing a continuous filament acrylic yarn has not been devised. Successful texturizing requires that the fiber be partially melted while the yarn is twisted. This is not possible with acrylic fibers because they do degrade rather than melt.

Staple lengths may vary from 25 to 150 mm, depending upon the spinning system that is used to convert fiber into spun yarn. The filament denier may vary from approximately 1.2 to 15 g/denier (or from 0.13 to 1.7 N/tex). Garments intended to have a very soft hand will be constructed of the low-denier fibers because of their low-bending modulus. Most typical acrylic apparel fabrics utilize filament deniers near 3 dpf, and some industrial and craft-knitting yarns use deniers of 5–6. Acrylic carpet yarns are usually made from 15-denier fiber to achieve the required stiffness and wear resistance.

The relationship between fiber structure, cross-sectional shape, and bending modulus (stiffness) was discussed in Section 12.5.2 and is also described by Morton and Hearle [353]. The influence of the cross-sectional shape is particularly important for acrylics. Fibers that have high aspect ratio cross sections, such as the dry-spun fibers with a dog-bone-shaped cross section, tend to be more compliant and softer in comparison with the wet-spun fibers, which have a more rounded cross section.

The tensile properties and boiling water shrinkage of seven commercial acrylic fibers are listed in Table 12.19. The fibers were chosen to represent different production processes, varied end uses, and the commercial denier range [6]. The tenacity (breaking strength) varies from 2.2 to 3.6 g/denier (or 0.19–0.32 N/tex), and the breaking elongation varies from 33 to 64%. The only apparent pattern among the fibers is the lower tenacity and higher elongation of the large-denier fibers, which suggests that the molecular orientation of the fiber may decrease with increasing denier.

The boiling water shrinkage of standard acrylic fibers is generally quite low, as seen in Table 12.19, but may be made intentionally high for special types of fibers. High-bulk fibers that have been stretched and cooled to retain a frozen-in strain are designed to shrink up to 30% in boiling water.

The tensile properties of a typical acrylic staple fiber are compared with other textile fibers in Table 12.20. The acrylics can be characterized as a moderately stiff fiber [354,355], and in comparison with the other two major synthetic fibers, nylon and polyester, acrylics have a lower breaking elongation and considerably lower work of rupture. Acrylics do not compete well with these fibers in applications where very high load-bearing properties are of paramount importance. They do, however, have a higher tenacity and work of rupture than wool. This is significant as acrylics have replaced wool in many markets. Van Krevelen [355] compares the complete stress–strain curves for many synthetic and natural fibers and characterizes the acrylics as wool-like. This author also gives the specific tenacity versus specific modulus for a range of conventional man-made fibers, including PAN fibers.

The shape of the stress–strain curve for acrylic fibers was analyzed by Rosenbaum [356,357]. The stress–strain curve may be divided into the three regions as shown in Figure 12.43. Region A is the Hookean region and covers approximately the first 3% extension up to the yield point. The stress in this region probably reflects the force required to perturb the intermolecular dipolar bonds. Region B is attributed to a short-range straightening out of individual molecules possibly arising from a transition from a pseudohelical conformation to a more extended planar zigzag conformation.

TABLE 12.19
Physical Properties of Selected Acrylic Fibers[a]

Fiber	Acrilan B-16	Orlon 42	Euroacril	Courtelle	Dow type 500	Acrilan B-96	Zefran 253A
Producer	Monsanto	DuPont	ANIC	Courtaulds	Dow Badische	Monsanto	Dow Badische
End use	Textile	Textile	Textile	Textile	Industrial	Carpet	Carpet
Tex	0.13	0.38	0.34	0.50	0.67	1.67	1.67
Denier	(1.2)	(3.4)	(3.1)	(4.5)	(6.0)	(15.0)	(15.0)
Tenacity (N/tex) (g/denier)	0.32 (3.6)	0.26 (3.0)	0.24 (2.7)	0.22 (2.5)	0.32 (3.5)	0.19 (2.2)	0.20 (2.3)
Elongation (%)	42	33	35	64	33	56	62
Relative knot tenacity (%)	90	82	91	93		80	82
Initial modulus(N/tex) (g/denier)	3.9 (44)	4.0 (45)	3.6 (41)	3.3 (37)		1.9 (21)	1.8 (20)
Boiling water shrinkage (%)	1.0	0.7	0.7	1.2		1.0	1.0

[a]Tensile properties measured at 65% RH and 21°C extension rate, 100%/min [6].

TABLE 12.20
Tensile Properties of Major Synthetic Fibers[a]

	Breaking strength (g/denier)		Tensile strength (psi/1000)	Breaking elongation (%)		Stiffness (g/denier)	Toughness (g cm)	Moisture regain 70°F, 65% RH
	Standard	Wet		Standard	Wet			
Acrylic (Acrilan, Monsanto)	2.2–2.3	1.8–2.4	30–40	40–55	40–46	5–7	0.4–0.5	1.5
Acrylic (Creslan, Cytec)	2.0–3.0	1.6–2.7	30–45	35–45	41–50	6–8	0.62	1.0–1.5
Modacrylic (SEF Plus, Monsanto)	1.7–2.6	1.5–2.4	29–45	45–60	45–65	3.8	0.5–2.5	
Polyester (Dacron, DuPont)	2.4–7.0	2.4–7.0	39–106	12–55	12–55	12–17	0.2–1.1	0.04
Nylon 6	3.5–7.2	—	62–98	30–90	42–100	17–20	0.64–0.78	2.8–5.0
Nylon 6,6	2.9–7.2	2.5–6.1	40–106	16–75	18–78	10–45	0.58–1.37	4.0–4.5
Olefin	3.5–4.5	3.5–4.5	41–52	70–100	70–100	20–30	1–3	0.01
Aramid (Kevlar, DuPont)	23	21.7	425	4.0	4.0	500	—	4.3
Aramid (Nomex, DuPont)	4.0–5.3	3.0–4.1	90	22–32	20–30	70–120	0.85	4.5

[a]Tests conducted in staple fiber at 20°C, 65% RH [354].

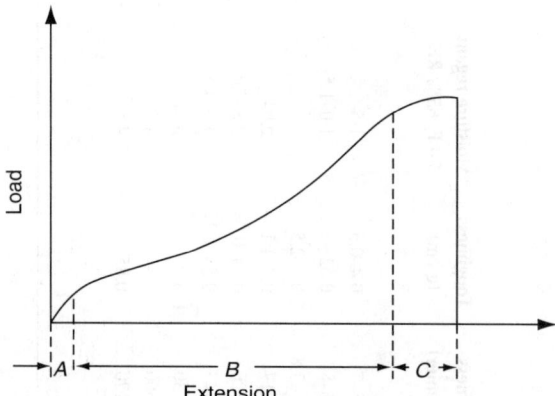

FIGURE 12.43 Generalized stress–strain curve for an acrylic fiber. Region A is Hookean behavior, region B is thought to be localized chain conformational changes, and region C, which occurs only when the fiber is above the glass transition temperature, is attributed to irreversible flow of chains. (From Rosenbaum, S., *J. Appl. Polym. Sci. 9*, 2071, 1965.)

Region C is only observed above T_g and is attributed to the irreversible flow of polymer chains and perhaps some rubber-like extension of chains that are not highly oriented. The temperature dependence of the stress–strain curves measured in water and silicone oil are compared in Figure 12.44 and Figure 12.45, respectively. The measurements in silicone oil are assumed to be characteristic of the dry fiber, since the oil does not plasticize the fiber. The disappearance of the yield point indicates that a transition from the glassy to the rubbery state has occurred near 115 and 70°C in the dry and wet measurements, respectively: this corroborates the strong plasticization effect of water that was discussed in Section 12.4.

The very significant reduction in modulus that occurs when the temperature in water exceeds T_g is the cause of the extremely poor hot–wet properties. In Table 12.21, the effects of water and temperature on the properties of common synthetic fibers are compared [358,359]. In Table 12.21, one can see how the properties change as, first, water is added to the dry fiber at 20°C, and, second, the water temperature is increased to 95°C. In the first case, the acrylics appear relatively

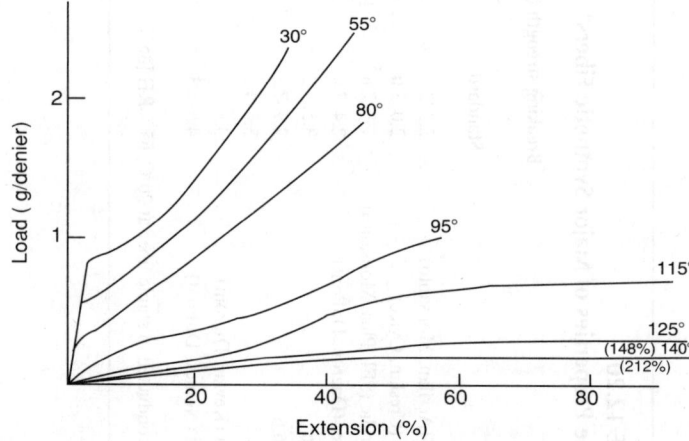

FIGURE 12.44 Effect of temperature on stress–strain curve for acrylic fiber in dry medium (silicone oil). (From Rosenbaum, S., *J. Appl. Polym. Sci. 9*, 2071, 1965.)

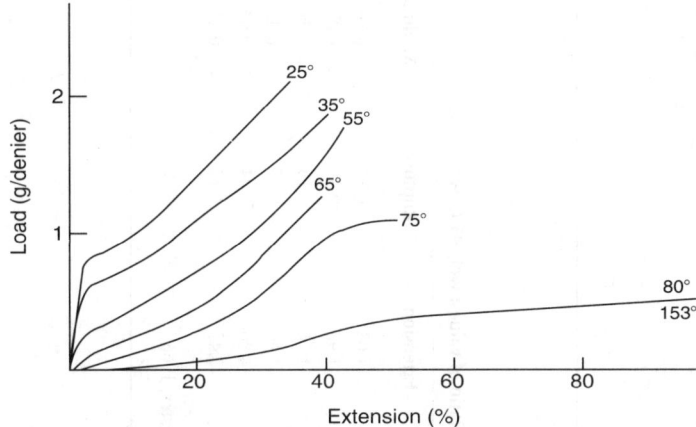

FIGURE 12.45 Effect of temperature on stress–strain curve for acrylic fiber in water. (From Rosenbaum, S., *J. Appl. Polym. Sci. 9*, 2071, 1965.)

impervious to the water. The only significant change is a relatively modest 15% decrease in tenacity. It compares well with the other petrochemical-based synthetics. One would conclude that the fiber is hydrophobic under these conditions. This is consistent with the low-equilibrium moisture uptake of the acrylic fiber, which is only 1–2% when measured at 20°C at 65% relative humidity [359]. Increasing the water temperature to 95°C reduces the modulus by a factor of 50, while the tenacity is reduced two thirds and the elongation increases fourfold. The low modulus of the acrylics under boiling water conditions results in a tendency for the fibers to lose bulk, shape, and resiliency during dyeing and other hot water treatments such as laundering.

The low hot–wet strength of acrylics can be attributed to the manner in which water plasticizes the unique laterally bonded acrylic fiber structure. The water lowers the T_g to approximately 70°C, but this is not sufficient to account for the extremely low modulus near the boil, since other fibers including nylon are also highly plasticized by water. However, these fibers contain a well-defined, stable, three-dimensional crystalline phase that is thought not to be penetrated by the water. Therefore, they can act to reinforce the fiber and limit the drop in modulus at temperatures above the T_g, where the amorphous phase has become rubbery. The crystalline phase in the acrylic fiber is highly imperfect (as was discussed in Section 12.4) and can probably be easily penetrated and plasticized by the water.

One way to improve the hot–wet modulus of acrylics is to introduce cross-links between the chains. Cross-linking is known to limit the decrease in modulus that occurs when a polymer is heated above T_g [360]. The success of attempts to do this will be covered in Section 12.6.4.

Acrylic fibers are less moisture absorbent than cotton, wool, and rayon and somewhat more moisture absorbent than nylon and polyester. Figure 12.46 shows that acrylic fibers swell approximately 5% when saturated with water at ambient conditions [361]. Cotton and wool, in contrast, swell by as much as 26% under similar conditions. This low moisture absorptivity makes acrylic fabrics relatively insensitive to changes in relative humidity. In one study, the dimensional change in going from 15% relative humidity to 90% was less than 0.5% for an Orlon fabric compared with approximately 2% for a similar wool fabric [362]. Rayon showed the greatest dimensional change, almost 5%, while polyester showed the least dimensional change. The Dacron polyester fiber changed only 0.2%.

One extremely important property of a textile fiber is the relative durability or wear resistance of a fabric made from the fiber. The mechanism of fabric wear resistance is very complicated because the nature of the wearing forces and the geometry of the fabric

TABLE 12.21
Effect of Moisture and Temperature on Tensile Properties

	Ratio of values: wet/65% RH				Ratio of values: wet, 95°C/wet, 20°C			
Fiber	Tenacity	Extension	Rupture	Modulus	Tenacity	Extension	Rupture	Modulus
Viscose rayon normal	0.50	1.58	0.69	0.03	0.90	1.03	0.89	0.80
Acetate	0.54	1.41	0.63	0.17	0.43	1.98	0.75	0.07
Nylon	0.80	1.05	0.87	0.82	0.79	1.76	1.19	0.21
Terylene (polyester fiber)	1.00	1.00	1.00	1.00	0.72	1.40	0.85	0.42
Orlon (acrylic fiber)	0.84	1.08	0.98	1.00	0.35	4.26	1.04	0.02
Polypropylene fiber	1.00	1.00	1.00	1.00	0.42	2.47	1.13	0.21

Source: From Farrow, B., *J. Text. Inst.*, 47, T58, 1956 and Ford, J.E., Ed., *Fiber Data Summaries*, Shirley Institute, Manchester, 1966.

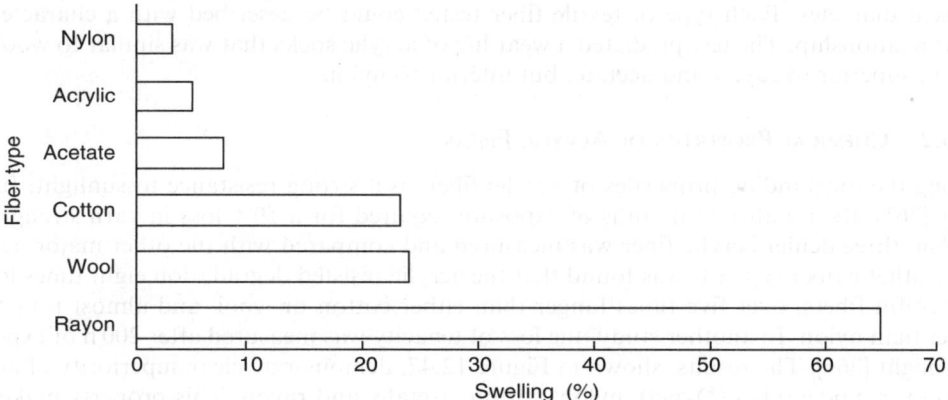

FIGURE 12.46 Cross-sectional swelling of synthetic fibers in water at 68°F (% increase in area). (From Morehead, F.F., *Text. Res. J. 17*, 96, 1947.)

construction are important in addition to the inherent durability of the individual filaments. Also, the latter is very difficult to measure because it involves determination of both transverse and longitudinal properties. Morton and Hearle [353] describe numerous mechanical tests that were developed to predict fabric durability. These tests are designed to subject the filaments to the types of forces that occur during actual fabric wear. Unfortunately, the predictive values of these tests leave much to be desired, albeit they can serve as useful aids for improving fiber properties. The performance of acrylic fiber is ranked against other textile fibers in Table 12.22. The acrylics as a class are much less durable than either nylon or polyester, but do compare favorably against other textile fabrics including wool, cotton, and rayon.

The nature of the mechanical forces that control fabric durability was investigated by Lefferdink and Briar [364]. They were able to show that weight loss due to abrasion did not correlate well with fabric durability as measured with actual wear tests. A fiber property that did correlate was the fatigue upon bending. A bending fatigue test was developed for single filaments where the log of the number of cycles to fail was plotted against the load per

TABLE 12.22
Performance Comparison of Synthetic Fibers in Abrasion Testing

Fiber	Walker abrader[a]	Stoll flex-abrasion[b]	Wet fabric[c]
Nylon	100	100	100
Polyester	>200	77	62
Wool	28, 13	29	—
Cotton	32, 5	44	—
Acrylic	18, 3	14	13
Rayon	6	9	4

Values are relative; nylon set to 100 in each case.
[a] Yarn-on-yarn abrasion. Yarn is twisted around a guide and then around itself.
[b] Yarns folded and rubbed over a bar.
[c] Same as b, but with a wet fabric.
Source: From Morton, W.E. and Hearle, J.W.S., *Physical Properties of Textile Fibers*, Wiley, New York, 1975, pp. 399–401.

filament diameter. Each type of textile fiber tested could be described with a characteristic linear relationship. The test predicted a wear life of acrylic socks that was similar to wool and cotton, superior to rayon and acetate, but inferior to nylon.

12.6.2 Chemical Properties of Acrylic Fibers

Among the outstanding properties of acrylic fibers is a strong resistance to sunlight. In one study [365], the number of months of exposure required for a 50% loss in yarn strength for Acrilan, three denier acrylic fiber was measured and compared with the other major natural and synthetic fiber types. It was found that the acrylic resisted degradation eight times longer than olefin fibers, over five times longer than either cotton or wool, and almost four times longer than nylon. In another study, the loss of tenacity was measured after 200 h of exposure to sunlight [366]. The results, shown in Figure 12.47, demonstrate clear superiority of acrylic fibers over modacrylics (Dynel), nylon, cotton, acetate, and rayon. This property makes the acrylics particularly useful for outdoor applications, such as in awnings, tents, and sandbags as well as upholstery for autos and outdoor furniture. Pigmented fibers with light-fast colors are particularly useful for outdoor applications. Acrylic fibers have excellent potential in draperies due to their good strength retention on exposure to light compared to other fibers commonly used in window coverings [367].

Acrylic fibers also have good heat stability. A detailed review of the thermal degradation of PAN has been published by Peters and Still [368]. Howell and Patil [369] determined the chromophores generated during exposure of acrylic fibers to light. The study included greige PAN and acrylic copolymer fibers and PAN fibers that had been dyed and finished. The effects of natural sunlight and accelerated exposure in a weatherometer were compared. IR spectral analysis revealed that greige PAN fibers exposed in the weatherometer developed various double-bonded species, including C=O, C=C, and C=N, but not carboxylic acid. The degradation was localized on or near the surface of the fiber. A similar pattern was observed for the dyed PAN fibers and the copolymers. The molecular weight changes in these weatherometer-degraded samples were analyzed using size-exclusion chromatography, and in every case M_n and M_w decreased substantially. Since the infrared analysis indicated that the degradation was located at or near the surface, there must be appreciable molecular weight reduction occurring since the molecular weight analysis samples the entire fiber.

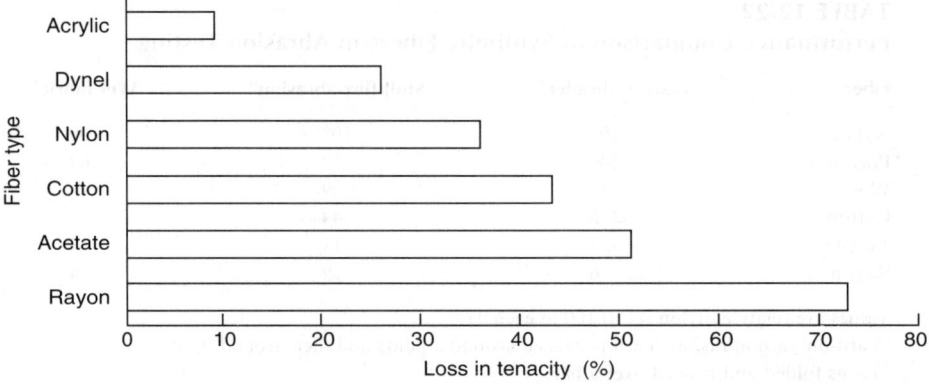

FIGURE 12.47 Loss of tenacity by fiber type after 200-h exposure in direct sunlight. (From Rugeley, E.W., Feild, Jr., T.A. and Fremon, G.H., *Ind. Eng. Chem. 40*, 1724, 1948.)

TABLE 12.23
The Stability of Orlon Fibers to Acids and Bases

Time	Temperature (°C)	Sulfuric acid (%)	Sodium hydroxide (%)
5 h	100	50	0.5
1 day	100	50	0.5
8 days	100	30	0.5
1 day	75	60	0.5
8 days	75	50	0.5
1 day	50	60	10
8 days	50	60	1.5
30 days	25	60	7

Data show the concentration, temperature, and time of exposure at which Orlon fibers begin to deteriorate.

Source: From Quig, J.B., *Papers Am. Assoc. Text. Tech.*, 4, 61, 1948; Quig, J.B., *Can. Text. J.*, 66(1), 42, 46, 1949; Quig, J.B., *Rayon Synth. Text.*, 30(2), 79, 1949; Quig, J.B., *Rayon Synth. Text.*, 30(3), 67, 1949; and Quig, J.B., *Rayon Synth. Text.*, 30(4), 91, 1949.

Sunlight induced similar spectral chemical changes to the fibers, but not to the same degree as those treated in the weatherometer.

Acrylic fibers are also resistant to all biological and most chemical agents. Acrylics are affected very little by weak acids, weak alkalis, organic solvents, oxidizing agents, and dry-cleaning solvents. These fibers are attacked only by strong base and highly polar organic solvents like DMAC, DMF, DMSO, and ethyl carbonate. Acrylic fibers tend to be much more susceptible to chemical attack by alkalis than acids or oxidizing agents. Data taken from Quig [370–374] shown in Table 12.23, for example, show that acrylic fibers are stable for up to 24 h at 100°C in 50% sulfuric acid, while these same fibers begin degrading with less than 0.5% sodium hydroxide at the same exposure time and temperature. The resistance of DuPont's acrylic Orlon fibers to oxidizing agents was compared with nylon, cotton, and acetate yarns [370–374]. The results, summarized in Table 12.24, show that the acrylic yarn was far superior to the others in retention of strength. In this study, the loss in yarn tenacity was measured as a function of bleaching time. After 300 min, the acrylic had lost 8% of its original tenacity whereas the nylon had lost 30%. Cotton and acetate yarns were even less resistant. Both of

TABLE 12.24
Yarn Tenacity Loss after Exposure to Bleaching Fluid

	Percent loss in yarn tenacity		
Fiber type	150 min	300 min	525 min
Orlon	7	8	9
Nylon	14	30	50
Cotton	42	65	98
Acetate	20	70	100

Source: From Quig, J.B., *Papers Am. Assoc. Text. Tech.*, 4, 61, 1948; Quig, J.B., *Can. Text. J.*, 66(1), 42, 46, 1949; Quig, J.B., *Rayon Synth. Text.*, 30(2), 79, 1949; Quig, J.B., *Rayon Synth. Text.*, 30(3), 67, 1949; and Quig, J.B., *Rayon Synth. Text.*, 30(4), 91, 1949.

these yarns lost 65–70% of their original tenacity. After 525 min of exposure, the cotton and acetate yarns had completely deteriorated, whereas the acrylic yarns retained approximately 92% of their original strength. Acrylic sandbags were first put into service by U.S. forces during the Vietnam War after testing showed their great superiority to jute burlap, cotton, and various synthetics. In addition to the superiority with respect to light stability, acrylic bags suffer no strength loss from fungal attack when buried in soil for up to 40 weeks, whereas cotton bags failed in 2 weeks and jute in 4 weeks [375].

Acrylic fibers discolor and decompose rather than melt when heated. Acrylic fibers, however, have very good color and heat stability at temperatures less than 230°C. In a study by American Cyanamid (using Federal Test Specification TT-P-141a Method 425.2), the yellowness of acrylic fabric was measured as a function of temperature [365]. Compared with a value of 0.00 for a pure-white body, the original fiber had a yellowness index of 0.04–0.10. After 30 min of exposure at 115°C, the yellowness increased only slightly from 0.11 to 0.17. After 6 h at 130°C, however, the yellowness increased from 0.38 to 0.41.

The excellent chemical resistance of acrylic fibers may stem from the unique laterally bonded structure discussed in Section 12.4.2. Dipole bonds formed between nitrile groups of adjacent chains must be broken before chemical attack, melting, or solvation can occur. In addition, the repulsive forces between adjacent nitrites resulted in a very stiff polymer backbone, which yields very little entropy gain when the bonds between adjacent chains are broken in solvation or melting. Therefore, relatively high temperatures are required for solvation and melting. Some specific data on the effects of chemicals are summarized in Table 12.25. In addition, a summary of the chemical properties of acrylic fibers, along with a comparison with other fiber types is given in Table 12.26.

12.6.3 FLAMMABILITY

Flammability and ignition behavior are important properties of acrylics and modacrylics. Fibers used in textiles must not ignite readily when placed in contact with a flame. In this respect, acrylic fibers compare favorably with the other natural and synthetic fibers currently on the market. There are, however, significant differences in the burning characteristics of the major fiber types. Cotton and rayon burn and form a char. Nylon, polyester, olefin, wool, and acrylics, on the other hand, burn and melt simultaneously. More rigorous standards are required for end uses such as carpets, sleepwear, draperies, and bedding. Fibers for these applications must also be self-extinguishing after removal from the ignition source. This is generally achieved in acrylonitrile-based fibers by incorporating halogen comonomers such as vinylidene chloride, vinyl chloride, and vinyl bromide. The burning characteristics of acrylic and modacrylic fibers are compared with those of other major commercial fibers in Table 12.27.

Another property used to compare the flammability of textile fibers is the limiting oxygen index (LOI). This measurement quantity describes the minimum oxygen content (%) in nitrogen necessary to sustain candle-like burning. Values of LOI, considered a measure of the intrinsic flammability of a fiber, are listed in Table 12.28 in order of decreasing flammability. Acrylic fibers, it can be seen, are similar in flammability to cotton. Modacrylics, on the other hand, are somewhat less flammable than any of the synthetics, except 100% PVC, and are substantially less flammable than cotton and wool.

Acrylics are also similar to cotton in smoke generation. National Bureau of Standards smoke cabinet results are shown in Figure 12.48 for acrylic, modacrylic, nylon, PVC, cotton, and wool [376]. These data show wool to give the highest smoke density while nylon and PVC give the lowest, under these burning conditions. The modacrylics are actually somewhat worse in smoke generation than acrylics. The thermal and flammability behavior of textile materials has been reviewed in detail by Rebenfeld et al. [377], Lewin et al. [378], and Lewin [379].

TABLE 12.25
Chemical Resistance of Acrylonitrile Fibers

	Concentration	Temperature	Time	Strength	Other	Fiber
Inorganic reagents						
Nitric acid	Fuming				Soluble	Acrilan
	40	90	8 h	Appreciable		Orlon
	40	40	8 week	Moderate		Orlon
	35	25	12 week	None		Acrilan
	10	90	2 week	Moderate		Orlon
Hydrochloric acid	37	50	72 h	45%		Orlon
	37	50	8 h	11%		Orlon
	37	25	9 days	Moderate		Orlon
Sodium hydroxide	50	50	20 h	0–6.6%		Dynel
	25	100	20 h	17%	Dark	Dynel
	5	80	4 h	Appreciable		Orlon
Ammonium hydroxide	28	25	20 h	None		Dynel
	5	100	48 h	100%	Brittle	Acrilan
	5	25	12 days		No change	Acrilan
Hydrogen peroxide	90	25	20 h	5–8%	Bleached	Dynel
Sodium hypochlorite	5.25	50	20 h	None		Dynel
Zinc chloride	50	75	24 h		Soluble	PAN
Organic reagents						
Acetic anydride	100	50	20 h	68%	Shrunken Stiffened Curled	Dynel
Acetic acid	100	75	48 h		Brown	Acrilan
Formic acid	100	50	20 h	5–9%	Bleached	Dynel
Aniline	10	25	12 days		No effect	Acrilan
	5	50	20 h	95%	Partly dissolved Discolored	

Source: From Ford, J.E., Ed., *Fiber Data Summaries*, Shirley Institute, Manchester, 1966; Quig, J.B., *Papers Am. Assoc. Text. Tech.*, 4, 61, 1948; Quig, J.B., *Can. Text. J.*, 66(1), 42, 46, 1949; Quig, J.B., *Rayon Synth. Text.*, 30(2), 79, 1949; Quig, J.B., *Rayon Synth. Text.*, 30(3), 67, 1949; and Quig, J.B., *Rayon Synth. Text.*, 30(4), 91, 1949.

12.6.4 MODIFICATIONS OF PROPERTIES

12.6.4.1 Modification of Handle

Many factors combine to determine the softness or handle of a fabric. For a given fabric construction the denier compliance, cross-sectional shape, degree of crimp, moisture absorption, and surface smoothness of the fibers all influence the softness of the final product. Very fine filament deniers are effective where good draping, anticrease properties, and softness of handle are desired [380]. Such yarns are expensive to produce and are more susceptible to abrasion than coarse-filament yarns. The desired softness of handle can also be achieved through modifying the fiber cross section. DeWitt Smith [381] has shown that a flattened cross section with a 3:1 ratio of principal axes has approximately one third the bending modulus of a round fiber with equal cross-sectional area. Wet-spun fibers generally have a cross section that is only slightly deviated from round. However, more rounded cross sections can be achieved by wet spinning by modifying the coagulation conditions and the spinneret

TABLE 12.26
Chemical Properties of Acrylic and Modacrylic Fibers

Chemical agent	Acrylic/modacrylic	Wool	Cotton	Nylon	Polyester	Polyolefin
Inorganic acids						
Weak	Resistant	Resistant	Hydrolyzed, hot	Fair	Resistant	Resistant
Strong	Dissolves in conc. HNO_3, H_2SO_4	Resistant	Hydrolyzed, oxidized	Poor	Resistant, but degrades in 96% H_2SO_4	Slowly oxidized
Alkalis						
Dilute	Resistant	Degrades easily	Resistant	Good	Resistant	Resistant
Concentrated	Degrades hot	Degrades easily	Swells	Good	Degrades hot	Resistant
Oxidizing agents	Resistant	Degrades	Degrades in conc. H_2SO_4	Resistant	Resistant	Moderately resistant
Reducing agents	Resistant hot	Degrades	Degrades in conc. bleaches			
Dry cleaning solvents	Insoluble					
Nonpolar solvents	Insoluble			Insoluble		
Polar solvents	Soluble in DMF, DMSO, DMAC, ethylene carbonate			Soluble in 90% formic acid		
Heat resistance	Degrades slowly above 200°C	Degrades above 150°C, chars 300°C	Excellent below 150°C, burns at 390°C	Degrades °C melts 250–255°C	Excellent melts 250°C	Degrades slowly and melts at 150–170°C
Stability in sunlight	Excellent	Fair degrades	Fair degrades	Good	Good	Must be stabilized
Resistance to insect	Resistant	Attacked by moths	Attacked by silverfish, not resistant-rotting	Resistant	Resistant	Resistant
Resistance to biological agents (mildew, fungus, bacteria, etc.)	Resistant	Resistant			Resistant	Resistant

TABLE 12.27
Burning Characteristics of Textile Fibers

Ignition behavior	Fiber type
Melts and burns	Nylon, polyester, acrylic, olefin, wool
Chars and burns	Cotton, rayon
Melts and self-extinguishing	Modacrylics (SEF, Verel, Kanekalon)
Chars and self-extinguishing	Treated cotton
Metal and nonburning	Telfon
Nonmelting and nonburning	Carbon, graphite, asbestos

orifices [382–387]. The rheological principles governing the formation of nonround filaments were discussed in Section 12.5.2.

Softness has also been achieved by increasing the bending compliance of the fibers. Grafting long-chain esters [388] and treating the fibers with various compounds [389–391] are techniques that have been used successfully. Fibers with a soft, wool-like handle have been produced by copolymerizing long-chain alkyl acrylates and alkyl methacrylates [392] and by blending conventional fibers with fibers containing hygroscopic comonomers, such as N,N-dimethylacrylamide [393] or derivatives of N-vinylpyrrolidone [394]. Techniques that simulate the scaled surface of wool by roughening the fiber surface have also been used successfully [395–397]. One route is the incorporation of insoluble particles [398]. Special techniques such as sheath–core spinning [399,400], conjugate spinning [401,402], and drawing in a bath containing sodium sulfate [403] have also been reported to give wool-like properties.

12.6.4.2 Wear Comfort

Cotton fabrics have long been considered a standard by which textile comfort properties are measured. Although many properties contribute to wear comfort [404], it is believed that the two dominant properties inherent to cotton are the ability to absorb moisture from the skin and softness. The second property is the result of the extremely fine denier of cotton fibers. Both properties can be achieved in acrylic fibers. However, fine-denier acrylics are achieved at the cost of reduced spinning productivity.

TABLE 12.28
Limiting Oxygen Index of Textile Fibers

Fiber	LOI	
Cotton	18.0	Most flammable
Acrylic	18.2	
Rayon	19.7	
Nylon	20.1	
Polyester	21.0	
Wool	25.0	
Modacrylic	27.0	
Verel modacrylic	33.0	
100% PVC	37.0	Least flammable

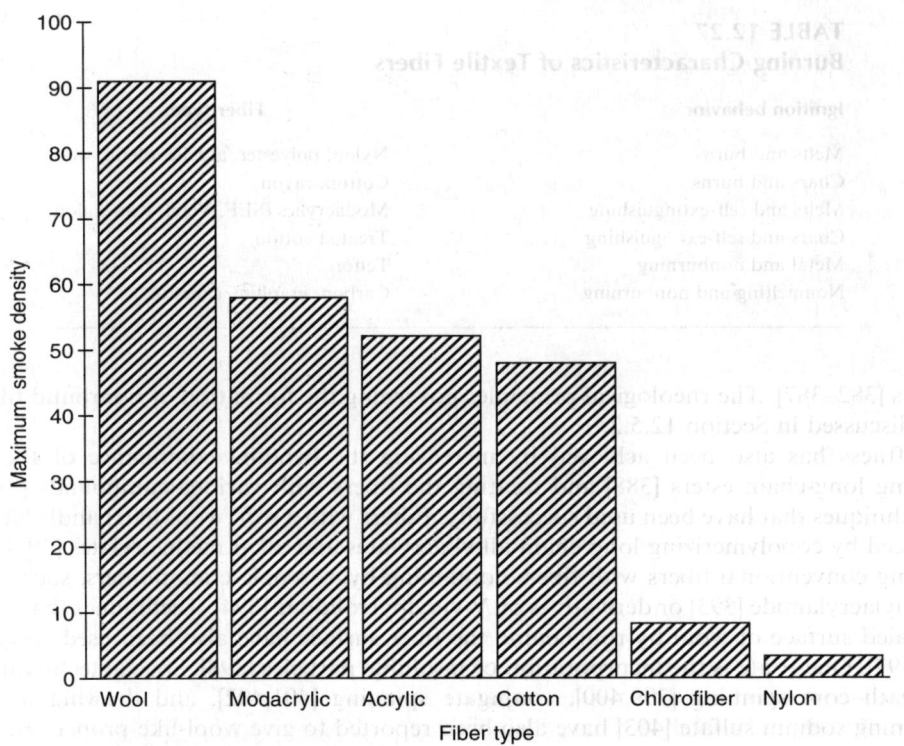

FIGURE 12.48 Maximum smoke density developed during controlled combustion of acrylic, modacrylic, and other fiber types. Results are from NBS smoke cabinet tests. (From *Fiber Producer*, p. 44, April, 1984.)

Many methods have been developed to increase the moisture-absorbing properties of acrylics. Acrylonitrile polymer can be modified by incorporating hydrophilic comonomers that decrease ultimate fiber density. Improved moisture retention can also be achieved by modifying the fiber-spinning process or by using postspinning treatments, such as modified finishes. An illustration of a polymer modification combined with a special spinning technique is Bayer's process for making porous, moisture-absorbent fibers [405]. This process uses copolymers containing carboxyl groups in the acid form. The polymer is dry-spun from a spinning dope containing a high-boiling glycol or alcohol derivative. The glycol or alcohol derivative is retained within the fibers in the dry-spinning tower, where a porous fiber morphology is established and subsequently extracted during the fiber-washing step. After spinning, the carboxyl groups are converted to the salt form, thereby enhancing their hydrophilic character. The resulting fibers have a unique porous sheath–core structure that allows water to diffuse into the fiber leaving the surface relatively dry. The fibers are said to rapidly absorb dye and still do not dye to deep shades like normal acrylics [406]. Fibers produced from this process can absorb large quantities of water without giving a clammy feel. Many variations of the sheath–core concept are described in the patent literature [407–411]. Dunova, an acrylic sheath–core fiber of this type marketed by Bayer, is a commercial example [412]. The subject of porous fiber structure has been reviewed by Bakerzyk [413].

12.6.4.3 Reduced Pilling

Acrylic fibers, like other fibers used in staple fabrics, develop small balls of fiber called "pills" as a result of abrasive action on the fabric surface [414]. Pills tend to build up more rapidly on acrylic fabrics than on comparable woolens, and this has been attributed to the relatively high flex life of acrylic fibers. With weaker fibers the pills are broken off and removed during cleaning operations, such as brushing, laundering, and, in the case of carpets, vacuuming. Many efforts have been made to overcome this problem. Examples include using highly twisted yarns, minimizing the number of short staple lengths, modifying the fiber cross section, and modifying the fiber surface to minimize tangling. The most effective methods are those designed to increase the likelihood that the pills will wear off. The required changes in fiber properties include reducing the fiber strength, introduction of defects in the fiber, brittleness, and lack of shear strength. All of these properties are easy to impart to the fibers, but increases in antipilling properties generally lead to corresponding decreases in the spinnability or ability to process the fibers. Thus, commercial antipilling processes depend on a balance between these factors.

Spinning conditions and polymer properties that give antipilling characteristics are described by Aotani and Ichimura [415]. The polymer can be modified to give more brittle fibers by decreasing the comonomer content, adding an ionic monomer, or inducing cross-linking during the spinning process. Wet-spinning processes can be modified by using low solvent concentration in the spin-bath and high spin-bath temperature to give brittle fibers with high void content. Low draw ratios result in low tensile strength. Drying the fibers under tension increases fiber brittleness and annealing the fibers at reduced temperatures reduces the fiber elongation to break. Skin-core fiber structures, described by Terada [416], have also been used to achieve anticrimping properties. Factors that cause pilling are discussed in a series of papers by Hirota [417].

Commercial processes have been patented by DuPont [418,419], Toray [420,424], Asahi [425,426], Mitsubishi Rayon [427], Courtaulds [428], and Hoechst [429]. The Hoechst antipilling fiber is marketed in various forms as Dolan 40, 41, and 44.

12.6.4.4 Improved Hot–Wet Properties

One of the deficiencies of acrylic fibers is their inability to retain their modules under hot–wet conditions. Knits and woven fabrics tend to lose their bulk and shape in dyeing and, to a more limited extent, in washing and drying cycles. This is a manifestation of the poor acrylic hot–wet properties discussed in Section 12.6.1. Under high humidity conditions, the fibers may lose their resilience, and wrinkles develop that are difficult to remove. The problem has been treated in a number of fundamental studies that have demonstrated that moisture lowers the wet glass transition temperature of acrylonitrile polymers [430] (see also Section 12.4.4).

Crimp imparted during spinning to enhance yarn bulk is lost when the yarn is exposed to conditions required to dyeing and laundering [431,432].

A number of polymer and fiber modifications have been devised to overcome this problem, although none has been successful enough to allow acrylics to compete in successfully easy-care apparel markets. The fibers may be treated with compounds such as ammonium sulfide [433], hydrazine derivatives [434,435], thiosemicarbazides [436], silicone oils [437], and emulsions of polysiloxane and polyepoxide [438]. Some success has been achieved by incorporating comonomers that increase the wet glass transition temperature of the polymer or make the copolymer more water-resistant [439–444]. Sheath–core fibers have been reported [445] in which the core polymer is stable under hot–wet conditions and the sheath polymer is used to compensate for deficiencies in dyeability.

Dope additives have also been reported. A patent issued to the American Cyanamid Co. [446], for example, describes a process in which a fine dispersion of a polycarbonate resin is incorporated into the spinning dope. The fibers produced in this manner contain fibrils of the polycarbonate resin that have the effect of improving the fiber hot–wet elasticity.

Many attempts to improve hot–wet properties by cross-linking have been reported [444,447]. Patents issued to Bayer [448–450], for example, claim that excellent properties under hot–wet conditions can be achieved by incorporating N-methylolvinyl monomers into the acrylic polymer. The fiber is spun from DMF and cross-linking is initiated by drying the fiber, tension-free, at 170°C.

The carpet market is one area in which acrylics have failed as a result of inadequate hot–wet stability. Although acrylic fibers produce a carpet of superior wool-like appearance and quality, the pile loses its resilience and bulk during dyeing and under high-humidity conditions. To compensate, acrylic carpets must be made in expensive, high-density constructions. As a result, acrylics have lost popularity as a carpet fiber since the peak year of 1968.

12.6.4.5 Improved Abrasion Resistance and Increased Fiber Density

The abrasion resistance of acrylic fibers is distinguished from fibrillation resistance in that fibrillation is primarily a surface-related phenomenon due to poor lateral cohesion. Surface fibrils are torn from the fiber surface causing a whitening effect in dyed fabrics due to scattering of light. True abrasion resistance relates more to fiber integrity and strength. It is directly related to fiber elasticity, toughness, and density. Abrasion resistance is generally improved by reducing the microvoid size, increasing the fiber density, and eliminating macrovoids. The effect of spinning conditions and polymer composition on the microstructure of acrylic fibers was reviewed in Section 12.5.

Void-free fibers have been produced by incorporating hydrophilic comonomers, such as sulfonated monomers, acrylamide derivatives, and hydroxyalkyl acrylates [451]. Polymer blends are also effective in reducing macrovoids. Examples include blends of hydrophilic polymers such as polyvinyl methyl formamide [452], poly N-vinylpyrrolidone, and acrylonitrile-dimethylacrylamide copolymer [453,454]. Dense fibers can be produced by incorporating comonomers, such as vinylidene chloride, with small molar volumes relative to acrylonitrile.

The spinning process also has a major effect on fiber density and abrasion resistance [455]. Modified wet-spinning processes have been reported [456,457]. Dry spinning, however, is known to produce a dense fiber structure than conventional wet spinning. In dry spinning, the fiber is formed by diffusion of solvent from the filament to the dry-spinning column. In contrast, wet-spun fibers are formed by counterdiffusion of solvent from the fiber filaments and diffusion of nonsolvent (usually water) into the filaments. The diffusion presence of water into the filaments precipitates the polymer, and is believed to cause macro- and microvoids within the filaments as well as a coarse and easily abraded skin. Wet-spinning techniques used to improve fiber density and abrasion resistance generally do so by approximating dry-spinning conditions in some way. Examples include modified spin bath, high solvent concentration, low spin-bath temperature, and additives to the dope or spin bath. Spin-bath additives include nondiffusing liquids, such as polyethylene glycol, and high-molecular-weight alcohols (e.g., t-butyl alcohol) [458]. Surface-active agents have been added to the dope as well as to the spin bath to increase fiber density [459]. Variations in wet spinning that approximate dry spinning are dry-jet wet spinning and spinning into a standard spin bath through an inert, low-density organic liquid dispersed on the spin-bath surface. Melt spinning of wet acrylonitrile copolymer in 120°C superheated steam is also reported to give dense, void-free fibers [460].

12.6.4.6 Improved Whiteness and Thermal Stability

Fiber whiteness and thermal stability can generally be improved by producing a polymer that is linear and free of conjugated or unstable chemical structures formed by side reactions during the polymerization process. The most straightforward approach is to maximize the probability that polymer radicals react with monomer molecules rather than with the polymer backbone itself. This is accomplished by maintaining the unreacted monomer concentration in the reactor as high as practical. Successful application of this technique has been accomplished by operating the polymerization process at low overall conversion of monomer to polymer [461,462]. Side reactions can also be minimized by avoiding comonomers having low reactivity with acrylonitrile and those comonomers that are known to undergo side reactions themselves. Vinyl acetate, for example, is known to undergo a branching reaction at the carbonyl group. The effect of branching on thermal stability, however, has not been established for acrylic fibers containing vinyl acetate. Polymerization reaction conditions are also important. High pH is known to cause hydrolysis of the nitrile groups in the polymer, and high-temperature reaction conditions tend to favor undesirable side reactions. Reaction temperatures above 50°C are normally avoided by using redox initiators, such as the persulfate–bisulfite–iron system described earlier. Additives used during polymerization include reducing agents such as organophosphorous compounds [463], sulfinic acid and other sulfur compounds [464], glycerol monoesters [465], and phenol-type antioxidants [466]. Comonomers [467,468] of amides and sulfonates and other comonomers with special functional groups have also been used effectively [469] and cross-linking of acrylonitrile and vinyl azetidinone has been reported [470]. Other additives for the reaction mixture [471–477] and for the spinning solution have also been studied [478–481]. Imperfect fiber color may be masked by using fluorescent brightening agents and blue-dye techniques [482–485].

A number of methods have been reported in the literature for improving the thermal stability of acrylic fibers. For example, impregnating wet-spun fibers in the gel state with salts of metals from group II in the periodic table of the elements [496] or sodium sulfide solution [487] have been reported. Also, dried fiber can be impregnated with sulfuric acid solutions of formamide [488], phosphates or borax [489], or organotin salts [490].

Other methods that have been reported include inducing the cyclization reaction by incorporating thioamide groups [491], copolymerized N-vinylpyrrolidone [492], or vinyl halogen monomers in copolymerized or grafted form [493]. Stabilizers based on maleic acid are also effective [494,495]. Reviews on the subject have been published by Brown [496] and Krcma [497].

12.6.4.7 Antistatic Fibers

Acrylic fibers present less of a problem with static and cling than most other synthetic fibers. Acrylics are naturally somewhat hydrophilic and tend to dissipate static charge more readily than other synthetic fiber types. Antistatic properties have been improved, however, by making the fibers hygroscopic with hydrolyzing and cross-linking agents [498,499].

Since static electricity is a surface phenomenon, effective antistatic surface treatments can be used as a means of dissipating static charge. Treatments that give an ionic surface character have been effective [500]. Surface treatments employing fatty acids [501], vinyl ether copolymers [502], and quaternary ammonium compounds [503] have all been reported. Other approaches are to incorporate metal or metallized fibers in combination with special conducting backings [504–510].

12.6.4.8 Antisoiling Fibers

Low dirt-absorbing fibers have been studied by incorporating paraffins into the spin bath [511,512]. The polymer composition has been modified by incorporating comonomers of fluorinated acrylates [513] and copolymers of fluorinated acrylamides [514] and other monomers [515]. Porous fibers with reduced staining tendencies have been made from acrylic copolymers containing methacrylate and sodium allylsulfonate [516]. Antisoiling properties can also be achieved by using finishes, either by treating the fiber during spinning, or by applying finishes directly to the fabric. The subject of soil abrasion and soil removal was reviewed by Jacobasch [517], and the effect of fiber properties on soiling resistance has been reviewed by Glubish and Kocher [518]. Chemical modifications of acrylic fiber-forming polymers and copolymers have been reviewed by Gabrielyan and Rogovin [519] and Ham [520].

12.7 ANALYSIS AND IDENTIFICATION

12.7.1 Fiber Identification

Identification of fibers is complicated by the presence of many generic types and modifications of both man-made fibers and common natural fibers. The major generic types of man-made fibers are summarized in Table 12.29. Visual and microscopic examination together with simple manual tests remain the primary methods of fiber identification, though many new sophisticated instrumental methods are available that are based on the chemical and physical property differences among the fibers. These methods are able to distinguish between closely related fibers that differ only in chemical composition or morphology.

Synthetic fibers are generally mixed with other fibers to achieve a desired balance of physical properties. Acrylic fiber staple may be blended with wool, cotton, polyester, rayon, and other fibers. Therefore, as a preliminary step in identification, the yarn or fabric must be separated into its constituent fibers. This will immediately establish if the fiber is a continuous filament or staple product. Staple length, brightness, and breaking strength, both wet and dry, are all useful tests that can be done during a cursory examination. A more critical identification can be made by a set of simple manual procedures based on burning, staining, solubility, density determination, and microscopic examination. The most important tests will be summarized below:

TABLE 12.29
Major Generic Types of Man-Made Fibers

Generic type	Key property or characteristic	Examples
Acetate	Derived from secondary cellulose acetate	Dicel
Acrylic	Contains 85% or more acrylonitrile	Acrilan Orlon Cashmilon Dolan
Chlorofiber	50% or more of poly(vinyl chloride) or poly(vinylidene chloride)	Saran Leavil
Modacrylic	50–85% acrylonitrile	SEF Kanekalon
Polyamide	Contains –CO–NH– repeat group	Ultron ICI Nylon
Polyester	85% or more of an ester of a diol and terephthalic acid	Dacron Trevira
Polyolefin	Polymer derived from an olefinic monomer—ethylene, propylene, etc.	Herculon Fibrite
Triacetate	Cellulose with at least 92% of acetylated hydroxyl groups	Tricel Arnel
Vinyal	Derived from poly(vinyl alcohol)	Vinylon
Viscose	Regenerated cellulose by viscose process	Evlan Fibro

1. *Burning*: Acrylics melt to form a black bead that is easily crushed in the fingers. There is usually no smell of acetic acid as with cellulose acetate. Many of the modacrylics, such as Dynel, do not form a bead in this test. In general, the modacrylics do not support combustion when removed from the source of ignition.
2. *Solubility*: Acrylic fibers are insoluble in acetone, but soluble in DMF, DMAC, and DMSO at room temperature. Modacrylics, such as Dynel and Verel, are exceptions because they may be soluble in acetone but insoluble in methanol. All of the modacrylics are soluble in butyrolactone at room temperature.
3. *Staining*: Two useful staining tests can be carried out to help identify various acrylic and modacrylic fibers by manufacturer and product type. These are the Meldrum and Sevron Orange staining tests. The Meldrum stain consists of 0.5 g of Lissamine Green SFS stain (Color Index Acid Green 5) and 0.5 g of Sevron Orange L (Color Index Basic Orange, aqueous volume of 100 ml). An undyed acrylic fiber test sample is immersed in the stain at 90°C. The sample is removed after 5 min and the color of the fiber is observed after washing and drying. The Sevron Orange staining test is carried out in a similar fashion with a second, undyed acrylic fiber sample, and the colors of the two fiber samples are compared with a standard chart of commercial acrylic fibers. Detailed information on these tests is compiled in a text published by the Textile Institute [521].
4. *Density*: Acrylics have a low specific gravity (1.12–1.20 g/cc) compared with all of the major natural fibers and most man-made fibers. Nylon has a similar specific gravity (1.14 g/cc) and the polyolefins have lower specific gravities (e.g., 0.90 g/cc for polypropylene). Again, the modacrylics and some acrylics with high levels of halogen comonomer of low molar volume are exceptions. Verel and Dynel, for example, have specific gravities of 1.37 and 1.31 g/cc, respectively.
5. *Microscopic examination*: All fibers have distinguishing features that allow either outright identification or classification into a narrower grouping for specialized analysis. Animal hair fibers, for example, have a characteristic scaled surface. In addition, many textile yarns are blends of two or more fiber types. A simple examination with a normal light microscope can establish this and allow the components of the yarn to be separated for more detailed evaluation. The major identifying characteristics are:

    a. Scaled surface indicates animal hair fiber.
    b. Convolutions, reversal zones, and lumen indicate cotton and seed hair fibers.
    c. Bundles of fibers indicate bast and leaf fibers.
    d. Smooth surface with or without striations indicates either silk or man-made fibers.
    e. Flat ribbon-like cross section indicates split film fibers.
    f. Voids and inclusions, such as delustering agents, pigments, flame-retardant agents, etc.
    g. Cross-sectional interfaces indicate bicomponent fibers.

The major cross-sectional shapes for acrylics and modacrylics are round, dog bone, kidney bean, elongated, crenelated, mushroom, and ribbon. Many of these shapes are illustrated in Figure 12.23 in Section 12.5.

12.7.2 Instrumental Analysis

It is very difficult to distinguish between the various acrylics and modacrylics using the simple tests described above, and one must turn to elemental analysis for identification. Specific atomic compositional data can be gained by determining the percentages of C, N, O, H, S, Br, Cl, Na, and K. In addition, the levels of many comonomers can be established using IR and UV spectroscopy. Also, manufacturers are able to identify their own products. To facilitate this, some manufacturers introduce a trace amount of a rare element as a built-in label, such as samarium.

General schemes for the identification of natural and synthetic fibers have been established by the Textile Institute [521] and by the American Association of Textile Chemists and Colorists [522]. A comprehensive treatment of burning, solvent, staining, microscopy, and density techniques is given by Wolfgang [523] and a general discussion of procedures for identifying man-made fibers is presented by Moncrieff [8].

12.7.2.1 Polymer Characterization

Many techniques are available for characterizing acrylic and modacrylic fibers. The properties of most concern are the dye site content, molecular weight average and molecular weight distribution, chemical composition, color and color stability, glass transition temperature, degree of crystalline order, and melting point. The dyeability of modern acrylic fibers is generally achieved by incorporating sulfate and sulfonate end groups onto the polymer chain. Dye site levels up to 50 meq/g can be incorporated as end groups derived from a redox initiator system, such as the commonly used combination of potassium persulfate–sodium bisulfite–ferrous sulfate. Higher dye site levels can be achieved through the use of a sulfonated vinyl monomer, such as SSS. The total dye site level can be measured by doing an elemental analysis for sulfur. The low levels of sulfur normally required for fiber dyeability can also be measured accurately by methods based on x-ray fluorescence. Other methods are available for fibers where nonsulfur acid dye sites are incorporated. Where weak acid groups, such as carboxylic acid, are present, the dye site content of the polymer can be measured using potentiometric titration. This method is time-consuming but very useful, where it is necessary to know the relative amounts of strong and weak acid groups in the polymer. Alternatively, the dye site level of the polymer may be determined directly by treating a suspension of the polymer with an aqueous solution of a basic dye standard. The amount of dye absorbed per gram of polymer can then be calculated from a photometric measurement of the residual dye in solution. Methods have been developed for measuring dye rates as well as the saturation dye level.

The average molecular weight of acrylic polymer can be measured by methods based on end-group analysis, osmotic pressure, light scattering, and dilute solution viscosity. The application of these methods to acrylic polymers has recently been reviewed by Frushour [524]. Knowledge of the dye site content of the polymer in cases where all dye sites are derived from end groups gives an interesting but not practical measure of the number average molecular weight of the polymer. If all polymer chains were formed by a kinetic sequence consisting of initiation by sulfate or sulfonate radicals, radical chain growth, and radical combination, then the number of dye sites per chain would be exactly 2.0. Thus, if the dye site level were 37 meq/g of polymer, then the number of miliequivalents of polymer per gram would be exactly 18.5. The number average molecular weight would be 54,000 and, assuming there is no comonomer, the degree of polymerization would be 1000. In practice, polymerization kinetics are much more complicated. Chain transfer reactions give polymer chains with just one dye site and branching reactions can give polymer chains with more than two dye sites. In addition, sulfone linkages can form within the polymer backbone and sulfate dye sites can be hydrolyzed in fiber processing to give further deviations from the expected number of dye sites per chain. At best, end-group analysis is useful as a means for estimating the number average molecular weight of the polymer. This method, however, is useful for linear condensation polymers, such as the polyamides and polyesters, in which the end groups consist of the easily measured −COOH, −NH$_2$, or −OH functional groups.

The colligative properties of dilute polymer solutions can be used to accurately measure M_n, the number average molecular weight of polymers [524,525]. This class of techniques is based on the thermodynamic principle that the lowering of the activity of a solvent by a solute is equal to the mole fraction of the solute. Thus the mole fraction of the polymer in solution can be measured by measuring the vapor pressure lowering, the boiling point elevation, freezing point depression, or osmotic pressure of the polymer solution relative to the pure solvent. Vapor pressure osmometry, the most widely used of these techniques, indirectly measures the vapor pressure lowering upon addition of a polymer solute. Though limited to M_n values of about 10,000 Da, vapor pressure osmometry has become very popular in recent years owing to its ease of use and the commercial availability of instruments.

Membrane osmometry also gives the M_n. The osmotic pressure is measured in a device that holds a solution of the polymer and the pure solvent in compartments separated by a semipermeable membrane. The membrane allows the solvent to diffuse between compartments while restricting the polymer to one compartment. Solvent molecules from the pure solvent compartment will diffuse into the polymer solution to dilute it and thus to equilibrate chemical potential of the pure solvents and chemical potential of the solvents in the polymer solution. However, the volume of the chamber is fixed so the additional solvent will create a hydrostatic pressure within the compartment containing the polymer solution. The osmotic pressure is determined by measuring the difference in the hydrostatic pressure head between the two chambers or by applying an external source of pressure to balance the osmotic pressure on the solution chamber. Although applicable to M_n values up to 500,000 Da, membrane osmometry is less popular than vapor pressure osmometry because it is limited to M_n values greater than 10,000 Da and the measurements are complicated by the difficulties of selecting good membranes and maintaining their performance.

While the polymer dye site level is closely related to the number average molecular weight, the weight average molecular weight has a greater impact on the rheological properties of the spinning dope and the mechanical properties of the ultimate fibers. These properties are influenced much more by the long-chain species in the polymer than the shorter ones. Short-chain molecules with ionic end groups or ionic comonomers may exert an influence on solution viscosity and gelation by acting as ionomers. In a solution, the ionic species are capable of coalescing into ionic domains to form at type of cross-linked structure with a higher solution viscosity than comparable nonionic polymers. The most widely used methods for measuring weight and viscosity averages molecular weight are light scattering and various methods based on dilute solution viscosity, respectively.

The basis for molecular weight determination by light scattering is the fact that the intensity of the light scattered from a single particle is proportional to the square of the mass of the particle. In a polydisperse sample, the larger molecules contribute proportionately more to the scattering intensity than the smaller ones. Thus, light scattering is ideally suited to studying macromolecules in the presence of smaller molecules. The resulting average molecular weight is a weight average. Experimentally, a beam of monochromatic polarized light is passed through the sample and the intensity of the scattered light is measured at a specific angle or series of angles after passing through a slit system and a second polarizer. The difference in refractive index between the polymer and solvent should be as high as possible and the solvent itself should exhibit low scattering.

The data are analyzed using a modified form of the Debye equation [526,527] and extrapolating to zero concentration and zero scattering angle. This method is suitable for macromolecules ranging in molecular weight from 10,000 to 10 million Da.

While osmometry and light-scattering techniques are still used to obtain true number and weight average molecular weights, simpler though less rigorous methods based on solution viscosity are commonly used whenever possible. The simplest method is to measure the viscosity of a solution of the polymer at a specified concentration and temperature. This may be done using either a Brookfield, capillary, or falling ball viscometer. The viscosity average molecular weight may be obtained from such measurements by extrapolating the reduced viscosity to zero polymer concentration. The intrinsic viscosity, thus derived, is then used in the well-known Staudinger equation [528] or Cleland–Stockmayer equation [529] to give the viscosity average molecular weight. A complicating factor in textile applications is the presence of ionic groups in the polymer chain. This becomes an important consideration when a sulfonated monomer is used or where there is a significant fraction of low-molecular-weight species present with ionic end groups. The ionic groups, depending on the size and distribution of the charge, cause an expansion of the hydrodynamic volume of the solvated polymer chains. This expansion effect increases considerably as the concentration of the polymer is reduced. Thus, the apparent molecular weight of the polymer increases rapidly upon extrapolating solution viscosity data to zero concentration. In practice, this effect is suppressed by adding a salt, such as lithium chloride, to the solution.

The dispersity of the polymer molecular weight has an important influence on spinning and fiber properties. The simplest measure of polydispersity is the ratio of the weight average molecular weight to the number average molecular weight, M_w/M_n. This parameter is commonly referred to as the polydispersity index of the polymer. Statistical analysis of the kinetics of free radical polymerization has shown that the polydispersity index of the polymer formed at any given instant is 1.5. In practice, however, variations in monomer concentration, imperfect mixing, multiple reaction phases, and the use of chain transfer agents tend to broaden the molecular weight distribution. In typical commercial polymers, the polydispersity index may vary from 1.5 to as much as 3.0.

Many times one would like to have a more detailed description of the molecular weight distribution, and this calls for chromatographic techniques that fraction the polymer's mass according to molecular weight. The most widely used technique is gel permeation chromatography (GPC) [525]. Typically a solution of about 0.5% polymer is injected into a column packed with a macroporous cross-linked polystyrene gel. As the polymer molecules pass through the column, they diffuse in and out of the pores of the gel. The gel pores may vary in size so that the larger molecules are excluded from a certain fraction of the pores. As a result of this pore size distribution, the smaller molecules are held up for a longer time within the pore structure. Therefore, a mathematical relationship can be established, by calibration, between the molecular weight of an eluted fraction and its elution volume. State-of-the-art GPC analysis may use UV light or refractometric detectors coupled with on-line molecular weight determination by low-angle light scattering. In addition, special solvents may be used to eliminate comonomer and ionic effects. These and other GPC techniques allow accurate determinations of molecular weight distribution for copolymers and for polymers containing high levels of ionic groups.

Copolymers may have a distribution in their copolymer composition as well as the molecular weight. This is especially pertinent for modacrylics, which may contain up to five different monomers. The distribution in composition is caused by variations in the composition of the monomer mixture during polymerization. This, in turn, may be attributed to imperfect mixing, unequal monomer reactivity, differences in monomer solubility, differences in interaction parameters for monomer–polymer–solvent combinations, and heterogeneity in the reaction mixture. Very little work is reported in the literature on this important topic because there is no simple universal technique for experimentally determining copolymer composition distributions. Topciev [530], however, developed the theoretical basis for fractionating copolymers using the Flory–Huggins theory of polymer solutions. When the polymer–solvent interaction parameters of the component homopolymers are equal, the fractionation takes place with respect to molecular weight only. Fractionation with respect to composition alone is not obtainable in practice, though by choosing a solvent system that maximizes the differences in interaction parameters of the various polymer species with the solution one can enhance the fractionation with respect to comonomer composition. Thus, by carrying out two series of fractionations or by combining GPC analysis with composition fractionation, it is possible to obtain separate molecular weight and composition distributions in binary copolymers. Glockner et al. [531] have reported two solvent–nonsolvent systems suitable for fractionating acrylonitrile-styrene copolymers, and Fritzsche et al. [532] have fractionated acrylonitrile–ethylacrylate copolymers using DMF–decaline [532].

Many different analytical methods are available for determining the overall chemical composition of acrylic and modacrylic polymers. The nitrile group of the acrylonitrile monomer has a strong IR absorbance, hence the monomer concentration can be determined with IR and Raman spectroscopy. Acrylonitrile content can also be determined using elemental analytical techniques for the nitrogen. Many of the usual comonomers found in acrylics can also be detected by their IR absorbances. Vinyl acetate, methyl acrylate, and methyl methacrylate, for example, can be detected by IR absorbance based on the carbonyl group. Sulfonated monomers, such as SSS, can be detected by the strong phenyl group absorbance in the UV spectrum. The levels of these monomers can also be measured by sulfur analysis using x-ray fluorescence, though the results may be in error due to the contribution of sulfur from the chain end groups, which is usually on the order of 20–50 meq/g. Halogen monomers can be measured quantitatively by pyrolyzing the polymer and analyzing the pyrolysis products by gas chromatography or halide titration. X-ray fluorescence alone may also be used to determine the level of specific halogens.

Other polymer properties that are often measured are whiteness, color, thermal stability, glass transition temperature (wet and dry), melting point (wet and dry), dope viscosity, and gelation characteristics. Techniques used for characterizing acrylic textile polymers are summarized in Table 12.30.

12.7.2.2 Fiber Characterization

In addition to characterizing the many properties introduced by the choice of monomers and the polymerization process, considerable characterization is also required to describe quantitatively the performance properties imparted by spinning and subsequent downstream processing. One endeavors to measure these performance properties and explain them in terms of the fiber structure; i.e., one determines the key structure–property relationships. Important fiber properties including hot–wet strength, abrasion resistance, crimp retention, and many others, including dyeability, can be rationalized in terms of the structural model for the fiber morphology. This general area was extensively discussed in Section 12.4.

TABLE 12.30
Common Analytical Techniques for Polymer Analysis

Required analysis	Analytical method
Comonomer content (vinyl acetate or methyl acrylate)	Infrared spectroscopy (nitrile and carbonyl)
End groups and dye sites (SO_4, HSO_3)	X-ray flourescence, strong acid titration
Molecular weight	Viscometry, size exclusion chromatography
Particle size	Sedimentation rate in solution
Dope color	Solution colorimetry
Residual acrylonitrile analysis	Gas chromatography
Residual acids	Microwave heating in water followed by titration
Metals in fiber	Acid digestion followed by ICP

ICP, Inductively coupled plasma.

TABLE 12.31
Methods and Instruments for Analysis of Acrylic and Modacrylic Fibers

Type of property	Method or instrument used
Specific gravity	Mercury density via pycnometer, density gradient column
Denier	Gravimetric measurement, vibrascope
Cross section	Optical microscopy
Moisture regain	Gravimetric measurement (70°F, 65% RH), thermogram method
Color—undyed	Colorimeter, visible spectroscopy in solution
Dyeability	Polymer dye site content, dye saturation value, dye diffusion rate
Microstructure	Scanning and transmission electron microscopy, mercury porosimetry
Mechanical properties	Instron (in United States)
Dynamic mechanical properties	Rheovibron, sonic modulus
Molecular orientation	Birefringence, x-ray diffraction
Melting transition, decomposition range, thermal effects	TGA, DTA, colorimetry
Luster/opacity	Clarity of fiber suspended in chlorobenzene measured using a photoelectric reflectance meter
Abrasion resistance	J.C. Penney abrasion test, Wyszenbeck, Stoll

The physical and mechanical properties of acrylic and modacrylic fibers are measured by standard test methods [533,534]. Many important physical properties are measured by simple gravimetric techniques. Denier, for example, can be measured exactly as it is defined, by determining the weight of 9000 m of a test filament. Fiber density is also measured gravimetrically, using a pycnometer, by weighing the amount of mercury displaced by a known weight of fiber. Moisture regain, a critical factor affecting wear comfort and static buildup, is measured by first weighing an oven-dry sample of yarn. Then the sample is equilibrated in a chamber at specified conditions (usually 70°F and 65% relative humidity) and weighed again. The difference in weights, expressed as a percentage of the dry weight, represents the equilibrium moisture content or moisture regain of the fiber. Dry-heat shrinkage and shrinkage in boiling water are measured by simply determining the difference in the length of a section of fiber after treatment at specified conditions.

The mechanical testing required to establish tensile properties, such as breaking elongation, tenacity, and modulus of elasticity, is carried out using devices like the Instron mechanical tester [535]. The fiber is gripped at the ends by two jaws, which are then pulled apart at a specified rate. The mechanical stress and strain on the fiber is recorded continuously until the fiber breaks. The result is a stress–strain curve for the fiber from which the breaking elongation, tenacity, and elastic modulus can be derived. Other properties, such as elastic recovery, stress relaxation, creep, cyclic loading effects, and hysteresis, can also be evaluated with this equipment.

A summary of the methods and equipment used for evaluating acrylic and modacrylic fiber properties in Table 12.31.

12.8 COMMERCIAL TEXTILE PRODUCTS

12.8.1 Standard Staple and Tow

Much of the growth in acrylic fibers usage has come from the replacement of wool. Acrylics have many of wool's desirable properties and are clearly superior in many areas where wool is deficient. Like wool, acrylic fibers are valued for their warmth, softness of hand, generous bulk and pile qualities, and ability to recover from stretching. At the same time, acrylics are less costly, more resistant to abrasion and chemical attack, and more stable toward degradation from light and heat. In addition, acrylics are not attacked by moths and biological agents and show very little of wool's tendency to felt.

The majority of acrylic fiber production is 3–5 denier staple and tow, furnished undyed in either bright or semidull luster. The major markets are in the apparel and home furnishings sectors. Within the apparel section, these fibers find extensive use in sweaters, and in single-knit jersey, double-knit, and warp-knit fabrics for a variety of knitted outerwear garments such as dresses, suits, and children's wear. Large markets for acrylics in the knit goods area are hand-knitting yarns, deep-pile fabrics, circular knit, fleece fabrics, half-hose, coarse-cut knitwear, and deep-pile fabrics for blankets.

As shown in Table 12.32, the applications for acrylic fibers depend on the needs, resulting from climate, economic conditions, etc., in each geographic region. China, for example, with its colder climate and developing economy, requires much of its acrylic for apparel. The countries with more advanced economies, such as the United States and Western Europe, require more acrylic for home furnishings. Prominent uses for acrylics in this area include area rugs, curtains, and upholstery. Acrylic carpets remain popular in Japan, but markets in the United States and Western Europe are dominated by nylon.

Until the early 1970s, acrylic fiber in 15–20 denier form was a major competitor in the carpet market. Its usage, compared to the total U.S. shipments of acrylic, is shown in Figure 12.49 [17]. Although the overall domestic shipments of acrylic fiber increased during most of the 1970s, the usage of acrylics in carpet has declined since peaking in 1972, as nylon began to dominate the carpet market. Strict flammability regulations put in effect during this period provided impetus for the development of flame-resistant acrylics and modacrylics. Fibers with vinyl chloride, vinyl bromide, or vinylidene chloride were developed to pass government tests, such as the tunnel test and the methenamine pill test. Today the usage of halogenated comonomers is restricted and the only commercially viable comonomer is vinylidene chloride.

Despite their decline in popularity, acrylic and modacrylic carpet fibers allow exceptional versatility in styling and color patterns. Acrylic carpets can also be superior to nylon esthetically, but dense and expensive constructions are required to match the pile height and durability of nylon carpets. In less dense constructions, acrylic carpets can develop wear patterns and pilling. Without the dense constructions acrylic carpet can also lose resilience and pile height in the dye bath or in service under hot, humid conditions.

Today, the carpet market is dominated by nylon-bulked continuous filament and staple fiber. A small market still exists in the United States and Europe, while in Japan acrylic carpets are still relatively popular. Numerous studies have been carried out to find ways of

TABLE 12.32
Synthetic Fiber Usage by Region and End Use

End use	China	United States	Japan	Western Europe
Home furnishings	20	50	36	44
Apparel	66	25	41	29
Medical and hygiene	1	4	2	4
Industrial	13	4	21	23

Source: *Int. Fiber J.*, excerpted from The Synthetic Fibres Market in China, Technon (U.K.) Ltd., October 1994, p. 44.

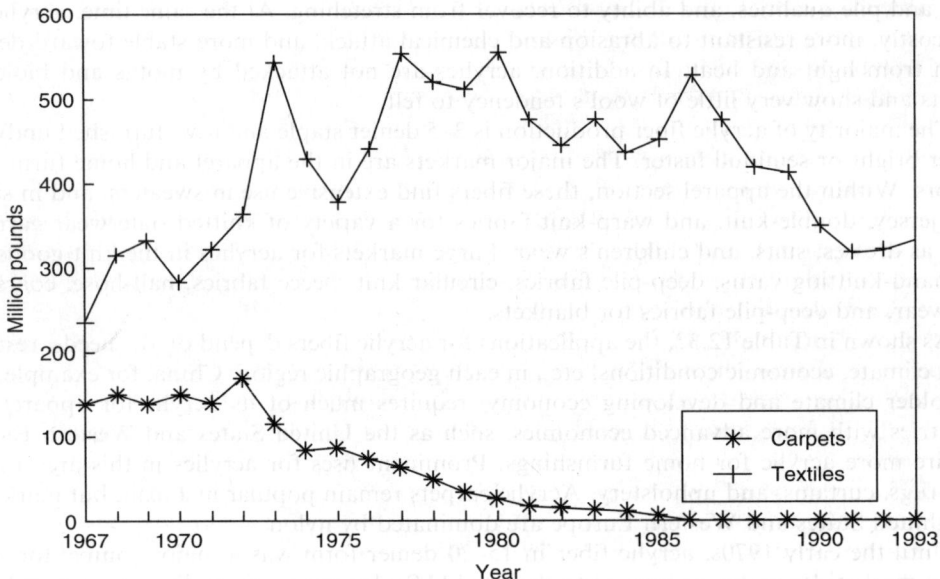

FIGURE 12.49 (Top) U.S. domestic acrylic fiber shipments; carpets versus textiles. (From Matzke, Jr., R.R., in *Acrylic Fiber Technology and Applications*, J.C. Masson, Ed., Marcel Dekker, New York, pp. 69–104, 1995.) (Bottom) U.S. total domestic acrylic fiber shipments.

improving the hot–wet and durability properties of acrylic carpet fibers. Abrasion-controlling properties, such as the fiber density and fibrillar structure, can be improved by using modified compositions and spinning processes. Producer-dyed fibers have been developed to eliminate the hot–wet performance problems that occur with dyeing. High-bulk blends are used to carpet body and minimize problems with hot–wet lean out. These consist of standard acrylic carpet fiber blended with high-shrinkage acrylic, with approximately 20–30% Superba shrinkage. Blends of staple and BCF nylon with high-shrinkage acrylic have also been developed to improve the performance of 100% nylon carpets. These blends, sold under the trademark Traffic Control, give improved tuft definition, less fuzz, and better floor performance than 100% nylon carpet [536,537].

In the apparel market, polyester in staple or filament form has a clear advantage because of its easy-care qualities and low cost. Although 100% polyester has limited appeal because of its less desirable handle, it is extremely popular in blends with cotton. The lack of a well-defined crystalline structure and low-wet T_g limit the attainment of the easy-care properties in acrylic fibers. However, textile researchers have devised methods that improve the hot–wet and easy-care properties of acrylics by cross-linking the oriented fiber molecules and by modifying the polymer composition or spinning process. These methods increase the cost of the fiber and consequently have not provided the technical breakthrough needed to compete on equal terms with low-cost polyester fiber. The cotton market itself would seem to be a natural target for acrylics. However, even though researchers have been able to match the wear-comfort qualities of cotton, consumer demand for genuine cotton remains the dominant market factor. A strong marketing effort may be required for acrylics to have a major impact as a cotton replacement.

Acrylics and modacrylics are also useful industrial fibers. Fibers low in comonomer content, such as Bayer's DOLAN 10, have exceptional resistance to chemicals and good dimensional stability under hot–wet conditions. These fibers are useful as industrial filters, battery separators, asbestos fiber replacement, hospital cubical curtains, office room dividers, uniform fabrics, and carbon fiber precursors. The excellent resistance of acrylic fibers to sunlight also makes them highly suitable for outdoor use. Typical applications, including modacrylics, are awnings, sandbags, tents, tarpaulins, covers for boats and swimming pools, cabanas, and duck for outdoor furniture [538].

A breakdown of U.S. domestic synthetic staple and tow fiber shipments by end use is given in Table 12.33. Organized by fiber type and application, these data show that acrylic fiber is strongest in hosiery, sweaters, craft yarns, pile fabrics, circular knits, flat knits, and blankets. Polyester, valued for its low cost and easy-care properties, has a strong hold on the fiberfil, nonwoven, flat-knit and broad-woven fabric markets. Nylon, valued for its durability and resiliency, dominates the carpet market. Olefin, a fiber type that appears to be growing in popularity, has great impact on the nonwovens market and has made inroads in the carpet face yarn markets versus nylon staple and filament. Table 12.34 shows the same type of data as Table 12.32, but arranged by distribution of end use for each type of synthetic staple and tow fiber.

Besides the standard staple and tow products, acrylic and modacrylic fibers are offered in many forms for specialized applications. Filament yarns are offered for outdoor applications, industrial fibers, and apparel fibers, where a fur-like or silky appearance is desired. Fibers with enhanced or modified properties are in great demand. Yarn bulk is enhanced by using bicomponent or biconstituent fibers (Wintuk, Monsanto). Pilling is reduced by producing fibers that are more brittle (Pil-Trol, Monsanto), and yarns with exceptionally soft handle are produced by using fine-denier fibers or fibers treated with special friction-reducing finishes (Fi-Lana, Monsanto). Abrasion resistance is enhanced by modifying the spinning conditions to produce a denser filament microstructure (Duraspun high-bulk abrasion resistant acrylic, Monsanto). Monsanto's Solara is one example of a modified acrylic fiber, designed for a specific end use. This fiber is currently used to make SAI Supersox, a high-performance, outdoor sock for hiking. Solara is a sort staple relaxed acrylic with enhanced abrasion resistance, good moisture transport properties, and sufficient bulk for comfort.

TABLE 12.33
Synthetic Staple and Tow Shipments by Fiber Type for 1995

	Acrylic and modacrylic	Nylon	Polyester	Olefin and Vinyon	Total in million pounds
Hosiery, sweaters, and craft yarns	62		38 = Total of others	101	
Pile fabrics	45		55		9
Circular knits and flat knits	17		83		521
Blankets	85		15		5
Other broadwoven	3		92	6	769
Carpet face yarns	<1	70	16	13	1091
Other acrylic markets	42	28	29	1	148
Fiberfil			100		442
Flock		100% = nylon + polyester			21
Nonwovens			51	49	547
Total shipments, %	7	23	57	13	
Millions of pounds	268	827	2100	458	3654

Source: From *Fiber Organon*, Fiber Economics Bureau, pub., March 1997.

TABLE 12.34
Distribution of Staple Fiber Products by Fiber Type for 1994

End use	Cellulose	Acrylic and modacrylic	Nylon	Polyester	Olefin and Vinyon
Hosiery, sweaters, and craft yarns		38			
Pile fabrics		6		0.3	
Circular knits and flat knits		28		22.6	
Nonwoven	37.7	0		12.6	57.2
Weaving	60.4	8.7		32.6	10.5
Fiberfil		0		19.4	
Carpet face yarns		<1	93.4	10.4	31.2
Other	1.9	19.6	6.6	2.1	1.1

Source: From *America's Textiles Int.*, September 1994, p. 224.

Acrylic fabrics and fabrics made from them can be colored using a variety of methods, dictated by the desired application. Basic dyes are the norm for acrylic fibers, but acid-dyeable fibers are made by copolymers of acrylonitrile with pyridines, tertiary amines, or quaternary ammonium salts. Cashmilon, produced by Asahi, is one example. Producer-dyed fibers offer a lower production cost to the mill and are useful for avoiding the hot–wet lean-out problems common with acrylic fibers. Fiber producers use several methods, gel-state dyeing, dope dying, and tow dyeing. The process works best with wet-spinning processes where, until the spun fiber is dried and collapsed, there is a sponge-like microstructure that accepts dye very efficiently. Pigmented fibers are offered where light-fastness is especially important. Among the obvious uses for pigmented fibers are outdoor fabrics, such as awnings, tents, and patio furniture. Pigments are generally added directly to the spinning solutions using dispersions in dilute polymer solutions. Methods of acrylic fiber dyeing are discussed in detail by Emsermann and Foppe [539] and product variants used for coloration are covered by Masson [540].

12.8.2 Acrylic Filament Yarns

Continuous filament acrylic yarns face stiff competition from nylon, polyester, and polyolefin fibers. Since they are more costly, acrylics have penetrated only those markets where they have a clear advantage in a critical property. Acrylics, for example, are clearly superior in resistance to sunlight. This makes continuous filament acrylics valuable in outdoor applications, such as convertible tops, tents, and awnings. Pigmented acrylics can be used in applications where maximum resistance to fading and loss of strength is needed. Modacrylic fibers may be used in applications, such as awnings, which require high resistance to flame. Acrylics with low comonomer content are highly resistant to chemical attack and thermal degradation. These properties make acrylics suitable for industrial filters and battery separators. In spite of their many desirable properties, acrylics have not significantly displaced continuous filament nylon, polyester, and polyolefin in the industrial fibers market.

As a filter medium, for example, acrylics lack sufficient abrasion resistance and dimensional stability at elevated temperatures to find universal acceptance. As a fiber for ropes and fish nets, acrylics excel in resistance to weathering but lack sufficient tensile strength for many applications. High-tenacity, continuous filament nylon, polyester, and UV-stabilized polyolefin are generally chosen where tensile strength is critical, even though the stability of these fibers toward sunlight and chemical exposure is inferior to acrylic fibers. Geotextiles is another large-volume market. Here acrylics have excellent properties but are too costly. A comparison of the major synthetic fibers in industrial and outdoor use has been published in *Textiles* [541] and similar information on acrylic and modacrylic fibers was published in *Industrial Fabric Products Review* [542].

In the general apparel markets, continuous filament acrylics and modacrylics find use as artificial fur or in fabrics with a fur-like appearance. These fibers are made in coarse deniers with special cross sections and surface modifications (e.g., surface inclusions or roughening) to simulate natural animal hairs. Uniform fabrics and silky fabrics are also produced from continuous filament acrylic and modacrylic yarns. In Japan, continuous filament yarns in very fine deniers are valued as a silk replacement. In this market, continuous filament is a premium product used in high-fashion dress fabrics, satins, and poplins or to produce a cloth suitable for surface raising to give a suede or fine velour effect. Fine-denier filaments may also be used to produce fabrics with very high-fiber density and low air permeability. These fabrics are useful for insulating material, for example, in quilted interliners for parkas and outerwear for skiing.

12.8.3 Fibers with High-Bulk and Pile Properties

High-bulk acrylic fibers are commonly made by blending high-shrinkage and low-shrinkage staple fibers. The high-shrinkage fiber can be spun using a hot-stretching process in which the fiber is drawn during washing, drying, or after drying. The low-shrinkage component is relatively unstretched or relaxed fiber. When the resulting yarn is allowed to relax, the high-shrink component causes the unstretched fiber to buckle and add bulk to the yarn. High-bulk acrylic yarns are also made by the turbo staple process in which tow is converted to staple by a stretch-break process that produces high-shrinkage staple of uneven cut lengths.

Another method of producing high-bulk yarns is with bicomponent fibers. There are a number of different types of bicomponent fibers on the market today. A true bicomponent fiber consists of filaments with high-shrinkage and low-shrinkage polymer types fused together along the longitudinal axis. In one type of bicomponent fiber, the copolymer on one side has a low comonomer level, while the other copolymer has a high comonomer level. When the fiber is heated, the polymer component with high comonomer content shrinks more than its counterpart, causing a helical crimp to develop. In the other major type of bicomponent, an ionic comonomer, such as SSS, is incorporated into one of the polymer components. During wet spinning, the ionic polymer component becomes more swollen along the longitudinal axis than the other polymer component. Thus, a helical crimp develops when this type of fiber is dried. This type of crimps is referred to as water-reversible crimp; the fiber loses much of its crimp when wet but regains it upon drying.

A variation of the bicomponent is known as the random bicomponent. This concept, described by Fitzgerald and Knudsen [543], is based on the fact that viscous spinning solutions can be merged without complete mixing. When passed through a standard spinnerette, bicomponent fibers are produced ranging in composition from 100% component A to 100% component B. This type of process gives a bicomponent fiber with good bulk development while yielding a substantial gain in productivity for the fiber producer.

Toray and Asahi have reported new types of bicomponent fibers that employ more than two functional components. Toray, for example, has announced the development of a water-repellent, antipilling, bicomponent fiber for the sport sweater market [544]. This product, which derives its water repellency by the addition of a fluorine-type resin, exhibits an exceptionally soft handle along with long-lasting water repellency. Asahi has developed a three-component fiber that is made by spinning one polymer as a continuous phase and intermittently spinning the remaining two polymers as dispersed phases [545]. When polymers of differing dyeability are used, this reportedly gives a fiber with a unique periodically varying shade. The subject of bicomponent acrylic fibers has been treated in more detail by Masson [540].

12.8.4 Flame-Resistant Fibers

Acrylics have reasonably good flame resistance compared with cotton and regenerated cellulosic fibers. In addition, acrylics exhibit more of a tendency to char when burning compared with polyesters and nylons, which form melts. A higher degree of flame resistance is required for certain end uses, such as children's sleepwear, blankets, carpets, and drapery fabrics. This can be achieved by copolymerizing acrylonitrile with halogen-containing monomers, such as vinyl chloride, vinyl bromide, and vinylidene chloride. Modacrylics are used where a high resistance to burning is required. In such fibers, the level of halogen-containing monomers may be over 50%, as in Dynel, which is a 40:60 acrylonitrile–vinyl chloride copolymer. Other modacrylics are Monsanto's SEF [546], Kanebo's Lufnen, Bayer's Dralon MA, and Kanegafuchi's Kanekalon. Flame-resistant acrylic fibers are also widely produced. These fibers, which by definition contain less than 15% comonomer, are made by copolymerizing or blending with halogen- or phosphorous-containing polymers [547,548] or by using dope additives, such as antimony compounds [549] or compounds of halogen or phosphorous. Monsanto's Type 90 carpet fiber is an example of a flame-resistant acrylic fiber.

The methods that have been used to enhance the flame resistance of acrylic and modacrylic fibers have been reviewed by Nametz [550]. Numerous reviews of flammability have been published [551–560], including the extensive work of Lewin et al. [378,379].

12.8.5 Specialized Products

A growing trend in the acrylic fibers industry is an effort to move to develop premium products based on proprietary technology. In the consumer markets specialty fibers, such as ultrafine denier fibers, are used for silk-like fabrics and superfine nonwoven fabric [561]. Hollow fibers for apparel [562] and coarse-denier fibers for wigs and simulated animal hair [563,564] are either available or remaining developed. Another important example of this trend is the so-called "functional fabrics." One of the most recent of these is the antimicrobial fabric [565–568,569,570]. One method of producing this type of fabric is by using microporous fibers [571]. Bayer's Donova, one example, is capable of absorbing large amounts of moisture. The void structure is treated with chemicals that retard the growth of bacteria and fungus. These fibers are valuable for apparel subjected to perspiration, such as socks, sportswear, underwear, and baby wear. Functional fabrics are also made from moisture-repellent fibers [572], moisture-absorbent, hydrophilic fibers [573], sheath-core fibers, surface modified moisture-absorbent fibers, and antistatic fibers [574,575].

Industrial applications include ion-exchange fibers, made by treating acrylic fiber with NH_2OH and NaOH to produce amidoxime fiber, used for removing uranium from seawater [576,577], and hydrazine-treated acrylic fibers have been used for removal of other heavy metals [578]. Fundamental studies of treatment methods have also appeared in the literature [579,580,581–584].

Metalized and semiconducting fibers have also been discussed [585–587]. Another key industrial application for acrylic fibers is for asbestos replacement, especially for concrete reinforcement and friction surfaces. An example of a concrete and mortar reinforcement fiber is Dolanit acrylic from Faserwerk Kelheim GmbH [588]. The subject has been reviewed by a number of authors [589–591]. Other acrylic fiber modifications and reviews of acrylic fiber modifications, in general, have been published by several authors [592–596]. A comprehensive summary of specialized acrylic and modacrylic fibers is given in Table 12.35. The major types,

Acrylic Fibers

TABLE 12.35
Summary of Specialized Acrylic Fibers and Yarns

Fiber or yarn	Application	Key technical elements	Commercial examples	Fiber producer
Filament	Silk replacement	Fine denier, high luster	Sumola Badiena Silpalon	Toray Asahi Mitsubishi Rayon
	Outdoor fabrics tents, awnings, etc.	Resistant to sunlight, light-fast-pigmented forms	Orlon 81	DuPont
High-bulk yarns	Wool replacement	Blend of high-shrinkage and low-shrinkage staple, e.g., by turbo staple process	High Shrink HIHI Mill Process	Toho Beslon
Bicomponent (heat sensitive)	Wool replacement, increased pile height, sweaters, craft yarns	Two polymers of different comonomer levels fused uniformly or randomly along filament	Bi-Loft	Monsanto
Bicomponent (water sensitive)	Wool replacement, rebulkable by consumer in normal clothes dryer	Two polymers of different ionomer level fused uniformly or randomly along filament	Paquel Remember Sayelle	Monsanto Monsanto DuPont
Soft handle	High-fashion apparel, velour effects, silk-like	Deniers from 0.1 to 1.2; special comonomers and finishes; modified cross sections	Fina Elfin Ultramere	Monsanto Asahi DuPont
Coarse denier	Wigs, carpets	Deniers from 50 to 60—wigs; deniers from 15 to 20—carpets and upholstery; usually modacrylic	Fina (wig) Velicren S.D. Acrilan 90	Monsanto Snia-Viscosa
Animal hair	Artificial fur, substitutes for mohair, alpaca, cashmere, angora	Modified cross section, special surface effects built in	MHD (mohair) NC (alpaca)	Asahi Asahi
Antipilling	No-pill sweaters and other knit goods	Spinning modified to produce brittle fibers	Corpilon Serista-K	Toray Asahi

continued

TABLE 12.35 (Continued)
Summary of Specialized Acrylic Fibers and Yarns

Fiber or yarn	Application	Key technical elements	Commercial examples	Fiber producer
Water absorbent	Cotton replacement, improved wear comfort	Porous fiber cross section; hydrophilic comonomers	Donova Aqualon ES	Bayer Kanebo
Soil resistant	Work clothes, uniforms, sportswear	Fluorocarbon surface treatments	Parasoil	Mitsubishi Rayon
Heat and chemical	Industrial filters, battery separators	Low comonomer content	Dolan 10	Hoechst
Antistatic	Underwear	Special comonomers and surface treatments	Parel Elekile	Toray Mitsubishi Rayon
Antimicrobial	Underwear, athletic socks, towels, etc.	Additives—usually to porous moisture absorbent fiber	Parasoil Libresh	Mitsubishi Rayon Toray
Water resistant	Raincoats, sportswear	Fluorine resin treatment	Toraymok Alpearl	Toray Asahi
High tenacity	Cement reinforcement, asbestos replacement, carbon and graphite fibers	Low comonomer, high orientation stretch-spinning process	Dolanit	Hoechst
Producer dyed	Light-fast outdoor fabrics, awnings, tents, car tops	Dyed in wet spinning during wash stage or pigmented dope	KCD Solara	Kanegafuchi Monsanto
Wear comfort	Knitted outerwear, cotton knits replacement	Blend of moisture absorbing and finedenier staple fibers	Comfort 12	DuPont
Flame-resistant	Children's sleepwear, home furnishings	Incorporate halogen monomers, such as vinylidene chloride and antimony compounds	SEF Kanekalon Lufnen Dralon MA	Monsanto Kanegafuchi Kanebo Bayer
Acid dyeable	Special color effects, cross-dyed knits	Cationic dye sites quaternary ammonium compounds	ACT Orlon 44	Toray DuPont

including high-bulk, bicomponent, and flame-resistant fibers, are described in detail below, while the technology employed in producing the more application-specific fibers is described in Section 12.6 and Section 12.9.

12.9 NEW PRODUCTS AND APPLICATIONS

12.9.1 Carbon and Graphite Fibers

Carbon and graphite fibers are the primary reinforcement materials used to achieve the exceptional stiffness and strength of lightweight-advanced composites. The term "carbon fiber" generally refers to a variety of filamentary products composed of more than 90% carbon and produced by the pyrolysis of acrylic fibers, mesophase pitch, and rayon. The diameter of the individual filaments in the fiber tow or threadline ranges from 5 to 10 μm. Approximately 95% of the precursor fiber used in the world is based on acrylic copolymers. Comonomers such as itaconic acid are added to reduce processing time and improve properties, though the precursors tend to have 90% acrylonitrile by weight, and in this way resemble the textile acrylic fibers.

Carbon and graphite fibers are valued for their unique combination of extremely high modulus, high breaking strength, and very low specific gravity. This is illustrated in Table 12.36, where the mechanical properties of the major commercial high-modulus fibers are compared.

Steel is the cheapest and most commonly used high-modulus fiber. Typically, steel fibers are made with a Young's modulus of approximately 21,000 kp/mm, far greater than fibers of glass (8000 kp/mm), nylon (350 kp/mm), polyester (350–1000 kp/mm), or Kevlar (4500–6000 kp/mm). Steel fibers, however, are very dense, 7.8 g/cm^3, and therefore, are not suitable for applications where light weight is critical. Carbon and graphite fibers, on the other hand, have a Young's modulus equivalent to steel at a density of only 1.8–2.0 g/cm^3 and the breaking strength (tenacity) of the carbon filters is actually higher than that of steel. Therefore, carbon and graphite fibers find wide use in applications such as sporting goods, aircraft, and space vehicles where high modulus and low weight are critical. The specific strength and stiffness of commercially available carbon fibers can be double that of other reinforcing fibers

TABLE 12.36
Comparison of High-Modulus Fibers

Fiber type	Density (g/cm^3)	Tenacity (kp/mm)	Extension at break (%)	Young's modulus (kp/mm)
Steel	7.8	35–120	10–28	21,000
Asbestos				
Glass	2.5	20–35	2–5	7,000–9,000
PAN (high modulus)	1.18	70–80 Cn/T	9–11	1,400–1,600 Cn/T
Dolanit 10		830–940 N/mm		16,500–19,000 N/mm
Nylon 6	1.14	35–70	20–60	25–350
Polyester	1.37	40–85	20–40	350–1,000
Kevlar 49	1.45	250–400	2–3	4,500–6,000
Boron	7.6	280–350	0.5–0.9	38,000–40,000
Carbon (PAN)	1.8–2.0	100–120	2–3	20,000–25,000
Carbon (pitch)	1.8–2.0	100–120	0.3–0.5	20,000–25,000
Graphite	1.8–2.0	100–200	0.3–0.5	40,000

such as Kevlar and S-glass, and they exceed metals by an order of magnitude. In addition to strength and stiffness, carbon fibers possess excellent fatigue strength, vibration damping characteristics, thermal resistance, and dimensional stability. Carbon filters also have good electrical, thermal conductivity and chemical inertness, except to oxidation [597]. A number of books have appeared on carbon fibers that cover the subject to a far greater extent than is possible here. These include, in order of appearance, Sittig [598], Delmonte [599], Donnet and Bansal [600], Fitzer [601], Watt and Perov [602], Dresselhaus et al. [603], Figueiredo et al. [604], Donnet and Bansal [605], and Peebles [606].

High-modulus carbon and graphite fibers are produced by the thermal decomposition of organic filaments at high temperature in a nonoxidizing atmosphere. In earlier days, a distinction was made between fibers heat-treated in the 1000–1500°C range, called carbon fibers, and those heat-treated above 2000°C, called graphite fibers. As the latter are not completely graphitic, they are designated as carbon fibers here. The term "graphitized" is sometimes used to distinguish fibers heat-treated above 2000°C and the term "carbonized" for those heat-treated below 2000°C. The more correct nomenclature is high-strength fibers for the lower treatment range and high-modulus fibers for the upper range [606].

Fibers spun from polyvinyl alcohol, polybenzimidazoles, polyamides, and aromatic polyamides have been used as carbon fiber precursors. However, at present, the most attractive precursors are made from acrylonitrile copolymers and pitch, and a small amount from rayon. Today more than 95% of the carbon fibers produced for advanced composite applications are based on acrylic precursors. Pitch-based precursors are generally the least expensive, but do not yield carbon fibers with an attractive combination of tenacity (breaking strength, modulus, and elongation as those made from a acrylic precursor fiber). The acrylic precursors provide a much higher carbon yield where compared to rayon, typically 55% versus 20% for rayon, and this translates directly into increased productivity.

In general, an increase in tensile strength will come at the expense of the modulus, but the breaking elongation will also be higher. Carbon fibers are classified by tensile modulus and strength. Tensile modulus classes range from low (<240 GPa), to standard (240 GPa), intermediate (280–300 GPa), high (350–500 GPa), and ultrahigh (500–1000 GPa). Typical physical and mechanical properties are presented in Table 12.37 [597]. Pitch-based carbon fibers can be described as stiffer and more brittle than PAN-based fibers; therefore, the latter must be used in applications where extensibility is a requirement.

TABLE 12.37
Mechanical and Physical Properties of Carbon Fibers

Property	PAN[a] standard grade	PAN-IM (intermediate modulus)	PAN-HM (high modulus)	PAN-UHM (ultrahigh modulus)	Pitch HM (high modulus)	Pitch UHM (ultrahigh modulus)
Tensile strength (Gpa)	3.5–4.8	4.1–7.0	1.7–4.7	1.7–3.9	2.4–3.0	2.2–3.3
Tensile modulus (Gpa)	230–240	280–300	350–480	500–600	380–520	550–827
Elongation at break (%)	1.5–2.1	1.5–2.4	0.4–1.4	0.3–0.7	0.4–0.6	0.27–0.5
Density (g/cm^3)	1.77–1.81	1.79–1.82	1.70–1.90	1.93–2.00	2.00–2.14	2.15–2.20

[a]PAN is a polyacrylonitrile-based fiber.
Source: From Venner, J.G., *Kirk–Othmer Encyclopediea of Chemical Technology*, 4th ed., Wiley-Interscience, New York, 1995.

Acrylic precursors are made using conventional wet-spinning manufacturing processes, except that increased stretch orientation is required to produce precursors with higher tenacity and modulus. Most manufacturers of carbon fibers also make the precursor fibers, and it is thought that many trade secrets are employed to optimize properties and manufacturing costs. A number of excellent review articles have been written on the conversion of acrylic fibers to carbon fibers, and include those by Goodhew et al. [607], Peebles [608], Henrici-Olive' and Olive' [609], Watt [610], Riggs [611], Thorne [612], Fitzer et al. [613], Fitzer [614], Domodaran et al. [615], Gupta et al. [616], Rajalingam and Radhakrishnan [617], Bashir [618], Grove et al. [619], as well as the books cited previously.

The first commercially feasible process for converting acrylic fibers to carbon fibers was developed by Walt, Phillips, and Johnson of the Royal Aircraft Establishment (RAE) in collaboration with the acrylic fiber producer, Courtaulds [621]. In the RAE process, the acrylic precursor is converted to carbon fiber in a two-step process [622]. Preoxidation or filament stabilization is carried out in the first stage. The precursor is heated in an oxygen atmosphere under tension at a temperature of approximately 200–250°C, well below its carbonizing temperature (approximately 800°C). At this temperature, the nitrile groups react with each other via a free radical addition process leading to the so-called ladder structure shown in reaction 12.34 [609,621–625].

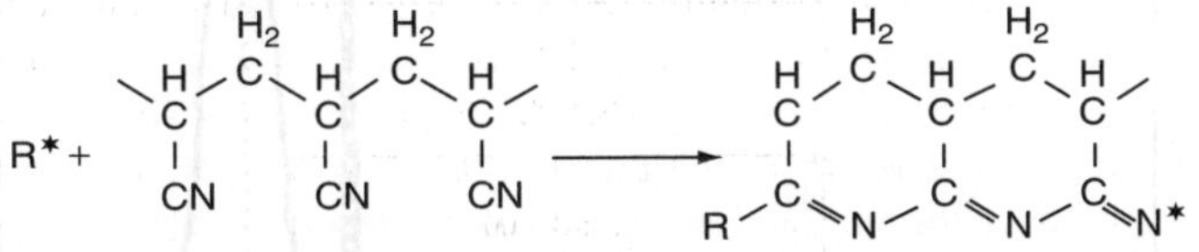

$$R^* + \text{(polyacrylonitrile chain with CN groups)} \longrightarrow \text{(ladder structure with C=N-C=N-C=N*)} \tag{12.34}$$

In addition, the dipole–dipole interactions between nitrile groups on adjacent chains brings these groups into a transition state favoring an intermolecular reaction of the nitrile groups analogous to the intramolecular reaction shown previously. This leads to the cross-linking reaction shown in reaction 12.35.

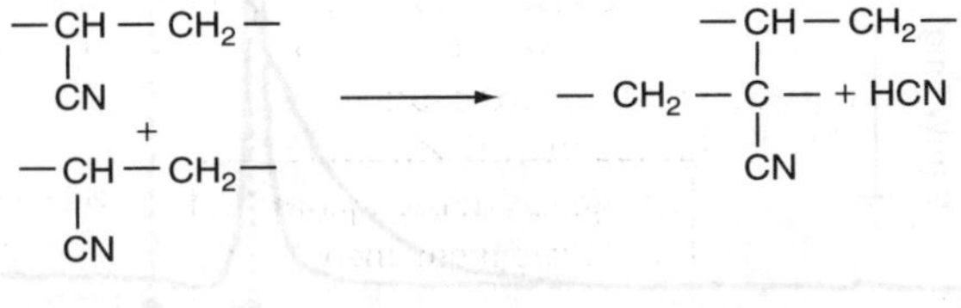

$$\begin{array}{c} -\text{CH}-\text{CH}_2- \\ | \\ \text{CN} \\ + \\ -\text{CH}-\text{CH}_2- \\ | \\ \text{CN} \end{array} \longrightarrow \begin{array}{c} -\text{CH}-\text{CH}_2- \\ | \\ -\text{CH}_2-\text{C}- \\ | \\ \text{CN} \end{array} + \text{HCN} \tag{12.35}$$

The combination of these two reactions stabilizes the carbon backbone of the polymer against melting, loss of molecular orientation, and carbon–carbon bond scission.

The stabilization process is a highly exothermic reaction that begins at approximately 200–230°C and peaks at 300–320°C. The temperature must be kept somewhat low in the early stages of the stabilization process. When the fiber is heated too rapidly, the heat released during cyclization can cause chain scission and fusion of filaments leading to the formation of a hard, brittle char. On the other hand, if the precursor is heated too slowly, the stabilization process may fail to go to completion. In this case, decomposition of the fiber core will occur when the fiber is heated to higher temperatures. The stabilization stage is the slowest part of the carbon fiber manufacturing process. Johnson et al. [622] state that as much as 24 h may be required for the preoxidation of a 2.5-denier fiber at 220°C.

This stabilization time may be shortened by using higher temperatures in combination with precursors that contain comonomers or additives that moderate the exothermic nitrile cyclization reactions [609]. The evolution of heat associated with the nitrile reaction can be conveniently followed using DSC. In Figure 12.50, the DSC traces for PAN homopolymer and a copolymer containing 5% methyl acrylate and 1% itaconic acid are compared using both an air and nitrogen environment [614]. The exotherms in both environments are much more narrow for the PAN homopolymer, which means that the evolution of heat will be very rapid and the fiber will be damaged by self-heating in the manner described previously. This evolution occurs over a wider temperature range for the copolymers and hence degradation due to self-heating can be controlled in the carbonization process.

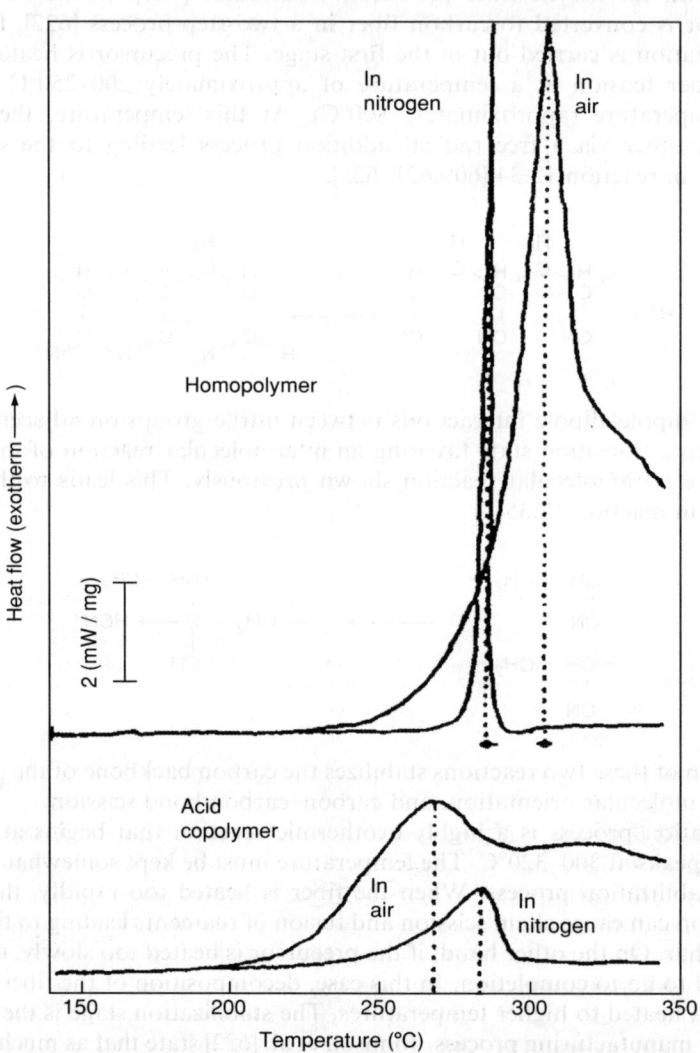

FIGURE 12.50 Comparison of the exotherms obtained using differential scanning calorimetry for a PAN homopolymer and a copolymer containing 6% methyl acrylate and 1% itaconic acid, both in nitrogen and in air, at a heating rate of 5°C/min. (From Fitzer, E., *Carbon 27*, 621, 1989.)

In the second stage, the fiber is carbonized by heating at a temperature of approximately 800–1400°C in a nonoxidizing atmosphere. Essentially all heteroatoms are expelled during this stage, leaving a fiber consisting of 99% carbon. The volatile products formed during carbonization are HCN, CO, CO_2, and CH_4. Thus, some carbon is also lost, giving a carbon fiber roughly 50% by weight of the original acrylic fiber. The gaseous products vary with the temperature of the carbonization. At 800°C, the primarily initial gaseous product is HCN [626]. This is believed to result from ring condensation, as shown in the reaction 12.36.

Above 1000°C, nitrogen becomes the primary gaseous product. This is believed to be the final step in converting the stabilized fiber to the desired carbon form.

The ultimate modulus of carbon fibers can be achieved by graphitizing in a third stage where the carbonized fiber is heat-treated at 1800–3000°C in a nonoxidizing atmosphere. The temperature of this third heat treatment is sufficient to induce bond rearrangement and development of graphite crystallites.

$$\text{(structure)} \longrightarrow \text{(structure)} + 2\text{HCN} \quad (12.36)$$

$$\text{(structure)} \longrightarrow \text{(structure)} + 2N_2 \quad (12.37)$$

The key structural features of the carbon and graphite produced by these heating processes are parallel planes of fused rings, as shown in Figure 12.51 [627]. The turbostratic structure is first achieved, at the carbon stage, and it can be considered to be a disordered form of graphite [597]. The formation of these graphite-like sheets, seen in Figure 12.51, is the key to the outstanding properties of carbon fibers. These sheets are aligned along the fiber axis so that tensile and compressive stresses are in the sheet direction and ultimately the load is supported by the rigid fused ring system, which is very strong and stiff. The strong sp^2 bonding within these graphite layer planes results in the highest absolute modulus and theoretical tensile strength of all known materials [628]. Transverse properties are significantly weaker than axial properties as graphite layers are held together by weak dispersive bonds resulting in low shear modulus and cross plane Young's modulus and strength.

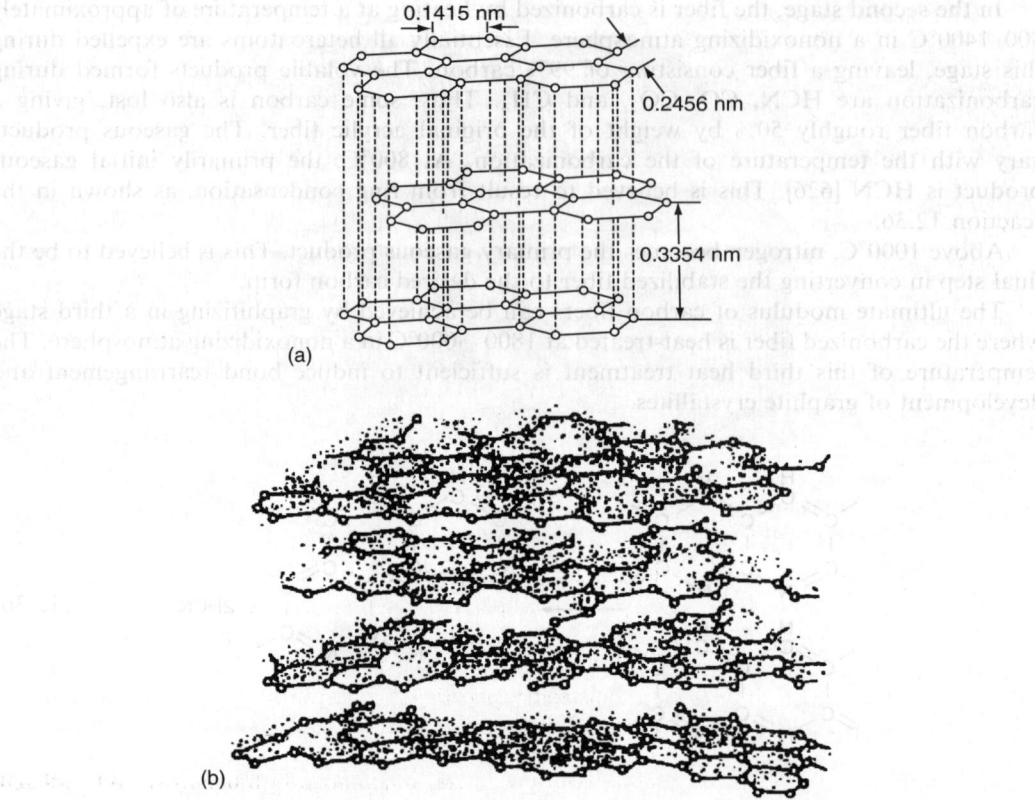

FIGURE 12.51 (a) Crystal structure of a graphite crystal. (b) Structure of turbostratic carbon. The two structures are not drawn to scale. The distance between layer planes in turbostratic carbon is greater than in graphite. (From Hoffman, W.P. et al., *J. Mater. Res. 6*, 1685, 1991.)

Maximum strength and modulus are achieved when the basal planes of the graphite yarns are oriented parallel to the longitudinal axis of the fiber. Since the polymer carbon backbone remains essentially intact during the conversion to graphite, orientation developed at any stage of the overall conversion process will remain to some degree in the final product. Thus, it is important to introduce as much orientation as possible during the precursor spinning process. One method frequently used to increase fiber orientation is heating the fiber at constant length during the stabilization, carbonization, or graphitization processes. The modulus of carbon fibers has been shown to be a strong function of the measured orientation of the precursor fiber, and this applies to PAN, pitch, and rayon-based fibers [629].

One model of a typical carbon fiber is shown in Figure 12.52, and it clearly shows the parallel arrangements of the sheets containing the fused rings, and the sheets are aligned along the fiber axis [630]. The figure also depicts some of the defect structures that particularly affect the transverse properties. Detailed models of the carbon fiber structure and elucidation of key structure–property relationships are included in the books and review articles mentioned previously. The influence of the precursor composition [631], spinning conditions [632,633], and structure [634,635] have been reported. The production of carbon fibers has been described by Morgan [636] and Fournel [637], and the chemistry of carbon fibers has been reviewed by Henrici-Olive' and Olive' [609].

Acrylic Fibers

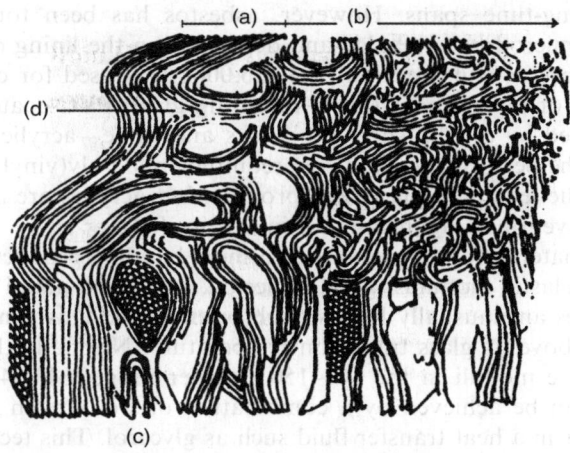

FIGURE 12.52 The combination of basic structural units into microdomains within a carbon fiber: (a) skin region; (b) core region; (c) a hairpin defect; and (d) a wedge disclination. (From Bennet, S.C. and Johnson, D.J. in *Society of Chemical Industry, London*, p. 377, 1978.)

Carbon fiber composites traditionally have been very expensive because the pyrolysis step must be done very slowly on relatively small tow bundles to avoid unwanted thermal degradation reactions that reduce the fiber properties. Therefore, the production rate in terms of pounds of fiber per hour is much lower in comparison to other common reinforcing fibers. However, progress has been made in recent years to develop grades of carbon fibers that are less expensive to produce and can be utilized in less demanding commercial applications. This will allow carbon composites to penetrate markets that are not considered high technology, such as automotive parts, construction, fireproof textiles, asbestos replacement, and other common materials applications. The market for carbon and graphite fibers is expected to grow rapidly as production costs fall. The pricing of carbon fibers depends on the precursor (PAN versus pitch), carbon fiber filament count (i.e., tow size), modulus, and product form (continuous filament, prepreg tape, chopped, or mat and paper). Estimated prices for four classes of PAN-based fibers in 1994 were as follows: standard modulus, 12–20 $/lb; intermediate modulus, 35–65 $/lb; high modulus, 75–125 $/lb, and ultrahigh modulus, 300–550 $/lb. These prices are based on fiber tows with a filament count of 12,000. Zoltek Corporation and Akzo each offer a 48,000–50,000 filament tow that sells for about 10–12 $/lb. Zoltek Corporation (St. Louis, MO) and RK Carbon Fibers Ltd. (Cheshire, U.K.) announced a new continuous carbon fiber-producing process using lower-cost, large-strand acrylic fibers; this new process should eventually reduce nongovernment-certified carbon fiber prices by as much as 25% [638]. Zoltek's long-range plan is to reach a 5 $/lb selling price while maintaining and improving margins. Growth is anticipated in the conductive and electronics market as well as in civil engineering, infrastructure, and automobiles. This is an example of a cost-sensitive customer, and meeting their costs will require processes utilizing large tows on the order of 200,000 filaments that the company can process. At a cost of $7.40 and below, carbon fiber becomes very competitive with steel in various automotive applications.

12.9.2 Asbestos Replacement Fibers and Other Composite Applications

Asbestos fiber is an excellent reinforcing fiber for cement because of its chemical and thermal inertness, fibrous structure, and high elastic modulus. Also, it has good adhesion to the cement, good dispersibility in the cement–water slurry, and excellent resistance to the alkaline environment over long-time spans. However, asbestos has been found to be extremely hazardous when the microdust is inhaled and deposited on the lining of the lungs. In 1981, 75% of the world asbestos consumption of 4,726,000 t was used for cement reinforcement [639]. A wide variety of replacement fibers, including glass, PVC-coated steel, carbon, and aramid, were evaluated by cement-sheet producers and three—acrylic, poly(vinyl alcohol), and cellulosic—met the performance and cost requirements. Poly(vinyl alcohol) has the best performance and is the leading replacement product, but acrylics are a strong contender in this area [640] and have found significant use in Europe [641].

As a reinforcing material for ambient-cured cement building products, acrylics offer three key properties: high elastic modulus, good adhesion, and good alkali resistance [642]. The high modulus requires an unusually high-stretch orientation. This can be accomplished by stretching the fiber above its glass transition temperature. Normally this is done in boiling water or steam to give moduli in the 100–150 g/denier range [643,644]. Alternatively, the stretch orientation can be achieved by a combination of wet stretch at 100°C and plastic stretch on hot rolls or in a heat transfer fluid such as glycerol. This technique is reported to give moduli as high as 200 g/denier [645,646].

Patents for acrylic asbestos replacement fibers have been obtained by Asahi [647–649]. Wuestfeld [650], REDCO [651], and Hoechst [652]. The Hoechst fiber, marketed under the trade name Dolanit (originally Dolan 10), is offered in two forms, as shown in Table 12.38 [653,654]. Similar products are available from Bayer (ATF 1055), Courtalds (Sekril), and Montefibre (Ricem).

Acrylic fibers, such as Dolanit, are blended in ambient-cured cement at a rate of 1–3%, compared with 9–15% by weight of asbestos. The flexural strength of cement sheets of acrylic-reinforced cement is equivalent to that of asbestos-reinforced cement and nearly double that of untreated cement, as shown in Table 12.39 [655].

This fiber is finding applications in reinforced concrete window boxes, wall cladding, roof tiles, short corrugated sheets, and drain pipes. Larger markets, such as reinforcement of plaster boards, cement floors, liquid concrete, and plaster, are sought. Other applications besides concrete reinforcement, such as adhesives, sealing compounds, and reinforced plastics, are developed. Acrylic fibers have already found use as a reinforcing fiber for molding resins made from acrylonitrile–styrene copolymer [656]. Today, only a few companies produce asbestos-free sheet and pipes using PAN or PVA fibers because the manufacturing costs are 30% higher compared to the recipes based on asbestos. The main use of PAN fiber is for partial replacement of asbestos in sheets to improve the productivity of the machine and the

TABLE 12.38
Properties of Dolanit Asbestos Replacement Fibers

Property	Dolanit 10	Dolanit VF
Staple length (mm)	6.0	1–24
Fineness (dtex)	3.0	1.5–25
Tenacity (CN/tex)	70–80	70–80
Elongation to break (%)	9–11	9–11
Initial modulus (CN/tex)	1400–1600	1400–1600
Wet strength (% of dry)	90–98	90–98

Source: From *High Perform. Text.*, 3(10), 3, 1983 and U.S. Patent 4,418,115.

TABLE 12.39
Flexural Strength of Acrylic Reinforced Cement

Additive	Flexural textile strength (N/mm)
None	8–10
Asbestos	24–40
2% Dolanit 10	25
2% Conventional acrylic	20

Source: From *High Perform. Text.*, 3(10), 3, 1983.

product quality and in pipes for total replacement of the crocidolite or blue asbestos, which is considered the most harmful and is banned in many countries.

Another application where acrylics have proven to be effective is in brake and clutch linings. Here thermal stability is a key property. Hoechst has developed an experimental heat-resistant acrylic fiber (VF 1003) that can be used in combination with glass fibers. The French company Valco is developing a friction material for brakes and clutches that is also based on the use of acrylic fibers [657]. Here the acrylic is used as a cross-linkable and fusible additive along with glass fibers, fillers, and binders. The fusible acrylic acts as a lubricant by melting when the friction material is rubbed. This allows the lining to compact thereby improving its wear resistance without affecting friction properties.

The use of fibers for battery plate reinforcing parallels the use in cement reinforcing. The purpose is to produce positive and negative plates that will not crack when drying or when they are handled during battery assembly. Later, during the battery operation the fibers prevent cracking and flaking caused by charging, discharging, and vibration. Modacrylic fibers are generally used in this application, although it is not clear that they are superior to acrylics for this purpose. The most important attribute of the fiber is resistance to hydrolysis in sulfuric acid solution, which is a characteristic of both acrylic and modacrylic fibers [658].

12.9.3 Fibers for Reverse Osmosis, Ion Exchange, and Filtration

Tubular fibers for reverse osmosis gas separations, ion exchange, ultrafiltration, and dialysis are a significant new application of acrylic fibers and other synthetics [659]. Commercial acrylic fibers have already been developed by Nippon Zeon [660–662], Asahi [663], and Rhone Poulenc. One such fiber, developed at the California Institute of Technology [664], is a hollow fiber made from copolymers of acrylonitrile and vinyl acetate or ethyl acrylate. The acrylonitrile-based hollow fiber acts as a substrate that is subsequently imbibed with liquid monomers that can be polymerized *in situ*. Monomers are used that form insoluble, cross-linked, ion-exchange resin particles embedded in the walls of the fiber. These fibers are useful for removing counterions from solution. Solid fibers alkylated with epichlorohydrin, having anion-exchange properties, are reported by Alikberova [665] for removing toxic cyanide compounds from air, and cation-exchange fibers, containing ionogenic carboxyl groups, are described by Dorokhina and Zharkova [666]. Polyampholyte ion-exchange fibers with high selectivity toward nonferrous, rare-earth, and heavy metals have also been synthesized, by modifying PAN fibers with hydroxylamine or polyethylene polyamine [667].

Acrylic fibers are used in filtration applications where aggressive environments, such as exposure to SO_2 and other acids, are encountered. The PAN homopolymer is preferred to a copolymer, since the PAN has a lower diffusion rate and less sensitivity to moisture [668].

TABLE 12.40
Maximum Use Temperatures for Various Fibers in Filtration Applications

Type of fiber	Maximum temperature (°C)	
	Moist	Dry
Glass	260	290
Poly(tetrafluoroethylene)	260	260
Aramid	180	230
Acrylic homopolymer (Dralon T)	40	140
Nylon 6,6	120	120
Copolymer acrylics	110	120
Polyester	100	150
Wool	90	110
Cotton	82	93
Polypropylene	93	107

Source: From Kneip, E., *Chemiefasern/Textilindustrie*, 32/84, June, E39, 1982.

Dralon T is supplied by Hoechst for this purpose. Both staple and filament types are sold. Acrylics are suitable as filter fabric in the midrange applications, based on the application temperatures shown in Table 12.40.

The operating temperature should be set approximately 15°C below these maximums to allow for short-term deviations. The homopolymer Dralon T has a 20–30°C advantage in maximum operating temperature over the acrylic copolymers. The resistance of Dralon T and Acrilan 16, an acrylic copolymer, to various chemicals is shown in Table 12.41.

12.9.4 Use as Moisture-Absorbent Synthetic Paper

Processes for making a water-absorbent synthetic paper with dimensional stability have been developed by several companies. In a process developed by Mitsubishi Rayon, acrylic fiber is rendered insoluble by hydrazine and then hydrolyzed with sodium hydroxide. The paper, formed from 100 parts fiber and 200 parts pulp, has a water absorption 38 times its own weight [671]. Processes for making hygroscopic fibers have also been prepared in the patent literature. These fibers are used in moisture-absorbing nonwovens for sanitary napkins, filters, and diapers. For example, in a process reported by Exlan the fibers are made by incorporating a hydrophilic comonomer, such as methylenebisacrylamide, then hydrolyzing the copolymer with caustic. The resulting copolymer, having hydroxyl groups thus incorporated, is then blended with a second hydrophilic acrylonitrile copolymer [672]. Developments in the use of PAN pulp and synthetic paper derived from it have been reviewed by Takeda [673] and Simionescu et al. [674].

12.9.5 Electrically Conducting Fibers

A process for making electrically conducting fibers based on treatment with zinc oxide has been reported by Kanebo [675] and copper ions by Exlan [676], Nippon Sanmo Dyeing Co. [677], and Teijin [678]. In the Exlan process, a copolymer of 86% acrylonitrile, 11% vinyl acetate, and 3% dimethylaminoethyl methacrylate was spun into fibers and then treated in a solution containing copper sulfate and $(NH_2OH)_2H_2SO_4$. The resulting fiber was then reduced in a solution containing $Zn(HSO_2CH_2O)_2$ to give fibers with a specific resistance of

TABLE 12.41
Strength Retention of Dralon T and Acrilan 16 after Exposure to Acids, Bases, and Other Chemicals

Chemical	Conc. (%)	Time (h)	Temp. (°C)	Acrilan (%)	Dralon
Acetic acid	5	10	99	97	
	40	10	21	91	
	50	15	100		Very good
	98	15	75		Very good
Acetone	100	1000	21	100	
Ammonia	28	15	20		Very good
Benzoic acid	3	10	99	94	
Chloroform	100	1000	21	100	
Ethyl ether	100	1000	21	100	
Ferric chloride	3	10	99	98	
Formic acid	5	10	99	98	
	10	15	75		Very good
	40	10	21	96	
	98	15	20		Good
Hydrochloric acid	10	10	21	100	
	10	15	75		Very good
	30	15	75		Good
	37	10	21	94	
Hydrogen peroxide[a]	0.3	10	21	100	
Hydrogen peroxide[b]	3	10	21	98	
Lactic acid	100	15	75		Very good
Nitric acid	7	10	21	100	
	10	10	21	94	
	10	15	75		Very good
	30	15	75		Satisfactory
	40	15	20		Good
Oxalic acid	5	10	21, 99	96	
Peracetic acid	2	10	21	95	
Phosphoric acid	10	15	75		Very good
Phosphoric acid	50	15	75		Good
	70	15	20		Good
Potassium hydroxide	10	15	20		Very good
	10	15	75		Good
Salicylic acid	3	10	99	98	
Sodium borate[c]	1	10	21	98	
Sodium carbonate	Dilute	15	20		Very good
	Saturated	15	100		Adequate
Sodium chlorite[d]	0.7	10	21	94	
Sodium hydroxide	10	15	20		Very good
	10	10	21	93	
	10	15	40		Good
	40	10	21	91	
	50	15	20		Good
Sodium hypochlorite[a]	0.4	10	21	98	

Continued

TABLE 12.41 (Continued)
Strength Retention of Dralon T and Acrilan 16 after Exposure to Acids, Bases, and Other Chemicals

Chemical	Conc. (%)	Time (h)	Temp. (°C)	Acrilan (%)	Dralon
Sulfuric acid	1	10	99	100	
	10	15	75		Very good
	30	15	75		Satisfactory
	60	15	75		Adequate

[a]pH 7.
[b]pH 6.
[c]pH 10.
[d]pH 4.

Source: From Kneip, E., *Chemiefasern/Textilindustrie*, 38/90, December T116, 1988 and Anon, *Industrial Uses of Acrilan Fiber*, TT-17 Monsanto Company Bulletin, August 1965, p. 6.

only 33 Ω/cm compared to 1010 Ω/cm without this treatment. Electrically conducting acrylic fibers are also made by spinning fibers containing carbon black [679,680] and other methods [681]. Electrically conducting fibers are useful in blends with fibers of other types to achieve antistatic properties in apparel fabrics and carpets. The process developed by Nippon Sanmo Dyeing Co., for example, is reportedly used by Asahi in Casmilon 2 denier staple fibers [682]. Okoniewski and Koprowska [683] studied PAN fibers treated with copper ions complexed with a sulfide ion. When blended at a rate of 3–4% with untreated acrylic or polyamide fibers, to make carpets and nonwovens, the resistivity of the end products was reduced by a factor of 10,000.

12.9.6 METALLIZED FIBERS

Metallized textiles are studied as a means to improve radar screens, making them more sensitive to small objects. The ability to be seen by radar is greatly increased by sheet-like materials containing thin layers of metallized fibers. Bayer has developed an acrylic filament yarn of 238 dtex, which gives radar reflectance of as much as 90%. The fiber is coated with a 0.02–2.5 μm layer of nickel by a currentless wet-chemical disposition process [684,685]. Other metallized fibers have been developed based on palladium [686].

12.9.7 OXIDIZED PAN

PANOX is PAN that has been oxidized; it is a trade name for a product made by R.K. Carbon Fibers in the U.K. A similar product is made by Stackpole in the United States. It is used as a flame-retardant textile material in braking systems, and as gland packings, but it primarily serves as a precursor for high-strength carbon fiber. In Section 12.9.1, we discussed the process by which acrylic fibers are converted to carbon fibers using the two-stage heating process. The first stage is a stabilization in air at approximately 200–300°C, followed by carbonization at over 1000°C. If the fibers are removed after the first step, i.e, after initiation of the nitrile cyclization reaction and oxidation reactions, then it still functions as a textile fiber, albeit with very specialized properties [687]. This is the PANOX filter. There is one major difference in processing between carbon and PANOX fibers; the latter is not restricted to continuous filament tows since PANOX is not intended for composites. PANOX may be processed as a stretch-broken tow or cut to stable, and hence the acrylic precursor may be a tow of 1.5 dpf, 480 K total denier.

The reported world tonnage in 1990 was 2000 t [688]. One of the major limitations to its use in visible applications is that it is available only in black. For protection against intense heat, however, it offers considerably more protection than conventional fire protection textile fibers. A PANOX-based fabric is reported to maintain a barrier against a 900°C flame for more than 5 min. In addition to its low flammability, it has an exceptionally low thermal conductivity [689].

12.10 FUTURE TRENDS

The acrylic fiber industry is mature. Traditionally, much of the appeal of acrylic fibers has been as a wool replacement. The wool market itself is a relatively small and declining market. Therefore, growth in acrylic volumes is very slow and controlled by a combination of personal income and population growth. Growth is strongest in the developing countries, especially countries like mainland China, where population growth is rapid and personal incomes are rising. Competition from other fiber types, particularly cotton, is not as strong in comparison to the developed countries. In China, for example, most of the land available for farming must be used to grow food, not fiber. These countries also have a cheap labor source. As a result acrylic fiber can be produced and finished garments can be made at a cost that makes exporting very profitable.

World production is summarized by fiber type in Table 12.42 [690]. The data show a decline in acrylic and nylon fiber production and a steadily increasing production of polyester fiber. Polyester, a mainstay of the low-cost and easy-care markets, has grown considerably in the developing countries. Recent advances in the recycle of polyester has put more cost pressure on markets where polyester has been excluded because of deficiencies in properties. An example is the carpet face yarn market where polyester has suffered from its lack of resiliency and tendency for matting. New, durable constructions are possible with lower cost polyester supplies. Olefin production, while relatively low compared to the others, is also growing steadily.

U.S. production of staple and tow for 1965–1996 is shown in Figure 12.53 [691]. The data clearly show the rise in polyester as the dominant fiber, beginning in 1965 and peaking by 1980. The data also show that acrylic enjoyed its best years in the late 1970s, with the decline in the carpet market starting in the early 1970s, followed by losses in knit markets beginning

TABLE 12.42
World Production of Man-Made Fibers by Type of Fiber (in Millions of Tons)

Year	Acrylic	Nylon	Polyester	Olefin and misc.
1970	21	40	34	5
1975	19	33	45	3
1980	19	30	47	4
1985	18	26	50	6
1990	15	24	53	8
1995	12	20	59	9

Source: From Akzo Nobel report on synthetic fibers, *Int. Fiber J.*, pp. 4–14, August 1996.

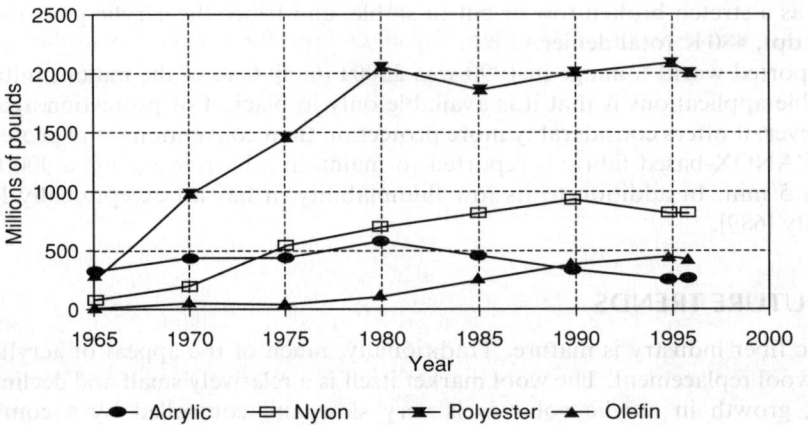

FIGURE 12.53 U.S. production of synthetic fiber staple and tow showing the relative quantities of acrylic fiber produced compared to nylon, polyester, and polyolefin. Acrylic production peaked in the late 1970s. One of the major factors was the decline of the acrylic carpet market, now dominated by nylon. (From *Manufactured Fiber Producer Handbook*, 1996; *Fiber Organon*, February 1997, Fiber Economics Bureau, pub.)

in the mid- to late 1980s. Total U.S. shipments, including imports and exports, are shown in Figure 12.54 [691].

Table 12.43 summarizes the shipments of U.S. synthetic staple and tow fiber by application for 1972–1995 [692]. Key points are the overall decline in the use of acrylic fibers as well as the decline of the carpet and craft yarn markets in particular. The category "other acrylic markets" includes upholstery and industrial fiber applications. This has improved

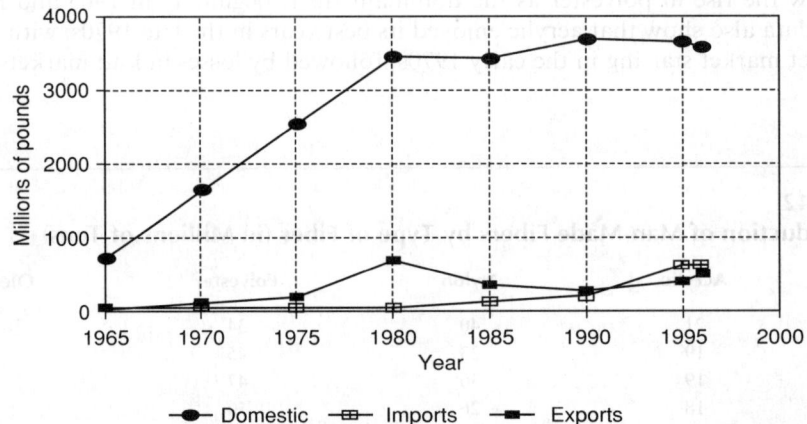

FIGURE 12.54 U.S. shipments of all synthetic fiber staple and tow, including imports and exports. Synthetic fiber staple and tow shipments, including exports, peaked in the late 1970s. Imports, however, have continued to increase since 1980. (From *Manufactured Fiber Producer Handbook*, 1996; *Fiber Organon*, February 1997, Fiber Economics Bureau, pub.)

TABLE 12.43
U.S. Synthetic Staple and Tow Shipments by Fiber Type for 1972–1995 (Millions of Pounds)

	1972	1975	1980	1983	1989	1992	1995
Hosiery, sweaters, and craft yarns	138	190	276	237	175	136	62
Pile fabrics	59	66	60	33	20	N/A	13
Other circular knits and flat knits	131	69	125	130	132	112	83
Blankets	24	35	42	27	30	N/A	28
Other broadwoven	32	13	42	33	41	38	—
Carpet face yarns	140	83	30	22	<10	<10	<10
Other acrylic markets	14	11	10	10	17	38	58
Total shipments	538	467	585	492	415	322	270

N/A: Not available.
Source: From *Fiber Organon*, Fiber Economics Bureau, pub., March 1997.

significantly as the acrylic industry in the United States seeks out higher value products to replace commodity production.

In the developed countries of Western Europe, Asia, and the United States, demand for internally produced fiber has decreased due to several factors. First, as mentioned above, finished goods from acrylic fibers are made at a low cost in inexpensive labor markets and imported for sale at very low prices compared to domestically produced goods. This trend reached a peak during 1985–1989. In 1990, dumping duties were imposed in the United States and the trend began to reverse. However, in 1994, the legality of these duties was questioned in the U.S. courts and the duties were dropped. The effects remain to be seen. In addition, fashion trends instilled in part by heavy advertising by the cotton industry have reduced the popularity of acrylic products. Cotton socks, for example, are still more popular than acrylic, in spite of the fact that acrylic socks offer excellent moisture transport and superior abrasion resistance. Acrylic athletic socks made from Monsanto's Duraspun wear-resistant fiber are promoted with some success. This trend toward cotton in sweaters, socks, and sweatshirts in the past 10–15 years may slow as crop acreage decreases and demand for clothing increases, especially in China.

Circular knit fabrics have declined in popularity from 195 million pounds in 1986 to less than 85 million pounds in 1995. Sweatshirts, leisure suits, and athletic warm-ups are now more popular in polyester and nylon fabrics. High-pile fabrics have also declined in popularity. The use of acrylics in products such as robes, sleepwear, fake furs, and coat liners has been supplanted by cheaper polyester fibers. Also, the number of domestic suppliers of high-pile and flame-retardant (FR) acrylic fibers has decreased, with Tennessee Eastman ceasing production of Verel (AN/VCl$_2$) in 1983, BASF in 1985, and DuPont in 1993. Imports of FR fibers are still strong, however, with Kanegafuchi supplying a wide range of deniers and cross sections needed for fake furs.

Federal regulations that restrict the use of vinyl bromide as a flame-retardant monomer will also have an impact as makers of modacrylic fibers are forced to find safe and economical alternatives. Changing lifestyles have also had an impact on the acrylic fiber market. Craft yarns, so popular in the 1970s and the 1980s, have declined and are examples of this trend. The rise in the number of two-wage-earner families had reduced the number of individuals with enough leisure time to enjoy craft knitting. Blankets have declined due to the decreased concern about energy cost, so strong in the late 1970s and the 1980s. Also, cotton has become a popular fiber for blankets. Another issue is that the amount of money spent on advertising acrylics is significantly higher for cotton, which is subsidized by a levy designated for this purpose. Also, the cost of acrylics is higher than the cost of polyester. This puts acrylic fibers at a disadvantage in markets where either of these two fibers can be used.

Several markets have improved, however, helping to offset some of these losses. Acrylic carpets, a market completely lost by 1985, has been resurrected by Monsanto in the 1990s. The inability to piece-dye acrylic carpet has been overcome by introducing producer-dyed fibers. Soiling has been reduced by surface treatments. In addition, Monsanto introduced innovative products, like Traffic Control carpets made from blends of high-shrinkage Acrilan fiber and standard nylon staple. The Traffic Control technology greatly improves the floor performance of residential staple carpets. Similar technology, combining high- and low-shrinkage acrylic fibers, is used in 100% acrylic carpets to yield high-bulk carpet yarns.

Fractional deniers from 0.50 to 0.90 dpf for socks and knit fashions, both 100% acrylic and in blends with cotton, wool, and rayon, are provided in the United States by Cytec, and by other companies in Europe and Japan. Microdeniers in the 0.05–0.10 dpf range are produced for suede and silk-like fabrics.

Acrylics continue to be strong and growing in the upholstery market, with total U.S. shipments at approximately 23 million pounds in 1995. A variety of fabrics including velvets and flat wovens can be produced with excellent color, hand, soiling, light fastness, and abrasion properties. Other outdoor applications, such as awnings, boat covers, and outdoor furniture, continue to enjoy the natural advantage that acrylics have in resistance to degradation in sunlight. Fabrics made from Monsanto's SEF and Pigmented SEF Plus fabrics make excellent products in this important market. Producer-dyed and pigmented fiber are excellent for outdoor use.

More functional fibers are produced today. These include antimicrobial and antifungal products sold by Cytec and Mitsubishi. Antistatic, conductive, and UV-blocking fibers for apparel and carpets are produced and sold in Japan. Industrial uses for acrylic fibers have been a mixture of gains and losses. The list includes awnings, boat covers, reinforcing fibers for cement in particular, sandbags, boat covers, battery separators, and pulp for friction linings, such as brakes and clutches. There will continue to be a strong market in this type of application because of the unique properties of acrylics, like resistance to degradation in sunlight, good flame resistance of modacrylics, and excellent resistance to chemical agents, such as acids.

A potentially large market is a precursor for carbon fibers. The latter fibers are used as the reinforcing element in advanced composites, and these fibers are produced by the controlled pyrolysis of fibers made from rayon, pitch, or acrylics. However, the most common precursor fiber is acrylic. The high price of carbon fiber-based advanced composites has limited the volume of acrylics going into this application, and special copolymer compositions are required. However, as lower performance and higher volume applications emerge, such as automobile bodies, recreational products, and materials of construction where light weight, chemical resistance, and the absence of corrosion is important, we will see a concomitant increase in acrylic fiber production.

In spite of pressure from inexpensive finished goods, the acrylic fiber industry in the United States depends heavily on exports. In 1992, exports accounted for approximately 28% of domestic shipments, surprisingly low compared to Western Europe and Japan. Rationalization of capacity, however, has left the United States with just two acrylic fiber producers, Monsanto and Cytec. However, in early 1997, Cytec announced the sale of its acrylic business to a subsidiary of Sterling Chemicals, Inc. Under the agreement of Cytec will continue to supply acrylonitrile to the business, located near Pensacola, Florida. Besides the withdrawal of DuPont in 1991, the last major producer. Mann Industries, discontinued operations in 1993 after initially acquiring the business from BASF in 1989. Expansion in low-wage countries is contributing to lower inflation and better profit margins. Thus, U.S. producers have little room to raise their domestic or export prices and the focus is on increasing productivity and cost cutting to remain competitive. Although corporate downsizing is not complete, the big gains from this have already been realized. Very good data are available for the U.S. market from the Fiber Economics Bureau [693].

For all the other developed countries, the product mix may vary from country to country but the world supply and demand situation follows essentially the same pattern as that in the United States. The developed countries of Western Europe and Japan depend heavily on exports to compensate for excess capacity and declining domestic demand. Market data can be obtained from several sources. Data for the European markets are available from CIRFS organization [694]. Data for the Japanese markets may be obtained from the Japan Chemical Fibres Association [695]. Acrylic fiber production by region is summarized in Figure 12.55 [690]. In Western Europe, there is nearly 1.8 billion pounds of fiber-producing capacity, with actual production in 1995 roughly 15% of the world's total. Only about half of that is consumed domestically. The rest must be exported from the region to maintain a profitable operation. Although there have been plant closures, mergers, and other business rationalization changes, this region is more resistant to downsizing because of the strong influence of industry socialization. Rationalization plans have been developed by the European Commission and production levels are often maintained in the face of losses in order to maintain jobs. Monsanto sold its facilities in Coleraine, Northern Ireland and Lingen, Germany. Rhone Poulenc withdrew from the business in 1984. About that same time, Anic fiber shut down its 90 million pound plant in Pisticci, Italy. Approximately 53 million pounds of this capacity was subsequently relocated to Mexico. Courtaulds closed their 180 million pound plant in Calais, France in 1990. In 1991–1992, Enichem closed two 75 million pound plants in Cesano and Villacidro. In 1993, Bayer announced cutbacks that eliminated 230 jobs, but did not reduce the production capacity of their 385 million pound plant in Dormagen. In 1994, the rationalization trend resulted in the merger of the acrylic fibers operations of Hoechst AG and Courtaulds. Traditionally, the wool replacement market has been strong in this region, especially since it is colder than in the United States. However, fashion trends have had an unfavorable impact and cheap imported apparel has cut into the domestic market. Also, acrylic carpets, once very popular in this region, have lost ground.

Japan has approximately 900 million pounds of capacity, with actual production in 1995 at approximately 8% of the worldwide production total [690]. Like Western Europe, Japan must export about half of their production to maintain profitable plant utilization. Japan has powerful trading companies that help keep exports at a high level. Nevertheless, rationalization is taking place in Japan. For example, Toray has cutback on exports; Toyobo (Japan Exlan) has cut 30 million pound capacity, and Mitsubishi Rayon has invested $100 million to develop a robotization system to cut costs. Joint research programs have been formed between fiber producers. The principal aim of these programs is to develop profitable technologies with minimal resources. Key objectives are the development of new process technologies to reduce conversion costs and the development of new high-value-added products. Process technologies of great interest are those that yield cost reduction, improved process control, and higher productivity. Opportunities are sought in energy conversion, high-yield polymerization, melt or emulsion spinning, high-speed spinning, robotization, process automation, multiend spinning, high-speed crimping, and improved staple technology [696,697].

The developing countries, in Eastern Europe, Asia, and Latin America, generally enjoy growing domestic markets, because of growing economies and populations. These countries export heavily, however, because they enjoy low labor costs, especially for apparel. Competition from competing fiber types and the effects of lifestyle and fashion trends are all much less

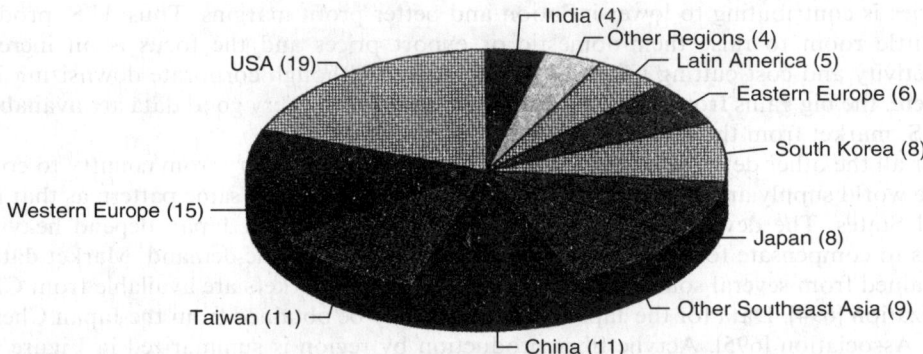

FIGURE 12.55 Worldwide acrylic fiber production for the year 1995. The three traditional acrylic fiber-producing regions of Japan, Western Europe, and the United States accounted for 42% of the total. However, large market shares are enjoyed by Taiwan, China, and South Korea. (From Akzo Nobel report on synthetic fibers, *Int. Fiber J.*, August 1996, pp. 4–14.)

important in these countries. Asian countries other than Japan, including China, South Korea, and Taiwan, account for 39% of the world production of acrylic fiber (see Figure 12.55). Growth of synthetic fiber production in China is expected to continue through the year 2000 [698,699]. A production level greater than 13% of the world's total is expected, moving China to second place in world ranking. Wool and cotton production does not satisfy demand. Therefore, acrylic fiber is in great demand. Production in 1993 was 150,000 t, yet imports of 222,000 t were required to meet demand. By the year 2002 production should reach 420,000 t, or 17% of the world's total. Even then imports of 110,000 t will be required to meet demand. These changes should have a significant effect on world trade. Today China imports 39% from Japan, 34% from East Asia, 14% from Western Europe, and 6% from the United States.

Today there are 55 acrylic fiber producers in 29 countries. These have a total production capacity of 6.4 billion pounds [700]. Rationalization is expected to continue with supply and demand gradually coming into balance in the future. Growth in both supply and demand is expected in the developing countries at a rate of about 1.8%, with most coming from China and the rest from Latin America and Eastern Europe. Generally the 1990s continue to be a period of reequilibration in world acrylic fiber markets. The larger, technology-oriented companies of the United States, Europe, and Japan are moving away from commodity markets where profit margins are small. These companies are attempting to use technology combined with strong patent estates to move into market areas with premium pricing. Fiber producers in the developing countries, on the other hand, are likely to capture market shares in commodity products where their natural advantage in labor cost allows them to compete most effectively.

Concern has been expressed over import–export trade agreements [701]. The GATT treaty, in the early 1960s, led to the Short and Long Term Cotton Agreements (SLTCA), which focused on cotton trade. In the 1970s, the multifiber agreement (MFA) was worked out to apply trade controls for synthetic fibers. The latest version, the Agreement on Textiles and Clothing (ATC), phases these pacts out by the year 2005. The concern is that the major exporting countries will prosper at the expense of the smaller ones. Countries that export less than existing quotas to the United States may be hurt the most, as more efficient suppliers increase their market share. A major player in this changing international trade balance is the North American Free Trade Agreement (NAFTA). This 1993 agreement removes trade

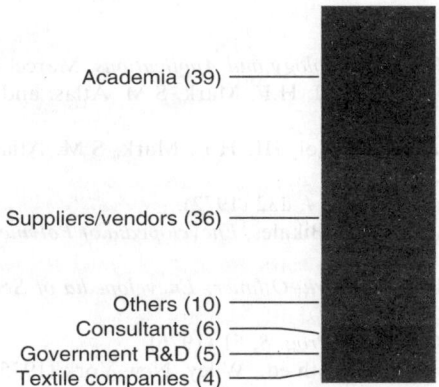

FIGURE 12.56 Analysis of the sources of technical presentations made to the AATCC ICE during 1987–1996. The most frequent source is academia, while the least frequent source is textile companies themselves. The corporate downsizing that has occurred has forced the textile companies to focus on proprietary product-oriented R&D. These companies now depend on the universities for fundamental research. (From Etters, J.N., *America's Textiles Int.*, November 1996, pp. 52–53.)

barriers between the United States, Canada, and Mexico. All three trading partners have shown positive export results in the fiber–textile–apparel markets [702]. As one of the major players in international trade agreements, such as MFA, are phased out.

In spite of the need for more research, a recent report on proceedings of the AATCC ICE over the period 1987–1996 showed that only 4% of technical presentations to that group originated from the textile companies [703,704]. Data from this report are shown in Figure 12.56. This is not surprising, however, because much research and development activity was eliminated during the drastic downsizing that has occurred over the past 20 years. The majority of research papers, 39%, originated from academia. The University of Georgia, North Carolina State University, Auburn University, Clemson, and Georgia Tech were singled out in this report as among the most active in this area and time period.

Products, developed by major acrylic fiber producers, fall into two broad categories: high-volume nontextile products and specialty products with high-value-added potential. New high-volume markets in nontextile markets include acrylics for asbestos replacement, in brakes, clutch linings, and cement reinforcement, and geotextiles applications, which are currently limited by cost factors. Products that have limited volume potential, but offer high-value-added opportunities, are carbon and graphite fiber precursors, fibers for reverse osmosis, ion-exchange fibers, and functional fibers, such as antimicrobial, antistatic, water-repellent, and highly reflectant fibers. Other product possibilities are those that fall into the category of acrylics like enhanced properties. Low-cost cotton-like properties, for example, could open up the huge cotton market to acrylic fibers. Acrylics like high modulus, high strength, and high toughness, similarly, could allow acrylics to gain greater penetration of the industrial fibers markets. In both cases, acrylic fibers already possess highly desirable properties for those respective markets. One glaring weakness in acrylic fibers is their lack of dimensional stability under hot–wet conditions. Though this appears to be an inherent limitation, an acrylic fiber with good hot–wet properties at competitive prices would gain penetration of the nylon carpet and easy-care apparel markets.

REFERENCES

1. J.C. Masson, Ed., *Acrylic Fiber Technology and Applications*, Marcel Dekker, New York (1995).
2. E. Cernia, *Man-Made Fibers*, Vol. III, H.F. Mark, S.M. Atlas, and E. Cernia, Eds., Interscience, New York (1968).
3. R.K. Kennedy, *Man-Made Fibers*, Vol. III, H.F. Mark, S.M. Atlas, and E. Cernia, Eds., Interscience, New York (1968).
4. P.H. Hobson, *Text. Chemist Colorist 4*, 232 (1972).
5. H.F. Mark, N.G. Gaylord, and N.B. Bikales, *Encyclopedia of Polymer Science and Technology*, Vol. I, Interscience, New York (1964).
6. P.H. Hobson and A.L. McPeters, *Kirk–Othmer: Encyclopedia of Science and Technology*, 3rd ed., Vol. I, Wiley, New York (1978).
7. D.J. Pyntun and J. Atkinson, *Text. Prog. 8*, 51 (1976).
8. R.W. Moncrieff, *Man-Made Fibers*, 6th ed., Wiley, New York (1975).
9. R.B. Beavers, *The Physical Properties of Polyacrylonitrile and Its Copolymers*, Macromolecular Reviews, Vol. III, Interscience, New York, 1968.
10. C. Moreau, *Ann. Chem. Phys. 2*, 186 (1893).
11. H. Rein, U.S. Patent 2,117,210 (May 10, 1938) and U.S. Patent 2,140,921 (December 20, 1938) (to I.G. Farbenindustrie).
12. R.C. Latham and R.C. Houtz, U.S. Patent 2,404,714 and U.S. Patent 2,404,720 (June 17, 1942) (to DuPont).
13. H. Rein, DPA 72,024 IVC/39b (April 13, 1942) (to I.G. Farbenindustrie).
14. E.W. Rugeley, T.A. Field, Jr., and G.H. Fremon, *Ind. Eng. Chem. 40*, 1724 (1948).
15. L.C. Shriver and G.H. Fremon, U.S. Patent 2,420,330 (May 13, 1947) (to Carbide and Carbon Chemicals Corp.).
16. H.R. Huff, *Fiber Producer*, p. 26 (April 1983).
17. R.R. Matzke, Jr., in *Acrylic Fiber Technology and Applications*, J.C. Masson, Ed., Marcel Dekker, New York, pp. 69–104 (1995).
18. D.J. Hadley, *Chem. Ind. 25* (February 1961).
19. O. Bayer and P. Kuntz, German Patent 728,767 (1939).
20. K. Sennewald, *Erdoel Kohle 12*, 364 (1959).
21. Belg. Patent 577,691 (1959) (to Distillers Co., Ltd.).
22. F.J. Bellringer, T. Bewley, and H.M. Stanely, U.S. Patent 2,691,037 (October 5, 1954) (to Distillers Co., Ltd.).
23. D.J. Hadley and C.A. Woodcock, U.S. Patent 3,009,943 (November 21, 1961) (to Distillers Co., Ltd.).
24. J.D. Idol, U.S. Patent 2,904,580 (September 9, 1959) (to Sohio Co.).
25. D.C. England, U.S. Patent 2,736,739 (February 28, 1956) (to E.I. DuPont de Nemours & Co., Inc.).
26. W.G. Johnson, U.S. Patent 2,980,726 (April 18, 1961) (to E.I. DuPont de Nemours & Co., Inc.).
27. F.M. Peng, *Encycl. Polym. Sci. Eng.*, J. Kroschwitz, Ed., *1*, 426 (1985).
28. K. Stueben, Acrylonitrile and related compounds, in *Vinyl and Diene Monomers*, Part I, E.C. Leonard, Ed., High Polymers Series Volume XXIV, Wiley-Interscience, New York, chap. 1, p. 181 (1970).
29. American Cyanamid Co., *The Chemistry of Acrylonitrile*, 2nd ed., American Cyanamid Publisher, New York (1959).
30. A.D. Jenkins, *Vinyl Polymerization*, Part I, G.E. Ham, Ed., Marcel Dekker, New York, chap. 6, pp. 369–400 (1967).
31. M.D. Goldfein and B.A. Zyubin, *Vysokomol. Soedin., Ser. A 32*(11), 2243 (1990).
32. J. Beniska and E. Staudner, *J. Polym. Sci., Polym. Symp.* No. 42, 429 (1973).
33. I. Czajlik et al., *Eur. Polym. J. 14*, 1059 (1978); *17*, 131 (1981).
34. I. Czajlik, T. Foldes-Bereznich, T. Tudos, and E. Vertes, *J. Macromol. Sci. Chem. Ed. 14*, 1243 (1980).
35. G. Vidotto, A. Grosatto-Arnaldi, and G. Talamini, *Makromol. Chem. 122*, 91 (1969).
36. G. Vidotto, S. Brugnaro, and G. Talamini, *Makromol. Chem. 140*, 263 (1970).

37. L.H. Peebles, Jr., *J. Polym. Sci. A 3*, 341 (1965); L.H. Peebles, Jr., *J. Polym. Sci. A 3*, 353, 361 (1965).
38. M. Yosida and K. Tonouchi, *Kobunshi Kagaku 20*, 545 (1963).
39. L.V. Smirnova, V.I. Zamelin, and Y.N. Kabanov, *Khim. Volokna 1*, 70 (1970).
40. J.C. Bevington, *Radical Polymerization*, Academic Press, New York (1961).
41. C.H. Bamford, W.C. Barb, A.D. Jenkins, and P.F. Onyon, *The Kinetics of Vinyl Polymerization by Radical Mechanisms*, Academic Press, New York (1958).
42. J. Brandrup and E.H. Immergut, *Polymer Handbook*, 2nd ed., Interscience, New York (1975).
43. W.M. Thomas, *Adv. Polym. Sci. 2*, 401 (1961).
44. N. Friis, D. Goosney, J.D. Wright, and A.E. Hamielic, *J. Appl. Polym. Sci. 18*, 1247 (1974).
45. J. Ulbricht, *Z. Phys. Chem.*, Leipzig, Ed., 5/6, 346 (1962).
46. L.H. Peebles, Jr., *J. Am. Chem. Soc. 80*, 5603 (1958).
47. C.H. Bamford and A.D. Jenkins, *Proc. R. Soc. London Ser. A 226*, 216 (1953); C.H. Bamford and A.D. Jenkins, *Proc. R. Soc. London Ser. A 228*, 220 (1955).
48. C.H. Bamford, A.D. Jenkins, M.C.R Symons, and M.G. Townsend, *J. Polym. Sci. 34*, 181 (1959).
49. A.D. Jenkins, *Vinyl Polymerization*, G. Ham, Ed., Marcel Dekker, New York, Part I, chap. 6 (1967).
50. L.H. Peebles, Jr., *Copolymerization*, Wiley Interscience, New York, chap. IX (1964).
51. K.E.J. Barret and H.R. Thomas, *Dispersion Polymerization in Organic Media*, K.E.J. Barret, Ed., Wiley, New York, chap. 4 (1975).
52. A. Guyot, *Makromol. Chem., Macromol. Symp. 10–11*, 461 (1987).
53. W.D. Harkins, *J. Am. Chem. Soc. 69*, 1428 (1947).
54. W.V. Smith and R.H. Ewart, *J. Chem. Phys. 16*, 592 (1948).
55. W.V. Smith, *J. Am. Chem. Soc. 70*, 3695 (1948); W.V. Smith, *J. Am. Chem. Soc. 71*, 4077 (1949).
56. J.L. Gardon, *J. Polym. Sci. A1*(6), 623 (1968); J.L. Gardon, *J. Polym. Sci. A1*(6), 643, 665, 687, 2853, 2859.
57. J.L. Gardon, *Br. Polym. J. 2*, 1 (1970).
58. R.M. Fitch and Lih-bin Shih, *Prog. Colloid Polym. Sci. 56*, 1 (1975).
59. D.G. King, S.J. McCarthy, B.A.W Coller, I.R. McKinnon, and I.R. Wilson, *Radiat. Phys. Chem. 36*(4), 595–600 (1990).
60. D.C. Blackley, *Emulsion Polymerization Theory and Practice*, Wiley, New York (1975).
61. D.R. Bassett and A.E. Hamielic, Eds., *Emulsion Polymers and Emulsion Polymerization*, ACS Symposium Ser. No. 165, American Chemical Society Publication, Washington, D.C. (1981).
62. I. Piirma, Ed., *Emulsion Polymerization*, Academic Press, New York (1982).
63. V.I. Eliseeva, S.S. Ivanchev, S.I. Kuchanov, and A.V. Lebedev, *Emulsion Polymerization and Its Applications in Industry*, Engl. Transl. Consultants Bureau, Division of Plenum Publishing Corporation, New York (1981).
64. K.W. Min and W.H. Ray, *J. Macromol. Sci. Rev. Macromol. Chem. C11*(2), 177 (1974).
65. L.H. Peebles, Jr., *J. Appl. Polym. Sci. 17*, 113 (1973).
66. L.H. Peebles, Jr., R.B. Thompson, Jr., J.R. Kirby, and M.E. Gibson, *J. Appl. Polym. Sci. 16*, 3341 (1972).
67. J.R. Ebdon, T.N. Huckerby, and T.C. Hunter, *Polymer 35*(2), 250 (1994).
68. J.R. Ebdon, T.N. Huckerby, and T.C. Hunter, *Polymer 35*(21), 4659 (1994).
69. S.R. Palit, T. Guha, R. Das, and R.S. Konar, *Encycl. Polym. Sci. Tech. 2*, 229 (1965).
70. W.M. Thomas, *Adv. Polym. Sci. 2*, 401 (1961).
71. J. Brandrup and E.H. Immergut, *Polymer Handbook*, 2nd ed. Interscience, New York (1975).
72. F.M. Peng, *J. Macromol. Sci., Chem. A22*(9), 1241 (1985).
73. F.R. Mayo and C. Walling, *Chem. Rev. 46*, 191 (1950).
74. F.R. Mayo and F.M. Lewis, *J. Am. Chem. Soc. 66*, 1594 (1944).
75. I. Skeist, *J. Am. Chem. Soc. 68*, 1781 (1946).
76. V.E. Meyer and G.G. Lowry, *J. Polym. Sci. A3*, 2843 (1965).
77. R.K.S Chan and V.E. Meyer, *J. Polym. Sci. C*(25), 11 (1968).
78. G.E. Ham, in *Copolymerization*, Wiley Interscience, New York, chap. I, pp. 1–64 (1964).
79. A. Valvassori and G. Sartori, *Adv. Polym. Sci. 5*, 28 (1967).

80. T. Alfrey, Jr., J.J. Bohrer, and H. Mark, Eds., *Copolymerization*, Wiley Interscience, New York (1952).
81. G.E. Ham, Ed., *Copolymerization*, Wiley Interscience, New York (1964).
82. G.E. Ham, *Encycl. Polym. Sci. Tech.* **4**, 165 (1966).
83. L.H. Peebles, Jr., *Copolymerization*, G.E. Ham, Ed., Wiley Interscience, New York, chap. IX, p. 525 (1964).
84. A.D. Jenkins and A. Ledwith, Eds., *Reactivity, Mechanism and Structure in Polymer Chemistry*, Wiley Interscience, New York (1974).
85. R.G. Fordyce and E.C. Chapin, *J. Am. Chem. Soc.* **69**, 581 (1947).
86. R.G. Fordyce, *J. Am. Chem. Soc.* **69**, 1903 (1947).
87. R.G. Fordyce and G.E. Ham, *J. Polym. Sci.* **3**(6), 891 (1948).
88. W.V. Smith, *J. Am. Chem. Soc.* **70**, 2177 (1948).
89. D.W. Ley and W.F. Fowler, Jr., *J. Polym. Sci. A2*, 1863 (1964).
90. W.F. Fowler, Jr., *Vinyl Polymerization*, Marcel Dekker, New York, Vol. 1, Part II, chap. 2, pp. 139–174 (1969).
91. V.I. Eliseeva, S.S. Ivanchev, S.I. Kuchanov, and A.V. Lebedev, *Emulsion Polymerization and Its Applications in Industry*, Engl. Transl. Consultants Bureau, Division of Plenum Publishing Corporation, New York (1981).
92. D.W. Cheape and W.R. Eberhardt, U.S. Patent 3,454,542 (July 8, 1969) (assigned to Monsanto Company).
93. S. Ito and H. Tamura, Jpn. Kokai Tokkyo Koho, 5 pp., JP 54029390 (March 5, 1979), Showa, JP 77-94742 (August 8, 1977).
94. S. Ito and K. Yoshida, *Kobunshi Ronbunshu* **40**(5), 307 (1983).
95. S. Ito, *Kobunshi Ronbunshu* **43**(1), 1 (1986).
96. S. Ito, *Kobunshi Ronbunshi* **41**(8), 445 (1984).
97. S. Ito, *Kobunshi Ronbunshu* **42**(12), 865 (1985).
98. S. Ito, *J. Appl. Polym. Sci.* **31**(3), 849 (1986).
99. S. Ito, Y. Kawai, and T. Oshita, *Kobunshi Ronbunshu* **43**(6), 345 (1986).
100. S. Ito and C. Okada, *Sen'i Gakkaishi* **42**(11), T618 (1986).
101. S. Ito, *Sen'i Gakkaishi* **43**(5), 236 (1987).
102. S. Ito, C. Okada, and K. Chiharu, *J. Appl. Polym. Sci.* **32**(3), 4001 (1986).
103. P. Melacini, R. Tedesco, L. Patron, and A. Moretti, Canadian Patent 911,650 (October 3, 1972) (assigned to Montedison Fibre).
104. U.S. Patent 3,787,365 (January 22, 1974); U.S. Patent 3,821,178 (June 28, 1974); U.S. Patent 3,879,360 (April 22, 1975). L. Patron et al. (assigned to Montefibre).
105. L.E. Alexander, *X-Ray Diffraction Methods in Polymer Science*, Wiley Interscience, New York, pp. 12–21 (1969).
106. R.J. Samuels, *Structured Polymer Properties*, Wiley, New York (1974).
107. P. Parrini and G. Crespi, *Encyclopedia of Polymer Science and Technology*, Vol. 13, Wiley Interscience, New York, p. 89 (1970).
108. G. Natta, P. Corradini, and P. Ganis, *J. Polym. Sci.* **58**, 1191 (1962).
109. J. Schaefer, *Macromolecules* **4**, 105 (1971).
110. F.A. Bovey and F.H. Winslow, *Macromolecules, An Introduction to Polymer Science*, Academic Press, New York, p. 223 (1979).
111. I. Pasquon and U. Giannini, Stereoregular linear polymers, in *Encyclopedia of Polymer Science and Engineering*, 2nd ed., Vol. 15, Wiley, New York, p. 632 (1989).
112. R. Chiang, *J. Polym. Sci. A3*, 2019 (1965).
113. R. Chiang, J.H. Rhodes, and V.F. Holland, *J. Polym. Sci. A3*, 479 (1965).
114. D.W. Van Krevelen, *Properties of Polymers*, Elsevier Publishing, New York, p. 43 (1972).
115. F.W. Billmeyer, Jr., *Textbook of Polymer Science*, 3rd ed., Wiley, New York, pp. 153–154 (1984).
116. G. Hinrici-Olive' and S. Olive', *Adv. Polym. Sci.* **32**, 128 (1979).
117. A.J. Hopfinger, *Conformational Properties of Macromolecules*, Academic Press, New York (1973).
118. W.R. Krigbaum and N. Tokita, *J. Polym. Sci.* **43**, 467 (1960).
119. R. de Daubney, C.W. Bunn, and C.J. Brown, *Proc. R. Soc. London* **226**, 531 (1954).

120. C.R. Bohn, J.R. Schaefgen, and W.O. Statton, *J. Polym. Sci.* 55, 531 (1961).
121. M. Saum, *J. Polym. Sci.* 42, 57 (1960).
122. R.N. Delmartino, *J. Polym. Sci.* 28, 1805 (1983).
123. P.H. Geil, *Polymer Single Crystals*, Interscience, New York, p. 21 (1963).
124. V.G. Holland, S.B. Mitchell, W.L. Hunter, and P.H. Lindenmeyer, *J. Polym. Sci.* 62, 145 (1962).
125. J.J. Klement and P.H. Geil, *J. Polym. Sci. A2*(6), 1381 (1968).
126. G.N. Patel and R.D. Patel, *J. Polym. Sci. A2*(8), 47 (1970).
127. R.M. Gohil, K.C. Patel, and R.D. Patel, *Makromol. Chem.* 169, 291 (1973).
128. R.M. Gohil, K.C. Patel, and R.D. Patel, *Colloid Polym. Sci.* 25, 559 (1976).
129. F. Kumamaru, T. Kajiyama, and M. Takayanagi, *J. Crystal Growth* 48(2), 202 (1980).
130. B.G. Colvin and P. Storr, *Eur. Polym. J.* 10, 337 (1974).
131. P. Corridini, G. Natta, P. Ganis, and P.A. Temussi, *J. Polym. Sci. C16*, 2477 (1967).
132. M. Takahashin and Y. Nukushima, *J. Polym. Sci.* 56, 519 (1962).
133. Z. Bashir, *J. Polym. Sci. Part B Polym. Phys.* 32, 1115 (1994).
134. Z. Bashir, *Polymer* 33, 4304 (1992).
135. I.R. Herbert, A. Tipping, and Z. Bashir, *J. Polym. Sci. Part B Polym. Phys.* 31, 1459 (1993).
136. Z. Bashir, S.P. Church, and D. Waldron, *Polymer* 35, 967 (1994).
137. D.M. Price and Z. Bashir, *Thermochim. Acta* 249, 351 (1995).
138. Z. Bashir, *J. Polym. Sci. Part B Polym. Phys.* 30, 1299 (1992).
139. P.H. Lindenmeyer and R. Hoseman, *J. Appl. Phys.* 34, 42 (1963).
140. X.D. Liu and W. Ruland, *Macromolecules* 26, 3030 (1993).
141. R.J. Hobson and A.H. Windle, *Macromolecules* 26, 6903 (1993).
142. P. Bell and J.H. Dumbleton, *Textile Res. J.* 41, 196 (1971).
143. G. Hinrichsen, *J. Polym. Sci.* 38, 303 (1972).
144. A.K. Gupta and R.P. Singhal, *J. Polym. Sci. Polym. Phys. Ed.* 21, 2243 (1983).
145. V.K. Matta, R.B. Mathur, and O.P. Bahl, *Carbon* 28(1), 241 (1990).
146. O.P. Bahl, R.B. Mathu, and T.L. Dhami, *Mat. Sci. Eng.* 13, 105 (1985).
147. Y. Imai, S. Minami, T. Yoshihara, Y. Joh, and H. Sato, *J. Polym. Sci. Part B, Polym. Lett.* 8, 291 (1970).
148. Y. Joh, *J. Polym. Sci., Polym. Chem. Ed.* 17, 4051 (1979).
149. G. Hinrichsen and H. Orth, *Polym. Lett.* 9, 529 (1971).
150. G. Hinrichsen, *Angew. Makromol. Chem.* 20, 121 (1974).
151. M.K. Jain and A. Abhiraman, *J. Mater. Sci.* 18, 1979 (1983).
152. S.B. Warner, D. Uhlmann, and L. Peebles. *J. Mater. Sci.* 10, 758 (1975).
153. S.B. Warner, *Polym. Lett.* 16, 287 (1978).
154. E.W. Fischer and S. Fakirov, *J. Mater. Sci.* 11, 1041 (1976).
155. A.K. Gupta, N. Chand, R. Singh, and A. Mansinga, *Eur. Polym. J.* 15, 129 (1979).
156. A.K. Gupta and N. Chand, *Eur. Polym. J.* 15, 899 (1979).
157. A.K. Gupta and N. Chand, *J. Polym. Sci. Polym. Phys. Ed.* 18, 1125 (1980).
158. A.K. Gupta, R.P. Singhal, and V.K. Agarwal, *Polymer* 22, 285 (1981).
159. A.K. Gupta and R.P. Singhal, *J. Appl. Polym. Sci.* 26, 3599 (1981).
160. A.K. Gupta and A.K. Maiti, *J. Appl. Polym. Sci.* 27, 2409 (1982).
161. A.K. Gupta, R.P. Singhal, and A.K. Maiti, *J. Appl. Polym. Sci.* 27, 4101 (1982).
162. A.K. Gupta, R.P. Singhal, and P. Bajaj, *J. Appl. Polym. Sci.* 28, 1167 (1983).
163. A.K. Gupta, R.P. Singhal, and V.K. Agarwal, *J. Appl. Polym. Sci.* 28, 2745 (1983).
164. M.L. Williams, R.F. Landel, and J.D. Ferry, *J. Am. Chem. Soc.* 77, 3701 (1955).
165. F. Beuche, *Physical Properties of Polymers*, Interscience, New York, chaps. 6–9 (1962).
166. K. Kulshreshtha, V.N. Garg, and Y.N. Sharma, *J. Appl. Polym. Sci.* 31, 1413 (1986).
167. B.G. Frushour, *Polym. Bull.* 4, 305 (1981).
168. R.D. Andrews, K. Miyachi, and R.S. Doshi, *J. Macromol. Sci. Phys.* B9, 281 (1974).
169. R. Miyachi and R.D. Andrews, *J. Appl. Polym. Sci. Appl. Polym. Symp.* 35, 127 (1974).
170. R.D. Andrews, R.C. Yen, and P. Changs, *J. Macromol. Sci. Phys.* B19, 729 (1981).
171. J. Grobelny, P. Tekely, and E. Turska, *Polymer* 22, 1649 (1981).
172. E. Turska and J. Grobelny, *Eur. Polym. J.* 19, 985 (1983).

173. J. Grobelny, M. Sokol, and E. Turska, *Polymer 25*, 1414 (1984).
174. M. Sokol, J. Grobelny, and E. Turska, *Polymer 28*, 843 (1987).
175. J. Grobelny, M. Sokol, and E. Turska, *Eur. Polym. J. 24*, 1195 (1988).
176. M. Lewin, H. Guttman, and Y. Naor, *J. Macromol. Sci. Chem. A25*(10–11), 1367 (1988).
177. H. Yasuda et al., Gel spinning processes, in *Advanced Fiber Spinning Processes*, P.T. Nakajima, Ed., Woodhead Publishing Limited, Cambridge, p. 174 (1994).
178. R.A. Allen, I.M. Ward, and Z. Bashir, *Polymer 35*, 2063 (1994).
179. N. Grassie and R. McGuchan, *Eur. Polym. J. 6*, 1277 (1970).
180. E.A. Turi, Ed., *Thermal Characteristics of Polymeric Materials*, Academic Press, New York (1981).
181. P.J. Flory, *Principles of Polymer Chemistry*, Cornell Press, Ithaca, NY, p. 568 (1953).
182. L. Mandelkern, *Chem. Rev. 56*, 903 (1956).
183. P.J. Flory, R.R. Garrett, S. Newman, and L. Mandelkern, *J. Polym. Sci. 12*, 94 (1954).
184. W.R. Krigbaum, *J. Polym. Sci. 28*, 213 (1958).
185. P.J. Flory, *Trans. Faraday Soc. 51*, 848 (1955).
186. R.K. Eby, *J. Appl. Phys. 34*, 2442 (1963).
187. P.E. Slade, *Thermochim. Acta 1*, 4559 (1970).
188. W. Berger, A. Heller, and H.-J. Adler, *Faserforschung Textiltechnik 24*, 484 (1973).
189. B.G. Frushour, *Polym. Bull. 7*, 1 (1982).
190. B.G. Frushour, *Polym. Bull. 11*, 375 (1984).
191. B.G. Min et al, *Polym. J. 24*, 841 (1992).
192. B.G. Min et al. *J. Appl. Polym. Sci. 46*, 1793 (1992).
193. B.G. Min et al. *J. Appl. Polym. Sci. 54*, 457 (1994).
194. J.D. Ferry, *Viscoelastic Properties of Polymers*, 2nd ed., Wiley, New York (1969).
195. R.M. Kimmel and R.D. Andrews, *J. Appl. Phys. 36*, 3063 (1965).
196. W.H. Howard, *J. Appl. Polym. Sci. 5*, 303 (1961).
197. S. Loshack and T.G Fox, *Bull. Am. Phys. Soc. 1*, 123 (1956).
198. B.G. Frushour, unpublished results.
199. D. Aiken, S.M. Burkinshaw, R. Cox, J. Catherall, R.E. Litchfield, D.M. Price, and N.G. Todd, *J. Appl. Polym. Sci. Appl. Polym. Symp. 47*, 263 (1991).
200. J.F. Bell and T. Murayama, *J. Appl. Polym. Sci. 12*, 1795 (1968).
201. Z. Gur-Arieh and W.C. Ingamells, *J. Soc. Dyers Colour 90*, 8 (1974).
202. T. Hori, H. Khang, T. Shimuzu, and H. Zollinger, *Text. Res. J. 58*, 227 (1988).
203. T. Murayama, *Dynamic Mechanical Analysis of Polymeric Materials*, Elsevier, New York (1978).
204. N.G. McCrum, B.E. Read, and G. Williams, *Molecular Theories of Relaxation in Anelastic and Dielectric Effects in Polymeric Solids*, Wiley, New York, chap. 5 (1967).
205. S. Minami, *J. Appl. Polym. Sci. Appl. Polym. Symp. 25*, 145 (1974).
206. S. Minami, H. Yamamori, and H. Sato, *Prog. Polym. Phys. Jpn. 14*, 379 (1971).
207. K. Schneider and K. Wolf, *Kolloid-Z 234*, 149 (1953).
208. G.R. Cotton and W.C. Schneider, *Kolloid-Z 192*, 16 (1963).
209. A. Minami, T. Yoshihara, and H. Sato, *Kobunshi Kagaku* (English ed.) *1*, 125 (1972).
210. S. Okajima, M. Ikeda, and T. Takeuchi, *J. Polym. Sci. A1*(6), 195 (1968).
211. S. Minami, T. Sakurai, T. Yoshihara, and H. Sato, *Prog. Polym. Phys. (Japan) 14*, 385 (1972).
212. A.S. Keyon and M.J. Rayford, *J. Appl. Polym. Sci. 23*, 717 (1979).
213. M.R. Padhye and A.V. Karandikar, *J. Appl. Polym. Sci. 33*, 1675 (1987).
214. M. Takayanagi, *Proceedings of the Fourth International Conference on Rheology*, Part I, Interscience, New York, p. 61 (1965).
215. L.E. Nielsen, *Mechanical Properties of Polymers*, Reinhold, New York, p. 163 (1962).
216. L.E. Nielsen, *J. Appl. Phys. 25*, 1209 (1954).
217. R.D. Andrews and R.M. Kimmel, *Polym. Lett. 3*, 167 (1965).
218. J.H. Dumbleton, T. Murayama, and J.P. Bell, *Kolloid-Z 228*, 54 (1968).
219. G.C. Capone, Wet-spinning technology, in *Acrylic Fiber Technology and Applications*, J.C. Masson, Ed., Marcel Dekker, New York, pp. 69–104 (1995).
220. B. von Falkai, Dry-spinning technology, in *Acrylic Fiber Technology and Applications*, J.C. Masson, Ed., Marcel Dekker, New York, pp. 105–166 (1995).

221. T. Tsurumi, Solution spinning, in *Advanced Fiber Spinning Technology*, T. Nakajima, Ed., Woodhead Publishing Limited, Cambridge, pp. 65–104 (1994).
222. S. Suzuki, *Kagaku (Kyoto) 23*, 315 (1953).
223. M. Takahashi and M. Wantanabe, *Kobunshi Kogaku 16*, 35 (1959).
224. M. Takahashi, *Kogyo Kagako Zasshi 63*, 2201 (1960).
225. D.R. Paul, *J. Appl. Polym. Sci. 11*, 439 (1967).
226. J. Bisschops, *J. Polym. Sci. 12*, 583 (1954).
227. J. Bisschops, *J. Polym. Sci. 17*, 89 (1955).
228. R. Jenny, Ciba-Geigy, private communications with Monsanto Company.
229. A. Cresswell and P.W. Cummings, U.S. Patent 2,605,246 (1952) (to American Cyanamid).
230. A. Mison and P. Tarbouriech, U.S. Patent 3,630,986 (December 28, 1971) (to Rhone-Poulenc).
231. A.A. Armstrong, U.S. Patent 3,969,305 (July 13, 1976) (to Monsanto).
232. R.S. Stoveken, U.S. Patent 3,010,932 (November 29, 1961) (to DuPont).
233. J.D. Chase and R.L. Potter, U.S. Patent 3,847,864 (November 12, 1974) (to American Cyanamid).
234. P.C. Yates, U.S. Patent 3,657,179 (April 18, 1972) (to DuPont).
235. B.M. Pettyjohn, U.S. Patent 3,383,350 (May 14, 1968) (to DuPont).
236. J.R. Kirby, U.S. Patent 3,784,511 (January 8, 1974) (to Monsanto).
237. J.C. Masson, U.S. Patent 3,784,512 (January 8, 1974) (to Monsanto).
238. Netherlands Pat. Appl. 65-1113 (February 28, 1966) (to Farbenfabriken Bayer).
239. H. Logemann, E. Roos, and C. Suling, U.S. Patent 3,436,364 (April 1, 1969) (to Farbenfabriken Bayer).
240. J.H. Hennes and C.R. Pfeifer, U.S. Patent 3,296,171 (January 3, 1967) (to Dow Chemical).
241. L. Cenzato, U.S. Patent 2,954,271 (September 27, 1960) (to DuPont).
242. W. Clapp and C.B. Mather, U.S. Patent 2,811,409 (October 29, 1957) (to Eastman Kodak).
243. A.L. McPeters and D.R. Paul, *Appl. Polym. Symp. 25*, 159 (1974).
244. D.R. Paul, *J. Appl. Polym. Sci. 12*, 2273 (1968).
245. W.E. Fitzgerald and J.P. Craig, *Appl. Polym. Symp. 6*, 67 (1967).
246. A. Ziabicki, *Sen'i Gakkaishi 26*, 156 (1970).
247. D.R. Paul, *J. Appl. Polym. Sci. 12*, 383 (1968).
248. V. Grobe and H. Heyer, *Faserforschung Textiltechnik 19*, 313 (1968).
249. Q. Baojin, P. Ding, and W. Zhenqiou, *Adv. Polym. Technol. 6*, 509 (1986).
250. Q. Jian, L. Zhaofeng, and P. Ding, *J. China Textile Univ.* No. 2, 15 (1986).
251. K. Terada, *Sen'i Gakkaishi* No. 8, 29 (1973).
252. C. Cohen, G. Tanny, and S. Prager, *J. Polym. Sci. Polym. Phys. Ed. 17*, 477 (1979).
253. A. Ziabicki, The Role of Phase and Structural Transitions in Fiber Spinning Processes, presented at Society of Fiber Science and Technology, Japan, 25th Anniversary, Tokyo-Osaka (1969).
254. K. Kawanishi, M. Komatsu, and T. Inoue, *Polymer 28*, 980 (1987).
255. J. Van Aartsen, *Eur. Polym. J. 6*, 919 (1970).
256. J. Van Aartsen and C. Cmolders, *Eur. Polym. J. 6*, 1105 (1970).
257. F. Nimts et al., *Khemicheskie Volokna* No. 3, 11 (1986).
258. G. Caneba and D. Soong, *Macromolecules 18*, 2538 (1985).
259. G. Caneba and D. Soong, *Macromolecules 18*, 2545 (1985).
260. M. Komatsu, T. Inoue, and K. Miyasaka, *J. Polym. Sci. Polym. Phys. Ed. 24*, 303 (1979).
261. V. Grobe and K. Meyer, *Faserforschung Textiltechnik 20*, 467 (1969).
262. J.P. Craig, J.P. Knudsen, and V.F. Holland, *Text. Res. J. 32*, 435 (1962).
263. A. Lulay, Apparel end uses, in *Acrylic Fiber Technology and Applications*, J.C. Masson, Ed., Marcel Dekker, New York, pp. 314–315 (1995).
264. W.J. Onions, E. Oxtoby, and P.P. Townend, *J. Text. Inst. 58*, 293 (1967).
265. *Manufacture's Journal*, Absorption Rate, Imbibition Test, EP-SAP 2301 (Basket Sink Test), June 18, 1980.
266. R.G. Quynn, *Text. Res. J. 33*, 21 (1963).
267. A. Gunier, *Ann. Phys. 12*, 161 (1939).
268. J.P. Knudsen, *Text. Res. J. 33*, 13 (1963).
269. K. Terada, *Sen'i-Gakkaishi 29*, 12, T451 (1973).

270. P.P. Singh, U.S. Patent 3,767,360 (October 23, 1973) (to DuPont).
271. W.W. Mosely, *J. Appl. Polym. Sci. 3*, 266 (1960).
272. C.R. Desper, *J. Macromol. Sci. Phys. B7(1)*, 105 (1973).
273. J.P. Bell and J.H. Dumbleton, *Text. Res. J. 41*, 196 (1971).
274. J.P. Knudsen and W.E. Fitzgerald, The Influence of Gel Network Mechanics on the Tensile Properties of Wet-Spun Fibers, presented at the Gordon Res. Conf. Textiles (1965).
275. Private discussion with J.P. Knudsen.
276. H. Terada, *Preprints of 20th Polymer Symposium*, Tokyo, 5D14, 755 (1971).
277. K. Terada, *Sen'i-Gakkaishi 30(8)*, T377 (1974).
278. K. Terada, *Sen'i-Gakkaishi 29(3)*, 78 (1973).
279. G.S. Kruglova, *Vysokomol. Soed. 17A(7)*, 1541 (1975).
280. D.J. Forrestal, *Faith, Hope and $5,000*, Simon & Schuster, New York, p. 121 (1977).
281. R.E. Harder and L.B. Ticknor, U.S. Patent 3,895,908 (July 22, 1975) (to Dow Badische).
282. P.H. Hobson, *Faserforsch. Textiltech. 22*, 80 (1971).
283. Y. Kobayashi, S. Okajima, and H. Kosuda, *J. Appl. Polym. Sci. 11*, 2525 (1967).
284. M. Sotton, A.M. Vialard, and C. Rabourdin, *Bull. Sci. ITF 2(7)*, 173 (1973).
285. C.R. Pfeifer and R.B. Hurley, *Appl. Polym Symp. 25*, 179 (1974).
286. W.E. Fitzgerald and J.P. Knudsen, *Text. Res. J. 37*, 447 (1967).
287. L.H. Belck and K.G. Siedschlag, Jr., U.S. Patent 3,039,525 (June 19, 1962) (to DuPont).
288. Y. Fujita, K. Shimoda, and K. Zoda, U.S. Patent 3,182,106 (May 4, 1965) (to American Cyanamid).
289. W.E. Fitzgerald and J.P. Knudsen, *Text. Res. J. 37*, 447 (1967).
290. F.B. Powell and B.K. Polk, U.S. Patent 3,295,552 (January 3, 1967) (to Monsanto).
291. A.L. Breen, U.S. Patent 3,038,236 (June 12, 1962) (to DuPont).
292. H.A. Hoffman, Jr., U.S. Patent 3,547,763 (December 15, 1970) (to DuPont).
293. H. Sekiguchi, N. Tsutsui, and T. Sumi, U.S. Patent 3,864,447 (February 4, 1975) (to Japan Exlan).
294. J.P. Knudsen, U.S. Patent 3,124,620 (March 10, 1964) (to Monsanto).
295. J.P. Knudsen, U.S. Patent 3,088,188 (May 7, 1963) (to Monsanto).
296. W. Dohrn, *Faserforsch. Textiltech. 25*, 28 (1974).
297. H. Takeda, U.S. Patent 3,758,659 (September 11, 1973) (to Toray).
298. R.M. Costa and F. Codignola, U.S. Patent 3,449,485 (June 10, 1969) (to Societa Italiana Resine).
299. M. Zwick, *Appl. Polym. Symp. 6*, 109–149 (1967).
300. F. Lieseberg, U.S. Patent 2,957,749 (October 25, 1960) (to Badische-Aniline).
301. P.A. Ucci, U.S. Patent 3,080,210 (March 5, 1963) (to Monsanto).
302. R.E. Vigneault, U.S. Patent 3,523,150 (August 4, 1970) (to Monsanto).
303. K. Kuratani and K. Fukushima, U.S. Patent 3,701,820 (October 31, 1972) (to Japan Exlan).
304. H. Yasuda, K. Ban, and Y. Ohta, Gel spinning processes, in *Advanced Fiber Spinning Technology*, T. Nakajima, Ed., Woodhead Publishing Limited, Cambridge, pp. 171–185 (1994).
305. Mitsubishi Rayon, Japan Patent (Laid Open) 89-104816; Japan Exlan and Tokyo Doseki, Japan Patent (Laid Open) 86-97415; U.S. Patent 4,657,004; Toray, Japan Patent (Laid Open) 88-812317.
306. Japan Pat. Appl. J5 0029-882 (March 25, 1975) (to Toho Beslon).
307. J.R. Caldwell, U.S. Patent 3,082,056 (March 19, 1963) (to Eastman Kodak).
308. G.A. Serad, Canadian Patent 907,771 (to Celanese).
309. S.L. Atureliya and Z. Bashir, *Polymer 34*, 5116 (1993).
310. T.C. Bohrer and A.E. Champ, U.S. Patent 3,655,857 (April 11, 1972) (to Celanese).
311. C.D. Cox, U.S. Patent 2,585,444 (to E.I. DuPont Company) (1952).
312. A. Goodman and M.S. Suwyn, U.S. Patent 3,896,204 (to E.I. DuPont Company) (1975).
313. R.A. Blickenstaff, U.S. Patent 3,984,601 (to E.I. DuPont Company) (1976).
314. H. Porosoff, U.S. Patent 4,163,770 (to American Cyanamid Co.) (1979).
315. U.S. Patent 3,402,231 (to Monsanto Company) (1964).
316. B. von Falkai, Dry-spinning technology, in *Acrylic Fiber Technology and Applications*, J.C. Masson, Ed., Marcel Dekker, New York, pp. 129–134 (1995).
317. Z. Bashir, *Polymer 35*, 967 (1994).
318. B.G. Min et al., *Polym. J. 24*, 841 (1992).
319. B.G. Min et al., *J. Appl. Polym. Sci. 54*, 457 (1994).

320. B.G. Min et al., *J. Appl. Polym. Sci. 46*, 1793 (1992).
321. C.D. Han and L. Segal, *J. Appl. Polym. Sci. 14*, 2973 (1970).
322. R.L. Ballman, *Rheo. Acta 4*, 137 (1965).
323. C.D. Han and L. Segal, *J. Appl. Polym. Sci. 14*, 2999 (1970).
324. J.D.I. McKelvey, *Polymer Processing*, Wiley, New York, chap. 3 (1962).
325. C.D. Han, *J. Appl. Polym. Sci. 15*, 1091(1971).
326. C.D. Han and J.Y. Park, *J. Appl. Polym. Sci. 17*, 187 (1973).
327. R.H. Peters, *Textile Chemistry. III. Physical Chemistry of Dyeing*, Elsevier, New York, pp. 549–580 (1975).
328. W.E. Morton and J.W.S Hearle, *Physical Properties of Textile Fibers*, Wiley, New York, pp. 177–179 (1975).
329. J.C. Masson, Product variants, in *Acrylic Fiber Technology and Applications*, J.C. Masson, Ed., Marcel Dekker, New York, p. 283 (1975).
330. D.F. Bittle and A.L. McPeters, Canadian Patent 954,256 (to Monsanto).
331. N.D. Sharma, P. Sharma, and R. Mehta, *Text. Asia 20*, 60 (1989).
332. G.B. Mullinax, New Developments in Colored Acrylic Fibers, Textile Research Institute presentation, Charlotte, NC (April 13, 1988).
333. Anon, Monsanto produced dyed fiber expansion widens color, denier range, Inside *Textiles 8*, 10 (1987).
334. D.E. Titheridge, *Chemiefasern/Textilindustrie* (English ed.), *32/84*, E26 (1982A).
335. G.E. Ham, *Text. Res. J. 24*, 604 (1954).
336. S. Rosenbaum, *J. Appl. Polym. Sci. 7*, 1225 (1963).
337. S. Rosenbaum, *Text. Res. J.* 899 (November 1963).
338. S. Rosenbaum, *Text. Res. J.* 52 (January 1964).
339. P. Bajaj and S.K. Munukutla, *Text. Res. Inst. 60*(2), 113 (1990).
340. S.R. Shukla, R.V. Hundekar, and A.N, Saligram, *Am. Dyestuff Rep.*, 41 (October 1991).
341. Z. Gur-Arieh and W.C. Ingamells, *J. Soc. Dyers Colour 90*, 8 (1974).
342. J.P. Bell and T. Murayama, *J. Appl. Polym. Sci. 12*, 1795 (1968).
343. Y. Bryant and T. Murayama, *J. Appl. Polym. Sci. 12*, 1795 (1968).
344. R.H. Peters and H. Wang, *J. Soc. Dyers Colour 98*, 432 (1982).
345. N.G. Morton and M.E. Kracht, *J. Soc. Dyers Colour 85*, 639 (1969).
346. C.L. Zimmerman and A.L. Cate, *Text. Chem. Col. 4*, 150 (1972).
347. H. Kellet, *J. Soc. Dyers Colour 84*, 257 (1978).
348. R.M. Rohner and H. Zollinger, *Text. Res. J. 56*, 1 (1986).
349. S. Chen, R.M. Rohner, and H. Zollinger, *Text. Res. J. 58*, 247 (1988).
350. T. Hori, H. Ahang, T. Shimizu, and H. Zollinger, *Text. Res. J. 58*, 227 (1988).
351. A. Lulay, Apparel end uses, in *Acrylic Fiber Technology and Applications*, J.C. Masson, Ed., Marcel Dekker, New York, pp. 315–318 (1995).
352. J.C. Masson, Home furnishings and industrial applications, in *Acrylic Fiber Technology and Applications*, J.C. Masson, Ed., Marcel Dekker, New York, pp. 341–369 (1995).
353. W.E. Morton and J.W.S Hearle, *Physical Properties of Textile Fibers*, Wiley, New York, pp. 399–401 (1975).
354. *Textile World*, Intertec Publishing, Overland Park, Kansas (August 1996) (insert).
355. D.W. Van Krevelen, *Properties of Polymers*, Elsevier, New York, pp. 425–428 (1990).
356. S. Rosenbaum, *J. Appl. Polym. Sci. 9*, 2071 (1965).
357. S. Rosenbaum, *J. Appl. Polym. Sci. 9*, 2085 (1965).
358. B. Farrow, *J. Text. Inst. 47*, T58 (1956).
359. J.E. Ford, Ed., *Fiber Data Summaries*, Shirley Institute, Manchester (1966).
360. L.E. Nielsen, *Mechanical Properties of Polymers*, Reinhold, New York (1962).
361. F.F. Morehead, *Text. Res. J. 17*, 96 (1947).
362. D.F. Holmes, *Am. Assoc. Tex. Technol. 7*, 85 (1952).
363. D.F. Holmes, *Am. Assoc. Tex. Technol. 353*, 431 (1952).
364. T.B. Lefferdink and H.P. Briar, *Text. Res. J. June*, 477 (1959).
365. G.H. Fremon, *Fibers from Synthetic Polymers*, R. Hill, Ed., Elsevier, New York, chap. 19 (1953).

366. E.W. Rugeley, T.A. Feild, Jr., and G.H. Fremon, *Ind. Eng. Chem. 40*, 1724 (1948).
367. J.C. Masson, Home furnishings and industrial applications, in *Acrylic Fiber Technology and Applications*, J.C. Masson, Ed., Marcel Dekker, New York, pp. 344–347 (1995).
368. R.H. Peters and R.H. Still, *Applied Fibre Science*, F. Happey, Ed., Academic Press, New York, Vol. 2, chap. 10, p. 321 (1979).
369. H.E. Howell and A.S. Patil, *J. Appl. Polym. Sci. 44*, 1523 (1992).
370. J.B. Quig, *Papers Am. Assoc. Text. Tech. 4*, 61 (1948).
371. J.B. Quig, *Can. Text. J. 66*(1), 42, 46 (1949).
372. J.B. Quig, *Rayon Synth. Text. 30*(2), 79 (1949).
373. J.B. Quig, *Rayon Synth. Text. 30*(3), 67 (1949)
374. J.B. Quig, *Rayon Synth. Text. 30*(4), 91 (1949)
375. J.C. Masson, Home furnishings and industrial applications, in *Acrylic FiberTechnology and Applications*, J.C. Masson, Ed., Marcel Dekker, New York, pp. 357–361 (1995).
376. *Fiber Producer*, p. 44 (April 1984).
377. L. Rebenfeld, B. Miller, and J.R. Martin, *Applied Fibre Science*, F. Happey, Ed., Academic Press, New York, Vol. 2, chap. 12, p. 465 (1979).
378. M. Lewin, E.M. Pearce, and S. Athas, Eds., *Flame Retardant, Polymeric Materials*, Vols. 1–3, Plenum Press, New York (1975, 1978, 1982).
379. M. Lewin, *Functional Finishes*, Part B, Vol. 20, M. Lewin and S.B. Sello, Eds., Marcel Dekker, New York, pp. 1–147 (1984).
380. Eur. Pat. Appl. 31,078 (July 1, 1981) (to Bayer AG.).
381. H. DeWitt Smith, *Proc. ASTM 44*, 543 (1944).
382. J.P. Bell and J.H. Dumbleton, *Text. Res. J. 41*(3), 196 (1971).
383. J.P. Knudsen, *Text. Res. J. 33*(1), 13 (1963).
384. Br. Patent 1,191,538 (May 13, 1970) (to Tokyo Rayon Co., Ltd.).
385. Br. Patent 1,271,334 (April 19, 1972) (to Toray Ind., Inc.).
386. A. Maranci, U.S. Patent 3,760,053 (September 18, 1973) (to American Cyanamid Co.).
387. Br. Patent 1,230,392 (April 28, 1971) (to Japan Exlan Co., Ltd.).
388. U.S. Patent 3,839,081 (August 22, 1973) (to American Cyanamid Co.).
389. U.S. Patent 3,844,825 (November 30, 1972) (to American Cyanamid Co.).
390. U.S. Patent 3,655,420 (May 6, 1970) (to E.I. DuPont de Nemours & Co., Inc.).
391. Br. Patent 1,380,422 (April 19, 1971) (to Ciba-Geigy Corp.).
392. Br. Patent 1,230,174 (April 24, 1967) (to Toray Ind., Inc.).
393. Dutch Pat. Appl. 7,010,775 (July 22, 1969) (to Asahi Chem. Ind., Co., Ltd.).
394. Japan Patent 33,289 (October 19,1979) (to Asahi Chem. Ind., Co., Ltd.).
395. Br. Patent 1,300,088 (December 20, 1972) (to Imperial Chem. Ind., Co., Ltd.).
396. U.S. Patent 3,679,355 (October 16, 1967) (to Asahi Chem. Ind., Co., Ltd.).
397. U.S. Patent 3,861,870 (May 4, 1973) (to Procter & Gamble).
398. Br. Patent 1,281,942 (July 31, 1968) (to Mitsubishi Rayon Co., Ltd.).
399. Japan Pat. Appl. 1,651 (1970) (to Toray Ind., Inc.).
400. Japan Patent 84,426 (June 25, 1980) (to Toray Ind., Inc.).
401. U.S. Patent 3,701,819 (August 16, 1968) (to Toho Beslon Co., Ltd.).
402. Japan Patent 33,290 (October 10, 1979) (to Asahi Chem. Ind., Co., Ltd.).
403. V. Sermadzhieva, I. Markovs, and P. Patronova, *Khim. Volokna 6*, 56 (1981).
404. K. Slater, *Text. Prog. 9*(4), 1 (1977).
405. V. Radlmann, U. Reinehr, and G. Nischk, U.S. Patent 4,123,200 (March 6, 1978) (to Bayer AG.).
406. Anonymous.
407. Japan Patent 158,321 (December 9, 1980) (to Mitsubishi Rayon Co., Ltd.).
408. Austrian Patent 363,168 (July 10, 1981) (to Chemiefaser Lenzing AG.).
409. U. Reinehr, T. Herbertz, and H.J. Jungverdorben, Ger. Offen. 2,706,522 (August 17, 1978) and Ger. Offen. 2,706,032 (August 17, 1978) (to Bayer AG.).
410. U.S. Patent 3,839,520 (April 10, 1972) (to American Cyanamid).
411. U.S. Patent 3,781,391 (January 24,1968) (to American Cyanamid).
412. W. Korner, G. Blankenstein, P. Dorsch, and U. Reinhrand, *Chemiefasern/Textilindustrie 29/81*, 452 (June 1979).

413. E. Bakerzyk, *Polimery 17*, 549 (1972).
414. A. Gintis and E.J. Mead, *Text. Res. J.*, 578 (July 1959).
415. H. Aotani and A. Ichimura, *Sen-i Gakkaishi 33*, 222 (1977).
416. H. Terada, *Sen-i Gakkaishi 29*, T-345 (1973).
417. T. Hirota, *Sen-i Seihn Shohi Kagaku 21*(5), 29 (1980).
418. J.F. Ryan, Jr., U.S. Patent 3,360,603 (February 2, 1971) (to E.I. DuPont de Nemours & Co., Inc.).
419. I.M. Jonkoff, U.S. Patent 3,618,307 (November 9, 1971) (to E.I. DuPont de Nemours & Co., Inc.).
420. M. Yoshitake, H. Autami, and M. Shimamura, Japan Patent 12,046 (May 1, 1970) (to Toray Ind., Inc.).
421. K. Saito, S. Kumura, and T. Sato, Japan Patent 13,958 (May 14, 1971) (to Toray Ind., Inc.).
422. A. Ogata et al., Japan Patent 9,130 (April 2, 1970) (to Tokyo Spinning Co., Ltd.).
423. K. Saito, S. Kumura, T. Nakamura, and T. Sato, Japan Patent 16,280 (June 15, 1970) (to Toray Ind., Inc.).
424. Japan Patent 66,177 (April 22, 1982) (to Toray Ind., Inc.).
425. Y. Honda, K. Yamamoto, and T. Iwasa, Japan Patent 126,321 (November 4, 1978) (to Asahi Chem. Ind., Co., Ltd.).
426. Japan Patent 25,410 (February 10, 1982) (to Asahi Chem. Ind., Co., Ltd.).
427. Br. Patent 1,309,051 (May 27, 1970) and Japan Patent 71,414 (May 4, 1982) (to Mitsubishi Rayon Co., Ltd.).
428. Br. Patent 1,263,082 (February 7, 1968) (to Courtaulds, Ltd.).
429. B. Sassenrath, *Chemiefasern/Textilindustrie, Anwendungstech. Text. Ind. 31*, 808 (1981).
430. J.F. Fuzek, *ACS Symp.* Ser. No. 127, 515 (1980).
431. U.S. Patent 3,624,195 (October 13, 1969) (to Asahi Chem. Ind., Co., Ltd.).
432. C.R. Pfeifer and H.B. Hurley, *Appl. Polym. Symp. 25*, 179 (1974).
433. L.M. Levites et al., *Khim. Volokna 15*(3), 21 (1973).
434. Jap. Pat. Appl. 32,432 (1971) (to Mitsubishi Rayon Co., Ltd.).
435. V.F. Androsov, K.I. Andreeva, V.S. Bondarenko, M.A. Zharkova, and T.A. Romanova, *Khim. Volokna 12*(2), 28 (1970); *Fiber Chem. USSR 1970*, 136 (1970).
436. M.A. Zharkova, G.I. Kudryartsev, I.F. Khudoshev, and T.A. Romanova, *Khim. Volokna 11*(2), 49 (1969); *Fiber Chem. USSR 1969*, 191 (1969).
437. Br. Patent 1,275,138 (December 19, 1969) (to Rolls-Royce, Ltd.).
438. French Patent 1,549,347 (December 21, 1966) (to E.I. DuPont de Nemours & Co., Inc.).
439. U.S. Patent 3,778,416 (January 7, 1921) (to Hoechst AG.).
440. U.S. Patent 3,626,049 (September 2, 1967) (to Tokyo Spinning Co., Ltd.).
441. Br. Patent 1,355,701 (May 2, 1970) (to Hoechst AG.).
442. U.S. Patent 3,809,685 (September 26, 1970) (to Asahi Chem. Ind., Co., Ltd.).
443. Swed. Patent 166,913 (April 21, 1969) (to Stockholms Superfostat Fabriks).
444. Br. Patent 1,227,410 (April 7, 1971) (to Tokyo Spinning Co., Ltd.).
445. Y. Yutaka and T. Iwasa, Japan Patent 59,4266 (May 14, 1979) (to Asahi Chem. Ind., Co., Ltd.).
446. K. Takeya, H. Suzuki, and N. Yamawaki, U.S. Patent 4,012,459 (March 15, 1977) (to American Cyanamid Co.).
447. A. Yamamoto et al., Japan Patent 12,474 (1971) (to Toyo Spinning Co., Ltd.).
448. G.D. Wolf et al., U.S. Patent 4,100,143 (July 11, 1978) (to Bayer AG.).
449. T. Neukam, U. Reinehr, F. Bentz, and G. Nischk, U.S. Patent, 4,059,556 (November 22, 1977) (to Bayer AG.).
450. T. Neukam, U. Reinehr, F. Bentz, and G. Nischk, Ger. Offen. 2,622,921 (December 1, 1977) (to Bayer AG.).
451. G. Palethorpe and L. Poole, Ger. Offen. 2,825,754 (December 21, 1978) (to Monsanto Co.)
452. Br. Patent 1,329,357 (September 25, 1970) (to Bayer AG.).
453. U.S. Patent 3,732,339 (March 26, 1970) (to FBy).
454. Br. Patent 1,348,044 (March 26, 1970) (to Bayer AG.).
455. Z. Choinski, B. Knetel, and K. Kamecka-Jedrzejczak, *Pregl. Wlok. 32*(6), 267 (1978).
456. J.P. Knudsen, U.S. Patent 3,124,620 (March 10, 1964) (to Monsanto Co.).
457. K. Weinrotter, H. Schmidt, H. Kraessig, G. Haslinger, and A. Verwaenger, Austrian Patent 364,069 (September 25, 1981) (to Chemiefaser Lenzing AG.).

458. H. Takeda et al., Japan Patent (June 1, 1979) (to Asahi Chem. Ind., Co., Ltd.).
459. G.S. Kruglova and E.P. Krasnov, *Khim. Volokna 13*(3), 18 (1971); *Fiber Chem. USSR 1971*, 255 (1971).
460. A. Sugiura, N. Kohashi, and M. Ozaki, Japan Patent 99,318 (August 20, 1977) (to Japan Exlan Co., Ltd.).
461. Y. Joh and T. Sugimori, U.S. Patent 3,813,372 (May 28, 1947) (to Mitsubishi Rayon Co., Ltd.).
462. Br. Patent 1,341,868 (December 28, 1973) (to Mitsubishi Rayon Co., Ltd.).
463. J.R. Kirby, U.S. Patent 3,784,511 (January 1, 1974) (to Monsanto Co.).
464. Br. Patent 1,360,669 (December 5, 1970) (to Bayer AG.).
465. H. Sekiguchi, K. Takeya, K. Tanghashi, and J. Tsuge, U.S. Patent 3,681,275 (August 1, 1972) (to Japan Exlan Co., Ltd.).
466. Br. Patent 1,349,669 (September 28, 1970) (to Mitsubishi Rayon Co., Ltd.).
467. Br. Patent 1,206,484 (December 5, 1966) and Br. Patent 1,178,913 (February 2, 1966) (to Bayer AG.).
468. Z.I. Burlyuk and A.N. Reshetova, *Proizvod. Sin. Volok*.
469. Br. Patent 1,355,701 (May 2, 1970) (to Hoechst AG.).
470. B. Huber, H.J. Kleiner, and H. Neumaier, U.S. Patent 4,052,551 (October 4, 1977) (to Hoechst AG.).
471. S. Hamada, M. Hoshima, Z. Izumi, H. Kitagawa, and H. Sakai, U.S. Patent 3,660,527 (May 1, 1972) (to Toray Ind., Inc.).
472. R.R. Lattime, Eur. Pat. Appl. 80,969 (December 11, 1982) (to Goodyear Tire & Rubber Co.).
473. S. Nakao, N. Numata, and T. Yamamoto, Japan Patent 29,872 (September 13, 1973); 29,873 (to Kanebo Co., Ltd.).
474. K. Nakao, M. Ohki, and T. Yamamoto, Japan Kokai Patent 73,979 (September 1, 1973) (to Kanegafuchi Co., Ltd.).
475. T. Ohfuka, K. Shirode, and Y. Ichikawa, Ger. Offen. 1,720,202 (September 21, 1972) (to Asahi Chem. Ind., Co., Ltd.).
476. S. Matsumura and C. Kanemitsu, Japan Patent 26,333 (July 30, 1971) (to Teijin Ltd.).
477. N. Yokouchi, T. Kawamura, and T. Tokitada, Japan Patent 38,561 (November 13, 1971) (to Mitsubishi Rayon Co., Ltd.).
478. Ger. Offen. 1,494,047 and 1,494,048 (November 2, 1962) (to Bayer AG.).
479. Ger. Offen. 1,946,008 (October 28, 1968) (to Engels).
480. Br. Patent 1,345,255 (Marcy 25, 1970) (to Bayer AG.).
481. Br. Patent 1,243,292 (September 13, 1968) (to Japan Exlan Co., Ltd.).
482. Y. Yamauchi, *Senryo To Yakuhin 15*(23), 75 (1970).
483. Y. Saki and H. Kato, *Sen-i Kako 25*(2), 88 (1973).
484. G. Eigenmann, L. Kaiser, and C. Leuthi, U.S. Patent 3,849,155 (November 19, 1974) and Br. Patent 1,313,332 (June 27, 1969) (to Ciba-Geigy Corp.).
485. Br. Patent 1,195,143 (September 2, 1967) (to Nippon Kayaku KK.).
486. U.S. Patent 3,682,004 (January 20, 1969) (to Courtaulds, Ltd.).
487. L.A. Yasnikov, E.A. Pakshver, and Y.V. Glazkovskii, *Khim. Volokna 14*(5), 11 (1972).
488. Z. Adamski, *Przegl. Wlok. 25*(3), 159 (1971).
489. A.H. Bruner and T.B. Truscott, U.S. Patent 3,558,765 (January 26, 1971) (to Monsanto Co.).
490. G. Palethorpe, U.S. Patent 3,642,628 (December 10, 1968) (to Monsanto Co.).
491. S.E. Shalabi et al., *Vysokomol. Soedin. Ser. A 16*, 1904 (1974).
492. I. Ismailov, A.M. Karimova, A.T. Dzhalilov, and M.A. Askarov, *Dokl. Akad. Nauk. Uzb. SSR 7*, 44 (1981).
493. T. Okada et al., *Nippon Genshiryoku Nempo JAERI 5026*, 35 (1970).
494. N. Grassie and R. McGuchan, *Eur. Polym. J. 6*, 1277 (1970).
495. J. Runge and W. Nelles, *Faserforsch. Textiltech. 22*, 1 (1971).
496. J.R. Brown, *Proc. R. Aust. Chem. Inst. 40*, 347 (1973).
497. L. Krcma, *Kem. Ind. 28*, 247 (1979).
498. T. Takashi, T. Ito, S. Nagai, S. Hiraoka, H. Mizuno, and Y. Kakumoto, Japan Patent 34,500 (March 13, 1979) (to Mitsubishi Rayon Co., Ltd.).

499. Br. Patent 1,202,793 (January 24, 1967) (to Toray Ind., Inc.).
500. P. Mallison, *J. Soc. Dyers Colour 90*, 67 (1974).
501. N.B. Zhurovska and B. Popov, *Khim. Ind. Sofia 43*, 436 (1971).
502. Y. Ichikawa, Y. Schichijo, Y. Sugaya, and T. Veda, U.S. Patent 3,770,494 (November 6, 1973) (to Asahi Chem. Ind., Co., Ltd.).
503. Br. Patent 1,301,029 (February 28, 1969) (to Mitsubishi Rayon Co., Ltd.).
504. P. Braid, *Can. Text. J. 91*(1), 27 (1974).
505. K.A. Reif, G. Lasch, A. Heinze, and W. Huber, *Tech. Textil. 15*, 108 (1972).
506. Br. Patent 1,213,248 (September 12, 1967) and Br. Patent 1,345,515 (January 9, 1970) (to Wissenschaftlich-Technische Zentrum Technische Textilien).
507. Br. Patent 1,247,940 (November 25, 1969) (to Dow Chemical Co.).
508. R. Zeidman, *Inustr. Textila 24*, 592 (1973); R. Zeidman, *Inustr. Textila 24*, 721 (1973).
509. A.S. Kryukova, T.V. Angarova, E.N. Cherkasova, and R.A. Lapina, *Statich. Elek. Polim., Sb. Dokl. Semin.*, 110 (1967).
510. C.M. Horikx, *Tex-Textilis 1*, 16 (1974).
511. H. Takeda, U.S. Patent 3,758,659 (September 11, 1973) (to Toray Ind., Inc.).
512. L. Lugo and C. Remi, U.S. Patent 3,608,043 (1967) (to Societa Italiana Resine S.P.A.).
513. Br. Patent 1,374,802 and Br. Patent 1,373,552 (August 20, 1970) (to American Cyanamid).
514. B. Sassenrath, *Cemiefasern/Text Anwendungstech. Text. Incl. 31*, 808 (1981).
515. Br. Patent 1,237,628 (June 12, 1968) (to American Cyanamid).
516. H.J. Jacobasch et al., German Patent (East) (October 31, 1979).
517. H.J. Jacobasch, *Tenside Deterg. 17*(3), 113 (1980).
518. P.A. Glubish and I.E. Kocher, *Khim. Teknol. (Kiev) 2*, 32 (1981).
519. G.A. Gabrielyan and Z.A. Rogovin, *Adv. Polym. Sci. 25*, 97 (1977).
520. G. Ham, *Text. Res. J. 24*, 597 (1954).
521. *Identification of Textile Materials*, 7th ed., The Textile Institute, Manchester (1975).
522. *Technical Manual of the AATCC*, Vol. 58, AATCC Test Methods 20-1980, Fiber Analysis: Qualitative and 20A-1982, Fiber Analysis: Quantitative (1982/1983).
523. W.G. Wolfgang, *Man-Made Textile Encyclopedia*, J.J. Press, Ed., Textile Publishers, Inc., chap. IV, p. 141 (1959).
524. B.G. Frushour, Acrylic polymer characterization in the solid state and in solution, in *Acrylic Fiber Technology and Applications*, J.C. Masson, Ed., Marcel Dekker, New York, pp. 242–253 (1995).
525. R.B. Seymour and C.E. Carrahar, Jr., *Polymer Chemistry, An Introduction*, 2nd ed., Marcel Dekker, New York, chap. 4 (1988).
526. P.J. Debye, *J. Appl. Phys. 17*, 392 (1944).
527. P.J. Debye, *J. Phys. Coll. Chem. 52*, 18 (1947).
528. H. Staudinger and W. Heuer, *Ber. Dtsch. Chem. Ges. 63*, 222 (1930).
529. R.L. Cleland and W.H. Stockmayer, *J. Polym. Sci. 18*, 473 (1955).
530. A.V. Topciev, A.D. Litmanovic, and V. Ja Stern, *Dokl. Akad. Nauk. SSSR 147*, 1389 (1962).
531. G. Glockner, F. Francuskiewicz, and K. Muller, *Plaste Kautschuk 18*, 564 (1971).
532. P. Fritzsche, P. Klug, and V. Grobe, *Faserforsch. Textiltech. 22*, 250 (1971).
533. R.H. Heidner and M.E. Gibson, *Encyclopedia of Industrial Chemical Analysis*, Vol. 4, F.D. Snell and C.L. Hilton, Eds., Wiley Interscience, New York, pp. 219–360 (1967).
534. G.C. East, *Text. Prog. 3*(1), 67 (1971).
535. G.A.M. Butterworth and N.J. Abbott, *Mechanical Testing of Polymeric Fibrous Materials*. Available from Instron Ltd., High Wycomb, U.K.
536. W. Baggett et al., U.S. Patents 4,839,211, 4,882,222 (assigned to Monsanto Co.).
537. M. Bheda, R.S. Knorr, and J. Yu, U.S. Pat. Appl. 08/458,152, filed 6/2/95 (Monsanto Co.).
538. J.C. Masson, *Acrylic Fiber Technology and Applications*, Marcel Dekker, New York, chap. 11, p. 341 (1995).
539. H. Emsermann and R. Foppe, in *Acrylic Fiber Technology and Applications*, J.C. Masson, Ed., Marcel Dekker, New York, chap. 9, p. 285 (1995).
540. J.C. Masson, *Acrylic Fiber Technology and Applications*, Marcel Dekker, New York, chap. 6, p. 167 (1995).

541. *Textiles* 4(3) (1981); *Textiles* 5(2) (1976); *Textiles* 6(1) (1977); *Textiles* 8(1) (1979); *Textiles* 10(1) (1981).
542. *Indust. Fabric Prod. Rev.* 59(10A), 1983.
543. W.E. Fitzgerald and J.P. Knudsen, *Text. Res. J.* 37, 447 (1967).
544. *Textile Daily*, April 8, 1983.
545. Japan Patent 137,219 (October 25, 1980) (to Asahi Chem. Ind., Co., Ltd.).
546. J.H. Gustafson, Conference and Exhibition, AATCC, pp. 170–174 (1987).
547. M. Banks, J.R. Ebdon, and M. Johnson, *Sch. Phys. Mater. Polym.* 35(16), 3470–1473 (1994).
548. M. Arai and K. Yoshida, Jpn. Kokai Tokkyo Koho, 5 pp., JP 04316616 A2 921109 Heisei.
549. X. Jiang, *Peop. Rep. China. Shanghai Huagong* 18(5), 18 (1993).
550. R.C. Nametz, *Ind. Eng. Chem.* 62, 41 (1970).
551. A.A. Vaidya and S. Chattopadhyay, *Tex. Dyer Print.* 10(8), 37 (1977).
552. C. Hsieh, *Hsin Hsien Wei* 20(11), 12 (1979).
553. M. Hatano and H. Ogawa, *Kagaku Keizoi* 27(7), 77 (1980).
554. A.A. Kumar, K.V. Vaidya, and K.V. Datye, *Man-Made Text. India* 24(1), 23 (1981).
555. B.B. Braddock and G.J. Brealey, *Rep. Prog. Appl. Chem.* 54, 527 (1969).
556. B.J.D. Torrance and K. Wade, *Rep. Prog. Appl. Chem.* 56, 77 (1971).
557. J.W. Lyons, *The Chemistry and Uses of Flame Retardants*, 1st ed., Wiley Interscience, New York (1970).
558. S.B. Sello, *J. Elastoplast.* 4, 5 (1972).
559. J.R. Brown, *Proc. R. Aust. Chem. Inst.* 40(12), 347 (1973).
560. M.M. Gauthier, R.D. Deanin, and C.J. Pope, *Polym. Plast. Technol. Eng.* 16(1), 39 (1981).
561. S. Emori, *Sen'I Gakkaishi* 50(2), 69 (1994) (to Asahi Chem. Ind., Co., Ltd.).
562. K. Wilson, *Text. Horizons* 5 (June 1984).
563. M. Senba and T. Hoshi, Jpn. Kokai Tokkyo Koho, 5 pp., JP 04034010 A2 920205 Heisei (to Asahi Chem. Ind., Co., Ltd.).
564. Japan Patent 28,373 (June 9, 1983) (to Kanebo).
565. U.S. Patent 258,332 (November 11, 1982) (to American Cyanamid Co.).
566. Japan Patent 212,519 (August 7, 1983) (to Kanebo).
567. Zh. A. Zgibneva, A.A. Geller, and B.E. Geller, *Khim. Volokna* 1, 69 (1971).
568. *Jpn. Text. News*, p. 32 (January 1984).
569. H. Intili, *J. Ind. Fabr.* 3(1), 4 (1984) (to Microban Prod. Co.).
570. M. Sato and M. Hoten, Jpn. Kokai Tokkyo Koho, 7 pp., JP 05140868 930608 Heisei JP 91-333956 91121 (to Toyo Boseki, Japan).
571. P. Bajaj, Pushpa, and M.S. Dara, *Indian J. Fibre Text. Res.* 19(2), 95 (1994).
572. T. Masui, S. Hagura, J. Fukui, S. Hayashi, and H. Ito, Jpn. Kokai Tokkyo Koho, 4 pp., JP 06116816 A2 940426 Heisei, JP 92-265101 921002.
573. S. Ito and K. Matsui, Japan Kokai, 2 pp., JP 49094921 740909 Showa.
574. Y. Mizukami, T. Matsumura, H. Yoshimura, and S. Kakegawa, Jpn. Kokai Tokkyo Koho, 4 pp., JP 06081216 940322 Heisei, JP 92-258982 920901.
575. Y. Fukui, H. Ito, and S. Minami, Jpn. Kokai Tokkyo Koho, 4 pp., JP 04077508 A2 920311 Heisei, JP 90-186591 900713.
576. T. Kato, T, Kago, K. Kusakabe, and S. Morooka, *J. Chem. Eng. Jpn.* 23(6), 755 (1990).
577. S. Morooka, T. Kato, M. Inada, T. Kago, and K. Kusakabe, *Ind. Eng. Chem. Res.* 30(1), 190–196 (1991).
578. B.W. Zhang, K. Fischer, D. Bieniek, and A. Kettrup, *React. Polym.* 24(1), 49 (1994).
579. N.V. Bystan, N.V. Voitova, G.V. Ivanova, L.Y. Emets, and A. Vollf, *Khim. Volokna* 4, 40 (1983).
580. S. Dorokhina and M.A. Zharkova, *Chim. Volokna* 1, 69 (1972).
581. E. Terlemezyan and S. Veleva, *Acta Polym.* 38(2), 119 (1987).
582. S. Veleva and E. Terlemezyan, *Dokl. Bulg. Akad. Nauk,* 41(7), 55 (1988).
583. S. Veleva, T. Terlemesyan, A. Arsov, and K. Dimov, *React. Kinet. Catal. Lett.* 25(3–4), 349 (1984).
584. J. Simitzis, *Angew. Makromol. Chem.* 228, 13 (1995).
585. F. Marchini and V. Massa, *Chemiefasern/Textilindustrie* 38(3), E26–E27, T5, T8, T10 (1988).
586. H. Onda, *Nippon Tenmo Senshoku K.K. Japan Kogyo Zairyo* 42(3), 47 (1994).

587. *High Perform. Text.* 3(10), 7 (1983).
588. H. Hahne, P. Huusen, and A.G. Hoechest, *Rev. Quim. Text.* 114, 76, 78, 80, 82, 84–85 (1993).
589. Z. Zheng and D. Feldman, *Prog. Polym. Sci.* 20(2), 185 (1995).
590. Z. Zheng and D. Feldman, *Prog. Polym. Sci.* 20(2), 185 (1995).
591. J.C. Masson, *Acrylic Fiber Technology and Applications*, Marcel Dekker, New York, chap. 6, pp. 183–186 (1995).
592. A. Gopalan, T. Vasudevan, P. Manisankar, C. Paruthimalkalaignan, and A. Ramasubramanian, *J. Appl. Polym. Sci.* 56(10), 1299 (1995).
593. A. Gopalan, T. Vasudevan, P. Manisankar, C. Paruthimalkalaignan, A. Ramasubramanian, and S.S. Hariharan, *J. Appl. Polym. Sci.* 56(13), 1715 (1995).
594. P. Bajaj, D.K. Paliwal, and A.K. Gupta, *Indian J. Fibre Text. Res.* 21(2), 143 (1996).
595. P. Bajaj and D.K. Paliwal, *J. Fibre Text. Res.* 16(1), 89 (1991).
596. P. Bajaj and K. Suurya, *JMS—Rev. Macromol. Chem. Phys.* C27(2), 181 (1987).
597. J.G. Venner, *Kirk–Othmer Encylopediea of Chemical Technology*, 4th ed., Wiley Interscience, New York (1995).
598. E. Sittig, Ed., *Carbon and Graphite Fibers: Manufacturing and Applications*, Noyes Data Corp., Park Ridge, NJ (1980).
599. J. Delmonte, *Technology of Carbon and Graphite Fiber Composites*, Van Nostrand Reinhold, New York (1981).
600. J.-B. Donnet and R. Bansal, *Carbon Fibers*, Marcel Dekker, New York (1984).
601. E. Fitzer, Ed., *Carbon Fibers and Their Composites*, Springer-Verlag, New York (1985).
602. W. Watt and B.V. Perov, Eds., *Handbook of Composites, Vol. 1, Strong Fibers*. Elsevier, New York (1985).
603. M.S. Dresselhaus et al., *Graphite Fibers and Filaments*, Springer-Verlag, New York (1988).
604. J.L. Figueiredo et al., Eds., *Carbon Fibers, Filaments, and Composites*, Kluwer Academic Publishers, Dordrecht, The Netherlands (1990).
605. J.-B. Donnet and R.C. Bansal, *Carbon Fibers*, 2nd ed., Marcel Dekker, New York (1990).
606. L.H. Peebles, *Carbon Fibers, Formation, Structure, and Properties*, CRC Press, Boca Raton, FL (1994).
607. P.J. Goodhew et al., *Mater. Sci. Eng.*, 17, 1 (1975).
608. L.H. Peebles, Jr., *Encyclopedia of Polymer Science and Technology*, Suppl. Vol. 1, Wiley, New York, p. 1 (1976).
609. G. Henrici-Olive' and S. Olive', *Adv. Polym. Sci.* 51, 1 (1983).
610. W. Watt, *Handbook of Composites. Vol. 1, Strong Fibers*, W. Watt and B.V. Perov, Eds., Elsevier, New York, p. 327 (1985).
611. J.P. Riggs, *Encyclopedia of Polymer Science and Technology*, Vol. 2, Wiley, New York, p. 640 (1985).
612. D.J. Thorne, *Handbook of Composites. Vol. 1, Strong Fibers*, W. Watt and B.V. Perov, Eds., Elsevier, New York, p. 475 (1985).
613. E. Fitzer et al., *Carbon* 24, 387 (1986).
614. E. Fitzer, *Carbon* 27, 621 (1989).
615. S. Domodaran et al., *J. Text. Inst.* (U.K.), Special Edition on Composites (1991).
616. A.K. Gupta et al., *J. Macromol. Sci. Rev. Macromol. Chem. Phys.* C31, 1 (1991).
617. P. Rajalingam and G. Radhakrishnan, *J. Macromol. Sci. Rev. Macromol. Chem. Phys.* C31, 301 (1991).
618. Z. Bashir, *Carbon* 29, 1081 (1991).
619. D. Grove et al., *Composite Applications: the Role of Matrix, Fiber, and Interface*, T. Vigo and B. Kinzig, Eds., VCH Publishers, Weinheim, Germany, chap. 5 (1992).
620. W. Watt, L.N. Phillips, and W. Johnson, Br. Patent 1,110,791 (April 24, 1964) (to National Research Development Corp.).
621. W. Watt, L.N. Phillips, and W. Johnson, Br. Patent 1,148,874 (June 15, 1966) (to National Research Development Corp.).
622. Johnson et al., U.S. Patent 3,412,062 (1968).
623. I.R. McCartney, *Mod. Plast.* 30, 118 (1953).

624. N. Grassie and I.C. McNeill, *J. Polym. Sci.* 27, 207 (1958).
625. N. Grassie and J.N. Hay, *J. Polym. Sci.* 56, 189 (1962).
626. A.K. Fiedler, E. Fitzer, and F. Rozploch, *Carbon 11*, 426 (1973).
627. W.P. Hoffman et al., *J. Mater. Res.* 6, 1685 (1991).
628. O.L. Blakslee et al., *J. Appl. Phys.* 41, 3373 (1970).
629. A. Shindo, Japan Patent 4405 (1962).
630. S.C. Bennet and D.J. Johnson, Strength–structure relationships in PAN-based carbon fibers, in *Society of Chemical Industry, London*, p. 377 (1978).
631. G. Henrici Olive' and S. Olive', *Polym. Bull. (Berl.)* 5(6), 229 (1980).
632. G. Radhakrishnan et al., *Leather Sci. (Madras)* 28(2), 27 (1981).
633. E. Fitzer and T. Mueller, *Ext. Abstr. Prog.-Bienn Conf. on Carbon 15*, 314 (1981).
634. O.P. Bahl, R. Mathur, and K. Kundra, *Fibre Sci. Technol.* 15(2), 147 (1981).
635. S. Chari, O. Bahl, and R. Mathur, *Fibre Sci. Technol.* 15(2), 153 (1981).
636. P.E. Morgan, *Text. Prog.* 8(1), 69 (1976).
637. F. Fournel, *Chemiefasern/Textilindustrie 32/84*, 433 (1982).
638. Anon, Zoltek Companies (September 27, 1996) Merrill Lynch & Co.
639. Caro Industrial Corp. Ltd., London (1982).
640. *Melliand Textilbericht 12*(2), 115 (1983).
641. J.C. Masson, in *Acrylic Fiber Technology and Applications*, J.C. Masson, Ed., Marcel Dekker, New York, chap. 11, p. 352 (1995).
642. *High Perform. Text.* 3(10), 3 (1983).
643. Br. Patent 2,018,188A (November 24, 1978) (to American Cyanamid).
644. H. Takeda, U.S. Patent 3,814,739 (June 4, 1974) (to Toray Ind., Inc.).
645. R. Moreton, Carbon Fibers: Their Composites and Applications, Paper No. 12, The Plastics Institute, London (1971).
646. Eur. Pat. Appl. EP 44,534 (January 27, 1982) (to Hoechst AG.).
647. Asahi Chem. Ind., Co., Ltd., Japan Patent 60,051 (May 6, 1980).
648. Asahi Chem. Ind., Co., Ltd., Japan Patent 98,025 (February 13, 1976).
649. Asahi Chem. Ind., Co., Ltd., Japan Patent 17,966 (February 20, 1981).
650. A. Wuestfeld, and C. Wuestfeld, German Patent DE 3,012,998 (October 15, 1981).
651. REDCO, S.A, Belg. Patent BE 889,260 (October 16, 1981).
652. A.G. Hoechst, Eur. Pat. Appl. EP 44,534 (January 27, 1982).
653. *High Perform. Text.* 3(10), 3 (1983).
654. U.S. Patent 4,418,115.
655. *High Perform. Text.* 3(10), 3 (1983).
656. Ube Cycon, Ltd., Japan Patent 112,909 (September 5, 1981).
657. M. LeLannou, Br. U.K. Pat. Appl. GB 2,083,062 (March 17, 1982).
658. J.C. Masson, *Acrylic Fiber Technology and Applications*, Marcel Dekker, New York, chap. 11, p. 357 (1995).
659. *High Perform. Text.* 2(10), 1 (1982).
660. J. Yasushi, N. Akihiko, N. Kaneko, and A. Fukutome, Japan Patent 6,627 (January 21, 1978) (to Nippon Zeon Co., Ltd.).
661. J. Yasushi, N. Akihiko, N. Kaneko, and A. Fukutome, Japan Patent 160,819 (December 19, 1979) (to Nippon Zeon Co., Ltd.).
662. J. Yasushi, N. Akihiko, N. Kaneko, and A. Fukutome, Japan Patent 30,427 (March 4, 1980) (Nippon Zeon Co., Ltd.).
663. Y. Uchida and T. Iwasa, Japan Patent 78,323 (July 22, 1978) (to Asahi Chem. Inc. Co.).
664. E. Klein, A. Rembaum, and S.P.S Yen, U.S. Patent 4,045,352 (August 30, 1977) (to California Inst. of Tech.).
665. L. Yu Alikberova, SU 000343, July 7, 1982.
666. I.S. Dorokhina and M.A. Zharkova, *Khim. Volokna 1*, 69 (1971).
667. N.V. Bytsan, N.V. Voitova, G.V. Ivanova, L.V. Emets, and L.A. Vol'F, *Khim. Volokna 4*, 40 (1982).
668. E. Kneip, *Chemiefasern/Textilindustrie, 32/84*, June, E39 (1982).

669. E. Kneip, *Chemiefasern/Textilindustrie, 38/90*, December T116 (1988).
670. Anon, *Industrial Uses of Acrilan Fiber*, TT-17 Monsanto Company Bulletin, p. 6 (August 1965).
671. Mitsubishi Rayon Co., Japan Patent 144,299 (November 10, 1981).
672. Jap. Exlan Co., Ltd., Japan Patent 154,514 (November 30,1982) and 118,938 (September 18, 1981).
673. H. Takeda, *Kami to Purasuchikku* 6(10), 1 (1978).
674. C.I. Simionescu, V. Rusan, and A. Liga, *Mater. Plast.* (*Bucharest*) *14*(2), 69 (1977).
675. Kanebo Synthetic Fibers, Ltd., Japan Patent 39,214 (March 4, 1982).
676. Japan Exlan Co., Ltd., Japan Patent 112,909 (September 5, 1981).
677. Nippon Sanmo Dyeing Co., U.S. Patent 4,336,028.
678. H. Tanaka and K. Tsunawaki, Eur. Pat. Appl. 14,944 (September 3, 1980); Japan Pat. Appl. 15,035 (February 14, 1979) (to Teijin, Ltd.).
679. Japan Exlan Co., Ltd., Japan Patent 4,713 (January 19, 1981).
680. Dow Badische Co., Belg. Patent 854,105 (August 16, 1977).
681. M.A. Magrupov, *Mekhanizmy Obraz. I. Svoistva Polimerov*, Tashkent, pp. 245–282 (1981), from ref. in *Zh. Khim.* Abstr. No. 7T4 (1982).
682. *High Perform. Text.* *3*(10), 7 (1983).
683. M. Okoniewski and J. Koprowska, *Textilveredlung* *17*(6), 269 (1982).
684. A.G. Bayer, U.S. Patent 4,320,403.
685. A.G. Bayer, U.S. Patent, 4,320,403, from ref. in *High Perform. Text.*
686. A.A. Nikitin, SU 968,876 (March 23, 1982).
687. J.C. Masson, Ed., *Acrylic Fiber Technology and Applications*, Marcel Dekker, New York, chap. 1, p. 361 (1995).
688. P.E. Morgan and F. Cognigni, *Text. Asia*, p. 61 (December 1990).
689. D. Ward, *Text. Month* (September 11, 1983).
690. Akzo Nobel report on synthetic fibers, *Int. Fiber J.*, pp. 4–14 (August 1996).
691. *Manufactured Fiber Producer Handbook* (1996); *Fiber Organon* (February 1997), Fiber Economics Bureau, pub.
692. *Fiber Organon* (March 1997), Fiber Economics Bureau, pub.
693. Fiber Economics Bureau, 101 Eisenhower Parkway, Roseland, NJ 07-68.
694. Comite International de la Rayonne et des Fibres Synthetiques, Paris, France; 1996 Yearbook CIFRS, Brussels, Belgium.
695. Japan Fibres Association, No. 1–11, Chou-ku, Tokyo 103, Japan.
696. *Jpn. Text. News*, p. 58 (February 1984).
697. *Jpn. Text. News*, p. 10 (March 1984).
698. J.E. Luke, Asia looms large in fiber's future, *Fiber World*, America's Textiles International (November 1996).
699. The synthetic fibers market in China 1993–2002, *Int. Fiber J.*, Technon (U.K.) Ltd. (October 1994).
700. Fiber Economics Bureau, World Directory of Fiber Producers, 1994–1995.
701. P.T. O'Day, *Int. Fiber J.*, 4 (February 1997).
702. J.E. Luke, *America's Textiles Int., Fiber World Suppl.*, pp. FW2–FW4 (January, 1997).
703. W.S. Perkins, *America's Textiles Int.*, pp. 50–53 (November 1996).
704. J.N. Etters, *America's Textiles Int.*, pp. 52–53 (November 1996).
705. *America's Textiles Int.*, p. 224 (September 1994).

13 Aramid Fibers

Vlodek Gabara, Jon D. Hartzler, Kiu-Seung Lee, David J. Rodini, and H.H. Yang

CONTENTS

13.1 Introduction 976
    13.1.1 Historical Perspective 976
    13.1.2 Aramid—Definition 977
    13.1.3 Examples of Compositions 977
13.2 Aramid Products: Forms and Properties 977
13.3 Producers of Aramid Products 978
13.4 Structure–Property Relationship 979
    13.4.1 Fine Structure 979
    13.4.2 Thermal Properties 980
    13.4.3 Solubility and Chemical Properties 981
    13.4.4 Fiber Mechanical Properties 981
    13.4.5 Films and Papers 984
13.5 Polymerization of Aromatic Polyamides 985
    13.5.1 Introduction 985
    13.5.2 Synthesis of Ingredients 986
        13.5.2.1 *m*-Phenylene Diamine 986
        13.5.2.2 *p*-Phenylene Diamine 987
        13.5.2.3 3,4'-Diaminodiphenyl Ether 988
        13.5.2.4 Diacid Chlorides 988
    13.5.3 Polymerization Fundamentals 989
        13.5.3.1 Reaction Mechanism 990
        13.5.3.2 Reaction Energetics 991
    13.5.4 Direct Polymerization by Catalysis 991
    13.5.5 Polymerization Methods 993
        13.5.5.1 Interfacial Polymerization 993
        13.5.5.2 Solution Polymerization 995
        13.5.5.3 Vapor-Phase Polymerization 999
        13.5.5.4 Plasticized Melt Polymerization 1000
13.6 Aramid Solutions 1001
    13.6.1 Isotropic Solutions 1001
        13.6.1.1 *m*-Aramid Solutions 1001
        13.6.1.2 *p*-Aramid Solutions 1001
    13.6.2 Anisotropic Solutions 1002
        13.6.2.1 Phase Behavior 1002
        13.6.2.2 Rheological Properties 1003

13.7 Preparation of Aramid Products.. 1003
   13.7.1 Fibers... 1003
      13.7.1.1 Dry Spinning... 1003
      13.7.1.2 Wet Spinning... 1005
      13.7.1.3 Dry-Jet Wet-Spinning... 1006
   13.7.2 Film.. 1009
   13.7.3 Fibrids.. 1010
   13.7.4 Pulp.. 1011
13.8 Applications 1012
   13.8.1 *m*-Aramid Fiber.. 1013
      13.8.1.1 Protective Apparel... 1013
      13.8.1.2 Thermal and Flame-Resistant Barriers 1014
      13.8.1.3 Elastomer Reinforcement.. 1015
      13.8.1.4 Filtration and Felts 1015
   13.8.2 *m*-Aramid Paper 1015
      13.8.2.1 Electrical.. 1015
      13.8.2.2 Core Structures.. 1016
      13.8.2.3 Miscellaneous 1017
   13.8.3 *p*-Aramid Fiber... 1017
      13.8.3.1 Armor... 1017
      13.8.3.2 Protective Apparel... 1018
      13.8.3.3 Tires and Mechanical Rubber Goods 1018
      13.8.3.4 Composites... 1019
      13.8.3.5 Optical and Electromechanical Cables 1019
      13.8.3.6 Ropes and Cables.. 1020
      13.8.3.7 Reinforced Thermoplastic Pipe 1020
      13.8.3.8 Civil Engineering.. 1021
   13.8.4 *p*-Aramid Paper 1021
      13.8.4.1 Core Structures.. 1021
      13.8.4.2 Printed Wiring Boards 1022
      13.8.4.3 *p*-Aramid Pulp... 1022
13.9 Conclusions and Direction.. 1024
References 1025

13.1 INTRODUCTION

13.1.1 HISTORICAL PERSPECTIVE

Development of aromatic polyamides had a very unique beginning. Its origin in an industrial corporation (DuPont) led to a combination of fundamental science, engineering, and applications research from the very early stages of the development. In 1948, with the commercialization of nylon fiber and the near-development of a polyester fiber, the management of the DuPont Technical Division launched very broad, long-range research programs with goals, among others, of developing very high-strength fibers and high-temperature-resistant fibers.

The first phase covered a decade from the early 1950s to the early 1960s. Clearly, materials with unusual properties are not easy to process and they would not have been possible without the development of low-temperature solution polymerization techniques by Paul Morgan's group at DuPont [1]. The next critical step was to understand factors governing solubility of these difficult to dissolve polymers. Beste and Stephens [2] elucidated the role of certain salts that help in obtaining good solutions of these polymers. This work culminated in the commercialization of Nomex, the first high-temperature-resistant, m-aramid fiber [3,4]. Starting in the early 1960s the work focused on new fibers with a performance superior to Nomex, and p-aramids became a logical choice. Stephanie Kwolek focused her initial work on the more tractable poly(1,4-benzamide) polymer and produced, in the mid-1960s, a fiber with a spectacular modulus of 400 gpd. After additional work, a yarn with 7.0 gpd tenacity and an unheard of modulus of 900 gpd was prepared. This fiber was known as fiber B. Subsequent work shifted to poly(p-phenylene terephthalamide) (PPTA). After significant effort by many in both polymerization and spinning areas, Herb Blades made a processing breakthrough by focusing on the air-gap spinning of concentrated solutions of high-molecular-weight PPTA polymer. The first PPTA fibers were produced by this process in early 1970, and by 1972 Kevlar[a] was introduced to the market place. This was clearly a significant achievement considering the novelty and complexity of the technology involved and the speed at which it was accomplished. In addition to the impressive blend of science and engineering required to commercialize Kevlar, this was also the first demonstration of fiber mechanical properties predicted by theoretical considerations developed as early as the mid-1930s. This provided a fundamental basis as well as an impetus to study and commercialize other materials with comparable properties.

13.1.2 Aramid—Definition

As alluded to in the introduction, properties of aromatic polyamides differ significantly from those of their aliphatic counterparts. This led the U.S. Federal Trade Commission to adopt the term "aramid" to designate fibers of the aromatic polyamide type in which at least 85% of the amide linkages are attached directly to two aromatic rings.

13.1.3 Examples of Compositions

The superior properties of these materials were the reason why significant research effort has been devoted to this chemistry. Yang [5] showed at least 100 different compositions in this area and that number has doubled during the past 15 years since Yang's book was published.

The early work by Sweeny, Kwolek, and others demonstrated that progress in this area of technology was the result of a constant trade-off between properties and processability. This is very likely the reason why after half-a-century of research only four compositions have reached commercial stage: poly(m-phenylene isophthalamide) (MPDI), PPTA, copoly(p-phenylene/3,4'-diphenyl ether terephthalamide) (ODA/PPTA), poly[5-amino-2-(p-aminophenyl)benzimidazole terephthalamide] (SVM), and its copolymers.

13.2 ARAMID PRODUCTS: FORMS AND PROPERTIES

The outstanding thermal and mechanical properties that can be derived from these compositions led to the exploration, as well as commercial realization, of various product forms. Currently these product forms include fibers, fibrids and pulps, films, papers, and particles.

---

[a]Kevlar—a registered trademark of E.I. DuPont de Nemours & Co., Inc., Wilmington, Delaware, USA.

poly (*m*-phenylene isophthalamide) **MPDI**

poly (*p*-phenylene terephthalamide) **PPTA**

copoly(*p*-phenylene/3,4'-diphenyl ether terephthalamide) **ODA/PPTA**

poly [5-amino-2-(*p*-aminophenyl)benzimidazole terephthalamide]–**SVM**

The largest commercial volume of these materials is in the form of fibers. Continuous filament yarns are preferred where very high mechanical properties are required and staple fiber is used for textile applications. The significant volumes involved in these applications led to the development of special spinning processes designed to produce these forms.

The excellent thermal properties of these materials led to high volume applications where these materials were used as binders or as short fiber reinforcing agents. This required the development of both fibrids and pulps. This chapter discusses both the processes of formation as well as the principles of applications of these forms.

Various nonwoven structures have been developed as well. The least important among sheet structures are films. There are two film products (see Section 13.3) based on *p*-aramids and none on *m*-aramids. The significant cost differential is the most likely reason for this situation. On the other hand a very large market has been developed for papers based on both *p*-aramids and *m*-aramids. In general, these papers are based on short fibers (floc) and a binder (fibrids), but other components have been explored as well. A very small market exists for particles other than fibrids and pulps.

13.3 PRODUCERS OF ARAMID PRODUCTS

The basic development and the first commercial introduction of these materials were done by DuPont, which continues to be the largest producer. *m*-Aramid fiber products (staple, continuous filament yarn, and floc) with the trademark Nomex[b] are produced by DuPont in the United States as well as Spain. The paper products come from the U.S. plant as well as from a facility in Japan. The only other major *m*-aramid producer is Teijin, with its fiber product Teijinconex[c] produced in Japan.

The situation is very similar on the *para* side of chemistry. The first and the largest producer—DuPont—has three facilities throughout the world. The largest one in the United States produces essentially all product forms except films. Fiber is also produced in Ireland and Japan. The other producer of *p*-aramids is Teijin Co., which produces two basic fibers: Twaron[d] based on PPTA and Technora[e] based on a copolymer. Twaron is produced in the Netherlands while Technora is manufactured in Japan.

A small amount of *p*-aramid fiber (Armos and Rusar) is produced in Russia. Both are copolymers based on diaminophenylbenzimidazole—a unique but expensive monomer.

There are two producers of *p*-aramid film. The first one was Toray with its Mictron[e] film based on a copolymer and Asahi with a product (Aramica[f]) based on PPTA homopolymer.

13.4 STRUCTURE–PROPERTY RELATIONSHIP

13.4.1 Fine Structure

In general aramid homopolymers crystallize with relative ease. PPTA is a highly crystalline material. Two structures have been identified for this polymer: the first was proposed by Northolt [6] and the second by Haraguchi [7]. Haraguchi [7] and Roche [8] proposed mechanisms for their formation. In both cases they proposed an interaction between the solution and the coagulation process. Roche proposed that to form the Haraguchi structure, PPTA solution has to crystallize into a crystal solvate [9] prior to the removal of sulfuric acid. After acid removal and drying the Haraguchi polymorph is formed. This is the less stable form and at an elevated temperature rearranges into the Northolt form. Coagulation of PPTA solution leads to the Northolt structure, according to Roche, and that is why all commercial fibers exhibit essentially the Northolt structure. Northolt [6] and later Tashiro [10] reported their estimates of the size of the orthorhombic unit cell. The values are listed in Table 13.1. Commercial fibers based on PPTA are highly crystalline. Estimates of the degree of crystallinity of Kevlar 29 are 68 to 85% and as high as 95% for Kevlar 49 [11,12].

In addition to crystallinity, PPTA fibers exhibit a larger scale organization. It has been proposed that PPTA fibers have an unusual radial orientation of pleated hydrogen-bonded sheets [13]. This unique morphology has a significant impact on the mechanical properties of the fibers.

MPDI has a triclinic unit cell and is significantly less crystalline than PPTA (Table 13.1). Savinov [14] proposed that crystallinity depends on the conditions of polymer precipitation from solution. Precipitation of polymer in water leads to a noncrystalline material while precipitation in water containing some solvent leads to a crystalline form. Krasnov [15] showed that increased fiber orientation leads to higher crystallinity. SVM, the Russian

---

[b]Nomex—a registered trademark of E.I. DuPont de Nemours & Co., Inc., Wilmington, Delaware, USA.
[c]Teijinconex, Technora—registered trademarks of Teijin, Ltd., Japan.
[d]Twaron—a registered trademark of Akzo Nobel, The Netherlands.
[e]Mictron—a registered trademark of Toray Co., Japan.
[f]Aramica—a registered trademark of Asahi Co., Japan.

TABLE 13.1
Crystallinity of Homopolymers

	PPTA	MPDI
Crystal system	Orthorhombic	Triclinic
Lattice constant		
a (Å)	7.80	5.27
b (Å)	5.19	5.25
c (Å)	12.9	11.3
α (degree)		111.5
β (degree)		111.4
γ (degree)	90	88.0
Number of chains in a unit cell	2	1
Density (g/cm^3)		
Calculated	1.50	1.45
Observed	1.43–1.45	1.38

Source: From Northolt, M.G.; *Eur. Polym. J.*, 10, 799, 1974; Haraguchi, K., Kajiyama, T., and Takayanagi, M.J., *J. Appl. Polym. Sci.*, 23, 915, 1979; Roche, E.J., Allen, S.R., Gabara, V., and Cox, B., *Polymer*, 30, 1776, 1989; Gardner, K.H., Matheson, R.R., Avakin, P., Chia, Y.T., and Gierke, T.D., *Polym. Prepr. (Am. Chem. Soc. Div. Polym. Chem.)*, 24(2), 469, 1983; Tashiro, K., Kobayashi, M., and Tadokoro, H., *Macromolecules*, 10(2), 413, 1977.

product based on poly[5-amino-2-(*p*-aminophenyl) benzimidazoleterephthalamide] is the only other commercial product based on a homopolymer. This material is noncrystalline, as might be expected, based on the structural irregularities that can arise from the orientation of repeat units in the polymer chain (cis–trans, head–tail).

Copolymers are noncrystalline materials. Blackwell has studied the fine structure of Technora fiber [16].

13.4.2 Thermal Properties

The search for materials with very good thermal properties was the original reason for research into aromatic polyamides. Bond dissociation energies of C—C and C—N bonds in aromatic polyamides are ~20% higher than those in aliphatic polyamides. This is the reason why the decomposition temperature of MPDI exceeds 450°C [17]. Conjugation between the amide group and the aromatic ring in PPTA increases chain rigidity as well as the decomposition temperature, which exceeds 550°C [17,18].

Obviously, hydrogen bonding and chain rigidity of these polymers translates to very high glass transition temperatures. Using low-molecular-weight polymers, Aharoni [19] measured glass transition temperatures of 272°C for MPDI and over 295°C for PPTA (which in this case had low crystallinity). Others have reported values as high as 492°C [20]. In most cases the measurement of T_g is difficult because PPTA is essentially 100% crystalline. As one would expect, these values are not strongly dependent on the molecular weight of the polymer above a DP of ~10 [21].

We have discussed above the crystalline nature of most of the fibers based on homopolymers. While information on melting of the crystalline phase of these polymers differs, all quoted melt temperatures are very high. For MPDI most values are similar to 435°C as determined by Takatsuka [17]. On the other hand, most authors report the decomposition temperature of PPTA to be lower than its melting point [17]. Chaudhuri [18] reported a value of 530°C. Table 13.2 summarizes some of the thermal properties of commercial aramid fibers [22,140–146].

The almost perfect orientation of p-aramid fibers is reflected in the anisotropic behavior of its thermal expansion coefficient. The linear expansion coefficient for these materials is negative (Table 13.2). Because the volumetric thermal expansion coefficient is not affected by orientation, the radial coefficient must increase as fiber orientation increases. The negative expansion coefficient of these materials has opened a whole field of applications in electronics (see section 13.8.4.2).

13.4.3 Solubility and Chemical Properties

The same structural characteristics that are responsible for the excellent thermal properties of these materials are responsible for their limited solubility as well as good chemical resistance. PPTA is soluble only in strong acids like H_2SO_4, HF, and methanesulfonic acid. Preparation of this polymer via solution polymerization in amide solvents is accompanied by polymer precipitation. As expected, based on its structure, MPDI is easier to solubilize then PPTA. It is soluble in neat amide solvents like N-methyl-2-pyrrolidone (NMP) and dimethylacetamide (DMAc), but adding salts like $CaCl_2$ or LiCl significantly enhances its solubility.

The significant rigidity of the PPTA chain (as discussed above) leads to the formation of anisotropic solutions when the solvent is good enough to reach a critical minimum solids concentration. The implications of this are discussed in greater detail later in this chapter.

It is well known that chemical properties differ significantly between crystalline and noncrystalline materials of the same composition. In general, aramids have very good chemical resistance as shown in Table 13.3. Obviously, the amide bond is subject to a hydrolytic attack by acids and bases. Exposure to very strong oxidizing agents results in a significant strength loss of these fibers. In addition to crystallinity, structure consolidation affects the rate of degradation of these materials.

The hydrophilicity of the amide group leads to a significant absorption of water by all aramids. While the chemistry is the driving factor, fiber structure also plays a very important role; for example, Kevlar 29 absorbs ~7% water, Kevlar 49 ~4%, and Kevlar 149 only 1%. Fukuda explored the relationship between fiber crystallinity and equilibrium moisture in great detail [23].

Because of their aromatic character, aramids absorb UV light, which results in an oxidative color change. Substantial exposure can lead to the loss of yarn tensile properties [24]. UV absorption by p-aramids is more pronounced than with m-aramids. In this case a self-screening phenomenon is observed, which makes thin structures more susceptible to degradation than thick ones. Very frequently p-aramids are covered with another material in the final application to protect them.

The high degree of aromaticity of these materials also provides significant flame resistance. All commercial aramids have a limited oxygen index in the range of 28–32%, which compares with ~20% for aliphatic polyamides (Table 13.2). The utilization of these properties is discussed in greater detail in the Applications section of this chapter.

13.4.4 Fiber Mechanical Properties

Typical properties of commercial aramid fibers are given in Table 13.4. While yarns of m-aramids have tensile properties that are no greater than those of aliphatic polyamides, they do retain useful mechanical properties at significantly higher temperatures. The high glass

TABLE 13.2
Thermal Properties of Aramid Fibers

Trade name polymer	Nomex MPDI	Teijinconex MPDI		Kevlar PPTA		Twaron PPTA		Technora ODA/PPTA
Fiber type	430	Std	HT	K-29	K-49	Std	HM	96
Property								
Specific heat (J/kg-K)	72	60	60	81	81	81	81	96
Thermal conductivity (W/m-K)	0.25	0.11	0.11	2.5	2.5			0.5
Coefficient of thermal expansion (cm/cm-°C)	1.8×10^{-5}	1.5×10^{-5}	1.5×10^{-5}	-4.0×10^{-6}	-4.9×10^{-6}	-3.5×10^{-6}	-3.5×10^{-6}	-6×10^{-6}
Heat of combustion (J/kg)	28×10^6			35×10^6	35×10^6			
Flammability LOI (%)	28	29–32	29–32	29	29	29	29	
Decomposition (in N_2) Temperature (°C)	400–420	400–430	400–430	520–540	520–540	520–540	520–540	500

Source: From DuPont Technical Guide for Kevlar Aramid Fiber, H-77848, 4/00; DuPont Technical Guide for Nomex Brand Aramid Fiber, H-52720, 7/01; Teijin Ltd, Teijinconex Heat Resistant Aramids Fiber 02.05; Teijin Ltd, High Tenacity Aramids Fibre: Technora TIE-05/87.5; Akzo Nobel, Twaron—Product Information:Yarns, Fibers and Pulp.

TABLE 13.3
Chemical Resistance of Aramid Fibers

Trade name Polymer		Nomex MPDI	Kevlar PPTA	Technora ODA/PPTA
Chemical	Time (h)/temp. (°C)	Percent	Strength	Retention
40% H$_2$SO$_4$	100/95			90
10% H$_2$SO$_4$	100/21	90–100	90–100	
10% H$_2$SO$_4$	1000/21		95	35*
10% HCl	1000/21	20–60	35	10*
10% HNO$_3$	100/21	60–80	90–100	20–60
10% NaOH	100/95			75
10% NaOH	1000/21	90–100	90	46
40% NaOH	1000/21	80–90	76	
28% NH$_4$OH	1000/21	90–100	90–100	65*
0.01% NaClO	1000/21	90–100		16
10% NaClO	100/95			55
0.4% H$_2$O$_2$	1000/21	90–100	56–75	
10% NaCl	1000/21	90–100		100*
100% Acetic acid	1000/21	90–100	90–100	90*
90% Formic acid	100/21	90–100	90–100	90–100
90% Formic acid	100/99	60–80	90–100	0–20
100% Acetone	1000/21	90–100	90–100	
100% Acetone	100/56	80–90	90–100	
100% Benzene	1000/21	90–100	90–100	100
100% Ethyl alcohol	1000/21	90–100	90–100	100
100% Ethyl alcohol	100/77	90–100		90–100
50% Ethylene glycol	1000/99	80–90	90–100	60–80
100% Gasoline	1000/21	90–100	90–100	90–100
100% Methyl alcohol	1000/21	90–100	90–100	90–100
100% Perchloroethylene	10/99		90–100	
100% Tetrahydrofuran	1000/21		90–100	

*Measurements made after 3 months (2200 h) exposure at room temperature.

Source: From DuPont Technical Guide for Kevlar Aramid Fiber, H-77848, 4/00; DuPont Technical Guide for Nomex Brand Aramid Fiber, H-52720, 7/01; Teijin Ltd., High Tenacity Aramids Fibre: Technora TIE-05/87.5.

TABLE 13.4
Properties of Aramid Fibers

Trade name Polymer	Nomex MPDI	Teijinconex MPDI		Kevlar PPTA		Twaron PPTA		Technora ODA/PPTA
Fiber type	430	std	HT	K-29	K-49	std	HM	
Density (g/cm^3)	1.38	1.38	1.38	1.44	1.44	1.44	1.45	1.39
Strength (Gpa)	0.59	0.61–0.68	0.73–0.86	2.9	3.0	2.9	2.9	3.4
Elongation (%)	31	35–45	20–30	3.6	2.4	3.6	2.5	4.6
Modulus (Gpa)	11.5	7.9–9.8	11.6–12.1	71	112	70	110	72

transition temperature leads to low (less than 1%) shrinkage at temperatures below 250°C. In general, mechanical properties of *m*-aramid fibers are developed on drawing (see below). This process produces fibers with a high degree of morphological homogeneity, which leads to very good fatigue properties.

The mechanical properties of *p*-aramid fibers have been the subject of much study. This is because these fibers were the first examples of organic materials with a very high level of both strength and stiffness. These materials are practical confirmation that nearly perfect orientation and full chain extension are required to achieve mechanical properties approaching those predicted for chemical bonds. In general, the mechanical properties reflect a significant anisotropy of these fibers—covalent bonds in the direction of the fiber axis with hydrogen bonding and van der Waals forces in the lateral direction.

Termonia has proposed a kinetic model for fiber strength [25–27]. His calculations suggest that molecular mass, its distribution, and intermolecular forces control fiber strength. Allen's work linked the failure mode of these fibers with their morphology very closely [16, 28–30]. He was able to show that fiber pleating is responsible for the fact that one needs to consider the asymptotic modulus (modulus close to the fiber breaking point) of these fibers rather than the initial modulus to explain mechanical properties. This interpretation confirmed a clear dependence of fiber strength on both local orientation (as measured by the asymptotic modulus) and secondary interactions (as measured by shear properties).

The use of *p*-aramids in composites has focused much research effort on the compressive properties of these fibers. Excellent tensile properties, approaching 80% of the theoretical modulus, and 30% of the theoretical strength are not matched by their compressive properties. PPTA fiber yields under compression at ~0.5% of strain. This is caused by a buckling phenomenon that is attributed to the relatively weak lateral properties of these highly anisotropic fibers. However, aramids with their hydrogen bonding have significantly better compressive strength than UHMWPE, which has extremely weak lateral properties. Allen [31] measured compressive strength by a recoil test and obtained 258 N/tex for Kevlar 49 compared to 7.5 N/tex for UHMWPE. Aramids also compare well with PBO, which has a compressive strength of 0.133 N/tex. All high strength organic fibers yield under compressive stress with formation of kink bands. However this, significant dislocation does not lead to major tensile strength loss. At a strain of 3% the loss is only ~10%.

This high degree of anisotropy of the *p*-aramids is reflected in fatigue properties. Tension–tension fatigue is very good. Wilfong [32] reported no failure after 10^7 cycles with loads at 60% of breaking strength. Compressive fatigue is not as good—especially at higher strains. At a strain of 0.5% no strength loss is observed even after 10^6 cycles but at a strain of 1% the strength loss begins at about 10^3 cycles [33].

Creep (long-term failure of fibers at loads below their breaking strength) is the final mechanical property for review. The kinetic model of fiber failure was applied by Termonia [25] to estimate creep behavior. His calculations suggest that the activation energy of covalent bond breaking controls the lifetime of materials. That is why UHMWPE fails after 2.5 min when strained to 50% of its breaking strain (measured at 1 sec). PPTA under the same conditions fails after 100 years. Lafitte [34] measured creep strain for Kevlar 29 at a load of 50% of its breaking strength and found a strain of 0.3% after 10^7 sec.

13.4.5 FILMS AND PAPERS

Although the primary focus of this chapter is on fibers, we have included some illustrations of sheet products based on this chemistry.

There are two examples of commercial *p*-aramid films. Toray produces a terpolymer film under the trade name Mictron, while Asahi introduced a PPTA homopolymer film called

TABLE 13.5
Properties of Aramid Films

Producer		Mictron toray		Aramica asahi	
Thickness (μm)		25		25	
Density (g/cm^3)		1.5		1.4	
Mechanical properties:					
Direction		machine	cross	machine	cross
Tensile strength	GPa	0.5		0.5	0.3
Elongation	%	60		15	25
Tensile modulus	GPa	13	9	19	10
Initial tear strength	Kg	—		25	
Long-term heat resistance	°C	180		~200	
Thermal expansion	(1/°C × 10^{-5})	0.1		0.2	
Moisture absorption					
At 75% RH and					
At room temp.	%	1.5		2.8	
Electrical properties:					
Dielectric constant at 1 KHz		—		4	
Dissipation factor at 1 KHz		—		0.02	
Volume resistivity	Ω/cm	5×10^{17}		1×10^{16}	
Surface resistivity	Ω/cm	—		1×10^{16}	
Dielectric strength	KV/mm	300		230	

Source: From Yasufuku, S., *IEEE Elec. Insu. Mag.*, 11(6), 27, 1995; Teijin Ltd., High Tenacity Aramids Fibre: Technora TIE-05/87.5; Asahi Chemical Industry America, Inc., Technical Brochure, Aramica Film, 1991; Akzo Nobel, Twron Product Information: Yarns, Fibers and Pulp.

Aramica. In both cases the product goal was a high strength, thin film for mass storage devices. Film properties are shown in Table 13.5.

Aramids papers are found in a much broader range of applications than films (see Applications section). Most papers are comprised of a composite structure of short fibers and a binder. Paper properties can be tailored by changing the composition and the processing conditions. Selected properties are illustrated in Table 13.6.

13.5 POLYMERIZATION OF AROMATIC POLYAMIDES

13.5.1 INTRODUCTION

We began this discussion with a description of the high melting point and difficult solubility of aromatic polyamides. Very clearly these properties present a significant challenge in their synthesis and fabrication.

First, the infusible nature of many of these polymers precludes the use of conventional bulk polymerization and melt processing techniques. Second, aromatic diamines are significantly less reactive than aliphatic diamines toward polyamidation. This requires the use of more reactive dicarboxylic acid intermediates or some activation mechanism to complete the polycondensation in a reasonable period of time. Some technological breakthroughs were necessary to make progress in the synthesis of aromatic polyamides. These came in the late

TABLE 13.6
Properties of Aramids Paper

Polymer		Nomex MPDI	Nomex MPDI	Nomex PPTA
Producer		DuPont	DuPont	DuPont
Type		410	410	N710
Thickness	μm	127	127	97
Density	g/cm^3	0.87	0.87	0.64
Mechanical properties:				
Direction		machine	cross	machine
Tensile strength	GPa	0.1	0.05	0.2
Elongation	%	16	13	1.5
Tensile modulus	GPa	—	—	5.4
Initial tear strength	Kg	3.3	1.6	—
Long-term heat resistance	°C	~200		~200
Thermal expansion	(1/°C × 10^{-5})	—		0.7
Moisture absorption				
At 55% RH and				
At room temp.	%	—		1.6
Electrical properties:				
Dielectric constant at 1 KHz		2.4		3.9[a]
Dissipation factor at 1 KHz		0.006		0.02[a]
Volume resistivity	Ω/cm	5 × 10^{16}		—
Surface resistivity	Ω/cm	1 × 10^{16}		—
Dielectric strength	KV/mm	25		82[a]

[a]Measurements made after three months (2200 hrs) exposure at room temperature.
Source: From E.I. DuPont de Nemours & Co., Inc, NOMEX Aramid Paper Type 410—Typical Properties, H-22368, 8/98; Magellan International; Hendren, G.L., Kirayoglu, B., Powell, D.J., and Weinhold, M., *Adv. Mater.*, 10(15), 1233, 1998; Yasufuku, S., *IEEE Elec. Insn. Mag.*, 11(6), 27, 1995.

1950s and the early 1960s when it was demonstrated that high molecular weight wholly aromatic polyamides could be prepared by low-temperature interfacial [35] and solution [36,37] polycondensation processes.

13.5.2 SYNTHESIS OF INGREDIENTS

It was also imperative to develop synthetic routes to high purity ingredients for these polymerizations to be successful. The syntheses of commercially important ingredients will be described here. Only one of several alternative routes will be illustrated. It must also be noted that the chemistry is constantly being modified to achieve less costly, more efficient and environmentally friendly processes.

13.5.2.1 *m*-Phenylene Diamine

The first step in *m*-phenylene diamine (MPD) synthesis is the nitration of benzene in 20% oleum (Equation 13.1). The nitration is a two-stage continuous process [38] replacing two protons on the benzene ring with two nitro groups by the catalytic action of sulfuric acid. The *m*-isomer is the dominant product.

Aramid Fibers

$$\text{C}_6\text{H}_6 + 2\,\text{HNO}_3 \xrightarrow{\text{H}_2\text{SO}_4} [\text{1,3-dinitrobenzene (Major)} + \text{1,2-dinitrobenzene (Minor)} + \text{1,4-dinitrobenzene (Minor)}] + 2\,\text{H}_2\text{O} \quad (13.1)$$

The resulting isomer mixture is washed with water and ammonia, centrifuged to remove acid and phenolic by-products and then catalytically hydrogenated [39]. MPD is isolated from the crude diamine mixture and purified by selective distillation (Equation 13.2)

$$\text{Isomeric mixture of dinitrobenzene} + 6\,\text{H}_2 \xrightarrow[\text{(Pt, Pd, Fe)}]{\text{Catalyst}} \text{MPD crude} + 4\,\text{H}_2\text{O} \quad (13.2)$$

13.5.2.2 p-Phenylene Diamine

The synthesis of *p*-phenylene diamine (PPD) starts with air oxidation of ammonia to form N_2O_3 (in equilibrium with NO and NO_2) (Equation 13.3)

$$4\,\text{NH}_3 + 6\,\text{O}_2 \xrightarrow[\text{1000°C}]{\text{Pt/Ru Catalyst}} 2\,\text{N}_2\text{O}_3 + 6\,\text{H}_2\text{O} \quad (13.3)$$

This mixture is then reacted with four moles of aniline to produce diphenyltriazine as follows:

$$\text{N}_2\text{O}_3 + 4\,\text{C}_6\text{H}_5\text{NH}_2 \rightleftharpoons 2\,\text{diphenyltriazine} + 3\,\text{H}_2\text{O} \quad (13.4)$$

Diphenyltriazine is rearranged to form a mixture of *p*- and *o*- aminoazobenzene using nitric acid as a catalyst

$$\text{diphenyltriazine} \underset{\text{Rearrangement}}{\overset{\text{HNO}_3}{\rightleftharpoons}} \text{p-aminoazobenzene} + \text{o-aminoazobenzene} \quad (13.5)$$

Finally, the aminoazobenzenes are hydrogenated to the corresponding phenylene diamines [40–42]. A mole of aniline is regenerated for every mole of phenylene diamine and is recycled. The phenylene diamine isomers are then separated, and the o-isomer is sold as an ingredient for the production of various fungicides.

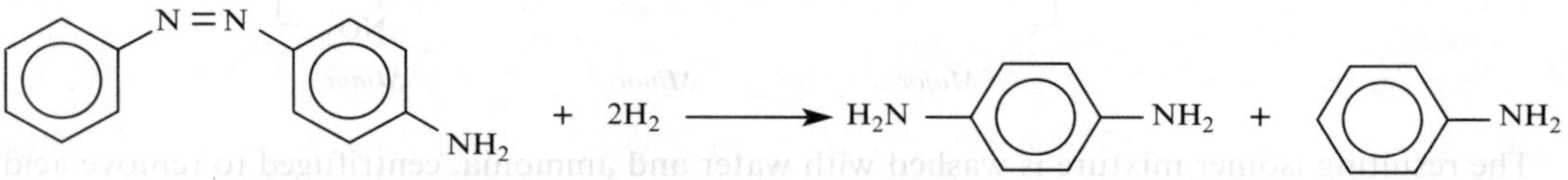

$$\text{PhN=N-C}_6\text{H}_4\text{-NH}_2 + 2\text{H}_2 \longrightarrow \text{H}_2\text{N-C}_6\text{H}_4\text{-NH}_2 + \text{C}_6\text{H}_5\text{-NH}_2 \quad (13.6a)$$

$$\text{PhN=N-C}_6\text{H}_4(\text{NH}_2) + 2\text{H}_2 \longrightarrow \text{H}_2\text{N-C}_6\text{H}_4\text{-NH}_2(\textit{o}) + \text{C}_6\text{H}_5\text{-NH}_2 \quad (13.6b)$$

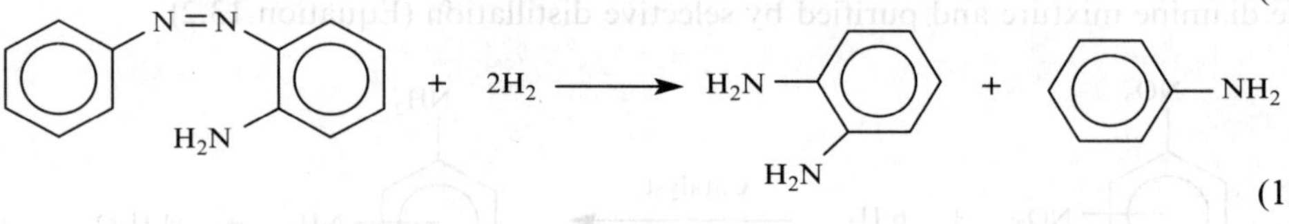

13.5.2.3 3,4'-Diaminodiphenyl Ether

The synthesis of 3,4'-diaminodiphenyl ether (3,4'-POP) is more complex than that of simple aromatic diamines such as MPD and PPD and hence this monomer is more expensive. Condensing 1,3-dinitrobenzene with 4-aminophenol using potassium carbonate in dimethylformamide (DMF) or DMAc produces 3,4'-POP. The resulting 3-nitro-4'-aminodiphenyl ether is then hydrogenated [42].

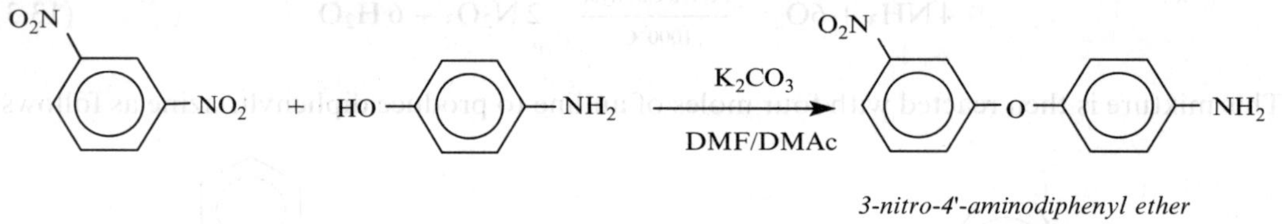

$$\text{O}_2\text{N-C}_6\text{H}_4\text{-NO}_2 + \text{HO-C}_6\text{H}_4\text{-NH}_2 \xrightarrow[\text{DMF/DMAc}]{\text{K}_2\text{CO}_3} \text{O}_2\text{N-C}_6\text{H}_4\text{-O-C}_6\text{H}_4\text{-NH}_2$$

3-nitro-4'-aminodiphenyl ether

(13.7)

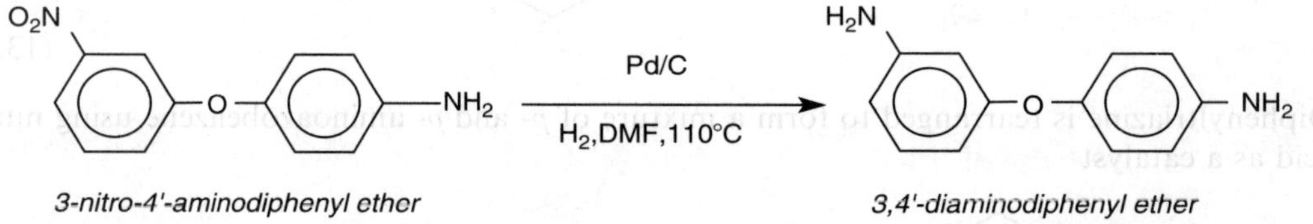

$$\text{O}_2\text{N-C}_6\text{H}_4\text{-O-C}_6\text{H}_4\text{-NH}_2 \xrightarrow[\text{H}_2,\text{DMF},110°\text{C}]{\text{Pd/C}} \text{H}_2\text{N-C}_6\text{H}_4\text{-O-C}_6\text{H}_4\text{-NH}_2$$

3-nitro-4'-aminodiphenyl ether   *3,4'-diaminodiphenyl ether*

(13.8)

A mixture of 4-aminophenol, 1-3-dinitrobenzene and K_2CO_3 in DMF was treated at 150°C for 4 h to give 96.3% 3-nitro-4'-aminodiphenyl ether. This was treated with Pd on C in DMF at 110°C and H_2(3 atm) for 5 h to give 98.0% 3,4'-diaminodiphenyl ether.

13.5.2.4 Diacid Chlorides

Terephthaloyl chloride (TCl) and isophthaloyl chloride (ICl) are produced by reacting the corresponding dicarboxylic acid with phosgene [43].

$$\text{HO-}\underset{\text{O}}{\overset{\text{O}}{\text{C}}}\text{-}\bigcirc\text{-}\underset{\text{O}}{\overset{\text{O}}{\text{C}}}\text{-OH} + 2\text{Cl-}\underset{\text{O}}{\overset{\text{O}}{\text{C}}}\text{-Cl} \xrightarrow{\text{DMF}} \text{Cl-}\underset{\text{O}}{\overset{\text{O}}{\text{C}}}\text{-}\bigcirc\text{-}\underset{\text{O}}{\overset{\text{O}}{\text{C}}}\text{-Cl} + 2\text{HCl} + 2\text{CO}_2$$

(13.9)

The reaction involves formation of a catalyst complex between DMF and phosgene, which then reacts with terephthalic acid.

$$\underset{\text{CH}_3}{\overset{\text{CH}_3}{\text{N}}}\text{-}\underset{\text{O}}{\overset{\text{O}}{\text{C}}}\text{-H} + \text{Cl-}\underset{\text{O}}{\overset{\text{O}}{\text{C}}}\text{-Cl} \longrightarrow \underset{\text{CH}_3}{\overset{\text{CH}_3}{\overset{\oplus}{\text{N}}}}\text{=}\underset{\text{H}}{\overset{\text{Cl}}{\text{C}}} \; \text{Cl}^{\ominus} + \text{CO}_2$$

"Complex"

(13.10)

$$\text{HO-}\underset{\text{O}}{\overset{\text{O}}{\text{C}}}\text{-}\bigcirc\text{-}\underset{\text{O}}{\overset{\text{O}}{\text{C}}}\text{-OH} + 2\,\underset{\text{CH}_3}{\overset{\text{CH}_3}{\overset{\oplus}{\text{N}}}}\text{=}\underset{\text{H}}{\overset{\text{Cl}}{\text{C}}}\,\text{Cl}^{\ominus} \xrightarrow{-2\text{HCl}} \text{Cl-}\underset{\text{O}}{\overset{\text{O}}{\text{C}}}\text{-}\bigcirc\text{-}\underset{\text{O}}{\overset{\text{O}}{\text{C}}}\text{-Cl} + 2\,\underset{\text{CH}_3}{\overset{\text{CH}_3}{\text{N}}}\text{-}\underset{\text{O}}{\overset{\text{O}}{\text{C}}}\text{-H}$$

"Complex"

(13.11)

The reaction is carried out in a slurry of TPA, DMF, and TCl with countercurrent injection of phosgene. The product, TCl, is degassed, heated to destroy the catalyst complex, and then distilled to remove impurities.

13.5.3 Polymerization Fundamentals

The usual methods for preparing aliphatic polyamides are not suitable for preparing high-molecular-weight aromatic polyamides because of the reduced reactivity of aromatic diamines and the high melting point of the resulting polymers. Polymerization of wholly aromatic polyamides is usually carried out in solution, instead of in bulk, using highly reactive diacid chlorides vs. diacids. The reaction is fast and takes place at a much lower temperature than conventional melt polymerizations. The synthesis is based on the familiar Schötten–Baumann reaction [44–49].

$$\text{RC-}\underset{\text{O}}{\overset{\text{O}}{\text{C}}}\text{-Cl} + \text{H-N}\underset{R_2}{\overset{R_1}{\diagdown}} \xrightarrow{\text{NaOH}} \text{R-}\underset{\text{O}}{\overset{\text{O}}{\text{C}}}\text{-N}\underset{R_2}{\overset{R_1}{\diagdown}} + \text{NaCl} + \text{H}_2\text{O}$$

(13.12)

Condensation polymers are formed if the complementary reagents are difunctional.

$$n\text{H}_2\text{N-R-NH}_2 + n\text{Cl-}\underset{\text{O}}{\overset{\text{O}}{\text{C}}}\text{-R'-}\underset{\text{O}}{\overset{\text{O}}{\text{C}}}\text{-Cl} \xrightarrow{2n\text{NaOH}} \left[\underset{\text{H}}{\overset{\text{H}}{\text{N}}}\text{-R-}\underset{\text{H}}{\overset{\text{H}}{\text{N}}}\text{-}\underset{\text{O}}{\overset{\text{O}}{\text{C}}}\text{-R'-}\underset{\text{O}}{\overset{\text{O}}{\text{C}}}\right]_n + 2n\text{NaCl} + 2n\text{H}_2\text{O}$$

(13.13)

A large amount of salt is generated in this reaction following neutralization of the by-product hydrochloric acid (HCl). The high salt concentration in the process stream requires the use of expensive corrosion resistant materials—one of the key contributors to the high cost of aramid fibers.

An alternative route to aromatic polyamides is referred to as a hydrogen transfer reaction [50]. This reaction between a diacid and diisocyanate is run at a low temperature to form an intermediate polymer that loses carbon dioxide on subsequent heating to form the aromatic polyamide (Equation 13.14).

$$nO=C=N-R_1-N=C=O + nHO-\overset{O}{\underset{\|}{C}}-R_2-\overset{O}{\underset{\|}{C}}-OH \xrightarrow{\text{H-transfer}} \left[\overset{O}{\underset{\|}{C}}-\overset{H}{\underset{|}{N}}-R_1-\overset{H}{\underset{|}{N}}-\overset{O}{\underset{\|}{C}}-O-\overset{O}{\underset{\|}{C}}-R_2-\overset{O}{\underset{\|}{C}}-O\right]_n$$

$$\left[\overset{O}{\underset{\|}{C}}-\overset{H}{\underset{|}{N}}-R_1-\overset{H}{\underset{|}{N}}-\overset{O}{\underset{\|}{C}}-O-\overset{O}{\underset{\|}{C}}-R_2-\overset{O}{\underset{\|}{C}}-O\right]_n \xrightarrow{\text{Heat}} \left[\overset{H}{\underset{|}{N}}-R_1-\overset{H}{\underset{|}{N}}-\overset{O}{\underset{\|}{C}}-R_2-\overset{O}{\underset{\|}{C}}\right]_n + 2n\ CO_2$$

(13.14)

This reaction is not widely used because of the higher cost of diisocyanates and the difficulty in eliminating all the carbon dioxide.

13.5.3.1 Reaction Mechanism

The first step in the condensation reaction is the attack of the amine nitrogen at the carbonyl carbon of the dicarboxylic acid. The local electron density at the aromatic amine nitrogen is greatly reduced by participation of the lone pair electrons with the aromatic π-cloud, whereas the local electron density of the aliphatic counterpart is enhanced by the inductive effect of aliphatic hydrocarbon. This leads to a significant difference in the polycondensation reaction rate between aromatic polyamides and aliphatic polyamides.

To compensate for reduced electron density at the amine nitrogen, the dicarboxylic acid is activated by increasing the partial positive charge at the carbonyl carbon. Halogen atoms (X) have proven to be effective because of their high electronegativity. An amide linkage is formed from the transition complex (Equation 13.15) by eliminating HX (Equation 13.16). Because the eliminated acid, HX, will react with the opposing amine to form a quaternary ammonium salt, it must be removed for the polymerization to continue. An organic amine, such as pyridine, is often used as an acid acceptor to regenerate the amine end from the quaternary salt (Equation 13.17). Polymerization solvents such as *N,N*-dimethyl acetamide (DMAc) and *N*-methylpyrrolidone (NMP) are sufficiently basic to function as acid acceptors as well.

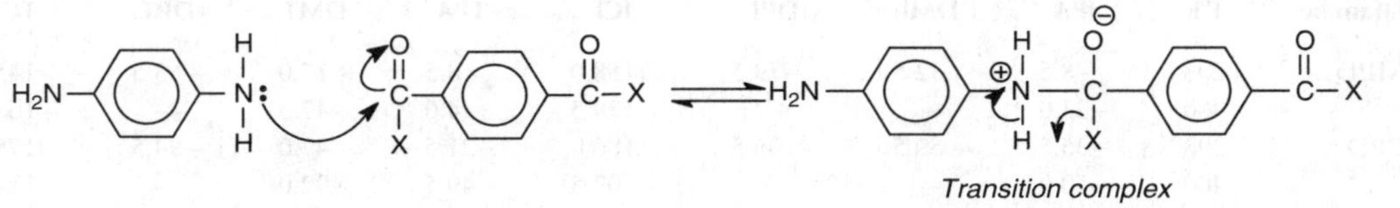

$$H_2N-\text{Ar}-\overset{H}{\underset{H}{N}}: \quad + \quad \overset{O}{\underset{X}{C}}-\text{Ar}-\overset{O}{C}-X \quad \rightleftharpoons \quad H_2N-\text{Ar}-\overset{H}{\underset{H}{\overset{\oplus}{N}}}-\overset{\overset{\ominus}{O}}{\underset{X}{C}}-\text{Ar}-\overset{O}{C}-X$$

Transition complex

(13.15)

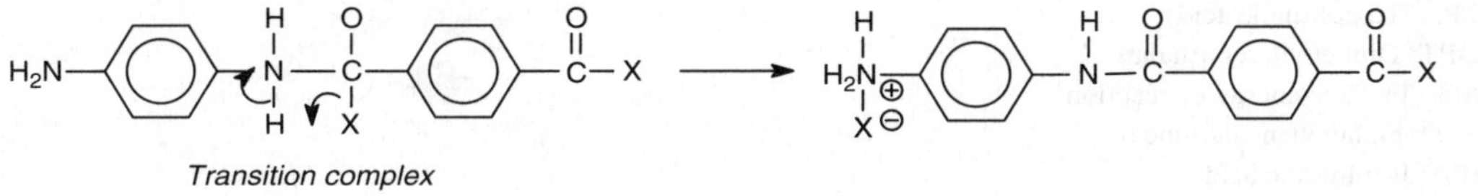

$$H_2N-\text{Ar}-\overset{H}{\underset{H}{N}}-\overset{\overset{\ominus}{O}}{\underset{X}{C}}-\text{Ar}-\overset{O}{C}-X \quad \longrightarrow \quad H_2\overset{\oplus}{N}-\text{Ar}-\overset{H}{N}-\overset{O}{C}-\text{Ar}-\overset{O}{C}-X \quad X^{\ominus}$$

Transition complex

(13.16)

Amine regeneration

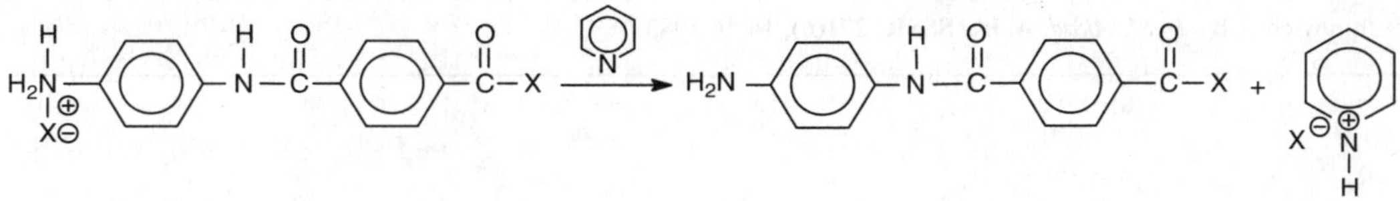

$$H_2\overset{\oplus}{\underset{X^\ominus}{N}}-\text{Ar}-\overset{H}{N}-\overset{O}{C}-\text{Ar}-\overset{O}{C}-X \quad + \quad \text{Py} \quad \longrightarrow \quad H_2N-\text{Ar}-\overset{H}{N}-\overset{O}{C}-\text{Ar}-\overset{O}{C}-X \quad + \quad X^\ominus\text{PyH}^\oplus$$

(13.17)

Factors that can limit the extent of the polymerization reaction include deactivation of chain-ends, stoichiometric imbalance of reagents, monofunctional impurities, and insufficient mobility of growing chain-ends. Some of these factors are used to control polymer molecular weight.

13.5.3.2 Reaction Energetics

As shown in Table 13.7, the free energy of reaction of aramid polymerizations is reported to be negative even with aromatic acid, ester, and diamine monomers. In spite of this driving force, the rate of reaction is extremely slow because of the high activation energy of the polymerization reaction [51].

13.5.4 DIRECT POLYMERIZATION BY CATALYSIS

Several different classes of catalysts, so-called condensing agents, have been reported in the literature [52–55] for the polycondensation reaction of aromatic diamines with aromatic diacids. This polycondensation is called "direct polymerization" because unmodified monomers can be used in the reaction. The condensing agents, which are generally derived from phosphorus or sulfur compounds, activate the dicarboxylic acid *in situ* during the

TABLE 13.7
Energetics of Aromatic Polycondensation

Diamine	T°K	ΔGr(T) (KJ/mole)				ΔGr(T) (KJ/mole)			
		IPA	DMI	DPI	ICl	TPA	DMT	DPT	TCl
MPD	298	−8.5	32.5	−79.5	−158.0	8.5	−17.0	−63.5	−145.0
	400	−23.0	—	—	−179.5	−4.0	−47.5	—	−168.5
PPD	298	−35.5	−59.5	−106.5	−186.0	−21.5	−47.0	−94.5	−175.0
	400	−50.0	—	—	−207.5	−49.5	−92.0	—	−214.0

MPD: *m*-Phenylenediamine
DPI: Diphenylisophthalate
DMI: Dimethyl isophthalate
TPA: Terephthalic acid
DPT: Diphenylterephthalate
ΔGr(T): Free energy of reaction
PPD: *p*-phenylenediamine
IPA: Isophthalic acid
ICl: Isophthaloyl chloride
DMT: Dimethylterephthalate
TCl: Terephthaloyl chloride

Source: From Hand, D.R., Hartert, R., and Bottger, C., Stab resistant and Anti-ballistic material. Method of making the same, U.S. Patent Application Publication U.S. 2004/0023580 A1, February 5, 2004; Karyakin, N.V. and Rabinovich, I.B., *Dokl. Akad. Nank.* SSSR, 271(6), 1429, 1983.

polymerization. The best-known route involves an N-P type intermediate as the activated complex. As an example, triphenyl phosphite is reacted with a carboxylic acid in the presence of a tertiary amine (e.g., pyridine) to form the N-phosphonium salt **5**, which gives the corresponding amide on aminolysis (Equation 13.18).

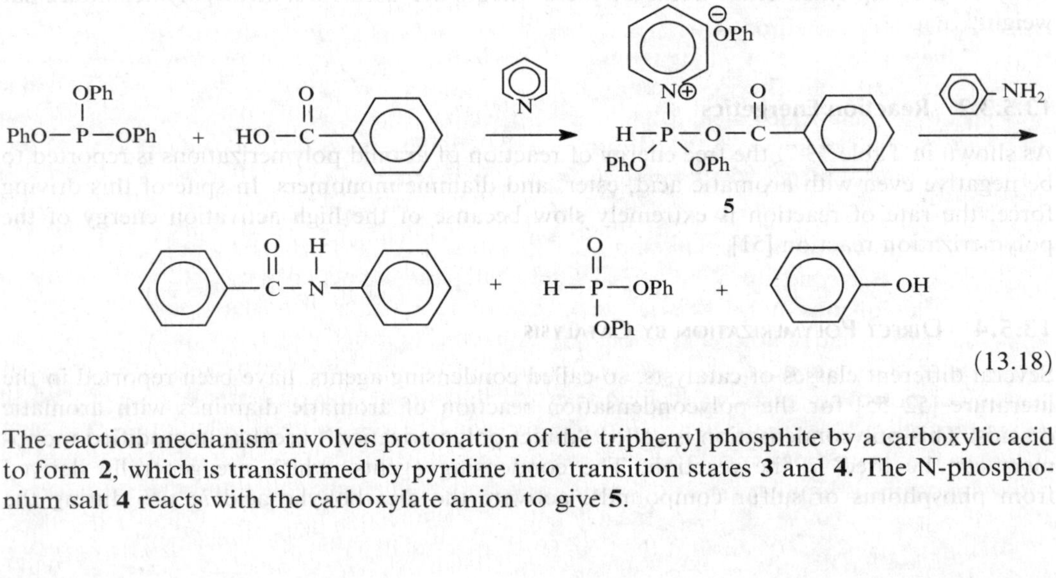

(13.18)

The reaction mechanism involves protonation of the triphenyl phosphite by a carboxylic acid to form **2**, which is transformed by pyridine into transition states **3** and **4**. The N-phosphonium salt **4** reacts with the carboxylate anion to give **5**.

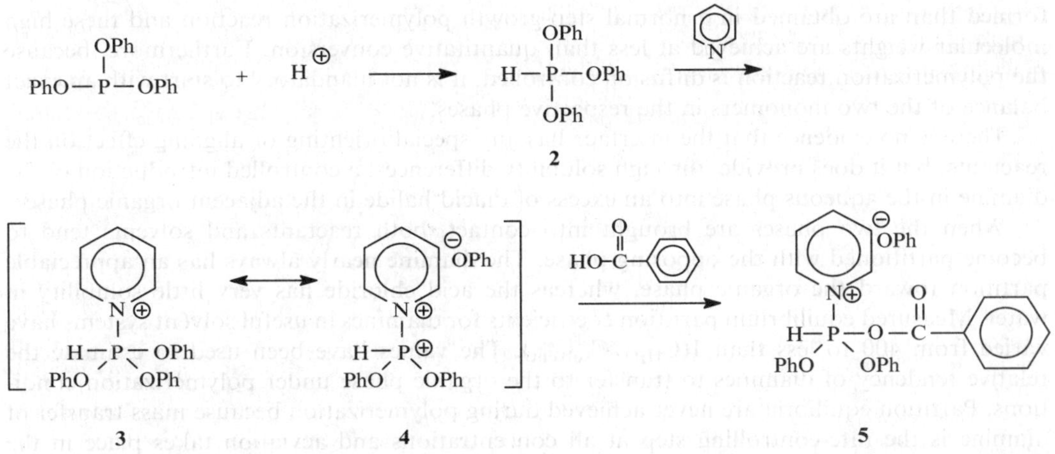

Aramid Fibers

$$\text{PhO-P(OPh)-OPh} + H^{\oplus} \longrightarrow H-\overset{\text{OPh}}{\underset{\text{OPh}}{P}}-\text{OPh} \xrightarrow{\text{Py}}$$

2

$$\left[\text{Py}^{\oplus}-\overset{H}{\underset{\text{PhO}}{P}}\!\!-\!\!\overset{\ominus}{\text{OPh}}\text{OPh} \longleftrightarrow \text{Py}^{\oplus}-\overset{H}{\underset{\text{PhO}}{P^{\oplus}}}\text{OPh} \right] \xrightarrow{\text{HO-C(O)-Ph}} \text{Py}^{\oplus}-\overset{\ominus\text{OPh}}{\underset{\text{PhO}}{\overset{|}{N}}}\overset{H}{\underset{\text{OPh}}{\overset{|}{P}}}-\text{O-C(O)-Ph}$$

3   **4**   **5**

(13.19)

In other words, the aromatic carboxylic acid is activated by the pyridinyl triphosphonate cation so that the weakly basic aromatic amine can effectively attack the carbonyl center. The reaction has not been utilized commercially because the costs of recovering and regenerating triphenylphosphite far outweigh the cost advantage of using unmodified diacids.

Similar activation mechanisms of the P-O-P type [56], C-O-P type [57], and N-S/C-O-S type [58], and reactions activated by silicon tetrachloride [59] and aromatic halo compounds such as picryl chloride have also been reported in the literature [60].

13.5.5 Polymerization Methods

The two principal methods used for the synthesis of aromatic polyamides are interfacial polymerization and solution polymerization. Vapor-phase polymerization and plasticized melt techniques have also been demonstrated but have not been adopted for practical use.

13.5.5.1 Interfacial Polymerization

In the interfacial method, the two fast-reacting intermediates are dissolved in a pair of immiscible liquids, one of which is preferably water. The water phase contains the diamine and any added alkali. The second phase consists of the diacid halide in an organic liquid such as carbon tetrachloride, dichloromethane, xylene, or hexane, etc. The two solutions are brought together with vigorous agitation and the reaction takes place at or near the interface of the two phases; hence, the name interfacial polymerization.

13.5.5.1.1 Reaction at the Interface
In interfacial polycondensation, the polymerization reaction occurs very close to the interface between the aqueous and organic layers generally just within the organic solvent layer that contains the diacid chloride [60,61]. The adjacent aqueous phase generally contains, in addition to the diamine, a basic reagent capable of neutralizing hydrogen chloride liberated in the reaction. The reaction rate is so fast that the polymerization reaction becomes "diffusion-controlled." As the polymerization proceeds, the diffusion of additional monomers through the formed polymer layer becomes increasingly difficult. As a result, the number of growing chains is limited. For this reason, polymers with much higher molecular weights are

formed than are obtained in a normal step-growth polymerization reaction and these high molecular weights are achieved at less than quantitative conversion. Furthermore, because the polymerization reaction is diffusion controlled, it is not mandatory to start with an exact balance of the two monomers in the respective phases.

There is no evidence that the interface has any special orienting or aligning effect on the reactants, but it does provide, through solubility differences, a controlled introduction of the diamine in the aqueous phase into an excess of diacid halide in the adjacent organic phase.

When the two phases are brought into contact, both reactants and solvents tend to become partitioned with the opposing phase. The diamine nearly always has an appreciable partition toward the organic phase, whereas the acid chloride has very little solubility in water. Measured equilibrium partition coefficients for diamines in useful solvent systems have varied from 400 to less than $1(C_{H_2O}/C_{solvent})$. The values have been used to estimate the relative tendency of diamines to transfer to the organic phase under polymerization conditions. Partition equilibria are never achieved during polymerization because mass transfer of diamine is the rate-controlling step at all concentrations and acylation takes place in the organic phase as rapidly as diamine is transferred.

13.5.5.1.2 Amine Acylation
At the onset of the polycondensation reaction, diamine monomer sees excess acid chloride and is presumably acylated at both ends. Ensuing diamine encounters a layer of acid chloride-terminated oligomer and some acid chloride. The reaction proceeds by an irreversible coupling of the oligomers by the diamine. The concentration and size of oligomers increase until a layer of high polymer is obtained. Thus, high polymer forms because of the high reaction rate and the increasing probability that the diamine will react with an acid chloride-terminated oligomer rather than with a free acid chloride monomer.

13.5.5.1.3 Acid Elimination
Hydrogen chloride, the product of the fast reaction between amine and acid chloride, diffuses to the aqueous phase. Any amine hydrochloride that might be formed is usually very insoluble in the organic phase but is soluble in the aqueous phase. Both hydrogen chloride and amine hydrochloride have to be neutralized in the aqueous phase with inorganic bases.

13.5.5.1.4 Major Variables
Variables affecting the polymerization include temperature, monomer ratio and concentration, impurities, additives, acid acceptor, and mode of addition. The polymerization of MPDI is used as a model for interfacial polymerization in the following discussion.

13.5.5.1.4.1 Temperature
Most interfacial polycondensation are initiated at ambient temperature. Because the reactions are rapid there is no need for heating and, in fact, cooling is frequently employed to control the temperature rise, especially on a larger scale [62–64]. Raising the temperature will change the solubility of both polymer and intermediates and will accelerate side reactions as well as the desired polymerization reaction.

13.5.5.1.4.2 Reactant Equivalence
The molecular weight of polymers made by interfacial polycondensation is far less sensitive to nonequivalence of reactants than that of polymers prepared by melt or solution methods for reasons already discussed—high reaction rate, diffusion control of monomers, and the nonequilibrium nature of the polymerization. The molecular weight of polymers precipitating as a coherent film from an unstirred interface is completely insensitive to the contents of the system as a whole, whereas the molecular weight of polymers from a stirred interface is generally more sensitive to reactant nonequivalence.

13.5.5.1.4.3 Impurities and Additives

Interfacial polymerization will tolerate the presence of impurities in the reactants that simply dilute the material and thereby produce nonequivalence of reactants. These diluents might be water or inert contaminants in the acid chloride. Reactive monofunctional species are harmful in either phase. To maximize molecular weight, it is essential to use high purity monomers. Molecular weight control can be achieved, if desired, with appropriate use of monofunctional reagents. Examples of impurities interfering with the interfacial polyamidation of MPDI are half hydrolyzed acid chloride, monoamide, partially oxidized amines, and reactive surfactants.

13.5.5.1.4.4 Acid Acceptors

Salts of basic diamines and strong acids are not sufficiently dissociated to permit the amine to react further. At least two moles of acid acceptor per mole of diamine are needed to maximize the yield of high polymer [65]. Of water soluble inorganic acid acceptors used in MPDI polymerizations in a water–DMeTMS solvent system, sodium carbonate appeared to be the most promising. Use of two equivalents of sodium carbonate gave white polymer with inherent viscosity of 2.48 in 100% yield. With 1.1 equivalents of sodium carbonate, white polymer with an inherent viscosity of 2.70 was obtained in 100% yield, while further reduction to one equivalent gave a polymer with an inherent viscosity of 1.97. Polymer with an inherent viscosity of 1.83 (98.5% yield) was obtained using two equivalents of sodium bicarbonate. Calcium hydroxide, potassium carbonate, and sodium hydroxide all gave polymers with lower inherent viscosity.

13.5.5.1.4.5 Reactant Addition

The mode of addition of reactants will also influence the reaction. Perhaps the best procedure would be to use a high-speed, low-volume mixer into which both solutions are charged simultaneously. In a typical batch polymerization process, rapid addition of the diacid chloride solution to a vigorously stirred diamine solution has given the best results. Rapid initial stirring appears to be an essential requirement for obtaining high-molecular-weight MPDI in water–DMeTMS. In two experiments employing rapid and slow stirring, respectively, in a Waring blender, polymers with inherent viscosity of 2.48 and 0.66 were obtained. In another experiment polymer obtained with initial low speed stirring for one minute followed by high speed stirring for an additional four minutes had a viscosity of only 0.41 [66].

These and other factors affecting the interfacial polycondensation reaction are discussed in more detail in P.W. Morgan's book entitled, *"Condensation Polymers,"* published by Interscience Publishers, John Wiley & Sons, 1965 [66].

13.5.5.2 Solution Polymerization

Solution polycondensation is carried out in an inert organic solvent.

Tertiary amines typically serve as the acid acceptor. The procedure generally starts with all the ingredients in solution but this is not always an essential requirement. The polymer may remain in solution or precipitate at any time.

13.5.5.2.1 Interfering Factors

Both physical and chemical factors can limit the polymerization reaction. Several effects that are classified as physical, even though they are physicochemical interactions, are the quality of stirring, precipitation of diamine salts, and precipitation of the polymer. Chemical factors include reactions with impurities and acid acceptors.

13.5.5.2.1.1 Impurities

The fast, low-temperature solution polymerization reactions are surprisingly tolerant of impurities but this tolerance varies considerably. The purity of the reactants and solvents must exceed the level required by the interfacial method. This is because all of the materials are in intimate proximity in a single-phase system.

Nonreactive impurities in the solvent are of minor significance except as they might depress the solubility of the polymer. Nonreactive impurities in the intermediates lead to an imbalance in the reactants thereby limiting molecular weight.

Reactive impurities are substances that can react with the monomers, the growing chain-ends, or the acid acceptor to terminate the polymerization prematurely. They can be introduced with the solvent or with the intermediates. The acid chloride may contain impurities originating in its synthesis or storage such as hydrogen chloride, thionyl chloride, phosphorus halides, or monoacid halides. The diamine may contain monoamines, water, or carbonates. It may degrade oxidatively in air or absorb moisture and carbon dioxide. The degree of interference caused by these impurities depends on both the quantity of the impurities as well the relative reaction rates of the desired polymerization vs. those of the impurities.

13.5.5.2.1.2 Solvent Reactivity

The solvent should not react with either the amine or the acid halide during the course of the polymerization. Solvent interference can be limited by minimizing the contact time between the monomer and the solvent; for example, the intermediates can be dissolved and allowed to react simultaneously. Alternatively, a small amount of nonreactive solvent can be used to dissolve one or both intermediates prior to polymerizing them in a more reactive medium.

13.5.5.2.1.3 Side Reactions with Acid Acceptors

Secondary amine acid acceptors can terminate chain growth by reacting with the diacid halide unless amine reactivity is minimized by steric effects. Reactions between a tertiary amine acid acceptor and the acid halide or certain solvents must also be avoided. An acid chloride and a tertiary amine can react to form a monoamide and an alkyl halide (Equation 13.20). This reaction is known to occur in fair yield at high temperatures and probably takes place to some extent at room temperature [67–69]. In the usual preparative method wherein diacid halide is added to a solution of diamine and a strongly basic acid acceptor, no difficulty is experienced if the polycondensation reaction is rapid. As the polycondensation reaction rate decreases, the potential for interference by side reactions increases. In a polymerization system, this would be a chain terminating reaction.

$$\text{PhC(=O)Cl} + \text{R}_3\text{N} \longrightarrow [\text{intermediate}] \longrightarrow \text{PhC(=O)NR}_2 + \text{R-Cl}$$

(13.20)

A reaction that can occur between an acid chloride and a tertiary amine in the presence of moisture is the formation of an acid anhydride (Equation 13.21).

$$2[\text{intermediate}] \longrightarrow \text{PhC(=O)-O-C(=O)Ph} + 2\text{R}_3\text{NH}^+\text{Cl}^-$$

(13.21)

An anhydride group in the polymer chain is a hydrolytically weak link and would likely be subject to cleavage on isolation of the polymer in water.

13.5.5.2.1.4 Diacylation
Diacylation of an amine by the acid halide leads to branched and network polymers. This side reaction has also been observed in interfacial polycondensation reactions [70].

13.5.5.2.2 Reaction Rates
Solution polycondensation employs the same reactions as used in interfacial polycondensation and similar reaction rates are involved. This means that the fastest reactions have rates on the order of 10^2–10^6 l/mole-sec. Polycondensations involving such reactions may be completed in a few minutes at room temperature.

13.5.5.2.3 Physical and Mechanical Effects
13.5.5.2.3.1 Temperature
Solution polycondensation reactions between diamines and diacid halides produce polymers with maximum molecular weight when carried out at room temperature or below. While reaction rates and polymer solubility would be expected to increase with increasing temperature, the rates of competitive side reactions will also increase.

13.5.5.2.3.2 Concentration
Solution polycondensation reactions have not shown any marked sensitivity to reactant concentration except as the concentration affects stirrability or temperature control. Lower concentrations are uneconomical and introduce relatively larger amounts of solvent impurities. Higher concentrations may yield unstirrable masses when the polymer or by-product salt precipitates, and the heat of the reaction is more difficult to control when reactants are mixed rapidly at high concentration.

13.5.5.2.3.3 Equivalence of Reactants and Mixing
Although both interfacial and solution polycondensation reactions show unusual insensitivity to nonequivalence of reactants, solution polycondensations are appreciably more sensitive to reactant balance.

Features common to both polymerization methods include: (1) use of fast reacting intermediates; (2) reaction irreversibility; (3) the reaction takes place essentially as fast as the contact of complementary reactants occurs; and (4) the growing polymer is in solution or highly swollen during the polymerization process. Unlike the interfacial process, the solution process has no interface to provide for the flow of one reactant into a higher concentration of the complementary reactant. It is this liquid–liquid interface that plays a significant role in attaining reactant balance in the interfacial process. The success of the solution process shows that an interfacial boundary, while helpful as a regulating device, is not essential for the formation of a high-molecular-weight polymer.

A key rationale for the insensitivity to nonequivalence of reactants in a single-phase system is that the rate of polymerization is often faster than the rate of mixing even in the absence of an interfacial boundary. It is presumed that in a solution polymerization system there are temporary interfaces or zones within which polymerization is proceeding independently of any potential effect of the ratio of the two reactants in the system as a whole. Thus, even a single drop of acid chloride solution in a large volume of diamine solution reacts rapidly with the local, or immediately surrounding diamine, before the droplet is dispersed. This leads to oligomers and polymer with higher molecular weight than would be obtained from a random reaction at the known reactant ratio. Further dropwise addition of one reactant continues this effect because each successive drop goes into a large system that consists in part of an active polymer with a higher than random degree of polymerization. Eventually as the system approaches equivalence and the concentration of reactive groups is reduced, there is a greater chance of a wider distribution of the increment of added reactant and the occurrence of random reaction [68]. Theoretical treatments of the effects of monomer ratio as well as side reactions have been described by Flory [71]. Kilkson has analyzed the problem of irreversible polymerization in both batch and steady-state reactors [72].

13.5.5.2.4 Acid Acceptors

Polycondensation reactions between diamines and diacid chlorides require the removal of the by-product hydrogen chloride. The acid acceptor need not be a basic substance but must retain the by-product acid in some way while the reaction proceeds. A variety of amines and some sterically hindered secondary amines have been used as acid acceptors in the solution preparation of polyamides. From an empirical point of view, the base strength of the acid acceptor should be about equal to or greater than the base strength of the terminal amine group at the end of an oligomer or polymer chain. The pKa scale in water is used for base strength. A different measure, E1/2, is used to quantify the base strength of amines in organic solvents. Hall [73,74] has defined the E1/2 of an amine as a potential (in millivolts) of solution at the half-titration point with perchloric acid and has shown that E1/2 is parallel to the pKa scale in water. Table 13.8 lists these values for some acid acceptors frequently used for solution polyamidation.

13.5.5.2.5 Solvent

The solvent has many roles. It dissolves the monomers and provides for their mixing and contact; it dissolves or swells the growing polymer so that the reaction is maintained; it carries the acid acceptor and facilitates the disposition of by-product salts; it influences the reaction rate by polarity or solvation effects; and it absorbs the heat of reaction.

The solvent should be inert and should ideally be able to dissolve the intermediates before the polymerization is started. A primary requisite for high polymer formation in all solution polycondensation reactions is that the solvent must be able to dissolve or swell the polymer sufficiently to permit the completion of the polymerization [75–77]. The solution polycondensation process requires a stronger polymer–solvent interaction than does the

TABLE 13.8
Basicity of Amine Acid Acceptors

		$E_{1/2}$ (mV)[a]	
Acid acceptor	pK$_a$	Ethyl acetate	Acetonitrile
tert-Butylamine	10.45	130	—
Diisobutylamine	10.59	207	—
Triethylamine	10.74	197	66
Tri-n-propylamine	10.70	228	—
Tri-n-butylamine	10.89	210	—
N-Ethylpiperidine	10.45	190	84
N-Ethylmorpholine	7.70	290	221
N,N-Diethyl-m-toluidine	7.24	—	—
N,N-Diethylaniline	6.56	467	425
Pyridine	5.26	—	—

[a] $E_{1/2}$ is the millivolt reading at the half-titration point at 25°C with perchloric acid as the titrant from the work of Hall.

Source: From Hall, H.K., J. Am. Chem. Soc., 79, 5439, 1957; Hall, H.K., J. Phys. Chem., 60, 63, 1956.

interfacial polycondensation method. The combination of solvent, diamine, and acid acceptor must be such that the diamine does not precipitate as a salt with limited solubility.

Although little is known about the effects of solvent polarity, viscosity, and specific gravity on these reactions, the reaction rate tends to increase with an increase in solvent polarity [78,79].

13.5.5.2.6 Solubilizing Aids

Occasionally, solubilizing aids or auxiliary solvents are added to boost the solvating power of the primary solvent. The polymerization of PPTA requires the presence of a solubilizing aid to obtain a high-molecular-weight polymer. Alkaline or alkaline earth metal halides such as $CaCl_2$ and LiCl are known to be effective solubilizing aids in substituted amide solvents such as NMP and DMAc. Solubilizing aids apparently increase the polarity of the solvent by complexing with the carbonyl group (Equation 13.22).

$$(13.22)$$

More recently, quaternary ammonium halides such as methyl tri-n-butyl ammonium chloride were used in the polymerization of PPTA in NMP [80]. Effective shielding of the ammonium cation by bulky alkyl groups stabilizes the ionized species in an organic medium so that it can facilitate the polarization of NMP (Equation 13.23).

$$(13.23)$$

13.5.5.2.7 Reactivity of Precipitated Polymer

In the solution polymerization of PPTA in NMP–$CaCl_2$ solvent, significant chain growth takes place after the polymer precipitates. At the beginning of the reaction, the polymerization proceeds in solution. As the molecular weight of the polymer increases, the viscosity of the solution increases rapidly to a gel point and eventually the polymer precipitates. At this stage, the molecular weight of the polymer is still very low (inherent viscosity ~2), but the polymerization continues in the precipitated state to an inherent viscosity of >6, in the absence of interfering contaminants such as water. This is a clear evidence that the chain-ends of the polymer are not deactivated on precipitation but retain enough mobility to react with the neighboring active groups. However, the rate of reaction becomes very slow after the polymer precipitates.

13.5.5.3 Vapor-Phase Polymerization

Vapor-phase polymerization has been described in the patent literature as an alternative route to aromatic polyamides from aromatic diamines and aromatic diacid chlorides [81]. The reaction is carried out in the gas phase by mixing vapors of the two monomers in the presence of an inert gas. The temperature at the reaction zone has to be higher than the glass transition temperature of the polymer to achieve segmental mobility of the growing polymer chain.

Polymer decomposition is minimal because the reaction time is very short. The polymer is deposited on removable inorganic or organic substrates maintained in the reaction zone.

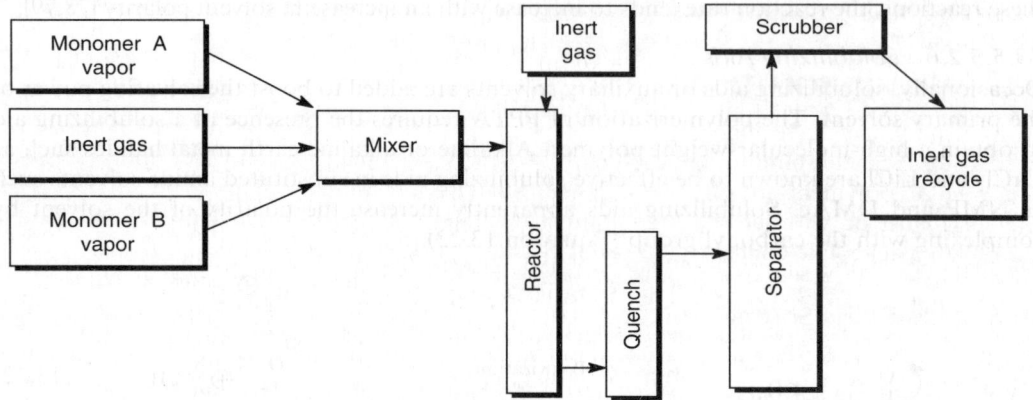

Schematic of the Vapor-Phase Polymerization Process

Vapors of two different monomers (A and B) together with a hot inert gas are fed to a mixer (such as a jet mixer, a simple short tube, or a combination of both) and then to the reactor inlet. Additional inert gas can be introduced as needed. The reactor effluent stream consisting of some polymer, possible oligomers, and by-product acid, is conducted through a quench chamber where the stream is cooled by a flow of relatively cold inert gas. The cooled stream is then led through a separator such as combination of a cyclone separator and filters to remove solid material. The filtered stream is then passed through a water scrubber to remove hydrogen halide and vented to the atmosphere or recycled.

Vapor-phase polycondensation has the distinct advantage of not having to use solvent and it makes possible the elimination of by-product HCl in the gas phase. However, the resulting polymers are usually highly branched due to the high reaction temperature required to maintain chain mobility. In addition, the stoichiometric balance of reagents is much more difficult to maintain than in the case of a condensed phase reaction.

13.5.5.4 Plasticized Melt Polymerization

Most aromatic polyamides cannot be made by a melt polymerization process because the polymer melt temperature exceeds the decomposition temperature. Singh developed a unique procedure for preparing certain aromatic polyamides by a melt process using an internal plasticizer generated *in-situ* during the polymerization [82]. The following reaction scheme was used to prepare aromatic polyamides in the absence of a solvent (Equation 13.24).

(13.24)

The melt polycondensation of isophthaloyl-*N,N*-bis (valerolactam) with *m*-phenylene diamine yielded the aromatic polyamide MPDI plasticized by liberated valerolactam. A small amount of valerolactam is polymerized to poly(valerolactam) during the polymerization, which the author claims can be minimized by adjusting the reaction parameters. It is proposed that the plasticizer can be removed by water extraction after the shaping process thereby recovering the infusible aromatic polyamide.

13.6 ARAMID SOLUTIONS

Aramid polymers have high melting points or melt with decomposition that makes fiber processing by melt spinning impractical [1][g]. Fibers are therefore spun from polymer solutions. These polymers not only do not melt but also are not easy to dissolve. Highly polar solvents, with or without the aid of inorganic salts such as lithium chloride or calcium chloride, or acids like concentrated sulfuric acid have to be used [88].

13.6.1 ISOTROPIC SOLUTIONS

Some aramids are processed from isotropic solutions. Flexible chain homo-polymers like MPDI can be dissolved in solvents like NMP and DMAc [88] to form such solutions but the degree of solubility can be further enhanced by copolymerization [83]. Isotropic solutions can be also obtained with p-aramids but in this case copolymerization is required to enhance solubility.

13.6.1.1 m-Aramid Solutions

As previously mentioned, DuPont and Teijin are the two major manufacturers of m-aramid fibers. Russian scientists also developed a commercial process for the manufacture of MPDI polymer and fiber under the trade name of Fenilon [84]. However, at this point Fenilon production has been suspended.

DuPont's m-aramid polymer, MPDI, is polymerized using essentially a 1:1 molar ratio of m-phenylenediamine and isophthaloyl chloride [85]. Patent literature indicates that the fiber, Nomex, is spun directly from the polymerization solution in DMAc, which contains calcium chloride. MPDI polymer solutions containing >3% by weight calcium chloride are quite stable [2].

Teijin's product, trademarked Teijinconex, is a 100/97/3 copolymer of MPD/ICl/TCl [83]. The polymer is prepared by interfacial polymerization, isolated and dissolved in NMP to form spin dopes of approximately 20% solids concentration [86]. The resulting isotropic solutions are stable at 100°C and are suitable for wet spinning. The solution has two solubility limits that include reversible and irreversible regions, as shown in Figure 13.1 [87]. If the irreversible limit is exceeded, the polymer becomes soluble only in sulfuric acid.

The Russian Fenilon process utilizes low-salt content MPDI solutions [89]. Most of the hydrochloric acid generated during the polymerization process is removed by treatment with ammonia. The resulting insoluble ammonium chloride is filtered from the polymerization solution. Residual HCl is likely neutralized with an organic base. The neutralized solution is suitable for wet spinning of fibers.

13.6.1.2 p-Aramid Solutions

p-Aramids are soluble in strong acids and in highly polar solvents in the presence of inorganic salts. They form isotropic solutions only at low polymer concentrations. Among commercial products, copolyamides from the SVM family as well as copoly(p-phenylene/3,4'-diaminodiphenylether terephthalamide) (Teijin's Technora base polymer) remain soluble in their polymerization mixture [90] and can be spun directly from that solution.

---

[g]Exception Teijinconex mono-filament process.

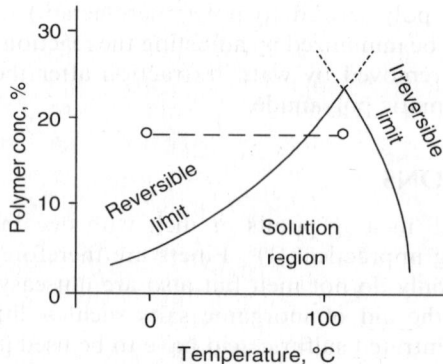

FIGURE 13.1 Stability of Teijinconex spin solution. (From Fujie, H., *Nikkyo Geppo*, 40, 8, 1987. With permission.)

13.6.2 ANISOTROPIC SOLUTIONS

13.6.2.1 Phase Behavior

A distinctive feature of semirigid polymers such as *p*-aramids is that their solutions develop molecular orientation under shear or extension with great ease. This results in a unique difference in properties in the direction of shear or extension vs. those perpendicular to the shear direction. There are two classes of materials that have this characteristic: lyotropic, which form anisotropic solutions; and thermotropic, which form anisotropic melts. As aramids do not melt we will focus here on lyotropic systems. Anisotropic solutions differ from isotropic solutions in many physical characteristics including light depolarization, rheological properties, phase behavior, and molecular orientation.

Observed structures of a lyotropic material are classified into three categories: nematic, smectic, and cholesteric. Nematic and cholesteric mesophases can be readily identified by microscopic examination. The existence of a smectic mesophase is not well defined and is only suggested in some cases. Solvent, solution concentration, polymer molecular weight, and temperature all affect the phase behavior of lyotropic polymer solutions. In general, the phase transition temperature of a lyotropic solution increases with increasing polymer molecular weight and concentration. It is often difficult to determine the critical concentration or transition temperature of a lyotropic polymer solution precisely. Some polymers even degrade below the nematic–isotropic transition temperature so that it is impossible to determine the transition temperatures. Phase behavior is also affected by the polymer molecular conformation and intermolecular interactions.

A good example of a lyotropic solution is that of PPTA in sulfuric acid. Figure 13.2 shows the viscosity–concentration relationship of a solution of PPTA of moderate molecular weight [91]. At low polymer concentrations, the solution viscosity increases with increasing concentration just like an isotropic solution of a flexible chain polymer. However, above a critical concentration of ~12%, the solution viscosity decreases abruptly with increasing concentration. This behavior is caused by the close packing of the rigid chain polymer molecules to form ordered domains. The solution viscosity reaches a minimum point at about 20% solids and then abruptly increases with additional solids. A solid phase will eventually appear when the solution becomes supersaturated. The anisotropic PPTA–H_2SO_4 solution exhibits liquid crystal behavior. It has the flow properties of a liquid and is crystal-like with the ability to depolarize cross-polarized light. When the solution is subjected

Aramid Fibers

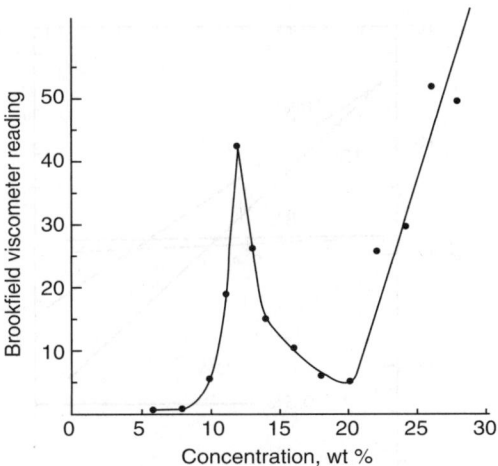

FIGURE 13.2 Bulk viscosity vs. concentration of PPTA–H_2SO_4 solution. (From Bair, T.I. and Morgan, P.W., U.S. Patent 3,673,143, 1972; U.S. Patent 3,817,941, 1974. With permission.)

to shear or elongational flow, the liquid crystal domains become aligned in the direction of flow to achieve a high degree of molecular orientation.

For fiber preparation, a lyotropic solution is best processed at a solids concentration near the minimum solution viscosity and at a temperature close to its anisotropic transition temperature (Figure 13.2). These conditions maximize solution ordering prior to spinning.

13.6.2.2 Rheological Properties

Lyotropic solutions generally exhibit viscoelastic behavior. They are pseudoplastic and exhibit shear thinning with increasing shear rate. For polymers of near-linear chain conformation, their lyotropic solutions are known to give less die swell and are less tractable than isotropic solutions. The PPTA–H_2SO_4 solution was the first to be used commercially and has been studied most extensively.

The rheological properties of PPTA–H_2SO_4 solutions have been studied by several investigators [92–97]. Figure 13.3 and Figure 13.4 show the relationship between shear viscosity, $\acute{\eta}$, and shear rate, γ, for Kevlar–H_2SO_4 solutions of various concentrations at 25 and 60°C, respectively. Figure 13.5 is a plot of shear viscosity vs. shear stress for PPTA solutions at 25°C [97]. The change in the slope of these curves between 8 and 10% solutions shows the effect of the isotropic–anisotropic phase transition. The viscosity–shear stress curves for 10 and 12% solutions tend to infinity, indicating the presence of a yield stress [94].

13.7 PREPARATION OF ARAMID PRODUCTS

13.7.1 Fibers

13.7.1.1 Dry Spinning

Solutions of m-aramid polymers are currently produced using dry-or-wet spinning processes. Processing steps after spinning can include drawing, drying, and heat treatment.

In the dry-spinning process, a solution of polymer is extruded through a spinneret that is mounted at the top of a heated column. As the solution is extruded in the presence of hot inert

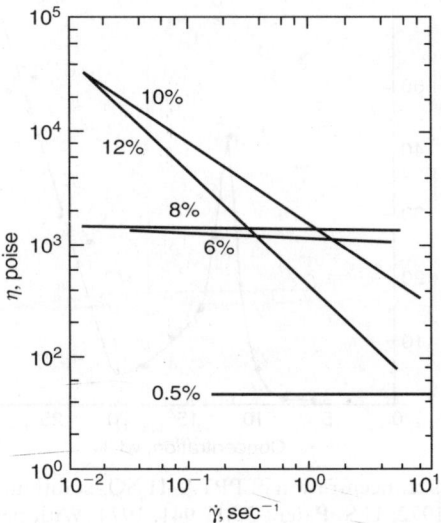

FIGURE 13.3 Shear viscosity vs. shear rate for re-dissolved Kevlar–H_2SO_4 solution at 25°C. (From Aoki, H., White, J.L., and Fellers, J.F., *J. Appl. Polym. Sci.*, 23, 2293, 1979. With permission.)

gas (or air), solvent evaporates from the incipient fiber. The temperature of the heated gases in the column is above the boiling point of the solvent. The solidified fiber is collected at the bottom of the column. The polymer solvent must be inert, stable at its boiling point, and a good solvent for the polymer. The heat of vaporization of the solvent must not be too high, it must have sufficient thermal resistance, low toxicity, a very low tendency to produce static charges, low risk of explosion, and be relatively easy to recover [98]. The dry-spinning process was initially developed for spinning acrylic fibers and was modified for spinning *m*-aramid polymer. DuPont developed processes for dry spinning Nomex from DMF and DMAc solutions [99]. The *m*-aramid polymer solution is disordered in the solution state. Some orientation is imparted during the extrusion of the solution through the spinneret capillary. The extent of fiber orientation tends to increase as the shear rate through the spinneret capillary is increased. Radial structural inhomogeneities are generally introduced during the solvent diffusion and evaporation stages of the dry-spinning process [10]. A skin

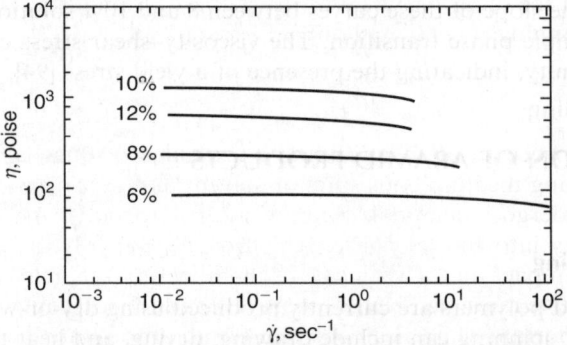

FIGURE 13.4 Shear viscosity vs. shear rate for re-dissolved Kevlar–H_2SO_4 solution at 60°C. (From Aoki, H., White, J.L., and Fellers, J.F., *J. Appl. Polym. Sci.*, 23, 2293, 1979. With permission.)

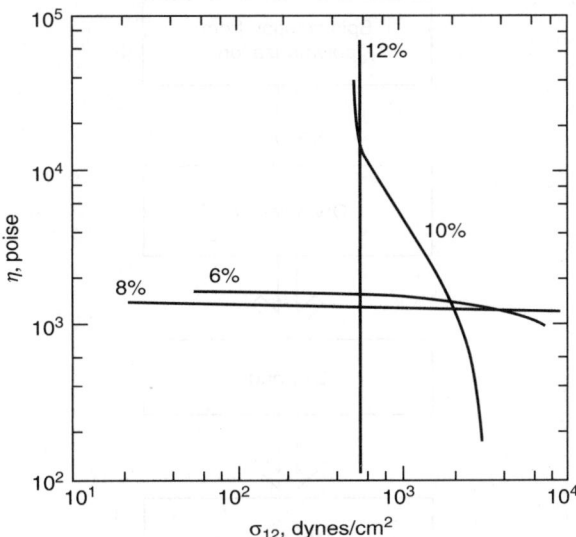

FIGURE 13.5 Shear viscosity vs. shear stress for re-dissolved Kevlar–H_2SO_4 solution at 25°C. (From Aoki, H., White, J.L., and Fellers, J.F., *J. Appl. Polym. Sci.*, 23, 2293, 1979. With permission.)

core structure forms because the outer skin of the fiber loses solvent faster than the inner core. As diffusion progresses, the loss of the solvent from the core through the solidified sheath reduces the mass of the core. This results in the sheath collapsing inward. Since the evaporation rate of the solvent in the sheath of the fiber is faster than the diffusion rate of solvent from the core of the fiber the cross section shape of the fiber can change from round to dogbone. *m*-Aramid fibers are spun at a spin stretch ratio of 1–20x, which is far lower than fibers processed from the melt, but this has little impact on fiber properties since there is very little orientation produced during this part of the process. The resulting *m*-aramid fibers at the bottom of the spin cell retain considerably more solvent (>20%) than dry spun acrylic fibers (<5%).

The as-spun fiber is then drawn to develop physical properties. Fiber drawing is generally done in a dilute water solution of solvent. The solvent partially plasticizes the fiber and facilitates drawing (3–5x). After the drawing step, the fibers are washed with water, dried, and crystallized by heating at a temperature above the polymer T_g (~275°C) [100,101]. Typical fiber properties are in the order of 0.6 GPa with an elongation to break of 30%. A schematic of the Nomex dry-spinning process is shown in Figure 13.6.

13.7.1.2 Wet Spinning

In the wet-spinning process polymer solution is extruded through a spinneret that is submerged in a coagulating medium consisting of solvent and nonsolvent. On coagulation, the spinning solution undergoes spinodal decomposition into polymer-rich and polymer-poor regions and ultimately into a solid phase. It is this polymer solvent–nonsolvent interaction that has the greatest impact on the structure of the fiber and the ultimate properties that can be achieved. The relative rates of solvent to nonsolvent diffusion control the process of phase separation [102]. Important variables controlling this process are polymer solids, solution composition and temperature, coagulating solution composition and temperature, the extrusion rate, and the residence time in the coagulating bath. Control of the size and character of

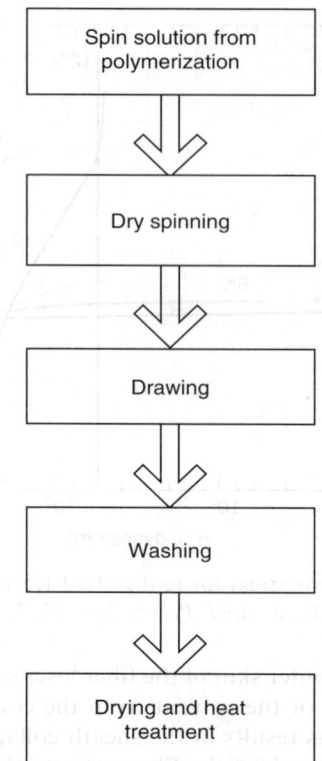

FIGURE 13.6 *m*-Aramid dry-spinning process.

the voids formed in such a process is key to achieving fibers with excellent mechanical properties [103].

A schematic of the Teijinconex wet-spinning process is shown in Figure 13.7 [87]. The process schematic for producing Fenilon is similar to that shown in Figure 13.7 with the exception that the polymer solution is spun directly from the polymerization process [104]. While the above processes require little or no inorganic salt content in the spinning solution, the process described by Tai et al. allows the use of salt-containing solutions [105].

13.7.1.3 Dry-Jet Wet-Spinning

Kwolek [106] demonstrated in her early work at DuPont that *p*-aramid fibers could be spun from amide and salt solutions using a conventional wet-spinning process. These solutions were typically of low concentration. The resulting fibers had low strength but high modulus after heat treatment. In later development, *p*-aramid fibers were spun from more concentrated solutions using dry-jet wet-spinning processes [107]. These solutions contained aramid polymer above a critical solids concentration and were anisotropic.

In 1970, Blades [108] discovered that high-strength, high-modulus fibers could be spun from anisotropic solutions of aramid polymers by dry-jet wet spinning (Figure 13.8). His process is shown schematically in Figure 13.8. The key feature of this process is that an anisotropic solution is extruded through an air gap between the spinneret and the coagulation bath. The coagulated filaments are washed, neutralized, and dried. This process

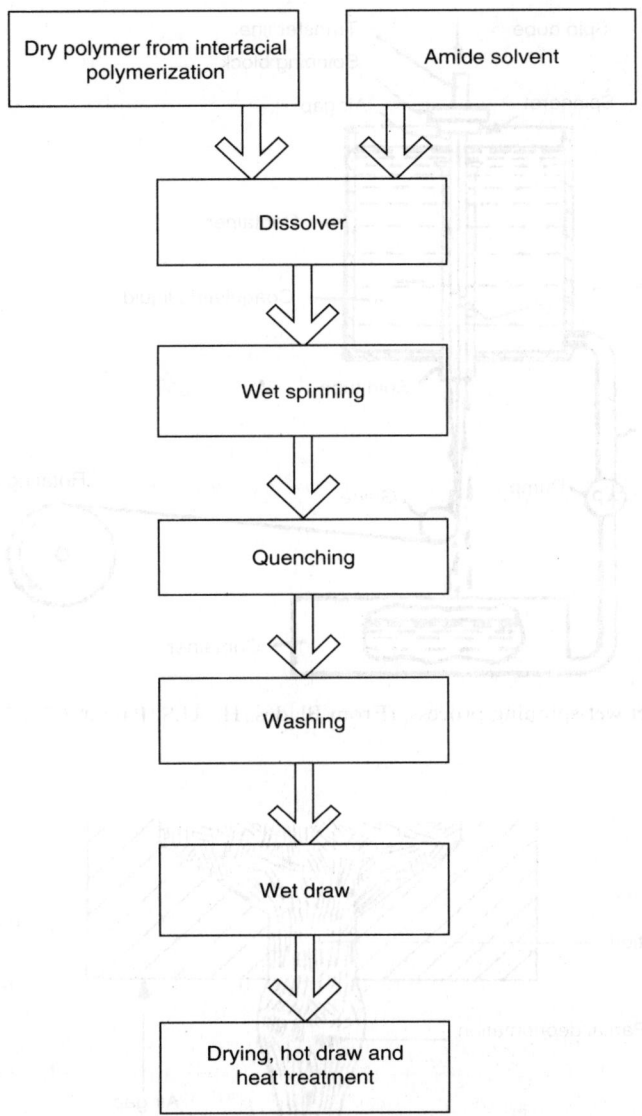

FIGURE 13.7 Teijinconex wet-spinning process. (From Fujie, H., *Nikkyo Geppo*, 40, 8, 1987.)

produces a fiber with tenacity and initial modulus 2–4 times that of a fiber prepared by a conventional wet-spinning process.

The mechanistic model of polymer molecular orientation in a dry-jet wet-spinning process is shown in Figure 13.9 [109]. Shear at the capillary wall causes the liquid crystalline domains to orient along the direction of flow when an anisotropic solution is extruded through a spinneret capillary. At the capillary exit, some deorientation of liquid crystalline domains occurs because of solution viscoelasticity. However, this deorientation is quickly overcome by threadline tension on the attenuating filament in the air gap. The attenuated filaments retain this highly oriented molecular structure on coagulation giving rise to highly crystalline, highly oriented fibers.

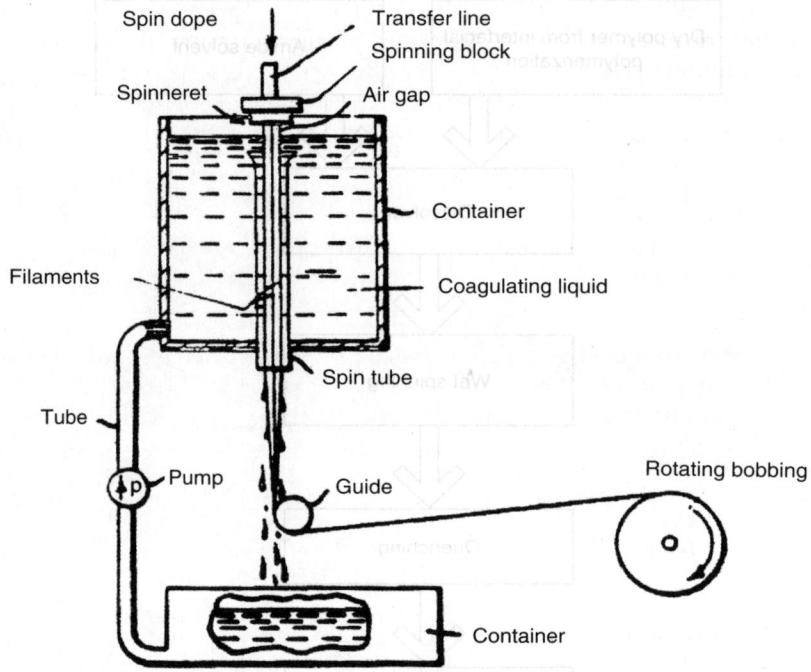

FIGURE 13.8 Dry-jet wet-spinning process. (From Blades, H., U.S. Patent 3,767,756, 1973.)

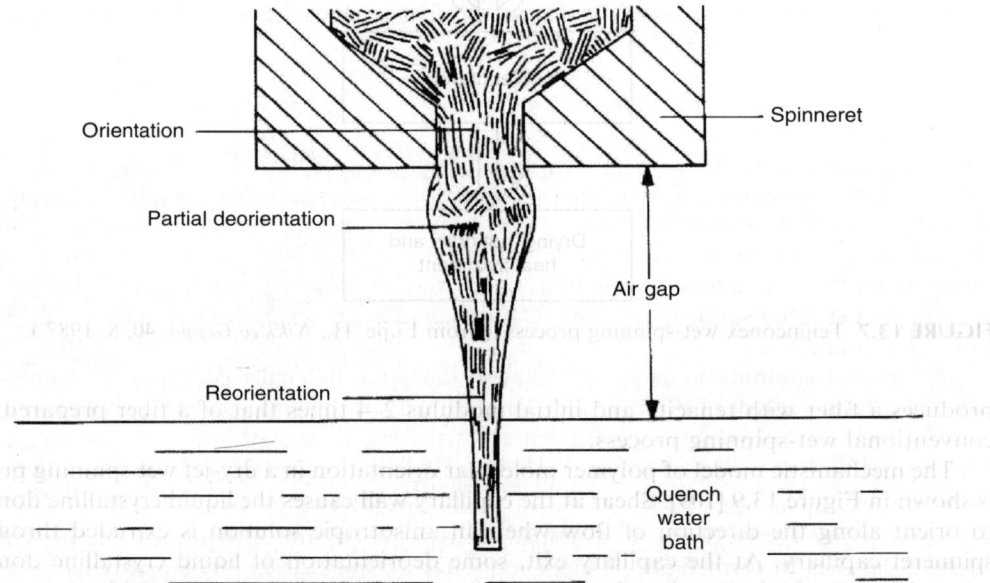

FIGURE 13.9 Molecular orientation during dry-jet wet spinning. (From Yang, H.H., Aramid fibers, in *Fibre Reinforcement for Composite Materials*, Bunsell, A.R., Ed., Elsevier, Amsterdam, 1988. With permission.)

The operating conditions for dry-jet wet spinning are proprietary for fiber producers and are therefore not revealed in detail. A review of the literature shows that the general conditions are as follows [110]:

Polymer molecular weight	5,000–35,000
Polymer inherent viscosity	3–20 dL/g
Spinning speed	>55 y/min (>50 m/min)
Number of filaments	10–1500
Spinneret hole diameter	0.002–0.004 in. (0.051–0.102 mm)
Filament size	1–6 denier/filament

The as-spun fiber from dry-jet wet spinning can be heat treated at high temperatures and high tension to increase its crystallinity and degree of crystalline orientation [111]. The heat treatment conditions are generally in the following ranges:

Temperature	250–550°C
Time	<10 min
Tension	5–50% of breaking strength

As discussed above, isotropic solutions are typically converted to fibers by a wet-spinning process. Ozawa [90] disclosed that the polymerization mixture of copoly(p-phenylene/3,4'-diaminodiphenylether terephthalamide) remained isotropic. He deviated from traditional spinning techniques and spun fiber from this solution using dry-jet wet spinning. Although as-spun fiber tensile properties were modest, high strength fiber was achieved with subsequent drawing. This fiber product was later commercialized as Technora aramid fiber by Teijin Ltd. The use of dry-jet wet spinning to prepare fibers from isotropic solutions has since been widely practiced.

The dry-jet wet-spinning process is unique in that the temperature of the spinning nozzle is different than that of the spin bath. In comparison, the spinning nozzle in a conventional wet-spinning process is immersed in the coagulation liquid and is therefore at the same temperature. This gives rise to several inherent limitations with the wet-spinning process. First, the coagulant temperature must exceed the freeze point of the spinning solution. Second, the spinning solution is exposed to the coagulant as soon as it exits the spinneret holes. This can limit attenuation of the incipient filament. The dry-jet wet-spinning method allows the use of a low temperature coagulant without concern for freezing the spin solution. The air gap permits the extruded solution to be more fully attenuated and to develop a higher degree of molecular orientation.

Dry-jet wet spinning is, however, a much more mechanically complicated process and requires careful control of both the air gap and the flow dynamics of the coagulant fluid.

13.7.2 Film

Aramid films have been in development since the late 1990s by several Japanese companies including Toray, Teijin, and Asahi. As with fibers, aramid solutions can be extruded through flat dies to form films. The conventional wet process can be employed to produce unidirectional and bi-oriented films from isotropic aramid solutions. Production of films from anisotropic solutions requires unique processes as shown by the example of PPTA film.

Forming films from anisotropic solution is extremely difficult because of the ease with which these solutions orient. Obviously once the films orient in the machine direction they are

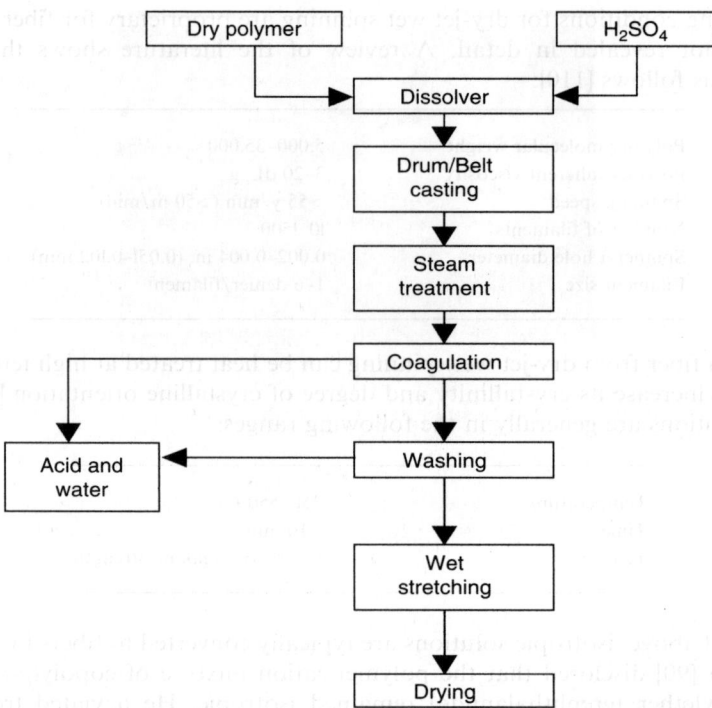

FIGURE 13.10 PPTA film process. (From Imanishi, T. and Muraoka S., U.S. Patent 4,752,643, June 21, 1988.)

very weak in the cross direction and, as a result, tend to fibrillate. Asahi has developed a process leading to a coherent film [112]. A schematic of this process is shown in Figure 13.10. An anisotropic PPTA solution in sulfuric acid is extruded through a die onto a drum or a belt where it is initially exposed to warm, humid air. Under these conditions, the solution reverts to an isotropic state as moisture is absorbed to reduce the effective acid concentration and raise the temperature. This is the critical step as it leads to the formation of an isotropic film. The structure is fixed on coagulation after which the film is washed with water to remove the remaining sulfuric acid. The wet film is biaxially stretched to develop mechanical properties in both directions and then dried. Finally, the film can be heat-treated to further improve properties.

13.7.3 Fibrids

Fibrids are film-like particles that are formed when—aramid solutions are precipitated in a nonsolvent under high shear [113,116]. The dimensions of as-formed fibrids are around 100 μm × 700 μm × 0.01 μm [113,114]. Fibrids have a high surface area, around 200–300 m²/g, and can function as a thixotrope or a reinforcing agent in composite, sealing, coating, and elastomer applications [114,115]. Fibrids are used primarily in aramid papers. Aramid papers are composed of a mixture of fibrids and short Nomex fibers referred to as floc (Figure 13.11). Fibrids serve as a binder for the short fibers and also improve the dielectric properties of high temperature, heat-resistant aramid papers (Figure 13.12) [115,116]. A process for making m-aramid papers is shown in Figure 13.13 [113].

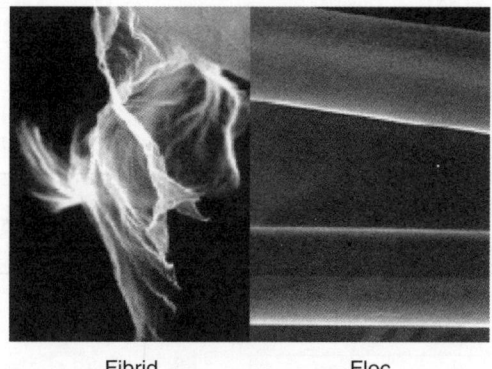

Fibrid   Floc

FIGURE 13.11 Photomicrographs of aramid fibrid and floc.

13.7.4 PULP

p-Aramid pulp is a highly fibrillated material that retains the key chemical and physical properties, low creep performance, and high temperature and wear resistance of the precursor *p*-aramid fiber. These characteristics make *p*-aramid pulp an excellent candidate to replace asbestos in friction products such as brake linings and clutch facings, gaskets, and industrial papers. The highly fibrillated structure of Kevlar pulp is characterized by a combination of high fibril aspect ratio (>100) and high specific surface area [116,117]. The fibrils can be attached to, or detached from, the core fiber.

Pulp is produced by passing a dilute slurry of short cut length, *p*-aramid fiber through one or more high shear refiners. The highly oriented, crystalline fiber is cut and readily split into fibrils of smaller diameter because of the relatively low compressive strength of the fiber. The refining process is controlled to produce a certain balance between the final fiber length and the degree of fibrillation or the degree of new surface generation. The optimum relationship between these two parameters is dictated by the process or product performance requirements of the specific end-use application. Water is removed from the resulting pulp slurry to produce a wet product or, with additional drying, a dry product. Wet pulp contains 50–70% moisture depending on the producer. Dry pulp contains 4–8% moisture. Handling of the pulp becomes difficult at lower moisture levels because of static problems.

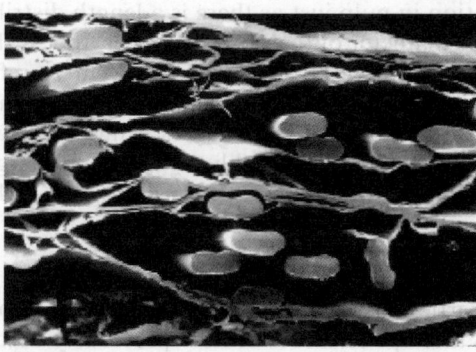

FIGURE 13.12 Photomicrograph of a cross section of Nomex Type 411 paper.

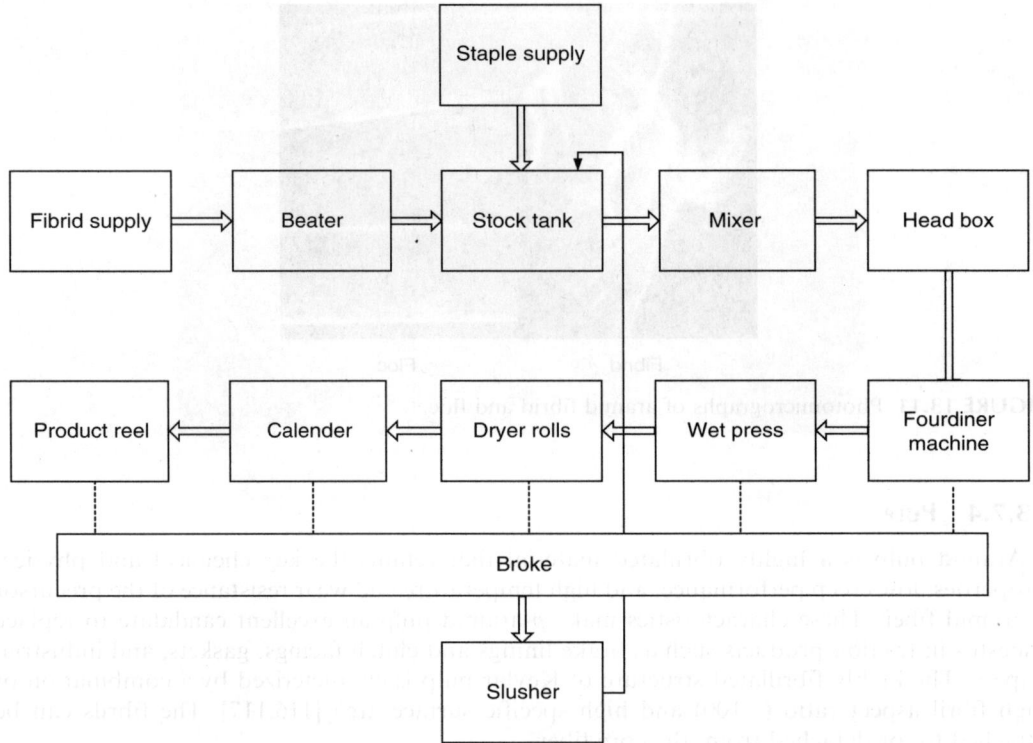

FIGURE 13.13 Process for making *m*-aramid papers.

Pulp is characterized in terms of fiber length, length distribution, and the degree of fibrillation. Absolute fiber (or fibril) length typically ranges from less than a millimeter to about 6 millimeters. The fiber length distribution is measured using a device such as a Kajaani 200 instrument and is reported in terms of a length-weighted average length ($\sum n_i l_i^2 / \sum n_i l_i$) or weight-weighted average length ($\sum n_i l_i^3 / \sum n_i l_i^2$). The length-weighted average length of typical commercial pulps is in the range of 0.6–1.1 mm. The degree of fibrillation is related to the specific surface area of the pulp or to the drainage rate of an aqueous pulp slurry determined by the Canadian Standard Freeness or Schopper–Riegler methods. There is a fiber–fibril diameter or width distribution in pulp just as there is a length distribution. The diameter will range from 12 to 15 μm, the diameter of the precursor fiber, to less than 1 μm for the smallest fibrils. Pulp specific surface area ranges from about 7 to 15 m²/g reflecting the breakdown of the initial fiber, with a surface area of about 0.2 m²/g, into a broad distribution of smaller diameter fibrils. Canadian Standard Freeness values range from about 100 ml for "high" surface area pulps to about 600 ml for less highly refined pulp merges. The highly fibrillated morphology characteristic of *p*-aramid pulp is shown in Figure 13.14.

13.8 APPLICATIONS

The broad range of properties of aramids is the main reason for their utility in diverse applications. Here we will attempt to illustrate how previously described properties of these fibers are exploited in their applications.

Aramid Fibers

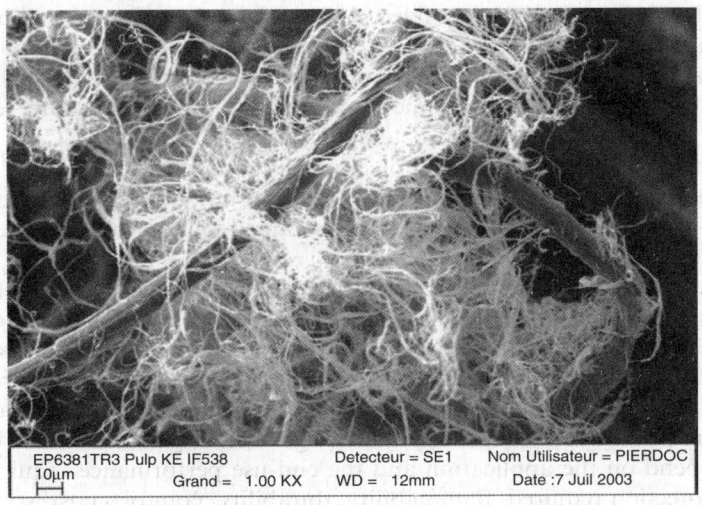

(Merge 1f538)

FIGURE 13.14 Scanning electron micrograph of Kevlar brand pulp.

13.8.1 *m*-Aramid Fiber

Many of the applications of *m*-aramid fibers are due to their unique combination of flame resistance with thermal and textile properties. Some applications also benefit from the fact that *m*-aramid fibers are available in colored form. In general, these fibers are very difficult to dye and thus most producers offer producer colored (pigmented) fiber. While pigments offer in general better UV stability, this approach limits the number of colors available. At this time only DuPont offers piece dyeable products. In general, dyeing of Nomex fibers requires the use of carriers, and dyeing technology is kept as proprietary information by dye houses.

In general, flammability as well as thermal properties are bulk properties of the material. When these properties are critical, compositions comprising 100% aramid fibers are used. Blends with nonaramid materials do come into play when other fiber properties or characteristics are desired.

13.8.1.1 Protective Apparel

Fabrics of *m*-aramids are widely used in thermal protective apparel because of their unique combination of thermal and textile properties. The fibers from which these fabrics are made are inherently flame resistant and do not melt or drip. A measure of the fiber's flammability is its limiting oxygen index (LOI), which is the concentration of oxygen in air that is required to support combustion once the material is ignited. Materials with an LOI > 21 are considered nonflammable. The inherent flame resistance of *m*- and *p*-aramids is essentially the same with LOI values of ~28–29. For apparel applications, *m*-aramids are generally preferred over *p*-aramids because the fabrics have a more comfortable, textile-like hand as a consequence of lower fiber modulus and higher elongation. Even though *m*-aramids fibers exhibit high glass transition temperature and high crystalline phase melting points (275°C and 425°C respectively) both glass transition temperature and melting temperature of the crystalline phase are high (275°C and 425°C respectively) in flame 100% *m*-aramid garments exhibit some shrinkage, which in turn can lead to fabric "break opening" and loss of protective barrier. Blends with *p*-aramids are often utilized to stabilize the protective garment against shrinkage and to reduce fabric "break-open" during flame exposure. At higher exposure to flame MPDI carbonizes and forms a tough char at a temperature of ~800°F (427°C). The intumescent nature of the char provides additional protection. Decomposition products on combustion will vary depending on the heating rate and the amount of oxygen present. In general, combustion by-products are similar to those obtained on burning wood, wool, cotton, polyester, and acrylic [118,121].

Both continuous filament and staple yarns are used in protective apparel fabrics. Typical filament deniers range from 0.85 to 2. Staple fiber length is 1.5–2 in. for processing on the cotton system. Yarns are available in dyeable and producer colored forms. Fabric forms include woven, knit, and nonwoven. The mechanical toughness of the fiber results in higher fabric strength than FRT cotton fabrics of even greater weight. Higher resistance to tear and abrasion also provides greater durability and longer useful garment life. Ultimately fabric selection will depend on the application and the end-use performance requirements such as the degree of protection required, flammability, durability, comfort, cost, style, etc.

m-Aramid fabrics are widely used in industrial, military, fire fighting, and auto racing applications. Chemical, petrochemical, and utility workers wear flame-resistant protective clothing where flash fire or electrical arc hazards exist. Military applications include flight suits and coveralls for combat vehicle and shipboard engineering crews. In firefighting apparel, *m*-aramids and blends with *p*-aramids find use in turnout gear, station uniforms, hood, gloves, and boots. The turnout is a three-component system (an outer shell, a moisture barrier, and a thermal barrier) designed to provide basic thermal protection in hot environments and in flashover conditions in addition to maximizing comfort and minimizing the potential for heat stress. Race car drivers and their crews wear clothing to protect themselves from flash fires resulting from crashes and pit accidents. The protective gear includes suits, underwear, socks, and gloves.

13.8.1.2 Thermal and Flame-Resistant Barriers

The same fiber properties that make *m*-aramids suitable for protective apparel applications find utility in thermal and flame-resistant barrier fabrics found in transportation (aircraft, train, and automobile) end-uses and in contract furnishings for hotels, offices, auditoriums, hospitals, and day care centers. Fabrics involved in aircraft and railroad car interior applications include upholstery, floor coverings, bulkheads, wall coverings, and blankets.

Fire-blocking materials increase the probability of safe egress of passengers from the cabin in a fire emergency. A fire-blocking fabric or thermal liner in aircraft seating provides a barrier between the flame source and, for example, a high fuel content polyurethane seat cushion. A typical construction would be a layer of a spunlaced fabric quilted to a woven *m*-aramid fabric to provide both durability and lightweight. The fire-block is designed to retard or delay ignition of the cushion once the flame has penetrated the outer upholstery fabric.

Because the fibers are inherently flame-resistant, there are no topical treatments that can wear off or be removed during routine laundering. The abrasion resistance and toughness of the fiber allows for easy maintenance of fabrics without concern for fading, cracking, or degradation.

Yarns can be dyed or are producer colored. This allows for the design of attractive interiors and at the same time, provides the safety of a flame-resistant material. The filament denier for these applications is higher than that of yarns for apparel fabrics and is generally in the range of 3–10.

13.8.1.3 Elastomer Reinforcement

There are a few elastomer reinforcement applications where *m*-aramid yarns are superior to *p*-aramid yarns. Continuous filament *m*-aramid yarn is used in a loose knit construction to reinforce automotive heater hose. Yarn on yarn abrasion resistance, and not strength, is key to performance in this application where the hose is exposed to significant thermal, impulse, and vibrational stresses. A second growing use is in the reinforcement of silicone elastomer hose for automobile turbochargers where *m*-aramid provides high thermal stability.

13.8.1.4 Filtration and Felts

Filter bags of *m*-aramid fiber felts are the material of choice in the bag houses of the hot mix asphalt (HMA) industry as well as in a variety of other applications. Bag houses are the preferred air cleaning system because they provide compliance with pollution codes and provide economic advantages over scrubbers. Bags can be manufactured from a variety of materials including Teflon®[h], fiberglass, polyester, and polyphenylene sulfide, but *m*-aramids are the most suitable for HMA plants. Key factors determining this include filtration performance, chemical resistance, tensile strength, durability, cost, temperature resistance, and combustibility [119,122].

Bags of Nomex fiber can withstand a continuous operating temperature of 400° F (204°C). Additionally the fiber remains dimensionally stable at this temperature—neither growing nor shrinking more than 1%. The common felt in the industry is a 14 oz/yd^2 felt made of 2 dpf fibers.

m-Aramid felts and fabrics are ideal for heavy-duty laundry textile covers used on calendars and ironing presses. These materials can meet the thermal stability requirements of calendars and presses operating at temperatures of up to 200° C. For equipment operating at lower temperatures (150–160° C), *m*-aramid fabrics provide greater reliability than lower cost polyester press covers whose use is still permissible at this temperature range. While heat resistance is the key criterion for covers, *m*-aramids also have the advantages of abrasion resistance, dimensional stability, and very good resistance to hydrolysis.

13.8.2 *m*-Aramid Paper

As we have mentioned earlier, *m*-aramid papers are produced exclusively by DuPont and thus most of the application data are based on Nomex papers.

13.8.2.1 Electrical

In the form of paper or pressboards, *m*-aramids provide an optimum balance of properties for use as electrical insulation in transformers, motor, generators, and other electrical equipment. Properly used, these materials can extend the life of an electrical equipment, reduce the frequency of premature failures, and protect against random electrical stress situations.

Papers and pressboards are made from two *m*-aramid forms—floc and fibrids. Floc is yarn cut to a short length. Floc retains the intrinsic properties of the yarn and gives the paper mechanical strength. Fibrids are microscopic film-like particles that provide dielectric strength and bind the floc particles together to give the sheet integrity.

Key properties are inherent dielectric strength, mechanical toughness, thermal stability, chemical compatibility, cryogenic capability, moisture insensitivity, and radiation resistance.

---

[h]Teflon®—a registered trademark of E.I. DuPont de Nemours & Co., Inc., Wilmington, Delaware, USA.

Depending on product type and thickness, densified products can withstand high short-term electrical stresses without further treatment with varnishes or resins. Densified products have good resistance to tear and abrasion and, in thin grades, are flexible. Electrical and mechanical properties are unaffected at temperatures up to 200°C. Useful properties are maintained for at least 10 years of continuous exposure at 220°C. Like *m*-aramid yarns, papers do not melt and do not support combustion. Products are compatible with all classes of varnishes and adhesives, transformer fluids, lubricating oils, and refrigerants. At the boiling point of nitrogen (77°K), selected types of Nomex paper and pressboards have tensile strengths exceeding values at room temperature. In equilibrium at 95% relative humidity, densified products retain 90% of their bone-dry dielectric strength. Products are unaffected by 800 megarads of ionizing radiation and retain useful electrical and mechanical properties after eight times this exposure [120,123].

Papers are available in many forms varying in thickness, degree of densification, and composition (additive type or floc to fibrid ratio). Pressboards, which differ from paper in thickness and rigidity, are likewise available in several thicknesses and degrees of densification. The product of choice will depend on many factors including end-use thermal and mechanical performance requirements, formability or ease of fabrication, and the desired degree of saturability.

Applications in transformers include conductor wrap, layer and barrier insulation, coil end filler, core tubes, section or phase insulation, lead and tap insulation, case insulation, and spacers. In motors and generators, the superior thermal properties of *m*-aramid products can enhance both performance and reliability. Their strength and resilience can also help extend the life of rotating equipment in severe operation conditions. Insulating parts where *m*-aramids are used in rotating equipment include conductor wrap, coil wrap, slot liners, wedges, phase insulation, end-laminations, pole pieces and coil supports, commutator V-rings, bushings, and lead insulation.

13.8.2.2 Core Structures

Core structures are more commonly referred to as honeycomb structures or cores. Cores of *m*-aramid honeycombs with carbon-fiber skins were first used in flooring panels of the British Aerospace VC-10 BOAC in the late 1960s. In 1970, Boeing's new generation aircraft, the 747, flew with a number of interior and exterior components fabricated with aramid core. Since then, aramid honeycomb cores have become a standard design material for flooring panels, fairings, radomes, rudders, elevators, cowlings, and thrust reversers. The primary purpose of core structures is to minimize weight while [121] maximizing stiffness. Lower weight translates to increased payloads and reduced fuel costs [124].

Aramid cores are made from paper (typically 1.5–4 mil in thickness) comprising *m*-aramid floc and fibrids, similar to the papers used in electrical applications discussed in the previous section. Adhesive node lines are printed on paper sheets that are then stacked, pressed, and heated to cure the adhesive. The resulting block is expanded. The adhesive-free areas form the hexagonal cells of the honeycomb configuration. The core is dipped several times in an epoxy or phenolic resin solution until the desired density and mechanical property levels are reached. The core is then cut into slices of the desired thickness. Face sheets are glued to each side of the core. The most common face sheet today is a composite of carbon fiber and epoxy resin.

Aramid cores have many attributes. *m*-Aramids have high thermal tolerance and are compatible with resins with cure temperatures to 400°F. Cores can be fabricated in a wide range of densities from 1.5 to 10 lb/ft^3. They have higher specific shear strength than foam cores and higher toughness, at equal density, than aluminum, glass, or foam cores. They have high wet strength and exhibit excellent creep and fatigue performance. Aramid cores do not corrode and do not promote galvanic action in contact with metals. They are easy to fabricate and the self-extinguishing character of m-aramids allows the structures to meet stringent flammability, smoke generation, and toxicity standards.

13.8.2.3 Miscellaneous

Tags and labels of m-aramid paper for in-process bar coding are used where high temperature stability and chemical resistance are required. In loudspeakers, m-aramid sheets are used for voice coil insulation and for the speaker cone itself. Bus bars in lithium ion batteries for portable telephones and computers are insulated with m-aramid paper. Photocopiers and laser printers that operate at high temperatures use cleaning rollers and webs made from m-aramid paper.

13.8.3 *p*-ARAMID FIBER

As m-aramid fibers are best known for their flame resistance, p-aramid fibers are universally recognized as the material of choice for ballistic protection. While p-aramids do play a critical role in this application we will attempt to show that their unusual properties are also suitable for a wide variety of other end-uses.

13.8.3.1 Armor

Aramid-based armor systems are designed to protect individuals and equipment against a variety of threats in both civilian and military environments. Handgun bullets and knives are the primary threats encountered in civilian law enforcement work. Military threats are more wide ranging and generally deal with higher velocity projectiles including rifle bullets, flechettes, and fragments from mortars, grenades, and mines. The design of the optimum protective system must take into consideration the nature of the threat and therefore civilian and military systems will necessarily differ. Armor systems can be roughly divided into soft and hard categories. Soft armor systems are assemblies of woven fabrics that are used to make bullet-resistant vests, flak jackets, and soft structures such as blankets, curtains, and liners. Hard or composite armor systems are used in helmets and in structures designed to protect vehicles, vessels, or shelters. These systems are made of multiple fabric layers impregnated with a vinyl ester or phenolic–polyvinylbutyral resin binder. Spall liners that are fitted inside armored military vehicles and protect against fragments resulting from hits by high velocity shells are a classical example of hard armor.

Beginning in the 1970s high strength fibers—particularly p-aramids—generally displaced glass and nylon as the preferred fibers for ballistic protection in soft armor. The evolution of vest design continues today with ever-increasing demands for greater ballistic protection, less weight, and greater comfort. Initial aramid-based vests of the 1970s had a weight of 1.26 lb/ft^2 compared to 1.3 for the incumbent nylon reinforced vests of the 1950s. Today's vest weighs even less, about 0.95 lb/ft^2, while providing greater ballistic protection. These advances have been made possible through the use of higher strength yarns with a broader range of deniers, achieved through spinning process modifications, and by optimizing the weave pattern of the reinforcing fabrics.

Vests providing ballistic protection do not necessarily provide adequate protection against threats from sharp implements such as knives. For civilian use, particularly in penal institutions, vests incorporating p-aramids have been designed that provide protection against penetration by knife, ice pick, and awl [122,123,125,126,127,128]. Designs that offer both ballistic and stab protection have also been claimed [124–130].

13.8.3.2 Protective Apparel

p-Aramid yarns are used in protective apparel where cut resistance, thermal resistance, or abrasion resistance is critical. Applications include gloves and sleeves for automotive, glass, steel and metal workers, chainsaw chaps and trousers for lumberjacks, and other apparel such as aprons and jackets. *p*-Aramid yarn does not support combustion and does not melt in contrast to competitive products made from nylon, polyester, and polyethylene. Gloves of *p*-aramids offer exceptional cut resistance and can substantially reduce the risk of hand and finger injuries in glass and metal handling operations.

Gloves are made primarily from spun yarns, although some are made from textured continuous filament yarns for applications where the tendency to form lint must be minimized. Yarn denier per filament can vary from 0.85 to 4.2 dpf with 2.25 dpf the predominant product. Generally, cut resistance increases as the denier is increased but dexterity is sacrificed. Gloves are made from 100% aramid yarns or from blends with other fibers, such as nylon or polyester, to reduce cost or to improve comfort or abrasion resistance. Yarns can also be spun with steel fibers to provide superior cut resistance. Most gloves are made of a knit construction although some are cut and sewn from woven fabric. Some *p*-aramid gloves are coated or "dotted" with elastomers to enhance grip; others have leather sewn over the palms and fingers to provide puncture resistance or to increase abrasion resistance.

p-Aramid gloves can be cleaned using conventional laundering or dry-cleaning processes with minimal impact on cut resistance. Unlike cotton, these gloves do not shrink when exposed to hot water or hot air. Overall cost per use can be reduced with cleaning and reuse, rather than disposal, of soiled items.

13.8.3.3 Tires and Mechanical Rubber Goods

p-Aramids are particularly well suited as reinforcing agents for belts of radial tires and for a variety of mechanical rubber goods because of their high strength and modulus, excellent dimensional stability, high temperature durability, and favorable strength to weight ratio. In spite of these attributes, lower cost steel wire continues to be the reinforcement of choice for passenger car tires. Nevertheless, aramid cords have slowly made inroads into tire applications since their introduction in the mid-1970s, particularly in the high performance arena where the performance to weight ratio is critical. Key performance criteria are speed capability, handling, and comfort. Additional factors that favor increasing aramid usage in automobile and truck tires are the ongoing efforts to reduce vehicle weight and to reduce rolling resistance to reduce energy consumption. Aramids also find use in aircraft, motorcycle, and bicycle tires where the performance attributes often outweigh cost. Typical yarn deniers for tire applications are 1000–3000 with a 1.5–2.25 dpf fiber. Product variants include so-called "adhesion activated" yarns that have a surface treatment that facilitates adhesion to the elastomer and can simplify subsequent tire cord and fabric processing steps by eliminating a dip-coating step [128,131].

Mechanical rubber goods include hoses, power transmission (PT) belts, and conveyor belts. Aramids compete with nylon, polyester, glass, and steel in these applications. Steel dominates the rubber hydraulic hose market and polyester is the reinforcement of choice in lower pressure thermoplastic hoses. Advantages of aramid vs. other textiles in hose applications include higher strength, which can lead to constructions with fewer plies and less weight, and better thermal stability, dimensional stability, and chemical resistance. When compared with steel, aramid will not corrode and can be fabricated into lower weight, more flexible hoses.

PT belts can be divided into two categories—v-belts and synchronous belts. Strength, dimensional stability, fatigue resistance, and adhesion are key reinforcement criteria. Polyester is the primary reinforcing fiber in v-belts where cost considerations are most important. Aramids can replace polyester in those applications where strength, shock loading, and dimensional stability requirements outweigh cost. Glass has been the primary reinforcing fiber in timing belts. However, aramid yarn is beginning to replace glass where higher fatigue performance is required to meet increasing demands for more durable, longer-lived belts.

In conveyor belts, as in hoses and PT belts, the superior performance potential of aramid reinforcement must be weighed against the higher material cost. Compared to steel, equivalent belt strength is achieved at one fifth the weight resulting in ease of handling, lower energy costs, and lower installation costs. Maintenance and repair costs are reduced because the fiber does not corrode. Personnel safety is enhanced by the absence of sparking potential. Aramid reinforced belts have higher strength and modulus than nylon or polyester belts and can be made thinner or constructed with fewer plies to lower belt weight, simplify handling, or increase section length by reducing the number of splices.

Yarns are available in high tenacity, high modulus, or high elongation versions to meet the performance requirements of specific end-uses.

13.8.3.4 Composites

p-Aramids are widely used in composite materials as the sole matrix-reinforcing agent or as a hybrid in combination with carbon or glass. Composite property balance will differ from application to application but the key requirement is cost-effective performance at reduced weight. Glass has lower strength and modulus and higher density than aramid or carbon but is the most widely used reinforcing fiber because of its low cost. Carbon fibers have the highest strength and modulus but the lowest elongation. Aramid fibers have a combination of high strength and modulus (although lower than carbon) with low density and high elongation that results in improved impact resistance. Composite structures are found in a host of applications including aerospace components, automobile parts, boats, sporting goods, protruded articles, and pressure vessels. In aircrafts, aramids are used in storage bins, air ducts, and a variety of core (honeycomb) structures. In general, aramid composites have demonstrated satisfactory performance in secondary aircraft structures. Aramid's high tensile strength lends itself well to the manufacture of canoes where weight can be reduced significantly while providing greater tear strength and puncture resistance than fiberglass composites. Hockey shafts, golf club shafts, fishing rods, skis, and tennis rackets have incorporated aramid composites. Fishing rods with unidirectional carbon fibers to provide longitudinal stiffness and aramid fibers woven to provide lateral stiffness yield a high performance rod that is both light weight and stable. In skis, aramid fibers dampen vibration for smoother, more comfortable skiing.

13.8.3.5 Optical and Electromechanical Cables

The primary function of *p*-aramid yarns in fiber optic and electromechanical cables is to protect the optic glass fiber and ductile power conductors from excessive loading or axial strain. *p*-Aramids are well suited to this task because of their high strength and modulus, low density, and resistance to creep. Yarn is used in two forms. Untwisted yarn is laid along the length of the cable to provide maximum modulus to resist stretching. Twisted yarn is inserted as a ripcord to provide maximum strength for tearing the protective sheathing when installing or repairing cable.

Initial usage as a reinforcing agent in ground cables has largely been replaced by less costly glass fiber that can provide the necessary strength and modulus where cable weight is not a critical factor. Aramid yarn is widely used in ADSS (all dielectric self-supporting) aerial cables where glass is unsuited because of its weight. Higher modulus aramid merges are used in this application to minimize cable sag and to prevent the cable from coming into close contact with neighboring electrical lines. Typical yarn deniers are 2840 and multiples thereof. More recent applications are in so-called premise cables that are used to connect devices within buildings. These cables provide more bandwidth, have lower power requirements, and are less costly to maintain than copper lines. Cable diameter is important in this application and therefore lower yarn deniers are used. These range from 380 to 1420. In addition to the attributes cited above, the aramid yarn is nonflammable, which allows the cable to pass mandated burn tests.

For electromechanical cables that are subject to fluctuating loads in use, tension–tension fatigue performance is key. For this application, aramids are superior to galvanized improved plow steel wire in fatigue resistance [129,132]. The high strength-per-unit weight of aramids also allows the cable designer to maximize payload or working length while retaining the ease of handling of a smaller and lighter system.

13.8.3.6 Ropes and Cables

Like fiber optic and electromechanical cables, p-aramids provide high strength and modulus and permit the design of cordage with high load carrying capability with smaller, lighter systems. Yarns are used in a variety of rope and cordage designs such as eight-strand plaited, single and double braids, parallel strands, and wire-lay construction. The choice of construction will depend on the balance of properties required for a specific application. Applications include mooring cables for ship, towlines, elevator cables, and deep-sea cables. Compared to heavy cables of steel wire, p-aramid cables provide equivalent strength at one fifth the weight and have a creep rate that approaches that of steel. Lower cable weight can be a significant factor in enhancing worker safety by reducing the potential for back injuries related to handling mooring lines. Unlike steel, aramid ropes will not corrode in an aqueous environment. Aramid ropes must be designed and handled in a way that minimizes the potential for severe internal or external abrasion and subsequent strength loss. This includes considerations of both rope construction and the appropriate sheave size for a given rope diameter.

A recent innovative machine-room-less traction elevator (ISIS) from ThyssenKrupp takes full advantage of the properties of p-aramid in the design of the hoist cable and associated traction sheaves [130–133]. The cable has three times the life of a steel rope, is smaller in size, and weighs 90% less than a steel rope at a comparable strength rating. The smaller size permits the use of smaller sheaves thereby decreasing torque requirements and operating costs. No lubrication is required because the inner strands are Teflon coated. Finally, the cable transmits less noise and provides a smoother, quieter ride.

Yarns are available in a variety of deniers and merge types that vary in the balance of tensile properties. Special finishes can be applied to increase lubricity, improve fatigue in wet applications, or provide better UV resistance. Ropes using Kevlar or Twaron are particularly useful for static applications or where maximum modulus is required. Technora-based ropes are suited for dynamic applications where resistance to fatigue is important.

13.8.3.7 Reinforced Thermoplastic Pipe

Reinforced thermoplastic pipe (RTP) is a relatively new composite product. At present there are four suppliers with products ranging in diameter from 4 to 10 in. and with pressure ratings up to 100 bars. The pipes are made in continuous lengths of polyethylene with *p*-aramid reinforcement [131,134]. Like the ISIS elevator example above, RTP takes full advantage of the intrinsic attributes of *p*-aramid fibers in the design of this new fluid transport system.

The oil industry is a major user of pipelines to transport oil and gas. In the oil field, flow lines connect individual wells to trunk lines that carry the crude to loading docks or to processing plants. Steel piping has traditionally been used for this application but the pipe is subject to corrosion from within or without over its lifetime. Leakage caused by corrosion is inevitable. Prior to the development of RTP, no suitable alternative to steel piping had been found. The pipeline operator has value for a system that can reduce installation and lifetime maintenance costs per unit length of pipe while meeting temperature and pressure requirements. RTP designs incorporating aramid reinforcement appear to have the necessary characteristics to replace steel piping in the flow line application.

Pipes are constructed with twisted cords to ensure the flexibility required to reel long lengths of pipe of relatively small diameter. The pipes are lightweight for ease of transportation and installation. Long lengths simplify installation and maintenance by reducing the number of couplings. Pipes are corrosion resistant, damage tolerant, and able to withstand high temperatures and pressures. Advantages of aramids over other reinforcement materials such as carbon or glass fiber include flexibility, ease of assembly, and damage tolerance during assembly.

13.8.3.8 Civil Engineering

Use of composite materials for concrete infrastructure repair that was initiated in the mid-1980s finally began to proliferate in the mid-1990s. Carbon and glass fiber reinforced epoxy resin composites have received the most interest. Aramid-based reinforcement has been viewed as a more specialty product for applications requiring high modulus and where the potential for electrical conductivity would preclude the use of carbon; for example, in Japan, aramid sheet is used for all tunnel repair. Product forms include dry fabrics or unidirectional sheets as well as pre-cured strips or bars. Fabrics or sheets are applied to a concrete surface that has been smoothed (by grinding or blasting) and wetted with a resin (usually epoxy). After air pockets are removed using rollers or flat, flexible squeegees, a second resin coat might be applied. The process is repeated for additional plies [132,135].

Reinforcement of concrete structures is important in earthquake prone areas such as Japan, Turkey, and Taiwan. Although steel plate is the primary material used to reinforce and repair concrete structures, higher priced fiber-based sheet structures offer advantages for small sites where ease of handling and corrosion resistance are important. The high strength, modulus, and damage tolerance of aramid-reinforced sheets makes the fiber especially suitable for protecting structures prone to seismic activity. The use of aramid sheet also simplifies the application process. Sheets are light in weight and can be easily handled without heavy machinery and can be applied in confined working spaces. Sheets are also flexible, so surface smoothing and corner rounding of columns are less critical than for carbon fiber sheets [133,136].

13.8.4 *p*-Aramid Paper

13.8.4.1 Core Structures

p-Aramid core structures are analogous to core structures based on *m*-aramids (Section 13.8.2.2) but the base paper uses stronger and stiffer *p*-aramid floc instead of *m*-aramid floc. In addition the component ratio of floc to fibrid is increased. This results in a more porous sheet structure that allows better penetration of the matrix resin in the dipping step. In addition to retaining all the attributes of *m*-aramid based cores, *p*-aramid cores have higher shear strength, higher modulus, and greater fatigue resistance at similar cell size and density. They also have higher hot–wet shear and compression properties than the *m*-analogues. *p*-Aramid cores also bring process advantages because of the lower thermal expansion coefficient and lower moisture-regain of the component fibers. This translates to improved dimensional stability and the ability to retain shape and dimensions throughout the fabrication and part consolidation process.

Because of their superior compression, shear, and fatigue properties, structures based on *p*-aramid cores allow even greater weight reduction than incumbent *m*-aramid cores. Recent commercial adoptions include flooring panels in weight critical programs such as the extended Airbus A-340 and the double deck Airbus A-380. *p*-Aramid cores have also replaced *m*-aramid cores in the elevators and rudders of these aircrafts [121,124], because of their superior hot–wet characteristics.

13.8.4.2 Printed Wiring Boards

Printed wiring boards (PWDs) made of *p*-aramid papers take advantage of the low axial coefficient of thermal expansion (CTE) of the fiber to restrain in-plane expansion of the impregnated resin when heat is applied to the composite laminate. Low CTE boards reduce the strain on solder joints of leadless ceramic chip carriers used in traditional avionics and military applications. In addition, low CTE laminates provide a reliable base for mounting new high-density chip packages where solder joint failure due to thermal cycling is a concern. These include the thin small outline package (TSOP) used for memory chips, the solder grid array (SGA) microprocessor package, and the high lead count ball grid array (BGA). Nonwoven aramid reinforcement is prepegged with epoxy resin on the same vertical path treaters that are used to process fine weave E-glass. At a resin loading of 45–55% by weight, the finished PWB has an in-plane CTE of 9–11 ppm/°C. [134,139].

13.8.4.3 *p*-Aramid Pulp

13.8.4.3.1 Brake Linings or Pads and Clutch Facings

Asbestos was the primary reinforcing agent used in friction materials before it was banned by Congressional legislation in 1978 for health reasons. Two classes of formulations were developed to replace asbestos: semimetallic and nonasbestos organic. Each has its own specific limitations and attributes. *p*-Aramid in the form of pulp is one of the few organic materials suited to the thermal demands of friction applications. Acrylic fiber in the form of pulp has also been used where temperature requirements are less severe. Pulp retains the strength, stiffness, and thermal properties of the precursor fiber and, in addition, provides surface area in the order of 7–15 m^2/g. This high surface area serves as a processing aid in certain manufacturing steps and also as a retention aid for multicomponent brake formulations. High fiber strength can lead to higher pad shear strength and increased resistance to cracking. Fiber thermal stability can influence the nature of the critical transfer layer that forms between the pad and the rotor. Brake formulations are optimized for a variety of performance characteristics such as wear, frictional behavior, and noise. Aramid pulp, at volume percentage levels of <1 to ~10, will influence each of these properties but overall performance is highly dependent on the combined performance of all of the components in the formulation.

Clutch facings are made from wet pulp and staple yarn. Friction papers for automatic transmissions are made from wet pulp that is formed into a sheet on a paper making machine and then impregnated with phenolic resin. Pulp provides strength in the initial paper making process and tensile strength in the final composite structure. The fibrillar pulp also influences sheet porosity. Sheet porosity is essential in this application to ensure adequate permeation of the transmission fluid to dissipate heat generated in service. The combined attributes of strength, heat and wear resistance, and durability that pulp brings to friction papers have become increasingly important as designers continue to reduce the number and size of plates in the transmission and, at the same time, auto manufacturers extend the warranty period for the transmission.

Manual transmissions use a clutch facing made from a resin impregnated wound structure composed of staple yarns. p-Aramid in yarn form provides more strength and durability than a pulp-based paper sheet in this more demanding application. Although aramid reinforced facings have sufficient thermal stability for this application, compositions based on glass and metal fibers dominate this market.

13.8.4.3.2 Gaskets

Like friction materials, asbestos was widely used in high temperature, high performance gaskets prior to the legislation in 1978. Asbestos was highly effective, very cheap, and comprised 80–85% of the weight of the gasket. Aramid pulp brought high strength and thermal stability to this end-use but the fiber cost was an order of magnitude higher than that of asbestos. To reduce this cost penalty, formulations with only 5–20% aramid and 60–80% inert fillers have been developed that provide goal performance in both compressed and beater-add type gaskets.

Compressed gaskets are made on a two-roll calendar from a mix of pulp, elastomer, fillers, curing agents, and toluene. Final gasket properties are very dependent on both the processing conditions and the specific gasket formulation. Tensile strength depends primarily on the amount and type (length and surface area) of pulp selected. Stress retention, compression and recovery, and sealability depend on a combination of factors including the relative amounts of fiber and elastomer, the type and particle size of the filler materials, as well as the mixing and calendaring conditions.

Beater-add gaskets are made in an aqueous paper making type process. Ingredients such as pulp, elastomer, fillers, curing agent, precipitation regulator, and precipitant are combined in water. The resultant slurry is laid down on a screen to drain the water and form a sheet that is then calendared and press cured. As with compressed gaskets, properties will depend on both process conditions and the relative amount and type of ingredients. Beater-add gaskets have been used primarily in cylinder head and other engine gaskets. Today, many auto manufacturers are replacing these beater-add gaskets with gaskets of multilayered steel.

13.8.4.3.3 Elastomer and Resin Reinforcement

p-Aramid in pulp and short fiber (1.5–6 mm length) forms is an effective reinforcing agent in both elastomer and thermoplastic resin matrices. Compared to traditional particulate reinforcing agents in elastomers such as carbon black and silica, aramid pulp provides superior reinforcement at much lower loadings. Advantages of pulp-reinforced elastomers include high low-strain modulus, property anisotropy, greater cut and abrasion resistance, improved wear performance and, in tire stocks, lower rolling resistance. These attributes are achieved, however, only when the pulp is fully dispersed in the rubber stock. Because high surface area pulp is rather difficult to open and wet out using standard rubber compounding processes, concentrated pulp masterbatches (Kevlar engineered elastomer and Rhenogran) have been formulated that allow compounders to more easily achieve adequate dispersion using standard mixing techniques. These masterbatches are available in a variety of elastomers including SBR, NBR, natural rubber, polychloroprene rubber, and EPDM [135,136,138,139].

Applications utilizing pulp and short fiber reinforcement include PT belts, tires, roll covers, hoses, and footwear. In v-belts, both wear resistance and durability increase. In synchronous timing belts, pulp placed in the tooth area increases modulus and reduces the propensity of tooth chunking or chipping. Pulp is used in high performance bicycle and motorcycle tires to improve handling characteristics and to increase puncture resistance. Use in several components of automobile tires continues to be investigated. In roll covers, improvements in tear and abrasion resistance are achieved without affecting compound hardness or processability.

Use of p-aramids in molded or extruded thermoplastic parts offers performance advantages over neat resins or glass-reinforced resins. Aramid reinforced parts exhibit improved mechanical and thermal properties and superior wear resistance with no abrasion to the counter surface. Because the fiber is not abrasive, there is less damage to processing equipment and machining of parts is simplified.

13.8.4.3.4 Sealants and Adhesives

Dry pulp is used as a thixotrope in sealants and adhesives to provide viscosity control at low cost. Viscosity is presumably built through the formation of physical networks of entangled fibrils of the high surface area pulp. Sag or run of applied sealants, adhesives, or coatings is thereby minimized. With shear, the viscosity of these fluids decreases as the networks break down, which facilitates application by spraying, brushing, or other means.

Compared to a common thixotrope such as fumed silica, pulp provides equivalent viscosity at less than one tenth the weight in a typical epoxy resin. In addition, fluid viscosity is unaffected by further processing (agitation) or aging—in contrast to fumed silica modified resin where viscosity drops and is not fully recovered under similar conditions. Pulp can also provide reinforcement in an adhesive matrix as shown by the significant increase in tensile strength, modulus, and tear strength of both a PVC plastisol adhesive and a silicone sealant on the addition of pulp [137,140].

13.9 CONCLUSIONS AND DIRECTION

This brief review of aramid fibers has summarized the very broad range of unusual functionalities that these products bring. While the chemistry plays an important role in defining the scope of applications for which these materials are suited, it is equally important that the final parts are designed to maximize the value of the inherent properties of these materials.

TABLE 13.9
Properties of High Performance Fibers

Fiber		Twaron HM	Carbon HS	PBO	M5 experimental	M5 target
Tenacity	GPa	3.2	3.5	5.5	5.3	9.5
Elongation	%	2.9	1.4	2.5	1.4	>2
E modulus	GPa	115	230	280	350	400–450
Compressive[a] strength	GPa	0.48	2.1	0.42	1.6	2
Compressive[a] strain	%	0.42	0.9	0.15	0.5	0.5
Density	g/cm^3	1.45	1.8	1.56	1.70	1.7
Onset of thermal degradation	°C	450	800	550	530	530
LOI	%	29	N/A	68	>50	>50

[a] In epoxy resin—3-point bending test.
Source: From Magellan International; Teijin Ltd., Teijinconex Heat Resistant Aramids Fiber 02.05.

It is very clear that these unusual properties are derived from structures that are quite different from those of incumbent materials; for example, to obtain very high strength and stiffness the polymer molecules must be perfectly oriented and fully extended, which leads to the highly anisotropic nature of the fibers. That is one of the major reasons why associated applications research efforts have gained such importance. The ultimate products have to be designed to take this anisotropy into account.

We hope that we were able to clearly exemplify the constant trade-off between functionality and processability that is an ongoing challenge with these advanced materials. The functionality that allows these materials to perform under extreme conditions has to be balanced against processability that allows them to be economically shaped into useful forms. This requirement is responsible for the fact that from hundreds of compositions evaluated in the laboratory only a handful are commercially viable.

The fundamental science of structure–property relationship developed as a result of work on aramids is being extended to other chemistries and offers the potential to develop materials with even more impressive properties (Table 13.9).

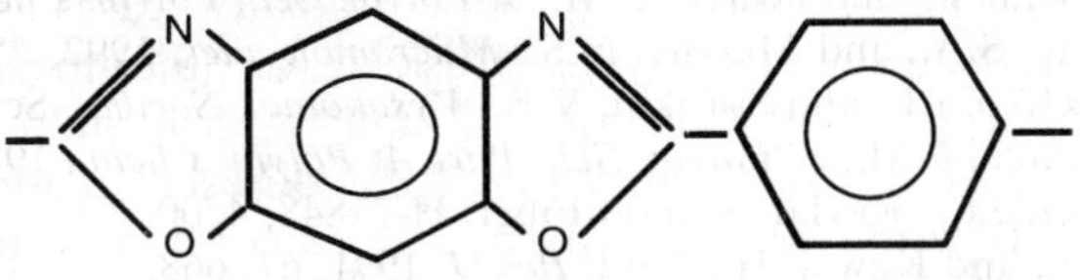

PBO

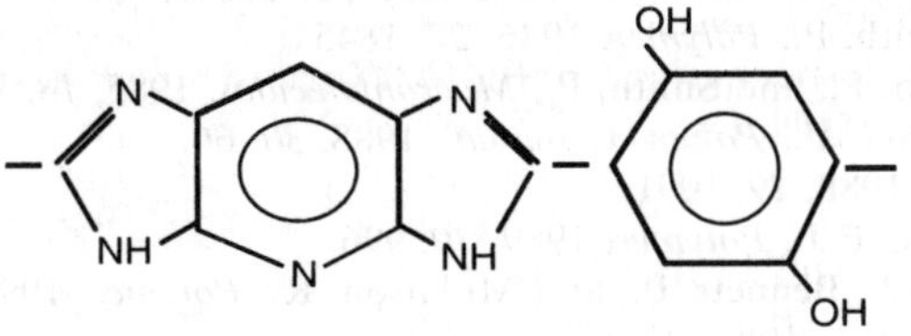

M5

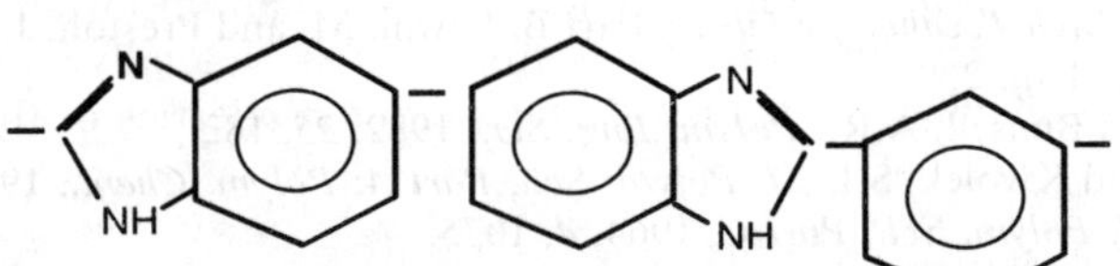

PBI

The structures shown above illustrate the movement to a higher level of aromatic "content" to obtain even better thermal and flame performance. In the case of PBO and M5, the structures are even more rigid than those of *p*-aramids and offer the potential for even greater properties. This is achieved at the expense of ease of processability and at a significantly higher cost. It is very clear that these compositions will not replace *p*-aramids but will likely be an important supplement to our "tool box" of solutions to problems that we face.

REFERENCES

1. Morgan, P.W., U.S. Patent 3,063,966.
2. Beste, L.F. and Stephens, C.W., U.S. Patent 3,068,188.
3. Hill, H.W., Kwolek, S.L., and Morgan, P.W., U.S. Patent 3,006,899, 1961.

4. Sweeny, W., U.S. Patent 3,287,324, 1966.
5. Yang, H.H., *Aromatic High-Strength Fibers*, John Wiley & Sons, New York, 1989, pp. 70–111.
6. Northolt, M.G., *Eur. Polym. J.*, 1974, *10*, 799.
7. Haraguchi, K., Kajiyama T., and Takayanagi, M.J., *J. Appl. Polym. Sci.*, 1979, *23*, 915.
8. Roche, E.J., Allen, S.R., Gabara, V., and Cox, B., *Polymer*, 1989, *30*, 1776.
9. Gardner, K.H., Matheson, R.R., Avakin, P., Chia, Y.T., and Gierke, T.D., *Polym. Prepr. (Am. Chem. Soc., Div. Polym. Chem.)*, 1983, *24*(2), 469.
10. Tashiro, K., Kobayashi, M., and Tadokoro, H., *Macromolecules*, 1977, *10*(2), 413.
11. Panar, M. et al., *J. Polym. Sci., Polym. Phys. Ed.*, 1983, *21*, 1955.
12. Hindeleh, A.M., Halim, N.A., and Ziq, K., *J. Macromol. Sci., Phys.*, 1984, *B23*(3), 289, 383.
13. Dobb, M.G., Johnson, D.J., and Saville, B.P., *J. Polym. Sci., Polym. Phys. Ed.*, 1977, *15*, 2201.
14. Savinov, V.M., Kuznetsov, G.A., Gerasimov, V.D., and Sokolov, L.B., *Vysokomol. Soedin., Ser. B: Kratk. Soobsc.*, 1967, *9*(8), 590.
15. Krasnov, E.P., Lavrov, B.B., Zakharov, V.S., Vorob'ev, E.A., Pantaev, V.A., and Gerasimova, L.S., *Khim. Volok.*, 1971, *1*, 48.
16. Blackwell, J., Cageao, R.A., and Biswas, A., *Macromolecules*, 1987, *20*, 667.
17. Takatsuka, R., Uno, K., Toda, F., and Iwakura, Y., *J. Polym. Sci.*, 1977, *15*, 1905.
18. Chaudhuri, B.A.K., Min, Y., and Pearce, E.M., *J. Polym. Sci., Polym. Chem. Ed.*, 1980, *18*, 2949.
19. Aharoni, S.M., Curran, S.A., and Murthy, N.S., *Macromolecules*, 1992, *25*(17), 4431.
20. Badaev, A.S., Perepechko, I.I., and Sorokin, V.E., *Vysokomol. Soedin.*, Ser. A, 1988, *30*(4), 874.
21. Khanna, Y.P. and Pearce, E.M., *J. Polym. Sci., Part A: Polym. Chem.*, 1986, *24*(9), 2377.
22. DuPont Technical Guide for Kevlar Aramid Fiber, H-77848, 4/00.
23. Fukuda, M., Ochi, M., and Kawai, H., *Text. Res. J.* 1991, *61*, 668.
24. DuPont Technical Guide for Kevlar Aramid Fiber, H-77848, 4/00.
25. Termonia, Y. and Smith, P., *Polymer*, 1986, *27*, 1845.
26. Termonia, Y., Meakin, P., and Smith, P., *Macromolecules*, 1985, *18*, 2246.
27. Smith, P. and Termonia, Y., *Polym. Commun.*, 1988, *30*, 66.
28. Allen, S.R., *Polymer*, 1988, *29*, 1091.
29. Allen, S.R. and Roche, E.J., *Polymer*, 1989, *30*, 996.
30. Allen, S.R., Roche, E.J., Bennett, B., and Molaison, R., *Polymer*, 1992, *33*, 1849.
31. Allen, S.R., *J. Mater. Sci.*, 1987, *22*, 857.
32. Wilfong, R.E. and Zimmerman, J., *J. Appl. Polym. Sci., Polym. Symp.*, 1977, *31*, 1.
33. Tanner D. et al., *High Technology Fibers*, Part B, Lewin, M. and Preston, J., Eds., Marcel Dekker, New York, 1989, chap. 2.
34. Lafitte, M.H. and Bunsell, A.R., *Polym. Eng. Sci.*, 1982, *25*, 182.
35. Morgan, P.W. and Kwolek, S.L., *J. Polym. Sci., Part A: Polym. Chem.*, 1964, *2*(1), 181.
36. Morgan, P.W., *J. Polym. Sci., Part C*, 1964, *4*, 1075.
37. Morgan, P.W. and Kwolek, S.L., *J. Polym. Sci., Part A: Polym. Chem.*, 1964, *2*(1), 209.
38. Olah, G. and Kuhn, I., *J. Am. Chem. Soc.*, 1936, *83*, 4571.
39. Kuhn, I., U.S. Patent 2,768,209, 1956.
40. Renne, A. and Zincke, T., *Ber. Dtsch. Chem. Ges.*, 1874, *7*, 869.
41. Quick, A.J., *J. Am. Chem. Soc.*, 1920, *42*, 1033.
42. Shimada, K., Yoshisato, E., Yoshitomi, T., and Matsumura, S., Jpn. Patent 07, 242, 606, 1995.
43. Christoph, F.J., Jr., Parker, S.H., and Seagraves, R.L., U.S. Patent 3,318,950, 1967.
44. Schötten, C., *Ber. Dtsch. Chem. Ges.*, 1882, *15*, 1947.
45. Schötten, C. and Baum, J., *Ber. Dtsch. Chem. Ges.*, 1884, *17*, 2548.
46. Schötten, C., *Ber. Dtsch. Chem. Ges.*, 1884, *17*, 2545.
47. Schötten, C., *Ber. Dtsch. Chem. Ges.*, 1888, *21*, 2238.
48. Schötten, C., *Ber. Dtsch. Chem. Ges.*, 1890, *23*, 3430.
49. Baumann, E., *Ber. Dtsch. Chem. Ges.*, 1882, *19*, 3218.
50. Osawa, M. and Jinno, M., Jpn. Patent 61,201,009, 1986.
51. Karyakin, N.V. and Rabinovich, I.B., *Dokl. Akad. Nauk. SSSR*, 1983, *271*, 1429.
52. Yamazaki, N., Higashi, F., and Iguchi, T., *J. Polym. Sci., Polym. Chem. Ed.*, 1975, *13*, 785.
53. Wu, G.C., Tanaka, H., Sanui, K., and Ogata, N., *Polym. J.*, 1982, *14*, 571.

54. Yamakawa, N. and Higashi, F., *Tetrahedron*, 1974, *30*, 1319.
55. Higashi, F., Murakami, T., and Taguchi, Y., *J. Polym. Sci., Polym. Chem. Ed.*, 1982, *20*, 103.
56. Ueda, M., *Kobunshi Ronbunshu*, 1986, *35*, 128.
57. Higashi, F., Hoshio, A., and Yamada, Y., *J. Polym. Sci., Polym. Chem. Ed.*, 1984, *22*, 2181.
58. Higashi, F., Mashimo, T., and Takahashi, I., *J. Polym. Sci., Polym. Chem. Ed.*, 1986, 24, 97.
59. Higashi, F., Akiyama, N., Takahashi, T., and Koyama, T., *J. Polym. Sci., Polym. Chem. Ed.*, 1984, *22*, 2181.
60. Morgan, P.W. and Kwolek, S.L., *J. Polym. Sci.*, 1959, *40*, 299.
61. Morgan, P.W., *Soc. Plastic Engrs. J.*, 1959, *15*, 485.
62. Beaman, R.G., Morgan, P.W., Koller, C.R., Wittbecker, E.L., and Magat, E.E., *J. Polym. Sci.*, 1959, *40*, 329.
63. Wittbecker E.L. and Morgan, P.W., *J. Polym. Sci.*, 1959, *40*, 289.
64. Magat, E.E., U.S. Patent 2,831,834, 1958.
65. Shanshoua, V.E. and Eareckson, W.M., *J. Polym. Sci.*, 1959, *40*, 343.
66. Morgan, P.W., *Condensation Polymers*, Wiley Interscience, New York, 1965.
67. Hess, O., *Ber. Dtsch. Chem. Ges.*, 1885, *18*, 685.
68. Tiffeneau, M. and Fuhrer, K., *Bull. Soc. Chim. Fr.*, 1914, *15*, 162.
69. Clarke, R.L., Mooradian, A., Lucas, P., and Slauson, T.J., *J. Am. Chem. Soc.*, 1949, *71*, 2821.
70. Morgan, P.W. and Kwolek, S.L., *J. Polym. Sci.*, 1962, *62*, 33.
71. Flory, P.J., *Priciples of Polymer Chemistry*, Cornell University Press, Ithaca, 1953.
72. Kilkson, H., *Ind. Eng. Chem. Fundam.*, 1964, *3*(4), 281.
73. Hall, H.K., *J. Am. Chem. Soc.*, 1957, *79*, 5439.
74. Hall, H.K., *J. Phys. Chem.*, 1956, *60*, 63.
75. Morgan, P.W. and Kwolek, S.L., *J. Polym. Sci.*, 1964, *A2*, 181.
76. Schnell, H., *J. Ind. Eng. Chem.*, 1959, *51*, 157.
77. Conix, A., *J. Ind. Eng. Chem.*, 1959, *51*, 147.
78. Frost, A.A. and Pearson, R.G., *Kinetics and Mechanisms*, John Wiley & Sons, New York, 1953, p. 1, 22, 130.
79. Pickles, N.J.T. and Hinshelwood, C.N., *J. Chem. Soc.*, 1936, 1353.
80. Lee, K.-S. and Hodge. J.D., U.S. Patent 4,684,409, 1987.
81. Shin, H., U.S. Patent 4,009,153, 1977.
82. Singh, G., Euro. Patent 366,316, 1990.
83. Ozawa, S., Fujie, H., and Aoki, A., Japanese Patent Application Publication, Kokoku No. 484,461, 1973.
84. Kudryavtev, G.I., *Khim. Volok.*, 4, 23–22, 1969.
85. E.I. DuPont de Nemours & Co., Inc., Belg. Patent 569,760, 1958; Hill, H.W., Kwolek, S.L., and Morgan, P.W., U.S. Patent 3,006,899, October 31, 1961; Kwolek, S.L., Morgan, P.W., and Sorenson, W.R., U.S. Patent 3,063,966, November 13, 1962.
86. Kouzai, K., Matsuda, K., Tabe, Y., Honda, H., and Mori, K., *Sen-i Gakkaishi*, 1992, *48*, 2, 55.
87. Fujie, H., *Nikkyo Geppo*, 1987, *40*, 8.
88. Preston, J., Polyamides, aromatic, in *Encyclopedia of Polymer Science and Technology*, 2nd ed., Wiley Interscience, New York, 1988.
89. Kudryavtev, G.I., *Khim. Volokuza*, 1969, *4*, 23.
90. Ozawa, S., Nakagawa, Y., Matsuda, K., Nishihara, T., and Yunoki, H., U.S. Patent 4,075,172, 1978.
91. Bair, T.I. and Morgan, P.W., U.S. Patent 3,673,143, 1972; U.S. Patent 3,817,941, 1974.
92. Hancock, T., Spruiell, J.E., and White, J.L., *J. Appl. Polym. Sci.*, 1977, *21*, 1227.
93. Aoki, H., Coffin, D.R., Hancock, T.A., Harwood, D., Lenk, R.S., Fellers, J.F., and White, J.L., *J. Polym. Sci., Polym. Symp.*, 1979, *65*, 29.
94. Papkov, S.P., Kulichikin, V.G., Kalymykovo, V.P., and Malkin, A.Y., *J. Polym. Sci.*, 1974, *12*, 1953.
95. Baird, D.G., *J. Appl. Polym. Sci.*, 1978, *22*, 2701.
96. Baird, D.G. and Bailman, R.L., *J. Rheol.*, 1979, *23*, 505.
97. Aoki, H., White, J.L., and Fellers, J.F., *J. Appl. Polym. Sci.*, 1979, *23*, 2293.
98. von Falkai, D., Dry-spinning technology, in *Acrylic Fiber Technology and Applications*, Masson, J.C., Ed., Marcel Dekker, Inc., New York, 1995.

99. Sweeney, W., U.S. Patent 3,287,324.
100. Kwolek, S.L., Morgan, P.W., and Sorenson, W.R., U.S. Patent 3,063,966, November 13, 1962.
101. Dierkes, G., Ingenieusch Textilwesen Krefeld, Krefeld, Fed. Rep. Ger. Spinner, Weber, *Textilveredlung*, 1969, *87*(1), 43.
102. Capone, G.J., Wet-spinning technology, in *Acrylic Fiber Technology and Applications*, Masson, J.C., Ed., Marcel Dekker, Inc., New York, 1995.
103. Cahn, J., *J. Chem. Phys.*, 1965, *42*, 93.
104. Kudryavtsev, G.I., USSR, *Khimicheskie Volokna*, 1969, *4*, 23.
105. Tai, T.M., Rodini, D.J., Masson, J.C., and Leonard, R.L., U.S. Patent 5,667,743, September 16, 1997.
106. Kwolek, S.L., U.S. Patent 3,671,542, 1972; U.S. Patent 3,819,587, 1974.
107. Kwolek, S.L. and Yang, H.H., History of aramid fibers, in *Manmade Fibers: Their Origin and Development*, Seymour, R.B. and Porter, R.S., Eds., Elsevier, London and New York, 1992.
109. Blades, H., U.S. Patent 3,767,756, 1973.
109. Yang, H.H., Aramid fibers, in *Fibre Reinforcement for Composite Materials*, Bunsell, A.R., Ed., Elsevier, Amsterdam, 1988.
110. Yang, H.H., Fiber spinning of anisotropic polymers, in *Advanced Fiber Spinning Technology*, Nakajima, T., Ed., Woodhead Publishing Ltd., Great Yarmouth, England, 1994.
111. Blades, H., U.S. Patent 3,869,429, 1975.
112. Imanishi, T. and Muraoka S., U.S. Patent 4,752,643, June 21, 1988.
113. Gross, G.C., U.S. Patent 3,756,908, September 4, 1973; Morgan, P.W., U.S. Patent 2,999,788.
114. Liang, R., Han, L., Doriswamy, D., and Gupta, R.K., 13th Proceedings of the International Congress of Rheology, Vol. 4, Cambridge, United Kingdom, August 20–25, 2000, pp. 136–138.
115. Gohlke, U. and Baum, E., *Inst. Polym. Chem.*, DAW, Teltow-Seehof, DDR-153; Ger. Dem. Rep. *Acta Polymerica*, 1979, *30*(3), 170.
116. Yang, H.H., *Kevlar Aramid Fiber*, John Wiley & Sons, England, 1993.
117. Morgan, P.W., U.S. Patent 2,999,788.
118. Gross, G.C., U.S. Patent 3,756,908, September 4, 1973.
119. Liang, R.,.Han, L., Doriswamy, D., and Gupta R.K., Proceedings of the 13th International Congress of Rheology, Vol. 4, Cambridge, United Kingdom, August 20–25, 2000, pp. 136–138.
120. Parrish, E., MaCartney, J., and Morgan, P.W., U.S. Patent 2,999,782.
121. DuPont Technical Bulletin H-52720, 2001, p. 8.
122. http://www.astecinc.com/literature/images/t_139.pdf (accessed May 2004).
123. DuPont Technical Bulletin H-50949, 2000.
124. Pinzelli, R. and Loken, H., Honeycomb cores: from NOMEX to KEVLAR aramid papers, JEC Composites, 2004, *8*, 133.
125. Chiou, M.J., Foy B.E., and Miner L.H., Penetration-Resistant Aramid Article, U.S. Patent 5,622,771, April 22, 1997.
126. Chiou, M.J., Knife-Stab-Resistant Composite, U.S. Patent 6,534,426 B1, March 18, 2003.
127. Chiou, M.J., Ren J., and Van Zijl, Penetration-Resistant Ballistic Article, U.S. Patent 6,133,169, October 17, 2000.
128. Chiou, M., Knife-Stab-Resistant Ballistic Article, U.S. Patent 6,475,936 B1, November 5, 2002.
129. Fels, A., Bottger, C., Polligkeit, W., Neu S., and Klingspor, C., Puncture and Bullet Proof Protective Clothing, U.S. Patent 6,656,570 B1, December 2, 2003.
130. Hand, D.R., Hartert, R., and Bottger, C., Stab Resistant and Anti-Ballistic Material. Method of Making the Same, U.S. Patent Application Publication US 2004/0023580 A1, February 5, 2004.
131. Willemsen, S., Weening W.E., and Steenbergen, A., Adhesive-coated multifilament yarn of an aromatic polyamide, a yarn package, a cord, a fabric, a reinforced object and a process for making said yarn, U.S. Patent 4,557,967, December 10, 1985.
132. DuPont Technical Bulletin H-50949, 3, 2002.
133. http://www.elevator-world.com/magazine/archive01/0402–001.shtml (accessed May 2004).
134. Frost S.R., Application of non-metallic materials for oil and gas transportation—Qualification issues, in *Composite Materials for Offshore Operations*—3, Third International Conference on Composite Materials for Offshore Operations (CM00-3), Houston, Texas, October 31–November

2, 2000; Wang, S.S., Williams, J.G., and Lo, K.H., Eds., University of Houston, CEAC, 2001, pp. 355–365.
135. Kliger, H.S., *SAMPE*, 2000, *36*(5), 18.
136. Sumida, A., Reinforcement, retrofit of concrete structures with aramid fiber, in *FRP Composites in Civil Engineering*, Proceedings of the International Conference on FRP Composites in Civil Engineering, Hong Kong, China, December 12–15, 2001, Teng, J.G., Ed., Elsevier, Amsterdam, 2001, p. 273.
137. Hendren, G.L., Kirayoglu, B., Powell, D.J., and Weinhold, M., *Adv. Mater.*, 1998, *10*(15), 1233.
138. Tsimpris, C.W. and Wartalski, J., *Rubber World*, 2001, *224*, 35.
139. Rheinchemie Product Bulletin.
140. DuPont Technical Bulletin H-67189, 1999.
141. DuPont Technical Guide for Nomex Brand Aramid Fiber, H-52720, 7/01.
142. Teijin Ltd., Teijinconex Heat Resistant Aramids Fiber 02.05.
143. Teijin Ltd., High Tenacity Aramids Fibre: Technora TIE-05/87.5.
144. Akzo Nobel, Twron—Product Information: Yarns, Fibers and Pulp.
145. Magellan International.
146. Yasufuku, S., *IEEE Elec. Insu. Mag.*, 1995, *11*(6), 27.
147. Asahi Chemical Industry America, Inc., Technical Brochure, Aramica Film, 1991.
148. E.I. DuPont de Nemours & Co., Inc., NOMEX Aramid Paper Type 410—Typical Properties, H-22368, 8/98.
149. Hendren, G.L., Kirayoglu, B., Powell, D.J., and Weinhold, M., *Adv. Mater.*, 1998, *10*(15), 1233.
150. Karyakin, N.V. and Rabinovich, I.B., *Dokl. Akad. Nauk. SSSR*, **1983**, *271*(6), 1429.

Index

A

Abaca fibers, uses, 459
Abaca plant, 456, 458
   stem stripping, 458
Abrasion, resistance, 803
Accessibility
   bromine method, 569
   solute exclusion, 583
Acetaldehyde, 816
Acetalization, bath, composition, 301
Acetalization polyvinylalcohol, 294
Acetic acid, 266
Acetylation, jute and kenaf, 425
Acetylene, 266, 816
Acetyl group, jute and kenaf, 412
Acetyl value, 784
Acrolein, 817
Acrylic fibers
   apparel section, 931
   asbestos replacement, 946
   basic properties, 906
   biocomponent fibers, 935, 936
   biological resistance, 915
   brake, clutch linings, 947
   chemical resistance, 916
   drying and collapsing, 888
   dyeing, 934
   electrically conducting, 950
   fabric durability, 913
   flammability, 916
   fluorinated comonomers, 924
   high bulk yarns, 935
   hot water treatment, 911
   hot-wet strength, 911
   identification-test, 925
   IR spectroscopy, 929
   mechanical testing (Instron), 930
   modulus, 910
   moisture-absorbing fibers, 920
   physical properties, 907
   polymer characterization, 926
      molecular weight, 926
   prediction of fabric durability, 913
   production, 954–956
   Raman spectroscopy, 929
   softness, 919
   specialized products, 936
   stress-strain curve, 910
   sunlight resistance, 914
   surface treatment, 923
   tensile properties, 907
   wear resistance, 911
   whiteness, 923
   x-ray fluorescence, 929
Acrylic and modacrylic fibers
   battery plate reinforcing parallels, 947
   filament, apparel markets, 935
   flame-resistant, 931, 936
   industrial fibers, 932, 933
   tubular, special uses, 947, 948
   uses, 933
Acrylic precursor fiber, 940
Acrylic-reinforced cement, 946
   uses, 946
Acrylics vs. polyester, 932
Acrylonitrile, 60, 427
   preparation, 816, 817
Acrylonitrile-vinylchloride copolymerization, 83
Activation energy
   ε-caprolactam polymerization, 47
   polyamidation, 46
N-Acyl lactam, 38
Adipic acid
   electrolytic coupling, 66
Adiponitrile, 67, 68
AH-salt, 70
   preparation, 73
Air filters, jute and kenaf, 437
Alkali cellulose, 716
   carbon dioxide, 710
   transition (scheme), 721
Alkali-spinning, polyamide, 300
Alkali treatment, 508, 509
Alkaline bath, polyamide, 304
1-Alkyl-3-methylimidazolium, 674
Alpaca, 371
Alternating copolymer, 830
Amide group, 113
Amidine group, 40
ε-Aminocaproic acid, 64

Amino capronitrile, 67
4-Aminophenol, 988
Ammonium sulfamate, 120
Ammonium sulfate, coagulation agent, 298
Ammonium thiocyanate/ammonia, 541, 671
Ammoxydation, 60
Amorphous domain, 840
Amorphous phase, polypropylene, 204
Amorphous state, polyamide, 68
Angora rabbit fibers, 371
Anionic (acidic) dyes, 900
Anionic polymerization, lactams, 38
Anisotropic nature, 15
Anisotropy, 17
   p-aramids, 984
Annealing, polypropylene, 227
Antimony-halogen finishes, 595
Antimony oxide, 120, 193
Antioxidant PG, 679
Antiparallel β-pleated sheet, 389
Aramid fibers, wet spinning, 1005, 1006
Aramid producers, 979
Aramids, 34
   compressive strength, 984
   concrete reinforcement, 1021
   creep, 984
   crystallinity, homopolymer, 980
   diacid and diisocyanate, 990
   "direct polymerization," 992, 993
   dry-jet wet spinning
      operating conditions, 1006–1010
   fiber skin-core structure, 1005
   fiber polymer molecules orientation, 1007
   filament, staple yarns, 1014
   interfacial polymerization, 993
   isotropic solution
      films bi-oriented, 1009, 1010
      films unidirectional, 1009, 1010
   liquid crystal behavior, 1002
   mechanical properties, 984
   papers, films, 985
   plasticized melt polymerization, 1000
   polymerization methods, 993
   Schötten–Baumann acylation, 989
   solubility, 1001
   solution polymerization, 995
      reaction conditions, 995–998
      side reactions, 996, 997
      solubilizing aids, 999
      solvents, 998
   solutions,
      reological properties, 1003
   uses, 1013
   UV light absorption, 981
   vapor-phase polymerization, 999
   water absorption, 981
m-Aramids
   core structure, 1015
   dry spinning, 1004
   electrical application, 1015, 1016
   filters, 1015
   flame resistance, 1013, 1014
   protective apparel, 1013, 1014
   spinning, 1001
p-Aramids, 1013, 1014
   fibers
      ballistic protection, 1017
      composite armor systems, 1017
      softarmor systems, 1017
   composite materials, 1019
   core structure, 1022
   fiber optics - electro cables protection, 1019
   pipe reinforcement, 1021
   printed wiring boards, 1022
   protective apparel, 1018
   pulp, 1011
      brake linings and clutch facings, 1022
      gaskets, 1023
      sealants and adhesives, 1024
   reinforcing agents, 1018
   ropes and cables, 1020
   spinning from reaction mixture, 1001
Atactic
   polypropylene, 150
   polystyrene, 249
Average molecular weights, 42
Average orientation, crystalline regions, 205
Avrami
   equation, 94
   parameter, 231

B

Banana fiber
   Barus effect, 165, 897
   uses, 459
Beckmann rearrangement, 63
Bending modulus, 802, 803
Bicomponent fibers, 23, 123, 192
   Polyvinylalcohol/polyvichloride, 311, 312;
      see also polychlal
Bicomponent spinning, 200
Biocompatibility, 18
Biodegradability polyvinylalcohol, 309
Bioengineered cotton, 525
Biotech cotton, 524
Birefringence
   factor, 100
   vegetable fibers, 491

Index 1033

Bis(4-aminocyclohexyl)methane, 59
1,3-Bis(hydroxymethyl)-4,5-dihydroxy-
　　　imidazolidinone-2, 592
Blends, polypropylene, 249
Block copolymers, 279
Borax, 545
Boric acid, 595
Bragg equation, 98
Branching, 20
　　radical polymerization, 271
Bulking, poly(ethylene terephthalate), 12
Bulk polymerization
　　acrylonitrile, 818, 822, 837
　　　　autocatalytic effect, 825
　　propylene, 156
Butadiene-1,3, carboxylation, 67
Byssinosis, 645

C
Calcium oxalate, 482
Calcium thiocyanate, 671
Callose-β-(1→3)-D-glucan, 534
Camel fibers, 371
Caprolactam, solid-phase polymerization, 72
Caprolactam sulfate, 64
ε-Caprolactone, 18
Capryllactam, 56
Carbon disulfide, 716
Carbon fibers, 940
　　carbonization, 943
　　pricing, 945
　　stabilization process, 941, 942
　　structure models, 944, 945
　　three-step process, 941
Carbon and graphite fibers, Young's
　　　modulus, 939
Carbonate process, 710
Cashgora goat fibers, 371
Cashmere, 371
Cationic (basic) dyes, 900
Cationic polymerization, lactams, 40
Cell membrane complex (CMC), interactions, 349
Cellobiose, 553, 556
　　conformational analysis, 556
　　hybrid modeling, 557
　　molecular dynamics, 556
　　residue, 547
Cellulose, 412, 477, 608, 775
Cellulose II, 564
Cellulose III and III_I, 564
Cellulose IV, 564
Cellulose
　　acetylation catalyst, 579
　　acid degradation, 603

in alkaline medium, 606, 607
amine-salt process, 709
anisotropic solutions, 669
anisotropic and isotropic solutions, 608
atomic force microscopy (AFM), 563
average degree of polymerization (DP), 719
biological deterioration, 608
birefringent solution 698
bromine accessibility method, 560–570;
　　see also Accesibility
carbamate, 703, 705
　　wet-spinning 708; see also Spinning process
carboxymethylation, 590
cellobiohydrolases, 616
cellulases, 616
conversion $I_\alpha \to I_\beta$, 562
conversion I→II, 565
crystal structures I–IV, 558
cyanoethylation, 591
degradation, 601
degree of polymerization, 601, 671
depolymerization, 779
determination of –CHO, carbonyl and
　　–COOH groups, 602, 603
distribution of molecular weights, 719
etherification, 590
experimental model, 553
flame retardants, 612, 613
β-(1→4)-D-glucan, 534
endo- and exo-1,4-β-glucanases, 116, 617
hydrogen bonds, 673, 776
kinetic study-pyrolyses, 611
liquid ammonia, 389
N-methylmorfolin N-oxide, 688–691
moisture content, 778
noncrystalline, 565
orientation primary hydroxyl
　　groups, 561
oxidative depolymerization, 614
peeling reaction, 605, 607
phosphoric acid, 698
photosensitized degradation, 614
polymorphic forms, 542, 668
pretreatment with $ZnCl_2$
purity, 777
pyrolysis, 610, 611
pyrolysis products, 610
reactivity, 777
scorch temperature, 613
sheet density, 778
smoldering combustion, 613
steam explosion-sodium hydroxide-water
　　thermal decomposition, 610
xanthate, 723

Cellulose acetate and triacetate, 779
  acetic acid recovery, 784
  acetic acid system, 781
  acetylation-batch type, 782
  cigarette filters, 806, 807
  crystallinity, 795
  die-swell, 790
  disperse dyes, 803, 804
  dope preheating, 790
  dry-spinning, 788; see also Dry-spinning fibers, 21
    heat resistance, 798
    heterogeneous system, 782
    melt spinning, 792
    moisture regain, 796
    tenacity, 798
  flake drying, 783
  flake washing, 783
  haze and color, 785
  homogeneous process, 782
  hydrolysis, 783
  imbibition, 798
  intrinsic viscosity, 785
  methylene chloride, 782
  miscellaneous applications, 807
  precipitation, 783
  pretreatment, 781
  properties, 793
  solubility, 784
  solvent recovery, 792
  spinneret design, 790
  spinning parameters, 791
  stain recovery, 802
  stress-strain curve, 799
  sulfate linkage, 779
  uses, 806, 807
  wet spinning, 792; see also Wet spinning
  wet tenacity, 801
Chain configuration, polyvinylacetate, 277
Chain radical, 266
Chain scission, 115, 601
Chain transfer, 820, 821
  backbiting, 281
  radical polymerization, 268
Change of free energy, 700
Char, 120
Characterization of polymers, methods, 44
"Chardonnet," "Basancon silk"
Chemical homogeneity, 830
Chemical resistance, 321
Cigarette tow, 792
Citrulline, 356
Clearing temperature, 699
Cleavage, glucosidic linkage of, 610

Coagulant, 870
Coagulation, 872, 873
  in polyglycoles, 892
Cocoon, 384
Coefficient, Huggins; Kraemer, 43
Coir, 472
  properties, 472
  uses, 474
"Cold drawing on stretching," 10
Colligative properties, 926, 927
Collodion, 713
Color stability, 187
Commingling jet effect, 87
Composites, jute, 433
Composition drift, copolymer composition, 832
Condensation polymers, 34
Condense-spinning, polyvinylchloride, 318
Condensing agents, 993
Conformational strain, lactams, 39
Conjugate spinning, 919
Continuous process, dispersion polymerization-acrylonitrile, 835, 836
Continuous stirred tank, 834
Controlled-rheology resin, polypropylene, 157
Copolymerization, 828
Copoly(p-phenylene/3,4'-diphenyl ether terephthalamide), 977, 978
Copoly(vinyl chloride-vinyl acetate), 313
Corona discharge, 619
Cortical cell, 333, 342
  distribution, 372
Cotton
  acetylation, 597
  antimicrobial resistence, 615
  average ordered fraction, 567, 573
  bale, 628
  biologically active conjugate, 599
  biopolishing, 618
  bleaching, 62
  cellulose modification, 617
  chemical properties, 584
  classification system, 635–638
  color, 631, 634
  corona treatment, 619
  I_α, I_β crystal structures, 562
  cuticle
  desizing, 620
  dust, 647
  enzyme-chemical modification, 619
  fabrics applications, 642, 644
  fiber
    classification, 628, 629
    length, 633

Index

morphology, 575
strength, 633
fineness, 622
formylation method, 571
fungicides, 616
general description, 524
grad, 630
inorganic acid esters, 599
linear density, 633
lumen, 543
maturity, 621, 622
maturity determination of, 622
maturity; indirect method, 622
mercerization, 542, 620; *see also* Mercerization
mineral acids, 510
naturally colored, 525
nep, 632
noncellulosic constituents, 537
organic acids, 538
pectin substrates, 538
pigments, 621
Pima and Upland grade standards, 631
pore structure, 583
production; consumption, markets, 639
quality parameters, 629
reactive deys, 621
scouring, 620
singeing, 620
sodium hydroxide treatment, 538
soluble sugars, 538
solvents for, 541, 542
tensile strength, 623
wax, 537
CR 144 light stabilzer, 181
Crack propagation, 445
Crimp, 85, 625, 891
bicomponent yarns, 891
texturing, 86
Cross-linkage
polyvinylacetate, 292
reaction, 592
Cross section, SEM, 875, 876
Crystal structure, unit cell polyamide, 284
α- and γ-Crystalline content, polyamides, 102
Crystalline domain, 840
Crystalline index, 209
polypropylene, 213
Crystalline orientation, 220
function polypropylene, 210
Crystalline phase, polypropylene, 204
Crystalline state, polyamides, 88
Crystallinity
change, 121
density, 101
polyvinylacetate, 265
polyvinylalcohol, 279
Crystallite size, 247
Crystallization
kinetics, 253
rate poly(ethyleneterephthalate), 8
PP-MM nanocomposite, 253
Cumen hydroperoxide, 838
Cuprammonium
process 675, 708, 709, 714
rayon, 714
Cupriethylene diamine (CUEN) hydroxide, 541
Cuticle, 532
cells, 333
Cyclic oligomers, polyamides, 71
Cyclization, caprolactam 51
Cyclohexane
carboxylic acid, 59
oxidation, 65
Cyclohexanone, 59
lactam production, 66
oxime, 59–62
cis- and *trans-*1,4-Cyclohexane-dimethanol, 15
Cyclohexenyl acetate, 64
Cysteic acid, 363, 370
Cystein, 357
Cystine, 357
content in fibers, 370
Cytoskeleton, 535

D

Deamination, polyamides, 114
Decarboxylation, polyamides, 114
Decomposition temperature, aramids, 98
Decortication, machine, 459, 469
Degradation(thermal),
poly(ethyleneterephthalate), 5
Degree
chain orientation, 904
crystallinity
coir, 487
polypropylene, 209
polyvinylacetate, 287
polymerization, 269, 819
cellulose, 488, 489
undegraded cotton, 776
substitution (DS), 778
swelling, polyvinylacetate, 292
Denier, 792
Denier per filament (dpf), 2, 792
Density, polyvinylacetate, 289
Desulfation, 784
Dew retting, 466
Dialysis membrane, 709

3,4'-Diaminodiphenylether, preparation (3,4'-POP), 988
Diammonium phosphate, 595
Dicotyledons, 410
1,4-Dicyanobutene-2, 69
Dielectric constant, 627
Diethyleneglycol in poly(ethyleneterephthalate), 5
Diethyl oxalate, 58
Differential scanning calorimetry (DSC), poly(ethyleneterephthalate), 5
Dilute alkali, 507
N,N-Dimethylacetamide (DMAc), 669, 919, 981, 999, 1001, 1004
N,N-Dimethylacetamide/lithium chloride, 541
Dimethylamin, 729
N,N-Dimethylformamide, 818, 1004
1,3-Dimethyl-2-imidazolidinone/lithium chloride, 541
Dimethylsulfoxide, 818
Dimethyl terephthalate, 3
1,3-Dinitrobenzene, 988
2-Dioxanone, 18
Disc refiner, 780
Disperse dyes, 900
Dispersion polymerization, 818, 826
Distribution, molecular weights, 817
Disulfide cross-links, reduction, 357
Dithiothreitol, 351
Dodecanedioic acid, 58, 59
ω-Dodecanolactam, 58
Donnan effect, 602
Double steeping process, 722
Drawing
 polyvinylacetate, 299
 polypropylene, 199
Draw ratio, 81
 Poly(ethyleneterephthalate), 11
Draw
 resonance, 170
 -twister, 199
Dry-spinning, 871 874
 Nylon-4, 56
 polyvinylacetate, 306
 polyvinylchloride, 313, 316
Dry vs. wet spinning, 871, 874
DSC, polypropylene, 233
Dye, 21
 diffusion, 903
 sites, 902
Dyeability, polypropylene, 147
Dyeing, 113, 620, 803
 characteristics, 804, 805
 ion-exchange equilibrium, 902
 nylons, 107
 package, piece, 900
 pore model, 905
 retarding agents, 904
 skein, stock, 900

E
Elastic recovery, 626
Electrical resistance, 628
 fibers, 493
Electrolytic coupling, catalysts, 66
Electronic structure, amide group, 88
Electrophoresis, 371
Electrospinning process, 710
Ellagic acid, 691
Elongation at break, 625
 poly(ethyleneterephthalate), 11
Emulsion polymerization, 818, 825
Emulsion spinning, 311
End groups, 46
 analysis, 926
Endocuticle, 333, 340, 341
Endotoxin, 646
End, uses, PPE, 181
Energy quencher, 181
Energy transfer agent, UV stability, 180
Entanglement density, poly(ethyleneterephthalate), 7
Enthalpy, fusion, 101, 854
Enzymes, 348, 618
 vegetable fibers, 512
Epicuticle, 337
Equilibrium constant, polycondensation, 36
Esterification, cellulose, 778
Ethylene, 266
Ethylenediamine, 709
Ethylene glycol, 3
Ethyleneoxide, 591
Exocuticle, 333, 340
Extraction, proteins, 351
Extensional viscosity, 897
Extent reaction, polycondensation, 41
Extrusion, high-temperature, 185

F
Fabric "break-open," 1014
False-twisting, 13, 86
 poly(ethyleneterephthalate), 12
False viscosity effect, 785
"Feel," "drape," "handle," 20
Fiber 66
Fiber
 cross section, 79
 elongation, polyamides, 81
 future trends, 951
 light stability, 807
 morphology, poly(ethyleneterephthalate), 3

properties vs. spinning conditions, 206, 207
reinforced cement (FRC), 326
ribbon width, 624
saturation point, 587
strength, 984
thermoplastic blends, 443
Fibrids, 1010
Fibrillation, Lyocell fibers, 697, 703;
    see also Lyocell fiber
Filament
    cross-section shape, 899
    diameter oscillation, polypropylene, 201
Filter value, 687
Flame retardance, polyvinylchloride, 321
Flame retardant, 119
    antimony trioxide, 596
    decabromodiphenyl ether, 596
    hexabromocyclododecane, 596
Flame suppression, 1,2,3,4-butanetetracarboxylic acid-citric acid, 597
Flaming combustion, 593
Flamstob NOR 116, 194
Flax
    plant, 463
    plant stem, 464
    uses, 467
1-Fluoro-2,4-dinitrobenzene, 361
Force to break, 624
Formaldehyde, 647
Formalization, polyvinylacetate fiber, 265
β-form polyamide, 91
Free, radical initiator, 266
Funiculus, 530

G

Gas-phase polymerization, propylene, 156
Gel
    dyeing, 900
    melting, 853
Gel permeation chromatography (GPC), 541
Gel
    spinning, polyethylene, 243
    state spinning, 892
Gelation, 872, 873
    polyacrylonitrile, 865
Geotextiles, 435; see also Juta and kenaf
Glass transition temperature, 861
    aramids, 981, 984
    B. mori silkworm cocoon
    Kodel
    nylon-6,T
    polyamides, 96, 73
    poly(ethyleneterephthalate), 3, 22

polypropylene, 160, 215
polyvinylacetate, 288
Glycolic acid, 18
1,4-β-Glycoside linkage, 720, 775
Glycosidic bond, 601
Gossypium
    G. aboreum, 524
    G. barbadense, 524
    G. herbaceum, 524
    G. hirsutum, 524
Grading, jute and kenaf, criteria, 422
Graphite fibers, 940
Graphitization, carbon fiber, 943
Guncotton, 599

H

"Hackling," 467
Half-cystin, 355
Hammer mill, 780
Hard armor, 18
Hard fibers, 455
Hard keratin IFs, 343, 346
"Head" to "head" addition, vinyl acetate, 277, 278
"Head" to "tail" insertion, polypropylene, 150
Heat-shrinkage, poly(ethyleneterephthalate), 11
Heat stabilizers, 869
Heat treatment, polyvinylacetate fibers, 264
Heavy-metal stains, wool, 343
Helical conformation, 366
Helical staple, three dimensional, 235, 237
Helix, 552
Hemicellulose, 412, 477, 507, 721, 776
Hemp, 470
    fiber extraction, 471
    fiber uses, 470, 471
Hemp stalks, retting, 471
Henequen, 462
Henequen fibers, uses, 462
Herbicides, 540
Hermans orientation factor, 487, 488
Hermans' RMS spiral angle, 420
"Heterofil" fibers, 23
Heterogeneous copolymerization, 833
Hexandiol-1,6, 67
Hexamethylenediamine
    from acrylonitrile, 69
    from butadiene, 69
High-modulus fibers, 15
High-performance cellulose fibers, 700
High-shrinkage polypropylene fibers, 247
    uses, 248
High-speed spinning, 84
High-strength synthetic fibers, 245
High-sulfur proteins, 354

High-velocity melt, 892
Hindered amine light stabilizers (HALS), 178
Holocellulose, 419
Hookean region, 366
Hook's law, 625
Hunter colorimeter, 785
HVI system, 639
Hydrated melts, 861
Hydrazine, 709
Hydrocellulose, 510, 544, 601, 604
Hydrogels, 397
Hydrogen bonding, 801, 720, 980, 984
Hydrogen bonds, 585
    polyamides, 88, 89, 91, 92
    polyvinylalcohol, 292, 300
Hydrogen peroxide, wool bleach, 358
ε-Hydroperoxy-ε-caprolactam, 117
Hydrophobic bond, 585
Hydrophobicity, wool, 338
Hydrophobic vs. hydrophilic, 793
Hydroxyl amine sulfate, 61
2-Hydroxybiphenyl, 22
3-Hydroxyvaleric acid, 18
Hypochlorite, 430

I

Ideal copolymerization, 830
Imide moiety, 38
Induction period, radical polymerization, 268
Inherent viscosity, poly(ethylenterephthalate), 4
Initiator, 36
Inner root sheath, protein composition, 356
Instability, polymer flows, 170
Interfacial polycondensation, reaction conditions, 994, 995
Interfacial polymerization, 35
Intermolecular bonds, bonds, 351
Intrachain disulfide bonds, 351
Intrinsic viscosity, 20, 42
    poly(ethyleneterephthalate), ionic liquid, 673
IR, wool, 336
isophthaloyl-N,N-bis(valerolactam), 1000
isophthaloyl chloride, preparation, 988
isotactic polypropylene, 150

J

Jet
    strech ratio, 872
    texturing, 85
Jute, 406
    bleaching process, 430
    charcoal, 449
    color fastness, 431
    composite with thermoplastic, 443
    cyanoethylation, 427
    dyed, 431
    presence of lignine, 429
    reaction with maleic anhydride, 449
    retting, 408
    woolenization, 432
Jute and kenaf
    chemical modification, 425
    composition, 414
    fiber quality, 423
    fiber structure, 409
    grading, 422
    inorganic matrix composites, 443
    light interaction, 430
    moisture effect, 430
    photochemical degradation, 428
    physical properties, 421
    reinforcing fillers, 444
    tensile properties
    uses, 423, 424, 433, 434
        filters, 437
        geotextiles, 435
        molded products, 439
        nonstructural composites, 438
        packaging, 439
        pulp and paper, 440
        sorbents, 438

K

Kenaf, 406
    acetylation, 444
    fiber structure, 409
    polypropylene composite, 444
    succinicanhydride modification, 448
Keratin proteins, 351
Kevlar pulp, 1011
Kinetic parameters, polyamides, 49
Kinetic scheme, radical polymerization, 818

L

Lactams, 36, 88
D-Lactic acid, 18
Lamellae, polypropylene, 214
Light
    induced degradation, 114
    stability, 869
    stabilizers, 118
Lignin, 412, 477
Lignin,
    coir, hemp, sisal, 479
    flax, 479
Limit oxygen index (LOI), 916

Limiting viscosity number, 281; see also
    Intrinsic viscosity
Linear polymer, β-D-glucopyranose, 536
Linen cross-linking, 511
Lint, linters, 527
Lipid moiety in CMC, 349
Liquid NH_3 treatment, 508, 509
LIST process, 682, 683, 684–687
Llama fibers, 371
Long air quench melt spinning, 196
Long-staple filter, 455
Long-term thermal stability, 178
Lorentz–Lorenz equation, 100
Low-sulfur proteins, 352
Lubricants, 19
Lubrication wheel, 789
Lyocell fibers
    alcohols as coagulants, 675, 696
    dry-jet wet spinning, 691
Lyocell process, 675, 679, 681, 692
Lyocell vs. viscose fibers, 694
Lysinoalanine, cross-links, 36

M

Macrofibrils, 343, 585
Macrovoids, 874
Mammalia fibers
    chemical composition, 370–372
    high-sulfur proteins, 353
Manila hemp, see Abaca
Mark-Houwink equation, 5, 43
Mass specific resistance, 628
Mechanical properties, polyamides, 107
Medulla, 347
Medullary index, 347
Melt-blowing process, 202
Melting, copolymers, 857–860
Melting point, polyamides, 79
Melt index, melt flow rate (MFR), 158
Melt spinning,
    liquid crystalline polymers, 16
    morphology development, 9
    polyamides, 78
    polypropylene, 196
    polyvinylacetate, 196
    polyvinylchloride, 313, 321
    semicrystalline polymers, 10
Melt temperature, polyamides, 94, 95
Membrane osmometry, 927
Mercaptoethanol, 351, 357
Mercerization, 508, 546, 587
Mercuric acetate, antimicrobial action, 513
Mercury porosimetry, 879, 888
Metallocene catalysts, 151, 153, 154

Metering pump, 196
2-Methyleneglutaronitrile, 70
Methyl ethyl ketone, 784
4-O-Methylglucuronoxylan, 786
N-Methylmorpholine-N-oxide, 541, 669, 676,
N-Methylmorpholine-N-oxide
    chemical properties, 678
    preparation, 676, 678
N-methylolpropionamide, 595
N-Methyl-2-pyrrolidone, 981, 991, 1001
Micelle, 825
Microfibers, 25
Microfibrillar structure, 109
Microfibrils, 214, 545, 585
Microtubules, 535
Microvoids, 877
Microvoid size, copolymer acrylonitrile-methyl
    acrylate, 887
Microvoid structure, 879
Miller indices, cellulose, 560
Mineral acids, pure cellulosic fiber, 510
Mineral acid treatment, vegetable fibers, 509
Miscelaneous applications, polypropylene, 148
Modacrylics, 817, 834
Modulus, 11
Modulus,
    polyamides, 84
    poly(ethyleneterephthalate), 3
    nylon-6, 111
Mohair, 371
Moisture uptake, polyethyleneterephthalate, 21
Molecular orientation, liquid crystalline
    polymers, 15
Molecular weight 817; see also Molecular
    weight distribution
Molecular weight, poly(ethyleneterephthalate), 5
Molecular weight distribution, 543, 928
Molecular weight distribution
    most probable, 282
    Nylon-6
    Poly(ethyleneterephthalate), 7
α-Monoclinic crystalline structure,
    polystyrene 244
Monomer reactivity ratios, 828, 829
Montmorillonite, 252
Morphology, acrylonitrile copolymers, 850
Morphology, swelling behavior, 851
Poly(m-phenylene isophthalamide) (MPDI),
    triclinic unit cell, 979

N

Nanocomposites
    polyamides, 74
    polypropylene, 251

Naphthalene-2,6-dicarboxylic acid, 15
"Neck," 10
Necking, polypropylene, 214
Newtonian fluid, 208
Nickel stabilizer, 187
Nitration method, determination
    DP cellulose, 601
Nitrocellulose, 713
Nitrosyl chloride, 62
Nitrosyl sulfuric acid, 65
Nomenclature polyamides, 35
Nylon-66, 33
Nylon-6, density, 92

O

Optical cables, 18
"Organic" cotton, 525, 526
Orientation, 199
Orientational functions, 205
Oriented fibrillar structure, 884
Orlon, 813
Orthocortex, 355
Osmium tetroxide, 343
Ovule, 530
Oxa-chromanol stabilizer (PBDP), 689
Oxycellulose, 601

P

Paper industry, 149
Parameters K and a, 43, 282; see also
    Mark–Houwink equation
Pectin, 475
Pentaerythritol, 20
Peracetic acid, 63, 358, 511
Perchloric acid, 779P
Performic acid, 358
Perlok spinning, polyvinylacetate, 325
Phase diagram, cellulose-liquid ammonia, 589
Phase separated fibers, polyvinylacetate, 309
Phase separation, 872, 873
Phase separation, mechanism, 874
Phenolic antioxidants, 176
Phenol, oxidation, 66
m-Phenylene diamine (MPD), preparation, 987
p-Phenylene diamine (PPD), preparation,
    987, 988
Phenylglycidyl ether, 19
Phosphoric acid, 674
Pigments, 147, 175, 184
Pigments
    color stability, 187
    effect on photostability, 189, 190
"Pilling," polyester, 19
"Pills," 921
Pinpicker, 780

Plasma modification, 193
Pleat-retaining properties,
    (poly(butylenes terephthalate), 14
Polyacrylonitrile
    branching, 822
    chain irregularity, 842
    degradation reaction, 852
    dynamic-mechanical properties, 862
    glass transition temperature, 861, 862
    melting, 852
    melting-like transition, 853
    nitrile dipolar interaction energy, 842
    single crystals, 846
    solvents, 864
    stereoregularity, 840, 841
    "two dimensional liquid crystalline like"
        structure, 840
    two-phase morphology, 847
    water system spinning, 893
    wet spinning, 869
Polyamides
    block copolymers, 73
    end uses, 124, 125
Poly[5-amino-2-(p-aminophenyl) benzimidazole
    terephthalamide] (SVM), 977
Poly(butylene terephthalate) (PBT), 2, 14
Polycaprolactam, 47
Polycaprolactam, high molecular weight, 72
Polychlal, 312
Polydispersity index, 44
Polyenantholactam, 56
Poly(ethylene adipate), 22
Plyethyleneglycols, 119
Poly(ethylene naphthlate) (PEN), 2
Polyethylene oxide (PEG), 21
Poly(ethyleneterephthalate) (PET), 2
Poly(ethyleneterephthalate)
    antimony trioxide (Sb_2O_3) catalyst, 4
    molecular weight, 4
    oligomers, 4
    physical properties, 14
Poly(hexamethylene adipamide), 45
Poly(3-hydroxybutyric acid), 18
Polylactams, 88
Poly(lactic acid) (PLA), 2, 25
Polymer blends, 249
Polymer blends, dyeing, 192
Polymer cleanliness, 5
Polymerization
    bulk, 273
    emulsion, 273
    solution, 273
    suspension, 273
Polymerization degree, polycondensation, 36;
    see also Polydispersity index

Polymerization kinetics, polyamides, 40
Polymerization mechanism, polypropylene, 151
Polymerization of acrylonitrile, inorganic salt, 818
Polymorphism, in polyamides, 90
Polynosic rayon, 761
Poly(m-phenylene isophthalamide), 977, 978
Poly(p-phenylene terephthalamide), 978
Polypropylene, 149
Polypropylene blends, 249
Polypropylene, isotactic, 142
Polypropylene, stability thermal and UV, 174
Poly(propylene terephthalate) (PPT), 2, 14, 24
Polysiloxane, finish, 24
Polyvinylacetate hydrolysis, autocatalytic reaction, 275
Polyvinylacetate, methanolysis-hydrolysis, 274
Polyvinylalcohol, tacticity, 280
Polyvinyl-cellulosic fibers, 615
Polyvinylchloride, 313
Polyvinylchloride/acetone-carbon disulfide solvent, 320
Polyvinylchloride fibers, applications, 325
Polyvinylchloride grafted polyacrylonitrile, 323
Polyvinylchloride/polyvinylalcohol fibers, 310
Pore size distribution, 879
Porosity, molecular probes, 583
Precipitation polymerization, 822
Pressley tester, 624
Pressure dyeing, 22
Primary crystallization, 121
Propagation reaction, polyamides, 38
1,3-Propanediol, 14
β-Propiolactam, 56
Propylene, 149, 817
Protofibrils, 345
Pseudocrystallinity, polyacrylonitrile, 840
Puckered ring, 775
Pump block; spin beam, polypropylene, 196
2-Pyrrolidinone, 56

R
Radical formation, in polypropylene, 175, 176
Ramie,
 degumming, 467, 469
 extraction, 468
 uses, 470
Raschig process, 61
Rate,
 crystallization, polypropylene, 231
 polycondensation, 37
 polymerization, 819
 polyethylene, 152

Rayon, 715
 cuprammonium, viscose, 758, 759
 fibers structure, 748
 process, 711
Reaction rate constant, polyamides, 48
Reactive dyeing, 593
Rectangular spinneret-pack assembly, 236
Recycling, polyester, 26
Redox initiation, 826
Relaxation, 890
Reprocessed silkworm silk, 397, 400
Residual water
 polyamides, 71
 polyvinylacetate, 300
Resilience, 626
 poly(butylene terephthalate) (PBT), 14
 polyvinylacetate, fibers, 302
Retting
 flax stalks, 462
 ultrasonic energy, 466
Rheological properties, polypropylene, 161–163
Rheovibron, 862
Rigidity, fiber, 627
Ring-opening polymerization, 34, 37
 caprolactam, 47

S
SAXS, microvoids, 879
Scherrer equation, 99
Schiff base, 115
Schötten–Baumann acylation, reaction mechanism, 990
Schweizer reagent, 708
Screw extrusion, 196
"Scutched flax" or "line," 467
Scutching, flax, 466
Secondary cellulose acetate, 780
Secondary wall, immature, 580
Self-bulking fiber, 24
Self-crimping yarn, 24
SEM, 251
 flax, 484
 polystyrene-polypropylene blend, 251
 ramie, 484
Semi batch polymerization, 834
Shear modulus
 liquid crystalline polymers, 18
 vegetable fibers, 502
Shear rate, 166, 1004
Sheat-core spinning, 919
Sheet film extrusion, polypropylene, 203
Shish-kebab structure, polypropylene, 204
Shrink-proof wool, chlorination, 358

Silica, 252
Silk I, II, III, 390, 391
　biomedical uses, 400
　crystalline forms, 390
　DSC, 388
　fibers, mechanical properties, 387
　formation, secondary structure, 393
　semicrystalline material, 389
　in vivo processing
Silkworm, *Bombyx mori*, 384
Silkworm cocoon silk, 385
　chemical composition, 388
　reprocessed, 395
Single-fiber, tensile tester, 624
Sisal fiber, uses, 460
Sisal plant, fiber extraction, 460, 462
β-Sitosterol-glucoside, 535
Size distribution of pores, polyacrylonitrile
　　copolymers, 881
Size exclusion chromatography (SEC), 928;
　　see also GPC
Slurry process, polypropylene
　　preparation, 155
Small angle x-ray analysis, 222
Smoldering combustion, 593
Sodium bisulfite, 358
　cellulosate, 720
　cellulose xanthate, 716
　chlorite, 477, 510, 511
　methallyl sulfonate, 817
　styrene sulfonate, 817
　sulfate
　　coagulation bath, 298
　sulfophenylmethallyl ether, 817
Soft ballistic protection, 18
Soft fibers, 455
Solution polymerization, 818–822
　chain transfer, 837
　spinning, 837
Sonic modulus, polymer chain 885
Speciality rayons, 763–764
Species of cotton, 524
Specific gravity, vegetable fibers, 489
Specific stress (tenacity), 624
Spherulites, 94
Spider silk, types, 386
Spin
　finish, 81
　pack, 197
Spinline stress, poly(ethylenterephthalate), 8
Spinnability, 893, 894, 899
Spinnaret, 540, 788
　polyamide, 78
　poly(ethyleneterephthalate), 6
　polypropylene, 197

Spinnarette, 714
Spinning, liquid crystal polyester, 16
Spinning process, 84
Spiral angle, jute and kenaf,
"Splittable pie" technique, 25
Spun-bonding process, 201
Stabilizer, n-propyl gallate, 689, 691
Stabilizers, 144
Stamicarbon, 61
Standard enthalpy ($\Delta H_P°$), polycondensation, 39
Standard entropy ($\Delta S_P°$), polycondensation, 39
Staple
　fiber, 12
　spinning line, 14
Static charge, poly(ethyleneterephthalate), 20
Stationary state, radical
　　polymerization, 267
Steeping, decrease in degree of
　　polymerization, 721
Stelometer, 624, 625
Step-growth polycondensation, 41
Stereospecific polymers, 141
Storage life, 706
Stress calculation, 206
"Stress in motion", 625
Stress relaxation, 227
Stress-stain
　characteristic, vegetable fibers, 496
　curve, 624
　polypropylene, 217
Structural changes, DSC technique, 859
Structure
　α-polyamides, 88, 89, 91, 92
　γ-polyamides, 88, 92
5-Sulfoisophthalic acid, 22
Sulfuric acid, 779
Sunn stalks, retting, 472
Sunn, uses, 471
Surgical sutures, 18
　polyvinylacetate, 263
Swelling, vegetable fibers, 491, 506
Synergises, 183

T
"Tail"-to-"tail" addition, vinyl acetate, 278
Take-up speed, 83, 84, 897
　polyamides, 80
　polypropylene, 198, 209
Tenacity, 11
　lignin content, 490, 500
　poly(ethyleneterephthalate) fibers, 12
　polypropylene fibers, 226
　polyvinyl acetate fibers, 290, 299
Tensile modulus, liquid crystalline
　　polymers, 16, 18

Index

Tensile properties, vegetable fibers-moisture content, 498
Tensile strain, 625
Tensile strength, polyamides, 109
Terephthalaldehyde, 302
Terephthalic acid, 3, 4
Terephthaloyl chloride (TCI), preparation, 988
Termination, by coupling and disproportionation, 267
Tetrakis(hydroxymethyl) phosponium chloride, 595
Tetramethylene diamine, 58
Tex, 792
Tex per filament, 792
Textiles, 806
Tex system, 624
Thermal insulation, 24, 236
Thermal stability, *B. mori* silkworm cocoon, 388
Thermo-oxidative degradation, 114
Thermotropic aromatic polyesters, 17
Thermotropic polyesters, 15, 16
Thiocyanates, 709
Thioglycolic acid, 343, 351, 357
Tire cord, 12
Titer, 86
Toughness, 627, 802
Tow bundle, 864
Transacylation, polyamides, 50
Transamidation, polyamides, 50
Transfer to polymer, radical polymerization, 269
Transgenic cotton, 524
Transmission electron microscopy (TEM), 341, 343, 349
  distribution microvoid size, 879
  mammalian fibers, 372, 373
Triboelectricity, 322
Trifluoroacetic acid, 670
Triphenyl phosphite, 992
Tris (2,3-dibromopropyl)phosphate, 595
Tuxying, 459
Two-phase, semicrystalline polymers, 839, 840
Two-stage, drawing, 239

U

Ultrasuede, 25
Urea, 705
Urea-formaldehyde resin, 513
U.S. HVI system, 629, 630
  global marketing system, 630
UV stability, polypropylene, 144
UV stabilizers, polypropylene, 142

V

Vapor pressure osmometry (VPO), 927
Vegetable fibers, chemical composition, 475–177
  dyeing, 514
  moisture absorption, 490
Vinyl acetate, polymerization, 282, 293
Vinyl alcohol, 262
Vinyl bromide, 817, 931
Vinyl chloride, 931
  mixed-gas method, 315
  low temperature polymerization, 316
  oxychlorination, 313
  suspension polymerization, 315
Vinylidene chloride, 817, 922, 931
N-Vinylpyrrolidone, 919
Vinyon N, 814
Viscometry, 42
Viscose rayon, 716
  aging
    hydrolysis, 722
    oxidation, 722
  all core structure, 752
  all skin structure, 752
  batch process
    description, 733
  belt xanthator, (CBX), 732, 737
  cellulose II, IV, 750
  classification of fibers, 744
  continuous process description, 733
  cross-section shape, 754–756
  filtration, 725, 793
  final processing, 742
  high wet-modulus, 718
  Hottenroth number, 726
  hydrogen sulfide formation, 729
  industrial yarn, 757–758
  Maurer–Buss Contisulf process, 738
  mixing, 725
  modal, 763
  modifiers, 752
    list of, 731
  modifiers-spin-bath, 729–731
  press weight ratio (PWR), 735
  production, 732
  properties, 756
  ripening, 725
    reaction involved,
  Salt index (SI), 726
  shredding, 722
  spinnerette, jet, 740
  spinning machine, 720, 740
  spinning, xanthate decomposition, 727
  steeping, 720

transxanthation, 728
uses, 760–763
xanthation, 723
x-ray diffraction pattern, 750
Viscosity average (M_v), 927, 928
Voids, 228

W

Waal's forces, polyamides, 88, 89
"Wash-and-wear," 793
Water-Quench melt spinning, polypropylene, 200
WAXS, orientational process, 885
Weissenburg number, 168
Wet-spinning, 696
    polyvinylacetate, 297
    polyvinylchloride, 320
Wet stretching, 883
White crumb, 716
Wide-line NMR measurement, 840
Winding, 87
Wool,
    acidic hydrolysis, 360
    alkaline hydrolysis, 359
    chlorination, 359
    β-conformation, α-helical conformation, 367
    cross-linking, 362
    equilibrium water content, 368, 369
    esterification, 362
    flammability, 364
    load-extension curve, 365
    Merino, 340
    photodegradation, 363
    reaction (amino acid side chains), 362
    water absorption, 368

Woolenization, stability of crimp, 433
Work of rupture, 624
Wrinkle resistance, 592
    polycarboxylic acid and catalysts, 598

X

Xanthation
    by-products, 724
    side-reaction, 724
X-ray diffraction, 205, 301, 840
    banana, 482, 488
    cellulose, 559
    coir, 487, 488
    flax, 484
    hemp, 486
    polyacrylonitrile, 843–845
    polyamides, 92, 98, 103
    polypropylene, 160
    polyvinylacetate, 263, 284
    ramie, 485
    sisal, 482
    wool, 345, 366
X-ray photoelectron spectroscopy, wool, 336

Y

Yak fibers, 373
Yield point, 625
Young's modulus, 626
    jute and kenaf, 421

Z

Ziegler–Natta catalyst, polypropylene, 153
Zirconium tetrafluoride, 153